Overview Animation: Life Cycle of a Retrovirus (Figure 4-49)

Focus Animation: Retroviral Gene Expression (Section 4.7)

Focus Animation: Retroviral Genome Integration (Section 4.7)

CHAPTER 5 MOLECULAR GENETIC TECHNIQUES

Focus Animation: Mitosis (Figure 5-3)

Focus Animation: Meiosis (Figure 5-3)

Technique Animation: Expression Cloning of Receptors (Section 5.2)

Technique Animation: Plasmid Cloning (Figure 5-14)

Technique Animation: Dideoxy Sequencing of DNA (Figure 5-21)

Technique Animation: Polymerase Chain Reaction (Figure 5-23)

Technique Animation: Synthesizing an Oligonucleotide Array (Figure 5-29)

Technique Animation: Screening for Patterns of Gene Expression (Figure 5-29)

Video: Microinjection of ES Cells into a Blastocyst (Figure 5-41)

Technique Animation: Creating a Transgenic Mouse (Figure 5-43)

Video: DNA Injected into a Pronucleus of a Mouse Zygote (Figure 5-43)

Podcast: RNA Interference (Figure 5-45)

CHAPTER 6 GENES, GENOMICS, AND CHROMOSOMES

Podcast: Eukaryotic Transcription Units (Figure 6-3)

Focus Animation: Retroviral Reverse Transcription (Figure 6-14)

Focus Animation: Telomere Replication (Figure 6-48)

Focus Animation: Telomere Replication (Figure 6-49)

CHAPTER 7 TRANSCRIPTIONAL CONTROL OF GENE EXPRESSION

Podcast: Assembly of the Pol II Preinitiation Complex (Figure 7-31)

Video: 3-D Model of an RNA Polymerase II Preinitation Compex (Figure 7-32)

Technique Animation: Yeast Two-Hybrid System (Figure 7-45)

Video: Hormone-Regulated Nuclear Translocation of the Glucocorticoid Receptor (Figure 7-49)

CHAPTER 8 POST-TRANSCRIPTIONAL GENE CONTROL

Overview Animation: Life Cycle of an mRNA (Figure 8-2)

Video: hnRNP A1 Nucleocytoplasmic Shuttling (Figure 8-4)

Podcast: Discovery of Introns (Figure 8-6)

Focus Animation: mRNA Splicing (Figure 8-9)

Focus Animation: mRNA Splicing (Figure 8-11)

Video: Dynamic Nature of Pre-mRNA Splicing Factor Movement in Living Cells (Figure 8-15)

CHAPTER 9 VISUALIZING, FRACTIONATING, AND CULTURING CELLS

Overview Animation: Protein Secretion (Figure 9-5)

Video: Three-Dimensional Model of a Golgi Complex (Figure 9-6)

Video: Three-Dimensional Model of a Mitochondrion (Figure 9-8)

Podcast: Light and Electron Microscopy (Figure 9-14)

Technique Animation: Preparing Monoclonal Antibodies (Figure 9-35)

CHAPTER 10 BIOMEMBRANE STRUCTURE

Podcast: Annular Phospholipids (Figure 10-17)

CHAPTER 11 TRANSMEMBRANE TRANSPORT OF IONS AND SMALL MOLECULES

Video: Frog Oocyte Expressing Aquaporin Bursts in Hypotonic Solution (Figure 11-7)

Overview Animation: Biological Energy Interconversions (Figure 11-12)

Overview Animation: Biological Energy Interconversions (Figure 11-24)

Overview Animation: Biological Energy Interconversions (Figure 11-25)

Podcast: The Two-Na^+/One-Leucine Symporter (Figure 11-26)

CHAPTER 12 CELLULAR ENERGETICS

Video: Mitochondrion Reconstructed by Electron Tomography (Figure 12-6)

Video: Mitochondrial Fusion and Fission (Figure 12-7)

Focus Animation: Electron Transport (Figure 12-16)

Focus Animation: Proton Translocating, Rotary F-ATPase (Figure 12-24)

Focus Animation: ATP Synthesis (Figure 12-25)

Podcast: ATP Synthesis (Figure 12-25)

Video: Rotation of Actin Filament Bound to ATP Synthase (Figure 12-26)

Focus Animation: Photosynthesis (Figure 12-33)

MOLECULAR CELL BIOLOGY

ABOUT THE AUTHORS

HARVEY LODISH is Professor of Biology and Professor of Bioengineering at the Massachusetts Institute of Technology and a member of the Whitehead Institute for Biomedical Research. Dr. Lodish is also a member of the National Academy of Sciences and the American Academy of Arts and Sciences and was President (2004) of the American Society for Cell Biology. He is well known for his work on cell-membrane physiology, particularly the biosynthesis of many cell-surface proteins, and on the cloning and functional analysis of several cell-surface receptor proteins, such as the erythropoietin and TGF-ß receptors. His laboratory also studies hematopoietic stem cells and has identified novel proteins that support their proliferation. Dr. Lodish teaches undergraduate and graduate courses in cell biology and biotechnology.

ARNOLD BERK is Professor of Microbiology, Immunology, and Molecular Genetics and a member of the Molecular Biology Institute at the University of California, Los Angeles. Dr. Berk is also a fellow of the American Academy of Arts and Sciences. He is one of the original discoverers of RNA splicing and of mechanisms for gene control in viruses. His laboratory studies the molecular interactions that regulate transcription initiation in mammalian cells, focusing in particular on transcription factors encoded by oncogenes and tumor suppressors. He teaches introductory courses in molecular biology and virology and an advanced course in cell biology of the nucleus.

CHRIS A. KAISER is Professor and Head of the Department of Biology at the Massachusetts Institute of Technology. His laboratory uses genetic and cell biological methods to understand the basic processes of how newly synthesized membrane and secretory proteins are folded and stored in the compartments of the secretory pathway. Dr. Kaiser is recognized as a top undergraduate educator at MIT, where he has taught genetics to undergraduates for many years.

MONTY KRIEGER is Whitehead Professor in the Department of Biology at the Massachusetts Institute of Technology. For his innovative teaching of undergraduate biology and human physiology as well as graduate cell-biology courses, he has received numerous awards. His laboratory has made contributions to our understanding of membrane trafficking through the Golgi apparatus and has cloned and characterized receptor proteins important for the movement of cholesterol into and out of cells, including the HDL receptor.

MATTHEW P. SCOTT is Professor of Developmental Biology, Genetics, and Bioengineering at Stanford University School of Medicine and Investigator at the Howard Hughes Medical Institute. He is a member of the National Academy of Sciences and the American Academy of Arts and Sciences and a past president of the Society for Developmental Biology. He is known for his work in developmental biology and genetics, particularly in areas of cell-cell signaling and homeobox genes and for discovering the roles of developmental regulators in cancer. Dr. Scott teaches cell and developmental biology to undergraduate students, development and disease mechanisms to medical students, and developmental biology to graduate students at Stanford University.

ANTHONY BRETSCHER is Professor of Cell Biology at Cornell University and Associate Director of the Cornell Institute of Cell and Molecular Biology. His laboratory is well known for identifying and characterizing new components of the actin cytoskeleton and elucidating the biological functions of those components in relation to cell polarity and membrane traffic. For this work, his laboratory exploits biochemical, genetic, and cell biological approaches in two model systems, vertebrate epithelial cells and the budding yeast. Dr. Bretscher teaches cell biology to graduate students at Cornell University.

HIDDE PLOEGH is Professor of Biology at the Massachusetts Institute of Technology and a member of the Whitehead Institute for Biomedical Research. One of the world's leading researchers in immune system behavior, Dr. Ploegh studies the various tactics that viruses employ to evade our immune responses and the ways our immune system distinguishes friend from foe. Dr. Ploegh teaches immunology to undergraduate students at Harvard University and MIT.

MOLECULAR CELL BIOLOGY

SIXTH EDITION

Harvey Lodish

Arnold Berk

Chris A. Kaiser

Monty Krieger

Matthew P. Scott

Anthony Bretscher

Hidde Ploegh

Paul Matsudaira

W. H. Freeman and Company

New York

Publisher: Sara Tenney
Executive Editor: Katherine Ahr
Developmental Editors: Matthew Tontonoz, Erica Pantages Frost, Elizabeth Rice
Associate Project Manager: Hannah Thonet
Assistant Editor: Nick Tymoczko
Asssociate Director of Marketing: Debbie Clare
Senior Project Editor: Mary Louise Byrd
Text Designer: Marsha Cohen
Page Makeup: Aptara, Inc.
Cover Design: Blake Logan
Illustration Coordinator: Susan Timmins
Illustrations: Network Graphics, Erica Beade, H. Adam Steinberg
Photo Editor: Cecilia Varas
Photo Researcher: Christina Micek
Production Coordinator: Susan Wein
Media and Supplements Editor: Hannah Thonet
Media Developers: Biostudio, Inc., Sumanas, Inc.
Composition: Aptara, Inc.
Manufacturing: RR Donnelley & Sons Company

About the cover: Mitotic PtK2 cells in late anaphase stained blue for DNA and green for tubulin. Courtesy of Torsten Wittman.

Library of Congress Cataloging-in-Publication Data
Molecular cell biology / Harvey Lodish . . . [et al.]. —6th ed.
p. cm.
Includes bibliographical references and index.
1. Cytology. 2. Molecular biology. I. Lodish, Harvey F.
QH581.2.M655 2007
571.6—dc22 2007006188
ISBN-13: 978-0-7167-7601-7
ISBN-10: 0-7167-7601-4

Printed in the United States of America

First printing

W. H. Freeman and Company
41 Madison Avenue, New York, NY 10010
Houndmills, Basingstoke
RG21 6XS, England

www.whfreeman.com

To our students and to our teachers,
from whom we continue to learn,
and to our families, for their support,
encouragement, and love

PREFACE

In writing the sixth edition of *Molecular Cell Biology* we have incorporated many of the spectacular advances made over the past four years in biomedical science, driven in part by new experimental technologies that have revolutionized many fields. High-velocity techniques for sequencing DNA, for example, have generated the complete sequence of dozens of eukaryotic genomes; these in turn have led to important discoveries about the organization of the human genome and regulation of gene expression, as well as novel insights into the evolution of life-forms and the functions of individual members of multiprotein families. New imaging techniques have generated profound revelations about cell organization and movement, and new molecular structures have greatly increased our understanding of life processes such as cell-cell signaling, photosynthesis, gene transcription, and chromatin structure.

New Author Team

Two new authors have been instrumental in refocusing this book toward these exciting new developments. **Anthony Bretscher** of Cornell University is known for identifying and characterizing new components of the actin cytoskeleton and elucidating their biological functions in relation to cell polarity and membrane traffic. **Hidde Ploegh,** of the Massachusetts Institute of Technology, has made major contributions to our understanding of immune system behavior, particularly in regard to the various tactics that viruses employ to evade our immune responses and the ways our immune system responds. Both authors are widely recognized for their research as well as their classroom teaching abilities.

We are grateful to Paul Matsudaira, Jim Darnell, Larry Zipursky, and David Baltimore for their exceptional contributions to the previous editions of *Molecular Cell Biology*. Much of their vision and insight is apparent at many places in this book.

Experimental Emphasis

The hallmark of *Molecular Cell Biology* has always been the use of experiments to teach students how we have learned what we know. A number of experimental organisms, from yeasts to worms to mice, are used throughout so the student can see how discoveries made with a "lower organism" can lead directly to insights even about human biology and disease. This experimental approach, evident in the text itself, has also been thoroughly integrated into the pedagogical framework. For example:

- Experimental Figures lead students through important experimental results.
- Classic Experiments essays focus on historically important and Nobel Prize–winning experiments.
- New and revised Analyze the Data problems at the end of each chapter require the student to synthesize real experimental data to answer a series of questions.
- Updated Perspectives for the Future essays explore potential applications of future discoveries and unanswered questions that lie ahead in research.

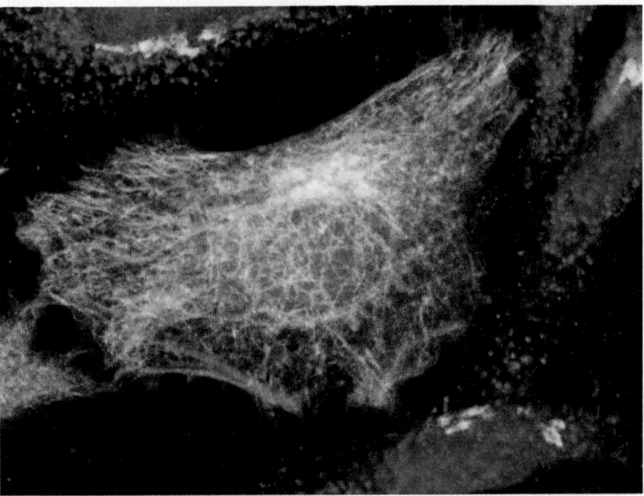

Fluorescence microscopy shows the location of DNA and multiple proteins within the same cell. [From B.N.G. Giepmans et al., 2006, *Science* **312**:217.]

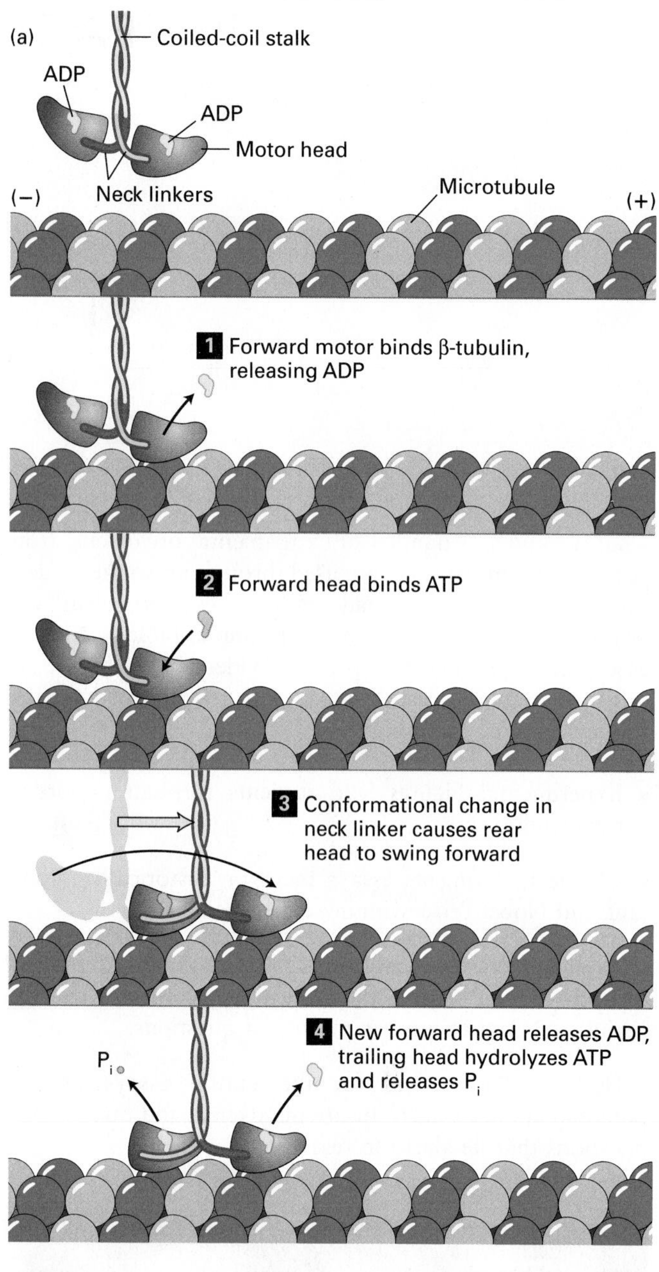
(a)
Coiled-coil stalk
ADP
ADP
Motor head
Neck linkers
Microtubule
(−)
(+)
1 Forward motor binds β-tubulin, releasing ADP
2 Forward head binds ATP
3 Conformational change in neck linker causes rear head to swing forward
4 New forward head releases ADP, trailing head hydrolyzes ATP and releases P_i
P_i

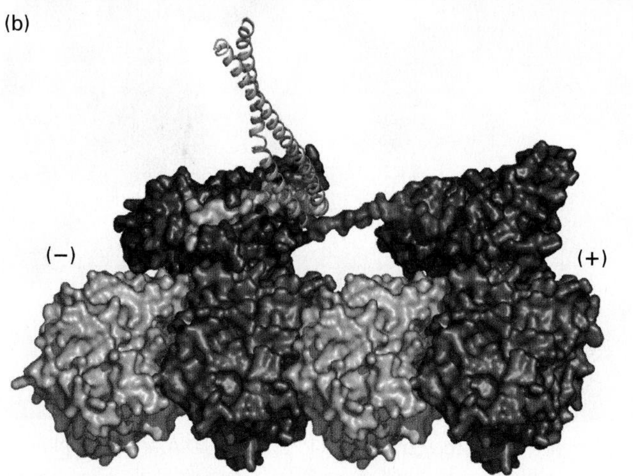
(b)
(−)
(+)

Figure 18-22 Kinesin-1 uses ATP to "walk" down a microtubule.

New Discoveries, New Methodologies

Methodological advances continue to expand and enrich our knowledge of molecular cell biology, and lead to new understanding. Following are just a selection of the new experimental methodologies and cutting-edge science introduced in this edition:

- Expanded coverage of proteomics, including organelle proteome profiling and advances in mass spectroscopy (Chapter 3)
- Expanded coverage of RNAi, including the use of shRNAs to inhibit any gene of interest in a cultured cell or organism (Chapters 5, 8)
- Updated discussions of chromatin, including structure and condensation (Chapter 6), control of gene expression by chromatin remodeling (Chapter 7), and chromatin-remodeling proteins and tumor development (Chapter 25)
- Evolution of chromosomes and the mitochondrion (Chapter 6)
- New molecular models, including pre-initiation complex and mediator complex (Chapter 7); annular phospholipids (Chapter 10); Ca^{2+} ATPase (Chapter 11); rhodopsin, transducin, and protein kinase A (Chapter 15); and myosin ATPase (Chapter 17)
- Latest advances in light and electron microscopy, including cryoelectron tomography (Chapter 9)
- Reactive oxygen species (ROS) (Chapter 12)
- Role of supercomplexes in electron transport (Chapter 12)
- Human epidermal growth factor receptors (HERs) and treatment of cancer (Chapter 16)
- Myosin ATPase cycle (Chapter 17)
- Kinesin-1 ATPase cycle (Chapter 18)
- Use of retrovirus infection for tracing cell lineage (Chapter 21)
- Axon guidance molecules (Chapter 23)
- Somatic gene rearrangement in immune cells (Chapter 24)
- Cancer stem cells (Chapter 25)
- Use of DNA microarray analysis in tumor typing (Chapter 25)

Increased Clarity and Accessibility

In writing the sixth edition we have taken a step back and asked what we, as experienced teachers of both undergraduate and graduate students, can do to make material that can be daunting for students more approachable. We have tried to balance our coverage of the latest scientific developments with students' need for clear explanations of the fundamentals, and we have intensified our commitment to making both the concepts and experiments more accessible.

In particular, we have improved the **chapter overviews,** making sure that they give students the big picture and provide a solid foundation for what is to come. We have also simplified the text and headings, removing jargon wherever possible. Clear **concept-oriented headings** now provide an easy-to-read roadmap of each chapter.

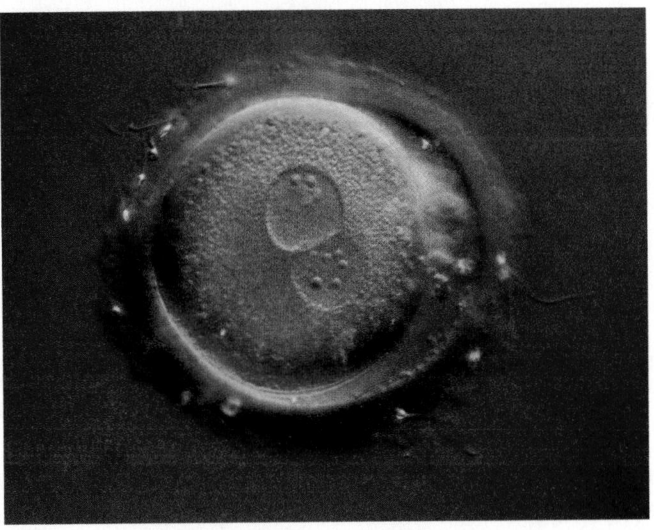

Fertilized human egg cell showing male and female pronuclei. [Courtesy of The Lennart Nilson Award Board.]

New Organization

In consultation with more than 100 professors of cell biology from around the country, we have revised the table of contents to unite related material and mirror the organization of most courses. We also streamlined and focused coverage of many topics by reorganizing several chapters, enabling us to make room for the addition of new material. Each chapter has an up-to-date list of references of landmark studies and comprehensive review articles that direct students and instructors to much additional information.

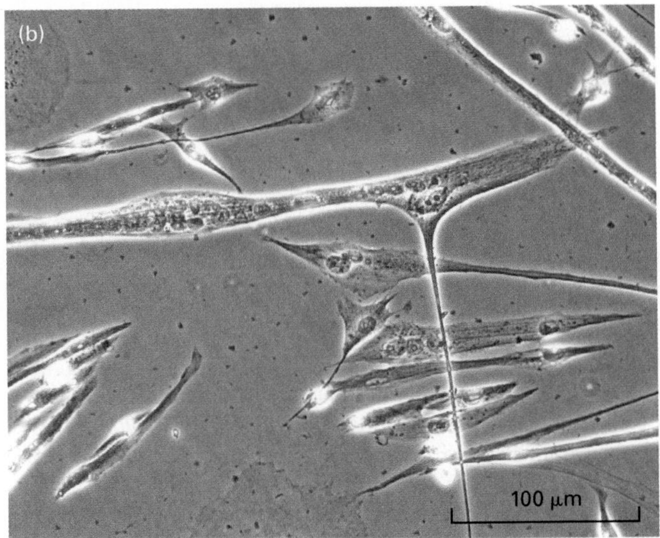

Experimental Figure 9-30 Differentiation in culture of primary mouse satellite (myoblast) cells into muscle cells. [Courtesy of C. Emerson and J. Chen, Boston Biomedical Research Institute.]

New Chapters

The sixth edition of *Molecular Cell Biology* includes three new chapters:

- "The Molecular Cell Biology of Development" (Chapter 22) presents the fundamentals of development from fertilization through pattern formation, with increased focus on mammalian development.
- "Nerve Cells" (Chapter 23) provides unified coverage of the cell biology of these specialized cells, including the latest advances in axonal path finding and sensory reception.
- "Immunology" (Chapter 24) applies principles of molecular cell biology to the vertebrate immune system.

Medical Relevance

Many advances in basic cellular and molecular biology have led to new treatments for cancer and other significant human diseases. These medical examples are woven throughout the chapters where appropriate to give students an appreciation for the clinical applications of the basic science they are learning. Many of these applications hinge on a detailed understanding of multiprotein complexes in cells–complexes that catalyze cell movements, regulate DNA transcription, coordinate metabolism, and connect cells to other cells and to proteins and carbohydrates in their extracellular environment. A complete list of the medical examples in the sixth edition can be found overleaf.

CLINICAL APPLICATIONS

This icon signals the start of a clinical application in the text. Additional briefer clinical correlations appear in the text as appropriate:

MEDIA AND SUPPLEMENTS

For Students

Companion Web Site
www.whfreeman.com/lodish6e

- **NEW: Podcasts** narrated by the authors give students a deeper understanding of key figures in the text and a sense of the thrill of discovery.

- **NEW:** Now available for your MP3 player or personal computer, more than 125 **animations and research videos** show the dynamic nature of key cellular processes and important experimental techniques. The animations were storyboarded by the textbook authors in conjunction with BioStudio, Inc., and programmed by Sumanas, Inc.

- **Classic Experiment** essays focus on classic groundbreaking experiments and explore the investigative process.

- **Online Quizzing** is provided, including multiple-choice and short answer questions.

Student Solutions Manual (ISBN:1-4292-0127-4), written by Brian Storrie, Eric A. Wong, Richard Walker, Glenda Gillaspy, and Jill Sible of Virginia Polytechnic Institute and State University and updated by Cindy Klevickis of James Madison University and Greg M. Kelly of the University of Western Ontario, contains complete worked-out solutions to all the end-of-chapter problems in the textbook.

NEW: eBook (ISBN: 1-4292-0955-0) New to the sixth edition, this customizable eBook fully integrates the complete contents of the text and its interactive media in a format that features a variety of helpful study tools, including full-text searching, note-taking, bookmarking, highlighting, and more. Easily accessible on any Internet-connected computer via a standard Web browser, the eBook enables students to take an active approach to their learning in an intuitive, easy-to-use format. Visit **http://ebooks.bfwpub.com** to learn more.

For Instructors

Companion Web Site
www.whfreeman.com/lodish6e

All the student resources, plus:

- All figures and tables from the book in **jpeg and layered PowerPoint** formats, which instructors can edit or project section by section, allowing students to follow underlying concepts. Optimized for lecture-hall presentation, including enhanced colors, enlarged labels, and boldface type.

- **Test Bank** in editable Microsoft Word format now featuring *new and revised questions* for every chapter. The test bank is written by Brian Storrie of the University of Arkansas for Medical Sciences and Eric A. Wong, Richard Walker, Glenda Gillaspy, and Jill Sible of Virginia Polytechnic Institute and State University and revised by Cindy Klevickis of James Madison University and Greg M. Kelly of the University of Ontario.

- Additional **Analyze the Data** problems are available in PDF format.

- **NEW:** Lecture-ready **Personal Response System "clicker" questions** are available as Microsoft Word files and Microsoft PowerPoint slides.

Instructor's Resource CD-ROM (ISBN: 1-4292-0126-6) includes all the instructor's resources from the Web site, including all the illustrations from the text, animations, videos, test bank files, clicker questions, and the solutions manual files.

Overhead Transparency Set (ISBN: 1-4292-0477-X) contains 250 key illustrations from the text, optimized for lecture-hall presentation.

eBook
http://ebooks.bfwpub.com

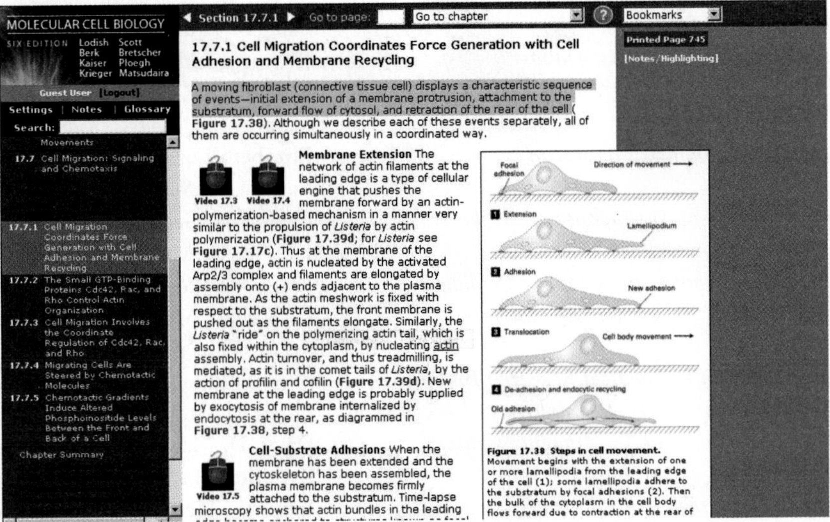

ACKNOWLEDGMENTS

In updating, revising and rewriting this book, we were given invaluable help by many colleagues. We thank the following people who generously gave of their time and expertise by making contributions to specific chapters in their areas of interest, providing us with detailed information about their courses, or by reading and commenting on one or more chapters:

Steven Ackerman, *University of Massachusetts, Boston*
Richard Adler, *University of Michigan, Dearborn*
Karen Aguirre, *Coastal Carolina University*
Jeff Bachant, *University of California, Riverside*
Kenneth Balazovich, *University of Michigan*
Ben A. Barres, *Stanford University*
Karen K. Bernd, *Davidson College*
Sanford Bernstein, *San Diego State University*
Doug Black, *Howard Hughes Medical Institute and University of California, Los Angeles*
Richard L. Blanton, *North Carolina State University*
Justin Blau, *New York University*
Steven Block, *Stanford University*
Jonathan E. Boyson, *University of Vermont*
Janet Braam, *Rice University*
Roger Bradley, *Montana State University*
William S. Bradshaw, *Brigham Young University*
Gregory G. Brown, *McGill University*
William J. Brown, *Cornell University*
Max M. Burger, *Friedrich Miescher Institute for Biomedical Research, Basel, Switzerland*
David Burgess, *Boston College*
Robin K. Cameron, *McMaster University*
W. Zacheus Cande, *University of California, Berkeley*
Steven A. Carr, *Broad Institute of Harvard University and Massachusetts Institute of Technology*
Alice Y. Cheung, *University of Massachusetts, Amherst*
Dennis O. Clegg, *University of California, Santa Barbara*
Paul Clifton, *Utah State University*
Randy W. Cohen, *California State University, Northridge*
Richard Dickerson, *University of California, Los Angeles*
Patrick J. DiMario, *Louisiana State University*
Santosh R. D'Mello, *University of Texas, Dallas*
Chris Doe, *HHMI and University of Oregon*
Robert S. Dotson, *Tulane University*
William Dowhan, *University of Texas–Houston Medical School*
Gerald B. Downes, *University of Massachusetts, Amherst*
Erastus C. Dudley, *Huntingdon College*
Susan Dutcher, *Washington University School of Medicine*
Matt Elrod-Erickson, *Middle Tennessee State University*
Susan Ely, *Cornell University*
Charles P. Emerson Jr., *Boston Biomedical Research Institute*
Irene M. Evans, *Rochester Institute of Technology*
James G. Evans, *Whitehead Institute Bio Imaging Center, Massachusetts Institute of Technology*
Marilyn Gist Farquhar, *University of California, San Diego*
Xavier Fernandez-Busquets, *Bioengineering Institute of Catalonia, Universitat de Barcelona, Spain*
Terrence G. Frey, *San Diego State University*
Margaret T. Fuller, *Stanford University School of Medicine*
Kendra J. Golden, *Whitman College*
David S. Goldfarb, *Rochester University*
Martha J. Grossel, *Connecticut College*
Lawrence I. Grossman, *Wayne State University School of Medicine*
Michael Grunstein, *University of California, Los Angeles, School of Medicine*
Barry M. Gumbiner, *University of Virginia*
Wei Guo, *University of Pennsylvania*
Leah Haimo, *University of California, Riverside*
Heidi E. Hamm, *Vanderbilt University Medical School*
Craig M. Hart, *Louisiana State University*
Merill B. Hille, *University of Washington*
Jerry E. Honts, *Drake University*
H. Robert Horvitz, *Massachusetts Institute of Technology*
Richard Hynes, *Massachusetts Institute of Technology and Howard Hughes Medical Institute*
Harry Itagaki, *Kenyon College*
Elizabeth R. Jamieson, *Smith College*
Marie A. Janicke, *State University of New York, Buffalo*
Bradley W. Jones, *University of Mississippi*
Mark Kainz, *Colgate University*
Naohiro Kato, *Louisiana State University*
Amy E. Keating, *Massachusetts Institute of Technology*
Charles H. Keith, *University of Georgia*
Thomas C. S. Keller III, *Florida State University*
Greg M. Kelly, *University of Western Ontario*
Stephen Kendall, *California State University, Fullerton*
Felipe Kierszenbaum, *Michigan State University*
Cindy Klevickis, *James Madison University*
Brian Kobilka, *Stanford University Medical School.*
Martina Koniger, *Wellesley University*
Catherine Koo, *Caldwell College*
Keith G. Kozminski, *University of Virginia*
Steven W. L'Hernault, *Emory University*
Douglas Lauffenburger, *Massachusetts Institute of Technology*
Robert J. Lefkowitz, *HHMI and Duke University Medical School*
R. L. Levine, *McGill University*
Fang Ju Lin, *Coastal Carolina University*
Elizabeth Lord, *University of California, Riverside*
Liqun Luo, *Stanford University*
Grant MacGregor, *University of California, Irvine*
Jennifer O. Manilay, *University of California, Merced*
Barry Margulies, *Towson University*
C. William McCurdy, *University of California, Davis, and Lawrence Berkeley National Laboratory*
Dennis W. McGee, *State University of New York, Binghamton*
James McGrath, *Rochester School of Medicine*
David D. McKemy, *University of Southern California*
Roderick MacKinnon, *Rockefeller University*
James A. McNew, *Rice University*

Stephanie Mel, *University of California, San Diego*
Ljiljana Milenkovic, *HHMI and Stanford University School of Medicine*
Elizabeth Miller, *Columbia University*
Pamela Mitchell, *Pennsylvania State University*
Vamsi Mootha, *Harvard Medical School*
Ben Murray, *California State University at Fullerton*
Byron K. Murray, *Brigham Young University*
Phillip Newmark, *University of Illinois, Urbana-Champaign*
Alan Nighorn, *University of Arizona*
James M. Ntambi, *University of Wisconsin, Madison*
Roel Nusse, *Stanford University*
David M. Ojcius, *University of California, Merced*
Anthony Oro, *Stanford University*
Nipam H. Patel, *HHMI and University of California, Berkeley*
Francesca Pignoni, *Harvard Medical School*
Dominic Poccia, *Amherst College*
Martin Privalsky, *University of California, Davis*
Kirsten Prüfer, *Louisiana State University*
David Pruyne, *Cornell University*
Tal Bachar Raveh, *Stanford University*
Mary K. Ritke, *University of Indianapolis*
Rajat Rohatgi, *Stanford University School of Medicine*
Robert D. Rosenberg, *Massachusetts Institute of Technology*
Anne G. Rosenwald, *Georgetown University*
Lorraine C. Santy, *Pennsylvania State University*
Jean Schaefer, *Washington University School of Medicine*
David Schneider, *Stanford University*
David A. Scicchitano, *New York University*
Donald Seto, *George Mason University*
Diane Shakes, *College of William & Mary*
Marlene Shaw, *University of Southern Indiana*
Eric A. Shelden, *Washington State University*
Morgan Sheng, *Massachusetts Institute of Technology*
Louis A. Sherman, *Purdue University*
Yan-Ting Elizabeth Shiu, *University of Utah*
Anu Singh-Cundy, *Western Washington University*
Frances M. Sladek, *University of California, Riverside*
Stephen T. Smale, *University of California, Los Angeles*
Gregory S. Smith, *Saint Louis University School of Medicine*
Susan Spencer, *Saint Louis University*
Veronica Stellmach, *Benedictine University*
Christine Sütterlin, *University of California, Irvine*
Salme Taagepera, *University of California, Irvine*
Susan S. Taylor, *HHMI and University of California, San Diego*
Paul Teesdale-Spittle, *Victoria University of Wellington, Wellington, New Zealand*
Robert M. Tombes, *Virginia Commonwealth University*
Vincent Tropepe, *University of Toronto*
Elizabeth Vallen, *Swarthmore College*
Amitabh Varshney, *University of Maryland*
Robert G. Van Buskirk, *State University of New York, Binghamton*
Volker V. Vogt, *Cornell University*
Claire Walczak, *Indiana University*
Angelika Amon, *HHMI and Massachusetts Institute of Technology*
Beverly Wendland, *Johns Hopkins University*
Gary M. Wessel, *Brown University*
Ding Xue, *University of Colorado, Boulder*
Michael B. Yaffe, *Massachusetts Institute of Technology*
P. Renee Yew, *The University of Texas Health Center at San Antonio*

We would also like to express our gratitude and appreciation to Leah Haimo of the University of California, Riverside, for her development of new Analyze the Data problems, to Cindy Klevickis of James Madison University and Greg M. Kelly of the University of Ontario for their authorship of excellent new Review the Concepts problems and Test Bank questions, and to Jill Sible of Virginia Polytechnic Institute and State University for her revision of the Online Quizzing problems. We are also grateful to Lisa Rezende of the University of Arizona for her development of the Classic Experiments and Podcasts.

This edition would not have been possible without the careful and committed collaboration of our publishing partners at W. H. Freeman and Company. We thank Kate Ahr, Mary Louise Byrd, Debbie Clare, Marsha Cohen, Blake Logan, Christina Micek, Bill O'Neal, Ruth Steyn, Karen Taschek, Sara Tenney, Hannah Thonet, Susan Timmins, Nick Tymoczko, Cecilia Varas, and Susan Wein for their labor and for their willingness to work overtime to produce a book that excels in every way.

In particular, we would like to acknowledge the talent and commitment of our text editors, Matthew Tontonoz, Erica Pantages Frost, and Elizabeth Rice. They are remarkable editors. Thank you for all you've done in this edition.

We are also indebted to H. Adam Steinberg for his pedagogical insight and his development of beautiful molecular models and Erica Beade of MBC Graphics (www.MBCGraphics.com) for her work in developing and enhancing the art program.

We would like to acknowledge those whose direct contributions to the fifth edition continue to influence in this edition; especially Sonia DiVittorio and Ruth Steyn.

Thanks to our own staff: Sally Bittancourt, Diane Bush, Mary Anne Donovan, Carol Eng, James Evans, George Kokkinogenis, Julie Knight, Guicky Waller, Nicki Watson, Rob Welsh, and members of the Scott laboratory.

Finally, special thanks to our families for inspiring us and for granting us the time it takes to work on such a book and to our mentors and advisers for encouraging us in our studies and teaching us much of what we know: *(Harvey Lodish)* my wife, Pamela; my children and grandchildren Heidi and Eric Steinert and Emma and Andrew Steinert; Martin Lodish, Kristin Schardt, and Sophia, Joshua, and Tobias Lodish; and Stephanie Lodish, Bruce Peabody, and Isaac and Violet Peabody; mentors Norton Zinder and Sydney Brenner; and also David Baltimore and Jim Darnell for collaborating on the first editions of this book; *(Arnold Berk)* my wife Sally, Jerry Berk, Shirley Berk, Angelina Smith, David Clayton, and Phil Sharp; *(Chris A. Kaiser)* my wife Kathy O'Neill; *(Monty Krieger)* Nancy Krieger, I. Jay Krieger, Mildred Krieger, Jonathan Krieger, and Joshua Krieger; (*Matthew P. Scott*) Margaret (Minx) Fuller, Lincoln Scott, Julia Scott, Peter Scott, Duscha Weisskopf, and adviser Mary Lou Pardue; *(Anthony Bretscher)* my wife Janice and daughters Heidi and Erika, and advisers A. Dale Kaiser and Klaus Weber; *(Hidde Ploegh)* my wife Anne Mahon.

CONTENTS IN BRIEF

CONTENTS

Part I Chemical and Molecular Foundations

Part III Cell Structure and Function

Part IV Cell Growth and Development

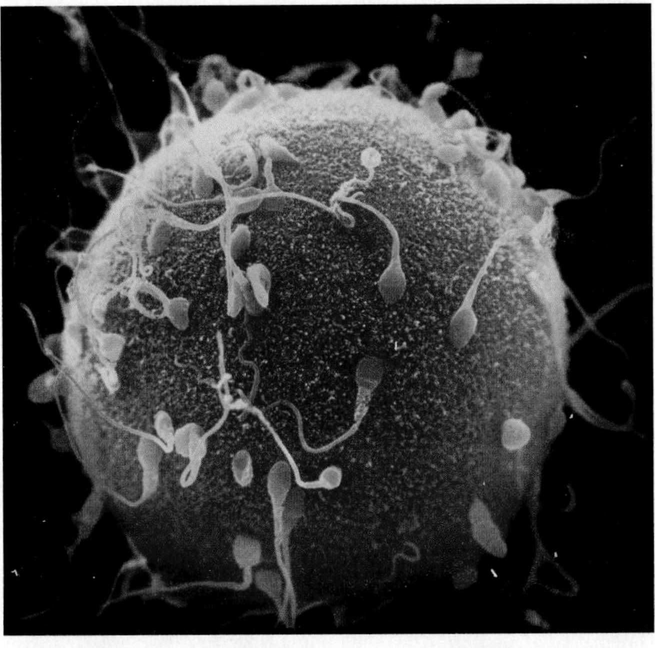

A single ~200 micrometer (μm) cell, the human egg, with sperm, which are also single cells. From the union of an egg and sperm will arise the 10 trillion cells of a human body. [Photo Researchers, Inc.]

CHAPTER

1 LIFE BEGINS WITH CELLS

Like ourselves, the individual cells that form our bodies can grow, reproduce, process information, respond to stimuli, and carry out an amazing array of chemical reactions. These abilities define life. We and other multicellular organisms contain billions or trillions of cells organized into complex structures, but many organisms consist of a single cell. Even simple unicellular organisms exhibit all the hallmark properties of life, indicating that the cell is the fundamental unit of life. As the twenty-first century opens, we face an explosion of new data about the components of cells, what structures they contain, how they touch and influence each other. Still, an immense amount remains to be learned, particularly about how information flows through cells and how they decide on the most appropriate ways to respond.

Molecular cell biology is a rich, integrative science that brings together biochemistry, biophysics, molecular biology, microscopy, genetics, physiology, computer science, and developmental biology. Each of these fields has its own emphasis and style of experimentation. In the following chapters, we will describe insights and experimental approaches drawn from all of these fields, gradually weaving the multifaceted story of the birth, life, and death of cells. We start in this prologue chapter by introducing the diversity of cells, their basic constituents and critical functions, and what we can learn from the various ways of studying cells.

1.1 The Diversity and Commonality of Cells

Cells come in an amazing variety of sizes and shapes (Figure 1-1). Some move rapidly and have fast-changing structures, as we can see in movies of amebas and rotifers. Others are largely stationary and structurally stable. Oxygen kills some cells but is an absolute requirement for others. Most cells in multicellular organisms are intimately involved with other cells. Although some unicellular organisms live in isolation, others form colonies or live in close association with other types of organisms, as do the bacteria that help plants to extract nitrogen from the air or the bacteria that live in our intestines and help us digest food. Despite these and numerous other differences, all cells share certain structural features and carry out many complicated processes in basically the same way. As the story of cells unfolds throughout this book, we will focus on the molecular basis of both the differences and similarities in the structure and function of various cells.

All Cells Are Prokaryotic or Eukaryotic

The biological universe consists of two types of cells—prokaryotic and eukaryotic. Eukaryotes include four kingdoms:

OUTLINE

▲ **FIGURE 1-1 Cells come in an astounding assortment of shapes and sizes.** Some of the morphological variety of cells is illustrated in these photographs. In addition to morphology, cells differ in their ability to move, internal organization (prokaryotic versus eukaryotic cells), and metabolic activities. (a) Eubacteria; note dividing cells. These are *Lactococcus lactis,* which are used to produce cheese such as Roquefort, Brie, and Camembert. (b) A mass of archaebacteria (*Methanosarcina*) that produce their energy by converting carbon dioxide and hydrogen gas to methane. Some species that live in the rumen of cattle give rise to >150 liters of methane gas/day. (c) Blood cells, shown in false color. The red blood cells are oxygen-bearing erythrocytes, the white blood cells (leukocytes) are part of the immune system and fight infection, and the green cells are platelets that provide substances to make blood clot at a wound. (d) Large single cells: fossilized dinosaur eggs. (e) A colonial single-celled green alga, *Volvox aureus.* The large spheres are made up of many individual cells, visible as blue or green dots. The yellow masses inside are daughter colonies, each made up of many cells. (f) A single Purkinje neuron of the cerebellum, an incredibly large cell that can form more than 100,000 connections with other cells through the branched network of dendrites. The cell was made visible by introduction of a fluorescent protein; the cell body is the bulb at the bottom. (g) Cells can form an epithelial sheet, as in the slice through intestine shown here. Each finger-like tower of cells, a villus, contains many cells in a continuous sheet. Nutrients are transferred from digested food through the epithelial sheet to the blood for transport to other parts of the body. New cells form continuously near the bases of the villi, and old cells are shed from the top. (h) Plant cells are fixed firmly in place in vascular plants, supported by a rigid cellulose skeleton. Spaces between the cells are joined into tubes for transport of water and food. [Part (a) Gary Gaugler/ Photo Researchers, Inc. Part (b) Ralph Robinson/ Visuals Unlimited, Inc. Part (c) NIH/Photo Researchers, Inc. Part (d) John D. Cunningham/Visuals Unlimited, Inc. Part (e) Carolina Biological/Visuals Unlimited, Inc. Part (f) Helen M. Blau, Stanford University. Part (g) Jeff Gordon, Washington University School of Medicine. Part (h) Richard Kessel and C. Shih/Visuals Unlimited, Inc.]

plants, animals, fungi, and protists. Prokaryotes include bacteria and archaea. Prokaryotic cells consist of a single closed compartment that is surrounded by the **plasma membrane,** lacks a defined **nucleus,** and has a relatively simple internal organization (Figure 1-2a). All **prokaryotes** consist of cells of this type. Bacteria, a numerous type of prokaryote, are single-celled organisms; the cyanobacteria, or blue-green algae, can be unicellular or be filamentous chains of cells. Although bacterial cells do not have membrane-bounded compartments, many proteins are precisely localized in their aqueous interior, or **cytosol,** indicating the presence of internal organization.

Although prokaryotes are small individually, they make up a huge part of the earth's biomass. A single *Escherichia coli* bacterium has a dry weight of about 25×10^{-14} g, yet bacteria account for an estimated 1–1.5 kg of the average human's weight. This represents more than 4×10^{17} individual organisms. The estimated number of bacteria on earth is 5×10^{30}, weighing a total of about 10^{12} kg. Prokaryotic cells have been found 7 miles deep in the ocean

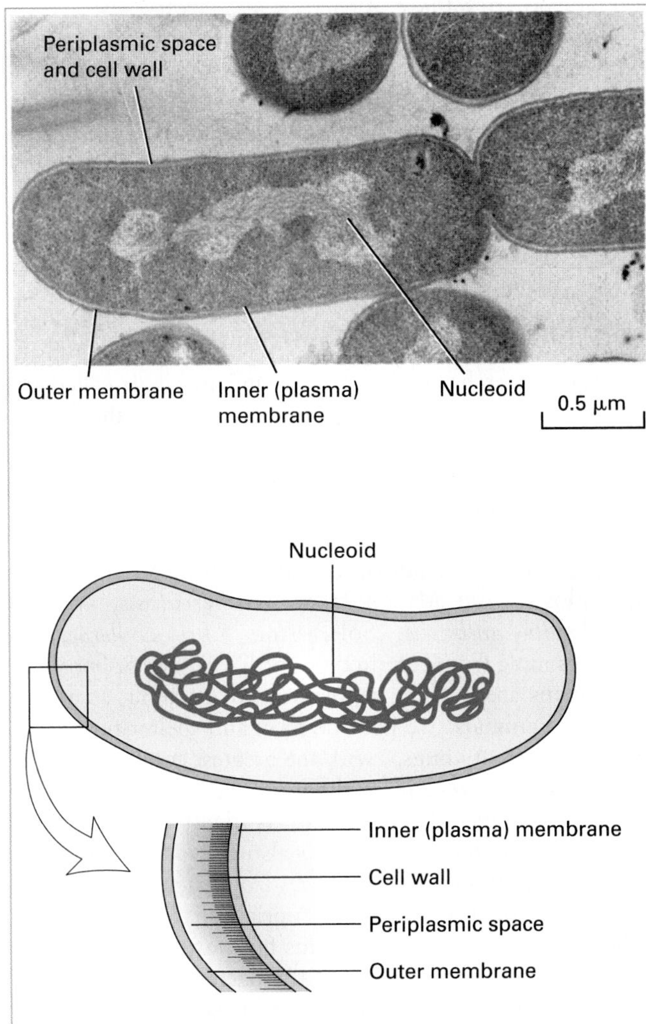

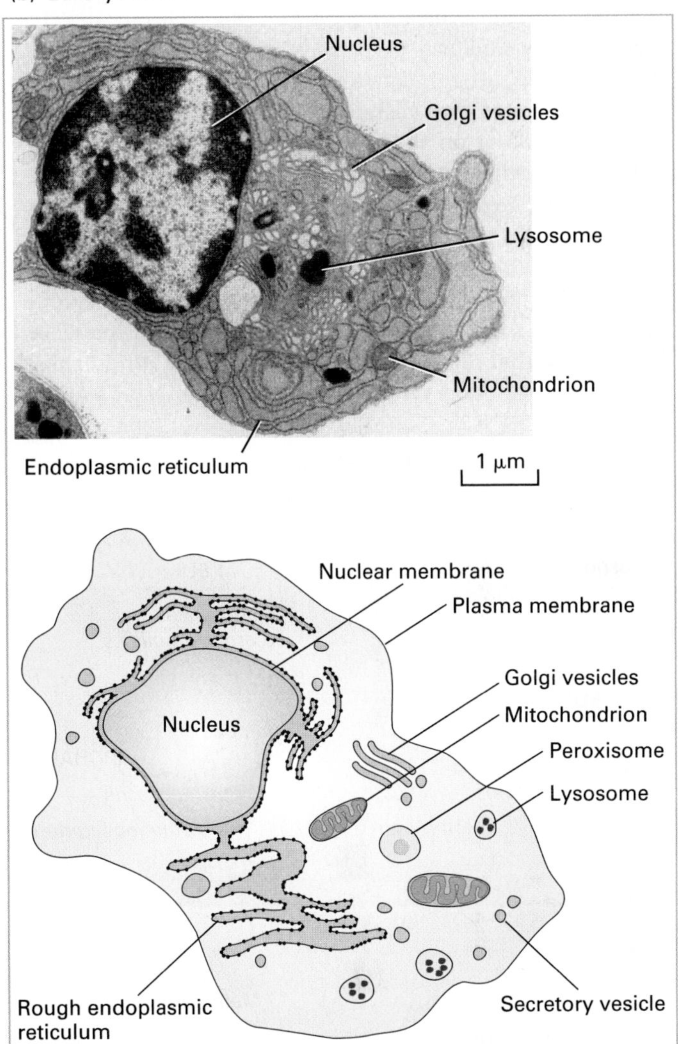

▲ **FIGURE 1-2 Prokaryotic cells have a simpler internal organization than eukaryotic cells.** (a) Electron micrograph of a thin section of *Escherichia coli,* a common intestinal bacterium. The nucleoid, consisting of the bacterial DNA, is not enclosed within a membrane. *E. coli* and some other bacteria are surrounded by two membranes separated by the periplasmic space. The thin cell wall is adjacent to the inner membrane. (b) Electron micrograph of a plasma cell, a type of white blood cell that secretes antibodies. Only a single membrane (the plasma membrane) surrounds the cell, but the interior contains many membrane-limited compartments, or organelles. The defining characteristic of eukaryotic cells is segregation of the cellular DNA within a defined nucleus, which is bounded by a double membrane. The outer nuclear membrane is continuous with the rough endoplasmic reticulum, a factory for assembling proteins. Golgi vesicles process and modify proteins, mitochondria generate energy, lysosomes digest cell materials to recycle them, peroxisomes process molecules using oxygen, and secretory vesicles carry cell materials to the surface to release them. [Part (a) courtesy of I. D. J. Burdett and R. G. E. Murray. Part (b) from P. C. Cross and K. L. Mercer, 1993, *Cell and Tissue Ultrastructure: A Functional Perspective,* W. H. Freeman and Company.]

and 40 miles up in the atmosphere; they are quite adaptable! The carbon stored in bacteria is nearly as much as the carbon stored in plants.

Eukaryotic cells, unlike prokaryotic cells, contain a defined membrane-bound nucleus and extensive internal membranes that enclose other compartments called **organelles** (Figure 1-2b). The region of the cell lying between the plasma membrane and the nucleus is the **cytoplasm,** comprising the cytosol (aqueous phase) and the organelles. **Eukaryotes** comprise all members of the plant and animal kingdoms, including the fungi, which exist in both multicellular forms (molds) and unicellular forms (yeasts), and the protozoans (*proto,* primitive; *zoan,* animal), which are exclusively unicellular. Eukaryotic cells are commonly about 10–100 μm across, generally much larger than bacteria. A typical human fibroblast, a connective tissue cell, might be about 15 μm across with a volume and dry weight some thousands of times those of an *E. coli* bacterial cell. An ameba, a single-celled protozoan, can be more than 0.5 mm long. An ostrich egg begins as a single cell that is even larger and is easily visible to the naked eye.

All cells are thought to have evolved from a common progenitor because the structures and molecules in all cells have so many similarities. In recent years, detailed analysis of the DNA sequences from a variety of prokaryotic organisms has revealed two distinct types: the **bacteria** and the **archaea.** Working on the assumption that organisms with more similar genes evolved from a common progenitor more recently than those with more dissimilar genes, researchers have developed the evolutionary lineage tree shown in Figure 1-3. According to this tree, the archaea and the eukaryotes diverged from bacteria billions of years ago before they diverged from each other. In addition to DNA sequence distinctions that define the three groups of organisms, archaea cell membranes have chemical properties that differ dramatically from those of bacteria and eukaryotes.

Many archaeans grow in unusual, often extreme, environments that may resemble the ancient conditions that existed when life first appeared on earth. For instance, halophiles ("salt loving") require high concentrations of salt to survive, and thermoacidophiles ("heat and acid loving") grow in hot (80° C) sulfur springs, where a pH of less than 2 is common. Still other archaeans live in oxygen-free milieus and generate methane (CH_4) by combining water with carbon dioxide.

Animals
Plants
Fungi
Ciliates
Euglena
Microsporidia
EUKARYOTA
Slime molds
Diplomonads (Giardia lamblia)
EUBACTERIA
E. coli
Sulfolobus
ARCHAEA
B. subtilus
Thermococcus
Thermotoga
Methanobacterium
Halococcus
Flavobacteria
Green sulfur bacteria
Halobacterium
Borrelia burgdorferi
Methanococcus jannaschii
Presumed common progenitor of all extant organisms
Presumed common progenitor of archaebacteria and eukaryotes

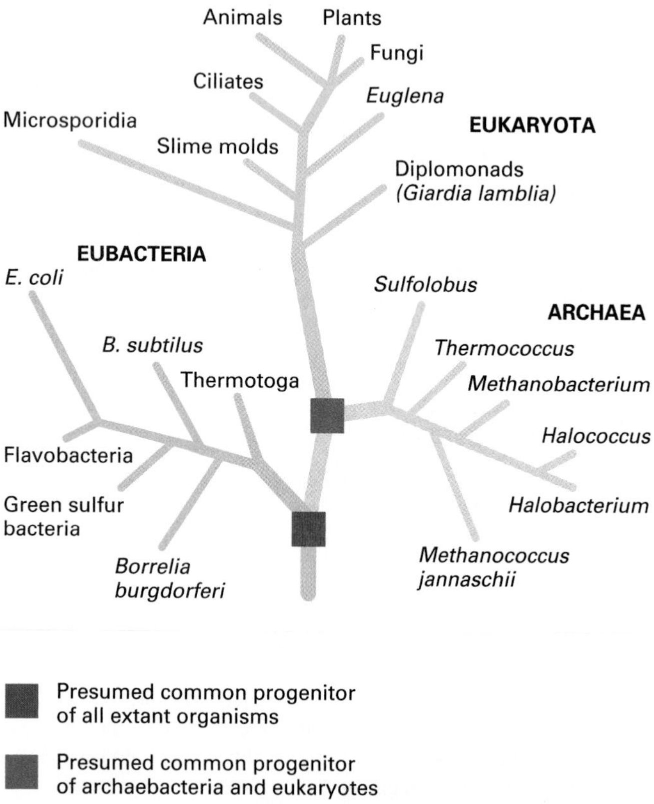

▲ **FIGURE 1-3 All organisms from simple bacteria to complex mammals probably evolved from a common, single-celled progenitor.** This family tree depicts the evolutionary relations among the three major lineages of organisms. The structure of the tree was initially ascertained from morphological criteria: creatures that look alike were put close together. More recently the sequences of DNA and proteins have been examined as a more information-rich criterion for assigning relationships. The greater the similarities in these macromolecular sequences, the more closely related organisms are thought to be. The trees based on morphological comparisons and the fossil record generally agree well with those based on molecular data. Although all organisms in the eubacterial and archaean lineages are prokaryotes, archaea are more similar to eukaryotes than to eubacteria ("true" bacteria) in some respects. For instance, archaean and eukaryotic genomes encode homologous histone proteins, which associate with DNA; in contrast, bacteria lack histones. Likewise, the RNA and protein components of archaean ribosomes are more like those in eukaryotes than those in bacteria.

Unicellular Organisms Help and Hurt Us

Bacteria and archaea, the most abundant single-celled organisms, are commonly 1–2 μm in size. Despite their small size and simple architecture, they are remarkable biochemical factories, converting simple chemicals into complex biological molecules. Bacteria are critical to the earth's ecology, but some cause major diseases: bubonic plague (Black Death) from *Yersinia pestis,* strep throat from *Streptomyces,* tuberculosis from *Mycobacterium tuberculosis,* anthrax from *Bacillus anthracis*, cholera from *Vibrio cholerae,* and food poisoning from certain types of *E. coli* and *Salmonella.*

Humans are walking repositories of bacteria, as are all plants and animals. We provide food and shelter for a staggering number of "bugs," with the greatest concentration in our intestines. In return for the food and shelter that allow them to reproduce, bacteria help us digest our food. One common gut bacterium, *E. coli,* is also a favorite experimental organism. In response to signals from bacteria such as *E. coli,* the intestinal cells form appropriate shapes to provide a niche where bacteria can live, thus facilitating proper digestion by the combined efforts of the bacterial and the intestinal cells. Conversely, exposure to intestinal cells changes the properties of the bacteria so that they participate more effectively in human digestion. Such communication and response is a common feature of cells.

The normal, peaceful mutualism of humans and bacteria is sometimes violated by one or both parties. When bacteria begin to grow where they are dangerous to us (e.g., in the bloodstream or in a wound), the cells of our immune system fight back, neutralizing or devouring the intruders. Powerful antibiotic medicines, which selectively poison prokaryotic cells, provide rapid assistance to our relatively slow-developing immune response. Understanding the molecular biology of bacterial cells leads to an understanding of how bacteria are normally poisoned by antibiotics, how they become resistant to antibiotics, and what processes or structures present in bacterial but not human cells might be usefully targeted by new drugs.

Like bacteria, protozoa are usually beneficial members of the food chain. They play key roles in the fertility of soil, controlling bacterial populations and excreting nitrogenous and phosphate compounds, and are key players in waste treatment systems—both natural and man-made. These unicellular eukaryotes are also critical parts of marine ecosystems, consuming large quantities of phytoplankton and harboring photosynthetic algae, which use sunlight to produce biologically useful energy forms and small fuel molecules.

Video: Plasmodium Sporozoite Entering and Exiting a Liver Cell

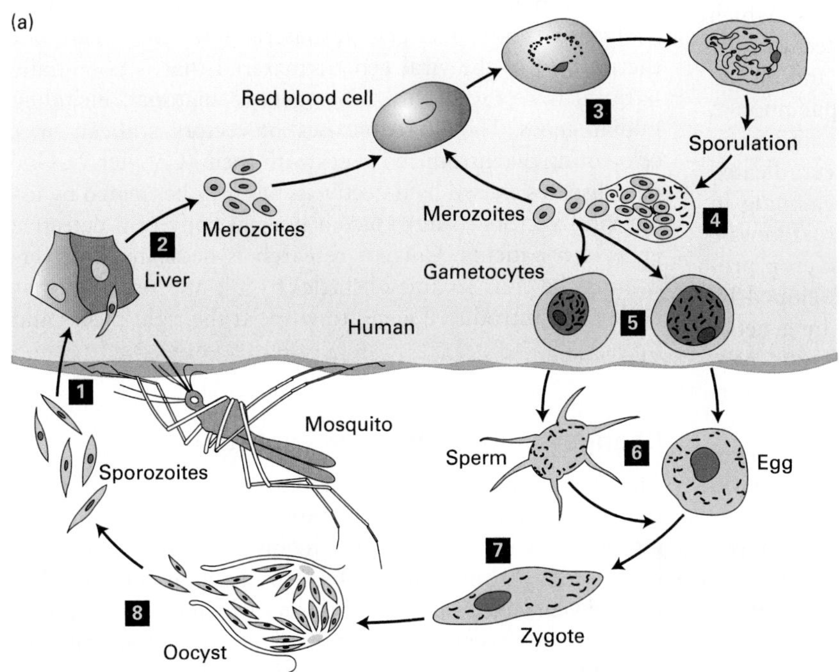

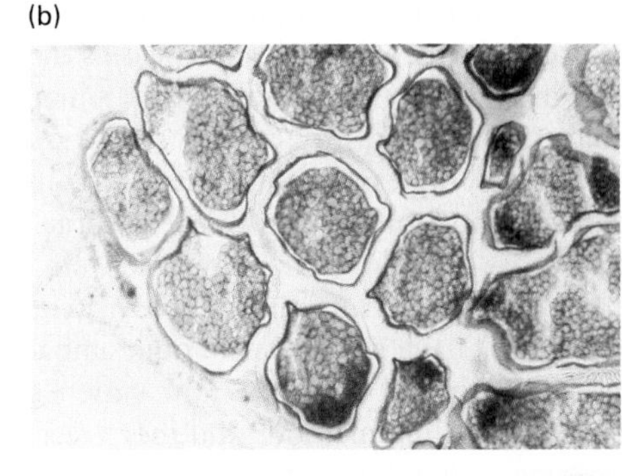

▲ **FIGURE 1-4 *Plasmodium* organisms, the parasites that cause malaria, are single-celled protozoans with a remarkable life cycle.** Many *Plasmodium* species are known, and they can infect a variety of animals, cycling between insect and vertebrate hosts. The four species that cause malaria in humans undergo several dramatic transformations within their human and mosquito hosts. (a) Diagram of the life cycle. Sporozoites enter a human host when an infected *Anopheles* mosquito bites a person **1**. They migrate to the liver, where they develop into merozoites, which are released into the blood **2**. Merozoites differ substantially from sporozoites, so this transformation is a metamorphosis (Greek, "to transform" or "many shapes"). Circulating merozoites invade red blood cells (RBCs) and reproduce within them **3**. Proteins produced by some *Plasmodium* species move to the surface of infected RBCs, causing the cells to adhere to the walls of blood vessels. This prevents infected RBCs from circulating to the spleen, where cells of the immune system would destroy the RBCs and the *Plasmodium* organisms they harbor. After growing and reproducing in RBCs for a period of time characteristic of each *Plasmodium* species, the merozoites suddenly burst forth in synchrony from large numbers of infected cells **4**. It is this event that brings on the fevers and shaking chills that are the well-known symptoms of malaria. Some of the released merozoites infect additional RBCs, creating a cycle of production and infection. Eventually, some merozoites develop into male and female gametocytes **5**, another metamorphosis. These cells, which contain half the usual number of chromosomes, cannot survive for long unless they are transferred in blood to an *Anopheles* mosquito. In the mosquito's stomach, the gametocytes are transformed into sperm or eggs (gametes), yet another metamorphosis marked by development of long hairlike flagella on the sperm **6**. Fusion of sperm and eggs generates zygotes **7**, which implant into the cells of the stomach wall and grow into oocysts, essentially factories for producing sporozoites. Rupture of an oocyst releases thousands of sporozoites **8**; these migrate to the salivary glands, setting the stage for infection of another human host. (b) Scanning electron micrograph of mature oocysts and emerging sporozoites. Oocysts abut the external surface of stomach wall cells and are encased within a membrane that protects them from the host immune system. [Part (b) courtesy of R. E. Sinden.]

However, some protozoa do give us grief: *Entamoeba histolytica* causes dysentery; *Trichomonas vaginalis*, vaginitis; and *Trypanosoma brucei*, sleeping sickness. Each year the deadliest of the protozoa, *Plasmodium falciparum* and related species, is the cause of more than 300 million new cases of malaria, a disease that kills 1.5 to 3 million people annually. These protozoans inhabit mammals and mosquitoes alternately, changing their morphology and behavior in response to signals in each of these environments. They also recognize receptors on the surfaces of the cells they infect. The complex life cycle of *Plasmodium* dramatically illustrates how a single cell can adapt to each new challenge it encounters (Figure 1-4). All of the transformations in cell type that occur during the *Plasmodium* life cycle are governed by instructions encoded in the genetic material of this parasite and triggered by environmental inputs.

The other group of single-celled eukaryotes, the yeasts, also have their good and bad points with respect to humans, as do their multicellular cousins, the molds. Yeasts and molds, which collectively constitute the fungi, have an important ecological role in breaking down plant and animal remains for reuse. They also make numerous antibiotics and are used in the manufacture of bread, beer, wine, and cheese. Not so pleasant are fungal diseases, which range from relatively innocuous skin infections such as jock itch and athlete's foot to life-threatening *Pneumocystis carinii* pneumonia, a common cause of death among AIDS patients.

Viruses Are the Ultimate Parasites

Not all microscopic pathogens are cells. The other most familiar disease-causing organisms are the **viruses,** which make use of the machinery inside the cells they infect to copy themselves. Virus-caused diseases are numerous and all too familiar: chickenpox, influenza, some types of pneumonia, polio, measles, rabies, hepatitis, the common cold, and many others. Smallpox, once a worldwide scourge, was eradicated by a decade-long global immunization effort beginning in the mid-1960s. Viral infections in plants (e.g., dwarf mosaic virus in corn) have a major economic impact on crop production. Planting of virus-resistant varieties, developed by traditional breeding methods and more recently by genetic-engineering techniques, can reduce crop losses significantly. Most viruses have a rather limited host range, infecting certain bacteria, plants, or animals (Figure 1-5).

Because viruses cannot grow or reproduce on their own, they are in this sense not considered to be alive. To survive, a virus must infect a host cell and take over its internal machinery to synthesize viral proteins and in some cases replicate the viral genetic material. When newly made viruses are released by budding from the cell membrane or when the infected cell bursts, the cycle starts anew. Viruses are much smaller than cells, on the order of 100 nanometer (nm) in diameter; in comparison, bacterial cells are usually >1000 nm (1 nm = 10^{-9} meters). A virus is typically composed of a protein coat that encloses a core containing the genetic material, which carries the information for producing more viruses (Chapter 4). The coat protects a virus from the environment and allows it to stick to, or enter, specific host cells. In some viruses, the protein coat is surrounded by an outer membrane-like envelope.

The ability of viruses to transport genetic material into cells and tissues represents a medical menace and a medical opportunity. Viral infections can be devastatingly destructive, causing cells to break open and tissues to fall apart. However, many methods for manipulating cells depend on using viruses to convey genetic material into cells. To do this, the portion of the viral genetic material that is potentially harmful is replaced with other genetic material, including human genes. The altered viruses, or **vectors,** still can enter cells toting the introduced genes with them (Chapter 9). One day, diseases caused by defective genes may be treated by using viral vectors to introduce a normal copy of a defective gene into patients. Current research is dedicated to overcoming the considerable obstacles to this approach, such as getting the introduced genes to work at the right places and times.

Changes in Cells Underlie Evolution

The most remarkable feature of organisms is their ability to reproduce. Biological reproduction, combined with continuing evolutionary selection for a highly functional body plan, is why today's horseshoe crabs look much as they did 300 million years ago, a time span during which entire mountain ranges have risen or fallen. The Teton Mountains in Wyoming, now about 14,000 feet high and still growing, did not exist a mere 10 million years ago. Yet horseshoe crabs, with a life span of about 19 years, have faithfully reproduced their ancient selves more than half a million times during that period. The common impression that biological structure is transient and geological structure is stable is the exact opposite of the truth. Despite the limited duration of our individual lives, reproduction gives us a potential for immortality that a mountain or a rock does not have.

Whereas some species have changed little over great periods of time, other organisms have changed dramatically

(a) T4 bacteriophage

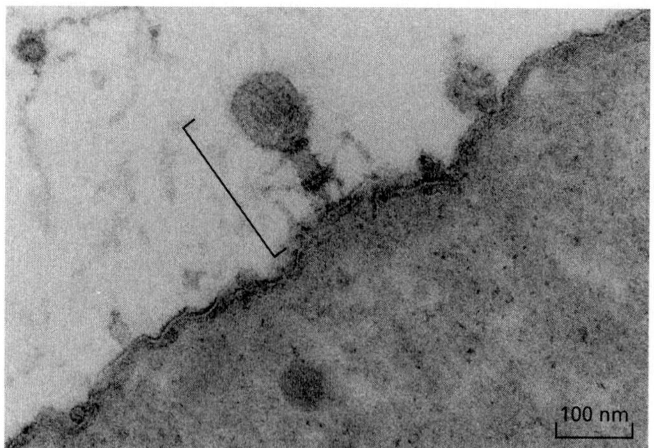

(b) Tobacco mosaic virus

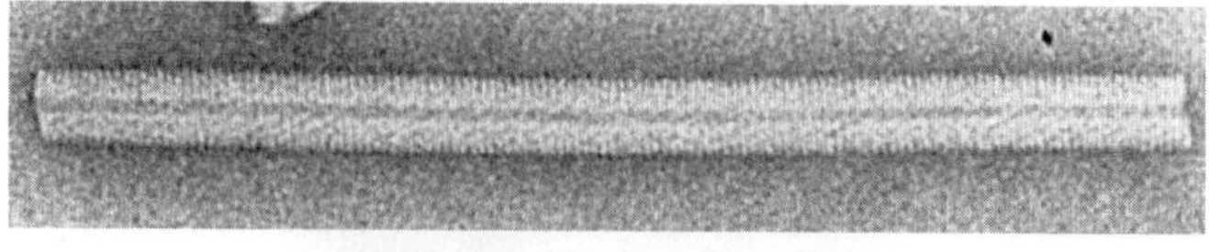

(c) Adenovirus

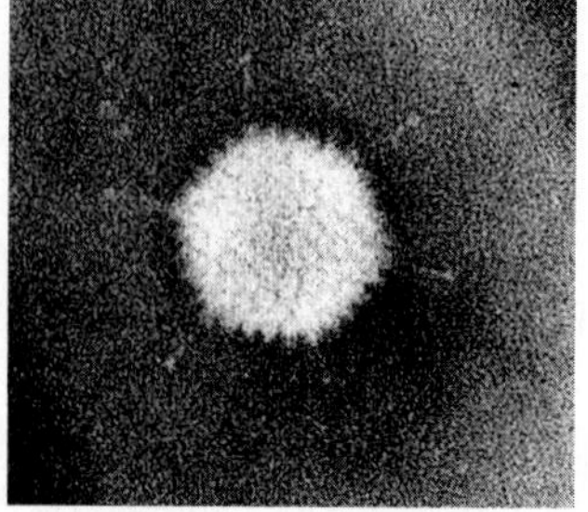

▲ **FIGURE 1-5 Viruses must infect a host cell to grow and reproduce.** These electron micrographs illustrate some of the structural variety exhibited by viruses. (a) T4 bacteriophage (bracket) attaches to a bacterial cell via a tail structure. Viruses that infect bacteria are called bacteriophages, or simply phages. (b) Tobacco mosaic virus causes a mottling of the leaves of infected tobacco plants and stunts their growth. (c) Adenovirus causes eye and respiratory tract infections in humans. This virus has an outer membranous envelope from which long glycoprotein spikes protrude. [Part (a) from A. Levine, 1991, *Viruses,* Scientific American Library, p. 20. Part (b) courtesy of R. C. Valentine. Part (c) courtesy of Robley C. Williams, University of California.]

during the same period. The changes came in response to pressures from the environment that caused increased survival of variant individuals. Both stasis and change are possible because the machinery of cells does an amazingly precise job of copying genetic material, yet its rare errors introduce some variation. If environmental conditions continue to select more or less the existing form, as in the case of horseshoe crabs, the species will change little. If a new variant has a survival advantage, perhaps because conditions have changed, it may persist and replace the old form.

Populations of bacteria exposed to antibiotics, for example, change their properties dramatically to escape and live. They do this because rare mutations, changes in the genetic material that allow antibiotic resistance, keep some cells alive while the cells without those mutations die. Most populations of any single species have a large repertoire of genetic alterations because there is a low but significant error rate in copying the genes. That error rate increases in the presence of radiation such as sunlight or certain chemical poisons. Current genome projects are exploring genetic variation among humans. "The" human genome sequence that has been determined already is just one version among billions. Understanding variation is essential to learn how we respond differently to certain infections or drugs and to exploring how our genetic heritage combines with our experience and learning to make each of us unique.

Underlying the reproduction of organisms is the copying of cells, something that must be precise in order to control the size, shape, and organization of animals and to prevent unwanted growth, such as cancer. The cell is a machine that can copy itself, unlike viruses, which cannot do so on their own. As we will see in Chapters 20 and 21, the cell cycle—from a single cell copying its own contents through division into two cells—is controlled by a series of elegant switches and cross-checking mechanisms. Reproducing cells accurately is a matter of life and death.

Even Single Cells Can Have Sex

If genetic material was never shared or exchanged, each individual would be the beginning of a new **clone** of individuals, and the members of a clone would share most of the same genetic strengths and weaknesses. Sex is a process of mingling genetic variation from two individuals, creating new individuals with a combination of properties unlike either parent and that may be beneficial for survival and reproduction. Each chromosome except the sex chromosomes is represented twice, one copy from the father and one from the mother. Since each pair of chromosomes trades pieces during the formation of eggs and sperm, new combinations of genes are created and inherited together in the next generation—variation is accelerated. The other benefit of having two copies of each chromosome is that a poorly functioning gene is backed up by the other copy.

The common yeast used to make bread and beer, *Saccharomyces cerevisiae,* appears fairly frequently in this book because it has proven to be a great experimental organism. Like many other unicellular organisms, yeasts have two mating types that are conceptually like the male and female gametes (eggs and sperm) of higher organisms. Two yeast cells of opposite mating type can fuse, or mate, to produce a third cell type containing the genetic material from each cell (Figure 1-6). Such sexual life cycles allow more rapid changes in genetic inheritance than would be possible without sex, resulting in valuable adaptations while quickly eliminating detrimental mutations. That, and not just Hollywood, is probably why sex is so ubiquitous.

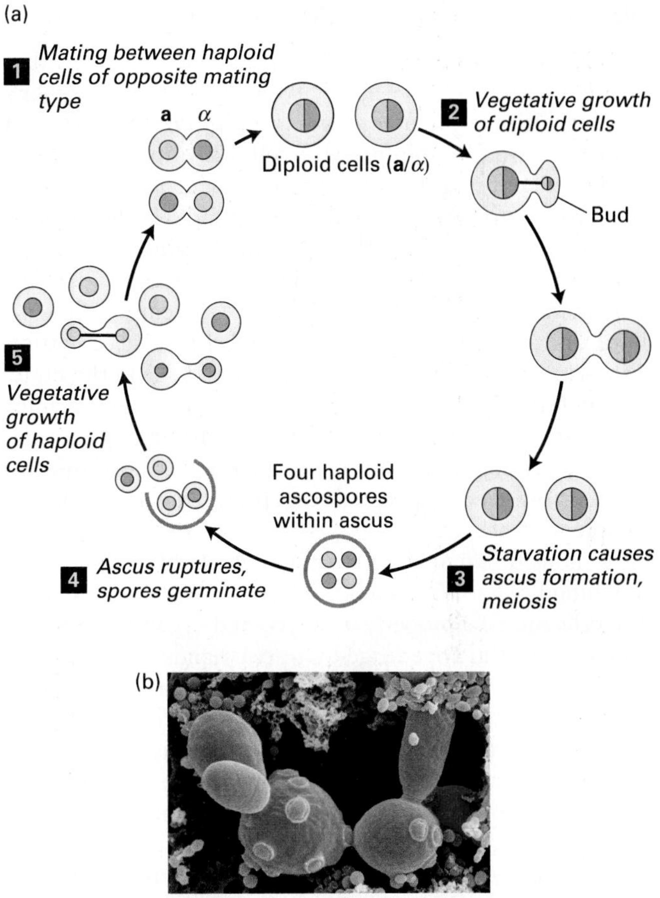

(a)

1 *Mating between haploid cells of opposite mating type*

a α

Diploid cells (**a**/α)

2 *Vegetative growth of diploid cells*

Bud

3 *Starvation causes ascus formation, meiosis*

Four haploid ascospores within ascus

4 *Ascus ruptures, spores germinate*

5 *Vegetative growth of haploid cells*

(b)

Budding (*S. cerevisiae*)

▲ **FIGURE 1-6 The yeast *Saccharomyces cerevisiae* reproduces sexually and asexually.** (a) Two cells that differ in mating type, called **a** and α, can mate to form an **a**/α cell **1**. The **a** and α cells are haploid, meaning they contain a single copy of each yeast chromosome, half the usual number. Mating yields a diploid **a**/α cell containing two copies of each chromosome. During vegetative growth, diploid cells multiply by mitotic budding, an asexual process **2**. Under starvation conditions, diploid cells undergo meiosis, a special type of cell division, to form haploid ascospores **3**. Rupture of an ascus releases four haploid spores, which can germinate into haploid cells **4**. These also can multiply asexually **5**. (b) Scanning electron micrograph of budding yeast cells. After each bud breaks free, a scar is left at the budding site, so the number of previous buds can be counted. The orange cells are bacteria. [Part (b) M. Abbey/Visuals Unlimited, Inc.]

We Develop from a Single Cell

In 1827, German physician Karl von Baer discovered that mammals grow from eggs that come from the mother's ovary. Fertilization of an egg by a sperm cell yields a **zygote,** a visually unimpressive cell 200 μm in diameter. Every human being begins as a zygote, which houses all the necessary instructions for building a human body containing about 100 trillion (10^{14}) cells, an amazing feat. **Development** begins with the fertilized egg cell dividing into two, four, then eight cells, forming the very early embryo (Figure 1-7). Continued cell proliferation and then **differentiation** into distinct cell types gives rise to every tissue in the body. One initial cell, the fertilized egg (zygote), generates hundreds of different kinds of cells that differ in contents, shape, size, color, mobility, and surface composition. We will see how genes and signals control cell diversification in Chapters 16 and 22.

Making different kinds of cells—muscle, skin, bone, neuron, blood cells—is not enough to produce the human body. The cells must be properly arranged and organized into tissues, organs, and appendages. Our two hands have the same kinds of cells, yet their different arrangements—in a mirror image—are critical for function. In addition, many cells exhibit distinct functional and/or structural asymmetries, a property often called **polarity.** From such polarized cells arise asymmetric, polarized tissues such as the lining of the intestines and structures such as hands and hearts. The features that make some cells polarized and how they arise also are covered in later chapters, including Chapter 21.

MEDIA CONNECTIONS

Video: Early Embryonic Development 

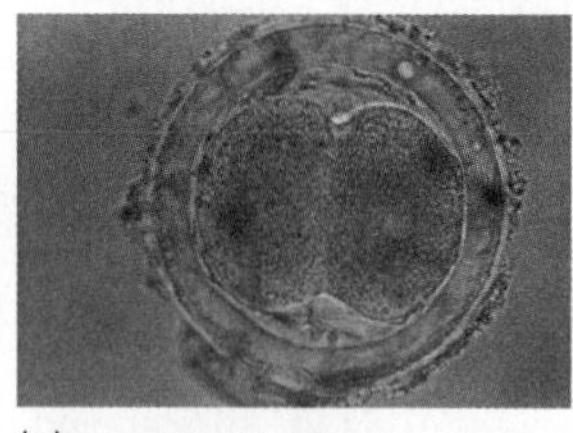
(a)

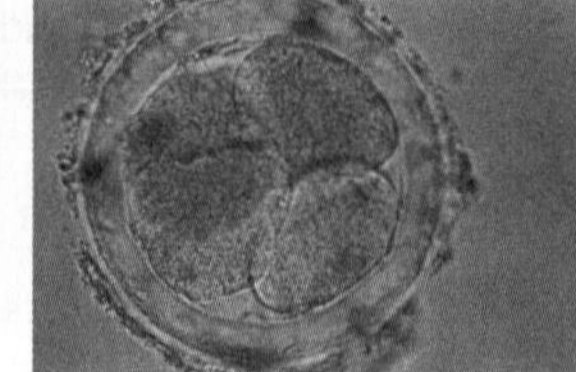
(b)

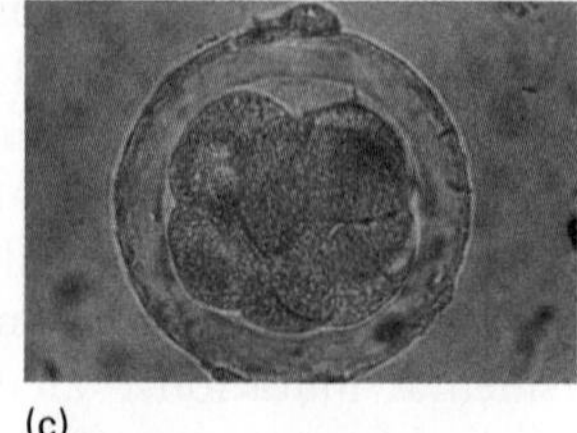
(c)

◄ **FIGURE 1-7 The first few cell divisions of a fertilized egg set the stage for all subsequent development.** A developing mouse embryo is shown at (a) the two-cell, (b) four-cell, and (c) eight-cell stages. The embryo is surrounded by supporting membranes. The corresponding steps in human development occur during the first few days after fertilization. [Claude Edelmann/Photo Researchers, Inc.]

Stem Cells, Fundamental to Forming Tissues and Organs, Offer Medical Opportunities

The biology of **stem cells,** cells that can give rise to specific cell types and tissues, has generated great interest. We can contrast stem cells to the simpler types of bacteria. When a bacterial *E. coli* cell divides, both daughter cells are pretty much equivalent in content, size, and shape. In some other bacteria and in many cases of eukaryotic cell division, the two daughters differ in important ways. Although both will have the same genetic material, the cells may differ in size, shape, and contents. The cells may have different *fates,* that is, they may become different types of differentiated cell. A division that produces two different daughter cells is sometimes described as an **asymmetric cell division.**

Stem-cell divisions are a special case of asymmetric division. One of the two daughter cells is identical to the parent cell; the other follows a path of differentiation, such as becoming a blood cell. The parent cell, called a **stem cell,** can go on reproducing itself at every division, at each division also producing another blood cell. Most tissues in our bodies form from stem cells. Blood, for example, is produced from stem cells that reside in the bone marrow and continue to produce new blood cells for our entire lives. This is the basis of the often successful bone marrow transplants that are used to treat cancer patients who have had their blood stem cells damaged by cancer treatments: what is being transplanted is stem cells. However, blood stem cells produce only more of themselves and blood cells, not other cell types. Thus each tissue must have its own stem cells, at least during the period of development when the tissue is formed. Stem cells for each tissue arise from even more capable stem cells that have the ability to form multiple stem cell types. The first stem cells are found in early embryos, where all the cells are capable of producing all cell types.

In mammals the ultimate stem cell is the fertilized egg, which produces early embryo cells capable of forming all the tissues of the body. This capability is illustrated by the formation of identical twins, which occur naturally when the mass of cells composing an early embryo divides into two parts, each of which develops and grows into an individual animal. This means that the cells cannot have divided up their embryo-forming duties prior to the time of embryo division. Each cell in an eight-cell-stage mouse embryo has the potential to give rise to any part of the entire animal. Cells with this capability are referred to as **embryonic stem (ES) cells.** As we will learn in Chapter 22, ES cells can be grown in the laboratory (cultured) and will develop into various types of differentiated cells under appropriate conditions.

The ability to make and manipulate mammalian embryos in the laboratory has led to new medical opportunities as well as various social and ethical concerns. In vitro fertilization, for instance, has allowed many otherwise infertile couples to have children. One technique involves extraction of nuclei from defective sperm incapable of normally fertilizing an egg, injection of the nuclei into eggs, and implantation of the resulting fertilized eggs into the mother.

In recent years, nuclei taken from cells of adult animals have been used to produce new animals. In this procedure, the nucleus is removed from a body cell (e.g., skin or blood cell) of a donor animal and introduced into an unfertilized mammalian egg that has been deprived of its own nucleus. In a step that has now been done with mice, cows, sheep, mules, and some other animals, the egg with its donor nucleus is implanted into a foster mother. The ability of such a donor nucleus to direct the development of an entire animal shows that all the information required for life is retained in the nuclei of some adult cells. Since all the cells in an animal produced in this way have the genes of the single original donor cell, the new animal is a genetic clone of the donor (Figure 1-8), though the animals may differ anyway due to their distinct environments and experiences. Repeating the process can give rise to many clones. Nuclei taken from ES cells work especially well, whereas nuclei from other parts of the body at later times in life work far less well. The majority of embryos produced by this technique do not survive due to birth defects, so the donor nuclei may not have all the needed information or the nuclei may be damaged by the cloning process. Even those animals that are born alive have abnormalities, including accelerated aging. The "rooting" of plants, in contrast, is a type of cloning that is readily accomplished by gardeners, farmers, and laboratory technicians.

Scientific interest in the cloning of humans is very limited. Virtually all scientists oppose it because of its high risk to the embryo (also, most people don't believe there is a critical shortage of twins and triplets). Of much greater scientific and medical interest is the ability to generate specific cell types starting from embryonic or adult stem cells. This procedure, **somatic cell nuclear transfer (SCNT)**, produces cells that are grown in culture and never turned into an embryo. The scientific interest in such cells comes from learning the signals that can unleash the potential of the genes to form a certain cell type. The medical interest comes from the possibility of treating the numerous diseases in which particular cell types are damaged or missing and of repairing wounds more completely. The cells may also be useful in culture to test the effects of drugs or other treatments. If the cells are produced using a donor nucleus from a patient, the properties of the cells may allow them to escape rejection by the patient's immune system, opening new possibilities for cell-transplant therapies.

▲ **FIGURE 1-8 Five genetically identical cloned sheep.** An early sheep embryo was divided into five groups of cells and each was separately implanted into a surrogate mother, much like the natural process of twinning. At an early stage the cells are able to adjust and form an entire animal; later in development the cells become progressively restricted and can no longer do so. An alternative way to clone animals is to replace the nuclei of multiple single-celled embryos with donor nuclei from cells of an adult sheep. Each embryo will be genetically identical to the adult from which the nucleus was obtained. Low percentages of embryos survive these procedures to give healthy animals, and the full impact of the techniques on the animals is not yet known. [Geoff Tompkinson/Science Photo Library/Photo Researchers, Inc.]

1.2 The Molecules of a Cell

Molecular cell biologists explore how all the remarkable properties of the cell arise from underlying molecular events: the assembly of large molecules, binding of large molecules to each other, catalytic effects that promote particular chemical reactions, and the deployment of information carried by giant molecules. Here we review the most important kinds of molecules that form the chemical foundations of cell structure and function.

Small Molecules Carry Energy, Transmit Signals, and Are Linked into Macromolecules

Much of the cell's content is a watery soup flavored with small molecules (e.g., simple sugars, amino acids, vitamins) and ions (e.g., sodium, chloride, calcium ions). The locations and concentrations of small molecules and ions within the cell are controlled by numerous proteins inserted in cellular membranes. These pumps, transporters, and ion channels move nearly all small molecules and ions into or out of the cell and its organelles (Chapter 11).

One of the best-known small molecules is **adenosine triphosphate (ATP)**, which stores readily available chemical energy in two of its chemical bonds (see Figure 2-31). When cells split apart these energy-rich bonds in ATP, the released energy can be harnessed to power an energy-requiring process such as muscle contraction or protein biosynthesis. To obtain energy for making ATP, cells break down food molecules. For instance, when sugar is degraded to carbon dioxide and water, the energy stored in the original chemical bonds is released and much of it can be "captured" in ATP (Chapter 12). Bacterial, plant, and animal cells can all make ATP by this process. In addition, plants and a few other organisms can harvest energy from sunlight to form ATP in **photosynthesis.**

Other small molecules act as signals both within and between cells; such signals direct numerous cellular activities (Chapters 15 and 16). The powerful effect on our bodies of a frightening event comes from the instantaneous flooding of the body with epinephrine, a small-molecule **hormone** that mobilizes the "fight-or-flight" response. The movements needed to fight or flee are triggered by nerve impulses that flow from the brain to our muscles with the aid of **neurotransmitters,** another type of small-molecule signal that we discuss in Chapter 23.

Certain small molecules (**monomers**) in the cellular soup can be joined to form **polymers** through repetition of a single type of chemical-linkage reaction (see Figure 2-1). Cells produce three types of large polymers, commonly called **macromolecules:** polysaccharides, proteins, and nucleic acids. Sugars, for example, are the monomers used to form **polysaccharides.** These macromolecules are critical structural components of plant cell walls and insect skeletons. A typical polysaccharide is a linear or branched chain of repeating identical sugar units. Such a chain carries information: the number of units. However, if the units are *not* identical, then the order and type of units carry additional information. As we will see in Chapter 6, some polysaccharides exhibit the greater informational complexity associated with a linear code made up of different units assembled in a particular order. But this property is most typical of the two other types of biological macromolecules—**proteins** and **nucleic acids.**

Proteins Give Cells Structure and Perform Most Cellular Tasks

The varied, intricate structures of proteins enable them to carry out numerous functions. Cells string together 20 different **amino acids** in a linear chain to form a protein (see Figure 2-14). Proteins commonly range in length from 100 to 1000 amino acids, but some are much shorter and others longer. We obtain amino acids either by synthesizing them from other molecules or by breaking down proteins that we eat. The "essential" amino acids, from a dietary standpoint, are the eight that we cannot synthesize and must obtain from food. Beans and corn together have all eight, making their combination particularly nutritious. Once a chain of amino acids is formed, it folds into a complex shape, conferring a distinctive three-dimensional structure and function on each protein (Figure 1-9).

Some proteins are similar to one another and therefore can be considered members of a **protein family.** A few hundred such families have been identified. Most proteins are designed to work in particular places within a cell or to be released into the extracellular (*extra,* "outside") space. Elaborate cellular pathways ensure that proteins are transported to their proper intracellular (*intra,* "within") locations or secreted (Chapters 13 and 14).

Proteins can serve as structural components of a cell, for example, by forming an internal skeleton (Chapters 10, 17, and 18). They can be sensors that change shape as temperature, ion concentrations, or other properties of the cell change. They can import and export substances across the plasma membrane (Chapter 11). They can be **enzymes,** causing chemical reactions to occur much more rapidly than they would without the aid of these protein **catalysts** (Chapter 3). They can bind to a specific gene, turning it on or off (Chapter 7). They can be extracellular signals, released from one cell to communicate with other cells, or intracellular signals, carrying information within the cell (Chapters 15 and 16). They can be motors that move other molecules around, burning chemical energy (ATP) to do so (Chapters 17 and 18).

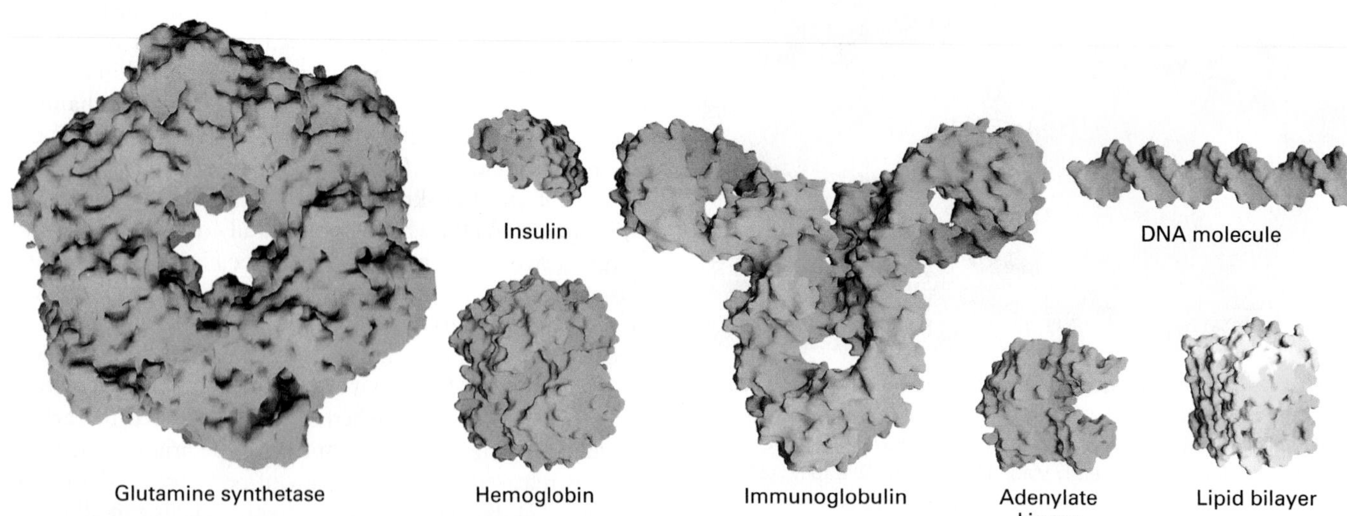

▲ **FIGURE 1-9 Proteins vary greatly in size, shape, and function.** These models of the water-accessible surface of some representative proteins are drawn to a common scale and reveal the numerous projections and crevices on the surface. Each protein has a defined three-dimensional shape (conformation) that is stabilized by numerous chemical interactions discussed in Chapters 2 and 3. The illustrated proteins include enzymes (glutamine synthetase and adenylate kinase), an antibody (immunoglobulin), a hormone (insulin), and the blood's oxygen carrier (hemoglobin). Models of a segment of the nucleic acid DNA and a small region of the lipid bilayer that forms cellular membranes (see Section 1.3) demonstrate the relative width of these structures compared with that of typical proteins. [Courtesy of Gareth White.]

How can 20 amino acids form all the different proteins needed to perform these varied tasks? It seems impossible at first glance. But if a "typical" protein is about 400 amino acids long, there are 20^{400} possible different protein sequences. Even assuming that many of these would be functionally equivalent, unstable, or otherwise discountable, the number of possible proteins is well along toward infinity.

Next we might ask how many protein molecules a cell needs to operate and maintain itself. To estimate this number, let's take a typical eukaryotic cell, such as a hepatocyte (liver cell). This cell, roughly a cube 15 μm (0.0015 cm) on a side, has a volume of 3.4×10^{-9} cm^3 (or milliliters). Assuming a cell density of 1.03 g/ml, the cell would weigh 3.5×10^{-9} g. Since protein accounts for approximately 20 percent of a cell's weight, the total weight of cellular protein is 7×10^{-10} g. The average yeast protein has a molecular weight of 52,700 (g/mol). Assuming this value is typical of eukaryotic proteins, we can calculate the total number of protein molecules per liver cell as about 7.9×10^9 from the total protein weight and Avogadro's number, the number of molecules per mole of any chemical compound (6.02×10^{23}). To carry this calculation one step further, consider that a liver cell contains about 10,000 different proteins; thus a cell contains close to a million molecules of each type of protein on average. In fact, the abundance of different proteins varies widely, from the quite rare insulin-binding receptor protein (20,000 molecules) to the abundant structural protein actin (5×10^8 molecules).

Nucleic Acids Carry Coded Information for Making Proteins at the Right Time and Place

The information about how, when, and where to produce each kind of protein is carried in the genetic material, a polymer called **deoxyribonucleic acid (DNA).** The three-dimensional structure of DNA consists of two long helical strands that are coiled around a common axis, forming a **double helix.** DNA strands are composed of monomers called **nucleotides;** these often are referred to as *bases* because their structures contain cyclic organic bases (Chapter 4).

Four different nucleotides, abbreviated A, T, C, and G, are joined end to end in a DNA strand, with the base parts projecting out from the helical backbone of the strand. Each DNA double helix has a simple construction: wherever one strand has an A, the other strand has a T, and each C is matched with a G (Figure 1-10). This **complementary** matching of the two strands is so strong that if complementary strands are separated, they will spontaneously zip back together in the right salt and temperature conditions. Such **nucleic acid hybridization** is extremely useful for detecting one strand using the other. For example, if one strand is purified and attached to a piece of paper, soaking the paper in a solution containing the other complementary strand will lead to zippering, even if the solution also contains many other DNA strands that do not match.

The genetic information carried by DNA resides in its sequence, the linear order of nucleotides along a strand. The information-bearing portion of DNA is divided into discrete functional units, the **genes,** which typically are 5000 to 100,000 nucleotides long. Most bacteria have a few thousand genes; humans, about 20,000–25,000. The genes that carry instructions for making proteins commonly contain two parts: a *coding region* that specifies the amino acid sequence of a protein and a *regulatory region* that controls when and in which cells the protein is made.

Cells use two processes in series to convert the coded information in DNA into proteins (Figure 1-11). In the first, called **transcription,** the coding region of a gene is copied into a single-stranded **ribonucleic acid (RNA)** version of the double-stranded DNA. A large enzyme, **RNA polymerase,** catalyzes the linkage of nucleotides into a RNA chain using DNA as a template. In eukaryotic cells, the initial RNA product is processed into a smaller **messenger RNA (mRNA)** molecule, which moves to the cytoplasm. Here the **ribosome,** an enormously complex molecular machine composed of both RNA and protein, carries out the second process, called **translation.** During translation, the ribosome assembles and links together amino acids in the precise order dictated by the mRNA sequence according to the nearly universal **genetic code.** We examine the cell components that carry out transcription and translation in detail in Chapter 4.

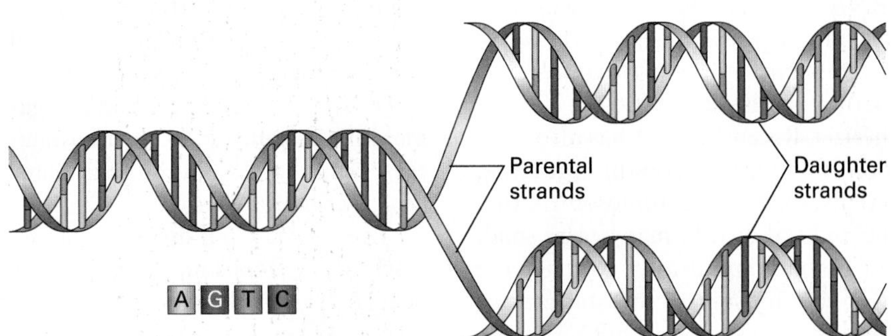

▲ **FIGURE 1-10 DNA consists of two complementary strands wound around each other to form a double helix.** *(Left)* The double helix is stabilized by weak hydrogen bonds between the A and T bases and between the C and G bases. *(Right)* During replication, the two strands are unwound and used as templates to produce complementary strands. The outcome is two copies of the original double helix, each containing one of the original strands and one new daughter (complementary) strand.

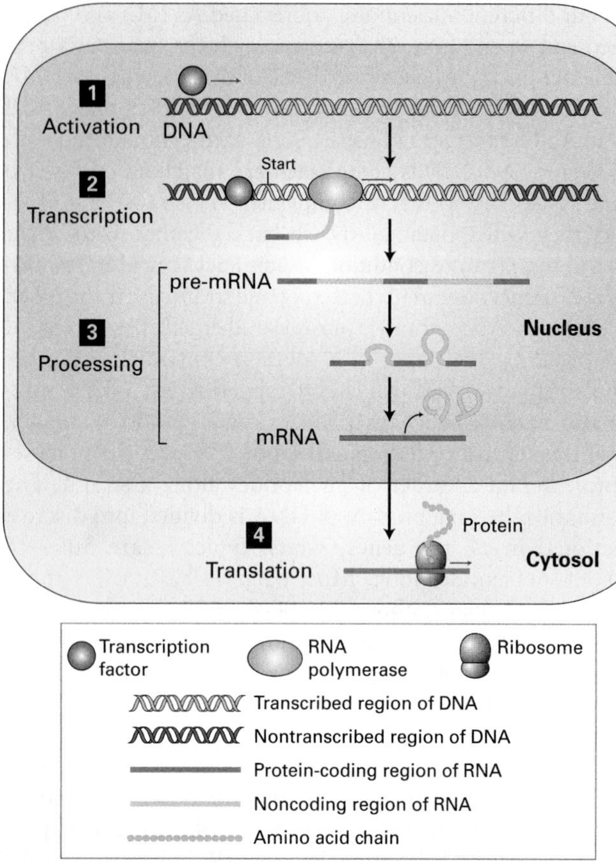

▲ **FIGURE 1-11 The coded information in DNA is converted into the amino acid sequences of proteins by a multistep process.** Step **1**: Transcription factors bind to the regulatory regions of the specific genes they control and activate them. Step **2**: Following assembly of a multiprotein initiation complex bound to the DNA, RNA polymerase begins transcription of an activated gene at a specific location, the start site. The polymerase moves along the DNA linking nucleotides into a single-stranded pre-mRNA transcript using one of the DNA strands as a template. Step **3**: The transcript is processed to remove noncoding sequences. Step **4**: In a eukaryotic cell, the mature messenger RNA (mRNA) moves to the cytoplasm, where it is bound by ribosomes that read its sequence and assemble a protein by chemically linking amino acids into a linear chain.

In addition to its role in transferring information from nucleus to cytoplasm, RNA can serve as a framework for building a molecular machine. For example, the ribosome has four RNA chains that team up with more than 50 proteins to make a remarkably precise and efficient mRNA reader and protein synthesizer. Recently, RNA has also been found to play a remarkably important role in regulating many aspects of gene activity, including chromosome structure and RNA processing and stability. In many cases small RNAs, 20–200 nucleotides long, specifically regulate the structure and function of chromosomes, the stability of larger RNA molecules, and the translation of mRNA molecules into protein.

All organisms have ways to control when and where their genes can be transcribed. For instance, nearly all the cells in our bodies contain the full set of human genes, but in each cell type only some of these genes are active, or turned on, and used to make proteins. That's why liver cells produce some proteins that are not produced by kidney cells and vice versa. Moreover, many cells can respond to external signals or changes in external conditions by turning specific genes on or off, thereby adapting their repertoire of proteins to meet current needs. Such control of gene activity depends on DNA-binding proteins called **transcription factors,** which bind to DNA and act as switches, either activating or repressing transcription of particular genes (Chapter 7).

Transcription factors are shaped so precisely that they are able to bind preferentially to the regulatory regions of just a few genes out of the thousands present in a cell's DNA. Typically a DNA-binding protein will recognize short DNA sequences about 6–12 base pairs long. A segment of DNA containing 10 base pairs can have 4^{10} possible sequences (1,048,576) since each position can be any of four nucleotides. Only a few copies of each such sequence will occur in the DNA of a cell, ensuring the specificity of gene activation and repression. Multiple copies of one type of transcription factor can coordinately regulate a set of genes if binding sites for that factor exist near each gene in the set. Transcription factors often work as multiprotein complexes, with more than one protein contributing its own DNA-binding specificity to selecting the regulated genes. In complex organisms, hundreds of different transcription factors are employed to form an exquisite control system that activates the right genes in the right cells at the right times. Small RNA molecules can have a dramatic effect on gene expression, regulating production and stability of gene transcripts. By some estimates small RNAs may regulate most or all genes, though the mechanisms and ubiquity of this type of regulation are still being explored.

The Genome Is Packaged into Chromosomes and Replicated During Cell Division

Most of the DNA in eukaryotic cells is located in the nucleus, extensively folded into the familiar structures we know as **chromosomes** (Chapter 6). Each chromosome contains a single linear DNA molecule associated with certain proteins. In prokaryotic cells, most or all of the genetic information resides in a single circular DNA molecule about a millimeter in length; this molecule lies, folded back on itself many times, in the central region of the cell (see Figure 1-2a). The **genome** of an organism comprises its entire complement of DNA. With the exception of eggs and sperm, every normal human cell has 46 chromosomes (Figure 1-12). Half of these, and thus half of the genes, can be traced back to Mom; the other half, to Dad.

Every time a cell divides, a large multiprotein replication machine, the replisome, separates the two strands of double-helical DNA in the chromosomes and uses each strand as a template to assemble nucleotides into a new complementary strand (see Figure 1-10). The outcome is a pair of double helices, each identical to the original. **DNA polymerase,** which is responsible for linking nucleotides into a DNA strand; the many other components of the replisome are described in

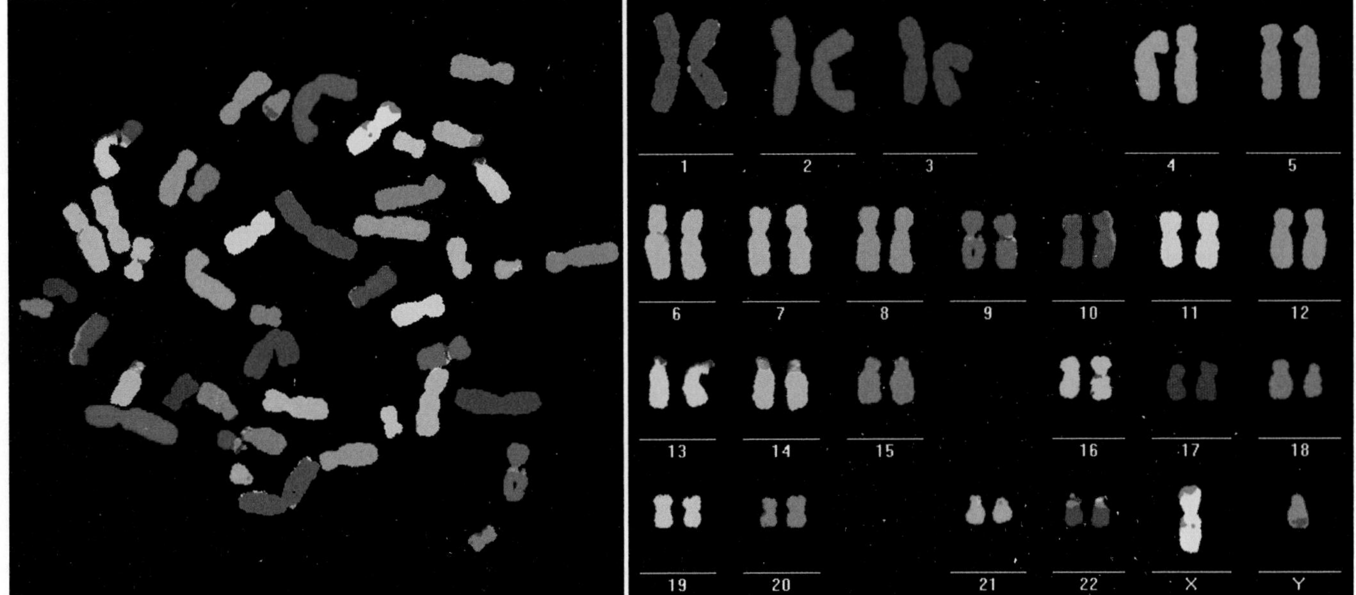

▲ **FIGURE 1-12 Chromosomes can be "painted" for easy identification.** A normal human has 23 pairs of morphologically distinct chromosomes; one member of each pair is inherited from the mother and the other member from the father. *(Left)* A chromosome spread from a human body cell midway through mitosis, when the chromosomes are fully condensed. This preparation was treated with fluorescent-labeled staining reagents that allow each of the 22 pairs and the X and Y chromosomes to appear in a different color when viewed in a fluorescence microscope. This technique of multiplex fluorescence in situ hybridization (M-FISH) sometimes is called chromosome painting (Chapter 6). *(Right)* Chromosomes from the preparation on the left arranged in pairs in descending order of size, an array called a karyotype. The presence of X and Y chromosomes identifies the sex of the individual as male. [Courtesy of M. R. Speicher.]

Chapter 4. The molecular design of DNA and the remarkable properties of the replisome ensure rapid, highly accurate copying. Many DNA polymerase molecules work in concert, each one copying part of a chromosome. The entire genome of fruit flies, about 1.2×10^8 nucleotides long, can be copied in three minutes! Because of the accuracy of DNA replication, nearly all the cells in our bodies carry the same genetic instructions, and we can inherit Mom's brown hair and Dad's blue eyes.

A rather dramatic example of gene control involves inactivation of an entire chromosome in human females. Women have two X chromosomes, whereas men have one X chromosome and one Y chromosome, which has different genes than the X chromosome. Yet the genes on the X chromosome must, for the most part, be equally active in female cells (XX) and male cells (XY). To achieve this balance, one of the X chromosomes in female cells is chemically modified and condensed into a very small mass called a Barr body, which is inactive and never transcribed.

Surprisingly, we inherit a small amount of genetic material entirely and uniquely from our mothers. This is the circular DNA present in mitochondria, the organelles in eukaryotic cells that synthesize ATP using the energy released by the breakdown of nutrients. Mitochondria contain multiple copies of their own DNA genomes, which code for some of the mitochondrial proteins (Chapter 6). Because each human inherits mitochondrial DNA only from his or her mother (it comes with the egg but not the sperm), the distinctive features of a particular mitochondrial DNA can be used to trace maternal history. Chloroplasts, the organelles that carry out photosynthesis in plants, also have their own circular genomes. Both mitochondria and chloroplasts are believed to be derived from **endosymbionts,** bacteria that took up residence inside eukaryotic cells in a mutually beneficial partnership. The mitochondrial and chloroplast circular DNAs appear to have originated as bacterial genomes, which also are usually circular, though the organelle genomes have lost most of the bacterial genes.

Mutations May Be Good, Bad, or Indifferent

Mistakes occasionally do occur spontaneously during DNA replication, causing changes in the sequence of nucleotides. Such changes, or **mutations,** also can arise from radiation that causes damage to the nucleotide chain or from chemical poisons, such as those in cigarette smoke, that lead to errors during the DNA-copying process (Chapter 25). Mutations come in various forms: a simple swap of one nucleotide for another; the deletion, insertion, or inversion of one to millions of nucleotides in the DNA of one chromosome; and translocation of a stretch of DNA from one chromosome to another.

In sexually reproducing animals such as ourselves, mutations can be inherited only if they are present in cells that potentially contribute to the formation of offspring. Such **germ-line cells** include eggs, sperm, and their precursor cells.

Body cells that do not contribute to offspring are called **somatic cells.** Mutations that occur in these cells never are inherited, although they may contribute to the onset of cancer. Plants have a less distinct division between somatic and germ-line cells, since many plant cells can function in both capacities.

Mutated genes that encode altered proteins or that cannot be controlled properly cause numerous inherited diseases. For example, sickle-cell disease is attributable to a single nucleotide substitution in the hemoglobin gene, which encodes the protein that carries oxygen in red blood cells. The single amino acid change caused by the sickle cell mutation reduces the ability of red blood cells to carry oxygen from the lungs to the tissues. Recent advances in detecting disease-causing mutations and in understanding how they affect cell functions offer exciting possibilities for reducing their often devastating effects.

Sequencing of the human genome has shown that a very large proportion of our DNA does not code for any RNA or have any discernible regulatory function, a quite unexpected finding. Mutations in these regions usually produce no immediate effects—good or bad. However, such "indifferent" mutations in nonfunctional DNA may have been a major player in evolution, leading to creation of new genes or new regulatory sequences for controlling already existing genes. For instance, since binding sites for transcription factors typically are only 10–12 nucleotides long, a few single-nucleotide mutations might convert a nonfunctional bit of DNA into a functional protein-binding regulatory site.

Much of the nonessential DNA in both eukaryotes and prokaryotes consists of highly repeated sequences that can move from one place in the genome to another. These **mobile DNA elements** can jump (transpose) into genes, most commonly damaging but sometimes activating them. Jumping generally occurs rarely enough to avoid endangering the host organism. Mobile elements, which were discovered first in plants, are responsible for leaf color variegation and the diverse beautiful color patterns of Indian corn kernels. By jumping in and out of genes that control pigmentation as plant development progresses, the mobile elements give rise to elaborate colored patterns. Mobile elements were later found in bacteria, in which they often carry and, unfortunately, disseminate genes for antibiotic resistance.

Now we understand that mobile elements have multiplied and slowly accumulated in genomes over evolutionary time, becoming a universal property of genomes in present-day organisms. They account for an astounding 45 percent of the human genome. Some of our own mobile DNA elements are copies—often highly mutated and damaged—of genomes from viruses that spend part of their life cycle as DNA segments inserted into host-cell DNA. Thus we carry in our chromosomes the genetic residues of infections acquired by our ancestors. Once viewed only as molecular parasites, mobile DNA elements are now thought to have contributed significantly to the evolution of higher organisms (Chapter 6).

1.3 The Work of Cells

In essence, any cell is simply a compartment with a watery interior that is separated from the external environment by a surface membrane (the plasma membrane) that prevents the free flow of molecules in and out. In addition, as we've noted, eukaryotic cells have extensive internal membranes that further subdivide the cell into various compartments, the organelles. Each compartment has contents and properties, such as specialized proteins or a certain pH, suited to its job. The plasma membrane and other cellular membranes are composed primarily of two layers of **phospholipid** molecules. These bipartite molecules have a "water-loving" (**hydrophilic**) end and a "water-hating" (**hydrophobic**) end. The two phospholipid layers of a membrane are oriented with all the hydrophilic ends directed toward the inner and outer surfaces and the hydrophobic ends buried within the interior (Figure 1-13). Smaller amounts of other lipids, such as cholesterol, and many kinds of proteins are inserted into the phospholipid framework. The lipid molecules and some proteins can float sidewise in the plane of the membrane, giving membranes a fluid character. This fluidity allows cells to change shape and even move. However, the attachment of some membrane proteins to other molecules inside or outside the cell restricts their lateral movement. We will learn more about membranes and how molecules cross them in Chapters 10 and 11.

The cytosol and the internal spaces of organelles differ from each other and from the cell exterior in terms of acidity, ionic composition, and protein contents. For example, the composition of salts inside the cell is often drastically different from what is outside. Because of these different "microclimates," each cell compartment has its own assigned

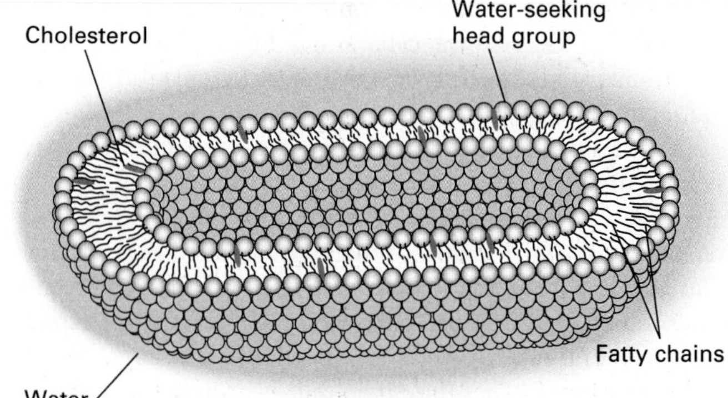

▲ **FIGURE 1-13 The watery interior of cells is surrounded by the plasma membrane, a two-layered shell of phospholipids.** The phospholipid molecules are oriented with their fatty acyl chains (black squiggly lines) facing inward and their water-seeking head groups (white spheres) facing outward. Thus both sides of the membrane are lined by head groups, mainly charged phosphates, adjacent to the watery spaces inside and outside the cell. All biological membranes have the same basic phospholipid bilayer structure. Cholesterol (red) and various proteins (not shown) are embedded in the bilayer. The interior space is actually much larger relative to the volume of the plasma membrane depicted here.

tasks in the overall work of the cell (Chapters 10, 12, and 13). The unique functions and microclimates of the various cell compartments are due largely to the proteins that reside in their membranes or interior.

We can think of the entire cell compartment as a factory dedicated to sustaining the well-being of the cell. Much cellular work is performed by molecular machines, some housed in the cytosol, some attached to the cytoskeleton, and some in various organelles. Here we quickly review the major tasks that cells carry out in their pursuit of the good life.

Cells Build and Degrade Numerous Molecules and Structures

As chemical factories, cells produce an enormous number of complex molecules from simple chemical building blocks. All of this synthetic work is powered by chemical energy extracted primarily from sugars and fats or sunlight, in the case of plant cells, and stored primarily in ATP, the universal "currency" of chemical energy (Figure 1-14). In animal and plant cells, most ATP is produced by large molecular machines located in two organelles, **mitochondria** and **chloroplasts.** Similar machines for generating ATP are located in the plasma membrane of bacterial cells. Both mitochondria and chloroplasts are thought to have originated as bacteria that took up residence inside eukaryotic cells and then became welcome collaborators (Chapter 12). Directly or indirectly, all of our food is created by plant cells using sunlight to build complex macromolecules during photosynthesis. Even underground oil supplies are derived from the decay of plant material.

Cells need to break down worn-out or obsolete parts into small molecules that can be discarded or recycled. This housekeeping task is assigned largely to **lysosomes,** organelles crammed with degradative enzymes. The interior of a lysosome has a pH of about 5.0, roughly 100 times more acidic than that of the surrounding cytosol. This aids in the breakdown of materials by lysosomal enzymes, which are specially designed to function at such a low pH. To create the low pH environment, proteins located in the lysosomal membrane pump hydrogen ions into the lysosome using energy supplied from ATP (Chapter 11). Lysosomes are assisted in the cell's cleanup work by **peroxisomes.** These small organelles are specialized for breaking down the lipid components of membranes and rendering various toxins harmless.

Most of the structural and functional properties of cells depend on proteins. Thus for cells to work properly, the numerous proteins composing the various working compartments must be transported from where they are made to their proper locations (Chapters 17 and 18). Some proteins are made on ribosomes that are free in the cytosol. Proteins secreted from the cell and most membrane proteins, however, are made on ribosomes associated with the **endoplasmic reticulum (ER).** This organelle produces, processes, and ships out both proteins and lipids. Protein chains produced on the ER move to the **Golgi complex,** where they are further modified before being forwarded to their final destinations. Proteins that travel in this way contain short sequences of amino acids or attached sugar chains (oligosaccharides) that serve as addresses for directing them to their correct destinations. These addresses work because they are recognized and bound by other proteins that do the sorting and shipping in various cell compartments.

Overview Animation: Biological Energy Interconversions

MEDIA CONNECTIONS

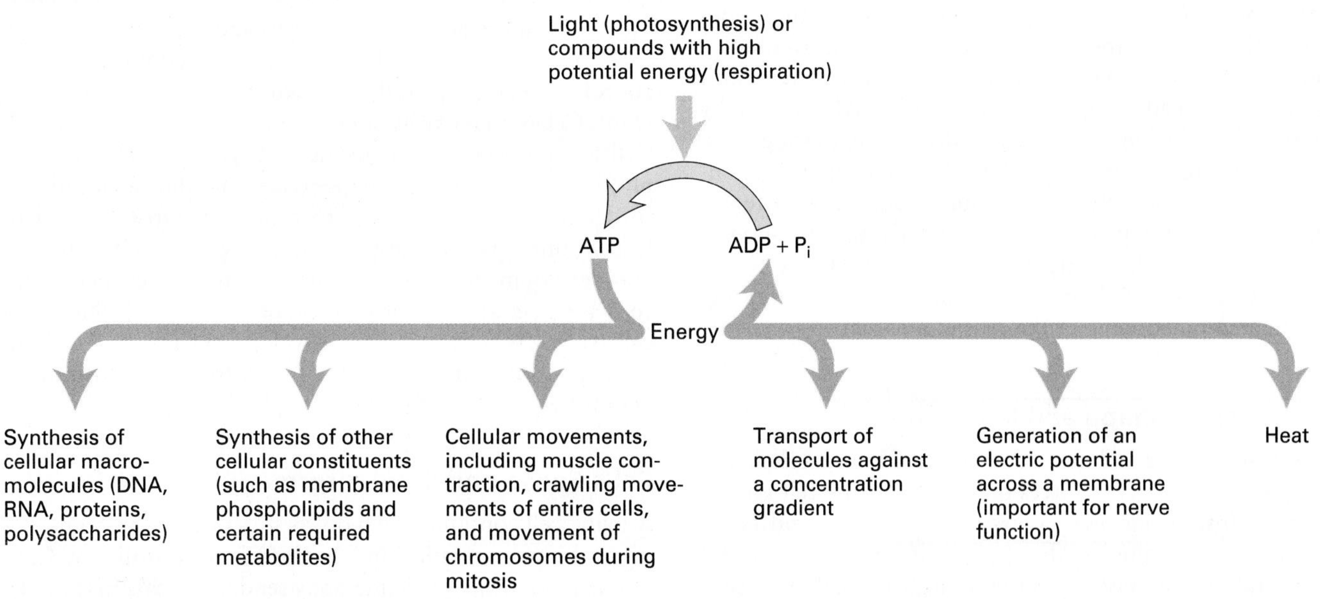

▲ **FIGURE 1-14 ATP is the most common molecule used by cells to capture and transfer energy.** ATP is formed from ADP and inorganic phosphate (P_i) by photosynthesis in plants and by the breakdown of sugars and fats in most cells. The energy released by the splitting (hydrolysis) of P_i from ATP drives many cellular processes.

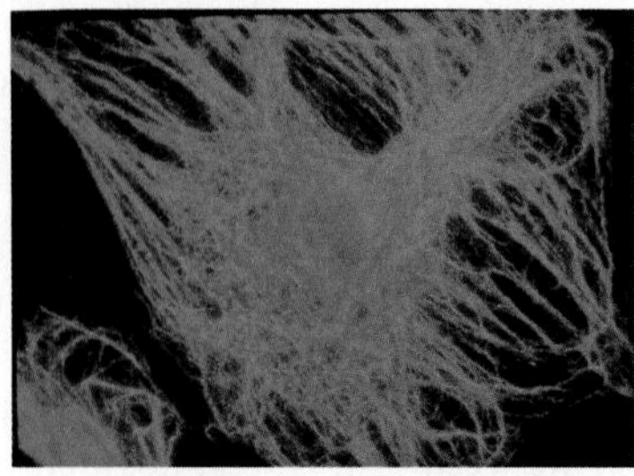

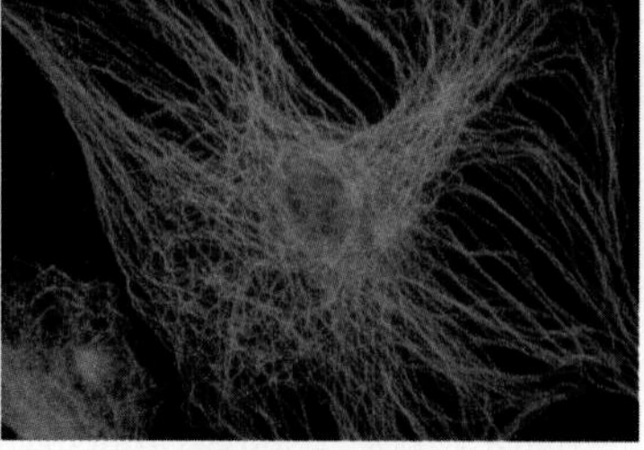

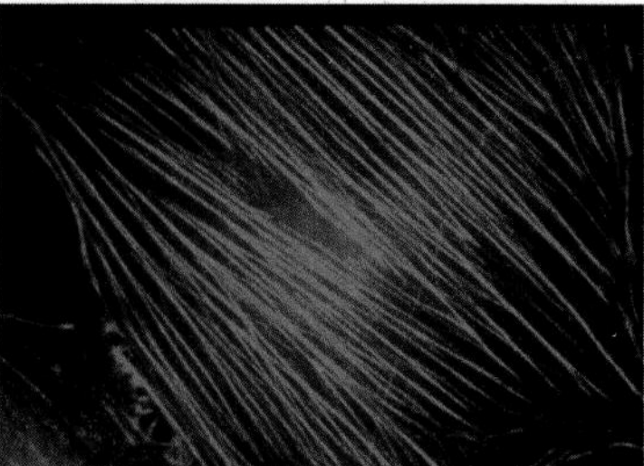

▲ **FIGURE 1-15 The three types of cytoskeletal filaments have characteristic distributions within cells.** Three views of the same cell. A cultured fibroblast was treated with three different antibody preparations. Each antibody binds specifically to the protein monomers forming one type of filament and is chemically linked to a differently colored fluorescent dye (green, blue, or red). Visualization of the stained cell in a fluorescence microscope reveals the location of filaments bound to a particular dye-antibody preparation. In this case, intermediate filaments are stained green; microtubules, blue; and microfilaments, red. All three fiber systems contribute to the shape and movements of cells. [Courtesy of V. Small.]

Animal Cells Produce Their Own External Environment and Glues

The simplest multicellular animals are single cells embedded in a jelly of proteins and polysaccharides called the **extracellular matrix.** Cells themselves produce and secrete these materials, thus creating their own immediate environment (Chapter 19). **Collagen,** the single most abundant protein in the animal kingdom, is a major component of the extracellular matrix in most tissues. In animals, the extracellular matrix cushions and lubricates cells. A specialized, especially tough matrix, the **basal lamina,** forms a supporting layer underlying sheetlike cell layers and helps prevent the cells from ripping apart.

The cells in animal tissues are "glued" together by **cell-adhesion molecules (CAMs)** embedded in their surface membranes. Some CAMs bind cells to one another; other types bind cells to the extracellular matrix, forming a cohesive unit. The cells of higher plants contain relatively few such molecules; instead, plants cells are rigidly tied together by extensive interlocking of the cell walls of neighboring cells. The cytosols of adjacent animal or plant cells often are connected by functionally similar but structurally different "bridges" called **gap junctions** in animals and **plasmodesmata** in plants. These structures allow cells to exchange small molecules including nutrients and signals, facilitating coordinated functioning of the cells in a tissue.

Cells Change Shape and Move

Cells change shape and move because their internal skeleton, the **cytoskeleton,** exerts forces on the rest of the cell and its contents. Just as our own bodies require both a rigid skeleton and a set of muscles that pull, cells have rigid skeletal fibers and protein motors that pull on them. Although cells are sometimes spherical, they more commonly have more elaborate shapes due to their internal skeletons and external attachments. Three types of protein filaments, organized into networks and bundles, form the cytoskeleton within animal cells (Figure 1-15). The cytoskeleton prevents the plasma membrane of animal cells from relaxing into a sphere (Chapter 10); it also functions in cell locomotion and the intracellular transport of vesicles, chromosomes, and macromolecules (Chapters 17 and 18). The cytoskeleton can be linked through the cell surface to the extracellular matrix or to the cytoskeleton of other cells, thus helping to form tissues (Chapter 19).

All cytoskeletal filaments are long polymers of protein subunits. Elaborate systems regulate the assembly and disassembly of the cytoskeleton, thereby controlling cell shape. In some cells the cytoskeleton is relatively stable, but in others it changes shape continuously. Shrinkage of the cytoskeleton in some parts of the cell and its growth in other parts can produce coordinated changes in shape that result in cell locomotion. For instance, a cell can send out an extension that attaches to a surface or to other cells and then retract the cell body from the other end. As this process continues due to coordinated changes in the cytoskeleton, the cell moves forward. Cells can move at rates on the order of 20 μm/second. Cell locomotion is used during embryonic development of multicellular animals to shape tissues and during adulthood to defend against infection, to transport nutrients, and to heal wounds. This process does not play a role in the growth and development of multicellular plants because new plant cells are generated by the division of existing cells that share cell walls. As a result, plant development involves cell enlargement but not movement of cells from one position to another.

Cells Sense and Send Information

A living cell continuously monitors its surroundings and adjusts its own activities and composition accordingly. Cells also communicate by deliberately sending signals that can be received and interpreted by other cells. Such signals are common not only within an individual organism but also between organisms. For instance, the odor of a pear signals a food source to us and other animals; consumption of the pear by an animal aids in distributing the pear's seeds. Everyone benefits! The signals employed by cells include simple small chemicals, gases, proteins, light, and mechanical movements. Cells possess numerous receptor proteins for detecting signals and elaborate pathways for transmitting them within the cell to evoke a response. At any time, a cell may be able to sense only some of the signals around it, and how a cell responds to a signal may change with time. In some cases, receiving one signal primes a cell to respond to a subsequent different signal in a particular way.

Both changes in the environment (e.g., an increase or decrease in a particular nutrient or the light level) and signals received from other cells represent external information that cells must process. The most rapid responses to such signals generally involve changes in the location or activity of preexisting proteins. For instance, soon after you eat a carbohydrate-rich meal, glucose pours into your bloodstream. The rise in blood glucose is sensed by β cells in the pancreas, which respond by releasing their stored supply of the protein hormone **insulin.** The circulating insulin signal causes glucose transporters in the cytoplasm of fat and muscle cells to move to the cell surface, where they begin importing glucose. Meanwhile, liver cells also are furiously taking in glucose via a different glucose transporter. In both liver and muscle cells, an intracellular signaling pathway triggered by binding of insulin to cell-surface receptors activates a key enzyme needed to make **glycogen,** a large glucose polymer (Figure 1-16a). The net result of these cell responses is that your blood glucose level falls and extra glucose is stored as glycogen, which your cells can use as a glucose source when you skip a meal to cram for a test.

The ability of cells to send and respond to signals is crucial to development. Many developmentally important signals are secreted proteins produced by specific cells at specific times and places in a developing organism. Often a receiving cell integrates multiple signals in deciding how to behave, for example, to differentiate into a particular tissue type, to extend a process, to die, to send back a confirming signal (yes, I'm here!), or to migrate.

The functions of about half the proteins in humans, roundworms, yeast, and several other eukaryotic organisms have been predicted based on analyses of genomic sequences (Chapter 6). Such analyses have revealed that at least 10–15 percent of the proteins in eukaryotes function as secreted extracellular signals, signal receptors, or intracellular **signal-transduction** proteins, which pass along a signal through a series of steps culminating in a particular cellular response (e.g., increased glycogen synthesis). Clearly, signaling and signal transduction are major activities of cells.

Cells Regulate Their Gene Expression to Meet Changing Needs

In addition to modulating the activities of existing proteins, cells often respond to changing circumstances and to signals from other cells by altering the amount or types of proteins they contain. **Gene expression,** the overall process of selectively reading and using genetic information, is commonly controlled at the level of transcription, the first step in the production of proteins. In this way cells can produce a particular mRNA only when the encoded protein is needed, thus minimizing wasted energy. Producing an mRNA is, however, only the first in a chain of regulated events that together determine whether an active protein product is produced from a particular gene.

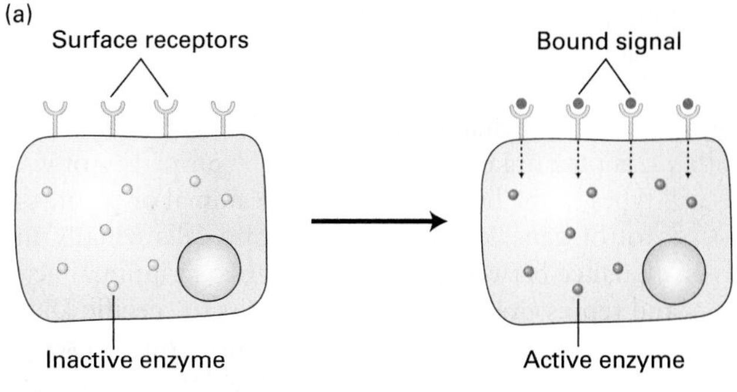

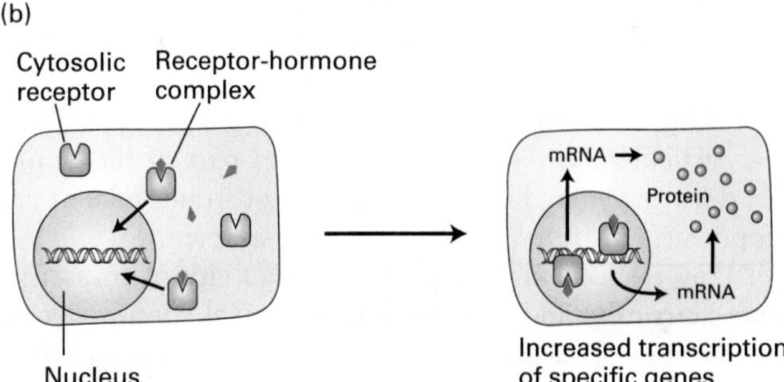

▲ **FIGURE 1-16 External signals commonly cause a change in the activity of preexisting proteins or in the amounts and types of proteins that cells produce.** (a) Binding of a hormone or other signaling molecule to its specific receptors can trigger an intracellular pathway that increases or decreases the activity of a preexisting protein. For example, binding of insulin to receptors in the plasma membrane of liver and muscle cells leads to activation of glycogen synthase, a key enzyme in the synthesis of glycogen from glucose. (b) The receptors for steroid hormones are located within cells, not on the cell surface. The hormone-receptor complexes activate transcription of specific target genes, leading to increased production of the encoded proteins. Many signals that bind to receptors on the cell surface also act, by more complex pathways, to modulate gene expression.

Transcriptional control of gene expression was first decisively demonstrated in the response of the gut bacterium *E. coli* to different sugar sources. *E. coli* cells prefer glucose as a sugar source, but they can survive on lactose in a pinch. These bacteria use both a DNA-binding *repressor* protein and a DNA-binding *activator* protein to change the rate of transcription of three genes needed to metabolize lactose depending on the relative amounts of glucose and lactose present (Chapter 4). Such dual positive/negative control of gene expression fine-tunes the bacterial cell's enzymatic equipment for the job at hand.

Like bacterial cells, unicellular eukaryotes may be subjected to widely varying environmental conditions that require extensive changes in cellular structures and function.

For instance, in starvation conditions yeast cells stop growing and form dormant spores (see Figure 1-4). In multicellular organisms, however, the environment around most cells is relatively constant. The major purpose of gene control in us and in other complex organisms is to tailor the properties of various cell types to the benefit of the entire animal or plant.

Control of gene activity in eukaryotic cells usually involves a balance between the actions of transcriptional activators and repressors. Binding of activators to specific DNA regulatory sequences called **enhancers** turns on transcription, and binding of repressors to other regulatory sequences called **silencers** turns off transcription. In Chapters 7 and 8, we take a close look at transcriptional activators and repressors and how they operate, as well as other mechanisms for controlling gene expression. In an extreme case, expression of a particular gene could occur only in part of the brain, only during evening hours, only during a certain stage of development, only after a large meal, and so forth.

Many external signals modify the activity of transcriptional activators and repressors that control specific genes. For example, lipid-soluble steroid hormones, such as estrogen and testosterone, can diffuse across the plasma membrane and bind to their specific receptors located in the cytoplasm or nucleus (Figure 1-16b). Hormone binding changes the shape of the receptor so that it can bind to specific enhancer sequences in the DNA, thus turning the receptor into a transcriptional activator. By this rather simple signal-transduction pathway, steroid hormones cause cells to change which genes they transcribe (Chapter 7). Since steroid hormones can circulate in the bloodstream, they can affect the properties of many or all cells in a temporally coordinated manner. Binding of many other hormones and of growth factors to receptors on the cell surface triggers different signal-transduction pathways that also lead to changes in the transcription of specific genes (Chapters 15 and 16). Although these pathways involve multiple components and are more complicated than those transducing steroid hormone signals, the general idea is the same.

Overview Animation: Life Cycle of a Cell

MEDIA CONNECTIONS

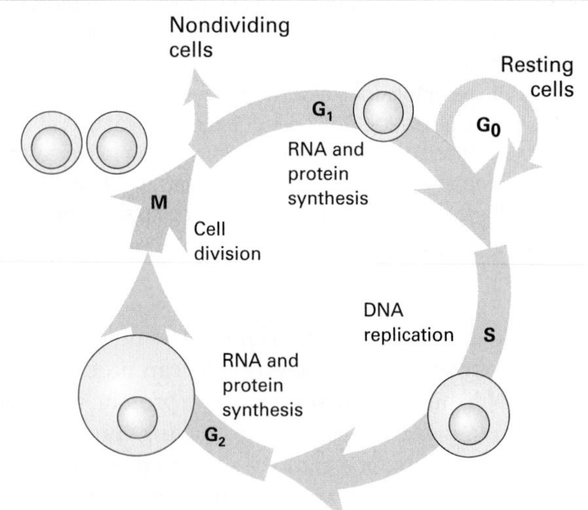

▲ **FIGURE 1-17 During growth, eukaryotic cells continually progress through the four stages of the cell cycle, generating new daughter cells.** In most proliferating cells, the four phases of the cell cycle proceed successively, taking from 10–20 hours depending on cell type and developmental state. During interphase, which consists of the G_1, S, and G_2 phases, the cell roughly doubles its mass. Replication of DNA during the S phase leaves the cell with four copies of each type of chromosome. In the mitotic (M) phase, the chromosomes are evenly partitioned to two daughter cells, and the cytoplasm divides roughly in half in most cases. Under certain conditions such as starvation or when a tissue has reached its final size, cells will stop cycling and remain in a waiting state called G_0. Most cells in G_0 can reenter the cycle if conditions change.

Cells Grow and Divide

As we have discussed, reproduction is at the heart of biology; rocks don't do it. The reproduction of organisms depends on the reproduction of cells. The simplest type of reproduction entails the division of a "parent" cell into two "daughter" cells. This occurs as part of the **cell cycle,** a series of events that prepares a cell to divide followed by the actual division process, called **mitosis.** The eukaryotic cell cycle commonly is represented as four stages (Figure 1-17). The chromosomes and the DNA they carry are copied during the **S (synthesis) phase.** The replicated chromosomes separate during the **M (mitotic) phase,** with each daughter cell getting a copy of each chromosome during cell division. The M and S phases are separated by two gap stages, the **G_1 phase** and **G_2 phase,** during which mRNAs and proteins are made. In single-celled organisms, both daughter cells often (though not always) resemble the parent cell. In multicellular organisms, stem cells can give rise to two different cells, one that resembles the parent cell and one that does not. Such asymmetric cell division is critical to the generation of different cell types in the body (Chapter 21).

During growth the cell cycle operates continuously, with newly formed daughter cells immediately embarking on their own path to mitosis. Under optimal conditions bacteria can divide to form two daughter cells once every 30 minutes. At this rate, in an hour one cell becomes four; in a day one cell becomes more than 10^{14}, which if dried would weigh about 25 grams. Under normal circumstances, however, growth cannot continue at this rate because the food supply becomes limiting.

Most eukaryotic cells take considerably longer than bacterial cells to grow and divide. Moreover, the cell cycle in adult plants and animals normally is highly regulated (Chapter 20). This tight control prevents imbalanced, excessive growth of tissues while ensuring that worn-out or damaged cells are replaced and that additional cells are formed in response to new circumstances or developmental needs. For instance, the proliferation of red blood cells increases substantially when a person ascends to a higher altitude and needs more capacity to capture oxygen. Some highly specialized cells in adult animals, such as nerve cells and striated muscle cells, rarely divide, if at all. The fundamental defect in cancer is loss of the ability to control the growth and division of cells. In Chapter 25, we examine the molecular and cellular events that lead to inappropriate, uncontrolled proliferation of cells.

Mitosis is an *asexual* process since the daughter cells carry the exact same genetic information as the parental cell. In *sexual* reproduction, fusion of two cells produces a third cell that contains genetic information from each parental cell. Since such fusions would cause an ever-increasing number of chromosomes, sexual reproductive cycles employ a special type of cell division, called **meiosis,** that reduces the number of chromosomes in preparation for fusion (see Figure 5-3). Cells with a full set of chromosomes are called **diploid** cells. During meiosis, a diploid cell replicates its chromosomes as usual for mitosis but then divides twice without copying the chromosomes in between. Each of the resulting four daughter cells, which have only half the full number of chromosomes, is said to be **haploid.**

Sexual reproduction occurs in animals and plants and even in unicellular organisms such as yeasts (see Figure 1-6). Animals spend considerable time and energy generating eggs and sperm, the haploid cells, called **gametes,** which are used for sexual reproduction. A human female will produce about half a million eggs in a lifetime, all of these cells forming before she is born; a young human male produces about 100 million sperm each day. Gametes are formed from diploid precursor germ-line cells, which in humans contain 46 chromosomes. In humans the X and Y chromosomes are called sex chromosomes because they determine whether an individual is male or female. In human diploid cells, the 44 remaining chromosomes, called **autosomes,** occur as pairs of 22 different kinds. Through meiosis, a man produces sperm that have 22 chromosomes plus either an X or a Y, and a woman produces ova (unfertilized eggs) with 22 chromosomes plus an X. Fusion of an egg and sperm (fertilization) yields a fertilized egg, the zygote, with 46 chromosomes, one pair of each of the 22 kinds and a pair of Xs in females or an X and a Y in males (Figure 1-18). Errors during meiosis can lead to disorders resulting from an abnormal number of chromosomes. These include Down's syndrome, caused by an extra chromosome 21, and Klinefelter's syndrome, caused by an extra X chromosome.

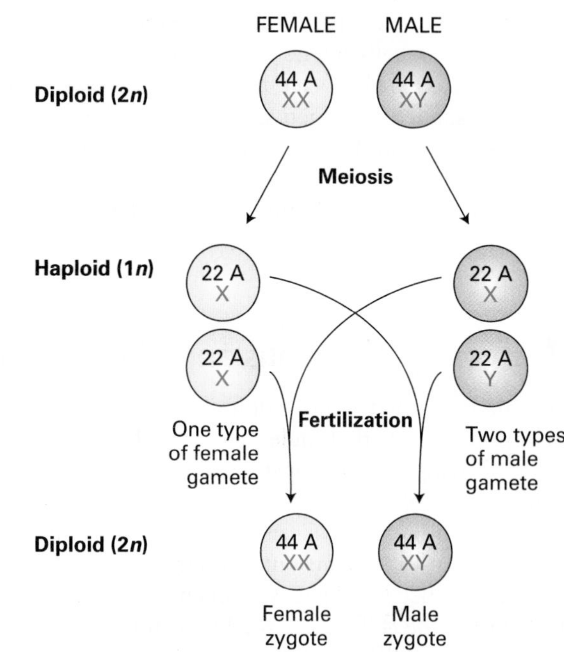

▲ **FIGURE 1-18 Dad made you a boy or girl.** In animals, meiosis of diploid precursor cells forms eggs and sperm (gametes). The male parent produces two types of sperm and determines the sex of the zygote. In humans, as shown here, X and Y are the sex chromosomes; the zygote must receive a Y chromosome from the male parent to develop into a male. A = autosomes (non-sex chromosomes).

Cells Die from Aggravated Assault or an Internal Program

When cells in multicellular organisms are badly damaged or infected with a virus, they die. Cell death resulting from such a traumatic event is messy and often releases potentially toxic cell constituents that can damage surrounding cells. Cells also may die when they fail to receive a life-maintaining signal or when they receive a death signal. In this type of programmed cell death, called **apoptosis,** a dying cell actually produces proteins necessary for self-destruction. Death by apoptosis avoids the release of potentially toxic cell constituents (Figure 1-19).

Programmed cell death is critical to the proper development and functioning of our bodies (Chapter 21). During fetal life, for instance, our hands initially develop with "webbing" between the fingers; the cells in the webbing subsequently die in an orderly and precise pattern that leaves the fingers and thumb free to play the piano. Nerve cells in the brain soon die if they do not make proper or useful electrical

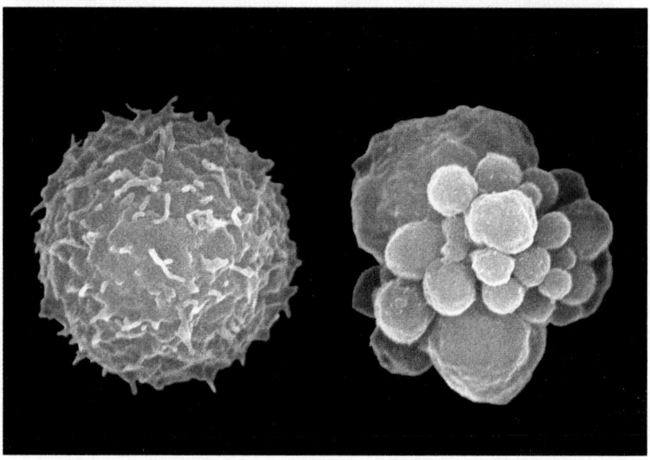

▲ **FIGURE 1-19 Apoptotic cells break apart without spewing forth cell constituents that might harm neighboring cells.** White blood cells normally look like the cell on the left. Cells undergoing programmed cell death (apoptosis), like the cell on the right, form numerous surface blebs that eventually are released. The cell is dying because it lacks certain growth signals. Apoptosis is important to eliminate virus-infected cells, to remove cells where they are not needed (like the webbing that disappears as fingers develop), and to destroy immune system cells that would react with our own bodies. [Gopal Murti/Visuals Unlimited, Inc.]

connections with other cells. Some developing **lymphocytes,** the immune-system cells intended to recognize foreign proteins and polysaccharides, have the ability to react against our own tissues. Such self-reactive lymphocytes become programmed to die before they fully mature. If these cells are not weeded out before reaching maturity, they can cause autoimmune diseases, in which our immune system destroys the very tissues it is meant to protect.

1.4 Investigating Cells and Their Parts

To build an integrated understanding of how the various molecular components that underlie cellular functions work together in a living cell, we must draw on various perspectives. Here, we look at how five disciplines—cell biology, biochemistry and biophysics, genetics, genomics, and developmental biology—can contribute to our knowledge of cell structure and function. The experimental approaches of each field probe the cell's inner workings in different ways, allowing us to ask different types of questions about cells and what they do. Cell division provides a good example to illustrate the role of different perspectives in analyzing a complex cellular process. Although we discuss the different disciplines separately for clarity, in practice most biologists use multiple approaches in concert. This is part of the fun of cell biology, putting together genetics with microscopy or enzymology with development.

The realm of biology ranges in scale more than a billion-fold (Figure 1-20). Beyond that, it's ecology and earth science at the "macro" end, chemistry and physics at the "micro" end. The visible plants and animals that surround us are measured in meters (10^0–10^2 m). By looking closely, we can see a biological world of millimeters (1 mm = 10^{-3} m) and even tenths of millimeters (10^{-4} m). Setting aside oddities like chicken eggs, most cells are 1–100 micrometers (1 μm = 10^{-6} m) long and thus clearly visible only when magnified. To see the structures within cells, we must go farther down the size scale to 10–100 nanometers (1 nm = 10^{-9} m).

Cell Biology Reveals the Size, Shape, Location, and Movements of Cell Components

The goal of cell biologists is to understand how a cell is able to control its own shape and surface properties, transport materials to the right locations, copy itself, and receive and send signals. Cell biologists use several types of microscopy to observe cells, while at the same time labeling specific cell components and altering them to see what happens. Analyses are generally done at the micrometer scale.

Actual observation of cells awaited development of the first, crude microscopes in the early 1600s. A compound

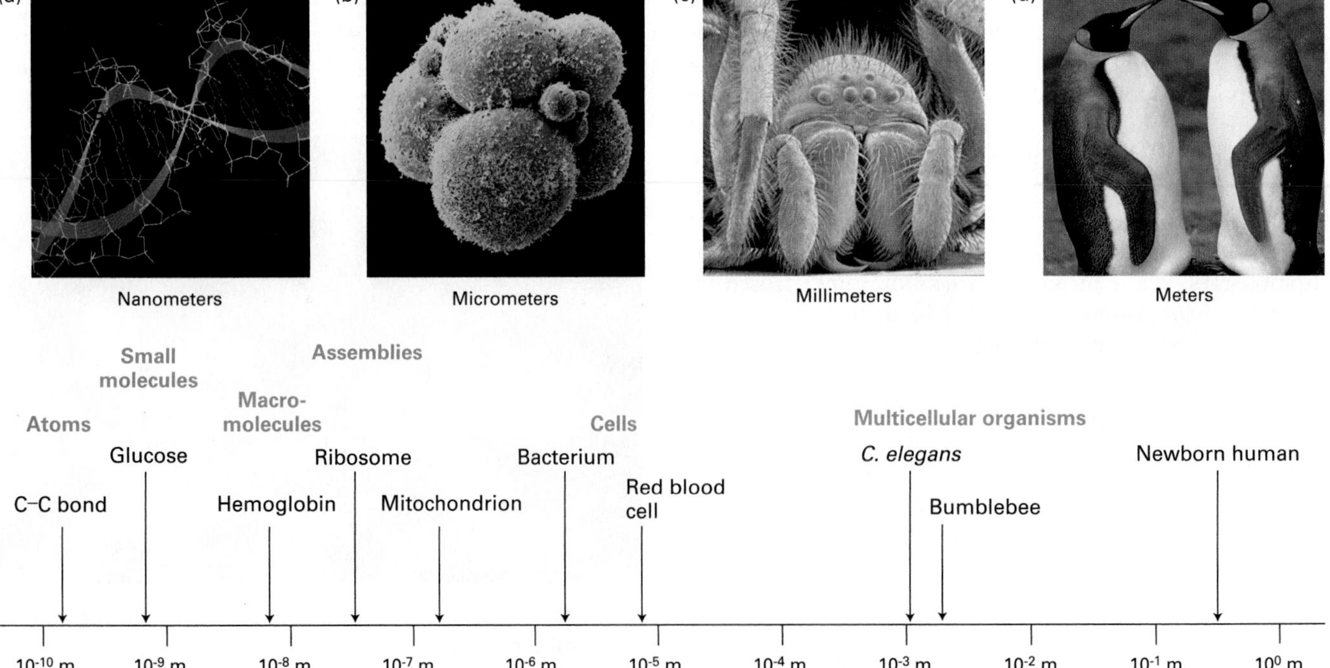

▲ **FIGURE 1-20 Biologists are interested in objects ranging in size from small molecules to the tallest trees.** A sampling of biological objects aligned on a logarithmic scale. (a) The DNA double helix has a diameter of about 2 nm. (b) Eight-cell-stage human embryo three days after fertilization, about 200 μm across. (c) A wolf spider, about 15 mm across. (d) Emperor penguins are about 1 m tall. [Part (a) Will and Deni McIntyre. Part (b) Yorgas Nikas/Photo Researchers, Inc. Part (c) Gary Gaugler/Visuals Unlimited, Inc. Part (d) Hugh S. Rose/Visuals Unlimited, Inc.]

microscope, the most useful type of light microscope, has two lenses. The total magnifying power is the product of the magnification by each lens. As better lenses were invented, the magnifying power and the ability to distinguish closely spaced objects, the **resolution,** increased greatly. Modern compound microscopes magnify the view about a thousandfold, so that a bacterium 1 micrometer (1 μm) long looks like it's a millimeter long. Objects about 0.2 μm apart can be discerned in these instruments.

Microscopy is most powerful when particular components of the cell are stained or labeled specifically, enabling them to be easily seen and located within the cell. A simple example is staining with dyes that bind specifically to DNA to visualize the chromosomes. Specific proteins can be detected by harnessing the binding specificity of **antibodies,** the proteins whose normal task is to help defend animals against infection and foreign substances. In general, each type of antibody binds to one protein or large polysaccharide and no other (Chapter 3). Purified antibodies can be chemically linked to a fluorescent molecule, which permits their detection in a special fluorescence microscope (Chapter 3). If a cell or tissue is treated with a detergent that partially dissolves cell membranes, fluorescent antibodies can drift in and bind to the specific protein they recognize. When the sample is viewed in the microscope, the bound fluorescent antibodies identify the location of the target protein (see Figure 1-15).

Better still is pinpointing proteins in living cells with intact membranes. One way of doing this is to introduce an engineered gene that codes for a hybrid protein: part of the hybrid protein is the cellular protein of interest; the other part is a protein that fluoresces when struck by ultraviolet light. A common fluorescent protein used for this purpose is *green fluorescent protein (GFP)*, a natural protein that makes some jellyfish colorful and fluorescent. GFP "tagging" could reveal, for instance, that a particular protein is first made on the endoplasmic reticulum and then is moved by the cell into the lysosomes. In this case, first the endoplasmic reticulum and later the lysosomes would glow in the dark.

Chromosomes are visible in the light microscope only during mitosis, when they become highly condensed. The extraordinary behavior of chromosomes during mitosis first was discovered using the improved compound microscopes of the late 1800s. About halfway through mitosis, the replicated chromosomes begin to move apart. Microtubules, one of the three types of cytoskeletal filaments, participate in this movement of chromosomes during mitosis. Fluorescent tagging of tubulin, the protein subunit that polymerizes to form microtubules, reveals structural details of cell division that otherwise could not be seen and allows observation of chromosome movement (Figure 1-21).

Electron microscopes use a focused beam of electrons instead of a beam of light. In transmission electron microscopy, specimens are cut into very thin sections and placed under a high vacuum, precluding examination of living cells. The resolution of transmission electron microscopes, about 0.1 nm, permits fine structural details to be distinguished, and their powerful magnification would make a 1-μm-long bacterial cell look like a soccer ball. Most of the organelles in eukaryotic cells and the double-layered structure of the plasma membrane were first observed with electron microscopes (Chapter 9). With new specialized electron microscopy techniques, three-dimensional models of organelles and large protein complexes can be constructed from multiple images. But to obtain a more detailed look at the individual macromolecules within cells, we must turn to techniques within the purview of biochemistry and biophysics.

Focus Animation: Mitosis

MEDIA CONNECTIONS

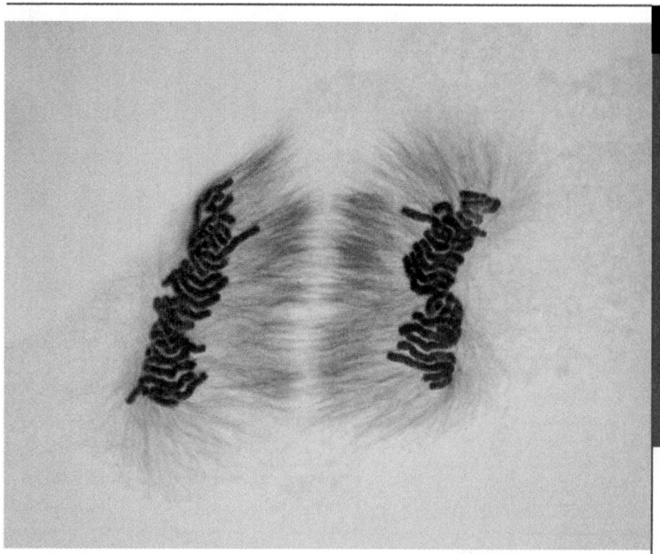

▲ **FIGURE 1-21 During the later stages of mitosis, microtubules (red) pull the replicated chromosomes (black) toward the ends of a dividing cell.** This plant cell is stained with a DNA-binding dye (ethidium) to reveal chromosomes and with fluorescent-tagged antibodies specific for tubulin to reveal microtubules. At this stage in mitosis, the two copies of each replicated chromosome (called chromatids) have separated and are moving away from each other. [Courtesy of Andrew Bajer.]

Biochemistry and Biophysics Reveal the Molecular Structure and Chemistry of Purified Cell Constituents

Biochemists extract the contents of cells and separate the constituents based on differences in their chemical or physical properties, a process called *fractionation.* The attention to individual molecules means operating at the nanometer scale. Of particular interest are proteins, the workhorses of many cellular processes. A typical fractionation scheme involves use of various separation techniques in a sequential fashion. These separation techniques commonly are based on differences in the size of molecules or the electric charge on their surface (Chapter 3). To purify a particular protein of interest, a purification scheme is designed so that each step yields a

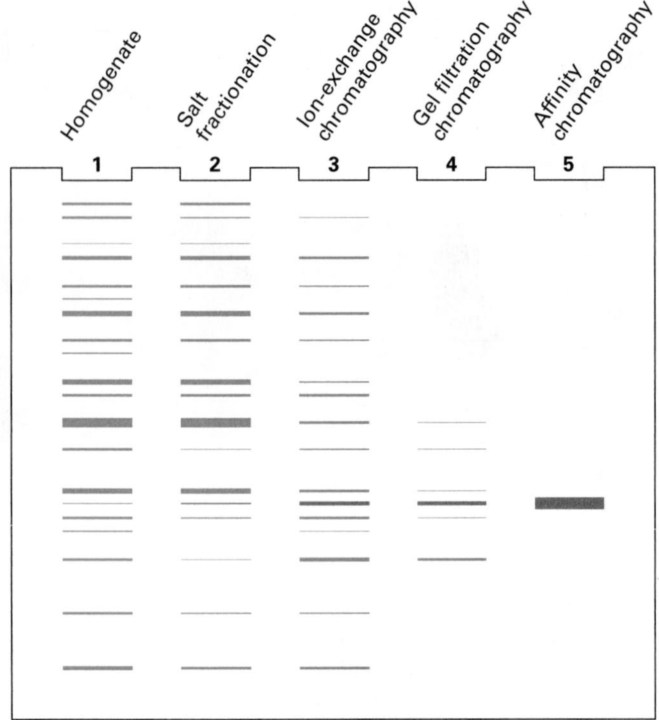

▲ **FIGURE 1-22 Biochemical purification of a protein from a cell extract often requires several separation techniques.** The purification can be followed by gel electrophoresis of the starting protein mixture and the fractions obtained from each purification step. In this procedure, a sample is applied to wells in the top of a gelatin-like slab and an electric field is applied. In the presence of appropriate salt and detergent concentrations, the proteins move through the fibers of the gel toward the anode, with larger proteins moving more slowly through the gel than smaller ones (see Figure 3-35). When the gel is stained, separated proteins are visible as distinct bands whose intensities are roughly proportional to the protein concentration. Shown here are schematic depictions of gels for the starting mixture of proteins (lane 1) and samples taken after each of several purification steps. In the first step, salt fractionation, proteins that precipitated with a certain amount of salt were redissolved; electrophoresis of this sample (lane 2) shows that it contains fewer proteins than the original mixture. The sample then was subjected in succession to three types of column chromatography that separate proteins by electrical charge, size, or binding affinity for a particular small molecule (see Figure 3-37). The final preparation is quite pure, as can be seen from the appearance of just one protein band in lane 5. [After J. Berg et al., 2002, *Biochemistry*, W. H. Freeman and Company, p. 87.]

preparation with fewer and fewer contaminating proteins, until finally only the protein of interest remains (Figure 1-22).

The initial purification of a protein of interest from a cell extract often is a tedious, time-consuming task. Once a small amount of purified protein is obtained, antibodies to it can be produced by methods discussed in Chapter 19. For a biochemist, antibodies are near-perfect tools for isolating larger amounts of a protein of interest for further analysis. In effect, antibodies can "pluck out" the protein they specifically recognize and bind from a semipure sample containing numerous different proteins. An increasingly common alternative is to engineer a gene that encodes a protein of interest with a small attached protein "tag," which can be used to pull out the protein from whole cell extracts.

Purification of a protein is a necessary prelude to studies on how it catalyzes a chemical reaction or carries out other functions and how its activity is regulated. Some enzymes are made of multiple protein chains (subunits) with one chain catalyzing a chemical reaction and other chains regulating when and where that reaction occurs. The molecular machines that perform many critical cell processes constitute even larger assemblies of proteins. By separating the individual proteins composing such assemblies, their individual catalytic or other activities can be assessed. For example, purification and study of the activity of the individual proteins composing the DNA replication machine provided clues about how they work together to replicate DNA during cell division (Chapter 4).

The folded, three-dimensional structure, or **conformation,** of a protein is vital to its function. To understand the relation between the function of a protein and its form, we need to know both what it does and its detailed structure. The most widely used method for determining the complex structures of proteins, DNA, and RNA is **x-ray crystallography,** one of the powerful tools of biophysics. Computer-assisted analysis of the data often permits the location of every atom in a large, complex molecule to be determined. The double-helix structure of DNA, which is key to its role in heredity, was first proposed based on x-ray crystallographic studies. Throughout this book you will encounter numerous examples of protein structures as we zero in on how proteins work.

Genetics Reveals the Consequences of Damaged Genes

Biochemical and crystallographic studies can tell us much about an individual protein, but they cannot prove that it is required for cell division or any other cell process. The importance of a protein is demonstrated most firmly if a mutation that prevents its synthesis or makes it nonfunctional adversely affects the process under study.

We define the **genotype** of an organism as its composition of genes; the term also is commonly used in reference to different versions of a single gene or a small number of genes of interest in an individual organism. A diploid organism generally carries two versions (**alleles**) of each gene, one derived from each parent. There are important exceptions, such as the genes on the X and Y chromosomes in males of some species, including our own. The **phenotype** is the visible outcome of a gene's action, such as blue eyes versus brown eyes or the shapes of peas. In the early days of genetics, the location and chemical identity of genes were unknown; only the observable characteristics, the phenotypes, could be followed. The concept that genes are like "beads" on a long "string," the chromosome, was proposed early in the 1900s based on genetic work with the fruit fly *Drosophila.*

In the classical genetics approach, mutants are isolated that lack the ability to do something a normal organism can do. Often large genetic "screens" are done to look for many different mutant individuals (e.g., fruit flies, yeast cells) that are unable to complete a certain process, such as cell division or muscle formation. In experimental organisms or cultured cells, mutations usually are produced by treatment with a **mutagen,** a chemical or physical agent that promotes mutations in a largely random fashion. But how can we isolate and maintain mutant organisms or cells that are defective in some process, such as cell division, that is necessary for survival? One way is to look for organisms with a **temperature-sensitive mutation.** These mutants are able to grow at one temperature, the *permissive* temperature, but not at another, usually higher temperature, the *nonpermissive* temperature. Normal cells can grow at either temperature. In most cases, a temperature-sensitive mutant produces an altered protein that works at the permissive temperature but unfolds and is nonfunctional at the nonpermissive temperature. Temperature-sensitive screens are readily done with viruses, bacteria, yeast, roundworms, and fruit flies.

By analyzing the effects of numerous different temperature-sensitive mutations that altered cell division, geneticists discovered all the genes necessary for cell division without knowing anything, initially, about which proteins they encode or how these proteins participate in the process. The great power of genetics is to reveal the existence and relevance of proteins without prior knowledge of their biochemical identity or molecular function. Eventually these "mutation-defined" genes were isolated and replicated (cloned) with **recombinant DNA** techniques discussed in Chapter 5. With the isolated genes in hand, the encoded proteins could be produced in the test tube or in engineered bacteria or cultured cells. Then the biochemists could investigate whether the proteins associate with other proteins or DNA or catalyze particular chemical reactions during cell division (Chapter 20).

The analysis of genome sequences from various organisms during the past decade has identified many previously unknown DNA regions that are likely to encode proteins (i.e., protein-coding genes). The general function of the protein encoded by a sequence-identified gene may be deduced by analogy with known proteins of similar sequence. Rather than randomly isolating mutations in novel genes, several techniques are now available for inactivating specific genes by engineering mutations into them or destroying their mRNA with interfering RNA molecules (Chapter 5). The effects of such deliberate gene-specific inactivation procedures provide information about the role of the encoded proteins in living organisms. This application of genetic techniques starts with a gene/protein sequence and ends up with a mutant phenotype; traditional genetics starts with a mutant phenotype and ends up with a gene/protein sequence.

Genomics Reveals Differences in the Structure and Expression of Entire Genomes

Biochemistry and genetics generally focus on one gene and its encoded protein at a time. While powerful, these traditional approaches do not give a comprehensive view of the structure and activity of an organism's genome, its entire set of genes. The field of **genomics** does just that, encompassing the molecular characterization of whole genomes and the determination of global patterns of gene expression. The recent completion of the genome sequences for more than 100 species of bacteria and several eukaryotes now permits comparisons of entire genomes from different species. The results provide overwhelming evidence of the molecular unity of life and the evolutionary processes that made us what we are (see Section 1.5). Genomics-based methods for comparing thousands of pieces of DNA from different individuals all at the same time are proving useful in tracing the history and migrations of plants and animals and in following the inheritance of diseases in human families.

DNA microarrays can simultaneously detect all the mRNAs present in a cell, thereby indicating which genes are being transcribed. Such global patterns of gene expression clearly show that liver cells transcribe a quite different set of genes than do white blood cells or skin cells. Changes in gene expression also can be monitored during a disease process, in response to drugs or other external signals, and during development. For instance, the identification of all the mRNAs present in cultured fibroblasts before, during, and after they divide has given us an overall view of transcriptional changes that occur during cell division (Figure 1-23). Cancer diagnosis is being transformed because previously indistinguishable cancer cells have distinct gene expression patterns and prognoses (Chapter 25). Similar studies with different organisms and cell types are revealing what is universal about the genes involved in cell division and what is specific to particular organisms. To find out which genes are directly regulated by a transcription factor, chromatin containing the protein of interest can be purified with an antibody and the associated DNA analyzed on microarrays, a procedure called chromatin immunoprecipitation.

The entire complement of proteins in a cell, its **proteome,** is controlled in part by changes in gene transcription. The regulated synthesis, processing, localization, and degradation of specific proteins also play roles in determining the proteome of a particular cell. Learning how proteins bind to other proteins, often in large, multiprotein complexes, is providing a comprehensive view of the molecular machines important for cell functioning. The field of proteomics will advance dramatically once high-throughput x-ray crystallography, currently under development, permits researchers to rapidly determine the structures of hundreds or thousands of proteins.

Developmental Biology Reveals Changes in the Properties of Cells as They Specialize

Another approach to viewing cells comes from studying how they change during development of a complex organism. Bacteria, algae, and unicellular eukaryotes (protozoans, yeasts) often, but by no means always, can work solo. The concerted actions of the trillions of cells that compose our bodies require an enormous amount of communication and division of labor. During the development of multicellular organisms, differentiation processes form hundreds of cell types, each specialized for a particular task: transmission of electric signals by neurons, transport of oxygen by red blood cells, destruction of infecting bacteria by macrophages, contraction by muscle cells, chemical processing by liver cells, and so on.

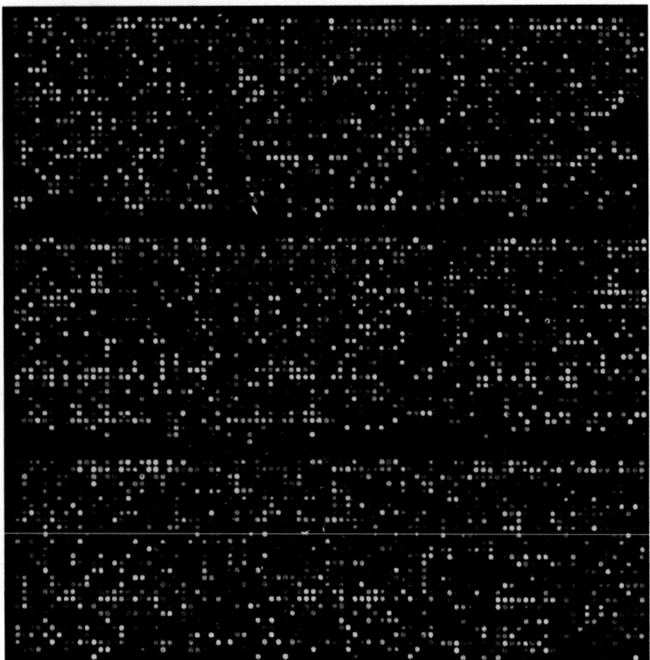

▲ **EXPERIMENTAL FIGURE 1-23 Microarray analysis of normal growing brain cells and brain tumor cells.** An experiment like this is a starting point for learning how tumor cells differ from normal cells. RNA was extracted from normal growing mouse brain cells from the cerebellum and from a tumor of the cerebellum. The RNA from the tumor was labeled with a red dye, and RNA from the normal, non-tumorous cerebellum was labeled with a green dye. The two RNA preparations were mixed and hybridized to a microarray containing thousands of spots of DNA. Each spot contains the DNA sequence of one gene. Unbound RNA was washed away and the microarray was exposed to UV light, which causes the dyes to fluoresce. Spots that are green have bound mostly normal cerebellum RNA, spots that are red have bound mostly tumor RNA, and spots that are yellow have bound roughly equal amounts of each. The faintly stained spots represent genes for which there is little RNA in either sample. The data indicate which genes have been transcribed in tumors, normal cerebellum, or both. Only part of the data is shown here. The entire data set requires analyzing the colors of more than 25,000 spots, all of which can be fitted onto one microscope slide. Precise measurements of color intensity are actually made by a spectrophotometer, but looking by eye shows that many genes are more highly expressed in normal or tumor cells. Some of these differences are the consequence of the change into tumor cells, but some may reveal gene expression changes that cause the tumors to form. In addition, proteins made exclusively in tumors, and perhaps necessary for uncontrolled growth, may be candidate targets for discovering anti-cancer drugs. [Courtesy of Tal Raveh and Matthew Scott, Stanford University School of Medicine.]

Many of the differences among differentiated cells are due to production of specific sets of proteins needed to carry out the unique functions of each cell type; that is, only a subset of an organism's genes is transcribed at any given time or in any given cell. Such **differential gene expression** at different times or in different cell types occurs in bacteria, fungi, plants, animals, and even viruses. Differential gene expression is readily apparent in an early fly embryo, in which all the cells look alike until they are stained to detect the proteins encoded by particular genes (Figure 1-24). Transcription can change within one cell type in response to an external signal or in accordance with a biological clock; some genes, for instance, undergo a daily cycle between low and high transcription rates.

Producing different kinds of cells is not enough to make an organism, any more than collecting all the parts of a truck in one pile gives you a truck. The various cell types must be organized and assembled into all the tissues and organs. Even more remarkable, these body parts must work almost immediately after their formation and continue working during the growth process. For instance, the human heart begins to beat when it is less than 3 mm long, when we are mere 23-day-old embryos, and continues beating as it grows into a fist-size muscle. From a few hundred cells to billions, and still ticking.

In the developing organism, cells grow and divide at some times and not others, they assemble and communicate, they prevent or repair errors in the developmental process, and they coordinate each tissue with others. In the adult organism, cell division largely stops in most organs. If part of an organ such as the liver is damaged or removed, cell division resumes until the organ is regenerated. The legend goes that Zeus punished Prometheus for giving humans fire by

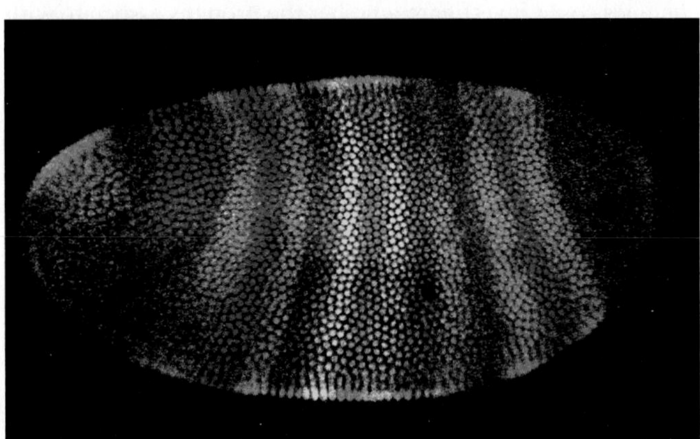

▲ **FIGURE 1-24 Differential gene expression can be detected in early fly embryos before cells are morphologically different.** An early *Drosophila* embryo has about 6000 cells covering its surface, most of which are indistinguishable by simple light microscopy. If the embryo is made permeable to antibodies with a detergent that partially dissolves membranes, the antibodies can find and bind to the proteins they recognize. In this embryo we see antibodies tagged with a fluorescent label bound to proteins that are in the nuclei; each small sphere corresponds to one nucleus. Three different antibodies were used, each specific for a different protein and each giving a distinct color (yellow, green, or blue) in a fluorescence microscope. The red color is added to highlight overlaps between the yellow and blue stains. The locations of the different proteins show that the cells are in fact different at this early stage, with particular genes turned on in specific stripes of cells. These genes control the subdivision of the body into repeating segments, like the black and yellow stripes of a hornet. [Courtesy of Sean Carroll, University of Wisconsin.]

chaining him to a rock and having an eagle eat his liver. The punishment was eternal because, as the Greeks evidently knew, the liver regenerates.

Developmental studies involve watching where, when, and how different kinds of cells form, discovering which signals trigger and coordinate developmental events, and understanding the differential gene action that underlies differentiation (Chapters 16 and 21). During development we can see cells change in their normal context of other cells. Cell biology, biochemistry, cell biology, genetics, and genomics approaches are all employed in studying cells during development.

Choosing the Right Experimental Organism for the Job

Our current understanding of the molecular functioning of cells rests on studies with viruses, bacteria, yeast, protozoa, slime molds, plants, frogs, sea urchins, worms, insects, fish, chickens, mice, and humans. For various reasons, some organisms are more appropriate than others for answering particular questions. Because of the evolutionary conservation of genes, proteins, organelles, cell types, and so forth, discoveries about biological structures and functions obtained with one experimental organism often apply to others. Thus researchers generally conduct studies with the organism that is most suitable for rapidly and completely answering the question being posed, knowing that the results obtained in one organism are likely to be broadly applicable. Figure 1-25 summarizes the typical experimental uses of various organisms whose genomes have been sequenced completely or nearly so. The availability of the genome sequences for these organisms makes them particularly useful for genetics and genomics studies.

Bacteria have several advantages as experimental organisms: they grow rapidly, possess elegant mechanisms for controlling gene activity, and have powerful genetics. This latter property relates to the small size of bacterial genomes, the ease of obtaining mutants, the availability of techniques for transferring genes into bacteria, an enormous wealth of knowledge about bacterial gene control and protein functions, and the relative simplicity of mapping genes relative to one another in the genome. Single-celled yeasts not only have some of the same advantages as bacteria but also possess the cell organization, marked by the presence of a nucleus and organelles, that is characteristic of all eukaryotes.

Studies of cells in specialized tissues make use of animal and plant "models," that is, experimental organisms with attributes typical of many others. Nerve cells and muscle cells, for instance, traditionally were studied in mammals or in creatures with especially large or accessible cells, such as the giant neural cells of the squid and sea hare or the flight muscles of birds. More recently, muscle and nerve development have been extensively studied in fruit flies (*Drosophila melanogaster*), roundworms (*Caenorhabditis elegans*), and zebrafish (*Danio rerio*), in which mutants can be readily isolated. Organisms with large-celled embryos that develop outside the mother (e.g., frogs, sea urchins, fish, and chickens) are extremely useful for tracing the fates of cells as they form different tissues and for making extracts for biochemical studies. For instance, a key protein in regulating mitosis was first identified in studies with frog and sea urchin embryos and subsequently purified from extracts (Chapter 20).

Using recombinant DNA techniques, researchers can engineer specific genes to contain mutations that inactivate or increase production of their encoded proteins. Such genes can be introduced into the embryos of worms, flies, frogs, sea urchins, chickens, mice, a variety of plants, and other organisms, permitting the effects of activating a gene abnormally or inhibiting a normal gene function to be assessed. This approach is being used extensively to produce mouse versions of human genetic diseases. Inactivating particular genes by introducing short pieces of interfering RNA is allowing quick tests of gene functions possible in many organisms. The expansion of genome projects to critically important disease organisms, such as malaria, and to creatures that span the evolutionary tree is bringing new options for medicine and new insights into how living organisms have diversified to take advantage of every possible ecological niche.

Mice have one enormous advantage over other experimental organisms: they are the closest to humans of any animal for which powerful genetic approaches are feasible. Engineered mouse genes carrying mutations similar to those associated with a particular inherited disease in humans can be introduced into mouse embryonic stem (ES) cells. These cells can be injected into an early embryo, which is then implanted into a pseudopregnant female mouse (a mouse treated with hormones to trigger physiological changes needed for pregnancy) (Chapter 5). If the mice that develop from the injected ES cells exhibit diseases similar to the human disease, then the link between the disease and mutations in a particular gene or genes is supported. Once mouse models of a human disease are available, further studies on the molecular defects causing the disease can be done and new treatments can be tested, thereby minimizing human exposure to untested treatments. Large-scale genetic screens are being done that take advantage of newly designed mutagenic transposons. The transposons allow efficient generation of mouse mutants and rapid identification of the gene that has been hit in each one.

A continuous unplanned genetic screen has been performed on human populations for millennia. What we mean is that all sorts of human variations have arisen and have been noticed, since they affect visible or noticeable human characteristics. Thousands of inherited traits have been identified and, more recently, mapped to locations on the chromosomes. Some of these traits are inherited propensities to get a disease; others are eye color or other minor characteristics. Genetic variations in virtually every aspect of cell biology can be found in human populations, allowing studies of normal and disease states and of variant cells in culture.

Less-common experimental organisms offer possibilities for exploring unique or exotic properties of cells and for studying standard properties of cells that are exaggerated in

MEDIA CONNECTIONS

(a)

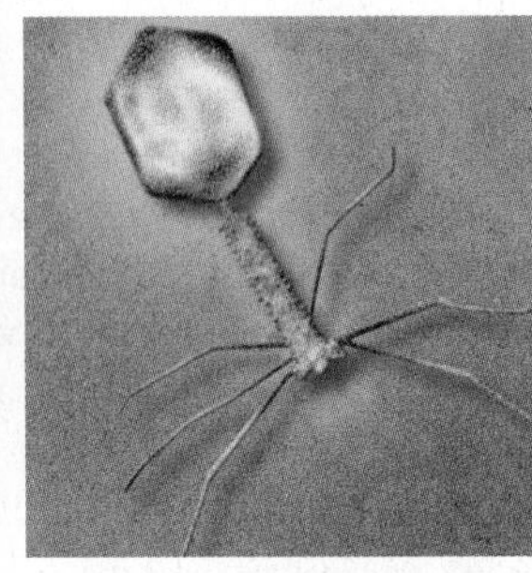

Viruses

Proteins involved in DNA, RNA, protein synthesis
Gene regulation
Cancer and control of cell proliferation
Transport of proteins and organelles inside cells
Infection and immunity
Possible gene therapy approaches

(b)

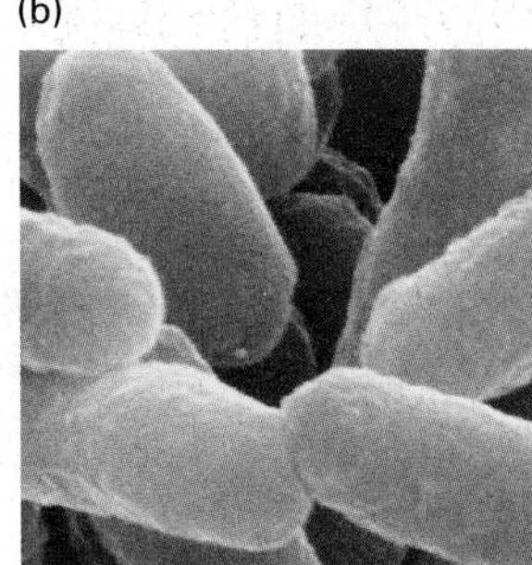

Bacteria

Proteins involved in DNA, RNA, protein synthesis, metabolism
Gene regulation
Targets for new antibiotics
Cell cycle
Signaling

(c)

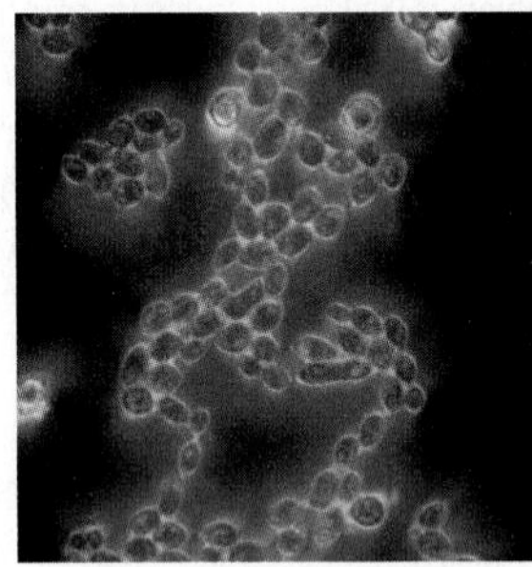

Yeast (*Saccharomyces cerevisiae*)

Control of cell cycle and cell division
Protein secretion and membrane biogenesis
Function of the cytoskeleton
Cell differentiation
Aging
Gene regulation and chromosome structure

(d)

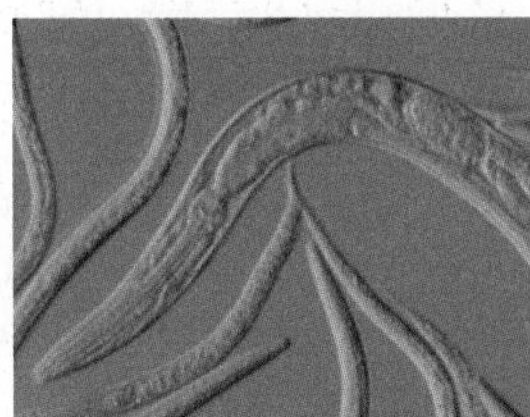

Roundworm (*Caenorhabditis elegans*)

Development of the body plan
Cell lineage
Formation and function of the nervous system
Control of programmed cell death
Cell proliferation and cancer genes
Aging
Behavior
Gene regulation and chromosome structure

(e)

Fruit fly (*Drosophila melanogaster*)

Development of the body plan
Generation of differentiated cell lineages
Formation of the nervous system, heart, and musculature
Programmed cell death
Genetic control of behavior
Cancer genes and control of cell proliferation
Control of cell polarization
Effects of drugs, alcohol, pesticides

(f)

Zebrafish

Development of vertebrate body tissues
Formation and function of brain and nervous system
Birth defects
Cancer

(g)

Mice, including cultured cells

Development of body tissues
Function of mammalian immune system
Formation and function of brain and nervous system
Models of cancers and other human diseases
Gene regulation and inheritance
Infectious disease

(h)

Plant (*Arabidopsis thaliana*)

Development and patterning of tissues
Genetics of cell biology
Agricultural applications
Physiology
Gene regulation
Immunity
Infectious disease

◀ **FIGURE 1-25 Each experimental organism used in cell biology has advantages for certain types of studies.** Viruses (a) and bacteria (b) have small genomes amenable to genetic dissection. Many insights into gene control initially came from studies with these organisms. The yeast *Saccharomyces cerevisiae* (c) has the cellular organization of a eukaryote but is a relatively simple single-celled organism that is easy to grow and to manipulate genetically. In the nematode worm *Caenorhabditis elegans* (d), which has a small number of cells arranged in a nearly identical way in every worm, the formation of each individual cell can be traced. The fruit fly *Drosophila melanogaster* (e), first used to discover the properties of chromosomes, has been especially valuable in identifying genes that control embryonic development. Many of these genes are evolutionarily conserved in humans. The zebrafish *Danio rerio* (f) is used for rapid genetic screens to identify genes that control development and organogenesis. Of the experimental animal systems, mice (*Mus musculus*) (g) are evolutionarily the closest to humans and have provided models for studying numerous human genetic and infectious diseases. The mustard-family weed *Arabidopsis thaliana*, sometimes described as the *Drosophila* of the plant kingdom, has been used for genetic screens to identify genes involved in nearly every aspect of plant life. Genome sequencing is completed for many viruses and bacterial species, the yeast *Saccharomyces cerevisiae,* the roundworm *C. elegans,* the fruit fly *D. melanogaster,* humans, and the plant *Arabidopsis thaliana.* It is mostly completed for mice and in progress for zebrafish. Other organisms, particularly frogs, sea urchins, chickens, and slime molds, continue to be immensely valuable for cell biology research. Increasingly, a wide variety of other species are used, especially for studies of evolution of cells and mechanisms. [Part (a) Visuals Unlimited, Inc. Part (b) Kari Lountmaa/Science Photo Library/Photo Researchers, Inc. Part (c) Scimat/Photo Researchers, Inc. Part (d) Photo Researchers, Inc. Part (e) Darwin Dale/Photo Researchers, Inc. Part (f) Inge Spence/Visuals Unlimited, Inc. Part (g) J. M. Labat/Jancana/Visuals Unlimited, Inc. Part (h) Darwin Dale/Photo Researchers, Inc.]

a useful fashion in a particular animal. For example, the ends of chromosomes, the **telomeres,** are extremely dilute in most cells. Human cells typically contain 92 telomeres (46 chromosomes $\times$ 2 ends per chromosome). In contrast, some protozoa with unusual "fragmented" chromosomes contain millions of telomeres per cell. Taking advantage of the unique properties of this well-chosen experimental organism has led to important recent discoveries about telomere structure.

The Most Successful Biological Studies Use Multiple Approaches

We have discussed five classes of approaches to biological problems: cell biology, biochemistry and biophysics, genetics, genomics, and developmental biology. Each has its own types of experiments, and most biological problems require more than one approach in order to reach a satisfying understanding of mechanism. Now we will survey how these approaches have been applied to the study of cell division to emphasize how important it is to use multiple types of experiments.

Cell division was viewed, and indeed discovered, by some of the earliest users of microscopes. More recently a variety of kinds of microscopy, including confocal and electron microscopy and time-lapse imaging (Chapter 9), have been used to characterize the steps of the cell cycle. Most biology begins with this sort of observation, defining the mysteries that must be tackled experimentally. Then manipulations begin. Antibodies were made against proteins that play critical roles in cell division both to detect proteins and in some cases to interfere with the functions of those proteins. Key proteins were fused to fluorescent proteins, starting with the jellyfish green fluorescent protein (GFP), so that key proteins could be followed in living cells. Questions about when and where proteins work could be addressed and their functions defined to an extent.

The apparatus of cell division, such as the mitotic spindle and other protein complexes, was purified and analyzed using the approaches of biochemistry and biophysics. Each protein had to be purified to find out whether it is part of a complex of proteins bound together as a machine, and the structures of key proteins were determined using x-ray crystallography and other methods (Chapter 3). Previously unknown enzymatic activities were detected in extracts using assays such as measuring the attachment of phosphate groups to cell division regulatory proteins by kinases, and then the relevant kinases could be purified.

Finding a novel protein in a complex of proteins involved in cell division makes it a good bet that the protein does something important, a sort of "guilt by association," but it does not provide *proof* that the protein matters. For that one must turn to genetics. Genetics can be used to identify mutants in the newly found protein. If cell division fails in a living organism when the protein is not working, you know the protein matters. Genetics is also a way to identify previously unknown genes and proteins since screens can be done (especially in bacteria and yeast, but also in more complex lab organisms) to look for all the genes that are needed for cell division. The newly discovered proteins can be incorporated into a complete picture of the mechanics of cell division machinery.

Genomics provides another way to look for working parts of the cell-division machine. Since it is often true that mRNAs and proteins are produced only when they are needed, using microarrays to look for all genes whose expression varies with the cell cycle is a powerful approach to identify candidates for cell division regulators.

Having identified new genes that are required for cell division, one must find out how these genes' protein products work. Simply knowing that a protein matters to cell division is not enough to understand the mechanism. Thus it is necessary to return to biochemical and biophysical approaches to work out the molecular biology and to cell biology to monitor protein locations and movements.

Finally, cell division does not happen in a vacuum; it happens in the context of the life cycle of the organism. To fully appreciate how the regulation works and is used, it is important to use the approaches of developmental biology to reveal when and where cell division normally happens and why. In these experiments the times and places of cell division during the development of the organism are monitored, and then the signals that stimulate or suppress cell division are identified and studied. Errors in developmental control of cell division, revealed by studying mutants, can cause organs and tissues to be the wrong size or cancer.

The same kinds of approaches that work for cell division can be applied to many other biological challenges, such as learning how muscles form or how they work or how the brain functions. Often it's good to use every tool in the toolkit.

1.5 A Genome Perspective on Evolution

Comprehensive studies of genes and proteins from many organisms are giving us an extraordinary documentation of the history of life. Nature is a laboratory that has been conducting experiments for billions of years, and some of the most successful genomes that emerged are still with us. We share with other eukaryotes thousands of individual proteins, hundreds of macromolecular machines, and most of our organelles, all as a result of our shared evolutionary history. New insights into molecular cell biology arising from genomics are leading to a fuller appreciation of the elegant molecular machines that arose during billions of years of genetic tinkering and evolutionary selection for the most efficient, precise designs. Due to alternative RNA splicing, the number of proteins vastly exceeds the number of genes, and the functions of many variant proteins and assemblies of proteins remain to be discovered. Once a more complete description of cells is in hand, we will be ready to fully investigate the rippling, flowing dynamics of living systems.

Metabolic Proteins, the Genetic Code, and Organelle Structures Are Nearly Universal

Even organisms that look incredibly different share many biochemical properties. For instance, the enzymes that catalyze degradation of sugars and many other simple chemical reactions in cells have similar structures and mechanisms in most living things. The genetic code whereby the nucleotide sequences of mRNA specifies the amino acid sequences of proteins can be read equally well by a bacterial cell and a human cell. Because of the universal nature of the genetic code, bacterial "factories" can be designed to manufacture growth factors, insulin, clotting factors, and other human proteins with therapeutic uses. The biochemical similarities among organisms also extend to the organelles found in eukaryotic cells. The basic structures and functions of these subcellular components are largely conserved in all eukaryotes.

Computer analysis of DNA sequence data, now available for numerous bacterial species and several eukaryotes, can locate protein-coding genes within genomes. With the aid of the genetic code, the amino acid sequences of proteins can be deduced from the corresponding gene sequences. Although simple conceptually, "finding" genes and deducing the amino acid sequences of their encoded proteins is complicated in practice because of the many noncoding regions in eukaryotic DNA (Chapter 5). Despite the difficulties and occasional ambiguities in analyzing DNA sequences, comparisons of the genomes from a wide range of organisms provide stunning, compelling evidence for the conservation of the molecular mechanisms that build and change organisms and for the common evolutionary history of all species.

Darwin's Ideas About the Evolution of Whole Animals Are Relevant to Genes

Darwin did not know that genes exist or how they change, but we do: the DNA replication machine makes an error, or a mutagen causes replacement of one nucleotide with another or breakage of a chromosome. Some changes in the genome are innocuous, some mildly harmful, some deadly; a very few are beneficial. Mutations can change the sequence of a gene in a way that modifies the activity of the encoded protein or alters when, where, and in what amounts the protein is produced in the body.

Gene-sequence changes that are harmful will be lost from a population of organisms because the affected individuals cannot survive as well as their relatives. This selection process is exactly what Darwin described without knowing the underlying mechanisms that cause organisms to vary. Thus the selection of whole organisms for survival is really a selection of genes, or more accurately sets of genes. A population of organisms often contains many variants that are all roughly equally well-suited to the prevailing conditions. When conditions change—a fire, a flood, loss of preferred food supply, climate shift—variants that are better able to adapt will survive, and those less suited to the new conditions will begin to die out. In this way, the genetic composition of a population of organisms can change over time.

Many Genes Controlling Development Are Remarkably Similar in Humans and Other Animals

As humans, we probably have a biased and somewhat exaggerated view of our status in the animal kingdom. Pride in our swollen forebrain and its associated mental capabilities may blind us to the remarkably sophisticated abilities of other species: navigation by birds, the sonar system of bats, homing by salmon, or the flight of a fly.

Despite all the evidence for evolutionary unity at the cellular and physiological levels, everyone expected that genes

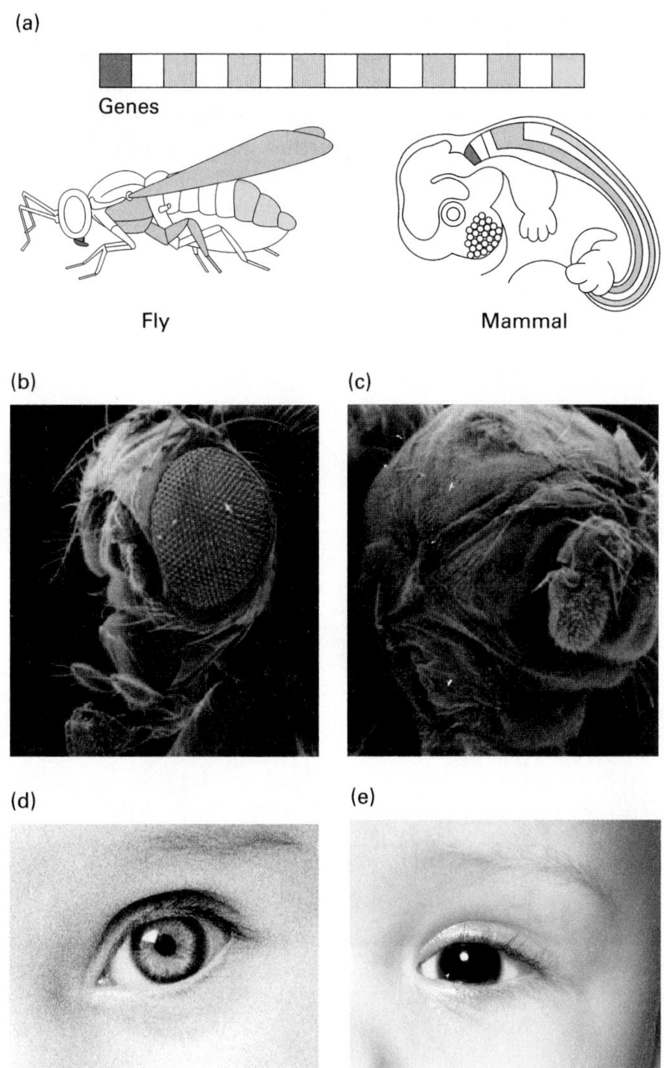

◀ **FIGURE 1-26 Similar genes, conserved during evolution, regulate many developmental processes in diverse animals.** Insects and mammals are estimated to have had a common ancestor about half a billion years ago. They share genes that control similar processes, such as growth of heart and eyes and organization of the body plan, indicating conservation of function from ancient times. (a) *Hox* genes are found in clusters on the chromosomes of most or all animals. *Hox* genes encode related proteins that control the activities of other genes. *Hox* genes direct the development of different segments along the head-to-tail axis of many animals, as indicated by corresponding colors. Each gene is activated (transcriptionally) in a specific region along the head-to-tail axis and controls the growth of tissues there. For example, in mice the *Hox* genes are responsible for the distinctive shapes of vertebrae. Mutations affecting *Hox* genes in flies cause body parts to form in the wrong locations, such as legs in lieu of antennae on the head. These genes provide a head-to-tail address and serve to direct formation of the right structures in the right places. (b) Development of the large compound eyes in fruit flies requires a gene called *eyeless* (named for the mutant phenotype). (c) Flies with inactivated *eyeless* genes lack eyes. (d) Normal human eyes require the human gene, called *Pax6,* that corresponds to *eyeless.* (e) People lacking adequate *Pax6* function have the genetic disease *aniridia,* a lack of irises in the eyes. *Pax6* and *eyeless* encode highly related proteins that regulate the activities of other genes and are descended from the same ancestral gene. [Parts (b) and (c) Andreas Hefti, Interdepartmental Electron Microscopy (IEM) Biocenter of the University of Basel. Part (d) © Simon Fraser/Photo Researchers, Inc. Part (e) Visuals Unlimited]

regulating animal development would differ greatly from one phylum to the next. After all, insects and sea urchins and mammals look *so* different. We must have many unique proteins to create a brain like ours . . . or must we? The fruits of research in developmental genetics during the past two decades reveal that insects and mammals, which have a common ancestor about half a billion years ago, possess many similar development-regulating genes (Figure 1-26). Indeed, a large number of these genes appear to be conserved in many and perhaps all animals. Remarkably, the developmental functions of the proteins encoded by these genes are also often preserved. For instance, certain proteins involved in eye development in insects are related to protein regulators of eye development in mammals. Same for development of the heart, gut, lungs, and capillaries and for placement of body parts along the head-to-tail and back-to-front body axes (Chapter 19).

This is not to say that all genes or proteins are evolutionarily conserved. Many striking examples exist of proteins that, as far as we can tell, are utterly absent from certain lineages of animals. Plants, not surprisingly, exhibit many such differences from animals after a billion-year separation in their evolution. Yet certain DNA-binding proteins differ between peas and cows at only two amino acids out of 102!

Human Medicine Is Informed by Research on Other Organisms

Mutations that occur in certain genes during the course of our lives contribute to formation of various human cancers. The normal, wild-type forms of such "cancer-causing" genes generally encode proteins that help regulate cell proliferation or death (Chapter 21). We also can inherit from our parents mutant copies of genes that cause all manner of genetic diseases, such as cystic fibrosis, muscular dystrophy, sickle cell anemia, and Huntington's disease. Happily we can also inherit genes that make us robustly resist disease. A remarkable number of genes associated with cancer and other human diseases are present in evolutionarily distant animals. For example, a recent study shows that more than three-quarters of the known human disease genes are related to genes found in the fruit fly *Drosophila*.

With the identification of human disease genes in other organisms, experimental studies in experimentally tractable organisms should lead to rapid progress in understanding the normal functions of the disease-related genes and what occurs when things go awry. Conversely, the disease states themselves constitute a genetic analysis with well-studied phenotypes. All the genes that can be altered to cause a certain disease may encode a group of functionally related proteins. Thus clues about the normal cellular functions of proteins come from human diseases and can be used to guide initial research into mechanism. For instance, genes initially identified because of their link to cancer in humans can be studied in the context of normal development in various model organisms, providing further insight about the functions of their protein products.

Polarized light microscopic image of crystals of ATP, whose hydrolysis is a primary source of energy that drives many cellular chemical reactions. [Dr. Arthur M. Siegelman/Visuals Unlimited.]

CHAPTER

2

CHEMICAL FOUNDATIONS

The life of a cell depends on thousands of chemical interactions and reactions exquisitely coordinated with one another in time and space and under the influence of the cell's genetic instructions and its environment. By understanding at a molecular level these interactions and reactions, we can begin to answer fundamental questions about cellular life: How does a cell extract critical nutrients and information from its environment? How does a cell convert the energy stored in nutrients into work (movement, synthesis of critical components)? How does a cell transform nutrients into the fundamental structures required for its survival (cell wall, nucleus, nucleic acids, proteins, cytoskeleton)? How does a cell link itself to other cells to form a tissue? How do cells communicate with one another so that a complex, efficiently functioning organism can develop and thrive? One of the goals of *Molecular Cell Biology* is to provide answers to these and other questions about the structure and function of cells and organisms in terms of the properties of individual molecules and ions.

For example, the properties of one such molecule, water, have controlled and continue to control the evolution, structure, and function of cells. You cannot understand biology without appreciating how the properties of water control the chemistry of life. Life first arose in a watery environment. Constituting 70–80 percent by weight of most cells, water is the most abundant molecule in biological systems. It is within this aqueous milieu that small molecules and ions, which make up about 7 percent of the weight of living matter, assemble into the larger macromolecules and macromolecular aggregates that make up a cell's machinery and architecture and so the remaining mass of organisms. These small molecules include amino acids (the building blocks of proteins), nucleotides (the building blocks of DNA and RNA), lipids (the building blocks of biomembranes), and sugars (the building blocks of starches and cellulose).

Many biomolecules (e.g., sugars) readily dissolve in water; these molecules are called **hydrophilic** (water liking). Others (e.g., cholesterol) are oily, fatlike substances that shun water; these are said to be **hydrophobic** (water fearing). Still other biomolecules (e.g., phospholipids) are a bit schizophrenic, containing both hydrophilic and hydrophobic regions; these molecule are said to be **amphipathic.** Phospholipids are used to build the flexible membranes that form the wall-like boundaries of cells and their internal organelles. The smooth functioning of cells, tissues, and organisms depends on all these molecules, from the smallest to the largest. Indeed, the chemistry of the simple proton (H^+) can be as important to the survival of a human cell as that of each gigantic, genetic-code-carrying DNA molecule (the mass of the DNA molecule in human

OUTLINE

(a) Molecular complementarity

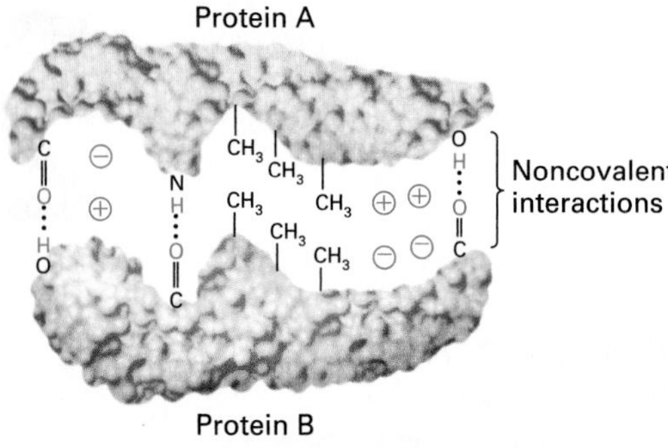

(b) Chemical building blocks

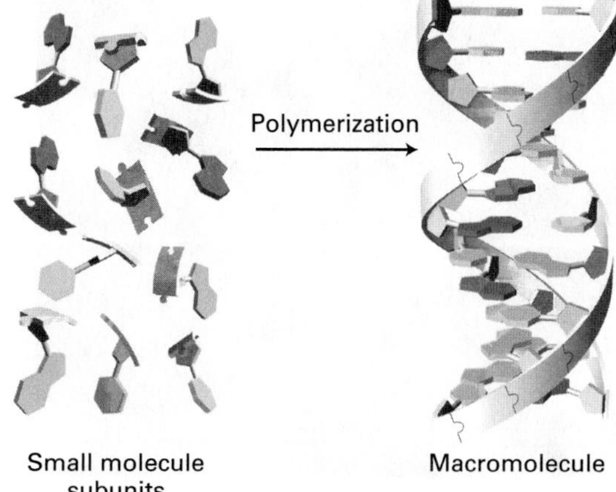

(c) Chemical equilibrium

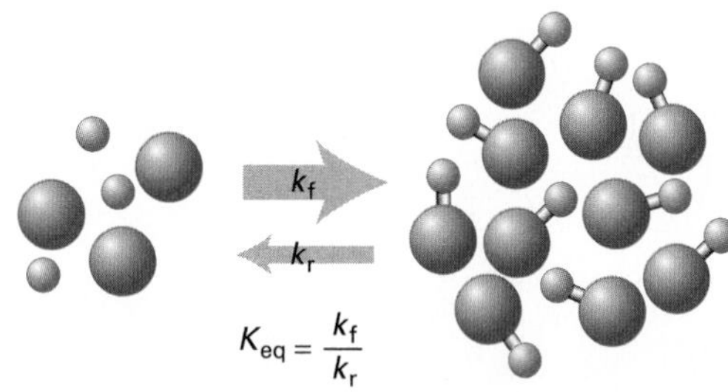

(d) Chemical bond energy

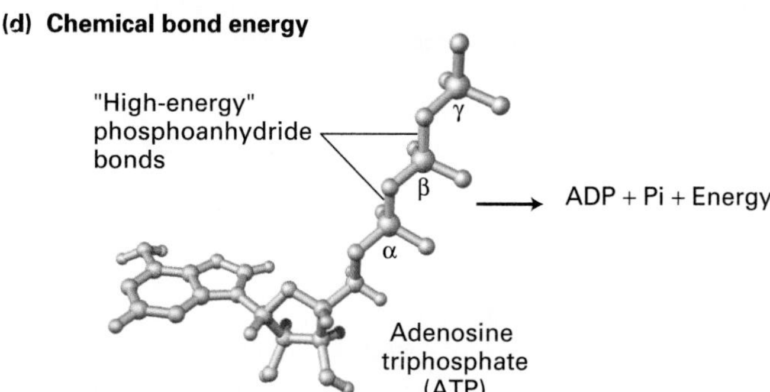

▲ **FIGURE 2-1 Chemistry of life: four key concepts.** (a) Molecular complementarity lies at the heart of all biomolecular interactions, as when two proteins with complementary shapes and chemical properties come together to form a tightly bound complex. (b) Small molecules serve as building blocks for larger structures. For example, to generate the information-carrying macromolecule DNA, four small nucleotide building blocks are covalently linked into long strings (polymers), which then wrap around each other to form the double helix. (c) Chemical reactions are reversible, and the distribution of the chemicals between starting reagents (*left*) and the products of the reactions (*right*) depends on the rate constants of the forward (k_f, upper arrow) and reverse (k_r, lower arrow) reactions. The ratio of these, K_{eq}, provides an informative measure of the relative amounts of products and reactants that will be present at equilibrium. (d) In many cases, the source of energy for chemical reactions in cells is the hydrolysis of the molecule ATP. This energy is released when a high-energy phosphoanhydride bond linking the β and γ phosphates in the ATP molecule (red) is broken by the addition of a water molecule, forming ADP and P_i.

chromosome 1 is 8.6×10^{10} times that of a proton!). The chemical interactions of all of these molecules, large and small, with water and with one another, define the nature of life.

Luckily, although many types of biomolecules interact and react in numerous and complex pathways to form functional cells and organisms, a relatively small number of chemical principles are necessary to understand cellular processes at the molecular level (Figure 2-1). In this chapter we review these key principles, some of which you already know well. We begin with the covalent bonds that connect atoms into a molecule and the noncovalent forces that stabilize groups of atoms into functional structures within and between molecules. We then consider the key properties of the basic chemical building blocks of macromolecules and macromolecular assemblies. After reviewing those aspects of chemical equilibrium that are most relevant to biological systems, we end the chapter with basic principles of biochemical energetics, including the central role of ATP (adenosine triphosphate) in capturing and transferring energy in cellular metabolism.

2.1 Covalent Bonds and Noncovalent Interactions

Strong and weak attractive forces between atoms are the "glue" that holds them together in individual molecules and permits interactions between different biomolecules. Strong forces form a **covalent bond** when two atoms share one pair of electrons ("single" bond) or multiple pairs of electrons ("double" bond, "triple" bond, etc.). The weak attractive forces of **noncovalent interactions** are equally important in determining the properties and functions of biomolecules such as proteins, nucleic acids, carbohydrates, and lipids. We will first review covalent bonds and then discuss the four

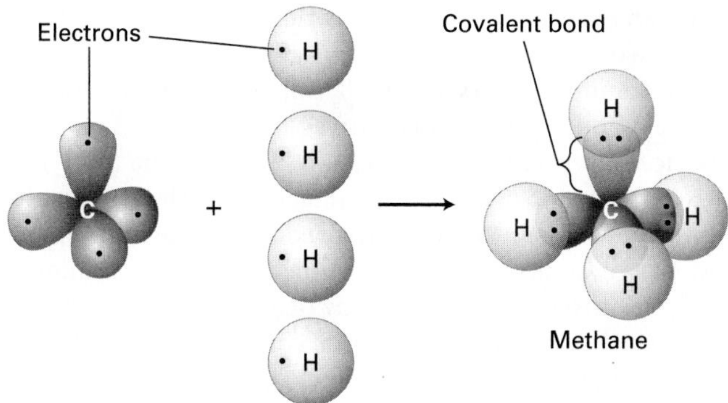

▲ **FIGURE 2-2 Covalent bonds form by the sharing of electrons.** Covalent bonds, the strong forces that hold atoms together into molecules, form when atoms share electrons from their outermost electron orbitals. Each atom forms a defined number and geometry of covalent bonds.

major types of noncovalent interactions: ionic bonds, hydrogen bonds, van der Waals interactions, and the hydrophobic effect.

The Electronic Structure of an Atom Determines the Number and Geometry of Covalent Bonds It Can Make

Hydrogen, oxygen, carbon, nitrogen, phosphorus, and sulfur are the most abundant elements in biological molecules. These atoms, which rarely exist as isolated entities, readily form covalent bonds, using electrons in the outermost electron orbitals surrounding their nuclei (Figure 2-2). As a rule, each type of atom forms a characteristic number of covalent bonds with other atoms, with a well-defined geometry determined by the atom's size and by both the distribution of electrons around the nucleus and the number of electrons that it can share. In some cases (e.g., carbon), the number of stable covalent bonds formed is fixed; in other cases (e.g., sulfur), different numbers of stable covalent bonds are possible.

All the biological building blocks are organized around the carbon atom, which normally forms four covalent bonds with three or four other atoms. As illustrated in Figure 2-3a for formaldehyde, carbon can bond to three atoms, all in a common plane. The carbon atom forms two typical single bonds with two atoms and a double bond (two shared electron pairs) with the third atom. In the absence of other constraints, atoms joined by a single bond generally can rotate freely about the bond axis, whereas those connected by a double bond cannot. The rigid planarity imposed by double bonds has enormous significance for the shapes and flexibility of biomolecules such as phospholipids, proteins, and nucleic acids.

Carbon can also bond to four rather than three atoms. As illustrated by the methane (CH_4) molecule, when carbon is bonded to four other atoms, the angle between any two bonds is 109.5° and the positions of bonded atoms define the four points of a tetrahedron (Figure 2-3b). This geometry defines the structures of many biomolecules. A carbon (or any other) atom bonded to four dissimilar atoms or groups in a nonplanar configuration is said to be asymmetric. The tetrahedral orientation of bonds formed by an **asymmetric carbon atom** can be arranged in three-dimensional space in two different ways, producing molecules that are mirror images of each other, a property called *chirality* (from the Greek word *cheir,* meaning "hand") (Figure 2-4). Such molecules are called *optical isomers,* or **stereoisomers.** Many molecules in cells contain at least one asymmetric carbon atom, often called a *chiral carbon* atom. The different stereoisomers of a molecule usually have completely different biological activities because the arrangement of atoms within their structures differs, yielding their unique abilities to interact and chemically react with other molecules.

Some drugs are mixtures of the stereoisomers of small molecules in which only one stereoisomer has the biological activity of interest. The use of a pure single stereoisomer of the chemical in place of the mixture can result in a more potent drug with reduced side effects. For example, one stereoisomer of the antidepressant drug citalopram (Celexa) is

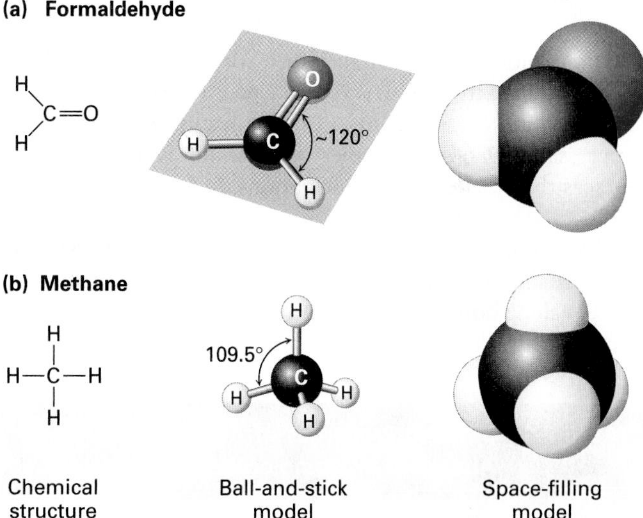

▲ **FIGURE 2-3 Geometry of bonds when carbon is covalently linked to three or four other atoms.** (a) A carbon atom can be bonded to three atoms, as in formaldehyde (CH_2O). In this case, the carbon-bonding electrons participate in two single bonds and one double bond, which all lie in the same plane. Unlike atoms connected by a single bond, which usually can rotate freely about the bond axis, those connected by a double bond cannot. (b) When a carbon atom forms four single bonds, as in methane (CH_4), the bonded atoms (all H in this case) are oriented in space in the form of a tetrahedron. The letter representation on the left clearly indicates the atomic composition of the molecule and the bonding pattern. The ball-and-stick model in the center illustrates the geometric arrangement of the atoms and bonds, but the diameters of the balls representing the atoms and their nonbonding electrons are unrealistically small compared with the bond lengths. The sizes of the electron clouds in the space-filling model on the right more accurately represent the structure in three dimensions.

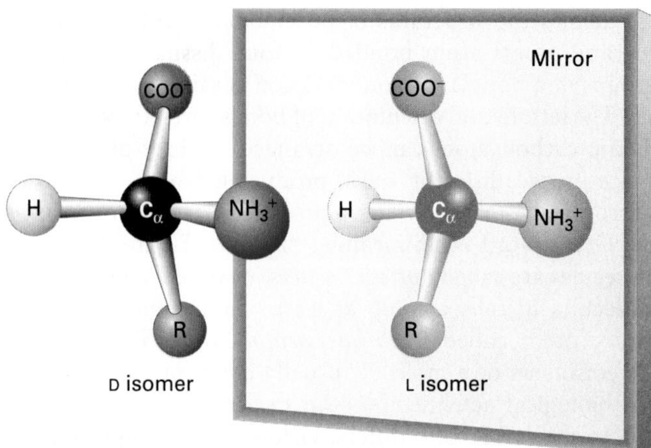

▲ **FIGURE 2-4 Stereoisomers.** Many molecules in cells contain at least one asymmetric carbon atom. The tetrahedral orientation of bonds formed by an asymmetric carbon atom can be arranged in three-dimensional space in two different ways, producing molecules that are mirror images, or stereoisomers, of each other. Shown here is the common structure of an amino acid, with its central asymmetric carbon and four attached groups, including the R group, discussed in Section 2.2. Amino acids can exist in two mirror-image forms, designated L and D. Although the chemical properties of such stereoisomers are identical, their biological activities are distinct. Only L amino acids are found in proteins.

170 times more potent than the other. Some stereoisomers have very different activities. Darvon is a pain reliever, whereas its stereoisomer, Novrad (*Darvon* spelled backward), is a cough suppressant. One stereoisomer of ketamine is an anesthetic, whereas the other causes hallucinations. ∎

The number of covalent bonds formed by other common atoms is shown in Table 2-1. A hydrogen atom forms only one covalent bond. An atom of oxygen usually forms only two covalent bonds but has two additional pairs of electrons that can participate in noncovalent interactions. Sulfur forms two covalent bonds in hydrogen sulfide (H_2S) but also can accommodate six covalent bonds, as in sulfuric acid (H_2SO_4) and its sulfate derivatives. Nitrogen and phosphorus each have five electrons to share. In ammonia (NH_3), the nitrogen atom forms three covalent bonds; the pair of electrons around the atom not involved in a covalent bond can take part in noncovalent interactions. In the ammonium ion (NH_4^+), nitrogen forms four covalent bonds, which have a tetrahedral geometry. Phosphorus commonly forms five covalent bonds, as in phosphoric acid (H_3PO_4) and its phosphate derivatives, which form the backbone of nucleic acids. Phosphate groups covalently attached to proteins play a key role in regulating the activity of many proteins, and the central molecule in cellular energetics, ATP, contains three phosphate groups (see Section 2.4). A summary of common covalent linkages and functional groups (portions of molecules that confer distinctive chemical properties) is provided in Table 2-2.

TABLE 2-1 Bonding Properties of Atoms Most Abundant in Biomolecules

ATOM AND OUTER ELECTRONS	USUAL NUMBER OF COVALENT BONDS	TYPICAL BOND GEOMETRY
H·	1	H
·Ö·	2	O
·S̈·	2, 4, or 6	S
·N̈·	3 or 4	N
·P̈·	5	P
·Ċ·	4	C

Electrons May Be Shared Equally or Unequally in Covalent Bonds

The extent of an atom's ability to attract an electron is called its *electronegativity*. In a bond between atoms with identical or similar electronegativities, the bonding electrons are essentially shared equally between the two atoms, as is the case for most C—C and C—H bonds. Such bonds are called **nonpolar.** In many molecules, the bonded atoms have different electronegativities, resulting in unequal sharing of the electrons. The bond between them is said to be **polar.**

One end of a polar bond has a partial negative charge (δ^-), and the other end has a partial positive charge (δ^+). In an O—H bond, for example, the greater electronegativity of the oxygen atom relative to hydrogen results in the electrons spending more time around the oxygen atom than the hydrogen. Thus the O—H bond possesses an *electric dipole,* a positive charge separated from an equal but opposite negative charge. The amount of δ^- charge on the oxygen atom of a O—H dipole is approximately 25 percent of that of an electron, with an equivalent δ^+ charge on the H atom. Because of its two O—H bonds that are not on exact opposite sides of the O atom, water molecules (H_2O) are dipoles (Figure 2-5) that form electrostatic, noncovalent interactions with one another and with other molecules. These interactions play a critical role in almost every biochemical interaction and so are fundamental to cell biology.

The polarity of the O=P double bond in H_3PO_4 results in a *resonance hybrid,* a structure between the two forms shown below in which nonbonding electrons are shown as pairs of dots:

H—O—P(=O)(—O—H)—O—H ⟷ H—O—P^+(—O^-)(—O—H)—O—H

TABLE 2-2 Common Functional Groups and Linkages in Biomolecules

FUNCTIONAL GROUPS			
—OH Hydroxyl (alcohol)	—C(=O)—R Acyl (triacylglycerol)	—C(=O)— Carbonyl (ketone)	—C(=O)—O^- Carboxyl (carboxylic acid)
—SH Sulfhydryl (Thiol)	—NH_2 or —$\overset{+}{N}H_3$ Amino (amines)	—O—P(=O)(O^-)—O^- Phosphate (phosphorylated molecule)	—O—P(=O)(O^-)—O—P(=O)(O^-)— Pyrophosphate (diphosphate)
LINKAGES			
—C—O—C(=O)— Ester	—C—O—C— Ether		—N—C(=O)— Amide

In the resonance hybrid on the right, one of the electrons from the P=O double bond has accumulated around the O atom, giving it a negative charge and leaving the P atom with a positive charge. These charges are important in noncovalent interactions.

Covalent Bonds Are Much Stronger and More Stable Than Noncovalent Interactions

Covalent bonds are very stable (i.e., considered to be strong) because the energies required to break them are much greater than the thermal energy available at room temperature (25 °C) or body temperature (37 °C). For example, the thermal energy at 25 °C is approximately 0.6 kilocalorie per mole (kcal/mol), whereas the energy required to break the carbon-carbon single bond (C—C) in ethane is about 140 times larger (Figure 2-6). Consequently, at room temperature (25 °C), fewer than 1 in 10^{12} ethane molecules is broken into a pair of $\cdot CH_3$ molecules, each containing an unpaired, nonbonding electron (called a radical).

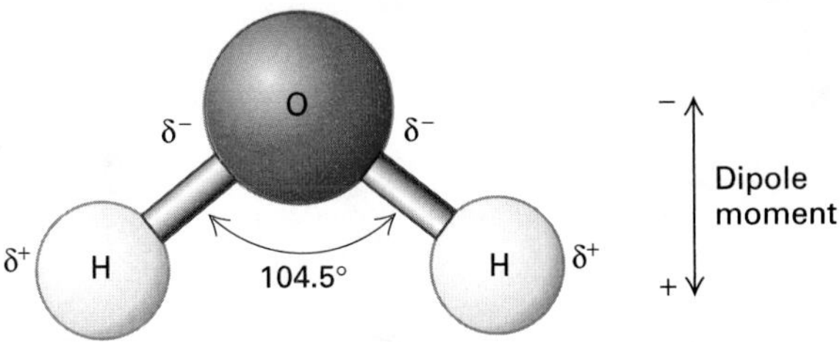

▲ **FIGURE 2-5 The dipole nature of a water molecule.** The symbol δ represents a partial charge (a weaker charge than the one on an electron or a proton). Because of the difference in the electronegativities of H and O, each of the polar H—O bonds in water is a dipole. The sizes and directions of the dipoles of each of the bonds determine the net distance and amount of charge separation, or **dipole moment,** of the molecule.

Covalent single bonds in biological molecules have energies similar to the energy of the C—C bond in ethane. Because more electrons are shared between atoms in double bonds, they require more energy to break than single bonds. For instance, it takes 84 kcal/mol to break a single C—O bond but 170 kcal/mol to break a C=O double bond. The most common double bonds in biological molecules are C=O, C=N, C=C, and P=O.

In contrast, the energy required to break noncovalent interactions is only 1–5 kcal/mol, much less than the bond energies of covalent bonds (see Figure 2-6). Indeed, noncovalent interactions are weak enough that they are constantly being formed and broken at room temperature. Although these interactions are weak and have a transient existence at physiological temperatures (25–37 °C), multiple noncovalent interactions can, as we will see, act together to produce highly stable and specific associations between different parts of a large molecule or between different macromolecules. Below, we review the four main types of noncovalent interactions and then consider their roles in the binding of biomolecules to one another and to other molecules.

► **FIGURE 2-6 Relative energies of covalent bonds and noncovalent interactions.** Bond energies are defined as the energy required to break a particular type of linkage. Covalent bonds, including those for single (C—C) and double (C=C) carbon-carbon bonds, are one to two powers of 10 stronger than noncovalent interactions. The latter are somewhat greater than the thermal energy of the environment at normal room temperature (25 °C). Many biological processes are coupled to the energy released during hydrolysis of a phosphoanhydride bond in ATP.

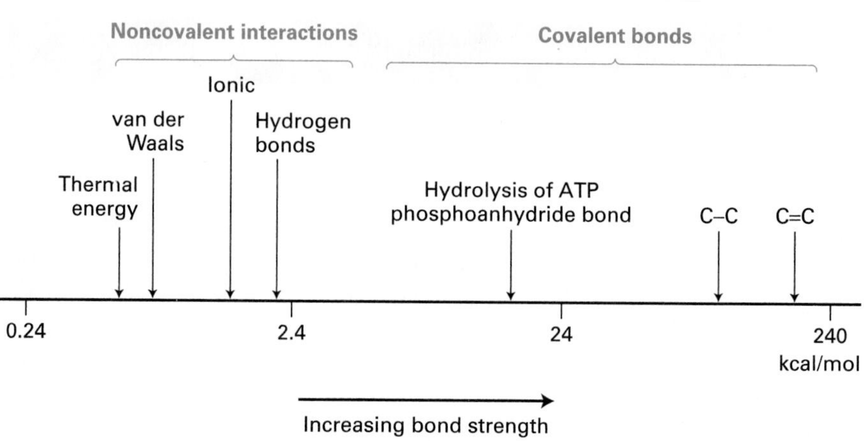

Ionic Interactions Are Attractions between Oppositely Charged Ions

Ionic interactions result from the attraction of a positively charged ion—a **cation**—for a negatively charged ion—an **anion.** In sodium chloride (NaCl), for example, the bonding electron contributed by the sodium atom is completely transferred to the chlorine atom. (Figure 2-7a). Unlike covalent bonds, ionic interactions do not have fixed or specific geometric orientations because the electrostatic field around an ion—its attraction for an opposite charge—is uniform in all directions. In solid NaCl, many ions pack tightly together in an alternating pattern to permit opposite charges to align and thus form a highly ordered crystalline array (salt crystals) (Figure 2-7b).

When solid salts dissolve in water, the ions separate from one another and are stabilized by their interactions with water molecules. In aqueous solutions, simple ions of biological significance, such as Na^+, K^+, Ca^{2+}, Mg^{2+}, and Cl^-, are hydrated, surrounded by a stable shell of water molecules held in place by ionic interactions between the central ion and the oppositely charged end of the water dipole (Figure 2-7c). Most ionic compounds dissolve readily in water because the energy of hydration, the energy released when ions tightly bind water molecules, is greater than the lattice energy that stabilizes the crystal structure. Parts or all of the aqueous *hydration shell* must be removed from ions when they directly interact with proteins. For example, water of hydration is lost when ions pass through protein pores in the cell membrane during nerve conduction.

The relative strength of the interaction between two ions, A^- and C^+, depends on the concentration of other ions in a solution. The higher the concentration of other ions (e.g., Na^+ and Cl^-), the more opportunities A^- and C^+ have to

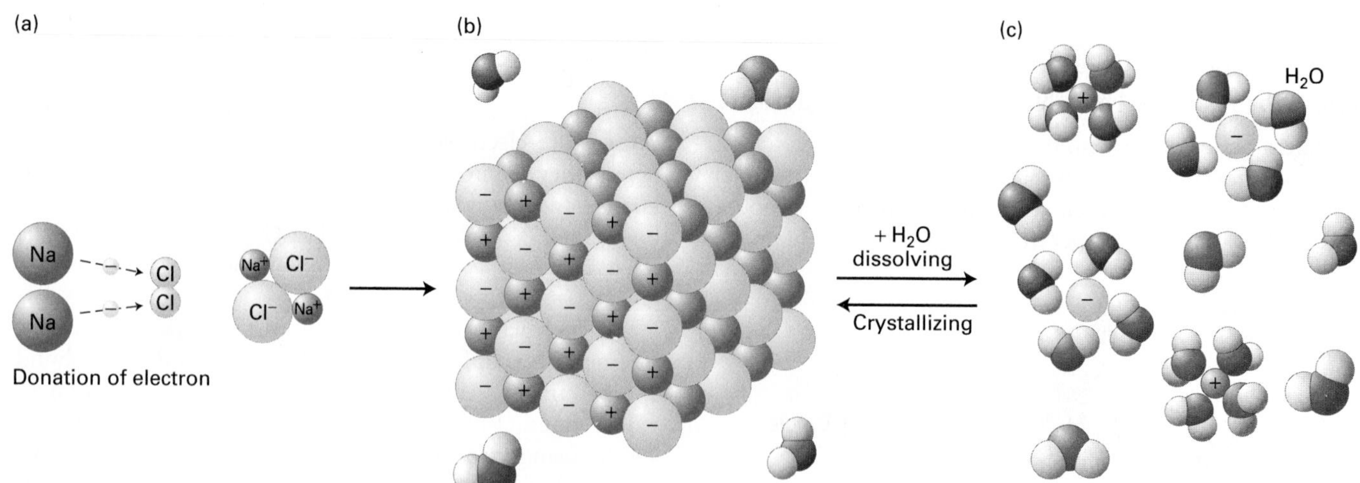

▲ **FIGURE 2-7 Electrostatic interactions of oppositely charged ions of salt (NaCl) in crystals and in aqueous solution.** (a) In crystalline table salt, sodium atoms are positively charged ions (Na^+) due to the loss of one electron each, whereas chloride atoms are correspondingly negatively charged (Cl^-) by gaining one electron each. (b) In solid form, ionic compounds form neatly ordered arrays, or crystals, of tightly packed ions in which the positive and negatively charged ions counterbalance each other. (c) When the crystals are dissolved in water, the ions separate and their charges, no longer balanced by immediately adjacent ions of opposite charge, are stabilized by interactions with polar water. Water molecules and the ions are held together by electrostatic interactions between the charges on the ion and the partial charges on the water's oxygen and hydrogen atoms. In aqueous solutions, all ions are surrounded by a hydration shell of water molecules.

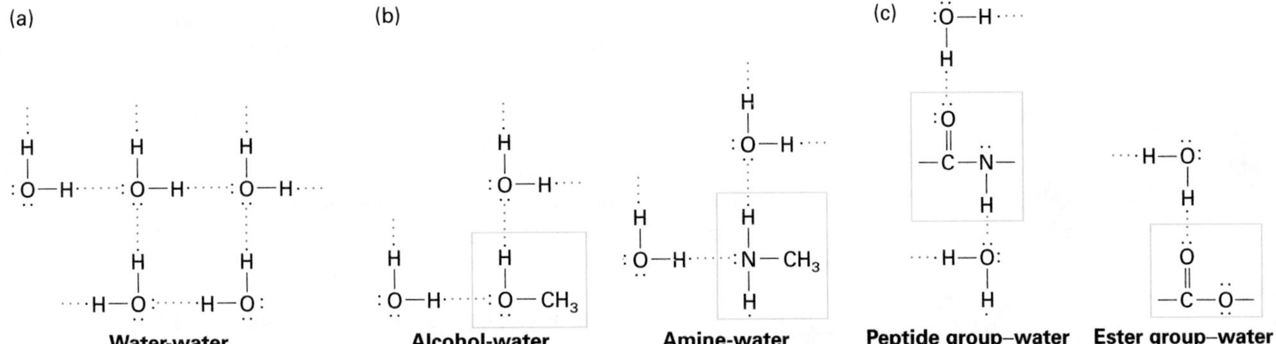

▲ **FIGURE 2-8 Hydrogen bonding of water with itself and with other compounds.** Each pair of nonbonding outer electrons in an oxygen or a nitrogen atom can accept a hydrogen atom in a hydrogen bond. The hydroxyl and the amino groups can also form hydrogen bonds with water. (a) In liquid water, each water molecule forms transient hydrogen bonds with several others, creating a dynamic network of hydrogen-bonded molecules. (b) Water also can form hydrogen bonds with alcohols and amines, accounting for the high solubility of these compounds. (c) The peptide group and ester group, which are present in many biomolecules, commonly participate in hydrogen bonds with water or polar groups in other molecules.

interact ionically with these other ions and thus the lower the energy required to break the interaction between A^- and C^+. As a result, increasing the concentrations of salts such as NaCl in a solution of biological molecules can weaken and even disrupt the ionic interactions holding the biomolecules together.

Hydrogen Bonds Determine the Water Solubility of Uncharged Molecules

A **hydrogen bond** is the interaction of a partially positively charged hydrogen atom in a molecular dipole (e.g., water) with unpaired electrons from another atom, either in the same (*intramolecular*) or different (*intermolecular*) molecule. Normally, a hydrogen atom forms a covalent bond with only one other atom. However, a hydrogen atom covalently bonded to an electronegative donor atom D may form an additional weak association, the hydrogen bond, with an acceptor atom A, which must have a nonbonding pair of electrons available for the interaction:

$$D^{\delta-}—H^{\delta+} + :A^{\delta-} \rightleftharpoons D^{\delta-}—\underbrace{H^{\delta+}\cdots\cdots:}_{\text{Hydrogen bond}}A^{\delta-}$$

The length of the covalent D—H bond is a bit longer than it would be if there were no hydrogen bond because the acceptor "pulls" the hydrogen away from the donor. An important feature of all hydrogen bonds is directionality. In the strongest hydrogen bonds, the donor atom, the hydrogen atom, and the acceptor atom all lie in a straight line. Nonlinear hydrogen bonds are weaker than linear ones; still, multiple nonlinear hydrogen bonds help to stabilize the three-dimensional structures of many proteins.

Hydrogen bonds are both longer and weaker than covalent bonds between the same atoms. In water, for example, the distance between the nuclei of the hydrogen and oxygen atoms of adjacent, hydrogen-bonded molecules is about 0.27 nm, about twice the length of the covalent O—H bonds within a single water molecule (Figure 2-8a). The strength of a hydrogen bond between water molecules (approximately 5 kcal/mol) is much weaker than a covalent O—H bond (roughly 110 kcal/mol), although it is greater than that for many other hydrogen bonds in biological molecules (1–2 kcal/ mol). The extensive hydrogen bonding between water molecules accounts for many of the key properties of this compound, including its unusually high melting and boiling points and its ability to interact with (e.g., dissolve) many other molecules.

The solubility of uncharged substances in an aqueous environment depends largely on their ability to form hydrogen bonds with water. For instance, the hydroxyl group (—OH) in an alcohol (XCH_2OH) and the amino group (—NH_2) in amines (XCH_2NH_2) can form several hydrogen bonds with water, enabling these molecules to dissolve in water to high concentrations (Figure 2-8b). In general, molecules with polar bonds that easily form hydrogen bonds with water, as well as charged molecules and ions that interact with the dipole in water, can readily dissolve in water; that is, they are hydrophilic (water liking). Many biological molecules contain, in addition to hydroxyl and amino groups, peptide and ester groups, which form hydrogen bonds with water via otherwise nonbonded electrons on their carbonyl oxygens (Figure 2-8c). X-ray crystallography combined with computational analysis permits an accurate depiction of the distribution of the outermost unbonded electrons of atoms as well as the electrons in covalent bonds, as illustrated in Figure 2-9.

Van der Waals Interactions Are Caused by Transient Dipoles

When any two atoms approach each other closely, they create a weak, nonspecific attractive force called a **van der Waals interaction.** These nonspecific interactions result from the momentary random fluctuations in the distribution of the electrons of any atom, which give rise to a transient

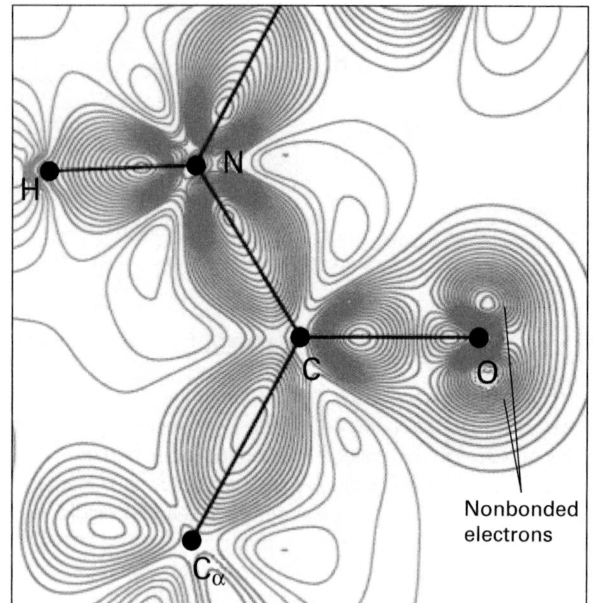

▲ **FIGURE 2-9 Distribution of bonding and outer nonbonding electrons in the peptide group.** Shown here is a peptide bond linking two amino acids within a protein called crambin. The black lines represent the covalent bonds between atoms. The red (negative) and blue (positive) lines represent contours of charge determined using x-ray crystallography and computational methods. The greater the number of contour lines, the higher the charge. The high density of red contour lines between atoms represents the covalent bonds (shared electron pairs). The two sets of red contour lines emanating from the oxygen (O) and not falling on a covalent bond (black line) represent the two pairs of nonbonded electrons on the oxygen that are available to participate in hydrogen bonding. The high density of blue contour lines near the hydrogen (H) bonded to nitrogen (N) represents a partial positive charge, indicating that this H can act as a donor in hydrogen bonding. [From C. Jelsch et al., 2000, *Proc. Nat'l. Acad. Sci.* USA **97**:3171. Courtesy of M. M. Teeter.]

unequal distribution of electrons. If two noncovalently bonded atoms are close enough together, electrons of one atom will perturb the electrons of the other. This perturbation generates a transient dipole in the second atom, and the two dipoles will attract each other weakly (Figure 2-10). Similarly, a polar covalent bond in one molecule will attract an oppositely oriented dipole in another.

Van der Waals interactions, involving either transiently induced or permanent electric dipoles, occur in all types of molecules, both polar and nonpolar. In particular, van der Waals interactions are responsible for the cohesion between nonpolar molecules such as heptane, $CH_3—(CH_2)_5—CH_3$, that cannot form hydrogen bonds or ionic interactions with other molecules. The strength of van der Waals interactions decreases rapidly with increasing distance; thus these noncovalent bonds can form only when atoms are quite close to one another. However, if atoms get too close together, they become repelled by the negative charges of their electrons. When the van der Waals attraction between two atoms exactly balances the repulsion between their two electron clouds, the atoms are said to be in van der Waals contact. The strength of the van der Waals interaction is about 1 kcal/mol, weaker than typical hydrogen bonds and only slightly higher than the average thermal energy of molecules at 25 °C. Thus multiple van der Waals interactions, a van der Waals interaction in conjunction with other noncovalent interactions, or both are required to significantly influence the stability of inter- and intramolecular contacts.

The Hydrophobic Effect Causes Nonpolar Molecules to Adhere to One Another

Because nonpolar molecules do not contain charged groups, possess a dipole moment, or become hydrated, they are insoluble or almost insoluble in water; that is, they are hydrophobic (water fearing). The covalent bonds between two carbon atoms and between carbon and hydrogen atoms are the most common nonpolar bonds in biological systems. **Hydrocarbons**—molecules made up only of carbon and hydrogen—are virtually insoluble in water. Large triacylglycerols (or triglycerides), which make up animal fats and vegetable oils, also are insoluble in water. As we will see later, the major portion of these molecules consists of long hydrocarbon chains. After being shaken in water, triacylglycerols form a separate phase. A familiar example is the separation of oil from the water-based vinegar in an oil-and-vinegar salad dressing.

Nonpolar molecules or nonpolar portions of molecules tend to aggregate in water owing to a phenomenon called the **hydrophobic effect.** Because water molecules cannot form hydrogen bonds with nonpolar substances, they tend to form "cages" of *relatively* rigid hydrogen-bonded pentagons and hexagons around nonpolar molecules

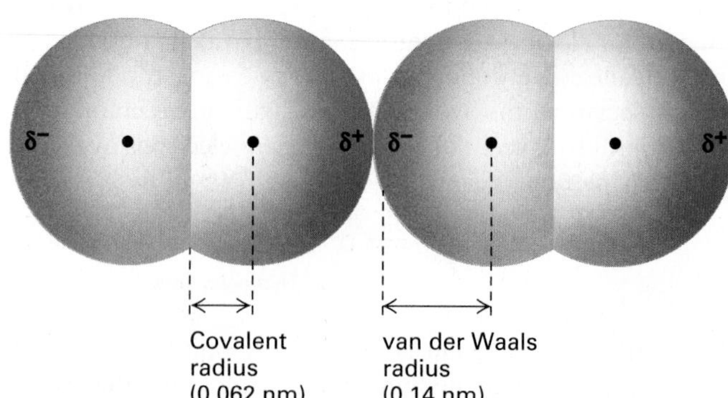

▲ **FIGURE 2-10 Two oxygen molecules in van der Waals contact.** In this model, red indicates negative charge and blue indicates positive charge. Transient dipoles in the electron clouds of all atoms give rise to weak attractive forces, called *van der Waals interactions.* Each type of atom has a characteristic van der Waals radius at which van der Waals interactions with other atoms are optimal. Because atoms repel one another if they are close enough together for their outer electrons to overlap without being shared in a covalent bond, the van der Waals radius is a measure of the size of the electron cloud surrounding an atom. The covalent radius indicated here is for the double bond of O=O; the single-bond covalent radius of oxygen is slightly longer.

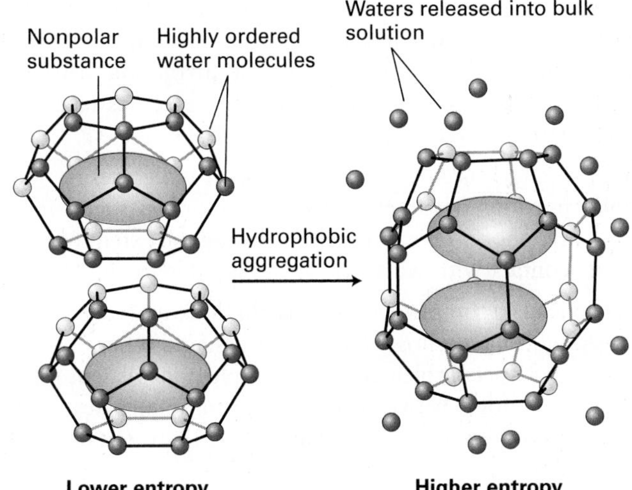

▲ **FIGURE 2-11 Schematic depiction of the hydrophobic effect.** Cages of water molecules that form around nonpolar molecules in solution are more ordered than water molecules in the surrounding bulk liquid. Aggregation of nonpolar molecules reduces the number of water molecules involved in highly ordered cages, resulting in a higher-entropy, more energetically favorable state (*right*) compared with the unaggregated state (*left*).

(Figure 2-11, *left*). This state is energetically unfavorable because it decreases the randomness (entropy) of the population of water molecules. (The role of entropy in chemical systems is discussed in a later section.) If nonpolar molecules in an aqueous environment aggregate with their hydrophobic surfaces facing each other, the hydrophobic surface area exposed to water is reduced (Figure 2-11, *right*). As a consequence, less water is needed to form the cages surrounding the nonpolar molecules, and entropy increases (an energetically more favorable state) relative to the unaggregated state. In a sense, then, water squeezes the nonpolar molecules into spontaneously forming aggregates. Rather than constituting an attractive force such as in hydrogen bonds, the hydrophobic effect results from an avoidance of an unstable state (extensive water cages around individual nonpolar molecules).

Nonpolar molecules can also associate, albeit weakly, through van der Waals interactions. The net result of the hydrophobic and van der Waals interactions is a very powerful tendency for hydrophobic molecules to interact with one another, not with water. Simply put, *like dissolves like.* Polar molecules dissolve in polar solvents such as water; nonpolar molecules dissolve in nonpolar solvents such as hexane.

Molecular Complementarity Mediated via Noncovalent Interactions Permits Tight, Highly Specific Binding of Biomolecules

Both inside and outside cells, ions and molecules are constantly bumping into one another. The greater the number of copies of any two types of molecules per unit volume (i.e., the higher their concentration), the more likely they are to encounter one another. When two molecules encounter each other, they most likely will simply bounce apart because the noncovalent interactions that would bind them together are weak and have a transient existence at physiological temperatures. However, molecules that exhibit **molecular complementarity,** a lock-and-key kind of fit between their shapes, charges, or other physical properties, can form multiple noncovalent interactions at close range. When two such structurally complementary molecules bump into each other, they can bind (stick) together.

Figure 2-12 illustrates how multiple, different weak bonds can bind two proteins together. Almost any other arrangement of the same groups on the two surfaces would not allow the molecules to bind so tightly. Such multiple, specific interactions between complementary regions within a protein molecule allow it to fold into a unique three-dimensional shape (Chapter 3) and hold the two chains of DNA together in a double helix (Chapter 4). Similar interactions underlie the association of groups of more than two molecules into multimolecular complexes, leading to formation of muscle fibers, to the gluelike associations between cells in solid tissues, and to numerous other cellular structures.

Depending on the number and strength of the noncovalent interactions between the two molecules and on their environment, their binding may be tight (strong) or loose (weak) and, as a consequence, either long lasting or transient. The higher the *affinity* of two molecules for each other, the better the molecular "fit" between them, the more noncovalent interactions can form, and the tighter they can bind

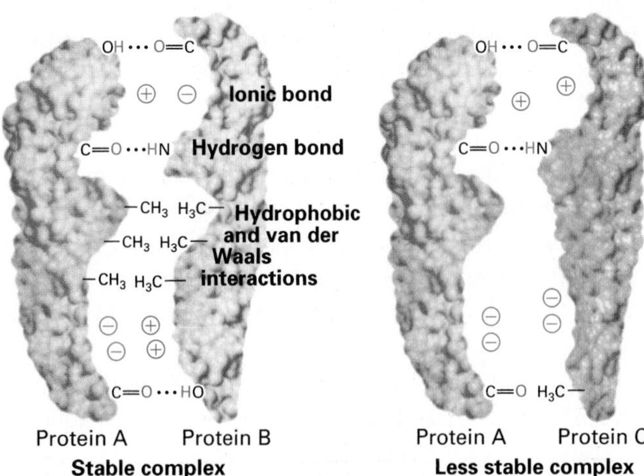

▲ **FIGURE 2-12 Molecular complementarity and the binding of proteins via multiple noncovalent interactions.** The complementary shapes, charges, polarity, and hydrophobicity of two protein surfaces permit multiple weak interactions, which in combination produce a strong interaction and tight binding. Because deviations from molecular complementarity substantially weaken binding, a particular surface region of any given biomolecule usually can bind tightly to only one or a very limited number of other molecules. The complementarity of the two protein molecules on the left permits them to bind much more tightly than the two noncomplementary proteins on the right.

together. An important quantitative measure of affinity is the binding dissociation constant K_d, described later.

As we discuss in Chapter 3, nearly all the chemical reactions that occur in cells also depend on the binding properties of enzymes. These proteins not only speed up, or catalyze, reactions but also do so with a high degree of *specificity,* a reflection of their ability to bind tightly to only one or a few related molecules. The specificity of intermolecular interactions and reactions, which depends on molecular complementarity, is essential for many processes critical to life.

KEY CONCEPTS OF SECTION 2.1

Covalent Bonds and Noncovalent Interactions

- Covalent bonds, which bind the atoms composing a molecule in a fixed orientation, consist of pairs of electrons shared by two atoms. They are stable in biological systems because the relatively high energies required to break them (50–200 kcal/mol) are much larger than the thermal kinetic energy available at room (25 °C) or body (37 °C) temperatures.
- Many molecules in cells contain at least one asymmetric carbon atom, which is bonded to four dissimilar atoms. Such molecules can exist as optical isomers (mirror images), designated D and L (see Figure 2-4), which have different biological activities. In biological systems, nearly all sugars are D isomers, whereas nearly all amino acids are L isomers.
- Electrons may be shared equally or unequally in covalent bonds. Atoms that differ in electronegativity form polar covalent bonds in which the bonding electrons are distributed unequally. One end of a polar bond has a partial positive charge and the other end has a partial negative charge (see Figure 2-5).
- Noncovalent interactions between atoms are considerably weaker than covalent bonds, with bond energies ranging from about 1–5 kcal/mol (see Figure 2-6).
- Four main types of noncovalent interactions occur in biological systems: ionic bonds, hydrogen bonds, van der Waals interactions, and interactions due to the hydrophobic effect.
- Ionic bonds result from the electrostatic attraction between the positive and negative charges of ions. In aqueous solutions, all cations and anions are surrounded by a shell of bound water molecules (see Figure 2-7c). Increasing the salt (e.g., NaCl) concentration of a solution can weaken the relative strength of and even break the ionic bonds between biomolecules.
- In a hydrogen bond, a hydrogen atom covalently bonded to an electronegative atom associates with an acceptor atom whose nonbonding electrons attract the hydrogen (see Figure 2-8).
- Weak and relatively nonspecific van der Waals interactions are created whenever any two atoms approach each other closely. They result from the attraction between transient dipoles associated with all molecules (see Figure 2-10).
- In an aqueous environment, nonpolar molecules or nonpolar portions of larger molecules are driven together by the hydrophobic effect, thereby reducing the extent of their direct contact with water molecules (see Figure 2-11).
- Molecular complementarity is the lock-and-key fit between molecules whose shapes, charges, and other physical properties are complementary. Multiple noncovalent interactions can form between complementary molecules, causing them to bind tightly (see Figure 2-12), but not between molecules that are not complementary.
- The high degree of binding specificity that results from molecular complementarity is one of the features that underlies intermolecular interactions and thus is essential for many processes critical to life.

2.2 Chemical Building Blocks of Cells

A common theme in biology is the construction of large molecules (**macromolecules**) and structures by the covalent or noncovalent association of many similar or identical smaller molecules. The three most abundant classes of the critically important biological macromolecules—**proteins, nucleic acids,** and **polysaccharides**—are all **polymers** composed of multiple covalently linked building block small molecules, or **monomers** (Figure 2-13). Proteins are linear polymers containing 10 to several thousand amino acids linked by **peptide bonds.** Nucleic acids are linear polymers containing hundreds to millions of nucleotides linked by **phosphodiester bonds.** Polysaccharides are linear or branched polymers of monosaccharides (sugars) such as glucose linked by **glycosidic bonds.** Although the actual mechanisms by which covalent bonds between monomers form are complex and will be discussed later, the formation of a covalent bond between two monomer molecules usually involves the net loss of a hydrogen (H) from one monomer and a hydroxyl (OH) from the other monomer—or the net loss of one water—and can therefore be thought of as a *dehydration reaction.* These bonds are stable under normal biological conditions (e.g., 37°C, neutral pH), and so these *biopolymers* are stable and can perform a wide variety of jobs in cells (store information, catalyze chemical reactions, serve as structural elements in defining cell shape and movement, etc.).

Macromolecular structures can also be assembled using noncovalent interactions. The macromolecular two-layered (bilayer) structure of cellular membranes is built up by the noncovalent assembly of many thousands of small molecules called phospholipids (see Figure 2-13). In this chapter, we will focus on the characteristics of the monomeric chemical

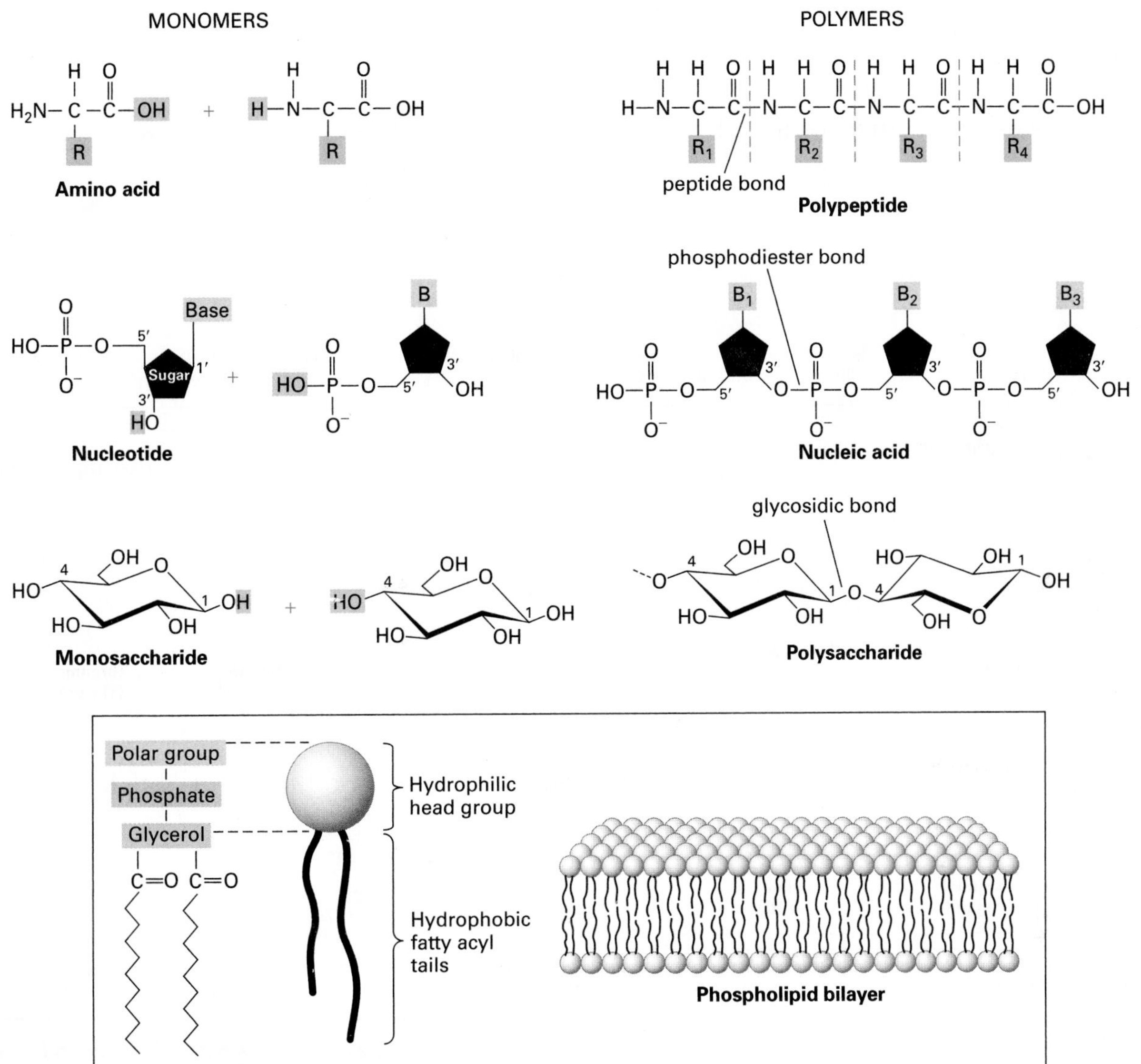

▲ **FIGURE 2-13 Overview of the cell's principal chemical building blocks.** *(Top)* The three major types of biological macromolecules are each assembled by the polymerization of multiple small molecules (monomers) of a particular type: proteins from amino acids (Chapter 3), nucleic acids from nucleotides (Chapter 4), and polysaccharides from monosaccharides (sugars). Each monomer is covalently linked into the polymer by a reaction whose net result is loss of a water molecule (dehydration). *(Bottom)* In contrast, phospholipid monomers noncovalently assemble into a bilayer structure, which forms the basis of all cellular membranes (Chapter 10).

building blocks—amino acids, nucleotides, sugars, and phospholipids. The structure, function, and assembly of proteins, nucleic acids, polysaccharides, and biomembranes are discussed in subsequent chapters.

Amino Acids Differing Only in Their Side Chains Compose Proteins

The monomeric building blocks of proteins are 20 **amino acids,** which when incorporated into a protein polymer are sometimes called **residues.** All amino acids have a characteristic structure consisting of a central **alpha (α) carbon atom** (C_α) bonded to four different chemical groups: an amino (NH_2) group, a carboxylic acid or carboxyl (COOH) group (hence the name *amino acid*), a hydrogen (H) atom, and one variable group, called a **side chain** or R group. Because the α carbon in all amino acids except glycine is asymmetric, these molecules can exist in two mirror-image forms called by convention the D (dextro) and the L (levo) isomers (see Figure 2-4). The two isomers cannot be interconverted (one made identical with the other) without breaking and then re-forming a chemical bond in one of them. With rare exceptions, only the L forms of amino acids are found in proteins.

To understand the three-dimensional structures and functions of proteins, discussed in detail in Chapter 3, you must be familiar with some of the distinctive properties of

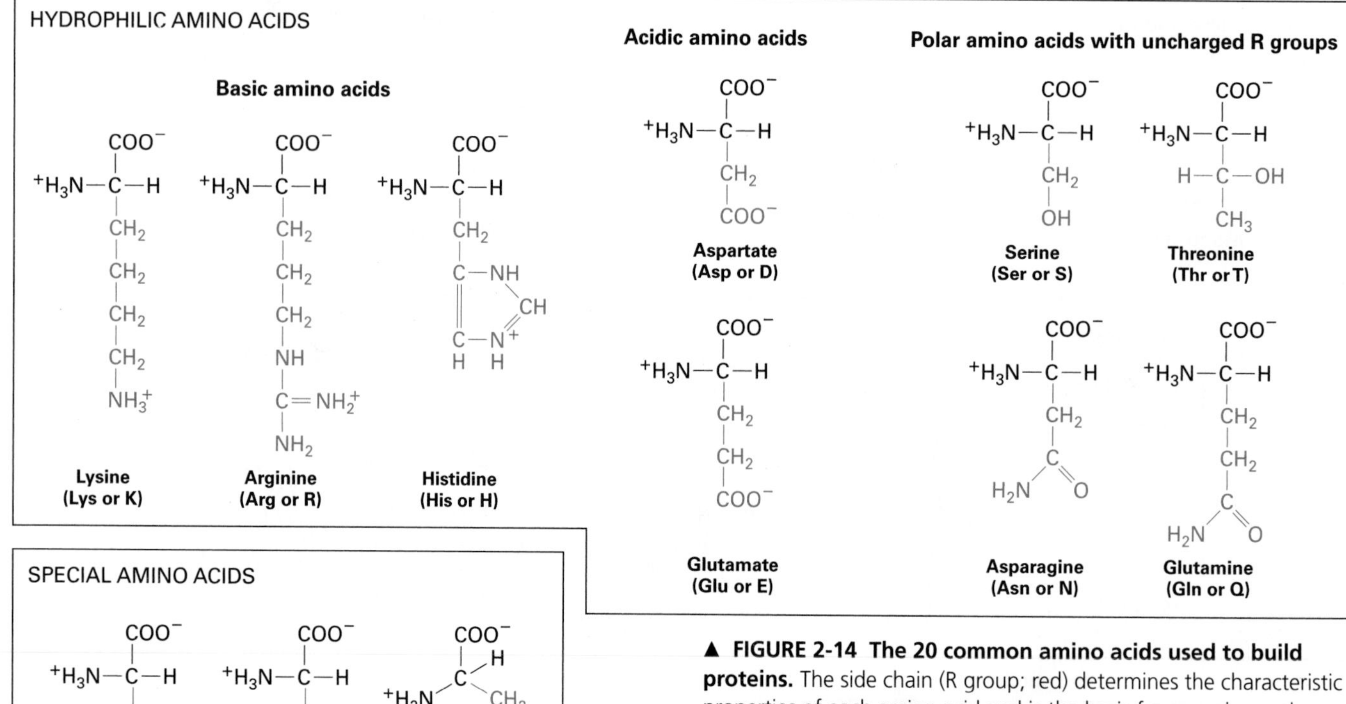

▲ **FIGURE 2-14 The 20 common amino acids used to build proteins.** The side chain (R group; red) determines the characteristic properties of each amino acid and is the basis for grouping amino acids into three main categories: hydrophobic, hydrophilic, and special. Shown are the ionized forms that exist at the pH (≈7) of the cytosol. In parentheses are the three-letter and one-letter abbreviations for each amino acid.

amino acids, which are determined in part by their side chains. You need not memorize the detailed structure of each type of side chain to understand how proteins work because amino acids can be classified into several broad categories based on the size, shape, charge, hydrophobicity (a measure of water solubility), and chemical reactivity of the side chains (Figure 2-14). However, you should be familiar with the general properties of each category.

Amino acids with nonpolar side chains are hydrophobic and so poorly soluble in water. The larger the nonpolar side chain, the more hydrophobic—less water soluble—the amino acid. The noncyclic side chains of *alanine, valine, leucine,* and *isoleucine* (called aliphatic), as well as *methionine,* consist entirely of hydrocarbons, except for the one sulfur atom in methionine, and all are nonpolar. *Phenylalanine, tyrosine,* and *tryptophan* have large, bulky aromatic side chains. In later chapters, we will see in detail how these hydrophobic side chains under the influence of the hydrophobic effect often pack in the interior of proteins or line the surfaces of proteins that are embedded within hydrophobic regions of biomembranes.

Amino acids with polar side chains are hydrophilic; the most hydrophilic of these amino acids is the subset with side chains that are charged (ionized) at the pH typical of biological fluids (≈7)—both inside and outside the cell (see Section 2.3). *Arginine* and *lysine* have positively charged side chains and are called basic amino acids; *aspartic acid* and *glutamic acid* have negatively charged side chains due to the carboxylic acid groups in their side chains (their charged forms are called *aspartate* and *glutamate*) and are called acidic. A fifth amino acid, *histidine,* has a side chain containing a ring with two nitrogens, called imidazole, which can shift from being positively charged to uncharged depending on small changes in the acidity of its environment:

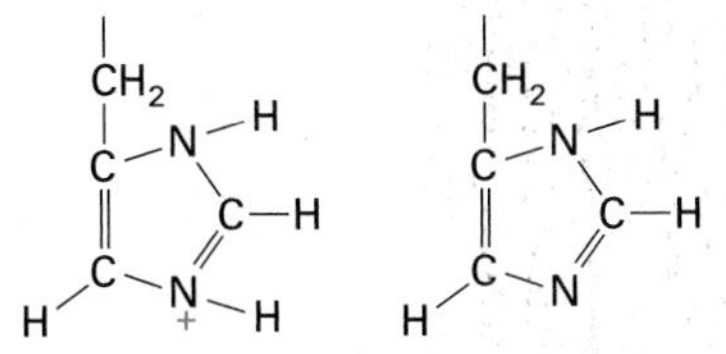

pH 5.8 pH 7.8

The activities of many proteins are modulated by shifts in environmental acidity through protonation or deprotonation of histidine side chains. *Asparagine* and *glutamine* are uncharged but have polar side chains containing amide groups with extensive hydrogen-bonding capacities. Similarly, *serine* and *threonine* are uncharged but have polar hydroxyl groups, which also participate in hydrogen bonds with other polar molecules.

Lastly, cysteine, glycine, and proline exhibit special roles in proteins because of the unique properties of their side chains. The side chain of *cysteine* contains a reactive *sulfhydryl group (—SH),* which can oxidize to form a covalent *disulfide bond (—S—S—)* to a second cysteine:

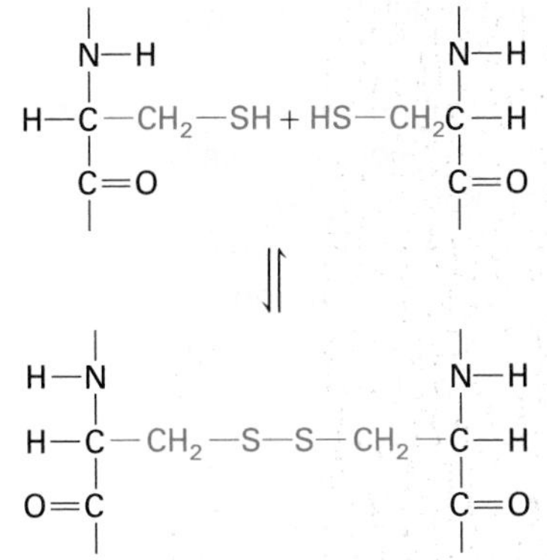

Regions within a single protein chain (intramolecular) or in separate chains (intermolecular) sometimes are cross-linked through disulfide bonds. Disulfide bonds stabilize the folded structure of some proteins. The smallest amino acid, *glycine,* has a single hydrogen atom as its R group. Its small size allows it to fit into tight spaces. Unlike the other common amino acids, the side chain of *proline* bends around to form a ring by covalently bonding to the nitrogen atom (amino group) attached to the C_α. As a result, proline is very rigid and creates a fixed kink in a protein chain, limiting how a protein can fold in the region of proline residues.

Some amino acids are more abundant in proteins than others. Cysteine, tryptophan, and methionine are rare amino acids: together they constitute approximately 5 percent of the amino acids in a protein. Four amino acids—leucine, serine, lysine, and glutamic acid—are the most abundant amino acids, totaling 32 percent of all the amino acid residues in a typical protein. However, the amino acid composition of proteins can vary widely from these values.

Although cells use the 20 amino acids shown in Figure 2-14 in the *initial* synthesis of proteins, analysis of cellular proteins reveals that they contain upward of 100 different amino acids. Chemical modifications of the amino acids account for this difference. Acetyl groups (CH_3CO) and a variety of other chemical groups can be added to specific internal amino acids after they are incorporated into proteins (Figure 2-15). An important modification is the addition of a phosphate (PO_4, phosphorylation) to hydroxyl groups in serine, threonine, and tyrosine residues. We will encounter numerous examples of proteins whose activity is regulated by reversible phosphorylation and dephosphorylation. Phosphorylation of nitrogen in the side chain of histidine is well known in bacteria, fungi, and plants but less studied—perhaps because of the relative instability of phosphorylated histidine—and apparently rare in mammals. The side chains of asparagine, serine, and threonine are sites for glycosylation, the attachment of linear and branched carbohydrate chains. Many secreted proteins and membrane proteins contain glycosylated residues. Other amino acid modifications found in selected proteins include the hydroxylation of proline and lysine residues in collagen (Chapter 19), the methylation of histidine residues in membrane receptors, and the γ carboxylation of glutamate in blood-clotting factors such as prothrombin.

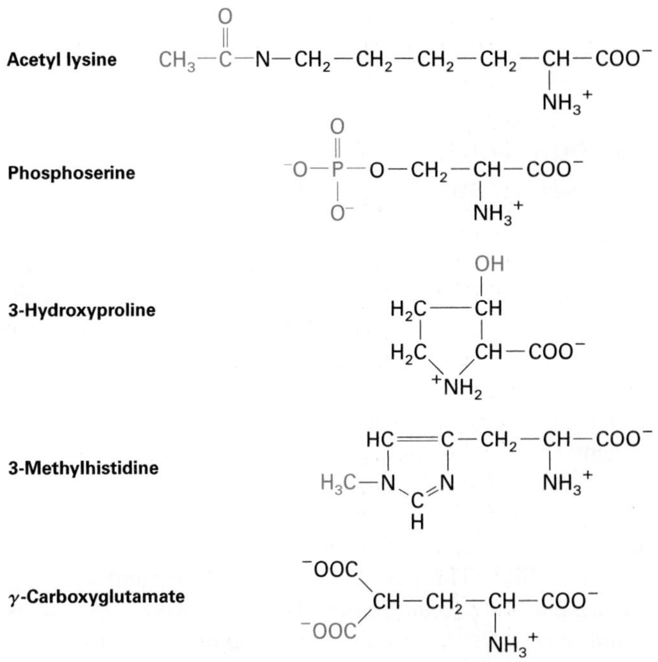

Acetylation, addition of an acetyl group, to the amino group of the N-terminal residue, is the most common form of amino acid chemical modification, affecting an estimated 80 percent of all proteins:

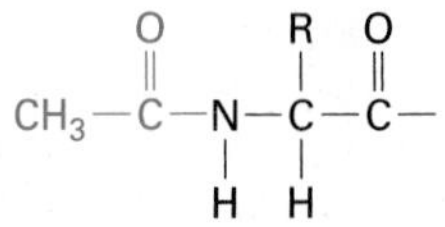

Acetylated N-terminus

Acetyl lysine

Phosphoserine

3-Hydroxyproline

3-Methylhistidine

γ-Carboxyglutamate

▲ **FIGURE 2-15 Common modifications of amino acid side chains in proteins.** These modified residues and numerous others are formed by addition of various chemical groups (red) to the amino acid side chains during or after synthesis of a polypeptide chain.

This modification may play an important role in controlling the life span of proteins within cells because nonacetylated proteins are rapidly degraded.

Five Different Nucleotides Are Used to Build Nucleic Acids

Two types of chemically similar nucleic acids, **DNA (deoxyribonucleic acid)** and **RNA (ribonucleic acid)**, are the principal genetic-information-carrying molecules of the cell. The monomers from which DNA and RNA polymers are built, called **nucleotides,** all have a common structure: a phosphate group linked by a phosphoester bond to a pentose (a five-carbon sugar molecule) that in turn is linked to a nitrogen- and carbon-containing ring structure commonly referred to as a base (Figure 2-16a). In RNA, the pentose is ribose; in DNA, it is deoxyribose that at position 2′ has a proton rather than the hydroxyl group at that site in ribose (Figure 2-16b). The bases *adenine, guanine,* and *cytosine* (Figure 2-17) are found in both DNA and RNA; *thymine* is found only in DNA, and *uracil* is found only in RNA.

Adenine and guanine are **purines,** which contain a pair of fused rings; cytosine, thymine, and uracil are **pyrimidines,** which contain a single ring (see Figure 2-17). The bases are often abbreviated A, G, C, T, and U, respectively; these same single-letter abbreviations are also commonly used to denote the entire nucleotides in nucleic acid polymers. In nucleotides, the 1′ carbon atom of the sugar (ribose or deoxyribose) is attached to the nitrogen at position 9 of a purine (N_9) or at position 1 of a pyrimidine (N_1). The acidic character of nucleotides is due to the phosphate group, which under normal intracellular conditions releases hydrogen ions (H^+), leaving the phosphate negatively charged (see Figure 2-16a). Most nucleic acids in cells are associated with proteins, which form ionic interactions with the negatively charged phosphates.

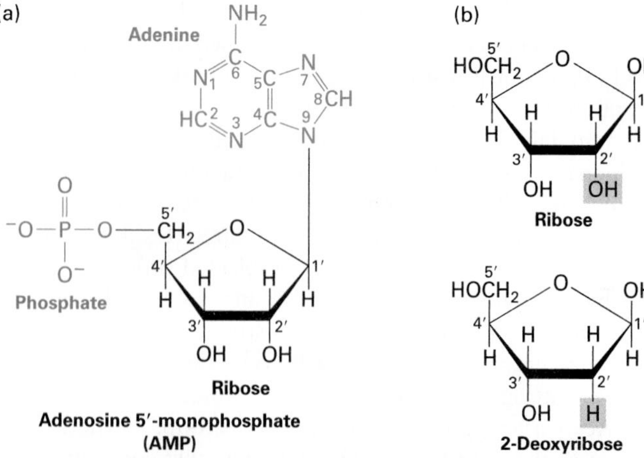

Cells and extracellular fluids in organisms contain small concentrations of **nucleosides,** combinations of a base and a sugar without a phosphate. Nucleotides are nucleosides that have one, two, or three phosphate groups esterified at the 5′ hydroxyl. Nucleoside monophosphates have a single esterified phosphate (see Figure 2-16a); nucleoside diphosphates contain a pyrophosphate group:

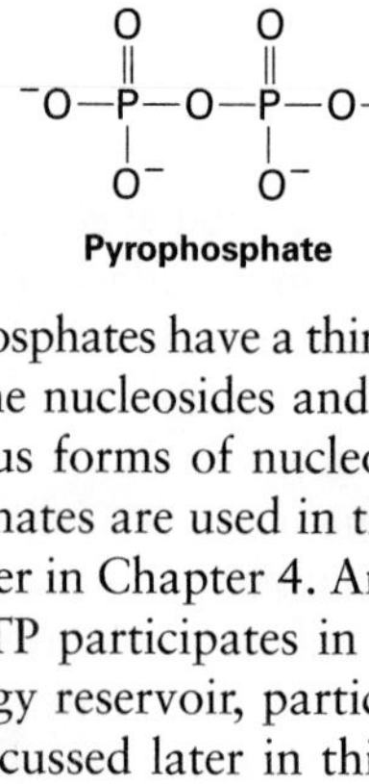

$$^{-}O-\overset{\overset{O}{\|}}{\underset{\underset{O^-}{|}}{P}}-O-\overset{\overset{O}{\|}}{\underset{\underset{O^-}{|}}{P}}-O-$$

Pyrophosphate

and nucleoside triphosphates have a third phosphate. Table 2-3 lists the names of the nucleosides and nucleotides in nucleic acids and the various forms of nucleoside phosphates. The nucleoside triphosphates are used in the synthesis of nucleic acids, which we cover in Chapter 4. Among their other functions in the cell, GTP participates in intracellular signaling and acts as an energy reservoir, particularly in protein synthesis, and ATP, discussed later in this chapter, is the most widely used biological energy carrier.

PURINES

Adenine (A) Guanine (G)

PYRIMIDINES

Uracil (U) Thymine (T) Cytosine (C)

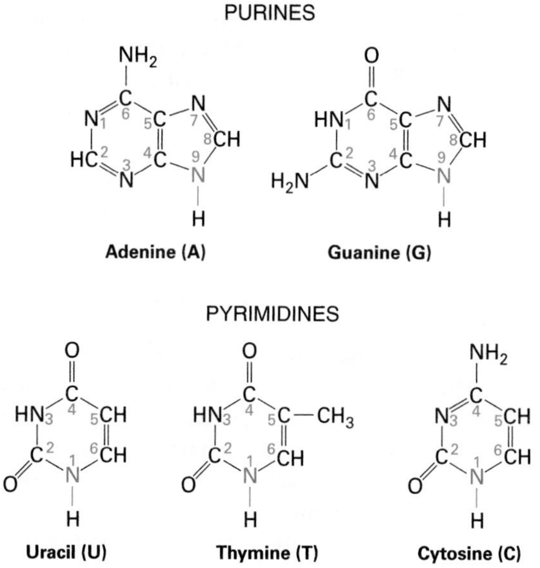

▲ **FIGURE 2-17 Chemical structures of the principal bases in nucleic acids.** In nucleic acids and nucleotides, nitrogen 9 of purines and nitrogen 1 of pyrimidines (red) are bonded to the 1′ carbon of ribose or deoxyribose. U is only in RNA, and T is only in DNA. Both RNA and DNA contain A, G, and C.

(a) Adenine; Phosphate; Ribose; Adenosine 5′-monophosphate (AMP)

(b) Ribose; 2-Deoxyribose

▲ **FIGURE 2-16 Common structure of nucleotides.** (a) Adenosine 5′-monophosphate (AMP), a nucleotide present in RNA. By convention, the carbon atoms of the pentose sugar in nucleotides are numbered with primes. In natural nucleotides, the 1′ carbon is joined by a β linkage to the base (in this case adenine); both the base (blue) and the phosphate on the 5′ hydroxyl (red) extend above the plane of the sugar ring. (b) Ribose and deoxyribose, the pentoses in RNA and DNA, respectively.

Monosaccharides Joined by Glycosidic Bonds Form Linear and Branched Polysaccharides

The building blocks of the polysaccharides are the simple sugars, or **monosaccharides.** Monosaccharides are **carbohydrates,** which are literally covalently bonded combinations of carbon and water in a one-to-one ratio $(CH_2O)_n$, where n equals 3, 4, 5, 6, or 7. **Hexoses** ($n = 6$) and **pentoses** ($n = 5$) are the most common monosaccharides. All monosaccharides

TABLE 2-3 Terminology of Nucleosides and Nucleotides

		PURINES		PYRIMIDINES	
BASES		ADENINE(A)	GUANINE(G)	CYTOSINE(C)	URACIL(U) THYMINE(T)
Nucleosides	in RNA	Adenosine	Guanosine	Cytidine	Uridine
	in DNA	Deoxyadenosine	Deoxyguanosine	Deoxycytidine	Deoxythymidine
Nucleotides	in RNA	Adenylate	Guanylate	Cytidylate	Uridylate
	in DNA	Deoxyadenylate	Deoxyguanylate	Deoxycytidylate	Deoxythymidylate
Nucleoside monophosphates		AMP	GMP	CMP	UMP
Nucleoside diphosphates		ADP	GDP	CDP	UDP
Nucleoside triphosphates		ATP	GTP	CTP	UTP
Deoxynucleoside mono-, di-, and triphosphates		dAMP, etc.	dGMP, etc.	dCMP, etc	dTMP, etc.

contain hydroxyl (—OH) groups and either an aldehyde or a keto group:

Aldehyde Keto

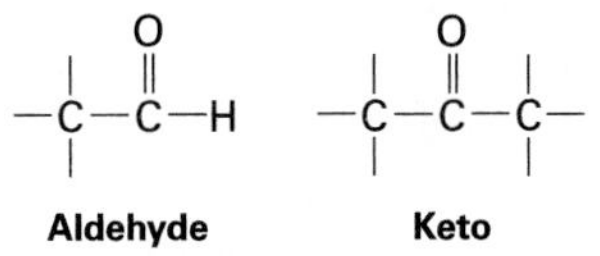

Many biologically important sugars are hexoses, including glucose, mannose, and galactose (Figure 2-18). Mannose is identical with glucose except that the orientation of the groups bonded to carbon 2 is reversed. Similarly, galactose, another hexose, differs from glucose only in the orientation of the groups attached to carbon 4. Interconversion of glucose and mannose or galactose requires the breaking and making of covalent bonds; such reactions are carried out by enzymes called *epimerases*.

D-Glucose ($C_6H_{12}O_6$) is the principal external source of energy for most cells in higher organisms and can exist in three different forms: a linear structure and two different hemiacetal ring structures (Figure 2-18a). If the aldehyde group on carbon 1 reacts with the hydroxyl group on carbon 5, the resulting hemiacetal, D-glucopyranose, contains a six-member ring. In the α anomer of D-glucopyranose, the hydroxyl group attached to carbon 1 points "downward" from the ring as shown in Figure 2-18a; in the β anomer, this hydroxyl points "upward." In aqueous solution the α and β anomers readily interconvert spontaneously; at equilibrium there is about one-third α anomer and two-thirds β, with very little of the open-chain form. Because enzymes can distinguish between the α and β anomers of D-glucose, these forms have distinct biological roles. Condensation of the

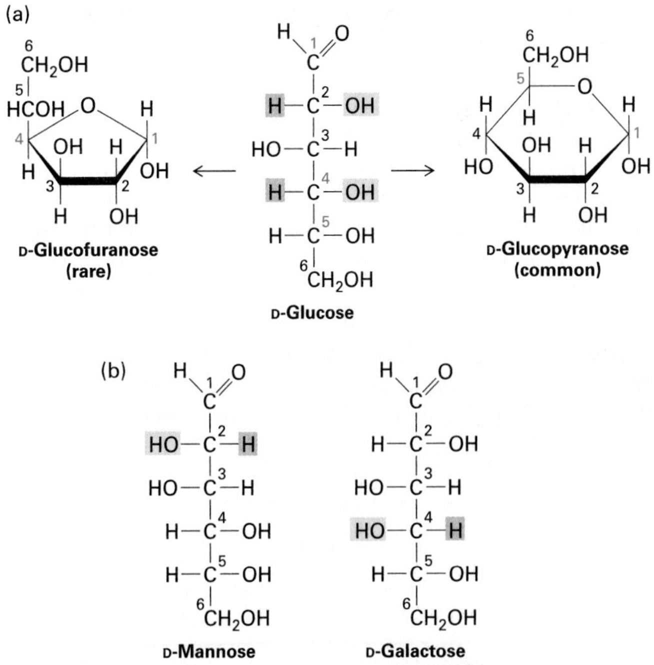

▲ **FIGURE 2-18 Chemical structures of hexoses.** All hexoses have the same chemical formula ($C_6H_{12}O_6$) and contain an aldehyde or a keto group. (a) The ring forms of D-glucose are generated from the linear molecule by reaction of the aldehyde at carbon 1 with the hydroxyl on carbon 5 or carbon 4. The three forms are readily interconvertible, although the pyranose form (*right*) predominates in biological systems. (b) In D-mannose and D-galactose, the configuration of the H (green) and OH (blue) bound to one carbon atom differs from that in glucose. These sugars, like glucose, exist primarily as pyranoses.

hydroxyl group on carbon 4 of the linear glucose with its aldehyde group results in the formation of D-glucofuranose, a hemiacetal containing a five-member ring. Although all three forms of D-glucose exist in biological systems, the pyranose form is by far the most abundant.

The pyranose ring in Figure 2-18a is depicted as planar. In fact, because of the tetrahedral geometry around carbon atoms, the most stable conformation of a pyranose ring has a nonplanar, chairlike shape. In this conformation, each bond from a ring carbon to a nonring atom (e.g., H or O) is either nearly perpendicular to the ring, referred to as axial (a), or nearly in the plane of the ring, referred to as equatorial (e):

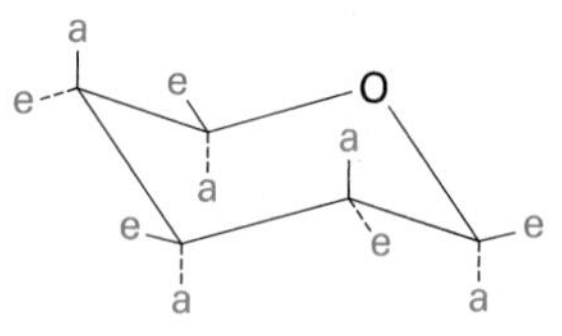

Pyranoses 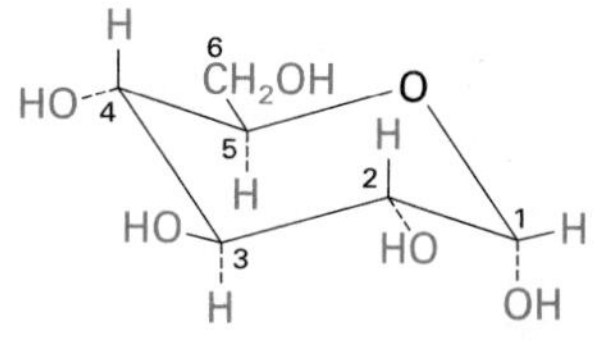 **α-D-Glucopyranose**

Disaccharides, formed from two monosaccharides, are the simplest polysaccharides. The disaccharide lactose, composed of galactose and glucose, is the major sugar in milk; the disaccharide sucrose, composed of glucose and fructose, is a principal product of plant photosynthesis and is refined into common table sugar (Figure 2-19).

Larger polysaccharides, containing dozens to hundreds of monosaccharide units, can function as reservoirs for glucose, as structural components, or as adhesives that help hold cells together in tissues. The most common storage carbohydrate in animal cells is **glycogen,** a very long, highly branched polymer of glucose. As much as 10 percent by weight of the liver can be glycogen. The primary storage carbohydrate in plant cells, **starch,** also is a glucose polymer. It occurs in an unbranched form (amylose) and lightly branched form (amylopectin). Both glycogen and starch are composed of the α anomer of glucose. In contrast, **cellulose,** the major constituent of plant cell walls (Chapter 19), is an unbranched polymer of the β anomer of glucose. Human digestive enzymes can hydrolyze the α glycosidic bonds in starch but not the β glycosidic bonds in cellulose. Many species of plants, bacteria, and molds produce cellulose-degrading enzymes. Cows and termites can break down cellulose because they harbor cellulose-degrading bacteria in their gut.

The enzymes that make the glycosidic bonds linking monosaccharides into polysaccharides are specific for the α or β anomer of one sugar and a particular hydroxyl group on the other. In principle, any two sugar molecules can be linked in a variety of ways because each monosaccharide has multiple hydroxyl groups that can participate in the formation of glycosidic bonds. Furthermore, any one monosaccharide has the potential of being linked to more than two other monosaccharides, thus generating a branch point and nonlinear polymers. Glycosidic bonds are usually formed between the growing polysaccharide chain and a covalently modified form of a monosaccharide. Such modifications include a phosphate (e.g., glucose 6-phosphate) or a nucleotide (e.g., UDP-galactose):

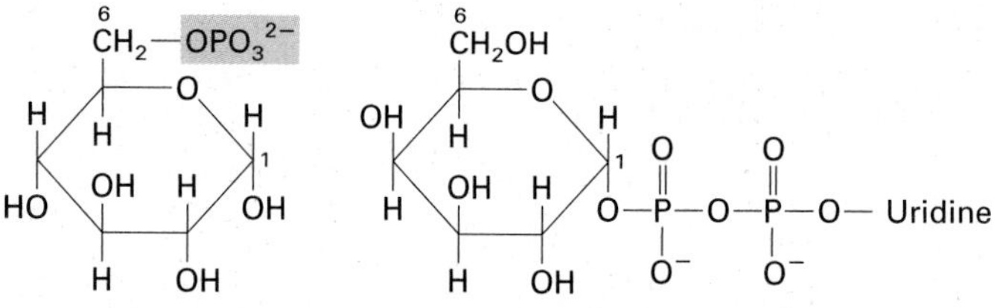

Glucose 6-phosphate **UDP-galactose**

The epimerase enzymes that interconvert different monosaccharides often do so using the nucleotide sugars rather than the unsubstituted sugars.

Many complex polysaccharides contain modified sugars that are covalently linked to various small groups, particularly amino, sulfate, and acetyl groups. Such modifications are abundant in **glycosaminoglycans,** major polysaccharide components of the extracellular matrix that we describe in Chapter 19.

Phospholipids Associate Noncovalently to Form the Basic Bilayer Structure of Biomembranes

Biomembranes are large flexible sheets that serve as the boundaries of cells and their intracellular organelles and

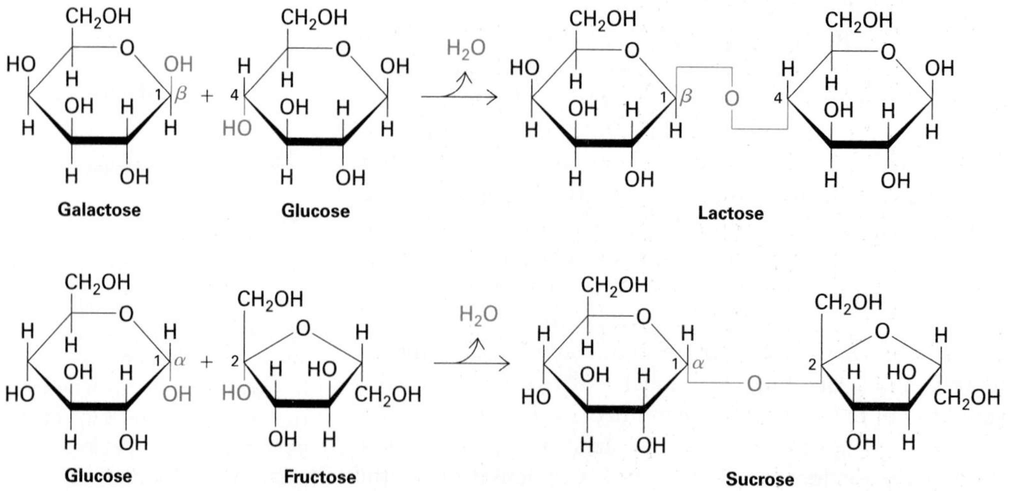

Galactose **Glucose** **Lactose**

Glucose **Fructose** **Sucrose**

▶ **FIGURE 2-19 Formation of the disaccharides lactose and sucrose.** In any glycosidic linkage, the anomeric carbon of one sugar molecule (in either the α or β conformation) is linked to a hydroxyl oxygen on another sugar molecule. The linkages are named accordingly; thus lactose contains a β(1 → 4) bond, and sucrose contains an α(1 → 2) bond.

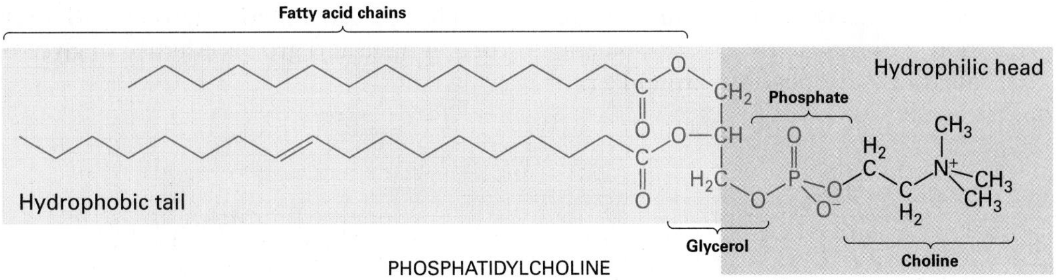

▲ **FIGURE 2-20 Phosphatidylcholine, a typical phosphoglyceride.** All phosphoglycerides are amphipathic phospholipids, having a hydrophobic tail (yellow) and a hydrophilic head (blue) in which glycerol is linked via a phosphate group to an alcohol. Either or both of the fatty acyl side chains in a phosphoglyceride may be saturated or unsaturated. In phosphatidic acid (red), the simplest phospholipid, the phosphate is not linked to an alcohol.

form the outer surfaces of some viruses. Membranes literally define what is a cell (the outer membrane and the contents within the membrane) and what is not (the extracellular space outside the membrane). Unlike the proteins, nucleic acids, and polysaccharides, membranes are assembled by the *noncovalent* association of their component building blocks. The primary building blocks of all biomembranes are **phospholipids,** whose physical properties are responsible for the formation of the sheetlike structure of membranes. The structures and functions of membranes, which include in addition to phospholipids a variety of other molecules (e.g., cholesterol, glycolipids, proteins), will be described in detail in Chapter 10.

Phospholipids consist of two long-chain, nonpolar fatty acid groups linked (usually by an ester bond) to small, highly polar groups, including a phosphate and a short organic molecule, such as glycerol (trihydroxy propanol) (Figure 2-20).

Fatty acids consist of a hydrocarbon chain attached to a carboxyl group (—COOH). Like glucose, fatty acids are an important energy source for many cells (Chapter 12). They differ in length, although the predominant fatty acids in cells have an even number of carbon atoms, usually 14, 16, 18, or 20. The major fatty acids in phospholipids are listed in Table 2-4. Fatty acids often are designated by the abbreviation Cx:y, where x is the number of carbons in the chain and y is the number of double bonds. Fatty acids containing 12 or more carbon atoms are nearly insoluble in aqueous solutions because of their long hydrophobic hydrocarbon chains.

Fatty acids with no carbon-carbon double bonds are said to be **saturated;** those with at least one double bond are **unsaturated.** Unsaturated fatty acids with more than one carbon-carbon double bond are referred to as **polyunsaturated.** Two "essential" polyunsaturated fatty acids, linoleic acid (C18:2) and linolenic acid (C18:3), cannot be synthesized

TABLE 2-4 Fatty Acids That Predominate in Phospholipids

COMMON NAME OF ACID (IONIZED FORM IN PARENTHESES)	ABBREVIATION	CHEMICAL FORMULA
SATURATED FATTY ACIDS		
Myristic (myristate)	C14:0	$CH_3(CH_2)_{12}COOH$
Palmitic (palmitate)	C16:0	$CH_3(CH_2)_{14}COOH$
Stearic (stearate)	C18:0	$CH_3(CH_2)_{16}COOH$
UNSATURATED FATTY ACIDS		
Oleic (oleate)	C18:1	$CH_3(CH_2)_7CH{=}CH(CH_2)_7COOH$
Linoleic (linoleate)	C18:2	$CH_3(CH_2)_4CH{=}CHCH_2CH{=}CH(CH_2)_7COOH$
Arachidonic (arachidonate)	C20:4	$CH_3(CH_2)_4(CH{=}CHCH_2)_3CH{=}CH(CH_2)_3COOH$

by mammals and must be supplied in their diet. Mammals can synthesize other common fatty acids. Two stereoisomeric configurations, cis and trans, are possible around each carbon-carbon double bond:

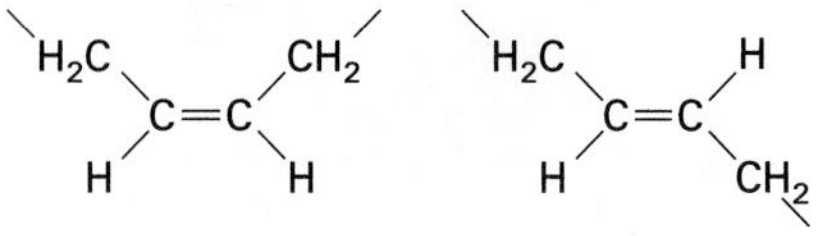

Cis

Trans

A cis double bond introduces a rigid kink in the otherwise flexible straight chain of a fatty acid (Figure 2-21). In general, the unsaturated fatty acids in biological systems contain only cis double bonds. Saturated fatty acids without the kink can pack together more tightly and so have higher melting points than unsaturated fatty acids (saturated fatty acids are usually solid rather than liquid at room temperature). The trans fatty acids (popularly called "trans fats") found in partially hydrogenated margarine (used to make solid margarine sticks) and other food products are not natural, arising from the catalytic process used for hydrogenation. Saturated and trans fatty acids have similar physical properties, and their consumption, relative to the consumption of unsaturated fats, is associated with increased plasma cholesterol levels.

Fatty acids can be covalently attached to another molecule by a type of dehydration reaction called *esterification,* in which the OH from the carboxyl group of the fatty acid and a H from a hydroxyl group on the other molecule are lost. In the combined molecule formed by this reaction, the portion derived from the fatty acid is called an *acyl group,* or *fatty acyl group.* This is illustrated by the most common form of phospholipids, **phosphoglycerides,** with two acyl groups attached to two of the hydroxyl groups of glycerol (see Figure 2-20) and **triacylglycerols, or triglycerides,** which contain three acyl groups esterfied to glycerol:

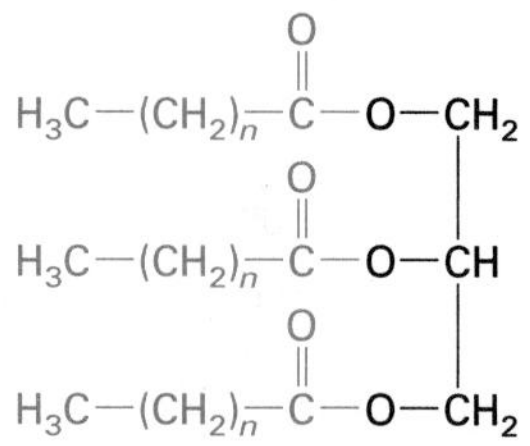

Triacylglycerol

Fatty acyl groups also can be covalently linked to another fatty molecule, cholesterol, to form cholesteryl esters.

Triglycerides and cholesteryl esters are extremely water-insoluble compounds in which fatty acids and cholesterol are either stored or transported. Triglycerides are the storage form of fatty acids in the fat cells of adipose tissue and are the principal components of dietary fats. Cholesteryl esters and triglycerides are transported between tissues through the bloodstream in specialized carriers called lipoproteins (Chapter 14).

In phosphoglycerides, one hydroxyl group of the glycerol is esterified to phosphate while the other two normally are esterified to fatty acids. The simplest phospholipid, phosphatidic acid, contains only these components. In most phospholipids found in membranes, the phosphate group is also esterified to a hydroxyl group on another hydrophilic compound. In phosphatidylcholine, for example, choline is attached to the phosphate (see Figure 2-20). The negative charge on the phosphate as well as the charged or polar groups esterified to it can interact strongly with water. The phosphate and its associated esterified group, the "head" group of a phospholipid, is hydrophilic, whereas the fatty acyl chains, the "tails," are hydrophobic. (Other common phosphoglycerides and associated head groups are shown in Table 2-5.) Molecules such as phospholipids that have both hydrophobic and

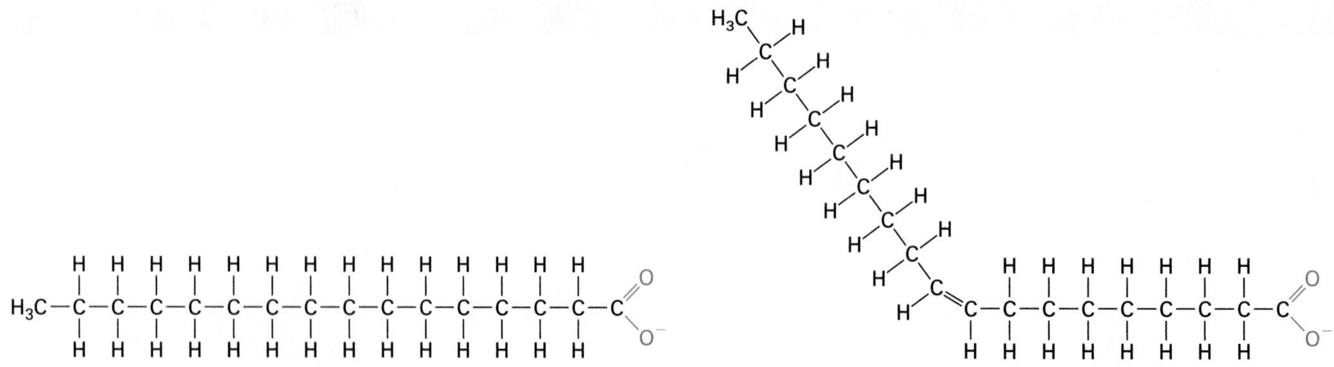

Palmitate
(ionized form of palmitic acid)

Oleate
(ionized form of oleic acid)

▲ **FIGURE 2-21 The effect of a double bond on the shape of fatty acids.** Shown are chemical structures of the ionized form of palmitic acid, a saturated fatty acid with 16 C atoms, and oleic acid, an unsaturated one with 18 C atoms. In saturated fatty acids, the hydrocarbon chain is often linear; the cis double bond in oleate creates a rigid kink in the hydrocarbon chain. [After L. Stryer, 1994, *Biochemistry,* 4th ed., W. H. Freeman and Company, p. 265.]

TABLE 2-5 Common Phosphoglycerides and Head Groups

COMMON PHOSPHOGLYCERIDES	HEAD GROUP
Phosphatidylcholine	Choline
Phosphatidylethanolamine	Ethanolamine
Phosphatidylserine	Serine
Phosphatidylinositol	Inositol

hydrophilic regions are called amphipathic. In Chapter 10, we will see how the amphipathic properties of phospholipids are responsible for the assembly of phospholipids into sheetlike bilayer biomembranes in which the fatty acyl tails point into the center of the sheet and the head groups point outward toward the aqueous environment (see Figure 2-13). ▮

KEY CONCEPTS OF SECTION 2.2

Chemical Building Blocks of Cells

- Three major biopolymers formed by polymerization reactions (net dehydration) of basic chemical building blocks are present in cells: proteins, composed of amino acids linked by peptide bonds; nucleic acids, composed of nucleotides linked by phosphodiester bonds; and polysaccharides, composed of monosaccharides (sugars) linked by glycosidic bonds (see Figure 2-13). Phospholipids, the fourth major chemical building block, assemble noncovalently into biomembranes.

- Differences in the size, shape, charge, hydrophobicity, and reactivity of the side chains of the 20 common amino acids determine the chemical and structural properties of proteins (see Figure 2-14).

- The bases in the nucleotides composing DNA and RNA are carbon- and nitrogen-containing rings attached to a pentose sugar. They form two groups: the purines—adenine (A) and guanine (G)—and the pyrimidines—cytosine (C), thymine (T), and uracil (U) (see Figure 2-17). A, G, T, and C are in DNA, and A, G, U, and C are in RNA.

- Glucose and other hexoses can exist in three forms: an open-chain linear structure, a six-member (pyranose) ring, and a five-member (furanose) ring (see Figure 2-18). In biological systems, the pyranose form of D-glucose predominates.

- Glycosidic bonds are formed between either the α or the β anomer of one sugar and a hydroxyl group on another sugar, leading to formation of disaccharides and other polysaccharides (see Figure 2-19).

- Phospholipids are amphipathic molecules with a hydrophobic tail (often two fatty acyl chains) connected by a small organic molecule (often glycerol) to a hydrophilic head (see Figure 2-20). The long hydrocarbon chain of a fatty acid may contain no carbon-carbon double bond (saturated) or one or more double bonds (unsaturated); a cis double bond bends the chain.

2.3 Chemical Equilibrium

We now shift our discussion to chemical reactions in which bonds, primarily covalent bonds in *reactant* chemicals, are broken and new bonds are formed to generate reaction *products*. At any one time, several hundred different kinds of chemical reactions are occurring simultaneously in every cell, and many chemicals can, in principle, undergo multiple chemical reactions. Both the *extent* to which reactions can proceed and the *rate* at which they take place determine the chemical composition of cells.

When reactants first mix together—before any products have been formed—their rate of reaction to form products (forward reaction) is determined in part by their initial concentrations, which determine the likelihood of reactants bumping into one another and reacting (Figure 2-22). As the reaction products accumulate, the concentration of each reactant decreases and so does the forward reaction rate. Meanwhile, some of the product molecules begin to participate in the reverse reaction, which re-forms the reactants (the ability of a reaction to go "backward" is called *microscopic reversibility*). This reverse reaction is slow at first but speeds up as the concentration of product increases. Eventually, the rates of the forward and reverse reactions become equal, so that the concentrations of reactants and products stop changing. The system is then said to be in **chemical equilibrium** (plural: *equilibria*).

At equilibrium, the ratio of products to reactants, called the **equilibrium constant,** is a fixed value that is independent of the rate at which the reaction occurs. The rate of a chemical reaction can be increased by a **catalyst,** which accelerates the chemical transformation (making and breaking of covalent bonds) but is not permanently changed during a reaction (see Section 2.4). In this section, we discuss several aspects of chemical equilibria; in the next section, we examine energy changes during reactions and their relationship to equilibria.

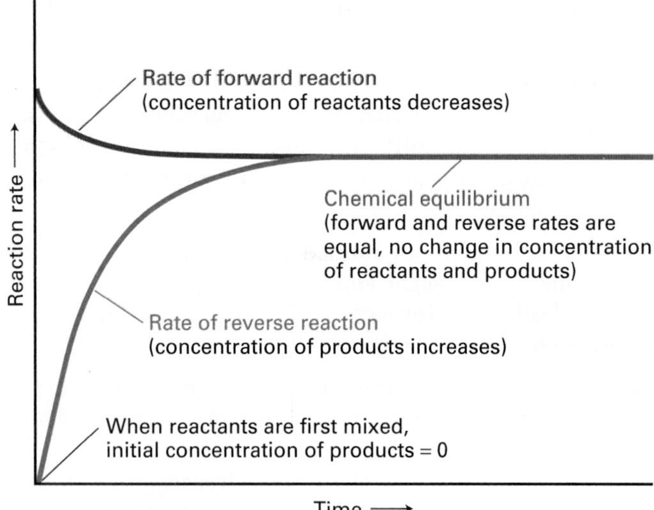

▲ **FIGURE 2-22 Time dependence of the rates of a chemical reaction.** The forward and reverse rates of a reaction depend in part on the initial concentrations of reactants and products. The net forward reaction rate slows as the concentration of reactants decreases, whereas the net reverse reaction rate increases as the concentration of products increases. At equilibrium, the rates of the forward and reverse reactions are equal and the concentrations of reactants and products remain constant.

Equilibrium Constants Reflect the Extent of a Chemical Reaction

The equilibrium constant K_{eq} depends on the nature of the reactants and products, the temperature, and the pressure (particularly in reactions involving gases). Under standard physical conditions (25 °C and 1 atm pressure for biological systems), the K_{eq} is always the same for a given reaction, whether or not a catalyst is present.

For the general reaction with three reactants and three products

$$aA + bB + cC \rightleftharpoons zZ + yY + xX \quad (2\text{-}1)$$

where capital letters represent particular molecules or atoms and lowercase letters represent the number of each in the reaction formula, the equilibrium constant is given by

$$K_{eq} = \frac{[X]^x [Y]^y [Z]^z}{[A]^a [B]^b [C]^c} \quad (2\text{-}2)$$

where brackets denote the concentrations of the molecules at equilibrium. The rate of the forward reaction (left to right in Equation (2-1) is

$$\text{Rate}_{\text{forward}} = k_f[A]^a[B]^b[C]^c$$

where k_f is the **rate constant** for the forward reaction. Similarly, the rate of the reverse reaction (right to left in Equation 2-1) is

$$\text{Rate}_{\text{reverse}} = k_r[X]^x[Y]^y[Z]^z$$

where k_r is the rate constant for the reverse reaction. At equilibrium the forward and reverse rates are equal, so $\text{Rate}_{\text{forward}}/\text{Rate}_{\text{reverse}} = 1$. By rearranging these equations, we can express the equilibrium constant as the ratio of the rate constants

$$K_{eq} = \frac{k_f}{k_r} \quad (2\text{-}3)$$

Chemical Reactions in Cells Are at Steady State

Under appropriate conditions and given sufficient time, individual biochemical reactions carried out in a test tube eventually will reach equilibrium. Within cells, however, many reactions are linked in pathways in which a product of one reaction serves as a reactant in another or is pumped out of the cell. In this more complex situation, when the rate of formation of a substance is equal to the rate of its consumption, the concentration of the substance remains constant, and the system of linked reactions for producing and consuming that substance is said to be in a **steady state** (Figure 2-23). One consequence of such linked reactions is that they prevent the accumulation of excess intermediates, protecting cells from the harmful effects of intermediates that have the potential of being toxic at high concentrations.

Dissociation Constants of Binding Reactions Reflect the Affinity of Interacting Molecules

The concept of equilibrium also applies to the binding of one molecule to another. Many important cellular processes depend on such binding "reactions," which involve the making and breaking of various noncovalent interactions rather than covalent bonds, as discussed above. A common example is the binding of a **ligand** (e.g., the hormone insulin or adrenaline) to its **receptor** on the surface of a cell forming a multimolecular assembly, or complex, that triggers a biological response. Another example is the binding of a protein to a specific sequence of base pairs in a molecule of DNA, which frequently causes the expression of a nearby gene to increase or decrease (Chapter 7). If the equilibrium constant for a binding reaction is

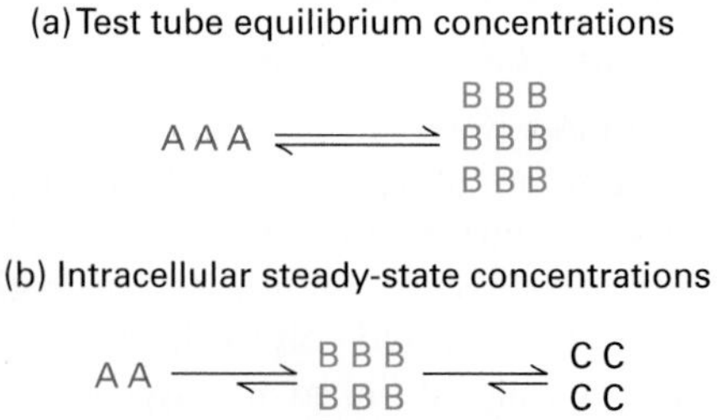

(a) Test tube equilibrium concentrations

AAA ⇌ BBB BBB BBB

(b) Intracellular steady-state concentrations

AA ⇌ BBB BBB ⇌ CC CC

▲ **FIGURE 2-23 Comparison of reactions at equilibrium and steady state.** (a) In the test tube, a biochemical reaction (A → B) eventually will reach equilibrium, in which the rates of the forward and reverse reactions are equal (as indicated by the reaction arrows of equal length). (b) In metabolic pathways within cells, the product B commonly would be consumed, in this example by conversion to C. A pathway of linked reactions is at steady state when the rate of formation of the intermediates (e.g., B) equals their rate of consumption. As indicated by the unequal length of the arrows, the individual reversible reactions constituting a metabolic pathway do not reach equilibrium. Moreover, the concentrations of the intermediates at steady state can differ from what they would be at equilibrium.

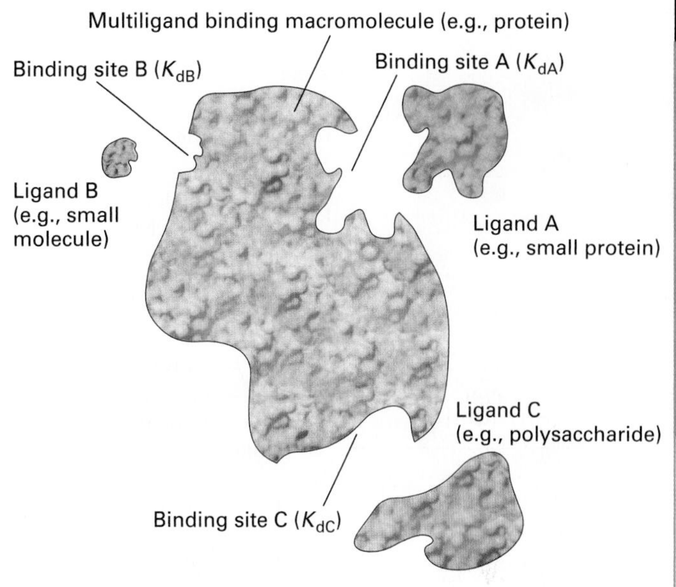

▶ **FIGURE 2-24 Macromolecules can have distinct binding sites for multiple ligands.** A large macromolecule (e.g., a protein, blue) with three distinct binding sites (A–C) is shown; each binding site exhibits molecular complementarity to three different binding partners (ligands A–C) with distinct dissociation constants (K_{dA-C}).

known, the intracellular stability of the resulting complex can be predicted. To illustrate the general approach for determining the concentration of noncovalently associated complexes, we will calculate the extent to which a protein (P) is bound to DNA (D) forming a protein-DNA complex (PD):

$$P + D \rightleftharpoons PD$$

Most commonly, binding reactions are described in terms of the **dissociation constant** K_d, which is the reciprocal of the equilibrium constant. For this binding reaction, the dissociation constant is given by

$$K_d = \frac{[P][D]}{[PD]} \quad (2\text{-}4)$$

Typical reactions in which a protein binds to a specific DNA sequence have a K_d of 10^{-10} M, where M symbolizes molarity, or moles per liter (mol/L). To relate the magnitude of this dissociation constant to the intracellular ratio of bound to unbound DNA, let's consider the simple example of a bacterial cell having a volume of 1.5×10^{-15} L and containing 1 molecule of DNA and 10 molecules of the DNA-binding protein P. In this case, given a K_d of 10^{-10} M, 99 percent of the time this specific DNA sequence will have a molecule of protein bound to it and 1 percent of the time it will not, even though the cell contains only 10 molecules of the protein! Clearly, P and D bind very tightly (have a high affinity), as reflected by the low value of the dissociation constant for their binding reaction. For protein-protein and protein-DNA binding, K_d values of $\leq 10^{-9}$ M (nanomolar) are considered to be tight, ~10^{-6} M (micromolar) modestly tight, and ~10^{-3} M (millimolar) relatively weak.

The large size of biological macromolecules, such as proteins, can result in the availability of multiple surfaces for complementary intermolecular interactions (Figure 2-24). As a consequence, many macromolecules have the capacity to bind several other molecules simultaneously. In some cases, these binding reactions are independent, with their own distinct K_d values that are constant. In other cases, binding of a molecule at one site on a macromolecule can change the three-dimensional shape of a distant site, thus altering the binding interactions of that distant site with some other molecule. This is an important mechanism by which one molecule can alter (regulate) the activity of a second molecule (e.g., a protein) by changing its capacity to interact with a third molecule. We examine this regulatory mechanism in more detail in Chapter 3.

Biological Fluids Have Characteristic pH Values

The solvent inside cells and in all extracellular fluids is water. An important characteristic of any aqueous solution is the concentration of positively charged hydrogen ions (H^+) and negatively charged hydroxyl ions (OH^-). Because these ions are the dissociation products of H_2O, they are constituents of all living systems, and they are liberated by many reactions that take place between organic molecules within cells. These ions also can be transported into or out of cells, as when highly acidic gastric juice is secreted by cells lining the walls of the stomach.

When a water molecule dissociates, one of its polar H—O bonds breaks. The resulting hydrogen ion, often referred to as a **proton**, has a short lifetime as a free ion and quickly combines with a water molecule to form a hydronium ion (H_3O^+). For convenience, however, we refer to the concentration of hydrogen ions in a solution, $[H^+]$, even though this really represents the concentration of hydronium ions, $[H_3O^+]$. Dissociation of H_2O generates one OH^- ion along with each H^+. The dissociation of water is a reversible reaction:

$$H_2O \rightleftharpoons H^+ + OH^-$$

At 25 °C, $[H^+][OH^-] = 10^{-14}$ M^2, so that in pure water, $[H^+] = [OH^-] = 10^{-7}$ M.

The concentration of hydrogen ions in a solution is expressed conventionally as its **pH**, defined as the negative log of the hydrogen ion concentration. The pH of pure water at 25 °C is 7:

$$pH = -\log[H^+] = \log\frac{1}{[H^+]} = \log\frac{1}{10^{-7}} = 7$$

It is important to keep in mind that a 1 unit difference in pH represents a tenfold difference in the concentration of protons. On the pH scale, 7.0 is considered neutral: pH values below 7.0 indicate acidic solutions (higher $[H^+]$), and values above 7.0 indicate basic (alkaline) solutions (Figure 2-25). For instance, gastric juice, which is rich in hydrochloric acid

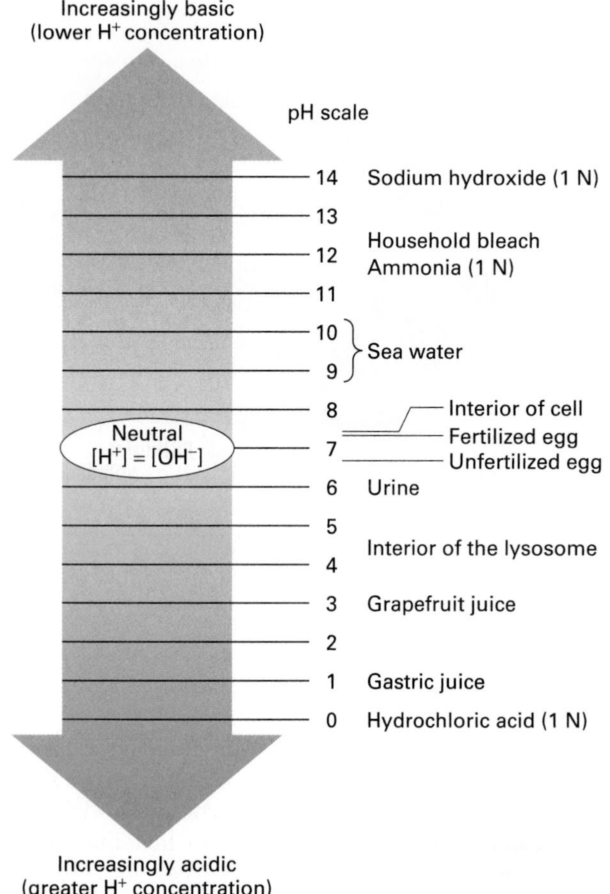

▲ **FIGURE 2-25 pH values of common solutions.** The pH of an aqueous solution is the negative log of the hydrogen ion concentration. The pH values for most intracellular and extracellular biological fluids are near 7 and are carefully regulated to permit the proper functioning of cells, organelles, and cellular secretions.

(HCl), has a pH of about 1. Its $[H^+]$ is roughly a millionfold greater than that of cytoplasm with a pH of about 7.2.

Although the cytosol of cells normally has a pH of about 7.2, the pH is much lower (about 4.5) in the interior of lysosomes, one type of organelle in eukaryotic cells (Chapter 9). The many degradative enzymes within lysosomes function optimally in an acidic environment, whereas their action is inhibited in the near neutral environment of the cytoplasm. This illustrates that maintenance of a specific pH is essential for proper functioning of some cellular structures. On the other hand, dramatic shifts in cellular pH may play an important role in controlling cellular activity. For example, the pH of the cytoplasm of an unfertilized egg of the sea urchin, an aquatic animal, is 6.6. Within 1 minute of fertilization, however, the pH rises to 7.2; that is, the $[H^+]$ decreases to about one-fourth its original value, a change necessary for subsequent growth and division of the egg.

Hydrogen Ions Are Released by Acids and Taken Up by Bases

In general, an **acid** is any molecule, ion, or chemical group that tends to release a hydrogen ion (H^+), such as hydrochloric acid (HCl) or the carboxyl group (—COOH), which tends to dissociate to form the negatively charged carboxylate ion (—COO^-). Likewise, a **base** is any molecule, ion, or chemical group that readily combines with a H^+, such as the hydroxyl ion (OH^-); ammonia (NH_3), which forms an ammonium ion (NH_4^+); or the amino group (—NH_2).

When acid is added to an aqueous solution, the $[H^+]$ increases (the pH goes down). Conversely, when a base is added to a solution, the $[H^+]$ decreases (the pH goes up). Because $[H^+][OH^-] = 10^{-14}M^2$, any increase in $[H^+]$ is coupled with a commensurate decrease in $[OH^-]$ and vice versa.

Many biological molecules contain both acidic and basic groups. For example, in neutral solutions (pH = 7.0), many amino acids exist predominantly in the doubly ionized form, in which the carboxyl group has lost a proton and the amino group has accepted one:

$$\begin{array}{c} NH_3^+ \\ | \\ H{-}C{-}COO^- \\ | \\ R \end{array}$$

where R represents the uncharged side chain. Such a molecule, containing an equal number of positive and negative ions, is called a *zwitterion*. Zwitterions, having no net charge, are neutral. At extreme pH values, only one of these two ionizable groups of an amino acid will be charged.

The dissociation reaction for an acid (or acid group in a larger molecule) HA can be written as $HA \rightleftharpoons H^+ + A^-$. The equilibrium constant for this reaction, denoted K_a (the subscript *a* stands for "acid"), is defined as $K_a = [H^+][A^-]/[HA]$. Taking the logarithm of both sides and rearranging the result yields a very useful relation between the equilibrium constant and pH:

$$pH = pK_a + \log \frac{[A^-]}{[HA]} \qquad (2\text{-}5)$$

where pK_a equals –log K_a.

From this expression, commonly known as the *Henderson-Hasselbalch equation*, it can be seen that the pK_a of any acid is equal to the pH at which half the molecules are dissociated and half are neutral (undissociated). This is because when $[A^-] = [HA]$, then $\log([A^-]/[HA]) = 0$, and thus pK_a = pH. The Henderson-Hasselbalch equation allows us to calculate the degree of dissociation of an acid if both the pH of the solution and the pK_a of the acid are known. Experimentally, by measuring the $[A^-]$ and [HA] as a function of the solution's pH, one can calculate the pK_a of the acid and thus the equilibrium constant K_a for the dissociation reaction (Figure 2-26).

Buffers Maintain the pH of Intracellular and Extracellular Fluids

A growing cell must maintain a constant pH in the cytoplasm of about 7.2–7.4 despite the metabolic production of many acids, such as lactic acid and carbon dioxide; the latter reacts with water to form carbonic acid (H_2CO_3). Cells have a reservoir of weak bases and weak acids, called **buffers,** which ensure that the cell's pH remains relatively constant

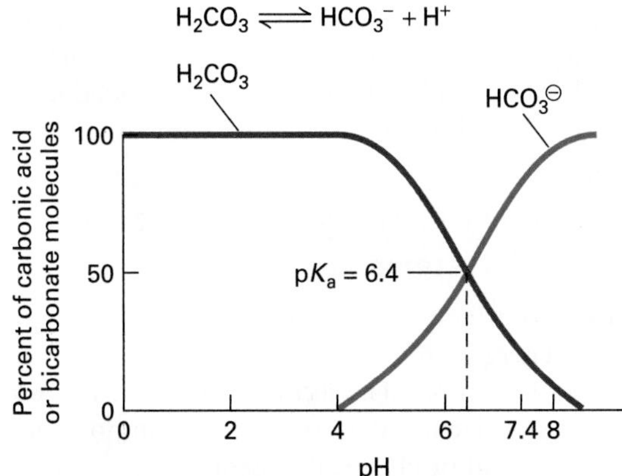

▲ **FIGURE 2-26 The relationship between pH, p*K*a, and the dissociation of an acid.** As the pH of a solution of carbonic acid rises from 0 to 8.5, the percentage of the compound in the undissociated, or un-ionized, form (H_2CO_3) decreases from 100 percent and that of the ionized form increases from 0 percent. When the pH (6.4) is equal to the acid's p*K*a, half of the carbonic acid has ionized. When the pH rises to above 8, virtually all of the acid has ionized to the bicarbonate form (HCO_3^-).

despite small fluctuations in the amounts of H^+ or OH^- being generated by metabolism or by the uptake or secretion of molecules and ions by the cell. Buffers do this by "soaking up" excess H^+ or OH^- when these ions are added to the cell or are produced by metabolism.

If additional acid (or base) is added to a buffered solution whose pH is equal to the pK_a of the buffer ([HA] = [A^-]), the pH of the solution changes, but it changes less than it would if the buffer had not been present. This is because protons released by the added acid are taken up by the ionized form of the buffer (A^-); likewise, hydroxyl ions generated by the addition of base are neutralized by protons released by the undissociated buffer (HA). The capacity of a substance to release hydrogen ions or take them up depends partly on the extent to which the substance has already taken up or released protons, which in turn depends on the pH of the solution relative to the pK_a of the substance. The ability of a buffer to minimize changes in pH, its *buffering capacity*, depends on the concentration of the buffer and the relationship between its pK_a value and the pH, which is expressed by the Henderson-Hasselbalch equation.

The titration curve for acetic acid shown in Figure 2-27 illustrates the effect of pH on the fraction of molecules in the un-ionized (HA) and ionized forms (A^-). At one pH unit below the pK_a of an acid, 91 percent of the molecules are in the HA form; at one pH unit above the pK_a, 91 percent are in the A^- form. At pH values more than one unit above or below the pK_a, the buffering capacity of weak acids and bases declines rapidly. In other words, the addition of the same number of moles of acid to a solution containing a mixture of HA and A^- that is at a pH near the pK_a will cause less of a pH change than it would if the HA and A^- were not present or if the pH were far from the pK_a value.

All biological systems contain one or more buffers. Phosphate ions, the ionized forms of phosphoric acid, are present in considerable quantities in cells and are an important factor

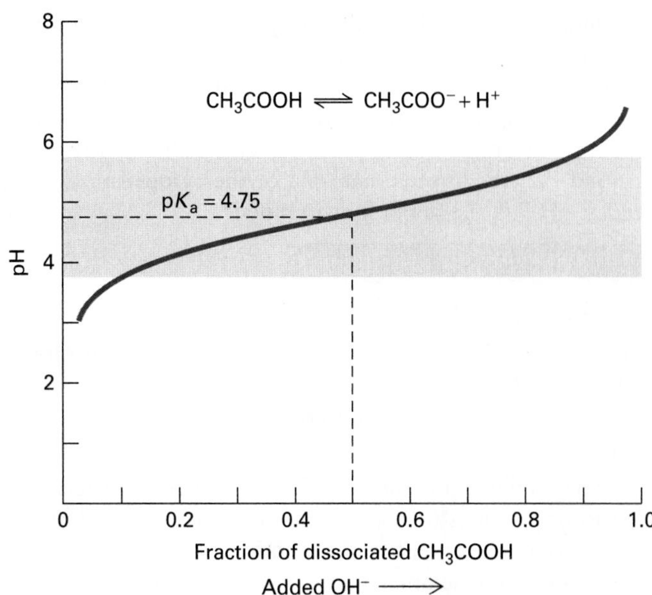

▲ **FIGURE 2-27 The titration curve of the buffer acetic acid (CH_3COOH).** The pK_a for the dissociation of acetic acid to hydrogen and acetate ions is 4.75. At this pH, half the acid molecules are dissociated. Because pH is measured on a logarithmic scale, the solution changes from 91 percent CH_3COOH at pH 3.75 to 9 percent CH_3COOH at pH 5.75. The acid has maximum buffering capacity in this pH range.

in maintaining, or buffering, the pH of the cytoplasm. Phosphoric acid (H_3PO_4) has three protons that are capable of dissociating, but they do not dissociate simultaneously. Loss of each proton can be described by a discrete dissociation reaction and pK_a, as shown in Figure 2-28. The titration curve

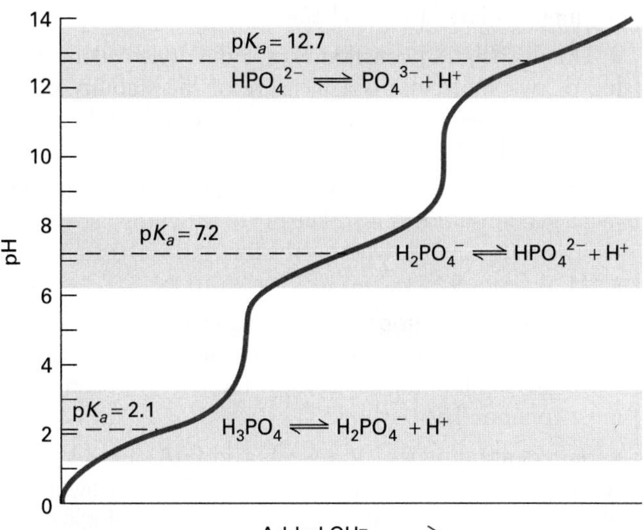

▲ **FIGURE 2-28 The titration curve of phosphoric acid (H_3PO_4), a common buffer in biological systems.** This biologically ubiquitous molecule has three hydrogen atoms that dissociate at different pH values; thus phosphoric acid has three pK_a values, as noted on the graph. The shaded areas denote the pH ranges—within one pH unit of the three pK_a values—where the buffering capacity of phosphoric acid is high. In these regions, the addition of acid (or base) will cause relatively small changes in the pH.

for phosphoric acid shows that the pK_a for the dissociation of the second proton is 7.2. Thus at pH 7.2, about 50 percent of cellular phosphate is $H_2PO_4^-$ and about 50 percent is HPO_4^{2-} according to the Henderson-Hasselbalch equation. For this reason, phosphate is an excellent buffer at pH values around 7.2, the approximate pH of the cytoplasm of cells, and at pH 7.4, the pH of human blood.

KEY CONCEPTS OF SECTION 2.3

Chemical Equilibrium

- A chemical reaction is at equilibrium when the rate of the forward reaction is equal to the rate of the reverse reaction (no net change in the concentration of the reactants or products).
- The equilibrium constant K_{eq} of a reaction reflects the ratio of products to reactants at equilibrium and thus is a measure of the extent of the reaction and the relative stabilities of the reactants and products.
- The K_{eq} depends on the temperature, pressure, and chemical properties of the reactants and products but is independent of the reaction rate and of the initial concentrations of reactants and products.
- For any reaction, the equilibrium constant K_{eq} equals the ratio of the forward rate constant to the reverse rate constant (k_f/k_r). The rates of conversion of reactants to products and vice versa depend on the rate constants and the concentrations of the reactants or products.
- Within cells, the linked reactions in metabolic pathways generally are at steady state, not equilibrium, at which rate of formation of the intermediates equals their rate of consumption (see Figure 2-23) and thus the concentrations of the intermediates are not changing.
- The dissociation constant K_d for the noncovalent binding of two molecules is a measure of the stability of the complex formed between the molecules (e.g., ligand-receptor or protein-DNA complexes).
- The pH is the negative logarithm of the concentration of hydrogen ions (–log [H+]). The pH of the cytoplasm is normally about 7.2–7.4, whereas the interior of lysosomes has a pH of about 4.5.
- Acids release protons (H^+) and bases bind them. In biological molecules, the carboxyl and phosphate groups are the most common acidic groups; the amino group is the most common basic group.
- Buffers are mixtures of a weak acid (HA) and its corresponding base form (A^-), which minimize the change in pH of a solution when acid or alkali is added. Biological systems use various buffers to maintain their pH within a very narrow range.

2.4 Biochemical Energetics

The production of energy, its storage, and its use are central to the economy of the cell. Energy may be defined as the ability to do work, a concept applicable to automobile engines and electric power plants in our day-to-day world and to cellular engines in the biological world. The energy associated with chemical bonds can be harnessed to support chemical work and the physical movements of cells.

Several Forms of Energy Are Important in Biological Systems

There are two principal forms of energy: kinetic and potential. **Kinetic energy** is the energy of movement—the motion of molecules, for example. The second form of energy, **potential energy,** or stored energy, is particularly important in the study of biological or chemical systems.

Thermal energy, or heat, is a form of kinetic energy—the energy of the motion of molecules. For heat to do work, it must flow from a region of higher temperature—where the average speed of molecular motion is greater—to one of lower temperature. Although differences in temperature can exist between the internal and external environments of cells, these thermal gradients do not usually serve as the source of energy for cellular activities. The thermal energy in warm-blooded animals, which have evolved a mechanism for thermoregulation, is used chiefly to maintain constant organismic temperatures. This is an important function because the rates of many cellular activities are temperature-dependent. For example, cooling mammalian cells from their normal body temperature of 37 °C to 4 °C can virtually "freeze" or stop many cellular processes (e.g., intracellular membrane movements).

Radiant energy is the kinetic energy of photons, or waves of light, and is critical to biology. Radiant energy can be converted to thermal energy, for instance when light is absorbed by molecules and the energy is converted to molecular motion. Radiant energy absorbed by molecules can also change the electronic structure of the molecules, moving electrons into higher-energy states (orbitals), whence it can later be recovered to perform work. For example, during photosynthesis, light energy absorbed by specialized molecules (e.g., chlorophyll) is subsequently converted into the energy of chemical bonds (Chapter 12).

Mechanical energy, a major form of kinetic energy in biology, usually results from the conversion of stored chemical energy. For example, changes in the lengths of cytoskeletal filaments generate forces that push or pull on membranes and organelles (Chapters 17 and 18).

Electric energy—the energy of moving electrons or other charged particles—is yet another major form of kinetic energy.

Several forms of potential energy are biologically significant. Central to biology is **chemical potential energy,** the energy stored in the bonds connecting atoms in molecules. Indeed, most of the biochemical reactions described in this book involve the making or breaking of at least one covalent chemical bond. We recognize this energy when chemicals undergo energy-releasing reactions. For example, the high potential energy in the covalent bonds of glucose can be released by controlled enzymatic combustion in cells (Chapter 12). This energy is harnessed by the cell to do many kinds of work.

A second biologically important form of potential energy is the energy in a **concentration gradient.** When the concentration of a substance on one side of a barrier, such as a membrane, is different from that on the other side, a concentration gradient exists. All cells form concentration gradients between their interior and the external fluids by selectively exchanging nutrients, waste products, and ions with their surroundings. Also, organelles within cells (e.g., mitochondria, lysosomes) frequently contain different concentrations of ions and other molecules; the concentration of protons within a lysosome, as we saw in the last section, is about 500 times that of the cytoplasm.

A third form of potential energy in cells is an **electric potential**—the energy of charge separation. For instance, there is a gradient of electric charge of ≈200,000 volts per cm across the plasma membrane of virtually all cells. We discuss how concentration gradients and the potential difference across cell membranes are generated and maintained in Chapter 11 and how they are converted to chemical potential energy in Chapter 12.

Cells Can Transform One Type of Energy into Another

According to the first law of thermodynamics, energy is neither created nor destroyed but can be converted from one form to another. (In nuclear reactions, mass is converted to energy, but this is irrelevant to biological systems.) In photosynthesis, for example, the radiant energy of light is transformed into the chemical potential energy of the covalent bonds between the atoms in a sucrose or starch molecule. In muscles and nerves, chemical potential energy stored in covalent bonds is transformed, respectively, into the kinetic energy of muscle contraction and the electric energy of nerve transmission. In all cells, potential energy, released by breaking certain chemical bonds, is used to generate potential energy in the form of concentration and electric potential gradients. Similarly, energy stored in chemical concentration gradients or electric potential gradients is used to synthesize chemical bonds or to transport molecules from one side of a membrane to another to generate a concentration gradient. The latter process occurs during the transport of nutrients such as glucose into certain cells and transport of many waste products out of cells.

Because all forms of energy are interconvertible, they can be expressed in the same units of measurement. Although the standard unit of energy is the joule, biochemists have traditionally used an alternative unit, the **calorie** (1 joule = 0.239 calorie). Throughout this book, we use the kilocalorie to measure energy changes (1 kcal = 1000 cal).

The Change in Free Energy Determines the Direction of a Chemical Reaction

Because biological systems are generally held at constant temperature and pressure, it is possible to predict the direction of a chemical reaction from the change in the **free energy** *G*, named after J. W. Gibbs, who showed that "all systems change in such a way that free energy [*G*] is minimized." In the case of a chemical reaction, reactants $\rightleftharpoons$ products, the free-energy change ΔG is given by

$$\Delta G = G_{\text{products}} - G_{\text{reactants}}$$

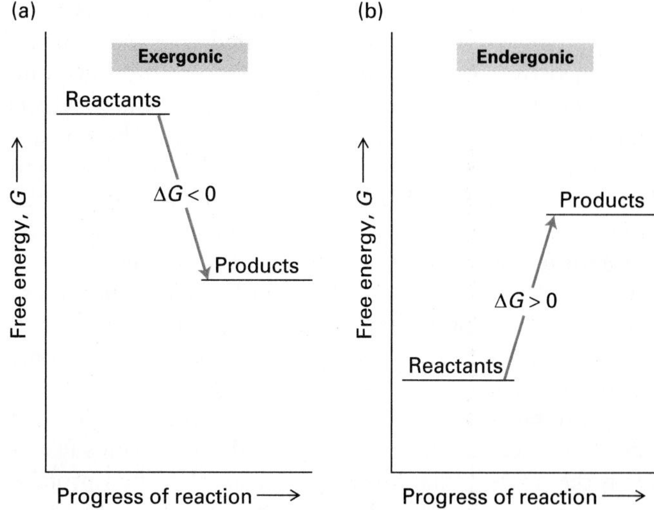

▲ **FIGURE 2-29 Changes in the free energy (ΔG) of exergonic and endergonic reactions.** (a) In exergonic reactions, the free energy of the products is lower than that of the reactants. Consequently, these reactions occur spontaneously and energy is released as the reactions proceed. (b) In endergonic reactions, the free energy of the products is greater than that of the reactants and these reactions do not occur spontaneously. An external source of energy must be supplied if the reactants are to be converted into products.

The relation of ΔG to the direction of any chemical reaction can be summarized in three statements:

- If ΔG is negative, the forward reaction will tend to occur spontaneously and energy usually will be released as the reaction takes place (**exergonic reaction**) (Figure 2-29).
- If ΔG is positive, the forward reaction will not occur spontaneously: energy will have to be added to the system in order to force the reactants to become products (**endergonic reaction**).
- If ΔG is zero, both forward and reverse reactions occur at equal rates and there will be no spontaneous conversion of reactants to products (or vice versa); the system is at equilibrium.

By convention, the standard free-energy change of a reaction $\Delta G^{\circ\prime}$ is the value of the change in free energy under the conditions of 298 K (25 °C), 1 atm pressure, pH 7.0 (as in pure water), and initial concentrations of 1 M for all reactants and products except protons, which are kept at 10^{-7} M (pH 7.0). Most biological reactions differ from standard conditions, particularly in the concentrations of reactants, which are normally less than 1 M.

The free energy of a chemical system can be defined as $G = H - TS$, where *H* is the bond energy, or **enthalpy,** of the system; *T* is its temperature in degrees Kelvin (K); and *S* is the **entropy,** a measure of its randomness or disorder. If temperature remains constant, a reaction proceeds spontaneously only if the free-energy change ΔG in the following equation is negative:

$$\Delta G = \Delta H - T\,\Delta S \tag{2-6}$$

In an **exothermic** reaction, the products contain less bond energy than the reactants, the liberated energy is usually converted to heat (the energy of molecular motion), and ΔH is negative. In an **endothermic** reaction, the products contain more bond energy than the reactants, heat is absorbed during the reaction, and ΔH is positive. The combined effects of the changes in the enthalpy and entropy determine if the ΔG for a reaction is positive or negative. An exothermic reaction ($\Delta H < 0$) in which entropy increases ($\Delta S > 0$) occurs spontaneously ($\Delta G < 0$). An endothermic reaction ($\Delta H > 0$) will occur spontaneously if ΔS increases enough so that the $T\,\Delta S$ term can overcome the positive ΔH.

Many biological reactions lead to an increase in order and thus a decrease in entropy ($\Delta S < 0$). An obvious example is the reaction that links amino acids to form a protein. A solution of protein molecules has a lower entropy than does a solution of the same amino acids unlinked because the free movement of any amino acid in a protein is restricted when it is bound into a long chain. Often cells compensate for decreases in entropy by "coupling" such synthetic, entropy-lowering reactions with independent reactions that have a very highly negative ΔG (see below). In this way cells can convert sources of energy in their environment into the building of highly organized structures and metabolic pathways that are essential for life.

The actual change in free energy ΔG during a reaction is influenced by temperature, pressure, and the initial concentrations of reactants and products and usually differs from $\Delta G^{\circ\prime}$. Most biological reactions—like others that take place in aqueous solutions—also are affected by the pH of the solution. We can estimate free-energy changes for different temperatures and initial concentrations using the equation

$$\Delta G = \Delta G^{\circ\prime} + RT \ln Q = \Delta G^{\circ\prime} + RT \ln \frac{[\text{products}]}{[\text{reactants}]} \quad (2\text{-}7)$$

where R is the gas constant of 1.987 cal/(degree·mol), T is the temperature (in degrees Kelvin), and Q is the *initial* ratio of products to reactants. For a reaction A + B $\rightleftharpoons$ C, in which two molecules combine to form a third, Q in Equation 2-7 equals [C]/[A][B]. In this case, an increase in the initial concentration of either [A] or [B] will result in a larger negative value for ΔG and thus drive the reaction toward more formation of C.

Regardless of the $\Delta G^{\circ\prime}$ for a particular biochemical reaction, it will proceed spontaneously within cells only if ΔG is negative, given the intracellular concentrations of reactants and products. For example, the conversion of glyceraldehyde 3-phosphate (G3P) to dihydroxyacetone phosphate (DHAP), two intermediates in the breakdown of glucose,

$$\text{G3P} \rightleftharpoons \text{DHAP}$$

has a $\Delta G^{\circ\prime}$ of −1840 cal/mol. If the initial concentrations of G3P and DHAP are equal, then $\Delta G = \Delta G^{\circ\prime}$ because $RT \ln 1 = 0$; in this situation, the reversible reaction G3P $\rightleftharpoons$ DHAP will proceed spontaneously in the direction of DHAP formation until equilibrium is reached. However, if the initial [DHAP] is 0.1 M and the initial [G3P] is 0.001 M, with other conditions standard, then Q in Equation 2-7 equals 0.1/0.001 = 100, giving a ΔG of +887 cal/mol. Under these conditions, the reaction will proceed in the direction of formation of G3P.

The ΔG for a reaction is independent of the reaction rate. Indeed, under usual physiological conditions, few if any of the biochemical reactions needed to sustain life would occur without some mechanism for increasing reaction rates. As we describe below and in more detail in Chapter 3, the rates of reactions in biological systems are usually determined by the activity of **enzymes**, the protein catalysts that accelerate the formation of products from reactants without altering the value of ΔG.

The $\Delta G^{\circ\prime}$ of a Reaction Can Be Calculated from Its K_{eq}

A chemical mixture at equilibrium is in a stable state of minimal free energy. For a system at equilibrium ($\Delta G = 0$, $Q = K_{eq}$), we can write

$$\Delta G^{\circ\prime} = -2.3RT \log K_{eq} = -1362 \log K_{eq} \quad (2\text{-}8)$$

under standard conditions (note the change to base 10 logarithms). Thus if we determine the concentrations of reactants and products at equilibrium (i.e., the K_{eq}), we can calculate the value of $\Delta G^{\circ\prime}$. For example, the K_{eq} for the interconversion of glyceraldehyde 3-phosphate to dihydroxyacetone phosphate (G3P $\rightleftharpoons$ DHAP) is 22.2 under standard conditions. Substituting this value into Equation 2-8, we can easily calculate the $\Delta G^{\circ\prime}$ for this reaction as −1840 cal/mol.

By rearranging Equation 2-8 and taking the antilogarithm, we obtain

$$K_{eq} = 10^{-(\Delta G^{\circ\prime}/2.3RT)} \quad (2\text{-}9)$$

From this expression, it is clear that if $\Delta G^{\circ\prime}$ is negative, the exponent will be positive and hence K_{eq} will be greater than 1. Therefore at equilibrium there will be more products than reactants; in other words, the formation of products from reactants is favored. Conversely, if $\Delta G^{\circ\prime}$ is positive, the exponent will be negative and K_{eq} will be less than 1.

The Rate of a Reaction Depends on the Activation Energy Necessary to Energize the Reactants into a Transition State

As a chemical reaction proceeds, reactants approach each other; some bonds begin to form while others begin to break. One way to think of the state of the molecules during this transition is that there are strains in the electronic configurations of the atoms and their bonds. In order for the collection of atoms to move from the relatively stable state of the reactants to this intermediate state during the reaction, an introduction of energy is necessary. This is illustrated in the reaction energy diagram in Figure 2-30. Thus the collection of atoms is transiently in a higher-energy state at some point during the course of the reaction. The state during a chemical reaction at which the system is at its highest energy level is called the **transition state** or **transition-**

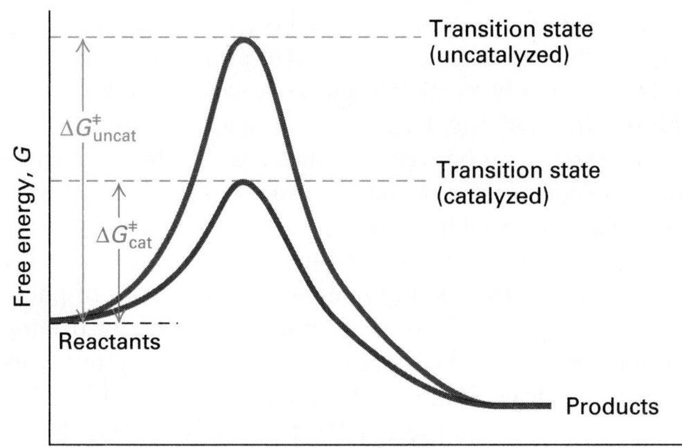

▲ **FIGURE 2-30 Activation energy of uncatalyzed and catalyzed chemical reactions.** This hypothetical reaction pathway (blue) depicts the changes in free energy G as a reaction proceeds. A reaction will take place spontaneously if the free energy (G) of the products is less than that of the reactants ($\Delta G < 0$). However, all chemical reactions proceed through one (shown here) or more high-energy transition states, and the rate of a reaction is inversely proportional to the activation energy ($\Delta G^{\ddagger}$), which is the difference in free energy between the reactants and the transition state. In a catalyzed reaction (red), the free energies of the reactants and products are unchanged but the free energy of the transition state is lowered, thus increasing the velocity of the reaction.

state intermediate. The energy needed to excite the reactants to this higher-energy state is called the **activation energy** of the reaction. The activation energy is usually represented by $\Delta G^{\ddagger}$, analogous to the representation of the change in Gibbs free energy (ΔG) already discussed. From the transition state, the collection of atoms can either release energy as the reaction products are formed or release energy as the atoms go "backward" and re-form the original reactants. The velocity (V) at which products are generated from reactants during the reaction under a given set of conditions (temperature, pressure, reactant concentrations) will depend on the concentration of material in the transition state, which in turn will depend on the activation energy and the characteristic rate constant (v) at which the transition state is converted to products. The higher the activation energy, the lower the fraction of reactants that reach the transition state and the slower the overall rate of the reaction. The relationship between the concentration of reactants, v, and V is

$$V = v\,[\text{reactants}] \times 10^{-(\Delta G^{\ddagger}/2.3RT)}$$

From this equation, we can see that lowering the activation energy—that is, decreasing the free energy of the transition state $\Delta G^{\ddagger}$—leads to an acceleration of the overall reaction rate V. A reduction in $\Delta G^{\ddagger}$ of 1.36 kcal/mol leads to a tenfold increase in the rate of the reaction, whereas a 2.72 kcal/mol reduction increases the rate 100-fold. Thus relatively small changes in $\Delta G^{\ddagger}$ can lead to large changes in the overall rate of the reaction.

Catalysts such as enzymes (Chapter 3) accelerate reaction rates by lowering the relative energy of the transition state and so the activation energy (see Figure 2-30). The relative energies of reactants and products will determine if a reaction is thermodynamically favorable (negative ΔG), whereas the activation energy will determine how rapidly products form (reaction kinetics). Thermodynamically favorable reactions will not occur if the activation energies are too high.

Life Depends on the Coupling of Unfavorable Chemical Reactions with Energetically Favorable Reactions

Many processes in cells are energetically unfavorable ($\Delta G > 0$) and will not proceed spontaneously. Examples include the synthesis of DNA from nucleotides and transport of a substance across the plasma membrane from a lower to a higher concentration. Cells can carry out an energy-requiring, or endergonic, reaction ($\Delta G_1 > 0$) by coupling it to an energy-releasing, or exergonic, reaction ($\Delta G_2 < 0$) if the sum of the two reactions has an overall net negative ΔG.

Suppose, for example, that the reaction $A \rightleftharpoons B + X$ has a ΔG of +5 kcal/mol and that the reaction $X \rightleftharpoons Y + Z$ has a ΔG of −10 kcal/mol:

$$\begin{array}{llll} (1) & A \rightleftharpoons B + X & \Delta G & = +5 \text{ kcal/mol} \\ (2) & X \rightleftharpoons Y + Z & \Delta G & = -10 \text{ kcal/mol} \\ \hline \text{Sum:} & A \rightleftharpoons B + Y + Z & \Delta G^{\circ\prime} & = -5 \text{ kcal/mol} \end{array}$$

In the absence of the second reaction, there would be much more A than B at equilibrium. However, because the conversion of X to Y + Z is such a favorable reaction, it will pull the first process toward the formation of B and the consumption of A. Energetically unfavorable reactions in cells often are coupled to the energy-releasing hydrolysis of ATP, as we discuss next.

Hydrolysis of ATP Releases Substantial Free Energy and Drives Many Cellular Processes

In almost all organisms, **adenosine triphosphate,** or **ATP,** is the most important molecule for capturing, transiently storing, and subsequently transferring energy to perform work (e.g., biosynthesis, mechanical motion). The useful energy in an ATP molecule is contained in **phosphoanhydride bonds,** which are covalent bonds formed from the condensation of two molecules of phosphate by the loss of water:

$$O^{-}\!-\!\overset{\overset{O}{\|}}{\underset{\underset{O^{-}}{|}}{P}}\!-\!OH + HO\!-\!\overset{\overset{O}{\|}}{\underset{\underset{O^{-}}{|}}{P}}\!-\!O^{-} \rightleftharpoons O^{-}\!-\!\overset{\overset{O}{\|}}{\underset{\underset{O^{-}}{|}}{P}}\!-\!O\!-\!\overset{\overset{O}{\|}}{\underset{\underset{O^{-}}{|}}{P}}\!-\!O^{-} + H_2O$$

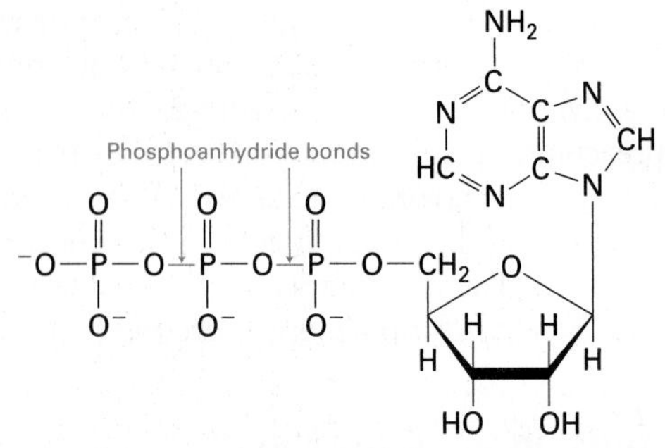

Phosphoanhydride bonds

Adenosine triphosphate (ATP)

▲ **FIGURE 2-31 Adenosine triphosphate (ATP).** The two phosphoanhydride bonds (red) in ATP, which link the three phosphate groups, each has a ΔG° of about -7.3 kcal/mol for hydrolysis. Hydrolysis of these bonds, especially the terminal one, is the source of energy that drives many energy-requiring reactions in biological systems.

An ATP molecule has two key phosphoanhydride (also called phosphodiester) bonds (Figure 2-31). Hydrolysis of a phosphoanhydride bond (~) in each of the following reactions has a highly negative $\Delta G^{\circ\prime}$ of about -7.3 kcal/mol:

$$\underset{\textbf{(ATP)}}{\text{Ap}{\sim}\text{p}{\sim}\text{p}} + H_2O \longrightarrow \underset{\textbf{(ADP)}}{\text{Ap}{\sim}\text{p}} + P_i + H^+$$

$$\underset{\textbf{(ATP)}}{\text{Ap}{\sim}\text{p}{\sim}\text{p}} + H_2O \longrightarrow \underset{\textbf{(AMP)}}{\text{Ap}} + PP_i + H^+$$

$$\underset{\textbf{(ADP)}}{\text{Ap}{\sim}\text{p}} + H_2O \longrightarrow \underset{\textbf{(AMP)}}{\text{Ap}} + P_i + H^+$$

In these reactions, P_i stands for inorganic phosphate (PO_4^{3-}) and PP_i for inorganic pyrophosphate, two phosphate groups linked by a phosphoanhydride bond. As the top two reactions show, the removal of a phosphate or a pyrophosphate group from ATP leaves adenosine diphosphate (ADP) or adenosine monophosphate (AMP), respectively.

A phosphoanhydride bond or other **high-energy bond** (commonly denoted by ~) is not intrinsically different from other covalent bonds. High-energy bonds simply release especially large amounts of energy when broken by addition of water (hydrolyzed). For instance, the $\Delta G^{\circ\prime}$ for hydrolysis of a phosphoanhydride bond in ATP (-7.3 kcal/mol) is more than three times the $\Delta G^{\circ\prime}$ for hydrolysis of the phosphoester bond (red) in glycerol 3-phosphate (-2.2 kcal/mol):

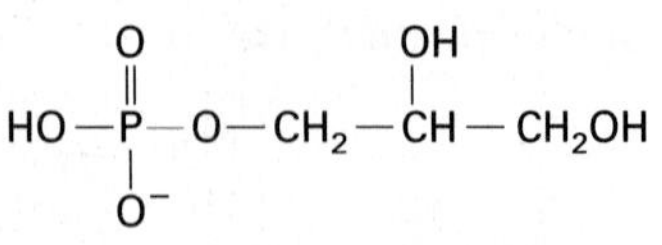

$$HO{-}\overset{\displaystyle O}{\overset{\|}{\underset{\displaystyle O^-}{\underset{|}{P}}}}{-}O{-}CH_2{-}\overset{\displaystyle OH}{\overset{|}{CH}}{-}CH_2OH$$

Glycerol 3-phosphate

A principal reason for this difference is that ATP and its hydrolysis products ADP and P_i are highly charged at neutral pH. During synthesis of ATP, a large input of energy is required to force the negative charges in ADP and P_i together. Conversely, much energy is released when ATP is hydrolyzed to ADP and P_i. In comparison, formation of the phosphoester bond between an uncharged hydroxyl in glycerol and P_i requires less energy, and less energy is released when this bond is hydrolyzed.

Cells have evolved protein-mediated mechanisms for transferring the free energy released by hydrolysis of phosphoanhydride bonds to other molecules, thereby driving reactions that would otherwise be energetically unfavorable. For example, if the ΔG for the reaction $B + C \rightarrow D$ is positive but less than the ΔG for hydrolysis of ATP, the reaction can be driven to the right by coupling it to hydrolysis of the terminal phosphoanhydride bond in ATP. In one common mechanism of such *energy coupling,* some of the energy stored in this phosphoanhydride bond is transferred to one of the reactants by breaking the bond in ATP and forming a covalent bond between the released phosphate group and one of the reactants. The phosphorylated intermediate generated in this way can then react with C to form $D + P_i$ in a reaction that has a negative ΔG:

$$B + ATP \rightarrow B{\sim}p + ADP$$

$$B{\sim}p + C \rightarrow D + P_i$$

The overall reaction

$$B + C + ATP \rightleftharpoons D + ADP + P_i$$

is energetically favorable ($\Delta G < 0$).

An alternative mechanism of energy coupling is to use the energy released by ATP hydrolysis to change the conformation of the molecule to an "energy-rich" stressed state. In turn, the energy stored as conformational stress can be released as the molecule "relaxes" back into its unstressed conformation. If this relaxation process can be mechanistically coupled to another reaction, the released energy can be harnessed to drive important cellular processes.

As with many biosynthetic reactions, transport of molecules into or out of the cell often has a positive ΔG and thus requires an input of energy to proceed. Such simple transport reactions do not *directly* involve the making or breaking of covalent bonds; thus the $\Delta G^{\circ\prime}$ is 0. In the case of a substance moving into a cell, Equation 2-7 becomes

$$\Delta G = RT \ln\frac{[C_{in}]}{[C_{out}]} \qquad (2\text{-}10)$$

where $[C_{in}]$ is the initial concentration of the substance inside the cell and $[C_{out}]$ is its concentration outside the cell. We can see from Equation 2-10 that ΔG is positive for transport of a substance into a cell against its concentration gradient (when $[C_{in}] > [C_{out}]$); the energy to drive such "uphill" transport often is supplied by the hydrolysis of

ATP. Conversely, when a substance moves down its concentration gradient ($[C_{out}] > [C_{in}]$), ΔG is negative. Such "downhill" transport releases energy that can be coupled to an energy-requiring reaction, say, the movement of another substance uphill across a membrane or the synthesis of ATP itself (see Chapters 11 and 12).

ATP Is Generated During Photosynthesis and Respiration

Clearly, to continue functioning, cells must constantly replenish their ATP supply. In nearly all cells, the initial energy source whose energy is ultimately transformed into the phosphoanhydride bonds of ATP and bonds in other compounds is sunlight. In **photosynthesis,** plants and certain microorganisms can trap the energy in light and use it to synthesize ATP from ADP and P_i. Much of the ATP produced in photosynthesis is hydrolyzed to provide energy for the conversion of carbon dioxide to six-carbon sugars, a process called **carbon fixation:**

$$6\,CO_2 + 6\,H_2O \xrightarrow{\text{ATP} \quad \text{ADP} + P_i} C_6H_{12}O_6 + 6\,O_2$$

In animals, the free energy in sugars and other molecules derived from food is released in the process of respiration. All synthesis of ATP in animal cells and in nonphotosynthetic microorganisms results from the chemical transformation of energy-rich compounds in the diet (e.g., glucose, starch). We discuss the mechanisms of photosynthesis and cellular respiration in Chapter 12.

The complete oxidation of glucose to yield carbon dioxide,

$$C_6H_{12}O_6 + 6\,O_2 \rightarrow 6\,CO_2 + 6\,H_2O$$

has a $\Delta G^{\circ\prime}$ of -686 kcal/mol and is the reverse of photosynthetic carbon fixation. Cells employ an elaborate set of protein-mediated reactions to couple the oxidation of 1 molecule of glucose to the synthesis of as many as 30 molecules of ATP from 30 molecules of ADP. This oxygen-dependent (**aerobic**) degradation (**catabolism**) of glucose is the major pathway for generating ATP in all animal cells, nonphotosynthetic plant cells, and many bacterial cells. Catabolism of fatty acids can also be an important source of ATP.

Light energy captured in photosynthesis is not the only source of chemical energy for all cells. Certain microorganisms that live in or around deep ocean vents, where adequate sunlight is unavailable, derive the energy for converting ADP and P_i into ATP from the oxidation of reduced inorganic compounds. These reduced compounds originate deep in the earth and are released at the vents.

NAD^+ and FAD Couple Many Biological Oxidation and Reduction Reactions

In many chemical reactions, electrons are transferred from one atom or molecule to another; this transfer may or may not accompany the formation of new chemical bonds or the release of energy that can be coupled to other reactions. The loss of electrons from an atom or a molecule is called **oxidation,** and the gain of electrons by an atom or a molecule is called **reduction.** Because electrons are neither created nor destroyed in a chemical reaction, if one atom or molecule is oxidized, another must be reduced. For example, oxygen draws electrons from Fe^{2+} (ferrous) ions to form Fe^{3+} (ferric) ions, a reaction that occurs as part of the process by which carbohydrates are degraded in mitochondria. Each oxygen atom receives two electrons, one from each of two Fe^{2+} ions:

$$2\,Fe^{2+} + \tfrac{1}{2}\,O_2 \rightarrow 2\,Fe^{3+} + O^{2-}$$

Thus Fe^{2+} is oxidized, and O_2 is reduced. Such reactions in which one molecule is reduced and another oxidized often are referred to as **redox reactions.** Oxygen is an electron acceptor in many redox reactions in cells under aerobic conditions.

Many biologically important oxidation and reduction reactions involve the removal or the addition of hydrogen atoms (protons plus electrons) rather than the transfer of isolated electrons on their own. The oxidation of succinate to fumarate, which also occurs in mitochondria, is an example (Figure 2-32). Protons are soluble in aqueous solutions (as H_3O^+), but electrons are not and must be transferred directly from one atom or molecule to another without a water-dissolved intermediate. In this type of oxidation reaction, electrons often are transferred to small electron-carrying molecules, sometimes referred to as coenzymes. The most common of these electron carriers are **NAD^+ (nicotinamide adenine dinucleotide),** which is reduced to NADH, and **FAD (flavin adenine dinucleotide),** which is reduced to $FADH_2$ (Figure 2-33). The reduced forms of these coenzymes can transfer protons and electrons to other molecules, thereby reducing them.

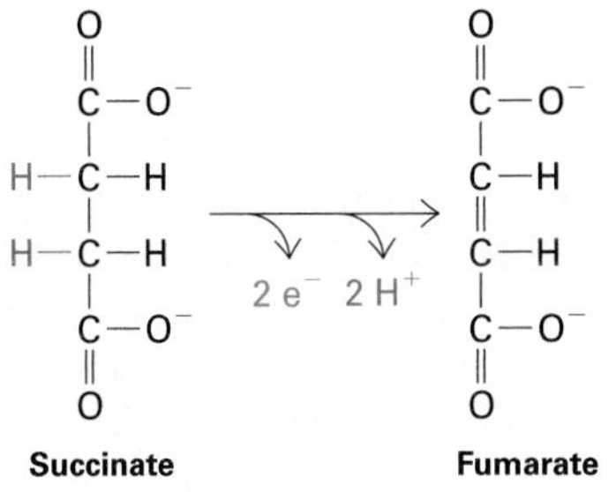

▲ **FIGURE 2-32 Conversion of succinate to fumarate.** In this oxidation reaction, which occurs in mitochondria as part of the citric acid cycle, succinate loses two electrons and two protons. These are transferred to FAD, reducing it to $FADH_2$.

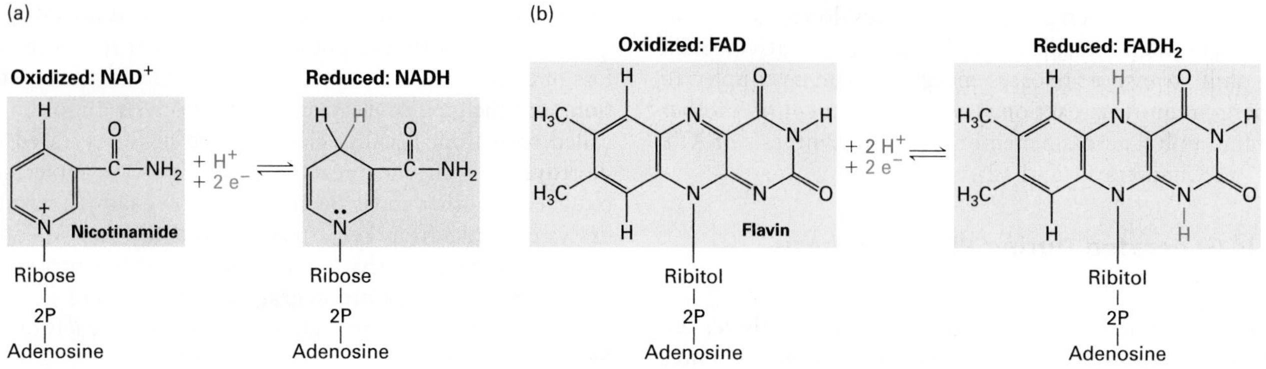

(a)

Oxidized: NAD^+ + H^+ + 2 e^- ⇌ Reduced: NADH

Nicotinamide — Ribose — 2P — Adenosine

$$NAD^+ + H^+ + 2\,e^- \rightleftharpoons NADH$$

(b)

Oxidized: FAD + 2 H^+ + 2 e^- ⇌ Reduced: $FADH_2$

Flavin — Ribitol — 2P — Adenosine

$$FAD + 2\,H^+ + 2\,e^- \rightleftharpoons FADH_2$$

▲ **FIGURE 2-33 The electron-carrying coenzymes NAD^+ and FAD.** (a) NAD^+ (nicotinamide adenine dinucleotide) is reduced to NADH by the addition of two electrons and one proton simultaneously. In many biological redox reactions, a pair of hydrogen atoms (two protons and two electrons) are removed from a molecule. In some cases, one of the protons and both electrons are transferred to NAD^+; the other proton is released into solution. (b) FAD (flavin adenine dinucleotide) is reduced to $FADH_2$ by the addition of two electrons and two protons, as occurs when succinate is converted to fumarate (see Figure 2-32). In this two-step reaction, addition of one electron together with one proton first generates a short-lived semiquinone intermediate (not shown), which then accepts a second electron and proton.

To describe redox reactions, such as the reaction of ferrous ion (Fe^{2+}) and oxygen (O_2), it is easiest to divide them into two half-reactions:

$$\textit{Oxidation of } Fe^{2+}\text{:}\quad 2\,Fe^{2+} \rightarrow 2\,Fe^{3+} + 2\,e^-$$
$$\textit{Reduction of } O_2\text{:}\quad 2\,e^- + \tfrac{1}{2}\,O_2 \rightarrow O^{2-}$$

In this case, the reduced oxygen (O^{2-}) readily reacts with two protons to form one water molecule (H_2O). The readiness with which an atom or a molecule *gains* an electron is its **reduction potential** E. The tendency to *lose* electrons, the **oxidation potential,** has the same magnitude but opposite sign as the reduction potential for the reverse reaction.

Reduction potentials are measured in volts (V) from an arbitrary zero point set at the reduction potential of the following half-reaction under standard conditions (25 °C, 1 atm, and reactants at 1 M):

$$H^+ + e^- \underset{\text{oxidation}}{\overset{\text{reduction}}{\rightleftharpoons}} \tfrac{1}{2}\,H_2$$

The value of E for a molecule or an atom under standard conditions is its standard reduction potential, E'_0. A molecule or an ion with a positive E'_0 has a higher affinity for electrons than the H^+ ion does under standard conditions. Conversely, a molecule or ion with a negative E'_0 has a lower affinity for electrons than the H^+ ion does under standard conditions. Like the values of $\Delta G^{\circ\prime}$, standard reduction potentials may differ somewhat from those found under the conditions in a cell because the concentrations of reactants in a cell are not 1 M.

In a redox reaction, electrons move spontaneously toward atoms or molecules having *more positive* reduction potentials. In other words, a compound having a more negative reduction potential can transfer electrons spontaneously to (i.e., reduce) a compound with a more positive reduction potential. In this type of reaction, the change in electric potential ΔE is the sum of the reduction and oxidation potentials for the two half-reactions. The ΔE for a redox reaction is related to the change in free energy ΔG by the following expression:

$$\Delta G\ (\text{cal/mol}) = -n\,(23{,}064)\,\Delta E\ (\text{volts}) \qquad (2\text{-}11)$$

where n is the number of electrons transferred. Note that a redox reaction with a positive ΔE value will have a negative ΔG and thus will tend to proceed spontaneously from left to right.

KEY CONCEPTS OF SECTION 2.4

Biochemical Energetics

- The change in free energy ΔG is the most useful measure for predicting the direction of chemical reactions in biological systems. Chemical reactions tend to proceed spontaneously in the direction for which ΔG is negative. The magnitude of ΔG is independent of the reaction rate.
- The chemical free-energy change $\Delta G^{\circ\prime}$ equals $-2.3\ RT \log K_{eq}$. Thus the value of $\Delta G^{\circ\prime}$ can be calculated from the experimentally determined concentrations of reactants and products at equilibrium.
- The rate of a reaction depends on the activation energy needed to energize reactants to a transition state. Catalysts such as enzymes speed up reactions by lowering the activation energy of the transition state.
- A chemical reaction having a positive ΔG can proceed if it is coupled with a reaction having a negative ΔG of larger magnitude.
- Many otherwise energetically unfavorable cellular processes are driven by the hydrolysis of phosphoanhydride bonds in ATP (see Figure 2-31).

- Directly or indirectly, light energy captured by photosynthesis in plants and photosynthetic bacteria is the ultimate source of chemical energy for almost all cells.
- An oxidation reaction (loss of electrons) is always coupled with a reduction reaction (gain of electrons).
- Biological oxidation and reduction reactions often are coupled by electron-carrying coenzymes such as NAD^+ and FAD (see Figure 2-33).
- Oxidation-reduction reactions with a positive ΔE have a negative ΔG and thus tend to proceed spontaneously.

Key Terms

acid *52*
α carbon atom (C_α) *41*
amino acids *41*
amphipathic *31*
base *52*
buffers *52*
chemical potential energy *54*
covalent bond *32*
dehydration reaction *40*
ΔG (free-energy change) *55*
disulfide bond *43*
endergonic *55*
endothermic *56*
energy coupling *58*
enthalpy (H) *55*
entropy (S) *55*
equilibrium constant *49*
exergonic *55*
exothermic *56*
fatty acids *47*
hydrogen bond *37*
hydrophilic *31*
hydrophobic *31*
hydrophobic effect *38*
ionic interactions *36*
molecular complementarity *39*
monosaccharides *44*
nucleosides *44*
nucleotides *44*
oxidation *59*
pH *51*
phosphoanhydride bonds *57*
phospholipid bilayers *41*
polar *34*
polymer *40*
redox reaction *59*
reduction *59*
saturated *47*
steady state *50*
stereoisomers *33*
unsaturated *47*
van der Waals interactions *37*

Review the Concepts

1. The gecko is a reptile with an amazing ability to climb smooth surfaces, including glass. Recent discoveries indicate that geckos stick to smooth surfaces via van der Waals interactions between septae on their feet and the smooth surface. How is this method of stickiness advantageous over covalent interactions? Given that van der Waals forces are among the weakest molecular interactions, how can the gecko's feet stick so effectively?

2. The K^+ channel is an example of a transmembrane protein (a protein that spans the phospholipid bilayer of the plasma membrane). What types of amino acids are likely to be found (a) lining the channel through which K^+ passes, (b) in contact with the hydrophobic core of the phospholipid bilayer containing fatty acyl groups, (c) in the cytosolic domain of the protein, and (d) in the extracellular domain of the protein?

3. V-M-Y-F-E-N: This is the single-letter amino acid abbreviation for a peptide. What is the net charge of this peptide at pH 7.0? An enzyme called a protein tyrosine kinase can attach phosphates to the hydroxyl groups of tyrosine. What is the net charge of the peptide at pH 7.0 after it has been phosphorylated by a tyrosine kinase? What is the likely source of phosphate utilized by the kinase for this reaction?

4. Disulfide bonds help to stabilize the three-dimensional structure of proteins. What amino acids are involved in the formation of disulfide bonds? Does the formation of a disulfide bond increase or decrease entropy (ΔS)?

5. In the 1960s, the drug thalidomide was prescribed to pregnant women to treat morning sickness. However, thalidomide caused severe limb defects in the children of some women who took the drug, and its use for morning sickness was discontinued. It is now known that thalidomide was administered as a mixture of two stereoisomeric compounds, one of which relieved morning sickness and the other of which was responsible for the birth defects. What are stereoisomers? Why might two such closely related compounds have such different physiologic effects?

6. Name the compound shown below.

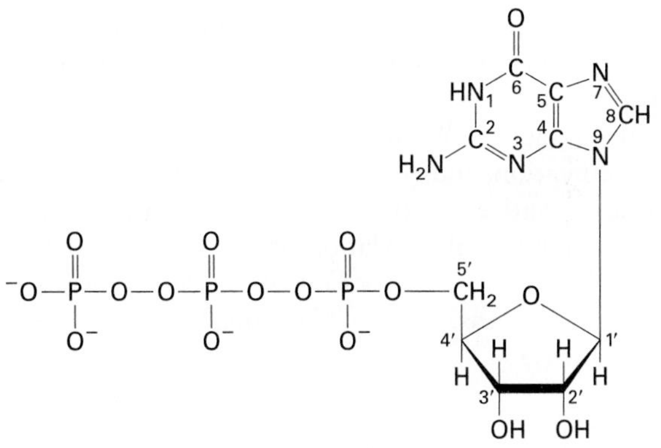

Is this nucleotide a component of DNA, RNA, or both? Name one other function of this compound.

7. The chemical basis of blood-group specificity resides in the carbohydrates displayed on the surface of red blood cells. Carbohydrates have the potential for great structural diversity. Indeed, the structural complexity of the oligosaccharides that can be formed from four sugars is greater than that for oligopeptides from four amino acids. What properties of carbohydrates make this great structural diversity possible?

8. Ammonia (NH_3) is a weak base that under acidic conditions becomes protonated to the ammonium ion in the following reaction:

$$NH_3 + H^+ \rightarrow NH_4^+$$

NH_3 freely permeates biological membranes, including those of lysosomes. The lysosome is a subcellular organelle with a pH of about 4.5–5.0; the pH of cytoplasm is ~7.0. What is the effect on the pH of the fluid content of lysosomes when cells are exposed to ammonia? *Note:* Protonated ammonia does not diffuse freely across membranes.

9. Consider the binding reaction L + R → LR, where L is a ligand and R is its receptor. When 1×10^{-3} M L is added to a solution containing 5×10^{-2} M R, 90% of the L binds to form LR. What is the K_{eq} of this reaction? How will the K_{eq} be affected by the addition of a protein that catalyzes this binding reaction? What is the K_d?

10. What is the ionization state of phosphoric acid in the cytoplasm? Why is phosphoric acid such a physiologically important compound?

11. The $\Delta G^{\circ\prime}$ for the reaction X + Y → XY is −1000 cal/mol. What is the ΔG at 25 °C (298 Kelvin) starting with 0.01 M each X, Y, and XY? Suggest two ways one could make this reaction energetically favorable.

12. According to health experts, saturated fatty acids, which come from animal fats, are a major factor contributing to coronary heart disease. What distinguishes a saturated fatty acid from an unsaturated fatty acid, and to what does the term *saturated* refer? Recently, trans unsaturated fatty acids, or trans fats, which raise total cholesterol levels in the body, have also been implicated in heart disease. How does the cis stereoisomer differ from the trans configuration, and what effect does the cis configuration have on the structure of the fatty acid chain?

13. Chemical modifications to amino acids contribute to the diversity and function of proteins. For instance, γ-carboxylation of specific amino acids is required to make some proteins biologically active. What particular amino acid undergoes this modification, and what is the biological relevance? Warfarin, a derivative of coumarin, which is present in many plants, inhibits γ-carboxylation of this amino acid and was used in the past as a rat poison. At present, it is also used clinically in humans. What patients might be prescribed warfarin and why?

References

Alberty, R. A., and R. J. Silbey. 2005. *Physical Chemistry,* 4th ed. Wiley.

Atkins, P., and J. de Paula. 2005. *The Elements of Physical Chemistry,* 4th ed. W. H. Freeman and Company.

Berg, J. M., J. L. Tymoczko, and L. Stryer. 2007. *Biochemistry,* 6th ed. W. H. Freeman and Company.

Cantor, P. R., and C. R. Schimmel. 1980. *Biophysical Chemistry.* W. H. Freeman and Company.

Davenport, H. W. 1974. *ABC of Acid-Base Chemistry,* 6th ed. University of Chicago Press.

Eisenberg, D., and D. Crothers. 1979. *Physical Chemistry with Applications to the Life Sciences.* Benjamin-Cummings.

Guyton, A. C., and J. E. Hall. 2000. *Textbook of Medical Physiology,* 10th ed. Saunders.

Hill, T. J. 1977. *Free Energy Transduction in Biology.* Academic Press.

Klotz, I. M. 1978. *Energy Changes in Biochemical Reactions.* Academic Press.

Murray, R. K., et al. 1999. *Harper's Biochemistry,* 25th ed. Lange.

Nicholls, D. G., and S. J. Ferguson. 1992. *Bioenergetics 2.* Academic Press.

Oxtoby, D., H. Gillis, and N. Nachtrieb. 2003. *Principles of Modern Chemistry,* 5th ed. Saunders.

Sharon, N. 1980. Carbohydrates. *Sci. Am.* **243**(5):90–116.

Tanford, C. 1980. *The Hydrophobic Effect: Formation of Micelles and Biological Membranes,* 2d ed. Wiley.

Tinoco, I., K. Sauer, and J. Wang. 2001. *Physical Chemistry—Principles and Applications in Biological Sciences,* 4th ed. Prentice Hall.

Van Holde, K., W. Johnson, and P. Ho. 1998. *Principles of Physical Biochemistry.* Prentice Hall.

Voet, D., and J. Voet. 2004. *Biochemistry,* 3d ed. Wiley.

Wood, W. B., et al. 1981. *Biochemistry: A Problems Approach,* 2d ed. Benjamin-Cummings.

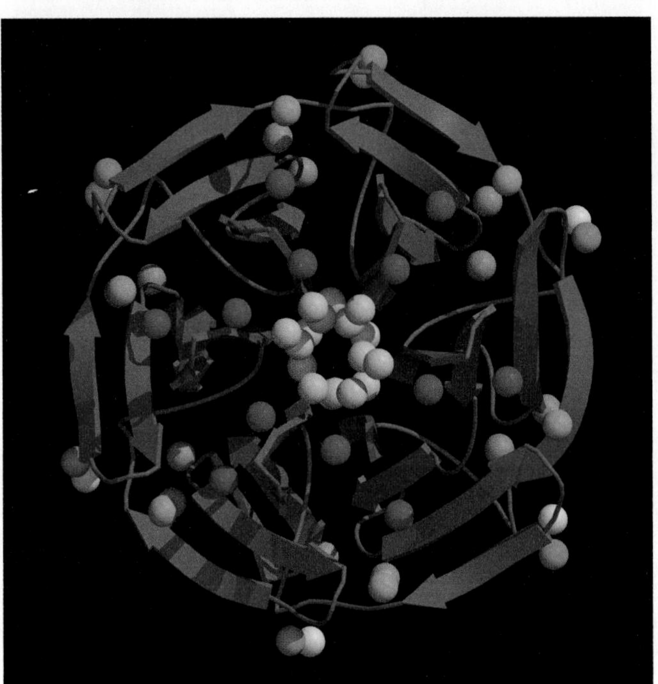

Ribbon diagram of a beta propeller domain from the human signaling protein Keapl. Ten water molecules (spheres) are bound to each of the six blades of the propeller. Many proteins are built from multiple, independently stable protein domains. [From L. J. Beamer, X. Li, C. A. Bottoms, and M. Hannink, 2005, *Acta Crystallogr. D: Biol. Crystallogr.* **61**(10):1335–1342.] Credit: Courtesy of Robert Huber, Martinsried.

CHAPTER

3

PROTEIN STRUCTURE AND FUNCTION

Proteins, which are polymers of amino acids, come in many sizes and shapes. Their three-dimensional diversity reflects underlying structural differences: principally variations in their lengths and amino acid sequences, and in some cases, differences also in the number of disulfide bonds or the attachment of small molecules or ions to their amino acid side chains. In general, the linear, unbranched polymer of amino acids composing any protein will fold into only one or a few closely related three-dimensional shapes—called **conformations.** The conformation of a protein together with the distinctive chemical properties of its amino acid side chains determines its function. As a consequence, proteins can perform a dazzling array of distinct functions inside and outside of cells that either are essential for life or provide selective evolutionary advantage to the cell or organism that contains them. It is, therefore, not surprising that characterizing the structures and activities of proteins is a fundamental prerequisite for understanding how cells work. Much of this textbook is devoted to examining how proteins act together to enable cells to live and function properly.

Many proteins can be grouped into just a few broad functional classes. *Structural proteins,* for example, determine the shapes of cells and their extracellular environments, and serve as guide wires or rails to direct the intracellular movement of molecules and organelles. They usually are formed by the assembly of multiple protein subunits into very large, long structures. *Scaffold proteins* bring other proteins together into ordered arrays to perform specific functions more efficiently than if those proteins were not assembled together. *Enzymes* are proteins that catalyze chemical reactions. *Membrane transport proteins* permit the flow of ions and molecules across cellular membranes. *Regulatory proteins* act as signals, sensors, and switches to control the activities of cells by altering the functions of other proteins and genes. These include *signaling proteins,* such as hormones and cell-surface receptors that transmit extracellular signals to the cell interior.

OUTLINE

Motor proteins are responsible for moving other proteins, organelles, cells—even whole organisms. Any one protein can be a member of more than one protein class, as is the case of some cell-surface signaling receptors that are both enzymes and regulator proteins because they transmit signals from outside to inside cells by catalyzing chemical reactions. To accomplish efficiently their diverse missions some proteins assemble into large complexes, often called *molecular machines*.

How do proteins mediate so many diverse functions? They do this by exploiting a few simple activities. Most fundamentally, proteins *bind*—to one another, to other macromolecules, such as DNA, and to small molecules and ions. In many cases such binding can induce a conformational change in the protein and thus influence its activity. Binding is based on molecular complementarity between a protein and its binding partner, as described in Chapter 2. A second key activity is enzymatic *catalysis*. Appropriate folding of a protein will place some amino acid side chains and carboxyl and amino groups of the backbone into positions that permit the catalysis of covalent bond rearrangements. A third activity involves *folding* into a channel or pore within a membrane through which molecules and ions flow. Although these are especially crucial protein activities, they are not the only ones. For example, fish that live in frigid waters—the Antarctic borchs and Arctic cods—have antifreeze proteins in their circulatory systems to prevent water crystallization at subzero temperatures.

A complete understanding of how proteins permit cells to live and thrive requires the identification and characterization of all the proteins used by a cell. In a sense, molecular cell biologists want to compile a complete protein 'parts list' and construct an all-inclusive "users manual" that describes how these proteins work. Compiling a comprehensive protein parts list has become feasible in recent years with the sequencing of entire **genomes**—complete sets of genes—of more and more organisms. From a computer analysis of genome sequences, researchers can deduce the number of amino acids and their sequence of most of the encoded proteins (Chapter 5). The term **proteome** was coined to refer to the entire protein complement of an organism. The human genome contains only 25,000 genes that encode proteins. However, variations in mRNA production (e.g., alternative splicing (Chapter 8)) and more than 100 types of protein modifications may generate hundreds of thousands of distinct human proteins. Remarkably, the human proteome is only about five times as large, comprising about 33,000 different proteins. By comparing protein sequences and structures of proteins of unknown function to those of known function, scientists can often deduce much about their functions. In the past, characterization of protein function by genetic, biochemical, or physiological methods often preceded the identification of particular proteins. In the modern genomic and proteomic era, a protein is usually identified prior to determining its function.

In this chapter, we begin our study of how the structure of a protein gives rise to its function, a theme that recurs throughout this book (Figure 3-1). The first section examines how chains of amino acid building blocks are arranged in a three-dimensional structural hierarchy. The next section discusses how proteins fold into these structures. We then turn to protein function, focusing on enzymes, the special class of proteins that catalyze chemical reactions. Various mechanisms that cells use to control the activities and life spans of proteins are covered in the next two sections. Next comes a section on commonly used techniques in the biologist's tool kit for isolating proteins and characterizing their properties. The chapter concludes with a discussion of the burgeoning field of proteomics.

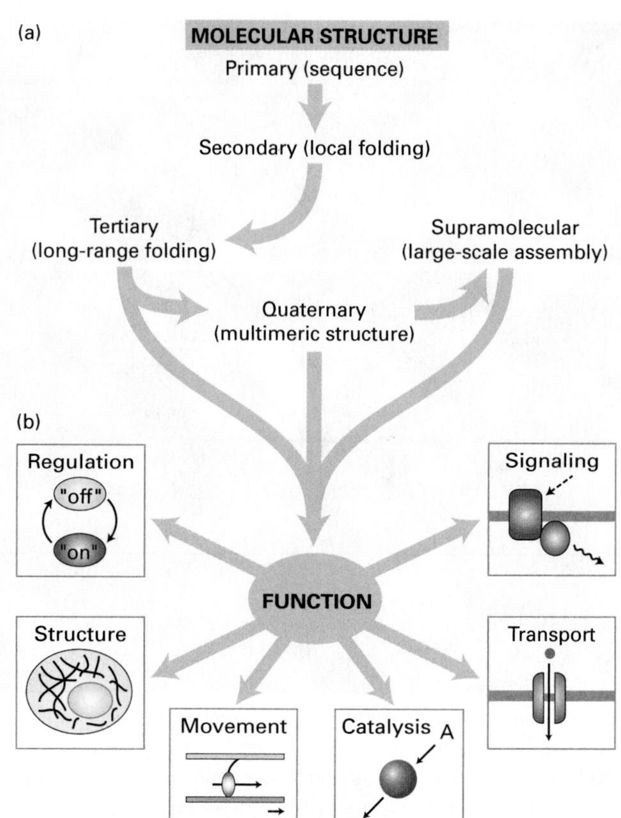

▲ **FIGURE 3-1 Overview of protein structure and function.** (a) Proteins are assembled according to a hierarchy of structures. A polypeptide's linear sequence of amino acids linked by peptide bonds (primary structure) folds into local helices or sheets (secondary structure) that pack into large (longer-range) complex three-dimensional structures (tertiary structure). Some individual polypeptides associate into multichain complexes (quaternary structure), which in some cases can be very large, consisting of tens to hundreds of subunits (supramolecular assemblies). (b) Protein function includes organization of the genome, other proteins, lipid bilayer membranes, and cytoplasm (structure); control of protein activity (regulation), monitoring of the environment and transmitting resultant information (signaling), flow of small molecules and ions across membranes (transport); catalysis of chemical reactions (via enzymes); and generation of force for movement (via motor proteins). These functions and others arise from specific binding interactions and conformational changes in the structure of a properly folded protein.

3.1 Hierarchical Structure of Proteins

A protein chain folds into a distinct three-dimensional shape that is stabilized by noncovalent interactions between regions in the linear sequence of amino acids. A key concept in

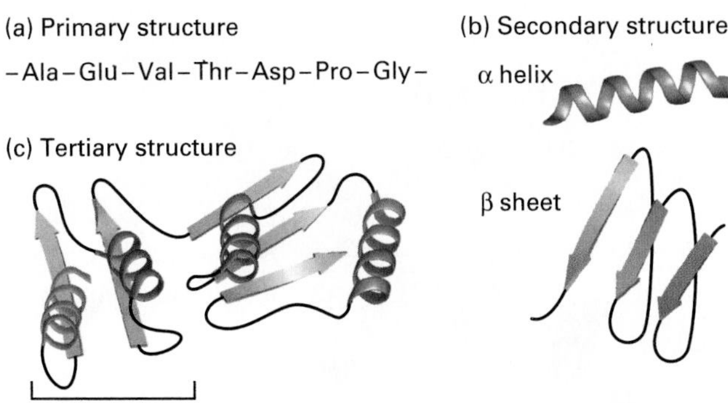

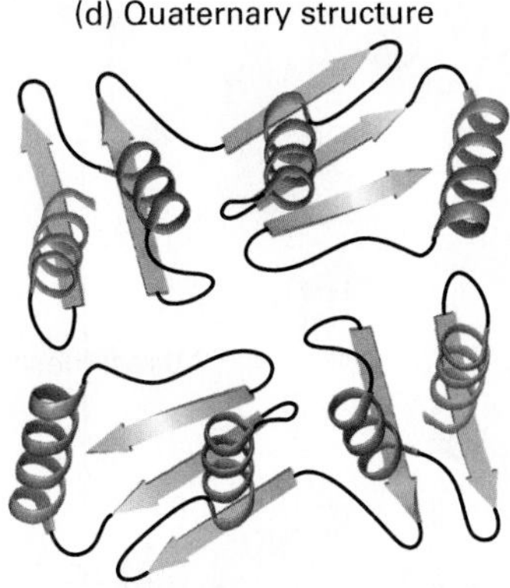

▲ **FIGURE 3-2 Four levels of protein hierarchy.** (a) The linear sequence of amino acids linked together by peptide bonds is the primary structure. (b) Folding of the polypeptide chain into local α helices or β sheets represents secondary structure. (c) Secondary structural elements together with various loops and turns in a single polypeptide chain pack into a larger independently stable structure, which may include distinct domains; this is tertiary structure. (d) Some individual polypeptides with their own tertiary structures can associate into a quaternary structure defining a multichain complex.

understanding how proteins work is that *function is derived from three-dimensional structure, and three-dimensional structure, which is determined primarily by noncovalent interactions between regions in the linear sequence of amino acids, is specified by amino acid sequence.* Indeed, principles relating biological structure and function initially were formulated by the biologists Johann von Goethe (1749–1832), Ernst Haeckel (1834–1919), and D'Arcy Thompson (1860–1948). They greatly influenced the school of "organic" architecture pioneered in the early twentieth century that is epitomized by the dicta "form follows function" (Louis Sullivan) and "form is function" (Frank Lloyd Wright). Here, we consider the architecture of proteins at four levels of organization: primary, secondary, tertiary, and quaternary (Figure 3-2).

The Primary Structure of a Protein Is Its Linear Arrangement of Amino Acids

As discussed in Chapter 2, proteins are constructed by the polymerization of 20 different types of amino acids. Individual amino acids are linked together in linear, unbranched chains by covalent amide bonds, called **peptide bonds,** with occasional disulfide bonds covalently linking side chains together. Peptide bond formation between the amino group of one amino acid and the carboxyl group of another results in the net release of a water molecule (dehydration) (Figure 3-3a). The repeated amide N, α carbon (C_α), carbonyl C and oxygen atoms of each amino acid residue form the backbone of a protein molecule from which the various side-chain groups project (Figure 3-3b, c). As a consequence of the peptide linkage, the backbone exhibits directionality because all the amino groups are located on the same side of the C_α atoms. Thus one end of a protein has a free (unlinked) amino group (the *N-terminus*), and the other end has a free carboxyl group (the *C-terminus*). The sequence of a protein chain is conventionally written with its

(a)

$$^{+}H_3N-C_\alpha H(R_1)-C(=O)-O^- \; + \; ^{+}H_3N-C_\alpha H(R_2)-C(=O)-O^- \longrightarrow \; ^{+}H_3N-C_\alpha H(R_1)-C(=O)-NH-C_\alpha H(R_2)-C(=O)-O^- + H_2O$$

Peptide bond

(b)

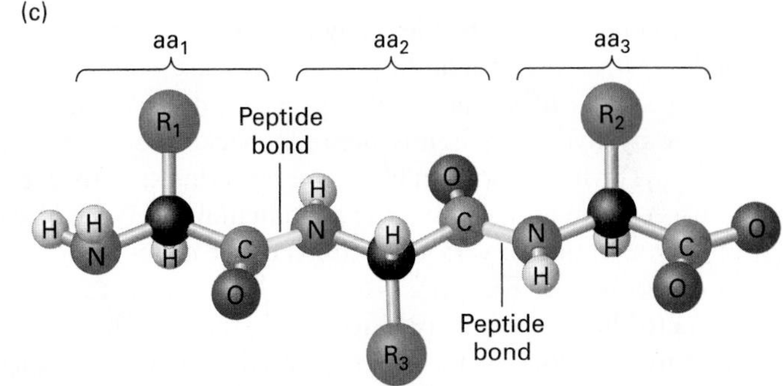

▲ **FIGURE 3-3 Structure of a polypeptide.** (a) Individual amino acids are linked together by peptide bonds, which form via reactions that result in a loss of water (dehydration). R_1, R_2, etc., represent the side chains ("R groups") of amino acids. (b) Linear polymers of peptide bond–linked amino acids are called *polypeptides,* which have a free amino end (N-terminus) and a free carboxyl end (C-terminus). (c) A ball-and-stick model shows peptide bonds (yellow) linking the amino nitrogen atom (blue) of one amino acid (aa) with the carbonyl carbon atom (gray) of an adjacent one in the chain. The R groups (green) extend from the α carbon atoms (black) of the amino acids. These side chains largely determine the distinct properties of individual proteins.

N-terminal amino acid on the left and its C-terminal amino acid on the right, and the amino acids are numbered sequentially starting from the amino terminus (number 1).

The **primary structure** of a protein is simply the linear arrangement, or sequence, of the amino acid residues that compose it. Many terms are used to denote the chains formed by the polymerization of amino acids. A short chain of amino acids linked by peptide bonds and having a defined sequence is called an **oligopeptide,** or just **peptide;** longer chains are referred to as **polypeptides.** Peptides generally contain fewer than 20–30 amino acid residues, whereas polypeptides are often 200–500 residues long. The longest protein described to date is the muscle protein titin with 26,926 residues. We generally reserve the term **protein** for a polypeptide (or complex of polypeptides) that has a well-defined three-dimensional structure. It is implied that proteins and peptides are the natural products of a cell.

The size of a protein or a polypeptide is reported as its mass in **daltons** (a dalton is 1 atomic mass unit) or as its molecular weight (MW), which is a dimensionless number. For example, a 10,000-MW protein has a mass of 10,000 daltons (Da), or 10 kilodaltons (kDa). In the penultimate section of this chapter, we will consider different methods for measuring the sizes and other physical characteristics of proteins. The known and predicted proteins encoded by the yeast genome have an average molecular weight of 52,728 and contain, on average, 466 amino acid residues. The average molecular weight of amino acids in proteins is 113, taking into account their average relative abundances. This value can be used to estimate the number of residues in a protein from its molecular weight or, conversely, its molecular weight from the number of residues.

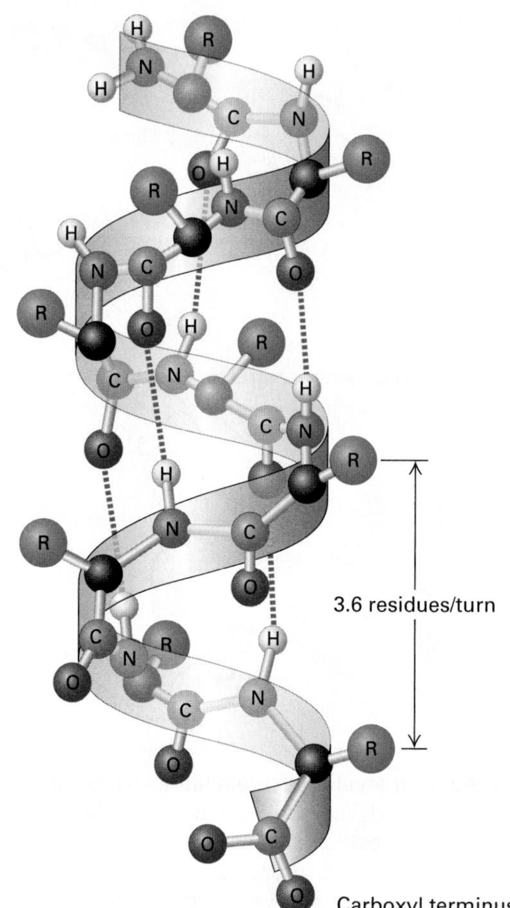

▲ **FIGURE 3-4 The α helix, a common secondary structure in proteins.** The polypeptide backbone (seen as a ribbon) is folded into a spiral that is held in place by hydrogen bonds between backbone oxygen and hydrogen atoms. Only hydrogens involved in bonding are shown. The outer surface of the helix is covered by the side-chain R groups (green).

Secondary Structures Are the Core Elements of Protein Architecture

The second level in the hierarchy of protein structure is **secondary structure.** Secondary structures are stable spatial arrangements of segments of a polypeptide chain held together by hydrogen bonds between backbone amide and carbonyl groups and often involving repeating structural patterns. A single polypeptide may contain multiple types of secondary structure in various portions of the chain, depending on its sequence. The principal secondary structures are the **alpha (α) helix,** the **beta (β) sheet,** and a short U-shaped **beta (β) turn.** Portions of the polypeptide that don't form these structures, but nevertheless have a well-defined, stable shape, are said to have an *irregular* structure. The term *random coil* applies to highly flexible portions of a polypeptide chain that have no fixed three-dimensional structure. In an average protein, 60 percent of the polypeptide chain exists as α helices and β sheets; the remainder of the molecule is in coils and turns. Thus, α helices and β sheets are the major internal supportive elements in most proteins. In this section, we explore the shapes of secondary structures and the forces that favor their formation. In later sections, we examine how linear arrays of secondary structure fold together into larger, more complex arrangements called tertiary structure.

The α Helix In a polypeptide segment folded into an α helix, the backbone forms a spiral structure in which the carbonyl oxygen atom of each peptide bond is hydrogen-bonded to the amide hydrogen atom of the amino acid four residues farther along the chain (in the direction of the C-terminus) (Figure 3-4). Within an α helix, all the backbone amino and carboxyl groups are hydrogen-bonded to one another, except at the very beginning and end of the helix. This periodic arrangement of bonds confers an amino-to-carboxy-terminal directionality on the helix because all the hydrogen bond acceptors (e.g., the carbonyl groups) have the same orientation (pointing in the downward direction in Figure 3-4) and results in a structure in which there is a complete turn of the spiral every 3.6 residues. An α helix 36 amino acids long has 10 turns of the helix and is 5.4 nm long (0.54 nm/turn).

The stable arrangement of hydrogen-bonded amino acids in the α helix holds the backbone in a straight, rodlike cylinder from which the side chains point outward. The relative hydrophobic or hydrophilic quality of a particular helix within a protein is determined entirely by the characteristics of the side chains, because all the polar amino and carboxyl groups of the peptide backbone are engaged in hydrogen bonding with one another in the helix. In water-soluble proteins, the hydrophilic helices tend to be found on the outside surfaces, where they can interact with the aqueous environment, whereas hydrophobic helices tend to be buried within the core of the folded protein. The amino acid proline is usually not found in α helices, because the covalent bonding of its amino group with a carbon in the side chain prevents its participation in stabilizing the backbone through normal hydrogen bonding. While the classic α helix is the most intrinsically stable, and most common helical form in proteins, there are variations, such as more tightly or loosely twisted helices. For example, in a specialized helix called a coiled coil (described several sections farther on), the helix is more tightly wound (3.5 residues and 0.51 nm per turn).

The β Sheet Another type of secondary structure, the β sheet, consists of laterally packed β strands. Each β strand is a short (5- to 8-residue), nearly fully extended polypeptide segment. Unlike in the α helix (where hydrogen bonding between the amino and carboxyl groups in the backbone occurs between nearly adjacent residues), hydrogen bonding in the β sheet occurs between backbone atoms in separate, but adjacent, β strands (Figure 3-5a). These distinct β strands may be either within a single polypeptide chain, with short or long loops between the β strand segments, or on different polypeptide chains. Figure 3-5b shows how two or more β strands align into adjacent rows, forming a nearly two-dimensional β pleated sheet (or simply *pleated sheet*), in which hydrogen bonds within the plane of the sheet hold the β strands together as the side chains stick out above and below the plane. Like α helices, β strands have a directionality defined by the orientation of the peptide bond. Therefore, in a pleated sheet, adjacent β strands can be oriented in the same (parallel) or opposite (antiparallel) directions with respect to each other. In some proteins, β sheets form the floor of a binding pocket or a hydrophobic core; in other proteins embedded in membranes the β sheets curve around and form a hydrophilic central pore through which ions and small molecules may flow (Chapter 11).

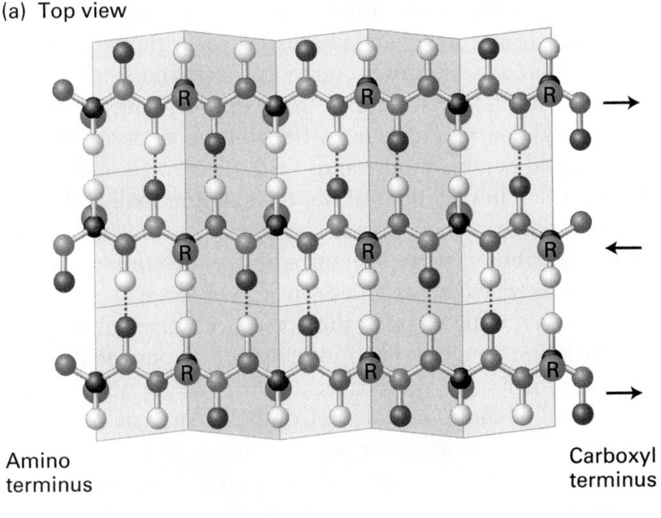

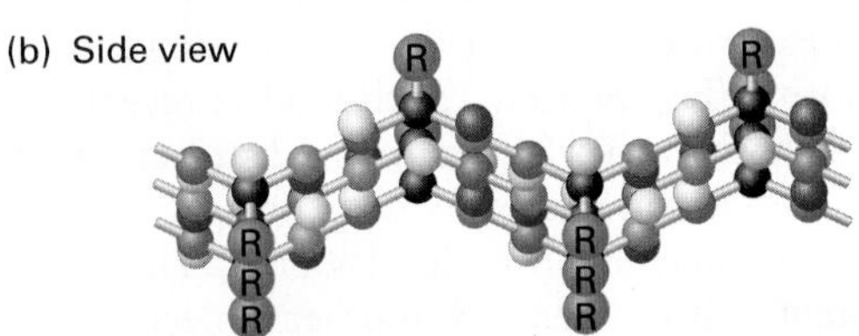

▲ **FIGURE 3-5 The β sheet, another common secondary structure in proteins.** (a) Top view of a simple three-stranded β sheet with antiparallel β strands. The stabilizing hydrogen bonds between the β strands are indicated by green dashed lines. (b) Side view of a β sheet. The projection of the R groups (green) above and below the plane of the sheet is obvious in this view. The fixed bond angles in the polypeptide backbone produce a pleated contour.

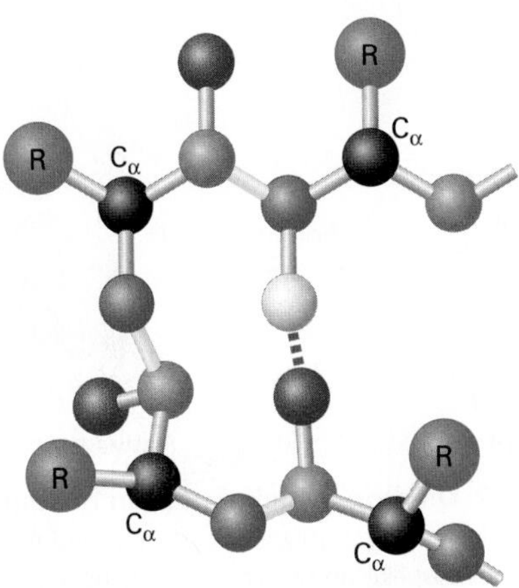

▲ **FIGURE 3-6 Structure of a β turn.** Composed of four residues, β turns reverse the direction of a polypeptide chain (≈180° U-turn). The C_α carbons of the first and fourth residues are usually <0.7 nm apart, and those residues are often linked by a hydrogen bond. β turns facilitate the folding of long polypeptides into compact structures.

β Turns Composed of four residues, β turns are located on the surface of a protein, forming sharp bends that reverse the direction of the polypeptide backbone, often toward the protein's interior. These short, U-shaped secondary structures are often stabilized by a hydrogen bond between their end residues (Figure 3-6). Glycine and proline are commonly present in turns. The lack of a large side chain in glycine and the presence of a built-in bend in proline allow the polypeptide backbone to fold into a tight U shape. β turns help large proteins to fold into highly compact structures. There are six types of well-defined turns, their detailed structures depending on the arrangement of H-bonding interactions. A polypeptide backbone also may contain longer bends, or loops. In contrast with tight β turns, which exhibit just a few well-defined conformations, longer loops can have many different conformations.

Overall Folding of a Polypeptide Chain Yields Its Tertiary Structure

Tertiary structure refers to the overall conformation of a polypeptide chain—that is, the three-dimensional arrangement of all its amino acid residues. In contrast with secondary

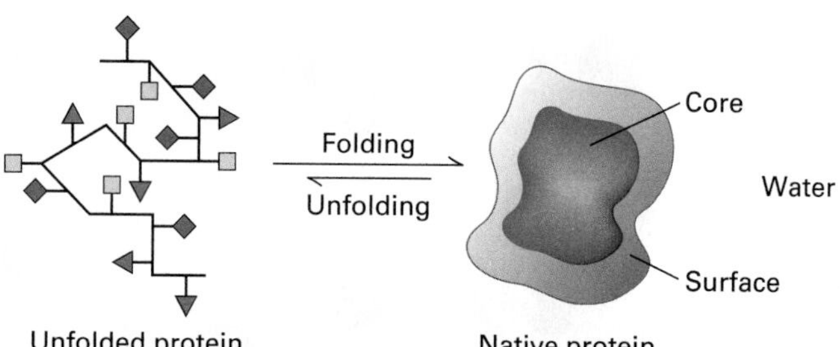

▲ **FIGURE 3-7 Oil drop model of protein folding.** The hydrophobic residues of a polypeptide chain tend to cluster together, somewhat like an oil drop, on the inside, or core, of a folded protein, driven away from the aqueous surroundings by the hydrophobic effect (Chapter 2). Charged and uncharged polar side chains appear on the protein's surface where they can form stabilizing interactions with surrounding water and ions.

structures, which are stabilized only by hydrogen bonds, tertiary structure is primarily stabilized by hydrophobic interactions between nonpolar side chains, together with hydrogen bonds between polar side chains and peptide bonds. These stabilizing forces compactly hold together elements of secondary structure—α helices, β strands, turns, and coils. Because the stabilizing interactions are weak, however, the tertiary structure of a protein is not rigidly fixed but undergoes continual, minute fluctuations, and some segments within the tertiary structure of a protein can be so very mobile they are considered to be disordered (that is, lacking well-defined, stable, three-dimensional structure). This variation in structure has important consequences for the function and regulation of proteins.

Chemical properties of amino acid side chains help define tertiary structure. Disulfide bonds between the side chains of cysteine residues in some proteins covalently link regions of proteins, thus restricting the mobility of proteins and increasing the stability of their tertiary structures. Amino acids with charged hydrophilic polar side chains tend to be on the outer surfaces of proteins; by interacting with water, they help to make proteins soluble in aqueous solutions and can form noncovalent interactions with other water-soluble molecules, including other proteins. In contrast, amino acids with hydrophobic nonpolar side chains are usually sequestered away from the water-facing surfaces of a protein, in many cases forming a water-insoluble central core (called the *oil drop model of globular proteins,* because of the relatively hydrophobic, or 'oily', core, Figure 3-7). Uncharged hydrophilic polar side chains are found on both the surface and inner core of proteins.

Proteins usually fall into one of three broad categories, based on their tertiary structure: fibrous proteins, globular proteins, and integral membrane proteins. *Fibrous proteins* are large, elongated, stiff molecules often composed of many tandem copies of a short sequence that forms a single repeating secondary structure (see the structure of collagen, the most abundant protein in mammals, in Chapter 19). Fibrous proteins, which often aggregate into large multiprotein fibers that do not readily dissolve in water, usually play a structural role or participate in cellular movements. *Globular proteins* are generally water-soluble, compactly folded structures, often but not exclusively spheroidal, that comprise a mixture of secondary structures (see the structure of myoglobin, below). *Integral membrane proteins* are embedded within the phospholipid bilayer of the membranes that serve as the walls of cells and organelles. The three broad categories of proteins noted here are not mutually exclusive—some proteins are made up of combinations of two or even all three of these categories.

Different Ways of Depicting the Conformation of Proteins Convey Different Types of Information

The simplest way to represent three-dimensional protein structure is to trace the course of the backbone atoms, sometimes only the C_α atoms, with a solid line (called a C_α trace, Figure 3-8a); the most complex model shows every atom (Figure 3-8b). The former shows the overall fold of the polypeptide chain without consideration of the amino acid side chains; the latter, a ball-and-stick model, details the interactions between side-chain atoms, including those that stabilize the protein's conformation and interact with other molecules, as well as the atoms of the backbone. Even though both views are useful, the elements of secondary structure are not always easily discerned in them. Another type of representation uses common shorthand symbols for depicting secondary structure—for example, coiled ribbons or solid cylinders for α helices, flat ribbons or arrows for β strands, and flexible thin strands for β turns, coils, and loops (Figure 3-8c). In variations of ribbon diagrams, ball-and-stick or space-filling models of side chains can be attached to the backbone ribbon, while ribbon and cylinder models make the secondary structures of a protein easy to see.

However, none of these three ways of representing protein structure conveys much information about the protein surface, which is of interest because it is where other molecules usually bind to a protein. Computer analysis can identify the surface atoms that are in contact with the watery environment. On this water-accessible surface, regions having a common chemical character (hydrophobicity or hydrophilicity) and electrical character (basic or acidic) can be indicated by coloring (Figure 3-8d). Such models reveal the topography of the protein surface and the distribution of charge, both important features of binding sites, as well as clefts in the surface where small molecules bind. This view represents a protein as it is "seen" by another molecule.

Structural Motifs Are Regular Combinations of Secondary and Tertiary Structures

Particular combinations of secondary and tertiary structures, called **structural motifs** or **folds,** appear often as segments within many different proteins. Structural motifs

(a) C_α backbone trace

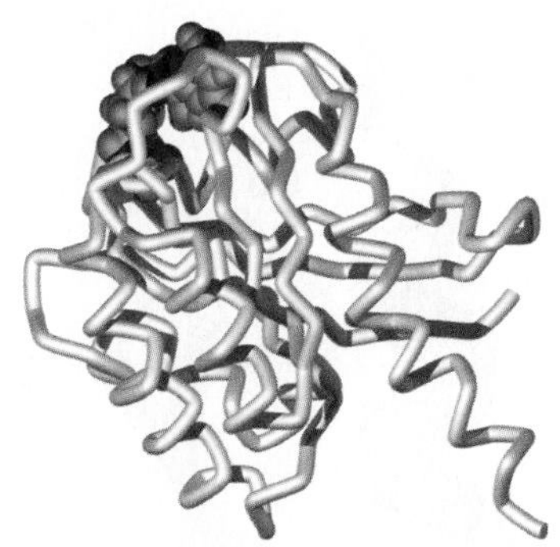

(b) Ball and stick

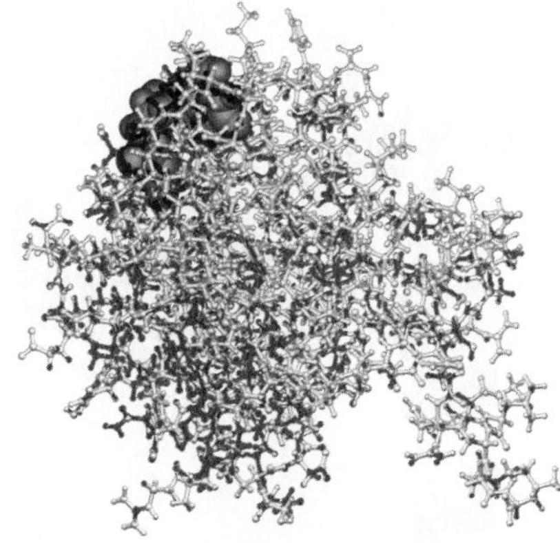

(c) Ribbons

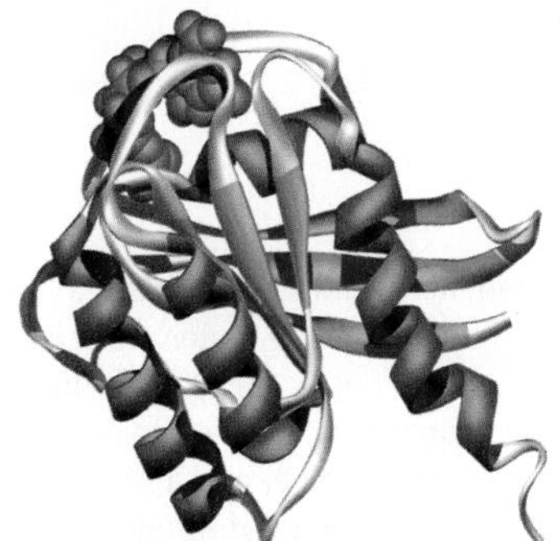

(d) Solvent-accessible surface

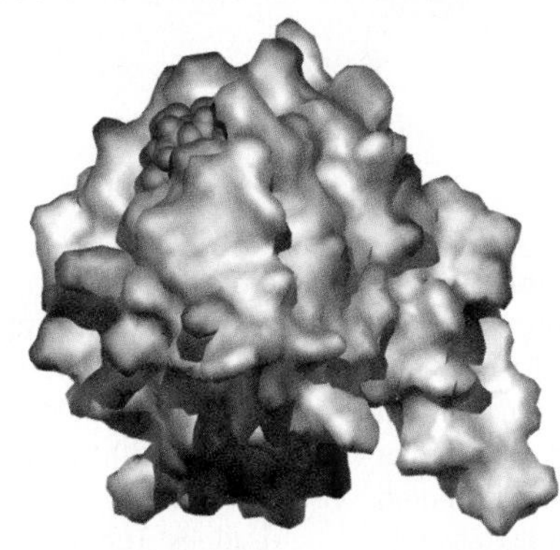

◀ **FIGURE 3-8 Four ways to visualize protein structure.** Ras, a monomeric guanine nucleotide–binding protein, is shown in all four panels, with guanosine diphosphate (GDP) always depicted in blue. (a) The C_α backbone trace demonstrates how the polypeptide is tightly packed into a small volume. (b) A ball-and-stick representation reveals the location of all atoms. (c) A ribbon representation emphasizes how β strands (light blue) and α helices (red) are organized in the protein. Note the turns and loops connecting pairs of helices and strands. (d) A model of the water-accessible surface reveals the numerous lumps, bumps, and crevices on the protein surface. Regions of positive charge are shaded purple; regions of negative charge are shaded red.

contribute to the global structure of the entire protein, and any particular structural motif often performs a common function in different proteins (e.g., binding to a particular small molecule or ion). The primary sequences responsible for any given structural motif may be very similar to one another. In other words, a common sequence motif can result in a common three-dimensional structural motif. However, it is possible for seemingly unrelated primary sequences to result in folding into a common structural motif. Conversely, it is possible that a commonly occurring sequence motif does not fold into a well-defined structural motif. Sometimes short sequence motifs that have an unusual abundance of a particular amino acid, e.g., proline or aspartate or glutamate, are called "domains"; however, these and other short contiguous segments are more appropriately called motifs than domains (which are defined below).

Many proteins, including fibrous proteins and DNA-regulating proteins called transcription factors (Chapter 7), assemble into dimers or trimers by using an α helix–based **coiled coil,** or heptad-repeat, structural motif. In this structural motif, α helices from two, three, or even four separate polypeptide chains coil about one another—resulting in a coil of coils, hence the name (Figure 3-9a). The individual helices bind tightly to one another because each helix has a strip of aliphatic (hydrophobic, but not aromatic) side chains (leucine valine, etc.) running along one side of the helix that interacts with a similar strip in the adjacent helix, thus sequestering the hydrophobic groups away from water and stabilizing the assembly of multiple independent helices. These hydrophobic strips are generated along only one side of the helix because the primary sequences of the helices exhibit a motif of repeating segments of seven amino acids (heptads) in which the sides chains of the first and fourth residues are aliphatic and the other side chains are often hydrophilic (Figure 3-9a). Because hydrophilic side chains extend from one side of the helix and hydrophobic side chains extend from the opposite side, the overall helical structure is **amphipathic.** Because leucine frequently appears in the fourth positions and the hydrophobic side chains merge together like the teeth of a zipper, these structural motifs are also called **leucine zippers.**

Many other structural motifs employ α helices. A common calcium-binding motif called the **EF hand** uses two short helices connected by a loop (Figure 3-9b). This structural motif found in more than 100 proteins is used for sensing the calcium levels in cells. The binding of a Ca^{2+} ion to oxygen atoms in conserved residues in the loop depends on the concentration of Ca^{2+} and often induces a conformational change in the protein, altering its activity. Thus, calcium concentrations can directly control proteins' structures and functions. Somewhat different helix-turn-helix and **basic helix-loop-helix** (bHLH) structural motifs are used for protein binding to DNA and consequently the regulation of gene activity. Yet another motif commonly found in proteins that bind RNA or DNA is the **zinc finger,** which contains three secondary structures—an α helix and two β strands

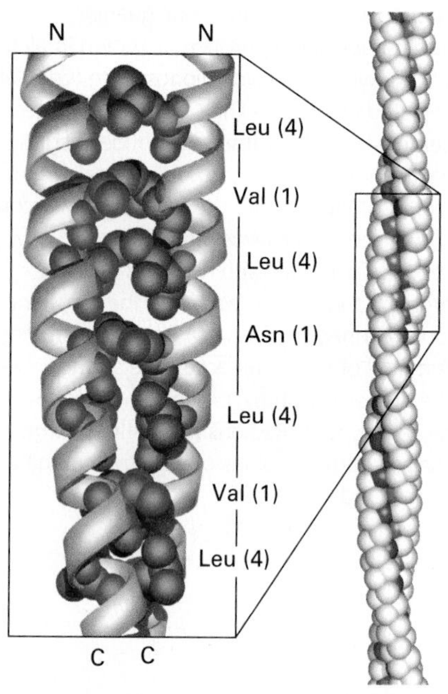

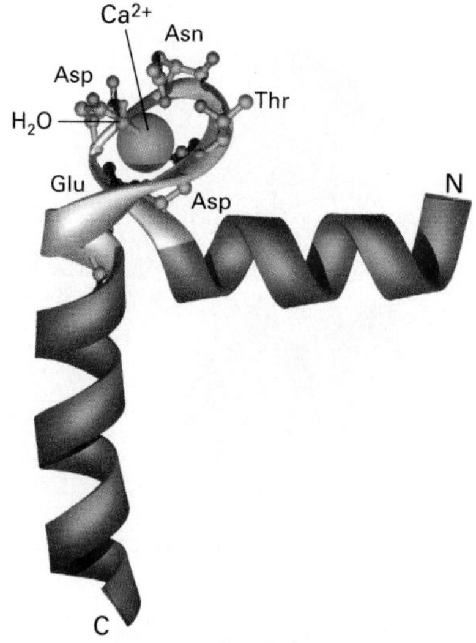

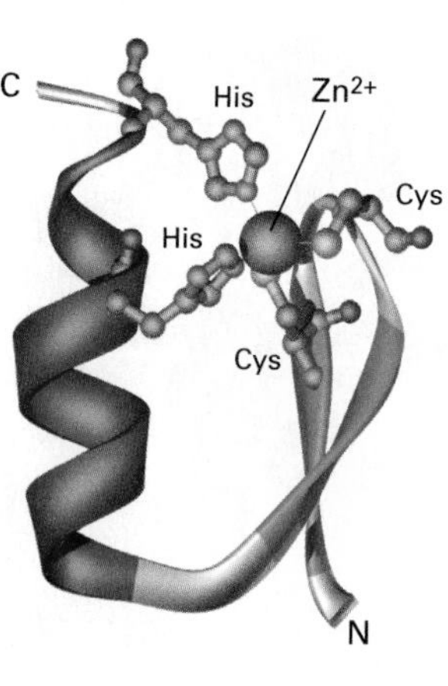

▲ **FIGURE 3-9 Motifs of protein secondary structure.** (a) The parallel two-stranded coiled-coil motif (*Left*) is characterized by two α helices wound around each other. Helix packing is stabilized by interactions between hydrophobic side chains (red and blue) present at regular intervals along each strand, and found along the seam of the intertwined helices. Each α helix exhibits a characteristic heptad repeat sequence with a hydrophobic residue often, but not always, at positions 1 and 4, as indicated. The coiled-coil nature of this structural motif is more apparent in long coiled coils (*Right* drawn at different scale). (b) An EF hand a type of helix-loop-helix motif, consists of two helices connected by a short loop in a specific conformation common to many proteins, including many calcium-binding and DNA-binding regulatory proteins. In calcium-binding proteins such as calmodulin, oxygen atoms from five residues in the acidic glutamate- and aspartate-rich loop and one water molecule form ionic bonds with a Ca^{2+} ion. (c) The zinc-finger motif is present in many DNA-binding proteins that help regulate transcription. A Zn^{2+} ion is held between a pair of β strands (blue) and a single α helix (red) by a pair of cysteine residues and a pair of histidine residues. The two invariant cysteine residues are usually at positions 3 and 6, and the two invariant histidine residues are at positions 20 and 24 in this 25-residue motif. [See A. Lewit-Bentley and S. Rety, 2000, EF-hand calcium-binding proteins, *Curr. Opin. Struc. Biol.* **10**:637–643; S. A. Wolfe, L. Nekludova, and C. O. Pabo, 2000, DNA recognition by Cys2His2 zinc finger proteins, *Ann. Rev. Biophys. Biomol. Struc.* **29**:183–212.]

with an antiparallel orientation—that form a fingerlike bundle held together by a zinc ion (Figure 3-9c).

We will encounter numerous additional motifs in later discussions of other proteins in this and other chapters. The presence of the same structural motif in different proteins with similar functions clearly indicates that these useful combinations of secondary structures have been conserved in evolution.

Structural and Functional Domains Are Modules of Tertiary Structure

Distinct regions of protein tertiary structure are often referred to as **domains.** There are three main classes of protein domains: functional, structural, and topological. A *functional domain* is a region of a protein that exhibits a particular activity characteristic of the protein, usually even when isolated from the rest of the protein. For instance, a particular region of a protein may be responsible for its catalytic activity (e.g., a kinase domain that covalently adds a phosphate group to another molecule) or binding ability (e.g., a DNA-binding domain or a membrane-binding domain). Functional domains are often identified experimentally by whittling down a protein to its smallest active fragment with the aid of **proteases,** enzymes that cleave one or more peptide bonds in a target polypeptide. Alternatively, the DNA encoding a protein can be modified so that when the modified DNA is used to generate a protein, only a particular region, or domain, of the full-length protein is made. Thus it is possible to determine if specific portions of a protein are responsible for particular activities exhibited by the protein. Indeed, functional domains are often also associated with corresponding structural domains.

A *structural domain* is a region ≈40 or more amino acids in length, arranged in a stable, distinct secondary or tertiary structure, that often can fold into its characteristic structure independently of the rest of the protein. As a consequence, distinct structural domains can be linked together—sometimes by short or long spacers—to form a large, multidomain protein. Each of the subunits in hemagglutinin, for example, contains

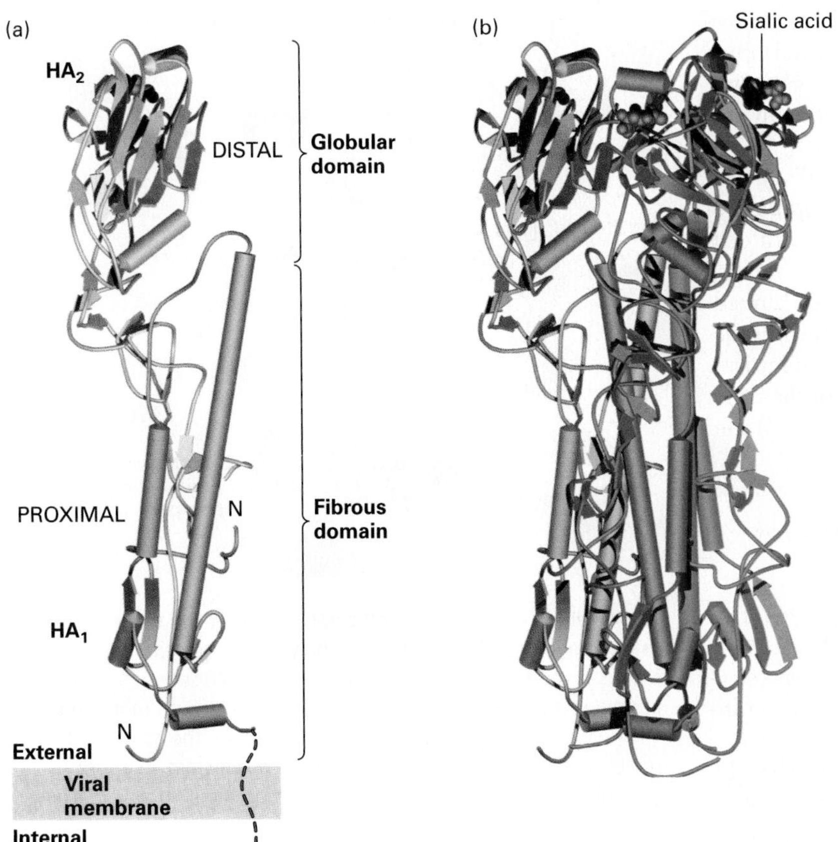

◀ **FIGURE 3-10 Tertiary and quaternary levels of structure.** The protein pictured here, hemagglutinin (HA), is found on the surface of influenza virus. This long, multimeric molecule has three identical subunits, each composed of two polypeptide chains, HA_1 and HA_2. (a) Tertiary structure of each HA subunit comprises the folding of its helices and strands into a compact structure that is 13.5 nm long and divided into two domains. The membrane-distal domain (silver) is folded into a globular conformation. The membrane-proximal domain (gold) has a fibrous, stemlike conformation owing to the alignment of two long α helices (cylinders) of HA_2 with β strands in HA_1. Short turns and longer loops, often at the surface of the molecule, connect the helices and strands in each chain. (b) Quaternary structure of HA is stabilized by lateral interactions between the long helices (cylinders) in the fibrous domains of the three subunits (gold, blue, and green), forming a triple-stranded coiled-coil stalk. Each of the distal globular domains in HA binds sialic acid (red) on the surface of target cells. Like many membrane proteins, HA contains several covalently linked carbohydrate chains (not shown).

a globular domain and a fibrous domain (Figure 3-10a). Like structural motifs (composed of secondary structures), structural domains (composed of secondary and tertiary structures) are incorporated as modules into different proteins. The modular approach to protein architecture is particularly easy to recognize in large proteins, which tend to be mosaics of different domains that confer distinct activities and thus can perform different functions simultaneously. Structural domains frequently are also functional domains in that they can have an activity independent of the rest of the protein. In Chapter 6 we consider the mechanism by which the gene segments that correspond to domains became shuffled in the course of evolution, resulting in their appearance in many proteins.

The epidermal growth factor (EGF) domain is a structural domain present in several proteins (Figure 3-11). EGF is a small, soluble peptide hormone that binds to cells in the embryo and in skin and connective tissue in adults, causing them to divide. It is generated by proteolytic (breaking of peptide bond) cleavage between repeated EGF domains in the EGF precursor protein, which is anchored in the cell membrane by a membrane-spanning domain. EGF domains with sequences similar to, but not identical with, those in the EGF peptide hormone are present in other proteins and can be liberated by proteolysis. These proteins include tissue plasminogen activator (TPA), a protease that is used to dissolve blood clots in heart attack victims; Neu protein, which takes part in embryonic differentiation; and Notch protein, a receptor protein in the plasma membrane that functions in developmentally important signaling (Chapter 16). Besides the EGF domain, these proteins have other domains in common with other proteins. For example, TPA possesses a trypsin domain, a functional domain in some proteases. It is estimated that there are about 1000 different types of structural domains in all proteins. Some of these are not very common, whereas others are found in many different proteins. Indeed, by some estimates only nine major types of domains account for as much as a third of all

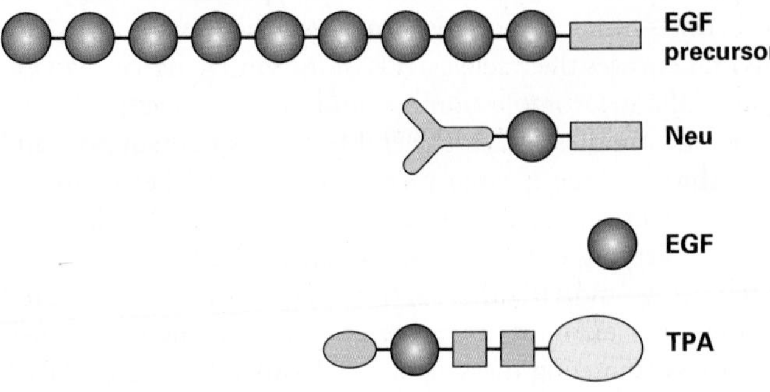

▲ **FIGURE 3-11 Modular nature of protein domains.** Epidermal growth factor (EGF) is generated by proteolytic cleavage of a precursor protein containing multiple EGF domains (green) and a membrane-spanning domain (blue). The EGF domain is also present in the Neu protein and in tissue plasminogen activator (TPA). These proteins also contain other widely distributed domains, indicated by shape and color. [Adapted from I. D. Campbell and P. Bork, 1993, *Curr. Opin. Struc. Biol.* **3**:385.]

the domains in all proteins. Structural domains can be recognized in proteins whose structures have been determined by x-ray crystallography or nuclear magnetic resonance (NMR) analysis or in images captured by electron microscopy.

Regions of proteins that are defined by their distinctive spatial relationships to the rest of the protein are *topological domains.* For example, some proteins associated with cell-surface membranes can have a portion extending inward into the cytoplasm (cytoplasmic domain), a portion embedded within the phospholipid bilayer membrane (membrane-spanning domain), and a portion extending outward into the extracellular space (extracellular domain). Each of these can comprise one or more structural motifs and structural and functional domains.

Proteins Associate into Multimeric Structures and Macromolecular Assemblies

Multimeric proteins consist of two or more polypeptide chains or subunits. A fourth level of structural organization, **quaternary structure,** describes the number (stoichiometry) and relative positions of the subunits in multimeric proteins. Hemagglutinin, for example, is a trimer of three identical subunits (homotrimer) held together by noncovalent bonds (Figure 3-10b). Other multimeric proteins can be composed of various numbers of identical (homomeric) or different (heteromeric) subunits (see the discussion of hemoglobin, below). Often, the individual monomeric subunits of a multimeric protein cannot function normally unless they are assembled into the multimeric protein. In some cases, assembly into a multimeric protein (oligomerization) permits proteins that act sequentially in a pathway to increase their efficiency of operation owing to their juxtaposition in space.

The highest level in the hierarchy of protein structure is the association of proteins into macromolecular assemblies. Typically, such structures are very large, in some cases exceeding 1 MDa in mass, approaching 30–300 nm in size, and containing tens to hundreds of polypeptide chains, and sometimes other biopolymers such as nucleic acids. The capsid that encases the nucleic acids of the viral genome is an example of a macromolecular assembly with a structural function. The bundles of cytoskeletal filaments that support and give shape to the plasma membrane are another example. Other macromolecular assemblies act as molecular machines, carrying out the most complex cellular processes by integrating individual functions into one coordinated process. For example, a transcriptional machine is responsible for synthesizing messenger RNA (mRNA) using a DNA template. This transcriptional machine, the operational details of which are discussed in Chapter 4, consists of RNA polymerase, itself a multimeric protein, and at least 50 additional components including general transcription factors, promoter-binding proteins, helicase, and other protein complexes (Figure 3-12). Ribosomes, also discussed in Chapter 4, are complex multiprotein and multi-nucleic acid machines that synthesize proteins.

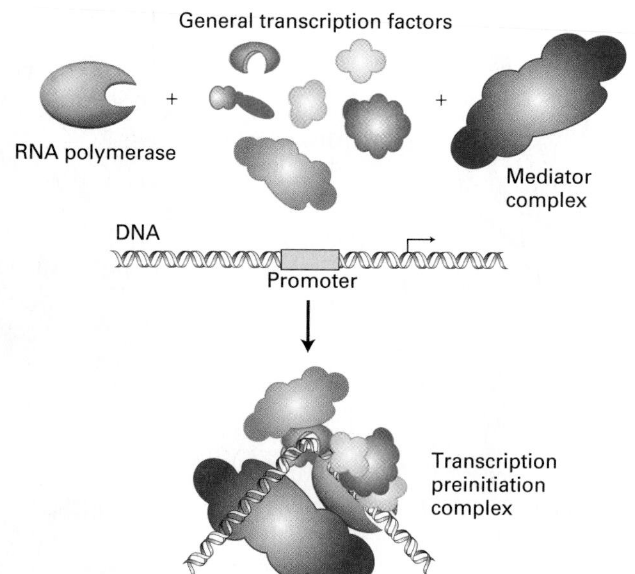

▲ **FIGURE 3-12 A macromolecular machine: the transcription-initiation complex.** The core RNA polymerase, general transcription factors, a mediator complex containing about 20 subunits, and other protein complexes not depicted here assemble at a promoter in DNA. The polymerase carries out transcription of DNA; the associated proteins are required for initial binding of polymerase to a specific promoter. The multiple components function together as a machine.

Members of Protein Families Have a Common Evolutionary Ancestor

Studies of myoglobin and hemoglobin, the oxygen-carrying proteins in muscle and red blood cells, respectively, provided early evidence that a protein's function derives from its three-dimensional structure, which in turn is specified by amino acid sequence. X-ray crystallographic analysis showed that the three-dimensional structures of myoglobin (a monomer) and the α and β subunits of hemoglobin (a $\alpha_2\beta_2$ tetramer) are remarkably similar. Sequencing of myoglobin and the hemoglobin subunits revealed that many identical or chemically similar residues are found in identical positions throughout the primary structures of both proteins. A mutation in the gene encoding the β chain that results in the substitution of a valine for a glutamic acid disturbs the folding and function of hemoglobin and causes sickle-cell anemia.

Similar comparisons between other proteins conclusively confirmed the relation between the amino acid sequence, three-dimensional structure, and function of proteins. Use of sequence comparisons to deduce protein function has expanded substantially in recent years as the genomes of more and more organisms have been sequenced.

The molecular revolution in biology during the last decades of the twentieth century also created a new scheme of biological classification based on similarities and differences in the amino acid sequences of proteins. Proteins that have a common ancestor are referred to as **homologs.** The main evidence for **homology** among proteins, and hence for their common ancestry, is similarity in their sequences or

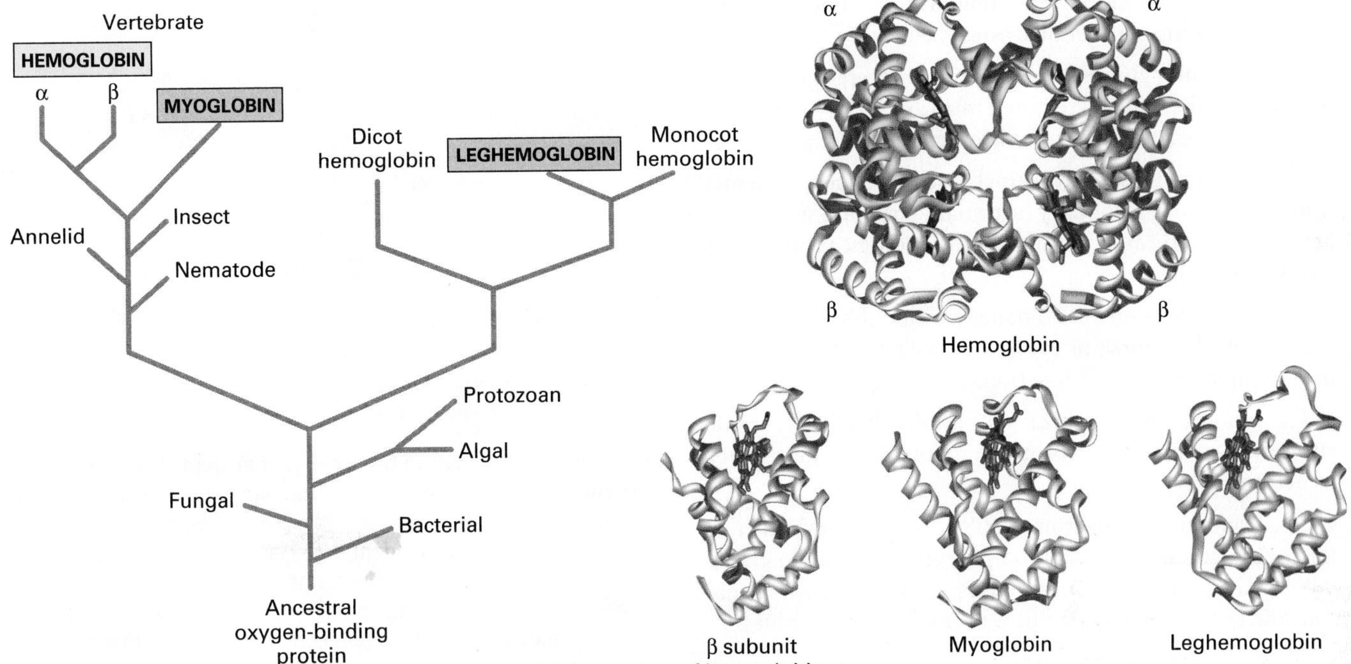

▲ **FIGURE 3-13 Evolution of the globin protein family.** *Left:* A primitive monomeric oxygen-binding globin is thought to be the ancestor of modern-day blood hemoglobins, muscle myoglobins, and plant leghemoglobins. Sequence comparisons have revealed that evolution of the globin proteins parallels the evolution of animals and plants. Major junctions occurred with the divergence of plant globins from animal globins and of myoglobin from hemoglobin. Later gene duplication gave rise to the α and β subunits of hemoglobin. *Right:* Hemoglobin is a tetramer of two α and two β subunits. The structural similarity of these subunits with leghemoglobin and myoglobin, both of which are monomers, is evident. A heme molecule (red) noncovalently associated with each globin polypeptide is directly responsible for oxygen-binding in these proteins. [(*Left*) Adapted from R. C. Hardison, 1996, *Proc. Nat'l Acad. Sci. USA* **93**:5675.]

structures. We can therefore describe homologous proteins as belonging to a "family" and can trace their lineage from comparisons of their sequences. The folded three-dimensional structures of homologous proteins are similar even if parts of their primary structure show little evidence of homology. Initially, proteins with relatively high sequence similarities (>50 percent exact matches, or "identities") and related functions or structures were defined as an evolutionarily related *family,* while a *superfamily* encompassed two or more families in which the interfamily sequences matched less well (≈30–40 percent identities) than within one family. It is generally thought that proteins with 30 percent sequence identity are likely to have similar three-dimensional structures; however, proteins with far less sequence matching can have very similar structures. Recently, revised definitions of *family* and *superfamily* have been proposed, in which a family comprises proteins with a clear evolutionary relationship (>30 percent identity or additional structural and functional information showing common descent but <30 percent identity), while a superfamily comprises proteins with only a probable common evolutionary origin (e.g., lower percent sequence identities). Often investigators consider proteins to constitute a common superfamily (have a common evolutionary origin) when they contain one or more common motifs or domains.

The kinship among homologous proteins is most easily visualized by a tree diagram based on sequence analyses. For example, the amino acid sequences of globins, the proteins of hemoglobin and myoglobin and their relatives from bacteria, plants, and animals, suggest that they evolved from an ancestral monomeric, oxygen-binding protein (Figure 3-13). With the passage of time, the gene for this ancestral protein slowly changed, initially diverging into lineages leading to animal and plant globins. Subsequent changes gave rise to myoglobin, the monomeric oxygen-storing protein in muscle, and to the α and β subunits of the tetrameric hemoglobin molecule ($\alpha_2\beta_2$) of the circulatory system.

KEY CONCEPTS OF SECTION 3.1

Hierarchical Structure of Proteins

- A protein is a linear polymer of amino acids linked together by peptide bonds. Various, mostly noncovalent, interactions between amino acids in the linear sequence stabilize a protein's specific folded three-dimensional structure, or conformation.
- The α helix, β strand and sheet, and β turn are the most prevalent elements of protein secondary structure. Secondary structures are stabilized by hydrogen bonds between atoms of the peptide backbone.
- Protein tertiary structure results from hydrophobic interactions between nonpolar side groups and hydrogen bonds between polar side groups and the polypeptide backbone.

These interactions stabilize folding of the secondary structure into a compact overall arrangement.

- Certain combinations of secondary structures give rise to different motifs, which are found in a variety of proteins and are often associated with specific functions (see Figure 3-9).
- Proteins often contain distinct domains, independently folded regions of secondary or tertiary structure with characteristic structural, functional, and topological properties (see Figure 3-10).
- The incorporation of domains as modules in different proteins in the course of evolution has generated diversity in protein structure and function.
- The number and organization of individual polypeptide subunits in multimeric proteins define their quaternary structure.
- Cells contain large macromolecular assemblies in which all the necessary participants in complex cellular processes (e.g., DNA, RNA, and protein synthesis; photosynthesis; signal transduction) are integrated to form molecular machines.
- Homologous proteins, which have similar sequences, structures, and functions, evolved from a common ancestor. They can be classified into families and superfamilies.

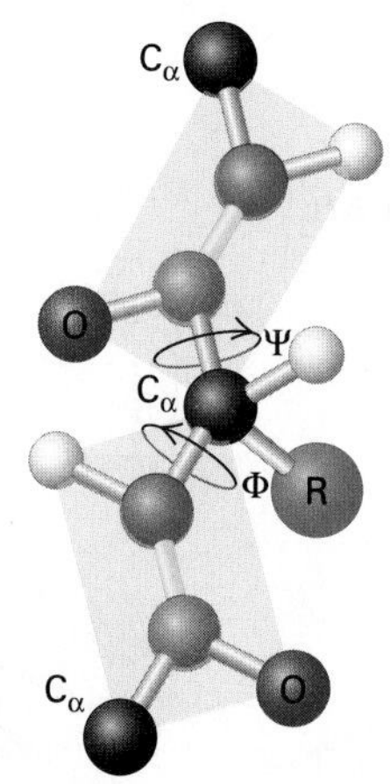

▲ **FIGURE 3-14 Rotation between planar peptide groups in proteins.** Rotation about the C_α–amino nitrogen bond (the ϕ angle) and the C_α–carbonyl carbon bond (the ψ angle) permits polypeptide backbones, in principle, to adopt a very large number of potential conformations. However steric restraints due to the structure of the polypeptide backbone and the properties of the amino acid side chains dramatically restrict the potential conformations that can be adopted by any given protein.

3.2 Protein Folding

As noted above, when it comes to biological structures such as proteins, "form follows function" and "form is function." Thus it is essential that when a polypeptide is synthesized with its particular primary structure (sequence), it folds into the proper three-dimension conformation with the appropriate secondary, tertiary, and possibly quaternary structure. A polypeptide chain is synthesized by a complex process called **translation** in which the assembly of amino acids in a particular sequence is dictated by **messenger RNA (mRNA)** and performed by a large protein–nucleic acid complex called a **ribosome.** The intricacies of translation are considered in Chapter 4. Here, we describe the key determinants of the proper folding of a *nascent* (newly formed or forming) polypeptide chain.

Planar Peptide Bonds Limit the Shapes into Which Proteins Can Fold

A critical structural feature of polypeptides that limits how the chain can fold is the planar peptide bond. Figure 3-3 illustrates the amide group in peptide bonds in a polypeptide chain. Because the peptide bond itself behaves partially like a double bond,

$$P_1-\overset{O}{\overset{\|}{C}}-\underset{H}{\underset{|}{N}}-P_2 \longleftrightarrow P_1-\overset{O^-}{\overset{|}{C}}=\underset{H}{\underset{|}{\overset{+}{N}}}-P_2$$

the carbonyl carbon and amide nitrogen and those atoms directly bonded to them must all lie in a fixed plane (Figure 3-14); there is no rotation possible about the peptide bond itself. As a consequence, the only flexibility in a polypeptide chain backbone, allowing it to adopt varying conformations (twists and turns to fold into different three-dimensional shapes), is rotation of the fixed planes of peptide bonds with respect to one another about two bonds—the C_α–amino nitrogen bond (rotational angle called ϕ) and the C_α–carbonyl carbon bond (rotational angle called ψ).

Yet a further constraint on the potential conformations that a polypeptide backbone chain can adopt is that only a limited number of ϕ and ψ angles are possible, because for most ϕ and ψ angles the backbone or side chain atoms would come too close to one another and thus the associated conformation would be highly unstable or even physically impossible to achieve.

Information Directing a Protein's Folding Is Encoded in Its Amino Acid Sequence

While the constraints of backbone bond angles seem very restrictive, any polypeptide chain containing only a few residues could, in principle, still fold into many conformations. For example, if the ϕ and ψ angles were limited to only eight combinations, an n-residue-long peptide would potentially have 8^n conformations—a very large number for even a small polypeptide of only 10 residues long (about 8.6 million possible conformations)! In general, however, any particular protein adopts only one or just a few very closely related characteristic functional conformations called the *native state;* for the vast majority of proteins, the native state is the most stably folded form of the molecule. In thermodynamic terms, the native state is usually the conformation with the lowest free energy. What features of proteins limit their folding from very many conformations to just one? The properties of the side chains (e.g., size, hydrophobicity, ability to form hydrogen and ionic bonds), together with their particular sequence along the polypeptide backbone, impose key restrictions. For example, a large side chain such as that of tryptophan might prevent (sterically block) one region of the chain from packing closely against another region, whereas a side chain with a positive charge such as arginine might attract a segment of the polypeptide that has a complementary negatively charged side chain (e.g., aspartic acid). Another example we have already discussed is the effect of the aliphatic side chains in heptad repeats on the formation of coiled coils. Thus, a polypeptide's primary structure determines its secondary, tertiary, and quaternary structure.

The initial evidence that the information necessary for a protein to fold properly is encoded in its sequence came from in vitro studies on the refolding of purified proteins. Various perturbations (such as thermal energy from heat, extremes of pH that alter the charges on amino acid side chains, and chemicals, called *denaturants,* such as urea or guanidine hydrochloride at concentrations of 6–8 M) can disrupt the weak noncovalent interactions that stabilize the native conformation of a protein, leading to its **denaturation.** Treatment with the reducing agents, such as β-mercaptoethanol, that break disulfide bonds can further destabilize disulfide-containing proteins. Under such unfolding or denaturing conditions, entropy increases when a population of uniformly folded molecules is destabilized and converted into a collection of many unfolded, or denatured, molecules that have many different non-native and biologically inactive conformations. As we have seen, there are very many possible non-native conformations (e.g., 8^n-1).

The spontaneous unfolding of proteins under denaturing conditions is not surprising, given the substantial increase in entropy. What is striking, however, is that when a pure sample of a single type of unfolded protein is shifted back to normal conditions (body temperature, normal pH levels, reduction in the concentration of denaturants by dilution or their removal), some denatured polypeptides can spontaneously renature (refold) into their native, biologically active states. This kind of refolding experiment, as well as studies that show synthetic proteins made chemically can fold properly, showed that sufficient information must be contained in the protein's primary sequence to direct correct refolding. Newly synthesized proteins appear to fold into their proper conformations just as denatured proteins do. The observed similarity in the folded, three-dimensional structures of proteins with similar amino acid sequences, noted in Section 3.1, provided additional evidence that the primary sequence also determines protein folding in vivo. It appears that formation of secondary structures and structural motifs occurs early in the folding process, followed by assembly of more compact and complex domains, which then associate into more complex tertiary and quaternary structures (Figure 3-15).

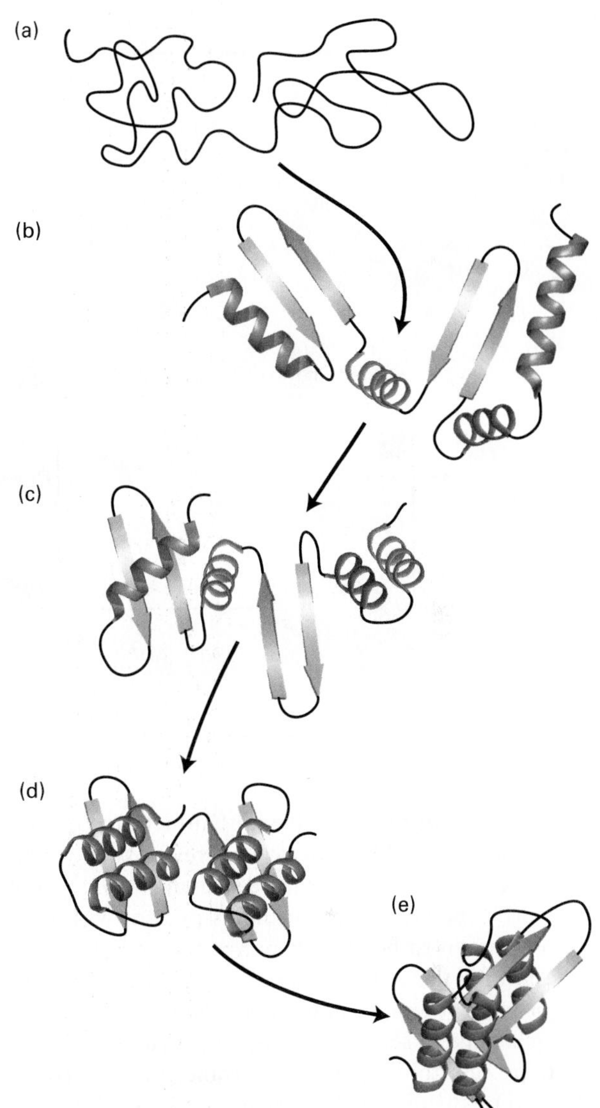

▲ **FIGURE 3-15 Hypothetical protein-folding pathway.** Folding of a monomeric protein follows the structural hierarchy of primary (a) → secondary (b–d) → tertiary (e) structure. Formation of small structural motifs (c) appears to precede formation of more stable domains (d) and the final tertiary structure (e).

Folding of Proteins in Vivo Is Promoted by Chaperones

The refolding of a denatured protein is presumed to mimic many aspects of the folding of a newly synthesized polypeptide. However the conditions inside a cell are not the same as those in test tubes used for in vitro refolding experiments with purified proteins. The presence of other biomolecules, including many other proteins at very high concentration, some of which are themselves nascent and in the process of folding, can potentially interfere with the autonomous, spontaneous folding of a protein. Furthermore, although protein folding into the native state can occur in vitro, this does not happen for all unfolded molecules in a timely fashion. Given such impediments, cells require a faster, more efficient mechanism for folding proteins into their correct shapes than sequence alone provides. Without such help, they would waste much energy in

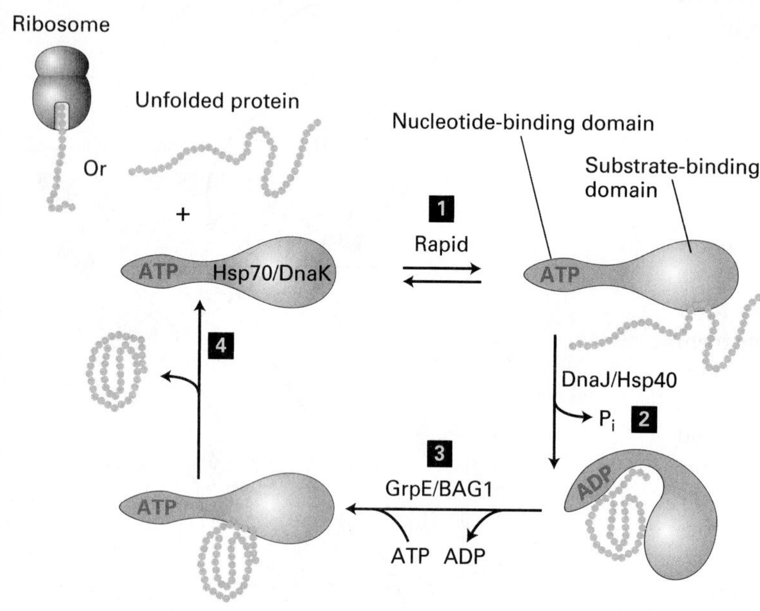

◀ **FIGURE 3-16 Chaperone-mediated protein folding.** Many proteins fold into their proper three-dimensional structures with the assistance of Hsp70-like proteins. These molecular chaperones transiently bind to a nascent polypeptide as it emerges from a ribosome or to proteins that have otherwise unfolded. In the Hsp70 cycle, a substrate unfolded protein binds in rapid equilibrium to the open conformation of the substrate-binding domain (SBD) of Hsp70, to which an ATP is bound in the nucleotide-binding domain (NBD) (step **1**). Accessory proteins (DnaJ/Hsp40) stimulate the hydrolysis of ATP and conformational change in Hsp70, resulting in the closed form, in which the substrate is locked into the SBD; here proper folding is facilitated (step **2**). Exchange of ATP for the bound ADP, stimulated by other accessory proteins (GrpE/BAG1), converts the Hsp70 back to the open form (step **3**), releasing the properly folded substrate (step **4**).

the synthesis of improperly folded, nonfunctional proteins, which would have to be destroyed to prevent their disrupting cell function. Cells clearly have such mechanisms, since more than 95 percent of the proteins present within cells have been shown to be in their native conformations. The explanation for the cell's remarkable efficiency in promoting the proper protein folding encoded in primary sequence is that cells make a set of proteins, called **chaperones,** that facilitate protein folding. The importance of chaperones is highlighted by the observations that they are evolutionarily conserved, they are found in all organisms from bacteria to humans, and some are highly homologous and use almost identical mechanisms to assist protein folding. Chaperones, which in eukaryotes are located in every cellular compartment and organelle, bind to the target proteins whose folding they will assist. Two general families of chaperones are recognized:

- **Molecular chaperones,** which bind and stabilize unfolded or partly folded proteins, thereby preventing these proteins from aggregating and being degraded
- **Chaperonins,** which form a small folding chamber into which an unfolded protein can be sequestered, giving it time and an appropriate environment to fold properly

One reason that chaperones are needed for intracellular protein folding is that they help prevent aggregation of unfolded proteins. Unfolded and partly folded proteins tend to aggregate into large, often water insoluble masses, from which it is extremely difficult for a protein to dissociate and then fold into its proper conformation. In part this aggregation is due to the exposure of hydrophobic side chains that have not yet had a chance to be buried in the inner core of the folded protein. These exposed hydrophobic side chains on different molecules will stick to one another owing to the hydrophobic effect (Chapter 2) and thus promote aggregation. When a newly synthesized molecule begins to fold, it is at risk of aggregating before it completes its proper folding. Molecular chaperones bind to the target polypeptide or sequester it from other partially or fully unfolded proteins, thereby preventing aggregation and thus giving the nascent protein time to fold properly.

Molecular Chaperones The heat-shock protein Hsp70 and its homologs (Hsp70 in the cytosol and mitochondrial matrix, BiP in the endoplasmic reticulum, and DnaK in bacteria) are molecular chaperones. They were first identified by their rapid appearance after a cell has been stressed by heat shock (*Hsp* stands for "*h*eat-*s*hock *p*rotein"). Hsp70 and its homologs are the major chaperones in all organisms. When bound to ATP, Hsp70-like proteins assume an open form in which an exposed hydrophobic pocket transiently binds to exposed hydrophobic regions of an incompletely folded or partially denatured target protein (Figure 3-16). Hydrolysis of the bound ATP causes the molecular chaperone to assume a closed form that appears to facilitate the target protein's folding, in part by preventing unfolded proteins from aggregating. The exchange of ATP for the protein-bound ADP causes a conformational change in the chaperone that releases the target protein.

Additional proteins, such as the co-chaperone Hsp40 in eukaryotes (DnaJ in bacteria), help increase efficiency of

Hsp70-mediated folding of many proteins by stimulating the hydrolysis of ATP by Hsp70/DnaK (see Figure 3-16). An additional protein called GrpE in bacteria (similar activity of BAG1 in mammals) also interacts with the Hsp70/DnaK, promoting the exchange of ATP for ADP. Multiple molecular chaperones are thought to bind all nascent polypeptide chains as they are being synthesized on ribosomes. In bacteria, 85 percent of the proteins are released from their chaperones and proceed to fold normally; an even higher percentage of proteins in eukaryotes follow this pathway.

Chaperonins The proper folding of a large variety of newly synthesized proteins also requires the assistance of another class of proteins, the chaperonins. These huge cylindrical macromolecular assemblies are formed from two rings of oligomers, which can exist in a "tight" peptide-binding state and a "relaxed" peptide-releasing state. The eukaryotic chaperonin TriC consists of eight subunits per ring. In the bacterial, mitochondrial, and chloroplast chaperonin, known as GroEL, each ring contains seven identical subunits (Figure 3-17a). The GroEL folding mechanism, which is better understood than TriC-mediated folding, serves as a general model (Figure 3-17b). A partly folded or misfolded polypeptide is inserted into the cavity of the barrel-like GroEL, where it binds to the inner wall and folds into its native conformation. In an ATP-dependent step, GroEL undergoes a conformational change and releases the folded protein, a process assisted by a co-chaperonin, GroES, which caps the ends of GroEL. The binding of ATP and the co-chaperonin GroES to one of the rings in the tight state of GroEL causes a twofold expansion of its cavity, shifting the equilibrium toward the relaxed peptide-folding state. There is a striking similarity between the capped-barrel design of GroEL/GroES, in which proteins are sequestered for folding, and the structure of the 26S proteasome that participates in protein degradation (discussed in Section 3.4).

Alternatively Folded Proteins Are Implicated in Diseases

As noted earlier, each protein normally folds into a single, energetically favorable conformation that is specified by its amino acid sequence. Recent evidence suggests, however, that a protein may fold into an alternative three-dimensional structure as the result of mutations, inappropriate covalent modifications made after the protein is synthesized, or other as-yet-unidentified reasons. Such "misfolding" not only leads to a loss of the normal function of the protein but often marks it for proteolytic degradation. However, when degradation isn't complete or doesn't keep pace with misfolding, the subsequent accumulation of the misfolded protein or its proteolytic fragments contributes to certain degenerative diseases characterized by the presence of insoluble protein plaques in various organs, including the liver and brain.

Some neurodegenerative diseases, including Alzheimer's disease and Parkinson's disease in humans and transmissible spongiform encephalopathy ("mad cow" disease) in cows and sheep, are marked by the formation of tangled filamentous plaques in a deteriorating brain (Figure 3-18). The *amyloid filaments* composing these structures derive from abundant natural proteins such as amyloid precursor protein, which is embedded in the plasma membrane; Tau, a microtubule-binding protein; and prion protein, an "infectious" protein. Influenced by unknown causes, these α helix–containing proteins or their proteolytic fragments fold into alternative β sheet–containing structures that polymerize into very stable filaments. Whether the extracellular deposits of these filaments or the soluble alternatively folded proteins are toxic to the cell is unclear.

Video: GroEl ATPase Cycle

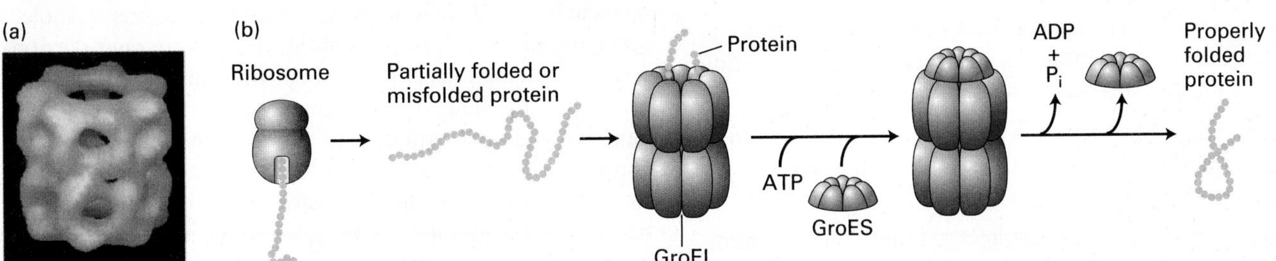

▲ **FIGURE 3-17 Chaperonin-mediated protein folding.** Proper folding of some proteins depends on chaperonins such as the prokaryotic GroEL. (a) GroEL is a hollow, barrel-shaped complex of 14 identical 60,000-MW subunits arranged in two stacked rings. (b) In the absence of ATP or presence of ADP, GroEL exists in a "tight" conformational state that binds partly folded or misfolded proteins. Binding of ATP shifts GroEL to a more open, "relaxed" state, which releases the folded protein. During this process, one end of GroEL is transiently blocked by the co-chaperonin GroES, an assembly of 10,000-MW subunits. [Part (a) from A. Roseman et al., 1996, *Cell* **87**:241; courtesy of H. Saibil.]

(a) (b)

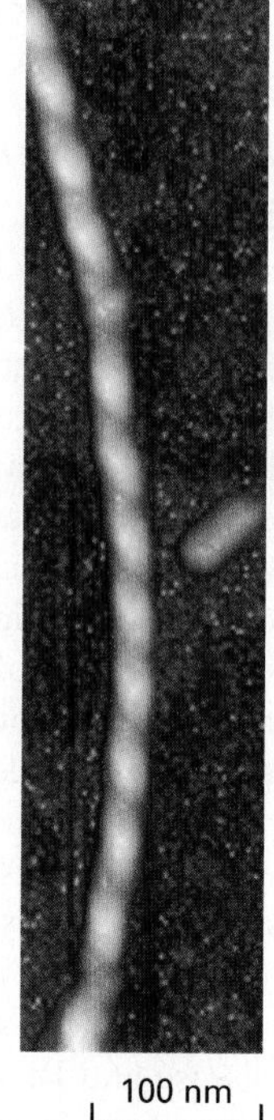

▲ **FIGURE 3-18 Alzheimer's disease is characterized by the formation of insoluble plaques composed of amyloid protein.** (a) At low resolution, an amyloid plaque in the brain of an Alzheimer's patient appears as a tangle of filaments. (b) The regular structure of filaments from plaques is revealed in the atomic force microscope. Proteolysis of the naturally occurring amyloid precursor protein yields a short fragment, called *β-amyloid protein,* that for unknown reasons changes from an α-helical to a β-sheet conformation. This alternative structure aggregates into the highly stable filaments (amyloid) found in plaques. Similar pathologic changes in other proteins cause other degenerative diseases. [Courtesy of K. Kosik.]

KEY CONCEPTS OF SECTION 3.2

Protein Folding

- The sequence of a protein determines its three-dimensional structure, which determines its function. In short, function derives from structure; structure derives from sequence.
- Because protein function derives from protein structure, newly synthesized proteins must fold into the correct shape to function properly.
- The planar structure of the peptide bond limits the number of conformations a polypeptide can have.
- The amino acid sequence of a protein dictates its folding into a specific three-dimensional conformation, the native state. Proteins will unfold, or denature, if treated under conditions that disrupt the noncovalent interactions stabilizing their three-dimensional structures.
- Protein folding in vivo occurs with assistance from chaperones, which bind to nascent polypeptides emerging from ribosomes and prevent their misfolding.
- Some neurodegenerative diseases are caused by aggregates of proteins that are stably folded in an alternative conformation.

3.3 Protein Function

Although proteins have many different shapes and sizes and mediate an extraordinarily diverse array of activities both inside and outside of cells, most of these diverse functions are based on the ability of proteins to engage in a common activity, the binding to themselves, other macromolecules, small molecules, and ions. Here we will describe some of the key features underlying protein binding, and then turn to look at one group of proteins, enzymes, in greater detail. The activities of the other functional classes of proteins (structural, scaffold, transport, regulatory, motor) will be described in other chapters.

Specific Binding of Ligands Underlies the Functions of Most Proteins

The molecule to which a protein binds is often called its **ligand.** In some cases ligand binding causes a change in the shape of a protein. Ligand-binding-driven conformational changes are integral to the mechanism of action of many proteins and are important in regulating protein activity.

Two properties of a protein characterize how it binds ligands. *Specificity* refers to the ability of a protein to bind one molecule in preference to other molecules. *Affinity* refers to the tightness or strength of binding, usually expressed as the dissociation constant (K_d). The K_d for a protein–ligand complex, which is the inverse of the equilibrium constant K_{eq} for the binding reaction, is the most common quantitative measure of affinity (Chapter 2). The stronger the interaction between a protein and ligand, the lower the value of K_d. Both the specificity and the affinity of a protein for a ligand depend on the structure of the *ligand-binding site.* For high-affinity and highly specific interactions to take place, the shape and chemical properties of the binding site must be complementary to that of the ligand molecule, a property termed **molecular complementarity.** As we saw in Chapter 2, molecular complementarity allows molecules to form multiple noncovalent interactions at close range and thus stick together.

One of the best-studied examples of protein–ligand binding, involving high affinity and exquisite specificity, is that of **antibodies** binding to **antigens.** Antibodies are proteins that circulate in the blood and are made by the immune system in response to the invasion by antigens, which are usually macromolecules present in infectious agents (e.g., a bacterium or a virus) or other foreign substances (e.g., proteins or polysaccharides in pollens). Different antibodies are generated in response to different antigens, and these antibodies have the remarkable characteristic of binding specifically to ("recognizing") a portion of the antigen, called an **epitope,** which initially induced the production of the antibody, and not to other molecules. Antibodies act as specific sensors for antigens, forming antibody–antigen complexes that initiate a cascade of protective reactions in cells of the immune system.

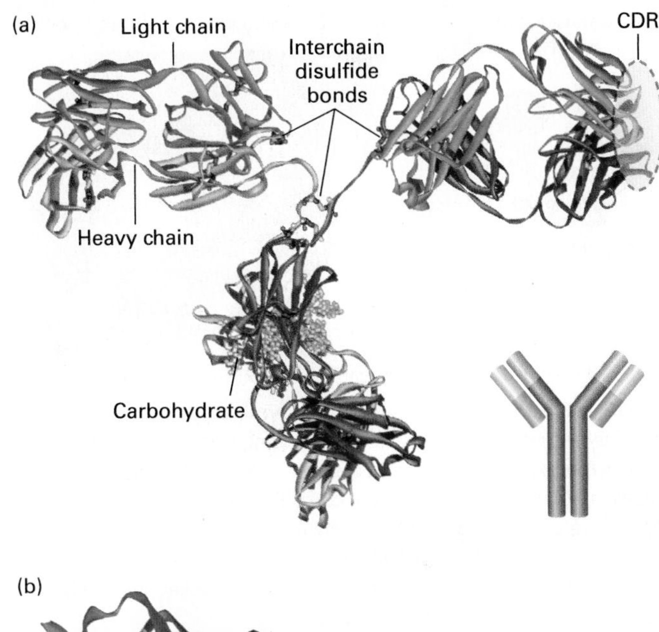

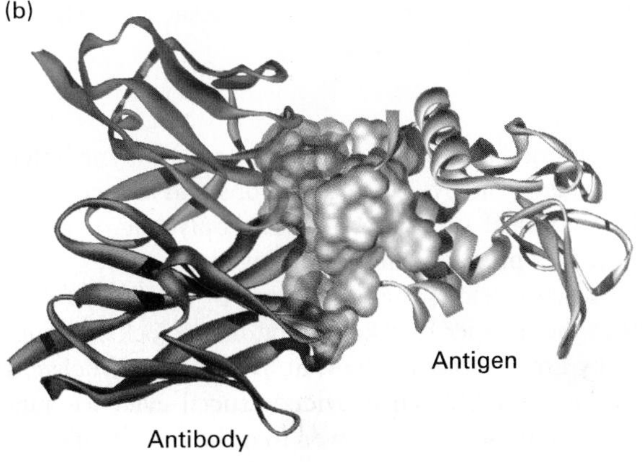

▲ **FIGURE 3-19 Protein–ligand binding of antibodies.** (a) Ribbon model of an antibody. Every antibody molecule of the immunoglobulin IgG class consists of two identical heavy chains (light and dark red) and two identical light chains (blue) covalently linked by disulfide bonds. The inset shows a diagram of the overall structure containing the two heavy and two light chains. (b) The hand-in-glove fit between an antibody and the site to which it binds (epitope) on its target antigen—in this case, chicken egg-white lysozyme. Regions where the two molecules make contact are shown as surfaces. The antibody contacts the antigen with residues from all its complementarity-determining regions (CDRs). In this view, the molecular complementarity of the antigen and antibody is especially apparent where "fingers" extending from the antigen surface are opposed to "clefts" in the antibody surface.

All antibodies are Y-shaped molecules formed from two identical heavy chains and two identical light chains (Figure 3-19a). Each arm of an antibody molecule contains a single light chain linked to a heavy chain by a disulfide bond. Near the end of each arm are six highly variable loops, called *complementarity-determining regions (CDRs),* which form the antigen-binding sites. The sequences of the six loops are highly variable among antibodies, generating unique complementary ligand-binding sites that make them specific for different epitopes (Figure 3-19b). The intimate contact between these two surfaces, stabilized by numerous noncovalent interactions, is responsible for the extremely precise binding specificity exhibited by an antibody.

The specificity of antibodies is so precise that they can distinguish between the cells of individual members of a species and in some cases can distinguish between proteins that differ by only a single amino acid. Because of their specificity and the ease with which they can be produced, antibodies are highly useful reagents used in many of the experiments discussed in subsequent chapters.

We will see many examples of protein–ligand binding throughout this book, including hormones binding to receptors (Chapter 15), regulatory molecules binding to DNA (Chapter 7), cell-adhesion molecules binding to extracellular matrix (Chapter 19), to name just a few. Next we will consider how the binding of one class of proteins, enzymes, to their ligands results in the catalysis of the chemical reactions essential for the survival and function of cells.

Enzymes Are Highly Efficient and Specific Catalysts

Proteins that catalyze chemical reactions, the making and breaking of covalent bonds, are called **enzymes,** and their ligands are called **substrates.** While not all proteins are enzymes, enzymes make up a large and very important class of protein—indeed, almost every chemical reaction in the cell is catalyzed by a specific enzyme. In many ways, enzymes are the cell's chemists, performing many of a cell's chemical reactions. (An additional form of catalytic macromolecule in cells is made from RNA. These RNAs are called **ribozymes.**)

Thousands of different types of enzymes, each of which catalyzes a single chemical reaction or set of closely related reactions, have been identified. Certain enzymes are found in the majority of cells because they catalyze the synthesis of common cellular products (e.g., proteins, nucleic acids, and phospholipids) or take part in the production of energy (e.g., by the conversion of glucose and oxygen into carbon dioxide and water). Other enzymes are present only in a particular type of cell because they catalyze chemical reactions unique to that cell type (e.g., the enzymes in nerve cells that convert tyrosine into dopamine, a neurotransmitter). Although most enzymes are located within cells, some are secreted and function at extracellular sites such as the blood, the digestive tract, or even outside the organism (e.g., toxic enzymes in the venom of poisonous snakes).

Like all **catalysts,** enzymes increase the rate of a reaction but do not affect the extent of a reaction, which is determined by the change in free energy ΔG between reactants and products, and are not themselves permanently changed as a consequence of the reaction they catalyze (Chapter 2). Enzymes increase the reaction rate by lowering the energy of the **transition state,** and therefore the **activation energy** (Figure 3-20). In the test tube, catalysts such as charcoal and platinum facilitate reactions but usually only at high temperatures or pressures, at extremes of high or low pH, or in organic solvents. As the cell's protein catalysts, however, enzymes must function effectively in aqueous environment at 37 °C and 1 atmosphere pressure, and at physiologic pH values, usually 6.5–7.5, but sometimes lower. Remarkably, enzymes exhibit immense catalytic power, accelerating the

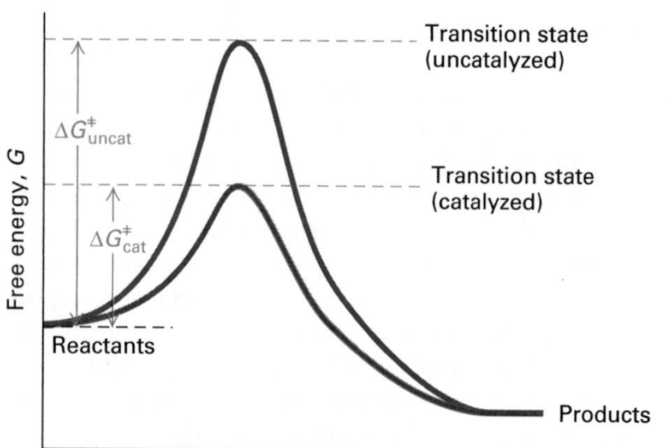

▲ **FIGURE 3-20 Effect of an enzyme on the activation energy of a chemical reaction.** This hypothetical reaction pathway depicts the changes in free energy *G* as a reaction proceeds. A reaction will take place spontaneously only if the total *G* of the products is less than that of the reactants (negative ΔG). However, all chemical reactions proceed through one or more high-energy transition states, and the rate of a reaction is inversely proportional to the activation energy ($\Delta G^{\ddagger}$), which is the difference in free energy between the reactants and the transition state (highest point along the pathway). Enzymes and other catalysts accelerate the rate of a reaction by reducing the free energy of the transition state and thus $\Delta G^{\ddagger}$.

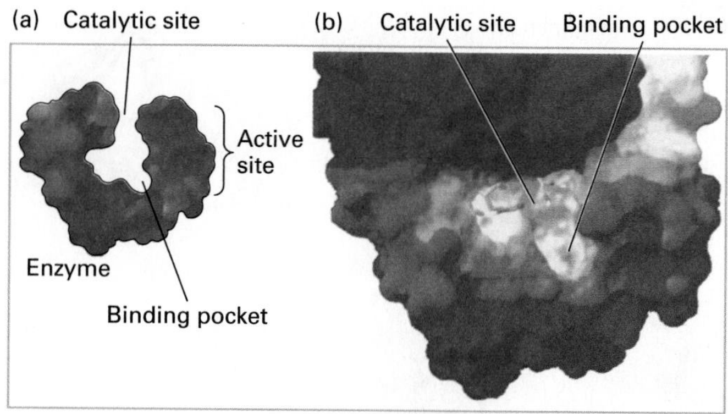

▲ **FIGURE 3-21 Active site of the enzyme trypsin.** (a) An enzyme's active site is composed of a binding pocket, which binds specifically to a substrate, and a catalytic site, which carries out catalysis. (b) A surface representation of the serine protease trypsin. Active site clefts containing the catalytic site (side chains of the catalytic triad Ser-195, Asp-102, and His-57 shown as stick figures) and the substrate side chain specificity binding pocket are clearly visible. [Part (b) courtesy of P. Teesdale-Spittle.]

rates of reactions 10^6–10^{12} times that of the corresponding uncatalyzed reactions under otherwise similar conditions.

An Enzyme's Active Site Binds Substrates and Carries Out Catalysis

Certain amino acids of an enzyme are particularly important in determining its specificity and catalytic power. In the native conformation of an enzyme, these side chains (which usually come from different parts of the linear sequence of the polypeptide) are brought into proximity, forming a cleft in the surface called the **active site** (Figure 3-21). Most active sites make up only a small fraction of the total protein, with the rest involved in folding of the polypeptide, regulation of the active site, and interactions with other molecules.

Active sites consist of two functionally important regions: the *substrate-binding site* that recognizes and binds the substrate (or substrates) and the *catalytic site* whose catalytic groups (amino acid side chains and backbone carbonyl and amino groups) mediate the chemical reaction once the substrate has bound. In some enzymes, the catalytic and substrate-binding sites overlap; in others, the two regions are structurally as well as functionally distinct.

The substrate-binding site is responsible for the remarkable specificity of enzymes—their ability to act selectively on one substrate or a small number of chemically similar substrates. The alteration of the structure of an enzyme's substrate by only one or a few atoms, or a subtle change in the geometry (e.g., stereochemistry) of the substrate, can result in a variant molecule that is no longer a substrate of the enzyme. As noted above, this specificity of enzymes is a consequence of the precise molecular complementarity between its substrate-binding site and the substrate, which is mediated by multiple weak noncovalent interactions and is very sensitive to the shapes of substrates. Usually only one or a few substrates can fit precisely into a binding site.

The idea that enzymes might function by binding to their substrates in the manner of a key fitting into a lock was suggested first by Emil Fischer in 1894. In 1913 Leonor Michaelis and Maud Leonora Menten provided crucial evidence supporting this hypothesis. They showed that the rate of an enzymatic reaction was proportional to the substrate concentration at low substrate concentrations, but that as the substrate concentrations increased, the rate reached a **maximal velocity** V_{max} and became substrate concentration–independent, with the value of V_{max} being directly proportional to the amount of enzyme present in the reaction mixture (Figure 3-22).

They deduced that this saturation at high substrate concentrations was due to the binding of substrate molecules (S) to a fixed and limited number of sites on the enzymes (E), and they called the bound species the enzyme-substrate (ES) complex. They proposed that the ES complex is in equilibrium with the unbound enzyme and substrate and is an intermediate step in the ultimately irreversible conversion of substrate to product (P) (Figure 3-23):

$$E + S \rightleftharpoons ES \rightarrow E + P$$

and that the rate V_0 of formation of product at a particular substrate concentration [S] is given by what is now called the *Michaelis-Menten equation:*

$$V_0 = V_{max}\frac{[S]}{[S] + K_m} \tag{3-1}$$

where the **Michaelis constant K_m**, a measure of the affinity of an enzyme for its substrate (see Figure 3-22), is the substrate concentration that yields a half-maximal reaction rate

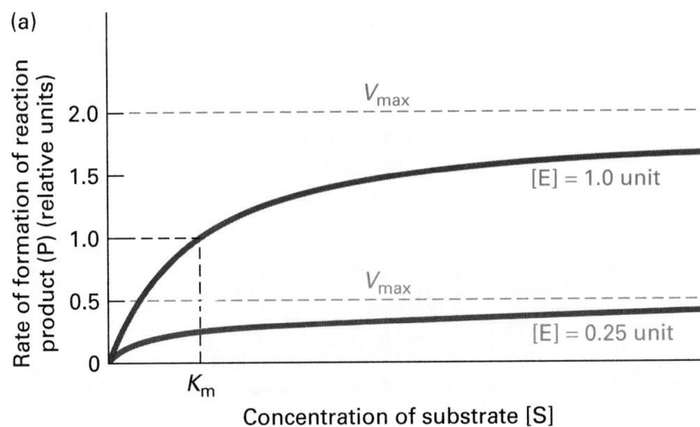

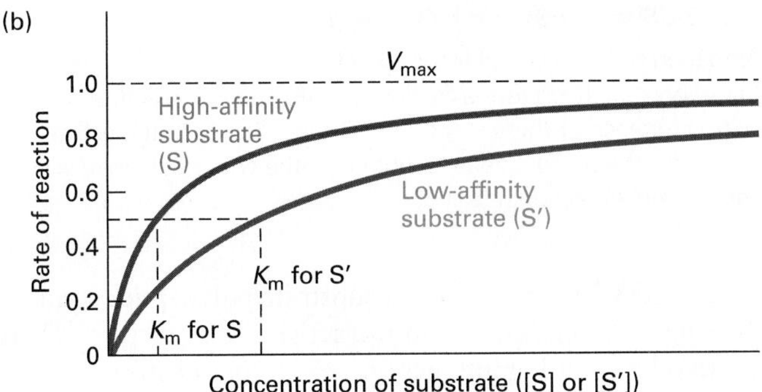

▲ **FIGURE 3-22 K_m and V_{max} for an enzyme-catalyzed reaction.** K_m and V_{max} are determined from analysis of the dependence of the initial reaction velocity on substrate concentration. The shape of these hypothetical kinetic curves is characteristic of a simple enzyme-catalyzed reaction in which one substrate (S) is converted into product (P). The initial velocity is measured immediately after addition of enzyme to substrate before the substrate concentration changes appreciably. (a) Plots of the initial velocity at two different concentrations of enzyme [E] as a function of substrate concentration [S]. The [S] that yields a half-maximal reaction rate is the Michaelis constant K_m, a measure of the affinity of E for turning S into P. Quadrupling the enzyme concentration causes a proportional increase in the reaction rate, and so the maximal velocity V_{max} is quadrupled; the K_m, however, is unaltered. (b) Plots of the initial velocity versus substrate concentration with a substrate S for which the enzyme has a high affinity and with a substrate S′ for which the enzyme has a lower affinity. Note that the V_{max} is the same with both substrates, because [E] is the same, but that K_m is higher for S′, the low-affinity substrate.

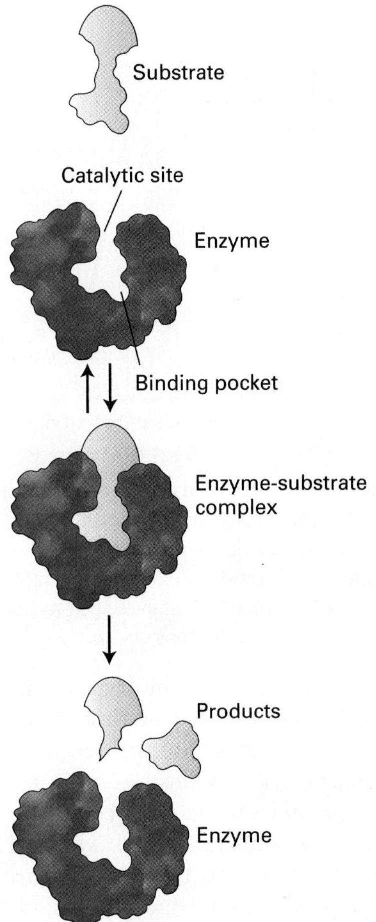

▲ **FIGURE 3-23 Schematic model of an enzyme's reaction mechanism.** Enzyme kinetics suggest that enzymes (E) bind substrate molecules (S) through a fixed and limited number of sites on the enzymes (the active sites). The bound species is known as an enzyme-substrate (ES) complex. The ES complex is in equilibrium with the unbound enzyme and substrate and is an intermediate step in the conversion of substrate to products (P).

(i.e., 1/2 V_{max}), and thus is analogous to the dissociation constant K_d (Chapter 2). The smaller the value of K_m, the more effective the enzyme is at making product from dilute solutions of substrate and the smaller the substrate concentration needed to reach half-maximal velocity. The concentrations of the various small molecules in a cell vary widely, as do the K_m values for the different enzymes that act on them. A good rule of thumb is that the intracellular concentration of a substrate is approximately the same as or somewhat greater than the K_m value of the enzyme to which it binds.

The rates of reaction at substrate saturation vary enormously among enzymes. The maximum number of substrate molecules converted to product at a single enzyme active site per second is called the *turnover number*, which can be less than 1 for very slow enzymes. The turnover number for carbonic anhydrase, one of the fastest enzymes, is 6×10^5 molecules/s.

Many enzymes catalyze the conversion of substrates to products by dividing the process into multiple, discrete chemical reactions that involve multiple, distinct enzyme substrate complexes (ES, ES′, ES″, etc.) generated prior to the final release of the products:

$$E + S \rightleftharpoons ES \rightleftharpoons ES' \rightleftharpoons ES'' \rightleftharpoons \ldots E + P$$

The energy profiles for such multistep reactions involve multiple hills and valleys (Figure 3-24), and methods have been developed to trap the intermediates in such reactions to learn more about the details of how enzymes catalyze reactions.

Serine Proteases Demonstrate How an Enzyme's Active Site Works

Serine proteases, a large family of proteolytic enzymes, are used throughout the biological world—to digest meals (the

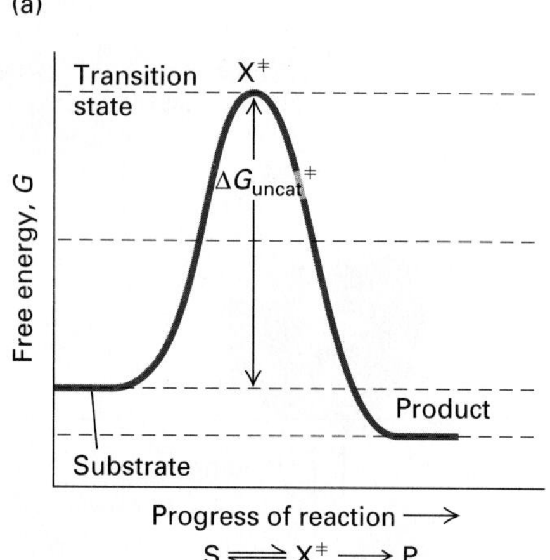

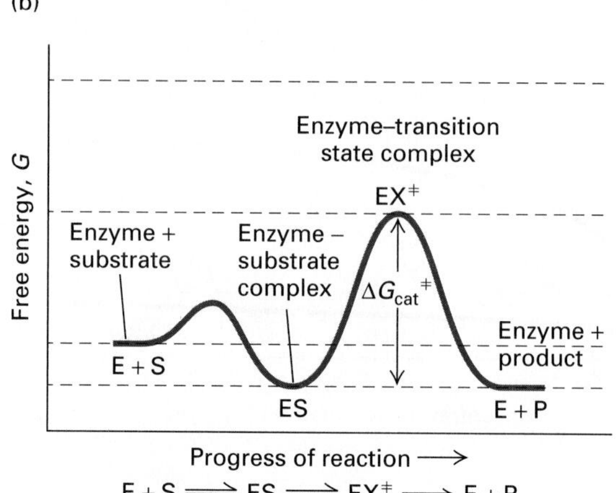

▲ **FIGURE 3-24 Free-energy reaction profiles of uncatalyzed and multistep enzyme-catalyzed reactions.** (a) The free-energy reaction profile of a hypothetical simple uncatalyzed reaction converting substrate (S) to product (P) via a single high-energy transition state. (b) Many enzymes catalyze such reactions by dividing the process into multiple discrete steps, in this case the initial formation of an ES complex followed by conversion via a single transition state ($EX^{\ddagger}$) to the free enzyme (E) and P. The activation energy for each of these steps is significantly less than the activation energy for the uncatalyzed reaction; thus the enzyme dramatically enhances the reaction rate.

pancreatic enzymes trypsin, chymotrypsin, and elastase), to control blood clotting (the enzyme thrombin), even to help silk moths chew their way out of their cocoons (cocoonase). This class of enzymes usefully illustrates how an enzyme's substrate-binding site and catalytic site cooperate in multistep reactions to convert substrates to products. Here we will consider how trypsin and its two evolutionarily closely related pancreatic proteases, chymotrypsin and elastase, catalyze cleavage of a peptide bond:

$$P_1{-}C({=}O){-}N(H){-}P_2 + H_2O \rightleftharpoons P_1{-}C(O)O^- + NH_3^+{-}P_2$$

where P_1 is the portion of the protein on the N-terminal side of the peptide bond, and P_2 is the portion on the C-terminal side. We first consider how serine proteases bind specifically to their substrates and then show in detail how catalysis takes place.

Figure 3-25a shows how a substrate polypeptide binds to the substrate-binding site in the active site of trypsin. There are two key binding interactions. First, the substrate and enzyme form hydrogen bonds that resemble a β sheet. Second, a key side chain of the substrate that determines which peptide in the substrate is to be cleaved extends into the enzyme's *side-chain-specificity binding pocket,* at the bottom of which resides the negatively charged side chain of the

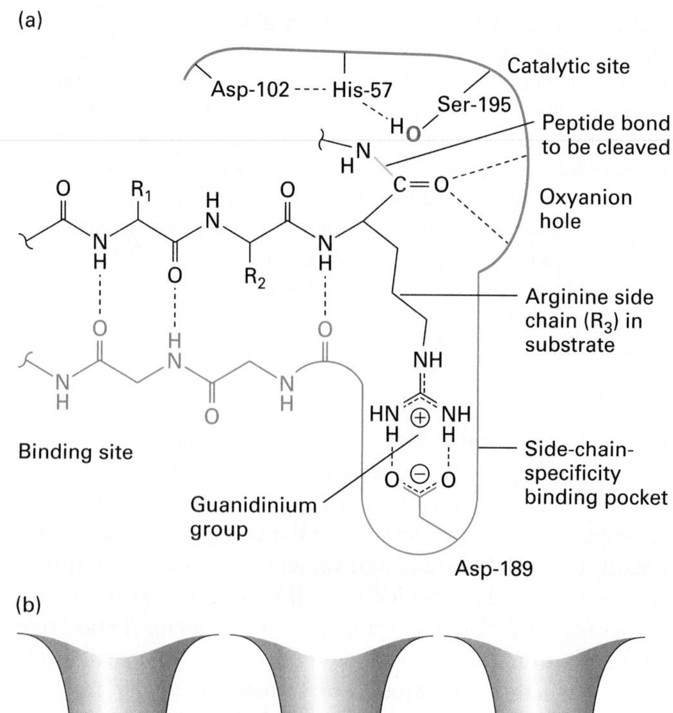

► **FIGURE 3-25 Substrate binding in the active site of typsin-like serine proteases.** (a) The active site of trypsin (blue molecule) with a bound substrate (black molecule). The substrate forms a two-stranded β sheet with the binding site, and the side chain of an arginine (R_3) in the substrate is bound in the side-chain-specificity binding pocket. Its positively charged guanidinium group is stabilized by the negative charge on the side chain of the enzyme's Asp-189. This binding aligns the peptide bond of the arginine appropriately for hydrolysis catalyzed by the enzyme's active-site catalytic triad (side chains of Ser-195, His-57 and Asp-102). (b) The amino acids lining the side-chain-specificity binding pocket determine its shape and charge, and thus its binding properties. Trypsin accommodates the positively charged side chains of arginine and lysine; chymotrypsin, large, hydrophobic side chains such as phenylalanine; and elastase, small side chains such as glycine and alanine. [Part (a) modified from J. J. Perona and C. S. Craik, 1997, *J. Biol. Chem.* **272**(48):29987–29990.]

enzyme's Asp-189. Trypsin has a marked preference for hydrolyzing proteins (black in Figure 3-25a) at the carboxyl side of a residue with a long positively charged side chain (arginine or lysine), because the side chain is stabilized in the specificity binding pocket by the negative Asp-189.

Slight differences in the structures of otherwise similar specificity pockets help explain the differing substrate specificities of the two related serine proteases: chymotrypsin prefers large aromatic groups (as in Phe, Tyr, Trp), and elastase prefers the small side chains of Gly and Ala (Figure 3-25b). The uncharged Ser-189 in chymotrypsin allows large, uncharged, hydrophobic side chains to bind stably in the pocket. The branched aliphatic side chains of valine and threonine in elastase replace glycines in the sides of the pocket in trypsin and thus prevent large side chains in substrates from binding, but allow stable binding of the short alanine or glycine side chain.

In the catalytic site, all three enzymes use the hydroxyl group on the side chain of a serine in position 195 to catalyze the hydrolysis of peptide bonds in protein substrates. A catalytic triad formed by the three side chains of Ser-195, His-57, Asp-102 participates in what is essentially a two-step reaction. Figure 3-26 shows how the catalytic triad cooperates in breaking the peptide bond, with Asp-102 and His-57 supporting the attack of the hydroxyl oxygen of Ser-195 on the carbonyl carbon in the substrate. This attack initially forms an unstable transition state with four groups attached to this carbon (tetrahedral intermediate). Breaking of the C—N peptide bond then releases one part of the protein (NH_3—P_2), while the other part remains covalently attached to the enzyme via an ester bond to the serine's oxygen, forming a relatively stable intermediate (the acyl enzyme). The subsequent replacement of this oxygen by one from water, in a reaction involving another unstable tetrahedral intermediate, leads to release of the final product (P_1—COOH). The tetrahedral intermediates are partially stabilized by hydrogen bonding from the enzyme's backbone amino groups in what is called the *oxyanion hole*. The large family of serine proteases and related enzymes with an active-site serine illustrates how an efficient reaction mechanism is used over and over by distinct enzymes to catalyze similar reactions.

The serine protease mechanism points out several general key features of enzymatic catalysis: (1) enzyme catalytic sites are designed to stabilize the binding of a transition state, thus lowering the activation energy and accelerating the overall

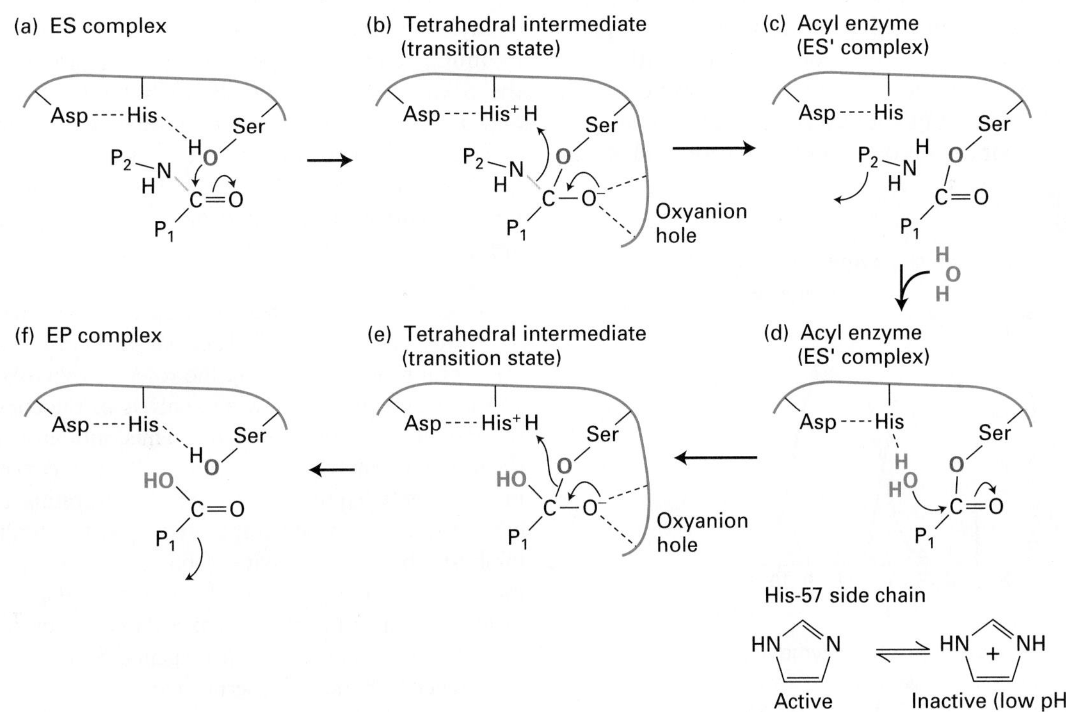

▲ FIGURE 3-26 Mechanism of serine protease–mediated hydrolysis of peptide bonds. The catalytic triad of Ser-195, His-57, and Asp-102 in the active sites of serine proteases employs a multistep mechanism to hydrolyze peptide bonds in target proteins. (a) After a polypeptide substrate binds to the active site (see Figure 3-23) forming an ES complex, the hydroxyl oxygen of Ser-195 attacks the carbonyl carbon of the substrate's targeted peptide bond (yellow). Movements of electrons are indicated by arrows. (b) This attack results in the formation of a transition state called the *tetrahedral intermediate,* in which the negative charge on the substrate's oxygen is stabilized by hydrogen bonds formed with the enzyme's *oxyanion hole.* (c) Additional electron movements result in the breaking of the peptide bond, release of one of the reaction products (NH_2-P_2), and formation of the acyl enzyme (ES′ complex). (d) An oxygen from a solvent water molecule then attacks the carbonyl carbon of the acyl enzyme. (e) This attack results in the formation of a second tetrahedral intermediate. (f) Additional electron movements result in the breaking of the Ser-195–substrate bond (formation of the EP complex) and release of the final reaction product (P_1-COOH). The side chain of His-57, which is held in the proper orientation by hydrogen bonding to the side chain of Asp-102, facilitates catalysis by withdrawing and donating protons throughout the reaction (*inset*). If the pH is too low and the side chain of His-57 is protonated, it cannot participate in catalysis and the enzyme is inactive.

reaction, (2) multiple side chains, together with the polypeptide backbone, carefully organized in three dimensions, work together to chemically transform substrate into product, often by multistep reactions, and (3) acid-base catalysis mediated by one or more amino acid side chains is often used by enzymes, as when the imidazole group of His-57 in serine proteases acts as a base to remove the hydrogen from Ser-195's hydroxyl group. As a consequence, often only a particular ionization state (protonated or nonprotonated) of one or more amino acid side chains in the catalytic site is compatible with catalysis, and thus the enzyme's activity is pH-dependent.

For example, the imidazole of His-57 in serine proteases, whose pK_a is ≈6.8, can help the Ser-195 hydroxyl attack the substrate only if it is not protonated. Thus, the activity of the protease is low at pH < 6.8, and the shape of the pH activity profile in the pH range 4–8 matches the titration of the His-57 side chain, which is governed by the Henderson-Hasselbalch equation, with an inflection near pH 6.8 (see Figure 3-27, *right,* and Chapter 2). The activity drops at higher pH values, generating a bell-shaped curve, because the proper folding of the protein is disrupted when the amino group at the protein's amino terminus is deprotonated ($pK_a \approx 9$); the conformation near the active site changes as a consequence.

The pH sensitivity of an enzyme's activity can be due to changes in the ionization of catalytic groups, groups that participate directly in substrate binding, or groups that influence the conformation of the protein. Pancreatic proteases evolved to function in the neutral or slightly basic conditions in the intestines; hence, their pH optima are ≈8. Proteases and other hydrolytic enzymes that function in acidic conditions must employ a different catalytic mechanism. This is the case for enzymes within the stomach (pH ≈ 1) such as the protease pepsin or those within lysosomes (pH ≈ 4.5), which play a key role in degrading macromolecules within cells (see Figure 3-27, *left*). Indeed, lysosomal hydrolases that degrade a wide variety of biomolecules (proteins, lipids, etc.) are relatively inactive at the pH in the cytosol (≈7), and that helps protect a cell from self-digestion should these enzymes escape the confines of the membrane-bound lysosome.

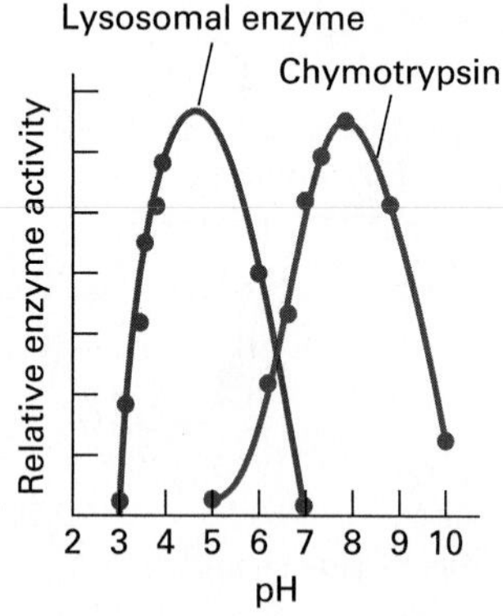

▲ **FIGURE 3-27 pH dependence of enzyme activity.** Ionizable (pH-titratable) groups in the active sites or elsewhere in enzymes often must be either protonated or deprotonated to permit proper substrate binding or catalysis or to permit the enzyme to adopt the correct conformation. Measurement of enzyme activity as a function of pH can be used to identify the pKa's of these groups. The pancreatic serine proteases, such as chymotrypsin (*right curve*), exhibit maximum activity around pH 8 because of titration of the active site His-57 (required for catalysis, $pK_a \approx 6.8$) and of the amino terminus of the protein (required for proper conformation, $pK_a \approx 9$). Many lysosomal hydrolases have evolved to exhibit a lower pH optimum (≈4.5, *left curve*) to match the low internal pH in lysosomes in which they function. [Adapted from P. Lozano, T. De Diego, and J. L. Iborra, 1997, *Eur. J. Biochem.* **248**(1):80–85, and W. A. Judice et al., 2004, *Eur. J. Biochem.* **271**(5):1046–1053.]

One key feature of enzymatic catalysis not seen in serine proteases, but found in many other enzymes, is a *cofactor,* or *prosthetic* (helper) *group*. This is a nonpolypeptide small molecule or ion (e.g., iron, zinc, copper, manganese) that is bound in the active site and plays an essential role in the reaction mechanism. Small organic prosthetic groups in enzymes are also called *coenzymes*. Some of these are chemically modified during the reaction and thus need to be replaced or regenerated after each reaction; others are not. Examples of the former include NAD^+ (nicotinamide adenine dinucleotide) and FAD (flavin adenine dinucleotide) (see Figure 2-33), whereas heme groups that bind oxygen in hemoglobin or transfer electrons in some cytochromes are examples of the latter (Figure 12-14). Thus, the chemistry catalyzed by enzymes is not restricted by the limited number of amino acids in polypeptide chains. Many vitamins—e.g., the B vitamins, thiamine (B_1), riboflavin (B_2), niacin (B_3), and pyridoxine (B_6), and vitamin C—which cannot be synthesized in higher animal cells, function as or are used to generate coenzymes. That is why supplements of vitamins must be added to the liquid medium in which animals cells are grown in the laboratory (Chapter 9).

Small molecules that can bind to active sites and disrupt the reactions are called *enzyme inhibitors*. Such inhibitors are useful tools for studying the roles of enzymes in cells and whole organisms by allowing analysis of the consequences of the loss of the enzyme's activity. Thus, inhibitors complement the use of mutations in genes for probing an enzyme's function in cells (see Chapter 5). However, interpreting results of inhibitor studies can be complicated if, as is often the case, the inhibitors block the activity of more than one protein. Small-molecule inhibition of protein activity is the basis for most drugs (e.g., aspirin inhibits enzymes called cyclooxygenases) and also for chemical warfare agents. Sarin and other nerve gases react with the active serine hydroxyl groups of both serine proteases and a related enzyme, acetylcholine esterase, which is a key enzyme in regulating nerve conduction (see Chapter 23). ▮

Enzymes in a Common Pathway Are Often Physically Associated with One Another

Enzymes taking part in a common metabolic process (e.g., the degradation of glucose to pyruvate) are generally located in the same cellular compartment (e.g., in the cytosol, at a membrane, or within a particular organelle). Within a compartment, products from one reaction can move by diffusion to the next enzyme in the pathway. However, diffusion

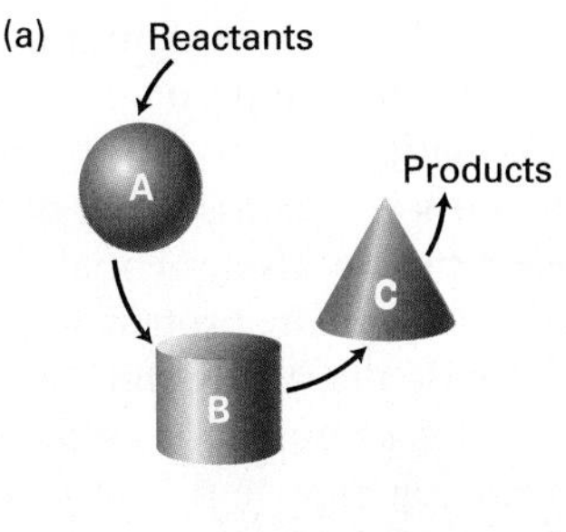

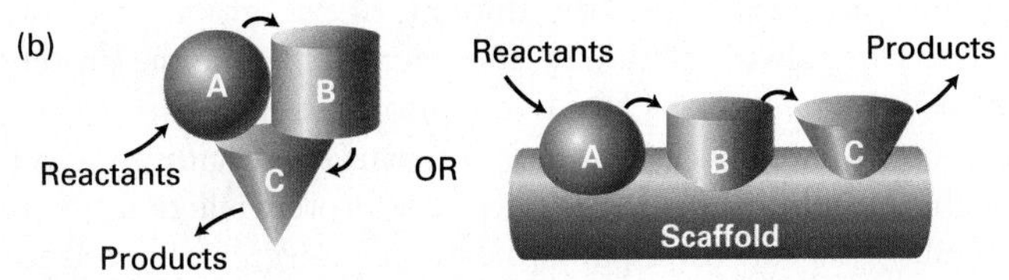

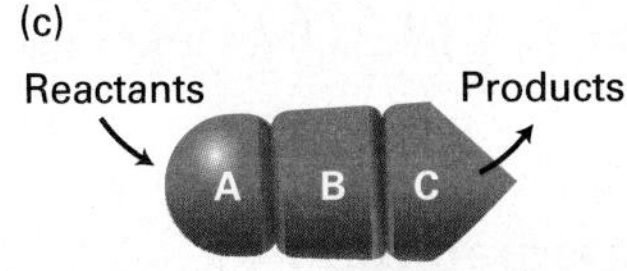

▲ **FIGURE 3-28 Assembly of enzymes into efficient multienzyme complexes.** In the hypothetical reaction pathways illustrated here the initial reactants are converted into final products by the sequential action of three enzymes: A, B, and C. (a) When the enzymes are free in solution or even constrained within the same cellular compartment, the intermediates in the reaction sequence must diffuse from one enzyme to the next, an inherently slow process. (b) Diffusion is greatly reduced or eliminated when the individual enzymes associate into multisubunit complexes, either by themselves or with the aid of a scaffold protein. (c) The closest integration of different catalytic activities occurs when the enzymes are fused at the genetic level, becoming domains in a single polypeptide chain.

entails random movement and can be a slow, relatively inefficient process for moving molecules between widely dispersed enzymes (Figure 3-28a). To overcome this impediment, cells have evolved mechanisms for bringing enzymes in a common pathway into close proximity.

In the simplest such mechanism, polypeptides with different catalytic activities cluster closely together as subunits of a multimeric enzyme or assemble on a common "scaffold" that holds them together (Figure 3-28b). This arrangement allows the products of one reaction to be channeled directly to the next enzyme in the pathway. In some cases, independent proteins have been fused together at the genetic level to create a single multidomain, multifunctional enzyme (Figure 3-28c).

Enzymes Called Molecular Motors Convert Energy into Motion

At the nanoscale of cells and molecules, movement is influenced by forces that differ from those in the macroscopic world. For example, the high protein concentration (200–300 mg/ml) of the cytoplasm prevents organelles and vesicles from diffusing faster than 100 μm/3 hours. Even a micrometer-sized bacterium experiences a drag force from water that stops its forward movement within a fraction of a nanometer when it stops actively swimming. To generate the forces necessary for many cellular movements, cells depend on specialized enzymes commonly called **molecular motors,** or **motor proteins.** These mechanochemical enzymes convert energy released by the hydrolysis of ATP or contained within ion gradients into a mechanical force, usually generating either linear or rotary motion.

From the observed activities of motor proteins, we can infer three general properties that they possess:

- The ability to transduce a source of energy, either ATP or an ion gradient, into linear or rotary movement
- The ability to bind and translocate along a substrate
- Net movement in a given direction

We will see many examples of such motors in subsequent chapters.

KEY CONCEPTS OF SECTION 3.3

Protein Function

- The functions of nearly all proteins depend on their ability to bind other molecules (ligands).
- The specificity of a protein for a particular ligand refers to the preferential binding of one or a few closely related ligands.
- The affinity of a protein for a particular ligand refers to the strength of binding, usually expressed as the dissociation constant K_d.
- Ligand-binding sites on proteins and the corresponding ligands themselves are chemically and spatially complementary.
- Enzymes are catalytic proteins that accelerate the rate of cellular reactions by lowering the activation energy and stabilizing transition-state intermediates (see Figure 3-20).
- An enzyme active site, which is usually only a small part of the protein, comprises two functional parts: a substrate-binding site and a catalytic site. The amino acids composing the active site are not necessarily adjacent in the amino acid sequence but are brought into proximity in the native conformation.
- The substrate-binding site is responsible for the exquisite specificity of enzymes owing to its molecular complementarity with the substrate and the transition state.
- The initial binding of substrates (S) to enzymes (E) results in the formation of an enzyme-substrate complex (ES), which then undergoes one or more reactions catalyzed by the catalytic groups in the active site until the products (P) are formed and diffuse away from the enzyme.
- From plots of reaction rate versus substrate concentration, two characteristic parameters of an enzyme can be determined: the Michaelis constant K_m, a rough measure of

the enzyme's affinity for converting substrate into product, and the maximal velocity V_{max}, a measure of its catalytic power (see Figure 3-22).

- The rates of enzyme-catalyzed reactions vary enormously, with the turnover numbers (number of substrate molecules converted to products at a single active site at substrate saturation) ranging between <1 to 6×10^5 molecules/s.
- Many enzymes catalyze the conversion of substrates to products by dividing the process into multiple discrete chemical reactions that involve multiple distinct enzyme substrate complexes (ES′, ES″, etc.).
- Serine proteases hydrolyze peptide bonds in protein substrates using as catalytic groups the side chains of Ser-195, His-57, and Asp-102.
- Amino acids lining the specificity binding pocket in the binding site of serine proteases determine the residue in a protein substrate that will be hydrolyzed and account for differences in the specificity of trypsin, chymotrypsin, and elastase.
- Enzymes often use acid-base catalysis mediated by one or more amino acid side chains, such as the imidazole group of His-57 in serine proteases, to catalyze reactions.
- The pH dependence of protonation of catalytic groups (pK_a) is often reflected in the pH-rate profile of the enzyme's activity. The pH sensitivity of an enzyme's activity can be due to changes in the ionization of catalytic groups, of groups that participate directly in substrate binding, or of groups that influence the conformation of the protein.
- In some enzymes, nonpolypeptide small molecules or ions, called *cofactors* or *prosthetic groups,* can bind to the active site and play an essential role in enzymatic catalysis. Small organic prosthetic groups in enzymes are also called *coenzymes;* vitamins, which cannot be synthesized in higher animal cells, function as or are used to generate coenzymes.
- Enzymes in a common pathway are located within specific cell compartments and may be further associated as domains of a monomeric protein, subunits of a multimeric protein, or components of a protein complex assembled on a common scaffold (see Figure 3-28).
- Motor proteins are mechanochemical enzymes that convert energy released by ATP hydrolysis into either linear or rotary movement.

3.4 Regulating Protein Function I: Protein Degradation

Most processes in cells do not take place independently of one another or at a constant rate. The activities of all proteins and other biomolecules are regulated to integrate their functions for optimal performance for survival. For example, the catalytic activity of enzymes is regulated so that the amount of reaction product is just sufficient to meet the needs of the cell. As a result, the steady-state concentrations of substrates and products will vary, depending on cellular conditions. Regulation of nonenzymatic proteins—the opening or closing of membrane channels or the assembly of a macromolecular complex, for example—is also essential.

In general, there are three ways to regulate protein activity. First, cells can increase or decrease the steady-state level of the protein by altering its rate of synthesis, its rate of degradation, or both. Second, cells can change the intrinsic activity, as distinct from the amount, of the protein (e.g., the affinity of substrate binding, the fraction of time the protein is in an active versus inactive conformation). Third, there can be a change in location or concentration within the cell of the protein itself, the target of the protein's activity (e.g., an enzyme's substrate), or some other molecule required for the protein's activity (e.g., an enzyme's cofactor). All three types of regulation play essential roles in the lives and functions of cells.

Regulated Synthesis and Degradation of Proteins Is a Fundamental Property of Cells

Control of Protein Synthesis The rate of synthesis of proteins is determined by the rate at which the DNA encoding the protein is converted to mRNA (transcription), the steady-state amount of the active mRNA in the cell, and the rate at which the mRNA is converted into newly synthesized protein (translation). These important pathways are described in detail in Chapter 4.

Control of Protein Degradation The life span of intracellular proteins varies from as short as a few minutes for mitotic cyclins, which help regulate passage through the mitotic stage of cell division, to as long as the age of an organism for proteins in the lens of the eye. Protein life span is controlled primarily by regulated protein degradation.

There are two especially important roles for protein degradation. First, degradation removes proteins that are potentially toxic, improperly folded or assembled, or damaged—including the products of mutated genes and proteins damaged by chemically active cell metabolites. Despite the existence of chaperone-mediated protein folding, it is estimated that as many as 30 percent of newly made proteins are rapidly degraded because they are misfolded, their assembly into complexes is defective, or they are otherwise unsuitable. Most other proteins are degraded more slowly, about 1–2 percent degradation per hour in mammalian cells. Second, the controlled destruction of otherwise normal proteins provides a powerful mechanism for maintaining the appropriate levels of the proteins and their activities, and for permitting rapid changes in these levels to help the cells respond to changing conditions.

Eukaryotic cells have several pathways for degrading proteins. One major pathway is degradation by enzymes within lysosomes, membrane-limited organelles whose acidic interior (pH $\approx$4.5) is filled with a host of hydrolytic enzymes. Lysosomal degradation is directed primarily toward aged or defective organelles of the cell—a process

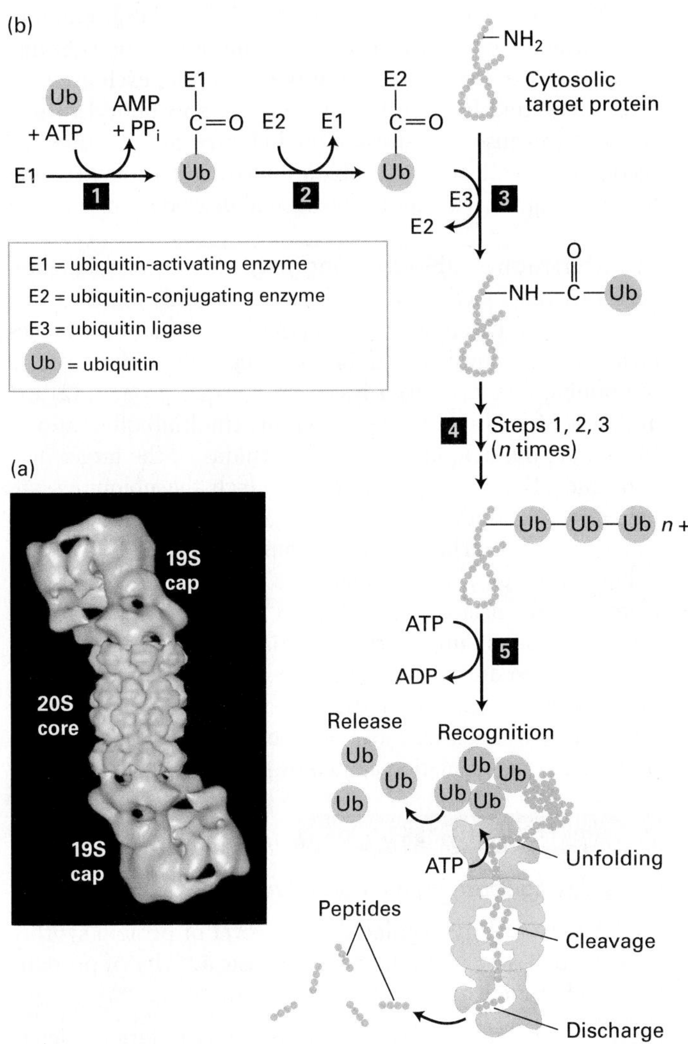

◀ FIGURE 3-29 Ubiquitin- and proteasome-mediated proteolysis. (a) Computer-generated image reveals that a proteasome has a cylindrical structure with a 19S cap (blue) at each end of a 20S core. Proteolysis of ubiquitin-tagged proteins occurs within the inner chamber of the core. (b) Proteins are targeted for proteasomal degradation by polyubiquitination. Enzyme E1 is activated by attachment of a ubiquitin (Ub) molecule (step **1**) and then transfers this Ub molecule to a cysteine residue in E2 (step **2**). Ubiquitin ligase (E3) transfers the bound Ub molecule on E2 to the side-chain —NH_2 of a lysine residue in a target protein (step **3**). Additional Ub molecules are added to the target protein by repeating steps **1**–**3**, forming a polyubiquitin chain (step **4**). The polyubiquitinylated target is recognized by the proteasome cap, which uses ATP hydrolysis to drive removal of the Ub groups, unfolding, and transfer of the unfolded protein into the proteolysis chamber in the core, from which the short peptide digestion fragments are later released (step **5**) [Part (a) from W. Baumeister et al., 1998, *Cell* **92**:357; courtesy of W. Baumeister.]

called autophagy (see Figure 9-2)—and toward extracellular proteins taken up by the cell. Lysosomes will be discussed at length in later chapters. Here we will focus on cytoplasmic protein degradation by proteasomes.

The Proteasome Is a Complex Molecular Machine Used to Degrade Proteins

Proteasomes are very large macromolecular machines consisting of ≈50 protein subunits and having a mass of 2–2.4 × 10^6 Da. They have a cylindrical, barrel-like catalytic core called the *20S proteasome* (where *S* is a Svedberg unit based on the sedimentation properties of the particle and is proportional to its size). Bound to the ends of this core are one or two cap complexes that regulate proteasomal activity. There are approximately 30,000 proteasomes in a typical mammalian cell. There are multiple forms of proteasomes. The best studied of these is the 26S proteasome (Figure 3-29a), which has a catalytic core approximately 14.8 nm tall and 11.3 nm in diameter and a 19S cap regulatory particle at each end. There are several distinct regulatory cap complexes with different activities. The 19S cap has 16–18 protein subunits, 6 of which can hydrolyze ATP (i.e., they are ATPases) to provide the energy needed to unfold protein substrates and selectively transfer them into the inner chamber of the proteasome. Genetic studies in yeast have shown that cells cannot survive without functional proteasomes, thus demonstrating their importance. Furthermore, proper proteasomal activity is so important that cells will expend as much as 30 percent of the energy needed to synthesize a protein to degrade it in a proteasome.

The proteasomal catalytic core comprises two inner rings, with six proteolytic active sites facing toward the inner chamber of the ≈1.7-nm-diameter barrel, and two outer rings that control substrate access. Proteasomes can degrade most proteins thoroughly because they have active sites (two each) that cleave after hydrophobic residues, acidic residues, and basic residues. Polypeptide substrates must enter the chamber via a regulated aperture at the center of the outer rings. In the 26S proteasome, the opening of the aperture, which is narrow and often allows the entry of only unfolded proteins, is controlled by ATPases in the 19S cap. The short peptide products of proteasomal digestion (2–24 residues long) exit the chamber and are further degraded rapidly by cytosolic peptidases, eventually being converted to individual amino acids. Some have quipped that a proteasome is a "cellular chamber of doom" in which proteins suffer a "death of a thousand cuts."

Inhibitors of proteasome function can be used therapeutically. Because of the global importance of proteasomes for cells, continuous, complete inhibition of proteasomes kills cells. However, partial, discontinuous proteasome inhibition has been introduced as an approach to cancer chemotherapy. To survive and grow, cells normally require the robust activity of a regulatory protein called $NF_{\kappa}B$, as well as other similar "pro-survival" proteins. In turn, $NF_{\kappa}B$ can function fully and promote survival only when its inhibitor, $I_{\kappa}B$, is disengaged and degraded by proteasomes (Chapter 16). Partial inhibition of proteasomal activity by a small-molecule inhibitor drug results in increased levels of $I_{\kappa}B$ and, consequently, reduced $NF_{\kappa}B$ activity (loss of pro-survival activity). Cells subsequently die by a mechanism called **apoptosis** (programmed cell death, Chapter 21). Because at least some types of tumor cells are more sensitive to being killed by proteasome inhibitors than normal cells are, *controlled* administration of proteasome inhibitors (at levels that kill the cancer cells but not normal cells) has proved to be an effective therapy for at least one type of lethal cancer, multiple myeloma. ■

Ubiquitin Marks Cytosolic Proteins for Degradation in Proteasomes

If proteasomes are to rapidly degrade only those proteins that are either defective or scheduled to be removed, they must be able to distinguish between those proteins that need to be degraded from most of the proteins that don't. To solve this problem, cells identify proteins that should be degraded by covalently attaching multiple copies of a 76-residue polypeptide called **ubiquitin** that is highly conserved from yeast to humans. A complex sensing system has evolved to determine which proteins are to be degraded, and then a three-step process is used to polyubiquitinylate the target proteins. The 19S regulatory cap of the 26S proteasome then recognizes the ubiquitin-labeled proteins, and unfolds and transports them into the proteasome for degradation. The ubiquitination process (Figure 3-29b) involves:

1. Activation of *ubiquitin-activating enzyme (E1)* by the addition of a ubiquitin molecule, a reaction that requires ATP

2. Transfer of this ubiquitin molecule to a cysteine residue in *ubiquitin-conjugating enzyme (E2)*

3. Formation of an isopeptide bond between the carboxyl terminus of the ubiquitin bound to E2 and the amino group of the side chain of a lysine residue in the target protein, a reaction catalyzed by *ubiquitin-protein ligase (E3)*. Subsequent ligase reactions covalently attach additional ubiquitins to the side chain of lysine 48 of the previously added ubiquitin to generate a linear polymer of ubiquitins, or a polyubiquitin-modified target protein.

Specificity of Degradation Targeting of specific proteins is primarily achieved through the substrate specificity of the E3 ligase. There are hundreds of E3 ligases in mammalian cells, ensuring that the wide variety of proteins to be polyubiquitinylated can be modified when necessary.

An example of the control of the activity of a key cellular protein by the ubiquitin-proteasome system is the regulated degradation of proteins called **cyclins**, which control the cell cycle (Chapter 20). Cyclins contain the internal sequence Arg-X-X-Leu-Gly-X-Ile-Gly-Asp/Asn (X can be any amino acid), which is recognized by specific ubiquitinylating enzyme complexes. At a specific time in the cell cycle, each cyclin is phosphorylated by a cyclin kinase. This phosphorylation is thought to cause a conformational change that exposes the recognition sequence to the ubiquitinylating enzymes, leading to polyubiquitination and proteasomal degradation.

Multifunctional Ubiquitin Tagging Some ubiquitination performs cell functions other than the degradation of a targeted protein. Examples of alternative ubiquitination schemes include (1) the covalent addition of a single ubiquitin molecule (monoubiquitination) to a lysine on a target protein, (2) the addition of multiple single ubiquitins (multiubiquitination), (3) linking the ubiquitin to the N-terminus of the target protein, and (4) polyubiquitination in which the ubiquitins are linked to one another via their Lys-63 residue instead of at the Lys-48 position. These modifications can influence the trafficking (sorting) of proteins within a cell (e.g., internalization from the cell surface), control DNA repair and regulation of transcription, and undoubtedly perform numerous other functions yet to be discovered. Cells also have a variety of deubiquitinylating enzymes that can remove ubiquitins from the target proteins and thus introduce the possibility in some cases of reversing the regulation caused by the initial ubiquitination.

KEY CONCEPTS OF SECTION 3.4

Regulating Protein Function I: Protein Degradation

- Proteins may be regulated at the level of protein synthesis, protein degradation, or the intrinsic activity of proteins through noncovalent or covalent interactions.
- The life span of intracellular proteins is largely determined by their susceptibility to proteolytic degradation.
- Many proteins are marked for destruction with a polyubiquitin tag and then degraded within proteasomes, large cylindrical complexes with multiple proteases in their interiors (see Figure 3-29).
- Variations in the nature of the covalent attachment of ubiquitin to proteins are involved in cellular functions other than proteasome-mediated degradation, such as changes in the location or activity of proteins.

3.5 Regulating Protein Function II: Noncovalent and Covalent Modifications

The intrinsic activities of proteins are modulated by both noncovalent and covalent changes in the protein. Noncovalent modifications usually involve the binding or dissociation of a molecule and a consequent change in the conformation of the protein. Often, in such cases, protein activation involves the release or rearrangement of an inhibitory subunit or domain.

Covalent modifications include hydrolysis of the polypeptide chain or addition of a molecule to the side chain of one or more residues or to the N- or C-terminus of the protein. Such modifications can cause a conformational change in the protein that can alter its activity (form is function). Covalent modifications can also modify the shape of a protein without changing the conformation of the polypeptide and its side chains, for example, by adding a charge or bulky group that can alter the ability of the protein to bind to other molecules. Lastly, covalent modifications can direct the protein to particular locations in a cell (e.g., the cytoplasmic surface of the plasma membrane).

Many noncovalent and covalent modifications are reversible, thus allowing the activity of an individual protein to be enhanced or suppressed multiple times during the lifetime of the protein. Others, such as proteolysis, are irreversible and can be superseded only by degradation of the modified protein and synthesis of a replacement. In the case of enzymes, these regulatory modifications alter K_m, V_{max}, or both. Nature has devised many different strategies for noncovalent and covalent regulation of activity. Here we discuss some common mechanisms for regulating protein function; additional examples will be described in other chapters.

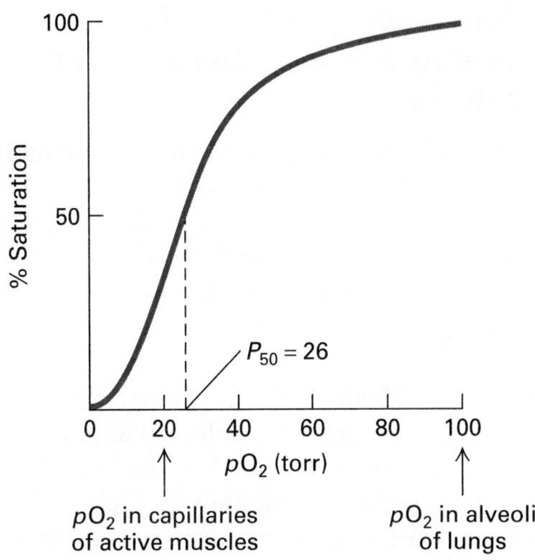

▲ **EXPERIMENTAL FIGURE 3-30 Hemoglobin binds oxygen cooperatively.** Each tetrameric hemoglobin protein has four oxygen-binding sites; at saturation all the sites are loaded with oxygen. The oxygen concentration is commonly measured as the partial pressure (pO_2). P_{50} is the pO_2 at which half the oxygen-binding sites at a given hemoglobin concentration are occupied; it is somewhat analogous to the K_m for an enzymatic reaction. The large change in the amount of oxygen bound over a small range of pO_2 values permits efficient unloading of oxygen in peripheral tissues such as muscle. The sigmoidal shape of a plot of percent saturation versus ligand concentration is indicative of cooperative binding. In the absence of cooperative binding, a binding curve is a hyperbola, similar to the curves in Figure 3-22. [Adapted from L. Stryer, 1995, *Biochemistry,* 4th ed., W. H. Freeman and Company.]

Noncovalent Binding Permits Allosteric, or Cooperative, Regulation of Proteins

One of the most important mechanisms for regulating protein function is through allosteric interactions. Broadly speaking, **allostery** (from the Greek "other shape") refers to any change in a protein's tertiary or quaternary structure, or in both, induced by the noncovalent binding of a ligand. When a ligand binds to one site (A) in a protein and induces a conformational change and associated change in activity of a different site (B), the ligand is called an *allosteric effector* of the protein, while site A is called an *allosteric binding site,* and the protein is called an *allosteric protein.* By definition, allosteric proteins have multiple binding sites for either a single type of ligand or for multiple different ligands. The allosteric change in activity can be positive or negative, i.e., can induce an increase or a decrease in protein activity. Allosteric regulation is particularly prevalent in multimeric enzymes and other proteins where conformational changes in one subunit are transmitted to an adjacent subunit. *Cooperativity* is a term often used synonymously with allostery, and usually refers to the influence (positive or negative) that the binding of a ligand at one site has on the binding of another molecule of the same type of ligand at a different site.

Hemoglobin presents a classic example of positive cooperative binding in that the binding of a single ligand, oxygen, increases the affinity of the binding of the next oxygen molecule. Each of the four subunits in hemoglobin contains one heme molecule. The heme groups are the oxygen-binding components of hemoglobin (see Figure 3-13). The binding of oxygen to the heme molecule in one of the four hemoglobin subunits induces a local conformational change whose effect spreads to the other subunits, lowering the K_m (increasing the affinity) for the binding of additional oxygen molecules to the remaining hemes and yielding a sigmoidal oxygen-binding curve (Figure 3-30). Because of the sigmoidal shape of the oxygen-saturation curve, it takes only a fourfold increase in oxygen concentration for the percent saturation of the oxygen binding sites in hemoglobin to go from 10 to 90 percent. Conversely, if there were no cooperativity and the shape of the curve was typical of that for Michaelis-Menten-type binding, it would take an 81-fold increase in oxygen concentration to accomplish the same increase in loading. This cooperativity permits hemoglobin to take up oxygen very efficiently in the lungs where the oxygen concentration is high, and unload it in tissues where the concentration is low. Thus, cooperativity amplifies the sensitivity of a system to concentration changes in its ligands, providing in many cases selective evolutionary advantage.

Negative cooperativity often involves the end product of a multistep biochemical pathway, which binds to and reduces the activity of an enzyme that catalyzes an early, rate-controlling step for that pathway. In this way excessive buildup of the product is prevented. This kind of regulation of a metabolic pathway is also called *end-product inhibition* or *feedback inhibition.*

Noncovalent Binding of Calcium and GTP Are Widely Used As Allosteric Switches to Control Protein Activity

Unlike oxygen, which causes graded allosteric changes in the activity of hemoglobin, other allosteric effectors act as switches, turning the activity of many different proteins on or off. Two important allosteric switches that we will encounter many times throughout this book are Ca^{2+} and GTP.

Ca^{2+}/Calmodulin-Mediated Switching The concentration of Ca^{2+} free in the cytosol (not bound to molecules other than water) is kept very low ($\approx 10^{-7}$ M) by specialized membrane transport proteins that continually pump Ca^{2+} out of the cytosol. However, as we learn in Chapter 11, the cytosolic Ca^{2+} concentration can increase from 10- to 100-fold when Ca^{2+}-permeable channels in the cell surface membranes open and allow extracellular Ca^{2+} to flow into the cell. This rise in cytosolic Ca^{2+} is sensed by specialized Ca^{2+}-binding proteins, which alter cellular behavior by turning other proteins on or off. The importance of extracellular Ca^{2+} for cell activity was first documented by S. Ringer in 1883, when he discovered that isolated rat hearts suspended in a NaCl solution made with 'hard' (Ca^{2+}-rich) London tap water contracted beautifully, whereas they beat poorly and stopped quickly if distilled water was used.

Many of the Ca^{2+}-binding proteins bind Ca^{2+} using the EF hand/helix-loop-helix structural motif discussed earlier (see Figure 3-9b). The prototype EF hand protein, **calmodulin,** is found in all eukaryotic cells and may exist as an individual monomeric protein or as a subunit of a multimeric protein. A dumbbell-shaped molecule, calmodulin contains four Ca^{2+}-binding EF hands with K_d's of $\approx 10^{-6}$ M. The binding of Ca^{2+} to calmodulin causes a conformational change that permits Ca^{2+}/calmodulin to bind to conserved sequences in various target proteins, thereby switching their activities on or off (Figure 3-31). Calmodulin and similar EF hand proteins thus function as *switch proteins,* acting in concert with changes in Ca^{2+} levels to modulate the activity of other proteins.

(a) Calmodulin without calcium

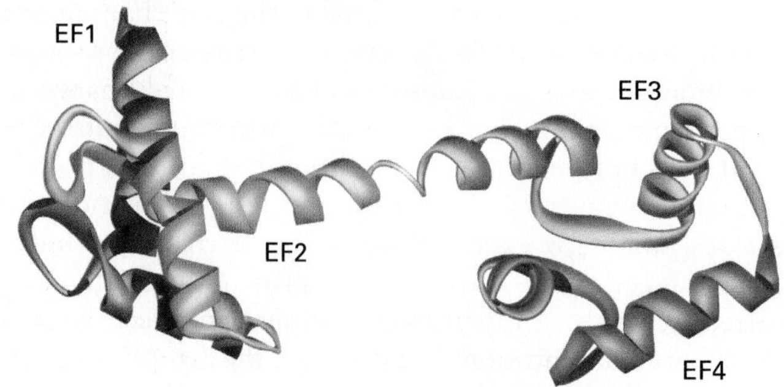

(b) Ca^{2+}/ calmodulin bound to target peptide

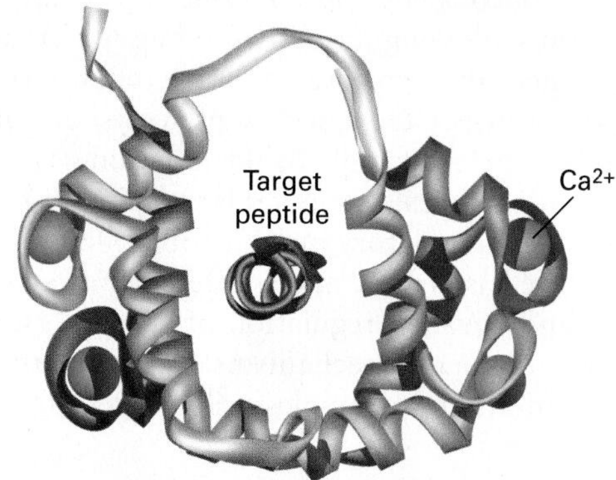

▲ **FIGURE 3-31 Conformational changes induced by Ca^{2+} binding to calmodulin.** Calmodulin is a widely distributed cytosolic protein that contains four Ca^{2+}-binding sites, one in each of its EF hands. Each EF hand has a helix-loop-helix motif. At cytosolic Ca^{2+} concentrations above about 5×10^{-7} M, binding of Ca^{2+} to calmodulin changes the protein's conformation from the dumbbell-shaped, unbound form (a) to one in which hydrophobic side chains become more exposed to solvent. The resulting Ca^{2+}/calmodulin can wrap around exposed helices of various target proteins (b), thereby altering their activity.

Switching Mediated by Guanine Nucleotide–Binding Proteins Another group of intracellular switch proteins constitutes the **GTPase superfamily.** As the name suggests, these proteins are enzymes, GTPases, that can hydrolyze GTP (guanosine triphosphate) to GDP (guanosine diphosphate). They include the monomeric Ras protein (see Figure 3-8) and the G_α subunit of the trimeric G proteins, both discussed at length in Chapter 15. Both Ras and G_α can bind to the plasma membrane, function in cell signaling, and play a key role in cell proliferation and differentiation. Other members of the GTPase superfamily function in protein synthesis, the transport of proteins between the nucleus and the cytoplasm, the formation of coated vesicles and their fusion with target membranes, and rearrangements of the actin cytoskeleton. The Hsp70 chaperone protein we encountered earlier is an example of an ATP/ADP switch, similar in many respects to a GTP/GDP switch.

All the GTPase switch proteins exist in two forms, or conformations (Figure 3-32): (1) an active ("on") form with bound GTP that modulates the activity of specific target proteins to which they bind and (2) an inactive ("off") form with bound GDP, which is generated by the relatively slow hydrolysis of the GTP bound to the active form. The amount of time any given GTPase switch remains active depends on the rate of its GTPase activity. Thus the GTPase activity acts as a timer to control this switch. Cells contain a variety of proteins that can modulate the baseline (or intrinsic) rate of GTPase activity for any given GTPase switch. For example, GTPase activity can be enhanced by specific GTPase-activating proteins, called *GAPs,* or depressed by other proteins acting as allosteric regulators. After the switch has been turned off (GTP hydrolysis), it can be turned back on by GTP exchange factor (GEF), which replaces the bound GDP with a different GTP molecule from the surrounding fluid. Thus, cells can control when the switch is turned on and how long the switch remains on. We examine the role of various GTPase switch proteins in regulating intracellular signaling and other processes in several later chapters.

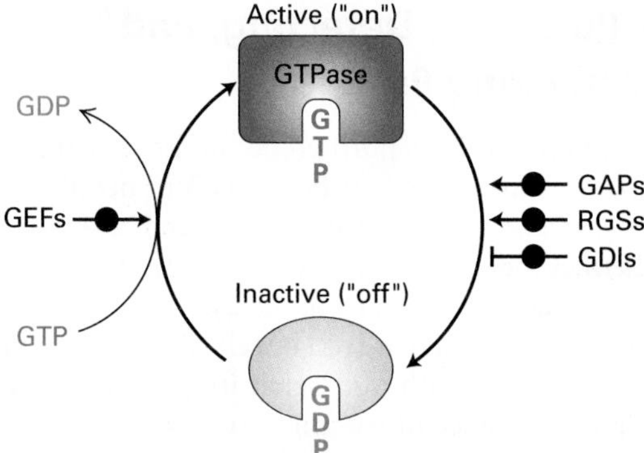

▲ **FIGURE 3-32 The GTPase switch.** Conversion of the active, GTP-bound GTPase into the inactive form by hydrolysis of GTP is accelerated by GAPs (GTPase-activating proteins) and RGSs (regulators of G protein signaling) and inhibited by GDIs (guanine nucleotide dissociation inhibitors). Reactivation by replacing GDP with GTP is promoted by GEFs (guanine nucleotide exchange factors).

Phosphorylation and Dephosphorylation Covalently Regulate Protein Activity

One of the most common mechanisms for regulating protein activity is *phosphorylation,* the addition of phosphate groups to hydroxyl groups on serine, threonine, or tyrosine residues. Protein **kinases** catalyze phosphorylation, and **phosphatases** catalyze *dephosphorylation.* The counteracting activities of kinases and phosphatases provide cells with a "switch" that can turn on or turn off the function of various proteins (Figure 3-33). Phosphorylation changes a protein's charge and generally leads to a conformational change; these effects can significantly alter ligand binding or other features of the protein, leading to an increase or decrease in its activity.

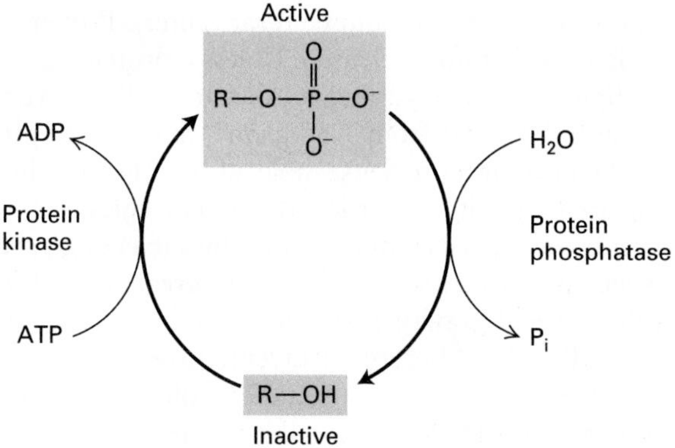

▲ **FIGURE 3-33 Regulation of protein activity by the kinase/phosphatase switch.** The cyclic phosphorylation and dephosphorylation of a protein is a common cellular mechanism for regulating protein activity. In this example, the target protein R is active (top) when phosphorylated and inactive (bottom) when dephosphorylated; some proteins have the opposite responses to phosphorylation.

Nearly 3 percent of all yeast proteins are protein kinases or phosphatases, indicating the importance of phosphorylation and dephosphorylation reactions even in simple cells. All classes of proteins—including structural proteins, scaffolds, enzymes, membrane channels, and signaling molecules—have members regulated by kinase/phosphatase switches. Different protein kinases and phosphatases are specific for different target proteins and can thus regulate a variety of cellular pathways, as discussed in later chapters. Some of these enzymes act on one or a few target proteins, whereas others have many targets. The latter are useful in integrating the activities of proteins that are coordinately controlled by a single kinase/phosphatase switch. Frequently, the target of the kinase (and phosphatase) is yet another kinase or phosphatase, creating a cascade effect. There are many examples of such kinase cascades, which permit amplification of a signal and many levels of fine-tuning control (see Chapter 15).

Proteolytic Cleavage Irreversibly Activates or Inactivates Some Proteins

Unlike phosphorylation, which is reversible, the activation or inactivation of protein function by proteolytic cleavage is an irreversible mechanism for regulating protein activity. For example, many polypeptide hormones, such as insulin, are synthesized as long precursors, and prior to secretion from cells some of their peptide bonds must be hydrolyzed for them to fold properly. In some cases, a single long precursor *prohormone* polypeptide can be cleaved into several distinct active hormones. To prevent the pancreatic serine proteases from inappropriately digesting proteins before they reach the small intestines, they are synthesized as *zymogens,* inactive precursor proteins. Cleavage of a peptide bond near the N-terminus of trypsinogen (the zymogen of trypsin) by a highly specific protease in the small intestine generates a new N-terminal residue (Ile-16), whose amino group can form an ionic bond with the carboxylic acid side chain of an internal aspartic acid. This causes a conformation change that opens the substrate-binding site, activating the enzyme. The active trypsin can then activate trypsinogen, chymotrypsinogen, and other zymogens. Similar, but more elaborate, protease cascades (one protease activating inactive precursors of others) that can amplify an initial signal play important roles in several systems, such as the blood-clotting cascade. The importance of carefully regulating such systems is clear—inappropriate clotting could fatally clog the circulatory system, while insufficient clotting could lead to uncontrolled bleeding.

An unusual and rare type of proteolytic processing, termed *protein self-splicing,* takes place in bacteria and some eukaryotes. This process is analogous to editing film: an internal segment of a polypeptide is removed and the ends of the polypeptide are rejoined (ligated). Unlike other forms of proteolytic processing, protein self-splicing is an autocatalytic process, which proceeds by itself without the participation of enzymes. The excised peptide appears to eliminate itself from the protein by a mechanism similar to that used in the processing of some RNA molecules (Chapter 8). In vertebrate cells, the processing of some proteins includes self-cleavage, but the subsequent ligation step is absent. One such protein is Hedgehog, a membrane-bound signaling molecule that is critical to a number of developmental processes (Chapter 16).

Higher-Order Regulation Includes Control of Protein Location and Concentration

All the regulatory mechanisms heretofore described affect a protein locally at its site of action, turning its activity on or off. Normal functioning of a cell, however, also requires the segregation of proteins to particular compartments such as the mitochondria, nucleus, and lysosomes. In regard to enzymes, compartmentation not only provides an opportunity for controlling the delivery of substrate or the exit of product, but also permits competing reactions to take place simultaneously in different parts of a cell. We describe the mechanisms that cells use to direct various proteins to different compartments in Chapters 12 and 13.

KEY CONCEPTS OF SECTION 3.5

Protein Regulation II: Noncovalent and Covalent Modifications

- In allostery, the noncovalent binding of one ligand molecule, the allosteric effector, induces a conformational change that alters a protein's activity or affinity for other ligands. The allosteric effector can be identical in structure to or different from the other ligands, whose binding it affects. The allosteric effector can be a substrate, activator, or inhibitor.
- In multimeric proteins, such as hemoglobin, that bind multiple identical ligand molecules (e.g., oxygen), the binding of one ligand molecule may increase or decrease the binding affinity for subsequent ligand molecules. This type of allostery is known as cooperativity.
- Several allosteric mechanisms act as switches, turning protein activity on and off in a reversible fashion.
- Two classes of intracellular switch proteins regulate a variety of cellular processes: (1) Ca^{2+}-binding proteins (e.g., calmodulin) and (2) members of the GTPase superfamily (e.g., Ras), which cycle between active GTP-bound and inactive GDP-bound forms (see Figure 3-32).
- The phosphorylation and dephosphorylation of hydroxyl groups on serine, threonine, or tyrosine residue side chains by protein kinases and phosphatases provide reversible on/off regulation of numerous proteins.
- Many types of covalent and noncovalent regulation are reversible, but some forms of regulation, like proteolytic cleavage, are irreversible.
- Higher-order regulation includes compartmentation of proteins and control of protein concentration.

3.6 Purifying, Detecting, and Characterizing Proteins

A protein often must be purified before its structure and the mechanism of its action can be studied in detail. However, because proteins vary in size, charge, and water-solubility, no single method can be used to isolate all proteins. To isolate one particular protein from the estimated 10,000 different proteins in a particular type of cell is a daunting task that requires methods both for separating proteins and for detecting the presence of specific proteins.

Any molecule, whether protein, carbohydrate, or nucleic acid, can be separated, or *resolved,* from other molecules on the basis of their differences in one or more physical or chemical characteristics. The larger and more numerous the differences between two proteins, the easier and more efficient their separation. The two most widely used characteristics for separating proteins are *size,* defined as either length or mass, and *binding affinity* for specific ligands. In this section, we briefly outline several important techniques for separating proteins; these separation techniques are also useful for the separation of nucleic acids and other biomolecules. (Specialized methods for removing membrane proteins from membranes are described in Chapter 10 after the unique properties of these proteins are discussed.) We then consider the use of radioactive compounds for tracking biological activity. Finally, we consider several techniques for characterizing a protein's mass, sequence, and three-dimensional structure.

Centrifugation Can Separate Particles and Molecules That Differ in Mass or Density

The first step in a typical protein purification scheme is centrifugation. The principle behind centrifugation is that two particles in suspension (cells, cell fragments, organelles, or molecules) with different masses or densities will settle to the bottom of a tube at different rates. Remember, mass is the weight of a sample (measured in grams), whereas density is the ratio of its weight to volume (grams/liter). Proteins vary greatly in mass but not in density. Unless a protein has an attached lipid or carbohydrate, its density will not vary by more than 15 percent from 1.37 g/cm^3, the average protein density. Heavier or more dense molecules settle, or sediment, more quickly than lighter or less dense molecules.

A centrifuge speeds sedimentation by subjecting particles in suspension to centrifugal forces as great as 1,000,000 times the force of gravity *g,* which can sediment particles as small as 10 kDa. Modern ultracentrifuges achieve these forces by reaching speeds of 150,000 revolutions per minute (rpm) or greater. However, small particles with masses of 5 kDa or less will not sediment uniformly even at such high speeds.

Centrifugation is used for two basic purposes: (1) as a preparative technique to separate one type of material from others and (2) as an analytical technique to measure physical properties (e.g., molecular weight, density, shape, and equilibrium binding constants) of macromolecules. The sedimentation

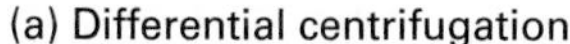

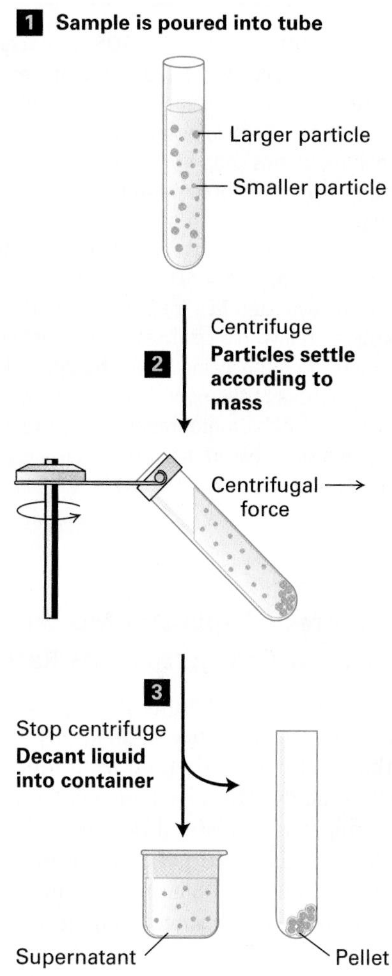

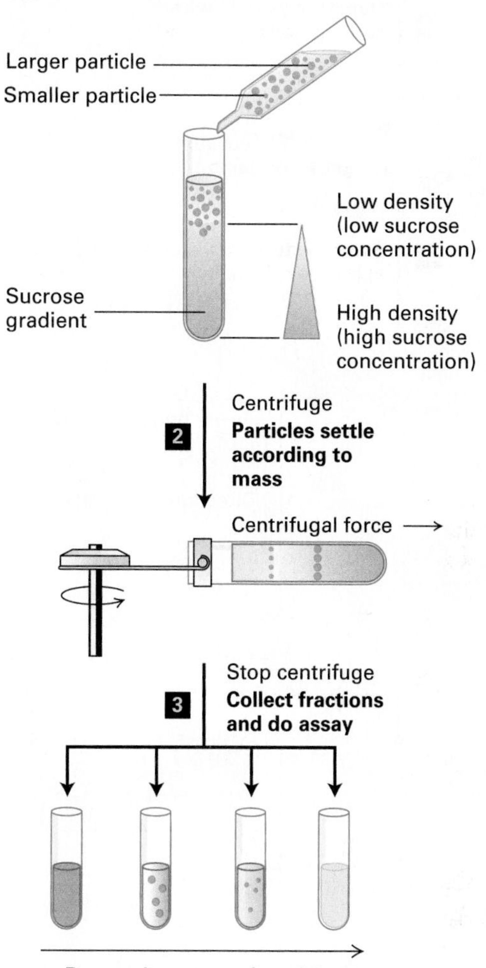

▲ EXPERIMENTAL FIGURE 3-34 **Centrifugation techniques separate particles that differ in mass or density.** (a) In differential centrifugation, a cell homogenate or other mixture is spun long enough to sediment the larger particles (e.g., cell organelles, cells), which collect as a pellet at the bottom of the tube (step 2). The smaller particles (e.g., soluble proteins, nucleic acids) remain in the liquid supernatant, which can be transferred to another tube (step 3). (b) In rate-zonal centrifugation, a mixture is spun (step 2) just long enough to separate molecules that differ in mass but may be similar in shape and density (e.g., globular proteins, RNA molecules) into discrete zones within a density gradient commonly formed by a concentrated sucrose solution. Fractions are removed from the bottom of the tube and subjected to testing (assayed).

constant s of a protein is a measure of its sedimentation rate. The sedimentation constant is commonly expressed in svedbergs (S), where a typical, large protein complex is about 3–5S, while a eukaryotic ribosome is 80S.

Differential Centrifugation The most common initial step in protein purification for cells or tissues is the separation of water-soluble proteins from insoluble cellular material by *differential centrifugation.* A starting mixture, commonly a cell homogenate (mechanically broken calls), is poured into a tube and spun at a rotor speed and for a period of time that forces cell organelles such as nuclei and large unbroken cells or large cell fragments to collect as a pellet at the bottom; the soluble proteins remain in the supernatant (Figure 3-34a). The supernatant fraction then is poured off, and either it or the pellet can be subjected to other purification methods to separate the many different proteins that they contain.

Rate-Zonal Centrifugation On the basis of differences in their masses, proteins can be separated by centrifugation through a solution of increasing density called a *density gradient.* A concentrated sucrose solution is commonly used to form density gradients. When a protein mixture is layered on top of a sucrose gradient in a tube and subjected to centrifugation, each protein in the mixture migrates down the tube at a rate controlled by the factors that affect the sedimentation constant. All the proteins start from a thin zone at the top of the tube and separate into bands, or zones (actually, disks), of proteins of

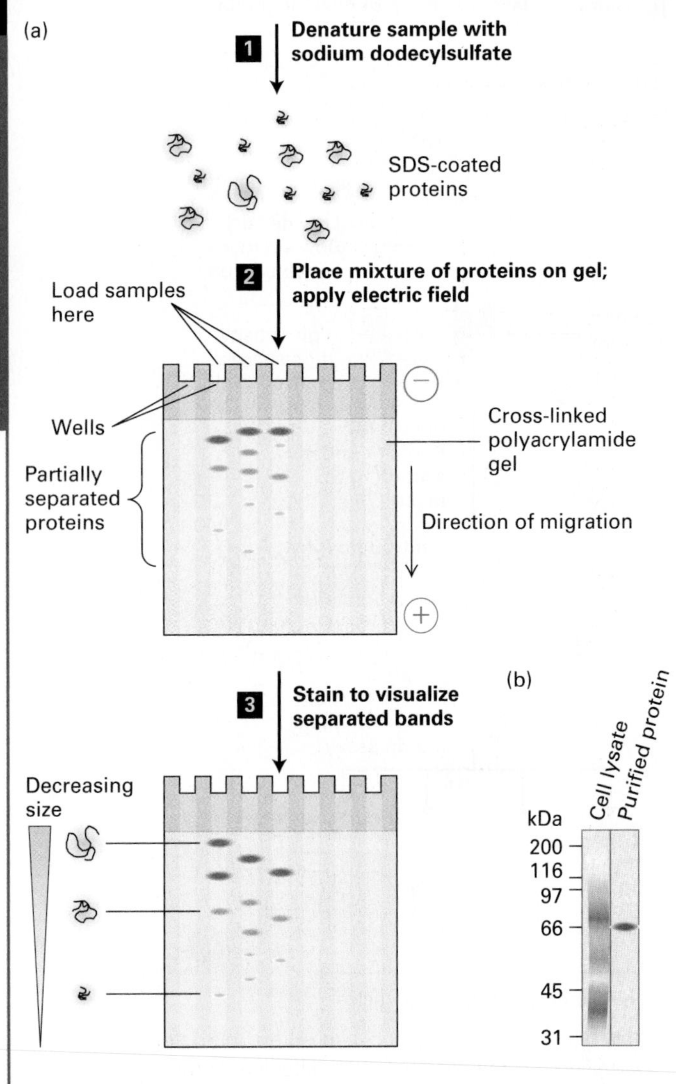

◀ EXPERIMENTAL FIGURE 3-35 **SDS-polyacrylamide gel electrophoresis (SDS-PAGE) separates proteins primarily on the basis of their masses.** (a) Initial treatment with SDS, a negatively charged detergent, dissociates multimeric proteins and denatures all the polypeptide chains (step **1**). During electrophoresis, the SDS-protein complexes migrate through the polyacrylamide gel (step **2**). Small complexes are able to move through the pores faster than larger ones. Thus the proteins separate into bands according to their sizes as they migrate. The separated protein bands are visualized by staining with a dye (step **3**). (b) Example of SDS-PAGE separation of all the proteins in a whole-cell lysate (detergent solublized cells): (*left*) the many separate stained proteins, appearing almost as a continuum; (*right*) a protein purified from the lysate by a single step of antibody-affinity chromatography. The proteins were visualized by staining with a silver-based dye. [Part (b) modified from B. Liu and M. Krieger, 2002, *J. Biol. Chem.* **277**(37):34125–34135.]

different masses. In this separation technique, called *rate-zonal centrifugation*, samples are centrifuged just long enough to separate the molecules of interest into discrete zones (Figure 3-34b). If a sample is centrifuged for too short a time, the different protein molecules will not separate sufficiently. If a sample is centrifuged much longer than necessary, all the proteins will end up in a pellet at the bottom of the tube.

Although the sedimentation rate is strongly influenced by particle mass, rate-zonal centrifugation is seldom effective in determining precise molecular weights because variations in shape also affect sedimentation rate. The exact effects of shape are hard to assess, especially for proteins or other molecules, such as single-stranded nucleic acid molecules, that can assume many complex shapes. Nevertheless, rate-zonal centrifugation has proved to be the most practical method for separating many different types of polymers and particles. A second density-gradient technique, called *equilibrium density-gradient centrifugation,* is used mainly to separate DNA, lipoproteins that carry lipids through the circulatory system, or organelles (see Figure 9-26).

Electrophoresis Separates Molecules on the Basis of Their Charge-to-Mass Ratio

Electrophoresis is a technique for separating molecules in a mixture under the influence of an applied electric field and is one of the most frequently used techniques to study proteins and nucleic acids. Dissolved molecules in an electric field move, or migrate, at a speed determined by their charge-to-mass (charge:mass) ratio. For example, if two molecules have the same mass and shape, the one with the greater net charge will move faster toward an electrode of the opposite polarity.

SDS-Polyacrylamide Gel Electrophoresis Because many proteins or nucleic acids that differ in size and shape have nearly identical charge:mass ratios, electrophoresis of these macromolecules in solution results in little or no separation of molecules of different lengths. However, successful separation of proteins and nucleic acids can be accomplished by electrophoresis in various gels (semisolid suspensions in water similar to the congealed gelatin found in desserts) rather than in a liquid solution. Electrophoretic separation of proteins is most commonly performed in polyacrylamide gels. When a mixture of proteins is placed in a gel and an electric current is applied, smaller proteins migrate faster through the gel than do larger proteins because the gel acts as a sieve, with smaller species able to maneuver more rapidly through the pores in the gel than larger species. The shape of a molecule can also influence its rate of migration (long asymmetric molecules migrate more slowly than spherical ones of the same mass).

Gels are cast between a pair of glass plates by polymerizing a solution of acrylamide monomers into polyacrylamide chains and simultaneously cross-linking the chains into a semisolid matrix. The pore size of a gel can be varied by adjusting the concentrations of polyacrylamide and the cross-linking reagent. The rate at which a protein moves through a gel is influenced by the gel's pore size and the strength of the electric field. By suitable adjustment of these parameters, proteins of widely varying sizes can be resolved (separated from one another) by polyacrylamide gel electrophoresis (PAGE).

In the most powerful technique for resolving protein mixtures, proteins are exposed to the ionic detergent SDS (sodium dodecylsulfate) before and during gel electrophoresis (Figure 3-35). SDS denatures proteins, in part because it binds to hydrophobic side chains, destabilizing the hydrophobic interactions in the core of a protein that contribute to its stable conformation. (SDS treatment is usually combined with heating in the presence of reducing agents that break disulfide bonds.) As a consequence, multimeric proteins dissociate into their subunits, and all polypeptide chains are forced into extended conformations with similar charge:mass ratios. SDS treatment thus eliminates the effect of differences in shape in native structures; therefore, chain length, which corresponds to mass, is the principal determinant of the migration rate of proteins in *SDS-polyacrylamide electrophoresis (SDS-PAGE)*. Even chains that differ in molecular weight by less than 10 percent can be resolved by this technique. Moreover, the molecular weight of a protein can be estimated by comparing the distance that it migrates through a gel with the distances that proteins of known molecular weight migrate (there is roughly a linear relationship between migration distance and the log of the molecular weight). Proteins within the gels can be extracted for further analysis (e.g., identification by methods described below).

Two-Dimensional Gel Electrophoresis Electrophoresis of all cellular proteins by SDS-PAGE can separate proteins having relatively large differences in mass but cannot readily resolve proteins having similar masses (e.g., a 41-kDa protein versus a 42-kDa protein). To separate proteins of similar masses, another physical characteristic must be exploited. Most commonly, this characteristic is electric charge, which is determined by the pH and the relative number of the protein's positively and negatively charged groups, which is in turn dependent on the pK_a's of the ionizable groups (see Chapter 2). Two unrelated proteins having similar masses are unlikely to have identical net charges because their sequences, and thus the number of acidic and basic residues, are different.

In two-dimensional electrophoresis, proteins are separated sequentially, first by their charges and then by their masses (Figure 3-36a). In the first step, a cell or tissue extract is fully denatured by high concentrations (8 M) of urea and then layered on a gel strip that contains a continuous pH gradient. SDS cannot be used, because its binding changes the charge of the protein. The gradient is formed by ampholytes, a mixture of polyanionic and polycationic molecules, that are cast into the gel, with the most acidic ampholyte at one end and the most basic ampholyte at the

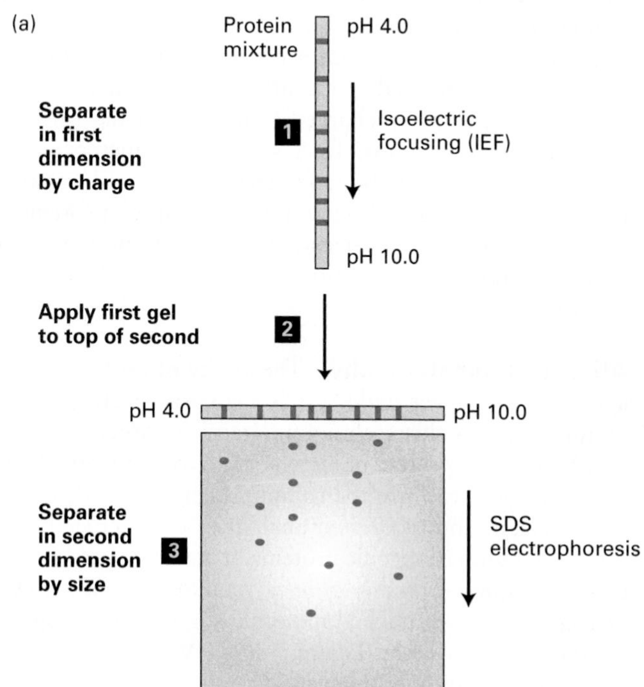

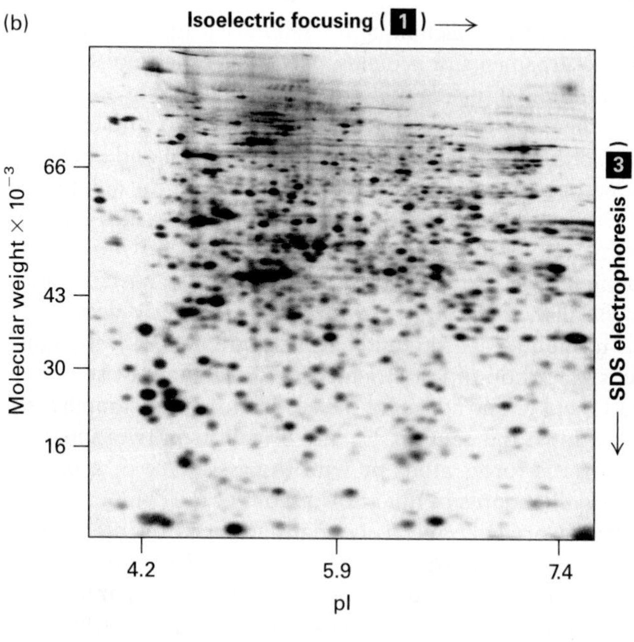

▲ EXPERIMENTAL FIGURE 3-36 **Two-dimensional gel electrophoresis separates proteins on the basis of charge and mass.** (a) In this technique, proteins are first separated into bands on the basis of their charges by isoelectric focusing (step 1). The resulting gel strip is applied to an SDS-polyacrylamide gel (step 2), and the proteins are separated into spots by mass (step 3). (b) In this two-dimensional gel of a protein extract from cultured cells, each spot represents a single polypeptide. Polypeptides can be detected by dyes, as here, or by other techniques such as autoradiography. Each polypeptide is characterized by its isoelectric point (pI) and molecular weight. [Part (b) courtesy of J. Celis.]

opposite end. A charged protein will migrate through the gradient until it reaches its **isoelectric point (pI)**, the pH at which the net charge of the protein is zero. This technique, called *isoelectric focusing (IEF)*, can resolve proteins that differ by only one charge unit. Proteins that have been separated on an IEF gel can then be separated in a second dimension on the basis of their molecular weights. To accomplish this separation, the IEF gel strip is placed lengthwise on one outside edge of a sheetlike (two-dimensional, or slab) polyacrylamide gel, this time saturated with SDS. When an electric field is imposed, the proteins will migrate from the IEF gel into the SDS slab gel and then separate according to their masses.

The sequential resolution of proteins by charge and mass can achieve excellent separation of cellular proteins (Figure 3-36b). For example, two-dimensional gels have been very useful in comparing the proteomes in undifferentiated and differentiated cells or in normal and cancer cells because as many as 1000 proteins can be resolved as individual spots simultaneously. Sophisticated methods have been developed to permit the comparison of complex patterns of proteins in two-dimensional gels from related, but distinct, samples (e.g., tissue from a normal versus a mutant individual) to permit identification of differences in the types or amounts of proteins in the samples (see section on proteomics, below).

Liquid Chromatography Resolves Proteins by Mass, Charge, or Binding Affinity

A third common technique for separating mixtures of proteins or fragments of proteins, as well as other molecules, is based on the principle that molecules dissolved in a solution can differentially interact (bind and dissociate) with a particular solid surface, depending on the physical and chemical properties of the molecule and the surface. If the solution is allowed to flow across the surface, then molecules that interact frequently with the surface will spend more time bound to the surface and thus flow past the surface more slowly than molecules that interact infrequently with it. In this technique, called **liquid chromatography** (LC), the sample is placed on top of a tightly packed column of spherical beads held within a glass or plastic cylinder. The sample then flows down the column, usually driven by gravitational or hydrostatic forces alone or with the assistance of a pump, and small aliquots of fluid flowing out of the column, called *fractions,* are collected sequentially for subsequent analysis for the presence of the proteins of interest. The nature of the beads in the column determines whether the separation of proteins depends on differences in mass, charge, or binding affinity.

Gel Filtration Chromatography Proteins that differ in mass can be separated on a column composed of porous beads made from polyacrylamide, dextran (a bacterial polysaccharide), or agarose (a seaweed derivative)—a technique called gel filtration chromatography. Although proteins flow around the spherical beads in gel filtration chromatography, they spend some time within the large depressions that cover a bead's surface. Because smaller proteins can penetrate into these depressions more readily than larger proteins can, they travel through a gel filtration column more slowly than larger proteins (Figure 3-37a). (In contrast, proteins migrate *through* the pores in an electrophoretic gel; thus smaller proteins move faster than larger ones.) The total volume of liquid required to elute (or separate and remove) a protein from a gel filtration column depends on its mass: the smaller the mass, the more time it is trapped on the beads, the greater the elution volume. By use of proteins of known mass as standards to calibrate the column, the elution volume can be used to estimate the mass of a protein in a mixture. A protein's shape as well as its mass can influence the elution volume.

Ion-Exchange Chromatography In ion-exchange chromatography, a second type of liquid chromatography, proteins are separated on the basis of differences in their charges. This technique makes use of specially modified beads whose surfaces are covered by amino groups or carboxyl groups and thus carry either a positive charge (NH_3^+) or a negative charge (COO^-) at neutral pH.

The proteins in a mixture carry various net charges at any given pH. When a solution of a protein mixture flows through a column of positively charged beads, only proteins with a net negative charge (acidic proteins) adhere to the beads; neutral and positively charged (basic) proteins flow unimpeded through the column (Figure 3-37b). The acidic proteins are then eluted selectively from the column by passing a solution of increasing concentrations of salt (a salt gradient) through the column. At low salt concentrations, protein molecules and beads are attracted by their opposite charges. At higher salt concentrations, negative salt ions bind to the positively charged beads, displacing the negatively charged proteins. In a gradient of increasing salt concentration, weakly bound proteins, those with relatively low charge, are eluted first and highly charged proteins are eluted last. Similarly, a negatively charged column can be used to retain and fractionate basic (positively charged) proteins.

Affinity Chromatography The ability of proteins to bind specifically to other molecules is the basis of affinity chromatography. In this technique, ligand or other molecules that bind to the protein of interest are covalently attached to the beads used to form the column. Ligands can be enzyme substrates, inhibitors or their analogues, or other small molecules that bind to specific proteins. In a widely used form of this technique—*antibody-affinity,* or *immunoaffinity, chromatography*—the attached molecule is an antibody specific for the desired protein (Figure 3-37c). (We discuss antibodies as tools to study proteins next).

An affinity column in principle will retain only those proteins that bind the molecule attached to the beads; the remaining proteins, regardless of their charges or masses, will pass through the column because they do not bind. However, if a retained protein is in turn bound to other molecules, forming a complex, then the entire complex is

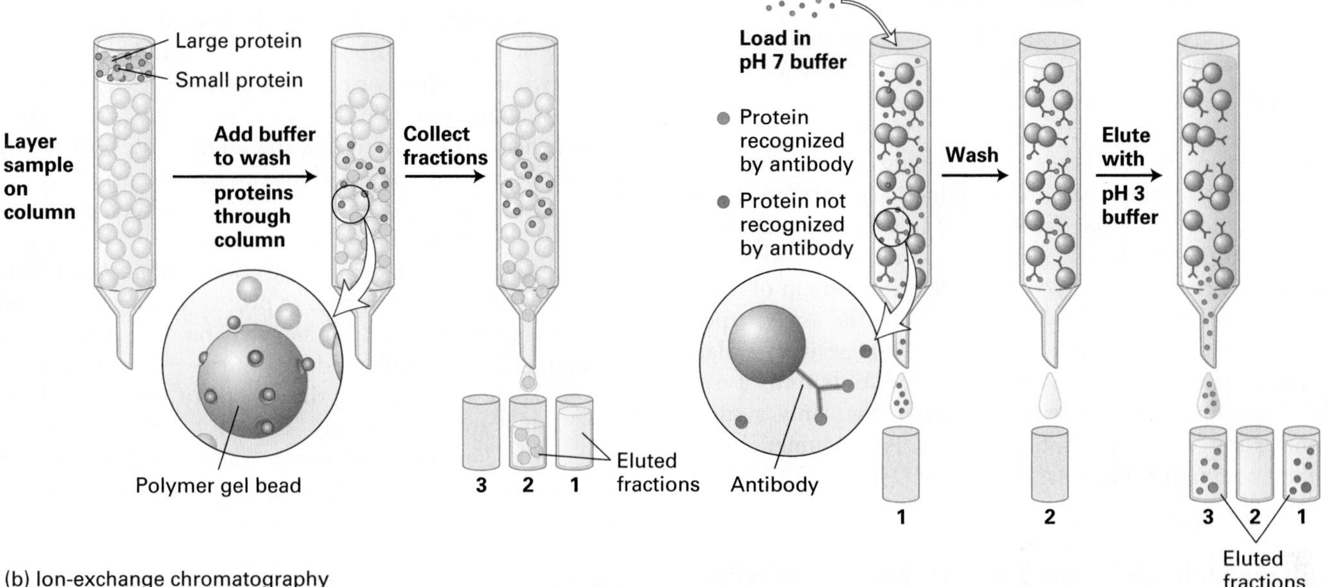

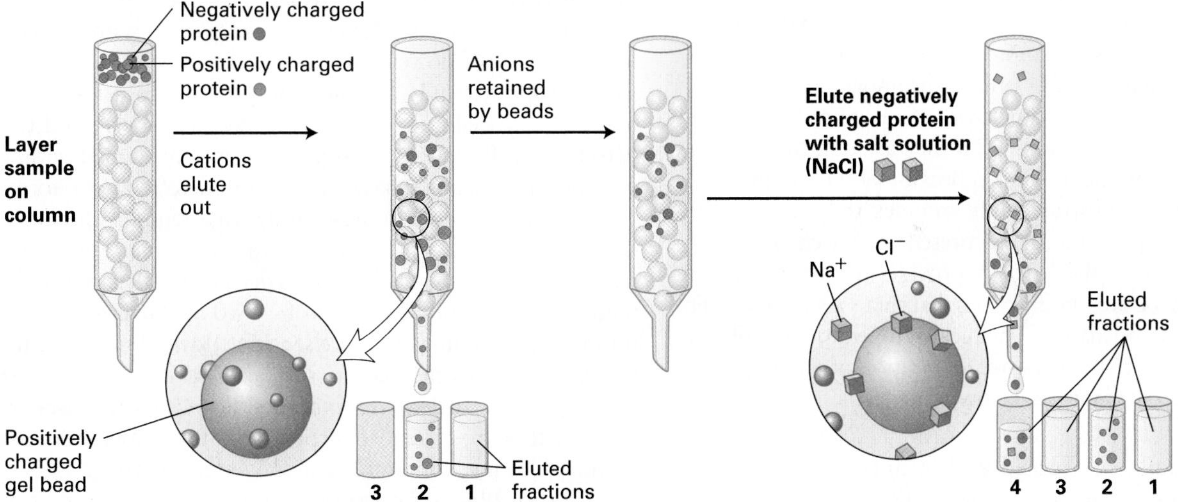

▲ EXPERIMENTAL FIGURE 3-37 **Three commonly used liquid chromatographic techniques separate proteins on the basis of mass, charge, or affinity for a specific binding partner.** (a) Gel filtration chromatography separates proteins that differ in size. A mixture of proteins is carefully layered on the top of a cylinder packed with porous beads. Smaller proteins travel through the column more slowly than larger proteins. Thus different proteins emerging in the eluate flowing out of the bottom of the column at different times (different elution volumes) can be collected in separate tubes, called *fractions.* (b) Ion-exchange chromatography separates proteins that differ in net charge in columns packed with special beads that carry either a positive charge (shown here) or a negative charge. Proteins having the same net charge as the beads are repelled and flow through the column, whereas proteins having the opposite charge bind to the beads more or less tightly, depending on their structures. Bound proteins—in this case, negatively charged—are subsequently eluted by passing a salt gradient (usually of NaCl or KCl) through the column. As the ions bind to the beads, they displace the proteins (more tightly bound proteins require higher salt concentration in order to be released). (c) In antibody-affinity chromatography, a mixture of proteins is passed through a column packed with beads to which a specific antibody is covalently attached. Only protein with high affinity for the antibody is retained by the column; all the nonbinding proteins flow through. After the column is washed, the bound protein is eluted with an acidic solution or some other solution that disrupts the antigen–antibody complexes; the released protein then flows out of the column and is collected.

retained on the column. The proteins bound to the affinity column are then eluted by adding an excess of a soluble form of the ligand or by changing the salt concentration or pH such that the binding to the molecule on the column is disrupted. The ability of this technique to separate particular proteins depends on the selection of appropriate binding partners that bind more tightly to the protein of interest than to other proteins.

Highly Specific Enzyme and Antibody Assays Can Detect Individual Proteins

The purification of a protein, or any other molecule, requires a specific assay that can detect the presence of the molecule of interest as it is separated from other molecules (e.g., in column or density-gradient fractions or gel bands or spots). An assay capitalizes on some highly distinctive characteristic of a protein: the ability to bind a particular ligand, to catalyze a particular reaction, or to be recognized by a specific antibody. An assay must also be simple and fast to minimize errors and the possibility that the protein of interest becomes denatured or degraded while the assay is performed. The goal of any purification scheme is to isolate sufficient amounts of a given protein for study; thus a useful assay must also be sensitive enough that only a small proportion of the available material is consumed by it. Many common protein assays require just 10^{-9} to 10^{-12} g of material.

Chromogenic and Light-Emitting Enzyme Reactions

Many assays are tailored to detect some functional aspect of a protein. For example, enzymatic activity assays are based on the ability to detect the loss of substrate or the formation of product. Some enzymatic assays utilize chromogenic substrates, which change color in the course of the reaction. (Some substrates are naturally chromogenic; if they are not, they can be linked to a chromogenic molecule.) Because of the specificity of an enzyme for its substrate, only samples that contain the enzyme will change color in the presence of a chromogenic substrate; the rate of the reaction provides a measure of the quantity of enzyme present. Enzymes that catalyze chromogenic reactions can also be fused or chemically linked to an antibody and used to "report" the presence or location of an antigen to which the antibody binds (see below).

Antibody Assays As noted earlier, antibodies have the distinctive characteristic of binding tightly and specifically to antigens. As a consequence, preparations of antibodies that recognize a protein antigen of interest can be generated and used to detect the presence of the protein, either in a complex mixture of other proteins (finding a needle in a haystack, as it were) or in a partially purified preparation of a particular protein. The tight binding of the antibody to its antigen, and thus the presence of the antigen, can be visualized by labeling the antibody with an enzyme, a fluorescent molecule, or radioactive isotopes. For example, luciferase, an enzyme present in fireflies and some bacteria, can be linked to an antibody. In the presence of ATP and the substrate luciferin, luciferase catalyzes a light-emitting reaction. In either case, after the antibody binds to the protein of interest (the antigen) and unbound antibody is washed away, substrates of the linked enzyme are added and the appearance of color or emitted light is monitored. The intensity is proportional to the amount of enzyme-linked antibody, and thus antigen, in the sample. A variation of this technique, particularly useful in detecting specific proteins within living cells, makes use of *green fluorescent protein (GFP),* a naturally fluorescent protein found in jellyfish (see Figure 9-12). Alternatively, after the first antibody binds to its target protein, a second, labeled antibody is used to bind to the complex of the first antibody and its target. This combination of two antibodies permits very high sensitivity in the detection of a target protein.

To generate the antibodies, the intact protein or a fragment of the protein is injected into an animal (usually a rabbit, mouse, or goat). Sometimes a short synthetic peptide of 10–15 residues based on the sequence of the protein is used as the antigen to induce antibody formation. A synthetic peptide, when coupled to a large protein carrier, can induce an animal to produce antibodies that bind to that portion (the epitope) of the full-sized, natural protein. Biosynthetically or chemically attaching the epitope to an unrelated protein is called *epitope tagging.* As we'll see throughout this book, antibodies generated using either peptide epitopes or intact proteins are extremely versatile reagents for isolating, detecting, and characterizing proteins.

Detecting Proteins in Gels Proteins embedded within a one- or two-dimensional gel usually are not visible. The two general approaches for detecting proteins in gels are either to label or stain the proteins while they are still within the gel or to electrophoretically transfer the proteins to a membrane made of nitrocellulose or polyvinylidene difluoride and then detect them. Proteins within gels are usually stained with an organic dye or a silver-based stain, both detected with normal visible light, or with a fluorescent dye that requires specialized detection equipment. Coomassie blue is the most commonly used organic dye, typically used to detect ≈1000 ng of protein, with a lower limit of detection of ≈4–10 ng. Silver staining or fluorescence staining are more sensitive (lower limit of ≈1 ng). Coomassie and other stains can also be used to visualize proteins after transfer to membranes; however, the most common method to visualize proteins in these membranes is immunoblotting, commonly called Western blotting.

Western blotting combines the resolving power of gel electrophoresis and the specificity of antibodies. This multistep procedure is commonly used to separate proteins and then identify a specific protein of interest. As shown in Figure 3-38, two different antibodies are used in this method, one that is specific for the desired protein and a second that binds to the first and is linked to an enzyme or other molecule that permits detection of the first antibody (and thus the protein of interest to which it binds). Enzymes to which the second antibody is attached can either generate a visible colored product or, by a process called *chemiluminescence,* produce light that can readily be recorded by film or a sensitive detector. The two different antibodies, sometimes called a "sandwich," are used to amplify the signals and improve sensitivity. If an antibody is not available, but the gene encoding the protein is available and can be used to express the protein, recombinant DNA methods (Chapter 5) can incorporate a small peptide epitope (epitope tagging) into the normal sequence of the protein that can be detected by a commercially available antibody to that epitope.

Technique Animation: Immunoblotting

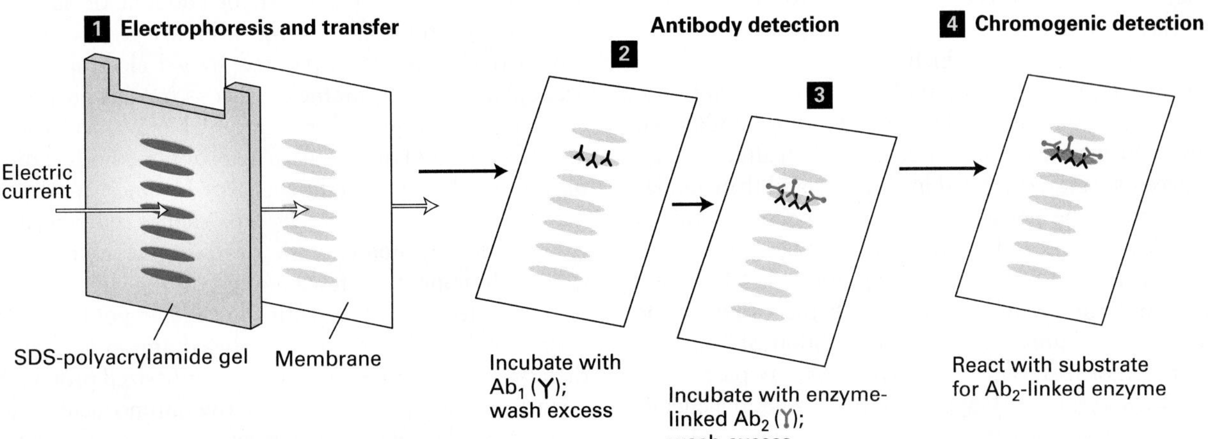

▲ EXPERIMENTAL FIGURE 3-38 **Western blotting (immunoblotting) combines several techniques to resolve and detect a specific protein.** Step **1**: After a protein mixture has been electrophoresed through an SDS gel, the separated bands (or spots, for a two-dimensional gel) are transferred (blotted) from the gel onto a porous membrane from which it is not readily removed. Step **2**: The membrane is flooded with a solution of antibody (Ab_1) specific for the desired protein and allowed to incubate for a while. Only the membrane-bound band containing this protein binds the antibody, forming a layer of antibody molecules (whose position cannot be seen at this point). Then the membrane is washed to remove unbound Ab_1. Step **3**: The membrane is incubated with a second antibody (Ab_2) that specifically recognizes and binds to the first Ab_1. This second antibody is covalently linked to either an enzyme (e.g., alkaline phosphatase, which can catalyze a chromogenic reaction), radioactive isotope, or some other substance whose presence can be detected with great sensitivity. Step **4**: Finally, the location and amount of bound Ab_2 are detected (e.g., by a deep-purple precipitate from chromogenic reaction), permitting the electrophoretic mobility (and therefore the mass) of the desired protein to be determined, as well as its quantity (based on band intensity).

Radioisotopes Are Indispensable Tools for Detecting Biological Molecules

A sensitive method for tracking a protein or other biological molecule is by detecting the radioactivity emitted from radioisotopes introduced into the molecule. At least one atom in a radiolabeled molecule is present in a radioactive form, called a **radioisotope.**

TABLE 3-1 Radioisotopes Commonly Used in Biological Research

ISOTOPE	HALF-LIFE
Phosphorus-32	14.3 days
Iodine-125	60.4 days
Sulfur-35	87.5 days
Tritium (hydrogen-3)	12.4 years
Carbon-14	5730.4 years

Radioisotopes Useful in Biological Research Hundreds of biological compounds (e.g., amino acids, nucleosides, and numerous metabolic intermediates) labeled with various radioisotopes are commercially available. These preparations vary considerably in their *specific activity,* which is the amount of radioactivity per unit of material, measured in disintegrations per minute (dpm) per millimole. The specific activity of a labeled compound depends on the probability of decay of the radioisotope, determined by its *half-life,* which is the time required for half the atoms to undergo radioactive decay. In general, the shorter the half-life of a radioisotope, the higher its specific activity (Table 3-1).

The specific activity of a labeled compound must be high enough that sufficient radioactivity is incorporated into molecules to be accurately detected. For example, methionine and cysteine labeled with sulfur-35 (^{35}S) are widely used to biosynthetically label cellular proteins because preparations of these amino acids with high specific activities ($>10^{15}$ dpm/mmol) are available. Likewise, commercial preparations of ^{3}H-labeled nucleic acid precursors have much higher specific activities than those of the corresponding ^{14}C-labeled preparations. In most experiments, the former are preferable because they allow RNA or DNA to be adequately labeled after a shorter time of incorporation or require a smaller cell sample. Various phosphate-containing compounds in which every phosphorus atom is the radioisotope phosphorus-32 are readily available. Because of their high specific activity, ^{32}P-labeled nucleotides are routinely used to label nucleic acids in cell-free systems.

Labeled compounds in which a radioisotope replaces atoms normally present in the molecule have virtually the same chemical properties as the corresponding nonlabeled compounds. Enzymes, for instance, generally cannot distinguish between substrates labeled in this way and their nonlabeled substrates. The presence of such radioactive atoms is indicated with the isotope in brackets (no hyphen) as a prefix (e.g., [^{3}H]leucine). In contrast, labeling almost all biomolecules (e.g., protein or nucleic acid) with the radioisotope iodine-125 (^{125}I) requires the covalent addition of ^{125}I to a molecule that normally does not have iodine as part of its structure. Because this labeling procedure modifies the chemical structure, the biological activity of the labeled molecule may differ somewhat from that of the nonlabeled form. The presence of such radioactive atoms is indicated with the isotope as a prefix with a hyphen (no bracket) (e.g., ^{125}I-trypsin). Standard methods for labeling proteins with ^{125}I result in covalent attachment of the ^{125}I primarily to the aromatic rings of tyrosine side chains (mono- and diiodotyrosine).

Labeling Experiments and Detection of Radiolabeled Molecules Whether labeled compounds are detected by **autoradiography,** a semiquantitative visual assay, or their radioactivity is measured in an appropriate "counter," a highly quantitative assay that can determine the amount of a radiolabeled compound in a sample, depends on the nature of the experiment. In some experiments, both types of detection are used.

In one use of autoradiography, a tissue, cell, or cell constituent is labeled with a radioactive compound, unassociated radioactive material is washed away, and the structure of the sample is stabilized either by chemically cross-linking the macromolecules ('fixation') or by freezing it. The sample is then overlaid with a photographic emulsion sensitive to radiation. Development of the emulsion yields small silver grains whose distribution corresponds to that of the radioactive material and is usually detected by microscopy. Autoradiographic studies of whole cells were crucial in determining the intracellular sites where various macromolecules are synthesized and the subsequent movements of these macromolecules within cells. Various techniques employing fluorescent microscopy, which we describe in Chapter 9, have largely supplanted autoradiography for studies of this type. However, autoradiography is commonly used in various assays for detecting specific isolated DNA or RNA sequences (Chapter 5).

Quantitative measurements of the amount of radioactivity in a labeled material are performed with several different instruments. A Geiger counter measures ions produced in a gas by the β particles or γ rays emitted from a radioisotope. In a scintillation counter, a radiolabeled sample is mixed with a liquid containing a fluorescent compound that emits a flash of light when it absorbs the energy of the β particles or γ rays released in the decay of the radioisotope; a phototube in the instrument detects and counts these light flashes. Phosphorimagers are used to detect radioactivity using a two-dimensional array detector, storing digital data on the number of decays in disintegrations per minute per small pixel of surface area. These instruments, which can be thought of as a kind of reusable electronic film, are commonly used to quantitate radioactive molecules separated by gel electrophoresis and are replacing photographic emulsions for this purpose.

A combination of labeling and biochemical techniques and of visual and quantitative detection methods is often employed in labeling experiments. For instance, to identify the major proteins synthesized by a particular cell type, a sample of the cells is incubated with a radioactive amino acid (e.g., [^{35}S]methionine) for a few minutes, during which time the labeled amino acid mixes with the cellular pool of unlabeled amino acids and some of the labeled amino acid is biosynthetically incorporated into newly synthesized protein. Subsequently unincorporated radioactive amino acid is washed away from the cells. The cells are harvested, the mixture of cellular proteins is extracted from the cells (for example, by a detergent solution), and then separated by gel electrophoresis; and the gel is subjected to autoradiography or phosphorimager analysis. The radioactive bands correspond to newly synthesized proteins, which have incorporated the radiolabeled amino acid. Alternatively, the proteins can be resolved by liquid chromatography, and the radioactivity in the eluted fractions can be determined quantitatively with a counter. To detect only one specific protein, rather than all the proteins biosynthetically labeled this way, a specific antibody to the protein of interest can be used to precipitate the protein away from the other proteins in the sample (immunoprecipitation). The precipitate is then solubilized in a detergent, separating the antibody from the protein, and the sample is subjected to SDS-PAGE followed by autoradiography. In this type of experiment, a fluorescent compound that is activated by the radiation is often infused into the gel so that the light emitted can be used to detect the presence of the labeled protein, either using film or a two-dimensional electronic detector.

Pulse-chase experiments are particularly useful for tracing changes in the intracellular location of proteins or the modification of a protein or metabolite over time. In this experimental protocol, a cell sample is exposed to a radiolabeled compound that can be incorporated or otherwise attached to a cellular molecule of interest—the "pulse"—for a brief period of time, then washed with buffer to remove the unincorporated label, and finally incubated with an unlabeled form of the compound—the "chase" (Figure 3-39). Samples taken periodically during the chase period are assayed to determine the location or chemical form of the radiolabel as a function of time. Often, pulse-chase experiments, in which the protein is detected by autoradiography after immunoprecipitation and SDS-PAGE, are used to follow the rate of synthesis, modification, and degradation of proteins by adding radioactive amino acid precursors during the pulse and then detecting the amounts and characteristics of the radioactive protein during the chase. One can thus observe postsynthetic modifications of the protein that change its electrophoretic mobility and the rate of degradation of a specific protein. A classic use of the pulse-chase technique was in studies to elucidate the pathway traversed by secreted proteins from their site of synthesis in the endoplasmic reticulum to the cell surface (Chapter 14).

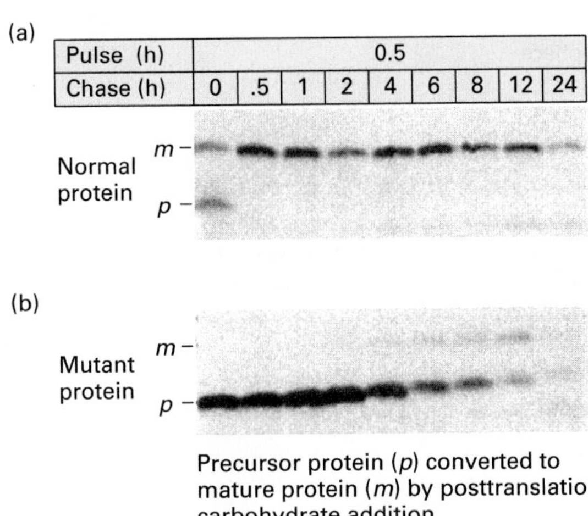

▲ EXPERIMENTAL FIGURE 3-39 **Pulse-chase experiments can track the pathway of protein modification or movement within cells.** (a) To follow the fate of a specific newly synthesized protein in a cell, cells were incubated with [^{35}S]methionine for 0.5 hr (the pulse) to label all newly synthesized proteins, and the radioactive amino acid not incorporated into the cells was then washed away. The cells were further incubated (the chase) for varying times up to 24 hours, and samples from each time of chase were subjected to immunoprecipitation to isolate one specific protein (here, the low-density lipoprotein receptor). SDS-PAGE of the immunoprecipitates followed by autoradiography permitted visualization of the one specific protein, which is initially synthesized as a small precursor (p) and then rapidly modified to a larger mature form (m) by addition of carbohydrates. About half of the labeled protein was converted from p to m during the pulse, the rest was converted after 0.5 hour of chase. The protein remains stable for 6–8 hours before it begins to be degraded (indicated by reduced band intensity). (b) The same experiment was performed in cells in which a mutant form of the protein is made. The mutant p form cannot be properly converted to the m form, and it is more quickly degraded than the normal protein. [Adapted from K. F. Kozarsky, H. A. Brush, and M. Krieger, 1986, J. Cell *Biol.* **102**(5):1567–1575.]

Mass Spectrometry Can Determine the Mass and Sequence of Proteins

Mass spectrometry (MS) is a powerful technique for characterizing proteins. MS is particularly useful in determining the mass of a protein or fragments of a protein. With such information in hand, it is also possible to determine part of or all the protein's sequence. This method permits the very highly accurate direct determination of the ratio of the mass (m) of a charged molecule (ion) to its charge (z), or m/z. Techniques are then used to deduce the absolute mass of the ion. There are four key features of all mass spectrometers. The first is an ion source, from which charge, usually in the form of protons, is transferred to the peptide or protein molecules. The formation of these ions occurs in the presence of a high electric field that then directs the charged molecular ions into the second key component, the mass analyzer. The mass analyzer, which is always in a high vacuum chamber, physically separates the ions on the basis of their differing mass-to-charge (m/z) ratios. The mass-separated ions are subsequently directed to strike a detector, the third key component, which provides a measure of the relative abundances of each of the ions in the sample. The fourth essential component is a computerized data system that is used to calibrate the instrument; acquire, store, and process the resulting data; and often direct the instrument automatically to collect additional specific types of data from the sample, based on the initial observations. This type of automated feedback is used for the tandem MS (MS/MS) peptide-sequencing methods described below.

The two most frequently used methods of generating ions of proteins and protein fragments are (1) matrix-assisted laser desorption/ionization (MALDI) and (2) electrospray (ES). In MALDI (Figure 3-40) the peptide or protein sample is mixed with a low-molecular-weight, UV-absorbing organic acid (the matrix) and then dried on a metal target. Energy from a laser ionizes and vaporizes the sample producing singly charged molecular ions from the constituent molecules. In ES (Figure 3-41a), the sample of peptides or proteins in solution is converted into a fine mist of tiny droplets by spraying through a narrow capillary at atmospheric pressure. The droplets are formed in the presence of a high electric field, rendering them highly charged. The droplets evaporate in their short flight (mm) to the entrance of the mass spectrometer's analyzer, forming multiply charged ions from the peptides and proteins. The gaseous ions are sampled into the analyzer region of the MS, where they are then accelerated by electric fields and separated by the mass analyzer on the basis of their m/z.

The two most frequently used mass analyzers are time-of-flight (TOF) instruments and ion traps. TOF instruments exploit the fact that the time it takes an ion to pass through the length of the analyzer before reaching the detector is

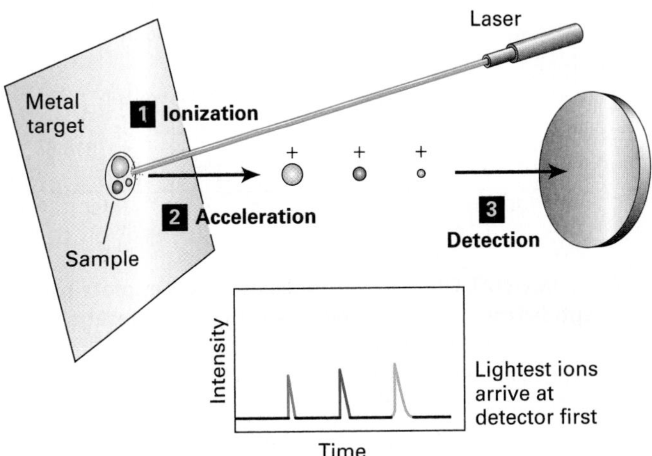

▲ EXPERIMENTAL FIGURE 3-40 **Molecular mass can be determined by matrix-assisted laser desorption/ionization time-of-flight (MALDI-TOF) mass spectrometry.** In a MALDI-TOF mass spectrometer, pulses of light from a laser ionize a protein or peptide mixture that is absorbed on a metal target (step 1). An electric field accelerates the ions in the sample toward the detector (steps 2 and 3). The time to the detector is proportional to the square root of the mass-to-charge (m/z) ratio. For ions having the same charge, the smaller ions move faster (shorter time to the detector). The molecular weight is calculated using the time of flight of a standard.

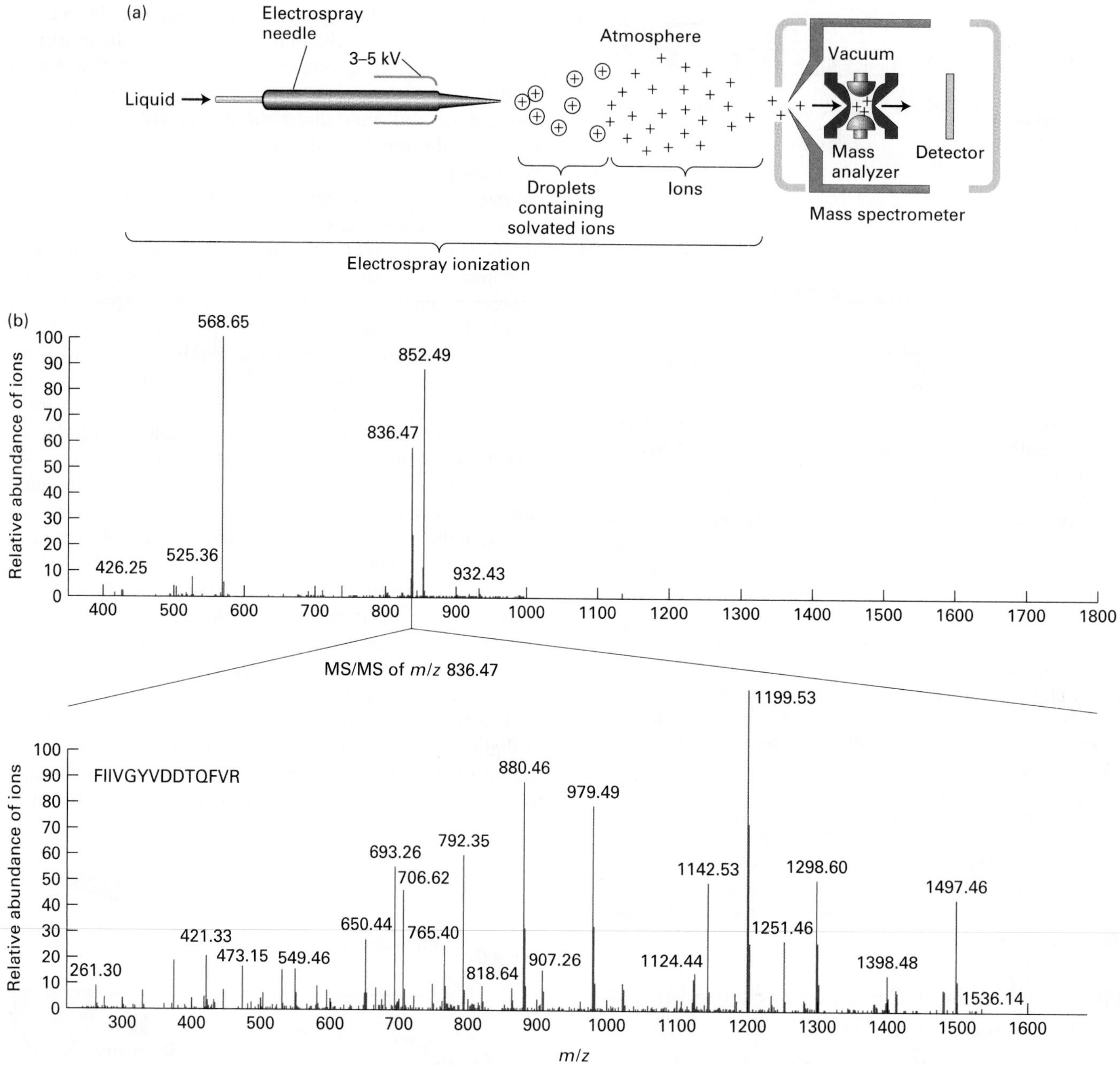

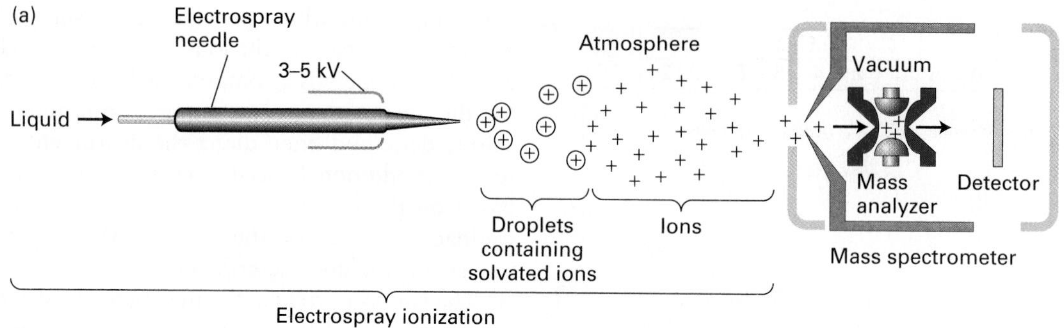

(b)

Relative abundance of ions

100 90 80 70 60 50 40 30 20 10 0

568.65 852.49 836.47 525.36 426.25 932.43

400 500 600 700 800 900 1000 1100 1200 1300 1400 1500 1600 1700 1800

MS/MS of *m/z* 836.47

Relative abundance of ions

100 90 80 70 60 50 40 30 20 10 0

FIIVGYVDDTQFVR

1199.53 880.46 979.49 792.35 693.26 706.62 1142.53 1298.60 1497.46 650.44 765.40 1251.46 421.33 473.15 549.46 907.26 1124.44 1398.48 261.30 818.64 1536.14

300 400 500 600 700 800 900 1000 1100 1200 1300 1400 1500 1600

m/z

▲ EXPERIMENTAL FIGURE 3-41 **Molecular mass of proteins and peptides can be determined by electrospray ionization ion-trap mass spectrometry.** (a) Electrospray (ES) ionization converts proteins and peptides in a solution into highly charged gaseous ions by passing the solution through a needle (forming the droplets) that has a high voltage across it (charging the droplets). Evaporation of the solvent produces gaseous ions that enter a mass spectrometer. The ions are analyzed by an ion-trap mass analyzer that then directs ions to the detector. (b) *Top panel:* Mass spectrum of a mixture of three major and several minor peptides is presented as the relative abundance of the ions striking the detector (y axis) as a function of the mass-to-charge (*m/z*) ratio (x axis). *Bottom panel:* In an MS/MS instrument (such as the ion trap shown in part (a), a specific peptide ion can be selected for fragmentation into smaller ions that are then analyzed and detected. The MS/MS spectrum (also called the product-ion spectrum) provides detailed structural information about the parent ion, including sequence information for peptides. Here, the ion with a *m/z* of 836.47 was selected, fragmented and the *m/z* mass spectrum of the product ions was measured. Note there is no longer an ion with an *m/z* of 836.47 present because it was fragmented. From the varying sizes of the product ions, the understanding that peptide bonds are often broken in such experiments, the known *m/z* values for individual amino acid fragments, and database information, the sequence of the peptide, FIIVGYVDDTQFVR, can be deduced. [Part (a) based on a figure from S. Carr; part (b), unpublished data from S. Carr.]

proportional to the square root of m/z (smaller ions move faster than larger ones with the same charge, see Figure 3-40). In ion-trap analyzers, tunable electric fields are used to capture, or 'trap,' ions with a specific m/z and to sequentially pass the trapped ions out of the analyzer onto the detector (see Figure 3-41a). By varying the electric fields, ions with a wide range of m/z values can be examined one by one, producing a mass spectrum, which is a graph of m/z (x axis) versus relative abundance (y axis) (Figure 3-41b, *top panel*).

In tandem, or MS/MS, instruments, any given parent ion in the original mass spectrum (Figure 3-41b, *top panel*) can be mass-selected, broken into smaller ions by collision with an inert gas, and then the m/z and relative abundances of the resulting fragment ions measured (Figure 3-41b, *bottom panel*), all within the same machine in about 0.1 s per selected parent ion. This second round of fragmentation and analysis permits the sequences of short peptides (<25 amino acids) to be determined, because collisional fragmentation occurs primarily at peptide bonds, so the differences in masses between ions correspond to the in-chain masses of the individual amino acids, permitting deduction of the sequence in conjunction with database sequence information (Figure 3-41b, *bottom panel*).

Mass spectrometry is highly sensitive, able to detect as little as 1×10^{-16} mol (100 attomoles) of a peptide or 10×10^{-15} mol (10 femtomoles) of a protein of 200,000 MW. Errors in mass measurement accuracy are dependent upon the specific mass analyzer used, but are typically ≈0.01 percent for peptides and 0.05–0.1 percent for proteins. As described in the proteomics section that follows, it is possible to use MS to analyze complex mixtures of proteins as well as purified proteins. Most commonly, protein samples are digested by proteases, and the peptide digestion products are subjected to analysis. An especially powerfully application of MS is to take a complex mixture of proteins from a biological specimen, digest it with trypsin or other proteases, partially separate the components using liquid chromatography (LC), and then transfer the solution flowing out of the chromatographic column directly into an ES tandem mass spectrometer. This technique, called *LC-MS/MS,* permits the nearly continuous analysis of a very complex mixture of proteins.

The abundances of ions determined by mass spectrometry in any given sample are relative, not absolute, values. Therefore, if one wants to use MS to compare the amounts of a particular protein in two different samples (e.g., from a normal versus a mutant organism), it is necessary to have an internal standard in the samples whose amounts do not differ between the two samples. One then determines the amounts of the protein of interest relative to that of the standard in each sample. This permits quantitatively accurate intersample comparisons of protein levels.

Protein Primary Structure Can Be Determined by Chemical Methods and from Gene Sequences

The classic method for determining the amino acid sequence of a protein is Edman degradation. In this procedure, the free amino group of the N-terminal amino acid of a polypeptide is labeled, and the labeled amino acid is then cleaved from the polypeptide and identified by high-pressure liquid chromatography. The polypeptide is left one residue shorter, with a new amino acid at the N-terminus. The cycle is repeated on the ever shortening polypeptide until all the residues have been identified.

Before about 1985, biologists commonly used the Edman chemical procedure for determining protein sequences. Now, however, complete protein sequences usually are determined primarily by analysis of genome sequences. The complete genomes of several organisms have already been sequenced, and the database of genome sequences from humans and numerous model organisms is expanding rapidly. As discussed in Chapter 5, the sequences of proteins can be deduced from DNA sequences that are predicted to encode proteins.

A powerful approach for determining the primary structure of an isolated protein combines MS and the use of sequence databases. First, the peptide "mass fingerprint" of the protein is obtained by MS. A *peptide mass fingerprint* is the list of the molecular weights of peptides that are generated from the protein by digestion with a specific protease, such as trypsin. The molecular weights of the parent protein and its proteolytic fragments are then used to search genome databases for any similarly sized protein with identical or similar peptide mass maps. Mass spectrometry can also be used to directly sequence peptides using MS/MS, as described above.

Protein Conformation Is Determined by Sophisticated Physical Methods

In this chapter, we have emphasized that protein function is dependent on protein structure. Thus, to figure out how a protein works, its three-dimensional structure must be known. Determining a protein's conformation requires sophisticated physical methods and complex analyses of the experimental data. We briefly describe three methods used to generate three-dimensional models of proteins.

X-ray Crystallography The use of **x-ray crystallography** to determine the three-dimensional structures of proteins was pioneered by Max Perutz and John Kendrew in the 1950s. In this technique, beams of x-rays are passed through a protein crystal in which millions of protein molecules are precisely aligned with one another in a rigid crystalline array. The wavelengths of x-rays are about 0.1–0.2 nm, short enough to determine the positions of individual atoms in the protein. The electrons in the atoms of the crystal scatter the x-rays, which produce a diffraction pattern of discrete spots when they are intercepted by photographic film or an electronic detector (Figure 3-42). Such patterns are extremely complex—composed of as many as 25,000 diffraction spots, or reflections, whose measured intensities vary depending on the distribution of the electrons, which is, in turn, determined by the atomic structure and three-dimensional conformation of the protein. Elaborate calculations and modifications of the protein (such as the binding of heavy metals) must be made to interpret the diffraction pattern and calculate the distribution of electrons (called the *electron density map*). With the three-dimensional electron density map in hand, one then "fits" a molecular model of the protein to match the electron density,

(a)

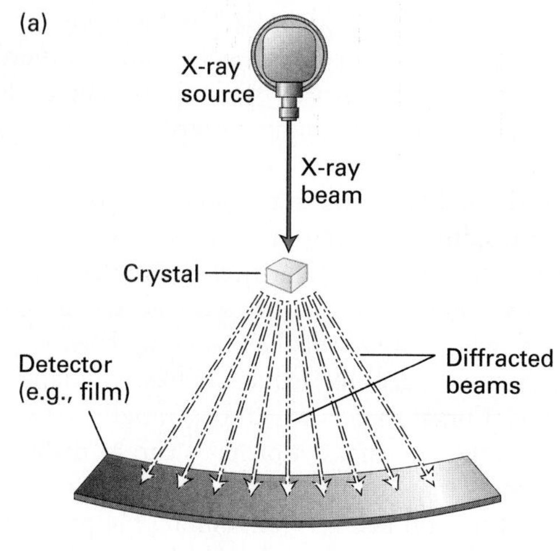

(b)

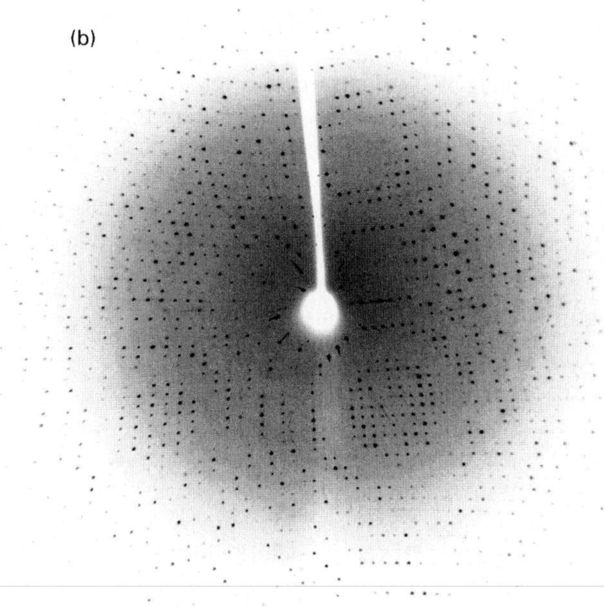

▲ **EXPERIMENTAL FIGURE 3-42 X-ray crystallography provides diffraction data from which the three-dimensional structure of a protein can be determined.** (a) Basic components of an x-ray crystallographic determination. When a narrow beam of x-rays strikes a crystal, part of it passes straight through and the rest is scattered (diffracted) in various directions. The intensity of the diffracted waves, which form periodic arrangements of diffraction spots, is recorded on an x-ray film or with a solid-state electronic detector. (b) X-ray diffraction pattern for a protein crystal collected on a solid-state detector. From complex analyses of patterns of spots like this one, the location of the atoms in a protein can be determined. [Part (a) adapted from L. Stryer, 1995, *Biochemistry,* 4th ed., W. H. Freeman and Company, p. 64; part (b) courtesy of J. Berger.]

and it is these models that one sees in the various diagrams of proteins throughout this book (e.g., Figure 3-8). The process is analogous to reconstructing the precise shape of a rock from the ripples that it creates when thrown into a pond. Although sometimes the structures of portions of the protein cannot be clearly defined, using x-ray crystallography researchers are systematically determining the structures of representative types of most proteins. To date, the detailed three-dimensional structures of more than 10,000 proteins have been established.

Cryoelectron Microscopy Although some proteins readily crystallize, obtaining crystals of others—particularly large multisubunit proteins and membrane-associated proteins—requires a time-consuming trial-and-error effort to find just the right conditions, if they can be found at all. (Growing crystals suitable for structural studies is as much an art as a science.) There are several ways to determine the structures of such difficult-to-crystallize proteins. One is cryoelectron microscopy. In this technique, a protein sample is rapidly frozen in liquid helium to preserve its structure and then examined in the frozen, hydrated state in a cryoelectron microscope. Pictures of the protein are taken at various angles and recorded on film using a low dose of electrons to prevent radiation-induced damage to the structure. Sophisticated computer programs analyze the images and reconstruct the protein's structure in three dimensions. Recent advances in cryoelectron microscopy permit researchers to generate molecular models that can help provide insight into how the protein functions. The use of cryoelectron microscopy and other types of electron microscopy for visualizing cell structures is discussed in Chapter 9.

NMR Spectroscopy The three-dimensional structures of small proteins containing as many as 200 amino acids can be studied with nuclear magnetic resonance (NMR) spectroscopy. In this technique, a concentrated protein solution is placed in a magnetic field, and the effects of different radio frequencies on the nuclear spin states of different atoms are measured. The spin state of any atom is influenced by neighboring atoms in adjacent residues, with closely spaced residues having a greater influence than distant residues. From the magnitude of the effect, the distances between residues can be calculated by a triangulation-like process; these distances are then used to generate a model of the three-dimensional structure of the protein.

Although NMR does not require the crystallization of a protein, a definite advantage, this technique is limited to proteins smaller than about 20 kDa. However, NMR analysis can also be applied to isolated protein domains, which can often be obtained as stable structures and tend to be small enough for this technique.

KEY CONCEPTS OF SECTION 3.6

Purifying, Detecting, and Characterizing Proteins

- Proteins can be separated from other cell components and from one another on the basis of differences in their physical and chemical properties.
- Various assays are used to detect and quantify proteins. Some assays use a light-producing reaction to generate a readily detected signal. Other assays produce an amplified colored signal with enzymes and chromogenic substrates.
- Centrifugation separates proteins on the basis of their rates of sedimentation, which are influenced by their masses and shapes.

- Electrophoresis separates proteins on the basis of their rates of movement in an applied electric field. SDS-polyacrylamide gel electrophoresis (PAGE) can resolve polypeptide chains differing in molecular weight by 10 percent or less (see Figure 3-35). Two-dimensional gel electrophoresis provides additional resolution by separating proteins first by charge (first dimension) and then by mass (second dimension).
- Liquid chromatography separates proteins on the basis of their rates of movement through a column packed with spherical beads. Proteins differing in mass are resolved on gel filtration columns; those differing in charge, on ion-exchange columns; and those differing in ligand-binding properties, on affinity columns, including antibody-based affinity chromatography (see Figure 3-37).
- Antibodies are powerful reagents used to detect, quantify, and isolate proteins.
- Immunoblotting, also called *Western blotting,* is a frequently used method to study specific proteins that exploits the high specificity and sensitivity of protein detection by antibodies and the high-resolution separation of proteins by SDS-PAGE.
- Radioisotopes play a key role in the study of proteins and other biomolecules. They can be incorporated into molecules without changing the chemical composition of the molecule, or as add-on tags. They can be used to help detect the synthesis, location, processing, and stability of proteins.
- Autoradiography is a semiquantitative technique for detecting radioactively labeled molecules in cells, tissues, or electrophoretic gels.
- Pulse-chase labeling can determine the intracellular fate of proteins and other metabolites (see Figure 3-39).
- Mass spectrometry is a very sensitive and highly precise method of detecting, identifying, and characterizing proteins and peptides.
- Three-dimensional structures of proteins are obtained by x-ray crystallography, cryoelectron microscopy, and NMR spectroscopy. X-ray crystallography provides the most detailed structures but requires protein crystallization. Cryoelectron microscopy is most useful for large protein complexes, which are difficult to crystallize. Only relatively small proteins are amenable to NMR analysis.

3.7 Proteomics

For most of the twentieth century, the study of proteins was restricted primarily to the analysis of individual proteins. For example, one would study an enzyme by determining its enzymatic activity (substrates, products, rate of reaction, requirement for cofactors, pH, etc.), its structure, and its mechanism of action. In some cases, the relationships between a few enzymes that participate in a metabolic pathway might also be studied. On a broader scale, the localization and activity of an enzyme would be examined in the context of a cell or tissue. The effects of mutations, diseases, or drugs on the expression and activity of the enzyme might also be the subject of investigation. This multipronged approach provided deep insight into the function and mechanisms of action of individual proteins or relatively small numbers of interacting proteins. However, this one-by-one approach to studying proteins does not efficiently provide insights into a global picture of what is happening in the proteome of a cell, tissue, or entire organism.

Proteomics Is the Study of All or a Large Subset of Proteins in a Biological System

The advent of genomics (sequencing of genomic DNA and its associated technologies, such as simultaneous analysis of the levels of all mRNAs in cells and tissues) clearly showed that a global, or systems, approach to biology could provide unique and highly valuable insights. Many scientists recognized that a global analysis of the proteins in biological systems had the potential for equally valuable contributions to our understanding. Thus, a new field was born—**proteomics.** Proteomics is the systematic study of the amounts, modifications, interactions, localization, and functions of all or subsets of proteins at the whole-organism, tissue, cellular, and subcellular levels.

A number of broad questions are addressed in proteomic studies:

- In a given sample (whole organism, tissue, cell, subcellular compartment), what fraction of the whole proteome is expressed (i.e., which proteins are present)?
- Of those proteins present in the sample, what are their relative abundances?
- What are the relative amounts of the different splice forms and chemically modified forms (e.g., phosphorylated, methylated, fatty acylated) of the proteins?
- Which proteins are present in large multiprotein complexes, and which proteins are in each complex? What are the functions of these complexes and how do they interact?
- When the state (e.g., growth rate, stage of cell cycle, differentiation, stress level) of a cell changes, do the proteins in the cell or secreted from the cell change in a characteristic (*fingerprint*-like) fashion? Which proteins change and how (relative amounts, modifications, splice forms, etc.)? (This is a form of *protein expression profiling* that complements the *transcriptional (mRNA) profiling* discussed in Chapter 7.)
- Can such fingerprint-like changes be used for diagnostic purposes? For example, do certain cancers or heart disease cause characteristic changes in blood proteins? Can the proteomic fingerprint help determine if a given cancer is resistant or sensitive to a particular chemotherapeutic drug? Proteomic fingerprints can also be the starting point for studies of the mechanisms underlying the change of state. Proteins (and other biomolecules) that show changes that are diagnostic of a particular state are called *biomarkers.*
- Can changes in the proteome help define targets for drugs or suggest mechanisms by which that drug might induce toxic side effects? If so, it might be possible to engineer modified versions of the drug with fewer side effects.

These are just a few of the questions that can be addressed using proteomics. The methods used to answer these questions are as diverse as the questions themselves, and their numbers are growing rapidly.

Advanced Techniques in Mass Spectrometry Are Critical to Proteomic Analysis

Advances in proteomics technologies (e.g., mass spectrometry) are having a profound effect on the types of questions that can be practically studied. For many years, two-dimensional gel electrophoresis has allowed researchers to separate, display, and characterize a mixture of proteins (Figure 3-36). The spots on a two-dimensional gel can be excised, the protein fragmented by proteolysis (e.g. by trypsin digestion), and the fragments identified by MS. An alternative to this two-dimensional gel method is *high throughput LC-MS/MS.* Figure 3-43 outlines the general LC-MS/MS approach in which a complex mixture of proteins is digested with a protease, the myriad resulting peptides are fractionated by LC into multiple, less complex fractions, which are slowly but continuously injected by electrospray ionization into a tandem mass spectrometer. The fractions are then sequentially subjected to multiple cycles of MS/MS until sequences of many of the peptides are determined and used to identify from databases the proteins in the original biological sample.

An example of the use of LC-MS/MS to identify many of the proteins in each organelle is seen in Figure 3-44. Cells from murine (mouse) liver tissue were mechanically broken to release the organelles, and the organelles were partially separated by density-gradient centrifugation. The locations of the organelles in the gradient were determined using immunoblotting with antibodies that recognize previously identified, organelle-specific proteins. Fractions from the gradient

Podcast: Use of Mass Spectrometry in Cell Biology 

MEDIA CONNECTIONS

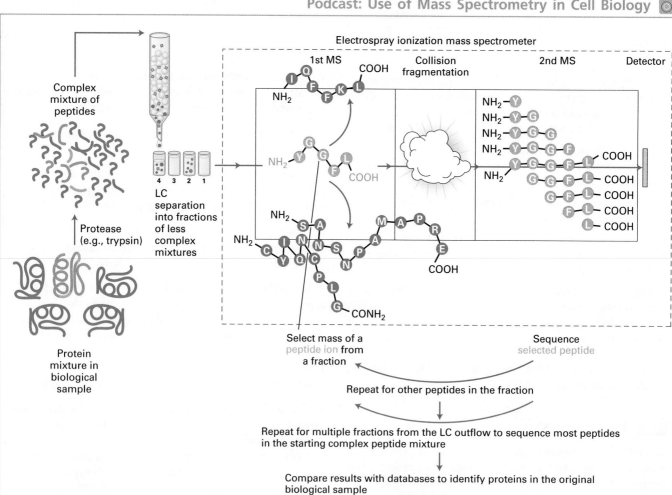

▲ EXPERIMENTAL FIGURE 3-43 **LC-MS/MS is used to identify the proteins in a complex biological sample.** A complex mixture of proteins in a biological sample (e.g., isolated preparation of Golgi organelles) is digested with a protease, the mixture of resulting peptides is fractionated by liquid chromatography (LC) into multiple, less complex, fractions, which are slowly but continuously injected by electrospray ionization into a tandem mass spectrometer. The fractions are then sequentially subjected to multiple cycles of MS/MS until masses and sequences of many of the peptides are determined and used to identify the proteins in the original biological sample through comparison with protein databases. [Based on a figure provided by S. Carr.]

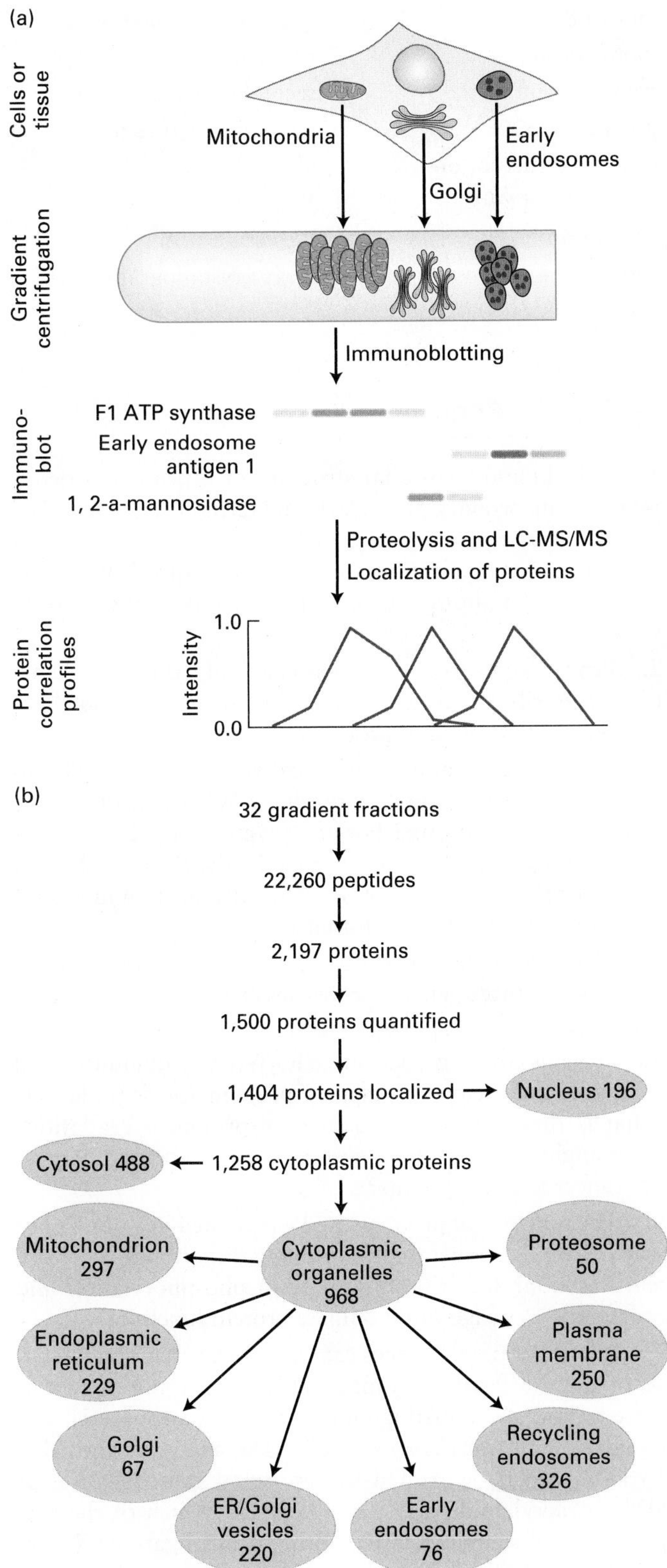

◀ EXPERIMENTAL FIGURE 3-44 **Density-gradient centrifugation and LC-MS/MS can be used to identify many of the proteins in organelles.** (a) The cells in liver tissue were mechanically broken to release the organelles, and the organelles were partially separated by density-gradient centrifugation. The locations of the organelles—which were spread out through the gradient and somewhat overlapped with one another—were determined using immunoblotting with antibodies that recognize previously identified, organelle-specific proteins. Fractions from the gradient were subjected to proteolysis and LC-MS/MS to identify the peptides, and hence the proteins, in each fraction. Comparisons with the locations of the organelles in the gradient (called *protein correlation profiling*) permitted assignment of many individual proteins to one or more organelles (organelle proteome identification). (b) The hierarchical breakdown of data derived from the procedures in part (a). Note that not all proteins identified could be assigned to organelles and some proteins were assigned to more than one organelle. [From L. J. Foster et al., 2006, *Cell* **125**(1):187–199.]

were subjected to LC-MS/MS to identify the proteins in each fraction, and the distributions in the gradient of many individual proteins were compared with the distributions of the organelles. This permitted assignment of many individual proteins to one or more organelles (organelle proteome profiling).

Proteomics combined with molecular genetics methods are currently being used to identify all protein complexes in a eukaryotic cell, the yeast *Saccharomyces cerevisiae*. Approximately 500 complexes have been identified, with an average of 4.9 distinct proteins per complex, and these in turn are involved in at least 400 complex-to-complex interactions. Such systematic proteomic studies are providing new insights into the organization of proteins within cells and how proteins work together to permit cells to live and function.

KEY CONCEPTS OF SECTION 3.7

Proteomics

- Proteomics is the systematic study of the amounts (and changes in the amounts), modifications, interactions, localization, and functions of all or subsets of all proteins in biological systems at the whole-organism, tissue, cellular, and subcellular levels.
- Proteomics provides insights into the fundamental organization of proteins within cells, and how this organization is influenced by the state of the cells (e.g., differentiation into distinct cell types; response to stress, disease, and drugs).
- A wide variety of methods are used for proteomic analyses, including two-dimensional gel electrophoresis, density-gradient centrifugation, and mass spectroscopy (MALDI-TOF and LC-MS/MS).
- Proteomics has helped begin to identify the proteomes of organelles ("organelle proteome profiling") and the organization of individual proteins into multiprotein complexes that interact in a complex network to support life and cellular function

Perspectives for the Future

Impressive expansion of the computational power of computers is at the core of advances in determining the three-dimensional structures of proteins. For example, vacuum tube computers running on programs punched on cards were used to solve the first protein structures on the basis of x-ray crystallography. In the future, researchers aim to predict the structures of proteins, using only amino acid sequences deduced from gene sequences. This computationally challenging problem requires supercomputers or large clusters of computers working in synchrony. Currently, only the structures of very small domains containing 100 residues or fewer can be predicted at a low resolution. However, continued developments in computing and models of protein folding, combined with large-scale efforts to solve the structures of all protein structural motifs by x-ray crystallography, will allow the prediction of the structures of larger proteins. With an exponentially expanding database of structurally defined motifs, domains, and proteins, scientists will be able to identify the motifs in an unknown protein, match the motif to the sequence, and use this to predict the three-dimensional structure of the entire protein.

New combined approaches will also help in determining high-resolution structures of molecular machines. Although these very large macromolecular assemblies usually are difficult to crystallize and thus to solve by x-ray crystallography, they can be imaged in a cryoelectron microscope at liquid helium temperatures and high electron energies. From millions of individual "particles," each representing a random view of the protein complex, the three-dimensional structure can be built. Because subunits of the complex may already be solved by crystallography, a composite structure consisting of the x-ray-derived subunit structures fit to the EM-derived model will be generated. An example of this approach for an electron-transport "supercomplex" is described in Chapter 12.

Methods for rapid structure determination combined with identification of novel substrates and inhibitors will help determine the structures of enzyme–substrate complexes and transition states, and thus help provide detailed information regarding the mechanisms of enzyme catalysis.

The rapid development of new technologies can be expected to help solve some of the still outstanding problems in proteomics. It should soon be possible to identify and sequence intact proteins in complex mixtures using MS techniques without first digesting the samples into peptides. An ongoing problem in proteomic analysis of complex mixtures is that it is difficult to detect and identify protein fragments from samples whose concentrations in the sample differ by more than 1000-fold: some samples, such as blood plasma, contain proteins whose concentrations vary over a 10^{11}-fold range. Routine analysis of specimens with such diverse concentrations should dramatically improve the mechanistic and diagnostic value of blood plasma proteomics.

Key Terms

α helix *66*
activation energy *79*
active site *80*
allostery *89*
amyloid filament *77*
autoradiography *100*
β sheet *66*
β turn *66*
chaperones *76*
conformations *63*
cooperativity *89*
domain *70*
electrophoresis *94*
homology *72*
K_m *80*
ligand *78*
liquid chromatography *96*
molecular machine *64*
motif *68*
motor proteins *85*
peptide bond *65*
polypeptides *66*
primary structure *66*
proteasomes *64*
protein *66*
proteome *64*
proteomics *105*
quaternary structure *72*
rate-zonal centrifugation *93*
secondary structure *66*
tertiary structure *67*
ubiquitin *88*
V_{max} *80*
Western blotting *98*
x-ray crystallography *103*

Review the Concepts

1. The three-dimensional structure of a protein is determined by its primary, secondary, and tertiary structures. Define the *primary, secondary,* and *tertiary structures.* What are some of the common secondary structures? What are the forces that hold together the secondary and tertiary structures?

2. Proper folding of proteins is essential for biological activity. Describe the roles of molecular chaperones and chaperonins in the folding of proteins.

3. Enzymes can catalyze chemical reactions. How do enzymes increase the rate of a reaction? What constitutes the active site of an enzyme? For an enzyme-catalyzed reaction, what are K_m and V_{max}? For enzyme X, the K_m for substrate A is 0.4 mM and for substrate B is 0.01 mM. Which substrate has a higher affinity for enzyme X?

4. Motor proteins convert energy into a mechanical force. Describe the three general properties characteristic of motor proteins.

5. Proteins are degraded in cells. What is ubiquitin, and what role does it play in tagging proteins for degradation? What is the role of proteasomes in protein degradation? How might proteasome inhibitors serve as chemotherapeutic (cancer-treating) agents?

6. The function of proteins can be regulated in a number of ways. What is cooperativity, and how does it influence protein function? Describe how protein phosphorylation and proteolytic cleavage can modulate protein function.

7. A number of techniques can separate proteins on the basis of their differences in mass. Describe the use of two of these techniques, centrifugation and gel electrophoresis. The blood proteins transferrin (MW 76 kDa) and lysozyme (MW 15 kDa) can be separated by rate-zonal centrifugation or SDS-polyacrylamide gel electrophoresis. Which of the two proteins will sediment faster during centrifugation? Which will migrate faster during electrophoresis?

8. Chromatography is an analytical method used to separate proteins. Describe the principles for separating proteins by gel filtration, ion-exchange, and affinity chromatography.

9. Various methods have been developed for detecting proteins. Describe how radioisotopes and autoradiography can be used for labeling and detecting proteins. How does Western blotting detect proteins?

10. Physical methods are often used to determine protein conformation. Describe how x-ray crystallography, cryo-electron microscopy, and NMR spectroscopy can be used to determine the shape of proteins.

11. Mass spectrometry is a powerful tool in proteomics. What are the four key features of a mass spectrometer? Describe briefly how MALDI and two-dimensional polyacrylamide gel electrophoresis (2D-PAGE) could be used to identify a protein expressed in cancer cells but not in normal healthy cells.

Analyze the Data

Proteomics involves the global analysis of protein expression. In one approach, all the proteins in control cells and treated cells are extracted and subsequently separated using two-dimensional gel electrophoresis. Typically, hundreds or thousands of protein spots are resolved and the steady-state levels of each protein are compared between control and treated cells. In the following example, only a few protein spots are shown for simplicity. Proteins are separated in the first dimension on the basis of charge by isoelectric focusing (pH 4–10) and then separated by size by SDS-polyacrylamide gel electrophoresis. Proteins are detected with a stain such as Coomassie blue and assigned numbers for identification.

a. Cells are treated with a drug ("+ Drug") or left untreated ("Control"), and then proteins are extracted and separated by two-dimensional gel electrophoresis. The stained gels are shown below. What do you conclude about the effect of the drug on the steady-state levels of proteins 1–7?

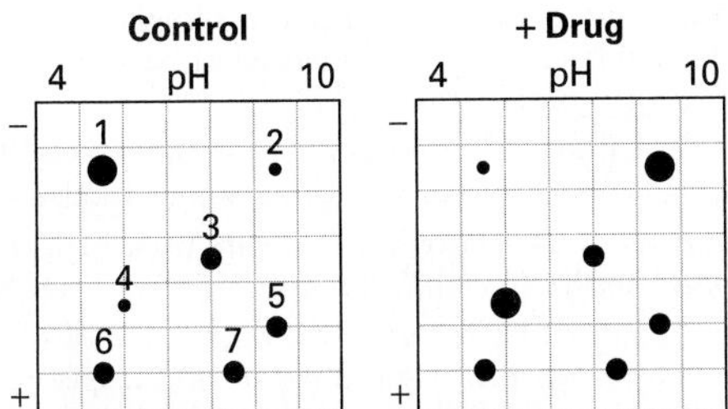

b. You suspect that the drug may be inducing a protein kinase and so repeat the experiment in part (a) in the presence of ^{32}P-labeled inorganic phosphate. In this experiment the two-dimensional gels are exposed to x-ray film to detect the presence of ^{32}P-labeled proteins. The x-ray films are shown below. What do you conclude from this experiment about the effect of the drug on proteins 1–7?

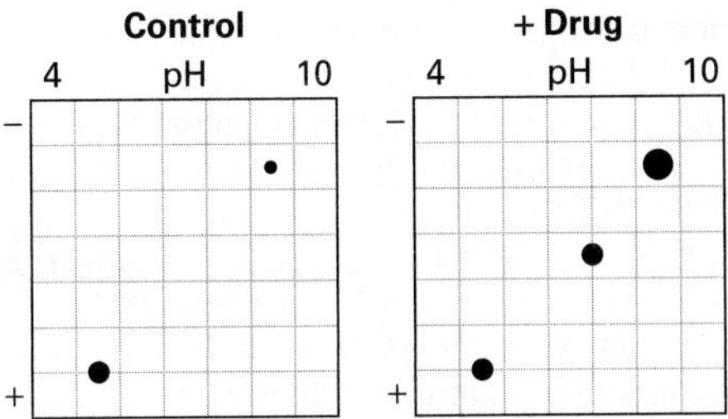

c. To determine the cellular localization of proteins 1–7, the cells from part (a) were separated into nuclear and cytoplasmic fractions by differential centrifugation. Two-dimensional gels were run, and the stained gels are shown below. What do you conclude about the cellular localization of proteins 1–7?

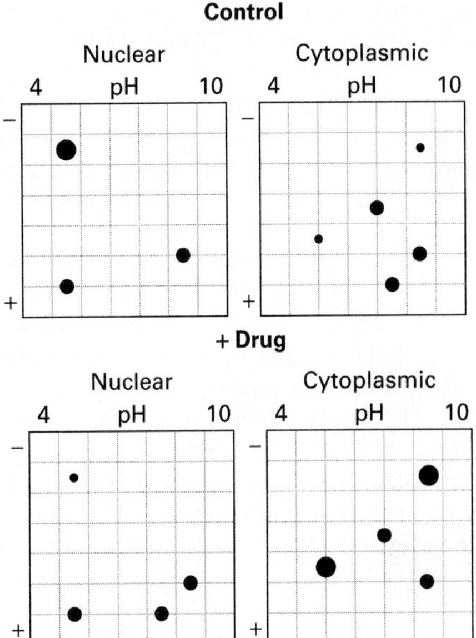

d. Summarize the overall properties of proteins 1–7, combining the data from parts (a), (b), and (c). Describe how you could determine the identity of any one of the proteins.

References

General References

Berg, J. M., J. L. Tymoczko, and L. Stryer. 2007. *Biochemistry,* 6th ed. W. H. Freeman and Company.

Nelson, D. L., and M. M. Cox. 2005. *Lehninger Principles of Biochemistry,* 4th ed. W. H. Freeman and Company.

Web Sites

Entry site into proteins, structures, genomes, and taxonomy: http://www.ncbi.nlm.nih.gov/Entrez/

The protein 3D structure database: http://www.rcsb.org/

Structural classifications of proteins: http://scop.berkeley.edu/

Sites containing general information about proteins: http://www.expasy.ch/; http://www.proweb.org/; http://scop.berkeley.edu/intro.html

PROSITE Database of protein families and domains: http://www.expasy.org/prosite/

Domain organization of proteins and large collection of multiple sequence alignments: http://www.sanger.ac.uk/Software/Pfam/

Hierarchical Structure of Proteins and Protein Folding

Branden, C., and J. Tooze. 1999. *Introduction to Protein Structure.* Garland.

Brodsky, J. L., and G. Chiosis. 2006. Hsp70 molecular chaperones: emerging roles in human disease and identification of small molecule modulators. *Curr. Top. Med. Chem.* **6**(11):1215–1225.

Bukau, B, J. Weissman, and A. Horwich. 2006. Molecular chaperones and protein quality control. *Cell* **125**(3):443–451.

Cohen, F. E. 1999. Protein misfolding and prion diseases. *J. Mol. Biol.* **293**:313–320.

Coulson, A. F., and J. Moult. 2002. A unifold, mesofold, and superfold model of protein fold use. *Proteins* **46**:61–71.

Daggett, V, and A. R. Fersht. 2003. Is there a unifying mechanism for protein folding? *Trends Biochem. Sci.* **28**(1):18–25.

Dobson, C. M. 1999. Protein misfolding, evolution, and disease. *Trends Biochem. Sci.* **24**:329–332.

Gimona, M. 2006. Protein linguistics—a grammar for modular protein assembly? *Nat. Rev. Mol. Cell Biol.* **7**(1):68–73.

Gough, J. 2006. Genomic scale sub-family assignment of protein domains. *Nucl. Acids Res.* **34**(13):3625–3633.

Koonin, E. V., Y. I. Wolf, and G. P. Karev. 2002. The structure of the protein universe and genome evolution. *Nature* **420**:218–223.

Lesk, A. M. 2001. *Introduction to Protein Architecture.* Oxford.

Lin, Z, and H. S. Rye. 2006. GroEL-mediated protein folding: making the impossible, possible. *Crit. Rev. Biochem. Mol. Biol.* **41**(4):211–239.

Orengo, C. A., D. T. Jones, and J. M. Thornton. 1994. Protein superfamilies and domain superfolds. *Nature* **372**: 631–634.

Patthy, L. 1999. *Protein Evolution.* Blackwell Science.

Rochet, J.-C., and P. T. Landsbury. 2000. Amyloid fibrillogenesis: themes and variations. *Curr. Opin. Struc. Biol.* **10**:60–68.

Vogel, C, and C. Chothia. 2006. Protein family expansions and biological complexity. *PLoS Comput. Biol.* **2**(5):e48.

Yaffe, M. B. 2006. "Bits" and pieces. *Sci STKE.* Jun 20 (340):pe28.

Young, J. C., et al. 2004. Pathways of chaperone-mediated protein folding in the cytosol. *Nat. Rev. Mol. Cell Biol.* **5**:781–791.

Protein Function

Dressler, D. H., and H. Potter. 1991. *Discovering Enzymes.* Scientific American Library.

Fersht, A. 1999. *Enzyme Structure and Mechanism,* 3d ed. W. H. Freeman and Company.

Jeffery, C. J. 2004. Molecular mechanisms for multitasking: recent crystal structures of moonlighting proteins. *Curr. Opin. Struc. Biol.* **14**(6):663–668.

Marnett, A. B., and C. S. Craik. 2005. Papa's got a brand new tag: advances in identification of proteases and their substrates. *Trends Biotechnol.* **23**(2):59–64.

Polgar, L. 2005. The catalytic triad of serine peptidases. *Cell Mol. Life Sci.* **62**(19–20):2161–2172.

Radisky, E. S., et al. 2006. Insights into the serine protease mechanism from atomic resolution structures of trypsin reaction intermediates. *Proc. Nat'l Acad. Sci. USA* **103**(18):6835–6840.

Schenone, M, B. C. Furie, and B. Furie. 2004. The blood coagulation cascade. *Curr. Opin. Hematol.* **11**(4):272–277.

Schramm, V. L. 2005. Enzymatic transition states and transition state analogues. *Curr. Opin. Struc. Biol.* **15**(6):604–613.

Regulating Protein Function I: Protein Degradation

Glickman, M. H., and A. Ciechanover. 2002 The ubiquitin-proteasome proteolytic pathway: destruction for the sake of construction. *Physiol. Rev.* **82**(2):373–428.

Goldberg, A. L. 2003. Protein degradation and protection against misfolded or damaged proteins. *Nature* **426**:895–899.

Goldberg, A. L, S. J. Elledge, and J. W. Harper. 2001. The cellular chamber of doom. *Sci. Am.* **284**(1):68–73.

Groll, M., and R. Huber. 2005. Purification, crystallization, and x-ray analysis of the yeast 20S proteasome. *Meth. Enzymol.* **398**:329–336.

Kisselev, A. F., A. Callard, and A. L. Goldberg. 2006. Importance of the different proteolytic sites of the proteasome and the efficacy of inhibitors varies with the protein substrate. *J. Biol. Chem.* **281**(13):8582–8590.

Rechsteiner, M., and C. P. Hill. 2005. Mobilizing the proteolytic machine: cell biological roles of proteasome activators and inhibitors. *Trends Cell Biol.* **15**(1):27–33.

Zhou, P. 2006. REGgamma: a shortcut to destruction. *Cell* **124**(2):256–257.

Zolk, O., C. Schenke, and A. Sarikas. 2006. The ubiquitin-proteasome system: focus on the heart. *Cardiovasc. Res.* **70**(3):410–421.

Regulating Protein Function II: Noncovalent and Covalent Modifications

Bellelli, A., et al. 2006. The allosteric properties of hemoglobin: insights from natural and site directed mutants. *Curr. Prot. Pep. Sci.* **7**(1):17–45.

Burack, W. R., and A. S. Shaw. 2000. Signal transduction: hanging on a scaffold. *Curr. Opin. Cell Biol.* **12**:211–216.

Halling, D. B, P. Aracena-Parks, and S. L. Hamilton. 2006. Regulation of voltage-gated Ca^{2+} channels by calmodulin. *Sci STKE* Jan 17 (318):er1.

Horovitz, A., et al. 2001. Review: allostery in chaperonins. *J. Struc. Biol.* **135**:104–114.

Kern, D., and E. R. Zuiderweg. 2003. The role of dynamics in allosteric regulation. *Curr. Opin. Struc. Biol.* **13**(6):748–757.

Lane, K. T., and L. S. Beese. 2006. Thematic review series: lipid posttranslational modifications. Structural biology of protein farnesyltransferase and geranylgeranyltransferase type I. *J. Lipid Res.* **47**(4):681–699.

Lim, W. A. 2002. The modular logic of signaling proteins: building allosteric switches from simple binding domains. *Curr. Opin. Struc. Biol.* **12**:61–68.

Martin, C., Y. Zhang. 2005. The diverse functions of histone lysine methylation. *Nat. Rev. Mol. Cell Biol.* **6**(11):838–849.

Sawyer, T. K., et al. 2005. Protein phosphorylation and signal transduction modulation: chemistry perspectives for small-molecule drug discovery. *Med. Chem.* **1**(3):293–319.

Xia, Z., and D. R. Storm. 2005. The role of calmodulin as a signal integrator for synaptic plasticity. *Nat. Rev. Neurosci.* **6**(4):267–276.

Yap, K. L., et al. 1999. Diversity of conformational states and changes within the EF-hand protein superfamily. *Proteins* **37**:499–507.

Purifying, Detecting, and Characterizing Proteins

Domon, B., and R. Aebersold. 2006. Mass spectrometry and protein analysis. *Science* **312**(5771):212–217.

Encarnacion, S., et al. 2005. Comparative proteomics using 2-D gel electrophoresis and mass spectrometry as tools to dissect stimulons and regulons in bacteria with sequenced or partially sequenced genomes. *Biol. Proc. Online* **7**:117–135.

Hames, B. D. *A Practical Approach.* Oxford University Press. A methods series that describes protein purification methods and assays.

Patton, W. F. 2002. Detection technologies in proteome analysis. *J. Chromatogr. B. Analyt. Technol. Biomed. Life Sci.* **771**(1–2):3–31.

White, I. R., et al. 2004. A statistical comparison of silver and SYPRO Ruby staining for proteomic analysis. *Electrophoresis* **25**(17):3048–3054.

Proteomics

Foster, L. J., et al. 2006. A mammalian organelle map by protein correlation profiling. *Cell* **125**(1):187–199.

Fu, Q., and J. E. Van Eyk. 2006. Proteomics and heart disease: identifying biomarkers of clinical utility. *Expert Rev. Proteomics* **3**(2):237–249.

Gavin. A. C., et al. 2006. Proteome survey reveals modularity of the yeast cell machinery. *Nature* **440**(7084):631–636.

Kislinger, T., et al. 2006. Global survey of organ and organelle protein expression in mouse: combined proteomic and transcriptomic profiling. *Cell* **125**(1):173–186.

Kolker, E, R. Higdon, and J. M. Hogan. 2006. Protein identification and expression analysis using mass spectrometry. *Trends Microbiol.* **14**(5):229–235.

Krogan, N. J., et al. 2006. Global landscape of protein complexes in the yeast *Saccharomyces cerevisiae. Nature* **440**(7084):637–643.

Ong, S. E., and M. Mann. 2005. Mass spectrometry-based proteomics turns quantitative. *Nat. Chem. Biol.* **1**(5):252–262.

Rifai, N., M. A. Gillette, and S. A. Carr. 2006. Protein biomarker discovery and validation: the long and uncertain path to clinical utility. *Nat. Biotech.* **24**(8):971–983.

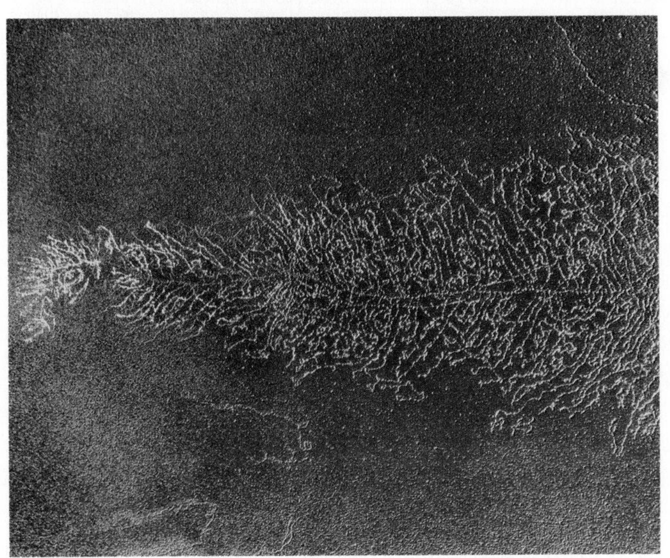

Colored transmission electron micrograph of one ribosomal RNA transcription unit from a *Xenopus* oocyte. Transcription proceeds from left to right, with nascent ribosomal ribonucleoprotein complexes (rRNPs) growing in length as each successive RNA polymerase I molecule moves along the DNA template at the center. In this preparation each rRNP is oriented either above or below the central strand of DNA being transcribed, so that the overall shape is similar to a feather. In the nucleolus of a living cell, the nascent rRNPs extend in all directions, like a bottlebrush. [Professor Oscar L. Miller/Science Photo Library.]

CHAPTER

4

BASIC MOLECULAR GENETIC MECHANISMS

The extraordinary versatility of proteins as molecular machines and switches, cellular catalysts, and components of cellular structures was described in Chapter 3. In this chapter we consider how proteins are made, as well as other cellular processes that are critical for the survival of an organism and its descendents. Our focus will be on the vital molecules known as **nucleic acids**, and how they ultimately are responsible for governing all cellular function. As introduced in Chapter 2, nucleic acids are linear polymers of four types of nucleotides (see Figures 2-13, 2-16 and 2-17). These macromolecules (1) contain in the precise sequence of their nucleotides the information for determining the amino acid sequence and hence the structure and function of all the proteins of a cell, (2) are critical functional components of the cellular macromolecular factories that select and align amino acids in the correct order as a polypeptide chain is being synthesized, and (3) catalyze a number of fundamental chemical reactions in cells, including formation of peptide bonds between amino acids during protein synthesis.

Deoxyribonucleic acid (DNA) is an informational molecule that contains in the sequence of its nucleotides the information required to build all the proteins of an organism, and hence the cells and tissues of that organism. It is ideally suited to perform this function on a molecular level. Chemically, it is extraordinarily stable under most terrestrial conditions, as exemplified by the ability to recover DNA sequence from fossils that are tens of thousands of years old. Because of this, and because of repair mechanisms that operate in living cells, the long polymers that make up a DNA molecule can be up to 10^9 nucleotides long. Virtually all the information required for the development of a fertilized human egg into an adult made of trillions of cells with specialized functions can be stored in the sequence of the four possible nucleotides that

OUTLINE

make up the $\approx 3 \times 10^9$ base pairs in the human genome. Because of the principles of base pairing discussed in this chapter, the information is readily copied with an error rate of <1 in 10^9 nucleotides per generation. The exact replication of this information in any species assures its genetic continuity from generation to generation and is critical to the normal development of an individual. DNA fulfills these functions so well that it is the vessel for genetic information in all forms of life known. (One exception is RNA viruses; however, these are limited to extremely short genomes because of the relative instability of RNA compared with DNA, as we will see.) The discovery that virtually all forms of life use DNA to encode their genetic information, and also use nearly the same nucleic acid sequence code to specify amino acid sequence, implies that all forms of life descended from a common ancestor on the basis of storage of information in nucleic acid sequence. This information is accessed and replicated by specific base pairing between bases. The information stored in DNA is arranged in hereditary units, now known as **genes,** which control identifiable traits of an organism. In the process of transcription, the information stored in DNA is copied into **ribonucleic acid (RNA),** which has three distinct roles in protein synthesis.

Portions of the DNA nucleotide sequence are copied into **messenger RNA (mRNA)** molecules that direct the synthesis of a specific protein. The nucleotide sequence of an mRNA molecule contains the information that specifies the correct order of amino acids during the synthesis of a protein. The remarkably accurate, stepwise assembly of amino acids into proteins occurs by **translation** of mRNA. In this process, the nucleotide sequence of an mRNA molecule is "read" by a second type of RNA called **transfer RNA (tRNA)** with the aid of a third type of RNA, **ribosomal RNA (rRNA),** and their associated proteins. As the correct amino acids are brought into sequence by tRNAs, they are linked by peptide bonds to make proteins. RNA synthesis is called **transcription** because the nucleotide sequence "language" of DNA is precisely copied, or *transcribed,* into the nucleotide sequence of an RNA molecule. Protein synthesis is referred to as *translation* because the nucleotide sequence language of DNA and RNA is *translated* into the amino acid sequence language of proteins.

Discovery of the structure of DNA in 1953 and subsequent elucidation of how DNA directs synthesis of RNA, which then directs assembly of proteins—the so-called *central dogma*—were monumental achievements marking the early days of molecular biology. However, the simplified representation of the central dogma as DNA → RNA → protein does not reflect the role of proteins in the synthesis of nucleic acids. Moreover, as discussed in Chapter 7, proteins are largely responsible for *regulating* **gene expression,** the entire process whereby the information encoded in DNA is decoded into a variety of proteins in the correct cells at the correct times in development. This regulation of gene expression allows hemoglobin to be expressed only in cells of the bone marrow (reticulocytes) destined to develop into circulating red blood cells (erythrocytes), and directs developing neurons to make the proper synapses (connections) with 10^{11} other developing neurons in the human brain. The fundamental molecular genetic processes of DNA replication, transcription, and translation must be carried out with extraordinary fidelity, speed, and accurate regulation in order for organisms as complex as prokaryotes and eukaryotes to develop normally. This is achieved by chemical processes that operate with extraordinary accuracy coupled with multiple layers of checkpoint, or surveillance, mechanisms that test whether critical steps in these processes have occurred correctly before the next step is initiated. The highly regulated expression of genes necessary for the development of a multicellular organism requires integrating information from signals sent by distant cells in the developing organism, as well as from neighboring cells, and an intrinsic developmental program determined by earlier steps in embryogenesis taken by that cell's progenitors. All this regulation is dependent on control sequences in the DNA that function with proteins called *transcription factors* to coordinate the expression of every gene. RNA sequences we discuss in Chapter 8 that regulate RNA processing and translation also are encoded in DNA originally. Thus nucleic acids function as the "brains and central nervous system" of the cell, while proteins carry out the functions they specify.

In this chapter, we first review the structures and properties of DNA and RNA, and explore how the different characteristics of each type of nucleic acid make them suited for their respective roles in the cell. In the next several sections we discuss the basic processes summarized in Figure 4-1: transcription of DNA into RNA precursors, processing of these precursors to make functional RNA molecules, translation of mRNAs into proteins, and the replication of DNA. After outlining the individual roles of mRNA, tRNA, and rRNA in protein synthesis, we present a detailed description of the components and biochemical steps in translation. We also consider the molecular problems involved in DNA replication and the complex cellular machinery for ensuring accurate copying of the genetic material. Along the way, we compare these processes in prokaryotes and eukaryotes. The next section describes how damage to DNA is repaired, and how regions of different DNA molecules are exchanged in the process of **recombination** to generate new combinations of traits in the individual organisms of a species. The final section of the chapter presents basic information about viruses, which, in addition to being significant pathogens, are important model organisms for studying macromolecular synthesis and other cellular processes. Viruses have relatively simple structures compared with cells, and small genomes that made them tractable for historic early studies of the basic processes of DNA replication, transcription, translation, recombination, and gene expression. Viruses continue to teach important lessons in molecular cell biology today and have been adapted as experimental tools for introducing any desired genes into cells, tools that are currently being tested for their effectiveness in human gene therapy.

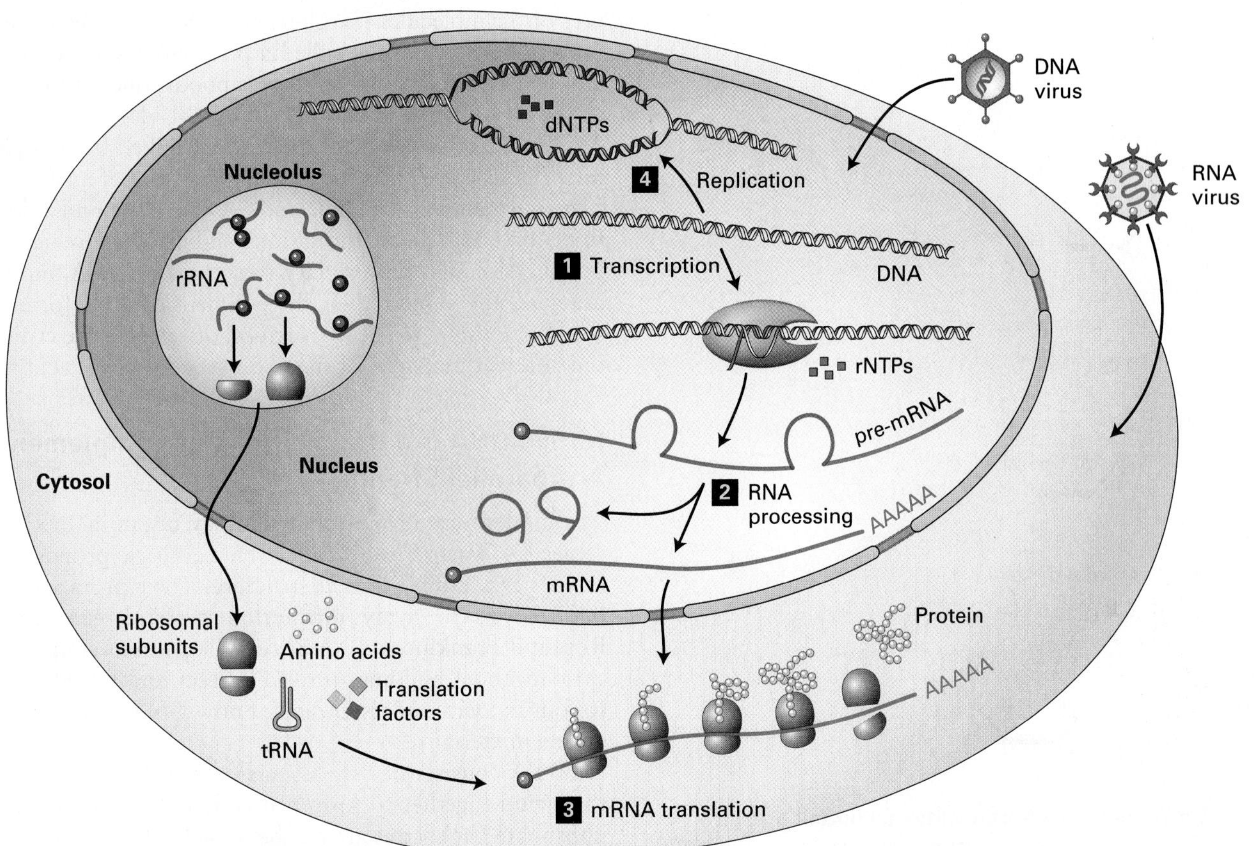

▲ **FIGURE 4-1 Overview of four basic molecular genetic processes.** In this chapter we cover the three processes that lead to production of proteins (1–3) and the process for replicating DNA (4). Because viruses utilize host-cell machinery, they have been important models for studying these processes. During transcription of a protein-coding gene by RNA polymerase (1), the four-base DNA code specifying the amino acid sequence of a protein is copied, or *transcribed,* into a precursor messenger RNA (pre-mRNA) by the polymerization of ribonucleoside triphosphate monomers (rNTPs). Removal of noncoding sequences and other modifications to the pre-mRNA (2), collectively known as *RNA processing*, produce a functional mRNA, which is transported to the cytoplasm. During *translation* (3), the four-base code of the mRNA is decoded into the 20–amino acid language of proteins. Ribosomes, the macromolecular machines that translate the mRNA code, are composed of two subunits assembled in the nucleolus from ribosomal RNAs (rRNAs) and multiple proteins (*left*). After transport to the cytoplasm, ribosomal subunits associate with an mRNA and carry out protein synthesis with the help of transfer RNAs (tRNAs) and various translation factors. During DNA replication (4), which occurs only in cells preparing to divide, deoxyribonucleoside triphosphate monomers (dNTPs) are polymerized to yield two identical copies of each chromosomal DNA molecule. Each daughter cell receives one of the identical copies.

4.1 Structure of Nucleic Acids

DNA and RNA are chemically very similar. The primary structures of both are linear **polymers** composed of **monomers** called **nucleotides.** Both function primarily as informational molecules, carrying information in the exact sequence of their nucleotides. Cellular RNAs range in length from fewer than one hundred to many thousands of nucleotides. Cellular DNA molecules can be as long as several hundred million nucleotides. These large DNA units in association with proteins can be stained with dyes and visualized in the light microscope as *chromosomes,* so named because of their stainability. Though chemically similar, DNA and RNA exhibit some very important differences. For example, RNA can also function as a catalytic molecule. As we'll see, it is the different and unique properties of DNA and RNA that make them each suited for their specific roles in the cell.

A Nucleic Acid Strand Is a Linear Polymer with End-to-End Directionality

In all organisms, DNA and RNA each comprise only four different nucleotides. Recall from Chapter 2 that all nucleotides consist of an organic base linked to a five-carbon sugar that has a phosphate group attached to carbon 5. In RNA, the sugar is ribose; in DNA, deoxyribose (see Figure 2-16). The nucleotides used in synthesis of DNA and RNA contain one of five different bases. The bases *adenine* (A) and *guanine* (G) are **purines,** which contain a pair of fused rings; the bases *cytosine* (C), *thymine* (T), and *uracil* (U) are

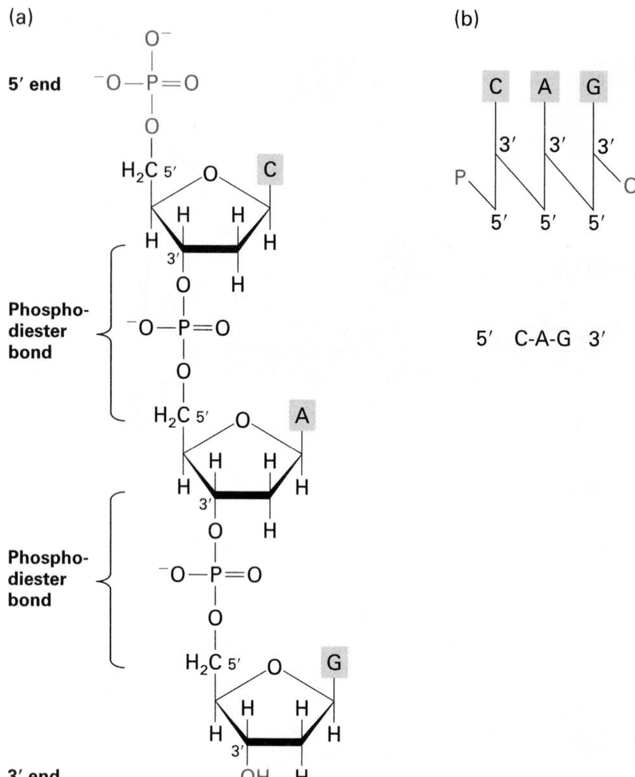

▲ **FIGURE 4-2 Chemical directionality of a nucleic acid strand.** Shown here are alternative representations of a single strand of DNA containing only three bases: cytosine (C), adenine (A), and guanine (G). (a) The chemical structure shows a hydroxyl group at the 3′ end and a phosphate group at the 5′ end. Note also that two phosphoester bonds link adjacent nucleotides; this two-bond linkage commonly is referred to as a *phosphodiester bond*. (b) In the "stick" diagram (*top*), the sugars are indicated as vertical lines and the phosphodiester bonds as slanting lines; the bases are denoted by their single-letter abbreviations. In the simplest representation (*bottom*), only the bases are indicated. By convention, a polynucleotide sequence is always written in the 5′→3′ direction (left to right) unless otherwise indicated.

pyrimidines, which contain a single ring (see Figure 2-17). Three of these bases—A, G, and C—are found in both DNA and RNA; however, T is found only in DNA, and U only in RNA. (Note that the single-letter abbreviations for these bases are also commonly used to denote the entire nucleotides in nucleic acid polymers.)

A single nucleic acid strand has a *backbone* composed of repeating pentose-phosphate units from which the purine and pyrimidine bases extend as side groups. Like a polypeptide, a nucleic acid strand has an end-to-end chemical orientation: the *5′ end* has a hydroxyl or phosphate group on the 5′ carbon of its terminal sugar; the *3′ end* usually has a hydroxyl group on the 3′ carbon of its terminal sugar (Figure 4-2). This directionality, plus the fact that synthesis proceeds 5′→3′, has given rise to the convention that polynucleotide sequences are written and read in the 5′→3′ direction (from left to right); for example, the sequence AUG is assumed to be (5′)AUG(3′). As we will see, the 5′→3′ directionality of a nucleic acid strand is an important property of the molecule. The chemical linkage between adjacent nucleotides, commonly called a **phosphodiester bond,** actually consists of two phosphoester bonds, one on the 5′ side of the phosphate and another on the 3′ side.

The linear sequence of nucleotides linked by phosphodiester bonds constitutes the primary structure of nucleic acids. Like polypeptides, polynucleotides can twist and fold into three-dimensional conformations stabilized by noncovalent bonds. Although the primary structures of DNA and RNA are generally similar, their three-dimensional conformations are quite different. These structural differences are critical to the different functions of the two types of nucleic acids.

Native DNA Is a Double Helix of Complementary Antiparallel Strands

The modern era of molecular biology began in 1953 when James D. Watson and Francis H. C. Crick proposed that DNA has a double-helical structure. Their proposal, based on analysis of x-ray diffraction patterns generated by Rosalind Franklin and Maurice Wilkins, and coupled with careful model building, proved correct and paved the way for our modern understanding of how DNA functions as the genetic material.

DNA consists of two associated polynucleotide strands that wind together to form a **double helix.** The two sugar-phosphate backbones are on the outside of the double helix, and the bases project into the interior. The adjoining bases in each strand stack on top of one another in parallel planes (Figure 4-3a). The orientation of the two strands is *antiparallel;* that is, their 5′→3′ directions are opposite. The strands are held in precise register by formation of **base pairs** between the two strands: A is paired with T through two hydrogen bonds; G is paired with C through three hydrogen bonds (Figure 4-3b). This base-pair complementarity is a consequence of the size, shape, and chemical composition of the bases. The presence of thousands of such hydrogen bonds in a DNA molecule contributes greatly to the stability of the double helix. Hydrophobic and van der Waals interactions between the stacked adjacent base pairs further stabilize the double-helical structure.

In natural DNA, A always hydrogen bonds with T and G with C, forming A·T and G·C base pairs, as shown in Figure 4-3b. These associations, always between a larger purine and smaller pyrimidine are often called *Watson-Crick base pairs.* Two polynucleotide strands, or regions thereof, in which all the nucleotides form such base pairs are said to be **complementary.** However, in theory and in synthetic DNAs, other base pairs can form. For example, guanine (a purine) could theoretically form hydrogen bonds with thymine (a pyrimidine), causing only a minor distortion in the helix. The space available in the helix also would allow pairing between the two pyrimidines cytosine and thymine. Although the nonstandard G·T and C·T base pairs are normally not found in DNA, G·U base pairs are quite common in double-helical regions that form within otherwise single-stranded RNA. Nonstandard base pairs do not occur naturally in duplex DNA because the DNA copying

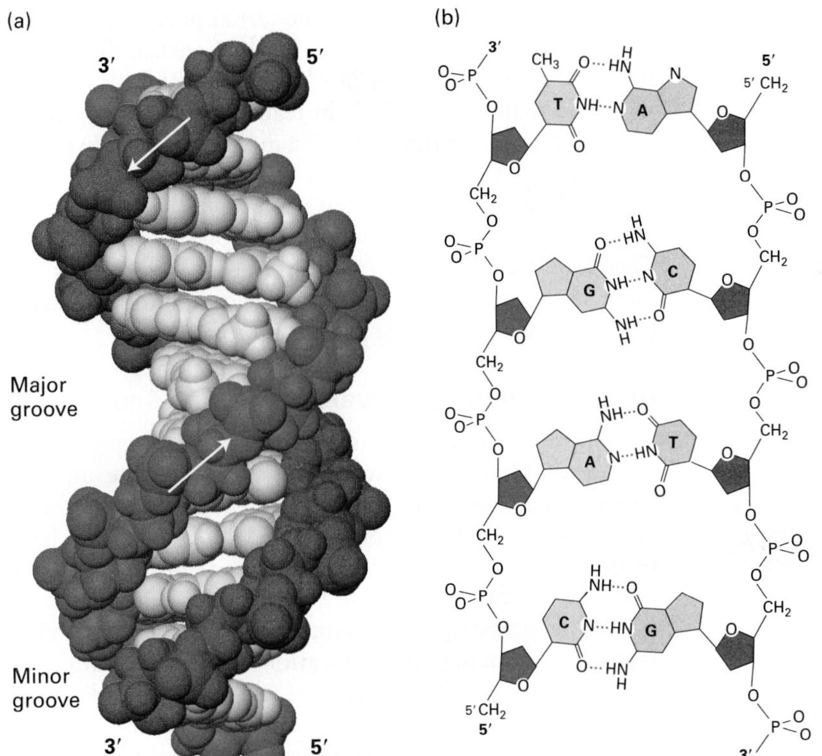

◀ **FIGURE 4-3 The DNA double helix.** (a) Space-filling model of B DNA, the most common form of DNA in cells. The bases (light shades) project inward from the sugar-phosphate backbones (dark red and blue) of each strand, but their edges are accessible through major and minor grooves. Arrows indicate the 5′→3′ direction of each strand. Hydrogen bonds between the bases are in the center of the structure. The major and minor grooves are lined by potential hydrogen bond donors and acceptors (highlighted in yellow). (b) Chemical structure of DNA double helix. This extended schematic shows the two sugar-phosphate backbones and hydrogen bonding between the Watson-Crick base pairs, A·T and G·C. [Part (a) from R. Wing et al., 1980, *Nature* **287**:755; part (b) from R. E. Dickerson, 1983, *Sci. Am.* **249**:94.]

enzyme, which is described later in this chapter, does not permit them.

Most DNA in cells is a *right-handed* helix. The x-ray diffraction pattern of DNA indicates that the stacked bases are regularly spaced 0.34 nm apart along the helix axis. The helix makes a complete turn every 3.4 nm; thus there are about 10.1 pairs per turn. This is referred to as the *B form* of DNA, the normal form present in most DNA stretches in cells. On the outside of B form DNA, the spaces between the intertwined strands form two helical grooves of different widths described as the *major* groove and the *minor* groove (see Figure 4-3a). As a consequence, the atoms on the edges of each base within these grooves are accessible from outside the helix, forming two types of binding surfaces. DNA-binding proteins can read the sequence of bases in duplex DNA by contacting atoms in either the major or the minor grooves.

Other structures of DNA have been described in addition to the major B form. Two of these are compared with B DNA in Figure 4-4. Under laboratory conditions in which most of the water is removed from DNA, the crystallographic structure of B DNA changes to the *A form,* which is wider and shorter than B DNA, with base pairs tilted rather than perpendicular to the helix axis. RNA-DNA and RNA-RNA helices exist in this form in cells and in vitro. Short DNA molecules composed of alternating purine-pyrimidine nucleotides (especially G's and C's) adopt an alternative left-handed helix configuration instead of the normal right-handed helix. This structure is called *Z DNA* because the bases seem to zigzag when viewed from the side. Some evidence suggests that Z DNA may occur in cells, although its function is unknown.

By far, the most important modifications in the structure of standard B form DNA come about as a result of protein binding to specific DNA sequences. Although the multitude of hydrogen and hydrophobic bonds between the bases provide stability to DNA, the double helix is flexible about its long axis. Unlike the α helix in proteins (see Figure 3-4),

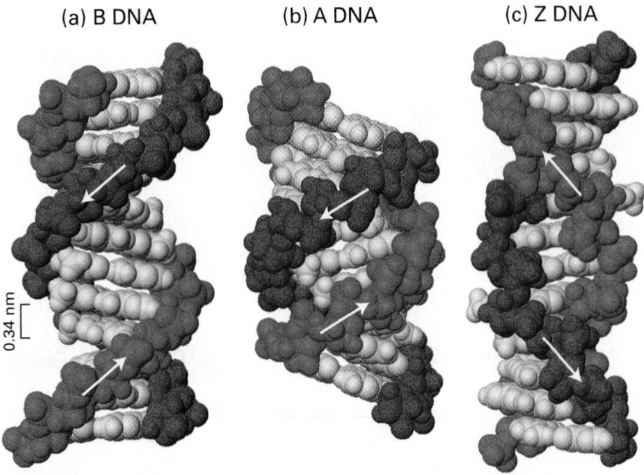

▲ **FIGURE 4-4 Models of various known DNA structures.** The sugar-phosphate backbones of the two strands, which are on the outside in all structures, are shown in red and blue; the bases (lighter shades) are oriented inward. (a) The B form of DNA has ≈10.1 base pairs per helical turn. Adjacent stacked base pairs are 0.34 nm apart. (b) The more compact A form of DNA has 11 base pairs per turn and exhibits a large tilt of the base pairs with respect to the helix axis. (c) Z DNA is a left-handed double helix.

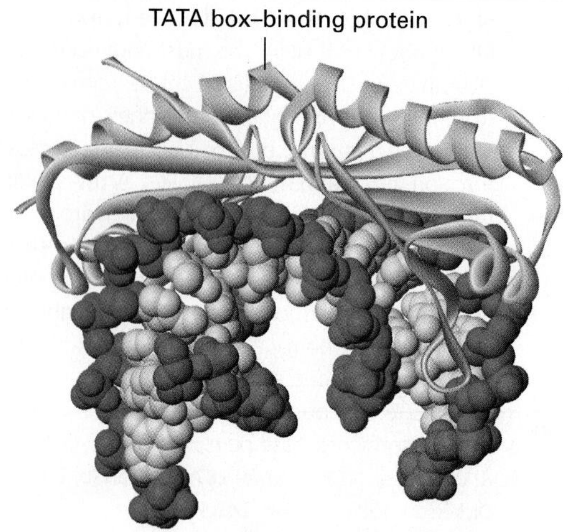

▲ **FIGURE 4-5 Protein interaction can bend DNA.** The conserved C-terminal domain of the TATA box–binding protein (TBP) binds to the minor groove of specific DNA sequences rich in A and T, untwisting and sharply bending the double helix. Transcription of most eukaryotic genes requires participation of TBP. [Adapted from D. B. Nikolov and S. K. Burley, 1997, *Proc. Nat'l Acad. Sci. USA* **94:**15.]

there are no hydrogen bonds parallel to the axis of the DNA helix. This property allows DNA to bend when complexed with a DNA-binding protein (Figure 4-5). Bending of DNA is critical to the dense packing of DNA in chromatin, the protein-DNA complex in which nuclear DNA occurs in eukaryotic cells (Chapter 6).

Why did DNA evolve to be the carrier of genetic information in cells as opposed to RNA? The hydrogen at the 2′ position in the deoxyribose of DNA makes it a far more stable molecule than RNA, which instead has a hydroxyl group at the 2′ position of ribose (See Figure 2-16). The 2′-hydroxyl groups in RNA participate in the slow, OH^--catalyzed hydrolysis of phosphodiester bonds at neutral pH (Figure 4-6). The absence of 2′-hydroxyl groups in DNA prevents this process. Therefore, the presence of deoxyribose in DNA makes it a more stable molecule—a characteristic critical to its function in the long-term storage of genetic information.

DNA Can Undergo Reversible Strand Separation

During replication and transcription of DNA, the strands of the double helix must separate to allow the internal edges of the bases to pair with the bases of the nucleotides being polymerized into new polynucleotide chains. In later sections, we describe the cellular mechanisms that separate and subsequently reassociate DNA strands during replication and transcription. Here we discuss fundamental factors influencing the separation and reassociation of DNA strands. These properties of DNA were elucidated by in vitro experiments.

The unwinding and separation of DNA strands, referred to as **denaturation,** or "melting," can be induced experimentally by increasing the temperature of a solution of DNA. As the thermal energy increases, the resulting increase in molecular motion eventually breaks the hydrogen bonds and other forces that stabilize the double helix;

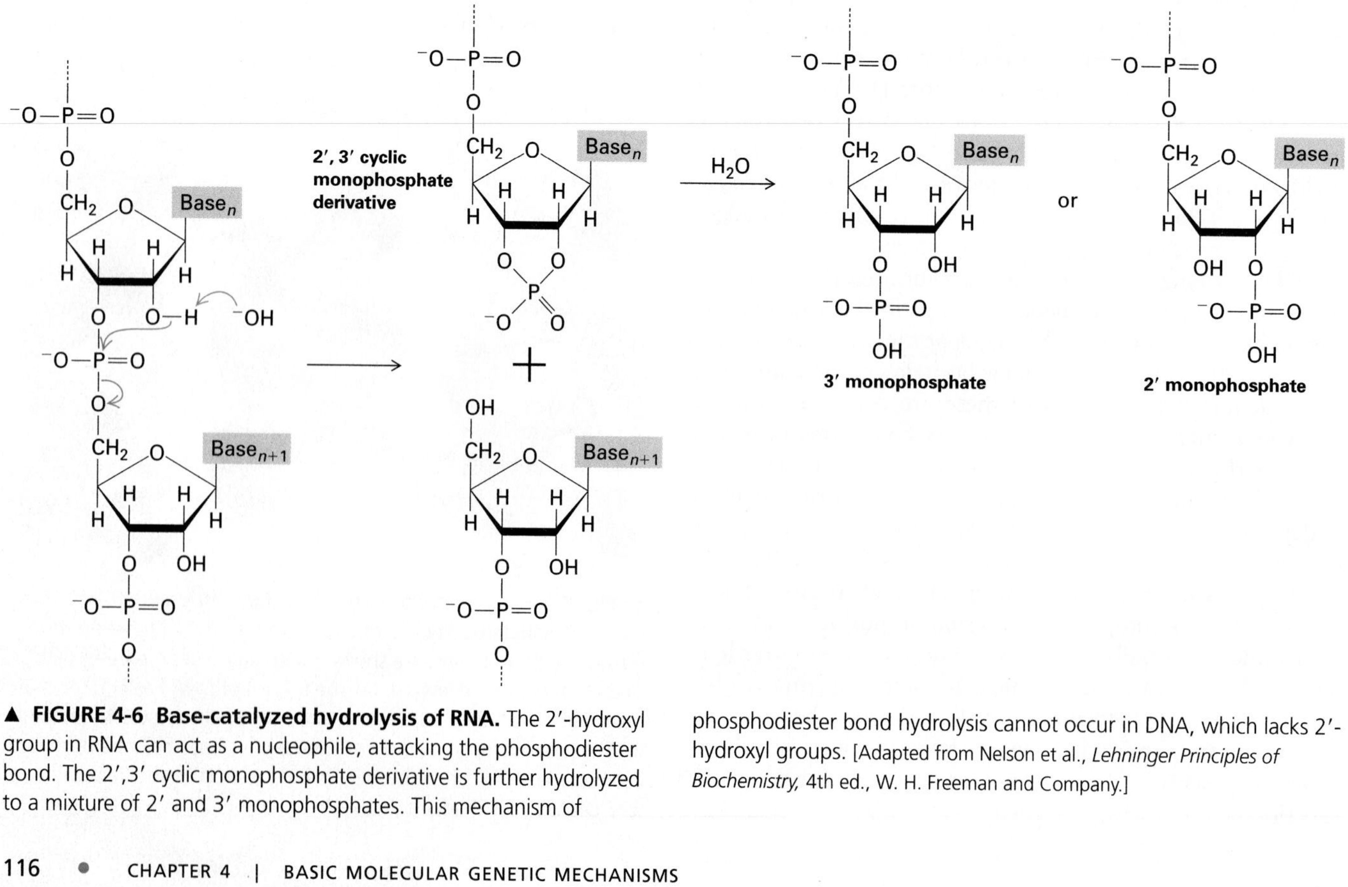

▲ **FIGURE 4-6 Base-catalyzed hydrolysis of RNA.** The 2′-hydroxyl group in RNA can act as a nucleophile, attacking the phosphodiester bond. The 2′,3′ cyclic monophosphate derivative is further hydrolyzed to a mixture of 2′ and 3′ monophosphates. This mechanism of phosphodiester bond hydrolysis cannot occur in DNA, which lacks 2′-hydroxyl groups. [Adapted from Nelson et al., *Lehninger Principles of Biochemistry,* 4th ed., W. H. Freeman and Company.]

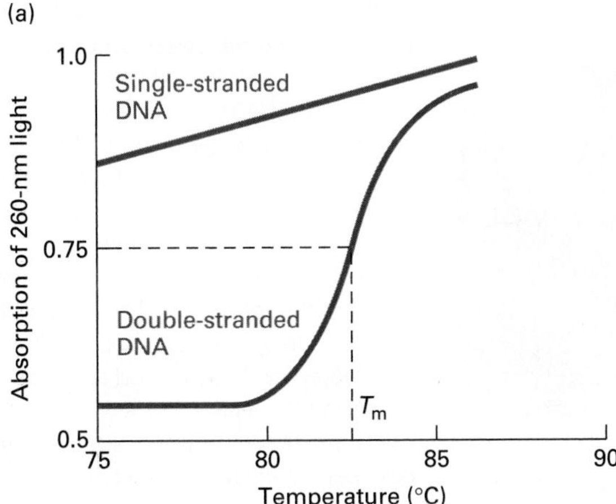

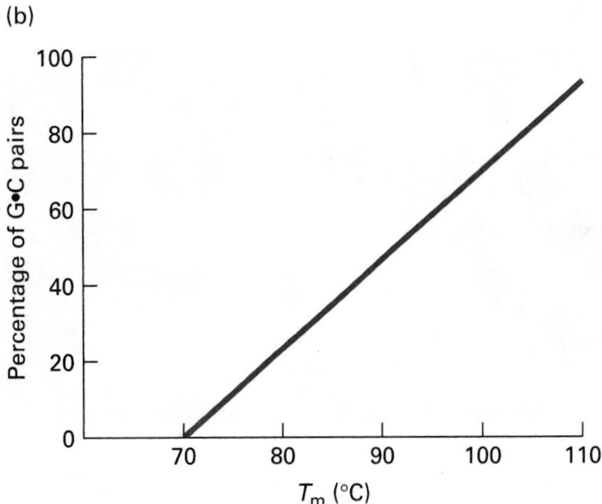

▲ **EXPERIMENTAL FIGURE 4-7 G·C content of DNA affects melting temperature.** The temperature at which DNA denatures increases with the proportion of G·C pairs. (a) Melting of doubled-stranded DNA can be monitored by the absorption of ultraviolet light at 260 nm. As regions of double-stranded DNA unpair, the absorption of light by those regions increases almost twofold. The temperature at which half the bases in a double-stranded DNA sample have denatured is denoted T_m (for "temperature of melting"). Light absorption by single-stranded DNA changes much less as the temperature is increased. (b) The T_m is a function of the G·C content of the DNA; the higher the G+C percentage, the greater the T_m.

the strands then separate, driven apart by the electrostatic repulsion of the negatively charged deoxyribose-phosphate backbone of each strand. Near the denaturation temperature, a small increase in temperature causes a rapid, near simultaneous loss of the multiple weak interactions holding the strands together along the entire length of the DNA molecules. Because the stacked base pairs in duplex DNA absorb less ultraviolet (UV) light than the unstacked bases in single-stranded DNA, this leads to an abrupt increase in the absorption of UV light, a phenomenon known as *hyperchromicity* (Figure 4-7a).

The *melting temperature* (T_m) at which DNA strands will separate depends on several factors. Molecules that contain a greater proportion of G·C pairs require higher temperatures to denature because the three hydrogen bonds in G·C pairs make these base pairs more stable than A·T pairs, which have only two hydrogen bonds. Indeed, the percentage of G·C base pairs in a DNA sample can be estimated from its T_m (Figure 4-7b). The ion concentration also influences the T_m because the negatively charged phosphate groups in the two strands are shielded by positively charged ions. When the ion concentration is low, this shielding is decreased, thus increasing the repulsive forces between the strands and reducing the T_m. Agents that destabilize hydrogen bonds, such as formamide or urea, also lower the T_m. Finally, extremes of pH denature DNA at low temperature. At low (acid) pH, the bases become protonated and thus positively charged, repelling one another. At high (alkaline) pH, the bases lose protons and become negatively charged, again repelling one another because of the similar charge. In cells, pH and temperature are, for the most part, maintained. These features of DNA separation are most useful for manipulating DNA in a laboratory setting.

The single-stranded DNA molecules that result from denaturation form random coils without an organized structure. Lowering the temperature, increasing the ion concentration, or neutralizing the pH causes the two complementary strands to reassociate into a perfect double helix. The extent of such *renaturation* is dependent on time, the DNA concentration, and the ionic concentration. Two DNA strands not related in sequence will remain as random coils and will not renature; most importantly, they will not inhibit complementary DNA partner strands from finding each other and renaturing. Denaturation and renaturation of DNA are the basis of nucleic acid **hybridization,** a powerful technique used to study the relatedness of two DNA samples and to detect and isolate specific DNA molecules in a mixture containing numerous different DNA sequences (see Figure 5-16).

Torsional Stress in DNA Is Relieved by Enzymes

Many prokaryotic genomic DNAs and many viral DNAs are circular molecules. Circular DNA molecules also occur in mitochondria, which are present in almost all eukaryotic cells, and in chloroplasts, which are present in plants and some unicellular eukaryotes.

Each of the two strands in a circular DNA molecule forms a closed structure without free ends. Localized unwinding of a circular DNA molecule, which occurs during DNA replication, induces torsional stress into the remaining portion of the molecule because the ends of the strands are not free to rotate. As a result, the DNA molecule twists back on itself, like a twisted rubber band, forming *supercoils* (Figure 4-8a). In other words, when part of the DNA helix is underwound, the remainder of the molecule becomes overwound. Bacterial and eukaryotic cells, however, contain *topoisomerase I,* which can relieve any torsional stress that develops in cellular DNA molecules during replication or other processes. This enzyme binds to DNA

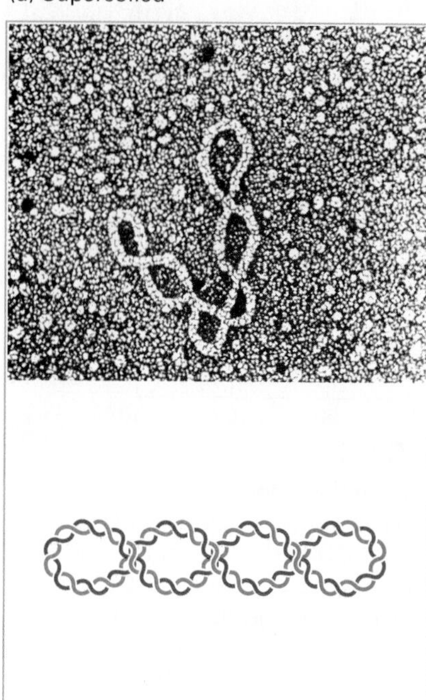

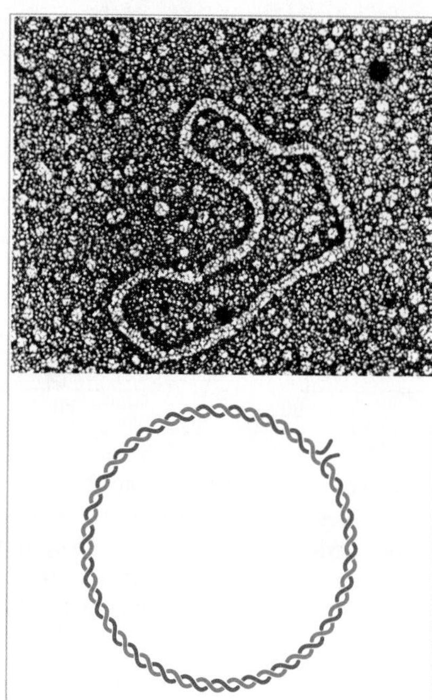

◀ **EXPERIMENTAL FIGURE 4-8 Topoisomerase I relieves torsional stress on DNA.** (a) Electron micrograph of SV40 viral DNA. When the circular DNA of the SV40 virus is isolated and separated from its associated protein, the DNA duplex is underwound and assumes the supercoiled configuration. (b) If a supercoiled DNA is nicked (i.e., one strand cleaved), the strands can rewind, leading to loss of a supercoil. Topoisomerase I catalyzes this reaction and also reseals the broken ends. All the supercoils in isolated SV40 DNA can be removed by the sequential action of this enzyme, producing the relaxed-circle conformation. For clarity, the shapes of the molecules at the bottom have been simplified.

at random sites and breaks a phosphodiester bond in one strand. Such a one-strand break in DNA is called a *nick*. The broken end then unwinds around the uncut strand, leading to loss of supercoils (Figure 4-8b). Finally, the same enzyme joins (ligates) the two ends of the broken strand. Another type of enzyme, *topoisomerase II*, makes breaks in both strands of a double-stranded DNA and then religates them. As a result, topoisomerase II can both relieve torsional stress and link together two circular DNA molecules as in the links of a chain.

Although eukaryotic nuclear DNA is linear, long loops of DNA are fixed in place within chromosomes (Chapter 6). Thus torsional stress and the consequent formation of supercoils also could occur during replication of nuclear DNA. As in bacterial cells, abundant topoisomerase I in eukaryotic nuclei relieves any torsional stress in nuclear DNA that would develop in the absence of this enzyme.

Different Types of RNA Exhibit Various Conformations Related to Their Functions

The primary structure of RNA is generally similar to that of DNA with two exceptions: the sugar component of RNA, ribose, has a hydroxyl group at the 2′ position (see Figure 2-16b), and thymine in DNA is replaced by uracil in RNA. The presence of thymine rather than uracil in DNA is important to the long-term stability of DNA because of its function in DNA repair (see Section 4.6). As noted earlier, the hydroxyl group on C_2 of ribose makes RNA more chemically labile than DNA. As a result of this lability, RNA is cleaved into mononucleotides by alkaline solution (see Figure 4-6), whereas DNA is not. The C_2 hydroxyl of RNA also provides a chemically reactive group that takes part in RNA-mediated catalysis. Like DNA, RNA is a long polynucleotide that can be double-stranded or single-stranded, linear or circular. It can also participate in a hybrid helix composed of one RNA strand and one DNA strand. As discussed above, RNA-RNA and RNA-DNA double helices have a compact conformation like the A form of DNA (see Figure 4-4b).

Unlike DNA, which exists primarily as a very long double helix, most cellular RNAs are single-stranded and exhibit a variety of conformations (Figure 4-9). Differences in the sizes and conformations of the various types of RNA permit them to carry out specific functions in a cell. The simplest secondary structures in single-stranded RNAs are formed by pairing of complementary bases within a linear sequence. "Hairpins" are formed by pairing of bases within $\approx$5–10 nucleotides of each other, and "stem-loops" by pairing of bases that are separated by $>$10 to several hundred nucleotides. These simple folds can cooperate to form more complicated tertiary structures, one of which is termed a "pseudoknot."

As discussed in detail later, tRNA molecules adopt a well-defined three-dimensional architecture in solution that is crucial in protein synthesis. Larger rRNA molecules also have locally well-defined three-dimensional structures, with more flexible links in between. Secondary and tertiary structures also have been recognized in mRNA, particularly near the ends of molecules. Clearly, then, RNA molecules are like proteins in that they have structured domains connected by less structured, flexible stretches.

The folded domains of RNA molecules not only are structurally analogous to the α helices and β strands found in proteins, but in some cases also have catalytic capacities.

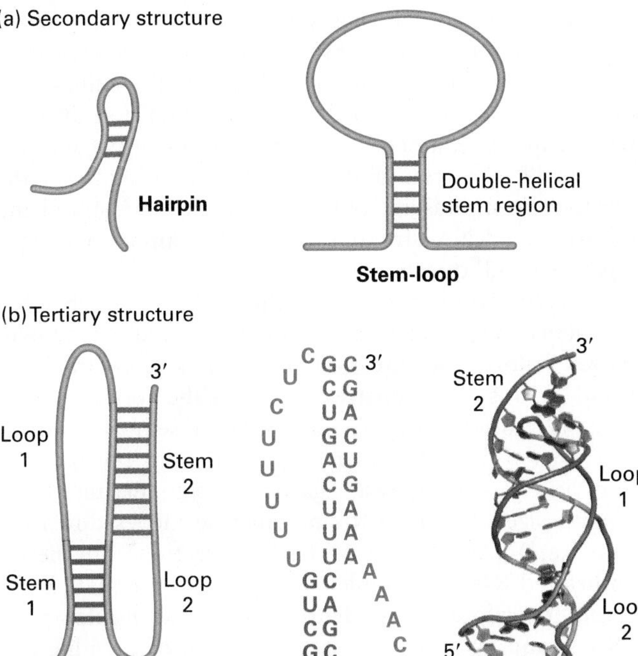

▲ **FIGURE 4-9 RNA secondary and tertiary structures.** (a) Stem-loops, hairpins, and other secondary structures can form by base pairing between distant complementary segments of an RNA molecule. In stem-loops, the single-stranded loop between the base-paired helical stem may be hundreds or even thousands of nucleotides long, whereas in hairpins, the short turn may contain as few as four nucleotides. (b) Pseudoknots, one type of RNA tertiary structure, are formed by interaction of secondary loops through base pairing between complementary bases. The structure shown forms the core domain of the human telomerase RNA. *Left:* Secondary-structure diagram with base-paired nucleotides in green and blue and single-stranded regions in red. *Middle:* Sequence of the telomerase RNA core domain, colored to correspond to the secondary-structure diagram at the left. *Right:* Diagram of the telomerase core domain structure determined by 2D-NMR, showing base-paired bases only and a tube for the sugar-phosphate backbone, colored to correspond to the diagrams to the left. [Part (b), *middle and right,* adapted from C. A. Theimer et al., 2005, *Mol. Cell* **17:**671.]

Such catalytic RNAs are called **ribozymes.** Although ribozymes usually are associated with proteins that stabilize the ribozyme structure, it is the RNA that acts as a catalyst. Some ribozymes can catalyze splicing, a remarkable process in which an internal RNA sequence is cut and removed, and the two resulting chains then ligated. This process occurs during formation of the majority of functional mRNA molecules in multicellular eukaryotes, and also occurs in single-celled eukaryotes such as yeast, bacteria, and archaea. Remarkably, some RNAs carry out *self-splicing,* with the catalytic activity residing in the sequence that is removed. The mechanisms of splicing and self-splicing are discussed in detail in Chapter 8. As noted later in this chapter, rRNA plays a catalytic role in the formation of peptide bonds during protein synthesis.

In this chapter, we focus on the functions of mRNA, tRNA, and rRNA in gene expression. In later chapters we will encounter other RNAs, often associated with proteins, that participate in other cell functions.

KEY CONCEPTS OF SECTION 4.1

Structure of Nucleic Acids

- Deoxyribonucleic acid (DNA), the genetic material, carries information to specify the amino acid sequences of proteins. It is transcribed into several types of ribonucleic acid (RNA), including messenger RNA (mRNA), transfer RNA (tRNA), and ribosomal RNA (rRNA), which function in protein synthesis (see Figure 4-1).
- Both DNA and RNA are long, unbranched polymers of nucleotides, which consist of a phosphorylated pentose linked to an organic base, either a purine or pyrimidine.
- The purines adenine (A) and guanine (G) and the pyrimidine cytosine (C) are present in both DNA and RNA. The pyrimidine thymine (T) present in DNA is replaced by the pyrimidine uracil (U) in RNA.
- Adjacent nucleotides in a polynucleotide are linked by phosphodiester bonds. The entire strand has a chemical directionality with 5′ and 3′ ends (see Figure 4-2).
- Natural DNA (B DNA) contains two complementary antiparallel polynucleotide strands wound together into a regular right-handed double helix with the bases on the inside and the two sugar-phosphate backbones on the outside (see Figure 4-3). Base pairing between the strands and hydrophobic interactions between adjacent base pairs stacked perpendicular to the helix axis stabilize this native structure.
- The bases in nucleic acids can interact via hydrogen bonds. The standard Watson-Crick base pairs are G·C, A·T (in DNA), and G·C, A·U (in RNA). Base pairing stabilizes the native three-dimensional structures of DNA and RNA.
- Binding of protein to DNA can deform its helical structure, causing local bending or unwinding of the DNA molecule.
- Heat causes the DNA strands to separate (denature). The melting temperature T_m of DNA increases with the percentage of G·C base pairs. Under suitable conditions, separated complementary nucleic acid strands will renature.
- Circular DNA molecules can be twisted on themselves, forming supercoils (see Figure 4-8). Enzymes called *topoisomerases* can relieve torsional stress and remove supercoils from circular DNA molecules. Long linear DNA can also experience torsional stress because long loops are fixed in place within chromosomes.
- Cellular RNAs are single-stranded polynucleotides, some of which form well-defined secondary and tertiary structures (see Figure 4-9). Some RNAs, called *ribozymes,* have catalytic activity.

4.2 Transcription of Protein-Coding Genes and Formation of Functional mRNA

The simplest definition of a gene is a "unit of DNA that contains the information to specify synthesis of a single polypeptide chain or functional RNA (such as a tRNA)." The DNA molecules of small viruses contain only a few genes, whereas the single DNA molecule in each of the chromosomes of higher animals and plants may contain several thousand genes. The vast majority of genes carry information to build protein molecules, and it is the RNA copies of such *protein-coding genes* that constitute the mRNA molecules of cells.

During synthesis of RNA, the four-base language of DNA containing A, G, C, and T is simply copied, or *transcribed,* into the four-base language of RNA, which is identical except that U replaces T. In contrast, during protein synthesis, the four-base language of DNA and RNA is *translated* into the 20–amino acid language of proteins. In this section, we focus on formation of functional mRNAs from protein-coding genes (see Figure 4-1, **1**). A similar process yields the precursors of rRNAs and tRNAs encoded by rRNA and tRNA genes; these precursors are then further modified to yield functional rRNAs and tRNAs. In addition, thousands of recently discovered **micro RNAs (miRNAs)** that function to regulate translation of specific target mRNAs and transcription of specific target genes are transcribed into precursors by RNA polymerases and processed into functional miRNAs. Transcription and processing of these other types of RNA is discussed in Chapter 8. Regulation of transcription allows distinct sets of genes to be expressed in the many different types of cells that make up a multicellular organism. It also allows different amounts of mRNA to be transcribed from different genes, resulting in differences in the amounts of the encoded proteins in a cell. Regulation of transcription is addressed in Chapter 7.

A Template DNA Strand Is Transcribed into a Complementary RNA Chain by RNA Polymerase

During transcription of DNA, one DNA strand acts as a *template,* determining the order in which ribonucleoside triphosphate (rNTP) monomers are polymerized to form a complementary RNA chain. Bases in the template DNA strand base-pair with complementary incoming rNTPs, which then are joined in a polymerization reaction catalyzed by **RNA polymerase.** Polymerization involves a nucleophilic attack by the 3′ oxygen in the growing RNA chain on the α phosphate of the next nucleotide precursor to be added, resulting in formation of a phosphodiester bond and release of pyrophosphate (PP_i). As a consequence of this mechanism, RNA molecules are always synthesized in the 5′→3′ direction (Figure 4-10a).

The energetics of the polymerization reaction strongly favor addition of ribonucleotides to the growing RNA chain because the high-energy bond between the α and β phosphate of rNTP monomers is replaced by the lower-energy phosphodiester bond between nucleotides. The equilibrium for the reaction is driven farther toward chain elongation by pyrophosphatase, an enzyme that catalyzes cleavage of the released PP_i into two molecules of inorganic phosphate. Like the two strands in DNA, the template DNA strand and the growing RNA strand that is base-paired to it have opposite 5′→3′ directionality.

By convention, the site on the DNA at which RNA polymerase begins transcription is numbered +1. **Downstream** denotes the direction in which a template DNA strand is transcribed; **upstream** denotes the opposite direction. Nucleotide positions in the DNA sequence downstream from a start site are indicated by a positive (+) sign; those upstream, by a negative (−) sign. Because RNA is synthesized 5′→3′, RNA polymerase moves down the template DNA strand in a 3′→5′ direction. The newly synthesized RNA is complementary to the template DNA strand; therefore, it is identical with the nontemplate DNA strand, except with uracil in place of thymine (see Figure 4-10b).

Stages in Transcription To carry out transcription, RNA polymerase performs several distinct functions, as depicted in Figure 4-11. During transcription *initiation,* RNA polymerase recognizes and binds to a specific site, called a **promoter,** in double-stranded DNA (step **1**). RNA polymerases require various protein factors, called general **transcription factors,** to help them locate promoters and initiate transcription. After binding to a promoter, RNA polymerase separates the DNA strands in order to make the bases in the template strand available for base pairing with the bases of the ribonucleoside triphosphates that it will polymerize together. RNA polymerases melt 12–14 base pairs of DNA around the transcription start site, which is located on the template strand within the promoter region (step **2**). This allows the template strand to enter the active site of the enzyme that catalyzes phosphodiester bond formation between ribonucleoside triphosphates that are complementary to the promoter template strand at the start site of transcription. The 12–14-base-pair region of melted DNA in the polymerase is known as the "transcription bubble." Transcription initiation is considered complete when the first two ribonucleotides of an RNA chain are linked by a phosphodiester bond (step **3**).

After several ribonucleotides have been polymerized, RNA polymerase dissociates from the promoter DNA and general transcription factors. During the stage of *strand elongation,* RNA polymerase moves along the template DNA one base at a time, opening the double-stranded DNA in front of its direction of movement and guiding the strands together so that they hybridize at the upstream end of the transcription bubble (Figure 4-11, step **4**). One ribonucleotide at a time is added to the 3′ end of the growing (*nascent*) RNA chain during strand elongation by the polymerase. The enzyme maintains a melted region of approximately 14 base pairs, the *transcription bubble.* Approximately eight nucleotides at

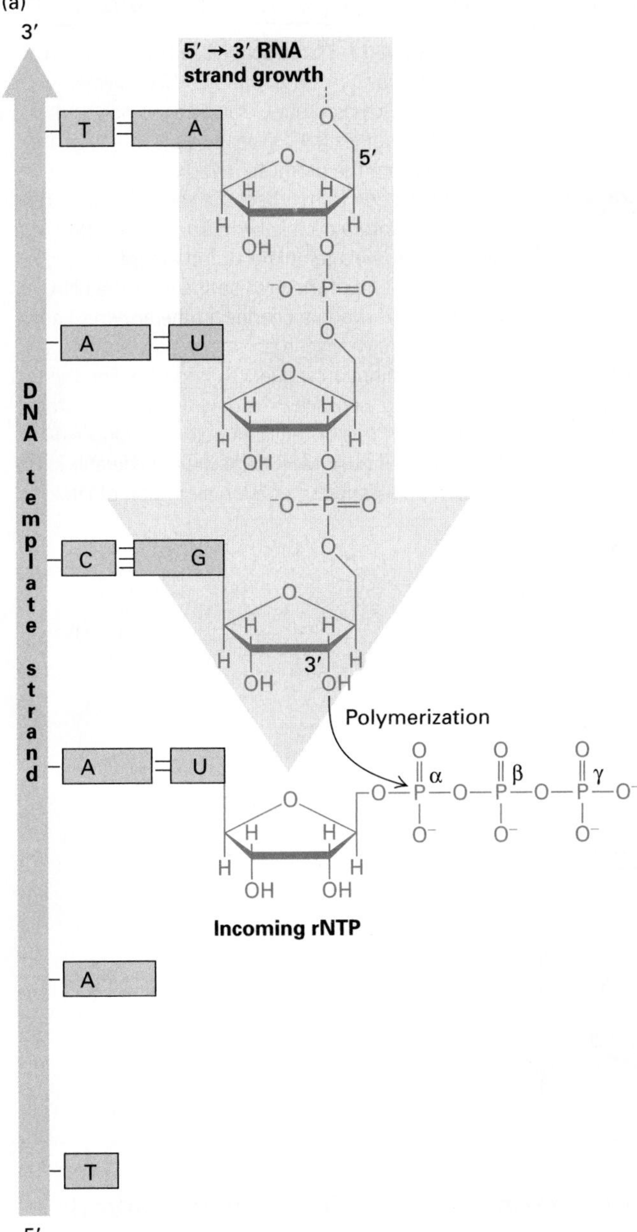

the 3′ end of the growing RNA strand remain base-paired to the template DNA strand in the transcription bubble. The elongation complex, comprising RNA polymerase, template DNA, and the growing (nascent) RNA strand, is extraordinarily stable. For example, RNA polymerase transcribes the longest known mammalian gene, containing about 2 million base pairs, without dissociating from the DNA template or releasing the nascent RNA. Since RNA synthesis occurs at a rate of about 1000 nucleotides per minute at 37 °C, the elongation complex must remain intact for more than 24 hours to assure continuous RNA synthesis of the pre-mRNA from this very long gene.

During transcription *termination*, the final stage in RNA synthesis, the completed RNA molecule, or **primary transcript**, is released from the RNA polymerase, and the polymerase dissociates from the template DNA (Figure 4-11, step **5**). Specific sequences in the template DNA signal the bound RNA polymerase to terminate transcription. Once released,

◀ **FIGURE 4-10 RNA is synthesized 5′→3′.** (a) Polymerization of ribonucleotides by RNA polymerase during transcription. The ribonucleotide to be added at the 3′ end of a growing RNA strand is specified by base-pairing between the next base in the template DNA strand and the complementary incoming ribonucleoside triphosphate (rNTP). A phosphodiester bond is formed when RNA polymerase catalyzes a reaction between the 3′ O of the growing strand and the α phosphate of a correctly base-paired rNTP. RNA strands always are synthesized in the 5′→3′ direction and are opposite in polarity to their template DNA strands. (b) Conventions for describing RNA transcription. *Top:* The DNA nucleotide where RNA polymerase begins transcription is designated +1. The direction the polymerase travels on the DNA is "downstream," and bases are marked with positive numbers. The opposite direction is "upstream," and bases are noted with negative numbers. Some important gene features lie upstream of the transcription start site, including the promoter sequence that localizes RNA polymerase to the gene. *(Bottom)* The DNA strand that is being transcribed is the template strand; its complement, the nontemplate strand. The RNA being synthesized is complementary to the template strand, and therefore identical with the nontemplate strand sequence, except with uracil in place of thymine. [Part (b) adapted from Griffiths et al., *Modern Genetic Analysis*, 2d ed., W. H. Freeman and Company.]

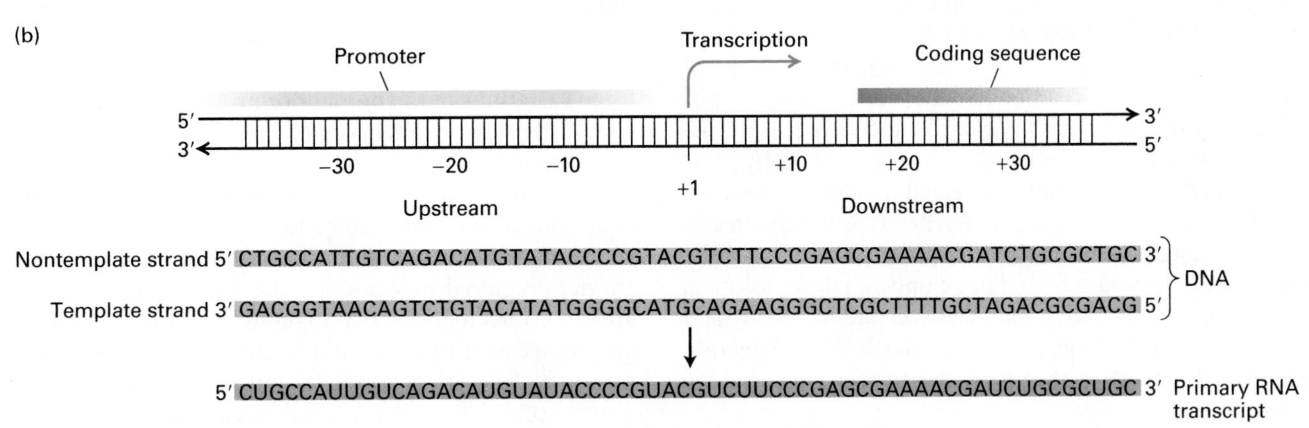

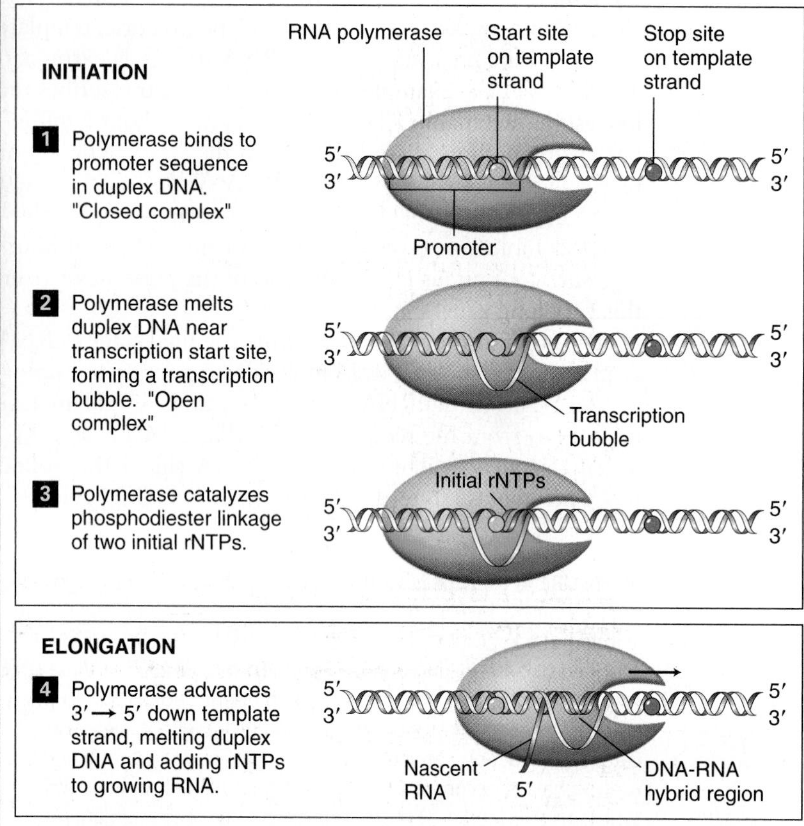

Initial rNTPs

4 Polymerase advances 3′→ 5′ down template strand, melting duplex DNA and adding rNTPs to growing RNA.

DNA-RNA hybrid region

◀ **FIGURE 4-11 Three stages in transcription.** During initiation of transcription, RNA polymerase forms a transcription bubble and begins polymerization of ribonucleotides (rNTPs) at the start site, which is located within the promoter region. Once a DNA region has been transcribed, the separated strands reassociate into a double helix. The nascent RNA is displaced from its template strand except at its 3′ end. The 5′ end of the RNA strand exits the RNA polymerase through a channel in the enzyme. Termination occurs when the polymerase encounters a specific termination sequence (stop site). See the text for details. For simplicity, the diagram depicts transcription of four turns of the DNA helix encoding ≈40 nucleotides of RNA. Most RNAs are considerably longer, requiring transcription of a longer region of DNA.

an RNA polymerase is free to transcribe the same gene again or another gene.

Structure of RNA Polymerases The RNA polymerases of bacteria, archaea, and eukaryotic cells are fundamentally similar in structure and function. Bacterial RNA polymerases are composed of two related large subunits (β' and β), two copies of a smaller subunit (α), and one copy of a fifth subunit (ω) that is not essential for transcription or cell viability but stabilizes the enzyme and assists in the assembly of its subunits. Archaeal and eukaryotic RNA polymerases have several additional small subunits associated with this core complex, which we describe in Chapter 7. Schematic diagrams of the transcription process generally show RNA polymerase bound to an unbent DNA molecule, as in Figure 4-11. However, X-ray crystallography and other studies of an elongating bacterial RNA polymerase indicate that the DNA bends at the transcription bubble (Figure 4-12).

Organization of Genes Differs in Prokaryotic and Eukaryotic DNA

Having outlined the process of transcription, we now briefly consider the large-scale arrangement of information in DNA and how this arrangement dictates the requirements for RNA synthesis so that information transfer goes smoothly. In recent years, sequencing of the entire **genomes** from several organisms has revealed not only large variations in the number of protein-coding genes but also differences in their organization in prokaryotes and eukaryotes.

The most common arrangement of protein-coding genes in all prokaryotes has a powerful and appealing logic: genes encoding proteins that function together, for example, the enzymes required to synthesize the amino acid tryptophan, are most often found in a contiguous array in the DNA. Such an arrangement of genes in a functional group is called an **operon** because it operates as a unit from a single promoter. Transcription of an operon produces a continuous strand of

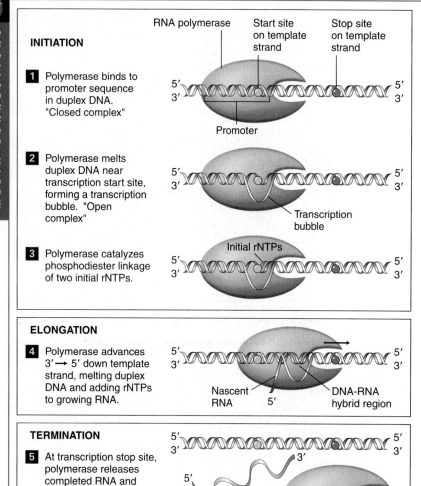

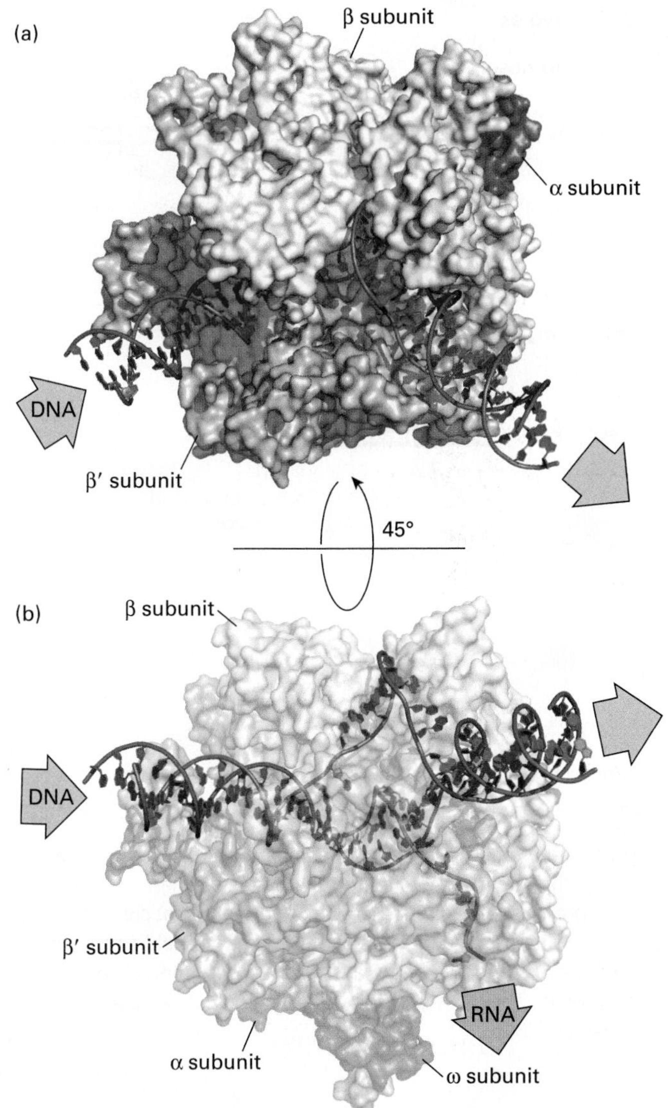

▲ **FIGURE 4-12 Bacterial RNA polymerase.** This structure corresponds to the polymerase molecule in the elongation phase (step **4**) of Figure 4-11. In these diagrams, transcription is proceeding in the leftward direction. Arrows indicate where downstream DNA enters the polymerase and upstream DNA exits at an angle from the downstream DNA; the coding strand is red, the noncoding strand blue, nascent RNA green. The RNA polymerase β′ subunit is gold, β is light gray, and the α subunit visible from this angle is brown. In (a) a space-filling model of the elongation complex is viewed from an angle that emphasizes the bend in the DNA as it passes through the polymerase. The elongation complex is rotated in (b) as shown, and proteins are made largely transparent to reveal the structure of the transcription bubble inside the polymerase that is not visible in the space-filling model. Nucleotides complementary to the template DNA are added to the 3′-end of the nascent RNA strand (at the left). The newly synthesized nascent RNA exits the polymerase at the bottom through a channel formed between the β and β′ subunits. The ω subunit and the other α subunit are visible from this angle. [Courtesy of Seth Darst; see N. Korzheva et al., 2000, *Science* **289,** 619–625, and N. Opalka et al., 2003, *Cell* **114:**335–345.]

mRNA that carries the message for a related series of proteins (Figure 4-13a). Each section of the mRNA represents the unit (or gene) that encodes one of the proteins in the series. This arrangement results in the *coordinate expression* of all the genes in the operon. Every time an RNA polymerase molecule initiates transcription at the promoter of the operon, all the genes of the operon are transcribed and translated. In prokaryotic DNA the genes are closely packed with very few noncoding gaps, and the DNA is transcribed directly into mRNA. Because DNA is not sequestered in a nucleus in prokaryotes, ribosomes have immediate access to the translation start sites in the mRNA as they emerge from the surface of the RNA polymerase. Consequently, translation of the mRNA begins even while the 3′ end of the mRNA is still being synthesized at the active site of the RNA polymerase.

This economic clustering of genes devoted to a single metabolic function does not occur in eukaryotes, even simple ones like yeasts, which can be metabolically similar to bacteria. Rather, eukaryotic genes encoding proteins that function together are most often physically separated in the DNA; indeed such genes usually are located on different chromosomes. Each gene is transcribed from its own promoter, producing one mRNA, which generally is translated to yield a single polypeptide (Figure 4-13b).

When researchers first compared the nucleotide sequences of eukaryotic mRNAs from multicellular organisms with the DNA sequences encoding them, they were surprised to find that the uninterrupted protein-coding sequence of a given mRNA was discontinuous in its corresponding section of DNA. They concluded that the eukaryotic gene existed in pieces of coding sequence, the **exons,** separated by non-protein-coding segments, the **introns.** This astonishing finding implied that the long initial primary transcript—the RNA copy of the entire transcribed DNA sequence—had to be clipped apart to remove the introns and then carefully stitched back together to produce eukaryotic mRNAs.

Although introns are common in multicellular eukaryotes, they are extremely rare in bacteria and archaea and uncommon in many unicellular eukaryotes such as baker's yeast. However, introns are present in the DNA of viruses that infect eukaryotic cells. Indeed, the presence of introns was first discovered in such viruses, whose DNA is transcribed by host-cell enzymes.

Eukaryotic Precursor mRNAs Are Processed to Form Functional mRNAs

In prokaryotic cells, which have no nuclei, translation of an mRNA into protein can begin from the 5′ end of the mRNA even while the 3′ end is still being synthesized by RNA polymerase. In other words, transcription and translation occur concurrently in prokaryotes. In eukaryotic cells, however, not only is the site of RNA synthesis—the nucleus—separated from the site of translation—the cytoplasm—but also the primary transcripts of protein-coding genes are precursor mRNAs (**pre-mRNAs**) that must undergo several

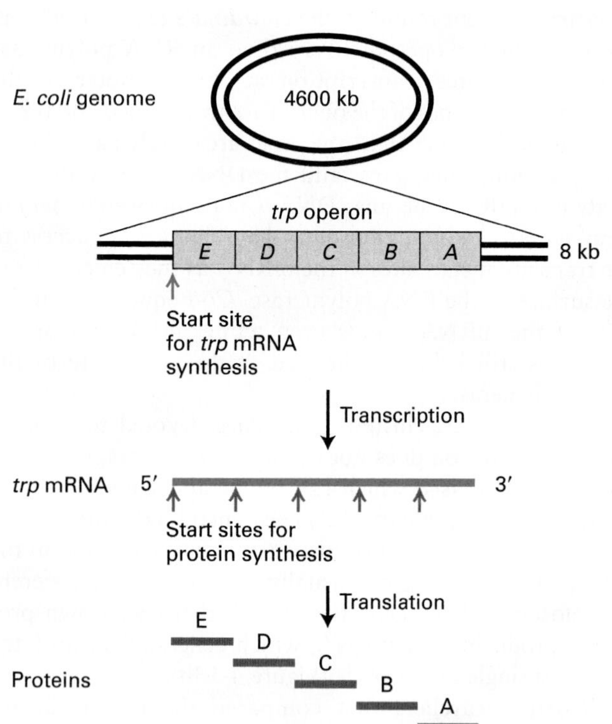

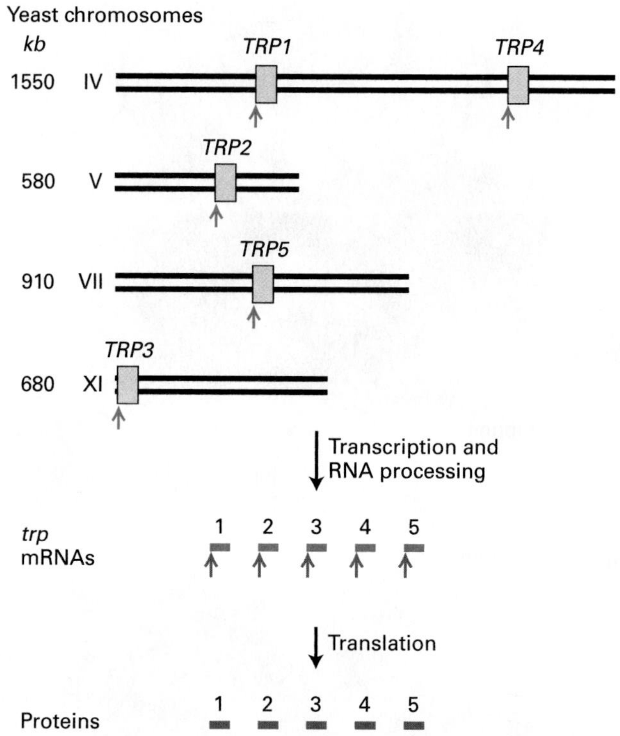

▲ **FIGURE 4-13 Gene organization in prokaryotes and eukaryotes.** (a) The tryptophan (*trp*) operon is a continuous segment of the *E. coli* chromosome, containing five genes (blue) that encode the enzymes necessary for the stepwise synthesis of tryptophan. The entire operon is transcribed from one promoter into one long continuous *trp* mRNA (red). Translation of this mRNA begins at five different start sites, yielding five proteins (green). The order of the genes in the bacterial genome parallels the sequential function of the encoded proteins in the tryptophan pathway. (b) The five genes encoding the enzymes required for tryptophan synthesis in yeast (*Saccharomyces cerevisiae*) are carried on four different chromosomes. Each gene is transcribed from its own promoter to yield a primary transcript that is processed into a functional mRNA encoding a single protein. The lengths of the various chromosomes are given in kilobases (10^3 bases).

modifications, collectively termed *RNA processing,* to yield a functional mRNA (see Figure 4-1, **2**). This mRNA then must be exported to the cytoplasm before it can be translated into protein. Thus transcription and translation cannot occur concurrently in eukaryotic cells.

All eukaryotic pre-mRNAs initially are modified at the two ends, and these modifications are retained in mRNAs. As the 5′ end of a nascent RNA chain emerges from the surface of RNA polymerase, it is immediately acted on by several enzymes that together synthesize the *5′ cap,* a 7-methylguanylate that is connected to the terminal nucleotide of the RNA by an unusual 5′,5′ triphosphate linkage (Figure 4-14). The cap protects an mRNA from enzymatic degradation and assists in its export to the cytoplasm. The cap also is bound by a protein factor required to begin translation in the cytoplasm.

Processing at the 3′ end of a pre-mRNA involves cleavage by an endonuclease to yield a free 3′-hydroxyl group to which a string of adenylic acid residues is added one at a time by an enzyme called *poly(A) polymerase.* The resulting *poly(A) tail* contains 100–250 bases, being shorter in yeasts and invertebrates than in vertebrates. Poly(A) polymerase is part of a complex of proteins that can locate and cleave a transcript at a specific site and then add the correct number of A residues, in a process that does not require a template.

The final step in the processing of many different eukaryotic mRNA molecules is **RNA splicing:** the internal cleavage of a transcript to excise the introns, followed by ligation of the coding exons. Figure 4-15 summarizes the basic steps in eukaryotic mRNA processing, using the β-globin gene as an example. We examine the cellular machinery for carrying out processing of mRNA, as well as tRNA and rRNA, in Chapter 8.

The functional eukaryotic mRNAs produced by RNA processing retain noncoding regions, referred to as 5′ and 3′ *untranslated regions (UTRs),* at each end. In mammalian mRNAs, the 5′ UTR may be a hundred or more nucleotides long, and the 3′ UTR may be several kilobases in length. Prokaryotic mRNAs also usually have 5′ and 3′ UTRs, but these are much shorter than those in eukaryotic mRNAs, generally containing fewer than 10 nucleotides.

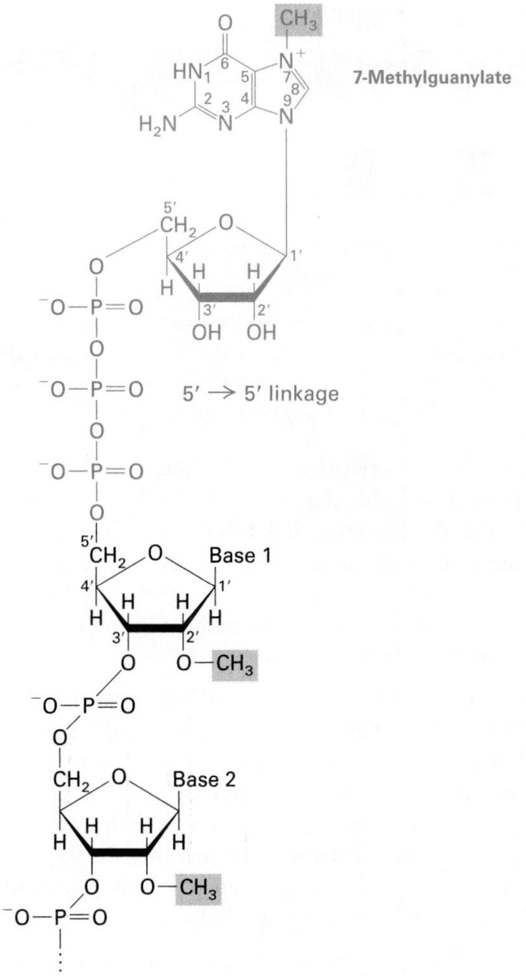

7-Methylguanylate

5′ → 5′ linkage

Base 1

Base 2

◀ **FIGURE 4-14 Structure of the 5′ methylated cap.** The distinguishing chemical features of the 5′ methylated cap on eukaryotic mRNA are (1) the 5′→5′ linkage of 7-methylguanylate to the initial nucleotide of the mRNA molecule and (2) the methyl group on the 2′ hydroxyl of the ribose of the first nucleotide (base 1). Both these features occur in all animal cells and in cells of higher plants; yeasts lack the methyl group on nucleotide 1. The ribose of the second nucleotide (base 2) also is methylated in vertebrates. [See A. J. Shatkin, 1976, *Cell* **9**:645.]

Alternative RNA Splicing Increases the Number of Proteins Expressed from a Single Eukaryotic Gene

In contrast to bacterial and archaeal genes, the vast majority of genes in higher, multicellular eukaryotes contain multiple introns. As noted in Chapter 3, many proteins from higher eukaryotes have a multidomain tertiary structure (see Figure 3-11). Individual repeated protein domains often are encoded by one exon or a small number of exons that code for identical or nearly identical amino acid sequences. Such repeated exons are thought to have evolved by the accidental multiple duplication of a length of DNA lying between two sites in adjacent introns, resulting in insertion of a string of repeated exons, separated by introns, between the original two introns. The presence of multiple introns in many eukaryotic genes permits expression of multiple, related proteins from a single gene by means of **alternative splicing.** In higher eukaryotes, alternative splicing is an important mechanism for production of different forms of a protein, called **isoforms,** by different types of cells.

Overview Animation: Life Cycle of an mRNA

MEDIA CONNECTIONS

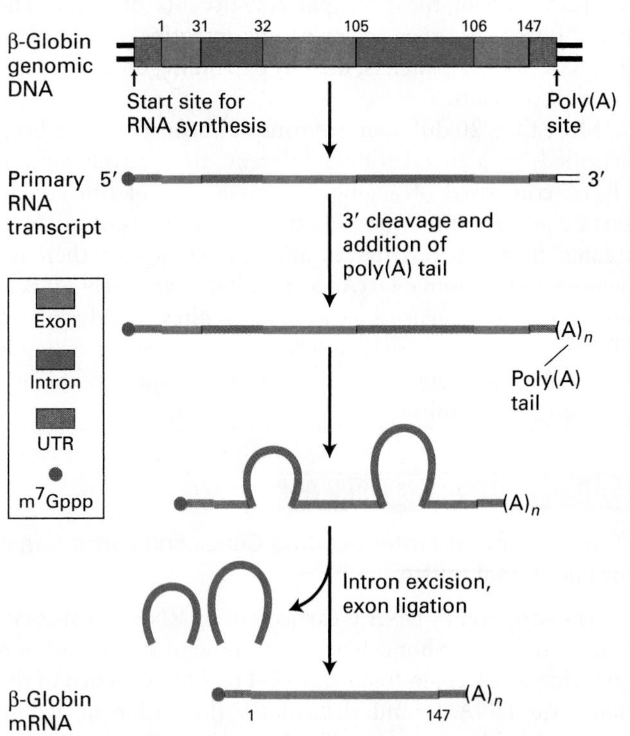

▶ **FIGURE 4-15 Overview of RNA processing.** RNA processing produces functional mRNA in eukaryotes. The β-globin gene contains three protein-coding exons (constituting the coding region, red) and two intervening noncoding introns (blue). The introns interrupt the protein-coding sequence between the codons for amino acids 31 and 32 and 105 and 106. Transcription of eukaryotic protein-coding genes starts before the sequence that encodes the first amino acid and extends beyond the sequence encoding the last amino acid, resulting in noncoding regions (gray) at the ends of the primary transcript. These untranslated regions (UTRs) are retained during processing. The 5′ cap (m^7Gppp) is added during formation of the primary RNA transcript, which extends beyond the poly(A) site. After cleavage at the poly(A) site and addition of multiple A residues to the 3′ end, splicing removes the introns and joins the exons. The small numbers refer to positions in the 147–amino acid sequence of β-globin.

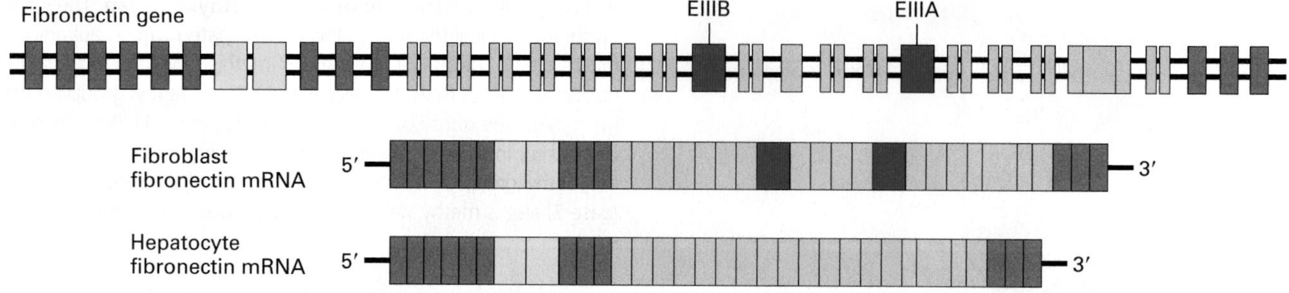

▲ **FIGURE 4-16 Alternative splicing.** The ≈75-kb fibronectin gene *(top)* contains multiple exons; splicing of fibronectin varies by cell type. The EIIIB and EIIIA exons (green) encode binding domains for specific proteins on the surface of fibroblasts. The fibronectin mRNA produced in fibroblasts includes the EIIIA and EIIIB exons, whereas these exons are spliced out of fibronectin mRNA in hepatocytes. In this diagram, introns (black lines) are not drawn to scale; most of them are much longer than any of the exons.

Fibronectin, a multidomain protein found in mammals, provides a good example of alternative splicing (Figure 4-16). Fibronectin is a long, adhesive protein secreted into the extracellular space that can bind other proteins together. What and where it binds depends on which domains are spliced together. The fibronectin gene contains numerous exons, grouped into several regions corresponding to specific domains of the protein. Fibroblasts produce fibronectin mRNAs that contain exons EIIIA and EIIIB; these exons encode amino acid sequences that bind tightly to proteins in the fibroblast plasma membrane. Consequently, this fibronectin isoform adheres fibroblasts to the extracellular matrix. Alternative splicing of the fibronectin primary transcript in hepatocytes, the major type of cell in the liver, yields mRNAs that lack the EIIIA and EIIIB exons. As a result, the fibronectin secreted by hepatocytes into the blood does not adhere tightly to fibroblasts or most other cell types, allowing it to circulate. During formation of blood clots, however, the fibrin-binding domains of hepatocyte fibronectin binds to fibrin, one of the principal constituents of clots. The bound fibronectin then interacts with integrins on the membranes of passing platelets, thereby expanding the clot by addition of platelets.

More than 20 different isoforms of fibronectin have been identified, each encoded by a different, alternatively spliced mRNA composed of a unique combination of fibronectin gene exons. Recent sequencing of large numbers of mRNAs isolated from various tissues and comparison of their sequences with genomic DNA has revealed that nearly 60 percent of all human genes are expressed as alternatively spliced mRNAs. Clearly, alternative RNA splicing greatly expands the number of proteins encoded by the genomes of higher, multicellular organisms.

KEY CONCEPTS OF SECTION 4.2

Transcription of Protein-Coding Genes and Formation of Functional mRNA

- Transcription of DNA is carried out by RNA polymerase, which adds one ribonucleotide at a time to the 3′ end of a growing RNA chain (see Figure 4-11). The sequence of the template DNA strand determines the order in which ribonucleotides are polymerized to form an RNA chain.
- During transcription initiation, RNA polymerase binds to a specific site in DNA (the promoter), locally melts the double-stranded DNA to reveal the unpaired template strand, and polymerizes the first two nucleotides complementary to the template strand. The melted region of 12–14 base pairs is known as the "transcription bubble."
- During strand elongation, RNA polymerase moves down the DNA, melting the DNA ahead of the polymerase, so that the template strand can enter the active site of the enzyme, and allowing the complementary DNA strands of the region just transcribed to reanneal behind it. The transcription bubble moves with the polymerase as the enzyme adds ribonucleotides complementary to the template strand to the 3′ end of the growing RNA chain.
- When RNA polymerase reaches a termination sequence in the DNA, the enzyme stops transcription, leading to release of the completed RNA and dissociation of the enzyme from the template DNA.
- In prokaryotic DNA, several protein-coding genes commonly are clustered into a functional region, an operon, which is transcribed from a single promoter into one mRNA encoding multiple proteins with related functions (see Figure 4-13a). Translation of a bacterial mRNA can begin before synthesis of the mRNA is complete.
- In eukaryotic DNA, each protein-coding gene is transcribed from its own promoter. The initial primary transcript very often contains noncoding regions (introns) interspersed among coding regions (exons).
- Eukaryotic primary transcripts must undergo RNA processing to yield functional RNAs. During processing, the ends of nearly all primary transcripts from protein-coding genes are modified by addition of a 5′ cap and 3′ poly(A) tail. Transcripts from genes containing introns undergo splicing, the removal of the introns and joining of the exons (see Figure 4-15).
- The individual domains of multidomain proteins found in higher eukaryotes are often encoded by individual exons or a small number of exons. Distinct isoforms of such proteins often are expressed in specific cell types as the result of alternative splicing of exons.

4.3 The Decoding of mRNA by tRNAs

Although DNA stores the information for protein synthesis and mRNA conveys the instructions encoded in DNA, most biological activities are carried out by proteins. As we saw in Chapter 3, the linear order of amino acids in each protein determines its three-dimensional structure and activity. For this reason, assembly of amino acids in their correct order, as encoded in DNA, is critical to production of functional proteins and hence the proper functioning of cells and organisms.

Translation is the whole process by which the nucleotide sequence of an mRNA is used as a template to join the amino acids in a polypeptide chain in the correct order (see Figure 4-1, **3**). In eukaryotic cells, protein synthesis occurs in the cytoplasm, where three types of RNA molecules come together to perform different but cooperative functions (Figure 4-17):

1. **Messenger RNA (mRNA)** carries the genetic information transcribed from DNA in a linear form. The mRNA is read in sets of three-nucleotide sequences, called **codons,** each of which specifies a particular amino acid.

2. **Transfer RNA (tRNA)** is the key to deciphering the codons in mRNA. Each type of amino acid has its own subset of tRNAs, which bind the amino acid and carry it to the growing end of a polypeptide chain when the next codon in the mRNA calls for it. The correct tRNA with its attached amino acid is selected at each step because each specific tRNA molecule contains a three-nucleotide sequence, an **anticodon,** that can base-pair with its complementary codon in the mRNA.

3. **Ribosomal RNA (rRNA)** associates with a set of proteins to form **ribosomes.** These complex structures, which physically move along an mRNA molecule, catalyze the assembly of amino acids into polypeptide chains. They also bind tRNAs and various accessory proteins necessary for protein synthesis. Ribosomes are composed of a large and a small subunit, each of which contains its own rRNA molecule or molecules.

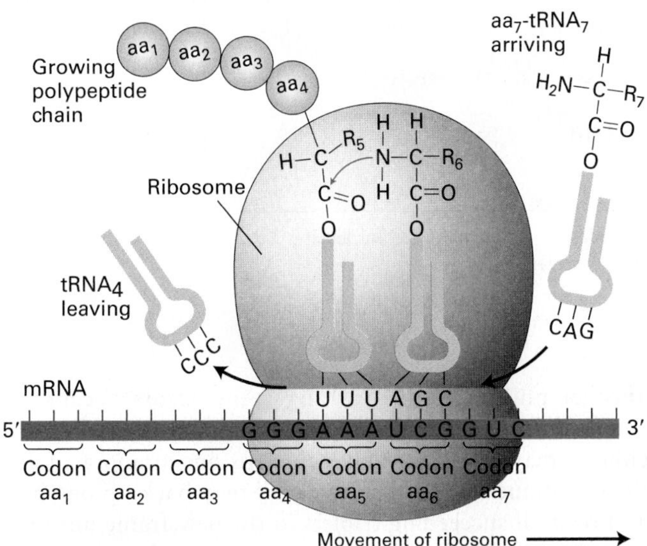

▲ **FIGURE 4-17 The three roles of RNA in protein synthesis.** Messenger RNA (mRNA) is translated into protein by the joint action of transfer RNA (tRNA) and the ribosome, which is composed of numerous proteins and two major ribosomal RNA (rRNA) molecules (not shown). Note the base pairing between tRNA anticodons and complementary codons in the mRNA. Formation of a peptide bond between the amino-group N on the incoming aa-tRNA and the carboxy-terminal C on the growing protein chain (green) is catalyzed by one of the rRNAs. aa = amino acid; R = side group. [Adapted from A. J. F. Griffiths et al., 1999, *Modern Genetic Analysis,* W. H. Freeman and Company.]

These three types of RNA participate in the synthesis of proteins in all organisms. Indeed, development of three functionally distinct RNAs was probably the molecular key to the origin of life. In this section, we focus on the decoding of mRNA by tRNA adaptors, and how the structure of each of these RNAs relates to its specific task. How they work together with rRNA, ribosomes, and other protein factors to synthesize proteins is detailed in the following section. Since translation is essential for protein synthesis, the two processes commonly are referred to interchangeably. However, the polypeptide chains resulting from translation undergo post-translational folding and often other changes (e.g., chemical modifications, association with other chains) that are required for production of mature, functional proteins (Chapter 3).

Messenger RNA Carries Information from DNA in a Three-Letter Genetic Code

As noted above, the **genetic code** used by cells is a *triplet* code, with every three-nucleotide sequence, or codon, being "read" from a specified starting point in the mRNA. Of the 64 possible codons in the genetic code, 61 specify individual amino acids and three are stop codons. Table 4-1 shows that most amino acids are encoded by more than one codon. Only two—methionine and tryptophan—have a single codon; at the other extreme, leucine, serine, and arginine are each specified by six different codons. The different codons for a given amino acid are said to be synonymous. The code itself is termed *degenerate,* meaning that a particular amino acid can be specified by multiple codons.

Synthesis of all polypeptide chains in prokaryotic and eukaryotic cells begins with the amino acid methionine. In bacteria, a specialized form of methionine is used with a formyl group linked to its amino group. In most mRNAs, the *start (initiator) codon* specifying this amino-terminal methionine is AUG. In a few bacterial mRNAs, GUG is used as the initiator codon, and CUG occasionally is used as an initiator codon for methionine in eukaryotes. The three codons UAA, UGA, and UAG do not specify amino acids but, rather, constitute *stop (termination) codons* that mark the carboxyl terminus of polypeptide chains in almost all cells. The sequence of codons that runs from a specific start

TABLE 4-1 The Genetic Code (Codons to Amino Acids)*

FIRST POSITION (5′ END)	SECOND POSITION U	C	A	G	THIRD POSITION (3′ END)
U	Phe	Ser	Tyr	Cys	U
	Phe	Ser	Tyr	Cys	C
	Leu	Ser	Stop	Stop	A
	Leu	Ser	Stop	Trp	G
C	Leu	Pro	His	Arg	U
	Leu	Pro	His	Arg	C
	Leu	Pro	Gln	Arg	A
	Leu (Met)*	Pro	Gln	Arg	G
A	Ile	Thr	Asn	Ser	U
	Ile	Thr	Asn	Ser	C
	Ile	Thr	Lys	Arg	A
	Met (Start)	Thr	Lys	Arg	G
G	Val	Ala	Asp	Gly	U
	Val	Ala	Asp	Gly	C
	Val	Ala	Glu	Gly	A
	Val (Met)*	Ala	Glu	Gly	G

*AUG is the most common initiator codon; GUG usually codes for valine and CUG for leucine, but, rarely, these codons can also code for methionine to initiate a protein chain.

codon to a stop codon is called a **reading frame.** This precise linear array of ribonucleotides in groups of three in mRNA specifies the precise linear sequence of amino acids in a polypeptide chain and also signals where synthesis of the chain starts and stops.

Because the genetic code is a non-overlapping triplet code without divisions between codons, a particular mRNA theoretically could be translated in three different reading frames. Indeed some mRNAs have been shown to contain overlapping information that can be translated in different reading frames, yielding different polypeptides (Figure 4-18). The vast majority of mRNAs, however, can be read in only one frame because stop codons encountered in the other two possible reading frames terminate translation before a functional protein is produced. Very rarely, another unusual coding arrangement occurs because of *frame-shifting.* In this case the protein-synthesizing machinery may read four nucleotides as one amino acid and then continue reading triplets, or it may back up one base and read all succeeding triplets in the new frame until termination of the chain occurs. Only a few dozen such instances are known.

The meaning of each codon is the same in most known organisms—a strong argument that life on earth evolved only once. In fact, the genetic code shown in Table 4-1 is known as the *universal code*. However, the genetic code has been found to differ for a few codons in many mitochondria, in ciliated protozoans, and in *Acetabularia,* a single-celled plant. As shown in Table 4-2, most of these changes involve reading of normal stop codons as amino acids, not an

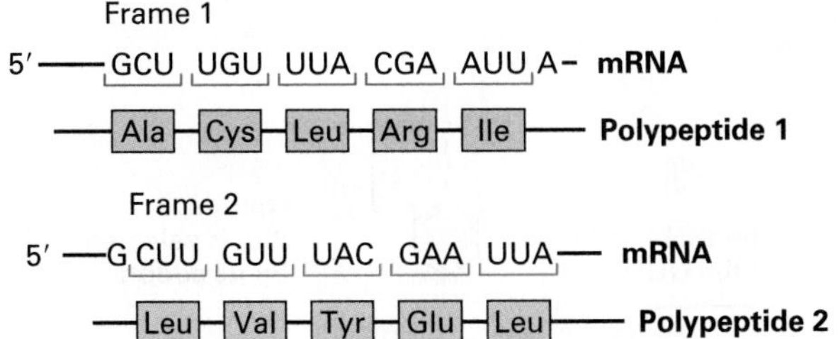

▲ **FIGURE 4-18 Multiple reading frames in an mRNA sequence.** If translation of the mRNA sequence shown begins at two different upstream start sites (not shown), then two overlapping reading frames are possible. In this example, the codons are shifted one base to the right in the lower frame. As a result, the same nucleotide sequence specifies different amino acids during translation. Although regions of sequence that are translated in two of the three possible reading frames are rare, there are examples in both prokaryotes and eukaryotes, and especially in their viruses, where the same sequence is used in two alternative mRNAs expressed from the same region of DNA, and the sequence is read in one reading frame in one mRNA and in an alternative reading frame in the other mRNA. There are even a few instances where the same short sequence is read in all three possible reading frames.

exchange of one amino acid for another. These exceptions to the universal code probably were later evolutionary developments; that is, at no single time was the code immutably fixed, although massive changes were not tolerated once a general code began to function early in evolution.

The Folded Structure of tRNA Promotes Its Decoding Functions

Translation, or decoding, of the four-nucleotide language of DNA and mRNA into the 20–amino acid language of proteins requires tRNAs and enzymes called *aminoacyl-tRNA synthetases*. To participate in protein synthesis, a tRNA molecule must become chemically linked to a particular amino acid via a high-energy bond, forming an **aminoacyl-tRNA** (Figure 4-19). The anticodon in the tRNA then base-pairs with a codon in mRNA so that the activated amino acid can be added to the growing polypeptide chain (see Figures 4-17 and 4-18).

Some 30–40 different tRNAs have been identified in bacterial cells and as many as 50–100 in animal and plant cells. Thus the number of tRNAs in most cells is more than the number of amino acids used in protein synthesis (20) and also differs from the number of amino acid codons in the genetic code (61). Consequently, many amino acids have more than one tRNA to which they can attach (explaining how there can be more tRNAs than amino acids); in addition, many tRNAs can pair with more than one codon (explaining how there can be more codons than tRNAs).

The function of tRNA molecules, which are 70–80 nucleotides long, depends on their precise three-dimensional structures. In solution, all tRNA molecules fold into a similar stem-loop arrangement that resembles a cloverleaf when drawn in two dimensions (Figure 4-20a). The four stems are short double helices stabilized by Watson-Crick base pairing; three of the four stems have loops containing seven or eight bases at their ends, while the remaining, unlooped stem contains the free 3′ and 5′ ends of the chain. The three nucleotides composing the anticodon are located at the center of the middle loop, in an accessible position that facilitates codon-anticodon base pairing. In all tRNAs, the 3′ end of the unlooped amino acid *acceptor stem* has the sequence CCA, which in most cases is added after synthesis and processing of the tRNA are complete. Several bases in most tRNAs also are modified after transcription, creating nonstandard nucleotides such as inosine, dihydrouridine, and pseudouridine. As we will see shortly, some of these modified bases are known to play an important role in protein synthesis. Viewed in three dimensions, the folded tRNA molecule has an L shape with the anticodon loop and acceptor stem forming the ends of the two arms (Figure 4-20b).

TABLE 4-2 Known Deviations from the Universal Genetic Code

CODON	UNIVERSAL CODE	UNUSUAL CODE*	OCCURRENCE
UGA	Stop	Trp	*Mycoplasma, Spiroplasma,* mitochondria of many species
CUG	Leu	Thr	Mitochondria in yeasts
UAA, UAG	Stop	Gln	*Acetabularia, Tetrahymena, Paramecium,* etc.
UGA	Stop	Cys	*Euplotes*

*Found in nuclear genes of the listed organisms and in mitochondrial genes as indicated.
SOURCE: S. Osawa et al., 1992, *Microbiol. Rev.* 56:229.

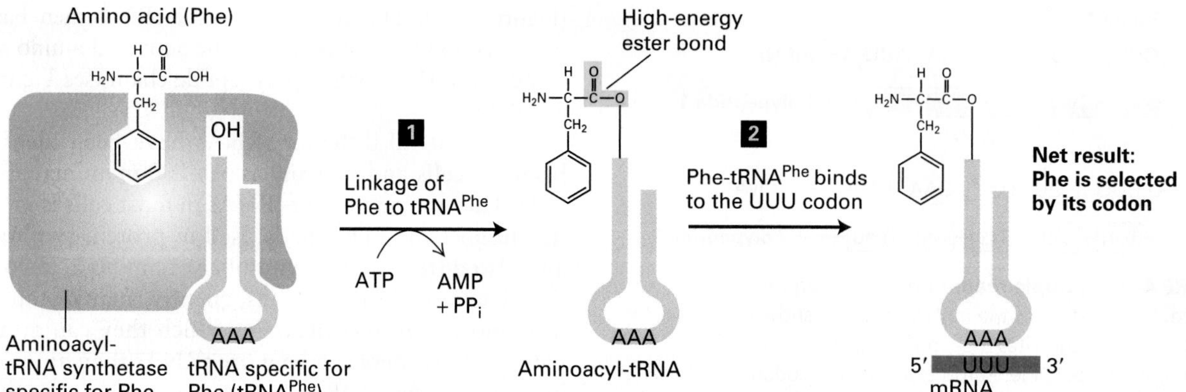

▲ **FIGURE 4-19 Decoding nucleic acid sequence into amino acid sequence.** The process for translating nucleic acid sequences in mRNA into amino acid sequences in proteins involves two steps. Step **1**: An aminoacyl-tRNA synthetase first couples a specific amino acid, via a high-energy ester bond (yellow), to either the 2′ or 3′ hydroxyl of the terminal adenosine in the corresponding tRNA. Step **2**: A three-base sequence in the tRNA (the anticodon) then base-pairs with a codon in the mRNA specifying the attached amino acid. If an error occurs in either step, the wrong amino acid may be incorporated into a polypeptide chain. Phe = phenylalanine.

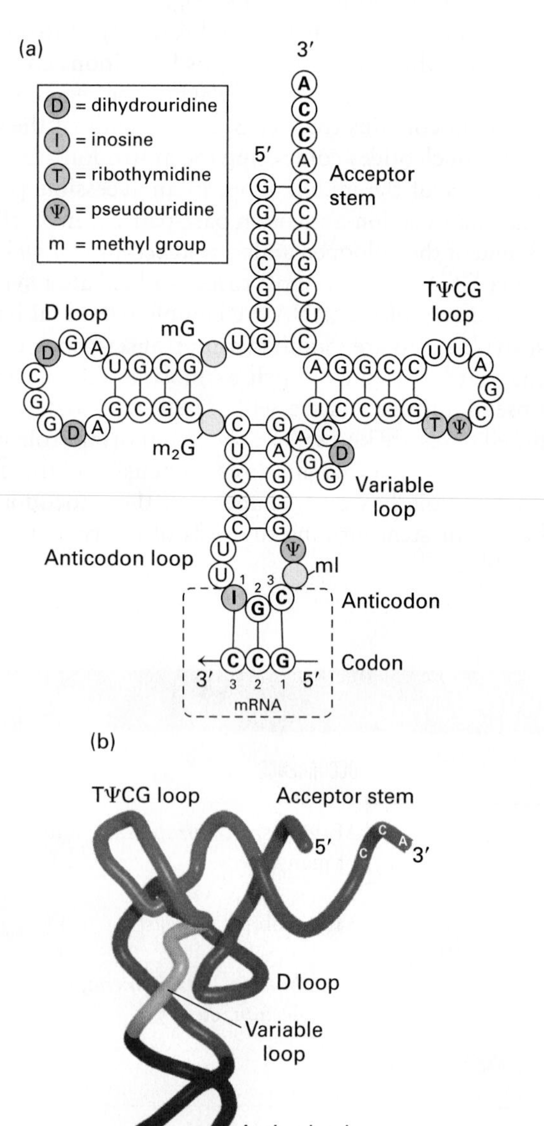

◀ **FIGURE 4-20 Structure of tRNAs.** (a) Although the exact nucleotide sequence varies among tRNAs, they all fold into four base-paired stems and three loops. The CCA sequence at the 3′ end also is found in all tRNAs. Attachment of an amino acid to the 3′ A yields an aminoacyl-tRNA. Some of the A, C, G, and U residues are modified post-transcriptionally in most tRNAs (see key). Dihydrouridine (D) is nearly always present in the D loop; likewise, ribothymidine (T) and pseudouridine (Ψ) are almost always present in the TΨCG loop. Yeast alanine tRNA, represented here, also contains other modified bases. The triplet at the tip of the anticodon loop base-pairs with the corresponding codon in mRNA. (b) Three-dimensional model of the generalized backbone of all tRNAs. Note the L shape of the molecule. [Part (a) see R. W. Holly et al., 1965, *Science* **147:**1462; part (b) from J. G. Arnez and D. Moras, 1997, *Trends Biochem. Sci.* **22:**211.]

Nonstandard Base Pairing Often Occurs Between Codons and Anticodons

If perfect Watson-Crick base pairing were demanded between codons and anticodons, cells would have to contain at least 61 different types of tRNAs, one for each codon that specifies an amino acid. As noted above, however, many cells contain fewer than 61 tRNAs. The explanation for the smaller number lies in the capability of a single tRNA anticodon to recognize more than one, but not necessarily every, codon corresponding to a given amino acid. This broader recognition can occur because of nonstandard pairing between bases in the so-called *wobble* position: that is, the third (3′) base in an mRNA codon and the corresponding first (5′) base in its tRNA anticodon.

The first and second bases of a codon almost always form standard Watson-Crick base pairs with the third and second bases, respectively, of the corresponding anticodon, but four nonstandard interactions can occur between bases in the wobble position. Particularly important is the G·U base pair, which structurally fits almost as well as the standard G·C pair. Thus, a given anticodon in tRNA with G in the first (wobble) position can base-pair with the two

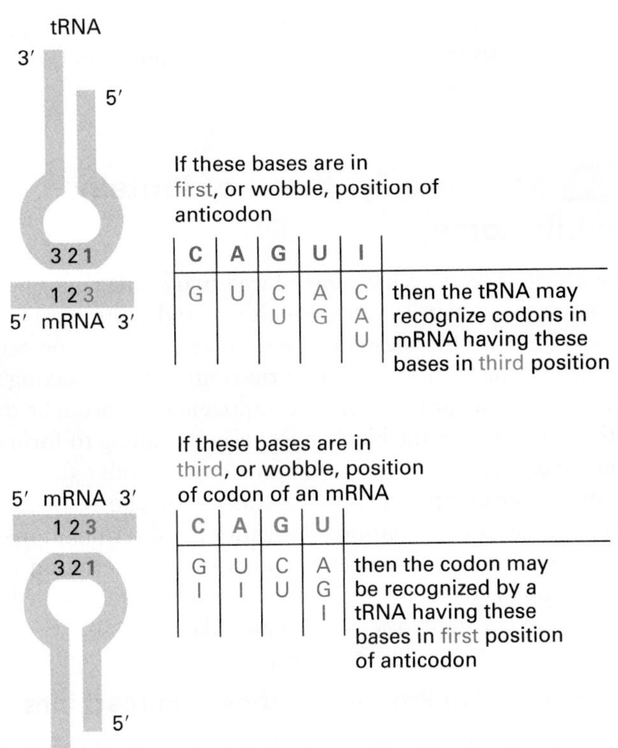
tRNA
3′
5′
If these bases are in first, or wobble, position of anticodon
3 2 1
1 2 3
5′ mRNA 3′
C A G U I
G U C U A G C A U
then the tRNA may recognize codons in mRNA having these bases in third position
If these bases are in third, or wobble, position of codon of an mRNA
5′ mRNA 3′
1 2 3
3 2 1
C A G U
G I U I C U A G I
then the codon may be recognized by a tRNA having these bases in first position of anticodon
5′
3′
tRNA

▲ **FIGURE 4-21 Nonstandard base pairing at the wobble position.** The base in the third (or wobble) position of an mRNA codon often forms a nonstandard base pair with the base in the first (or wobble) position of a tRNA anticodon. Wobble pairing allows a tRNA to recognize more than one mRNA codon *(top)*; conversely, it allows a codon to be recognized by more than one kind of tRNA *(bottom)*, although each tRNA will bear the same amino acid. Note that a tRNA with I (inosine) in the wobble position can "read" (become paired with) three different codons, and a tRNA with G or U in the wobble position can read two codons. Although A is theoretically possible in the wobble position of the anticodon, it is almost never found in nature.

corresponding codons that have either pyrimidine (C or U) in the third position (Figure 4-21). For example, the phenylalanine codons UUU and UUC (5′→3′) are both recognized by the tRNA that has GAA (5′→3′) as the anticodon. In fact, any two codons of the type NNPyr (N = any base; Pyr = pyrimidine) encode a single amino acid and are decoded by a single tRNA with G in the first (wobble) position of the anticodon.

Although adenine rarely is found in the anticodon wobble position, many tRNAs in plants and animals contain inosine (I), a deaminated product of adenine, at this position. Inosine can form nonstandard base pairs with A, C, and U. A tRNA with inosine in the wobble position thus can recognize the corresponding mRNA codons with A, C, or U in the third (wobble) position (see Figure 4-21). For this reason, inosine-containing tRNAs are heavily employed in translation of the synonymous codons that specify a single amino acid. For example, four of the six codons for leucine (CUA, CUC, CUU, and UUA) are all recognized by the same tRNA with the anticodon 3′-GAI-5′; the inosine in the wobble position forms nonstandard base pairs with the third base in the four codons. In the case of the UUA codon, a nonstandard G·U pair also forms between position 3 of the anticodon and position 1 of the codon.

Amino Acids Become Activated When Covalently Linked to tRNAs

Recognition of the codon or codons specifying a given amino acid by a particular tRNA is actually the second step in decoding the genetic message. The first step, attachment of the appropriate amino acid to a tRNA, is catalyzed by a specific *aminoacyl-tRNA synthetase.* Each of the 20 different synthetases recognizes *one* amino acid and *all* its compatible, or *cognate,* tRNAs. These coupling enzymes link an amino acid to the free 2′ or 3′ hydroxyl of the adenosine at the 3′ terminus of tRNA molecules by an ATP-requiring reaction. In this reaction, the amino acid is linked to the tRNA by a high-energy bond and is thus said to be *activated.* The energy of this bond subsequently drives formation of the peptide bonds linking adjacent amino acids in a growing polypeptide chain. The equilibrium of the aminoacylation reaction is driven further toward activation of the amino acid by hydrolysis of the high-energy phosphoanhydride bond in the released pyrophosphate (see Figure 4-19).

Aminoacyl-tRNA synthetases recognize their cognate tRNAs by interacting primarily with the anticodon loop and acceptor stem, although interactions with other regions of a tRNA also contribute to recognition in some cases. Also, specific bases in incorrect tRNAs that are structurally similar to a cognate tRNA will inhibit charging of the incorrect tRNA. Thus, recognition of the correct tRNA depends on both positive interactions and the absence of negative interactions. Still, because some amino acids are so similar structurally, aminoacyl-tRNA synthetases sometimes make mistakes. These are corrected, however, by the enzymes themselves, which have a *proofreading* activity that checks the fit in their amino acid–binding pocket. If the wrong amino acid becomes attached to a tRNA, the bound synthetase catalyzes removal of the amino acid from the tRNA. This crucial function helps guarantee that a tRNA delivers the correct amino acid to the protein-synthesizing machinery. The overall error rate for translation in *E. coli* is very low, approximately 1 per 50,000 codons, evidence of both the fidelity of tRNA recognition and the importance of proofreading by aminoacyl-tRNA synthetases.

KEY CONCEPTS OF SECTION 4.3

The Decoding of mRNA by tRNAs

- Genetic information is transcribed from DNA into mRNA in the form of an overlapping, degenerate triplet code.
- Each amino acid is encoded by one or more three-nucleotide sequences (codons) in mRNA. Each codon

specifies one amino acid, but most amino acids are encoded by multiple codons (see Table 4-1).

- The AUG codon for methionine is the most common start codon, specifying the amino acid at the NH_2-terminus of a protein chain. Three codons (UAA, UAG, UGA) function as stop codons and specify no amino acids.
- A reading frame, the uninterrupted sequence of codons in mRNA from a specific start codon to a stop codon, is translated into the linear sequence of amino acids in a polypeptide chain.
- Decoding of the nucleotide sequence in mRNA into the amino acid sequence of proteins depends on tRNAs and aminoacyl-tRNA synthetases.
- All tRNAs have a similar three-dimensional structure that includes an acceptor arm for attachment of a specific amino acid and a stem-loop with a three-base anticodon sequence at its ends (see Figure 4-20). The anticodon can base-pair with its corresponding codon in mRNA.
- Because of nonstandard interactions, a tRNA may base-pair with more than one mRNA codon; conversely, a particular codon may base-pair with multiple tRNAs. In each case, however, only the proper amino acid is inserted into a growing polypeptide chain.
- Each of the 20 aminoacyl-tRNA synthetases recognizes a single amino acid and covalently links it to a cognate tRNA, forming an aminoacyl-tRNA (see Figure 4-19). This reaction activates the amino acid, so it can participate in peptide bond formation.

4.4 Stepwise Synthesis of Proteins on Ribosomes

The previous sections have introduced two of the major participants in protein synthesis—mRNA and aminoacylated tRNA. Here we first describe the third key player in protein synthesis—the rRNA-containing ribosome—before taking a detailed look at how all three components are brought together to carry out the biochemical events leading to formation of polypeptide chains on ribosomes. Similar to transcription, the complex process of translation can be divided into three stages—initiation, elongation, and termination—which we consider in order. We focus our description on translation in eukaryotic cells, but the mechanism of translation is fundamentally the same in all cells.

Ribosomes Are Protein-Synthesizing Machines

If the many components that participate in translating mRNA had to interact in free solution, the likelihood of simultaneous collisions occurring would be so low that the rate of amino acid polymerization would be very slow. The

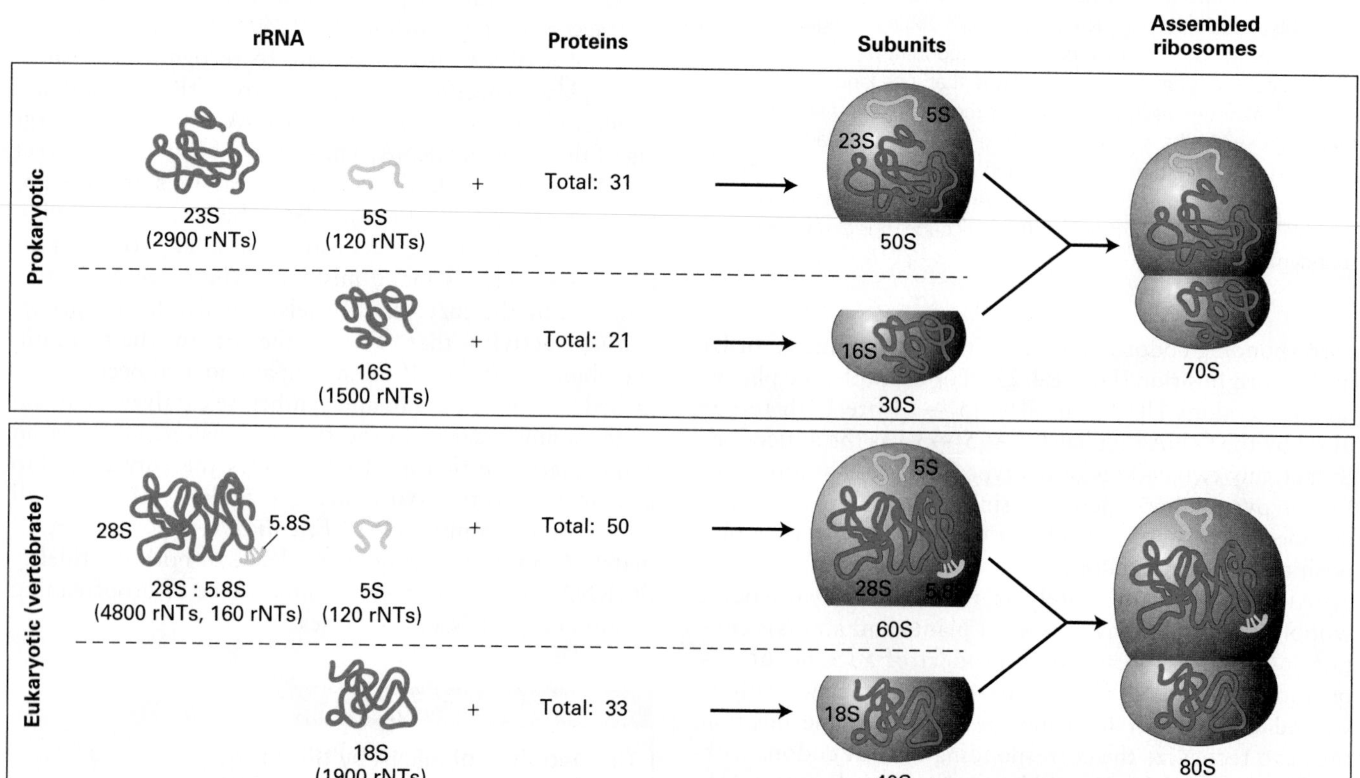

▲ **FIGURE 4-22 Prokaryotic and eukaryotic ribosome components.** In all cells, each ribosome consists of a large and a small subunit. The two subunits contain rRNAs (red) of different lengths, as well as a different set of proteins. All ribosomes contain two major rRNA molecules (23S and 16S rRNA in bacteria; 28S and 18S rRNA in vertebrates) and a 5S rRNA. The large subunit of vertebrate ribosomes also contains a 5.8S rRNA base-paired to the 28S rRNA. The number of ribonucleotides (rNTs) in each rRNA type is indicated.

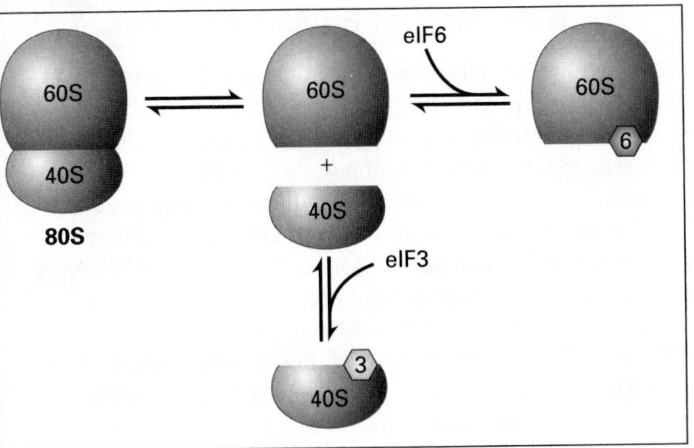

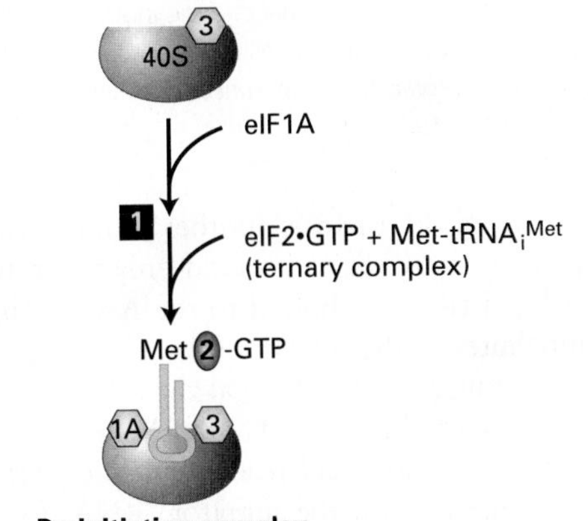

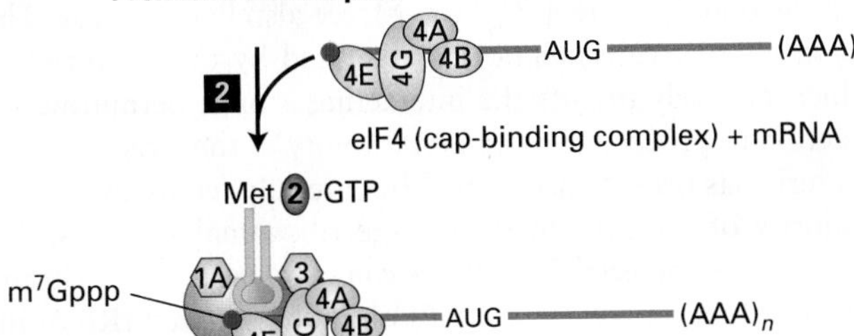

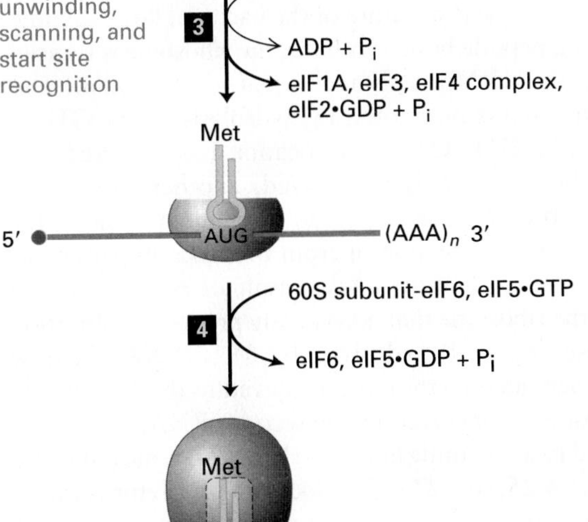

◀ **FIGURE 4-24 Initiation of translation in eukaryotes.** *Inset:* When a ribosome dissociates at the termination of translation, the 40S and 60S subunits associate with initiation factors eIF3 and eIF6, respectively, forming complexes that can initiate another round of translation. Steps **1** and **2**: Sequential addition of the indicated components to the 40S subunit–eIF3 complex forms the initiation complex. Step **3**: Scanning of the mRNA by the associated initiation complex leads to positioning of the small subunit and bound Met-tRNA$_i^{Met}$ at the start codon. Step **4**: Association of the large subunit (60S) forms an 80S ribosome ready to translate the mRNA. Two initiation factors, eIF2 (step **1**) and eIF5 (step **4**) are GTP-binding proteins, whose bound GTP is hydrolyzed during translation initiation. The precise time at which particular initiation factors are released is not yet well characterized. See the text for a more detailed discussion. [Adapted from R. Mendez and J. D. Richter, 2001, *Nature Rev. Mol.* Cell Biol. **2**:521.]

start codon by 4–7 nucleotides. Base pairing between this sequence in the mRNA, called the Shine-Dalgarno sequence after its discoverers, and the small rRNA places the small ribosomal subunit in the proper position for initiation. Next, f-Met-tRNA$_i^{Met}$ and initiation factors comparable to eIF1A, eIF2, and eIF3 associate with the small subunit, followed by association of the large subunit to form the complete bacterial ribosome by a mechanism similar to that in eukaryotes.

During Chain Elongation Each Incoming Aminoacyl-tRNA Moves Through Three Ribosomal Sites

The correctly positioned ribosome–Met-tRNA$_i^{Met}$ complex is now ready to begin the task of stepwise addition of amino acids by the in-frame translation of the mRNA. As is the case with initiation, a set of special proteins, termed translation **elongation factors (EFs)**, is required to carry out this process of chain elongation. The key steps in elongation are entry of each succeeding aminoacyl-tRNA with an anticodon complementary to the next codon, formation of a peptide bond, and the movement, or *translocation,* of the ribosome one codon at a time along the mRNA.

At the completion of translation initiation, as noted already, Met-tRNA$_i^{Met}$ is bound to the P site on the assembled 80S ribosome (Figure 4-25, *top*). This region of the ribosome is called the *P* site because the tRNA chemically linked to the growing poly*p*eptide chain is located here. The second aminoacyl-tRNA is brought into the ribosome as a ternary complex in association with EF1α·GTP and becomes bound to the *A* site, so named because it is where *a*minoacylated tRNAs bind (step **1**). EF1α·GTP bound to various aminoacyl-tRNAs diffuse into the A site, but the next step in translation proceeds only when the tRNA anticodon base-pairs with the second codon in the coding region. When that occurs properly, the GTP in the associated EF1α·GTP is hydrolyzed. The hydrolysis of GTP promotes a conformational change in the ribosome that leads to tight binding of the aminoacyl-tRNA in the A site and release of the resulting EF1α·GDP complex (step **2**). This conformational change also positions the aminoacylated 3′ end of the tRNA in the A site in close proximity to the 3′ end of the Met-tRNA$_i^{Met}$

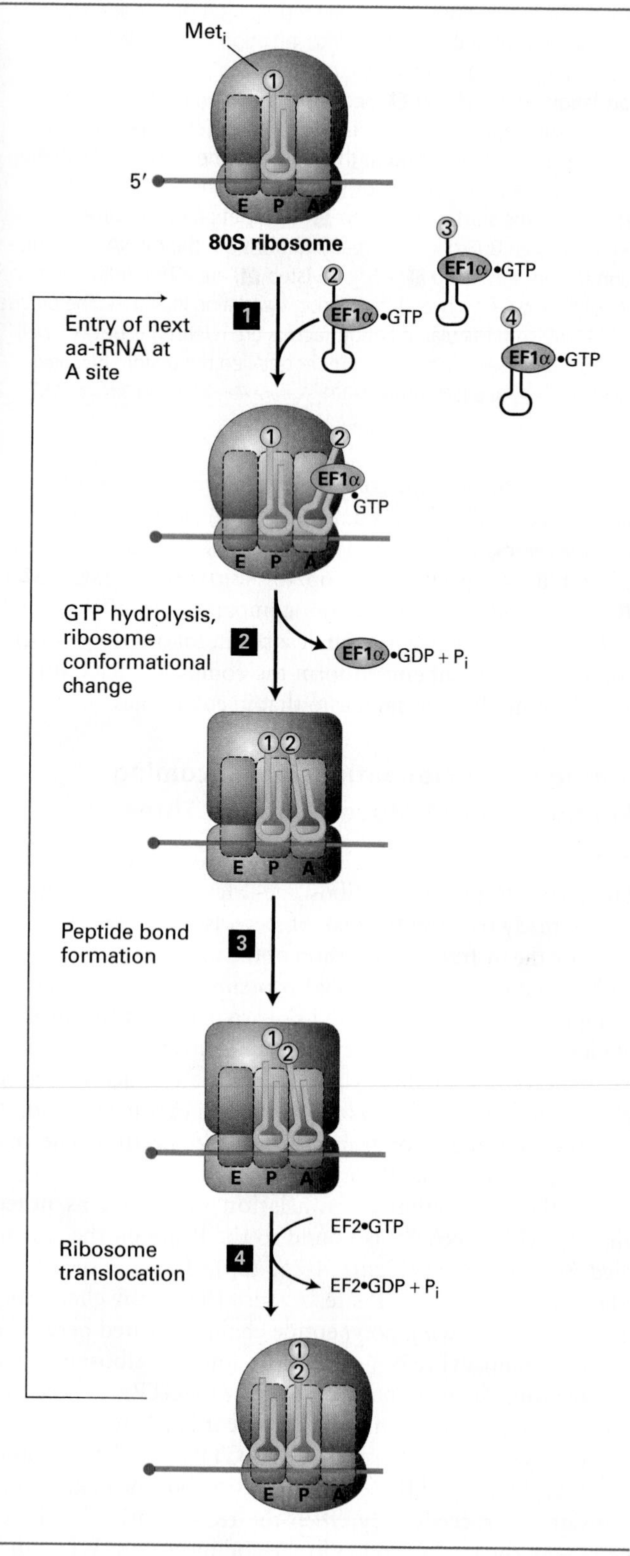

◀ **FIGURE 4-25 Peptidyl chain elongation in eukaryotes.** Once the 80S ribosome with Met-tRNA$_i^{Met}$ in the ribosome P site is assembled *(top)*, a ternary complex bearing the second amino acid (aa_2) coded by the mRNA binds to the A site (step **1**). Following a conformational change in the ribosome induced by hydrolysis of GTP in EF1α·GTP (step **2**), the large rRNA catalyzes peptide bond formation between Met$_i$ and aa_2 (step **3**). Hydrolysis of GTP in EF2·GTP causes another conformational change in the ribosome that results in its translocation one codon along the mRNA and shifts the unacylated tRNA$_i^{Met}$ to the E site and the tRNA with the bound peptide to the P site (step **4**). The cycle can begin again with binding of a ternary complex bearing aa_3 to the now open A site. In the second and subsequent elongation cycles, the tRNA at the E site is ejected during step **2** as a result of the conformational change induced by hydrolysis of GTP in EF1α·GTP. [Adapted from K. H. Nierhaus et al., 2000, in R. A. Garrett et al., eds., *The Ribosome: Structure, Function, Antibiotics, and Cellular Interactions,* ASM Press, p. 319.]

in the P site. GTP hydrolysis, and hence tight binding, does not occur if the anticodon of the incoming aminoacyl-tRNA cannot base-pair with the codon at the A site. In this case, the ternary complex diffuses away, leaving an empty A site that can associate with other aminoacyl-tRNA–EF1α·GTP complexes until a correctly base-paired tRNA is bound. Thus, GTP hydrolysis by EF1α is another proofreading step that allows protein synthesis to proceed only when the correct aminoacylated tRNA is bound to the A site. This phenomenon contributes to the fidelity of protein synthesis.

With the initiating Met-tRNA$_i^{Met}$ at the P site and the second aminoacyl-tRNA tightly bound at the A site, the α amino group of the second amino acid reacts with the "activated" (ester-linked) methionine on the initiator tRNA, forming a peptide bond (Figure 4-25, step **3**; see also Figure 4-17). This *peptidyltransferase reaction* is catalyzed by the large rRNA, which precisely orients the interacting atoms, permitting the reaction to proceed. The catalytic ability of the large rRNA in bacteria has been demonstrated by carefully removing the vast majority of the protein from large ribosomal subunits. The nearly pure bacterial 23S rRNA can catalyze a peptidyltransferase reaction between analogs of aminoacylated-tRNA and peptidyl-tRNA. Further support for the catalytic role of large rRNA in protein synthesis comes from crystallographic studies showing that no proteins lie near the site of peptide bond synthesis in the crystal structure of the bacterial large subunit.

Following peptide bond synthesis, the ribosome is translocated along the mRNA a distance equal to one codon. This translocation step is monitored by hydrolysis of the GTP in eukaryotic EF2·GTP. Once translocation has occurred correctly, the bound GTP is hydrolyzed, another irreversible process that prevents the ribosome from moving along the RNA in the wrong direction or from translocating an incorrect number of nucleotides. As a result of conformational changes in the ribosome that accompany proper translocation and the resulting GTP hydrolysis by EF2, tRNA$_i^{Met}$, now without its activated methionine, is moved to the *E* (*e*xit) site on the ribosome; concurrently, the second tRNA, now covalently bound to a dipeptide (a peptidyl-tRNA), is moved to the P site (Figure 4-25, step **4**). Translocation thus returns the ribosome conformation to a state in which the A site is open and able to accept another aminoacylated tRNA complexed with EF1α·GTP, beginning another cycle of chain elongation.

Repetition of the elongation cycle depicted in Figure 4-25 adds amino acids one at a time to the C-terminus of the

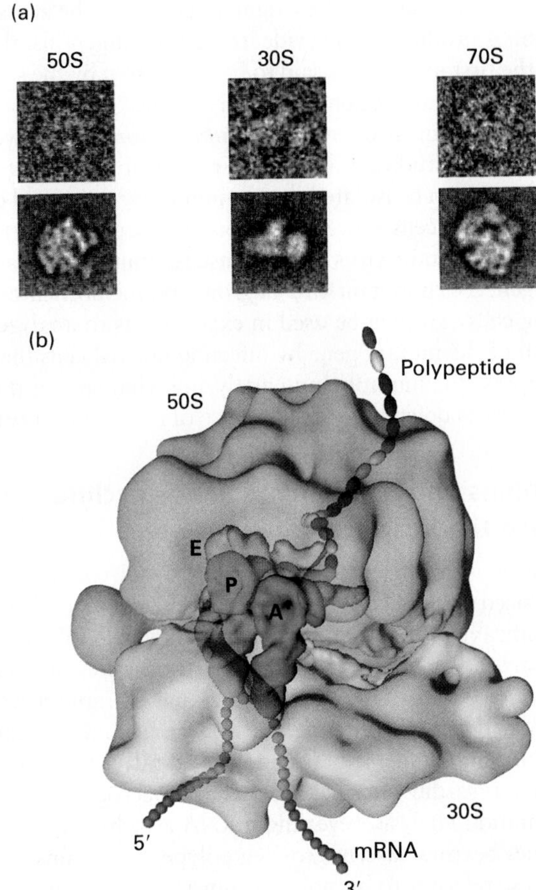

▲ **FIGURE 4-26 Low-resolution model of *E. coli* 70S ribosome.** (a) Top panels show cryoelectron microscopic images of *E. coli* 70S ribosomes and 50S and 30S subunits. Bottom panels show computer-derived averages of many dozens of images in the same orientation. (b) Model of a 70S ribosome based on the computer-derived images and on chemical cross-linking studies. Three tRNAs are superimposed on the A (pink), P (green), and E (yellow) sites. The nascent polypeptide chain is buried in a tunnel in the large ribosomal subunit that begins close to the acceptor stem of the tRNA in the P site. [See I. S. Gabashvili et al., 2000, *Cell* **100**:537; courtesy of J. Frank.]

growing polypeptide as directed by the mRNA sequence, until a stop codon is encountered. In subsequent cycles, the conformational change that occurs in step **2** ejects the unacylated tRNA from the E site. As the nascent polypeptide chain becomes longer, it threads through a channel in the large ribosomal subunit, exiting at a position opposite the side that interacts with the small subunit (Figure 4-26).

In the absence of the ribosome, the three-base-pair RNA-RNA hybrid between the tRNA anticodons and the mRNA codons in the A and P sites would not be stable; RNA-RNA duplexes between separate RNA molecules must be considerably longer to be stable under physiological conditions. However, multiple interactions between the large and small rRNAs and general domains of tRNAs (e.g., the D and TΨCG loops, see Figure 4-20) stabilize the tRNAs in the A and P sites, while other RNA-RNA interactions sense correct codon-anticodon base pairing, assuring that the genetic code is read properly. Then, interactions between rRNAs and the general domains of all tRNAs result in the movement of the tRNAs between the A, P, and E sites as the ribosome translocates along the mRNA one three-nucleotide codon at a time.

Translation Is Terminated by Release Factors When a Stop Codon Is Reached

The final stage of translation, like initiation and elongation, requires highly specific molecular signals that decide the fate of the mRNA–ribosome–peptidyl-tRNA complex. Two types of specific protein **release factors** (RFs) have been discovered. Eukaryotic eRF1, whose shape is similar to that of tRNAs, apparently acts by binding to the ribosomal A site and recognizing stop codons directly. Like some of the initiation and elongation factors discussed previously, the second eukaryotic release factor, eRF3, is a GTP-binding protein. The eRF3·GTP acts in concert with eRF1 to promote cleavage of the peptidyl-tRNA, thus releasing the completed protein chain (Figure 4-27). Bacteria have two release factors (RF1 and RF2) that are functionally analogous to eRF1 and

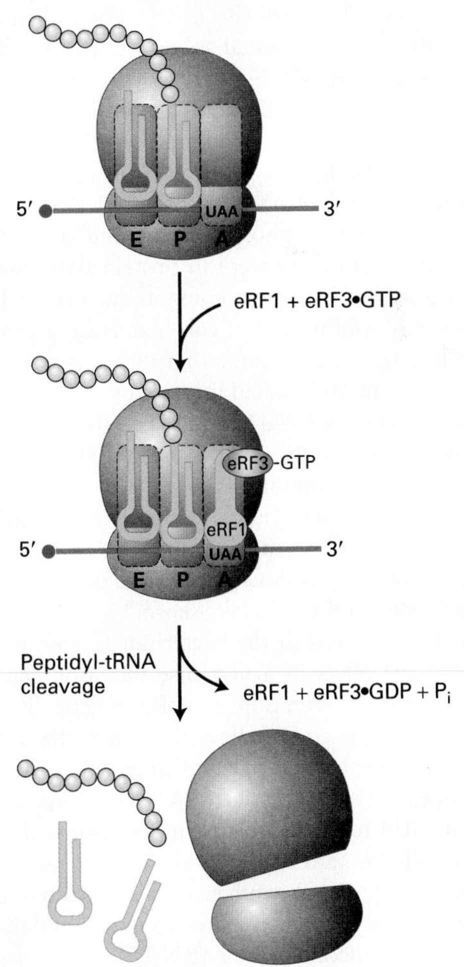

▲ **FIGURE 4-27 Termination of translation in eukaryotes.** When a ribosome bearing a nascent protein chain reaches a stop codon (UAA, UGA, UAG), release factor eRF1 enters the ribosomal complex, probably at or near the A site together with eRF3·GTP. Hydrolysis of the bound GTP is accompanied by cleavage of the peptide chain from the tRNA in the P site and release of the tRNAs and the two ribosomal subunits.

a GTP-binding factor (RF3) that is analogous to eRF3. Once again, the eRF3 GTPase monitors the correct recognition of a stop codon by eRF1. The peptidyl-tRNA bond of the tRNA in the P site is not cleaved, terminating translation, until one of the three stop codons is correctly recognized by eRF1, another example of a proofreading step in protein synthesis.

After its release from the ribosome, a newly synthesized protein folds into its native three-dimensional conformation, a process facilitated by other proteins called **chaperones** (Chapter 3). Additional release factors then promote dissociation of the ribosome, freeing the subunits, mRNA, and terminal tRNA for another round of translation.

We can now see that one or more GTP-binding proteins participate in each stage of translation. These proteins belong to the **GTPase superfamily** of switch proteins that cycle between a GTP-bound active form and GDP-bound inactive form (see Figure 3-32). Hydrolysis of the bound GTP causes a conformational change in the GTPase itself and other associated proteins that are critical to various complex molecular processes. In translation initiation, for instance, hydrolysis of eIF2·GTP to eIF2·GDP prevents further scanning of the mRNA once the start site is encountered and allows binding of the large ribosomal subunit to the small subunit (see Figure 4-24, step **3**). Similarly, hydrolysis of EF2·GTP to EF2·GDP during chain elongation leads to correct translocation of the ribosome along the mRNA (see Figure 4-25, step **4**), and hydrolysis of eRF3·GTP to eRF3·GDP assures correct termination of translation. Since hydrolysis of the high-energy β–γ phosphoester bond of GTP is irreversible, coupling of these steps in protein synthesis to GTP hydrolysis prevents them from going in the reverse direction.

One kind of **mutation** that can inactivate a gene in any organism is a base-pair change that converts a codon normally encoding an amino acid into a stop codon, e.g., UAC (encoding tyrosine) $\rightarrow$ UAG (stop). When this occurs early in the reading frame, the resulting truncated protein usually is nonfunctional. Such mutations are called *nonsense* mutations because when the genetic code equating each triplet codon sequence with a single amino acid was being deciphered, the three stop codons were found not to encode any amino acid—they did not "make sense."

In genetic studies with the bacterium *E. coli,* it was discovered that the effect of a nonsense mutation can be suppressed by a second mutation in a tRNA gene. This occurs when the sequence encoding the anticodon in the tRNA gene is changed to a triplet that is complementary to the original stop codon, e.g., a mutation in $tRNA^{Tyr}$ that changes its anticodon from GUA to CUA, which can base-pair with the UAG stop codon. The mutant tRNA can still be recognized by the tyrosine aminoacyl-tRNA synthetase and coupled to tyrosine. Cells with both the original nonsense mutation and the second mutation in the anticodon of the $tRNA^{Tyr}$ gene consequently can insert a tyrosine at the position of the mutant stop codon, allowing protein synthesis to continue past the original nonsense mutation. The resulting protein can function normally. This mechanism of suppression is not highly efficient, so that translation of normal mRNAs with a UAG stop codon terminates at the normal position in most instances. If enough of the protein encoded by the original gene with the nonsense mutation is produced to provide its essential functions, the effect of the first mutation is said to be *suppressed* by the second mutation in the anticodon of the tRNA gene.

This mechanism of *nonsense suppression* is a powerful tool in genetic studies in bacteria. For example, mutant bacterial viruses can be isolated that cannot grow in normal cells, but can grow in cells expressing a nonsense-suppressing tRNA because the mutant virus has a nonsense mutation in an essential gene. Such mutant viruses grown on the nonsense-suppressing cells can then be used in experiments to analyze the function of the mutant gene by infecting normal cells that do not suppress the mutation and analyzing what step in the viral life cycle is defective in the absence of the mutant protein.

Polysomes and Rapid Ribosome Recycling Increase the Efficiency of Translation

Translation of a single eukaryotic mRNA molecule to yield a typical-sized protein takes one to two minutes. Two phenomena significantly increase the overall rate at which cells can synthesize a protein: (1) the simultaneous translation of a single mRNA molecule by multiple ribosomes and (2) rapid recycling of ribosomal subunits after they disengage from the 3′ end of an mRNA. Simultaneous translation of an mRNA by multiple ribosomes is readily observable in electron micrographs and by sedimentation analysis, revealing mRNA attached to multiple ribosomes bearing nascent growing polypeptide chains. These structures, referred to as **polyribosomes** or *polysomes,* were seen to be circular in electron micrographs of some tissues. Subsequent studies with yeast cells explained the circular shape of polyribosomes and suggested the mechanism by which ribosomes recycle efficiently.

These studies revealed that multiple copies of a cytosolic protein found in all eukaryotic cells, *poly(A)-binding protein I (PABPI),* can interact with both an mRNA poly(A) tail and the 4G subunit of yeast eIF4. Recall that the 4E subunit of yeast eIF4 binds to the 5′ end of an mRNA. As a result of these interactions, the two ends of an mRNA molecule can be bridged by the intervening proteins, forming a "circular" mRNA (Figure 4-28a). Because the two ends of a polysome are relatively close together, ribosomal subunits that disengage from the 3′ end are positioned near the 5′ end, facilitating re-initiation by the interaction of the 40S subunit with eIF4 bound to the 5′ cap. The circular pathway depicted in Figure 4-28b, which may operate in many eukaryotic cells, would enhance ribosome recycling and thus increase the efficiency of protein synthesis.

KEY CONCEPTS OF SECTION 4.4

Stepwise Synthesis of Proteins on Ribosomes

- Both prokaryotic and eukaryotic ribosomes—the large ribonucleoprotein complexes on which translation occurs—consist of a small and a large subunit (see Figure 4-22). Each subunit contains numerous different proteins and one major rRNA molecule (small or large). The large subunit also contains one accessory 5S rRNA in bacteria and two accessory rRNAs in eukaryotes (5S and 5.8S in vertebrates).

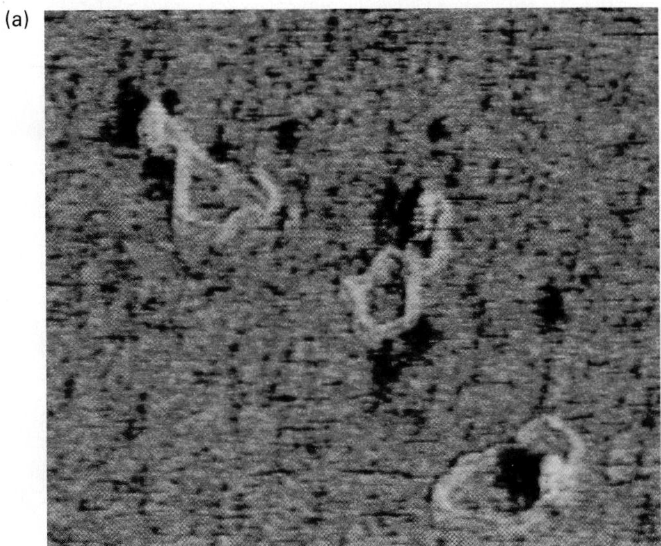

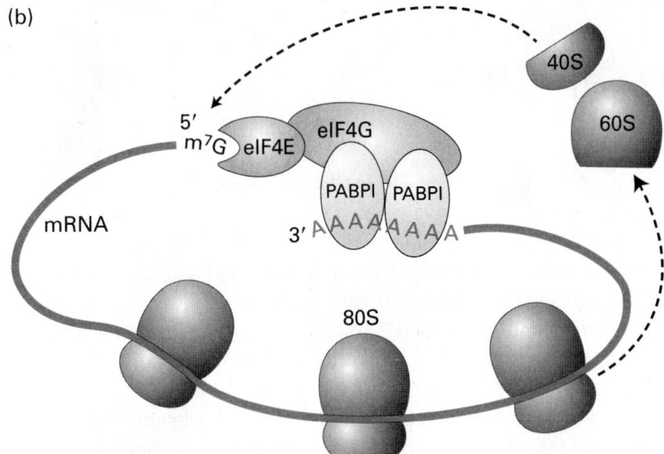

▲ EXPERIMENTAL FIGURE 4-28 **The circular structure of mRNA increases translation efficiency.** Eukaryotic mRNA forms a circular structure owing to interactions of three proteins. (a) In the presence of purified poly(A)-binding protein I (PABPI), eIF4E, and eIF4G, eukaryotic mRNAs form circular structures, visible in this atomic force micrograph. In these structures, protein-protein and protein-mRNA interactions form a bridge between the 5′ and 3′ ends of the mRNA. (b) Model of protein synthesis on circular polysomes and recycling of ribosomal subunits. Multiple individual ribosomes can simultaneously translate a eukaryotic mRNA, shown here in circular form stabilized by interactions between proteins bound at the 3′ and 5′ ends. When a ribosome completes translation and dissociates from the 3′ end, the separated subunits can rapidly find the nearby 5′ cap (m^7G) and initiate another round of synthesis. [Part (a) courtesy of A. Sachs.]

- Analogous rRNAs from many different species fold into quite similar three-dimensional structures containing numerous stem-loops and binding sites for proteins, mRNA, and tRNAs. Much smaller ribosomal proteins are associated with the periphery of the rRNAs.
- Of the two methionine tRNAs found in all cells, only one ($tRNA_i^{Met}$) functions in initiation of translation.
- Each stage of translation—initiation, chain elongation, and termination—requires specific protein factors, including GTP-binding proteins that hydrolyze their bound GTP to GDP when a step has been completed successfully.
- During initiation, the ribosomal subunits assemble near the translation start site in an mRNA molecule with the tRNA carrying the amino-terminal methionine (Met-$tRNA_i^{Met}$) base-paired with the start codon (Figure 4-24).
- Chain elongation entails a repetitive four-step cycle: (1) loose binding of an incoming aminoacyl-tRNA to the A site on the ribosome, (2) tight binding of the correct aminoacyl-tRNA to the A site accompanied by release of the previously used tRNA from the E site, (3) transfer of the growing peptidyl chain to the incoming amino acid catalyzed by the large rRNA, and (4) translocation of the ribosome to the next codon, thereby moving the peptidyl-tRNA in the A site to the P site and the now unacylated tRNA in the P site to the E site (see Figure 4-25).
- In each cycle of chain elongation, the ribosome undergoes two conformational changes monitored by GTP-binding proteins. The first (involving EF1α) permits tight binding of the incoming aminoacyl-tRNA to the A site and ejection of a tRNA from the E site, and the second (involving EF2) leads to translocation.
- Termination of translation is carried out by two types of termination factors: those that recognize stop codons and those that promote hydrolysis of peptidyl-tRNA (see Figure 4-27). Once again, correct recognition of a stop codon is monitored by a GTPase (eRF3).
- The efficiency of protein synthesis is increased by the simultaneous translation of a single mRNA by multiple ribosomes, forming a polyribosome, or polysome. In eukaryotic cells, protein-mediated interactions bring the two ends of a polyribosome close together, thereby promoting the rapid recycling of ribosomal subunits, which further increases the efficiency of protein synthesis (see Figure 4-28b).

4.5 DNA Replication

Now that we have seen how genetic information encoded in the nucleotide sequence of DNA is translated into the proteins that perform most cell functions, we can appreciate the necessity for precisely copying DNA sequences during DNA replication, in preparation for cell division (see Figure 4-1, **4**). The regular pairing of bases in the double-helical DNA structure suggested to Watson and Crick that new DNA strands are synthesized by using the existing (*parental*) strands as templates in the formation of new, *daughter* strands complementary to the parental strands.

This base-pairing template model theoretically could proceed either by a *conservative* or a *semiconservative* mechanism. In a conservative mechanism, the two daughter strands would form a new double-stranded (*duplex*) DNA molecule, and the parental duplex would remain intact. In a semiconservative mechanism, the parental strands are permanently separated, and each forms a duplex molecule with the daughter strand base-paired to it. Definitive evidence

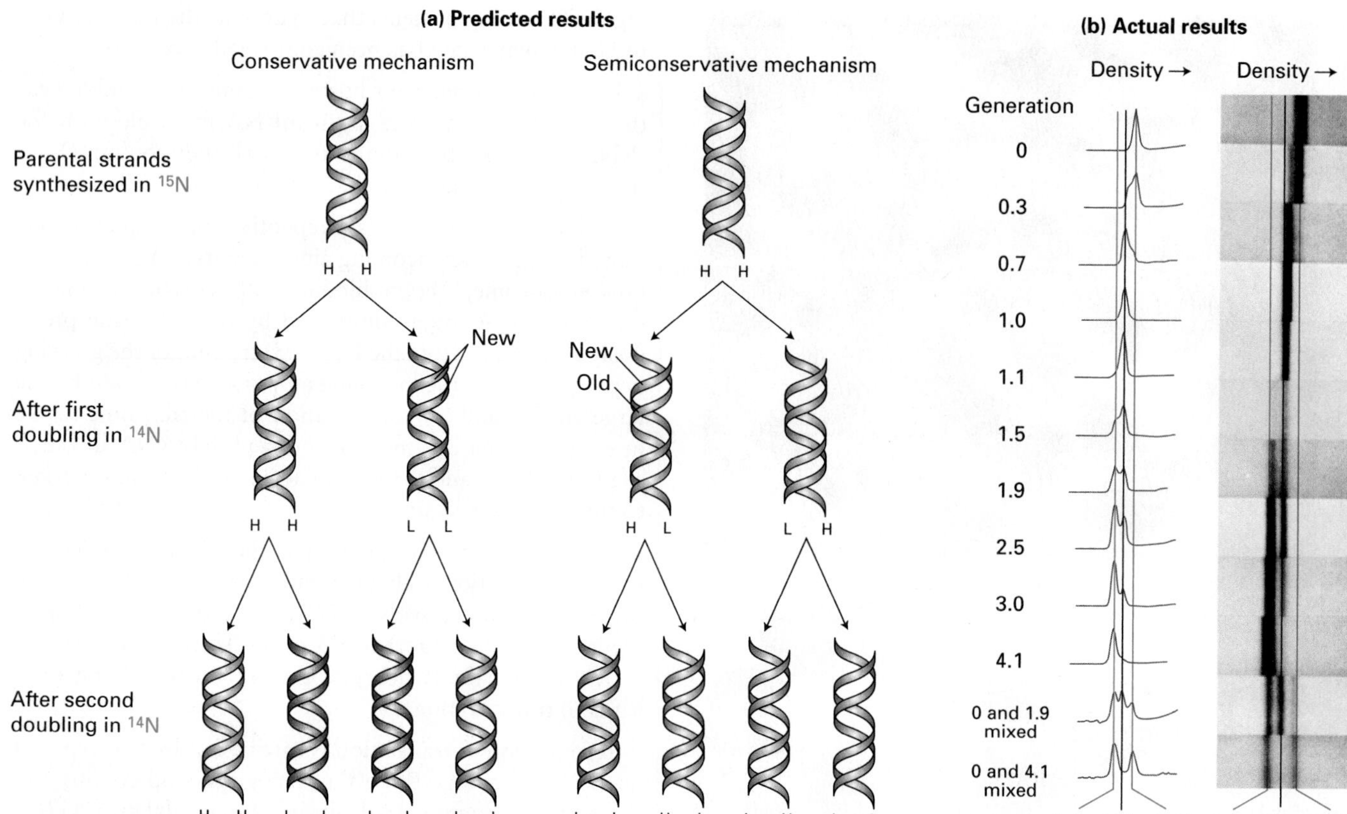

▲ **EXPERIMENTAL FIGURE 4-29 The Meselson-Stahl experiment shows replication to be semiconservative.** This experiment showed that DNA replicates by a semiconservative mechanism. *E. coli* cells initially were grown in a medium containing ammonium salts prepared with "heavy" nitrogen (^{15}N) until all the cellular DNA was labeled. After the cells were transferred to a medium containing the normal "light" isotope (^{14}N), samples were removed periodically from the cultures and the DNA in each sample was analyzed by equilibrium density-gradient centrifugation, a procedure that separates macromolecules on the basis of their density. This technique can separate heavy-heavy (H-H), light-light (L-L), and heavy-light (H-L) duplexes into distinct bands. (a) Expected composition of daughter duplex molecules synthesized from ^{15}N-labeled DNA after *E. coli* cells are shifted to ^{14}N-containing medium if DNA replication occurs by a conservative or semiconservative mechanism. Parental heavy (H) strands are in red; light (L) strands synthesized after shift to ^{14}N-containing medium are in blue. Note that the conservative mechanism never generates H-L DNA and that the semiconservative mechanism never generates H-H DNA but does generate H-L DNA during the second and subsequent doublings. With additional replication cycles, the ^{15}N-labeled (H) strands from the original DNA are diluted, so that the vast bulk of the DNA would consist of L-L duplexes with either mechanism. (b) Actual banding patterns of DNA subjected to equilibrium density-gradient centrifugation before and after shifting ^{15}N-labeled *E. coli* cells to ^{14}N-containing medium. DNA bands were visualized under UV light and photographed. The traces on the left are a measure of the density of the photographic signal, and hence the DNA concentration, along the length of the centrifuge cells from left to right. The number of generations (*far left*) following the shift to ^{14}N-containing medium was determined by counting the concentration of *E. coli* cells in the culture. This value corresponds to the number of DNA replication cycles that had occurred at the time each sample was taken. After one generation of growth, all the extracted DNA had the density of H-L DNA. After 1.9 generations, approximately half the DNA had the density of H-L DNA; the other half had the density of L-L DNA. With additional generations, a larger and larger fraction of the extracted DNA consisted of L-L duplexes; H-H duplexes never appeared. These results match the predicted pattern for the semiconservative replication mechanism depicted in (a). The bottom two centrifuge cells contained mixtures of H-H DNA and DNA isolated at 1.9 and 4.1 generations in order to clearly show the positions of H-H, H-L, and L-L DNA in the density gradient. [Part (b) from M. Meselson and F. W. Stahl, 1958, *Proc. Nat'l Acad. Sci. USA* **44:**671.]

that duplex DNA is replicated by a semiconservative mechanism came from a now classic experiment conducted by M. Meselson and W. F. Stahl, outlined in Figure 4-29.

Copying of a DNA template strand into a complementary strand thus is a common feature of DNA replication, transcription of DNA into RNA, and, as we will see later in this chapter, DNA repair and recombination. In all cases, the information in the template in the form of the specific sequence of nucleotides is preserved. In some viruses, single-stranded RNA molecules function as templates for synthesis of complementary RNA or DNA strands. However, the vast preponderance of RNA and DNA in cells is synthesized from preexisting duplex DNA.

DNA Polymerases Require a Primer to Initiate Replication

Analogous to RNA, DNA is synthesized from deoxynucleoside 5′-triphosphate precursors (dNTPs). Also like RNA synthesis, DNA synthesis always proceeds in the 5′→3′ direction because chain growth results from formation of a phosphoester bond between the 3′ oxygen of a growing strand and the α phosphate of a dNTP (see Figure 4-10a). As discussed earlier, an RNA polymerase can find an appropriate transcription start site on duplex DNA and initiate the synthesis of an RNA complementary to the template DNA strand (see Figure 4-11). In contrast, **DNA polymerases** cannot initiate chain synthesis de novo; instead, they require a short, preexisting RNA or DNA strand, called a **primer,** to begin chain growth. With a primer base-paired to the template strand, a DNA polymerase adds deoxynucleotides to the free hydroxyl group at the 3′ end of the primer as directed by the sequence of the template strand:

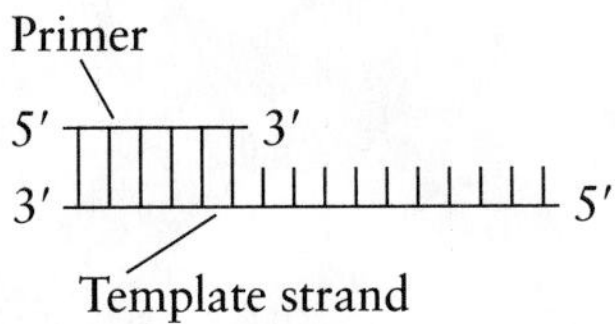

When RNA is the primer, the daughter strand that is formed is RNA at the 5′ end and DNA at the 3′ end.

Duplex DNA Is Unwound and Daughter Strands Are Formed at the DNA Replication Fork

In order for duplex DNA to function as a template during replication, the two intertwined strands must be unwound, or melted, to make the bases available for base pairing with the bases of the dNTPs that are polymerized into the newly synthesized daughter strands. This unwinding of the parental DNA strands is by specific **helicases,** beginning at unique segments in a DNA molecule called *replication origins,* or simply *origins*. The nucleotide sequences of origins from different organisms vary greatly, although they usually contain A·T-rich sequences. Once helicases have unwound the parental DNA at an origin, a specialized RNA polymerase called **primase** forms a short RNA primer complementary to the unwound template strands. The primer, still base-paired to its complementary DNA strand, is then elongated by a DNA polymerase, thereby forming a new daughter strand.

The DNA region at which all these proteins come together to carry out synthesis of daughter strands is called the **replication fork,** or growing fork. As replication proceeds, the growing fork and associated proteins move away from the origin. As noted earlier, local unwinding of duplex DNA produces torsional stress, which is relieved by topoisomerase I. In order for DNA polymerases to move along and copy a duplex DNA, helicase must sequentially unwind the duplex and topoisomerase must remove the supercoils that form.

A major complication in the operation of a DNA replication fork arises from two properties: the two strands of the parental DNA duplex are antiparallel, and DNA polymerases (like RNA polymerases) can add nucleotides to the growing new strands only in the 5′→3′ direction. Synthesis of one daughter strand, called the **leading strand,** can proceed continuously from a single RNA primer in the 5′→3′ direction, *the same direction as movement of the replication fork* (Figure 4-30). The problem comes in synthesis of the other daughter strand, called the **lagging strand.**

Because growth of the lagging strand must occur in the 5′→3′ direction, copying of its template strand must somehow occur in the *opposite* direction from the movement of the replication fork. A cell accomplishes this feat by synthesizing a new primer every few hundred bases or so on the second parental strand, as more of the strand is exposed by unwinding. Each of these primers, base-paired to their template strand, is elongated in the 5′→3′ direction, forming discontinuous segments called **Okazaki fragments** after their discoverer Reiji Okazaki (see Figure 4-30). The RNA primer of each Okazaki fragment is removed and replaced by DNA chain growth from the neighboring Okazaki fragment; finally an enzyme called *DNA ligase* joins the adjacent fragments.

Several Proteins Participate in DNA Replication

Detailed understanding of the eukaryotic proteins that participate in DNA replication has come largely from studies with small viral DNAs, particularly SV40 DNA, the circular

Focus Animation: Nucleotide Polymerization by DNA Polymerase

MEDIA CONNECTIONS

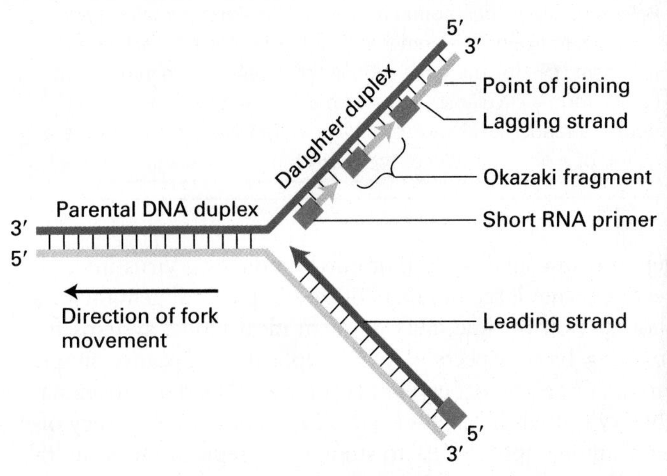

▶ **FIGURE 4-30 Leading-strand and lagging-strand DNA synthesis.** Nucleotides are added by a DNA polymerase to each growing daughter strand in the 5′→3′ direction (indicated by arrowheads). The leading strand is synthesized continuously from a single RNA primer (red) at its 5′ end. The lagging strand is synthesized discontinuously from multiple RNA primers that are formed periodically as each new region of the parental duplex is unwound. Elongation of these primers initially produces Okazaki fragments. As each growing fragment approaches the previous primer, the primer is removed and the fragments are ligated. Repetition of this process eventually results in synthesis of the entire lagging strand.

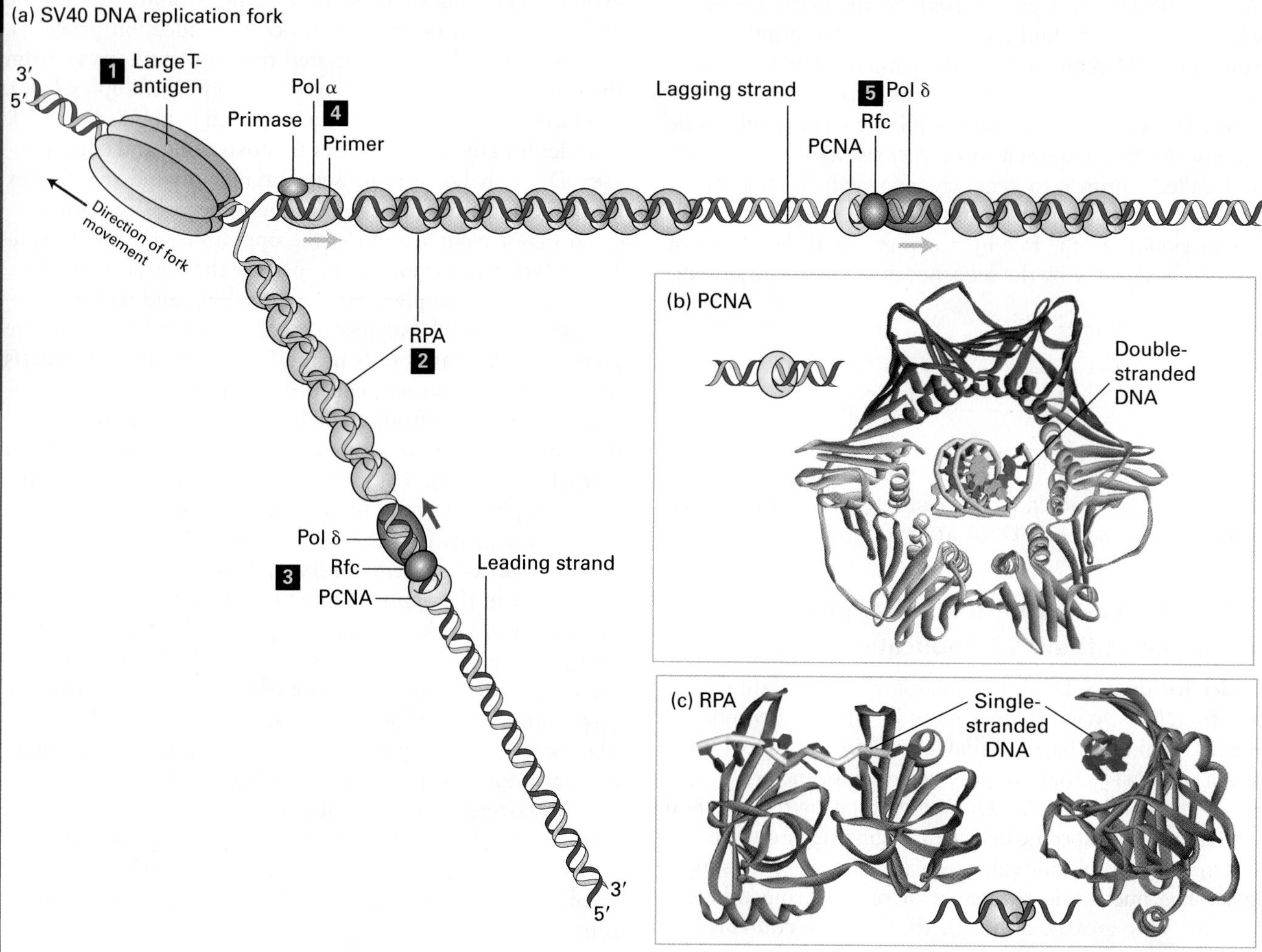

MEDIA CONNECTIONS

Focus Animation: Coordination of Leading- and Lagging-Strand Synthesis

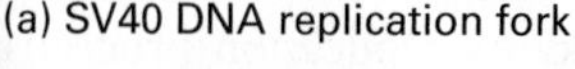

▲ **FIGURE 4-31 Model of an SV40 DNA replication fork.** (a) A hexamer of large T-antigen (**1**), a viral protein, functions as a helicase to unwind the parental DNA strands. Single-strand regions of the parental template unwound by large T-antigen are bound by multiple copies of the heterotrimeric protein RPA (**2**). The leading strand is synthesized by a complex of DNA polymerase δ (Pol δ), PCNA, and Rfc (**3**). Primers for lagging-strand synthesis (red, RNA; light blue, DNA) are synthesized by a complex of DNA polymerase α (Pol α) and primase (**4**). The 3′ end of each primer synthesized by Pol α–primase is then bound by a PCNA-Rfc–Pol δ complex, which proceeds to extend the primer and synthesize most of each Okazaki fragment (**5**). (b) The three subunits of PCNA, shown in different colors, form a circular structure with a central hole through which double-stranded DNA passes. A diagram of DNA is shown in the center of a ribbon model of the PCNA trimer. The diagram at the upper left shows the icon representing PCNA bound to DNA in part a. (c) The large subunit of RPA contains two domains that bind single-stranded DNA. On the left, the structure determined for the two DNA-binding domains of the large subunit bound to single-stranded DNA is shown with the DNA backbone (white backbone with blue bases) parallel to the plane of the page. Note that the single DNA strand is extended with the bases exposed, an optimal conformation for replication by a DNA polymerase. On the right, the view is down the length of the single DNA strand, revealing how RPA β strands wrap around the DNA. The diagram at bottom center shows the icon representing heterotrimeric RPA bound to single-stranded DNA in part (a). [Part (a) adapted from S. J. Flint et al., 2000, *Virology: Molecular Biology, Pathogenesis, and Control,* ASM Press; part (b) after J. M. Gulbis et al., 1996, *Cell* **87:**297; and part (c) after A. Bochkarev et al., 1997, *Nature* **385:**176.]

genome of a small virus that infects monkeys. Virus-infected cells replicate large numbers of the simple viral genome in a short period of time, making them ideal model systems for studying basic aspects of DNA replication. Because simple viruses like SV40 depend largely on the DNA replication machinery of their host cells (in this case monkey cells), they offer a unique opportunity to study DNA replication of multiple identical small DNA molecules by cellular proteins. Figure 4-31 depicts the multiple proteins that coordinate copying of SV40 DNA at a replication fork. The assembled proteins at a replication fork further illustrate the concept of molecular machines introduced in Chapter 3. These multicomponent complexes permit the cell to carry out an ordered sequence of events that accomplish essential cell functions.

The molecular machine that replicates SV40 DNA contains only one viral protein. All other proteins involved in

Focus Animation: Bidirectional Replication of DNA

SV40 DNA replication are provided by the host cell. This viral protein, *large T-antigen*, forms a hexamer that unwinds the parental strands at a replication fork. Primers for leading and lagging daughter-strand DNA are synthesized by a complex of primase, which synthesizes a short RNA primer, and *DNA polymerase α* (Pol α), which extends the RNA primer with deoxynucleotides, forming a mixed RNA-DNA primer.

The primer is extended into daughter-strand DNA by *DNA polymerase δ* (Pol δ), which is less likely to make errors during copying of the template strand than is Pol α because of its proofreading mechanism (see Section 4.6 below). Pol δ forms a complex with *Rfc* (*r*eplication *f*actor C) and *PCNA* (*p*roliferating *c*ell *n*uclear *a*ntigen), which displaces the primase–Pol α complex following primer synthesis. As illustrated in Figure 4-31b, PCNA is a homotrimeric protein that has a central hole through which the daughter duplex DNA passes, thereby preventing the PCNA-Rfc–Pol δ complex from dissociating from the template. Pol δ is the main polymerase used by eukaryotes for elongating DNA strands during replication.

After parental DNA is separated into single-stranded templates at the replication fork, it is bound by multiple copies of RPA (*r*eplication *p*rotein *A*), a heterotrimeric protein (Figure 4-31c). Binding of RPA maintains the template in a uniform conformation optimal for copying by DNA polymerases. Bound RPA proteins are dislodged from the parental strands by Pol α and Pol δ as they synthesize the complementary strands base-paired with the parental strands.

Several eukaryotic proteins that function in DNA replication are not depicted in Figure 4-31. DNA polymerase ε also contributes to the synthesis of cellular chromosomal DNA, though its exact role is uncertain. Still other specialized DNA polymerases are involved in repair of mismatches and damaged lesions in DNA (see Section 4.6). A topoisomerase associates with the parental DNA ahead of the helicase to remove torsional stress introduced by the unwinding of the parental strands. Ribonuclease H and FEN I remove the ribonucleotides at the 5′ ends of Okazaki fragments; these are replaced by deoxynucleotides added by DNA polymerase δ as it extends the upstream Okazaki fragment. Successive Okazaki fragments are coupled by DNA ligase through standard 5′→3′ phosphoester bonds. Replication of a linear DNA molecule presents a special problem at the ends of the molecule since the 5′-most RNA primers of the lagging strands cannot be replaced by DNA by this mechanism. In most eukaryotes, this problem is solved by the RNA-protein complex called *telomerase* that carries its own template as discussed in Chapter 6, *Genes, Genomics, and Chromosomes.*

DNA Replication Usually Occurs Bidirectionally from Each Origin

As indicated in Figures 4-30 and 4-31, both parental DNA strands that are exposed by local unwinding at a replication fork are copied into a daughter strand. In theory, DNA replication from a single origin could involve one replication fork that moves in one direction. Alternatively, two replication forks might assemble at a single origin and then move in opposite directions, leading to *bidirectional growth* of both daughter strands. Several types of experiments, including the one shown in Figure 4-32, provided early evidence in support of bidirectional strand growth.

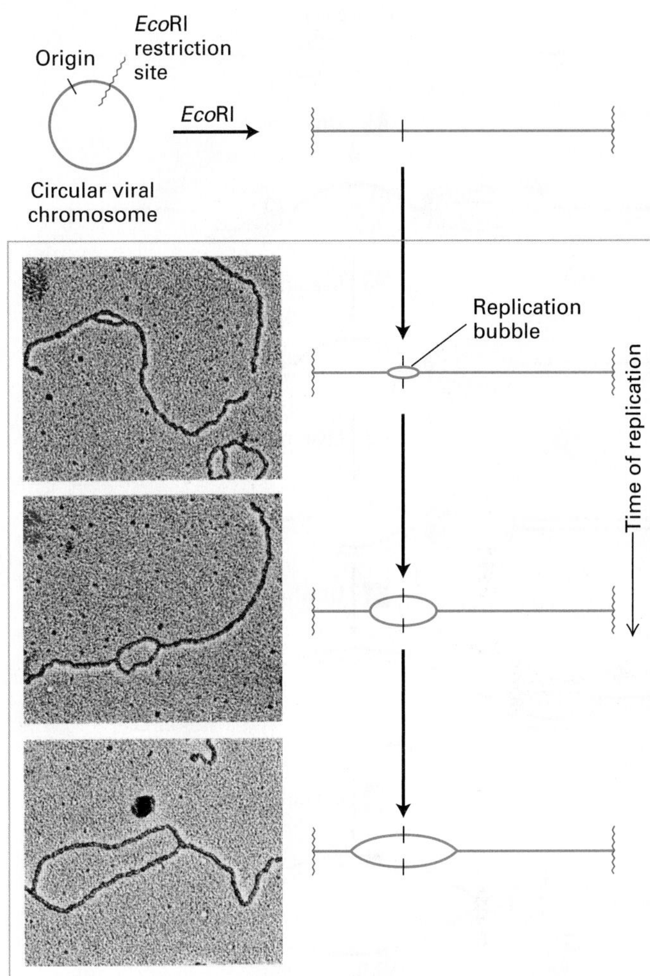

▲ **EXPERIMENTAL FIGURE 4-32 Electron microscopy proves bidirectional replication of SV40 DNA.** Electron microscopy of replicating SV40 DNA indicates bidirectional growth of DNA strands from an origin. The replicating viral DNA from SV40-infected cells was cut by the restriction enzyme *Eco*RI, which recognizes one site in the circular DNA. This was done to provide a landmark for a specific sequence in the SV40 genome: the *Eco*RI recognition sequence is now easily recognized as the ends of linear DNA molecules visualized by electron microscopy. Electron micrographs of *Eco*RI-cut replicating SV40 DNA molecules showed a collection of cut molecules with increasingly longer replication "bubbles," whose centers are a constant distance from each end of the cut molecules. This finding is consistent with chain growth in two directions from a common origin located at the center of a bubble, as illustrated in the corresponding diagrams. [See G. C. Fareed et al., 1972, *J. Virol.* **10**:484; photographs courtesy of N. P. Salzman.]

The general consensus is that all prokaryotic and eukaryotic cells employ a bidirectional mechanism of DNA

MEDIA CONNECTIONS

MEDIA CONNECTIONS

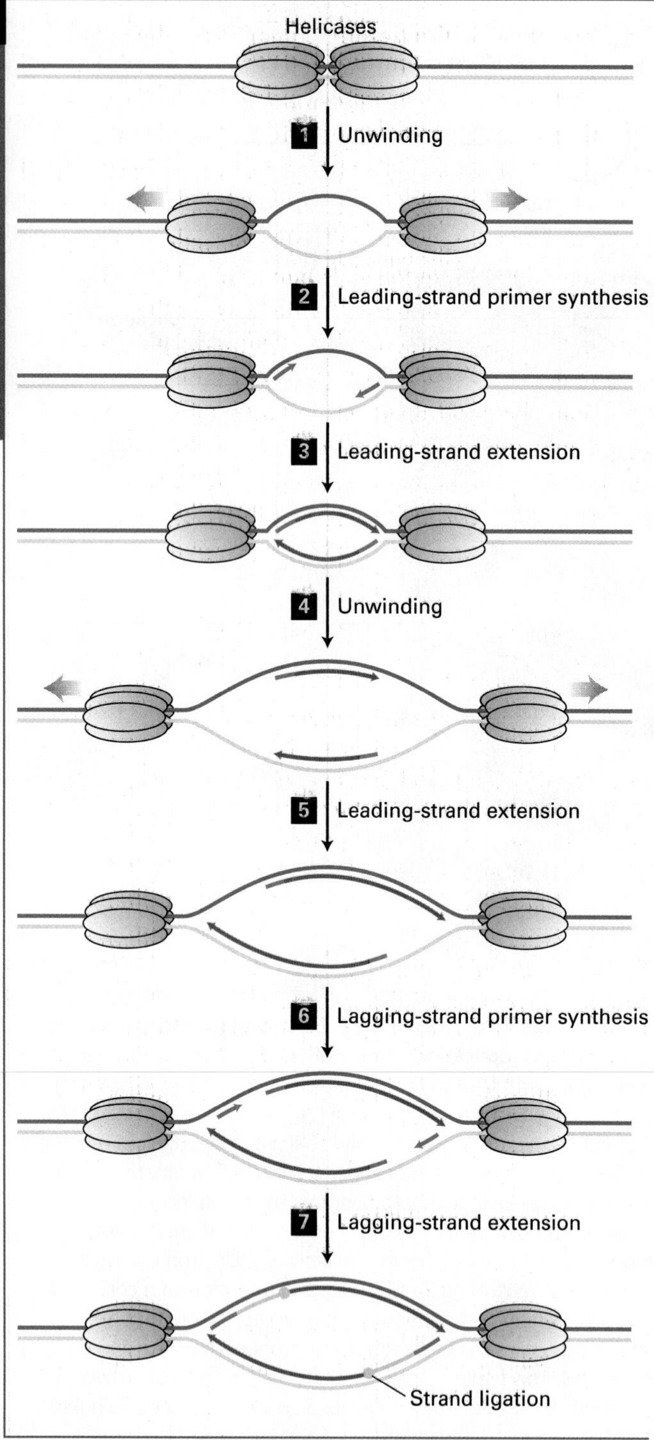

◄ **FIGURE 4-33 Bidirectional mechanism of DNA replication.** The left replication fork here is comparable to the replication fork diagrammed in Figure 4-31, which also shows proteins other than large T-antigen. *Top:* Two large T-antigen hexameric helicases first bind at the replication origin in opposite orientations. Step **1**: Using energy provided from ATP hydrolysis, the helicases move in opposite directions, unwinding the parental DNA and generating single-strand templates that are bound by RPA proteins. Step **2**: Primase–Pol α complexes synthesize short primers (red) base-paired to each of the separated parental strands. Step **3**: PCNA-Rfc–Pol δ complexes replace the primase–Pol α complexes and extend the short primers, generating the leading strands (dark green) at each replication fork. Step **4**: The helicases further unwind the parental strands, and RPA proteins bind to the newly exposed single-strand regions. Step **5**: PCNA-Rfc–Pol δ complexes extend the leading strands further. Step **6**: Primase–Pol α complexes synthesize primers for lagging-strand synthesis at each replication fork. Step **7**: PCNA-Rfc–Pol δ complexes displace the primase–Pol α complexes and extend the lagging-strand Okazaki fragments (light green), which eventually are ligated to the 5′ ends of the leading strands. The position where ligation occurs is represented by a circle. Replication continues by further unwinding of the parental strands and synthesis of leading and lagging strands as in Steps **4**–**7**. Although depicted as individual steps for clarity, unwinding and synthesis of leading and lagging strands occur concurrently.

replication. In the case of SV40 DNA, replication is initiated by binding of two large T-antigen hexameric helicases to the single SV40 origin and assembly of other proteins to form two replication forks. These then move away from the SV40 origin in opposite directions with leading- and lagging-strand synthesis occurring at both forks. As shown in Figure 4-33, the left replication fork extends DNA synthesis in the leftward direction; similarly, the right replication fork extends DNA synthesis in the rightward direction.

Unlike SV40 DNA, eukaryotic chromosomal DNA molecules contain multiple replication origins separated by tens to hundreds of kilobases. A six-subunit protein called *ORC*, for *o*rigin *r*ecognition *c*omplex, binds to each origin and associates with other proteins required to load cellular hexameric helicases composed of six homologous *MCM* proteins (for *m*ini*c*hromosome *m*aintenance, the genetic screen initially used to identify the genes encoding them). Two opposed MCM helicases separate the parental strands at an origin, with RPA proteins binding to the resulting single-stranded DNA. Synthesis of primers and subsequent steps in replication of cellular DNA are thought to be analogous to those in SV40 DNA replication (see Figures 4-31 and 4-33).

Replication of cellular DNA and other events leading to proliferation of cells are tightly regulated, so that the appropriate numbers of cells constituting each tissue are produced during development and throughout the life of an organism. As in transcription of most genes, control of the initiation step is the primary mechanism for regulating cellular DNA replication. Activation of MCM helicase activity, which is required to initiate cellular DNA replication, is regulated by specific protein kinases called ***S-phase* cyclin-dependent kinases.** Other cyclin-dependent kinases regulate additional aspects of cell proliferation, including the complex process of mitosis by which a eukaryotic cell divides into two daughter cells. We discuss the various regulatory mechanisms that determine the rate of cell division in Chapter 20.

KEY CONCEPTS OF SECTION 4.5

DNA Replication

- Each strand in a parental duplex DNA acts as a template for synthesis of a daughter strand and remains base-paired to the new strand, forming a daughter duplex (using a semiconservative mechanism). New strands are formed in the 5′→3′ direction.

- Replication begins at a sequence called an origin. Each eukaryotic chromosomal DNA molecule contains multiple replication origins.
- DNA polymerases, unlike RNA polymerases, cannot unwind the strands of duplex DNA and cannot initiate synthesis of new strands complementary to the template strands.
- At a replication fork, one daughter strand (the leading strand) is elongated continuously. The other daughter strand (the lagging strand) is formed as a series of discontinuous Okazaki fragments from primers synthesized every few hundred nucleotides (Figure 4-30).
- The ribonucleotides at the 5′ end of each Okazaki fragment are removed and replaced by elongation of the 3′ end of the next Okazaki fragment. Finally, adjacent Okazaki fragments are joined by DNA ligase.
- Helicases use energy from ATP hydrolysis to separate the parental (template) DNA strands. Primase synthesizes a short RNA primer, which remains base-paired to the template DNA. This initially is extended at the 3′ end by DNA polymerase α (Pol α), resulting in a short (5′)RNA-(3′)DNA daughter strand.
- Most of the DNA in eukaryotic cells is synthesized by Pol δ, which takes over from Pol α and continues elongation of the daughter strand in the 5′→3′ direction. Pol δ remains stably associated with the template by binding to Rfc protein, which in turn binds to PCNA, a trimeric protein that encircles the daughter duplex DNA (see Figure 4-31).
- DNA replication generally occurs by a bidirectional mechanism in which two replication forks form at an origin and move in opposite directions, with both template strands being copied at each fork (see Figure 4-33).
- Synthesis of eukaryotic DNA in vivo is regulated by controlling the activity of the MCM helicases that initiate DNA replication at multiple origins spaced along chromosomal DNA.

4.6 DNA Repair and Recombination

Damage to DNA is unavoidable and arises in many ways. DNA damage can be caused by spontaneous cleavage of chemical bonds in DNA, by environmental agents such as ultraviolet and ionizing radiation, and by reaction with genotoxic chemicals that are by-products of normal cellular metabolism or occur in the environment. A change in the normal DNA sequence, called a **mutation**, can occur during replication when a DNA polymerase inserts the wrong nucleotide as it reads a damaged template. Mutations also occur at a low frequency as the result of copying errors introduced by DNA polymerases when they replicate an undamaged template. If such mutations were left uncorrected, cells might accumulate so many mutations that they could no longer function properly. In addition, the DNA in germ cells might incur too many mutations for viable offspring to be formed. Thus the prevention of DNA sequence errors in all types of cells is important for survival, and several cellular mechanisms for repairing damaged DNA and correcting sequence errors have evolved. One mechanism for repairing double-stranded DNA breaks, by a process called **recombination,** is also used by eukaryotic cells to generate new combinations of maternal and paternal genes on each chromosome through the exchange of segments of the chromosomes during the production of germ cells (e.g., sperm and eggs).

Significantly, defects in DNA repair mechanisms and cancer are closely related. When repair mechanisms are compromised, mutations accumulate in the cell's DNA. If these mutations affect genes that are normally involved in the careful regulation of cell division, cells can begin to divide uncontrollably, leading to tumor formation, and cancer. Chapter 25 outlines in detail how cancer arises from defects in DNA repair. We will encounter a few examples in this section, as well, as we first consider the ways in which DNA integrity can be compromised, and then discuss the repair mechanisms that cells have evolved to ensure the fidelity of this very important molecule.

DNA Polymerases Introduce Copying Errors and Also Correct Them

The first line of defense in preventing mutations is DNA polymerase itself. Occasionally, when replicative DNA polymerases progress along the template DNA, an incorrect nucleotide is added to the growing 3′ end of the daughter strand (see Figure 4-31). *E. coli* DNA polymerases, for instance, introduce about 1 incorrect nucleotide per 10^4 polymerized nucleotides. Yet the measured mutation rate in bacterial cells is much lower: about 1 mistake in 10^9 nucleotides incorporated into a growing strand. This remarkable accuracy is largely due to *proofreading* by *E. coli* DNA polymerases.

Proofreading depends on a *3′→5′ exonuclease activity* of some DNA polymerases. When an incorrect base is incorporated during DNA synthesis, base-pairing between the 3′ nucleotide of the nascent strand and the template strand does not occur. As a result, the polymerase pauses, then transfers the 3′ end of the growing chain to its exonuclease site, where the incorrect mispaired base is removed (Figure 4-34). Then the 3′ end is transferred back to the polymerase site, where this region is copied correctly. Like the *E. coli* DNA polymerases, two eukaryotic DNA polymerases, δ and ε, used for replication of most chromosomal DNA in animal cells, also have proofreading activity. It seems likely that proofreading is indispensable for all cells to avoid excessive mutations.

Chemical and Radiation Damage to DNA Can Lead to Mutations

DNA is continually subjected to a barrage of damaging chemical reactions; estimates of the number of DNA damage events in a single human cell range from 10^4 to 10^6 per day! Even if DNA were not exposed to damaging chemicals, certain aspects of DNA structure are inherently unstable.

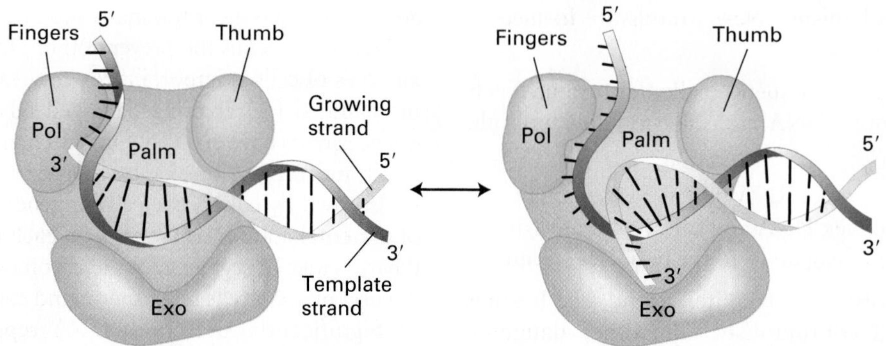

▲ **FIGURE 4-34 Proofreading by DNA polymerase.** All DNA polymerases have a similar three-dimensional structure, which resembles a half-opened right hand. The "fingers" bind the single-stranded segment of the template strand, and the polymerase catalytic activity (Pol) lies in the junction between the fingers and palm. As long as the correct nucleotides are added to the 3′ end of the growing strand, it remains in the polymerase site. Incorporation of an incorrect base at the 3′ end causes melting of the newly formed end of the duplex. As a result, the polymerase pauses, and the 3′ end of the growing strand is transferred to the 3′→5′ exonuclease site (Exo) about 3 nm away, where the mispaired base and probably other bases are removed. Subsequently, the 3′ end flips back into the polymerase site and elongation resumes. [Adapted from C. M. Joyce and T. T. Steitz, 1995, *J. Bacteriol.* **177:**6321, and S. Bell and T. Baker, 1998, *Cell* **92:**295.]

For example, the bond connecting a purine base to deoxyribose is prone to hydrolysis at low rate under physiological conditions, leaving a sugar without an attached base. Thus coding information is lost, and this can lead to a mutation during DNA replication. Normal cellular reactions, including the movement of electrons along the electron-transport chain in mitochondria and lipid oxidation in peroxisomes, produce several chemicals that react with and damage DNA, including hydroxyl radicals and superoxide (O_2^-). These too can cause mutations, including those that lead to cancers.

Many spontaneous mutations are **point mutations,** which involve a change in a single base pair in the DNA sequence. One of the most frequent point mutations comes from *deamination* of a cytosine (C) base, which converts it into a uracil (U) base. In addition, the common modified base 5-methylcytosine forms thymine when it is deaminated. If these alterations are not corrected before the DNA is replicated, the cell will use the strand containing U or T as template to form a U·A or T·A base pair, thus creating a permanent change to the DNA sequence (Figure 4-35).

Radiation from the environment can also have dramatic consequences for DNA. High-energy *ionizing radiation* such as x-rays and gamma rays cause double-stranded breaks in DNA. *UV radiation* found in sunlight causes distortions in

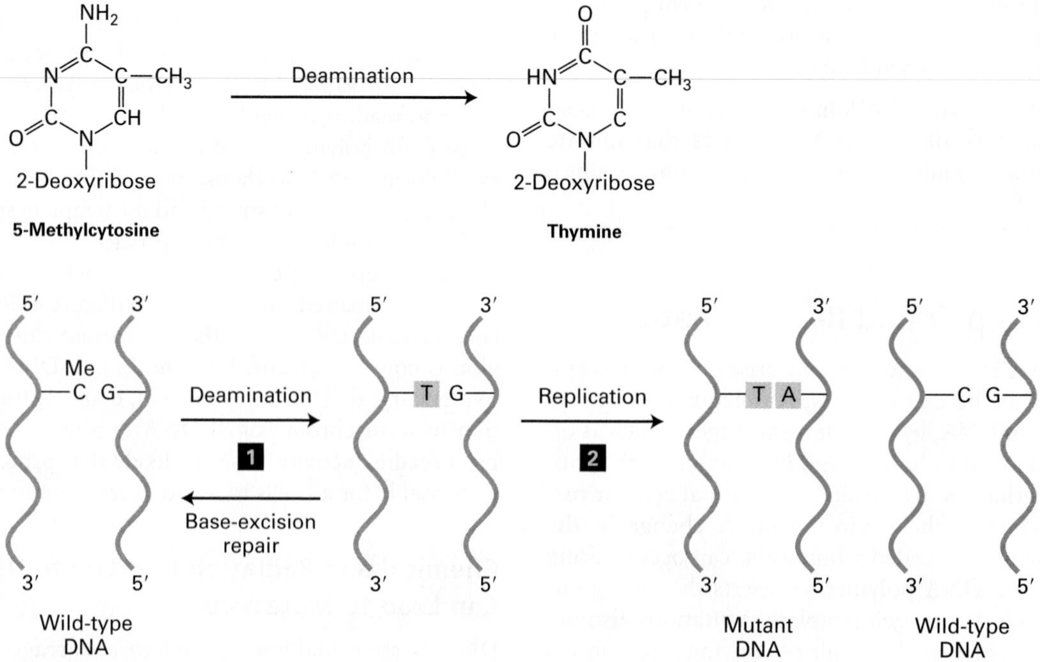

2-Deoxyribose
5-Methylcytosine
Deamination
2-Deoxyribose
Thymine

Wild-type DNA
Deamination
1
Base-excision repair
Replication
2
Mutant DNA
Wild-type DNA

▲ **FIGURE 4-35 Deamination leads to point mutations.** A spontaneous point mutation can form by deamination of 5-methylcytosine (C) to form thymine (T). If the resulting T·G base pair is not restored to the normal C·G base pair by base excision-repair mechanisms (step 1), it will lead to a permanent change in sequence (i.e., a mutation) following DNA replication (step 2). After one round of replication, one daughter DNA molecule will have the mutant T·A base pair and the other will have the wild-type C·G base pair.

the DNA double helix that interfere with proper replication and transcription.

High-Fidelity DNA Excision-Repair Systems Recognize and Repair Damage

In addition to proofreading, cells have other repair systems for preventing mutations due to copying errors, spontaneous mutation, and exposure to chemicals and radiation. Several DNA **excision-repair systems** that normally operate with a high degree of accuracy have been well studied. These systems were first elucidated through a combination of genetic and biochemical studies in *E. coli*. Homologs of the key bacterial proteins exist in eukaryotes from yeast to humans, indicating that these error-free mechanisms arose early in evolution to protect DNA integrity. Each of these systems functions in a similar manner—a segment of the damaged DNA strand is excised, and the gap is filled by DNA polymerase and ligase using the complementary DNA strand as template.

We will now turn to a closer look at some of the mechanisms of DNA repair, ranging from repair of single base mutations to repair of DNA broken across both strands. Some of these accomplish their repairs with great accuracy; others are less precise.

Base Excision Repairs T·G Mismatches and Damaged Bases

In humans, the most common type of point mutation is a C to T, which is caused by deamination of 5-methyl C to T (see Figure 4-35). The conceptual problem with *base excision repair* in this case is determining which is the normal and which is the mutant DNA strand, and repairing the latter so that it is properly base-paired with the normal strand. Since a G·T mismatch is almost invariably caused by chemical conversion of C to U or 5-methyl C to T, the repair system evolved to remove the T and replace it with a C (Figure 4-36).

The G·T mismatch is recognized by a DNA glycosylase that flips the thymine base out of the helix and then hydrolyzes the bond that connects it to the sugar-phosphate DNA backbone. Following this initial incision, an apurinic (AP) endonuclease cuts the DNA strand near the abasic site. The deoxyribose phosphate lacking the base is then removed and replaced with a C by a specialized repair DNA polymerase that reads the G in the template strand. As mentioned earlier, this repair must take place prior to DNA replication because the incorrect base in this pair, T, occurs naturally in normal DNA. Consequently, it would be able to engage in normal Watson-Crick base pairing during replication, generating a stable point mutation that is now unable to be recognized by repair mechanisms (see Figure 4-35, step **2**).

Human cells contain a battery of glycosylases, each of which is specific for a different set of chemically modified DNA bases. For example, one removes 8-oxyguanine, an oxidized form of guanine, allowing its replacement by an undamaged G, and others remove bases modified by alkylating agents. The resulting nucleotide lacking a base is then replaced

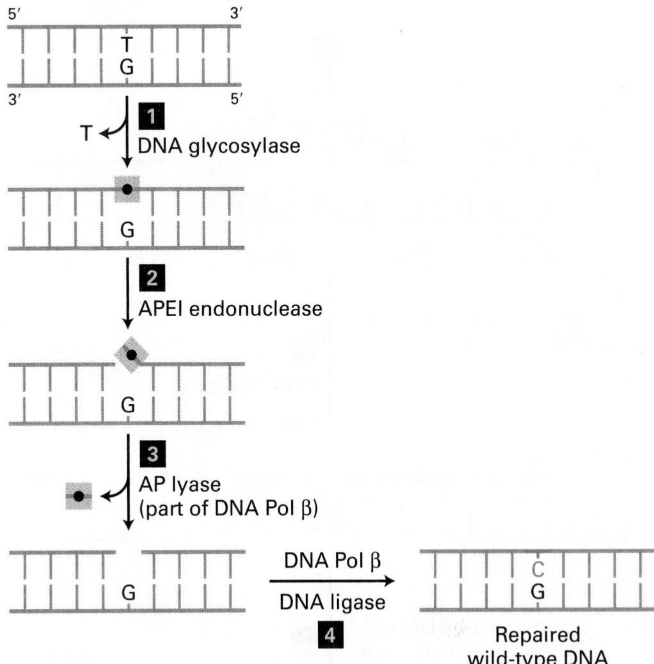

▲ **FIGURE 4-36 Base excision repair of a T·G mismatch.** A DNA glycosylase specific for G·T mismatches, usually formed by deamination of 5-mC residues (see Figure 4-35), flips the thymine base out of the helix and then cuts it away from the sugar-phosphate DNA backbone (step **1**), leaving just the deoxyribose (black dot). An endonuclease specific for the resultant baseless site [apurinic endonuclease I (APE1)] then cuts the DNA backbone (step **2**), and the deoxyribose phosphate is removed by an endonuclease, apurinic lyase (AP lyase), associated with DNA polymerase β, a specialized DNA polymerase used in repair (step **3**). The gap is then filled in by DNA Pol β and sealed by DNA ligase (step **4**), restoring the original G·C base pair. [After O. Schärer, 2003, *Angewandte Chemie* **42**:2946.]

by the repair mechanism just discussed. A similar mechanism functions in the repair of lesions resulting from *depurination*, the loss of a guanine or adenine base from DNA resulting from hydrolysis of the glycosylic bond between deoxyribose and the base. Depurination occurs spontaneously and is fairly common in mammals. The resulting abasic sites, if left unrepaired, generate mutations during DNA replication because they cannot specify the appropriate paired base.

Mismatch Excision Repairs Other Mismatches and Small Insertions and Deletions

Another process, also conserved from bacteria to man, principally eliminates base-pair mismatches and insertions or deletions of one or a few nucleotides that are accidentally introduced by DNA polymerases during replication. As with base excision repair of a T in a T·G mismatch, the conceptual problem with *mismatch excision repair* is determining which is the normal and which is the mutant DNA strand, and repairing the latter. How this happens in human cells is not known with certainty. It is thought that the proteins that bind to the mismatched segment of DNA distinguish the template

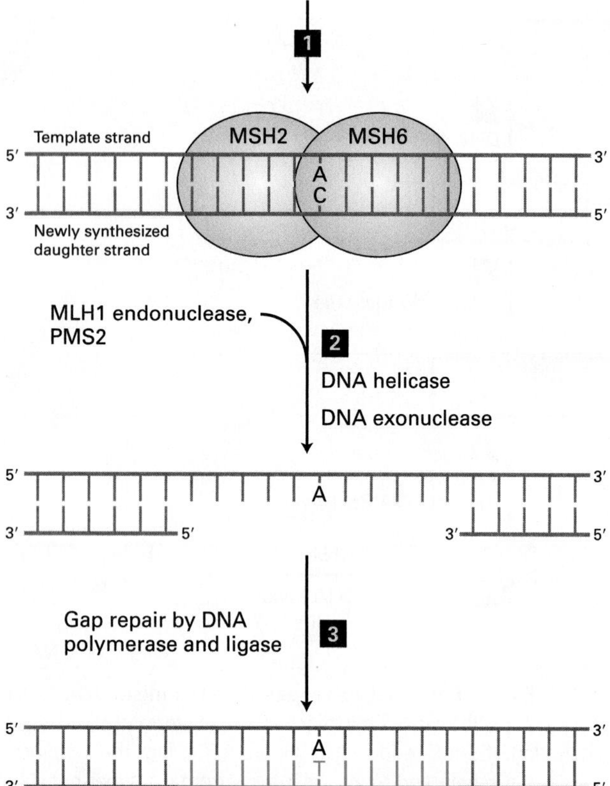

▲ **FIGURE 4-37 Mismatch excision repair in human cells.** The mismatch excision-repair pathway corrects errors introduced during replication. A complex of the MSH2 and MSH6 proteins (bacterial *MutS h*omologs 2 and 6) binds to a mispaired segment of DNA in such a way as to distinguish between the template and newly synthesized daughter strands (step **1**). This triggers binding of MLH1 and PMS2 (both homologs of bacterial MutL). The resulting DNA-protein complex then binds an endonuclease that cuts the newly synthesized daughter strand. Next a DNA helicase unwinds the helix, and an exonuclease removes several nucleotides from the cut end of the daughter strand, including the mismatched base (step **2**). Finally, as with base excision repair, the gap is then filled in by a DNA polymerase (Pol δ, in this case) and sealed by DNA ligase (step **3**).

and daughter strands; then the mispaired segment of the daughter strand—the one with the replication error—is excised and repaired to become an exact complement of the template strand (Figure 4-37). In contrast to base excision repair, mismatch excision repair occurs after DNA replication.

Predisposition to a colon cancer known as *hereditary nonpolyposis colorectal cancer* results from an inherited loss-of-function mutation in one copy of either the *MLH1* or the *MSH2* gene. The MSH2 and MLH1 proteins are essential for DNA mismatch repair (see Figure 4-37). Cells with at least one functional copy of each of these genes exhibit normal mismatch repair. However, tumor cells frequently arise from those cells that have experienced a random mutation in the second copy; when both copies of one gene are not functional, the mismatch-repair system is lost. Inactivating mutations in these genes are also common in noninherited forms of colon cancer. ▮

Nucleotide Excision Repairs Chemical Adducts That Distort Normal DNA Shape

Cells use *nucleotide excision repair* to fix DNA regions containing chemically modified bases, often called *chemical adducts,* that distort the normal shape of DNA locally. A key to this type of repair is the ability of certain proteins to slide along the surface of a double-stranded DNA molecule looking for bulges or other irregularities in the shape of the double helix. For example, this mechanism repairs *thymine-thymine dimers,* a common type of damage caused by UV light (Figure 4-38); these dimers interfere with both replication and transcription of DNA.

Figure 4-39 illustrates how the nucleotide excision-repair system repairs damaged DNA. Some 30 proteins are involved in this repair process, the first of which were identified through a study of the defects in DNA repair in cultured cells from individuals with xeroderma pigmentosum, a hereditary disease associated with a predisposition to cancer. Individuals with this disease frequently develop the skin cancers called *melanomas* and *squamous cell carcinomas* if their skin is exposed to the UV rays in sunlight. Cells of affected patients lack a functional nucleotide excision-repair system. Mutations in any of at least seven different genes, called *XP-A* through *XP-G,*

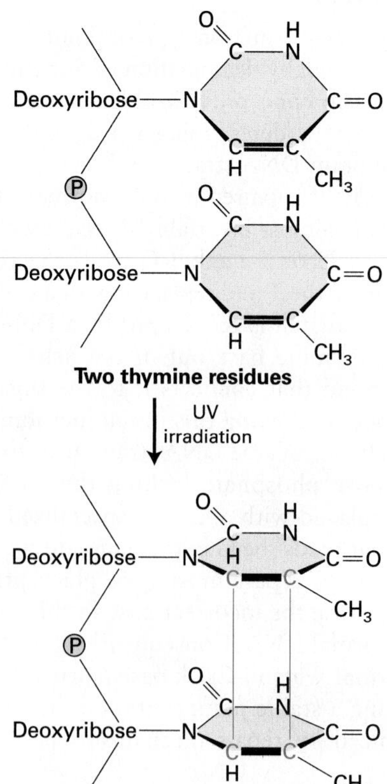

▲ **FIGURE 4-38 Formation of thymine-thymine dimers.** The most common type of DNA damage caused by UV irradiation, thymine-thymine dimers can be repaired by an excision-repair mechanism.

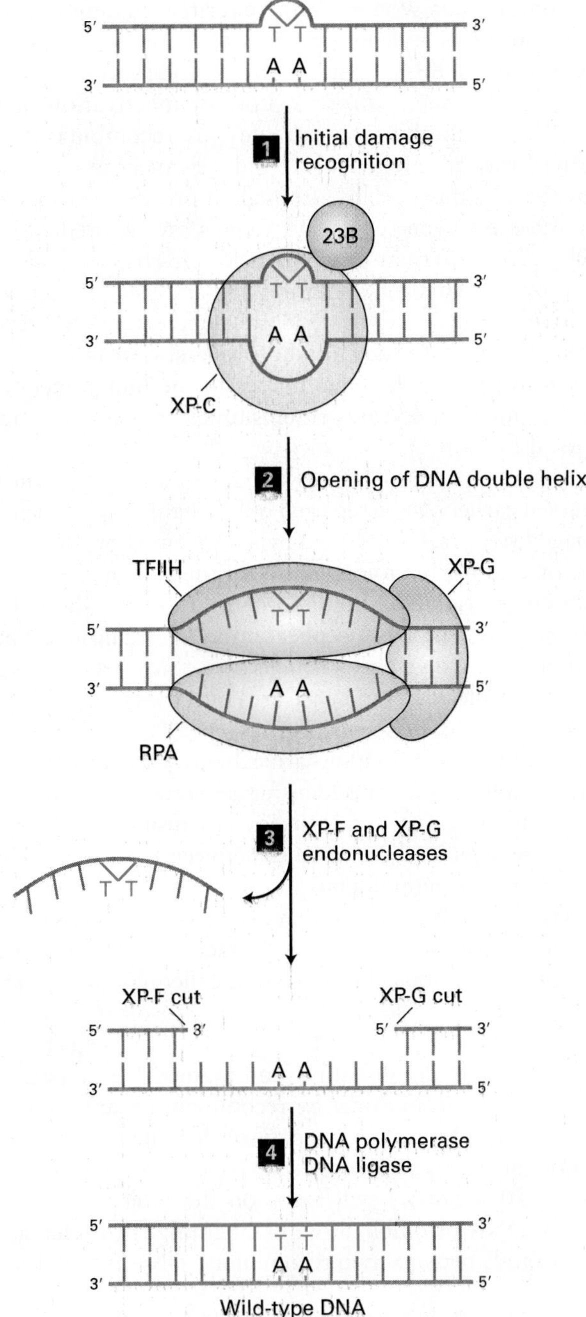

▲ **FIGURE 4-39 Nucleotide excision repair in human cells.** A DNA lesion that causes distortion of the double helix, such as a thymine dimer, is initially recognized by a complex of the XP-C (*x*eroderma *p*igmentosum *C* protein) and 23B proteins (step **1**). This complex then recruits transcription factor TFIIH, whose helicase subunits, powered by ATP hydrolysis, partially unwind the double helix. XP-G and RPA proteins then bind to the complex and further unwind and destabilize the helix until a bubble of ≈25 bases is formed (step **2**). Then XP-G (now acting as an endonuclease) and XP-F, a second endonuclease, cut the damaged strand at points 24–32 bases apart on each side of the lesion (step **3**). This releases the DNA fragment with the damaged bases, which is degraded to mononucleotides. Finally the gap is filled by DNA polymerase exactly as in DNA replication, and the remaining nick is sealed by DNA ligase (step **4**). [Adapted from J. Hoeijmakers, 2001, *Nature* **411**:366, and O. Schärer, 2003, *Angewandte Chemie* **42**:2946.]

lead to inactivation of this repair system and cause xeroderma pigmentosum; all produce the same phenotype and have the same consequences. The roles of most of these XP proteins in nucleotide excision repair are now well understood (see Figure 4-39). ■

Remarkably, five polypeptide subunits of TFIIH, a general transcription factor required for transcription of all genes (see Figure 7-31), are also required for nucleotide excision repair in eukaryotic cells. Two of these subunits have homology to helicases, as shown in Figure 4-39. In transcription, the helicase activity of TFIIH unwinds the DNA helix at the start site, allowing RNA polymerase to initiate (see Figure 7-31). It appears that nature has used a similar protein assembly in two different cellular processes that require helicase activity.

The use of shared subunits in transcription and DNA repair may help explain the observation that DNA damage in higher eukaryotes is repaired at a much faster rate in regions of the genome being actively transcribed than in nontranscribed regions—so-called *transcription-coupled repair.* Since only a small fraction of the genome is transcribed in any one cell in higher eukaryotes, transcription-coupled repair efficiently directs repair efforts to the most critical regions. In this system, if an RNA polymerase becomes stalled at a lesion on DNA (e.g., a thymine-thymine dimer), a small protein, CSB, is recruited to the RNA polymerase; this triggers opening of the DNA helix at that point, recruitment of TFIIH, and the reactions of steps **2** through **4** depicted in Figure 4-39.

Two Systems Utilize Recombination to Repair Double-Strand Breaks in DNA

Ionizing radiation (e.g. x- and γ-radiation) and some anticancer drugs cause double-strand breaks in DNA. These are particularly severe lesions because incorrect rejoining of double strands of DNA can lead to gross chromosomal rearrangements that can affect the functioning of genes. For example, incorrect joining could create a "hybrid" gene that codes for the N-terminal portion of one amino acid sequence fused to the C-terminal portion of a completely different amino acid sequence; or a chromosomal rearrangement could bring the promoter of one gene into close proximity to the coding region of another gene, changing the level or cell type in which that gene is expressed.

Two systems have evolved to repair double-strand breaks. The predominant mechanism for repairing double-strand breaks in multicellular organisms, *nonhomologous end-joining,* involves rejoining the nonhomologous ends of two DNA molecules. This process is error-prone; even if the joined DNA fragments come from the same chromosome, the repair process invariably results in loss of several base pairs at the joining point (Figure 4-40). Formation of such a possibly mutagenic deletion is one example of how repair of DNA damage can introduce mutations. The second mechanism, *homologous recombination,* is discussed as part of the next section.

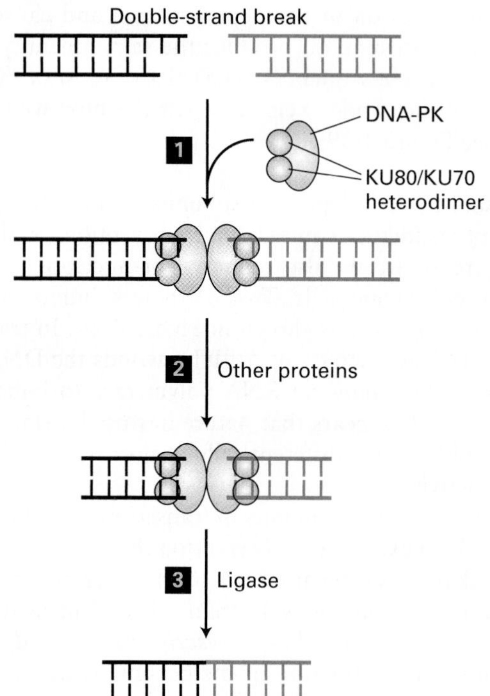

▲ **FIGURE 4-40 Nonhomologous end-joining.** When sister chromatids are not available to help repair double-strand breaks, nucleotide sequences are butted together that were not apposed in the unbroken DNA. These DNA ends are usually from the same chromosome locus, and when linked together, several base pairs are lost. Occasionally, ends from different chromosomes are accidentally joined together. A complex of two proteins, Ku and DNA-dependent protein kinase, binds to the ends of a double-strand break (step **1**). After formation of a synapse, the ends are further processed by nucleases, resulting in removal of a few bases (step **2**), and the two double-stranded molecules are ligated together (step **3**). As a result, the double-strand break is repaired, but several base pairs at the site of the break are removed. [Adapted from G. Chu, 1997, *J. Biol. Chem.* **272**:24097; M. Lieber et al., 1997, *Curr. Opin. Genet. Devel.* **7**:99; and D. van Gant et al., 2001, *Nature Rev. Genet.* **2**:196.]

Since movement of DNA within the protein-dense nucleus is fairly minimal, the correct ends are generally rejoined together, albeit with loss of base pairs. However, occasionally broken ends from different chromosomes are joined together, leading to translocation of pieces of DNA from one chromosome to another. Such translocations may generate chimeric genes that can have drastic effects on normal cell function, such as uncontrollable cell growth, which is the hallmark of cancer. The devastating effects of double-strand breaks make this the "most unkindest cut of all," to borrow a phrase from Shakespeare's *Julius Caesar.*

Homologous Recombination Can Repair DNA Damage and Generate Genetic Diversity

At one time homologous recombination was thought to be a minor repair process in human cells. This changed when it was realized that several human cancers are potentiated by inherited mutations in genes essential for **homologous recombination repair** (see Table 25-1). For example, the vast majority of women with inherited susceptibility to breast cancer have a mutation in one allele of either the BRCA-1 or the BRCA-2 genes that encode proteins participating in this repair process. Loss or inactivation of the second allele inhibits the homologous recombination repair pathway and thus tends to induce cancer in mammary or ovarian epithelial cells, although at present it is not clear why these estrogen-responsive tissues are favored sites of carcinogenesis. Yeasts can repair double-strand breaks induced by γ-irradiation. Isolation and analysis of radiation-sensitive (*RAD*) mutants that are deficient in this repair system facilitated study of the process. Virtually all the yeast Rad proteins have homologs in the human genome, and the human and yeast proteins function in an essentially identical fashion.

A variety of DNA lesions not repaired by mechanisms discussed earlier can be repaired by mechanisms in which the damaged sequence is replaced by a segment copied from the same or a highly homologous DNA sequence on the homologous chromosome of diploid organisms, or the sister chromosome following DNA replication in all organisms. These mechanisms involve an exchange of strands between separate DNA molecules and hence are collectively referred to as *DNA recombination* mechanisms.

In addition to providing a mechanism for DNA repair, similar recombination mechanisms generate genetic diversity among the individuals of a species by causing the exchange of large regions of chromosomes between the maternal and paternal pair of homologous chromosomes during the special type of cellular division that generates germ cells (sperm and eggs), **meiosis** (Figure 5-3). In fact the exchange of regions of homologous chromosomes, called **crossing over,** is required for proper segregation of chromosomes during the first meiotic cell division. Meiosis and the consequences of generating new combinations of maternal and paternal genes on one chromosome by recombination are discussed further in Chapter 5. The mechanisms leading to proper segregation of chromosomes during meiosis are discussed in Chapter 20. Here we will focus on the molecular mechanisms of DNA recombination, highlighting the exchange of DNA strands between two recombining DNA molecules.

Repair of a Collapsed Replication Fork An example of recombinational DNA repair is the repair of a "collapsed" replication fork. If a break in the phosphodiester backbone of one DNA strand is not repaired before a replication fork passes, the replicated portions of the daughter chromosomes become separated when the replication helicase reaches the "nick" in the parental DNA strand because there are no covalent bonds between the two fragments of the parental strand on either side of the nick. This process is called *replication fork collapse* (Figure 4-41, step **1**). If it is not repaired, it is generally lethal to at least one daughter cell following cell division because of the loss of genetic information between the nick and the end of the chromosome. The recombination process that repairs the resulting double-stranded break and regenerates a replication fork involves multiple enzymes and other proteins, only some of which are mentioned here.

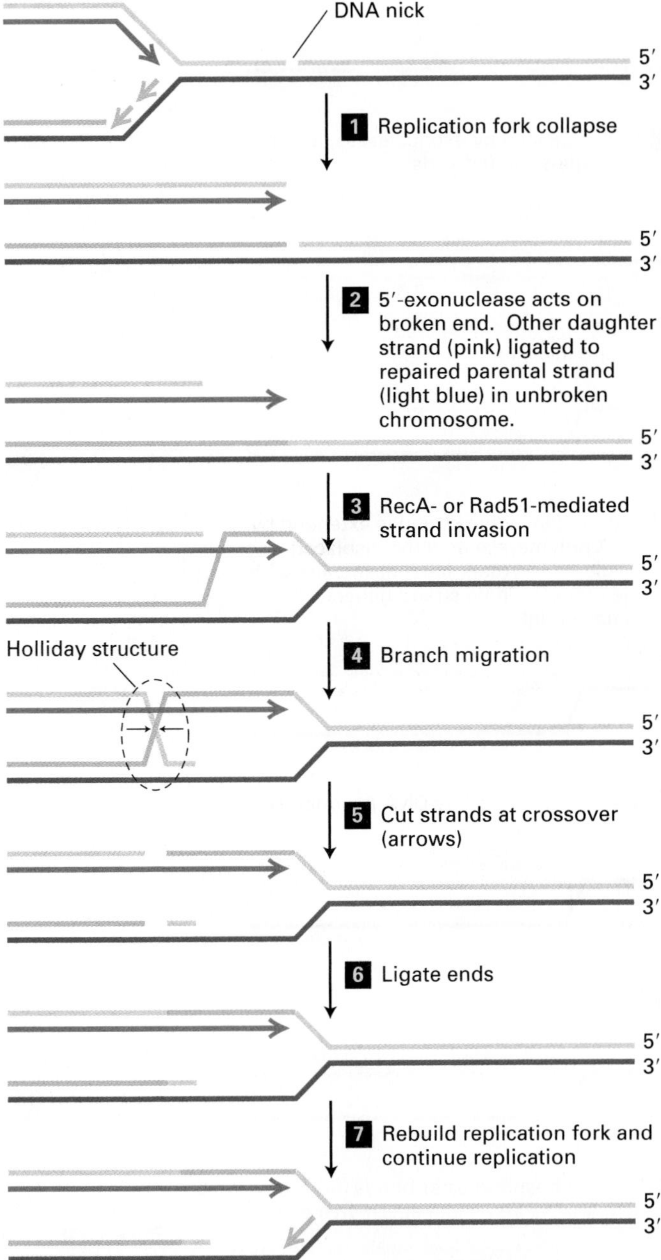

▲ **FIGURE 4-41 Recombinational repair of a collapsed replication fork.** Parental strands are light and dark blue. The leading daughter strand is dark red, and the lagging daughter strand pink. Diagonal lines in step 3 and beyond represent a single phosphodiester bond from the DNA strand of the corresponding color. Small black arrows following step 4 represent cleavage of the phosphodiester bonds at the crossover of DNA strands in the Holliday structure. See http://www.sdsc.edu/journals/mbb/ruva.html for an animation of branch migration catalyzed by *E. coli* proteins RuvA and RuvB. See http://engels.genetics.wisc.edu/Holliday/holliday3D.html for an animation of the Holliday structure and its resolution. See the text for a discussion. [Adapted from D. L. Nelson and M. M. Cox, 2005, *Lehninger Principles of Biochemistry*, 4th ed., W. H. Freeman and Company.]

The first step in the repair of the double-strand break is exonucleolytic digestion of the strand (light blue) that has its 5′ end at the break, leaving a portion of the other (dark red) strand (the one with the 3′ end at the break) single-stranded (Figure 4-41, step 2). The lagging nascent strand (pink) created on the unnicked, homologous, parent strand is ligated to the unreplicated portion of the parent chromosome, as shown in Figure 4-41, step 2). A critical protein required for the next step is RecA in bacteria, or the homologous Rad51 in *S. cerevisiae* and other eukaryotes. Multiple RecA/Rad51 molecules bind to the single-stranded DNA (now considered the invading strand) and catalyze its hybridization to a perfectly or nearly perfectly complementary sequence in another, homologous, double-stranded DNA molecule, either the ligated molecule created after fork collapse (as shown in the figure) or the other homologous chromosome in diploid organisms. The other strand (dark blue) of this target double-stranded DNA (the strand not base-pairing with the invading strand) is displaced as a single-stranded loop of DNA along the region of hybridization between its complement and the invading strand (refer to Figure 4-41, step 3). The RecA/Rad51-catalyzed invasion of a duplex DNA by a single-stranded complement of one of the strands is key to the recombination process. Since no base pairs are lost or gained in this process, called *strand invasion*, it does not require an input of energy.

Next, the hybrid region between target DNA (pink) and the invading strand (dark red) is extended in the direction away from the break by proteins that use energy from ATP hydrolysis. This process is called *branch migration* (Figure 4-41, step 4) because the point at which the target DNA strand (pink) crosses from one complementary strand (dark blue) to its complement in the broken DNA molecule (dark red), is called a *branch* in the DNA structure. In this diagram, the diagonal lines represent only one phosphodiester bond. Molecular modeling and other studies show that the first base on either side of the branch is base-paired to a complementary nucleotide. As this branch *migrates* to the left, the number of base pairs remains constant; one new base pair formed with the (red) invading strand is matched by the loss of one base pair with the parental (dark blue) strand.

When the region of hybridization extends beyond the 5′ end of the broken strand (light blue), this broken parental DNA strand becomes increasingly single-stranded, as its complement, the invading (dark red) strand, base-pairs instead with the target (pink) DNA strand. This single-stranded (light blue) parental strand then base-pairs with the complementary region of the other parental strand (dark blue) that has likewise become single-stranded as the branch migrates to the left (Figure 4-41, step 4). The resulting structure is called a *Holliday structure*, after the geneticist who first proposed it as an intermediate in genetic recombination. Again, the diagonal lines in the diagram following step 4 represent single phosphodiester bonds (not a stretch of DNA), and all bases in the Holliday structure are base-paired to complementary bases in the parental strands. Cleavage of the phosphodiester bonds that cross over from one parental strand to the other and ligation of the 5′ and 3′ ends base-paired to the same parental strands (steps 5 and 6) result in the generation of a structure similar to a replication fork. Rebinding of replication fork proteins results in extension of the leading

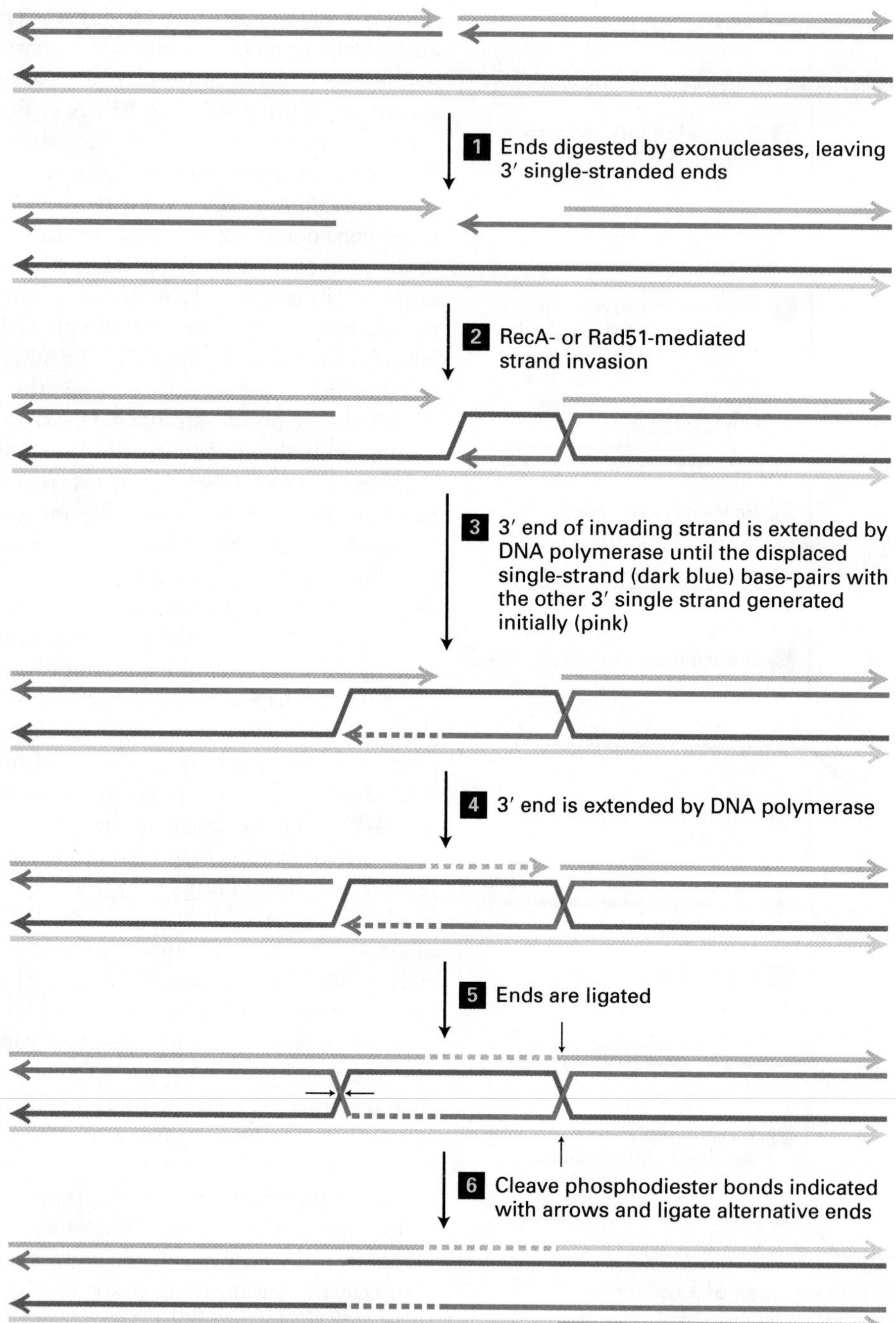

1 Ends digested by exonucleases, leaving 3′ single-stranded ends
2 RecA- or Rad51-mediated strand invasion
3 3′ end of invading strand is extended by DNA polymerase until the displaced single-strand (dark blue) base-pairs with the other 3′ single strand generated initially (pink)
4 3′ end is extended by DNA polymerase
5 Ends are ligated
6 Cleave phosphodiester bonds indicated with arrows and ligate alternative ends

▲ **FIGURE 4-42 Double-strand DNA break repair by homologous recombination.** For simplicity, each DNA double helix is represented by two parallel lines with the polarities of the strands indicated by arrowheads at their 3′ ends. The upper molecule has a double-strand break. Note that in the diagram of the upper DNA molecule the strand with its 3′ end at the right is on the top, while in the diagram of the lower DNA molecule this strand is drawn on the bottom. See the text for discussion. [After T. L. Orr-Weaver and J. W. Szostak, 1985, *Microbiol. Rev.* **49:**33.]

strand past the point of the original strand break and reinitiation of lagging-strand synthesis (step **7**), thus regenerating a replication fork. The overall process allows the undamaged upper strand in the lower molecule following step **2** (pink/light blue) to serve as template for extension of the leading (dark red) strand in step **7**.

Double-Stranded DNA Break Repair by Homologous Recombination A similar mechanism called *homologous recombination* can repair a double-strand break in a chromosome and can also exchange large segments of two double-stranded DNA molecules (Figure 4-42). Homologous recombination is also dependent on strand invasion catalyzed principally by RecA in bacteria and Rad51 in eukaryotes (steps **1** and **2**). The 3′ end of the invading DNA strand is then extended by a DNA polymerase, displacing the parental strand as an enlarging single-stranded loop of DNA (dark blue, step **3**). When DNA synthesis extends sufficiently far, the displaced parental strand (dark blue) that is complementary to the 3′ single-stranded region generated at the other broken end of DNA (the pink single-stranded region on the left following step **1**), the complementary sequences base-pair (step **3**). This 3′ end (pink) is then extended by a DNA polymerase, using the displaced single-stranded loop of parental DNA (dark blue) as template (step **4**).

Next, the two 3′ ends generated by DNA synthesis are ligated (step **5**) to the 5′ ends generated in step **1** by 5′-exonuclease digestion of the broken ends. This generates two Holliday structures in the paired molecules (step **5**). Branch migration of these Holliday structures can occur in either direction (not diagrammed). Finally, cleavage of the strands at the positions shown by the arrows, and ligation of the alternative 5′ and 3′ ends at each cleaved Holliday structure generates two recombinant chromosomes that each contain the DNA of one parental DNA molecule (pink and red strands) on one side of the break point and the DNA of the other parental DNA molecule (light and dark blue) on the other side of the break point (step **6**). Each chromosome contains a third region, located in the immediate vicinity of the initial break point, that forms a *heteroduplex;* here one strand from one parent is base-paired to the complementary strand of the other parent (pink or red base-paired to dark or light blue). Base-pair mismatches between the two parental strands are usually repaired by repair mechanisms discussed above to generate a complementary base pair. In the process, sequence differences between the two parents are lost, a process referred to as *gene conversion.*

Figure 4-43 diagrams how cleavage of one or the other pair of strands at the four-way strand junction in the Holliday structure generates parental or recombinant molecules. This process, called *resolution* of the Holliday structure, separates DNA molecules initially joined by RecA/Rad51-catalyzed strand invasion. Each Holliday structure in the intermediate following step **5** of Figure 4-42 can be cleaved and religated in the two possible ways shown by the two sets of small black arrows. Consequently, there are four possible products of the recombination process. Two of these regenerate the parental chromosomes (with the exception of the heteroduplex region at the break point that is repaired into the sequence of one parent or the other (gene conversion), and two generate recombinant chromosomes as shown in Figure 4-42.

Meiotic Recombination Meiosis is the specialized form of cell division in eukaryotes that generates haploid germ cells (e.g., sperm and eggs) from a diploid cell (Figure 20-38). At least one recombination occurs between the paternal and maternal homologous chromosomes before the first meiotic cell division. Recombination is initiated by an enzyme that makes a double-stranded break in the DNA of one chromosome at any one of a very large number of sites. The process diagrammed in Figure 4-42 is then followed. The entire process from cleavage of the DNA of one chromosome through resolution of the Holliday structures is repeated until at least one recombination, also called a *crossover,* occurs between one pair of each of the homologous chromosomes. As mentioned earlier, the resulting link between homologous chromosomes is required for their proper segregation during the first meiotic cell division (see Chapter 20). As a consequence, every germ cell contains multiple recombinant chromosomes made of large segments of either the maternal or the paternal chromosome.

KEY CONCEPTS OF SECTION 4.6

DNA Repair and Recombination

- Changes in the DNA sequence result from copying errors and the effects of various physical and chemical agents.
- Many copying errors that occur during DNA replication are corrected by the proofreading function of DNA polymerases that can recognize incorrect (mispaired) bases at the 3′ end of the growing strand and then remove them by an inherent 3′→5′ exonuclease activity (see Figure 4-34).

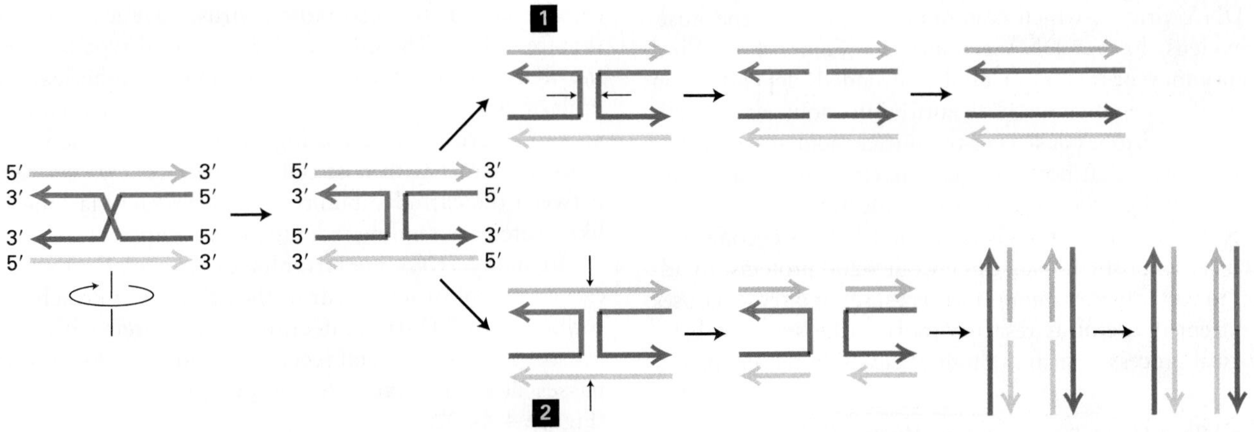

▲ **FIGURE 4-43 Alternative resolution of a Holliday structure.** Diagonal and vertical lines represent a single phosphodiester bond. It is simplest to diagram the process by rotating the diagram of the bottom molecule 180° so that the top and bottom molecules have the same strand orientations. Cutting the bonds as shown in **1** and ligating the ends as indicated regenerates the original chromosomes. Cutting the strands as shown in **2** and religating as shown at the bottom generates recombinant chromosomes. See http://engels.genetics.wisc.edu/Holliday/holliday3D.html for a three-dimensional animation of the Holliday structure and its resolution.

- Eukaryotic cells have three excision-repair systems for correcting mispaired bases and for removing UV-induced thymine-thymine dimers or large chemical adducts from DNA. Base excision repair, mismatch repair, and nucleotide excision repair operate with high accuracy and generally do not introduce errors.
- Repair of double-strand breaks by the nonhomologous end-joining pathway can link segments of DNA from different chromosomes, possibly forming an oncogenic translocation. The repair mechanism also produces a small deletion, even when segments from the same chromosome are joined.
- Inherited defects in the nucleotide excision-repair pathway, as in individuals with xeroderma pigmentosum, predispose them to skin cancer. Inherited colon cancer frequently is associated with mutant forms of proteins essential for the mismatch repair pathway. Defects in repair by homologous recombination are associated with inheritance of one mutant allele of the BRCA-1 or BRCA-2 gene and result in predisposition to breast and uterine cancer.
- Error-free repair of double-strand breaks in DNA is accomplished by homologous recombination using the undamaged sister chromatid as a template. This process can lead to recombination of parental chromosomes and is exploited by eukaryotes to generate genetic diversity by recombination of paternal and maternal chromosomes in developing germ cells.

4.7 Viruses: Parasites of the Cellular Genetic System

Viruses are obligate, intracellular parasites. They cannot reproduce by themselves and must commandeer a host cell's machinery to synthesize viral proteins and in some cases to replicate the viral genome. RNA viruses, which usually replicate in the host-cell cytoplasm, have an RNA genome, and DNA viruses, which commonly replicate in the host-cell nucleus, have a DNA genome (see Figure 4-1). Viral genomes may be single- or double-stranded, depending on the specific type of virus. The entire infectious virus particle, called a **virion,** consists of the nucleic acid and an outer shell of protein that both protects the viral nucleic acid and functions in the process of host-cell infection. The simplest viruses contain only enough RNA or DNA to encode four proteins; the most complex can encode ≈200 proteins. In addition to their obvious importance as causes of disease, viruses are extremely useful as research tools in the study of basic biological processes, such as those discussed in this chapter.

Most Viral Host Ranges Are Narrow

The surface of a virion contains many copies of one type of protein that binds specifically to multiple copies of a receptor protein on a host cell. This interaction determines the *host range*—the group of cell types that a virus can infect—and begins the infection process. Most viruses have a rather limited host range.

A virus that infects only bacteria is called a **bacteriophage,** or simply a *phage*. Viruses that infect animal or plant cells are referred to generally as *animal viruses* or *plant viruses*. A few viruses can grow in both plants or animals and the insects that feed on them. The highly mobile insects serve as vectors for transferring such viruses between susceptible plant or animal hosts. Wide host ranges are also characteristic of some strictly animal viruses, such as vesicular stomatitis virus, which grows in insect vectors and in many different types of mammals. Most animal viruses, however, do not cross phyla, and some (e.g., poliovirus) infect only closely related species such as primates. The host-cell range of some animal viruses is further restricted to a limited number of cell types because only these cells have appropriate surface receptors to which the virions can attach.

Viral Capsids Are Regular Arrays of One or a Few Types of Protein

The nucleic acid of a virion is enclosed within a protein coat, or **capsid,** composed of multiple copies of one protein or a few different proteins, each of which is encoded by a single viral gene. Because of this structure, a virus is able to encode all the information for making a relatively large capsid in a small number of genes. This efficient use of genetic information is important, since only a limited amount of DNA or RNA, and therefore a limited number of genes, can fit into a virion capsid. A capsid plus the enclosed nucleic acid is called a **nucleocapsid.**

Nature has found two basic ways of arranging the multiple capsid protein subunits and the viral genome into a nucleocapsid. In some viruses, multiple copies of a single coat protein form a *helical* structure that encloses and protects the viral RNA or DNA, which runs in a helical groove within the protein tube. Viruses with such a helical nucleocapsid, such as tobacco mosaic virus, have a rodlike shape (Figure 4-44a). The other major structural type is based on the *icosahedron,* a solid, approximately spherical object built of 20 identical faces, each of which is an equilateral triangle (Figure 4-44b). During infection, some icosahedral viruses interact with host cell-surface receptors via clefts in between the capsid subunits; others interact via long fiberlike proteins extending from the nucleocapsid.

In many DNA bacteriophages, the viral DNA is located within an icosahedral "head" that is attached to a rodlike "tail." During infection, viral proteins at the tip of the tail bind to host-cell receptors, and then the viral DNA passes down the tail into the cytoplasm of the host cell (Figure 4-44c).

In some viruses, the symmetrically arranged nucleocapsid is covered by an external membrane, or **envelope,** which consists mainly of a phospholipid bilayer but also contains one or two types of virus-encoded glycoproteins (Figure 4-44d).

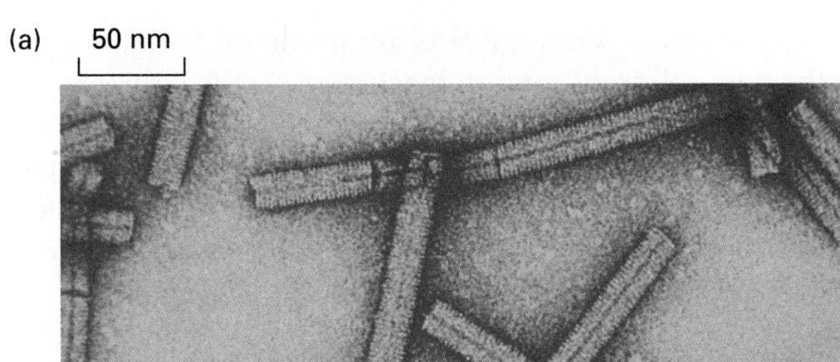

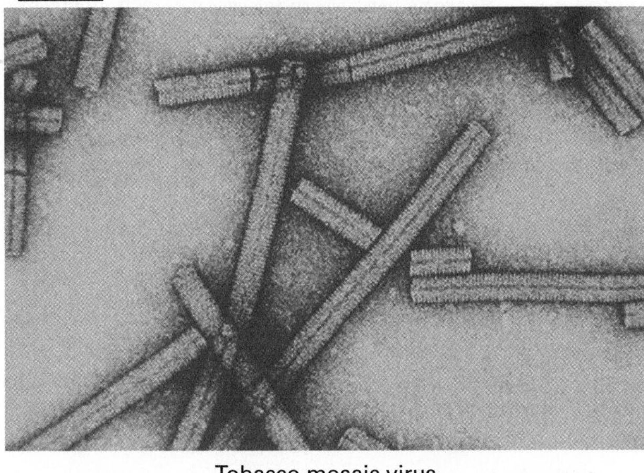

Tobacco mosaic virus

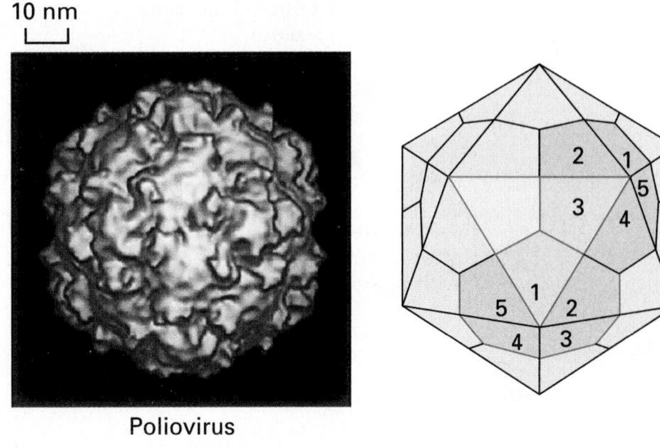

Poliovirus

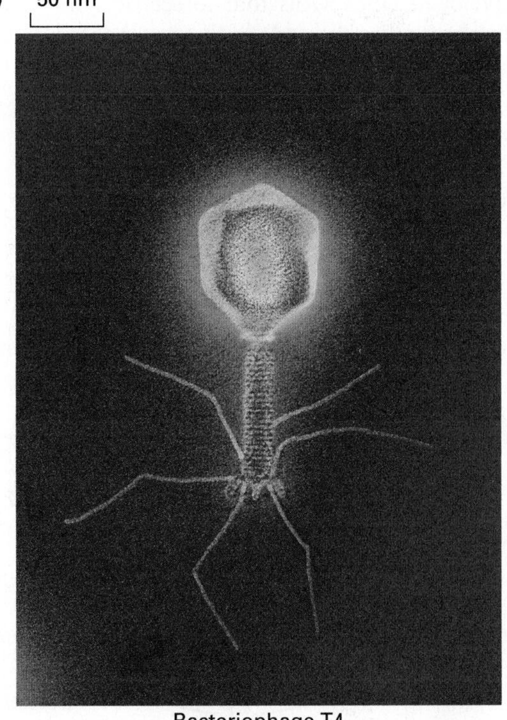

Bacteriophage T4

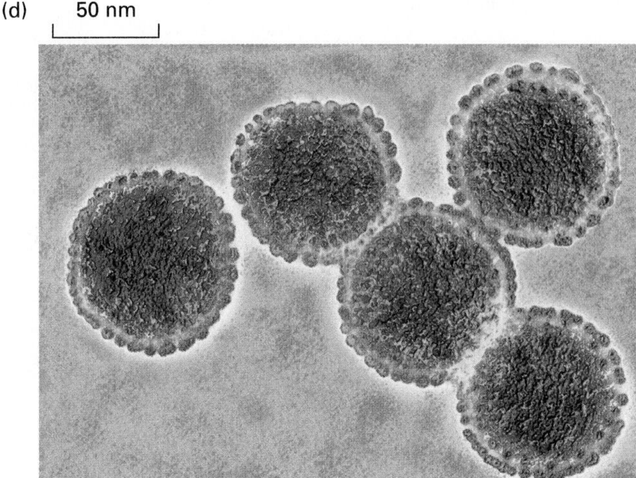

Avian influenza virus

◀ **FIGURE 4-44 Virion structures.** (a) Helical tobacco mosaic virus. (b) Small icosahedral virus. The example shown is poliovirus. (c) Bacteriophage T4. (d) Influenza virus, an example of an enveloped virus. [Part (a) O. Bradfute, Peter Arnold/Science Photo Library; part (b) courtesy of T. S. Baker; part (c) Department of Microbiology, Biozentrum/Science Photo Library; part (d) James Cavallini/Photo Researchers, Inc.]

The phospholipids in the viral envelope are similar to those in the plasma membrane of an infected host cell. The viral envelope is, in fact, derived by budding from that membrane, but contains mainly viral glycoproteins, as we discuss shortly.

Viruses Can Be Cloned and Counted in Plaque Assays

The number of infectious viral particles in a sample can be quantified by a **plaque assay.** This assay is performed by culturing a dilute sample of viral particles on a plate covered with host cells and then counting the number of local lesions, called *plaques,* that develop (Figure 4-45). A plaque develops on the plate wherever a single virion initially infects a single cell. The virus replicates in this initial host cell and then lyses (ruptures) the cell, releasing many progeny virions that infect the neighboring cells on the plate. After a few such cycles of infection, enough cells are lysed to produce a visible clear area, the plaque, in the layer of remaining uninfected cells.

Since all the progeny virions in a plaque are derived from a single parental virus, they constitute a virus **clone.** This type of plaque assay is in standard use for bacterial and animal viruses. Plant viruses can be assayed similarly by counting local lesions on plant leaves inoculated with viruses. Analysis of viral mutants, which are commonly isolated by plaque assays, has contributed extensively to current understanding of molecular cellular processes.

(a)

Confluent layer of susceptible host cells growing on surface of a plate

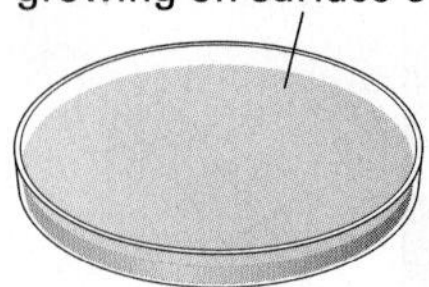

Add dilute suspension containing virus; after infection, cover layer of cells with agar; incubate

Plaque

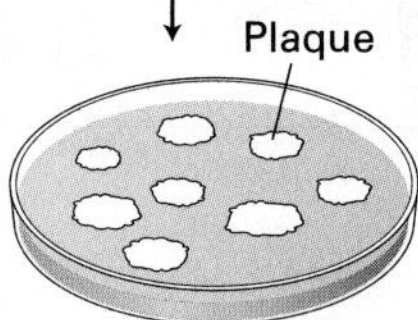

Each plaque represents cell lysis initiated by one viral particle (agar restricts movement so that virus can infect only contiguous cells)

(b)

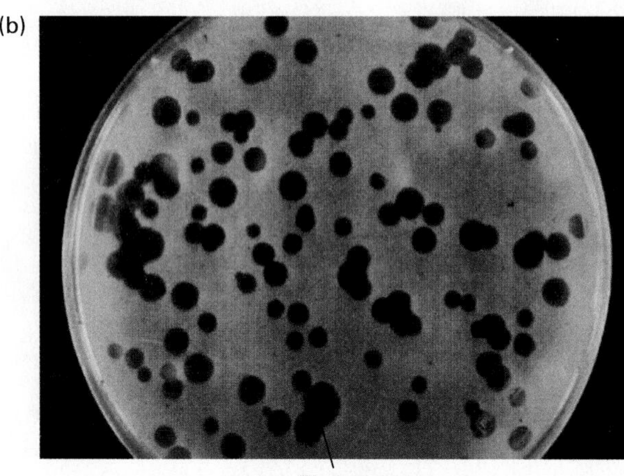

▲ **EXPERIMENTAL FIGURE 4-45 The plaque assay determines the number of infectious particles in a viral suspension.** (a) Each lesion, or plaque, which develops where a single virion initially infected a single cell, constitutes a pure viral clone. (b) Plaques on a lawn of *Pseudomonas fluorescens* bacteria made by bacteriophage ϕS1. [Part (b) Courtesy of Dr. Pierre Rossi, Ecole Polytechnique Fédérale de Lausanne (LBE-EPFL).]

Lytic Viral Growth Cycles Lead to the Death of Host Cells

Although details vary among different types of viruses, those that exhibit a *lytic cycle* of growth proceed through the following general stages:

1. *Adsorption*—Virion interacts with a host cell by binding of multiple copies of capsid protein to specific receptors on the cell surface.

2. Penetration—Viral genome crosses the plasma membrane. For some viruses, viral proteins packaged inside the capsid also enter the host cell.

3. *Replication*—Viral mRNAs are produced with the aid of the host-cell transcription machinery (DNA viruses) or by viral enzymes (RNA viruses). For both types of viruses, viral mRNAs are translated by the host-cell translation machinery. Production of multiple copies of the viral genome is carried out either by viral proteins alone or with the help of host-cell proteins.

4. *Assembly*—Viral proteins and replicated genomes associate to form progeny virions.

5. *Release*—Infected cell either ruptures suddenly (**lysis**), releasing all the newly formed virions at once, or disintegrates gradually, with slow release of virions. Both cases lead to the death of the infected cell.

Figure 4-46 illustrates the lytic cycle for T4 bacteriophage, a nonenveloped DNA virus that infects *E. coli.* Viral capsid proteins generally are made in large amounts because many

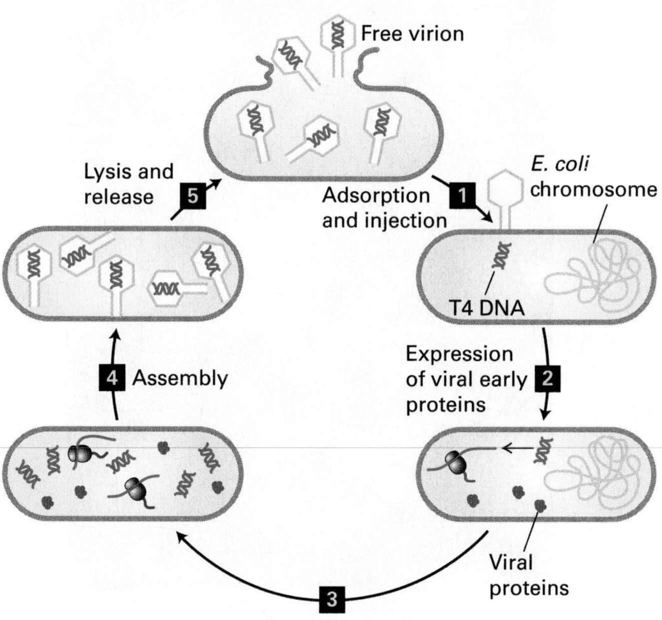

▲ **FIGURE 4-46 Lytic replication cycle of a nonenveloped, bacterial virus.** *E. coli* bacteriophage T4 has a double-stranded DNA genome and lacks a membrane envelope. After viral coat proteins at the tip of the tail in T4 interact with specific receptor proteins on the exterior of the host cell, the viral genome is injected into the host (step 1). Host-cell enzymes then transcribe viral "early" genes into mRNAs and subsequently translate these into viral "early" proteins (step 2). The early proteins replicate the viral DNA and induce expression of viral "late" proteins by host-cell enzymes (step 3). The viral late proteins include capsid and assembly proteins and enzymes that degrade the host-cell DNA, supplying nucleotides for synthesis of more viral DNA. Progeny virions are assembled in the cell (step 4) and released (step 5) when viral proteins lyse the cell. Newly liberated viruses initiate another cycle of infection in other host cells.

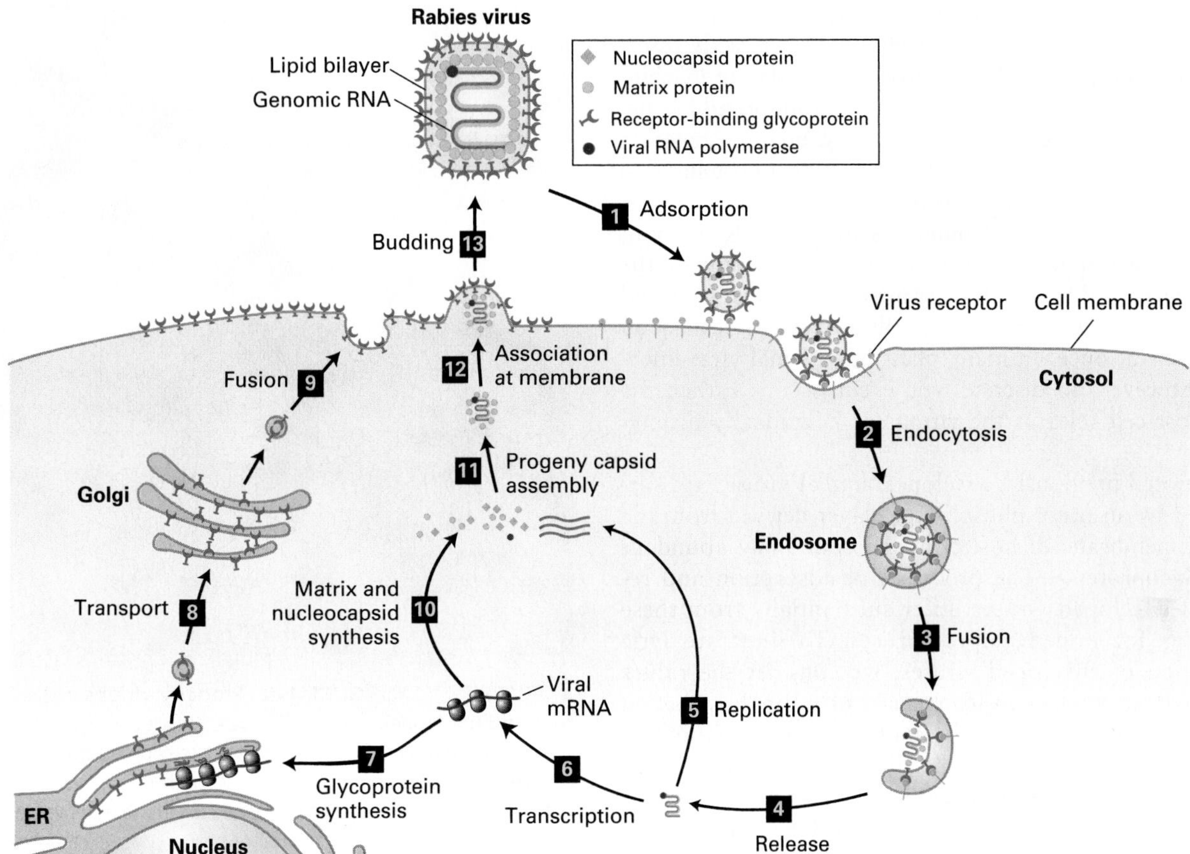

▲ **FIGURE 4-47 Lytic replication cycle of an enveloped, animal virus.** Rabies virus is an enveloped virus with a single-stranded RNA genome. The structural components of this virus are depicted at the top. After a virion adsorbs to multiple copies of a specific host membrane protein (step **1**), the cell engulfs it in an endosome (step **2**). A cellular protein in the endosome membrane pumps H^+ ions from the cytosol into the endosome interior. The resulting decrease in endosomal pH induces a conformational change in the viral glycoprotein, leading to fusion of the viral envelope with the endosomal lipid bilayer membrane and release of the nucleocapsid into the cytosol (steps **3** and **4**). Viral RNA polymerase uses ribonucleoside triphosphates in the cytosol to replicate the viral RNA genome (step **5**) and to synthesize viral mRNAs (step **6**). One of the viral mRNAs encodes the viral transmembrane glycoprotein, which is inserted into the membrane of the endoplasmic reticulum (ER) as it is synthesized on ER-bound ribosomes (step **7**). Carbohydrate is added to the large folded domain inside the ER lumen and is modified as the membrane and the associated glycoprotein pass through the Golgi apparatus (step **8**). Vesicles with mature glycoprotein fuse with the host plasma membrane, depositing viral glycoprotein on the cell surface with the large receptor-binding domain outside the cell (step **9**). Meanwhile, other viral mRNAs are translated on host-cell ribosomes into nucleocapsid protein, matrix protein, and viral RNA polymerase (step **10**). These proteins are assembled with replicated viral genomic RNA (bright red) into progeny nucleocapsids (step **11**), which then associate with the cytosolic domain of viral transmembrane glycoproteins in the plasma membrane (step **12**). The plasma membrane is folded around the nucleocapsid, forming a "bud" that eventually is released (step **13**).

copies of them are required for the assembly of each progeny virion. In each infected cell, about 100–200 T4 progeny virions are produced and released by lysis.

The lytic cycle is somewhat more complicated for DNA viruses that infect eukaryotic cells. In most such viruses, the DNA genome is transported (with some associated proteins) into the cell nucleus. Once inside the nucleus, the viral DNA is transcribed into RNA by the host's transcription machinery. Processing of the viral RNA primary transcript by host-cell enzymes yields viral mRNA, which is transported to the cytoplasm and translated into viral proteins by host-cell ribosomes, tRNA, and translation factors. The viral proteins are then transported back into the nucleus, where some of them either replicate the viral DNA directly or direct cellular proteins to replicate the viral DNA, as in the case of SV40 discussed earlier. Assembly of the capsid proteins with the newly replicated viral DNA occurs in the nucleus, yielding thousands to hundreds of thousands of progeny virions.

Most plant and animal viruses with an RNA genome do not require nuclear functions for lytic replication. In some of these viruses, a virus-encoded enzyme that enters the host during penetration transcribes the genomic RNA into mRNAs in the cell cytoplasm. The mRNA is directly translated into viral proteins by the host-cell translation machinery. One or more of these proteins then produces additional copies of the viral RNA genome. Finally, progeny genomes are assembled with newly synthesized capsid proteins into progeny virions in the cytoplasm.

After the synthesis of hundreds to hundreds of thousands of new virions has been completed, depending on the type of virus and host cell, most infected bacterial cells and some infected plant and animal cells are lysed, releasing all the virions at once. In many plant and animal viral infections, however, no discrete lytic event occurs; rather, the dead host cell releases the virions as it gradually disintegrates.

As noted previously, enveloped animal viruses are surrounded by an outer phospholipid layer derived from the plasma membrane of host cells and containing abundant viral glycoproteins. The processes of adsorption and release of enveloped viruses differ substantially from these processes for nonenveloped viruses. To illustrate lytic replication of enveloped viruses, we consider the rabies virus, whose nucleocapsid consists of a single-stranded RNA genome surrounded by multiple copies of nucleocapsid protein. Like most other lytic RNA viruses, rabies virions are replicated in the cytoplasm and do not require host-cell nuclear enzymes. As shown in Figure 4-47, a rabies virion is adsorbed to a host cell by binding to a specific cell-surface receptor molecule and then enters the cell by endocytosis. Progeny virions are released from a host cell by *budding* from the host-cell plasma membrane. Budding virions are clearly visible in electron micrographs of infected cells, as illustrated in Figure 4-48. Many tens of thousands of progeny virions bud from an infected host cell before it dies.

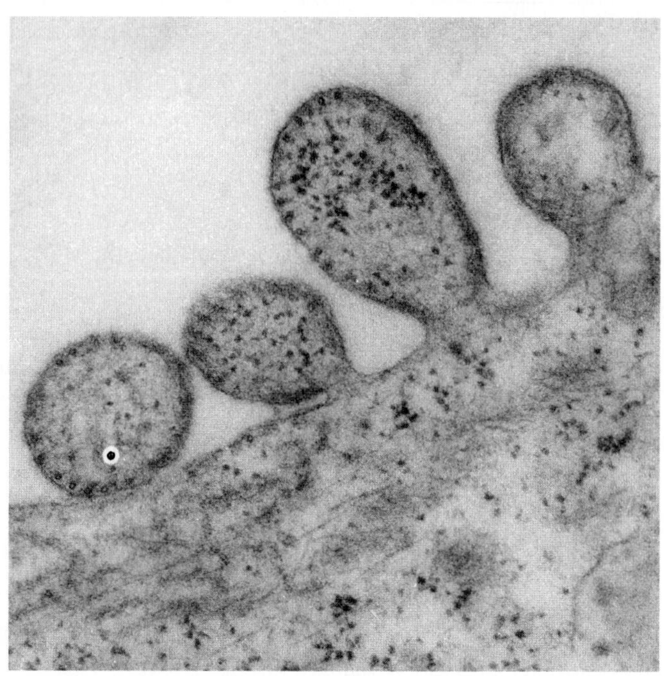

▲ EXPERIMENTAL FIGURE 4-48 **Progeny virions are released by budding.** Progeny virions of enveloped viruses are released by budding from infected cells. In this transmission electron micrograph of a cell infected with measles virus, virion buds are clearly visible protruding from the cell surface. Measles virus is an enveloped RNA virus with a helical nucleocapsid, like rabies virus, and replicates as illustrated in Figure 4-47. [From A. Levine, 1991, *Viruses,* Scientific American Library, p. 22.]

Viral DNA Is Integrated into the Host-Cell Genome in Some Nonlytic Viral Growth Cycles

Some bacterial viruses, called *temperate phages,* can establish a nonlytic association with their host cells that does not kill the cell. For example, when bacteriophage infects *E. coli,* the viral DNA may be integrated into the host-cell chromosome rather than being replicated. The integrated viral DNA, called a *prophage,* is replicated as part of the cell's DNA from one host-cell generation to the next. This phenomenon is referred to as **lysogeny.** Under certain conditions, the prophage DNA is activated, leading to its excision from the host-cell chromosome and entrance into the lytic cycle, with subsequent production and release of progeny virions.

The genomes of a number of animal viruses also can integrate into the host-cell genome. One of the most important are the **retroviruses,** which are enveloped viruses with a genome consisting of two identical strands of RNA. These viruses are known as *retroviruses* because their RNA genome acts as a template for formation of a DNA molecule—the opposite flow of genetic information compared with the more common transcription of DNA into RNA. In the retroviral life cycle (Figure 4-49), a viral enzyme called **reverse transcriptase** initially copies the viral RNA genome into single-stranded DNA complementary to the virion RNA; the same enzyme then catalyzes synthesis of a complementary DNA strand. (This complex reaction is detailed in Chapter 6 when we consider closely related intracellular parasites called *retrotransposons.*) The resulting double-stranded DNA is integrated into the chromosomal DNA of the infected cell. Finally, the integrated DNA, called a *provirus,* is transcribed by the cell's own machinery into RNA, which either is translated into viral proteins or is packaged within virion coat proteins to form progeny virions that are released by budding from the host-cell membrane. Because most retroviruses do not kill their host cells, infected cells can replicate, producing daughter cells with integrated proviral DNA. These daughter cells continue to transcribe the proviral DNA and bud progeny virions.

Some retroviruses contain cancer-causing genes (**oncogenes**), and cells infected by such retroviruses areoncogenically transformed into tumor cells. Studies of oncogenic retroviruses (mostly viruses of birds and mice) have revealed a great deal about the processes that lead to transformation of a normal cell into a cancer cell (Chapter 25).

Overview Animation: Life Cycle of a Retrovirus

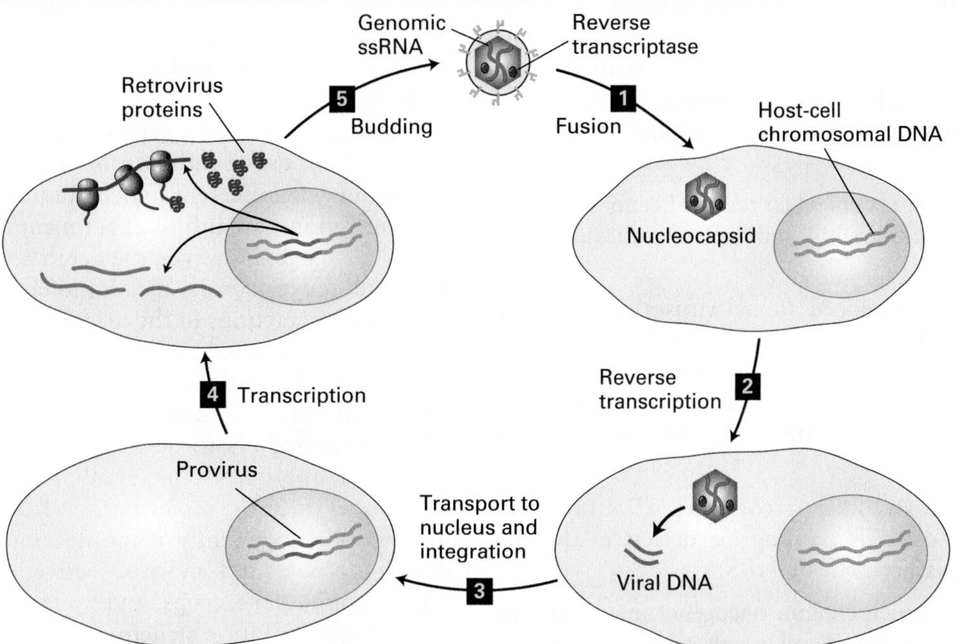

▲ **FIGURE 4-49 Retroviral life cycle.** Retroviruses have a genome of two identical copies of single-stranded RNA and an outer envelope. Step **1**: After viral glycoproteins in the envelope interact with a specific host-cell membrane protein, the retroviral envelope fuses directly with the plasma membrane, allowing entry of the nucleocapsid into the cytoplasm of the cell. Step **2**: Viral reverse transcriptase and other proteins copy the viral ssRNA genome into a double-stranded DNA. Step **3**: The viral dsDNA is imported into the nucleus and integrated into one of many possible sites in the host-cell chromosomal DNA. For simplicity, only one host-cell chromosome is depicted. Step **4**: The integrated viral DNA (provirus) is transcribed by the host-cell RNA polymerase, generating mRNAs (dark red) and genomic RNA molecules (bright red). The host-cell machinery translates the viral mRNAs into glycoproteins and nucleocapsid proteins. Step **5**: Progeny virions then assemble and are released by budding as illustrated in Figure 4-47.

Among the known human retroviruses are human T-cell lymphotrophic virus (HTLV), which causes a form of leukemia, and human immunodeficiency virus (HIV), which causes acquired immune deficiency syndrome (AIDS). Both of these viruses can infect only specific cell types, primarily certain cells of the immune system and, in the case of HIV, some central nervous system neurons and glial cells. Only these cells have cell-surface receptors that interact with viral envelope proteins, accounting for the host-cell specificity of these viruses. Unlike most other retroviruses, HIV eventually kills its host cells. The eventual death of large numbers of immune-system cells results in the defective immune response characteristic of AIDS.

Some DNA viruses also can integrate into a host-cell chromosome. One example is the human papillomaviruses (HPVs), which most commonly cause warts and other benign skin lesions. The genomes of certain HPV serotypes, however, occasionally integrate into the chromosomal DNA of infected cervical epithelial cells, initiating development of cervical cancer. Routine Pap smears can detect cells in the early stages of the transformation process initiated by HPV integration, permitting effective treatment. ▮

KEY CONCEPTS OF SECTION 4.7

Viruses: Parasites of the Cellular Genetic System

- Viruses are small parasites that can replicate only in host cells. Viral genomes may be either DNA (DNA viruses) or RNA (RNA viruses) and either single- or double-stranded.
- The capsid, which surrounds the viral genome, is composed of multiple copies of one or a small number of virus-encoded proteins. Some viruses also have an outer envelope, which is similar to the plasma membrane but contains viral transmembrane proteins.
- Most animal and plant DNA viruses require host-cell nuclear enzymes to carry out transcription of the viral genome into mRNA and production of progeny genomes. In contrast, most RNA viruses encode enzymes that can transcribe the RNA genome into viral mRNA and produce new copies of the RNA genome.
- Host-cell ribosomes, tRNAs, and translation factors are used in the synthesis of all viral proteins in infected cells.

■ Lytic viral infection entails adsorption, penetration, synthesis of viral proteins and progeny genomes (replication), assembly of progeny virions, and release of hundreds to thousands of virions, leading to death of the host cell (see Figure 4-46). Release of enveloped viruses occurs by budding through the host-cell plasma membrane (see Figure 4-47).

■ Nonlytic infection occurs when the viral genome is integrated into the host-cell DNA and generally does not lead to cell death.

■ Retroviruses are enveloped animal viruses containing a single-stranded RNA genome. After a host cell is penetrated, reverse transcriptase, a viral enzyme carried in the virion, converts the viral RNA genome into double-stranded DNA, which integrates into chromosomal DNA (see Figure 4-49).

■ Unlike infection by other retroviruses, HIV infection eventually kills host cells, causing the defects in the immune response characteristic of AIDS.

■ Tumor viruses, which contain oncogenes, may have an RNA genome (e.g., human T-cell lymphotrophic virus) or a DNA genome (e.g., human papillomaviruses). In the case of these viruses, integration of the viral genome into a host-cell chromosome can cause transformation of the cell into a tumor cell.

Perspectives for the Future

In this chapter we first reviewed the basic structure of DNA and RNA and then described fundamental aspects of the transcription of DNA by RNA polymerases. RNA polymerases are discussed in greater detail in Chapter 7, along with additional factors required for transcription initiation in eukaryotic cells and interactions with regulatory transcription factors that control transcription initiation in both bacterial and eukaryotic cells. Next, we discussed the genetic code and the participation of tRNA and the protein-synthesizing machine, the ribosome, in decoding the information in mRNA to allow accurate assembly of protein chains. Mechanisms that regulate protein synthesis are considered further in Chapter 8. Then, we considered the molecular details underlying the accurate replication of DNA required for cell division. Chapter 20 covers the mechanisms that regulate when a cell replicates its DNA and that coordinate DNA replication with the complex process of mitosis that distributes the daughter DNA molecules equally to each daughter cell. The next section addressed mechanisms for repairing damage to DNA, including recombination mechanisms that also lead to the generation of genetic diversity among individuals of a species. This genetic recombination contributes to the diversity of traits subjected to natural selection during the evolution of contemporary species. In Chapter 20, we discuss the mechanisms that segregate chromosomes into haploid germ cells, a process that requires recombination between maternal and paternal chromosomes. Finally, we discussed viruses, parasites of the cellular molecular genetic system and important model systems and useful tools for studying multiple aspects of molecular cell biology.

The basic molecular genetic processes discussed in this chapter form the foundation of contemporary molecular cell biology. Our current understanding of these processes is grounded in a wealth of experimental results and is not likely to change. However, the depth of our understanding will continue to increase as additional details of the structures and interactions of the macromolecular machines involved are uncovered. The determination in recent years of the three-dimensional structures of RNA polymerases, ribosomal subunits, and DNA replication proteins has allowed researchers to design ever more penetrating experimental approaches for revealing how these macromolecules operate at the molecular level. The detailed level of understanding currently being developed may allow the design of new and more effective drugs for treating illnesses of humans, crops, and livestock. For example, the recent high-resolution structures of ribosomes are providing insights into the mechanism by which antibiotics inhibit bacterial protein synthesis without affecting the function of mammalian ribosomes. This new knowledge may allow the design of even more effective antibiotics. Similarly, detailed understanding of the mechanisms regulating transcription of specific human genes may lead to therapeutic strategies that can reduce or prevent inappropriate immune responses that lead to multiple sclerosis and arthritis, the inappropriate cell division that is the hallmark of cancer, and other pathological processes.

Much of current biological research is focused on discovering how molecular interactions endow cells with decision-making capacity and their special properties. For this reason several of the following chapters describe current knowledge about how such interactions regulate transcription and protein synthesis in multicellular organisms and how such regulation endows cells with the capacity to become specialized and grow into complicated organs. Other chapters deal with how protein-protein interactions underlie the construction of specialized organelles in cells, and how they determine cell shape and movement. The rapid advances in molecular cell biology in recent years hold promise that in the not too distant future we will understand how the regulation of specialized cell function, shape, and mobility coupled with regulated cell replication and cell death (apoptosis) lead to the growth of complex organisms like flowering plants and human beings.

Key Terms

anticodon *127*
codons *127*
complementary *114*
crossing over *150*
deamination *146*
depurination *147*

Review the Concepts

1. What are Watson-Crick base pairs? Why are they important?

2. TATA box–binding protein binds to the minor groove of DNA, resulting in the bending of the DNA helix (see Figure 4-5). What property of DNA allows the TATA box–binding protein to recognize the DNA helix?

3. Preparing plasmid (double-stranded, circular) DNA for sequencing involves annealing a complementary, short, single-stranded oligonucleotide DNA primer to one strand of the plasmid template. This is routinely accomplished by heating the plasmid DNA and primer to 90 °C and then slowly bringing the temperature down to 25 °C. Why does this protocol work?

4. What difference between RNA and DNA helps to explain the greater stability of DNA? What implications does this have for the function of DNA?

5. What are the major differences in the synthesis and structure of prokaryotic and eukaryotic mRNAs?

6. While investigating the function of a specific growth factor receptor gene from humans, researchers found that two types of proteins are synthesized from this gene. A larger protein containing a membrane-spanning domain functions to recognize growth factors at the cell surface, stimulating a specific downstream signaling pathway. In contrast, a related, smaller protein is secreted from the cell and functions to bind available growth factor circulating in the blood, thus inhibiting the downstream signaling pathway. Speculate on how the cell synthesizes these disparate proteins.

7. The transcription of many bacterial genes relies on functional groups called *operons*, such as the tryptophan operon (Figure 4-13a). What is an operon? What advantages are there to having genes arranged in an operon, compared with the arrangement in eukaryotes?

8. Contrast how selection of the translational start site occurs on bacterial, eukaryotic, and poliovirus mRNAs.

9. What is the evidence that the 23S rRNA in the large rRNA subunit has a peptidyltransferase activity?

10. How would a mutation in the poly(A)-binding protein I gene affect translation? How would an electron micrograph of polyribosomes from such a mutant differ from the normal pattern?

11. What characteristic of DNA results in the requirement that some DNA synthesis is discontinuous? How are Okazaki fragments and DNA ligase utilized by the cell?

12. Eukaryotes have repair systems that prevent mutations due to copying errors and exposure to mutagens. What are the three excision-repair systems found in eukaryotes, and which one is responsible for correcting thymine-thymine dimers that form as a result of UV light damage to DNA?

13. DNA-repair systems are responsible for maintaining genomic fidelity in normal cells despite the high frequency with which mutational events occur. What type of DNA mutation is generated by (a) UV irradiation and (b) ionizing radiation? Describe the system responsible for repairing each of these types of mutations in mammalian cells. Postulate why a loss of function in one or more DNA-repair systems typifies many cancers.

14. What is the name given to the process that can repair DNA damage *and* generate genetic diversity? Briefly describe the similarities and differences of the two processes.

15. The genome of a retrovirus can integrate into the host-cell genome. What gene is unique to retroviruses, and why is the protein encoded by this gene absolutely necessary for maintaining the retroviral life cycle? A number of retroviruses can infect certain human cells. List two of them, briefly describe the medical implications resulting from these infections, and describe why only certain cells are infected.

Analyze the Data

Protein synthesis in eukaryotes normally begins at the first AUG codon in the mRNA. Sometimes, however, the ribosomes do not begin protein synthesis at this first AUG but scan past it (leaky scanning), and protein synthesis begins instead at an internal AUG. In order to understand what features of an mRNA affect efficiency of initiation at the first AUG, studies have been undertaken in which the synthesis of chloramphenicol acetyltransferase was examined. Translation of its message can give rise to a protein referred to as preCAT or give rise to a slightly smaller protein, CAT (see M. Kozak, 2005, *Gene* **361**:13). The two proteins differ in that CAT lacks several amino acids found at the

N-terminus of preCAT. CAT is not derived by cleavage of preCAT but, instead, by initiation of translation of the mRNA at an internal AUG:

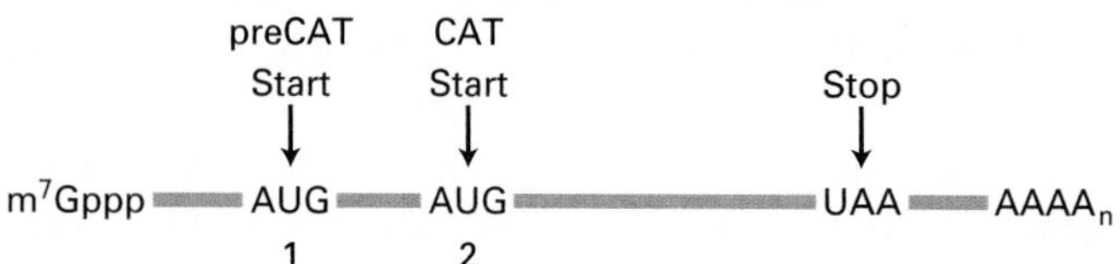

a. Results from a number of studies have given rise to the hypothesis that the sequence (−3)ACC**AUGG**(+4), in which the start codon AUG is shown in boldface, provides an optimal context for initiation of protein synthesis and ensures that ribosomes do not scan past this first AUG to begin initiation instead at a downstream AUG. In the numbering scheme used here, the A of the AUG initiation is designated (+1); bases 5′ of this are given negative numbers [so that the first base of this sequence is (−3)], and bases 3′ to the (+1) A are given positive numbers [so that the last base of this sequence is (+4)]. To test the hypothesis that the start site sequence (−3)ACC**AUGG**(+4) prevents leaky scanning, the chloramphenicol acetyltransferase mRNA sequence was modified and the resulting effects on translation assessed. In the following figure, the sequence (red) surrounding the first AUG codon (black) of the mRNA that gives rise to the synthesis of preCAT is shown above lane 3. Modification of this message is shown above the other gel lanes (altered nucleotides are in blue), and the completed proteins generated from each modified message appear as bands on the SDS-polyacrylamide gel below. The intensity of each band is an indication of the amount of that protein synthesized. Analyze the alterations to the wild-type sequence, and describe how they affect translation. Are the positions of some nucleotides more important than others? Do the data shown in this figure provide support for the hypothesis that the context in which the first AUG is present affects efficiency of translation from this site? Is ACCAUGG an optimal context for initiation from the first AUG?

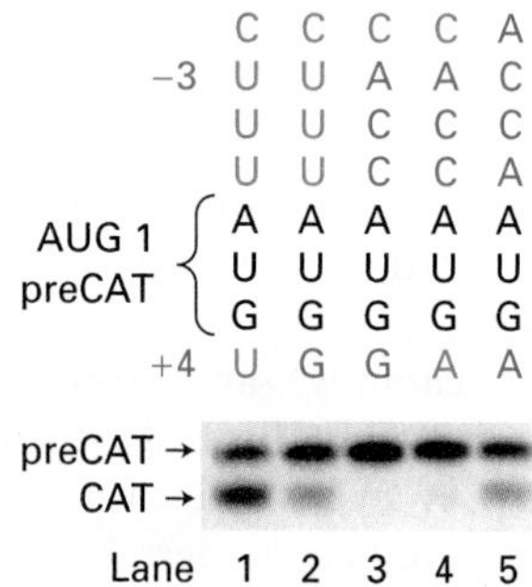

b. What are some additional alterations to this message, other than those shown in the figure, that would further elucidate the importance of the ACCAUGG sequence as an optimal context for synthesis of preCAT rather than CAT? How would you further examine whether A at the (−3) position and G at the (+4) position are the most important nucleotides to provide context for the AUG start?

c. A mutation causing a severe blood disease has been found in a single family (see T. Matthes et al., 2004, *Blood* **104**:2181). The mutation, shown in red in the figure below, has been mapped to the 5′ untranslated region of the gene encoding hepcidin and has been found to alter the gene's mRNA. The shaded regions indicate the coding sequence of the normal and mutant genes. No hepcidin is produced from the altered mRNA, and lack of hepcidin results in the disease. Can you provide a reasonable explanation for the lack of synthesis of hepcidin in the family members who have inherited this mutation? What can you deduce about the importance of the context in which the start site for initiation of protein synthesis occurs in this case?

Start hepcidin

.....GCA**G**UGGGACAGCCAGACAGACGGCACGAUGGCACUG..... No

.....GCAAUGGGACAGCCAGACAGACGGCACGAUGGCACU........ Mu

References

Structure of Nucleic Acids

Arnott, S. 2006. Historical article: DNA polymorphism and the early history of the double helix. *Trends Biochem. Sci.* **31**:349–354.

Berger, J. M., and J. C. Wang. 1996. Recent developments in DNA topoisomerase II structure and mechanism. *Curr. Opin. Struc. Biol.* **6**:84–90.

Dickerson, R. E. 1983. The DNA helix and how it is read. *Sci. Am.* **249**:94–111.

Dickerson, R. E., and H. L. Ng. 2001. DNA structure from A to B. *Proc. Nat'l Acad. Sci. USA.* **98**:6986–69888.

Doudna, J. A., and T. R. Cech. 2002. The chemical repertoire of natural ribozymes. *Nature* **418**:222–228.

Kornberg, A., and T. A. Baker. 2005. *DNA Replication.* University Science, chap. 1. A good summary of the principles of DNA structure.

Lilley, D. M. 2005. Structure, folding and mechanisms of ribozymes. *Curr. Opin. Struc. Biol.* **15**:313–323.

Vicens, Q., and T. R. Cech. 2005. Atomic level architecture of group I introns revealed. *Trends Biochem. Sci.* **31**:41–51.

Wang, J. C. 1980. Superhelical DNA. *Trends Biochem. Sci.* **5**:219–221.

Wigley, D. B. 1995. Structure and mechanism of DNA topoisomerases. *Ann. Rev. Biophys. Biomol. Struc.* **24**:185–208.

Transcription of Protein-Coding Genes and Formation of Functional mRNA

Brenner, S., F. Jacob, and M. Meselson. 1961. An unstable intermediate carrying information from genes to ribosomes for protein synthesis. *Nature* **190**:576–581.

Murakami, K. S., and S. A. Darst. 2003. Bacterial RNA polymerases: the whole story. *Curr. Opin. Struc. Biol.* **13**:31–39.

Okamoto K., Y. Sugino, and M. Nomura. 1962. Synthesis and turnover of phage messenger RNA in E. coli infected with bacteriophage T4 in the presence of chloromycetin. *J. Mol. Biol.* **5**:527–534.

Steitz, T. A. 2006. Visualizing polynucleotide polymerase machines at work. *EMBO J.* **25**:3458–3468.

The Decoding of mRNA by tRNAs

Alexander, R. W., and P. Schimmel. 2001. Domain-domain communication in aminoacyl-tRNA synthetases. *Prog. Nucl. Acid Res. Mol. Biol.* **69**:317–349.

Hatfield, D. L., and V. N. Gladyshev. 2002. How selenium has altered our understanding of the genetic code. *Mol. Cell Biol.* **22**:3565–3576.

Hoagland, M. B., et al. 1958. A soluble ribonucleic acid intermediate in protein synthesis. *J. Biol. Chem.* **231**:241–257.

Ibba, M., and D. Soll. 2004. Aminoacyl-tRNAs: setting the limits of the genetic code. *Genes Dev.* **18**:731–738.

Khorana, G. H., et al. 1966. Polynucleotide synthesis and the genetic code. *Cold Spring Harbor Symp. Quant. Biol.* **31**:39–49.

Nakanishi, K., and O. Nureki. 2005. Recent progress of structural biology of tRNA processing and modification. *Mol. Cells* **19**:157–166.

Nirenberg, M., et al. 1966. The RNA code in protein synthesis. *Cold Spring Harbor Symp. Quant. Biol.* **31**:11–24.

Rich, A., and S.-H. Kim. 1978. The three-dimensional structure of transfer RNA. *Sci. Am.* **240**(1):52–62 (offprint 1377).

Stepwise Synthesis of Proteins on Ribosomes

Abbott, C. M., and C. G. Proud. 2004. Translation factors: in sickness and in health. *Trends Biochem. Sci.* **29**:25–131.

Auerbach, T., A. Bashan, and A. Yonath. 2004. Ribosomal antibiotics: structural basis for resistance, synergism and selectivity. *Trends Biotechnol.* **22**:570–576.

Frank, J., et al. 2005. The role of tRNA as a molecular spring in decoding, accommodation, and peptidyl transfer. *FEBS Lett.* **579**:959–962.

Ganoza, M. C., M. C. Kiel, and H. Aoki. 2002. Evolutionary conservation of reactions in translation. *Microbiol. Mol. Biol. Rev.* **66**:460–485.

Gualerzi, C. O., et al. 2001. Initiation factors in the early events of mRNA translation in bacteria. *Cold Spring Harbor Symp. Quant. Biol.* **66**:363–376.

Hellen, C. U., and P. Sarnow. 2001. Internal ribosome entry sites in eukaryotic mRNA molecules. *Genet. Devel.* **15**:1593–1612.

Kahvejian, A., G. Roy, and N. Sonenberg. 2001. The mRNA closed-loop model: the function of PABP and PABP-interacting proteins in mRNA translation. *Cold Spring Harbor Symp. Quant. Biol.* **66**:293–300.

Kapp, L. D., and J. R. Lorsch. 2004. The molecular mechanics of eukaryotic translation. *Ann. Rev. Biochem.* **73**:657–704.

Mitra, K., and J. Frank. 2006. Ribosome dynamics: insights from atomic structure modeling into cryo-electron microscopy maps. *Ann. Rev. Biophys. Biomol. Struc.* **35**:299–317.

Noller, H. F. 2005. RNA structure: reading the ribosome. *Science* **309**:1508–1514.

Noller, H. F., et al. 2002. Translocation of tRNA during protein synthesis. *FEBS Lett.* **514**:11–16.

Polacek, N., and A. S. Mankin. 2005. The ribosomal peptidyl transferase center: structure, function, evolution, inhibition. *Crit. Rev. Biochem. Mol. Biol.* **40**:285–311.

Preiss, T., and M. W. Hentze. 2003. Starting the protein synthesis machine: eukaryotic translation initiation. *BioEssays* **25**:1201–1211.

Richter, J. D., and N. Sonenberg. 2005. Regulation of cap-dependent translation by eIF4E inhibitory proteins. *Nature* **433**:477–480.

Scheper, G. C., C. G. Proud, and M. S. van der Knaap. 2006. Defective translation initiation causes vanishing of cerebral white matter. *Trends Mol. Med.* **12**:159–166.

Sonenberg, N. 2006. *Translational Control in Biology and Medicine* (CSH monograph). Cold Spring Harbor Press.

Sonenberg, N., and T. E. Dever. 2003. Eukaryotic translation initiation factors and regulators. *Curr. Opin. Struc. Biol.* **13**:56–63.

Steitz, T. A. 2005. On the structural basis of peptide-bond formation and antibiotic resistance from atomic structures of the large ribosomal subunit. *FEBS Lett.* **579**:955–958.

Taylor, S. S., N. M. Haste, and G. Ghosh. 2005. PKR and eIF2α: integration of kinase dimerization, activation, and substrate docking. *Cell* **122**:823–825.

DNA Replication

Brautigam, C. A, and T. A. Steitz. 1998. Structural and functional insights provided by crystal structures of DNA polymerases and their substrate complexes. *Curr. Opin. Struc. Biol.* **8**:54–63.

Bullock, P. A. 1997. The initiation of simian virus 40 DNA replication in vitro. *Crit. Rev. Biochem. Mol. Biol.* **32**:503–568.

DePamphilis, M. L., ed. 2006 *DNA Replication and Human Disease*. Cold Spring Harbor Laboratory Press.

Kornberg, A., and T. A. Baker. 2005. *DNA Replication.* University Science.

Langston, L. D., and M. O'Donnell. 2006. DNA replication: keep moving and don't mind the gap. *Mol. Cells* **23**:155–160.

Mendez, J., and B. Stillman. 2003. Perpetuating the double helix: molecular machines at eukaryotic DNA replication origins. *BioEssays* **25**:1158–1167.

O'Donnell, M. 2006. Replisome architecture and dynamics in *Escherichia coli. J. Biol. Chem.* **281**:10653–10656.

Sclafani, R. A., R. J. Fletcher, and X. S. Chen. 2004. Two heads are better than one: regulation of DNA replication by hexameric helicases. *Genes Dev.* **18**:2039–2045.

DNA Repair and Recombination

Andressoo, J. O., and J. H. Hoeijmakers. 2005. Transcription-coupled repair and premature aging. *Mutat. Res.* **577**:179–194.

Barnes, D. E., and T. Lindahl. 2004. Repair and genetic consequences of endogenous DNA base damage in mammalian cells. *Ann. Rev. Genet.* **38**:445–476.

Bell, C. E. 2005. Structure and mechanism of *Escherichia coli* RecA ATPase. *Mol. Microbiol.* **58**:358–366.

Friedberg, E. C., et al. 2006. DNA repair: from molecular mechanism to human disease. *DNA Repair* **5**:986–996.

Haber, J. E. 2000. Partners and pathways repairing a double-strand break. *Trends Genet.* **16**:259–264.

Jiricny, J. 2006. The multifaceted mismatch-repair system. *Nat. Rev. Mol. Cell Biol.* **7**:335–346.

Khuu, P. A., et al. 2006. The stacked-X DNA Holliday junction and protein recognition. *J. Mol. Recog.* **19**:234–242.

Lilley, D. M., and R. M. Clegg. 1993. The structure of the four-way junction in DNA. *Ann. Rev. Biophys. Biomol. Struc.* **22**:299–328.

Mirchandani, K. D., and A. D. D'Andrea. 2006. The Fanconi anemia/BRCA pathway: a coordinator of cross-link repair. *Exp. Cell Res.* **312**:2647–2653.

Mitchell, J. R., J. H. Hoeijmakers, and L. J. Niedernhofer. 2003. Divide and conquer: nucleotide excision repair battles cancer and aging. *Curr. Opin. Cell Biol.***15**:232–240.

Orr-Weaver, T. L., and J. W. Szostak. 1985. Fungal recombination. *Microbiol. Rev.* **49**:33–58.

Shin, D. S., et al. 2004. Structure and function of the double strand break repair machinery. *DNA Repair* **3**:863–873.

Wood, R. D., M. Mitchell, and T. Lindahl. Human DNA repair genes. *Mutat. Res.* 577:275–283.

Yoshida, K., and Y. Miki. 2004. Role of BRCA1 and BRCA2 as regulators of DNA repair, transcription, and cell cycle in response to DNA damage. *Cancer Sci.* 95:866–871.

Viruses: Parasites of the Cellular Genetic System

Flint, S. J., et al. 2000. *Principles of Virology: Molecular Biology, Pathogenesis, and Control.* ASM Press.

Hull, R. 2002. *Mathews' Plant Virology*. Academic Press.

Klug, A. 1999. The tobacco mosaic virus particle: structure and assembly. *Phil. Trans. R. Soc. Lond. B Biol. Sci.* 354:531–535.

Knipe, D. M., and P. M. Howley, eds. 2001. *Fields Virology.* Lippincott Williams & Wilkins.

Kornberg, A., and T. A. Baker. 1992. *DNA Replication,* 2d ed. W. H. Freeman and Company. Good summary of bacteriophage molecular biology.

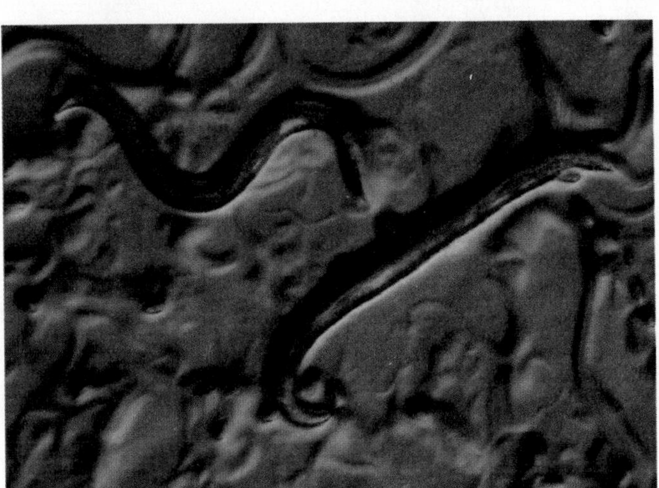

RNA interference (RNAi) can be used to silence most genes in the *C. elegans* genome. The transgenic worm on the right (marked by a GFP reporter in head neurons) expresses dsRNA to the muscle gene *unc-15*, resulting in the potent degradation of the *unc-15* mRNA and leading to complete paralysis of the worm. In contrast, the wild-type worm on the left exhibits the typical sinusoidal body movement. [Courtesy of John Kim.]

CHAPTER

5

MOLECULAR GENETIC TECHNIQUES

In previous chapters, we were introduced to the variety of tasks that proteins perform in biological systems. Indeed, the very field of molecular cell biology seeks to understand the molecular mechanisms of individual proteins and how groups of proteins work together to perform their biological functions. In studying a newly discovered protein, cell biologists usually begin by asking three questions about it: what is its function, where is it located, and what is its structure? To answer these questions, investigators employ three tools: the gene that encodes the protein, a mutant cell line or organism that lacks the function of the protein, and a source of the purified protein for biochemical studies. In this chapter we consider various aspects of two basic experimental strategies for obtaining all three tools (Figure 5-1).

The first strategy, often referred to as *classical genetics*, begins with isolation of a mutant that appears to be defective in some process of interest. Genetic methods then are used to identify and isolate the affected gene. The isolated gene can be manipulated to produce large quantities of the protein for biochemical experiments and to design probes for studies of where and when the encoded protein is expressed in an organism. The second strategy follows essentially the same steps as the classical approach but in reverse order, beginning with isolation of an interesting protein or its identification based on analysis of an organism's genomic sequence. Once the corresponding gene has been isolated, the gene can be altered and then reinserted into an organism. In both strategies, by examining the phenotypic consequences of mutations that inactivate a particular gene, geneticists are able to connect knowledge about the sequence, structure, and biochemical activity of the encoded protein to its function in the context of a living cell or multicellular organism.

An important component in both strategies for studying a protein and its biological function is isolation of the corresponding gene. Thus we discuss various techniques by which researchers can isolate, sequence, and manipulate specific regions of an organism's DNA. Next we introduce a variety of techniques that are commonly used to analyze where and when a particular gene is expressed and where in the cell its protein is localized. In some cases, knowledge of protein function can lead to significant medical advances, and the first step in developing treatments for an inherited disease is to identify and isolate the affected gene, which we describe here. Finally we discuss techniques that abolish normal protein function in order to analyze the role of the protein in the cell.

OUTLINE

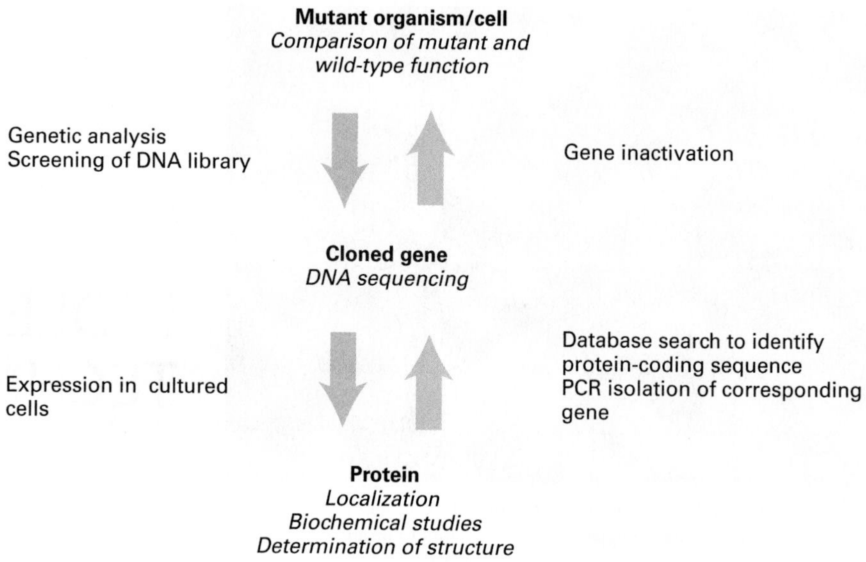

▶ **FIGURE 5-1 Overview of two strategies for relating the function, location, and structure of gene products.** A mutant organism is the starting point for the classical genetic strategy (green arrows). The reverse strategy (orange arrows) usually begins with identification of a protein-coding sequence by analysis of genome sequence databases. In both strategies, the actual gene is isolated either from a DNA library or by specific amplification of the gene sequence from genomic DNA. Once a cloned gene is isolated, it can be used to produce the encoded protein in bacterial or eukaryotic expression systems. Alternatively, a cloned gene can be inactivated by one of various techniques and used to generate mutant cells or organisms.

5.1 Genetic Analysis of Mutations to Identify and Study Genes

As described in Chapter 4, the information encoded in the DNA sequence of genes specifies the sequence—and therefore the structure and function—of every protein molecule in a cell. The power of genetics as a tool for studying cells and organisms lies in the ability of researchers to selectively alter every copy of just one type of protein in a cell by making a change in the gene for that protein. Genetic analyses of mutants defective in a particular process can reveal (a) new genes required for the process to occur, (b) the order in which gene products act in the process, and (c) whether the proteins encoded by different genes interact with one another. Before seeing how genetic studies of this type can provide insights into the mechanism of complicated cellular or developmental process, we first explain some basic genetic terms used throughout our discussion.

The different forms, or variants, of a gene are referred to as **alleles.** Geneticists commonly refer to the numerous naturally occurring genetic variants that exist in populations, particularly human populations, as alleles. The term **mutation** usually is reserved for instances in which an allele is known to have been newly formed, such as after treatment of an experimental organism with a **mutagen,** an agent that causes a heritable change in the DNA sequence.

Strictly speaking, the particular set of alleles for all the genes carried by an individual is its **genotype.** However, this term also is used in a more restricted sense to denote just the alleles of the particular gene or genes under examination. For experimental organisms, the term **wild type** often is used to designate a standard genotype for use as a reference in breeding experiments. Thus the normal, nonmutant allele will usually be designated as the wild type. Because of the enormous naturally occurring allelic variation that exists in human populations, the term *wild type* usually denotes an allele that is present at a much higher frequency than any of the other possible alternatives.

Geneticists draw an important distinction between the *genotype* and the **phenotype** of an organism. The phenotype refers to all the physical attributes or traits of an individual that are the consequence of a given genotype. In practice, however, the term *phenotype* often is used to denote the physical consequences that result from just the alleles that are under experimental study. Readily observable phenotypic characteristics are critical in the genetic analysis of mutations.

Recessive and Dominant Mutant Alleles Generally Have Opposite Effects on Gene Function

A fundamental genetic difference between experimental organisms is whether their cells carry a single set of chromosomes or two copies of each chromosome. The former are referred to as **haploid;** the latter, as **diploid.** Complex multicellular organisms (e.g., fruit flies, mice, humans) are diploid, whereas many simple unicellular organisms are haploid. Some organisms, notably the yeast *Saccharomyces cerevisiae,* can exist in either haploid or diploid states. Many cancer cells and the normal cells of some organisms, both plants and animals, carry more than two copies of each chromosome. However, our discussion of genetic techniques and analysis relates to diploid organisms, including diploid yeasts.

Although many different alleles of a gene might occur in different organisms in a population, any individual diploid organism will carry two copies of each gene and thus at most can have two different alleles. An individual with two different alleles is **heterozygous** for a gene, whereas an individual that carries two identical alleles is **homozygous** for a gene. A **recessive** mutant allele is defined as one in which both alleles must be mutant in order for the mutant phenotype to be observed; that is, the individual must be homozygous for the mutant allele to show the mutant phenotype. In contrast, the phenotypic consequences of a **dominant** mutant allele can be

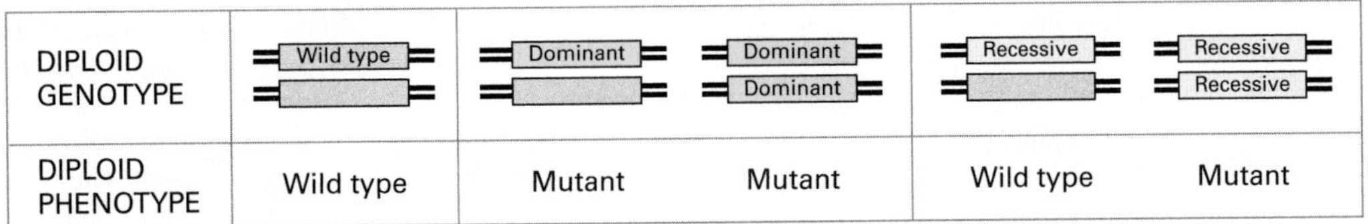

▲ **FIGURE 5-2 Effects of dominant and recessive mutant alleles on phenotype in diploid organisms.** A single copy of a dominant allele is sufficient to produce a mutant phenotype, whereas both copies of a recessive allele must be present to cause a mutant phenotype. Recessive mutations usually cause a loss of function; dominant mutations usually cause a gain of function or an altered function.

observed in a heterozygous individual carrying one mutant and one wild-type allele (Figure 5-2).

Whether a mutant allele is recessive or dominant provides valuable information about the function of the affected gene and the nature of the causative mutation. Recessive alleles usually result from a mutation that inactivates the affected gene, leading to a partial or complete *loss of function.* Such recessive mutations may remove part of the gene or the entire gene from the chromosome, disrupt expression of the gene, or alter the structure of the encoded protein, thereby altering its function. Conversely, dominant alleles are often the consequence of a mutation that causes some kind of *gain of function.* Such dominant mutations may increase the activity of the encoded protein, confer a new function on it, or lead to its inappropriate spatial or temporal pattern of expression.

Dominant mutations in certain genes, however, are associated with a loss of function. For instance, some genes are *haploinsufficient,* meaning that both alleles are required for normal function. Removing or inactivating a single allele in such a gene leads to a mutant phenotype. In other rare instances a dominant mutation in one allele may lead to a structural change in the protein that interferes with the function of the wild-type protein encoded by the other allele. This type of mutation, referred to as a *dominant-negative,* produces a phenotype similar to that obtained from a loss-of-function mutation.

Some alleles can exhibit both recessive and dominant properties. In such cases, statements about whether an allele is dominant or recessive must specify the phenotype. For example, the allele of the hemoglobin gene in humans designated Hb^s has more than one phenotypic consequence. Individuals who are homozygous for this allele (Hb^s/Hb^s) have the debilitating disease sickle-cell anemia, but heterozygous individuals (Hb^s/Hb^a) do not have the disease. Therefore, Hb^s is *recessive* for the trait of sickle-cell disease. On the other hand, heterozygous (Hb^s/Hb^a) individuals are more resistant to malaria than homozygous (Hb^a/Hb^a) individuals, revealing that Hb^s is *dominant* for the trait of malaria resistance. ▮

A commonly used agent for inducing mutations (mutagenesis) in experimental organisms is ethylmethane sulfonate (EMS). Although this mutagen can alter DNA sequences in several ways, one of its most common effects is to chemically modify guanine bases in DNA, ultimately leading to the conversion of a G·C base pair into an A·T base pair. Such an alteration in the sequence of a gene, which involves only a single base pair, is known as a **point mutation.** A *silent* point mutation causes no change in the amino acid sequence or activity of a gene's encoded protein. However, observable phenotypic consequences due to changes in a protein's activity can arise from point mutations that result in substitution of one amino acid for another (*missense* mutation), introduction of a premature stop codon (*nonsense* mutation), or a change in the reading frame of a gene (*frameshift* mutation). Because alterations in the DNA sequence leading to a decrease in protein activity are much more likely than alterations leading to an increase or qualitative change in protein activity, mutagenesis usually produces many more recessive mutations than dominant mutations.

Segregation of Mutations in Breeding Experiments Reveals Their Dominance or Recessivity

Geneticists exploit the normal life cycle of an organism to test for the dominance or recessivity of alleles. To see how this is done, we need first to review the type of cell division that gives rise to **gametes** (sperm and egg cells in higher plants and animals). Whereas the body (somatic) cells of most multicellular organisms divide by **mitosis,** the **germ cells** that give rise to gametes undergo **meiosis.** Like somatic cells, premeiotic germ cells are diploid, containing two **homologs** of each morphological type of chromosome. The two homologs constituting each pair of **homologous chromosomes** are descended from different parents, and thus their genes may exist in different allelic forms. Figure 5-3 depicts the major events in mitotic and meiotic cell division. In mitosis DNA replication is always followed by cell division, yielding two diploid daughter cells. In meiosis *one* round of DNA replication is followed by *two* separate cell divisions, yielding four haploid ($1n$) cells that contain only one chromosome of each homologous pair. The apportionment, or **segregation,** of the replicated homologous chromosomes to daughter cells during the first meiotic division is random; that is, maternally and paternally derived homologs segregate independently, yielding daughter cells with different mixes of paternal and maternal chromosomes.

As a way to avoid unwanted complexity, geneticists usually strive to begin breeding experiments with strains that are homozygous for the genes under examination. In such *true-breeding* strains, every individual will receive the same allele from each parent and therefore the composition of alleles will not change from one generation to the next. When a true-breeding mutant strain is mated to a true-breeding wild-type strain, all the first filial (F_1) progeny will be heterozygous

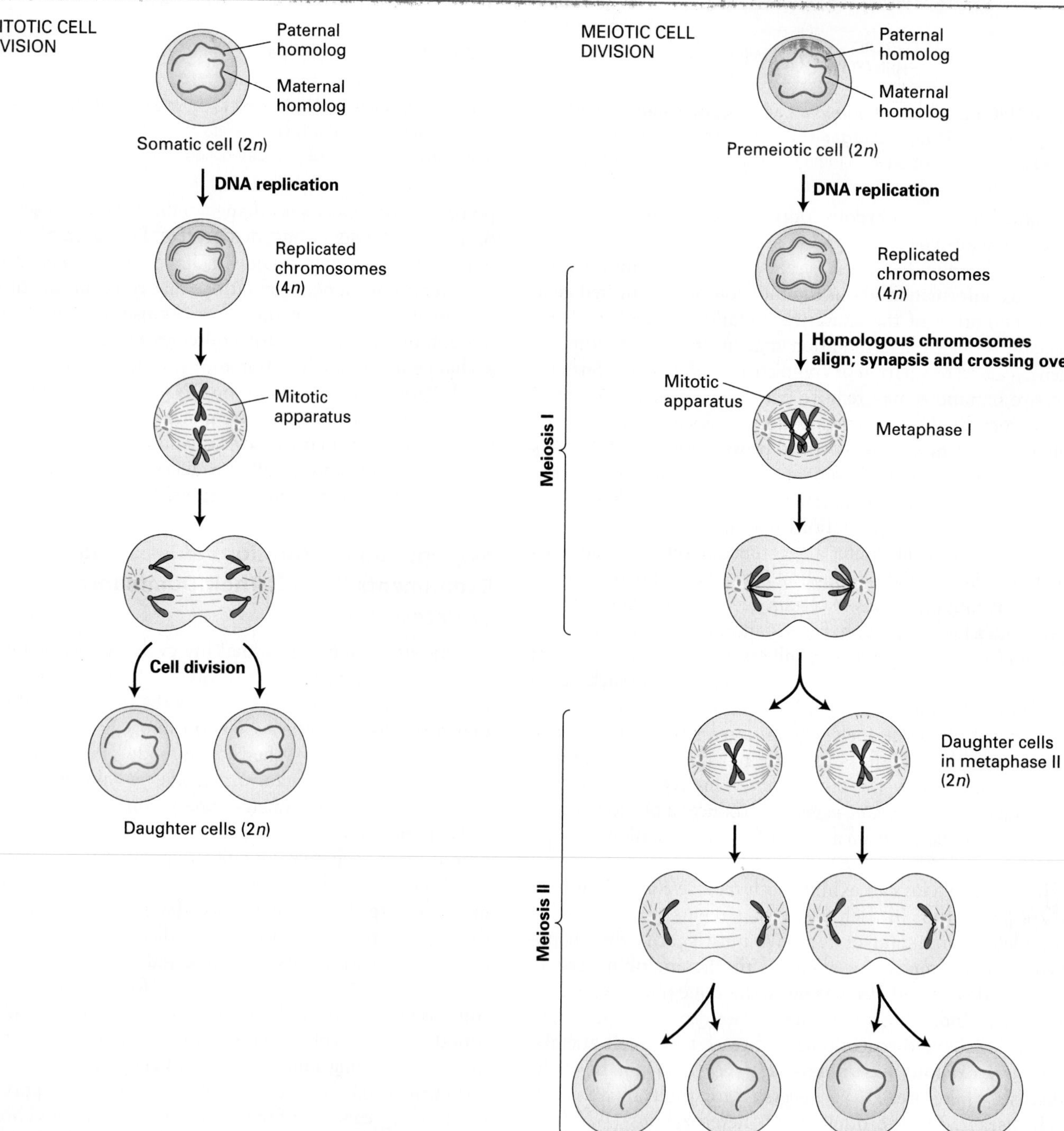

▲ **FIGURE 5-3 Comparison of mitosis and meiosis.** Both somatic cells and premeiotic germ cells have two copies of each chromosome (2n), one maternal and one paternal. In mitosis, the replicated chromosomes, each composed of two sister chromatids, align at the cell center in such a way that both daughter cells receive a maternal and paternal homolog of each morphological type of chromosome. During the first *meiotic* division, however, each replicated chromosome pairs with its homologous partner at the cell center; this pairing off is referred to as *synapsis*, and crossing over between homologous chromosomes is evident at this stage. One replicated chromosome of each morphological type then goes into each daughter cell. The resulting cells undergo a second division without intervening DNA replication, with the sister chromatids of each morphological type being apportioned to the daughter cells. In the second meiotic division the alignment of chromatids and their equal segregation into daughter cells is the same as in mitotic division. The alignment of pairs of homologous chromosome in metaphase I is random with respect to other chromosome pairs, resulting in a mix of paternally and maternally derived chromosomes in each daughter cell.

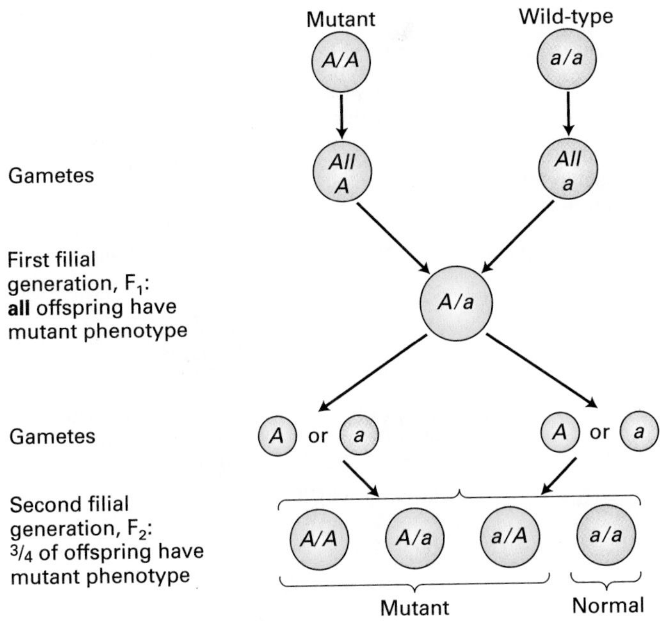

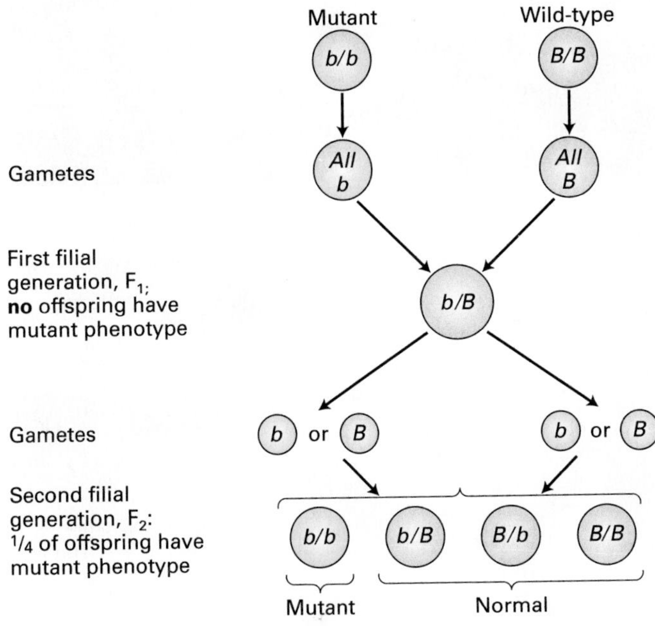

▲ **FIGURE 5-4 Segregation patterns of dominant and recessive mutations in crosses between true-breeding strains of diploid organisms.** All the offspring in the first (F_1) generation are heterozygous. If the mutant allele is dominant, the F_1 offspring will exhibit the mutant phenotype, as in part (a). If the mutant allele is recessive, the F_1 offspring will exhibit the wild-type phenotype, as in part (b). Crossing of the F_1 heterozygotes among themselves also produces different segregation ratios for dominant and recessive mutant alleles in the F_2 generation.

(Figure 5-4). If the F_1 progeny exhibit the mutant trait, then the mutant allele is dominant; if the F_1 progeny exhibit the wild-type trait, then the mutant is recessive. Further crossing between F_1 individuals will also reveal different patterns of inheritance according to whether the mutation is dominant or recessive. When F_1 individuals that are heterozygous for a dominant allele are crossed among themselves, three-fourths of the resulting F_2 progeny will exhibit the mutant trait. In contrast, when F_1 individuals that are heterozygous for a recessive allele are crossed among themselves, only one-fourth of the resulting F_2 progeny will exhibit the mutant trait.

As noted earlier, the yeast *S. cerevisiae,* an important experimental organism, can exist in either a haploid or a diploid state. In these unicellular eukaryotes, crosses between haploid cells can determine whether a mutant allele is dominant or recessive. Haploid yeast cells, which carry one copy of each chromosome, can be of two different mating types known as **a** and α. Haploid cells of opposite mating type can mate to produce **a**/α diploids, which carry two copies of each chromosome. If a new mutation with an observable phenotype is isolated in a haploid strain, the mutant strain can be mated to a wild-type strain of the opposite mating type to produce **a**/α diploids that are heterozygous for the mutant allele. If these diploids exhibit the mutant trait, then the mutant allele is dominant, but if the diploids appear as wild-type, then the mutant allele is recessive. When **a**/α diploids are placed under starvation conditions, the cells undergo meiosis, giving rise to a tetrad of four haploid spores, two of type **a** and two of type α. Sporulation of a heterozygous diploid cell yields two spores carrying the mutant allele and two carrying the wild-type allele (Figure 5-5). Under

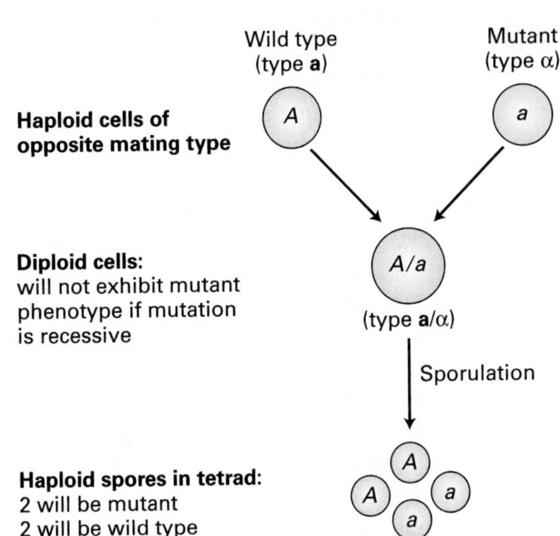

▲ **FIGURE 5-5 Segregation of alleles in yeast.** Haploid *Saccharomyces* cells of opposite mating type (i.e., one of mating type α and one of mating type **a**) can mate to produce an **a**/α diploid. If one haploid carries a dominant wild-type allele and the other carries a recessive mutant allele of the same gene, the resulting heterozygous diploid will express the dominant trait. Under certain conditions, a diploid cell will form a tetrad of four haploid spores. Two of the spores in the tetrad will express the recessive trait and two will express the dominant trait.

appropriate conditions, yeast spores will germinate, producing vegetative haploid strains of both mating types.

Conditional Mutations Can Be Used to Study Essential Genes in Yeast

The procedures used to identify and isolate mutants, referred to as *genetic screens,* depend on whether the experimental organism is haploid or diploid and, if the latter, whether the mutation is recessive or dominant. Genes that encode proteins essential for life are among the most interesting and important ones to study. Since phenotypic expression of mutations in essential genes leads to death of the individual, ingenious genetic screens are needed to isolate and maintain organisms with a lethal mutation.

In haploid yeast cells, essential genes can be studied through the use of *conditional mutations.* Among the most common conditional mutations are **temperature-sensitive mutations,** which can be isolated in bacteria and lower eukaryotes but not in warm-blooded eukaryotes. For instance, a single missense mutation may cause the resulting mutant protein to have reduced thermal stability such that the protein is fully functional at one temperature (e.g., 23 °C) but begins to denature and is inactive at another temperature (e.g., 36 °C), whereas the normal protein would be fully stable and functional at both temperatures. A temperature at which the mutant phenotype is observed is called *nonpermissive;* a *permissive* temperature is one at which the mutant phenotype is not observed even though the mutant allele is present. Thus mutant strains can be maintained at a permissive temperature and then subcultured at a nonpermissive temperature for analysis of the mutant phenotype.

An example of a particularly important screen for temperature-sensitive mutants in the yeast *S. cerevisiae* comes from the studies of L. H. Hartwell and colleagues in the late 1960s and early 1970s. They set out to identify genes important in regulation of the **cell cycle** during which a cell synthesizes proteins, replicates its DNA, and then undergoes mitotic cell division, with each daughter cell receiving a copy of each chromosome. Exponential growth of a single yeast cell for 20–30 cell divisions forms a visible yeast colony on solid agar medium. Since mutants with a complete block in the cell cycle would not be able to form a colony, conditional mutants were required to study mutations that affect this basic cell process. To screen for such mutants, the researchers first identified mutagenized yeast cells that could grow normally at 23 °C but that could not form a colony when placed at 36 °C (Figure 5-6a).

Once temperature-sensitive mutants were isolated, further analysis revealed that some indeed were defective in cell division. In *S. cerevisiae,* cell division occurs through a budding process, and the size of the bud, which is easily visualized by light microscopy, indicates a cell's position in the cell cycle. Each of the mutants that could not grow at 36 °C was examined by microscopy after several hours at the nonpermissive temperature. Examination of many different temperature-sensitive mutants revealed that about 1 percent exhibited a distinct block in the cell cycle. These mutants were therefore designated *cdc* (*c*ell-*d*ivision *c*ycle) mutants. Importantly,

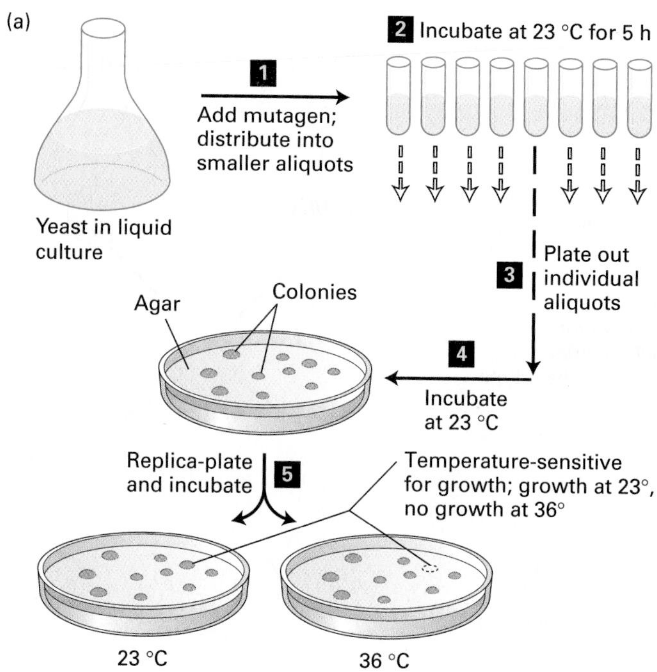

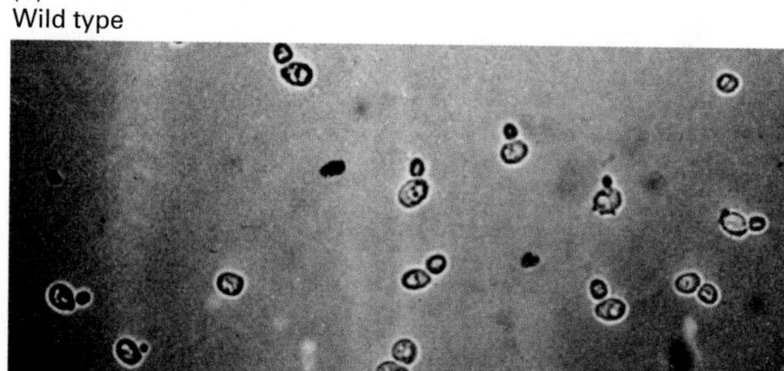

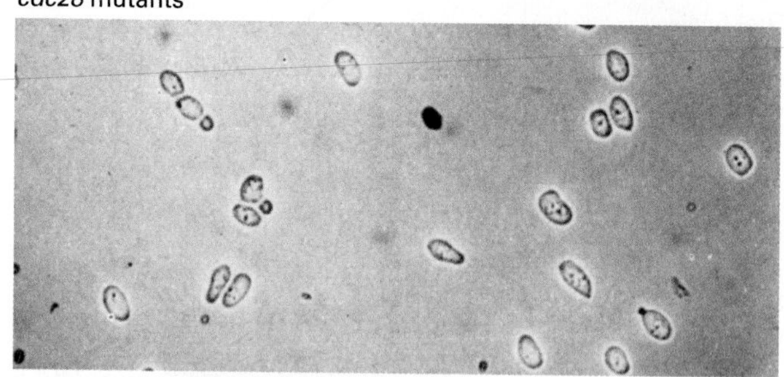

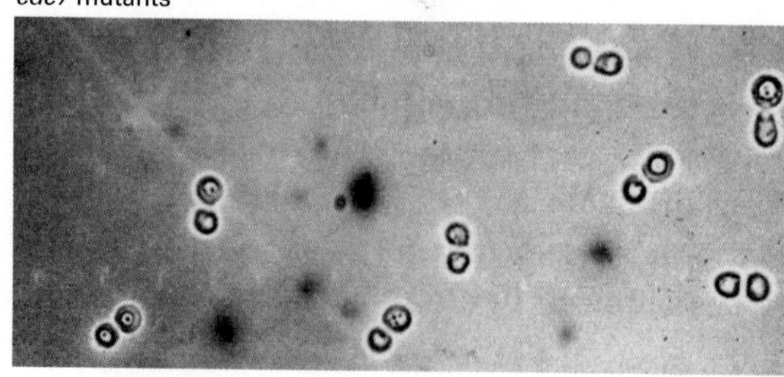

◀ **EXPERIMENTAL FIGURE 5-6 Haploid yeasts carrying temperature-sensitive lethal mutations are maintained at permissive temperature and analyzed at nonpermissive temperature.** (a) Genetic screen for temperature-sensitive cell-division cycle (*cdc*) mutants in yeast. Yeasts that grow and form colonies at 23 °C (permissive temperature) but not at 36 °C (nonpermissive temperature) may carry a lethal mutation that blocks cell division. (b) Assay of temperature-sensitive colonies for blocks at specific stages in the cell cycle. Shown here are micrographs of wild-type yeast and two different temperature-sensitive mutants after incubation at the nonpermissive temperature for 6 h. Wild-type cells, which continue to grow, can be seen with all different sizes of buds, reflecting different stages of the cell cycle. In contrast, cells in the lower two micrographs exhibit a block at a specific stage in the cell cycle. The *cdc28* mutants arrest at a point before emergence of a new bud and therefore appear as unbudded cells. The *cdc7* mutants, which arrest just before separation of the mother cell and bud (emerging daughter cell), appear as cells with large buds. [Part (a) see L. H. Hartwell, 1967, *J. Bacteriol.* **93**:1662; part (b) from L. M. Hereford and L. H. Hartwell, 1974, *J. Mol. Biol.* **84**:445.]

these yeast mutants did not simply fail to grow, as they might if they carried a mutation affecting general cellular metabolism. Rather, at the nonpermissive temperature, the mutants of interest grew normally for part of the cell cycle but then arrested at a particular stage of the cell cycle, so that many cells at this stage were seen (Figure 5-6b). Most *cdc* mutations in yeast are recessive; that is, when haploid *cdc* strains are mated to wild-type haploids, the resulting heterozygous diploids are neither temperature-sensitive nor defective in cell division.

Recessive Lethal Mutations in Diploids Can Be Identified by Inbreeding and Maintained in Heterozygotes

In diploid organisms, phenotypes resulting from recessive mutations can be observed only in individuals homozygous for the mutant alleles. Since mutagenesis in a diploid organism typically changes only one allele of a gene, yielding heterozygous mutants, genetic screens must include inbreeding steps to generate progeny that are homozygous for the mutant alleles. The geneticist H. Muller developed a general and efficient procedure for carrying out such inbreeding experiments in the fruit fly *Drosophila*. Recessive lethal mutations in *Drosophila* and other diploid organisms can be maintained in heterozygous individuals and their phenotypic consequences analyzed in homozygotes.

The Muller approach was used to great effect by C. Nüsslein-Volhard and E. Wieschaus, who systematically screened for recessive lethal mutations affecting embryogenesis in *Drosophila*. Dead homozygous embryos carrying recessive lethal mutations identified by this screen were examined under the microscope for specific morphological defects in the embryos. Current understanding of the molecular mechanisms underlying development of multicellular organisms is based, in large part, on the detailed picture of embryonic development revealed by characterization of these *Drosophila* mutants. We will discuss some of the fundamental discoveries based on these genetic studies in Chapter 22.

Complementation Tests Determine Whether Different Recessive Mutations Are in the Same Gene

In the genetic approach to studying a particular cellular process, researchers often isolate multiple recessive mutations that produce the same phenotype. A common test for determining whether these mutations are in the same gene or in different genes exploits the phenomenon of **genetic complementation,** that is, the restoration of the wild-type phenotype by mating of two different mutants. If two recessive mutations, *a* and *b,* are in the *same* gene, then a diploid organism heterozygous for both mutations (i.e., carrying one *a* allele and one *b* allele) will exhibit the mutant phenotype because neither allele provides a functional copy of the gene. In contrast, if mutation *a* and *b* are in *separate* genes, then heterozygotes carrying a single copy of each mutant allele will not exhibit the mutant phenotype because a wild-type allele of each gene will also be present. In this case, the mutations are said to *complement* each other. Complementation analysis cannot be performed on dominant mutants, because the phenotype conferred by the mutant allele is displayed even in the presence of a wild-type allele of the gene.

Complementation analysis of a set of mutants exhibiting the same phenotype can distinguish the individual genes in a set of functionally related genes, all of which must function to produce a given phenotypic trait. For example, the screen for *cdc* mutations in *Saccharomyces* described previously yielded many recessive temperature-sensitive mutants that appeared arrested at the same cell-cycle stage. To determine how many genes were affected by these mutations, Hartwell and his colleagues performed complementation tests on all of the pair-wise combinations of *cdc* mutants following the general protocol outlined in Figure 5-7. These tests identified more than 20 different *CDC* genes. The subsequent molecular characterization of the *CDC* genes and their encoded proteins, as described in detail in Chapter 20, has provided a framework for understanding how cell division is regulated in organisms ranging from yeast to humans.

Double Mutants Are Useful in Assessing the Order in Which Proteins Function

Based on careful analysis of mutant phenotypes associated with a particular cellular process, researchers often can deduce the order in which a set of genes and their protein products function. Two general types of processes are amenable to such analysis: (a) biosynthetic pathways in which a precursor material is converted via one or more intermediates to a final product and (b) signaling pathways that regulate other processes and involve the flow of information rather than chemical intermediates.

Ordering of Biosynthetic Pathways A simple example of the first type of process is the biosynthesis of a metabolite such as the amino acid tryptophan in bacteria. In this case, each of the enzymes required for synthesis of tryptophan catalyzes the conversion of one of the intermediates in the pathway to the

▶ **EXPERIMENTAL FIGURE 5-7**

Complementation analysis determines whether recessive mutations are in the same or different genes. Complementation tests in yeast are performed by mating haploid **a** and α cells carrying different recessive mutations to produce diploid cells. In the analysis of *cdc* mutations, pairs of different haploid temperature-sensitive *cdc* strains were systematically mated and the resulting diploids tested for growth at the permissive and nonpermissive temperatures. In this hypothetical example, the *cdcX* and *cdcY* mutants complement each other and thus have mutations in different genes, whereas the *cdcX* and *cdcZ* mutants have mutations in the same gene.

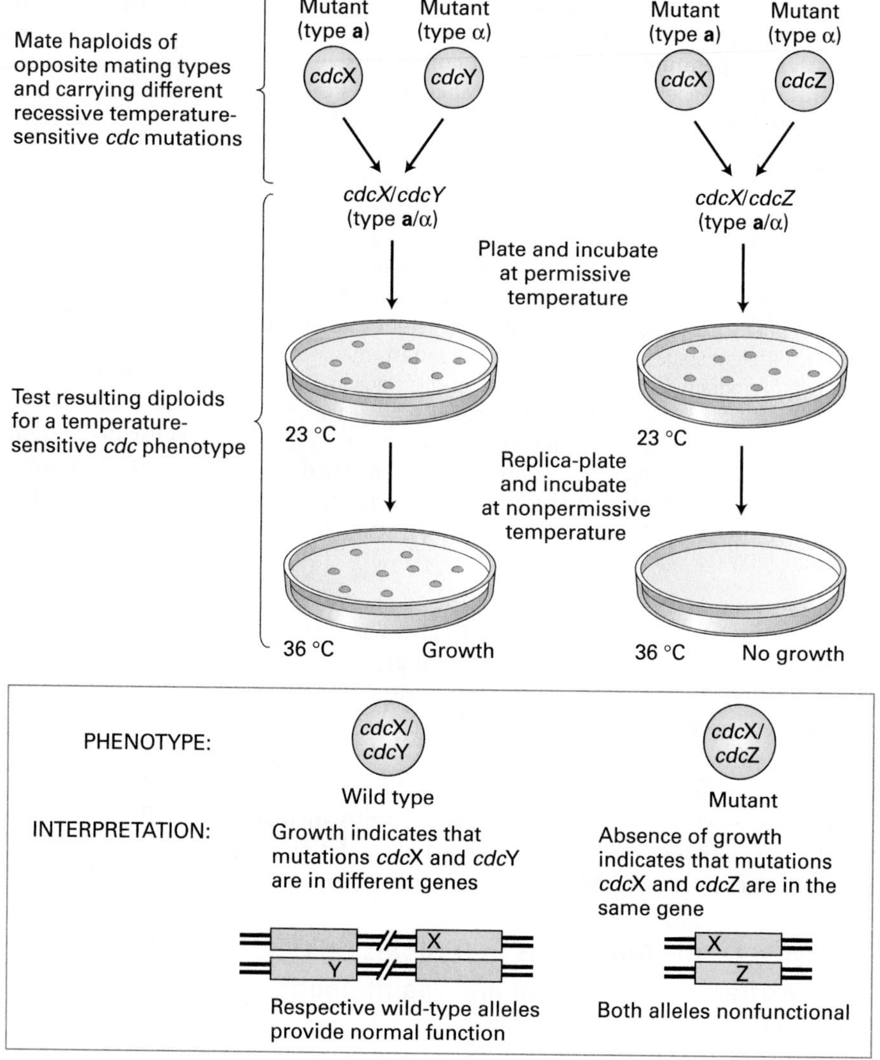

next. In *E. coli,* the genes encoding these enzymes lie adjacent to one another in the genome, constituting the *trp* operon (see Figure 4-13a). The order of action of the different genes for these enzymes, hence the order of the biochemical reactions in the pathway, initially was deduced from the types of intermediate compounds that accumulated in each mutant. In the case of complex synthetic pathways, however, phenotypic analysis of mutants defective in a single step may give ambiguous results that do not permit conclusive ordering of the steps. Double mutants defective in two steps in the pathway are particularly useful in ordering such pathways (Figure 5-8a).

In Chapter 14 we discuss the classic use of the double-mutant strategy to help elucidate the secretory pathway. In this pathway proteins to be secreted from the cell move from their site of synthesis on the rough endoplasmic reticulum (ER) to the Golgi complex, then to secretory vesicles, and finally to the cell surface.

Ordering of Signaling Pathways As we learn in later chapters, expression of many eukaryotic genes is regulated by signaling pathways that are initiated by extracellular hormones, growth factors, or other signals. Such signaling pathways may include numerous components, and double-mutant analysis often can provide insight into the functions and interactions of these components. The only prerequisite for obtaining useful information from this type of analysis is that the two mutations must have opposite effects on the output of the same regulated pathway. Most commonly, one mutation represses expression of a particular reporter gene even when the signal is present, while another mutation results in reporter gene expression even when the signal is absent (i.e., constitutive expression). As illustrated in Figure 5-8b, two simple regulatory mechanisms are consistent with such single mutants, but the double-mutant phenotype can distinguish between them. This general approach has enabled geneticists to delineate many of the key steps in a variety of different regulatory pathways, setting the stage for more specific biochemical assays.

Note that this technique differs from complementation analysis just described in that when testing two recessive mutations, the double mutant created is *homozygous* for both mutations. Furthermore, dominant mutants can be subjected to double-mutant analysis.

(a) Analysis of a biosynthetic pathway

A mutation in A accumulates intermediate 1.

A mutation in B accumulates intermediate 2.

PHENOTYPE OF DOUBLE MUTANT: A double mutation in A and B accumulates intermediate 1.

INTERPRETATION: *The reaction catalyzed by A precedes the reaction catalyzed by B.*

1 —A→ 2 —B→ 3

(b) Analysis of a signaling pathway

A mutation in A gives repressed reporter expression.

A mutation in B gives constitutive reporter expression.

PHENOTYPE OF DOUBLE MUTANT: A double mutation in A and B gives repressed reporter expression.

INTERPRETATION: *A positively regulates reporter expression and is negatively regulated by B.*

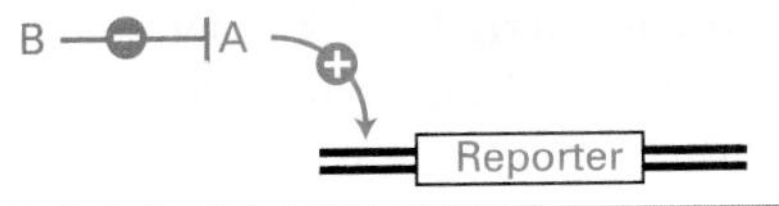

PHENOTYPE OF DOUBLE MUTANT: A double mutation in A and B gives constitutive reporter expression.

INTERPRETATION: *B negatively regulates reporter expression and is negatively regulated by A.*

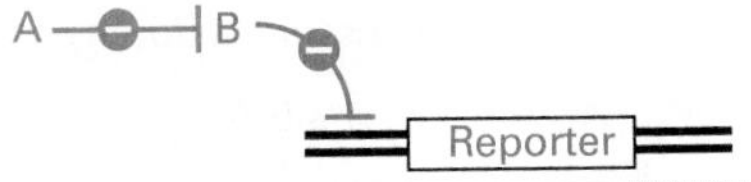

▲ **FIGURE 5-8 Analysis of double mutants often can order the steps in biosynthetic or signaling pathways.** When mutations in two different genes affect the same cellular process but have distinctly different phenotypes, the phenotype of the double mutant can often reveal the order in which the two genes must function. (a) In the case of mutations that affect the same biosynthetic pathway, a double mutant will accumulate the intermediate immediately preceding the step catalyzed by the protein that acts earlier in the wild-type organism. (b) Double-mutant analysis of a signaling pathway is possible if two mutations have opposite effects on expression of a reporter gene. In this case, the observed phenotype of the double mutant provides information about the order in which the proteins act and whether they are positive or negative regulators.

Genetic Suppression and Synthetic Lethality Can Reveal Interacting or Redundant Proteins

Two other types of genetic analysis can provide additional clues about how proteins that function in the same cellular process may interact with one another in the living cell. Both of these methods, which are applicable in many experimental organisms, involve the use of double mutants in which the phenotypic effects of one mutation are changed by the presence of a second mutation.

Suppressor Mutations The first type of analysis is based on *genetic suppression.* To understand this phenomenon, suppose that point mutations lead to structural changes in one protein (A) that disrupt its ability to associate with another protein (B) involved in the same cellular process. Similarly, mutations in protein B lead to small structural changes that inhibit its ability to interact with protein A. Assume, furthermore, that the normal functioning of proteins A and B depends on their interacting. In theory, a specific structural change in protein A might be suppressed by compensatory changes in protein B, allowing the mutant proteins to interact. In the rare cases in which such **suppressor mutations** occur, strains carrying both mutant alleles would be normal, whereas strains carrying only one or the other mutant allele would have a mutant phenotype (Figure 5-9a).

The observation of genetic suppression in yeast strains carrying a mutant actin allele (*act1-1*) and a second mutation (*sac6*) in another gene provided early evidence for a

(a) Suppression

Genotype	*AB*	*aB*	*Ab*	*ab*
Phenotype	Wild type	Mutant	Mutant	Suppressed mutant

INTERPRETATION

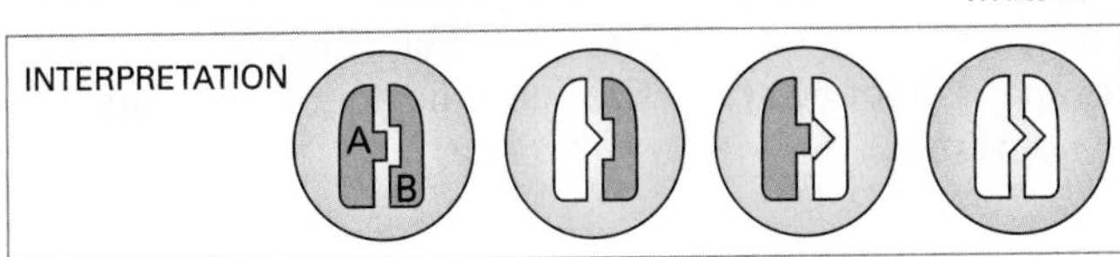

(b) Synthetic lethality 1

Genotype	*AB*	*aB*	*Ab*	*ab*
Phenotype	Wild type	Partial defect	Partial defect	Severe defect

INTERPRETATION

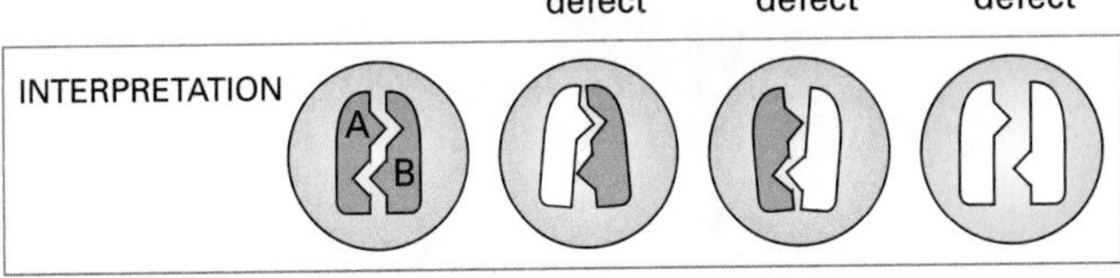

(c) Synthetic lethality 2

Genotype	*AB*	*aB*	*Ab*	*ab*
Phenotype	Wild type	Wild type	Wild type	Mutant

INTERPRETATION

Precursor (A, B) → Product; Precursor (a, B); Precursor (A, b); Precursor (a, b)

▲ **FIGURE 5-9 Mutations that result in genetic suppression or synthetic lethality reveal interacting or redundant proteins.** (a) Observation that double mutants with two defective proteins (A and B) have a wild-type phenotype but that single mutants give a mutant phenotype indicates that the function of each protein depends on interaction with the other. (b) Observation that double mutants have a more severe phenotypic defect than single mutants also is evidence that two proteins (e.g., subunits of a heterodimer) must interact to function normally. (c) Observation that a double mutant is nonviable but that the corresponding single mutants have the wild-type phenotype indicates that two proteins function in redundant pathways to produce an essential product.

direct interaction in vivo between the proteins encoded by the two genes. Later biochemical studies showed that these two proteins—Act1 and Sac6—do indeed interact in the construction of functional actin structures within the cell.

Synthetic Lethal Mutations Another phenomenon, called *synthetic lethality,* produces a phenotypic effect opposite to that of suppression. In this case, the deleterious effect of one mutation is greatly exacerbated (rather than suppressed) by a second mutation in a related gene. One situation in which such **synthetic lethal mutations** can occur is illustrated in Figure 5-9b. In this example, a heterodimeric protein is partially, but not completely, inactivated by mutations in either one of the nonidentical subunits. However, in double mutants carrying specific mutations in the genes encoding both subunits, little interaction between subunits occurs, resulting in severe phenotypic effects. Synthetic lethal mutations also can reveal nonessential genes whose encoded proteins function in redundant pathways for producing an essential cell component. As depicted in Figure 5-9c, if either pathway alone is inactivated by a mutation, the other pathway will be able to supply the needed product. However, if both pathways are inactivated at the same time, the essential product cannot be synthesized and the double mutants will be nonviable.

Genes Can Be Identified by Their Map Position on the Chromosome

The preceding discussion of genetic analysis illustrates how a geneticist can gain insight into gene function by observing the phenotypic effects produced by joining together different combinations of mutant alleles in the same cell or organism. For example, combinations of different alleles of the same gene in a diploid can be used to determine whether a mutation is dominant or recessive or whether two different recessive mutations are in the same gene. Furthermore, combinations of mutations in different genes can be used to determine the order of gene function in a pathway or to identify functional relationships between genes such as suppression and synthetic enhancement. Generally speaking, all these methods can be viewed as analytical tests based on *gene function.* We will now consider a fundamentally different type of genetic analysis based on *gene position.* Studies designed to determine the position of a gene on a chromosome, often referred to as **genetic mapping** studies, can be used to identify the gene affected by a particular mutation or to determine whether two mutations are in the same gene.

In many organisms genetic mapping studies rely on exchanges of genetic information that occur during meiosis. As discussed in Chapter 4 and as shown in Figure 5-10a,

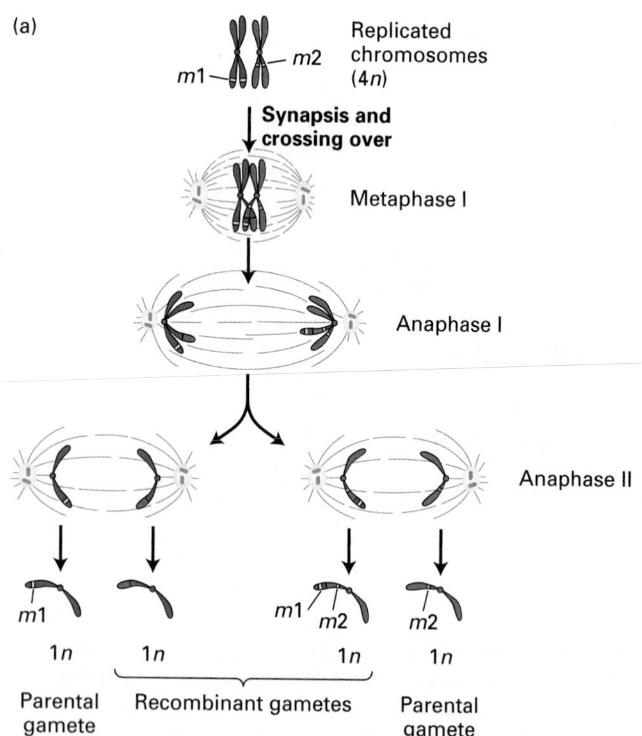

(b) Consider two linked genes *A* and *B* with recessive alleles a and b.

Cross of two mutants to construct a doubly heterozygous strain:

$$A/A\ b/b \times a/a\ B/B$$

$$\downarrow$$

$$\frac{A \quad b}{a \quad B}$$

Cross of double heterozygote to test strain:

$$\frac{A \quad b}{a \quad B} \times \frac{a \quad b}{a \quad b}$$

$$\downarrow$$

$$\underbrace{\frac{A \quad b}{a \quad b} \quad \frac{a \quad B}{a \quad b}}_{\text{Parental types}} \qquad \underbrace{\frac{A \quad B}{a \quad b} \quad \frac{a \quad b}{a \quad b}}_{\text{Recombinant types}}$$

Genetic distance between *A* and *B* can be determined from frequency of parental and recombinant gametes:

$$\text{Genetic distance in cM} = 100 \times \frac{\text{recombinant gametes}}{\text{total gametes}}$$

▲ **FIGURE 5-10 Recombination during meiosis can be used to map the position of genes.** (a) Shown is an individual that carries two mutations, designated *m1* (yellow) and *m2* (green), that are on the maternal and paternal versions of the same chromosome. If crossing over occurs at an interval between *m1* and *m2* before the first meiotic division, then two recombinant gametes are produced; one carries both *m1* and *m2,* whereas the other carries neither mutation. The longer the distance between two mutations on a chromatid, the more likely they are to be separated by recombination and the greater the proportion of recombinant gametes produced. (b) In a typical mapping experiment, a strain that is heterozygous for two different genes is constructed. The frequency of parental or recombinant gametes produced by this strain can be determined from the phenotypes of the progeny in a testcross to a homozygous recessive strain. The genetic map distance in centimorgans (cM) is given as the percent of the gametes that are recombinant.

genetic **recombination** takes place before the first meiotic cell division in germ cells, when the replicated chromosomes of each homologous pair align with each other. At this time, homologous DNA sequences on maternally and paternally derived chromatids can exchange with each other, a process known as **crossing over.** We now know that the resulting crossovers between homologous chromosomes provide structural links that are important for the proper segregation of pairs of homologous chromatids to opposite poles during the first meiotic cell division (for discussion see Chapter 20).

Consider two different mutations, one inherited from each parent, that are located close to one another on the same chromosome. Two different types of gametes can be produced according to whether a crossover occurs between the mutations during meiosis. If no crossover occurs between them, gametes known as *parental types*, which contain either one or the other mutation, will be produced. In contrast, if a crossover occurs between the two mutations, gametes known as *recombinant types* will be produced. In this example recombinant chromosomes would contain either both mutations, or neither of them. The sites of recombination occur more or less at random along the length of chromosomes; thus the closer together two genes are, the less likely that recombination will occur between them during meiosis. In other words, *the less frequently recombination occurs between two genes on the same chromosome, the more tightly they are linked and the closer together they are.* Two genes that are sufficiently close together such that there are significantly fewer recombinant gametes produced than parental gametes are considered to be *genetically linked.*

The technique of recombinational mapping was devised in 1911 by A. Sturtevant while he was an undergraduate working in the laboratory of T. H. Morgan at Columbia University. Originally used in studies on *Drosophila,* this technique is still used today to assess the distance between two genetic loci on the same chromosome in many experimental organisms. A typical experiment designed to determine the map distance between two genetic positions would involve two steps. In the first step, a strain is constructed that carries a different mutation at each position, or **locus.** In the second step, the progeny of this strain are assessed to determine the relative frequency of inheritance of parental or recombinant types. A typical way to determine the frequency of recombination between two genes is to cross one diploid parent heterozygous at each of the genetic loci to another parent homozygous for each gene. For such a cross, the proportion of recombinant progeny is readily determined because recombinant phenotypes will differ from the parental phenotypes. By convention, one *genetic map unit* is defined as the distance between two positions along a chromosome that results in one recombinant individual in 100 total progeny. The distance corresponding to this 1 percent recombination frequency is called a *centimorgan* (cM) in honor of Sturtevant's mentor, Morgan (see Figure 5-10b).

A complete discussion of the methods of genetic mapping experiments is beyond the scope of this introductory discussion; however, two features of measuring distances by recombination mapping need particular emphasis. First, the frequency of genetic exchange between two loci is strictly proportional to the physical distance in base pairs separating them only for loci that are relatively close together (say, less than about 10 cM). For loci that are farther apart than this, a distance measured by the frequency of genetic exchange tends to underestimate the physical distance because of the possibility of two or more crossovers occurring within an interval. In the limiting case in which the number of recombinant types will equal the number of parental types, the two loci under consideration could be far apart on the same chromosome or they could be on different chromosomes, and in such cases the loci are considered to be *unlinked.*

A second important concept needed for interpretation of genetic mapping experiments in different types of organisms is that although genetic distance is defined in the same way for different organisms, the relationship between recombination frequency (i.e., genetic map distance) and physical distance varies between organisms. For example, a 1 percent recombination frequency (i.e., a genetic distance of 1 cM) represents a physical distance of about 2.8 kilobases in yeast compared with a distance of about 400 kilobases in *Drosophila* and about 780 kilobases in humans.

One of the chief uses of genetic mapping studies is to locate the gene that is affected by a mutation of interest. The presence of many different already mapped genetic traits, or **genetic markers**, distributed along the length of a chromosome permits the position of an unmapped mutation to be determined by assessing its segregation with respect to these marker genes during meiosis. Thus the more markers that are available, the more precisely a mutation can be mapped. In Section 5.4 we will see how the genes affected in inherited human diseases can be identified using such methods. A second general use of mapping experiments is to determine whether two different mutations are in the same gene. If two mutations are in the same gene, they will exhibit tight **linkage** in mapping experiments, but if they are in different genes, they will usually be unlinked or exhibit weak linkage.

KEY CONCEPTS OF SECTION 5.1

Genetic Analysis of Mutations to Identify and Study Genes

- Diploid organisms carry two copies (alleles) of each gene, whereas haploid organisms carry only one copy.
- Recessive mutations lead to a loss of function, which is masked if a normal allele of the gene is present. For the mutant phenotype to occur, both alleles must carry the mutation.
- Dominant mutations lead to a mutant phenotype in the presence of a normal allele of the gene. The phenotypes associated with dominant mutations often represent a gain of function but in the case of some genes result from a loss of function.
- In meiosis, a diploid cell undergoes one DNA replication and two cell divisions, yielding four haploid cells in which maternal and paternal alleles are randomly assorted (see Figure 5-3).

- Dominant and recessive mutations exhibit characteristic segregation patterns in genetic crosses (see Figure 5-4).
- In haploid yeast, temperature-sensitive mutations are particularly useful for identifying and studying genes essential to survival.
- The number of functionally related genes involved in a process can be defined by complementation analysis (see Figure 5-7).
- The order in which genes function in a signaling pathway can be deduced from the phenotype of double mutants defective in two steps in the affected process.
- Functionally significant interactions between proteins can be deduced from the phenotypic effects of allele-specific suppressor mutations or synthetic lethal mutations.
- Genetic mapping experiments make use of crossing over between homologous chromosomes during meiosis to measure the genetic distance between two different mutations on the same chromosome.

5.2 DNA Cloning and Characterization

Detailed studies of the structure and function of a gene at the molecular level require large quantities of the individual gene in pure form. A variety of techniques, often referred to as *recombinant DNA technology,* are used in **DNA cloning,** which permits researchers to prepare large numbers of identical DNA molecules. **Recombinant DNA** is simply any DNA molecule composed of sequences derived from different sources.

The key to cloning a DNA fragment of interest is to link it to a **vector** DNA molecule that can replicate within a host cell. After a single recombinant DNA molecule, composed of a vector plus an inserted DNA fragment, is introduced into a host cell, the inserted DNA is replicated along with the vector, generating a large number of identical DNA molecules. The basic scheme can be summarized as follows:

Vector + DNA fragment

↓

Recombinant DNA

↓

Replication of recombinant DNA within host cells

↓

Isolation, sequencing, and manipulation of purified DNA fragment

Although investigators have devised numerous experimental variations, this flow diagram indicates the essential steps in DNA cloning. In this section, we first describe methods for isolating a specific sequence of DNA from a sea of other DNA sequences. This process often involves cutting the genome into fragments and then placing each fragment in a vector so that the entire collection can be propagated as recombinant molecules in separate host cells. While many different types of vectors exist, our discussion will mainly focus on plasmid vectors in *E. coli* host cells, which are commonly used. Various techniques can then be employed to identify the sequence of interest from this collection of DNA fragments, known as a DNA library. Once a specific DNA fragment is isolated, it is typically characterized by determining the exact sequence of nucleotides in the molecule. We end with a discussion of the polymerase chain reaction (PCR). This powerful and versatile technique can be used in many ways to generate large quantities of a specific sequence and otherwise manipulate DNA in the laboratory. The various uses of cloned DNA fragments are discussed in subsequent sections.

Restriction Enzymes and DNA Ligases Allow Insertion of DNA Fragments into Cloning Vectors

A major objective of DNA cloning is to obtain discrete, small regions of an organism's DNA that constitute specific genes. In addition, only relatively small DNA molecules can be cloned in any of the available vectors. For these reasons, the very long DNA molecules that compose an organism's genome must be cleaved into fragments that can be inserted into the vector DNA. Two types of enzymes—**restriction enzymes** and **DNA ligases**—facilitate production of such recombinant DNA molecules.

Cutting DNA Molecules into Small Fragments Restriction enzymes are endonucleases produced by bacteria that typically recognize specific 4- to 8-bp sequences, called *restriction sites,* and then cleave both DNA strands at this site. Restriction sites commonly are short *palindromic* sequences; that is, the restriction-site sequence is the same on each DNA strand when read in the 5′→3′ direction (Figure 5-11).

For each restriction enzyme, bacteria also produce a *modification enzyme,* which protects a bacterium's own DNA from cleavage by modifying it at or near each potential cleavage site. The modification enzyme adds a methyl group to one or two bases, usually within the restriction site. When

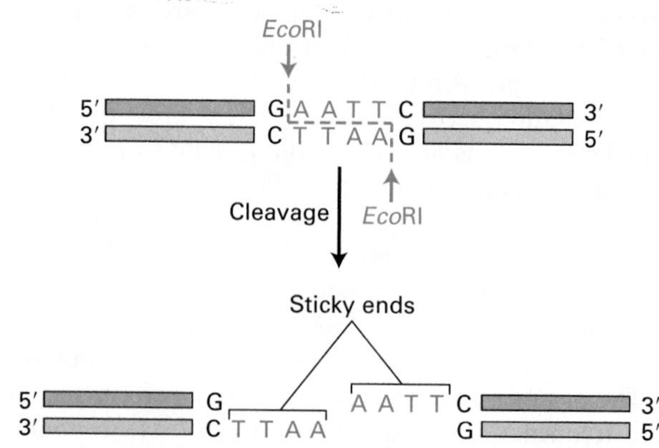

▲ **FIGURE 5-11 Cleavage of DNA by the restriction enzyme *Eco*RI.** This restriction enzyme from *E. coli* makes staggered cuts at the specific 6-bp palindromic sequence shown, yielding fragments with single-stranded, complementary "sticky" ends. Many other restriction enzymes also produce fragments with sticky ends.

a methyl group is present there, the restriction endonuclease is prevented from cutting the DNA. Together with the restriction endonuclease, the methylating enzyme forms a restriction-modification system that protects the host DNA while it destroys incoming foreign DNA (e.g., bacteriophage DNA or DNA taken up during transformation) by cleaving it at all the restriction sites in the DNA.

Many restriction enzymes make staggered cuts in the two DNA strands at their recognition site, generating fragments that have a single-stranded "tail" at both ends, *sticky ends* (see Figure 5-11). The tails on the fragments generated at a given restriction site are complementary to those on all other fragments generated by the same restriction enzyme. At room temperature, these single-stranded regions can transiently base-pair with those on other DNA fragments generated with the same restriction enzyme. A few restriction enzymes, such as *Alu*I and *Sma*I, cleave both DNA strands at the same point within the restriction site, generating fragments with "blunt" (flush) ends in which all the nucleotides at the fragment ends are base-paired to nucleotides in the complementary strand.

The DNA isolated from an individual organism has a specific sequence, which purely by chance will contain a specific set of restriction sites. Thus a given restriction enzyme will cut the DNA from a particular source into a reproducible set of fragments called **restriction fragments.** The frequency with which a restriction enzyme cuts DNA, and thus the average size of the resulting restriction fragments, depends largely on the length of the recognition site. For example, a restriction enzyme that recognizes a 4-bp site will cleave DNA an average of once every 4^4, or 256, base pairs, whereas an enzyme that recognizes an 8-bp sequence will cleave DNA an average of once every 4^8 base pairs (≈65 kb). Restriction enzymes have been purified from several hundred different species of bacteria, allowing DNA molecules to be cut at a large number of different sequences corresponding to the recognition sites of these enzymes (see Table 5-1).

Inserting DNA Fragments into Vectors DNA fragments with either sticky ends or blunt ends can be inserted into vector DNA with the aid of DNA ligases. During normal DNA replication, DNA ligase catalyzes the end-to-end joining (ligation) of

TABLE 5-1 Selected Restriction Enzymes and Their Recognition Sequences

ENZYME	SOURCE MICROORGANISM	RECOGNITION SITE*	ENDS PRODUCED
*Bam*HI	*Bacillus amyloliquefaciens*	↓ -G-G-A-T-C-C- -C-C-T-A-G-G- ↑	Sticky
*Sau*3A	*Staphylococcus aureus*	↓ -G-A-T-C- -C-T-A-G- ↑	Sticky
*Eco*RI	*Escherichia coli*	↓ -G-A-A-T-T-C- -C-T-T-A-A-G ↑	Sticky
*Hin*dIII	*Haemophilus influenzae*	↓ -A-A-G-C-T-T- -T-T-C-G-A-A- ↑	Sticky
*Sma*I	*Serratia marcescens*	↓ -C-C-C-G-G-G- -G-G-G-C-C-C- ↑	Blunt
*Not*I	*Nocardia otitidis-caviarum*	↓ -G-C-G-G-C-C-G-C- -C-G-C-C-G-G-C-G- ↑	Sticky

*Many of these recognition sequences are included in a common polylinker sequence (see Figure 5-13).

short fragments of DNA called **Okazaki fragments.** For purposes of DNA cloning, purified DNA ligase is used to covalently join the ends of a restriction fragment and vector DNA that have complementary ends (Figure 5-12). The vector DNA and restriction fragment are covalently ligated together through the standard 3′→5′ phosphodiester bonds of DNA. In addition to ligating complementary sticky ends, the DNA ligase from bacteriophage T4 can ligate any two blunt DNA ends. However, blunt-end ligation is inherently inefficient and requires a higher concentration of both DNA and DNA ligase than does ligation of sticky ends.

E. coli Plasmid Vectors Are Suitable for Cloning Isolated DNA Fragments

Plasmids are circular, double-stranded DNA (dsDNA) molecules that are separate from a cell's chromosomal DNA. These extrachromosomal DNAs, which occur naturally in bacteria and in lower eukaryotic cells (e.g., yeast), exist in a parasitic or symbiotic relationship with their host cell. Like the host-cell chromosomal DNA, plasmid DNA is duplicated before every cell division. During cell division, copies of the plasmid DNA segregate to each daughter cell, assuring continued propagation of the plasmid through successive generations of the host cell.

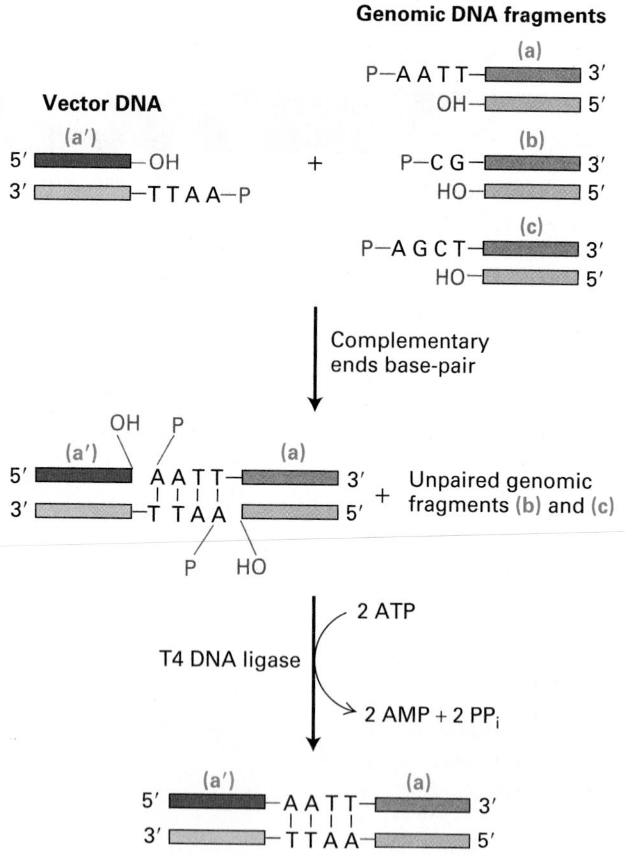

▲ **FIGURE 5-12 Ligation of restriction fragments with complementary sticky ends.** In this example, vector DNA cut with *Eco*RI is mixed with a sample containing restriction fragments produced by cleaving genomic DNA with several different restriction enzymes. The short base sequences composing the sticky ends of each fragment type are shown. The sticky end on the cut vector DNA (a′) base-pairs only with the complementary sticky ends on the *EcoRI* fragment (a) in the genomic sample. The adjacent 3′ hydroxyl and 5′ phosphate groups (red) on the base-paired fragments then are covalently joined (ligated) by T4 DNA ligase.

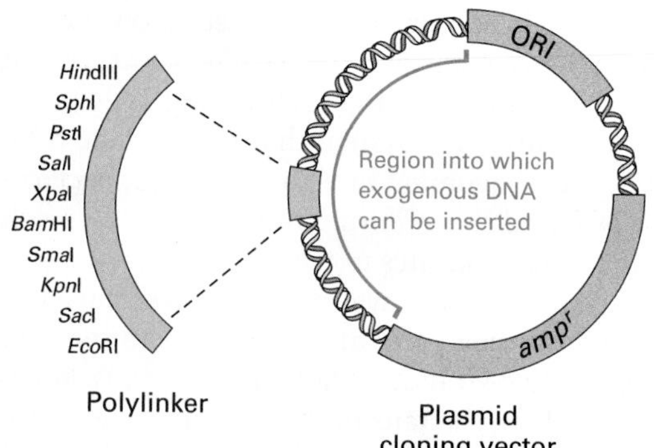

▲ **FIGURE 5-13 Basic components of a plasmid cloning vector that can replicate within an *E. coli* cell.** Plasmid vectors contain a selectable gene such as *amp*^{r}, which encodes the enzyme β-lactamase and confers resistance to ampicillin. Exogenous DNA can be inserted into the bracketed region without disturbing the ability of the plasmid to replicate or express the *amp*^{r} gene. Plasmid vectors also contain a replication origin (ORI) sequence where DNA replication is initiated by host-cell enzymes. Inclusion of a synthetic polylinker containing the recognition sequences for several different restriction enzymes increases the versatility of a plasmid vector. The vector is designed so that each site in the polylinker is unique on the plasmid.

The plasmids most commonly used in recombinant DNA technology are those that replicate in *E. coli.* Investigators have engineered these plasmids to optimize their use as vectors in DNA cloning. For instance, removal of unneeded portions from naturally occurring *E. coli* plasmids yields plasmid vectors, (≈1.2–3 kb in circumferential length, that contain three regions essential for DNA cloning: a replication origin; a marker that permits selection, usually a drug-resistance gene; and a region in which exogenous DNA fragments can be inserted (Figure 5-13). Host-cell enzymes replicate a plasmid beginning at the replication origin (ORI), a specific DNA sequence of 50–100 base pairs. Once DNA replication is initiated at the ORI, it continues around the circular plasmid regardless of its nucleotide sequence. Thus any DNA sequence inserted into such a plasmid is replicated along with the rest of the plasmid DNA.

Figure 5-14 outlines the general procedure for cloning a DNA fragment using *E. coli* plasmid vectors. When *E. coli* cells are mixed with recombinant vector DNA under certain conditions, a small fraction of the cells will take up the plasmid DNA, a process known as **transformation.** Typically, 1 cell in about 10,000 incorporates a *single* plasmid DNA molecule and thus becomes transformed. After plasmid vectors are incubated with *E. coli,* those cells that take up the plasmid can be easily selected from the much larger number of cells. For instance, if the plasmid carries a gene that confers resistance to the antibiotic ampicillin,

Technique Animation: Plasmid Cloning

MEDIA CONNECTIONS

▶ **EXPERIMENTAL FIGURE 5-14 DNA cloning in a plasmid vector permits amplification of a DNA fragment.** A fragment of DNA to be cloned is first inserted into a plasmid vector containing an ampicillin-resistance gene (amp^r), such as that shown in Figure 5-13. Only the few cells transformed by incorporation of a plasmid molecule will survive on ampicillin-containing medium. In transformed cells, the plasmid DNA replicates and segregates into daughter cells, resulting in formation of an ampicillin-resistant colony.

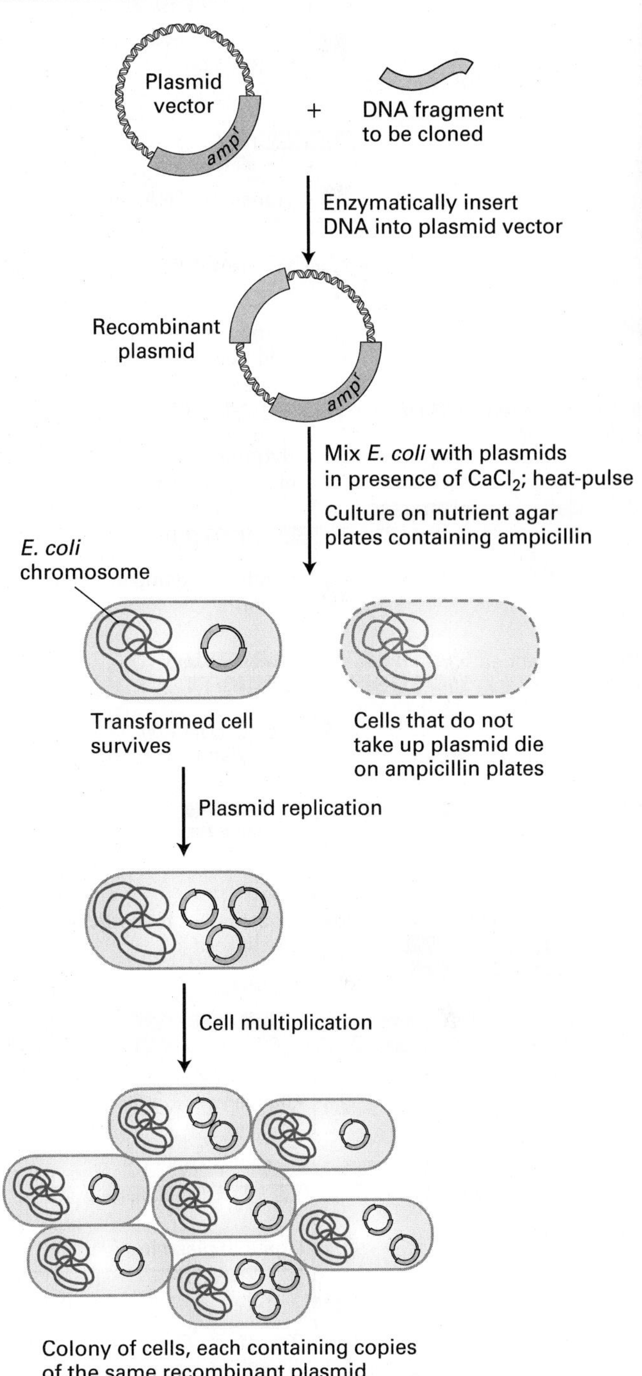

transformed cells can be selected by growing them in an ampicillin-containing medium.

DNA fragments from a few base pairs up to ≈10 kb commonly are inserted into plasmid vectors. When a recombinant plasmid with an inserted DNA fragment transforms an *E. coli* cell, all the antibiotic-resistant progeny cells that arise from the initial transformed cell will contain plasmids with the same inserted DNA. The inserted DNA is replicated along with the rest of the plasmid DNA and segregates to daughter cells as the colony grows. In this way, the initial fragment of DNA is replicated in the colony of cells into a large number of identical copies. Since all the cells in a colony arise from a single transformed parental cell, they constitute a **clone** of cells, and the initial fragment of DNA inserted into the parental plasmid is referred to as *cloned DNA* or a *DNA clone*.

The versatility of an *E. coli* plasmid vector is increased by the addition of a *polylinker*, a synthetically generated sequence containing one copy of several different restriction sites that are not present elsewhere in the plasmid sequence (see Figure 5-13). When such a vector is treated with a restriction enzyme that recognizes a restriction site in the polylinker, the vector is cut only once within the polylinker. Subsequently any DNA fragment of appropriate length produced with the same restriction enzyme can be inserted into the cut plasmid with DNA ligase. Plasmids containing a polylinker permit a researcher to use the same plasmid vector when cloning DNA fragments generated with different restriction enzymes, which simplifies experimental procedures.

For some purposes, such as the isolation and manipulation of large segments of the human genome, it is desirable to clone DNA segments as large as several megabases [1 megabase (Mb) = 1 million nucleotides). For this purpose specialized plasmid vectors known as *BACs* (*b*acterial *a*rtificial *c*hromosomes) have been developed. One type of BAC uses a replication origin derived from an endogenous plasmid of *E. coli* known as the *F factor*. The F factor and cloning vectors derived from it can be stably maintained at a single copy per *E. coli* cell even when they contain inserted sequences of up to about 2 Mb. Production of BAC libraries requires special methods for the isolation, ligation, and transformation of large segments of DNA because segments of DNA larger than about 20 kb are highly vulnerable to mechanical breakage by even standard manipulations such as pipetting.

cDNA Libraries Represent the Sequences of Protein-Coding Genes

A collection of DNA molecules each cloned into a vector molecule is known as a **DNA library.** When genomic DNA from a particular organism is the source of the starting DNA, the set of clones that collectively represent all the DNA sequences in the genome is known as a *genomic*

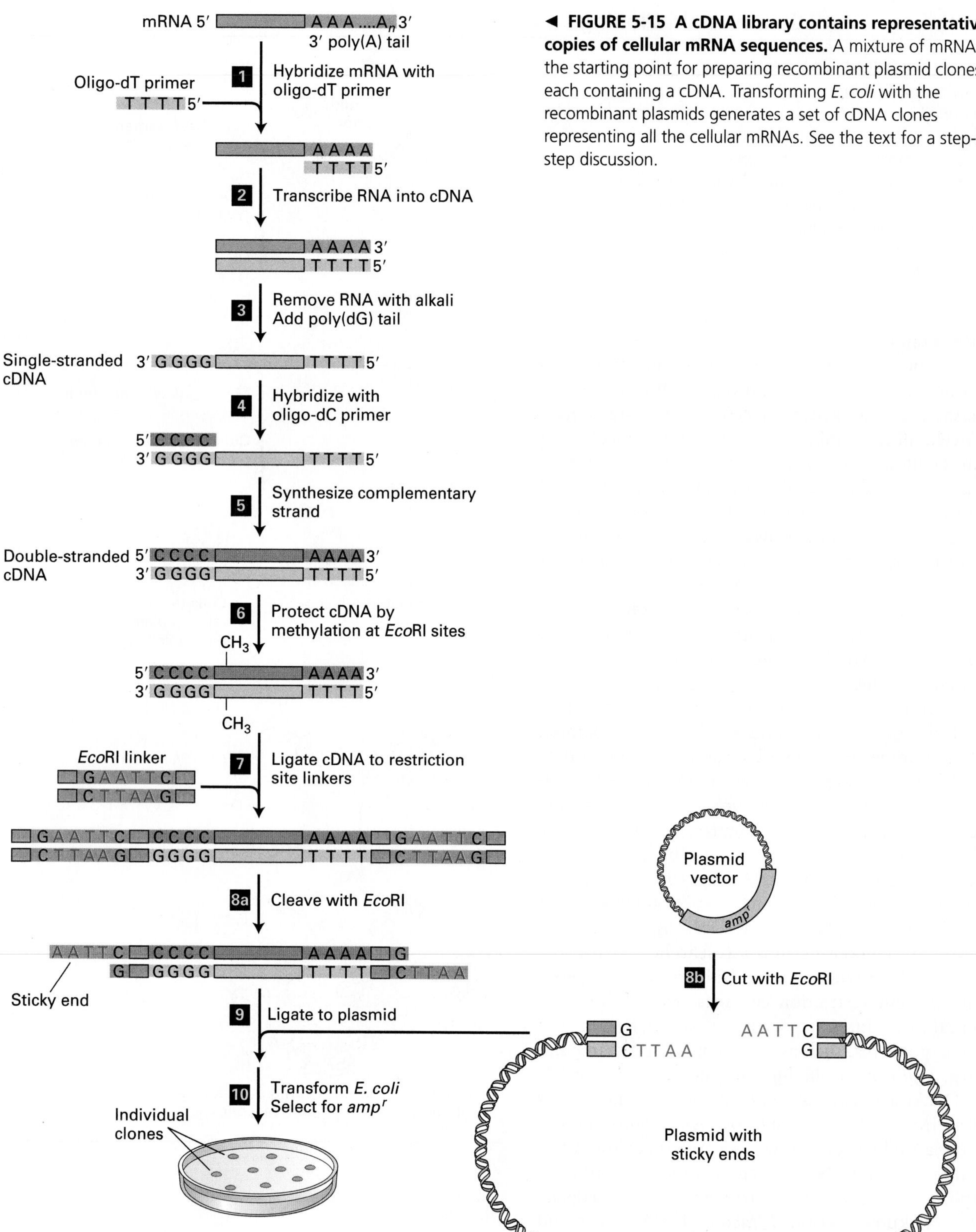

◀ **FIGURE 5-15 A cDNA library contains representative copies of cellular mRNA sequences.** A mixture of mRNAs is the starting point for preparing recombinant plasmid clones each containing a cDNA. Transforming *E. coli* with the recombinant plasmids generates a set of cDNA clones representing all the cellular mRNAs. See the text for a step-by-step discussion.

library. Such genomic libraries are ideal for representing the genetic content of relatively simple organisms such as bacteria or yeast, but present certain experimental difficulties for higher eukaryotes. First, the genes from such organisms usually contain extensive intron sequences and therefore can be too large to be inserted intact into plasmid vectors. As a result, the sequences of individual genes are broken apart and carried in more than one clone. Moreover, the presence of introns and long intergenic regions in genomic DNA often makes it difficult to identify the important parts of a gene that actually encode protein sequences. For example, only about 1.5 percent of the human genome actually represents protein-coding gene sequences. Thus for many studies, cellular mRNAs, which lack the noncoding regions present in genomic DNA, are a more useful starting material for generating a DNA library. In this approach, DNA copies of mRNAs, called **complementary DNAs (cDNAs)**, are synthesized and cloned into plasmid vectors. A large collection of the resulting cDNA clones, representing all the mRNAs expressed in a cell type, is called a *cDNA library*.

cDNAs Prepared by Reverse Transcription of Cellular mRNAs Can Be Cloned to Generate cDNA Libraries

The first step in preparing a cDNA library is to isolate the total mRNA from the cell type or tissue of interest. Because of their poly(A) tails, mRNAs are easily separated from the much more prevalent rRNAs and tRNAs present in a cell extract by use of a column to which short strings of thymidylate (oligo-dTs) are linked to the matrix. The general procedure for preparing a cDNA library from a mixture of cellular mRNAs is outlined in Figure 5-15. The enzyme **reverse transcriptase**, which is found in retroviruses, is used to synthesize a strand of DNA complementary to each mRNA molecule, starting from an oligo-dT primer (steps 1 and 2). The resulting cDNA-mRNA hybrid molecules are converted in several steps to double-stranded cDNA molecules corresponding to all the mRNA molecules in the original preparation (steps 3–5). Each double-stranded cDNA contains an oligo-dC·oligo-dG double-stranded region at one end and an oligo-dT·oligo-dA double-stranded region at the other end. Methylation of the cDNA protects it from subsequent restriction enzyme cleavage (step 6).

To prepare double-stranded cDNAs for cloning, short double-stranded DNA molecules containing the recognition site for a particular restriction enzyme are ligated to both ends of the cDNAs using DNA ligase from bacteriophage T4 (Figure 5-15, step 7). As noted earlier, this ligase can join "blunt-ended" double-stranded DNA molecules lacking sticky ends. The resulting molecules are then treated with the restriction enzyme specific for the attached linker, generating cDNA molecules with sticky ends at each end (step 8a). In a separate procedure, plasmid DNA first is treated with the same restriction enzyme to produce the appropriate sticky ends (step 8b).

The vector and the collection of cDNAs, all containing complementary sticky ends, then are mixed and joined covalently by DNA ligase (Figure 5-15, step 9). The resulting DNA molecules are transformed into *E. coli* cells to generate individual clones; each clone carrying a cDNA derived from a single mRNA.

Because different genes are transcribed at very different rates, cDNA clones corresponding to abundantly transcribed genes will be represented many times in a cDNA library, whereas cDNAs corresponding to infrequently transcribed genes will be extremely rare or not present at all. This property is advantageous if an investigator is interested in a gene that is transcribed at a high rate in a particular cell type. In this case, a cDNA library prepared from mRNAs expressed in that cell type will be enriched in the cDNA of interest, facilitating isolation of clones carrying that cDNA from the library. However, to have a reasonable chance of including clones corresponding to slowly transcribed genes, mammalian cDNA libraries must contain 10^6–10^7 individual recombinant clones.

DNA Libraries Can Be Screened by Hybridization to an Oligonucleotide Probe

Both genomic and cDNA libraries of various organisms contain hundreds of thousands to upwards of a million individual clones in the case of higher eukaryotes. Two general approaches are available for screening libraries to identify clones carrying a gene or other DNA region of interest: (1) detection with oligonucleotide **probes** that bind to the clone of interest and (2) detection based on expression of the encoded protein. Here we describe the first method; an example of the second method is presented in the next section.

The basis for screening with oligonucleotide probes is **hybridization**, the ability of complementary single-stranded DNA or RNA molecules to associate (hybridize) specifically with each other via base pairing. As discussed in Chapter 4, double-stranded (duplex) DNA can be denatured (melted) into single strands by heating in a dilute salt solution. If the temperature then is lowered and the ion concentration raised, complementary single strands will reassociate (hybridize) into duplexes. In a mixture of nucleic acids, only complementary single strands (or strands containing complementary regions) will reassociate; moreover, the extent of their reassociation is virtually unaffected by the presence of noncomplementary strands. As we will see later in this chapter, the ability to identify a particular DNA or RNA sequence within a highly complex mixture of molecules through nucleic acid hybridization is the basis for many techniques employed to study gene expression.

The steps involved in screening an *E. coli* plasmid cDNA library are depicted in Figure 5-16. First, the DNA to be screened must be attached to a solid support. A *replica* of the petri dish containing a large number of individual *E. coli* clones is reproduced on the surface of a nitrocellulose membrane. The DNA on the membrane is denatured, and the membrane is then incubated in a solution containing a radioactively labeled probe specific for the recombinant DNA containing the fragment of interest. Under hybridization conditions (near neutral pH, 40–65 °C, 0.3–0.6 M NaCl),

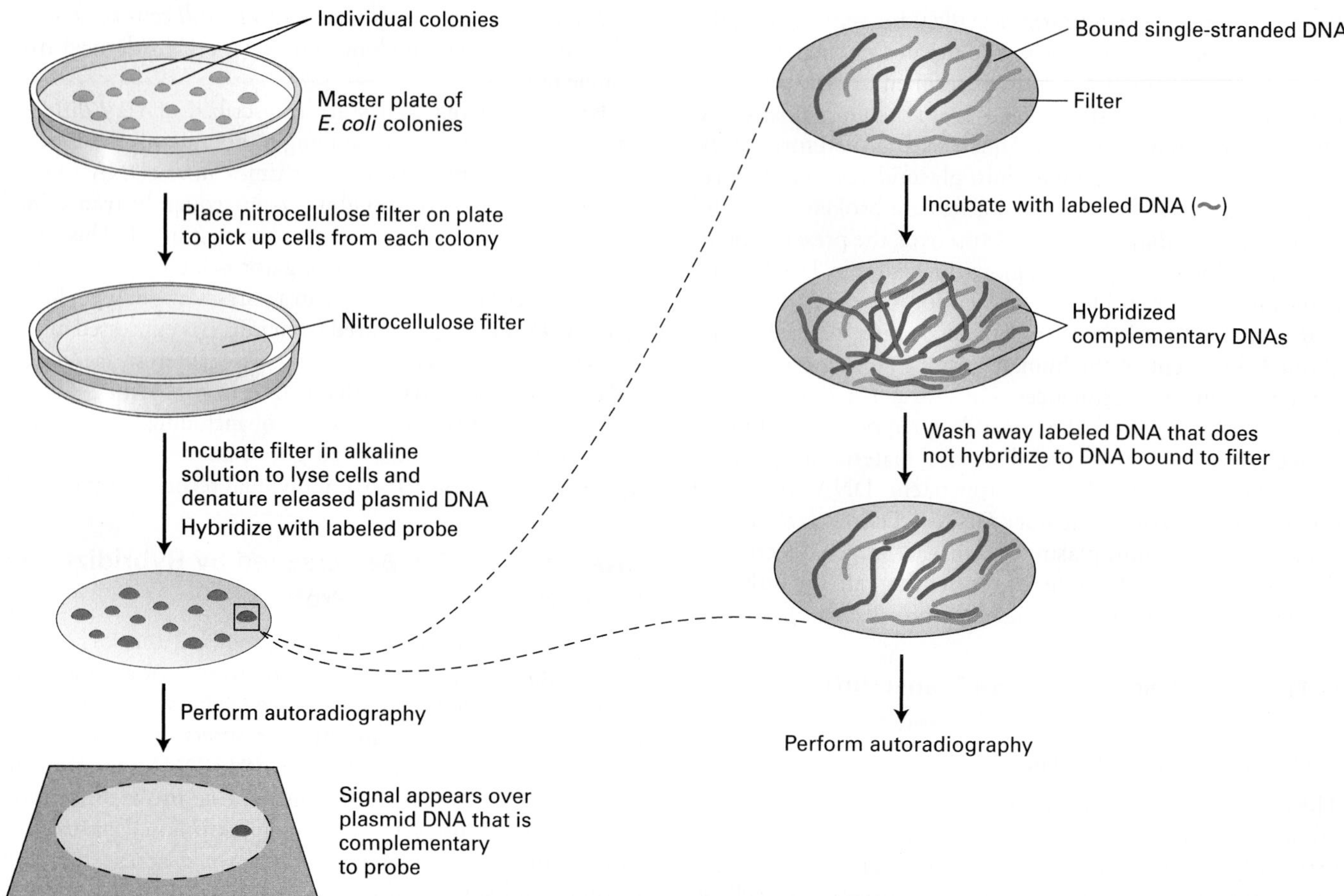

▲ EXPERIMENTAL FIGURE 5-16 **cDNA libraries can be screened with a radiolabeled probe to identify a clone of interest.** The appearance of a spot on the autoradiogram indicates the presence of a recombinant clone containing DNA complementary to the probe. The position of the spot on the autoradiogram is the mirror image of the position of that particular clone on the original petri dish (although for ease of comparison, it is not shown reversed here). Aligning the autoradiogram with the original petri dish will locate the corresponding clone from which *E. coli* cells can be recovered.

this labeled probe hybridizes to any complementary nucleic acid strands bound to the membrane. Any excess probe that does not hybridize is washed away, and the labeled hybrids are detected by autoradiography of the filter. This technique can be used to screen both genomic and cDNA libraries, but is most commonly used to isolate specific cDNAs.

Clearly, identification of specific clones by the membrane-hybridization technique depends on the availability of complementary radiolabeled probes. For an oligonucleotide to be useful as a probe, it must be long enough for its sequence to occur uniquely in the clone of interest and not in any other clones. For most purposes, this condition is satisfied by oligonucleotides containing about 20 nucleotides. This is because a specific 20-nucleotide sequence occurs once in every 4^{20} ($\approx 10^{12}$) nucleotides. Since all genomes are much smaller ($\approx 3 \times 10^9$ nucleotides for humans), a specific 20-nucleotide sequence in a genome usually occurs only once. With automated instruments now available, researchers can program the chemical synthesis of oligonucleotides of specific sequence up to about 100 nucleotides long. Longer probes can be prepared by the polymerase chain reaction (PCR), a widely used technique for amplifying specific DNA sequences that is described later.

How might an investigator design an oligonucleotide probe to identify a clone encoding a particular protein? It helps if all or a portion of the amino acid sequence of the protein is known. Thanks to the availability of the complete genomic sequences for humans and some important model organisms such as the mouse, *Drosophila,* and the roundworm *Caenorhabditis elegans,* a researcher can use an appropriate computer program to search the genomic sequence database for the coding sequence that corresponds to the amino acid sequence of the protein under study. If a match is found, then a single, unique DNA probe based on this known genomic sequence will hybridize perfectly with the clone encoding the protein of interest.

Yeast Genomic Libraries Can Be Constructed with Shuttle Vectors and Screened by Functional Complementation

In some cases a DNA library can be screened for the ability to express a functional protein that complements a recessive

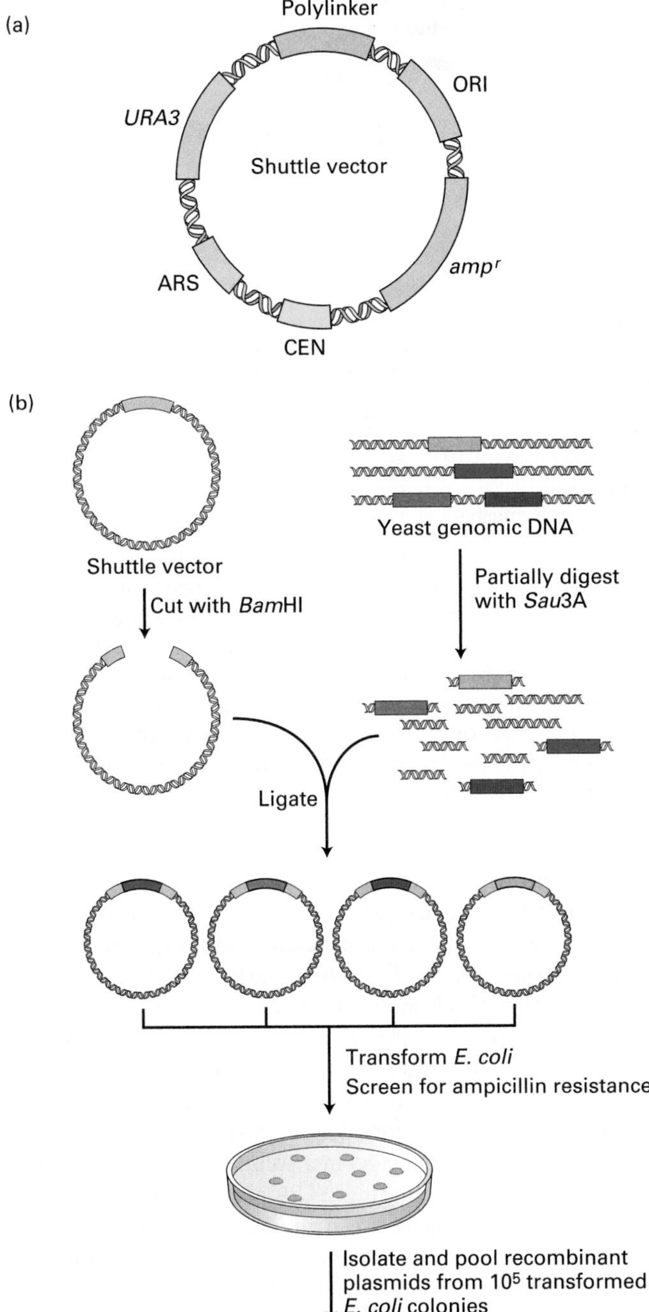

◀ EXPERIMENTAL FIGURE 5-17 **A yeast genomic library can be constructed in a plasmid shuttle vector that can replicate in yeast and *E. coli*.** (a) Components of a typical plasmid shuttle vector for cloning *Saccharomyces* genes. The presence of a yeast origin of DNA replication (ARS) and a yeast centromere (CEN) allows stable replication and segregation in yeast. Also included is a yeast selectable marker such as *URA3*, which allows a $ura3^-$ mutant to grow on medium lacking uracil. Finally, the vector contains sequences for replication and selection in *E. coli* (ORI and amp^r) and a polylinker for easy insertion of yeast DNA fragments. (b) Typical protocol for constructing a yeast genomic library. Partial digestion of total yeast genomic DNA with *Sau*3A is adjusted to generate fragments with an average size of about 10 kb. The vector is prepared to accept the genomic fragments by digestion with *Bam*HI, which produces the same sticky ends as *Sau*3A. Each transformed clone of *E. coli* that grows after selection for ampicillin resistance contains a single type of yeast DNA fragment.

mutation. Such a screening strategy would be an efficient way to isolate a cloned gene that corresponds to an interesting recessive mutation identified in an experimental organism. To illustrate this method, referred to as **functional complementation,** we describe how yeast genes cloned in special *E. coli* plasmids can be introduced into mutant yeast cells to identify the wild-type gene that is defective in the mutant strain.

Libraries constructed for the purpose of screening among yeast gene sequences usually are constructed from genomic DNA rather than cDNA. Because *Saccharomyces* genes do not contain multiple introns, they are sufficiently compact that the entire sequence of a gene can be included in a genomic DNA fragment inserted into a plasmid vector. To construct a plasmid genomic library that is to be screened by functional complementation in yeast cells, the plasmid vector must be capable of replication in both *E. coli* cells and yeast cells. This type of vector, capable of propagation in two different hosts, is called a **shuttle vector.** The structure of a typical yeast shuttle vector is shown in Figure 5-17a. This vector contains the basic elements that permit cloning of DNA fragments in *E. coli*. In addition, the shuttle vector contains an autonomously replicating sequence (ARS), which functions as an origin for DNA replication in yeast; a yeast centromere (called CEN), which allows faithful segregation of the plasmid during yeast cell division; and a yeast gene encoding an enzyme for uracil synthesis (*URA3*), which serves as a selectable marker in an appropriate yeast mutant.

To increase the probability that all regions of the yeast genome are successfully cloned and represented in the plasmid library, the genomic DNA usually is only partially digested to yield overlapping restriction fragments of ≈10 kb. These fragments are then ligated into the shuttle vector in which the polylinker has been cleaved with a restriction enzyme that produces sticky ends complementary to those on the yeast DNA fragments (Figure 5-17b). Because the 10-kb restriction fragments of yeast DNA are incorporated into the shuttle vectors randomly, at least 10^5 *E. coli* colonies, each containing a particular recombinant shuttle vector, are necessary to assure that each region of yeast DNA has a high probability of being represented in the library at least once.

Figure 5-18 outlines how such a yeast genomic library can be screened to isolate the wild-type gene corresponding to one of the temperature-sensitive *cdc* mutations mentioned earlier in this chapter. The starting yeast strain is a double mutant that requires uracil for growth due to a *ura3* mutation and is temperature-sensitive due to a *cdc28* mutation identified by its phenotype (see Figure 5-6). Recombinant plasmids isolated from the yeast genomic library are mixed with yeast cells under conditions that promote transformation of the cells with foreign DNA. Since transformed yeast cells carry a plasmid-borne copy of the wild-type *URA3*

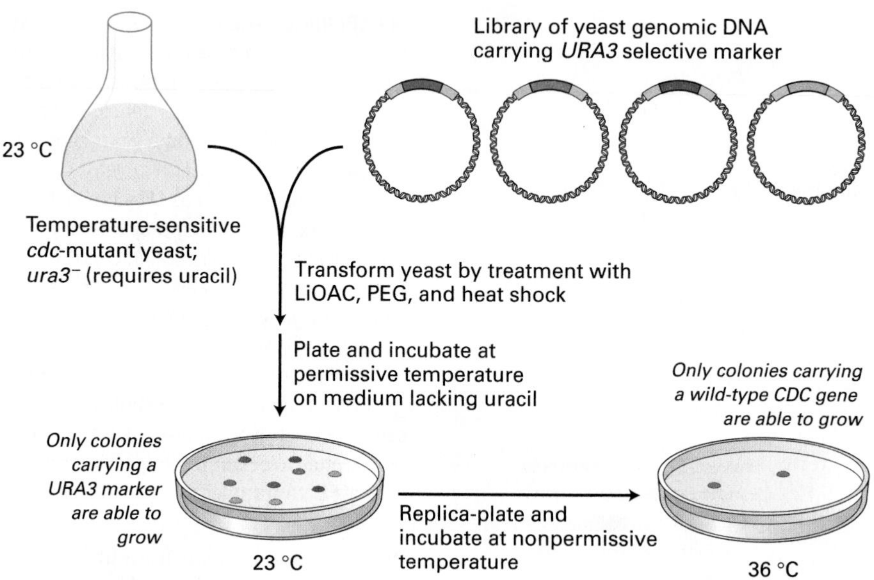
Library of yeast genomic DNA carrying URA3 selective marker
23 °C
Temperature-sensitive cdc-mutant yeast; ura3⁻ (requires uracil)
Transform yeast by treatment with LiOAC, PEG, and heat shock
Plate and incubate at permissive temperature on medium lacking uracil
Only colonies carrying a wild-type CDC gene are able to grow
Only colonies carrying a URA3 marker are able to grow
23 °C
Replica-plate and incubate at nonpermissive temperature
36 °C

▲ EXPERIMENTAL FIGURE 5-18 **Screening of a yeast genomic library by functional complementation can identify clones carrying the normal form of a mutant yeast gene.** In this example, a wild-type *CDC* gene is isolated by complementation of a *cdc* yeast mutant. The *Saccharomyces* strain used for screening the yeast library carries *ura3*$^-$ and a temperature-sensitive *cdc* mutation. This mutant strain is grown and maintained at a permissive temperature (23 °C). Pooled recombinant plasmids prepared as shown in Figure 5-17 are incubated with the mutant yeast cells under conditions that promote transformation. The relatively few transformed yeast cells, which contain recombinant plasmid DNA, can grow in the absence of uracil at 23 °C. When transformed yeast colonies are replica-plated and placed at 36 °C (a nonpermissive temperature), only clones carrying a library plasmid that contains the wild-type copy of the *CDC* gene will survive. LiOAC = lithium acetate; PEG = polyethylene glycol.

gene, they can be selected by their ability to grow in the absence of uracil. Typically, about 20 petri dishes, each containing about 500 yeast transformants, are sufficient to represent the entire yeast genome. This collection of yeast transformants can be maintained at 23°C, a temperature permissive for growth of the *cdc28* mutant. The entire collection on 20 plates is then transferred to replica plates, which are placed at 36 °C, a nonpermissive temperature for *cdc* mutants. Yeast colonies that carry recombinant plasmids expressing a wild-type copy of the *CDC28* gene will be able to grow at 36 °C. Once temperature-resistant yeast colonies have been identified, plasmid DNA can be extracted from the cultured yeast cells and analyzed by subcloning and DNA sequencing, topics we take up next.

Gel Electrophoresis Allows Separation of Vector DNA from Cloned Fragments

In order to manipulate or sequence a cloned DNA fragment, it sometimes must first be separated from the vector DNA. This can be accomplished by cutting the recombinant DNA clone with the same restriction enzyme used to produce the recombinant vectors originally. The cloned DNA and vector DNA then are subjected to gel **electrophoresis,** a powerful method for separating DNA molecules of different size (Figure 5-19).

Near neutral pH, DNA molecules carry a large negative charge and therefore move toward the positive electrode during gel electrophoresis. Because the gel matrix restricts random diffusion of the molecules, molecules of the same length migrate together as a band whose width equals that of the well into which the original DNA mixture was placed at the start of the electrophoretic run. Smaller molecules move through the gel matrix more readily than larger molecules, so that molecules of different length migrate as distinct bands. Smaller DNA molecules from about 10 to 2000 nucleotides can be separated electrophoretically on *polyacrylamide gels,* and larger molecules from about 200 nucleotides to more than 20 kb on *agarose gels.*

A common method for visualizing separated DNA bands on a gel is to incubate the gel in a solution containing the fluorescent dye ethidium bromide. This planar molecule binds to DNA by intercalating between the base pairs. Binding concentrates ethidium in the DNA and also increases its intrinsic fluorescence. As a result, when the gel is illuminated with ultraviolet light, the regions of the gel containing DNA fluoresce much more brightly than the regions of the gel without DNA.

Once a cloned DNA fragment, especially a long one, has been separated from vector DNA, it often is treated with various restriction enzymes to yield smaller fragments. After separation by gel electrophoresis, all or some of these smaller fragments can be ligated individually into a plasmid vector and cloned in *E. coli* by the usual procedure. This process, known as *subcloning,* is an important step in rearranging parts of genes into useful new configurations. For instance, an investigator who wants to change the conditions under which a gene is expressed might use subcloning to replace the normal promoter associated with a cloned gene with a DNA segment containing a different promoter. Subcloning also can be used to obtain cloned DNA fragments that are of an appropriate length for determining the nucleotide sequence.

(a) 

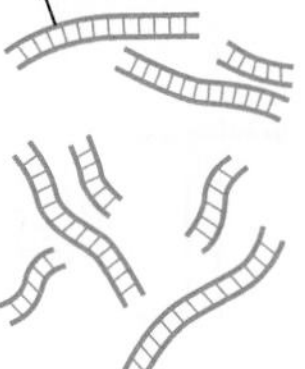

Place mixture in the well of an agarose or polyacrylamide gel. Apply electric field

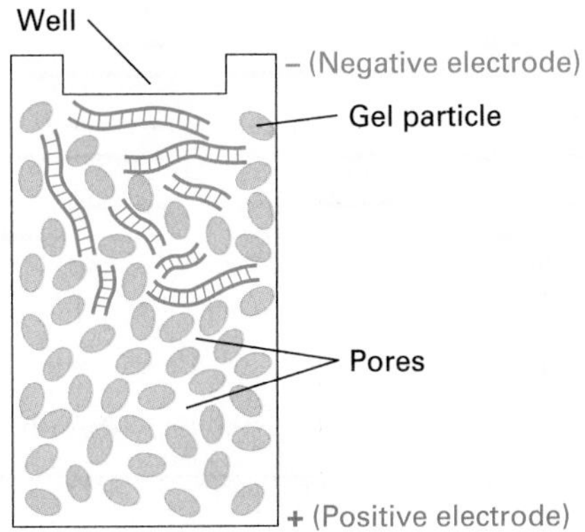

Molecules move through pores in gel at a rate inversely proportional to their chain length

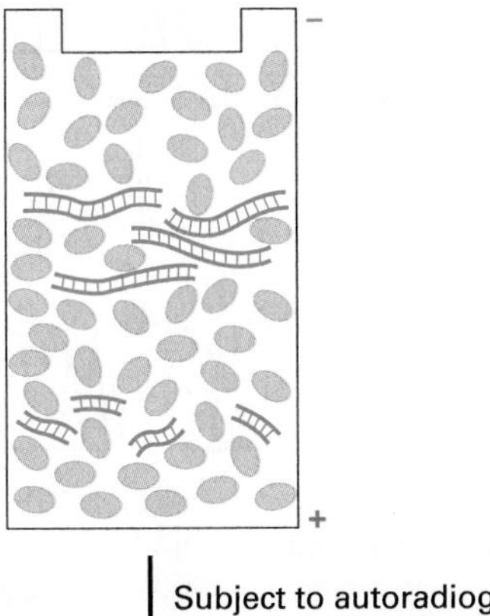

Subject to autoradiography or incubate with fluorescent dye

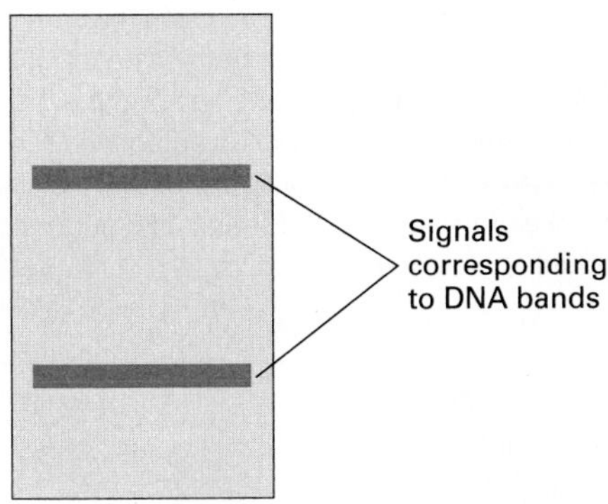

(b) 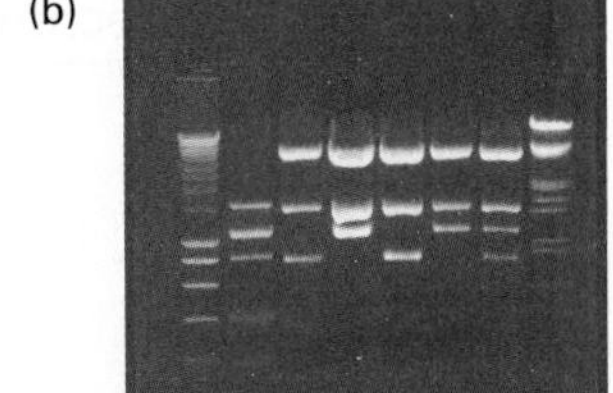

◀ EXPERIMENTAL FIGURE 5-19 **Gel electrophoresis separates DNA molecules of different lengths.** (a) A gel is prepared by pouring a liquid containing either melted agarose or unpolymerized acrylamide between two glass plates a few millimeters apart. As the agarose solidifies or the acrylamide polymerizes into polyacrylamide, a gel matrix (orange ovals) forms consisting of long, tangled chains of polymers. The dimensions of the interconnecting channels, or pores, depend on the concentration of the agarose or acrylamide used to form the gel. The separated bands can be visualized by autoradiography (if the fragments are radiolabeled) or by addition of a fluorescent dye (e.g., ethidium bromide) that binds to DNA. (b) A photograph of a gel stained with ethidium bromide (EtBr). EtBr binds to DNA and fluoresces under UV light. The bands in the far left and far right lanes are known as DNA ladders—DNA fragments of known size that serve as a reference for determining the length of the DNA fragments in the experimental sample. [Part (b) Science Photo Library.]

Deoxyribonucleoside triphosphate (dNTP)

Dideoxyribonucleoside triphosphate (ddNTP)

▲ **FIGURE 5-20 Structures of deoxyribonucleoside triphosphate (dNTP) and dideoxyribonucleoside triphosphate (ddNTP).** Incorporation of a ddNTP residue into a growing DNA strand terminates elongation at that point.

MEDIA CONNECTIONS

(a)

5′ TAGCTGACTC3′
3′ ATCGACTGAGTCAAGAACTATTGGGCTTAA . . .

DNA polymerase
+ dATP, dGTP, dCTP, dTTP
+ ddGTP in low concentration

5′ TAGCTGACTCAG3′
3′ ATCGACTGAGTCAAGAACTATTGGGCTTAA . . .
+
5′ TAGCTGACTCAGTTCTTG3′
3′ ATCGACTGAGTCAAGAACTATTGGGCTTAA . . .
+
5′ TAGCTGACTCAGTTCTTGATAACCCG3′
3′ ATCGACTGAGTCAAGAACTATTGGGCTTAA . . .

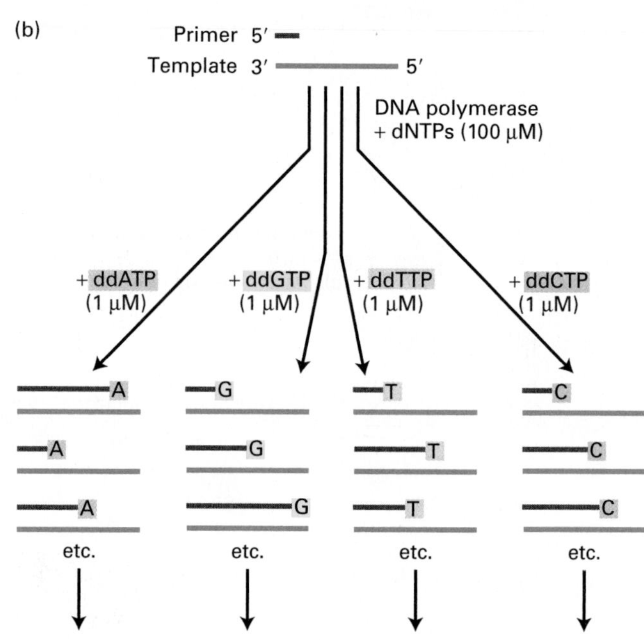

(c)

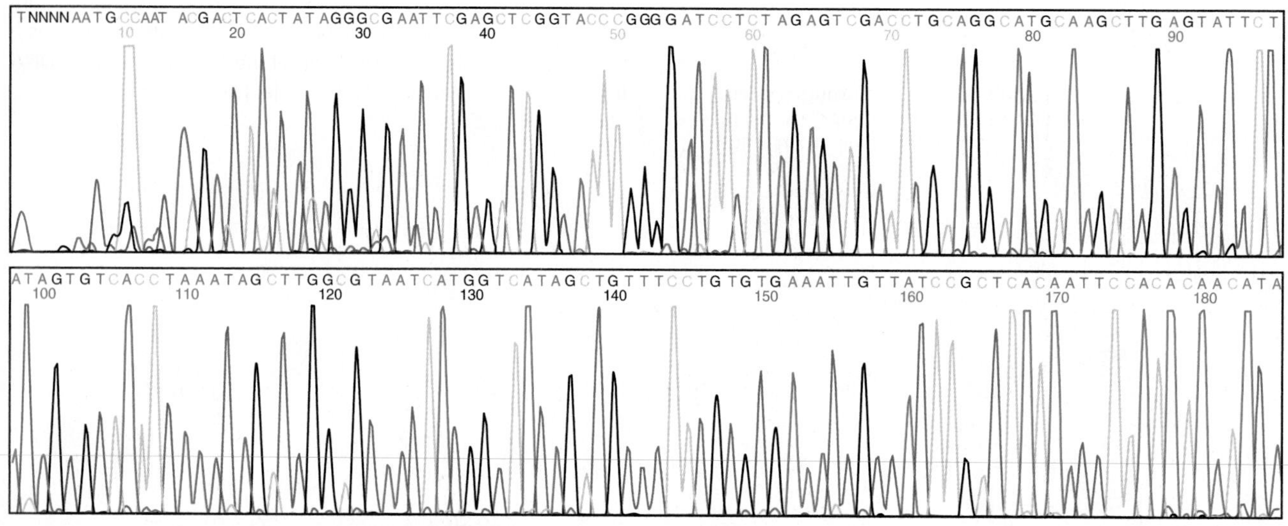

▲ **EXPERIMENTAL FIGURE 5-21 Cloned DNAs can be sequenced by the Sanger method, using fluorescent-tagged dideoxyribonucleoside triphosphates (ddNTPs).** (a) A single (template) strand of the DNA to be sequenced (blue letters) is hybridized to a synthetic deoxyribonucleotide primer (black letters). The primer is elongated in a reaction mixture containing the four normal deoxyribonucleoside triphosphates plus a relatively small amount of one of the four dideoxyribonucleoside triphosphates. In this example, ddGTP (yellow) is present. Because of the relatively low concentration of ddGTP, incorporation of a ddGTP, and thus chain termination, occurs at a given position in the sequence only about 1 percent of the time. Eventually the reaction mixture will contain a mixture of prematurely terminated (truncated) daughter fragments ending at every occurrence of ddGTP. (b) To obtain the complete sequence of a template DNA, four separate reactions are performed, each with a different dideoxyribonucleoside triphosphate (ddNTP). The ddNTP that terminates each truncated fragment can be identified by use of ddNTPs tagged with four different fluorescent dyes (indicated by colored highlights). (c) In an automated sequencing machine, the four reaction mixtures are subjected to gel electrophoresis, and the order of appearance of each of the four different fluorescent dyes at the end of the gel is recorded. Shown here is a sample printout from an automated sequencer from which the sequence of the original template DNA can be deduced from the sequence of the synthesized strand. N = nucleotide that cannot be assigned. [Part (c) from Griffiths et al., Figure 14-27.]

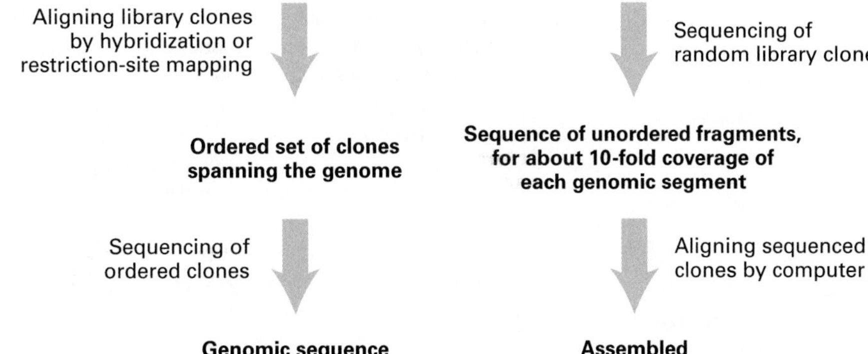

▶ **FIGURE 5-22 Two Strategies for Assembling Whole Genome Sequences.** One method depends on isolating and assembling a set of cloned DNA segments that span the genome. This can be done by matching cloned segments by hybridization or by alignment of restriction site maps. The DNA sequence of the ordered clones can then be assembled into a complete genomic sequence. The alternative method depends on the relative ease of automated DNA sequencing and bypasses the laborious step of ordering the library. By sequencing enough random library clones so that each segment of the genome is represented from 3 to 10 times it is possible to reconstruct the genomic sequence by computer alignment of the very large number of sequence fragments.

Cloned DNA Molecules Are Sequenced Rapidly by the Dideoxy Chain-Termination Method

The complete characterization of any cloned DNA fragment requires determination of its nucleotide sequence. F. Sanger and his colleagues developed the method now most commonly used to determine the exact nucleotide sequence of DNA fragments up to ≈500 nucleotides long. The basic idea behind this method is to synthesize from the DNA fragment to be sequenced a set of daughter strands that are labeled at one end and differ in length by one nucleotide. Separation of the truncated daughter strands by gel electrophoresis can then establish the nucleotide sequence of the original DNA fragment.

Synthesis of truncated daughter stands is accomplished by use of 2′,3′-dideoxyribonucleoside triphosphates (ddNTPs). These molecules, in contrast to normal deoxyribonucleotides (dNTPs), lack a 3′ hydroxyl group (Figure 5-20). Although ddNTPs can be incorporated into a growing DNA chain by DNA polymerase, once incorporated they cannot form a phosphodiester bond with the next incoming nucleotide triphosphate. Thus incorporation of a ddNTP terminates chain synthesis, resulting in a daughter strand truncated at specific positions corresponding to the base complementary to the added ddNTP on the template strand.

Sequencing using the Sanger *dideoxy chain-termination method* is usually carried out using an automated DNA sequencing machine. The reaction begins by denaturing a double-stranded DNA fragment to generate template strands for in vitro DNA synthesis. A synthetic oligodeoxynucleotide is used as the primer for the polymerization reaction that contains a low concentration of each of the four ddNTPs in addition to higher concentrations of the normal dNTPs. The ddNTPs are randomly incorporated at the positions of the corresponding dNTP, causing termination of polymerization at those positions in the sequence (Figure 5-21a). Inclusion of fluorescent tags of different colors on each of the four ddNTPs allows each set of truncated daughter fragments to be distinguished by their corresponding fluorescent label (Figure 5-21b). For example, all truncated fragments that end with a G would fluoresce one color (e.g., yellow), and those ending with an A would fluoresce another color (e.g., red), regardless of their lengths. The mixtures of truncated daughter fragments from each of the four reactions are subjected to electrophoresis on special polyacrylamide gels that can separate single-stranded DNA molecules differing in length by only 1 nucleotide. A fluorescence detector that can distinguish the four fluorescent tags is located at the end of the gel. The sequence of the original DNA template strand can be determined from the order in which different labeled fragments migrate past the fluorescence detector (Figure 5-21c).

In order to sequence a long continuous region of genomic DNA or even the entire genome of an organism, researchers usually employ one of the strategies outlined in Figure 5-22. The first method requires the isolation of a collection of cloned DNA fragments whose sequences overlap. Once the sequence of one of these fragments is determined, oligonucleotides based on that sequence can be chemically synthesized for use as primers in sequencing the adjacent overlapping fragments. In this way, the sequence of a long stretch of DNA is determined incrementally by sequencing of the overlapping cloned DNA fragments that compose it. A second method, which is called *whole genome shotgun sequencing,* bypasses the time-consuming step of isolating an ordered collection of DNA segments that span the genome. This method involves simply sequencing random clones from a genomic library. A total number of clones are chosen for sequencing so that on average each segment of the genome is sequenced about 10 times. This degree of coverage ensures that each segment of the genome is sequenced more than once. The entire genomic sequence is then assembled using a computer algorithm that aligns all the sequences, using their regions of overlap. Whole genome shotgun sequencing is the fastest and most cost-effective method for sequencing long stretches of DNA, and most genomes, including the human genome, have been sequenced by this method.

MEDIA CONNECTIONS

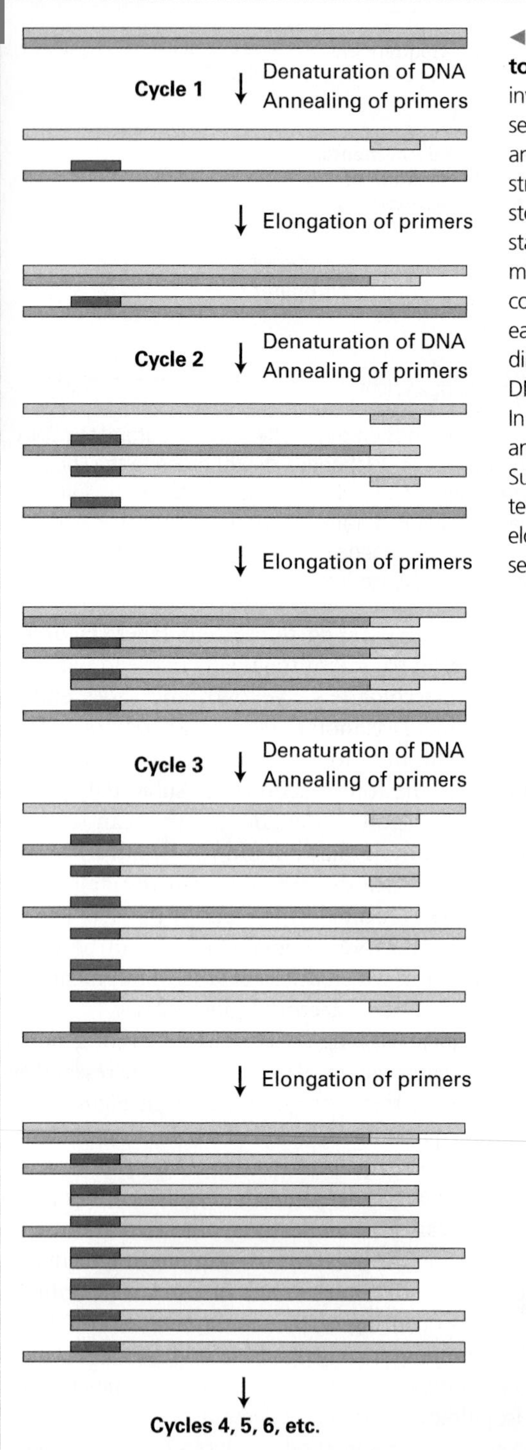

◀ EXPERIMENTAL FIGURE 5-23 **The polymerase chain reaction (PCR) is widely used to amplify DNA regions of known sequences.** To amplify a specific region of DNA, an investigator will chemically synthesize two different oligonucleotide primers complementary to sequences of approximately 18 bases flanking the region of interest (designated as light blue and dark blue bars). The complete reaction is composed of a complex mixture of double-stranded DNA (usually genomic DNA containing the target sequence of interest), a stoichiometric excess of both primers, the four deoxynucleoside triphosphates, and a heat-stable DNA polymerase known as Taq *polymerase*. During each PCR cycle, the reaction mixture is first heated to separate the strands and then cooled to allow the primers to bind to complementary sequences flanking the region to be amplified. *Taq* polymerase then extends each primer from its 3′ end, generating newly synthesized strands that extend in the 3′ direction to the 5′ end of the template strand. During the third cycle, two double-stranded DNA molecules are generated equal in length to the sequence of the region to be amplified. In each successive cycle the target segment, which will anneal to the primers, is duplicated, and will eventually vastly outnumber all other DNA segments in the reaction mixture. Successive PCR cycles can be automated by cycling the reaction for timed intervals at high temperature for DNA melting and at a defined lower temperature for the annealing and elongation portions of the cycle. A reaction that cycles 20 times will amplify the specific target sequence 1-million-fold.

The Polymerase Chain Reaction Amplifies a Specific DNA Sequence from a Complex Mixture

If the nucleotide sequences at the ends of a particular DNA region are known, the intervening fragment can be amplified directly by the **polymerase chain reaction (PCR)**. Here we describe the basic PCR technique and three situations in which it is used.

The PCR depends on the ability to alternately denature (melt) double-stranded DNA molecules and hybridize complementary single strands in a controlled fashion. As outlined in Figure 5-23, a typical PCR procedure begins by heat-denaturation of a DNA sample into single strands. Next, two synthetic oligonucleotides complementary to the 3′ ends of the target DNA segment of interest are added in great excess to the denatured DNA, and the temperature is lowered to 50–60 °C. These specific oligonucleotides, which are at a very high concentration, will hybridize with their complementary sequences in the DNA sample, whereas the long strands of the sample DNA remain apart because of their low concentration. The hybridized oligonucleotides then serve as primers for DNA chain synthesis in the presence of deoxynucleotides (dNTPs) and a temperature-resistant DNA polymerase such as that from *Thermus aquaticus* (a bacterium that lives in hot springs). This enzyme, called Taq *polymerase,* can remain active even after being heated to 95 °C and can extend the primers at temperatures up to 72 °C. When synthesis is complete, the whole mixture is then heated to 95 °C to denature the newly formed DNA duplexes. After the temperature is lowered again, another cycle of synthesis takes place because excess primer is still present. Repeated cycles of denaturation (heating) followed by hybridization and synthesis (cooling) quickly amplify the sequence of interest. At each cycle, the number of copies of the sequence between the primer sites is doubled; therefore, the desired sequence increases exponentially—about a million-fold after 20 cycles—whereas all other sequences in the original DNA sample remain unamplified.

Direct Isolation of a Specific Segment of Genomic DNA

For organisms in which all or most of the genome has been sequenced, PCR amplification starting with the total genomic DNA often is the easiest way to obtain a specific DNA region of interest for cloning. In this application, the two oligonucleotide primers are designed to hybridize to sequences flanking the genomic region of interest and to include sequences

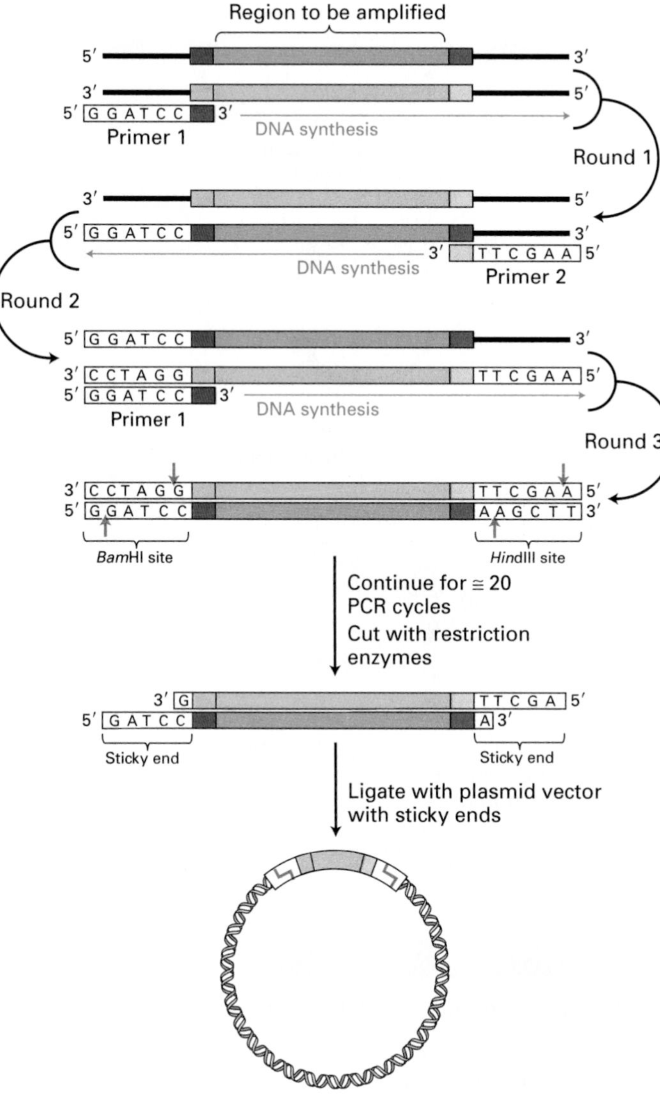

◀ **EXPERIMENTAL FIGURE 5-24 A specific target region in total genomic DNA can be amplified by PCR for use in cloning.** Each primer for PCR is complementary to one end of the target sequence and includes the recognition sequence for a restriction enzyme that does not have a site within the target region. In this example, primer 1 contains a *Bam*HI sequence, whereas primer 2 contains a *Hin*dIII sequence. (Note that for clarity, in any round, amplification of only one of the two strands is shown, the one in brackets.) After amplification, the target segments are treated with appropriate restriction enzymes, generating fragments with sticky ends. These can be incorporated into complementary plasmid vectors and cloned in *E. coli* by the usual procedure (see Figure 5-13).

that are recognized by specific restriction enzymes (Figure 5-24). After amplification of the desired target sequence for about 20 PCR cycles, cleavage with the appropriate restriction enzymes produces sticky ends that allow efficient ligation of the fragment into a plasmid vector cleaved by the same restriction enzymes in the polylinker. The resulting recombinant plasmids, all carrying the identical genomic DNA segment, can then be cloned in *E. coli* cells. With certain refinements of the PCR, even DNA segments greater than 10 kb in length can be amplified and cloned in this way.

Note that this method does not involve cloning of large numbers of restriction fragments derived from genomic DNA and their subsequent screening to identify the specific fragment of interest. In effect, the PCR method inverts this traditional approach and thus avoids its most tedious aspects. The PCR method is useful for isolating gene sequences to be manipulated in a variety of useful ways described later. In addition the PCR method can be used to isolate gene sequences from mutant organisms to determine how they differ from the wild type.

A variation on the PCR method allows PCR amplification of a specific cDNA sequence from cellular mRNAs. This method, known as *reverse transcriptase–PCR (RT-PCR)*, begins with the same procedure described previously for isolation of cDNA from a collection of cellular mRNAs. Typically, an oligo-dT primer, which will hybridize to the 3′ poly(A) tail of the mRNA, is used as the primer for the first strand of cDNA synthesis by reverse transcriptase. A specific cDNA can then be isolated from this complex mixture of cDNAs by PCR amplification using two oligonucleotide primers designed to match sequences at the 5′ and 3′ ends of the corresponding mRNA. As described previously, these primers could be designed to include restriction sites to facilitate the insertion of amplified cDNA into a suitable plasmid vector.

Preparation of Probes Earlier we discussed how oligonucleotide probes for hybridization assays can be chemically synthesized. Preparation of such probes by PCR amplification requires chemical synthesis of only two relatively short primers corresponding to the two ends of the target sequence. The starting sample for PCR amplification of the target sequence can be a preparation of genomic DNA, or a preparation of cDNA synthesized from the total cellular mRNA. To generate a radiolabeled product from PCR, ^{32}P-labeled dNTPs are included during the last several amplification cycles. Because probes prepared by PCR are relatively long and have many radioactive ^{32}P atoms incorporated into them, these probes usually give a stronger and more specific signal than chemically synthesized probes.

Tagging of Genes by Insertion Mutations Another useful application of the PCR is to amplify a "tagged" gene from the genomic DNA of a mutant strain. This approach is a simpler method for identifying genes associated with a particular mutant phenotype than screening of a library by functional complementation (see Figure 5-18).

The key to this use of the PCR is the ability to produce mutations by insertion of a known DNA sequence into the genome of an experimental organism. Such insertion mutations can be generated by use of **mobile DNA elements,** which can move (or transpose) from one chromosomal site to another. As discussed in more detail in Chapter 6, these DNA sequences occur naturally in the genomes of most organisms and may give rise to loss-of-function mutations if they transpose into a protein-coding region.

For example, researchers have modified a *Drosophila* mobile DNA element, known as the *P element,* to optimize

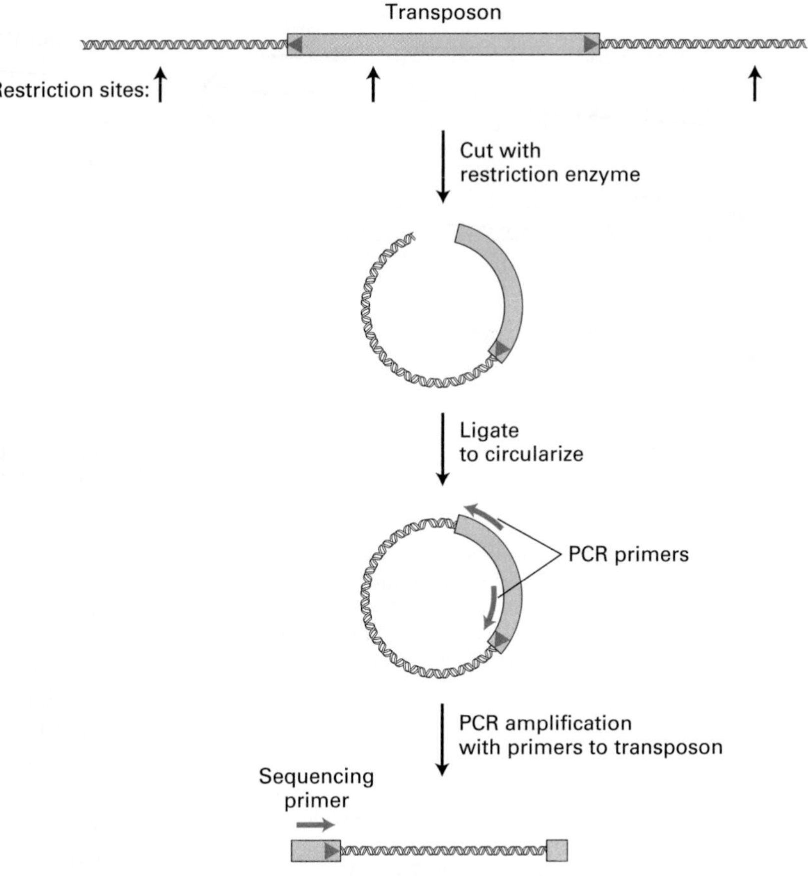

▶ **EXPERIMENTAL FIGURE 5-25 The genomic sequence at the insertion site of a transposon is revealed by PCR amplification and sequencing.** To obtain the DNA sequence of the insertion site of a P-element transposon it is necessary to PCR-amplify the junction between known transposon sequences and unknown flanking chromosomal sequences. One method to achieve this is to cleave genomic DNA with a restriction enzyme that cleaves once within the transposon sequence. Ligation of the resulting restriction fragments will generate circular DNA molecules. By using appropriately designed DNA primers that match transposon sequences it is possible to PCR-amplify the desired junction fragment. Finally, a DNA sequencing reaction (see Figure 5-21) is performed using the PCR-amplified fragment as a template and an oligonucleotide primer that matches sequences near the end of the transposon, to obtain the sequence of the junction between the transposon and chromosome.

its use in the experimental generation of insertion mutations. Once it has been demonstrated that insertion of a P element causes a mutation with an interesting phenotype, the genomic sequences adjacent to the insertion site can be amplified by a variation of the standard PCR protocol that uses synthetic primers complementary to the known P-element sequence but that allows unknown neighboring sequences to be amplified. One such method, depicted in Figure 5-25, begins by cleaving *Drosophila* genomic DNA containing a P-element insertion with a restriction enzyme that cleaves once within the P-element DNA. The collection of cleaved DNA fragments treated with DNA ligase yields circular molecules, some of which will contain P-element DNA. The chromosomal region flanking the P element can then be amplified by PCR using primers that match P-element sequences and are elongated in opposite directions. The sequence of the resulting amplified fragment can then be determined using a third DNA primer. The crucial sequence for identifying the site of P-element insertion is the junction between the end of the P-element and genomic sequences. Overall, this approach avoids the cloning of large numbers of DNA fragments and their screening to detect a cloned DNA corresponding to a mutated gene of interest.

Similar methods have been applied to other organisms for which insertion mutations can be generated using either mobile DNA elements or viruses with sequenced genomes that can insert randomly into the genome.

KEY CONCEPTS OF SECTION 5.2

DNA Cloning and Characterization

- In DNA cloning, recombinant DNA molecules are formed in vitro by inserting DNA fragments into vector DNA molecules. The recombinant DNA molecules are then introduced into host cells, where they replicate, producing large numbers of recombinant DNA molecules.
- Restriction enzymes (endonucleases) typically cut DNA at specific 4- to 8-bp palindromic sequences, producing defined fragments that often have self-complementary single-stranded tails (sticky ends).
- Two restriction fragments with complementary ends can be joined with DNA ligase to form a recombinant DNA molecule (see Figure 5-12).
- *E. coli* cloning vectors are small circular DNA molecules (plasmids) that include three functional regions: an origin of replication, a drug-resistance gene, and a site where a DNA fragment can be inserted. Transformed cells carrying a vector grow into colonies on the selection medium (see Figure 5-13).
- A cDNA library is a set of cDNA clones prepared from the mRNAs isolated from a particular type of tissue. A genomic library is a set of clones carrying restriction fragments produced by cleavage of the entire genome.
- In cDNA cloning, expressed mRNAs are reverse-transcribed into complementary DNAs, or cDNAs. By a series of reactions, single-stranded cDNAs are converted into

double-stranded DNAs, which can then be ligated into a plasmid vector (see Figure 5-15).

- A particular cloned DNA fragment within a library can be detected by hybridization to a radiolabeled oligonucleotide whose sequence is complementary to a portion of the fragment (see Figure 5-16).
- Shuttle vectors that replicate in both yeast and *E. coli* can be used to construct a yeast genomic library. Specific genes can be isolated by their ability to complement the corresponding mutant genes in yeast cells (see Figure 5-17).
- Long cloned DNA fragments often are cleaved with restriction enzymes, producing smaller fragments that are then separated by gel electrophoresis and subcloned in plasmid vectors prior to sequencing or experimental manipulation.
- DNA fragments up to about 500 nucleotides long are sequenced in automated instruments based on the Sanger (dideoxy chain-termination) method (see Figure 5-21).
- Whole genome sequences can be assembled from the sequences of a large number of overlapping clones from a genomic library (see Figure 5-22).
- The polymerase chain reaction (PCR) permits exponential amplification of a specific segment of DNA from just a single initial template DNA molecule if the sequence flanking the DNA region to be amplified is known (see Figure 5-23).
- PCR is a highly versatile method that can be programmed to amplify a specific genomic DNA sequence, a cDNA, or a sequence at the junction between a transposable element and flanking chromosomal sequences.

5.3 Using Cloned DNA Fragments to Study Gene Expression

In the last section we described the basic techniques for using recombinant DNA technology to isolate specific DNA clones, and ways in which the clones can be further characterized. Now we consider how an isolated DNA clone can be used to study gene expression. We discuss several widely used general techniques that rely on nucleic acid hybridization to elucidate when and where genes are expressed, as well as methods for generating large quantities of protein and otherwise manipulating amino acid sequences to determine their expression patterns, structure, and function. More specific applications of all these basic techniques are examined in the following sections.

Hybridization Techniques Permit Detection of Specific DNA Fragments and mRNAs

Two very sensitive methods for detecting a particular DNA or RNA sequence within a complex mixture combine separation by gel electrophoresis and hybridization with a complementary radiolabeled DNA probe. A third method involves hybridizing labeled probes directly onto a prepared tissue sample. We will encounter references to all three of these techniques, which have numerous applications, in other chapters.

Southern Blotting The first hybridization technique to detect DNA fragments of a specific sequence is known as **Southern blotting** after its originator E. M. Southern. This technique is capable of detecting a single specific restriction fragment in the highly complex mixture of fragments produced by cleavage of the entire human genome with a restriction enzyme. When such a complex mixture is subjected to gel electrophoresis, so many different fragments of nearly the same length are present it is not possible to resolve any particular DNA fragments as a discrete band on the gel. Nevertheless it is possible to identify a particular fragment migrating as a band on the gel by its ability to hybridize to a specific DNA probe. To accomplish this, the restriction fragments present in the gel are denatured with alkali and transferred onto a nitrocellulose filter or nylon membrane by blotting (Figure 5-26). This procedure preserves the distribution of the fragments in the gel, creating a replica of the gel on the filter. (The blot is used because probes do not readily diffuse into the original gel.) The filter then is incubated under hybridization conditions with a specific radiolabeled DNA probe, which usually is

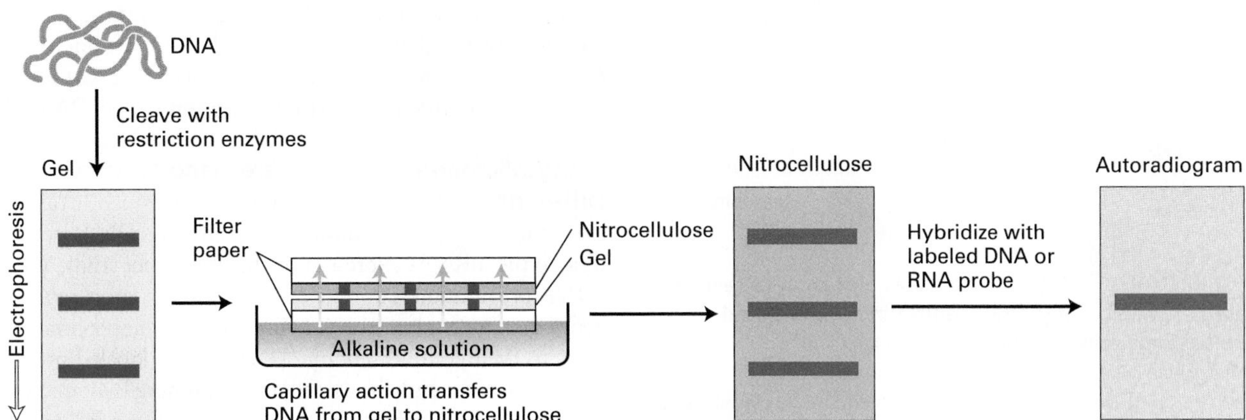

▲ EXPERIMENTAL FIGURE 5-26 **Southern blot technique can detect a specific DNA fragment in a complex mixture of restriction fragments.** The diagram depicts three different restriction fragments in the gel, but the procedure can be applied to a mixture of millions of DNA fragments. Only fragments that hybridize to a labeled probe will give a signal on an autoradiogram. A similar technique called *Northern blotting* detects specific mRNAs within a mixture. [See E. M. Southern, 1975, *J. Mol. Biol.* **98**:508.]

generated from a cloned restriction fragment. The DNA restriction fragment that is complementary to the probe hybridizes, and its location on the filter can be revealed by autoradiography.

Northern Blotting One of the most basic ways to characterize a cloned gene is to determine when and where in an organism the gene is expressed. Expression of a particular gene can be followed by assaying for the corresponding mRNA by **Northern blotting,** named, in a play on words, after the related method of Southern blotting. An RNA sample, often the total cellular RNA, is denatured by treatment with an agent such as formaldehyde that disrupts the hydrogen bonds between base pairs, ensuring that all the RNA molecules have an unfolded, linear conformation. The individual RNAs are separated according to size by gel electrophoresis and transferred to a nitrocellulose filter to which the extended denatured RNAs adhere. As in Southern blotting, the filter then is exposed to a labeled DNA probe that is complementary to the gene of interest; finally, the labeled filter is subjected to autoradiography. Because the amount of a specific RNA in a sample can be estimated from a Northern blot, the procedure is widely used to compare the amounts of a particular mRNA in cells under different conditions (Figure 5-27).

In Situ Hybridization Northern blotting requires extracting the mRNA from a cell or mixture of cells, which means that the cells are removed from their normal location within an organism or tissue. As a result, the location of a cell and its relation to its neighbors is lost. To retain such positional information in precise studies of gene expression, a whole or sectioned tissue or even a whole permeabilized embryo may be subjected to **in situ hybridization** to detect the mRNA encoded by a particular gene. This technique allows gene transcription to be monitored in both time and space (Figure 5-28).

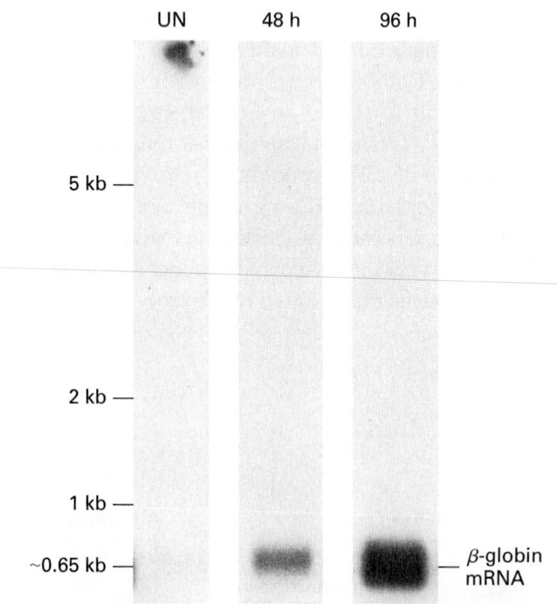

▲ EXPERIMENTAL FIGURE 5-27 **Northern blot analysis reveals increased expression of β-globin mRNA in differentiated erythroleukemia cells.** The total mRNA in extracts of erythroleukemia cells that were growing but uninduced and in cells induced to stop growing and allowed to differentiate for 48 hours or 96 hours was analyzed by Northern blotting for β-globin mRNA. The density of a band is proportional to the amount of mRNA present. The β-globin mRNA is barely detectable in uninduced cells (UN lane) but increases more than 1000-fold by 96 hours after differentiation is induced. [Courtesy of L. Kole.]

DNA Microarrays Can Be Used to Evaluate the Expression of Many Genes at One Time

Monitoring the expression of thousands of genes simultaneously is possible with **DNA microarray** analysis, another technique based on the concept of nucleic acid hybridization. A DNA microarray consists of an organized array of thousands of individual, closely packed gene-specific sequences attached to the surface of a glass microscope slide. By coupling microarray analysis with the results from genome sequencing projects, researchers can analyze the global patterns of gene expression of an organism during specific physiological responses or developmental processes.

Preparation of DNA Microarrays In one method for preparing microarrays, an ≈1-kb portion of the coding region of each gene analyzed is individually amplified by the PCR. A robotic device is used to apply each amplified DNA sample to the surface of a glass microscope slide, which then is chemically processed to permanently attach the DNA sequences to the glass surface and to denature them. A typical array might contain ≈6000 spots of DNA in a 2 × 2– cm grid.

In an alternative method, multiple DNA oligonucleotides, usually at least 20 nucleotides in length, are synthesized from an initial nucleotide that is covalently bound to the surface of a glass slide. The synthesis of an oligonucleotide of specific sequence can be programmed in a small region on the surface of the slide. Several oligonucleotide sequences from a single gene are thus synthesized in neighboring regions of the slide to analyze expression of that gene. With this method, oligonucleotides representing thousands of genes can be produced on a single glass slide. Because the methods for constructing these arrays of synthetic oligonucleotides were adapted from methods for manufacturing microscopic integrated circuits used in computers, these types of oligonucleotide microarrays are often called *DNA chips*.

Using Microarrays to Compare Gene Expression under Different Conditions The initial step in a microarray expression study is to prepare fluorescently labeled cDNAs corresponding to the mRNAs expressed by the cells under study. When the cDNA preparation is applied to a microarray, spots representing genes that are expressed will hybridize under appropriate conditions to their complementary cDNAs in the labeled probe mix, and can subsequently be detected in a scanning laser microscope.

Figure 5-29 depicts how this method can be applied to examine the changes in gene expression observed after starved human fibroblasts are transferred to a rich, serum-containing, growth medium. In this type of experiment, the separate cDNA preparations from starved and serum-grown

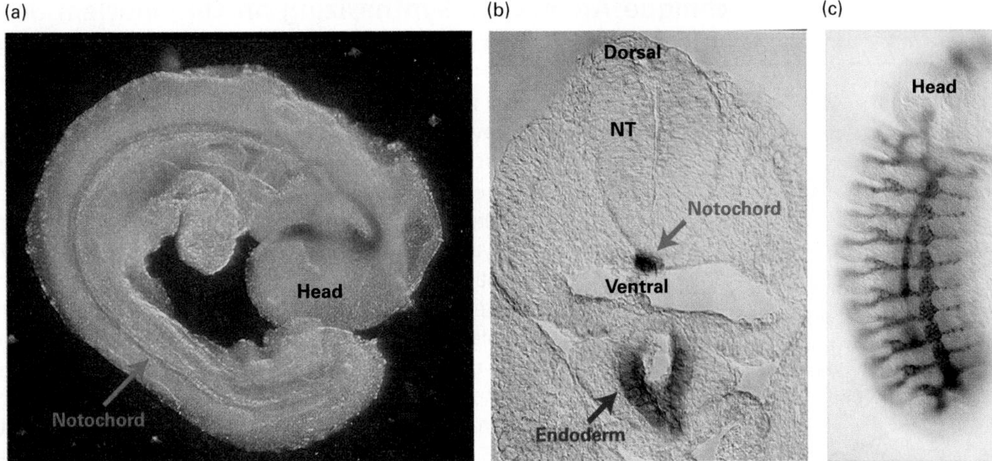

▲ EXPERIMENTAL FIGURE 5-28 **In situ hybridization can detect activity of specific genes in whole and sectioned embryos.** The specimen is permeabilized by treatment with detergent and a protease to expose the mRNA to the probe. A DNA or RNA probe, specific for the mRNA of interest, is made with nucleotide analogs containing chemical groups that can be recognized by antibodies. After the permeabilized specimen has been incubated with the probe under conditions that promote hybridization, the excess probe is removed with a series of washes. The specimen is then incubated in a solution containing an antibody that binds to the probe. This antibody is covalently joined to a reporter enzyme (e.g., horseradish peroxidase or alkaline phosphatase) that produces a colored reaction product. After excess antibody has been removed, substrate for the reporter enzyme is added. A colored precipitate forms where the probe has hybridized to the mRNA being detected. (a) A whole mouse embryo at about 10 days of development probed for *Sonic hedgehog* mRNA. The stain marks the notochord (red arrow), a rod of mesoderm running along the future spinal cord. (b) A section of a mouse embryo similar to that in part (a). The dorsal/ventral axis of the neural tube (NT) can be seen, with the *Sonic hedgehog*–expressing notochord (red arrow) below it and the endoderm (blue arrow) still farther ventral. (c) A whole *Drosophila* embryo probed for an mRNA produced during trachea development. The repeating pattern of body segments is visible. Anterior (head) is up; ventral is to the left. [Courtesy of L. Milenkovic and M. P. Scott.]

fibroblasts are labeled with differently colored fluorescent dyes. A DNA array comprising 8600 mammalian genes then is incubated with a mixture containing equal amounts of the two cDNA preparations under hybridization conditions. After unhybridized cDNA is washed away, the intensity of green and red fluorescence at each DNA spot is measured using a fluorescence microscope and stored in computer files under the name of each gene according to its known position on the slide. The relative intensities of red and green fluorescence signals at each spot are a measure of the relative level of expression of that gene in response to serum. Genes that are not transcribed under these growth conditions give no detectable signal. Genes that are transcribed at the same level under both conditions will hybridize equally to both red and green–labeled cDNA preparations. Microarray analysis of gene expression in fibroblasts showed that transcription of about 500 of the 8600 genes examined changed substantially after addition of serum.

Cluster Analysis of Multiple Expression Experiments Identifies Co-regulated Genes

Firm conclusions rarely can be drawn from a single microarray experiment about whether genes that exhibit similar changes in expression are co-regulated and hence likely to be closely related functionally. For example, many of the observed differences in gene expression just described in fibroblasts could be indirect consequences of the many different changes in cell physiology that occur when cells are transferred from one medium to another. In other words, genes that appear to be co-regulated in a single microarray expression experiment may undergo changes in expression for very different reasons and may actually have very different biological functions. A solution to this problem is to combine the information from a set of expression array experiments to find genes that are similarly regulated under a variety of conditions or over a period of time.

This more informative use of multiple expression array experiments is illustrated by examining the relative expression of the 8600 genes at different times after serum addition, generating more than 10^4 individual pieces of data. A computer program, related to the one used to determine the relatedness of different protein sequences, can organize these data and cluster genes that show similar expression over the time course after serum addition. Remarkably, such *cluster analysis* groups sets of genes whose encoded proteins participate in a common cellular process, such as cholesterol biosynthesis or the cell cycle (Figure 5-30).

In the future, microarray analysis will be a powerful diagnostic tool in medicine. For instance, particular sets of mRNAs have been found to distinguish tumors with a poor prognosis from those with a good prognosis. Previously indistinguishable disease variations are now detectable. Analysis of tumor biopsies for these distinguishing mRNAs

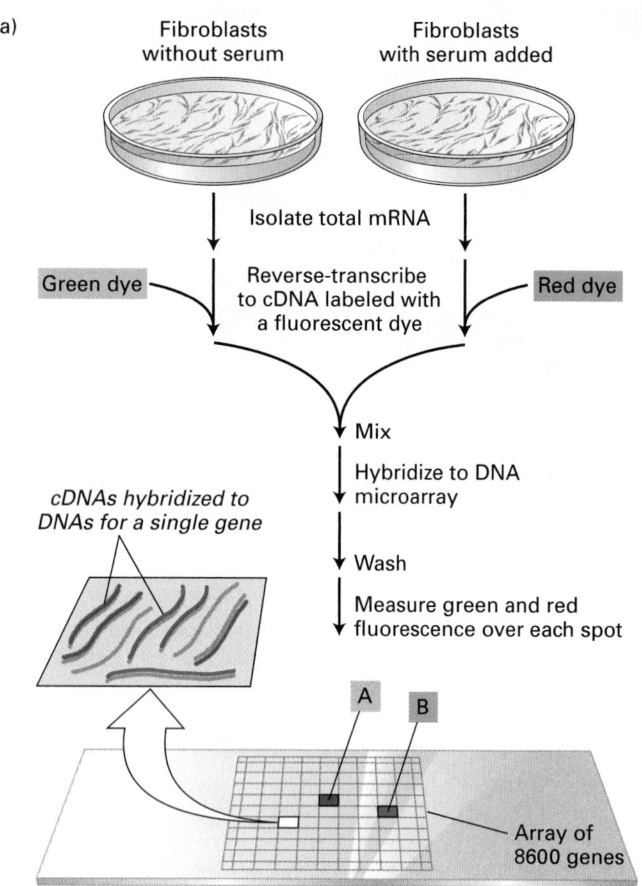
(a)
Fibroblasts without serum
Fibroblasts with serum added
Isolate total mRNA
Green dye
Reverse-transcribe to cDNA labeled with a fluorescent dye
Red dye
Mix
Hybridize to DNA microarray
cDNAs hybridized to DNAs for a single gene
Wash
Measure green and red fluorescence over each spot
A
B
Array of 8600 genes
A If a spot is green, expression of that gene decreases in cells after serum addition
B If a spot is red, expression of that gene increases in cells after serum addition

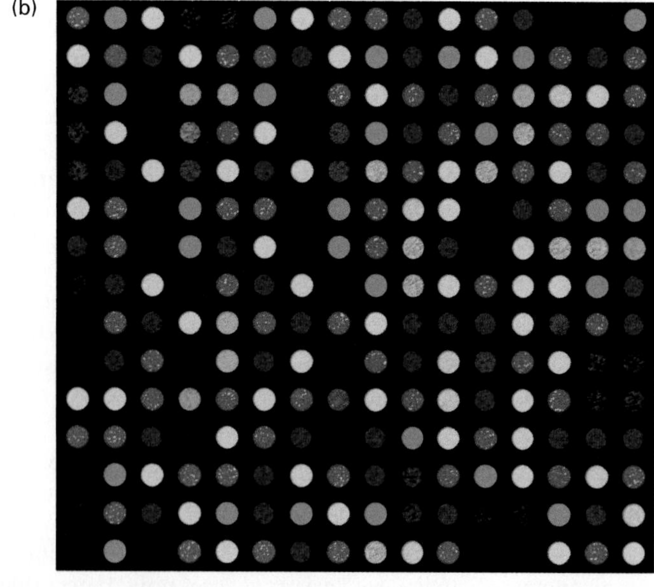
(b)

◀ EXPERIMENTAL FIGURE 5-29 **DNA microarray analysis can reveal differences in gene expression in fibroblasts under different experimental conditions.** (a) In this example, cDNA prepared from mRNA isolated from fibroblasts either starved for serum or after serum addition is labeled with different fluorescent dyes. A microarray composed of DNA spots representing 8600 mammalian genes is exposed to an equal mixture of the two cDNA preparations under hybridization conditions. The ratio of the intensities of red and green fluorescence over each spot, detected with a scanning confocal laser microscope, indicates the relative expression of each gene in response to serum. (b) A micrograph of a small segment of an actual DNA microarray. Each spot in this 16 × 16 array contains DNA from a different gene hybridized to control and experimental cDNA samples labeled with red and green fluorescent dyes. (A yellow spot indicates equal hybridization of green and red fluorescence, indicating no change in gene expression). [Part (b) Alfred Pasieka/Photo Researchers, Inc.]

will help physicians to select the most appropriate treatment. As more patterns of gene expression characteristic of various diseased tissues are recognized, the diagnostic use of DNA microarrays will be extended to other conditions. ▮

E. coli Expression Systems Can Produce Large Quantities of Proteins from Cloned Genes

Many protein hormones and other signaling or regulatory proteins are normally expressed at very low concentrations, precluding their isolation and purification in large quantities by standard biochemical techniques. Widespread therapeutic use of such proteins, as well as basic research on their structure and functions, depends on efficient procedures for producing them in large amounts at reasonable cost. Recombinant DNA techniques that turn *E. coli* cells into factories for synthesizing low-abundance proteins now are used to commercially produce granulocyte colony-stimulating factor (G-CSF), insulin, growth hormone, and other human proteins with therapeutic uses. For example, G-CSF stimulates the production of granulocytes, the phagocytic white blood cells critical to defense against bacterial infections. Administration of G-CSF to cancer patients helps offset the reduction in granulocyte production caused by chemotherapeutic agents, thereby protecting patients against serious infection while they are receiving chemotherapy. ▮

The first step in producing large amounts of a low-abundance protein is to obtain a cDNA clone encoding the full-length protein by methods discussed previously. The second step is to engineer plasmid vectors that will express large amounts of the encoded protein when it is inserted into *E. coli* cells. The key to designing such **expression vectors** is inclusion of a **promoter,** a DNA sequence from which transcription of the cDNA can begin. Consider, for example, the

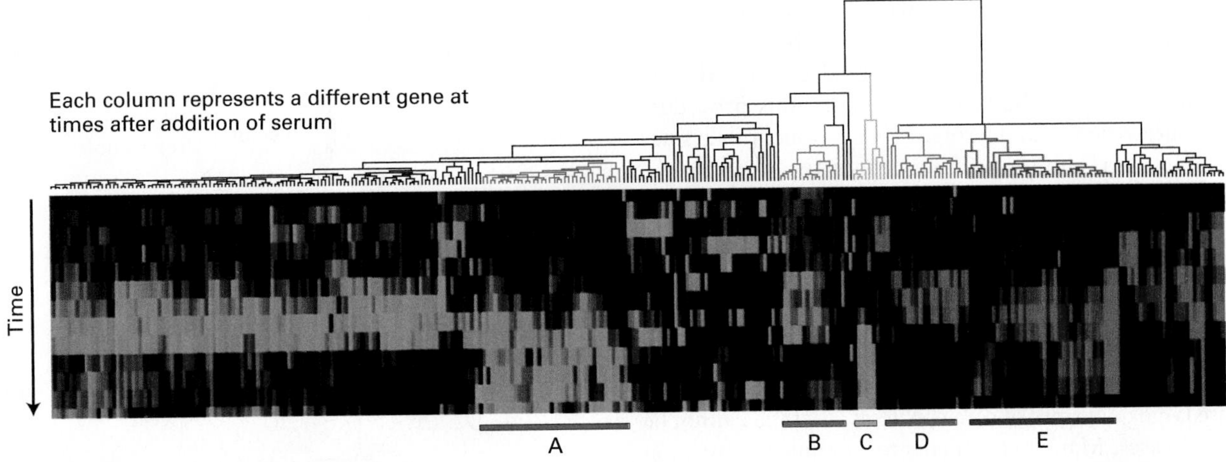

▲ EXPERIMENTAL FIGURE 5-30 **Cluster analysis of data from multiple microarray expression experiments can identify co-regulated genes.** The expression of 8600 mammalian genes was detected by microarray analysis at time intervals over a 24-hour period after serum-starved fibroblasts were provided with serum. The cluster diagram shown here is based on a computer algorithm that groups genes showing similar changes in expression compared with a serum-starved control sample over time. Each column of colored boxes represents a single gene, and each row represents a time point. A red box indicates an increase in expression relative to the control; a green box, a decrease in expression; and a black box, no significant change in expression. The "tree" diagram at the top shows how the expression patterns for individual genes can be organized in a hierarchical fashion to group together the genes with the greatest similarity in their patterns of expression over time. Five clusters of coordinately regulated genes were identified in this experiment, as indicated by the bars at the bottom. Each cluster contains multiple genes whose encoded proteins function in a particular cellular process: cholesterol biosynthesis (A), the cell cycle (B), the immediate-early response (C), signaling and angiogenesis (D), and wound healing and tissue remodeling (E). [Courtesy of Michael B. Eisen, Lawrence Berkeley National Laboratory.]

relatively simple system for expressing G-CSF shown in Figure 5-31. In this case, G-CSF is expressed in *E. coli* transformed with plasmid vectors that contain the *lac* promoter adjacent to the cloned cDNA encoding G-CSF. Transcription from the *lac* promoter occurs at high rates only when lactose, or a lactose analog such as isopropylthiogalactoside (IPTG), is added to the culture medium. Even larger quantities of a desired protein can be produced in more complicated *E. coli* expression systems.

To aid in purification of a eukaryotic protein produced in an *E. coli* expression system, researchers often modify the cDNA encoding the recombinant protein to facilitate its separation from endogenous *E. coli* proteins. A commonly used modification of this type is to add a short nucleotide sequence to the end of the cDNA, so that the expressed protein will have six histidine residues at the C-terminus. Proteins modified in this way bind tightly to an affinity matrix

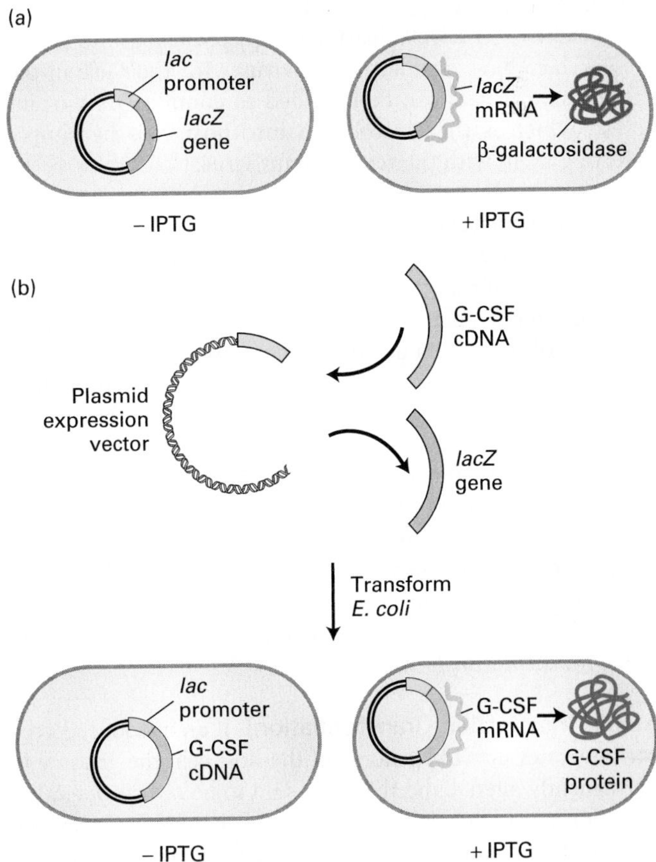

▶ EXPERIMENTAL FIGURE 5-31 **Some eukaryotic proteins can be produced in *E. coli* cells from plasmid vectors containing the *lac* promoter.** (a) The plasmid expression vector contains a fragment of the *E. coli* chromosome containing the *lac* promoter and the neighboring *lacZ* gene. In the presence of the lactose analog IPTG, RNA polymerase normally transcribes the *lacZ* gene, producing *lacZ* mRNA, which is translated into the encoded protein, β-galactosidase. (b) The *lacZ* gene can be cut out of the expression vector with restriction enzymes and replaced by a cloned cDNA, in this case one encoding granulocyte colony-stimulating factor (G-CSF). When the resulting plasmid is transformed into *E. coli* cells, addition of IPTG and subsequent transcription from the *lac* promoter produce G-CSF mRNA, which is translated into G-CSF protein.

that contains chelated nickel atoms, whereas most *E. coli* proteins will not bind to such a matrix. The bound proteins can be released from the nickel atoms by decreasing the pH of the surrounding medium. In most cases, this procedure yields a pure recombinant protein that is functional, since addition of short amino acid sequences to either the C-terminus or the N-terminus of a protein usually does not interfere with the protein's biochemical activity.

Plasmid Expression Vectors Can Be Designed for Use in Animal Cells

While bacterial expression systems can be used successfully to create large quantities of some proteins, bacteria cannot be used in all cases. Many experiments to examine the function of a protein in an appropriate cellular context require expression of a genetically modified protein in cultured animal cells. Genes are cloned into specialized eukaryotic expression vectors and are introduced into cultured animal cells by a process called **transfection.** Two common methods for transfecting animal cells differ in whether the recombinant vector DNA is or is not integrated into the host-cell genomic DNA.

In both methods, cultured animal cells must be treated to facilitate their initial uptake of the recombinant plasmid vector. This can be done by exposing cells to a preparation of lipids that penetrate the plasma membrane, increasing its permeability to DNA. Alternatively, subjecting cells to a brief electric shock of several thousand volts, a technique known as *electroporation,* makes them transiently permeable to DNA. Usually the plasmid DNA is added in sufficient concentration to ensure that a large proportion of the cultured cells will receive at least one copy of the plasmid DNA. Researchers have also harnessed viruses for their use in the laboratory; viruses can be modified to contain DNA of interest, which is then introduced into host cells by simply infecting them with the recombinant virus.

Transient Transfection The simplest of the two expression methods, called *transient transfection*, employs a vector similar to the yeast shuttle vectors described previously. For use in mammalian cells, plasmid vectors are engineered also to carry an origin of replication derived from a virus that infects mammalian cells, a strong promoter recognized by mammalian RNA polymerase, and the cloned cDNA encoding the protein to be expressed adjacent to the promoter (Figure 5-32a). Once such a plasmid vector enters a mammalian cell, the viral origin of replication allows it to replicate efficiently, generating numerous plasmids from which the protein is expressed. However, during cell division such plasmids are not faithfully segregated into both daughter cells and in time a substantial fraction of the cells in a culture will not contain a plasmid, hence the name *transient transfection*.

Stable Transfection (Transformation) If an introduced vector integrates into the genome of the host cell, the genome is permanently altered and the cell is said to be *transformed*. Integration most likely is accomplished by mammalian enzymes that normally function in DNA repair and recombination.

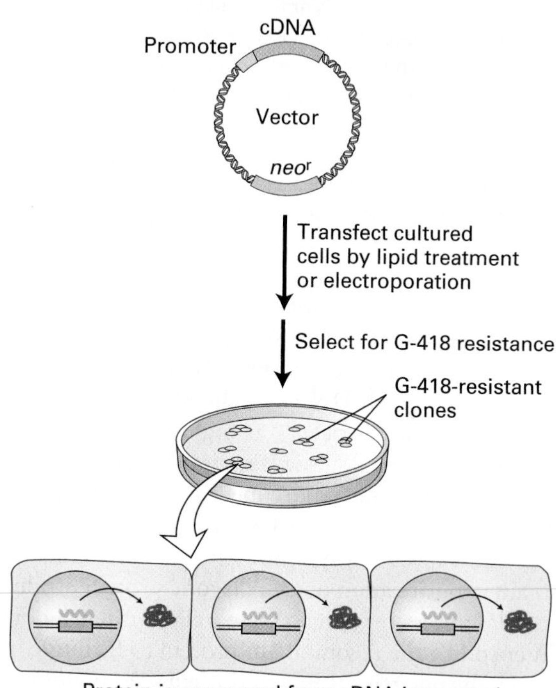

▲ EXPERIMENTAL FIGURE 5-32 **Transient and stable transfection with specially designed plasmid vectors permit expression of cloned genes in cultured animal cells.** Both methods employ plasmid vectors that contain the usual elements—ORI, selectable marker (e.g., *amp*^{r}), and polylinker—that permit propagation in *E. coli* and insertion of a cloned cDNA with an adjacent animal promoter. For simplicity, these elements are not depicted. (a) In transient transfection, the plasmid vector contains an origin of replication for a virus that can replicate in the cultured animal cells. Since the vector is not incorporated into the genome of the cultured cells, production of the cDNA-encoded protein continues only for a limited time. (b) In stable transfection, the vector carries a selectable marker such as *neo*^{r}, which confers resistance to G-418. The relatively few transfected animal cells that integrate the exogenous DNA into their genomes are selected on medium containing G-418. Because the vector is integrated into the genome, these stably transfected, or transformed, cells will continue to produce the cDNA-encoded protein as long as the culture is maintained. See the text for discussion.

▶ **EXPERIMENTAL FIGURE 5-33**
Retroviral vectors can be used for efficient integration of cloned genes into the mammalian genome. See the text for discussion.

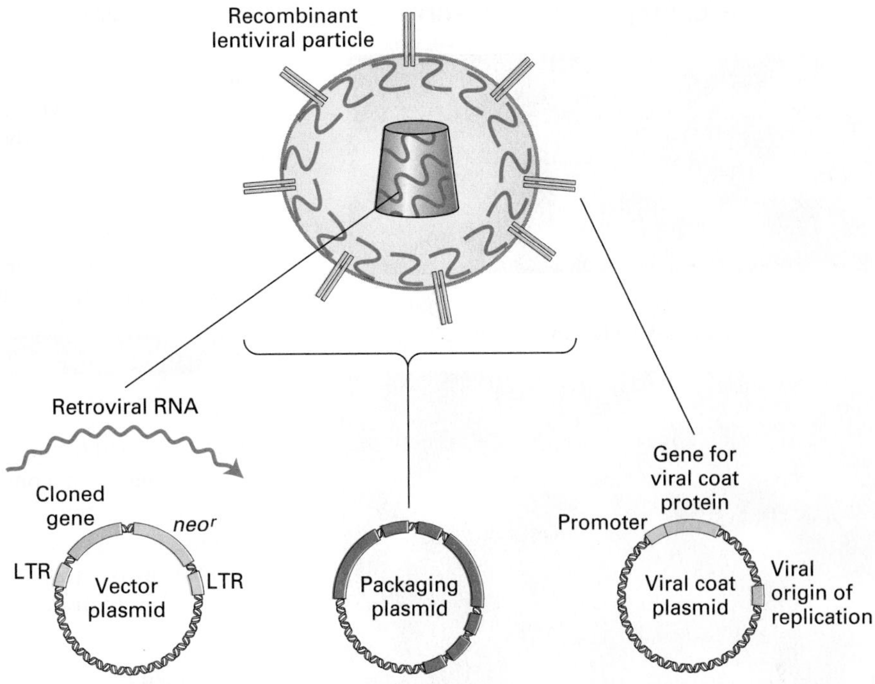

Because integration occurs at random sites in the genome, individual transformed clones resistant to G-418 will often differ in their rates of transcribing the inserted cDNA. A commonly used selectable marker is the gene for neomycin phosphotransferase (designated *neo*^{r}), which confers resistance to a toxic compound chemically related to neomycin known as G-418. The basic procedure for expressing a cloned cDNA by *stable transfection* is outlined in Figure 5-32b. Only those cells that have integrated the expression vector into the host chromosome will survive and give rise to a clone in the presence of a high concentration of G-418. Because integration occurs at random sites in the genome, individual transformed clones resistant to G-418 will differ in their rates of transcribing the inserted cDNA. Therefore, the stable transfectants usually are screened to identify those that produce the protein of interest at the highest levels.

Retroviral Expression Systems Researchers have exploited the basic mechanisms used by viruses for introduction of genetic material into animal cells and subsequent insertion into chromosomal DNA to greatly increase the efficiency by which a modified gene can be stably expressed in animal cells. An example of one such viral expression is derived from a class of retroviruses known as *lentiviruses*. As shown in Figure 5-33, three different plasmids, introduced into cells by transient transfection, are used to produce recombinant lentivirus particles suitable for efficient introduction of a cloned gene into target animal cells. The first plasmid, known as the *vector plasmid* contains a cloned gene of interest next to a selectable marker such as *neo*^{r} flanked by lentivirus LTR sequences. As described in Chapter 6, viral LTR sequences direct synthesis of a viral RNA molecule that on introduction into a virally infected target cell can be copied into DNA by reverse transcription and then be integrated into chromosomal DNA. A second plasmid known as the packaging plasmid carries all the viral genes, except for the major viral envelope protein, necessary for packaging LTR-containing viral RNA into a functional lentivirus particle. The final plasmid allows expression of a viral envelope protein that when incorporated into a recombinant lentivirus will allow the resulting hybrid virus particles to infect a desired target cell type. A common envelope protein used in this context is the glycoprotein of the vesicular stomatitis virus (VSV-G protein), which can readily replace the normal lentivirus envelope protein on the surface of completed virus particles, and will allow the resulting virus particles to infect a wide variety of mammalian cell types, including hematopoietic stem cells, neurons, muscle and liver cells. After cell infection the cloned gene flanked by viral LTR sequences is reverse-transcribed into DNA, which is transported into the nucleus and then integrated into the host genome. If necessary, as in the case for stable transfection, cells with a stably integrated cloned gene and *neo*^{r} marker can be selected for by resistance to G-418.

Gene and Protein Tagging Expression vectors can provide a way to study the expression and intracellular localization of eukaryotic proteins. This method often relies on using a reporter protein such as *green fluorescent protein (GFP)* that can conveniently be detected in cells. Here we describe two ways to create a hybrid gene that connect expression of the reporter protein to the protein of interest. When the hybrid gene is reintroduced into cells either by transfection with a plasmid expression vector containing the modified gene or by creation of a transgenic animal as described in Section 5.5, the expression of the reporter protein can be used to determine where and when a gene is expressed. This method provides similar data to in situ hybridization experiments described previously, but often with greater resolution and sensitivity.

(a) Promoter-fusion; ODR10 promoter fused to GFP

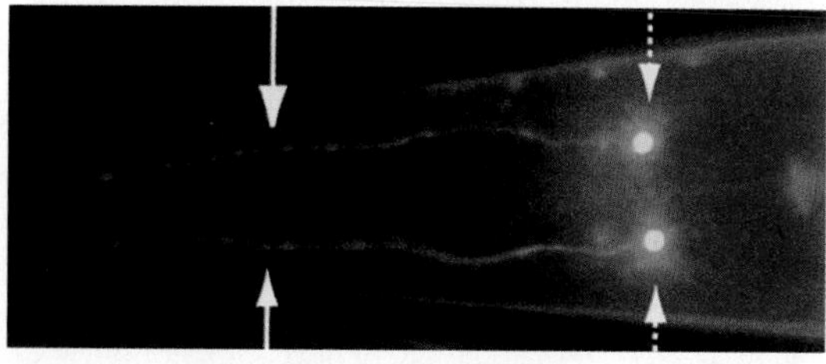

(b) Protein-fusion; ODR10-GFP fusion protein

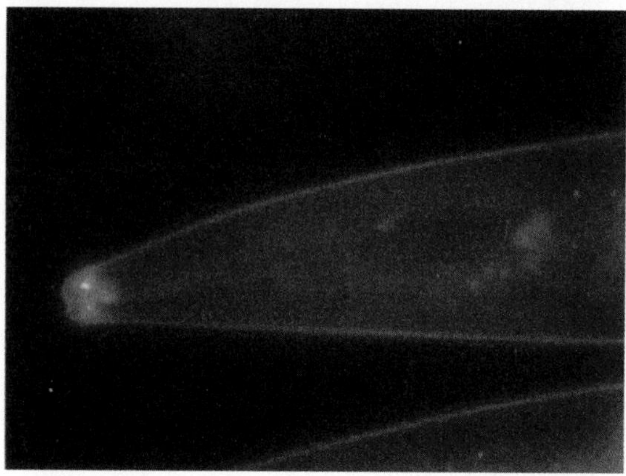

▲ EXPERIMENTAL FIGURE 5-34 **Gene and protein tagging facilitate cellular localization of proteins expressed from cloned genes.** In this experiment, the gene encoding a chemical odorant receptor, Odr10, of *C. elegans* was fused to the gene sequence for green fluorescent protein (GFP). (a) A promoter-fusion was generated by linking GFP to the promoter and the first four amino acid codons of Odr10. This protein is expressed in the cytoplasm of specific sensory neurons in the head of *C. elegans*. Note that the cell body (dashed arrow) and sensory dendrites (solid arrow) are fluorescently labeled. (b) A protein-fusion was constructed by linking GFP to the end of the full-length Odr10 coding sequence. In this case the Odr10-GFP fusion protein is targeted to the membrane at the tip of the sensory neurons and is only apparent at the distal end of the sensory cilia. The observed distribution can be inferred to reflect the normal location of Odr10 protein in specific neurons. [P. Sengupta et al., 1996, *Cell* **84**:899 (derived from Figures 4 and 5).]

Figure 5-34 illustrates the use of two different types of GFP-tagging experiments to study the expression of an odorant receptor protein in *C. elegans*. When the promoter for the odorant receptor is linked directly to the coding sequence of GFP in a configuration usually known as a *promoter-fusion*, GFP is expressed in specific neurons, filling the cytoplasm of those neurons. In contrast, when the hybrid gene is constructed by linking GFP to the coding sequence of the receptor, the resulting *protein-fusion* can be localized by GFP fluorescence at the distal cilia in sensory neurons, the site at which the receptor protein is normally located.

An alternative to GFP tagging for detecting the intracellular location of a protein is to modify the gene of interest by fusing it with a short DNA sequence that encodes a short stretch of amino acids recognized by a known monoclonal antibody. Such a short peptide that can be bound by an antibody is called an **epitope**; hence, this method is known as *epitope tagging*. After transfection with a plasmid expression vector containing the modified gene, the expressed epitope-tagged form of the protein can be detected by immunofluorescence-labeling of the cells with the monoclonal antibody specific for the epitope. The choice of whether to use a short epitope or GFP to tag a given protein often depends on what types of modification a cloned gene can tolerate and still remain functional.

KEY CONCEPTS OF SECTION 5.3

Using Cloned DNA Fragments to Study Gene Expression

- Southern blotting can detect a single, specific DNA fragment within a complex mixture by combining gel electrophoresis, transfer (blotting) of the separated bands to a filter, and hybridization with a complementary radiolabeled DNA probe (see Figure 5-26). The similar technique of Northern blotting detects a specific RNA within a mixture.
- The presence and distribution of specific mRNAs can be detected in living cells by in situ hybridization.
- DNA microarray analysis simultaneously detects the relative level of expression of thousands of genes in different types of cells or in the same cells under different conditions (see Figure 5-29).
- Cluster analysis of the data from multiple microarray expression experiments can identify genes that are similarly regulated under various conditions. Such co-regulated genes commonly encode proteins that have biologically related functions.
- Expression vectors derived from plasmids allow the production of abundant amounts of a protein from a cloned gene.
- Eukaryotic expression vectors can be used to express cloned genes in yeast or mammalian cells. An important application of these methods is the tagging of proteins with GFP or an epitope for antibody detection.

5.4 Identifying and Locating Human Disease Genes

Inherited human diseases are the phenotypic consequence of defective human genes. Table 5-2 lists several of the most commonly occurring inherited diseases. Although a "disease" gene may result from a new mutation that arose in the preceding generation, most cases of inherited diseases are caused by preexisting mutant alleles that have been passed from one generation to the next for many generations. ▮

Nowadays, the typical first step in deciphering the underlying cause for any inherited human disease is to identify the affected gene and its encoded protein. Comparison of the sequences of a disease gene and its product with those of genes and proteins whose sequence and function are known can

TABLE 5-2 Common Inherited Human Diseases

DISEASE	MOLECULAR AND CELLUAR DEFECT	INCIDENCE
AUTOSOMAL RECESSIVE		
Sickle-cell anemia	Abnormal hemoglobin causes deformation of red blood cells, which can become lodged in capillaries; also confers resistance to malaria.	1/625 of sub-Saharan African origin
Cystic fibrosis	Defective chloride channel (CFTR) in epithelial cells leads to excessive mucus in lungs.	1/2500 of European origin
Phenylketonuria (PKU)	Defective enzyme in phenylalanine metabolism (tyrosine hydroxylase) results in excess phenylalanine, leading to mental retardation, unless restricted by diet.	1/10,000 of European origin
Tay-Sachs disease	Defective hexosaminidase enzyme leads to accumulation of excess sphingolipids in the lysosomes of neurons, impairing neural development.	1/1000 eastern European Jews
AUTOSOMAL DOMINANT		
Huntington's disease	Defective neural protein (huntingtin) may assemble into aggregates causing damage to neural tissue.	1/10,000 of European origin
Hypercholesterolemia	Defective LDL receptor leads to excessive cholesterol in blood and early heart attacks.	1/122 French Canadians
X-LINKED RECESSIVE		
Duchenne muscular dystrophy (DMD)	Defective cytoskeletal protein dystrophin leads to impaired muscle function.	1/3500 males
Hemophilia A	Defective blood clotting factor VIII leads to uncontrolled bleeding.	1–2/10,000 males

provide clues to the molecular and cellular cause of the disease. Historically, researchers have used whatever phenotypic clues might be relevant to make guesses about the molecular basis of inherited diseases. An early example of successful guesswork was the hypothesis that sickle-cell anemia, known to be a disease of blood cells, might be caused by defective hemoglobin. This idea led to identification of a specific amino acid substitution in hemoglobin that causes polymerization of the defective hemoglobin molecules, causing the sickle-like deformation of red blood cells in individuals who have inherited two copies of the Hb^s allele for sickle-cell hemoglobin.

Most often, however, the genes responsible for inherited diseases must be found without any prior knowledge or reasonable hypotheses about the nature of the affected gene or its encoded protein. In this section, we will see how human geneticists can find the gene responsible for an inherited disease by following the segregation of the disease in families. The segregation of the disease can be correlated with the segregation of many other genetic markers, eventually leading to identification of the chromosomal position of the affected gene. This information, along with knowledge of the sequence of the human genome, can ultimately allow the affected gene and the disease-causing mutations to be pinpointed.

Many Inherited Diseases Show One of Three Major Patterns of Inheritance

Human genetic diseases that result from mutation in one specific gene exhibit several inheritance patterns depending on the nature and chromosomal location of the alleles that cause them. One characteristic pattern is that exhibited by a dominant allele in an **autosome** (that is, one of the 22 human chromosomes that is not a sex chromosome). Because an *autosomal dominant* allele is expressed in the heterozygote, usually at least one of the parents of an affected individual will also have the disease. It is often the case that the diseases caused by dominant alleles appear later in life after the reproductive age. If this were not the case, natural selection would have eliminated the allele during human evolution. An example of an

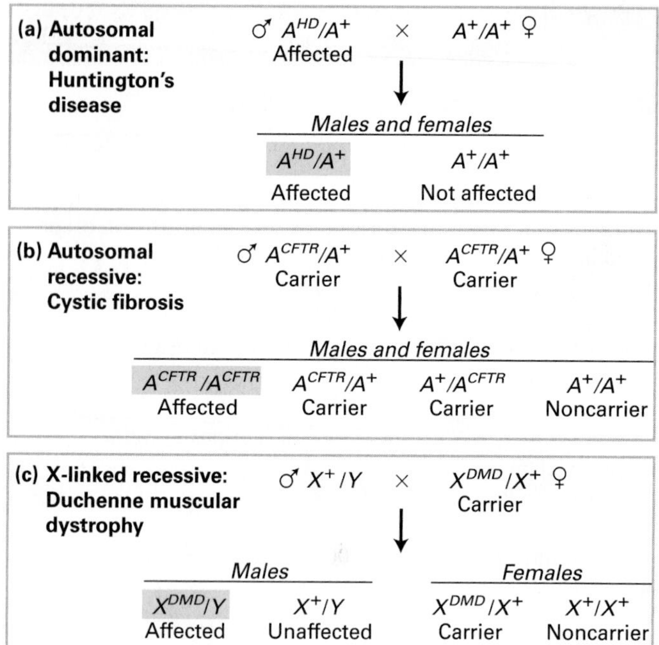

▲ **FIGURE 5-35 Three common inheritance patterns for human genetic diseases.** Wild-type autosomal (A) and sex chromosomes (X and Y) are indicated by superscript plus signs. (a) In an autosomal dominant disorder such as Huntington's disease, only one mutant allele is needed to confer the disease. If either parent is heterozygous for the mutant *HD* allele, his or her children have a 50 percent chance of inheriting the mutant allele and getting the disease. (b) In an autosomal recessive disorder such as cystic fibrosis, two mutant alleles must be present to confer the disease. Both parents must be heterozygous carriers of the mutant *CFTR* gene for their children to be at risk of being affected or being carriers. (c) An X-linked recessive disease such as Duchenne muscular dystrophy is caused by a recessive mutation on the X chromosome and exhibits the typical sex-linked segregation pattern. Males born to mothers heterozygous for a mutant *DMD* allele have a 50 percent chance of inheriting the mutant allele and being affected. Females born to heterozygous mothers have a 50 percent chance of being carriers.

autosomal dominant disease is Huntington's disease, a neural degenerative disease that generally strikes in mid- to late life. If either parent carries a mutant *HD* allele, each of his or her children (regardless of sex) has a 50 percent chance of inheriting the mutant allele and being affected (Figure 5-35a).

A recessive allele in an autosome exhibits a quite different segregation pattern. For an *autosomal recessive* allele, both parents must be heterozygous *carriers* of the allele in order for their children to be at risk of being affected with the disease. Each child of heterozygous parents has a 25 percent chance of receiving both recessive alleles and thus being affected, a 50 percent chance of receiving one normal and one mutant allele and thus being a carrier, and a 25 percent chance of receiving two normal alleles. A clear example of an autosomal recessive disease is cystic fibrosis, which results from a defective chloride-channel gene known as *CFTR* (Figure 5-35b). Related individuals (e.g., first or second cousins) have a relatively high probability of being carriers for the same recessive alleles. Thus children born to related parents are much more likely than those born to unrelated parents to be homozygous for, and therefore affected by, an autosomal recessive disorder.

The third common pattern of inheritance is that of an *X-linked recessive* allele. A recessive allele on the X chromosome will most often be expressed in males, who receive only one X chromosome from their mother, but not in females, who receive an X chromosome from both their mother and their father. This leads to a distinctive sex-linked segregation pattern where the disease is exhibited much more frequently in males than in females. For example, Duchenne muscular dystrophy (DMD), a muscle degenerative disease that specifically affects males, is caused by a recessive allele on the X chromosome. DMD exhibits the typical sex-linked segregation pattern in which mothers who are heterozygous and therefore phenotypically normal can act as carriers, transmitting the DMD allele, and therefore the disease, to 50 percent of their male progeny (Figure 5-35c).

DNA Polymorphisms Are Used in Linkage-Mapping Human Mutations

Once the mode of inheritance has been determined, the next step in determining the position of a disease allele is to genetically map its position with respect to known genetic markers using the basic principle of genetic linkage as described in Section 5.1. The presence of many different already mapped genetic traits, or markers, distributed along the length of a chromosome facilitates the mapping of a new mutation by assessing its possible linkage to these marker genes in appropriate crosses. The more markers that are available, the more precisely a mutation can be mapped. The density of genetic markers needed for a high-resolution human genetic map is about one marker every 5 centimorgans (cM) (as discussed previously, one genetic map unit, or centimorgan, is defined as the distance between two positions along a chromosome that results in one recombinant individual in 100 progeny). Thus a high-resolution genetic map requires 25 or so genetic markers of known position spread along the length of each human chromosome.

In the experimental organisms commonly used in genetic studies, numerous markers with easily detectable phenotypes are readily available for genetic mapping of mutations. This is not the case for mapping genes whose mutant alleles are associated with inherited diseases in humans. However, recombinant DNA technology has made available a wealth of useful **DNA-based molecular markers.** Because most of the human genome does not code for protein, a large amount of sequence variation exists between individuals. Indeed, it has been estimated that nucleotide differences between unrelated individuals can be detected on an average of every 10^3 nucleotides. If these variations in DNA sequence, referred to as *DNA polymorphisms,* can be followed from one generation to the next, they can serve as genetic markers for linkage studies. Currently, a panel of as many as 10^4 different known polymorphisms whose locations have been mapped in the human genome is used for genetic linkage studies in humans.

Restriction fragment length polymorphisms (**RFLPs**) were the first type of molecular markers used in linkage studies. RFLPs arise because mutations can create or destroy the sites recognized by specific restriction enzymes that happen to lie in human DNA, leading to variations between individuals in the length of restriction fragments produced from identical regions of the genome. Differences in the sizes of restriction fragments between individuals can be detected by Southern blotting with a probe specific for a region of DNA known to contain an RFLP (Figure 5-36a). The segregation and meiotic recombination of such DNA polymorphisms can be followed like typical genetic markers. Figure 5-36b illustrates how RFLP analysis of a family can detect the segregation of an RFLP that can be used to test for statistically significant linkage to the allele for an inherited disease or some other human trait of interest.

The amassed genomic sequence information from different humans has led to identification of other useful DNA polymorphisms in recent years. *Single-nucleotide polymorphisms* (*SNPs*) constitute the most abundant type and are therefore useful for constructing genetic maps of maximum resolution. Another useful type of DNA polymorphism consists of a variable number of repetitions of a one- two-, or three-base sequence. Such polymorphisms, known as simple sequence repeats (SSRs) or *microsatellites*, presumably are formed by recombination or a slippage mechanism of either the template or newly synthesized strands during DNA replication. A useful property of SSRs is that different individuals will often have different numbers of repeats. The existence of multiple versions of an SSR makes it more likely to produce an informative segregation pattern in a given pedigree and therefore be of more general use in mapping the positions of disease genes. If an SNP or SSR alters a restriction site, it can be detected by RFLP analysis. More commonly, however, these polymorphisms do not alter restriction fragments and must be detected by PCR amplification and DNA sequencing.

Linkage Studies Can Map Disease Genes with a Resolution of About 1 Centimorgan

Without going into all the technical considerations, let's see how the allele conferring a particular dominant trait (e.g., familial hypercholesterolemia) might be mapped. The first step is to obtain DNA samples from all the members of a family containing individuals that exhibit the disease. The DNA from each affected and unaffected individual then is analyzed to determine the identity of a large number of known DNA polymorphisms (either SSR or SNP markers can be used). The segregation pattern of each DNA polymorphism within the family is then compared with the segregation of the disease under study to find those polymorphisms that tend to segregate along with the disease. Finally, computer analysis of the segregation data is used to

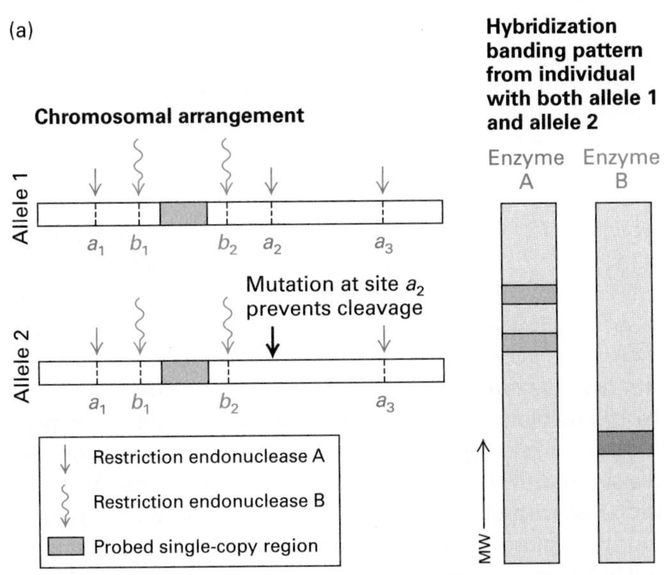

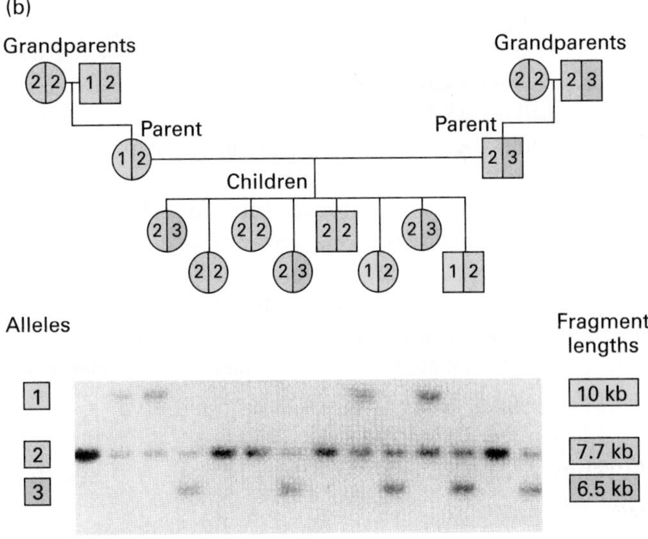

▲ **EXPERIMENTAL FIGURE 5-36 Restriction fragment length polymorphisms (RFLPs) can be followed like genetic markers.** (a) In the two homologous chromosomes shown, DNA is treated with two different restriction enzymes (A and B), which cut DNA at different sequences (*a* and *b*). The resulting fragments are subjected to Southern blot analysis (see Figure 5-26) with a radioactive probe that binds to the indicated DNA region (green) to detect the fragments. Since no differences between the two homologous chromosomes occur in the sequences recognized by the B enzyme, only one fragment is recognized by the probe, as indicated by a single hybridization band. However, treatment with enzyme A produces radiographically distinct fragments of two different lengths (a_1–a_2 and a_1–a_3), and two bands are seen, indicating that a mutation has caused the loss of one of the *a* sites in one of the two chromosomes. (b) Pedigree based on RFLP analysis of the DNA from a region known to be present on chromosome 5. The DNA samples were cut with the restriction enzyme *TaqI* and analyzed by Southern blotting. In this family, this region of the genome exists in three allelic forms characterized by *TaqI* sites spaced 10, 7.7, or 6.5 kb apart. Each individual has two alleles; some contain allele 2 (7.7 kb) on both chromosomes, and others are heterozygous at this site. Circles indicate females; squares indicate males. The gel lanes are in the same order as the subjects in the pedigree above and are generally aligned below them. [After H. Donis-Keller et al., 1987, *Cell* **51**:319.]

calculate the likelihood of linkage between each DNA polymorphism and the disease-causing allele.

In practice, segregation data are collected from different families exhibiting the same disease and pooled. The more families exhibiting a particular disease that can be examined, the greater the statistical significance of evidence for linkage that can be obtained and the greater the precision with which the distance can be measured between a linked DNA polymorphism and a disease allele. Most family studies have a maximum of about 100 individuals in which linkage between a disease gene and a panel of DNA polymorphisms can be tested. This number of individuals sets the practical upper limit on the resolution of such a mapping study to about 1 centimorgan, or a physical distance of about 7.5×10^5 base pairs.

A phenomenon called *linkage disequilibrium* is the basis for an alternative strategy, which in some cases can afford a higher degree of resolution in mapping studies. This approach depends on the particular circumstance in which a genetic disease commonly found in a particular population results from a single mutation that occurred many generations in the past. The DNA polymorphisms carried by this ancestral chromosome are collectively known as the *haplotype* of that chromosome. As the disease allele is passed from one generation to the next, only the polymorphisms that are closest to the disease gene will not be separated from it by recombination. After many generations the region that contains the disease gene will be evident because this will be the only region of the chromosome that will carry the haplotype of the ancestral chromosome conserved through many generations (Figure 5-37). By assessing the distribution of specific markers in all the affected individuals in a population, geneticists can identify DNA markers tightly associated with the disease, thus localizing the disease-associated gene to a relatively small region. Under ideal circumstances linkage disequilibrium studies can improve the resolution of mapping studies to less than 0.1 centimorgan. The resolving power of this method comes from the ability to determine whether a polymorphism and the disease allele were ever separated by a meiotic recombination event at any time since the disease allele first appeared on the ancestral chromosome—in some cases this can amount to finding markers that are so closely linked to the disease gene that even after hundreds of meioses they have never been separated by recombination.

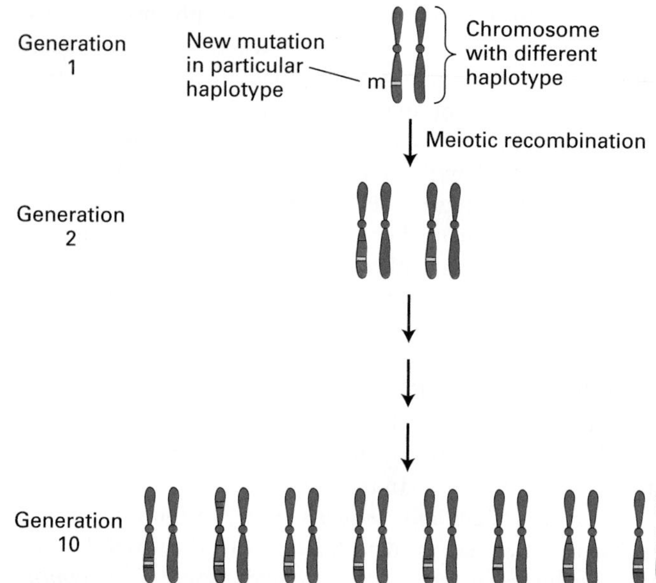

▲ **FIGURE 5-37 Linkage disequilibrium studies of human populations can be used to map genes at high resolution.** A new disease mutation will arise in the context of an ancestral chromosome among a set of polymorphisms known as the *haplotype* (indicated by pink shading). After many generations, chromosomes that carry the disease mutation will also carry segments of the ancestral haplotype that have not been separated from the disease mutation by recombination. The blue segments of these chromosomes represent general haplotypes derived from the general population and not from the ancestral haplotype in which the mutation originally arose. This phenomenon is known as *linkage disequilibrium*. The position of the disease mutation can be located by scanning chromosomes containing the disease mutation for highly conserved polymorphisms corresponding to the ancestral haplotype.

Further Analysis Is Needed to Locate a Disease Gene in Cloned DNA

Although linkage mapping can usually locate a human disease gene to a region containing about 10^5 base pairs, as many as 10 different genes may be located in a region of this size. The ultimate objective of a mapping study is to locate the gene within a cloned segment of DNA and then to determine the nucleotide sequence of this fragment. The relative scales of a chromosomal genetic map and physical maps corresponding to ordered sets of plasmid clones and the nucleotide sequence are shown in Figure 5-38.

One strategy for further localizing a disease gene within the genome is to identify mRNA encoded by DNA in the region of the gene under study. Comparison of gene expression in tissues from normal and affected individuals may suggest tissues in which a particular disease gene normally is expressed. For instance, a mutation that phenotypically affects muscle, but no other tissue, might be in a gene that is expressed only in muscle tissue. The expression of mRNA in both normal and affected individuals generally is determined by Northern blotting or in situ hybridization of labeled DNA or RNA to tissue sections. Northern blots, in situ hybridization, or microarray experiments permit comparison of both the level of expression and the size of mRNAs in mutant and wild-type tissues. Although the sensitivity of in situ hybridization is lower than that of Northern blot analysis, it can be very helpful in identifying an mRNA that is expressed at low levels in a given tissue but at very high levels in a subclass of cells within that tissue. An mRNA that is altered or missing in various individuals affected with a disease compared with wild-type individuals would be an excellent candidate for encoding the protein whose disrupted function causes that disease.

In many cases, point mutations that give rise to disease-causing alleles may result in no detectable change in the level of expression or electrophoretic mobility of mRNAs. Thus if comparison of the mRNAs expressed in normal and affected individuals reveals no detectable differences in the candidate mRNAs, a search for point mutations in the

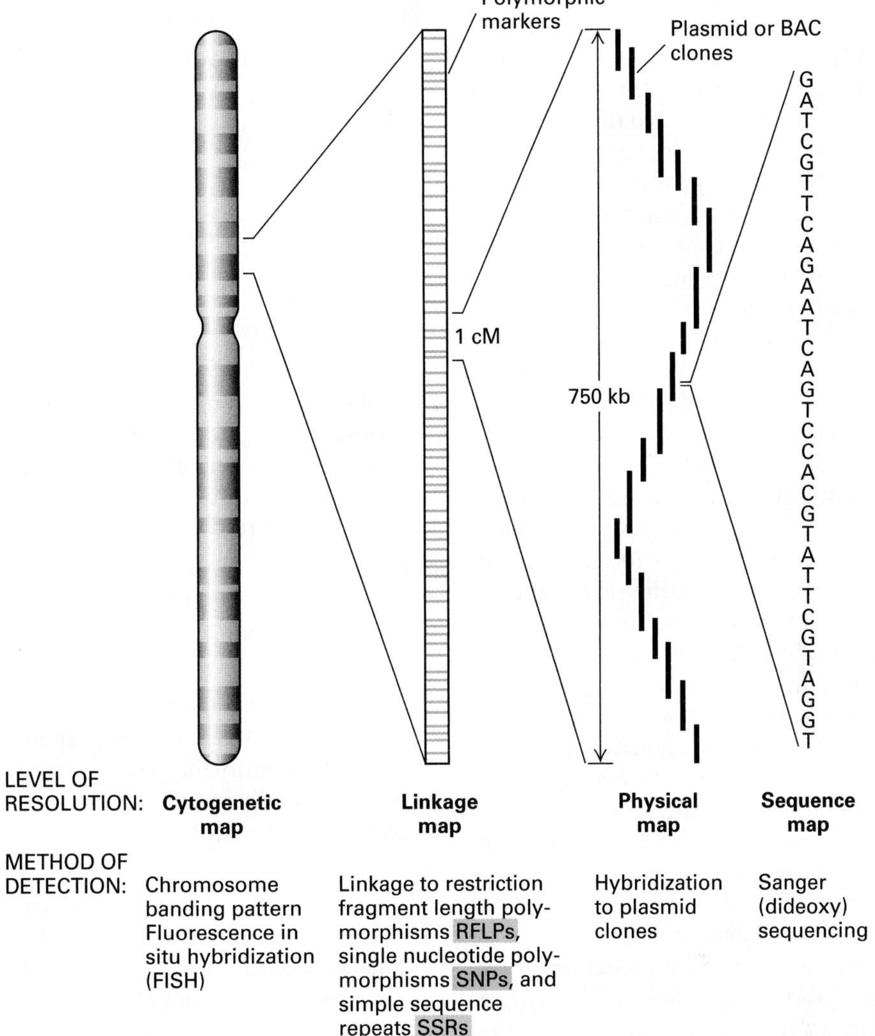

▲ **FIGURE 5-38 The relationship between the genetic and physical maps of a human chromosome.** The diagram depicts a human chromosome analyzed at different levels of detail. The chromosome as a whole can be viewed in the light microscope when it is in a condensed state that occurs at metaphase, and the approximate location of specific sequences can be determined by fluorescence *in situ* hybridization (FISH). At the next level of detail, genetic traits can be mapped relative to DNA-based genetic markers. Local segments of the chromosome can be analyzed at the level of DNA sequences identified by Southern hybridization or PCR. Finally, important genetic differences can be most precisely defined by differences in the nucleotide sequence of the chromosomal DNA. [Adapted from L. Hartwell et al., 2003, *Genetics: From Genes to Genomes*, 2d ed., McGraw Hill.]

DNA regions encoding the mRNAs is undertaken. Now that highly efficient methods for sequencing DNA are available, researchers frequently determine the sequence of candidate regions of DNA isolated from affected individuals to identify point mutations. The overall strategy is to search for a coding sequence that consistently shows possibly deleterious alterations in DNA from individuals that exhibit the disease. A limitation of this approach is that the region near the affected gene may carry naturally occurring polymorphisms unrelated to the gene of interest. Such polymorphisms, not functionally related to the disease, can lead to misidentification of the DNA fragment carrying the gene of interest. For this reason, the more mutant alleles available for analysis, the more likely that a gene will be correctly identified.

Many Inherited Diseases Result from Multiple Genetic Defects

Most of the inherited human diseases that are now understood at the molecular level are *monogenetic traits;* that is, a clearly discernible disease state is produced by a defect in a single gene. Monogenic diseases caused by mutation in one specific gene exhibit one of the characteristic inheritance patterns shown in Figure 5-35. The genes associated with most of the common monogenic diseases have already been mapped using DNA-based markers as described previously.

However, many other inherited diseases show more complicated patterns of inheritance, making the identification of the underlying genetic cause much more difficult.

One type of added complexity that is frequently encountered is *genetic heterogeneity.* In such cases, mutations in any one of multiple different genes can cause the same disease. For example, retinitis pigmentosa, which is characterized by degeneration of the retina usually leading to blindness, can be caused by mutations in any one of more than 60 different genes. In human linkage studies, data from multiple families usually must be combined to determine whether a statistically significant linkage exists between a disease gene and known molecular markers. Genetic heterogeneity such as that exhibited by retinitis pigmentosa can confound such an approach because any statistical trend in the mapping data from one family tends to be canceled out by the data obtained from another family with an unrelated causative gene.

Human geneticists used two different approaches to identify the many genes associated with retinitis pigmentosa. The first approach relied on mapping studies in exceptionally large single families that contained a sufficient number of affected individuals to provide statistically significant evidence for linkage between known DNA polymorphisms and a single causative gene. The genes identified in such studies showed that several of the mutations that cause retinitis pigmentosa lie within genes that encode abundant proteins of the retina. Following up on this clue, geneticists concentrated their attention on those genes that are highly expressed in the retina when screening other individuals with retinitis pigmentosa. This approach of using additional information to focus screening efforts on a subset of candidate genes led to identification of additional rare causative mutations in many different genes encoding retinal proteins.

A further complication in the genetic dissection of human diseases is posed by diabetes, heart disease, obesity, predisposition to cancer, and a variety of mental disorders that have at least some heritable properties. These and many other diseases can be considered to be *polygenic traits* in the sense that alleles of multiple genes, acting together within an individual, contribute to both the occurrence and the severity of disease. A systematic solution to the problem of mapping complex polygenic traits in humans does not yet exist. Future progress may come from development of refined diagnostic methods that can distinguish the different forms of diseases resulting from multiple causes.

Models of human disease in experimental organisms may also contribute to unraveling the genetics of complex traits such as obesity or diabetes. For instance, large-scale controlled breeding experiments in mice can identify mouse genes associated with diseases analogous to those in humans. The human orthologs of the mouse genes identified in such studies would be likely candidates for involvement in the corresponding human disease. DNA from human populations then could be examined to determine if particular alleles of the candidate genes show a tendency to be present in individuals affected with the disease but absent from unaffected individuals. This "candidate gene" approach is currently being used intensively to search for genes that may contribute to the major polygenic diseases in humans.

KEY CONCEPTS OF SECTION 5.4

Identifying and Locating Human Disease Genes

- Inherited diseases and other traits in humans show three major patterns of inheritance: autosomal dominant, autosomal recessive, and X-linked recessive (see Figure 5-35).
- Genes for human diseases and other traits can be mapped by determining their cosegregation during meiosis with markers whose locations in the genome are known. The closer a gene is to a particular marker, the more likely they are to cosegregate.
- Mapping of human genes with great precision requires thousands of molecular markers distributed along the chromosomes. The most useful markers are differences in the DNA sequence (polymorphisms) between individuals in noncoding regions of the genome.
- DNA polymorphisms useful in mapping human genes include restriction fragment length polymorphisms (RFLPs), single-nucleotide polymorphisms (SNPs), and simple sequence repeats (SSRs).
- Linkage mapping often can locate a human disease gene to a chromosomal region that includes as many as 10 genes. To identify the gene of interest within this candidate region typically requires expression analysis and comparison of DNA sequences between wild-type and disease-affected individuals.
- Some inherited diseases can result from mutations in different genes in different individuals (genetic heterogeneity). The occurrence and severity of other diseases depend on the presence of mutant alleles of multiple genes in the same individuals (polygenic traits). Mapping of the genes associated with such diseases is particularly difficult because the occurrence of the disease cannot readily be correlated to a single chromosomal locus.

5.5 Inactivating the Function of Specific Genes in Eukaryotes

The elucidation of DNA and protein sequences in recent years has led to identification of many genes, using sequence patterns in genomic DNA and the sequence similarity of the encoded proteins with proteins of known function. As discussed in Chapter 6, the general functions of proteins identified by sequence searches may be predicted by analogy with known proteins. However, the precise in vivo roles of such "new" proteins may be unclear in the absence of mutant forms of the corresponding genes. In this section, we describe several ways for disrupting the normal function of a specific gene in the genome of an organism. Analysis of the resulting mutant phenotype often helps reveal the in vivo function of the normal gene and its encoded protein.

Three basic approaches underlie these gene-inactivation techniques: (1) replacing a normal gene with other sequences, (2) introducing an allele whose encoded protein inhibits functioning of the expressed normal protein, and (3) promoting destruction of the mRNA expressed from a gene. The normal endogenous gene is modified in techniques based on the first approach but is not modified in the other approaches.

Normal Yeast Genes Can Be Replaced with Mutant Alleles by Homologous Recombination

Modifying the genome of the yeast *S. cerevisiae* is particularly easy for two reasons: yeast cells readily take up exogenous DNA under certain conditions, and the introduced DNA is efficiently exchanged for the homologous chromosomal site in the recipient cell. This specific, targeted recombination of identical stretches of DNA allows any gene in yeast chromosomes to be replaced with a mutant allele. (As we discuss in Section 5.1, recombination between homologous chromosomes also occurs naturally during meiosis.)

In one popular method for disrupting yeast genes in this fashion, the PCR is used to generate a *disruption construct* containing a selectable marker that subsequently is transfected into yeast cells. As shown in Figure 5-39a, primers for PCR amplification of the selectable marker are designed to include about 20 nucleotides identical with sequences flanking the yeast gene to be replaced. The resulting amplified construct comprises the selectable marker (e.g., the *kanMX* gene, which like *neo*^{r} confers resistance to G-418) flanked by about 20 base pairs that match the ends of the target yeast gene. Transformed diploid yeast cells in which one of the two copies of the target endogenous gene has been replaced by the disruption construct are identified by their resistance to G-418 or other selectable phenotype. These heterozygous diploid yeast cells generally grow normally regardless of the function of the target gene, but half the haploid spores derived from these cells will carry only the disrupted allele (Figure 5-39b). If a gene is essential for viability, then spores carrying a disrupted allele will not survive.

Disruption of yeast genes by this method is proving particularly useful in assessing the role of proteins identified by analysis of the entire genomic DNA sequence (see Chapter 6). A large consortium of scientists has replaced each of the approximately 6000 genes identified by this analysis with the *kanMX* disruption construct and determined which gene disruptions lead to nonviable haploid spores. These analyses have shown that about 4500 of the 6000 yeast genes are not required for viability, an unexpectedly large number of apparently nonessential genes. In some cases, disruption of a particular gene may give rise to subtle defects that do not compromise the viability of yeast cells growing under laboratory conditions. Alternatively, cells carrying a disrupted gene may be viable because of operation of backup or compensatory pathways. To investigate this possibility, yeast geneticists currently are searching for synthetic lethal mutations that might reveal nonessential genes with redundant functions (see Figure 5-9c).

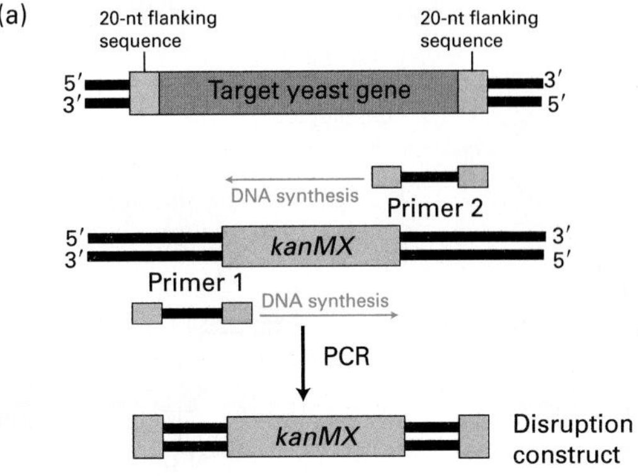

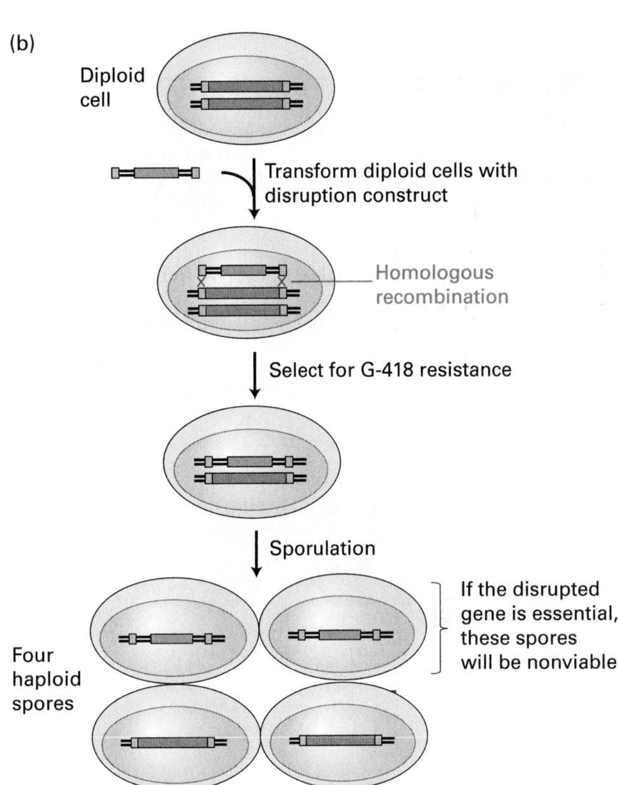

▲ **EXPERIMENTAL FIGURE 5-39 Homologous recombination with transfected disruption constructs can inactivate specific target genes in yeast.** (a) A suitable construct for disrupting a target gene can be prepared by the PCR. The two primers designed for this purpose each contain a sequence of about 20 nucleotides (nt) that is homologous to one end of the target yeast gene as well as sequences needed to amplify a segment of DNA carrying a selectable marker gene such as *kanMX,* which confers resistance to G-418. (b) When recipient diploid *Saccharomyces* cells are transformed with the gene disruption construct, homologous recombination between the ends of the construct and the corresponding chromosomal sequences will integrate the *kanMX* gene into the chromosome, replacing the target gene sequence. The recombinant diploid cells will grow on a medium containing G-418, whereas nontransformed cells will not. If the target gene is essential for viability, half the haploid spores that form after sporulation of recombinant diploid cells will be nonviable.

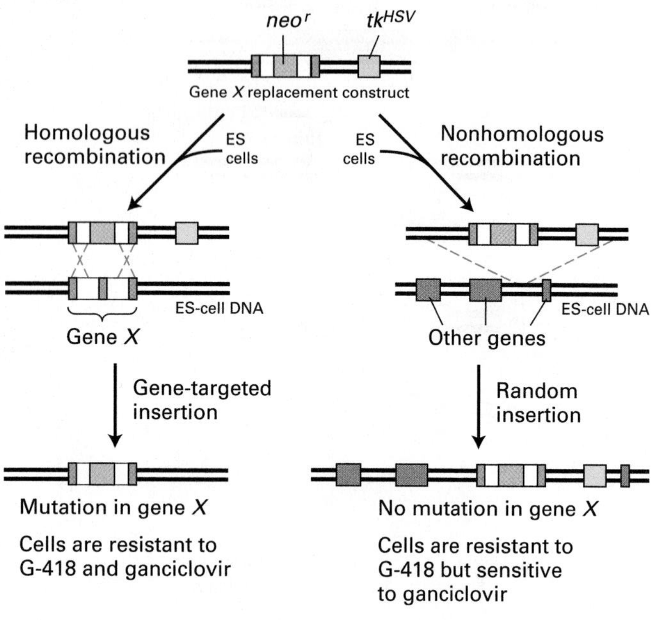

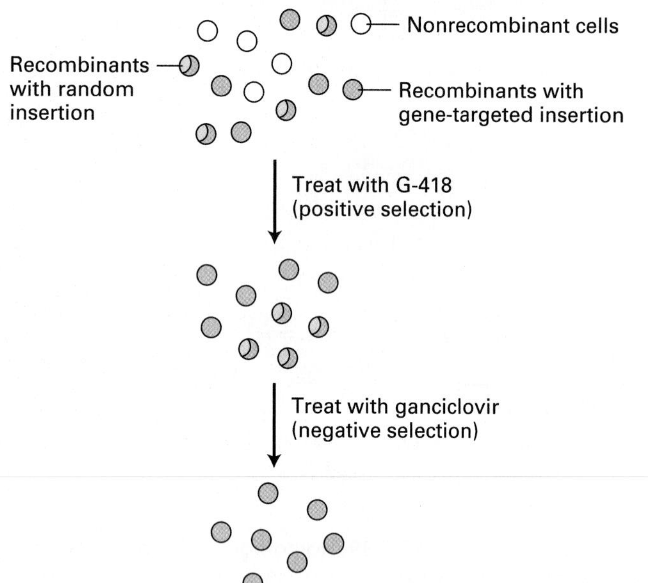

◄ **EXPERIMENTAL FIGURE 5-40 Isolation of mouse ES cells with a gene-targeted disruption is the first stage in production of knockout mice.** (a) When exogenous DNA is introduced into embryonic stem (ES) cells, random insertion via nonhomologous recombination occurs much more frequently than gene-targeted insertion via homologous recombination. Recombinant cells in which one allele of gene *X* (orange and white) is disrupted can be obtained by using a recombinant vector that carries gene *X* disrupted with *neo*^{r} (green), which confers resistance to G-418, and, outside the region of homology, *tk*^{HSV} (yellow), the thymidine kinase gene from herpes simplex virus. The viral thymidine kinase, unlike the endogenous mouse enzyme, can convert the nucleotide analog ganciclovir into the monophosphate form; this is then modified to the triphosphate form, which inhibits cellular DNA replication in ES cells. Thus ganciclovir is cytotoxic for recombinant ES cells carrying the *tk*^{HSV} gene. Nonhomologous insertion includes the *tk*^{HSV} gene, whereas homologous insertion does not; therefore, only cells with nonhomologous insertion are sensitive to ganciclovir.
(b) Recombinant cells are selected by treatment with G-418, since cells that fail to pick up DNA or integrate it into their genome are sensitive to this cytotoxic compound. The surviving recombinant cells are treated with ganciclovir. Only cells with a targeted disruption in gene *X*, and therefore lacking the *tk*^{HSV} gene and its accompanying cytotoxicity, will survive. [See S. L. Mansour et al., 1988, *Nature* **336**:348.]

Transcription of Genes Ligated to a Regulated Promoter Can Be Controlled Experimentally

Although disruption of an essential gene required for cell growth will yield nonviable spores, this method provides little information about what the encoded protein actually does in cells. To learn more about how a specific gene contributes to cell growth and viability, investigators must be able to selectively inactivate the gene in a population of growing cells. One method for doing this employs a regulated promoter to selectively shut off transcription of an essential gene.

A useful promoter for this purpose is the yeast *GAL1* promoter, which is active in cells grown on galactose but completely inactive in cells grown on glucose. In this approach, the coding sequence of an essential gene (*X*) ligated to the *GAL1* promoter is inserted into a yeast shuttle vector (see Figure 5-17a). The recombinant vector then is introduced into haploid yeast cells in which gene *X* has been disrupted. Haploid cells that are transformed will grow on galactose medium, since the normal copy of gene *X* on the vector is expressed in the presence of galactose. When the cells are transferred to a glucose-containing medium, gene *X* no longer is transcribed; as the cells divide, the amount of the encoded protein X gradually declines, eventually reaching a state of depletion that mimics a complete loss-of-function mutation. The observed changes in the phenotype of these cells after the shift to glucose medium may suggest which cell processes depend on the protein encoded by the essential gene *X*.

In an early application of this method, researchers explored the function of cytosolic *Hsc70* genes in yeast. Haploid cells with a disruption in all four redundant *Hsc70* genes were nonviable, unless the cells carried a vector containing a copy of the *Hsc70* gene that could be expressed from the *GAL1* promoter on galactose medium. On transfer to glucose, the vector-carrying cells eventually stopped growing because of insufficient Hsc70 activity. Careful examination of these dying cells revealed that their secretory proteins could no longer enter the endoplasmic reticulum (ER). This study provided the first evidence for the unexpected role of Hsc70 protein in translocation of secretory proteins into the ER, a process examined in detail in Chapter 13.

Video: Microinjection of ES Cells into a Blastocyst

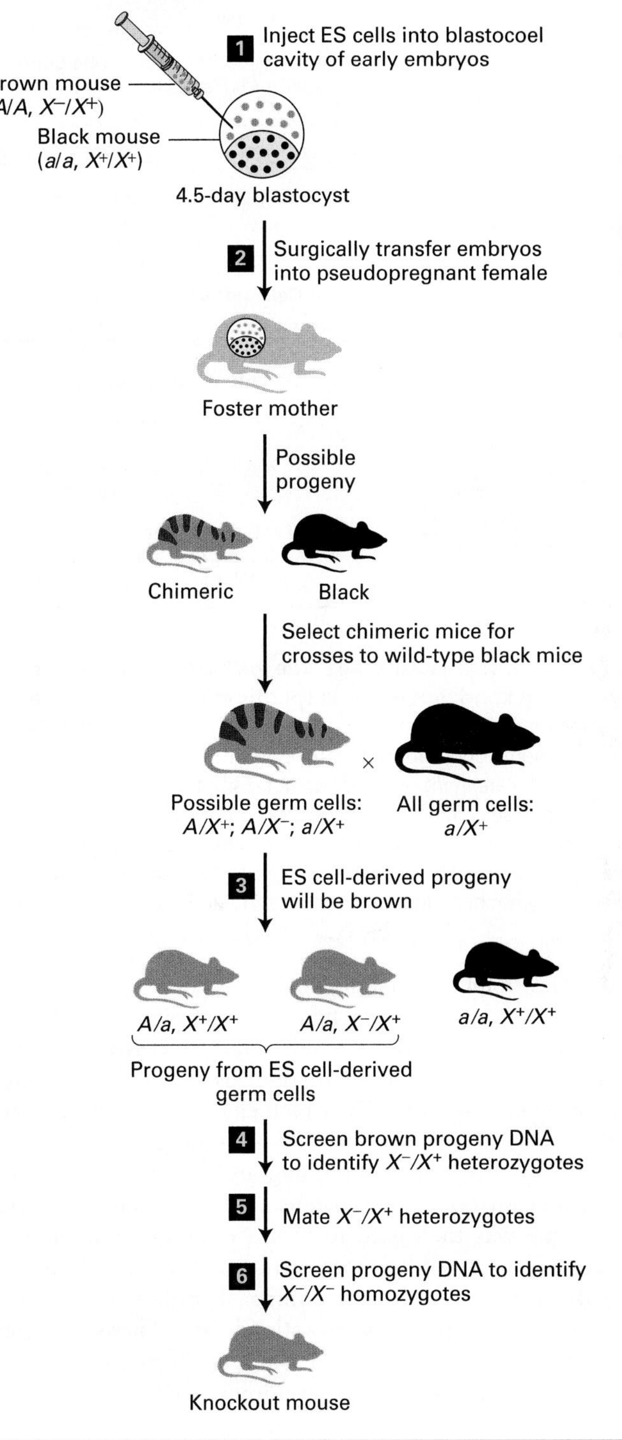

▶ EXPERIMENTAL FIGURE 5-41 **ES cells heterozygous for a disrupted gene are used to produce gene-targeted knockout mice.** Step 1: Embryonic stem (ES) cells heterozygous for a knockout mutation in a gene of interest (*X*) and homozygous for a dominant allele of a marker gene (here, brown coat color, *A*) are transplanted into the blastocoel cavity of 4.5-day embryos that are homozygous for a recessive allele of the marker (here, black coat color, *a*). Step 2: The early embryos then are implanted into a pseudopregnant female. Those progeny containing ES-derived cells are chimeras, indicated by their mixed black and brown coats. Step 3: Chimeric mice then are backcrossed to black mice; brown progeny from this mating have ES-derived cells in their germ line. Steps 4–6: Analysis of DNA isolated from a small amount of tail tissue can identify brown mice heterozygous for the knockout allele. Intercrossing of these mice produces some individuals homozygous for the disrupted allele, that is, knockout mice. [Adapted from M. R. Capecchi, 1989, *Trends Genet.* **5**:70.]

Specific Genes Can Be Permanently Inactivated in the Germ Line of Mice

Many of the methods for disrupting genes in yeast can be applied to genes of higher eukaryotes. These altered genes can be introduced into the **germ line** via homologous recombination to produce animals with a **gene knockout,** or simply "knockout." Knockout mice in which a specific gene is disrupted are a powerful experimental system for studying mammalian development, behavior, and physiology. They also are useful in studying the molecular basis of certain human genetic diseases.

Gene-targeted knockout mice are generated by a two-stage procedure. In the first stage, a DNA construct containing a disrupted allele of a particular target gene is introduced into **embryonic stem (ES) cells.** These cells, which are derived from the blastocyst, can be grown in culture through many generations (see Figure 21-7). In a small fraction of transfected cells, the introduced DNA undergoes homologous recombination with the target gene, although recombination at nonhomologous chromosomal sites occurs much more frequently. To select for cells in which homologous gene-targeted insertion occurs, the recombinant DNA construct introduced into ES cells needs to include two selectable marker genes (Figure 5-40). One of these genes (neo^r), which confers G-418 resistance, is inserted within the target gene (*X*), thereby disrupting it. The other selectable gene, the thymidine kinase gene from herpes simplex virus (tk^{HSV}), confers sensitivity to ganciclovir, a cytotoxic nucleotide analog; it is inserted into the construct outside the target-gene sequence. Only ES cells that undergo homologous recombination, and therefore do not incorporate tk^{HSV}, can survive in the presence of both G-418 and ganciclovir. In these cells one allele of gene *X* will be disrupted.

In the second stage in production of knockout mice, ES cells heterozygous for a knockout mutation in gene *X* are injected into a recipient wild-type mouse blastocyst, which subsequently is transferred into a surrogate pseudopregnant female mouse (Figure 5-41). The resulting progeny will be **chimeras,** containing tissues derived from both the transplanted ES cells and the host cells. If the ES cells also are homozygous for a visible marker trait (e.g., coat color), then chimeric progeny in which the ES cells survived and proliferated can be identified easily. Chimeric mice are then mated with mice homozygous for another allele of the marker trait to determine if the knockout mutation is incorporated into the germ line. Finally, mating of mice, each heterozygous for the knockout allele, will produce progeny homozygous for the knockout mutation.

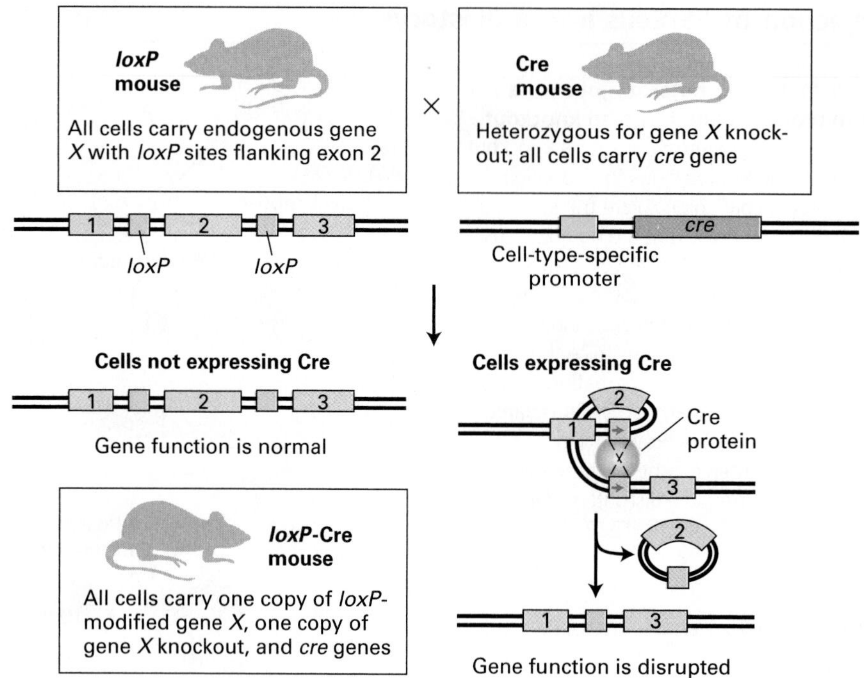

▲ **EXPERIMENTAL FIGURE 5-42 The *loxP*-Cre recombination system can knock out genes in specific cell types.** A *loxP* site is inserted on each side of the essential exon 2 of the target gene *X* (blue) by homologous recombination, producing a *loxP* mouse. Since the *loxP* sites are in introns, they do not disrupt the function of *X*. The Cre mouse carries one gene *X* knockout allele and an introduced *cre* gene (orange) from bacteriophage P1 linked to a cell-type-specific promoter (yellow). The *cre* gene is incorporated into the mouse genome by nonhomologous recombination and does not affect the function of other genes. In the *loxP*-Cre mice that result from crossing, Cre protein is produced only in those cells in which the promoter is active. Thus these are the only cells in which recombination between the *loxP* sites catalyzed by Cre occurs, leading to deletion of exon 2. Since the other allele is a constitutive gene *X* knockout, deletion between the *loxP* sites results in complete loss of function of gene *X* in all cells expressing Cre. By using different promoters, researchers can study the effects of knocking out gene *X* in various types of cells.

Development of knockout mice that mimic certain human diseases can be illustrated by cystic fibrosis. By methods discussed in Section 5.4, the recessive mutation that causes this disease eventually was shown to be located in a gene known as *CFTR*, which encodes a chloride channel. Using the cloned wild-type human *CFTR* gene, researchers isolated the homologous mouse gene and subsequently introduced mutations in it. The gene-knockout technique was then used to produce homozygous mutant mice, which showed symptoms (i.e., a phenotype), including disturbances to the functioning of epithelial cells, similar to those of humans with cystic fibrosis. These knockout mice are currently being used as a model system for studying this genetic disease and developing effective therapies. ▮

Somatic Cell Recombination Can Inactivate Genes in Specific Tissues

Investigators often are interested in examining the effects of knockout mutations in a particular tissue of the mouse, at a specific stage in development, or both. However, mice carrying a germ-line knockout may have defects in numerous tissues or die before the developmental stage of interest. To address this problem, mouse geneticists have devised a clever technique to inactivate target genes in specific types of somatic cells or at particular times during development.

This technique employs site-specific DNA recombination sites (called loxP *sites*) and the enzyme Cre that catalyzes recombination between them. The loxP-*Cre recombination system* is derived from bacteriophage P1, but this site-specific recombination system also functions when placed in mouse cells. An essential feature of this technique is that expression of Cre is controlled by a cell-type-specific promoter. In *loxP*-Cre mice generated by the procedure depicted in Figure 5-42, inactivation of the gene of interest (*X*) occurs only in cells in which the promoter controlling the *cre* gene is active.

An early application of this technique provided strong evidence that a particular neurotransmitter receptor is important for learning and memory. Previous pharmacological and physiological studies had indicated that normal learning requires the NMDA class of glutamate receptors in the hippocampus, a region of the brain. But mice in which the gene encoding an NMDA receptor subunit was knocked out died neonatally, precluding analysis of the receptor's role in learning. Following the protocol in Figure 5-42, researchers generated mice in which the receptor subunit gene was inactivated in the hippocampus but expressed in other tissues. These mice survived to adulthood and showed

 Technique Animation: Creating a Transgenic Mouse

Video: DNA injected into a Pronucleus of a Mouse Zygote

Inject foreign DNA into one of the pronuclei
Pronuclei
Fertilized mouse egg prior to fusion of male and female pronuclei
Transfer injected eggs into foster mother
About 10–30% of offspring will contain foreign DNA in chromosomes of all their tissues and germ line
Breed mice expressing foreign DNA to propagate DNA in germ line

▶ EXPERIMENTAL FIGURE 5-43 **Transgenic mice are produced by random integration of a foreign gene into the mouse germ line.** Foreign DNA injected into one of the two pronuclei (the male and female haploid nuclei contributed by the parents) has a good chance of being randomly integrated into the chromosomes of the diploid zygote. Because a transgene is integrated into the recipient genome by nonhomologous recombination, it does not disrupt endogenous genes. [See R. L. Brinster et al., 1981, *Cell* **27**:223.]

learning and memory defects, confirming a role for these receptors in the ability of mice to encode their experiences into memory.

Dominant-Negative Alleles Can Functionally Inhibit Some Genes

In diploid organisms, as noted in Section 5.1, the phenotypic effect of a recessive allele is expressed only in homozygous individuals, whereas dominant alleles are expressed in heterozygotes. Thus an individual must carry two copies of a recessive allele but only one copy of a dominant allele to exhibit the corresponding phenotypes. We have seen how strains of mice that are homozygous for a given recessive knockout mutation can be produced by crossing individuals that are heterozygous for the same knockout mutation (see Figure 5-41). For experiments with cultured animal cells, however, it is usually difficult to disrupt both copies of a gene in order to produce a mutant phenotype. Moreover, the difficulty in producing strains with both copies of a gene mutated is often compounded by the presence of related genes of similar function that must also be inactivated in order to reveal an observable phenotype.

For certain genes, the difficulties in producing homozygous knockout mutants can be avoided by use of an allele carrying a **dominant-negative** mutation. These alleles are genetically dominant; that is, they produce a mutant phenotype even in cells carrying a wild-type copy of the gene. However, unlike other types of dominant alleles, dominant-negative alleles produce a phenotype equivalent to that of a loss-of-function mutation.

Useful dominant-negative alleles have been identified for a variety of genes and can be introduced into cultured cells by transfection or into the germ line of mice or other organisms. In both cases, the introduced gene is integrated into the genome by nonhomologous recombination. Such randomly inserted genes are called **transgenes;** the cells or organisms carrying them are referred to as **transgenic.** Transgenes carrying a dominant-negative allele usually are engineered so that the allele is controlled by a regulated promoter, allowing expression of the mutant protein in different tissues at different times. As noted above, the random integration of exogenous DNA via nonhomologous recombination occurs at a much higher frequency than insertion via homologous recombination. Because of this phenomenon, the production of transgenic mice is an efficient and straightforward process (Figure 5-43).

Among the genes that can be functionally inactivated by introduction of a dominant-negative allele are those encoding small (monomeric) GTP-binding proteins belonging to the GTPase superfamily. As we will examine in several later chapters, these proteins (e.g., Ras, Rac, and Rab) act as intracellular switches. Conversion of the small GTPases from an inactive GDP-bound state to an active GTP-bound state depends on their interacting with a corresponding guanine nucleotide exchange factor (GEF). A mutant small GTPase that permanently binds to the GEF protein will block conversion of endogenous wild-type small GTPases to the active GTP-bound state, thereby inhibiting them from performing their switching function (Figure 5-44).

(a) Cells expressing only wild-type alleles of a small GTPase

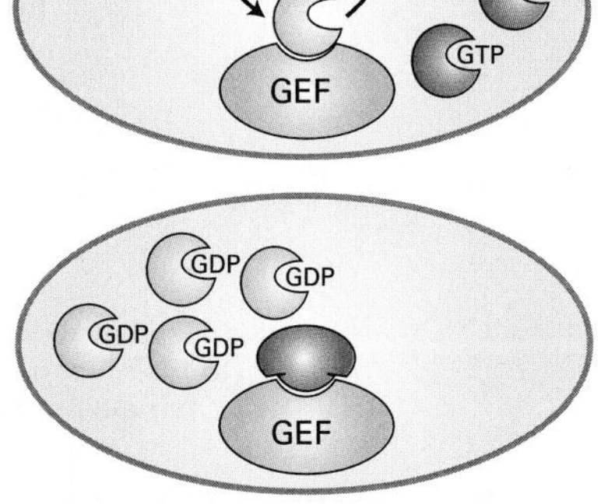

(b) Cells expressing both wild-type alleles and a dominant-negative allele

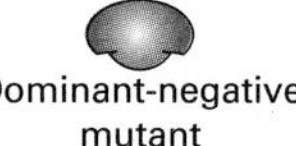

▲ **FIGURE 5-44 Inactivation of the function of a wild-type GTPase by the action of a dominant-negative mutant allele.** (a) Small (monomeric) GTPases (purple) are activated by their interaction with a guanine nucleotide exchange factor (GEF), which catalyzes the exchange of GDP for GTP. (b) Introduction of a dominant-negative allele of a small GTPase gene into cultured cells or transgenic animals leads to expression of a mutant GTPase that binds to and inactivates the GEF. As a result, endogenous wild-type copies of the same small GTPase are trapped in the inactive GDP-bound state. A single dominant-negative allele thus causes a loss-of-function phenotype in heterozygotes similar to that seen in homozygotes carrying two recessive loss-of-function alleles.

RNA Interference Causes Gene Inactivation by Destroying the Corresponding mRNA

A recently discovered phenomenon known as **RNA interference (RNAi)** is perhaps the most straightforward method to inhibit the function of specific genes. This approach is technically simpler than the methods described above for disrupting genes. First observed in the roundworm *C. elegans,* RNAi refers to the ability of double-stranded RNA to block expression of its corresponding single-stranded mRNA but not that of mRNAs with a different sequence.

As described in Chapter 8, the phenomenon of RNAi rests on the general ability of eukaryotic cells to cleave double-stranded RNA into short (23-nt) double-stranded segments known as small inhibitory RNA (siRNA). The RNA endonuclease that catalyzes this reaction, known as Dicer, is found in all metazoans but not in simpler eukaryotes such as yeast. The siRNA molecules, in turn, can cause cleavage of mRNA molecules of matching sequence, in a reaction catalyzed by a protein complex known as *RISC.* RISC mediates recognition and hybridization between one strand of the siRNA and its complementary sequence on the target mRNA; subsequently, specific nucleases in the RISC complex then cleave the mRNA/siRNA hybrid. This model accounts for the specificity of RNAi, since it depends on base pairing, and for its potency in silencing gene function, since the complementary mRNA is permanently destroyed by nucleolytic degradation. The normal function of both Dicer and RISC is to allow for gene regulation by small endogenous RNA molecules known as **micro RNAs miRNAs.**

Researchers exploit the micro RNA pathway for intentional silencing of a gene of interest by using either of two general methods for generating siRNAs of defined sequence. In the first method a double-stranded RNA corresponding to the target gene sequence is produced by in vitro transcription of both sense and antisense copies of this sequence (Figure 5-45a). This dsRNA is injected into the gonad of an adult worm, where it is converted to siRNA by Dicer in the developing embryos. In conjunction with the RISC complex, the siRNA molecules cause the corresponding mRNA molecules to be destroyed rapidly. The resulting worms display a phenotype similar to the one that would result from disruption of the corresponding gene itself. In some cases, entry of just a few molecules of a particular dsRNA into a cell is sufficient to inactivate many copies of the corresponding mRNA. Figure 5-45b illustrates the ability of an injected dsRNA to interfere with production of the corresponding endogenous mRNA in *C. elegans* embryos. In this experiment, the mRNA levels in embryos were determined by incubating the embryos with a fluorescently labeled probe specific for the mRNA of interest. This technique, in situ hybridization, is useful in assaying expression of a particular mRNA in cells and tissue sections.

The second method is to produce a specific double-stranded RNA in vivo. An efficient way to do this is to express a synthetic gene that is designed to contain tandem segments of both sense and anti-sense sequences corresponding to the target gene (Figure 5-45c). When this gene is transcribed, a double-stranded RNA "hairpin" structure forms, known as *small hairpin RNA*, or shRNA. The shRNA will then be cleaved by Dicer to form siRNA molecules. The lentiviral expression vectors are particularly useful for introducing synthetic genes for the expression of shRNA constructs into animal cells.

Both RNAi methods lend themselves to systematic studies to inactivate each of the known genes in an organism and to observe what goes wrong. For example, in initial studies with *C. elegans,* RNA interference with 16,700 genes (about 86 percent of the genome) yielded 1722 visibly abnormal phenotypes. The genes whose functional inactivation causes particular abnormal phenotypes can be grouped into sets; each member of a set presumably controls the same signals or events. The regulatory relations between the genes in the set—for example, the genes that control muscle development—can then be worked out.

Other organisms in which RNAi-mediated gene inactivation has been successful include *Drosophila,* many kinds of plants, zebra fish, the frog *Xenopus,* and mice, and are now the subjects of large-scale RNAi screens. For example, lentiviral vectors have been designed to inactivate by RNAi more than 10,000 different genes expressed in cultured mammalian cells. The function of the inactivated genes can be inferred from defects in growth or morphology of cell clones transfected with lentiviral vectors.

Podcast: RNA Interference

(a) In vitro production of double-stranded RNA

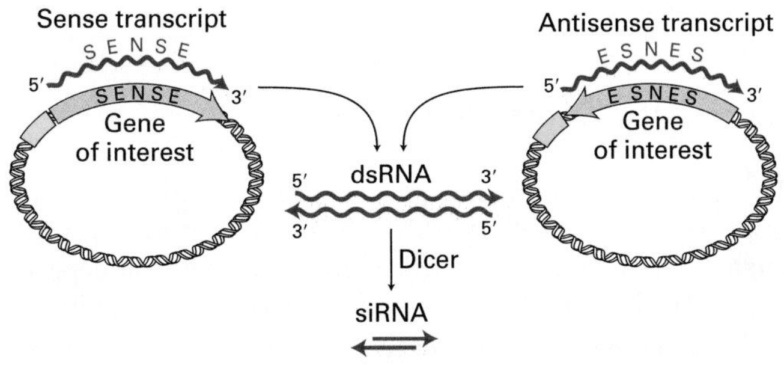

(b)

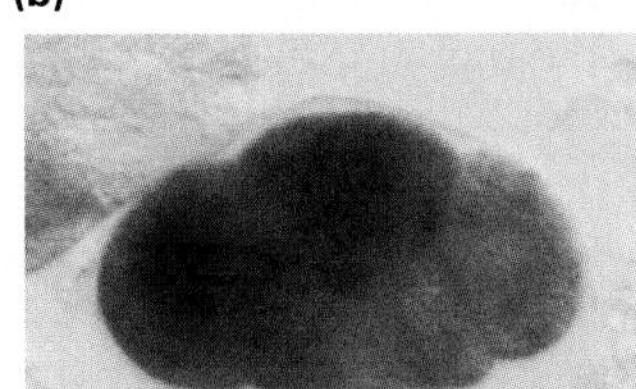

Noninjected

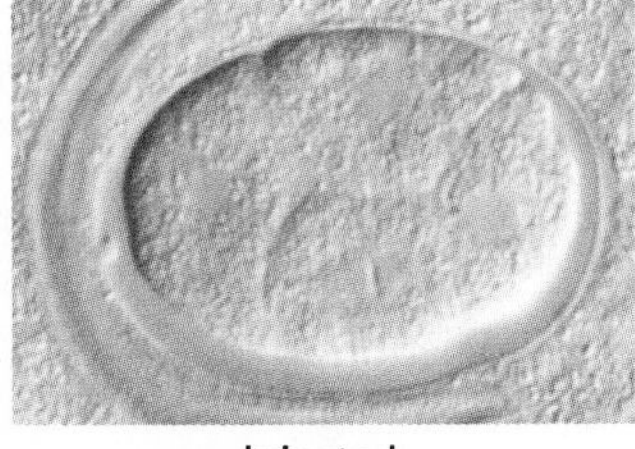

Injected

(c) In vivo production of double-stranded RNA

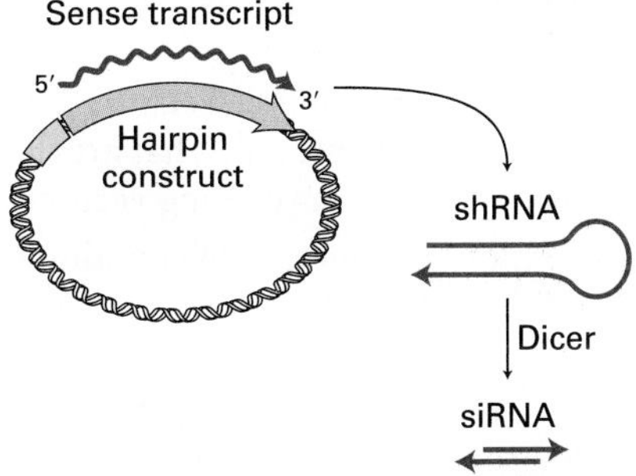

◀ EXPERIMENTAL FIGURE 5-45 **RNA interference (RNAi) can functionally inactivate genes in *C. elegans* and other organisms.** (a) In vitro production of double-stranded RNA (dsRNA) for RNAi of a specific target gene. The coding sequence of the gene, derived from either a cDNA clone or a segment of genomic DNA, is placed in two orientations in a plasmid vector adjacent to a strong promoter. Transcription of both constructs in vitro using RNA polymerase and ribonucleoside triphosphates yields many RNA copies in the sense orientation (identical with the mRNA sequence) or complementary antisense orientation. Under suitable conditions, these complementary RNA molecules will hybridize to form dsRNA. When the dsRNA is injected into cells, it is cleaved by Dicer into siRNAs. (b) Inhibition of *mex3* RNA expression in worm embryos by RNAi (see the text for the mechanism). (*Left*) Expression of *mex3* RNA in embryos was assayed by in situ hybridization with a probe specific for this mRNA, that is, linked to an enzyme that produces a colored (purple) product. (*Right*) The embryo derived from a worm injected with double-stranded *mex3* mRNA produces little or no endogenous *mex3* mRNA, as indicated by the absence of color. Each four-cell-stage embryo is ≈50 μm in length. (c) In vivo production of double-stranded RNA occurs via an engineered plasmid introduced directly into cells. The synthetic gene construct is a tandem arrangement of both sense and antisense sequences of the target gene. When it is transcribed, double-stranded small hairpin RNA forms (shRNA). The shRNA is cleaved by Dicer to form siRNA. [Part (b) from A. Fire et al., 1998, *Nature* **391**:806.]

KEY CONCEPTS OF SECTION 5.5

Inactivating the Function of Specific Genes in Eukaryotes

- Once a gene has been cloned, important clues about its normal function in vivo can be deduced from the observed phenotypic effects of mutating the gene.
- Genes can be disrupted in yeast by inserting a selectable marker gene into one allele of a wild-type gene via homologous recombination, producing a heterozygous mutant. When such a heterozygote is sporulated, disruption of an essential gene will produce two nonviable haploid spores (Figure 5-39).
- A yeast gene can be inactivated in a controlled manner by using the *GAL1* promoter to shut off transcription of a gene when cells are transferred to glucose medium.
- In mice, modified genes can be incorporated into the germ line at their original genomic location by homologous recombination, producing knockouts (see Figures 5-40 and 5-41). Mouse knockouts can provide models for human genetic diseases such as cystic fibrosis.
- The *loxP*-Cre recombination system permits production of mice in which a gene is knocked out in a specific tissue.
- In the production of transgenic cells or organisms, exogenous DNA is integrated into the host genome by nonhomologous recombination (see Figure 5-43). Introduction of a dominant-negative allele in this way can functionally inactivate a gene without altering its sequence.
- In many organisms, including the roundworm *C. elegans*, double-stranded RNA triggers destruction of the all the mRNA molecules with the same sequence (see Figure 5-45). This phenomenon, known as *RNAi* (RNA interference), provides a specific and potent means of functionally inactivating genes without altering their structure.

Perspectives for the Future

As the examples in this chapter and throughout the book illustrate, genetic analysis is the foundation of our understanding of many fundamental processes in cell biology. By examining the phenotypic consequences of mutations that inactivate a particular gene, geneticists are able to connect knowledge about the sequence, structure, and biochemical activity of the encoded protein to its function in the context of a living cell or multicellular organism. The classical approach to making these connections in both humans and simpler, experimentally accessible organisms has been to identify new mutations of interest based on their phenotypes and then to isolate the affected gene and its protein product.

Although scientists continue to use this classical genetic approach to dissect fundamental cellular processes and biochemical pathways, the availability of complete genomic sequence information for most of the common experimental organisms has fundamentally changed the way genetic experiments are conducted. Using various computational methods, scientists have identified the protein-coding gene sequences in most experimental organisms including *E. coli*, yeast, *C. elegans*, *Drosophila*, *Arabidopsis*, mouse, and humans. The gene sequences, in turn, reveal the primary amino acid sequence of the encoded protein products, providing us with a nearly complete list of the proteins found in each of the major experimental organisms.

The approach taken by most researchers has thus shifted from discovering new genes and proteins to discovering the functions of genes and proteins whose sequences are already known. Once an interesting gene has been identified, genomic sequence information greatly speeds subsequent genetic manipulations of the gene, including its designed inactivation, to learn more about its function. Already sets of vectors for RNAi inactivation of most defined genes in the nematode *C. elegans* now allow efficient genetic screens to be performed in this multicellular organism. These methods are now being applied to large collections of genes in cultured mammalian cells and in the near future either RNAi or knockout methods will have been used to inactivate every gene in the mouse.

In the past, a scientist might spend many years studying only a single gene, but nowadays scientists commonly study whole sets of genes at once. For example, with DNA microarrays the level of expression of all genes in an organism can be measured almost as easily as the expression of a single gene. One of the great challenges facing geneticists in the twenty-first century will be to exploit the vast amount of available data on the function and regulation of individual genes to understand how groups of genes are organized to form complex biochemical pathways and regulatory networks.

Key Terms

alleles *166*
clone *179*
complementary DNAs (cDNAs) *181*
complementation *183*
DNA cloning *176*
DNA library *179*
DNA microarray *192*
dominant *166*
gene knockout *207*
genomics *179*
genotype *166*
heterozygous *166*
homozygous *166*
hybridization *181*
in situ hybridization *192*
linkage *175*
mutation *166*
Northern blotting *192*
phenotype *166*
plasmids *178*
polymerase chain reaction (PCR) *188*
probes *181*
recessive *166*
recombinant DNA *176*
recombination *175*
restriction enzymes *176*
RNA interference (RNAi) *210*
segregation *167*
Southern blotting *191*
temperature-sensitive mutations *170*
transfection *196*
transformation *178*
transgenes *209*
vector *176*

Review the Concepts

1. Genetic mutations can provide insights into the mechanisms of complex cellular or developmental processes. How might your analysis of a genetic mutation be different depending on whether a particular mutation is recessive or dominant?

2. Give an example of how and why temperature-sensitive mutations might be used to study the function of essential genes.

3. Describe how complementation analysis can be used to reveal whether two mutations are in the same or in different genes. Explain why complementation analysis will not work with dominant mutations.

4. Compare the different uses of suppressor and synthetic lethal mutations in genetic analysis.

5. Restriction enzymes and DNA ligase play essential roles in DNA cloning. How is it that a bacterium that produces a restriction enzyme does not cut its own DNA? Describe some general features of restriction enzyme sites. What are the three types of DNA ends that can be generated after cutting DNA with restriction enzymes? What reaction is catalyzed by DNA ligase?

6. Bacterial plasmids often serve as cloning vectors. Describe the essential features of a plasmid vector. What are the advantages and applications of plasmids as cloning vectors?

7. A DNA library is a collection of clones, each containing a different fragment of DNA, inserted into a cloning vector. What is the difference between a cDNA and a genomic DNA library? How can you use hybridization or expression to screen a library for a specific gene? How many different oligonucleotide primers would need to be synthesized as probes to screen a library for the gene encoding the peptide Met-Pro-Glu-Phe-Tyr?

8. In 1993, Kerry Mullis won the Nobel prize in chemistry for his invention of the PCR process. Describe the three steps in each cycle of a PCR reaction. Why was the discovery of a thermostable DNA polymerase (e.g., *Taq* polymerase) so important for the development of PCR?

9. Southern and Northern blotting are powerful tools in molecular biology based on hybridization of nucleic acids. How are these techniques the same? How do they differ? Give some specific applications for each blotting technique.

10. A number of foreign proteins have been expressed in bacterial and mammalian cells. Describe the essential features of a recombinant plasmid that are required for expression of a foreign gene. How can you modify the foreign protein to facilitate its purification? What is the advantage of expressing a protein in mammalian cells versus bacteria?

11. What is a DNA microarray? How are DNA microarrays used for studying gene expression? How do experiments with microarrays differ from Northern blotting experiments?

12. In determining the identity of the protein that corresponds to a newly discovered gene, it often helps to know the pattern of tissue expression for that gene. For example, researchers have found that a gene called *SERPINA6* is expressed in the liver, kidney, and pancreas but not in other tissues. What techniques might researchers use to find out which tissues express a particular gene?

13. DNA polymorphisms can be used as DNA markers. Describe the differences between RFLP, SNP, and SSR polymorphisms. How can these markers be used for DNA mapping studies?

14. How can linkage disequilibrium mapping sometimes provide a much higher resolution of gene location than classical linkage mapping?

15. Genetic linkage studies can usually only roughly locate the chromosomal position of a "disease" gene. How can expression analysis and DNA sequence analysis help locate a disease gene within the region identified by linkage mapping?

16. Gene targeting by siRNA techniques exploits a normal micro RNA pathway that is found in all metazoans but not in simpler eukaryotes such as yeast. What are the roles of Dicer and RISC in this pathway?

17. The ability to selectively modify the genome in the mouse has revolutionized mouse genetics. Outline the procedure for generating a knockout mouse at a specific genetic locus. How can the *loxP*-Cre system be used to conditionally knock out a gene? What is an important medical application of knockout mice?

18. Two methods for functionally inactivating a gene without altering the gene sequence are by dominant-negative mutations and RNA interference (RNAi). Describe how each method can inhibit expression of a gene.

Analyze the Data

A culture of yeast that requires uracil for growth ($ura3^-$) was mutagenized, and two mutant colonies, X and Y, have been isolated. Mating type **a** cells of mutant X are mated with mating type α cells of mutant Y to form diploid cells. The parental ($ura3^-$), X, Y, and diploid cells are streaked onto agar plates containing uracil and incubated at 23 °C or 32 °C. Cell growth was monitored by the formation of colonies on the culture plates as shown in the figure below.

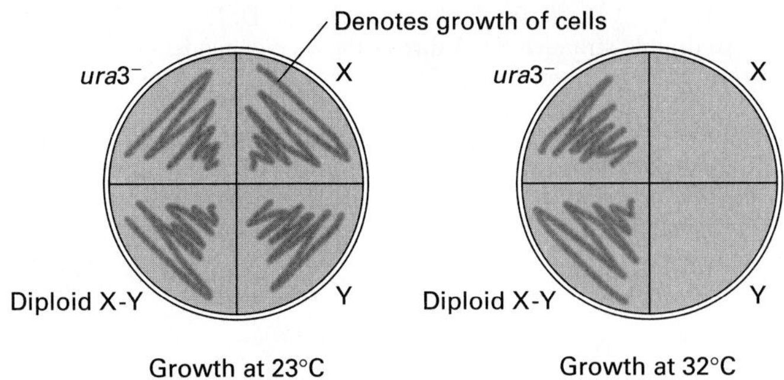

a. What can be deduced about mutants X and Y from the data provided?

b. A wild-type yeast cDNA library, prepared in a plasmid that contains the wild-type $URA3^+$ gene, is used to transform X cells, which are then cultured as indicated. Each black spot below represents a single clone growing on a petri plate. What are the molecular differences between the clones growing on the two plates? How can these results be used to identify the gene encoding X?

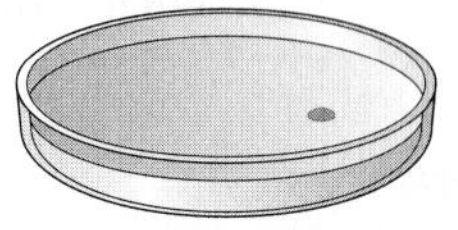

c. DNA is extracted from the parental cells, from X cells, and from Y cells and digested with a restriction enzyme. The digests are analyzed by Southern blot analysis, shown at the left below, using a probe obtained from the gene encoding X. In addition, PCR primers are used to amplify the gene encoding X in both the parental and the X cells. The primers are complementary to regions of DNA just external to the gene encoding X. The PCR results are shown in the gel at the right. What can be deduced about the mutation in the X gene from these data?

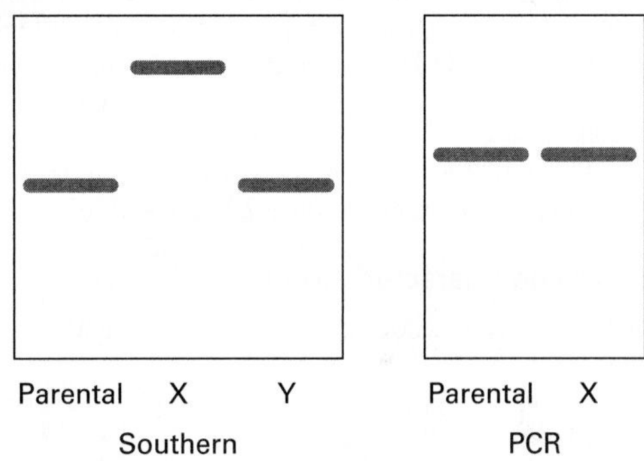

d. A construct of the wild-type gene X is engineered to encode a fusion protein in which the green fluorescent protein (GFP) is present at the N-terminus (GFP-X) or the C-terminus (X-GFP) of protein X. Both constructs, present on a $URA3^+$ plasmid, are used to transform X cells grown in the absence of uracil. The transformants are then monitored for growth at 32 °C, shown below at the left. At the right are typical fluorescent images of X-GFP and GFP-X cells grown at 23 °C in which green denotes the presence of green fluorescent protein. What is a reasonable explanation for growth of GFP-X but not X-GFP cells at 32 °C?

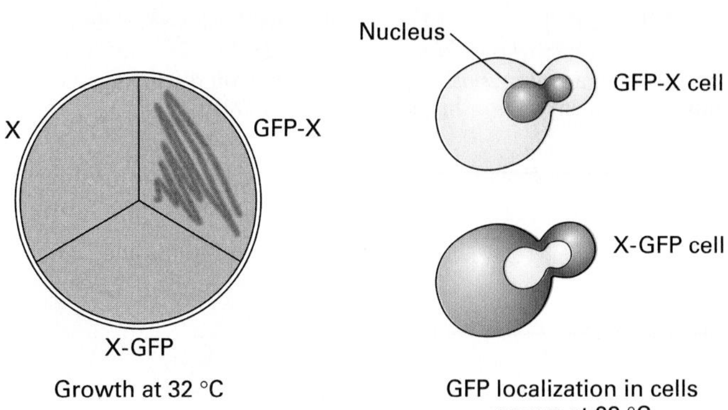

e. Haploid offspring of the diploid cells from part (a) above are generated. XY double mutants constitute 1/4 of these offspring. Haploid X cells, Y cells, and XY cells in liquid culture are synchronized at a stage just prior to budding and then are shifted from 23 °C to 32 °C. Examination of the cells 24 hours later reveals that X cells are arrested with small buds, Y cells are arrested with large buds, and XY cells are arrested with small buds. What is the relationship between X and Y?

References

Genetic Analysis of Mutations to Identify and Study Genes

Adams, A. E. M., D. Botstein, and D. B. Drubin. 1989. A yeast actin-binding protein is encoded by *sac6*, a gene found by suppression of an actin mutation. *Science* **243**:231.

Griffiths, A. G. F., et al. 2000. *An Introduction to Genetic Analysis,* 7th ed. W. H. Freeman and Company.

Guarente, L. 1993. Synthetic enhancement in gene interaction: a genetic tool comes of age. *Trends Genet.* **9**:362–366.

Hartwell, L. H. 1967. Macromolecular synthesis of temperature-sensitive mutants of yeast. *J. Bacteriol.* **93**:1662.

Hartwell, L. H. 1974. Genetic control of the cell division cycle in yeast. *Science* **183**:46.

Nüsslein-Volhard, C., and E. Wieschaus. 1980. Mutations affecting segment number and polarity in *Drosophila. Nature* **287**:795–801.

Simon, M. A., et al. 1991. Ras1 and a putative guanine nucleotide exchange factor perform crucial steps in signaling by the sevenless protein tyrosine kinase. *Cell* **67**:701–716.

Tong, A. H., et al. 2001. Systematic genetic analysis with ordered arrays of yeast deletion mutants. *Science* **294**:2364–2368.

DNA Cloning and Characterization

Ausubel, F. M., et al. 2002. *Current Protocols in Molecular Biology.* Wiley.

Gubler, U., and B. J. Hoffman. 1983. A simple and very efficient method for generating cDNA libraries. *Gene* **25**:263–289.

Han, J. H., C. Stratowa, and W. J. Rutter. 1987. Isolation of full-length putative rat lysophospholipase cDNA using improved methods for mRNA isolation and cDNA cloning. *Biochem.* **26**:1617–1632.

Itakura, K., J. J. Rossi, and R. B. Wallace. 1984. Synthesis and use of synthetic oligonucleotides. *Ann. Rev. Biochem.* **53**:323–356.

Maniatis, T., et al. 1978. The isolation of structural genes from libraries of eucaryotic DNA. *Cell* **15**:687–701.

Nasmyth, K. A., and S. I. Reed. 1980. Isolation of genes by complementation in yeast: molecular cloning of a cell-cycle gene. *Proc. Nat'l Acad. Sci. USA* **77**:2119–2123.

Nathans, D., and H. O. Smith. 1975. Restriction endonucleases in the analysis and restructuring of DNA molecules. *Ann. Rev. Biochem.* **44**:273–293.

Roberts, R. J., and D. Macelis. 1997. REBASE—restriction enzymes and methylases. *Nucl. Acids Res.* **25**:248–262. Information on accessing a continuously updated database on restriction and modification enzymes at http://www.neb.com/rebase.

Using Cloned DNA Fragments to Study Gene Expression

Andrews, A. T. 1986. *Electrophoresis,* 2d ed. Oxford University Press.

Erlich, H., ed. 1992. *PCR Technology: Principles and Applications for DNA Amplification.* W. H. Freeman and Company.

Pellicer, A., M. Wigler, R. Axel, and S. Silverstein. 1978. The transfer and stable integration of the HSV thymidine kinase gene into mouse cells. *Cell* **41**:133–141.

Saiki, R. K., et al. 1988. Primer-directed enzymatic amplification of DNA with a thermostable DNA polymerase. *Science* **239**:487–491.

Sanger, F. 1981. Determination of nucleotide sequences in DNA. *Science* **214**:1205–1210.

Souza, L. M., et al. 1986. Recombinant human granulocyte-colony stimulating factor: effects on normal and leukemic myeloid cells. *Science* **232**:61–65.

Wahl, G. M., J. L. Meinkoth, and A. R. Kimmel. 1987. Northern and Southern blots. *Meth. Enzymol.* **152**:572–581.

Wallace, R. B., et al. 1981. The use of synthetic oligonucleotides as hybridization probes. II: Hybridization of oligonucleotides of mixed sequence to rabbit β-globin DNA. *Nucl. Acids Res.* **9**:879–887.

Identifying and Locating Human Disease Genes

Botstein, D., et al. 1980. Construction of a genetic linkage map in man using restriction fragment length polymorphisms. *Am. J. Genet.* **32**:314–331.

Donis-Keller, H., et al. 1987. A genetic linkage map of the human genome. *Cell* **51**:319–337.

Hartwell, et al. 2000. *Genetics: From Genes to Genomes.* McGraw-Hill.

Hastbacka, T., et al. 1994. The diastrophic dysplasia gene encodes a novel sulfate transporter: positional cloning by fine-structure linkage disequilibrium mapping. *Cell* **78**:1073.

Orita, M., et al. 1989. Rapid and sensitive detection of point mutations and DNA polymorphisms using the polymerase chain reaction. *Genomics* **5**:874.

Tabor, H. K., N. J. Risch, and R. M. Myers. 2002. Opinion: candidate-gene approaches for studying complex genetic traits: practical considerations. *Nat. Rev. Genet.* **3**:391–397.

Inactivating the Function of Specific Genes in Eukaryotes

Capecchi, M. R. 1989. Altering the genome by homologous recombination. *Science* **244**:1288–1292.

Deshaies, R. J., et al. 1988. A subfamily of stress proteins facilitates translocation of secretory and mitochondrial precursor polypeptides. *Nature* **332**:800–805.

Fire, A., et al. 1998. Potent and specific genetic interference by double-stranded RNA in *Caenorhabditis elegans. Nature* **391**:806–811.

Gu, H., et al. 1994. Deletion of a DNA polymerase beta gene segment in T cells using cell type-specific gene targeting. *Science* **265**:103–106.

Zamore, P. D., T. Tuschl, P. A. Sharp, and D. P. Bartel. 2000. RNAi: double-stranded RNA directs the ATP-dependent cleavage of mRNA at 21 to 23 nucleotide intervals. *Cell* **101**:25–33.

Zimmer, A. 1992. Manipulating the genome by homologous recombination in embryonic stem cells. *Ann. Rev. Neurosci.* **15**:115

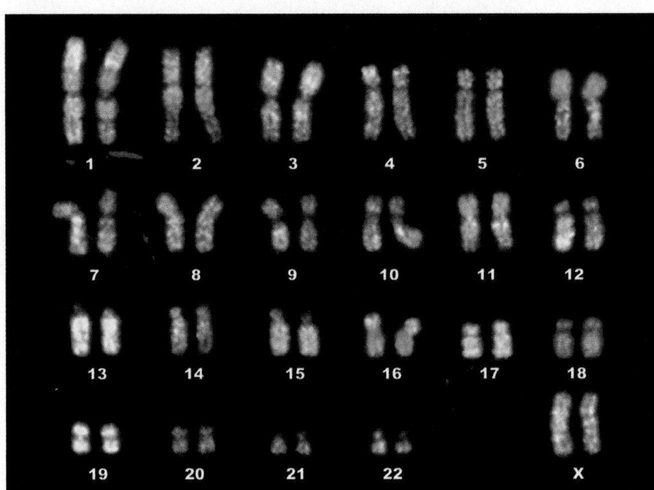

These brightly colored RxFISH-painted chromosomes are both beautiful and useful in revealing chromosome anomalies and in comparing karyotypes of different species. [© Department of Clinical Cytogenetics, Addenbrookes Hospital/Photo Researchers, Inc.]

CHAPTER

6

GENES, GENOMICS, AND CHROMOSOMES

In previous chapters we learned how the structure and composition of proteins allow them to perform a wide variety of cellular functions. We also examined another vital component of cells, the nucleic acids, and the process by which information encoded in the sequence of DNA is translated into protein. In this chapter, our focus again is on DNA and proteins as we consider the characteristics of eukaryotic nuclear and organellar genomes: the features of genes and the other DNA sequence that comprise the genome, and how this DNA is structured and organized by proteins within the cell.

By the beginning of the twenty-first century, molecular biologists had completed sequencing the entire genomes of hundreds of viruses, scores of bacteria, and one unicellular eukaryote, the budding yeast *S. cerevisiae*. In addition, the vast majority of the genome sequence is also known for the fission yeast *S. pombe,* and several multicellular eukaryotes including the roundworm *C. elegans,* the fruit fly *D. melanogaster,* mice, and humans. Detailed analysis of these sequencing data has revealed insights into genome organization and gene function. It has allowed researchers to identify previously unknown genes and to estimate the total number of protein-coding genes encoded in each genome. Comparisons between gene sequences often provide insight into possible functions of newly identified genes. Comparisons of genome sequence and organization between species also help us understand the evolution of organisms.

Surprisingly, DNA sequencing revealed that a large portion of the genomes of higher eukaryotes does not encode mRNAs or any other RNAs required by the organism. Remarkably, such noncoding DNA constitutes ≈98.5 percent of human chromosomal DNA! The noncoding DNA in multicellular organisms contains many regions that are similar but not identical. Variations within some stretches of this *repetitious DNA* between individuals are so great that every person can be distinguished by a DNA "fingerprint" based on these sequence variations. Moreover, some repetitious DNA sequences are not found in the same positions in the genomes of different individuals of the same species. At one time, all noncoding DNA was collectively termed "junk DNA" and was considered to serve no purpose. We now understand the evolutionary basis of all this extra DNA, and the variation in location of certain sequences between

OUTLINE

individuals. Cellular genomes harbor symbiotic **transposable (mobile) DNA elements,** sequences that can copy themselves and move throughout the genome. Although transposable DNA elements seem to have little function in the life cycle of an individual organism, over evolutionary time they have shaped our genomes and contributed to the rapid evolution of multicellular organisms.

In higher eukaryotes, DNA regions encoding proteins or functional RNAs—that is, **genes**—lie amidst this expanse of apparently nonfunctional DNA. In addition to the nonfunctional DNA *between* genes, noncoding **introns** are common *within* genes of multicellular plants and animals. Sequencing of the same protein-coding gene in a variety of eukaryotic species has shown that evolutionary pressure selects for maintenance of relatively similar sequences in the coding regions, or **exons.** In contrast, wide sequence variation, even including total loss, occurs among introns, suggesting that most intron sequences have little functional significance. However, as we shall see, although most of the DNA sequence of introns is not functional, the existence of introns has favored the evolution of multidomain proteins that are common in higher eukaryotes. It also allowed the rapid evolution of proteins with new combinations of functional domains.

Mitochondria and chloroplasts also contain DNA that encodes proteins essential to the function of these vital organelles. We shall see that mitochondrial and chloroplast DNAs are evolutionary remnants of the origins of these organelles. Comparison of DNA sequences between different classes of bacteria and mitochondrial and chloroplast genomes has revealed that these organelles evolved from intracellular bacteria that developed symbiotic relationships with ancient eukaryotic cells.

The sheer length of cellular DNA is a significant problem with which cells must contend. The DNA in a single human cell, which measures about 2 meters in total length, must be contained within cells with diameters of less than 10 μm, a compaction ratio of greater than 10^5 to 1. In relative terms, if a cell were 1 centimeter in diameter, the length of DNA packed into its nucleus would be about 2 kilometers! Specialized eukaryotic proteins associated with nuclear DNA exquisitely fold and organize the DNA so that it fits into nuclei. And yet at the same time, any given portion of this highly compacted DNA can be accessed readily for transcription, DNA replication, and repair of DNA damage without the long DNA molecules becoming tangled or broken. Furthermore, the integrity of DNA must be maintained during the process of cell division when it is partitioned into daughter cells. In eukaryotes, the complex of DNA and the proteins that organize it, called **chromatin,** can be visualized as individual **chromosomes** during mitosis (see chapter opening figure). As we will see in this and the following chapter, the organization of DNA into chromatin allows a mechanism for regulation of gene expression that is not available in bacteria.

In the first five sections of this chapter, we provide an overview of the landscape of eukaryotic genes and genomes. First we discuss the structure of eukaryotic genes and the complexities that arise in higher organisms from the processing of mRNA precursors into alternatively spliced mRNAs. Next we discuss the main classes of eukaryotic DNA including the special properties of transposable DNA elements and how they shaped

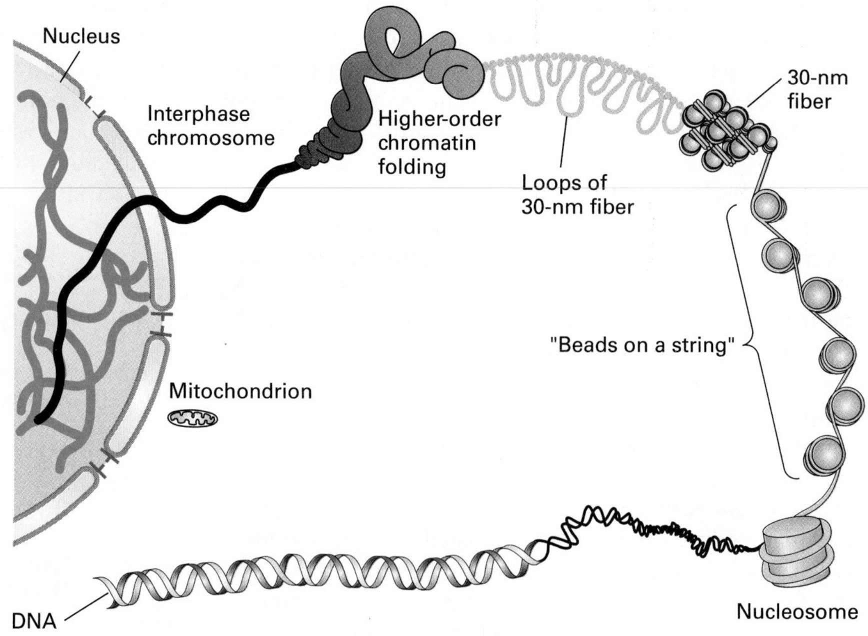

▶ **FIGURE 6-1 Overview of the structure of genes and chromosomes.** DNA of higher eukaryotes consists of unique and repeated sequences. Only ≈1.5 percent of human DNA encodes proteins and functional RNAs and the regulatory sequences that control their expression; the remainder is merely introns within genes and intergenic DNA between genes. Much of the intergenic DNA, ≈45 percent in humans, is derived from transposable (mobile) DNA elements, genetic symbionts that have contributed to the evolution of contemporary genomes. Each chromosome consists of a single, long molecule of DNA up to ≈280 Mb in humans, organized into increasing levels of condensation by the histone and nonhistone proteins with which it is intricately complexed. Much smaller DNA molecules are localized in mitochondria and chloroplasts.

contemporary genomes. We then consider organelle DNA and how it differs from nuclear DNA. This background prepares us to discuss **genomics**, computer-based methods for analyzing and interpreting vast amounts of sequence data. The final two sections of the chapter address how DNA is physically organized in eukaryotic cells. We consider the packaging of DNA and histone proteins into compact complexes (nucleosomes) that are the fundamental building blocks of chromatin, the large-scale structure of chromosomes, and the functional elements required for chromosome duplication and segregation. Figure 6-1 provides an overview of these interrelated subjects. The understanding of genes, genomics, and chromosomes gained in this chapter will prepare us to explore current knowledge about how the synthesis and concentration of each protein and functional RNA in a cell is regulated in the following two chapters.

6.1 Eukaryotic Gene Structure

In molecular terms, a gene commonly is defined as *the entire nucleic acid sequence that is necessary for the synthesis of a functional gene product (polypeptide or RNA).* According to this definition, a gene includes more than the nucleotides encoding an amino acid sequence or a functional RNA, referred to as the *coding region.* A gene also includes all the DNA sequences required for synthesis of a particular RNA transcript, no matter where those sequences are located in relation to the coding region. For example, in eukaryotic genes, transcription-control regions known as **enhancers** can lie 50 kb or more from the coding region. As we learned in Chapter 4, other critical noncoding regions in eukaryotic genes include the promoter, as well as sequences that specify 3′ cleavage and polyadenylation, known as *poly(A) sites,* and splicing of primary RNA transcripts, known as *splice sites* (see Figure 4-15). Mutations in these sequences, which control transcription initiation and RNA processing, affect the normal expression and function of RNAs, producing distinct phenotypes in mutant organisms. We examine these various control elements of genes in greater detail in Chapters 7 and 8.

Although most genes are transcribed into mRNAs, which encode proteins, some DNA sequences are transcribed into RNAs that do not encode proteins (e.g., tRNAs and rRNAs described in Chapter 4 and micro RNAs that regulate mRNA stability and translation discussed in Chapter 8). Because the DNA that encodes tRNAs, rRNAs and micro RNAs can cause specific phenotypes when mutated, these DNA regions generally are referred to as tRNA, rRNA and micro RNA *genes,* even though the final products of these genes are RNA molecules and not proteins.

In this section, we will examine the structure of genes in bacteria and eukaryotes and discuss how their respective gene structures influence gene expression and evolution.

Most Eukaryotic Genes Contain Introns and Produce mRNAs Encoding Single Proteins

As discussed in Chapter 4, many bacterial mRNAs (e.g., the mRNA encoded by the *trp* operon) include the coding region for several proteins that function together in a biological process. Such mRNAs are said to be *polycistronic.* (A *cistron* is a genetic unit encoding a single polypeptide.) In contrast, most eukaryotic mRNAs are *monocistronic;* that is, each mRNA molecule encodes a single protein. This difference between polycistronic and monocistronic mRNAs correlates with a fundamental difference in their translation.

Within a bacterial polycistronic mRNA a ribosome-binding site is located near the start site for each of the protein-coding regions, or cistrons, in the mRNA. Translation initiation can begin at any of these multiple internal sites, producing multiple proteins (see Figure 4-13a). In most eukaryotic mRNAs, however, the 5′-cap structure directs ribosome binding, and translation begins at the closest AUG start codon (see Figure 4-13b). As a result, translation begins only at this site. In many cases, the primary transcripts of eukaryotic protein-coding genes are processed into a single type of mRNA, which is translated to give a single type of polypeptide (see Figure 4-15).

Unlike bacterial and yeast genes, which generally lack introns, most genes in multicellular animals and plants contain introns, which are removed during RNA processing in the nucleus before the fully processed mRNA is exported to the cytosol for translation. In many cases, the introns in a gene are considerably longer than the exons. Although many introns are ≈90 bp long, the median intron length in human genes is 3.3 kb. Some, however, are much longer: the longest known human intron is 17,106 bp, and lies within *titan,* a gene encoding a structural protein in muscle cells. In comparison, most human exons contain only 50–200 base pairs. The typical human gene encoding an average-size protein is ≈50,000 bp long, but more than 95 percent of that sequence is present in introns and flanking noncoding 5′ and 3′ regions.

Many large proteins in higher organisms that have repeated domains are encoded by genes consisting of repeats of similar exons separated by introns of variable length. An example of this is fibronectin, a component of the extracellular matrix. The fibronectin gene contains multiple copies of five types of exons (see Figure 4-16). Such genes evolved by tandem duplication of the DNA encoding the repeated exon, probably by unequal crossing over during meiosis as shown in Figure 6-2a.

Simple and Complex Transcription Units Are Found in Eukaryotic Genomes

The cluster of genes that form a bacterial operon comprises a single **transcription unit** that is transcribed from a specific promoter in the DNA sequence to a termination site, producing a single primary transcript. In other words, genes and transcription units often are distinguishable in prokaryotes since a single transcription unit contains several genes when they are part of an operon. In contrast, most eukaryotic genes are expressed from separate transcription units, and each mRNA is translated into a single protein. Eukaryotic transcription units, however, are classified into two types, depending on the fate of the primary transcript.

The primary transcript produced from a *simple* transcription unit is processed to yield a single type of mRNA, encoding

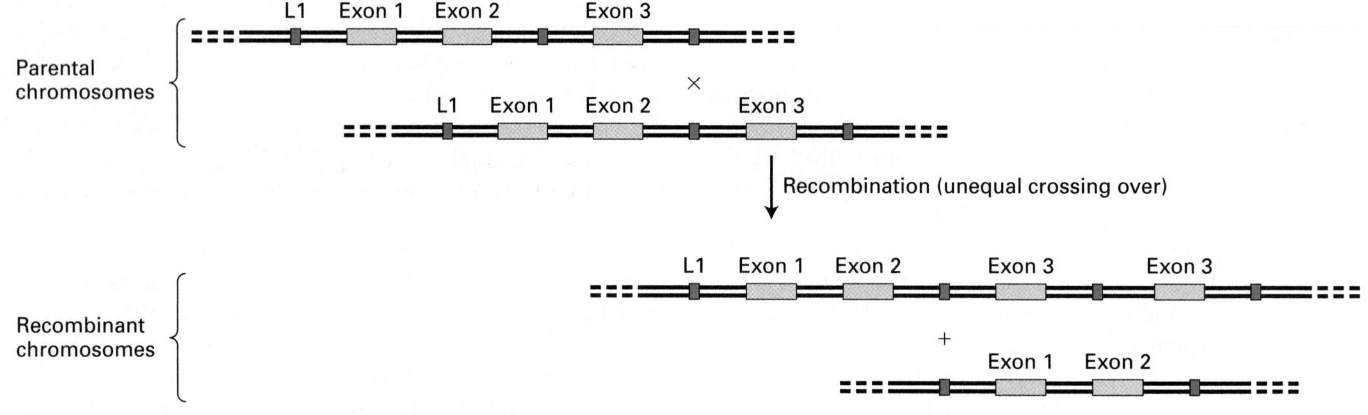

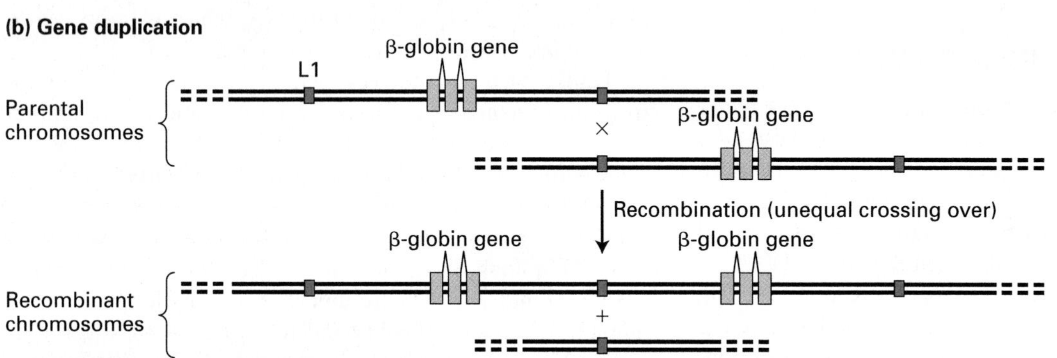

▲ **FIGURE 6-2 Exon and gene duplication.** (a) Exon duplication results from unequal crossing over during meiosis. Each parental chromosome contains one ancestral gene containing three exons (numbered 1–3) and two introns. Homologous noncoding L1 repeated sequences lie 5′ and 3′ of the gene, and also in the intron between exons 2 and 3. As discussed in Section 6.3, L1 sequences have been repeatedly transposed to new sites in the genome over the course of human evolution, so that all chromosomes are peppered with them. The parental chromosomes are shown displaced relative to each other, so that the L1 sequences are aligned. Homologous recombination between L1 sequences as shown would generate one recombinant chromosome in which the gene has four exons (two copies of exon 3) and one chromosome in which the gene is missing exon 3. (b) Unequal crossing over between L1 sequences also can generate duplications of entire genes. In this example, each parental chromosome contains one ancestral β-globin gene, and one of the recombinant chromosomes contains two duplicated β-globin genes. Subsequent independent mutations in the duplicated genes could lead to slight changes in sequence that might result in slightly different functional properties of the encoded proteins. Unequal crossing over also can result from rare recombinations between unrelated sequences. Note that the scale in part (b) is much larger than in part (a). [Part (b) see D. H. A. Fitch et al., 1991, *Proc. Nat'l. Acad. Sci. USA* **88:**7396.]

a single protein. Mutations in exons, introns, and transcription-control regions all may influence expression of the protein encoded by a simple transcription unit (Figure 6-3a).

In the case of *complex* transcription units, which are quite common in multicellular organisms, the primary RNA transcript can be processed in more than one way, leading to formation of mRNAs containing different exons. Each alternate mRNA, however, is monocistronic, being translated into a single polypeptide, with translation usually initiating at the first AUG in the mRNA. Multiple mRNAs can arise from a primary transcript in three ways, as shown in Figure 6-3b.

Examples of all three types of alternative RNA processing occur in the genes that regulate sexual differentiation in *Drosophila* (see Figure 8-16). Commonly, one mRNA is produced from a complex transcription unit in some cell types, and a different mRNA is made in other cell types. For example, **alternative splicing** of the primary fibronectin transcript in fibroblasts and hepatocytes determines whether or not the secreted protein includes domains that adhere to cell surfaces (see Figure 4-16). The phenomenon of alternative splicing greatly expands the number of proteins encoded in the genomes of higher organisms. It is estimated that ≈60 percent of human genes are contained within complex transcription units that give rise to alternatively spliced mRNAs encoding proteins with distinct functions, as for the fibroblast and hepatocyte forms of fibronectin.

The relationship between a mutation and a gene is not always straightforward when it comes to complex transcription units. A mutation in the control region or in an exon shared by alternatively spliced mRNAs will affect all the alternative proteins encoded by a given complex transcription unit. On the other hand, mutations in an exon present in only one of the alternative mRNAs will affect only the protein encoded by that mRNA. As explained in

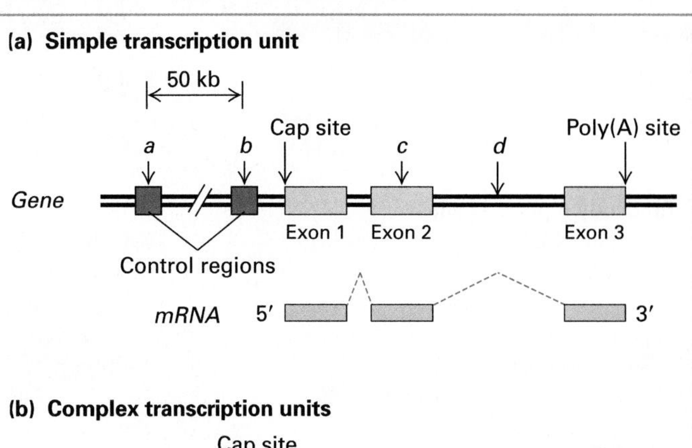

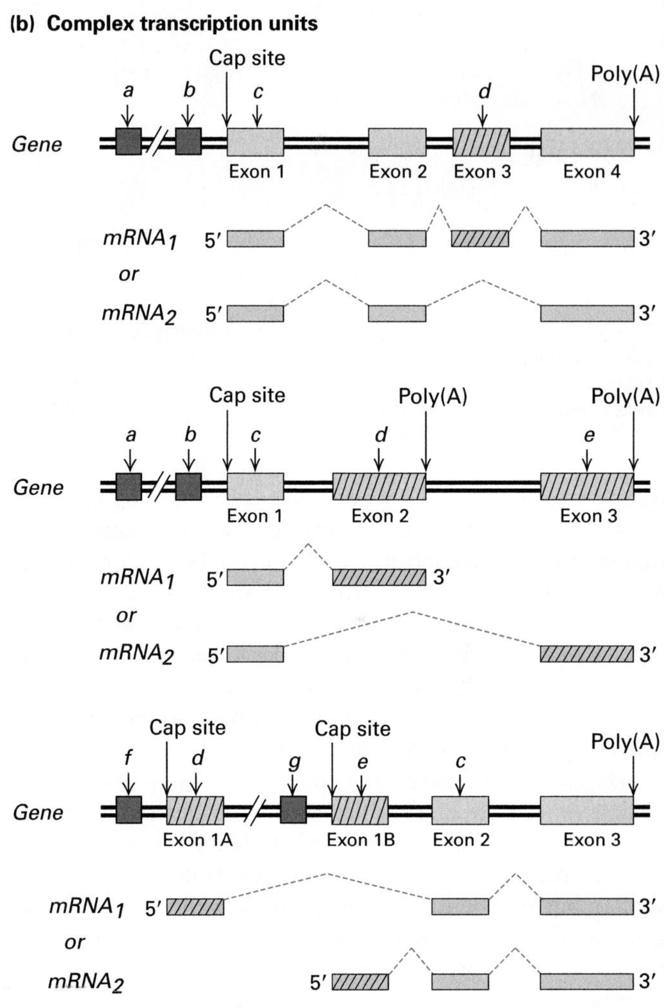

▶ **FIGURE 6-3 Simple and complex eukaryotic transcription units.** (a) A simple transcription unit includes a region that encodes one protein, extending from the 5′ cap site to the 3′ poly(A) site, and associated control regions. Introns lie between exons (light blue rectangles) and are removed during processing of the primary transcripts (dashed red lines); thus they do not occur in the functional monocistronic mRNA. Mutations in a transcription-control region (*a, b*) may reduce or prevent transcription, thus reducing or eliminating synthesis of the encoded protein. A mutation within an exon (*c*) may result in an abnormal protein with diminished activity. A mutation within an intron (*d*) that introduces a new splice site results in an abnormally spliced mRNA encoding a nonfunctional protein. (b) Complex transcription units produce primary transcripts that can be processed in alternative ways. (*Top*) If a primary transcript contains alternative splice sites, it can be processed into mRNAs with the same 5′ and 3′ exons but different internal exons. (*Middle*) If a primary transcript has two poly(A) sites, it can be processed into mRNAs with alternative 3′ exons. (*Bottom*) If alternative promoters (*f* or *g*) are active in different cell types, $mRNA_1$, produced in a cell type in which *f* is activated, has a different first exon (1A) than $mRNA_2$ has, which is produced in a cell type in which *g* is activated (and where exon 1B is used). Mutations in control regions (*a* and *b*) and those designated *c* within exons shared by the alternative mRNAs affect the proteins encoded by both alternatively processed mRNAs. In contrast, mutations (designated *d* and *e*) within exons unique to one of the alternatively processed mRNAs affect only the protein translated from that mRNA. For genes that are transcribed from different promoters in different cell types *(bottom)*, mutations in different control regions (*f* and *g*) affect expression only in the cell type in which that control region is active.

Chapter 5, **genetic complementation** tests commonly are used to determine if two mutations are in the same or different genes (see Figure 5-7). However, in the complex transcription unit shown in Figure 6-3b (*middle*), mutations *d* and *e* would complement each other in a genetic complementation test, even though they occur in the same gene. This is because a chromosome with mutation *d* can express a normal protein encoded by $mRNA_2$ and a chromosome with mutation *e* can express a normal protein encoded by $mRNA_1$. Both mRNAs produced from this gene would be present in a diploid cell carrying both mutations, generating both protein products and hence a wild-type phenotype. However, a chromosome with mutation *c* in an exon common to both mRNAs would not complement either mutation *d* or *e*. In other words, mutation *c* would be in the same complementation groups as mutations *d* and *e*, even though *d* and *e* themselves would not be in the same complementation group! Given these complications with the genetic definition of a gene, the genomic definition outlined at the beginning of this section is commonly used. In the case of protein-coding genes, a gene is the DNA sequence transcribed into a pre-mRNA precursor, equivalent to a transcription unit, plus any other regulatory elements required for synthesis of the primary transcript. The various proteins encoded by the alternatively spliced mRNAs expressed from one gene are called **isoforms.**

Protein-Coding Genes May Be Solitary or Belong to a Gene Family

The nucleotide sequences within chromosomal DNA can be classified on the basis of structure and function, as shown in Table 6-1. We will examine the properties of each class, beginning with protein-coding genes, which comprise two groups.

In multicellular organisms, roughly 25–50 percent of the protein-coding genes are represented only once in the haploid genome and thus are termed *solitary* genes. A well-studied example of a solitary protein-coding gene is the chicken

TABLE 6-1 Major Classes of Nuclear Eukaryotic DNA and Their Representation in the Human Genome

CLASS	LENGTH	COPY NUMBER IN HUMAN GENOME	FRACTION OF HUMAN GENOME (%)
Protein-coding genes	0.5–2200 kb	≈25,000	≈55* (1.8)†
Tandemly repeated genes			
U2 snRNA	6.1 kb‡	≈20	<0.001
rRNAs	43 kb‡	≈300	0.4
Repetitious DNA			
Simple-sequence DNA	1–500 bp	Variable	≈6
Interspersed repeats (mobile DNA elements)			
DNA transposons	2–3 kb	300,000	3
LTR retrotransposons	6–11 kb	440,000	8
Non-LTR retrotransposons			
LINEs	6–8 kb	860,000	21
SINEs	100–400 bp	1,600,000	13
Processed pseudogenes	Variable	1–≈100	≈0.4
Unclassified spacer DNA§	Variable	n.a.	≈25

* Complete transcription units including introns.
† Transcription units not including introns. Protein-coding regions (exons) total 1.1% of the genome.
‡ Length of tandemly repeated sequence.
§ Sequences between transcription units that are not repeated in the genome; n.a. = not applicable.
SOURCE: International Human Genome Sequencing Consortium, 2001, *Nature* **409:**860 and 2004, *Nature* **431:**931.

lysozyme gene. The 15-kb DNA sequence encoding chicken lysozyme constitutes a simple transcription unit containing four exons and three introns. The flanking regions, extending for about 20 kb upstream and downstream from the transcription unit, do not encode any detectable mRNAs. Lysozyme, an enzyme that cleaves the polysaccharides in bacterial cell walls, is an abundant component of chicken egg-white protein and also is found in human tears. Its activity helps to keep the surface of the eye and the chicken egg sterile.

Duplicated genes constitute the second group of protein-coding genes. These are genes with close but nonidentical sequences that often are located within 5–50 kb of one another. A set of duplicated genes that encode proteins with similar but nonidentical amino acid sequences is called a **gene family;** the encoded, closely related, homologous proteins constitute a **protein family.** A few protein families, such as protein kinases, vertebrate immunoglobulins, and olfactory receptors include hundreds of members. Most protein families, however, include from just a few to 30 or so members; common examples are cytoskeletal proteins, the myosin heavy chain, and the α- and β-globins in vertebrates.

The genes encoding the β-like globins are a good example of a gene family. As shown in Figure 6-4a, the β-like globin gene family contains five functional genes designated β, δ, A_γ, G_γ, and ε; the encoded polypeptides are similarly designated. Two identical β-like globin polypeptides combine with two identical α-globin polypeptides (encoded by another gene family) and four small heme groups to form a hemoglobin molecule (see Figure 3-13). All the hemoglobins formed from the different β-like globins carry oxygen in the blood, but they exhibit somewhat different properties that are suited to specific roles in human physiology. For example, hemoglobins containing either the A_γ or G_γ polypeptides are expressed only during fetal life. Because these fetal hemoglobins have a higher affinity for oxygen than adult hemoglobins, they can effectively extract oxygen from the maternal circulation in the placenta. The lower oxygen affinity of adult hemoglobins, which are expressed after birth, permits better release of oxygen to the tissues, especially muscles, which have a high demand for oxygen during exercise.

The different β-globin genes arose by duplication of an ancestral gene, most likely as the result of an "unequal crossover" during meiotic recombination in a developing germ cell (egg or sperm) (Figure 6-2b). Over evolutionary time the two copies of the gene that resulted accumulated random mutations; beneficial mutations that conferred some refinement in the basic oxygen-carrying function of hemoglobin were retained by natural selection, resulting in *sequence drift.* Repeated gene duplications and subsequent sequence drift are thought to have generated the contemporary globin-like genes observed in humans and other mammals today.

Two regions in the human β-like globin gene cluster contain nonfunctional sequences, called **pseudogenes,** similar to those of the functional β-like globin genes (see Figure 6-4a). Sequence analysis shows that these pseudogenes have the same apparent exon–intron structure as the functional β-like globin genes, suggesting that they also arose by duplication of the same ancestral gene. However, there was little selective pressure to maintain the function of these genes. Consequently

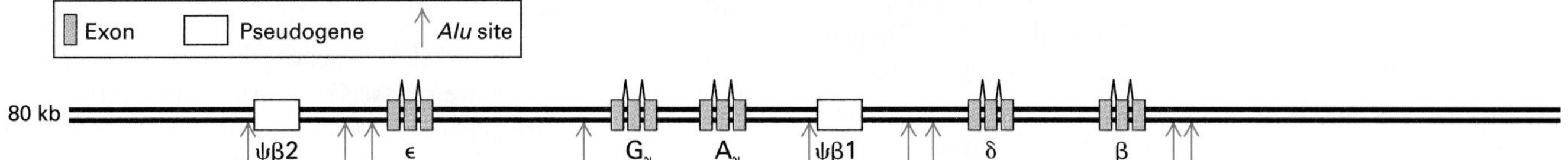

▲ **FIGURE 6-4 Structure of β-globin gene cluster and comparison of gene density in higher and lower eukaryotes.** (a) In the diagram of the β-globin gene cluster on human chromosome 11, the green boxes represent exons of β-globin–related genes. Exons spliced together to form one mRNA are connected by caret-like spikes. The human β-globin gene cluster contains two pseudogenes (white); these regions are related to the functional globin-type genes but are not transcribed. Each red arrow indicates the location of an *Alu* sequence, an ≈300-bp noncoding repeated sequence that is abundant in the human genome. (b) In the diagram of yeast DNA from chromosome III, the green boxes indicate open reading frames. Most of these potential protein-coding sequences are functional genes without introns. Note the much higher proportion of noncoding-to-coding sequences in the human DNA than in the yeast DNA. [Part (a), see F. S. Collins and S. M. Weissman, 1984, *Prog. Nucl. Acid Res. Mol. Biol.* **31**:315. Part (b), see S.G. Oliver et al., 1992, *Nature* **357**:28.]

sequence drift during evolution generated sequences that either terminate translation or block mRNA processing, rendering such regions nonfunctional. Because such pseudogenes are not deleterious, they remain in the genome and mark the location of a gene duplication that occurred in one of our ancestors.

Duplications of segments of a chromosome (called *segmental duplication*) occurred fairly often during the evolution of multicellular plants and animals. As a result, a large fraction of the genes in these organisms today have been duplicated, allowing the process of sequence drift to generate gene families and pseudogenes. The extent of sequence divergence between duplicated copies of the genome and characterization of the homologous genome sequences in related organisms allow an estimate of the time in evolutionary history when the duplication occurred. For example, the human fetal γ-globin genes (G_γ and A_γ) evolved following the duplication of a 5.5-kb region in the β-globin locus that included the single γ-globin gene in the common ancestor of caterrhine primates (old world monkeys, apes, and humans) and platyrrhine primates (new world monkeys) about 50 million years ago.

Although members of gene families that arose relatively recently during evolution are often found near each other on the same chromosome, as for genes of the human β-globin locus, members of gene families may also be found on different chromosomes in the same organism. This is the case for the human α-globin genes, which were separated from the β-globin genes by an ancient chromosomal translocation. Both the α- and β-globin genes evolved from a single ancestral globin gene that was duplicated (see Figure 6-2b) to generate the predecessors of the contemporary α- and β-globin genes in mammals. Both the primordial α- and β-globin genes then underwent further duplications to generate the different genes of the α- and β-globin gene clusters found in mammals today.

Several different gene families encode the various proteins that make up the cytoskeleton. These proteins are present in varying amounts in almost all cells. In vertebrates, the major cytoskeletal proteins are the actins, tubulins, and intermediate filament proteins like the keratins discussed in Chapters 17, 18 and 19. We examine the origin of one such family, the tubulin family, in Section 6.5. Although the physiological rationale for the cytoskeletal protein families is not as obvious as it is for the globins, the different members of a family probably have similar but subtly different functions suited to the particular type of cell in which they are expressed.

Heavily Used Gene Products Are Encoded by Multiple Copies of Genes

In vertebrates and invertebrates, the genes encoding ribosomal RNAs and some other nonprotein-coding RNAs such as those involved in RNA splicing occur as *tandemly repeated arrays*. These are distinguished from the duplicated genes of gene families in that the multiple tandemly repeated genes encode identical or nearly identical proteins or functional RNAs. Most often copies of a sequence appear one after the other, in a head-to-tail fashion, over a long stretch of DNA. Within a tandem array of rRNA genes, each copy is nearly exactly like all the others. Although the transcribed portions of rRNA genes are the same in a given individual, the nontranscribed spacer regions between the transcribed regions can vary.

These tandemly repeated RNA genes are needed to meet the great cellular demand for their transcripts. To understand why, consider that a fixed maximal number of RNA copies can be produced from a single gene during one cell generation when the gene is fully loaded with RNA polymerase molecules. If more RNA is required than can be transcribed from one gene, multiple copies of the gene are necessary. For example, during early embryonic development in humans, many embryonic cells have a doubling time of ≈24 hours and contain 5–10 million ribosomes. To produce enough rRNA to form this many ribosomes, an embryonic human cell needs at least 100 copies of the large and small subunit rRNA genes, and most of these must be close to maximally active for the cell to divide every 24 hours. That is, multiple RNA polymerases must be transcribing each rRNA gene at the same time (see Figure 8-33). Indeed, all eukaryotes, including yeasts, contain 100 or more copies of the genes encoding 5S rRNA and the large and small subunit rRNAs.

Multiple copies of tRNA genes and the genes encoding the histone proteins also occur. As we'll see later in this chapter, histones bind and organize nuclear DNA. Just as the cell requires multiple rRNA and tRNA genes to support efficient translation, multiple copies of the histone genes are required to produce sufficient histone protein to bind the large amount of nuclear DNA. While tRNA and histone genes often occur in clusters, they generally do not occur in tandem arrays in the human genome.

Nonprotein-Coding Genes Encode Functional RNAs

In addition to rRNA and tRNA genes, there are hundreds of additional genes that are transcribed into nonprotein-coding RNAs, some with various known functions, and many whose functions are not yet known. For example, **small nuclear RNAs (snRNAs)** function in RNA splicing, and **small nucleolar RNAs (snoRNAs)** function in rRNA processing and base modification in the nucleolus. The RNase P RNA functions in tRNA processing, and a large family (≈1000) of short **micro RNAs (miRNAs)** regulate the stability and translation of specific mRNAs. The functions of these nonprotein-coding RNAs are discussed in Chapter 8. An RNA found in telomerase (Chapter 4) functions in maintaining the sequence at the ends of chromosomes, and the 7SL RNA functions in the import of secreted proteins and most membrane proteins into the endoplasmic reticulum (Chapter 13). These and other nonprotein-coding RNAs encoded in the human genome and their functions, when known, are listed in Table 6-2.

TABLE 6-2 Known Nonprotein-Coding RNAs and Their Functions

RNA	NUMBER OF GENES IN HUMAN GENOME	FUNCTION
rRNAs	≈300	Protein synthesis
tRNAs	≈500	Protein synthesis
snRNAs	≈80	mRNA splicing
U7 snRNA	1	Histone mRNA 3′ processing
snoRNAs	≈85	Pre-rRNA processing and rRNA modification
miRNAs	≈1000	Regulation of gene expression
Xist	1	X-chromosome inactivation
7SK	1	Transcription control
RNAse P	1	tRNA 5′ processing
7SL RNA	3	Protein secretion (component of signal recognition particle, SRP)
RNAse MRP	1	rRNA processing
Telomerase RNA	1	Template for addition of telomeres
Vault RNAs	3	Components of Vault ribonucleoproteins (RNPs), function unknown
hY1, hY3, hY4, hY5	≈30	Components of Ro ribonucleoproteins (RNPs), function unknown
H19	1	Unknown

SOURCE: International Human Genome Sequencing Consortium, 2001, *Nature* **409**:860, and P. D. Zamore and B. Haley, 2005, *Science* **309**:1519.

KEY CONCEPTS OF SECTION 6.1

Eukaryotic Gene Structure

- In molecular terms, a gene is the entire DNA sequence required for synthesis of a functional protein or RNA molecule. In addition to the coding regions (exons), a gene includes control regions and sometimes introns.
- A simple eukaryotic transcription unit produces a single monocistronic mRNA, which is translated into a single protein.
- A complex eukaryotic transcription unit is transcribed into a primary transcript that can be processed into two or more different monocistronic mRNAs depending on the choice of splice sites or polyadenylation sites. A complex transcription unit with alternate promoters also generates two or more different mRNAs (see Figure 6-3b).
- Many complex transcription units (e.g., the fibronectin gene) express one mRNA in one cell type and an alternate mRNA in a different cell type.
- About half the protein-coding genes in vertebrate genomic DNA are solitary genes, each occurring only once in the haploid genome. The remainder are duplicated genes, which arose by duplication of an ancestral gene and subsequent independent mutations (see Figure 6-2b). The proteins encoded by a gene family have homologous but nonidentical amino acid sequences and exhibit similar but slightly different properties.
- In invertebrates and vertebrates, rRNAs are encoded by multiple copies of genes located in tandem arrays in genomic DNA. Multiple copies of tRNA and histone genes also occur, often in clusters, but not generally in tandem arrays.
- Many genes also encode functional RNAs that are not translated into protein but nonetheless perform significant functions, such as rRNA, tRNA and snRNAs. Among these are micro RNAs, possibly up to 1000 in humans, whose biological significance in regulating gene expression has only recently been appreciated.

6.2 Chromosomal Organization of Genes and Noncoding DNA

Having reviewed the relationship between transcription units and genes, we now consider the organization of genes on chromosomes and the relationship of noncoding DNA sequences to coding sequences.

Genomes of Many Organisms Contain Much Nonfunctional DNA

Comparisons of the total chromosomal DNA per cell in various species first suggested that much of the DNA in certain organisms does not encode RNA or have any apparent regulatory function. For example, yeasts, fruit flies, chickens, and humans have successively more DNA in their haploid chromosome sets (12; 180; 1300; and 3300 Mb, respectively), in keeping with what we perceive to be the increasing complexity of these organisms. Yet the vertebrates with the greatest amount of DNA per cell are amphibians, which are surely less complex than humans in their structure and behavior. Even more surprising, the unicellular protozoan species *Amoeba dubia* has 200 times more DNA per cell than humans. Many plant species also have considerably more DNA per cell than humans have. For example, tulips have 10 times as much DNA per cell as humans. The DNA content per cell also varies considerably between closely related species. All insects or all amphibians would appear to be similarly complex, but the amount of haploid DNA in species within each of these phylogenetic classes varies by a factor of 100.

Detailed sequencing and identification of exons in chromosomal DNA have provided direct evidence that the genomes of higher eukaryotes contain large amounts of noncoding DNA. For instance, only a small portion of the β-globin gene cluster of humans, about 80 kb long, encodes protein (see Figure 6-4a). In contrast, a typical 80-kb stretch of DNA from the yeast *S. cerevisiae,* a single-celled eukaryote, contains many closely spaced protein-coding sequences without introns and relatively much less noncoding DNA (see Figure 6-4b). Moreover, compared with other regions of vertebrate DNA, the β-globin gene cluster is unusually rich in protein-coding sequences, and the introns in globin genes are considerably shorter than those in many human genes.

The density of genes varies greatly in different regions of human chromosomal DNA, from "gene-rich" regions, such as the β-globin cluster, to large gene-poor "gene deserts." Of the 96 percent of human genomic DNA that has been sequenced, only ≈1.5 percent corresponds to protein-coding sequences (exons). We learned in the previous section that the intron sequences of genes are often significantly longer than the exon sequences. Approximately one-third of human genomic DNA is thought to be transcribed into pre-mRNA precursors or nonprotein-coding RNAs in one cell or another, but some 95 percent of this sequence is intronic, and thus removed by RNA splicing. This amounts to a large fraction of the total genome. The remaining two-thirds of human DNA is noncoding DNA between genes as well as regions of repeated DNA sequences that make up the centromeres and telomeres of the human chromosomes. Consequently, ≈98.5 percent of human DNA is noncoding.

Different selective pressures during evolution may account, at least in part, for the remarkable difference in the amount of nonfunctional DNA in unicellular and multicellular organisms. For example, microorganisms must compete for limited amounts of nutrients in their environment, and metabolic economy thus is a critical characteristic. Since synthesis of nonfunctional (i.e., noncoding) DNA requires time, nutrients and energy, presumably there was selective pressure to lose nonfunctional DNA during the evolution of microorganisms. On the other hand, natural selection in vertebrates depends largely on their behavior. The energy invested in DNA synthesis is trivial compared with the metabolic energy required for the movement of muscles; thus there was little selective pressure to eliminate nonfunctional DNA in vertebrates.

Most Simple-Sequence DNAs Are Concentrated in Specific Chromosomal Locations

Besides duplicated protein-coding genes and tandemly repeated genes, eukaryotic cells contain multiple copies of other DNA sequences in the genome, generally referred to as repetitious DNA (see Table 6-1). Of the two main types of repetitious DNA, the less prevalent is **simple-sequence DNA,** or **satellite DNA,** which constitutes about 6 percent of the human genome and is composed of perfect or nearly perfect repeats of relatively short sequences. The more common type of repetitious DNA, collectively called *interspersed repeats,* is composed of much longer sequences. These sequences, consisting of several types of transposable elements, are discussed in Section 6.3.

The length of each repeat in simple-sequence DNA can range from 1 to 500 base pairs. Simple-sequence DNAs in which the repeats contain 1–13 base pairs are often called *microsatellites.* Most microsatellite DNA has a repeat length of 1–4 base pairs and usually occurs in tandem repeats of 150 repeats or fewer. Microsatellites are thought to have originated by "backward slippage" of a daughter strand on its template strand during DNA replication so that the same short sequence is copied twice (Figure 6-5).

Microsatellites occasionally occur within transcription units. Some individuals are born with a larger number of repeats in specific genes than observed in the general population, presumably because of daughter-strand slippage during DNA replication in a germ cell from which they developed. Such expanded microsatellites have been found to cause at least 14 different types of neuromuscular diseases, depending on the gene in which they occur. In some cases expanded microsatellites behave like a recessive mutation because they interfere with just the function or expression of the encoded gene. But in the more common types of diseases associated with expanded microsatellite repeats, such as *myotonic dystrophy* and *spinocerebellar ataxia,* the expanded repeats behave like dominant mutations because they interfere with *all* RNA processing in the muscle cells and neurons where the affected genes are expressed. For example, in patients with myotonic dystrophy, transcripts of the *DMPK* gene contain between 100–4000 repeats of the sequence CUG in the 3′ untranslated region, compared to 50–100 repeats in normal individuals. The extended stretch of CUG repeats in affected individuals forms long RNA hairpins (see Figure 4-9) that interfere with normal RNA processing and export of transcripts from the nucleus to the cytosol. The double-stranded (ds) regions of these long RNA hairpins bind nuclear dsRNA-binding proteins, interfering with their normal function in regulating the alternative splicing of other specific pre-mRNAs essential for normal muscle and nerve cell function. ▮

Most satellite DNA is composed of repeats of 14–500 base pairs in tandem arrays of 20–100 kb. In situ hybridization studies with metaphase chromosomes have localized these simple-sequence DNAs to specific chromosomal regions. Much of this DNA lies near centromeres, the discrete chromosomal regions that attach to spindle microtubules during mitosis and meiosis (Figure 6-6). Experiments in the fission yeast *Schizosaccharomyces pombe* indicate that these sequences are required to form a specialized chromatin structure called *centromeric heterochromatin,* necessary for the proper segregation of chromosomes to daughter cells during mitosis. Simple-sequence DNA is also found in long tandem repeats at the ends of chromosomes, the telomeres, where they function to maintain chromosome ends and prevent their joining to the ends of other DNA molecules, as discussed further in the last section of this chapter.

(a) Normal replication

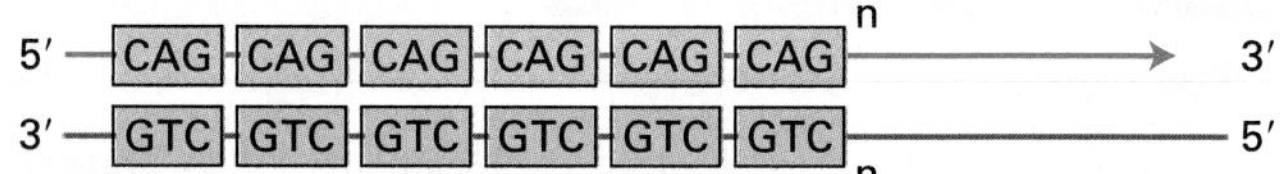

(b) Backward slippage

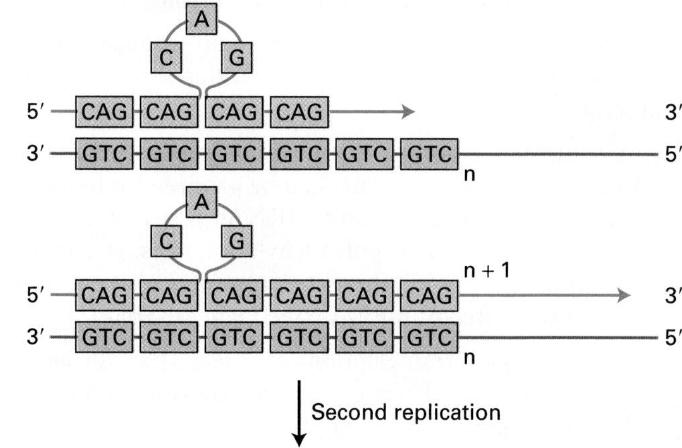

(c) Second replication

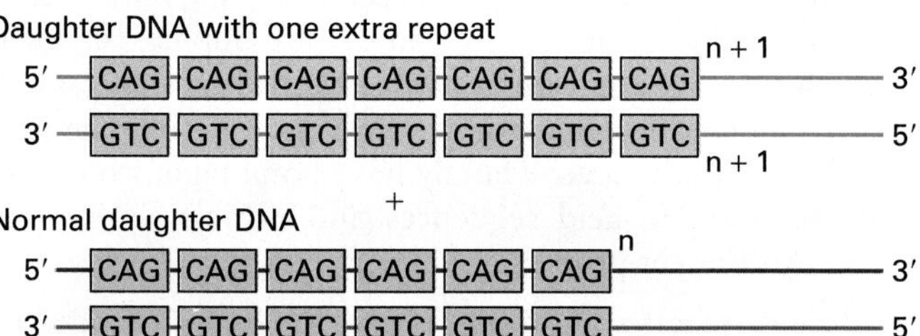

▲ **FIGURE 6-5 Generation of microsatellite repeats by backward slippage of the nascent daughter strand during DNA replication.** If during replication (a), the nascent daughter strand "slips" backward relative to the template strand by one repeat, one new copy of the repeat is added to the daughter strand when DNA replication continues (b). This extra copy of the repeat forms a single-stranded loop in the daughter strand of the daughter duplex DNA molecule. If this single-stranded loop is not removed by DNA repair proteins before the next round of DNA replication (c), the extra copy of the repeat is added to one of the double-stranded daughter DNA molecules; the other daughter molecule will be normal.

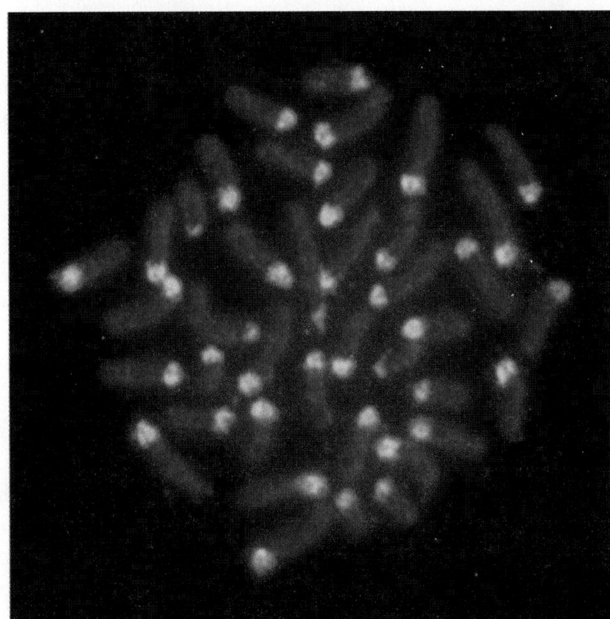

▲ **EXPERIMENTAL FIGURE 6-6 Simple-sequence DNA is localized at the centromere in mouse chromosomes.** Purified simple-sequence DNA from mouse cells was copied in vitro using *E. coli* DNA polymerase I and fluorescently labeled dNTPs to generate a fluorescently labeled DNA probe for mouse simple-sequence DNA. Chromosomes from cultured mouse cells were fixed and denatured on a microscope slide, and then the chromosomal DNA was hybridized in situ to the labeled probe (blue-green). The slide was also stained with DAPI, a DNA-binding dye, to visualize the full-length of the chromosomes (dark blue). Fluorescence microscopy shows that the simple-sequence probe hybridizes primarily to one end of the telocentric mouse chromosomes (i.e., chromosomes in which the centromeres are located near one end). [Courtesy of Sabine Mal, Ph.D., Manitoba Institute of Cell Biology, Canada.]

DNA Fingerprinting Depends on Differences in Length of Simple-Sequence DNAs

Within a species, the nucleotide sequences of the repeat units composing simple-sequence DNA tandem arrays are highly conserved among individuals. In contrast, the *number* of repeats, and thus the length of simple-sequence tandem arrays containing the same repeat unit, is quite variable among individuals. These differences in length are thought to result from unequal crossing over within regions of simple-sequence DNA during meiosis. As a consequence of this unequal crossing over, the lengths of some tandem arrays are unique in each individual.

In humans and other mammals, some of the simple-sequence DNA exists in relatively short 1- to 5-kb regions made up of 20–50 repeat units, each containing ≈14–100 base pairs. These regions are called *minisatellites.* Even slight differences in the total lengths of various minisatellites from different individuals can be detected by the polymerase chain reaction (PCR), using a mixture of several primers that hybridize to unique sequences flanking multiple minisatellites. These DNA *polymorphisms* (i.e., differences in sequence between individuals of the same species) form the basis of *DNA fingerprinting,* which is superior to conventional fingerprinting for identifying individuals (Figure 6-7). The use of PCR methods allows analysis of minute amounts of DNA, and individuals can be distinguished more precisely and reliably than by conventional fingerprinting.

Unclassified Spacer DNA Occupies a Significant Portion of the Genome

As Table 6.1 shows, ≈25 percent of human DNA lies between transcription units and is not repeated anywhere else in the genome. Much of this DNA probably arose from ancient transposable elements that accumulated so many mutations over evolutionary time they can no longer be recognized as having arisen from this source (Section 6.3).

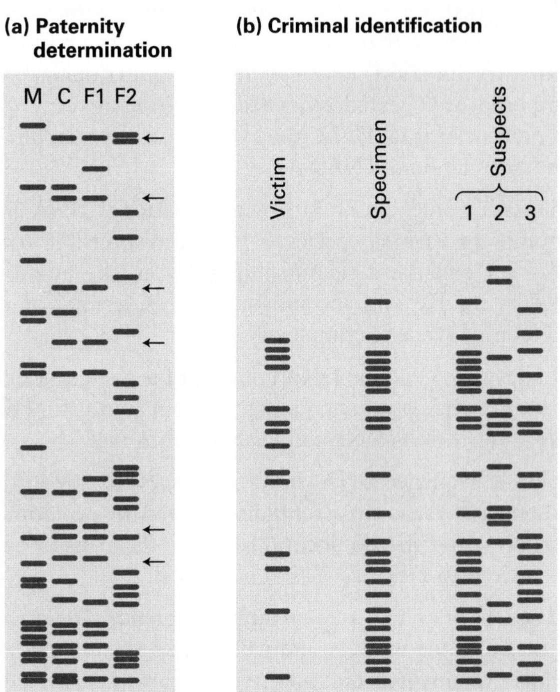

▲ **EXPERIMENTAL FIGURE 6-7 DNA fingerprinting is used to identify individuals in paternity cases and criminal investigations.** In DNA fingerprinting, a single PCR reaction is performed on template DNA from an individual using several sets of primers to unique sequences flanking the minisatellite repeat sequences. Gel electrophoresis of the PCR products generates a DNA fingerprint for the individual: that is, a set of minisatellites of different repeat lengths and hence mobility in the gel. (a) In this analysis of paternity, lane M shows the products using the mother's DNA as template in the PCR reaction; C, using the child's DNA; and F1 and F2 using DNA from two potential fathers. The child has minisatellite repeat lengths inherited from either the mother or F1, demonstrating that F1 is the father. Arrows indicate PCR products from F1, but not F2, found in the child's DNA. (b) In these DNA fingerprints of a specimen isolated from a rape victim and three men suspected of the crime, it is clear that minisatellite repeat lengths in the specimen match those of suspect 1. The victim's DNA was included in the analysis to ensure that the specimen DNA was not contaminated with DNA from the victim. [From T. Strachan and A. P. Read, *Human Molecular Genetics 2*, 1999, John Wiley & Sons.]

Transcription-control regions on the order of 50–200 base pairs in length that help to regulate transcription from distant promoters also occur in these long stretches of unclassified spacer DNA. In some cases, sequences of this seemingly nonfunctional DNA are nonetheless conserved during evolution, indicating that they may perform a significant function that is not yet understood. For example, they may contribute to the structures of chromosomes discussed in Section 6.7.

KEY CONCEPTS OF SECTION 6.2

Chromosomal Organization of Genes and Noncoding DNA

- In the genomes of prokaryotes and most lower eukaryotes, which contain few nonfunctional sequences, coding regions are densely arrayed along the genomic DNA.
- In contrast, vertebrate and higher plant genomes contain many sequences that do not code for RNAs or have any regulatory function. Much of this nonfunctional DNA is composed of repeated sequences. In humans, only about 1.5 percent of total DNA (the exons) actually encodes proteins or functional RNAs.
- Variation in the amount of nonfunctional DNA in the genomes of various species is largely responsible for the lack of a consistent relationship between the amount of DNA in the haploid chromosomes of an animal or plant and its phylogenetic complexity.
- Eukaryotic genomic DNA consists of three major classes of sequences: genes encoding proteins and functional RNAs, repetitious DNA, and spacer DNA (see Table 6-1).
- Simple-sequence DNA, short sequences repeated in long tandem arrays, is preferentially located in centromeres, telomeres, and specific locations within the arms of particular chromosomes.
- The length of a particular simple-sequence tandem array is quite variable between individuals in a species, probably because of unequal crossing over during meiosis. Differences in the lengths of some simple-sequence tandem arrays form the basis for DNA fingerprinting (see Figure 6-7).

6.3 Transposable (Mobile) DNA Elements

Interspersed repeats, the second type of repetitious DNA in eukaryotic genomes, is composed of a very large number of copies of relatively few sequence families (see Table 6-1). Also known as *moderately repeated DNA*, or *intermediate-repeat DNA*, these sequences are interspersed throughout mammalian genomes and make up ≈25–50 percent of mammalian DNA (≈45 percent of human DNA).

Because interspersed repeats have the unique ability to "move" in the genome, they are collectively referred to as transposable DNA elements or mobile DNA elements (we use these terms interchangeably). Although transposable DNA elements originally were discovered in eukaryotes, they also are found in prokaryotes. The process by which these sequences are copied and inserted into a new site in the genome is called **transposition.** Transposable DNA elements are essentially molecular symbionts that in most cases appear to have no specific function in the biology of their host organisms, but exist only to maintain themselves. For this reason, Francis Crick referred to them as "selfish DNA."

When transposition occurs in germ cells, the transposed sequences at their new sites are passed on to succeeding generations. In this way, mobile elements have multiplied and slowly accumulated in eukaryotic genomes over evolutionary time. Since mobile elements are eliminated from eukaryotic genomes very slowly, they now constitute a significant portion of the genomes of many eukaryotes.

Not only are mobile elements the source for much of the DNA in our genomes, but they also provided a second mechanism, in addition to meiotic recombination, for bringing about chromosomal DNA rearrangements during evolution (see Figure 6-2). This is because during transposition of a particular mobile element, adjacent DNA sometimes also is mobilized. Transpositions occur rarely: in humans, about one new germ-line transposition for every eight individuals. Since 98.5 percent of our DNA is noncoding, most transpositions have no deleterious effects. But over time they played an essential part in the evolution of genes having multiple exons and of genes whose expression is restricted to specific cell types or developmental periods. In other words, although transposable elements probably evolved as cellular symbionts, they have played a central role in the evolution of complex, multicellular organisms.

Transposition also may occur within a somatic cell; in this case the transposed sequence is transmitted only to the daughter cells derived from that cell. In rare cases, such somatic-cell transposition may lead to a somatic-cell mutation with detrimental phenotypic effects, for example, the inactivation of a tumor-suppressor gene (Chapter 25). In this section, we first describe the structure and transposition mechanisms of the major types of transposable DNA elements and then consider their likely role in evolution.

Movement of Mobile Elements Involves a DNA or an RNA Intermediate

Barbara McClintock discovered the first mobile elements while doing classical genetic experiments in maize (corn) during the 1940s. She characterized genetic entities that could move into and back out of genes, changing the phenotype of corn kernels. Her theories were very controversial until similar mobile elements were discovered in bacteria, where they were characterized as specific DNA sequences, and the molecular basis of their transposition was deciphered.

As research on mobile elements progressed, they were found to fall into two categories: (1) those that transpose directly as DNA and (2) those that transpose via an RNA intermediate transcribed from the mobile element by an RNA polymerase and then converted back into double-stranded DNA by a **reverse**

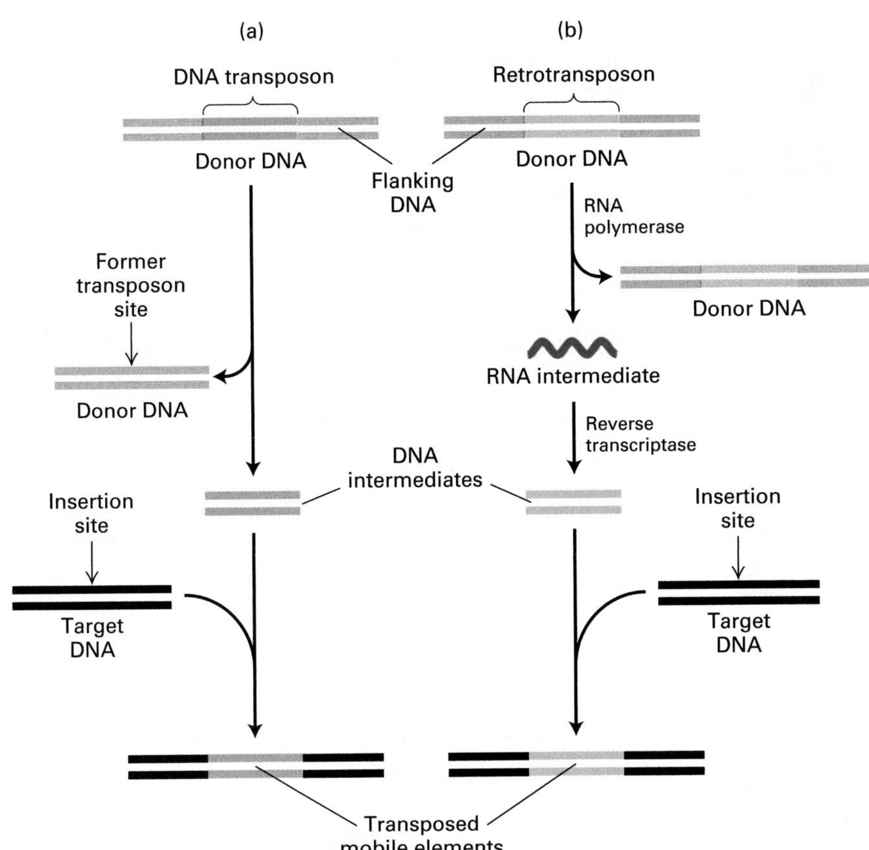

◀ **FIGURE 6-8 Two major classes of mobile DNA elements.** (a) Eukaryotic DNA transposons (orange) move via a DNA intermediate, which is excised from the donor site. (b) Retrotransposons (green) are first transcribed into an RNA molecule, which then is reverse-transcribed into double-stranded DNA. In both cases, the double-stranded DNA intermediate is integrated into the target-site DNA to complete movement. Thus DNA transposons move by a cut-and-paste mechanism, whereas retrotransposons move by a copy-and-paste mechanism.

transcriptase (Figure 6-8). Mobile elements that transpose directly as DNA are generally referred to as **DNA transposons,** or simply **transposons.** Eukaryotic DNA transposons excise themselves from one place in the genome, leaving that site and moving to another. Mobile elements that transpose to new sites in the genome via an RNA intermediate are called **retrotransposons.**

Retrotransposons make an RNA copy of themselves and introduce this new copy into another site in the genome, while also remaining at their original location. The movement of retrotransposons is analogous to the infectious process of retroviruses. Indeed, retroviruses can be thought of as retrotransposons that evolved genes encoding viral coats, thus allowing them to transpose between cells. Retrotransposons can be further classified on the basis of their specific mechanism of transposition. To summarize, DNA transposons can be thought of as transposing by a "cut-and-paste" mechanism, while retrotransposons move by a "copy-and-paste" mechanism in which the copy is an RNA intermediate.

DNA Transposons Are Present in Prokaryotes and Eukaryotes

Most mobile elements in bacteria transpose directly as DNA. In contrast, most mobile elements in eukaryotes are retrotransposons, but eukaryotic DNA transposons also occur. Indeed, the original mobile elements discovered by Barbara McClintock are DNA transposons.

Bacterial Insertion Sequences The first molecular understanding of mobile elements came from the study of certain *E. coli* mutations caused by the spontaneous insertion of a DNA sequence, ≈1–2 kb long, into the middle of a gene. These inserted stretches of DNA are called *insertion sequences,* or *IS elements.* So far, more than 20 different IS elements have been found in *E. coli* and other bacteria.

Transposition of an IS element is a very rare event, occurring in only one in 10^5–10^7 cells per generation, depending on the IS element. Often, transpositions inactivate essential genes, killing the host cell and the IS elements it carries. Therefore, higher rates of transposition would probably result in too great a mutation rate for the host organism to survive. However, since IS elements transpose more or less randomly, some transposed sequences enter nonessential regions of the genome (e.g., regions between genes), allowing the cell to survive. At a very low rate of transposition, most host cells survive and therefore propagate the symbiotic IS element. IS elements also can insert into plasmids or lysogenic viruses, and thus be transferred to other cells. In this way, IS elements can transpose into the chromosomes of virgin cells.

The general structure of IS elements is diagrammed in Figure 6-9. An *inverted repeat* of ≈50 base pairs is invariably present at each end of an insertion sequence. In an inverted repeat, the $5' \rightarrow 3'$ sequence on one strand is repeated on the other strand, such as:

$$\begin{array}{l} 5'\ \overrightarrow{\text{GAGC}}\text{——GCTC}\ 3' \\ 3'\ \text{CTCG——}\underleftarrow{\text{CGAG}}\ 5' \end{array}$$

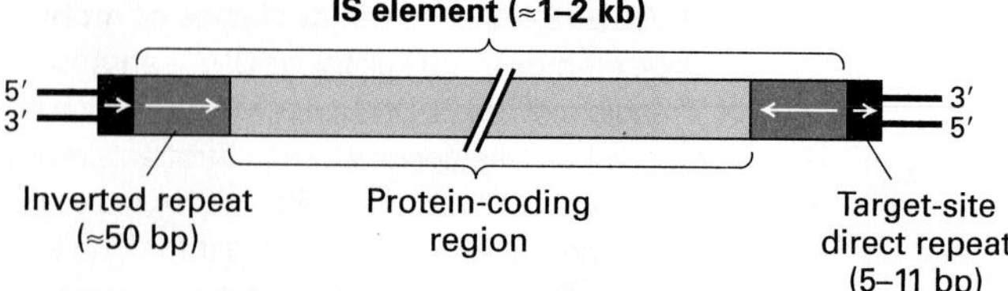

▲ **FIGURE 6-9 General structure of bacterial IS elements.** The relatively large central region of an IS element, which encodes one or two enzymes required for transposition, is flanked by an inverted repeat at each end. The sequences of the inverted repeats are nearly identical, but they are oriented in opposite directions. The inverted-repeat sequence is characteristic of a particular IS element. The 5′ and 3′ short *direct* (as opposed to *inverted*) repeats are not transposed with the insertion element; rather, they are insertion-site sequences that become duplicated, with one copy at each end, during insertion of a mobile element. The length of the direct repeats is constant for a given IS element, but their sequence depends on the site of insertion and therefore varies with each transposition of the IS element. Arrows indicate sequence orientation. The regions in this diagram are not to scale; the coding region makes up most of the length of an IS element.

Between the inverted repeats is a region that encodes *transposase,* an enzyme required for transposition of the IS element to a new site. The transposase is expressed very rarely, accounting for the very low frequency of transposition. An important hallmark of IS elements is the presence of a short *direct-repeat sequence,* containing 5–11 base pairs, immediately adjacent to both ends of the inserted element. The *length* of the direct repeat is characteristic of each type of IS element, but its *sequence* depends on the target site where a particular copy of the IS element inserted. When the sequence of a mutated gene containing an IS element is compared with the wild-type gene sequence, only one copy of the short direct-repeat sequence is found in the wild-type gene. Duplication of this target-site sequence to create the second direct repeat adjacent to an IS element occurs during the insertion process.

As depicted in Figure 6-10, transposition of an IS element occurs by a "cut-and-paste" mechanism. Transposase performs three functions in this process: it (1) precisely excises the IS element in the donor DNA, (2) makes staggered cuts in a short sequence in the target DNA, and (3) ligates the 3′ termini of the IS element to the 5′ ends of the cut donor DNA. Finally, a host-cell DNA polymerase fills in the single-stranded gaps, generating the short direct repeats that flank IS elements, and DNA ligase joins the free ends.

Eukaryotic DNA Transposons McClintock's original discovery of mobile elements came from observation of spontaneous mutations in maize that affect production of enzymes required to make anthocyanin, a purple pigment in maize kernels. Mutant kernels are white, and wild-type kernels are purple. One class of these mutations is revertible at high frequency, whereas a second class of mutations does not revert unless they occur in the presence of the first class of mutations. McClintock called the agent responsible for the first class of mutations the *activator (Ac) element* and those responsible for the second class *dissociation (Ds) elements* because they also tended to be associated with chromosome breaks.

Many years after McClintock's pioneering discoveries, cloning and sequencing revealed that Ac elements are equivalent to bacterial IS elements. Like IS elements, they contain inverted terminal repeat sequences that flank the coding region for a transposase, which recognizes the terminal repeats and catalyzes transposition to a new site in DNA. Ds elements

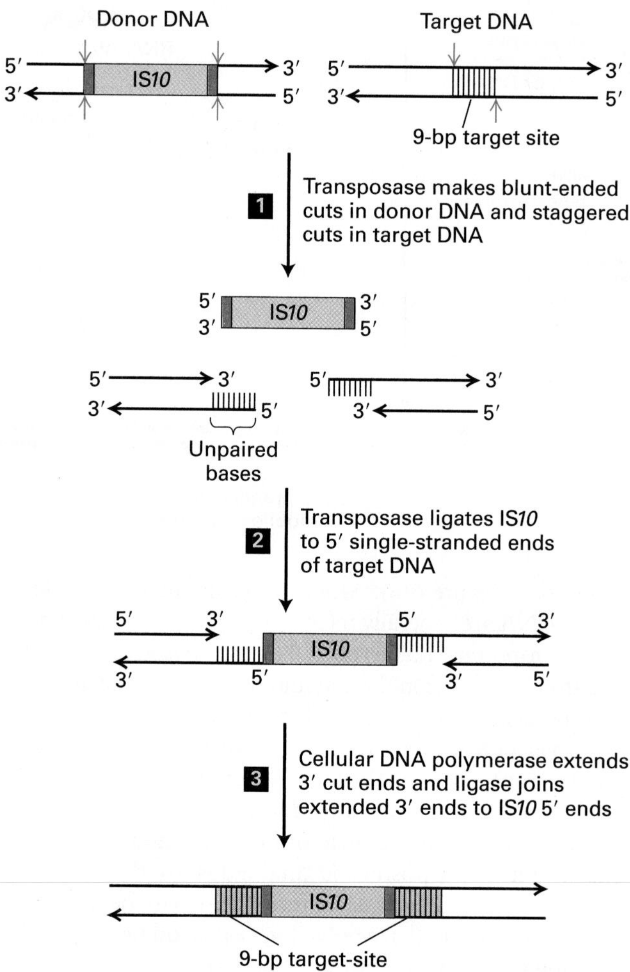

▲ **FIGURE 6-10 Model for transposition of bacterial insertion sequences.** Step **1**: Transposase, which is encoded by the IS element (IS*10* in this example), cleaves both strands of the donor DNA next to the inverted repeats (dark red), excising the IS*10* element. At a largely random target site, transposase makes staggered cuts in the target DNA. In the case of IS*10,* the two cuts are 9 bp apart. Step **2**: Ligation of the 3′ ends of the excised IS element to the staggered sites in the target DNA also is catalyzed by transposase. Step **3**: The 9-bp gaps of single-stranded DNA left in the resulting intermediate are filled in by a cellular DNA polymerase; finally cellular DNA ligase forms the 3′→5′ phosphodiester bonds between the 3′ ends of the extended target DNA strands and the 5′ ends of the IS*10* strands. This process results in duplication of the target-site sequence on each side of the inserted IS element. Note that the length of the target site and IS*10* are not to scale. [See H. W. Benjamin and N. Kleckner, 1989, *Cell* **59**:373, and 1992, *Proc. Nat'l. Acad. Sci. USA* **89**:4648.]

are deleted forms of the Ac element in which a portion of the sequence encoding transposase is missing. Because it does not encode a functional transposase, a Ds element cannot move by itself. However, in plants that carry the Ac element and thus express a functional transposase, Ds elements can be transposed because they retain the inverted terminal repeats recognized by the transposase.

Since McClintock's early work on mobile elements in corn, transposons have been identified in other eukaryotes. For instance, approximately half of all the spontaneous mutations observed in *Drosophila* are due to the insertion of mobile elements. Although most of the mobile elements in *Drosophila* function as retrotransposons, at least one—the *P element*—functions as a DNA transposon, moving by a mechanism similar to that used by bacterial insertion sequences. Current methods for constructing transgenic *Drosophila* depend on engineered, high-level expression of the P-element transposase and use of the P-element inverted terminal repeats as targets for transposition, as discussed in Chapter 5 (see Figure 5-25).

DNA transposition by the cut-and-paste mechanism can result in an increase in the copy number of a transposon if it occurs during the S phase of the cell cycle, when DNA synthesis occurs. An increase in the copy number happens when the donor DNA is from one of the two daughter DNA molecules in a region of a chromosome that has replicated and the target DNA is in the region that has not yet replicated. When DNA replication is complete at the end of the S phase, the target DNA in its new location is also replicated, resulting in a net increase in the total number of these transposons in the cell (Figure 6-11). When such a transposition occurs during the S phase preceding meiosis, one of the four germ cells produced contains the extra copy of the transposon. Repetition of this process over evolutionary time has resulted in the accumulation of large numbers of DNA transposons in the genomes of some organisms. Human DNA contains about 300,000 copies of full-length and deleted DNA transposons, amounting to ≈3 percent of human DNA. As we will see shortly, this mechanism can lead to the transposition of genomic DNA as well as the transposon itself.

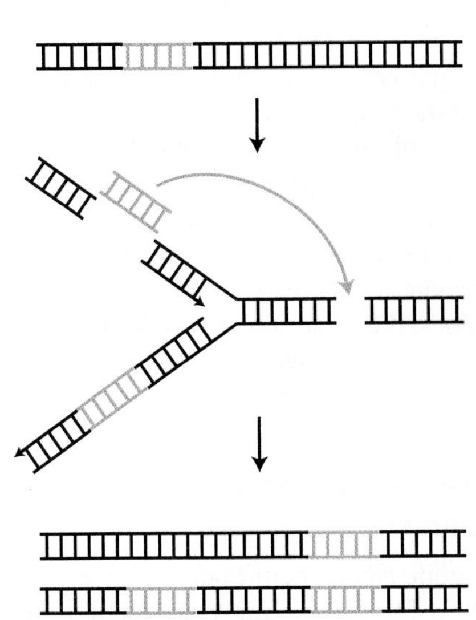

One copy of transposon before S phase

S phase: DNA replication and DNA transposition

After S phase, one daughter molecule has two copies of the transposon

▲ **FIGURE 6-11 Mechanism for increasing the copy number of DNA transposons.** In this model, a DNA transposon that moves by a cut-and-paste mechanism (see Figure 6-10) transposes during the S phase from a region of the chromosome that has replicated to a region that has not yet replicated. If this occurs, one of the two daughter chromosomes will have a net increase of one transposon insertion once chromosomal replication is completed.

LTR Retrotransposons Behave Like Intracellular Retroviruses

The genomes of all eukaryotes studied from yeast to humans contain retrotransposons, mobile DNA elements that transpose through an RNA intermediate utilizing a reverse transcriptase (see Figure 6-8b). These mobile elements are divided into two major categories, those containing and those lacking **long terminal repeats (LTRs).** LTR retrotransposons, which we discuss here, are common in yeast (e.g., Ty elements) and in *Drosophila* (e.g., *copia* elements). Although less abundant in mammals than non-LTR retrotransposons, LTR retrotransposons nonetheless constitute ≈8 percent of human genomic DNA. In mammals, retrotransposons lacking LTRs are the most common type of mobile element; these are described in the next section.

The general structure of LTR retrotransposons found in eukaryotes is depicted in Figure 6-12. In addition to short 5′ and 3′ direct repeats typical of all transposons, these retrotransposons are marked by the presence of LTRs flanking the central protein-coding region. These *l*ong direct *t*erminal *r*epeats, containing ≈250–600 base pairs depending on the type of LTR retrotransposon, are characteristic of integrated retroviral DNA and are critical to the life cycle of retroviruses. In addition to sharing LTRs with retroviruses, LTR retrotransposons encode all the proteins of the most common type of retroviruses, except for the envelope proteins. Lacking these envelope proteins, LTR retrotransposons cannot bud from their host cell and infect other cells; however, they can transpose to new sites in the DNA

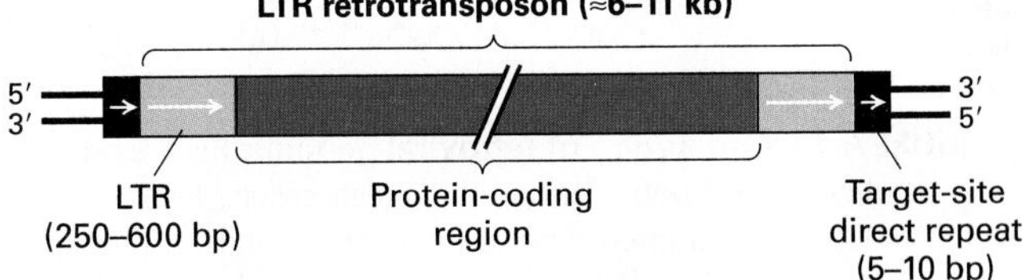

▲ **FIGURE 6-12 General structure of eukaryotic LTR retrotransposons.** The central protein-coding region is flanked by two long terminal repeats (LTRs), which are element-specific direct repeats. Like other mobile elements, integrated retrotransposons have short target-site direct repeats at each end. Note that the different regions are not drawn to scale. The protein-coding region constitutes 80 percent or more of a retrotransposon and encodes reverse transcriptase, integrase, and other retroviral proteins.

of their host cell. Because of their clear relationship with retroviruses, this class of retrotransposons are often called *retrovirus-like elements*.

A key step in the retroviral life cycle is formation of retroviral genomic RNA from integrated retroviral DNA (see Figure 4-49). This process serves as a model for generation of the RNA intermediate during transposition of LTR retrotransposons. As depicted in Figure 6-13, the leftward retroviral LTR functions as a promoter that directs host-cell RNA polymerase to initiate transcription at the 5′ nucleotide of the R sequence. After the entire downstream retroviral DNA has been transcribed, the RNA sequence corresponding to the rightward LTR directs host-cell RNA-processing enzymes to cleave the primary transcript and add a poly(A) tail at the 3′ end of the R sequence. The resulting retroviral RNA genome, which lacks a complete LTR, exits the nucleus and is packaged into a virion that buds from the host cell.

After a retrovirus infects a cell, reverse transcription of its RNA genome by the retrovirus-encoded reverse transcriptase yields a double-stranded DNA containing complete LTRs (Figure 6-14). This DNA synthesis takes place in the cytosol. The double-stranded DNA with an LTR at each end is then transported into the nucleus in a complex with integrase, another enzyme encoded by retroviruses. Retroviral integrases are closely related to the transposases encoded by DNA transposons and use a similar mechanism to insert the double-stranded retroviral DNA into the host-cell genome. In this process, short direct repeats of the target-site sequence are generated at either end of the inserted viral DNA sequence. Although the mechanism of reverse transcription is complex, it is a critical aspect of the retrovirus life cycle. The process generates the complete 5′ LTR that functions as a promoter for initiation of transcription precisely at the 5′ nucleotide of the R sequence, while the complete 3′ LTR functions as a poly(A) site leading to polyadenylation precisely at the 3′ nucleotide of the R sequence. Consequently, no nucleotides are lost from a LTR retrotransposon as it undergoes successive rounds of insertion, transcription, reverse transcription and re-insertion at a new site.

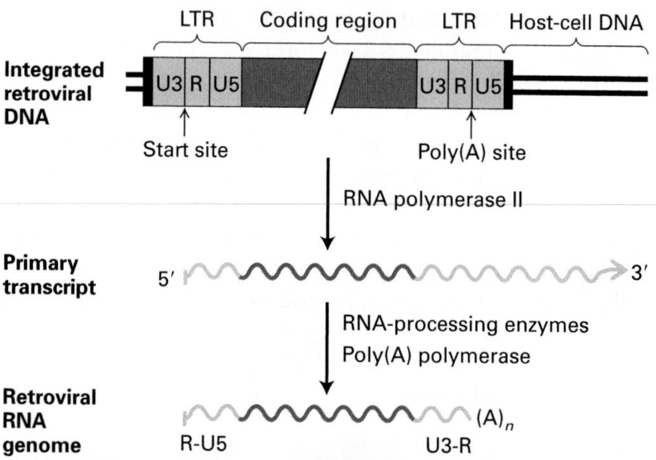

▲ **FIGURE 6-13 Generation of retroviral genomic RNA from integrated retroviral DNA.** The left LTR directs cellular RNA polymerase to initiate transcription at the first nucleotide of the left R region. The resulting primary transcript extends beyond the right LTR. The right LTR, now present in the RNA primary transcript, directs cellular enzymes to cleave the primary transcript at the last nucleotide of the right R region and to add a poly(A) tail, yielding a retroviral RNA genome with the structure shown at the top of Figure 6-14. A similar mechanism is thought to generate the RNA intermediate during transposition of retrotransposons. The short direct-repeat sequences (black) of target-site DNA are generated during integration of the retroviral DNA into the host-cell genome.

► **FIGURE 6-14 Model for reverse transcription of retroviral genomic RNA into DNA.** In this model, a complicated series of nine events generates a double-stranded DNA copy of the single-stranded RNA genome of a retrovirus (*top*). The genomic RNA is packaged in the virion with a retrovirus-specific cellular tRNA hybridized to a complementary sequence near its 5′ end called the *primer-binding site* (PBS). The retroviral RNA has a short direct-repeat terminal sequence (R) at each end. The overall reaction is carried out by reverse transcriptase, which catalyzes polymerization of deoxyribonucleotides. RNaseH digests the RNA strand in a DNA–RNA hybrid. The entire process yields a double-stranded DNA molecule that is longer than the template RNA and has a long terminal repeat (LTR) at each end. The different regions are not shown to scale. The PBS and R regions are actually much shorter than the U5 and U3 regions, and the central coding region is very much longer than the other regions. [See E. Gilboa et al., 1979, *Cell* **18**:93.]

As noted above, LTR retrotransposons encode reverse transcriptase and integrase. By analogy with retroviruses, these mobile elements move by a "copy-and-paste" mechanism whereby reverse transcriptase converts an RNA copy of a donor element into DNA, which is inserted into a target site by integrase. The experiments depicted in Figure 6-15 provided strong evidence for the role of an RNA intermediate in transposition of Ty elements.

The most common LTR retrotransposons in humans are called *ERVs*, for *e*ndogenous *r*etro*v*iruses. Most of the 443,000 ERV-related DNA sequences in the human genome consist only of isolated LTRs. These are derived from full-length proviral DNA by homologous recombination between the two LTRs, resulting in deletion of the internal retroviral sequences. Isolated LTRs such as these cannot be transposed to a new position in the genome, but recombination between homologous LTRs at different positions in the genome have likely contributed to the chromosomal DNA rearrangements leading to gene and exon duplications, the evolution of proteins with new combinations of exons, and, as we will see in Chapter 7, the evolution of complex control of gene expression.

Non-LTR Retrotransposons Transpose by a Distinct Mechanism

The most abundant mobile elements in mammals are retrotransposons that lack LTRs, sometimes called *nonviral retrotransposons*. These moderately repeated DNA sequences form two classes in mammalian genomes: **LINEs** (long interspersed elements) and **SINEs** (short interspersed elements). In humans, full-length LINEs are ≈6 kb long, and SINEs are ≈300 bp long (see Table 6-1). Repeated sequences with the characteristics of LINEs have been observed in protozoans, insects, and plants, but for unknown reasons they

Focus Animation: Retroviral Reverse Transcription

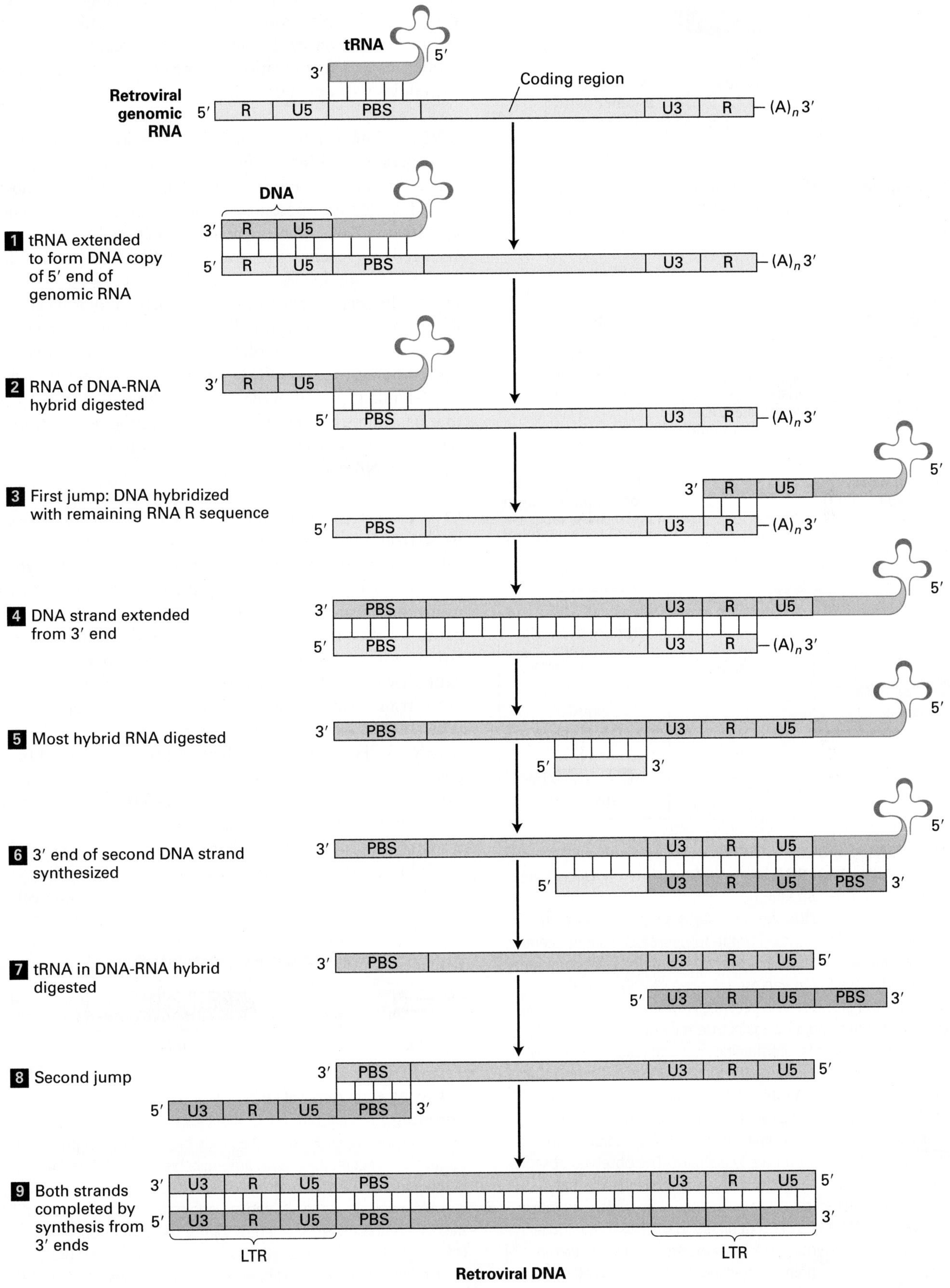

MEDIA CONNECTIONS

Retroviral genomic RNA

5′ R | U5 | PBS | Coding region | U3 | R —(A)$_n$ 3′

tRNA 3′ 5′

1 tRNA extended to form DNA copy of 5′ end of genomic RNA

DNA

3′ R | U5

5′ R | U5 | PBS | U3 | R —(A)$_n$ 3′

2 RNA of DNA-RNA hybrid digested

3′ R | U5

5′ PBS | U3 | R —(A)$_n$ 3′

3 First jump: DNA hybridized with remaining RNA R sequence

3′ R | U5 5′

5′ PBS | U3 | R —(A)$_n$ 3′

4 DNA strand extended from 3′ end

3′ PBS | U3 | R | U5 5′

5′ PBS | U3 | R —(A)$_n$ 3′

5 Most hybrid RNA digested

3′ PBS | U3 | R | U5 5′

5′ 3′

6 3′ end of second DNA strand synthesized

3′ PBS | U3 | R | U5 5′

5′ U3 | R | U5 | PBS 3′

7 tRNA in DNA-RNA hybrid digested

3′ PBS | U3 | R | U5 5′

5′ U3 | R | U5 | PBS 3′

8 Second jump

3′ PBS | U3 | R | U5 5′

5′ U3 | R | U5 | PBS 3′

9 Both strands completed by synthesis from 3′ ends

3′ U3 | R | U5 | PBS | U3 | R | U5 5′

5′ U3 | R | U5 | PBS | 3′

LTR LTR

Retroviral DNA

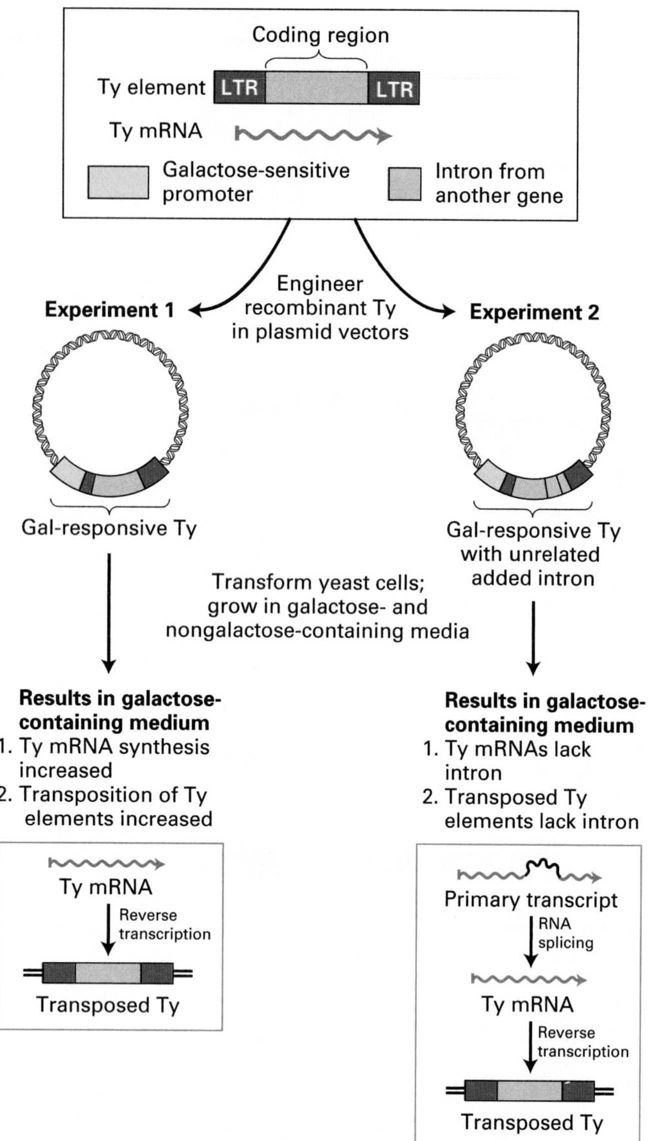

▲ EXPERIMENTAL FIGURE 6-15 **The yeast Ty element transposes through an RNA intermediate.** When yeast cells are transformed with a Ty-containing plasmid, the Ty element can transpose to new sites, although normally this occurs at a low rate. Using the elements diagrammed at the top, researchers engineered two different recombinant plasmid vectors containing recombinant Ty elements adjacent to a galactose-sensitive promoter. These plasmids were transformed into yeast cells, which were grown in a galactose-containing and a nongalactose medium. In experiment 1, growth of cells in galactose-containing medium resulted in many more transpositions than in nongalactose medium, indicating that transcription into an mRNA intermediate is required for Ty transposition. In experiment 2, an intron from an unrelated yeast gene was inserted into the putative protein-coding region of the recombinant galactose-responsive Ty element. The observed absence of the intron in transposed Ty elements is strong evidence that transposition involves an mRNA intermediate from which the intron was removed by RNA splicing, as depicted in the box in the lower right. In contrast, eukaryotic DNA transposons, like the Ac element of maize, contain introns within the transposase gene, indicating that they do not transpose via an RNA intermediate. [See J. Boeke et al., 1985, *Cell* **40**:491.]

are particularly abundant in the genomes of mammals. SINEs also are found primarily in mammalian DNA. Large numbers of LINEs and SINEs in higher eukaryotes have accumulated over evolutionary time by repeated copying of sequences at a few positions in the genome and insertion of the copies into new positions.

LINEs Human DNA contains three major families of LINE sequences that are similar in their mechanism of transposition, but differ in their sequences: L1, L2, and L3. Only members of the L1 family transpose in the contemporary human genome. Apparently there are no remaining functional copies of L2 or L3. LINE sequences are present at ≈900,000 sites in the human genome, accounting for a staggering 21 percent of total human DNA. The general structure of a complete LINE is diagrammed in Figure 6-16. LINEs usually are flanked by short direct repeats, the hallmark of mobile elements, and contain two long open reading frames (ORFs). ORF1, ≈1 kb long, encodes an RNA-binding protein. ORF2, ≈4 kb long, encodes a protein that has a long region of homology with the reverse transcriptases of retroviruses and LTR retrotransposons, but also exhibits DNA endonuclease activity.

Evidence for the mobility of L1 elements first came from analysis of DNA cloned from humans with certain genetic diseases such as hemophilia and myotonic dystrophy. DNA from these patients was found to carry mutations resulting from insertion of an L1 element into a gene, whereas no such element occurred within this gene in either parent. About 1 in 600 mutations that cause significant disease in humans are due to L1 transpositions or SINE transpositions that are catalyzed by L1-encoded proteins. Later experiments similar to those just described with yeast Ty elements (see Figure 6-15) confirmed that L1 elements transpose through an RNA intermediate. In these experiments, an intron was introduced into a cloned mouse L1 element, and the recombinant L1 element was stably transformed into cultured hamster cells. After several cell doublings, a PCR-amplified fragment corresponding to the L1 element but lacking the inserted intron was detected in

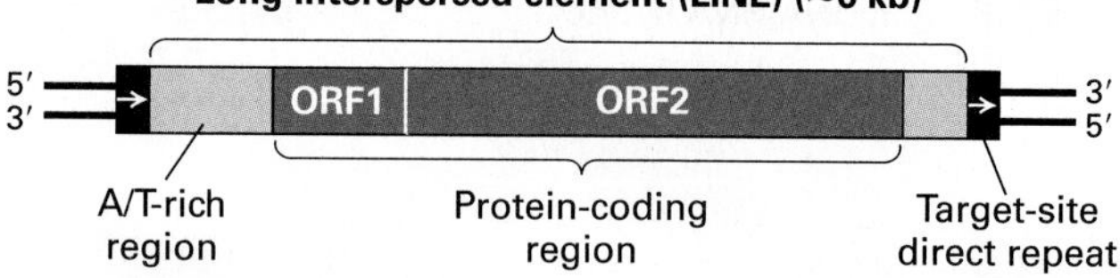

▲ **FIGURE 6-16 General structure of a LINE, a non-LTR retrotransposon.** Mammalian DNA carries two classes of non-LTR retrotransposons, LINES and SINES (not shown). The length of the target-site direct repeats varies among copies of a LINE at different sites in the genome. Although the full-length L1 sequence is ≈6 kb long, variable amounts of the left end are absent at over 90 percent of the sites where this mobile element is found. The shorter open reading frame (ORF1), ≈1 kb in length, encodes an RNA-binding protein. The longer ORF2, ≈4 kb in length, encodes a bifunctional protein with reverse transcriptase and DNA endonuclease activity. Note that LINEs lack the long terminal repeats found in LTR retrotransposons.

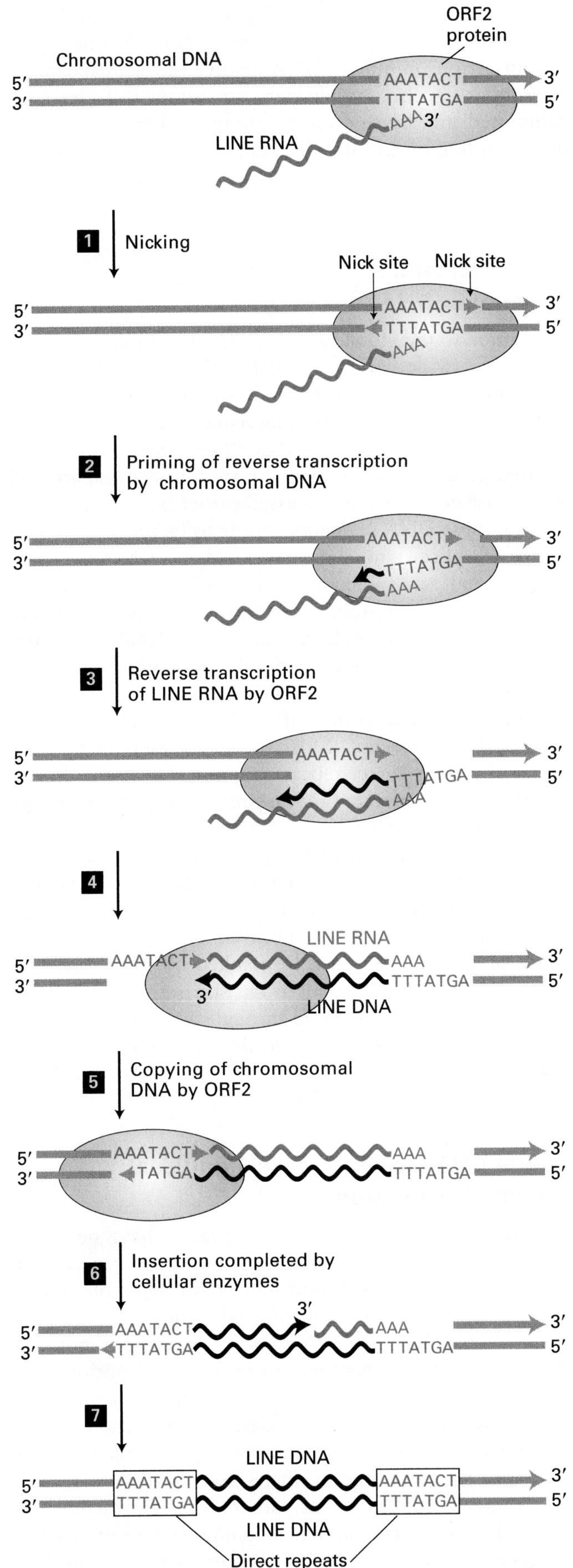

◀ **FIGURE 6-17 Proposed mechanism of LINE reverse transcription and integration.** After synthesis of ORF1 and ORF2 in the cytosol, a complex of LINE RNA (red), multiple copies of ORF1 and one copy of ORF2 bound to the Poly(A) tail moves into the nucleus. Only ORF2 protein, a reverse transcriptase, is represented. Newly synthesized LINE DNA is shown in black. Step **1**: ORF2 makes staggered nicks in chromosomal DNA on either side of any accessible A/T-rich sequence in the genome. Step **2**: Reverse transcription of LINE RNA by ORF2 is primed by the single-stranded T-rich sequence generated by the nick in the bottom strand, which hybridizes to the LINE poly(A) tail. Steps **3**–**5**: ORF2 reverse-transcribes the LINE RNA and then continues this new DNA strand, switching to the single-stranded region of the upper chromosomal strand as a template. Step **6**: Cellular enzymes then hydrolyze the RNA and extend the 3′ end of the chromosomal DNA top strand, replacing the LINE RNA strand with DNA. Step **7**: Finally, the 5′ and 3′ ends of the DNA strands are ligated, completing the insertion. The last two steps probably are catalyzed by the same cellular enzymes that remove RNA primers and ligate Okazaki fragments during DNA replication (see Figure 4-30). [Adapted from D. D. Luan et al., 1993, *Cell* **72**:595.]

the cells. This finding strongly suggests that over time the recombinant L1 element containing the inserted intron had transposed to new sites in the hamster genome through an RNA intermediate that underwent RNA splicing to remove the intron. ∎

Since LINEs do not contain LTRs, their mechanism of transposition through an RNA intermediate differs from that of LTR retrotransposons. ORF1 and ORF2 proteins are translated from a LINE RNA. In vitro studies indicate that transcription by RNA polymerase is directed by promoter sequences at the left end of integrated LINE DNA. LINE RNA is polyadenylated by the same post-transcriptional mechanism that polyadenylates other mRNAs. The LINE RNA then is exported into the cytosol, where it is translated into ORF1 and ORF2 proteins. Multiple copies of ORF1 protein then bind to the LINE RNA, and ORF2 protein binds to the poly(A) tail. The LINE RNA is then transported back into the nucleus as a complex with ORF1 and ORF2 proteins, and is reverse-transcribed into LINE DNA in the nucleus by ORF2. The mechanism involves staggered cleavage of cellular DNA at the insertion site followed by priming of reverse transcription by the resulting cleaved cellular DNA as detailed in Figure 6-17. The complete process results in insertion of a copy of the original LINE retrotransposon into a new site in chromosomal DNA. A short direct repeat is generated at the insertion site because of the initial staggered cleavage of the two chromosomal DNA strands.

As noted already, the DNA form of an LTR retrotransposon is synthesized from its RNA form in the cytosol (see Figure 6-14) and then transported into the nucleus where it is integrated into chromosomal DNA by a retrotransposon-encoded integrase. In contrast, the DNA form of a non-LTR retrotransposon is synthesized in the nucleus. The synthesis of the first strand of the non-LTR retroviral DNA by ORF2, a reverse transcriptase, is primed by the 3′-end of cleaved chromosomal DNA, which base pairs with the poly(A) tail of the non-LTR retroviral RNA (see Figure 6-17, step **2**). Since its synthesis is primed by the cut end of a cleaved chromosome, and synthesis of the other strand of the non-LTR retrotransposon DNA is primed by the 3′-end of chromosomal DNA on the other side of the initial cut in chromosomal DNA (step **6**), the mechanism of synthesis results in integration of the non-LTR retrotransposon DNA. There is no need for an integrase to insert the non-LTR retrotransposon DNA.

The vast majority of LINEs in the human genome are truncated at their 5′ end, suggesting that reverse transcription terminated before completion, and the resulting fragments extending variable distances from the poly(A) tail were inserted. Because of this shortening, the average size of LINE elements is only about 900 base pairs, even though the full-length sequence is ≈6 kb long. Truncated LINE elements, once formed, probably are not further transposed because they lack a promoter for formation of the RNA intermediate in transposition. In addition to the fact that most L1 insertions are truncated, nearly all the full-length elements contain stop codons and frameshift mutations in ORF1 and ORF2; these mutations probably have accumulated in most LINE sequences over evolutionary time. As a result, only ≈0.01 percent of the LINE sequences in the human genome are full-length with intact open reading frames for ORF1 and ORF2, representing ≈60–100 in total number.

SINEs The second most abundant class of mobile elements in the human genome, SINEs constitute ≈13 percent of total human DNA. Varying in length from about 100 to 400 base pairs, these retrotransposons do not encode protein, but most contain a 3′ A/T-rich sequence similar to that in LINEs. SINEs are transcribed by the same nuclear RNA polymerase that transcribes genes encoding tRNAs, 5S rRNAs, and other small stable RNAs. Most likely, the ORF1 and ORF2 proteins expressed from full-length LINEs mediate reverse transcription and integration of SINEs by the mechanism depicted in Figure 6-17. Consequently, SINEs can be viewed as parasites of the LINE symbionts, competing with LINE RNAs for binding and reverse transcription/integration by LINE encoded ORF1 and ORF2.

SINEs occur at about 1.6 million sites in the human genome. Of these, ≈1.1 million are *Alu elements,* so named because most of them contain a single recognition site for the restriction enzyme *Alu*I. *Alu* elements exhibit considerable sequence homology with and probably evolved from 7SL RNA, a cytosolic RNA in a ribonucleoprotein complex called the signal recognition particle. This abundant cytosolic ribonucleoprotein particle aids in targeting certain polypeptides to the membranes of the endoplasmic reticulum (Chapter 13). *Alu* elements are scattered throughout the human genome at sites where their insertion has not disrupted gene expression: between genes, within introns, and in the 3′ untranslated regions of some mRNAs. For instance, nine *Alu* elements are located within the human β-globin gene cluster (see Figure 6-4a). Of the one new germ-line non-LTR retrotransposition that is estimated to occur in about every eight individuals, ≈40 percent involve L1 and 60 percent involve SINEs, of which ≈90 percent are *Alu* elements. (Note that nearly all new insertions in human DNA involve retrotransposons.)

Similar to other mobile elements, most SINEs have accumulated mutations from the time of their insertion in the germ line of an ancient ancestor of modern humans. Like LINEs, many SINEs also are truncated at their 5′ end.

Other Retrotransposed RNAs Are Found in Genomic DNA

In addition to the mobile elements listed in Table 6-1, DNA copies of a wide variety of mRNAs appear to have integrated into chromosomal DNA. Since these sequences lack introns and do not have flanking sequences similar to those of the functional gene copies, they clearly are not simply duplicated genes that have drifted into nonfunctionality and become pseudogenes, as discussed earlier (see Figure 6-4a). Instead, these DNA segments appear to be retrotransposed copies of spliced and polyadenylated mRNA. Compared with normal genes encoding mRNAs, these inserted segments generally contain multiple mutations, which are thought to have accumulated since their mRNAs were first reverse-transcribed and randomly integrated into the genome of a germ cell in an ancient ancestor. These nonfunctional genomic copies of mRNAs are referred to as *processed pseudogenes.* Most processed pseudogenes are flanked by short direct repeats, supporting the hypothesis that they were generated by rare retrotransposition events involving cellular mRNAs.

Other interspersed repeats representing partial or mutant copies of genes encoding small nuclear RNAs (snRNAs) and tRNAs are found in mammalian genomes. Like processed pseudogenes derived from mRNAs, these nonfunctional copies of small RNA genes are flanked by short direct repeats and most likely result from rare retrotransposition events that have accumulated through the course of evolution. Enzymes expressed from a LINE are thought to have carried out all these retrotransposition events involving mRNAs, snRNAs, and tRNAs.

Mobile DNA Elements Have Significantly Influenced Evolution

Although mobile DNA elements appear to have no direct function other than to maintain their own existence, their presence has had a profound impact on the evolution of modern-day organisms. As mentioned earlier, about half the spontaneous mutations in *Drosophila* result from insertion of a mobile DNA element into or near a transcription unit. In mammals, mobile elements cause a much smaller proportion of spontaneous mutations: ≈10 percent in mice, and only 0.1–0.2 percent in humans. Still, mobile elements have been found in mutant alleles associated with several human genetic diseases. For example, insertions into the clotting factor IX gene cause hemophilia, and insertions into the gene encoding the muscle protein dystrophin lead to myotonic dystrophy, commonly known as Duchenne muscular dystrophy. The genes encoding factor IX and dystrophin are both on the X chromosome. Because the male genome has only one copy of the X chromosome, transposition insertions into these genes predominantly affect males.

In lineages leading to higher eukaryotes, homologous recombination between mobile DNA elements dispersed throughout ancestral genomes may have generated gene duplications and other DNA rearrangements during evolution (see Figure 6-2b). For instance, cloning and sequencing of the β-globin gene cluster from various primate species has provided strong evidence that the human G_γ and A_γ genes arose from an unequal homologous crossover between two L1 sequences flanking an ancestral globin gene. Subsequent divergence of such duplicated genes could lead to acquisition of distinct, beneficial functions associated with each member of a gene family. Unequal crossing over between mobile elements located within introns of a particular gene could lead to the duplication of exons within that gene (see Figure 6-2a). This process most likely influenced the evolution of genes that contain multiple copies of similar exons encoding similar protein domains, such as the fibronectin gene (see Figure 4-16).

Some evidence suggests that during the evolution of higher eukaryotes, recombination between mobile DNA elements (e.g., *Alu* elements) in introns of *two separate* genes also occurred, generating new genes made from novel combinations of preexisting exons (Figure 6-18). This evolutionary process, termed **exon shuffling,** may have occurred during evolution of the genes encoding tissue plasminogen activator, the Neu receptor, and epidermal growth factor, which all contain an EGF domain (see Figure 3-11). In this case, exon shuffling presumably resulted in insertion of an EGF domain–encoding exon into an intron of the ancestral form of each of these genes.

Both DNA transposons and LINE retrotransposons have been shown to occasionally carry unrelated flanking sequences when they transpose to new sites by the mechanisms diagrammed in Figure 6-19. These mechanisms likely also contributed to exon shuffling during the evolution of contemporary genes.

In addition to causing changes in coding sequences in the genome, recombination between mobile elements and transposition of DNA adjacent to DNA transposons and retrotransposons likely played a significant role in the evolution of regulatory sequences that control gene expression. As noted earlier, eukaryotic genes have transcription-control regions called enhancers that can operate over distances of tens of thousands of base pairs. Transcription of many genes is controlled through the combined effects of several enhancer elements. Insertion of mobile elements near such transcription-control regions probably contributed to the evolution of new combinations of enhancer sequences. These in turn control which specific genes are expressed in particular cell types and the amount of the encoded protein produced in modern organisms, as we discuss in the next chapter.

These considerations suggest that the early view of mobile DNA elements as completely selfish molecular parasites misses the mark. Rather, they have contributed profoundly to the evolution of higher organisms by promoting (1) the generation of gene families via gene duplication, (2) the creation of new genes via shuffling of preexisting exons, and (3) formation of more complex regulatory regions that provide multifaceted control of gene expression. Today, researchers are attempting to harness transposition mechanisms for inserting therapeutic genes into patients as a form of gene therapy.

KEY CONCEPTS OF SECTION 6.3

Transposable (Mobile) DNA Elements

- Transposable DNA elements are moderately repeated sequences interspersed at multiple sites throughout the genomes of higher eukaryotes. They are present less frequently in prokaryotic genomes.
- DNA transposons move to new sites directly as DNA; retrotransposons are first transcribed into an RNA copy of the element, which then is reverse-transcribed into DNA (see Figure 6-8).
- A common feature of all mobile elements is the presence of short direct repeats flanking the sequence.
- Enzymes encoded by transposons themselves catalyze insertion of these sequences at new sites in genomic DNA.
- Although DNA transposons, similar in structure to bacterial IS elements, occur in eukaryotes (e.g., the *Drosophila* P element), retrotransposons generally are much more abundant, especially in vertebrates.
- LTR retrotransposons are flanked by long terminal repeats (LTRs), similar to those in retroviral DNA; like retroviruses, they encode reverse transcriptase and integrase. They move in the genome by being transcribed into RNA, which then undergoes reverse transcription in

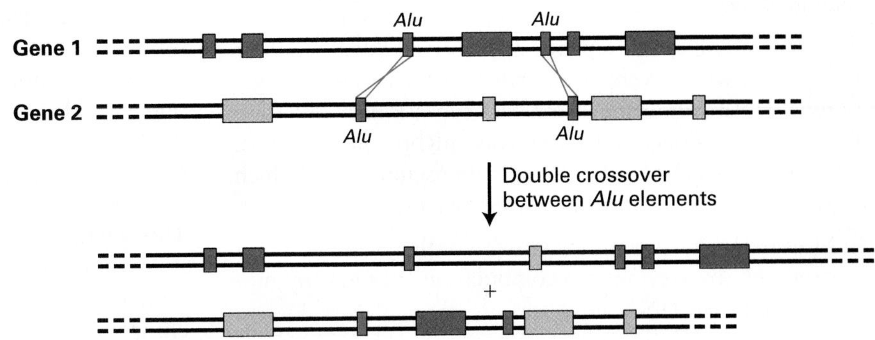

▶ **FIGURE 6-18 Exon shuffling via recombination between homologous interspersed repeats.** Recombination between interspersed repeats in the introns of separate genes produces transcription units with a new combination of exons (green and blue). In the example shown here, a double crossover between two sets of *Alu* elements, (the most abundant SINEs in humans) results in an exchange of exons between the two genes.

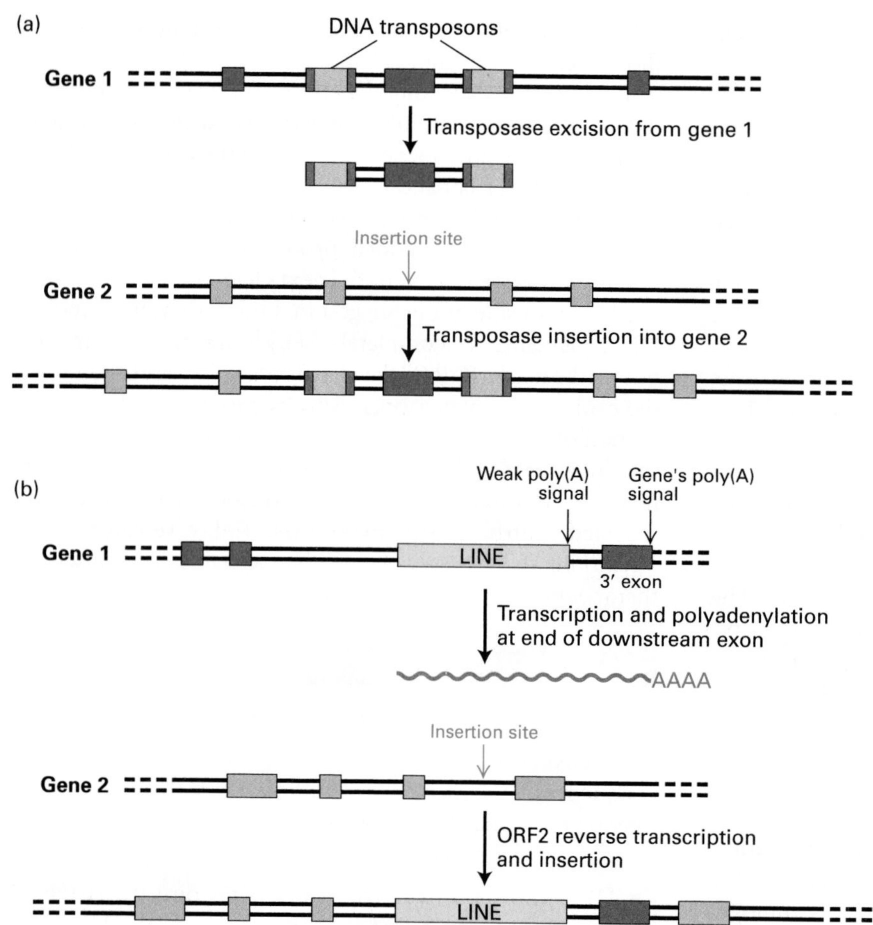

◀ **FIGURE 6-19 Exon shuffling via transposition of a DNA transposon or LINE retrotransposon.** (a) Transposition of an exon (blue) flanked by homologous DNA transposons into an intron of a second gene. As we saw in Figure 6-10, step **1**, transposase can recognize and cleave the DNA at the ends of the transposon inverted repeats. In gene 1, if the transposase cleaves at the left end of the transposon on the left and at the right end of the transposon on the right, it can transpose all the intervening DNA, including the exon from gene 1, to a new site in an intron of gene 2. The net result is an insertion of the exon from gene 1 into gene 2. (b) Integration of an exon into another gene via transposition of a LINE containing a weak poly(A) signal. If such a LINE is in the 3′-most intron of gene 1, transcription of the LINE into an RNA intermediate may continue beyond its own poly(A) site and extend into the 3′ exon, transcribing the cleavage and polyadenylation site of gene 1 itself. This RNA can then be reverse-transcribed and integrated by the LINE ORF2 protein (see Figure 6-17) into an intron on gene 2, introducing a new 3′ exon (from gene 1) into gene 2.

the cytosol, nuclear import of the resulting DNA with LTRs, and integration into a host-cell chromosome (see Figure 6-14).

- Non-LTR retrotransposons, including long interspersed elements (LINEs) and short interspersed elements (SINEs), lack LTRs and have an A/T-rich stretch at one end. They are thought to move by a nonviral retrotransposition mechanism mediated by LINE-encoded proteins involving priming of reverse transcription by chromosomal DNA (see Figure 6-17).
- SINE sequences exhibit extensive homology with small cellular RNAs and are transcribed by the same RNA polymerase. *Alu* elements, the most common SINEs in humans, are ≈300-bp sequences found scattered throughout the human genome.
- Some interspersed repeats are derived from cellular RNAs that were reverse-transcribed and inserted into genomic DNA at some time in evolutionary history. Processed pseudogenes derived from mRNAs lack introns, a feature that distinguishes them from pseudogenes, which arose by sequence drift of duplicated genes.
- Mobile DNA elements most likely influenced evolution significantly by serving as recombination sites and by mobilizing adjacent DNA sequences.

6.4 Organelle DNAs

Although the vast majority of DNA in most eukaryotes is found in the nucleus, some DNA is present within the mitochondria of animals, plants, and fungi and within the chloroplasts of plants. These organelles are the main cellular sites for ATP formation, during oxidative phosphorylation in mitochondria and photosynthesis in chloroplasts (Chapter 12). Many lines of evidence indicate that mitochondria and chloroplasts evolved from bacteria that were endocytosed into ancestral cells containing a eukaryotic nucleus, forming **endosymbionts** (Figure 6-20). Over evolutionary time, most of the bacterial genes were lost from organellar DNAs. Some, such as genes encoding proteins involved in nucleotide, lipid and amino acid biosynthesis, were lost because their functions were provided by genes in the nucleus of the host cell. Other genes encoding components of the present-day organelles were transferred to the nucleus. However, mitochondria and chloroplasts in today's eukaryotes retain DNAs encoding some proteins essential for organellar function, as well as the ribosomal and transfer RNAs required for synthesis of these proteins. Thus eukaryotic cells have multiple genetic systems: a predominant nuclear system and secondary systems with their own DNA, ribosomes, and tRNAs in mitochondria and chloroplasts.

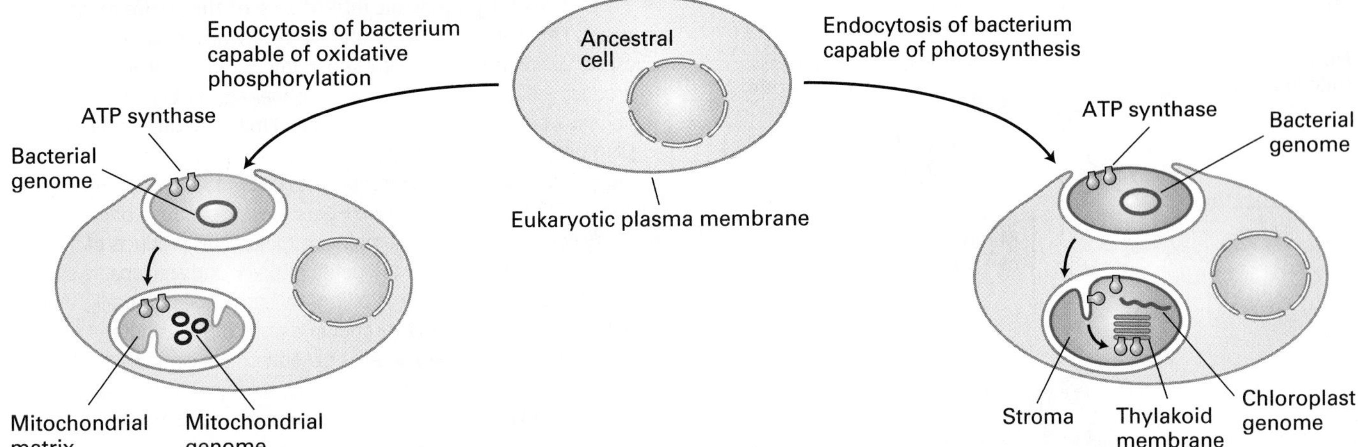

▲ **FIGURE 6-20 Model for endosymbiotic origin of mitochondria and chloroplasts.** Endocytosis of a bacterium by an ancestral eukaryotic cell would generate an organelle with two membranes, the outer membrane derived from the eukaryotic plasma membrane (gray) and the inner one from the bacterial plasma membrane (red). Proteins localized to the ancestral bacterial membrane retain their orientation, such that the portion of the protein once facing the extracellular space now faces the intermembrane space. Budding of vesicles from the inner chloroplast membrane, such as occurs during development of chloroplasts in contemporary plants, would generate the thylakoid membranes of chloroplasts. The organellar DNAs are indicated.

Mitochondria Contain Multiple mtDNA Molecules

Individual mitochondria are large enough to be seen under the light microscope, and even the mitochondrial DNA (mtDNA) can be detected by fluorescence microscopy. The mtDNA is located in the interior of the mitochondrion, the region known as the matrix (see Figure 12-6). As judged by the number of yellow fluorescent "dots" of mtDNA, a *Euglena gracilis* cell contains at least 30 mtDNA molecules (Figure 6-21).

Replication of mtDNA and division of the mitochondrial network can be followed in living cells using time-lapse microscopy. Such studies show that in most organisms mtDNA replicates throughout interphase. At mitosis each daughter cell receives approximately the same number of mitochondria, but since there is no mechanism for apportioning exactly equal numbers of mitochondria to the daughter cells, some cells contain more mtDNA than others. By isolating mitochondria from cells and analyzing the DNA extracted from them, it can be seen that each mitochondrion contains multiple mtDNA molecules. Thus the total amount of mtDNA in a cell depends on the number of mitochondria, the size of the mtDNA, and the number of mtDNA molecules per mitochondrion. Each of these parameters varies greatly between different cell types.

mtDNA Is Inherited Cytoplasmically

Studies of mutants in yeasts and other single-celled organisms first indicated that mitochondria exhibit *cytoplasmic inheritance* and thus must contain their own genetic system (Figure 6-22). For instance, *petite* yeast mutants exhibit structurally abnormal mitochondria and are incapable of oxidative phosphorylation. As a result, petite cells grow more slowly than wild-type yeasts and form smaller colonies. Genetic crosses between different (haploid) yeast strains showed that the *petite* mutation does not segregate with any known nuclear gene or chromosome. In later studies, most petite mutants were found to contain deletions of mtDNA.

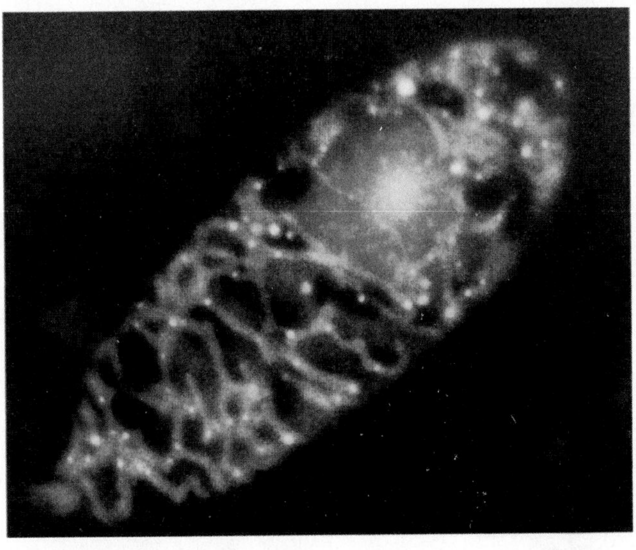

▲ EXPERIMENTAL FIGURE 6-21 **Dual staining reveals the multiple mitochondrial DNA molecules in a growing *Euglena gracilis* cell.** Cells were treated with a mixture of two dyes: ethidium bromide, which binds to DNA and emits a red fluorescence, and DiOC6, which is incorporated specifically into mitochondria and emits a green fluorescence. Thus the nucleus emits a red fluorescence, and areas rich in mitochondrial DNA fluoresce yellow—a combination of red DNA and green mitochondrial fluorescence. [From Y. Hayashi and K. Ueda, 1989, *J. Cell Sci.* **93:**565.]

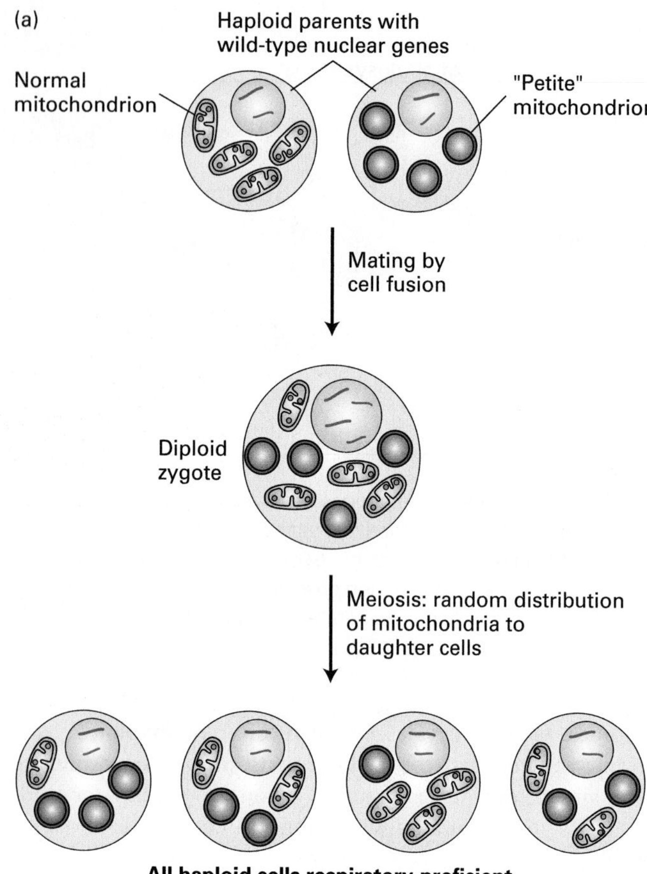

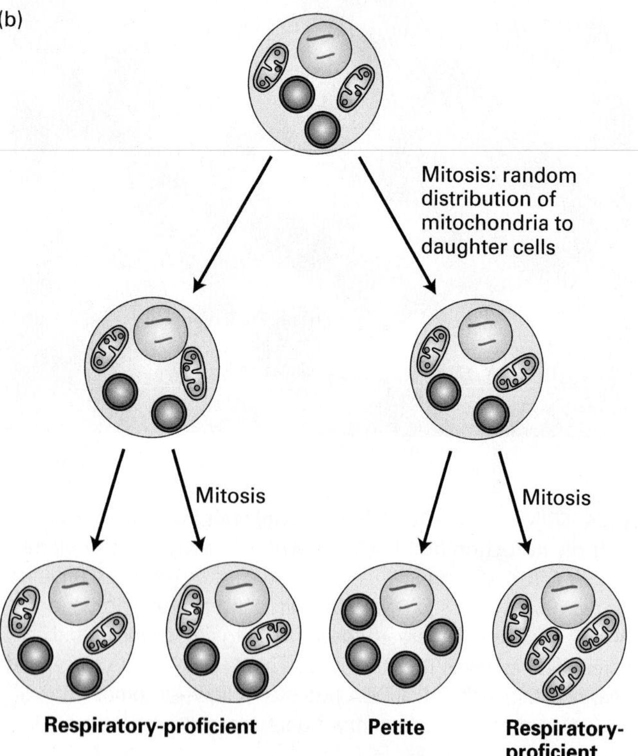

◀ **FIGURE 6-22 Cytoplasmic inheritance of the *petite* mutation in yeast.** Petite-strain mitochondria are defective in oxidative phosphorylation owing to a deletion in mtDNA. (a) Haploid cells fuse to produce a diploid cell that undergoes meiosis, during which random segregation of parental chromosomes and mitochondria containing mtDNA occurs. Note that alleles for genes in nuclear DNA (represented by large and small nuclear chromosomes colored red and blue) segregate 2:2 during meiosis (see Figure 5-5). In contrast, since yeast normally contain ≈50 mtDNA molecules per cell, all products of meiosis usually contain both normal and petite mtDNAs and are capable of respiration. (b) As these haploid cells grow and divide mitotically, the cytoplasm (including the mitochondria) is randomly distributed to the daughter cells. Occasionally, a cell is generated that contains only defective petite mtDNA and yields a petite colony. Thus formation of such petite cells is independent of any nuclear genetic marker.

In the mating by fusion of haploid yeast cells, both parents contribute equally to the cytoplasm of the resulting diploid; thus inheritance of mitochondria is biparental. In mammals and most other multicellular organisms, however, the sperm contributes little (if any) cytoplasm to the zygote, and virtually all the mitochondria in the embryo are derived from those in the egg, not the sperm. Studies in mice have shown that 99.99 percent of mtDNA is maternally inherited, but a small part (0.01 percent) is inherited from the male parent. In higher plants, mtDNA is inherited exclusively in a uniparental fashion through the female parent (egg), not the male (pollen).

The Size, Structure, and Coding Capacity of mtDNA Vary Considerably Between Organisms

Surprisingly, the size of the mtDNA, the number and nature of the proteins it encodes, and even the mitochondrial genetic code itself vary greatly between different organisms. The mtDNAs of most multicellular animals are ≈16-kb circular molecules that encode intron-less genes compactly arranged on both DNA strands. Vertebrate mtDNAs encode the two rRNAs found in mitochondrial ribosomes, the 22 tRNAs used to translate mitochondrial mRNAs, and 13 proteins involved in electron transport and ATP synthesis (Chapter 12). The smallest mitochondrial genomes known are in *Plasmodium*, single-celled obligate intracellular parasites that cause malaria in humans. *Plasmodium* mtDNAs are only ≈6 kb, encoding five proteins and the mitochondrial rRNAs.

The entire mitochondrial genomes from a number of different metazoan organisms (i.e., multicellular animals) have now been cloned and sequenced, and mtDNAs from all these sources encode essential mitochondrial proteins (Figure 6-23). All proteins encoded by mtDNA are synthesized on mitochondrial ribosomes. Most mitochondria-synthesized polypeptides identified thus far are subunits of multimeric complexes used in electron transport, ATP synthesis, or insertion of proteins into the inner mitochondrial membrane or intermembrane space. However, most of the proteins localized in mitochondria, such as those involved in the processes listed at the top of Figure 6-23, are encoded by nuclear genes, synthesized on cytosolic ribosomes, and imported into the organelle by processes discussed in Chapter 13.

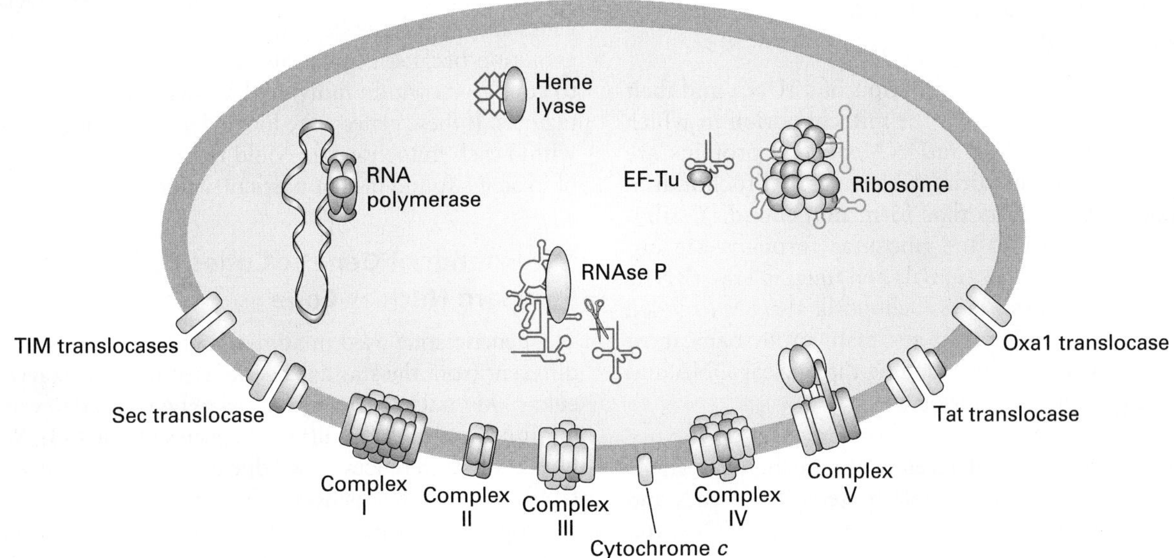

▲ **FIGURE 6-23 Proteins encoded in mitochondrial DNA and their involvement in mitochondrial processes.** Only the mitochondrial matrix and inner membrane are depicted. Most mitochondrial components are encoded by the nucleus (blue); mitochondrial processes carried out by exclusively nucleus-encoded components are listed at the top. Mitochondrial components shown in pink are encoded by mtDNA in some eukaryotes but by the nuclear genome in other eukaryotes. The relatively few components invariably encoded in mtDNA are shown in orange. Complexes I–V are involved in electron transport and oxidative phosphorylation. TIM, Sec, Tat, and Oxa1 translocases are involved in protein import and export, and insertion of proteins into the inner membrane (Chapter 13). RNase P is a ribozyme that processes the 5′-end of tRNAs (Chapter 8). It should be noted that the majority of eukaryotes have a multi-subunit complex I as depicted here, with three subunits invariantly encoded by mtDNA. However, in a few organisms (e.g., *Saccharomyces, Schizosaccharomyces,* and *Plasmodium*), this complex is replaced by a nucleus-encoded, single-polypeptide enzyme. For more details on mitochondrial metabolism and transport, see Chapters 12 and 13. [Adapted from G. Burger et al., 2003, *Trends Genet.* **19:**709.]

In contrast to metazoan mtDNAs, plant mtDNAs are many times larger, and most of the DNA does not encode protein. For instance, the mtDNA in the important model plant *Arabidopsis thaliana* is 366,924 base pairs, and the largest known mtDNA is ≈2 Mb, found in cucurbit plants (e.g., melon and cucumber). Most plant mtDNA consists of long introns, pseudogenes, mobile DNA elements restricted to the mitochondrial compartment, and pieces of foreign (chloroplast, nuclear and viral) DNA that were probably inserted into plant mitochondrial genomes during their evolution. Duplicated sequences also contribute to the greater length of plant mtDNAs.

Differences in the number of genes encoded by the mtDNA from various organisms most likely reflect the movement of DNA between mitochondria and the nucleus during evolution. Direct evidence for this movement comes from the observation that several proteins encoded by mtDNA in some species are encoded by nuclear DNA in other, closely related species. The most striking example of this phenomenon involves the *cox II* gene, which encodes subunit 2 of cytochrome *c* oxidase, which constitutes complex IV in the mitochondrial electron transport chain (see Figure 12-16). This gene is found in mtDNA in all multicellular plants studied except for certain related species of legumes, including the mung bean and soybeans, in which the *cox II* gene is nuclear. The *cox II* gene is completely missing from mung bean mtDNA, but a defective *cox II* pseudogene that has accumulated many mutations can still be recognized in soybean mtDNA.

Many RNA transcripts of plant mitochondrial genes are edited, mainly by the enzyme-catalyzed conversion of selected C residues to U, and occasionally U to C. (RNA editing is discussed in Chapter 8.) The nuclear *cox II* gene of mung bean corresponds more closely to the edited cox II RNA transcripts than to the mitochondrial *cox II* genes found in other legumes. These observations are strong evidence that the *cox II* gene moved from the mitochondrion to the nucleus during mung bean evolution by a process that involved an RNA intermediate. Presumably this movement involved a reverse-transcription mechanism similar to that by which processed pseudogenes are generated in the nuclear genome from nucleus-encoded mRNAs.

In addition to the large differences in the sizes of mtDNAs in different eukaryotes, the structure of the mtDNA also varies greatly. As mentioned above, mtDNA in most animals is a circular molecule ≈16 kb. However, the mtDNA of many organisms such as the protist *Tetrahymena* exists as linear head-to-tail concatomers of repeating sequence. In the most extreme examples, the mtDNA of the protist *Amoebidium parasiticum* is composed of several hundred distinct short linear molecules. And the mtDNA of *Trypanosoma* is comprised of multiple *maxicircles* concatenated (interlocked) to thousands of *minicircles* encoding *guide RNAs* involved in editing the sequence of the mitochondrial mRNAs encoded in the maxicircles.

Products of Mitochondrial Genes Are Not Exported

As far as is known, all RNA transcripts of mtDNA and their translation products remain in the mitochondrion in which they are produced, and all mtDNA-encoded proteins are synthesized on mitochondrial ribosomes. Mitochondrial DNA encodes the rRNAs that form mitochondrial ribosomes, although most of the ribosomal proteins are imported from the cytosol. In animals and fungi, all the tRNAs used for protein synthesis in mitochondria also are encoded by mtDNAs. However, in plants and many protozoans, most mitochondrial tRNAs are encoded by the nuclear DNA and imported into the mitochondrion.

Reflecting the bacterial ancestry of mitochondria, mitochondrial ribosomes resemble bacterial ribosomes and differ from eukaryotic cytosolic ribosomes in their RNA and protein compositions, their size, and their sensitivity to certain antibiotics (see Figure 4-22). For instance, chloramphenicol blocks protein synthesis by bacterial and mitochondrial ribosomes from most organisms, but cycloheximide does not. This sensitivity of mitochondrial ribosomes to the important aminoglycoside class of antibiotics that includes chloramphenicol is the main cause of the toxicity that these antibiotics can cause. Conversely, cytosolic ribosomes are sensitive to cycloheximide and resistant to chloramphenicol. ▮

Mitochondria Evolved from a Single Endosymbiotic Event Involving a *Rickettsia*-like Bacterium

Analysis of the mtDNA sequences from various eukaryotes, including single-celled protists that diverged from other eukaryotes early in evolution, provides strong support for the idea that the mitochondrion had a single origin. Mitochondria most likely arose from a bacterial symbiont whose closest contemporary relatives are in the *Rickettsiaceae* group. Bacteria in this group are obligate intracellular parasites. Thus, the ancestor of the mitochondrion probably also had an intracellular life style, putting it in a good location for evolving into an intracellular symbiont. The mtDNA with the largest number of encoded genes so far found is in the protist species *Reclinomonas americana.* All other mtDNAs have a subset of the *R. americana* genes, strongly implying that they evolved from a common ancestor with *R. americana,* losing different groups of mitochondrial genes by deletion and/or transfer to the nucleus over time.

In organisms whose mtDNA includes only a limited number of genes, the same set of mitochondrial genes are retained, independent of the phyla that includes these organisms (see Figure 6-23, orange proteins). One hypothesis for why these genes were never successfully transferred to the nuclear genome is that their encoded polypeptides are too hydrophobic to cross the outer mitochondrial membrane, and therefore would not be imported back into the mitochondria if they were synthesized in the cytosol. Similarly, the large size of rRNAs may interfere with their transport from the nucleus through the cytosol into mitochondria. Alternatively, these genes may not have been transferred to the nucleus during evolution because regulation of their expression in response to conditions within individual mitochondria may be advantageous. If these genes were located in the nucleus, conditions within each mitochondria could not influence the expression of proteins found in that particular mitochondrion.

Mitochondrial Genetic Codes Differ from the Standard Nuclear Code

The genetic code used in animal and fungal mitochondria is different from the standard code used in all prokaryotic and eukaryotic nuclear genes; remarkably, the code even differs in mitochondria from different species (Table 6-3). Why and how these differences arose during evolution is mysterious. UGA, for example, is normally a stop codon, but is read as tryptophan by human and fungal mitochondrial translation systems; however, in plant mitochondria, UGA is still recognized as a stop codon. AGA and AGG, the standard nuclear codons for arginine, also code for arginine in fungal and plant mtDNA, but they are stop codons in mammalian mtDNA and serine codons in *Drosophila* mtDNA.

As shown in Table 6-3, plant mitochondria appear to utilize the standard genetic code. However, comparisons of the amino acid sequences of plant mitochondrial proteins with the nucleotide sequences of plant mtDNAs suggested that CGG could code for *either* arginine (the "standard" amino acid) or tryptophan. This apparent nonspecificity of the plant mitochondrial code is explained by editing of mitochondrial RNA transcripts, which can convert cytosine residues to uracil residues. If a CGG sequence is edited to UGG, the codon specifies tryptophan, the standard amino acid for UGG, whereas unedited CGG codons encode the standard arginine. Thus the translation system in plant mitochondria does utilize the standard genetic code. ▮

Mutations in Mitochondrial DNA Cause Several Genetic Diseases in Humans

The severity of disease caused by a mutation in mtDNA depends on the nature of the mutation and on the proportion of mutant and wild-type mtDNAs present in a particular cell type. Generally, when mutations in mtDNA are found, cells contain mixtures of wild-type and mutant mtDNAs—a condition known as *heteroplasmy.* Each time a mammalian somatic or germ-line cell divides, the mutant and wild-type mtDNAs segregate randomly into the daughter cells, as occurs in yeast cells (see Figure 6-22b). Thus, the mtDNA genotype, which fluctuates from one generation and from one cell division to the next, can drift toward predominantly wild-type or predominantly mutant mtDNAs. Since all enzymes required for the replication and growth of mammalian mitochondria, such as the mitochondrial DNA and

special case of chromatin condensation, the inactivation of X chromosomes in female mammals.

Histone Acetylation Histone-tail lysines undergo reversible acetylation and deacetylation by enzymes that act on specific lysines in the N-termini. In the acetylated form, the positive charge of the lysine ε-amino group is neutralized. As mentioned above, lysine 16 in histone H4 is particularly important for the folding of the 30-nm fiber because it interacts with a negatively charged patch on the surface of the neighboring nucleosome in the fiber. Consequently, when H4 lysine 16 is acetylated, the chromatin tends to form the less condensed "beads-on-a-string" conformation conducive for transcription and replication.

Histone acetylation at other sites in H4 and in other histones (see Figure 6-31b) is correlated with increased sensitivity of chromatin DNA to digestion by nucleases. This phenomenon can be demonstrated by digesting isolated nuclei with DNase I. Following digestion, the DNA is completely separated from chromatin protein, digested to completion with a restriction enzyme, and analyzed by Southern blotting. An intact gene treated with a restriction enzyme yields fragments of characteristic sizes. If a gene is exposed first to DNase, it is cleaved at random sites within the boundaries of the restriction enzyme cut sites. Consequently, any Southern blot bands normally seen with that gene will be lost. This method has been used to show that the transcriptionally inactive β-globin gene in non-erythroid cells, where it is associated with relatively unacetylated histones, is much more resistant to DNase I than is the active, transcribed β-globin gene in erythroid precursor cells, where it is associated with acetylated histones (Figure 6-32). These results indicate that the chromatin structure of nontranscribed DNA is more condensed, and therefore more protected from DNase digestion, than that of transcribed DNA. In condensed chromatin, the DNA is largely inaccessible to DNase I because of its close association with histones and other chromatin-associated proteins that bind to unacetylated histone tails. In contrast, actively transcribed DNA is much more accessible to DNase I digestion because it is present in the extended, beads-on-a-string form of chromatin.

Genetic studies in yeast indicate that *histone acetylases (HATs)*, which acetylate specific lysine residues in histones, are required for the full activation of transcription of a number of genes. Consequently, the control of acetylation of histone N-termini in specific chromosomal regions is thought to contribute to the transcriptional control of gene expression by mechanisms described below and in the next chapter. Just as genes in condensed, folded regions of chromatin are less accessible to exogenously added DNase I than genes in decondensed, extended regions of chromatin, RNA polymerase and other proteins required for transcription are also inhibited from interacting with DNA in condensed chromatin.

Other Histone Modifications As shown in Figure 6-31b, histone tails in chromatin can undergo a variety of other covalent modifications at specific amino acids. Lysine ε-amino groups can be methylated, a process that prevents acetylation, thus maintaining their positive charge. Moreover, lysine ε-amino groups can be methylated once, twice, or three times. Arginine side chains can also be methylated. Serine and threonine side chains can be reversibly phosphorylated, introducing a negative charge. Finally, a single 76-amino-acid ubiquitin molecule can be reversibly added to a lysine in

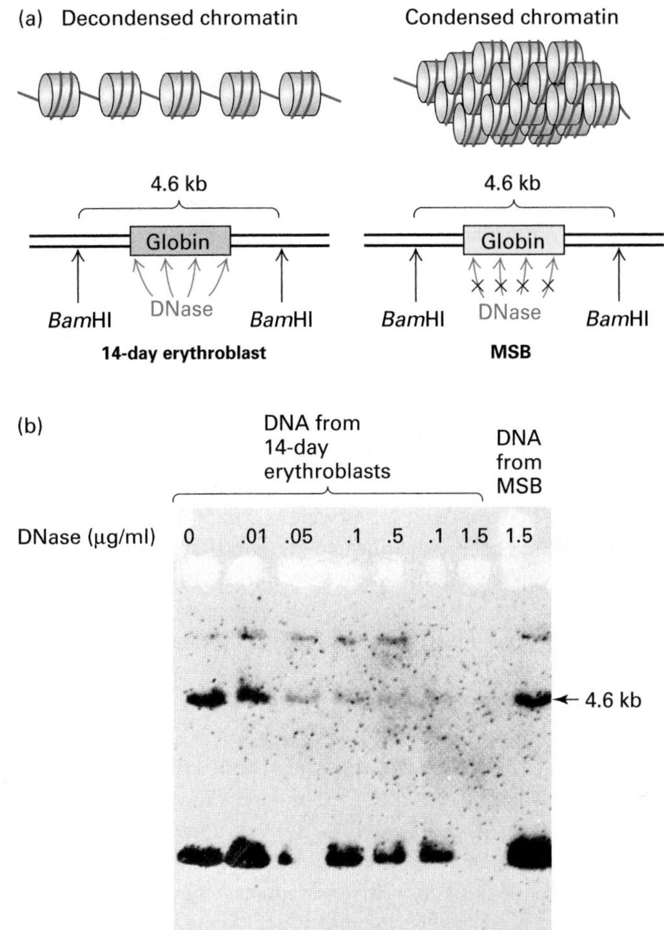

▲ EXPERIMENTAL FIGURE 6-32 **Nontranscribed genes are less susceptible to DNase I digestion than active genes.** Chick embryo erythroblasts at 14 days actively synthesize globin, whereas cultured undifferentiated MSB cells do not. (a) Nuclei from each type of cell were isolated and exposed to increasing concentrations of DNase I. The nuclear DNA was then extracted and treated with the restriction enzyme *Bam*HI, which cleaves the DNA around the globin sequence and normally releases a 4.6-kb globin fragment. (b) The DNase I- and *Bam*HI-digested DNA was subjected to Southern blot analysis with a probe of labeled cloned adult globin DNA, which hybridizes to the 4.6-kb *Bam*HI fragment. If the globin gene is susceptible to the initial DNase digestion, it would be cleaved repeatedly and would not be expected to show this fragment. As seen in the Southern blot, the transcriptionally active DNA from the 14-day globin-synthesizing cells was sensitive to DNase I digestion, indicated by the absence of the 4.6-kb band at higher nuclease concentrations. In contrast, the inactive DNA from MSB cells was resistant to digestion. These results suggest that the inactive DNA is in a more condensed form of chromatin in which the globin gene is shielded from DNase digestion. [See J. Stalder et al., 1980, *Cell* **19**:973; photograph courtesy of H. Weintraub.]

the C-terminal tails of H2A and H2B. Recall that addition of multiple linked ubiquitin molecules to a protein can mark it for degradation by the proteasome (see Figure 3-29b). In this case, however, the addition of a single ubiquitin molecule does not affect the stability of a histone, although it does influence chromatin structure.

As mentioned previously, it is the precise combination of modified amino acids in histone tails that helps control the condensation, or compaction, of chromatin and its ability to be transcribed, replicated, and repaired. This can be illustrated by comparing the specific modifications observed in highly condensed chromatin, known as **heterochromatin,** with those in less condensed chromatin, known as **euchromatin** (Figure 6-33a). Heterochromatin does not fully decondense following mitosis, remaining in a compacted state during interphase. It is typically found at the centromeres and telomeres of chromosomes, as well as some other discrete locations. When cells are subjected to dyes that bind DNA, regions of heterochromatin stain very darkly. In contrast, areas of euchromatin, which are in a less compacted state during interphase, stain lightly with DNA dyes. Typically, most transcribed regions of DNA are found in euchromatin, while heterochromatin remains transcriptionally inactive.

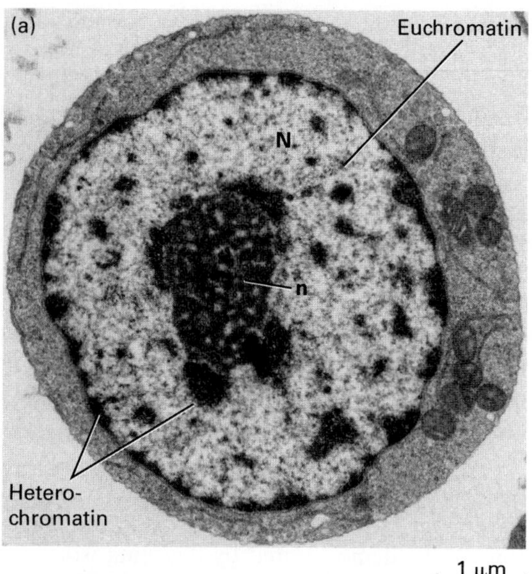

(b)

Heterochromatin (inactive/condensed)

Me_3 (K9)
H3 ARTKQTARKSTGGKAPRKQLATKAARKSAPAT
9

Me_3 (K27)
H3 ARTKQTARKSTGGKAPRKQLATKAARKSAPAT
27

Euchromatin (active/open)

Ac (K9), P (S10), Ac (K14)
H3 ARTKQTARKSTGGKAPRKQLATKAARKSAPAT
9 10 14

Me_3 (K4), Ac (K14)
H3 ARTKQTARKSTGGKAPRKQLATKAARKSAPAT
4 14

▲ **FIGURE 6-33 Heterochromatin versus euchromatin.** (a) In this electron micrograph of a bone marrow stem cell, the dark-staining areas in the nucleus (N) outside the nucleolus (n) are heterochromatin. The light-staining, whitish areas are euchromatin. (b) The modifications of histone N-terminal tails in heterochromatin and euchromatin differ, as illustrated here for histone H3. Note in particular that histone tails are generally much more extensively acetylated in euchromatin compared with heterochromatin. Heterochromatin is much more condensed (thus less accessible to proteins) and is much less transcriptionally active than is euchromatin. [Part (a) P. C. Cross and K. L. Mercer, 1993, *Cell and Tissue Ultrastructure,* W. H. Freeman and Company, p. 165. Part (b) adapted from T. Jenuwein and C. D. Allis, 2001, *Science* **293:**1074.]

Reading the Histone Code The histone code is "read" by proteins that bind to the modified histone tails and in turn promote condensation or decondensation of chromatin, forming "closed" or "open" chromatin structures. Higher eukaryotes express a number of proteins containing a so-called *chromodomain* that binds to histone tails when they are methylated at specific lysines. One example is *heterochromatin protein 1 (HP1).* In addition to histones, HP1 is one of the major proteins associated with heterochromatin. The HP1 chromodomain binds the H3 N-terminal tail only when it is tri-methylated at lysine 9 (Figure 6-33b). HP1 also contains a second domain called a *chromoshadow domain* because it is frequently found in proteins that contain a chromodomain. The chromoshadow domain binds to other chromoshadow domains. Consequently, chromatin containing H3 tri-methylated at lysine 9 ($H3K9Me_3$) is assembled into a condensed chromatin structure by HP1, although the structure of this chromatin is not well understood (Figure 6-34a).

In addition to binding to itself, the chromoshadow domain also binds the enzyme that methylates H3 lysine 9, H3K9 *histone methyl transferase (HMT).* As a consequence, nucleosomes adjacent to a region of HP1-containing heterochromatin also become methylated at lysine 9 (Figure 6-34b). This creates a binding site for another HP1 that can bind the H3K9 histone methyl transferase, resulting in "spreading" of the heterochromatin structure along the chromosome until a *boundary element* is encountered that blocks further spreading. Boundary elements so far characterized are generally regions in chromatin where several nonhistone proteins bind to DNA, possibly blocking histone methylation on the other side of the boundary.

Significantly, the model of heterochromatin formation in Figure 6-34b provides an explanation for how heterochromatic regions of a chromosome are re-established following DNA replication during the S phase of the cell cycle. When DNA in heterochromatin is replicated, the histone octamers that are tri-methylated at H3 lysine 9 become distributed to both daughter chromosomes along with an equal number of newly assembled histone octamers. The H3K9 histone methyl transferase associated with the H3K9 tri-methylated nucleosomes methylate lysine 9 of the newly assembled nucleosomes, regenerating the heterochromatin in both daughter chromosomes.

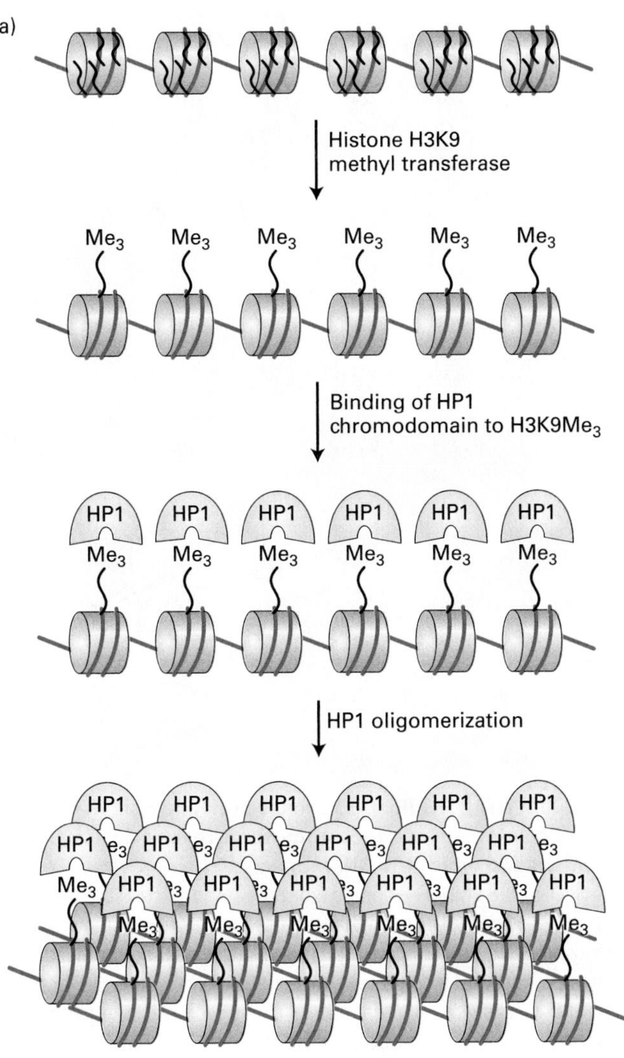

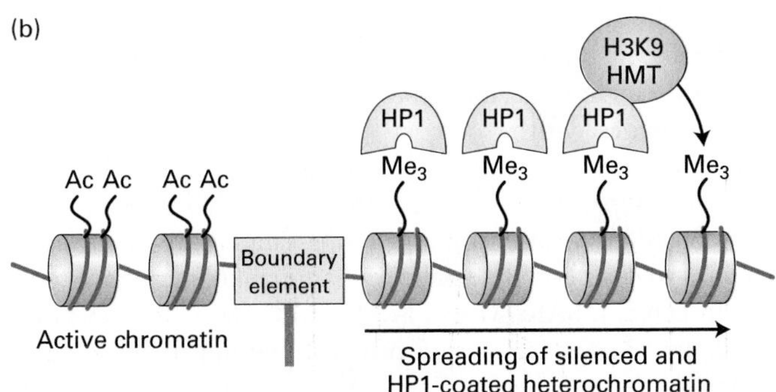

◄ **FIGURE 6-34 Model for the formation of heterochromatin by binding of HP1 to histone H3 tri-methylated at lysine 9.** (a) HP1 contributes to the condensation of heterochromatin by binding to histone H3 N-terminal tails tri-methylated at lysine 9 (H3K9Me$_3$), followed by association of histone-bound HP1 molecules with each other. (b) Heterochromatin condensation can spread along a chromosome because HP1 binds a histone methyltransferase (HMT) that methylates lysine 9 of histone H3. This creates a binding site for HP1 on the neighboring nucleosome. The spreading process continues until a "boundary element" is encountered. [Part (a) adapted from G. Thiel et al., 2004, *Eur. J. Biochem.* **271:**2855. Part (b) adapted from A.J. Bannister et al., 2001, *Nature* **410:**120.]

Other protein domains associate with histone-tail modifications typical of euchromatin. For example, the *bromodomain* binds to acetylated histone tails and therefore is associated with transcriptionally active chromatin. TFIID, a protein involved in transcription, contains two closely spaced bromodomains, which probably help TFIID to associate with transcriptionally active chromatin (i.e., euchromatin). This protein also has histone acetylase activity, which may maintain the chromatin in a hyperacetylated state conducive to transcription.

In summary, multiple types of covalent modifications of histone tails can influence chromatin structure by altering nucleosome–nucleosome interactions and interactions with additional proteins that participate in or regulate processes such as transcription and DNA replication. The mechanisms and molecular processes governing chromatin modifications that regulate transcription are discussed in greater detail in the next chapter.

X-Chromosome Inactivation in Mammalian Females

One important case of heterochromatin formation that correlates with gene inactivation in mammals is the random inactivation and condensation of one of the two female sex chromosomes (the X chromosomes) in virtually all the diploid cells of adult females. Inactivation of one X chromosome in females results in *dosage compensation,* a process that generates equal expression of genes on the sex chromosome in males and females. The inactive X appears as heterochromatin in interphase cells. It is visible as a dark-staining, peripheral nuclear structure called the *Barr body,* named after its discoverer.

Each female mammal has two X chromosomes, one contributed by the egg from which they developed (X_m) and one contributed by the sperm (X_p). Early during embryologic development, random inactivation of either the X_m or the X_p chromosome occurs in each cell. In the female embryo, about half the cells have an inactive X_m, and the other half have an inactive X_p. All subsequent daughter cells maintain the same inactive X chromosomes as their parent cells. As a result, the adult female is a mosaic of clones, some expressing the genes from the X_m and the rest expressing the genes from the X_p. Histones associated with the inactive X chromosome have post-translational modifications characteristic of other regions of heterochromatin: hypoacetylation of lysines, di- and tri-methylation of histone H3 lysine 9, tri-methylation of H3 lysine 27, and a lack of methylation at histone H3 lysine 4 (see Figure 6-33b). X-chromosome inactivation at an early stage in embryonic development is controlled by the X-inactivation center, a complex locus on the X chromosome that determines which of the two X chromosomes will be inactivated and in which cells. The X-inactivation center also contains the *Xist* gene, which encodes a remarkable RNA that coats only the X chromosome it was transcribed from, thereby triggering silencing of the chromosome.

Although the mechanism of X-chromosome inactivation is not fully understood, it involves several processes including the action of *Polycomb* protein complexes that are discussed further in Chapter 7. One subunit of the Polycomb complex contains a chromodomain that binds to histone H3 tails when they are tri-methylated at lysine 27. The Polycomb complex also contains a histone methyl transferase specific for H3 lysine 27. This finding helps to explain how the X-inactivation process spreads along large regions of the X chromosome and how it is maintained through DNA replication, similar to heterochromatization by the binding of HP1 to histone H3 tails methylated at lysine 9 (see Figure 6-34b).

X-chromosome inactivation is an **epigenetic** process: that is, a process that affects the expression of specific genes and is inherited by daughter cells, but is not the result of a change in DNA sequence. Instead, the activity of genes on the X chromosome in female mammals is controlled by chromatin structure rather than the nucleotide sequence of the underlying DNA. And the inactivated X chromosome (either X_m or X_p) is maintained as the inactive chromosome in the progeny of all future cell divisions because the histones are modified in a specific, repressing manner that is faithfully inherited through each cell division.

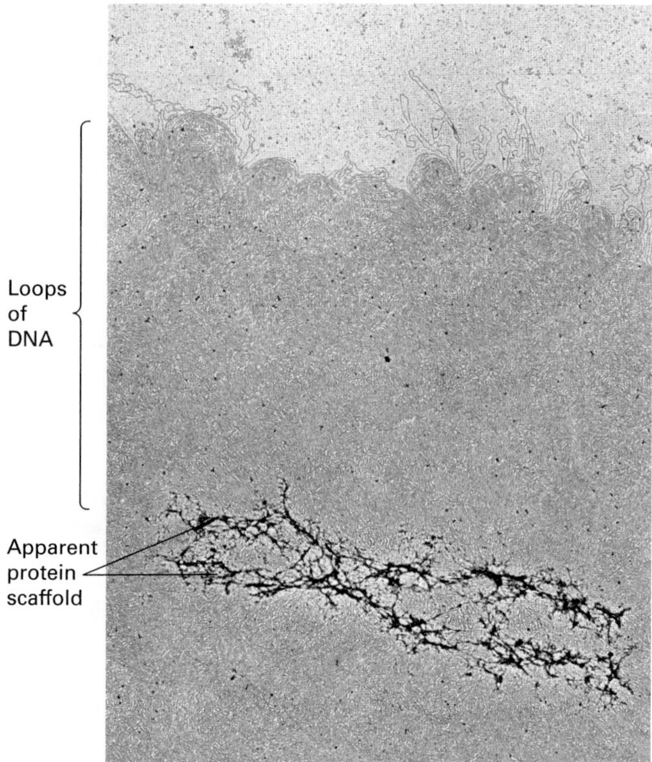

▲ EXPERIMENTAL FIGURE 6-35 **An electron micrograph of a histone-depleted metaphase chromosome reveals an apparent scaffold around which the DNA appears to be organized.** Long loops of DNA are visible extending from the nonhistone protein "scaffold" (the dark structure) that reflects the shape of the metaphase chromosome. However, recent studies indicate that these nonhistone proteins do not form a continuous structure solely responsible for determining the shape of a metaphase chromosome as one might expect for a true scaffold structure. The chromosome was prepared from HeLa cells by treatment with a mild detergent. [From J. R. Paulson and U. K. Laemmli, 1977, *Cell* **12:**817. Copyright 1977 MIT.]

Nonhistone Proteins Provide a Structural Scaffold for Long Chromatin Loops

Although histones are the predominant proteins in chromatin, less abundant, nonhistone chromatin-associated proteins, and even the DNA molecule itself, are also crucial to chromosome structure. Electron micrographs of histone-depleted metaphase chromosomes from HeLa cells reveal long loops of DNA anchored to what appears to be a protein *chromosome scaffold* composed of nonhistone proteins (Figure 6-35). Although this chromosome scaffold has the shape of the metaphase chromosome, recent results indicate that it is not protein alone that gives a metaphase chromosome its structure.

Micromechanical studies of large metaphase chromosomes from newts in the presence of proteases or nucleases indicate that DNA, not protein, is responsible for the mechanical integrity of a metaphase chromosome when it is pulled from its ends. These results are inconsistent with a continuous protein scaffold at the chromosome axis. Rather, the integrity of chromosome structure requires the complete chromatin complex of DNA, histone octamers, and nonhistone chromatin-associated proteins.

In situ hybridization experiments with several different fluorescent-labeled probes to the DNA of one chromosome in human interphase cells support a model in which chromatin is arranged in large loops. In these experiments, some probe sequences separated by millions of base pairs in linear DNA appeared reproducibly very close to one another in interphase nuclei from different cells of the same type (Figure 6-36). These closely spaced probe sites are postulated to lie close to regions of chromatin, called *scaffold-associated regions (SARs)* or *matrix-attachment regions (MARs)*, that are located at the bases of the DNA loops observed in histone-depleted metaphase chromosomes (see Figure 6-35). SARs/MARs have been mapped by digesting histone-depleted chromosomes with restriction enzymes and then recovering the fragments that remain associated with the histone-depleted preparation. The measured distances between probes are consistent with chromatin loops ranging in size from 1 million to 4 million base pairs in mammalian interphase cells.

In general, SARs/MARs are found between transcription units, and genes are located primarily within the chromatin loops. As discussed below, the loops are tethered at their bases by a mechanism that does not break the duplex DNA molecule, which extends the entire length of the chromosome. Evidence indicates that SARs/MARs may affect transcription of neighboring genes. Experiments with transgenic mice indicate that in some cases SARs/MARs are required for high-level expression of genes in the vicinity of SARs/MARs. And in *Drosophila,* some SARs/MARs function as **insulators,** that is, DNA sequences of tens to hundreds of base pairs that insulate transcription units from each other. Proteins regulating transcription of one gene cannot influence the transcription of a neighboring gene that is separated from it by an insulator.

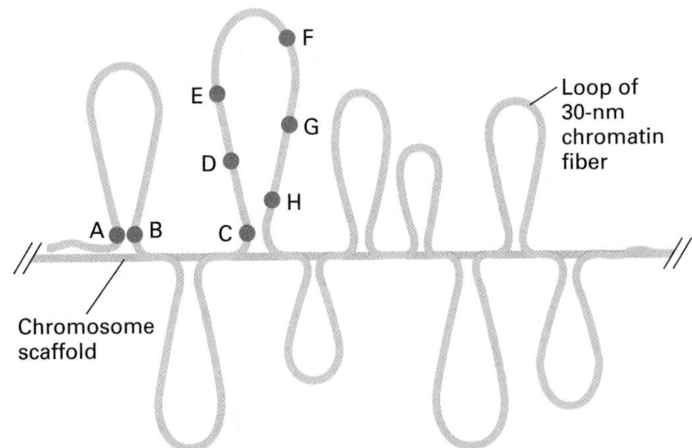

▲ EXPERIMENTAL FIGURE 6-36 **Fluorescent-labeled probes hybridized to interphase chromosomes demonstrate chromatin loops and permit their measurement.** In situ hybridization of interphase cells was carried out with several different probes specific for sequences separated by known distances in linear, cloned DNA. Lettered red circles represent probes. Measurement of the distances between different hybridized probes, which could be distinguished by their color, showed that some sequences (e.g., A, B, and C), separated from one another by millions of base pairs, appear located near one another within nuclei. For some sets of sequences, the measured distances in nuclei between one probe (e.g., C) and sequences successively farther away initially appear to increase (e.g., D, E, and F) and then appear to decrease (e.g., G and H). [Adapted from H. Yokota et al., 1995, *J. Cell Biol.* **130:**1239.]

Individual interphase chromosomes, which are less condensed than metaphase chromosomes, cannot be resolved by standard microscopy or electron microscopy. Nonetheless, the chromatin of one chromosome in interphase cells is not spread throughout the nucleus. Rather, interphase chromatin is organized into *chromosome territories*. As illustrated in Figure 6-37, in situ hybridization of interphase nuclei with chromosome-specific fluorescent-labeled probes shows that the probes are visualized within restricted regions of the nucleus rather than appearing throughout the nucleus. Use of probes specific for different chromosomes shows that there is little overlap between chromosomes in interphase nuclei. However, the precise positions of chromosomes are not reproducible between cells.

Ringlike Structure of SMC Protein Complexes Characterization of proteins associated with metaphase chromosomes identified a small family of proteins called *structural maintenance of chromosome proteins*, or **SMC proteins.** These nonhistone proteins are critical for maintaining the morphological structure of chromosomes. In yeast with mutations in certain SMC proteins chromosome condensation during the prophase period of mitosis does not occur. Mutants with defects in other SMC proteins fail to properly associate daughter chromatids following DNA replication in the S phase. As a result, chromosomes do not properly segregate to daughter cells during mitosis. Related SMC proteins are required for proper segregation of chromosomes in bacteria and archaea, indicating that this is an ancient class of proteins vital to chromosome structure and segregation in all kingdoms of life.

An SMC monomer contains two globular domains, a head domain and hinge domain, that are separated by a very long coiled-coil domain. The head domain is formed from the N- and C-termini of the polypeptide, which are folded together in the native protein structure. The hinge domain forms where the polypeptide folds back on itself. The hinge domain of one monomer binds to the hinge domain of a second monomer, forming a roughly U-shaped dimeric complex (Figure 6-38a). The head domains of the monomers have ATPase activity and are linked by members of another small protein family called *kleisins*.

Interphase Chromosome Structure Studies have indicated that SMC proteins can link two circular DNA molecules by a mechanism that does not require direct protein–DNA binding. Rather, the two DNA molecules are topologically linked and can be separated by either cleavage of the SMC complex with a protease, or cleavage of one of the circular DNA molecules by a restriction enzyme. These results, combined with the U-shaped structure of an SMC complex, suggest that an SMC complex can link two 30-nm chromatin fibers by encircling both of them as depicted in Figure 6-38b. Using the technique of *chromatin immunoprecipitation*, discussed in the next chapter, researchers have demonstrated that SMC proteins in yeast interphase cells associate with chromatin primarily at regions between genes. It is possible that ringlike SMC protein complexes are "pushed" into these regions by RNA polymerases transcribing the regions of chromatin between them.

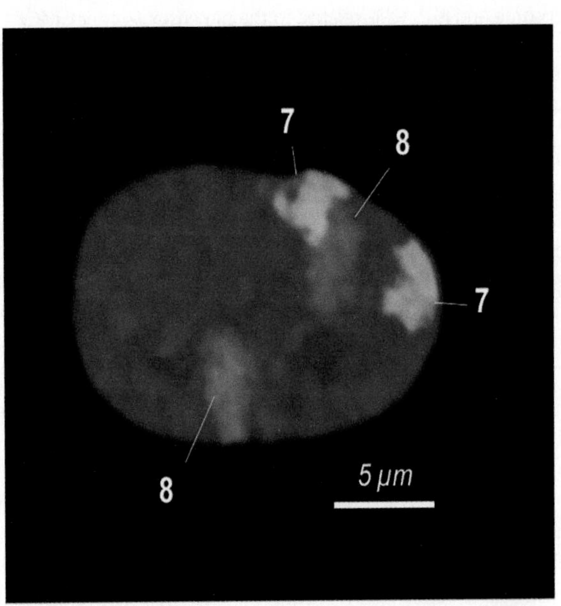

▲ EXPERIMENTAL FIGURE 6-37 **During interphase, human chromosomes remain in specific territories in the nucleus.** Fixed interphase human fibroblasts were hybridized in situ to fluorescently labeled probes specific for sequences along the full length of human chromosomes 7 (cyan) and 8 (purple). DNA was stained blue with DAPI. In this diploid cell, each of the two chromosome 7s and two chromosome 8s can be seen to be restricted to a territory or domain within the nucleus, rather than stretching throughout the entire nucleus. [Courtesy of Drs. I. Solovei and T. Cremer.]

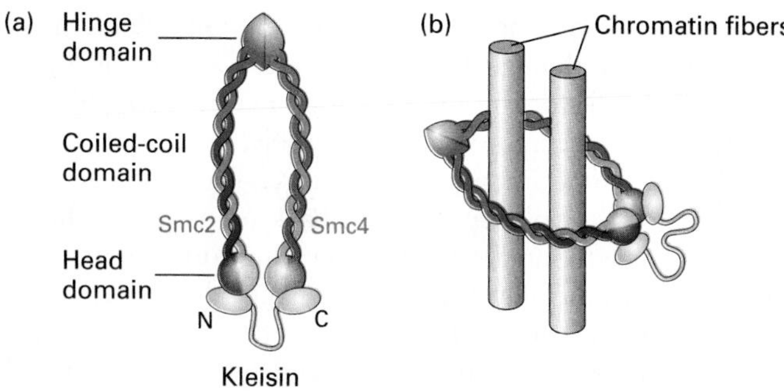

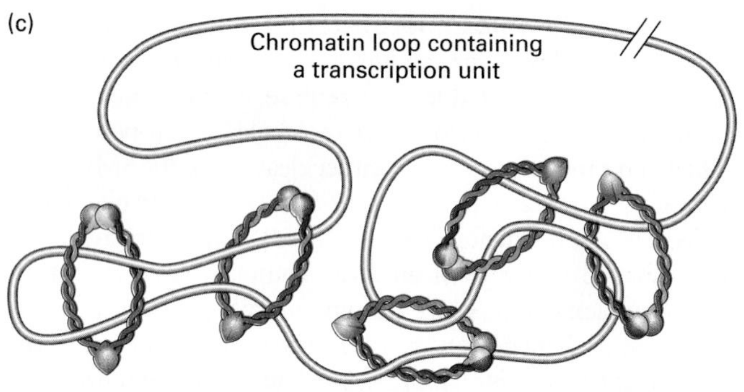

▲ **FIGURE 6-38 Models of SMC complexes and their association with 30-nm chromatin fibers in interphase cells.** (a) An SMC protein complex consists of two monomers, SMC2 (blue) and SMC4 (red), whose hinge domains associate. The head domains, which have ATPase activity, are linked by a kleisin protein, forming a ringlike structure. (b) The ringlike SMC complex topologically links two chromatin fibers (gray cylinders). The cylinder diameter represents the diameter of a nucleosome and is to scale relative to the dimensions of the SMC complex. (c) Loops of transcriptionally active chromatin may be tethered at their base by several SMC complexes, forming a topological knot. [Adapted from K. Nasmyth and C.H. Haering, 2005, *Ann. Rev. Biochem.* **74:**595.]

Based on these various lines of evidence, a recent model proposes that the long loops of chromatin detected in interphase chromosomes (see Figure 6-36) are tethered at the base of each loop by several SMC complexes (Figure 6-38c). These topological knots of SMC proteins and chromatin at the base of each loop are probably linked together in some way to produce the apparent protein scaffold shape visualized in histone-depleted metaphase chromosomes (see Figure 6-35). The linkage may require additional types of proteins, or may result from linkage of SMC complexes alone. In either case, the model in Figure 6-38c can explain why cleavage of the DNA at a relatively small number of sites leads to rapid dissolution of chromosome structure, whereas protease cleavage has only a minor effect on chromosome structure until most of the protein is digested: When the DNA is cut anywhere in a chromatin loop, the broken ends can slip through the SMC protein rings, "untying" the topological knots that constrain the loops of chromatin. In contrast, most of the individual rings of SMC proteins must be broken before the topological constraints holding the base of the loops together is released.

Metaphase Chromosome Structure Condensation of chromosomes during prophase may involve the formation of many more loops of chromatin, so that the length of each loop is greatly reduced compared with that in interphase cells. However, the folding of chromatin in metaphase chromosomes is not well understood. Microscopic analysis of mammalian chromosomes as they condense during prophase indicates that the 30-nm fiber folds into a 100- to 130-nm fiber called a *chromonema* fiber. As depicted in Figure 6-39, a chromonema fiber then folds into a structure with a diameter of 200–250-nm called a *middle prophase chromatid,* which then folds into the 500- to 750-nm diameter chromatids observed during metaphase.

Additional Nonhistone Proteins Regulate Transcription and Replication

The total mass of the histones associated with DNA in chromatin is about equal to that of the DNA. Interphase chromatin and metaphase chromosomes also contain small amounts of a complex set of other proteins. For instance, hundreds to thousands of different **transcription factors** are associated with interphase chromatin. The structure and function of these critical nonhistone proteins, which help regulate transcription, are examined in Chapter 7. Other low-abundance nonhistone proteins associated with chromatin regulate DNA replication during the eukaryotic cell cycle (Chapter 20).

A few other nonhistone DNA-binding proteins are present in much larger amounts than the transcription or replication factors. Some of these exhibit high mobility during electrophoretic separation and thus have been designated

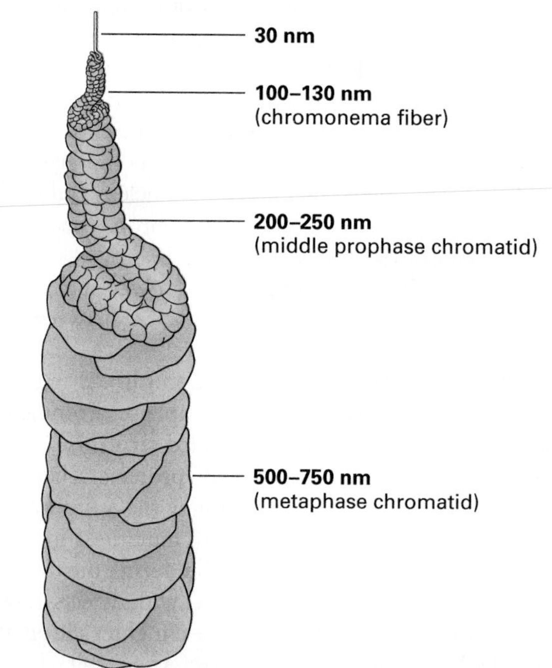

▲ **FIGURE 6-39 Model for the folding of the 30-nm chromatin fiber in a metaphase chromosome.** This drawing depicts the sequential folding of a single 30-nm fiber into a single chromatid of a metaphase chromosome. [Adapted from N. Kireeva et al., 2004, *J. Cell Biol.* **166:**775.]

HMG (high-mobility group) proteins. When genes encoding the most abundant HMG proteins are deleted from yeast cells, normal transcription is disturbed in most genes examined. Some HMG proteins have been found to bind to DNA cooperatively with transcription factors that bind to specific DNA sequences, stabilizing multiprotein complexes that regulate transcription of a neighboring gene.

KEY CONCEPTS OF SECTION 6.6

Structural Organization of Eukaryotic Chromosomes

- In eukaryotic cells, DNA is associated with about an equal mass of histone proteins in a highly condensed nucleoprotein complex called chromatin. The building block of chromatin is the nucleosome, consisting of a histone octamer around which is wrapped 147 bp of DNA (see Figure 6-29).
- The chromatin in transcriptionally inactive regions of DNA within cells is thought to exist in a condensed, 30-nm fiber form and higher-order structures built from it (see Figures 6-30b and 6-39).
- The chromatin in transcriptionally active regions of DNA within cells is thought to exist in an open, extended form (see Figure 6-28a).
- The histone H4 tails, particularly H4 lysine 16, are required for beads-on-a-string chromatin (the 10-nm chromatin fiber) to fold into a 30-nm fiber.
- Histone tails can be modified by acetylation, methylation, phosphorylation, and monoubiquitination (see Figure 6-31). These modifications influence chromatin structure by regulating the binding of histone tails to other less abundant chromatin-associated proteins.
- The reversible acetylation and deacetylation of lysine residues in the N-termini of the core histones regulates chromatin condensation. Proteins involved in transcription, replication, and repair, and enzymes like DNaseI can more easily access chromatin with hyperacetylated histone tails (euchromatin) than chromatin with hypoacetylated histone tails (heterochromatin).
- When metaphase chromosomes decondense during interphase, areas of heterochromatin remain much more condensed than regions of euchromatin.
- Heterochromatin protein 1 (HP1) uses a chromodomain to bind to histone H3 tri-methylated on lysine 9. The chromoshadow domain of HP1 also associates with itself and with the histone methyl transferase that methylates H3 lysine 9. These interactions cause condensation of the 30-nm chromatin fiber and spreading of the heterochromatic structure along the chromosome until a boundary element is encountered (see Figure 6-34).
- One X chromosome in nearly every cell of mammalian females is highly condensed heterochromatin, resulting in repression of expression of nearly all genes on the inactive chromosome. This inactivation results in dosage compensation so that genes on the X chromosome are expressed at the same level in both males and females.
- Each eukaryotic chromosome contains a single DNA molecule packaged into nucleosomes and folded into a 30-nm chromatin fiber, which is associated with a protein scaffold made up in part of structural maintenance of chromosome (SMC) proteins at sites between transcription units (see Figure 6-38c). Additional folding of the scaffold further compacts the structure into the highly condensed form of metaphase chromosomes (see Figure 6-39).

6.7 Morphology and Functional Elements of Eukaryotic Chromosomes

Having examined the detailed structural organization of chromosomes in the previous section, we now view them from a more global perspective. Early microscopic observations on the number and size of chromosomes and their staining patterns led to the discovery of many important general characteristics of chromosome structure. Researchers subsequently identified specific chromosomal regions critical to their replication and segregation to daughter cells during cell division. In this section we discuss these functional elements of chromosomes and consider how chromosomes evolved through rare rearrangements of ancestral chromosomes.

Chromosome Number, Size, and Shape at Metaphase Are Species-Specific

As noted previously, in nondividing cells individual chromosomes are not visible, even with the aid of histologic stains for DNA (e.g., Feulgen or Giemsa stains) or electron microscopy. During mitosis and meiosis, however, the chromosomes condense and become visible in the light microscope. Therefore, almost all cytogenetic work (i.e., studies of chromosome morphology) has been done with condensed metaphase chromosomes obtained from dividing cells—either somatic cells in mitosis or dividing gametes during meiosis.

The condensation of metaphase chromosomes probably results from several orders of folding of 30-nm chromatin fibers (see Figure 6-39). At the time of mitosis, cells have already progressed through the S phase of the cell cycle and have replicated their DNA. Consequently, the chromosomes that become visible during metaphase are *duplicated* structures. Each metaphase chromosome consists of two sister **chromatids,** which are linked at a constricted region, the centromere (Figure 6-40). The number, sizes, and shapes of the metaphase chromosomes constitute the **karyotype,** which is distinctive for each species. In most organisms, all cells have the same karyotype. However, species that appear quite similar can have very different karyotypes, indicating that similar genetic potential can be organized on chromosomes in very different ways. For example, two species of small deer—the Indian muntjac and Reeves muntjac—contain about the same total amount of genomic DNA. In one species, this DNA is organized into 22 pairs of homologous **autosomes** and two physically separate sex chromosomes. In contrast, the other species contains the smallest number of chromosomes in any mammal, only three pairs of autosomes;

▲ **FIGURE 6-40 Typical metaphase chromosome.** As seen in this scanning electron micrograph, each chromosome has replicated and comprises two chromatids, each containing one of two identical DNA molecules. The centromere, where chromatids are attached at a constriction, is required for their separation late in mitosis. Special telomere sequences at the ends function in preventing chromosome shortening. [Andrew Syred/Photo Researchers, Inc.]

one sex chromosome is physically separate, but the other is joined to the end of one autosome.

During Metaphase, Chromosomes Can Be Distinguished by Banding Patterns and Chromosome Painting

Certain dyes selectively stain some regions of metaphase chromosomes more intensely than other regions, producing characteristic banding patterns that are specific for individual chromosomes. The regularity of chromosomal bands serve as useful visible landmarks along the length of each chromosome and can help to distinguish chromosomes of similar size and shape.

G bands are produced when metaphase chromosomes are subjected briefly to mild heat or proteolysis and then stained with Giemsa reagent, a permanent DNA dye (Figure 6-41). G bands correspond to large regions of the human genome that have an unusually low G + C content. Treatment of chromosomes with a hot alkaline solution before staining with Giemsa reagent produces *R bands* in a pattern that is approximately the reverse of the G-band pattern. The distinctive banding patterns of each chromosome permit cytologists to identify specific parts of a chromosome and to locate the sites of chromosomal breaks and translocations (Figure 6-42a). In addition, cloned DNA probes that have hybridized to specific sequences in the chromosomes can be located in particular bands.

The method of *spectral karyotyping* or *chromosome painting* greatly simplifies differentiating chromosomes of similar size and shape. This technique, a variation of **fluorescence in situ hybridization (FISH)**, makes use of probes specific for sites scattered along the length of each chromosome. The probes are labeled with several different fluorescent dyes with distinct excitation and emission wavelengths. Probes specific for each chromosome are labeled with a predetermined fraction of each of the dyes. After the probes are hybridized to chromosomes and the excess removed, the sample is observed with a fluorescent microscope in which a detector determines the fraction of each dye present at each fluorescing position in the microscopic field. This information is conveyed to a computer, and a special program assigns a false color image to each type of chromosome. A related technique called *multicolor FISH* can detect chromosomal translocations (Figure 6-42b). The much more detailed analysis possible with these techniques permits detection of chromosomal translocations that banding analysis does not reveal. The photograph at the beginning of the chapter illustrates the use of multicolor FISH in preparing the karyotype of a human female.

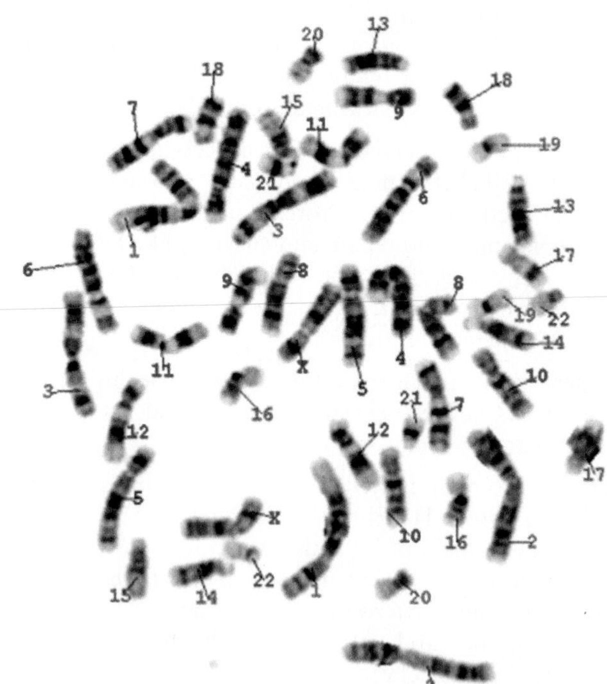

▲ EXPERIMENTAL FIGURE 6-41 **G bands produced with Giemsa stains are useful markers for identifying specific chromosomes.** Shown here are the chromosomes from a human male that were subjected to brief proteolytic treatment and then staining with Giemsa reagent. The resulting dark bands at characteristic places are distinctive for each chromosome. [Courtesy of Sabine Mal, Ph.D., Manitoba Institute of Cell Biology, Canada]

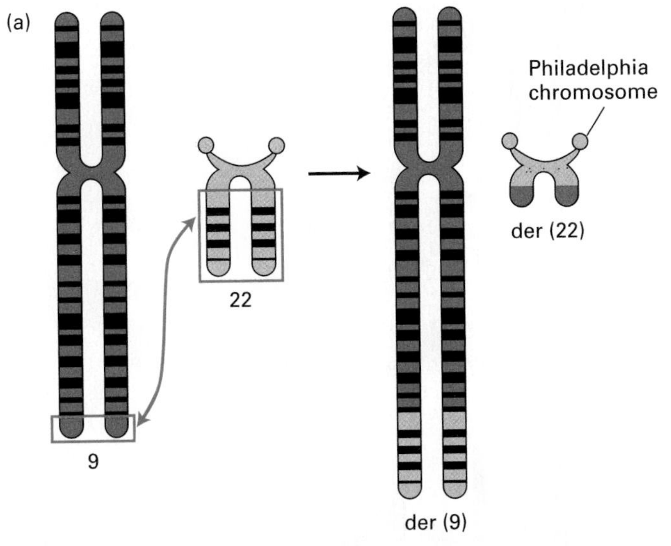

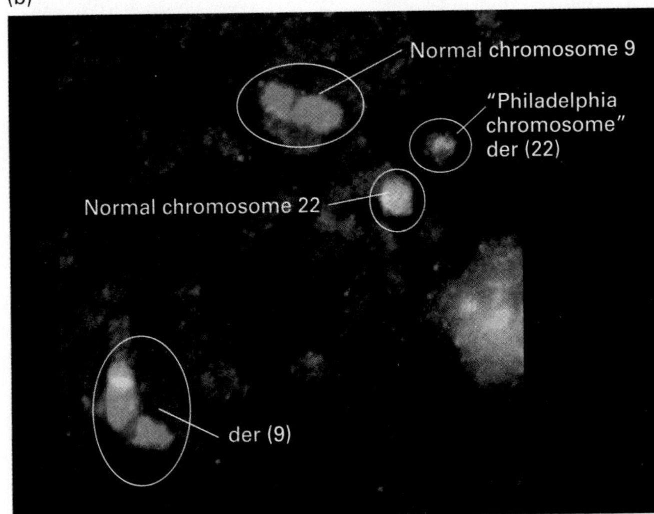

▲ EXPERIMENTAL FIGURE 6-42 **Chromosomal translocations can be analyzed using banding patterns and multicolor FISH.** Characteristic chromosomal translocations are associated with certain genetic disorders and specific types of cancers. For example, in nearly all patients with chronic myelogenous leukemia, the leukemic cells contain the Philadelphia chromosome, a shortened chromosome 22 [der (22)], and an abnormally long chromosome 9 [der (9)] ("der" stands for derivative). These result from a translocation between normal chromosomes 9 and 22. This translocation can be detected by classical banding analysis (a) and by multicolor FISH (b). [Part (a) from J. Kuby, 1997, *Immunology,* 3d ed., W. H. Freeman and Company, p. 578. Part (b) courtesy of J. Rowley and R. Espinosa.]

Chromosome Painting and DNA Sequencing Reveal the Evolution of Chromosomes

Analysis of chromosomes from different species has provided considerable insight about how chromosomes evolved. For example, hybridization of chromosome paint probes for chromosome 16 of the tree shrew (*Tupia belangeri*) to tree shrew metaphase chromosomes revealed the two copies of chromosome 16, as expected (Figure 6-43a). However, when the same chromosome paint probes were hybridized to human metaphase chromosomes, most of the probes hybridized to the long arm of chromosome 10 (Figure 6-43b). Further, when multiple probes from the long arm of human chromosome 10 with different fluorescent dye labels were hybridized to human chromosome 10 and tree shrew metaphase chromosomes, tree shrew sequences homologous to each of these probes were found along tree shrew chromosome 16 in the same order that they occur on human chromosome 10.

These results indicate that during the evolution of humans and tree shrews from a common ancestor that lived ≈85 million years ago, a long, continuous DNA sequence on one of the ancestral chromosomes became chromosome 16 in tree shrews, but evolved into the long arm of chromosome 10 in humans. The phenomenon of genes occurring in the same order on a chromosome in two different species is referred to as conserved **synteny** (derived from Latin for "on the same ribbon"). The presence of two or more genes in a common chromosomal region in two or more species indicates a conserved syntenic segment.

The relationships between the chromosomes of many primates have been determined by cross-species hybridizations of chromosome paint probes as shown for human and tree shrew in Figure 6-43a and b. From these relationships and higher resolution analyses of regions of synteny by DNA sequencing and other methods, it has been possible to propose the karyotype of the common ancestor of all primates based on the minimum number of chromosomal rearrangements necessary to generate the regions of synteny in chromosomes of contemporary primates.

Human chromosomes are thought to have derived from a common primate ancestor with 23 autosomes plus the X and Y sex chromosomes by several different mechanisms (Figure 6-43c). Some human chromosomes were derived without large scale rearrangements of chromosome structure. Others are thought to have evolved by breakage of an ancestral chromosome into two chromosomes or, conversely, by fusion of two ancestral chromosomes. Still other human chromosomes appear to have been generated by exchanges of parts of the arms of distinct chromosomes, that is, by reciprocal translocation involving two ancestral chromosomes. Analysis of regions of conserved synteny between the chromosomes of many mammals indicates that chromosomal rearrangements such as breakage, fusion and translocations occurred rarely in mammalian evolution, about once every five million years. When such chromosomal rearrangements did occur, they very likely contributed to the evolution of new species that cannot interbreed with the species from which they evolved.

Chromosomal rearrangements similar to those inferred for the primate lineage have been inferred for other groups of related organisms, including invertebrate, plant, and fungi lineages. The excellent agreement between predictions of evolutionary relationships based on analysis of syntenic regions of chromosomes from organisms with related anatomical structure (i.e., among mammals, among insects with similar body organization, among similar plants, etc.) and the evolutionary relationships based on the fossil record and on the extent of divergence of DNA sequences for homologous genes is a strong argument for the validity of evolution as the process that generated the diversity of contemporary organisms.

(a)

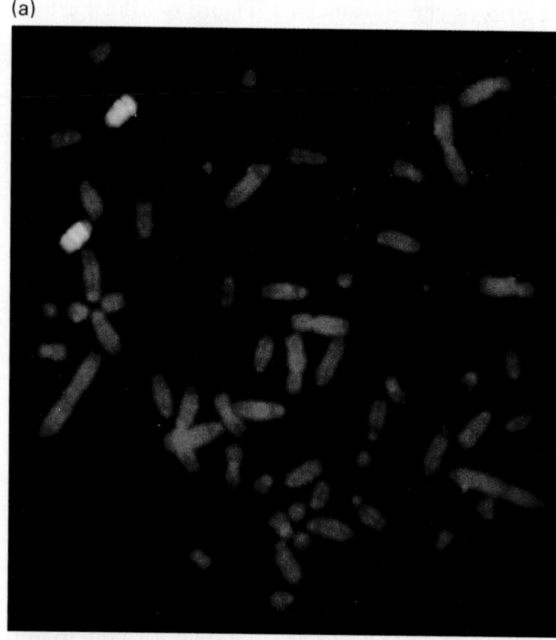

(b)

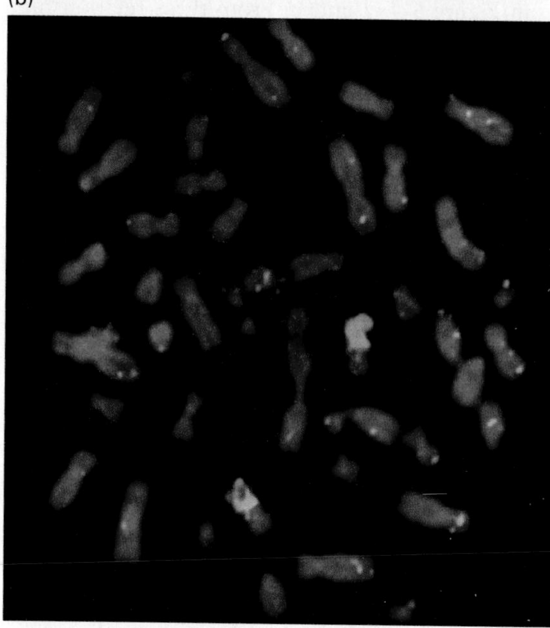

(c)

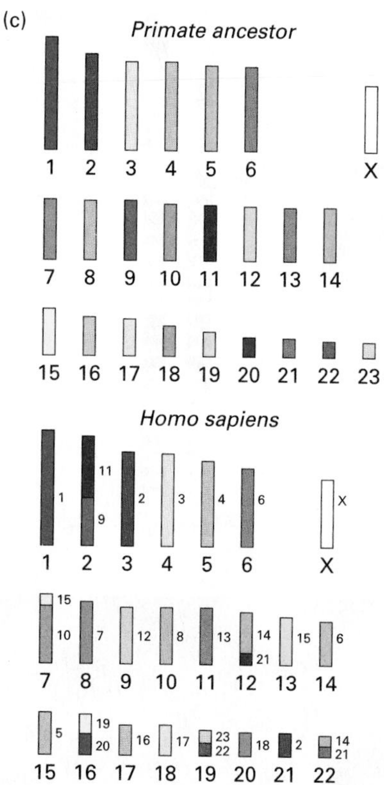

▲ **EXPERIMENTAL FIGURE 6-43 Evolution of primate chromosomes.** (a) Chromosome paint probes for chromosome 16 of the tree shrew (*T. belangeri,* a primate-like animal distantly related to humans) were hybridized (yellow) to tree shrew metaphase chromosomes (red). These probes "painted" the entirety of both copies of chromosome 16. (b) The same tree shrew chromosome 16 paint probes were hybridized to human metaphase chromosomes. These probes were largely localized to the long arms of the two chromosome 10s. (c) Proposed evolution of human chromosomes (*bottom*) from the chromosomes of the common ancestor of all primates (*top*). The proposed common primate ancestor chromosomes are numbered according to their sizes, with each chromosome represented by a different color. The human chromosomes are also numbered according to their relative sizes with colors taken from the colors of the proposed common primate ancestor chromosomes from which they were derived. Small numbers to the right of the colored regions of the human chromosomes indicate the number of the ancestral chromosome from which the region was derived. Human chromosomes were derived from the proposed chromosomes of the common primate ancestor in several ways: without significant rearrangements (e.g., human chromosome 1); by fusion (e.g., human chromosome 2 by fusion of ancestral chromosomes 9 and 11); breakage (e.g., human chromosomes 14 and 15 by breakage of ancestral chromosome 5), and chromosomal translocations (e.g., human chromosomes 12 and 22 by a reciprocal translocation between ancestral chromosomes 14 and 21). [Parts (a) and (b) courtesy of Professor Dr. Johannes Weinberg, Institute for Human Genetics and Anthropology, University of Munich. Part (c) courtesy of Lutz Froenicke Ph.D., School of Veterinary Medicine, University of California, Davis.]

Interphase Polytene Chromosomes Arise by DNA Amplification

The larval salivary glands of *Drosophila* species and other dipteran insects contain enlarged interphase chromosomes that are visible in the light microscope. When fixed and stained, these **polytene chromosomes** are characterized by a large number of reproducible, well-demarcated bands that have been assigned standardized numbers (Figure 6-44a). The highly reproducible banding pattern seen in *Drosophila* salivary gland chromosomes provides an extremely powerful method for locating specific DNA sequences along the lengths of the chromosomes in this species. For example, the chromosomal location of a cloned DNA sequence can be accurately determined by hybridizing a labeled sample of the cloned DNA to polytene chromosomes prepared from larval salivary glands (Figure 6-44b). Chromosomal translocations and inversions also are readily detectable in polytene chromosomes, and specific chromosomal proteins can be localized on interphase polytene chromosomes by immunostaining with specific antibodies raised against them (see Figure 7-11). Insect polytene chromosomes offer one of the only experimental

▶ EXPERIMENTAL FIGURE 6-44 **Banding on *Drosophila* polytene salivary gland chromosomes and in situ hybridization are used together to localize gene sequences.** (a) In this light micrograph of *Drosophila melanogaster* larval salivary gland chromosomes, four chromosomes can be observed (X, 2, 3, and 4), with a total of approximately 5000 distinguishable bands. The banding pattern results from reproducible packing of DNA and protein within each amplified site along the chromosome. Dark bands are regions of more highly compacted chromatin. The centromeres of all four chromosomes often appear fused at the chromocenter. The tips of chromosomes 2 and 3 are labeled (L = left arm; R = right arm), as is the tip of the X chromosome. (b) A particular DNA sequence can be mapped on *Drosophila* salivary gland chromosomes by in situ hybridization. This photomicrograph shows a portion of a chromosome that was hybridized with a cloned DNA sequence labeled with biotin-derivatized nucleotides. Hybridization is detected with the biotin-binding protein avidin that is covalently bound to the enzyme alkaline phosphatase. On addition of a soluble substrate, the enzyme catalyzes a reaction that results in formation of an insoluble colored precipitate at the site of hybridization (asterisk). Since the very reproducible banding patterns are characteristic of each *Drosophila* polytene chromosome, the hybridized sequence can be located on a particular chromosome. The numbers indicate major bands. Bands between those indicated are designated with numbers and letters (not shown). [Part (a) courtesy of J. Gall. Part (b) courtesy of F. Pignoni.]

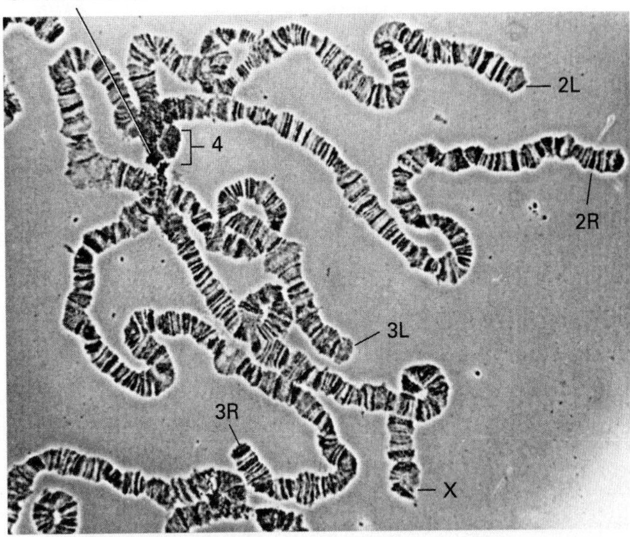

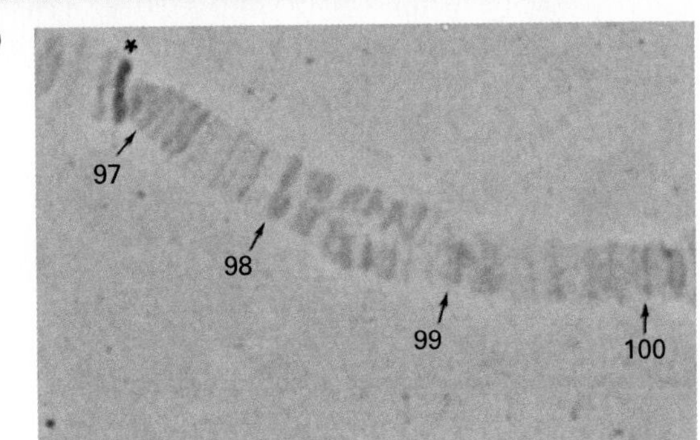

systems in all of nature where such immuno-localization studies on decondensed interphase chromosomes are possible.

A generalized amplification of DNA gives rise to the polytene chromosomes found in the salivary glands of *Drosophila.* This process, termed *polytenization,* occurs when the DNA repeatedly replicates everywhere except at the telomeres and centromere, but the daughter chromosomes do not separate. The result is an enlarged chromosome composed of many parallel copies of itself (Figure 6-45). The amplification of chromosomal DNA greatly increases gene copy number, presumably to supply sufficient mRNA for protein synthesis in the massive salivary gland cells. Although the bands seen in G-banded human metaphase chromosomes probably represent very long folded or compacted stretches of DNA containing about 10^7 base pairs, the bands in *Drosophila* polytene chromosomes represent much shorter stretches of only 50,000–100,000 base pairs.

Three Functional Elements Are Required for Replication and Stable Inheritance of Chromosomes

Although chromosomes differ in length and number between species, cytogenetic studies have shown that they all behave similarly at the time of cell division. Moreover, any eukaryotic chromosome must contain three functional elements in order to replicate and segregate correctly: (1) **replication origins** at which DNA polymerases and other proteins initiate synthesis of DNA (see Figures 4-31 and 4-33); (2) the **centromere**, the constricted region required for proper segregation of daughter chromosomes; and (3) the two ends, or **telomeres.** The yeast transformation studies depicted in Figure 6-46 demonstrated the functions of these three chromosomal elements and established their importance for chromosome function.

As discussed in Chapter 4, replication of DNA begins from sites that are scattered throughout eukaryotic chromosomes. The yeast genome contains many ≈100-bp sequences, called *autonomously replicating sequences (ARSs),* that act as replication origins. The observation that insertion of an ARS into a circular plasmid allows the plasmid to replicate in yeast cells provided the first functional identification of origin sequences in eukaryotic DNA (see Figure 6-46a).

Even though circular ARS-containing plasmids can replicate in yeast cells, only about 5–20 percent of progeny cells contain the plasmid because mitotic segregation of the plasmids is faulty. However, plasmids that also carry a CEN sequence, derived from the centromeres of yeast chromosomes,

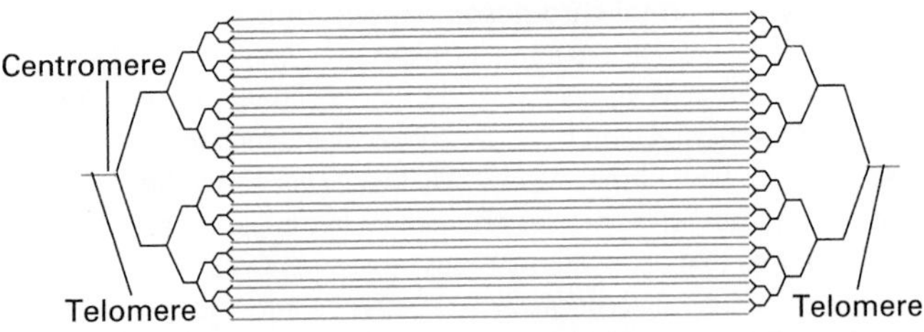

▲ **FIGURE 6-45 The pattern of generalized DNA amplification of one polytene chromosome during five replications.** Double-stranded DNA is represented by a single line. During polytenization, telomere and centromere DNA are not amplified and the daughter chromosomes do not separate. In salivary gland polytene chromosomes, each parental chromosome undergoes ≈10 replications (2^{10} = 1024 strands). [Adapted from C. D. Laird et al., 1973, *Cold Spring Harbor Symp. Quant. Biol.* **38**:311.]

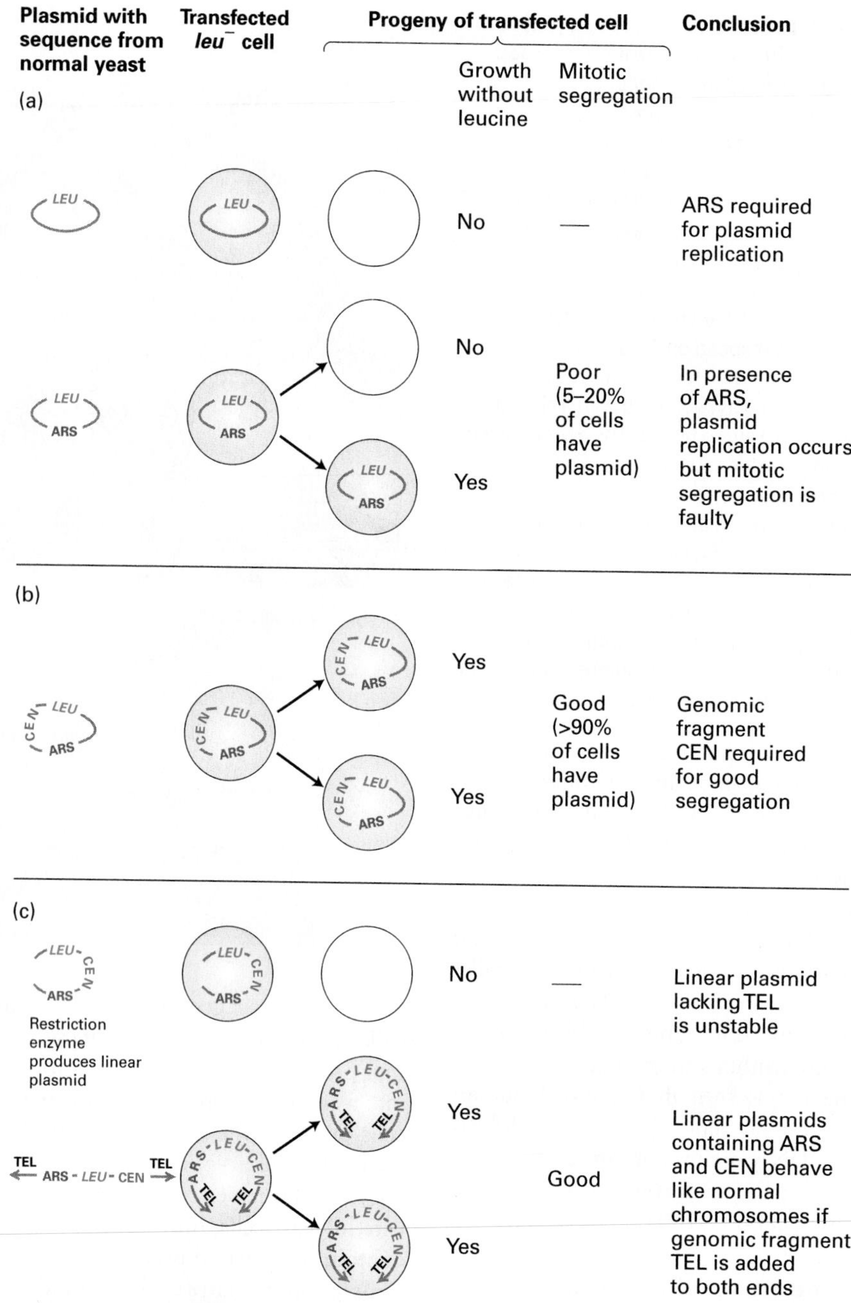

▲ EXPERIMENTAL FIGURE 6-46 **Yeast transfection experiments identify the functional chromosomal elements necessary for normal chromosome replication and segregation.** In these experiments, plasmids containing the *LEU* gene from normal yeast cells are constructed and introduced into *leu*$^-$ cells by transfection. If the plasmid is maintained in the *leu*$^-$ cells, they are transformed to *LEU*$^+$ by the *LEU* gene on the plasmid and can form colonies on medium lacking leucine. (a) Sequences that allow autonomous replication (ARS) of a plasmid were identified because their insertion into a plasmid vector containing a cloned *LEU* gene resulted in a high frequency of transformation to *LEU*$^+$. However, even plasmids with ARS exhibit poor segregation during mitosis, and therefore do not appear in each of the daughter cells. (b) When randomly broken pieces of genomic yeast DNA are inserted into plasmids containing ARS and *LEU*, some of the subsequently transfected cells produce large colonies, indicating that a high rate of mitotic segregation among their plasmids is facilitating the continuous growth of daughter cells. The DNA recovered from plasmids in these large colonies contains yeast centromere (CEN) sequences. (c) When *leu*$^-$ yeast cells are transfected with linearized plasmids containing *LEU*, ARS, and CEN, no colonies grow. Addition of telomere (TEL) sequences to the ends of the linear DNA gives the linearized plasmids the ability to replicate as new chromosomes that behave very much like a normal chromosome in both mitosis and meiosis. [See A. W. Murray and J. W. Szostak, 1983, *Nature* **305:**89, and L. Clarke and J. Carbon, 1985, *Ann. Rev. Genet.* **19:**29.]

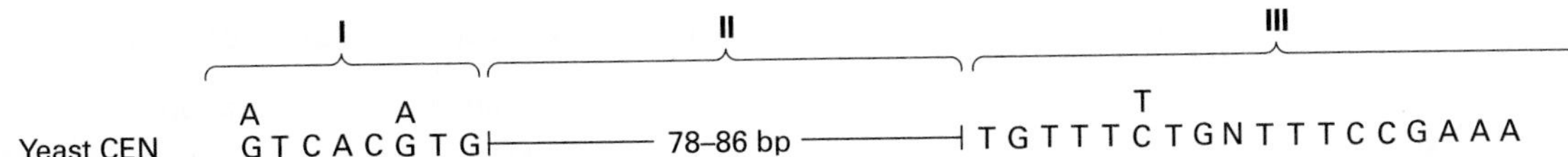

▲ **FIGURE 6-47 Yeast centromere (CEN) sequence.** The consensus yeast CEN sequence shown here, based on analysis of centromeres from 10 different *S. cerevisiae* chromosomes, includes three conserved regions. Region II, although variable in sequence, is fairly constant in length and is rich in A and T residues. Yeast chromosomes are quite short and their CEN sequences are simpler than those in other eukaryotes. [See L. Clarke and J. Carbon, 1985, *Ann. Rev. Genet.* **19:**29.]

segregate equally or nearly so to both mother and daughter cells during mitosis (see Figure 6-46b).

If circular plasmids containing an ARS and CEN sequence are cut once with a restriction enzyme, the resulting linear plasmids do not produce LEU^+ colonies unless they contain special telomeric (TEL) sequences ligated to their ends (see Figure 6-46c). The first successful experiments involving transfection of yeast cells with linear plasmids were achieved by using the ends of a DNA molecule that was known to replicate as a linear molecule in the ciliated protozoan *Tetrahymena.* During part of the life cycle of *Tetrahymena,* much of the nuclear DNA is repeatedly copied in short pieces to form a so-called *macronucleus.* One of these repeated fragments was identified as a dimer of ribosomal DNA, the ends of which contained a repeated sequence $(G_4T_2)_n$. When a section of this repeated TEL sequence was ligated to the ends of linear yeast plasmids containing ARS and CEN, replication and good segregation of the linear plasmids occurred.

Centromere Sequences Vary Greatly in Length

Once the yeast centromere regions that confer mitotic segregation were cloned, their sequences could be determined and compared, revealing three regions (I, II, and III) that are conserved among different chromosomes (Figure 6-47). Short, fairly well conserved nucleotide sequences are present in regions I and III. Although region II seems to have a fairly constant length, it contains no definite consensus sequence; however, it is rich in A and T residues. Regions I and III are bound by proteins that interact with a set of more than 30 other proteins, which in turn bind to microtubules. As a result of these interactions, each of the *S. cerevisiae* chromosomes becomes attached to one microtubule of the spindle apparatus during mitosis. Region II is bound to a nucleosome that has a variant form of histone H3 replacing the usual H3. Centromeres from all eukaryotes similarly are bound by nucleosomes with this specialized, centromere-specific form of histone H3, called CENP-A in humans, that is essential for centromere function. *S. cerevisiae* has by far the simplest centromere sequence known in nature.

In the fission yeast *S. pombe,* centromeres are ≈40 kb in length and are composed of repeated copies of sequences similar to those in *S. cerevisiae* centromeres. Multiple copies of proteins homologous to those that interact with *S. cerevisiae* centromeres bind to these complex *S. pombe* centromeres and in turn bind the much longer *S. pombe* chromosomes to several microtubules of the mitotic spindle apparatus. In plants and animals, centromeres are megabases in length and are composed of multiple repeats of simple-sequence DNA. In humans, centromeres contain 2- to 4-megabase arrays of a 171-bp simple-sequence DNA called *alphoid* DNA that is bound by nucleosomes containing the CENP-A histone H3 variant, as well as other repeated simple-sequence DNA.

In higher eukaryotes, a complex protein structure called the **kinetochore** assembles at centromeres and associates with multiple mitotic spindle fibers during mitosis. Homologs of most of the centromeric proteins found in the yeasts occur in humans and other higher eukaryotes and are thought to be components of kinetochores. The role of the centromere and proteins that bind to it in the segregation of sister chromatids during mitosis is described in Chapters 18 and 20.

Addition of Telomeric Sequences by Telomerase Prevents Shortening of Chromosomes

Sequencing of telomeres from multiple organisms, including humans, has shown that most are repetitive oligomers with a high G content in the strand with its 3′ end at the end of the chromosome. The telomere repeat sequence in humans and other vertebrates is TTAGGG. These simple sequences are repeated at the very termini of chromosomes for a total of a few hundred base pairs in yeasts and protozoans, and a few thousand base pairs in vertebrates. The 3′ end of the G-rich strand extends 12–16 nucleotides beyond the 5′ end of the complementary C-rich strand. This region is bound by specific proteins that protect the ends of linear chromosomes from attack by exonucleases.

The need for a specialized region at the ends of eukaryotic chromosomes is apparent when we consider that all known DNA polymerases elongate DNA chains at the 3′ end, and all require an RNA or DNA primer. As the replication fork approaches the end of a linear chromosome, synthesis of the leading strand continues to the end of the DNA template strand, completing one daughter DNA double helix. However, because the lagging-strand template is copied in a discontinuous fashion, it cannot be replicated in its entirety (Figure 6-48). When the final RNA primer is removed, there is no upstream strand onto which DNA polymerase can build to fill the resulting gap. Without some special mechanism, the daughter DNA strand resulting from lagging-strand synthesis would be shortened at each cell division.

The problem of telomere shortening is solved by an enzyme that adds telomeric (TEL) sequences to the ends of each chromosome. The enzyme is a protein–RNA complex called *telomere terminal transferase,* or *telomerase.* Because the sequence of the telomerase-associated RNA, as we will see, serves as the template for addition of deoxyribonucleotides to the ends of telomeres, the source of the enzyme and not the source of the telomeric DNA primer determines the sequence added. This was proved by transforming *Tetrahymena* with a mutated form of the gene encoding the telomerase-associated RNA. The resulting telomerase added a DNA sequence complementary to the

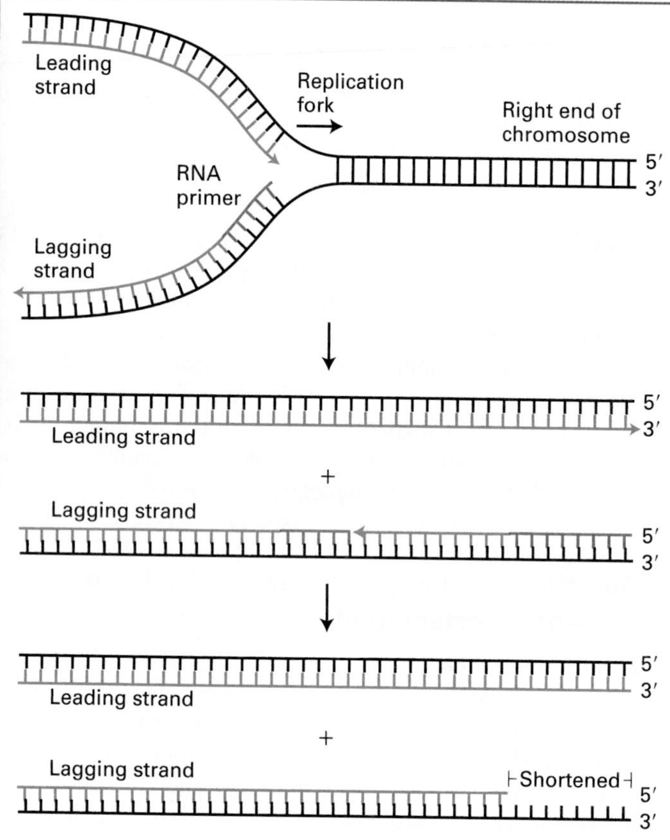

◀ **FIGURE 6-48 Standard DNA replication leads to loss of DNA at the 5′ end of each strand of a linear DNA molecule.** Replication of the right end of a linear DNA is shown; the same process occurs at the left end (shown by inverting the figure). As the replication fork approaches the end of the parental DNA molecule, the leading strand can be synthesized all the way to the end of the parental template strand without the loss of deoxyribonucleotides. However, since synthesis of the lagging strand requires RNA primers, the right end of the lagging daughter DNA strand would remain as ribonucleotides which cannot serve as the template for a replicative DNA polymerase. Alternative mechanisms must be utilized by cells (and viruses with linear DNA genomes) to prevent successive shortening of the lagging strand with each round of replication.

mutated RNA sequence to the ends of telomeric primers. Thus telomerase is a specialized form of a reverse transcriptase that carries its own internal RNA template to direct DNA synthesis.

Figure 6-49 depicts how telomerase, by reverse transcription of its associated RNA, elongates the 3′ end of the single-stranded DNA at the end of the G-rich strand mentioned above. Cells from knockout mice that cannot produce the telomerase-associated RNA exhibit no telomerase activity, and their telomeres shorten successively with each cell generation. Such mice can breed and reproduce normally for three generations before the long telomere repeats become substantially eroded. Then, the absence of telomere DNA results in adverse effects, including fusion of chromosome termini and chromosomal loss. By the fourth generation, the reproductive potential of these knockout mice declines, and they cannot produce offspring after the sixth generation.

The human genes expressing the telomerase protein and the telomerase-associated RNA are active in germ cells and stem cells, but are turned off in most cells of adult tissues that replicate only a limited number of times, or will never replicate again (such cells are called *postmitotic*). However, these genes are activated in most human cancer cells, where telomerase is required for the multiple cell divisions necessary to form a tumor. This phenomenon has stimulated a search for inhibitors of human telomerase as potential therapeutic agents for treating cancer.

While telomerase prevents telomere shortening in most eukaryotes, some organisms use alternative strategies. *Drosophila* species maintain telomere lengths by the regulated insertion of non-LTR retrotransposons into telomeres. This is one of the few instances in which a mobile element has a specific function in its host organism. ■

KEY CONCEPTS OF SECTION 6.7

Morphology and Functional Elements of Eukaryotic Chromosomes

- During metaphase, eukaryotic chromosomes become sufficiently condensed that they can be visualized individually in the light microscope.
- The chromosomal karyotype is characteristic of each species. Closely related species can have dramatically different karyotypes, indicating that similar genetic information can be organized on chromosomes in different ways.
- Banding analysis and chromosome painting are used to identify the different human metaphase chromosomes and to detect translocations and deletions (see Figure 6-42).
- Analysis of chromosomal rearrangements and regions of conserved synteny between related species allows scientists to make predictions about the evolution of chromosomes (see Figure 6-43c). The evolutionary relationships between organisms indicated by these studies are consistent with proposed evolutionary relationships based on the fossil record and DNA sequence analysis.
- The highly reproducible banding patterns of polytene chromosomes make it possible to localize cloned *Drosophila* DNA on a *Drosophila* chromosome by in situ hybridization (see Figure 6-44) and to visualize chromosomal deletions and rearrangements as changes in the normal pattern of bands.
- Three types of DNA sequences are required for a long linear DNA molecule to function as a chromosome: a replication origin, called ARS in yeast; a centromere (CEN) sequence; and two telomere (TEL) sequences at the ends of the DNA (see Figure 6-46).
- Telomerase, a protein–RNA complex, has a special reverse transcriptase activity that completes replication of telomeres during DNA synthesis (see Figure 6-49). In the absence of telomerase, the daughter DNA strand resulting from lagging-strand synthesis would be shortened at each cell division in most eukaryotes (see Figure 6-48).

Focus Animation: Telomere Replication

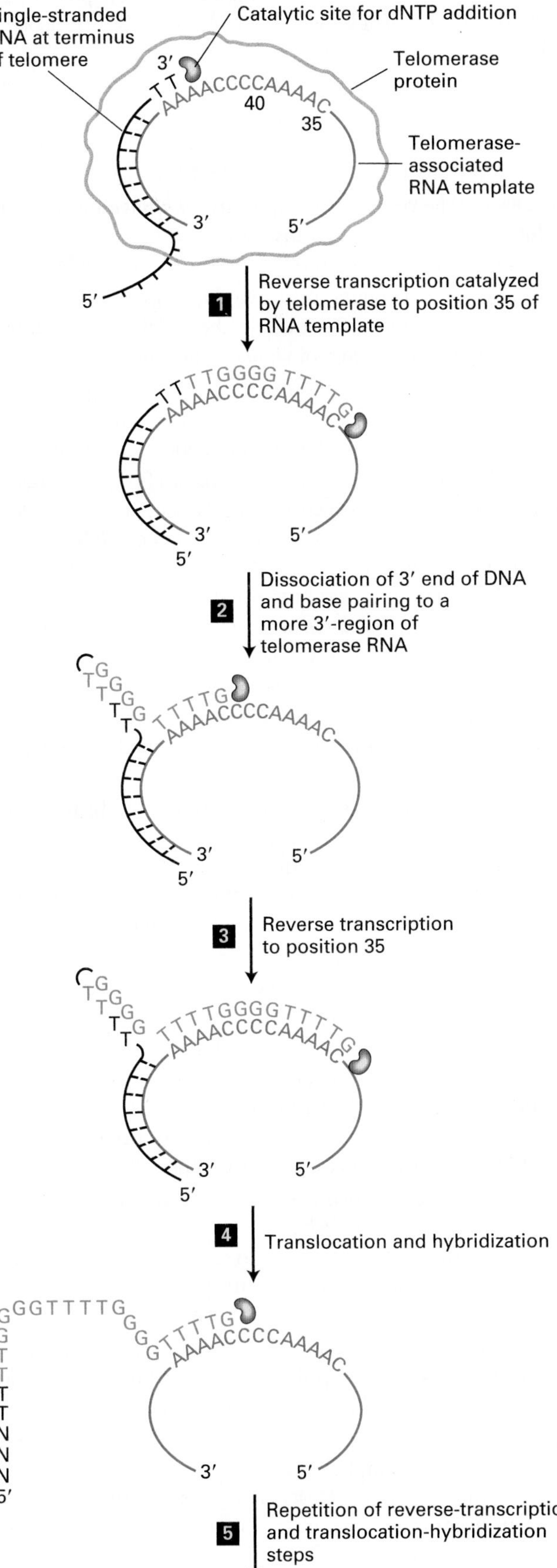

◀ **FIGURE 6-49 Mechanism of action of telomerase.** The single-stranded 3′ terminus of a telomere is extended by telomerase, counteracting the inability of the DNA replication mechanism to synthesize the extreme terminus of linear DNA. Telomerase elongates this single-stranded end by a reiterative reverse-transcription mechanism. The action of the telomerase from the protozoan *Oxytricha,* which adds a T_4G_4 repeat unit, is depicted; other telomerases add slightly different sequences. The telomerase contains an RNA template (red) that base-pairs to the 3′ end of the lagging-strand template. The telomerase catalytic site (green) then adds deoxyribonucleotides (blue) using the RNA molecule as a template; this reverse transcription proceeds to position 35 of the RNA template (step **1**). The strands of the resulting DNA–RNA duplex are then thought to slip relative to each other, leading to displacement of a single-stranded region of the telomeric DNA strand and to uncovering of part of the RNA template sequence (step **2**). The lagging-strand telomeric sequence is again extended to position 35 by telomerase, and the DNA–RNA duplex undergoes translocation and hybridization as before (steps **3** and **4**). Telomerases can add multiple repeats by repetition of steps **3** and **4**. DNA polymerase α-primase can prime synthesis of new Okazaki fragments on this extended template strand. The net result prevents shortening of the lagging strand at each cycle of DNA replication. [Adapted from D. Shippen-Lentz and E. H. Blackburn, 1990, *Nature* **247:**550.]

MEDIA CONNECTIONS

Perspectives for the Future

The human genome sequence is a goldmine for new discoveries in molecular cell biology, in identifying new proteins that may be the basis of effective therapies of human diseases, and for understanding early human history and evolution. However, finding new genes is like finding a needle in a haystack because only ≈1.5 percent of the sequence encodes proteins or functional RNA. Identification of genes in bacterial genome sequences is relatively simple because of the scarcity of introns; simply searching for long open reading frames free of stop codons identifies most genes. In contrast, the search for human genes is complicated by the structure of human genes, most of which are composed of multiple, relatively short exons separated by much longer, noncoding introns. Identification of complex transcription units by analysis of genomic DNA sequences alone is extremely challenging. Future improvements in bioinformatic methods for gene identification, and characterization of cDNA copies of mRNAs isolated from the hundreds of human cell types, will likely lead to the discovery of new proteins, to a better understanding of biological processes, and may lead to applications in medicine and agriculture.

We have seen that although most transposons do not function directly in cellular processes, they have helped to shape modern genomes by promoting gene duplications, exon shuffling, the generation of new combinations of transcription-control sequences, and other aspects of contemporary genomes. They also have the potential to teach us about our own history and origins, because L1 and *Alu* retrotransposons have inserted into new sites in individuals throughout our history. Large numbers of these interspersed repeats are polymorphic within populations, occurring at a particular site in some individuals and not others. Individuals sharing an insertion at a particular site descended from a common ancestor that developed from an egg or sperm in which that insertion occurred. The time elapsed from the initial insertion can be estimated by the differences in sequences of the element that arose from the accumulation of random mutations. Analysis of retrotransposon polymorphisms will undoubtedly add immensely to our understanding both of human migrations since *Homo sapiens* first evolved, as well as the history of contemporary populations.

Key Terms

Barr body *253*
bioinformatics *243*
centromere *262*
chromatid *257*
chromatin *216*
cytoplasmic inheritance *237*
DNA transposons *227*
epigenetic *254*
euchromatin *252*
exon shuffling *235*
fluorescence in situ hybridization (FISH) *258*
gene family *220*
genomics *216*
heterochromatin *252*
histones *247*
histone code *250*
insulator *254*
karyotype *257*
LINEs *230*
long terminal repeats (LTRs) *229*
matrix-associated regions (MARs) *254*
monocistronic *245*
nucleosome *248*
open reading frame (ORF) *244*
polytene chromosome *261*
protein family *220*
pseudogene *220*
retrotransposons *227*
repetitious DNA *215*
scaffold-associated regions (SARs) *254*
simple-sequence (satellite) DNA *224*
SINEs *230*
SMC proteins *255*
synteny *259*
telomere *262*
transcription unit *217*
transposable DNA element *216*

Review the Concepts

1. Genes can be transcribed into mRNA for protein-coding genes or RNA for genes such as ribosomal or transfer RNAs. Define a gene. Describe how a complex transcription unit can be alternatively processed to generate a variety of mRNAs and ultimately proteins.

2. Sequencing of the human genome has revealed much about the organization of genes. Describe the differences between solitary genes, gene families, pseudogenes, and tandemly repeated genes.

3. Much of the human genome consists of repetitious DNA. Describe the difference between microsatellite and minisatellite DNA. How is this repetitious DNA useful for identifying individuals by the technique of DNA fingerprinting?

4. Mobile DNA elements that can move or transpose to a new site directly as DNA are called DNA transposons. Describe the mechanism by which a bacterial DNA transposon, called an insertion sequence, can transpose.

5. Retrotransposons are a class of mobile elements that transpose via a RNA intermediate. Contrast the mechanism of transposition between retrotransposons that contain long terminal repeats (LTRs) and those that lack LTRs.

6. Discuss the role that transposons may have played in the evolution of modern organisms. What is exon shuffling? What role do transposons play in the process of exon shuffling?

7. Mitochondria contain their own DNA molecules. Describe the types of genes encoded in the mitochondrial genome. How do the mitochondrial genomes of plants, fungi, and animals differ?

8. Mitochondria and chloroplasts are thought to have evolved from symbiotic bacteria present in nucleated cells. Review the experimental evidence that supports this hypothesis.

9. Why is screening of sequence databases for genes based on the presence of ORFs (open reading frames) more useful for bacterial genomes than for eukaryotic genomes? What are paralogous and orthologous genes? What are some of the explanations for the finding that humans are a much more complex organism than the roundworm *C. elegans,* yet have only fewer than one and a half as many genes (25,000 versus 18,000)?

10. The DNA in a cell associates with proteins to form chromatin. What is a nucleosome? What role do histones play in nucleosomes? How are nucleosomes arranged in condensed 30-nm fibers?

11. What post-translation modifications of histones are associated with transcribed genes (euchromatin) and with repressed genes (heterochromatin)? What protein is associated with heterochromatin in most eukaryotes? How does this affect heterochromatin formation over a region of a chromosome?

12. Describe the general organization of a eukaryotic chromosome. What structural role do scaffold-associated regions (SARs) or matrix attachment regions (MARs) play? Where are genes primarily located relative to chromosome structure?

13. FISH is a powerful diagnostic tool readily used by cytogeneticists. What is FISH? Briefly describe how it is used to characterize chromosomal translocations associated with certain genetic disorders and specific types of cancers.

14. Metaphase chromosomes can be identified by characteristic banding patterns. What are G bands and R bands? What is chromosome painting, and how is this technique useful? How can chromosome paint probes be used to analyze the evolution of mammalian chromosomes?

15. Certain organisms contain cells that possess polytene chromosomes. What are polytene chromosomes, where are they found and what function do they serve?

16. Replication and segregation of eukaryotic chromosomes require three functional elements: replication origins, a centromere, and telomeres. Describe how these three elements function. What is the role of telomerase in maintaining chromosome structure?

Analyze the Data

To determine if gene transfer from an organelle genome to the nucleus can be observed in the laboratory, a chloroplast transformation vector was constructed that contained two selectable antibiotic-resistance markers each with its own promoter: the spectinomycin-resistance gene and the kanamycin-resistance gene (see S. Stegemann et al., 2003, *Proc. Nat'l. Acad. Sci. USA* **100**:8828–8833). The spectinomycin-resistance gene was controlled by a chloroplast promoter, yielding a chloroplast-specific selectable marker. Plants grown on spectinomycin are white unless they express the spectinomycin-resistance gene in the chloroplast. The kanamycin-resistance gene, inserted into the plasmid adjacent to the spectinomycin-resistance gene, was under the control of a strong nuclear promoter. Transgenic, spectinomycin-resistant tobacco plants were selected following transformation with this plasmid by identifying green plants grown on medium with spectinomycin. These plants contain the two antibiotic-resistance genes inserted into the chloroplast genome by a recombination event; however, kanamycin resistance is not expressed because it is under the control of a nuclear promoter. These spectinomycin-resistant plants were grown for multiple generations and used in the following studies.

a. Leaves from the spectinomycin-resistant transgenic plants were placed in a plant regeneration medium containing kanamycin. Some of the leaf cells were resistant to kanamycin, and grew into kanamycin-resistant plants. Pollen (paternal) from kanamycin-resistant plants was used to pollinate wild-type (nontransgenic) plants. In tobacco, no chloroplasts are inherited from pollen. The resulting seeds were germinated on media with and without kanamycin. Half of the resulting seedlings were kanamycin resistant. When these kanamycin-resistant plants were allowed to self-pollinate, the offspring exhibited a 3:1 ratio of kanamycin-resistant to sensitive phenotypes. What can be deduced from these data about the location of the kanamycin-resistance gene?

b. To determine if transfer of the kanamycin-resistance gene to the nucleus was mediated via DNA or an RNA intermediate, DNA was extracted from 10 seedling plants germinated from seeds produced by a wild-type plant pollinated with a kanamycin-resistant plant. The 10 seedling plants, numbered 1–10 in the corresponding gel lanes in the figure below, consist of 5 kanamycin-resistant (+) and 5 kanamycin-sensitive (−) plants. Each DNA sample was subjected to PCR analysis using primers to amplify the kanamycin-resistance gene (gel at left) or the spectinomycin-resistance gene (gel at right). The lane marked M shows molecular weight markers. What does the correspondence between the presence or absence of PCR products generated in the same plant with both sets of primers suggest about the mode of transfer of the kanamycin gene to the nucleus?

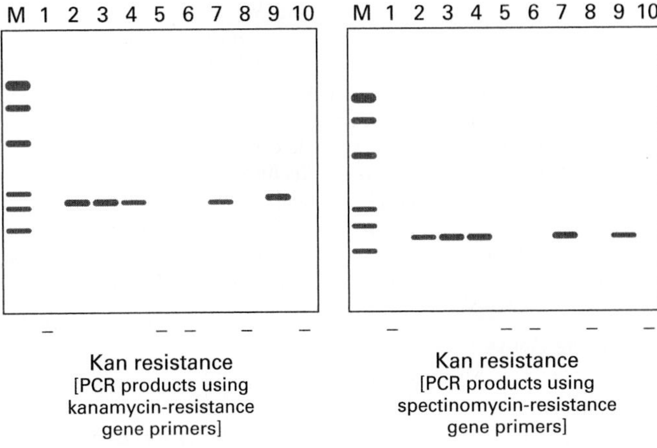

c. When the original transgenic plants, which were selected on spectinomycin but not on kanamycin, were used to pollinate wild-type plants, none of the offspring were kanamycin resistant. What can be deduced from these observations?

References

Eukaryotic Gene Structure

Blencowe, B. J. 2006. Alternative splicing: new insights from global analyses. *Cell* **126**:37–47.

Do, J. H, and D. K. Choi. 2006. Computational approaches to gene prediction. *J. Microbiol.* **44**:137–144.

Landry, J. R, D. L. Mager, and B. T. Wilhelm. 2003. Complex controls: the role of alternative promoters in mammalian genomes. *Trends Genet.* **19**:640–648.

Suzuki, Y., and S. Sugano. 2006. Transcriptome analyses of human genes and applications for proteome analyses. *Curr. Protein Peptide Sci.* **7**:147–163.

Chromosomal Organization of Genes and Noncoding DNA

International Human Genome Sequencing Consortium. 2004. Finishing the euchromatic sequence of the human genome. *Nature* **431**:931 – 945.

Paterson, A.H. 2006. Leafing through the genomes of our major crop plants: strategies for capturing unique information. *Nature Rev. Genet.* **7**:174–184.

Pearson, C. E, K. Nichol Edamura, and J. D. Cleary. 2005. Repeat instability: mechanisms of dynamic mutations. *Nature Rev. Genet.* **6**:729–742.

Ranum, L. P., and T. A. Cooper. 2006. RNA-mediated neuromuscular disorders. *Annu. Rev. Neurosci.* **29**:259–277.

Transposable (Mobile) DNA Elements

Feschotte, C., N. Jiang, and S. R. Wessler. 2002. Plant transposable elements: where genetics meets genomics. *Nature Rev. Genet.* **3**:329–341.

Gray, Y. H. 2000. It takes two transposons to tango: transposable-element-mediated chromosomal rearrangements. *Trends Genet.* **16**:461–468.

Jones, R. N. 2005. McClintock's controlling elements: the full story. *Cytogenet. Genome Res.* **109**:90–103.

Kazazian, H. H., Jr. 1999. An estimated frequency of endogenous insertional mutations in humans. *Nature Genet.* **22**:130.

Kazazian, H.H., Jr. 2004. Mobile elements: drivers of genome evolution. *Science* **303**:1626–1632.

Kleckner, N., et al. 1996. Tn10 and IS10 transposition and chromosome rearrangements: mechanism and regulation in vivo and in vitro. *Curr. Topics Microbiol. Immunol.* **204**:49-82.

Mahillon, J., and M. Chandler. 1998. Insertion sequences. *Microbiol. Mol. Biol. Rev.* **62**:725–774.

Morgante, M. 2006. Plant genome organisation and diversity: the year of the junk! *Curr. Opin. Biotechnol.* **17**:168–173.

Ostertag, E. M., and H. H. Kazazian, Jr. 2001. Biology of mammalian L1 retrotransposons. *Ann. Rev. Genet.* **35**:501–538.

Steiniger-White M, I. Rayment, and W. S. Reznikoff. 2004. Structure/function insights into Tn5 transposition. *Curr. Opin. Struc. Biol.* **14**:50–57.

Organelle DNAs

Bendich, A. J. 2004. Circular chloroplast chromosomes: the grand illusion. *Plant Cell* **16**:1661–1666.

Chan, D. C. 2006. Mitochondria: dynamic organelles in disease, aging, and development. *Cell* **125**:1241–1252.

Clayton, D. A. 2000. Transcription and replication of mitochondrial DNA. *Hum. Reprod.* **2**(Suppl.):11–17.

Daniell, H., M. S. Khan, and L. Allison. 2002. Milestones in chloroplast genetic engineering: an environmentally friendly era in biotechnology. *Trends Plant Sci.* **7**:84–91.

Gray, M. W., G. Burger, and B. F. Lang. 2001. The origin and early evolution of mitochondria. *Genome Biol.* **2**(Reviews):1018.1–1018.5.

Shutt, T. E., and M. W. Gray. 2006. Bacteriophage origins of mitochondrial replication and transcription proteins. *Trends Genet.* **22**:90–95.

Sugiura, M., T. Hirose, and M. Sugita. 1998. Evolution and mechanism of translation in chloroplasts. *Ann. Rev. Genet.* **32**:437–459.

Genomics: Genome-wide Analysis of Gene Structure and Expression

BLAST Information can be found at: http://www.ncbi.nlm.nih.gov/Education/BLASTinfo/information3.htm

Binnewies, T. T., et al. 2006. Ten years of bacterial genome sequencing: comparative-genomics-based discoveries. *Funct. Integr. Genomics* **6**:165–185.

Celniker, S. E., and G. M. Rubin. 2003. The *Drosophila melanogaster* genome. *Annu. Rev. Genomics Hum. Genet.* **4**:89–117.

Chimpanzee Sequencing and Analysis Consortium. 2005. Initial sequence of the chimpanzee genome and comparison with the human genome. *Nature* **437**:69-87.

Gebhardt, C., R. Schmidt, and K. Schneider. 2005. Plant genome analysis: the state of the art. *Int'l. Rev. Cytol.* **247**:223–284.

Gene Ontology Consortium. 2006. The Gene Ontology (GO) project in 2006. *Nucleic Acids Res.* **34**:D322-D326.

Guigo, R., et al. 2006. EGASP: the human ENCODE genome annotation assessment project. *Genome Biol.* **7**(Suppl 1):S2.1–S2.31.

Harris, T. W., and L. D. Stein. 2006. WormBase: methods for data mining and comparative genomics. *Methods Mol. Biol.* **351**:31–50.

International HapMap Consortium. 2003. The international HapMap project. *Nature* **426**:789–796.

International Human Genome Sequencing Consortium. 2004. Finishing the euchromatic sequence of the human genome. *Nature* **431**:931–945.

Lander, E. S., et al. 2001. Initial sequencing and analysis of the human genome. *Nature* **409**:860–921.

Ness, S. A. 2006. Basic microarray analysis: strategies for successful experiments. *Methods Mol. Biol.* **316**:13–33.

Shianna, K. V., and H. F. Willard. 2006. In search of normality. *Nature* **444**:428–429.

Sjoblom, T. 2006. The consensus coding sequences of human breast and colorectal cancers. *Science* **314**:268–274.

Waterston, R. H., et al. 2002. Initial sequencing and comparative analysis of the mouse genome. *Nature* **420**:520–562.

Windsor, A. J., and T. Mitchell-Olds. 2006. Comparative genomics as a tool for gene discovery. *Curr. Opin. Biotechnol.* **17**:161–167.

Structural Organization of Eukaryotic Chromosomes

Carroll, C. W., and A. F. Straight. 2006. Centromere formation: from epigenetics to self-assembly. *Trends Cell Biol.* **16**:70–78.

Dehghani, H., G. Dellaire, and D. P. Bazett-Jones. 2005. Organization of chromatin in the interphase mammalian cell. *Micron* **36**:95-108.

Dillon, N. 2004. Heterochromatin structure and function. *Biol. Cell.* 2004. **96**:631–637.

Henikoff, S., and Y. Dalal. 2005. Centromeric chromatin: what makes it unique? *Curr. Opin. Genet. Dev.* **15**:177–184.

Horn, P. J., and C. L. Peterson. 2002. Molecular biology. Chromatin higher order folding—wrapping up transcription. *Science* **297**:1824–1827.

Luger, K. 2003. Structure and dynamic behavior of nucleosomes. *Curr. Opin. Genet. Dev.* **13**:127–135.

Luger, K., and T. J. Richmond. 1998. The histone tails of the nucleosome. *Curr. Opin. Genet. Dev.* **8**:140–146.

Margueron, R., P. Trojer, and D. Reinberg. 2005. The key to development: interpreting the histone code? *Curr. Opin. Genet. Dev.* **15**:163–176.

McBryant, S. J., V. H. Adams, and J. C. Hansen. 2006. Chromatin architectural proteins. *Chromosome Res.* **14**:39–51.

Nasmyth, K., and C. H. Haering. 2005. The structure and function of SMC and kleisin complexes. *Annu. Rev. Biochem.* **74**:595–648.

Sarma, K., and D. Reinberg. 2005. Histone variants meet their match. *Nature Rev. Mol. Cell Biol.* **6**:139–149.

Schalch, T., et al. 2005. X-ray structure of a tetranucleosome and its implications for the chromatin fibre. *Nature* **436**:138–141.

Woodcock, C. L. 2006. Chromatin architecture. *Curr. Opin. Struc. Biol.* **16**:213–220.

Woodcock, C. L., A. I. Skoultchi, and Y. Fan. 2006. Role of linker histone in chromatin structure and function: H1 stoichiometry and nucleosome repeat length. *Chromosome Res.* **14**:17–25.

Morphology and Functional Elements of Eukaryotic Chromosomes

Armanios, M., and C. W. Greider. 2005. Telomerase and cancer stem cells. *Cold Spring Harb. Symp. Quant. Biol.* **70**:205–208.

Belmont, A. S. 2002. Mitotic chromosome scaffold structure: New approaches to an old controversy. *Proc. Nat'l. Acad. Sci. USA* **99**:15855–15857.

Belmont, A. S., et al. 1999. Large-scale chromatin structure and function. *Curr. Opin. Cell Biol.* **11**:307–311.

Blackburn, E. H. 2005. Telomeres and telomerase: their mechanisms of action and the effects of altering their functions. *FEBS Lett.* **579**:859–862.

Carroll, C. W., and A. F. Straight. 2006. Centromere formation: from epigenetics to self-assembly. *Trends Cell. Biol.* **16**:70–78.

Cvetic, C., and J. C. Walter. 2005. Eukaryotic origins of DNA replication: could you please be more specific? *Semin. Cell Dev. Biol.* **16**:343–353.

de Lange, T. 2006. Lasker laurels for telomerase. *Cell* **126**:1017–1020.

Froenicke, L. 2005. Origins of primate chromosomes as delineated by Zoo-FISH and alignments of human and mouse draft genome sequences. *Cytogenet Genome Res.* **108**:122–138.

Gassmann, R., et al. 2004. Mitotic chromosome formation and the condensin paradox. *Exp. Cell Res.* **296**:35–42.

MacAlpine, D. M, and S. P. Bell. 2005. A genomic view of eukaryotic DNA replication. *Chromosome Res.* **13**:309–326.

Malik, H. S. 2006. A hitchhiker's guide to survival finally makes CENs. *J. Cell Biol.* **174**:747–749.

Marko, J. F., and M. G. Poirier. 2003. Micromechanics of chromatin and chromosomes. *Biochem. Cell Biol.* **81**:209–220.

Robinson, N. P., and S. D. Bell. 2005. Origins of DNA replication in the three domains of life. *FEBS J.* **272**:3757–3766.

Warburton, P. E. 2004. Chromosomal dynamics of human neocentromere formation. *Chromosome Res.* **12**:617–626.

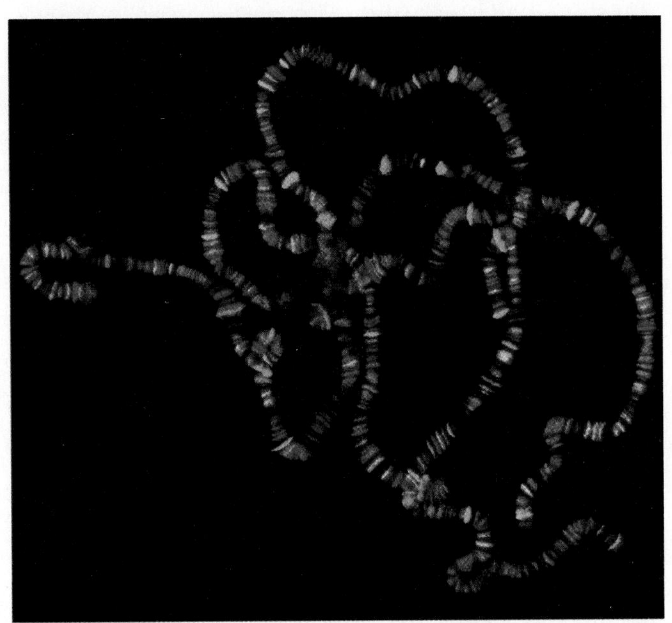

Drosophila polytene chromosomes stained with antibodies against a chromatin remodeling ATPase called Kismet (blue), RNA polymerase II with low CTD phosphorylation (red), and RNA polymerase II with high CTD phosphorylation (green). [Courtesy of John Tamkun; see S. Srinivasan et al., 2005, *Development* **132:**1623.]

CHAPTER

7 TRANSCRIPTIONAL CONTROL OF GENE EXPRESSION

In previous chapters we have seen that the properties and functions of each cell type are determined by the proteins it contains. In this and the next chapter, we consider how the kinds and amounts of the various proteins produced by a particular cell type in a multicellular organism are regulated. This regulation of **gene expression** is the fundamental process that controls the development of a multicellular organism such as ourselves from a single fertilized egg cell into the thousands of cell types from which we are made. When gene expression goes awry, cellular properties are altered, a process that all too often leads to the development of cancer. As discussed further in Chapter 25, genes encoding proteins that restrain cell growth are abnormally repressed in cancer cells, whereas genes encoding proteins that promote cell growth and replication are inappropriately activated in cancer cells. Abnormalities in gene expression also result in developmental defects such as cleft pallet, tetrology of Fallot (a common, serious developmental defect of the heart that can be treated surgically), and many others. Regulation of gene expression also plays a vital role in bacteria and other single-celled microorganisms, where it allows cells to adjust their enzymatic machinery and structural components in response to their changing nutritional and physical environment. Consequently, to understand how microorganisms respond to their environment and how multicellular organisms normally develop, as well as how pathological abnormalities of gene expression occur, it is essential to understand the molecular interactions that control protein production.

The basic steps in gene expression, i.e., the entire process whereby the information encoded in a particular gene is decoded into a particular protein, are reviewed in Chapter 4. Synthesis of mRNA requires that an **RNA polymerase** initiate transcription, polymerize ribonucleoside triphosphates complementary to the DNA coding strand, and then terminate transcription (see Figure 4-11). In prokaryotes, ribosomes and translation-initiation factors have immediate access to newly formed RNA transcripts, which function as mRNA

OUTLINE

without further modification. In eukaryotes, however, the initial RNA transcript is subjected to processing that yields a functional mRNA (see Figure 4-15). The mRNA then is transported from its site of synthesis in the nucleus to the cytoplasm, where it is translated into protein with the aid of ribosomes, tRNAs, and translation factors (see Figure 4-25).

Theoretically, regulation at any one of the various steps in gene expression outlined above could lead to *differential* production of proteins in different cell types or developmental stages or in response to external conditions. Although examples of regulation at each step in gene expression have been found, control of transcription initiation—the first step—is the most important mechanism for determining whether most genes are expressed and how much of the encoded mRNAs and, consequently, proteins are produced. The molecular mechanisms that regulate transcription initiation are critical to numerous biological phenomena, including the development of a multicellular organism from a single fertilized egg cell as mentioned above, the immune responses that protect us from pathogenic microorganisms, and neurological processes such as learning and memory. When these regulatory mechanisms controlling transcription function improperly, pathological processes may occur. For example, reduced activity of the *Pax6* gene causes *aniridia,* failure to develop an iris. Pax6 is a **transcription factor** that normally regulates transcription of genes involved in eye development (see Figure 1-26e). In other organisms, mutations of transcription factors cause an extra pair of wings to develop in *Drosophila* (see Figure 22-31), change the structure of flowers in plants (see Figure 22-36), and are responsible for multiple other developmental abnormalities.

Transcription is a complex process involving many layers of regulation. In this chapter, we focus on the molecular events that determine when transcription of a gene occurs. First, we consider the relatively basic mechanisms of gene expression in bacteria, where **repressor** and **activator** proteins recognize and bind to specific regions of DNA to control the transcription of a nearby gene. The remainder of the chapter focuses on eukaryotic transcription regulation and how the basic tenets of bacterial regulation are applied in more complex ways in higher organisms. Figure 7-1 provides an overview of eukaryotic gene regulation and the processes outlined in this chapter. We discuss how specific DNA sequences function as **transcription-control regions** by serving as the binding sites for transcription factors (repressors and activators) and how the RNA polymerases responsible for transcription bind to **promoter** sequences initiate the synthesis of an RNA molecule complementary to template DNA. Next, we consider how activators and repressors influence transcription through interactions with large, multiprotein complexes. Some of these multiprotein complexes modify chromatin condensation, altering access of chromosomal DNA to transcription factors and RNA polymerases. Other complexes influence the rate at which RNA polymerase binds to DNA at the site of transcription initiation, as well as the frequency of initiation. Then we discuss how transcription of specific genes can be specified by particular combinations of the ≈2,000 transcription factors encoded in the human genome, giving rise to cell-type-specific gene expression. We also further consider the various ways in which the activities of transcription factors themselves are controlled to ensure genes are expressed only at the right time and in the right place. Finally, we'll examine control of transcription elongation and termination and the transcription of nonprotein-coding RNAs. RNA processing and various post-transcriptional mechanisms for controlling eukaryotic gene expression are covered in the next chapter. Subsequent chapters, particularly Chapters 16 and 22, provide examples of how transcription is regulated by interactions between cells and how the resulting **gene control** contributes to the development and function of specific types of cells in multicellular organisms.

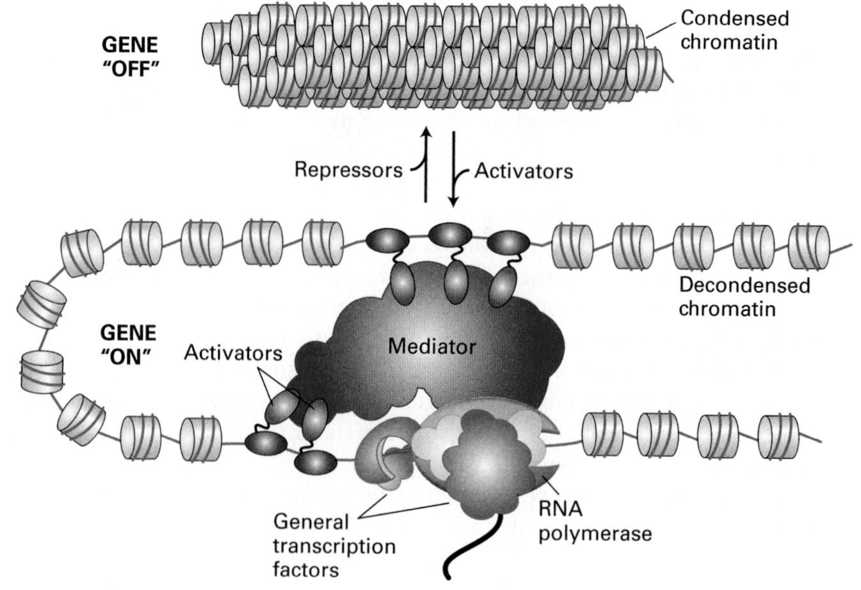

▶ **FIGURE 7-1 Overview of eukaryotic transcription control.** Activator proteins bind to specific DNA control elements in chromatin and interact with multiprotein co-activator machines, such as mediator, to decondense chromatin and assemble RNA polymerase and general transcription factors on promoters. Inactive genes are assembled into regions of condensed chromatin that inhibit RNA polymerases and their associated general transcription factors (GTFs) from interacting with promoters. Alternatively, repressor proteins bind to other control elements to inhibit initiation by RNA polymerase and interact with multiprotein co-repressor complexes to condense chromatin.

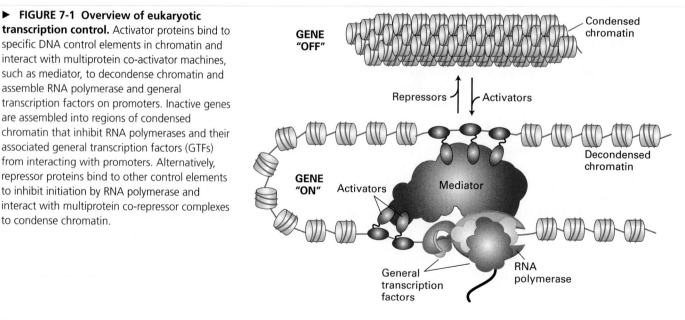

7.1 Control of Gene Expression in Bacteria

Since the structure and function of a cell are determined by the proteins it contains, the control of gene expression is a fundamental aspect of molecular cell biology. Most commonly, the "decision" to initiate transcription of the gene encoding a particular protein is the major mechanism for controlling production of the encoded protein in a cell. By controlling transcription initiation, a cell can regulate which proteins it produces and how rapidly. When transcription of a gene is *repressed,* the corresponding mRNA and encoded protein or proteins are synthesized at low rates. Conversely, when transcription of a gene is *activated,* both the mRNA and encoded protein or proteins are produced at much higher rates.

In most bacteria and other single-celled organisms, gene expression is highly regulated in order to adjust the cell's enzymatic machinery and structural components to changes in the nutritional and physical environment. Thus, at any given time, a bacterial cell normally synthesizes only those proteins of its entire proteome required for survival under the particular conditions. Here we describe the basic features of transcription control in bacteria, using the *lac* operon and the glutamine synthetase gene in *E. coli* as our primary examples. Many of the same processes, as well as others, are involved in eukaryotic transcription control, which is the subject of the remainder of this chapter.

Transcription Initiation by Bacterial RNA Polymerase Requires Association with a Sigma Factor

In *E. coli,* about half the genes are clustered into **operons**, each of which encodes enzymes involved in a particular metabolic pathway or proteins that interact to form one multisubunit protein. For instance, the *trp* operon discussed in Chapter 4 encodes five enzymes needed in the biosynthesis of tryptophan (see Figure 4-13). Similarly, the *lac* operon encodes three enzymes required for the metabolism of lactose, a sugar present in milk. Since a bacterial operon is transcribed from one start site into a single mRNA, all the genes within an operon are **coordinately regulated**; that is, they are all activated or repressed to the same extent.

Transcription of operons, as well as of isolated genes, is controlled by an interplay between RNA polymerase and specific repressor and activator proteins. In order to initiate transcription, however, *E. coli* RNA polymerase must be associated with one of a small number of *σ (sigma) factors.* The most common one in eubacterial cells is σ^{70}. σ^{70} binds to RNA polymerase and to promoter DNA sequences, bringing the RNA polymerase enzyme to a promoter. σ^{70} recognizes and binds to both a six-base pair sequence centered at −10 and a seven-base pair sequence centered at −35 from the +1 transcription start. Consequently, the −10 plus the −35 sequence constitute a promoter for *E. coli* RNA polymerase associated with σ^{70} (see Figure 4-10b). Although the promoter sequences contacted by σ^{70} are located at −35 and −10, *E. coli* RNA polymerase binds to the promoter region DNA from ≈−50 to ≈+20 through interactions with DNA that do not depend on the sequence. σ^{70} also assists the RNA polymerase in separating the DNA strands at the transcription start site and inserting the coding strand into the active site of the polymerase so that transcription starts at +1 (see Figure 4-11, step 2). The optimal σ^{70}-RNA polymerase promoter sequence, determined as the consensus sequence of multiple strong promoters, is:

–35 region —— –10 region
TTGACAT——15–17 bp——TATAAT

The size of the font indicates the importance of that base at this position. The sequence shows the strand of DNA that has the same 5′→3′ orientation as the transcribed RNA (i.e., the nontemplate strand). However, the σ^{70}-RNA polymerase initially binds to double-stranded DNA. After the polymerase transcribes a few tens of base pairs, σ^{70} is released. Thus σ^{70} acts as an *initiation factor* required for transcription initiation but not for RNA-strand elongation once initiation has taken place.

Initiation of *lac* Operon Transcription Can Be Repressed and Activated

When *E. coli* is in an environment that lacks lactose, synthesis of *lac* mRNA is repressed so that cellular energy is not wasted synthesizing enzymes the cells cannot use. In an environment containing both lactose and glucose, *E. coli* cells preferentially metabolize glucose, the central molecule of carbohydrate metabolism. Lactose is metabolized at a high rate only when lactose is present and glucose is largely depleted from the medium. This metabolic adjustment is achieved by repressing transcription of the *lac* operon until lactose is present and allowing synthesis of only low levels of *lac* mRNA until the cytosolic concentration of glucose falls to low levels. Transcription of the *lac* operon under different conditions is controlled by *lac* repressor and *c*atabolite *a*ctivator *p*rotein (CAP) (also called CRP for *c*atabolite *r*eceptor *p*rotein), each of which binds to a specific DNA sequence in the *lac* transcription-control region (Figure 7-2, *top*).

For transcription of the *lac* operon to begin, the σ^{70} subunit of the RNA polymerase must bind to the *lac* promoter at the −35 and −10 promoter sequences. When no lactose is present, the *lac* repressor binds to a sequence called the *lac* **operator,** which overlaps the transcription start site. Therefore, *lac* repressor bound to the operator site blocks binding and hence transcription initiation by RNA polymerase (Figure 7-2a). When lactose is present, it binds to specific binding sites in each subunit of the tetrameric *lac* repressor, causing a conformational change in the protein that makes it dissociate from the *lac* operator. As a result, the polymerase can bind to the promoter and initiate transcription of the *lac* operon. However, when glucose also is present, the rate of transcription initiation (i.e., the number of times per minute different RNA polymerase molecules initiate transcription) is very low, resulting in synthesis of only low levels of *lac* mRNA and the proteins encoded in the *lac* operon (Figure 7-2b). The frequency of transcription initiation is low because

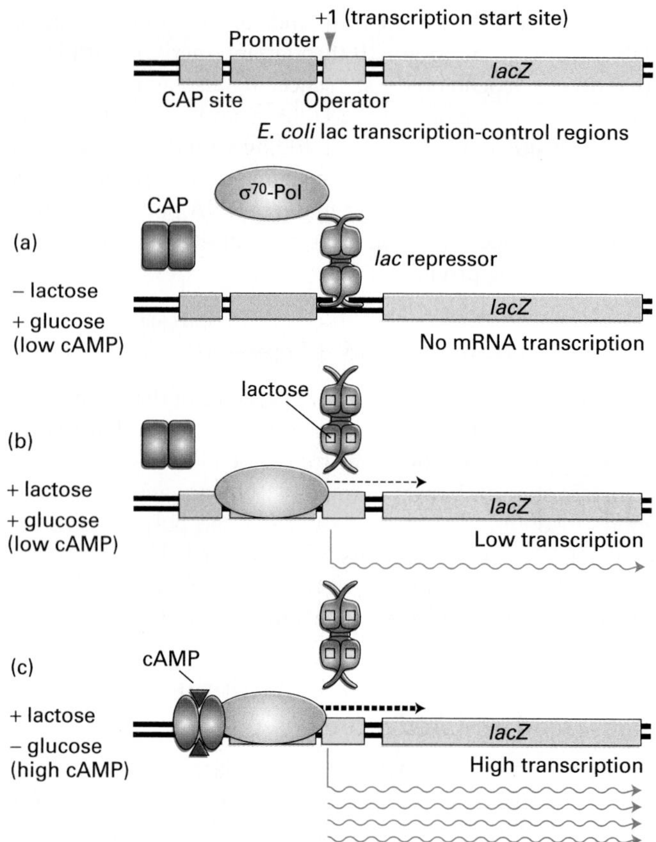

▲ **FIGURE 7-2 Regulation of transcription from the *lac* operon of *E. coli*.** *(Top)* The transcription-control region, composed of ≈100 base pairs, includes three protein-binding regions: the CAP site, which binds catabolite activator protein; the *lac* promoter, which binds the σ^{70}–RNA polymerase complex; and the *lac* operator, which binds *lac* repressor. The *lacZ* gene, the first of three genes in the operon, is shown to the right. (a) In the absence of lactose, very little *lac* mRNA is produced because the *lac* repressor binds to the operator, inhibiting transcription initiation by σ^{70}–RNA polymerase. (b) In the presence of glucose and lactose, *lac* repressor binds lactose and dissociates from the operator, allowing σ^{70}–RNA polymerase to initiate transcription at a low rate. (c) Maximal transcription of the *lac* operon occurs in the presence of lactose and absence of glucose. In this situation, cAMP increases in response to the low glucose concentration and forms the CAP-cAMP complex, which binds to the CAP site, where it interacts with RNA polymerase to stimulate the rate of transcription initiation.

the −35 and −10 sequences in the *lac* promoter differ from the ideal σ^{70}-binding sequences shown previously.

Once glucose is depleted from the media and the intracellular glucose concentration falls, *E. coli* cells respond by synthesizing cyclic AMP, or cAMP. As the concentration of cAMP increases, it binds to a site in each subunit of the dimeric CAP protein, causing a conformational change that allows the protein to bind to the CAP site in the *lac* transcription-control region. The bound CAP-cAMP complex interacts with the polymerase bound to the promoter, greatly stimulating the rate of transcription initiation. This activation leads to synthesis of high levels of *lac* mRNA and subsequently of the enzymes encoded by the *lac* operon (Figure 7-2c).

In fact, the *lac* operon is more complex than depicted in the simplified model of Figure 7-2. The tetrameric *lac* repressor actually binds to two sites simultaneously, one at the primary operator (*lacO1*) that overlaps the region of DNA bound by RNA polymerase at the promoter and at one of two secondary operators centered at +412 (*lacO2*) and −82 (*lacO3*) (Figure 7-3). The *lac* repressor tetramer is a dimer of dimers. Each dimer binds to one operator. Simultaneous binding of the tetrameric *lac* repressor to the primary *lac* operator *O1* and one of the two secondary operators is possible because DNA is quite flexible, as we saw in the wrapping of DNA around the surface of a histone octomer in the nucleosomes of eukaryotes (Figure 6-29). These secondary operators function to increase the local concentration of *lac* repressor in the micro-vicinity of the primary operator where repressor binding blocks RNA polymerase binding. Since the equilibrium of binding reactions depends on the concentrations of the binding partners, the resulting increased local concentration of *lac* repressor in the vicinity of *O1* increases repressor binding to *O1*. There are approximately 10 *lac* repressor tetramers per *E. coli* cell. Because of binding to *O2* and *O3*, there is nearly always a *lac* repressor tetramer much closer to *O1* than would otherwise be the case if the 10 repressors were diffusing randomly through the cell. If both *O2* and *O3* are mutated so that the *lac* repressor no longer binds to them with high affinity, repression at the *lac* promoter is reduced by a factor of 70. Mutation of only *O2* or only *O3* reduces repression twofold, indicating that either one of these secondary operators provides most of the stimulation of repression.

Although the promoters for different *E. coli* genes exhibit considerable homology, their exact sequences differ. The promoter sequence determines the intrinsic rate at which an RNA polymerase–σ complex initiates transcription of a gene in the absence of a repressor or activator protein. Promoters that support a high rate of transcription initiation have −10 and −35 sequences similar to the ideal promoter shown previously and are called *strong promoters*. Those that support a low rate of transcription initiation differ from this ideal sequence and are called *weak promoters*. The *lac* operon, for instance, has a weak promoter. Its sequence differs from the consensus strong promoter at several positions. This low intrinsic rate of initiation is further reduced by the *lac* repressor and substantially increased by the cAMP-CAP activator.

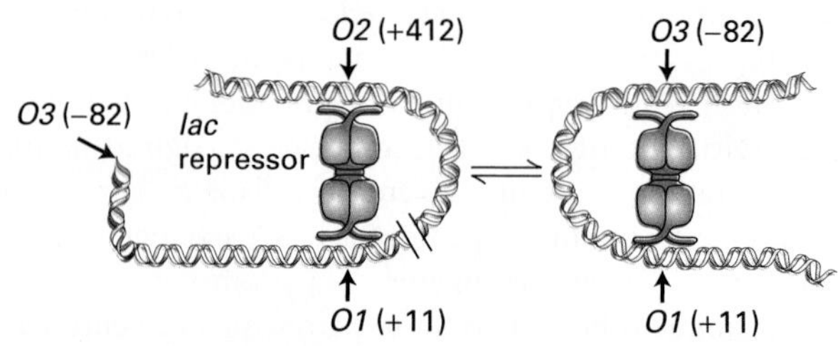

▲ **FIGURE 7-3 *Lac* repressor-operator interactions.** The tetrameric *lac* repressor binds to the primary *lac* operator (*O1*) and one of two secondary operators (*O2* or *O3*) simultaneously. The two structures are in equilibrium. [Adapted from B. Muller-Hill, 1998, *Curr. Op. Microbiol.* **1**:145.]

Small Molecules Regulate Expression of Many Bacterial Genes via DNA-Binding Repressors and Activators

Transcription of most *E. coli* genes is regulated by processes similar to those described for the *lac* operon, although the detailed interactions differ at each promoter. The general mechanism involves a specific repressor that binds to the operator region of a gene or operon, thereby blocking transcription initiation. A small molecule ligand (or ligands) binds to the repressor, controlling its DNA-binding activity and consequently the rate of transcription as appropriate for the needs of the cell. As for the *lac* operon, many eubacterial transcription-control regions contain one or more secondary operators that contribute to the level of repression.

Specific activator proteins, such as CAP in the *lac* operon, also control transcription of a subset of bacterial genes that have binding sites for the activator. Like CAP, other activators bind to DNA together with RNA polymerase, stimulating transcription from a specific promoter. The DNA-binding activity of an activator can be modulated in response to cellular needs by binding specific small molecule ligands (e.g., cAMP) or by post-translational modifications, such as phosphorylation, that alter the conformation of the activator.

Transcription Initiation from Some Promoters Requires Alternative Sigma Factors

Most *E. coli* promoters interact with σ^{70}-RNA polymerase, the major initiating form of the bacterial enzyme. Transcription of certain groups of genes, however, is initiated by *E. coli* RNA polymerases containing one of several alternative sigma factors that recognize different consensus promoter sequences than σ^{70} does (Table 7-1). These alternative σ-factors are required for the transcription of sets of genes with related functions such as those involved in the response to heat shock or nutrient deprivation, motility, or sporulation in gram-positive eubacteria. In *E. coli* there are six alternative σ-factors in addition to the major "housekeeping" σ-factor, σ^{70}. The genome of the gram-positive, sporulating bacterium *Streptomyces coelicolor* encodes 63 σ-factors, the current record, based on sequence analysis of

TABLE 7-1 Sigma Factors of *E. Coli*

		PROMOTER CONSENSUS	
SIGMA FACTOR	PROMOTERS RECOGNIZED	−35 REGION	−10 REGION
σ^{70}	Housekeeping genes, most genes in exponentially replicating cells	TTGACA	TATAAT
σ^{S}	Stationary-phase genes and general stress response	TTGACA	TATAAT
σ^{32}	Induced by unfolded proteins in the cytoplasm; genes encoding chaperones that refold unfolded proteins and protease systems leading to the degradation of unfolded proteins in the cytoplasm	TCTCNCCCTTGAA	CCCCATNTA
σ^{E}	Activated by unfolded proteins in the periplasmic space and cell membrane; genes encoding proteins that restore integrity to the cellular envelope	GAACTT	TCTGA
σ^{F}	Genes involved in flagellum assembly	CTAAA	CCGATAT
FecI	Genes required for iron uptake	TTGGAAA	GTAATG
		−24 REGION	−12 REGION
σ^{54}	Genes for nitrogen metabolism and other functions	CTGGNA	TTGCA

SOURCES: C. A. Gross, M. Lonetto, and R. Losick, 1992, in *Transcriptional Regulation*, S. L. McKnight and K. R. Yamamoto, eds., Cold Spring Harbor Laboratory Press; D. N. Arnosti and M. J. Chamberlin, 1989, *Proc. Nat'l. Acad. Sci. USA* **86**:830; K. Tanaka et al., 1993, *Proc. Nat'l. Acad. Sci., USA* **90**:3511; C. Dartigalongue et al., 2001, *J. Biol. Chem.* **276**:20866; A. Angerer and V. Braun, 1998, *Arch. Microbiol.* **169**:483.

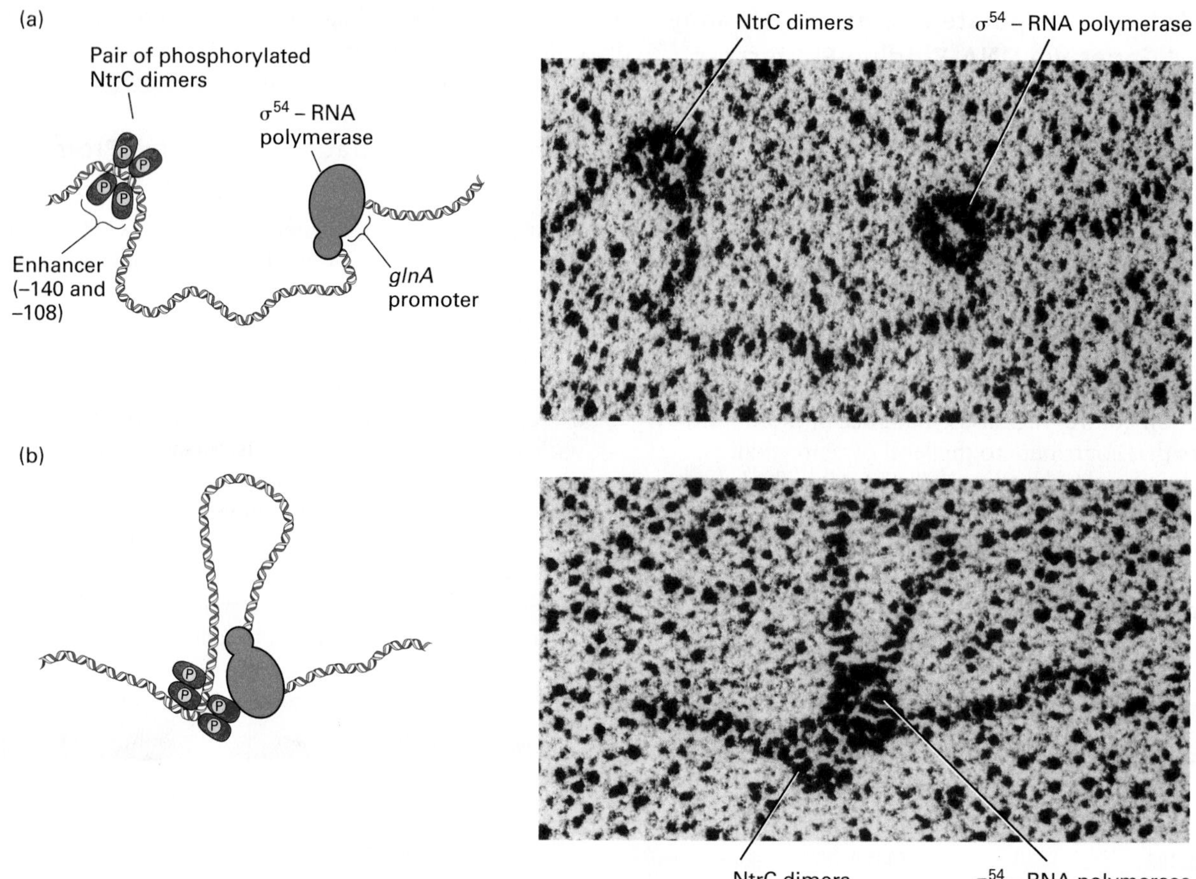

▲ EXPERIMENTAL FIGURE 7-4 **DNA looping permits interaction of bound NtrC and σ^{54}–RNA polymerase.** (a) Drawing (*left*) and electron micrograph (*right*) of DNA restriction fragment with phosphorylated NtrC dimers binding to the enhancer region near one end and σ^{54}–RNA polymerase bound to the *glnA* promoter near the other end. (b) Drawing (*left*) and electron micrograph (*right*) of the same fragment preparation showing NtrC dimers and σ^{54}–RNA polymerase binding to each other with the intervening DNA forming a loop between them. [Micrographs from W. Su et al., 1990, *Proc. Nat'l. Acad. Sci. USA* **87**:5505; courtesy of S. Kustu.]

nearly 100 eubacterial genomes. Most are structurally and functionally related to σ^{70}. But one class is unrelated, represented in *E. coli* by σ^{54}. Transcription initiation by RNA polymerases containing σ^{70}-like factors is regulated by repressors and activators that bind to DNA near the region where the polymerase binds, similar to initiation by σ^{70}-RNA polymerase itself.

Transcription by σ^{54}-RNA Polymerase Is Controlled by Activators That Bind Far from the Promoter

The sequence of one *E. coli* sigma factor, σ^{54}, is distinctly different from that of all the σ^{70}-like factors. Transcription of genes by RNA polymerases containing σ^{54} is regulated solely by activators whose binding sites in DNA, referred to as **enhancers,** generally are located 80–160 base pairs upstream from the start site. Even when enhancers are moved more than a kilobase away from a start site, σ^{54}-activators can activate transcription.

The best-characterized σ^{54}-activator—the NtrC protein (nitrogen regulatory protein C)—stimulates transcription of the *glnA* gene. *glnA* encodes the enzyme glutamine synthetase, which synthesizes the amino acid glutamine from glutamic acid and ammonia. The σ^{54}-RNA polymerase binds to the *glnA* promoter but does not melt the DNA strands and initiate transcription until it is activated by NtrC, a dimeric protein. NtrC, in turn, is regulated by a protein kinase called NtrB. In response to low levels of glutamine, NtrB phosphorylates dimeric NtrC, which then binds to an enhancer upstream of the *glnA* promoter. Enhancer-bound phosphorylated NtrC then stimulates the σ^{54}-polymerase bound at the promoter to separate the DNA strands and initiate transcription.

Electron microscopy studies have shown that phosphorylated NtrC bound at enhancers and σ^{54}-polymerase bound at the promoter directly interact, forming a loop in the DNA between the binding sites (Figure 7-4). As discussed later in this chapter, this activation mechanism resembles the predominant mechanism of transcriptional activation in eukaryotes.

NtrC has ATPase activity, and ATP hydrolysis is required for activation of bound σ^{54}-polymerase by phosphorylated NtrC. Evidence for this is that mutants with an NtrC defective in ATP hydrolysis are invariably defective in stimulating the σ^{54}-polymerase to melt the DNA strands at the transcription start site. It is postulated that ATP hydrolysis supplies the energy required for melting the DNA strands. In contrast, the σ^{70}-polymerase does not require ATP hydrolysis to separate the strands at a start site.

Many Bacterial Responses Are Controlled by Two-Component Regulatory Systems

As we have just seen, control of the *E. coli glnA* gene depends on two proteins, NtrC and NtrB. Such two-component regulatory systems control many responses of bacteria to changes in their environment. Another example involves the *E. coli* proteins PhoR and PhoB, which regulate transcription in response to the concentration of free phosphate. PhoR is a transmembrane protein, located in the inner (plasma) membrane, whose periplasmic domain binds phosphate with moderate affinity and whose cytosolic domain has protein kinase activity; PhoB is a cytosolic protein.

Large protein pores in the *E. coli* outer membrane allow ions to diffuse freely between the external environment and the periplasmic space. Consequently, when the phosphate concentration in the environment falls, it also falls in the periplasmic space, causing phosphate to dissociate from the PhoR periplasmic domain, as depicted in Figure 7-5. This causes a conformational change in the PhoR cytoplasmic domain that activates its protein kinase activity. The activated PhoR initially transfers a γ-phosphate from ATP to a histidine (H) side chain in the PhoR kinase domain itself. The same phosphate is then transferred to a specific aspartic acid (D) side chain in PhoB, converting PhoB from an inactive to an active transcriptional activator. Phosphorylated, active PhoB then induces transcription from several genes that help the cell cope with low phosphate conditions.

Many other bacterial responses are regulated by two proteins with homology to PhoR and PhoB. In each of these regulatory systems, one protein, called a *sensor,* contains a transmitter domain homologous to the PhoR protein kinase domain. The transmitter domain of the sensor protein is regulated by a second unique protein domain (e.g., the periplasmic domain of PhoR) that senses environmental changes. The second protein, called a *response regulator,* contains a receiver domain homologous to the region of PhoB that is phosphorylated by activated PhoR. The receiver domain of the response regulator is associated with a second domain that determines the protein's function. The activity of this second functional domain is regulated by phosphorylation of the receiver domain. Although all transmitter domains are homologous (as are receiver domains), the transmitter domain of a specific sensor protein will phosphorylate only the receiver domains of specific response regulators, allowing specific responses to different environmental changes. Note that NtrB and NtrC, discussed above, function as sensor and response regulator proteins, respectively, in the two-component regulatory system that controls transcription of *glnA*. Similar two-component histidyl-aspartyl phosphorelay regulatory systems are also found in plants.

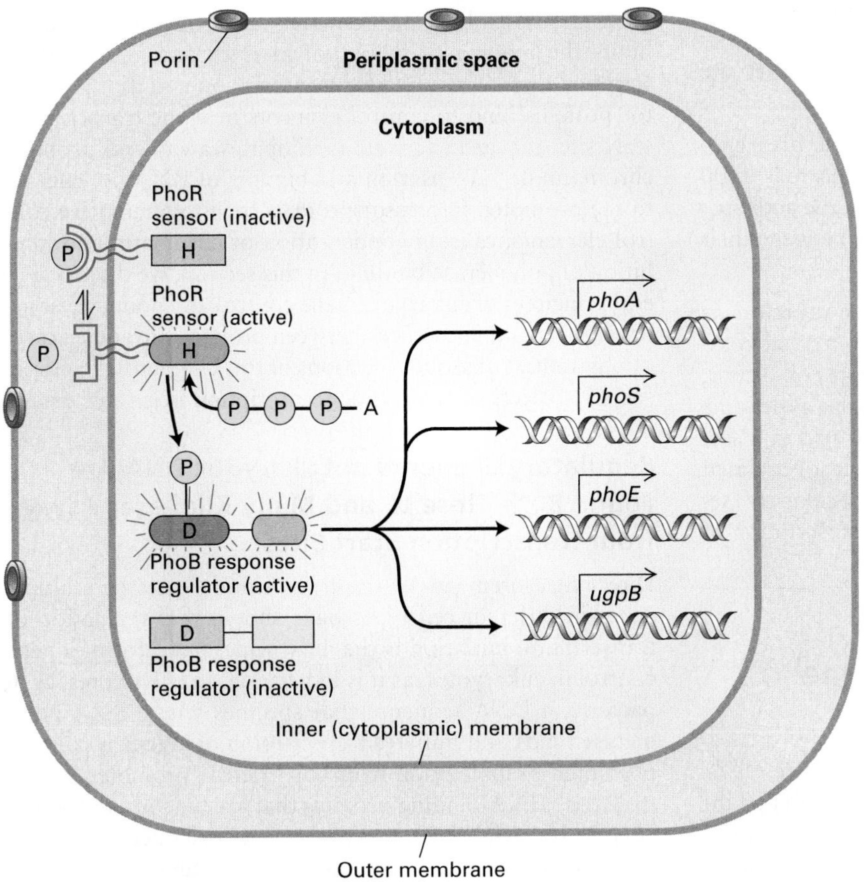

◀ FIGURE 7-5 The PhoR/PhoB two-component regulatory system in *E. coli.* In response to low phosphate concentrations in the environment and periplasmic space, a phosphate ion dissociates from the periplasmic domain of the inactive sensor protein PhoR. This causes a conformational change that activates a protein kinase transmitter domain in the cytosolic region of PhoR. The activated transmitter domain transfers an ATP γ phosphate to a conserved histidine (H) in the transmitter domain. This phosphate is then transferred to an aspartic acid (D) in the receiver domain of the response regulator PhoB. Several PhoB proteins can be phosphorylated by one activated PhoR. Phosphorylated PhoB proteins then activate transcription from genes encoding proteins that help the cell to respond to low phosphate, including *phoA, phoS, phoE,* and *ugpB.*

KEY CONCEPTS OF SECTION 7.1

Control of Gene Expression in Bacteria

- Gene expression in both prokaryotes and eukaryotes is regulated primarily by mechanisms that control the initiation of transcription.
- The first step in the initiation of transcription in *E. coli* is binding of the σ subunit complexed with an RNA polymerase to a promoter.
- The nucleotide sequence of a promoter determines its strength, that is, how frequently different RNA polymerase molecules can bind and initiate transcription per minute.
- Repressors are proteins that bind to operator sequences, which overlap or lie adjacent to promoters. Binding of a repressor to an operator inhibits transcription initiation.
- The DNA-binding activity of most bacterial repressors is modulated by small molecule ligands. This allows bacterial cells to regulate transcription of specific genes in response to changes in the concentration of various nutrients in the environment and metabolites in the cytoplasm.
- The *lac* operon and some other bacterial genes also are regulated by activator proteins that bind next to promoters and increase the rate of transcription initiation by RNA polymerase.
- The major sigma factor in *E. coli* is σ^{70}, but several other less abundant sigma factors are also found, each recognizing different consensus promoter sequences.
- Transcription initiation by all *E. coli* RNA polymerases, except those containing σ^{54}, can be regulated by repressors and activators that bind near the transcription start site (see Figure 7-2).
- Genes transcribed by σ^{54}–RNA polymerase are regulated by activators that bind to enhancers located $\approx$100 base pairs upstream from the start site. When the activator and σ^{54}–RNA polymerase interact, the DNA between their binding sites forms a loop (see Figure 7-4).
- In two-component regulatory systems, one protein acts as a sensor, monitoring the level of nutrients or other components in the environment. Under appropriate conditions, the γ-phosphate of an ATP is transferred first to a histidine in the sensor protein and then to an aspartic acid in a second protein, the response regulator. The phosphorylated response regulator then binds to DNA regulatory sequences, thereby stimulating or repressing transcription of specific genes (see Figure 7-5).

7.2 Overview of Eukaryotic Gene Control and RNA Polymerases

In bacteria, gene control serves mainly to allow a single cell to adjust to changes in its environment so that its growth and division can be optimized. In multicellular organisms, environmental changes also induce changes in gene expression. An example is the response to low oxygen (hypoxia) in which a specific set of genes is rapidly induced that help the cell survive under the hypoxic conditions. These include secreted angiogenic proteins that stimulate the growth and penetration of new capillaries into the surrounding tissue. However, the most characteristic and biologically far-reaching purpose of gene control in multicellular organisms is execution of the genetic program that underlies embryological development. Generation of the many different cell types that collectively form a multicellular organism depends on the right genes being activated in the right cells at the right time during the developmental period.

In most cases, once a developmental step has been taken by a cell, it is not reversed. Thus these decisions are fundamentally different from the reversible activation and repression of bacterial genes in response to environmental conditions. In executing their genetic programs, many differentiated cells (e.g., skin cells, red blood cells, and antibody-producing cells) march down a pathway to final cell death, leaving no progeny behind. The fixed patterns of gene control leading to differentiation serve the needs of the whole organism and not the survival of an individual cell. Despite the differences in the purposes of gene control in bacteria and eukaryotes, two key features of transcription control first discovered in bacteria and described in the previous section also apply to eukaryotic cells. First, protein-binding regulatory DNA sequences, or control elements, are associated with genes. Second, specific proteins that bind to a gene's regulatory sequences determine where transcription will start and either activate or repress its transcription. As represented in Figure 7-1, in multicellular eukaryotes, inactive genes are assembled into condensed chromatin, which inhibits the binding of RNA polymerases and general transcription factors required for transcription initiation. Activator proteins bind to control elements near the transcription start site of a gene as well as kilobases away and promote chromatin decondensation and binding of RNA polymerase to the promoter. Repressor proteins bind to alternative control elements, causing condensation of chromatin and inhibition of polymerase binding. In this section, we discuss general principles of eukaryotic gene control and point out some similarities and differences between prokaryotic and eukaryotic systems. Subsequent sections of this chapter will address specific aspects of eukaryotic transcription in greater detail.

Regulatory Elements in Eukaryotic DNA Are Found Both Close to and Many Kilobases Away from Transcription Start Sites

Direct measurements of the transcription rates of multiple genes in different cell types have shown that regulation of transcription initiation is the most widespread form of gene control in eukaryotes, as it is in bacteria. In eukaryotes, as in bacteria, a DNA sequence that specifies where RNA polymerase binds and initiates transcription of a gene is called a promoter. Transcription from a particular promoter is controlled by DNA-binding proteins that are functionally equivalent to bacterial repressors and activators. Since these transcriptional regulatory proteins can often function either to

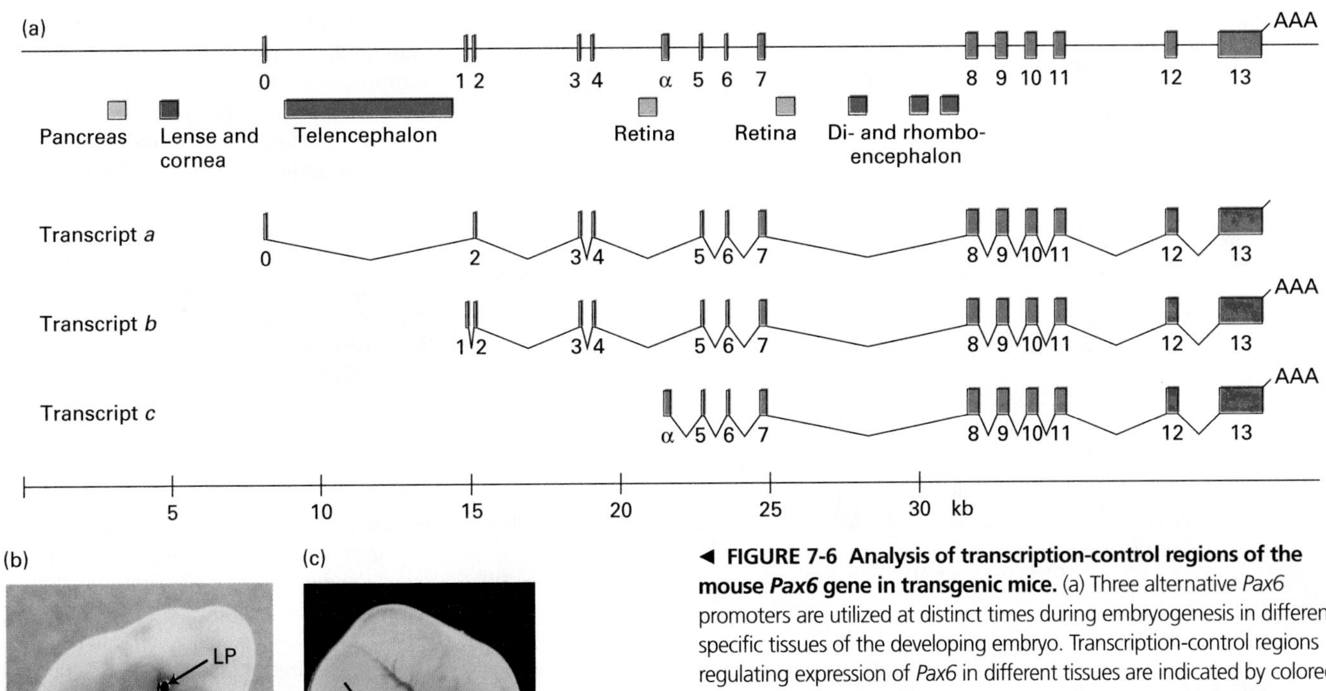

◄ **FIGURE 7-6 Analysis of transcription-control regions of the mouse *Pax6* gene in transgenic mice.** (a) Three alternative *Pax6* promoters are utilized at distinct times during embryogenesis in different specific tissues of the developing embryo. Transcription-control regions regulating expression of *Pax6* in different tissues are indicated by colored rectangles. The telencephalon-specific control region in intron 1 between exons 0 and 1 has not been mapped to high resolution. The other control regions shown are ≈200–500 base pairs in length. (b) β-galactosidase expressed in tissues of a mouse embryo with a β-galactosidase reporter transgene 10.5 days after fertilization. The genome of the mouse embryo contained a transgene with 8 kb of DNA upstream from exon 0 fused to the β-galactosidase coding region. Lense pit (LP) is the tissue that will develop into the lense of the eye. Expression was also observed in tissue that will develop into the pancreas (P). (c) β-galactosidase expression in a 13.5-day embryo with a β-galactosidase reporter gene under control of the sequence in part (a) between exons 4 and 5 marked Retina. Arrow points to nasal and temporal regions of the developing retina. *Pax6* transcription-control regions have also been found ≈17 kb downstream from the 3′ exon in an intron of the neighboring gene. [Part (a) adapted from B. Kammendal et al., 1999, *Dev. Biol.* **205:**79; parts (b) and (c): Courtesy of Peter Gruss]

activate or to repress transcription depending on their association with other proteins, they are more generally called transcription factors. The DNA control elements in eukaryotic genomes that bind transcription factors often are located much farther from the promoter they regulate than is the case in prokaryotic genomes. In some cases, transcription factors that regulate expression of protein-coding genes in higher eukaryotes bind at regulatory sites tens of thousands of base pairs either **upstream** (opposite to the direction of transcription) or **downstream** (in the same direction as transcription) from the promoter. As a result of this arrangement, transcription of a single gene may be regulated by binding of multiple transcription factors to alternative control elements, directing expression of the same gene in different types of cells and at different times during development.

For example, several separate transcription-control DNA sequences regulate expression of the mammalian gene encoding the transcription factor *Pax6*. Pax6 protein is required for development of the eye, certain regions of the brain and spinal cord, and the cells in the pancreas that secrete hormones such as insulin. Heterozygous humans with only one functional *Pax6* gene are born with *aniridia,* a lack of irises in the eyes (Figure 1-26). The *Pax6* gene is expressed from at least three alternative promoters that function in different cell types and at different times during embryogenesis (Figure 7-6a). When transgenic mice are prepared (see Figure 5-43) containing a *β-galactosidase* **reporter gene** fused to 8 kb of DNA upstream from *Pax6* exon 0, β-galactosidase is observed in the developing lense, cornea, and pancreas of the embryo halfway through gestation (Figure 7-6b). Analysis of transgenic mice with smaller fragments of DNA from this region allowed the mapping of separate transcription-control regions regulating transcription in the pancreas and in the lense and cornea. Transgenic mice with other reporter gene constructs revealed additional transcription-control regions (Figure 7-6a). These controlled transcription in the developing retina and different regions of the brain (encephalon). Some of these transcription-control regions are in introns between exons 4 and 5 and between exons 7 and 8. For example, a reporter gene under control of the region labeled *retina* in Figure 7-6a between exons 4 and 5 led to reporter gene expression specifically in the retina (Figure 7-6c).

Control regions for many genes are found several hundreds of kilobases away from the coding exons of the gene. One method for identifying such distant control regions is to

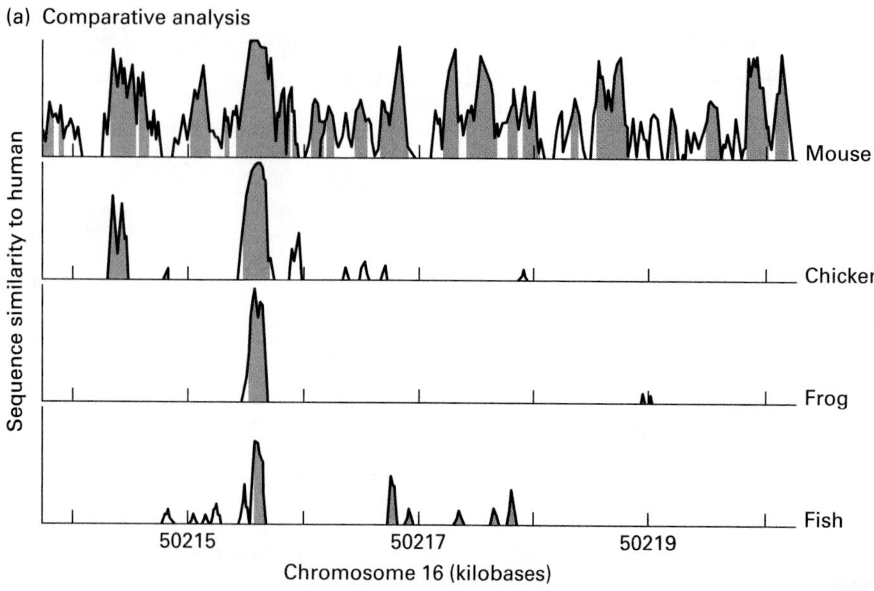

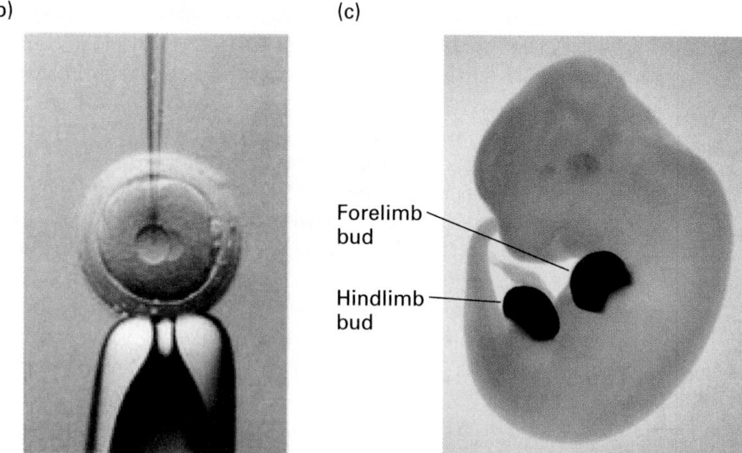

◄ **EXPERIMENTAL FIGURE 7-7 The human *SALL1* gene enhancer activates expression of a reporter gene in limb buds of the developing mouse embryo.** (a) Graphic representation of the conservation of DNA sequence in a region of the human genome (from 50214–50220.5 kb of the chromosome 16 sequence) ≈500 kb downstream from the *SALL1* gene encoding a zinc finger transcription repressor. A region of ≈500 bp of non-coding sequence is conserved from fish to human. 900 bp including this conserved region were inserted into a plasmid next to the coding region for *E. coli* β-galactosidase. (b) The plasmid was microinjected into a pronucleus of a fertilized mouse egg and implanted in the uterus of a pseudo-pregnant mouse to generate a transgenic mouse embryo with the "reporter gene" on the injected plasmid incorporated into its genome (see Figure 5-43). (c) After 11.5 days of development when limb buds develop, the fixed and permeabilized embryo was incubated in X-gal which is converted into an insoluble intensely blue compound by β-galactosidase. The ≈ 900 bp region of human DNA contained an enhancer that stimulated strong transcription of the β-galactosidase reporter gene in limb buds specifically. [From the VISTA Enhancer Browser http://enhancer. lbl.gov; Parts (b) and (c): Courtesy of Len A. Pennacchio, Joint Genome Institute, Lawrence Berkeley National Laboratory.]

compare the sequences of distantly related organisms. Transcription control regions for a conserved gene are also often conserved and can be recognized in the background of nonfunctional sequence that diverges during evolution. For example, there is a human DNA sequence ≈500 kilobases downstream of the *SALL1* gene that is highly conserved in mice, frogs, and fish (Figure 7-7a). This gene encodes a transcription repressor required for normal development of the lower intestine, kidneys, limbs, and ears. When transgenic mice were generated containing this conserved DNA sequence linked to a β-galactosidase reporter gene (Figure 7-7b), the transgenic embryos expressed a very high level of the β-galactosidase reporter gene specifically in the developing limb buds (Figure 7-7c). Human patients with deletions in this region of the genome develop with limb abnormalities. These results indicate that this conserved region directs transcription of the *SALL1* gene in the developing limb. Presumably, other enhancers control expression of this gene in other types of cells where it functions in the normal development of the lower intestine, kidneys and ears.

Three Eukaryotic Polymerases Catalyze Formation of Different RNAs

The nuclei of all eukaryotic cells examined so far (e.g., vertebrate, *Drosophila,* yeast, and plant cells) contain three different RNA polymerases, designated I, II, and III. These enzymes are eluted at different salt concentrations during ion-exchange chromatography, reflecting the polymerases' various net charges. The three polymerases also differ in their sensitivity to α-amanitin, a poisonous cyclic octapeptide produced by some mushrooms (Figure 7-8). RNA polymerase I is very insensitive to α-amanitin, but RNA polymerase II is very sensitive—the drug binds near the active site of the enzyme and inhibits translocation of the enzyme along the DNA template. RNA polymerase III has intermediate sensitivity.

Each eukaryotic RNA polymerase catalyzes transcription of genes encoding different classes of RNA (Table 7-2). *RNA polymerase I,* located in the nucleolus, transcribes genes encoding precursor rRNA (**pre-rRNA**), which is processed into 28S, 5.8S, and 18S rRNAs. *RNA polymerase III* transcribes genes encoding tRNAs, 5S rRNA, and an array of small,

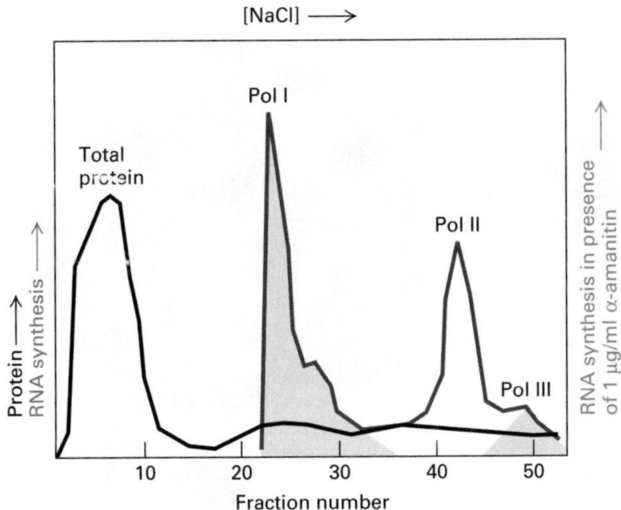

▲ EXPERIMENTAL FIGURE 7-8 **Column chromatography separates and identifies the three eukaryotic RNA polymerases, each with its own sensitivity to α-amanitin.** A protein extract from the nuclei of cultured eukaryotic cells is passed through a DEAE Sephadex column and adsorbed protein eluted (black curve) with a solution of constantly increasing NaCl concentration. Three fractions from the eluate subsequently showed RNA polymerase activity (red curve). At a concentration of 1 μg/ml, α-amanitin inhibits polymerase II activity but has no effect on polymerases I and III (green shading). Polymerase III is inhibited by 10 μg/ml of α-amanitin, whereas polymerase I is unaffected even at this higher concentration. [See R. G. Roeder, 1974, *J. Biol. Chem.* **249**:241.]

stable RNAs, including one involved in RNA splicing (U6) and the RNA component of the signal-recognition particle (SRP) involved in directing nascent proteins to the endoplasmic reticulum (Chapter 13). *RNA polymerase II* transcribes all protein-coding genes; that is, it functions in production of mRNAs. RNA polymerase II also produces four of the five small nuclear RNAs that take part in RNA splicing.

Each of the three eukaryotic RNA polymerases is more complex than *E. coli* RNA polymerase, although their structures are similar (Figure 7-9a, b). All three contain two large subunits and 10–14 smaller subunits, some of which are common between two or all three of the polymerases. The best-characterized eukaryotic RNA polymerases are from the yeast *Saccharomyces cerevisiae*. Each of the yeast genes encoding the polymerase subunits has been cloned and sequenced and the effects of gene-knockout mutations have been characterized. In addition, the three-dimensional structure of yeast RNA polymerase II has been determined (Figure 7-9b, c). The three nuclear RNA polymerases from all eukaryotes so far examined are very similar to those of yeast.

The two large subunits (RPB1 and RPB2) of all three eukaryotic RNA polymerases are related to each other and are similar to the *E. coli* β′ and β subunits, respectively (Figure 7-10). Each of the eukaryotic polymerases also contains an ω-like and two nonidentical α-like subunits. The extensive similarity in the structures of these core subunits in RNA polymerases from various sources indicates that this enzyme arose early in evolution and was largely conserved. This seems logical for an enzyme catalyzing a process so basic as copying RNA from DNA.

In addition to their core subunits related to the *E. coli* RNA polymerase subunits, all three yeast RNA polymerases contain four additional small subunits, common to them but not to the bacterial RNA polymerase. Finally, each eukaryotic nuclear RNA polymerase has several enzyme-specific subunits that are not present in the other two nuclear RNA polymerases. Gene-knockout experiments in yeast indicate that most of these subunits are essential for cell viability. Disruption of the few polymerase subunit genes that are not absolutely essential for viability (subunits 4 and 7) nevertheless results in very poorly growing cells. Thus it seems likely that all the subunits are necessary for eukaryotic RNA polymerases to function normally.

The Largest Subunit in RNA Polymerase II Has an Essential Carboxyl-Terminal Repeat

The carboxyl end of the largest subunit of RNA polymerase II (RPB1) contains a stretch of seven amino acids that is nearly precisely repeated multiple times. Neither RNA polymerase I

TABLE 7-2 Classes of RNA Transcribed by the Three Eukaryotic Nuclear RNA Polymerases and Their Functions

POLYMERASE	RNA TRANSCRIBED	RNA FUNCTION
RNA polymerase I	Pre r-RNA (28S, 18S, 5.8S rRNAs)	Ribosome components, protein synthesis
RNA polymerase II	mRNA snRNAs miRNAs	Encodes protein RNA Splicing Post-transcriptional gene control
RNA polymerase III	tRNAs 5S rRNA snRNA U6 7S RNA Other stable short RNAs	Protein synthesis Ribosome component, protein synthesis RNA Splicing Signal-recognition particle for insertion of polypeptides into the endoplasmic reticulum Various functions, unknown for many

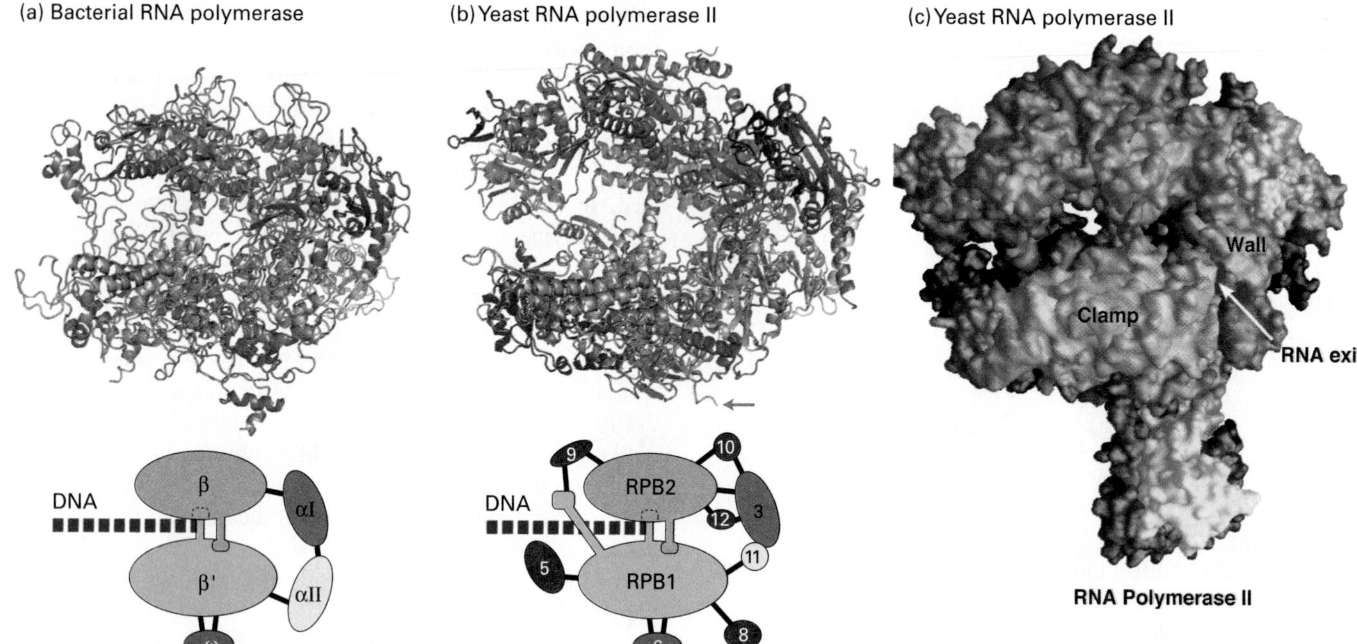

▲ **FIGURE 7-9 Comparison of three-dimensional structures of bacterial and eukaryotic RNA polymerases.** (a, b) These C_α trace models are based on x-ray crystallographic analysis of RNA polymerase from the bacterium *T. aquaticus* and core RNA polymerase II from *S. cerevisiae.* (a) The five subunits of the bacterial enzyme are distinguished by color. Only the N-terminal domains of the α subunits are included in this model. (b) Ten of the 12 subunits constituting yeast RNA polymerase II are shown in this model. Subunits that are similar in conformation to those in the bacterial enzyme are shown in the same colors. The C-terminal domain of the large subunit RPB1 was not observed in the crystal structure, but it is known to extend from the position marked with a red arrow. (RPB is the abbreviation for "*R*NA *p*olymerase *B*," which is an alternative way of referring to RNA polymerase II.) (c) Space-filling model of yeast RNA polymerase including subunits 4 and 7. These subunits extend from the core portion of the enzyme shown in (b) near the region of the C-terminal domain of the large subunit. Clamp is a domain of RPB1 that swings on a hinge when RNA is in the exit channel to close over upstream DNA so that the polymerase cannot release from the template until transcription has terminated. Wall is the domain of RPB2 that forces DNA entering the jaws of the polymerase from the left to bend before it exits the polymerase. The RNA exit channel is indicated. [Part (a) based on crystal structures from G. Zhang et al., 1999, *Cell* **98:**811. Part (b) from P. Cramer et al., 2001, *Science* **292:**1863. Part (c) from K. J. Armache et al., 2003, *PNAS* **100:**6964, and D. A. Bushnell and R. D. Kornberg, 2003, *PNAS* **100:**6969; colored model courtesy of Steven Hahn.]

nor III contains these repeating units. This heptapeptide repeat, with a consensus sequence of Tyr-Ser-Pro-Thr-Ser-Pro-Ser, is known as the *carboxyl-terminal domain (CTD).* Yeast RNA polymerase II contains 26 or more repeats, vertebrate enzymes have 52 repeats, and an intermediate number of repeats occur in RNA polymerase II from nearly all other eukaryotes. The CTD is critical for viability, and at least 10 copies of the repeat must be present for yeast to survive.

In vitro experiments with model promoters first showed that RNA polymerase II molecules that initiate transcription have an unphosphorylated CTD. Once the polymerase initiates transcription and begins to move away from the promoter, many of the serine and some tyrosine residues in the CTD are phosphorylated. Analysis of polytene chromosomes from *Drosophila* salivary glands prepared just before molting of the larva, a time of active transcription, indicate that the CTD also is phosphorylated during in vivo transcription. The large chromosomal "puffs" induced at this time in development are regions where the genome is very actively transcribed. Staining with antibodies specific for the phosphorylated or unphosphorylated CTD demonstrated that RNA polymerase II associated with the highly transcribed puffed regions contains a phosphorylated CTD (Figure 7-11).

RNA Polymerase II Initiates Transcription at DNA Sequences Corresponding to the 5′ Cap of mRNAs

In vitro transcription experiments involving RNA polymerase II, a protein extract prepared from the nuclei of cultured cells, and DNA templates containing sequences encoding the 5′ ends of mRNAs for a number of abundantly expressed genes revealed that the transcripts produced always contained a cap structure at their 5′ ends identical with that present at the 5′ end of nearly all eukaryotic mRNAs (see Figure 4-14). In these experiments, the 5′ cap is added to the 5′ end of the nascent RNA by enzymes in the nuclear extract, which can only add a cap to an RNA that has a 5′ tri- or diphosphate. Because a 5′ end generated by cleavage of a longer RNA would have a 5′ monophosphate, it would not be capped. Consequently, researchers concluded that the capped nucleotides generated in the in vitro transcription reactions must have been the nucleotides with which transcription was initiated. Sequence analysis revealed that for a given gene, the sequence at the 5′ end of the RNA transcripts produced in vitro is the same as that at the 5′ end of the mRNAs isolated from cells, confirming that the capped

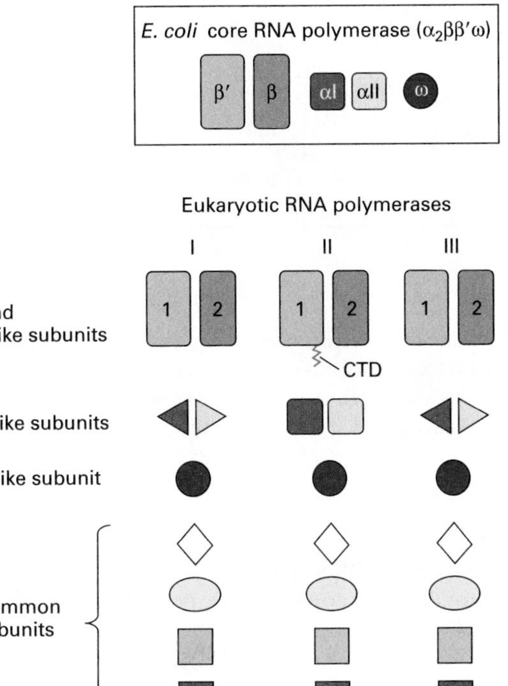

▲ **FIGURE 7-10 Schematic representation of the subunit structure of the *E. coli* RNA core polymerase and yeast nuclear RNA polymerases.** All three yeast polymerases have five core subunits homologous to the β, β′, two α, and ω subunits of *E. coli* RNA polymerase. The largest subunit (RPB1) of RNA polymerase II also contains an essential C-terminal domain (CTD). RNA polymerases I and III contain the same two nonidentical α-like subunits, whereas RNA polymerase II contains two other nonidentical α-like subunits. All three polymerases share the same ω-like subunit and four other common subunits. In addition, each yeast polymerase contains three to seven unique smaller subunits.

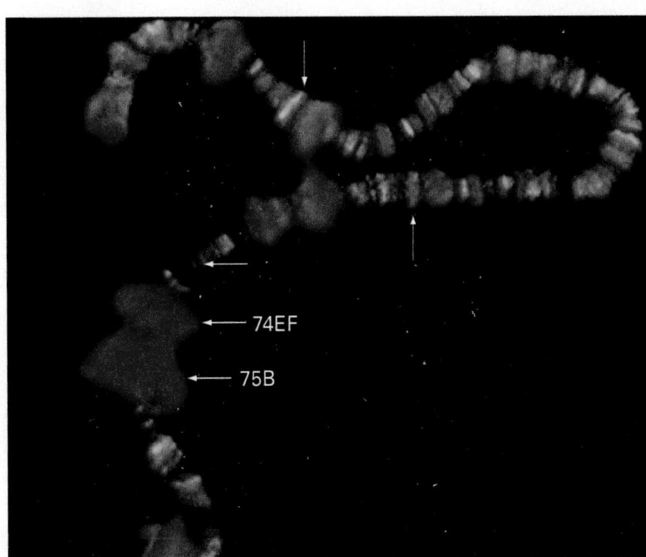

▲ EXPERIMENTAL FIGURE 7-11 **Antibody staining demonstrates that the carboxyl-terminal domain (CTD) of RNA polymerase II is phosphorylated during in vivo transcription.** Salivary gland polytene chromosomes were prepared from *Drosophila* larvae just before molting. The preparation was treated with a rabbit antibody specific for phosphorylated CTD and with a goat antibody specific for unphosphorylated CTD. The preparation then was stained with fluorescein-labeled anti-goat antibody (green) and rhodamine-labeled anti-rabbit antibody (red). Thus polymerase molecules with an unphosphorylated CTD stain green, and those with a phosphorylated CTD stain red. The molting hormone ecdysone induces very high rates of transcription in the puffed regions labeled 74EF and 75B; note that only phosphorylated CTD is present in these regions. Smaller puffed regions transcribed at high rates also are visible. Nonpuffed sites that stain red (up arrow) or green (horizontal arrow) also are indicated, as is a site staining both red and green, producing a yellow color (down arrow). [From J. R. Weeks et al., 1993, *Genes Dev.* **7**:2329; courtesy of J. R. Weeks and A. L. Greenleaf.]

nucleotide of eukaryotic mRNAs coincides with the transcription start site. Today, the transcription start site for a newly characterized mRNA generally is determined simply by identifying the DNA sequence encoding the 5′ end of the mRNA.

KEY CONCEPTS OF SECTION 7.2

Overview of Eukaryotic Gene Control and RNA Polymerases

- The primary purpose of gene control in multicellular organisms is the execution of precise developmental decisions so that the proper genes are expressed in the proper cells during development and cellular differentiation.
- Transcriptional control is the primary means of regulating gene expression in eukaryotes, as it is in bacteria.
- In eukaryotic genomes, DNA transcription-control elements may be located many kilobases away from the promoter they regulate. Different control regions can control transcription of the same gene in different cell types.
- Eukaryotes contain three types of nuclear RNA polymerases. All three contain two large and three smaller core subunits with homology to the β′, β, α, and ω subunits of *E. coli* RNA polymerase, as well several additional small subunits (see Figure 7-10).
- RNA polymerase I synthesizes only pre-rRNA. RNA polymerase II synthesizes mRNAs and some of the small nuclear RNAs that participate in mRNA splicing. RNA polymerase III synthesizes tRNAs, 5S rRNA, and several other relatively short, stable RNAs (see Table 7-2).
- The carboxyl-terminal domain (CTD) in the largest subunit of RNA polymerase II becomes phosphorylated during transcription initiation and remains phosphorylated as the enzyme transcribes the template.
- RNA polymerase II initiates transcription of genes at the nucleotide in the DNA template that corresponds to the 5′ nucleotide that is capped in the encoded mRNA.

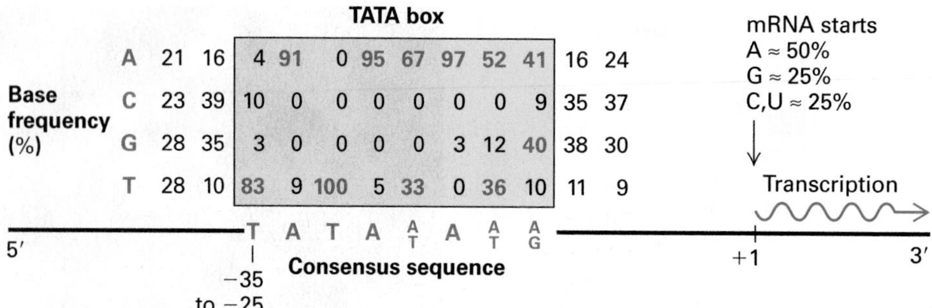

▲ **FIGURE 7-12 Determination of consensus TATA box sequence.** The nucleotide sequences upstream of the start site in 900 different eukaryotic protein-coding genes were aligned to maximize homology in the region from −35 to −25. The tabulated numbers are the percentage frequency of each base at each position. Maximum homology occurs over an eight-base region, referred to as the *TATA box,* whose consensus sequence is shown at the bottom. The initial base in mRNAs encoded by genes containing a TATA box most frequently is an A. [See P. Bucher, 1990, *J. Mol. Biol.* **212:**563.]

7.3 Regulatory Sequences in Protein-Coding Genes

As noted in the previous section, expression of eukaryotic protein-coding genes is regulated by multiple protein-binding DNA sequences, generically referred to as transcription-control regions. These include promoters and other types of control elements located near transcription start sites, as well as sequences located far from the genes they regulate. In this section, we take a closer look at the properties of various control elements found in eukaryotic protein-coding genes and some techniques used to identify them.

The TATA Box, Initiators, and CpG Islands Function as Promoters in Eukaryotic DNA

The first genes to be sequenced and studied through in vitro transcription systems were viral genes and cellular protein-coding genes that are very actively transcribed either at particular times of the cell cycle or in specific differentiated cell types. In all these highly transcribed genes, a conserved sequence called the **TATA box** was found ≈25–35 base pairs upstream of the start site (Figure 7-12). Mutagenesis studies have shown that a single-base change in this nucleotide sequence drastically decreases in vitro transcription by RNA polymerase II of genes adjacent to a TATA box. In most cases, sequence changes between the TATA box and start site do not significantly affect the transcription rate. If the base pairs between the TATA box and the normal start site are deleted, transcription of the altered, shortened template begins at a new site ≈25 base pairs downstream from the TATA box. Consequently, the TATA box acts similarly to an *E. coli* promoter to position RNA polymerase II for transcription initiation (see Figure 4-12).

Instead of a TATA box, some eukaryotic genes contain an alternative promoter element called an *initiator.* Most naturally occurring initiator elements have a cytosine (C) at the −1 position and an adenine (A) residue at the transcription start site (+1). Directed mutagenesis of mammalian genes with an initiator-containing promoter revealed that the nucleotide sequence immediately surrounding the start site determines the strength of such promoters. Unlike the conserved TATA box sequence, however, only an extremely degenerate initiator consensus sequence has been defined:

$$(5') \text{ Y-Y-A}^{+1}\text{-N-T/A-Y-Y-Y } (3')$$

where A^{+1} is the base at which transcription starts, Y is a pyrimidine (C or T), N is any of the four bases, and T/A is T or A at position +3.

Transcription of genes with promoters containing a TATA box or initiator element begins at a well-defined initiation site. However, transcription of many protein-coding genes has been shown to begin at any one of multiple possible sites over an extended region, often 20–200 base pairs in length. As a result, such genes give rise to mRNAs with multiple alternative 5′ ends. These genes, which generally are transcribed at low rates (e.g., genes encoding the enzymes required for basic metabolic processes required in all cells, often called "housekeeping genes"), do not contain a TATA box or an initiator. Most genes of this type contain a CG-rich stretch of 20–50 nucleotides within ≈100 base pairs upstream of the start-site region. The dinucleotide CG is statistically underrepresented in vertebrate DNAs, and the presence of a CG-rich region, or *CpG island,* just upstream from a start site is a distinctly nonrandom distribution. For this reason, the presence of a CpG island in genomic DNA suggests that it may contain a transcription-initiation region.

Promoter-Proximal Elements Help Regulate Eukaryotic Genes

Recombinant DNA techniques have been used to systematically mutate the nucleotide sequences of various eukaryotic genes in order to identify transcription-control regions. For example, alternative transcription-control elements regulate expression of the mammalian gene that encodes transthyretin (*TTR*), which transports thyroid hormone in blood and the cerebrospinal fluid that surrounds the brain and spinal cord. Transthyretin is expressed in hepatocytes, which synthesize and secrete most of the blood serum proteins, and in choroid plexus cells in the brain, which secrete cerebrospinal fluid and its constituent proteins. The control

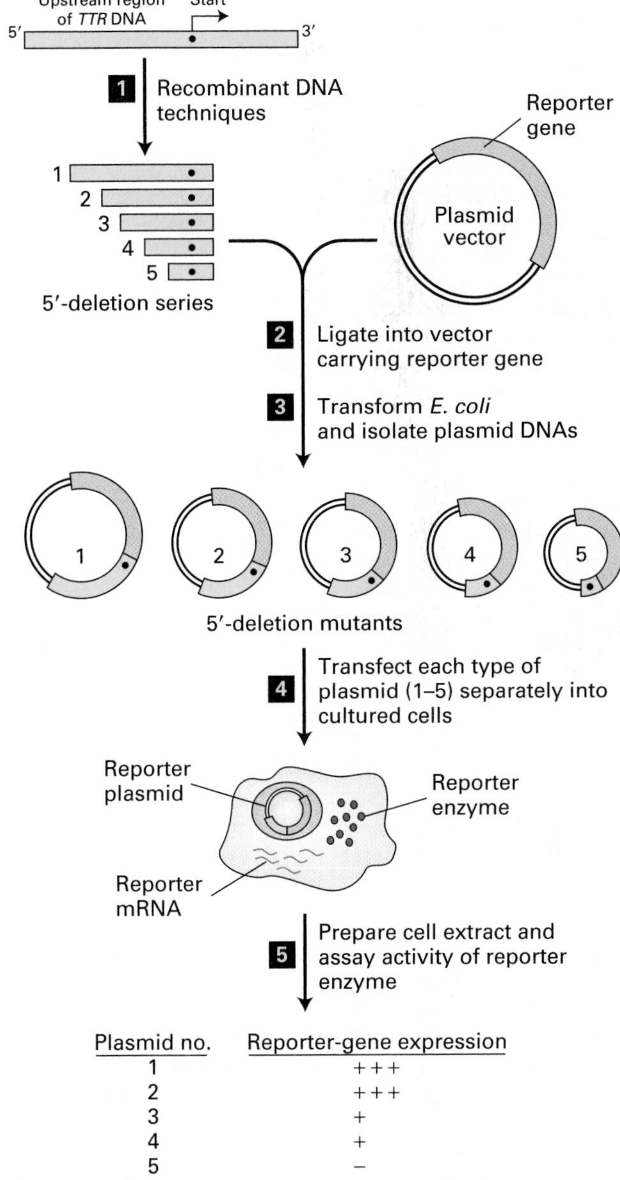

Plasmid no.	Reporter-gene expression
1	+++
2	+++
3	+
4	+
5	−

▲ EXPERIMENTAL FIGURE 7-13 **5′-Deletion analysis can identify transcription-control sequences in DNA upstream of a eukaryotic gene.** Step **1**: Recombinant DNA techniques are used to prepare a series of DNA fragments that extend from the 5′-untranslated region of a gene various distances upstream. Step **2**: The DNA fragments are ligated into a reporter plasmid upstream of an easily assayed reporter gene. Step **3**: The DNA is transformed into *E. coli* to isolate plasmids with deletions of various lengths 5′ to the transcription start site. Step **4**: Each plasmid is then transfected into cultured cells (or used to prepare transgenic organisms) and expression of the reporter gene is assayed (step **5**). The results of this hypothetical example (*bottom*) indicate that the test fragment contains two control elements. The 5′ end of one lies between deletions 2 and 3; the 5′ end of the other lies between deletions 4 and 5.

elements required for transcription of the *TTR* gene were identified by the procedure outlined in Figure 7-13. In this experimental approach, DNA fragments with varying extents of sequence upstream of a start site are cloned in front of a reporter gene in a bacterial plasmid using recombinant DNA techniques. Reporter genes express enzymes that are easily assayed in cell extracts. Commonly used reporter genes include the *E. coli lacZ* gene encoding β-galactosidase; the firefly gene encoding luciferase, which converts energy from ATP hydrolysis into light; and the jellyfish gene encoding green fluorescent protein (GFP).

By constructing and analyzing a *5′-deletion series* upstream of the *TTR* gene, researchers identified two control elements that stimulate reporter-gene expression in cultured hepatocytes but not in other cell types. One region mapped between the transcription start site and ≈200 base pairs upstream of the start site; the other mapped between ≈1.85 and 2.01 kb upstream of the *TTR* gene transcription start site. Further studies demonstrated that alternative DNA sequences control *TTR* transcription in choroid plexus cells, demonstrating again that alternative control elements often regulate transcription of eukaryotic genes in different cell types.

By now, hundreds of eukaryotic genes have been analyzed, and scores of transcription-control regions have been identified. These control elements, together with the TATA box or initiator, often are referred to as the promoter of the gene they regulate. However, we prefer to reserve the term *promoter* for the TATA box or initiator sequences that determine the initiation site on the template. We use the term **promoter-proximal elements** for control regions lying within 100–200 base pairs upstream of the start site. In some cases, promoter-proximal elements are cell-type specific; that is, they function only in specific differentiated cell types. In eukaryotes the term *enhancer* is used to refer to transcription-control regions greater than ≈200 bp from the transcription start site.

Once a transcription-control region has been identified, analysis of *linker scanning mutations* can pinpoint the sequences within the regulatory region that function to control transcription. In this approach, a set of constructs with contiguous overlapping mutations are assayed for their effect on expression of a reporter gene or production of a specific mRNA (Figure 7-14a). One of the first uses of this type of analysis identified promoter-proximal elements of the thymidine kinase (*tk*) gene from herpes simplex virus (HSV). The results demonstrated that the DNA region upstream of the HSV *tk* gene contains three separate transcription-control sequences: a TATA box in the interval from −32 to −16 and two other control elements farther upstream (Figure 7-14b).

To test the spacing constraints on control elements in the HSV *tk* promoter region identified by analysis of linker scanning mutations, researchers prepared and assayed constructs containing small deletions and insertions between the elements. Changes in spacing between the promoter and promoter-proximal control elements of 20 nucleotides or fewer had little effect. However, insertions of 30 to 50 base pairs between a promoter-proximal element and the TATA box was equivalent to deleting the element. Similar analyses of other eukaryotic promoters have also indicated that considerable flexibility in the spacing between promoter-proximal elements is generally tolerated, but separations of several tens of base pairs may decrease transcription.

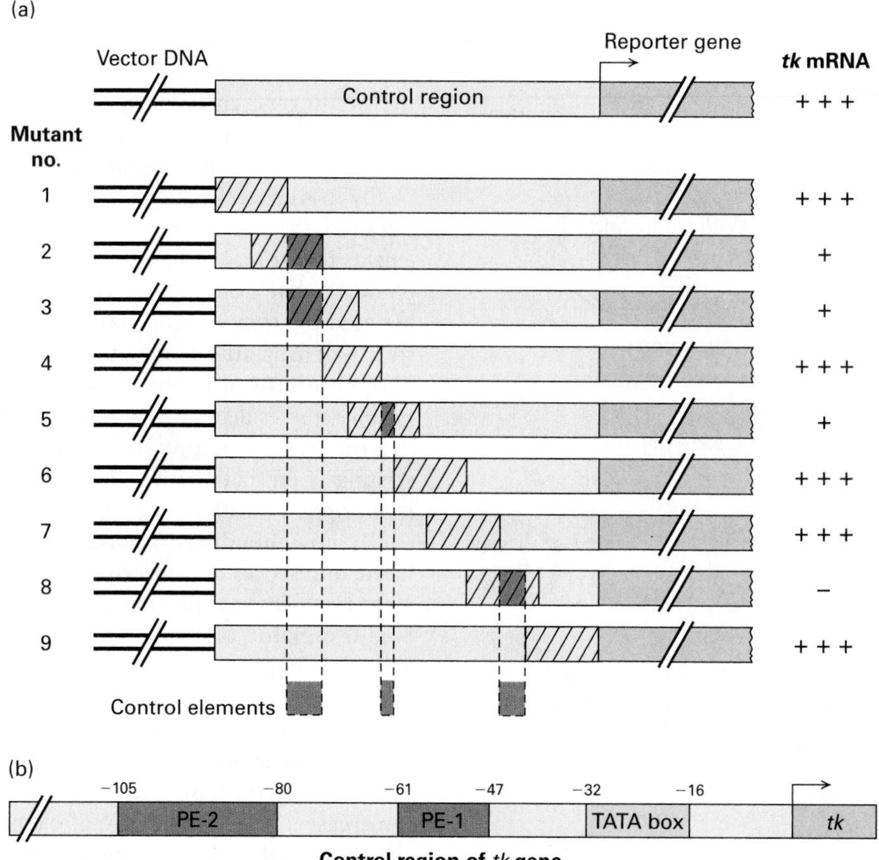

▲ **EXPERIMENTAL FIGURE 7-14 Linker scanning mutations identify transcription-control elements.** (a) A region of eukaryotic DNA (tan) that supports high-level expression of a reporter gene (light purple) is cloned in a plasmid vector as diagrammed at the top. Overlapping linker scanning (LS) mutations (crosshatch) are introduced from one end of the region being analyzed to the other. These mutations result from scrambling the nucleotide sequence in a short stretch of the DNA. After the mutant plasmids are transfected separately into cultured cells, the activity of the reporter-gene product is assayed. In the hypothetical example shown here, LS mutations 1, 4, 6, 7, and 9 have little or no effect on expression of the reporter gene, indicating that the regions altered in these mutants contain no control elements. Reporter-gene expression is significantly reduced in mutants 2, 3, 5, and 8, indicating that control elements (brown) lie in the intervals shown at the bottom. (b) Analysis of LS mutations in the transcription-control region of the thymidine kinase (*tk*) gene from herpes simplex virus (HSV) identified a TATA box and two promoter-proximal elements (PE-1 and PE-2). [Part (b) see S. L. McKnight and R. Kingsbury, 1982, *Science* **217**:316.]

Distant Enhancers Often Stimulate Transcription by RNA Polymerase II

As noted earlier, transcription from many eukaryotic promoters can be stimulated by control elements located thousands of base pairs away from the start site. Such long-distance transcription-control elements, referred to as enhancers, are common in eukaryotic genomes but fairly rare in bacterial genomes. The first enhancer to be discovered that stimulates transcription of eukaryotic genes was in a 366-bp fragment of the simian virus 40 (SV40) genome (Figure 7-15). Further analysis of this region of SV40 DNA revealed that an ≈100-bp sequence lying ≈100 base pairs upstream of the SV40 early transcription start site was responsible for its ability to enhance transcription. In SV40, this enhancer sequence functions to stimulate transcription from viral promoters. The SV40 enhancer, however, stimulates transcription from all mammalian promoters that have been tested when it is inserted in either orientation anywhere on a plasmid carrying the test promoter, even when it is thousands of base pairs from the start site. An extensive linker scanning mutational analysis of the SV40 enhancer indicated that it is composed of multiple individual elements, each of which contributes to the total activity of the enhancer. As discussed later, each of these regulatory elements is a protein-binding site.

Soon after discovery of the SV40 enhancer, enhancers were identified in other viral genomes and in eukaryotic cellular DNA. Some of these control elements are located 50 or more kilobases from the promoter they control. Analyses of many different eukaryotic cellular enhancers have shown that they can occur upstream from a promoter, downstream from a promoter within an intron, or even downstream from the final exon of a gene, as in the case of the *Pax6* gene (see Figure 7-6). Like promoter-proximal elements, many enhancers are cell-type specific. For example, an enhancer controlling *Pax6* expression in the retina has been characterized in the intron between exons 4 and 5 (see Figure 7-6). Analyses of the effects of deletions and linker scanning mutations in cellular enhancers have shown that, like the SV40

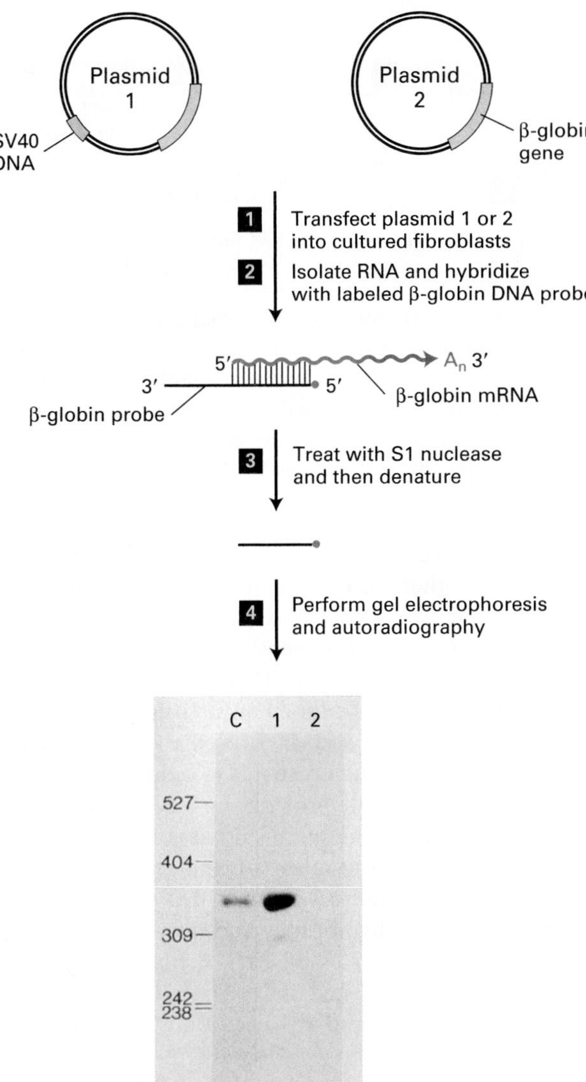

▲ **EXPERIMENTAL FIGURE 7-15** **Plasmids containing a particular SV40 DNA fragment showed marked increase in mRNA production compared with plasmids lacking this enhancer.** Plasmids containing the β-globin gene with or without a 366-bp fragment of SV40 DNA were constructed. These plasmids were transfected into cultured cells, and any resulting RNA was hybridized to a β-globin DNA probe (steps **1** and **2**).The amount of β-globin mRNA synthesized by cells transfected with one or the other plasmid was assayed by the S1 nuclease–protection method (step **3**). The restriction fragment probe, generated from a β-globin cDNA clone, was complementary to the 5′ end of β-globin mRNA. The 5′ end of the probe was labeled with ^{32}P (red dot). Hybridization of β-globin mRNA to the probe protected an ≈340-nucleotide fragment of the probe from digestion by S1 nuclease, which digests single-stranded DNA but not DNA in an RNA-DNA hybrid. Autoradiography of electrophoresed S1-protected fragments (step **4**) revealed that cells transfected with plasmid 1 (lane 1) produced much more β-globin mRNA than those transfected with plasmid 2 (lane 2). Lane C is a control assay of β-globin mRNA isolated from reticulocytes, which actively synthesize β-globin. These results show that the SV40 DNA fragment in plasmid 1 contains an element, the enhancer, that greatly stimulates synthesis of β-globin mRNA. [Adapted from J. Banerji et al., 1981, *Cell* **27**:299.]

enhancer, they generally are composed of multiple elements that contribute to the overall activity of the enhancer.

Most Eukaryotic Genes Are Regulated by Multiple Transcription-Control Elements

Initially, enhancers and promoter-proximal elements were thought to be distinct types of transcription-control elements. However, as more enhancers and promoter-proximal elements were analyzed, the distinctions between them became less clear. For example, both types of element generally can stimulate transcription even when inverted, and both types often are cell-type specific. The general consensus now is that a spectrum of control elements regulates transcription by RNA polymerase II. At one extreme are enhancers, which can stimulate transcription from a promoter tens of thousands of base pairs away (e.g., the SV40 enhancer). At the other extreme are promoter-proximal elements, such as the upstream elements controlling the HSV *tk* gene, which lose their influence when moved an additional 30–50 base pairs farther from the promoter. Researchers have identified a large number of transcription-control elements that can stimulate transcription from distances between these two extremes.

Figure 7-16a summarizes the locations of transcription-control sequences for a hypothetical mammalian gene. The start site at which transcription initiates encodes the first (5′) nucleotide of the first exon of an mRNA, the nucleotide that is capped. For many genes, especially those encoding abundantly expressed proteins, a TATA box located approximately 25–35 base pairs upstream from the start site directs RNA polymerase II to begin transcription at the proper nucleotide. Promoter-proximal elements, which are relatively short (≈10 base pairs), are located within the first ≈200 base pairs upstream of the start site. Enhancers, in contrast, usually are about 50–200 base pairs long and are composed of multiple elements of ≈10 base pairs. Enhancers may be located up to 50 kilobases or more upstream or downstream from the start site or within an intron. Many mammalian genes are controlled by more than one enhancer region.

The *S. cerevisiae* genome contains regulatory elements called **upstream activating sequences (UASs)**, which function similarly to enhancers and promoter-proximal elements in higher eukaryotes. Most yeast genes contain only one UAS, which generally lies within a few hundred base pairs of the start site. In addition, *S. cerevisiae* genes contain a TATA box ≈90 base pairs upstream from the transcription start site (Figure 7-16b).

KEY CONCEPTS OF SECTION 7.3

Regulatory Sequences in Protein-Coding Genes

- Expression of eukaryotic protein-coding genes generally is regulated through multiple protein-binding control regions that are located close to or distant from the start site (Figure 7-16).

- Promoters direct binding of RNA polymerase II to DNA, determine the site of transcription initiation, and influence transcription rate.

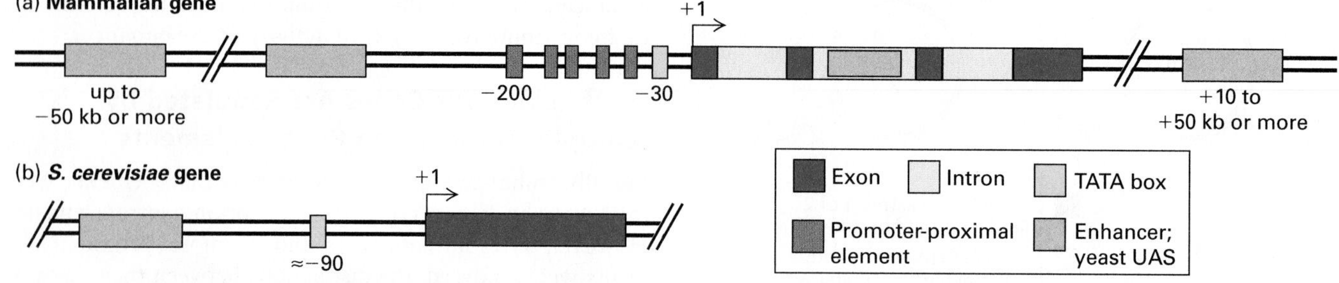

▲ **FIGURE 7-16 General organization of control elements that regulate gene expression in multicellular eukaryotes and yeast.** (a) Genes of multicellular organisms contain both promoter-proximal elements and enhancers, as well as a TATA box or other promoter element. The promoter elements position RNA polymerase II to initiate transcription at the start site and influence the rate of transcription. Enhancers may be either upstream or downstream and as far away as 50 kb or more from the transcription start site. In some cases, enhancers lie within introns. For some genes, promoter-proximal elements occur downstream from the start site as well as upstream. (b) Most *S. cerevisiae* genes contain only one regulatory region, called an *upstream activating sequence (UAS),* and a TATA box, which is ≈90 base pairs upstream from the start site.

- Three principal types of promoter sequences have been identified in eukaryotic DNA. The TATA box, the most common, is prevalent in highly transcribed genes. Initiator promoters are found in some genes, and CpG islands are characteristic of genes transcribed at a low rate.
- Promoter-proximal elements occur within ≈200 base pairs upstream of a start site. Several such elements, containing ≈10 base pairs, may help regulate a particular gene.
- Enhancers, which contain multiple short control elements, may be located from 200 base pairs to tens of kilobases upstream or downstream from a promoter, within an intron, or downstream from the final exon of a gene.
- Promoter-proximal elements and enhancers often are cell-type specific, functioning only in specific differentiated cell types.

7.4 Activators and Repressors of Transcription

The various transcription-control elements found in eukaryotic DNA are binding sites for regulatory proteins. The simplest eukaryotic cells encode hundreds of transcription factors, and the human genome encodes over 2000. The transcription of each gene in the genome is independently regulated by combinations of *specific transcription factors* that bind to its transcription-control regions. The number of possible combinations of this many transcription factors is astronomical, sufficient to generate unique controls for every gene encoded in the genome. In this section, we discuss the identification, purification, and structures of these transcription factors, which function to activate or repress expression of eukaryotic protein-coding genes.

Footprinting and Gel-Shift Assays Detect Protein-DNA Interactions

In yeast, *Drosophila,* and other genetically tractable eukaryotes, numerous genes encoding transcriptional activators and repressors have been identified by classical genetic analyses like those described in Chapter 5. However, in mammals and other vertebrates, which are less amenable to such genetic analysis, most transcription factors have been detected initially and subsequently purified by biochemical techniques. In this approach, a DNA regulatory element that has been identified by the kinds of mutational analyses described in the previous section is used to identify *cognate* proteins that bind specifically to it. Two common techniques for detecting such cognate proteins are DNase I footprinting and the electrophoretic mobility shift assay.

DNase I footprinting takes advantage of the fact that when a protein is bound to a region of DNA, it protects that DNA sequence from digestion by nucleases. As illustrated in Figure 7-17, when samples of a DNA fragment that is labeled at one end are digested under carefully controlled conditions in the presence and absence of a DNA-binding protein and then denatured, electrophoresed, and the resulting gel subjected to autoradiography, the region protected by the bound protein appears as a gap, or "footprint," in the array of bands resulting from digestion in the absence of protein. When footprinting is performed with a DNA fragment containing a known DNA control element, the appearance of a footprint indicates the presence of a transcription factor that binds that control element in the protein sample being assayed. Footprinting also identifies the specific DNA sequence to which the transcription factor binds.

The *electrophoretic mobility shift assay (EMSA),* also called the *gel-shift* or *band-shift* assay, is more useful than the footprinting assay for quantitative analysis of DNA-binding proteins. In general, the electrophoretic mobility of a DNA fragment is reduced when it is complexed to protein, causing a shift in the location of the fragment band. This assay can be used to detect a transcription factor in protein fractions incubated with a radiolabeled DNA fragment containing a known control element (Figure 7-18).

In the biochemical isolation of a transcription factor, an extract of cell nuclei commonly is subjected sequentially to several types of column chromatography (Chapter 3). Fractions eluted from the columns are assayed by DNase I footprinting or EMSA using DNA fragments containing an identified regulatory element (see Figures 7-17 and 7-18).

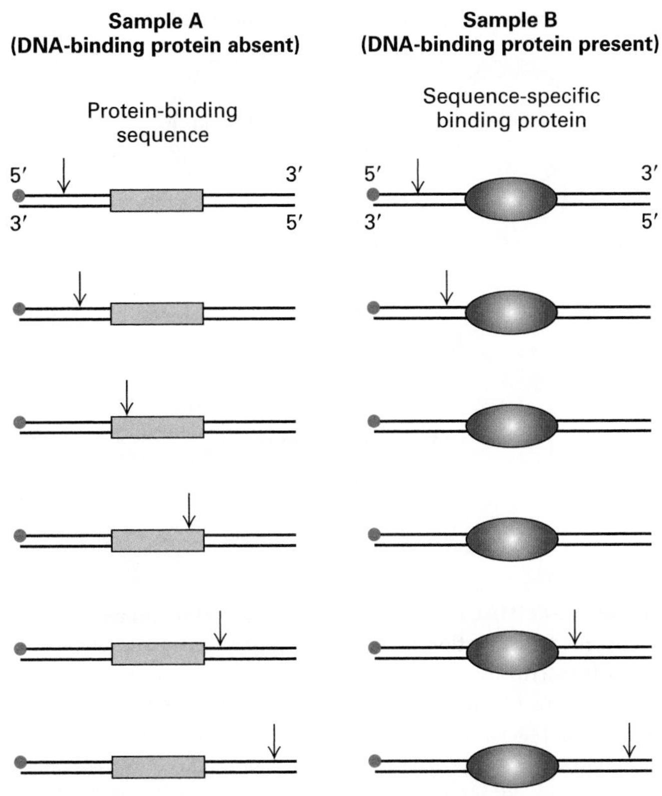

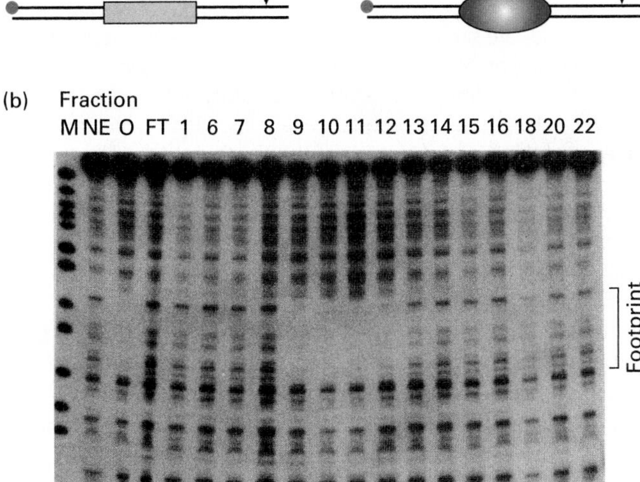

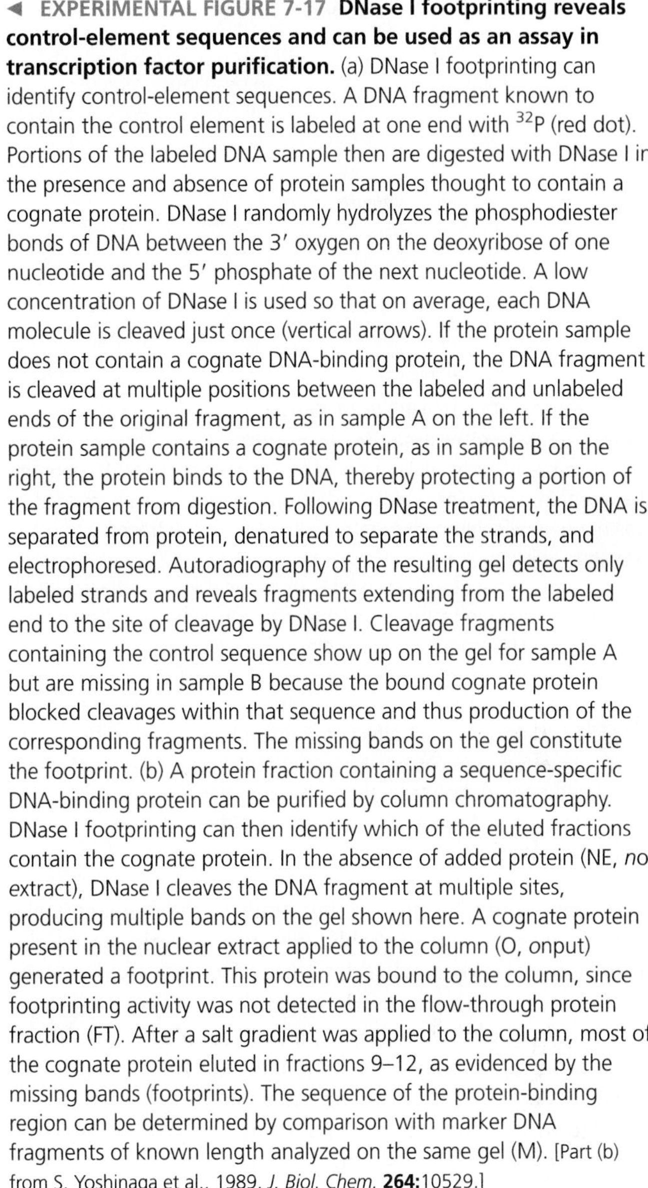

◀ EXPERIMENTAL FIGURE 7-17 **DNase I footprinting reveals control-element sequences and can be used as an assay in transcription factor purification.** (a) DNase I footprinting can identify control-element sequences. A DNA fragment known to contain the control element is labeled at one end with ^{32}P (red dot). Portions of the labeled DNA sample then are digested with DNase I in the presence and absence of protein samples thought to contain a cognate protein. DNase I randomly hydrolyzes the phosphodiester bonds of DNA between the 3′ oxygen on the deoxyribose of one nucleotide and the 5′ phosphate of the next nucleotide. A low concentration of DNase I is used so that on average, each DNA molecule is cleaved just once (vertical arrows). If the protein sample does not contain a cognate DNA-binding protein, the DNA fragment is cleaved at multiple positions between the labeled and unlabeled ends of the original fragment, as in sample A on the left. If the protein sample contains a cognate protein, as in sample B on the right, the protein binds to the DNA, thereby protecting a portion of the fragment from digestion. Following DNase treatment, the DNA is separated from protein, denatured to separate the strands, and electrophoresed. Autoradiography of the resulting gel detects only labeled strands and reveals fragments extending from the labeled end to the site of cleavage by DNase I. Cleavage fragments containing the control sequence show up on the gel for sample A but are missing in sample B because the bound cognate protein blocked cleavages within that sequence and thus production of the corresponding fragments. The missing bands on the gel constitute the footprint. (b) A protein fraction containing a sequence-specific DNA-binding protein can be purified by column chromatography. DNase I footprinting can then identify which of the eluted fractions contain the cognate protein. In the absence of added protein (NE, *no* extract), DNase I cleaves the DNA fragment at multiple sites, producing multiple bands on the gel shown here. A cognate protein present in the nuclear extract applied to the column (O, onput) generated a footprint. This protein was bound to the column, since footprinting activity was not detected in the flow-through protein fraction (FT). After a salt gradient was applied to the column, most of the cognate protein eluted in fractions 9–12, as evidenced by the missing bands (footprints). The sequence of the protein-binding region can be determined by comparison with marker DNA fragments of known length analyzed on the same gel (M). [Part (b) from S. Yoshinaga et al., 1989, *J. Biol. Chem.* **264:**10529.]

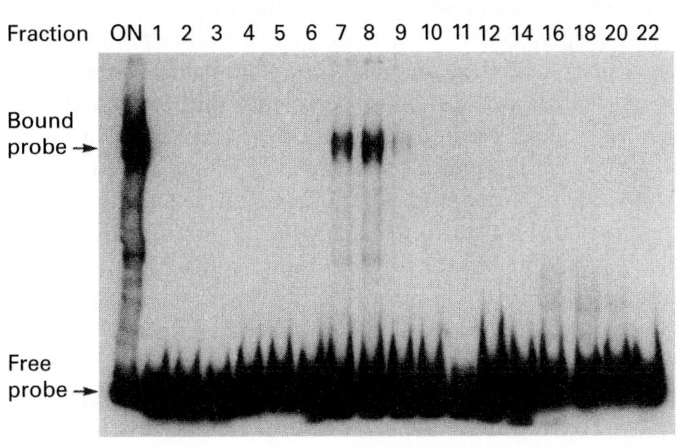

◀ EXPERIMENTAL FIGURE 7-18 **Electrophoretic mobility shift assay can be used to detect transcription factors during purification.** In this example, protein fractions separated by column chromatography were assayed for their ability to bind to a radiolabeled DNA-fragment probe containing a known regulatory element. After an aliquot of the protein sample was loaded onto the column (ON) and successive column fractions (numbers) were incubated with the labeled probe, the samples were electrophoresed under conditions that do not disrupt protein-DNA interactions. The free probe not bound to protein migrated to the bottom of the gel. A protein in the preparation applied to the column and in fractions 7 and 8 bound to the probe, forming a DNA-protein complex that migrated more slowly than the free probe. These fractions therefore likely contain the regulatory protein being sought. [From S. Yoshinaga et al., 1989, *J. Biol. Chem.* **264:**10529.]

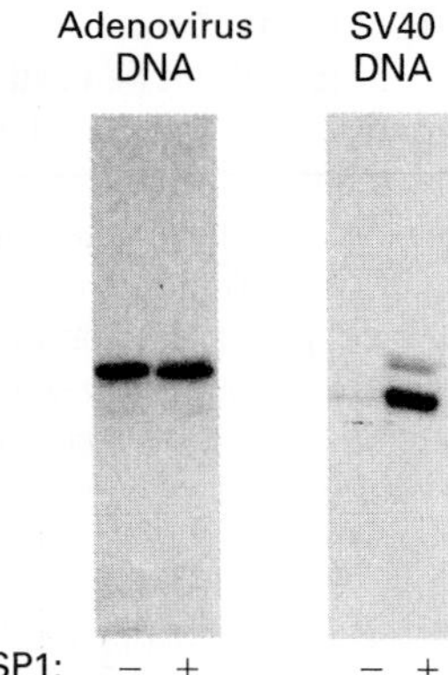

▲ **EXPERIMENTAL FIGURE 7-19 Transcription factors can be identified by in vitro assay for transcription activity.** SP1 was identified based on its ability to bind to a region of the SV40 genome that contains six copies of a GC-rich promoter-proximal element and was purified by column chromatography. To test the transcription-activating ability of purified SP1, it was incubated in vitro with template DNA, a protein fraction containing RNA polymerase II and associated general transcription factors, and labeled ribonucleoside triphosphates. The labeled RNA products were subjected to electrophoresis and autoradiography. Shown here are autoradiograms from assays with adenovirus and SV40 DNA in the absence (−) and presence (+) of SP1. SP1 had no significant effect on transcription from the adenovirus promoter, which contains no SP1-binding sites. In contrast, SP1 stimulated transcription from the SV40 promoter about tenfold. [Adapted from M. R. Briggs et al., 1986, *Science* **234:**47.]

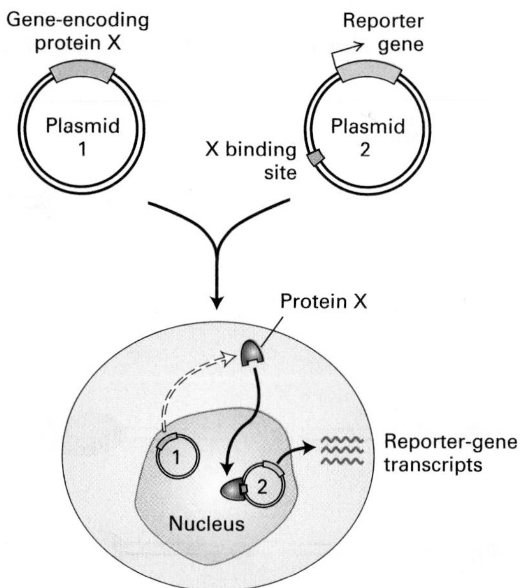

▲ **EXPERIMENTAL FIGURE 7-20 In vivo transfection assay measures transcription activity to evaluate proteins believed to be transcription factors.** The assay system requires two plasmids. One plasmid contains the gene encoding the putative transcription factor (protein X). The second plasmid contains a reporter gene (e.g., *lacZ*) and one or more binding sites for protein X. Both plasmids are simultaneously introduced into cells that lack the gene encoding protein X. The production of reporter-gene RNA transcripts is measured; alternatively, the activity of the encoded protein can be assayed. If reporter-gene transcription is greater in the presence of the X-encoding plasmid than in its absence, then the protein is an activator; if transcription is less, then it is a repressor. By use of plasmids encoding a mutated or rearranged transcription factor, important domains of the protein can be identified.

Fractions containing protein that binds to the regulatory element in these assays probably contain a putative transcription factor. A powerful technique commonly used for the final step in purifying transcription factors is *sequence-specific DNA affinity chromatography,* a particular type of affinity chromatography in which long DNA strands containing multiple copies of the transcription factor–binding site are coupled to a column matrix. As a final test that an isolated protein is in fact a transcription factor, its ability to modulate transcription of a template containing the corresponding protein-binding sites is assayed in an in vitro transcription reaction. Figure 7-19 shows the results of such an assay for SP1, a transcription factor that binds to GC-rich sequences, thereby activating transcription from nearby promoters.

Once a transcription factor is isolated and purified, its partial amino acid sequence can be determined and used to clone the gene or cDNA encoding it, as outlined in Chapter 5. The isolated gene can then be used to test the ability of the encoded protein to activate or repress transcription in an in vivo transfection assay (Figure 7-20).

Activators Are Modular Proteins Composed of Distinct Functional Domains and Promote Transcription

Studies with a yeast transcription activator called GAL4 provided early insight into the domain structure of transcription factors. The gene encoding the GAL4 protein, which promotes expression of enzymes needed to metabolize galactose, was identified by complementation analysis of *gal4* mutants (Chapter 5). Directed mutagenesis studies like those described previously identified UASs for the genes activated by GAL4. Each of these UASs was found to contain one or more copies of a related 17-bp sequence called UAS_{GAL}. DNase I footprinting assays with recombinant GAL4 protein produced in *E. coli* from the yeast *GAL4* gene showed that GAL4 protein binds to UAS_{GAL} sequences. When a copy of UAS_{GAL} was cloned upstream of a TATA box followed by a *lacZ* reporter gene, expression of *lacZ* was activated in galactose media in wild-type cells but not in *gal4* mutants. These results showed that UAS_{GAL} is a transcription-control element activated by the GAL4 protein in galactose media.

A remarkable set of experiments with *gal4* deletion mutants demonstrated that the GAL4 transcription factor is composed of separable functional domains: an N-terminal *DNA-binding domain,* which binds to specific DNA sequences, and a C-terminal *activation domain,* which interacts with other proteins to stimulate transcription from a nearby promoter (Figure 7-21). When the N-terminal DNA-binding domain of GAL4 was fused directly to various portions of its own C-terminal region, the resulting truncated proteins retained the ability to stimulate expression of a reporter gene in an in vivo assay like that

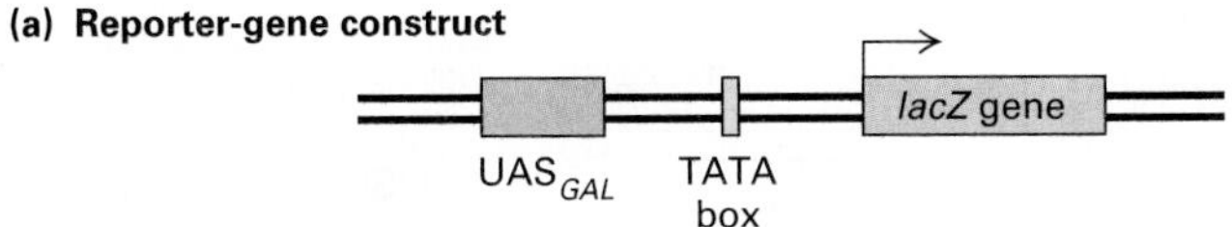

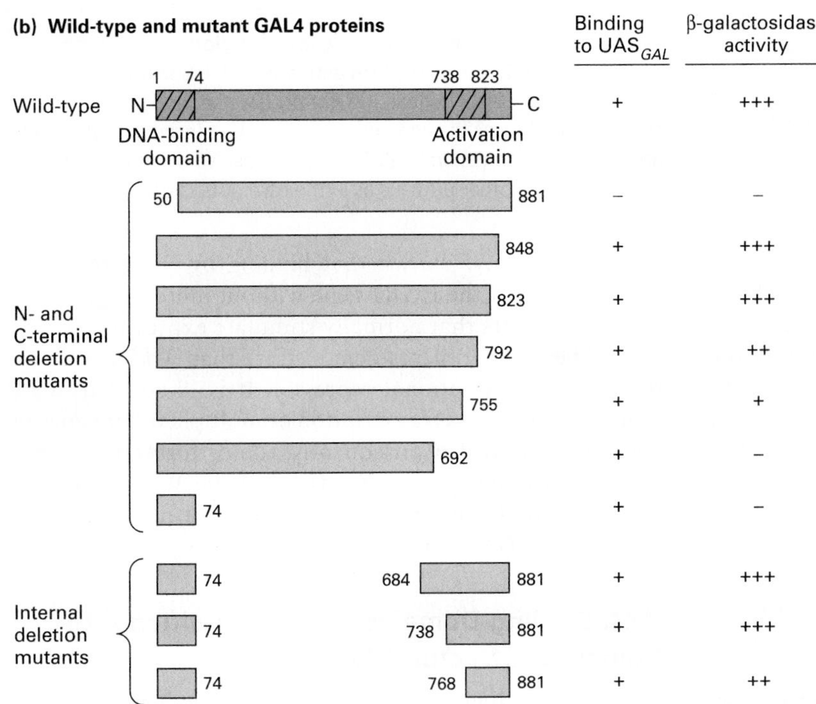

◄ **EXPERIMENTAL FIGURE 7-21 Deletion mutants of the *GAL4* gene in yeast with a UAS_{GAL} reporter-gene construct demonstrate the separate functional domains in an activator.** (a) Diagram of DNA construct containing a *lacZ* reporter gene and TATA box ligated to UAS_{GAL}, a regulatory element that contains several GAL4-binding sites. The reporter-gene construct and DNA encoding wild-type or mutant (deleted) GAL4 were simultaneously introduced into mutant (*gal4*) yeast cells, and the activity of β-galactosidase expressed from *lacZ* was assayed. Activity will be high if the introduced *GAL4* DNA encodes a functional protein. (b) Schematic diagrams of wild-type GAL4 and various mutant forms. Small numbers refer to positions in the wild-type sequence. Deletion of 50 amino acids from the N-terminal end destroyed the ability of GAL4 to bind to UAS_{GAL} and to stimulate expression of β-galactosidase from the reporter gene. Proteins with extensive deletions from the C-terminal end still bound to UAS_{GAL}. These results localize the DNA-binding domain to the N-terminal end of GAL4. The ability to activate β-galactosidase expression was not entirely eliminated unless somewhere between 126 and 189 or more amino acids were deleted from the C-terminal end. Thus the activation domain lies in the C-terminal region of GAL4. Proteins with internal deletions (*bottom*) also were able to stimulate expression of β-galactosidase, indicating that the central region of GAL4 is not crucial for its function in this assay. [See J. Ma and M. Ptashne, 1987, *Cell* **48:**847; I. A. Hope and K. Struhl, 1986, *Cell* **46:**885; and R. Brent and M. Ptashne, 1985, *Cell* **43:**729.]

depicted in Figure 7-20. Thus the internal portion of the protein is not required for functioning of GAL4 as a transcription factor. Similar experiments with another yeast transcription factor, GCN4, which regulates genes required for synthesis of many amino acids, indicated that it contains an ≈60-aa DNA-binding domain at its C-terminus and an ≈20-aa activation domain near the middle of its sequence.

Further evidence for the existence of distinct activation domains in GAL4 and GCN4 came from experiments in which their activation domains were fused to a DNA-binding domain from an entirely unrelated *E. coli* DNA-binding protein. When these fusion proteins were assayed in vivo, they activated transcription of a reporter gene containing the cognate site for the *E. coli* protein. Thus functional transcription factors can be constructed from entirely novel combinations of prokaryotic and eukaryotic elements.

Studies such as these have now been carried out with many eukaryotic activators. The structural model of eukaryotic activators that has emerged from these studies is a modular one in which one or more activation domains are connected to a sequence-specific DNA-binding domain through flexible protein domains (Figure 7-22). In some cases, amino acids included in the DNA-binding domain also contribute to transcriptional activation. As discussed in a later section, activation domains are thought to function by binding other proteins involved in transcription. The presence of flexible domains connecting the DNA-binding domains to activation

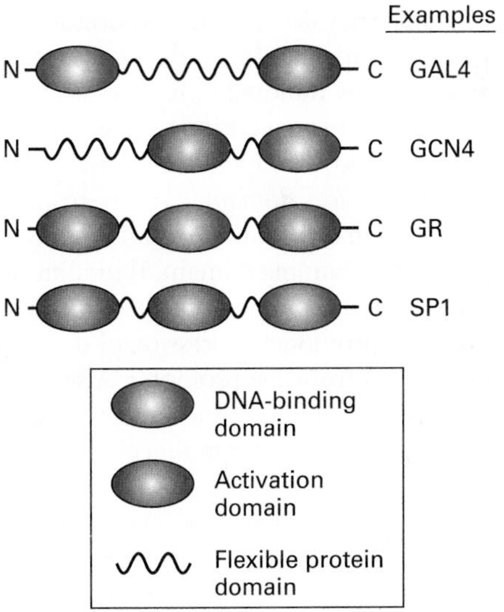

▲ **FIGURE 7-22 Schematic diagrams illustrating the modular structure of eukaryotic transcription activators.** These transcription factors may contain more than one activation domain but rarely contain more than one DNA-binding domain. GAL4 and GCN4 are yeast transcription activators. The glucocorticoid receptor (GR) promotes transcription of target genes when certain hormones are bound to the C-terminal activation domain. SP1 binds to GC-rich promoter elements in a large number of mammalian genes.

domains may explain why alterations in the spacing between control elements are so well tolerated in eukaryotic control regions. Thus even when the positions of transcription factors bound to DNA are shifted relative to each other, their activation domains may still be able to interact because they are attached to their DNA-binding domains through flexible protein regions.

Repressors Inhibit Transcription and Are the Functional Converse of Activators

Eukaryotic transcription is regulated by repressors as well as activators. For example, geneticists have identified mutations in yeast that result in continuously high expression of certain genes. This type of unregulated, abnormally high expression is called **constitutive** expression and results from the inactivation of a repressor that normally inhibits the transcription of these genes. Similarly, mutants of *Drosophila* and *Caenorhabditis elegans* have been isolated that are defective in embryonic development because they express genes in embryonic cells where those genes are normally repressed. The mutations in these mutants inactivate repressors, leading to abnormal development.

Repressor-binding sites in DNA have been identified by systematic linker scanning mutation analysis similar to that depicted in Figure 7-14. In this type of analysis, mutation of an activator-binding site leads to decreased expression of the linked reporter gene, whereas mutation of a repressor-binding site leads to increased expression of a reporter gene. Repressor proteins that bind such sites can be purified and assayed using the same biochemical techniques described earlier for activator proteins.

Eukaryotic transcription repressors are the functional converse of activators. They can inhibit transcription from a gene they do not normally regulate when their cognate binding sites are placed within a few hundred base pairs of the gene's start site. Like activators, most eukaryotic repressors are modular proteins that have two functional domains: a DNA-binding domain and a *repression domain.* Similar to activation domains, repression domains continue to function when fused to another type of DNA-binding domain. If binding sites for this second DNA-binding domain are inserted within a few hundred base pairs of a promoter, expression of the fusion protein inhibits transcription from the promoter. Also like activation domains, repression domains function by interacting with other proteins, as discussed later in this chapter.

The absence of appropriate repressor activity can have devastating consequences. For instance, the protein encoded by the *Wilms' tumor (WT1)* gene is a repressor that is expressed preferentially in the developing kidney. Children who inherit mutations in both the maternal and paternal *WT1* genes, so that they produce no functional WT1 protein, invariably develop kidney tumors early in life. The WT1 protein binds to the control region of the gene encoding EGR1, which itself is a transcription factor as well (Figure 7-23). This gene, like many other eukaryotic genes, is subject to both repression and activation.

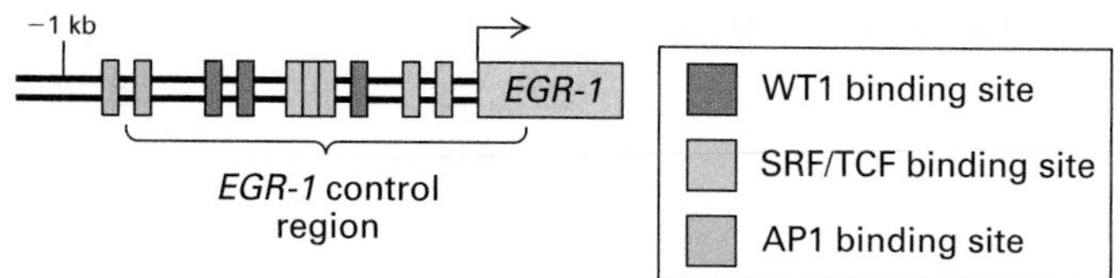

▲ **FIGURE 7-23 Diagram of the control region of the gene encoding EGR-1, a transcription activator.** The binding sites for WT1, a eukaryotic repressor protein, do not overlap the binding sites for the activator AP1 or the composite binding site for the activators SRF and TCF. Thus repression by WT1 does not involve direct interference with binding of other proteins as in the case of bacterial repressors.

Experiments have shown that binding by WT1 represses transcription of the *EGR1* gene without inhibiting binding of the activators that normally stimulate expression of this gene. These experiments demonstrate that WT1 can function as a transcriptional repressor. It is likely that WT1 functions to repress transcription of multiple other genes in addition to *EGR1*. Consequently, tumor formation in patients with homozygous mutations of the *WT1* gene may result in part from abnormal activation of multiple genes such as the *EGR1* gene. ▮

DNA-Binding Domains Can Be Classified into Numerous Structural Types

The DNA-binding domains of eukaryotic activators and repressors contain a variety of structural motifs that bind specific DNA sequences. The ability of DNA-binding proteins to bind to specific DNA sequences commonly results from noncovalent interactions between atoms in an α helix in the DNA-binding domain and atoms on the edges of the bases within a major groove in the DNA. Interactions with sugar-phosphate backbone atoms and, in some cases, with atoms in a DNA minor groove also contribute to binding.

The principles of specific protein-DNA interactions were first discovered during the study of bacterial repressors. Many bacterial repressors are dimeric proteins in which an α helix from each monomer inserts into a major groove in the DNA helix (Figure 7-24). This α helix is referred to as the *recognition helix* or *sequence-reading helix* because most of the amino acid side chains that contact DNA extend from this helix. The recognition helix that protrudes from the surface of bacterial repressors to enter the DNA major groove and make multiple, specific interactions with atoms in the DNA is usually supported in the protein structure in part by hydrophobic interactions with a second α helix just N-terminal to it. This structural element, which is present in many bacterial repressors, is called a *helix-turn-helix* motif.

Many additional motifs that can present an α helix to the major groove of DNA are found in eukaryotic transcription factors, which often are classified according to the type of DNA-binding domain they contain. Because most of these motifs have characteristic consensus amino acid sequences, newly characterized transcription factors frequently can be classified once the corresponding genes or cDNAs are cloned and sequenced. The genomes of higher eukaryotes encode

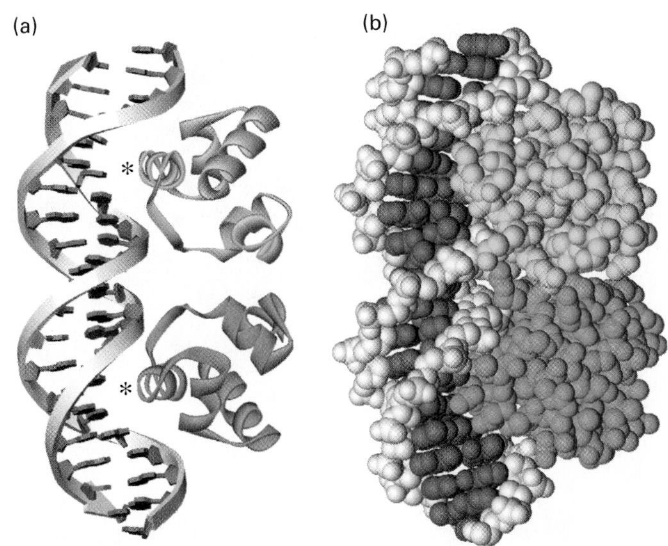

▲ **FIGURE 7-24 Interaction of bacteriophage 434 repressor with DNA.** (a) Ribbon diagram of 434 repressor bound to its specific operator DNA. Repressor monomers are in yellow and green. The recognition helices are indicated by asterisks. A space-filling model of the repressor-operator complex (b) shows how the protein interacts intimately with one side of the DNA molecule over a length of 1.5 turns. [Adapted from A. K. Aggarwal et al., 1988, *Science* **242:**899.]

dozens of classes of DNA-binding domains and hundreds to thousands of transcription factors. The human genome, for instance, encodes ≈2000 transcription factors.

Here we introduce several common classes of DNA-binding proteins whose three-dimensional structures have been determined. In all these examples and many other transcription factors, at least one α helix is inserted into a major groove of DNA. However, some transcription factors contain alternative structural motifs (e.g., β strands and loops) that interact with DNA.

Homeodomain Proteins Many eukaryotic transcription factors that function during development contain a conserved 60-residue DNA-binding motif, called a **homeodomain** that is similar to the helix-turn-helix motif of bacterial repressors. These transcription factors were first identified in *Drosophila* mutants in which one body part was transformed into another during development (Chapter 22). The conserved homeodomain sequence has also been found in vertebrate transcription factors, including those that have similar master-control functions in human development.

Zinc-Finger Proteins A number of different eukaryotic proteins have regions that fold around a central Zn^{2+} ion, producing a compact domain from a relatively short length of the polypeptide chain (Figure 7-25a). Termed a **zinc finger,** this structural motif was first recognized in DNA-binding domains but now is known to occur also in proteins that do not bind to DNA. Here we describe two of the several classes of zinc-finger motifs that have been identified in eukaryotic transcription factors.

The *C_2H_2 zinc finger* is the most common DNA-binding motif encoded in the human genome and the genomes of most other multicellular animals. It is also common in multicellular plants but is not the dominant type of DNA-binding domain in plants as it is in animals. This motif has a 23- to 26-residue consensus sequence containing two conserved cysteine (C) and two conserved histidine (H) residues, whose side chains bind one Zn^{2+} ion (Figure 3-9c). The name "zinc finger" was coined because a two-dimensional diagram of the structure resembles a finger. When the three-dimensional structure was solved, it became clear that the binding of the Zn^{2+} ion by the two cysteine and two histidine residues folds the relatively short polypeptide sequence into a compact domain, which can insert its α helix into the major groove of DNA. Many transcription factors contain multiple C_2H_2 zinc fingers, which interact with successive groups of base pairs, within the major groove, as the protein wraps around the DNA double helix (Figure 7-25b).

A second type of zinc-finger structure, designated the *C_4 zinc finger* (because it has four conserved cysteines in contact with the Zn^{2+}), is found in ≈50 human transcription factors. The first members of this class were identified as specific intracellular high-affinity binding proteins, or "receptors," for steroid hormones, leading to the name *steroid receptor superfamily.* Because similar intracellular receptors for nonsteroid hormones subsequently were found, these transcription factors are now commonly called **nuclear receptors.** The characteristic feature of C_4 zinc fingers is the presence of two groups of four critical cysteines, one toward each end of the 55- or 56-residue domain. Although the C_4 zinc finger initially was named by analogy with the C_2H_2 zinc finger, the three-dimensional structures of proteins containing these DNA-binding motifs later were found to be quite distinct. A particularly important difference between the two is that C_2H_2 zinc-finger proteins generally contain three or more repeating finger units and bind as monomers, whereas C_4 zinc-finger proteins generally contain only two finger units and generally bind to DNA as homodimers or heterodimers. Homodimers of C_4 zinc-finger DNA-binding domains have twofold rotational symmetry (Figure 7-25c). Consequently, homodimeric nuclear receptors bind to consensus DNA sequences that are inverted repeats.

Leucine-Zipper Proteins Another structural motif present in the DNA-binding domains of a large class of transcription factors contains the hydrophobic amino acid leucine at every seventh position in the sequence. These proteins bind to DNA as dimers, and mutagenesis of the leucines showed that they were required for dimerization. Consequently, the name **leucine zipper** was coined to denote this structural motif.

The DNA-binding domain of the yeast GCN4 transcription factor mentioned earlier is a leucine-zipper domain. X-ray crystallographic analysis of complexes between DNA and the GCN4 DNA-binding domain has shown that the dimeric protein contains two extended α helices that "grip" the DNA molecule, much like a pair of scissors, at two adjacent major grooves separated by about half a turn of the double helix

► **FIGURE 7-25 Zinc-Finger Proteins.** (a) This two-dimensional depiction of C_2H_2 zinc fingers illustrates the shape that gave this motif its name. Individual amino acids are in purple; the amino acids that contact the Zn^{+2} ion are in blue. (b) GL1 is a monomeric protein that contains five C_2H_2 zinc fingers. α-Helices are shown as cylinders, Zn^{+2} ions as spheres. Finger 1 does not interact with DNA, whereas the other four fingers do. (c) The glucocorticoid receptor is a homodimeric C_4 zinc-finger protein. α-Helices are shown as purple ribbons, β-strands as green arrows, Zn^{+2} ions as spheres. Two α helices (darker shade), one in each monomer, interact with the DNA. Like all C_4 zinc-finger homodimers, this transcription factor has twofold rotational symmetry; the center of symmetry is shown by the yellow ellipse. In contrast, heterodimeric nuclear receptors do not exhibit rotational symmetry. [See N. P. Pavletich and C. O. Pabo, 1993, *Science* **261:**1701, and B. F. Luisi et al., 1991, *Nature* **352:**497.]

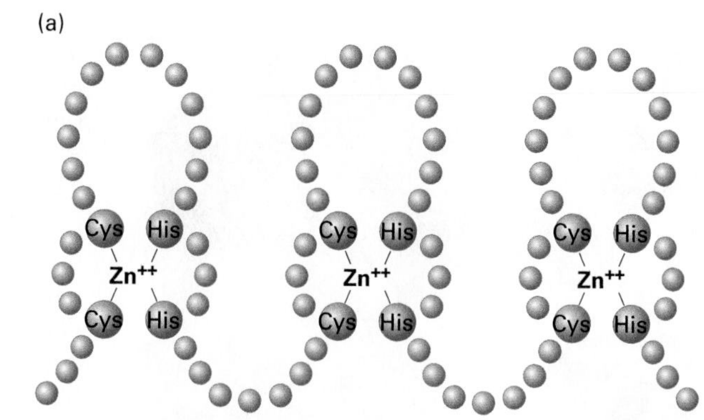

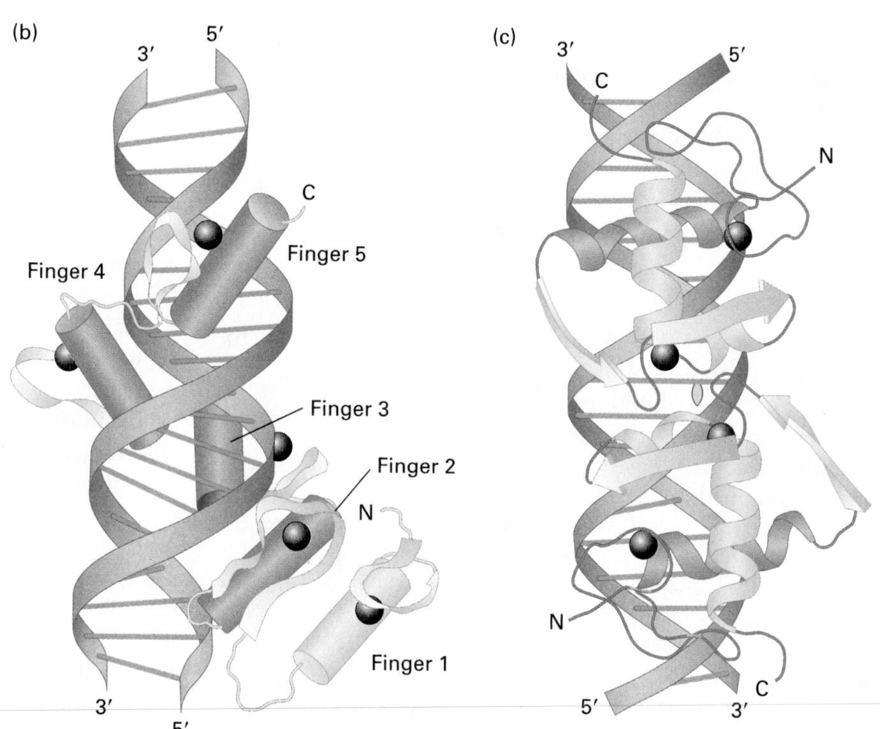

(Figure 7-26a). The portions of the α helices contacting the DNA include positively charged (basic) residues that interact with phosphates in the DNA backbone and additional residues that interact with specific bases in the major groove.

GCN4 forms dimers via hydrophobic interactions between the C-terminal regions of the α helices, forming a **coiled-coil** structure. This structure is common in proteins containing amphipathic α helices in which hydrophobic amino acid residues are regularly spaced alternately three or four positions apart in the sequence, forming a stripe down one side of the α helix. These hydrophobic stripes make up the interacting surfaces between the α-helical monomers in a coiled-coil dimer (see Figure 3-9a).

Although the first leucine-zipper transcription factors to be analyzed contained leucine residues at every seventh position in the dimerization region, additional DNA-binding proteins containing other hydrophobic amino acids in these positions subsequently were identified. Like leucine-zipper proteins, they form dimers containing a C-terminal coiled-coil dimerization region and an N-terminal DNA-binding domain. The term *basic zipper (bZIP)* now is frequently used to refer to all proteins with these common structural features. Many basic-zipper transcription factors are heterodimers of two different polypeptide chains, each containing one basic-zipper domain.

Basic Helix-Loop-Helix (bHLH) Proteins The DNA-binding domain of another class of dimeric transcription factors contains a structural motif very similar to the basic-zipper motif except that a nonhelical loop of the polypeptide chain separates two α-helical regions in each monomer (Figure 7-26b). Termed a **basic helix-loop-helix (bHLH)**, this motif was predicted from the amino acid sequences of these proteins, which contain an N-terminal α helix with basic residues that interact

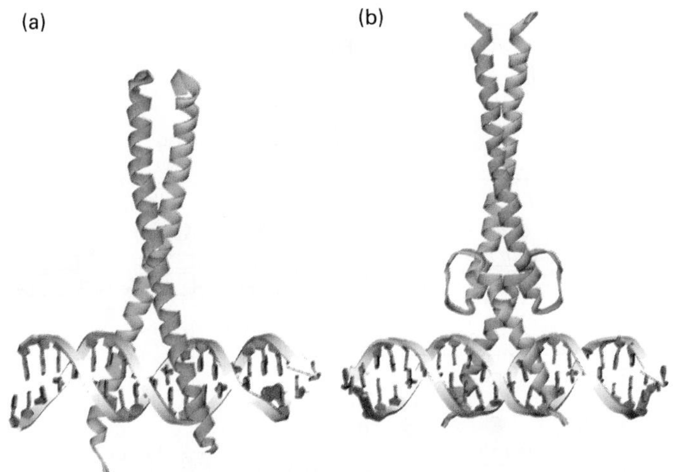

▲ **FIGURE 7-26 Interaction of homodimeric leucine-zipper and basic helix-loop-helix (bHLH) proteins with DNA.** (a) In leucine-zipper proteins, basic residues in the extended α-helical regions of the monomers interact with the DNA backbone at adjacent major grooves. The coiled-coil dimerization domain is stabilized by hydrophobic interactions between the monomers. (b) In bHLH proteins, the DNA-binding helices at the bottom (N-termini of the monomers) are separated by nonhelical loops from a leucine-zipper-like region containing a coiled-coil dimerization domain. [Part (a) see T. E. Ellenberger et al., 1992, *Cell* **71:**1223; part (b) see A. R. Ferre-D'Amare et al., 1993, *Nature* **363:**38.]

with DNA, a middle loop region, and a C-terminal region with hydrophobic amino acids spaced at intervals characteristic of an amphipathic α helix. As with basic-zipper proteins, different bHLH proteins can form heterodimers.

Structurally Diverse Activation and Repression Domains Regulate Transcription

Experiments with fusion proteins composed of the GAL4 DNA-binding domain and random segments of *E. coli* proteins demonstrated that a diverse group of amino acid sequences can function as activation domains, ≈1 percent of all *E. coli* sequences, even though they evolved to perform other functions. Many transcription factors contain activation domains marked by an unusually high percentage of particular amino acids. GAL4, GCN4, and most other yeast transcription factors, for instance, have activation domains that are rich in acidic amino acids (aspartic and glutamic acids). These so-called *acidic activation domains* generally are capable of stimulating transcription in nearly all types of eukaryotic cells—fungal, animal, and plant cells. Activation domains from some *Drosophila* and mammalian transcription factors are glutamine-rich, and some are proline-rich; still others are rich in the closely related amino acids serine and threonine, both of which have hydroxyl groups. However, some strong activation domains are not particularly rich in any specific amino acid.

Biophysical studies indicate that acidic activation domains have an unstructured, random-coil conformation. These domains stimulate transcription when they are bound to a protein *co-activator.* The interaction with a co-activator causes the activation domain to assume a more structured α-helical conformation in the activation domain–co-activator complex. A well-studied example of a transcription factor with an acidic activation domain is the mammalian CREB protein, which is phosphorylated in response to increased levels of cAMP. This regulated phosphorylation is required for CREB to bind to its co-activator CBP (CREB *b*inding *p*rotein), resulting in the transcription of genes whose control regions contain a CREB-binding site (see Figure 16-31). When the phosphorylated random coil activation domain of CREB interacts with CBP, it undergoes a conformational change to form two α helices linked by a short loop that wrap around the interacting domain of CBP.

Some activation domains are larger and more highly structured than acidic activation domains. For example, the ligand-binding domains of nuclear receptors function as activation domains when they bind their specific ligand (Figure 7-27). Binding of ligand induces a large conformational change that allows the ligand-binding domain with bound hormone to interact with a short α helix in nuclear-receptor co-activators; the resulting complex then can activate transcription of genes whose control regions bind the nuclear receptor.

Thus the acidic activation domain in CREB and the ligand-binding activation domains in nuclear receptors represent two structural extremes. The CREB acidic activation

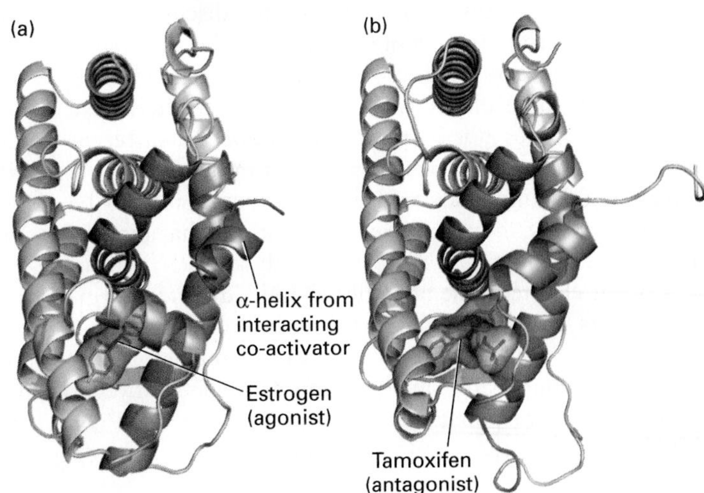

► **FIGURE 7-27 Effect of ligand binding on conformation of the estrogen receptor.** Only the ligand-binding domain of the receptor is shown. (a) When estrogen is bound to the domain, the green α helix interacts with the ligand, generating a hydrophobic groove in the ligand-binding domain (dark brown helices) that binds an amphipathic α helix in a co-activator subunit (blue). (b) The conformation of the estrogen receptor in the absence of hormone is thought to be stabilized by binding of the estrogen antagonist tamoxifen. In this conformation, the green helix of the receptor folds into a conformation that interacts with the co-activator binding groove of the active receptor, sterically blocking binding of co-activators. [From A. K. Shiau et al., 1998, *Cell* **95:**927.]

▶ **FIGURE 7-28 Combinatorial possibilities due to formation of heterodimeric transcription factors.** (a) In some heterodimeric transcription factors, each monomer recognizes the same DNA sequence. In the hypothetical example shown, transcription factors A, B, and C can all interact with one another, creating six different alternative combinations of activation domains that can all bind at the same site. Each composite binding site is divided into two half-sites, and each heterodimeric factor contains the activation domains of its two constituent monomers. (b) When transcription factor monomers recognize different DNA sequences, alternative combinations of the three factors bind to six different DNA sequences (sites 1–6), each with a unique combination of activation domains. (c) Expression of an inhibitory factor (red) that interacts only with factor A inhibits binding; hence transcriptional activation at sites 1, 4, and 5 is inhibited, but activation at sites 2, 3, and 6 is unaffected.

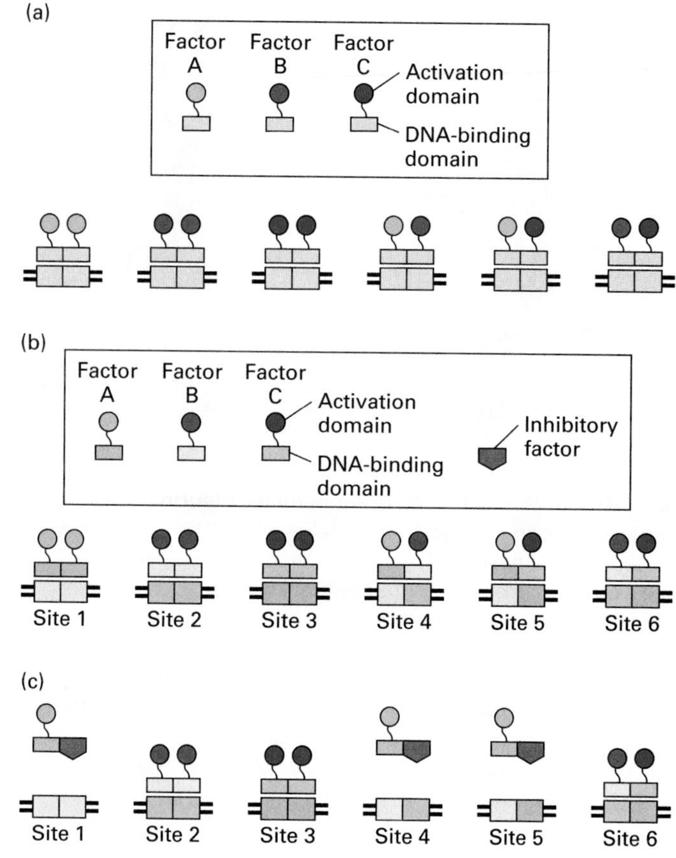

domain is a random coil that folds into two α-helices when it binds to the surface of a globular domain in a co-activator. In contrast, the nuclear-receptor ligand-binding activation domain is a structured globular domain that interacts with a short α helix in a co-activator, which probably is a random coil before it is bound. In both cases, however, specific protein-protein interactions between co-activators and the activation domains permit the transcription factors to stimulate gene expression.

Currently, less is known about the structure of repression domains. The globular ligand-binding domains of some nuclear receptors function as repression domains in the absence of their specific hormone ligand. Like activation domains, repression domains may be relatively short, comprising 15 or fewer amino acids. Biochemical and genetic studies indicate that repression domains also mediate protein-protein interactions and bind to *co-repressor* proteins, forming a complex that inhibits transcription initiation by mechanisms that are discussed later in the chapter.

Transcription Factor Interactions Increase Gene-Control Options

Two types of DNA-binding proteins discussed previously—basic-zipper proteins and bHLH proteins—often exist in alternative heterodimeric combinations of monomers. Other classes of transcription factors not discussed here also form heterodimeric proteins. In some heterodimeric transcription factors, each monomer recognizes the same sequence. In these proteins, the formation of alternative heterodimers does not increase the number of different sites on which the monomers can act but rather allows the activation domains associated with each monomer to be brought together in alternative combinations that bind to the same site (Figure 7-28a). As we will see later, and in subsequent chapters, the activities of individual transcription factors can be regulated by multiple mechanisms. Consequently, a single bZIP or bHLH DNA regulatory element in the control region of a gene may elicit different transcriptional responses depending on which bZIP or bHLH monomers that bind to that site are expressed in a particular cell at a particular time and how their activities are regulated.

In some heterodimeric transcription factors, however, each monomer has a different DNA-binding specificity. The resulting combinatorial possibilities increase the number of potential DNA sequences that a family of transcription factors can bind. Three different factor monomers theoretically could combine to form six homo- and heterodimeric factors, as illustrated in Figure 7-28b. Four different factor monomers could form a total of 10 dimeric factors; five monomers, 16 dimeric factors; and so forth. In addition, inhibitory factors are known that bind to some basic-zipper and bHLH monomers, thereby blocking their binding to DNA. When these inhibitory factors are expressed, they repress transcriptional activation by the factors with which they interact (Figure 7-28c). The rules governing the interactions of members of a heterodimeric transcription factor class are complex. This combinatorial complexity expands both the number of DNA sites from which these factors can activate transcription and the ways in which they can be regulated.

Similar combinatorial transcriptional regulation is achieved through the interaction of structurally unrelated transcription factors bound to closely spaced binding sites in DNA. An example is the interaction of two transcription factors, NFAT and AP1, which bind to neighboring

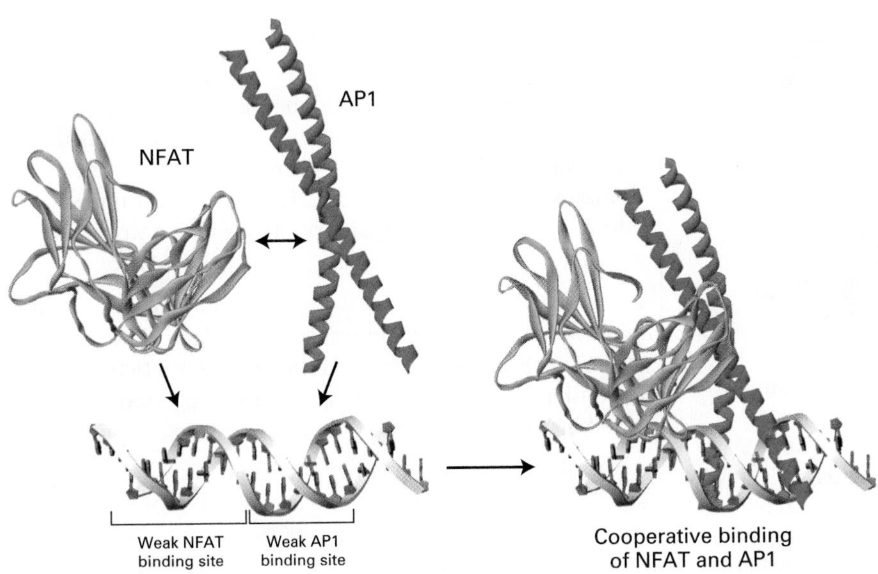

◄ **FIGURE 7-29 Cooperative binding of two unrelated transcription factors to neighboring sites in a composite control element.** By themselves, both monomeric NFAT and heterodimeric AP1 transcription factors have low affinity for their respective binding sites in the *IL-2* promoter-proximal region. Protein-protein interactions between NFAT and AP1 add to the overall stability of the NFAT-AP1-DNA complex, so that the two proteins bind to the composite site cooperatively. [See L. Chen et al., 1998, *Nature* **392**:42.]

sites in a composite promoter-proximal element regulating the gene encoding interleukon-2 (IL-2). Expression of the *IL-2* gene is critical to the immune response, but abnormal expression of IL-2 can lead to autoimmune diseases such as rheumatoid arthritis. Neither NFAT nor AP1 binds to its site in the *IL-2* control region in the absence of the other. The affinities of the factors for these particular DNA sequences are too low for the individual factors to form a stable complex with DNA. However, when both NFAT and AP1 are present, protein-protein interactions between them stabilize the DNA ternary complex composed of NFAT, AP1, and DNA (Figure 7-29). Such *cooperative DNA binding* of various transcription factors results in considerable combinatorial complexity of transcription control. As a result, the approximately 2000 transcription factors encoded in the human genome can bind to DNA through a much larger number of cooperative interactions, resulting in unique transcriptional control for each of the several tens of thousands of human genes. In the case of *IL-2*, transcription occurs only when both NFAT is activated, resulting in its transport from the cytoplasm to the nucleus, and the two subunits of AP1 are synthesized. These events are controlled by distinct signal transduction pathways (Chapters 15 and 16), allowing stringent control of IL-2 expression.

Cooperative binding by NFAT and AP1 occurs only when their weak binding sites are positioned quite close to each other in DNA. The sites must be located at a precise distance from each other for effective binding. Recent studies have shown that the requirements for cooperative binding are not so stringent in the case of some other transcription factors and control regions. For example, the *EGR-1* control region contains a composite binding site to which the SRF and TCF transcription factors bind cooperatively (see Figure 7-23). Because a TCF has a long, flexible domain that interacts with SRF, the two proteins can bind cooperatively when their individual sites in DNA are separated by any distance up to 10 base pairs or are inverted relative to each other.

Multiprotein Complexes Form on Enhancers

As noted previously, enhancers generally range in length from about 50 to 200 base pairs and include binding sites for several transcription factors. The multiple transcription factors that bind to a single enhancer are thought to interact. Analysis of the ≈70-bp enhancer that regulates expression of β-interferon, an important protein in defense against viral infections in humans, provides a good example of such transcription-factor interactions. The β-interferon enhancer contains four control elements that bind four different transcription factors simultaneously. In the presence of a small, abundant protein associated with chromatin called *HMGI*, binding of the transcription factors is highly cooperative, similar to the binding of NFAT and AP1 to the composite promoter-proximal site in the *IL-2* control region (Figure 7-29). This cooperative binding produces a multiprotein complex on the β-interferon enhancer DNA (Figure 7-30). The term **enhancesome** has been coined to describe such large nucleoprotein complexes that assemble from transcription factors as they bind cooperatively to their multiple binding sites in an enhancer.

HMGI binds to the minor groove of DNA regardless of the sequence and, as a result, bends the DNA molecule sharply. This bending of the enhancer DNA permits bound transcription factors to interact properly. The inherently weak, noncovalent protein-protein interactions between transcription factors are strengthened by their binding to neighboring DNA sites, which keeps the proteins at very high relative concentrations.

Because of the presence of flexible regions connecting the DNA-binding domains and activation or repression

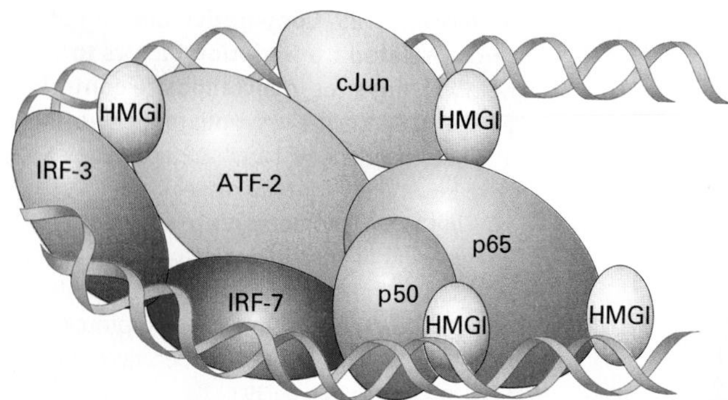

▲ **FIGURE 7-30 Model of the enhancesome that forms on the β-interferon enhancer.** Two monomeric transcription factors, IRF-3 and IRF-7, and two heterodimeric factors, Jun/ATF-2 and p50/ p65 (NF-κB), bind to the four control elements in this enhancer. Cooperative binding of these transcription factors is facilitated by HMGI, which binds to the minor groove of DNA and also interacts directly with the dimeric factors. Bending of the enhancer sequence resulting from HMGI binding is critical to formation of an enhancesome. Different DNA-bending proteins act similarly at other enhancers. [Adapted from D. Thanos and T. Maniatis, 1995, *Cell* **83:**1091, and M. A. Wathel et al., 1998, *Mol. Cell* **1:**507.]

domains in transcription factors (see Figure 7-22) and the ability of bound proteins to bend DNA, considerable leeway in the spacing between regulatory elements in transcription-control regions is permissible. This property probably contributed to rapid evolution of gene control in eukaryotes. Transposition of DNA sequences and recombination between repeated sequences over evolutionary time likely created new combinations of control elements that were subjected to natural selection and retained if they proved beneficial. The latitude in spacing between regulatory elements probably allowed many more functional combinations to be subjected to this evolutionary experimentation than would be the case if constraints on the spacing between regulatory elements were strict, as for most genes in bacteria.

KEY CONCEPTS OF SECTION 7.4

Activators and Repressors of Transcription

- Transcription factors, which stimulate or repress transcription, bind to promoter-proximal regulatory elements and enhancers in eukaryotic DNA.
- Transcription activators and repressors are generally modular proteins containing a single DNA-binding domain and one or a few activation domains (for activators) or repression domains (for repressors). The different domains frequently are linked through flexible polypeptide regions (see Figure 7-22).
- Among the most common structural motifs found in the DNA-binding domains of eukaryotic transcription factors are the C_2H_2 zinc finger, homeodomain, basic helix-loop-helix (bHLH), and basic zipper (leucine zipper). All these and many other DNA-binding motifs contain one or more α helices that interact with major grooves in their cognate site in DNA.
- Activation and repression domains in transcription factors exhibit a variety of amino acid sequences and three-dimensional structures. In general, these functional domains interact with co-activators or co-repressors, which are critical to the ability of transcription factors to modulate gene expression.
- The transcription-control regions of most genes contain binding sites for multiple transcription factors. Transcription of such genes varies depending on the particular repertoire of transcription factors that are expressed and activated in a particular cell at a particular time.
- Combinatorial complexity in transcription control results from alternative combinations of monomers that form heterodimeric transcription factors (see Figure 7-28) and from cooperative binding of transcription factors to composite control sites (see Figure 7-29).
- Cooperative binding of multiple activators to nearby sites in an enhancer forms a multiprotein complex called an enhancesome (see Figure 7-30). Assembly of enhancesomes often requires small proteins that bind to the DNA minor groove and bend the DNA sharply, allowing bound proteins on either side of the bend to interact more readily.

7.5 Transcription Initiation by RNA Polymerase II

In previous sections, many of the eukaryotic proteins and DNA sequences that participate in transcription and its control have been introduced. In this section, we focus on assembly of *transcription preinitiation complexes.* The preinitiation complex is an association of RNA polymerase II and several protein initiation factors that assemble together at the start site and begin to unwind the DNA in preparation for transcription of the gene. Recall that this eukaryotic RNA polymerase II catalyzes synthesis of mRNAs and a few small nuclear RNAs (snRNAs). Specific activator and repressor proteins regulate the mechanisms that control the assembly of Pol II transcription preinitiation complexes—and hence the rate of transcription of protein-coding genes—and are considered in the next section.

General Transcription Factors Position RNA Polymerase II at Start Sites and Assist in Initiation

In vitro transcription by purified RNA polymerase II requires the addition of several initiation factors that are separated from the polymerase during purification. These initiation factors, which position polymerase molecules at transcription start sites and help to melt the DNA strands so that the template strand can enter the active site of the enzyme, are called *general transcription factors.* In contrast to the transcription

Podcast: Assembly of the Pol II Preinitiation Complex

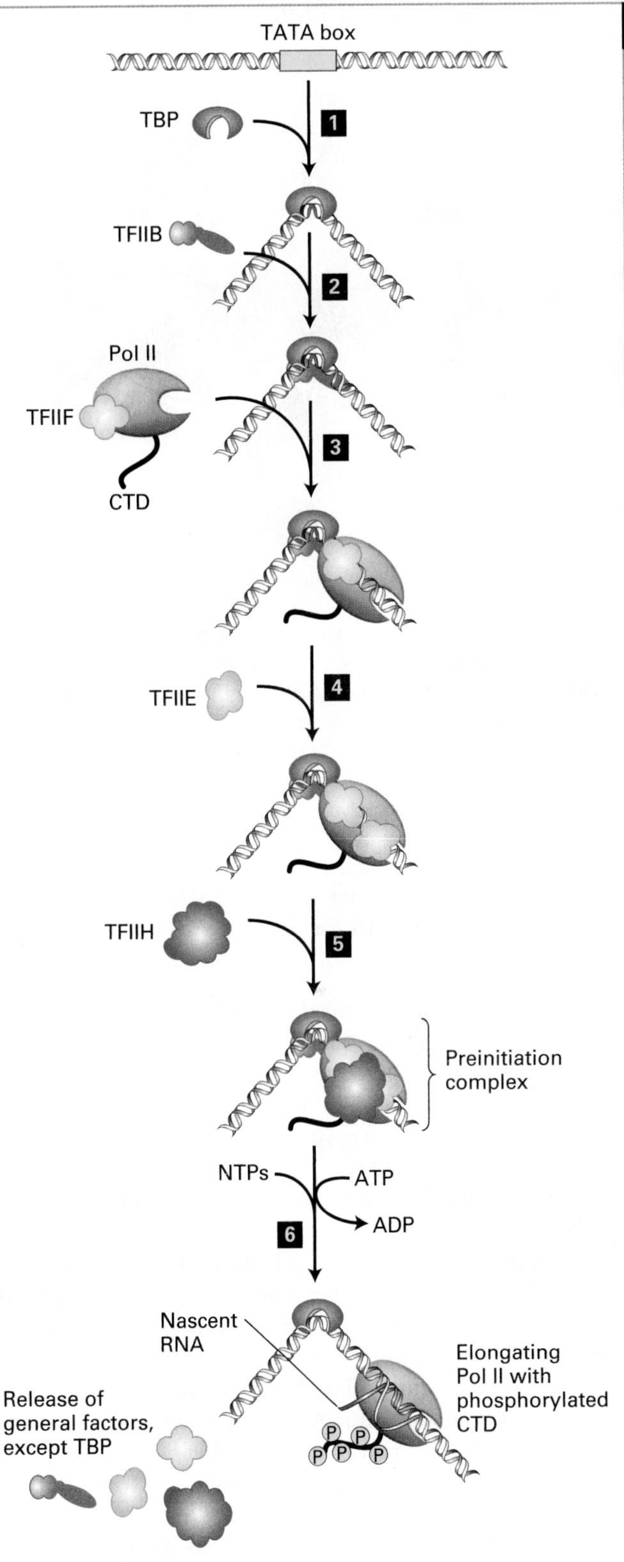

▶ **FIGURE 7-31 In vitro assembly of RNA polymerase II preinitiation complex.** The indicated general transcription factors and purified RNA polymerase II (Pol II) bind sequentially to TATA-box DNA to form a preinitiation complex. ATP hydrolysis then provides the energy for unwinding of DNA at the start site by a TFIIH subunit. As Pol II initiates transcription in the resulting open complex, the polymerase moves away from the promoter and its CTD becomes phosphorylated. In vitro, the general transcription factors (except for TBP) dissociate from the TBP-promoter complex, but it is not yet known which factors remain associated with promoter regions following each round of transcription initiation in vivo.

factors discussed in the previous section, which bind to specific sites in a limited number of genes, general transcription factors are required for synthesis of RNA from most genes.

The general transcription factors that assist Pol II in initiation of transcription from most TATA-box promoters in vitro have been isolated and characterized. These proteins are designated *TFIIA, TFIIB,* etc., and most are multimeric proteins. The largest is TFIID, which consists of a single 38-kDa *TATA-box-binding protein (TBP)* and 13 TBP-associated factors (TAFs). General transcription factors with similar activities have been isolated from cultured human cells, rat liver, *Drosophila* embryos, and yeast. The genes encoding these proteins in yeast have been sequenced as part of the complete yeast genome sequence, and many of the cDNAs encoding human and *Drosophila* Pol II general transcription factors have been cloned and sequenced. In all cases, equivalent general transcription factors from different eukaryotes are highly conserved.

Sequential Assembly of Proteins Forms the Pol II Transcription Preinitiation Complex in Vitro

The complex of RNA polymerase II and its general transcription factors bound to a promoter and ready to initiate transcription is called a *preinitiation complex.* Understanding how a preinitiation complex is assembled on a promoter is important because, in principle, each step in the process can be controlled, allowing complex regulation of the overall process of transcription initiation. Detailed biochemical studies using DNase I footprinting and electrophoretic mobility shift assays were used to determine the order in which Pol II and general transcription factors bound to TATA-box promoters. Because the complete, multisubunit TFIID is difficult to purify, researchers used only the isolated TBP component of this general transcription factor in these experiments. Pol II can initiate transcription in vitro in the absence of the other TFIID subunits.

Figure 7-31 summarizes our current understanding of the stepwise assembly of the Pol II transcription preinitiation complex in vitro. TBP is the first protein to bind to a TATA-box promoter. All eukaryotic TBPs analyzed to date have very similar C-terminal domains of 180 residues. The sequence of this region is 80 percent identical in the yeast and human proteins, and most differences are conservative substitutions. This conserved C-terminal domain functions as well as the full-length protein does for in vitro transcription. (The N-terminal domain of TBP, which varies greatly in sequence and length among different eukaryotes, functions in the Pol II–catalyzed transcription of genes encoding snRNAs.) TBP is a monomer that folds into a saddle-shape structure; the two halves of the molecule exhibit an overall dyad symmetry but are not identical. Like the HMGI and

Video: 3D Model of an RNA Polymerase II Preinitiation Complex

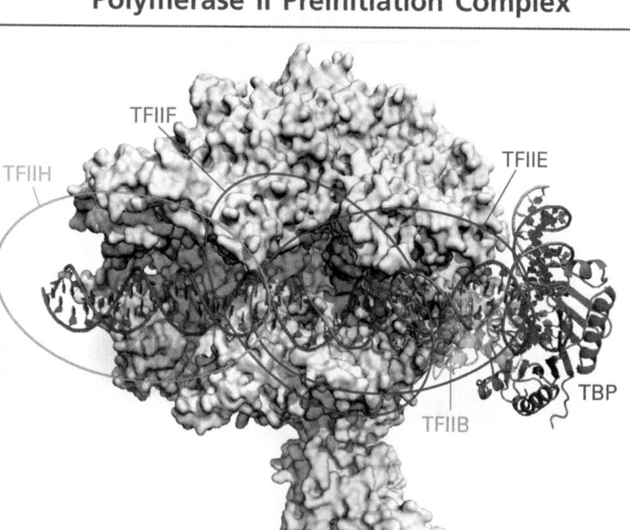

▲ **FIGURE 7-32 Model for the structure of an RNA polymerase II preinitiation complex.** Yeast RNA polymerase II is shown as a space-filling model with the direction of transcription to the left. The template strand of DNA is shown in dark blue and the nontemplate strand in red. The start site of transcription is shown as a space-filling red and dark blue base pair. TBP and TFIIB are shown as green and yellow worm traces of the polypeptide backbone. Structures for TFIIE, F, and H have not been determined to high resolution. Their approximate positions lying over the DNA in the preinitiation complex are shown by ellipses for TFIIE (light blue), TFIIF (red), and TFIIH (orange). [Adapted from G. Miller and S. Hahn, 2006, *Nature Struct. Biol.*, **13:**603.]

other DNA-bending proteins that participate in formation of enhancesomes, TBP interacts with the minor groove in DNA, bending the helix considerably (see Figure 4-5). The DNA-binding surface of TBP is conserved in all eukaryotes, explaining the high conservation of the TATA-box promoter element (see Figure 7-12).

Once TBP has bound to the TATA box, TFIIB can bind. TFIIB is a monomeric protein, slightly smaller than TBP. The C-terminal domain of TFIIB makes contact with both TBP and DNA on either side of the TATA box, while its N-terminal domain extends toward the transcription start site. Following TFIIB binding, a preformed complex of tetrameric TFIIF and Pol II binds, positioning the polymerase over the start site. At most promoters, two more general transcription factors must bind before the DNA duplex can be separated to expose the template strand. First to bind is tetrameric TFIIE, creating a docking site for TFIIH, another multimeric factor containing ten subunits. Binding of TFIIH completes assembly of the transcription preinitiation complex in vitro (Figure 7-31). Figure 7-32 shows a current model for the structure of a preinitiation complex.

The **helicase** activity of one of the TFIIH subunits uses energy from ATP hydrolysis to unwind the DNA duplex at the start site, allowing Pol II to form an *open* complex in which the DNA duplex surrounding the start site is melted and the template strand is bound at the polymerase active site. If the remaining ribonucleoside triphosphates are present, Pol II begins transcribing the template strand. As the polymerase transcribes away from the promoter region, another subunit of TFIIH phosphorylates the Pol II CTD at multiple sites (see Figure 7-31). In the minimal in vitro transcription assay containing only these general transcription factors and purified RNA polymerase II, TBP remains bound to the TATA box as the polymerase transcribes away from the promoter region, but the other general transcription factors dissociate.

The first subunits of TFIIH to be cloned from humans were identified because mutations in the genes encoding them cause defects in the repair of damaged DNA. In normal individuals, when a transcribing RNA polymerase becomes stalled at a region of damaged template DNA, a subcomplex of TFIIH is thought to recognize the stalled polymerase and then recruit other proteins that function with TFIIH in repairing the damaged DNA region. In patients with mutant forms of TFIIH subunits, such repair of damaged DNA in transcriptionally active genes is impaired. As a result, affected individuals have extreme skin sensitivity to sunlight (a common cause of DNA damage) and exhibit a high incidence of cancer. Depending on the severity of the defect in TFIIH function, these individuals may suffer from diseases such as xeroderma pigmentosum and Cockayne's syndrome (Chapter 25). ▮

In Vivo Transcription Initiation by Pol II Requires Additional Proteins

Although the general transcription factors discussed above allow Pol II to initiate transcription in vitro, another general transcription factor, TFIIA, is required for initiation by Pol II in vivo. Purified TFIIA forms a complex with TBP and TATA-box DNA. X-ray crystallography of this complex shows that TFIIA interacts with the side of TBP that is upstream from the direction of transcription. Biochemical experiments suggest that in cells of higher eukaryotes, TFIIA and TFIID, with its multiple TAF subunits, bind first to TATA-box DNA and then the other general transcription factors subsequently bind as indicated in Figure 7-31.

The TAF subunits of TFIID appear to play a role in initiating transcription from promoters that lack a TATA box. For instance, some TAF subunits contact the initiator element in promoters where it occurs, probably explaining how such sequences can replace a TATA box. Additional TFIID TAF subunits can bind to a consensus sequence A/G-G-A/T-C-G-T-G centered ≈30 base pairs downstream from the transcription start site in many genes that lack a TATA-box promoter. Because of its position, this regulatory sequence is called the *downstream promoter element (DPE)*. The DPE facilitates transcription of TATA-less genes that contain it by increasing TFIID binding.

In addition to general transcription factors, specific activators and repressors regulate transcription of genes by Pol II. In the next section we will examine how these regulatory proteins influence Pol II transcription initiation.

KEY CONCEPTS OF SECTION 7.5

Transcription Initiation by RNA Polymerase II

- Transcription of protein-coding genes by Pol II can be initiated in vitro by sequential binding of the following in the indicated order: TBP, which binds to TATA-box DNA; TFIIB; a complex of Pol II and TFIIF; TFIIE; and finally TFIIH (see Figure 7-31).
- The helicase activity of a TFIIH subunit separates the template strands at the start site in most promoters, a process that requires hydrolysis of ATP. As Pol II begins transcribing away from the start site, its CTD is phosphorylated by another TFIIH subunit.
- In vivo transcription initiation by Pol II also requires TFIIA and, in metazoans, a complete TFIID protein, including its multiple TAF subunits as well as the TBP subunit.

7.6 Molecular Mechanisms of Transcription Repression and Activation

The repressors and activators that bind to specific sites in DNA and regulate expression of the associated protein-coding genes do so by two general mechanisms. First, these regulatory proteins act in concert with other proteins to modulate chromatin structure, inhibiting or stimulating the ability of general transcription factors to bind to promoters. Recall from Chapter 6 that the DNA in eukaryotic cells is not free but is associated with a roughly equal mass of protein in the form of **chromatin.** The basic structural unit of chromatin is the **nucleosome,** which is composed of ≈147 base pairs of DNA wrapped tightly around a disk-shaped core of **histone** proteins. Residues within the N-terminal region of each histone, and the C-terminal regions of histones H2A and H2B, called *histone tails,* extend from the surface of the nucleosome and can be reversibly modified (see Figure 6-31b). Such modifications, especially the acetylation of histone H3 and H4 tails, influence the relative condensation of chromatin and thus its accessibility to proteins required for transcription initiation. In addition to their role in such chromatin-mediated transcriptional control, activators and repressors interact with a large multiprotein complex called the *mediator of transcription complex,* or simply **mediator.** This complex in turn binds to Pol II and directly regulates assembly of transcription preinitiation complexes.

In this section, we review current understanding of how repressors and activators control chromatin structure and preinitiation complex assembly. In the next section of the chapter, we discuss how the concentrations and activities of activators and repressors themselves are controlled, so that gene expression is precisely attuned to the needs of the cell and organism.

Formation of Heterochromatin Silences Gene Expression at Telomeres, Near Centromeres, and in Other Regions

For many years it has been clear that inactive genes in eukaryotic cells are often associated with **heterochromatin,** regions of chromatin that are more highly condensed and stain more darkly with DNA dyes than **euchromatin,** where most transcribed genes are located (see Figure 6-33a). Regions of chromosomes near the centromeres and telomeres and additional specific regions that vary in different cell types are organized into heterochromatin. The DNA in heterochromatin is less accessible to externally added proteins than DNA in euchromatin and consequently is often referred to as "closed" chromatin. For instance, in an experiment described in Chapter 6, the DNA of inactive genes was found to be far more resistant to digestion by DNase I than the DNA of transcribed genes (see Figure 6-32).

Chromatin-Mediated Repression in Yeast Study of DNA regions in *S. cerevisiae* that behave like the heterochromatin of higher eukaryotes provided early insight into the *chromatin-mediated repression* of transcription. This yeast can grow either as haploid or diploid cells. Haploid cells exhibit one of two possible mating types, called **a** and α. Cells of different mating type can "mate," or fuse, to generate a diploid cell (see Figure 1-6). When a haploid cell divides by budding, the larger "mother" cell switches its mating type (see Figure 21-27). Genetic and molecular analyses have revealed that three genetic loci on yeast chromosome III control the mating type of yeast cells (Figure 7-33). Only the central mating-type locus, termed *MAT,* is actively transcribed. How the proteins encoded at the *MAT locus* determine whether a cell has the **a** or α phenotype is explained in Chapter 21. The two additional loci, termed *HML* and *HMR,* near the left and right telomere, respectively, contain "silent" (nontranscribed) copies of the **a** or α genes. These sequences are transferred alternately from *HML*α or *HMR***a** into the *MAT* locus by a type of nonreciprocal recombination between sister chromatids during cell division. When the *MAT* locus contains the DNA sequence from *HML*α, the cells behave as α cells. When the *MAT* locus contains the DNA sequence from *HMR***a,** the cells behave like **a** cells.

Our interest here is how transcription of the silent mating-type loci at *HML* and *HMR* is repressed. If the genes at these loci are expressed, as they are in yeast mutants with defects in the repressing mechanism, both **a** and α proteins are expressed, causing the cells to behave like diploid cells, which cannot mate. The promoters and UASs controlling transcription of the **a** and α genes lie near the center of the DNA sequence that is transferred and are identical whether the sequences are at the *MAT* locus or at one of the silent loci. This indicates that the function of the transcription factors that interact with these sequences must somehow be blocked at *HML* and *HMR* but not at the *MAT* locus. This repression of the silent loci depends on **silencer sequences** located next to the region of transferred DNA at *HML* and *HMR* (Figure 7-33). If the silencer is deleted, the adjacent silent locus is transcribed. Remarkably, any gene placed near the yeast mating-type silencer sequence by recombinant DNA techniques is repressed, or "silenced," even a tRNA gene transcribed by RNA polymerase III, which uses a different set of general transcription factors than RNA polymerase II uses.

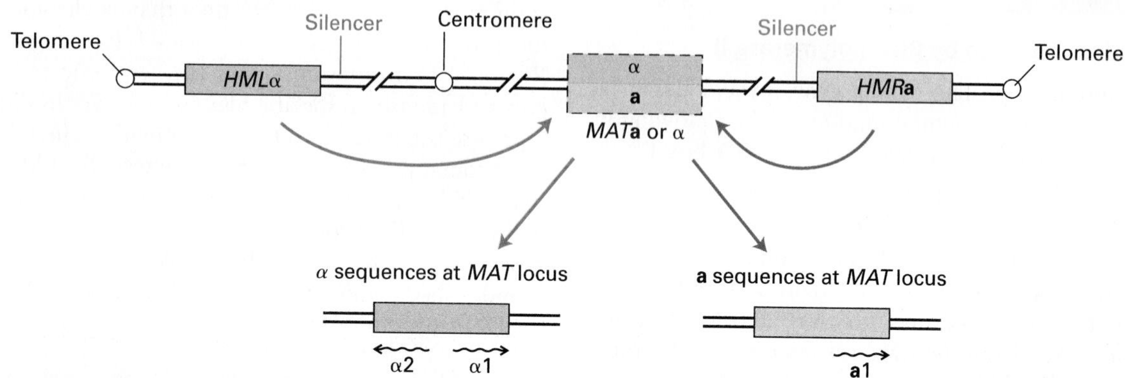

▲ **FIGURE 7-33 Arrangement of mating-type loci on chromosome III in the yeast *S. cerevisiae.*** Silent (unexpressed) mating-type genes (either **a** or α, depending on the strain) are located at the *HML* locus. The opposite mating-type genes are present at the silent *HMR* locus. When the α or **a** sequences are present at the *MAT* locus, they can be transcribed into mRNAs whose encoded proteins specify the mating-type phenotype of the cell. The silencer sequences near *HML* and *HMR* bind proteins that are critical for repression of these silent loci. Haploid cells can switch mating types in a process that transfers the DNA sequence from *HML* or *HMR* to the transcriptionally active *MAT* locus.

Several lines of evidence indicate that repression of the *HML* and *HMR* loci results from a condensed chromatin structure that sterically blocks transcription factors from interacting with the DNA. In one telling experiment, the gene encoding an *E. coli* enzyme that methylates adenine residues in GATC sequences was introduced into yeast cells under the control of a yeast promoter so that the enzyme was expressed. Researchers found that GATC sequences within the *MAT* locus and most other regions of the genome in these cells were methylated, but not those within the *HML* and *HMR* loci. These results indicate that the DNA of the silent loci is inaccessible to the *E. coli* methylase and presumably to proteins in general, including transcription factors and RNA polymerase. Similar experiments conducted with various yeast histone mutants indicated that specific interactions involving the histone tails of H3 and H4 are required for formation of a fully repressed chromatin structure. Other studies have shown that the telomeres of every yeast chromosome also behave like silencer sequences. For instance, when a gene is placed within a few kilobases of any yeast telomere, its expression is repressed. In addition, this repression is relieved by the same mutations in the H3 and H4 histone tails that interfere with repression at the silent mating-type loci.

Genetic studies led to identification of several proteins, RAP1 and three SIR proteins, that are required for repression of the silent mating-type loci and the telomeres in yeast. RAP1 was found to bind within the DNA silencer sequences associated with *HML* and *HMR* and to a sequence that is repeated multiple times at each yeast chromosome telomere. Further biochemical studies showed that the SIR2 protein is a *histone deacetylase;* it removes acetyl groups on lysines of the histone tails. Also, the RAP1, and SIR2, 3, and 4 proteins bind to one another, and SIR3 and SIR4 bind to the N-terminal tails of histones H3 and H4 that are maintained in a largely unacetylated state by the deacetylase activity of SIR2. Several experiments using fluorescence confocal

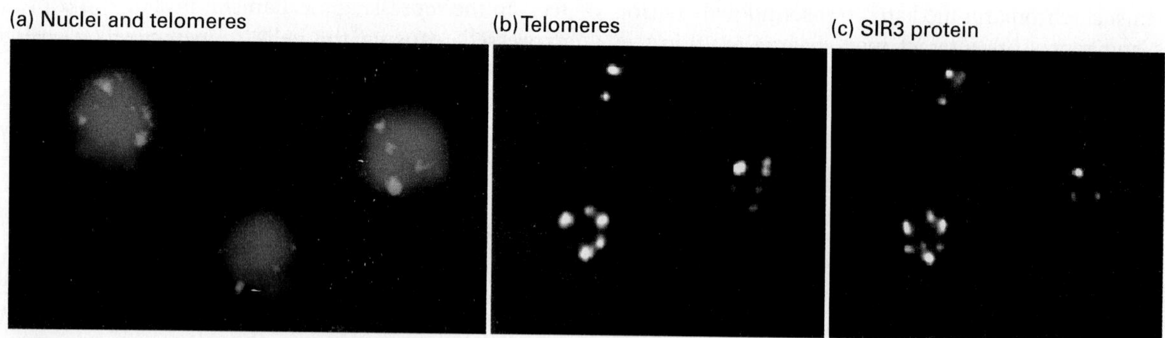

▲ **EXPERIMENTAL FIGURE 7-34 Antibody and DNA probes colocalize SIR3 protein with telomeric heterochromatin in yeast nuclei.** (a) Confocal micrograph 0.3 μm thick through three diploid yeast cells, each containing 68 telomeres. Telomeres were labeled by hybridization to a fluorescent telomere-specific probe (yellow). DNA was stained red to reveal the nuclei. The 68 telomeres coalesce into a much smaller number of regions near the nuclear periphery. (b, c) Confocal micrographs of yeast cells labeled with a telomere-specific hybridization probe (b) and a fluorescent-labeled antibody specific for SIR3 (c). Note that SIR3 is localized in the repressed telomeric heterochromatin. Similar experiments with RAP1, SIR2, and SIR4 have shown that these proteins also colocalize with the repressed telomeric heterochromatin. [From M. Gotta et al., 1996, *J. Cell Biol.* **134:**1349; courtesy of M. Gotta, T. Laroche, and S. M. Gasser.]

microscopy of yeast cells either stained with fluorescent-labeled antibody to any one of the SIR proteins or RAP1 or hybridized to a labeled telomere-specific DNA probe revealed that these proteins form large, condensed telomeric nucleoprotein structures resembling the heterochromatin found in higher eukaryotes (Figure 7-34). Figure 7-35 depicts a model for the chromatin-mediated silencing at yeast telomeres based on these and other studies. Formation of heterochromatin at telomeres is nucleated by multiple RAP1 proteins bound to repeated sequences in a nucleosome-free region at the extreme end of a telomere. A network of protein-protein interactions involving telomere-bound RAP1, three SIR proteins (2, 3, and 4), and hypoacetylated histones H3 and H4 creates a stable, higher-order nucleoprotein complex that includes several telomeres and in which the DNA is largely inaccessible to external proteins. One additional protein, SIR1, is also required for silencing of the silent mating-type loci. It binds to the silencer regions associated with *HML* and *HMR* together with RAP1 and other proteins to initiate assembly of a similar multiprotein silencing complex that encompasses *HML* and *HMR*.

An important feature of this model is the dependence of silencing on *hypoacetylation* of the histone tails. This was shown in experiments with yeast mutants expressing histones in which lysines in histone N-termini were substituted with arginines or glycines. Arginine is positively charged like lysine but cannot be acetylated. Glycine, on the other hand, is neutral and simulates the absence of a lysine. Recall from Chapter 6 that acetylation of lysines neutralizes their positive charge and eliminates their interaction with DNA phosphate groups, reducing chromatin condensation. Repression at telomeres and at the silent mating-type loci was defective in the mutants with glycine substitutions but not in mutants with arginine substitutions. Further, acetylation of H3 and H4 lysines interferes with binding by SIR3 and SIR4 and consequently prevents repression at the silent loci and telomeres.

Chromatin Condensation in Higher Eukaryotes In higher eukaryotes, a similar process leads to the formation of condensed heterochromatin at centromeres and telomeres, as well as repression of genes at internal chromosome positions that differ depending on the cell type. But in

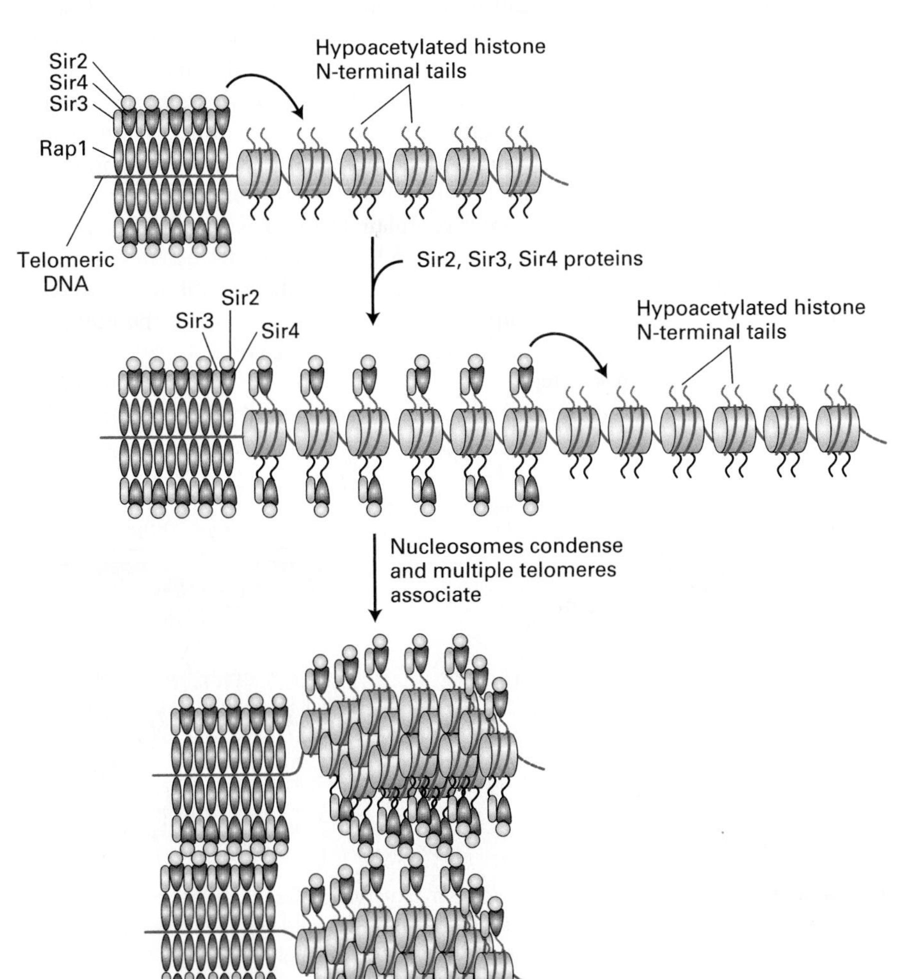

◀ **FIGURE 7-35 Schematic model of silencing mechanism at yeast telomeres.** *(Top)* Multiple copies of RAP1 bind to a simple repeated sequence at each telomere region that lacks nucleosomes. SIR3 and SIR4 bind to RAP1, and SIR2 binds to SIR4. SIR2 is a histone deacetylase that deacetylates the tails on the histones neighboring the repeated RAP1 binding site. *(Middle)* The hypoacetylated histone tails are also binding sites for SIR3 and SIR4, which in turn binds additional SIR2, deacetylating neighboring histones. Repetition of this process results in spreading of the region of hypoacetylated histones with associated SIR2, SIR3, and SIR4. *(Bottom)* Interactions between complexes of SIR2, SIR3, and SIR4 cause the chromatin to condense and several telomeres to associate, as shown in Figure 7-34. The higher-order chromatin structure generated sterically blocks other proteins from interacting with the underlying DNA. [Adapted from M. Grunstein, 1997, *Curr. Opin. Cell Biol.* **9**:383.]

multicellular organisms, methylation of specific lysine residues in histone H3 contribute to chromatin condensation in addition to the deacetylation of histone lysines. We learned in Chapter 6 how HP1 (heterochromatin protein 1) promotes chromatin condensation by binding to nucleosomes that are methylated at lysine 9 of histone H3. Because HP1 also binds to a *histone methyl transferase* that methylates H3 lysine 9 on neighboring nucleosomes, additional HP1 is recruited to the area. Further associations between the individual HP1 molecules themselves remodel chromatin into a more compact structure (see Figure 6-34).

Another kind of chromatin structure essential for repression of genes in specific cell types in multicellular organisms involves a set of proteins collectively known as *Polycomb proteins,* after a phenotype observed in *Drosophila* when they are mutated. The Polycomb repression mechanism is essential for maintaining the repression of genes in specific types of cells and in all of the subsequent cells that develop from them throughout the life of an organism. Important genes regulated by Polycomb proteins include the **Hox genes,** encoding master regulatory transcription factors. As discussed in Chapter 22, different combinations of Hox transcription factors help to direct the development of specific tissues and organs in a developing embryo. Early during embryogenesis, expression of Hox genes is controlled by typical activator and repressor proteins. However, the expression of these activators and repressors stops at an early point in embryogenesis. Correct expression of the Hox genes is then maintained throughout the remainder of embryogenesis and on into adult life by the Polycomb proteins. Complexes of the Polycomb proteins maintain repression of the Hox genes in cells in which they were initially repressed and in all future descendents of those cells. *Trithorax proteins* perform the opposing function to Polycomb proteins, maintaining the expression of the Hox genes that were expressed in a specific cell early in embryogenesis and in all the subsequent descendents of that cell.

Remarkably, virtually all cells in the developing embryo and adult express a similar set of Polycomb and Trithorax proteins and all cells contain the same set of Hox genes. Yet only the Hox genes in cells where they were initially repressed in early embryogenesis remain repressed, even though the same Hox genes in other cells remain active in the presence of the same Polycomb proteins. Consequently, as in the case of the yeast silent-mating-type loci, the expression of Hox genes is regulated by a process that involves more than simply specific DNA sequences interacting with proteins that diffuse through the nucleoplasm. The reason is the same collection of Polycomb and Trithorax proteins and the same Hox DNA sequences in all cells result in the expression of specific Hox genes in cells comprising the anterior of an embryo and their repression in cells located in the posterior of the embryo.

A current model for repression by Polycomb proteins is depicted in Figure 7-36. Most Polycomb proteins are subunits of one of two multiprotein complexes, PRC1 and PRC2. PRC2 is thought to act initially by associating with specific repressors bound to their cognate DNA sequences early in embryogenesis. The PRC2 complex contains a subunit with a *SET* domain, the enzymatically active domain of several histone methyl transferases. This SET domain methylates histone H3 on lysine 27. The PRC1 complex then binds the methylated nucleosomes through dimeric Pc subunits each containing a binding domain (called a *chromodomain*) specific for methylated H3 lysine 27. Binding of the dimeric Pc to neighboring nucleosomes is proposed to condense the chromatin into a structure that inhibits transcription. PRC2 or another methyltransferase is postulated to associate with PRC1, maintaining methylation of H3 lysine 27 in nucleosomes in the region. This results in association of the chromatin with PRC1 and PRC2 complexes even after expression of the initial repressor proteins in Figure 7-36a has ceased.

A key feature of Polycomb repression is its maintenance in daughter cells through successive cell divisions for the life

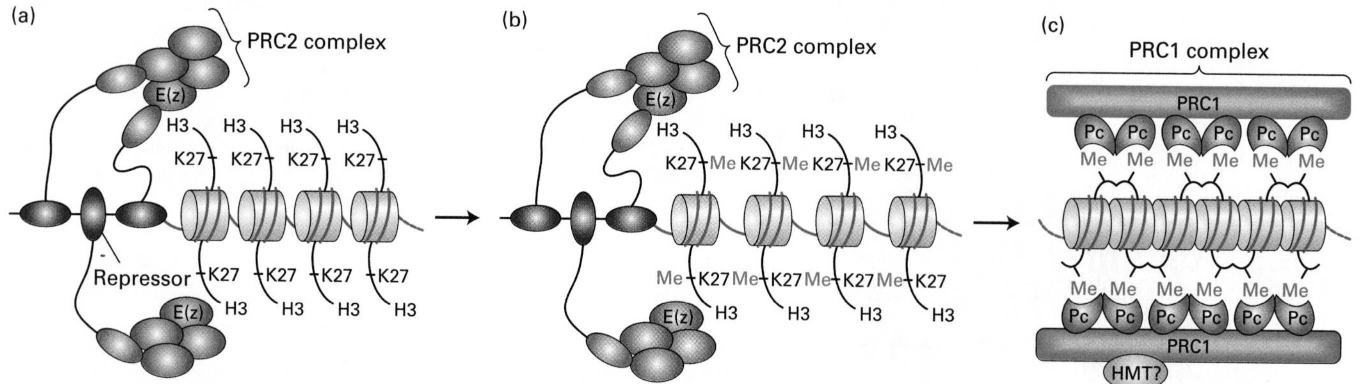

▲ FIGURE 7-36 Model for repression by Polycomb complexes. (a) During early embryogenesis repressors associate with the PRC2 complex. (b) This results in methylation (Me) of neighboring nucleosomes on histone H3 lysine 27 (K27) by the SET-domain-containing subunit E(z). (c) The PRC1 complexes bind nucleosomes methylated at H3 lysine 27 through a dimeric, chromodomain-containing subunit Pc. The PRC1 complex condenses the chromatin into a repressed chromatin structure. PRC2 complexes or another histone methyl transferase associates with PRC1 complexes to maintain H3 lysine 27 methylation of neighboring nucleosomes (not shown). As a consequence, PRC1 association with the region is maintained when expression of the repressor proteins in (a) ceases. [Modified from A. H. Lund and M. van Lohuizen, 2004, *Curr. Op. Cell Biol.* **16:**239.]

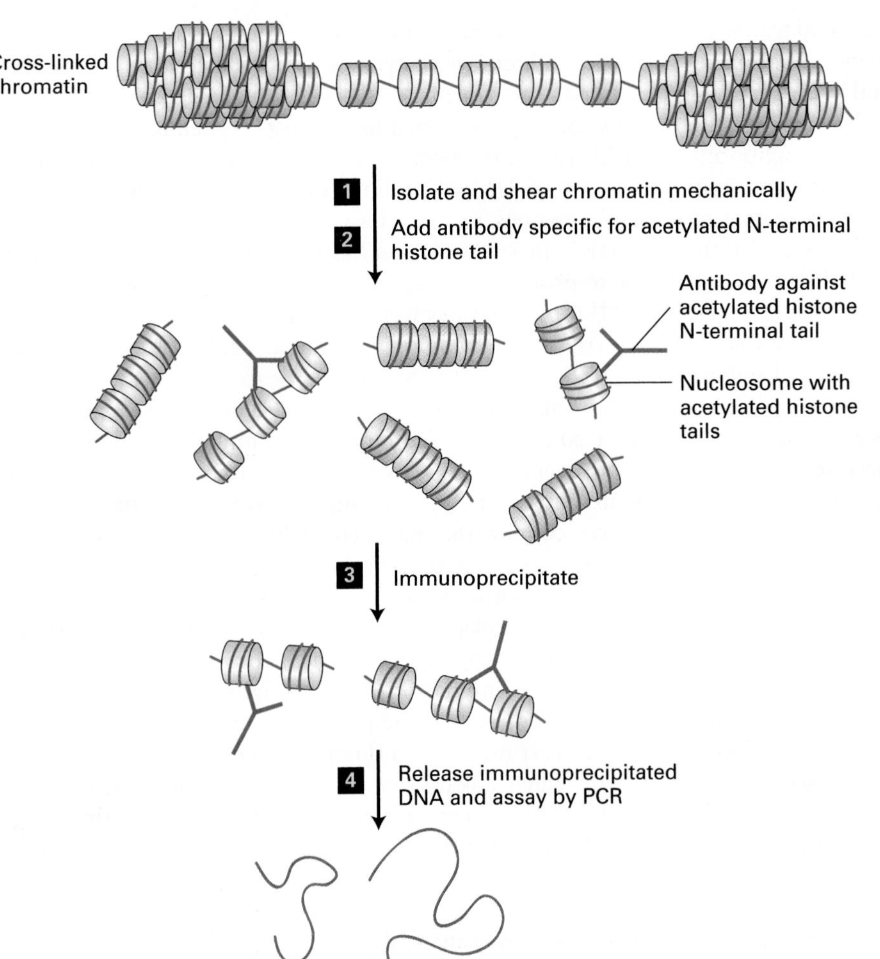
Cross-linked chromatin
1
Isolate and shear chromatin mechanically
2
Add antibody specific for acetylated N-terminal histone tail
Antibody against acetylated histone N-terminal tail
Nucleosome with acetylated histone tails
3
Immunoprecipitate
4
Release immunoprecipitated DNA and assay by PCR

◀ EXPERIMENTAL FIGURE 7-37 **The chromatin immunoprecipitation method can reveal the acetylation state of histones in chromatin.** Histones are lightly cross-linked to DNA in vivo using a cell-permeable, reversible, chemical cross-linking agent. Nucleosomes with acetylated histone tails are shown in green. Step **1**: Cross-linked chromatin is then isolated and sheared to an average length of two to three nucleosomes. Step **2**: An antibody against a particular acetylated histone tail sequence is added, and (step **3**) bound nucleosomes are immunoprecipitated. Step **4**: DNA in the immunoprecipitated chromatin fragments is released by reversing the cross-link and then is quantitated using a sensitive PCR method. The method can be used to analyze the in vivo association of any protein with a specific sequence of DNA by using an antibody against the protein of interest in step **2**. [See S. E. Rundlett et al., 1998, *Nature* **392:**831.]

of an organism ($\approx$100 years for some vertebrates, 2,000 years for a sugar cone pine!). This stability of Hox gene expression state is thought to result from the distribution of nucleosomes with methylated H3 lysine 27 to both daughter DNA molecules immediately following DNA replication. Association of PRC1 complexes with these H3 lysine 27-methylated nucleosomes and methylation of new nucleosomes assembled on the replicated DNA on H3 lysine 27 would allow the complete PRC1 complex to be reestablished over the same region of chromatin in both replicated daughter chromosomes. Although Polycomb repression may involve additional mechanisms as well, this model can explain how repression of specific genes can be maintained in all daughter cells derived from an initial cell where the gene was repressed by a transiently expressed set of repressors.

Alternatively, a complex of Trithorax proteins includes a histone methyl transferase that methylates histone H3 lysine 4, a histone methylation associated with the promoters of actively transcribed genes. This histone modification is thought to create a binding site for histone acetylase and chromatin remodeling complexes that promote transcription and prevent methylation of histone H3 at lysine 9, preventing the binding of HP1, and at lysine 27, preventing the binding of the PRC1 repressing complex. Nucleosomes marked with histone H3 lysine 4 methylation also are thought to be distributed to both daughter DNA molecules during DNA replication. Binding of Trithorax complexes to nucleosomes with the histone H3 lysine 4 methylation mark may cause the same methylation at unmodified histones incorporated into the daughter chromatin, leading to perpetuation of the chromatin mark in this region. In this way, inheritance of the expression status of Hox genes and other genes regulated by the Polycomb/Trithorax system is templated through chromatin replication by post-translational modifications on histones rather than DNA sequence. This type of inheritance through modifications of chromatin structure rather than modification of DNA sequence is referred to as **epigenetic** inheritance.

Repressors Can Direct Histone Deacetylation and Methylation at Specific Genes

The importance of *histone deacetylation* and *methylation* in chromatin-mediated gene repression has been further supported by studies of eukaryotic repressors that regulate genes at internal chromosomal positions. These proteins are now known to act in part by causing deacetylation of histone tails in nucleosomes that bind to the TATA box and promoter-proximal region of the genes they repress. In vitro studies have shown that when promoter DNA is assembled onto a nucleosome with unacetylated histones, the general transcription factors cannot bind to the TATA box and initiation region. In unacetylated histones, the N-terminal lysines are positively charged and interact strongly with DNA phosphates. The unacetylated histone tails also interact with neighboring histone octamers, favoring the folding of chromatin into condensed, higher-order structures whose precise conformation is not well understood. The net effect is that general transcription factors cannot assemble into a preinitiation complex on a promoter associated with hypoacetylated histones. In contrast, binding of general transcription factors is repressed much less by histones with hyperacetylated tails in which the positively charged lysines are neutralized and electrostatic interactions with DNA phosphates are eliminated.

The connection between histone deacetylation and repression of transcription at specific yeast promoters became clearer when the cDNA encoding a human *histone deacetylase* was found to have high homology to the yeast *RPD3* gene, known to be required for the normal repression of a number of yeast genes. Further work showed that RPD3 protein has histone deacetylase activity. The ability of RPD3 to deacetylate histones at a number of promoters depends on two other proteins: UME6, a repressor that binds to a specific upstream regulatory sequence (URS1), and SIN3, which is part of a large, multiprotein complex that also contains RPD3. SIN3 also binds to the repression domain of UME6, thus positioning the RPD3 histone deacetylase in the complex so it can interact with nearby promoter-associated nucleosomes and remove acetyl groups from histone tail lysines. Additional experiments, using the *chromatin immunoprecipitation* technique outlined in Figure 7-37, demonstrated that in wild-type yeast, one or two nucleosomes in the immediate vicinity of UME6-binding sites are hypoacetylated. These DNA regions include the promoters of genes repressed by UME6. In *sin3* and *rpd3* deletion mutants, not only were these promoters derepressed, but the nucleosomes near the UME6-binding sites were hyperacetylated.

All these findings provide considerable support for the model of repressor-directed deacetylation shown in Figure 7-38a. In this model, the SIN3-RPD3 complex functions as a *co-repressor.* Co-repressor complexes containing histone deacetylases also have been found associated with many repressors from mammalian cells. Some of these complexes contain the mammalian homolog of SIN3 (mSin3), which interacts with the repression domain of the repressor, as in yeast. Other histone deacetylase complexes identified in mammalian cells appear to contain additional or different repressor-binding proteins. These various repressor and co-repressor combinations are thought to mediate histone deacetylation at specific promoters by a mechanism similar to the yeast mechanism (Figure 7-38a).

In higher eukaryotes, some co-repressor complexes also contain histone methyl transferase subunits that methylate histone H3 at lysine 9, generating a binding site for HP1 protein,

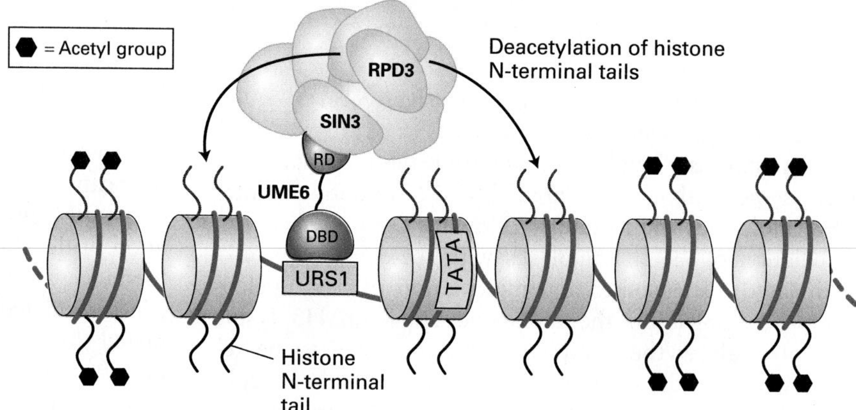

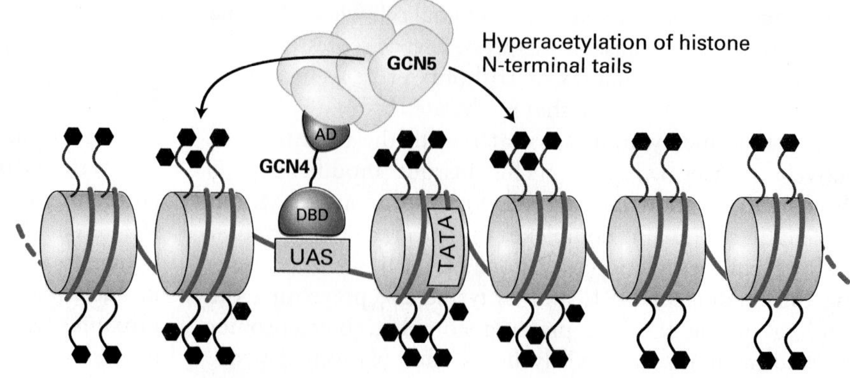

▶ **FIGURE 7-38 Proposed mechanism of histone deacetylation and hyperacetylation in yeast transcription control.** (a) Repressor-directed deacetylation of histone N-terminal tails. The DNA-binding domain (DBD) of the repressor UME6 interacts with a specific upstream control element (URS1) of the genes it regulates. The UME6 repression domain (RD) binds SIN3, a subunit of a multiprotein complex that includes RPD3, a histone deacetylase. Deacetylation of histone N-terminal tails on nucleosomes in the region of the UME6-binding site inhibits binding of general transcription factors at the TATA box, thereby repressing gene expression. (b) Activator-directed hyperacetylation of histone N-terminal tails. The DNA-binding domain of the activator GCN4 interacts with specific upstream activating sequences (UAS) of the genes it regulates. The GCN4 activation domain (AD) then interacts with a multiprotein histone acetylase complex that includes the GCN5 catalytic subunit. Subsequent hyperacetylation of histone N-terminal tails on nucleosomes in the vicinity of the GCN4-binding site facilitates access of the general transcription factors required for initiation. Repression and activation of many genes in higher eukaryotes occurs by similar mechanisms.

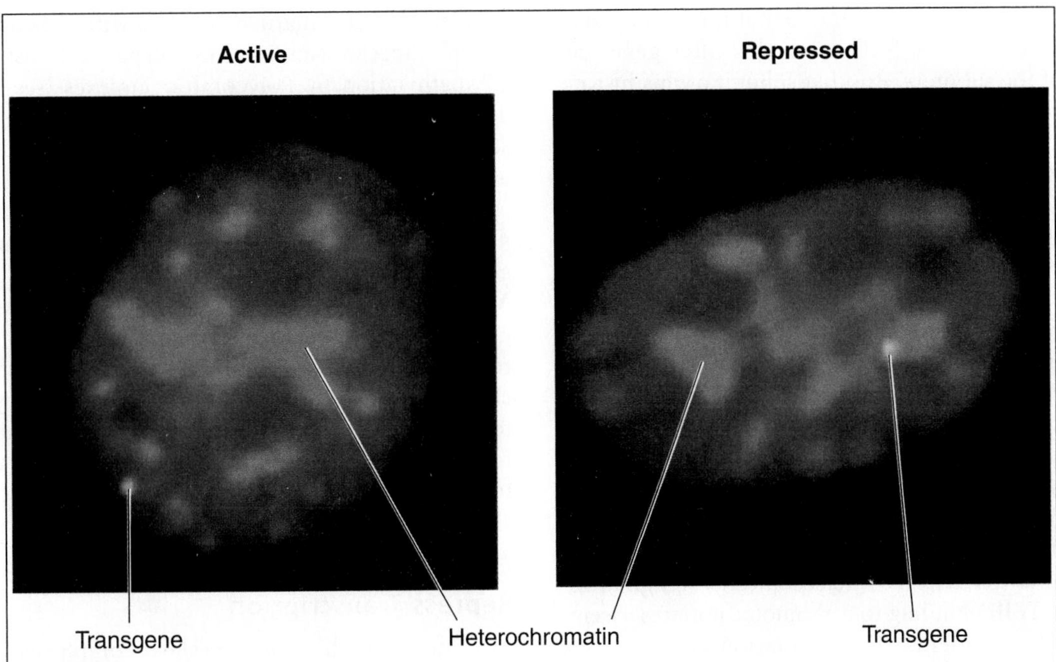

▲ **FIGURE 7-39 Association of a repressed transgene with heterochromatin.** Mouse fibroblasts were stably transformed with a transgene with binding sites for an engineered repressor. The repressor was a fusion between a DNA-binding domain, a repression domain that interacts with the KAP1 co-repressor complex, and the ligand-binding domain of a nuclear receptor that allows the nuclear import of the fusion protein to be controlled experimentally (see Figure 7-49). DNA was stained blue with the intercalating dye DAPI. Brighter-staining regions are regions of heterochromatin, where the DNA concentration is higher than in euchromatin. The transgene was detected by hybridization of a fluorescently labeled complementary probe (green). When the recombinant repressor was retained in the cytoplasm, the transgene was transcribed (*left*) and was associated with euchromatin in most cells. When hormone was added so that the recombinant repressor entered the nucleus, the transgene was repressed (*right*) and associated with heterochromatin. Chromatin immunoprecipitation assays (see Figure 7-37) showed that the repressed gene was associated with histone H3 methylated at lysine 9 and HP1, whereas the active gene was not. [Courtesy of Frank Rauscher from Ayyanathan et al., 2003 *Genes and Dev.* **17:**1855.]

as discussed earlier. For example, the KAP1 co-repressor complex functions with a class of more than 200 zinc-finger transcription factors encoded in the human genome. This co-repressor complex includes an H3 lysine 9 methyl transferase that methylates nucleosomes over the promoter region of repressed genes, leading to HP1 binding and repression of transcription. An integrated transgene in cultured mouse fibroblasts that was repressed through the action of the KAP1 co-repressor associated with heterochromatin in most cells, whereas the active form of the same transgene associated with euchromatin (Figure 7-39). Chromatin immunoprecipitation assays (see Figure 7-37) showed that the repressed gene was associated with histone H3 methylated at lysine 9 and HP1, whereas the active gene was not.

Interestingly, in addition to methylation of histone proteins, methylation of the DNA sequence itself can also be a trigger for chromatin condensation. The discovery of mSin3-containing histone deacetylase complexes provided an explanation for earlier observations that in vertebrates transcriptionally inactive DNA regions often contain the modified cytidine residue *5-methylcytidine* (mC) followed immediately by a G, whereas transcriptionally active DNA regions contain fewer mC residues. DNA containing 5-methylcytidine has been found to bind a specific protein that in turn interacts specifically with mSin3. This finding suggests that association of mSin3-containing co-repressors with methylated sites in DNA leads to deacetylation of histones in neighboring nucleosomes, making these regions inaccessible to general transcription factors and Pol II and hence transcriptionally inactive.

Activators Can Direct Histone Acetylation and Methylation at Specific Genes

Just as repressors function through co-repressors that bind to their repression domains, the activation domains of DNA-binding activators function by binding multisubunit *co-activator* complexes. One of the first co-activator complexes to be characterized was the yeast *SAGA* complex, which functions with the GCN4 activator protein described in Section 7.4. Early genetic studies indicated that full activity of the GCN4 activator required a protein called GCN5. The clue to GCN5's function came from biochemical studies of a *histone acetylase* purified from the protozoan *Tetrahymena*, the first histone acetylase to be purified. Sequence analysis revealed homology between the *Tetrahymena* protein and yeast GCN5, which was soon shown to have histone acetylase activity as well. Further genetic and biochemical studies revealed that GCN5 is one subunit of a multiprotein co-activator complex, named the SAGA complex after genes encoding some of the subunits. Another subunit of this histone acetylase complex binds to activation domains in multiple yeast activator proteins, including GCN4. The model shown in Figure 7-38b is consistent with the observation that nucleosomes near the promoter region of a gene regulated by the GCN4 activator are specifically hyperacetylated compared to most histones in the cell. This activator-directed hyperacetylation of nucleosomes near a promoter region opens the chromatin structure so as to facilitate the binding of other proteins required for transcription initiation. The chromatin structure is less condensed compared to most chromatin, as indicated by its sensitivity to digestion with nucleases in isolated nuclei. Also, the acetylation of specific histone lysines generates binding sites for proteins with *bromodomains* that bind them. For example, a subunit of the general transcription factor TFIID contains two bromodomains that bind to acetylated nucleosomes with high affinity. Recall that TFIID binding to a promoter initiates assembly of an RNA polymerase II preinitiation complex (see Figure 7-31). Nucleosomes at promoter regions of virtually all active genes are hyperacetylated.

A similar activation mechanism operates in higher eukaryotes. Mammalian cells contain multisubunit histone acetylase co-activator complexes homologous to the yeast SAGA complex. They also express two related ≈400-kDa, multidomain proteins called *CBP* and *P300,* which are thought to function similarly. As noted earlier, one domain of CBP binds the phosphorylated acidic activation domain in the CREB transcription factor. Other domains of CBP interact with different activation domains in other activators. Yet another domain of CBP has histone acetylase activity, and another CBP domain associates with additional multisubunit histone acetylase complexes. CREB and many other mammalian activators are thought to function in part by directing CBP and the associated histone acetylase complex to specific nucleosomes, where they acetylate histone tails, facilitating the interaction of general transcription factors with promoter DNA. In addition, the largest TFIID subunit also has histone acetylase activity and may function to maintain histone tail hyperacetylation in promoter regions.

Methylation of histone H3 lysines 9 or 27 results in transcriptional repression mediated by binding proteins of the HP1 or Polycomb class, respectively, as discussed above. In contrast, H3 lysine 4 methylation is observed in the promoter regions of active genes through the targeting of methyl transferases specific for H3 lysine 4. One example of this involves the *Trithorax* complex proteins. As discussed earlier, Trithorax complex proteins maintain expression of Hox genes in appropriate cells, just as Polycomb proteins maintain repression of the same Hox genes in other cells. Trithorax proteins, including one with a SET domain that methylates lysines, assemble into a multiprotein complex that tri-methylates H3 lysine 4. H3 tails tri-methylated at lysine 4 then serve as a binding site for another subunit of the Trithorax complex so that the methyl transferase subunit of the Trithorax complex can maintain H3 lysine 4 in the methylated state in chromatin associated with the complex. This is a similar mechanism to the maintenance of histone H3 lysine 27 methylation by Polycomb complexes (see Figure 7-36). The H3 amino-terminal tail tri-methylated on lysine 4 also serves as a binding site for co-activator complexes. For example, SAGA-like histone acetylase complexes also contain a domain that binds specifically to tri-methylated H3 lysine 4. This results in acetylation of histone tail lysines, thereby generating a chromatin structure conducive to transcription. Many genes in multicellular organisms in addition to Hox genes are expressed in lineage-specific expression programs regulated by Trithorax and Polycomb proteins. This is most clearly revealed by staining *Drosophila* salivary gland polytene chromosomes with antibodies to Polycomb and Trithorax proteins. This experiment reveals binding of these proteins to more than 100 sites on fly chromosomes in these cells.

Chromatin-Remodeling Factors Help Activate or Repress Transcription

In addition to histone acetylase complexes, multiprotein *chromatin-remodeling complexes* also are required for activation at many promoters. The first of these characterized was the yeast SWI/SNF chromatin-remodeling complex. One of the SWI/SNF subunits has homology to DNA helicases, enzymes that use energy from ATP hydrolysis to disrupt interactions between base-paired nucleic acids or between nucleic acids and proteins. In vitro, the SWI/SNF complex is thought to pump or push DNA into the nucleosome so that DNA bound to the surface of the histone octomer transiently dissociates from the surface and translocates, causing the nucleosomes to "slide" along the DNA. The net result of such chromatin remodeling is to facilitate the binding of transcription factors to specific DNA sequences in chromatin. Many activation domains bind to chromatin-remodeling complexes, and this binding stimulates in vitro transcription from chromatin templates (DNA bound to nucleosomes). Thus the SWI/SNF complex represents another type of co-activator complex. The experiment shown in Figure 7-40 dramatically demonstrates how an activation domain can cause decondensation of a region of chromatin. This is thought to result from the interaction of the activation domain with chromatin-remodeling and histone acetylase complexes.

Chromatin-remodeling complexes are required for many processes involving DNA in eukaryotic cells, including transcription control, DNA replication, recombination, and repair. Several types of chromatin-remodeling complexes are found in eukaryotic cells, all with homologous DNA helicase domains. SWI/SNF complexes and related chromatin-remodeling complexes in multicellular organisms contain subunits with bromodomains that bind to acetylated histone tails. Consequently, SWI/SNF complexes remain associated with activated, acetylated regions of chromatin, presumably maintaining them in a decondensed conformation. Some chromatin-remodeling complexes contain subunits that bind to histone H3 methylated on lysine 4, contributing to transcriptional activation by Trithorax proteins. Surprisingly, chromatin-remodeling complexes can also participate in

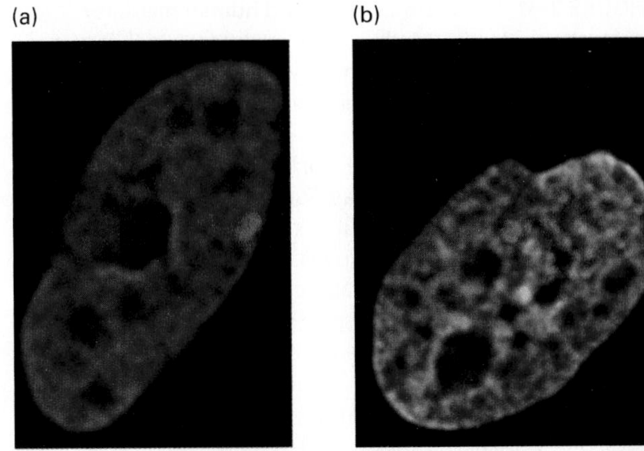

▲ EXPERIMENTAL FIGURE 7-40 **Expression of fusion proteins demonstrates chromatin decondensation in response to an activation domain.** A cultured hamster cell line was engineered to contain multiple copies of a tandem array of *E. coli lac operator* sequences integrated into a chromosome in a region of heterochromatin. (a) When an expression vector for the *lac* repressor was transfected into these cells, *lac* repressors bound to the *lac* operator sites could be visualized in a region of condensed chromatin using an antibody against the *lac* repressor (red). DNA was visualized by staining with DAPI (blue), revealing the nucleus. (b) When an expression vector for the *lac* repressor fused to an activation domain was transfected into these cells, staining as in (a) revealed that the activation domain causes this region of chromatin to decondense into a thinner chromatin fiber that fills a much larger volume of the nucleus. Bar = 1 μm. [Courtesy of Andrew S. Belmont, 1999, *J. Cell Biol.* **145:**1341.]

transcriptional repression. These chromatin-remodeling complexes bind to transcription repression domains of repressors and contribute to repression, presumably, by folding chromatin into condensed structures. Much remains to be learned about how this important class of proteins alters chromatin structure to influence gene expression and other processes.

Histone Modifications Vary Greatly in Their Stabilities

Pulse-chase radiolabeling experiments have shown that acetyl groups on histone lysines turn over rapidly, whereas methyl groups are much more stable. The acetylation state at a specific histone lysine on a particular nucleosome results from a dynamic equilibrium between acetylation and deacetylation by histone acetylases and histone deacetylases, respectively. Acetylation of histones in a localized region of chromatin predominates when local DNA-bound activators transiently bind histone acetylase complexes. De-acetylation predominates when repressors transiently bind histone deacetylase complexes. In addition to these processes, localized to relatively short lengths of chromatin, which include promoters and other transcription-control regions, histone acetylases and deacetylases also function globally on all euchromatin, constantly removing and replacing histone lysine acetyl groups.

In contrast to acetyl groups, methyl groups on histone lysines are much more stable and turn over much less rapidly than acetyl groups. Histone lysine methyl groups can be removed by recently discovered *histone lysine demethylases.* But the resulting turnover of histone lysine methyl groups is much slower than the turnover of histone lysine acetyl groups.

Multiple other post-translational modifications have been characterized on histones, many of them summarized in Figure 6-31b. These all have the potential to positively or negatively regulate the binding of proteins that interact with the chromatin fiber to regulate transcription and other processes. A picture of chromatin is emerging in which histone tails extending as random coils from the chromatin fiber are post-translationally modified to generate one of many possible combinations of modifications that regulate transcription and other processes by regulating the binding of a large number of different protein complexes. Some of these modifications, like histone lysine acetylation, are rapidly reversible, whereas others, like histone lysine methylation, can be templated through chromatin replication, generating epigenetic inheritance in addition to inheritance of DNA sequence.

The Mediator Complex Forms a Molecular Bridge Between Activation Domains and Pol II

Now let's shift our attention from how activators and repressors control chromatin structure to the other mechanism of gene regulation outlined in the introduction to this section—regulation of the assembly of transcription preinitiation complexes.

Interaction of activators with the multiprotein mediator complex (Figure 7-41) directly assists in assembly of Pol II preinitiation complexes. Some of the ≈30 mediator subunits bind to RNA polymerase II, and other mediator subunits bind to activation domains in various activator proteins. Thus mediator can form a molecular bridge between an activator bound to its cognate site in DNA and Pol II at a promoter. In addition, one of the mediator subunits has histone acetylase activity and may function to maintain a promoter region in a hyperacetylated state.

Experiments with temperature-sensitive yeast mutants indicate that some mediator subunits are required for transcription of virtually all yeast genes. These subunits most likely help maintain the overall structure of the mediator complex or bind to Pol II and therefore are required for activation by all activators. In contrast, other mediator subunits are required for normal activation or repression of specific subsets of genes. DNA microarray analysis of yeast gene expression in mutants with defects in these mediator subunits indicates that each such subunit influences transcription of ≈3–10 percent of all genes to the extent that its deletion either increases or decreases mRNA expression by a factor of twofold or more (see Figure 5-29 for DNA microarray technique). These mediator subunits are thought to interact with specific activation domains; thus when one subunit is defective, transcription of genes regulated by activators that bind to that subunit is severely depressed, but transcription of other genes is unaffected. Consistent with this explanation are binding studies showing that some activation domains do indeed interact with specific mediator subunits. However, recent studies suggest that most activation domains may interact with more than one mediator subunit.

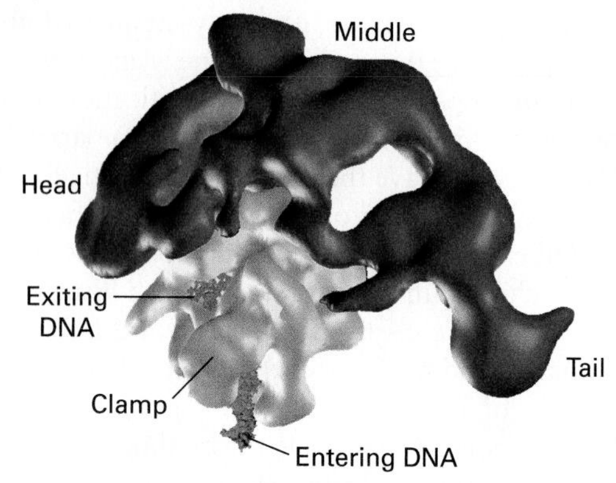

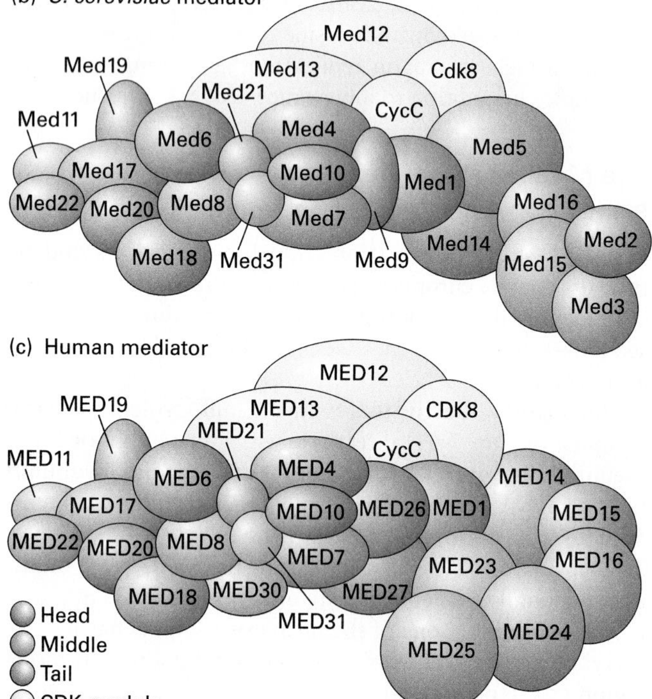

◀ FIGURE 7-41 Structure of yeast and human mediator complexes. (a) Reconstructed image of mediator from *S. cerevisiae* bound to Pol II. Multiple electron microscopy images were aligned and computer-processed to produce this average image in which the three-dimensional Pol II structure (light orange) is shown associated with the yeast mediator complex (dark blue). (b) Diagrammatic representation of mediator subunits from *S. cerevisiae*. Subunits shown in the same color are thought to form a module. Mutations in one subunit of a module may inhibit association of other subunits in the same module with the rest of the complex. (c) Diagrammatic representation of human mediator subunits. The relative position of each human mediator subunit is arbitrary except for the subunits that are homologous to *S. cerevisiae* mediator subunits. [Part (a) from S. Hahn, 2004, *Nat. Struct. Mol. Biol.* **11**:394, based on J. Davis et al., 2002, *Mol. Cell* **10**:409. Part (b) from B. Guglielmi et al., 2004, *Nuc. Acids Res.* **32**:5379. Part (c) adapted from S. Malik and R. G. Roeder, 2005, *Trends Biochem. Sci.* **30**:256.]

Large mediator complexes, isolated from yeast and from cultured mammalian cells, are required for mammalian activators to stimulate transcription by Pol II in vitro. Since genes encoding homologs of mediator subunits are found in the genomes of *C. elegans, Drosophila,* and plants, it appears that most multicellular organisms have homologous mediator complexes. About half of the metazoan (multicellular animals) mediator subunits are clearly homologous to yeast mediator subunits (Figure 7-41b, c). But the remaining subunits, which appear to be distinct from any yeast proteins, may interact with activation domains that are not found in yeast.

The various experimental results indicating that individual mediator subunits bind to specific activation domains suggest that multiple activators influence transcription from a single promoter by interacting with a mediator complex simultaneously (Figure 7-42). Activators bound at enhancers or promoter-proximal elements can interact with mediator associated with a promoter because chromatin, like DNA, is flexible and can form a loop bringing the regulatory regions and the promoter close together, as observed for the *E. coli* NtrC activator and σ^{54}-RNA polymerase (see Figure 7-4). The multiprotein nucleoprotein complexes that form on eukaryotic promoters may comprise as many as 100 polypeptides with a total mass of ≈3 megadaltons (MDa), as large as a ribosome.

Transcription of Many Genes Requires Ordered Binding and Function of Activators and Co-activators

We can now extend the model of Pol II transcription initiation in Figure 7-31 to take into account the role of activators and co-activators. These accessory proteins function not only to make genes within nucleosomal DNA accessible to general transcription factors and Pol II but also directly recruit Pol II to promoter regions.

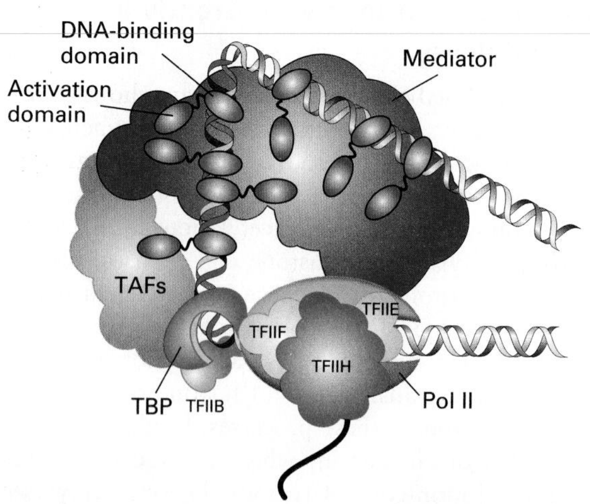

▲ FIGURE 7-42 Model of several DNA-bound activators interacting with a single mediator complex. The ability of different mediator subunits to interact with specific activation domains may contribute to the integration of signals from several activators at a single promoter. See the text for discussion.

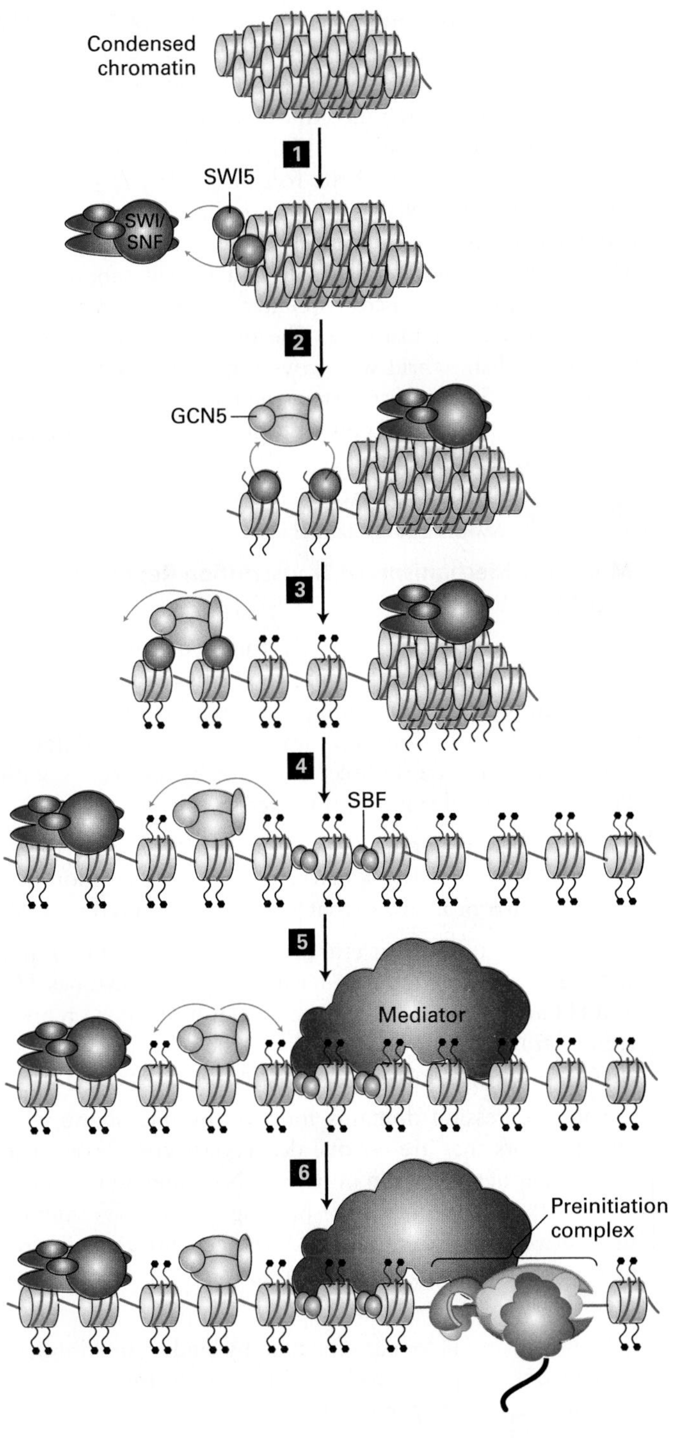

◀ **FIGURE 7-43 Ordered binding and interaction of activators and co-activators leading to transcription of the yeast *HO* gene.** Step **1**: Initially, the *HO* gene is packaged into condensed chromatin. Activation begins when the SWI5 activator binds to enhancer sites 1200–1400 base pairs upstream of the start site and interacts with the SWI/SNF chromatin-remodeling complex. Step **2**: The SWI/SNF complex acts to decondense the chromatin, thereby exposing histone tails. Step **3**: A GCN5-containing histone acetylase complex associates with bound SWI5 and acetylates histone tails in the *HO* locus as SWI/SNF continues to decondense adjacent chromatin. Step **4**: SWI5 is released from the DNA, but the SWI/SNF and GCN5 complexes remain associated with the *HO* control region because of subunits of both complexes that bind acetylated histone tails through bromodomains. Their action allows the SBF activator to bind several sites in the promoter-proximal region. Step **5**: SBF then binds the mediator complex. Step **6**: Subsequent binding of Pol II and general transcription factors results in assembly of a transcription preinitiation complex whose components are detailed in Figure 7-42. [Adapted from C. J. Fry and C. L. Peterson, 2001, *Curr. Biol.* **11:**R185. See also M. P. Cosma et al., 1999, *Cell* **97:**299, and M. P. Cosma et al., 2001, *Mol. Cell* **7:**1213.]

Recent studies have analyzed the order in which activators bind to a transcription-control region and interact with co-activators as a gene is induced. Such studies show that assembly of preinitiation complexes depends on multiple protein-DNA and protein-protein interactions, as illustrated in Figure 7-43, which depicts activation of the yeast *HO* gene. This gene encodes a sequence-specific nuclease that initiates mating-type switching in haploid yeast cells (see Figure 7-33). Activation of the *HO* gene begins with binding of the SWI5 activator to an upstream enhancer. Bound SWI5 then interacts with the SWI/SNF chromatin-remodeling complex and the GCN5-containing SAGA histone acetylase complex. Once the chromatin in the *HO* control region is decondensed and hyperacetylated, a second activator, SBF, can bind to several sites in the promoter-proximal region. Subsequent binding of the mediator complex by SBF then leads to assembly of the transcription preinitiation complex containing Pol II and the general transcription factors shown in Figure 7-31.

We can now see that the assembly of a preinitiation complex and stimulation of transcription at a promoter results from the interaction of several activators with various multiprotein co-activator complexes. These include chromatin-remodeling complexes, histone acetylase complexes, and a mediator complex. Although much remains to be learned about these processes, it is clear that the net result of these multiple molecular events is that activation of transcription at a promoter depends on highly cooperative interactions initiated by several activators. This allows genes to be regulated in a cell-type-specific manner by specific combinations of transcription factors.

The *TTR* gene, which encodes transthyretin in mammals, is a good example of this. As noted earlier, transthyretin is expressed in hepatocytes and in choroid plexus cells. Transcription of the *TTR* gene in hepatocytes is controlled by at least five different transcriptional activators (Figure 7-44). Even though three of these activators—HNF4, C/EBP, and AP1—are also expressed in cells of the intestine and kidney, *TTR* transcription does not occur in these cells, because all five activators are required and HNF1 and HNF3 are missing in kidney and intestine cells. Other hepatocyte-specific enhancers and promoter-proximal regions that regulate

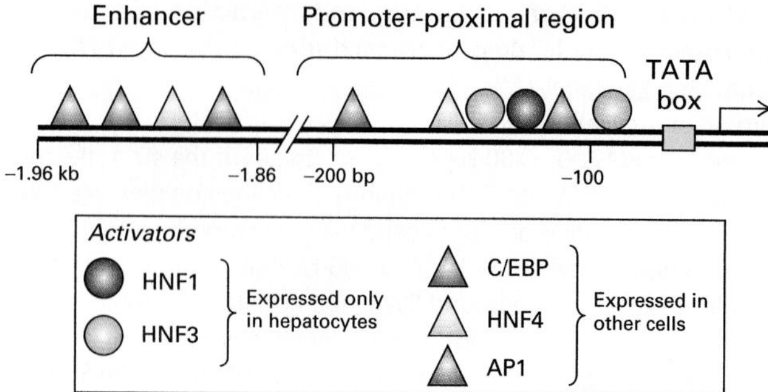

▲ **FIGURE 7-44 Transcription-control region of the mouse transthyretin (*TTR*) gene.** Binding sites for the five activators required for transcription of *TTR* in hepatocytes are indicated. The complete set of activators is expressed at the required concentrations to stimulate transcription only in hepatocytes. A different set of activators stimulates transcription in choroid plexus cells. [See R. Costa et al., 1989, *Mol. Cell Biol.* **9**:1415, and K. Xanthopoulus et al., 1989, *Proc. Nat'l. Acad. Sci. USA* **86**:4117.]

additional genes expressed only in hepatocytes contain binding sites for other specific combinations of transcription factors found only in these cells, together with those expressed ubiquitously.

The Yeast Two-Hybrid System Exploits Activator Flexibility to Detect cDNAs That Encode Interacting Proteins

A powerful molecular genetic method called the *yeast two-hybrid system* exploits the flexibility in activator structures to identify genes whose products bind to a specific protein of interest. Because of the importance of protein-protein interactions in virtually every biological process, the yeast two-hybrid system is used widely in biological research.

This method employs a yeast vector for expressing a DNA-binding domain and flexible linker region without the associated activation domain, such as the deleted GAL4 containing amino acids 1–692 (see Figure 7-21). A cDNA sequence encoding a protein or protein domain of interest, called the *bait domain,* is fused in frame to the flexible linker region so that the vector will express a hybrid protein composed of the DNA-binding domain, linker region, and bait domain (Figure 7-45a, *left*). A cDNA library is cloned into multiple copies of a second yeast vector that encodes a strong activation domain and flexible linker to produce a vector library expressing multiple hybrid proteins, each containing a different *fish domain* (Figure 7-45a, *right*).

The bait vector and library of fish vectors are then transfected into engineered yeast cells in which the only copy of a gene required for histidine synthesis (*HIS*) is under control of a UAS with binding sites for the DNA-binding domain of the hybrid bait protein. Transcription of the *HIS* gene requires activation by proteins bound to the UAS. Transformed cells that express the bait hybrid and an *interacting* fish hybrid will be able to activate transcription of the *HIS* gene (Figure 7-45b). This system works because of the flexibility in the spacing between the DNA-binding and activation domains of eukaryotic activators.

A two-step selection process is used (Figure 7-45c). The bait vector also expresses a wild-type *TRP* gene, and the hybrid vector expresses a wild-type *LEU* gene. Transfected cells are first grown in a medium that lacks tryptophan and leucine but contains histidine. Only cells that have taken up the bait vector and one of the fish plasmids will survive in this medium. The cells that survive then are plated on a medium that lacks histidine. Those cells expressing a fish hybrid that does not bind to the bait hybrid cannot transcribe the *HIS* gene and consequently will not form a colony on medium lacking histidine. The few cells that express a bait-binding fish hybrid will grow and form colonies in the absence of histidine. Recovery of the fish vectors from these colonies yields cDNAs encoding protein domains that interact with the bait domain.

KEY CONCEPTS OF SECTION 7.6

Molecular Mechanisms of Transcription Repression and Activation

- Eukaryotic transcription activators and repressors exert their effects largely by binding to multisubunit co-activators or co-repressors that influence assembly of Pol II transcription preinitiation complexes either by modulating chromatin structure (indirect effect) or by interacting with Pol II and general transcription factors (direct effect).
- The DNA in condensed regions of chromatin (heterochromatin) is relatively inaccessible to transcription factors and other proteins, so that gene expression is repressed.
- The interactions of several proteins with each other and with the hypoacetylated N-terminal tails of histones H3 and H4 are responsible for the chromatin-mediated repression of transcription that occurs in the telomeres and the silent mating-type loci in *S. cerevisiae* (see Figure 7-35).
- Some repression domains function by interacting with co-repressors that are histone deacetylase complexes. The subsequent deacetylation of histone N-terminal tails in nucleosomes near the repressor-binding site inhibits interaction between the promoter DNA and general transcription factors, thereby repressing transcription initiation (see Figure 7-38a).
- Some activation domains function by binding multiprotein co-activator complexes such as histone acetylase complexes. The subsequent hyperacetylation of histone N-terminal tails in nucleosomes near the activator-binding site facilitates interactions between the promoter DNA and general transcription factors, thereby stimulating transcription initiation (see Figure 7-38b).
- SWI/SNF chromatin-remodeling factors constitute another type of co-activator. These multisubunit complexes can transiently dissociate DNA from histone cores in an ATP-dependent reaction and may also decondense regions of chromatin, thereby promoting the binding of DNA-binding proteins needed for initiation to occur at some promoters.
- Mediator, another type of co-activator, is an ≈30-subunit complex that forms a molecular bridge between

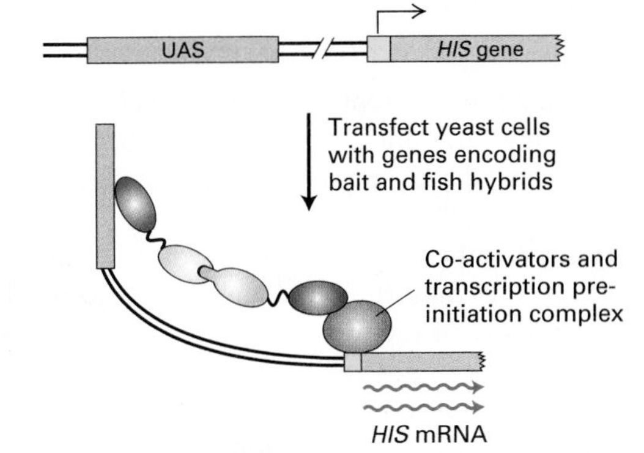

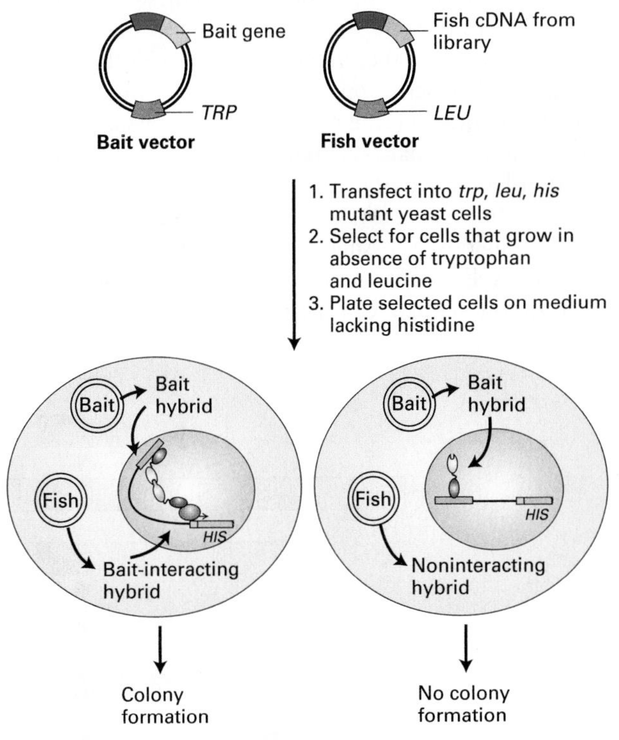

▶ EXPERIMENTAL FIGURE 7-45 **The yeast two-hybrid system provides a way of screening a cDNA library for clones encoding proteins that interact with a specific protein of interest.** This is a common technique for screening a cDNA library for clones encoding proteins that interact with a specific protein of interest. (a) Two vectors are constructed containing genes that encode hybrid (chimeric) proteins. In one vector (*left*), the coding sequence for the DNA-binding domain of a transcription factor is fused to the sequences for a known protein, referred to as the "bait" domain (light blue). The second vector (*right*) expresses an activation domain fused to a "fish" domain (green) that interacts with the bait domain. (b) If yeast cells are transformed with vectors expressing both hybrids, the bait and fish portions of the chimeric proteins interact to produce a functional transcriptional activator. In this example, the activator promotes transcription of a *HIS* gene. One end of this protein complex binds to the upstream activating sequence (UAS) of the *HIS3* gene; the other end, consisting of the activation domain, stimulates assembly of the transcription preinitiation complex (orange) at the promoter (yellow). (c) To screen a cDNA library for clones encoding proteins that interact with a particular bait protein of interest, the library is cloned into the vector encoding the activation domain so that hybrid proteins are expressed. The bait vector and fish vectors contain wild-type selectable genes (e.g., a *TRP* or *LEU* gene). The only transformed cells that survive the indicated selection scheme are those that express the bait hybrid and a fish hybrid that interacts with it. See the text for discussion. [See S. Fields and O. Song, 1989, *Nature* **340**:245.]

activation domains and RNA polymerase II by binding directly to the polymerase and activation domains. By binding to several different activators simultaneously, mediator probably helps integrate the effects of multiple activators on a single promoter (see Figure 7-42).

- Activators bound to a distant enhancer can interact with transcription factors bound to a promoter because DNA is flexible and the intervening DNA can form a large loop.
- The highly cooperative assembly of preinitiation complexes in vivo generally requires several activators. A cell must produce the specific set of activators required for transcription of a particular gene in order to express that gene.
- The yeast two-hybrid system is widely used to detect cDNAs encoding protein domains that bind to a specific protein of interest (see Figure 7-45).

7.7 Regulation of Transcription-Factor Activity

We have seen in the preceding discussion how combinations of activators and repressors that bind to specific DNA regulatory sequences control transcription of eukaryotic genes. Whether or not a specific gene in a multicellular organism is expressed in a particular cell at a particular time is largely a consequence of the nuclear concentrations and activities of the transcription factors that interact with the regulatory sequences of that gene. Which transcription factors are expressed in a particular cell type, and the amounts produced, are determined by multiple regulatory interactions between transcription-factor genes that occur during the development and differentiation of a particular cell type. In

► **FIGURE 7-46 Examples of hormones that bind to nuclear receptors.** These and related lipid-soluble hormones bind to receptors located in the cytosol or nucleus. The ligand-receptor complex functions as a transcription activator.

Cortisol

Retinoic acid

Thyroxine

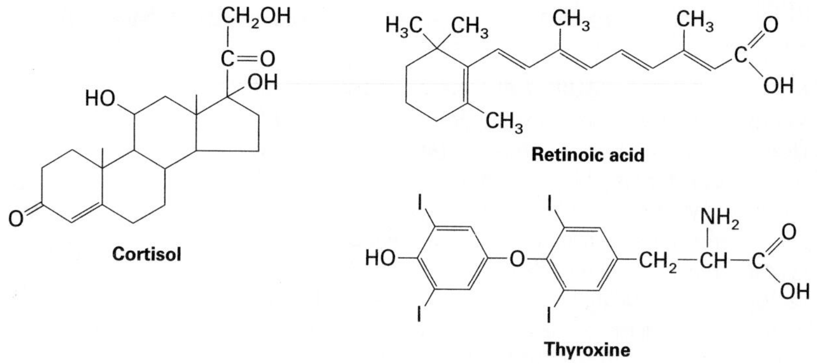

Chapters 16 and 22, we present examples of such regulatory interactions during development and discuss the principles of development and differentiation that have emerged from these examples.

In addition to controlling the expression of hundreds to thousands of specific transcription factors, cells also regulate the activities of many of the transcription factors expressed in a particular cell type. For example, transcription factors are often regulated in response to extracellular signals. Interactions between the extracellular domains of transmembrane receptor proteins on the surface of the cell and specific protein ligands for these receptors activate protein domains associated with the intracellular domains of these transmembrane proteins, transducing the signal received on the outside of the cell to a signal on the inside of the cell that eventually reaches transcription factors in the nucleus. In Chapter 16, we describe the major types of cell-surface receptors and intracellular signaling pathways that regulate transcription-factor activity.

In this section, we discuss the second major group of extracellular signals, the small, lipid-soluble hormones—including many different steroid hormones, retinoids, and thyroid hormones—that can diffuse through plasma and nuclear membranes and interact directly with the transcription factors they control (Figure 7-46). As noted earlier, the intracellular receptors for most of these lipid-soluble hormones, which constitute the *nuclear-receptor superfamily*, function as transcription activators when bound to their ligands.

All Nuclear Receptors Share a Common Domain Structure

Cloning and sequencing of the genes encoding various nuclear receptors revealed a remarkable conservation in their amino acid sequences and three functional regions (Figure 7-47). All the nuclear receptors have a unique N-terminal region of variable length (100–500 amino acids). Portions of this variable region function as activation domains in some nuclear receptors. The DNA-binding domain maps near the center of the primary sequence and has a repeat of the C_4 zinc-finger motif. The hormone-binding domain, located near the C-terminal end, contains a hormone-dependent activation domain. In some nuclear receptors, the hormone-binding domain functions as a repression domain in the absence of ligand.

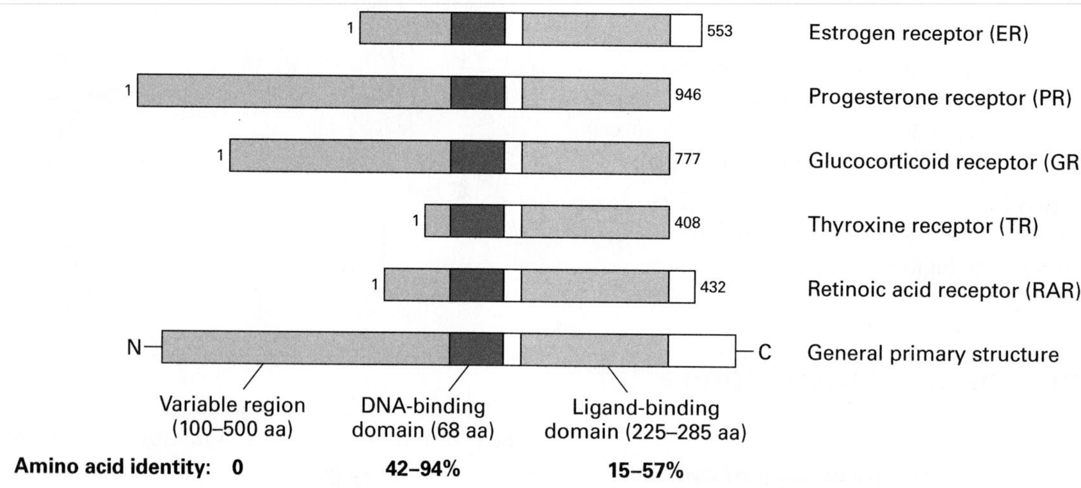

▲ **FIGURE 7-47 General design of transcription factors in the nuclear-receptor superfamily.** The centrally located DNA-binding domain exhibits considerable sequence homology among different receptors and contains two copies of the C_4 zinc-finger motif. The C-terminal hormone-binding domain exhibits somewhat less homology. The N-terminal regions in various receptors vary in length, have unique sequences, and may contain one or more activation domains. [See R. M. Evans, 1988, *Science* **240:**889.]

Nuclear-Receptor Response Elements Contain Inverted or Direct Repeats

The characteristic nucleotide sequences of the DNA sites, called *response elements,* that bind several nuclear receptors have been determined. The sequences of the consensus response elements for the glucocorticoid and estrogen receptors are 6-bp inverted repeats separated by any three base pairs (Figure 7-48a, b). This finding suggested that the cognate steroid hormone receptors would bind to DNA as symmetrical dimers, as was later shown from the x-ray crystallographic analysis of the homodimeric glucocorticoid receptor's C_4 zinc-finger DNA-binding domain (see Figure 7-25c).

Some nuclear-receptor response elements, such as those for the receptors that bind vitamin D_3, thyroid hormone, and retinoic acid, are direct repeats of the same sequence recognized by the estrogen receptor, separated by three to five base pairs (Figure 7-48c–e). The specificity for responding to these different hormones by binding distinct receptors is determined by the spacing between the repeats. The receptors that bind to such direct-repeat response elements do so as heterodimers with a common nuclear-receptor monomer called RXR. The vitamin D_3 response element, for example, is bound by the RXR-VDR heterodimer, and the retinoic acid response element is bound by RXR-RAR. The monomers composing these heterodimers interact with each other in such a way that the two DNA-binding domains lie in the same rather than inverted orientation, allowing the RXR heterodimers to bind to direct repeats of the binding site for each monomer. In contrast, the monomers in homodimeric nuclear receptors (e.g., GRE and ERE) have an inverted orientation.

(a) **GRE**
5′ AGAACA(N)$_3$TGTTCT 3′
3′ TCTTGT(N)$_3$ACAAGA 5′

(b) **ERE**
5′ AGGTCA(N)$_3$TGACCT 3′
3′ TCCAGT(N)$_3$ACTGGA 5′

(c) **VDRE**
5′ AGGTCA(N)$_3$AGGTCA 3′
3′ TCCAGT(N)$_3$TCCAGT 5′

(d) **TRE**
5′ AGGTCA(N)$_4$AGGTCA 3′
3′ TCCAGT(N)$_4$TCCAGT 5′

(e) **RARE**
5′ AGGTCA(N)$_5$AGGTCA 3′
3′ TCCAGT(N)$_5$TCCAGT 5′

▲ **FIGURE 7-48 Consensus sequences of DNA response elements that bind three nuclear receptors.** The response elements for the glucocorticoid receptor (GRE) and estrogen receptor (ERE) contain inverted repeats that bind these homodimeric proteins. The response elements for heterodimeric receptors contain a common direct repeat separated by three to five base pairs for the vitamin D_3 receptor (VDRE), thyroid hormone receptor (TRE), and retinoic acid receptor (RARE). The repeat sequences are indicated by red arrows. [See K. Umesono et al., 1991, *Cell* **65**:1255, and A. M. Naar et al., 1991, *Cell* **65**:1267.]

Hormone Binding to a Nuclear Receptor Regulates Its Activity as a Transcription Factor

The mechanism whereby hormone binding controls the activity of nuclear receptors differs for heterodimeric and homodimeric receptors. Heterodimeric nuclear receptors (e.g., RXR-VDR, RXR-TR, and RXR-RAR) are located exclusively in the nucleus. In the absence of their hormone ligand, they repress transcription when bound to their cognate sites in DNA. They do so by directing histone deacetylation at nearby nucleosomes by the mechanism described earlier (see Figure 7-38a). In the ligand-bound conformation, heterodimeric nuclear receptors containing RXR can direct hyperacetylation of histones in nearby nucleosomes, thereby reversing the repressing effects of the free ligand-binding domain. In the presence of ligand, ligand-binding domains of nuclear receptors also bind mediator, stimulating preinitiation complex assembly.

In contrast to heterodimeric nuclear receptors, homodimeric receptors are found in the cytoplasm in the absence of their ligands. Hormone binding to these receptors leads to their translocation to the nucleus. The hormone-dependent translocation of the homodimeric glucocorticoid receptor (GR) was demonstrated in the transfection experiments shown in Figure 7-49. The GR hormone-binding domain alone mediates this transport. Subsequent studies showed that, in the absence of hormone, GR is anchored in the cytoplasm as a large protein aggregate complexed with inhibitor proteins, including Hsp90, a protein related to Hsp70, the major heat-shock chaperone in eukaryotic cells. As long as the receptor is confined to the cytoplasm, it cannot interact with target genes and hence cannot activate transcription. Hormone binding to a homodimeric nuclear receptor releases the inhibitor proteins, allowing the receptor to enter the nucleus, where it can bind to response elements associated with target genes (Figure 7-50). Once the receptor with bound hormone binds to a response element, it activates transcription by interacting with chromatin-remodeling and histone acetylase complexes and mediator.

KEY CONCEPTS OF SECTION 7.7

Regulation of Transcription-Factor Activity

- The activities of many transcription factors are indirectly regulated by binding of extracellular proteins and peptides to cell-surface receptors. These receptors activate intracellular signal-transduction pathways that regulate specific transcription factors through a variety of mechanisms discussed in Chapter 16.
- Nuclear receptors constitute a superfamily of dimeric C_4 zinc-finger transcription factors that bind lipid-soluble hormones and interact with specific response elements in DNA (see Figure 7-47).
- Hormone binding to nuclear receptors induces conformational changes that modify their interactions with other proteins.

MEDIA CONNECTIONS

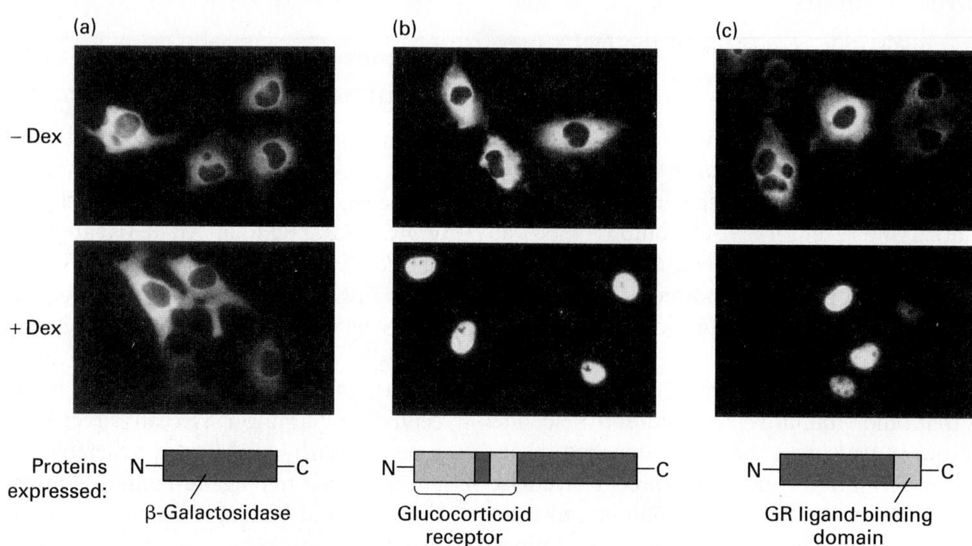

▲ EXPERIMENTAL FIGURE 7-49 **Fusion proteins from expression vectors demonstrate that the hormone-binding domain of the glucocorticoid receptor (GR) mediates translocation to the nucleus in the presence of hormone.** Cultured animal cells were transfected with expression vectors encoding the proteins diagrammed at the bottom. Immunofluorescence with a labeled antibody specific for β-galactosidase was used to detect the expressed proteins in transfected cells. (a) In cells that expressed β-galactosidase alone, the enzyme was localized to the cytoplasm in the presence and absence of the glucocorticoid hormone dexamethasone (Dex). (b) In cells that expressed a fusion protein consisting of β-galactosidase and the entire glucocorticoid receptor (GR), the fusion protein was present in the cytoplasm in the absence of hormone but was transported to the nucleus in the presence of hormone. (c) Cells that expressed a fusion protein composed of β-galactosidase and just the GR ligand-binding domain (light purple) also exhibited hormone-dependent transport of the fusion protein to the nucleus. [From D. Picard and K. R. Yamamoto, 1987, *EMBO J.* **6**:3333; courtesy of the authors.]

- Heterodimeric nuclear receptors (e.g., those for retinoids, vitamin D, and thyroid hormone) are found only in the nucleus. In the absence of hormone, they repress transcription of target genes with the corresponding response element. When bound to their ligands, they activate transcription.
- Steroid hormone receptors are homodimeric nuclear receptors. In the absence of hormone, they are trapped in the cytoplasm by inhibitor proteins. When bound to their ligands, they can translocate to the nucleus and activate transcription of target genes (see Figure 7-50).

7.8 Regulated Elongation and Termination of Transcription

In eukaryotes, the mechanisms for terminating transcription differ for each of the three RNA polymerases. Transcription of pre-rRNA genes by RNA polymerase I is terminated by a mechanism that requires a polymerase-specific termination factor. This DNA-binding protein binds to a specific DNA sequence downstream of the transcription unit. Efficient termination requires that the termination factor bind to the template DNA in the correct orientation. Purified RNA

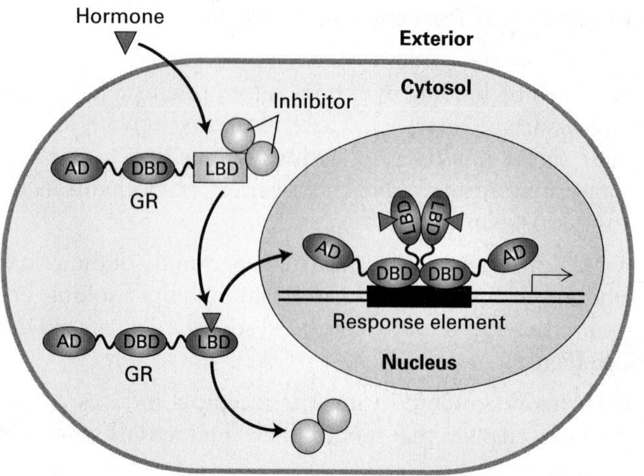

◄ **FIGURE 7-50 Model of hormone-dependent gene activation by a homodimeric nuclear receptor.** In the absence of hormone, the receptor is kept in the cytoplasm by interaction between its ligand-binding domain (LBD) and inhibitor proteins. When hormone is present, it diffuses through the plasma membrane and binds to the ligand-binding domain, causing a conformational change that releases the receptor from the inhibitor proteins. The receptor with bound ligand is then translocated into the nucleus, where its DNA-binding domain (DBD) binds to response elements, allowing the ligand-binding domain and an additional activation domain (AD) at the N-terminus to stimulate transcription of target genes.

polymerase III terminates after polymerizing a series of U residues. The deoxy$(A)_n$-ribo$(U)_n$ DNA-RNA hybrid that results when a stretch of U's are synthesized is particularly unstable compared with all other base-paired sequences. The ease with which this hybrid can be melted probably contributes to the mechanism of termination by RNA polymerase III.

In most mammalian protein-coding genes transcribed by RNA polymerase II, once the polymerase has transcribed beyond about 50 bases, further elongation is highly processive and does not terminate until after a sequence is transcribed that directs cleavage and polyadenylation of the RNA at the sequence that forms the 3′ end of the encoded mRNA. RNA polymerase II then can terminate at multiple sites located over a distance of 0.5–2 kb beyond this poly(A) addition site. Experiments with mutant genes show that termination is coupled to the process that cleaves and polyadenylates the 3′ end of a transcript, which is discussed in the next chapter. Biochemical and chromatin immunoprecipitation experiments suggest that the protein complex that cleaves and polyadenylates the nascent mRNA transcript at specific sequences associates with the phosphorylated carboxyl-terminal domain (CTD) of RNA polymerase II following initiation (see Figure 7-31). This cleavage/polyadenylation complex may suppress termination by RNA polymerase II until the sequence signaling cleavage and polyadenylation is transcribed by the polymerase.

Although transcription termination is unregulated for most genes, for some specific genes, a choice is made between elongation and termination or pausing within a few tens of bases from the transcription start site. This choice between elongation and termination or pausing can be regulated; thus expression of the encoded protein is controlled not only by transcription initiation but also by control of transcription elongation early in the transcription unit. We discuss two examples of such regulation next.

Transcription of the HIV Genome Is Regulated by an Antitermination Mechanism

Currently, transcription of the human immunodeficiency virus (HIV) genome by RNA polymerase II provides the best-understood example of regulated transcription termination in eukaryotes. Efficient expression of HIV genes requires a small viral protein encoded at the *tat* locus. Cells infected with *tat*$^-$ mutants produce short viral transcripts that hybridize to restriction fragments containing promoter-proximal regions of the HIV DNA but not to restriction fragments farther downstream from the promoter. In contrast, cells infected with wild-type HIV synthesize long viral transcripts that hybridize to restriction fragments throughout the single HIV transcription unit. Thus Tat protein functions as an *antitermination factor,* permitting RNA polymerase II to read through a transcriptional block. Since antitermination by Tat protein is required for HIV replication, further understanding of this gene-control mechanism may offer possibilities for designing effective therapies for acquired immunodeficiency syndrome (AIDS).

Tat is a sequence-specific RNA-binding protein. It binds to the RNA copy of a sequence called TAR, which is located near the 5′ end of the HIV transcript. The TAR sequence folds into an RNA hairpin with a bulge in the middle of the stem (Figure 7-51). TAR contains two binding sites: one that interacts with Tat and one that interacts with a cellular protein called *cyclin T.* As depicted in Figure 7-51, the HIV Tat protein and cellular cyclin T each bind to TAR RNA and also interact directly with each other so that they bind cooperatively, much like the cooperative binding of DNA-binding transcription factors (see Figure 7-29). Interaction of cyclin T with a protein kinase called CDK9 activates the kinase, whose substrate is the CTD of RNA polymerase II. In vitro transcription studies using a specific inhibitor of CDK9 suggest that RNA polymerase II molecules that initiate transcription on the HIV promoter terminate after transcribing ≈50 bases unless the CTD is hyperphosphorylated by CDK9. Cooperative binding of cyclin T and Tat to the TAR sequence at the 5′ end of the HIV transcript positions CDK9 so that it can phosphorylate the CTD, thereby preventing termination and permitting the polymerase to continue chain elongation.

Several additional cellular proteins, including Spt4 and Spt5 and the NELF complex, participate in the process by which HIV Tat controls elongation versus termination (Figure 7-51). Experiments with the specific inhibitor of CDK9 mentioned above and with *spt4* and *spt5* yeast mutants indicate that these cellular proteins are required for transcription elongation beyond ≈50 bases for most cellular genes. But for most genes, these proteins appear to function constitutively, that is, without being regulated. As discussed in Chapter 8, RNA polymerase II pausing instigated by Spt4/5 and NELF is thought to delay elongation until mRNA processing factors associate with the phosphorylated CTD. Further phosphorylation of the CTD by cyclin T–CDK9 (also known

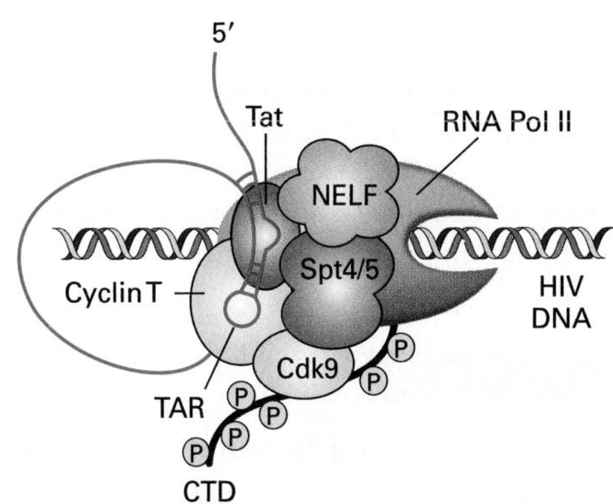

▲ **FIGURE 7-51 Model of antitermination complex composed of HIV Tat protein and several cellular proteins.** The TAR element in the HIV transcript contains sequences recognized by Tat and the cellular protein cyclin T. Cyclin T activates and helps position the protein kinase CDK9 near its substrate, the CTD of RNA polymerase II. CTD phosphorylation prevents termination and allows transcription to continue. Cellular proteins Spt4 and Spt5 and the NELF complex are also involved in regulating HIV transcript termination. [See P. Wei et al., 1998, *Cell* **92**:451; T. Wada et al., 1998, *Genes Dev.* **12**:357; and Y. Yamaguchi et al., 1999, *Cell* **97**:41.]

as pTEFb) appears to reverse this pause and allow elongation to continue. Currently, it is not clear why this process is not constitutive for the HIV promoter, where cooperative binding of HIV Tat and cyclin T to the TAR RNA sequence is required for CDK9 activation and efficient elongation.

Promoter-Proximal Pausing of RNA Polymerase II Occurs in Some Rapidly Induced Genes

The *heat-shock genes* (e.g., *hsp70*) illustrate another mechanism for regulating RNA chain elongation in eukaryotes. During transcription of these genes, RNA polymerase II pauses after transcribing ≈25 nucleotides but does not terminate transcription (as it does when transcribing the HIV genome in the absence of Tat protein). The paused polymerase remains associated with the nascent RNA and template DNA and then, after a few minutes, continues transcription of the gene. As the first polymerase transcribes away from the promoter region, another RNA polymerase II binds to the promoter, initiates transcription, and pauses after transcribing ≈25 nucleotides, waiting several minutes before the process is repeated. When heat shock occurs, the heat-shock transcription factor (HSTF) is activated. Subsequent binding of activated HSTF to specific sites in the promoter-proximal region of heat-shock genes stimulates the paused polymerase to continue chain elongation and promotes rapid re-initiation by additional RNA polymerase II molecules, leading to many transcription initiations per minute.

The pausing during transcription of heat-shock genes initially was discovered in *Drosophila,* but a similar mechanism has been shown to occur in human cells. Heat-shock genes are induced by intracellular conditions that denature proteins (such as elevated temperature, "heat shock"). Some encode proteins that are relatively resistant to denaturing conditions and act to protect other proteins from denaturation; others are chaperonins that refold denatured proteins (Chapter 3). The mechanism of transcriptional control that evolved to regulate expression of these genes permits a rapid response: these genes are always paused in a state of suspended transcription and therefore, when an emergency arises, require no time to remodel and acetylate chromatin over the promoter and assemble a transcription preinitiation complex.

KEY CONCEPTS OF SECTION 7.8

Regulated Elongation and Termination of Transcription

- Different mechanisms of transcription termination are employed by each of the eukaryotic nuclear RNA polymerases. Transcription of most protein-coding genes is not terminated until an RNA sequence is synthesized that specifies a site of RNA cleavage and polyadenylation.
- Transcription of the HIV genome by RNA polymerase II is regulated by an antitermination mechanism that requires cooperative binding by the virus-encoded Tat protein and cyclin T to the *TAR* sequence near the 5′ end of the HIV RNA.
- During transcription of *Drosophila* heat-shock genes, RNA polymerase II pauses within the downstream promoter-proximal region; this interruption in transcription is released when the HSTF transcription factor is activated, resulting in very rapid transcription of the heat-shock genes in response to the accumulation of denatured proteins.

7.9 Other Eukaryotic Transcription Systems

We conclude this chapter with a brief discussion of transcription initiation by the other two eukaryotic nuclear RNA polymerases, Pol I and Pol III, and by the distinct polymerases that transcribe mitochondrial and chloroplast DNA. Although these systems, particularly their regulation, are less thoroughly understood than transcription by RNA polymerase II, they are equally as fundamental to the life of eukaryotic cells.

Transcription Initiation by Pol I and Pol III Is Analogous to That by Pol II

The formation of transcription-initiation complexes involving Pol I and Pol III is similar in some respects to assembly of Pol II initiation complexes (see Figure 7-31). However, each of the three eukaryotic nuclear RNA polymerases requires its own polymerase-specific general transcription factors and recognizes different DNA control elements. Moreover, neither Pol I nor Pol III requires ATP hydrolysis to initiate transcription, whereas Pol II does.

Transcription initiation by Pol I, which synthesizes pre-rRNA, and by Pol III, which synthesizes tRNAs, 5S rRNA, and other short, stable RNAs (see Table 7-2), has been characterized most extensively in *S. cerevisiae* using both biochemical and genetic approaches. It is clear that synthesis of tRNAs and of rRNAs, which are incorporated into ribosomes, is tightly coupled to the rate of cell growth and proliferation. However, much remains to be learned about how transcription initiation by Pol I and Pol III is regulated so that synthesis of pre-rRNA, 5S rRNA, and tRNAs is coordinated with the growth and replication of cells.

Initiation by Pol I The regulatory elements directing Pol I initiation are similarly located relative to the transcription start site in both yeast and mammals. A *core element* spanning the transcription start site from −40 to +5 is essential for Pol I transcription. An additional *upstream element* extending from roughly −155 to −60 stimulates in vitro Pol I transcription tenfold.

Assembly of a fully active Pol I initiation complex begins with binding of a multimeric upstream activating factor (UAF) to the upstream element (Figure 7-52). Two of the six subunits composing UAF are histones, which probably participate in DNA binding. Next, a trimeric core factor binds to the core element together with TBP, which makes contact with both the bound UAF and the core factor. Finally, a preformed complex of Pol I and Rrn3p associates with the bound proteins, positioning Pol I near the start site. In human cells, TBP is stably

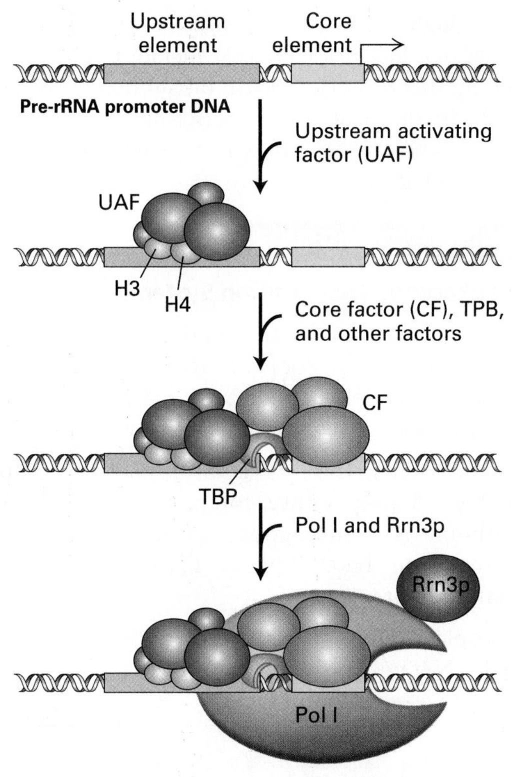

▲ **FIGURE 7-52 In vitro assembly of the yeast Pol I transcription initiation complex.** UAF and CF, both multimeric general transcription factors, bind to the upstream element (UE) and core element, respectively, in the promoter DNA. TBP and a monomeric factor (Rrn3p) associated with RNA polymerase I (Pol I) also participate in forming the initiation complex. [Adapted from N. Nomura, 1998, in R. M. Paule, *Transcription of Ribosomal RNA Genes by Eukaryotic RNA Polymerase I*, Landes Bioscience, pp. 157–172.]

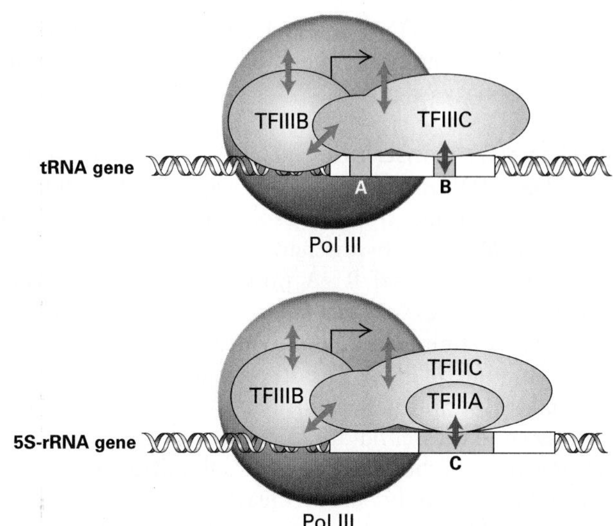

▲ **FIGURE 7-53 Transcription-control elements in genes transcribed by RNA polymerase III.** Both tRNA and 5S-rRNA genes contain internal promoter elements (yellow) located downstream from the start site and named A, B, and C boxes, as indicated. Assembly of transcription initiation complexes on these genes begins with the binding of Pol-III-specific general transcription factors TFIIIA, TFIIIB, and TFIIIC to these control elements. Green arrows indicate strong, sequence-specific protein-DNA interactions. Blue arrows indicate interactions between general transcription factors. Purple arrows indicate interactions between general transcription factors and Pol III. [From L. Schramm and N. Hernandez, 2002, *Genes Dev.* **16**:2593.]

bound to three other polypeptides, forming an initiation factor called *SL1* that binds to the core promoter element and is functionally equivalent to yeast core factor plus TBP.

Initiation by Pol III Unlike protein-coding genes and pre-rRNA genes, the promoter regions of tRNA and 5S-rRNA genes lie entirely within the transcribed sequence (Figure 7-53). Two such *internal* promoter elements, termed the *A box* and *B box*, are present in all tRNA genes. These highly conserved sequences not only function as promoters but also encode two invariant portions of eukaryotic tRNAs that are required for protein synthesis. In 5S-rRNA genes, a single internal control region, the *C box*, acts as a promoter.

Three general transcription factors are required for Pol III to initiate transcription of tRNA and 5S-rRNA genes in vitro. Two multimeric factors, TFIIIC and TFIIIB, participate in initiation at both tRNA and 5S-rRNA promoters; a third factor, TFIIIA, is required for initiation at 5S-rRNA promoters. As with assembly of Pol I and Pol II initiation complexes, the Pol III general transcription factors bind to promoter DNA in a defined sequence.

The N-terminal half of one TFIIIB subunit, called *BRF* (for TFII*B*-*r*elated *f*actor), is similar in sequence to TFIIB (a Pol II factor). This similarity suggests that BRF and TFIIB perform a similar function in initiation, namely, to direct the polymerase to the correct start site. Once TFIIIB has bound to either a tRNA or 5S-rRNA gene, Pol III can bind and initiate transcription in the presence of ribonucleoside triphosphates. The BRF subunit of TFIIIB interacts specifically with one of the polymerase subunits unique to Pol III, accounting for initiation by this specific nuclear RNA polymerase.

Another of the three subunits composing TFIIIB is TBP, which we can now see is a component of a general transcription factor for all three eukaryotic nuclear RNA polymerases. The finding that TBP participates in transcription initiation by Pol I and Pol III was surprising, since the promoters recognized by these enzymes often do not contain TATA boxes. Nonetheless, recent studies indicate that the TBP subunit of TFIIIB interacts with DNA similarly to the way it interacts with TATA boxes.

Mitochondrial and Chloroplast DNAs Are Transcribed by Organelle-Specific RNA Polymerases

As discussed in Chapter 6, mitochondria and chloroplasts probably evolved from bacteria that were endocytosed into ancestral cells containing a eukaryotic nucleus. In modern-day eukaryotes, both organelles contain distinct DNAs that encode some of the proteins essential to their specific functions. Interestingly, the RNA polymerases that transcribe mitochondrial (mt) DNA and chloroplast DNA are similar to polymerases from bacteria and bacteriophages, reflecting their evolutionary origins.

Mitochondrial Transcription The RNA polymerase that transcribes mtDNA is encoded in nuclear DNA. After synthesis of the enzyme in the cytosol, it is imported into the mitochondrial matrix by mechanisms described in Chapter 13. The mitochondrial RNA polymerases from *S. cerevisiae* and the frog *Xenopus laevis* both consist of a large subunit with ribonucleotide-polymerizing activity and a small B subunit (TFBM). Another matrix protein, mitochondrial transcription factor A (TFAM), binds to mtDNA promoters and is essential for initiating transcription at the start sites used in the cell. The large subunit of yeast mitochondrial RNA polymerase clearly is related to the monomeric RNA polymerases of bacteriophage T7 and similar bacteriophages. However, the mitochondrial enzyme is functionally distinct from the bacteriophage enzyme in its dependence on two other polypeptides for transcription from the proper start sites.

The promoter sequences recognized by mitochondrial RNA polymerases include the transcription start site. These promoter sequences, which are rich in A residues, have been characterized in the mtDNA from yeast, plants, and animals. The circular, human mitochondrial genome contains two related 15-bp promoter sequences, one for the transcription of each strand. Each strand is transcribed in its entirety; the long primary transcripts are then processed to yield mitochondrial mRNAs, rRNAs, and tRNAs. A second promoter appears to be responsible for transcribing additional copies of the rRNAs. Currently, there is relatively little understanding of how transcription of the mitochondrial genome is regulated to coordinate the production of the few mitochondrial proteins it encodes with synthesis and import of the thousands of nuclear DNA-encoded proteins that comprise the mitochondria.

Chloroplast Transcription Chloroplast DNA is transcribed by two types of RNA polymerases, one multisubunit protein similar to bacterial RNA polymerases and one similar to the single subunit enzymes of bacteriophage and mitochondria. The core subunits of the bacterial-type enzyme, α, β, β', and ω subunits, are encoded in the chloroplast DNAs of higher plants, whereas six σ^{70}-like σ factors are encoded in the nuclear DNA of higher plants. This is another example of the transfer of genes from organellar genomes to nuclear genomes during evolution. In this case, genes encoding the regulatory transcription initiation factors have been transferred to the nucleus, where the control of their transcription by nuclear RNA polymerase II likely indirectly controls the expression of sets of chloroplast genes. The bacterial-like chloroplast RNA polymerase is called the plastid polymerase since its catalytic core is encoded by the chloroplast genome. Most chloroplast genes are transcribed by these enzymes and have −35 and −10 control regions similar to promoters in cyanobacteria, from which they evolved. The chloroplast T7–like RNA polymerase is also encoded in the nuclear genome of higher plants. It transcribes a different set of chloroplast genes. Curiously, this includes genes encoding subunits of the bacterial-like multisubunit plastid polymerase. As for mitochondrial transcription, currently relatively little is known about how transcription of chloroplast DNA is regulated.

KEY CONCEPTS OF SECTION 7.9

Other Eukaryotic Transcription Systems

- The process of transcription initiation by Pol I and Pol III is similar to that by Pol II but requires different general transcription factors, is directed by different promoter elements, and does not require ATP hydrolysis.
- Mitochondrial DNA is transcribed by a nuclear-encoded RNA polymerase composed of two subunits. One subunit is homologous to the monomeric RNA polymerase from bacteriophage T7; the other resembles bacterial σ factors.
- Chloroplast DNA is transcribed by a chloroplast-encoded RNA polymerase homologous to bacterial RNA polymerases, except that it lacks a σ factor.

Perspectives for the Future

A great deal has been learned in recent years about transcription control in eukaryotes. Genes encoding about 2000 activators and repressors can be recognized in the human genome. We now have a glimpse of how the astronomical number of possible combinations of these transcription factors can generate the complexity of gene control required to produce organisms as remarkable as those we see around us. But very much remains to be understood. Although we now have some understanding of what processes turn a gene on and off, we have very little understanding of how the frequency of transcription is controlled in order to provide a cell with the appropriate amounts of its various proteins. In a red blood cell precursor, for example, the globin genes are transcribed at a far greater rate than the genes encoding the enzymes of intermediary metabolism (the so-called housekeeping genes). How are the vast differences in the frequency of transcription initiation at various genes achieved? What happens to the multiple interactions between activation domains, co-activator complexes, general transcription factors, and RNA polymerase II when the polymerase initiates transcription and transcribes away from the promoter region? Do these completely dissociate at promoters that are transcribed infrequently, so that the combination of multiple factors required for transcription must be reassembled anew for each round of transcription? Do complexes of activators with their multiple interacting co-activators remain assembled at promoters from which re-initiation takes place at a high rate, so that the entire assembly does not have to be reconstructed each time a polymerase initiates?

Much remains to be learned about the structure of chromatin and how that structure influences transcription. What additional components besides HP1 and methylated histone H3 lysine 9 are required to direct certain regions of chromatin to form heterochromatin, where transcription is repressed? Precisely how is the structure of chromatin changed by activators and repressors, and how does this promote or inhibit transcription? Once chromatin-remodeling complexes and histone acetylase complexes become associated with a promoter region, how do they remain associated? Current models suggest that certain subunits of these complexes associate with modified histone tails so that the combination of binding to a specific histone tail modification plus modification of neighboring histone tails in the same way results in retention of the modifying complex at an activated promoter region. In some cases, this type of assembly mechanism causes the complexes to spread along the length of a chromatin fiber. What controls when such complexes spread and how far they will spread?

Single activation domains have been discovered to interact with several co-activator complexes. Are these interactions transient, so that the same activation domain can interact with several co-activators sequentially? Is a specific order of co-activator interaction required? How does the interaction of activation domains with mediator stimulate transcription? Do these interactions simply stimulate the assembly of a preinitiation complex, or do they also influence the rate at which RNA polymerase II initiates transcription from an assembled preinitiation complex?

Transcriptional activation is a highly cooperative process so that genes expressed in a specific type of cell are expressed only when the complete set of activators that control that gene are expressed and activated. As mentioned earlier, some of the transcription factors that control expression of the *TTR* gene in the liver are also expressed in intestinal and kidney cells. Yet the *TTR* gene is not expressed in these other tissues, since its transcription requires two additional transcription factors expressed only in the liver. What mechanisms account for this highly cooperative action of transcription factors that is critical to cell-type-specific gene expression?

A thorough understanding of normal development and of abnormal processes associated with disease will require answers to these and many related questions. As further understanding of the principles of transcription control are discovered, applications of the knowledge will likely be made. This understanding may allow fine control of the expression of therapeutic genes introduced by gene therapy vectors as they are developed. Detailed understanding of the molecular interactions that regulate transcription may provide new targets for the development of therapeutic drugs that inhibit or stimulate the expression of specific genes. A more complete understanding of the mechanisms of transcriptional control may allow improved engineering of crops with desirable characteristics. Certainly, further advances in the area of transcription control will help to satisfy our desire to understand how complex organisms such as ourselves develop and function.

Key Terms

activation domain *288*
activators *271*
antitermination factor *315*
carboxyl-terminal domain (CTD) *280*
chromatin-mediated repression *299*
co-activator *293*
co-repressor *294*
DNase I footprinting *286*
enhancers *274*
enhancesome *295*
general transcription factors *296*
heat-shock genes *316*
histone deacetylation *300*
leucine zipper *291*
MAT locus (in yeast) *299*
mediator *299*
nuclear receptors *291*
promoter *276*
promoter-proximal elements *283*
repression domain *290*
repressors *271*
RNA polymerase II *279*
silencer sequences *299*
specific transcription factors *286*
TATA box *282*
TATA box-binding protein (TBP) *297*
upstream activating sequences (UASs) *285*
yeast two-hybrid system *310*
zinc finger *291*

Review the Concepts

1. Describe the molecular events that occur at the *lac* operon when *E. coli* cells are shifted from a glucose-containing medium to a lactose-containing medium.

2. The concentration of free phosphate affects transcription of some *E. coli* genes. Describe the mechanism for this.

3. What types of genes are transcribed by RNA polymerases I, II, and III? Design an experiment to determine whether a specific gene is transcribed by RNA polymerase II.

4. The CTD of the largest subunit of RNA polymerase II can be phosphorylated at multiple serine residues. What are the conditions that lead to the phosphorylated versus unphosphorylated RNA polymerase II CTD?

5. What do TATA boxes, initiators, and CpG islands have in common? Which was the first of these to be identified? Why?

6. Describe the methods used to identify the location of DNA-control elements in promoter proximal regions of genes.

7. What is the difference between a promoter-proximal element and a distal enhancer?

8. Describe the methods used to identify the location of DNA-binding proteins in the regulatory regions of genes.

9. Describe the structural features of transcriptional activator and repressor proteins.

10. What happens to transcription of the *EGR-1* gene in patients with Wilm's tumor? Why?

11. Using CREB and nuclear receptors as examples, compare and contrast the structural changes that take place when these transcription factors bind to their co-activators.

12. What general transcription factors associate with an RNA polymerase II promoter in addition to the polymerase? In what order do they bind in vitro? What structural change occurs in the DNA when an "open" transcription-initiation complex is formed?

13. Expression of recombinant proteins in yeast is an important tool for biotechnology companies that produce new drugs for human use. In an attempt to get a new gene *X* expressed in yeast, a researcher has integrated gene *X* into the yeast genome near a telomere. Will this strategy result in good expression of gene *X*? Why or why not? Would the outcome of this experiment differ if the experiment had been performed in a yeast line containing mutations in the H3 or H4 histone tails?

14. You have isolated a new protein called STICKY. You can predict from comparisons with other known proteins that STICKY contains a bHLH domain and a Sin3-interacting domain. Predict the function of STICKY and rationale for the importance of these domains in STICKY function.

15. The yeast two-hybrid method is a powerful molecular genetic method to identify a protein(s) that interacts with a known protein or protein domain. You have isolated the glucocorticoid receptor (GR) and have evidence that it is a modular protein containing an activation domain, a DNA-binding domain, and a second ligand-binding activation domain. Further analysis reveals that in pituitary cells, the protein is anchored in the cytoplasm in the absence of its hormone ligand, a result leading you to speculate that it binds to other inhibitory proteins. Describe how a two-hybrid analysis could be used to identify the protein(s) GR interacts with. How would you specifically identify the domain in the GR that binds the inhibitor(s)?

16. Some heat-shock genes encode proteins that act rapidly to protect other proteins from harsh conditions. Describe the mechanism that has evolved to regulate the expression of such genes.

Analyze the Data

In eukaryotes, the three RNA polymerases, Pol I, II and III, each transcribe unique genes required for the synthesis of ribosomes: 25S and 18S rRNAs (Pol I), 5S rRNA (Pol III), and mRNAs for ribosomal proteins (Pol II). Researchers have long speculated that the activities of the three RNA polymerases are coordinately regulated according to the demand for ribosome synthesis: high in replicating cells in rich nutrient conditions and low when nutrients are scarce. To determine if the activities of the three polymerases are coordinated, Laferte and colleagues engineered a strain of yeast to be partially resistant to the inhibition of cell growth by the drug rapamycin (2006, *Genes Dev.* **20**:2030–2040). As discussed in Chapter 8, rapamycin inhibits a protein kinase (called TOR for *t*arget *o*f *r*apamycin) that regulates the overall rate of protein synthesis and ribosome synthesis. When TOR is inhibited by rapamycin, the transcription of rRNAs by Pol I and Pol III and ribosomal protein mRNAs by RNA polymerase II are all rapidly repressed. Part of the inhibition of Pol I rRNA synthesis results from the dissociation of the Pol I transcription factor Rrn3 from Pol I (see Figure 7-52). In the strain constructed by Laferte and colleagues, the wild-type *Rrn3* gene and the wild-type *A43* gene, encoding the Pol I subunit to which Rrn3 binds, were replaced with a gene encoding a fusion protein of the A43 Pol I subunit with Rrn3. The idea was that the covalent fusion of the two proteins would prevent the Rrn3 dissociation from Pol I otherwise caused by rapamycin treatment. The resulting CARA (*c*onstitutive *a*ssociation of *R*rn3 and *A*43) strain was found to be partially resistant to rapamycin. In the absence of rapamycin, the CARA strain grew at the same rate and had equal numbers of ribosomes as do wild-type cells.

a. To analyze rRNA transcription by Pol I, total RNA was isolated from rapidly growing wild-type (WT) and CARA cells at various times following the addition of rapamycin. The concentration of the 35S rRNA precursor transcribed by Pol I (see Figure 8-35) was assayed by the primer-extension method. Since the 5' end of the 35S rRNA precursor is degraded during the processing of 25S and 18S rRNA, this method measures the relatively short-lived pre-rRNA precursor. This is an indirect measure of the rate of rRNA transcription by Pol I. The results of this primer extension assay are shown below. How does the CARA Pol I–Rrn3 fusion affect the response of Pol I transcription to rapamycin?

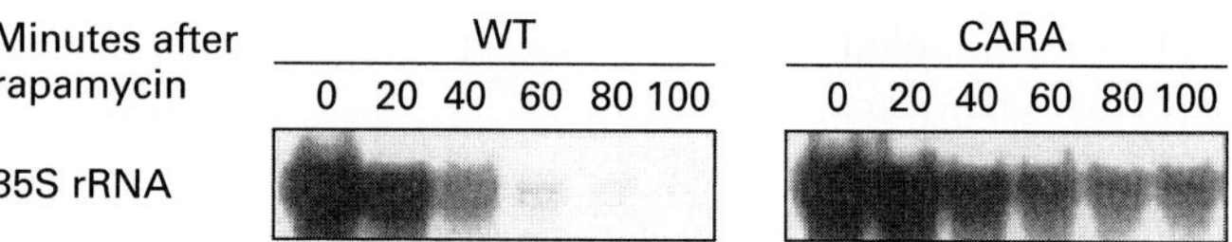

b. The concentrations of four mRNAs encoding ribosomal proteins, RPL30, RPS6a, RPL7a, and RPL5, and the mRNA for actin (*ACT1*), a protein present in the cytoskeleton, were assessed in wild-type and CARA cells by Northern blotting at various times after addition of rapamycin to rapidly growing cells (upper autoradiograms). 5S rRNA transcription was assayed by pulse labeling rapidly growing WT and CARA cells with ^{3}H uracil (for 20 minutes) at various times

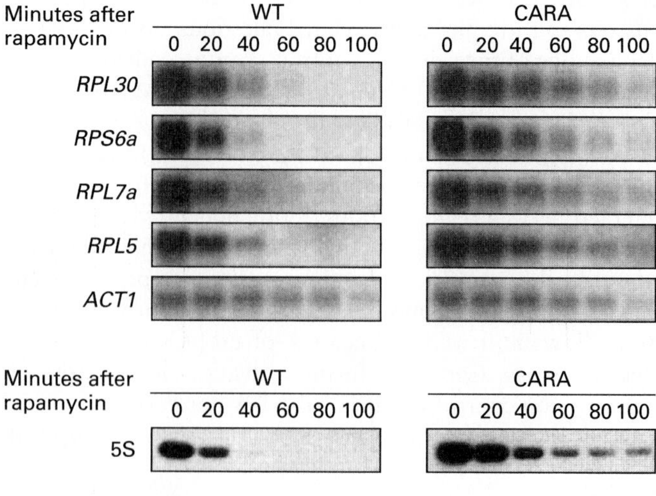

after addition of rapamycin to the media. Total cellular RNA was isolated and subjected to gel electrophoresis and autoradiography. The lower autoradiogram shows the region of the gel containing 5S rRNA. Based on these data, what can be concluded about the influence of Pol I transcription on the transcription of ribosomal protein genes by Pol II and 5S rRNA by Pol III?

c. To determine if the difference in behavior of wild-type and CARA cells can be observed under normal physiological conditions (i.e., without drug treatment), cells were subjected to a shift in their food source, from nutrient-rich media to nutrient-poor media. Under these conditions, in wild-type cells, the TOR protein kinase becomes inactive. Consequently, shifting cells from nutrient-rich media to nutrient-poor media should result in a normal physiological response that is equivalent to treating cells with rapamycin, which inhibits TOR. To determine how the CARA fusion protein affected the response to this media shift, RNA was extracted from wild-type and CARA cells and used to probe microarrays containing all yeast open reading frames. The extent of RNA hybridization with the arrays was quantified and is expressed in the graphs as $\log_2$ of the ratio of CARA-cell RNA concentration to wild-type-cell RNA concentration for each open reading frame. A value of zero indicates that the two strains of yeast exhibit the same level of expression for those specific RNAs. A value of 1 indicates that the CARA cells contain twice as much of that particular RNA as do wild-type cells. The graphs below show the number of open reading frames (y axis) that have values for $\log_2$ of this ratio, indicated by the x axis. The results of hybridization to open reading frames encoding mRNAs for ribosomal proteins are shown by black bars, those for all other mRNAs by white bars. The graph on the left gives results for cells grown in nutrient-rich medium, the graph on the right for cells shifted to nutrient-poor medium for 90 minutes. What do these data suggest about the regulation of ribosomal protein gene transcription by Pol II?

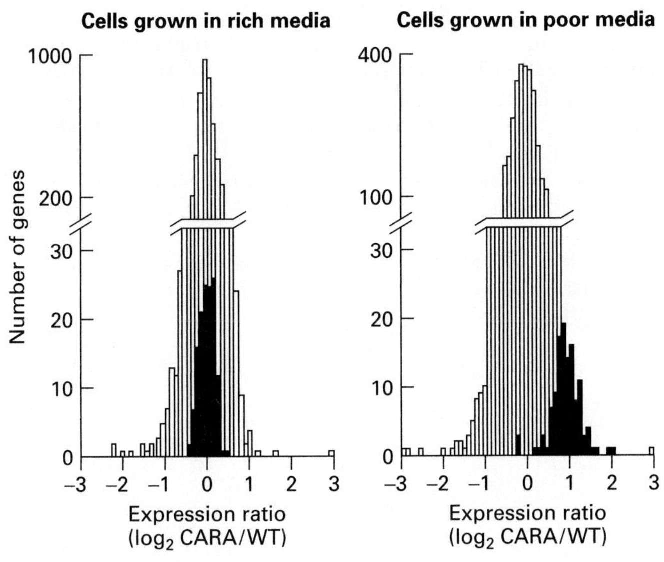

References

Control of Gene Expression in Bacteria

Borukhov, S., and J. Lee. 2005. RNA polymerase structure and function at lac operon. *C. R. Biol.* **328**:576–587.

Halford, S. E., and J. F. Marko. 2004. How do site-specific DNA-binding proteins find their targets? *Nucleic Acids Res.* **32**:3040–3052.

Lawson, C. L., et al. 2004. Catabolite activator protein: DNA binding and transcription activation. *Curr. Opin. Struct. Biol.* **14**:10–20.

Lewis, M. 2005. The lac repressor. *C. R. Biol.* **328**:521–548.

Muller-Hill, B. 1998. Some repressors of bacterial transcription. *Curr. Opin. Microbiol.* **1**:145–151.

Murakami, K. S., and S. A. Darst. 2003. Bacterial RNA polymerases: the whole story. *Curr. Opin. Struct. Biol.* **13**:31–39.

Wigneshweraraj, S. R., et al. 2005. The second paradigm for activation of transcription. *Prog. Nucleic Acid Res. Mol. Biol.* **79**:339–369.

Overview of Eukaryotic Gene Control and RNA Polymerases

Boeger, H., et al.. 2005. Structural basis of eukaryotic gene transcription. *FEBS Lett.* **579**:899–903.

Regulatory Sequences in Protein-Coding Genes

Brenner, S., et al. 2002. Conserved regulation of the lymphocyte-specific expression of lck in the Fugu and mammals. *Proc. Natl. Acad. Sci. USA* **99**:2936–2941.

Dean, A. 2006. On a chromosome far, far away: LCRs and gene expression. *Trends Genet.* **22**:38–45.

Gaszner, M., and G. Felsenfeld. 2006. Insulators: exploiting transcriptional and epigenetic mechanisms. *Nat. Rev. Genet.* **7**:703–713.

Kleinjan, D. A., and V. van Heyningen. 2005. Long-range control of gene expression: emerging mechanisms and disruption in disease. *Am. J. Hum. Genet.* **76**:8–32.

Nobrega, M. A., I. Ovcharenko, V. Afzal, and E. M. Rubin. 2003. Scanning human gene deserts for long-range enhancers. *Science* **302**:413.

Smale, S. T., and J. T. Kadonaga. 2003. The RNA polymerase II core promoter. *Annu. Rev. Biochem.* **72**:449–479.

Activators and Repressors of Transcription

Brivanlou, A. H., and J. E. Darnell, Jr. 2002. Signal transduction and the control of gene expression. *Science* **295**:813–818.

Broun, P. 2004. Transcription factors as tools for metabolic engineering in plants. *Curr. Opin. Plant Biol.* **7**:202–209.

Garvie, C. W., and C. Wolberger. 2001. Recognition of specific DNA sequences. *Mol. Cell.* **8**:937–946.

Kadonaga, J. T. 2004. Regulation of RNA polymerase II transcription by sequence-specific DNA binding factors. *Cell* **116**:247–257.

Luscombe, N. M., et al. 2000. An overview of the structures of protein-DNA complexes. *Genome Biol.* **1**:1–37.

Marmorstein, R., and M. X. Fitzgerald. 2003. Modulation of DNA-binding domains for sequence-specific DNA recognition. *Gene* **304**:1–12.

Riechmann, J. L., et al. 2000. *Arabidopsis* transcription factors: genome-wide comparative analysis among eukaryotes. *Science* **290**:2105–2110.

Tupler, R., G. Perini, and M. R. Green. 2001. Expressing the human genome. *Nature* **409**:832–833.

Transcription Initiation by RNA Polymerase II

Hahn, S. 2004. Structure and mechanism of the RNA polymerase II transcription machinery. *Nat. Struct. Mol. Biol.* **11**:394–403.

Molecular Mechanisms of Transcription Repression and Activation

Courey, A. J., and S. Jia. 2001. Transcriptional repression: the long and the short of it. *Genes Dev.* **15**:2786–2796.

Green, M. R. 2005. Eukaryotic transcription activation: right on target. *Mol. Cell* **18**:399–402.

Horn, P. J., and C. L. Peterson. 2006. Heterochromatin assembly: a new twist on an old model. *Chromosome Res.* **14**:83–94.

Klose, R. J., and A. P. Bird. 2006. Genomic DNA methylation: the mark and its mediators. *Trends Biochem. Sci.* **31**:89–97.

Kornberg, R. D. 2005. Mediator and the mechanism of transcriptional activation. *Trends Biochem. Sci.* **30**:235–239.

Levine, S. S., I. F. King, and R. E. Kingston. 2004. Division of labor in polycomb group repression. *Trends Biochem. Sci.* **29**:478–485.

Lonard, D. M., and B. W. O'Malley. 2006. The expanding cosmos of nuclear receptor coactivators. *Cell* **125**:411–414.

Lund, A. H., and M. van Lohuizen. 2004. Polycomb complexes and silencing mechanisms. *Curr. Opin. Cell Biol.* **16**:239–246.

Malik, S., and R. G. Roeder. 2005. Dynamic regulation of pol II transcription by the mammalian Mediator complex. *Trends Biochem. Sci.* **30**:256–263.

Margueron, R., P. Trojer, and D. Reinberg. 2005. The key to development: interpreting the histone code? *Curr. Opin. Genet. Dev.* **15**:163–176.

Martin, C., and Y. Zhang. 2005. The diverse functions of histone lysine methylation. *Nat. Rev. Mol. Cell Biol.* **6**:838–849.

Mohrmann, L., and C. P. Verrijzer. 2005. Composition and functional specificity of SWI2/SNF2 class chromatin remodeling complexes. *Biochim. Biophys. Acta* **1681**:59–73.

Roeder, R. G. 2005. Transcriptional regulation and the role of diverse coactivators in animal cells. *FEBS Lett.* **579**:909–915.

Smith, C. L., and C. L. Peterson. 2005. ATP-dependent chromatin remodeling. *Curr. Top. Dev. Biol.* **65**:115–148.

Spiegelman, B. M., and R. Heinrich. 2004. Biological control through regulated transcriptional coactivators. *Cell* **119**:157–167.

Struhl, K. 2005. Transcriptional activation: mediator can act after preinitiation complex formation. *Mol. Cell* **17**:752–754.

Taatjes, D. J., M. T. Marr, and R. Tjian. 2004. Regulatory diversity among metazoan co-activator complexes. *Nat. Rev. Mol. Cell Biol.* **5**:403–410.

Thiel, G., M. Lietz, and M. Hohl. 2004. How mammalian transcriptional repressors work. *Eur. J. Biochem.* **271**:2855–2862.

Trojer, P., and D. Reinberg. 2006. Histone lysine demethylases and their impact on epigenetics. *Cell* **125**:213–217.

Villard, J. 2004. Transcription regulation and human diseases. *Swiss Med. Wkly.* **134**:571–579.

Wang, W. 2003. The SWI/SNF family of ATP-dependent chromatin remodelers: similar mechanisms for diverse functions. *Curr. Top. Microbiol. Immunol.* **274**:143–169.

Regulation of Transcription-Factor Activity

Boisvert, F. M., C. A. Chenard, and S. Richard. 2005. Protein interfaces in signaling regulated by arginine methylation. *Sci. STKE***271**:re2.

Brahimi-Horn, C., N. Mazure, and J. Pouyssegur. 2005. Signalling via the hypoxia-inducible factor-1alpha requires multiple posttranslational modifications. *Cell. Signal.* **17**:1–9.

Gardner, K. H., and M. Montminy. 2005. Can you hear me now? Regulating transcriptional activators by phosphorylation. *Sci. STKE* **301**:pe44.

Glozak, M. A., N. Sengupta, X. Zhang, and E. Seto. 2005. Acetylation and deacetylation of non-histone proteins. *Gene* **363**:15–23.

Khidekel, N., and L. C. Hsieh-Wilson. 2004. A "molecular switchboard"—covalent modifications to proteins and their impact on transcription. *Org. Biomol. Chem.* **7**:1–7.

Moehren, U., M. Eckey, and A. Baniahmad. 2004. Gene repression by nuclear hormone receptors. *Essays Biochem.* **40**:89–104.

Picard, D. 2006. Chaperoning steroid hormone action. *Trends Endocrinol. Metab.* **17**:229–235.

Rosenfeld, M. G., V. V. Lunyak, and C. K. Glass. 2006. Sensors and signals: a coactivator/corepressor/epigenetic code for integrating signal-dependent programs of transcriptional response. *Genes Dev.* 2006 **20**:1405–1428.

Regulated Elongation and Termination of Transcription

Peterlin, B. M., and D. H. Price. 2006. Controlling the elongation phase of transcription with P-TEFb. *Mol. Cell* **23**:297–305.

Saunders, A., L. J. Core, and J. T. Lis. 2006. Breaking barriers to transcription elongation. *Nat. Rev. Mol. Cell Biol.* **7**:557–567.

Winston, F. 2001. Control of eukaryotic transcription elongation. *Genome Biol.* **2**:reviews1006.

Yamada, T., et al. 2006. P-TEFb-mediated phosphorylation of hSpt5 C-terminal repeats is critical for processive transcription elongation. *Mol. Cell* **21**:227–237.

Other Eukaryotic Transcription Systems

Gaspari, M., N. G. Larsson, and C. M. Gustafsson. 2004. The transcription machinery in mammalian mitochondria. *Biochim. Biophys. Acta* **1659**:148–152.

Grummt, I. 2003. Life on a planet of its own: regulation of RNA polymerase I transcription in the nucleolus. *Genes Dev.* **17**:1691–1702.

Kanamaru, K., and K. Tanaka. 2004. Roles of chloroplast RNA polymerase sigma factors in chloroplast development and stress response in higher plants. *Biosci. Biotechnol. Biochem.* **68**:2215–2223.

Nomura, M. 2001. Ribosomal RNA genes, RNA polymerases, nucleolar structures, and synthesis of rRNA in the yeast *Saccharomyces cerevisiae. Cold Spring Harb. Symp. Quant. Biol.* **66**:555–565.

Raska, I., et al. 2004. The nucleolus and transcription of ribosomal genes. *Biol. Cell.* **96**:579–594.

Schramm, L., and N. Hernandez. 2002. Recruitment of RNA polymerase III to its target promoters. *Genes Dev.* **16**:2593–2620.

Shiina, T., Y. Tsunoyama, Y. Nakahira, and M. S. Khan. 2005. Plastid RNA polymerases, promoters, and transcription regulators in higher plants. *Int. Rev. Cytol.* **244**:1–68.

Shutt, T. E., and M. W. Gray. 2005. Bacteriophage origins of mitochondrial replication and transcription proteins. *Trends Genet.* **22**:90–95.

White, R. J. 2004. RNA polymerases I and III, growth control and cancer. *Nat. Rev. Mol. Cell Biol.* **6**:69–78.

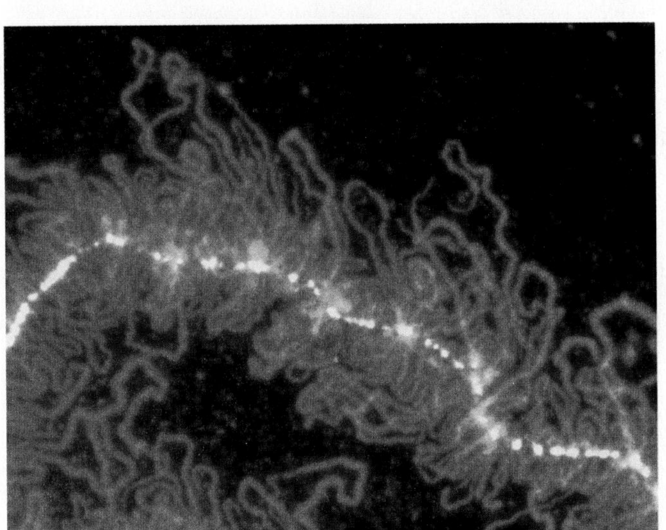

Portion of a "lampbrush chromosome" from an oocyte of the newt *Nophthalmus viridescens;* hnRNP protein associated with nascent RNA transcripts fluoresces red after staining with a monoclonal antibody. [Courtesty of M. Roth and J. Gall.]

CHAPTER

8

POST-TRANSCRIPTIONAL GENE CONTROL

In the previous chapter, we saw that most genes are regulated at the first step in gene expression, namely, the initiation of transcription. However, once transcription has been initiated, synthesis of the encoded RNA requires that RNA polymerase transcribe the entire gene and not terminate prematurely. Moreover, the initial **primary transcripts** produced from eukaryotic genes must undergo various processing reactions to yield the corresponding functional RNAs. For mRNAs, the 5′ cap structure necessary for translation must be added (see Figure 4-14), introns must be spliced out of pre-mRNAs, and the 3′ end must be polyadenylated (see Figure 4-15). Once formed in the nucleus, mature, functional RNAs are exported to the cytoplasm as components of ribonucleoproteins. Both processing of RNAs and their export from the nucleus offer opportunities for further regulating gene expression after the initiation of transcription.

Recently, the vast amount of sequence data on human cDNAs has revealed that ~60 percent of human genes give rise to alternatively spliced mRNAs. These alternatively spliced mRNAs encode related proteins with differences in sequences limited to specific functional domains. In many cases, alternative RNA splicing is regulated to meet the need for a specific protein isoform in a specific cell type. Given the complexity of pre-mRNA splicing, it is not surprising that mistakes are occasionally made, giving rise to mRNA precursors with improperly spliced exons. However, eukaryotic cells have evolved *RNA surveillance* mechanisms that prevent the transport of incorrectly processed RNAs to the cytoplasm or lead to their degradation if they are transported.

Additional control of gene expression can occur in the cytoplasm. In the case of protein-coding genes, for instance, the amount of protein produced depends on the stability of the corresponding mRNAs in the cytoplasm and the rate of their translation. For example, during an immune response, lymphocytes communicate by secreting polypeptide hormones called cytokines that signal neighboring lymphocytes through cytokine receptors that span their plasma membranes (Chapter 24). It is important for lymphocytes to synthesize and secrete cytokines in short bursts. This is possible because cytokine mRNAs are extremely unstable. Consequently, the concentration of the mRNA in the cytoplasm falls rapidly once its synthesis is stopped. In contrast, mRNAs encoding proteins required in large amounts that function over long periods, such as ribosomal proteins, are extremely stable so that multiple polypeptides are transcribed from each mRNA.

In addition to regulation of pre-mRNA processing, nuclear export, and translation, the cellular locations of some mRNAs are regulated so that newly synthesized protein is concentrated

OUTLINE

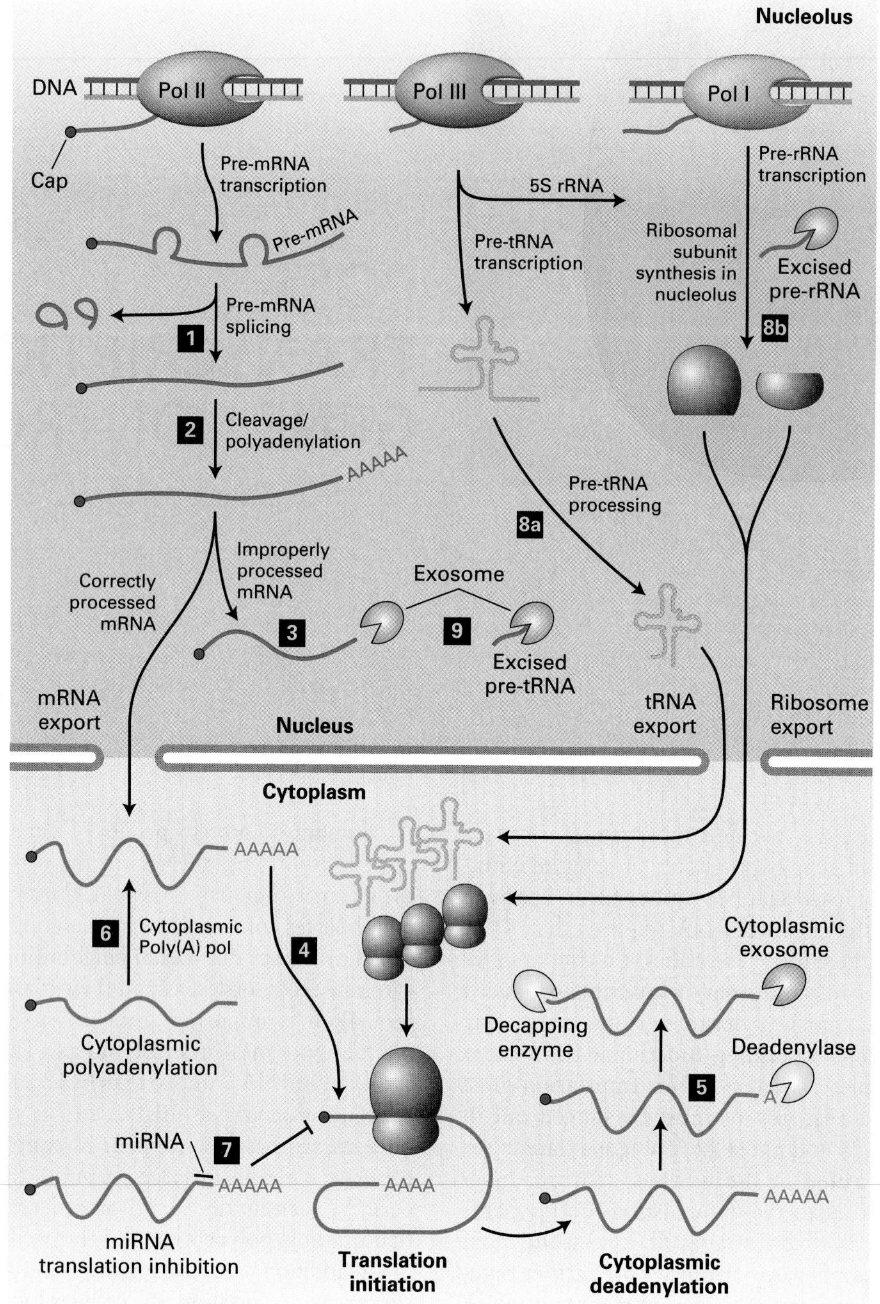

▲ **FIGURE 8-1 Overview of RNA processing and post-transcriptional gene control.** Nearly all cytoplasmic RNAs are processed from primary transcripts in the nucleus before they are exported to the cytoplasm. For protein coding genes transcribed by RNA polymerase II, gene control can be exerted through **1** the choice of alternative exons during pre-mRNA splicing and the **2** choice of alternative poly(A) sites. Improperly processed mRNAs are blocked from export to the cytoplasm and degraded **3** by a large complex called the exosome that contains multiple ribonucleases. Once exported to the cytoplasm, **4** translation initiation factors bind to the mRNA 5′-cap cooperatively with poly(A)-binding protein I bound to the poly(A) tail and initiate translation (see Figure 4-28). **5** mRNA is degraded in the cytoplasm by de-adenylation and decapping followed by degradation by cytoplasmic exosomes. The degradation rate of each mRNA is controlled, thereby regulating the mRNA concentration and, consequently, the amount of protein translated. Some mRNAs are synthesized without long poly(A) tails. Their translation is regulated by **6** controlling the synthesis of a long poly(A) tail by a cytoplasmic poly(A) polymerase. **7** Translation is also regulated by other mechanisms including miRNAs. When expressed, these ~22 nucleotide RNAs inhibit translation of mRNAs to which they hybridize, usually in the 3′-untranslated region. tRNAs and rRNAs are also synthesized as precursor RNAs that must be **8** processed before they are functional. Regions of precursors cleaved from the mature RNAs are degraded by nuclear exosomes **9**. [Adapted from Houseley *et. al. Nat Rev Mol Cell Biol.* (2006) **7:**529.]

where it is needed. Particularly striking examples of this occur in the nervous systems of multicellular animals. Some neurons in the human brain make more than 1000 separate synapses with other neurons. During the process of learning, synapses that fire more frequently than others increase in size many times, even though other synapses made by the same neuron do not. This can occur because mRNAs encoding proteins critical for formation of an enlarged synapse are stored at all synapses. Then the translation of these localized, stored mRNAs is regulated at each synapse independently by the frequency at which it initiates translation. In this way, synthesis of synapse-associated proteins can be regulated independently at each of the many synapses made by the same neuron.

Another type of gene regulation that has recently come to light involves micro RNAs (miRNAs), which regulate the stability and translation of specific target mRNAs in multicellular animals and plants. Analyses of these short miRNAs in various human tissues indicate that there are ≈1000 miRNAs expressed in the multiple types of human cells. Although some have been recently discovered to function through inhibition of target gene expression in the appropriate tissue and at the appropriate time in development, the functions of the vast majority of human miRNAs are unknown and are the subject of a growing new area of research. If most miRNAs do indeed have significant functions, miRNA genes constitute a significant fraction of the ≈25,000 human genes. A closely related process called RNA interference (RNAi) leads to the degradation of viral RNAs in infected cells and the degradation of transposon-encoded RNAs in most eukaryotes. This is of tremendous significance to biological researchers because it is possible to design **short interfering** RNAs (siRNA) to inhibit the translation of specific mRNAs experimentally by a process called RNA knockdown. This makes it possible to inhibit the function of any desired gene, even in organisms that are not amenable to classic genetic methods for isolating mutants.

We refer to all the mechanisms that regulate gene expression following transcription as *post-transcriptional gene control* (Figure 8-1). Since the stability and translation rate of an mRNA contribute to the amount of protein expressed from a gene, these post-transcriptional processes are important components of **gene control**. Indeed, the protein output of a gene is regulated at every step in the life of an mRNA from the initiation of its synthesis to its degradation. Thus genetic regulatory processes act on RNA as well as DNA. In this chapter, we consider the events that occur in the processing of mRNA following transcription initiation and the various mechanisms that are known to regulate these events. In the last section, we briefly discuss the processing of primary transcripts produced from genes encoding rRNAs and tRNAs.

8.1 Processing of Eukaryotic Pre-mRNA

In this section, we take a closer look at how eukaryotic cells convert the initial primary transcript synthesized by RNA polymerase II into a functional mRNA. Three major events occur during the process: *5′ capping, 3′ cleavage/polyadenylation,* and *RNA splicing* (Figure 8-2). Adding these specific modifications to the 5′ and 3′ ends of the pre-mRNA is important to protect it from enzymes that quickly digest uncapped RNAs generated by RNA processing, such as spliced-out introns and RNA transcribed downstream from a polyadenylation site. The 5′-cap and 3′-poly(A) tail distinguish pre-mRNA molecules from the many other kinds of RNAs in the nucleus. Pre-mRNA molecules are bound by nuclear proteins that function in mRNA export to the cytoplasm. After mRNAs are exported to the cytoplasm, they are bound by another set of cytoplasmic proteins that stimulate translation and are critical for mRNA stability in the cytoplasm. Furthermore, introns must be removed to generate the correct coding region of the mRNA. In higher eukaryotes, alternative splicing is intricately regulated in order to substitute different functional domains into proteins, producing a considerable expansion of the proteome of these organisms, which include ourselves.

The pre-mRNA processing events of capping, splicing, and polyadenylation occur in the nucleus as the nascent mRNA precursor is being transcribed. Thus pre-mRNA processing is *co-transcriptional.* As the RNA emerges from the surface of RNA polymerase II, its 5′ end is immediately modified by the addition of the 5′-cap structure found on all mRNAs (see Figure 4-14). As the nascent pre-mRNA continues to emerge from the surface of the polymerase, it is immediately bound by members of a complex group of RNA-binding proteins that assist in RNA splicing and export of the fully processed mRNA through nuclear pore complexes into the cytoplasm. Some of these proteins remain associated with the mRNA in the cytoplasm, but most either remain in the nucleus or shuttle back into the nucleus shortly after the mRNA is exported to the cytoplasm. Cytoplasmic RNA-binding proteins are exchanged for the nuclear ones. Consequently, mRNAs never occur as free RNA molecules in the cell but are always associated with protein as ***ribonucleoprotein* (RNP) complexes,** first as nascent *pre-mRNPs* that are capped and spliced as they are transcribed. Then, following cleavage and polyadenylation, they are referred to as nuclear *mRNPs.* Following the exchange of proteins that accompanies export to the cytoplasm, they are called *cytoplasmic mRNPs.* Although we frequently refer to pre-mRNAs and mRNAs, it is important to remember that they are always associated with proteins as RNP complexes.

The 5′ Cap Is Added to Nascent RNAs Shortly After Transcription Initiation

As the nascent RNA transcript emerges from the RNA channel of RNA polymerase II and reaches a length of ≈25 nucleotides, a protective cap composed of 7-methylguanosine and methylated riboses is added to the 5′ end of eukaryotic mRNAs (see Figure 4-14). The *5′ cap* marks RNA molecules as mRNA precursors and serves to protect them from RNA-digesting enzymes (5′-exoribonucleases) in the nucleus and cytoplasm. This initial step in RNA processing is catalyzed by a dimeric capping enzyme, which associates with the

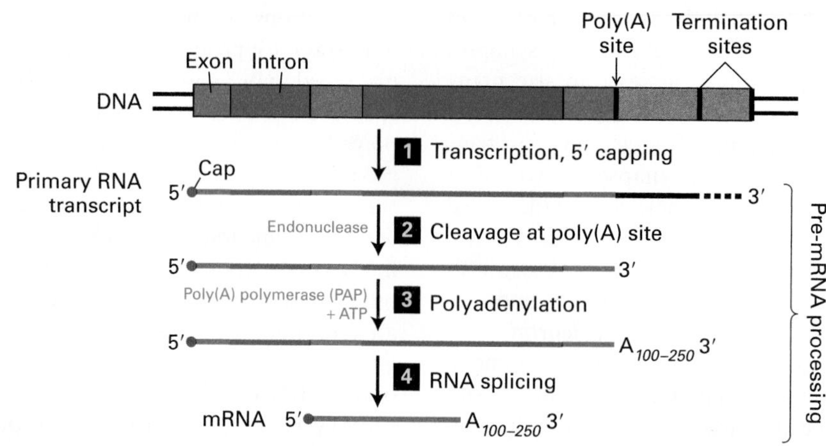

▲ **FIGURE 8-2 Overview of mRNA processing in eukaryotes.** Shortly after RNA polymerase II initiates transcription at the first nucleotide of the first exon of a gene, the 5′ end of the nascent RNA is capped with 7-methylguanylate (step **1**). Transcription by RNA polymerase II terminates at any one of multiple termination sites downstream from the poly(A) site, which is located at the 3′ end of the final exon. After the primary transcript is cleaved at the poly(A) site (step **2**), a string of adenosine (A) residues is added (step **3**). The poly(A) tail contains ≈250 A residues in mammals, ≈150 in insects, and ≈100 in yeasts. For short primary transcripts with few introns, splicing (step **4**) usually follows cleavage and polyadenylation, as shown. For large genes with multiple introns, introns often are spliced out of the nascent RNA during its transcription, i.e., before transcription of the gene is complete. Note that the 5′ cap and sequence adjacent to the poly(A) tail are retained in mature mRNAs.

phosphorylated carboxyl-terminal domain (CTD) of RNA polymerase II. Recall that the CTD becomes phosphorylated by the TFIIH general transcription factor during transcription initiation (see Figure 7-31). Binding to the phosphorylated CTD stimulates the activity of the capping enzymes so that it is focused on RNAs with a 5′-triphosphate that emerge from RNA polymerase II and not on RNA polymerases I or III that do not have a CTD. This is important because pre-mRNA synthesis accounts for less than 10 percent of the total RNA synthesized in replicating cells. About 80 percent of RNA synthesis is preribosomal RNA transcribed by RNA polymerase I, and about 15 percent is transcribed by RNA polymerase III, including 5S rRNA, tRNAs, and other stable small RNAs. The two mechanisms of (1) capping enzyme binding to initiated RNA polymerase II specifically through its unique, phosphorylated CTD and (2) activation of capping enzyme activity by binding to the phosphorylated CTD result in specific capping of the minor fraction of RNAs transcribed by RNA polymerase II.

One subunit of the capping enzyme removes the γ-phosphate from the 5′ end of the nascent RNA (Figure 8-3). Another domain of this subunit transfers the GMP moiety from GTP to the 5′-diphosphate of the nascent transcript, creating the unusual guanosine 5′-5′-triphosphate structure. In the final steps, separate enzymes transfer methyl groups from *S*-adenosylmethionine to the N_7 position of the guanine and the 2′ oxygens of riboses at the 5′ end of the nascent RNA.

Considerable evidence indicates that capping of the nascent transcript is coupled to elongation by RNA polymerase II so that all of its transcripts are capped during the earliest phase of elongation. In one model, during the initial phase of transcription the polymerase elongates the nascent transcript very slowly due to the action of elongation factors NELF (*n*egative *el*ongation *f*actor) and Spt4/5, as occurs at the HIV LTR promoter (see Figure 7-51). Once the 5′-end of the nascent RNA is capped, phosphorylation of the RNA polymerase CTD and Spt5 by the cyclin T-Cdk9 protein kinase is stimulated, causing the release of NELF. This allows RNA polymerase II to enter into a faster mode of elongation that rapidly transcribes away from the promoter. The net effect of this mechanism is that the polymerase waits for the nascent RNA to be capped before elongating at a rapid rate. HIV appears to have exploited this mechanism to add an additional level of transcription control through the Tat protein and the TAR element, which are required to bring the cyclin T-Cdk9 kinase to the elongation complex that forms at the HIV LTR promoter (see Figure 7-51). Currently, it is not understood what distinguishes transcription from the HIV LTR promoter from other promoters where Tat protein is not required to bring in the cyclin T-Cdk9 protein kinase.

A Diverse Set of Proteins with Conserved RNA-Binding Domains Associate with Pre-mRNAs

As noted earlier, nascent RNA transcripts from protein-coding genes and mRNA processing intermediates, collectively referred to as **pre-mRNA,** do not exist as free RNA molecules in the nuclei of eukaryotic cells. From the time nascent transcripts first emerge from RNA polymerase II until mature mRNAs are transported into the cytoplasm, the RNA molecules are associated with an abundant set of nuclear proteins.

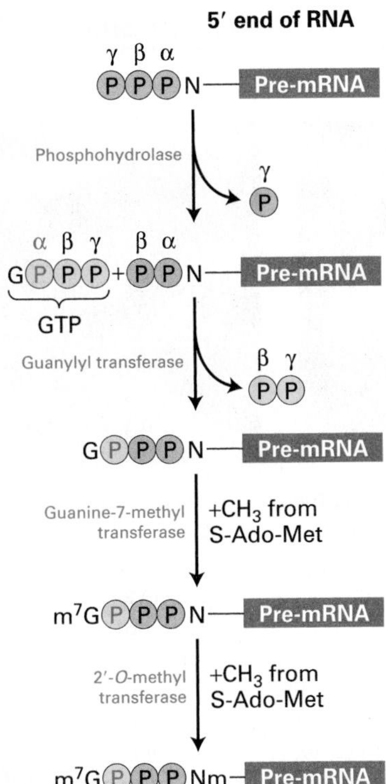

▲ **FIGURE 8-3 Synthesis of 5′-cap on eukaryotic mRNAs.** The 5′ end of a nascent RNA contains a 5′-triphosphate from the initiating NTP. The γ-phosphate is removed in the first step of capping, while the remaining α- and β-phosphates (orange) remain associated with the cap. The third phosphate of the 5-5′−triphosphate bond is derived from the α phosphate of the GTP that donates the guanine. The methyl donor for methylation of the cap guanine and the first one or two riboses of the mRNA is *S*-adenosyl methionine. [From S. Venkatesan and B. Moss, 1982, *Proc. Natl. Acad. Sci. USA* **79**:304.]

These are the major protein components of *heterogeneous ribonucleoprotein particles (hnRNPs),* which contain *heterogeneous nuclear RNA (hnRNA),* a collective term referring to pre-mRNA and other nuclear RNAs of various sizes. These hnRNP proteins contribute to further steps in RNA processing, including splicing and polyadenylation and export through nuclear pore complexes to the cytoplasm.

Researchers identified hnRNP proteins by first exposing cultured cells to high-dose UV irradiation, which causes covalent cross-links to form between RNA bases and closely associated proteins. Chromatography of nuclear extracts from treated cells on an oligo-dT cellulose column, which binds RNAs with a poly(A) tail, was used to recover the proteins that had become cross-linked to nuclear polyadenylated RNA. Subsequent treatment of cell extracts from unirradiated cells with monoclonal antibodies specific for the major proteins identified by this cross-linking technique revealed a complex set of abundant hnRNP proteins ranging in size from ≈30 to ≈120 kDa.

Like transcription factors, most hnRNP proteins have a modular structure. They contain one or more *RNA-binding domains* and at least one other domain that interacts with other proteins. Several different RNA-binding motifs have been identified by constructing deletions of hnRNP proteins and testing their ability to bind RNA.

Functions of hnRNP Proteins The association of pre-mRNAs with hnRNP proteins prevents formation of short secondary structures dependent on base pairing of complementary regions, thereby making the pre-mRNAs accessible for interaction with other RNA molecules or proteins. Pre-mRNAs associated with hnRNP proteins present a more uniform substrate for subsequent processing steps than would free, unbound pre-mRNAs, where each mRNA forms a unique secondary structure due to its specific sequence.

Binding studies with purified hnRNP proteins indicate that different hnRNP proteins associate with different regions of a newly made pre-mRNA molecule. For example, the hnRNP proteins A1, C, and D bind preferentially to the pyrimidine-rich sequences at the 3′ ends of introns (see Figure 8-7). Some hnRNP proteins interact with the RNA sequences that specify RNA splicing or cleavage/polyadenylation and contribute to the structure recognized by RNA-processing factors. Finally, cell-fusion experiments have shown that some hnRNP proteins remain localized in the nucleus, whereas others cycle in and out of the cytoplasm, suggesting that they function in the transport of mRNA (Figure 8-4).

Conserved RNA-Binding Motifs The *RNA recognition motif (RRM)*, also called the RNP motif and the RNA-binding domain (RBD), is the most common RNA-binding domain in hnRNP proteins. This ≈80-residue domain, which occurs in many other RNA-binding proteins, contains two highly conserved sequences (RNP1 and RNP2) that are found across organisms ranging from yeast to human—indicating that like many DNA-binding domains, they evolved early in eukaryotic evolution.

Structural analyses have shown that the RRM domain consists of a four-stranded β sheet flanked on one side by two α helices. To interact with the negatively charged RNA phosphates, the β sheet forms a positively charged surface. The conserved RNP1 and RNP2 sequences lie side by side on the two central β strands, and their side chains make multiple contacts with a single-stranded region of RNA that lies across the surface of the β sheet (Figure 8-5).

The 45-residue *KH motif* is found in the hnRNP K protein and several other RNA-binding proteins. The three-dimensional structure of representative KH domains is similar to that of the RRM domain but smaller, consisting of a three-stranded β sheet supported from one side by two α helices. Nonetheless, the KH domain interacts with RNA much differently than does the RRM domain. RNA binds to the KH domain by interacting with a hydrophobic surface formed by the α helices and one β strand. The *RGG box*, another RNA-binding motif found in hnRNP proteins, contains five Arg-Gly-Gly (RGG) repeats with several interspersed aromatic amino acids. Although the structure of this motif has not yet been determined, its arginine-rich nature is similar to the RNA-binding domains of the HIV Tat protein. KH domains and RGG repeats are often interspersed in two or more sets in a single RNA-binding protein.

(a)

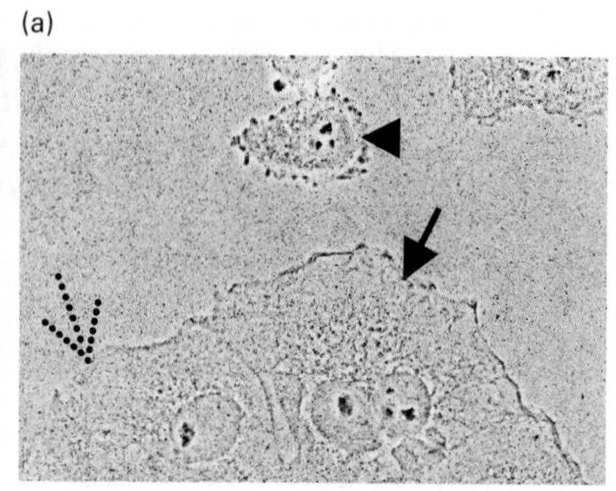

(b)

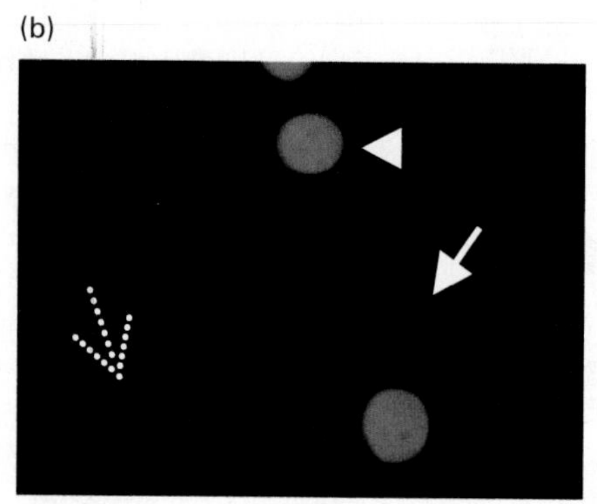

(c)

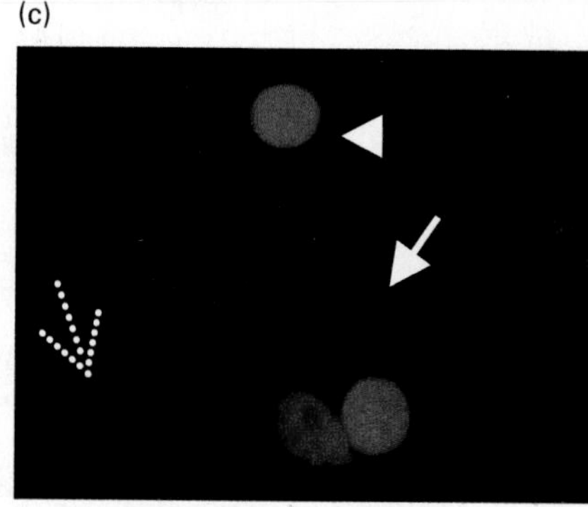

▲ **FIGURE 8-4 Human hnRNP A1 protein can cycle in and out of the cytoplasm, but human hnRNP C protein cannot.** Cultured HeLa cells and *Xenopus* cells were fused by treatment with polyethylene glycol, producing heterokaryons containing nuclei from each cell type. The hybrid cells were treated with cycloheximide immediately after fusion to prevent protein synthesis. After 2 hours, the cells were fixed and stained with fluorescent-labeled antibodies specific for human hnRNP C and A1 proteins. These antibodies do not bind to the homologous *Xenopus* proteins. (a) A fixed preparation viewed by phase-contrast microscopy includes unfused HeLa cells (arrowhead) and *Xenopus* cells (dotted arrow), as well as fused heterokaryons (solid arrow). In the heterokaryon in this micrograph, the round HeLa-cell nucleus is to the right of the oval-shaped *Xenopus* nucleus. (b, c) When the same preparation was viewed by fluorescence microscopy, the stained hnRNP C protein appeared green and the stained hnRNP A1 protein appeared red. Note that the unfused *Xenopus* cell on the left is unstained, confirming that the antibodies are specific for the human proteins. In the heterokaryon, hnRNP C protein appears only in the HeLa-cell nucleus (b), whereas the A1 protein appears in both nucleus (c). Since protein synthesis was blocked after cell fusion, some of the human hnRNP A1 protein must have left the HeLa-cell nucleus, moved through the cytoplasm, and entered the *Xenopus* nucleus in the heterokaryon. [See S. Pinol-Roma and G. Dreyfuss, 1992, *Nature* **355:**730; courtesy of G. Dreyfuss.]

(a) RNA recognition motif (RRM)

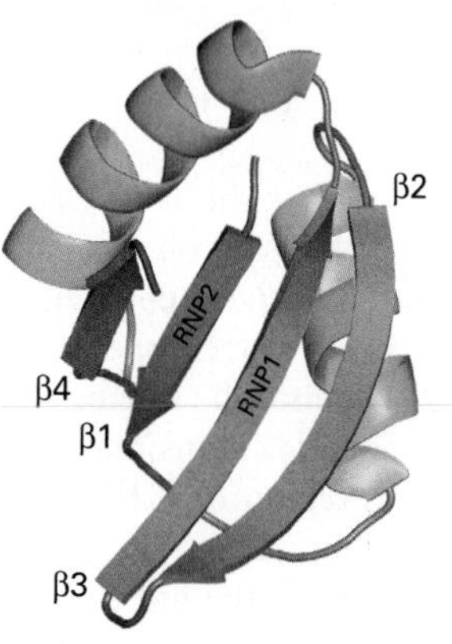

(b) Sex-lethal (Sxl) RRM domains

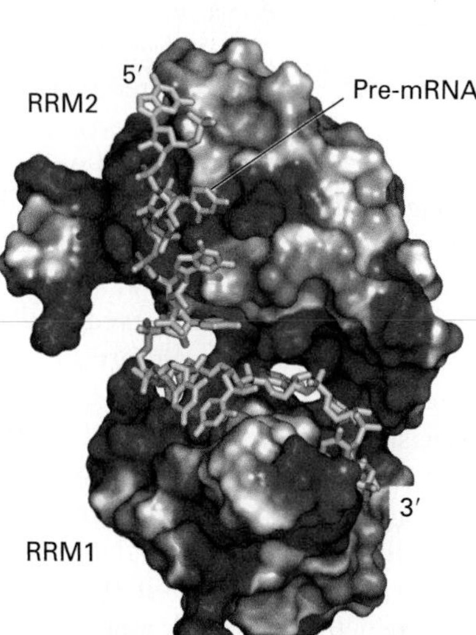

(c) Polypyrimidine tract binding protein (PTB)

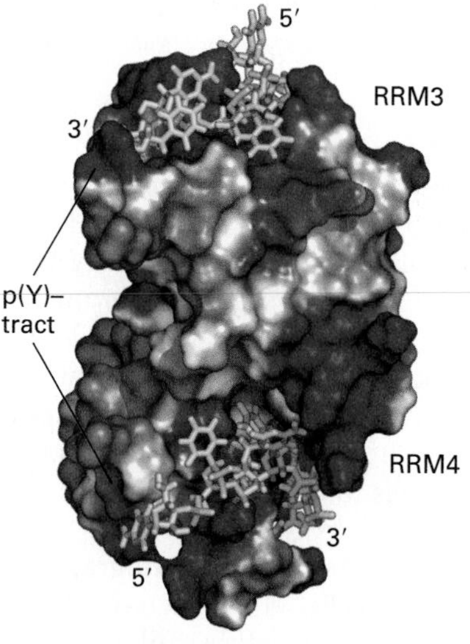

▲ **FIGURE 8-5 Structure of the RRM domain and its interaction with RNA.** (a) Diagram of the RRM domain showing the two α helices (green) and four β strands (red) that characterize this motif. The conserved RNP1 and RNP2 regions are located in the two central β strands. (b) Surface representation of the two RRM domains in *Drosophila* Sex-lethal (Sxl) protein, which bind a nine-base sequence in transformer pre-mRNA (yellow). The two RRMs are oriented like the two parts of an open pair of castanets, with the β sheet of RRM1 facing upward and the β sheet of RRM2 facing downward. Positively charged regions in Sxl protein are shown in shades of blue; negatively charged regions, in shades of red. The pre-mRNA is bound to the surfaces of the positively charged β sheets, making most of its contacts with the RNP1 and RNP2 regions of each RRM. (c) Strikingly different orientation of RRM domains in a different hnRNP, the polypyrimidine tract binding (PTB) protein, illustrating that RRMs are oriented in different relative positions in different hnRNPs; colors are as in (b). Polypyrimidine (p(Y)) single stranded RNA is bound to the upward (RRM3) and downward (RRM4) facing β-sheets. RNA is shown in yellow. [Part (a) adapted from K. Nagai et al., 1995, *Trends Biochem. Sci.* **20:**235. Part (b) after N. Harada et al., 1999, *Nature* **398:**579. Part (c) after F. C. Oberstrass et al., 2006, *Science* **309:**2054.]

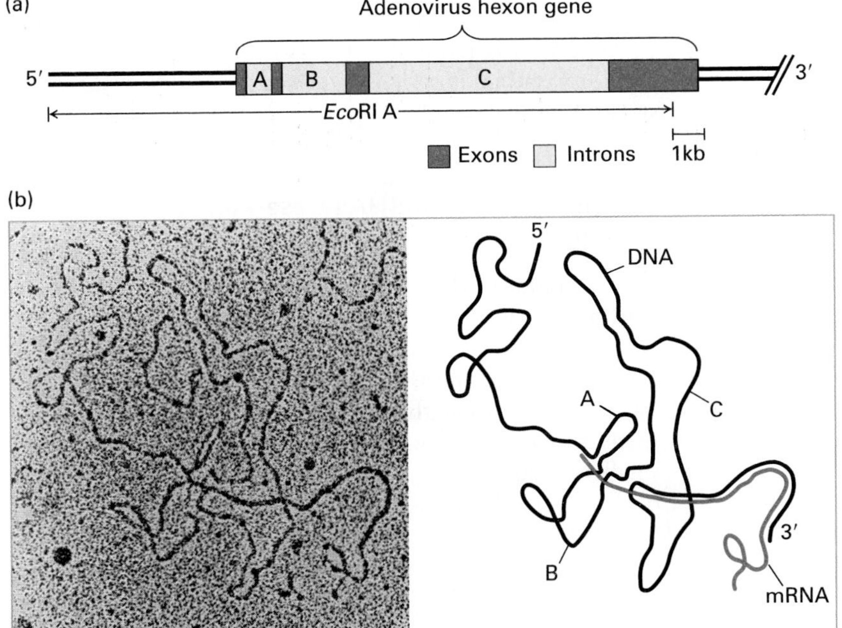

◄ **EXPERIMENTAL FIGURE 8-6 Electron microscopy of mRNA-template DNA hybrids shows that introns are spliced out during pre-mRNA processing.** (a) Diagram of the *Eco*RI A fragment of adenovirus DNA, which extends from the left end of the genome to just before the end of the final exon of the hexon gene. The gene consists of three short exons and one long (≈3.5 kb) exon separated by three introns of ≈1, 2.5, and 9 kb. (b) Electron micrograph (*left*) and schematic drawing (*right*) of hybrid between an *Eco*RI A fragment and hexon mRNA. The loops marked A, B, and C correspond to the introns indicated in (a). Since these intron sequences in the viral genomic DNA are not present in mature hexon mRNA, they loop out between the exon sequences that hybridize to their complementary sequences in the mRNA. [Micrograph from S. M. Berget et al., 1977, *Proc. Nat'l. Acad. Sci. USA* **74:**3171; courtesy of P. A. Sharp.]

Splicing Occurs at Short, Conserved Sequences in Pre-mRNAs via Two Transesterification Reactions

During formation of a mature, functional mRNA, the **introns** are removed and **exons** are spliced together. For short transcription units, **RNA splicing** usually follows cleavage and polyadenylation of the 3′ end of the primary transcript, as depicted in Figure 8-2. However, for long transcription units containing multiple exons, splicing of exons in the nascent RNA begins before transcription of the gene is complete.

Early evidence that introns are removed during splicing came from electron microscopy of RNA-DNA hybrids between adenovirus DNA and the mRNA encoding hexon, a major virion capsid protein (Figure 8-6). Other studies revealed nuclear viral RNAs that were colinear with the viral DNA (primary transcripts) and RNAs with one or two of the introns removed (processing intermediates). These results, together with the findings that the 5′ cap and 3′ poly(A) tail at each end of long mRNA precursors are retained in shorter mature cytoplasmic mRNAs, led to the realization that introns are removed from primary transcripts as exons are spliced together.

The location of *splice sites*—that is, exon-intron junctions—in a pre-mRNA can be determined by comparing the sequence of genomic DNA with that of the cDNA prepared from the corresponding mRNA (see Figure 5-15). Sequences that are present in the genomic DNA but absent from the cDNA represent introns and indicate the positions of splice sites. Such analysis of a large number of different mRNAs revealed moderately conserved, short consensus sequences at the splice sites flanking introns in eukaryotic pre-mRNAs; in higher organisms, a pyrimidine-rich region just upstream of the 3′ splice site also is common (Figure 8-7). Studies of mutant genes with deletions introduced into introns have shown that much of the center portion of introns can be removed without affecting splicing; generally only 30–40

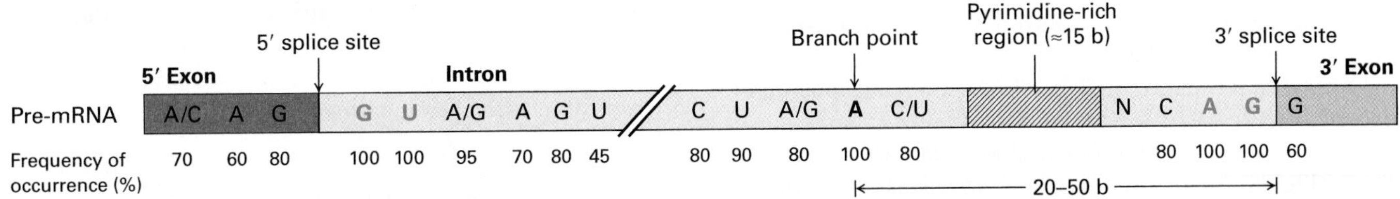

▲ **FIGURE 8-7 Consensus sequences around splice sites in vertebrate pre-mRNAs.** The only nearly invariant bases are the 5′ GU and the 3′ AG of the intron (blue), although the flanking bases indicated are found at frequencies higher than expected based on a random distribution. A pyrimidine-rich region (hatch marked) near the 3′ end of the intron is found in most cases. The branch-point adenosine, also invariant, usually is 20–50 bases from the 3′ splice site. The central region of the intron, which may range from 40 bases to 50 kilobases in length, generally is unnecessary for splicing to occur. [See R. A. Padgett et al., 1986, *Ann. Rev. Biochem.* **55:**1119, and E. B. Keller and W. A. Noon, 1984, *Proc. Nat'l. Acad. Sci. USA* **81:**7417.]

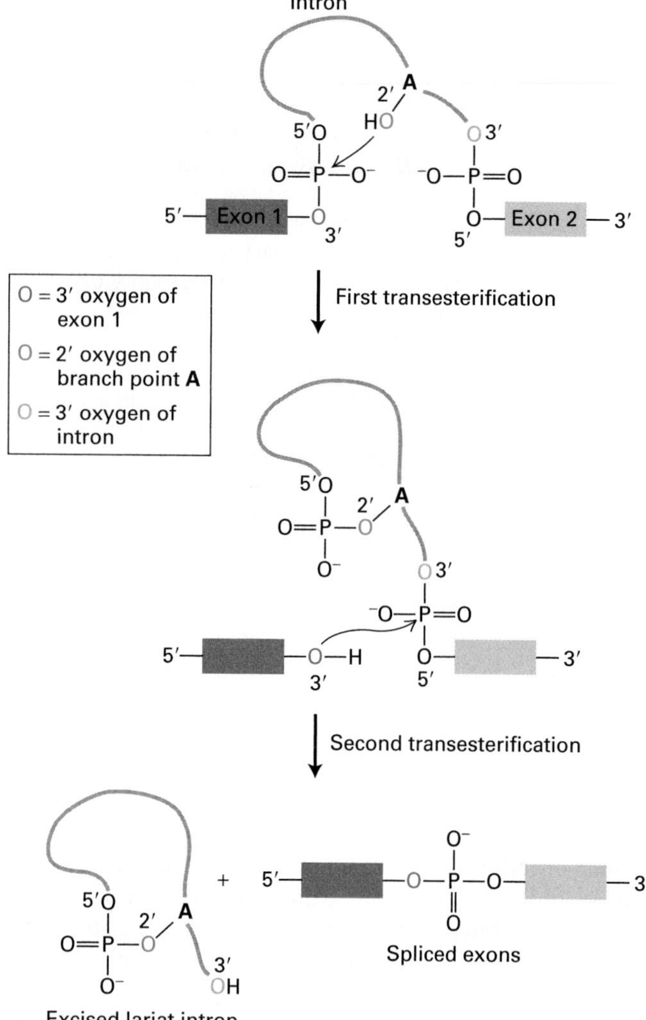

▲ **FIGURE 8-8 Two transesterification reactions that result in splicing of exons in pre-mRNA.** In the first reaction, the ester bond between the 5′ phosphorus of the intron and the 3′ oxygen (dark red) of exon 1 is exchanged for an ester bond with the 2′ oxygen (blue) of the branch-site **A** residue. In the second reaction, the ester bond between the 5′ phosphorus of exon 2 and the 3′ oxygen (orange) of the intron is exchanged for an ester bond with the 3′ oxygen of exon 1, releasing the intron as a lariat structure and joining the two exons. Arrows show where activated hydroxyl oxygens react with phosphorus atoms.

nucleotides at each end of an intron are necessary for splicing to occur at normal rates.

Analysis of the intermediates formed during splicing of pre-mRNAs in vitro led to the discovery that splicing of exons proceeds via two sequential *transesterification reactions* (Figure 8-8). Introns are removed as a lariat-like structure in which the 5′ G of the intron is joined in an unusual 2′,5′-phosphodiester bond to an adenosine near the 3′ end of the intron. This A residue is called *branch point A* because it forms an RNA branch in the lariat structure. In each transesterification reaction, one phosphoester bond is exchanged for another. Since the number of phosphoester bonds in the molecule is not changed in either reaction, no energy is consumed. The net result of these two reactions is that two exons are ligated and the intervening intron is released as a branched lariat structure.

During Splicing, snRNAs Base-Pair with Pre-mRNA

Splicing requires the presence of both **small nuclear RNAs (snRNAs)**, important for base pairing with the pre-mRNA, and associated proteins. Five U-rich snRNAs, designated U1, U2, U4, U5, and U6, participate in pre-mRNA splicing. Ranging in length from 107 to 210 nucleotides, these snRNAs are associated with 6 to 10 proteins in the many small nuclear ribonucleoprotein particles (snRNPs) in the nuclei of eukaryotic cells.

Definitive evidence for the role of U1 snRNA in splicing came from experiments indicating that base pairing between the 5′ splice site of a pre-mRNA and the 5′ region of U1 snRNA is required for RNA splicing (Figure 8-9a). In vitro experiments showed that a synthetic oligonucleotide that hybridizes with the 5′-end region of U1 snRNA blocks RNA splicing. In vivo experiments showed that base-pairing-disrupting mutations in the pre-mRNA 5′ splice block RNA splicing. However, splicing can be restored by expression of a mutant U1 snRNA with a compensating mutation that restores base pairing to the mutant pre-mRNA 5′ splice site (Figure 8-9b). Involvement of U2 snRNA in splicing initially was suspected when it was found to have an internal sequence that is largely complementary to the consensus sequence flanking the branch point in pre-mRNAs (see Figure 8-7). Compensating mutation experiments, similar to those conducted with U1 snRNA and 5′ splice sites, demonstrated that base pairing between U2 snRNA and the branch-point sequence in pre-mRNA also is critical to splicing.

Figure 8-9a illustrates the general structures of the U1 and U2 snRNAs and how they base-pair with pre-mRNA during splicing. Significantly, branch-point A itself, which is not base-paired to U2 snRNA, "bulges out" (Figure 8-10a), allowing its 2′ hydroxyl to participate in the first transesterification reaction of RNA splicing (see Figure 8-8).

Similar studies with other snRNAs demonstrated that base pairing between them also occurs during splicing. Moreover, rearrangements in these RNA-RNA interactions are critical in the splicing pathway, as we describe next.

Spliceosomes, Assembled from snRNPs and a Pre-mRNA, Carry Out Splicing

The five splicing snRNPs and other proteins involved in splicing assemble on a pre-mRNA, forming a large ribonucleoprotein

Focus Animation: mRNA Splicing

(a)

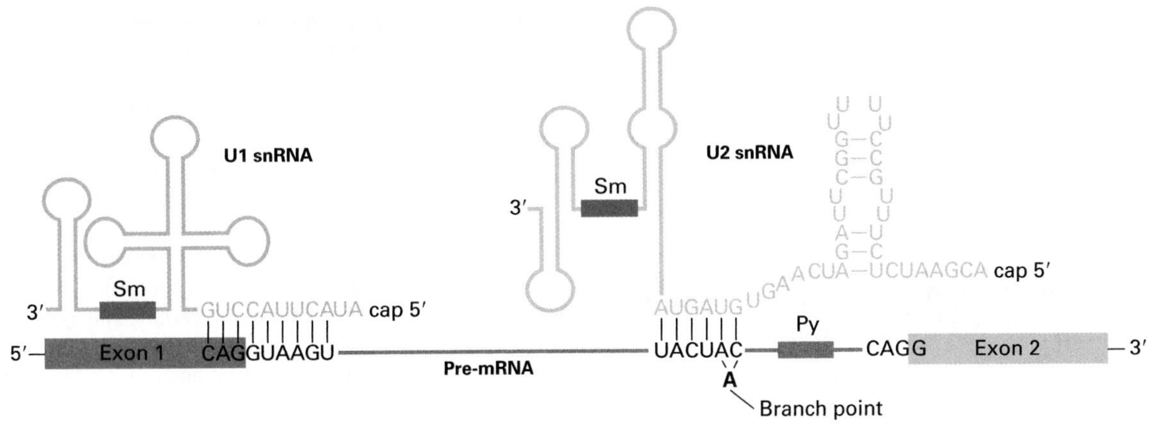

(b)

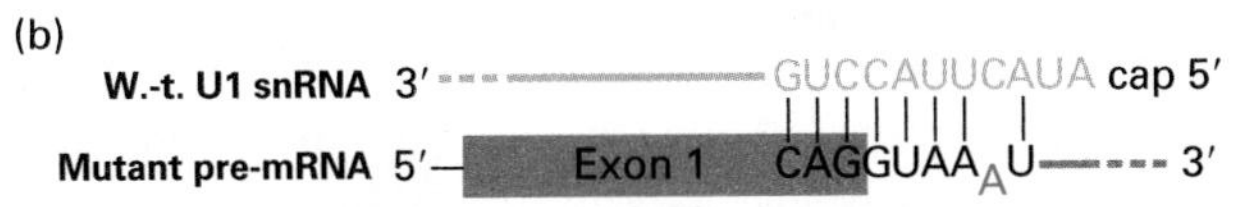

Mutation in pre-mRNA 5′ splice site blocks splicing

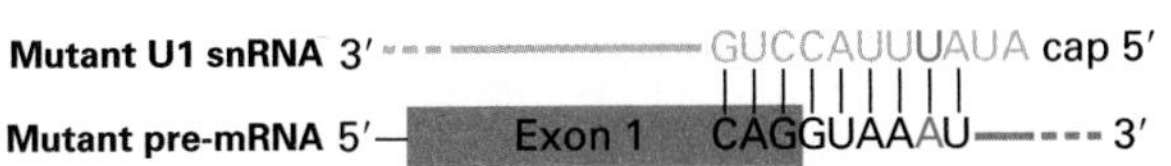

Compensatory mutation in U1 restores splicing

▲ **FIGURE 8-9 Base pairing between pre-mRNA, U1 snRNA, and U2 snRNA early in the splicing process.** (a) In this diagram, secondary structures in the snRNAs that are not altered during splicing are depicted schematically. The yeast branch-point sequence is shown here. Note that U2 snRNA base-pairs with a sequence that includes the branch-point A, although this residue is not base-paired. The purple rectangles represent sequences that bind snRNP proteins recognized by anti-Sm antibodies. For unknown reasons, antisera from patients with the autoimmune disease systemic lupus erythematosus (SLE) contain antibodies to snRNP proteins, which have been useful in characterizing components of the splicing reaction. (b) Only the 5′ ends of U1 snRNAs and 5′ splice sites in pre-mRNAs are shown. *(Left)* A mutation (A) in a pre-mRNA splice site that interferes with base pairing to the 5′ end of U1 snRNA blocks splicing. Expression of a U1 snRNA with a compensating mutation (U) that restores base pairing also restores splicing of the mutant pre-mRNA. [Part (a) adapted from M. J. Moore et al., 1993, in R. Gesteland and J. Atkins, eds., *The RNA World*, Cold Spring Harbor Press, pp. 303–357. Part (b) see Y. Zhuang and A. M. Weiner, 1986, Cell **46**:827.]

Video: Dynamic Nature of Pre-mRNA Splicing Factor Movement in Living Cells

MEDIA CONNECTIONS

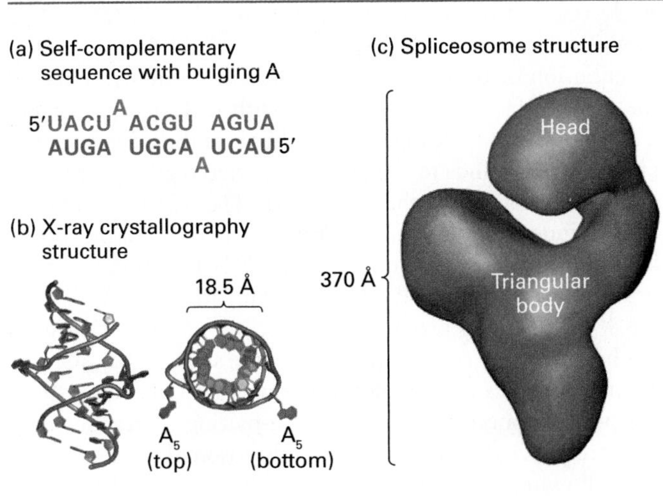

◄ **FIGURE 8-10 Structures of a bulged A in an RNA-RNA helix and an intermediate in the splicing process.** (a) The structure of an RNA duplex with the sequence shown, containing bulged A residues (red) at position 5 in the RNA helix, was determined by x-ray crystallography. (b) The bulged A residues extend from the side of an A-form RNA-RNA helix. The phosphate backbone of one strand is shown in green; the other strand in blue. The structure on the right is turned 90 degrees for a view down the axis of the helix. (c) 40 Å resolution structure of a spliceosomal splicing intermediate containing U2, U4, U5, and U6 snRNPs, determined by cryoelectron microscopy and image reconstruction. The U4/U6/U5 tri-snRNP complex has a similar structure to the triangular body of this complex, suggesting that these snRNPs are at the bottom of the structure shown here and that the head is composed largely of U2snRNP. [Parts (a) and (b) from J. A. Berglund et al., 2001, *RNA* **7**:682. Part (c) from D. Boehringer et al., 2004, *Nat. Struct. Mol. Biol.* **11**:463. See also H. Stark and R. Luhrmann, 2006, *Annu. Rev. Biophys. Biomol. Struct.* **35**:435.]

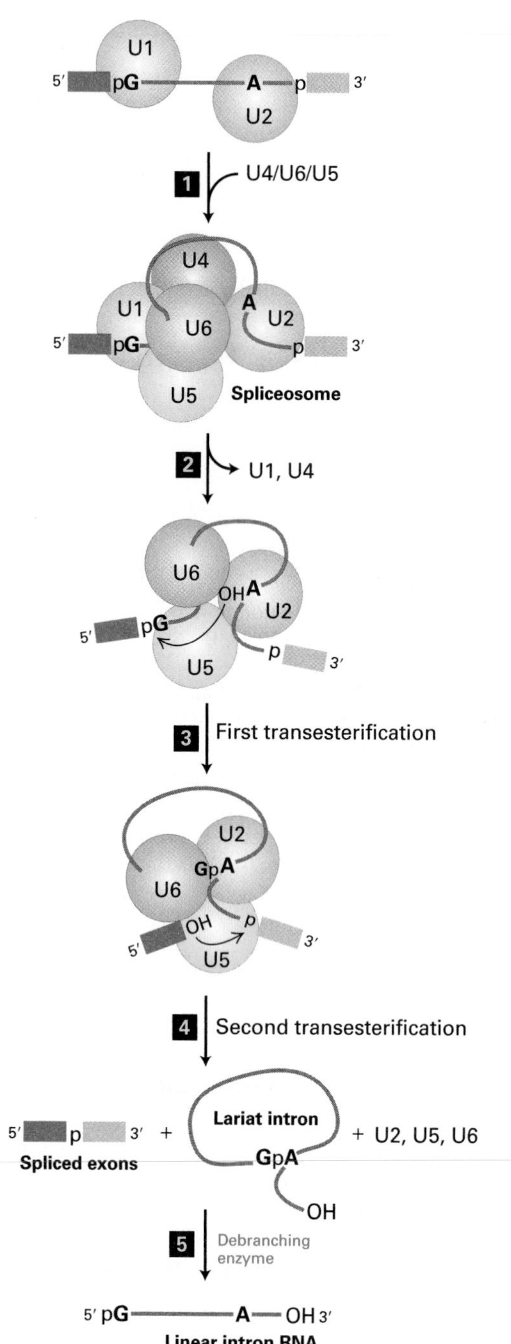

◄ **FIGURE 8-11 Model of spliceosome-mediated splicing of pre-mRNA.** Step **1**: After U1 and U2 snRNPs associate with the pre-mRNA, via base-pairing interactions shown in Figure 8-9, a trimeric snRNP complex of U4, U5, and U6 joins the initial complex to form the spliceosome. Step **2**: Rearrangements of base-pairing interactions between snRNAs converts the spliceosome into a catalytically active conformation and destabilizes the U1 and U4 snRNPs, which are released. Step **3**: The catalytic core, thought to be formed by U6 and U2, then catalyzes the first transesterification reaction, forming the intermediate containing a 2′,5′-phosphodiester bond shown in Figure 8-8. Step **4**: Following further rearrangements between the snRNPs, the second transesterification reaction joins the two exons by a standard 3′,5′-phosphodiester bond and releases the intron as a lariat structure and the remaining snRNPs. Step **5**: The excised lariat intron is converted into a linear RNA by a debranching enzyme. [Adapted from T. Villa et al., 2002, *Cell* **109:**149.]

complex called a **spliceosome** (Figure 8-11). The spliceosome is a large ribonuclear protein complex with a mass similar to that of a ribosome. Assembly of a spliceosome begins with the base pairing of the snRNAs of the U1 and U2 snRNPs to the pre-mRNA (see Figure 8-9). Extensive base pairing between the snRNAs in the U4 and U6 snRNPs forms a complex that associates with U5 snRNP. The U4/U6/U5 complex then associates with the previously formed U1/U2/pre-mRNA complex to generate a spliceosome.

After formation of the spliceosome, extensive rearrangements in the pairing of snRNAs and the pre-mRNA lead to release of first the U1 and then the U4 snRNPs. Figure 8-10b shows a cryoelectron-microscopy structure of an intermediate in the splicing process lacking U1snRNP. A further rearrangement of spliceosomal components occurs with the loss of U4snRNP. This generates a complex that catalyzes the first transesterification reaction that forms the 2′,5′-phosphodiester bond between the 2′ hydroxyl on branch point A and the phosphate at the 5′ end of the intron (Figure 8-11). Following another rearrangement of the snRNPs, the second transesterification reaction ligates the two exons in a standard 3′,5′-phosphodiester bond, releasing the intron as a lariat structure associated with the snRNPs. This final intron-snRNP complex rapidly dissociates, and the individual snRNPs released can participate in a new cycle of splicing. The excised intron is then rapidly degraded by a debranching enzyme and other nuclear RNases discussed later.

As mentioned above, a spliceosome is roughly the size of a ribosome and is composed of about 100 proteins, collectively referred to as splicing factors, in addition to the five snRNPs. This makes RNA splicing comparable in complexity to initiation of transcription and protein synthesis. Some of the splicing factors are associated with snRNPs, but others are not. For instance, the 65-kD subunit of the U2-associated factor (U2AF) binds to the pyrimidine-rich region near the 3′ end of introns and to the U2 snRNP. The 35-kD subunit of U2AF binds to the AG dinucleotide at the 3′ end of the intron and also interacts with the larger U2AF subunit bound nearby. These two U2AF subunits act together to help specify the 3′ splice site by promoting interaction of U2 snRNP with the branch point (see Figure 8-9). Some splicing factors also exhibit sequence homologies to known RNA helicases; these are probably necessary for the base-pairing rearrangements that occur in snRNAs during the spliceosomal splicing cycle.

Following RNA splicing, a specific set of hnRNP proteins remain bound to the spliced RNA approximately 20 nucleotides 5′ to each exon-exon junction, thus forming an *exon-junction complex*. One of the hnRNP proteins associated with the exon junction complex is the *R*NA *e*xport *f*actor (REF), which functions in the export of fully processed

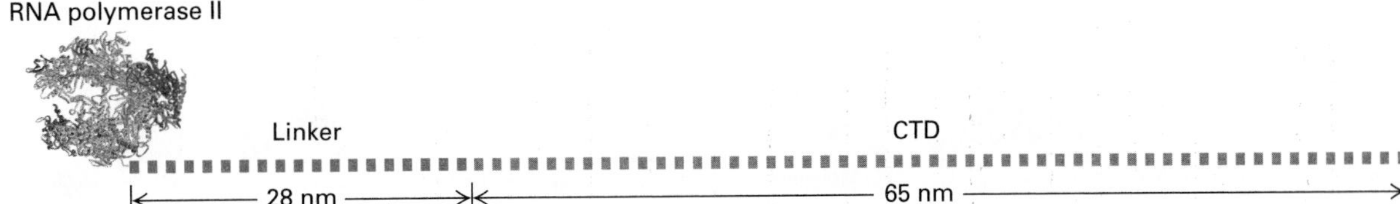

▲ **FIGURE 8-12 Schematic diagram of RNA polymerase II with the CTD extended.** The length of the fully extended yeast RNA polymerase II carboxyl-terminal domain (CTD) and the linker region that connects it to the polymerase is shown relative to the globular domain of the polymerase. The CTD of mammalian RNA polymerase II is twice as long. In its extended form, the CTD can associate with multiple RNA-processing factors simultaneously. [From P. Cramer, D. A. Bushnell, and R. D. Kornberg, 2001, *Science* **292:**1863.]

mRNPs from the nucleus to the cytoplasm, as discussed in Section 8.3. Other proteins associated with the exon junction complex function in a quality-control mechanism that leads to the degradation of improperly spliced mRNAs, known as nonsense-mediated decay (Section 8.4).

A small fraction of pre-mRNAs (≈1% in humans) contain introns whose splice sites do not conform to the standard consensus sequence. This class of introns begins with AU and ends with AC rather than following the usual "GU-AG rule" (see Figure 8-7). Splicing of this special class of introns appears to occur via a splicing cycle analogous to that shown in Figure 8-11, except that four novel, low-abundance snRNPs, together with the standard U5 snRNP, are involved.

Nearly all functional mRNAs in vertebrate, insect, and plant cells are derived from a single molecule of the corresponding pre-mRNA by removal of internal introns and splicing of exons. However, in two types of protozoans—trypanosomes and euglenoids—mRNAs are constructed by splicing together separate RNA molecules. This process, referred to as *trans-splicing,* is also used in the synthesis of 10–15 percent of the mRNAs in the nematode (roundworm) *Caenorhabditis elegans,* an important model organism for studying embryonic development. Trans-splicing is carried out by snRNPs by a process similar to the splicing of exons in a single pre-mRNA.

Chain Elongation by RNA Polymerase II Is Coupled to the Presence of RNA-Processing Factors

How is RNA processing efficiently coupled with the transcription of a pre-mRNA? The key lies in the long carboxyl-terminal domain (CTD) of RNA polymerase II, which, as discussed in Chapter 7, is composed of multiple repeats of a seven-residue (heptapeptide) sequence. When fully extended, the CTD domain in the yeast enzyme is about 65 nm long (Figure 8-12); the CTD in human RNA polymerase II is about twice as long. The remarkable length of the CTD apparently allows multiple proteins to associate simultaneously with a single RNA polymerase II molecule. For instance, as mentioned earlier, the enzymes that add the 5′ cap to nascent transcripts associate with the phosphorylated CTD shortly after transcription initiation. In addition, RNA splicing and polyadenylation factors have been found to associate with the phosphorylated CTD. As a consequence, these processing factors are present at high local concentrations when splice sites and poly(A) signals are transcribed by the polymerase, enhancing the rate and specificity of RNA processing.

Recent results indicate that the association of RNA splicing factors with phosphorylated CTD also stimulates transcription elongation. Thus chain elongation is coupled to the binding of RNA-processing factors to the phosphorylated CTD. This mechanism may ensure that a pre-mRNA is not synthesized unless the machinery for processing it is properly positioned.

SR Proteins Contribute to Exon Definition in Long Pre-mRNAs

The average length of an exon in the human genome is ≈150 bases, whereas the average length of an intron is much longer (≈3500 bases). The longest introns contain upward of 500 kb! Because the sequences of 5′ and 3′ splice sites and branch points are so degenerate, multiple copies are likely to occur randomly in long introns. Consequently, additional sequence information is required to define the exons that should be spliced together in higher organisms with long introns.

The information for defining the splice sites that demarcate exons is encoded within the sequences of exons. A family of RNA-binding proteins, the *SR proteins,* interact with sequences within exons called *exonic splicing enhancers.* SR proteins are a subset of the hnRNP proteins discussed earlier and so contain one or more RRM RNA-binding domains. They also contain several protein-protein interaction domains rich in serine (S) and arginine (R) residues. When bound to exonic splicing enhancers, SR proteins mediate the cooperative binding of U1 snRNP to a true 5′ splice site and U2 snRNP to a branch point through a network of protein-protein interactions that span across an exon (Figure 8-13). The complex of SR proteins, snRNPs, and other splicing factors (e.g., U2AF) that assemble across an exon, which has been called a **cross-exon recognition complex,** permits precise specification of exons in long pre-mRNAs.

In the transcription units of higher organisms with long introns, exons not only encode the amino acid sequences of different portions of a protein but also contain binding sites for SR proteins. Mutations that interfere with the binding of an SR protein to an exonic splicing enhancer, even if they do not change the encoded amino acid sequence, would prevent formation of the cross-exon recognition complex. As a result, the affected exon is "skipped" during splicing and is not included in the final processed mRNA. The truncated mRNA produced in this case is either degraded or

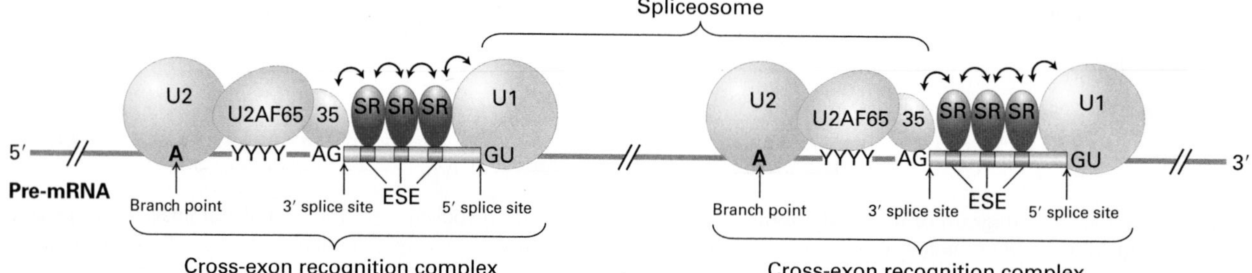

▲ **FIGURE 8-13 Exon recognition through cooperative binding of SR proteins and splicing factors to pre-mRNA.** The correct 5′ GU and 3′ AG splice sites are recognized by splicing factors on the basis of their proximity to exons. The exons contain exonic splicing enhancers (ESEs) that are binding sites for SR proteins. When bound to ESEs, the SR proteins interact with one another and promote the cooperative binding of the U1 snRNP to the 5′ splice site of the downstream intron, the U2 snRNP to the branch point of the upstream intron, the 65- and 35-kD subunits of U2AF to the pyrimidine-rich region and AG 3′ splice site of the upstream intron, and other splicing factors (not shown). The resulting RNA-protein cross-exon recognition complex spans an exon and activates the correct splice sites for RNA splicing. Note that the U1 and U2 snRNPs in this unit are not part of the same spliceosome. The U2 snRNP on the right forms a spliceosome with the U1 snRNP bound to the 5′ end of the same intron. The U1 snRNP shown on the right forms a spliceosome with the U2 snRNP bound to the branch point of the downstream intron (not shown), and the U2 snRNP on the left forms a spliceosome with a U1 snRNP bound to the 5′ splice site of the upstream intron (not shown). Double-headed arrows indicate protein-protein interactions. [Adapted from T. Maniatis, 2002, *Nature* **418:**236; see also S. M. Berget, 1995, *J. Biol. Chem.* **270:**2411.]

translated into a mutant, abnormally functioning protein. Recent studies have implicated this type of mutation in human genetic diseases. For example, *spinal muscle atrophy* is one of the most common genetic causes of childhood mortality. This disease results from mutations in a region of the genome containing two closely related genes, *SMN1* and *SMN2*, that arose by gene duplication. *SMN2* encodes a protein identical with *SMN1*. SMN2 is expressed at much lower level because a silent mutation in one exon interferes with the binding of an SR protein. This leads to exon skipping in most of the *SMN2* mRNAs. The homologous *SMN* gene in the mouse, where there is only a single copy, is essential for cell viability. Spinal muscle atrophy in humans results from homozygous mutations that inactivate *SMN1*. The low level of protein translated from the small fraction of *SMN2* mRNAs that are correctly spliced is sufficient to maintain cell viability during embryogenesis and fetal development, but it is not sufficient to maintain viability of spinal cord motor neurons in childhood, resulting in their death and the associated disease.

Approximately 15 percent of the single-base-pair mutations that cause human genetic diseases interfere with proper exon definition. Some of these mutations occur in 5′ or 3′ splice sites, often resulting in the use of nearby alternative "cryptic" splice sites present in the normal gene sequence. In the absence of the normal splice site, the cross-exon recognition complex recognizes these alternative sites. Other mutations that cause abnormal splicing result in a new consensus splice site sequence that becomes recognized in place of the normal splice site. Finally, some mutations can interfere with the binding of specific SR proteins to pre-mRNAs. These mutations inhibit splicing at normal splice sites, as in the case of the *SMN2* gene, and thus lead to exon skipping. ■

Self-Splicing Group II Introns Provide Clues to the Evolution of snRNAs

Under certain nonphysiological in vitro conditions, pure preparations of some RNA transcripts slowly splice out introns in the absence of any protein. This observation led to the recognition that some introns are *self-splicing*. Two types of self-splicing introns have been discovered: *group I introns*, present in nuclear rRNA genes of protozoans, and *group II introns*, present in protein-coding genes and some rRNA and tRNA genes in mitochondria and chloroplasts of plants and fungi. Discovery of the catalytic activity of self-splicing introns revolutionized concepts about the functions of RNA. As discussed in Chapter 4, RNA is now thought to catalyze peptide-bond formation during protein synthesis in ribosomes. Here we discuss the probable role of group II introns, now found only in mitochondrial and chloroplast DNA, in the evolution of snRNAs; the functioning of group I introns is considered in the later section on rRNA processing.

Even though their precise sequences are not highly conserved, all group II introns fold into a conserved, complex secondary structure containing numerous stem loops (Figure 8-14a). Self-splicing by a group II intron occurs via two transesterification reactions, involving intermediates and products analogous to those found in nuclear pre-mRNA splicing. The mechanistic similarities between group II intron self-splicing and spliceosomal splicing led to the hypothesis that snRNAs function analogously to the stem loops in the secondary structure of group II introns. According to this hypothesis, snRNAs interact with 5′ and 3′ splice sites of pre-mRNAs and with each other to produce a three-dimensional RNA structure functionally analogous to that of group II self-splicing introns (Figure 8-14b).

An extension of this hypothesis is that introns in ancient pre-mRNAs evolved from group II self-splicing introns through the progressive loss of internal RNA structures, which concurrently evolved into trans-acting snRNAs that perform the same functions. Support for this type of evolutionary model comes from experiments with group II intron mutants in which domain V and part of domain I are deleted. RNA transcripts containing such mutant introns are defective in self-splicing, but when RNA molecules equivalent

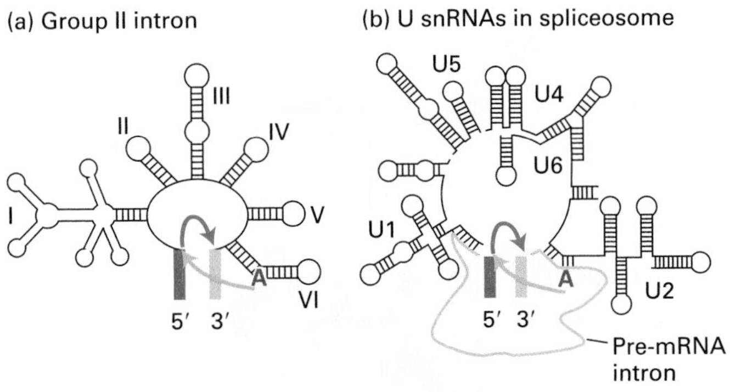

(a) Group II intron (b) U snRNAs in spliceosome

▲ **FIGURE 8-14 Comparison of group II self-splicing introns and the spliceosome.** The schematic diagrams comparing the secondary structures of (a) group II self-splicing introns and (b) U snRNAs present in the spliceosome. The first transesterification reaction is indicated by light green arrows; the second reaction, by blue arrows. The branch-point A is boldfaced. The similarity in these structures suggests that the spliceosomal snRNAs evolved from group II introns, with the trans-acting snRNAs being functionally analogous to the corresponding domains in group II introns. The colored bars flanking the introns in (a) and (b) represent exons. [Adapted from P. A. Sharp, 1991, *Science* **254**:663.]

to the deleted regions are added to the in vitro reaction, self-splicing occurs. This finding demonstrates that these domains in group II introns can be trans-acting, like snRNAs.

The similarity in the mechanisms of group II intron self-splicing and spliceosomal splicing of pre-mRNAs also suggests that the splicing reaction is catalyzed by the snRNA, not the protein, components of spliceosomes. Although group II introns can self-splice in vitro at elevated temperatures and Mg^{2+} concentrations, under in vivo conditions proteins called *maturases*, which bind to group II intron RNA, are required for rapid splicing. Maturases are thought to stabilize the precise three-dimensional interactions of the intron RNA required to catalyze the two splicing transesterification reactions. By analogy, snRNP proteins in spliceosomes are thought to stabilize the precise geometry of snRNAs and intron nucleotides required to catalyze pre-mRNA splicing.

The evolution of snRNAs may have been an important step in the rapid evolution of higher eukaryotes. As internal intron sequences were lost and their functions in RNA splicing supplanted by trans-acting snRNAs, the remaining intron sequences would be free to diverge. This in turn likely facilitated the evolution of new genes through **exon shuffling** since there are few constraints on the sequence of new introns generated in the process (see Figures 6-18 and 6-19). It also permitted the increase in protein diversity that results from alternative RNA splicing and an additional level of gene control resulting from regulated RNA splicing.

3′ Cleavage and Polyadenylation of Pre-mRNAs Are Tightly Coupled

In eukaryotic cells, all mRNAs, except histone mRNAs, have a 3′ *poly(A) tail.* (The major histone mRNAs are transcribed from repeated genes at prodigious levels in replicating cells during the S phase. They undergo a special form of 3′-end processing that involves cleavage but not polyadenylation. Specialized RNA-binding proteins that help to regulate histone mRNA translation bind to the 3′ end generated by this specialized system.) Early studies of pulse-labeled adenovirus and SV40 RNA demonstrated that the viral primary transcripts extend beyond the site from which the poly(A) tail extends. These results suggested that A residues are added to a 3′ hydroxyl generated by endonucleolytic cleavage of a longer transcript, but the predicted downstream RNA fragments never were detected in vivo, presumably because of their rapid degradation. However, detection of both predicted cleavage products was observed in in vitro processing reactions performed with nuclear extracts of cultured human cells. The cleavage/polyadenylation process and degradation of the RNA downstream of the cleavage site occurs much more slowly in these in vitro reactions, simplifying detection of the downstream cleavage product.

Early sequencing of cDNA clones from animal cells showed that nearly all mRNAs contain the sequence AAUAAA 10–35 nucleotides *upstream* from the poly(A) tail (Figure 8-15). Polyadenylation of RNA transcripts is virtually eliminated when the corresponding sequence in the template DNA is mutated to any other sequence except one encoding a closely related sequence (AUUAAA). The unprocessed RNA transcripts produced from such mutant templates do not accumulate in nuclei but are rapidly degraded. Further mutagenesis studies revealed that a second signal downstream from the cleavage site is required for efficient cleavage and polyadenylation of most pre-mRNAs in animal cells. This downstream signal is not a specific sequence but rather a GU-rich or simply a U-rich region within ≈50 nucleotides of the cleavage site.

Identification and purification of the proteins required for cleavage and polyadenylation of pre-mRNA have led to the model shown in Figure 8-15. According to this model, a 360-kDa cleavage and polyadenylation specificity factor (CPSF), composed of four different polypeptides, first forms an unstable complex with the upstream AAUAAA poly(A) signal. Then at least three additional proteins bind to the CPSF-RNA complex: a 200-kDa heterotrimer called *cleavage stimulatory factor (CStF),* which interacts with the G/U-rich sequence; a 150-kDa heterotrimer called *cleavage factor I (CFI);* and a second, poorly characterized cleavage factor (CFII). Finally, a *poly(A) polymerase (PAP)* binds to the complex *before* cleavage can occur. This requirement for PAP binding links cleavage and polyadenylation, so that the free 3′ end generated is rapidly polyadenylated and no essential information is lost to exonuclease degradation of an unprotected 3′ end.

Assembly of the large, multiprotein **cleavage/polyadenylation complex** around the AU-rich poly(A) signal in a pre-mRNA is analogous in many ways to formation of the transcription-preinitiation complex at the AT-rich TATA box of a template DNA molecule (see Figure 7-31). In both cases, multiprotein complexes assemble cooperatively through a network of specific protein–nucleic acid and protein-protein interactions.

Following cleavage at the poly(A) site, polyadenylation proceeds in two phases. Addition of the first 12 or so A

▶ **FIGURE 8-15 Model for cleavage and polyadenylation of pre-mRNAs in mammalian cells.** Cleavage and polyadenylation specificity factor (CPSF) binds to the upstream AAUAAA poly(A) signal. CStF interacts with a downstream GU- or U-rich sequence and with bound CPSF, forming a loop in the RNA; binding of CFI and CFII help stabilize the complex. Binding of poly(A) polymerase (PAP) then stimulates cleavage at a poly(A) site, which usually is 10–35 nucleotides 3′ of the upstream poly(A) signal. The cleavage factors are released, as is the downstream RNA cleavage product, which is rapidly degraded. Bound PAP then adds ≈12 A residues at a slow rate to the 3′-hydroxyl group generated by the cleavage reaction. Binding of poly(A)-binding protein II (PABPII) to the initial short poly(A) tail accelerates the rate of addition by PAP. After 200–250 A residues have been added, PABPII signals PAP to stop polymerization.

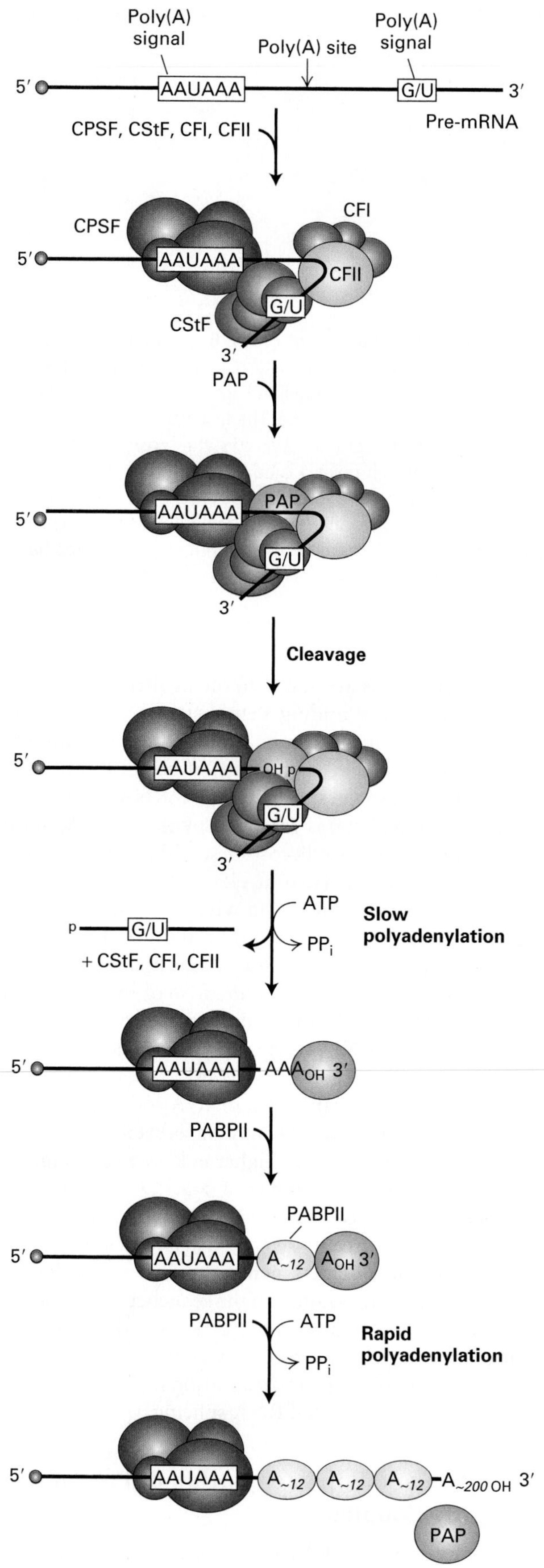

residues occurs slowly, followed by rapid addition of up to 200–250 more A residues. The rapid phase requires the binding of multiple copies of a *poly(A)-binding protein* containing the RRM motif. This protein is designated *PABPII* to distinguish it from the poly(A)-binding protein present in the cytoplasm. PABPII binds to the short A tail initially added by PAP, stimulating the rate of polymerization of additional A residues by PAP, resulting in the fast phase of polyadenylation (Figure 8-15). PABPII is also responsible for signaling poly(A) polymerase to terminate polymerization when the poly(A) tail reaches a length of 200–250 residues, although the mechanism for controlling the length of the tail is not yet understood. Binding of PABP to the poly(A) tail is essential for mRNA export into the cytoplasm.

Nuclear Exonucleases Degrade RNA That Is Processed Out of Pre-mRNAs

Because the human genome contains long introns, only ≈5 percent of the nucleotides that are polymerized by RNA polymerase II during transcription are retained in mature, processed mRNAs. Although this process appears inefficient, it probably evolved in multicellular organisms because the process of exon shuffling facilitated the evolution of new genes in organisms with long introns (Chapter 6). The introns that are spliced out and the region downstream from the cleavage and polyadenylation site are degraded by nuclear exoribonucleases that hydrolyze one base at a time from either the 5′ or 3′ end of a RNA strand.

As mentioned earlier, the 2′,5′-phosphodiester bond in excised introns is hydrolyzed by a debranching enzyme, yielding a linear molecule with unprotected ends that can be attacked by exonucleases (see Figure 8-11). The predominant nuclear decay pathway is 3′ → 5′ hydrolysis by 11 exonucleases that associate with one another in a large protein complex called the **exosome.** Other proteins in the complex include RNA helicases that disrupt base pairing and RNA-protein interactions that would otherwise impede the exonucleases. Exosomes also function in the cytoplasm, as discussed later. In addition, the exosome appears to degrade pre-mRNAs that have not been properly spliced or polyadenylated. It is not yet clear how the exosome recognizes improperly processed pre-mRNAs. But in yeast cells with temperature-sensitive mutant poly(A) polymerase (Figure 8-15), pre-mRNAs are retained at their sites of transcription at the nonpermissive temperature. These abnormally processed pre-mRNAs are released in cells with a second mutation in a subunit of the exosome found only in nuclear and not cytoplasmic exosomes (PM-Scl; 100 kD in humans). Also, exosomes are found concentrated at sites of transcription in *Drosophila* polytene chromosomes, where they are associated with RNA polymerase II elongation factors. These results suggest that the exosome participates in an as yet poorly understood quality-control mechanism that recognizes aberrantly processed pre-mRNAs, preventing their export to the cytoplasm and ultimately leading to their degradation.

To avoid being degraded by nuclear exonucleases, nascent transcripts, pre-mRNA processing intermediates, and mature mRNAs in the nucleus must have their ends protected. As discussed above, the 5′ end of a nascent transcript is protected by addition of the 5′ cap structure as soon as the 5′ end emerges from the polymerase. The 5′ cap is protected because it is bound by a *nuclear cap-binding complex,* which protects it from 5′ exonucleases and also functions in export of mRNA to the cytoplasm. The 3′ end of a nascent transcript lies within the RNA polymerase and thus is inaccessible to exonucleases (see Figure 4-12). As discussed previously, the free 3′ end generated by cleavage of a pre-mRNA downstream from the poly(A) signal is rapidly polyadenylated by the poly(A) polymerase associated with the other 3′ processing factors, and the resulting poly(A) tail is bound by PABPII (Figure 8-15). This tight coupling of cleavage and polyadenylation protects the 3′ end from exonuclease attack.

KEY CONCEPTS OF SECTION 8.1

Processing of Eukaryotic Pre-mRNA

- In the nucleus of eukaryotic cells, pre-mRNAs are associated with hnRNP proteins and processed by 5′ capping, 3′ cleavage and polyadenylation, and splicing before being transported to the cytoplasm (see Figure 8-2).
- Shortly after transcription initiation, a capping enzyme associated with the phosphorylated CTD of RNA polymerase II adds the 5′ cap to the nascent transcript. Other RNA processing factors involved in RNA splicing, 3′ cleavage, and polyadenylation also associate with the phosphorylated CTD, increasing the rate of transcription elongation. Consequently, transcription does not proceed at a high rate until RNA processing factors become associated with the CTD, where they are poised to interact with the nascent pre-mRNA as it emerges from the surface of the polymerase.
- Five different snRNPs interact via base pairing with one another and with pre-mRNA to form the spliceosome (see Figure 8-11). This very large ribonucleoprotein complex catalyzes two transesterification reactions that join two exons and remove the intron as a lariat structure, which is subsequently degraded (see Figure 8-8).
- SR proteins that bind to exonic splicing enhancer sequences in exons are critical in defining exons in the large pre-mRNAs of higher organisms. A network of interactions between SR proteins, snRNPs, and splicing factors forms a cross-exon recognition complex that specifies correct splice sites (see Figure 8-13).
- The snRNAs in the spliceosome are thought to have an overall tertiary structure similar to that of group II self-splicing introns.
- For long transcription units in higher organisms, splicing of exons usually begins as the pre-mRNA is still being formed. Cleavage and polyadenylation to form the 3′ end of the mRNA occur after the poly(A) site is transcribed.
- In most protein-coding genes, a conserved AAUAAA poly(A) signal lies slightly upstream from a poly(A) site where cleavage and polyadenylation occur. A GU- or U-rich sequence downstream from the poly(A) site contributes to the efficiency of cleavage and polyadenylation.
- A multiprotein complex that includes poly(A) polymerase (PAP) carries out the cleavage and polyadenylation of a pre-mRNA. A nuclear poly(A)-binding protein, PABPII, stimulates addition of A residues by PAP and stops addition once the poly(A) tail reaches 200–250 residues (see Figure 8-15).
- Excised introns and RNA downstream from the cleavage/poly(A) site are degraded primarily by exosomes, multiprotein complexes that contain eleven 3′ → 5′ exonucleases as well as RNA helicases. Exosomes also degrade improperly processed pre-mRNAs.

8.2 Regulation of Pre-mRNA Processing

Now that we've seen how pre-mRNAs are processed into mature, functional mRNAs, we consider how regulation of this process can contribute to gene control. Recall from Chapter 6 that higher eukaryotes contain both simple and complex transcription units encoded in their DNA. The primary transcripts produced from the former contain one poly(A) site and exhibit only one pattern of RNA splicing, even if multiple introns are present; thus simple transcription units encode a single mRNA. In contrast, the primary transcripts produced from complex transcription units ($\approx$60% of all human transcription units) can be processed in alternative ways to yield different mRNAs that encode distinct proteins (see Figure 6-3).

Alternative Splicing Is the Primary Mechanism for Regulating mRNA Processing

The discovery that a large fraction of transcription units in higher organisms encode alternatively spliced mRNAs and that differently spliced mRNAs are expressed in different cell types revealed that regulation of RNA splicing is an important gene-control mechanism in higher eukaryotes. Although many examples of cleavage at alternative poly(A) sites in pre-mRNAs are known, **alternative splicing** of different exons is the more common mechanism for expressing different

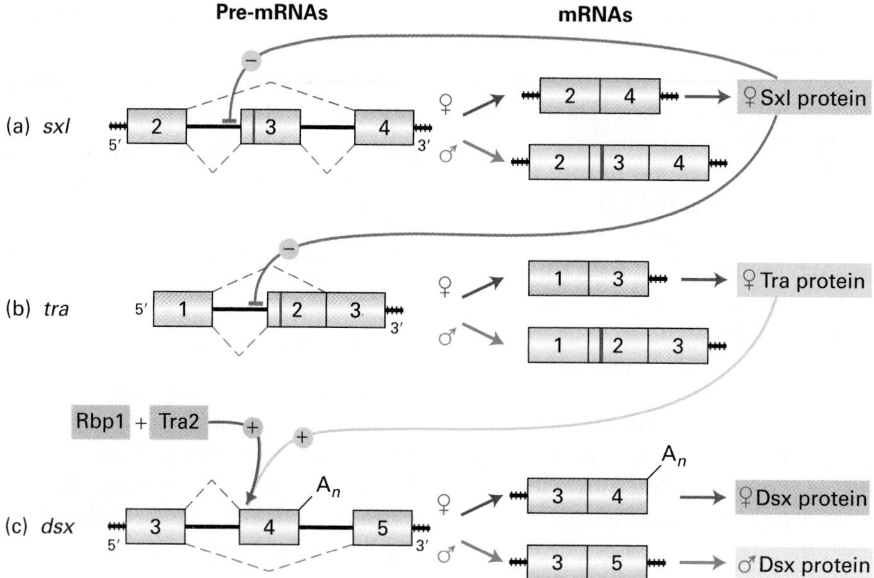

▲ **FIGURE 8-16 Cascade of regulated splicing that controls sex determination in *Drosophila* embryos.** For clarity, only the exons (boxes) and introns (black lines) where regulated splicing occurs are shown. Splicing is indicated by red dashed lines above (female) and blue dashed lines below (male) the pre-mRNAs. Vertical red lines in exons indicate in-frame stop codons, which prevent synthesis of functional protein. Only female embryos produce functional Sxl protein, which *represses* splicing between exons 2 and 3 in *sxl* pre-mRNA (a) and between exons 1 and 2 in *tra* pre-mRNA (b). (c) In contrast, the cooperative binding of Tra protein and two SR proteins, Rbp1 and Tra2, *activates* splicing between exons 3 and 4 and cleavage/polyadenylation A_n at the 3′ end of exon 4 in *dsx* pre-mRNA in female embryos. In male embryos, which lack functional Tra, the SR proteins do not bind to exon 4, and consequently exon 3 is spliced to exon 5. The distinct Dsx proteins produced in female and male embryos as the result of this cascade of regulated splicing repress transcription of genes required for sexual differentiation of the opposite sex. [Adapted from M. J. Moore et al., 1993, in R. Gesteland and J. Atkins, eds., *The RNA World,* Cold Spring Harbor Press, pp. 303–357.]

proteins from one complex transcription unit. In Chapter 4, for example, we mentioned that fibroblasts produce one type of the extracellular protein fibronectin, whereas hepatocytes produce another type. Both fibronectin **isoforms** are encoded by the same transcription unit, which is spliced differently in the two cell types to yield two different mRNAs (see Figure 4-16). In other cases, alternative processing may occur simultaneously in the same cell type in response to different developmental or environmental signals. First we discuss one of the best-understood examples of regulated RNA processing and then briefly consider the consequences of RNA splicing in the development of the nervous system.

A Cascade of Regulated RNA Splicing Controls Drosophila Sexual Differentiation

One of the earliest examples of regulated alternative splicing of pre-mRNA came from studies of sexual differentiation in *Drosophila*. Genes required for normal *Drosophila* sexual differentiation were first characterized by isolating *Drosophila* mutants defective in the process. When the proteins encoded by the wild-type genes were characterized biochemically, two of them were found to regulate a cascade of alternative RNA splicing in *Drosophila* embryos. More recent research has provided insight into how these proteins regulate RNA processing and ultimately lead to the creation of two different sex-specific transcriptional repressors that suppress the development of characteristics of the opposite sex.

The Sxl protein, encoded by the *sex-lethal* gene, is the first protein to act in the cascade (Figure 8-16). The Sxl protein is present only in female embryos. Early in development, the gene is transcribed from a promoter that functions only in female embryos. Later in development, this female-specific promoter is shut off and another promoter for *sex-lethal* becomes active in both male and female embryos. However, in the absence of early Sxl protein, the *sex-lethal* pre-mRNA in male embryos is spliced to produce an mRNA that contains a stop codon early in the sequence. The net result is that male embryos produce no functional Sxl protein either early or later in development.

In contrast, the Sxl protein expressed in early female embryos directs splicing of the *sex-lethal* pre-mRNA so that a functional *sex-lethal* mRNA is produced (Figure 8-16a). Sxl accomplishes this by binding to a sequence in the pre-mRNA near the 3′ end of the intron between exon 2 and exon 3, thereby blocking the proper association of U2AF and U2 snRNP. As a consequence, the U1 snRNP bound to the 3′ end of exon 2 assembles into a spliceosome with U2 snRNP bound to the branch point at the 3′ end of the intron between exons 3 and 4, leading to splicing of exon 2 to 4 and skipping of exon 3. The resulting female-specific *sex-lethal* mRNA is translated into functional Sxl protein, which reinforces its own expression in female embryos by continuing to cause skipping of exon 3. The absence of Sxl protein in male embryos allows the inclusion of exon 3 and, consequently, of the stop codon that prevents translation of functional Sxl protein.

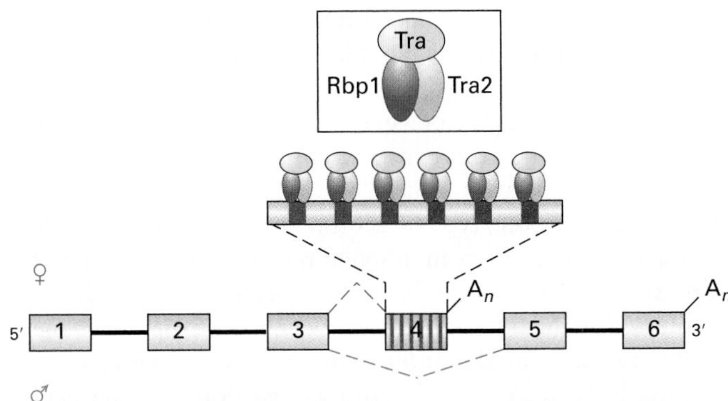

▲ **FIGURE 8-17 Model of splicing activation by Tra protein and the SR proteins Rbp1 and Tra2.** In female *Drosophila* embryos, splicing of exons 3 and 4 in *dsx* pre-mRNA is activated by binding of Tra/Tra2/Rbp1 complexes to six sites in exon 4. Because Rbp1 and Tra2 cannot bind to the pre-mRNA in the absence of Tra, exon 4 is skipped in male embryos. See the text for discussion. A_n = polyadenylation. [Adapted from T. Maniatis and B. Tasic, 2002, *Nature* **418**:236.]

Sxl protein also regulates alternative RNA splicing of the *transformer* gene pre-mRNA (Figure 8-16b). In male embryos, where no Sxl is expressed, exon 1 is spliced to exon 2, which contains a stop codon that prevents synthesis of a functional *transformer* protein. In female embryos, however, binding of Sxl protein to the 3′ end of the intron between exons 1 and 2 blocks binding of U2AF at this site. The interaction of Sxl with *transformer* pre-mRNA is mediated by two RRM domains in the protein (see Figure 8-5). When Sxl is bound, U2AF binds to a lower-affinity site farther 3′ in the pre-mRNA; as a result exon 1 is spliced to this alternative 3′ splice site, eliminating exon 2 with its stop codon. The resulting female-specific *transformer* mRNA, which contains additional constitutively spliced exons, is translated into functional Transformer (Tra) protein.

Finally, Tra protein regulates the alternative processing of pre-mRNA transcribed from the *double-sex* gene (Figure 8-16c). In female embryos, a complex of Tra and two constitutively expressed proteins, Rbp1 and Tra2, directs splicing of exon 3 to exon 4 and also promotes cleavage/polyadenylation at the alternative poly(A) site at the 3′ end of exon 4—leading to a short, female-specific version of the Dsx protein. In male embryos, which produce no Tra protein, exon 4 is skipped, so that exon 3 is spliced to exon 5. Exon 5 is constitutively spliced to exon 6, which is polyadenylated at its 3′ end—leading to a longer, male-specific version of the Dsx protein. As a result of the cascade of regulated RNA processing depicted in Figure 8-16, different Dsx proteins are expressed in male and female embryos. The male Dsx protein is a transcriptional repressor that inhibits the expression of genes required for female development. Conversely, the female Dsx protein represses transcription of genes required for male development.

Figure 8-17 illustrates how the Tra/Tra2/Rbp1 complex is thought to interact with *double-sex (dsx)* pre-mRNA. Rbp1 and Tra2 are SR proteins, but they do not interact with exon 4 in the absence of the Tra protein. Tra protein interacts with Rbp1 and Tra2, resulting in the cooperative binding of all three proteins to six exonic splicing enhancers in exon 4. The bound Tra2 and Rbp1 proteins then promote the binding of U2AF and U2 snRNP to the 3′ end of the intron between exons 3 and 4, just as other SR proteins do for constitutively spliced exons (see Figure 8-13). The Tra/Tra2/Rbp1 complexes may also enhance binding of the cleavage/polyadenylation complex to the 3′ end of exon 4.

Splicing Repressors and Activators Control Splicing at Alternative Sites

As is evident from Figure 8-16, the *Drosophila* Sxl protein and Tra protein have opposite effects: Sxl prevents splicing, causing exons to be skipped, whereas Tra promotes splicing. The action of similar proteins may explain the cell-type-specific expression of fibronectin isoforms in humans. For instance, an Sxl-like splicing repressor expressed in hepatocytes might bind to splice sites for the EIIIA and EIIIB exons in the fibronectin pre-mRNA, causing them to be skipped during RNA splicing (see Figure 4-16). Alternatively, a Tra-like splicing activator expressed in fibroblasts might activate the splice sites associated with the fibronectin EIIIA and EIIIB exons, leading to inclusion of these exons in the mature mRNA. Experimental examination in some systems has revealed that inclusion of an exon in some cell types versus skipping of the same exon in other cell types results from the combined influence of several splicing repressors and enhancers.

Alternative splicing of exons is especially common in the nervous system, generating multiple isoforms of many proteins required for neuronal development and function in both vertebrates and invertebrates. The primary transcripts from these genes often show complex splicing patterns that can generate several different mRNAs, with different spliced forms expressed in different anatomical locations within the central nervous system. We consider two remarkable examples that illustrate the critical role of this process in neural function.

Expression of K^+-Channel Proteins in Vertebrate Hair Cells In the inner ear of vertebrates, individual "hair cells," which are ciliated neurons, respond most strongly to a specific frequency of sound. Cells tuned to low frequency ($\approx$50 Hz) are found at one end of the tubular cochlea that makes up the inner ear; cells responding to high frequency ($\approx$5000 Hz) are found at the other end (Figure 8-18a). Cells in between the ends respond to a gradient of frequencies between these extremes. One component in the tuning of hair cells in reptiles and birds is the opening of K^+ ion channels in response to increased intracellular Ca^{2+} concentrations. The Ca^{2+} concentration at which the channel opens determines the frequency with which the membrane potential oscillates and hence the frequency to which the cell is tuned.

The gene encoding this Ca^{2+}-activated K^+ channel is expressed as multiple, alternatively spliced mRNAs. The various proteins encoded by these alternative mRNAs open at different Ca^{2+} concentrations. Hair cells with different response frequencies express different isoforms of the channel protein depending on their position along the length of the cochlea (see Figure 23-30). The sequence variation in the protein is very complex: there are at least eight regions in the mRNA where alternative exons are utilized, permitting the expression of 576 possible isoforms (Figure 8-18b). PCR analysis of mRNAs from

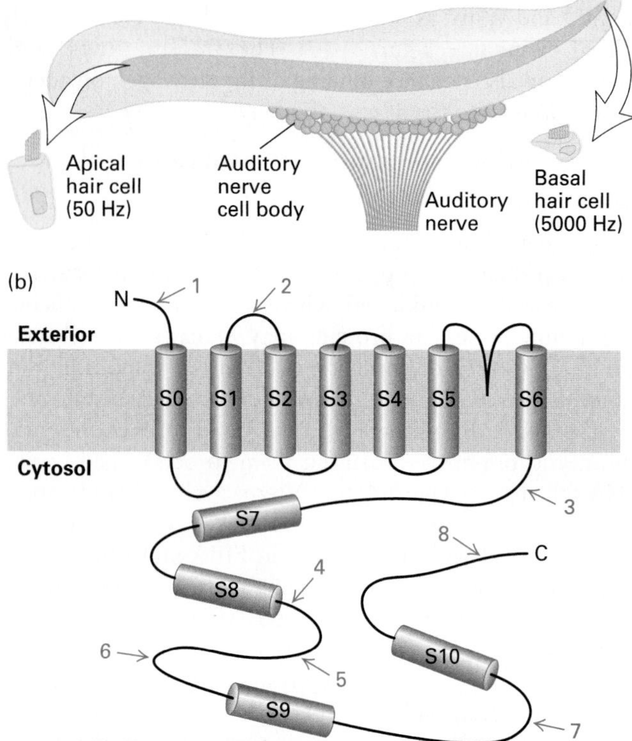

▲ **FIGURE 8-18 Role of alternative splicing in the perception of sounds of different frequency.** (a) The chicken cochlea, a 5-mm-long tube, contains an epithelium of auditory hair cells that are tuned to a gradient of vibrational frequencies from 50 Hz at the apical end (*left*) to 5000 Hz at the basal end (*right*). (b) The Ca^{2+}-activated K^+ channel contains seven transmembrane α helices (S0–S6), which associate to form the channel. The cytosolic domain, which includes four hydrophobic regions (S7–S10), regulates opening of the channel in response to Ca^{2+}. Isoforms of the channel, encoded by alternatively spliced mRNAs produced from the same primary transcript, open at different Ca^{2+} concentrations and thus respond to different frequencies. Red numbers refer to regions where alternative splicing produces different amino acid sequences in the various isoforms. [Adapted from K. P. Rosenblatt et al., 1997, *Neuron* **19:**1061.]

individual hair cells has shown that each hair cell expresses a mixture of different alternative Ca^{2+}-activated K^+-channel mRNAs, with different forms predominating in different cells according to their position along the cochlea. This remarkable arrangement suggests that splicing of the Ca^{2+}-activated K^+-channel pre-mRNA is regulated in response to extracellular signals that inform the cell of its position along the cochlea.

Other studies demonstrated that splicing at one of the alternative splice sites in the Ca^{2+}-activated K^+-channel pre-mRNA in the rat is suppressed when a specific protein kinase is activated by neuron depolarization in response to synaptic activity from interacting neurons. This observation raises the possibility that a splicing repressor specific for this site may be activated when it is phosphorylated by this protein kinase, whose activity in turn is regulated by synaptic activity. Since hnRNP and SR proteins are extensively modified by phosphorylation and other post-translational modifications, it seems likely that complex regulation of alternative RNA splicing through post-translational modifications of splicing factors plays a significant role in modulating neuron function.

Many examples of genes similar to the cochlear K^+-channel neurons have been observed in vertebrate neurons; alternatively spliced mRNAs co-expressed from a specific gene in one type of neuron are expressed at different relative concentrations in different regions of the central nervous system. Expansions in the number of microsatellite repeats within the transcribed regions of genes expressed in neurons can cause an alteration in the relative concentrations of alternatively spliced mRNAs transcribed from multiple genes. In Chapter 6, we discussed how backward slippage during DNA replication can lead to expansion of a microsatellite repeat (see Figure 6-5). At least 14 different types of neurological disease result from expansion of microsatellite regions within transcription units expressed in neurons. The resulting long regions of repeated simple sequences in nuclear RNAs of these neurons results in the abnormalities in the relative concentrations of alternatively spliced mRNAs. For example, the most common of these types of diseases, myotonic dystrophy, is characterized by paralysis, cognitive impairment, and personality and behavior disorders. Myotonic dystrophy results from increased copies of either CUG repeats in one transcript in some patients or CCUG repeats in another transcript in other patients. When the number of these repeats increases to 10 or more times the normal number of repeats in these genes, abnormalities are observed in the level of two hnRNP proteins that bind to these repeated sequences. This probably results because these hnRNPs are bound by the abnormally high concentration of this RNA sequence in the nuclei of neurons in such patients. The abnormal concentrations of these hnRNP proteins are thought to lead to alterations in the rate of splicing of different alternative splice sites in multiple pre-mRNAs normally regulated by these hnRNP proteins. ▮

Expression of *Dscam* Isoforms in *Drosophila* Retinal Neurons The most extreme example of regulated alternative RNA processing yet uncovered occurs in expression of the *Dscam* gene in *Drosophila.* Mutations in this gene interfere with the normal synaptic connections made between axons and dendrites during fly development. Analysis of the *Dscam* gene showed that it contains 95 alternatively spliced exons that could be spliced to generate over 38,000 possible isoforms! Recent results have shown that *Drosophila* mutants with a version of the gene that can be spliced in only about 22,000 different ways have specific defects in connectivity between neurons. These results indicate that expression of most of the possible *Dscam* isoforms through regulated RNA splicing helps to specify the tens of millions of different specific synaptic connections between neurons in the *Drosophila* brain. In other words, the correct wiring of neurons in the brain requires regulated RNA splicing.

RNA Editing Alters the Sequences of Some Pre-mRNAs

In the mid-1980s, sequencing of numerous cDNA clones and corresponding genomic DNAs from multiple organisms led

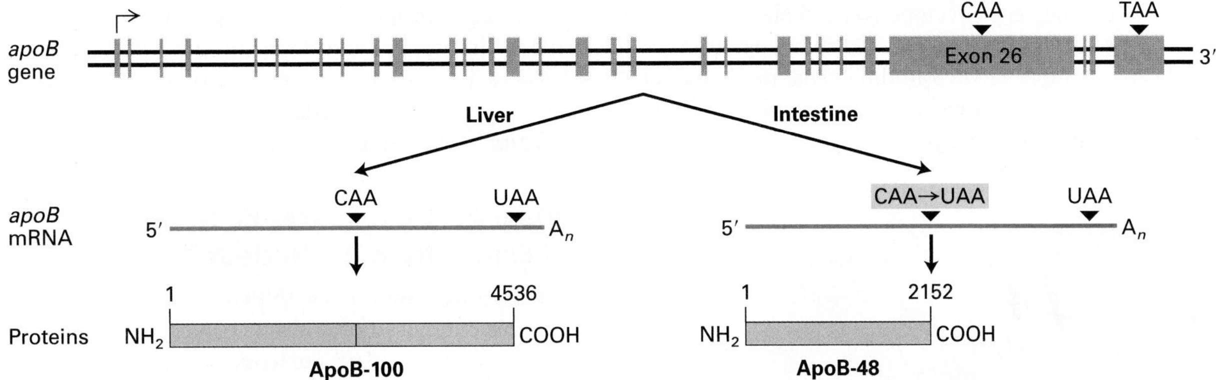

▲ **FIGURE 8-19 RNA editing of *apo-B* pre-mRNA.** The *apoB* mRNA produced in the liver has the same sequence as the exons in the primary transcript. This mRNA is translated into apoB-100, which has two functional domains: an N-terminal domain (green) that associates with lipids and a C-terminal domain (orange) that binds to LDL receptors on cell membranes. In the *apo-B* mRNA produced in the intestine, the CAA codon in exon 26 is edited to a UAA stop codon. As a result, intestinal cells produce apoB-48, which corresponds to the N-terminal domain of apoB-100. [Adapted from P. Hodges and J. Scott, 1992, *Trends Biochem. Sci.* **17**:77.]

to the unexpected discovery of another type of pre-mRNA processing. In this type of processing, called **RNA editing,** the sequence of a pre-mRNA is altered; as a result, the sequence of the corresponding mature mRNA differs from the exons encoding it in genomic DNA.

RNA editing is widespread in the mitochondria of protozoans and plants and also in chloroplasts. In the mitochondria of certain pathogenic trypanosomes, more than half the sequence of some mRNAs is altered from the sequence of the corresponding primary transcripts. Additions and deletions of specific numbers of U's follows templates provided by base-paired short "guide" RNAs. These RNAs are encoded by thousands of mini-mitochondrial DNA circles catenated to many fewer large mitochondrial DNA molecules. The reason for this baroque mechanism for encoding mitochondrial proteins in such protozoans is not clear. But this system does represent a potential target for drugs to inhibit the complex processing enzymes essential to the microbe that do not exist in the cells of their human or other vertebrate hosts.

In higher eukaryotes, RNA editing is much rarer, and thus far, only single-base changes have been observed. Such minor editing, however, turns out to have significant functional consequences in some cases. An important example of RNA editing in mammals involves the *apoB* gene. This gene encodes two alternative forms of a serum protein central to the uptake and transport of cholesterol. Consequently, it is important in the pathogenic processes that lead to *atherosclerosis,* the arterial disease that is the major cause of death in the developed world. The *apoB* gene expresses both the serum protein apolipoprotein B-100 (apoB-100) in hepatocytes, the major cell type in the liver, and apoB-48, expressed in intestinal epithelial cells. The ≈240-kDa apoB-48 corresponds to the N-terminal region of the ≈500-kDa apoB-100. As we detail in Chapter 10, both apoB proteins are components of large lipoprotein complexes that transport lipids in the serum. However, only low-density lipoprotein (LDL) complexes, which contain apoB-100 on their surface, deliver cholesterol to body tissues by binding to the LDL receptor present on all cells.

The cell-type-specific expression of the two forms of apoB results from editing of *apoB* pre-mRNA so as to change the nucleotide at position 6666 in the sequence from a C to a U. This alteration, which occurs only in intestinal cells, converts a CAA codon for glutamine to a UAA stop codon, leading to synthesis of the shorter apoB-48 (Figure 8-19). Studies with the partially purified enzyme that performs the post-transcriptional deamination of C_{6666} to U shows that it can recognize and edit an RNA as short as 26 nucleotides with the sequence surrounding C_{6666} in the *apoB* primary transcript.

KEY CONCEPTS OF SECTION 8.2

Regulation of Pre-mRNA Processing

- Because of alternative splicing of primary transcripts, the use of alternative promoters, and cleavage at different poly(A) sites, different mRNAs may be expressed from the same gene in different cell types or at different developmental stages (see Figure 6-3 and Figure 8-16).
- Alternative splicing can be regulated by RNA-binding proteins that bind to specific sequences near regulated splice sites. Splicing repressors may sterically block the binding of splicing factors to specific sites in pre-mRNAs or inhibit their function. Splicing activators enhance splicing by interacting with splicing factors, thus promoting their association with a regulated splice site.
- In RNA editing the nucleotide sequence of a pre-mRNA is altered in the nucleus. In vertebrates, this process is fairly rare and entails deamination of a single base in the mRNA sequence, resulting in a change in the amino acid specified by the corresponding codon and production of a functionally different protein (see Figure 8-19).

8.3 Transport of mRNA Across the Nuclear Envelope

Fully processed mRNAs in the nucleus remain bound by hnRNP proteins in complexes now referred to as *nuclear mRNPs.* Before an mRNA can be translated into its encoded protein, it must be exported out of the nucleus into the cytoplasm. The **nuclear envelope** is a double membrane that separates the nucleus from the cytoplasm (see Figure 9-1). Like the plasma membrane surrounding cells, each nuclear membrane consists of a water-impermeable phospholipid bilayer and multiple associated proteins. mRNPs and other macromolecules including tRNAs and ribosomal subunits traverse the nuclear envelope through *nuclear pores.* This section will focus on the export of mRNPs through the nuclear pore and the mechanisms that allow some level of regulation of this step.

(a)
Nucleoplasm
Inner nuclear membrane
Nuclear envelope
Outter nuclear membrane
Cytoplasm
Nuclear basket
Lumen
Central transporter
Cytoplasmic filaments

(b)

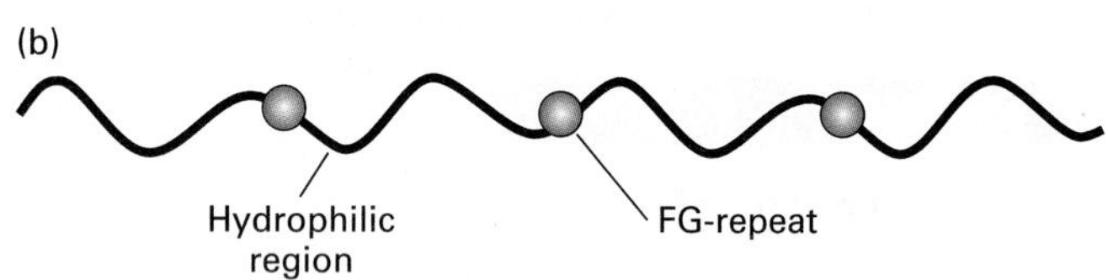

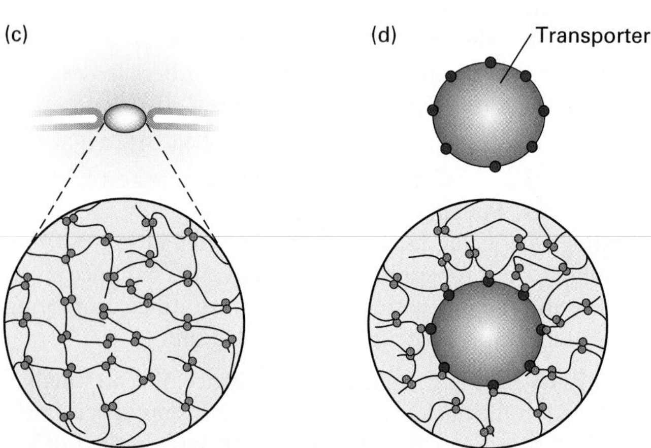

▲ **FIGURE 8-20 Model of transporter passage through an NPC.** (a) Diagram of NPC structure (b) The FG-domains of FG-nucleoporins are thought to have an extended conformation with long hydrophilic regions of polypeptide separated by short FG-hydrophobic domains. (c) FG-repeats are thought to associate with each other reversibly and rapidly, forming a constantly re-arranging molecular sieve that allows diffusion of small water-soluble molecules through it. However, macromolecules are too big to fit through the channels in the molecular sieve. (d) Nuclear transporters have hydrophobic regions on their surface (dark blue dots) that bind reversibly to the FG-domains in the FG-nucleoporins. As a consequence, they can penetrate the molecular sieve in the NPC central channel and diffuse in and out of the nucleus. [From K. Ribbeck and D. Gorlich, 2001, *EMBO J.* **20:**1320.]

Nuclear Pore Complexes Control Import and Export from the Nucleus

Nuclear pore complexes (NPCs) are large, complicated structures composed of multiple copies of approximately 30 different proteins called **nucleoporins.** Embedded in the nuclear envelope, NPCs are roughly octagonal in shape with eight filaments extending into the nucleoplasm and another eight filaments extending into the cytoplasm. (Figure 8-20a; see also Figure 13-32 for a more detailed discussion of NPC structure). The filaments extending from the nuclear face of the NPC are joined at their ends by a terminal ring forming a structure called the nuclear basket. A special class of nucleoporins called **FG-nucleoporins** line the central channel through the NPC. FG-nucleoporins contain long stretches of hydrophilic amino acid sequences punctuated by hydrophobic *FG-repeats,* short sequences rich in hydrophobic phenylalanine (F) and glycine (G).

Water, ions, metabolites, and small globular proteins up to ≈40 kDa can diffuse through the water-filled channel in the nuclear pore complex. However, FG-nucleoporins in the central channel form a barrier restricting the diffusion of macromolecules between the cytoplasm and nucleus. These larger molecules must be selectively transported across the nuclear envelope with the assistance of soluble transporter proteins that bind them and also interact with FG-repeats of FG-nucleoporins.

In one current model of the nuclear pore complex central channel, FG-repeat domains of FG-nucleoporins form random coil regions of polypeptide that extend into the channel (Figure 8-20b). The FG-repeat domains of different FG-nucleoporin molecules are proposed to associate with each other reversibly, forming a constantly rearranging molecular meshwork that acts like a sieve (Figure 8-20c). Small molecules can diffuse through the spaces between FG-repeat domains. But large proteins and ribonucleoproteins, including mRNPs, are too large to pass between the protein filaments that form the molecular sieve and consequently cannot diffuse through the nuclear pore complex.

Nuclear transport proteins bind reversibly to the hydrophobic FG-regions of FG-nucleoporins. These interactions are thought to involve multiple surfaces of the transporters, allowing the proteins to diffuse through the central channel (Figure 8-20d).

mRNPs are transported through the NPC by the *mRNP exporter.* The mRNP exporter is a heterodimer consisting of a large subunit, called *n*uclear export *f*actor 1 (NXF1) or TAP, and a small subunit, *n*uclear export *t*ransporter 1 (Nxt1). TAP binds nuclear mRNPs through associations with both RNA and other proteins in the mRNP complex. One of the most important of these is REF (*R*NA *e*xport *f*actor), a component of the exon junction complexes discussed

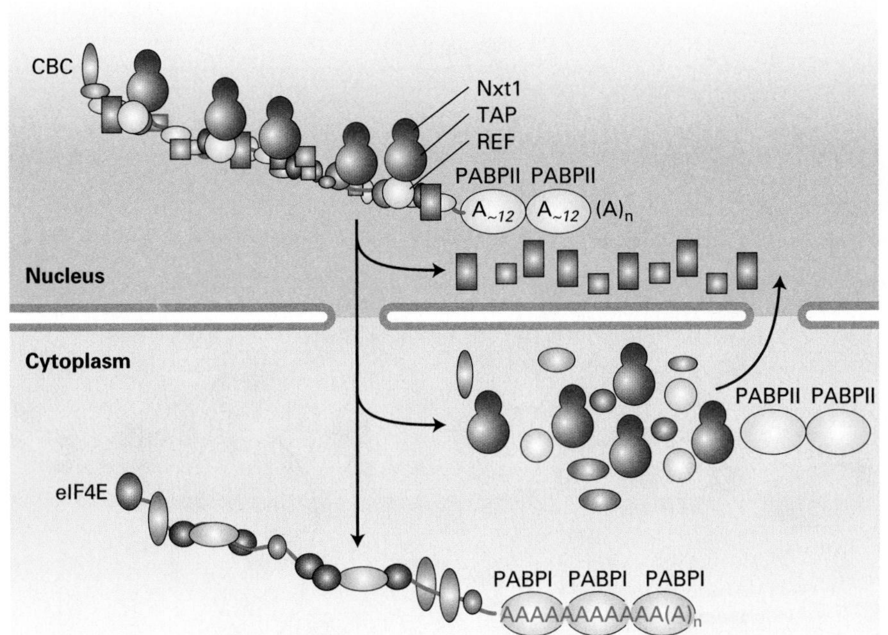

◀ **FIGURE 8-21 Remodeling of mRNPs during nuclear export.** Some mRNP proteins (rectangles) dissociate from nuclear mRNP complexes before export through an NPC. Some (ovals) are exported through the NPC associated with the mRNP but dissociate in the cytoplasm and are shuttled back into the nucleus through an NPC. In the cytoplasm, translation initiation factor eIF4E replaces CBC bound to the 5′ cap and PABPI replaces PABPII.

earlier, bound approximately 20 nucleotides 5′ to each exon-exon junction (Figure 8-21). The TAP/Nxt1 mRNP exporter also associates with SR proteins bound to exonic splicing enhancers. Thus SR proteins associated with exons function to direct both the splicing of pre-mRNAs and the export of fully processed mRNAs through NPCs to the cytoplasm. mRNPs are probably bound along their length by several TAP/Nxt1 mRNP exporters, which both interact with the FG-domains of FG-nucleoporins to facilitate export of mRNPs through the NPC central channel (Figure 8-20). The filaments that extend from the nuclear and cytoplasmic faces of the NPC also assist in mRNP export. Gle2, an adapter protein that reversibly binds both TAP and a nucleoporin in the nuclear basket, brings nuclear mRNPs to the pore in preparation for export. A nucleoporin in the cytoplasmic filaments of the NPC binds an RNA helicase that is proposed to function in the dissociation of TAP/Nxt1 and other hnRNP proteins from the mRNP as it reaches the cytoplasm.

In a process called *mRNP remodeling,* the proteins associated with an mRNA in the nuclear mRNP complex are exchanged for a different set of proteins as the mRNP is transported through the NPC. Some nuclear mRNP proteins dissociate early in tranport, remaining in the nucleus to bind to newly synthesized nascent pre-mRNA. Other nuclear mRNP proteins remain with the mRNP complex as it traverses the pore and do not dissociate from the mRNP until the complex reaches the cytoplasm. Proteins in this category include the TAP/Nxt1 mRNP exporter, CBC bound to the 5′ cap, and PABPII bound to the poly(A) tail. They dissociate from the mRNP on the cytoplasmic side of the NPC through the action of the RNA helicase that associates with cytoplasmic NPC filaments, as discussed above. These proteins are then imported back into the nucleus as discussed for other nuclear proteins in Chapter 13, where they can function in the export of another mRNP. In the cytoplasm, the cap-binding translation initiation factor eIF4E replaces CBC bound to the 5′ cap of nuclear mRNPs. In vertebrates, the nuclear poly(A)-binding protein PABPII is replaced with the cytoplasmic poly(A)-binding protein PABPI (so named because it was discovered before PABPII). (Only a single PABP is found in budding yeast, in both the nucleus and the cytoplasm.)

Yeast SR Protein Recent results suggest that the direction of mRNP export from the nucleus into the cytoplasm is controlled by phosphorylation and dephosphorylation of mRNP adapter proteins such as REF that assist in the binding of the TAP/Nxt1 mRNP exporter to mRNPs. In one case, a yeast SR protein (Npl3) functions as an adapter protein that promotes the binding of the yeast mRNP exporter (Figure 8-22). The SR-protein initially binds to nascent pre-mRNA in its phosphorylated form. When 3′-cleavage and polyadenylation are completed, the adapter protein is dephosphorylated by a specific nuclear protein phosphatase essential for mRNP export. Only the dephosphorylated adapter protein can bind the mRNP exporter, thereby coupling mRNP export to correct polyadenylation. This is one form of mRNA "quality control." If the nascent mRNP is not correctly processed, it is not recognized by the phosphatase that dephosphorylates Npl3. Consequently, it is not bound by the mRNA exporter and not exported from the nucleus. Instead it is degraded by exosomes, the multiprotein complexes that degrade unprotected RNAs in the nucleus and cytoplasm (see Figure 8-1).

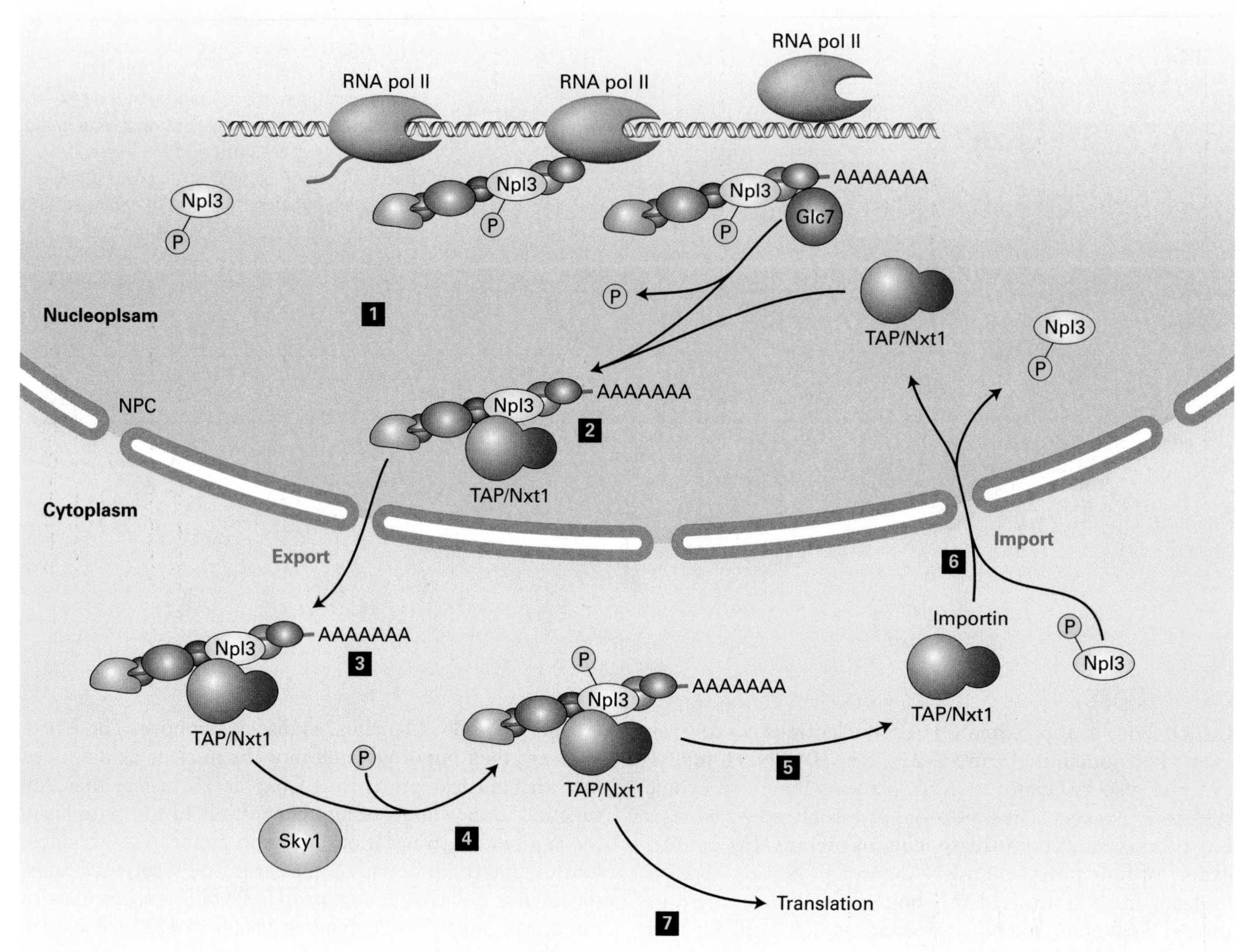

▲ **FIGURE 8-22 Reversible phosphorylation and direction of mRNP nuclear export.** Step **1**: The yeast SR protein Npl3 binds nascent pre-mRNAs in its phosphorylated form. Step **2**. When polyadenylation has occurred successfully, the Glc7 nuclear phosphatase essential for mRNP export dephosphorylates Npl3, promoting the binding of the yeast mRNP exporter, TAP/Nxt1. Step **3**. The mRNP exporter allows diffusion of the mRNP complex through the central channel of the nuclear core complex (NPC). Step **4**. The cytoplasmic protein kinase Sky1 phosphorylates Npl3 in the cytoplasm, causing **5** dissociation of the mRNP exporter, and phosphorylated Npl3 probably through the action of an RNA helicase associated with NPC cytoplasmic filaments. **6** The mRNA transporter and phosphorylated Npl3 are transported back into the nucleus through NPCs. **7** Transported mRNA is available for translation in the cytoplasm. [From E. Izaurralde, 2004, *Nat. Struct. Mol. Biol.* **11:**210–212. See W. Gilbert and C. Guthrie, 2004, *Mol. Cell* **13:**201–212.]

Following export to the cytoplasm, the Npl3 SR protein is phosphorylated by a specific protein kinase restricted to the cytoplasm. This causes it to dissociate from the mRNP, along with the mRNP exporter. In this way, dephosphorylation of adapter mRNP proteins in the nucleus once RNA processing is complete and their phosphorylation in the cytoplasm results in a higher concentration of mRNP exporter–mRNP complexes in the nucleus, where they form, and a lower concentration of these complexes in the cytoplasm, where they dissociate. As a result, the direction of mRNP export may be driven by simple diffusion down a concentration gradient of the transport-competent mRNP exporter-mRNP complex across the NPC from high in the nucleus to low in the cytoplasm.

Nuclear Export of Balbiani Ring mRNPs The larval salivary glands of the insect *Chironomous tentans* provide a good model system for electron microscopy studies of the

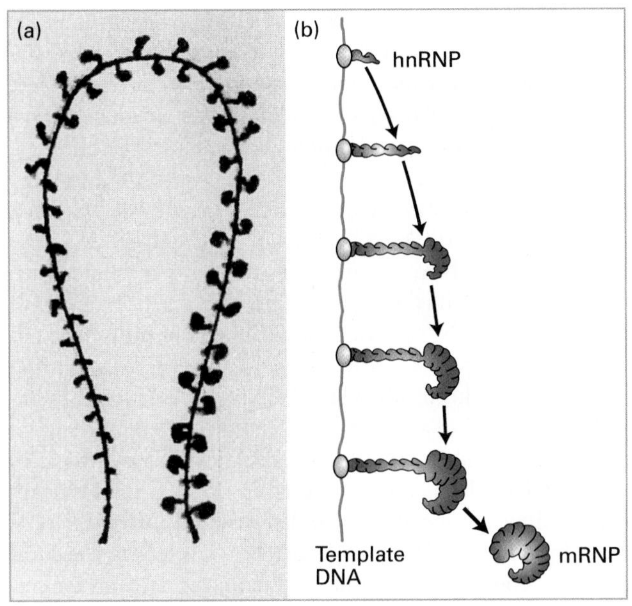

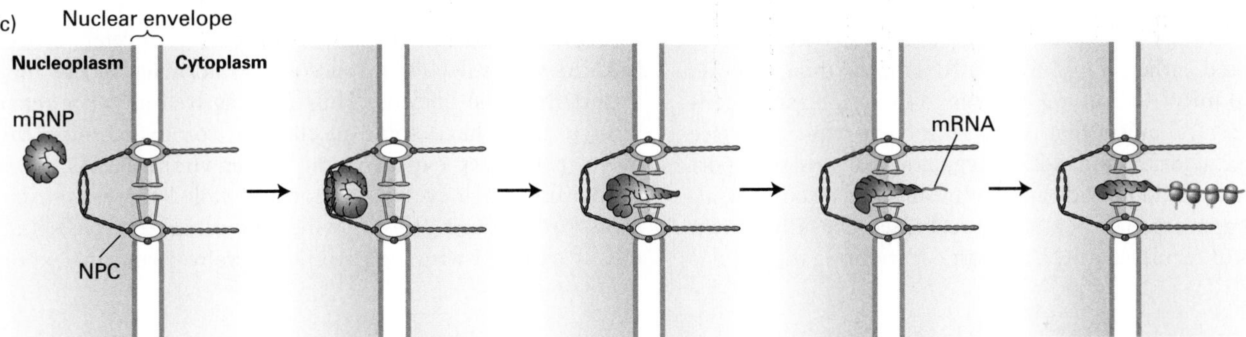

◀ **FIGURE 8-23 Formation of heterogeneous ribonucleoprotein particles (hnRNPs) and export of mRNPs from the nucleus.** (a) Model of a single chromatin transcription loop and assembly of Balbiani ring (BR) mRNP in *Chironomous tentans.* Nascent RNA transcripts produced from the template DNA rapidly associate with proteins, forming hnRNPs. The gradual increase in size of the hnRNPs reflects the increasing length of RNA transcripts at greater distances from the transcription start site. The model was reconstructed from electron micrographs of serial thin sections of salivary gland cells. (b) Schematic diagram of the biogenesis of hnRNPs. Following processing of the pre-mRNA, the resulting ribonucleoprotein particle is referred to as an mRNP. (c) Model for the transport of BR mRNPs through the nuclear pore complex (NPC) based on electron microscopic studies. Note that the curved mRNPs appear to uncoil as they pass through nuclear pores. As the mRNA enters the cytoplasm, it rapidly associates with ribosomes, indicating that the 5′ end passes through the NPC first. [Part (a) from C. Erricson et al., 1989, *Cell* **56:**631; courtesy of B. Daneholt. Parts (b) and (c) adapted from B. Daneholt, 1997, *Cell* **88:**585. See also B. Daneholt, 2001, *Proc. Nat'l. Acad. Sci. USA* **98:**7012.]

formation of hnRNPs and their export through NPCs. In these larvae, genes in large chromosomal puffs called Balbiani rings are abundantly transcribed into nascent pre-mRNAs that associate with hnRNP proteins and are processed into coiled mRNPs with an mRNA of ≈75 kb (Figure 8-23a, b). These giant mRNAs encode large glue proteins that adhere the developing larvae to a leaf. After processing of the pre-mRNA in Balbiani ring hnRNPs, the resulting mRNPs move through nuclear pores to the cytoplasm. Electron micrographs of sections of these cells show mRNPs that appear to uncoil during their passage through nuclear pores and then bind to ribosomes as they enter the cytoplasm. This uncoiling is probably a consequence of the remodeling of mRNPs as the result of phosphorylation of mRNP proteins by cytoplasmic kinases and the action of an RNA helicase associated with NPC cytoplasmic filaments, as discussed in the previous section. The observation that mRNPs become associated with ribosomes during transport indicates that the 5′ end leads the way through the nuclear pore complex. Detailed electron microscopic studies of the transport of Balbiani ring mRNPs through nuclear pore complexes led to the model depicted in Figure 8-23c.

Pre-mRNAs in Spliceosomes Are Not Exported from the Nucleus

It is critical that only fully processed mature mRNAs be exported from the nucleus because translation of incompletely processed pre-mRNAs containing introns would produce defective proteins that might interfere with the functioning of the cell. To prevent this, pre-mRNAs associated with snRNPs in spliceosomes usually are prevented from being transported to the cytoplasm.

In one type of experiment demonstrating this restriction, a gene encoding a pre-mRNA with a single intron that normally is spliced out was mutated to introduce deviations from the consensus splice-site sequences. Mutation of either the 5′ or the 3′ invariant splice-site bases at the ends of the intron resulted in pre-mRNAs that were bound by snRNPs to form spliceosomes; however, RNA splicing was blocked, and the pre-mRNA was retained in the nucleus. In contrast, mutation of *both* the 5′ and 3′ splice sites in the same pre-mRNA resulted in export of the unspliced pre-mRNA, although less efficiently than for the spliced mRNA. When both splice sites were mutated, the pre-mRNAs were not efficiently bound by snRNPs, and, consequently, their export was not blocked.

Recent studies in yeast have shown that a nuclear protein that associates with a nucleoporin in the NPC nuclear basket is required to retain pre-mRNAs associated with snRNPs in the nucleus. If either this protein or the nucleoporin to which it binds is deleted, unspliced pre-mRNAs are exported.

Many cases of thalassemia, an inherited disease that results in abnormally low levels of globin proteins, are due to mutations in globin-gene splice sites that decrease the efficiency of splicing but do not prevent association of the pre-mRNA with snRNPs. The resulting unspliced globin pre-mRNAs are retained in reticulocyte nuclei and are rapidly degraded. ▮

HIV Rev Protein Regulates the Transport of Unspliced Viral mRNAs

As discussed earlier, transport of mRNPs containing mature, functional mRNAs from the nucleus to the cytoplasm entails a complex mechanism that is crucial to gene expression (see Figures 8-21, 8-22, and 8-23). Regulation of this transport theoretically could provide another means of gene control, although it appears to be relatively rare. Indeed, the only known examples of regulated mRNA export occur during the cellular response to conditions (e.g., heat shock) that cause protein denaturation or during viral infection when virus-induced alterations in nuclear transport maximize viral replication. Here we describe the regulation of mRNP export mediated by a protein encoded by human immunodeficiency virus (HIV).

A retrovirus, HIV integrates a DNA copy of its RNA genome into the host-cell DNA (see Figure 4-49). The integrated viral DNA, or provirus, contains a single transcription unit, which is transcribed into a single primary transcript by cellular RNA polymerase II. The HIV transcript can be spliced in alternative ways to yield three classes of mRNAs: a 9-kb unspliced mRNA; ≈4-kb mRNAs formed by removal of one intron; and ≈2-kb mRNAs formed by removal of two or more introns (Figure 8-24). After their synthesis in the host-cell nucleus, all three classes of HIV mRNAs are transported to the cytoplasm and translated into viral proteins; some of the 9-kb unspliced RNA is used as the viral genome in progeny virions that bud from the cell surface.

Since the 9-kb and 4-kb HIV mRNAs contain splice sites, they can be viewed as incompletely spliced mRNAs. However, as discussed earlier, association of such incompletely spliced mRNAs with snRNPs in spliceosomes normally blocks their export from the nucleus. Thus HIV, as well as other retroviruses, must have some mechanism for overcoming this block, permitting export of the longer viral mRNAs. Some retroviruses have evolved a sequence called the *constitutive transport element (CTE)*, which binds to the TAP-Nxt1 mRNP exporter with high affinity, thereby permitting export

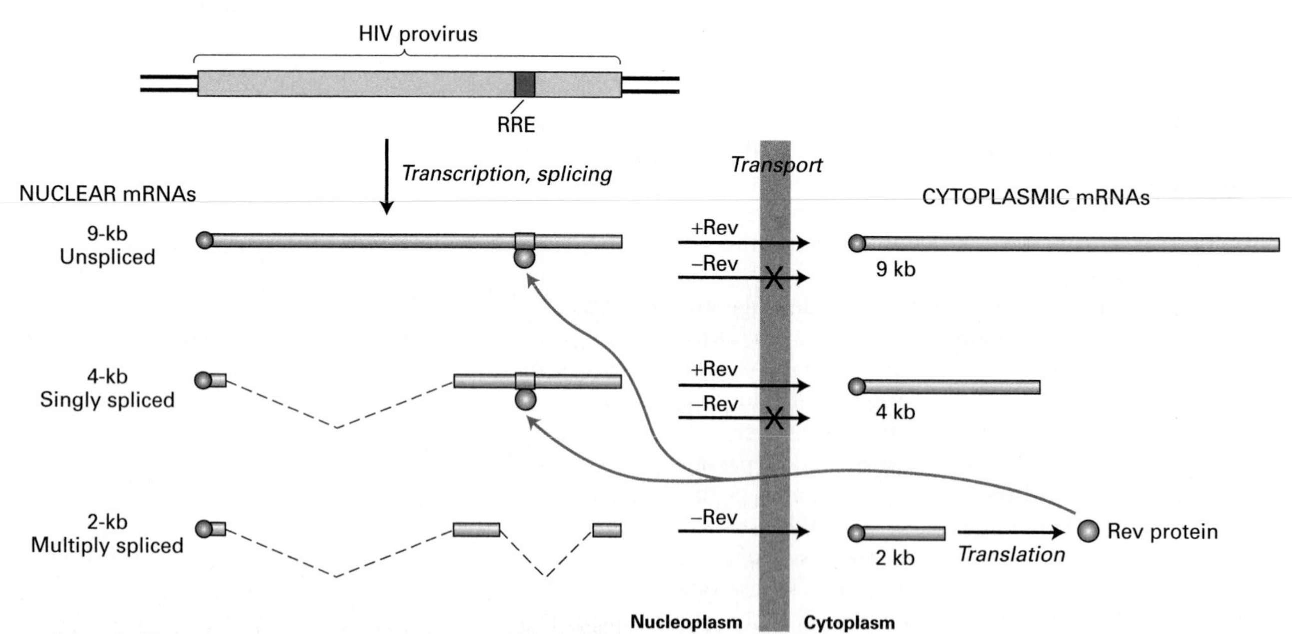

▲ **FIGURE 8-24 Transport of HIV mRNAs from the nucleus to the cytoplasm.** The HIV genome, which contains several coding regions, is transcribed into a single 9-kb primary transcript. Several ≈4-kb mRNAs result from alternative splicing out of any one of several introns (dashed lines) and several ≈2-kb mRNAs from splicing out of two or more alternative introns. After transport to the cytoplasm, the various RNA species are translated into different viral proteins. Rev protein, encoded by a 2-kb mRNA, interacts with the Rev-response element (RRE) in the unspliced and singly spliced mRNAs, stimulating their transport to the cytoplasm. [Adapted from B. R. Cullen and M. H. Malim, 1991, *Trends Biochem. Sci.* **16:**346.]

of unspliced retroviral RNA into the cytoplasm. HIV solved the problem differently.

Studies with HIV mutants showed that transport of unspliced 9-kb and singly spliced 4-kb viral mRNAs from the nucleus to the cytoplasm requires the virus-encoded Rev protein. Subsequent biochemical experiments demonstrated that Rev binds to a specific Rev-response element (RRE) present in HIV RNA. In cells infected with HIV mutants lacking the RRE, unspliced and singly spliced viral mRNAs remain in the nucleus, demonstrating that the RRE is required for Rev-mediated stimulation of nuclear export. Rev contains a leucine-rich nuclear export signal that interacts with transporter exportinl. As discussed in Chapter 13, this results in export of the unspliced and singly spliced HIV mRNAs through the nuclear pore complex. ▮

KEY CONCEPTS OF SECTION 8.3

Transport of mRNA Across the Nuclear Envelope

- Most mRNPs are exported from the nucleus by a heterodimeric mRNP exporter that interacts with FG-repeats of FG-nucleoporins (see Figure 8-20). The direction of transport (nucleus → cytoplasm) may result from dissociation of the exporter-mRNP complex in the cytoplasm by phosphorylation of mRNP proteins by cytoplasmic kinases and the action of an RNA helicase associated with the cytoplasmic filaments of the nuclear pore complexes (see Figure 8-20).
- The mRNP exporter binds to most mRNAs cooperatively with SR proteins bound to exons and REF associated with exon-junction complexes that bind to mRNAs following RNA splicing, and to additional mRNP proteins.
- Pre-mRNAs bound by a spliceosome normally are not exported from the nucleus, ensuring that only fully processed, functional mRNAs reach the cytoplasm for translation.

8.4 Cytoplasmic Mechanisms of Post-transcriptional Control

Before proceeding, let's quickly review the steps in gene expression at which control is exerted. We saw in the previous chapter that regulation of transcription initiation is the principal mechanism for controlling the expression of genes. In preceding sections of this chapter, we also learned that the expression of protein isoforms is controlled by regulating alternative RNA splicing. Although nuclear export of fully and correctly processed mRNPs to the cytoplasm is rarely regulated, the export of improperly processed or aberrantly remodeled pre-mRNPs is prevented, and such abnormal transcripts are degraded by the exosome. However, retroviruses, including HIV, have evolved mechanisms that permit pre-mRNAs that retain splice sites to be exported and translated (see Figure 8-24).

In this section we consider other mechanisms of post-transcriptional control that contribute to regulating the expression of some genes. Most of these mechanisms operate in the cytoplasm, controlling the stability or localization of mRNA or its translation into protein. We begin by discussing two recently discovered and related mechanisms of gene control that provide powerful new techniques for manipulating the expression of specific genes for experimental and therapeutic purposes. These mechanisms are controlled by short, ≈21 nucleotide, single-stranded RNAs called **micro RNAs (miRNAs)** and **short interfering (siRNAs)**. Both base-pair with specific target mRNAs, either inhibiting their translation (miRNAs) or causing their degradation (siRNAs). Humans express ≈1,000 miRNAs. Most of these are expressed in specific cell types at particular times during embryogenesis and after birth. Many miRNAs can target more than one mRNA. Consequently, these newly discovered mechanisms contribute significantly to the regulation of gene expression. siRNAs, involved in the process called RNA interference, are also an important cellular defense against viral infection and excessive transposition by transposons.

Micro RNAs Repress Translation of Specific mRNAs

Micro RNAs (miRNAs) were first discovered during analysis of mutations in the *lin-4* and *let-7* genes of the nematode *C. elegans,* which influence development of the organism. Cloning and analysis of wild-type *lin-4* and *let-7* revealed that they encode no protein products but rather RNAs only 21 and 22 nucleotides long, respectively. The RNAs hybridize to the 3′ untranslated regions of specific target mRNAs. For example, the *lin-4* miRNA, which is expressed early in embryogenesis, hybridizes to the 3′ untranslated regions of both the *lin-14* and *lin-28* mRNAs in the cytoplasm, thereby repressing their translation by a mechanism discussed below. Expression of *lin-4* miRNA ceases later in development, allowing translation of newly synthesized *lin-14* and *lin-28* mRNAs at that time. Expression of *let-7* miRNA occurs at comparable times during embryogenesis of all bilaterally symmetric animals. The role of *lin-4* and *let-7* miRNAs in coordinating the timing of early developmental events in *C. elegans* is discussed in Chapter 22. Here we focus on what is currently understood about how miRNAs repress translation.

miRNA regulation of translation appears to be widespread in all multicellular plants and animals. In the past few years, small RNAs of 20–26 nucleotides have been isolated, cloned, and sequenced from various tissues of multiple model organisms. Recent estimates suggest the expression of one-third of all human genes is regulated by ≈1,000 human miRNAs isolated from various tissues. The potential for regulation of multiple mRNAs by one miRNA is great because base pairing between the miRNA and the sequence in the 3′-ends of mRNAs that they regulate need not be perfect (Figure 8-25). In fact, considerable experimentation with

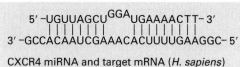

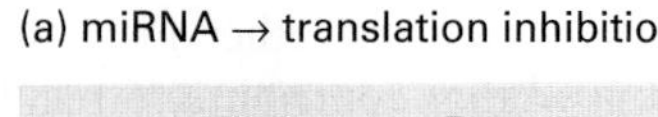

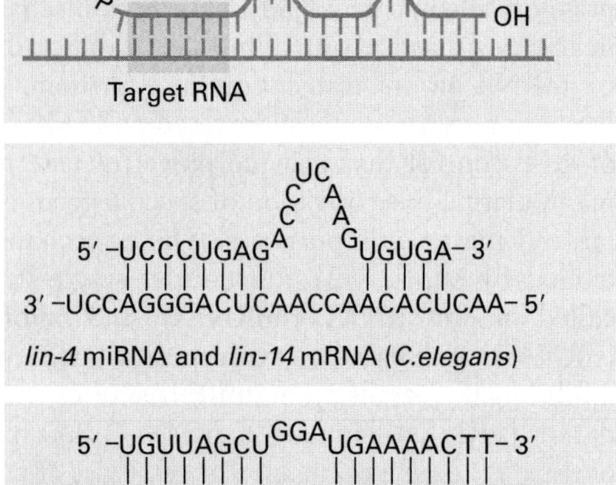

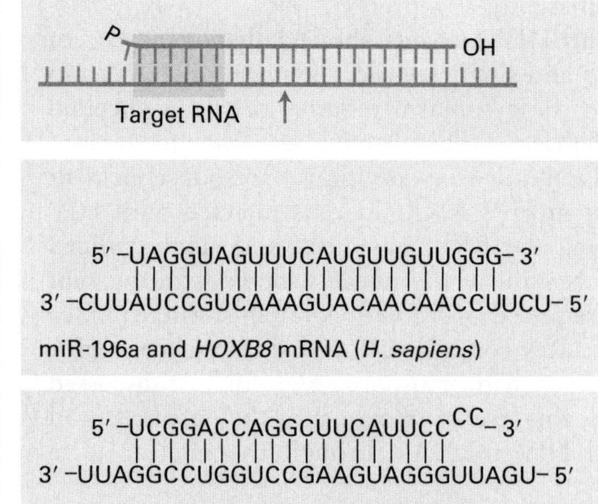

▲ **FIGURE 8-25 Base pairing with target RNAs distinguishes miRNA and siRNA.** (a) miRNAs hybridize imperfectly with their target mRNAs, repressing translation of the mRNA. Nucleotides 2 to 7 of an miRNA (highlighted blue) are the most critical for targeting it to a specific mRNA. (b) siRNA hybridizes perfectly with its target mRNA, causing cleavage of the mRNA at the position indicated by the red arrow, triggering its rapid degradation. [Adapted from P. D. Zamore and B. Haley, 2005, *Science* **309:**1519.]

synthetic miRNAs has shown that complementarity between the six or seven 5′ nucleotides of an miRNA and its target mRNA 3′ untranslated region are most critical for target mRNA selection.

Most miRNAs are processed from RNA polymerase II transcripts of several hundred to thousands of nucleotides in length called pri (for *pri*mary transcript)-miRNAs (Figure 8-26). Pri-miRNAs can contain the sequence of one or more miRNAs. miRNAs are also processed out of some excised introns and from 3′ untranslated regions of some pre-mRNAs. Within these long transcripts are sequences that fold into hairpin structures of ≈70 nucleotides in length with imperfect base pairing in the stem. A nuclear RNase specific for double-stranded RNA called *Drosha* acts with a nuclear double-stranded RNA-binding protein called DGCR8 in humans (Pasha in *Drosophila*) and cleaves the hairpin region out of the long precursor RNA, generating a pre-miRNA. Pre-miRNAs are recognized and bound by a specific nuclear transporter, *Exportin5,* which interacts with the FG-domains of nucleoporins, allowing the complex to diffuse through the inner channel of the nuclear pore complex, as discussed above (see Figure 8-20), and in Chapter 13. Once in the cytoplasm, a cytoplasmic double-stranded RNA-specific RNase called *Dicer* acts with a cytoplasmic double-stranded RNA-binding protein called *TRBP* in humans (for *T*ar *b*inding *p*rotein; called *Loquocious* in *Drosophila*) to further process the pre-miRNA into a double-stranded miRNA. The double-stranded miRNA is approximately two turns of an A-form RNA helix in length, with strands 21–23 nucleotides long and two unpaired 3′-nucleotides at each end. Finally, one of the two strands is selected for assembly into a mature ***R****NA-**i**nduced* ***s****ilencing* ***c****omplex* **(RISC)** containing a single-stranded mature miRNA bound by a multidomain *Argonaute* protein, a member of a protein family with a recognizable conserved sequence. Several Argonaute proteins are expressed in some organisms, especially plants, and are found in distinct RISC complexes with different functions.

The miRNA-RISC complexes associate with target mRNPs by base pairing between the Argonaute-bound mature miRNA and complementary regions in the 3′-untranslated regions (3′-UTRs) of target mRNAs (see Figure 8-25). Inhibition of target mRNA translation requires the binding of two or more RISC complexes to distinct complementary regions in the target mRNA 3′-UTR. It has been suggested that this may allow combinatorial regulation of mRNA translation by separately regulating the transcription of two or more different pri-miRNAs, which are processed to miRNAs that are required in combination to suppress the translation of a specific target mRNA.

The binding of several RISC complexes to an mRNA inhibit translation initiation by a mechanism currently being analyzed. Recent discoveries showed that binding of RISC complexes causes the bound mRNPs to associate with dense cytoplasmic domains many times the size of a ribosome called cytoplasmic RNA-processing bodies, or simply P bodies. **P bodies,** which will be described in greater detail below, are sites of RNA degradation that contain no ribosomes or translation factors, potentially explaining the inhibition of translation. The association with P bodies may also explain why expression of an miRNA often decreases the stability of a targeted mRNA.

As mentioned earlier, approximately 1000 different human miRNAs have been observed, many of them expressed only in specific cell types. Determining the function of these miRNAs is currently a highly active area of research.

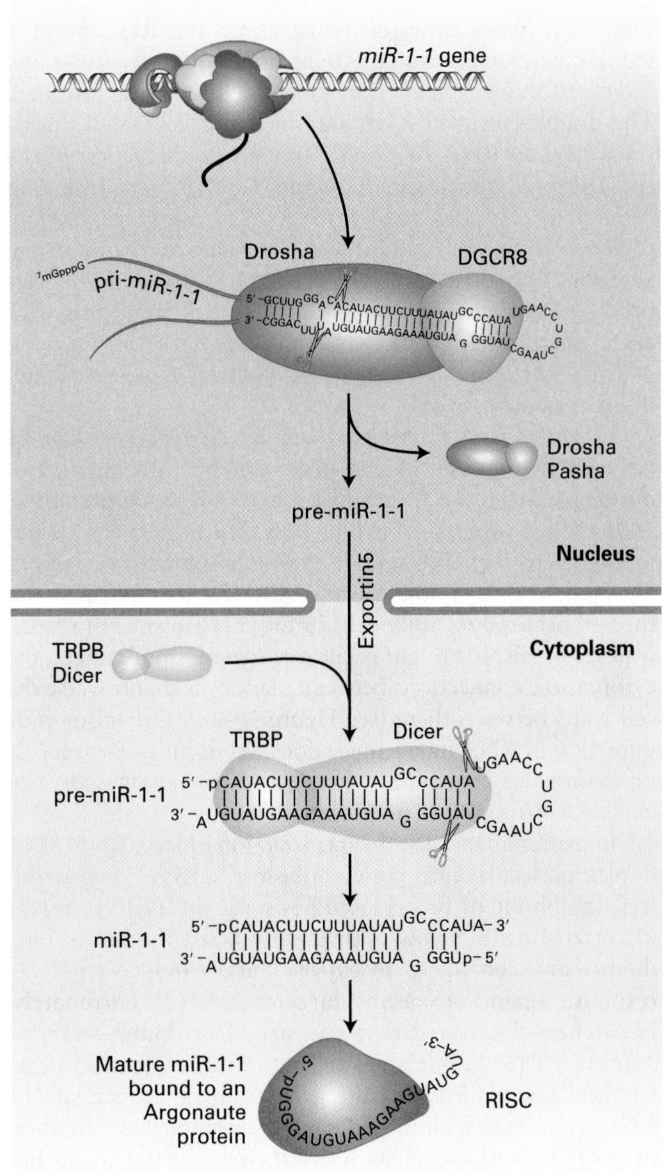

▲ **FIGURE 8-26 miRNA processing.** This diagram shows transcription and processing of the miR-1-1 miRNA. The primary miRNA transcript (pri-miRNA) is transcribed by RNA polymerase II. The nuclear double-stranded specific endoribonuclease Drosha with its partner double-stranded RNA-binding protein DGCR8 (Pasha in *Drosophila)* make the initial cleavages in the pri-miRNA, generating a ≈70 nucleotide pre-miRNA that is exported to the cytoplasm by nuclear transporter Exportin 5. The pre-miRNA is further processed in the cytoplasm to a double-stranded miRNA with a two-base single-stranded 3′ end by Dicer in conjunction with the DS RNA-binding protein TRBP (Loquatious in *Drosophila*). Finally, one of the two strands is incorporated into an RISC complex, where it is bound by an Argonaute protein. [Adapted from P. D. Zamore and B. Haley, 2005, *Science* **309:**1519.]

In one example, a specific miRNA called miR-133 is induced when myoblasts differentiate into muscle cells. miR-133 suppresses the translation of PTB, a regulatory splicing factor that functions similarly to Sxl in *Drosophila* (see Figure 8-16). It binds to the 3′ splice site region in the pre-mRNAs of many genes, leading to exon skipping or use of alternative 3′ splice sites. When miR-133 is expressed in differentiating myoblasts, the PTB concentration falls without a significant decrease in the concentration of PTB mRNA. As a result, alternative isoforms of multiple proteins important for muscle-cell function are expressed in the differentiated cells.

Other examples of miRNA regulation in various organisms are being discovered at a rapid pace. Knocking out the *dicer* gene eliminates the generation of miRNA in mammals. This causes embryonic death early in mouse development. However, when *dicer* is knocked out only in limb primordia, the influence of miRNA on the development of the nonessential limbs can be observed (Figure 8-27). Although all major cell types differentiated and fundamental aspects of limb patterning were maintained, development was abnormal—demonstrating the importance of miRNAs in regulating the proper level of translation of multiple mRNAs. Of the ≈1,000 human miRNAs, 53 appear to be unique to primates. It seems likely that new miRNAs arose readily during evolution by the duplication of a pri-miRNA gene followed by mutation of bases encoding the mature miRNA. miRNAs are particularly abundant in plants—more than 1.5 million distinct miRNAs have been characterized in *Arabidopsis thaliana!*

RNA Interference Induces Degradation of Precisely Complementary mRNAs

RNA interference (RNAi) was discovered unexpectedly during attempts to experimentally manipulate the expression of specific genes. Researchers tried to inhibit the expression of a gene in *C. elegans* by microinjecting a single-stranded, complementary RNA that would hybridize to the encoded mRNA and prevent its translation, a method called antisense inhibition. But in control experiments, perfectly base-paired double-stranded RNA a few hundred base pairs long was

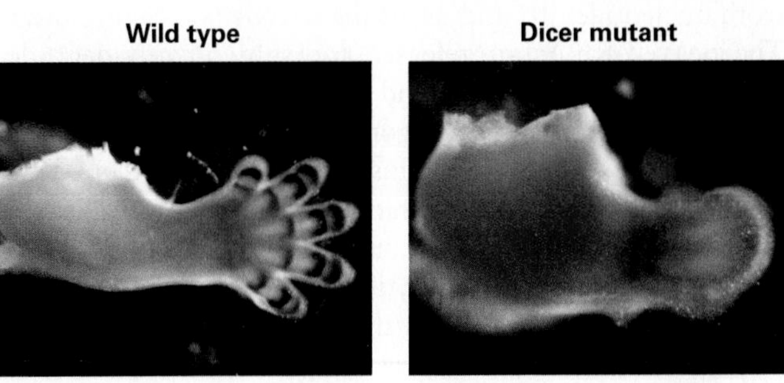

▲ EXPERIMENTAL FIGURE 8-27 **miRNA function in limb development.** Micrographs comparing normal (*left*) and Dicer knockout (*right*) limbs of embryonic development day-13 mouse embryos immunostained for the Gd5 protein, a marker of joint formation. Dicer is knocked out in developing mouse embryos by conditional expression of Cre to induce deletion of the Dicer gene only in these cells (see Figure 5-42) [From B. D. Harfe et al., 2005, *Proc. Natl. Acad. Sci. USA* **102:**10898.]

much more effective at inhibiting expression of the gene than the antisense strand alone. Similar inhibition of gene expression by an introduced double-stranded RNA soon was observed in plants. In each case, the double-stranded RNA induced degradation of all cellular RNAs containing a sequence that was exactly the same as one strand of the double-stranded RNA. Because of the specificity of RNA interference in targeting mRNAs for destruction, it has become a powerful experimental tool for studying gene function (see Figure 5-45).

Subsequent biochemical studies with extracts of *Drosophila* embryos showed that a long double-stranded RNA that mediates interference is initially processed into a double-stranded short interfering RNA (siRNA). The strands in siRNA contain 21–23 nucleotides hybridized to each other so that the two bases at the 3′ end of each strand are single-stranded. Further studies revealed that the cytoplasmic double-stranded RNA-specific ribonuclease that cleaves long double-stranded RNA into siRNAs is the same Dicer enzyme involved in processing pre-miRNAs after their nuclear export to the cytoplasm (see Figure 8-27). This discovery led to the realization that RNA interference and miRNA-mediated translational repression are related processes. In both cases, the mature short single-stranded RNA, either mature siRNA or mature miRNA, is assembled into RISC complexes in which the short RNAs are bound by an Argonaute protein. What distinguishes a RISC complex containing an siRNA from one containing an miRNA is that the siRNA base-pairs extensively with its target RNA and induces its cleavage, whereas a RISC complex associated with an miRNA recognizes its target through imperfect base-pairing and results in inhibition of translation.

The Argonaute protein appears to be responsible for cleavage of target RNA; one domain of the Argonaute protein is homologous to RNase H enzymes that degrade the RNA of an RNA-DNA hybrid (see Figure 6-14). When the 5′ end of the short RNA of a RISC complex base-pairs precisely with a target mRNA over a distance of one turn of an RNA helix (10-12 base pairs), this domain of Argonaute cleaves the phosphodiester bond of the target RNA across from nucleotides 10 and 11 of the siRNA (see Figure 8-25). The cleaved RNAs are released and subsequently degraded by cytoplasmic exosomes and 5′ exoribonucleases. If base pairing is not perfect, the Argonaute domain does not cleave or release the target mRNA. Instead, if several miRNA-RISC complexes associate with a target mRNA, its translation is inhibited and the mRNA becomes associated with P bodies, where, as mentioned earlier, it is probably degraded by a different and slower mechanism than the degradation pathway initiated by RISC cleavage of a perfectly complementary target RNA.

When double-stranded RNA is introduced into the cytoplasm of eukaryotic cells, it enters the pathway for assembly of siRNAs into a RISC complex because it is recognized by the cytoplasmic Dicer enzyme and TRBP double-stranded RNA binding protein that process pre-miRNAs (see Figure 8-26). This process of RNA interference is believed to be an ancient cellular defense against certain viruses and mobile genetic elements in both plants and animals. Plants with mutations in the genes encoding Dicer and RISC proteins exhibit increased sensitivity to infection by RNA viruses and increased movement of **transposons** within their genomes. The double-stranded RNA intermediates generated during replication of RNA viruses are thought to be recognized by the Dicer ribonuclease, inducing a RNAi response that ultimately degrades viral mRNAs. During transposition, transposons are inserted into cellular genes in a random orientation, and their transcription from different promoters produces complementary RNAs that can hybridize with each other, initiating the RNAi system that then interferes with the expression of transposon proteins required for additional transpositions.

In plants and *C. elegans* the RNAi response can be induced in all cells of the organism by introduction of double-stranded RNA into just a few cells. Such organism-wide induction requires production of a protein that is homologous to the RNA replicases of RNA viruses. This finding suggests that double-stranded siRNAs are replicated and then transferred to other cells in these organisms. In plants, transfer of siRNAs might occur through **plasmodesmata,** the cytoplasmic connections between plant cells that traverse the cell walls between them (see Figure 19-38). Organism-wide induction of RNA interference does not occur in *Drosophila* or mammals, presumably because their genomes do not encode RNA replicase homologs.

In mammalian cells, the introduction of long RNA-RNA duplex molecules into the cytoplasm results in the generalized inhibition of protein synthesis via the PKR pathway, discussed further below. This greatly limits the use of long double-stranded RNAs to experimentally induce an RNAi response against a specific targeted mRNA. Fortunately, researchers discovered that one strand of double-stranded siRNAs of 21–23 nucleotides in length with two-base 3′ single-stranded regions leads to the generation of mature siRNA RISC complexes without inducing the generalized inhibition of protein synthesis. This has allowed researchers to use synthetic double-stranded siRNAs to "knock down" the expression of specific genes in human cells as well as in other mammals. This method of **siRNA knockdown** is now widely used in studies of diverse processes, including the RNAi pathway itself.

RNAi Inhibition of Transcription In plants and the fission yeast *Schizosaccharomyces pombe,* double-stranded RNA also induces the formation of **heterochromatin** on genes with the same sequence as the double-stranded RNA, inhibiting their transcription. Nuclear proteins homologous to cytoplasmic Dicer and Argonaute proteins generate nuclear siRNA complexes composed of different proteins from the cytoplasmic RISC complexes. These nuclear siRNA complexes are thought to be targeted to specific genes by base pairing with nascent pre-mRNAs during their transcription. This interaction induces the methylation of histone H3 at lysine 9, generating a binding site for HP1 proteins and the subsequent assembly of heterochromatin, as discussed in Chapter 6 (see Figure 6-34). In plants, the DNA in these heterochromatic regions also is methylated, contributing to the formation of heterochromatin. Components of the RNAi system are also required for the formation of heterochromatin at centromeres and the proper function of centromeres in *S. pombe,* plants, and cultured mammalian cells. Centromeres from most organisms contain highly repetitive DNA sequences. Consequently, the RNAi system that leads to heterochromatization of repeated genes in *S. pombe* and plants may be exploited generally by most eukaryotes for the proper formation of the DNA-protein kinetochore complex formed at centromeres and critical for cell division (Chapter 20).

Cytoplasmic Polyadenylation Promotes Translation of Some mRNAs

In addition to repression of translation by miRNAs, other protein-mediated translational controls help regulate expression of some genes. Regulatory sequences, or elements, in mRNAs that interact with specific proteins to control translation generally are present in the untranslated region (UTR) at the 3′ or 5′ end of an mRNA. Here we discuss a type of protein-mediated translational control involving 3′ regulatory elements. A different mechanism involving RNA-binding proteins that interact with 5′ regulatory elements is discussed later.

Translation of many eukaryotic mRNAs is regulated by sequence-specific RNA-binding proteins that bind cooperatively to neighboring sites in 3′ UTRs. This allows them to function in a combinatorial manner, similar to the cooperative binding of transcription factors to regulatory sites in an enhancer or promoter region. In most cases studied, translation is repressed by protein binding to 3′ regulatory elements and regulation results from derepression at the appropriate time or place in a cell or developing embryo. The mechanism of such repression is best understood for mRNAs that must undergo *cytoplasmic polyadenylation* before they can be translated.

Cytoplasmic polyadenylation is a critical aspect of gene expression in the early embryo of animals. The egg cells (oocytes) of multicellular animals contain many mRNAs, encoding numerous different proteins, that are not translated until after the egg is fertilized by a sperm cell. Some of these "stored" mRNAs have a short poly(A) tail, consisting of only ≈20–40 A residues, to which just a few molecules of cytoplasmic poly(A)-binding protein (PABPI) can bind. As discussed in Chapter 4, multiple PABPI molecules bound to the long poly(A) tail of an mRNA interact with the eIF4G initiation factor, thereby stabilizing the interaction of the mRNA 5′ cap with eIF4E, which is required for translation initiation (see Figure 4-28b). Because this stabilization cannot occur with mRNAs that have short poly(A) tails, such mRNAs stored in oocytes are not translated efficiently. At the appropriate time during oocyte maturation or after fertilization of an egg cell, usually in response to an external signal, approximately 150 A residues are added to the short poly(A) tails on these mRNAs in the cytoplasm, stimulating their translation.

Recent studies with mRNAs stored in *Xenopus* oocytes have helped elucidate the mechanism of this type of translational control. Experiments in which short-tailed mRNAs are injected into oocytes have shown that two sequences in their 3′ UTR are required for their polyadenylation in the cytoplasm: the AAUAAA poly(A) signal that is also required for the nuclear polyadenylation of pre-mRNAs and one or more copies of an upstream U-rich *cytoplasmic polyadenylation element (CPE).* This regulatory element is bound by a highly conserved *CPE-binding protein (CPEB)* that contains an RRM domain and a zinc-finger domain.

According to the current model, in the absence of a stimulatory signal, CPEB bound to the U-rich CPE interacts with the protein Maskin, which in turn binds to eIF4E associated with the mRNA 5′ cap (Figure 8-28, *left*). As a result, eIF4E cannot interact with other initiation factors and the 40S

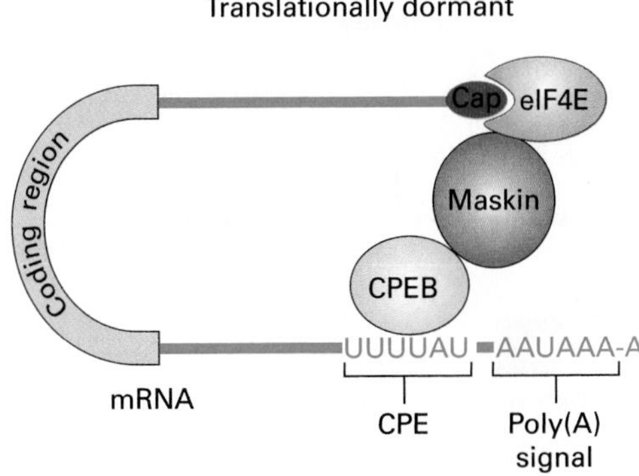

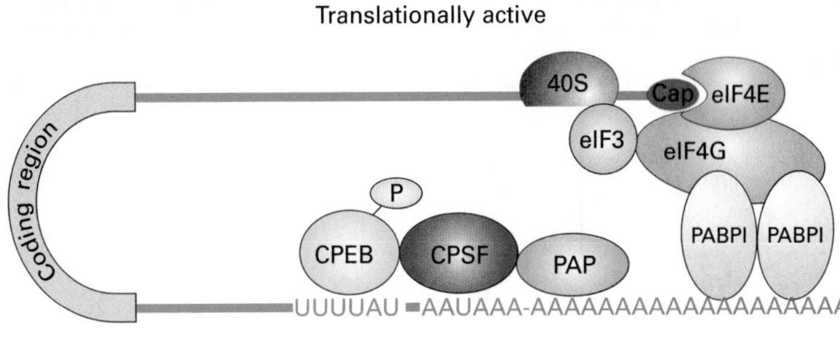

▲ **FIGURE 8-28 Model for control of cytoplasmic polyadenylation and translation initiation.** *(Left)* In immature oocytes, mRNAs containing the U-rich cytoplasmic polyadenylation element (CPE) have short poly(A) tails. CPE-binding protein (CPEB) mediates repression of translation through the interactions depicted, which prevent assembly of an initiation complex at the 5′ end of the mRNA. *(Right)* Hormone stimulation of oocytes activates a protein kinase that phosphorylates CPEB, causing it to release Maskin. The cleavage and polyadenylation specificity factor (CPSF) then binds to the poly(A) site, interacting with both bound CPEB and the cytoplasmic form of poly(A) polymerase (PAP). After the poly(A) tail is lengthened, multiple copies of the cytoplasmic poly(A)-binding protein I (PABPI) can bind to it and interact with eIF4G, which functions with other initiation factors to bind the 40S ribosome subunit and initiate translation. [Adapted from R. Mendez and J. D. Richter, 2001, *Nature Rev. Mol. Cell Biol.* **2:**521.]

ribosomal subunit, so translation initiation is blocked. During oocyte maturation, a specific CPEB serine is phosphorylated, causing Maskin to dissociate from the complex. This allows cytoplasmic forms of the cleavage and polyadenylation specificity factor (CPSF) and poly(A) polymerase to bind to the mRNA cooperatively with CPEB. Once the poly(A) polymerase catalyzes the addition of A residues, PABPI can bind to the lengthened poly(A) tail, leading to the stabilized interaction of all the participants needed to initiate translation (Figure 8-28, *right;* see also Figure 4-28). In the case of *Xenopus* oocyte maturation, the protein kinase that phosphorylates CPEB is activated in response to the hormone progesterone. Thus timing of the translation of stored mRNAs encoding proteins needed for oocyte maturation is regulated by this external signal.

Considerable evidence indicates that a similar mechanism of translational control plays a role in learning and memory. In the central nervous system, the axons from a thousand or so neurons can make connections (synapses) with the dendrites of a single postsynaptic neuron (Figure 23-23). When one of these axons is stimulated, the postsynaptic neuron "remembers" which one of these thousands of synapses was stimulated. The next time that synapse is stimulated, the strength of the response triggered in the postsynaptic cell differs from the first time. This change in response has been shown to result largely from the translational activation of mRNAs stored in the region of the synapse, leading to the local synthesis of new proteins that increase the size and alter the neurophysiological characteristics of the synapse. The finding that CPEB is present in neuronal dendrites has led to the proposal that cytoplasmic polyadenylation stimulates translation of specific mRNAs in dendrites, much as it does in oocytes. In this case, presumably, synaptic activity (rather than a hormone) is the signal that induces phosphorylation of CPEB and subsequent activation of translation.

Degradation of mRNAs in the Cytoplasm Occurs by Several Mechanisms

The concentration of an mRNA is a function of both its rate of synthesis and its rate of degradation. For this reason, if two genes are transcribed at the same rate, the steady-state concentration of the corresponding mRNA that is more stable will be higher than the concentration of the other. The stability of an mRNA also determines how rapidly synthesis of the encoded protein can be shut down. For a stable mRNA, synthesis of the encoded protein persists long after transcription of the gene is repressed. Most bacterial mRNAs are unstable, decaying exponentially with a typical half-life of a few minutes. For this reason, a bacterial cell can rapidly adjust the synthesis of proteins to accommodate changes in the cellular environment. Most cells in multicellular organisms, on the other hand, exist in a fairly constant environment and carry out a specific set of functions over periods of days to months or even the lifetime of the organism (nerve cells, for example). Accordingly, most mRNAs of higher eukaryotes have half-lives of many hours.

However, some proteins in eukaryotic cells are required only for short periods and must be expressed in bursts. For example, as discussed in the chapter introduction, certain signaling molecules called **cytokines,** which are involved in the immune response of mammals, are synthesized and secreted in short bursts. Similarly, many of the transcription factors that regulate the onset of the S phase of the cell cycle, such as c-Fos and c-Jun, are synthesized for brief periods only (Chapter 20). Expression of such proteins occurs in short bursts because transcription of their genes can be rapidly turned on and off, and their mRNAs have unusually short half-lives, on the order of 30 minutes or less.

Cytoplasmic mRNAs are degraded by one of the three pathways shown in Figure 8-29. For most mRNAs, the *deadenylation-dependent pathway* is followed: the length

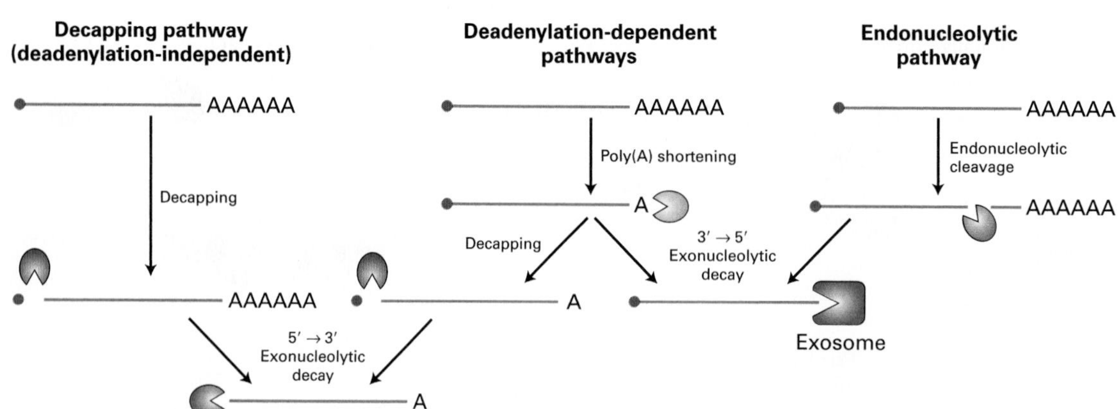

▲ **FIGURE 8-29 Pathways for degradation of eukaryotic mRNAs.** In the deadenylation-dependent (*middle*) pathways, the poly(A) tail is progressively shortened by a deadenylase (orange) until it reaches a length of 20 or fewer A residues, at which point the interaction with PABPI is destabilized, leading to weakened interactions between the 5′ cap and translation-initiation factors. The deadenylated mRNA then may either (1) be decapped and degraded by a 5′ → 3′ exonuclease or (2) be degraded by a 3′ → 5′ exonuclease in cytoplasmic exosomes. Some mRNAs (*right*) are cleaved internally by an endonuclease and the fragments degraded by an exosome. Other mRNAs (*left*) are decapped before they are deadenylated and then degraded by a 5′ → 3′ exonuclease. [Adapted from M. Tucker and R. Parker, 2000, *Ann. Rev. Biochem.* **69**:571.]

of the poly(A) tail gradually decreases with time through the action of a deadenylating nuclease. When it is shortened sufficiently, PABPI molecules can no longer bind and stabilize the interaction of the 5′ cap and translation initiation factors (see Figure 4-28b). The exposed cap then is removed by a decapping enzyme (Dcp1/Dcp2 in *S. cerevisiae*), and the unprotected mRNA is degraded by a 5′ → 3′ exonuclease (Xrn1 in *S. cerevisiae*). Removal of the poly(A) tail also makes mRNAs susceptible to degradation by cytoplasmic exosomes containing 3′ → 5′ exonucleases. The 5′ → 3′ exonucleases predominate in yeast, and the 3′ → 5′ exosome predominates in mammalian cells. The decapping enzymes and 5′ → 3′ exonuclease are concentrated in the P bodies, regions of the cytoplasm of unusually high density (see Figure 8-31).

Some mRNAs are degraded primarily by a deadenylation-independent *decapping pathway* (see Figure 8-29). This is because certain sequences at the 5′ end of an mRNA seem to make the cap sensitive to the decapping enzyme. For these mRNAs, the rate at which they are decapped controls the rate at which they are degraded because once the 5′ cap is removed, the RNA is rapidly hydrolyzed by the 5′ → 3′ exonuclease.

The rate of mRNA deadenylation varies inversely with the frequency of translation initiation for an mRNA: the higher the frequency of initiation, the slower the rate of deadenylation. This relation probably is due to the reciprocal interactions between translation initiation factors bound at the 5′ cap and PABPI bound to the poly(A) tail. For an mRNA that is translated at a high rate, initiation factors are bound to the cap much of the time, stabilizing the binding of PABPI and thereby protecting the poly(A) tail from the deadenylation exonuclease.

Many short-lived mRNAs in mammalian cells contain multiple, sometimes overlapping copies of the sequence AUUUA in their 3′-untranslated region. Specific RNA-binding proteins have been found that both bind to these 3′ AU-rich sequences and also interact with a deadenylating enzyme and with the exosome. This causes rapid deadenylation and subsequent 3′ → 5′ degradation of these mRNAs. In this mechanism, the rate of mRNA degradation is uncoupled from the frequency of translation. Thus mRNAs containing the AUUUA sequence can be translated at high frequency yet also be degraded rapidly, allowing the encoded proteins to be expressed in short bursts.

As shown in Figure 8-29, some mRNAs are degraded by an *endonucleolytic pathway* that does not involve decapping or significant deadenylation. One example of this type of pathway is the RNAi pathway discussed above (see Figure 8-25). Each siRNA-RISC complex can degrade thousands of targeted RNA molecules. The fragments generated by internal cleavage then are degraded by exonucleases.

P Bodies As mentioned above, P bodies are sites of translational repression of mRNAs bound by miRNA-RISC complexes. They are also the major sites of mRNA degradation in the cytoplasm. These dense regions of cytoplasm contain the decapping enzyme (Dcp1/Dcp2 in yeast), activators of decapping (Dhh, Pat1, Lsm1-7 in yeast), the major 5′ → 3′ exonuclease (Xrn1), as well as densely associated mRNAs. P bodies are dynamic structures that grow and shrink in size depending on the rate at which mRNPs associate with them, the rate at which mRNAs are degraded, and the rate at which mRNPs exit P bodies and reenter the pool of translated mRNPs.

Protein Synthesis Can Be Globally Regulated

Like proteins involved in other processes, translation initiation factors and ribosomal proteins can be regulated by post-translational modifications such as phosphorylation. Such mechanisms affect the translation rate of most mRNAs and hence the overall rate of cellular protein synthesis.

TOR Pathway The TOR pathway was discovered through research into the mechanism of action of rapamycin, an antibiotic produced by a strain of *Streptomyces* bacteria, useful for suppressing the immune response in organ transplant patients. The *target of rapamycin (TOR)* was identified by isolating yeast mutants resistant to rapamycin inhibition of cell growth. TOR is a large (~2400 amino acid residue) protein kinase that regulates several cellular processes in yeast cells in response to nutritional status. In multicellular eukaryotes, *metazoan TOR (mTOR)* also responds to multiple signals from cell-surface-signaling proteins to coordinate cell growth with developmental programs as well as nutritional status.

Current understanding of the mTOR pathway is summarized in Figure 8-30. Active mTOR stimulates the overall rate of protein synthesis by phosphorylating two critical proteins that regulate translation directly. mTOR also activates transcription factors that control expression of ribosomal components, tRNAs, and translation factors, further activating protein synthesis and cell growth.

Recall that the first step in translation of a eukaryotic mRNA is binding of the eIF4 initiation complex to the 5′ cap via its eIF4E cap-binding subunit (see Figure 4-24). The concentration of active eIF4E is regulated by a small family of homologous *eIF4E-binding proteins (4E-BPs)* that inhibit the interaction of eIF4E with mRNA 5′ caps. 4E-BPs are direct targets of mTOR. When phosphorylated by mTOR, 4E-BPs release eIF4E, stimulating translation initiation. mTOR also phosphorylates and activates another protein kinase that phosphorylates the small ribosomal subunit protein S6 (S6K) and probably additional substrates, leading to a further increase in the rate of protein synthesis.

Translation of a specific subset of mRNAs that have a string of pyrimidines in their 5′ untranslated regions (called TOP mRNAs for *t*ract of *o*ligo*p*yrimidine) is stimulated particularly strongly by mTOR. The 5′ TOP mRNAs encode ribosomal proteins and translation elongation factors. mTOR also activates the RNA polymerase I transcription factor TIF1A, stimulating transcription of the large rRNA precursor (see Section 8.5 below). mTOR also activates transcription

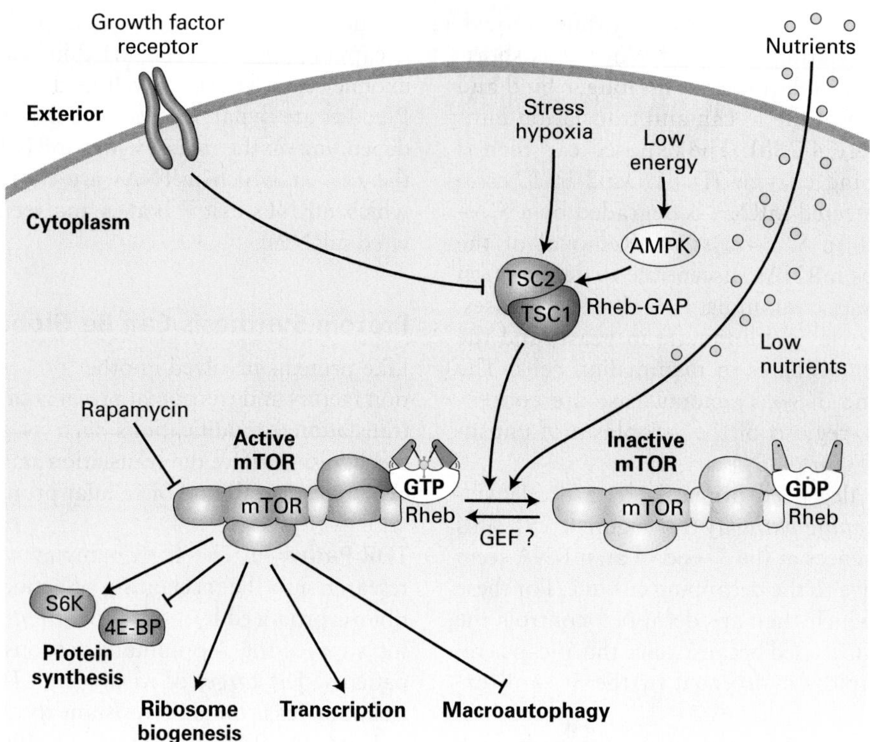

▲ **FIGURE 8-30 mTOR pathway.** mTOR is an active protein kinase when bound by a complex of Rheb and an associated GTP (*lower left*). In contrast, mTOR is inactive when bound by a complex of Rheb associated with GDP (*lower right*). When active, the TSC1/TSC2 Rheb-GTPase activating protein (Rheb-GAP) causes hydrolysis of Rheb-bound GTP to GDP, thereby inactivating mTOR. The TSC1/TSC2 Rheb-GAP is activated (arrows) by phosphorylation by AMP kinase (AMPK) when cellular energy charge is low and by other cellular stress responses. Signal-transduction pathways activated by cell-surface growth factor receptors lead to phosphorylation of inactivating sites on TSC1/TSC2, inhibiting its GAP activity. Consequently, they leave a higher fraction of cellular Rheb in the GTP conformation that activates mTOR protein kinase activity. Low nutrient concentration also regulates Rheb GTPase activity by a mechanism that does not require TSC1/TSC2. Active mTOR phosphorylates 4E-BP, causing it to release eIF4E, stimulating translation initiation. It also phosphorylates and activates S6 kinase (S6K), which in turn phosphorylates ribosomal proteins, stimulating translation. Activated mTOR also activates transcription factors for RNA polymerases I, II, and III, leading to synthesis and assembly of ribosomes, tRNAs, and translation factors. In the absence of mTOR activity, all of these processes are inhibited. In contrast, activated mTOR inhibits macroautophagy, which is stimulated in cells with inactive mTOR. [Adapted from S. Wullschleger et al., 2006, *Cell* **124:**471.]

by RNA polymerase III, although the mechanism is not clear. In addition, mTOR activates two RNA polymerase II activators that stimulate transcription of ribosomal protein and translation factor genes. Finally, mTOR stimulates processing of the rRNA precursor (Section 8.5). As a consequence of phosphorylation of these several mTOR substrates, the synthesis and assembly of ribosomes as well as the synthesis of translation factors and tRNAs are greatly increased. Alternatively, when mTOR kinase activity is inhibited, these substrates become dephosphorylated, greatly decreasing the rate of protein synthesis and the production of ribosomes, translation factors, and tRNAs, thus halting cell growth.

mTOR activity is regulated by a **monomeric small G protein** in the Ras protein family called Rheb. Like other small G proteins, Rheb is in its active conformation when it is bound to GTP. Rheb · GTP binds the mTOR complex, activating mTOR kinase activity, probably by inducing a conformation change in its kinase domain. Rheb is in turn regulated by a heterodimer composed of subunits TSC1 and TSC2, named for their involvement in the medical syndrome *t*uberous *s*clerosis *c*omplex, as discussed below. In the active conformation, the TSC1/TSC2 heterodimer functions as a GTPase activating protein for Rheb (Rheb-GAP), causing hydrolysis of the Rheb-bound GTP to GDP. This converts Rheb to its GDP-bound conformation, which binds to the mTOR complex and inhibits its kinase activity. Finally, the activity of the TSC1/TSC2 Rheb-GAP is regulated by several inputs, allowing the cell to integrate different cellular signaling pathways to control the overall rate of protein synthesis. Signaling from cell-surface growth factor receptors leads to phosphorylation of TSC1/TSC2 at inhibitory sites, causing an increase in Rheb·GTP and activation of mTOR kinase activity. This type of regulation through cell-surface receptors links the control of cell growth to developmental processes controlled by cell-cell interactions.

mTOR activity also is regulated in response to nutritional status. When energy from nutrients is not sufficient for cell growth, the resulting fall in the ratio of ATP to AMP concentrations is detected by the AMP kinase. The activated AMP kinase phosphorylates TSC1/TSC2 at activating sites, stimulating its Rheb-GAP activity and consequently inhibiting mTOR kinase activity and the global rate of translation. Hypoxia and other cellular stresses also activate the TSC1/TSC2 Rheb-GAP. Finally, the concentration of nutrients in the extracellular space also regulates Rheb by an unknown mechanism that does not require the TSC1/TSC2 complex.

In addition to regulating the global rate of cellular protein synthesis and the production of ribosomes, tRNAs, and translation factors, mTOR regulates at least one other process involved in the response to low levels of nutrients: macroautophagy. Starved cells degrade cytoplasmic constituents, including whole organelles, to supply energy and precursors for essential cellular processes. During this process a large, double-membrane structure engulfs a region of cytoplasm to form an autophagosome, which then fuses with a lysosome where the entrapped proteins, lipids, and other macromolecules are degraded, completing the process of macroautophagy. Activated mTOR inhibits macroautophagy in growing cells when nutrients are plentiful. Macroautophagy is stimulated when mTOR activity falls in nutrient-deprived cells.

Genes encoding components of the mTOR pathway are mutated in many human cancers, resulting in cell growth in the absence of normal growth signals. TSC1 and TSC2 (Figure 8-30) were initially identified because one or the other of the proteins is mutant in a rare human genetic syndrome: *t*uberous *s*clerosis *c*omplex. Patients with this disorder develop benign tumors in multiple tissues. The disease results because inactivation of either TSC1 or TSC2 eliminates the Rheb-GAP activity of the TSC1/TSC2 heterodimer, resulting in an abnormally high and unregulated level of Rheb · GTP and the resulting high, unregulated activity of mTOR. Mutations in components of cell-surface receptor signal-transduction pathways that lead to inhibition of TSC1/TSC2 Rheb-GAP activity are also common in human tumors and contribute to cell growth and replication in the absence of normal signals for growth and proliferation.

High mTOR protein kinase activity in tumors correlates with a poor clinical prognosis. Consequently, mTOR inhibitors are currently in clinical trials to test their effectiveness for treating cancers in conjunction with other modes of therapy. Rapamycin and other structurally related mTOR inhibitors are potent suppressors of the immune response because they inhibit activation and replication of T lymphocytes in response to foreign antigens (Chapter 24). Several viruses encode proteins that activate mTOR early after viral infection. The resulting stimulation of translation has an obvious selective advantage for these cellular parasites. ▮

eIF2 Kinases eIF2 kinases also regulate the global rate of cellular protein synthesis. Figure 4-24 summarizes the steps in translation initiation. Translation initiation factor eIF2 brings the charged initiator tRNA to the small ribosome subunit P site. eIF2 is a **trimeric G protein** and consequently exists in either a GTP-bound or a GDP-bound conformation. Only the GTP-bound form of eIF2 is able to bind the charged initiator tRNA and associate with the small ribosomal subunit. The small subunit with bound initiation factors and charged initiator tRNA then interacts with the eIF4 complex bound to the 5′ cap of an mRNA via its eIF4E subunit. The small ribosomal subunit then scans down the mRNA in the 3′ direction until it reaches an AUG initiation codon that can base-pair with the initiator tRNA in its P site. When this occurs, the GTP bound by eIF2 is hydrolyzed to GDP and the resulting eIF2·GDP complex is released. GTP hydrolysis results in an irreversible "proofreading" step that prepares the small ribosomal subunit to associate with the large subunit only when an initiator tRNA is properly bound in the P site and is properly base-paired with the AUG start codon. Before eIF2 can participate in another round of initiation, its bound GDP must be replaced with a GTP. This process is catalyzed by the translation initiation factor eIF2B, a guanine nucleotide exchange factor (GEF) specific for eIF2.

A mechanism for inhibiting general protein synthesis in stressed cells involves phosphorylation of the eIF2 α subunit at a specific serine. Phosphorylation at this site does not interfere with eIF2 function in protein synthesis directly. Rather, phosphorylated eIF2 has very high affinity for the eIF2 guanine nucleotide exchange factor, eIF2B, which cannot release the phosphorylated eIF2 and consequently is blocked from catalyzing GTP exchange of additional eIF2 factors. Since there is an excess of eIF2 over eIF2B, phosphorylation of a fraction of eIF2 results in inhibition of all the cellular eIF2B. The remaining eIF2 accumulates in its GDP-bound form, which cannot participate in protein synthesis, thereby inhibiting nearly all cellular protein synthesis. However, some mRNAs have 5′ regions that allow translation initiation at the low eIF2-GTP concentration that results from eIF2 phosphorylation. These mRNAs include those for chaperone proteins that function to refold cellular proteins denatured as the result of cellular stress, additional proteins that help the cell to cope with stress, and transcription factors that activate transcription of the genes encoding these stress-induced proteins.

Human cells contain four eIF2 kinases that phosphorylate the same inhibitory eIF2α serine. Each of these is regulated by a different type of cellular stress, inhibiting protein synthesis and allowing cells to divert the large fraction of cellular resources devoted to protein synthesis in growing cells for use in responding to the stress.

The GCN2 (general control non-derepressible 2) eIF2-kinase is activated by binding uncharged tRNAs. The concentration of uncharged tRNAs increases when cells are starved for amino acids, activating GCN2 eIF2-kinase activity and greatly inhibiting protein synthesis.

PEK (pancreatic eIF2α kinase) is activated when proteins translocated into the endoplasmic reticulum (ER) do not fold properly because of abnormalities in the ER lumen environment. Inducers include abnormal carbohydrate concentration because this inhibits the glycosylation of many ER proteins and inactivating mutations in an ER chaperone required for proper folding of many ER proteins (Chapters 13 and 14).

Heme-regulated inhibitor (HRI) is activated in developing red blood cells when the supply of heme prosthetic group is too low to accommodate the rate of globin protein synthesis. This negative feedback loop lowers the rate of globin protein synthesis until it matches the rate of heme synthesis. HR1 is also activated in other types of cells in response to oxidative stress or heat shock.

Finally, protein kinase RNA activated (PKR) is activated by double-stranded RNAs longer than ≈30 base pairs. Under normal circumstances in mammalian cells, such double-stranded RNAs are produced only during a viral infection. Long regions of double-stranded RNA are generated in replication intermediates of RNA viruses or from hybridization of complementary regions of RNA transcribed from both strands of DNA virus genomes. Inhibition of protein synthesis prevents the production of progeny virions, protecting neighboring cells from infection. Interestingly, adenoviruses evolved a defense against PKR: they express prodigious amounts of an ≈160-nucleotide virus-associated (VA) RNA with long double-stranded hairpin regions. VA RNA is transcribed by RNA polymerase III and exported from the nucleus by Exportin5, the exportin for pre-miRNAs (see Figure 8-27). VA RNA binds to PKR with high affinity, inhibiting its protein kinase activity and preventing the inhibition of protein synthesis observed in cells infected with a mutant adenovirus from which the VA gene was deleted.

Sequence-Specific RNA-Binding Proteins Control Specific mRNA Translation

In contrast to global mRNA regulation, mechanisms have also evolved for controlling the translation of certain specific mRNAs. This is usually done by sequence-specific RNA-binding proteins that bind to a particular sequence or RNA structure in the mRNA. When binding is in the 5′ untranslated region (5′-UTR) of an mRNA, the ribosome's ability to scan to the first initiation codon is blocked, inhibiting translation initiation. Binding in other regions can either promote or inhibit mRNA degradation.

Control of intracellular iron concentration by the *iron response element–binding protein (IRE-BP)* is an elegant example of a single protein that regulates the translation of one mRNA and the degradation of another. Precise regulation of cellular iron ion concentration is critical to the cell. Multiple enzymes and proteins contain Fe^{2+} as a cofactor, such as enzymes of the Krebs cycle (see Figure 12-10) and electron-carrying proteins involved in the generation of ATP by mitochondria and chloroplasts (Chapter 12). On the other hand, excess Fe^{2+} generates free radicals that react with and damage cellular macromolecules. When intracellular iron stores are low, a dual-control system operates to increase the level of cellular iron; when iron is in excess, the system operates to prevent accumulation of toxic levels of free ions.

One component in this system is the regulation of the production of ferritin, an intracellular iron-binding protein that binds and stores excess cellular iron. The 5′ untranslated region of ferritin mRNA contains iron-response elements (IREs) that have a stem-loop structure. The IRE-binding protein (IRE-BP) recognizes five specific bases in the IRE loop and the duplex nature of the stem. At low iron concentrations, IRE-BP is in an active conformation that binds to the IREs (Figure 8-31a). The bound IRE-BP blocks the small ribosomal subunit from scanning for the AUG start codon (see Figure 4-24), thereby inhibiting translation initiation. The resulting decrease in ferritin means less iron is complexed with the ferritin and is therefore available to iron-requiring enzymes. At high iron concentrations, IRE-BP is in an inactive conformation that does not bind to the 5′ IREs, so translation initiation can proceed. The newly synthesized ferritin then binds free iron ions, preventing their accumulation to harmful levels.

The other part of this regulatory system controls the import of iron into cells. In vertebrates, ingested iron is carried through the circulatory system bound to a protein called transferrin. After binding to the transferrin receptor (TfR) in the plasma membrane, the transferrin-iron complex is brought into cells by receptor-mediated endocytosis (Chapter 14). The 3′ untranslated region of TfR mRNA contains IREs whose stems have AU-rich destabilizing sequences (Figure 8-31b). At high iron concentrations, when the IRE-BP is in the inactive, nonbinding conformation, these AU-rich sequences promote degradation of TfR mRNA by the same mechanism that leads to rapid degradation of other short-lived mRNAs, as described previously. The resulting decrease in production of the transferrin receptor quickly reduces iron import, thus protecting the cell from excess iron. At low iron concentrations, however, IRE-BP can bind to the 3′ IREs in TfR mRNA. The bound IRE-BP blocks recognition of the destabilizing AU-rich sequences by the proteins that would otherwise rapidly degrade the mRNAs. As a result, production of the transferrin receptor increases and more iron is brought into the cell.

Other regulated RNA-binding proteins may also function to control mRNA translation or degradation, much like the dual-acting IRE-BP. For example, a heme-sensitive RNA-binding protein controls translation of the mRNA encoding aminolevulinate (ALA) synthase, a key enzyme in the synthesis of heme. In vitro studies have shown that the mRNA

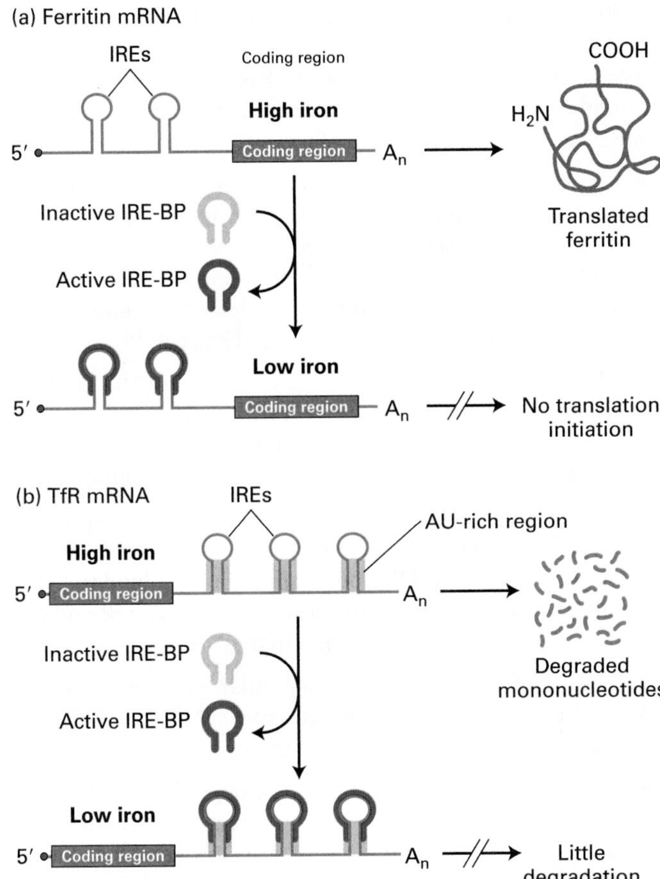

▲ **FIGURE 8-31 Iron-dependent regulation of mRNA translation and degradation.** The iron response element–binding protein (IRE-BP) controls translation of ferritin mRNA (a) and degradation of transferrin-receptor (TfR) mRNA (b). At low intracellular iron concentrations IRE-BP binds to iron-response elements (IREs) in the 5′ or 3′ untranslated region of these mRNAs. At high iron concentrations, IRE-BP undergoes a conformational change and cannot bind either mRNA. The dual control by IRE-BP precisely regulates the level of free iron ions within cells. See the text for discussion.

encoding the milk protein casein is stabilized by the hormone prolactin and rapidly degraded in its absence.

Surveillance Mechanisms Prevent Translation of Improperly Processed mRNAs

Translation of an improperly processed mRNA could lead to production of an abnormal protein that interferes with the gene's normal function. (This effect is equivalent to that resulting from dominant-negative mutations, discussed in Chapter 5 (Figure 5-44), although the cause is different.) Several mechanisms collectively termed **mRNA surveillance** help cells avoid the translation of improperly processed mRNA molecules. We have previously mentioned two such surveillance mechanisms: the recognition of improperly processed pre-mRNAs in the nucleus and their degradation by the exosome and the general restriction against nuclear export of incompletely spliced pre-mRNAs that remain associated with a spliceosome.

Another mechanism called nonsense-mediated decay causes degradation of mRNAs in which one or more exons have been incorrectly skipped during splicing. Such exon skipping often will alter the open reading frame of the mRNA 3′ to the improper exon junction, resulting in introduction of out-of-frame missense mutations and an incorrect stop codon. For nearly all properly spliced mRNAs, the stop codon is in the last exon. The process of *nonsense-mediated decay (NMD)* results in the rapid degradation of mRNAs with stop codons that occur before the last splice junction in the mRNA since in most cases, such mRNAs arise from errors in RNA splicing.

A search for possible molecular signals that might indicate the positions of splice junctions in a processed mRNA led to the discovery of exon-junction complexes. As noted already, these complexes of several proteins (including Y14, Magoh, eIF4IIIA, Upf3, and REF), bound ≈20 nucleotides 5′ to an exon-exon junction following RNA splicing, stimulate export of mRNPs from the nucleus by interacting with the mRNA exporter (see Figure 8-21). Analysis of yeast mutants indicate that one of the proteins in exon-junction complexes (Upf3) functions in nonsense-mediated decay. In the cytoplasm, this component of exon-junction complexes interacts with a protein (Upf1) that causes the mRNA to associate with P bodies, repressing translation of the mRNA. An additional protein (Upf2) then associates with the mRNP complex and binds a P-body associated deadenylase that rapidly removes the poly(A) tail from an associated mRNA, leading to its rapid decapping and degradation by the P body associated 5′ → 3′ exonuclease (see Figure 8-29). In the case of properly spliced mRNAs, the exon-junction complexes are thought to be dislodged from the mRNA by passage of the first "pioneer" ribosome to translate the mRNA, thereby protecting the mRNA from degradation. However, for mRNAs with a stop codon before the final exon junction, one or more exon-junction complexes remain associated with the mRNA, resulting in nonsense-mediated decay.

Localization of mRNAs Permits Production of Proteins at Specific Regions Within the Cytoplasm

Many cellular processes depend on localization of particular proteins to specific structures or regions of the cell. In later chapters we examine how some proteins are transported *after* their synthesis to their proper cellular location. Alternatively, protein localization can be achieved by localization of mRNAs to specific regions of the cell cytoplasm in which their encoded proteins function. In most cases examined thus far, such mRNA localization is specified by sequences in the 3′ untranslated region of the mRNA.

As mentioned earlier, in neurons, localization of specific mRNAs at synapses far from the nucleus in the cell body plays an essential function in learning and memory (Figure 8-32). Some of these mRNAs are initially synthesized with short poly(A) tails that do not allow translation initiation. Synaptic activity can stimulate their polyadenylation in the region of the synapse, activating the translation of encoded proteins that increase the size and alter the neurophysiological

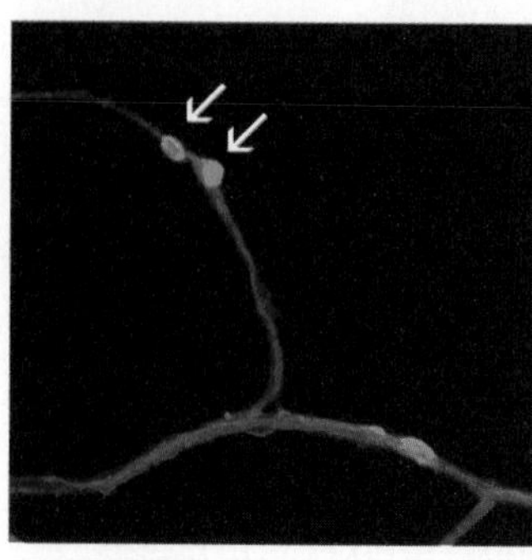
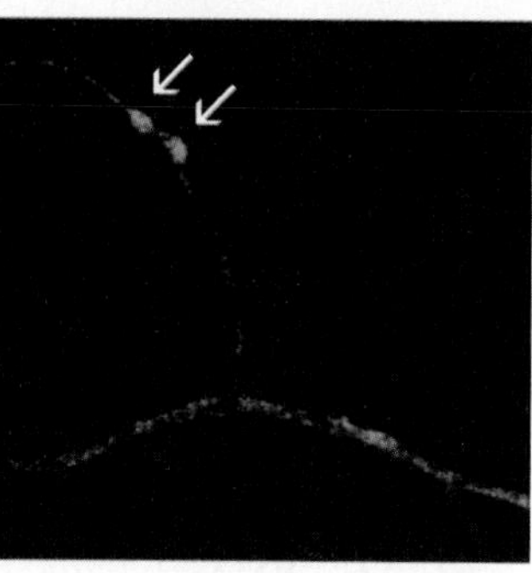

▲ **EXPERIMENTAL FIGURE 8-32 A specific neuronal mRNA localizes to synapses.** Sensory neurons from the sea slug *Aplysia californica*—among the largest neurons in the animal kingdom—were cultured with target motor neurons so that processes from the sensory neuron formed synapses with processes from the motor neuron. The micrograph at the left shows motor neuron processes visualized with a blue fluorescent dye. GFP-VAMP (green) was expressed in sensory neurons and marks the location of synapses formed between sensory and motor neuron processes (arrows). The micrograph on the right shows red fluorescence from in situ hybridization of an antisensorin mRNA probe. Sensorin is a neurotransmitter expressed by the sensory neuron only; sensory neuron processes are not otherwise visualized in this preparation, but they lie adjacent to the motor neuron processes. The in situ hybridization results indicate that sensorin mRNA is localized to synapses. [From V. Lyles, Y. Zhao, and K. C. Martin, 2006, *Neuron* **49**:323.]

properties of this one synapse out of hundreds to thousands of synapses made by a neuron.

Cultured myoblasts localize β-actin mRNA to the leading edge of the cell, where newly synthesized β-actin is polymerized into an actin cytoskeletal network that pushes the cell membrane forward as it assembles under it (Chapter 17). Treatment of cultured myoblasts with cytochalasin D, which disrupts actin microfilaments, leads to rapid delocalization of β-actin mRNA, indicating that cytoskeletal actin microfilaments participate in the localization process. Disruption of other cytoskeletal components, however, does not alter the localization of actin mRNAs. Other types of evidence also implicate the actin cytoskeleton in mRNA localization. Presumably certain RNA-binding proteins interact with specific sequences in the 3′ untranslated regions of an mRNA and with specific components of microfilaments, including motor proteins that move cargo along the length of microfilaments. The best-understood example of this mechanism of mRNA localization occurs during cell division in *S. cerevisiae,* as we describe in Chapter 21.

KEY CONCEPTS OF SECTION 8.4

Cytoplasmic Mechanisms of Post-transcriptional Control

- Translation can be repressed by micro-RNAs (miRNAs), which form imperfect hybrids with sequences in the 3′ untranslated region (UTR) of specific target mRNAs.
- The related phenomenon of RNA interference, which probably evolved as an early defense system against viruses and transposons, leads to degradation of mRNAs that form perfect hybrids with short interfering RNAs (siRNAs).
- Both miRNAs and siRNAs contain 21–23 nucleotides, are generated from longer precursor molecules, and are assembled into a multiprotein RNA-induced silencing complex (RISC) that either represses translation of target mRNAs or cleaves them (see Figures 8-26 and 8-27).
- Cytoplasmic polyadenylation is required for translation of mRNAs with a short poly(A) tail. Binding of a specific protein to regulatory elements in their 3′ UTRs represses translation of these mRNAs. Phosphorylation of this RNA-binding protein, induced by an external signal, leads to lengthening of the 3′ poly(A) tail and translation (see Figure 8-29).
- Most mRNAs are degraded as the result of the gradual shortening of their poly(A) tail (deadenylation) followed by exosome-mediated 3′ → 5′ digestion or removal of the 5′ cap and digestion by a 5′ → 3′ exonuclease (see Figure 8-30).
- Eukaryotic mRNAs encoding proteins that are expressed in short bursts generally have repeated copies of an AU-rich sequence in their 3′ UTR. Specific proteins that bind to these elements also interact with the deadenylating enzyme and cytoplasmic exosomes, promoting rapid RNA degradation.
- Binding of various proteins to regulatory elements in the 3′ or 5′ UTRs of mRNAs regulates the translation or degradation of many mRNAs in the cytoplasm.
- Translation of ferritin mRNA and degradation of transferrin receptor (TfR) mRNA are both regulated by the same iron-sensitive RNA-binding protein. At low iron concentrations, this protein is in a conformation that binds to specific elements in the mRNAs, inhibiting ferritin mRNA translation or degradation of TfR mRNA (see Figure 8-32). This dual control precisely regulates the iron level within cells.
- Nonsense-mediated decay and other mRNA surveillance mechanisms prevent the translation of improperly processed mRNAs encoding abnormal proteins that might interfere with functioning of the corresponding normal proteins.
- Some mRNAs are directed to specific subcellular locations by sequences usually found in the 3′ UTR, leading to localization of the encoded proteins.

8.5 Processing of rRNA and tRNA

Approximately 80 percent of the total RNA in rapidly growing mammalian cells (e.g., cultured HeLa cells) is rRNA, and 15 percent is tRNA; protein-coding mRNA thus constitutes only a small portion of the total RNA. The primary transcripts produced from most rRNA genes and from tRNA genes, like pre-mRNAs, are extensively processed to yield the mature, functional forms of these RNAs.

The ribosome is a highly evolved, complex structure (see Figure 4-23), optimized for its function in protein synthesis. Eukaryotic cells devote an extraordinary number of proteins and other RNAs to the construction of these complex molecular machines. Their synthesis requires the function and coordination of all three nuclear RNA polymerases. The 28S and 5.8S rRNAs associated with the large ribosomal subunit and the single 18S rRNA of the small subunit are transcribed by

RNA polymerase I. The 5S rRNA of the large subunit is transcribed by RNA polymerase III, and the mRNAs encoding the ribosomal proteins are transcribed by RNA polymerase II. In addition to the four rRNAs and ≈70 ribosomal proteins, at least 150 other RNAs and proteins interact transiently with the two ribosomal subunits during their assembly through a series of coordinated steps. Furthermore, multiple specific bases and riboses of the mature rRNAs are modified to optimize their function in protein synthesis. Although most of the steps in ribosomal subunit synthesis and assembly occur in the **nucleolus** (a subcompartment of the nucleus not bounded by a membrane), some occur in the nucleoplasm during passage from the nucleolus to nuclear pore complexes. A quality-control step occurs before nuclear export so that only fully functional subunits are exported to the cytoplasm, where the final steps of ribosome subunit maturation occur. tRNAs also are processed from precursor primary transcripts in the nucleus and modified extensively before they are exported to the cytoplasm and used in protein synthesis. First we'll discuss the processing and modification of rRNA and the assembly and nuclear export of ribosomes. Then we'll consider the processing and modification of tRNAs.

Pre-rRNA Genes Function as Nucleolar Organizers and Are Similar in All Eukaryotes

The 28S and 5.8S rRNAs associated with the large (60S) ribosomal subunit and the 18S rRNA associated with the small (40S) ribosomal subunit in higher eukaryotes (and the functionally equivalent rRNAs in all other eukaryotes) are encoded by a single type of **pre-rRNA** transcription unit. In human cells, transcription by RNA polymerase I yields a 45S (≈13.7 kb) primary transcript (pre-rRNA), which is processed into the mature 28S, 18S, and 5.8S rRNAs found in cytoplasmic ribosomes. Sequencing of the DNA encoding pre-rRNA from many species showed that this DNA shares several properties in all eukaryotes. First, the pre-rRNA genes are arranged in long tandem arrays separated by nontranscribed spacer regions ranging in length from ≈2 kb in frogs to ≈30 kb in humans (Figure 8-33). Second, the genomic regions corresponding to the three mature rRNAs are always arranged in the same 5′ → 3′ order: 18S, 5.8S, and 28S. Third, in all eukaryotic cells (and even in bacteria), the pre-rRNA gene codes for regions that are removed during processing and rapidly degraded. These regions probably contribute to proper folding of the rRNAs but are not required once the folding has occurred. The general structure of pre-rRNAs is diagrammed in Figure 8-34.

The synthesis and most of the processing of pre-rRNA occurs in the *nucleolus*. When pre-rRNA genes initially were identified in the nucleolus by in situ hybridization, it was not known whether any other DNA was required to form the nucleolus. Subsequent experiments with transgenic *Drosophila* strains demonstrated that a single complete pre-rRNA transcription unit induces formation of a small nucleolus. Thus a single pre-rRNA gene is sufficient to be a *nucleolar organizer*, and all the other components of the ribosome diffuse to the newly formed pre-rRNA. The structure

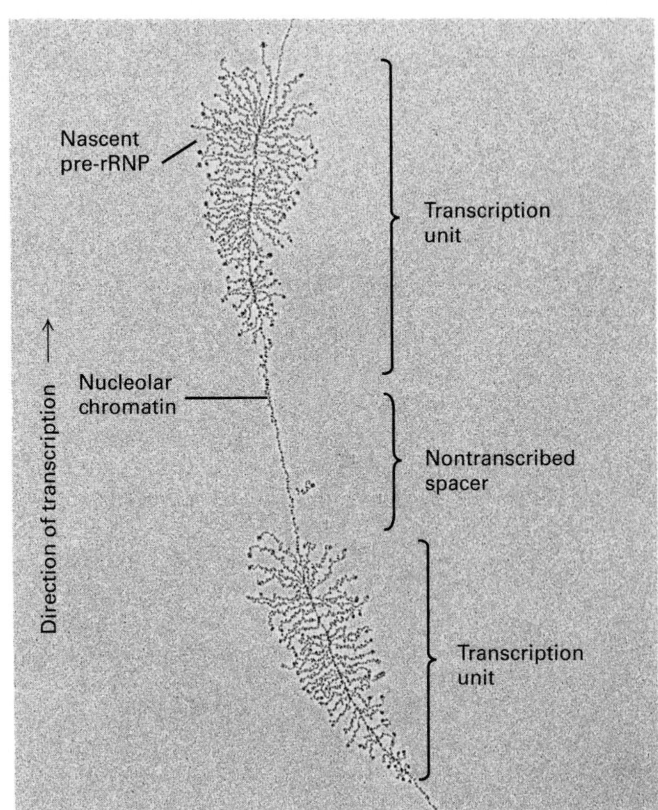

▲ EXPERIMENTAL FIGURE 8-33 **Electron micrograph of pre-rRNA transcription units from the nucleolus of a frog oocyte.** Each "feather" represents multiple pre-rRNA molecules associated with protein in a pre-ribonucleoprotein complex (pre-RNP) emerging from a transcription unit. Note the dense "knob" at the 5′-end of each nascent pre-RNP thought to be a processome. Pre-rRNA transcription units are arranged in tandem, separated by nontranscribed spacer regions of nucleolar chromatin. [Courtesy of Y. Osheim and O. J. Miller, Jr.]

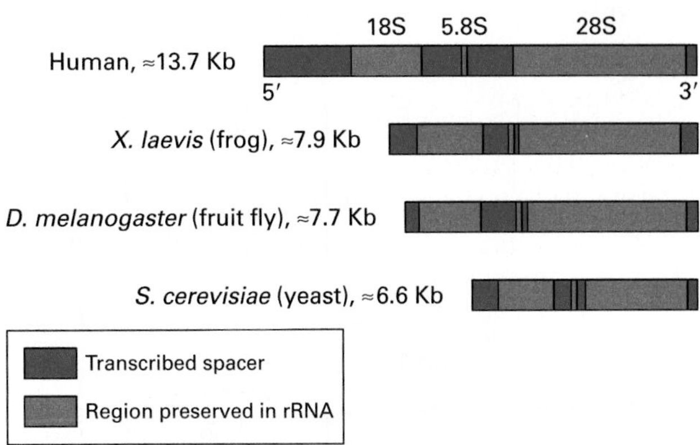

▲ FIGURE 8-34 **General structure of eukaryotic pre-rRNA transcription units.** The three coding regions (red) encode the 18S, 5.8S, and 28S rRNAs found in ribosomes of higher eukaryotes or their equivalents in other species. The order of these coding regions in the genome is always 5′ → 3′. Variations in the lengths of the transcribed spacer regions (blue) account for the major difference in the lengths of pre-rRNA transcription units in different organisms.

of the nucleolus observed by light and electron microscopy results from the processing of pre-RNA and the assembly of ribosomal subunits.

Small Nucleolar RNAs Assist in Processing Pre-rRNAs

Ribosomal subunit assembly, maturation, and export to the cytoplasm are best understood in the yeast *S. cerevisiae*. However, nearly all the proteins and RNAs involved are highly conserved in multicellular eukaryotes, where the fundamental aspects of ribosome biosynthesis are likely to be the same. As for pre-mRNAs, nascent pre-rRNA transcripts are immediately bound by proteins, forming preribosomal ribonucleoprotein particles (pre-rRNPs). For reasons not yet known, cleavage of the pre-rRNA does not begin until transcription of the pre-rRNA is nearly complete. In yeast, it takes approximately six minutes for a pre-rRNA to be transcribed. Once transcription is complete, the rRNA is cleaved, and bases and riboses are modified in about 10 seconds. In a rapidly growing yeast cell, ≈40 pairs of ribosomal subunits are synthesized, processed, and transported to the cytoplasm every second. This extremely high rate of ribosome synthesis in the face of the seemingly long period required to transcribe a pre-rRNA is possible because pre-rRNA genes are packed with RNA polymerase I molecules transcribing the same gene simultaneously (see Figure 8-33) and because there are 100–200 such genes on chromosome XII, the yeast nucleolar organizer.

The primary transcript of ≈7 kb is cut in a series of cleavage and exonucleolytic steps that ultimately yield the mature rRNAs found in ribosomes (Figure 8-35). During processing,

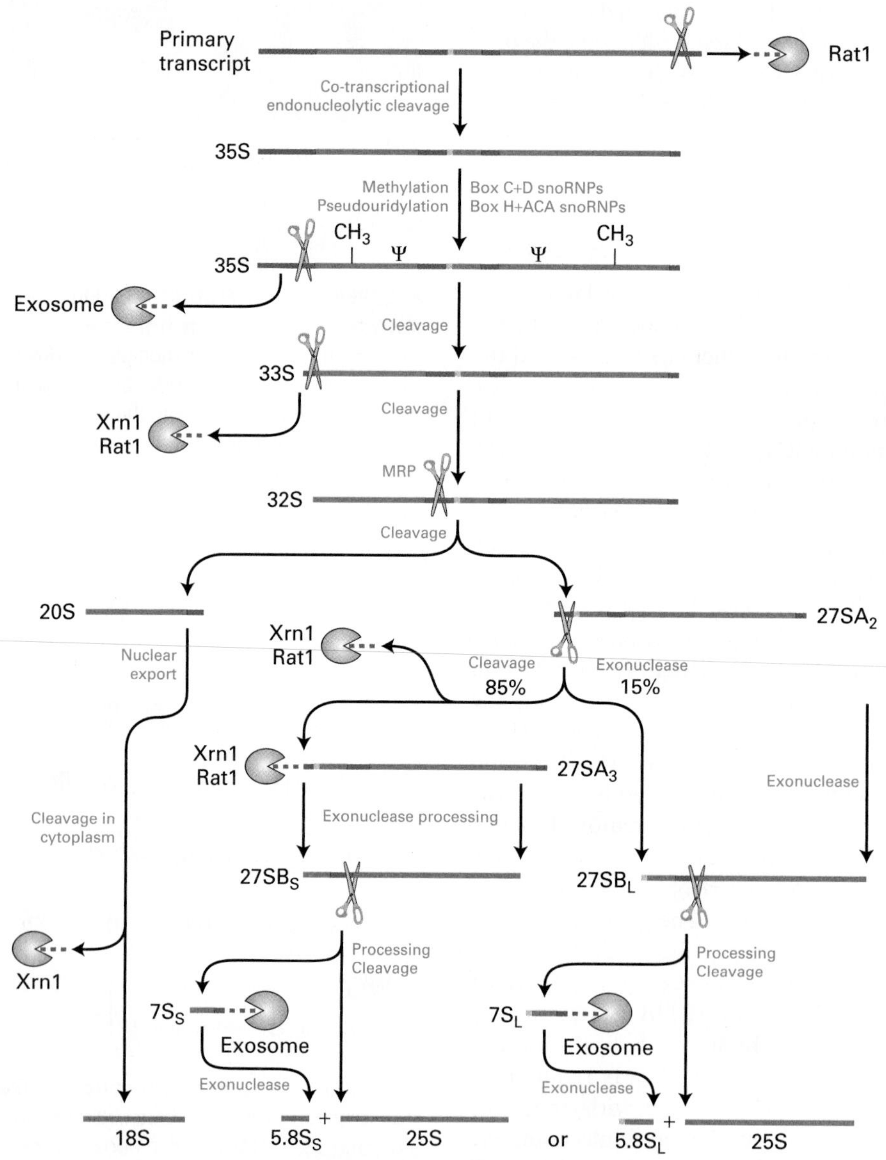

▲ **FIGURE 8-35 rRNA processing.** Endoribonucleases that make internal cleavages are represented as scissors. Exoribonucleases that digest from one end, either 5′ or 3′, are shown as Pac-Men. Most 2′-*O*-ribose methylation (CH_3) and generation of pseudouridines in the rRNAs occurs following the initial cleavage at the 3′ end, before the initial cleavage at the 5′ end. Proteins and snoRNPs known to participate in these steps are indicated. [From J. Venema and D. Tollervey, 1999, *Ann. Rev. Genetics* **33**:261.]

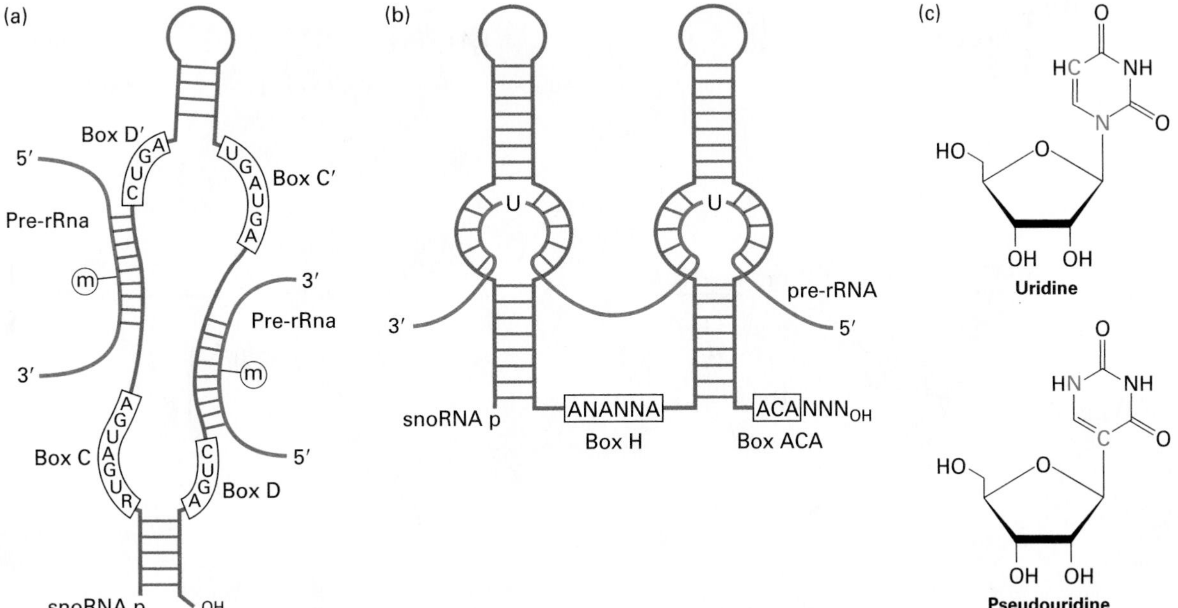

▲ **FIGURE 8-36 SnoRNP-directed modification of pre-rRNA.** (a) A snoRNA called box C+D snoRNA is involved in ribose 2′-*O*-methylation. Sequences in this snoRNA hybridize to two different regions in the pre-rRNA, directing methylation at the indicated sites. (b) Box H+ACA snoRNAs fold into two stem loops with internal single-stranded bulges in the stems. Pre-rRNA hybridizes to the single-stranded bulges, demarcating a site of pseudouridylation. (c) Conversion from uridine to pseudouridine directed by the box H+ACA snoRNAs of part (b). [Part (a) from T. Kiss, 2001, *EMBO J.* **20:**3617. Part (b) from U. T. Meier, 2005, *Chromosoma* **114:**1.]

pre-rRNA also is extensively modified, mostly by methylation of the 2′-hydroxyl group of specific riboses and conversion of specific uridine residues to pseudouridine. These post-transcriptional modifications of rRNA are probably important for protein synthesis because they are highly conserved. Virtually all of these modifications occur in the most conserved core structure of the ribosome, which is directly involved in protein synthesis. The positions of the specific sites of 2′-O-methylation and pseudouridine formation are determined by approximately 150 different small nucleolus-restricted RNA species, called **small nucleolar RNAs (snoRNAs),** which hybridize transiently to pre-rRNA molecules. Like the snRNAs that function in pre-mRNA processing, snoRNAs associate with proteins, forming ribonucleoprotein particles called snoRNPs. One class of more than 40 snoRNPs (box C+D snoRNAs) positions a methyltransferase enzyme near methylation sites in the pre-mRNA. The multiple different box C+D snoRNAs direct methylation at multiple sites through a similar mechanism. They share common sequence and structural features and are bound by a common set of proteins. One or two regions of each of these snoRNAs are precisely complementary to sites on the pre-rRNA and direct the methyltransferase to specific riboses in the hybrid region (Figure 8-36a). A second major class of snoRNPs (box H+ACA snoRNAs) positions the enzyme that converts uridine to pseudouridine. This conversion involves rotation of the pyrimidine ring (Figure 8-36c). Bases on either side of the modified uridine in the pre-rRNA base-pair with bases in the bulge of a stem in the H+ACA snoRNAs, leaving the modified uridine bulged out of the helical double-stranded region, like the branch point A bulges out in pre-mRNA spliceosomal splicing (see Figure 8-10). Other modifications of pre-rRNA nucleotides, such as adenine dimethylation, are carried out by specific proteins without the assistance of guiding snoRNAs.

The U3snoRNA is assembled into a large snoRNP containing ≈25 proteins called the "processome," which specifies cleavage at site A_0, the initial cut near the 5′ end of the pre-rRNA (see Figure 8-35). U3snoRNA base-pairs with an upstream region of the pre-rRNA to specify the location of the cleavage. The processome is thought to form the "5′ knob" visible in electron micrographs of pre-rRNPs (see Figure 8-33). Base pairing of other snoRNPs specify additional cleavage reactions that remove transcribed spacer regions. The first cleavage to initiate processing of the 5.8S and 25S rRNAs of the large subunit is performed by RNAse MRP, a complex of nine proteins with an RNA. Once cleaved from pre-rRNAs, these sequences are degraded by the same exosome-associated 3′ → 5′ nuclear exonucleases that degrade introns spliced from pre-mRNAs. Nuclear 5′ → 3′ exoribonucleases (Rat1; Xrn1) also remove some regions of 5′ spacer.

Some snoRNAs are expressed from their own promoters by RNA polymerase II or III. Remarkably, however, the large majority of snoRNAs are processed from spliced-out introns of genes encoding functional mRNAs for proteins involved in ribosome synthesis or translation. Some snoRNAs are processed from introns spliced from apparently nonfunctional mRNAs. The genes encoding these mRNAs seem to exist only to express snoRNAs from excised introns.

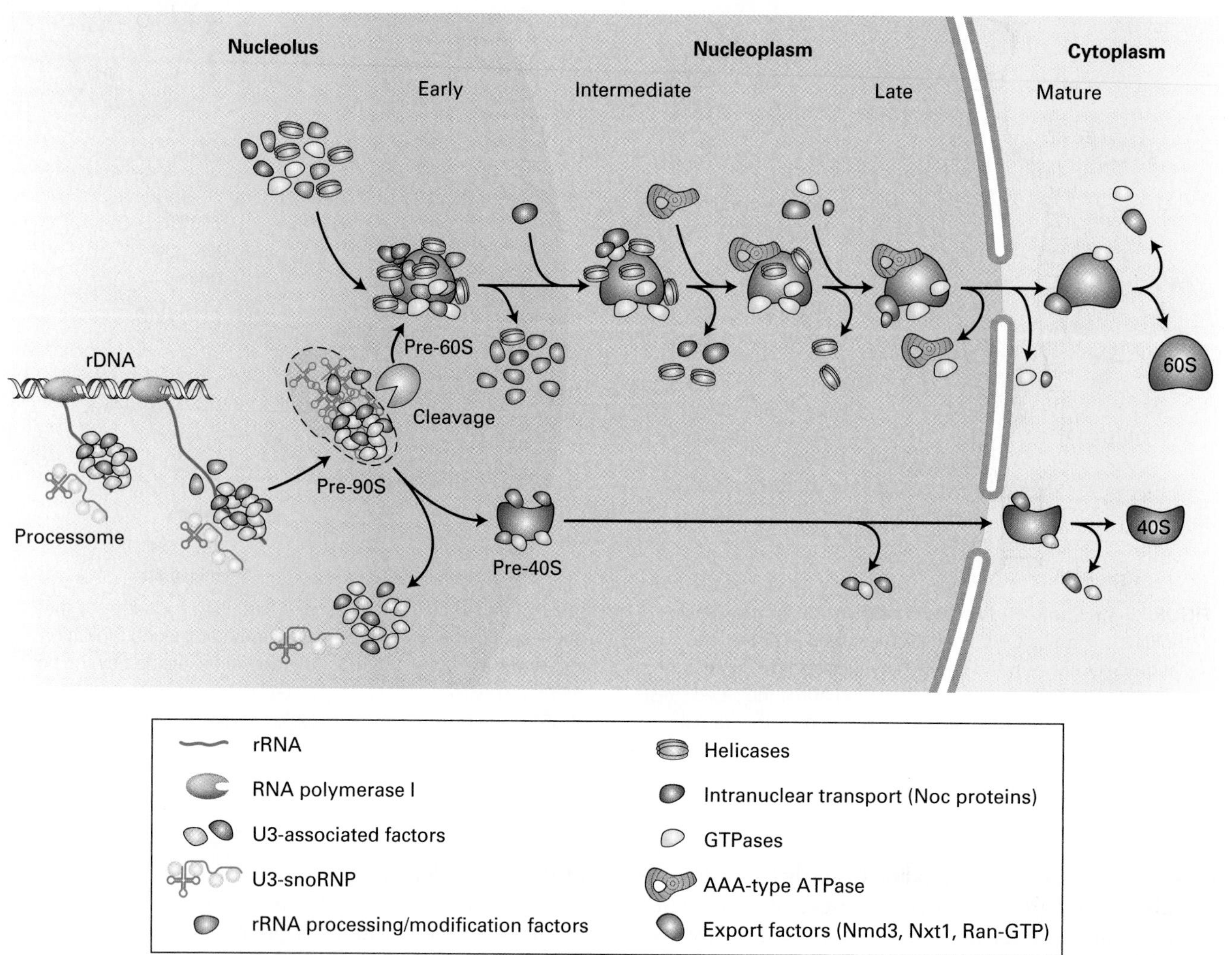

▲ **FIGURE 8-37 Ribosomal subunit assembly.** Ribosomal proteins and RNAs in the maturing small and large ribosomal subunits are depicted in blue, with a shape similar to the icons for the mature subunits in the cytoplasm. Other factors that associate transiently with the maturing subunits are depicted in different colors, as shown in the key. [From H. Tschochner and E. Hurt, 2003, *Trends Cell Biol.* **13:**255.]

Unlike pre-rRNA genes, 5S rRNA genes are transcribed by RNA polymerase III in the nucleoplasm outside the nucleolus. With only minor additional processing to remove nucleotides at the 3′ end, 5S rRNA diffuses to the nucleolus, where it assembles with the pre-rRNA precursor and remains associated with the region that is cleaved into the precursor of the large ribosomal subunit.

Most of the ribosomal proteins of the small 40S ribosomal subunit associate with the nascent pre-mRNA during transcription (Figure 8-37). Cleavage of the full-length pre-rRNA in the 90S RNP precursor releases a pre-40S particle that requires only a few more remodeling steps before it is transported to the cytoplasm. Once the pre-40S particle leaves the nucleolus, it traverses the nucleoplasm quickly and is exported through nuclear pore complexes (NPCs), as discussed below. Final maturation of the small ribosomal subunit occurs in the cytoplasm: exonucleolytic processing of the 20S rRNA into mature small subunit 18S rRNA by the cytoplasmic 5′ → 3′ exoribonuclease Xrn1 and the dimethylation of two adjacent adenines near the 3′ end of 18S rRNA by the cytoplasmic enzyme Dim1.

In contrast to the pre-40S particle, the precursor of the large subunit requires considerably more remodeling through many more transient interactions with nonribosomal proteins before it is sufficiently mature for export to the cytoplasm. Consequently, it takes a considerably longer period for the maturing 60S subunit to exit the nucleus than for the 40S subunit (30 minutes compared to 5 minutes for export of the small subunit in cultured human cells). Multiple presumptive RNA helicases and small G proteins are associated with the maturing pre-60S subunits. Some RNA helicases are necessary to dislodge the snoRNPs that base-pair perfectly with pre-rRNA over up to 30 base pairs. Other RNA helicases may function in the disruption of protein-RNA interactions. The requirement for so many GTPases suggests that there are many quality-control checkpoints in the assembly and remodeling of the large subunit RNP, where one step must be completed before a GTPase is activated to allow the next step to proceed. Members of the **AAA ATPase family** are also bound transiently. This class of proteins is often involved in large molecular movements and may be required to fold the complex, large rRNA into the proper conformation. Some steps in 60S subunit maturation occur in the nucleoplasm, during passage from the nucleolus to nuclear pore complexes (see Figure 8-37). Much remains to be learned about the complex, fascinating, and essential remodeling processes that occur during formation of the ribosomal subunits.

The large ribosomal subunit is one of the largest structures to pass through nuclear pore complexes. Maturation of the large subunit in the nucleoplasm leads to the generation of binding sites for a nuclear export adapter called Nmd3. Nmd3 is bound by the nuclear tranporter Exportin1 (also called Crm1). This is another quality-control step since only correctly assembled subunits can bind Nmd3 and be exported. The small subunit of the mRNP exporter (Nxt1) also becomes associated with the nearly mature large ribosomal subunit. These nuclear transporters interact with FG-domains of FG-nucleoporins. This mechanism allows penetration of the molecular meshwork that makes up the central channel of the NPC (see Figure 8-20). Several specific nucleoporins without FG-domains are also required for ribosomal subunit export and may have additional functions specific for this task. The dimensions of ribosomal subunits (≈25–30 nm in diameter) and the central channel of the NPC are comparable, so passage may not require distortion of either the ribosomal subunit or the channel. Final maturation of the large subunit in the cytoplasm includes removal of these export factors. As for the export of most macromolecules from the nucleus, including tRNAs and pre-miRNAs (but not most mRNPs), ribosome subunit export requires the function of a small G protein called Ran, as discussed in Chapter 13.

Self-Splicing Group I Introns Were the First Examples of Catalytic RNA

During the 1970s, the pre-rRNA genes of the protozoan *Tetrahymena thermophila* were discovered to contain an intron. Careful searches failed to uncover even one pre-rRNA gene without the extra sequence, indicating that splicing is required to produce mature rRNA in these organisms. In 1982, in vitro studies showing that the pre-rRNA was spliced at the correct sites in the absence of any protein provided the first indication that RNA can function as a catalyst, like enzymes.

A whole raft of self-splicing sequences subsequently were found in pre-rRNAs from other single-celled organisms, in mitochondrial and chloroplast pre-rRNAs, in several pre-mRNAs from certain *E. coli* bacteriophages, and in some bacterial tRNA primary transcripts. The self-splicing sequences in all these precursors, referred to as *group I introns,* use guanosine as a cofactor and can fold by internal base pairing to juxtapose closely the two exons that must be joined. As discussed earlier, certain mitochondrial and chloroplast pre-mRNAs and tRNAs contain a second type of self-splicing intron, designated group II.

The splicing mechanisms used by group I introns, group II introns, and spliceosomes are generally similar, involving two transesterification reactions, which require no input of energy (Figure 8-38). Structural studies of the group I intron from *Tetrahymena* pre-rRNA combined with mutational and biochemical experiments have revealed that the RNA folds into a precise three-dimensional structure that, like protein enzymes, contains deep grooves for binding substrates and solvent-inaccessible regions that function in catalysis. The group I intron functions like a metalloenzyme to precisely orient the atoms that participate in the two transesterification reactions adjacent to catalytic Mg^{2+} ions. Considerable evidence now indicates that splicing by group II introns and by snRNAs in the spliceosome also involves bound catalytic Mg^{2+} ions. In both the group I and II self-splicing introns and probably in the spliceosome, RNA functions as a **ribozyme,** an RNA sequence with catalytic ability.

Pre-tRNAs Undergo Extensive Modification in the Nucleus

Mature cytosolic tRNAs, which average 75–80 nucleotides in length, are produced from larger precursors (pre-tRNAs) synthesized by RNA polymerase III in the nucleoplasm. Mature tRNAs also contain numerous modified bases that are not present in tRNA primary transcripts. Cleavage and base modification occur during processing of all pre-tRNAs; some pre-tRNAs also are spliced during processing. All of these processing and modification events occur in the nucleus.

A 5′ sequence of variable length that is absent from mature tRNAs is present in all pre-tRNAs (Figure 8-39). This occurs because the 5′ end of mature tRNAs is generated by an endonucleolytic cleavage specified by the tRNA three-dimensional structure rather than the start site of transcription. These extra 5′ nucleotides are removed by ribonuclease P (RNase P), a ribonucleoprotein endonuclease. Studies with *E. coli* RNase P indicate that at high Mg^{2+} concentrations, the RNA component alone can recognize and cleave *E. coli* pre-tRNAs. The RNase P polypeptide increases the rate of cleavage by the RNA, allowing it to proceed at physiological Mg^{2+} concentrations. A comparable RNase P functions in eukaryotes.

About 10 percent of the bases in pre-tRNAs are modified enzymatically during processing. Three classes of base modifications occur (Figure 8-39): (1) U residues at the 3′ end of pre-tRNA are replaced with a CCA sequence. The CCA sequence is found at the 3′ end of all tRNAs and is required for their charging by aminoacyl-tRNA synthetases during protein synthesis. This step in tRNA synthesis likely functions as a quality-control point since only properly folded tRNAs are recognized by the CCA addition enzyme. (2) Addition of methyl and isopentenyl groups to the heterocyclic ring of purine bases and methylation of the 2′-OH group in the ribose of any residue. (3) Conversion of specific uridines to dihydrouridine, pseudouridine, or ribothymidine residues. The functions of these base and ribose modifications are not well understood, but since they are highly conserved, they probably have a positive influence on protein synthesis.

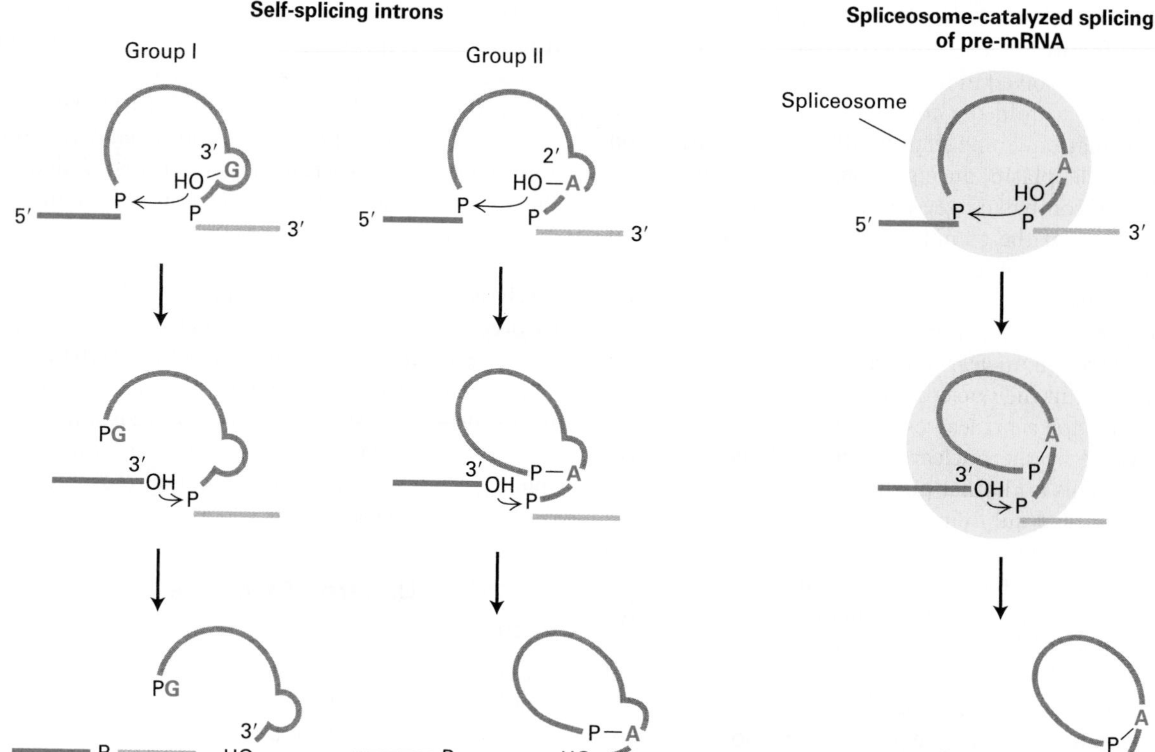

▲ **FIGURE 8-38 Splicing mechanisms in group I and group II self-splicing introns and spliceosome-catalyzed splicing of pre-mRNA.** The intron is shown in gray, the exons to be joined in red. In group I introns, a guanosine cofactor (G) that is not part of the RNA chain associates with the active site. The 3′-hydroxyl group of this guanosine participates in a transesterification reaction with the phosphate at the 5′ end of the intron; this reaction is analogous to that involving the 2′-hydroxyl groups of branch-site As in group II introns and pre-mRNA introns spliced in spliceosomes (see Figure 8-8). The subsequent transesterification that links the 5′ and 3′ exons is similar in all three splicing mechanisms. Note that spliced-out group I introns are linear structures, unlike the branched intron products in the other two cases. [Adapted from P. A. Sharp, 1987, *Science* **235:**769.]

As shown in Figure 8-39, the pre-tRNA expressed from the yeast tyrosine tRNA ($tRNA^{Tyr}$) gene contains a 14-base intron that is not present in mature $tRNA^{Tyr}$. Some other eukaryotic tRNA genes and some archaeal tRNA genes also contain introns. The introns in nuclear pre-tRNAs are shorter than those in pre-mRNAs and lack the consensus splice-site sequences found in pre-mRNAs (see Figure 8-7). Pre-tRNA introns also are clearly distinct from the much longer self-splicing group I and group II introns found in chloroplast and mitochondrial pre-rRNAs. The mechanism of pre-tRNA splicing differs in three fundamental ways from the mechanisms utilized by self-splicing introns and spliceosomes (see Figure 8-38). First, splicing of pre-tRNAs is catalyzed by proteins, not by RNAs. Second, a pre-tRNA intron is excised in one step that entails simultaneous cleavage at both ends of the intron. Finally, hydrolysis of GTP and ATP is required to join the two tRNA halves generated by cleavage on either side of the intron.

After pre-tRNAs are processed in the nucleoplasm, the mature tRNAs are transported to the cytoplasm through nuclear pore complexes by exportin-t, as discussed previously. In the cytoplasm, tRNAs are passed between aminoacyl-tRNA synthetases, elongation factors, and ribosomes during protein synthesis (Chapter 4). Thus tRNAs generally are associated with proteins and spend little time free in the cell, as is also the case for mRNAs and rRNAs.

Nuclear Bodies Are Functionally Specialized Nuclear Domains

High-resolution visualization of plant- and animal-cell nuclei by electron microscopy and subsequent staining with fluorescently labeled antibodies has revealed domains in nuclei in addition to chromosome territories and nucleoli. These specialized nuclear domains, called **nuclear bodies,** are not surrounded by membranes but are nonetheless regions of high concentrations of specific proteins and RNAs that form distinct, roughly spherical structures within the nucleus. The most prominent nuclear bodies are nucleoli, the sites of ribosomal subunit synthesis and assembly discussed earlier. Several other types of nuclear bodies also have been described in structural studies.

Experiments with fluorescently labeled nuclear proteins have shown that the nucleus is a highly dynamic environment, with rapid diffusion of proteins through the nucleoplasm. Proteins associated with nuclear bodies are often also observed at lower concentration in the nucleoplasm outside

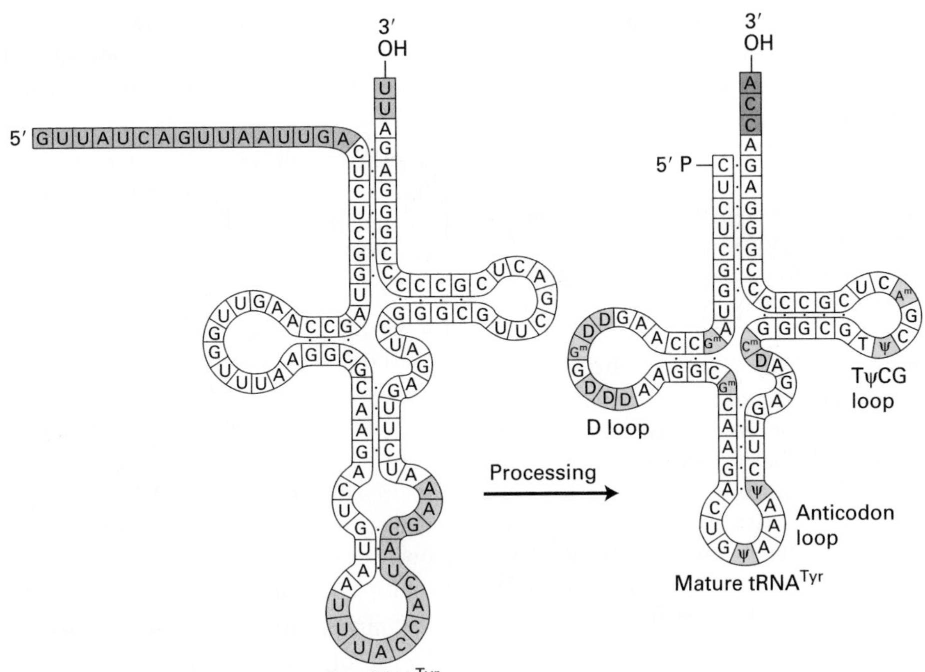

▲ **FIGURE 8-39 Changes that occur during the processing of tyrosine pre-tRNA.** A 14–nucleotide intron (blue) in the anticodon loop is removed by splicing. A 16-nucleotide sequence (green) at the 5′ end is cleaved by RNase P. U residues at the 3′ end are replaced by the CCA sequence (red) found in all mature tRNAs. Numerous bases in the stem loops are converted to characteristic modified bases (yellow). Not all pre-tRNAs contain introns that are spliced out during processing, but they all undergo the other types of changes shown here. D = dihydrouridine; ψ = pseudouridine.

the nuclear bodies, and fluorescence studies indicate that they diffuse in and out of the nuclear bodies. Based on these measurements of molecular mobility in living cells, nuclear bodies can be mathematically modeled as the expected steady state for diffusing proteins that interact with sufficient affinity to form self-organized regions of high concentrations of specific proteins but with low enough affinity for each other to be able to diffuse in and out of the structure. Electron micrographs show these structures appear to be a heterogeneous, spongelike network of interacting components. We discuss a few of these nuclear bodies here as examples of these nuclear domains.

Cajal Bodies Cajal bodies are ≈0.2–1 μm spherical structures that have been observed in large nuclei for more than a century (Figure 8-40). Current research indicates that like nucleoli, Cajal bodies are centers of RNP-complex assembly for spliceosomal snRNPs and other RNPs. Like rRNAs, snRNAs undergo specific modifications, such as the conversion of specific uridine residues to pseudouridine and addition of methyl groups to the 2′-hydroxyl groups of specific riboses. These post-transcriptional modifications are important for the proper assembly and function of snRNPs in pre-mRNA splicing. These modifications occur in Cajal bodies, where they are directed by a class of snoRNA-like guide RNA molecules called *scaRNAs* (small Cajal body–*a*ssociated RNAs). There is also evidence that the Cajal body is the site of reassembly of the U4/U6/U5 tri-snRNP complexes required for pre-mRNA splicing from the free U4, U5, and U6 snRNPs released during the removal of each intron (see Figure 8-11). Since Cajal bodies also contain a high concentration of the *U7snRNP* involved in the specialized 3′-end processing of the major histone mRNAs, it is likely that this process also occurs in Cajal bodies, as may the assembly of the telomerase RNP.

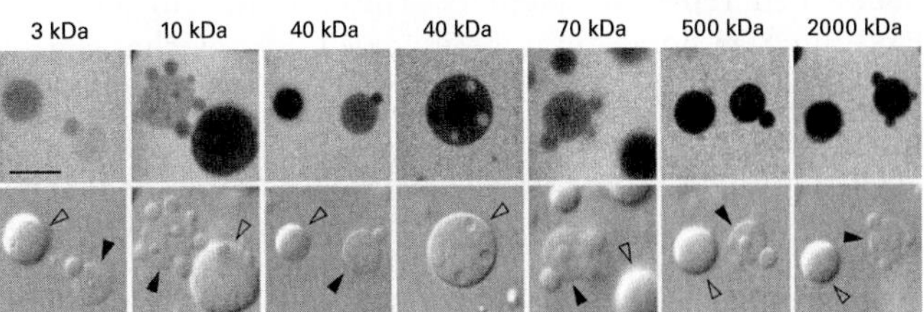

▲ **FIGURE 8-40 Nuclear bodies are differentially permeable to molecules in the bulk nucleoplasm.** Each pair of panels shows a single area through a living *Xenopus* oocyte nucleus, which was previously injected with fluorescent dextran of the indicated molecular mass (3–2000 kDa). Each section of the upper panel is a confocal image in which the intensity of fluorescence is a measure of dextran concentration (i.e., darker areas show regions where dextran has been excluded). Each section of the lower panel is a differential interference contrast image of the same field. Open arrowheads indicate nucleoli, closed arrowheads Cajal bodies (CBs) with attached nuclear speckles which are much larger in *Xenopus* oocytes than in most somatic cells. Dextrans of low molecular mass (e.g., 3 kDa) almost completely penetrated CBs but were excluded more from nuclear speckles and nucleoli. Exclusion of dextran increased with molecular mass. Bar, 10 μm. [From K. E. Handwerger, 2005, *Mol. Biol. Cell* **16:**202.]

Nuclear Speckles Nuclear speckles were observed with fluorescently labeled antibodies to snRNP proteins and other proteins involved in pre-mRNA splicing as ≈25–50 irregular, amorphous structures 0.5–2 μm in diameter that are distributed through the nucleoplasm of vertebrate cells. Since speckles are not located at sites of co-transcriptional pre-mRNA splicing, which are associated closely with chromatin, they are thought to be regions of storage of snRNPs and proteins involved in pre-mRNA splicing that are released into the nucleoplasm when required.

Promyelocytic Leukemia (PML) Nuclear Bodies The *PML* gene was originally discovered when chromosomal translocations within the gene were observed in the leukemic cells of patients with the rare disease promyelolcytic leukemia (PML). When antibodies specific for the PML protein were used in immunofluorescence microscopy studies, the protein was found to localize to ≈10–30 roughly spherical regions 0.3–1 μm in diameter in the nuclei of mammalian cells. Multiple functions have been proposed for PML nuclear bodies, but a consensus is emerging that they function as sites for the assembly and modification of protein complexes involved in DNA repair and the induction of apoptosis. For example, the important p53 tumor suppressor protein appears to be post-translationally modified by phosphorylation and acetylation in PML nuclear bodies in response to DNA damage, increasing its ability to activate the expression of DNA-damage response genes.

Recent results indicate that PML nuclear bodies are also sites of protein post-translational modification through the addition of a small, ubiquitin-like protein called *SUMO1* (*s*mall *u*biquitin-like *mo*iety-1), which can control the activity and subcellular localization of the modified protein. Many transcriptional repressors are sumoylated, and mutation of their site of sumoylation reduces their repression activity. These results suggest that PML nuclear bodies may be involved in a mechanism of transcriptional repression that remains to be studied and understood.

In addition to PML nuclear bodies, the first nuclear bodies to be observed, the nucleoli may have specialized regions of substructure that are dedicated to functions other than ribosome biogenesis. There is evidence that immature SRP ribonucleoprotein complexes involved in protein secretion and ER membrane insertion (Chapter 13) are assembled in nucleoli and then exported to the cytoplasm, where their final maturation takes place.

KEY CONCEPTS OF SECTION 8.5

Processing of rRNA and tRNA

- A large precursor pre-rRNA (45S in humans) synthesized by RNA polymerase I undergoes cleavage, exonucleolytic digestion, and base modifications to yield mature 28S, 18S, and 5.8S rRNAs, which associate with ribosomal proteins into ribosomal subunits.
- Synthesis and processing of pre-rRNA occur in the nucleolus. The 5S rRNA component of the large ribosomal subunit is synthesized in the nucleoplasm by RNA polymerase III.
- ≈150 snoRNAs associated with proteins in snoRNPs base-pair with specific sites in pre-RNA, directing ribose methylation, modification of uridine to pseudouridine, and cleavage at specific sites during rRNA processing in the nucleolus.
- Group I and group II self-splicing introns and probably snRNAs in spliceosomes all function as ribozymes, or catalytically active RNA sequences, that carry out splicing by analogous transesterification reactions requiring bound Mg^{+2} ions (see Figure 8-38).
- Pre-tRNAs synthesized by RNA polymerase III in the nucleoplasm are processed by removal of the 5′-end sequence, addition of CCA to the 3′ end, and modification of multiple internal bases (see Figure 8-39).
- Some pre-tRNAs contain a short intron that is removed by a protein-catalyzed mechanism distinct from the splicing of pre-mRNA and self-splicing introns.
- All species of RNA molecules are associated with proteins in various types of ribonucleoprotein particles both in the nucleus and after export to the cytoplasm.
- Nuclear bodies are functionally specialized regions in the nucleus where interacting proteins form self-organized structures. Many of these, like the nucleolus, are regions of assembly of RNP complexes.

Perspectives for the Future

In this and the previous chapter, we have seen that in eukaryotic cells, mRNAs are synthesized and processed in the nucleus, transported through nuclear pore complexes to the cytoplasm, and then, in some cases, transported to specific areas of the cytoplasm before being translated by ribosomes. Each of these fundamental processes is carried out by complex macromolecular machines composed of scores of proteins and in many cases RNAs as well. The complexity of these macromolecular machines ensures accuracy in finding promoters and splice sites in the long length of DNA and RNA sequences and provides various avenues for regulating synthesis of a polypeptide chain. Much remains to be learned about the structure, operation, and regulation of such complex machines as spliceosomes and the cleavage/polyadenylation apparatus.

Recent examples of the regulation of pre-mRNA splicing raise the question of how extracellular signals might control such events, especially in the nervous system of vertebrates. A case in point is the remarkable situation in the chick inner ear, where multiple isoforms of the Ca^{2+}-activated K^{+} channel called *Slo* are produced by alternative RNA splicing. Cell-cell interactions appear to inform cells of their position in the cochlea, leading to alternative splicing of Slo pre-mRNA. The challenging task facing researchers is to discover how such cell-cell interactions regulate the activity of RNA-processing factors.

The mechanism of mRNP transport through nuclear pore complexes poses many intriguing questions. Future research will likely reveal additional activities of hnRNP and nuclear mRNP proteins and clarify their mechanisms of action. For instance, there is a small gene family encoding proteins homologous to the large subunit of the mRNA exporter. What are the functions of these related proteins? Do they participate in the transport of overlapping sets of mRNPs? Some hnRNP proteins contain nuclear-retention signals that prevent nuclear export when fused to hnRNP proteins with nuclear-export signals (NESs). How are these hnRNP proteins selectively removed from processed mRNAs in the nucleus, allowing the mRNAs to be transported to the cytoplasm?

The localization of certain mRNAs to specific subcellular locations is fundamental to the development of multicellular organisms. As discussed in Chapter 22, during development an individual cell frequently divides into daughter cells that function differently from each other. In the language of developmental biology, the two daughter cells are said to have different developmental fates. In many cases, this difference in developmental fate results from the localization of an mRNA to one region of the cell before mitosis so that after cell division, it is present in one daughter cell and not the other. Much exciting work remains to be done to fully understand the molecular mechanisms controlling mRNA localization that are critical for the normal development of multicellular organisms.

Some of the most exciting and unanticipated discoveries in molecular cell biology in recent years concern the existence and function of miRNAs and the process of RNA interference. RNA interference (RNAi) provides molecular cell biologists with a powerful method for studying gene function. The discovery of ≈1000 miRNAs in humans and other organisms suggests that multiple significant examples of translational control by this mechanism await to be characterized. Recent studies in *S. pombe* and plants link similar short nuclear RNAs to the control of DNA methylation and the formation of heterochromatin. Will similar processes control gene expression through the assembly of heterochromatin in humans and other animals? What other regulatory processes might be directed by other kinds of small RNAs? Since control by these mechanisms depends on base pairing between miRNAs and target mRNAs or genes, genomic and bioinformatic methods will probably suggest genes that may be controlled by these mechanisms. What other processes in addition to translation control, mRNA degradation, and heterochromatin assembly might be controlled by miRNAs?

These are just a few of the fascinating questions concerning RNA processing, post-transcriptional control, and nuclear transport that will challenge molecular cell biologists in the coming decades. The astounding discoveries of entirely unanticipated mechanisms of gene control by miRNAs remind us that many more surprises are likely in the future.

Key Terms

5′ cap *325*
alternative splicing *337*
cleavage/polyadenylation complex *335*
cross-exon recognition complex *333*
Dicer *348*
Drosha *348*
exosome *336*
exportin *347*
FG-nucleoporins *342*
group I introns *363*
group II introns *334*
importin *344*
iron-response element–binding protein (IRE-BP) *356*
micro RNAs (miRNAs) *347*
mRNP exporter *343*
mRNA surveillance *357*
nuclear pore complex (NPC) *342*
poly(A) tail *335*
pre-mRNA *326*
pre-rRNA *359*
ribozyme *363*
RNA editing *341*
RNA-induced silencing complex (RISC) *348*
RNA interference (RNAi) *349*
RNA splicing *329*
short interfering RNAs (siRNA) *347*
siRNA knockdown *351*
small nuclear RNAs (snRNAs) *330*
small nucleolar RNAs (snoRNAs) *361*
spliceosome *332*
SR proteins *333*

Review the Concepts

1. Describe three types of post-transcriptional regulation of protein-coding genes.
2. You are investigating the transcriptional regulation of genes from a recently discovered eukaryotic organism. You find that this organism has three RNA polymerase–like enzymes and that genes encoding proteins are transcribed using an RNA polymerase II–like enzyme. When examining the full-length mRNAs produced, you find that there is no 5′ structure (i.e., cap). Are you surprised? Why/why not?
3. What is the evidence that transcription termination by RNA polymerase II is coupled to polyadenylation?
4. It has been suggested that manipulation of HIV antitermination might provide for effective therapies in combating AIDS. What effect would a mutation in the TAR sequence that abolishes Tat binding have on HIV transcription after HIV infection and why? A mutation in Cdk9 that abolishes activity?
5. Describe how the discovery that introns are removed during splicing was made. How are the locations of exon-intron junctions predicted?
6. What is the difference between hnRNAs, snRNAs, miRNAs, siRNAs, and snoRNAs?
7. What are the mechanistic similarities between group II intron self-splicing and spliceosomal splicing? What is the evidence that there may be an evolutionary relationship between the two?

8. Where do researchers believe most transcription and RNA-processing events occur? What is the evidence to support this?

9. You obtain the sequence of a gene containing 10 exons, 9 introns, and a 3′ UTR containing a polyadenylation consensus sequence. The fifth intron also contains a polyadenylation site. To test whether both polyadenylation sites are used, you isolate mRNA and find a longer transcript from muscle tissue and a shorter mRNA transcript from all other tissues. Speculate about the mechanism involved in the production of these two transcripts.

10. RNA editing is a common process occurring in the mitochondria of trypanosomes and plants, in chloroplasts, and in rare cases in higher eukaryotes. What is RNA editing and what benefit does it serve in the documented example with apoB in humans?

11. A protein complex in the nucleus is responsible for transporting mRNA molecules into the cytoplasm. Describe the proteins that form this exporter and what two protein groups are likely behind the mechanism involved in the directional movement of the mRNP and complex into the cytosol.

12. RNA knockdown has become a powerful tool in the arsenal of methods to deregulate gene expression. Briefly describe how gene expression can be knocked down. What effect would introducing siRNAs to TSC1 have on human cells?

13. Speculate about why plants deficient in Dicer activity show increased sensitivity to infection by RNA viruses.

14. What is the evidence that some mRNAs are directed to accumulate in specific subcellular locations?

Analyze the Data

Most humans are infected with herpes simplex virus-1 (HSV-1), the causative agent of cold sores. The HSV-1 genome encodes about 100 genes, most of which are expressed in infected host cells at the site of oral sores. The infectious process involves replication of viral DNA, transcription and translation of viral genes, assembly of new viral particles, and death of the host cell as the viral progeny are released. Unlike most other types of viruses, herpesvirus also has a latent phase, in which the virus remains hidden in neurons. These latently infected neurons are the source of active infections, causing cold sores when latency is overcome.

Interestingly, only a single viral transcript is expressed during latency. This transcript, *LAT* (*l*atency-*a*ssociated *t*ranscript), does not encode a protein, and neurons infected with mutant HSV-1 lacking the *LAT* gene undergo cell death by apoptosis at a rate twice that of cells infected with wild-type HSV-1. To determine if *LAT* functions to block apoptosis by encoding a miRNA, the following studies were done (see Gupta et al., 2006, *Nature* **442**:82–85).

a. A cell line was transfected with an expression vector that expresses a Pst-Msu (part b, diagram below) fragment of the *LAT* gene. The percentage of these transfected cells that then undergo drug-induced cell death was compared to that of control cells. The experiment was repeated in cells in which Dicer expression was knocked down using Dicer siRNA. The data obtained are shown in the graph below. Why did the scientists who conducted this study examine the effects of silencing Dicer? Under what conditions does the *LAT* gene protect the cells from apoptosis?

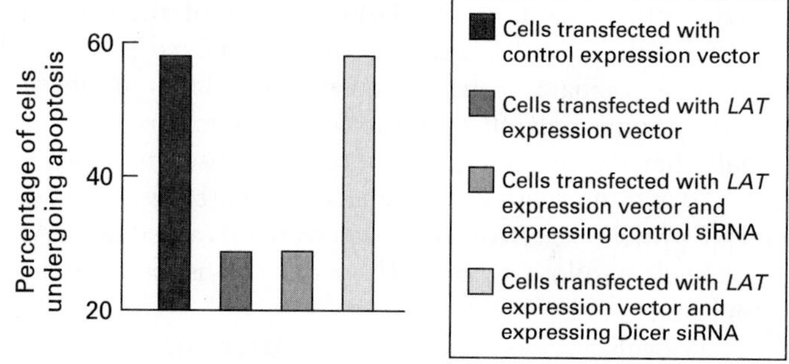

b. Cells were transfected with an expression vector expressing the Pst-Msu fragment of the *LAT* gene from which the region between the two Sty restriction sites was deleted (ΔSty; diagram below). When these cells were induced to undergo apoptosis, they died at the same rate as did non-transfected cells. In additional studies, cells were transfected with an expression vector expressing the Sty-Sty region of the *LAT* gene. These cells exhibited the same resistance to apoptosis as did cells transfected with the Pst-Msu fragment. What can be deduced from these findings about the region of the *LAT* gene required to protect cells from apoptosis?

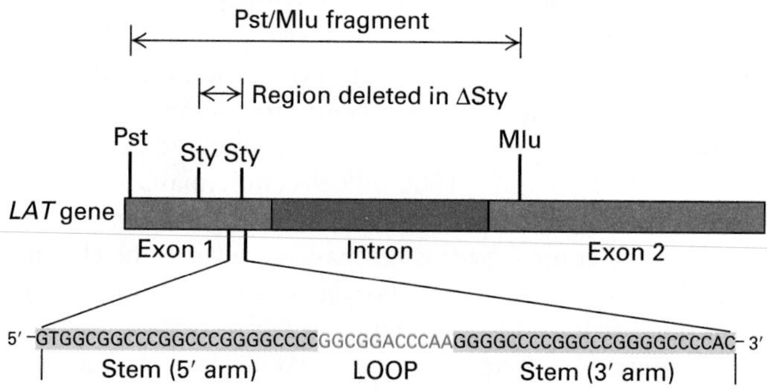

c. RNA encoded within the Sty-Sty region is predicted to form a stem loop (part b, diagram above). Northern blot analysis was performed on total-cell RNA isolated from control cells (mock), cells infected with wild-type HSV-1, cells infected with an HSV-1 deletion mutant from which the sequence between the two Sty sites in the *LAT* gene was deleted (ΔSty), and cells infected with a rescued ΔSty virus into which the deleted region was re-inserted into the viral genome (StyR). The probe used for the Northern blot was the labeled 3′ stem region of the *LAT* RNA in the Sty-Sty region, as diagrammed in part (b). The RNAs recognized by this probe were either ≈55 nucleotides or 20 nucleotides, as shown in the Northern blot below. Why were two different-size RNAs detected? When a second probe was used that was the labeled 5′ stem region of the RNA sequence shown in part (b), only the ≈55-nucleotide RNA was detected. What can you deduce about the processing of RNA expressed from the *LAT* gene? What enzyme likely produced the ≈55 nucleotide RNA? In what part of the cell? What enzyme likely produced the 20-nucleotide RNA? In what part of the cell?

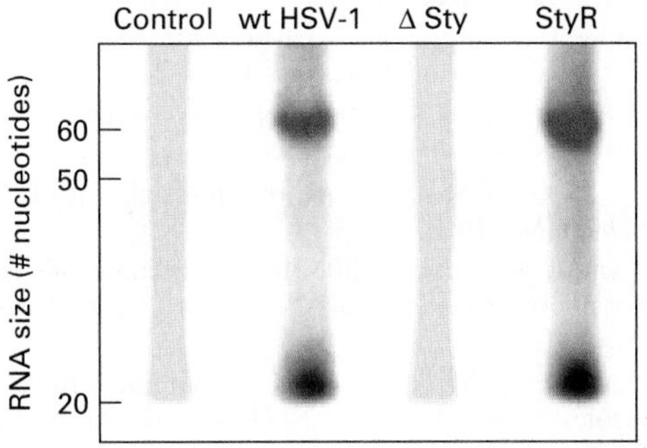

d. TGF-β mRNA encodes a protein, transforming growth factor β, that inhibits cell growth and induces apoptosis. The 3′ untranslated region (3′-UTR) of TGF-β mRNA can form an imperfect duplex with RNA encoded by the 5′ stem region of the *LAT* Sty-Sty domain (miR-LAT), as shown below. In what way might the expression levels of TGF-β differ in cells infected with wild-type HSV-1 compared to uninfected cells? What can you infer about latent HSV-1 infections from these studies?

```
                                         C
                            miR-LAT     / \
   C-C-G-G-G-G-C-C-C-G-G-C-C-C-G-G   G-G-U 5'
   | | | | | | | | | | | | | | | |   | | |
5' C-G-C-C-C-C-G-G G-C-C-C-G-G-C-C-C-C-A
                  | |
                  CAG        TGF-β-3′UTR
```

References

Processing of Eukaryotic Pre-mRNA

Bentley, D. L. 2005. Rules of engagement: co-transcriptional recruitment of pre-mRNA processing factors. *Curr. Opin. Cell Biol.* **17**:251–256.

Buratowski, S. 2005. Connections between mRNA 3′ end processing and transcription termination. *Curr. Opin. Cell Biol.* **17**:257–261.

Gu, M., and C. D. Lima. 2005. Processing the message: structural insights into capping and decapping mRNA. *Curr. Opin. Struct. Biol.* **15**:99–106.

Houseley, J., J. LaCava, and D. Tollervey. 2006. RNA-quality control by the exosome. *Nat. Rev. Mol. Cell Biol.* **7**:29–539.

Hsieh, J., A. J. Andrews, and C. A. Fierke. 2004. Roles of protein subunits in RNA-protein complexes: lessons from ribonuclease P. *Biopolymers* **73**:79–89.

Lambowitz, A. M., and S. Zimmerly. 2004. Mobile group II introns. *Annu. Rev. Genet.* **38**:1–35.

Lehmann, K., and U. Schmidt. 2003. Group II introns: structure and catalytic versatility of large natural ribozymes. *Crit. Rev. Biochem. Mol. Biol.* **38**:249–303.

Moore, M. J. 2005. From birth to death: the complex lives of eukaryotic mRNAs. *Science* **309**:1514–1518.

Sharp, P. A. 2005. The discovery of split genes and RNA splicing. *Trends Biochem. Sci.* **30**:279–281.

Shatkin, A. J., and J. L. Manley. 2000. The ends of the affair: capping and polyadenylation. *Nature Struct. Biol.* **7**:838–842.

Soller, M. 2006. Pre-messenger RNA processing and its regulation: a genomic perspective. *Cell Mol. Life Sci.* **63**:796–819.

Valadkhan, S. 2005. snRNAs as the catalysts of pre-mRNA splicing. *Curr. Opin. Chem. Biol.* **9**:603–608.

Villa, T., J. A. Pleiss, and C. Guthrie. 2002. Spliceosomal snRNAs: $Mg(2^+)$-dependent chemistry at the catalytic core? *Cell* **109**:149–152.

Regulation of Pre-mRNA Processing

Black, D. L. 2003. Mechanisms of alternative pre-mRNA splicing. *Ann. Rev. Biochem.* **72**:291–336.

Blencowe, B. J. 2006. Alternative splicing: new insights from global analyses. *Cell* **126**:37–47.

Buratti, E., M. Baralle, and F. E. Baralle. 2006. Defective splicing, disease and therapy: searching for master checkpoints in exon definition. *Nucleic Acids Res.* **34**:3494–3510.

Lee, C., and Q. Wang. 2005. Bioinformatics analysis of alternative splicing. *Brief Bioinform.* **6**:23–33.

Licatalosi, D. D., and R. B. Darnell. 2006. Splicing regulation in neurologic disease. *Neuron* **52**:93–101.

Maniatis, T., and B. Tasic. 2002. Alternative pre-mRNA splicing and proteome expansion in metazoans. *Nature* **418**:236–243.

Sanford, J. R., J. Ellis, and J. F. Caceres. 2005. Multiple roles of arginine/serine-rich splicing factors in RNA processing. *Biochem. Soc. Trans.* **33**:443–446.

Xing, Y., and C. Lee. 2006. Alternative splicing and RNA selection pressure—evolutionary consequences for eukaryotic genomes. *Nat. Rev. Genet.* **7**:499–509.

Transport of mRNA Across the Nuclear Envelope

Cole, C. N., and J. J. Scarcelli. 2006. Transport of messenger RNA from the nucleus to the cytoplasm. *Curr. Opin. Cell Biol.* **18**:299–306.

Darzacq, X., R. H. Singer, and Y. Shav-Tal. 2005. Dynamics of transcription and mRNA export. *Curr. Opin. Cell Biol.* **17**:332–339.

Dimaano, C., and K. S. Ullman. 2004. Nucleocytoplasmic transport: integrating mRNA production and turnover with export through the nuclear pore. *Mol. Cell Biol.* **24**:3069–3076.

Fried, H., and U. Kutay. 2003. Nucleocytoplasmic transport: taking an inventory. *Cell Mol. Life Sci.* **60**:1659–1688.

Huang, Y., and J. A. Steitz. 2005. SRprises along a messenger's journey. *Mol. Cell.* **17**:613–615.

Izaurralde, E. 2004. Directing mRNA export. *Nat. Struct. Mol. Biol.* **11**:210–212.

Kuersten, S., and E. B. Goodwin. 2005. Linking nuclear mRNP assembly and cytoplasmic destiny. *Biol. Cell.* **97**:469–478.

Lim, R. Y., and B. Fahrenkrog. 2006. The nuclear pore complex up close. *Curr. Opin. Cell Biol.* **18**:342–347.

Maco, B., B. Fahrenkrog, N. P. Huang, and U. Aebi. 2006. Nuclear pore complex structure and plasticity revealed by electron and atomic force microscopy. *Methods Mol. Biol.* **322**:273–288.

Reed, R., and H. Cheng. 2005. TREX, SR proteins and export of mRNA. *Curr. Opin. Cell Biol.* **17**:269–273.

Ribbeck, K., and D. Gorlich. 2001. Kinetic analysis of translocation through nuclear pore complexes. *EMBO J.* **20**:1320–1330.

Rodriguez, M. S., C. Dargemont, and F. Stutz. 2004. Nuclear export of RNA. *Biol. Cell* **96**:639–655.

Saguez, C., J. R. Olesen, and T. H. Jensen. 2005. Formation of export-competent mRNP: escaping nuclear destruction. *Curr. Opin. Cell Biol.* **17**:287–293.

Sommer, P., and U. Nehrbass. 2005. Quality control of messenger ribonucleoprotein particles in the nucleus and at the pore. *Curr. Opin. Cell Biol.* **17**:294–301.

Tran, E. J., and S. R. Wente. 2006. Dynamic nuclear pore complexes: life on the edge. *Cell* **125**:1041–1053.

Cytoplasmic Mechanisms of Post-transcriptional Control

Almeida, R., and R. C. Allshire. 2005. RNA silencing and genome regulation. *Trends Cell Biol.* **15**:251–258.

Ambros, V. 2004. The functions of animal microRNAs. *Nature* **431**:350–355.

Anant, S., V. Blanc, and N. O. Davidson. 2003. Molecular regulation, evolutionary, and functional adaptations associated with C to U editing of mammalian apolipoproteinB mRNA. *Prog. Nucleic Acid Res. Mol. Biol.* **75**:1–41.

Baulcombe, D. 2005. RNA silencing. *Trends Biochem. Sci.* **30**:290–293.

Chan, S. W., and I. R. Henderson, and S. E. Jacobsen. 2005. Gardening the genome: DNA methylation in *Arabidopsis thaliana*. *Nat. Rev. Genet.* **6**:351–360.

Chen, C. Z., and H. F. Lodish. 2005. MicroRNAs as regulators of mammalian hematopoiesis. *Semin. Immunol.* **17**:155–165.

Engels, B. M., and G. Hutvagner. 2006. Principles and effects of microRNA-mediated post-transcriptional gene regulation. *Oncogene* **25**:6163–6169.

Filipowicz, W. 2005. RNAi: the nuts and bolts of the RISC machine. *Cell* **122**:17–20.

Grewal, S. I., and J. C. Rice. 2004. Regulation of heterochromatin by histone methylation and small RNAs. *Curr. Opin. Cell Biol.* **16**:230–238.

Hirokawa, N. 2006. mRNA transport in dendrites: RNA granules, motors, and tracks. *J. Neurosci.* **26**:7139–7142.

Huang, Y. S., and J. D. Richter. 2004. Regulation of local mRNA translation. *Curr. Opin. Cell Biol.* **16**:308–313.

Kidner, C. A., and R. A. Martienssen. 2005. The developmental role of microRNA in plants. *Curr. Opin. Plant Biol.* **8**:38–44.

Lejeune, F., and L. E. Maquat. 2005. Mechanistic links between nonsense-mediated mRNA decay and pre-mRNA splicing in mammalian cells. *Curr. Opin. Cell Biol.* **17**:309–315.

Maas, S., A. Rich, and K. Nishikura. 2003. A-to-I RNA editing: recent news and residual mysteries. *J. Biol. Chem.* **278**:1391–1394.

Mangus, D. A., M. C. Evans, and A. Jacobson. 2003. Poly(A)-binding proteins: multifunctional scaffolds for the post-transcriptional control of gene expression. *Genome Biol.* **4**:223.

Matzke, M., et al. 2004. Genetic analysis of RNA-mediated transcriptional gene silencing. *Biochim. Biophys. Acta* **1677**:129–141.

Mello, C. C., and D. Conte Jr. 2004. Revealing the world of RNA interference. *Nature* **431**:338–342. C. C. Mello's Nobel Prize lecture can be viewed at http://nobelprize.org/nobel_prizes/medicine/laureates/2006/announcement.html

Pak, J., and A. Fire. 2006. Distinct populations of primary and secondary effectors during RNAi in *C. elegans*. *Science* Nov. 23, 2006 (E-pub.). A. Z. Fire's Nobel Prize lecture can be viewed at http://nobelprize.org/nobel_prizes/medicine/laureates/2006/announcement.html

Parker, R., and H. Song. 2004. The enzymes and control of eukaryotic mRNA turnover. *Nat. Struct. Mol. Biol.* **11**:121–127.

Richter, J. D., and N. Sonenberg. 2005. Regulation of cap-dependent translation by eIF4E inhibitory proteins. *Nature* **433**:477–480.

Ruvkun, G. B. 2004. The tiny RNA world. *Harvey Lect.* **99**:1–21.

Schmauss, C. 2003. Serotonin 2C receptors: suicide, serotonin, and runaway RNA editing. *Neuroscientist* **9**:237–242.

Shikanai, T. 2006. RNA editing in plant organelles: machinery, physiological function and evolution. *Cell Mol. Life Sci.* **63**:698–708.

Simpson, L., S. Sbicego, and R. Aphasizhev. 2003. Uridine insertion/deletion RNA editing in trypanosome mitochondria: a complex business. *RNA* **9**:265–276.

Sontheimer, E. J., and R. W. Carthew. 2005. Silence from within: endogenous siRNAs and miRNAs. *Cell* **122**:9–12.

St. Johnston, D. 2005. Moving messages: the intracellular localization of mRNAs. *Nat. Rev. Mol. Cell Biol.* **6**:363–375.

Tang, G. 2005. siRNA and miRNA: an insight into RISCs. *Trends Biochem. Sci.* **30**:106–114.

Ule, J., and R. B. Darnell. 2006. RNA binding proteins and the regulation of neuronal synaptic plasticity. *Curr. Opin. Neurobiol.* **16**:102–110.

Valencia-Sanchez, M. A., J. Liu, G. J. Hannon, and R. Parker. 2006. Control of translation and mRNA degradation by miRNAs and siRNAs. *Genes Dev.* **20**:515–524.

Verdel, A., and D. Moazed. 2005. RNAi-directed assembly of heterochromatin in fission yeast. *FEBS Lett.* **579**:5872–5878.

Wullschleger, S., R. Loewith, and M. N. Hall. 2006. TOR signaling in growth and metabolism. *Cell* **124**:471–484.

Zamore, P. D., and B. Haley. 2005. Ribo-gnome: the big world of small RNAs. *Science* **309**:1519–1524.

Processing of rRNA and tRNA

Anderson, J. T. 2005. RNA turnover: unexpected consequences of being tailed. *Curr. Biol.* **15**:R635–R638.

Decatur, W. A., and M. J. Fournier. 2003. RNA-guided nucleotide modification of ribosomal and other RNAs. *J. Biol. Chem.* **278**:695–698.

Dez, C., and D. Tollervey. 2004. Ribosome synthesis meets the cell cycle. *Curr. Opin. Microbiol.* **7**:631–637.

Evans, D., S. M. Marquez, and N. R. Pace. 2006. RNase P: interface of the RNA and protein worlds. *Trends Biochem. Sci.* **31**:333–341.

Fatica, A., and D. Tollervey. 2002. Making ribosomes. *Curr. Opin. Cell Biol.* **14**:313–318.

Handwerger, K. E., and J. G. Gall. 2006. Subnuclear organelles: new insights into form and function. *Trends Cell Biol.* **16**:19–26.

Hopper, A. K., and E. M. Phizicky. 2003. tRNA transfers to the limelight. *Genes Dev.* **17**:162–180.

Mayer, C., and I. Grummt. 2006. Ribosome biogenesis and cell growth: mTOR coordinates transcription by all three classes of nuclear RNA polymerases. *Oncogene* 2006 **25**:6384–6391.

Stahley, M. R., and S. A. Strobel. 2006. RNA splicing: group I intron crystal structures reveal the basis of splice site selection and metal ion catalysis. *Curr. Opin. Struct. Biol.* **16**:319–326.

Tschochner, H., and E. Hurt. 2003. Pre-ribosomes on the road from the nucleolus to the cytoplasm. *Trends Cell Biol.* **13**:255–263.

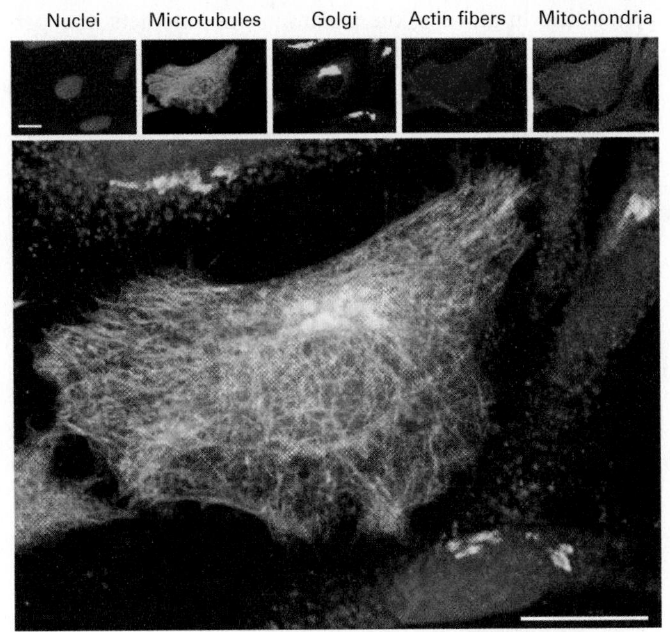

Fluorescence microscopy shows the location of DNA and multiple proteins within the same cell. Here, fluorescent tagging and staining techniques using different fluorescent molecules reveal the cytoskeletal proteins α-tubulin (green) and actin (red), DNA (blue), the Golgi complex (yellow), and mitochondria (purple). The images along the top are false-colored images of each structure stained individually. The larger image merges these separate images to depict the full cell. Scale bars, 20 μm. [From B. N. G. Giepmans, et al., 2006, *Science* **312:**217.]

CHAPTER

9

VISUALIZING, FRACTIONATING, AND CULTURING CELLS

Almost 200 years ago Matthias Schleiden and Theodore Schwann used a primitive light microscope to show that individual cells constitute the fundamental unit of life, and light microscopy has been an increasingly important research tool for biologists ever since. Sophisticated light microscopes developed in the last two decades now enable cell biologists to reveal the myriad movements of cells ranging from the translocation of chromosomes and vesicles to cell crawling and swimming.

Electron microscopy provides a much higher resolution of cell ultrastructure than light microscopy, but the technology requires that the cell be fixed and sectioned and thus all cell movements are frozen in time. Electron microscopy revealed that all eukaryotic cells—be they of fungal, plant, or animal origin—are divided into similar multiple membrane-limited compartments termed **organelles.** In the first section of this chapter we describe the basic structures and functions of the major organelles in animal and plant cells.

In the second and third sections of this chapter we discuss many modern techniques in light and electron microscopy that are suitable for detecting and imaging particular structural features of the cell. Developments in both light and electron microscopy, together with those for generating monoclonal antibodies, have enabled modern cell biologists to detect specific proteins in fixed cells, thus providing a static image of their location within cells, as illustrated in the chapter opening figure. Such studies led to the important concept that the membranes and interior spaces of each type of organelle contain a unique group of proteins that are essential for the organelle to carry out its unique functions. In addition, we see how certain *chimeric proteins*—consisting of a protein of interest covalently linked to a naturally fluorescent protein—enable biologists to image movements of individual proteins in live cells. Introduction of digital systems also has resulted in the improved quality of microscopic images, as well as digital storage and retrieval. Digital algorithms also permit three-dimensional reconstructions of cell components from two-dimensional images, and allow visualization and quantification of specific proteins and other molecules in cells.

OUTLINE

Parallel developments in subcellular fractionation have enabled cell biologists to isolate individual organelles to a high degree of purity. These techniques, detailed in the fourth section of this chapter, continue to provide important information about the protein composition and biochemical function of organelles. For example, the use of proteomic approaches including mass spectrometry to determine the identity of all of the major proteins present in preparations of purified mitochondria has revealed many novel functions for this organelle.

Many technical constraints hamper studies on cells in intact animals and plants. One alternative is the use of intact organs that are removed from animals and treated to maintain their physiologic integrity and function. However, the organization of organs, even isolated ones, is sufficiently complex to pose numerous problems for research. Thus molecular cell biologists often conduct experimental studies on cells isolated from an organism. In the fifth section of this chapter we learn how to isolate certain types of cells to high purity from a complex mixture of cells.

In many cases, isolated cells can be maintained in the laboratory under conditions that permit their survival and growth, a procedure known as *culturing*. Cultured cells have several advantages over intact organisms for cell biology research. Cells of a single specific type can be grown in culture, experimental conditions can be better controlled, and in many cases a single cell can be readily grown into a colony of many identical cells. The resulting strain of cells, which is genetically homogeneous, is called a **clone.** In culture, certain lines of undifferentiated mammalian cells can be induced to differentiate over a period of days to a specific type of cell such as muscle or nerve when switched to a different culture medium. As we learn in the last section of this chapter, such lines of cells provide powerful tools for understanding how specific types of differentiated cells are formed in the body.

9.1 Organelles of the Eukaryotic Cell

The major organelles in animal and plant cells are depicted in Figure 9-1. Unique proteins in the interior and membranes of each type of organelle largely determine its specific functional characteristics, which are examined in more detail in later chapters. Those organelles bounded by a single membrane are covered first, followed by the three types that have a double membrane—the nucleus, mitochondrion, and chloroplast. The internal organization of organelles, and the structure of the plasma membrane, is organized by the fibrous cytoskeleton; Chapters 17 and 18 discuss these fibers in detail.

The Plasma Membrane Has Many Common Functions in All Cells

Although the lipid composition of a membrane largely determines its physical characteristics, its complement of proteins is primarily responsible for a membrane's functional properties. In all cells, the plasma membrane acts as a permeability barrier that prevents the entry of unwanted materials from the extracellular milieu and the exit of needed metabolites. Specific **membrane transport proteins** in the plasma membrane permit the passage of nutrients into the cell and metabolic wastes out of it; others function to maintain the proper ionic composition and pH (≈7.2) of the **cytosol,** the aqueous portion of the cytoplasm excluding organelles, membranes, and insoluble cytoskeletal components. The structure and function of proteins that make the plasma membrane selectively permeable to different molecules are discussed in Chapters 10 and 11.

Unlike animal cells, most bacterial, and all fungal and plant cells are surrounded by a rigid cell wall and lack the extracellular matrix found in animal tissues. The plasma membrane is intimately engaged in the assembly of cell walls, which in plants are built primarily of **cellulose.** The cell wall prevents the swelling or shrinking of a cell that would otherwise occur when it is placed in a medium that is less concentrated (**hypotonic**) or more concentrated (**hypertonic**), respectively, than the cell interior. For this reason, cells surrounded by a wall can grow in media having an osmotic strength much less than that of the cytosol. The properties, function, and formation of the plant cell wall are covered in Chapter 19.

In addition to these universal functions, the plasma membrane has other crucial roles in multicellular organisms. Few of the cells in multicellular plants and animals exist as isolated entities; rather, groups of cells with related specializations combine to form tissues. In animal cells, specialized areas of the plasma membrane, called **cell junctions,** contain proteins and glycolipids that form specific structures between cells that strengthen tissues and allow the exchange of metabolites between cells. Certain plasma membrane proteins anchor cells to components of the **extracellular matrix,** the mixture of fibrous proteins and polysaccharides that provides a bedding on which sheets of epithelial cells or small glands lie. We examine both of these membrane functions in Chapter 19. Still other proteins in the plasma membrane act as anchoring points for many of the cytoskeletal fibers that permeate the cytosol, imparting shape and strength to cells (Chapters 17 and 18).

The plasma membranes of many types of eukaryotic cells also contain proteins that function as **receptors** by binding specific signaling molecules, such as hormones, growth factors, and neurotransmitters, leading to various cellular responses. These proteins, which are critical for cell development and functioning, are described in several later chapters. Finally, peripheral cytosolic proteins that are recruited to the membrane surface function as enzymes, intracellular signal transducers, and structural proteins for stabilizing the membrane.

Endosomes Take Up Soluble Macromolecules from the Cell Exterior

Although transport proteins in the plasma membrane mediate the movement of ions and small molecules into and out

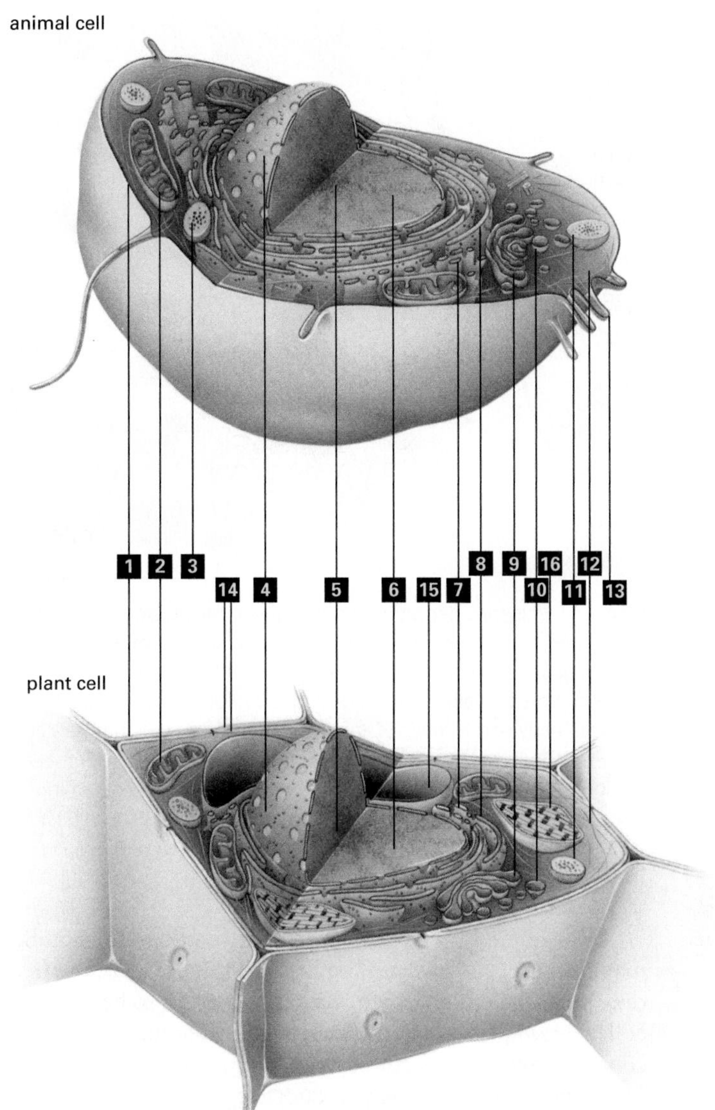

1 Plasma membrane controls movement of molecules in and out of the cell and functions in cell-cell signaling and cell adhesion.

2 Mitochondria, which are surrounded by a double membrane, generate ATP by oxidation of glucose and fatty acids.

3 Lysosomes, which have an acidic lumen, degrade material internalized by the cell and worn-out cellular membranes and organelles.

4 Nuclear envelope, a double membrane, encloses the contents of the nucleus; the outer nuclear membrane is continuous with the rough ER.

5 Nucleolus is a nuclear subcompartment where most of the cell's rRNA is synthesized.

6 Nucleus is filled with chromatin composed of DNA and proteins; site of mRNA and tRNA synthesis.

7 Smooth endoplasmic reticulum (ER) synthesizes lipids and detoxifies certain hydrophobic compounds.

8 Rough endoplasmic reticulum (ER) functions in the synthesis, processing, and sorting of secreted proteins, lysosomal proteins, and certain membrane proteins.

9 Golgi complex processes and sorts secreted proteins, lysosomal proteins, and membrane proteins synthesized on the rough ER.

10 Secretory vesicles store secreted proteins and fuse with the plasma membrane to release their contents.

11 Peroxisomes detoxify various molecules and also break down fatty acids to produce acetyl groups for biosynthesis.

12 Cytoskeletal fibers form networks and bundles that support cellular membranes, help organize organelles, and participate in cell movement.

13 Microvilli increase surface area for absorption of nutrients from surrounding medium.

14 Cell wall, composed largely of cellulose, helps maintain the cell's shape and provides protection against mechanical stress.

15 Vacuole stores water, ions, and nutrients, degrades macromolecules, and functions in cell elongation during growth.

16 Chloroplasts, which carry out photosynthesis, are surrounded by a double membrane and contain a network of internal membrane-bounded sacs.

▲ **FIGURE 9-1 Schematic overview of a "typical" animal cell (top) and plant cell (bottom) and their major substructures.** Not every cell will contain all the organelles, granules, and fibrous structures shown here, and other substructures can be present in some. Cells also differ considerably in shape and in the prominence of various organelles and substructures.

of the cell across the lipid bilayer, proteins and some other soluble macromolecules in the extracellular milieu are internalized by **endocytosis.** In this process, a segment of the plasma membrane invaginates into a "coated pit," whose cytosolic face is lined by a specific set of proteins; several types of endocytosis, each involving different sets of proteins, are known. In **receptor-mediated endocytosis**, for example, certain receptor proteins in the plasma membrane bind macromolecules in the cell exterior and then become incorporated into the invaginating coated pit. The pit pinches from the membrane into a small membrane-bounded vesicle that contains extracellular material—both soluble and bound to receptors—and is delivered to an early **endosome,** a sorting station of membrane-limited tubules and vesicles (Figure 9-2a, b). From this compartment, some membrane proteins are recycled back to the plasma membrane; other membrane proteins are transported to a late endosome where further sorting takes place. The endocytic pathway ends when a late endosome delivers its membrane and internal contents—including material from the extracellular solution—to lysosomes for degradation. The entire endocytic pathway is described in some detail in Chapter 14.

Lysosomes Are Acidic Organelles That Contain a Battery of Degradative Enzymes

Lysosomes provide an excellent example of the ability of intracellular membranes to form closed compartments in which the composition of the **lumen** (the aqueous interior of the compartment) differs substantially from that of the surrounding cytosol. Found exclusively in animal cells, **lysosomes** are responsible for degrading certain components that have become obsolete for the cell or organism. (The vacuoles in plant and fungal cells have many of the same functions as

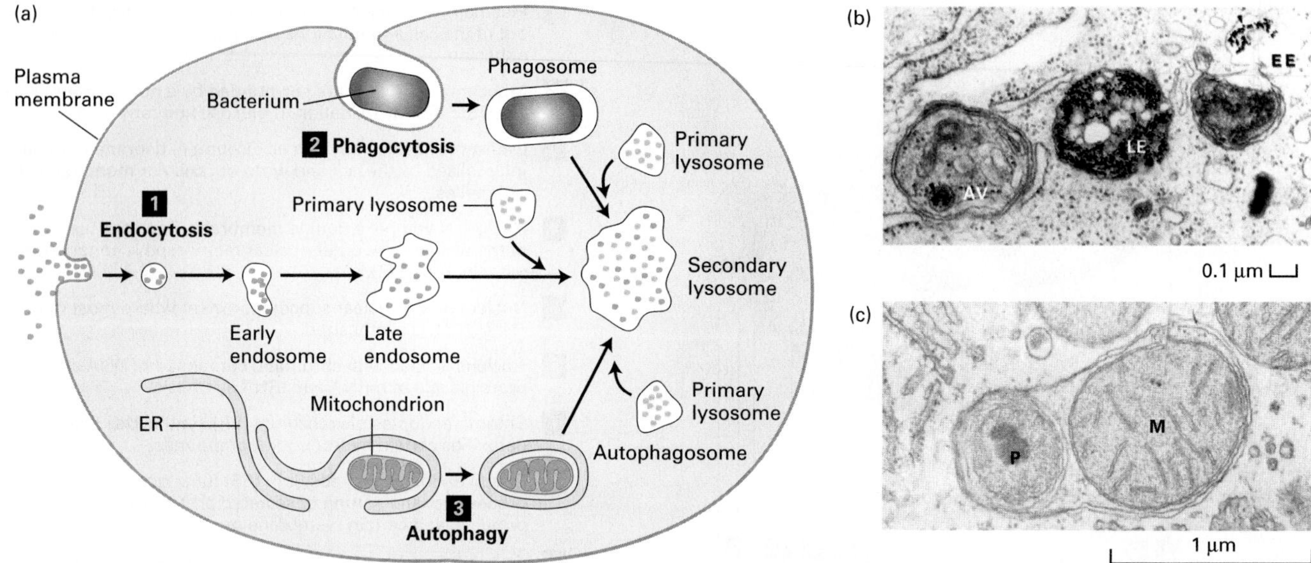

▲ **FIGURE 9-2 Cellular structures that participate in delivering materials to lysosomes.** (a) Schematic overview of three pathways by which materials are moved to lysosomes. Soluble macromolecules are taken into the cell by invagination of coated pits in the plasma membrane and delivered to lysosomes through the endocytic pathway **1**. Whole cells and other large, insoluble particles move from the cell surface to lysosomes through the phagocytic pathway **2**. Worn-out organelles and bulk cytoplasm are delivered to lysosomes through the autophagic pathway **3**. Within the acidic lumen of lysosomes, hydrolytic enzymes degrade proteins, nucleic acids, and other large molecules. (b) An electron micrograph of a section of a cultured mammalian cell that had taken up small gold particles coated with the egg protein ovalbumin. Gold-labeled ovalbumin (black spots) is found in early endosomes (EE) and late endosomes (LE), but very little is present in autophagosomes (AV). (c) Electron micrograph of a section of a rat liver cell showing a secondary lysosome containing fragments of a mitochondrion (M) and a peroxisome (P). [Part (b) from T. E. Tjelle et al., 1996, *J. Cell Sci.* **109**:2905. Part (c) courtesy of D. Friend.]

animal lysosomes.) The process by which an aged organelle is degraded in a lysosome is called *autophagy* ("eating oneself"). Materials are taken into a cell not only by endocytosis but also by **phagocytosis,** a process in which large, insoluble particles (e.g., bacteria) are enveloped by the plasma membrane and internalized (see Figure 9-2a). In both cases, the internalized material may be degraded in lysosomes.

Lysosomes contain a group of enzymes that degrade polymers, releasing their monomeric subunits. For example, nucleases degrade RNA and DNA to mononucleotides; proteases degrade a variety of proteins and peptides to amino acids; phosphatases remove phosphate groups from mononucleotides, phospholipids, and other compounds; still other enzymes degrade complex polysaccharides and glycolipids into smaller units. All the lysosomal enzymes work most efficiently at acid pH values and collectively are termed *acid hydrolases.* Two types of transport proteins in the lysosomal membrane work together to pump H^+ and Cl^- ions (HCl) from the cytosol across the membrane, thereby acidifying the lumen (see Figure 11-13). The acid pH helps to denature proteins, making them accessible to the action of the lysosomal hydrolases, which themselves are resistant to acid denaturation. Lysosomal enzymes are poorly active at the neutral pH of cells and most extracellular fluids. Thus, if a lysosome releases its enzymes into the cytosol, where the pH is between 7.0 and 7.3, they cause little degradation of cytosolic components. Cytosolic and nuclear proteins generally are not degraded in lysosomes but rather in proteasomes, large multiprotein complexes in the cytosol (see Figure 3-29).

Lysosomes vary in size and shape, and several hundred may be present in a typical animal cell (Figure 9-3). In effect, they function as sites where various materials to be degraded collect. *Primary lysosomes* are roughly spherical and do not contain obvious particulate or membrane debris. *Secondary lysosomes,* which are larger and irregularly shaped, result from the fusion of primary lysosomes with late endosomes and other membrane-bounded organelles and vesicles. They contain particles or membranes in the process of being digested (Figure 9-2c).

Several human diseases are caused by defects in specific lysosomal enzymes because their substrates accumulate inside the organelle. For example, *Tay-Sachs disease* is caused by a defect in one enzyme catalyzing a step in the lysosomal breakdown of gangliosides. The resulting accumulation of these glycolipids, especially in nerve cells, has devastating consequences. The symptoms of this inherited disease are usually evident before the age of 1. Affected children commonly become demented and blind by age 2 and die before their third birthday. Nerve cells from such children are greatly enlarged with swollen lipid-filled lysosomes. ▮

Peroxisomes Degrade Fatty Acids and Toxic Compounds

All animal cells (except erythrocytes) and many plant cells contain **peroxisomes,** a class of roughly spherical organelles, 0.2–1.0 μm in diameter (Figure 9-4). Peroxisomes contain

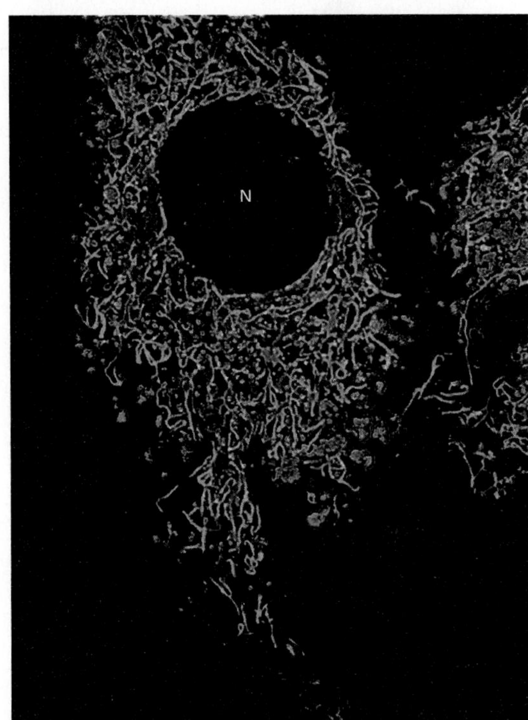

▲ **FIGURE 9-3 Location of lysosomes and mitochondria in a cultured living bovine pulmonary artery endothelial cell.** The cell was stained with a green-fluorescing dye that is specifically bound to mitochondria, and a red-fluorescing dye that is specifically incorporated into lysosomes. The image was sharpened using a deconvolution computer program discussed later in the chapter. N = nucleus. [Courtesy Invitrogen/ Molecular Probes Inc.]

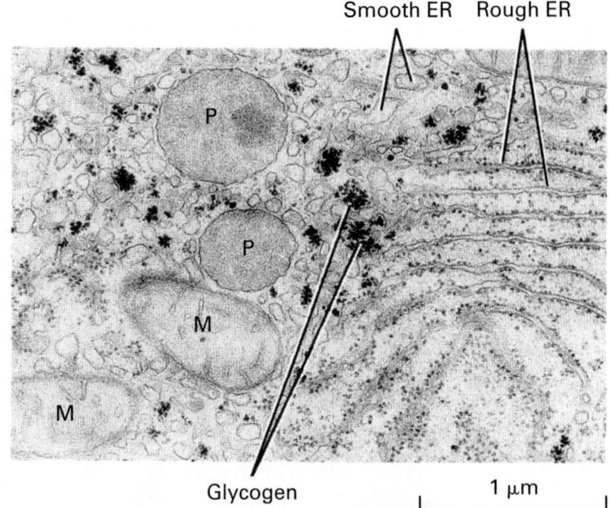

▲ **FIGURE 9-4 Electron micrograph showing various organelles in a rat liver cell.** Two peroxisomes (P) lie in close proximity to mitochondria (M) and the rough and smooth endoplasmic reticulum (ER). Also visible are accumulations of glycogen, a polysaccharide that is the primary glucose-storage molecule in animals. [Courtesy of P. Lazarow.]

several *oxidases*—enzymes that use molecular oxygen to oxidize organic substances, in the process forming hydrogen peroxide (H_2O_2), a corrosive substance. Peroxisomes also contain copious amounts of the enzyme *catalase,* which degrades hydrogen peroxide to yield water and oxygen:

$$2H_2O_2 \xrightarrow{\text{catalase}} 2H_2O + O_2$$

In contrast with the oxidation of fatty acids in mitochondria, which produces CO_2 and is coupled to the generation of ATP, peroxisomal oxidation of fatty acids yields acetyl groups and is not linked to ATP formation (see Figure 12-12). The energy released during peroxisomal oxidation is converted into heat, and the acetyl groups are transported into the cytosol, where they are used in the synthesis of cholesterol and other metabolites. In most eukaryotic cells, the peroxisome is the principal organelle in which fatty acids are oxidized, thereby generating precursors for important biosynthetic pathways. Particularly in liver and kidney cells, various toxic molecules that enter the bloodstream also are degraded in peroxisomes, producing harmless products.

Plant seeds contain *glyoxisomes,* small organelles that oxidize stored lipids as a source of carbon and energy for growth. They are similar to peroxisomes and contain many of the same types of enzymes as well as additional ones used to convert fatty acids into glucose precursors. ∎

The Endoplasmic Reticulum Is a Network of Interconnected Internal Membranes

Generally, the largest membrane in a eukaryotic cell encloses the **endoplasmic reticulum (ER)**—an extensive network of closed, flattened membrane-bounded sacs called **cisternae** (see Figure 9-1). The endoplasmic reticulum has a number of functions in the cell but is particularly important in the synthesis of lipids, membrane proteins, and secreted proteins. The *smooth endoplasmic reticulum* is smooth because it lacks ribosomes. In contrast, the cytosolic face of the *rough endoplasmic reticulum* is studded with ribosomes.

The Smooth Endoplasmic Reticulum The synthesis of fatty acids and phospholipids takes place in the smooth ER. Although many cells have very little smooth ER, this organelle is abundant in hepatocytes. Enzymes in the smooth ER of the liver also modify or detoxify hydrophobic chemicals such as pesticides and carcinogens by chemically converting them into more water-soluble, conjugated products that can be excreted from the body. High doses of such compounds result in a large proliferation of the smooth ER in liver cells.

The Rough Endoplasmic Reticulum Cytoplasmic ribosomes bound to the rough ER synthesize certain membrane and organelle proteins and virtually all proteins to be secreted from the cell (Chapter 13). Sequences in the nascent polypeptide chain growing on the ribosome sequentially bind to proteins in the rough ER membrane and thus localize the ribosome to the ER. As the growing polypeptide emerges from the ribosome, it passes through the rough ER membrane, with the help of specific translocation proteins in the membrane, into the

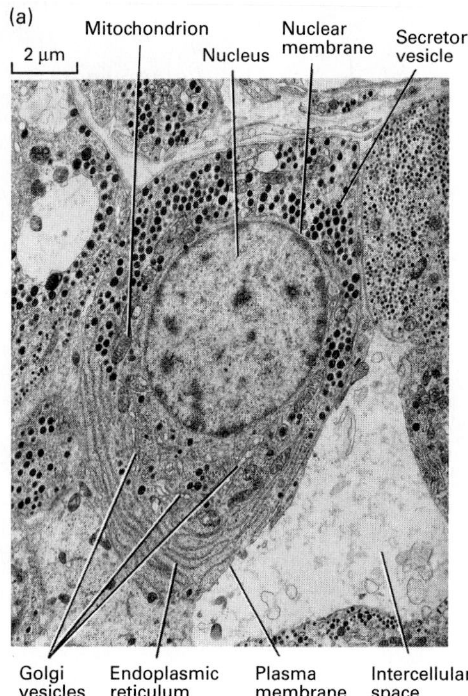

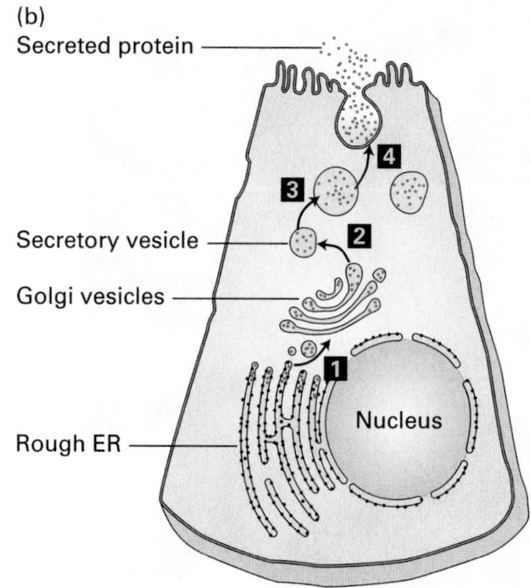

▲ **FIGURE 9-5 Characteristic features of cells specialized to secrete large amounts of particular proteins (e.g., hormones, antibodies).** (a) Electron micrograph of a thin section of a hormone-secreting cell from the rat pituitary. The basal end of the cell (*bottom*) is filled with abundant rough ER and Golgi sacs, where polypeptide hormones are synthesized and packaged. At the opposite apical end of the cell (*top*) are numerous secretory vesicles, which contain recently made hormones eventually to be secreted. (b) Diagram of a typical secretory cell tracing the pathway followed by a protein (small red dots) to be secreted. Immediately after their synthesis on ribosomes (small black dots) of the rough ER, secreted proteins are found in the lumen of the rough ER. Transport vesicles bud off and carry these proteins to the Golgi complex (**1**), where the proteins are concentrated and packaged into immature secretory vesicles (**2**). These vesicles then coalesce to form larger mature secretory vesicles that lose water to the cytosol, leaving an almost crystalline mixture of secreted proteins in the lumen (**3**). After these vesicles accumulate under the apical surface, they fuse with the plasma membrane and release their contents (exocytosis) in response to appropriate hormonal or nerve stimulation (**4**). [Part (a) courtesy of Biophoto Associates.]

interior space, or lumen. There the protein folds, assisted by folding catalysts termed **chaperones.** Secreted proteins are modified in several ways by enzymes localized in the ER lumen, including covalent addition of sugars (*glycosylation*) and formation of disulfide bonds. Newly made membrane proteins remain associated with the rough ER membrane, and proteins to be secreted accumulate in the lumen of the organelle.

All eukaryotic cells contain a discernible amount of rough ER because it is needed for the synthesis of plasma membrane proteins and secreted proteins that comprise the extracellular matrix. Phospholipids are also synthesized in association with the rough ER. Rough ER is particularly abundant in specialized cells that produce an abundance of specific proteins to be secreted. For example, plasma cells produce antibodies, pancreatic acinar cells synthesize digestive enzymes, and cells in the pancreatic islets of Langerhans produce the polypeptide hormones insulin and glucagon. In these secretory cells and others, a large part of the cytosol is filled with rough ER and secretory vesicles (Figure 9-5).

The Golgi Complex Processes and Sorts Secreted and Membrane Proteins

Several minutes after proteins are synthesized in the rough ER, most of them leave the organelle within small membrane-bounded transport vesicles. These vesicles, which bud from regions of the rough ER not coated with ribosomes, carry the proteins to another membrane-limited organelle, the **Golgi complex** (see Figure 9-5), named after the Italian microscopist Camillio Golgi.

Three-dimensional reconstructions from serial sections of a Golgi complex reveal this organelle to be a series of flattened membrane vesicles or sacs (cisternae), surrounded by a number of more or less spherical membrane-limited vesicles (Figure 9-6). The stack of Golgi cisternae has three defined regions—the *cis,* the *medial,* and the *trans.* Transport vesicles from the rough ER fuse with the *cis* region of the Golgi complex, where they deposit their protein contents. As detailed in Chapter 14, these proteins then progress from the *cis* to the *medial* to the *trans* region. Within each region are different luminal enzymes that modify proteins to be secreted

 Video: Three-Dimensional Model of a Golgi Complex

MEDIA CONNECTIONS

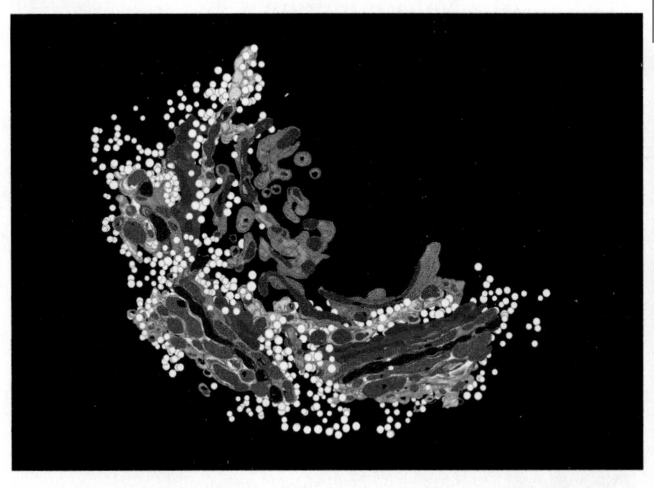

▶ **FIGURE 9-6 Model of the Golgi complex based on three-dimensional reconstruction of electron microscopy images.** Transport vesicles (white spheres) that have budded off the rough ER fuse with the *cis* membranes (light blue) of the Golgi complex. By mechanisms described in Chapter 14, proteins move from the *cis* region to the *medial* region and finally to the *trans* region of the Golgi complex. Eventually, vesicles bud off the *trans*-Golgi membranes (orange and red); some move to the cell surface and others move to lysosomes. The Golgi complex, like the rough endoplasmic reticulum, is especially prominent in secretory cells. [From B. J. Marsh et al., 2001, *Proc Nat'l. Acad. Sci USA* **98**:2399.]

and membrane proteins differently, depending on their structures and their final destinations.

After proteins to be secreted and membrane proteins are modified in the Golgi complex, they are transported out of the complex by a second set of vesicles, which bud from the *trans* side of the Golgi complex. Some vesicles carry membrane proteins destined for the plasma membrane or soluble proteins to be released into the extracellular medium; others carry soluble or membrane proteins to lysosomes or other organelles. Continuous budding of vesicles in this manner would result in accumulation of phospholipids at the plasma membrane, but endocytic vesicles (see Figure 9-2) return plasma membrane phospholipids to lysosomes or to the Golgi. Yet other vesicles bud from Golgi vesicles and fuse with earlier Golgi vesicles or with the rough ER. How intracellular transport vesicles "know" with which membranes to fuse and where to deliver their contents is also discussed in Chapter 14.

Plant Vacuoles Store Small Molecules and Enable a Cell to Elongate Rapidly

Most plant cells contain at least one membrane-limited internal vacuole. The number and size of vacuoles depend on both the type of cell and its stage of development; a single vacuole may occupy as much as 80 percent of a mature plant cell (Figure 9-7). A variety of transport proteins in the vacuolar membrane allow plant cells to accumulate and store water, ions, and nutrients (e.g., sucrose, amino acids) within vacuoles (Chapter 11). Like a lysosome, the lumen of a vacuole contains a battery of degradative enzymes and has an acidic pH, which is maintained by similar transport proteins in the vacuolar membrane. Thus plant vacuoles may also have a degradative function similar to that of lysosomes in animal cells. Similar storage vacuoles are found in green algae and many microorganisms such as fungi.

Like most cellular membranes, the vacuolar membrane is permeable to water but is poorly permeable to the small molecules stored within it. Because the solute concentration is much higher in the vacuole lumen than in the cytosol or extracellular fluids, water tends to move by osmotic flow into vacuoles. This influx of water, which causes the vacuole to expand, creates hydrostatic pressure, or *turgor*, inside the cell. This pressure is balanced by the mechanical

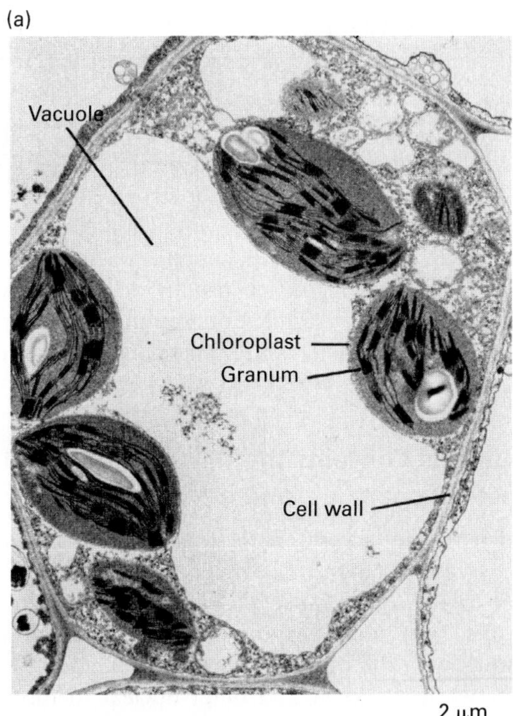

▲ **FIGURE 9-7 Electron micrograph of a thin section of a leaf cell showing a prominent vacuole.** In this cell, the single large vacuole occupies much of the cell volume. Parts of five chloroplasts and the cell wall also are visible. Note the internal subcompartments in the chloroplasts. [Courtesy of Biophoto Associates/Myron C. Ledbetter/Brookhaven National Laboratory.]

Video: Three-Dimensional Model of a Mitochondrion

MEDIA CONNECTIONS

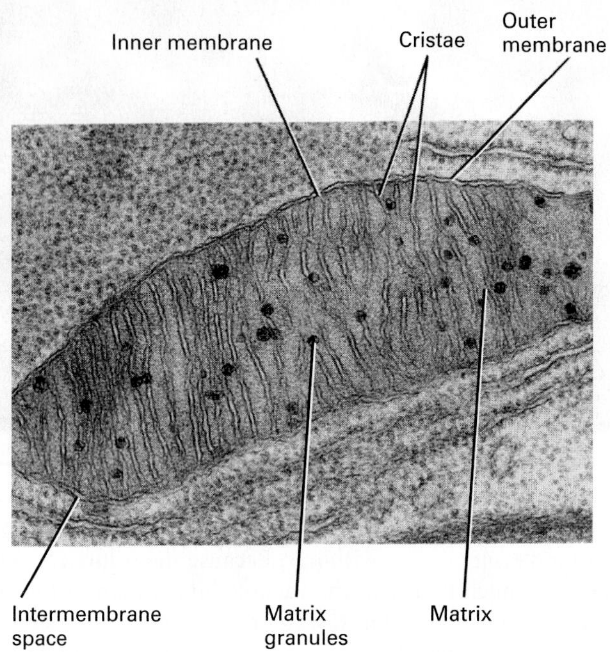

▲ **FIGURE 9-8 Electron micrograph of a mitochondrion.** Most ATP production in nonphotosynthetic cells takes place in mitochondria. The inner membrane, which surrounds the matrix space, has many infoldings, called cristae. Small calcium-containing matrix granules also are evident. [From D. W. Fawcett, 1981, *The Cell,* 2d ed., Saunders, p. 421.]

resistance of the cellulose-containing cell walls that surround plant cells. Most plant cells have a turgor of 5–20 atmospheres (atm); their cell walls must be strong enough to react to this pressure in a controlled way. Unlike animal cells, plant cells can elongate extremely rapidly, at rates of 20–75 μm/h. This elongation, which usually accompanies plant growth, occurs when a segment of the somewhat elastic cell wall stretches under the pressure created by water taken into the vacuole. ▮

The Nucleus Contains the DNA Genome, RNA Synthetic Apparatus, and a Fibrous Matrix

The **nucleus,** the largest organelle in animal cells, is surrounded by two membranes, each containing many different types of proteins. The inner nuclear membrane defines the nucleus itself. In most cells, the outer nuclear membrane is continuous with the rough endoplasmic reticulum, and the space between the inner and outer nuclear membranes is continuous with the lumen of the rough endoplasmic reticulum (see Figure 9-1). The two nuclear membranes appear to fuse at **nuclear pores,** the ringlike complexes composed of specific proteins termed **nucleoporins** through which material moves between the nucleus and the cytosol. The structure of nuclear pores and the regulated transport of material through them are detailed in Chapter 8.

Most of the cell's ribosomal RNA is synthesized in the **nucleolus,** a subcompartment of the nucleus that is not bounded by a phospholipid membrane. Ribosomal proteins, like all nuclear-encoded proteins, are made in the cytosol. Most enter the nucleus via nuclear pores and are added to ribosomal RNAs within the nucleolus. The finished or partly finished ribosomal subunits, exit through a nuclear pore into the cytosol for use in protein synthesis (Chapter 4). In mature erythrocytes from nonmammalian vertebrates and in other types of "resting" cells, the nucleus is inactive or dormant and minimal synthesis of DNA and RNA takes place. Similarly, mRNA and tRNA is synthesized in the nucleus; following extensive processing within the nucleus, particles containing these RNAs also exit the nucleus into the cytosol through nuclear pores.

The packaging of nuclear DNA into chromosomes is described in Chapter 6. Only during cell division are individual chromosomes visible by light microscopy. In the electron microscope, the nonnucleolar regions of the nucleus, called the *nucleoplasm,* can be seen to have dark- and light-staining areas. The dark areas, which are often closely associated with the nuclear membrane, contain condensed concentrated DNA, called **heterochromatin** (see Figure 6-33a). Fibrous proteins called **lamins** form a two-dimensional network along the inner surface of the inner membrane, giving it shape and apparently binding DNA to it. The breakdown of this network occurs early in cell division, as we detail in Chapter 20.

Mitochondria Are the Principal Sites of ATP Production in Aerobic Nonphotosynthetic Cells

Most eukaryotic cells contain many **mitochondria** (sing., mitochondrion), which occupy up to 25 percent of the volume of the cytoplasm (see Figure 9-3). These complex organelles, the main sites of ATP production during aerobic metabolism, are generally exceeded in size only by the nucleus, vacuoles, and chloroplasts (which also produce ATP).

The two membranes that bound a mitochondrion differ in composition and function. The outer membrane, composed of about half lipid and half protein, contains porin proteins (see Figure 10-18) that render the membrane permeable to molecules having molecular weights as high as 10,000. The inner membrane is much less permeable; its surface area is greatly increased by a large number of infoldings, or *cristae,* that protrude into the *matrix,* or central space (Figure 9-8).

In nonphotosynthetic cells, the principal fuels for ATP synthesis are fatty acids and glucose. The complete aerobic degradation of glucose to CO_2 and H_2O is coupled to the synthesis of about 30 molecules of ATP (the exact number is in question). In eukaryotic cells, the initial stages of glucose degradation take place in the cytosol, where 2 ATP molecules per glucose molecule are generated. The terminal stages of oxidation and the coupled synthesis of ATP are carried out by enzymes in the mitochondrial matrix and inner membrane (Chapter 12). As many as 28 ATP molecules per glucose molecule are generated in mitochondria. Similarly, virtually all the ATP formed in the oxidation of fatty acids to CO_2 is generated in mitochondria. Thus mitochondria can be regarded as the "power plants" of the cell.

Chloroplasts Contain Internal Compartments in Which Photosynthesis Takes Place

Except for vacuoles, **chloroplasts** are the largest and the most characteristic organelles in the cells of plants and green algae. They can be as long as 10 μm and are typically 0.5–2 μm thick, but they vary in size and shape in different cells, especially among the algae. In addition to the double membrane that bounds a chloroplast, this organelle also contains an extensive internal system of interconnected membrane-limited sacs called **thylakoids,** which are flattened to form disks (Figure 9-9). Thylakoids often form stacks called *grana* and are embedded in a matrix space, the *stroma*. The thylakoid membranes contain green pigments (**chlorophylls**) and other pigments that absorb light, as well as enzymes that generate ATP during photosynthesis. Some of the ATP is used to convert CO_2 into three-carbon intermediates by enzymes located in the stroma; the intermediates are then exported to the cytosol and converted into sugars. ▌

The molecular mechanisms by which ATP is formed in mitochondria and chloroplasts are very similar, as explained in Chapter 12. Chloroplasts and mitochondria have other features in common: both often migrate from place to place within cells, and they contain their own DNA, which encodes some of the key organellar proteins (Chapter 6). The proteins encoded by mitochondrial or chloroplast DNA are synthesized on ribosomes within the organelles. However, most of the proteins in each organelle are encoded in nuclear DNA and are synthesized in the cytosol; these proteins are then incorporated into the organelles by processes described in Chapter 13.

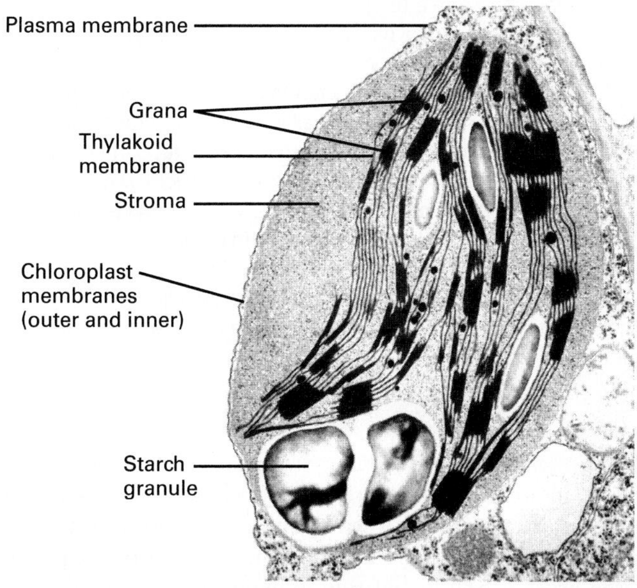

▲ **FIGURE 9-9 Electron micrograph of a plant chloroplast.** The internal membrane vesicles (thylakoids) are fused into stacks (grana), which reside in a matrix (the stroma). All the chlorophyll in the cell is contained in the thylakoid membranes, where the light-induced production of ATP takes place during photosynthesis. [Courtesy of Biophoto Associates/M. C. Ledbetter/Brookhaven National Laboratory.]

KEY CONCEPTS OF SECTION 9.1

Organelles of the Eukaryotic Cell

- All eukaryotic cells contain a nucleus as well as numerous other organelles in their cytoplasm (see Figure 9-1).
- The nucleus, mitochondrion, and chloroplast are bounded by two bilayer membranes separated by an intermembrane space. All other organelles are surrounded by single membranes.
- The plasma membrane acts as a permeability barrier. It contains a multitude of proteins that transport nutrients and waste molecules or that bind components of the extracellular matrix or, in plants, the cell wall. The plasma membranes of many eukaryotic cells also contain receptor proteins that bind specific signaling molecules.
- Endosomes internalize plasma membrane proteins and soluble materials from the extracellular medium, and sort the internalized materials back to the membrane or to lysosomes for degradation.
- Lysosomes have an acidic interior and contain various hydrolases that degrade worn-out or unneeded cellular components and some ingested materials (see Figure 9-2).
- Peroxisomes are small organelles containing enzymes that oxidize various organic compounds without the production of ATP. By-products of oxidation are used in biosynthetic reactions.
- Secreted proteins and membrane proteins are synthesized on the rough endoplasmic reticulum, a network of flattened membrane-bounded sacs studded with ribosomes.
- Proteins synthesized on the rough ER first move to the Golgi complex, where they are processed and sorted for transport to the cell surface or other destinations (see Figure 9-5).
- Plant cells contain one or more large vacuoles, which are storage sites for ions and nutrients. Osmotic flow of water into vacuoles generates turgor pressure that pushes the plasma membrane against the cell wall.
- The nucleus houses the genome of a cell. The inner and outer nuclear membranes are fused at numerous nuclear pores, through which materials pass between the nucleus and the cytosol. The outer nuclear membrane is continuous with that of the rough endoplasmic reticulum.
- Mitochondria have a highly permeable outer membrane and an inner membrane that is extensively folded. Enzymes in the inner mitochondrial membrane and central matrix space carry out the terminal stages of sugar and lipid oxidation coupled to ATP synthesis.
- Chloroplasts contain a complex system of thylakoid membranes in their interiors. These membranes contain the pigments and enzymes that absorb light and produce ATP during photosynthesis.

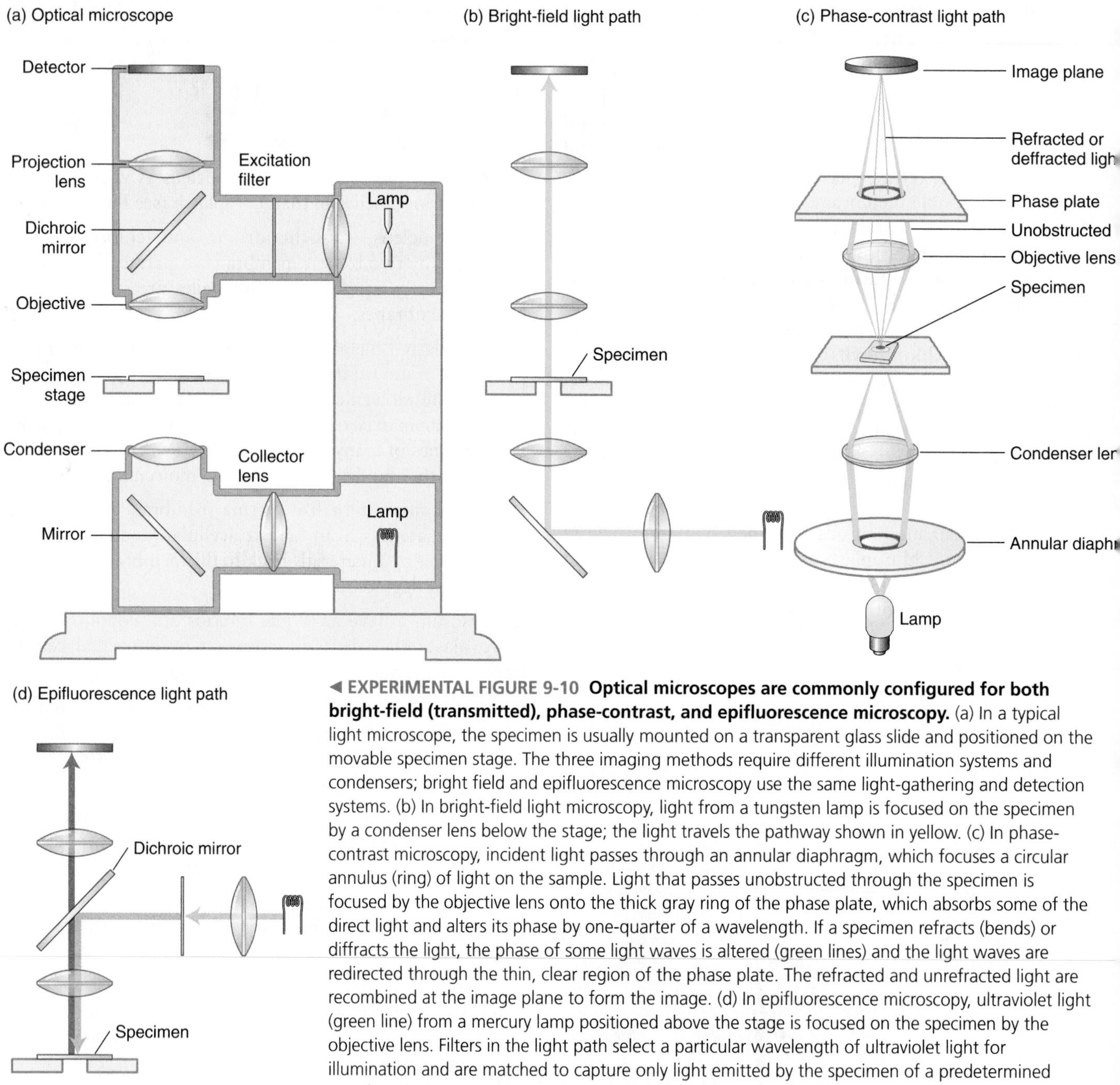

(a) Optical microscope

Detector

Projection lens

Excitation filter

Lamp

Dichroic mirror

Objective

Specimen stage

Condenser

Collector lens

Lamp

Mirror

(b) Bright-field light path

Specimen

(c) Phase-contrast light path

Image plane

Refracted or deffracted ligh

Phase plate

Unobstructed

Objective lens

Specimen

Condenser ler

Annular diaph

Lamp

(d) Epifluorescence light path

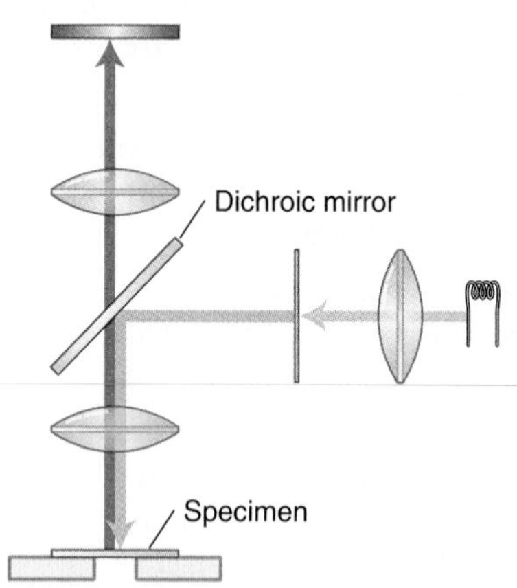

◄ EXPERIMENTAL FIGURE 9-10 **Optical microscopes are commonly configured for both bright-field (transmitted), phase-contrast, and epifluorescence microscopy.** (a) In a typical light microscope, the specimen is usually mounted on a transparent glass slide and positioned on the movable specimen stage. The three imaging methods require different illumination systems and condensers; bright field and epifluorescence microscopy use the same light-gathering and detection systems. (b) In bright-field light microscopy, light from a tungsten lamp is focused on the specimen by a condenser lens below the stage; the light travels the pathway shown in yellow. (c) In phase-contrast microscopy, incident light passes through an annular diaphragm, which focuses a circular annulus (ring) of light on the sample. Light that passes unobstructed through the specimen is focused by the objective lens onto the thick gray ring of the phase plate, which absorbs some of the direct light and alters its phase by one-quarter of a wavelength. If a specimen refracts (bends) or diffracts the light, the phase of some light waves is altered (green lines) and the light waves are redirected through the thin, clear region of the phase plate. The refracted and unrefracted light are recombined at the image plane to form the image. (d) In epifluorescence microscopy, ultraviolet light (green line) from a mercury lamp positioned above the stage is focused on the specimen by the objective lens. Filters in the light path select a particular wavelength of ultraviolet light for illumination and are matched to capture only light emitted by the specimen of a predetermined wavelength that is longer than that of the incident light (red line).

9.2 Light Microscopy: Visualizing Cell Structure and Localizing Proteins Within Cells

For many years, light microscopy has been an essential part of virtually all research on eukaryotic cells. Basic light microscopes are used to enumerate the number of cells under study, and simple staining procedures enable the study of live cells. More specialized techniques in light microscopy enable the investigator to visualize movements such as the crawling of cells along a substrate, extension of nerve axons, and movements of chromosomes and organelles within cells. Using techniques of recombinant DNA, researchers can express, in a cell or in an organism, a **chimera** of any desired protein of interest linked to a naturally fluorescent protein. Usually a chimera exhibits the normal function of the desired protein, and its location in living cells can be observed over time. Using immunofluorescence, researchers can determine the location of specific proteins in fixed cells, as well as any localization changes in response to changes in the cell's environment. Finally, many microscopic images of the same cell can be stored in a computer data base; digital reconstructions permits three-dimensional reconstructions of cell components from two-dimensional images, and a determination of whether two or more proteins are found in the same subcellular compartment.

The Resolution of the Light Microscope Is About 0.2 μm

All microscopes produce a magnified image of a small object, but the nature of the image depends on the type of microscope employed and on the way in which the specimen is prepared. The compound microscope, used in conventional *bright-field light microscopy,* contains several lenses that magnify the image of a specimen under study (Figure 9-10a, b). The total magnification is a product of the magnification of the individual lenses: if the *objective lens,* the lens closest to the specimen, magnifies 100-fold (a 100× lens, the maximum usually employed) and the *projection lens,* sometimes called the ocular or eyepiece, magnifies 10-fold, the final magnification recorded by the human eye or on film will be 1000-fold.

However, the most important property of any microscope is not its magnification but its resolving power, or **resolution**—the ability to distinguish between two very closely positioned objects. Merely enlarging the image of a specimen accomplishes nothing if the image is blurred. The resolution of a microscope lens is numerically equivalent to D, the minimum distance between two distinguishable objects. The smaller the value of D, the better the resolution. The value of D is given by the equation

$$D = \frac{0.61\lambda}{N \sin\alpha} \qquad (9\text{-}1)$$

where α is the angular aperture, or half-angle, of the cone of light entering the objective lens from the specimen; N is the refractive index of the medium between the specimen and the objective lens (i.e., the relative velocity of light in the medium compared with the velocity in air); and λ is the wavelength of the incident light. Resolution is improved by using shorter wavelengths of light (decreasing the value of λ) or gathering more light (increasing either N or α). Note that the magnification is not part of this equation.

Owing to limitations on the values of α, λ, and N based on the physical properties of light, the *limit of resolution* of a light microscope using visible light is about 0.2 μm (200 nm). No matter how many times the image is magnified, a conventional light microscope can never resolve objects that are less than ≈0.2 μm apart or reveal details smaller than ≈0.2 μm in size. But exotic new light microscopes described below can resolve two fluorescent objects even if they are as close together as ≈20 nm.

A conventional light microscope can be used to track the location of a small bead of known size to a precision of only a few nanometers. If we know the precise size and shape of an object—say, a 5-nm sphere of gold that is attached to an antibody in turn bound to a cell-surface protein—and if we use a video camera to record the microscopic image as a digital image, then a computer can calculate the position of the center of the object to within a few nanometers. In this way, computer algorithms can be used to make observations at a more precise level—in this case the movement of a cell-surface protein labeled with the gold-tagged antibody—than would be possible based on the light microscope's resolution alone. This technique has also been used to measure nanometer-size steps as molecules and vesicles move along cytoskeletal filaments (see Figures 17-21, 17-26, and 17-27).

Phase-Contrast and Differential Interference Contrast Microscopy Visualize Unstained Living Cells

Two common methods for imaging live cells and unstained tissues generate contrast by taking advantage of differences in the refractive index and thickness of cellular materials. These methods, called *phase-contrast microscopy* and *differential interference contrast (DIC) microscopy* (or Nomarski interference microscopy), produce images that differ in appearance and reveal different features of cell architecture. Figure 9-11 compares images of live, cultured cells obtained with these two methods and standard bright-field microscopy.

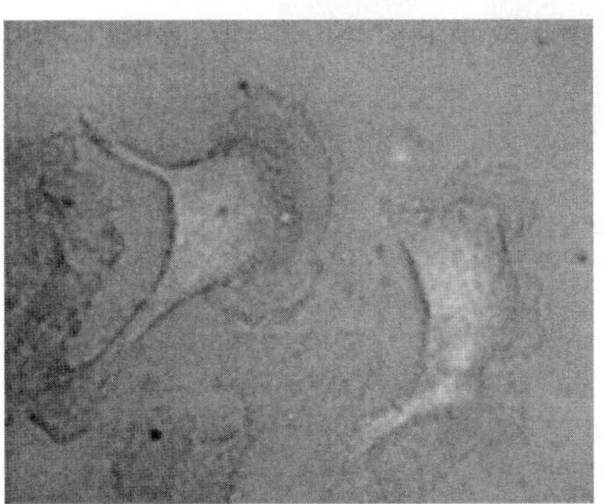

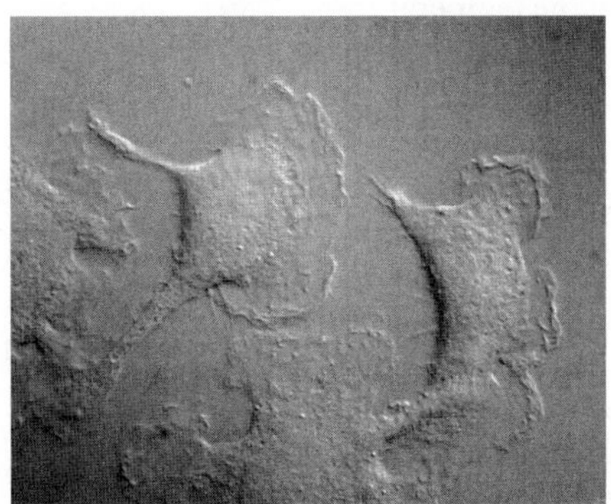

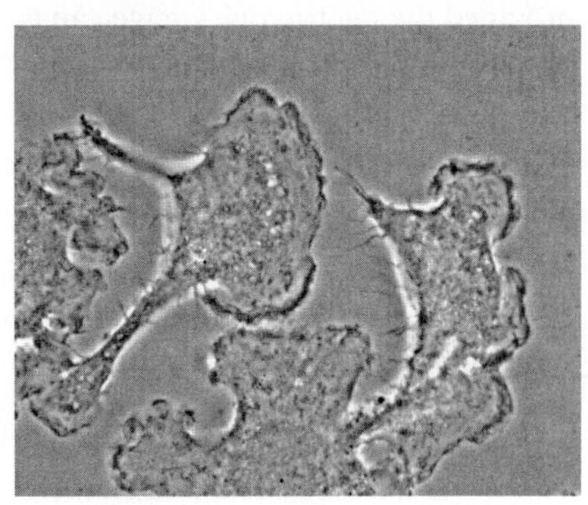

▲ EXPERIMENTAL FIGURE 9-11 **Live cells can be visualized by microscopy techniques that generate contrast by interference.** These micrographs show live, cultured macrophage cells viewed by bright-field microscopy (*left*), differential interference contrast (DIC) microscopy (*middle*), and phase-contrast microscopy (*right*). In a phase-contrast image, cells are surrounded by alternating dark and light bands; in-focus and out-of-focus details are simultaneously imaged in a phase-contrast microscope. In a DIC image, cells appear in pseudorelief. Because only a narrow in-focus region is imaged, a DIC image is an optical slice through the object. [Courtesy of N. Watson and J. Evans.]

Phase-contrast microscopy (Figure 9-10c) generates an image in which the degree of darkness or brightness of a region of the sample depends on the *refractive index* of that region. Light moves more slowly in a medium of higher refractive index. Thus, a beam of light is refracted (bent) once as it passes from air into a transparent object and again when it departs. Consequently, part of a light wave that passes through a specimen will be refracted and will be out of phase (out of synchrony) with the part of the wave that does not pass through the specimen. How much their phases will differ depends on the difference in refractive index along the two paths and on the thickness of the specimen. If the two parts of the light wave are recombined, the resultant light will be brighter if they are in phase and less bright if they are out of phase. The refracted and unrefracted light are recombined at the image plane to form the image. Phase-contrast microscopy is suitable for observing single cells or thin cell layers, but not thick tissues. It is particularly useful for examining the location and movement of larger organelles in live cells.

DIC microscopy is based on interference between polarized light and is the method of choice for visualizing extremely small details and thick objects. Contrast is generated by differences in the index of refraction of the object and its surrounding medium. In DIC images, objects appear to cast a shadow to one side. The "shadow" primarily represents a difference in the refractive index of a specimen rather than its topography. DIC microscopy easily defines the outlines of large organelles, such as the nucleus and vacuole. In addition to having a "relief"-like appearance, a DIC image is a thin *optical section,* or slice, through the object. Thus details of the nucleus in thick specimens (e.g., an intact *Caenorhabditis elegans* roundworm; see Figure 21-4) can be observed in a series of such optical sections, and the three-dimensional structure of the object can be reconstructed by combining the individual DIC images.

Both phase-contrast and DIC microscopy can be used in *time-lapse microscopy,* in which the same cell is photographed at regular intervals over periods of several hours. This procedure allows the observer to study cell movement, provided the microscope's stage can control the temperature of the specimen and the gas environment.

Fluorescence Microscopy Can Localize and Quantify Specific Molecules in Live Cells

Perhaps the most versatile and powerful technique for localizing proteins within a cell by light microscopy is **fluorescent staining** of cells and observation by *fluorescence microscopy.* A chemical is said to be fluorescent if it absorbs light at one wavelength (the excitation wavelength) and emits light (fluoresces) at a specific and longer wavelength. In modern fluorescence microscopes, only fluorescent light emitted by the sample is used to form an image; light of the exciting wavelength induces the fluorescence but is then not allowed to pass the filters placed between the objective lens and the eye or camera (Figure 9-10d).

Expression of Fluorescent Proteins in Live Cells and Organisms A naturally fluorescent protein found in the jellyfish *Aequorea victoria* can be exploited to visualize live cells and specific proteins within them. This 238-residue protein, called *green fluorescent protein (GFP),* contains a serine, tyrosine, and glycine sequence whose side chains spontaneously cyclize to form a green-fluorescing chromophore. With the use of recombinant DNA techniques discussed in Chapter 5, the *GFP* gene can be introduced into living cultured cells or into specific cells of an entire animal. Cells containing the introduced gene will produce GFP and thus emit a green fluorescence when irradiated; this GFP fluorescence can be used to localize the cells within a tissue or even a whole organism, as illustrated in Figure 9-12 for a primitive metazoan organism, the Cnidairian *Hydra vulgaris.*

In a particularly useful application of GFP, a cellular protein of interest is "tagged" with GFP to localize it. In this technique, the gene for GFP is fused to the gene for a particular cellular protein, producing a recombinant DNA encoding one long chimeric protein that contains the entirety of both proteins. GFP fusion proteins frequently retain normal function, even with the large GFP polypeptide appended. Thus GFP fusions are extremely useful

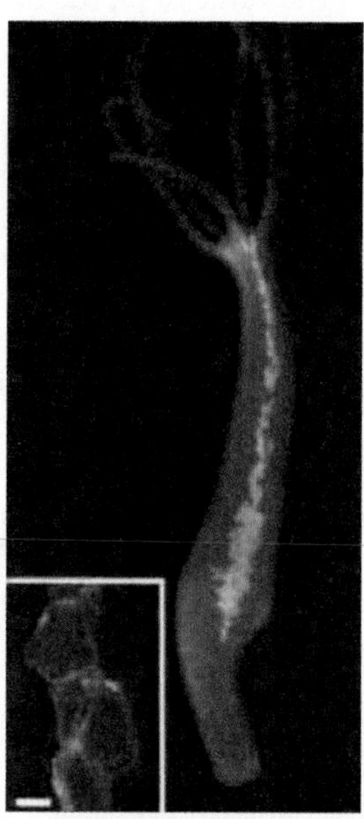

▲ EXPERIMENTAL FIGURE 9-12 **Transgenic hydra *H. vulgaris* expressing green fluorescent protein (GFP).** A recombinant plasmid in which the promoter of the *H. vulgaris β-actin* gene drives expression of the *GFP* gene was microinjected into embryos at the two- to eight-cell stage. In certain hydra lines, the plasmid integrated into the genome of one or a few adjacent cells that then underwent division. The GFP-producing cells (green) were visualized by fluorescence microscopy. In the specimen shown here, GFP was confined to certain internal endodermal epithelial cells. (*Inset*) A patch of GFP^+ endodermal epithelial cells visualized by confocal microscopy. Scale bar, 20 μm. With both imaging techniques, the GFP-producing cells can be followed during growth and differentiation of the organism. [From J. Wittlieb et al., 2006, *Proc. Nat'l Acad. Sci. USA* **103**:6208.]

tools for looking at the normal function and trafficking of native proteins. Cells in which this recombinant DNA has been introduced will synthesize the chimeric protein whose green fluorescence reveals the subcellular location of the protein of interest. This GFP-tagging technique, for example, has been used to visualize the expression and distribution of lamin A, a protein that lines the inner, or intranuclear, surface of the inner nuclear membrane; it also forms tubule-like structures that protrude into the nucleus (Figure 9-13).

In a variation of this technique, a cDNA encoding a protein of interest is modified at its C-terminus by the addition of a segment that encodes four cysteine residues in a defined sequence (Cys-Cys-Xaa-Xaa-Cys-Cys). Following expression of the recombinant protein in cultured cells, a chemically modified red-fluorescing dye (ReAsH) is added to the culture; this dye forms stable covalent bonds with the four cysteines in the tetracysteine sequence. Because this sequence is not found in natural proteins, the dye binds to no other cellular proteins. After the cells are washed in buffers to remove unattached dye, the protein of interest can be visualized by fluorescence microscopy on either live or fixed cells. The subcellular localization of the tagged protein in the same cells can also be visualized at a higher resolution by electron microscopy. Figure 9-14 shows the use of this fluorescent-tagging technique to localize **connexin,** a plasma membrane protein that is found in **gap junctions.** These cell-surface structures permit the rapid diffusion of small, water-soluble molecules between the cytoplasms of two adjacent cells (Chapter 19). Another example of this tagging technique is shown in the chapter opening figure. In this case, HeLa cells were transfected with a cDNAs encoding a β actin with a tetracysteine tag and then stained with ReAsH dye to label the β actin.

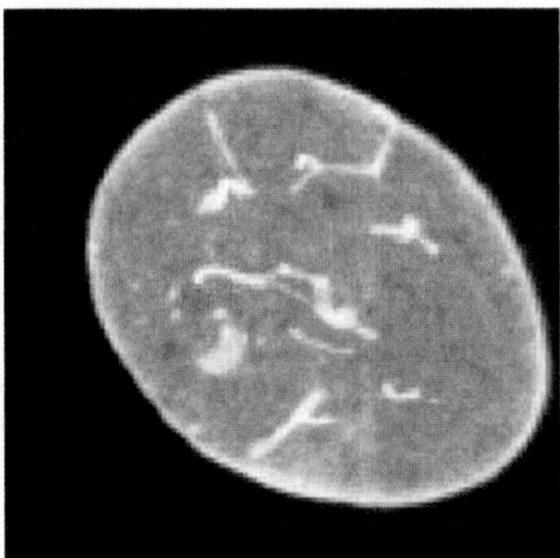

▲ EXPERIMENTAL FIGURE 9-13 **An optical section of a living CHO-K1 cell expressing a recombinant lamin A–GFP chimeric protein.** Shown here is the oval-shaped nucleus with lamin A (white) lining the inner nuclear membrane. Lamin A is also found in intra- and transnuclear tubule-like structures. The rest of the cell does not contain any lamin–GFP fusion protein and thus is black. [From J. L. V. Broers et. al., 1999, *J. Cell Sci.* **112:**3463.]

Podcast: Light and Electron Microscopy

MEDIA CONNECTIONS

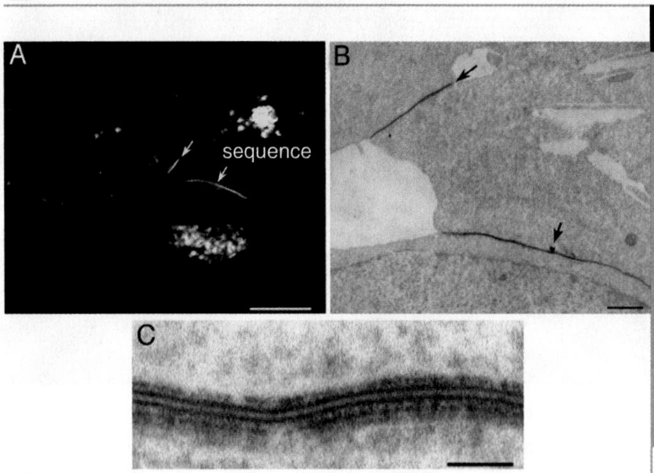

▲ EXPERIMENTAL FIGURE 9-14 **Detection of tetracysteine-tagged Connexin43, a gap-junction protein, by both light and electron microscopy.** The cDNA encoding Connexin43 (C×43) was extended at the C-terminus with a sequence that encodes a short peptide containing the Cys-Cys-Xaa-Xaa-Cys-Cys sequence. The recombinant cDNA was expressed in HeLa cells, which were then stained with ReAsH, a red-fluorescing dye that covalently binds only to proteins with this specific tetracysteine (TC) sequence. (a) Confocal image reveals two segments of the plasma membranes of adjacent cells that are connected by multiple gap junctions that contain the C×43-TC protein (bright lines indicated by white arrows). The cells were subsequently fixed with 2% glutaraldehyde, treated with the dye diaminobenzidine, and subjected to intense illumination. Under these conditions, the ReAsH dye undergoes a photoconversion that results in formation of a dense precipitate, which can be seen, after the cells are sectioned, in the electron microscope. (b) In this electron micrograph of the same area shown in (a) after photoconversion, the dark lines (black arrows) represent the gap junctions. (c) High-magnification view of the section depicted in panel (b) reveals many C×43-containing gap junction proteins in the plasma membranes of the two adjacent cells (dark stain). This image highlights the high resolution obtainable with the tetracysteine labeling method. Bar in (a), 10 μm; in (b), 1 μm; and in (c), 100 nm. [From G. Gaietta et al., 2002, *Science* **296:**503; see also B. N. G. Giepmans et al., 2006, *Science* Vol. 312, page 207.]

Determination of Intracellular Ca^{2+} and H^+ Levels with Ion-Sensitive Fluorescent Dyes The concentration of Ca^{2+} or H^+ within live cells can be measured with the aid of fluorescent dyes, or *fluorochromes,* whose fluorescence depends on the concentration of these ions. As discussed in later chapters, intracellular Ca^{2+} and H^+ concentrations have pronounced effects on many cellular processes. For instance, many hormones and other stimuli cause a rise in cytosolic Ca^{2+} from the resting level of about 10^{-7} M to 10^{-6} M, which induces various cellular responses including the contraction of muscle.

The fluorescent dye *fura-2,* which is sensitive to Ca^{2+}, contains five carboxylate groups that form ester linkages with ethanol. The resulting fura-2 ester is lipophilic and can

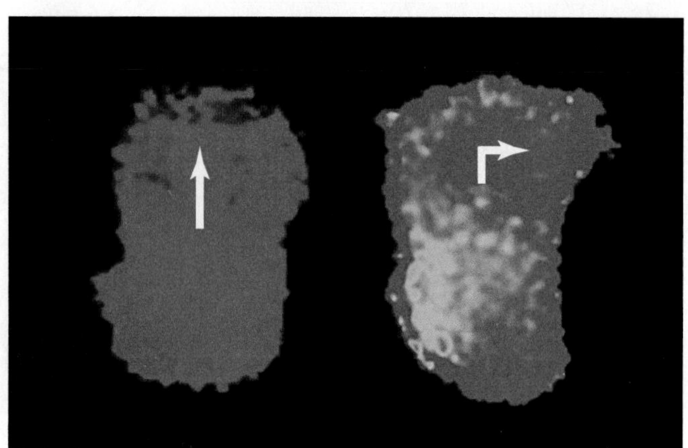

▲ **EXPERIMENTAL FIGURE 9-15 Fura-2, a Ca^{2+}-sensitive fluorochrome, can be used to monitor the relative concentrations of cytosolic Ca^{2+} in different regions of live cells.** (*Left*) In a moving leukocyte, a Ca^{2+} gradient is established. The highest levels (green) are at the rear of the cell, where cortical contractions take place, and the lowest levels (blue) are at the cell front, where actin undergoes polymerization. (*Right*) When a pipette filled with chemotactic molecules placed to the side of the cell induces the cell to turn, the Ca^{2+} concentration momentarily increases throughout the cytoplasm and a new gradient is established. The gradient is oriented such that the region of lowest Ca^{2+} (blue) lies in the direction that the cell will turn, whereas a region of high Ca^{2+} (yellow) always forms at the site that will become the rear of the cell. [From R. A. Brundage et al., 1991, *Science* **254:**703; courtesy of F. Fay.]

diffuse from the medium across the plasma membrane into cells. Within the cytosol, esterases hydrolyze fura-2 ester, yielding fura-2 whose free carboxylate groups render the molecule nonlipophilic and thus unable to cross cellular membranes, so it remains in the cytosol. Inside cells, each fura-2 molecule can bind a single Ca^{2+} ion but no other cellular cation. This binding, which is proportional to the cytosolic Ca^{2+} concentration over a certain range, increases the fluorescence of fura-2 at one particular wavelength. At a second wavelength, the fluorescence of fura-2 is the same whether or not Ca^{2+} is bound and provides a measure of the total amount of fura-2 in a region of the cell. By examining cells continuously in the fluorescence microscope and measuring rapid changes in the ratio of fura-2 fluorescence at these two wavelengths, one can quantify rapid changes in the fraction of fura-2 that has a bound Ca^{2+} ion and thus in the concentration of cytosolic Ca^{2+} (Figure 9-15).

Fluorescent dyes (e.g., SNARF-1) that are sensitive to the H^+ concentration can similarly be used to monitor the cytosolic pH of living cells. Other useful probes consist of a fluorochrome linked to a weak base that is only partially protonated at neutral pH and that can freely permeate cell membranes. In acidic organelles, however, these probes become protonated; because the protonated probes cannot recross the organelle membrane, they accumulate in the lumen in concentrations many fold greater than in the cytosol. Thus this type of fluorescent dye can be used to specifically stain lysosomes in living cells, as Figure 9-3 shows.

Imaging Subcellular Details Often Requires that the Samples Be Fixed, Sectioned, and Stained

Live cells and tissues generally lack compounds that absorb light and are thus nearly invisible in a light microscope. Although such specimens can be visualized by the special techniques we just discussed, these methods do not reveal the fine details of structure. Microscopy of live cells also requires that cells be housed in special glass-faced chambers, called culture chambers, that can be mounted on a microscope stage. For these reasons, cells are often fixed, sectioned, and stained to reveal subcellular structures.

Specimens for light and electron microscopy are commonly fixed with a solution containing chemicals that cross-link most proteins and nucleic acids. Formaldehyde, a common fixative, cross-links amino groups on adjacent molecules; these covalent bonds stabilize protein–protein and protein–nucleic acid interactions and render the molecules insoluble and stable for subsequent procedures. After fixation, a sample used for light microscopy is usually embedded in paraffin and cut into sections 0.5–50 μm thick (Figure 9-16). For electron microscopy samples are imbedded in liquid plastic and, after hardening, sections 50–100 nm thick are cut. Alternatively, the sample can be frozen without

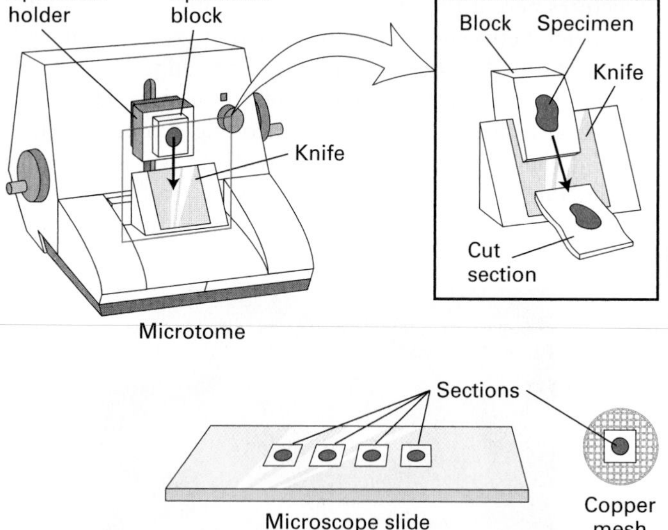

▲ **EXPERIMENTAL FIGURE 9-16 Tissues for microscopy are commonly fixed, embedded in a solid medium, and cut into thin sections.** A fixed tissue is dehydrated by soaking in a series of alcohol–water solutions, ending with an organic solvent compatible with the embedding medium. To embed the tissue for sectioning, the tissue is placed in liquid paraffin for light microscopy or in liquid plastic for electron microscopy. After the block containing the specimen has hardened, it is mounted on the arm of a microtome and slices are cut with a knife. Typical sections cut for light microscopy are 0.5–50 μm thick. Those cut for electron microscopy are generally 50–100 nm thick. The sections are collected either on microscope slides (light microscopy) or copper mesh grids (electron microscopy) and stained with an appropriate agent.

prior fixation and then sectioned; such treatment preserves the activity of enzymes for later detection by cytochemical reagents. Cultured cells growing on glass coverslips, as described later, are thin enough so they can be fixed in situ and visualized by light microscopy without the need for sectioning.

A final step in preparing a specimen for light microscopy is to stain it so as to visualize the main structural features of the cell or tissue. Many chemical stains bind to molecules that have specific features. For example, *hematoxylin* binds to basic amino acids (lysine and arginine) on many different kinds of proteins, whereas *eosin* binds to acidic molecules (such as DNA and side chains of aspartate and glutamate). Because of their different binding properties, these dyes stain various cell types sufficiently differently that they are distinguishable visually. If an enzyme catalyzes a reaction that produces a colored or otherwise visible precipitate from a colorless precursor, the enzyme may be detected in cell sections by their colored reaction products. Such staining techniques, although once quite common, have been largely replaced by other techniques for visualizing particular proteins, as discussed next.

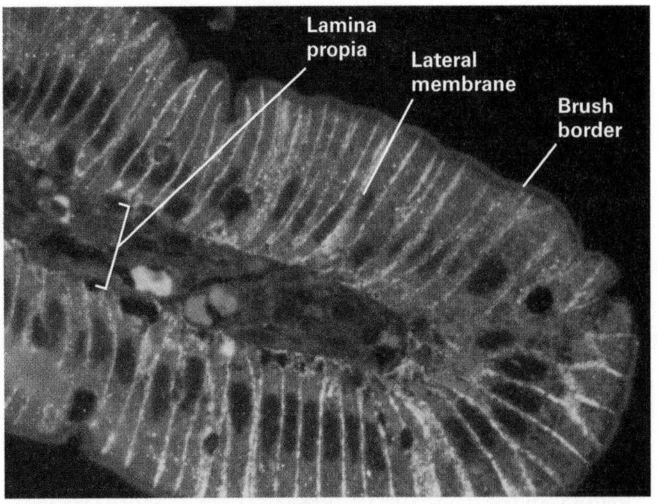

▲ EXPERIMENTAL FIGURE 9-17 **A specific protein can be localized in fixed tissue sections by immunofluorescence microscopy.** A section of the rat intestinal wall was stained with Evans blue, which generates a nonspecific red fluorescence, and with a yellow green–fluorescing antibody specific for GLUT2, a glucose transport protein. As evident from this fluorescence micrograph, GLUT2 is present in the basal and lateral sides of the intestinal cells but is absent from the brush border, composed of closely packed microvilli on the apical surface facing the intestinal lumen. Capillaries run through the lamina propria, a loose connective tissue beneath the epithelial layer. [See B. Thorens et al., 1990, *Am. J. Physio.* **259:**C279; courtesy of B. Thorens.]

Immunofluorescence Microscopy Can Detect Specific Proteins in Fixed Cells

The common chemical dyes just mentioned stain nucleic acids or broad classes of proteins, but investigators often want to detect the presence and location of specific proteins. A widely used method for this purpose employs specific antibodies that are detected by fluorescence. In one version of this technology, the antibody—generally a monoclonal antibody—is covalently linked to a fluorochrome. Commonly used fluorochromes include rhodamine and Texas red, which emit red light; Cy3, which emits orange light; and fluorescein, which emits green light. These fluorochromes can be chemically coupled to purified antibodies specific for almost any desired macromolecule. When a fluorochrome–antibody complex is added to a permeabilized cell or tissue section, the complex will bind to the corresponding antigens, which then light up when illuminated by the exciting wavelength, a technique called *immunofluorescence microscopy* (Figure 9-17). Staining a specimen with different dyes that fluoresce at different wavelengths allows multiple proteins as well as DNA to be localized within the same cell (see chapter opening figure).

In a variation of the immunofluorescence technique, a monoclonal or polyclonal antibody is applied to the fixed tissue section, followed by a second fluorochrome-tagged antibody that binds to the common (Fc) segment of the first antibody. For example, a "second" antibody (called "goat anti-rabbit") is prepared by immunizing a goat with the Fc segment that is common to all rabbit IgG antibodies; when coupled to a fluorochrome, this second antibody preparation will detect any rabbit antibody used to stain a tissue or cell. Because several goat anti-rabbit antibody molecules can bind to a single rabbit antibody molecule in a section, the fluorescence is often brighter than if just a protein-specific antibody directly coupled to a fluorochrome is used.

In another widely used version of this technology, cells are transfected with a cDNA encoding a recombinant protein that has appended, generally to one end, a short sequence of amino acids called an *epitope tag*. Two commonly used epitope tags are called FLAG, encoding the amino acid sequence DYKDDDDK (single-letter code), and myc, encoding the sequence EQKLISEEDL. Commercial fluorochrome-coupled monoclonal anti-epitope antibodies can then be used to detect the recombinant protein in the cell.

New types of fluorescence microscopy, such as one with the exotic name of SPED (stimulated emission depletion), enable two fluorescent objects to be resolved even if they are as close together as ≈20 nm, well below the ≈200-nm resolution limit for standard light microscopy. For example, nerve cells contain a class of membrane-lined vesicles, termed synaptic vesicles, that are too small (≈40 nm in diameter) and too densely packed to be resolved by available fluorescence microscopes. However, STED can resolve individual vesicles in nerve cells. This technique should also enable investigators to detect single fluorescent protein molecules in the membranes of purified organelles.

Confocal and Deconvolution Microscopy Enable Visualization of Three-Dimensional Objects

Conventional fluorescence microscopy has two major limitations. First, the fluorescent light emitted by a sample comes from molecules above and below the plane of focus; thus the observer sees a blurred image caused by the superposition of fluorescent images from molecules at many depths in the cell. The blurring effect makes it difficult to determine the actual molecular arrangements (Figure 9-18a). Second, to visualize thick specimens, consecutive (serial) sections must be prepared, imaged, and aligned to reconstruct structures in thick tissues. Two approaches can be used to avoid the preparation of serial sections and to obtain high-resolution three-dimensional information.

One approach, called *confocal microcopy* differs from conventional fluorescence microscopy by the use of a pinhole located in front of the detector that blocks light not originating from that focal plane. The resulting images do not contain blurs from structures above and below the current position of the focal plane (Figure 9-18b). The majority of confocal microscopes use a laser as the source of illumination; lasers provide a defined excitation wavelength and because of their focused energy are often well suited to penetrating thick specimens. Since a laser is focused on a single point on the specimen, it must be scanned across and down to build an image. The intensity of light from these in-focus areas is recorded by a photomultiplier tube, and the image is stored in a computer. The scanning process can often take a few minutes and so is not well-suited to imaging fast biological processes in living samples. A revolutionary newly developed variation of laser-scanning confocal microscopy is Nipkow confocal microscopy. This method uses a spinning disk with multiple pinholes that split the laser into hundreds of beamlets, increasing the rate at which the image is collected by a factor of 10 or more.

Images of the highest possible resolution currently available can be obtained by deconvolution, a computationally intensive mathematical process whereby blurred objects are sharpened. Similar algorithms are used by astronomers to sharpen images of distant stars. In *deconvolution microscopy,* a series of images of an object are taken at different focal planes with a conventional fluorescence microscope or confocal microscope. A separate series of images are made from a test slide containing tiny fluorescent beads smaller (e.g., 0.2 μm in diameter) than the resolution of the microscope. Each bead represents a pinpoint of light that becomes a blurred object because of the imperfect optics of the microscope; from these images a *point spread function* is determined that enables the investigator to calculate the distribution of fluorescent point sources that generated the "blur" in the object image. In other words, the light generating the blurred sample image from adjacent focal planes is reassigned to the correct focal plane via iterative comparison with the point spread function. Images restored by deconvolution display impressive detail without any blurring as illustrated in Figure 9-19.

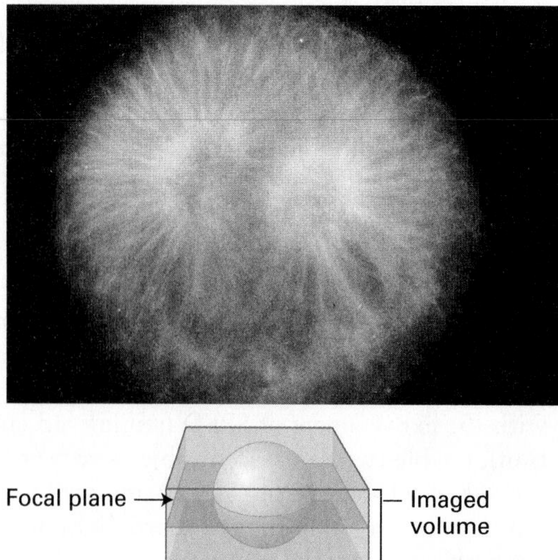

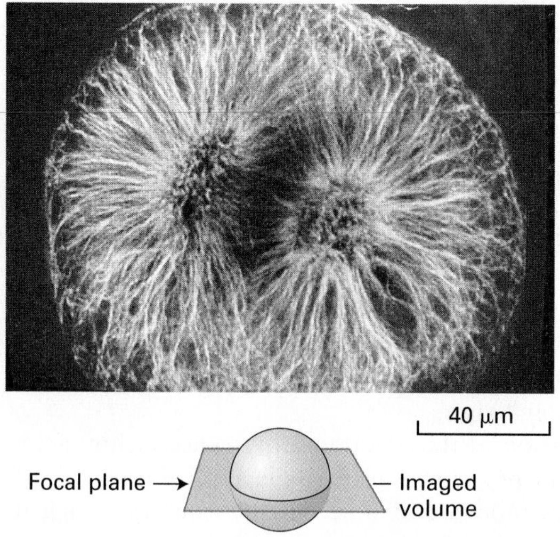

▲ **EXPERIMENTAL FIGURE 9-18 Confocal microscopy produces an in-focus optical section through thick cells.** A mitotic fertilized egg from a sea urchin (*Psammechinus*) was lysed with a detergent, exposed to an anti-tubulin antibody, and then exposed to a fluorescein-tagged antibody that binds to the anti-tubulin antibody. (a) When viewed by conventional fluorescence microscopy, the mitotic spindle is blurred. This blurring occurs because background fluorescence is detected from tubulin above and below the focal plane as depicted in the sketch. (b) The confocal microscopic image is sharp, particularly in the center of the mitotic spindle. In this case, fluorescence is detected only from molecules in the focal plane, generating a very thin optical section. [Micrographs from J. G. White et al., 1987, *J. Cell Biol.* **104**:41.]

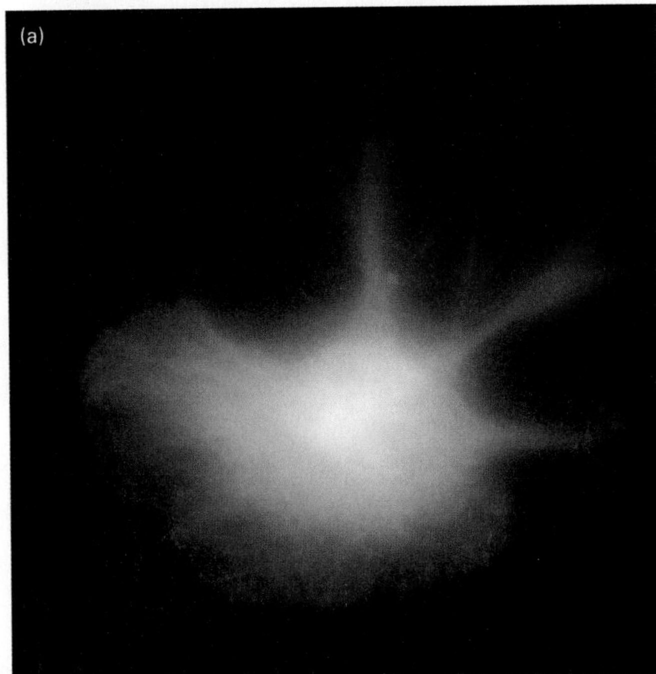

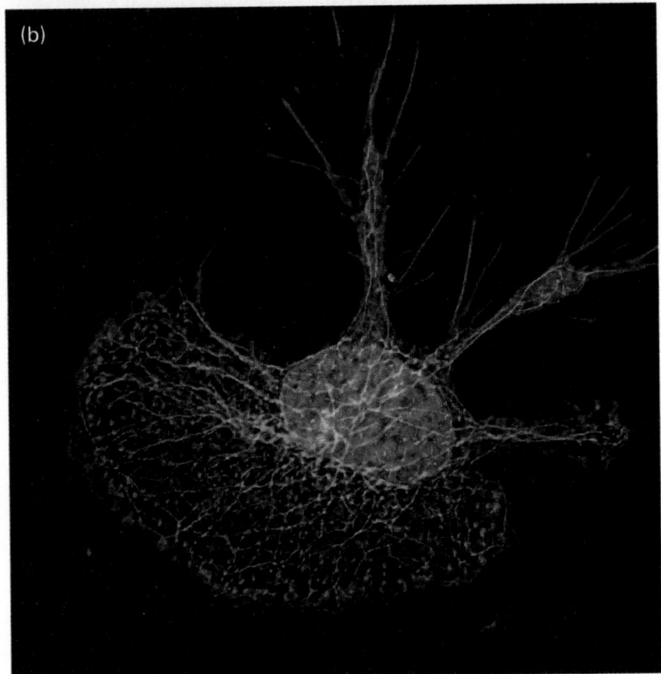

▲ EXPERIMENTAL FIGURE 9-19 **Deconvolution fluorescence microscopy yields high-resolution optical sections that can be reconstructed into one three-dimensional image.** A macrophage cell was stained with fluorochrome-labeled reagents specific for DNA (blue), microtubules (green), and actin microfilaments (red). The series of fluorescent images obtained at consecutive focal planes (optical sections) through the cell were recombined in three dimensions. (a) In this three-dimensional reconstruction of the raw images, the DNA, microtubules, and actin appear as diffuse zones in the cell. (b) After application of the deconvolution algorithm to the images, the fibrillar organization of microtubules and the localization of actin to adhesions become readily visible in the reconstruction. [Courtesy of J. Evans, Whitehead Institute.]

Graphics and Informatics Have Transformed Modern Microscopy

In the past decade, digital cameras have largely replaced optical cameras to record microscopy images. Digital images can be stored in a computer and manipulated by conventional photographic software as well as by specialized algorithms. As mentioned above, deconvolution algorithms can sharpen an image by restoring out-of-focus photons to their origin. Other computer algorithms, which previously required supercomputing facilities, enable visualization of intricate three-dimensional structures reconstructed on desktop computers (see Figure 9-19). Informatics including image analysis algorithms and statistical approaches allow quantitation of shapes, movements and signal intensities within objects such as cells or organelles.

KEY CONCEPTS OF SECTION 9.2

Light Microscopy: Visualizing Cell Structure and Localizing Proteins Within Cells

- The limit of resolution of a light microscope is about 200 nm.
- Because cells and tissues are almost transparent, various types of stains and optical techniques are used to generate sufficient contrast for imaging.
- Phase-contrast and differential interference contrast (DIC) microscopy are used to view the details of live, unstained cells and to monitor cell movement (see Figure 9-11).
- When proteins tagged with naturally occurring green fluorescent protein (GFP) or its variants are expressed in live cells, they can be visualized in a fluorescence microscope (see Figure 9-12).
- With the use of dyes whose fluorescence is proportional to the concentration of Ca^{2+} or H^+ ions, fluorescence microscopy can measure the local concentration of Ca^{2+} ions and intracellular pH in living cells.
- In immunofluorescence microscopy, specific proteins and organelles in fixed cells are stained with fluorochrome-labeled monoclonal antibodies. Multiple proteins can be localized in the same sample by staining with antibodies labeled with different fluorochromes (see chapter opening figure).
- Confocal microscopy and deconvolution microscopy use different methods to optically section a specimen, thereby reducing the blurring due to out-of-focus fluorescence light (see Figures 9-18 and 9-19). Both methods provide much sharper images, particularly of thick specimens, than does standard fluorescence or light microscopy.
- Advances in computation and graphics enable measurements to be extracted from microscope images.

9.3 Electron Microscopy: Methods and Applications

Electron microscopy of cells and tissues provides a much higher resolution of cell ultrastructure than can be obtained by light microscopy. However, the technology requires that the cell be fixed and sectioned, and thus living cells cannot be imaged. Researchers weigh these kinds of trade-offs and select from a variety of methods to produce images that answer their questions. For example, *immunoelectron microscopy* can be used to localize specific proteins at high resolution within cells. Two proteins, but generally not more, can be detected simultaneously, though the procedure is technically challenging. By comparison, fluorescence microscopy can be used to detect four or more proteins at the same time (see chapter opening figure) albeit at a lower resolution. Preparations of certain purified particles, such as viruses and ribosomes, can be visualized by electron microscopy without prior fixation or staining if the sample is frozen in liquid nitrogen and maintained in the frozen state while being observed.

Resolution of Transmission Electron Microscopy Is Vastly Greater Than That of Light Microscopy

The fundamental principles of electron microscopy are similar to those of light microscopy; the major difference is that electromagnetic lenses focus a high-velocity electron beam instead of the visible light used by optical lenses. In the *transmission electron microscope (TEM),* electrons are emitted from a filament and accelerated in an electric field. A condenser lens focuses the electron beam onto the sample; objective and projector lenses focus the electrons that pass through the specimen and project them onto a viewing screen or other detector (Figure 9-20, *left*). Because atoms in air absorb electrons, the entire tube between the electron source and the detector is maintained under an ultrahigh vacuum.

The short wavelength of electrons means that the limit of resolution for a transmission electron microscope is theoretically 0.005 nm (less than the diameter of a single atom), or 40,000 times better than the resolution of a light microscope, and 2 million times better than that of the unaided human eye. However, the effective resolution of the transmission electron microscope in the study of biological systems is considerably less than this ideal. Under optimal conditions, a resolution of 0.10 nm can be obtained with transmission electron microscopes, about 2000 times better than the best resolution of light microscopes. Several examples of cells and subcellular structures imaged by TEM were included in Section 9.1.

Because TEM requires very thin, fixed sections (about 50 nm), only a small part of a cell can be observed in any one section. Sectioned specimens are prepared in a manner similar to that used for light microscopy, by using a knife capable of producing sections 50–100 nm in thickness (see Figure 9-16). The generation of the image depends on differential scattering of the incident electrons by molecules in the preparation. Without staining, the beam of electrons passes through a specimen uniformly, and so the entire sample appears uniformly bright with little differentiation of components. To obtain useful images by TEM, samples are commonly stained with heavy metals such as lead and uranium, during or just after the fixation step. Metal-stained areas appear dark on a micrograph because the metals scatter (diffract) most of the incident electrons; scattered electrons are not focused by the electromagnetic

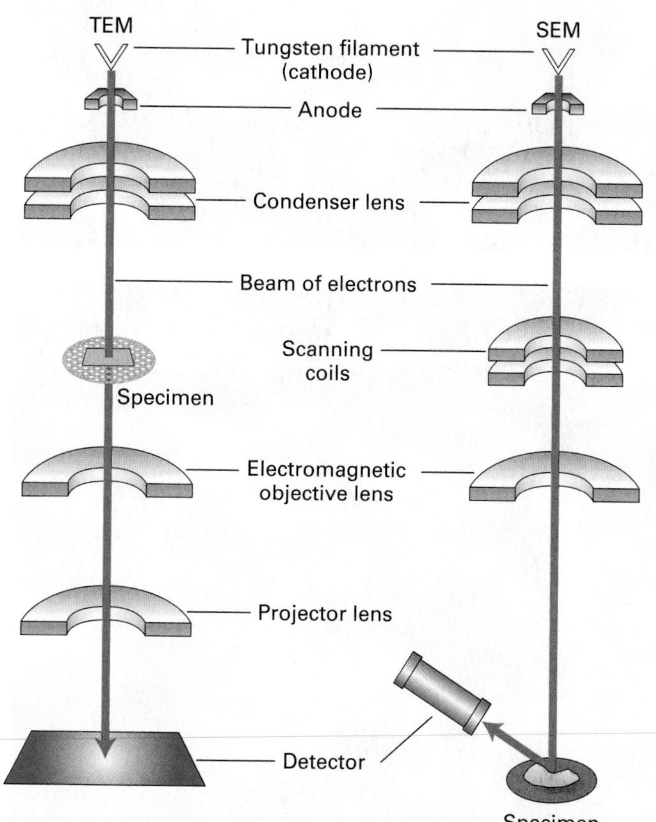

▲ EXPERIMENTAL FIGURE 9-20 **In electron microscopy, images are formed from electrons that pass through a specimen or are scattered from a metal-coated specimen.** In a transmission electron microscope (TEM), electrons are extracted from a heated filament, accelerated by an electric field, and focused on the specimen by a magnetic condenser lens. Electrons that pass through the specimen are focused by a series of magnetic objective and projector lenses to form a magnified image of the specimen on a detector, which may be a fluorescent viewing screen, photographic film, or a charged-couple-device (CCD) camera. In a scanning electron microscope (SEM), electrons are focused by condensor and objective lenses on a metal-coated specimen. Scanning coils move the beam across the specimen, and electrons scattered from the metal are collected by a photomultiplier tube detector. In both types of microscopes, because electrons are easily scattered by air molecules, the entire column is maintained at a very high vacuum.

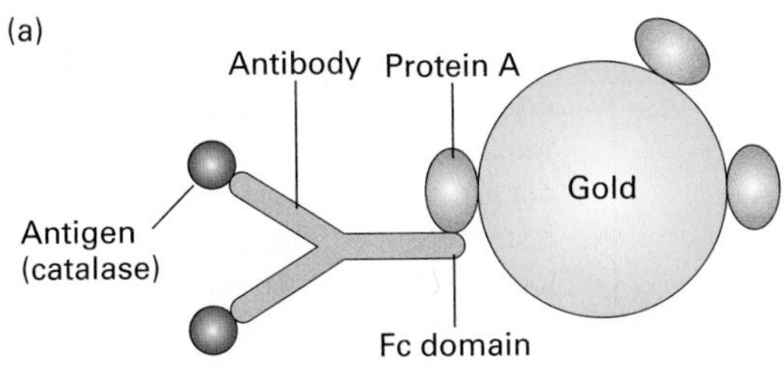

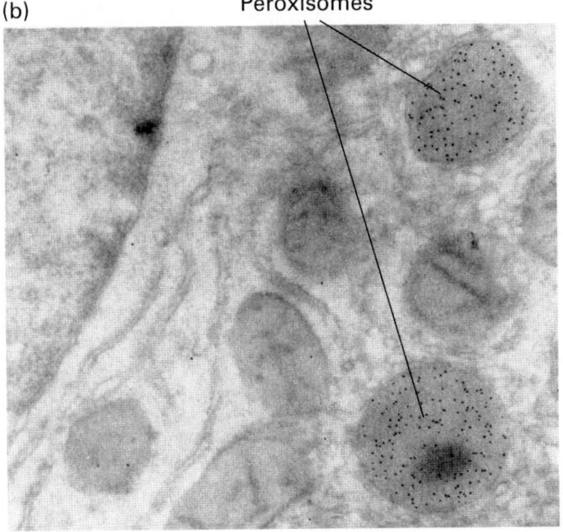

◀ EXPERIMENTAL FIGURE 9-21 **Gold particles coated with protein A are used to detect an antibody-bound protein by transmission electron microscopy.** (a) First antibodies are allowed to interact with their specific antigen (e.g., catalase) in a section of fixed tissue. Then the section is treated with electron-dense gold particles coated with protein A from the bacterium *S. aureus*. Binding of the bound protein A to the Fc domains of the antibody molecules makes the location of the target protein, catalase in this case, visible in the electron microscope. (b) A slice of liver tissue was fixed with glutaraldehyde, sectioned, and then treated as described in part (a) to localize catalase. The gold particles (black dots) indicating the presence of catalase are located exclusively in peroxisomes. [From H. J. Geuze et al., 1981, *J. Cell Biol.* **89**:653. Reproduced from the *Journal of Cell Biology* by copyright permission of The Rockefeller University Press.]

lenses and do not contribute to the image. Areas that take up less stain appear lighter. Osmium tetroxide preferentially stains certain cellular components, such as membranes (see Figure 10-6).

Specific proteins can be detected in thin sections using protein-specific antibodies. In this technique the cells are lightly fixed (to avoid denaturing epitopes on the desired protein) and then frozen at the temperature of liquid nitrogen and sectioned. After thawing, an antibody is applied to the section, which then is treated with electron-dense gold particles coated with protein A, a bacterial protein that binds the Fc segment of all antibody molecules (Figure 9-21). Because the gold particles diffract incident electrons, they appear as dark dots.

Cryoelectron Microscopy Allows Visualization of Particles Without Fixation or Staining

Standard transmission electron microscopy cannot be used to study live cells because they are generally too vulnerable to the required conditions and preparatory techniques. In particular, the absence of water causes macromolecules to become denatured and nonfunctional. However, hydrated, unfixed, and unstained biological specimens can be viewed directly in a transmission electron microscope if the sample is frozen. In this technique of *cryoelectron microscopy,* an aqueous suspension of a sample is applied to a grid in an extremely thin film, frozen in liquid nitrogen, and maintained in this state by means of a special mount. The frozen sample then is placed in the electron microscope. The very low temperature (−196 °C) keeps water from evaporating, even in a vacuum. Thus, the sample can be observed in detail in its native, hydrated state without fixing or heavy metal staining, although cells and some viruses subjected to this treatment will be killed. By computer-based averaging of hundreds of images, a three-dimensional model can be generated almost to atomic resolution. For example, this method has been used to generate models of ribosomes (see Figure 4-26), the muscle calcium pump discussed in Chapter 11, and other large proteins that are difficult to crystallize.

Many viruses contain coats, or capsids, that contain multiple copies of one or a few proteins arranged in a symmetric array. In a cryoelectron microscope, images of these particles can be viewed from a number of angles. A computer analysis of multiple images can make use of the symmetry of the particle to the calculate three-dimensional structure of the capsid to about 5-nm resolution. Examples of such images are shown in Figure 4-44.

An extension of this technique, *cryoelectron tomography,* allows researchers to determine the three-dimensional architecture of organelles or even whole cells embedded in ice, that is, in a state close to life. In this technique, the specimen holder is tilted in small increments around the axis perpendicular to the electron beam; thus images of the object viewed from different directions are obtained (Figure 9-22a, b). The images are then merged computationally into a three-dimensional reconstruction termed a *tomogram* (Figure 9-22c). A disadvantage of cryoelectron tomography is that the samples must be relatively thin, about 200 nm; this is much thinner than the samples (200 μm thick) that can be studied by confocal light microscopy.

(a)

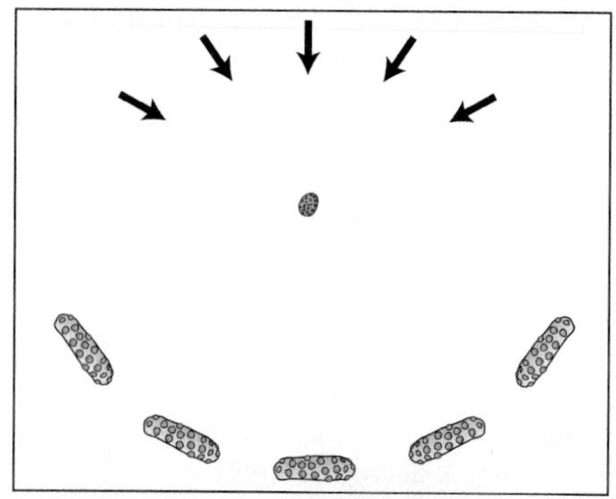

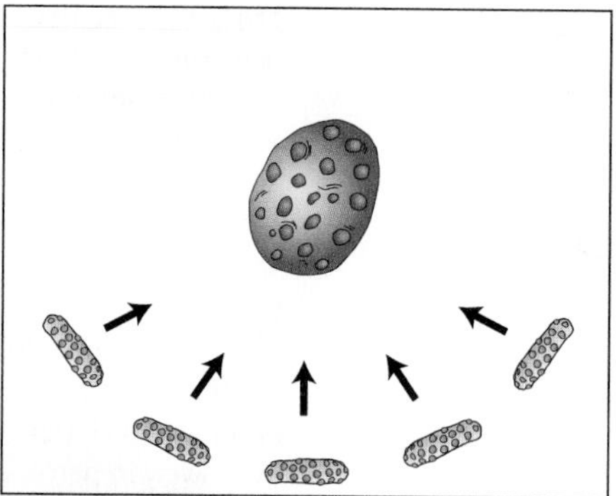

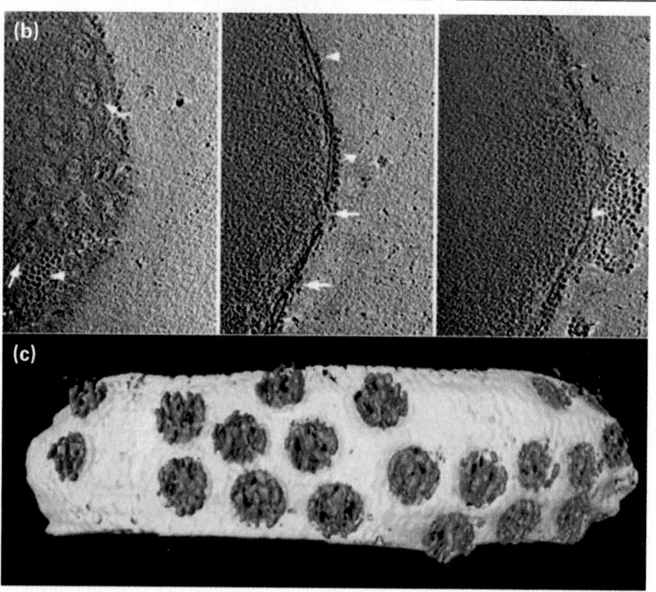

◀ **EXPERIMENTAL FIGURE 9-22 Structure of the nuclear pore complex (NPC) by cryoelectron tomography.** (a) In electron tomography, a semicircular series of two-dimensional projection images is recorded from the three-dimensional specimen that is located at the center; the specimen is tilted while the electron optics and detector remain stationary. The three-dimensional structure is computed from the individual two-dimensional images that are obtained when the object is imaged by electrons coming from different directions (arrows in left panel.) These individual images are used to generate a 3-dimensional image of the object (arrows, right panel) (b) Isolated nuclei from the cellular slime mold *Dictyostelium discoideum* were quick-frozen in liquid nitrogen and maintained in this state as the sample was observed in the electron microscope. The panel shows three sequential tilted images. Different orientations of NPCs (arrows) are shown in top-view (*left and center*) and side-view (*right*). Ribosomes connected to the outer nuclear membrane are visible, as is a patch of rough ER (arrowheads). (c) Computer-generated surface-rendered representation of a segment of the nuclear envelope membrane (yellow) studded with NPCs (blue). [Part (a) after S. Nickell et al., 2006, *Nature Rev. Mol. Cell. Biol.* 7:225. Parts (b) and (c) from M. Beck et al., 2004, *Science* **306:**1387.]

Electron Microscopy of Metal-Coated Specimens Can Reveal Surface Features of Cells and Their Components

Information about the shapes of purified viruses, fibers, enzymes, and other subcellular particles can be obtained with TEM by using *metal shadowing*. In this preparative technique, a thin layer of metal, such as platinum, is evaporated on a fixed or rapidly frozen biological sample (Figure 9-23a). Treatment with acid and bleach dissolves away the cell, leaving a metal replica that is viewed in a transmission electron microscope (Figure 9-23b).

Alternatively, the *scanning electron microscope* (SEM) allows investigators to view the surfaces of unsectioned metal-coated specimens. An intense electron beam inside the microscope scans rapidly over the sample. Molecules in the coating are excited and release secondary electrons that are focused onto a scintillation detector; the resulting signal is displayed on a cathode-ray tube much like a conventional television (see Figure 9-20, *right*). The resulting scanning electron micrograph has a three-dimensional appearance because the number of secondary electrons produced by any one point on the sample depends on the angle of the electron beam in relation to the surface (Figure 9-24). The resolving power of scanning electron microscopes, which is limited by the thickness of the metal coating, is only about 10 nm, much less than that of transmission instruments.

KEY CONCEPTS OF SECTION 9.3

Electron Microscopy: Methods and Applications

- Specimens for transmission electron microscopy (TEM) generally must be fixed, dehydrated, embedded, sectioned, and then stained with electron-dense heavy metals.
- Cryoelectron microscopy allows examination of hydrated, unfixed, and unstained biological specimens directly in a transmission electron microscope; samples are frozen in liquid nitrogen and maintained in this state.
- Surface details of objects can be revealed by transmission electron microscopy of metal-coated specimens.
- Scanning electron microscopy (SEM) of metal-coated unsectioned cells or tissues produces images that appear to be three-dimensional.

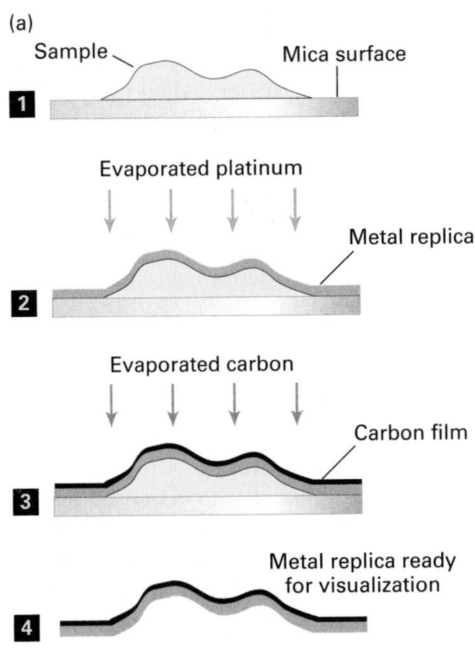

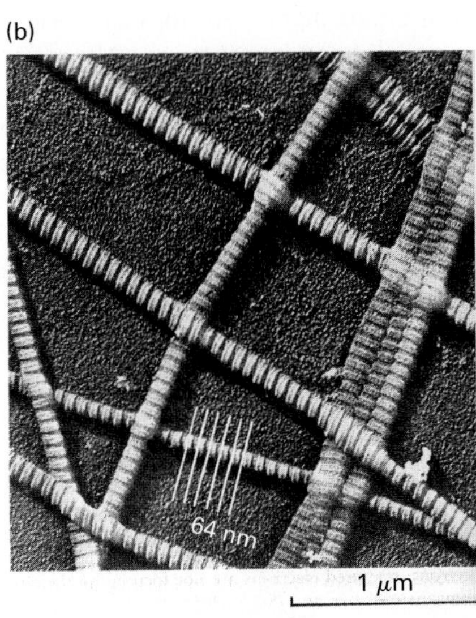

▲ EXPERIMENTAL FIGURE 9-23 **Metal shadowing makes surface details on very small objects visible by transmission electron microscopy.** (a) The sample is spread on a mica surface and then dried in a vacuum evaporator (1). The sample grid is coated with a thin film of a heavy metal, such as platinum or gold, evaporated from an electrically heated metal filament (2). To stabilize the replica, the specimen is then coated with a carbon film evaporated from an overhead electrode (3). The biological material is then dissolved by acid and bleach (4), which is viewed in a TEM. In electron micrographs of such preparations, the carbon-coated areas appear light—the reverse of micrographs of simple metal-stained preparations in which the areas of heaviest metal staining appear the darkest. (b) A platinum-shadowed replica of the substructural fibers of calfskin collagen, the major structural protein of tendons, bone, and similar tissues. The fibers are about 200 nm thick; a characteristic 64-nm repeated pattern (white parallel lines) is visible along the length of each fiber. [Courtesy of R. Bruns.]

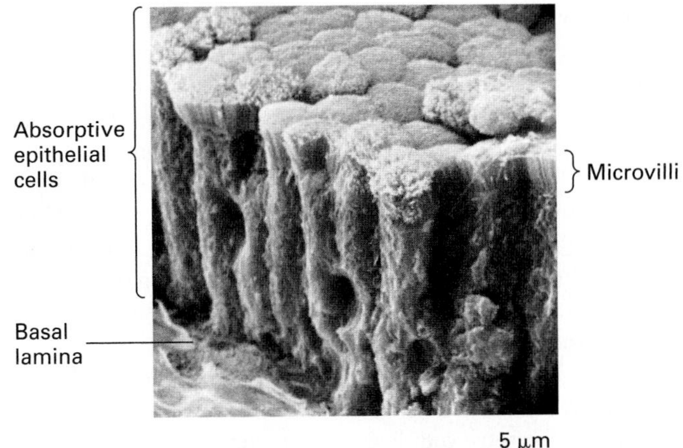

▲ EXPERIMENTAL FIGURE 9-24 **Scanning electron microscopy (SEM) produces a three-dimensional image of the surface of an unsectioned specimen.** Shown here is an SEM image of the epithelium lining the lumen of the intestine. Abundant fingerlike microvilli extend from the lumen-facing surface of each cell. The basal lamina beneath the epithelium helps support and anchor it to the underlying connective tissue. Compare this image of intestinal cells with that in Figure 9-17, a fluorescence micrograph. [From R. Kessel and R. Kardon, 1979, *Tissues and Organs: A Text-Atlas of Scanning Electron Microscopy,* W. H. Freeman and Company, p. 176.]

9.4 Purification of Cell Organelles

Many studies on cell structure and function require samples of a particular type of subcellular organelle. As one example, a recent proteomic study on mitochondria purified from mouse brain, heart, kidney, and liver revealed 591 mitochondrial proteins, including 163 proteins not previously associated with this organelle. Several proteins were found in mitochondria only in specific cell types. Determining the functions associated with these newly identified mitochondrial proteins is a major objective of current research on this organelle. In this section, we describe several commonly used techniques for separating different organelles.

Disruption of Cells Releases Their Organelles and Other Contents

The initial step in purifying subcellular structures is to release the cell's contents by rupturing the plasma membrane and the cell wall, if present. First, the cells are suspended in a solution of appropriate pH and salt content, usually isotonic sucrose (0.25 M) or a combination of salts similar in composition to those in the cell's interior. Many cells can then be broken by stirring the cell suspension in a high-speed blender or by exposing it to ultrahigh-frequency sound (*sonication*). Plasma membranes can also be sheared by special pressurized tissue homogenizers in which the cells are forced through a very narrow space between a plunger and the vessel wall; the pressure of being forced between the wall of the vessel and the plunger ruptures the cell.

Recall that water flows into cells when they are placed in a hypotonic solution, that is, one with a lower concentration of ions and small molecules than found inside the cell. This osmotic flow can be used to cause cells to swell, weakening the plasma membrane and facilitating its rupture. Generally, the cell solution is kept at 0 °C to best preserve enzymes and other constituents after their release from the stabilizing forces of the cell.

Disrupting the cell produces a mix of suspended cellular components, the homogenate, from which the desired organelles can be retrieved. Because rat liver contains an abundance of a single cell type, this tissue has been used in many classic studies of cell organelles. However, the same isolation principles apply to virtually all cells and tissues, and modifications of these cell-fractionation techniques can be used to separate and purify any desired components.

Centrifugation Can Separate Many Types of Organelles

In Chapter 3, we considered the principles of centrifugation and the uses of centrifugation techniques for separating proteins and nucleic acids. Similar approaches are used for separating and purifying the various organelles, which differ in both size and density and thus undergo sedimentation at different rates.

Most cell-fractionation procedures begin with *differential centrifugation* of a filtered cell homogenate at increasingly higher speeds (Figure 9-25). After centrifugation at each speed for an appropriate time, the liquid that remains at the top of the vessel, called the supernatant, is poured off and centrifuged at higher speed. The pelleted fractions obtained by differential centrifugation generally contain a mixture of organelles, although nuclei and viral particles can sometimes be purified completely by this procedure.

An impure organelle fraction obtained by differential centrifugation can be further purified by *equilibrium density-gradient centrifugation,* which separates cellular components according to their density. After the fraction is resuspended, it is layered on top of a solution that contains a gradient of a dense nonionic substance (e.g., sucrose or glycerol). The tube is centrifuged at a high speed (about 40,000 rpm) for several hours, allowing each particle to migrate to an equilibrium position where the density of the surrounding liquid is equal to the density of the particle (Figure 9-26). The different layers of liquid are then recovered by pumping out the contents of the centrifuge tube through a narrow piece of tubing.

Because each organelle has unique morphological features, the purity of organelle preparations can be assessed by examination in an electron microscope. Alternatively, organelle-specific marker molecules can be quantified. For example, the protein cytochrome *c* is present only in mitochondria; so the presence of this protein in a fraction of lysosomes would indicate its contamination by mitochondria. Similarly, catalase is present only in peroxisomes; acid phosphatase, only in lysosomes; and ribosomes, only in the rough endoplasmic reticulum or the cytosol.

▶ EXPERIMENTAL FIGURE 9-25 Differential centrifugation is a common first step in fractionating a cell homogenate. The homogenate resulting from disrupting cells is usually filtered to remove unbroken cells and then centrifuged at a fairly low speed to selectively pellet the nucleus—the largest organelle. The undeposited material (the supernatant) is next centrifuged at a higher speed to sediment the mitochondria, chloroplasts, lysosomes, and peroxisomes. Subsequent centrifugation in the ultracentrifuge at 100,000g for 60 minutes results in deposition of the plasma membrane, fragments of the endoplasmic reticulum, and large polyribosomes. The recovery of ribosomal subunits, small polyribosomes, and particles such as complexes of enzymes requires additional centrifugation at still higher speeds. Only the cytosol—the soluble aqueous part of the cytoplasm—remains in the supernatant after centrifugation at 300,000g for 2 hours.

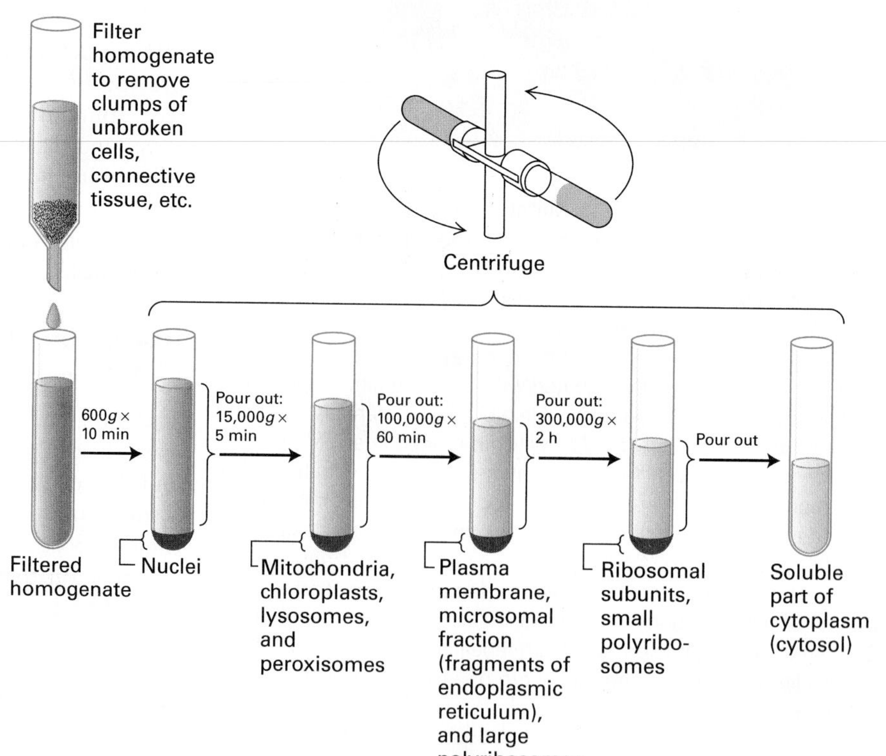

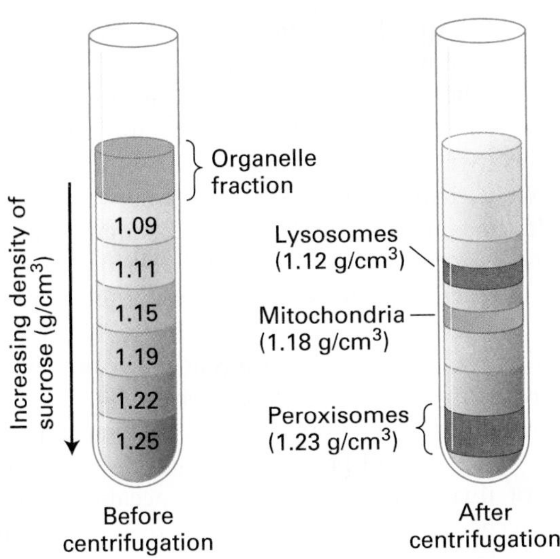

◀ EXPERIMENTAL FIGURE 9-26 **A mixed organelle fraction can be further separated by equilibrium density-gradient centrifugation.** In this example, utilizing rat liver, material in the pellet from centrifugation at 15,000g (see Figure 9-25) is resuspended and layered on a gradient of increasingly more dense sucrose solutions in a centrifuge tube. During centrifugation for several hours, each organelle migrates to its appropriate equilibrium density and remains there. To obtain a good separation of lysosomes from mitochondria, the liver is perfused with a solution containing a small amount of detergent before the tissue is disrupted. During this perfusion period, detergent is taken into the cells by endocytosis and transferred to the lysosomes, making them less dense than they would normally be and permitting a "clean" separation of lysosomes from mitochondria.

Organelle-Specific Antibodies Are Useful in Preparing Highly Purified Organelles

Cell fractions remaining after differential and equilibrium density-gradient centrifugation usually contain more than one type of organelle. Monoclonal antibodies for various organelle-specific membrane proteins are a powerful tool for further purifying such fractions. One example is the purification of vesicles whose outer surface is covered with the protein **clathrin**; these coated vesicles, which participate in formation of endosomes from the plasma membrane (see Figure 9-2), are discussed in detail in Chapter 14. An antibody to clathrin, bound to a bacterial carrier, can selectively bind these vesicles in a crude preparation of membranes, and the whole antibody complex can then be isolated by low-speed centrifugation (Figure 9-27). A related technique uses tiny metallic beads coated with specific antibodies. Organelles that bind to the antibodies, and are thus linked to the metallic beads, are recovered from

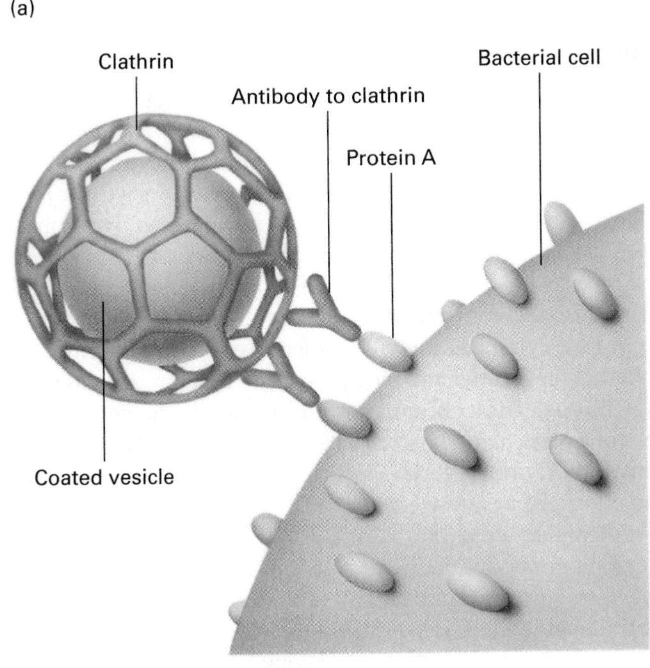

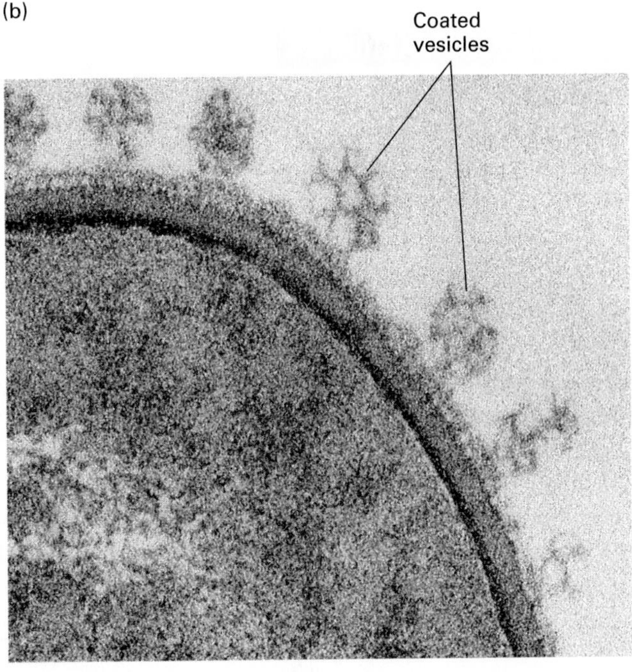

▲ EXPERIMENTAL FIGURE 9-27 **Small coated vesicles can be purified by binding of antibody specific for a vesicle surface protein and linkage to bacterial cells.** In this example, a suspension of membranes from rat liver is incubated with an antibody specific for clathrin, a protein that coats the outer surface of certain cytosolic vesicles. To this mixture is added a suspension of *Staphylococcus aureus* bacteria whose surface membrane contains protein A, which binds to the Fc constant region of antibodies. (a) Interaction of protein A with antibodies bound to clathrin-coated vesicles links the vesicles to the bacterial cells. The vesicle–bacteria complexes can then be recovered by low-speed centrifugation. (b) A thin-section electron micrograph reveals clathrin-coated vesicles bound to an *S. aureus* cell. [See E. Merisko et al., 1982, *J. Cell Biol.* 93:846. Micrograph courtesy of G. Palade.]

the preparation by adhesion to a small magnet on the side of the test tube.

All cells contain a dozen or more different types of small membrane-limited vesicles of about the same size (50–100 nm in diameter) and density, which makes them difficult to separate from one another by centrifugation techniques. Immunological techniques are particularly useful for purifying specific classes of such vesicles. Fat and muscle cells, for instance, contain a particular glucose transporter (GLUT4) that is localized to the membrane of one of these types of vesicle. When insulin is added to the cells, these vesicles fuse with the cell-surface membrane and increase the number of glucose transporters able to take up glucose from the blood. As we will see in Chapter 15, this process is critical to maintaining the appropriate concentration of sugar in the blood. The GLUT4-containing vesicles can be purified by using an antibody that binds to a segment of the GLUT4 protein that faces the cytosol. Likewise, the various transport vesicles discussed in Chapter 14 are characterized by unique surface proteins that permit their separation with the aid of specific antibodies.

A variation of this technique is employed when no antibody specific for the organelle under study is available. A gene encoding an organelle-specific membrane protein is modified by the addition of a segment encoding an epitope tag; the tag is placed on a segment of the protein that faces the cytosol. Following stable expression of the recombinant protein in the cell under study, an anti-epitope monoclonal antibody (described above) can be used to purify the organelle.

KEY CONCEPTS OF SECTION 9.4

Purification of Cell Organelles

- Disruption of cells by vigorous homogenization, sonication, or other techniques releases their organelles. Swelling of cells in a hypotonic solution weakens the plasma membrane, making it easier to rupture.
- Sequential differential centrifugation of a cell homogenate yields fractions of partly purified organelles that differ in mass and density (see Figure 9-25).
- Equilibrium density-gradient centrifugation, which separates cellular components according to their densities, can further purify cell fractions obtained by differential centrifugation (see Figure 9-26).
- Immunological techniques using antibodies against organelle-specific membrane proteins are particularly useful in purifying organelles and vesicles of similar sizes and densities (see Figure 9-27).

9.5 Isolation, Culture, and Differentiation of Metazoan Cells

Most animal and plant tissues contain multiple types of cells, but biochemical and molecular investigations are best accomplished on homogenous cell populations. In the first part of this section we describe a powerful instrument, the **fluorescence-activated cell sorter (FACS)**, in which cells that express specific cell-surface proteins can be separated from a complex mixture of cells. With this instrument, an investigator can isolate homogenous populations of specific cell types for study. Next we discuss techniques for culturing *primary cells*—that is, maintaining cells directly isolated from an animal in the laboratory under conditions that permit their survival and growth for at least several divisions.

Certain types of primary cells, especially those from embryos, can undergo differentiation in culture. As an example, we will describe how muscle cell precursors grown in culture can differentiate and form apparently normal muscle cells, providing a good system for studying this developmental process. Although many types of primary cells undergo only a limited number of divisions in culture, some accumulate cancer-causing (oncogenic) mutations that allow them to be cultured indefinitely. In many cases a single cell can be readily grown into a colony of identical cells, a process called *cell cloning.* Because these cells are genetically homogeneous, they are particularly suitable for many types of biochemical and genetic studies. Certain cloned cells can undergo differentiation into specific cells types such as adipocytes (fat-storing cells), nerve, or muscle, allowing studies on the mechanism of cell differentiation to be conducted on homogenous cell populations. Many times in this chapter we have shown how monoclonal antibodies facilitate cell biological experiments; at the end of this section we describe how special cultured cells are used to generate these antibodies.

Flow Cytometry Separates Different Cell Types

Some cell types differ sufficiently in density that they can be separated on the basis of this physical property. White blood cells (leukocytes) and red blood cells (erythrocytes), for instance, have very different densities because erythrocytes have no nucleus; thus these cells can be separated by equilibrium density centrifugation (described above). Because most cell types cannot be differentiated so easily, other techniques such as flow cytometry must be used to separate them.

A flow cytometer identifies different cells by measuring the light that they scatter and the fluorescence that they emit as they flow through a laser beam; thus it can quantify the numbers of cells of a particular type from a mixture. Indeed, a fluorescence-activated cell sorter (FACS), which is based on flow cytometry, can select one or a few cells from thousands of other cells and sort them into a separate culture dish (Figure 9-28). For example, if an antibody specific to a certain cell-surface molecule is linked to a fluorescent dye, any cell bearing this molecule will bind the antibody and will then be separated from other cells when it fluoresces in the FACS. Having been sorted from other cells, the selected cells can be grown in culture.

The FACS procedure is commonly used to purify the different types of white blood cells, each of which bears on its surface one or more distinctive proteins and will thus bind monoclonal antibodies specific for that protein. Only the T cells of the immune system, for instance, have both CD3 and Thy1.2 proteins on their surfaces. The presence of these surface proteins allows T cells to be separated easily from other

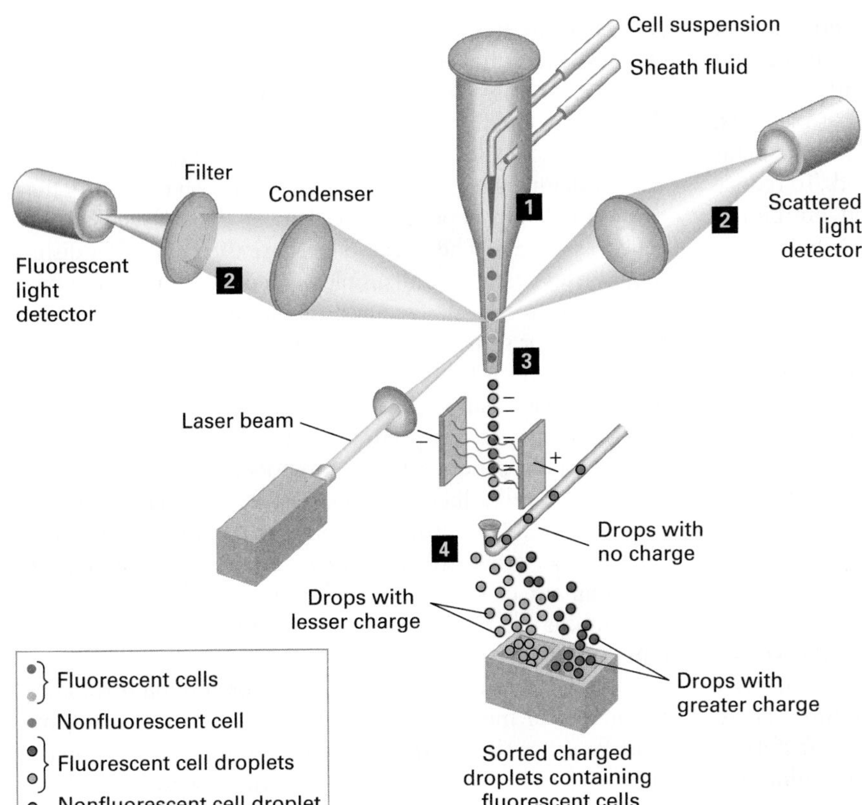

◀ EXPERIMENTAL FIGURE 9-28 **Fluorescence-activated cell sorter (FACS) separates cells that are labeled differentially with a fluorescent reagent.** Step **1**: A concentrated suspension of labeled cells is mixed with a buffer (the sheath fluid) so that the cells pass single-file through a laser light beam. Step **2**: Both the fluorescent light emitted and the light scattered by each cell are measured; from measurements of the scattered light, the size and shape of the cell can be determined. Step **3**: The suspension is then forced through a nozzle, which forms tiny droplets containing at most a single cell. At the time of formation, each droplet is given a negative electric charge proportional to the amount of fluorescence of its cell. Step **4**: Droplets with no charge and those with different electric charges are separated by an electric field and collected. Because it takes only milliseconds to sort each droplet, as many as 10 million cells per hour can pass through the machine. [Adapted from D. R. Parks and L. A. Herzenberg, 1982, *Meth. Cell Biol.* **26**:283.]

types of blood cells or spleen cells (Figure 9-29). Small magnetic beads may be used in a variation of this process that does not involve flow cytometry. The beads are coated with a monoclonal antibody specific for a surface protein such as CD3 or Thy1.2. Only cells with these proteins will stick to the beads and can be recovered from the preparation by adhesion to a small magnet on the side of the test tube.

Other uses of flow cytometry include the measurement of a cell's DNA and RNA content and the determination of its general shape and size. The FACS can make simultaneous measurements of the size of a cell (from the amount of scattered light) and the amount of DNA that it contains (from the amount of fluorescence emitted from a DNA-binding dye). Measurements of the DNA content of individual cells are used to follow replication of DNA as the cells progress through the cell cycle (Chapter 20).

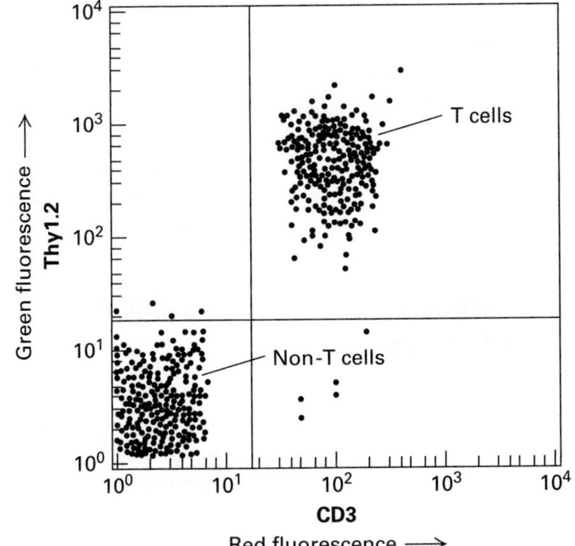

◀ EXPERIMENTAL FIGURE 9-29 **T cells bound to fluorescence-tagged antibodies to two cell-surface proteins are separated from other white blood cells by FACS.** Spleen cells from a mouse were treated with a red fluorescent monoclonal antibody specific for the CD3 cell-surface protein and with a green fluorescent monoclonal antibody specific for a second cell-surface protein, Thy1.2. As the cells were passed through a FACS machine, the intensity of the green and red fluorescence emitted by each cell was recorded. Each dot represents a single cell. This plot of the green fluorescence (vertical axis) versus red fluorescence (horizontal axis) for thousands of spleen cells shows that about half of them—the T cells—express both CD3 and Thy1.2 proteins on their surfaces (upper-right quadrant). The remaining cells, which exhibit low fluorescence (lower-left quadrant), express only background levels of these proteins and are other types of white blood cells. Note the logarithmic scale on both axes. [Courtesy of Chengcheng Zhang, Whitehead Institute.]

Culture of Animal Cells Requires Nutrient-Rich Media and Special Solid Surfaces

In contrast with most bacterial cells, which can be cultured quite easily, animal cells require many specialized nutrients and often specially coated dishes for successful culturing. To permit the survival and normal function of cultured tissues or cells, the temperature, pH, ionic strength, and access to essential nutrients must simulate as closely as possible the conditions within an intact organism. Isolated animal cells are typically placed in a nutrient-rich liquid, called the culture medium, within specially treated plastic dishes or flasks. The cultures are kept in incubators in which the temperature, atmosphere, and humidity can be controlled. To reduce the chances of bacterial or fungal contamination, antibiotics are often added to the culture medium. To further guard against contamination, investigators usually transfer cells between dishes, add reagents to the culture medium, and otherwise manipulate the specimens within special cabinets containing circulating air that is filtered to remove microorganisms and other airborne contaminants.

Media for culturing animal cells must supply the nine amino acids that cannot be synthesized by adult vertebrate animal cells. In addition, most cultured cells require three other amino acids that are synthesized only by specialized cells in intact animals. The other necessary components of a medium for culturing animal cells are vitamins, various salts, fatty acids, glucose, and serum—the fluid remaining after the noncellular part of blood (plasma) has been allowed to clot. Serum contains various protein factors that are needed for the proliferation of mammalian cells in culture. These factors include the polypeptide hormone insulin; transferrin, which supplies iron in a bioaccessible form; and numerous growth factors. In addition, certain cell types require specialized protein growth factors not present in serum. For instance, progenitors of red blood cells require erythropoietin, and T lymphocytes require interleukin 2 (Chapter 16). A few mammalian cell types can be grown in a chemically defined, serum-free medium containing amino acids, glucose, vitamins, and salts plus certain trace minerals, specific protein growth factors, and other components.

Unlike bacterial and yeast cells, which can be grown in suspension, most animal cells will grow only on a solid surface. This requirement highlights the importance of cell-surface proteins, called **cell-adhesion molecules (CAMs)**, that normally bind cells to other cells or proteins such as collagen or fibronectin in the extracellular matrix surrounding many animal cells (Chapter 19). Many types of cells can attach to and grow on glass or on specially treated plastics. Many cells secrete extracellular matrix (ECM) components, which adhere to these surfaces, and help the cells attach to the culture dish and grow. A single cell cultured on a glass or a plastic dish proliferates to form a visible mass, or *colony,* containing thousands of genetically identical cells in 4–14 days, depending on the growth rate. Some specialized blood cells and tumor cells can be maintained or grown in suspension as single cells.

Primary Cell Cultures Can Be Used to Study Cell Differentiation

Normal animal tissues (e.g., skin, kidney, liver) or whole embryos are commonly used to establish *primary cell cultures.* To prepare tissue cells for a primary culture, the cell–cell and cell–matrix interactions must be broken. To do so, tissue fragments are treated with a combination of a protease (e.g., trypsin, the collagen-hydrolyzing enzyme collagenase, or both) and a divalent cation chelator (e.g., EDTA) that depletes the medium of usable Ca^{2+} and Mg^{2+}. Many cell-adhesion molecules require calcium and are thus inactivated when calcium is removed; other cell-adhesion molecules that are not calcium dependent need to be proteolyzed for the cells to separate. The released cells are then placed in dishes in a nutrient-rich, serum-supplemented medium, where they can adhere to the surface and one another. The same protease/chelator solution is used to remove adherent cells from a culture dish for biochemical studies or subculturing (transfer to another dish).

Fibroblasts are the predominant cells in connective tissue and normally produce ECM components such as collagen that bind to cell-adhesion molecules, thereby anchoring cells to a surface. In culture, fibroblasts usually divide more rapidly than other cells in a tissue, eventually becoming the predominant cell type in a primary culture unless special precautions are taken to remove them when isolating other types of cells.

Several regions of vertebrate embryos contain *myoblasts,* precursors of muscle cells that both divide rapidly and migrate to other sites in the embryo. At some point myoblasts align with each other, cease division, and fuse to form a large multinucleate cell called a *myotube.* Concomitant with fusion, the cells induce synthesis of dozens of muscle-specific proteins necessary for further muscle development and function. Similar mononucleated cells, termed *satellite cells,* are found in adult muscle and can fuse to form multinucleated myotubes and simultaneously induce synthesis of muscle-specific proteins such as myosin heavy chain. As Figure 9-30 shows, this developmental process can be recapitulated in culture. When placed in the appropriate culture medium, primary mouse satellite cells fuse (Figures 9-30a, b) and differentiate into muscle cells (Figures 9-30c, d). This system has enabled investigators to elucidate the role of specific transcription factors and cell-adhesion molecules in controlling muscle development.

Certain cells from blood, spleen, or bone marrow adhere poorly, if at all, to a culture dish but nonetheless grow well in culture. In the body, such *nonadherent* cells are held in suspension (in the blood) or they are loosely adherent (in the bone marrow and spleen). Because some of these cells represent immature stages in the development of differentiated blood cells, they are very useful for studying normal blood cell differentiation and the abnormal development of leukemias.

Primary Cell Cultures and Cell Strains Have a Finite Life Span

When cells removed from an embryo or an adult animal are cultured, most of the adherent ones will divide a finite number of times and then cease growing (cell senescence). For instance, human fetal fibroblasts divide about 50 times

Undifferentiated cells

(a)

100 μm

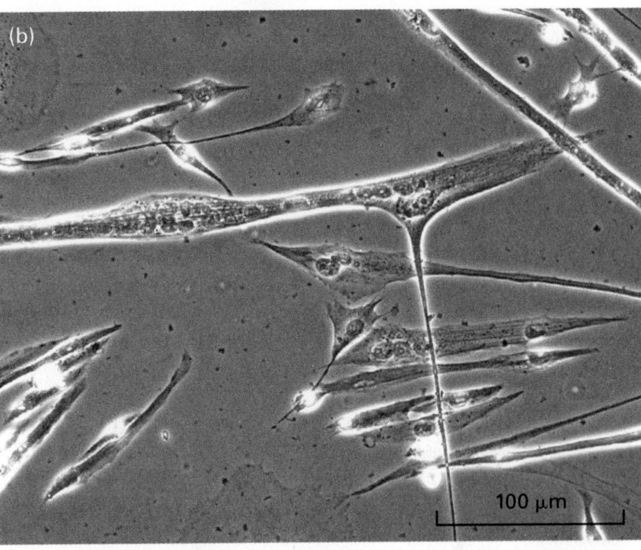

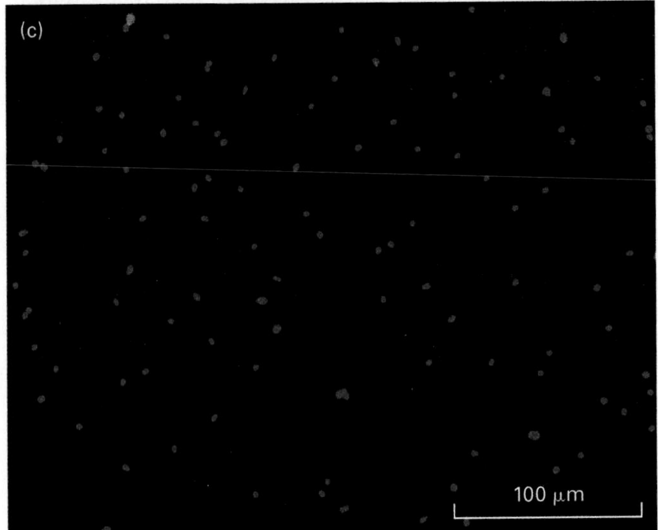

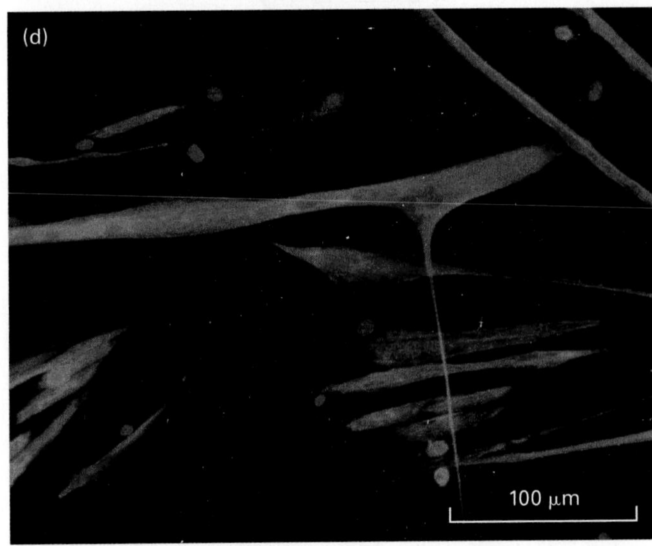

▲ EXPERIMENTAL FIGURE 9-30 **Differentiation in culture of primary mouse satellite (myoblast) cells into muscle cells.** Primary mouse satellite cells were cultured under nondifferentiating proliferation conditions (a,c) or differentiation conditions (b,d). (a,b) Phase-contrast images show formation of multinucleated syncytia during muscle differentiation. (c,d) Staining with a red-fluorescing antibody specific for myosin heavy chain reveals that this muscle-specific protein is induced during differentiation. Staining with Hoechst dye (blue) detects nuclei. Bar = 100 μm. [Courtesy Charles Emerson and Jennifer Chen, Boston Biomedical Research Institute.]

before they cease growth (Figure 9-31a). Starting with 10^6 cells, 50 doublings can produce $10^6 \times 2^{50}$, or more than 10^{20} cells, which is equivalent to the weight of about 1000 people. Normally, only a very small fraction of these cells are used in any one experiment. Thus, even though its lifetime is limited, a single culture, if carefully maintained, can be studied through many generations. Such a lineage of cells originating from one initial primary culture is called a **cell strain.**

Research with cell strains is simplified by the ability to freeze and successfully thaw them at a later time for experimental analysis. Cell strains can be frozen in a state of suspended animation and stored for extended periods at liquid nitrogen temperature, provided that a preservative that prevents the formation of damaging ice crystals is used. Although some cells do not survive thawing, many do survive and resume growth.

Transformed Cells Can Grow Indefinitely in Culture

To be able to clone individual cells, modify cell behavior, or select mutants, biologists often want to maintain cell

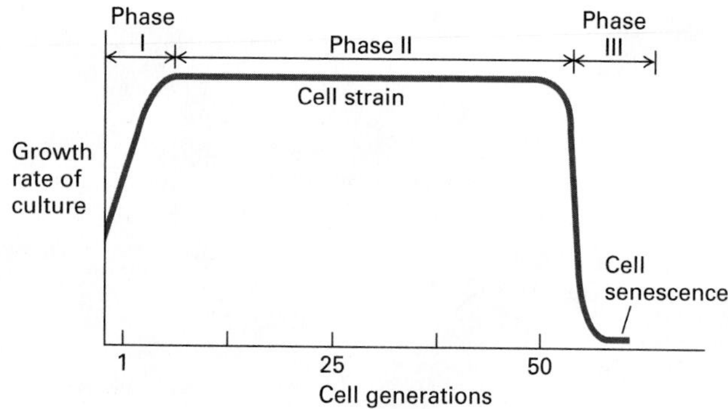

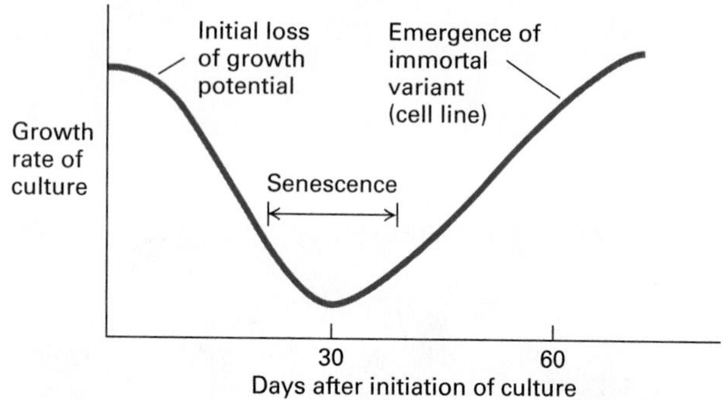

▲ **FIGURE 9-31 Stages in the establishment of a cell culture.** (a) When cells isolated from human tissue are initially cultured, some cells die and others (mainly fibroblasts) start to grow; overall, the growth rate increases (phase I). If the remaining cells are harvested, diluted, and replated into dishes again and again, the cell strain continues to divide at a constant rate for about 50 cell generations (phase II), after which the growth rate falls rapidly. In the ensuing period (phase III), all the cells in the culture stop growing (senescence). (b) In a culture prepared from mouse or other rodent cells, initial cell death (not shown) is coupled with the emergence of healthy growing cells. As these dividing cells are diluted and allowed to continue growth, they soon begin to lose growth potential, and most stop growing (i.e., the culture goes into senescence). Very rare cells undergo oncogenic mutations that allow them to survive and continue dividing until their progeny overgrow the culture. These cells constitute a cell line, which will grow indefinitely if it is appropriately diluted and fed with nutrients. Such cells are said to be immortal.

cultures for many more than 100 doublings. Such prolonged growth is exhibited by cells derived from some tumors. In addition, rare cells in a population of primary cells may undergo spontaneous oncogenic mutations, leading to oncogenic **transformation** (Chapter 25). Such cells, said to be oncogenically transformed or simply *transformed,* are able to grow indefinitely. A culture of cells with an indefinite life span is considered immortal and is called a **cell line.**

The HeLa cell line, the first human cell line, was originally obtained in 1952 from a malignant tumor (carcinoma) of the uterine cervix. Although primary cell cultures of normal human cells rarely undergo transformation into a cell line, rodent cells commonly do. After rodent cells are grown in culture for several generations, the culture goes into senescence (Figure 9-31b). During this period, most of the cells stop growing, but often a rapidly dividing transformed cell arises spontaneously and takes over, or overgrows, the culture. A cell line derived from such a transformed variant will grow indefinitely if provided with the necessary nutrients.

Regardless of the source, cells in immortalized lines often have chromosomes with abnormal DNA sequences. In addition, the number of chromosomes in such cells is usually greater than that in the normal cell from which they arose, and the chromosome number expands and contracts as the cells continue to divide in culture. A noteworthy exception is the Chinese hamster ovary (CHO) line and its derivatives, which have fewer chromosomes than their hamster progenitors. Cells with an abnormal number of chromosomes are said to be *aneuploid.*

Some Cell Lines Undergo Differentiation in Culture

Most cell lines have lost some or many of the functions characteristic of the differentiated cells from which they were derived. Such relatively undifferentiated cells are poor models for investigating the normal functions of specific cell types. Better in this regard are several more-differentiated cell lines that exhibit many properties of normal nontransformed cells. These lines include the human liver tumor (hepatoma) HepG2 line, which synthesizes most of the serum proteins made by normal liver cells (hepatocytes) and have been employed in studies identifying transcription factors that regulate synthesis of liver proteins.

More useful for cell and developmental biologists are cell lines that grow without acquiring characteristics of a differentiated cell, yet can undergo differentiation into a particular cell type when placed in a different culture medium. Because large numbers of such cells can be induced to undergo synchronized differentiation, they are often used in biochemical and molecular studies.

One example is a line of transformed mouse myoblasts termed C2C12 cells. Derived from adult mouse muscle, these cells divide rapidly and induce none of the principal muscle-specific proteins when grown in media rich in growth factors. When the culture is placed in a medium with low concentrations of growth factors, the cells stop dividing. Like primary myoblasts, the cells then align with each other, cease division, fuse to form a large multinucleate myotubes, and induce synthesis of muscle-specific proteins (Figure 9-32). Because one can introduce or knock-down expression of any desired gene and observe the effects on differentiation, these cells have been particularly valuable in uncovering the roles of many transcription factors in muscle development (Chapter 22).

Similarly, the murine 3T3-L1 preadipocyte cell line grows with fibroblast-like morphology in culture. When switched to a medium containing insulin and dexamethasone (a glucocorticoid steroid hormone), 3T3-L1 cells undergo synchronized differentiation into adipocytes, as shown both by accumulation of intracellular lipids (Figure 9-33a–d) and induction of adipocyte-specific mRNAs (Figure 9-33e). Because primary adipocytes do not divide in culture, these cell lines are widely used in biochemical, molecular, and cell biological studies on the development and function of adipose cells.

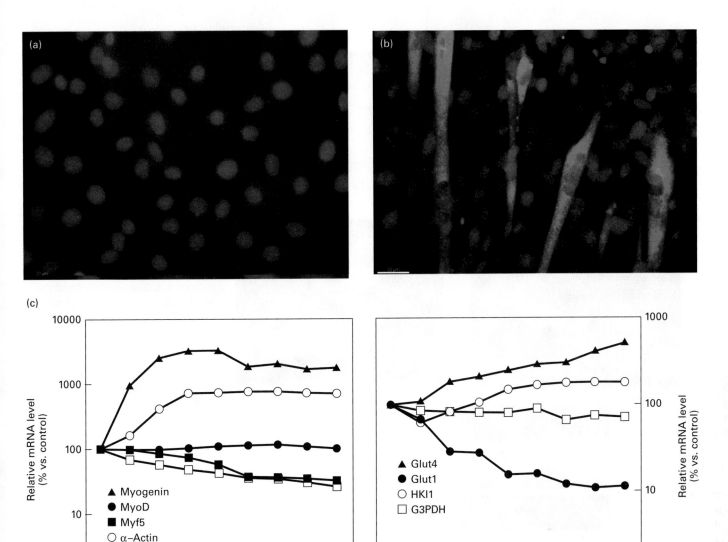

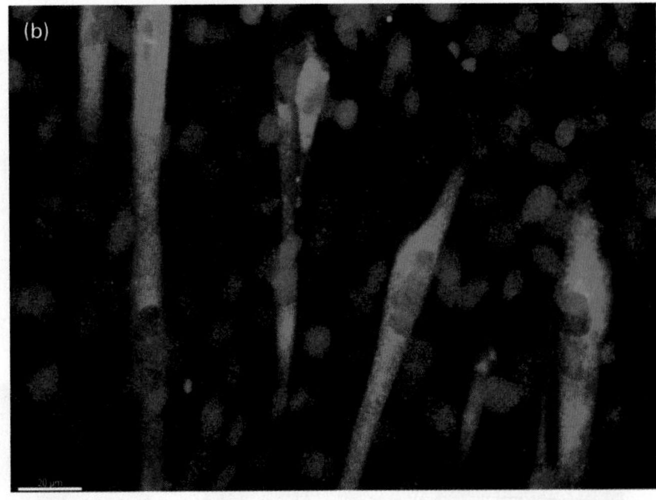

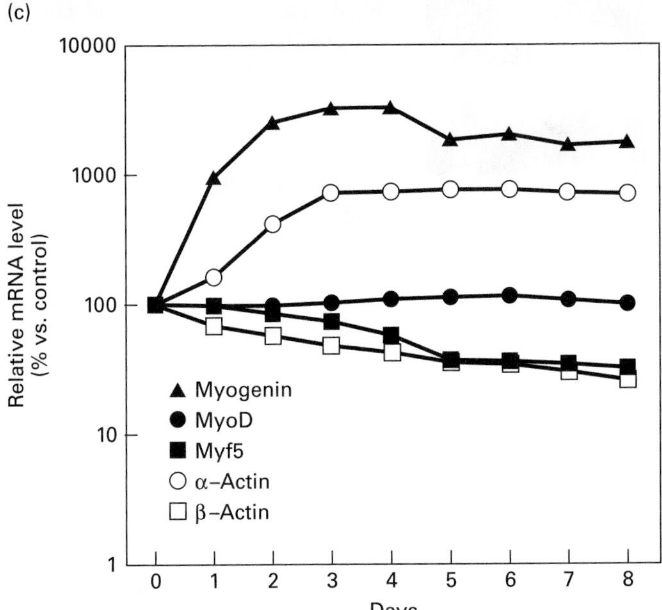

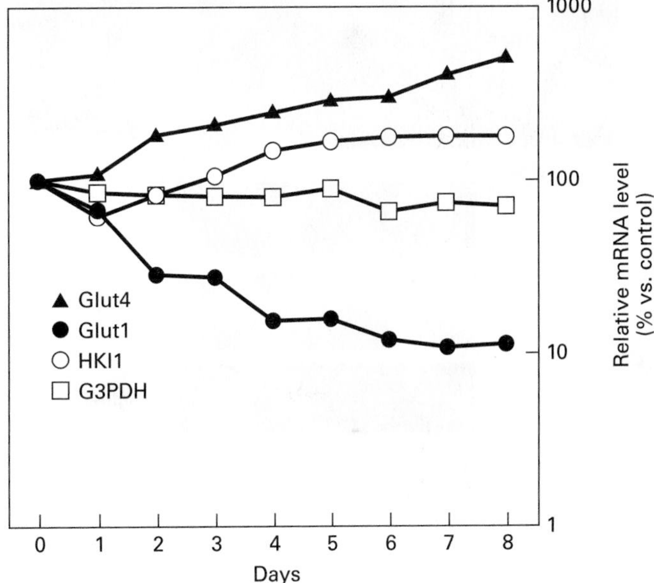

▲ EXPERIMENTAL FIGURE 9-32 **Differentiation in culture of C2C12 cells, a line of transformed mouse myoblasts.** (a) C2C12 cells proliferate but do not differentiate in a medium containing fetal calf serum, which contains high concentrations of mitogenic growth factors. (b) After placing C2C12 cells in a medium with horse serum, which contains lower concentrations of such growth factors, many of the cells fused to form multinucleate syncytia that express the muscle-specific myosin heavy chain, detected with a green-fluorescing antibody specific for this protein. Staining of DNA with blue DAPI dye reveals the nuclei. Bar denotes 20 μm. (c) During differentiation of cultured C2C12 cells the levels of certain mRNAs increase markedly, as shown in these graphs. For instance, the mRNAs for myogenin (a muscle-specific transcription factor), α-actin (major component of contractile filaments), and the GLUT4 glucose transporter (found only in muscle and fat cells) increased 5 to 50 fold. In contrast, mRNAs encoding proteins specific for growing cells, such as the GLUT1 glucose transporter and ß-actin, were downregulated. Other mRNAs, such as the glycolytic enzyme glyceraldehyde 3 phosphate dehydrogenase (G3PDH), were unaffected. Individual mRNAs were measured by using reverse transcriptase to copy total cellular mRNA into DNA, followed by quantitative polymerase chain reaction (PCR) amplification of specific cDNAs. The results were normalized to the amount present in growing cells. [Parts (a) and (b) courtesy James Evans and Prakash Rao, Whitehead Institute. Part (c) adapted from T. Shimokawa et al., 1998, *Biochem. Biophys. Res. Comm.* **246:**287.]

As detailed in Chapter 19, many cell types function only when closely linked to other cells. Key examples are the sheet-like layers of epithelial tissue, called **epithelia** (sing., **epithelium**), which cover the external and internal surfaces of organs. Typically, the distinct surfaces of a polarized epithelial cell are called the **apical** (top), **basal** (base or bottom), and **lateral** (side) surfaces (see Figure 19-8). The basal surface usually contacts an underlying extracellular matrix called the **basal lamina,** whose composition and function are discussed in Section 19.3. Several types of cell junctions interconnect adjacent epithelial cells and anchor them to the basal lamina. Cultured cells called *Madin-Darby canine kidney (MDCK) cells* are often used to study the formation and function of epithelial cells. When grown in specialized containers, these

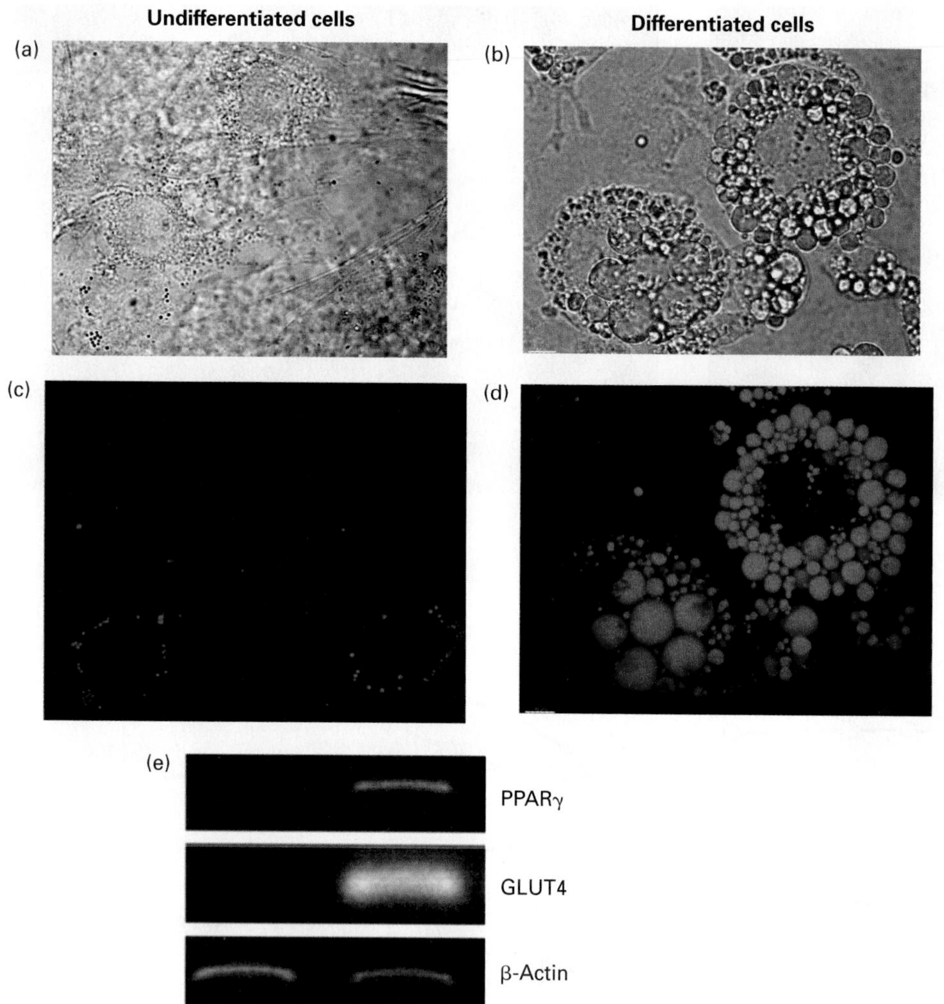

◀ **EXPERIMENTAL FIGURE 9-33 A line of 3T3L1 preadipocytes can differentiate into adipocytes and express adipocyte-specific mRNAs in culture.** Fibroblast-like 3T3-L1 preadipocytes growing in standard serum-containing proliferation medium (a,c) were switched to a differentiation medium, containing insulin, the steroid hormone dexamethasone, and isobutylmethylxanthine, an inhibitor of cAMP phosphodiesterase, for 8 days (b,d). (a,b) Differential interference contrast (DIC) microscopy reveals the considerable morphological changes in the cells during differentiation. (c,d) Staining with Oil Red O reveals droplets of triglycerides (red) in differentiated but not undifferentiated cells. (e) Northern blot analysis demonstrates expression of two key adipocyte genes, encoding the transcription factor PPARγ and the insulin-responsive glucose transporter GLUT4 in the differentiated 3T3-L1 cells (*right lane*) but not in the undifferentiated 3T3-L1 preadipocytes (*left lane*). β-Actin was used as a control to show loading of equal amounts of RNA in the two gel lanes. [Part a-d courtesy James Evans and Huangming Xie, Whitehead Institute. Part (e) from N. L Harvey et al., 2005, *Nature Gen.* **37:**1072.]

cells form a continuous one-cell-thick sheet (monolayer) of polarized kidney-like epithelial cells (Figure 9-34). The importance of many proteins in forming interconnecting cell junctions in MDCK cells was revealed by inhibiting their production with small inhibitory RNAs (siRNAs), a phenomenon called RNA interference (see Figure 5-45).

A major disadvantage of cultured cells is that they are not in their normal environment and hence their activities are not regulated by the other cells and tissues as they would be in an intact organism. For example, insulin produced by the pancreas has an enormous effect on liver glucose metabolism; however, this normal regulatory mechanism does not operate in a purified population of liver cells (called hepatocytes) grown in culture. In addition, as already described, the three-dimensional distribution of cells and extracellular matrix around a cell influences its shape and behavior. Because the immediate environment of cultured cells differs radically from this "normal" environment, their properties may be affected in various ways. Thus care must always be exercised in drawing conclusions about the normal properties of cells in complex tissues and organisms only on the basis of experiments with isolated, cultured cells.

Hybrid Cells Called Hybridomas Produce Abundant Monoclonal Antibodies

In addition to serving as research models for studies on cell function, cultured cells can be converted into "factories" for producing specific proteins. In Chapter 5, we described how this is done by introducing genes encoding insulin, growth factors, and other therapeutically useful proteins into bacterial or eukaryotic cells (see Figures 5-31 and 5-32). Here we consider the use of special cultured cells to generate monoclonal antibodies, which we have seen are experimental tools widely used in many aspects of cell biological research. Increasingly, they are being used for diagnostic and therapeutic purposes in medicine, as we discuss in later chapters.

To understand the challenge of generating monoclonal antibodies, we need to briefly review how mammals produce antibodies; more detail is provided in Chapter 24. Each

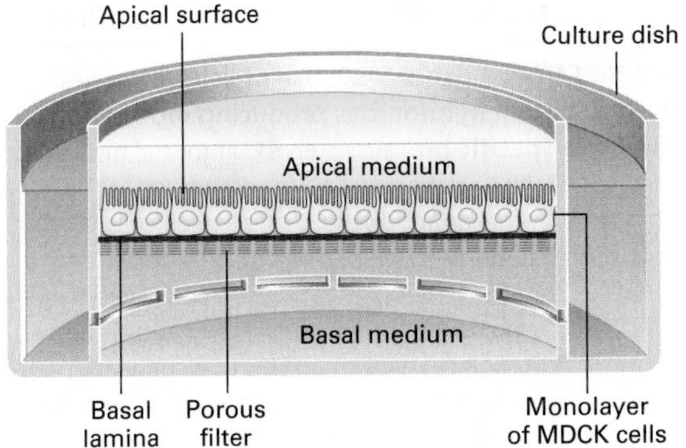

▲ EXPERIMENTAL FIGURE 9-34 **Madin-Darby canine kidney (MDCK) cells grown in specialized containers provide a useful experimental system for studying epithelial cells.** MDCK cells form a polarized epithelium when grown on a porous membrane filter coated on one side with collagen and other components of the basal lamina. With the use of the special culture dish shown here, the medium on each side of the filter (apical and basal sides of the monolayer) can be experimentally manipulated and the movement of molecules across the layer monitored. Several cell junctions that interconnect the cells form only if the growth medium contains sufficient Ca^{2+}.

normal antibody-producing B lymphocyte in a mammal is capable of producing a single type of antibody that can bind to a specific chemical structure (called a determinant or **epitope**) on an antigen molecule. If an animal is injected with an antigen, many B lymphocytes that make antibodies recognizing that antigen are stimulated to grow and secrete the antibodies. Each antigen-activated B lymphocyte forms a clone of cells in the spleen or lymph nodes, with each cell of the clone producing the identical antibody—that is, a monoclonal antibody. Because most natural antigens contain multiple epitopes, exposure of an animal to an antigen usually stimulates the formation of multiple different B-lymphocyte clones, each producing a different antibody. The resulting mixture of antibodies that recognize different epitopes on the same antigen is said to be *polyclonal.* Such polyclonal antibodies circulate in the blood and can be isolated as a group.

Although polyclonal antibodies are useful for a variety of experiments, monoclonal antibodies are required for many types of experiments and medical applications. Unfortunately, the biochemical purification of any one type of monoclonal antibody from blood is not feasible for two main reasons: the concentration of any given antibody is quite low, and all antibodies have the same basic molecular architecture (see Figure 3-19).

Because of their limited life span, primary cultures of normal B lymphocytes are of limited usefulness for the production of monoclonal antibodies. Thus the first step in producing a monoclonal antibody is to generate immortal, antibody-producing cells. This immortality is achieved by fusing normal B lymphocytes from an immunized animal with transformed, immortal lymphocytes called *myeloma cells* that themselves synthesize neither the heavy (H) nor the light (L) polypeptides that constitute all antibodies (see Figure 3-19). During cell fusion, the plasma membranes of two cells fuse together, allowing their cytosols and organelles to intermingle. Treatment with certain viral glycoproteins or the chemical polyethylene glycol promotes cell fusion. Some of the fused cells undergo division, and their nuclei eventually coalesce, producing viable *hybrid cells* with a single nucleus that contains chromosomes from both "parents." The fusion of two cells that are genetically different can yield a hybrid cell with novel characteristics. For instance, the fusion of a myeloma cell with a normal antibody-producing cell from a rat or mouse spleen yields a hybrid that proliferates into a clone called a **hybridoma.** Like myeloma cells, hybridoma cells grow rapidly and are immortal. Each hybridoma produces the monoclonal antibody encoded by its B-lymphocyte parent.

The second step in this procedure for producing monoclonal antibody is to separate, or select, the hybridoma cells from the unfused parental cells and the self-fused cells generated by the fusion reaction. This selection is usually performed by incubating the mixture of cells in a special culture medium, called *selection medium,* that permits the growth of only the hybridoma cells because of their novel characteristics. If the myeloma cells used for the fusion carry a mutation that blocks a metabolic pathway, a selection medium can be used that is lethal to them and not their lymphocyte fusion partners that do not have the mutation. In the immortal hybrid cells, the functional gene from the lymphocyte can supply the missing gene product, and thus the hybridoma cells will be able to grow in the selection medium. Because the lymphocytes used in the fusion are not immortalized and do not divide rapidly, only the hybridoma cells will proliferate rapidly in the selection medium and can thus be readily isolated from the initial mixture of cells.

Figure 9-35 depicts the general procedure for generating and selecting hybridomas. In this case, normal B lymphocytes in a sample of spleen cells are fused with myeloma cells that cannot grow in *HAT medium,* a common selection medium. Only the myeloma–lymphocyte hybrids can survive and grow for an extended period in HAT medium for reasons described shortly. Thus, this selection medium permits the separation of hybridoma cells from both types of parental cells and any self-fused cells. Finally, each selected hybridoma clone is then tested for the production of the desired antibody; any clone producing that antibody is then grown in large cultures, from which a substantial quantity of pure monoclonal antibody can be obtained.

Monoclonal antibodies are commonly employed in affinity chromatography to isolate and purify proteins from complex mixtures (see Figure 3-37c). They can also be used to label and thus locate a particular protein in specific cells of an organ and within cultured cells with the use of immunofluorescence microscopy techniques or in specific cell fractions with the use of immunoblotting (see Figure 3-38). Monoclonal antibodies also have become important diagnostic and therapeutic tools in medicine. For example, monoclonal antibodies that bind to and inactivate toxins secreted by bacterial pathogens are used to treat diseases. Other monoclonal antibodies are specific for cell-surface proteins expressed by certain types of tumor cells. Several of these anti-tumor antibodies are widely used in cancer therapy, including monoclonal antibody against a mutant form of the Her2 receptor that is overexpressed in some breast cancers (see Chapter 16, Figure 16-18).

MEDIA CONNECTIONS

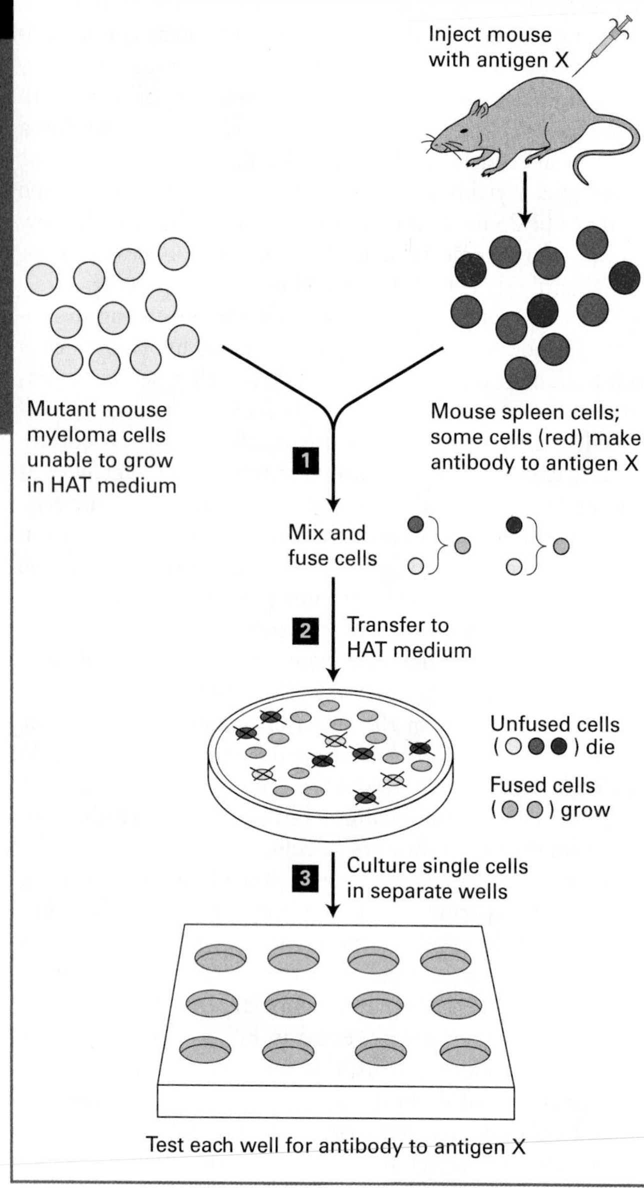
Inject mouse
with antigen X
Mutant mouse
myeloma cells
unable to grow
in HAT medium
Mouse spleen cells;
some cells (red) make
antibody to antigen X
1
Mix and
fuse cells
2
Transfer to
HAT medium
Unfused cells
( ) die
Fused cells
( ) grow
3
Culture single cells
in separate wells
Test each well for antibody to antigen X

◀ EXPERIMENTAL FIGURE 9-35 **Use of cell fusion and selection to obtain hybridomas producing monoclonal antibody to a specific protein.** Step **1**: Immortal myeloma cells that lack HGPRT, an enzyme required for growth on HAT medium, are fused with normal antibody-producing spleen cells from an animal that was immunized with antigen X. The spleen cells can make HGPRT. Step **2**: When plated on HAT medium, unfused and self-fused cells do not grow: the mutant myeloma cells because they cannot make purines through an HGPRT-dependent metabolic "salvage" pathway (see Figure 9-36), and the spleen cells because they have a limited life span in culture. Thus only fused cells formed from a myeloma cell *and* a spleen cell survive on HAT medium, proliferating into clones called hybridomas. Each hybridoma produces a single antibody. Step **3**: Testing of individual clones identifies those that recognize antigen X. After a hybridoma that produces a desired antibody has been identified, the clone can be cultured to yield large amounts of that antibody.

HAT Medium Is Commonly Used to Isolate Hybrid Cells

The principles underlying HAT selection are important not only for understanding how hybridoma cells are isolated but also for understanding several other frequently used selection methods, including selection of the embryonic stem (ES) cells used in generating knockout mice (see Figure 5-40). HAT medium contains *h*ypoxanthine (a purine), *a*minopterin, and *t*hymidine. Most animal cells can synthesize the purine and pyrimidine nucleotides from simpler carbon and nitrogen compounds (Figure 9-36, *top*). The folic acid antagonists amethopterin and aminopterin block these biochemical pathways; they interfere with the donation of methyl and formyl groups by tetrahydrofolic acid in the early stages of the synthesis of glycine, purine nucleoside monophosphates, and thymidine monophosphate. These drugs are called *antifolates* because they block reactions of tetrahydrofolate, an active form of folic acid.

Many cells, however, are resistant to antifolates because they contain enzymes that can synthesize the necessary nucleotides by a different route from purine bases and thymidine (Figure 9-36, *bottom*). Two key enzymes in these *nucleotide salvage pathways* are thymidine kinase (TK) and hypoxanthine-guanine phosphoribosyl transferase (HGPRT). Cells that produce these enzymes can grow on HAT medium, which supplies a salvageable purine and thymidine, whereas those lacking either of these enzymes cannot.

Cells with a TK mutation that prevents the production of the functional TK enzyme can be isolated because such cells are resistant to the otherwise toxic thymidine analog 5-bromodeoxyuridine. Cells containing TK convert this compound into 5-bromodeoxyuridine monophosphate, which is then converted into a nucleoside triphosphate by other enzymes. The triphosphate analog is incorporated by DNA polymerase into DNA, where it exerts its toxic effects. This pathway is blocked in TK^- mutants, and thus they are resistant to the toxic effects of 5-bromodeoxyuridine. Similarly, cells lacking the HGPRT enzyme, such as the $HGPRT^-$ myeloma cell lines used in producing hybridomas, can be isolated because they are resistant to the otherwise toxic guanine analog 6-thioguanine.

Normal cells can grow in HAT medium because even though the aminopterin in the medium blocks de novo synthesis of purines and TMP, the thymidine in the medium is transported into the cell and converted into TMP by TK and the hypoxanthine is transported and converted into usable purines by HGPRT. On the other hand, neither TK^- nor $HGPRT^-$ cells can grow in HAT medium because each lacks an enzyme of the salvage pathway. However, hybrids formed by the fusion of these two mutants will carry a normal *TK* gene from the $HGPRT^-$ parent and a normal *HGPRT* gene from the TK^- parent. The hybrids will thus produce both functional salvage-pathway enzymes and will grow on HAT medium.

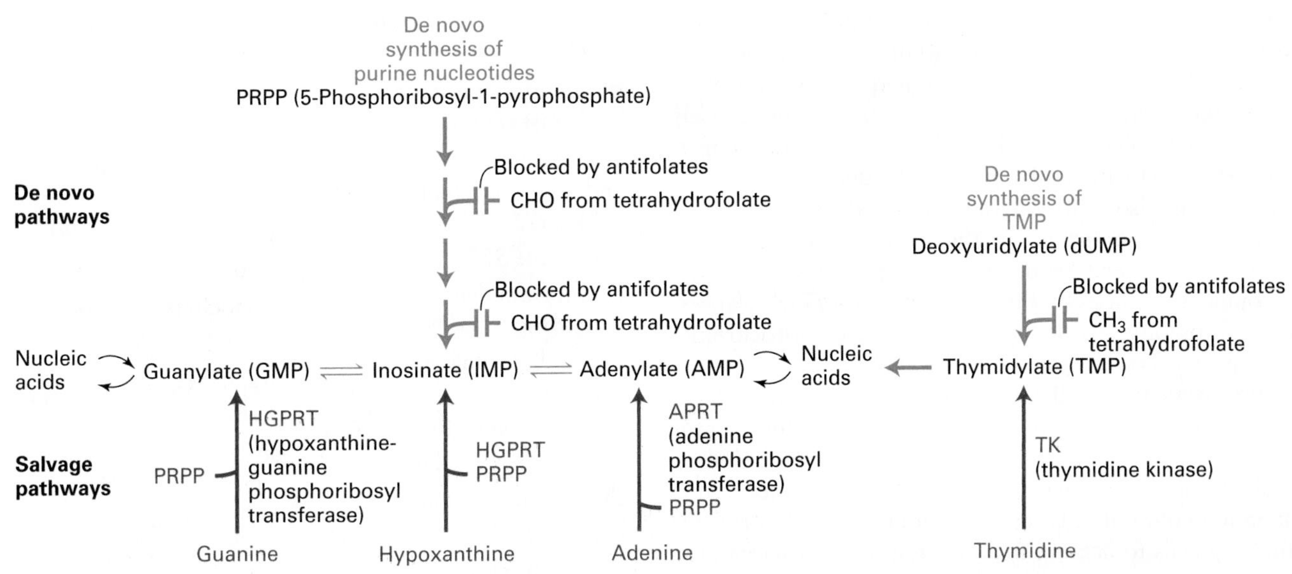

▲ **FIGURE 9-36 De novo and salvage pathways for nucleotide synthesis.** Animal cells can synthesize purine nucleotides (AMP, GMP, IMP) and thymidylate (TMP) from simpler compounds by de novo pathways (blue). Several of these reactions require the transfer of a methyl or formyl ("CHO") group from an activated form of tetrahydrofolate (e.g., N^5,N^{10}-methylenetetrahydrofolate), as shown in the upper part of the diagram. Antifolates, such as aminopterin and amethopterin, block the reactivation of tetrahydrofolate, preventing purine and thymidylate synthesis. Many animal cells can also use salvage pathways (red) to incorporate purine bases or nucleosides and thymidine. If these precursors are present in the medium, normal cells will grow even in the presence of antifolates. However, cultured cells lacking one of the enzymes—HGPRT, APRT, or TK—of the salvage pathways will not survive in media containing antifolates.

KEY CONCEPTS OF SECTION 9.5

Isolation, Culture, and Differentiation of Metazoan Cells

- Flow cytometry can identify different cells on the basis of the light that they scatter and the fluorescence that they emit. The fluorescence-activated cell sorter (FACS) is useful in separating different types of cells (see Figures 9-28 and 9-29).
- Growth of vertebrate cells in culture requires rich media containing essential amino acids, vitamins, fatty acids, and peptide or protein growth factors; the last are frequently provided by serum.
- Most cultured vertebrate cells will grow only when attached to a negatively charged substratum that is coated with components of the extracellular matrix.
- Primary cells, which are derived directly from animal tissue, have limited growth potential in culture and may give rise to a cell strain (see Figure 9-31). Some primary cells can undergo differentiation into a specific type of cell.
- Transformed cells, which are derived from animal tumors or arise spontaneously by transformation of primary cells, grow indefinitely in culture, forming cell lines.
- Certain cell lines can be induced to undergo differentiation in culture to generate muscle, adipose, epithelial, and other types of cells and are widely used in cell biological studies (see Figures 9-30, 9-32, and 9-33).
- The fusion of an immortal myeloma cell and a single B lymphocyte yields a hybrid cell that can proliferate indefinitely, forming a clone called a hybridoma (see Figure 9-35). Because each individual B lymphocyte produces antibodies specific for one antigenic determinant (epitope), a hybridoma produces only the monoclonal antibody synthesized by its original B-lymphocyte parental cell.
- HAT medium is commonly used to isolate hybridoma cells and other types of hybrid cells.

Perspectives for the Future

Light microscopy will continue to be a major tool in cell biology, providing images that relate to both the interactions among proteins and the movements, or mechanics, involved in various cell processes. The use of more fluorescent labels and tags will allow visualization of five or six different types of molecules simultaneously. With more labeled proteins, the complex interactions among proteins and organelles inside cells will become better understood.

Very recent advances in light microscopy are opening entire new areas for investigation. As an example, the two-photon fluorescence microscope enables visualization of fluorescent proteins (e.g., GFP expressed from a reporter gene) in thick tissue samples. With this technology the properties of individual cells can assessed in tissues inside living animals.

Although the limit of resolution of a light microscope using visible light is about 0.2 μm (200 nm), new types of fluorescence microscopy, such as SPED (stimulated emission depletion), can resolve two fluorescent objects as close together as ≈20 nm. For example, we noted that synaptic vesicles are too small (≈40 nm in diameter) and too densely packed to be resolved by available fluorescence microscopes. However, STED can resolve these individual vesicles. This technique will also enable investigators to detect single fluorescent protein molecules in the membranes of purified organelles. Another type of fluorescence microscopy under development, termed *total internal reflection (TIR) fluorescence microscopy,* enables detection of single fluorochrome-tagged proteins on the surface of living cells.

Improvements in cell culture technology will allow both primary cells and cultured lines to be studied in more natural contexts, not on glass cover slips but in three-dimensional gels of extracellular matrix molecules. This technique will permit cultured cells such as liver and several hormone-producing cells to achieve and maintain their differentiated state for days, enabling many types of experiments to be performed. Bioengineers also are fabricating artificial tissues based on a synthetic three-dimensional architecture incorporating layers of different cells. Eventually such artificial tissues will provide replacements for defective tissues in sick, injured, or aging individuals.

In parallel, investigators will use advanced microfabrication techniques to culture minute numbers of cells in microliter volumes on a glass slide consisting of microfabricated wells and channels. With this technique, reagents in nanoliter volumes can be introduced and exposed to selected parts of individual cells; the responses of the cells can then be detected by light microscopy and analyzed by powerful image-processing software. In these types of studies cells can be screened rapidly with millions of different chemical compounds, thus facilitating the discovery of new drugs, detection of subtle phenotypes of mutant cells (e.g., tumor cells), and development of comprehensive models of cellular processes. Thus advances in bioengineering will make major contributions not only to our understanding of cell and tissue function but also to the quality of human health.

Finally, the electron microscope will become the dominant instrument for studying the structure of multiprotein machines in vitro and in situ. Tomographic methods applied to single cells and molecules combined with automated reconstruction methods will generate models of protein-based structures that cannot be determined by x-ray crystallography. High resolution three-dimensional models of molecules in cells will help explain the intricate biochemical interactions among proteins.

Key Terms

autophagy *374*
bright-field light microscopy *381*
cell line *398*
cell strain *397*
chimeric proteins *371*
chloroplast *379*
cryoelectron microscopy *389*
culturing *372*
cytosol *372*
differential centrifugation *392*
differential interference contrast (DIC) microscopy *381*
endoplasmic reticulum (ER) *375*
endosome *373*
equilibrium density-gradient centrifugation *392*
extracellular matrix (ECM) *372*
fluorescence-activated cell sorter (FACS) *394*
fluorescent staining *382*
gap junctions *383*
Golgi complex *376*
HAT medium *401*
hybridoma *401*
immunofluorescence microscopy *385*
lumen *373*
lysosome *373*
membrane transport protein *372*
metal shadowing *390*
mitochondrion *378*
monoclonal antibody *394*
nucleolus *378*
organelles *371*
peroxisome *374*
phase-contrast microscopy *381*
resolution *381*
scanning electron microscope (SEM) *390*
transmission electron microscope (TEM) *388*

Review the Concepts

1. One of the defining features of eukaryotic cells is the presence of organelles. What are the major organelles of eukaryotic cells, and what is the function of each? What is the cytosol? What cellular processes occur within the cytosol?

2. The internalization of proteins, soluble macromolecules and large particles is a fundamental process required by all cells. The size of the molecule often dictates how it gets internalized. Describe how phagocytosis differs from endocytosis and what generally happens to the internalized material.

3. Cell organelles such as mitochondria, chloroplasts, and the Golgi apparatus each have unique structures. How is the structure of each organelle related to its function?

4. Much of what we know about cellular function depends on experiments utilizing specific cells and specific parts (e.g., organelles) of cells. What techniques do scientists commonly use to isolate cells and organelles from complex mixtures, and how do these techniques work?

5. Both light and electron microscopy are commonly used to visualize cells, cell structures, and the location of specific molecules. Explain why a scientist may choose one or the other microscopy technique for use in research.

6. The magnification possible with any type of microscope is an important property, but its resolution, the ability to distinguish between two very closely apposed objects, is even more critical. Describe why the resolving power of a microscope is more important for seeing finer details than its magnification. What is the formula to describe the resolution of a microscope lens and what are the limitations placed on the values in the formula?

7. Why are chemical stains required for visualizing cells and tissues with the basic light microscope? What advantage do fluorescent dyes and fluorescence microscopy provide in

comparison to the chemical dyes used to stain specimens for light microscopy? What advantages do confocal scanning microscopy and deconvolution microscopy provide in comparison to conventional fluorescence microscopy?

8. In certain electron microscopy methods, the specimen is not directly imaged. How do these methods provide information about cellular structure, and what types of structures do they visualize?

9. What is the difference between a cell strain, a cell line, and a clone?

10. Explain why certain cell lines are used to study the differentiation and function of muscle or fat cells.

11. Explain why the process of cell fusion is necessary to produce monoclonal antibodies used for research.

Analyze the Data

Mouse liver cells were homogenized and the homogenate subjected to equilibrium density-gradient centrifugation with sucrose gradients. Fractions obtained from these gradients were assayed for *marker molecules* (i.e., molecules that are limited to specific organelles). The results of these assays are shown in the figure. The marker molecules have the following functions: Cytochrome oxidase is an enzyme involved in the process by which ATP is formed in the complete aerobic degradation of glucose or fatty acids; ribosomal RNA forms part of the protein-synthesizing ribosomes; catalase catalyzes decomposition of hydrogen peroxide; acid phosphatase hydrolyzes monophosphoric esters at acid pH; cytidylyl transferase is involved in phospholipid biosynthesis; and amino acid permease aids in transport of amino acids across membranes.

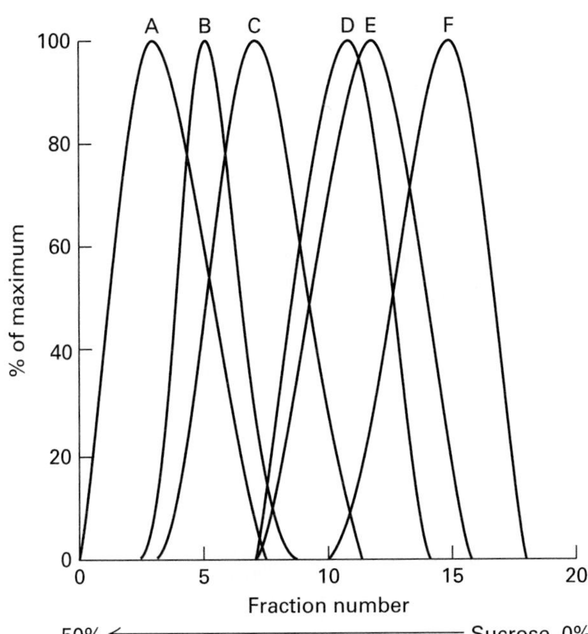

Curve A = cytochrome oxidase
Curve B = ribosomal RNA
Curve C = catalase
Curve D = acid phosphatase
Curve E = cytidylyl transferase
Curve F = amino acid permease

a. Name the marker molecule and give the number of the fraction that is *most* enriched for each of the following cell components: lysosomes; peroxisomes; mitochondria; plasma membrane; rough endoplasmic reticulum; smooth endoplasmic reticulum.

b. Is the rough endoplasmic reticulum more or less dense than the smooth endoplasmic reticulum? Why?

c. Describe an alternative approach by which you could identify which fraction was enriched for which organelle.

d. How would addition of a detergent to the homogenate, which disrupts membranes by solubilizing their lipid and protein components, affect the equilibrium density-gradient results?

References

Organelles of the Eukaryotic Cell

Bainton, D. 1981. The discovery of lysosomes. *J. Cell Biol.* **91**:66s–76s.

Cuervo, A. M., and J. F. Dice. 1998. Lysosomes: a meeting point of proteins, chaperones, and proteases. *J. Mol. Med.* **76**:6–12.

de Duve, C. 1996. The peroxisome in retrospect. *Ann. NY Acad. Sci.* **804**:1–10.

Foster, L. J., et al. 2006. A mammalian organelle map by protein correlation profiling. *Cell* **125**:187–199.

Holtzman, E. 1989. *Lysosomes.* Plenum Press.

Lamond, A., and W. Earnshaw. 1998. Structure and function in the nucleus. *Science* **280**:547–553.

Mootha, V.K., et al. 2003. Integrated analysis of protein composition, tissue diversity, and gene regulation in mouse mitochondria. *Cell* **115**:629–640.

Palade, G. 1975. Intracellular aspects of the process of protein synthesis. *Science* **189**:347–358. The Nobel Prize lecture of a pioneer in the study of cellular organelles.

Wanders, R., and H. R. Waterham. 2006. Biochemistry of mammalian peroxisomes revisited. Ann. Rev. Biochem. **75**:295–332.

Light Microscopy: Visualizing Cell Structure and Localizing Proteins Within Cells

Chen, X., M. Velliste, and R. F. Murphy. 2006. Automated interpretation of subcellular patterns in fluorescence microscope images for location proteomics. *Cytometry* (Part A) **69A**: 631–640.

Egner, A., and S. Hell. 2005. Fluorescence microscopy with super-resolved optical sections. *Trends Cell Biol.* **15**:207–215.

Gaietta, G., et al. 2002. Multicolor and electron microscopic imaging of connexin trafficking. *Science* **296**:503–507.

Giepmans, B. N. G., et al. 2006. The fluorescent toolbox for assessing protein location and function. *Science* **312**:217–224.

Gilroy, S. 1997. Fluorescence microscopy of living plant cells. *Ann. Rev. Plant Physiol. Plant Mol. Biol.* **48**:165–190.

Inoué, S., and K. Spring. 1997. *Video Microscopy,* 2d ed. Plenum Press.

Matsumoto, B., ed. 2002. *Methods in Cell Biology.* Vol. 70: *Cell Biological Applications of Confocal Microscopy.* Academic Press.

Misteli, T., and D. L. Spector. 1997. Applications of the green fluorescent protein in cell biology and biotechnology. *Nature Biotech.* **15**:961–964.

Pepperkok, R., and Ellenberg, J. 2006. High-throughput fluorescence microscopy for systems biology. *Nature Rev. Mol. Cell Biol.* AOP, published online July 19 2006 (doi:10.1038/nrm1979).

Sako, Y., S. Minoguchi, and T. Yanagida. 2000. Single-molecule imaging of EGFR signalling on the surface of living cells. *Nature Cell Biol.* **2**:168–172.

Simon, S., and J. Jaiswal. 2004. Potentials and pitfalls of fluorescent quantum dots for biological imaging. *Trends Cell Biol.* **14**:497–504.

Sluder, G., and D. Wolf, eds. 1998. *Methods in Cell Biology.* Vol. 56: *Video Microscopy.* Academic Press.

So, P.T.C., et al. 2000. Two-photon excitation fluorescence microscopy. *Ann. Rev. Biomed. Eng.* **2**:399–429.

Willig, K. I., et al. 2006. STED microscopy reveals that synaptotagmin remains clustered after synaptic vesicle exocytosis. *Nature* **440**:935–939.

Electron Microscopy: Methods and Applications

Beck, M., et al. 2004. Nuclear pore complex structure and dynamics revealed by cryoelectron tomography. *Science* **306**:1387–1390.

Frey, T. G., G. A. Perkins, and M. H. Ellisman. 2006. Electron tomography of membrane-bound cellular organelles. *Ann. Rev. Biophy. Biomol. Struc.* **35**:199–224.

Hyatt, M. A. *Principles and Techniques of Electron Microscopy,* 4th ed. 2000. Cambridge University Press.

Koster, A., and J. Klumperman. 2003. Electron microscopy in cell biology: integrating structure and function. *Nature Rev. Mol. Cell Biol.* 4:SS6–SS10.

Medalia, O., et al. 2002. Macromolecular architecture in eukaryotic cells visualized by cryoelectron tomography. *Science* **298**:1209–1213.

Nickell, S., et al. 2006. A visual approach to proteomics. *Nature Rev. Mol. Cell Biol.* **7**:225–230.

Purification of Cell Organelles

Battye, F. L., and K. Shortman. 1991. Flow cytometry and cell-separation procedures. *Curr. Opin. Immunol.* **3**:238–241.

de Duve, C. 1975. Exploring cells with a centrifuge. *Science* **189**:186–194. The Nobel Prize lecture of a pioneer in the study of cellular organelles.

de Duve, C., and H. Beaufay. 1981. A short history of tissue fractionation. *J. Cell Biol.* **91**:293s–299s.

Howell, K. E., E. Devaney, and J. Gruenberg. 1989. Subcellular fractionation of tissue culture cells. *Trends Biochem. Sci.* **14**:44–48.

Ormerod, M. G., ed. 1990. *Flow Cytometry: A Practical Approach.* IRL Press.

Rickwood, D. 1992. *Preparative Centrifugation: A Practical Approach.* IRL Press.

Isolation, Culture, and Differentiation of Metazoan Cells

Bissell, M. J., A. Rizki, and I. S. Mian. 2003. Tissue architecture: the ultimate regulator of breast epithelial function. *Curr. Opin. Cell Biol.* **15**:753–762.

Davis, J. M., ed. 1994. *Basic Cell Culture: A Practical Approach.* IRL Press.

Edwards, B., et al. 2004. Flow cytometry for high-throughput, high-content screening. *Curr. Opin. Chem. Biol.* **8**:392–398.

Goding, J. W. 1996. *Monoclonal Antibodies: Principles and Practice. Production and Application of Monoclonal Antibodies in Cell Biology, Biochemistry, and Immunology,* 3d ed. Academic Press.

Griffith, L. G., and M. A. Swartz. 2006. Capturing complex 3D tissue physiology in vitro. *Nature Rev. Mol. Cell. Biol.* **7**:211–224.

Krutzik, P., et al. 2004. Analysis of protein phosphorylation and cellular signaling events by flow cytometry: techniques and clinical applications. *Clin. Immunol.* **110**:206 – 221.

Paszek, M. J., and V. M. Weaver. 2004. The tension mounts: mechanics meets morphogenesis and malignancy. J. Mammary Gland Biol. Neoplasia **9**:325–342.

Shaw, A. J., ed. 1996. *Epithelial Cell Culture.* IRL Press.

Tyson, C. A., and J. A. Frazier, eds. 1993. *Methods in Toxicology. Vol. I (Part A): In Vitro Biological Systems.* Academic Press. Describes methods for growing many types of primary cells in culture.

CLASSIC EXPERIMENT 9.1

SEPARATING ORGANELLES

Beaufay et al., 1964, *Biochem J.* **92**:191

In the 1950s and 1960s scientists used two techniques to study cell organelles: microscopy and fractionation. Christian de Duve was at the forefront of cell fractionation. In the early 1950s he used centrifugation to distinguish a new organelle, the lysosome, from previously characterized fractions: the nucleus, the mitochondrial-rich fraction, and the microsomes. Soon thereafter he used equilibrium-density centrifugation to uncover yet another organelle.

Background

Eukaryotic cells are highly organized and composed of cell structures known as organelles that perform specific functions. Although microscopy has allowed biologists to describe the location and appearance of various organelles, it is of limited use in uncovering an organelle's function. To do this, cell biologists have relied on a technique known as cell fractionation. Here, cells are broken open, and the cellular components are separated on the basis of size, mass, and density using a variety of centrifugation techniques. Scientists could then isolate and analyze cell components of different densities, called *fractions.* Using this method, biologists had divided the cell into four fractions: nuclei, mitochondrial-rich fraction, microsomes, and cell sap.

De Duve was a biochemist interested in the subcellular locations of metabolic enzymes. He had already completed a large body of work on the fractionation of liver cells, in which he had determined the subcellular location of numerous enzymes. By locating these enzymes in specific cell fractions, he could begin to elucidate the function of the organelle. He noted that his work was guided by two hypotheses: the "postulate of biochemical homogeneity" and "the postulate of single location." In short, these hypotheses propose that the entire composition of a subcellular population will contain the same enzymes and that each enzyme is located at a discrete site within the cell. Armed with these hypotheses and the powerful tool of centrifugation, de Duve further subdivided the mitochondrial-rich fraction. First, he identified the light mitochondrial fraction, which is made up of hydrolytic enzymes that are now known to compose the lysosome. Then, in a series of experiments described here, he identified another discrete subcellular fraction, which he called the **peroxisome,** within the mitochondrial-rich fraction.

The Experiment

De Duve studied the distribution of enzymes in rat liver cells. Highly active in energy metabolism, the liver contains a number of useful enzymes to study. To look for the presence of various enzymes during the fractionation, de Duve relied on known tests, called enzyme assays, for enzyme activity. To retain maximum enzyme activity, he had to take precautions, which included performing all fractionation steps at 0°C because heat denatures protein, compromising enzyme activity.

De Duve used rate-zonal centrifugation to separate cellular components by successive centrifugation steps. He removed the rat's liver and broke it apart by homogenization. The crude preparation of homogenized cells was then subjected to relatively low-speed centrifugation. This initial step separated the cell nucleus, which collects as sediment at the bottom of the tube, from the cytoplasmic extract, which remains in the supernatant. Next, de Duve further subdivided the cytoplasmic extract into heavy mitochondrial fraction, light mitochondrial fraction, and microsomal fraction. He accomplished separating the cytoplasm by employing successive centrifugation steps of increasing force. At each step he collected and stored the fractions for subsequent enzyme analysis. Once the fractionation was complete, de Duve performed enzyme assays to determine the subcellular distribution of each enzyme. He then graphically plotted the distribution of the enzyme throughout the cell. As had been shown previously, the activity of cytochrome oxidase, an important enzyme in the electron transfer system, was found primarily in the heavy mitochondrial fractions. The microsomal fraction was shown to contain another previously characterized enzyme, glucose-6-phosphatase. The light mitochondrial fraction, which is made up of the lysosome, showed the characteristic acid phosphatase activity. Unexpectedly, de Duve observed a fourth pattern when he assayed uricase activity. Rather than following the pattern of the reference enzymes, uricase activity was sharply concentrated within the light mitochondrial fraction. This sharp concentration, in contrast to the broad distribution, suggested to de Duve that the uricase might be secluded in another subcellular population separate from the lysosomal enzymes.

To test this theory, de Duve employed a technique known as equilibrium density-gradient centrifugation, which separates macromolecules on the basis of density. Equilibrium density-gradient centrifugation can be performed using a number of different

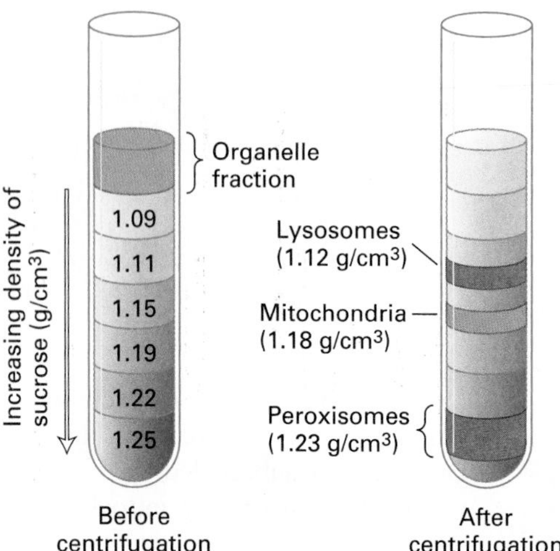

▲ **FIGURE 1 Schematic depiction of the separation of the lysosomes, mitochondria, and peroxisomes by equilibrium density centrifugation.** The mitochondrial-rich fraction from rate-zonal centrifugation was separated in a sucrose gradient, and the organelles were separated on the basis of density. [From Lodish et al., *Molecular Cell Biology,* 3d ed., W. H. Freeman and Company, p. 166.]

gradients, including sucrose and glycogen. In addition, the gradient can be made up in either water or "heavy water," which contains the hydrogen isotope deuterium in place of hydrogen. In his experiment de Duve separated the mitochondrial-rich fraction prepared by rate-zonal centrifugation in each of these different gradients (Figure 1). If uricase were part of a separate subcellular compartment, it would separate from the lysosomal enzymes in each gradient tested. De Duve performed the fractionations in this series of gradients, then performed enzyme assays as before. In each case, he found uricase in a separate population than the lysosomal enzyme acid phosphatase and the mitochondrial enzyme cytochrome oxidase (Figure 2). By repeatedly observing uricase activity in a distinct fraction from the activity of the lysosomal and mitochondrial enzymes, de Duve concluded that uricase was part of a separate organelle. The experiment also showed that two other enzymes, catalase and D–amino acid oxidase, segregated into the same fractions as uricase. Because each of these enzymes either produced or used hydrogen peroxide, de Duve proposed that this fraction represented an organelle responsible for the peroxide metabolism and dubbed it the **peroxisome.**

Discussion

De Duve's work on cellular fractionation provided an insight into the function of cell structures as he sought to map the location of known enzymes. Examining the inventory of enzymes in a given cell fraction gave him clues to its function. His careful work result in the uncovering of two organelle the lysosome and the **peroxisome.** H work also provided important clues the organelles' function. The lysosom where de Duve found so many pote tially destructive enzymes, is no known to be an important site fo degradation of biomolecules. The **pe oxisome** has been shown to be the si of fatty acid and amino acid oxidatio reactions that produce a large amou of hydrogen peroxide. In 1974, Duve received the Nobel Prize fo Physiology and Medicine in recogn tion of his pioneering work.

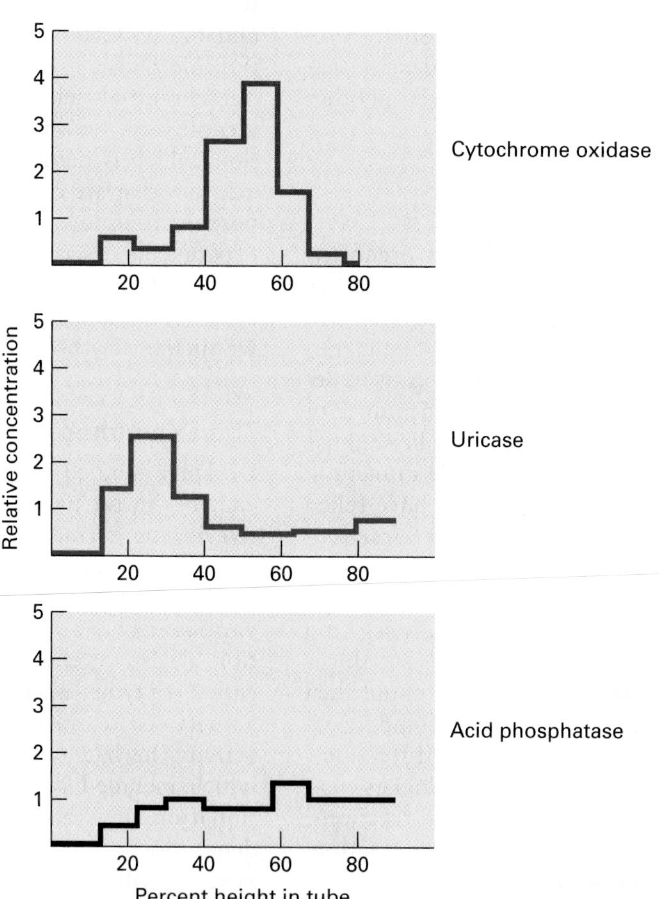

▲ **FIGURE 2 Graphical representation of the enzyme analysis of products from a sucrose gradient.** The mitochondrial-rich fraction was separated as depicted in Figure 10.1, and then enzyme assays were performed. The relative concentration of active enzyme is plotted on the y axis; the height in the tube is plotted on the x axis. The peak activities of cytochrome oxidase (*top*) and acid phosphatase (*bottom*) are observed near the top of the tube. The peak activity of uricase (*middle*) migrates to the bottom of the tube. [Adapted fron Beaufay et al., 1964, *Biochem J.* **92:**191.]

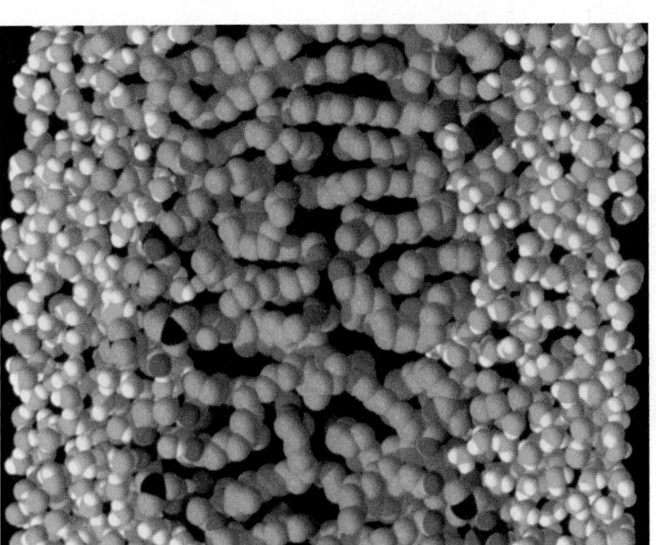

Molecular model of a phospholipid bilayer surrounded by water, as determined by molecular dynamics calculations. © National Institutes of Health/Photo Researchers, Inc.

CHAPTER

10

BIOMEMBRANE STRUCTURE

Membranes participate in many aspects of cell structure and function. The **plasma membrane** defines the cell and separates the inside from the outside. In eukaryotes, membranes also define the intracellular organelles such as the nucleus and lysosome. These biomembranes all have the same basic architecture—a phospholipid bilayer—but they are not static; their function is not to prevent all exchange across a border. Each cellular membrane has its own set of proteins that allow it to carry out its multitude of specific functions (Figure 10-1).

Prokaryotes, the simplest and smallest cells, are about 1–2 μm in length and are surrounded by a single plasma membrane; in most cases they contain no internal membrane-limited subcompartments (see Figure 1-2a). However, this single plasma membrane contains hundreds of different types of proteins that are integral to the function of the cell. Some of these proteins catalyze ATP synthesis and initiation of DNA replication, for instance. Others represent the many types of **membrane transport proteins** that enable specific ions, sugars, amino acids, and vitamins to cross the otherwise impermeable phospholipid bilayer to enter the cell and that allow specific metabolic products to exit.

In eukaryotic cells the plasma membrane (Figure 10-2) is not a site for ATP generation or DNA synthesis. Eukaryotic plasma membranes are studded with a multitude of membrane transport proteins that allow selective import and export of small molecules and ions. **Receptors** in the plasma membrane are proteins that allow the cell to recognize chemical signals—many sent by neighboring cells—present in its environment and adjust its metabolism or, especially during development, its pattern of gene expression in response. Other specialized plasma membrane proteins allow the cell to adhere to other cells and to components in the surrounding fibrous extracellular matrix. Many plasma membrane proteins bind components of the **cytoskeleton,** the dense network of protein filaments that permeates the cytosol and mechanically supports cellular membranes, interactions that are essential for the cell to assume its specific shape and for many types of cell movements.

The plasma membrane bends, folds, and flexes in three dimensions. Some segments bleb inward, incorporating components from the extracellular medium into intracellular vesicles (see Figure 9-2). Viruses such as HIV bud outward from the cell-surface membrane, enveloping themselves with a bit of the plasma membrane that contains virus-specific proteins (Figure 10-3).

OUTLINE

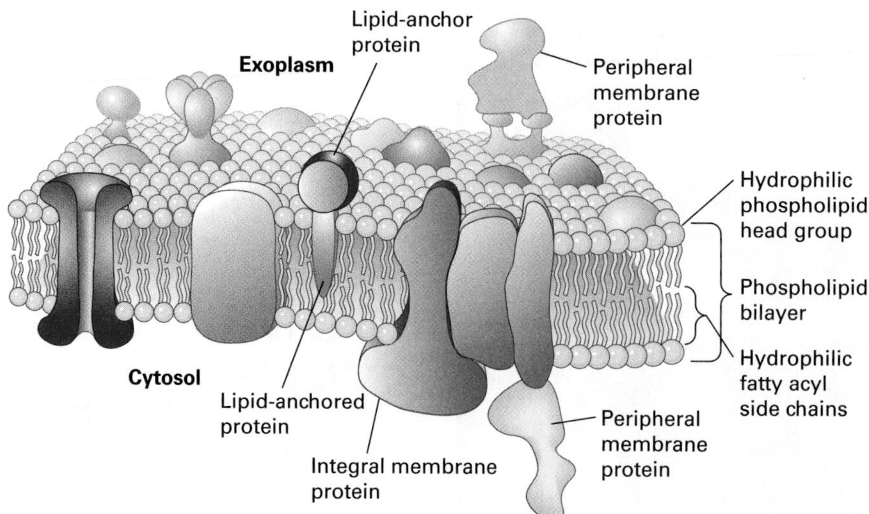

▲ **FIGURE 10-1 Fluid mosaic model of biomembranes.** A bilayer of phospholipids ~3 nm thick provides the basic architecture of all cellular membranes; membrane proteins give each cellular membrane its unique set of functions. Integral (transmembrane) proteins span the bilayer and often form dimers and higher-order oligomers. Lipid-anchored proteins are tethered to one leaflet by a long covalently attached hydrocarbon chain. Peripheral proteins associate with the membrane primarily by specific noncovalent interactions with integral proteins or membrane lipids. [After D. Engelman, 2005, *Nature* **438:**578–80.]

Each organelle in a eukaryotic cell (Figures 1-2, 10-2 and 10-4) contains a unique complement of proteins—some embedded in its membrane(s), others contained in its aqueous interior space, the **lumen**—that enable it to carry out its characteristic cellular functions. For instance, the internal pH of the lysosome, an organelle that contains many degradative enzymes, is about 5, compared to pH 7.2 of the **cytosol,** the aqueous part of the cytoplasm. The lysosome membrane contains ATP-powered H^+ pumps that use the energy of hydrolysis of ATP phosphoanhydride bonds to pump protons from the cytosol into the lumen of the lysosome, lowering its pH.

We begin our examination of biomembranes by considering their lipid components. These not only affect membrane shape and function but also play important roles in anchoring proteins to the membrane, modifying membrane protein activities, and transducing signals to the cytoplasm. We then consider the structure of membrane proteins. Many, such as transport proteins and receptors, have large segments that are imbedded in the hydrocarbon core of the

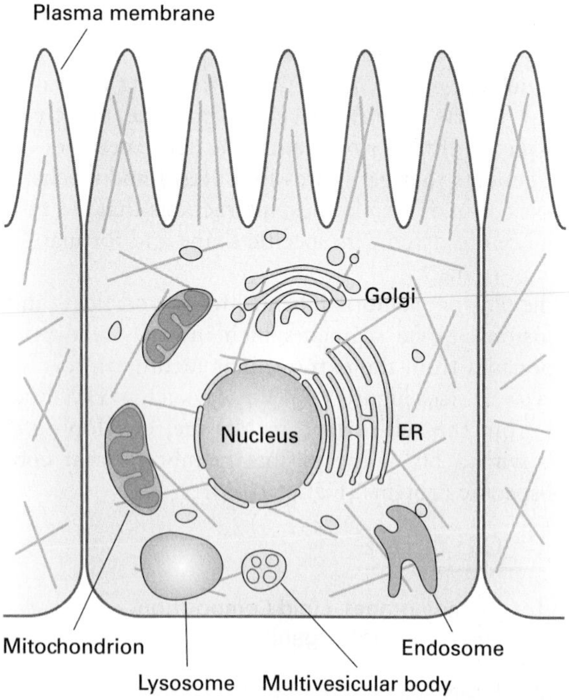

Plasma membrane
Golgi
Nucleus
ER
Mitochondrion
Endosome
Lysosome
Multivesicular body

▲ **FIGURE 10-2 Cellular membranes** The plasma membrane defines the exterior of the cell and controls the movement of molecules between the cytosol and the extracellular medium. Different types of organelles and smaller vesicles enclosed within their own distinctive membranes carry out special functions such as gene expression, energy production, membrane synthesis, and intracellular transport.

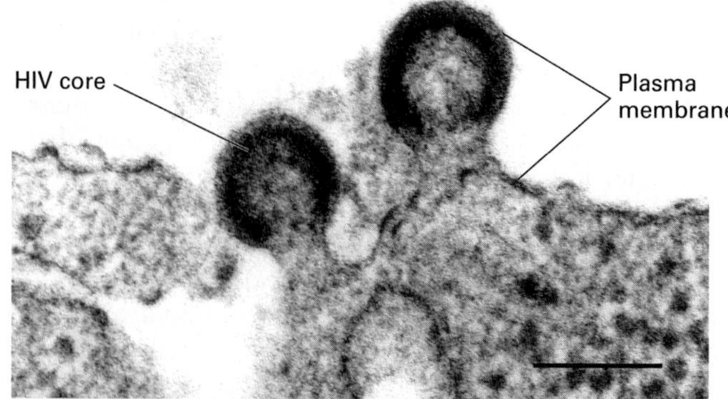

▲ **FIGURE 10-3 Eukaryotic cell membranes are dynamic structures.** An electron micrograph of the plasma membrane of an HIV-infected cell, showing HIV particles budding into the culture medium. As the virus core buds from the cell, it becomes enveloped by a membrane derived from the cell's plasma membrane that contains specific viral proteins. [From W. Sundquist and U. von Schwedler, University of Utah.]

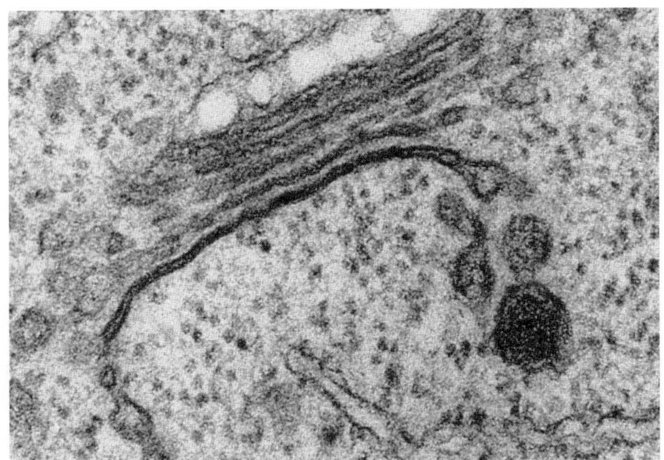

▲ **FIGURE 10-4 Stacked membranes of the Golgi complex.** Note the irregular shape and curvature of these membranes. [From C. Hopkins and J. Burden, Imperial College London.]

phospholipid bilayer, and we will focus on the principal classes of such membrane proteins. Finally, we consider how phospholipids and cholesterol are synthesized in cells and distributed to the many membranes and organelles. Cholesterol is an essential component of the plasma membrane of all animal cells but is toxic to the organism if present in excess.

10.1 Biomembranes: Lipid Composition and Structural Organization

In Chapter 2 we learned that phosphoglycerides are the principal building blocks of most biomembranes. Like the two other principal classes of membrane lipids, sphingolipids and cholesterol (Figure 10-5), phosphoglycerides are **amphipathic** molecules—they consist of two segments with very different chemical properties. In phosphoglycerides and sphingolipids the hydrocarbon "tails" of the fatty acid side chains are **hydrophobic** and partition away from water, whereas the "head groups" are strongly **hydrophilic,** or water loving, and tend to interact with water molecules. Steroids such as cholesterol, in contrast, are mostly hydrophobic except for one hydrophilic hydroxyl group. All three types of phospholipids have the necessary qualities to form membranes and play different roles in the function of the cell.

Phospholipids Spontaneously Form Bilayers

The amphipathic nature of phospholipids, which governs their interactions, is critical to the structure of biomembranes. When a suspension of phospholipids is mechanically dispersed in aqueous solution, the phospholipids aggregate into one of three forms: spherical **micelles** and **liposomes** and sheetlike **phospholipid bilayers,** which are two molecules thick (Figure 10-6). The type of structure formed by a pure phospholipid or a mixture of phospholipids depends on several factors, including the length of the fatty acyl chains, their degree of saturation, and temperature. In all three structures, the hydrophobic effect causes the fatty acyl chains to aggregate and exclude water molecules from the "core." Micelles are rarely formed from natural phosphoglycerides, whose fatty acyl chains generally are too bulky to fit into the interior of a micelle. However, micelles are formed if one of the two fatty acyl chains is removed from the phosphoglyceride by hydrolysis, forming a lysophospholipid. In aqueous solution, common detergents and soaps form micelles that behave as tiny ball bearings, thus giving soap solutions their slippery feel and lubricating properties.

Under suitable conditions, phospholipids of the composition present in cells spontaneously form symmetric phospholipid bilayers. Each phospholipid layer in this lamellar structure is called a *leaflet.* The hydrophobic fatty acyl chains in each leaflet minimize contact with water by aligning themselves tightly together in the center of the bilayer, forming a hydrophobic core that is about 3–4 nm thick (Figure 10-6b). The close packing of these nonpolar tails is stabilized by van der Waals interactions between these hydrocarbon chains. Ionic and hydrogen bonds stabilize the interaction of the phospholipid polar head groups with one another and with water. Electron microscopy of thin membrane sections of cells stained with osmium tetroxide, which binds strongly to the polar head groups of phospholipids, reveals the bilayer structure (Figure 10-6a). A cross section of a single membrane stained with osmium tetroxide looks like a railroad track: two thin dark lines (the stained head group complexes) with a uniform light space of about 2 nm between them (the hydrophobic tails).

A phospholipid bilayer can be of almost unlimited size—from micrometers (μm) to millimeters (mm) in length or width—and can contain tens of millions of phospholipid molecules. Because of their hydrophobic core, bilayers are virtually impermeable to salts, sugars, and most other small hydrophilic molecules. The phospholipid bilayer is the basic structural unit of nearly all biological membranes; although they contain other molecules (e.g., cholesterol, glycolipids, proteins), biomembranes have a hydrophobic core that separates two aqueous solutions and acts as a permeability barrier.

Phospholipid Bilayers Form a Sealed Compartment Surrounding an Internal Aqueous Space

Phospholipid bilayers can be generated in the laboratory by simple experimental manipulations; these utilize either chemically pure phospholipids or lipid mixtures of the composition

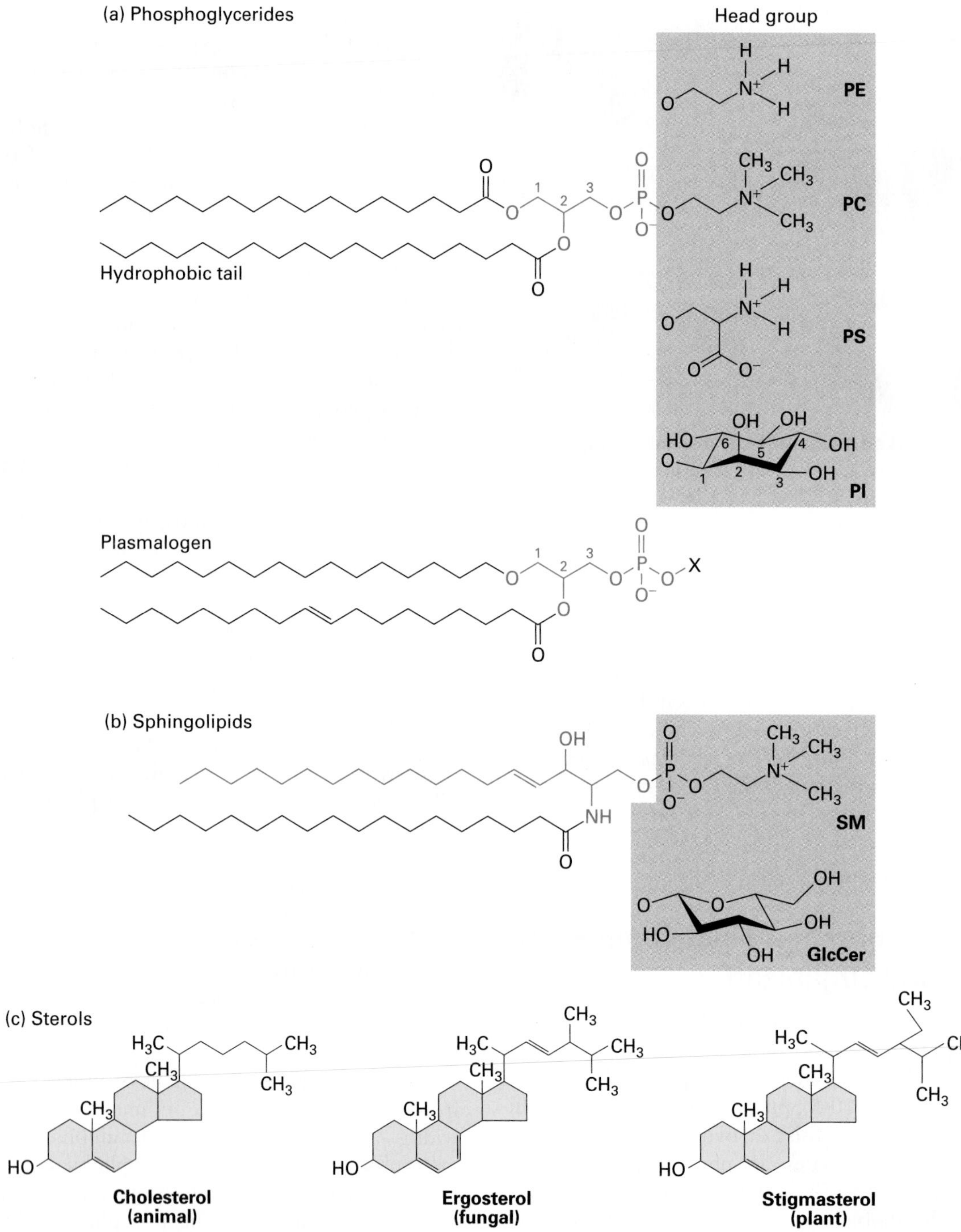

(a) Phosphoglycerides

Head group

PE

PC

Hydrophobic tail

PS

PI

Plasmalogen

(b) Sphingolipids

SM

GlcCer

(c) Sterols

Cholesterol (animal)

Ergosterol (fungal)

Stigmasterol (plant)

▲ **FIGURE 10-5 Three classes of membrane lipids.** (a) Most phosphoglycerides are derivatives of glycerol 3-phosphate (red), which contains two esterified fatty acyl chains that constitute the hydrophobic "tail" and a polar "head group" esterified to the phosphate. The fatty acids can vary in length and be saturated (no double bonds) or unsaturated (one, two, or three double bonds). In phosphatidylcholine (PC), the head group is choline. Also shown are the molecules attached to the phosphate group in three other common phosphoglycerides: phosphatidylethanolamine (PE), phosphatidylserine (PS), and phosphatidylinositol (PI). Plasmalogens contain one fatty acyl chain attached to glycerol by an ester linkage and one attached by an ether linkage; these contain similar head groups as phosphoglycerides. (b) Sphingolipids are derivatives of sphingosine (red), an amino alcohol with a long hydrocarbon chain. Various fatty acyl chains are connected to sphingosine by an amide bond. The sphingomyelins (SM), which contain a phosphocholine head group, are phospholipids. Other sphingolipids are glycolipids in which a single sugar residue or branched oligosaccharide is attached to the sphingosine backbone. For instance, the simple glycolipid glucosylcerebroside (GlcCer) has a glucose head group. (c) The major sterols in animals (cholesterol), fungi (ergosterol), and plants (stigmasterol) differ slightly in structure, but all serve as key components of cellular membranes. The basic structure of steroids is a four-ring hydrocarbon (yellow). Like other membrane lipids, sterols are amphipathic. The single hydroxyl group is equivalent to the polar head group in other lipids; the conjugated ring and short hydrocarbon chain form the hydrophobic tail. [See H. Sprong et al., 2001, *Nature Rev. Mol. Cell Biol.* **2**:504.]

(a) Membrane bilayer

Exterior

Cytosol

(b) Polar head groups

Hydrophobic tails

Polar head groups

(c)

Micelle

Liposome

▲ **FIGURE 10-6 The bilayer structure of biomembranes.** (a) Electron micrograph of a thin section through an erythrocyte membrane stained with osmium tetroxide. The characteristic "railroad track" appearance of the membrane indicates the presence of two polar layers, consistent with the bilayer structure for phospholipid membranes. (b) Schematic interpretation of the phospholipid bilayer in which polar groups face outward to shield the hydrophobic fatty acyl tails from water. The hydrophobic effect and van der Waals interactions between the fatty acyl tails drive the assembly of the bilayer (Chapter 2). (c) Cross-sectional views of two other structures formed by dispersal of phospholipids in water. A spherical micelle has a hydrophobic interior composed entirely of fatty acyl chains; a spherical liposome consists of a phospholipid bilayer surrounding an aqueous center. [Part (a) courtesy of J. D. Robertson.]

found in cell membranes (Figure 10-7). Studies on these bilayers have shown that they possess three important properties. First, the hydrophobic core is an impermeable barrier that prevents the diffusion of water-soluble (hydrophilic) solutes across the membrane. Importantly, this simple barrier function is modulated by the presence of membrane proteins that mediate the transport of specific molecules across this otherwise impermeable bilayer. The second property of the bilayer is its stability. The bilayer structure is maintained by hydrophobic and van der Waals interactions between the lipid chains. Even though the exterior aqueous environment can vary widely in ionic strength and pH, the bilayer has the strength to retain its characteristic architecture. Third, all phospholipid bilayers can spontaneously form sealed closed compartments where the aqueous space on the inside is

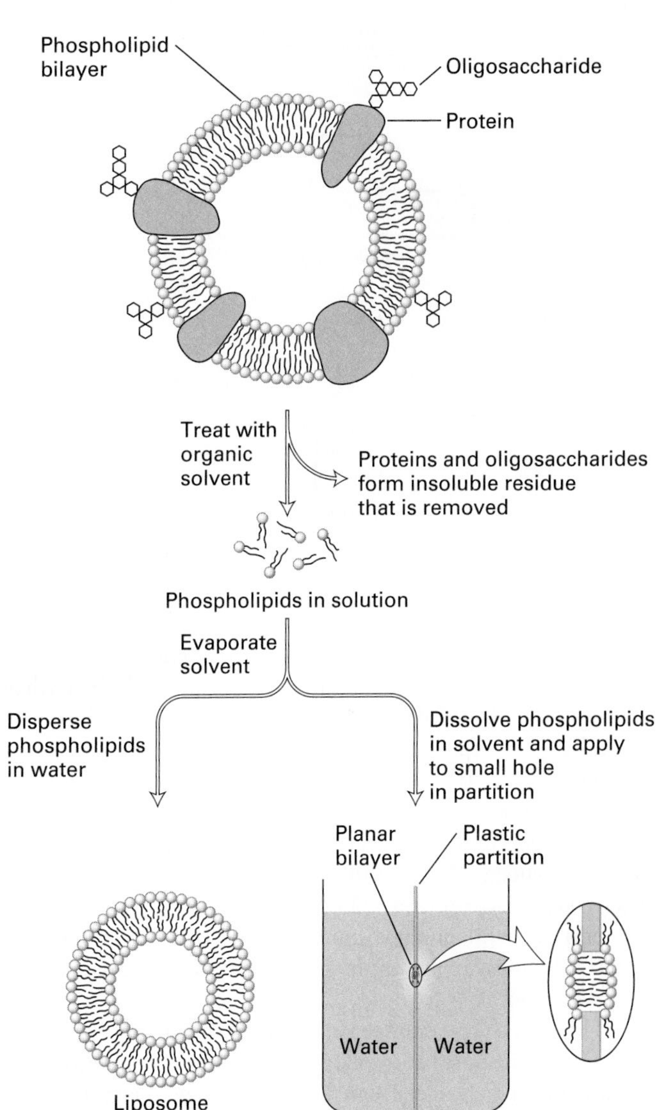

▲ EXPERIMENTAL FIGURE 10-7 **Formation and study of pure phospholipid bilayers.** *(Top)* A preparation of biological membranes is treated with an organic solvent, such as a mixture of chloroform and methanol (3:1), which selectively solubilizes the phospholipids and cholesterol. Proteins and carbohydrates remain in an insoluble residue. The solvent is removed by evaporation. *(Bottom left)* If the lipids are mechanically dispersed in water, they spontaneously form a liposome, shown in cross section, with an internal aqueous compartment. *(Bottom right)* A planar bilayer, also shown in cross section, can form over a small hole in a partition separating two aqueous phases; such bilayers can be used to study the movement of solutes from one solution to the other through the membrane.

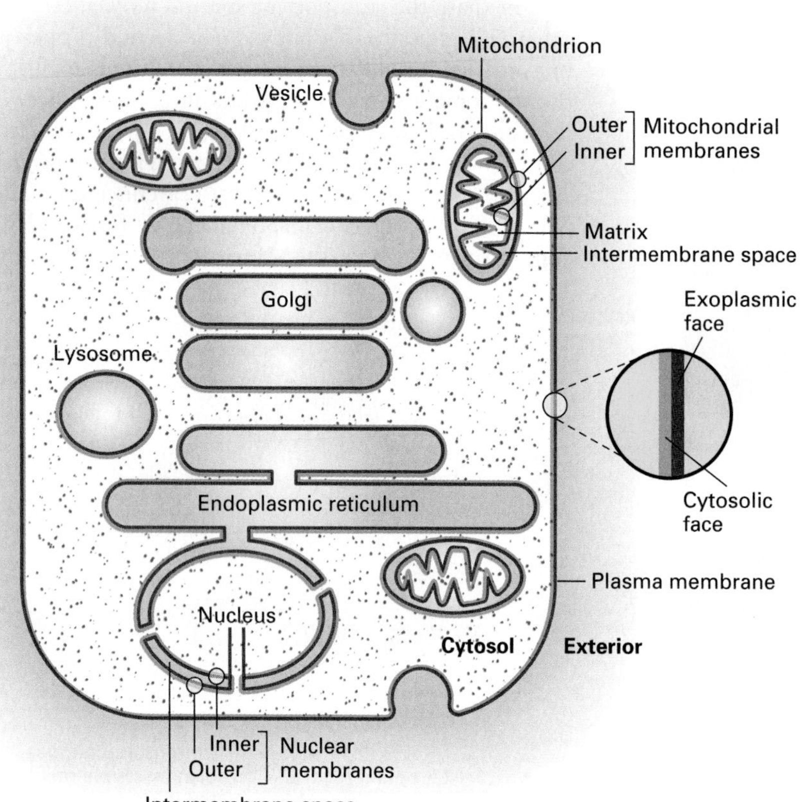

◄ **FIGURE 10-8 The faces of cellular membranes.** The plasma membrane, a single bilayer membrane, encloses the cell. In this highly schematic representation, internal cytosol (green stipple) and external environment (purple) define the cytosolic (red) and exoplasmic (black) faces of the bilayer. Vesicles and some organelles have a single membrane and their internal aqueous space (purple) is topologically equivalent to the outside of the cell. Three organelles—the nucleus, mitochondrion, and chloroplast (which is not shown)—are enclosed by two membranes separated by a small intermembrane space. The exoplasmic faces of the inner and outer membranes around these organelles border the intermembrane space between them. For simplicity, the hydrophobic membrane interior is not indicated in this diagram.

separated from that on the outside. An "edge" of a phospholipid bilayer, as depicted in Figure 10-6b with the hydrocarbon core of the bilayer exposed to an aqueous solution, is unstable; the exposed fatty acyl side chains would be in an energetically much more stable state if they were not adjacent to water molecules but surrounded by other fatty acyl chains (hydrophobic effect; Chapter 2). Thus in aqueous solution, sheets of phospholipid bilayers spontaneously seal their edges, forming a spherical bilayer that encloses an aqueous central compartment. The liposome depicted in Figure 10-6 is an example of such a structure viewed in cross section.

This physical chemical property of a phospholipid bilayer has important implications for cellular membranes: no membrane in a cell can have an "edge" with exposed hydrocarbon fatty acyl chains. All membranes form closed compartments, similar in basic architecture to liposomes. Because all cellular membranes enclose an entire cell or an internal compartment, they have an *internal face* (the surface oriented toward the interior of the compartment) and an *external face* (the surface presented to the environment).

More commonly, we designate the two surfaces of a cellular membrane as the **cytosolic face** and the **exoplasmic face.** This nomenclature is useful in highlighting the topological equivalence of the faces in different membranes, as diagrammed in Figures 10-8 and 10-9. For example, the

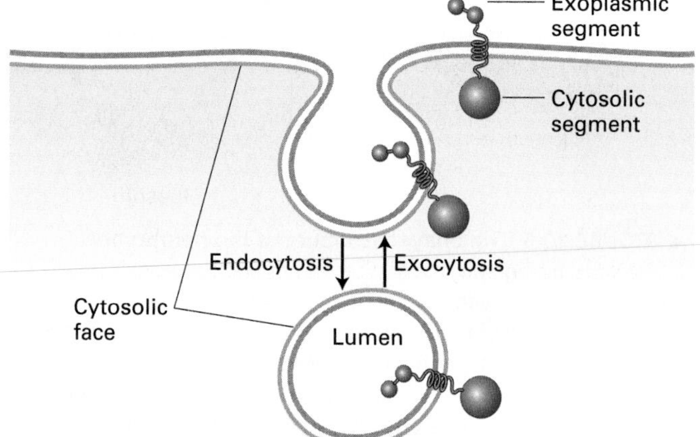

▲ **FIGURE 10-9 Faces of cellular membranes are conserved during membrane budding and fusion.** Red membrane surfaces are cytosolic faces; black are exoplasmic faces. During endocytosis a segment of the plasma membrane buds inward toward the cytosol and eventually pinches off a separate vesicle. During this process the cytosolic face of the plasma membrane remains facing the cytosol and the exoplasmic face of the new vesicle membrane faces the lumen. During exocytosis an intracellular vesicle fuses with the plasma membrane, and the lumen of the vesicle (exoplasmic face) connects with the extracellular medium. Proteins that span the membrane retain their asymmetric orientation during vesicle budding and fusion events; in particular the same segment always faces the cytosol.

exoplasmic face of the plasma membrane is directed away from the cytosol, toward the extracellular space or external environment, and defines the outer limit of the cell. The cytosolic face of the plasma membrane faces the cytosol. Similarly for organelles and vesicles surrounded by a single membrane, the cytosolic face faces the cytosol. The exoplasmic face is always directed away from the cytosol and in this case is on the inside of the organelle in contact with the internal aqueous space, or lumen. That the inside, or lumen, of these vesicles is topologically equivalent to the extracellular space is most easily understood for vesicles that arise by invagination (endocytosis) of the plasma membrane. This process results in the external face of the plasma membrane becoming the internal face of the vesicle membrane, and in the vesicle the cytosolic face of the plasma membrane remains facing the cytosol (see Figure 10-9).

Two distinct membranes surround three organelles—the nucleus, mitochondrion, and chloroplast; the exoplasmic surface of each membrane faces the space between the two membranes. This can perhaps best be understood by reference to the *endosymbiont hypothesis,* discussed in Chapter 1, which posits that mitochondria and chloroplasts arose early in the evolution of eukaryotic cells by the engulfment of bacteria capable of oxidative phosphorylation or photosynthesis, respectively (see Figure 6-20). Thus the exoplasmic face of the mitochondrial inner membrane, derived from the exoplasmic face of the ancestral bacterial plasma membrane, as well as the exoplasmic face of the outer mitochondrial membrane, derived from the exoplasmic face of the ancestral plasma membrane, both face the intermembrane space.

Natural membranes from different cell types exhibit a variety of shapes, which complement a cell's function (Figure 10-10; see Figures 10-3 and 10-4). The smooth, flexible surface of the erythrocyte plasma membrane allows the cell to squeeze through narrow blood capillaries. Some cells have a long, slender extension of the plasma membrane, called a **cilium** or **flagellum,** which beats in a whiplike manner. This motion causes fluid to flow across the surface of a sheet of cells or a sperm cell to swim toward an egg. Membranes are dynamic structures; Figure 10-3 shows an HIV virus budding from the surface of a human cell. In infected cells, specific viral proteins become inserted into the plasma membrane, and segments of the plasma membrane envelop the viral core, or nucleocapsid, that contains the viral RNA genome as the virus buds from the cell. The membrane-coated virus then pinches off from the plasma membrane and is released into the surrounding medium. Internal cellular membranes such as the Golgi complex (see Figure 10-4) are constantly budding off membrane vesicles into the cytosol. These then fuse with other membranes to transport the luminal contents from one organelle to another (Chapter 14).

(a)

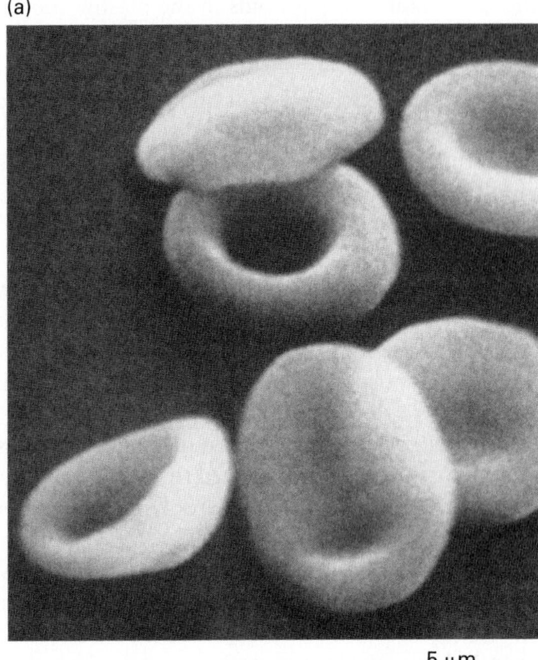

(b)

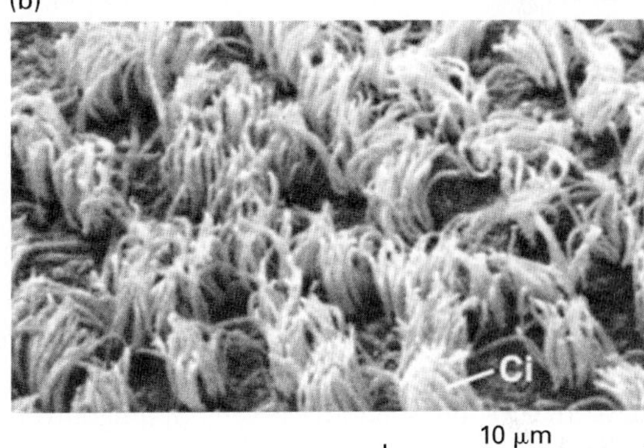

▲ **FIGURE 10-10 Variation in biomembranes in different cell types.** (a) A smooth, flexible membrane covers the surface of the discoid erythrocyte cell as seen in this scanning electron micrograph. (b) Tufts of cilia (Ci) project from the ependymal cells that line the brain ventricles. [Part (a) Copyright © Omi Kron/Photo Researchers, Inc. Part (b) from R. G. Kessel and R. H. Kardon, 1979, *Tissues and Organs: A Text-Atlas of Scanning Electron Microscopy,* W. H. Freeman and Company.]

Biomembranes Contain Three Principal Classes of Lipids

As noted above, a typical biomembrane is assembled from three classes of amphipathic lipids: phosphoglycerides, sphingolipids, and steroids, which differ in their chemical structures, abundance, and functions in the membrane.

Phosphoglycerides, the most abundant class of lipids in most membranes, are derivatives of glycerol 3-phosphate (see Figure 10-5a). A typical phosphoglyceride molecule consists of a hydrophobic tail composed of two fatty acyl chains esterified to the two hydroxyl groups in glycerol phosphate and a polar head group attached to the phosphate group. The two fatty acyl chains may differ in the number of carbons that they contain (commonly 16 or 18) and their degree of saturation (0, 1, or 2 double bonds). A phosphoglyceride is classified according to the nature of its head group. In phosphatidylcholines, the most abundant phospholipids in the plasma membrane, the head group consists of choline, a positively charged alcohol, esterified to the negatively charged phosphate. In other phosphoglycerides, an OH-containing molecule such as ethanolamine, serine, and the sugar derivative inositol is linked to the phosphate group. The negatively charged phosphate group and the positively charged groups or the hydroxyl groups on the head group interact strongly with water. At neutral pH, some phosphoglycerides (e.g., phosphatidylcholine and phosphatidylethanolamine) carry no net electric charge, whereas others (e.g., phosphatidylinositol and phosphatidylserine) carry a single net negative charge. Nonetheless, the polar head groups in all phospholipids can pack together into the characteristic bilayer structure.

The *plasmalogens* are a group of phosphoglycerides that contain one fatty acyl chain attached to carbon 2 of glycerol by an ester linkage and one long hydrocarbon chain attached to carbon 1 of glycerol by an ether (C—O—C) rather than an ester linkage. The abundance of plasmalogens varies among tissues and species but is especially high in human brain and heart tissue. The additional chemical stability of the ether linkage in plasmalogens, compared to the ester linkage, or the subtle differences in their three-dimensional structure compared with that of other phosphoglycerides may have as yet unrecognized physiologic significance.

A second class of membrane lipid is the **sphingolipids.** All of these compounds are derived from sphingosine, an amino alcohol with a long hydrocarbon chain, and contain a long-chain fatty acid attached in an amide linkage to the sphingosine amino group. Like phosphoglycerides, sphingolipids have a phosphate-based polar head. In sphingomyelin, the most abundant sphingolipid, phosphocholine is attached to the terminal hydroxyl group of sphingosine (see Figure 10-5b). Thus sphingomyelin is a phospholipid, and its overall structure is quite similar to that of phosphatidylcholine. Sphingomyelins are similar in shape to phosphoglycerides and can form mixed bilayers with them. Other sphingolipids are amphipathic **glycolipids** whose polar head groups are sugars that are not linked via a phosphate group. Glucosylcerebroside, the simplest glycosphingolipid, contains a single glucose unit attached to sphingosine. In the complex glycosphingolipids called *gangliosides,* one or two branched sugar chains (oligosaccharides) containing sialic acid groups are attached to sphingosine. Glycolipids constitute 2–10 percent of the total lipid in plasma membranes; they are most abundant in nervous tissue.

Cholesterol and its analogs constitute the third important class of membrane lipids, the **steroids.** The basic structure of steroids is a four-ring hydrocarbon. The structures of the principal yeast sterol (ergosterol) and plant phytosterols (e.g., stigmasterol) differ slightly from that of cholesterol, the major animal sterol (see Figure 10-5c). The small differences in the biosynthetic pathways of fungal and animal sterols and in their structures are the basis of most antifungal drugs currently in use. Cholesterol, like the two other sterols, has a hydroxyl substituent on one ring. Although cholesterol is almost entirely hydrocarbon in composition, it is amphipathic because its hydroxyl group can interact with water. Cholesterol is especially abundant in the plasma membranes of mammalian cells but is absent from most prokaryotic and all plant cells. As much as 30–50 percent of the lipids in plant plasma membranes consist of certain steroids unique to plants. Between 50 and 90 percent of the cholesterol in most mammalian cells is present in the plasma membrane and associated vesicles. Cholesterol and other sterols are too hydrophobic to form a bilayer structure on their own. Instead, at concentrations found in natural membranes, these sterols must intercalate between phospholipid molecules to be incorporated into biomembranes.

In addition to its structural role in membranes, cholesterol is the precursor for several important bioactive molecules. They include *bile acids,* which are made in the liver and help emulsify dietary fats for digestion and absorption in the intestines; steroid hormones produced by endocrine cells (e.g., adrenal gland, ovary, testes); and vitamin D produced in the skin and kidneys. Another critical function of cholesterol is its covalent addition to Hedgehog protein, a key signaling molecule in embryonic development (Chapter 16).

Most Lipids and Many Proteins Are Laterally Mobile in Biomembranes

In the two-dimensional plane of a bilayer, thermal motion permits lipid molecules to rotate freely around their long axes and to diffuse laterally within each leaflet. Because such movements are lateral or rotational, the fatty acyl chains remain in the hydrophobic interior of the bilayer. In both natural and artificial membranes, a typical lipid molecule

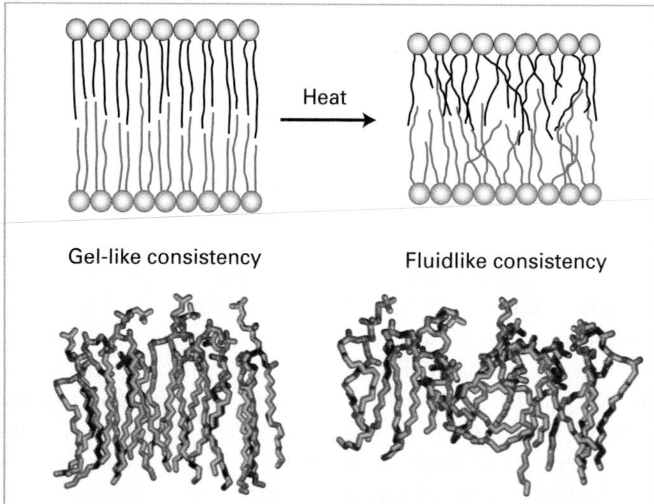

▲ **FIGURE 10-11 Gel and fluid forms of the phospholipid bilayer.** *(Top)* Depiction of gel-to-fluid transition. Phospholipids with long saturated fatty acyl chains tend to assemble into a highly ordered, gel-like bilayer in which there is little overlap of the nonpolar tails in the two leaflets. Heat disorders the nonpolar tails and induces a transition from a gel to a fluid within a temperature range of only a few degrees. As the chains become disordered, the bilayer also decreases in thickness. *(Bottom)* Molecular models of phospholipid monolayers in gel and fluid states, as determined by molecular dynamics calculations. [Bottom based on H. Heller et al., 1993, *J. Phys. Chem.* **97:**8343.]

exchanges places with its neighbors in a leaflet about 10^7 times per second and diffuses several micrometers per second at 37 °C. These diffusion rates indicate that the viscosity of the bilayer is 100 times as great as that of water—about the same as the viscosity of olive oil. Even though lipids diffuse more slowly in the bilayer than in an aqueous solvent, a membrane lipid could diffuse the length of a typical bacterial cell (1 μm) in only 1 second and the length of an animal cell in about 20 seconds. When fluid artificial pure phospholipid membranes are cooled below 37 °C, the lipids can undergo a *phase transition* from a liquidlike (fluid) state to a gel-like (semisolid) state, analogous to the liquid-solid transition when liquid water freezes (Figure 10-11). Below the phase transition temperature, the rate of diffusion of the lipids drops precipitously. At usual physiologic temperatures, the hydrophobic interior of natural membranes generally has a low viscosity and a fluidlike consistency, in contrast to the gel-like consistency observed at lower temperatures.

In pure membrane bilayers, phospholipids and sphingolipids rotate and move laterally, but they do not spontaneously migrate, or flip-flop, from one leaflet to the other. The energetic barrier is too high; migration would require moving the polar head group from its aqueous environment through the hydrocarbon core of the bilayer to the aqueous solution on the other side. Special membrane proteins discussed in Chapter 11 are required for membrane lipids and other polar molecules to flip from one leaflet to the other.

The lateral movements of specific plasma-membrane proteins and lipids can be quantified by a technique called *fluorescence recovery after photobleaching (FRAP)*. Phospholipids containing a fluorescent substituent are used to monitor lipid movement. For proteins, a fragment of a monoclonal antibody that is specific for the exoplasmic domain of the desired protein and that has only a single antigen-binding site is tagged with a fluorescent dye. With this method, described in Figure 10-12, the rate at which

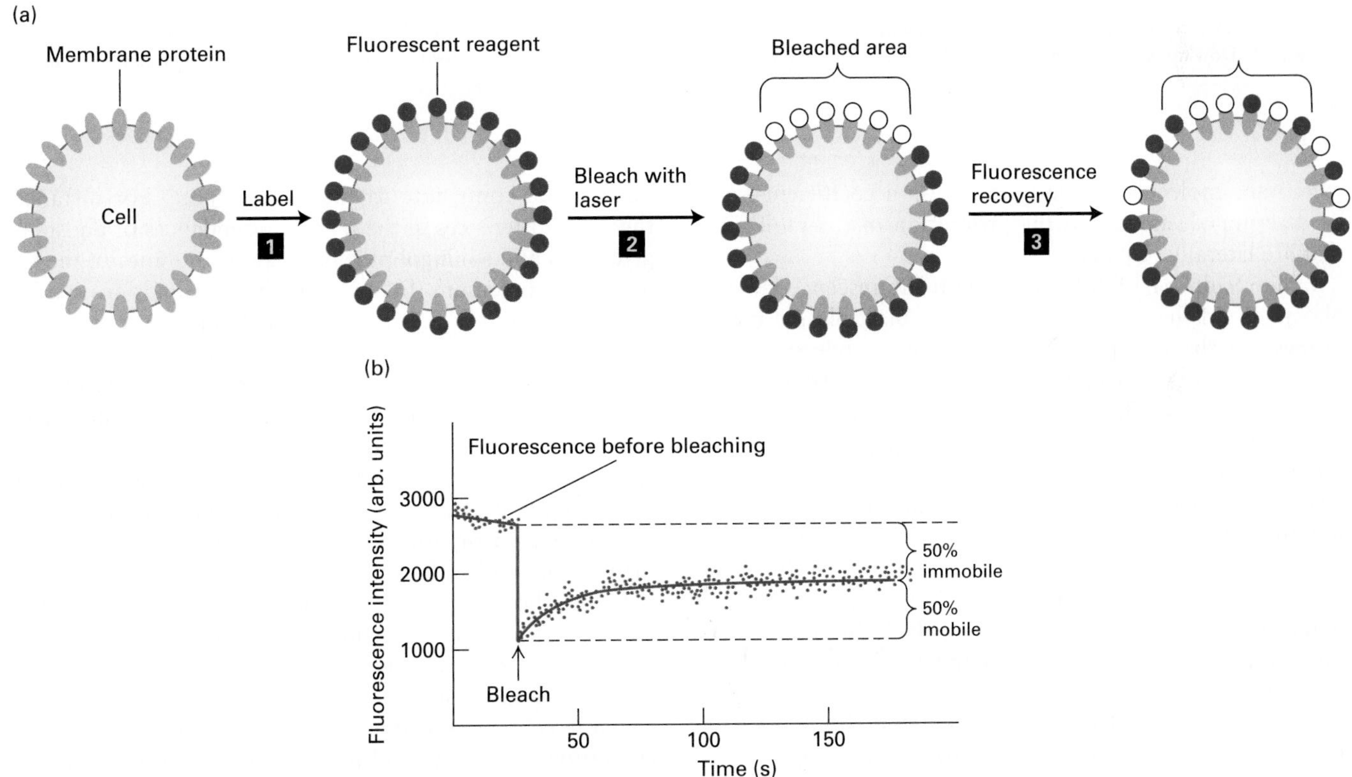

▲ EXPERIMENTAL FIGURE 10-12 **Fluorescence recovery after photobleaching (FRAP) experiments can quantify the lateral movement of proteins and lipids within the plasma membrane.** (a) Experimental protocol. Step (**1**): Cells are first labeled with a fluorescent reagent that binds uniformly to a specific membrane lipid or protein. Step (**2**): A laser light is then focused on a small area of the surface, irreversibly bleaching the bound reagent and thus reducing the fluorescence in the illuminated area. Step (**3**): In time, the fluorescence of the bleached patch increases as unbleached fluorescent surface molecules diffuse into it and bleached ones diffuse outward. The extent of recovery of fluorescence in the bleached patch is proportional to the fraction of labeled molecules that are mobile in the membrane. (b) Results of FRAP experiment with human hepatoma cells treated with a fluorescent antibody specific for the asialoglycoprotein receptor protein. The finding that 50 percent of the fluorescence returned to the bleached area indicates that 50 percent of the receptor molecules in the illuminated membrane patch were mobile and 50 percent were immobile. Because the rate of fluorescence recovery is proportional to the rate at which labeled molecules move into the bleached region, the diffusion coefficient of a protein or lipid in the membrane can be calculated from such data. [See Y. I. Henis et al., 1990, *J. Cell Biol.* **111**:1409.]

TABLE 10-1 Major Lipid Components of Selected Biomembranes

	COMPOSITION (MOL %)			
SOURCE/LOCATION	PC	PE + PS	SM	CHOLESTEROL
Plasma membrane (human erythrocytes)	21	29	21	26
Myelin membrane (human neurons)	16	37	13	34
Plasma membrane (*E. coli*)	0	85	0	0
Endoplasmic reticulum membrane (rat)	54	26	5	7
Golgi membrane (rat)	45	20	13	13
Inner mitochondrial membrane (rat)	45	45	2	7
Outer mitochondrial membrane (rat)	34	46	2	11
Primary leaflet location	Exoplasmic	Cytosolic	Exoplasmic	Both

PC = phosphatidylcholine; PE = phosphatidylethanolamine; PS = phosphatidylserine; SM = sphingomyelin.
SOURCE: W. Dowhan and M. Bogdanov, 2002, in D. E. Vance and J. E. Vance, eds., *Biochemistry of Lipids, Lipoproteins, and Membranes*, Elsevier.

membrane molecules move—the diffusion coefficient—can be determined, as well as the proportion of the molecules that are laterally mobile.

The results of FRAP studies with fluorescence-labeled phospholipids have shown that in fibroblast plasma membranes, all the phospholipids are freely mobile over distances of about 0.5 μm, but most cannot diffuse over much longer distances. These findings suggest that protein-rich regions of the plasma membrane about 1 μm in diameter separate lipid-rich regions containing the bulk of the membrane phospholipid. Phospholipids are free to diffuse within such regions but not from one lipid-rich region to an adjacent one. Furthermore, the rate of lateral diffusion of lipids in the plasma membrane is nearly an order of magnitude slower than in pure phospholipid bilayers: diffusion constants of 10^{-8} cm^2/s and 10^{-7} cm^2/s are characteristic of the plasma membrane and a lipid bilayer, respectively. This difference suggests that lipids may be tightly but not irreversibly bound to certain integral proteins in some membranes, as indeed has recently been demonstrated (see Figure 10-17 below).

Lipid Composition Influences the Physical Properties of Membranes

A typical cell contains myriad types of membranes, each with unique properties bestowed by its particular mix of lipids and proteins. The data in Table 10-1 illustrate the variation in lipid composition in different biomembranes. Several phenomena contribute to these differences. For instance, there are differences in the relative abundances of phosphoglycerides and sphingolipids between membranes in the endoplasmic reticulum (ER), where phospholipids are synthesized, and the Golgi, where sphingolipids are synthesized. The proportion of sphingomyelin as a percentage of total membrane lipid phosphorus is about six times as high in Golgi membranes as it is in ER membranes. In other cases, the movement of membranes from one cellular compartment to another can selectively enrich certain membranes in lipids such as cholesterol. In responding to differing environments throughout an organism, different types of cells generate membranes with differing lipid compositions. In the cells that line the intestinal tract, for example, the membranes that face the harsh environment in which dietary nutrients are digested have a sphingolipid-to-phosphoglyceride-to-cholesterol ratio of 1:1:1 rather than the 0.5:1.5:1 ratio found in cells subject to less stress. The relatively high concentration of sphingolipid in this intestinal membrane may increase its stability because of extensive hydrogen bonding by the free —OH group in the sphingosine moiety (see Figure 10-5).

The degree of bilayer fluidity depends on the lipid composition, structure of the phospholipid hydrophobic tails, and temperature. As already noted, van der Waals interactions and the hydrophobic effect cause the nonpolar tails of phospholipids to aggregate. Long, saturated fatty acyl chains have the greatest tendency to aggregate, packing tightly together into a gel-like state. Phospholipids with

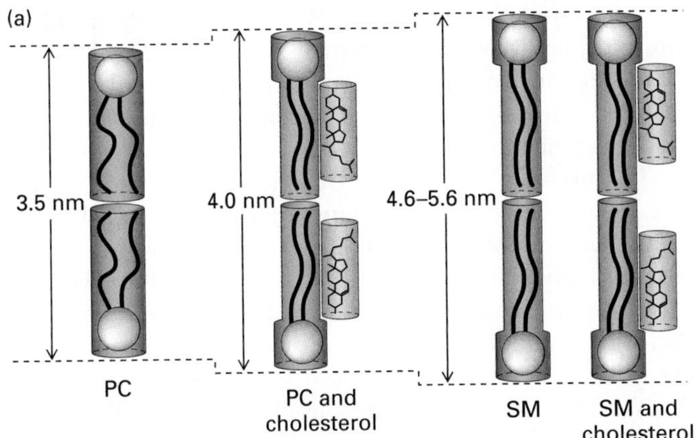

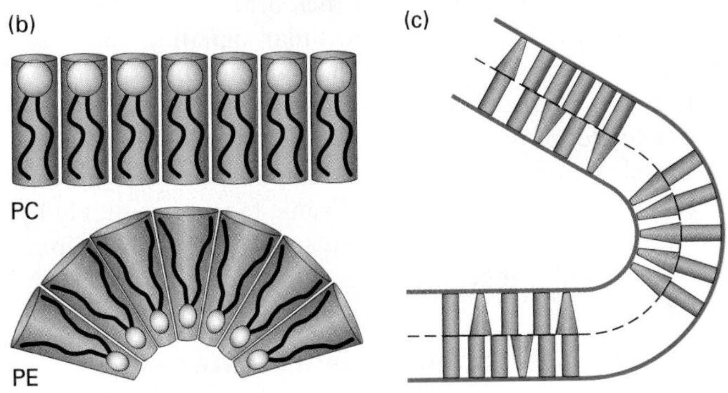

◀ **FIGURE 10-13 Effect of lipid composition on bilayer thickness and curvature.** (a) A pure sphingomyelin (SM) bilayer is thicker than one formed from a phosphoglyceride such as phosphatidylcholine (PC). Cholesterol has a lipid-ordering effect on phosphoglyceride bilayers that increases their thickness but does not affect the thickness of the more ordered SM bilayer. (b) Phospholipids such as PC have a cylindrical shape and form more or less flat monolayers, whereas those with smaller head groups such as phosphatidylethanolamine (PE) have a conical shape. (c) A bilayer enriched with PC in the exoplasmic leaflet and with PE in the cytosolic face, as in many plasma membranes, would have a natural curvature. [Adapted from H. Sprong et al., 2001, *Nature Rev. Mol. Cell Biol.* **2**:504.]

short fatty acyl chains, which have less surface area and therefore fewer van der Waals interactions, form more fluid bilayers. Likewise, the kinks in cis- unsaturated fatty acyl chains (Chapter 2) result in their forming less stable van der Waals interactions with other lipids, and hence more fluid bilayers, than do straight saturated chains, which can pack more tightly together.

Cholesterol is important in maintaining the appropriate fluidity of natural membranes, a property that appears to be essential for normal cell growth and reproduction. Cholesterol restricts the random movement of phospholipid head groups at the outer surfaces of the leaflets, but its effect on the movement of long phospholipid tails depends on concentration. At cholesterol concentrations present in the plasma membrane, the interaction of the steroid ring with the long hydrophobic tails of phospholipids tends to immobilize these lipids and thus decrease biomembrane fluidity. At lower cholesterol concentrations, however, the steroid ring separates and disperses phospholipid tails, causing the inner regions of the membrane to become slightly more fluid.

The lipid composition of a bilayer also influences its thickness, which in turn may influence the distribution of other membrane components, such as proteins, in a particular membrane. The results of biophysical studies on artificial membranes demonstrate that sphingomyelin associates into a more gel-like and thicker bilayer than phospholipids do (Figure 10-13a). Cholesterol and other molecules that decrease membrane fluidity also increase membrane thickness. Because sphingomyelin tails are already optimally stabilized, the addition of cholesterol has no effect on the thickness of a sphingomyelin bilayer.

Another property dependent on the lipid composition of a bilayer is its curvature, which depends on the relative sizes of the polar head groups and nonpolar tails of its constituent phospholipids. Lipids with long tails and large head groups are cylindrical in shape; those with small head groups are cone shaped (Figure 10-13b). As a result, bilayers composed of cylindrical lipids are relatively flat, whereas those containing large numbers of cone-shaped lipids form curved bilayers (Figure 10-13c). This effect of lipid composition on bilayer curvature may play a role in the formation of highly curved membranes, such as sites of viral budding (see Figure 10-3) and formation of internal vesicles from the plasma membrane (see Figure 10-9) and in specialized stable membrane structures such as microvilli. Several proteins bind to the surface of phospholipid bilayers and cause the membrane to curve; such proteins are important in formation of transport vesicles that bud from a donor membrane (Chapter 14).

Lipid Composition Is Different in the Exoplasmic and Cytosolic Leaflets

A characteristic of all membranes is an asymmetry in lipid composition across the bilayer. Although most phospholipids are present in both membrane leaflets, they are commonly more abundant in one or the other leaflet. For instance, in plasma membranes from human erythrocytes and certain canine kidney cells grown in culture, almost all the sphingomyelin and phosphatidylcholine, both of which form less fluid bilayers, are found in the exoplasmic leaflet. In contrast, phosphatidylethanolamine, phosphatidylserine, and phosphatidylinositol, which form more fluid bilayers, are preferentially located in the cytosolic leaflet. This segregation of lipids across the bilayer

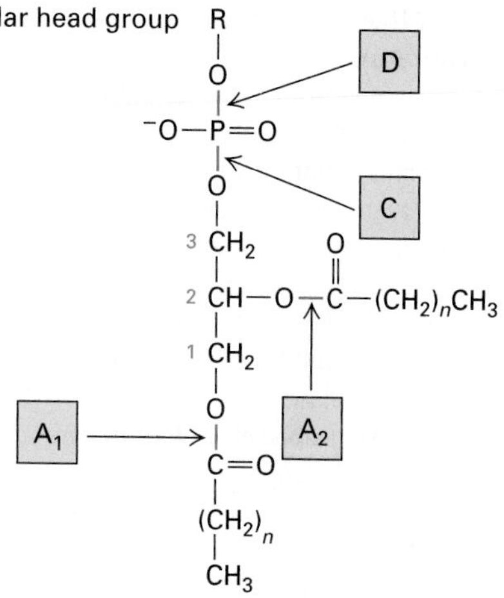

Polar head group R
O
D
$^{-}O-P=O$
O
C
3 CH_2 O
2 CH—O—C—$(CH_2)_nCH_3$
1 CH_2
O
A_1 A_2
C=O
$(CH_2)_n$
CH_3

▲ **FIGURE 10-14 Specificity of phospholipases.** Each type of phospholipase cleaves one of the susceptible bonds shown in red. The glycerol carbon atoms are indicated by small numbers. In intact cells, only phospholipids in the exoplasmic leaflet of the plasma membrane are cleaved by phospholipases in the surrounding medium. Phospholipase C, a cytosolic enzyme, cleaves certain phospholipids in the cytosolic leaflet of the plasma membrane.

may influence membrane curvature (see Figure 10-13c). Unlike particular phospholipids, cholesterol is relatively evenly distributed in both leaflets of cellular membranes. The relative abundance of a particular phospholipid in the two leaflets of a plasma membrane can be determined experimentally on the basis of the susceptibility of phospholipids to hydrolysis by **phospholipases,** enzymes that cleave various bonds in the hydrophilic ends of phospholipids (Figure 10-14). When added to the external medium, phospholipases cannot cross the membrane, and thus they cleave off the head groups of only those lipids present in the exoplasmic face; phospholipids in the cytosolic leaflet are resistant to hydrolysis because the enzymes cannot penetrate to the cytosolic face of the plasma membrane.

How the asymmetric distribution of phospholipids in membrane leaflets arises is still unclear. As noted, in pure bilayers phospholipids do not spontaneously migrate, or flip-flop, from one leaflet to the other. To a first approximation, the asymmetry in phospholipid distribution results from synthesis of these lipids in the endoplasmic reticulum and Golgi. Sphingomyelin is synthesized on the luminal (exoplasmic) face of the Golgi, which becomes the exoplasmic face of the plasma membrane. In contrast, phosphoglycerides are synthesized on the cytosolic face of the ER membrane, which is topologically identical with the cytosolic face of the plasma membrane (see Figure 10-8). Clearly, this explanation does not account for the preferential location of phosphatidylcholine in the exoplasmic leaflet. Movement of this phosphoglyceride, and perhaps others, from one leaflet to the other in some natural membranes is catalyzed by ATP-powered transport proteins called **flippases,** which are discussed in Chapter 11.

The preferential location of lipids on one face of the bilayer is necessary for a variety of membrane-based functions. For example, the head groups of all phosphorylated forms of phosphatidylinositol (see Figure 10-5; PI) face the cytosol. Stimulation of many cell-surface receptors by their corresponding hormone results in activation of the cytosolic enzyme phospholipase C, which can then hydrolyze the bond connecting the phosphoinositols to the diacylglycerol. As we will see in Chapter 15, both water soluble phosphoinositols and membrane-embedded diacylglycerol participate in intracellular signaling pathways that affect many aspects of cellular metabolism. Phosphatidylserine also is normally most abundant in the cytosolic leaflet of the plasma membrane. In the initial stages of platelet stimulation by serum, phosphatidylserine is briefly translocated to the exoplasmic face, presumably by a flippase enzyme, where it activates enzymes participating in blood clotting.

Cholesterol and Sphingolipids Cluster with Specific Proteins in Membrane Microdomains

Membrane lipids are not randomly distributed (evenly mixed) in each leaflet of a bilayer. One hint that lipids may be organized within the leaflets was the discovery that the lipids remaining after the extraction of plasma membranes with nonionic detergents predominantly contain two species: cholesterol and sphingomyelin. Because these two lipids are found in more ordered, less fluid bilayers, researchers hypothesized that they form microdomains, termed **lipid rafts,** surrounded by other, more fluid phospholipids that are more readily extracted by detergents. Some biochemical and microscopic evidence supports the existence of lipid rafts, which in natural membranes are typically 50 nm in diameter. Rafts can be disrupted by methyl-β-cyclodextrin, which specifically extracts cholesterol out of membranes, or by antibiotics, such as filipin, that sequester cholesterol into aggregates within the membrane. Such findings indicate the importance of cholesterol in maintaining the integrity of these rafts. As judged by their presence in the complex of cholesterol and sphingomyelin remaining after detergent extraction, lipid rafts in plasma membranes are thought to be enriched for a subset of plasma membrane proteins, including those that participate in sensing extracellular signals and transmitting them into the cytosol. Thus by bringing many key proteins into close proximity these lipid-protein complexes may facilitate signaling by cell-surface receptors and the subsequent activation of cytosolic events. However, much remains to be learned about the structure and biological function of lipid rafts.

KEY CONCEPTS OF SECTION 10.1

Biomembranes: Lipid Composition and Structural Organization

- The eukaryotic cell is demarcated from the external environment by the plasma membrane and organized into membrane-limited internal compartments (organelles and vesicles).
- The phospholipid bilayer, the basic structural unit of all biomembranes, is a two-dimensional lipid sheet with hydrophilic faces and a hydrophobic core, which is impermeable to water-soluble molecules and ions (see Figure 10-6).
- The primary lipid components of biomembranes are phosphoglycerides, sphingolipids, and sterols such as cholesterol (see Figure 10-5).
- Most lipids and many proteins are laterally mobile in biomembranes.
- Membranes can undergo phase transitions from the fluid to gel-like states depending on the temperature and composition of the membrane.
- Different cellular membranes vary in lipid composition (see Table 10-1). Phospholipids and sphingolipids are asymmetrically distributed in the two leaflets of the bilayer, whereas cholesterol is fairly evenly distributed in both leaflets.
- Natural biomembranes generally have a viscous consistency with fluidlike properties. In general, membrane fluidity is decreased by sphingolipids and cholesterol and increased by phosphoglycerides. The lipid composition of a membrane also influences its thickness and curvature (see Figure 10-13).
- Lipid rafts are microdomains containing cholesterol, sphingolipids, and certain membrane proteins that form in the plane of the bilayer. These aggregates might facilitate signaling by certain plasma-membrane receptors.

10.2 Biomembranes: Protein Components and Basic Functions

Membrane proteins are defined by their location within or at the surface of a phospholipid bilayer. Although every biological membrane has the same basic bilayer structure, the proteins associated with a particular membrane are responsible for its distinctive activities. The kinds and amounts of proteins associated with biomembranes vary depending on cell type and subcellular location. For example, the inner mitochondrial membrane is 76 percent protein; the myelin membrane that surrounds nerve axons, only 18 percent. The high phospholipid content of myelin allows it to electrically insulate the nerve from its environment, as we discuss in Chapter 23. The importance of membrane proteins is suggested from the finding that approximately a third of all yeast genes encode a membrane protein. The relative abundance of genes for membrane proteins is greater in multicellular organisms in which membrane proteins have additional functions in cell adhesion.

The lipid bilayer presents a distinctive two-dimensional hydrophobic environment for membrane proteins. Some proteins contain segments that are imbedded within the hydrophobic core of the phospholipid bilayer; other proteins are associated with the exoplasmic or cytosolic leaflet of the bilayer. Protein domains on the extracellular surface of the plasma membrane generally bind to extracellular molecules, including external signaling proteins, ions, and small metabolites (e.g., glucose, fatty acids), as well as proteins on other cells or in the external environment. Segments of the protein within the plasma membrane have a variety of functions, including those that form channels and pores in transport proteins that move molecules and ions into and out of cells. Domains lying along the cytosolic face of the plasma membrane have a wide range of functions, from anchoring cytoskeletal proteins to the membrane to triggering intracellular signaling pathways.

In many cases, the function of a membrane protein and the topology of its polypeptide chain in the membrane can be predicted on the basis of its similarity with other well-characterized proteins. In this section, we examine the characteristic structural features of membrane proteins and some of their basic functions. We will describe the structures of several proteins to help you get a feel for the way membrane proteins interact with membranes. More complete characterization of the properties of various types of membrane proteins is presented in later chapters that focus on their structures and activities in the context of their cellular functions.

Proteins Interact with Membranes in Three Different Ways

Membrane proteins can be classified into three categories—integral, lipid-anchored, and peripheral—on the basis of the nature of the membrane–protein interactions (see Figure 10-1). **Integral membrane proteins,** also called *transmembrane proteins,* span a phospholipid bilayer and comprise three segments. The cytosolic and exoplasmic domains have hydrophilic exterior surfaces that interact with the aqueous solutions on the cytosolic and exoplasmic faces of the membrane. These domains resemble segments of other water-soluble proteins in their amino acid composition and structure. In contrast, the membrane-spanning segments usually contain many hydrophobic amino acids whose side chains protrude outward and interact with the hydrophobic hydrocarbon core of the phospholipid bilayer. In all transmembrane proteins examined to date, the membrane-spanning domains consist of one or more α helices or of multiple β strands. Because they must be inserted into membranes, the ribosomal synthesis and posttranslational processing of integral membrane proteins differs from that of soluble cytosolic proteins and is discussed separately in Chapters 13 and 14.

Lipid-anchored membrane proteins are bound covalently to one or more lipid molecules. The hydrophobic segment of the attached lipid is embedded in one leaflet of the membrane and anchors the protein to the membrane. The polypeptide chain itself does not enter the phospholipid bilayer.

Peripheral membrane proteins do not directly contact the hydrophobic core of the phospholipid bilayer. Instead they are bound to the membrane either indirectly by interactions with integral or lipid-anchored membrane proteins or directly by interactions with lipid head groups. Peripheral proteins can be bound to either the cytosolic or the exoplasmic face of the plasma membrane. In addition to these proteins, which are closely associated with the bilayer, cytoskeletal filaments can be more loosely associated with the cytosolic face, usually through one or more peripheral (adapter) proteins. Such associations with the cytoskeleton provide support for various cellular membranes, helping to determine cell shape and mechanical properties, and play a role in the two-way communication between the cell interior and the exterior, as we learn in Chapter 17. Finally, peripheral proteins on the outer surface of the plasma membrane and the exoplasmic domains of integral membrane proteins are often attached to components of the extracellular matrix or to the cell wall surrounding bacterial and plant cells, providing a crucial interface between the cell and its environment.

Most Transmembrane Proteins Have Membrane-Spanning α Helices

Soluble proteins exhibit hundreds of distinct localized folded structures, or motifs (see Figure 3-9). In comparison, the repertoire of folded structures in the transmembrane domains of integral membrane proteins is quite limited, with the hydrophobic α helix predominating. Proteins containing membrane-spanning α-helical domains are stably embedded in membranes because of energetically favorable hydrophobic and van der Waals interactions of the hydrophobic side chains in the domain with specific lipids and probably also by ionic interactions with the polar head groups of the phospholipids.

A single α-helical domain is sufficient to incorporate an integral membrane protein into a membrane. However, many proteins have more than one transmembrane α helix. Typically, a membrane-embedded α helix is composed of a continuous segment of 20–25 hydrophobic (uncharged) amino acids (see Figure 2-14). The predicted length of such an α helix (3.75 nm) is just sufficient to span the hydrocarbon core of a phospholipid bilayer. In many membrane proteins, these helices are perpendicular to the plane of the membrane, whereas in others, the helices traverse the membrane at an oblique angle. The hydrophobic side chains protrude outward from the helix (Figure 10-15) and form van

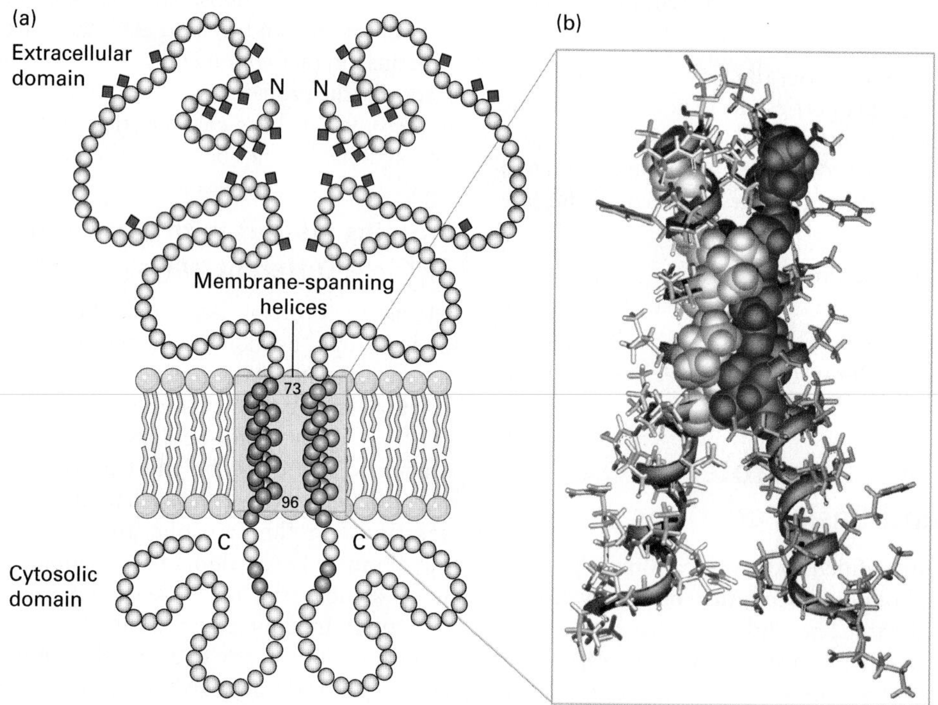

▲ **FIGURE 10-15 Structure of glycophorin A, a typical single-pass transmembrane protein.** (a) Diagram of dimeric glycophorin showing major sequence features and its relation to the membrane. The single 23-residue membrane-spanning α helix in each monomer is composed of amino acids with hydrophobic (uncharged) side chains (red and green spheres). By binding negatively charged phospholipid head groups, the positively charged arginine and lysine residues (blue spheres) near the cytosolic side of the helix help anchor glycophorin in the membrane. Both the extracellular and the cytosolic domains are rich in charged residues and polar uncharged residues; the extracellular domain is heavily glycosylated, with the carbohydrate side chains (green diamonds) attached to specific serine, threonine, and asparagine residues. (b) Molecular model of the transmembrane domain of dimeric glycophorin corresponding to residues 73–96. The hydrophobic side chains of the α helix in one monomer are shown in pink; those in the other monomer, in green. Residues depicted as space-filling structures participate in intermonomer van der Waals interactions that stabilize the coiled-coil dimer. [Part (b) adapted from K. R. MacKenzie et al., 1997, *Science* **276:**131.]

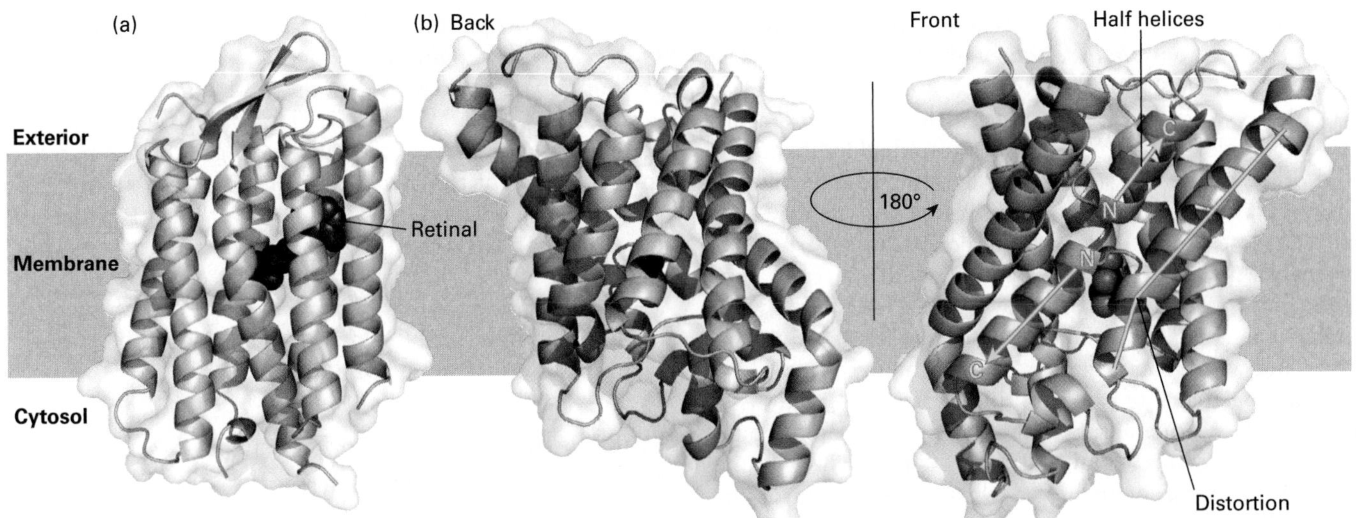

▲ **FIGURE 10-16 Structural models of two multipass membrane proteins.** (a) Bacteriorhodopsin, a photoreceptor in certain bacteria. The seven hydrophobic α helices in bacteriorhodopsin traverse the lipid bilayer roughly perpendicular to the plane of the membrane. A retinal molecule (black) covalently attached to one helix absorbs light. The large class of G protein–coupled receptors in eukaryotic cells also has seven membrane-spanning α helices; their three-dimensional structure is thought to be similar to that of bacteriorhodopsin. (b) Two views of the glycerol channel Glpf, rotated 180° with respect to each other along an axis perpendicular to the plane of the membrane. Note several membrane-spanning α helices that are at oblique angles, the two helices that penetrate only halfway through the membrane (purple with yellow arrows), and one long membrane-spanning helix with a "break" or distortion in the middle (purple with yellow line). The glycerol molecule in the hydrophilic "core" is colored red. The structure was approximately positioned in the hydrocarbon core of the membrane by finding the most hydrophobic 3 mm slab of the protein perpendicular to the membrane plane. [Part (a) After H. Luecke et al., 1999, *J. Mol. Biol.* **291:**899. Part (b) after J. Bowie, 2005, *Nature* **438:**581–589, and D. Fu et. al., 2000, *Science* **290:**481–486.]

der Waals interactions with the fatty acyl chains in the bilayer. In contrast, the hydrophilic amide peptide bonds are in the interior of the α helix (see Figure 3-4); each carbonyl (C=O) group forms a hydrogen bond with the amide hydrogen atom of the amino acid four residues toward the C-terminus of the helix. These polar groups are shielded from the hydrophobic interior of the membrane.

To help you get a better sense of the structures of proteins with α-helical domains, we will briefly discuss three different kinds of such proteins: glycophorin A, G protein-coupled receptors, and aquaporins (water/glycerol channels).

Glycophorin A, the major protein in the erythrocyte plasma membrane, is a representative *single-pass* transmembrane protein, which contains only one membrane-spanning α helix (Figure 10-15). The transmembrane helix of one glycophorin A polypeptide associates with the corresponding transmembrane helix in a second glycophorin A to form a coiled-coil dimer (Figure 10-15b). Such interactions of membrane-spanning α helices are a common mechanism for creating dimeric membrane proteins, and many membrane proteins form oligomers (two or more polypeptides bound together noncovalently) by interactions between their membrane-spanning helices.

A large and important group of integral proteins is defined by the presence of seven membrane-spanning α helices; this includes the large family of G protein–coupled cell-surface receptors discussed in Chapter 15. One such *multipass* transmembrane protein whose structure is known in molecular detail is bacteriorhodopsin, a protein found in the membrane of certain photosynthetic bacteria; it illustrates the general structure of all these proteins (Figure 10-16a). Absorption of light by the retinal group covalently attached to this protein causes a conformational change in the protein that results in the pumping of protons from the cytosol across the bacterial membrane to the extracellular space. The proton concentration gradient thus generated across the membrane is used to synthesize ATP (Chapter 12). In the high-resolution structure of bacteriorhodopsin the positions of all the individual amino acids, retinal, and the surrounding lipids are clearly defined. As might be expected, virtually all of the amino acids on the exterior of the membrane-spanning segments of bacteriorhodopsin are hydrophobic, permitting energetically favorable interactions with the hydrocarbon core of the surrounding lipid bilayer.

The **aquaporins** are a large family of highly conserved proteins that transport water, glycerol, and other hydrophilic molecules across biomembranes. They illustrate several aspects of the structure of multipass transmembrane proteins. Aquaporins are tetramers of four identical subunits. Each of the four subunits has six membrane-spanning α helixes, some of which traverse the membrane at oblique angles rather than perpendicularly. Because the aquaporins have similar structures, we will focus on one that has an especially well-defined structure determined by x-ray diffraction studies (Figure 10-16b). This aquaporin has one long transmembrane helix with a bend in the middle, and more strikingly, there are two α helixes that penetrate only *halfway* through the membrane. The N-termini of these helices face each other (yellow N's in the figure), and together they span the membrane at an oblique angle. Thus some membrane-embedded helices—and other nonhelical structures we will encounter later—do not traverse the entire bilayer. As we will see in Chapter 11, these short helixes in aquaporins form part of the glycerol/water-selective pore in the middle of each subunit. This highlights the considerable diversity in the ways membrane-spanning α helices interact with the lipid bilayer and with other segments of the protein.

The specificity of phospholipid-protein interactions is revealed in great detail in the structure of a different aquaporin, aquaporin 0 (Figure 10-17). Aquaporin 0 is the most abundant protein in the plasma membrane of the fiber cells that make up the bulk of the lens of the mammalian eye. Like other aquaporins, it is a tetramer of identical subunits. The protein's surface is not covered by a set of uniform binding sites for phospholipid molecules. Instead, fatty acyl side chains pack tightly against the irregular hydrophobic outer surface of the protein; some of the fatty acyl chains are straight, in the *all-trans* conformation (Chapter 2), whereas others are kinked in order to interact with bulky hydrophilic side chains on the surface of the protein. Some of the lipid head groups are parallel to the surface of the membrane, as is the case in purified phospholipid bilayers. Others, however, are oriented almost at right angles to the plane of the membrane. Thus there can be specific interactions between phospholipids and membrane-spanning proteins, and the function of many membrane proteins can be affected by the specific types of phospholipid present in the bilayer.

MEDIA CONNECTIONS

Podcast: Annular Phospholipids

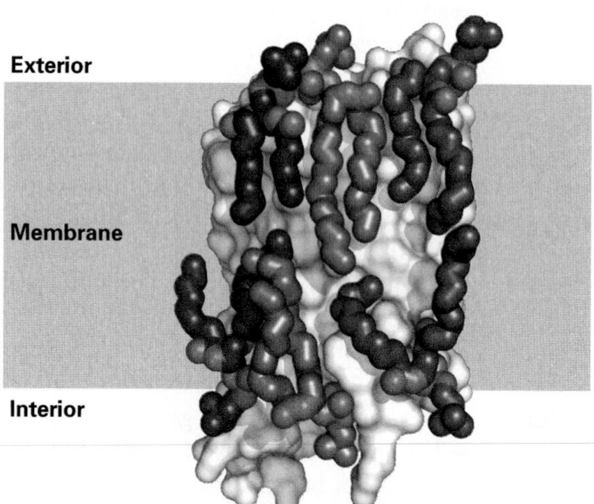

▲ **FIGURE 10-17 Annular phospholipids.** Side view of the three-dimensional structure of one subunit of the lens-specific aquaporin 0 homotetramer, crystallized in the presence of the phospholipid dimyristoylphosphatidylcholine, a phospholipid with 14 carbon-saturated fatty acyl chains. Note the lipid molecules forming a bilayer shell around the protein. The protein is shown as a surface plot (the lighter background molecule). The lipid molecules are shown in space-fill format; the polar lipid head groups (grey and red) and the lipid fatty acyl chains (black and grey) form a bilayer with almost uniform thickness around the protein. Presumably, in the membrane, lipid fatty acyl chains will cover the whole of the hydrophobic surface of the protein; only the most ordered of the lipid molecules will be resolved in the crystallographic structure. [After A. Lee, 2005, *Nature* **438**:569–570, and T. Gonen et. al., 2005, *Nature* **438**:633–688.]

Multiple β Strands in Porins Form Membrane-Spanning "Barrels"

The **porins** are a class of transmembrane proteins whose structure differs radically from that of other integral proteins based on α-helical transmembrane domains. Several types of porins are found in the outer membrane of gram-negative bacteria such as *E. coli* and in the outer membranes of mitochondria and chloroplasts. The outer membrane protects an intestinal bacterium from harmful agents (e.g., antibiotics, bile salts, and proteases) but permits the uptake and disposal of small hydrophilic molecules, including nutrients and waste products. Different types of porins in the outer membrane of an *E. coli* cell provide channels for the passage of specific types of disaccharides or other small molecules as well as of ions such as phosphate. The amino acid sequences of porins contain none of the long hydrophobic segments typical of integral proteins with α-helical membrane-spanning domains. X-ray crystallography has revealed that porins are trimers of identical subunits. In each subunit, 16 β strands form a sheet that twists into a barrel-shaped structure with a pore in the center (Figure 10-18). Unlike a typical water-soluble globular protein, a porin has a hydrophilic interior and a hydrophobic exterior; in this sense, porins are inside out. In a porin monomer, the outward-facing side groups on each of the β strands are hydrophobic and form a nonpolar ribbonlike band that encircles the outside of the barrel. This hydrophobic band interacts with the fatty acyl groups of the membrane lipids or with other porin monomers. The side groups facing the inside of a porin monomer are predominantly hydrophilic; they line the pore through which small water-soluble molecules cross the membrane. (Note that the aquaporins discussed above, despite their name, are not porins and contain multiple transmembrane α helices.)

Covalently Attached Hydrocarbon Chains Anchor Some Proteins to Membranes

In eukaryotic cells, several types of covalently attached lipids anchor some otherwise typically water soluble proteins to one or the other leaflet of the plasma membrane or other cellular membranes. In these lipid-anchored proteins, the lipid

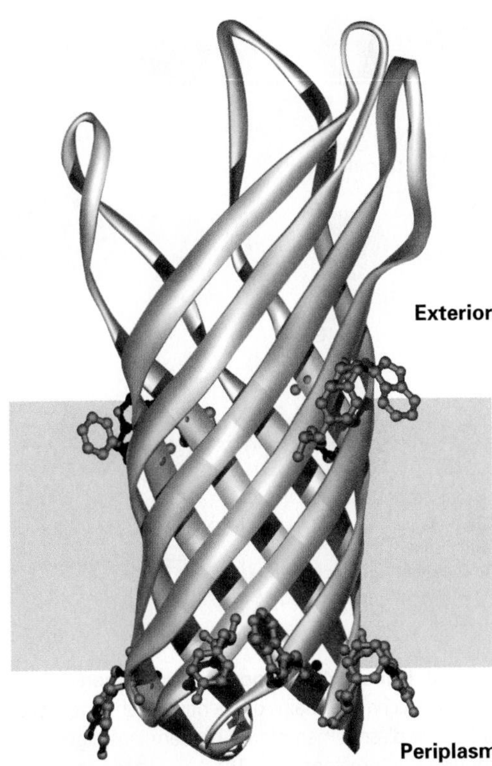

▲ **FIGURE 10-18 Structural model of one subunit of OmpX, a porin found in the outer membrane of *E. coli*.** All porins are trimeric transmembrane proteins. Each subunit is barrel shaped, with β strands forming the wall and a transmembrane pore in the center. A band of aliphatic (hydrophobic and noncyclic) side chains (yellow) and a border of aromatic (ring-containing) side chains (red) position the protein in the bilayer. [After G. E. Schulz, 2000, *Curr. Opin. Struc. Biol.* **10**:443.]

hydrocarbon chains are embedded in the bilayer, but the protein itself does not enter the bilayer. The anchors used to insert proteins at the cytosolic face are not used for the exoplasmic face and vice versa.

One group of cytosolic proteins are anchored to the cytosolic face of a membrane by a fatty acyl group (e.g., myristate or palmitate) covalently attached to an N-terminal glycine residue (Figure 10-19a). Retention of such proteins at the membrane by the N-terminal acyl anchor, called *acylation*, may play an important role in a membrane-associated function. For example, v-Src, a mutant form of a cellular tyrosine kinase, induces abnormal cellular growth that can lead to cancer but only when it has a myristylated N-terminus.

A second group of cytosolic proteins are anchored to membranes by a hydrocarbon chain attached to a cysteine residue at or near the C-terminus (Figure 10-19b). Some of these chains are *prenyl anchors*, built from 5-carbon isoprene units, which, as detailed in the following section, are also used in the synthesis of cholesterol. In these proteins, a 15-carbon farnesyl or 20-carbon geranylgeranyl group is bound through a thioether bond to the —SH group of a C-terminal cysteine residue. In some cases, a second geranylgeranyl group or a fatty acyl palmitate group is linked to a nearby cysteine residue. The additional hydrocarbon anchor is thought to reinforce the attachment of the protein to the membrane. For example, Ras, a GTPase superfamily protein that functions in intracellular signaling (Chapter 16), is recruited to the cytosolic face of the plasma membrane by such a double anchor. Rab proteins, which also belong to the GTPase superfamily, are similarly bound to the cytosolic surface of intracellular vesicles by prenyl anchors; these proteins are required for the fusion of vesicles with their target membranes (Chapter 14).

Some cell-surface proteins and specialized proteins with distinctive covalently attached polysaccharides called

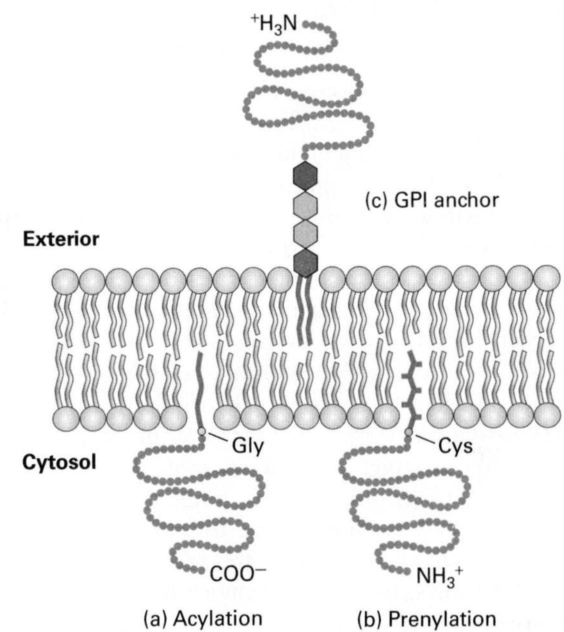

▲ **FIGURE 10-19 Anchoring of plasma-membrane proteins to the bilayer by covalently linked hydrocarbon groups.** (a) Cytosolic proteins such as v-Src are associated with the plasma membrane through a single fatty acyl chain attached to the N-terminal glycine (Gly) residue of the polypeptide. Myristate (C_{14}) and palmitate (C_{16}) are common acyl anchors. (b) Other cytosolic proteins (e.g., Ras and Rab proteins) are anchored to the membrane by prenylation of one or two cysteine (Cys) residues, at or near the C-terminus. The anchors are farnesyl (C_{15}) and geranylgeranyl (C_{20}) groups, both of which are unsaturated. (c) The lipid anchor on the exoplasmic surface of the plasma membrane is glycosylphosphatidylinositol (GPI). The phosphatidylinositol part (red) of this anchor contains two fatty acyl chains that extend into the bilayer. The phosphoethanolamine unit (purple) in the anchor links it to the protein. The two green hexagons represent sugar units, which vary in number, nature, and arrangement in different GPI anchors. The complete structure of a yeast GPI anchor is shown in Figure 13-14. [Adapted from H. Sprong et al., 2001, *Nature Rev. Mol. Cell Biol.* **2**:504.]

proteoglycans (Chapter 19) are bound to the exoplasmic face of the plasma membrane by a third type of anchor group, glycosylphosphatidylinositol (GPI). The exact structures of *GPI anchors* vary greatly in different cell types, but they always contain phosphatidylinositol (PI), whose two fatty acyl chains extend into the lipid bilayer just like those of typical membrane phospholipids; phosphoethanolamine, which covalently links the anchor to the C-terminus of a protein; and several sugar residues (Figure 10-19c). Therefore GPI anchors are glycolipids. The GPI anchor is both necessary and sufficient for binding proteins to the membrane. For instance, the enzyme phospholipase C cleaves the phosphate-glycerol bond in phospholipids and in GPI anchors (see Figure 10-14). Treatment of cells with phospholipase C releases GPI-anchored proteins such as Thy-1 and placental alkaline phosphatase (PLAP) from the cell surface.

All Transmembrane Proteins and Glycolipids Are Asymmetrically Oriented in the Bilayer

Every type of transmembrane protein has a specific orientation, known as its *topology*, with respect to the membrane faces. Its cytosolic segments are always facing the cytoplasm, and exoplasmic segments always face the opposite side of the membrane. This asymmetry in protein orientation confers different properties on the two membrane faces. The orientation of different types of transmembrane proteins is established during their synthesis, as we describe in Chapter 13. Membrane proteins have never been observed to flip-flop across a membrane; such movement, requiring a transient movement of hydrophilic amino acid residues through the hydrophobic interior of the membrane, would be energetically unfavorable. Accordingly, the asymmetric topology of a transmembrane protein, which is established during its biosynthetic insertion into a membrane, is maintained throughout the protein's lifetime. As Figure 10-9 shows, membrane proteins retain their asymmetric orientation in the membrane during membrane budding and fusion events; the same segment always faces the cytosol and the same segment is always exposed to the exoplasmic face.

Many transmembrane proteins contain carbohydrate chains covalently linked to serine, threonine, or asparagine side chains of the polypeptide. Such transmembrane **glycoproteins** are always oriented so that all carbohydrate chains are in the exoplasmic domain (see Figure 10-15). Likewise, glycolipids, in which a carbohydrate chain is attached to the glycerol or sphingosine backbone, are always located in the exoplasmic leaflet with the carbohydrate chain protruding from the membrane surface. The biosynthetic basis for the asymmetric glycosylation of proteins is described in Chapter 14. Both glycoproteins and glycolipids are especially abundant in the plasma membranes of eukaryotic cells; they are absent from the inner mitochondrial membrane, chloroplast lamellae, and several other intracellular membranes. Because the carbohydrate chains of glycoproteins and glycolipids in the plasma membrane

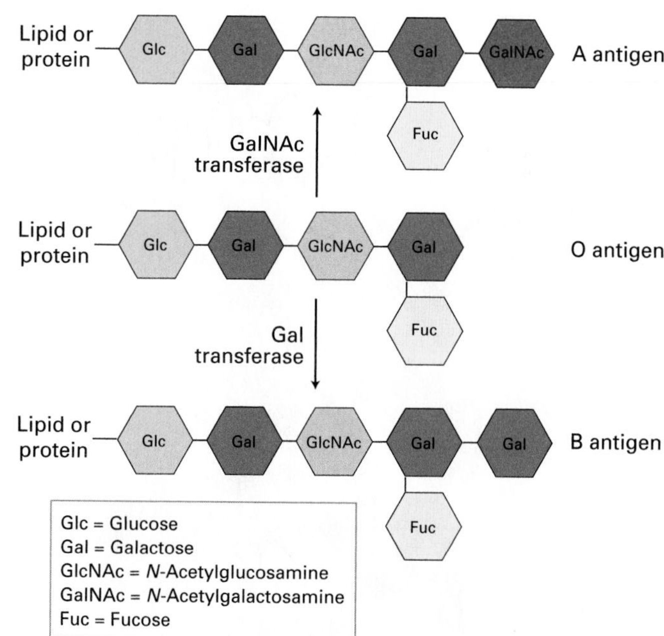

▲ **FIGURE 10-20 Human ABO blood group antigens.** These antigens are oligosaccharide chains covalently attached to glycolipids or glycoproteins in the plasma membrane. The terminal oligosaccharide sugars distinguish the three antigens. The presence or absence of the glycosyltransferases that add galactose (Gal) or *N*-acetylgalactosamine (GalNAc) to O antigen determine a person's blood type.

extend into the extracellular space, they are available to interact with components of the extracellular matrix as well as **lectins** (proteins that bind specific sugars), growth factors, and antibodies.

One important consequence of such interactions is illustrated by the A, B, and O blood group antigens. These three structurally related oligosaccharide components of certain glycoproteins and glycolipids are expressed on the surfaces of human red blood cells and many other cell types (Figure 10-20). All humans have the enzymes for synthesizing O antigen. Persons with type A blood also have a glycosyltransferase enzyme that adds an extra modified monosaccharide called *N*-acetylgalactosamine to O antigen to form A antigen. Those with type B blood have a different transferase that adds an extra galactose to O antigen to form B antigen. People with both transferases produce both A and B antigen (AB blood type); those who lack these transferases produce O antigen only (O blood type).

People whose erythrocytes lack the A antigen, the B antigen, or both on their surface normally have antibodies against the missing antigen(s) in their serum. Thus if a type A or O person receives a transfusion of type B blood, antibodies against the B antigen will bind to the introduced red cells and trigger their destruction. To prevent such harmful reactions, blood group typing and appropriate matching of blood donors and recipients are required in all transfusions (Table 10-2).

TABLE 10-2 ABO Blood Groups

BLOOD GROUP	ANTIGENS ON RBCS*	SERUM ANTIBODIES	CAN RECEIVE BLOOD TYPES
A	A	Anti-B	A and O
B	B	Anti-A	B and O
AB	A and B	None	All
O	O	Anti-A and anti-B	O

*See Figure 10-20 for antigen structures.

Lipid-Binding Motifs Help Target Peripheral Proteins to the Membrane

Many water-soluble enzymes use membrane phospholipids as their substrates and thus must bind to membrane surfaces. As exemplified by the phospholipases, many such enzymes initially bind to the polar head groups of membrane phospholipids to carry out their catalytic functions. As noted earlier, phospholipases hydrolyze various bonds in the head groups of phospholipids (see Figure 10-14). These enzymes have an important role in the degradation of damaged or aged cell membranes and also are active components in many snake venoms. The mechanism of action of phospholipase A_2 illustrates how such water-soluble enzymes can reversibly interact with membranes and catalyze reactions at the interface of an aqueous solution and lipid surface (interfacial chemistry). When this enzyme is in aqueous solution, its Ca^{2+}-containing active site is buried in a channel lined with hydrophobic amino acids. The enzyme binds with greatest affinity to bilayers composed of negatively charged phospholipids (e.g., phosphotidylserine). This finding suggests that the rim of positively charged lysine and arginine residues around the entrance to the catalytic channel is particularly important in interfacial binding (Figure 10-21a). Binding induces a small conformational change in phospholipase A_2 that strengthens its binding to the phospholipid heads and opens the hydrophobic channel. As a phospholipid molecule moves from the bilayer into the channel, the enzyme-bound Ca^{2+} binds to the phosphate in the head group, thereby positioning the ester bond to be cleaved in the catalytic site (Figure 10-21b).

Proteins Can Be Removed from Membranes by Detergents or High-Salt Solutions

Membrane proteins are often difficult to purify and study, mostly because of their tight association with membrane lipids and other membrane proteins. These tasks are greatly facilitated by *detergents,* amphipathic molecules that disrupt membranes by intercalating into phospholipid bilayers and solubilize lipids and many membrane proteins. The hydrophobic part of a detergent molecule is attracted to hydrocarbons and mingles with them readily; the hydrophilic part is strongly attracted to water. Some detergents are natural products, but most are synthetic molecules developed for cleaning and for dispersing mixtures of oil and

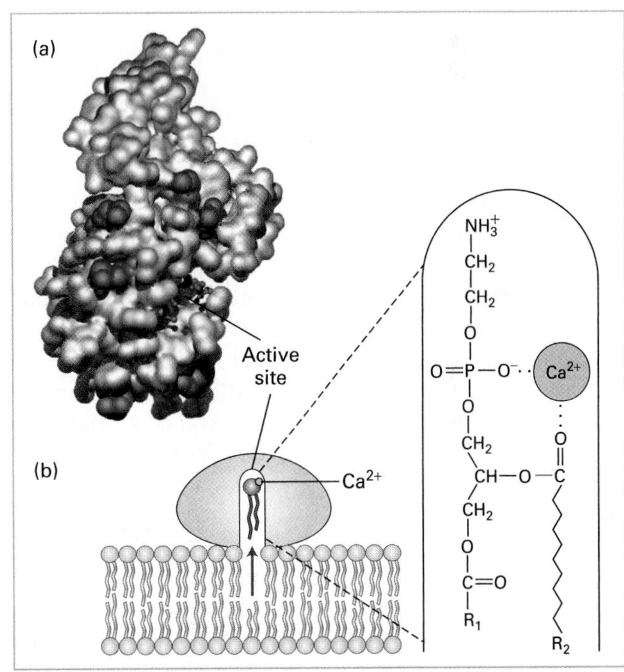

▲ **FIGURE 10-21 Interfacial binding surface and mechanism of action of phospholipase A_2.** (a) A structural model of the enzyme showing the surface that interacts with a membrane. This interfacial binding surface contains a rim of positively charged arginine and lysine residues, shown in blue surrounding the cavity of the catalytic active site, in which a substrate lipid (red stick structure) is bound. (b) Diagram of catalysis by phospholipase A_2. When docked on a model lipid membrane, positively charged residues of the interfacial binding site bind to negatively charged polar groups at the membrane surface. This binding triggers a small conformational change, opening a channel lined with hydrophobic amino acids that leads from the bilayer to the catalytic site. As a phospholipid moves into the channel, an enzyme-bound Ca^{2+} ion (green) binds to the head group, positioning the ester bond to be cleaved (red) next to the catalytic site. [Part (a) adapted from M. H. Gelb et al., 1999, *Curr. Opin. Struc. Biol.* **9**:428. Part (b), see D. Blow, 1991, *Nature* **351**:444.]

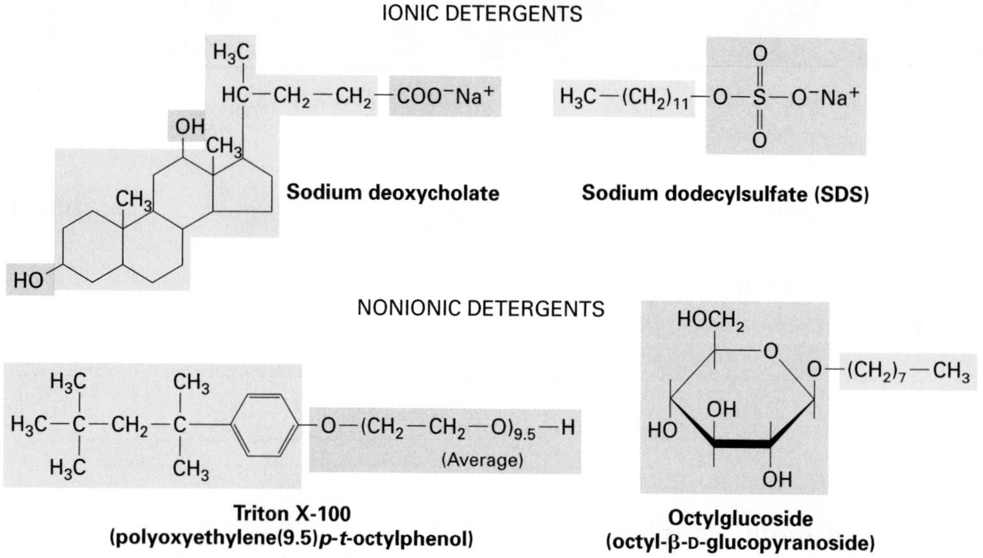

IONIC DETERGENTS

Sodium deoxycholate

Sodium dodecylsulfate (SDS)

NONIONIC DETERGENTS

Triton X-100 (polyoxyethylene(9.5)*p*-*t*-octylphenol)

Octylglucoside (octyl-β-D-glucopyranoside)

▲ **FIGURE 10-22 Structures of four common detergents.** The hydrophobic part of each molecule is shown in yellow; the hydrophilic part, in blue. The bile salt sodium deoxycholate is a natural product; the others are synthetic. Although ionic detergents commonly cause denaturation of proteins, nonionic detergents do not and are thus useful in solubilizing integral membrane proteins.

water (Figure 10-22). Ionic detergents, such as sodium deoxycholate and sodium dodecylsulfate (SDS), contain a charged group; nonionic detergents, such as Triton X-100 and octylglucoside, lack a charged group. At very low concentrations, detergents dissolve in pure water as isolated molecules. As the concentration increases, the molecules begin to form micelles—small, spherical aggregates in which hydrophilic parts of the molecules face outward and the hydrophobic parts cluster in the center (see Figure 10-6(c)). The *critical micelle concentration* (CMC) at which micelles form is characteristic of each detergent and is a function of the structures of its hydrophobic and hydrophilic parts.

Ionic detergents bind to the exposed hydrophobic regions of membrane proteins as well as to the hydrophobic cores of water-soluble proteins. Because of their charge, these detergents also disrupt ionic and hydrogen bonds. At high concentrations, for example, sodium dodecylsulfate completely denatures proteins by binding to every side chain, a property that is exploited in SDS gel electrophoresis (see Figure 3-35). *Nonionic detergents* generally do not denature proteins and are thus useful in extracting proteins from membranes before the proteins are purified. These detergents act in different ways at different concentrations. At high concentrations (above the CMC), they solubilize biological membranes by forming mixed micelles of detergent, phospholipid, and integral membrane proteins (Figure 10-23). At low concentrations (below the CMC), these detergents bind to the hydrophobic regions of most integral membrane proteins, making them soluble in aqueous solution.

Treatment of cultured cells with a buffered salt solution containing a nonionic detergent such as Triton X-100 extracts water-soluble proteins as well as integral membrane proteins. As noted earlier, the exoplasmic and cytosolic domains of integral membrane proteins are generally hydrophilic and soluble in water. The membrane-spanning domains, however, are rich in hydrophobic and uncharged residues (see Figure 10-15). When separated from membranes, these exposed hydrophobic segments tend to interact with one another, causing the protein molecules to aggregate and precipitate from aqueous solutions. The hydrophobic parts of nonionic detergent molecules preferentially bind to the hydrophobic segments of transmembrane proteins, preventing protein aggregation and allowing the proteins to remain in the aqueous solution. Detergent-solubilized transmembrane proteins can then be purified by affinity chromatography and other techniques used in purifying water-soluble proteins (Chapter 3).

As discussed previously, most peripheral proteins are bound to specific transmembrane proteins or membrane phospholipids by ionic or other weak interactions. Generally, peripheral proteins can be removed from the membrane by solutions of high ionic strength (high salt concentrations), which disrupt ionic bonds, or by chemicals that bind divalent cations such as Mg^{2+}. Unlike integral proteins, most peripheral proteins are soluble in aqueous solution and need not be solubilized by nonionic detergents.

KEY CONCEPTS OF SECTION 10.2

Biomembranes: Protein Components and Basic Functions

- Biological membranes usually contain both integral (transmembrane) proteins and peripheral membrane proteins, which do not enter the hydrophobic core of the bilayer (see Figure 10-1).
- Most integral membrane proteins contain one or more membrane-spanning hydrophobic α helices bracketed by

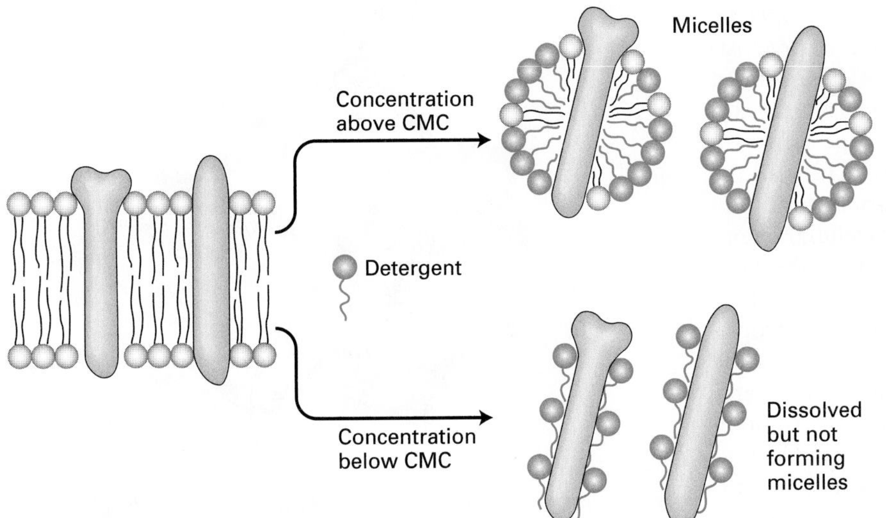

◀ **FIGURE 10-23 Solubilization of integral membrane proteins by nonionic detergents.** At a concentration higher than its critical micelle concentration (CMC), a detergent solubilizes lipids and integral membrane proteins, forming mixed micelles containing detergent, protein, and lipid molecules. At concentrations below the CMC, nonionic detergents (e.g., octylglucoside, Triton X-100) can dissolve membrane proteins without forming micelles by coating the membrane-spanning regions.

hydrophilic domains that extend into the aqueous solutions surrounding the cytosolic and exoplasmic faces of the membrane (see Figures 10-15 and 10-16).

- Fatty acyl side chains as well as the polar heads of membrane lipids groups pack tightly and irregularly around the hydrophobic segments of integral membrane proteins.
- All transmembrane proteins and glycolipids are asymmetrically oriented in the bilayer; invariably carbohydrate chains are attached to the exoplasmic surface of the protein.
- The porins, unlike other integral proteins, contain membrane-spanning β sheets that form a barrel-like channel through the bilayer (see Figure 10-18).
- Long-chain lipids attached to certain amino acids anchor some proteins to one or the other membrane leaflet (see Figure 10-19).
- The binding of a water-soluble enzyme (e.g., a phospholipase, kinase, or phosphatase) to a membrane surface brings the enzyme close to its substrate and in some cases activates it. Such interfacial binding is often due to the attraction between positive charges on basic residues in the protein and negative charges on phospholipid head groups in the bilayer.
- Transmembrane proteins are selectively solubilized and purified with the use of nonionic detergents.

10.3 Phospholipids, Sphingolipids, and Cholesterol: Synthesis and Intracellular Movement

In this section, we consider some of the special challenges that a cell faces in synthesizing and transporting lipids, which are poorly soluble in the aqueous interior of cells. One focus will be on fatty acids, the precursors of the phospholipids that are the structural backbone of cellular membranes. Fatty acids are oxidized in mitochondria to release energy for cellular functions (Chapter 12) and are stored and transported through the bloodstream primarily in the form of triglycerides, in which three fatty acids are esterified to a glycerol molecule. We will also review the synthesis of cholesterol, both because it is an important membrane component and because it is a precursor for steroid hormones such as testosterone and estrogen; vitamin D, which helps controls calcium metabolism; and other biologically active lipids.

$H_3C{-}(CH_2)_n{-}C(=O){-}O{-}CH_2$
$H_3C{-}(CH_2)_n{-}C(=O){-}O{-}CH$
$H_3C{-}(CH_2)_n{-}C(=O){-}O{-}CH_2$

Triacylglycerol

We focus our discussion of lipid biosynthesis and movement on the major lipids found in cellular membranes and their precursors. In lipid biosynthesis, water-soluble precursors are assembled into membrane-associated intermediates that are then converted into membrane lipid products. The movement of lipids, especially membrane components, between different organelles is critical for maintaining the proper composition and properties of membranes and overall cell structure, but our understanding of such intracellular lipid transport is still rudimentary.

A fundamental principle of membrane biosynthesis is that cells synthesize new membranes only by the expansion of existing membranes. Although some early steps in the synthesis of membrane lipids take place in the cytoplasm, the final steps are catalyzed by enzymes bound to preexisting cellular membranes, and the products are incorporated into the membranes as they are generated. Evidence for this phenomenon is seen when cells are briefly exposed to radioactive precursors (e.g., phosphate or fatty acids): all the phospholipids and sphingolipids incorporating these precursor substances are associated with intracellular membranes; as expected from the hydrophobicity of the fatty acyl chains, none are found free in the cytosol. After they are formed, membrane lipids must be distributed appropriately both in leaflets of a given membrane and among the independent membranes of different organelles in eukaryotic cells. Here, we focus on the synthesis and distribution of phospholipids, sphingolipids, and cholesterol; in Chapter 13 we discuss how membrane proteins are inserted into cell membranes.

Fatty Acids Synthesis Is Mediated by Several Important Enzymes

Fatty acids are key components of both phospholipids and sphingolipids; they also anchor some proteins to cellular membranes (see Figure 10-19). Thus the regulation of fatty acid synthesis plays a key role in the regulation of membrane synthesis as a whole. The major fatty acids in phospholipids contain 14, 16, 18, or 20 carbon atoms and include both saturated and unsaturated chains.

Fatty acids are synthesized from the two-carbon building block acetate, CH_3COO^-. In cells, both acetate and the intermediates in fatty acid biosynthesis are esterified to the large water-soluble molecule coenzyme A (CoA), as exemplified by the structure of **acetyl CoA:**

$$\underbrace{H_3C-\overset{O}{\overset{\|}{C}}}_{\text{Acetyl}}-S-(CH_2)_2-\overset{H}{\overset{|}{N}}-\underset{O}{\underset{\|}{C}}-(CH_2)_2-\overset{H}{\overset{|}{N}}-\underset{O}{\underset{\|}{C}}-\overset{H}{\overset{|}{\underset{OH}{\underset{|}{C}}}}-\overset{CH_3}{\overset{|}{\underset{CH_3}{\underset{|}{C}}}}-CH_2-O-\overset{O}{\overset{\|}{\underset{O^-}{\underset{|}{P}}}}-O-\overset{O}{\overset{\|}{\underset{O^-}{\underset{|}{P}}}}-O-\underset{\text{Phosphate}}{\underset{|}{\text{Ribose}}}-\text{Adenine}$$

Coenzyme A (CoA)

Acetyl CoA is an important intermediate in the metabolism of glucose, fatty acids, and many amino acids, as detailed in Chapter 12. It also contributes acetyl groups in many biosynthetic pathways. **Saturated** fatty acids (no double bonds) containing 14 or 16 carbon atoms are made from acetyl CoA by two enzymes, *acetyl-CoA carboxylase* and *fatty acid synthase.* In animal cells, these enzymes are found in the cytosol; in plants, they are found in chloroplasts. Palmitoyl CoA (16 carbon fatty acyl group linked to CoA) can be elongated to 18–24 carbons by the sequential addition of two-carbon units in the endoplasmic reticulum (ER) or sometimes in the mitochondrion. Desaturase enzymes, also located in the ER, introduce double bonds at specific positions in some fatty acids, yielding **unsaturated** fatty acids. Oleyl CoA (oleate linked to CoA, see Table 2-4), for example, is formed by removal of two H atoms from stearyl CoA. In contrast to free fatty acids, fatty acyl CoA derivatives are soluble in aqueous solutions because of the hydrophilicity of the CoA segment.

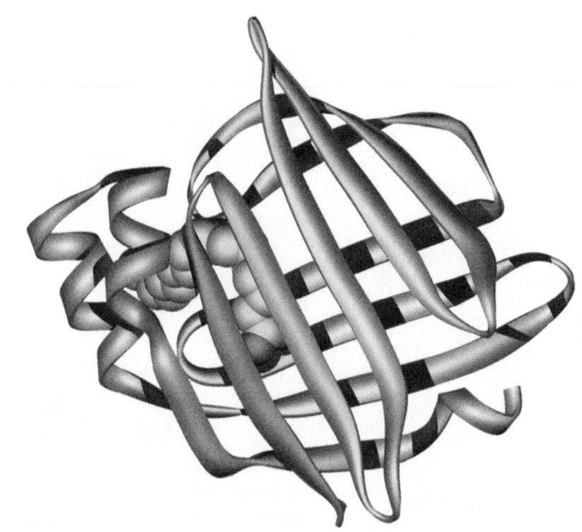

▲ **FIGURE 10-24 Binding of a fatty acid to the hydrophobic pocket of a fatty-acid-binding protein (FABP).** The crystal structure of adipocyte FABP (ribbon diagram) reveals that the hydrophobic binding pocket is generated from two β sheets that are nearly at right angles to each other, forming a clam-shell-like structure. A fatty acid (carbons yellow; oxygens red) interacts noncovalently with hydrophobic amino acid residues within this pocket. [See A. Reese-Wagoner et al., 1999, *Biochim. Biophys. Acta* **23:**1441(2–3):106–116.]

Small Cytosolic Proteins Facilitate Movement of Fatty Acids

In order to be transported through the cell cytoplasm free, or unesterified, fatty acids (those unlinked to a CoA), commonly are bound by *fatty-acid-binding proteins* (FABPs), which belong to a group of small cytosolic proteins that facilitate the intracellular movement of many lipids. These proteins contain a hydrophobic pocket lined by β sheets (Figure 10-24). A long-chain fatty acid can fit into this pocket and interact noncovalently with the surrounding protein.

The expression of cellular FABPs is regulated coordinately with cellular requirements for the uptake and release of fatty acids. Thus FABP levels are high in active muscles that are using fatty acids for generation of ATP and in adipocytes (fat-storing cells) when they are either taking up fatty acids to be stored as triglycerides or releasing fatty acids for use by other cells. The importance of FABPs in fatty acid metabolism is highlighted by the observations that they can compose as much as 5 percent of all cytosolic proteins in the liver and that genetic inactivation of cardiac muscle FABP converts the heart from a muscle that primarily burns fatty acids for energy into one that primarily burns glucose.

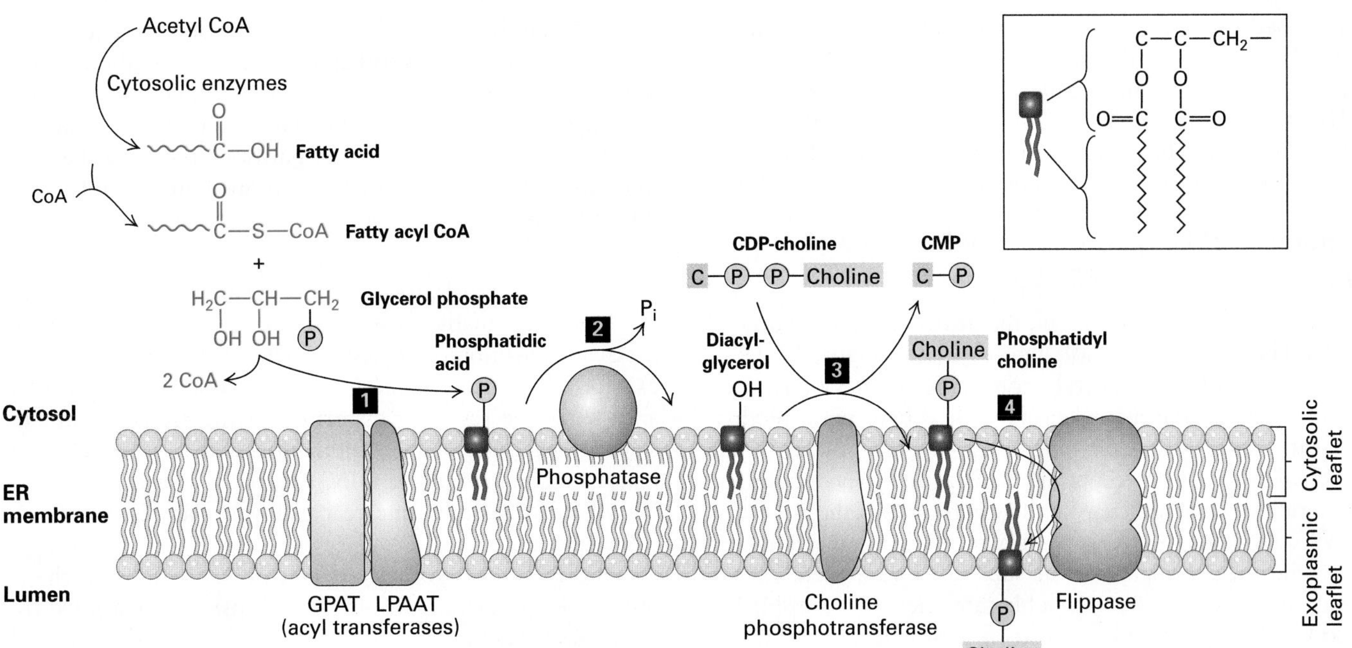

▲ **FIGURE 10-25 Phospholipid synthesis.** Because phospholipids are amphipathic molecules, the last stages of their multistep synthesis take place at the interface between a membrane and the cytosol and are catalyzed by membrane-associated enzymes. Step (**1**): Two fatty acids from fatty acyl CoA are esterified to the phosphorylated glycerol backbone, forming phosphatidic acid, whose two long hydrocarbon chains anchor the molecule to the membrane. Step (**2**): A phosphatase converts phosphatidic acid into diacylglycerol. Step (**3**): A polar head group (e.g., phosphorylcholine) is transferred from cytosine diphosphocholine (CDP-choline) to the exposed hydroxyl group. Step (**4**): Flippase proteins catalyze the movement of phospholipids from the cytosolic leaflet in which they are initially formed to the exoplasmic leaflet.

Incorporation of Fatty Acids into Membrane Lipids Takes Place on Organelle Membranes

Fatty acids are not directly incorporated into phospholipids; rather, in eukaryotic cells they are first converted into CoA esters. The subsequent synthesis of many *diacyl glycerophospholipids* from fatty acyl CoAs, glycerol 3-phosphate, and polar head-group precursors is carried out by enzymes associated with the cytosolic face of the ER membrane, usually the smooth ER, in animal cells (Figure 10-25); the ER is described in detail in Chapters 9 and 13. Mitochondria synthesize some of their own membrane lipids and import others.

Sphingolipids are derivatives of sphingosine, an amino alcohol that contains a long, unsaturated hydrocarbon chain (Figure 10-5). Sphingosine is made in the ER, beginning with the coupling of a palmitoyl group from palmitoyl CoA to serine; the subsequent addition of a second fatty acyl group to form *N*-acyl sphingosine (ceramide) also takes place in the ER. The later addition of a polar head group to ceramide in the Golgi yields *sphingomyelin,* whose head group is phosphorylcholine, and various *glycosphingolipids,* in which the head group may be a monosaccharide or a more complex oligosaccharide (see Figure 10-5b). Some sphingolipid synthesis can also take place in mitochondria. In addition to serving as the backbone for sphingolipids, ceramide and its metabolic products are important signaling molecules that can influence cell growth, proliferation, endocytosis, resistance to stress, and apoptosis.

After their synthesis is completed in the Golgi, sphingolipids are transported to other cellular compartments through vesicle-mediated mechanisms similar to those discussed in Chapter 14. In contrast, phospholipids, as well as cholesterol, can move between organelles by different mechanisms, described below.

Flippases Move Phospholipids from One Membrane Leaflet to the Opposite Leaflet

Even though phospholipids are initially incorporated into the cytosolic leaflet of the ER membrane, various phospholipids are asymmetrically distributed in the two leaflets of the ER membrane and of other cellular membranes. As noted above, phospholipids spontaneously flip-flop from one leaflet to the other only very slowly. For the ER membrane to expand by growth of both leaflets and have asymmetrically distributed phospholipids, its phospholipid components must be able to rapidly and selectively flip-flop from one membrane leaflet to the other. Although the mechanisms employed to generate and maintain membrane phospholipid asymmetry are not well understood, it is clear that flippases play a key role. As described in Chapter 11, these integral membrane proteins use the energy of ATP hydrolysis to facilitate the movement of phospholipid molecules from one leaflet to the other (see Figure 11-16).

The usual asymmetric distribution of phospholipids in membrane leaflets is broken down as cells (e.g., red blood cells) become senescent or undergo apoptosis. For instance, phosphatidylserine and phosphatidylethanolamine are preferentially located in the cytosolic leaflet of cellular membranes.

Increased exposure of these anionic phospholipids on the exoplasmic face of the plasma membrane appears to serve as a signal for scavenger cells to remove and destroy old or dying cells. Annexin V, a protein that specifically binds to anionic phospholipids, can be fluorescently labeled and used to detect apoptotic cells in cultured cells and in tissues.

Cholesterol Is Synthesized by Enzymes in the Cytosol and ER Membrane

Next we focus on cholesterol, the principal sterol in animal cells. The first steps of cholesterol synthesis (Figure 10-26)—conversion of three acetyl groups linked to CoA (acetyl CoA) forming the 6-carbon molecule β-hydroxy-β-methylglutaryl linked to CoA (HMG-CoA)—take place in the cytosol. The conversion of HMG CoA into mevalonate, the key rate-controlling step in cholesterol biosynthesis, is catalyzed by *HMG-CoA reductase,* an ER integral membrane protein, even though both its substrate and its product are water soluble. The water-soluble catalytic domain of HMG-CoA reductase extends into the cytosol, but its eight transmembrane α helices firmly embed the enzyme in the ER membrane. Five of the transmembrane α helices compose the so-called *sterol-sensing domain* and regulate enzyme stability. When levels of cholesterol in the ER membrane are high, binding of cholesterol to this domain causes the protein to bind to two other integral ER membrane proteins, Insig-1 and Insig-2. This in turn induces ubiquitination (see Figure 3-29) of HMG-CoA reductase and its degradation by the proteosome pathway, reducing the production of mevalonate, the key intermediate in cholesterol biosynthesis.

Atherosclerosis, frequently called cholesterol-dependent clogging of the arteries, is characterized by the progressive deposition of cholesterol and other lipids, cells, and extracellular matrix material in the inner layer of the wall of an artery. The resulting distortion of the artery's wall can lead, either alone or in combination with a blood clot, to major blockage of blood flow. Atherosclerosis accounts for 75 percent of deaths due to cardiovascular disease in the United States.

Cholesterol is synthesized mainly in the liver. Perhaps the most successful anti-atherosclerosis medications are the *statins*. These drugs bind to HMG-CoA reductase and directly inhibit its activity, thereby lowering cholesterol biosynthesis. As a consequence, the amount of low-density lipoproteins (see Figure 14-27)—the small, membrane-enveloped particles containing cholesterol esterified to fatty

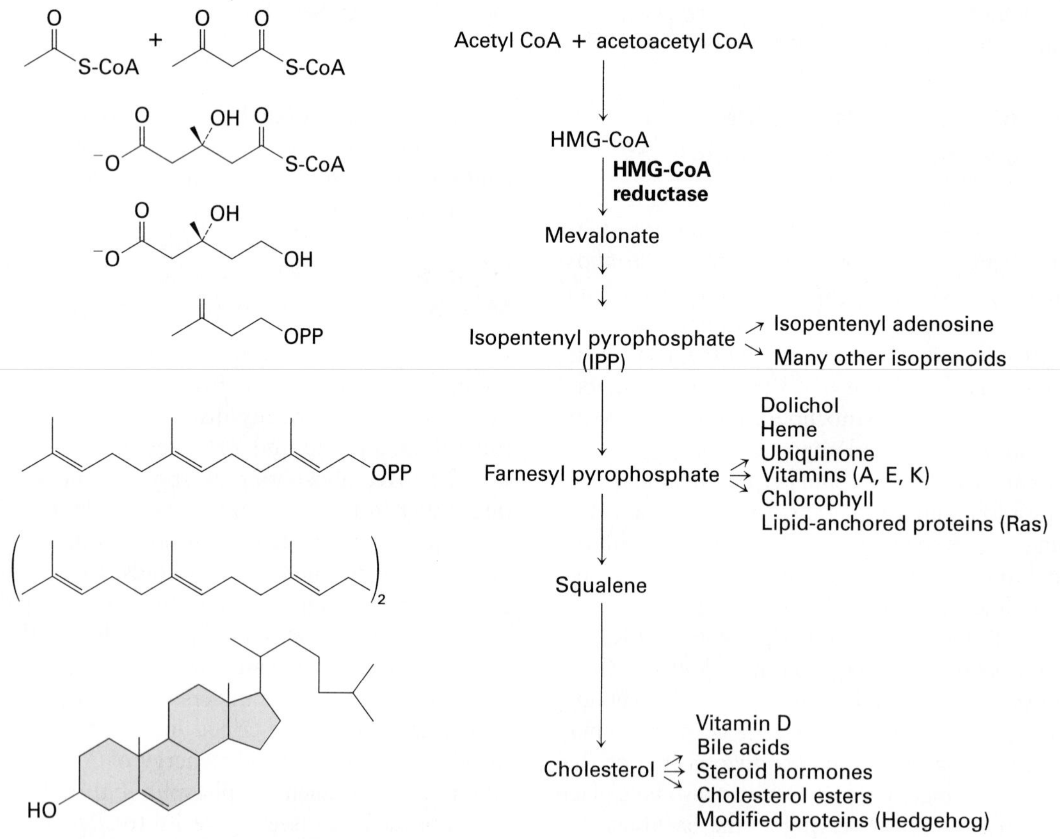

Acetyl CoA + acetoacetyl CoA
↓
HMG-CoA
↓ **HMG-CoA reductase**
Mevalonate
↓
↓
Isopentenyl pyrophosphate (IPP) → Isopentenyl adenosine; Many other isoprenoids
↓
Farnesyl pyrophosphate → Dolichol; Heme; Ubiquinone; Vitamins (A, E, K); Chlorophyll; Lipid-anchored proteins (Ras)
↓
Squalene
↓
Cholesterol → Vitamin D; Bile acids; Steroid hormones; Cholesterol esters; Modified proteins (Hedgehog)

▲ **FIGURE 10-26 Cholesterol biosynthetic pathway.** The regulated rate-controlling step in cholesterol biosynthesis is the conversion of β-hydroxy-β-methylglutaryl CoA (HMG-CoA) into mevalonic acid by HMG-CoA reductase, an ER-membrane protein. Mevalonate is then converted into isopentenyl pyrophosphate (IPP), which has the basic five-carbon isoprenoid structure. IPP can be converted into cholesterol and into many other lipids, often through the polyisoprenoid intermediates shown here. Some of the numerous compounds derived from isoprenoid intermediates and cholesterol itself are indicated.

acids that often and rightly are called "bad cholesterol"—drops in the blood, reducing the formation of atherosclerotic plaques. ■

Mevalonate, the 6-carbon product formed by HMG-CoA reductase, is converted in several steps into the 5-carbon isoprenoid compound isopentenyl pyrophosphate (IPP) and its stereoisomer, dimethylallyl pyrophosphate (DMPP) (see Figure 10-26). These reactions are catalyzed by cytosolic enzymes, as are the subsequent reactions in the cholesterol synthesis pathway, in which six IPP units condense to yield squalene, a branched-chain 30-carbon intermediate. Enzymes bound to the ER membrane catalyze the multiple reactions that convert squalene into cholesterol in mammals or into related sterols in other species. One of the intermediates in this pathway, farnesyl pyrophosphate, is the precursor of the prenyl lipid that anchors Ras and related proteins to the cytosolic surface of the plasma membrane (see Figure 10-19) as well as other important biomolecules (see Figure 10-26).

Cholesterol and Phospholipids Are Transported Between Organelles by Several Mechanisms

As already noted, the final steps in the synthesis of cholesterol and phospholipids take place primarily in the ER, although some of these membrane lipids (plasmalogens) are produced in mitochondria and peroxisomes. Thus the plasma membrane and the membranes bounding other organelles must obtain these lipids by means of one or more intracellular transport processes. Membrane lipids accompany both soluble and membrane proteins during the secretory pathway described in Chapter 14; membrane vesicles bud from the ER and fuse with membranes in the Golgi complex, and other membrane vesicles bud from the Golgi complex and fuse with the plasma membrane (Figure 10-27a). However, several lines of evidence suggest that there is substantial interorganelle movement of cholesterol and phospholipids through other mechanisms. For example, chemical inhibitors of the classic secretory pathway and mutations that impede vesicular traffic in this pathway do not prevent cholesterol or phospholipid transport between membranes.

A second mechanism entails direct protein-mediated contact of ER or ER-derived membranes with membranes of other organelles (Figure 10-27b). In the third mechanism, small lipid-transfer proteins facilitate the exchange of phospholipids or cholesterol between different membranes (Figure 10-27c). Although such transfer proteins have been identified in assays in vitro, their role in intracellular movements of most phospholipids is not well defined. For instance, mice with a knockout mutation in the gene encoding the phosphatidylcholine-transfer protein appear to be normal in most respects, indicating that this protein is not essential for cellular phospholipid metabolism.

As noted earlier, the lipid compositions of different organelle membranes vary considerably (see Table 10-1). Some of these differences are due to different sites of synthesis. For example, a phospholipid called cardiolipin, which is localized to the mitochondrial membrane, is made only in mitochondria and little is transferred to other organelles. Differential transport of lipids also plays a role in determining the lipid compositions of different cellular membranes. For instance, even though cholesterol is made in the ER, the cholesterol concentration (cholesterol-to-phospholipid molar ratio) is ~1.5–13-fold higher in the plasma membrane than in other organelles (ER, Golgi, mitochondrion, lysosome). Although the mechanisms responsible for establishing and maintaining these differences are not well understood, we have seen that the distinctive lipid composition of each membrane has a major influence on its physical and biological properties.

KEY CONCEPTS OF SECTION 10.3

Phospholipids, Sphingolipids, and Cholesterol: Synthesis and Intracellular Movement

- Saturated and unsaturated fatty acids of various chain lengths are components of phospholipids, sphingolipids, and triglycerides.

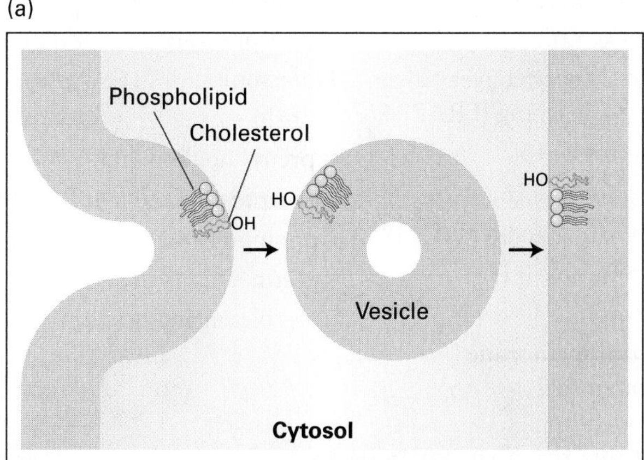

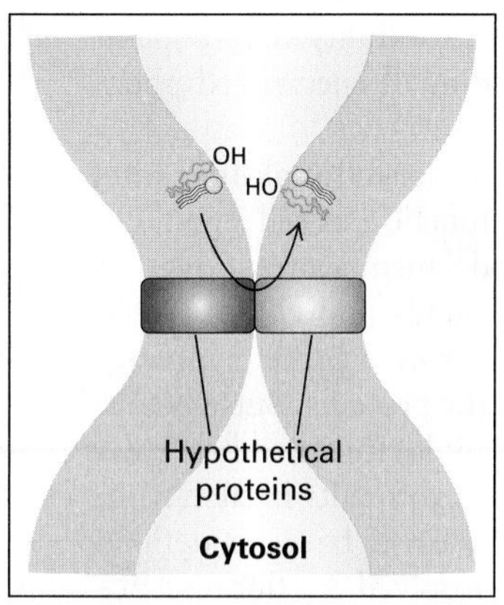

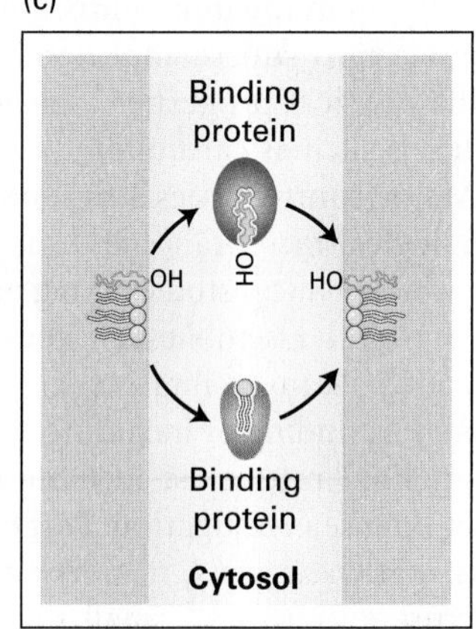

▲ **FIGURE 10-27 Proposed mechanisms of transport of cholesterol and phospholipids between membranes.** In mechanism (a), vesicles transfer lipids between membranes. In mechanism (b), lipid transfer is a consequence of direct contact between membranes that is mediated by membrane-embedded proteins. In mechanism (c), transfer is mediated by small, soluble lipid-transfer proteins. [Adapted from F. R. Maxfield and D. Wustner, 2002, *J. Clin. Invest.* **110:**891.]

- Fatty acids are synthesized from acetyl CoA by water-soluble enzymes and modified by elongation and desaturation in the endoplasmic reticulum (ER).
- The final steps in the synthesis of glycerophospholipids, plasmalogens, and sphingolipids are catalyzed by membrane-associated enzymes primarily on the cytosolic face of the ER (see Figure 10-25).
- Each type of lipid is initially incorporated into the preexisting membranes on which it is made.
- Most membrane phospholipids are preferentially distributed in either the exoplasmic or the cytosolic leaflet. This asymmetry results in part from the action of phospholipid flippases.
- The initial steps in cholesterol biosynthesis take place in the cytosol, whereas the last steps are catalyzed by enzymes associated with the ER membrane.
- The rate-controlling step in cholesterol biosynthesis is catalyzed by HMG-CoA reductase, whose transmembrane segments are embedded in the ER membrane and contain a sterol-sensing domain.
- Considerable evidence indicates that Golgi-independent vesicular transport, direct protein-mediated contacts between different membranes, soluble protein carriers, or all three may account for some inter-organelle transport of cholesterol and phospholipids (see Figure 10-27).

Perspectives for the Future

One fundamental question in lipid biology concerns the generation, maintenance, and function of the asymmetric distribution of lipids within the leaflets of one membrane and the variation in lipid composition among the membranes of different organelles. What are the mechanisms underlying this complexity, and why is such complexity needed? We already know that certain lipids can specifically interact with and influence the activity of some proteins. For example, the large multimeric proteins that participate in oxidative phosphorylation in the inner mitochondrial membrane appear to assemble into supercomplexes whose stability may depend on the physical properties and binding of specialized phospholipids such as cardiolipin.

Major controversies surround the existence of lipid rafts in biological membranes and their function in cell signaling. Many biochemical studies using model membranes show that stable lateral assemblies of sphingolipids and cholesterol—lipid rafts—can facilitate selective protein-protein interactions by excluding or including specific proteins. But whether or not lipid rafts exist in natural biological membranes is under intense investigation. Part of the difficulty in such studies is that these rafts, if indeed they exist, have no defined structure and are too small to be resolved by fluorescence microscopy. Proving their existence in cells will require development of new biophysical and microscopic tools.

Despite considerable progress in our understanding of the cellular metabolism and movement of lipids, the mechanisms for transporting cholesterol and phospholipids between organelle membranes remain poorly characterized. In particular, we lack a detailed understanding of how various transport proteins move lipids from one membrane leaflet to another (flippase activity) and into and out of cells. Such understanding will undoubtedly require a determination of many high-resolution structures of these molecules, their capture in various stages of the transport process, and careful kinetic and other biophysical analyses of their function, similar to the approaches discussed in Chapter 11 for elucidating the operation of ion channels and ATP-powered pumps.

Recent advances in solubilizing and crystallizing integral membrane proteins have led to the delineation of the molecular structures of many important types of proteins, such as ion channels, ATP-powered ion pumps, and aquaporins, as we will see in Chapter 11. However, many important classes of membrane proteins have proven recalcitrant to even these new approaches. For example, we lack the structure of any protein that transports glucose into a eukaryotic cell. As we will learn in Chapters 15 and 16, many classes of receptors span the plasma membrane with one or more α helixes. Perhaps surprisingly, we lack the molecular structure of the transmembrane segment of any eukaryotic cell-surface receptor, and so many aspects of the function of these proteins are still mysterious. Elucidating the molecular structures of these and many other types of membrane proteins will clarify many aspects of molecular cell biology.

Key Terms

acylation *425*
bile acid *416*
cholesterol *416*
critical micelle concentration (CMC) *428*
cytosolic face *414*
detergent *427*
exoplasmic face *414*
external face *414*
flippase *420*
fluorescence recovery after photobleaching (FRAP) *417*
glycolipid *416*
GPI anchor *426*
HMG-CoA reductase *432*
hydrophobic *411*
hydrophilic *411*
integral membrane protein *421*
internal face *414*
leaflet *411*
lipid raft *420*
liposome *411*
lumen *410*
micelle *411*
peripheral membrane protein *422*
phosphoglyceride *415*
phospholipid bilayer *411*
plasmologen *416*
porin *423*
prenyl anchor *425*
receptor proteins *409*
sphingolipid *416*
statin *432*
sterol-sensing domain *432*

Review the Concepts

1. When viewed by electron microscopy, the lipid bilayer is often described as looking like a railroad track. Explain how the structure of the bilayer creates this image.

2. Biomembranes contain many different types of lipid molecules. What are the three main types of lipid molecules found in biomembranes? How are the three types similar, and how are they different?

3. Proteins may be bound to the exoplasmic or cytosolic face of the plasma membrane by way of covalently attached lipids. What are the three types of lipid anchors responsible for tethering proteins to the plasma membrane bilayer, and which type is used by cell-surface proteins that face the external medium and by glycosylated proteoglycans?

4. Lipid bilayers are considered to be two-dimensional fluids; what does this mean? What drives the movement of lipid molecules and proteins within the bilayer? How can such movement be measured? What factors affect the degree of membrane fluidity?

5. Phospholipid biosynthesis at the interface between the endoplasmic reticulum (ER) and the cytosol presents a number of challenges that must be solved by the cell. Explain how each of the following is handled.

a. The substrates for phospholipid biosynthesis are all water soluble, yet the end products are not.

b. The immediate site of incorporation of all newly synthesized phospholipids is the cytosolic leaflet of the ER membrane, yet phospholipids must be incorporated into both leaflets.

c. Many membrane systems in the cell, for example, the plasma membrane, are unable to synthesize their own phospholipids, yet these membranes must also expand if the cell is to grow and divide.

6. Fatty acids must associate with lipid chaperones in order to move within the cell. Why are these chaperones needed, and what is the name given to a group of proteins that are responsible for this intracellular trafficking of fatty acids? What is the key distinguishing feature of these proteins that allows fatty acids to move within the cell?

7. What are the common fatty acid chains in glycerophospholipids, and why do these fatty acid chains differ in their number of carbon atoms by multiples of 2?

8. The biosynthesis of cholesterol is a highly regulated process. What is the key regulated enzyme in cholesterol biosynthesis? This enzyme is subject to feedback inhibition. What is feedback inhibition? How does this enzyme sense cholesterol levels in a cell?

9. It is evident that one function of cholesterol is structural because it is the most common single lipid molecule in the plasma membrane of mammals. Yet cholesterol may also have other functions. What aspects of cholesterol and its metabolism lead to the conclusion that cholesterol is a multifunctional membrane lipid?

10. Phospholipids and cholesterol must be transported from their site of synthesis to various membrane systems within cells. One way of doing this is through vesicular transport, as is the case for many proteins in the secretory pathway. However, most phospholipid and cholesterol membrane-to-membrane transport in cells is not by Golgi-mediated vesicular transport. What is the evidence for this statement? What appear to be the major mechanisms for phospholipid and cholesterol transport?

11. Explain the following statement: The structure of all biomembranes depends on the chemical properties of phospholipids, whereas the function of each specific biomembrane depends on the specific proteins associated with that membrane.

12. Name the three groups into which membrane-associated proteins may be classified. Explain the mechanism by which each group associates with a biomembrane.

13. Although both faces of a biomembrane are composed of the same general types of macromolecules, principally lipids and proteins, the two faces of the bilayer are not identical. What accounts for the asymmetry between the two faces?

Analyze the Data

The behavior of receptor X (XR), a transmembrane protein present in the plasma membrane of mammalian cells, is being investigated. The protein has been engineered as a fusion protein containing the green fluorescent protein (GFP) at its N-terminus. GFP-XR is a functional protein and can replace XR in cells.

a. Cells expressing GFP-XR or artificial lipid vesicles (liposomes) containing GFP-XR are subjected to fluorescence recovery after photobleaching (FRAP). The intensity of the fluorescence of a small spot on the surface of the cells (solid line) or on the surface of the liposomes (dashed line) is measured prior to and following laser bleaching (arrow). The data are shown below.

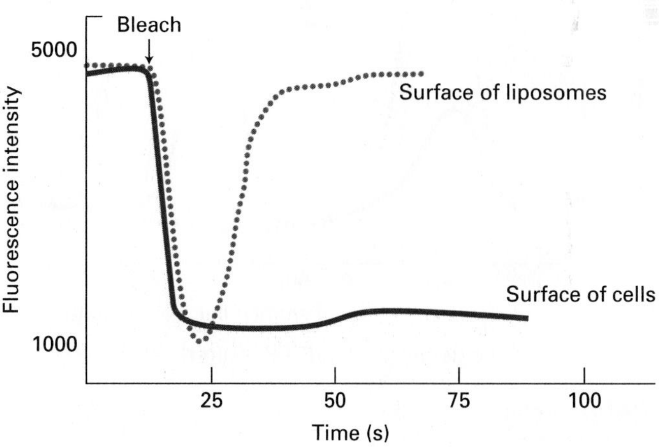

What explanation could account for the differing behavior of GFP-XR in liposomes versus in the plasma membrane of a cell?

b. Tiny gold particles can be attached to individual molecules and their movement then followed in a light microscope by single-particle tracking. This method allows one to observe the behavior of individual proteins in a membrane. The tracks generated during a 5-second observational period by a gold particle attached to XR present in a cell (left) or in a liposome (middle) or to XR adhered to a microscope slide (right) are shown below.

XR present on a microscope slide

What additional information do these data provide beyond what can be determined from the FRAP data?

c. Fluorescence resonance energy transfer (FRET) is a technique by which a fluorescent molecule, following its excitation with the appropriate wavelength of light, can transfer its emission energy to and excite a nearby different fluorescent molecule (see Figure 15-14). Cyan fluorescent protein (CFP) and yellow fluorescent protein (YFP) are related to GFP but fluoresce at cyan and yellow wavelengths rather than at green. If CFP is excited with the appropriate wavelength of light and a YFP molecule is very near, then energy can be transferred from CFP emission and used to excite YFP, as indicated by a loss of emission of cyan fluorescence and an increase in emission of yellow fluorescence. CFP-XR and YFP-XR are expressed together in a cell line or are both incorporated into liposomes. The number of molecules of YFP-XR and CFP-XR per cm^2 of membrane is equivalent in the cells and the liposomes. The cells and liposomes are then irradiated with a wavelength of light that causes CFP but not YFP to fluoresce. The amount of cyan (CFP) and yellow (YFP) fluorescence emitted by the cells (solid line) or liposomes (dashed line) is then monitored, as shown below.

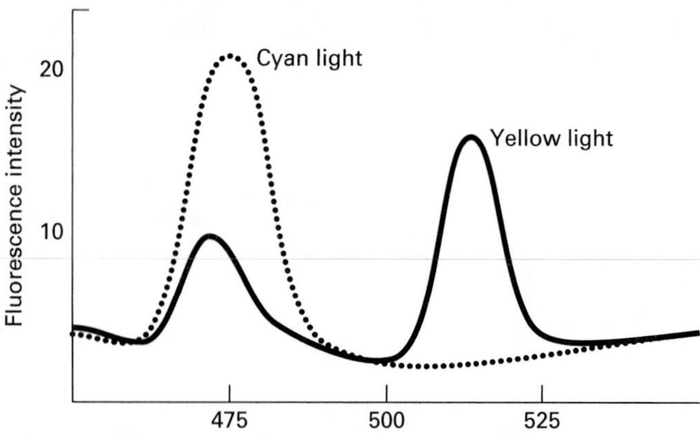

What can be deduced about XR from these data?

References

10.1 Biomembranes: Lipid Composition and Structural Organization

McMahon, H., and J. L. Gallop. 2005. Membrane curvature and mechanisms of dynamic cell membrane remodeling. *Nature* **438**:590–596.

Mukherjee, S., and F. R. Maxfield. 2004. Membrane domains. *Annu. Rev. Cell Dev. Biol.* **20**:839–866.

Simons, K., and D. Toomre. 2000. Lipid rafts and signal transduction. *Nature Rev. Mol. Cell Biol.* **1**:31–41.

Simons, K., and W. L. C. Vaz. Model systems, lipid rafts, and cell membranes. 2004. *Annu. Rev. Biophys. Biomolec. Struct.* **33**:269–295.

Tamm, L. K., V. K. Kiessling, and M. L. Wagner. 2001. Membrane dynamics. *Encyclopedia of Life Sciences.* Nature Publishing Group.

Vance, D. E., and J. E. Vance. 2002. *Biochemistry of Lipids, Lipoproteins, and Membranes,* 4th ed. Elsevier.

Van Meer, G. 2006 Cellular lipidomics. *EMBO J.* **24**:3159–3165.

Yeager, P. L. 2001. Lipids. *Encyclopedia of Life Sciences.* Nature Publishing Group.

Zimmerberg, J., and M. M. Kozlov. 2006. How proteins produce cellular membrane curvature. *Nature Rev. Mol. Cell Biol.* **7**:9–19.

10.2 Biomembranes: Protein Components and Basic Functions

Bowie, J. Solving the membrane protein folding problem. 2005. *Nature* **438**:581–589.

Cullen, P. J., G. E. Cozier, G. Banting, and H. Mellor. 2001. Modular phosphoinositide-binding domains: their role in signalling and membrane trafficking. *Curr. Biol.* **11**:R882–R893.

Engelman, D. Membranes are more mosaic than fluid. 2005. *Nature* **438**:578–580.

Lanyi, J. K., and H. Luecke. 2001. Bacteriorhodopsin. *Curr. Opin. Struc. Biol.* **11**:415–519.

Lee, A. G. A greasy grip. 2005. *Nature* **438**:569–570.

MacKenzie, K. R., J. H. Prestegard, and D. M. Engelman. 1997. A transmembrane helix dimer: structure and implications. *Science* **276**:131–133.

McIntosh, T. J., and S. A. Simon. 2006. Roles of bilayer material properties in function and distribution of membrane proteins *Ann. Rev. Biophys. Biomolec. Struct.* **35**:177–198.

Schulz, G. E. 2000. β-Barrel membrane proteins. *Curr. Opin. Struc. Biol.* **10**:443–447.

10.3 Phospholipids, Sphingolipids, and Cholesterol: Synthesis and Intracellular Movement

Bloch, K. 1965. The biological synthesis of cholesterol. *Science* **150**:19–28.

Daleke, D. L., and J. V. Lyles. 2000. Identification and purification of aminophospholipid flippases. *Biochim. Biophys. Acta* **1486**:108–127.

Futerman, A., and H. Riezman. 2005. The inns and outs of sphingolipid synthesis. *Trends Cell Biol.* **15**:312–318.

Hajri, T., and N. A. Abumrad. 2002. Fatty acid transport across membranes: relevance to nutrition and metabolic pathology. *Ann. Rev. Nutr.* **22**:383–415.

Henneberry, A. L., M. M. Wright, and C. R. McMaster. 2002. The major sites of cellular phospholipid synthesis and molecular determinants of fatty acid and lipid head group specificity. *Mol. Biol. Cell* **13**:3148–3161.

Holthuis, J. C. M., and T. P. Levine. 2005. Lipid traffic: floppy drives and a superhighway *Nature Rev. Molec. Cell Biol.* **6**:209–220.

Ioannou, Y. A. 2001. Multidrug permeases and subcellular cholesterol transport. *Nature Rev. Mol. Cell Biol.* **2**:657–668.

Kent, C. 1995. Eukaryotic phospholipid biosynthesis. *Ann. Rev. Biochem.* **64**:315–343.

Maxfield, F. R., and I. Tabas. 2005. Role of cholesterol and lipid organization in disease. *Nature* **438**:612–621.

Stahl, A., R. E. Gimeno, L. A. Tartaglia, and H. F. Lodish. 2001. Fatty acid transport proteins: a current view of a growing family. *Trends Endocrinol. Metab.* **12**(6):266–273.

van Meer, G., and H. Sprong. 2004. Membrane lipids and vesicular traffic. *Curr. Opin. Cell Biol.* **16**:373–378.

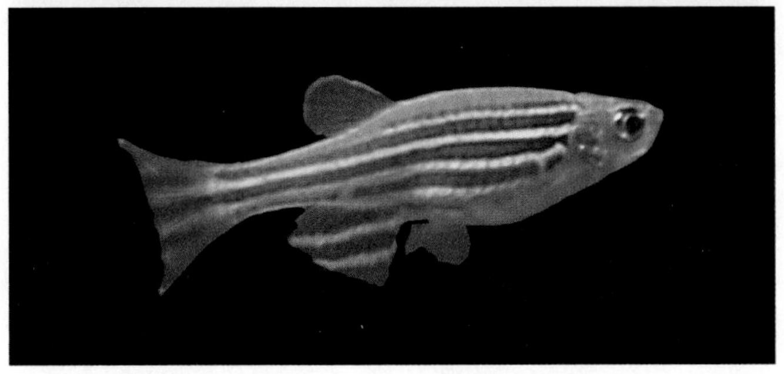

A study of mutant zebrafish with pale stripes led to the identification of a sodium/calcium transporter that regulates the darkness of human skin. © Christina Micek

CHAPTER

11

TRANSMEMBRANE TRANSPORT OF IONS AND SMALL MOLECULES

In all cells, the plasma membrane forms the permeability barrier that separates the cytoplasm from the exterior environment, thus topologically and biochemically defining a cell and distinguishing it from its surroundings. As a permeability barrier, the plasma membrane allows import of essential nutrients, ensures that metabolic intermediates remain in the cell where they belong, and enables waste products to leave the cell. By preventing the unimpeded movement of molecules into and out of cells, the plasma membrane maintains essential differences between the composition of the extracellular fluid and that of the cytoplasm; for example, the concentration of NaCl in the blood and extracellular fluids of animals is generally above 150 mM, similar to that of the seawater in which all cells are thought to have evolved, whereas the Na^+ concentration in the cytoplasm is tenfold lower. In contrast, the potassium ion concentration is higher in the cytoplasm than outside.

The plasma membrane, like all cellular membranes, is based on a bilayer of phospholipids into which proteins and other lipids are embedded. If the plasma membrane were a pure phospholipid bilayer, it would be an excellent chemical barrier, impermeable to virtually all ions, amino acids, sugars, and other water-soluble molecules that must selectively enter or leave a cell. In fact, only a few gases and uncharged small molecules can readily diffuse across a pure phospholipid membrane (Figure 11-1). However, plasma membranes must not only serve as barriers but also must play another, almost contradictory role. They must permit the selective transport of material and information between the cell's interior and exterior spaces, which often involves regulated entry or exit of many diverse small biomolecules, ions, and water. Plasma and other membranes serve as both barriers and conduits.

Integral membrane proteins called **transport proteins,** embedded in the plasma membrane and other membranes within cells by multiple transmembrane domains, permit the controlled and selective transport of molecules and ions across the membrane. In some cases, this involves molecules moving from a higher to a lower concentration, a thermodynamically spontaneous process that does not require the input of energy. Examples include the movement of water or glucose from the blood into body cells.

In other cases, molecules must be moved "uphill" across a membrane—against a concentration gradient—a thermodynamically unfavorable process that can only occur when an external source of energy is available. Examples include

OUTLINE

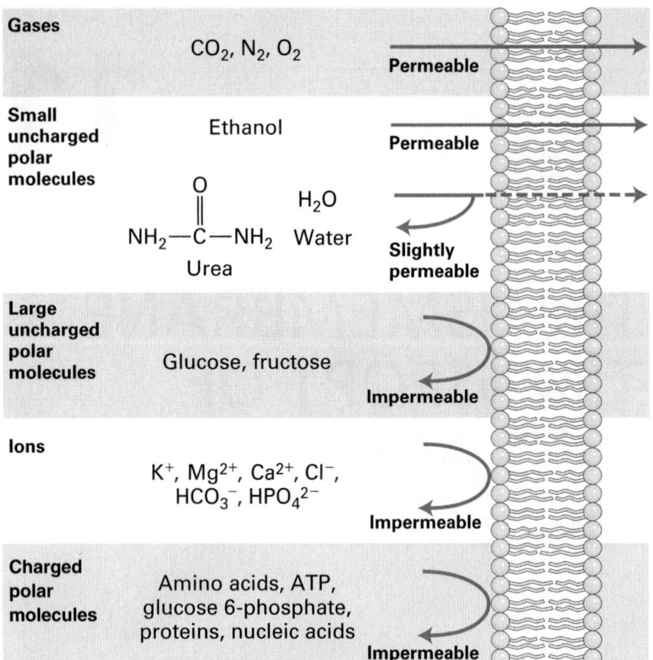

▲ **FIGURE 11-1 Relative permeability of a pure phospholipid bilayer to various molecules.** A bilayer is permeable to small hydrophobic molecules and small uncharged polar molecules, slightly permeable to water and urea, and essentially impermeable to ions and to large polar molecules.

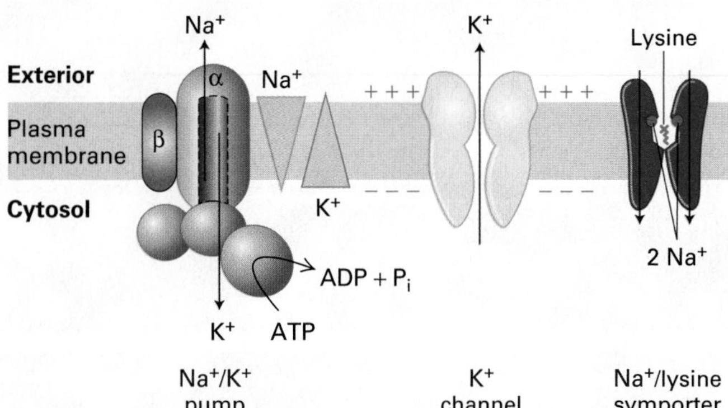

▲ **FIGURE 11-2 Multiple transport proteins function together in the plasma membrane of metazoan cells.** Gradients are indicated by triangles with the tip pointing toward lower concentration. The **Na^+/K^+ ATPase** in the plasma membrane uses energy released by ATP hydrolysis to pump Na^+ out of the cell and K^+ inward; this creates a concentration gradient of Na^+ that is greater out than in and one of K^+ that is greater in than out. Movement of positively charged K^+ ions out of the cell through membrane **K^+ channel** proteins creates an electric potential across the plasma membrane—the cytosolic face is negative with respect to the extracellular face. A **sodium-lysine transporter, a typical sodium–amino acid transporter,** moves Na^+ ions together with lysine from the extracellular medium into the cell. "Uphill" movement of the amino acid is powered by "downhill" movement of Na^+ ions, powered both by the out-greater-than-in Na^+ concentration gradient and by the negative potential on the inside of the cell membrane, which attracts the positively charged Na^+ ions. The ultimate source of the energy to power amino acid uptake comes from the ATP hydrolyzed by the Na^+/K^+ ATPase, since this pump creates both the Na^+ ion concentration gradient and, via the K^+ channels, the membrane potential, which together power influx of Na^+ ions.

concentrating protons within lysosomes to generate a low pH in the lumen. Often this energy is provided by mechanistically coupling the energy-releasing hydrolysis of the terminal phosphoanhydride bond in ATP; such transporters are called **ATP-powered pumps.** When such pumps transport ions, such as Na^+ and K^+, they generate across the membrane an electrical gradient, or potential, as well as a chemical concentration gradient. The energy stored in such gradients can subsequently be used to do work or convey information. Other transport proteins couple the movement of one molecule or ion against its concentration gradient with the movement of another down its gradient, using the energy released by the downhill movement of one molecule or ion to thermodynamically drive the uphill movement of another.

Frequently, several different types of transport proteins work in concert to achieve a physiological function. An example is seen in Figure 11-2, where an ATP-powered transporter pumps Na^+ out of the cell and K^+ ions inward; this pump establishes the opposing concentration gradients of Na^+ and K^+ ions across the plasma membrane. The human genome encodes hundreds of different types of transport proteins that use the energy stored in the Na^+ concentration gradient and the electric potential to transport a wide variety of molecules into cells against their concentration gradients.

We begin our discussion by reviewing some of the general principles of transport across membranes and distinguishing between three major classes of transport proteins. In subsequent sections, we describe the structure and operation of specific examples of each class and show how members of families of homologous transport proteins have different properties that enable different cell types to function appropriately. We also explain how subcellular membranes encompass specific combinations of transport proteins that enable cells to carry out essential physiological processes, including the maintenance of cytosolic pH, the accumulation of sucrose and salts in plant cell vacuoles, and the directed flow of water in both plants and animals. Epithelial cells, such as those lining the small intestine, transport ions, sugars, and other small molecules and water from one side to the other. We will see how this understanding has led to the development of sports drinks and also therapies for cholera.

11.1 Overview of Membrane Transport

Many different proteins contribute to the transport of ions and small molecules across membranes. As we learn about different transport processes in this chapter, we will see how different kinds of membrane-embedded proteins accomplish the task of moving molecules in different ways.

Only Small Hydrophobic Molecules Cross Membranes by Simple Diffusion

As we saw above, only gases, such as O_2 and CO_2, and small uncharged polar molecules, such as urea and ethanol, can readily move by **simple diffusion** across an artificial membrane composed of pure phospholipid or of phospholipid and cholesterol (see Figure 11-1). Such molecules also can diffuse across cellular membranes without the aid of transport proteins. No metabolic energy is expended because movement is from a high to a low concentration of the molecule, down its chemical concentration gradient. As noted in Chapter 2, such transport reactions are spontaneous because they have a positive ΔS value (increase in entropy) and thus a negative ΔG (decrease in free energy).

The relative diffusion rate of any substance across a pure phospholipid bilayer is proportional to its concentration gradient across the bilayer and to its hydrophobicity and size; the movement of charged molecules is also affected by any electric potential across the membrane. When a phospholipid bilayer separates two aqueous compartments, membrane permeability can be easily determined by adding a small amount of radioactive material to one compartment and measuring its rate of appearance in the other compartment. The greater the concentration gradient of the substance, the faster its rate of movement across a bilayer.

The hydrophobicity of a substance is measured by its partition coefficient K, the equilibrium constant for its partition between oil and water. The higher a substance's partition coefficient, the more lipid soluble it is. The first and rate-limiting step in transport by simple diffusion is movement of a molecule from the aqueous solution into the hydrophobic interior of the phospholipid bilayer, which resembles oil in its chemical properties. This is the reason that the more hydrophobic a molecule is, the faster it diffuses across a pure phospholipid bilayer. For example, diethylurea, with an ethyl group (CH_3CH_2—) attached to each nitrogen atom of urea, has a K of 0.01, whereas urea has a K of 0.0002 (see Figure 11-1). Diethylurea, which is 50 times (0.01/0.0002) more hydrophobic than urea, will therefore diffuse through phospholipid bilayer membranes about 50 times faster than urea. Similarly, fatty acids with longer hydrocarbon chains are more hydrophobic than those with shorter chains and will diffuse more rapidly across a pure phospholipid bilayer at all concentrations.

If a transported substance carries a net charge, its movement is influenced by both its concentration gradient and the **membrane potential,** the electric potential (voltage) across the membrane. The combination of these two forces, called the **electrochemical gradient,** determines the energetically favorable direction of transport of a charged molecule across a membrane. The electric potential that exists across most cellular membranes results from a small imbalance in the concentration of positively and negatively charged ions on the two sides of the membrane. We discuss how this ionic imbalance, and resulting potential, arise and are maintained in Sections 11.4 and 11.5.

Membrane Proteins Mediate Transport of Most Molecules and All Ions Across Biomembranes

As is evident from Figure 11-1, very few molecules and no ions can cross a pure phospholipid bilayer at appreciable rates by simple diffusion. Thus transport of most molecules into and out of cells requires the assistance of specialized membrane proteins. Even transport of molecules with relatively large partition coefficients (e.g., urea and certain gases such as CO_2) is frequently accelerated by specific proteins because their transport by simple diffusion usually is not sufficiently rapid to meet cellular needs.

All transport proteins are transmembrane proteins containing multiple membrane-spanning segments that generally are α helices. By forming a protein-lined pathway across the membrane, transport proteins are thought to allow movement of hydrophilic substances without their coming into contact with the hydrophobic interior of the membrane. Here we introduce the various types of transport proteins covered in this chapter (Figure 11-3).

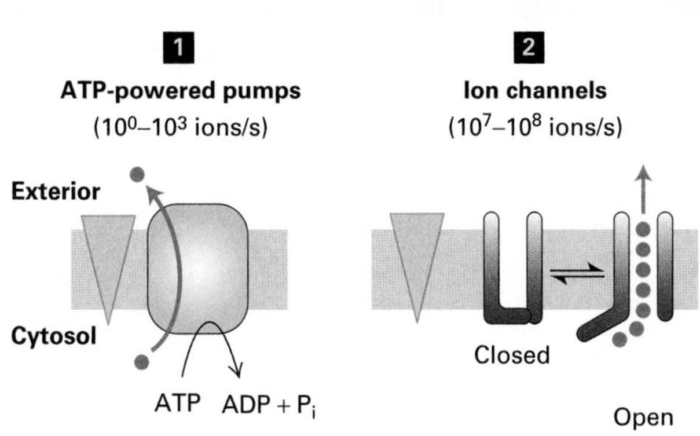

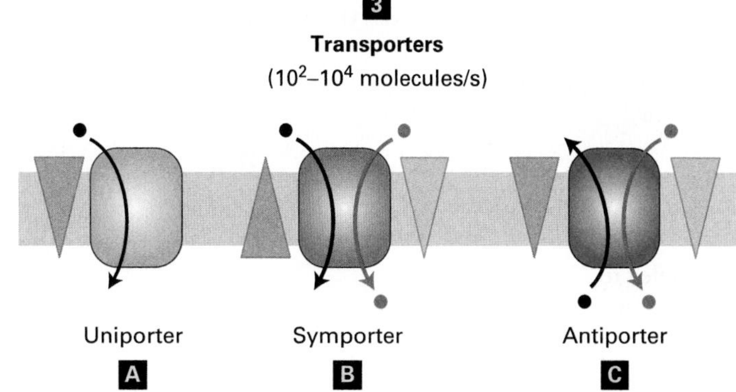

▲ **FIGURE 11-3 Overview of membrane transport proteins.** Gradients are indicated by triangles with the tip pointing toward lower concentration, electrical potential, or both. **1** Pumps utilize the energy released by ATP hydrolysis to power movement of specific ions (red circles) or small molecules against their electrochemical gradient. **2** Channels permit movement of specific ions (or water) down their electrochemical gradient. **3** Transporters, which fall into three groups, facilitate movement of specific small molecules or ions. Uniporters transport a single type of molecule down its concentration gradient **3A**. Cotransport proteins (symporters, **3B**, and antiporters, **3C**) catalyze the movement of one molecule *against* its concentration gradient (black circles), driven by movement of one or more ions down an electrochemical gradient (red circles). Differences in the mechanisms of transport by these three major classes of proteins account for their varying rates of solute movement.

ATP-powered pumps (or simply *pumps*) are ATPases that use the energy of ATP hydrolysis to move ions or small molecules across a membrane *against* a chemical concentration gradient, an electric potential, or both. This process, referred to as **active transport,** is an example of a coupled chemical reaction (Chapter 2). In this case, transport of ions or small molecules "uphill" against an electrochemical gradient, which requires energy, is coupled to the hydrolysis of ATP, which releases energy. The overall reaction—ATP hydrolysis and the "uphill" movement of ions or small molecules—is energetically favorable. The Na^+/K^+ pump shown in our overview figure (see Figure 11-2) is an example of an ATP-powered pump.

Channel proteins transport water, specific ions, or hydrophilic small molecules *down* their concentration or electric potential gradients via **facilitated transport** (or facilitated diffusion), the protein-assisted movement of a substance down its concentration gradient. Channel proteins form a hydrophilic passageway across the membrane through which multiple water molecules or ions move simultaneously, single file, at a very rapid rate. Some channels are open much of the time; these are referred to as *nongated* channels. Most ion channels, however, open only in response to specific chemical or electric signals; these are referred to as *gated* channels. The plasma membrane K^+ channel in Figure 11-2 is an example of a nongated ion channel. Channel proteins, like all transport proteins, are very selective for the type of molecule they transport.

Transporters (also called *carriers*) move a wide variety of ions and molecules across cell membranes. Three types of transporters have been identified. *Uniporters* transport a single type of molecule *down* its concentration gradient via facilitated diffusion. Glucose and amino acids cross the plasma membrane into most mammalian cells with the aid of uniporters. In contrast, *antiporters* and *symporters* couple the movement of one type of ion or molecule *against* its concentration gradient with the movement of one or more different ions *down* its concentration gradient, in the same (symporter) or different (antiporter) directions. These proteins often are called *cotransporters,* referring to their ability to transport two or more different solutes simultaneously. In Figure 11-2, lysine is moved into the cell via a Na^+/lysine symporter.

Like ATP pumps, cotransporters mediate coupled reactions in which an energetically unfavorable reaction (i.e., uphill movement of one type of molecule) is coupled to an energetically favorable reaction, the downhill movement of another. Note, however, that the nature of the energy-supplying reaction driving active transport by these two classes of proteins differs. ATP pumps use energy from hydrolysis of ATP, whereas cotransporters use the energy stored in an electrochemical gradient. This latter process sometimes is referred to as *secondary* active transport.

Table 11-1 summarizes the four mechanisms by which small molecules and ions are transported across cellular membranes. Conformational changes are essential to the function of all transport proteins. ATP-powered pumps and transporters undergo a cycle of conformational change exposing a binding site (or sites) to one side of the membrane in one conformation and to the other side in a second conformation. Because each such cycle results in movement of only one (or a few) substrate molecules, these proteins are characterized by relatively slow rates of transport ranging from 10^0 to 10^4 ions or molecules per second (see Figure 11-3). Ion channels shuttle between a closed state and an open state, but many ions can pass through an open channel without any further conformational change. For this reason, channels are characterized by very fast rates of transport, up

TABLE 11-1 Mechanisms for Transporting Ions and Small Molecules across Cell Membranes

PROPERTY	SIMPLE DIFFUSION	FACILITATED TRANSPORT	ACTIVE TRANSPORT	COTRANSPORT*
Requires specific protein	–	+	+	+
Solute transported against its gradient	–	–	+	+
Coupled to ATP hydrolysis	–	–	+	–
Driven by movement of a cotransported ion down its gradient	–	–	–	+
Examples of molecules transported	O_2, CO_2, steroid hormones, many drugs	Glucose and amino acids (uniporters); ions and water (channels)	Ions, small hydrophilic molecules, lipids (ATP-powered pumps)	Glucose and amino acids (symporters); various ions and sucrose (antiporters)

*Also called *secondary active transport.*

to 10^8 ions per second. We will begin with the simplest transport proteins, the molecules responsible for the transport of glucose and water, before moving on to progressively more complex transport molecules.

KEY CONCEPTS OF SECTION 11.1

Overview of Membrane Transport

- The plasma membrane regulates the traffic of molecules into and out of the cell.
- With the exception of gases (e.g., O_2 and CO_2) and small hydrophobic molecules, most molecules cannot diffuse across a pure phospholipid bilayer at rates sufficient to meet cellular needs.
- Three classes of transmembrane proteins mediate transport of ions, sugars, amino acids, and other metabolites across cell membranes: ATP-powered pumps, channels, and transporters (see Figure 11-3).
- In active transport, a transport protein couples movement of a substrate against its concentration gradient to ATP hydrolysis.
- In facilitated diffusion, a transport protein assists in the movement of a specific substrate (molecule or ion) down its concentration gradient.
- In secondary active transport, or cotransport, a transport protein couples movement of a substrate against its concentration gradient to the movement of a second substrate down its concentration gradient (see Table 11-1).

11.2 Uniport Transport of Glucose and Water

Most animal cells utilize glucose as a source for ATP production; they employ a glucose uniporter to take up glucose from the blood or other extracellular fluid down its concentration gradient. Many cells utilize membrane transport proteins called aquaporins to increase the rate of water movement across their surface membranes. Here, we discuss the structure and function of these and other uniport proteins.

Several Features Distinguish Uniport Transport from Simple Diffusion

The protein-mediated transport of glucose and other small hydrophilic molecules across a membrane, a process known as **uniport,** exhibits the following distinguishing properties:

1. The rate of facilitated diffusion by uniporters is far higher than simple diffusion through a pure phospholipid bilayer.

2. Because the transported molecules never enter the hydrophobic core of the phospholipid bilayer, the partition coefficient K is irrelevant.

3. Transport occurs via a limited number of uniporter molecules *rather than throughout the phospholipid bilayer.* Consequently, there is a maximum transport rate V_{max} that is achieved when the concentration gradient across the membrane is very large and each uniporter is working at its maximal rate.

4. Transport is specific. Each uniporter transports only a single species of molecule or a single group of closely related molecules. A measure of the affinity of a transporter for its substrate is K_m, which is the concentration of substrate at which transport is half maximal.

These properties also apply to transport mediated by the other classes of proteins depicted in Figure 11-3.

One of the best-understood uniporters is the glucose transporter *GLUT1* found in the plasma membrane of erythrocytes. The properties of GLUT1 and many other transport proteins from mature erythrocytes have been extensively studied. These cells, which have no nucleus or other internal organelles, are essentially "bags" of hemoglobin containing relatively few other intracellular proteins and a single membrane, the plasma membrane (see Figure 10-10a). Because the erythrocyte plasma membrane can be isolated in high purity, isolating and purifying a transport protein from mature erythrocytes is a fairly straightforward procedure.

Figure 11-4 shows that glucose uptake by erythrocytes and liver cells exhibits kinetics characteristic of a simple

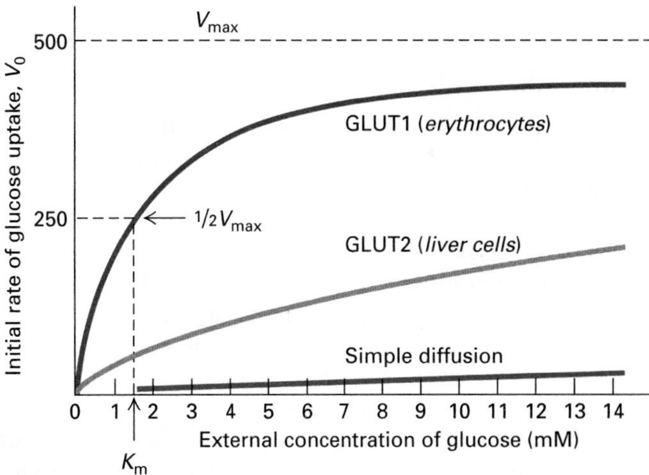

▲ EXPERIMENTAL FIGURE 11-4 **Cellular uptake of glucose mediated by GLUT proteins exhibits simple enzyme kinetics and greatly exceeds the calculated rate of glucose entry solely by simple diffusion.** The initial rate of glucose uptake (measured as micromoles per milliliter of cells per hour) in the first few seconds is plotted against increasing glucose concentration in the extracellular medium. In this experiment, the initial concentration of glucose in the cells is always zero. Both GLUT1, expressed by erythrocytes, and GLUT2, expressed by liver cells, greatly increase the rate of glucose uptake (burgundy and tan curves) compared with that associated with simple diffusion (blue curve) at all external concentrations. Like enzyme-catalyzed reactions, GLUT-facilitated uptake of glucose exhibits a maximum rate (V_{max}). The K_m is the concentration at which the rate of glucose uptake is half maximal. GLUT2, with a K_m of about 20 mM, has a much lower affinity for glucose than GLUT1, with a K_m of about 1.5 mM.

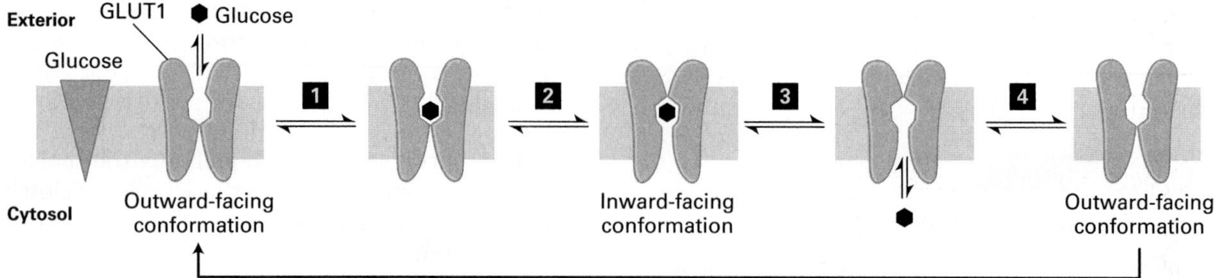

▲ **FIGURE 11-5 Model of uniport transport by GLUT1.** In one conformation, the glucose-binding site faces outward; in the other, the binding site faces inward. Binding of glucose to the outward-facing site (step **1**) triggers a conformational change in the transporter such that the binding site now faces inward toward the cytosol (step **2**). Glucose then is released to the inside of the cell (step **3**). Finally, the transporter undergoes the reverse conformational change, regenerating the outward-facing binding site (step **4**). If the concentration of glucose is higher inside the cell than outside, the cycle will work in reverse (step **4** → step **1**), resulting in net movement of glucose from inside to out. The actual conformational changes are probably smaller than those depicted here.

enzyme-catalyzed reaction involving a single substrate. The kinetics of transport reactions mediated by other types of proteins are more complicated than for uniporters. Nonetheless, all protein-assisted transport reactions occur faster than allowed by simple diffusion, are substrate-specific as reflected in lower K_m values for some substrates than others, and exhibit a maximal rate (V_{max}).

GLUT1 Uniporter Transports Glucose into Most Mammalian Cells

Most mammalian cells use blood glucose as the major source of cellular energy and express GLUT1. Since the glucose concentration usually is higher in the extracellular medium (blood, in the case of erythrocytes) than in the cell, GLUT1 generally catalyzes the net import of glucose from the extracellular medium into the cell. Under this condition, V_{max} is achieved at high external glucose concentrations.

Like other uniporters, GLUT1 alternates between two conformational states: in one, a glucose-binding site faces the outside of the membrane; in the other, a glucose-binding site faces the inside. Figure 11-5 depicts the sequence of events occurring during the unidirectional transport of glucose from the cell exterior inward to the cytosol. GLUT1 also can catalyze the net export of glucose from the cytosol to the extracellular medium when the glucose concentration is higher inside the cell than outside.

The kinetics of the unidirectional transport of glucose from the outside of a cell inward via GLUT1 can be described by the same type of equation used to describe a simple enzyme-catalyzed chemical reaction. For simplicity, let's assume that the substrate glucose, S, is present initially only on the outside of the membrane. In this case, we can write

$$S_{out} + \text{GLUT1} \overset{K_m}{\rightleftharpoons} S_{out} - \text{GLUT1} \overset{V_{max}}{\rightleftharpoons} S_{in} + \text{GLUT1}$$

where S_{out} − GLUT1 represents GLUT1 in the outward-facing conformation with a bound glucose. This equation is similar to the one describing the path of a simple enzyme-catalyzed reaction where the protein binds a single substrate and then transforms it into a different molecule. Here, however, no chemical modification occurs to the GLUT1-bound sugar; rather, it is moved across a cellular membrane. Nonetheless, the kinetics of this transport reaction are similar to those of simple enzyme-catalyzed reactions, and we can use the same derivation as that of the Michaelis-Menten equation in Chapter 3 to derive the following expression for v, the initial transport rate for S into the cell catalyzed by GLUT1:

$$v = \frac{V_{max}}{1 + \dfrac{K_m}{C}} \qquad (11\text{-}1)$$

where C is the concentration of S_{out} (initially, the concentration of $S_{in} = 0$). V_{max}, the rate of transport when all molecules of GLUT1 contain a bound S, occurs at an infinitely high S_{out} concentration. The lower the value of K_m, the more tightly the substrate binds to the transporter and the greater the transport rate at a fixed concentration of substrate. Equation 11-1 describes the curve for glucose uptake by erythrocytes shown in Figure 11-4 as well as similar curves for other uniporters.

For GLUT1 in the erythrocyte membrane, the K_m for glucose transport is 1.5 mM; at this concentration, roughly half the transporters with outward-facing binding sites would have a bound glucose and transport would occur at 50 percent of the maximal rate. Since blood glucose is normally 5 mM, the erythrocyte glucose transporter usually is functioning at 77 percent of the maximal rate, as can be seen from Equation 11-1. GLUT1 and the very similar GLUT3 are expressed by erythrocytes and other cells that need to take up glucose from the blood continuously at high rates;

the rate of glucose uptake by such cells will remain high regardless of small changes in the concentration of blood glucose.

In addition to glucose, the isomeric sugars D-mannose and D-galactose, which differ from D-glucose in the configuration at only one carbon atom, are transported by GLUT1 at measurable rates. However, the K_m for glucose (1.5 mM) is much lower than the K_m for D-mannose (20 mM) or D-galactose (30 mM). Thus GLUT1 is quite specific, having a much higher affinity (indicated by a lower K_m) for the normal substrate D-glucose than for other substrates.

GLUT1 accounts for 2 percent of the protein in the plasma membrane of erythrocytes. After glucose is transported into the erythrocyte, it is rapidly phosphorylated, forming glucose 6-phosphate, which cannot leave the cell. Because this reaction, the first step in the metabolism of glucose (see Figure 12-3), is rapid and occurs at a constant rate, the intracellular concentration of glucose is kept low even when glucose is imported from the medium. Consequently the concentration gradient of glucose outside greater than inside the cell is maintained at a sufficiently high ratio to support import of additional glucose molecules and maintain a constant rate of glucose metabolism.

The Human Genome Encodes a Family of Sugar-Transporting GLUT Proteins

The human genome encodes at least 12 highly homologous **GLUT proteins,** GLUT1–GLUT12, that are all thought to contain 12 membrane-spanning α helices, suggesting that they evolved from a single ancestral transport protein. Although no three-dimensional structure of GLUT1 is available, detailed biochemical studies have shown that the amino acid residues in the transmembrane α helices are predominantly hydrophobic; several helices, however, bear amino acid residues (e.g., serine, threonine, asparagine, and glutamine) whose side chains can form hydrogen bonds with the hydroxyl groups on glucose. These residues are thought to form the inward-facing and outward-facing glucose-binding sites in the interior of the protein (see Figure 11-5).

The structures of all GLUT isoforms are thought to be quite similar, and all transport sugars. Nonetheless, their differential expression in various cell types and isoform-specific functional properties enable different body cells to regulate glucose metabolism independently and at the same time maintain a constant concentration of glucose in the blood. For instance, GLUT3 is found in neuronal cells of the brain. Neurons depend on a constant influx of glucose for metabolism, and the low K_m of GLUT3 for glucose, like that of GLUT1, ensures that these cells incorporate glucose from brain extracellular fluids at a high and constant rate.

GLUT2, expressed in liver and the insulin-secreting β cells of the pancreas, has a K_m of ≈20 mM, about 13 times higher than the K_m of GLUT1. As a result, when blood glucose rises from its basal level of 5 mM to 10 mM or so after a meal, the rate of glucose influx will almost double in GLUT2-expressing cells, whereas it will increase only slightly in GLUT1-expressing cells (see Figure 11-4). In liver, the "excess" glucose brought into the cell is stored as the polymer glycogen. In islet β cells, the rise in glucose triggers secretion of the hormone insulin, which in turn lowers blood glucose by increasing glucose uptake and metabolism in muscle and by inhibiting glucose production in liver (see Figure 15-32).

Another GLUT isoform, GLUT4, is expressed only in fat and muscle cells, the cells that respond to insulin by increasing their uptake of glucose, thereby removing glucose from the blood. In the absence of insulin, GLUT4 is found in intracellular membranes, not on the plasma membrane, and is unable to facilitate glucose uptake. By a process detailed in Chapter 15, insulin causes these GLUT4-rich internal membranes to fuse with the plasma membrane, increasing the number of GLUT4 molecules on the cell surface and thus the rate of glucose uptake. A defect in this process, one principal mechanism by which insulin lowers blood glucose, is one of the causes of adult onset, or type II, diabetes, a disease marked by continuously high blood glucose.

Transport Proteins Can Be Enriched Within Artificial Membranes and Cells

Although transport proteins can be isolated from membranes and purified, the functional properties of these proteins can be studied only when they are associated with a membrane. Most cellular membranes contain many different types of transport proteins but a relatively low concentration of any particular one, making functional studies of a single protein difficult. To facilitate such studies, researchers use two approaches for enriching a transport protein of interest so that it predominates in the membrane.

In one common approach, a specific transport protein is extracted and purified; the purified protein then is reincorporated into pure phospholipid bilayer membranes, such as liposomes (see Figure 10-6). For example, all of the integral proteins of the erythrocyte membrane can be solubilized by a nonionic detergent, such as octylglucoside. The glucose uniporter GLUT1 can be purified by antibody affinity chromatography (Chapter 3) on a column containing a specific monoclonal antibody and then incorporated into liposomes made of pure phospholipids.

Alternatively, the gene encoding a specific transport protein can be expressed at high levels in a cell type that normally does not express it. The difference in transport of a substance by the transfected and by control nontransfected cells will be due to the expressed transport protein. In these systems, the functional properties of the various membrane proteins can be examined without ambiguity. As an example, overexpressing GLUT1 in lines of cultured fibroblasts increases severalfold their rate of uptake of glucose, and expression of mutant GLUT1 proteins with specific amino acid alterations can identify residues important for substrate binding.

Osmotic Pressure Causes Water to Move Across Membranes

Movement of water in and out of cells is an important feature of the life of both plants and animals. **Aquaporins** are a family of membrane proteins that allow water and a few other small uncharged molecules, such as glycerol, to cross biomembranes. But before discussing these transport proteins, we need to review osmosis, the force that powers the movement of water.

Water tends to move across a semipermeable membrane from a solution of low solute concentration to one of high concentration, a process termed **osmosis,** or osmotic flow. In other words, since solutions with a high concentration of dissolved solute have a lower concentration of water, water will spontaneously move from a solution of high water concentration to one of lower. In effect, osmosis is equivalent to "diffusion" of water. Osmotic pressure is defined as the hydrostatic pressure required to stop the net flow of water across a membrane separating solutions of different compositions (Figure 11-6). In this context, the "membrane" may be a layer of cells or a plasma membrane that is permeable to water but not to the solutes. The osmotic pressure is directly proportional to the difference in the concentration of the total number of solute molecules on each side of the membrane. For example, a 0.5 M NaCl solution is actually 0.5 M Na^+ ions and 0.5 M Cl^- ions and has the same osmotic pressure as a 1 M solution of glucose or sucrose.

The movement of water across the plasma membrane also determines the volume of individual cells, which must be regulated to avoid damage to the cell. Osmotic pressure is the force powering the movement of water in biological systems.

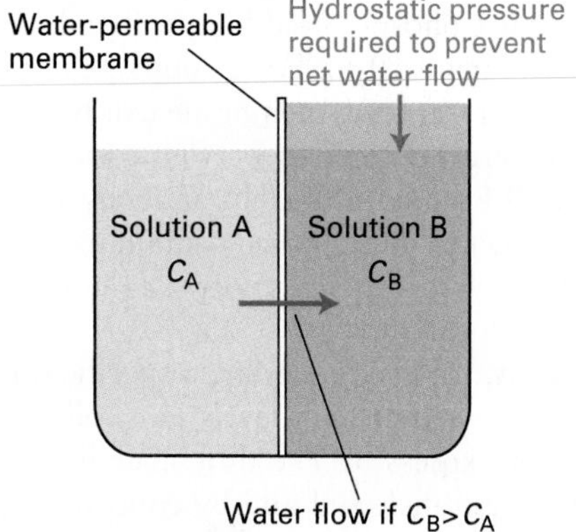

▲ **FIGURE 11-6 Osmotic pressure.** Solutions A and B are separated by a membrane that is permeable to water but impermeable to all solutes. If C_B (the total concentration of solutes in solution B) is greater than C_A, water will tend to flow across the membrane from solution A to solution B. The osmotic pressure π between the solutions is the hydrostatic pressure that would have to be applied to solution B to prevent this water flow. From the van't Hoff equation, osmotic pressure is given by $\pi = RT(C_B - C_A)$, where R is the gas constant and T is the absolute temperature.

In higher plants, water and minerals are absorbed from the soil by the roots and move up the plant through conducting tubes (the xylem); water loss from the plant, mainly by evaporation from the leaves, drives these movements of water. Unlike animal cells, plant, algal, fungal, and bacterial cells are surrounded by rigid cell walls, which resist the expansion of the volume of the cell when the intracellular osmotic pressure increases. Without such a wall, animal cells expand when internal osmotic pressure increases—if that pressure rises too much, the cells will burst like overextended balloons. Because of the cell wall in plants, the osmotic influx of water that occurs when such cells are placed in a hypotonic solution (even pure water) leads to an increase in intracellular pressure but not in cell volume. In plant cells, the concentration of solutes (e.g., sugars and salts) usually is higher in the vacuole (see Figure 9-7) than in the cytosol, which in turn has a higher solute concentration than the extracellular space. The osmotic pressure, called *turgor pressure,* generated from the entry of water into the cytosol and then into the vacuole pushes the cytosol and the plasma membrane against the resistant cell wall. Plant cells can harness this pressure to help them grow. Cell elongation during growth occurs by a hormone-induced localized loosening of a defined region of the cell wall, followed by influx of water into the vacuole, increasing its size and thus the size of the cell. ▮

Although most protozoans (like animal cells) do not have a rigid cell wall, many contain a contractile vacuole that permits them to avoid osmotic lysis. A contractile vacuole takes up water from the cytosol and, unlike a plant vacuole, periodically discharges its contents through fusion with the plasma membrane. Thus even though water continuously enters the protozoan cell by osmotic flow, the contractile vacuole prevents too much water from accumulating in the cell and swelling it to the bursting point.

Aquaporins Increase the Water Permeability of Cell Membranes

Small changes in extracellular osmotic strength cause most animal cells to swell or shrink rapidly. When placed in a **hypotonic** solution (i.e., one in which the concentration of solutes is *lower* than in the cytosol), animal cells swell owing to the osmotic flow of water inward. Conversely, when placed in a **hypertonic** solution (i.e., one in which the concentration of solutes is *higher* than in the cytosol), animal cells shrink as cytosolic water leaves the cell by osmotic flow. Consequently, cultured animal cells must be maintained in an **isotonic** medium, which has a solute concentration and thus osmotic strength identical with that of the cell cytosol.

In contrast, frog oocytes and eggs do not swell when placed in pond water of very low osmotic strength, even though their internal salt (mainly KCl) concentration is comparable to that of other cells ($\approx$150 mM KCl). These observations were what first led investigators to suspect that the plasma membranes of

Video: Frog Oocyte Expressing Aquaporin Bursts in Hypotonic Solution

0.5 min

1.5 min

2.5 min

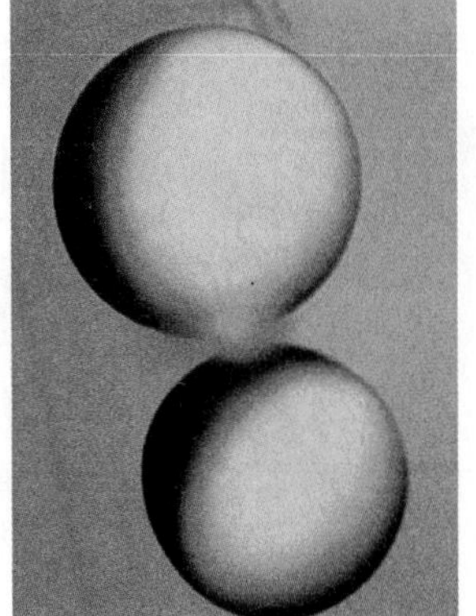

3.5 min

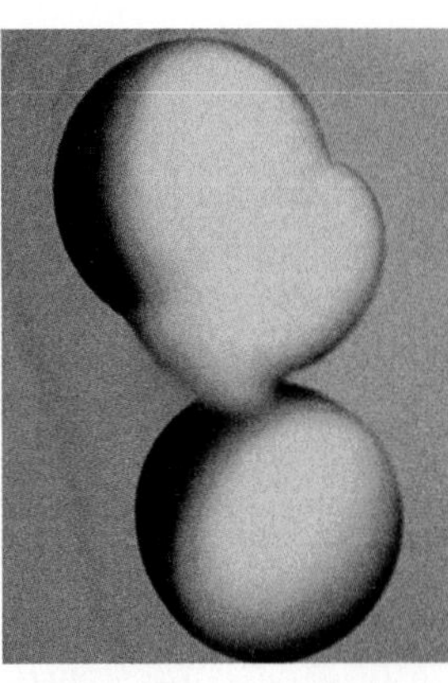

▲ EXPERIMENTAL FIGURE 11-7 **Expression of aquaporin by frog oocytes increases their permeability to water.** Frog oocytes, which normally are impermeable to water and do not express an aquaporin protein, were microinjected with mRNA encoding aquaporin. These photographs show control oocytes (bottom cell in each panel) and microinjected oocytes (top cell in each panel) at the indicated times after transfer from an isotonic salt solution (0.1 mM) to a hypotonic salt solution (0.035 M). The volume of the control oocytes remained unchanged because they are poorly permeable to water. In contrast, the microinjected oocytes expressing aquaporin swelled and then burst because of an osmotic influx of water, indicating that aquaporin is a water-channel protein. [Courtesy of Gregory M. Preston and Peter Agre, Johns Hopkins University School of Medicine. See L. S. King, D. Kozono, and P. Agre., 2004, *Nat. Rev. Mol. Cell Biol.* **5**:687–98.]

erythrocytes and other cell types, but not frog oocytes, contain water-channel proteins that accelerate the osmotic flow of water. The experimental results shown in Figure 11-7 demonstrate that an aquaporin in the erythrocyte plasma membrane functions as a water channel.

In its functional form, aquaporin is a tetramer of identical 28-kDa subunits (Figure 11-8a). Each subunit contains six membrane-spanning α helices that form a central pore through which water moves (Figure 11-8b, c). At its center, the $\approx$2-nm-long water-selective gate, or pore, is only 0.28 nm in diameter—only slightly larger than the diameter of a water molecule. The molecular sieving properties of the constriction are determined by several conserved hydrophilic amino acid residues whose side-chain and carbonyl groups extend into the middle of the channel. Several water molecules move simultaneously through the channel, each of which sequentially forms specific hydrogen bonds and displaces another water molecule downstream. The formation of hydrogen bonds between the oxygen atom of water and the amino groups of two amino acid side chains ensures that only water passes through the channel; even protons cannot pass through, allowing ionic gradients to be maintained across membranes even when water is flowing across.

Mammals express a family of aquaporins; 11 are known in humans. Aquaporin 1 is expressed in abundance in erythrocytes, and the homologous aquaporin 2 is found in the kidney epithelial cells that resorb water from the urine, thus controlling the amount of water in the body. The activity of aquaporin 2 is regulated by vasopressin, also called antiduretic hormone. The regulation of the activity of aquaporin 2 in resting kidney cells resembles that of GLUT4 in fat and muscle in that when its activity is not required, when the cells are in their resting state and water is excreted to form urine, aquaporin 2 is localized to intracellular vesicle membranes, not to the plasma membrane, and so is unable to catalyze water import into the cell. When the polypeptide hormone vasopressin binds to the cell-surface vasopressin receptor, it activates a signaling pathway (detailed in Chapter 15) that causes these aquaporin 2-containing vesicles to fuse with the plasma membrane, increasing the rate of water uptake and its return into the circulation instead of the urine. Inactivating mutations in either the vasopressin or the aquaporin 2 gene cause *diabetes insipidus,* a disease marked by excretion of large volumes of dilute urine. This finding establishes the etiology of the disease and demonstrates that the level of aquaporin 2 is rate limiting for water resorption from urine being formed by the kidney. ▮

Other members of the aquaporin family transport hydroxyl-containing molecules such as glycerol rather than water. Human aquaporin 3, for instance, transports glycerol and is similar in amino acid sequence and structure to the *Escherichia coli* glycerol transport protein *GlpF.*

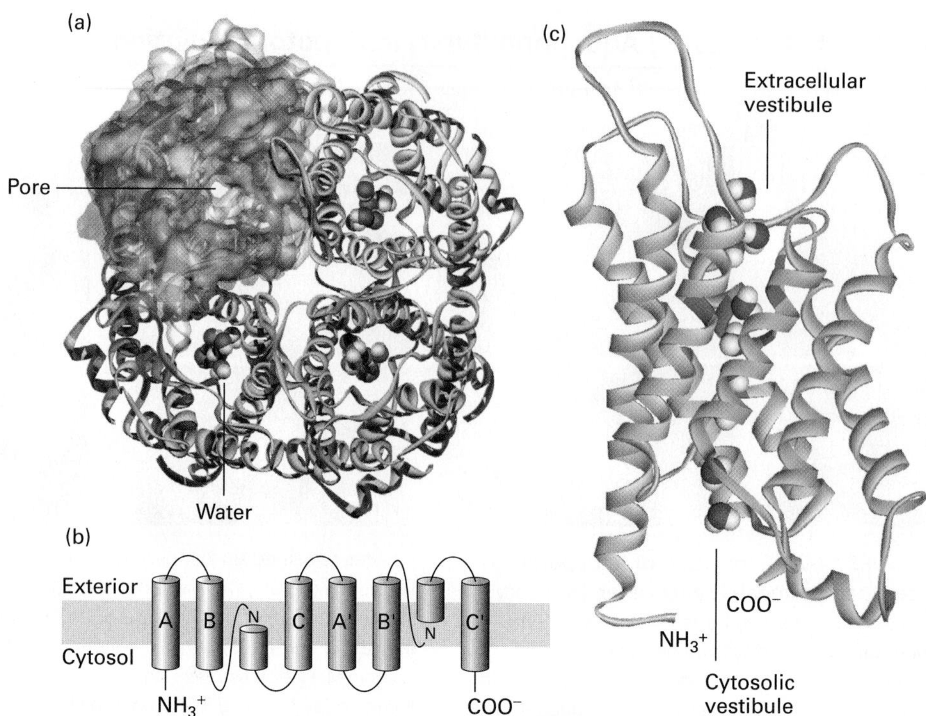

▲ **FIGURE 11-8 Structure of the water-channel protein aquaporin.** (a) Structural model of the tetrameric protein comprising four identical subunits. Each subunit forms a water channel, as seen in this view looking down on the protein from the exoplasmic side. One of the monomers is shown with a molecular surface in which the pore entrance can be seen. (b) Schematic diagram of the topology of a single aquaporin subunit in relation to the membrane. Three pairs of homologous transmembrane α helices (A and A′, B and B′, and C and C′) are oriented in the opposite direction with respect to the membrane and are connected by two hydrophilic loops containing short non-membrane-spanning helices and conserved asparagine (N) residues. The loops bend into the cavity formed by the six transmembrane helices, meeting in the middle to form part of the water-selective gate. (c) Side view of the pore in a single aquaporin subunit in which several water molecules (red oxygens and white hydrogens) are seen within the 2-nm-long water-selective gate that separates the water filled cytosolic and extracellular vestibules. The gate contains highly conserved arginine and histidine residues, as well as the two asparagine residues (blue) whose side chains form hydrogen bonds with transported waters. (Key gate residues are highlighted in blue.) Transported waters also form hydrogen bonds to the main-chain carbonyl group of a cysteine residue. The arrangement of these hydrogen bonds and the narrow pore diameter of 0.28 nm prevent passage of protons (i.e., H_3O^+) or other ions. [After H. Sui et al., 2001, *Nature* **414**:872. See also T. Zeuthen, 2001, *Trends Biochem. Sci.* **26**:77, and K. Murata et al., 2000, *Nature* **407**:599.]

KEY CONCEPTS OF SECTION 11.2

Uniport Transport of Glucose and Water

- Protein-catalyzed transport of a solute across a membrane occurs much faster than passive diffusion, exhibits a V_{max} when the limited number of transporter molecules are saturated with substrate, and is highly specific for substrate (see Figure 11-4).
- Uniport proteins, such as the glucose transporters (GLUTs), are thought to shuttle between two conformational states, one in which the substrate-binding site faces outward and one in which the binding site faces inward (see Figure 11-5).
- All members of the GLUT protein family transport sugars and have similar structures. Differences in their K_m values, expression in different cell types, and substrate specificities are important for proper sugar metabolism in the body.
- Two common experimental systems for studying the functions of transport proteins are liposomes containing a purified transport protein and cells transfected with the gene encoding a particular transport protein.
- Most biological membranes are semipermeable, more permeable to water than to ions or most other solutes. Water moves by osmosis across membranes from a solution of lower solute concentration to one of higher solute concentration.
- The rigid cell wall surrounding plant cells prevents their swelling and leads to generation of turgor pressure in response to the osmotic influx of water.
- In response to the entry of water, protozoans maintain their normal cell volume by extruding water from contractile vacuoles.
- Aquaporins are water-channel proteins that specifically increase the permeability of biomembranes for water (see Figure 11-8).
- Aquaporin 2 in the plasma membrane of certain kidney cells is essential for resorption of water from urine being formed; the absence of aquaporin 2 leads to the medical condition diabetes insipidus.

11.3 ATP-Powered Pumps and the Intracellular Ionic Environment

In previous sections, we focused on transport proteins that move molecules down their concentrations gradients. Here we focus our attention on a major class of proteins—the ATP-powered pumps—that use the energy released by hydrolysis of the terminal phosphoanhydride bond of ATP to transport ions and various small molecules across membranes against their concentration gradients. All ATP-powered pumps are transmembrane proteins with one or more binding sites for ATP located on subunits or segments of the protein that face the cytosol. Although these proteins commonly are called *ATPases,* they normally do not hydrolyze ATP into ADP and P_i unless ions or other molecules are simultaneously transported. Because of this tight coupling between ATP hydrolysis and transport, the energy stored in the phosphoanhydride bond is not dissipated but rather used to move ions or other molecules uphill against an electrochemical gradient. As Figure 11-2 illustrates, these concentration gradients are used by other transport proteins to power uphill movement of yet other types of molecules.

Different Classes of Pumps Exhibit Characteristic Structural and Functional Properties

The general structures of the four classes of ATP-powered pumps are depicted in Figure 11-9, with specific examples in each class listed below the figure. Note that the members of three of the classes (P, F, and V) only transport ions, whereas members of the ABC superfamily primarily transport small molecules such as amino acids and sugars.

All *P-class ion pumps* possess two identical catalytic α subunits, each of which contains an ATP-binding site. Most also have two smaller β subunits that usually have regulatory functions. During transport, at least one of the α

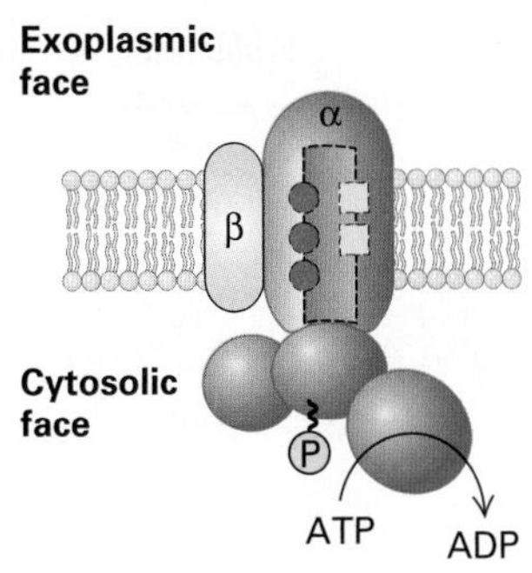

P-class pumps

Plasma membrane of plants, fungi, bacteria (H^+ pump)

Plasma membrane of higher eukaryotes (Na^+/K^+ pump)

Apical plasma membrane of mammalian stomach (H^+/K^+ pump)

Plasma membrane of all eukaryotic cells (Ca^{2+} pump)

Sarcoplasmic reticulum membrane in muscle cells (Ca^{2+} pump)

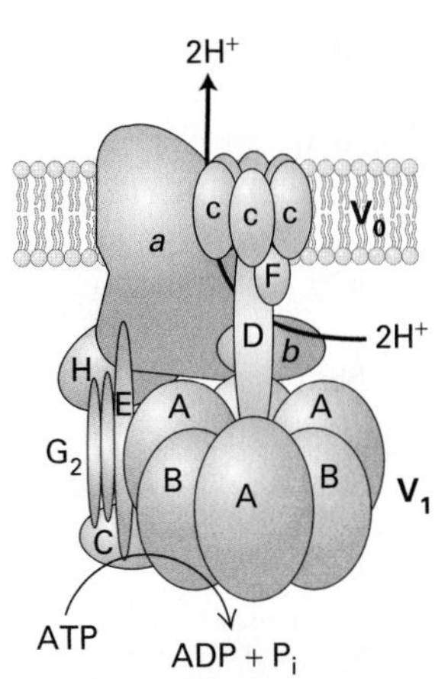

V-class proton pumps

Vacuolar membranes in plants, yeast, other fungi

Endosomal and lysosomal membranes in animal cells

Plasma membrane of osteoclasts and some kidney tubule cells

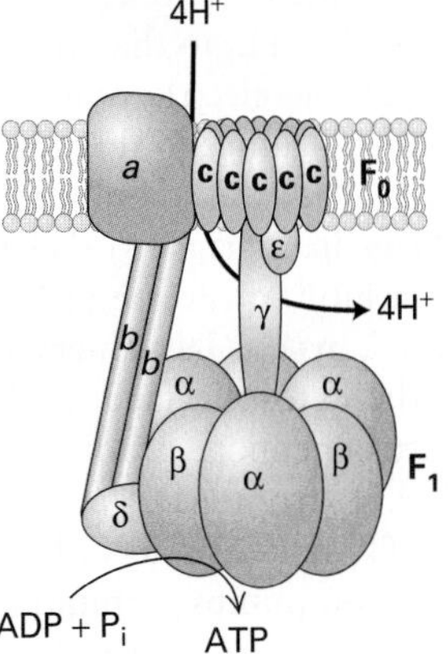

F-class proton pumps

Bacterial plasma membrane

Inner mitochondrial membrane

Thylakoid membrane of chloroplast

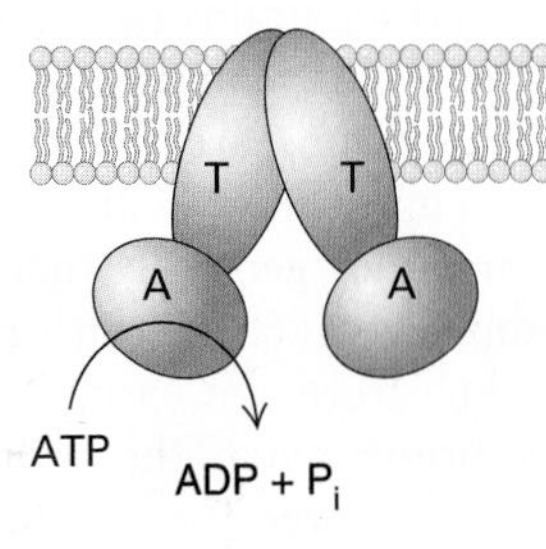

ABC superfamily

Bacterial plasma membranes (amino acid, sugar, and peptide transporters)

Mammalian plasma membranes (transporters of phospholipids, small lipophilic drugs, cholesterol, other small molecules)

▲ **FIGURE 11-9 The four classes of ATP-powered transport proteins.** The location of specific pumps are indicated below each class. P-class pumps are composed of two catalytic α subunits, which become phosphorylated as part of the transport cycle. Two β subunits, present in some of these pumps, may regulate transport. Only one α and β subunit are depicted. V-class and F-class pumps do not form phosphoprotein intermediates and transport only protons. Their structures are similar and contain similar proteins, but none of their subunits are related to those of P-class pumps. V-class pumps couple ATP hydrolysis to transport of protons against a concentration gradient, whereas F-class pumps normally operate in the reverse direction to utilize energy in a proton concentration or electrochemical gradient to synthesize ATP. All members of the large ABC superfamily of proteins contain two transmembrane (T) domains and two cytosolic ATP-binding (A) domains, which couple ATP hydrolysis to solute movement. These core domains are present as separate subunits in some ABC proteins (depicted here) but are fused into a single polypeptide in other ABC proteins. [See T. Nishi and M. Forgac, 2002, *Nature Rev. Mol. Cell Biol.* **3**:94; C. Toyoshima et al., 2000, *Nature* **405**:647; D. McIntosh, 2000, *Nature Struc. Biol.* **7**:532; and T. Elston, H. Wang, and G. Oster, 1998, *Nature* **391**:510.]

subunits is phosphorylated (hence the name "P" class), and the transported ions move through the phosphorylated subunit. The amino acid sequence around the phosphorylated residue is homologous in different pumps. This class includes the Na^+/K^+ ATPase in the plasma membrane, which generates the low cytosolic Na^+ and high cytosolic K^+ concentrations typical of animal cells (See Figure 11-2). Certain Ca^{2+} ATPases pump Ca^{2+} ions out of the cytosol into the external medium; others pump Ca^{2+} from the cytosol into the endoplasmic reticulum or into the specialized ER called the **sarcoplasmic reticulum,** which is found in muscle cells. Another member of the P class, found in acid-secreting cells of the mammalian stomach, transports protons (H^+ ions) out of and K^+ ions into the cell.

The structures of *V-class* and *F-class ion pumps* are similar to one another but unrelated to, and more complicated than, P-class pumps. V- and F-class pumps contain several different transmembrane and cytosolic subunits. All known V and F pumps transport only protons and do so in a process that does not involve a phosphoprotein intermediate. V-class pumps generally function to maintain the low pH of plant vacuoles and of lysosomes and other acidic vesicles in animal cells by pumping protons from the cytosolic to the exoplasmic face of the membrane against a proton electrochemical gradient. The H^+ pump that generates and maintains the plasma membrane electric potential in plant, fungal, and bacterial cells also belongs to this class. F-class pumps are found in bacterial plasma membranes and in mitochondria and chloroplasts. In contrast to V-class pumps, they generally function as a kind of reverse proton pump, in which the energy released by the movement of protons from the exoplasmic to the cytosolic face of the membrane *down* the proton electrochemical gradient is used to power the synthesis of ATP from ADP and P_i. Because of their importance in ATP synthesis in chloroplasts and mitochondria, F-class proton pumps, commonly called ATP synthases, are treated separately in Chapter 12.

The final class of ATP-powered pumps is a large family of multiple members that are more diverse in function than those of the other classes. Referred to as the **ABC** (*A*TP-*b*inding *c*assette) **superfamily,** this class includes several hundred different transport proteins found in organisms ranging from bacteria to humans. As detailed below, some of these transport proteins were first identified as multidrug-resistance proteins that, when overexpressed in cancer cells, export anticancer drugs and render tumors resistant to their action. Each ABC protein is specific for a single substrate or group of related substrates, which may be ions, sugars, amino acids, phospholipids, cholesterol, peptides, polysaccharides, or even proteins. All ABC transport proteins share a structural organization consisting of four "core" domains: two transmembrane (T) domains, forming the passageway through which transported molecules cross the membrane, and two cytosolic ATP-binding (A) domains. In some ABC proteins, mostly those in bacteria, the core domains are present in four separate polypeptides; in others, the core domains are fused into one or two multidomain polypeptides.

ATP-Powered Ion Pumps Generate and Maintain Ionic Gradients Across Cellular Membranes

The specific ionic composition of the cytosol usually differs greatly from that of the surrounding extracellular fluid. In virtually all cells—including microbial, plant, and animal cells—the cytosolic pH is kept near 7.2 *regardless* of the extracellular pH. In the most extreme case there is a millionfold difference in H^+ concentration between the pH of the cytosol of the epithelial cells lining the stomach and the pH of the stomach lumen. Also, the cytosolic concentration of K^+ is much higher than that of Na^+. In both invertebrates and vertebrates the concentration of K^+ is 20–40 times higher in cells than in the blood, while the concentration of Na^+ is 8–12 times lower in cells than in the blood (see Figure 11-2 and Table 11-2). Some Ca^{2+} in the cytosol is bound to the negatively charged groups in ATP and other

TABLE 11-2 Typical Intracellular and Extracellular Ion Concentrations

ION	CELL (mM)	BLOOD (mM)
SQUID AXON (INVERTEBRATE)*		
K^+	400	20
Na^+	50	440
Cl^-	40–150	560
Ca^{2+}	0.0003	10
X^{-}†	300–400	5–10
MAMMALIAN CELL (VERTEBRATE)		
K^+	139	4
Na^+	12	145
Cl^-	4	116
HCO_3^-	12	29
X^-	138	9
Mg^{2+}	0.8	1.5
Ca^{2+}	< 0.0002	1.8

*The large nerve axon of the squid has been widely used in studies of the mechanism of conduction of electric impulses.

†X^- represents proteins, which have a net negative charge at the neutral pH of blood and cells.

molecules, but it is the concentration of free, unbound Ca^{2+} that is critical to its functions in signaling pathways and muscle contraction. The concentration of free Ca^{2+} in the cytosol is generally less than 0.2 micromolar (2×10^{-7} M), a thousand or more times lower than that in the blood. Plant cells and many microorganisms maintain similarly high cytosolic concentrations of K^+ and low concentrations of Ca^{2+} and Na^+ even if the cells are cultured in very dilute salt solutions.

The ion pumps discussed in this section are largely responsible for establishing and maintaining the usual ionic gradients across the plasma and intracellular membranes. In carrying out this task, cells expend considerable energy. For example, up to 25 percent of the ATP produced by nerve and kidney cells is used for ion transport, and human erythrocytes consume up to 50 percent of their available ATP for this purpose; in both cases, most of this ATP is used to power the Na^+/K^+ pump. The resultant Na^+ and K^+ gradients in nerve cells are essential for their ability to conduct electric signals rapidly and efficiently, as we detail in Chapter 23. Certain enzymes required for protein synthesis in all cells require a high K^+ concentration and are inhibited by high concentrations of Na^+; these would cease to function without the operation of the Na^+/K^+ pump. In cells treated with poisons that inhibit the production of ATP (e.g., 2,4-dinitrophenol in aerobic cells), the pumping stops and the ion concentrations inside the cell gradually approach those of the exterior environment as ions spontaneously move through channels in the plasma membrane down their electrochemical gradients. Eventually treated cells die: partly because protein synthesis requires a high concentration of K^+ ions and partly because in the absence of a Na^+ gradient across the cell membrane, a cell cannot import certain nutrients such as amino acids. Studies on the effects of such poisons provided early evidence for the existence and significance of ion pumps.

Muscle Relaxation Depends on Ca^{2+} ATPases That Pump Ca^{2+} from the Cytosol into the Sarcoplasmic Reticulum

In skeletal muscle cells, Ca^{2+} ions are concentrated and stored in the sarcoplasmic reticulum (SR); release of stored Ca^{2+} ions from the SR lumen via ion channels into the cytosol causes contraction, as discussed in Chapter 17. A P-class Ca^{2+} ATPase located in the SR membrane of skeletal muscle pumps Ca^{2+} from the cytosol back into the lumen of the SR, thereby inducing muscle relaxation. Because this *muscle calcium pump* constitutes more than 80 percent of the integral protein in SR membranes, it is easily purified from other membrane proteins and has been studied extensively. Determination of the three-dimensional structure of this protein in several conformational states that represent different steps in the pumping process has revealed much about its mechanism of action.

In the cytosol of muscle cells, the free Ca^{2+} concentration ranges from 10^{-7} M (resting cells) to more than 10^{-6} M (contracting cells), whereas the *total* Ca^{2+} concentration in the SR lumen can be as high as 10^{-2} M. Two soluble proteins in the lumen of the SR vesicles bind Ca^{2+} and serve as a reservoir for intracellular Ca^{2+}, thereby reducing the concentration of free Ca^{2+} ions in the SR vesicles and consequently the energy needed to pump Ca^{2+} ions into the SR from the cytosol. The activity of the muscle Ca^{2+} ATPase increases as the free Ca^{2+} concentration in the cytosol rises. In skeletal muscle cells the calcium pump in the SR membrane works in concert with a similar Ca^{2+} pump located in the plasma membrane to ensure that the cytosolic concentration of free Ca^{2+} in resting muscle remains below 1 μM.

The current model for the mechanism of action of the Ca^{2+} ATPase in the SR membrane involves multiple conformational states. For simplicity, we group these into E1 states, in which the two binding sites for Ca^{2+}, located in the center of the membrane-spanning domain, face the cytosol, and E2 states, in which these binding sites face the exoplasmic face of the membrane, pointing into the lumen of the SR. Coupling of ATP hydrolysis with ion pumping involves several conformational changes in the protein that must occur in a defined order, as shown in Figure 11-10. When the protein is in the E1 conformation, two Ca^{2+} ions bind to two high-affinity binding sites accessible from the cytosolic side; even though the Ca^{2+} concentration is very low (see Table 11-2), calcium ions still fill these sites. Next, an ATP binds to a site on the cytosolic surface (step 1). The bound ATP is hydrolyzed to ADP in a reaction that requires Mg^{2+}, and the liberated phosphate is transferred to a specific aspartate residue in the protein, forming the high-energy acyl phosphate bond denoted by E1 ~ P (step 2). The protein then undergoes a conformational change that generates E2, in which the affinity of the two Ca^{2+}-binding sites is reduced (Figure 11-11a) and in which these sites are now accessible to the SR lumen (step 3). The free energy of hydrolysis of the aspartyl-phosphate bond in E1 ~ P is greater than that in E2–P, and this reduction in free energy of the aspartyl-phosphate bond can be said to power the E1 → E2 conformational change. The Ca^{2+} ions spontaneously dissociate from the low-affinity sites to enter the SR lumen, because even though the Ca^{2+} concentration there is higher than in the cytosol, it is lower than the K_d for Ca^{2+} binding in the low-affinity state (step 4). Finally the aspartyl-phosphate bond is hydrolyzed (step 5); dephosphorylation powers the E2 → E1 conformational change (step 6), and E1 is ready to transport two more Ca^{2+} ions. Thus hydrolysis of a phosphoanhydride bond in ATP is being used to pump Ca^{2+} against its concentration gradient into the SR lumen.

Much structural and biophysical evidence supports the model depicted in Figure 11-10. For instance, the muscle calcium pump has been isolated with phosphate linked to the key aspartate residue, and spectroscopic studies have detected slight alterations in protein conformation during the E1 → E2 conversion. The two phosphorylated states can also be distinguished biochemically; addition of ADP to phosphorylated E1 results in synthesis of ATP, the reverse of step 1 in Figure 11-10, whereas addition of ADP to phosphorylated E2 does not. Each principal conformational

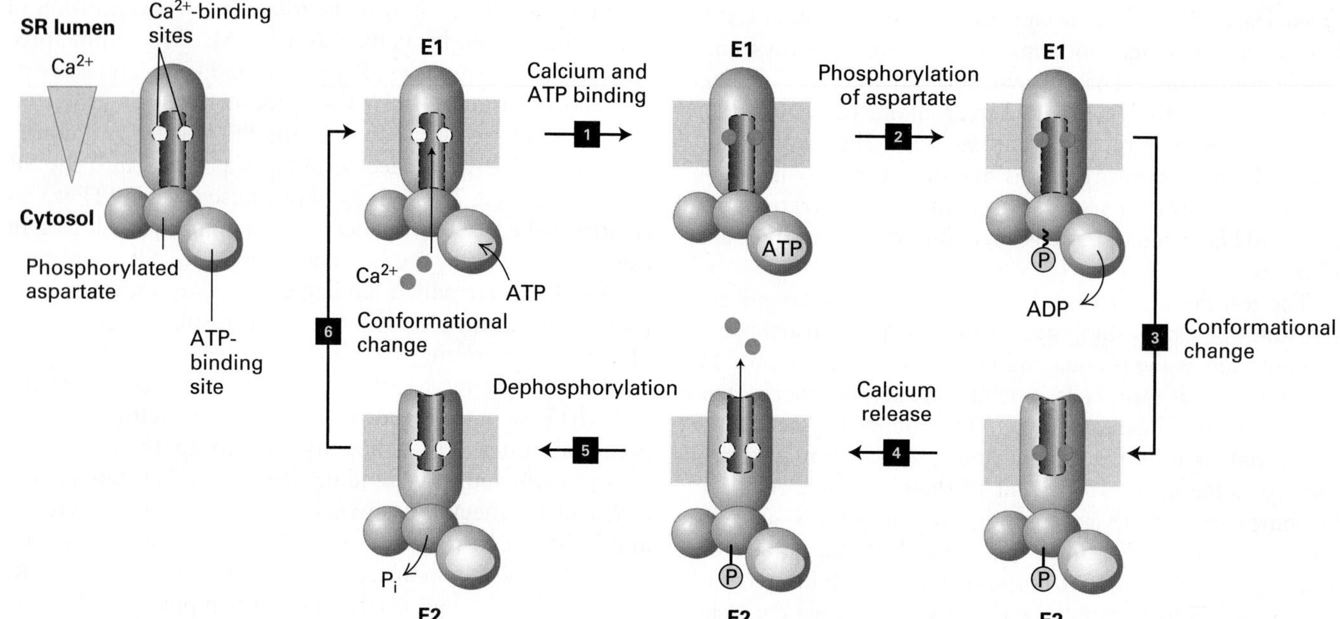

▲ **FIGURE 11-10 Operational model of the Ca^{2+} ATPase in the SR membrane of skeletal muscle cells.** Only one of the two catalytic α subunits of this P-class pump is depicted. E1 and E2 are alternative conformations of the protein in which the Ca^{2+}-binding sites are accessible to the cytosolic and exoplasmic faces, respectively. An ordered sequence of steps (1–6), as diagrammed here, is essential for coupling ATP hydrolysis and the transport of Ca^{2+} ions across the membrane. In the figure, ~P indicates a high-energy aspartyl phosphate bond; −P indicates a low-energy bond. Because the affinity of Ca^{2+} for the cytosolic-facing binding sites in E1 is a thousandfold greater than the affinity of Ca^{2+} for the exoplasmic-facing sites in E2, this pump transports Ca^{2+} unidirectionally from the cytosol to the SR lumen. See the text and Figure 11-11 for more details. [See C. Toyoshima and G. Inesi, 2004, *Ann. Rev. Biochem.* **73**:269–92.]

state of the reaction cycle can also be characterized by a different susceptibility to various proteolytic enzymes such as trypsin.

As seen in the three-dimensional structure of the Ca^{2+} pump in the E1 state, the 10 membrane-spanning α helices in the catalytic subunit form the passageway through which Ca^{2+} ions move, and amino acids in four of these helices form the two high-affinity E1 Ca^{2+}-binding sites (Figure 11-11a, *left*). One site is formed out of negatively charged oxygen atoms from the carboxyl groups (COO^-) of glutamate and aspartate side chains, as well as from water molecules. The other site is formed from side- and main-chain oxygen atoms. Thus most of the water molecules that normally surround a Ca^{2+} ion in aqueous solution are replaced by oxygen atoms attached to the protein. In contrast, in the E2 state (Figure 11-a, *right*), several of these binding side chains have moved fractions of a nanometer and are unable to interact with Ca^{2+} ions, accounting for the low affinity of the E2 state for Ca^{2+} ions.

The cytoplasmic part of the Ca^{2+} pump consists of three domains that are well separated from each other in the E1 state (Figure 11-11b). Each of these domains is connected to the membrane-spanning helices by short segments of amino acids, and movements of these cytosolic domains will cause corresponding movements of attached membrane-spanning α helices. The phosphorylated residue, Asp 351, is located on the P domain, and the adenosine moiety of ATP binds to the N domain. Following ATP and Ca^{2+} binding, the N domain moves until the γ phosphate of the bound ATP becomes adjacent to the aspartate on the P domain that is to receive the phosphate. Although the details of these conformational changes are not yet clear, this movement is thought to be transmitted by scissorlike motions of the connecting segments to movements of several membrane-spanning α helices. These changes are especially apparent in the four helices that contain the two Ca^{2+}-binding sites: the changes prevent the bound Ca^{2+} ions from dissociating back into the cytosol but enable them to dissociate into the exoplasmic medium. These changes also weaken the binding of the two Ca^{2+} ions, as can be seen by comparing the structures in Figure 11-11a. This weakening enables the bound ions to dissociate into the exoplasmic space, here, the SR lumen. After the Ca^{2+} ions dissociate and the aspartyl phosphate is hydrolyzed, the protein reverts back to the E1 conformation.

All P-class ion pumps, regardless of which ion they transport, are phosphorylated on a highly conserved aspartate residue during the transport process. Thus the operational model in Figure 11-11 is generally applicable to all of these ATP-powered ion pumps. In addition, the catalytic α subunits of all the P pumps examined to date have a similar molecular weight and, as deduced from their amino acid sequences derived from cDNA clones, have a similar arrangement of transmembrane α helices and cytosol-facing A, P, and N domains (see Figure 11-10). These

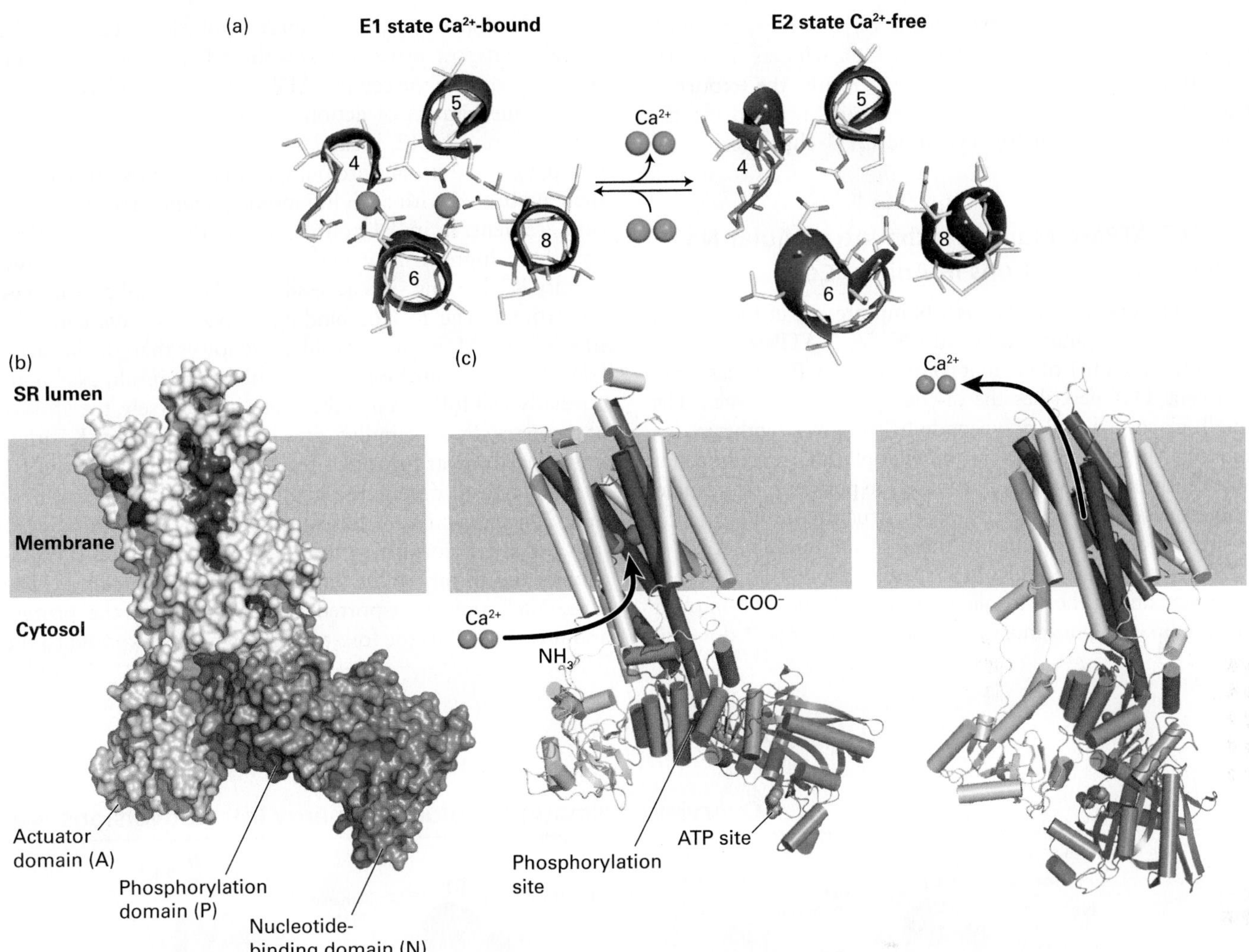

▲ **FIGURE 11-11 Structure of the catalytic α subunit of the muscle Ca^{2+} ATPase.** (a) Ca^{2+}-binding sites in the low-affinity E2 state (*right*), without bound ions, and in the E1 state (*left*), with two bound calcium ions. Side chains of key amino acids are white, and the oxygen atoms on the glutamate and aspartate side chains are red. In the high-affinity E1 conformation, Ca^{2+} ions bind at two sites between helices 4, 5, 6, and 8 inside the membrane. One site is formed out of negatively charged oxygen atoms from glutamate and aspartate side chains and of water molecules (not shown), and the other is formed out of side- and main-chain oxygen atoms. Seven oxygen atoms surround the Ca^{2+} ion in both sites. (b) Three-dimensional model of the protein in the E1 state based on the structure determined by x-ray crystallography. There are 10 transmembrane α helices, four of which (purple) contain residues that participate in Ca^{2+} binding. The cytosolic segment forms three domains: the nucleotide-binding domain N (blue), the phosphorylation domain P (green), and the actuator domain A (beige) that connects two of the membrane-spanning helices. (c) Models of the pump in the E1 state (left) and E2 state (right). Note the differences between the E1 and E2 states in the conformations of the nucleotide-binding and actuator domains; these movements power the conformational changes of the membrane-spanning α helices (purple) that constitute the Ca^{2+}-binding sites, converting them from one in which the Ca^{2+}-binding sites are accessible to the cytosolic face (E1 state) to one in which the now loosely bound calcium ions are accessible to the exoplasmic face (E2 state). [Adapted from C. Toyoshima and H. Nomura, 2002, *Nature*, **418**:605–11; C. Toyoshima and G. Inesi, 2004, *Ann. Rev. Biochem.* **73**:269–92; and E. Gouaux and R. MacKinnon, 2005, *Science* **310**:1461.]

findings strongly suggest that all such proteins evolved from a common precursor although they now transport different ions.

Calmodulin Regulates the Plasma Membrane Ca^{2+} Pumps That Control Cytosolic Ca^{2+} Concentrations

As we explain in Chapter 15, small increases in the concentration of free Ca^{2+} ions in the cytosol trigger a variety of cellular responses. In order for Ca^{2+} to function in intracellular signaling, the concentration of Ca^{2+} ions free in the cytosol usually must be kept below 0.1–0.2 μM. Animal, yeast, and probably plant cells express plasma membrane Ca^{2+} ATPases that transport Ca^{2+} out of the cell against its electrochemical gradient. The catalytic α subunit of these P-class pumps is similar in structure and sequence to the α subunit of the muscle SR Ca^{2+} pump.

The activity of plasma membrane Ca^{2+} ATPases is regulated by **calmodulin**, a cytosolic Ca^{2+}-binding protein (see Figure 3-31). A rise in cytosolic Ca^{2+} induces the binding of Ca^{2+} ions to calmodulin, which triggers allosteric activation of the Ca^{2+} ATPase. As a result, the export of Ca^{2+} ions from the cell accelerates, quickly restoring the low concentration of free cytosolic Ca^{2+} characteristic of the resting cell.

Na^+/K^+ ATPase Maintains the Intracellular Na^+ and K^+ Concentrations in Animal Cells

A second important P-class ion pump present in the plasma membrane of all animal cells is the **Na^+/K^+ ATPase.** This ion pump is a tetramer of subunit composition $\alpha_2\beta_2$. (Classic Experiment 11.1 describes the discovery of this enzyme.) The small, glycosylated β polypeptide helps newly synthesized α subunits to fold properly in the endoplasmic reticulum but apparently is not involved directly in ion pumping. The amino acid sequence and predicted tertiary structure of the catalytic α subunit are very similar to those of the muscle SR Ca^{2+} ATPase (see Figure 11-11). In particular, the Na^+/K^+ ATPase has segments on the cytosolic face that link the ATP-binding N domain, the phosphorylated aspartate on the P domain, and the A domain to the membrane-embedded segment. The overall transport process moves three Na^+ ions *out* of and two K^+ ions *into* the cell per ATP molecule hydrolyzed.

The mechanism of action of the Na^+/K^+ ATPase, outlined in Figure 11-12, is similar to that of the muscle calcium pump, except that ions are pumped in *both* directions across the membrane, with each ion moving *against* its concentration gradient. In its E1 conformation, the Na^+/K^+ ATPase has three high-affinity Na^+-binding sites and two low-affinity K^+-binding sites accessible to the cytosolic surface of the protein. The K_m for binding of Na^+ to these cytosolic sites is 0.6 mM, a value considerably lower than the intracellular Na^+ concentration of ≈12 mM; as a result, Na^+ ions normally will fully occupy these sites. Conversely, the affinity of the cytosolic K^+-binding sites is low enough that K^+ ions, transported inward through the protein, dissociate from E1 into the cytosol despite the high intracellular K^+ concentration. During the E1 → E2 transition, the three bound Na^+ ions become accessible to the exoplasmic face, and simultaneously the affinity of the three Na^+-binding sites drops. The three Na^+ ions, transported outward through the protein and now bound to the low-affinity Na^+ sites exposed to the

Overview Animation: Biological Energy Interconversions

MEDIA CONNECTIONS

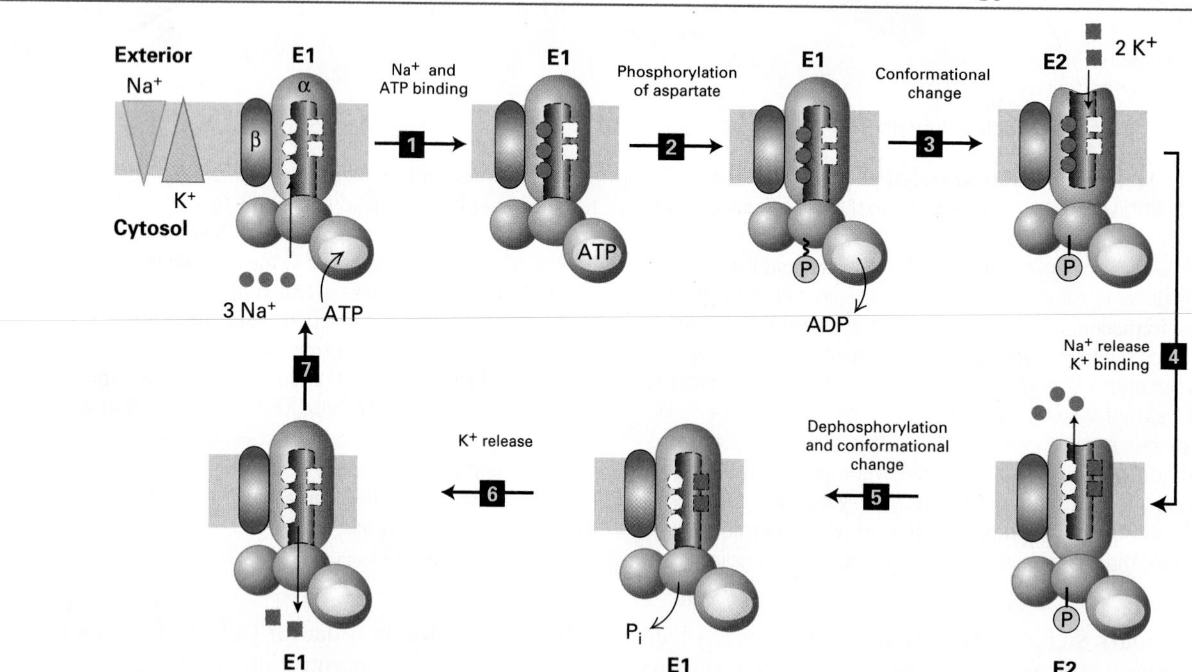

▲ **FIGURE 11-12 Operational model of the Na^+/K^+ ATPase in the plasma membrane.** Only one of the two catalytic α subunits of this P-class pump is depicted. It is not known whether just one or both subunits in a single ATPase molecule transport ions. Ion pumping by the Na^+/K^+ ATPase involves phosphorylation, dephosphorylation, and conformational changes similar to those in the muscle Ca^{2+} ATPase (see Figure 11-10). In this case, hydrolysis of the E2–P intermediate powers the E2 → E1 conformational change and concomitant transport of two K^+ ions inward. Na^+ ions are indicated by red circles; K^+ ions, by purple squares; high-energy acyl phosphate bond, by ~P; low-energy phosphoester bond, by –P. [See K. Sweadner and C. Donnet, 2001, *Biochem. J.* **356**:6875, for details of the structure of the α subunit.]

exoplasmic face, dissociate one at a time into the extracellular medium despite the high extracellular Na^+ concentration. Transition to the E2 conformation also generates two high-affinity K^+ sites accessible to the exoplasmic face. Because the K_m for K^+ binding to these sites (0.2 mM) is lower than the extracellular K^+ concentration (4 mM), these sites will fill with K^+ ions as the Na^+ ions dissociate. Similarly, during the E2 $\rightarrow$ E1 transition, the two bound K^+ ions are transported inward and then released into the cytosol.

Certain drugs (e.g., ouabain and digoxin) bind to the exoplasmic domain of the plasma membrane Na^+/K^+ ATPase and specifically inhibit its ATPase activity. The resulting disruption in the Na^+/K^+ balance of cells is strong evidence for the critical role of this ion pump in maintaining the normal K^+ and Na^+ ion concentration gradients.

V-Class H^+ ATPases Maintain the Acidity of Lysosomes and Vacuoles

All V-class ATPases transport only H^+ ions. These proton pumps, present in the membranes of lysosomes, endosomes, and plant vacuoles, function to acidify the lumen of these organelles. The pH of the lysosomal lumen can be measured precisely in living cells by use of particles labeled with a pH-sensitive fluorescent dye. When these particles are added to the extracellular fluid, the cells engulf and internalize them (phagocytosis; see Chapter 17), ultimately transporting them into lysosomes. The lysosomal pH can be calculated from the spectrum of the fluorescence emitted. Maintenance of the 100-fold or more proton gradient between the lysosomal lumen (pH $\approx$4.5–5.0) and the cytosol (pH $\approx$7.0) depends on a V-class ATPase and thus ATP production by the cell. The low lysosomal pH is necessary for optimal function of the many proteases, nucleases, and other hydrolytic enzymes in the lumen; on the other hand, a cytosolic pH of 5 would disrupt the functions of many proteins optimized to act at pH 7 and lead to death of the cell.

The ATP-powered proton pumps in lysosomal and vacuolar membranes have been isolated, purified, and incorporated into liposomes. As shown in Figure 11-9 (center), these V-class proton pumps contain two discrete domains: a cytosolic hydrophilic domain (V_1) and a transmembrane domain (V_0) with multiple subunits forming each domain. Binding and hydrolysis of ATP by the B subunits in V_1 provide the energy for pumping of H^+ ions through the proton-conducting channel formed by the c and a subunits in V_0. Unlike P-class ion pumps, V-class proton pumps are not phosphorylated and dephosphorylated during proton transport. The structurally similar F-class proton pumps, which we describe in Chapter 12, normally operate in the "reverse" direction to generate ATP rather than pump protons; their mechanism of action is understood in great detail.

Pumping of relatively few protons is required to acidify an intracellular vesicle. To understand why, recall that a solution of pH 4 has an H^+ ion concentration of 10^{-4} moles per liter, or 10^{-7} moles of H^+ ions per milliliter. Since there are 6.02×10^{23} atoms of H per mole (Avogadro's number), then a milliliter of a pH 4 solution contains 6.02×10^{16} H^+ ions. Thus at pH 4, a primary spherical lysosome with a volume of 4.18×10^{-15} ml (diameter of 0.2 μm) will contain just 252 protons. At pH 7, the same organelle would have only an average of only 0.2 protons in its lumen, and thus pumping of only approximately 250 proteins is necessary for lysosome acidification.

By themselves, ATP-powered proton pumps cannot acidify the lumen of an organelle (or the extracellular space) because these pumps are *electrogenic;* that is, a net movement of electric charge occurs during transport. Pumping of just a few protons causes a buildup of positively charged H^+ ions on the exoplasmic (inside) face of the organelle membrane. For each H^+ pumped across, a negative ion (e.g., OH^- or Cl^-) will be "left behind" on the cytosolic face, causing a buildup of negatively charged ions there. These oppositely charged ions attract each other on opposite faces of the membrane, generating a charge separation, or electric potential, across the membrane. As more and more protons are pumped, the excess of positive charges on the exoplasmic face repels other H^+ ions, soon preventing pumping of additional protons long before a significant transmembrane H^+ concentration gradient is established (Figure 11-13). In fact, this is the way that P-class H^+

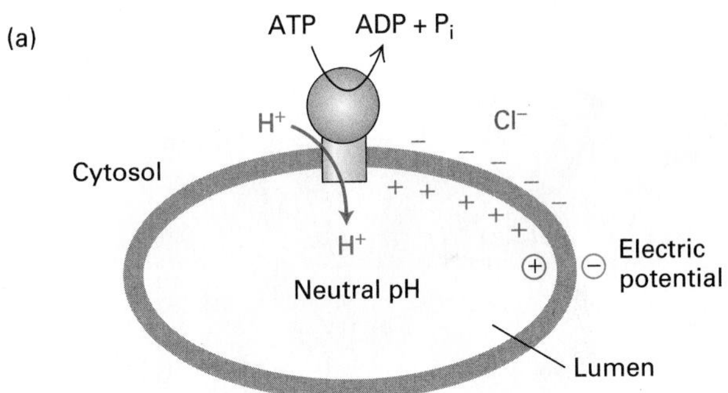

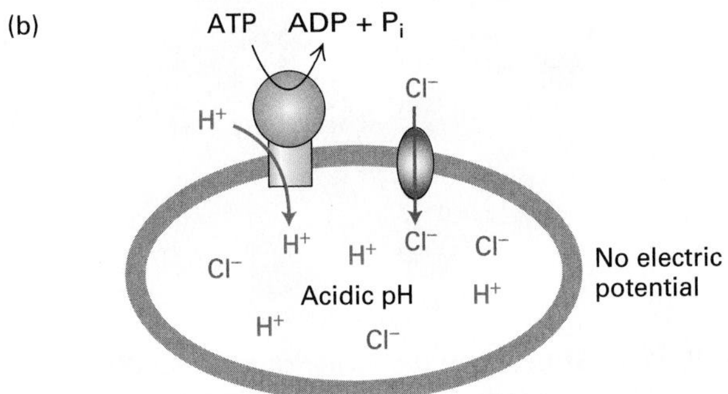

▲ **FIGURE 11-13 Effect of V-class H^+ pumps on H^+ concentration gradients and electric potential gradients across cellular membranes.** (a) If an intracellular organelle contains only V-class pumps, proton pumping generates an electric potential across the membrane, cytosol- facing side negative and lumenal-side positive but no significant change in the intraluminal pH. (b) If the organelle membrane also contains Cl^- channels, anions passively follow the pumped protons, resulting in an accumulation of H^+ and Cl^- ions in the lumen (low luminal pH) but no electric potential across the membrane.

pumps generate a cytosol-negative potential across plant and yeast plasma membranes.

In order for an organelle lumen or an extracellular space (e.g., the lumen of the stomach) to become acidic, movement of protons must be accompanied either by (1) movement of an equal number of anions (e.g., Cl^-) in the same direction or by (2) movement of equal numbers of a different cation in the opposite direction. The first process occurs in lysosomes and plant vacuoles, whose membranes contain V-class H^+ ATPases and anion channels through which accompanying Cl^- ions move (Figure 11-13b). The second process occurs in the lining of the stomach, which contains a P-class H^+/K^+ ATPase that is not electrogenic and pumps one H^+ outward and one K^+ inward. Operation of this pump is discussed later in the chapter.

Bacterial Permeases Are ABC Proteins That Import a Variety of Nutrients from the Environment

As noted earlier, all members of the very large and diverse ABC superfamily of transport proteins contain two transmembrane (T) domains and two cytosolic ATP-binding (A) domains (Figure 11-14). The T domains, each built of 10 membrane-spanning α helices, form the pathway through which the transported substance (substrate) crosses the membrane and determine the substrate specificity of each ABC protein. The sequences of the A domains are approximately 30–40 percent homologous in all members of this superfamily, indicating a common evolutionary origin. Some ABC proteins also contain an additional exoplasmic substrate-binding subunit or regulatory subunit.

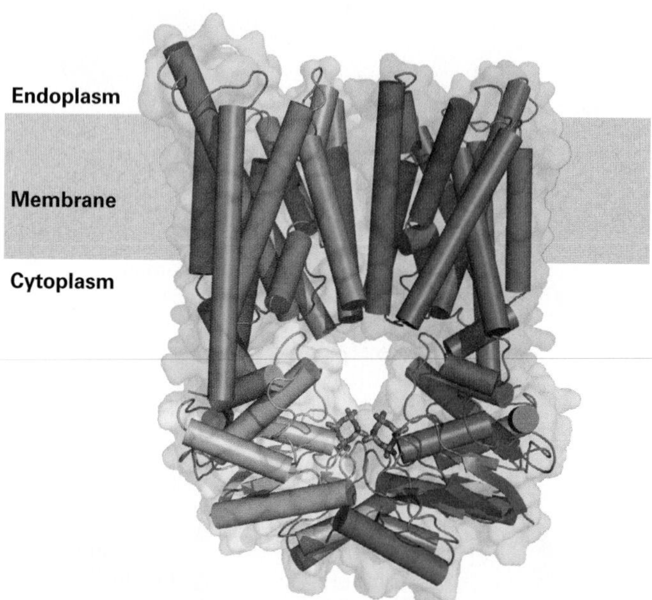

▲ **FIGURE 11-14 Structure of the *Escherichia coli* BtuCD protein, an ABC transporter mediating vitamin B_{12} uptake.** The complete transporter is assembled from four subunits, two identical membrane-spanning subunits (green), and two ATP-binding subunits (blue). Molecules of cyclotetravanadate, an analog of the phosphate groups in ATP, are located in the ATP-binding sites and are depicted in ball-and-stick. Approximate boundaries of the membrane bilayer are indicated by the gray shaded area, with the external surface at the top and the cytoplasm at the bottom. [After K. Locher et. al., 2002, *Science* **296**:1091–98.]

The plasma membrane of many bacteria contains numerous *permeases* that belong to the ABC superfamily. These proteins use the energy released by hydrolysis of ATP to transport specific amino acids, sugars, vitamins, and even peptides into the cell. Since bacteria frequently grow in soil or pond water where the concentration of nutrients is low, these ABC transport proteins enable the cells to import nutrients against substantial concentration gradients. Bacterial permeases generally are inducible; that is, the quantity of a transport protein in the cell membrane is regulated by both the concentration of the nutrient in the medium and the metabolic needs of the cell.

In the *E. coli* vitamin B_{12} permease, a typical bacterial ABC protein whose structure is known in molecular detail (see Figure 11-18), the two transmembrane domains and two cytosolic ATP-binding domains are formed by four separate subunits. Gram-negative bacteria, such as *E. coli*, contain, besides the plasma membrane, an outer membrane also built of a phospholipid bilayer (see Figure 1-2). This outer membrane contains porin proteins (see Figure 10-18) that render it highly permeable to most small molecules, including amino acids and vitamins. Thus these molecules can enter into the periplasmic space in between the plasma and outer membranes (see Figure 1-2). A soluble vitaminB_{12}-binding protein, located in the periplasmic space, binds vitamin B_{12} tightly and directs it to the T subunits of the permease, through which the vitamin crosses the plasma membrane powered by ATP hydrolysis. Precisely how transport occurs is not known, but it is thought that B_{12} binding is signaled to the nucleotide hydrolysis sites where the affinity for ATP increases, a prerequisite for a productive transport cycle. The two ATP-binding domains then carry out a highly cooperative ATP-binding and hydrolysis reaction that is concurrent with a substantial conformational change in the membrane-spanning segments; this change somehow allows the bound substrate to pass between the two membrane-spanning domains into the cell cytoplasm. Then the transporter returns to the resting state through the dissociation of ADP and inorganic phosphate.

Mutant *E. coli* cells that are defective in any of the B_{12} permease subunits or the soluble periplasmic-binding protein are unable to transport B_{12} into the cell but are able to transport other molecules such as amino acids whose uptake is facilitated by other transport proteins. Such genetic analyses provide strong evidence that these permeases function to transport specific solutes into bacterial cells.

The Approximately 50 Mammalian ABC Transporters Play Diverse and Important Roles in Cell and Organ Physiology

Discovery of the first eukaryotic ABC protein to be recognized came from studies on tumor cells and cultured cells that exhibited resistance to several drugs with unrelated chemical structures. Such cells eventually were shown to express elevated levels of a *multidrug-resistance (MDR) transport protein* known as *MDR1*. This protein uses the energy derived from ATP hydrolysis to *export* a large variety of drugs from the cytosol to the extracellular medium. The *Mdr1* gene is frequently amplified in multidrug-resistant cells, resulting in a large overproduction of the MDR1 protein.

Most drugs transported by MDR1 are small hydrophobic molecules that diffuse from the medium across the plasma membrane, unaided by transport proteins, into the cell cytosol, where they block various cellular functions. Two such drugs are colchicine and vinblastine, which block assembly of microtubules (Chapter 18). ATP-powered export of such drugs by MDR1 reduces their concentration in the cytosol. As a result, a much higher extracellular drug concentration is required to kill cells that express MDR1 than those that do not. That MDR1 is an ATP-powered small-molecule pump has been demonstrated with liposomes containing the purified protein. The ATPase activity of these liposomes is enhanced by different drugs in a dose-dependent manner corresponding to their ability to be transported by MDR1.

About 50 different mammalian ABC transport proteins are now recognized (Table 11-3). Several are expressed in abundance in the liver, intestines, and kidney—sites where natural toxic and waste products are removed from the body. Substrates for these ABC proteins include sugars, amino acids, cholesterol, bile acids, phospholipids, peptides, proteins, toxins, and foreign substances. The normal function of MDR1 most likely is to transport various natural and metabolic toxins into the bile or intestinal lumen or to the urine being formed in the kidney. During the course of its evolution, MDR1 appears to have acquired the ability to transport drugs whose structures are similar to those of these endogenous toxins. Tumors derived from MDR-expressing cell types, such as hepatomas (liver cancers), frequently are resistant to virtually all chemotherapeutic agents and are thus difficult to treat, presumably because the tumors exhibit increased expression of MDR1 or the related MDR2.

Several human genetic diseases are associated with defective ABC proteins. The best studied is cystic fibrosis (CF), caused by a mutation in the gene encoding the *cystic fibrosis transmembrane regulator (CFTR)*. This Cl^- transport protein is expressed in the apical plasma membranes of epithelial cells in the lung, sweat glands, pancreas, and other tissues. For instance, CFTR protein is important for resorption of Cl^- into cells of sweat glands, and babies with cystic fibrosis, if licked, often taste "salty." An increase in cyclic AMP (cAMP), a small intracellular signaling molecule, causes

TABLE 11-3 Selected Human ABC Proteins

PROTEIN	TISSUE EXPRESSION	FUNCTION	DISEASE CAUSED BY DEFECTIVE PROTEIN
ABCB1 (MDR1)	Adrenal, kidney, brain	Exports lipophilic drugs	
ABCB4 (MDR2)	Liver	Exports phosphatidylcholine into bile	
ABCB11	Liver	Exports bile salts into bile	
CFTR	Exocrine tissue	Transports Cl ions	Cystic fibrosis
ABCDl	Ubiquitous in peroxisomal membrane	Influences activity of peroxisomal enzyme that oxidizes very long chain fatty acids	Adrenoleukodystrophy (ADL)
ABCG5/8	Liver, instestine	Exports cholesterol and other sterols	β-Sitosterolemia
ABCA1	Ubiquitous	Exports cholesterol and phospholipid for uptake into high-density lipoprotein (HDL)	Tangier's disease

phosphorylation of CFTR and stimulates Cl^- transport by such cells from normal individuals but not from CF individuals, who have a defective CFTR protein. (The role of cAMP in numerous signaling pathways is covered in Chapter 15.) The sequence and predicted structure of the CFTR protein, based on analysis of the cloned gene, are very similar to those of the MDR1 protein except for the presence of an additional domain, the regulatory (R) domain, on the cytosolic face. Moreover, the Cl^--transport activity of CFTR protein is enhanced by the binding of ATP, but it is uncertain whether or not CFTR is actually an ATP-powered Cl^- pump. ∎

Certain ABC Proteins "Flip" Phospholipids and Other Lipid-Soluble Substrates from One Membrane Leaflet to the Opposite Leaflet

The substrates of mammalian MDR1 are primarily planar, lipid-soluble molecules with one or more positive charges; they all compete with one another for transport by MDR1, suggesting that they bind to the same site or sites on the protein. In contrast to bacterial ABC proteins, all four domains of mammalian MDR1 are fused into a single 170,000-MW protein. The recently determined three-dimensional structure of a homologous *E. coli* lipid-transport protein reveals that the molecule is V shaped, with the apex in the membrane and the arms containing the ATP-binding sites protruding into the cytosol (Figure 11-15).

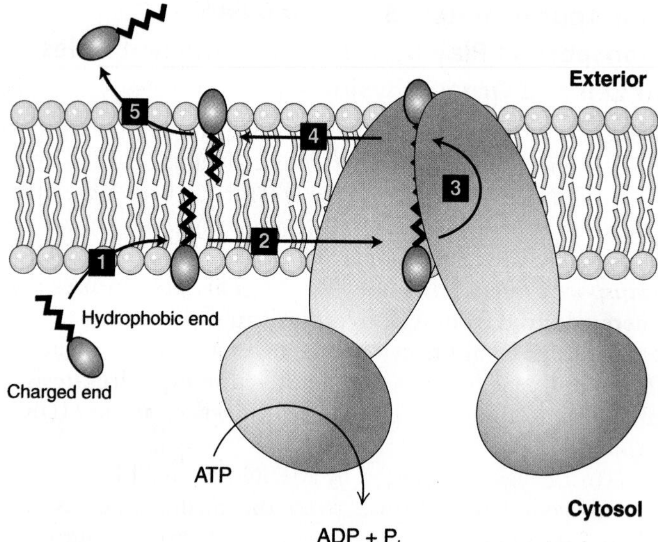

▲ **FIGURE 11-15 Flippase model of transport by MDR1 and similar ABC proteins.** Step **1**: The hydrophobic portion (black) of a substrate molecule moves spontaneously from the cytosol into the cytosolic-facing leaflet of the lipid bilayer, while the charged end (red) remains in the cytosol. Step **2**: The substrate diffuses laterally until it encounters and binds to a site on the MDR1 protein within the bilayer. Step **3**: The protein then "flips" the charged substrate molecule into the exoplasmic leaflet, an energetically unfavorable reaction involving conformational changes in the membrane-spanning domains that are powered by the coupled hydrolysis of ATP by the cytosolic domains. Steps **4** and **5**: Once in the exoplasmic face, the substrate again can diffuse laterally in the membrane and ultimately moves into the aqueous phase on the outside of the cell. [Adapted from P. Borst, N. Zelcer, and A. van Helvoort, 2000, *Biochim. Biophys. Acta* **1486**:128.]

Although the mechanism of transport by MDR1 and similar ABC proteins has not been definitively demonstrated, a likely candidate is the *flippase model* depicted in Figure 11-15. According to this model, MDR1 "flips" an amphiphilic substrate molecule from the cytosolic to the exoplasmic leaflet, an energetically unfavorable reaction powered by the coupled ATPase activity of the protein.

Support for the flippase model of transport by MDR1 comes from ABCB4 (originally called MDR2), a homologous protein present in the region of the liver-cell plasma membrane that faces the bile duct. ABCB4 moves phosphatidylcholine from the cytosolic to the exoplasmic leaflet of the plasma membrane for subsequent release into the bile in combination with cholesterol and bile acids, which themselves are transported by other ABC family members. Several other ABC superfamily members participate in the cellular export of various lipids (see Table 11-3).

ABCB4 was first suspected of having phospholipid flippase activity because mice with homozygous loss-of-function mutations in the *ABCB4* gene exhibited defects in the secretion of phosphatidylcholine into bile. To determine directly if ABCB4 was in fact a flippase, researchers performed experiments on a homogeneous population of purified vesicles with ABCB4 in the membrane and with the cytosolic face directed outward. These vesicles were obtained by introducing cDNA encoding mammalian ABCB4 into a temperature-sensitive yeast *sec* mutant. At lower, permissive temperatures where the *sec* protein is functional, the ABCB4 protein expressed by the transfected cells moves normally through the secretory pathway to the cell surface (Chapter 14). At higher nonpermissive temperatures, however, the sec protein is nonfunctional and secretory vesicles cannot fuse with the plasma membrane, as they do in wild-type cells; so vesicles containing ABCB4 and other yeast proteins accumulate in the cells. After purifying these secretory vesicles, investigators labeled them in vitro with a fluorescent phosphatidylcholine derivative. The fluorescence-quenching assay outlined in Figure 11-16 was used to demonstrate that the vesicles containing ABCB4 exhibited an ATP-dependent flippase activity, whereas those without ABCB4 did not.

KEY CONCEPTS OF SECTION 11.3

ATP-Powered Pumps and the Intracellular Ionic Environment

- Four classes of transmembrane proteins couple the energy-releasing hydrolysis of ATP with the energy-requiring transport of substances against their concentration gradient: P-, V-, and F-class pumps and ABC proteins (see Figure 11-9).
- The combined action of P-class Na^+/K^+ ATPases in the plasma membrane and homologous Ca^{2+} ATPases in the

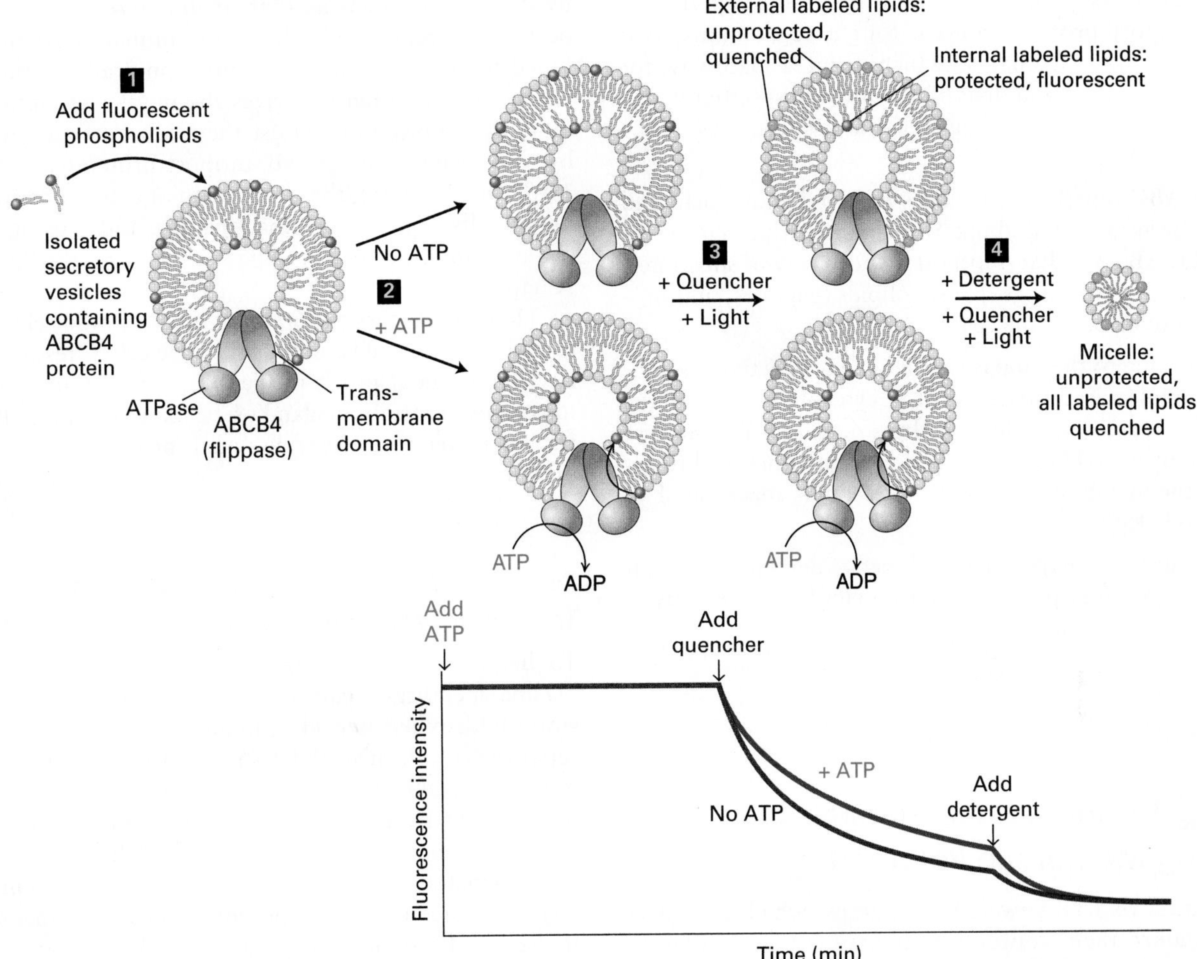

▲ **EXPERIMENTAL FIGURE 11-16 In vitro fluorescence-quenching assay can detect phospholipid flippase activity of ABCB4.** A homogeneous population of secretory vesicles containing ABCB4 protein was purified from a yeast *sec* mutant transfected with the *ABCB4* gene. Step **1**: Synthetic phospholipids containing a fluorescently modified head group (blue) were incorporated primarily into the outer, cytosolic leaflets of the purified vesicles. Step **2**: If ABCB4 acted as a flippase, then on addition of ATP to the outside of the vesicles, a small fraction of the outward-facing labeled phospholipids would be flipped to the inside leaflet. Step **3**: Flipping was detected by adding a membrane-impermeable quenching compound called dithionite to the medium surrounding the vesicles. Dithionite reacts with the fluorescent head group, destroying its ability to fluoresce (gray). In the presence of the quencher, only labeled phospholipid in the protected environment on the inner leaflet will fluoresce. Subsequent to the addition of the quenching agent, the total fluorescence decreases with time until it plateaus at the point at which all external fluorescence is quenched and only the internal phospholipid fluorescence can be detected. The observation of greater fluorescence (less quenching) in the presence of ATP than in its absence indicates that ABCB4 has flipped some of the labeled phospholipid to the inside. Step **4**: Addition of detergent to the vesicles generates micelles and makes all fluorescent lipids accessible to the quenching agent and lowers the fluorescence to baseline values. [Adapted from S. Ruetz and P. Gros, 1994, *Cell* **77**:1071.]

plasma membrane or sarcoplasmic reticulum creates the usual ion milieu of animal cells: high K^+, low Ca^{2+}, and low Na^+ in the cytosol; low K^+, high Ca^{2+}, and high Na^+ in the extracellular fluid.

- In P-class pumps, phosphorylation of the α (catalytic) subunit and a change in conformational states are essential for coupling ATP hydrolysis to transport of H^+, Na^+, K^+, or Ca^{2+} ions (see Figures 11-10, 11-11, and 11-12).
- V- and F-class ATPases, which transport protons exclusively, are large, multisubunit complexes with a proton-conducting channel in the transmembrane domain and ATP-binding sites in the cytosolic domain.
- V-class H^+ pumps in animal lysosomal and endosomal membranes and plant vacuole membranes are responsible for maintaining a lower pH inside the organelles than in the surrounding cytosol (see Figure 11-13b).

- All members of the large and diverse ABC superfamily of transport proteins contain four core domains: two transmembrane domains, which form a pathway for solute movement and determine substrate specificity, and two cytosolic ATP-binding domains (see Figures 11-14 and 11-15).
- The ABC superfamily includes bacterial amino acid and sugar permeases and about 50 mammalian proteins (e.g., MDR1, ABCA1) that transport a wide array of substrates, including toxins, drugs, phospholipids, peptides, and proteins, into or out of the cell.
- According to the flippase model of MDR activity, a substrate molecule diffuses into the cytosolic leaflet of the plasma membrane, then is flipped to the exoplasmic leaflet in an ATP-powered process, and finally diffuses from the membrane into the extracellular space (see Figure 11-15).
- Biochemical experiments directly demonstrate that ABCB4 (MDR2) possesses phospholipid flippase activity (see Figure 11-16).

11.4 Nongated Ion Channels and the Resting Membrane Potential

In addition to ATP-powered ion pumps, which transport ions *against* their concentration gradients, the plasma membrane contains channel proteins that allow the principal cellular ions (Na^+, K^+, Ca^{2+}, and Cl^-) to move through them at different rates *down* their concentration gradients. Ion concentration gradients generated by pumps and selective movements of ions through channels constitute the principal mechanism by which a difference in voltage, or electric potential, is generated across the plasma membrane. In other words, ATP-powered ion pumps generate differences in ion concentrations across the plasma membrane, and ion channels utilize these concentration gradients to generate an electric potential across the membrane (see Figure 11-2).

In all cells the magnitude of this electric potential generally is ≈70 millivolts (mV) with the *inside* of the cell membrane always *negative* with respect to the outside. This value does not seem like much until we consider that the thickness of the plasma membrane is only ≈3.5 nm. Thus the voltage gradient across the plasma membrane is 0.07 V per 3.5×10^{-7} cm, or 200,000 volts per centimeter! (To appreciate what this means, consider that high-voltage transmission lines for electricity utilize gradients of about 200,000 volts per kilometer.)

The ionic gradients and electric potential across the plasma membrane play crucial roles in many biological processes. As noted previously, a rise in the cytosolic Ca^{2+} concentration is an important regulatory signal, initiating contraction in muscle cells and triggering protein secretion by many cells, such as that of digestive enzymes in exocrine pancreatic cells. In many animal cells, the combined force of the Na^+ concentration gradient and membrane electric potential drives the uptake of amino acids and other molecules against their concentration gradient by ion-linked symport and antiport proteins (see Section 11.5). Furthermore, the conduction of action potentials by nerve cells depends on the opening and closing of ion channels in response to changes in the membrane potential (Chapter 23).

Here, we discuss the origin of the membrane electric potential in resting cells, often called the cell's "resting potential"; how ion channels mediate the selective movement of ions across a membrane; and useful experimental techniques for characterizing the functional properties of channel proteins.

Selective Movement of Ions Creates a Transmembrane Electric Potential Difference

To help explain how an electric potential across the plasma membrane can arise, we first consider a set of simplified experimental systems in which a membrane separates a 150 mM NaCl/15 mM KCl solution (similar to the extracellular medium surrounding metazoan cells) on the right from a 15 mM NaCl/150 mM KCl solution (similar to that of the cytosol) on the left. A potentiometer (voltmeter) is connected to both solutions to measure any difference in electric potential across the membrane. If the membrane is impermeable to all ions, no ions will flow across it. There will be no difference in voltage, or electric potential gradient across the membrane, as shown in Figure 11-17a.

Now suppose that the membrane contains Na^+-channel proteins that accommodate Na^+ ions but exclude K^+ and Cl^- ions (Figure 11-17b). Na^+ ions then tend to move down their concentration gradient from the right side to the left, leaving an excess of negative Cl^- ions compared with Na^+ ions on the right side and generating an excess of positive Na^+ ions compared with Cl^- ions on the left side. The excess Na^+ on the left and Cl^- on the right remain near the respective surfaces of the membrane because the excess positive charges on one side of the membrane are attracted to the excess negative charges on the other side. The resulting separation of charge across the membrane constitutes an electric potential, or voltage, with the left (cytosolic) side of the membrane having excess positive charge with respect to the right.

As more and more Na^+ ions move through channels across the membrane, the magnitude of this charge difference (i.e., voltage) increases. However, continued right-to-left movement of the Na^+ ions eventually is inhibited by the mutual repulsion between the excess positive (Na^+) charges accumulated on the left side of the membrane and by the attraction of Na^+ ions to the excess negative charges built up on the right side. The system soon reaches an equilibrium point at which the two opposing factors that determine the

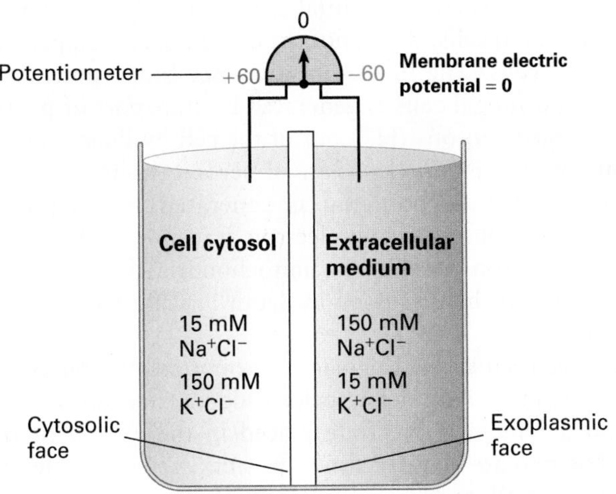

◀ EXPERIMENTAL FIGURE 11-17 **Generation of a transmembrane electric potential (voltage) depends on the selective movement of ions across a semipermeable membrane.** In this experimental system, a membrane separates a 15 mM NaCl/150 mM KCl solution (*left*) from a 150 mM NaCl/15 mM KCl solution (*right*); these ion concentrations are similar to those in cytosol and blood, respectively. If the membrane separating the two solutions is impermeable to all ions (a), no ions can move across the membrane and no difference in electric potential is registered on the potentiometer connecting the two solutions. If the membrane is selectively permeable only to Na^+ (b) or to K^+ (c), then diffusion of ions through their respective channels leads to a separation of charge across the membrane. At equilibrium, the membrane potential caused by the charge separation becomes equal to the Nernst potential E_{Na} or E_K registered on the potentiometer. See the text for further explanation.

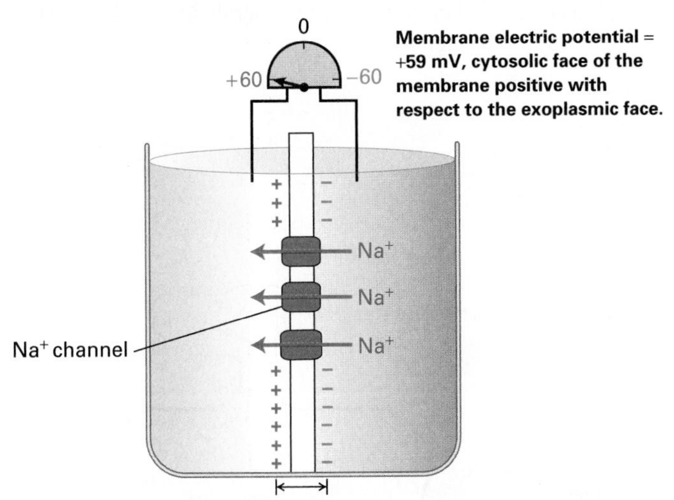

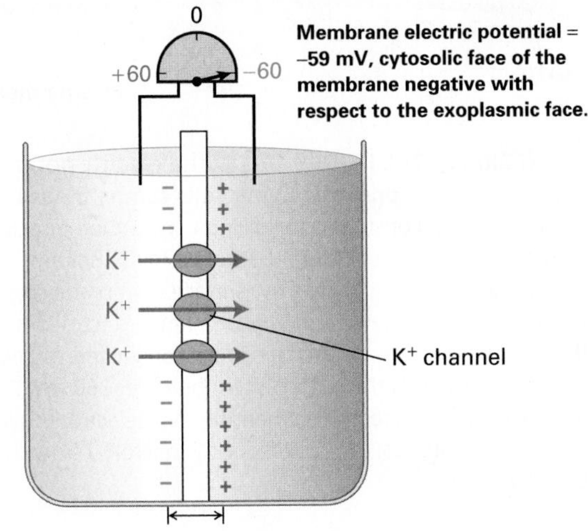

movement of Na^+ ions—the membrane electric potential and the ion concentration gradient—balance each other out. At equilibrium, no net movement of Na^+ ions occurs across the membrane. Thus this semipermeable membrane, like all biological membranes, acts as a *capacitor*—a device consisting of a thin sheet of nonconducting material (the hydrophobic interior) surrounded on both sides by electrically conducting material (the polar phospholipid head groups and the ions in the surrounding aqueous solution) that can store positive charges on one side and negative charges on the other.

If a membrane is permeable only to Na^+ ions, then at equilibrium the measured electric potential across the membrane equals the sodium equilibrium potential in volts, E_{Na}. The magnitude of E_{Na} is given by the *Nernst equation,* which is derived from basic principles of physical chemistry:

$$E_{Na} = \frac{RT}{ZF} \ln \frac{[Na_r]}{[Na_l]} \tag{11-2}$$

where R (the gas constant) = 1.987 cal/(degree · mol), or 8.28 joules/(degree · mol); T (the absolute temperature in degrees Kelvin) = 293 K at 20 °C; Z (the charge, also called the valency) here equal to +1; F (the Faraday constant) = 23,062 cal/(mol · V), or 96,000 coulombs/(mol · V); and $[Na_l]$ and $[Na_r]$ are the Na^+ concentrations on the left and right sides, respectively, at equilibrium. By convention the potential is expressed as the *cytosolic* face of the membrane relative to the *exoplasmic* face, and the equation is written with the concentration of ion in the extracellular solution (here the right side of the membrane) placed in the numerator and that of the cytosol in the denominator.

At 20 °C, Equation 11-2 reduces to

$$E_{Na} = 0.059 \log_{10} \frac{[Na_r]}{[Na_l]} \tag{11-3}$$

If $[Na_r]/[Na_l] = 10$, a tenfold ratio of concentrations as in Figure 11-17b, then $E_{Na} = +0.059$ V (or +59 mV), with the left, cytosolic side positive with respect to the exoplasmic right side.

If the membrane is permeable only to K^+ ions and not to Na^+ or Cl^- ions, then a similar equation describes the potassium equilibrium potential E_K:

$$E_{Na} = 0.059 \log_{10} \frac{[K_r]}{[K_l]} \quad (11\text{-}4)$$

The *magnitude* of the membrane electric potential is the same (59 mV for a tenfold difference in ion concentrations), except that the left, cytosolic side is now *negative* with respect to the right (Figure 11-17b), opposite to the polarity obtained across a membrane selectively permeable to Na^+ ions.

The Membrane Potential in Animal Cells Depends Largely on Potassium Ion Movements Through Open Resting K^+ Channels

The plasma membranes of animal cells contain many open K^+ channels but few open Na^+, Cl^-, or Ca^{2+} channels. As a result, the major ionic movement across the plasma membrane is that of K^+ from the *inside outward,* powered by the K^+ concentration gradient, leaving an excess of *negative* charge on the inside and creating an excess of *positive* charge on the outside, similar to the experimental system shown in Figure 11-17c. This outward flow of K^+ ions through these channels, called **resting K^+ channels,** is the major determinant of the inside-negative membrane potential. Like all channels, these alternate between an open and a closed state, but since their opening and closing are not affected by the membrane potential or by small signaling molecules, these channels are called *nongated.* The various gated channels discussed in Chapter 23 open only in response to specific ligands or to changes in membrane potential.

Quantitatively, the usual resting membrane potential of −70 mV is close to the potassium equilibrium potential, calculated from the Nernst equation and the K^+ concentrations in cells and surrounding media depicted in Table 11-2. Usually the potential is lower than that calculated from the Nernst equation because of the presence of a few open Na^+ channels. These open Na^+ channels allow the net *inward* flow of Na^+ ions, making the cytosolic face of the plasma membrane more positive, that is, less negative, than predicted by the Nernst equation for K^+. The K^+ concentration gradient that drives the flow of ions through resting K^+ channels is generated by the Na^+/K^+ ATPase described previously (see Figures 11-2 and 11-12). In the absence of this pump, or when it is inhibited, the K^+ concentration gradient cannot be maintained, and eventually the magnitude of the membrane potential falls to zero.

Although resting K^+ channels play the dominant role in generating the electric potential across the plasma membrane of animal cells, this is not the case in bacterial, plant, and fungal cells. The inside-negative membrane potential in plant and fungal cells is generated by transport of positively charged protons (H^+) out of the cell by P-class proton pumps (see Figure 11-13a). In aerobic bacterial cells the inside negative potential is generated by outward pumping of protons during electron transport, a process similar to proton pumping in mitochondrial inner membranes that will be discussed in detail in Chapter 12 (see Figure 12-16).

The potential across the plasma membrane of large cells can be measured with a microelectrode inserted inside the cell and a reference electrode placed in the extracellular fluid. The two are connected to a potentiometer capable of measuring small potential differences (Figure 11-18). The potential across the surface membrane of most animal cells generally does not vary with time. In contrast, neurons and muscle cells—the principal types of electrically active cells—undergo controlled changes in their membrane potential that we discuss in Chapter 23.

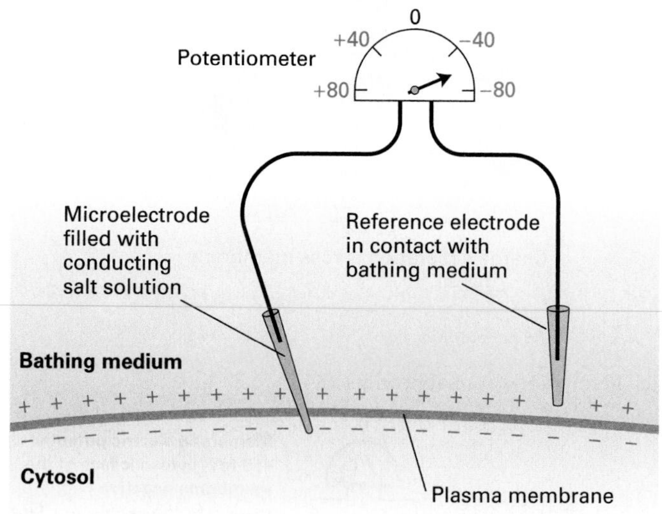

▲ **EXPERIMENTAL FIGURE 11-18 The electric potential across the plasma membrane of living cells can be measured.** A microelectrode, constructed by filling a glass tube of extremely small diameter with a conducting fluid such as a KCl solution, is inserted into a cell in such a way that the surface membrane seals itself around the tip of the electrode. A reference electrode is placed in the bathing medium. A potentiometer connecting the two electrodes registers the potential, in this case −60 mV. A potential difference is registered only when the microelectrode is inserted into the cell; no potential is registered if the microelectrode is in the bathing fluid.

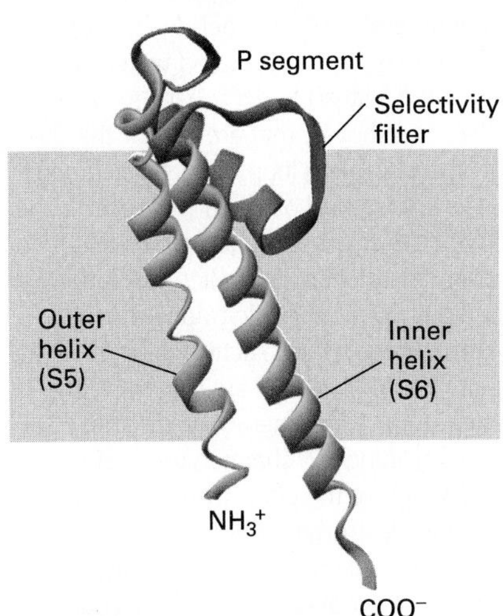

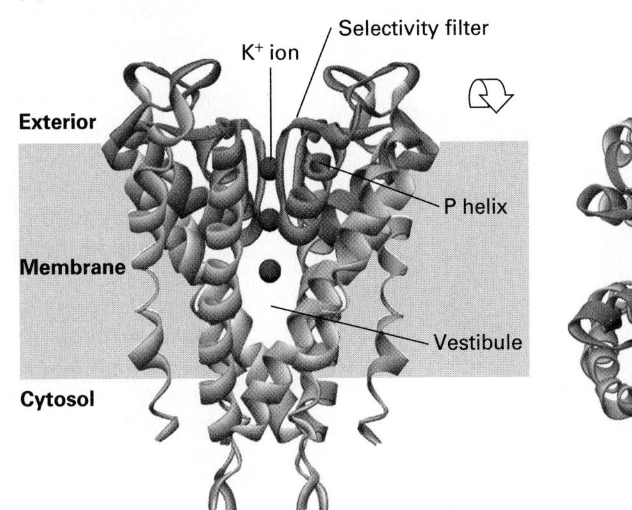

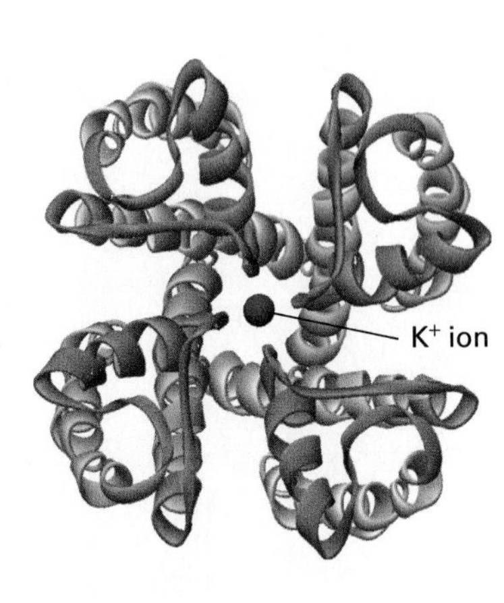

▲ **FIGURE 11-19 Structure of resting K$^+$ channel from the bacterium *Streptomyces lividans*.** All K$^+$ channel proteins are tetramers comprising four identical subunits, each containing two conserved membrane-spanning α helices, called by convention S5 and S6 (yellow), and a shorter P, or pore segment (pink). (a) One of the subunits, viewed from the side, with key structural features indicated. (b) The complete tetrameric channel viewed from the side (*left*) and the top, or extracellular, end (*right*). The P segments are located near the exoplasmic surface and connect the S5 and S6 α helices; they consist of a nonhelical "turret," which lines the upper part of the pore; a short α helix; and an extended loop that protrudes into the narrowest part of the pore and forms the ion-selectivity filter. This filter allows K$^+$ (purple spheres) but not other ions to pass. Below the filter is the central cavity or vestibule lined by the inner, or S6 α, helixes. The subunits in gated K$^+$ channels, which open and close in response to specific stimuli, contain additional transmembrane helices not shown here; these are discussed in Chapter 23. [See Y. Zhou et al., 2001, *Nature* **414**:43.]

Ion Channels Contain a Selectivity Filter Formed from Conserved Transmembrane Segments

All ion channels exhibit specificity for particular ions: K$^+$ channels allow K$^+$ but not closely related Na$^+$ ions to enter, whereas Na$^+$ channels admit Na$^+$ but not K$^+$. Determination of the three-dimensional structure of a bacterial K$^+$ channel first revealed how this exquisite ion selectivity is achieved. Comparisons of the sequences of other K$^+$, Na$^+$, and Ca^{2+} channels established that all such proteins share a common structure and probably evolved from a single type of channel protein.

Like all other K$^+$ channels, bacterial K$^+$ channels are built of four identical subunits symmetrically arranged around a central pore (Figure 11-19). Each subunit contains two membrane-spanning α helices (S5 and S6) and a short P (pore domain) segment that partly penetrates the membrane bilayer. In the tetrameric K$^+$ channel, the eight transmembrane α helices (two from each subunit) form an "inverted tepee," generating a water-filled cavity called the *vestibule* in the central portion of the channel that extends halfway through the membrane. Four extended loops that are part of the four P segments form the actual *ion-selectivity filter* in the narrow part of the pore near the exoplasmic surface, above the vestibule.

Several related pieces of evidence support the role of P segments in ion selection. First, the amino acid sequence of the P segment is highly homologous in all known K$^+$ channels and different from that in other ion channels. Second, mutation of certain amino acids in this segment alters the ability of a K$^+$ channel to distinguish Na$^+$ from K$^+$. Finally, replacing the P segment of a bacterial K$^+$ channel with the homologous segment from a mammalian K$^+$ channel yields a chimeric protein that exhibits normal selectivity for K$^+$ over other ions. Thus all K$^+$ channels are thought to use the same mechanism to distinguish K$^+$ over other ions.

Na$^+$ ions are smaller than K$^+$ ions. How, then, can a channel protein exclude Na$^+$ ions, yet allow passage of K$^+$? The ability of the ion-selectivity filter in K$^+$ channels to select K$^+$ over Na$^+$ is due mainly to backbone carbonyl oxygens on residues located in a Gly-Tyr-Gly sequence that is found in an analogous position in the P segment in every known K$^+$ channel. As a K$^+$ ion enters the narrow selectivity filter, it loses its water of hydration but becomes bound in the same geometry to eight backbone carbonyl oxygens, two from the extended loop in each of the four P segments lining the channel (Figure 11-20a, left). Thus little energy is required to strip off the eight waters of hydration of a K$^+$ ion, and as a result, a relatively low activation energy is required for passage of K$^+$ ions

(a) K^+ and Na^+ ions in the pore of a K^+ channel (top view)

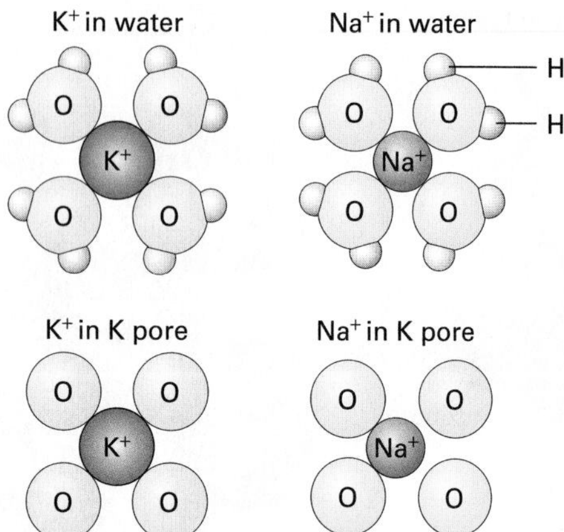

(b) K^+ ions in the pore of a K^+ channel (side view)

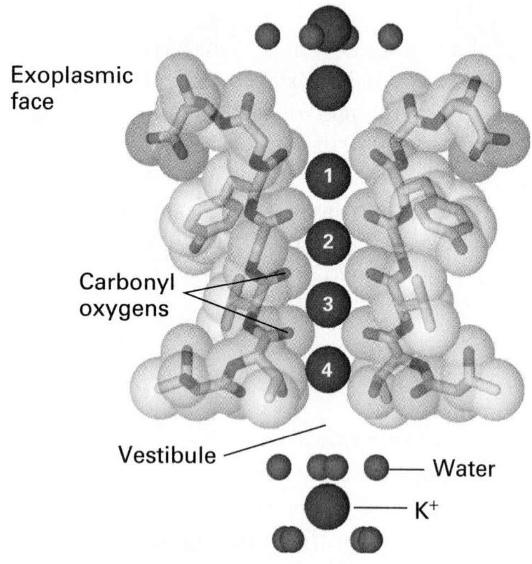

(c) Ion movement through selectivity filter

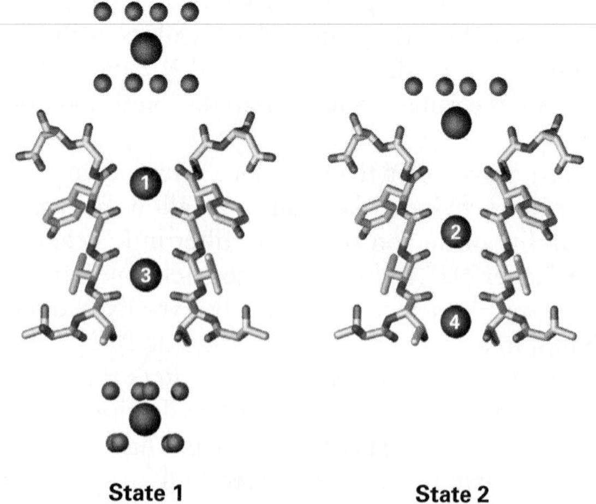

◄ **FIGURE 11-20 Mechanism of ion selectivity and transport in resting K^+ channels.** (a) Schematic diagrams of K^+ and Na^+ ions hydrated in solution and in the pore of a K^+ channel. As K^+ ions pass through the selectivity filter, they lose their bound water molecules and become coordinated instead to eight backbone carbonyl oxygens, four of which are shown, that are part of the conserved amino acids in the channel-lining loop of each P segment. The smaller Na^+ ions with their tighter shell of water molecules cannot perfectly coordinate with the channel oxygen atoms and therefore pass through the channel only rarely. (b) High-resolution electron density map obtained from x-ray crystallography showing K^+ ions (purple spheres) passing through the selectivity filter. Only two of the diagonally opposed channel subunits are shown. Within the selectivity filter each unhydrated K^+ ion interacts with eight carbonyl oxygen atoms (red sticks) lining the channel, two from each of the four subunits, as if to mimic the eight waters of hydration. (c) Interpretation of the electron density map showing the two alternating states by which K^+ ions move through the channel. In state 1, moving from the exoplasmic side of the channel inward, one sees a hydrated K^+ ion with its eight bound water molecules, K^+ ions at positions 1 and 3 within the selectivity filter, and a fully hydrated K^+ ion within the vestibule. During K^+ movement each ion in state 1 moves one step inward, forming state 2. Thus in state 2 the K^+ ion on the exoplasmic side of the channel has lost four of its eight waters, the ion at position 1 in state 1 has moved to position 2, and the ion at position 3 in state 1 has moved to position 4. In going from state 2 to state 1, the K^+ at position 4 moves into the vestibule and picks up eight water molecules, while another hydrated K^+ ion moves into the channel opening and the other K^+ ions move down one step. [Part (a) adapted from C. Armstrong, 1998, *Science* **280**:56. Parts (b) and (c) adapted from Y. Zhou et al., 2001, *Nature* **414**:43.]

through the channel. A dehydrated Na^+ ion is too small to bind to all eight carbonyl oxygens that line the selectivity filter with the same geometry as a Na^+ ion surrounded by its normal eight water molecules. As a result, Na^+ ions would "prefer" to remain in water rather than enter the selectivity filter, and thus the activation energy for passage of Na^+ ions is relatively high (Figure 11-20a, right). This difference in activation energies favors passage of K^+ ions over Na^+ by a factor of 1,000. Like Na, the dehydrated Ca^{2+} ion is smaller than the dehydrated K^+ ion and cannot interact properly with the O atoms in the selectivity filter. Also, more energy is required to strip the waters of hydration from Ca^{2+} than from K^+.

Recent x-ray crystallographic studies reveal that both when open and closed, the channel contains K^+ ions within the selectivity filter; without these ions the channel probably would collapse. The K^+ ions are thought to be present either at positions 1 and 3 or at 2 and 4, each surrounded by eight carbonyl oxygen atoms (Figure 11-20b and c). Several K^+ ions move simultaneously through the channel such that when the ion on the exoplasmic face that has been partially stripped of its water of hydration moves into position 1, the ion at position 2 jumps to position 3 and the one at position 4 exits the channel (Figure 11-20c).

Although the amino acid sequences of the P segment in Na^+ and K^+ channels differ somewhat, they are similar

▶ **EXPERIMENTAL FIGURE 11-21 Current flow through individual ion channels can be measured by patch-clamping technique.** (a) Basic experimental arrangement for measuring current flow through individual ion channels in the plasma membrane of a living cell. The patch electrode, filled with a current-conducting saline solution, is applied, with a slight suction, to the plasma membrane. The 0.5μm-diameter tip covers a region that contains only one or a few ion channels. The second electrode is inserted through the membrane into the cytosol. A recording device measures current flow only through the channels in the patch of plasma membrane. (b) Photomicrograph of the cell body of a cultured neuron and the tip of a patch pipette touching the cell membrane. (c) Different patch-clamping configurations. Isolated, detached patches are the best configurations for studying the effects on channels of different ion concentrations and solutes such as extracellular hormones and intracellular second messengers (e.g., cAMP). [Part (b) from B. Sakmann, 1992, *Neuron* **8**:613 (Nobel lecture); also published in E. Neher and B. Sakmann, 1992, *Sci. Am.* **266**(3):44. Part (c) adapted from B. Hille, 1992, *Ion Channels of Excitable Membranes,* 2d ed., Sinauer Associates, p. 89.]

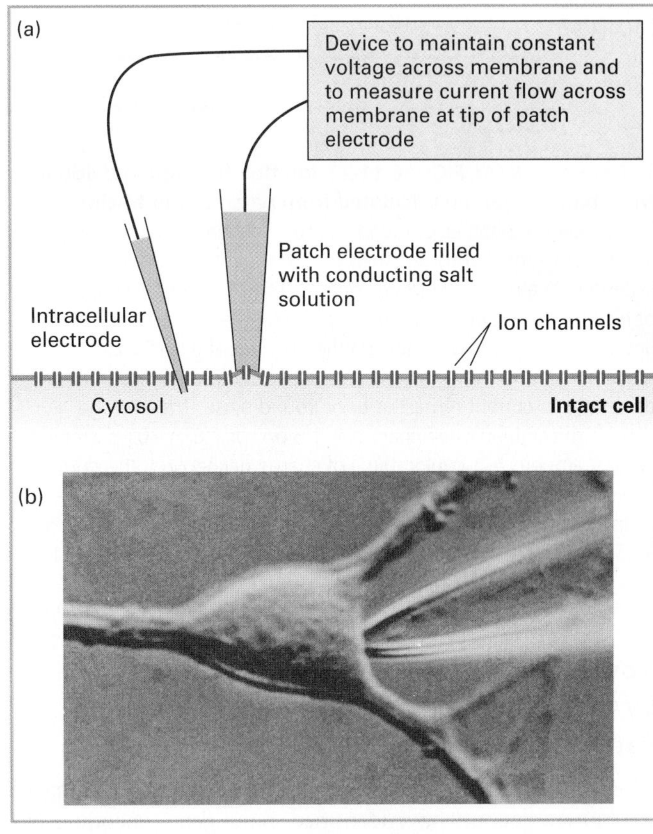

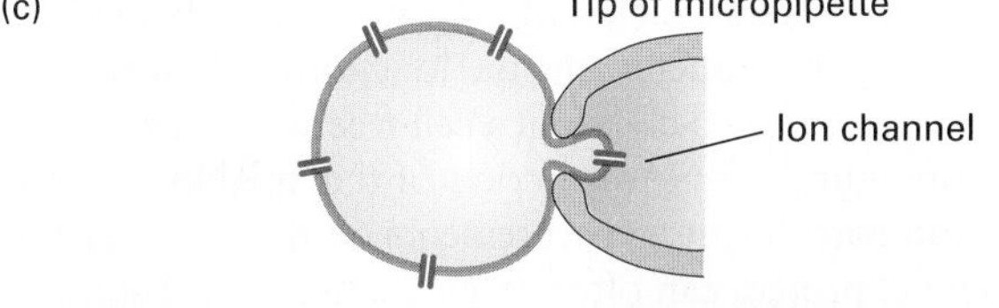

On-cell patch measures indirect effect of extracellular solutes on channels within membrane patch on intact cell

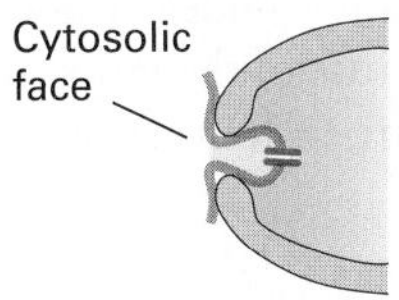

Inside-out detached patch measures effects of intracellular solutes on channels within isolated patch

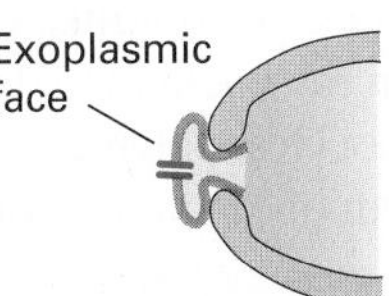

Outside-out detached patch measures effects of extracellular solutes on channels within isolated patch

enough to suggest that the general structure of the ion-selectivity filters are comparable in both types of channels. Presumably the diameter of the filter in Na^+ channels is small enough that it permits dehydrated Na^+ ions to bind to the backbone carbonyl oxygens but excludes the larger K^+ ions from entering, but as yet no three-dimensional structure of a Na^+ channel is available.

Patch Clamps Permit Measurement of Ion Movements Through Single Channels

The technique of **patch clamping** enables workers to investigate the opening, closing, regulation, and ion conductance of a *single* ion channel. In this technique, the inward or outward movement of ions across a patch of membrane is quantified from the amount of electric current needed to maintain the membrane potential at a particular "clamped" value (Figure 11-21a and b). To preserve electroneutrality and to keep the membrane potential constant, the entry of each positive ion (e.g., a Na^+ ion) into the cell through a channel in the patch of membrane is balanced by the addition of an electron into the cytosol through a microelectrode inserted into the cytosol; an electronic device measures the numbers of electrons (current) required to counterbalance the inflow of ions through the membrane channels. Conversely, the exit of each positive ion from the cell (e.g., a K^+ ion) is balanced by the withdrawal of an electron from the cytosol. The patch-clamping technique can be employed on whole cells or isolated membrane patches to measure the effects of different substances and ion concentrations on ion flow (Figure 11-21c).

The patch-clamp tracings in Figure 11-22 illustrate the use of this technique to study the properties of voltage-gated Na^+ channels in the plasma membrane of muscle cells. As we discuss in Chapter 23, these channels normally are closed in resting muscle cells and open following nervous stimulation. Patches of muscle membrane, each containing one Na^+ channel, were clamped at a voltage slightly less than the resting membrane potential. Under these circumstances, transient pulses of current cross the membrane as individual Na^+ channels open and then close. Each channel is either fully open or completely closed. From such tracings, it is possible to determine the time that a channel is open and the ion flux through it. For the channels measured in Figure 11-22, the flux is about 10 million Na^+ ions per channel per second, a typical value for ion channels. Replacement of the NaCl within the patch pipette (corresponding to the outside of the cell) with KCl or choline chloride abolishes current through the channels, confirming that they conduct only Na^+ ions, not K^+ or other ions.

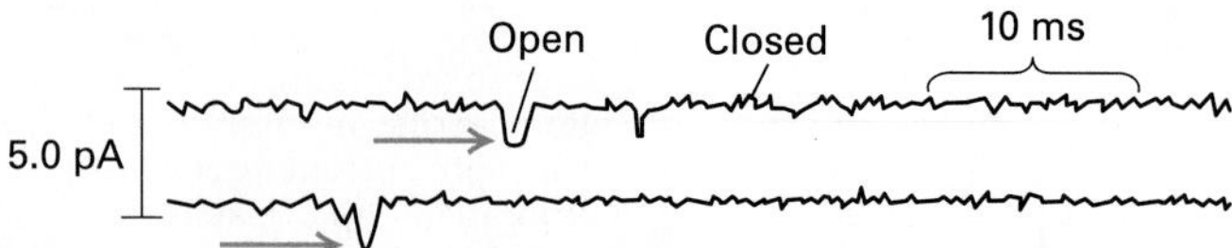

▲ EXPERIMENTAL FIGURE 11-22 **Ion flux through individual Na^+ channels can be calculated from patch clamp tracings.** Two inside-out patches of muscle plasma membrane were clamped at a potential of slightly less than that of the resting membrane potential. The patch electrode contained NaCl. The transient pulses of electric current in picoamperes (pA), recorded as large downward deviations (blue arrows), indicate the opening of a Na^+ channel and movement of Na^+ ions inward across the membrane. The smaller deviations in current represent background noise. The average current through an open channel is 1.6 pA, or 1.6×10^{-12} amperes. Since 1 ampere = 1 coulomb (C) of charge per second, this current is equivalent to the movement of about 9900 Na^+ ions per channel per millisecond: $(1.6 \times 10^{-12}$ C/s$)(10^{-3}$ s/ms$)(6 \times 10^{23}$ molecules/mol$) \div$ 96,500 C/mol. [See F. J. Sigworth and E. Neher, 1980, *Nature* **287**:447.]

1 Microinject mRNA encoding channel protein of interest

mRNA

Plasma membrane

2 Incubate 24–48 h for synthesis and movement of channel protein to plasma membrane

Newly synthesized channel protein

3 Measure channel-protein activity by patch-clamping technique

Patch electrode

▲ EXPERIMENTAL FIGURE 11-23 **Oocyte expression assay is useful in comparing the function of normal and mutant forms of a channel protein.** A follicular frog oocyte is first treated with collagenase to remove the surrounding follicle cells, leaving a denuded oocyte, which is microinjected with mRNA encoding the channel protein under study. [Adapted from T. P. Smith, 1988, *Trends Neurosci.* **11**:250.]

Novel Ion Channels Can Be Characterized by a Combination of Oocyte Expression and Patch Clamping

Cloning of human-disease-causing genes and sequencing of the human genome have identified many genes encoding putative channel proteins, including 67 putative K^+ channel proteins. One way of characterizing the function of these proteins is to transcribe a cloned cDNA in a cell-free system to produce the corresponding mRNA. Injection of this mRNA into frog oocytes and patch clamp measurements on the newly synthesized channel protein can often reveal its function (Figure 11-23). This experimental approach is especially useful because frog oocytes normally do not express any channel proteins on their surface membrane, so only the channel under study is present in the membrane. In addition, because of the large size of frog oocytes, patch-clamping studies are technically easier to perform on them than on smaller cells.

Na^+ Entry into Mammalian Cells Has a Negative Change in Free Energy (ΔG)

As mentioned earlier, two forces govern the movement of ions across selectively permeable membranes: the voltage and the ion concentration gradient across the membrane. The sum of these forces, which may act in the same direction or in opposite directions, constitutes the electrochemical gradient. To calculate the free-energy change ΔG corresponding to the transport of any ion across a membrane, we need to consider the independent contributions from each of the forces to the electrochemical gradient.

For example, when Na^+ moves from outside to inside the cell, the free-energy change generated from the Na^+ concentration gradient is given by

$$\Delta G_c = RT \ln \frac{[Na_{in}]}{[Na_{out}]} \qquad (11\text{-}5)$$

At the concentrations of Na_{in} and Na_{out} shown in Figure 11-24, which are typical for many mammalian cells, ΔG_c, the change in free energy due to the concentration gradient, is -1.45 kcal for transport of 1 mol of Na^+ ions from outside to inside the cell, assuming there is no membrane electric potential. Note the free energy is negative, indicating spontaneous movement of Na^+ into the cell.

The free-energy change generated from the membrane electric potential is given by

$$\Delta G_m = FE \qquad (11\text{-}6)$$

where F is the Faraday constant and E is the membrane electric potential. If $E = -70$ mV, then ΔG_m, the free-energy change due to the membrane potential, is -1.61 kcal for transport of 1 mol of Na^+ ions from outside to inside the cell, assuming there is no Na^+ concentration gradient. Since both forces do in fact act on Na^+ ions, the total ΔG is the sum of the two partial values:

$$\Delta G = \Delta G_c + \Delta G_m = (-1.45) + (-1.61) = -3.06 \text{ kcal/mole}$$

In this example, the Na^+ concentration gradient and the membrane electric potential contribute almost equally to the total ΔG for transport of Na^+ ions. Since ΔG is <0, the inward movement of Na^+ ions is thermodynamically favored. As discussed in the next section, certain cotransport proteins use the inward movement of Na^+ to power the uphill movement of other ions and several types of small molecules into

 Overview Animation: Biological Energy Interconversions

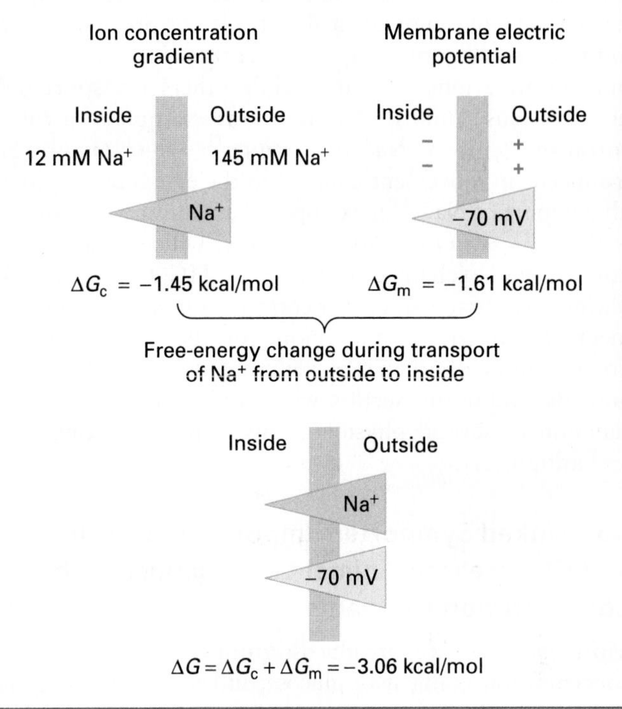

▶ **FIGURE 11-24 Transmembrane forces acting on Na^+ ions.** As with all ions, the movement of Na^+ ions across the plasma membrane is governed by the sum of two separate forces—the ion concentration gradient and the membrane electric potential. At the internal and external Na^+ concentrations typical of mammalian cells, these forces usually act in the same direction, making the inward movement of Na^+ ions energetically favorable.

or out of animal cells. The rapid, energetically favorable movement of Na^+ ions through gated Na^+ channels also is critical in generating action potentials in nerve and muscle cells, as we discuss in Chapter 23.

KEY CONCEPTS OF SECTION 11.4

Nongated Ion Channels and the Resting Membrane Potential

- An inside-negative electric potential (voltage) of 50–70 mV exists across the plasma membrane of all cells.
- In animal cells, the membrane potential is generated primarily by movement of cytosolic K^+ ions through resting K^+ channels to the external medium. Unlike the more common gated ion channels, which open only in response to various signals, these nongated K^+ channels are usually open.
- In plants and fungi, the membrane potential is maintained by the ATP-driven pumping of protons from the cytosol to the exterior of the cell.
- K^+ channels are assembled from four identical subunits, each of which has at least two conserved membrane-spanning α helices and a nonhelical P segment that lines the ion pore and forms the selectivity filter (see Figure 11-19).
- The ion specificity of K^+ channel proteins is due mainly to coordination of the selected ion with the carbonyl oxygen atoms of specific amino acids in the P segments, thus lowering the activation energy for passage of the selected K^+ compared with other ions (see Figure 11-20).
- Patch-clamping techniques, which permit measurement of ion movements through single channels, are used to determine the ion conductivity of a channel and the effect of various signals on its activity (see Figure 11-21).
- Recombinant DNA techniques and patch clamping allow the expression and functional characterization of channel proteins in frog oocytes (see Figure 11-23).
- The electrochemical gradient across a semipermeable membrane determines the direction of ion movement through channel proteins. The two forces constituting the electrochemical gradient, the membrane electric potential and the ion concentration gradient, may act in the same or opposite directions (see Figure 11-24).

11.5 Cotransport by Symporters and Antiporters

In previous sections we saw how ATP-powered pumps generate ion concentration gradients across cell membranes and how ion channel proteins use these gradients to establish an electric potential across these membranes. In this section we see how cotransporters use the energy stored in the electric potential and concentration gradients of Na^+ or H^+ ions to power the uphill movement of another substance, which may be a small organic molecule such as glucose or an amino acid or a different ion (see Figure 11-2). For instance, the energetically favored movement of a Na^+ ion (the cotransported ion) into a cell across the plasma membrane, driven both by its concentration gradient and by the transmembrane voltage gradient (see Figure 11-24), can be coupled to movement of the transported molecule (e.g., glucose or lysine) against its concentration gradient. An important feature of such **cotransport** is that neither molecule can move alone; movement of both molecules together is obligatory, or *coupled.*

Cotransporters share common features with uniporters such as the GLUT proteins. The two types of transporters exhibit certain structural similarities, operate at equivalent rates, and undergo cyclical conformational changes during transport of their substrates. They differ in that uniporters can only accelerate thermodynamically favorable transport down a concentration gradient, whereas cotransporters can harness the energy of a coupled favorable reaction to actively transport molecules against a concentration gradient.

When the transported molecule and cotransported ion move in the same direction, the process is called **symport;** when they move in opposite directions, the process is called **antiport** (see Figure 11-3). Some cotransporters transport only positive ions (cations), while others transport only negative ions (anions). An important example of a cation cotransporter is the *Na^+/H^+ antiporter,* which exports H^+ from cells in movement coupled to the energetically favorable import of Na^+. An example of an anion cotransporter is the *AE1 anion antiporter protein,* which catalyzes the one-for-one exchange of Cl^- and HCO_3^- across the plasma membrane. Yet other cotransporters mediate movement of both cations and anions together. Cotransporters are present in all organisms, including bacteria, plants, and animals, and in this section we describe the operation and function of several physiologically important symporters and antiporters.

Na^+-Linked Symporters Import Amino Acids and Glucose into Animal Cells Against High Concentration Gradients

Most body cells import glucose from the blood *down* the concentration gradient of glucose, utilizing GLUT proteins to facilitate this transport. However, certain cells, such as those lining the small intestine and the kidney tubules, need to import glucose from the intestinal lumen or forming urine against a very large concentration gradient. Such cells utilize a *two-Na^+/one-glucose symporter,* a protein that couples import of one glucose molecule to the import of two Na^+ ions:

$$2\ Na^+_{out} + \text{glucose}_{out} \rightleftharpoons 2\ Na^+_{in} + \text{glucose}_{in}$$

Quantitatively, the free-energy change for the symport transport of two Na^+ ions and one glucose molecule can be written

$$\Delta G = RT\ln\frac{[\text{glucose}_{in}]}{[\text{glucose}_{out}]} + 2RT\ln\frac{[Na^+_{in}]}{[Na^+_{out}]} + 2FE \quad (11\text{-}7)$$

Thus the ΔG for the overall reaction is the sum of the free-energy changes generated by the glucose concentration gradient (1 molecule transported), the Na^+ concentration gradient (2 Na^+ ions transported), and the membrane potential (2 Na^+ ions transported). As illustrated in Figure 11-24, the free energy released by movement of Na^+ into mammalian cells down its electrochemical gradient has a free-energy change ΔG of about -3 kcal per mole of Na^+ transported. Thus the ΔG for transport of two moles of Na^+ inward is about -6 kcal. This negative free energy of sodium import is coupled to the uphill transport of glucose, a process with a positive ΔG. We can calculate the glucose concentration gradient, inside greater than outside, against which glucose can be transported by realizing that at equilibrium for sodium-coupled glucose import, $\Delta G = 0$. By substituting the values for sodium import into Equation 11-7 and setting $\Delta G = 0$, we see that

$$0 = RT\ln\frac{[\text{glucose}_{in}]}{[\text{glucose}_{out}]} - 6\text{kcal}$$

and we can calculate that at equilibrium, the ratio of glucose_{in}/ glucose_{out} = $\approx$30,000. Thus the inward flow of two moles of Na^+ can generate an intracellular glucose concentration that is $\approx$30,000 times greater than the exterior concentration. If only one Na^+ ion were imported (ΔG of -3 kcal/mol) per glucose molecule, then the available energy could generate a glucose concentration gradient (inside > outside) of only about 170-fold. Thus by coupling the transport of two Na^+ ions to the transport of one glucose, the two-Na^+/one-glucose symporter permits cells to accumulate a very high concentration of glucose relative to the external

MEDIA CONNECTIONS

Oveview Animation: Biological Energy Interconversions

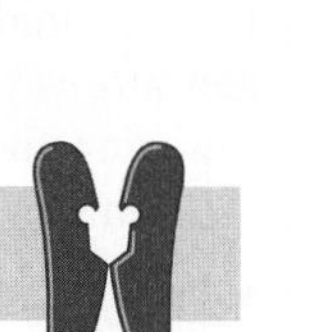

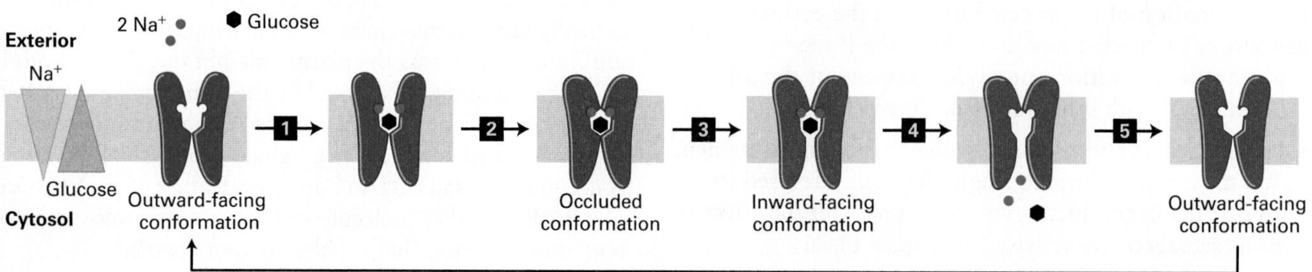

▲ **FIGURE 11-25 Operational model for the two-Na^+/one-glucose symporter.** Simultaneous binding of Na^+ and glucose to the conformation with outward-facing binding sites (step **1**) causes a conformational change in the protein such that the bound substrates are transiently occluded, unable to dissociate into either medium (step **2**). In step **3** the protein assumes a third conformation with inward-facing sites. Dissociation of the bound Na^+ and glucose into the cytosol (step **4**) allows the protein to revert to its original outward-facing conformation (step **5**), ready to transport additional substrate. [See E. Wright et. al., 2004, *Physiology* **19**:370 for details on the structure and function of this and related transporters.]

concentration. This means that glucose present even at very low concentrations in the lumen of the intestine or in the forming urine can be efficiently transported into the lining cells and not lost from the body.

The two-Na^+/one-glucose symporter is thought to contain 14 transmembrane α helices with both its N- and C-termini extending into the cytosol. A truncated recombinant protein consisting of only the five C-terminal transmembrane α helices can transport glucose independently of Na^+ across the plasma membrane, *down* its concentration gradient. This portion of the molecule thus functions as a glucose uniporter. The N-terminal portion of the protein, including helices 1–9, is required to couple Na^+ binding and influx to the transport of glucose against a concentration gradient.

Figure 11-25 depicts the current model of transport by Na^+/glucose symporters. This model entails conformational changes in the protein analogous to those that occur in uniport transporters, such as GLUT1, which do not require a cotransported ion (see Figure 11-5). Binding of all substrates to their sites on the extracellular domain is required before the protein undergoes the conformational change that transitions the substrate-binding sites from outward to inward facing; this ensures that inward transport of glucose and Na^+ ions are coupled.

Bacterial Symporter Structure Reveals the Mechanism of Substrate Binding

No three-dimensional structure of a mammalian sodium symporter has been determined, but the structures of several homologous bacterial sodium–amino acid transporters have provided considerable information about symport function. The bacterial two-Na^+/one-leucine symporter shown in Figure 11-26a consists of 12 membrane-spanning α helices. Two of the helices (numbers 1 and 6) have non-helical segments in the middle of the membrane that form part of the leucine-binding site.

Podcast: The Two-Na^+/One-Leucine Symporter

MEDIA CONNECTIONS

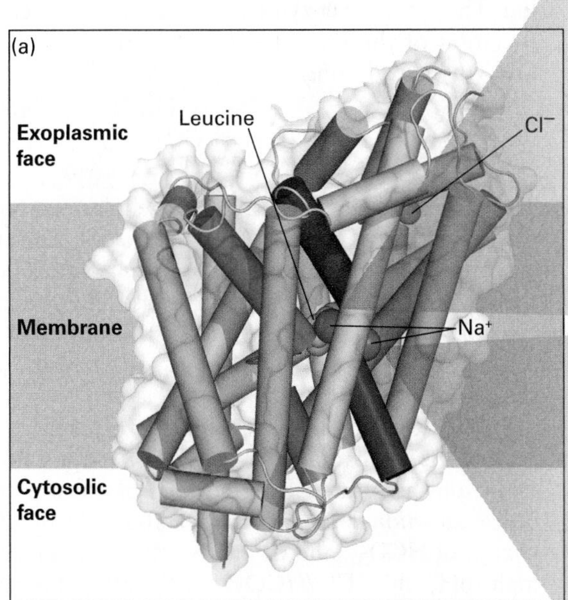

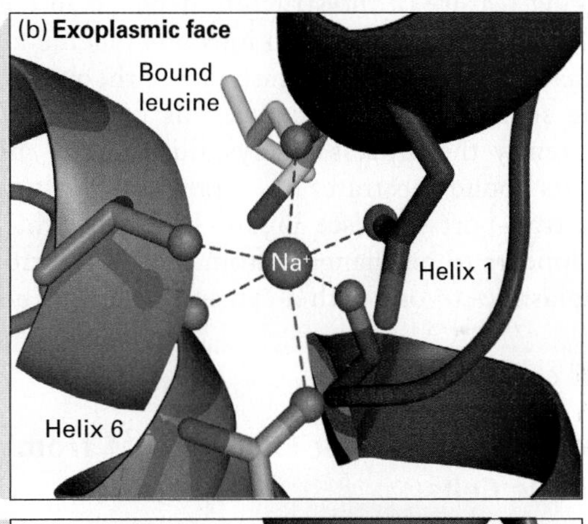

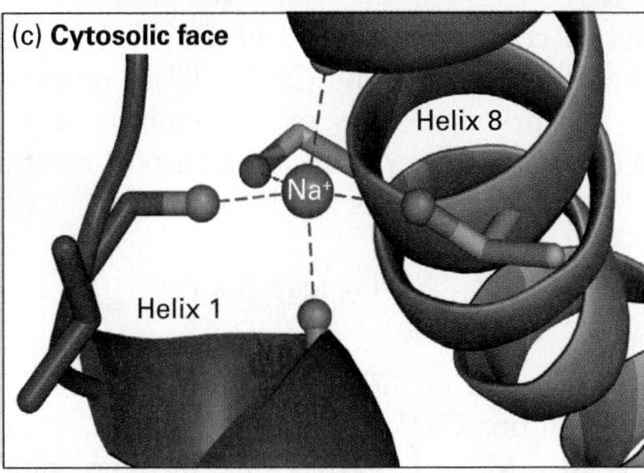

▲ **FIGURE 11-26 Three-dimensional structure of the two-Na^+-/one-leucine symporter from the bacterium *Aquifex aeolicus*.** (a) The bound L-leucine, two sodium ions, and a chloride ion are shown as CPK models in yellow, purple and green, respectively. The three membrane-spanning α helices that bind the Na^+ or leucine are colored brown, blue, and orange. (b) Binding of the two sodium ions to carbonyl main-chain oxygen atoms or carboxyl side-chain oxygens (red) that are part of helices 1 (brown), 6 (blue), or 8 (orange). It is important that one of the sodium ions (top) is also bound to the carboxyl group of the transported leucine (yellow). [From A. Yamashita et al., 2005, *Nature* **437**:215; see also D. Yernool et al., 2004, *Nature* **431**:811 for details on the structure and function of this and related transporters.]

Amino acid residues involved in binding the leucine and the two sodium ions are located in the middle of the membrane-spanning segment (as depicted for the two-Na^+/one-glucose symporter in Figure 11-25) and are close together in three-dimensional space. This demonstrates that the coupling of substrate and ion transport in these transporters is the consequence of direct or nearly direct physical interactions of the substrates. Indeed, one of the sodium ions (number 1 in Figure 11-26b) is bound to the carboxyl group of the transported leucine, indicating how binding of sodium and leucine are coupled. Each of the two sodium ions is bound to six oxygen atoms. Sodium 1, for example, is also bound to carbonyl oxygens of several transporter amino acids as well as to carbonyl oxygens and the hydroxyl oxygen of one threonine. Equally importantly, there are no water molecules surrounding either of the bound sodium atoms, as is the case for K^+ ions in potassium channels (see Figure 11-20). Thus as the sodium ions lose their water of hydration in binding to the transporter, they bind to six oxygen atoms with a similar geometry. This reduces the activation energy for binding of sodium ions and prevents other ions, such as potassium, from binding in place of sodium.

One striking feature of the structure depicted in Figure 11-26 is that the bound sodium ions and leucine are *occluded*—that is, they cannot diffuse out of the protein to either the surrounding extracellular or cytoplasmic media. Apparently the process of crystallization of this protein with its bound substrates has "trapped" it into an intermediate transport step (see Figure 11-25) in which the protein appears to be changing from a conformation with an exoplasmic- to one with a cytosolic-facing binding site.

Na^+-Linked Ca^{2+} Antiporter Exports Ca^{2+} from Cardiac Muscle Cells

In cardiac muscle cells a *three-Na^+/one-Ca^{2+} antiporter,* rather than the plasma membrane Ca^{2+} ATPase discussed earlier, plays the principal role in maintaining a low concentration of Ca^{2+} in the cytosol. The transport reaction mediated by this *cation antiporter* can be written

$$3\ Na^+_{out} + Ca^{2+}_{in} \rightleftharpoons 3\ Na^+_{in} + Ca^{2+}_{out}$$

Note that the movement of three Na^+ ions is required to power the export of one Ca^{2+} ion from the cytosol, with a $[Ca^{2+}]$ of $\approx 2 \times 10^{-7}$ M, to the extracellular medium, with a $[Ca^{2+}]$ of 2×10^{-3} M, a gradient of some 10,000-fold. In all muscle cells, a rise in the cytosolic Ca^{2+} concentration in cardiac muscle triggers contraction; by lowering cytosolic Ca^{2+}, operation of the Na^+/Ca^{2+} antiporter reduces the strength of heart muscle contraction.

The Na^+/K^+ ATPase in the plasma membrane of cardiac muscle cells, as in other body cells, creates the Na^+ concentration gradient necessary for export of Ca^{2+} by the Na^+-linked Ca^{2+} antiporter. As mentioned earlier, inhibition of the Na^+/K^+ ATPase by the drugs ouabain and digoxin lowers the cytosolic K^+ concentration and, more relevant here, simultaneously increases cytosolic Na^+. The resulting reduced Na^+ electrochemical gradient across the membrane causes the Na^+-linked Ca^{2+} antiporter to function less efficiently. As a result, fewer Ca^{2+} ions are exported and the cytosolic Ca^{2+} concentration increases, causing the muscle to contract more strongly. Because of their ability to increase the force of heart muscle contractions, drugs such as ouabain and digoxin that inhibit the Na^+/K^+ ATPase are widely used in the treatment of congestive heart failure. ▮

Several Cotransporters Regulate Cytosolic pH

The anaerobic metabolism of glucose yields lactic acid, and aerobic metabolism yields CO_2, which adds water to form carbonic acid (H_2CO_3). These weak acids dissociate, yielding H^+ ions (protons); if these excess protons were not removed from cells, the cytosolic pH would drop precipitously, endangering cellular functions. Two types of cotransport proteins help remove some of the "excess" protons generated during metabolism in animal cells. One is a *$Na^+HCO_3^-/Cl^-$ antiporter,* which imports one Na^+ ion together with one HCO_3^-, in exchange for export of one Cl^- ion. The cytosolic enzyme *carbonic anhydrase* catalyzes dissociation of the imported HCO_3^- ions into CO_2 and an OH^- (hydroxyl) ion:

$$HCO_3^- \rightleftharpoons CO_2 + OH^-$$

The OH^- ions combine with intracellular protons, forming water, and the CO_2 diffuses out of the cell. Thus the overall action of this transporter is to *consume* cytosolic H^+ ions, thereby *raising* the cytosolic pH. Also important in raising cytosolic pH is a *Na^+/H^+ antiporter,* which couples entry of one Na^+ ion into the cell down its concentration gradient to the export of one H^+ ion.

Under certain circumstances, the cytosolic pH can rise beyond the normal range of 7.2–7.5. To cope with the excess OH^- ions associated with elevated pH, many animal cells utilize an *anion antiporter* that catalyzes the one-for-one exchange of HCO_3^- and Cl^- across the plasma membrane. At high pH, this *Cl^-/HCO_3^- antiporter* exports HCO_3^- (which can be viewed as a "complex" of OH^- and CO_2) in exchange for import of Cl^-, thus lowering the cytosolic pH. The import of Cl^- down its concentration gradient ($Cl^-_{medium} > Cl^-_{cytosol}$) powers the transport.

The activity of all three of these antiport proteins depends on pH, providing cells with a finely tuned mechanism for controlling the cytosolic pH. The two antiporters that operate to increase cytosolic pH are activated when the pH of the cytosol falls. Similarly, a rise in pH above 7.2 stimulates the Cl^-/HCO_3^- antiporter, leading to a more rapid export of HCO_3^- and decrease in the cytosolic pH. In this manner, the cytosolic pH of growing cells is maintained very close to pH 7.4.

A Putative Cation Exchange Protein Plays a Key Role in Evolution of Human Skin Pigmentation

Sequencing of the human, mouse, and rat genomes indicates the presence of hundreds of putative transport proteins, but the functions of most of these are as yet unknown. A particularly interesting human transporter called *SLC24A5* emerged from a study of zebrafish that had abnormal skin color; in fish homozygous for the *golden* mutation, the eponymous black horizontal stripes were very pale (Figure 11-27a and b). Microscopy showed that the mutant fish had a much lower amount of the black pigment called *melanin,* and melanin vesicles, called *melanosomes,* were much smaller and paler than normal (Figure 11-27 c and d). Positional cloning of the *golden* gene demonstrated that it encodes a putative cation exchange protein termed SLC24A5. Immunofluorescence studies showed that the protein is found in intracellular membranes, likely in the membrane of the melanosome or its precursor, but the ions transported by SLC24A5 are not yet known. However, the amino acid sequence of the SLAC24A5 protein is closest to that of several sodium/calcium antiporters, so the protein is likely a sodium/calcium antiporter.

Most strikingly, investigators showed that the human version of SLC24A5 is highly similar in sequence to the zebrafish protein; when the human protein is expressed in mutant *golden* zebrafish, it complements the mutant phenotype and the fish have normal black stripes. The most evolutionarily conserved form of the gene, or allele, most similar to the wild-type zebrafish gene predominates in dark-skinned African and East Asian human populations. In contrast, a version, or variant allele, of the SLC24A5 gene with a single amino acid change that is thought to encode a less active protein is found in virtually all people of European origin. Studies of the allele frequencies in admixed populations indicate that different forms, or polymorphisms, in just this cation transporter play a key role in determining the darkness of human skin color. Clearly much needs to be learned about the role of this transporter in cell physiology and how a single point mutation in this gene accounts for the large differences in skin pigmentation characteristic of individuals of European, African, and Asian origin.

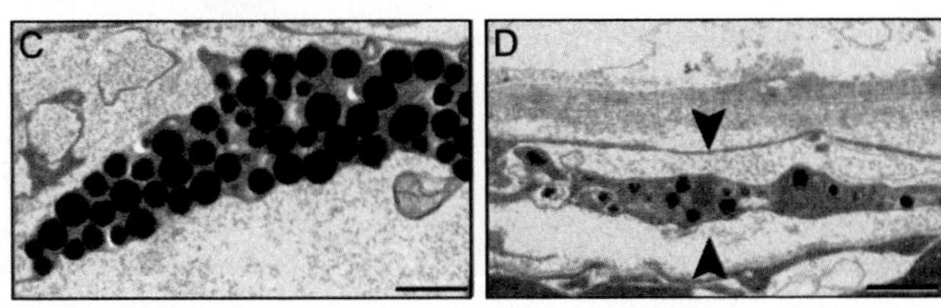

▲ **FIGURE 11-27 Zebrafish mutations in the gene encoding the cation exchanger SLC24A5 cause the *golden* skin pigment phenotype.** Lateral views of adult wild-type (a) and *golden* (b) zebrafish. Insets show melanophores (arrowheads). Scale bars, 5 mm (inset, 0.5 mm). *Golden* mutants have melanophores that are on average smaller, paler, and more transparent than normal. Transmission electron micrographs of skin melanophores from wild-type (c) and *golden* (d) larvae show that *golden* skin melanophores (arrowheads show edges) are thinner and contain fewer melanosomes than normal. Scale bars in (c) and (d), 1000 nm. [From R. L. Lamason et al., 2005, *Science* **310**:1782.]

Numerous Transport Proteins Enable Plant Vacuoles to Accumulate Metabolites and Ions

The lumen of plant vacuoles is much more acidic (pH 3–6) than is the cytosol (pH 7.5). The acidity of vacuoles is maintained by a V-class ATP-powered proton pump (see Figure 11-9) and by a pyrophosphate-powered pump that is unique to plants. Both of these pumps, located in the vacuolar membrane, import H^+ ions into the vacuolar lumen against a concentration gradient. The vacuolar membrane also contains Cl^- and NO_3^- channels that transport these anions from the cytosol into the vacuole. Entry of these anions against their concentration gradients is driven by the inside-positive potential generated by the H^+ pumps. The combined operation of these proton pumps and anion channels produces an inside-positive electric potential of about 20 mV across the vacuolar membrane and also a substantial pH gradient (Figure 11-28).

The proton electrochemical gradient across the plant vacuole membrane is used in much the same way as the Na^+ electrochemical gradient across the animal-cell plasma membrane: to power the selective uptake or extrusion of ions and small molecules by various antiporters. In the leaf, for example, excess sucrose generated during photosynthesis in the day is stored in the vacuole; during the night, the stored sucrose moves into the cytoplasm and is metabolized to CO_2 and H_2O with concomitant generation of ATP from ADP and P_i. A *proton/sucrose antiporter* in the vacuolar membrane operates to accumulate sucrose in plant vacuoles. The inward movement of sucrose is powered by the outward movement of H^+, which is favored by its concentration

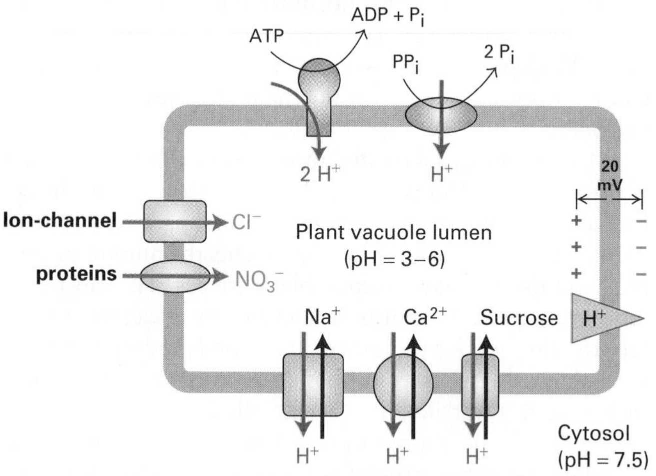

▲ **FIGURE 11-28 Concentration of ions and sucrose by the plant vacuole.** The vacuolar membrane contains two types of proton pumps (orange): a V-class H^+ ATPase (*left*) and a pyrophosphate-hydrolyzing proton pump (*right*) that differs from all other ion transport proteins and probably is unique to plants. These pumps generate a low luminal pH as well as an inside-positive electric potential across the vacuolar membrane owing to the inward pumping of H^+ ions. The inside-positive potential powers the movement of Cl^- and NO_3^- from the cytosol through separate channel proteins (purple). Proton antiporters (green), powered by the H^+ gradient, accumulate Na^+, Ca^{2+}, and sucrose inside the vacuole. [After P. Rea and D. Sanders, 1987, *Physiol. Plant* **71**:131; J. M. Maathuis and D. Sanders, 1992, *Curr. Opin. Cell Biol.* **4**:661; and P. A. Rea et al., 1992, *Trends Biochem. Sci.* **17**:348.]

gradient (lumen > cytosol) and by the cytosolic-negative potential across the vacuolar membrane (see Figure 11-28). Uptake of Ca^{2+} and Na^+ into the vacuole from the cytosol against their concentration gradients is similarly mediated by proton antiporters.

Understanding of the transport proteins in plant vacuolar membranes has the potential for increasing agricultural production in high-salt (NaCl) soils, which are found throughout the world. Because most agriculturally useful crops cannot grow in such saline soils, agricultural scientists have long sought to develop salt-tolerant plants by traditional breeding methods. With the availability of the cloned gene encoding the vacuolar Na^+/H^+ antiporter, researchers can now produce transgenic plants that overexpress this transport protein, leading to increased sequestration of Na^+ in the vacuole. For instance, transgenic tomato plants that overexpress the vacuolar Na^+/H^+ antiporter can grow, flower, and produce fruit in the presence of soil NaCl concentrations that kill wild-type plants. Interestingly, although the leaves of these transgenic tomato plants accumulate large amounts of salt, the fruit has a very low salt content. ▮

KEY CONCEPTS OF SECTION 11.5

Cotransport by Symporters and Antiporters

- Cotransporters use the energy released by movement of an ion (usually H^+ or Na^+) down its electrochemical gradient to power the import or export of a small molecule or different ion against its concentration gradient.
- The cells lining the small intestine and kidney tubules express symport proteins that couple the energetically favorable entry of Na^+ to the import of glucose and amino acids against their concentration gradients (see Figure 11-25).
- The molecular structure of a bacterial Na^+–amino acid symporter reveals how binding of Na^+ and leucine are coupled and provides a snapshot of an occluded transport intermediate in which the bound substrates cannot diffuse out of the protein.
- In cardiac muscle cells, the export of Ca^{2+} is coupled to and powered by the import of Na^+ by a cation antiporter, which transports 3 Na^+ ions inward for each Ca^{2+} ion exported.
- As judged by mutations in zebrafish and polymorphisms in humans, the presumed sodium/calcium cotransporter SLC24A5 plays a major role in forming melanin granules and in regulating the darkness of human skin pigmentation.
- Two cotransporters that are activated at low pH help maintain the cytosolic pH in animal cells very close to 7.4 despite metabolic production of carbonic and lactic acids. One, a Na^+/H^+ antiporter, exports excess protons. The other, a $Na^+HCO_3^-/Cl^-$ cotransporter, imports HCO_3^-, which dissociates in the cytosol to yield pH-raising OH^- ions.
- A Cl^-/HCO_3^- antiporter that is activated at high pH functions to export HCO_3^- when the cytosolic pH rises above normal and causes a decrease in pH.
- Uptake of sucrose, Na^+, Ca^{2+}, and other substances into plant vacuoles is carried out by proton antiporters in the vacuolar membrane. Ion channels and proton pumps in the membrane are critical in generating a large enough proton concentration gradient to power accumulation of ions and metabolites in vacuoles by these proton antiporters (see Figure 11-28).

11.6 Transepithelial Transport

Previous sections illustrated how several types of transporters function together to carry out important cell functions (see Figure 11-2). Here, we extend this concept by focusing on the transport of several types of molecules and ions across the sheetlike layers of epithelial cells that cover most external and internal surfaces of body organs. Like all epithelial cells, an intestinal epithelial cell is said to be **polarized** because its plasma membrane is organized into at least two discrete regions. Typically, the surface that faces the outside of the organism, here the lumen of the intestine, is called the **apical**, or top, surface, and the surface that faces the inside of the organism is called the **basolateral** surface (see Figure 19-9).

Specialized regions of the epithelial-cell plasma membrane, called **cell junctions**, connect the cells and provide strength and rigidity to the cell sheet (see Figure 19-9 for details). One of these types of cell junctions—the **tight junction**—is of particular interest here since tight junctions prevent many water-soluble substances on one side of an epithelium from moving across to the other side through the extracellular space between cells. For this reason, absorption of nutrients from the intestinal lumen into the blood occurs by the two-stage process called *transcellular transport:* import of molecules through the plasma membrane on the apical side of intestinal epithelial cells and their export through the plasma membrane on the blood-facing (basolateral, or serosal) side (Figure 11-29). The apical portion of the plasma membrane, which faces the intestinal lumen, is specialized for absorption of sugars, amino acids, and other molecules that are produced from food by various digestive enzymes. Numerous fingerlike projections (100 nm in diameter) called **microvilli** greatly increase the area of the apical surface and so the number of transport proteins it can contain, enhancing the cell's absorptive capacity.

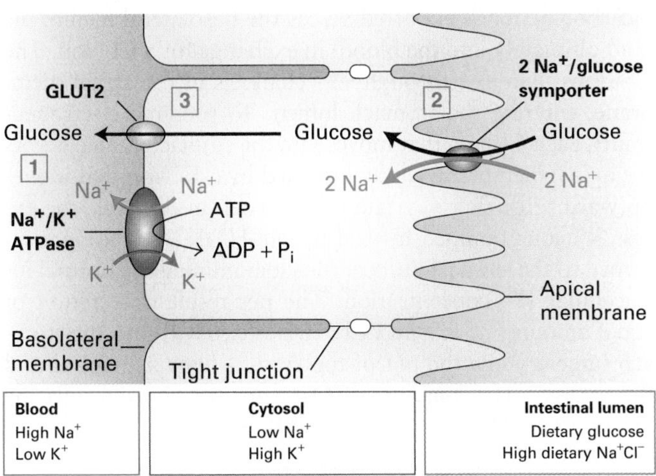

▲ **FIGURE 11-29 Transcellular transport of glucose from the intestinal lumen into the blood.** The Na^+/K^+ ATPase in the basolateral surface membrane generates Na^+ and K^+ concentration gradients (step 1). The outward movement of K^+ ions through nongated K^+ channels (not shown) generates an inside-negative membrane potential. Both the Na^+ concentration gradient and the membrane potential are used to drive the uptake of glucose from the intestinal lumen by the two-Na^+/one-glucose symporter located in the apical surface membrane (step 2). Glucose leaves the cell via facilitated diffusion catalyzed by GLUT2, a glucose uniporter located in the basolateral membrane (step 3).

Multiple Transport Proteins Are Needed to Move Glucose and Amino Acids Across Epithelia

Figure 11-29 depicts the proteins that mediate absorption of glucose from the intestinal lumen into the blood and illustrates the important concept that different types of proteins are localized to the apical and basolateral membranes of epithelial cells. In the first stage of this process, a two-Na^+/one-glucose symporter located in microvillar membranes imports glucose, against its concentration gradient, from the intestinal lumen across the apical surface of the epithelial cells. As noted above, this symporter couples the energetically unfavorable inward movement of one glucose molecule to the energetically favorable inward transport of two Na^+ ions (see Figure 11-25). In the steady state, all the Na^+ ions transported from the intestinal lumen into the cell during Na^+/glucose symport, or the similar process of Na^+/amino acid symport, are pumped out across the basolateral membrane, which faces the underlying tissue. Thus the low intracellular Na^+ concentration is maintained. The Na^+/K^+ ATPase that accomplishes this is found exclusively in the basolateral membrane of intestinal epithelial cells. The coordinated operation of these two transport proteins allows uphill movement of glucose and amino acids from the intestine into the cell. This first stage in transcellular transport ultimately is powered by ATP hydrolysis by the Na^+/K^+ ATPase.

In the second stage, glucose and amino acids concentrated inside intestinal cells by symporters are exported down their concentration gradients into the blood via uniport proteins in the basolateral membrane. In the case of glucose, this movement is mediated by GLUT2 (see Figure 11-29). As noted earlier, this GLUT isoform has a relatively low affinity for glucose but increases its rate of transport substantially when the glucose gradient across the membrane rises (see Figure 11-4).

The net result of this two-stage process is movement of Na^+ ions, glucose, and amino acids from the intestinal lumen across the intestinal epithelium into the extracellular medium that surrounds the basolateral surface of intestinal epithelial cells. Tight junctions between the epithelial cells prevent these molecules from diffusing back into the intestinal lumen, and eventually they move into the blood. The increased osmotic pressure created by transcellular transport of salt, glucose, and amino acids across the intestinal epithelium draws water from the intestinal lumen into the extracellular medium that surrounds the basolateral surface. In a sense, salts, glucose, and amino acids "carry" the water along with them.

Simple Rehydration Therapy Depends on the Osmotic Gradient Created by Absorption of Glucose and Na^+

An understanding of osmosis and the intestinal absorption of salt and glucose forms the basis for a simple therapy that saves millions of lives each year, particularly in less-developed countries. In these countries, cholera and other intestinal pathogens are major causes of death of young children. A toxin released by the bacteria activates chloride secretion by the intestinal epithelial cells into the lumen; water follows osmotically, and the resultant massive loss of water causes diarrhea, dehydration, and ultimately death. A cure demands not only killing the bacteria with antibiotics but also *rehydration*—replacement of the water that is lost from the blood and other tissues.

Simply drinking water does not help, because it is excreted from the gastrointestinal tract almost as soon as it enters. However, as we have just learned, the coordinated transport of glucose and Na^+ across the intestinal epithelium creates a transepithelial osmotic gradient, forcing movement of water from the intestinal lumen across the cell layer and ultimately into the blood. Thus giving affected children a solution of sugar and salt to drink (but not sugar or salt alone) causes the osmotic flow of water into the blood from the intestinal lumen and leads to rehydration. Similar sugar-salt solutions are the basis of popular drinks used by athletes to get sugar as well as water into the body quickly and efficiently. ■

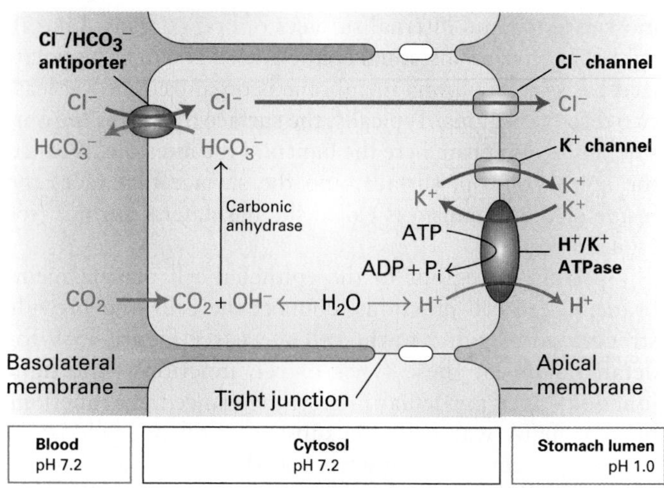

▲ **FIGURE 11-30 Acidification of the stomach lumen by parietal cells in the gastric lining.** The apical membrane of parietal cells contains an H^+/K^+ ATPase (a P-class pump) as well as Cl^- and K^+ channel proteins. Note the cyclic K^+ transport across the apical membrane: K^+ ions are pumped inward by the H^+/K^+ ATPase and exit via a K^+ channel. The basolateral membrane contains an anion antiporter that exchanges HCO_3^- and Cl^- ions. The combined operation of these four different transport proteins and carbonic anhydrase acidifies the stomach lumen while maintaining the neutral pH and electroneutrality of the cytosol.

Parietal Cells Acidify the Stomach Contents While Maintaining a Neutral Cytosolic pH

The mammalian stomach contains a 0.1 M solution of hydrochloric acid (HCl). This strongly acidic medium kills many ingested pathogens and denatures many ingested proteins before they are degraded by proteolytic enzymes (e.g., pepsin) that function at acidic pH. Hydrochloric acid is secreted into the stomach by specialized epithelial cells called *parietal cells* (also known as *oxyntic cells*) in the gastric lining. These cells contain a *H^+/K^+ ATPase* in their apical membrane, which faces the stomach lumen and generates a millionfold H^+ concentration gradient: pH = 1.0 in the stomach lumen versus pH = 7.0 in the cell cytosol. This transport protein is a P-class ATP-powered ion pump similar in structure and function to the plasma membrane Na^+/K^+ ATPase discussed earlier. The numerous mitochondria in parietal cells produce abundant ATP for use by the H^+/K^+ ATPase.

If parietal cells simply exported H^+ ions in exchange for K^+ ions, the loss of protons would lead to a rise in the concentration of OH^- ions in the cytosol and thus a marked increase in cytosolic pH. (Recall that $[H^+] \times [OH^-]$ always is a constant, 10^{-14} M^2.) Parietal cells avoid this rise in cytosolic pH in conjunction with acidification of the stomach lumen by using Cl^-/HCO_3^- antiporters in the basolateral membrane to export the "excess" OH^- ions from the cytosol to the blood. As noted earlier, this anion antiporter is activated at high cytosolic pH.

The overall process by which parietal cells acidify the stomach lumen is illustrated in Figure 11-30. In a reaction catalyzed by carbonic anhydrase, the "excess" cytosolic OH^- combines with CO_2 that diffuses in from the blood, forming HCO_3^-. Catalyzed by the basolateral anion antiporter, this bicarbonate ion is exported across the basolateral membrane (and ultimately into the blood) in exchange for a Cl^- ion. The Cl^- ions then exit through Cl^- channels in the apical membrane, entering the stomach lumen. To preserve electroneutrality, each Cl^- ion that moves into the stomach lumen across the apical membrane is accompanied by a K^+ ion that moves outward through a separate K^+ channel. In this way, the excess K^+ ions pumped inward by the H^+/K^+ ATPase are returned to the stomach lumen, thus maintaining the normal intracellular K^+ concentration. The net result is secretion of equal amounts of H^+ and Cl^- ions (i.e., HCl) into the stomach lumen, while the pH of the cytosol remains neutral and the excess OH^- ions, as HCO_3^-, are transported into the blood.

KEY CONCEPTS OF SECTION 11.6

Transepithelial Transport

■ The apical and basolateral plasma membrane domains of epithelial cells contain different transport proteins and carry out quite different transport processes.

■ In the intestinal epithelial cell, the coordinated operation of Na^+-linked symporters in the apical membrane with Na^+/K^+ ATPases and uniporters in the basolateral membrane mediates transcellular transport of amino acids and glucose from the intestinal lumen to the blood (see Figure 11-29).

■ The combined action of carbonic anhydrase and four different transport proteins permits parietal cells in the stomach lining to secrete HCl into the lumen while maintaining their cytosolic pH near neutrality (see Figure 11-30).

Perspectives for the Future

In this chapter, we have explained the action of specific membrane transport proteins and their impact on certain aspects of human physiology; such a molecular physiology approach has many medical applications. Even today, specific inhibitors or activators of channels, pumps, and transporters constitute the largest single class of drugs. For instance, an inhibitor of the gastric H^+/K^+ ATPase that acidifies the stomach is the most widely used drug for treating stomach ulcers and gastric reflux syndrome. Inhibitors of channel proteins in the kidney are widely used to control hypertension (high blood pressure); by blocking resorption of water from forming urine into the blood, these drugs reduce blood volume and thus blood pressure. Calcium-channel blockers are widely employed to control the intensity of contraction of the heart. Drugs that inhibit a particular potassium channel in β islet cells enhance secretion of insulin (see Figure 15-32) and are widely used to treat adult-onset (type II) diabetes.

With the completion of the human genome project, we are positioned to learn the sequences of all human membrane transport proteins. Already we know that mutations in many of them cause disease—cystic fibrosis, due to mutations in CFTR, is one example. This exploding basic knowledge will enable researchers to identify new types of compounds that inhibit or activate just one of these membrane transport proteins and not its homologs. An important challenge, however, is to understand the role of an individual transport protein in each of the several tissues in which it is expressed.

Another major challenge is to understand how each channel, transporter, and pump is regulated to meet the needs of the cell. Like other cellular proteins, many of these proteins undergo reversible phosphorylation, ubiquitination, and other covalent modifications that affect their activity, but in the vast majority of cases, we do not understand how this regulation affects cellular function. Many channels, transporters, and pumps normally reside on intracellular membranes, not on the plasma membrane, and move to the plasma membrane only when a particular hormone is present. The addition of insulin to muscle, for instance, causes the GLUT4 glucose transporter to move from intracellular membranes to the plasma membrane, increasing the rate of glucose uptake. We noted earlier that the addition of vasopressin to certain kidney cells similarly causes an aquaporin to traffic to the plasma membrane, increasing the rate of water transport. But despite much research, the underlying cellular mechanisms by which hormones stimulate the movement of transport proteins to and from the plasma membrane remain obscure.

Key Terms

ABC superfamily *448*
active transport *440*
antiport *466*
aquaporins *444*
cotransport *465*
electrochemical gradient *439*
facilitated transport *440*
F-class pumps *448*
flippase model *456*
gated channel *440*
GLUT proteins *443*
hypertonic *444*
hypotonic *444*
isotonic *444*
membrane potential *439*
microvilli *471*
Na^+/K^+ ATPase *452*
patch clamping *463*
P-class pumps 448
resting K^+ channels *460*
simple diffusion *439*
symport *466*
tight junctions *471*
transcellular transport *471*
uniport *441*
V-class pumps *448*

Review the Concepts

1. The basic structural unit of a biomembrane is the phospholipid bilayer. Acetic acid and ethanol are each composed of two carbons, hydrogen and oxygen, and both enter cells by passive diffusion. At pH 7, one is much more membrane permeable than the other. Which is more permeable and why? Predict how the permeability of each is altered when the pH is reduced to 1.0, a value typical of the stomach.

2. Uniporters and ion channels support facilitated diffusion across biomembranes. Although both are examples of facilitated diffusion, the rates of ion movement via an ion channel are roughly 10^4- to 10^5-fold faster than that of molecules via a uniporter. What key mechanistic difference results in this large difference in transport rate?

3. Name the three classes of transporters. Explain which of these classes is able to move glucose or bicarbonate (HCO_3^-), for example, against an electrochemical gradient. In the case of bicarbonate, but not glucose, the ΔG of the transport process has two terms. What are these two terms, and why does the second not apply to glucose? Why are cotransporters often referred to as examples of secondary active transport?

4. GLUT1, found in the plasma membrane of erythrocytes, is a classic example of a uniporter. Design a set of experiments to prove that GLUT1 is indeed a glucose-specific uniporter rather than a galactose- or mannose-specific uniporter. Glucose is a 6-carbon sugar while ribose is a 5-carbon sugar. Despite this smaller size, ribose is not efficiently transported by GLUT1. How can this be explained?

5. Name the four classes of ATP-powered pumps that produce active transport of ions and molecules. Indicate which of these classes transport ions only and which transport primarily small molecules. The initial discovery of one class of these ATP-powered pumps came from studying not the transport of a natural substrate but rather artificial substrates used as cancer chemotherapy drugs. What do investigators now think are common examples of the natural substrates of this particular class of ATP-powered pumps?

6. Genome sequencing projects continue, and the complete genome sequences for an increasing number of organisms are known. How does this information allow us to state the total number of transporters or pumps of a given type in either mice or humans? Many of the sequence-identified transporters or pumps are "orphan" proteins, in the sense that their natural substrate or physiological role is not known. How can this be, and how might one establish the physiological role of an orphan protein?

7. As we saw in the section Perspectives for the Future, specific inhibitors or activators of channels, pumps, and transporters constitute the largest single class of drugs produced by the pharmaceutical industry. Skeletal muscle contraction is caused by elevation of Ca^{2+} concentration in the cytosol. What is the expected effect on muscle contraction of selective drug inhibition of sarcoplasmic reticulum (SR) P-class Ca^{2+} ATPase?

8. The membrane potential in animal cells, but not in plants, depends largely on resting K^+ channels. How do these channels contribute to the resting potential? Why are these channels considered to be nongated channels? How do these channels achieve selectivity for K^+ versus Na^+?

9. Patch clamping can be used to measure the conductance properties of individual ion channels. Describe how patch clamping can be used to determine whether or not the gene coding for a putative K^+ channel actually codes for a K^+ or Na^+ channel.

10. Plants use the proton electrochemical gradient across the vacuole membrane to power the accumulation of salts and sugars in the organelle. This creates a hypertonic situation. Why does this not result in the plant cell bursting? How does the plasma membrane Na^+/K^+ ATPase allow animal cells to avoid osmotic lysis even under isotonic conditions?

11. In the case of the bacterial sodium-leucine transporter, what is the key distinguishing feature about the bound sodium ions that ensures that other ions, particularly K^+, do not bind? Describe the symport process by which cells lining the small intestine import glucose. What ion is responsible for the transport, and what two particular features facilitate the energetically favored movement of this ion across the plasma membrane?

12. Sequencing several genomes, including that of the zebrafish, has revealed a number of receptors and transporters. In one case, positional cloning of the zebrafish *golden* gene identified a putative cation exchange protein called SLC24A5. When the *golden* gene is mutated, the normally black horizontal stripes are pale or golden in appearance because the amount of black pigment or melanin is greatly reduced. What is the name of the melanin-containing vesicles present in fish and humans? Design an experiment to identify where the protein encoded by the *golden* gene is expressed and localized in the zebrafish. What is the term used when a mutant gene and its phenotype, for instance *golden* in zebrafish, is rescued by expressing an orthologous gene from another animal?

13. Movement of glucose from one side to the other side of the intestinal epithelium is a major example of transcellular transport. How does the Na^+/K^+ ATPase power the process? Why are tight junctions essential for the process? Rehydration supplements such as sport drinks include a sugar and a salt. Why are both important to rehydration?

Analyze the Data

Imagine that you are investigating the transepithelial transport of radioactive glucose. Intestinal epithelial cells are grown in culture to form a complete sheet so that the fluid bathing the apical domain of the cells (the apical medium) is completely separated from the fluid bathing the basolateral domain of the cells (the basolateral medium). Radioactive (^{14}C-labeled) glucose is added to the apical medium, and the appearance of radioactivity in the basolateral medium is monitored in terms of counts per milliliter (cpm/ml), a measure of radioactivity per unit volume.

Treatment 1: The apical and basolateral media each contain 150 mM Na^+ (curve 1).

Treatment 2: The apical medium contains 1 mM Na^+, and the basolateral medium contains 150 mM Na^+ (curve 2).

Treatment 3: The apical medium contains 150 mM Na^+, and the basolateral medium contains 1 mM Na^+ (curve 3).

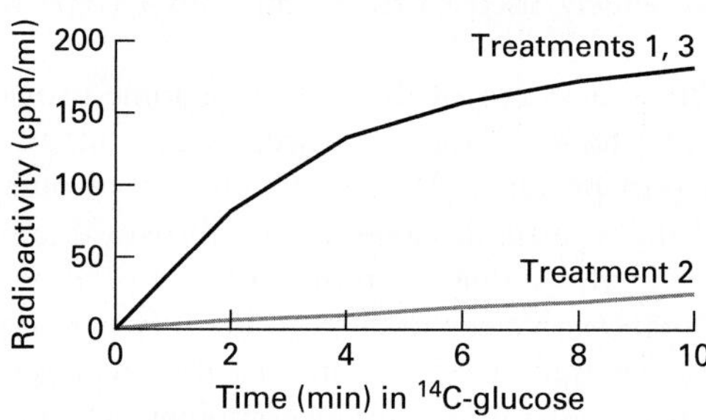

a. What is a likely explanation for the different results obtained in treatments 1 and 3 versus treatment 2?

In additional studies, the drug ouabain, which inhibits Na^+/K^+ ATPases, is included as noted.

Treatment 4: The apical and basolateral media contain 150 mM Na^+ and the apical media contains ouabain (curve 4)

Treatment 5: The apical and basolateral media contain 150 mM Na^+ and the basolateral media contains ouabain (curve 5)

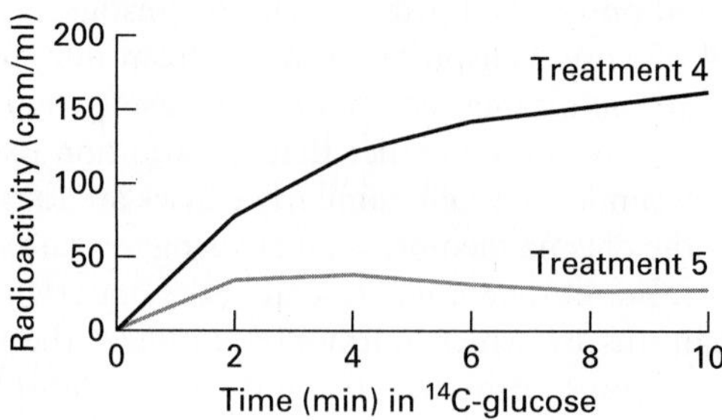

b. What is a likely explanation for the different results obtained in treatment 4 versus treatment 5?

c. A population of epithelial cells used in the above studies has been engineered to express GLUT1 rather than GLUT2 in their basolateral membrane. These engineered cells appear to be much less robust than the parental cells and do not survive long in culture. What is a reasonable explanation for this finding?

References

Uniport Transport of Glucose and Water

Engel, A., Y. Fujiyoshi, and P. Agre. 2000. The importance of aquaporin water channel protein structures. *EMBO J.* **19**:800–806.

Hedfalk, K., et al. 2006. Aquaporin gating. *Curr. Opinion Structural Biology* **16**:1–10.

Hruz, P. W., and M. M. Mueckler. 2001. Structural analysis of the GLUT1 facilitative glucose transporter (review). *Mol. Memb. Biol.* **18**:183–193.

King, L. S., D. Kozono, and P. Agre. 2004. From structure to disease: the evolving tale of aquaporin biology. *Nat. Rev. Mol. Cell Biol.* **5**:687–698.

Maurel, C. 1997. Aquaporins and water permeability of plant membranes. *Ann. Rev. Plant Physiol. Plant Mol. Biol.* **48**:399–430.

Mueckler, M. 1994. Facilitative glucose transporters. *Eur. J. Biochem.* **219**:713–725.

Schafer, J. A. 2004. Renal water reabsorption: a physiologic retrospective in a molecular era. *Kidney Int. Suppl.* **91**:S20-27.

Schultz, S. G. 2001. Epithelial water absorption: osmosis or cotransport? *Proc. Nat'l. Acad. Sci. USA* **98**:3628–3630.

Wang, Y., K. Schulten, and E. Tajkhorshid. 2005. What makes an aquaporin a glycerol channel? A comparative study of AqpZ and GlpF structure. *Structure* **13**:1107–1118.

Verkman, A. S. 2005. Novel roles of aquaporins revealed by phenotypic analysis of knockout mice. *Rev. Physiol. Biochem. Pharmacol.* **155**:31–55.

ATP-Powered Pumps and the Intracellular Ionic Environment

Borst, P., N. Zelcer, and A. van Helvoort. 2000. ABC transporters in lipid transport. *Biochim. Biophys. Acta* **1486**:128–144.

Guerini, D., L. Coletto, and E. Carafoli. 2005. Exporting calcium from cells. *Cell Calcium* **38**:281–289.

Davidson, A. L., and J. Chen. 2004. ATP-binding cassette transporters in bacteria. *Ann. Rev. Biochem.* 73:241–268.

Davies, J., F. Chen, and Y. Ioannou. 2000. Transmembrane molecular pump activity of Niemann-Pick C1 protein. *Science* **290**:2295–2298.

Gottesman, M. M. 2002. Mechanisms of cancer drug resistance. *Ann. Rev. Med.* **53**:615–627.

Gottesman, M. M., and V. Ling. 2006. The molecular basis of multidrug resistance in cancer: the early years of P-glycoprotein research. *FEBS Lett.* **580**:998–1009.

Inoue, T., S. Wilkens, and M. Forgac. 2003. Subunit structure, function, and arrangement in the yeast and coated vesicle V-ATPases. *J. Bioenerg. Biomembr.* **35**:291–299.

Jencks, W. P. 1995. The mechanism of coupling chemical and physical reactions by the calcium ATPase of sarcoplasmic reticulum and other coupled vectorial systems. *Biosci. Rept.* **15**:283–287.

Kühlbrandt, W. 2004. Biology, structure and mechanism of p-type ATPases. *Nature Rev. Mol. Cell Biol.* **5**:282–295.

K. P. Locher, A. Lee, and D. C. Rees. 2002. The *E. coli* BtuCD structure: a framework for ABC transporter architecture and mechanism. *Science* **296**:1091.

K. Obara, et al. 2005. Structural role of countertransport revealed in Ca^{2+} pump crystal structure in the absence of Ca^{2+} PNAS. **102**:14489–14496.

Ostedgaard, L. S., O. Baldursson, and M. J. Welsh. 2001. Regulation of the cystic fibrosis transmembrane conductance regulator Cl^- channel by its R domain. *J. Biol. Chem.* **276**:7689–7692.

Raggers, R. J., et al. 2000. Lipid traffic: the ABC of transbilayer movement. *Traffic* **1**:226–234.

Rea, P. A., et al. 1992. Vacuolar H^+-translocating pyrophosphatases: a new category of ion translocase. *Trends Biochem. Sci.* **17**:348–353.

Riordan, J. 2005. Assembly of functional CFTR chloride channels. *Ann. Rev. Physiol.* **67**:701–718.

Toyoshima, C., and G. Inesi. 2004. Structural basis of ion pumping by Ca^2-ATPase of the sarcoplasmic reticulum. *Ann. Rev. Biochem.* **73**:269.

Verkman, A. S., G. L. Lukacs, and L. J. Galietta. 2006. CFTR chloride channel drug discovery—inhibitors as antidiarrheals and activators for therapy of cystic fibrosis. *Curr. Pharm. Des.* **12**:2235-2247.

Nongated Ion Channels and the Resting Membrane Potential

Clapham, D. 1999. Unlocking family secrets: K^+ channel transmembrane domains. *Cell* **97**:547–550.

Cooper, E. C., and L. Y. Jan. 1999. Ion channel genes and human neurological disease: recent progress, prospects, and challenges. *Proc. Nat'l. Acad. Sci. USA* **96**:4759–4766.

Dutzler, R., et al. 2002. X-ray structure of a ClC chloride channel at 3.0 Å reveals the molecular basis of anion selectivity. *Nature* **415**:287–294.

Gouaux, E., and R. Mackinnon. 2005. Principles of selective ion transport in channels and pumps. *Science* **310**:1461–1465.

Hille, B. 2001. *Ion Channels of Excitable Membranes,* 3d ed. Sinauer Associates.

Jentsch, T., M. Poët, J. Fuhrmann, and A. Zdebik 2005. Physiological functions of CLC Cl^- channels gleaned from human genetic disease and mouse models. *Ann. Rev. Physiol.* **67**:779–807.

Jiang, Y., et al. 2003. X-ray structure of a voltage-dependent K^+ channel. *Nature* **423**:33–41.

MacKinnon, R. 2004. Potassium channels and the atomic basis of selective ion conduction. Nobel Lecture reprinted in *Biosci. Rep.* **24**:75–100.

Montell, C., L. Birnbaumer, and V. Flockerzi. 2002. The TRP channels, a remarkably functional family. *Cell* **108**:595–598.

Neher, E. 1992. Ion channels for communication between and within cells. Nobel Lecture reprinted in *Neuron* **8**:605–612 and *Science* **256**:498–502.

Neher, E., and B. Sakmann. 1992. The patch clamp technique. *Sci. Am.* **266**(3):28–35.

Roux, B. 2005. Ion conduction and selectivity in K^+ channels. 2005. *Ann. Rev. Biophys. Biomol. Struct.* **34**:153–171.

Zhou, Y., J. Morais-Cabral, A. Kaufman, and R. MacKinnon. 2001. Chemistry of ion coordination and hydration revealed by a K^+ channel–Fab complex at 2 Å resolution. *Nature* **414**:43–48.

Cotransport by Symporters and Antiporters

Alper, S. L., M. N. Chernova, and A. K. Stewart. 2001. Regulation of Na^+-independent Cl^-/HCO_3^- exchangers by pH. *J. Pancreas* **2**:171–175.

Barkla, B., and O. Pantoja. 1996. Physiology of ion transport across the tonoplast of higher plants. *Ann. Rev. Plant Physiol. Plant Mol. Biol.* **47**:159–184.

Orlowski, J., and S. Grinstein. 2004. Diversity of the mammalian sodium/proton exchanger SLC9 gene family. *Pflugers Arch.* **447**:549–565.

Wakabayashi, S., M. Shigekawa, and J. Pouyssegur. 1997. Molecular physiology of vertebrate Na^+/H^+ exchangers. *Physiol. Rev.* **77**:51–74.

Wright, E. M. 2001. Renal Na($^+$)-glucose cotransporters. *Am. J. Physiol. Renal Physiol.* **280**:F10–F18.

Wright, E. M., and D. D. Loo. 2000. Coupling between Na^+, sugar, and water transport across the intestine. *Ann. NY Acad. Sci.* **915**:54–66.

A. Yamashita, et al. 2005. Crystal structure of a bacterial homologue of Na^+/Cl^--dependent neurotransmitter transporters. *Nature* **437**:215.

Zhang, H-X., and E. Blumwald. 2001. Transgenic salt-tolerant tomato plants accumulate salt in foliage but not in fruit. *Nature Biotech.* **19**:765–769.

Transepithelial Transport

Furuse, M., and S. Tsukita. 2006. Claudins in occluding junctions of humans and flies. *Trends Cell Biol.* **16**:181–188.

Goodenough, D. A. 1999. Plugging the leaks. *Proc. Nat'l. Acad. Sci. USA* **96**:319.

Mitic, L., and J. Anderson. 1998. Molecular architecture of tight junctions. *Ann. Rev. Physiol.* **60**:121–142.

Rao, M. 2004. Oral rehydration therapy: new explanations for an old remedy. *Ann. Rev. Physiol.* **66**:385–417.

CLASSIC EXPERIMENT 11.1

STUMBLING UPON ACTIVE TRANSPORT

J. Skou, 1957, *Biochem. Biophys. Acta* **23**:394

In the mid-1950s Jens Skou was a young physician researching the effects of local anesthetics on isolated lipid bilayers. He needed an easily assayed membrane-associated enzyme to use as a marker in his studies. What he discovered was an enzyme critical to the maintenance of membrane potential, the Na^+/K^+ ATPase, a molecular pump that catalyzes active transport.

Background

During the 1950s many researchers around the world were actively investigating the physiology of the cell membrane, which plays a role in a number of biological processes. It was well known that the concentration of many ions differs inside and outside the cell. For example, the cell maintains a lower intracellular sodium (Na^+) concentration and higher intracellular potassium (K^+) concentration than is found outside the cell. Somehow the membrane can regulate intracellular salt concentrations. Additionally, movement of ions across cell membranes had been observed, suggesting that some sort of transport is system is present. To maintain normal intracellular Na^+ and K^+ concentrations, the transport system could not rely on passive diffusion because both ions must move across the membrane against their concentration gradients. This energy-requiring process was termed active transport.

At the time of Skou's experiments, the mechanism of active transport was still unclear. Surprisingly, Skou had no intention of helping to clarify the field. He found the Na^+/K^+ ATPase completely by accident in his search for an abundant, easily measured enzyme activity associated with lipid membranes. A recent study had shown that membranes derived from squid axons contained a membrane-associated enzyme that could hydrolyze ATP. Thinking that this would be an ideal enzyme for his purposes, Skou set out to isolate such an ATPase from a more readily available source, crab leg neurons. It was during his characterization of this enzyme that he discovered the protein's function.

The Experiment

Since the original goal of his study was to characterize the ATPase for use in subsequent studies, Skou wanted to know under what experimental condition its activity was both robust and reproducible. As often is the case with the characterization of a new enzyme, this requires careful titration of the various components of the reaction. Before this can be done, one must be sure the system is free from outside sources of contamination.

In order to study the influence of various cations, including three that are critical for the reaction—Na^+, K^+, and Mg^{2+}—Skou had to make sure that no contaminating ions were brought into the reaction from another source. Therefore all buffers used in the purification of the enzyme were prepared from salts that did not contain these cations. An additional source of contaminating cations was the ATP substrate, which contains three phosphate groups, giving it an overall negative charge. Because stock solutions of ATP often included a cation to balance the charge, Skou converted the ATP used in his reactions to the acid form so that balancing cations would not affect the experiments. Once he had a well-controlled environment, he could characterize the enzyme activity. These precautions were fundamental to his discovery.

Skou first showed that his enzyme could indeed catalyze the cleavage of ATP into ADP and inorganic phosphate. He then moved on to look for the optimal conditions for this activity by varying the pH of the reaction, and the concentrations of salts and other cofactors, which bring cations into the reaction. He could easily determine a pH optimum as well as an optimal concentration of Mg^{2+}, but optimizing Na^+ and K^+ proved to be more difficult. Regardless of the amount of K^+ added to the reaction, the enzyme was inactive without Na^+. Similarly, without K^+, Skou observed only a low-level ATPase activity that did not increase with increasing amounts of Na^+.

These results suggested that the enzyme required both Na^+ and K^+ for optimal activity. To demonstrate that this was the case, Skou performed a series of experiments in which he measured the enzyme activity as he varied both the Na^+ and K^+ concentrations in the reaction (Figure 1). Although both cations clearly were required for significant activity, something interesting occurred at high concentrations of each cation. At the optimal concentration of Na^+ and K^+, the ATPase activity reached a peak. Once at that peak, further increasing the concentration did not affect the ATPase activity. Na^+ thus behaved like a classic enzyme substrate, with increasing input leading to increased activity until a saturating concentration was achieved, at which the activity plateaued. K^+, on the other hand, behaved differently. When the K^+ concentration was increased beyond the optimum, ATPase activity declined. Thus while K^+ was required for optimal activity, at high concentrations it inhibited the enzyme. Skou reasoned that the enzyme must have

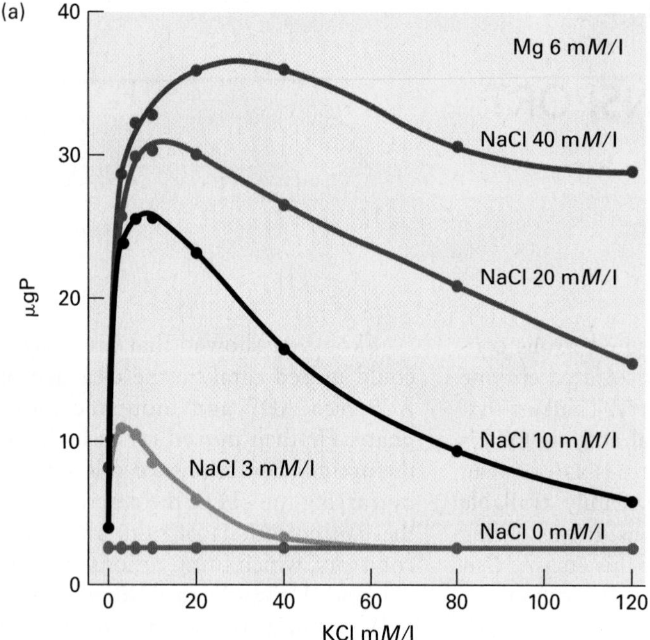

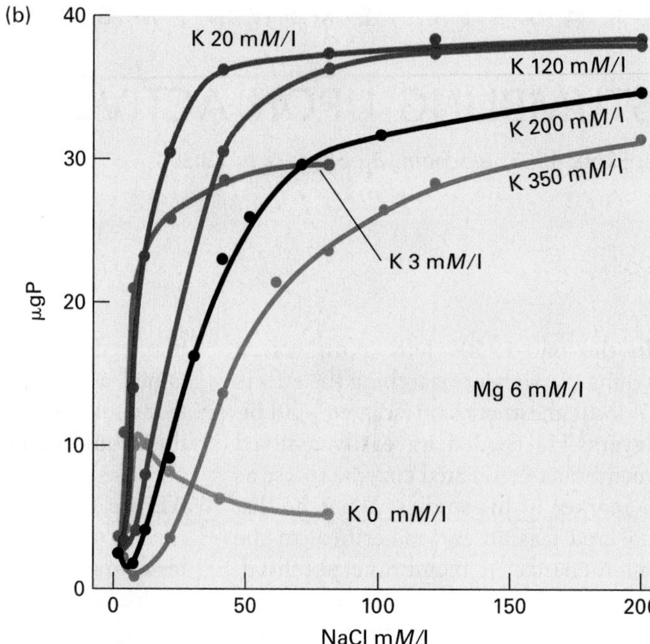

▲ **FIGURE 1 Demonstration of the dependence of the Na^+/K^+ ATPase activity on the concentration of each ion.** The graph on the left shows that increasing K^+ leads to an inhibition of the ATPase activity. The graph on the right shows that with increasing Na^+, the enzyme activity increases up to a peak and then levels out. This graph also demonstrates the dependence of the activity on low levels of K^+. [Adapted from J. Skou, 1957, *Biochem. Biophys. Acta* **23**:394.]

separate binding sites for Na^+ and K^+. For optimal ATPase activity, both must be filled. However, at high concentrations K^+ could compete for the Na^+-binding site, leading to enzyme inhibition. He hypothesized that this enzyme was involved in active transport, that is, the pumping of Na^+ out of the cell, coupled to the import of K^+ into the cell. Later studies would prove that this enzyme was indeed the pump that catalyzed active transport. This finding was so exciting that Skou devoted his subsequent research to studying the enzyme, never using it as a marker, as he initially intended.

Discussion

Skou's finding that a membrane ATPase used both Na^+ and K^+ as substrates was the first step in understanding active transport on a molecular level. How did Skou know to test both Na^+ and K^+? In his Nobel lecture in 1997, he explained that in his first attempts at characterizing the ATPase, he took no precautions to avoid the use of buffers and ATP stock solutions that contained Na^+ and K^+. Pondering the puzzling and unreproducible results that he obtained led to the realization that contaminating salts might be influencing the reaction. When he repeated the experiments, this time avoiding contamination by Na and K at all stages, he obtained clear-cut, reproducible results.

The discovery of the Na^+/K^+ ATPase had an enormous impact on membrane biology, leading to a better understanding of the membrane potential. The generation and disruption of membrane potential forms the basis of many biological processes, including neurotransmission and the coupling of chemical and electrical energy. For this fundamental discovery, Skou was awarded the Nobel Prize for Chemistry in 1997.

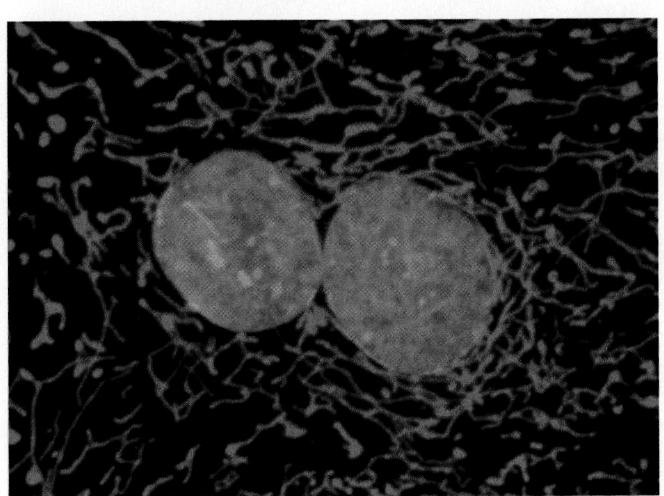

Immunofluorescence micrograph showing the intertwined network of mitochondria (red) in a cell from the ovary of the Himalayan Tahr mountain goat. The unusual twin nuclei in this cell are stained blue. [Courtesy of M. Davidson]

CHAPTER

12

CELLULAR ENERGETICS

From the growth and division of a cell to the beating of a heart to the electrical activity of a neuron that underlies thinking, life requires energy. Cells are complex systems in which a multitude of chemical reactions and transport processes are coordinately regulated in time and space. Cells cannot generate and maintain their highly organized structures and conduct extensive metabolism (e.g., carbohydrate synthesis) without material and energy from their environments. This chapter describes the molecular mechanisms by which cells use sunlight or chemical nutrients as sources of energy, with a special focus on how cells convert these external sources of energy into a biologically universal, intracellular, chemical energy carrier, **adenosine triphosphate,** or **ATP** (Figure 12-1). ATP, found in all types of organisms and presumably present in the earliest life-forms, is generated from ADP and inorganic phosphate (HPO_4^{2-}, often abbreviated as P_i). Cells use the energy released during hydrolysis of the terminal high-energy phosphoanhydride bond in ATP (see Figure 2-31) to power many otherwise energetically unfavorable processes. Examples include the synthesis of proteins from amino acids and of nucleic acids from nucleotides (Chapter 4), transport of molecules against a concentration gradient by ATP-powered pumps (Chapter 11), contraction of muscle (Chapter 17), and beating of cilia (Chapter 18).

The energy to drive ATP synthesis from ADP (ΔG = 7.3 kcal/mol) is produced primarily by two processes: **aerobic oxidation,** which occurs in **mitochondria** in nearly all eukaryotic cells (Figure 12-1, *top*), and **photosynthesis,** which occurs in **chloroplasts** only in leaf cells of plants (Figure 12-1, *bottom*) and certain single-celled organisms, such as cyanobacteria. Two additional processes, glycolysis and the citric acid cycle, are also important direct or indirect sources of ATP in both animal and plant cells.

In aerobic oxidation, breakdown products of sugars (carbohydrates) and fatty acids (hydrocarbons)—both derived in animals from the digestion of food—are converted by oxidation with O_2 to carbon dioxide and water. The energy released from this overall reaction is transformed into the chemical energy of phosphoanhydride bonds in ATP. This is analogous to burning wood (carbohydrates) or oil (hydrocarbons) to generate heat in furnaces or motion in automobile engines: both consume O_2 and generate carbon dioxide and water. The key difference is that cells break the overall reaction down into many intermediate steps. This permits the amount of energy released in any given step to

OUTLINE

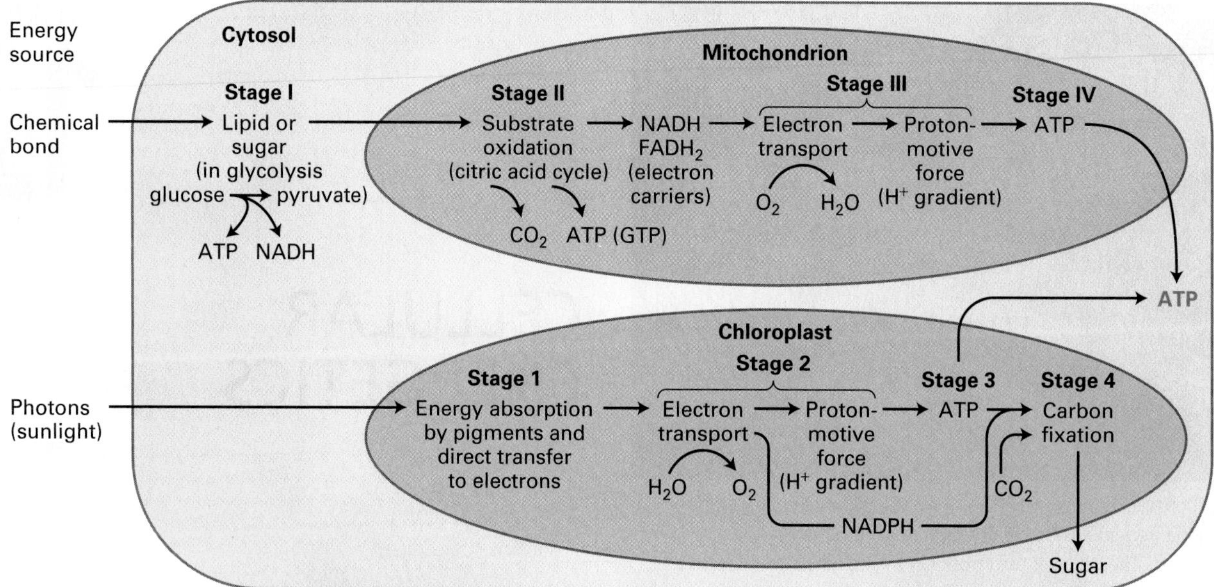

▲ **FIGURE 12-1 Overview of aerobic oxidation and photosynthesis.** Eukaryotic cells use two fundamental mechanisms to convert external sources of energy into ATP. *(Top)* In aerobic oxidation, "fuel" molecules (primarily sugars and fatty acids) undergo preliminary processing in the cytosol, e.g., breakdown of glucose to pyruvate (**stage I**), and are then transferred into mitochondria, where they are converted by oxidation with O_2 to carbon dioxide and water (**stages II** and **III**) and ATP is generated (**stage IV**). *(Bottom)* In photosynthesis, which occurs in chloroplasts, the radiant energy of light is absorbed by specialized pigments (**stage 1**); the absorbed energy is used to both oxidize water to O_2 and establish conditions (**stage 2**) necessary for the generation of ATP (**stage 3**) and carbohydrates from CO_2 (carbon fixation, **stage 4**). Both mechanisms involve the production of reduced high-energy electron carriers (NADH, NADPH, $FADH_2$) and movement of electrons down an electrical potential in an electron transport chain through specialized membranes. Energy from these electrons is released and captured as a proton-motive force (proton electrochemical gradient) that is then used to drive ATP synthesis. Bacteria utilize comparable processes.

match closely the amount of energy required for the next intermediate stage of the process. If there were not a close match, excess released energy would be lost as heat (which would be very inefficient) or not enough energy would be released to drive the next step in the process (which would be ineffective).

In photosynthesis, the radiant energy of light is absorbed by pigments such as chlorophyll and used to make ATP and carbohydrates (primarily sucrose and starch). Unlike aerobic oxidation, which uses carbohydrates and O_2 to generate CO_2, photosynthesis uses CO_2 as a substrate and generates O_2 and carbohydrates as products.

This reciprocal relationship between aerobic oxidation in mitochondria and photosynthesis in chloroplasts underlies a profound symbiotic relationship between photosynthetic and nonphotosynthetic organisms and is responsible for much of the life on earth. The oxygen generated during photosynthesis is the source of virtually all the oxygen in the air, and the carbohydrates produced are the ultimate source of energy for virtually all nonphotosynthetic organisms. (An exception is bacteria living in deep ocean vents—and the organisms that feed on them—that obtain energy for converting CO_2 into carbohydrates by oxidation of geologically generated reduced inorganic compounds released by the vents.)

At first glance, it would seem that the molecular mechanisms underlying the reciprocal processes of photosynthesis and aerobic oxidation have little in common. However, a revolutionary discovery in cell biology established that bacteria, mitochondria, and chloroplasts all use the same mechanism, known as **chemiosmosis,** to generate ATP from ADP and P_i. In chemiosmosis (also known as chemiosmotic coupling), a proton electrochemical gradient is generated across a membrane, driven by energy released as electrons travel through an **electron transport chain.** The energy stored in this gradient, called the **proton-motive force**, is used directly to power the synthesis of ATP and other energy-requiring processes (Figure 12-2). In this chapter, we explore the molecular mechanisms of the two processes that share this central mechanism, focusing first on aerobic oxidation and then on photosynthesis.

12.1 First Steps of Glucose and Fatty Acid Catabolism: Glycolysis and the Citric Acid Cycle

In an automobile engine, hydrocarbon fuel is oxidatively and explosively converted in an essentially one step process to mechanical work (i.e., driving a piston). The process is

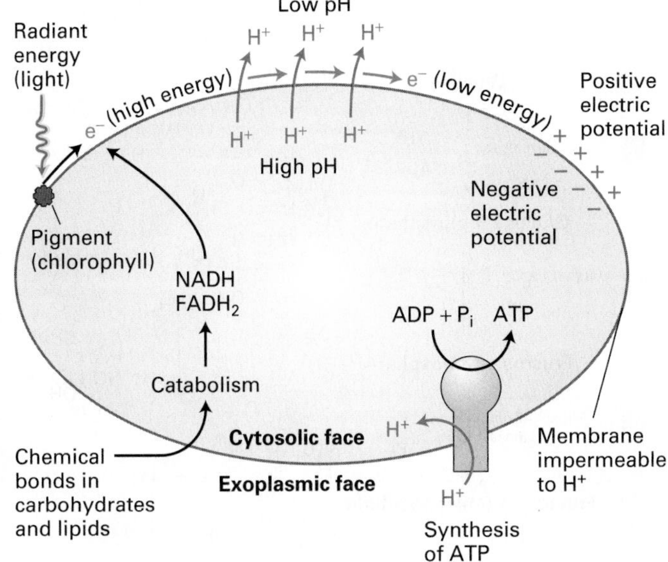

▲ **FIGURE 12-2 Proton-motive force.** Transmembrane proton concentration and electrical (voltage) gradients, collectively called the *proton-motive force,* are generated during aerobic oxidation and photosynthesis in eukaryotes and prokaryotes (bacteria). High-energy electrons generated by light absorption by pigments (e.g., chlorophyll) or held in the reduced form of electron carriers (e.g., NADH, $FADH_2$) made during the catabolism of sugars and lipids pass down an electron transport chain (blue arrows), releasing energy throughout the process. The energy is recovered by coupling its release to pumping protons across the membrane (red arrows), generating the proton-motive force. In chemiosmotic coupling, the energy released when protons flow down the gradient through ATP synthase drives the synthesis of ATP. The proton-motive force can also power transport of metabolites across the membrane against their concentration gradient and rotation of bacterial flagella.

relatively inefficient in that both substantial amounts of the chemical energy stored in the fuel are wasted as they are converted to unused heat and substantial amounts of fuel are only partially oxidized and released as carbonaceous, sometimes toxic, exhaust. In the competition to survive, organisms cannot afford to squander their sometimes limited energy sources on an equivalently inefficient process. Cells have evolved incredibly efficient mechanisms for hydrocarbon (fatty acid) and carbohydrate (sugar) combustion coupled to ATP synthesis. That mechanism is aerobic oxidation. Each stage of fuel conversion to energy comprises multiple steps that are catalyzed or mediated by specific proteins. This strategy provides the following advantages:

- By dividing the process into multiple steps that generate several energy-carrying intermediates, bond energy is efficiently channeled into the synthesis of ATP and energy lost as heat is reduced.

- Different fuels are reduced to common intermediates that can then share subsequent pathways for combustion and ATP synthesis.

- Since total energy stored in the bonds of the initial fuel molecules is substantially greater than that required to drive the synthesis of a single ATP molecule (~7.3 kcal/mole), many ATP molecules are produced.

In our discussion of aerobic oxidation, we will be tracing the fate of the two main energy-producing digestive products of food: sugars (principally glucose) and fatty acids. Under certain conditions amino acids also feed into these metabolic pathways.

The complete aerobic oxidation of each molecule of glucose yields 6 molecules of CO_2 and the energy released is coupled to the synthesis of as many as 30 molecules of ATP. The overall reaction is

$$C_6H_{12}O_6 + 6\ O_2 + 30\ Pi^{2-} + 30\ ADP^{3-} + 30\ H^+ \rightarrow 6\ CO_2 + 30\ ATP^{4-} + 36\ H_2O,\ \Delta G = 686\ \text{kcal/mol}$$

Glucose oxidation in eukaryotes takes place in four stages (see Figure 12-1):

I. Conversion in the cytosol of one 6-carbon glucose molecule to two 3-carbon pyruvate molecules (**glycolysis**)

II. Pyruvate oxidation to CO_2 in the mitochondrion via a 2-carbon acetyl CoA intermediate (**citric acid cycle**)

III. **Electron transport** to generate a proton-motive force

IV. ATP synthesis in the mitochondrion (**oxidative phosphorylation**)

In this section, we discuss stages I and II: the biochemical pathways that break down glucose and fatty acids to CO_2, generating some ATP and high-energy electrons in the process; the fate of the released electrons (stage III) is described in the next section.

During Glycolysis (Stage I), Cytosolic Enzymes Convert Glucose to Pyruvate

Glycolysis occurs in the cytosol in both eukaryotes and prokaryotes and does not require molecular oxygen; thus it is called *anaerobic* glucose **catabolism** (biological breakdown of complex to simpler substances). A set of 10 water-soluble cytosolic enzymes catalyze the reactions constituting the *glycolytic pathway* (*glyco,* "sweet"; *lysis,* "split"), in which one molecule of glucose is converted to two molecules of pyruvate (Figure 12-3). All the reaction intermediates produced by these enzymes are water soluble, phosphorylated compounds called *metabolic intermediates.* In addition to chemically converting one glucose molecule into these intermediates and the two pyruvates, these enzymatic reactions generate four ATP molecules by phosphorylation of four ADPs (reactions 7 and 10), a process called **substrate-level phosphorylation** (to distinguish it from the oxidative phosphorylation that generates ATP in the third stage of aerobic oxidation). Unlike later stages of

▶ **FIGURE 12-3 The glycolytic pathway.** Glucose is degraded to pyruvate. Two reactions consume ATP, forming ADP and phosphorylated sugars (red), two generate ATP from ADP by substrate-level phosphorylation (green), and one yields NADH by reduction of NAD^+ (yellow). Note that all the intermediates between glucose and pyruvate are phosphorylated compounds. Reactions 1, 3, and 10, with single arrows, are essentially irreversible (large negative ΔG values) under ordinary conditions in cells.

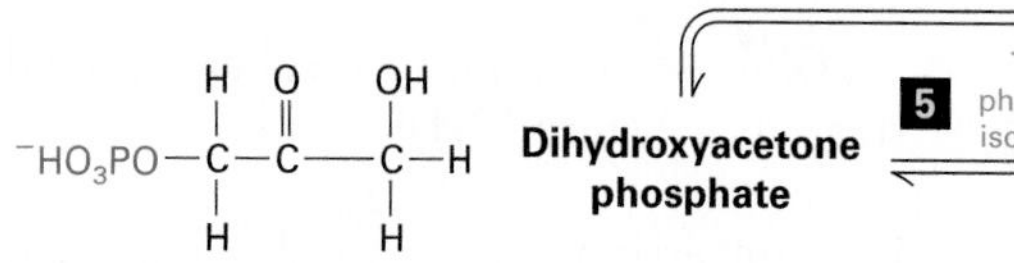

Glucose

1 Hexokinase ATP → ADP

Glucose 6-phosphate

2 Phosphoglucose isomerase

Fructose 6-phosphate

3 Phosphofructo-kinase-1 ATP → ADP

Fructose 1,6-bisphosphate

4 Aldolase

Dihydroxyacetone phosphate

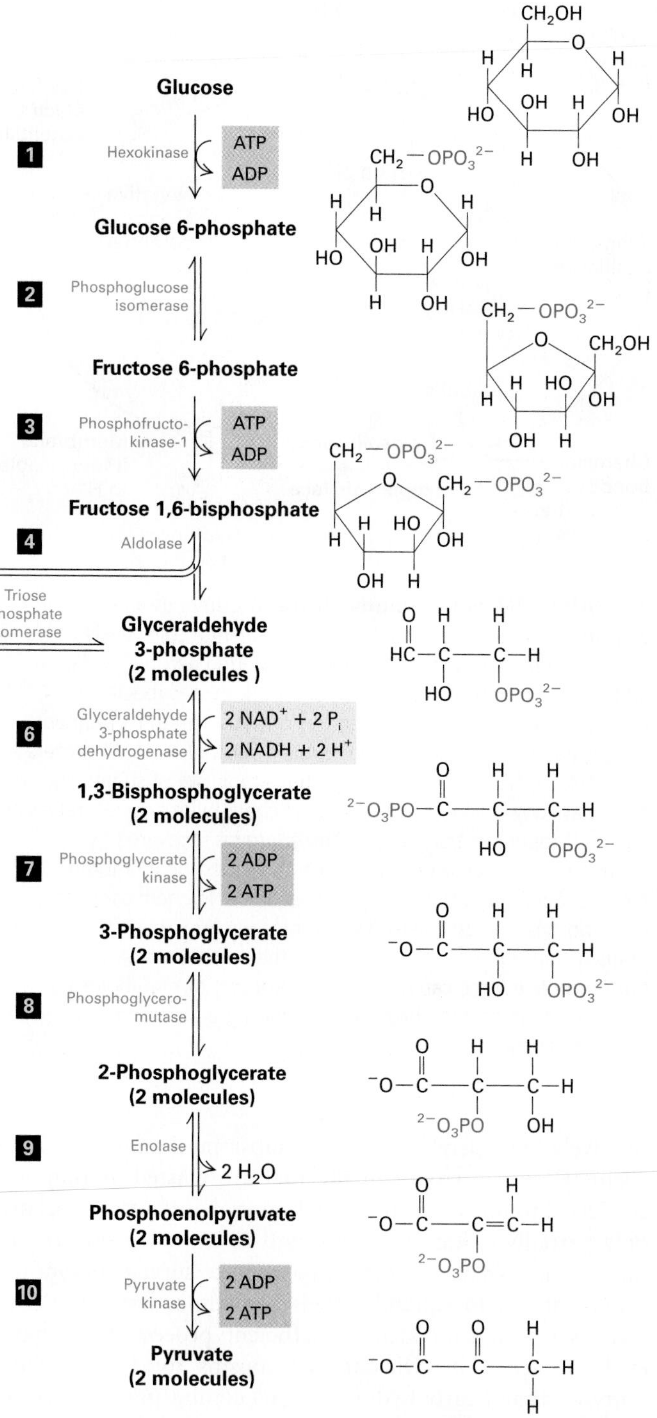

5 Triose phosphate isomerase

Glyceraldehyde 3-phosphate (2 molecules)

6 Glyceraldehyde 3-phosphate dehydrogenase $2\ NAD^+ + 2\ P_i$ → $2\ NADH + 2\ H^+$

1,3-Bisphosphoglycerate (2 molecules)

7 Phosphoglycerate kinase 2 ADP → 2 ATP

3-Phosphoglycerate (2 molecules)

8 Phosphoglyceromutase

2-Phosphoglycerate (2 molecules)

9 Enolase → $2\ H_2O$

Phosphoenolpyruvate (2 molecules)

10 Pyruvate kinase 2 ADP → 2 ATP

Pyruvate (2 molecules)

ATP formation in mitochondria and chloroplasts, a proton-motive force is not involved in substrate-level phosphorylation. However, substrate-level phosphorylation requires the addition (in reactions 1 and 3) of two phosphates from two ATPs. These can be thought of as "pump priming" reactions, which introduce a little energy up front in order to effectively recover more energy downstream. Thus glycolysis yields a net of only two ATP molecules per glucose molecule.

The balanced chemical equation for the conversion of glucose to pyruvate shows that four hydrogen atoms (four protons and four electrons) are also released:

$$\underset{\text{Glucose}}{C_6H_{12}O_6} \longrightarrow 2\ \underset{\text{Pyruvate}}{CH_3-\overset{O}{\overset{\|}{C}}-\overset{O}{\overset{\|}{C}}-OH} + 4\,H^+ + 4\,e^-$$

(For convenience, we show pyruvate here in its un-ionized form, pyruvic acid, although at physiological pH it would be largely dissociated.) All four electrons and two of the four protons are transferred (Figure 12-3, reaction 6) to two molecules of the oxidized form of **nicotinamide adenine dinucleotide (NAD^+)** to produce the reduced form, NADH (see Figure 2-33):

$$2\,H^+ + 4\,e^- + 2\,NAD^+ \rightarrow 2\,NADH$$

Later we will see that the energy carried by the electrons in NADH and an analogous carrier $FADH_2$, the reduced form of **flavin adenine dinucleotide (FAD)**, can be used to make additional ATPs via the electron transport chain.

The overall chemical equation for this first stage of glucose metabolism is

$$C_6H_{12}O_6 + 2\,NAD^+ + 2\,ADP^{3-} + 2\,Pi^{2-} \rightarrow 2\,C_3H_4O_3 + 2\,NADH + 2\,ATP^{4-}$$

After glycolysis, only a fraction of the energy available in glucose has been extracted and converted to ATP and NADH. The rest remains in the covalent bonds of the two pyruvate molecules. The ability to efficiently convert the energy in pyruvate to ATP depends on molecular oxygen. As we will see, in the presence of oxygen (**aerobic** conditions), energy conversion is highly efficient. In its absence (**anaerobic** conditions), the process is much less efficient.

The Rate of Glycolysis Is Adjusted to Meet the Cell's Need for ATP

Enzyme-catalyzed reactions and metabolic pathways are regulated by cells so as to produce the needed amounts of metabolites but not an excess. The primary function of the oxidation of glucose to CO_2 via the glycolytic pathway is to produce NADH and $FADH_2$, whose oxidation in mitochondria generates ATP. Operation of the glycolytic pathway (stage I), as well as the citric acid cycle (stage II), is continuously regulated, primarily by **allosteric** mechanisms, to meet the cell's need for ATP (see Chapter 3 for general principles of allosteric control).

Three allosterically controlled glycolytic enzymes play a key role in regulating the entire glycolytic pathway (Figure 12-3). *Hexokinase* (step **1**) is inhibited by its reaction product, glucose 6-phosphate. *Pyruvate kinase* (step **10**) is inhibited by ATP, so glycolysis slows down if too much ATP is present. The third enzyme, *phosphofructokinase-1* (step **3**), is the principal rate-limiting enzyme of the glycolytic pathway. Emblematic of its critical role in regulating the rate of glycolysis, this enzyme is allosterically controlled by several molecules (Figure 12-4).

For example, phosphofructokinase-1 is allosterically *inhibited* by ATP and allosterically *activated* by AMP. As a result, the rate of glycolysis is very sensitive to the cell's energy charge, reflected in the ATP:AMP ratio. The allosteric inhibition of phosphofructokinase-1 by ATP may seem unusual, since ATP is also a substrate of this enzyme. But the affinity of the substrate-binding site for ATP is much higher (has a lower K_m) than that of the allosteric site. Thus at low concentrations, ATP binds to the catalytic but not to the inhibitory allosteric site, and enzymatic catalysis proceeds at near maximal rates. At high concentrations, ATP also binds to the allosteric site, inducing a conformational change that reduces the affinity of the enzyme for the other substrate, fructose 6-phosphate, and thus reduces the rate of this reaction and the overall rate of glycolysis.

Another important allosteric activator of phosphofructokinase-1 is *fructose 2,6-bisphosphate.* This metabolite is formed from fructose 6-phosphate by an enzyme called *phosphofructokinase-2.* Fructose 6-phosphate accelerates the formation of fructose 2,6-bisphosphate, which in turn activates phosphofructokinase-1. This type of control is known as *feed-forward activation,* in which the abundance of a metabolite (here, fructose 6-phosphate) induces an acceleration in its subsequent metabolism. Fructose 2,6-bisphosphate allosterically activates phosphofructokinase-1 in liver cells by decreasing the inhibitory effect of high ATP and by increasing the affinity of phosphofructokinase-1 for one of its substrates, fructose 6-phosphate.

The three glycolytic enzymes that are regulated by allostery catalyze reactions with large negative $\Delta G^{\circ\prime}$ values—reactions that are essentially irreversible under ordinary conditions. These enzymes thus are particularly suitable for regulating the entire glycolytic pathway. Additional control is exerted by glyceraldehyde 3-phosphate dehydrogenase, which catalyzes the reduction of NAD^+ to NADH (see Figure 12-3, step **6**). If cytosolic NADH builds up owing to a slowdown in mitochondrial oxidation, this reaction becomes thermodynamically less favorable.

Glucose metabolism is controlled differently in various mammalian tissues to meet the metabolic needs of the organism as a whole. During periods of carbohydrate starvation, for instance, it is necessary for the liver to release glucose into the bloodstream. To do this, the liver converts

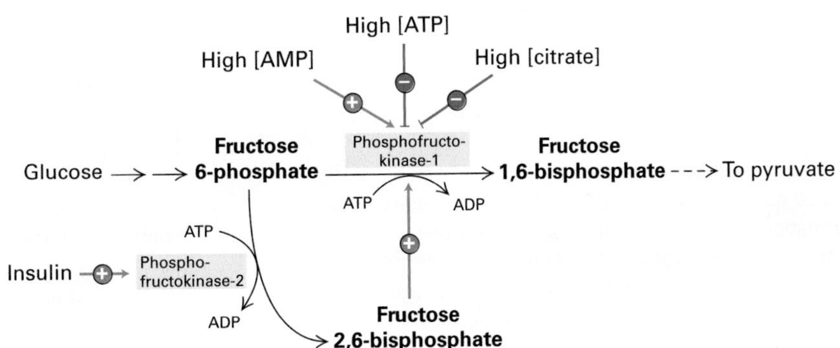

▲ **FIGURE 12-4 Allosteric regulation of glucose metabolism.** The key regulatory enzyme in glycolysis, phosphofructokinase-1, is allosterically activated by AMP and fructose 2,6-bisphosphate, which are elevated when the cell's energy stores are low. The enzyme is inhibited by ATP (when energy stores are high) and citrate, both of which are elevated when the cell is actively oxidizing glucose to CO_2. Later we will see that citrate is generated during stage II of glucose oxidation. Phosphofructokinase-2 (PFK2) is a bifunctional enzyme: its kinase activity forms fructose 2,6-bisphosphate from fructose 6-phosphate, and its phosphatase activity catalyzes the reverse reaction. Insulin, which is released by the pancreas when blood glucose levels are high, promotes PFK2 kinase activity and thus stimulates glycolysis. At low blood glucose, glucagon is released by the pancreas and promotes PFK2 phosphatase activity in the liver, indirectly slowing down glycolysis.

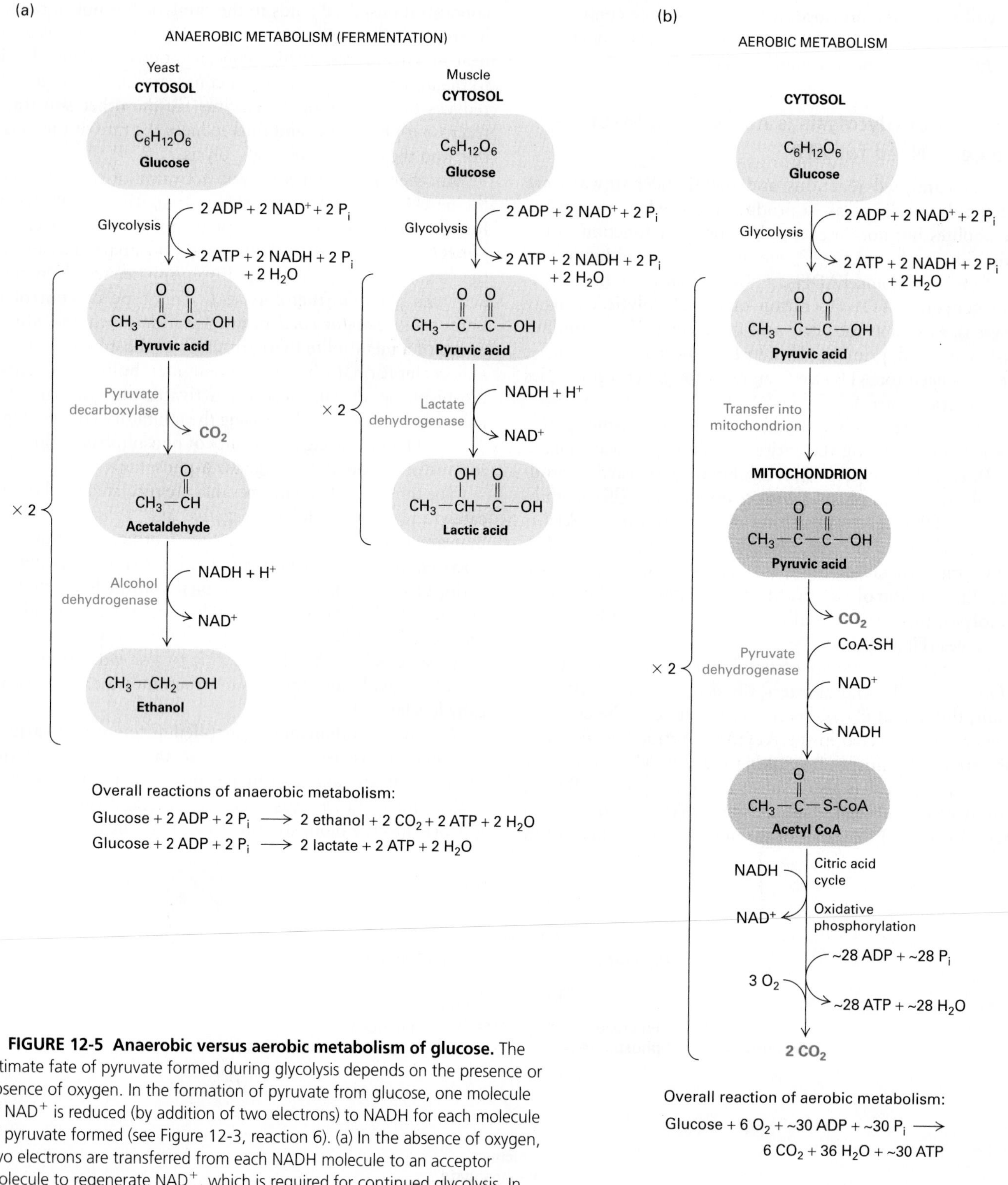

(a)

ANAEROBIC METABOLISM (FERMENTATION)

Yeast
CYTOSOL

$C_6H_{12}O_6$
Glucose

Glycolysis
2 ADP + 2 NAD$^+$ + 2 P$_i$
2 ATP + 2 NADH + 2 P$_i$ + 2 H_2O

$CH_3-C(=O)-C(=O)-OH$
Pyruvic acid

Pyruvate decarboxylase
CO_2

$CH_3-CH(=O)$
Acetaldehyde

Alcohol dehydrogenase
NADH + H$^+$
NAD$^+$

CH_3-CH_2-OH
Ethanol

× 2

Muscle
CYTOSOL

$C_6H_{12}O_6$
Glucose

Glycolysis
2 ADP + 2 NAD$^+$ + 2 P$_i$
2 ATP + 2 NADH + 2 P$_i$ + 2 H_2O

$CH_3-C(=O)-C(=O)-OH$
Pyruvic acid

Lactate dehydrogenase
NADH + H$^+$
NAD$^+$

$CH_3-CH(OH)-C(=O)-OH$
Lactic acid

× 2

Overall reactions of anaerobic metabolism:

$$\text{Glucose} + 2\ \text{ADP} + 2\ \text{P}_i \longrightarrow 2\ \text{ethanol} + 2\ CO_2 + 2\ \text{ATP} + 2\ H_2O$$

$$\text{Glucose} + 2\ \text{ADP} + 2\ \text{P}_i \longrightarrow 2\ \text{lactate} + 2\ \text{ATP} + 2\ H_2O$$

(b)

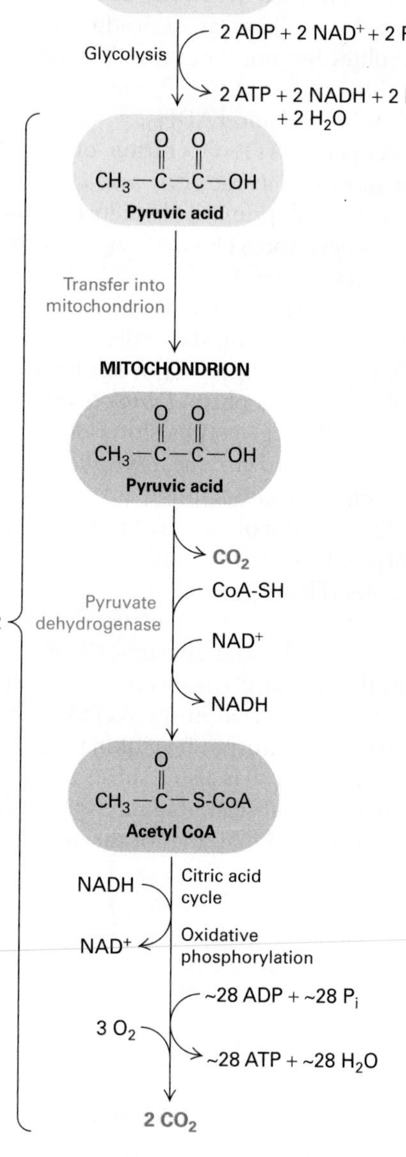

Overall reaction of aerobic metabolism:

$$\text{Glucose} + 6\ O_2 + {\sim}30\ \text{ADP} + {\sim}30\ \text{P}_i \longrightarrow 6\ CO_2 + 36\ H_2O + {\sim}30\ \text{ATP}$$

▲ **FIGURE 12-5 Anaerobic versus aerobic metabolism of glucose.** The ultimate fate of pyruvate formed during glycolysis depends on the presence or absence of oxygen. In the formation of pyruvate from glucose, one molecule of NAD$^+$ is reduced (by addition of two electrons) to NADH for each molecule of pyruvate formed (see Figure 12-3, reaction 6). (a) In the absence of oxygen, two electrons are transferred from each NADH molecule to an acceptor molecule to regenerate NAD$^+$, which is required for continued glycolysis. In yeasts (*left*), acetaldehyde is the electron acceptor and ethanol is the product. This process is called *alcoholic fermentation.* When oxygen is limiting in muscle cells (*right*), NADH reduces pyruvate to form lactic acid, regenerating NAD$^+$. (b) In the presence of oxygen, pyruvate is transported into mitochondria. First it is converted by pyruvate dehydrogenase into one molecule of CO_2 and one of acetic acid, the latter linked to coenzyme A (CoA-SH) to form acetyl CoA, concomitant with reduction of one molecule of NAD$^+$ to NADH. Further metabolism of acetyl CoA and NADH generates approximately an additional 28 molecules of ATP per glucose molecule oxidized.

the polymer glycogen, a storage form of glucose (Chapter 2), directly to glucose 6-phosphate (without involvement of hexokinase, step **1**). Under these conditions, there is a reduction in fructose 2,6-bisphosphate levels and decreased phosphofructokinase-1 activity (Figure 12-4). As a result, glucose 6-phosphate derived from glycogen is not metabolized to pyruvate; rather, it is converted to glucose by a phosphatase and released into the blood to nourish the brain and red blood cells, which depend primarily on glucose as an energy fuel. In all cases, the activity of these regulated enzymes is controlled by the level of small-molecule metabolites, generally by allosteric interactions, or by hormone-mediated phosphorylation and dephosphorylation reactions (Chapter 15 gives a more detailed discussion of hormonal control of glucose metabolism in liver and muscle).

Glucose Is Fermented Under Anaerobic Conditions

Many eukaryotes are *obligate aerobes:* they grow only in the presence of molecular oxygen and metabolize glucose (or related sugars) completely to CO_2, with the concomitant production of a large amount of ATP. Most eukaryotes, however, can generate some ATP by anaerobic metabolism. A few eukaryotes are *facultative anaerobes:* they grow in either the presence or the absence of oxygen. For example, annelids, mollusks, and some yeasts can live and grow for days without oxygen.

In the absence of oxygen, yeasts convert the pyruvate produced by glycolysis to one molecule each of ethanol and CO_2; in these reactions two NADH molecules are oxidized to NAD^+ for each two pyruvates converted to ethanol, thereby regenerating the supply of NAD^+ (Figure 12-5a, *left*). This anaerobic degradation of glucose, called *fermentation,* is the basis of beer and wine production.

Oxygen deprivation can also affect glucose metabolism in animals. During prolonged contraction of mammalian skeletal muscle cells—for example, during exercise—oxygen within the muscle tissue can become limited and glucose catabolism is limited to glycolysis (stage I). As a consequence, muscle cells convert the pyruvate from glycolysis to two molecules of lactic acid by a reduction reaction that also oxidizes two NADHs to NAD^+s (Figure 12-5a, right). Although the lactic acid is released from the muscle into the blood, if the contractions are sufficiently rapid and strong, the lactic acid can transiently accumulate in that tissue and contribute to muscle and joint pain during exercise. Once it is secreted into the blood, some of the lactic acid passes into the liver, where it is reoxidized to pyruvate and either further metabolized to CO_2 aerobically or converted back to glucose. Much lactate is metabolized to CO_2 by the heart, which is highly perfused by blood and can continue aerobic metabolism at times when exercising, oxygen-poor skeletal muscles secrete lactate. Lactic acid bacteria (the organisms that spoil milk) and other prokaryotes also generate ATP by the fermentation of glucose to lactate.

Under Aerobic Conditions, Mitochondria Efficiently Oxidize Pyruvate and Generate ATP (Stages II–IV)

In the presence of oxygen, pyruvate formed by glycolysis is transported into mitochondria, where it is oxidized by O_2 to CO_2 and H_2O via a series of oxidation reactions. The overall process by which cells use O_2 and produce CO_2 is collectively termed *cellular respiration* (Figure 12-5b). Reactions in the mitochondria (stages II–IV) generate an estimated 28 additional ATP molecules per original glucose molecule, far outstripping the ATP yield from anaerobic glucose metabolism.

Oxygen-producing photosynthetic cyanobacteria appeared about 2.7 billion years ago. The subsequent buildup in the earth's atmosphere of sufficient oxygen during the next approximately billion years opened the way for organisms to evolve the very efficient aerobic oxidation pathway, which in turn permitted the evolution, especially during what is called the Cambrian explosion, of large and complex body forms and associated metabolic activities. In effect, mitochondria are ATP-generating factories, taking full advantage of this plentiful oxygen. We first describe their structure and then the reactions they employ to degrade pyruvate.

Mitochondria Are Dynamic Organelles with Two Structurally and Functionally Distinct Membranes

Mitochondria (Figure 12-6) are among the larger organelles in the cell. A mitochondrion is about the size of an *E. coli* bacterium, which is not surprising, because bacteria are thought to be the evolutionary precursors of mitochondria (see Chapter 6 and the discussion of endosymbiont hypothesis, below). Most eukaryotic cells contain many mitochondria, collectively occupying as much as 25 percent of the volume of the cytoplasm. The numbers of mitochondria in a cell, hundreds to thousands in mammalian cells, are regulated to match the cell's requirements for ATP (e.g., stomach cells, which use a lot of ATP for acid secretion, have many mitochondria). Analysis of fluorescently labeled mitochondria in living cells has shown that mitochondria are highly dynamic. They undergo frequent fusions and fissions that generate tubular, sometimes branched networks (Figure 12-7), which may account for the wide variety of mitochondrial morphologies seen in different types of cells. Fusions and fissions apparently play a functional role as well because genetic disruptions in GTPase superfamily genes required for these dynamic processes can disrupt function, such as maintenance of proper inner membrane electrical potential, and cause human disease, such as the neuromuscular disease Charcot-Marie-Tooth subtype 2A.

The details of mitochondrial structure can be observed with electron microscopy (see Figure 9-8). Mitochondria have two distinct kinds of concentrically related membranes. The outer membrane defines the smooth outer perimeter of the mitochondrion. The inner membrane has numerous

Video: Mitochondrion Reconstructed by Electron Tomography

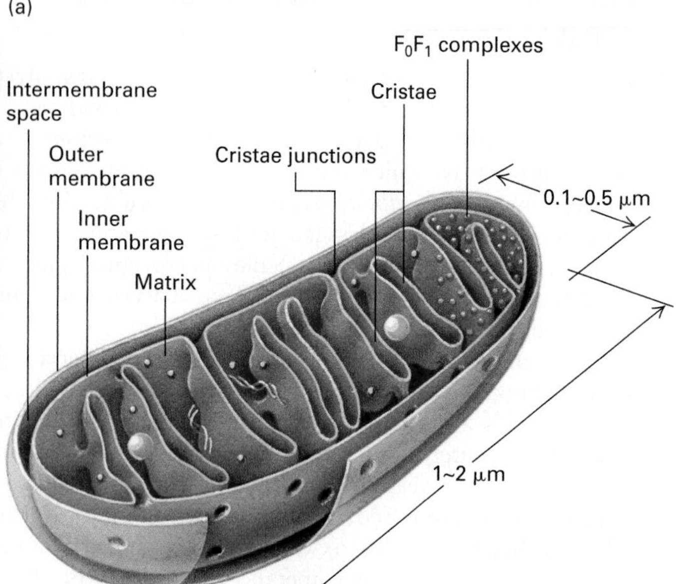

▲ **FIGURE 12-6 Internal structure of a mitochondrion.** (a) Schematic diagram showing the principal membranes and compartments. The cristae form sheets and tubes by invagination of the inner membrane and connect to the inner membrane through relatively small uniform tubular structures called *crista junctions.* The intermembrane space appears continuous with the lumen of each crista. The F_0F_1 complexes (small red spheres), which synthesize ATP, are intramembrane particles that protrude from the cristae and inner membrane into the matrix. The matrix contains the mitochondrial DNA (blue strand), ribosomes (small blue spheres), and granules (large yellow spheres). (b) Computer-generated model of a section of a mitochondrion from chicken brain. This model is based on a three-dimensional electron microscopic image calculated from a series of two-dimensional electron micrographs recorded at regular intervals. This technique is analogous to a three-dimensional x-ray tomogram or CAT scan used in medical imaging. Note the tightly packed cristae (yellow-green), the inner membrane (light blue), and the outer membrane (dark blue). [Part (a) courtesy of T. Frey; part (b) from T. Frey and C. Mannella, 2000, *Trends Biochem. Sci.* **25:**319.]

Video: Mitochondrial Fusion and Fission

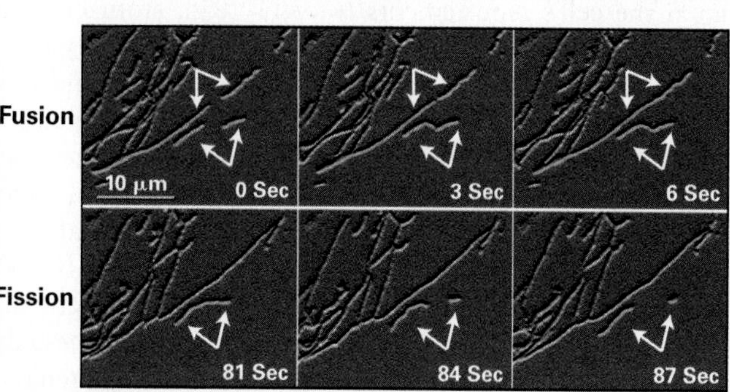

▲ EXPERIMENTAL FIGURE 12-7 **Mitochondria undergo rapid fusion and fission in living cells.** Mitochondria labeled with a fluorescent protein in a living normal murine embryonic fibroblast were observed using time-lapse fluorescence microscopy. Several mitochondria undergoing fusion (*top*) or fission (*bottom*) are artificially highlighted in blue and with arrows. [Modified from D. C. Chan, 2006, *Cell* **125**(7):1241–1252.]

invaginations called *cristae* (see Figure 12-6). These membranes topologically define two submitochondrial compartments: the *intermembrane space,* between the outer and inner membranes, and the *matrix,* or central compartment, which forms the lumen within the inner membrane. When individual mitochondria fuse, each of their distinct compartments intermixes (e.g., matrix with matrix, inner membrane with inner membrane). Fractionation and purification of these membranes and compartments have made it possible to determine their protein, DNA, and phospholipid compositions and to localize each enzyme-catalyzed reaction to a specific membrane or compartment. About 1000 polypeptides are required to make and maintain mitochondria and permit them to function. Only a small number of these—13 in humans—are encoded by mitochondrial DNA genes, while the remaining proteins are encoded by nuclear genes (Chapter 6).

The most abundant protein in the outer membrane is mitochondrial **porin,** a transmembrane channel protein similar in structure to bacterial porins (see Figure 10-18). Ions and most small molecules (up to about 5000 Da) can readily pass through these channel proteins when they are open. Although there may be metabolic regulation of the opening of mitochondrial porins and thus the flow of metabolites across the outer membrane, the inner membrane with its cristae are the major permeability barriers between the cytosol and the mitochondrial matrix, limiting the rate of mitochondrial oxidation.

Protein constitutes 76 percent of the total weight of the inner membrane—a higher fraction than in any other cellular membrane. Many of these proteins are key participants in cellular respiration. They include ATP synthase, proteins responsible for electron transport, and a wide variety of transport proteins that permit the movement of metabolites between the cytosol and the mitochondrial matrix. The human genome encodes 48 members of a family of mitochondrial transport proteins. One of these is called the ADP/ATP carrier, an antiporter that moves newly synthesized ATP out of the matrix and into the inner membrane space (and subsequently the cytosol) in exchange for ADP originating from the cytosol. Without this essential antiporter, the energy trapped in the chemical bonds in mitochondrial ATP would not be available to the rest of the cell.

The invaginating cristae greatly expand the surface area of the inner mitochondrial membrane (see Figure 12-6), enhancing its capacity to generate ATP. In typical liver mitochondria, for example, the area of the inner membrane, including cristae, is about five times that of the outer membrane. In fact, the total area of all inner mitochondrial membranes in liver cells is about 17 times that of the plasma membrane. The mitochondria in heart and skeletal muscles contain three times as many cristae as are found in typical liver mitochondria—presumably reflecting the greater demand for ATP by muscle cells.

Note that plants have mitochondria and perform cellular respiration as well. In plants, stored carbohydrates, mostly in the form of starch, are hydrolyzed to glucose. Glycolysis then produces pyruvate, which is transported into mitochondria, as in animal cells. Mitochondrial oxidation of pyruvate and concomitant formation of ATP occur in photosynthetic cells during dark periods when photosynthesis is not possible and in roots and other nonphotosynthetic tissues at all times.

The mitochondrial inner membrane, cristae, and matrix are the sites of most reactions involving the oxidation of pyruvate and fatty acids to CO_2 and H_2O and the coupled synthesis of ATP from ADP and P_i, with each reaction occurring in a discrete membrane or space in the mitochondrion (Figure 12-8).

The last three of the four stages of glucose oxidation are

- **Stage II.** Conversion of pyruvate to acetyl CoA, followed by oxidation to CO_2 in the citric acid cycle. These oxidations are coupled to reduction of NAD^+ to NADH and of FAD to $FADH_2$. (Fatty acid oxidation follows a similar route, with conversion of fatty acyl CoA to acetyl CoA.) Most of the reactions occur in or on the membrane facing the matrix.

- **Stage III.** Electron transfer from NADH and $FADH_2$ to O_2 via an electron transport chain within the inner membrane, which generates a proton-motive force across that membrane.

- **Stage IV.** Harnessing the energy of the proton-motive force for ATP synthesis in the mitochondrial inner membrane. Stages III and IV are together called oxidative phosphorylation.

In Stage II, Pyruvate Is Oxidized to CO_2 and High-Energy Electrons Stored in Reduced Coenzymes

Pyruvate formed during glycolysis in stage I in the cytosol is transported into the mitochondrial matrix (Figure 12-8). Stage II metabolism accomplishes three things: (1) it converts the 3-carbon pyruvate to three molecules of CO_2; (2) it generates high-energy **electron carriers** (NADH and $FADH_2$) that will be used for electron transport (stage III); and (3) it generates a GTP molecule, which is then converted to ATP:

$$\text{GTP} + \text{ADP} \rightleftharpoons \text{GDP} + \text{ATP}$$

Stage II can be subdivided into two distinct parts: (1) the generation of **acetyl CoA** plus one molecule of CO_2 and one NADH and (2) the conversion of acetyl CoA to two molecules of CO_2 and the high-energy intermediates NADH (3 molecules), $FADH_2$, and GTP.

Generation of Acetyl CoA Within the mitochondrial matrix, pyruvate reacts with coenzyme A, forming CO_2 and acetyl CoA and NADH (Figure 12-8). This reaction, catalyzed by *pyruvate dehydrogenase*, is highly exergonic ($\Delta G^{\circ\prime} = -8.0$ kcal/mol) and essentially irreversible.

Acetyl CoA (Figure 12-9) plays a central role in the oxidation of fatty acids and amino acids. In addition, it is an intermediate in numerous biosynthetic reactions, including transfer of an acetyl group to histone proteins and many mammalian proteins, and synthesis of lipids such as cholesterol. In respiring mitochondria, however, the acetyl group of acetyl CoA is almost always oxidized to CO_2 via the citric acid cycle.

Citric Acid Cycle Nine sequential reactions operate in a cycle to oxidize acetyl CoA to CO_2. The cycle is referred to by several names: the citric acid cycle, the tricarboxylic acid (or TCA) cycle, and the Krebs cycle. The net result is that for each acetyl group entering the cycle as acetyl CoA, two molecules of CO_2, three of NADH, and one each of $FADH_2$ and GTP are produced.

As shown in Figure 12-10, the cycle begins with condensation of the two-carbon acetyl group from acetyl CoA with the four-carbon molecule *oxaloacetate* to yield the six-carbon *citric acid* (hence the name citric acid cycle). In both reactions 4 and 5, a CO_2 molecule is released and NAD^+ is reduced to NADH. Reduction of NAD^+ to NADH also occurs during reaction 9; thus three NADHs

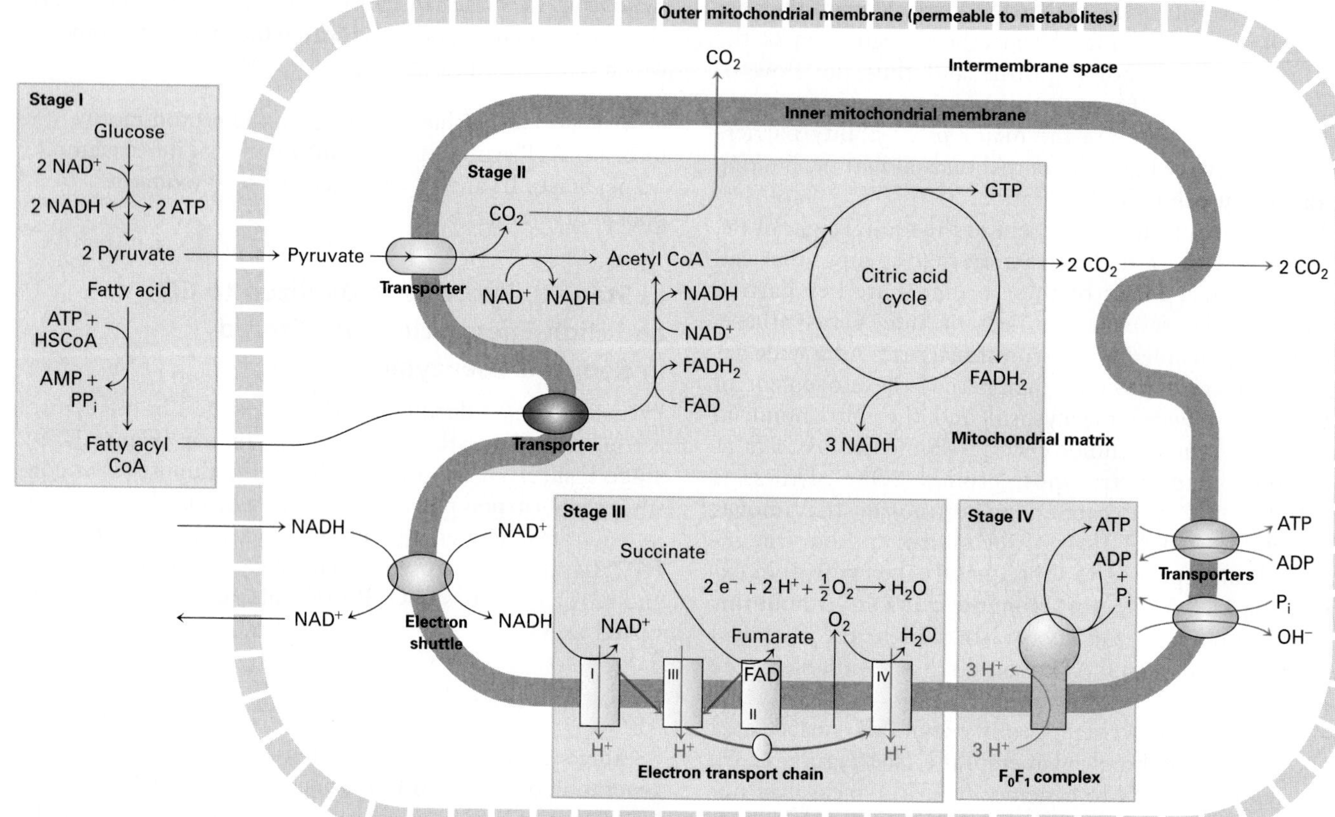

▲ FIGURE 12-8 Summary of aerobic oxidation of glucose and fatty acids. Stage I: In the cytosol, glucose is converted to pyruvate (glycolysis) and fatty acid to fatty acyl CoA. Pyruvate and fatty acyl CoA then move into the mitochondrion. Mitochondrial porins make the outer membrane permeable to these metabolites, but specific transport proteins (colored ovals) in the inner membrane are required to import pyruvate (yellow) and fatty acids (blue) into the matrix. Fatty acyl groups are transferred from fatty acyl CoA to an intermediate carrier, transported across the inner membrane (blue oval), and then reattached to CoA on the matrix side. **Stage II:** In the mitochondrial matrix, pyruvate and fatty acyl CoA are converted to acetyl CoA and then oxidized, releasing CO_2. Pyruvate is converted to acetyl CoA with the formation of NADH and CO_2; two carbons from fatty acyl CoA are converted to acetyl CoA with the formation of $FADH_2$ and NADH. Oxidation of acetyl CoA in the citric acid cycle generates NADH and $FADH_2$, GTP, and CO_2. **Stage III:** Electron transport reduces oxygen to water and generates a proton-motive force. Electrons (blue) from reduced coenzymes are transferred via electron-transport complexes (blue boxes) to O_2 concomitant with transport of H^+ ions (red) from the matrix to the intermembrane space, generating the proton-motive force. Electrons from NADH flow directly from complex I to complex III, bypassing complex II. Electrons from $FADH_2$ flow directly from complex II to complex III, bypassing complex I. **Stage IV:** ATP synthase, the F_0F_1 complex (orange), harnesses the proton-motive force to synthesize ATP in the matrix. Antiporter proteins (purple and green ovals) transport ADP and P_i into the matrix and export hydroxyl groups and ATP. NADH generated in the cytosol is not transported directly to the matrix because the inner membrane is impermeable to NAD^+ and NADH; instead, a shuttle system (red) transports electrons from cytosolic NADH to NAD^+ in the matrix. O_2 diffuses into the matrix, and CO_2 diffuses out.

are generated per turn of the cycle. In reaction 7, two electrons and two protons are transferred to FAD, yielding the reduced form of this coenzyme, $FADH_2$. Reaction 7 is distinctive because it not only is an intrinsic part of the citric acid cycle (stage II), but also it is catalyzed by a membrane-attached enzyme that is an intrinsic part of the electron transport chain (stage III). In reaction 6, hydrolysis of the high-energy thioester bond in succinyl CoA is coupled to synthesis of one GTP by substrate-level phosphorylation. (Because GTP and ATP are interconvertible, this can be

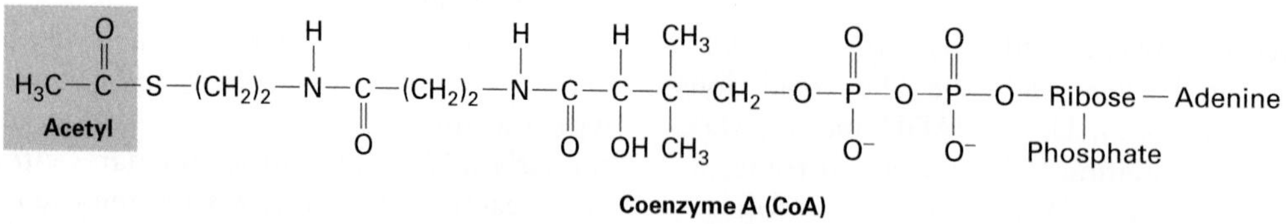

$H_3C-C(=O)-S-(CH_2)_2-NH-C(=O)-(CH_2)_2-NH-C(=O)-CH(OH)-C(CH_3)_2-CH_2-O-P(=O)(O^-)-O-P(=O)(O^-)-O-$ Ribose — Adenine (Ribose bears Phosphate)

Acetyl

Coenzyme A (CoA)

▲ FIGURE 12-9 The structure of acetyl CoA. This compound is an important intermediate in the aerobic oxidation of pyruvate, fatty acids, and many amino acids. It also contributes acetyl groups in many biosynthetic pathways.

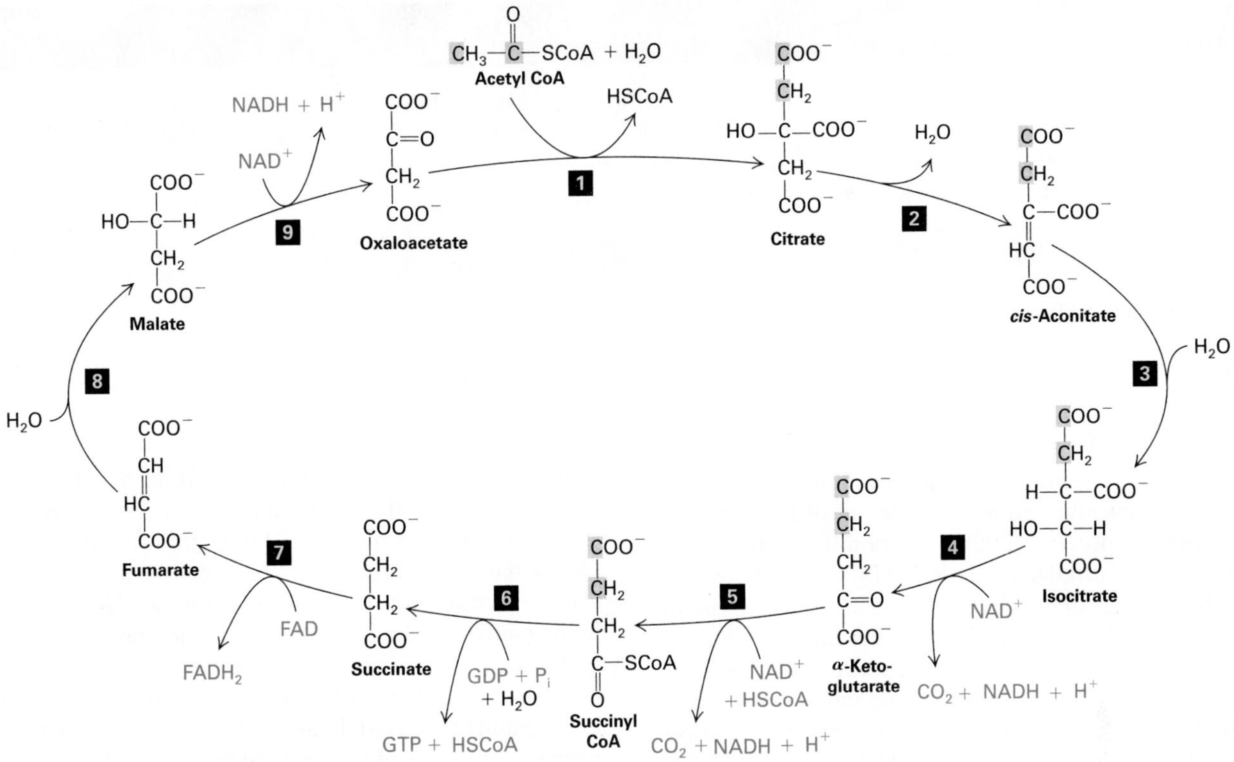

▲ **FIGURE 12-10 The citric acid cycle.** Acetyl CoA is metabolized to CO_2 and the high-energy electron carriers NADH and $FADH_2$. In reaction 1, a two-carbon acetyl residue from acetyl CoA condenses with the four-carbon molecule oxaloacetate to form the six-carbon citrate. In the remaining reactions (2–9) each molecule of citrate is eventually converted back to oxaloacetate, losing two CO_2 molecules in the process. In each turn of the cycle, four pairs of electrons are removed from carbon atoms, forming three molecules of NADH, one molecule of $FADH_2$, and one molecule of GTP. The two carbon atoms that enter the cycle with acetyl CoA are highlighted in blue through succinyl CoA. In succinate and fumarate, which are symmetric molecules, they can no longer be specifically denoted. Isotope-labeling studies have shown that these carbon atoms are *not* lost in the turn of the cycle in which they enter; on average, one will be lost as CO_2 during the next turn of the cycle and the other in subsequent turns.

considered an ATP-generating step.) Reaction 9 regenerates oxaloacetate, so the cycle can begin again. Note that molecular O_2 does not participate in the citric acid cycle.

Most enzymes and small molecules involved in the citric acid cycle are soluble in the aqueous mitochondrial matrix. These include CoA, acetyl CoA, succinyl CoA, NAD^+, and NADH, as well as most of the eight cycle enzymes. *Succinate dehydrogenase* (reaction 7), however, is a component of an integral membrane protein in the inner membrane, with its active site facing the matrix. When mitochondria are disrupted by gentle ultrasonic vibration or osmotic lysis, non-membrane-bound enzymes in the citric acid cycle are released as very large multiprotein complexes. Within such complexes the reaction product of one enzyme is thought to pass directly to the next enzyme without diffusing through the solution. However, much work is needed to determine the structures of these large enzyme complexes as they exist in the cell.

Since glycolysis of one glucose molecule generates two acetyl CoA molecules, the reactions in the glycolytic pathway and citric acid cycle produce six CO_2 molecules, 10 NADH molecules, and two $FADH_2$ molecules per glucose molecule (Table 12-1). Although these reactions also generate four high-energy phosphoanhydride bonds in the form of two ATP and two GTP molecules, this represents only a small fraction of the available energy released in the complete aerobic oxidation of glucose. The remaining energy is stored as high-energy electrons in the reduced coenzymes NADH and $FADH_2$. The goal of stages III and IV is to recover this energy in the form of ATP.

Transporters in the Inner Mitochondrial Membrane Help Maintain Appropriate Cytosolic and Matrix Concentrations of NAD^+ and NADH

In the cytosol NAD^+ is required for step 6 of glycolysis (see Figure 12-3), and in the mitochondrial matrix NAD^+ is required for conversion of pyruvate to acetyl CoA and for three steps in the citric acid cycle (4, 5, and 9 in Figure 12-10). In each case, NADH is a product of the reaction. For ongoing glycolysis and oxidation of pyruvate, NAD^+ must be regenerated by oxidation of NADH. (Similarly, the $FADH_2$ generated in stage II reactions must be reoxidized to FAD if FAD-dependent reactions are to continue.) As we will see in the next section, the stage III electron transport chain *within* the mitochondrion converts NADH to NAD^+ and $FADH_2$ to FAD as it reduces O_2 to water and converts the energy stored in the high-energy electrons in the reduced

TABLE 12-1 Net Result of the Glycolytic Pathway and the Citric Acid Cycle				
REACTION	CO_2 MOLECULES PRODUCED	NAD^+ MOLECULES REDUCED TO NADH	FAD MOLECULES REDUCED TO $FADH_2$	ATP (OR GTP)
1 glucose molecule to 2 pyruvate molecules	0	2	0	2
2 pyruvates to 2 acetyl CoA molecules	2	2	0	0
2 acetyl CoA to 4 CO_2 molecules	4	6	2	2
Total	6	10	2	4

forms of these molecules into a proton-motive force. Even though O_2 is not involved in any reaction of the citric acid cycle, in the absence of O_2, this cycle soon stops operating as the intramitochondrial supplies of NAD^+ and FAD dwindle due to the inability of the electron transport chain to oxidize NADH and $FADH_2$. These observations raise the question of how a supply of NAD^+ in the cytosol is regenerated.

If the NADH from the cytosol could move into the mitochondrial matrix and be oxidized by the electron transport chain and if the NAD^+ product could be transported back into the cytosol, regeneration of cytosolic NAD^+ would be simple. However, the inner mitochondrial membrane is impermeable to NADH. To bypass this problem and permit the electrons from cytosolic NADH to be transferred *indirectly* to O_2 via the electron transport chain, cells use several *electron shuttles* to transfer electrons from cytosolic NADH to NAD^+ in the matrix. Operation of the most widespread shuttle—the *malate-aspartate shuttle*—is depicted in Figure 12-11. For every complete "turn" of the cycle, there is no overall change in the numbers of NADH and NAD^+ molecules or the intermediates aspartate or malate used by the shuttle. However, in the cytosol, NADH is oxidized to NAD^+, which can be used

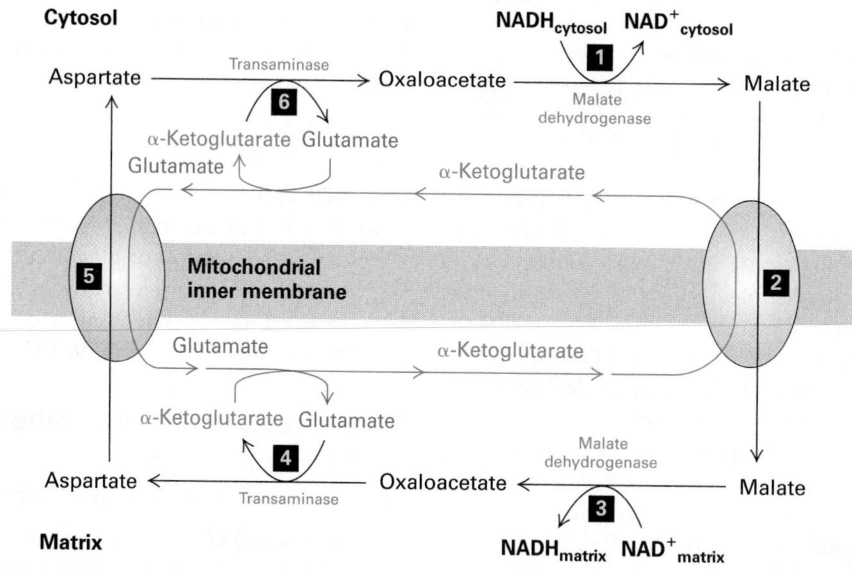

▲ **FIGURE 12-11 The malate shuttle.** This cyclical series of reactions transfers electrons from NADH in the cytosol (intermembrane space) across the inner mitochondrial membrane, which is impermeable to NADH itself, to NAD^+ in the matrix. The net result is the replacement of cytosolic NADH with NAD^+ and matrix NAD^+ with NADH. Step **1**: Cytosolic malate dehydrogenase transfers electrons from cytosolic NADH to oxaloacetate, forming malate. Step **2**: An antiporter (blue oval) in the inner mitochondrial membrane transports malate into the matrix in exchange for α-ketoglutarate. Step **3**: Mitochondrial malate dehydrogenase converts malate back to oxaloacetate, reducing NAD^+ in the matrix to NADH in the process. Step **4**: Oxaloacetate, which cannot directly cross the inner membrane, is converted to aspartate by addition of an amino group from glutamate. In this transaminase-catalyzed reaction in the matrix, glutamate is converted to α-ketoglutarate. Step **5**: A second antiporter (red oval) exports aspartate to the cytosol in exchange for glutamate. Step **6**: A cytosolic transaminase converts aspartate to oxaloacetate and α-ketoglutarate to glutamate, completing the cycle. The blue arrows reflect the movement of the α-ketoglutarate, the red arrows the movement of glutamate, and the black arrows that of aspartate/malate. It is noteworthy that, as aspartate and malate cycle clockwise, glutamate and α-ketoglutarate cycle in the opposite direction.

for glycolysis, and in the matrix, NAD^+ is reduced to NADH, which can be used to generate ATP via stages III and IV:

$$NADH_{cytosol} + NAD^+_{matrix} \rightarrow NAD^+_{cytosol} + NADH_{matrix}$$

Mitochondrial Oxidation of Fatty Acids Generates ATP

Up to now, we have focused mainly on the oxidation of carbohydrates, namely glucose, for ATP generation. Fatty acids are another important source of cellular energy. Cells can take up either glucose or fatty acids from the extracellular space with the help of specific transporter proteins (Chapter 11). Should a cell not need to immediately burn these molecules, it can store them as a polymer of glucose called glycogen (especially in muscle or liver) or as a trimer of fatty acids covalently linked to glycerol, called a **triacylglycerol** or **triglyceride.** In some cells, excess glucose is converted into fatty acids and then triacylglycerols for storage. However, unlike microorganisms, animals are unable to convert fatty acids to glucose. When the cells need to burn these energy stores to make ATP (e.g., when a resting muscle begins to do work), enzymes break down glycogen to glucose or hydrolyze triacylglycerols to fatty acids, which are then oxidized to generate ATP:

$$\begin{array}{l} CH_3-(CH_2)_n-\overset{O}{\overset{\|}{C}}-O-CH_2 \\ CH_3-(CH_2)_n-\overset{O}{\overset{\|}{C}}-O-CH \\ CH_3-(CH_2)_n-\overset{O}{\overset{\|}{C}}-O-CH_2 \end{array} + 3\,H_2O \longrightarrow 3\,CH_3-(CH_2)_n-\overset{O}{\overset{\|}{C}}-OH + \begin{array}{l} HO-CH_2 \\ HO-CH \\ HO-CH_2 \end{array}$$

Triacylglycerol; Fatty acid; Glycerol

Fatty acids are the major energy source for many tissues, particularly adult heart muscle. In humans, in fact, the oxidation of fats is quantitatively more important than the oxidation of glucose as a source of ATP. The oxidation of 1 g of triacylglyceride to CO_2 generates about six times as much ATP as does the oxidation of 1 g of hydrated glycogen. Thus triglycerides are more efficient than carbohydrates for storage of energy, in part because they are stored in anhydrous form and can yield more energy when oxidized and also because they are intrinsically more reduced (have more hydrogens) than carbohydrates. In mammals, the primary site of storage of triacylglycerides is fat (adipose) tissue, whereas the primary sites for glycogen storage are muscle and the liver.

Just as there are four stages in the oxidation of glucose, there are four stages in the oxidation of fatty acids. To optimize the efficiency of ATP generation, part of stage II (citric acid cycle oxidation of acetyl CoA) and all of stages III and IV of fatty acid oxidation are identical to those of glucose oxidation. The differences lie in the cytosolic stage I and the first part of the mitochondrial stage II. In stage I, fatty acids are converted to a fatty acyl CoA in the cytosol in a reaction coupled to the hydrolysis of ATP to AMP and PP_i (inorganic pyrophosphate) (see Figure 12-8):

$$\underset{\text{Fatty acid}}{R-\overset{O}{\overset{\|}{C}}-O^-} + HSCoA + ATP \longrightarrow \underset{\text{Fatty acyl CoA}}{R-\overset{O}{\overset{\|}{C}}-SCoA} + AMP + PP_i$$

Subsequent hydrolysis of PP_i to two molecules of P_i drives this reaction to completion. To transfer the fatty acyl group into the mitochondrial matrix, it is covalently transferred to a molecule called carnitine, moved across the inner mitochondrial membrane by an acylcarnitine transporter protein (see Figure 12-8, blue oval), and then on the matrix side, the fatty acyl group is released from carnitine and reattached to another CoA molecule. The activity of the acyl carnitine transporter is regulated to prevent oxidation of fatty acids when cells have adequate energy (ATP) supplies.

In the first part of stage II, each molecule of a fatty acyl CoA in the mitochondrion is oxidized in a cyclical sequence of four reactions in which all the carbon atoms are converted two at a time to acetyl CoA with generation of $FADH_2$ and NADH (Figure 12-12a). For example, mitochondrial oxidation of each molecule of the 18-carbon stearic acid, $CH_3(CH_2)_{16}COOH$, yields nine molecules of acetyl CoA and eight molecules each of NADH and $FADH_2$. In the second part of stage II, as with acetyl CoA generated from pyruvate, these acetyl groups enter the citric acid cycle and are oxidized to CO_2. As will be described in detail in the next section, the reduced NADH and $FADH_2$ with their high-energy electrons from stage II will be used in stage III to generate a proton-motive force that in turn is used in stage IV to power ATP synthesis.

Peroxisomal Oxidation of Fatty Acids Generates No ATP

Mitochondrial oxidation of fatty acids is the major source of ATP in mammalian liver cells, and biochemists at one time believed this was true in all cell types. However, rats treated with clofibrate, a drug that affects many features of lipid metabolism, were found to exhibit an increased rate of fatty acid oxidation and a large increase in the number of peroxisomes in their liver cells. This finding suggested that peroxisomes, as well as mitochondria, can oxidize fatty acids. These small organelles, ≈0.2–1 μm in diameter, are lined by a single membrane (see Figure 9-4). They are present in all mammalian cells except erythrocytes and are also found in plant cells, yeasts, and probably most other eukaryotic cells.

Mitochondria preferentially oxidize short-(fewer than 8 carbons in the fatty acyl chain, or $<C_8$), medium-(C_8–C_{12}),

► **FIGURE 12-12 Oxidation of fatty acids in mitochondria and peroxisomes.** In both mitochondrial oxidation (a) and peroxisomal oxidation (b), fatty acids are converted to acetyl CoA by a series of four enzyme-catalyzed reactions (shown down the center of the figure). A fatty acyl CoA molecule is converted to acetyl CoA and a fatty acyl CoA shortened by two carbon atoms. Concomitantly, one FAD molecule is reduced to $FADH_2$ and one NAD^+ molecule is reduced to NADH. The cycle is repeated on the shortened acyl CoA until fatty acids with an even number of carbon atoms are completely converted to acetyl CoA. In mitochondria, electrons from $FADH_2$ and NADH enter the electron transport chain and ultimately are used to generate ATP; the acetyl CoA generated is oxidized in the citric acid cycle, resulting in release of CO_2 and ultimately the synthesis of additional ATP. Because peroxisomes lack the electron transport complexes composing the electron transport chain and the enzymes of the citric acid cycle, oxidation of fatty acids in these organelles yields no ATP. [Adapted from D. L. Nelson and M. M. Cox, *Lehninger Principles of Biochemistry,* 3d ed., 2000, Worth Publishers.]

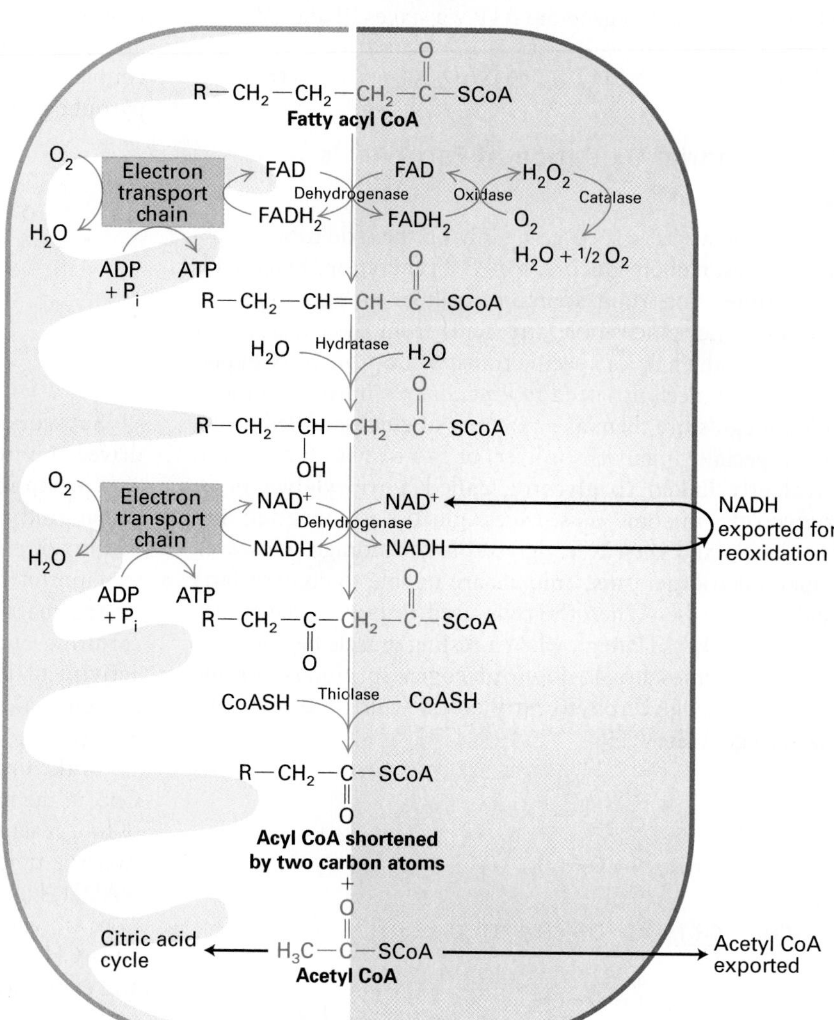

and long-(C_{14}–C_{20}) chain fatty acids; whereas peroxisomes preferentially oxidize very long-chain fatty acids (VLCFAs, $>C_{20}$), which cannot be oxidized by mitochondria. Most dietary fatty acids have long chains, which means they are oxidized mostly in mitochondria. In contrast to mitochondrial oxidation of fatty acids, which is coupled to generation of ATP, peroxisomal oxidation of fatty acids is not linked to ATP formation, and energy is released as heat.

The reaction pathway by which fatty acids are degraded to acetyl CoA in peroxisomes is similar to that used in mitochondria (Figure 12-12b). However, peroxisomes lack an electron transport chain, and electrons from the $FADH_2$ produced during the oxidation of fatty acids are immediately transferred to O_2 by *oxidases*, regenerating FAD and forming hydrogen peroxide (H_2O_2). In addition to oxidases, peroxisomes contain abundant *catalase,* which quickly decomposes the H_2O_2, a highly cytotoxic metabolite. NADH produced during oxidation of fatty acids is exported and reoxidized in the cytosol; there is no need for a malate/aspartate shuttle here. Peroxisomes also lack the citric acid cycle, so acetyl CoA generated during peroxisomal degradation of fatty acids cannot be oxidized further; instead it is transported into the cytosol for use in the synthesis of cholesterol (Chapter 10) and other metabolites.

KEY CONCEPTS OF SECTION 12.1

First Steps of Glucose and Fatty Acid Catabolism: Glycolysis and the Citric Acid Cycle

- In a process known as aerobic oxidation, cells convert the energy released by the oxidation ("burning") of glucose or fatty acids into the terminal phosphoanhydride bond of ATP.
- The complete aerobic oxidation of each molecule of glucose produces six molecules of CO_2 and approximately 30 ATP molecules. The entire process, which starts in the cytosol and moves into the mitochondrion, can be divided into four stages: (I) glycolysis to pyruvate in the cytosol, (II) pyruvate oxidation to CO_2 in the mitochondrion, (III) electron transport to generate a proton-motive force together with conversion of molecular oxygen to water, and (IV) ATP synthesis.
- The mitochondrion has two distinct membranes (outer and inner) and two distinct subcompartments (intermembrane

space between the two membranes and the matrix surrounded by the inner membrane). Aerobic oxidation occurs in the mitochondrial matrix and on the inner mitochondrial membrane.

- Each turn of the citric acid cycle releases two molecules of CO_2 and generates three NADH molecules, one $FADH_2$ molecule, and one GTP.
- In glycolysis (stage I), cytosolic enzymes convert glucose to two molecules of pyruvate and generate two molecules each of NADH and ATP.
- The rate of glucose oxidation via glycolysis and the citric acid cycle is regulated by the inhibition or stimulation of several enzymes, depending on the cell's need for ATP. Glucose is stored (as glycogen or fat) when ATP is abundant.
- Some of the energy released in the early stages of oxidation is temporarily stored in the reduced coenzymes NADH or $FADH_2$, which carry high-energy electrons that subsequently drive the electron transport chain (stage III).
- In the absence of oxygen (anaerobic conditions), cells can metabolize pyruvate to lactate or (in the case of yeast) to ethanol and CO_2, in the process converting NADH back to NAD^+, which is necessary for continued glycolysis. In aerobic conditions (presence of oxygen), pyruvate is transported into the mitochondrion, where stages II through IV occur.
- In stage II, the three-carbon pyruvate molecule is first oxidized to generate one molecule each of CO_2, NADH, and acetyl CoA. The acetyl CoA is then oxidized to CO_2 by the citric acid cycle.
- Neither glycolysis (stage I) nor the citric acid cycle (stage II) directly use molecular oxygen (O_2).
- The malate/aspartate shuttle regenerates the supply of cytosolic NAD^+ necessary for continued glycolysis.
- Like glucose oxidation, the oxidation of fatty acids takes place in four stages. In stage I, fatty acids are converted to fatty acyl CoA in the cytosol. In stage II, the fatty acyl CoA is first converted into multiple acetyl CoA molecules with generation of NADH and $FADH_2$. Then, as in glucose oxidation, the acetyl CoA enters the citric acid cycle. Stages III and IV are identical for fatty acid and glucose oxidation.
- In most eukaryotic cells, oxidation of short- to long-chain fatty acids occurs in mitochondria with production of ATP, whereas oxidation of very long chain fatty acids occurs primarily in peroxisomes and is not linked to ATP production; the released energy is converted to heat.

12.2 The Electron Transport Chain and Generation of the Proton-Motive Force

Most of the energy released during the oxidation of glucose and fatty acids to CO_2 (stages I and II) is converted into high-energy electrons in the reduced coenzymes NADH and $FADH_2$. We now turn to stage III, in which the energy transiently stored in the coenzymes is converted by an electron transport chain, also known as the **respiratory chain** into the proton-motive force. We first describe the logic and components of the electron transport chain and the pumping of protons across the mitochondrial inner membrane. We conclude the section with a discussion of the magnitude of the proton-motive force produced by electron transport and proton pumping. In the following section, we describe stage IV, focusing on the structure of the ATP synthase and how it uses the proton-motive force to synthesize ATP.

Stepwise Electron Transport Efficiently Releases the Energy Stored in NADH and $FADH_2$

During electron transport, electrons are released from NADH and $FADH_2$ and eventually transferred to O_2, forming H_2O according to the following overall reactions:

$$\text{NADH} + \text{H}^+ + \tfrac{1}{2}\,\text{O}_2 \longrightarrow \text{NAD}^+ + \text{H}_2\text{O},\quad \Delta G = -52.6\ \text{kcal/mol}$$

$$\text{FADH}_2 + \tfrac{1}{2}\,\text{O}_2 \longrightarrow \text{FAD} + \text{H}_2\text{O},\quad \Delta G = -43.4\ \text{kcal/mol}$$

Recall that the conversion of 1 glucose molecule to CO_2 via the glycolytic pathway and citric acid cycle yields 10 NADH and 2 $FADH_2$ molecules (see Table 12-1). Oxidation of these reduced coenzymes has a total $\Delta G^{\circ\prime}$ of -613 kcal/mol $[10(-52.6) + 2(-43.4)]$. Thus of the potential free energy present in the chemical bonds of glucose (-686 kcal/mol), about 90 percent is conserved in the reduced coenzymes. Why should there be two different coenzymes, NADH and $FADH_2$? Although many of the reactions involved in glucose and fatty acid oxidation are sufficiently energetic to reduce NAD^+, several are not, so those reactions are coupled to reduction of FAD, which requires less energy.

The energy carried in the reduced coenzymes can be released by oxidizing them. The biochemical challenge faced by the mitochondrion is to transfer, as efficiently as possible, the energy released by this oxidation into the energy in the terminal phosphoanhydride bond in ATP.

$$\text{P}_i^{2-} + \text{H}^+ + \text{ADP}^{3-} \rightarrow \text{ATP}^{4-} + \text{H}_2\text{O},\quad \Delta G = +7.3\ \text{kcal/mol}$$

A relatively simple one-to-one reaction in which reduction of one coenzyme molecule and synthesis of one ATP occurs would be terribly inefficient, because the $\Delta G^{\circ\prime}$ for ATP generation from ADP and P_i is substantially less than for the coenzyme oxidation and much energy would be lost as heat. To efficiently recover the energy, the mitochondrion first converts the energy of coenzyme oxidation into a proton-motive force using a series of electron carriers, all but one of which are integral components of the inner membrane.

Electron Transport in Mitochondria Is Coupled to Proton Pumping

At several sites during electron transport from NADH and $FADH_2$ to O_2, protons from the mitochondrial matrix are

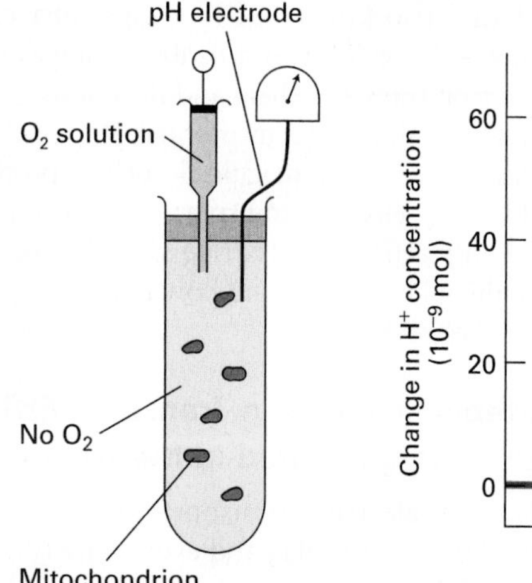

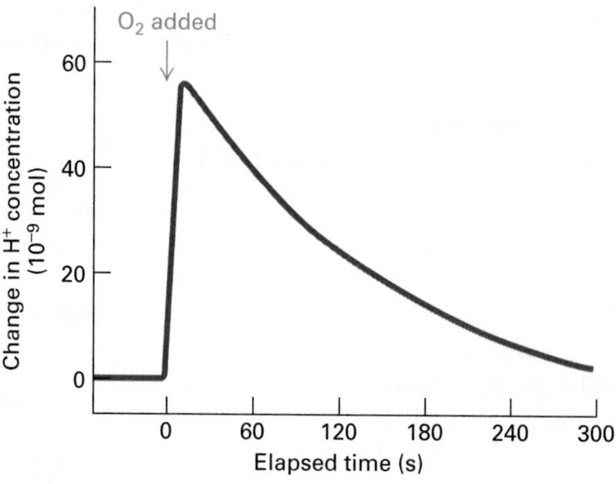

▲ EXPERIMENTAL FIGURE 12-13 **Electron transfer from NADH to O_2 is coupled to proton transport across the mitochondrial membrane.** If NADH is added to a suspension of mitochondria depleted of O_2, no NADH is oxidized. When a small amount of O_2 is added to the system (arrow), there is a sharp rise in the concentration of protons in the surrounding medium outside the mitochondria (decrease in pH). Thus the oxidation of NADH by O_2 is coupled to the movement of protons out of the matrix. Once the O_2 is depleted, the excess protons slowly move back into the mitochondria (powering the synthesis of ATP) and the pH of the extracellular medium returns to its initial value.

pumped across the inner membrane; this generates proton concentration and electrical gradients across the inner membrane (see Figure 12-2). This pumping causes the pH of the mitochondrial matrix to become higher (i.e., the H^+ concentration is lower) than that of the intermembrane space and cytosol. An electric potential across the inner membrane also results from the pumping of H^+ outward from the matrix, which becomes negative with respect to the intermembrane space. Thus free energy released during the oxidation of NADH or $FADH_2$ is stored both as an electric potential and a proton concentration gradient—collectively, the proton-motive force—across the inner membrane. As we will see, the movement of protons back across the inner membrane, driven by this force, is coupled to the synthesis of ATP from ADP and P_i by ATP synthase.

The synthesis of ATP from ADP and P_i, driven by the transfer of electrons from NADH or $FADH_2$ to O_2, is the major source of ATP in aerobic nonphotosynthetic cells. Much evidence shows that in mitochondria and bacteria this process of oxidative phosphorylation depends on generation of a proton-motive force across the inner membrane (mitochondria) or bacterial plasma membrane, with electron transport, proton pumping, and ATP formation occurring simultaneously. In the laboratory, for instance, addition of O_2 and an oxidizable substrate such as pyruvate or succinate to isolated intact mitochondria results in a net synthesis of ATP if the inner mitochondrial membrane is intact. In the presence of minute amounts of detergents that make the membrane leaky, electron transport and the oxidation of these metabolites by O_2 still occurs. However, under these conditions no ATP is made, because the proton leak prevents the maintenance of the transmembrane proton concentration gradient and the membrane electric potential.

The coupling between electron transport from NADH (or $FADH_2$) to O_2 and proton transport across the inner mitochondrial membrane can be demonstrated experimentally with isolated, intact mitochondria (Figure 12-13). As soon as O_2 is added to a suspension of mitochondria in an otherwise O_2-free solution that contains NADH, the medium outside the mitochondria transiently becomes more acidic (increased proton concentration), because the mitochondrial outer membrane is freely permeable to protons. (Remember that malate/aspartate and other shuttles can convert the NADH in the solution into NADH in the matrix.) Once the O_2 is depleted by its reduction, the excess protons in the medium slowly leak back into the matrix. From analysis of the measured pH change in such experiments, one can calculate that about 10 protons are transported out of the matrix for every electron pair transferred from NADH to O_2.

To obtain numbers for $FADH_2$, the above experiment can be repeated, but with succinate instead of NADH as the substrate. (Recall that oxidation of succinate to fumarate in the citric acid cycle generates $FADH_2$; see Figure 12-10). The amount of succinate added can be adjusted so that the amount of $FADH_2$ generated is equivalent to the amount of NADH in the first experiment. As in the first experiment, addition of oxygen causes the medium outside the mitochondria to become acidic, but less so than with NADH. This is not surprising because electrons in $FADH_2$ have less potential energy (43.4 kcal/mol) than electrons in NADH (52.6 kcal/mole), and thus it drives the translocation of fewer protons from the matrix and a smaller change in pH.

Electrons Flow from $FADH_2$ and NADH to O_2 Through Four Multiprotein Complexes

We now examine more closely the energetically favored movement of electrons from NADH and $FADH_2$ to the final electron acceptor, O_2. For simplicity, we will focus our discussion on NADH. In respiring mitochondria, each NADH molecule releases two electrons to the electron transport chain; these electrons ultimately reduce one oxygen atom (half of an O_2 molecule), forming one molecule of water:

$$NADH \rightarrow NAD^+ + H^+ + 2\,e^-$$
$$2\,e^- + 2\,H^+ + {}^1/_2\,O_2 \rightarrow H_2O$$

As electrons move from NADH to O_2, their potential declines by 1.14 V, which corresponds to 26.2 kcal/mol of electrons transferred, or ≈53 kcal/mol for a pair of electrons. As noted earlier, much of this energy is conserved in the proton-motive force generated across the inner mitochondrial membrane.

There are four large multiprotein complexes in the electron transport chain that span the inner mitochondrial membrane: *NADH-CoQ reductase* (**complex I,** >40 subunits), *succinate-CoQ reductase* (**complex II,** 4 subunits), *CoQH$_2$–cytochrome c reductase* (**complex III,** 11 subunits), and *cytochrome c oxidase* (**complex IV,** 13 subunits). Electrons from NADH flow from complex I to III to IV, bypassing complex II; electrons from $FADH_2$ flow from complex II to III to IV, bypassing complex I (see Figure 12-8).

Each complex contains several *prosthetic groups* that participate in moving electrons. These small nonpeptide organic molecules or metal ions are tightly and specifically associated with the multiprotein complexes (Table 12-2).

TABLE 12-2 Electron-Carrying Prosthetic Groups in the Respiratory Chain

PROTEIN COMPONENT	PROSTHETIC GROUPS*
NADH-CoQ reductase (complex I)	FMN Fe-S
Succinate-CoQ reductase (complex II)	FAD Fe-S
$CoQH_2$–cytochrome *c* reductase (complex III)	Heme b_L Heme b_H Fe-S Heme c_1
Cytochrome *c*	Heme *c*
Cytochrome *c* oxidase (complex IV)	Cu_a^{2+} Heme *a* Cu_b^{2+} Heme a_3

*Not included is coenzyme Q, an electron carrier that is not permanently bound to a protein complex.
SOURCE: J. W. De Pierre and L. Ernster, 1977, *Ann. Rev. Biochem.* **46**:201.

Heme and the Cytochromes Several types of *heme,* an iron-containing prosthetic group similar to that in hemoglobin and myoglobin (Figure 12-14a), are tightly bound (covalently or noncovalently) to a set of mitochondrial proteins called **cytochromes.** Each cytochrome is designated by a letter, such as *a, b, c,* or c_1. Electron flow through the cytochromes occurs by oxidation and reduction of the Fe atom in the center of the heme molecule:

$$Fe^{3+} + e^- \rightleftharpoons Fe^{2+}$$

Because the heme ring in cytochromes consists of alternating double- and single-bonded atoms, a large number of resonance hybrid forms exist. These allow the extra electron delivered to the cytochrome to be delocalized throughout the heme carbon and nitrogen atoms as well as the Fe ion.

The various cytochromes have slightly different heme groups and surrounding atoms (called axial ligands), which

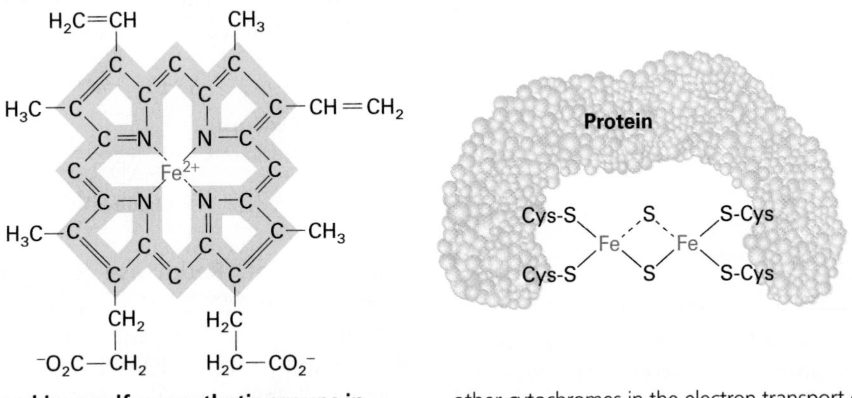

(a)

H₂C=CH CH₃ H₃C CH=CH₂ N Fe²⁺ H₃C CH₃ CH₂ H₂C ⁻O₂C—CH₂ H₂C—CO₂⁻

(b)

▲ **FIGURE 12-14 Heme and iron-sulfur prosthetic groups in the electron transport chain.** (a) Heme portion of cytochromes b_L and b_H, which are components of $CoQH_2$–cytochrome *c* reductase (complex III). The same porphyrin ring (yellow) is present in all hemes. The chemical substituents attached to the porphyrin ring differ in the other cytochromes in the electron transport chain. All hemes accept and release one electron at a time. (b) Dimeric iron-sulfur cluster (Fe-S). Each Fe atom is bonded to four S atoms: two are inorganic sulfur and two are in cysteine side chains of the associated protein. All Fe-S clusters accept and release one electron at a time.

generate different environments for the Fe ion. Therefore, each cytochrome has a different reduction potential, or tendency to accept an electron—an important property dictating the unidirectional "downhill" electron flow along the chain. Just as water spontaneously flows downhill from a higher to lower potential energy state—but not uphill—so too do electrons flow in only one direction from one heme (or other prosthetic group) to another due to their differing reduction potentials. All the cytochromes, except cytochrome *c*, are components of integral membrane multiprotein complexes in the inner mitochondrial membrane.

Iron-Sulfur Clusters *Iron-sulfur clusters* are nonheme, iron-containing prosthetic groups consisting of Fe atoms bonded both to inorganic S atoms and to S atoms on cysteine residues in a protein (Figure 12-14b). Some Fe atoms in the cluster bear a +2 charge; others have a +3 charge. However, the net charge of each Fe atom is actually between +2 and +3, because electrons in their outermost orbitals together with the extra electron delivered via the transport chain are dispersed among the Fe atoms and move rapidly from one atom to another. Iron-sulfur clusters accept and release electrons one at a time.

Coenzyme Q (CoQ) *Coenzyme Q* (CoQ), also called *ubiquinone*, is the only small-molecule electron carrier in the chain that is not a protein-bound prosthetic group (Figure 12-15). It is a carrier of both protons and electrons. The oxidized quinone form of CoQ can accept a single electron to form a semiquinone, a charged free radical denoted by $CoQ^{-}\cdot$. Addition of a second electron and two protons (thus a total of two hydrogen atoms) to $CoQ^{-}\cdot$ forms dihydroubiquinone ($CoQH_2$), the fully reduced form. Both CoQ and $CoQH_2$ are soluble in phospholipids and diffuse freely in the hydrophobic center of the inner mitochondrial membrane. This is how it participates in the electron transport chain—carrying electrons and protons between the complexes.

As shown in Figure 12-16, CoQ accepts electrons released from NADH-CoQ reductase (complex I) or succinate-CoQ reductase (complex II) and donates them to $CoQH_2$–cytochrome *c* reductase (complex III). Importantly, reduction and oxidation of CoQ are coupled to pumping of protons. Whenever CoQ accepts electrons, it does so at a binding site on the matrix (also called the cytosolic) face of the protein complex, always picking up protons from the medium there. Whenever $CoQH_2$ releases its electrons, it does so at a site on the intermembrane space (also called the exoplasmic) side of the protein complex, releasing protons into the intermembrane (or exoplasmic) fluid. Thus transport of each pair of electrons by CoQ is obligatorily coupled to movement of two protons from the matrix to the intermembrane space fluid.

Ubiquinone (CoQ) (oxidized form)

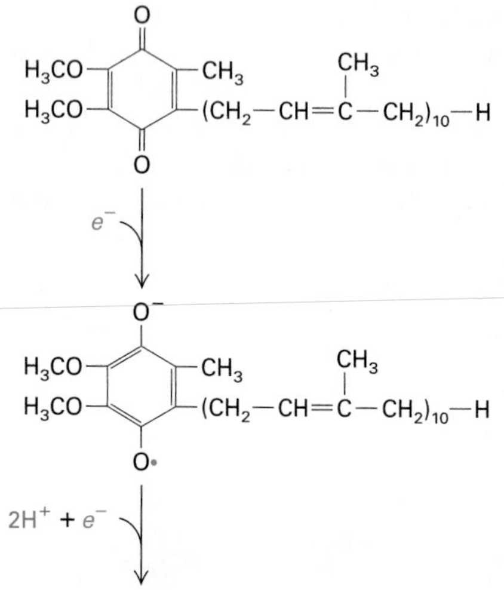

Semiquinone ($CoQ^{\bar{\cdot}}$) (free radical)

Dihydroquinone ($CoQH_2$) (fully reduced form)

▲ **FIGURE 12-15 Oxidized and reduced forms of coenzyme Q (CoQ), which can carry two protons and two electrons.** Because of its long hydrocarbon "tail" of isoprene units, CoQ, also called ubiquinone, is soluble in the hydrophobic core of phospholipid bilayers and is very mobile. Reduction of CoQ to the fully reduced form, QH_2 (dihydroquinone), occurs in two steps with a half-reduced free-radical intermediate, called semiquinone.

NADH-CoQ Reductase (Complex I) Electrons are transferred from NADH to CoQ by NADH-CoQ reductase (Figure 12-16). In bacteria the mass of this complex is about 500 kDa (~14 subunits), whereas for the L-shaped eukaryotic complex it is 1 MDa (14 central and as many as 32 accessory subunits). NAD^+ is exclusively a two-electron carrier: it accepts or releases a pair of electrons simultaneously. In NADH-CoQ reductase (complex I), electrons first flow from NADH to FMN (flavin mononucleotide), a cofactor related to FAD, then to an iron-sulfur cluster, and finally to CoQ. FMN, like FAD, can accept two electrons but does so one electron at a time.

Each transported electron undergoes a drop in potential of ≈360 mV, equivalent to a $\Delta G^{\circ\prime}$ of −16.6 kcal/mol for the two electrons transported. Much of this released energy is used to transport four protons across the inner membrane per molecule of NADH oxidized by the complex I. Those four protons are distinct from the two protons transferred to the CoQ in the chemical reaction shown above.

The overall reaction catalyzed by this complex is

$$\underset{\text{(Reduced)}}{\text{NADH}} + \underset{\text{(Oxidized)}}{\text{CoQ}} + 6\,H^+_{in} \rightarrow \underset{\text{(Oxidized)}}{\text{NAD}^+} + H^+_{in} + \underset{\text{(Reduced)}}{\text{CoQH}_2} + 4\,H^+_{out}$$

Succinate-CoQ Reductase (Complex II) Succinate dehydrogenase, the enzyme that oxidizes a molecule of succinate to fumarate in the citric acid cycle (and in the process generates the reduced coenzyme $FADH_2$), is one of the four subunits of complex II. In this way the citric acid cycle is physically as well as functionally linked to the electron transport chain. The two electrons released in conversion of succinate to fumarate are transferred first to FAD in succinate dehydrogenase, then to iron-sulfur clusters—regenerating FAD—and finally to CoQ,

Animation: Electron Transport

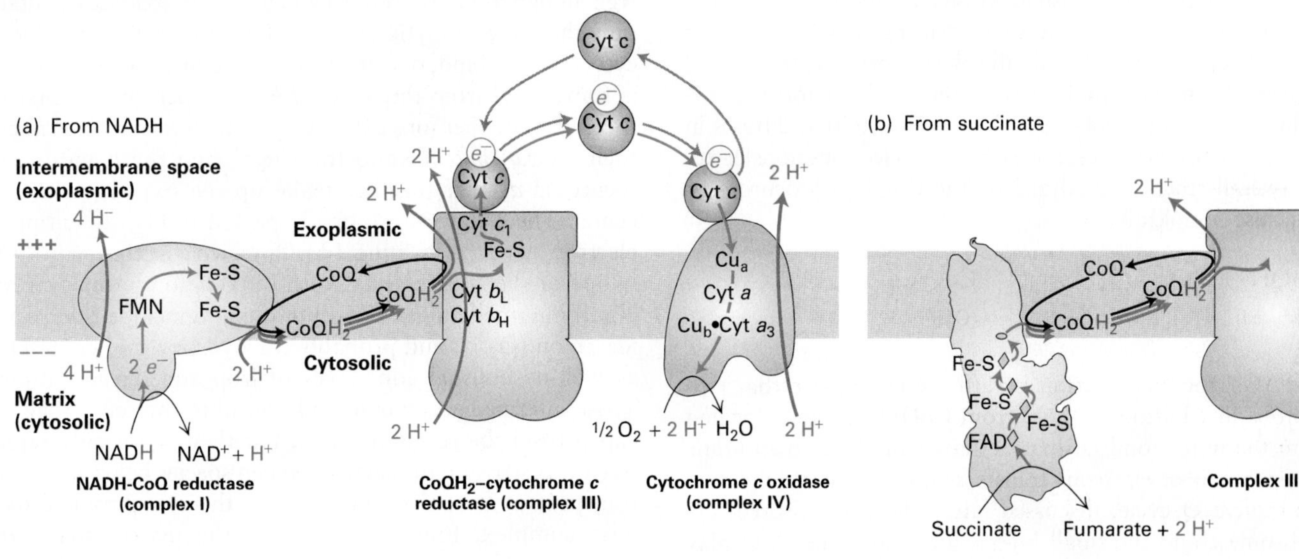

▲ **FIGURE 12-16 Multiprotein complexes and mobile electron carriers of the electron transport chain.** Electrons (blue arrows) flow through four major multiprotein complexes (I–IV). Electron movement between complexes is mediated either by the lipid-soluble molecule coenzyme Q (CoQ, oxidized form; $CoQH_2$, reduced form) or the water-soluble protein cytochrome *c* (cyt *c*). The multiple protein complexes use the energy released from passing electrons to pump protons from the matrix to the intramembrane space (red arrows). (a) Pathway from NADH. From NADH electrons flow to complex I, then III and then IV. A total of 10 protons are translocated per pair of electrons that flow from NADH to O_2. The protons released into the matrix space during oxidation of NADH by complex I are consumed in the formation of water from O_2 by complex IV, resulting in no net proton translocation from these reactions. (b) Pathway from succinate. Electrons flow from succinate (via $FADH_2$) in complex II to complex III and then IV. Electrons released during oxidation of succinate to fumarate in complex II are used to reduce CoQ to $CoQH_2$ without translocating additional protons. The remainder of electron transport from $CoQH_2$ proceeds by the same pathway as for the NADH pathway in (a). Thus for every pair of electrons transported from succinate to O_2, six protons are translocated by complexes III and IV.

which binds to a cleft on the matrix side of the transmembrane portions of complex II (Figure 12-16).

The overall reaction catalyzed by this complex is

$$\underset{\text{(Reduced)}}{\text{Succinate}} + \underset{\text{(Oxidized)}}{\text{CoQ}} \rightarrow \underset{\text{(Oxidized)}}{\text{fumarate}} + \underset{\text{(Reduced)}}{\text{CoQH}_2}$$

Although the $\Delta G^{\circ\prime}$ for this reaction is negative, the released energy is insufficient for proton pumping in addition to reduction of CoQ to form $CoQH_2$, which then dissociates from complex II. Thus no protons are translocated directly across the membrane by the succinate-CoQ reductase complex, and no proton-motive force is generated in this part of the respiratory chain. Shortly we will see how the protons and electrons in the $CoQH_2$ molecules generated by complex I and complex II contribute to the generation of the proton-motive force.

Complex II generates $CoQH_2$ from succinate via FAD/$FADH_2$-mediated redox reactions. Another set of proteins in the matrix and inner mitochondrial membrane performs a comparable set of FAD/$FADH_2$-mediated redox reactions to generate $CoQH_2$ from fatty acyl CoA. *Fatty acyl–CoA dehydrogenase,* which is a water-soluble enzyme, catalyzes the first step of the oxidation of fatty acyl CoA in the mitochondrial matrix (see Figure 12-12). There are several fatty acyl–CoA dehydrogenase enzymes with specificities for fatty acyl chains of different lengths. These enzymes mediate the initial step in a four-step process that removes two carbons from the fatty acyl group by oxidizing the carbon in the β position of the fatty acyl chain (thus the entire process is often referred to as β-oxidation). These reactions generate acetyl CoA, which in turn enters the citric acid cycle. They also generate an $FADH_2$ intermediate and NADH. The $FADH_2$ generated remains bound to the enzyme during the redox reaction, as is the case for complex II. A water-soluble protein called *electron transfer flavoprotein (ETF)* transfers the high-energy electrons from the $FADH_2$ in the acyl-CoA dehydrogenase to *electron transfer flavoprotein:ubiquinone oxidoreductase (ETF:QO)*, a membrane protein that reduces CoQ to $CoQH_2$ in the inner membrane. This $CoQH_2$ intermixes in the membrane with the other $CoQH_2$ molecules generated by complexes I and II.

$CoQH_2$–Cytochrome *c* Reductase (Complex III) A $CoQH_2$ generated either by complex I or complex II (or ETF:QO) donates two electrons to $CoQH_2$–cytochrome *c* reductase (complex III), regenerating oxidized CoQ. Concomitantly it releases into the intermembrane space two protons previously picked up on the matrix face, generating part of the proton-motive force

(Figure 12-16). Within complex III, the released electrons first are transferred to an iron-sulfur cluster within complex III and then to cytochrome c_1 or to two *b*-type cytochromes (b_L and b_H, see Q cycle below). Finally, the two electrons are transferred sequentially to two molecules of the oxidized form of cytochrome *c*, a water-soluble peripheral protein that diffuses in the intermembrane space. For each pair of electrons transferred, the overall reaction catalyzed by the $CoQH_2$–cytochrome *c* reductase complex is

$$\underset{\text{(Reduced)}}{CoQH_2} + \underset{\text{(Oxidized)}}{2\ Cyt\,c^{3+}} + 2\ H^+_{in} \rightarrow \underset{\text{(Oxidized)}}{CoQ} + 4\ H^+_{out} + \underset{\text{(Reduced)}}{2\ Cyt\,c^{2+}}$$

The $\Delta G^{\circ\prime}$ for this reaction is sufficiently negative that two protons in addition to those from $CoQH_2$ are translocated from the mitochondrial matrix across the inner membrane for each pair of electrons transferred; this involves the proton-motive Q cycle, discussed later. The heme protein cytochrome c and the small lipid-soluble molecule CoQ play similar roles in the electron transport chain in that they both serve as mobile electron shuttles, transferring electrons (and thus energy) between the complexes of the electron transport chain.

Cytochrome *c* Oxidase (Complex IV) Cytochrome *c*, after being reduced by $CoQH_2$–cytochrome *c* reductase (complex III), is reoxidized as it transports electrons, one at a time, to cytochrome *c* oxidase (complex IV) (Figure 12-16). Mitochondrial cytochrome *c* oxidases contain 13 different subunits, but the catalytic core of the enzyme consists of only three subunits. The function of the remaining subunits is less well understood. Bacterial cytochrome *c* oxidases contain only the three catalytic subunits. Four molecules of reduced cytochrome *c* bind, one at a time, to the oxidase. An electron is transferred from the heme of each cytochrome *c*, first to the pair of copper ions called Cu_a^{2+}, then to the heme *in* cytochrome *a*, and next to the Cu_b^{2+} and the heme in cytochrome a_3 that together make up the oxygen reduction center. The electrons are finally passed to O_2, the ultimate electron acceptor, yielding 4 H_2O, which together with CO_2 is one of the end products of the oxidation pathway. Proposed intermediates in oxygen reduction include the peroxide anion (O_2^{2-}) and probably the hydroxyl radical (OH·), as well as unusual complexes of iron and oxygen atoms. These intermediates would be harmful to the cell if they escaped from the reaction center, but they do so only rarely (see the discussion of reactive oxygen species below). During transport of four electrons through the cytochrome *c* oxidase complex, four protons from the matrix space are translocated across the membrane. However, the mechanism by which these protons are translocated is not known.

For each four electrons transferred, the overall reaction catalyzed by cytochrome *c* oxidase is

$$\underset{\text{(Reduced)}}{4\ Cyt\,c^{2+}} + 8\ H^+_{in} + O_2 \rightarrow \underset{\text{(Oxidized)}}{4\ Cyt\,c^{3+}} + 2\ H_2O + 4\ H^+_{out}$$

The poison cyanide, used as a chemical warfare agent, by spies to commit suicide when captured, in gas chambers to execute prisoners, and by the Nazis (Zyklon B

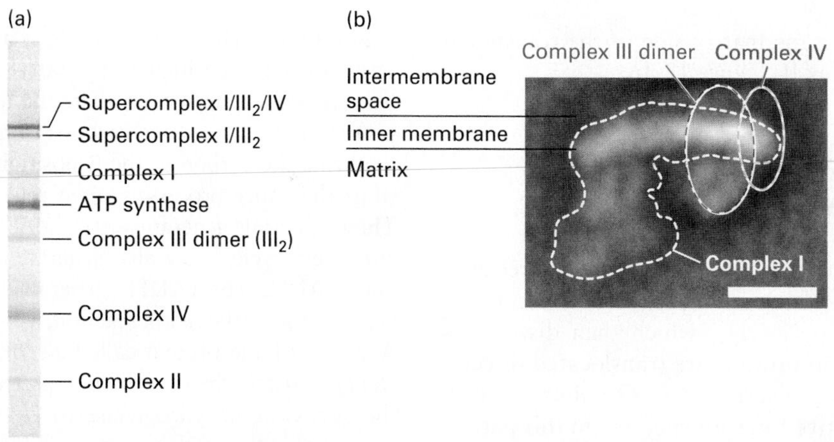

▲ **EXPERIMENTAL FIGURE 12-17 Electrophoresis and electron microscopic imaging identifies an electron transport chain supercomplex containing complexes I, III, and IV.** (a) Membrane proteins in isolated bovine heart mitochondria were solubilized with a detergent, and the complexes and supercomplexes were separated by gel electrophoresis using the blue native (BN)-PAGE method. Each blue-stained band within the gel represents the indicated protein complex or supercomplex, with III_2 representing a dimer of complex III. Intensity of the blue stain is approximately proportional to the amount of complex or supercomplex present. (b) Supercomplex I/III_2/IV was extracted from the gel, and the particles were negatively stained with 1% uranyl acetate and visualized by transmission electron microscopy. Images of 228 particles were combined at a resolution of ~3.4 nm to generate an averaged image of the complex viewed from the side in the plane of the membrane. Approximate locations of the complex III dimer and complex IV are indicated by dashed ovals; the outline of complex I is also indicated by a dashed line (white). Scale bar is 10 nm. [Adapted from E. Schafer et al., 2006, *J. Biol. Chem.* **281**(22):15370–15375.]

gas) for the mass murder of Jews and others, is toxic because it binds to the heme a_3 in mitochondrial cytochrome *c* oxidase (complex IV), inhibiting cellular respiration and therefore production of ATP. Cyanide is one of many toxic small molecules that interfere with energy production in mitochondria. ▮

Electron Transport Supercomplexes Over 50 years ago Britton Chance proposed that electron transport complexes might assemble into large supercomplexes. Doing so would bring the complexes into close and highly organized proximity, which might improve the speed and efficiency of the overall process. However, an alternative view holding that the complexes behaved as independent entities diffusing freely in the inner membrane became the dominant paradigm. During the past several years, genetic, biochemical, and biophysical studies have provided very strong evidence for the existence of electron transport chain supercomplexes. These studies involved relatively new gel electrophoretic methods called blue native (BN)-PAGE and colorless native (CN)-PAGE, which permit separation of very large macromolecular protein complexes, and electron microscopic analysis of their three-dimensional structures. One such supercomplex contains one copy of complex I, a dimer of complex III (III_2), and one or more copies of complex IV (Figure 12-17). The unique phospholipid *cardiolipin* (diphosphatidyl glycerol)

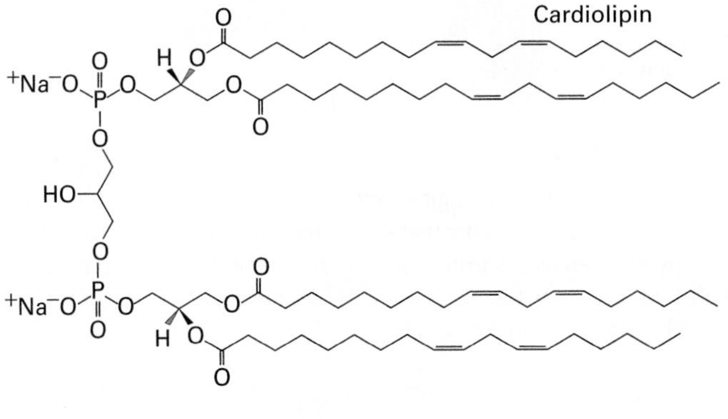

Cardiolipin

appears to play an important role in the assembly and function of these supercomplexes. Generally not observed in other membranes of eukaryotic cells, cardiolipin has been observed to bind to integral membrane proteins of the inner membrane (e.g., complex II). Genetic and biochemical studies in yeast mutants in which cardiolipin synthesis is blocked have established that cardiolipin contributes to the formation and activity of mitochondrial supercomplexes, and thus it has been called the glue that holds together the electron transport chain, though the precise mechanism remains to be defined. In addition, there is evidence that cardiolipin may influence the inner membrane's binding and permeability to protons and consequently the proton-motive force.

Reduction Potentials of Electron Carriers Favor Electron Flow from NADH to O_2

As we saw in Chapter 2, the **reduction potential** *E* for a partial reduction reaction

$$\text{Oxidized molecule} + e^- \rightleftharpoons \text{reduced molecule}$$

is a measure of the equilibrium constant of that partial reaction. With the exception of the *b* cytochromes in the $CoQH_2$–cytochrome *c* reductase complex, the standard reduction potential $E^{\circ\prime}$ of the electron carriers in the mitochondrial respiratory chain increases steadily from NADH to O_2. For instance, for the partial reaction

$$NAD^+ + H^+ + 2\,e^- \rightleftharpoons NADH$$

the value of the standard reduction potential is −320 mV, which is equivalent to a $\Delta G^{\circ\prime}$ of +14.8 kcal/mol for transfer of two electrons. Thus this partial reaction tends to proceed toward the left, that is, toward the oxidation of NADH to NAD^+.

By contrast, the standard reduction potential for the partial reaction

$$\text{Cytochrome } c_{ox}\,(Fe^{3+}) + e^- \rightleftharpoons \text{cytochrome } c_{red}\,(Fe^{2+})$$

is +220 mV ($\Delta G^{\circ\prime} = -5.1$ kcal/mol) for transfer of one electron. Thus this partial reaction tends to proceed toward the right, that is, toward the reduction of cytochrome *c* (Fe^{3+}) to cytochrome *c* (Fe^{2+}).

The final reaction in the respiratory chain, the reduction of O_2 to H_2O

$$2\,H^+ + {}^1/_2\,O_2 + 2\,e^- \rightarrow H_2O$$

has a standard reduction potential of +816 mV ($\Delta G^{\circ\prime} = -37.8$ kcal/mol for transfer of two electrons), the most positive in the whole series; thus this reaction also tends to proceed toward the right.

As illustrated in Figure 12-18, the steady increase in $E^{\circ\prime}$ values, and the corresponding decrease in $\Delta G^{\circ\prime}$ values, of the carriers in the electron transport chain favors the flow of electrons from NADH and $FADH_2$ (generated from succinate) to oxygen.

Experiments Using Purified Complexes Established the Stoichiometry of Proton Pumping

The multiprotein complexes responsible for proton pumping coupled to electron transport have been identified by selectively extracting mitochondrial membranes with detergents, isolating each of the complexes in nearly pure form, and then preparing artificial phospholipid vesicles (liposomes) containing each complex. When an appropriate electron donor and electron acceptor are added to such

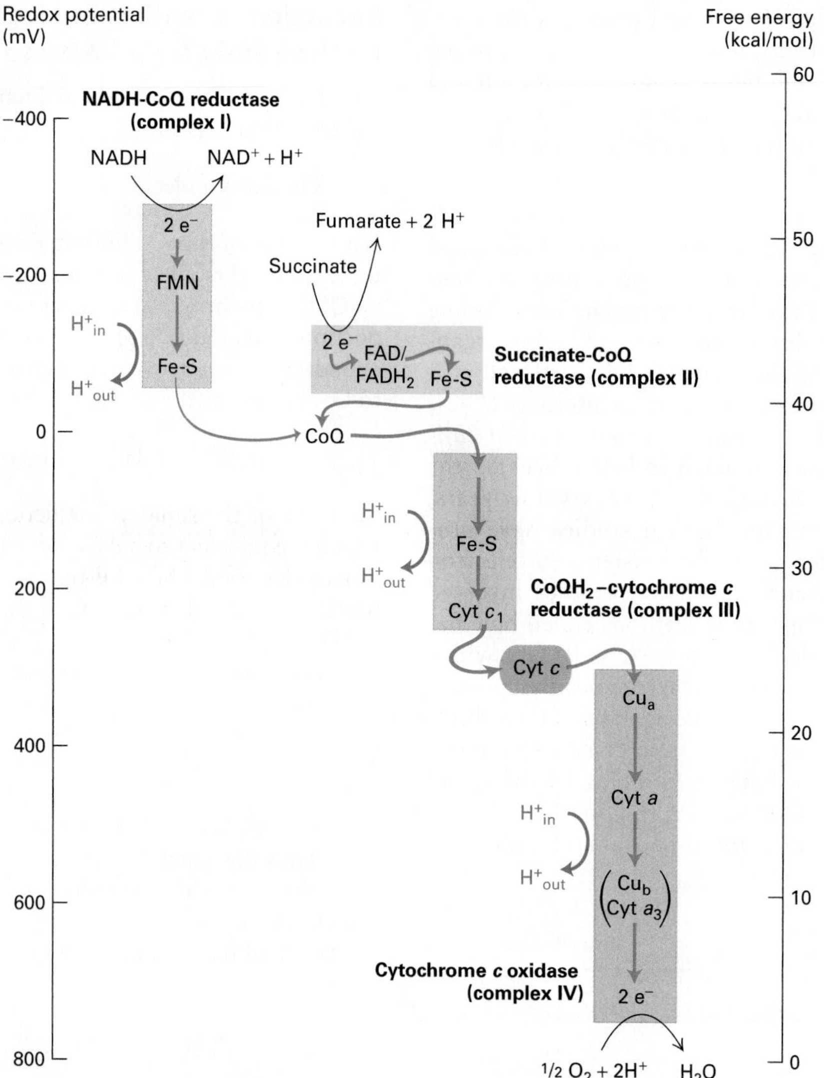

▲ **FIGURE 12-18 Changes in redox potential and free energy during stepwise flow of electrons through the respiratory chain.** Blue arrows indicate electron flow; red arrows, translocation of protons across the inner mitochondrial membrane. Electrons pass through the multiprotein complexes from those at a lower reduction potential to those with a higher (more positive) potential (left scale), with a corresponding reduction in free energy (right scale). The energy released as electrons flow through three of the complexes is sufficient to power the pumping of H^+ ions across the membrane, establishing a proton-motive force.

liposomes, a change in pH of the medium will occur if the embedded complex transports protons (Figure 12-19). Studies of this type indicate that NADH-CoQ reductase (complex I) translocates four protons per pair of electrons transported, whereas cytochrome *c* oxidase (complex IV) translocates two protons per electron pair transported (or, equivalently, for every two molecules of cytochrome *c* oxidized).

Current evidence suggests that a total of 10 protons are transported from the matrix space across the inner mitochondrial membrane for every electron pair that is transferred from NADH to O_2 (see Figure 12-16). Because succinate-CoQ reductase (complex II) does not transport protons and complex I is bypassed when the electrons come from succinate-derived $FADH_2$, only six protons are transported across the membrane for every electron pair that is transferred from this $FADH_2$ to O_2.

The Q Cycle Increases the Number of Protons Translocated as Electrons Flow Through Complex III

Experiments such as the one depicted in Figure 12-19 have shown that four protons are translocated across the membrane per electron pair transported from $CoQH_2$ through $CoQH_2$–cytochrome *c* reductase (complex III). Thus this complex transports two protons per electron transferred, whereas cytochrome *c* oxidase (complex IV) transports only one proton per electron transferred. An evolutionarily conserved mechanism, called the *Q cycle,* accounts for the

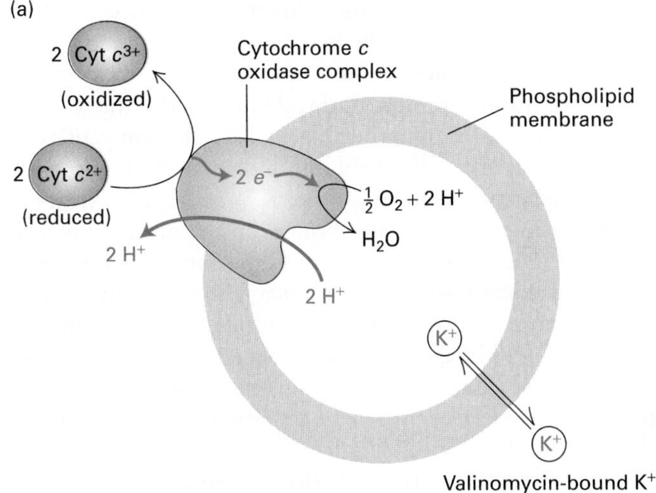

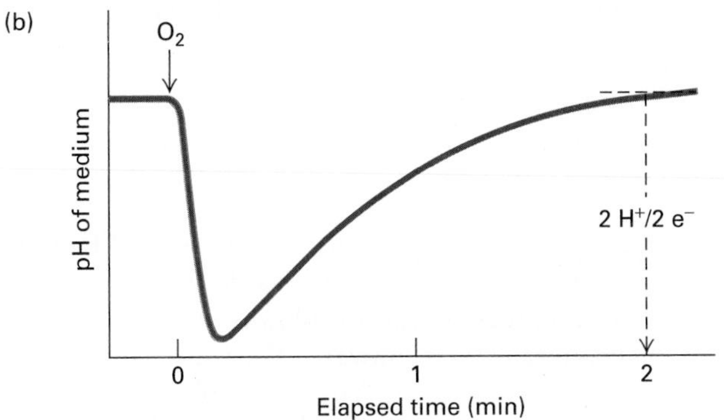

▲ EXPERIMENTAL FIGURE 12-19 **Electron transfer from reduced cytochrome *c* to O_2 via cytochrome c oxidase (complex IV) is coupled to proton transport.** The oxidase complex is incorporated into liposomes with the binding site for cytochrome *c* positioned on the outer surface. (a) When O_2 and reduced cytochrome *c* are added, electrons are transferred to O_2 to form H_2O and protons are transported from the inside to the medium outside of the vesicles. A drug called valinomycin was added to the medium to dissipate the voltage gradient generated by the translocation of H^+, which would otherwise reduce the number of protons moved across the membrane. (b) Monitoring of the medium's pH reveals a sharp drop in pH following addition of O_2. As the reduced cytochrome *c* becomes fully oxidized, protons leak back into the vesicles, and the pH of the medium returns to its initial value. Measurements show that two protons are transported per O atom reduced. Two electrons are needed to reduce one O atom, but cytochrome *c* transfers only one electron; thus two molecules of Cyt c^{2+} are oxidized for each O reduced. [Adapted from B. Reynafarje et al., 1986, *J. Biol. Chem.* **261**:8254.]

two-for-one transport of protons and electrons by complex III (Figure 12-20).

The substrate for complex III, $CoQH_2$, is generated by several enzymes, including NADH-CoQ reductase (complex I) and succinate-CoQ reductase (complex II), *electron transfer flavoprotein:ubiquinone oxidoreductase (ETF:QO, during β-oxidation)*, and, as we shall see, by complex III itself.

In one turn of the Q cycle, two molecules of $CoQH_2$ are oxidized to CoQ at the Q_o site and release a total of four protons into the intermembrane space, but one molecule of $CoQH_2$ is regenerated from CoQ at the Q_i site (see Figure 12-20, *bottom*). Thus the *net* result of the Q cycle is that four protons are translocated to the intermembrane space for every two electrons transported through the $CoQH_2$–cytochrome *c* reductase complex and accepted by two molecules of cytochrome *c*. The translocated protons are all derived from $CoQH_2$, which

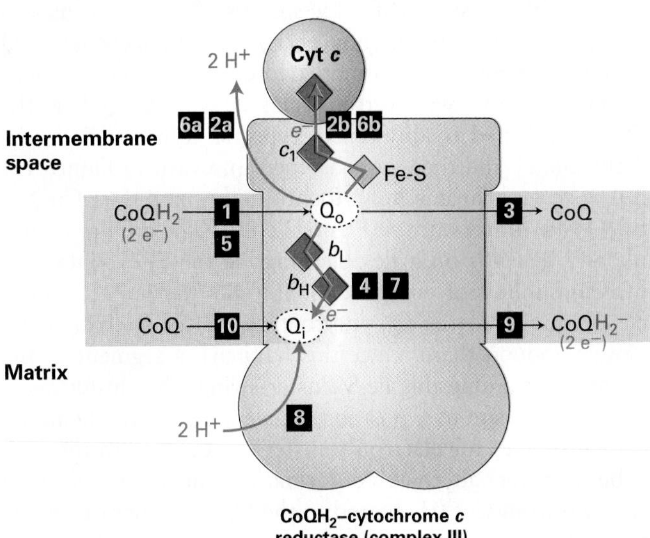

At Q_o site: $2\,CoQH_2 + 2\,Cyt\ c^{3+} \longrightarrow$ (4 H^+, 4 e^-)
$2\,CoQ + 2\,Cyt\ c^{2+} + 2\,e^- + 4\,H^+_{(outside)}$ (2 e^-)

At Q_i site: $CoQ + 2\,e^- + 2\,H^+_{(matrix\ side)} \longrightarrow CoQH_2$ (2 H^+, 2 e^-)

Net Q cycle (sum of reactions at Q_o and Q_i):
$CoQH_2 + 2\,Cyt\ c^{3+} + 2\,H^+_{(matrix\ side)} \longrightarrow$ (2 H^+, 2 e^-)
$CoQ + 2\,Cyt\ c^{2+} + 4\,H^+_{(outside)}$ (2 e^-)

Per 2 e^- transferred through complex III to cytochrome *c*, 4 H^+ released to the intermembrane space

▲ **FIGURE 12-20 The Q cycle.** The Q cycle begins when a molecule from the combined pool of reduced $CoQH_2$ in the membrane binds to the Q_o site on the *intermembrane space (outer) side* of the transmembrane portion of complex III (step **1**). There, $CoQH_2$ releases two protons into the intermembrane space (step **2a**) and two electrons and the resulting CoQ dissociate (step **3**). One of the electrons is transported, via an iron-sulfur protein and cytochrome c_1, directly to cytochrome *c* (step **2b**). (Recall that each cytochrome *c* shuttles one electron from complex III to complex IV.) The other electron moves through cytochromes b_L and b_H and partially reduces an oxidized CoQ molecule bound to the second, Q_i, site on the *matrix (inner) side* of the complex, forming a CoQ semiquinone anion, $Q^{-}\cdot$ (step **4**). The process is repeated with the binding of a second $CoQH_2$ at the Q_o site (step **5**), proton release (step **6a**), reduction of another cytochrome *c* (step **6b**), and addition of the other electron to the $Q^{-}\cdot$ bound at the Q_i site (step **7**). There, the addition of two protons from the matrix yields a fully reduced $CoQH_2$ molecule at the Q_i site, which then dissociates (steps **8** and **9**), freeing the Q_i to bind a new molecule of CoQ (step **10**) and begin the Q cycle over again. [Adapted from B. Trumpower, 1990, *J. Biol. Chem.* **265**:11409, and E. Darrouzet et al., 2001, *Trends Biochem. Sci.* **26**:445.]

obtained its protons from the matrix, as a consequence of the reduction of CoQ catalyzed by either NADH-CoQ reductase (complex I) or by $CoQH_2$–cytochrome *c* reductase (complex III) (see Figure 12-16). Although seemingly cumbersome, the Q cycle optimizes the numbers of protons pumped per pair of electrons moving through complex III. The Q cycle is found in all plants and animals as well as in bacteria. Its formation at a very early stage of cellular evolution was likely essential for the success of all life-forms as a way of converting the potential energy in reduced coenzyme Q into the maximum proton-motive force across a membrane.

How are the two electrons released from $CoQH_2$ at the Q_o site directed to different acceptors, either to Fe-S, cytochrome c_1 and then cytochrome *c* (upward in Figure 12-20) or to cytochrome b_L, cytochrome b_H, and then CoQ at the Q_i site (downward in Figure 12-20)? The answer is simple and depends on a flexible hinge in the Fe-S-containing protein subunit of complex III. Initially the Fe-S cluster is close enough to the Q_o site to pick up an electron from $CoQH_2$ bound there. Once this happens, a segment of the protein containing this Fe-S cluster swings the cluster away from the Q_o site to a position near enough to the heme on cytochrome c_1 for electron transfer to occur. With the Fe-S subunit in this alternate conformation, the second electron released from $CoQH_2$ bound to the Q_o site cannot move to the Fe-S cluster—it is too far away, so it takes an alternative path open to it via a somewhat less thermodynamically favored route to cytochrome b_L.

The Proton-Motive Force in Mitochondria Is Due Largely to a Voltage Gradient Across the Inner Membrane

One result of the electron transport chain is the generation of the proton-motive force (pmf), which is the sum of a transmembrane proton concentration (pH) gradient and electric potential, or voltage, gradient. It has been possible to determine experimentally the relative contribution of the two components to the total pmf. The relative contributions depend on the permeability of the membrane to ions other than H^+. A significant voltage gradient can develop only if the membrane is poorly permeable to other cations and to anions. Otherwise, anions would leak across from the matrix to the intermembrane space along with the protons and prevent a voltage gradient from forming. Similarly, cations leaking across from the intermembrane space to the matrix (exchange of like charge) would also short-circuit voltage gradient formation. Indeed, the inner mitochondrial membrane is poorly permeable to other ions. Thus proton pumping generates a voltage gradient that makes it energetically difficult for additional protons to move across because of charge repulsion. As a consequence, proton pumping by the electron transport chain establishes a robust voltage gradient in the context of a rather small pH gradient.

Because mitochondria are much too small to be impaled with electrodes, the electric potential and pH gradient across the inner mitochondrial membrane cannot be determined by direct measurement. Nevertheless, it has been possible to develop methods to measure indirectly these critical values. The electric potential can be measured by adding radioactive $^{42}K^+$ ions and a trace amount of valinomycin to a suspension of respiring mitochondria. Although the inner membrane is normally impermeable to K^+, valinomycin is an *ionophore*, a small lipid-soluble molecule that selectively binds a specific ion (in this case, K^+) and carries it across otherwise impermeable membranes. In the presence of valinomycin, $^{42}K^+$ equilibrates across the inner membrane of isolated mitochondria in accordance with the electric potential: the more negative the matrix side of the membrane, the more $^{42}K^+$ will be attracted to and accumulate in the matrix.

At equilibrium, the measured concentration of radioactive K^+ ions in the matrix, $[K_{in}]$, is about 500 times greater than that in the surrounding medium, $[K_{out}]$. Substitution of this value into the Nernst equation (Chapter 11) shows that the electric potential *E* (in mV) across the inner membrane in respiring mitochondria is −160 mV, with the matrix (inside) negative:

$$E = -59 \log \frac{[K_{in}]}{[K_{out}]} = -59 \log 500 = -160 \text{ mV}$$

Researchers can measure the matrix (inside) pH by trapping pH-sensitive fluorescent dyes inside vesicles formed from the inner mitochondrial membrane, with the matrix side of the membrane facing inward. They also can measure the pH outside of the vesicles (equivalent to the intermembrane space) and thus determine the pH gradient (ΔpH), which turns out to be ~1 pH unit. Since a difference of one pH unit represents a tenfold difference in H^+ concentration, according to the Nernst equation a pH gradient of one unit across a membrane is equivalent to an electric potential of 59 mV at 20 °C. Thus, knowing the voltage and pH gradients, we can calculate the proton-motive force, pmf, as

$$\text{pmf} = \Psi - \left(\frac{RT}{F} \times \Delta\text{pH}\right) = \Psi - 59\ \Delta\text{pH}$$

where *R* is the gas constant of 1.987 cal/(degree·mol), *T* is the temperature (in degrees Kelvin), *F* is the Faraday constant [23,062 cal/(V·mol)], and Ψ is the transmembrane electric potential; Ψ and pmf are measured in millivolts. The electric potential Ψ across the inner membrane is −160 mV (negative inside matrix) and ΔpH is equivalent to ≈60 mV. Thus the total pmf is −220 mV, with the transmembrane electric potential responsible for about 73 percent of the total.

Toxic By-products of Electron Transport Can Damage Cells

About 1–2 percent of the oxygen metabolized by aerobic organisms, rather than being converted to water, is partially reduced to the superoxide anion radical (O_2^-). Superoxide is unstable in aqueous biological liquids, breaking down into especially toxic hydrogen peroxide (H_2O_2) and then hydroxyl radicals. These and other *reactive oxygen species* (ROS), which contribute to what is often called *cellular oxidative stress,* can be highly toxic, because they chemically modify proteins, DNA, and unsaturated fatty acyl groups in membrane lipids, thus interfering with normal function. Indeed, ROS are purposefully generated by body defense cells (e.g., macrophages) to kill pathogens. In humans, excessive or inappropriate generation of ROS has been implicated in many diverse diseases, including heart failure, neurodegenerative diseases, alcohol-induced liver disease, diabetes, and aging.

Although ROS can be generated by a number of metabolic pathways, the major source of ROS appears to be the electron transport chain, in particular mechanisms coupled to complexes I and III. The semiquinone form of ubiquinone, $CoQ^{-}\cdot$ (see Figure 12-15), an intermediate form of CoQ generated in the Q cycle, may play a particularly important role in superoxide generation.

To help protect against ROS toxicity, mitochondria have evolved several defense mechanisms, including the use of enzymes that inactivate superoxide first by converting it to H_2O_2 (Mn-containing superoxide dismutase) and then to H_2O (glutathione peroxidase, which also detoxifies the lipid hydroperoxide products formed when ROS react with unsaturated fatty acyl groups). Cardiac mitochondria also have catalase (normally only found in peroxisomes) to help breakdown H_2O_2. This is not surprising, because the most oxygen-consuming organ in mammals is the heart. In addition, the small molecule antioxidants α-lipoic acid and vitamin E help protect the mitochondrion from ROS. ▌

KEY CONCEPTS OF SECTION 12.2

Electron Transport and Generation of the Proton-Motive Force

- By the end of the citric acid cycle (stage II), much of the energy originally present in the covalent bonds of glucose and fatty acids is converted into high-energy electrons in the reduced coenzymes NADH and $FADH_2$. The energy from these electrons is used to generate the proton-motive force.

- In the mitochondrion, the proton-motive force is generated by coupling electron flow (from NADH and $FADH_2$ to O_2) to the uphill transport of protons from the matrix across the inner membrane to the intermembrane space. This process together with the synthesis of ATP from ADP and P_i driven by the proton-motive force is called oxidative phosphorylation.

- The flow of electrons from $FADH_2$ and NADH to O_2 is directed through multiprotein complexes. The four major complexes are NADH-CoQ reductase (complex I), succinate-CoQ reductase (complex II), $CoQH_2$–cytochrome *c* reductase (complex III), and cytochrome *c* oxidase (complex IV).

- Each complex contains one or more electron-carrying prosthetic groups: iron-sulfur clusters, flavins, heme groups, and copper ions (see Table 12-2). Cytochrome *c*, which contains heme, and coenzyme Q (CoQ), a lipid-soluble small molecule, are mobile carriers that shuttle electrons between the complexes.

- Complexes I, III, and IV pump protons from the matrix into the intermembrane space. Complexes I and II reduce CoQ to $CoQH_2$, which carries protons and high-energy electrons to complex III. The heme protein cytochrome *c* carries electrons from complex III to complex IV, which uses them to pump protons and reduce molecular oxygen to water.

- The high-energy electrons from NADH enter the electron transport chain through complex I, whereas the high-energy electrons from $FADH_2$ (derived from succinate in the citric acid cycle) enter the electron transport chain through complex II. Additional electrons derived from $FADH_2$ by the initial step of fatty acyl–CoA β-oxidation increase the supply of $CoQH_2$ available for electron transport.

- Within the inner membrane, electron transport complexes assemble into supercomplexes held together by cardiolipin, a specialized phospholipid. Supercomplex formation may enhance the speed and efficiency of generation of the proton-motive force.

- Each electron carrier accepts an electron or electron pair from a carrier with a less positive reduction potential and transfers the electron to a carrier with a more positive reduction potential. Thus the reduction potentials of electron carriers favor unidirectional electron flow from NADH and $FADH_2$ to O_2 (see Figure 12-18).

- The Q cycle allows four protons (rather than two) to be translocated per pair of electrons moving through complex III (see Figure 12-20).

- A total of 10 H^+ ions are translocated from the matrix across the inner membrane per electron pair flowing from NADH to O_2 (see Figure 12-16), whereas 6 H^+ ions are translocated per electron pair flowing from $FADH_2$ to O_2.

- The proton-motive force is due largely to a voltage gradient across the inner membrane produced by proton pumping; the pH gradient plays a quantitatively less important role.

- Reactive oxygen species (ROS) are toxic by-products of the electron transport chain that can modify and damage proteins, DNA, and lipids. Specific enzymes (e.g., glutathione peroxidase, catalase) and small molecule antioxidants (e.g., vitamin E) help protect against ROS-induced damage.

12.3 Harnessing the Proton-Motive Force for Energy-Requiring Processes

The hypothesis that a proton-motive force across the inner mitochondrial membrane is the immediate source of energy for ATP synthesis was proposed in 1961 by Peter Mitchell. Virtually all researchers studying oxidative phosphorylation and photosynthesis initially rejected his *chemiosmotic hypothesis*. They favored a mechanism similar to the then well-elucidated substrate-level phosphorylation in glycolysis, in which chemical transformation of a substrate molecule (i.e., phosphoenolpyruvate) is directly coupled to ATP synthesis. Despite intense efforts by a large number of investigators, however, compelling evidence for such a direct mechanism was never observed.

Definitive evidence supporting Mitchell's hypothesis depended on development of techniques to purify and reconstitute organelle membranes and membrane proteins. The experiment with vesicles made from chloroplast thylakoid membranes (described in detail below) that contain **ATP synthase,** outlined in Figure 12-21, was one of several demonstrating that this protein is an ATP-generating enzyme and that ATP generation is dependent on proton movement down an electrochemical gradient. It turns out that the protons actually move *through* the ATP synthase as they traverse the membrane!

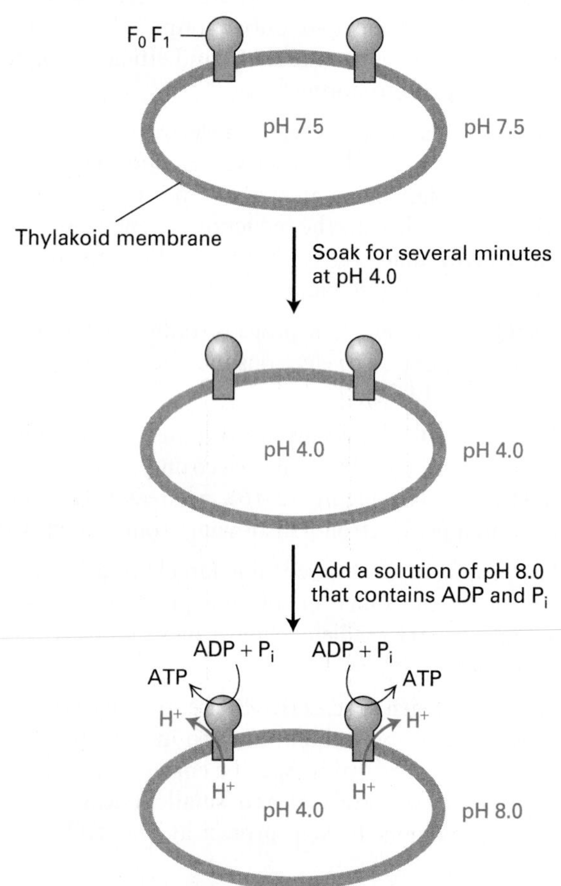

▲ EXPERIMENTAL FIGURE 12-21 **Synthesis of ATP by F_0F_1 depends on a pH gradient across the membrane.** Isolated chloroplast thylakoid vesicles containing F_0F_1 particles were equilibrated in the dark with a buffered solution at pH 4.0. When the pH in the thylakoid lumen became 4.0, the vesicles were rapidly mixed with a solution at pH 8.0 containing ADP and P_i. A burst of ATP synthesis accompanied the transmembrane movement of protons driven by the 10,000-fold H^+ concentration gradient (10^{-4} M versus 10^{-8} M). In similar experiments using "inside-out" preparations of mitochondrial membrane vesicles, an artificially generated membrane electric potential also resulted in ATP synthesis.

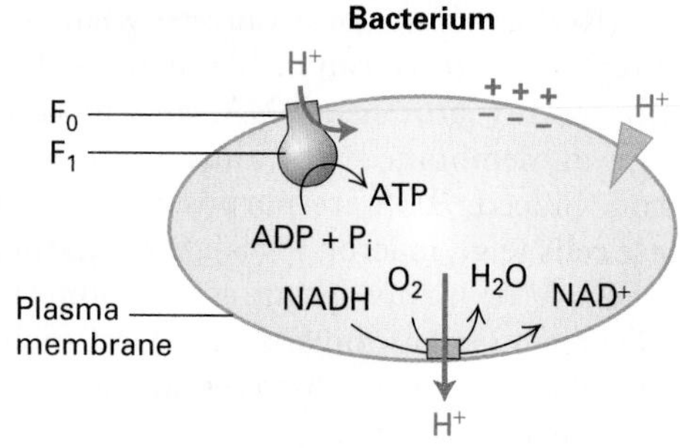

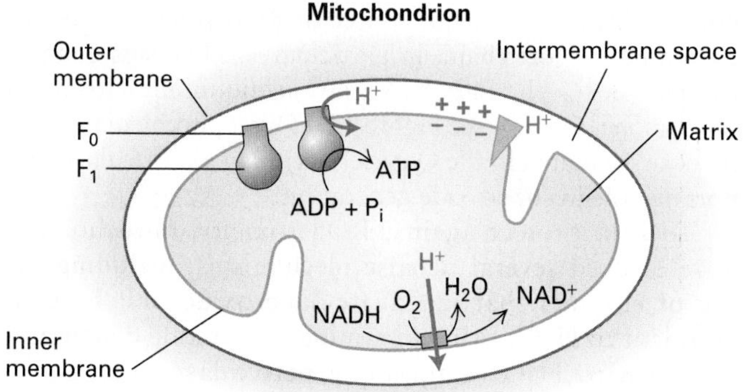

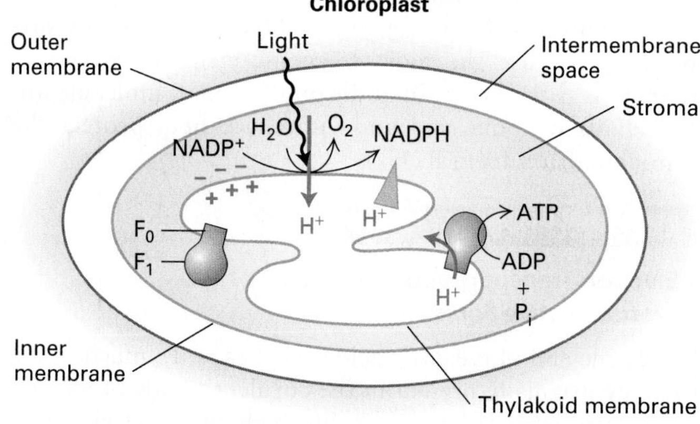

▲ FIGURE 12-22 **Chemiosmosis in bacteria, mitochondria, and chloroplasts.** The membrane surface facing a shaded area is a cytosolic face; the surface facing an unshaded, white area is an exoplasmic face. Note that the cytosolic face of the bacterial plasma membrane, the matrix face of the inner mitochondrial membrane, and the stromal face of the thylakoid membrane are all equivalent. During electron transport, protons are always pumped from the cytosolic face to the exoplasmic face, creating a proton concentration gradient (exoplasmic face > cytosolic face) and an electric potential (negative cytosolic face and positive exoplasmic face) across the membrane. During the synthesis of ATP, protons flow in the reverse direction (down their electrochemical gradient) through ATP synthase (F_0F_1 complex), which protrudes in a knob at the cytosolic face in all cases.

As we shall see, the ATP synthase is a multiprotein complex that can be subdivided into two subcomplexes called F_0 (containing the transmembrane portions of the complex) and F_1 (containing the globular portions of the complex that sit above the membrane and point toward the matrix space in mitochondria). Thus the ATP synthase is often also called the F_0F_1 complex; we will use the terms interchangeably.

The Mechanism of ATP Synthesis Is Shared Among Bacteria, Mitochondria, and Chloroplasts

Although bacteria lack any internal membranes, aerobic bacteria nonetheless carry out oxidative phosphorylation by the same processes that occur in eukaryotic mitochondria. Enzymes that catalyze the reactions of both the glycolytic pathway and the citric acid cycle are present in the cytosol of bacteria; enzymes that oxidize NADH to NAD^+ and transfer the electrons to the ultimate acceptor O_2 reside in the bacterial plasma membrane.

The movement of electrons through these membrane carriers is coupled to the pumping of protons out of the cell. The movement of protons back into the cell, down their concentration gradient, is coupled to the synthesis of ATP. This general process is similar for bacteria and eukaryotes (in both mitochondrial and chloroplasts) (Figure 12-22). The bacterial ATP synthases (F_0F_1 complex) are essentially identical in structure and function to the mitochondrial and chloroplast ATP synthases but are simpler to purify and study.

A primitive aerobic bacterium was probably the progenitor of mitochondria in eukaryotic cells (Figure 12-23). According to this *endosymbiont hypothesis*, the inner mitochondrial membrane would be derived from the bacterial plasma membrane with its cytosolic face pointing toward what became the matrix space of the mitochondrion. Similarly, in plants the progenitor's plasma membrane became the chloroplast's thylakoid membrane and its cytosolic face pointed toward what became the stromal space of the chloroplast (described in detail below).

In all cases, ATP synthase is positioned with the globular F_1 domain, which catalyzes ATP synthesis, on the cytosolic face of the membrane, so ATP is always formed on the cytosolic face of the membrane (see Figure 12-22). Protons always flow through ATP synthase from the exoplasmic to the cytosolic face of the membrane, which in the mitochondrion is from the intermembrane to the matrix space. This flow is driven by the proton motive force. Invariably, the cytosolic face has a negative electric potential relative to the exoplasmic face.

In addition to ATP synthesis, the proton-motive force across the bacterial plasma membrane is used to power other processes, including the uptake of nutrients such as sugars (using proton/sugar symporters) and the rotation of bacterial flagella. Chemiosmotic coupling thus illustrates an important principle introduced in our discussion of active transport in Chapter 11: *the membrane potential, the concentration gradients of protons (and other ions) across a membrane, and the phosphoanhydride bonds in ATP are equivalent and interconvertible forms of chemical potential energy.* Indeed, ATP synthesis through ATP synthase can be thought of as active transport in reverse.

ATP Synthase Comprises Two Multiprotein Complexes Termed F_0 and F_1

With general acceptance of Mitchell's chemiosmotic mechanism, researchers turned their attention to the structure and operation of the F_0F_1 complex. The F_0F_1 complex, or ATP synthase, has two principal components, F_0 and F_1, both of which are

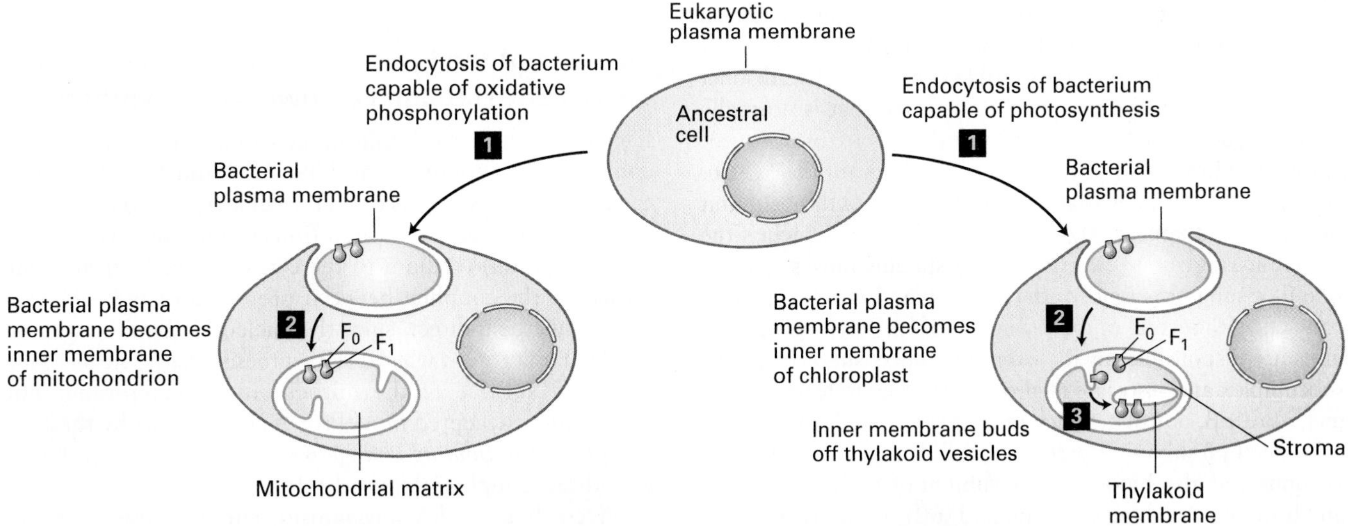

▲ **FIGURE 12-23 Endosymbiont hypothesis for the evolutionary origin of mitochondria and chloroplasts.** Endocytosis of a bacterium by an ancestral eukaryotic cell (step 1) would generate an organelle with two membranes, the outer membrane derived from the eukaryotic plasma membrane and the inner one from the bacterial membrane (step 2). The F_1 subunit of ATP synthase, localized to the cytosolic face of the bacterial membrane, would then face the matrix of the evolving mitochondrion (*left*) or chloroplast (*right*). Budding of vesicles from the inner chloroplast membrane, such as occurs during development of chloroplasts in contemporary plants, would generate the thylakoid membranes with the F_1 subunit remaining on the cytosolic face, facing the chloroplast stroma (step 3). Membrane surfaces facing a shaded area are cytosolic faces; surfaces facing an unshaded area are exoplasmic faces.

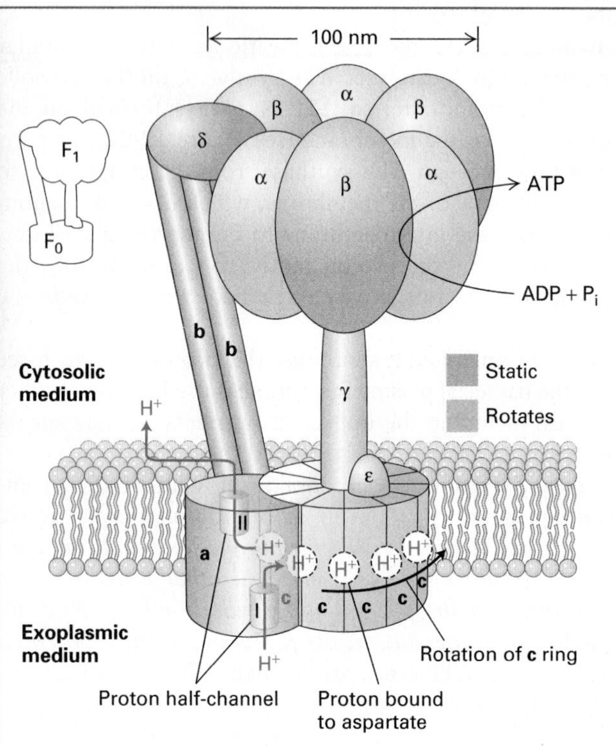

◀ **FIGURE 12-24 Structure and function of ATP synthase (the F_0F_1 complex) in the bacterial plasma membrane.** The F_0 membrane-embedded portion of ATP synthase is built of three integral membrane proteins: one copy of **a,** two copies of **b,** and on average 10 copies of **c** arranged in a ring in the plane of the membrane. Two proton half-channels lie at the interface between the **a** subunit and the **c** ring. Half-channel I allows protons to move one at a time from the exoplasmic medium and bind to aspartate-61 in the center of a **c** subunit near the middle of the membrane. Half-channel II (after rotation of the **c** ring) permits protons to dissociate from the aspartate and move into the cytosolic medium. The F_1 portion contains three copies each of subunits α and β that form a hexamer resting atop the single rod-shaped γ subunit, which is inserted into the **c** ring of F_0. The ε subunit is rigidly attached to the γ subunit and also to several of the **c** subunits. The δ subunit permanently links one of the α subunits in the F_1 complex to the **b** subunit of F_0. Thus the F_0 **a** and **b** subunits and the F_1 δ subunit and $(\alpha\beta)_3$ hexamer form a rigid structure anchored in the membrane (orange). During proton flow, the **c** ring and the attached F_1 ε and γ subunits rotate as a unit (green), causing conformation changes in the F_1 β subunits, leading to ATP synthesis. [Adapted from M. J. Schnitzer, 2001, *Nature* **410:**878, and P. D. Boyer, 1999, *Nature* **402:**247.]

multimeric proteins (Figure 12-24). The F_0 component contains three types of integral membrane proteins, designated **a**, **b**, and **c**. In bacteria and in yeast mitochondria the most common subunit composition is $a_1b_2c_{10}$, but F_0 complexes in animal mitochondria have 12 **c** subunits and those in chloroplasts have 14. In all cases the **c** subunits form a doughnut-shaped ring in the plane of the membrane. The **a** and two **b** subunits are rigidly linked to one another but not to the ring of **c** subunits, a critical feature of the protein to which we will return shortly.

The F_1 portion is a water-soluble complex of five distinct polypeptides with the composition $\alpha_3\beta_3\gamma\delta\epsilon$ that is normally firmly bound to the F_0 subcomplex at the surface of the membrane. The lower end of the rodlike γ subunit of the F_1 subcomplex is a coiled coil that fits into the center of the c-subunit ring of F_0 and appears rigidly attached to it. Thus when the c-subunit ring rotates, the rodlike γ subunit moves with it. The F_1 ε subunit is rigidly attached to γ and also forms tight contacts with several of the **c** subunits of F_0. The α and β subunits are responsible for the overall globular shape of the F_1 subcomplex and associate in alternating order to form a hexamer, αβαβαβ, or $(\alpha\beta)_3$, which rests atop the single long γ subunit. The F_1 δ subunit is permanently linked to one of the F_1 α subunits and also binds to the **b** subunit of F_0. Thus the F_0 **a** and **b** subunits and the δ subunit and $(\alpha\beta)_3$ hexamer of the F_1 complex form a rigid structure anchored in the membrane. The rodlike **b** subunits form a "stator" that prevents the $(\alpha\beta)_3$ hexamer from moving while it rests on the γ subunit, whose rotation together with the **c** subunits of F_0 plays an essential role in the ATP synthesis mechanism described below.

When ATP synthase is embedded in a membrane, the F_1 component forms a knob that protrudes from the cytosolic (in the mitochondrion this is the matrix) face. Because F_1 separated from membranes is capable of catalyzing ATP hydrolysis (ATP conversion to ADP plus P_i) in the absence of the F_0 component, it has been called the F_1 ATPase; however, its function in the cells is the reverse, to synthesize ATP. ATP hydrolysis is a spontaneous process ($\Delta G < 0$); thus energy is required to drive the ATPase in "reverse" and generate ATP.

Rotation of the F_1 γ Subunit, Driven by Proton Movement Through F_0, Powers ATP Synthesis

Each of the three β subunits in the globular F_1 portion of the complete F_0F_1 complex can bind ADP and P_i and catalyze the endergonic synthesis of ATP when coupled to the flow of protons from the exoplasmic (intermembrane space in the mitochondrion) medium to the cytosolic (matrix) medium. However, the coupling between proton flow and ATP synthesis must be indirect, since the nucleotide-binding sites on the β subunits of F_1, where ATP synthesis occurs, are 9–10 nm from the surface of the mitochondrial membrane. The most widely accepted model for ATP synthesis by the F_0F_1 complex—the *binding-change mechanism*—posits just such an indirect coupling (Figure 12-25).

According to this mechanism, energy released by the "downhill" movement of protons through F_0 directly powers rotation of the c-subunit ring together with its attached γ and ε subunits (see Figure 12-24). The γ subunit acts as a cam, or nonsymmetrical rotating shaft, whose rotation within center of the static $(\alpha\beta)_3$ hexamer of F_1 causes it to push sequentially against each of the β subunits and thus cause cyclical changes in their conformations between three

Animation: ATP Synthesis

Podcast: ATP Synthesis

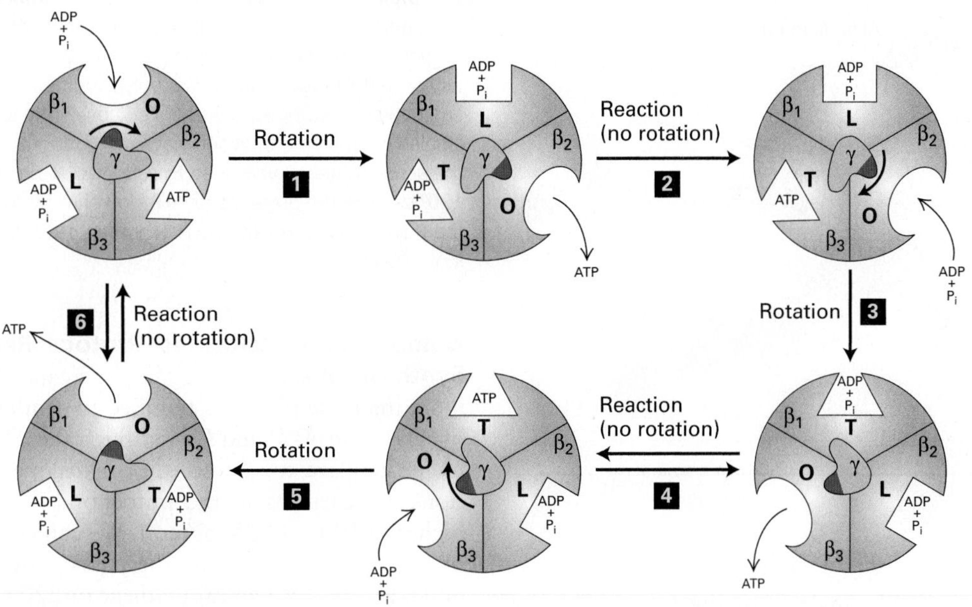

▲ **FIGURE 12-25 The binding-change mechanism of ATP synthesis from ADP and P_i.** This view is looking up at F_1 from the membrane surface (see Figure 12-24). As the γ subunit rotates by 120° in the center, each of the otherwise identical F_1 β subunits alternates between three conformational states (O, open with oval representation of the binding site; L, loose with a rectangular binding site; T, tight with a triangular site) that differ in their binding affinities for ATP, ADP, and P_i. The cycle begins (upper left) when ADP and P_i bind loosely to one of the three β subunits (here, arbitrarily designated β_1) whose nucleotide-binding site is in the O (open) conformation. Proton flux through the F_0 portion of the protein powers a 120° rotation of the γ subunit (relative to the fixed β subunits) (step **1**). This causes the rotating γ subunit, which is asymmetric, to push differentially against the β subunits, resulting in a conformational change and an increase in the binding affinity of the β_1 subunit for ADP and P_i (from O $\rightarrow$ L), an increase in the binding affinity of the β_3 subunit for ADP and P_i that were previously bound (from L $\rightarrow$ T), and a decrease in the binding affinity of the β_2 subunit for a previously bound ATP (from T $\rightarrow$ O), causing release of the bound ATP. Step **2**: Without additional rotation the ADP and P_i in the T site (here the β_3 subunit) form ATP, a reaction that does not require an input of additional energy due to the special environment in the active site of the T state. At the same time a new ADP and P_i bind loosely to the unoccupied O site on β_2. Step **3**: Proton flux powers another 120° rotation of the γ subunit, consequent conformational changes in the binding sites (L $\rightarrow$T, O$\rightarrow$ L, T$\rightarrow$ O), and release of ATP from β_3. Step **4**: Without additional rotation the ADP and P_i in the T site of β_1 form ATP, and additional ADP and P_i bind to the unoccupied O site on β_3. The process continues with rotation (step **5**) and ATP formation (step **6**) until the cycle is complete, with three ATPs having been produced for every 360° rotation of γ. [Adapted from P. Boyer, 1989, *FASEB J.* **3:**2164; Y. Zhou et al., 1997, *Proc. Nat'l. Acad. Sci. USA* **94:**10583; and M. Yoshida, E. Muneyuki, and T. Hisabori, 2001, *Nat. Rev. Mol. Cell Biol.* **2:**669–677.]

different states. As schematically depicted in a view of the bottom of the $(\alpha\beta)_3$ hexamer's globular structure in Figure 12-25, rotation of the γ subunit relative to the fixed $(a\beta)_3$ hexamer causes the nucleotide-binding site of each β subunit to cycle through three conformational states in the following order:

1. An O (open) state that binds ATP very poorly and ADP and P_i weakly
2. An L (loose) state that binds ADP and P_i more strongly but cannot bind ATP
3. A T (tight) state that binds ADP and P_i so tightly that they spontaneously react and form ATP

In the T state the ATP produced is bound so tightly that it cannot readily dissociate from the site—it is trapped until another rotation of the γ subunit returns that β subunit to the O state, thereby releasing ATP and beginning the cycle again. ATP or ADP also binds to regulatory or allosteric sites on the three α subunits; this binding modifies the rate of ATP synthesis according to the level of ATP and ADP in the matrix, but is not directly involved in synthesis of ATP from ADP and P_i.

Several types of evidence support the binding-change mechanism. First, biochemical studies showed that one of the three β subunits on isolated F_1 particles can tightly bind ADP and P_i and then form ATP, which remains tightly bound. The measured ΔG for this reaction is near zero, indicating that once ADP and P_i are bound to the T state of a β subunit, they spontaneously form ATP. Importantly, dissociation of the bound ATP from the β subunit on isolated F_1 particles occurs extremely slowly. This finding suggested that dissociation of ATP would have to be powered by a conformational change in the β subunit, which in turn would be caused by proton movement.

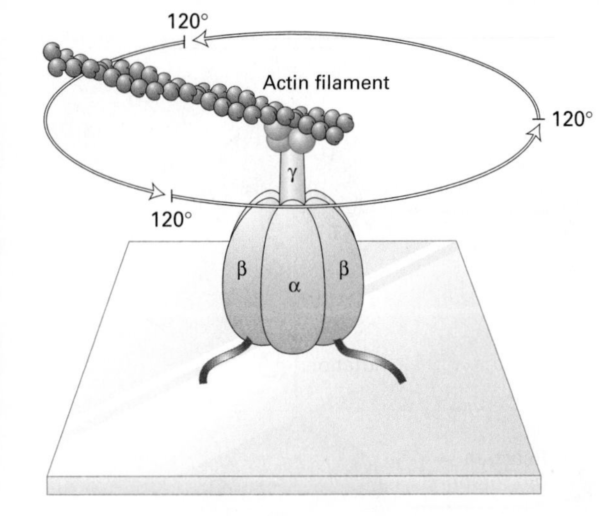

◀ **EXPERIMENTAL FIGURE 12-26 The γ subunit of the F_1 complex rotates relative to the $(\alpha\beta)_3$ hexamer.** F_1 complexes were engineered that contained β subunits with an additional His_6 sequence, which causes them to adhere to a glass plate coated with a metal reagent that binds polyhistidine. The γ subunit in the engineered F_1 complexes was linked covalently to a fluorescently labeled actin filament. When viewed in a fluorescence microscope, the actin filaments were seen to rotate counterclockwise in discrete 120° steps in the presence of ATP, powered by ATP hydrolysis by the β subunits. [Adapted from H. Noji et al., 1997, *Nature* **386:**299, and R. Yasuda et al., 1998, *Cell* **93:**1117.]

X-ray crystallographic analysis of the $(\alpha\beta)_3$ hexamer yielded a striking conclusion: although the three β subunits are identical in sequence and overall structure, the ADP/ATP-binding sites have different conformations in each subunit. The most reasonable conclusion was that the three β subunits cycle in an energy-dependent reaction between three conformational states (O, L, T), in which the nucleotide-binding site has substantially different structures.

In other studies, intact F_0F_1 complexes were treated with chemical cross-linking agents that covalently linked the γ and ε subunits and the c-subunit ring. The observation that such treated complexes could synthesize ATP or use ATP to power proton pumping indicates that the cross-linked proteins normally rotate together.

Finally, rotation of the γ subunit relative to the fixed $(\alpha\beta)_3$ hexamer, as proposed in the binding-change mechanism, was observed directly in the clever experiment depicted in Figure 12-26. In one modification of this experiment in which tiny gold particles, rather than an actin filament, were attached to the γ subunit, rotation rates of 134 revolutions per second were observed. Hydrolysis of 3 ATPs, which you recall is the reverse reaction catalyzed by the same enzyme, is thought to power one revolution; this result is close to the experimentally determined rate of ATP hydrolysis by F_0F_1 complexes: about 400 ATPs per second. In a related experiment, a γ subunit linked to an ε subunit and a ring of c subunits was seen to rotate relative to the fixed $(\alpha\beta)_3$ hexamer. Rotation of the γ subunit in these experiments was powered by ATP hydrolysis, the reverse of the normal process in which proton movement through the F_0 complex drives rotation of the γ subunit. These observations established that the γ subunit, along with the attached c ring and ε subunit, does indeed rotate, thereby driving the conformational changes in the β subunits that are required for binding of ADP and P_i, followed by synthesis and subsequent release of ATP.

Number of Translocated Protons Required for ATP Synthesis A simple calculation indicates that the passage of more than one proton is required to synthesize one molecule of ATP from ADP and P_i. Although the ΔG for this reaction under standard conditions is +7.3 kcal/mol, at the concentrations of reactants in the mitochondrion, ΔG is probably higher (+10 to +12 kcal/mol). We can calculate the amount of free energy released by the passage of 1 mol of protons down an electrochemical gradient of 220 mV (0.22 V) from the Nernst equation, setting $n = 1$ and measuring ΔE in volts:

$$\begin{aligned}\Delta G(\text{cal/mol}) &= -nF\Delta E = -(23{,}062\ \text{cal}\cdot\text{V}^{-1}\cdot\text{mol}^{-1})\Delta E \\ &= (23{,}062\ \text{cal}\cdot\text{V}^{-1}\cdot\text{mol}^{-1})(0.22\ \text{V}) \\ &= -5074\ \text{cal/mol, or } -5.1\ \text{kcal/mol}\end{aligned}$$

Since the downhill movement of 1 mol of protons releases just over 5 kcal of free energy, the passage of at least two protons is required for synthesis of each molecule of ATP from ADP and P_i.

Proton Movement Through F_0 and Rotation of the c Ring Each copy of subunit **c** contains two membrane-spanning α helices that form a hairpin-like structure. An aspartate residue, Asp61, in the center of one of these helices is thought to participate in proton movement. Chemical modification of this aspartate by the poison dicyclohexylcarbodiimide or its mutation to alanine specifically blocks proton movement through F_0. According to one current model, two proton half-channels, I and II, lie at the interface between the **a** subunit and **c** ring (see Figure 12-24). Protons are thought to move one at a time through half-channel I from the exoplasmic medium and bind to the carboxylate side chain on Asp61 of one **c** subunit. Binding of a proton to this aspartate would result in a conformational change in the **c** subunit, causing it to move relative to the fixed **a** subunit or equivalently to rotate in the membrane plane. This rotation would bring the adjacent **c** subunit, with its ionized aspartyl side chain, into channel I, thereby allowing it to receive the next proton and subsequently move relative to the **a** subunit. Continued rotation of the **c** ring, due to binding of protons to additional **c** subunits, eventually would align the first **c** subunit containing a protonated Asp61 with the second half-channel (II), which is connected to the cytosol. A positively charged side chain of Arg210 in the **a** subunit has been proposed to interact with the negatively charged Asp61 and facilitate movement of the **c** subunits and proton translocation. Once this occurs, the proton on the aspartyl residue could dissociate (forming ionized aspartate) and move into the cytosolic medium.

Since the γ subunit of F_1 is tightly attached to the **c** ring of F_0, rotation of the **c** ring associated with proton movement causes rotation of the γ subunit. According to the binding-change mechanism, a 120° rotation of γ powers synthesis of one ATP (see Figure 12-25). Thus complete rotation of the **c** ring by 360° would generate three ATPs. In *E. coli,* where the F_0 composition is $\mathbf{a}_1\mathbf{b}_2\mathbf{c}_{10}$, movement of 10 protons drives one complete rotation and thus synthesis of three ATPs. This value is consistent with experimental data on proton flux during ATP synthesis, providing indirect support for the model coupling proton movement to **c**-ring rotation depicted in Figure 12-24. The F_0 from chloroplasts contains 14 **c** subunits per ring, and movement of 14 protons would be needed for synthesis of three ATPs. Why these otherwise similar F_0F_1 complexes have evolved to have different H^+:ATP ratios is not clear.

ATP-ADP Exchange Across the Inner Mitochondrial Membrane Is Powered by the Proton-Motive Force

In addition to powering ATP synthesis, the proton-motive force across the inner mitochondrial membrane powers the exchange of ATP formed by oxidative phosphorylation inside the mitochondrion for ADP and P_i in the cytosol. This exchange, which is required to supply ADP and P_i substrate for oxidative phosphorylation to continue, is mediated by two proteins in the inner membrane: a *phosphate transporter* (HPO_4^{2-}/OH^- antiporter) that mediates the import of one HPO_4^{2-} coupled to the export of one OH^- and an *ATP/ADP antiporter* (Figure 12-27).

The ATP/ADP antiporter allows one molecule of ADP to enter only if one molecule of ATP exits simultaneously. The ATP/ADP antiporter, a dimer of two 30,000-Da subunits, makes up 10–15 percent of the protein in the inner membrane, so it is one of the more abundant mitochondrial proteins. Functioning of the two antiporters together produces an influx of one ADP^{3-} and one P_i^{2-} and efflux of one ATP^{4-} together with one OH^-. Each OH^- transported outward combines with a proton, translocated during electron transport to the intermembrane space, to form H_2O. This drives the overall reaction in the direction of ATP export and ADP and P_i import.

Because some of the protons translocated out of the mitochondrion during electron transport provide the power (by combining with the exported OH^-) for the ATP-ADP exchange, fewer protons are available for ATP synthesis. It is estimated that for every four protons translocated out, three are used to synthesize one ATP molecule and one is used to power the export of ATP from the mitochondrion in exchange for ADP and P_i. This expenditure of energy from the proton concentration gradient to export ATP from the mitochondrion in exchange for ADP and P_i ensures a high ratio of ATP to ADP in the cytosol, where hydrolysis of the high-energy phosphoanhydride bond of ATP is utilized to power many energy-requiring reactions.

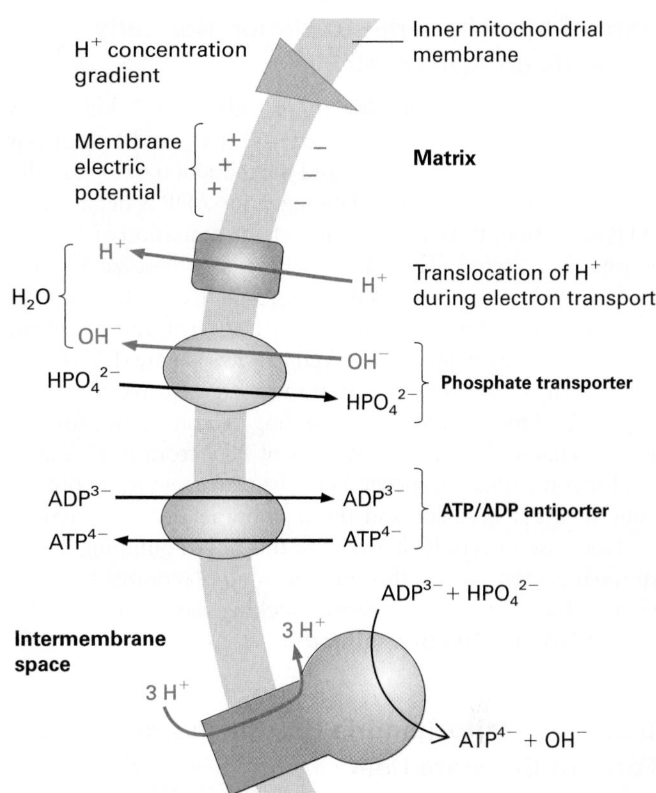

▲ **FIGURE 12-27 The phosphate and ATP/ADP transport system in the inner mitochondrial membrane.** The coordinated action of two antiporters (purple and green) results in the uptake of one ADP^{3-} and one HPO_4^{2-} in exchange for one ATP^{4-} and one hydroxyl, powered by the outward translocation of one proton (mediated by the proteins of the electron transport chain, blue) during electron transport. The outer membrane is not shown here because it is permeable to molecules smaller than 5000 Da.

Studies of what turned out to be ATP/ADP antiporter activity were first recorded about 2000 years ago, when Dioscorides described a poisonous herb from the thistle *Atractylis gummifera,* found commonly in the Mediterranean region. The same agent is found in the traditional Zulu multipurpose herbal remedy *impila (Callilepis laureola).* In Zulu *impila* means "health," although it has been associated with numerous poisonings. In 1962 the active agent in the herb, atractyloside, which inhibits the ATP/ADP antiporter, was shown to inhibit oxidative phosphorylation of extramitochondrial ADP but not intramitochondrial ADP. This demonstrated the importance of the ATP/ADP antiporter and has provided a powerful tool to study the mechanism by which this transporter functions.

Dioscorides (~AD 40–90) lived near Tarsus, at the time a province of Rome in southeastern Asia Minor in what is now Turkey. His five-volume *De Materia Medica (The Materials of Medicine)* "on the preparation, properties, and testing of drugs" described the medicinal properties of about 1000 natural products and 4740 medicinal usages of them. For approximately 1600 years it was the basic reference in medicine from northern Europe to the Indian Ocean, comparable to today's *Physicians' Desk Reference* as a guide for using drugs. ▮

Rate of Mitochondrial Oxidation Normally Depends on ADP Levels

If intact isolated mitochondria are provided with NADH (or a source of $FADH_2$ such as succinate) plus O_2 and P_i, but not ADP, the oxidation of NADH and the reduction of O_2 rapidly cease, because the amount of endogenous ADP is depleted by ATP formation. If ADP is then added, the oxidation of NADH is rapidly restored. Thus mitochondria can oxidize $FADH_2$ and NADH only as long as there is a source of ADP and P_i to generate ATP. This phenomenon, termed **respiratory control,** occurs because oxidation of NADH and succinate ($FADH_2$) is obligatorily coupled to proton transport across the inner mitochondrial membrane. If the resulting proton-motive force is not dissipated during the synthesis of ATP from ADP and P_i (or for some other purpose), both the transmembrane proton concentration gradient and the membrane electric potential will increase to very high levels. At this point, pumping of additional protons across the inner membrane requires so much energy that it eventually ceases, blocking the coupled oxidation of NADH and other substrates.

Brown-Fat Mitochondria Use the Proton-Motive Force to Generate Heat

Brown-fat tissue, whose color is due to the presence of abundant mitochondria, is specialized for the generation of heat. In contrast, *white-fat tissue* is specialized for the storage of fat and contains relatively few mitochondria.

The inner membrane of brown-fat mitochondria contains *thermogenin,* a protein that functions as a natural **uncoupler** of oxidative phosphorylation and generation of a proton-motive force. Thermogenin, or UCP1, is one of several uncoupling proteins (UCPs) found in most eukaryotes (but not in fermentative yeasts). Thermogenin dissipates the proton-motive force by rendering the inner mitochondrial membrane permeable to protons. As a consequence the energy released by NADH oxidation in the electron transport chain is converted to heat. Thermogenin is a proton transporter, not a proton channel, and shuttles protons across the membrane at a rate that is a millionfold slower than that of typical ion channels (see Figure 11-3). Thermogenin is similar in sequence to the mitochondrial ATP/ADP transporter, as are many other mitochondrial transporter proteins that compose the ATP/ADP transporter family. Certain small-molecule poisons also function as uncouplers by rendering the inner mitochondrial membrane permeable to protons. One example is the lipid-soluble chemical 2,4-dinitrophenol (DNP), which can reversibly bind to and release protons and shuttle them across the inner membrane from the intermembrane space into the matrix.

Environmental conditions regulate the amount of thermogenin in brown-fat mitochondria. For instance, during the adaptation of rats to cold, the ability of their tissues to generate heat is increased by the induction of thermogenin synthesis. In cold-adapted animals, thermogenin may constitute up to 15 percent of the total protein in the inner mitochondrial membrane.

Adult humans have little brown fat, but human infants have a great deal. In the newborn, thermogenesis by brown-fat mitochondria is vital to survival, as it also is in hibernating mammals. In fur seals and other animals naturally acclimated to the cold, muscle-cell mitochondria contain thermogenin; as a result, much of the proton-motive force is used for generating heat, thereby maintaining body temperature.

KEY CONCEPTS OF SECTION 12.3

Harnessing the Proton-Motive Force for Energy-Requiring Processes

- Peter Mitchell proposed the chemiosmotic hypothesis that a proton-motive force across the inner mitochondrial membrane is the direct source of energy for ATP synthesis.
- Bacteria, mitochondria, and chloroplasts all use the same chemiosmotic mechanism and a similar ATP synthase to generate ATP.
- ATP synthase (the F_0F_1 complex) catalyzes ATP synthesis as protons flow through the inner mitochondrial membrane (plasma membrane in bacteria) down their electrochemical proton gradient.
- F_0 contains a ring of 10–14 **c** subunits that is rigidly linked to the rod-shaped γ subunit and ϵ subunit of F_1. Together they rotate during ATP synthesis. Resting atop the γ subunit is the hexameric knob of F_1 $[(\alpha\beta)_3]$, which protrudes into the mitochondrial matrix (cytosol in bacteria). The three β subunits are the sites of ATP synthesis (see Figure 12-24).
- Movement of protons across the membrane via two half-channels at the interface of the F_0 **a** subunit and the **c** ring powers rotation of the **c** ring with its attached F_1 ϵ and γ subunits.
- Rotation of the F_1 γ subunit, which is inserted in the center of the nonrotating $(\alpha\beta)_3$ hexamer and operates like a camshaft, leads to changes in the conformation of the nucleotide-binding sites in the three F_1 β subunits (see Figure 12-25). By means of this binding-change mechanism, the β subunits bind ADP and P_i, condense them to form ATP, and then release the ATP. Three ATPs are made for each revolution made by the assembly of **c**, γ, and ϵ subunits.
- The proton-motive force also powers the uptake of P_i and ADP from the cytosol in exchange for mitochondrial ATP and OH^-, thus reducing some of the energy available for ATP synthesis. The ATP/ADP antiporter that participates in this exchange is one of the most abundant proteins in the inner mitochondrial membrane.
- Continued mitochondrial oxidation of NADH and the reduction of O_2 are dependent on sufficient ADP being present in the matrix. This phenomenon, termed respiratory control, is an important mechanism for coordinating oxidation and ATP synthesis in mitochondria.
- In brown fat, the inner mitochondrial membrane contains the uncoupler protein thermogenin, a proton trans-

porter that dissipates the proton-motive force into heat. Certain chemicals also function as uncouplers (e.g., DNP) and have the same effect, uncoupling oxidative phosphorylation from electron transport.

12.4 Photosynthesis and Light-Absorbing Pigments

We now shift our attention to photosynthesis, the second main process for synthesizing ATP. Photosynthesis in plants occurs in chloroplasts, large organelles found mainly in leaf cells. The principal end products generated from carbon dioxide and water are two carbohydrates that are polymers of hexose (six-carbon) sugars: sucrose, a glucose-fructose disaccharide (see Figure 2-19), and leaf **starch,** a large insoluble glucose polymer that is the primary storage carbohydrate in higher plants (Figure 12-28). Leaf starch is synthesized and stored in the chloroplast. Sucrose is synthesized in the leaf cytosol from three-carbon precursors generated in the chloroplast; it is transported to nonphotosynthetic (nongreen) plant tissues (e.g., roots and seeds), which metabolize sucrose for energy by the pathways described in the previous sections. Photosynthesis in plants, as well as in eukaryotic single-celled algae and in several photosynthetic bacteria (e.g., the cyanobacteria and prochlorophytes), also generates oxygen. The overall reaction of oxygen-generating photosynthesis,

$$6\,CO_2 + 6\,H_2O \rightarrow 6\,O_2 + C_6H_{12}O_6$$

is the reverse of the overall reaction by which carbohydrates are oxidized to CO_2 and H_2O. In effect, photosynthesis in chloroplasts produces energy-rich sugars that are broken down and harvested for energy by mitochondria during the process of cellular respiration.

Although green and purple bacteria also carry out photosynthesis, they use a process that does not generate oxygen. As discussed in Section 12.5, detailed analysis of the photosynthetic system in these bacteria has provided insights about the first stages in the more common process of oxygen-generating photosynthesis. In this section, we provide an overview of the stages in oxygen-generating photosynthesis and introduce the main components, including the **chlorophylls,** the principal light-absorbing pigments. ■

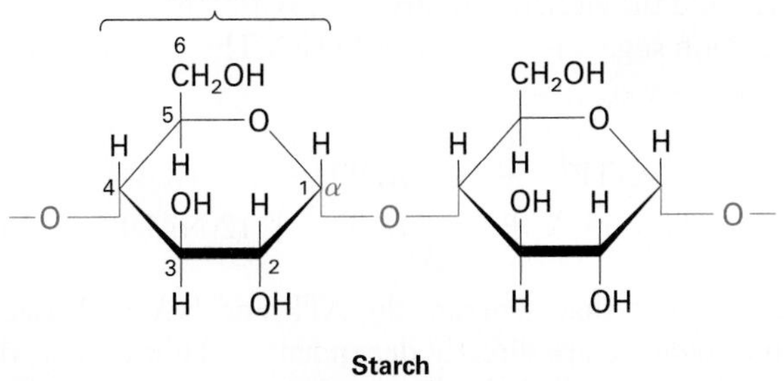

▲ **FIGURE 12-28 Structure of starch.** This large glucose polymer and the disaccharide sucrose (see Figure 2-19) are the principal end products of photosynthesis. Both are built of six-carbon sugars (hexoses).

Thylakoid Membranes in Chloroplasts Are the Sites of Photosynthesis in Plants

Chloroplasts are lens shaped with an approximate diameter of 5 μm and a width of ~2.5 μm, bounded by two membranes, which do not contain chlorophyll and do not participate directly in photosynthesis (Figure 12-29). As in mitochondria, the outer membrane of chloroplasts contains porins and thus is permeable to metabolites of small molecular weight. The inner membrane forms a permeability barrier that contains transport proteins for regulating the movement of metabolites into and out of the organelle.

Unlike mitochondria, chloroplasts contain a third membrane—the *thylakoid membrane*—on which photosynthesis occurs. The chloroplast thylakoid membrane is believed to constitute a single sheet that forms numerous small, interconnected flattened structures, the **thylakoids,** which commonly are arranged in stacks termed *grana* (Figure 12-29). The spaces within all the thylakoids constitute a single continuous compartment, the *thylakoid lumen.* The thylakoid membrane contains a number of integral membrane proteins to which are bound several important prosthetic groups and light-absorbing pigments, most notably chlorophyll. Carbohydrate synthesis occurs in the *stroma,* the soluble phase between the thylakoid membrane and the inner membrane. In photosynthetic bacteria extensive invaginations of the plasma membrane form a set of internal membranes, also termed thylakoid membranes, where photosynthesis occurs.

Three of the Four Stages in Photosynthesis Occur Only During Illumination

The photosynthetic process in plants can be divided into four stages (Figure 12-30), each localized to a defined area of the chloroplast: (1) absorption of light, generation of a high energy electron and formation of O_2 from H_2O; (2) electron transport leading to reduction of $NADP^+$ to NADPH, and generation of a proton-motive force; (3) synthesis of ATP; and (4) conversion of CO_2 into carbohydrates, commonly referred to as **carbon fixation.** All four stages of photosynthesis are tightly coupled and controlled so as to produce the amount of carbohydrate required by the plant. All the reactions in stages 1–3 are catalyzed by multiprotein complexes in the thylakoid membrane. The generation of a pmf and the use of the pmf to synthesize ATP resemble stages III and IV of mitochondrial oxidative phosphorylation. The enzymes that incorporate CO_2 into chemical intermediates and then convert them to starch are soluble constituents of the chloroplast stroma; the enzymes that form sucrose from three-carbon intermediates are in the cytosol.

Stage 1: Absorption of Light The initial step in photosynthesis is the absorption of light by chlorophylls attached to proteins in the thylakoid membranes. Like the heme component of cytochromes, chlorophylls consist of a porphyrin ring attached to a long hydrocarbon side chain

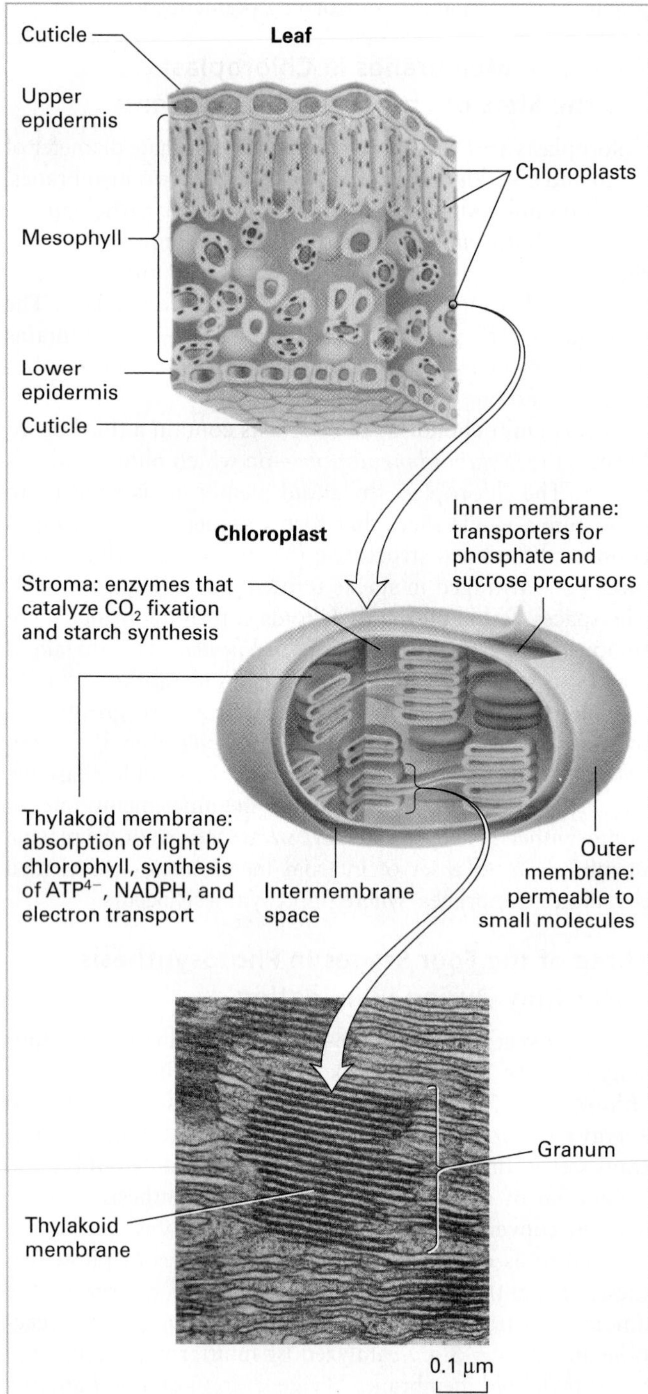

▲ **FIGURE 12-29 Cellular structure of a leaf and chloroplast.** Like mitochondria, plant chloroplasts are bounded by a double membrane separated by an intermembrane space. Photosynthesis occurs on a third membrane, the thylakoid membrane, which is surrounded by the inner membrane and forms a series of flattened vesicles (thylakoids) that enclose a single interconnected *luminal* space. The green color of plants is due to the green color of chlorophyll, all of which is localized to the thylakoid membrane. A granum is a stack of adjacent thylakoids. The stroma is the space enclosed by the inner membrane and surrounding the thylakoids. [Photomicrograph courtesy of Katherine Esau, University of California, Davis.]

(Figure 12-31). In contrast to the hemes (see Figure 12-14), chlorophylls contain a central Mg^{2+} ion (rather than Fe) and have an additional five-membered ring. The energy of the absorbed light ultimately is used to remove electrons from a donor (water in green plants), forming oxygen:

$$2\,H_2O \xrightarrow{\text{light}} O_2 + 4\,H^+ + 4\,e^-$$

The electrons are transferred to a *primary electron acceptor,* a quinone designated Q, which is similar to CoQ in mitochondria. In plants the oxidation of water takes place in a multiprotein complex called *photosystem II (PSII).*

Stage 2: Electron Transport and Generation of a Proton-Motive Force Electrons move from the quinone primary electron acceptor through a series of electron carriers until they reach the ultimate electron acceptor, usually the oxidized form of **nicotinamide adenine dinucleotide phosphate ($NADP^+$)**, reducing it to NADPH. $NADP^+$ is identical in structure with NAD^+ except for the presence of an additional phosphate group. Both molecules gain and lose electrons in the same way (see Figure 2-33). In plants the reduction of $NADP^+$ takes place in a complex called *photosystem I (PSI).* The transport of electrons in the thylakoid membrane is coupled to the movement of protons from the stroma to the thylakoid lumen, forming a pH gradient across the membrane ($pH_{lumen} < pH_{stroma}$). This process is analogous to generation of a proton-motive force across the inner mitochondrial membrane and in bacterial membranes during electron transport (see Figure 12-22).

Thus the overall reaction of stages 1 and 2 can be summarized as

$$2\,H_2O + 2\,NADP^+ \xrightarrow{\text{light}} 2\,H^+ + 2\,NADPH + O_2$$

Stage 3: Synthesis of ATP Protons move down their concentration gradient from the thylakoid lumen to the stroma through the F_0F_1 complex (ATP synthase), which couples proton movement to the synthesis of ATP from ADP and P_i. The chloroplast ATP synthase works similarly to the synthases of mitochondria and bacteria (see Figure 12-25).

Stage 4: Carbon Fixation The ATP and NADPH generated by the second and third stages of photosynthesis provide the energy and the electrons to drive the synthesis of polymers of six-carbon sugars from CO_2 and H_2O. The overall chemical equation is written as

$$6\,CO_2 + 18\,ATP^{4-} + 12\,NADPH + 12\,H_2O \rightarrow$$
$$C_6H_{12}O_6 + 18\,ADP^{3-} + 18\,P_i^{2-} + 12\,NADP^+ + 6\,H^+$$

The reactions that generate the ATP and NADPH used in carbon fixation are directly dependent on light energy; thus stages 1–3 are called the *light reactions* of photosynthesis. The reactions in stage 4 are indirectly dependent on light energy; they are sometimes called the *dark reactions* of photosynthesis because they can occur in the dark, utilizing the supplies of ATP and NADPH generated by light energy.

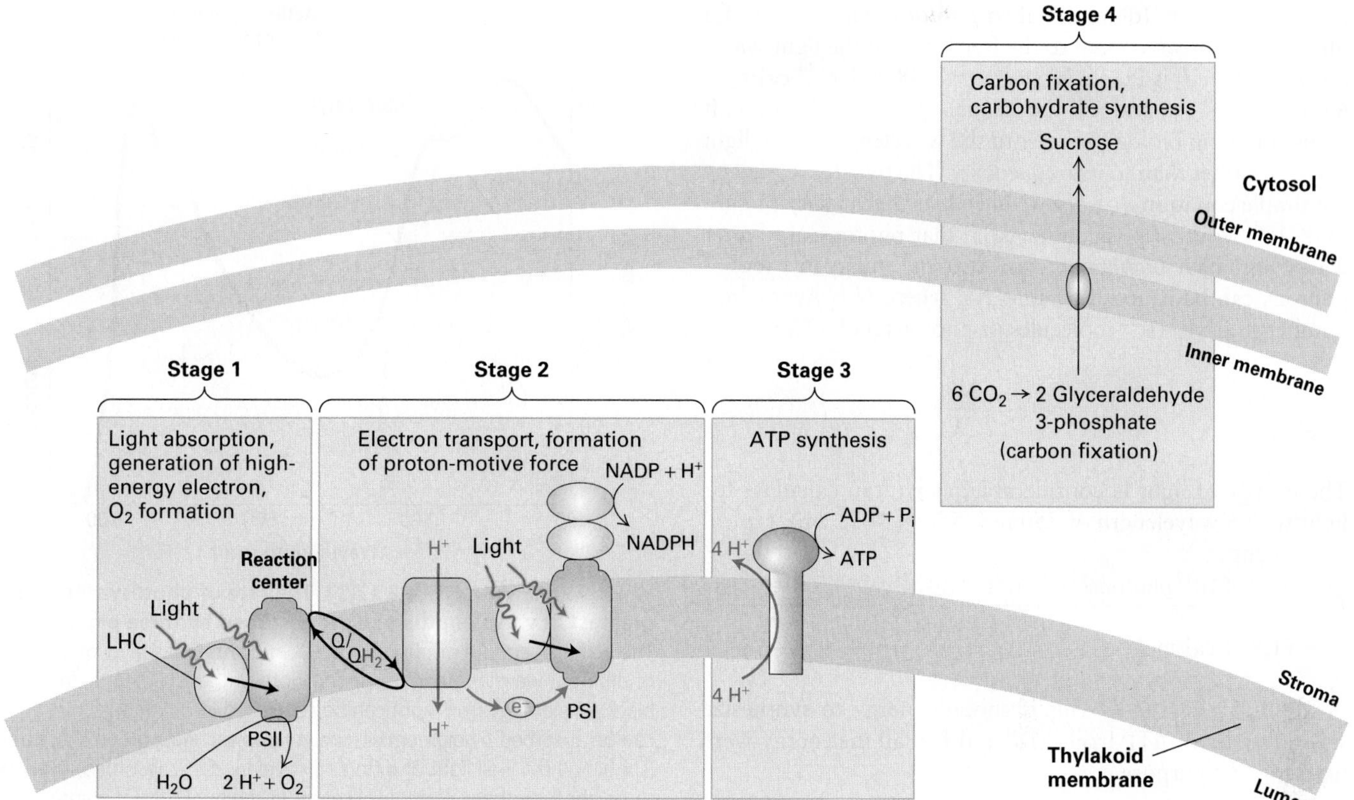

▲ **FIGURE 12-30 Overview of the four stages of photosynthesis.** In **stage 1,** light is absorbed by light-harvesting complexes (LHC) and reaction center of photosystem II (PSII). The LHCs transfer the absorbed energy to the reaction centers, which use it, or the energy absorbed by directly from a photon, to oxidize water to molecular oxygen and generate high-energy electrons. In **stage 2,** these electrons move down an electron transport chain, which uses either lipid-soluble (Q/QH_2) or water-soluble (plastocyanin, PC) electron carriers to shuttle electrons between multiple protein complexes. As electrons move down the chain, they release energy that the complexes use to generate a proton-motive force and, after additional energy is introduced by absorption of light in photosystem I (PSI), to synthesize the high-energy electron carrier NADPH. In **stage 3,** movement of proteins powered by the proton-motive force drives the synthesis of ATP by an F_0F_1 ATP synthase. **Stages 1–3** in plants take place in the thylakoid membrane of the chloroplast. In **stage 4,** in the chloroplast stroma, the energy stored in NADPH and ATP is used to convert CO_2 initially into three-carbon molecules (glyceraldehyde 3-phosphate), a process known as carbon fixation. These molecules are then transported to the cytosol of the cell for conversion to hexose sugars in the form of sucrose. Glyceraldehyde 3-phosphate is also used to make starch within the chloroplast.

However, the reactions in stage 4 are not confined to the dark; in fact, they occur primarily during illumination.

Each Photon of Light Has a Defined Amount of Energy

Quantum mechanics established that light, a form of electromagnetic radiation, has properties of both waves and particles. When light interacts with matter, it behaves as discrete

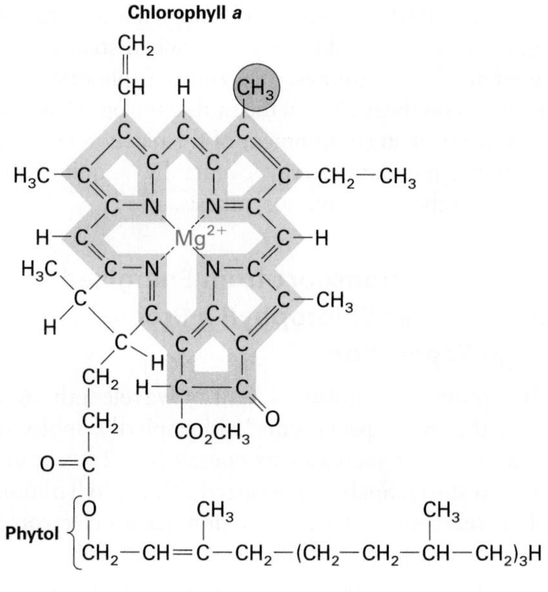

Chlorophyll *a*

Phytol

◀ **FIGURE 12-31 Structure of chlorophyll *a*, the principal pigment that traps light energy.** Electrons are delocalized among three of chlorophylls *a*'s four central rings (yellow) and the atoms that interconnect them. In chlorophyll, a Mg^{2+} ion, rather than the Fe^{3+} ion found in heme, sits at the center of the porphyrin ring and an additional five-membered ring (blue) is present; otherwise, the structure of chlorophyll is similar to that of heme, found in molecules such as hemoglobin and cytochromes (see Figure 12-14a). The hydrocarbon phytol "tail" facilitates binding of chlorophyll to hydrophobic regions of chlorophyll-binding proteins. The CH_3 group (green) is replaced by a formaldehyde (CHO) group in chlorophyll *b*.

packets of energy (quanta) called *photons*. The energy of a photon, ϵ, is proportional to the frequency of the light wave: $\epsilon = h\gamma$, where h is Planck's constant (1.58×10^{-34} cal·s, or 6.63×10^{-34} J·s) and γ is the frequency of the light wave. It is customary in biology to refer to the wavelength of the light wave, λ, rather than to its frequency γ. The two are related by the simple equation $\gamma = c \div \lambda$, where c is the velocity of light (3×10^{10} cm/s in a vacuum). Note that photons of *shorter* wavelength have *higher* energies. Also, the energy in 1 mol of photons can be denoted by $E = N\epsilon$, where N is Avogadro's number (6.02×10^{23} molecules or photons/mol). Thus

$$E = Nh\gamma = \frac{Nhc}{\lambda}$$

The energy of light is considerable, as we can calculate for light with a wavelength of 550 nm (550×10^{-7} cm), typical of sunlight:

$$E = \frac{(6.02 \times 10^{23}\text{photons/mol})(1.58 \times 10^{-34}\text{cal}\cdot\text{s})(3 \times 10^{10}\text{cm/s})}{550 \times 10^{-7}\text{ cm}}$$
$$= 51{,}881 \text{ cal/mol}$$

or about 52 kcal/mol. This is enough energy to synthesize several moles of ATP from ADP and P_i if all the energy were used for this purpose.

Photosystems Comprise a Reaction Center and Associated Light-Harvesting Complexes

The absorption of light energy and its conversion into chemical energy occurs in multiprotein complexes called **photosystems.** Found in all photosynthetic organisms, both eukaryotic and prokaryotic, photosystems consist of two closely linked components: a *reaction center,* where the primary events of photosynthesis occur, and an antenna complex consisting of numerous protein complexes, including internal antenna within the photosystem proper and external antenna made up of specialized proteins termed *light-harvesting complexes (LHCs),* which capture light energy and transmit it to the reaction center (see Figure 12-30).

Both reaction centers and antennas contain tightly bound light-absorbing pigment molecules. Chlorophyll *a* is the principal pigment involved in photosynthesis, being present in both reaction centers and antennas. In addition to chlorophyll *a,* antennas contain other light-absorbing pigments: *chlorophyll b* in vascular plants and *carotenoids* in both plants and photosynthetic bacteria. Carotenoids consist of long branched hydrocarbon chains with alternating single and double bonds; they are similar in structure to the visual pigment retinal, which absorbs light in the eye. The presence of various antenna pigments, which absorb light at different wavelengths, greatly extends the range of light that can be absorbed and used for photosynthesis.

One of the strongest pieces of evidence for the involvement of chlorophylls and carotenoids in photosynthesis is that the absorption spectrum of these pigments is similar to the action spectrum of photosynthesis (Figure 12-32). The latter is a measure of the relative ability of light of different wavelengths to support photosynthesis.

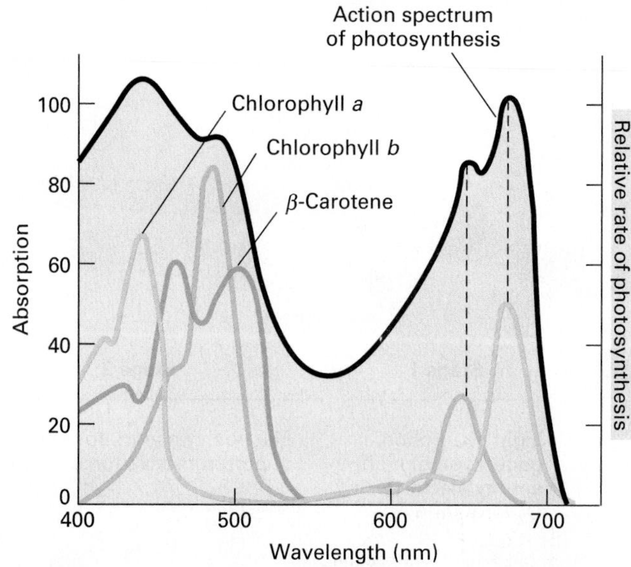

▲ EXPERIMENTAL FIGURE 12-32 **The rate of photosynthesis is greatest at wavelengths of light absorbed by three pigments.** The action spectrum of photosynthesis in plants (the ability of light of different wavelengths to support photosynthesis) is shown in black. The energy from light can be converted into ATP only if it can be absorbed by pigments in the chloroplast. Absorption spectra (showing how well light of different wavelengths is absorbed) for three photosynthetic pigments present in the antennas of plant photosystems are shown in color. Comparison of the action spectrum with the individual absorption spectra suggests that photosynthesis at 680 nm is primarily due to light absorbed by chlorophyll *a;* at 650 nm, to light absorbed by chlorophyll *b;* and at shorter wavelengths, to light absorbed by chlorophylls *a* and *b* and by carotenoid pigments, including β-carotene.

When chlorophyll *a* (or any other molecule) absorbs visible light, the absorbed light energy raises electrons in the chlorophyll *a* to a higher-energy (excited) state. This state differs from the ground (unexcited) state largely in the distribution of the electrons around the C and N atoms of the porphyrin ring. Excited states are unstable, and the electrons return to the ground state by one of several competing processes. For chlorophyll *a* molecules dissolved in organic solvents such as ethanol, the principal reactions that dissipate the excited-state energy are the emission of light (fluorescence and phosphorescence) and thermal emission (heat). When the same chlorophyll *a* is bound in the unique protein environment of the reaction center, dissipation of excited-state energy occurs by a quite different process that is the key to photosynthesis.

Photoelectron Transport from Energized Reaction-Center Chlorophyll *a* Produces a Charge Separation

The absorption of a photon of light of wavelength ≈680 nm by one of the two "special-pair" chlorophyll *a* molecules in the reaction center increases its energy by 42 kcal/mol (the first excited state). Such an energized chlorophyll *a* molecule in a plant reaction center rapidly donates an electron to an intermediate acceptor, and the electron is rapidly passed on

Animation: Photosynthesis

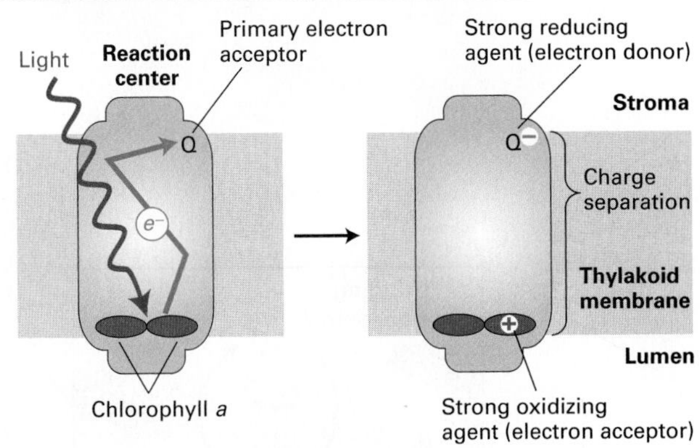

◀ **FIGURE 12-33 Photoelectron transport, the primary event in photosynthesis.** After absorption of a photon of light, one of the excited special pair of chlorophyll *a* molecules in the reaction center (*left*) donates via several intermediates an electron to a loosely bound acceptor molecule, the quinone Q, on the stromal surface of the thylakoid membrane, creating an essentially irreversible charge separation across the membrane (*right*). The electron cannot easily return through the reaction center to neutralize the positively charged chlorophyll *a*. In plants the oxidation of water to molecular oxygen takes place in a multiprotein complex called photosystem II. The complex photosystem I uses a similar photoelectron transport pathway, but instead of oxidizing water, it reduces the electron carrier $NADP^+$.

MEDIA CONNECTIONS

to the primary electron acceptor, quinone Q, near the stromal surface of the thylakoid membrane (Figure 12-33). This light-driven electron transfer, called **photoelectron transport,** depends on the unique environment of both the chlorophylls and the acceptor within the reaction center. Photoelectron transport, which occurs nearly every time a photon is absorbed, leaves a positive charge on the chlorophyll *a* close to the luminal surface of the thylakoid membrane (opposite side from the stroma) and generates a reduced, negatively charged acceptor (Q^-) near the stromal surface.

The Q^- produced by photoelectron transport is a powerful reducing agent with a strong tendency to transfer an electron to another molecule, ultimately to $NADP^+$. The positively charged chlorophyll a^+, a strong oxidizing agent, attracts an electron from an electron donor on the luminal surface to regenerate the original chlorophyll *a*. In plants, the oxidizing power of four chlorophyll a^+ molecules is used, by way of intermediates, to remove four electrons from 2 H_2O molecules bound to a site on the luminal surface to form O_2:

$$2\ H_2O + 4\ \text{chlorophyll}\ a^+ \rightarrow 4\ H^+ + O_2 + 4\ \text{chlorophyll}\ a$$

These potent biological reductants and oxidants provide all the energy needed to drive all subsequent reactions of photosynthesis: electron transport (stage 2), ATP synthesis (stage 3), and CO_2 fixation (stage 4).

Chlorophyll *a* also absorbs light at discrete wavelengths shorter than 680 nm (see Figure 12-32). Such absorption raises the molecule into one of several excited states, whose energies are higher than that of the first excited state described above, which decay by releasing energy within 10^{-12} seconds (1 picosecond, ps) to the lower-energy first excited state with loss of the extra energy as heat. Because photoelectron transport and the resulting charge separation occur only from the first excited state of the reaction-center chlorophyll *a*, the quantum yield—the amount of photosynthesis per absorbed photon—is the same for all wavelengths of visible light shorter (and therefore of higher energy) than 680 nm. How closely the wavelength of light matches the absorption spectra of the pigments will determine how likely it is that the photon will be absorbed. Once absorbed, the photon's exact wavelength is not critical, provided it is energetic enough to push the chlorophyll into the first excited state.

Internal Antenna and Light-Harvesting Complexes Increase the Efficiency of Photosynthesis

Although chlorophyll *a* molecules within a reaction center that are involved directly with charge separation and electron transfer are capable of directly absorbing light and initiating photosynthesis, they most commonly are energized indirectly by energy transferred to them from other light-absorbing and energy-transferring pigments. These other pigments, which include many other chlorophyll molecules, are involved with absorption of photons and passing the energy to the chlorophyll *a* molecules in the reaction center. Some are bound to protein subunits that are considered to be intrinsic components of the photosystem and thus are called internal antennas; others are bound to proteins complexes that bind to but are distinct from the photosystem core proteins and are called light-harvesting complexes (LHCs). Even at the maximum light intensity encountered by photosynthetic organisms (tropical noontime sunlight), each reaction-center chlorophyll *a* molecule absorbs only about one photon per second, which is not enough to support photosynthesis sufficient for the needs of the plant. The involvement of internal antenna and LHCs greatly increases the efficiency of photosynthesis, especially at more typical light intensities, by increasing absorption of 680-nm light and by extending the range of wavelengths of light that can be absorbed by other antenna pigments.

Photons can be absorbed by any of the pigment molecules in internal antennas or an LHC. The absorbed energy is then rapidly transferred (in $<10^{-9}$ seconds) to one of the two "special-pair" chlorophyll *a* molecules in the associated reaction center, where it promotes the primary photosynthetic charge separation (Figure 12-33). Photosystem core proteins and LHC proteins maintain the pigment molecules in the precise orientation and position optimal for light absorption and energy transfer, thereby maximizing the very rapid and efficient *resonance transfer* of energy from antenna pigments to

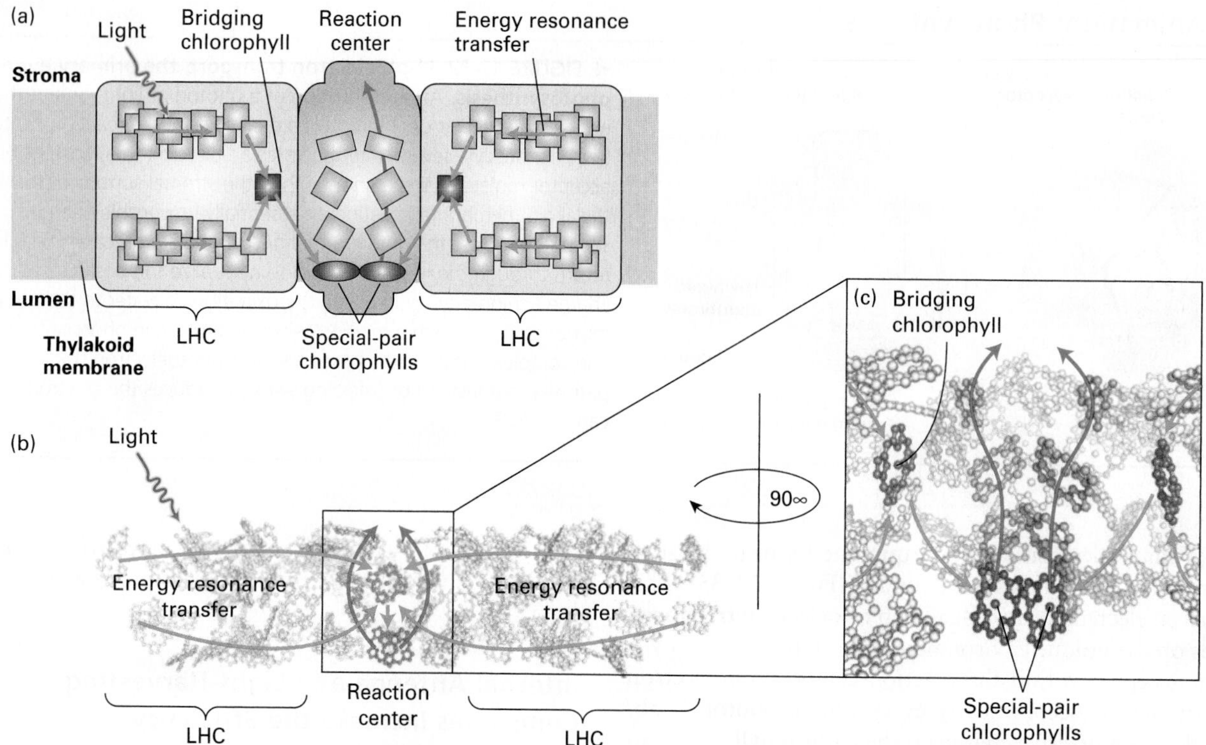

▲ **FIGURE 12-34 Light-harvesting complexes and photosystems in cyanobacteria and plants.** (a) Diagram of the membrane of a cyanobacterium, in which the multiprotein light-harvesting complex (LHC) contains 90 chlorophyll molecules (green) and 31 other small molecules, all held in a specific geometric arrangement for optimal light absorption and energy transfer. Of the six chlorophyll molecules in the reaction center, two constitute the special-pair chlorophylls (ovals, dark green) that can initiate photoelectron transport when excited (blue arrow). Resonance transfer of energy (red arrows) rapidly funnels energy from absorbed light to one of two "bridging" chlorophylls (squares, dark green) and thence to chlorophylls in the reaction center. (b) Three-dimensional organization of the photosystem I (PSI) with its LHCs of *Pisum sativum* (garden pea) as determined by x-ray crystallography and as seen from the plane of the membrane. Only the chlorophylls together with the reaction center electron carriers are shown. (c) Expanded view of the reaction center from (b) rotated 90° about a vertical axis. [Part (a) adapted from W. Kühlbrandt, 2001, *Nature* **411:**896, and P. Jordan et al., 2001, *Nature* **411:**909. Parts (b and c) based on the structural determination by A. Ben-Sham et al., 2003, *Nature* **426:**630.]

reaction-center chlorophylls. Resonance energy transfer does not involve the transfer of an electron. Studies on one of the two photosystems in cyanobacteria, which are similar to those in higher plants, suggest that energy from absorbed light is funneled first to a "bridging" chlorophyll in each LHC and then to the special pair of reaction-center chlorophylls (Figure 12-34a). Surprisingly, however, the molecular structures of LHCs from plants and cyanobacteria are completely different from those in green and purple bacteria, even though both types contain carotenoids and chlorophylls in a clustered arrangement within the membrane. Figure 12-34b shows the distribution of the chlorophyll pigments in the plant photosystem I from *Pisum sativum* (garden pea) together with its peripheral LHCI antenna. The large number of internal and LHC antenna chlorophylls surround the core reaction center to permit efficient transfer of absorbed light energy to the special chlorophylls in the reaction center.

Although LHC antenna chlorophylls can transfer light energy absorbed from a photon, they cannot release an electron. As we've seen already, this function resides in the two reaction-center chlorophylls. To understand their electron-releasing ability, we examine the structure and function of the reaction center in bacterial and plant photosystems in the next section.

KEY CONCEPTS OF SECTION 12.4

Photosynthetic Stages and Light-Absorbing Pigments

- The principal end products of photosynthesis in plants are molecular oxygen and polymers of six-carbon sugars (starch and sucrose).

- The light-capturing and ATP-generating reactions of photosynthesis occur in the thylakoid membrane located within chloroplasts. The permeable outer membrane and inner membrane surrounding chloroplasts do not participate directly in photosynthesis (see Figure 12-29).

- There are four stages in photosynthesis: (1) absorption of light, generation of a high energy electron and formation of O_2 from H_2O; (2) electron transport leading to reduction of $NADP^+$ to NADPH, and to generation of a proton-motive force; (3) synthesis of ATP; and (4) conversion of CO_2 into carbohydrates (carbon fixation).

- In stage 1 of photosynthesis, light energy is absorbed by one of two "special-pair" chlorophyll *a* molecules bound to reaction-center proteins in the thylakoid membrane. The energized chlorophylls donate via intermediates an electron to a quinone on the opposite side of the membrane, creating a charge separation (see Figure 12-33). In green plants, the positively charged chlorophylls then remove electrons from water, forming molecular oxygen (O_2).

- In stage 2, electrons are transported from the reduced quinone via carriers in the thylakoid membrane until they reach the ultimate electron acceptor, usually $NADP^+$, reducing it to NADPH. Electron transport is coupled to movement of protons across the membrane from the stroma to the thylakoid lumen, forming a pH gradient (proton-motive force) across the thylakoid membrane.

- In stage 3, movement of protons down their electrochemical gradient through F_0F_1 complexes (ATP synthase) powers the synthesis of ATP from ADP and P_i.

- In stage 4, the NADPH and ATP generated in stages 2 and 3 provide the energy and the electrons to drive the fixation of CO_2, which results in the synthesis of carbohydrates. These reactions occur in the thylakoid stroma and cytosol.

- Associated with each reaction center are multiple internal antenna and light-harvesting complexes (LHCs), which contain chlorophylls *a* and *b*, carotenoids, and other pigments that absorb light at multiple wavelengths. Energy, but not an electron, is transferred from the internal antenna and LHC chlorophyll molecules to reaction-center chlorophylls by resonance energy transfer (see Figure 12-34).

12.5 Molecular Analysis of Photosystems

As noted in the previous section, photosynthesis in the green and purple bacteria does not generate oxygen, whereas photosynthesis in cyanobacteria, algae, and higher plants does.* This difference is attributable to the presence of two types of photosystem (PS) in the latter organisms: PSI reduces $NADP^+$ to NADPH, and PSII forms O_2 from H_2O. In contrast, the green and purple bacteria have only one type of photosystem, which cannot form O_2. We first discuss the simpler photosystem of purple bacteria and then consider the more complicated photosynthetic machinery in chloroplasts.

The Single Photosystem of Purple Bacteria Generates a Proton-Motive Force but No O_2

The three-dimensional structures of the photosynthetic reaction centers have been determined, permitting scientists to trace in detail the paths of electrons during and after the absorption of light. Similar proteins and pigments compose photosystems I and II of plants and photosynthetic bacteria.

The reaction center of purple bacteria contains three protein subunits (L, M, and H) located in the plasma membrane (Figure 12-35). Bound to these proteins are the prosthetic groups that absorb light and transport electrons during photosynthesis. The prosthetic groups include a "special pair" of bacteriochlorophyll *a* molecules equivalent to the reaction-center chlorophyll *a* molecules in plants, as well as several other pigments and two quinones, termed Q_A and Q_B, that are structurally similar to mitochondrial ubiquinone.

Initial Charge Separation The mechanism of charge separation in the photosystem of purple bacteria is identical with that in plants outlined earlier; that is, energy from absorbed light is used to strip an electron from a reaction-center bacteriochlorophyll *a* molecule and transfer it, via several different pigments, to the primary electron acceptor Q_B, which is loosely bound to a site on the cytosolic membrane face.

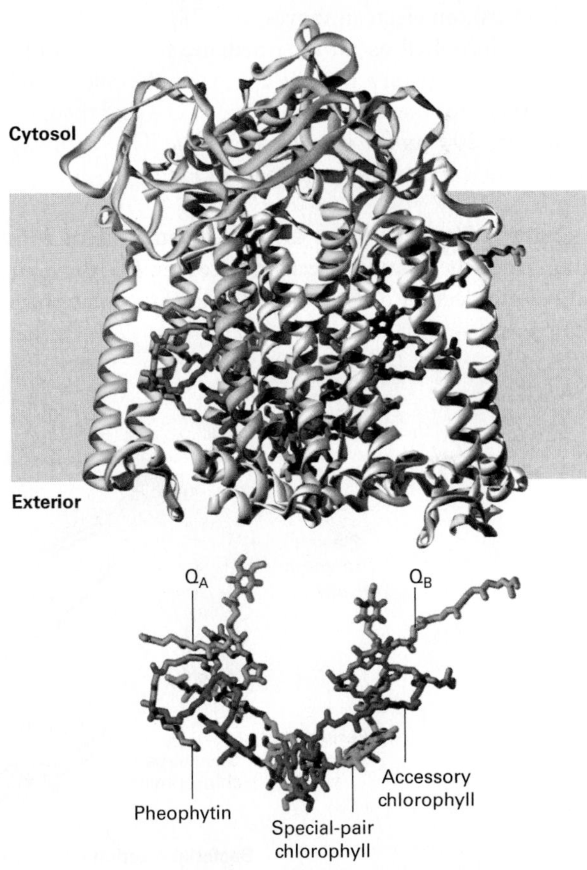

▲ **FIGURE 12-35 Three-dimensional structure of the photosynthetic reaction center from the purple bacterium *Rhodobacter spheroides*.** *(Top)* The L subunit (yellow) and M subunit (gray) each form five transmembrane α helices and have a very similar structure overall; the H subunit (light blue) is anchored to the membrane by a single transmembrane α helix. A fourth subunit (not shown) is a peripheral protein that binds to the exoplasmic segments of the other subunits. *(Bottom)* Within each reaction center, but not easily distinguished in the top image, is a special pair of bacteriochlorophyll *a* molecules (green), capable of initiating photoelectron transport; two accessory chlorophylls (purple); two pheophytins (dark blue), and two quinones, Q_A and Q_B (orange). Q_B is the primary electron acceptor during photosynthesis. [After M. H. Stowell et al., 1997, *Science* **276:**812.]

*A very different type of bacterial photosynthesis, which occurs only in certain archaebacteria, is not discussed here because it is very different from photosynthesis in higher plants. In this type of photosynthesis, the plasma-membrane protein bacteriorhodopsin pumps one proton from the cytosol to the extracellular space for every quantum of light absorbed.

The chlorophyll thereby acquires a positive charge, and Q_B acquires a negative charge. To determine the pathway traversed by electrons through the bacterial reaction center, researchers exploited the fact that each pigment absorbs light of only certain wavelengths, and its absorption spectrum changes when it possesses an extra electron. Because these electron movements are completed in less than 1 millisecond (ms), a special technique called *picosecond absorption spectroscopy* is required to monitor the changes in the absorption spectra of the various pigments as a function of time shortly after the absorption of a light photon.

When a preparation of bacterial membrane vesicles is exposed to an intense pulse of laser light lasting less than 1 ps, each reaction center absorbs one photon (Figure 12-36). Light absorbed by the chlorophyll *a* molecules in each reaction center converts them to the excited state, and the subsequent electron transfer processes are synchronized in all reaction centers in the experimental sample. Within 4×10^{-12} seconds (4 ps), an electron moves, possibly via the accessory bacterial chlorophyll as an intermediate, to the pheophytin molecules (Ph), leaving a positive charge on the chlorophyll *a*. It takes 200 ps for the electron to move to Q_A, and then, in the slowest step, 200 μs for it to move to Q_B. This pathway of electron flow is traced in the left part of Figure 12-36.

Subsequent Electron Flow and Coupled Proton Movement After the primary electron acceptor, Q_B, in the bacterial reaction center accepts one electron, forming $Q_B^{-}\cdot$, it accepts a second electron from the same reaction-center chlorophyll following its re-excitation (e.g., by absorption of a second photon or transfer of energy from antenna molecules). The quinone then binds two protons from the cytosol, forming the reduced quinone (QH_2), which is released from the reaction center (Figure 12-36). QH_2 diffuses within the bacterial membrane to the Q_o site on the exoplasmic face of a cytochrome bc_1 electron transport complex similar in structure to complex III in mitochondria. There it releases its two protons into the periplasmic space (the space between the plasma membrane and the bacterial cell wall). This process moves protons from the cytosol to the outside of the cell, generating a proton-motive force across the plasma membrane. Simultaneously, QH_2 releases its two electrons, which move through the cytochrome bc_1 complex exactly as depicted for the mitochondrial complex III ($CoQH_2$–cytochrome *c* reductase) in Figure 12-20. The Q cycle in the bacterial reaction center, like the Q cycle in mitochondria, pumps additional protons from the cytosol to the intermembrane space, thereby increasing the proton-motive force.

The acceptor for electrons transferred through the cytochrome bc_1 complex is a soluble cytochrome, a one-electron carrier, in the periplasmic space, which is reduced from the Fe^{3+} to the Fe^{2+} state. The reduced cytochrome (analogous to cytochrome *c* in mitochondria) then diffuses to a reaction center, where it releases its electron to a positively charged chlorophyll a^{+}, returning that chlorophyll to the uncharged ground state and the cytochrome to the Fe^{3+} state. This *cyclic* electron

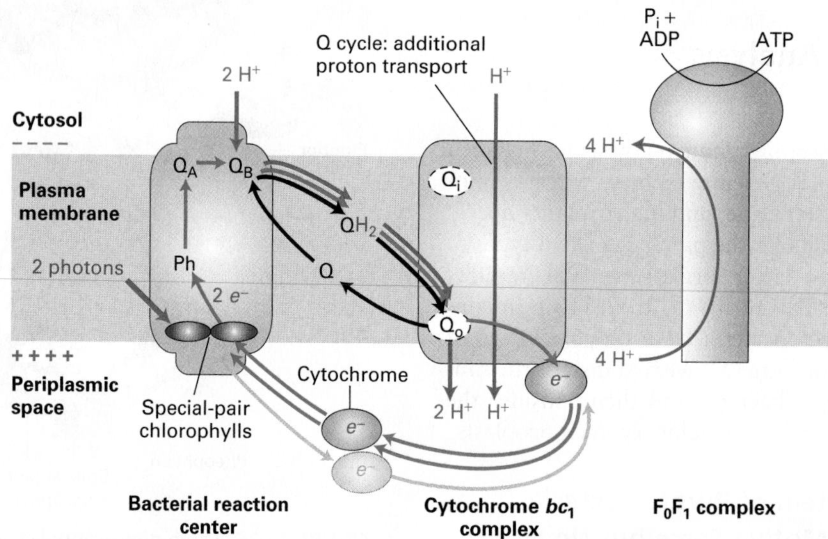

▲ **FIGURE 12-36 Cyclic electron flow in the single photosystem of purple bacteria.** Cyclic electron flow generates a proton-motive force but no O_2. Blue arrows indicate flow of electrons; red arrows indicate proton movement. *(Left)* Energy absorbed directly from light or funneled from an associated LHC (not illustrated here) energizes one of the special-pair chlorophylls in the reaction center. Photoelectron transport from the energized chlorophyll, via accessory chlorophyll, pheophytin (Ph), and quinone A (Q_A), to quinone B (Q_B) forms the semiquinone $Q^{-}\cdot$ and leaves a positive charge on the chlorophyll. Following absorption of a second photon and transfer of a second electron to the semiquinone, the quinone rapidly picks up two protons from the cytosol to form QH_2. *(Center)* After diffusing through the membrane and binding to the Q_o site on the exoplasmic face of the cytochrome bc_1 complex, QH_2 donates two electrons and simultaneously gives up two protons to the external medium in the periplasmic space, generating a proton electrochemical gradient (proton-motive force). Electrons are transported back to the reaction-center chlorophyll via a soluble cytochrome, which diffuses in the periplasmic space. Note the cyclic path (blue) of electrons. Operation of a Q cycle in the cytochrome bc_1 complex pumps additional protons across the membrane to the external medium, as in mitochondria. [Adapted from J. Deisenhofer and H. Michael, 1991, *Ann. Rev. Cell Biol.* **7**:1.]

flow generates no oxygen and no reduced coenzymes, but it has generated a proton-motive force.

Electrons also can flow through the single photosystem of purple bacteria via a *linear* (noncyclic) pathway. In this case, instead of the electron removed from reaction-center chlorophylls by light moving to the cytochrome bc_1 complex and then cycling back again via the water-soluble cytochrome to the reaction center, the electron removed from reaction-center chlorophyll is eventually transferred to NAD^+ (rather than $NADP^+$ as in plants), forming NADH. As a consequence, an electron from a different source must be returned to the special-pair chlorophylls if additional light energy is to be harvested by photoelectron transport. In bacteria using a linear pathway, the electrons to do this come from the oxidation of either hydrogen sulfide (H_2S)

$$H_2S \rightarrow S + 2\,H^+ + 2\,e^-$$

to form elemental sulfur (S) or hydrogen gas (H_2):

$$H_2 \rightarrow 2\,H^+ + 2\,e^-$$

These electrons are used to reduce a cytochrome, which in turn passes an electron to the special-pair chlorophylls in the reaction center to bring the oxidized reaction-center chlorophyll *a* back to its ground state. Overall, the linear pathway results in the light-mediated oxidation of H_2S (or H_2) and the reduction of NAD^+ to NADH. Since H_2O is not the electron donor, no O_2 is formed.

Both the cyclic and linear pathways of electron flow in the bacterial photosystem generate a proton-motive force. As in other systems, this proton-motive force is used by the F_0F_1 complex located in the bacterial plasma membrane to synthesize ATP and also to transport molecules across the membrane against a concentration gradient.

Chloroplasts Contain Two Functionally and Spatially Distinct Photosystems

In the 1940s, biophysicist R. Emerson discovered that the rate of plant photosynthesis generated by light of wavelength 700 nm can be greatly enhanced by adding light of shorter wavelength (higher energy). He found that a combination of light at, say, 600 and 700 nm supports a greater rate of photosynthesis than the sum of the rates for the two separate wavelengths. This so-called *Emerson effect* led researchers to conclude that photosynthesis in plants involves the interaction of two separate photosystems, referred to as *PSI* and *PSII*. PSI is driven by light of wavelength 700 nm or less; PSII, only by shorter-wavelength light (<680 nm).

In chloroplasts, the special-pair reaction-center chlorophylls that initiate photoelectron transport in PSI and PSII differ in their light-absorption maxima because of differences in their protein environments. For this reason, these chlorophylls are often denoted P_{680} (PSII) and P_{700} (PSI). Like a bacterial reaction center, each chloroplast reaction center is associated with multiple internal antenna and light-harvesting complexes (LHCs); the LHCs associated with PSII and PSI contain different proteins.

The two photosystems also are distributed differently in thylakoid membranes: PSII primarily in stacked regions (grana, see Figure 12-29) and PSI primarily in unstacked regions. The stacking of the thylakoid membranes may be due to the binding properties of the proteins in PSII. Evidence for this distribution came from studies in which thylakoid membranes were gently fragmented into vesicles by ultrasound. Stacked and unstacked thylakoid vesicles were then fractionated by density-gradient centrifugation. The stacked fractions contained primarily PSII protein and the unstacked fraction PSI.

Finally, and most importantly, the two chloroplast photosystems differ significantly in their functions (Figure 12-37): only PSII splits water to form oxygen, whereas only PSI transfers electrons to the final electron acceptor, $NADP^+$. Photosynthesis in chloroplasts can follow a linear or cyclic pathway, again like green and purple bacteria. The linear pathway, which we discuss first, can support carbon fixation as well as ATP synthesis. In contrast, the cyclic pathway supports only ATP synthesis and generates no reduced NADPH for use in carbon fixation. Photosynthetic algae and cyanobacteria contain two photosystems analogous to those in chloroplasts.

Linear Electron Flow Through Both Plant Photosystems, PSII and PSI, Generates a Proton-Motive Force, O_2, and NADPH

Linear electron flow in chloroplasts involves PSII and PSI in an obligate series in which electrons are transferred from H_2O to $NADP^+$. The process begins with absorption of a photon by PSII, causing an electron to move from a P_{680} chlorophyll *a* to an acceptor plastoquinone (Q_B) on the stromal surface (Figure 12-37). The resulting oxidized $P_{680}{}^+$ strips one electron from the relatively unwilling donor H_2O, forming an intermediate in O_2 formation and a proton, which remains in the thylakoid lumen and contributes to the proton-motive force. After P_{680} absorbs a second photon, the semiquinone $Q^{-}\cdot$ accepts a second electron and picks up two protons from the stromal space, generating QH_2. After diffusing in the membrane, QH_2 binds to the Q_o site on a cytochrome *bf* complex (analogous to bacterial cytochrome bc_1 complex and mitochondrial complex III). As in these systems, a Q cycle operates, thereby increasing the proton-motive force generated by electron transport. After the cytochrome *bf* complex accepts electrons from QH_2, it transfers them, one at a time, to the Cu^{2+} form of the soluble electron carrier plastocyanin (analogous to bacterial cytochrome *c*), reducing it to the Cu^+ form. Reduced plastocyanin then diffuses in the thylakoid lumen, carrying the electron to PSI.

Absorption of a photon by PSI leads to removal of an electron from the reaction-center chlorophyll *a*, P_{700} (Figure 12-37). The resulting oxidized $P_{700}{}^+$ is reduced by an electron passed from the PSII reaction center via the cytochrome *bf* complex and plastocyanin. Again, this is analogous to situation in mitochondria, where cytochrome *c* acts as a single electron shuttle from complex III to complex IV (see Figure 12-16). The electron taken up at the luminal surface by P_{700} energized by photon absorption moves within PSI via several carriers to the stromal surface of the thylakoid membrane,

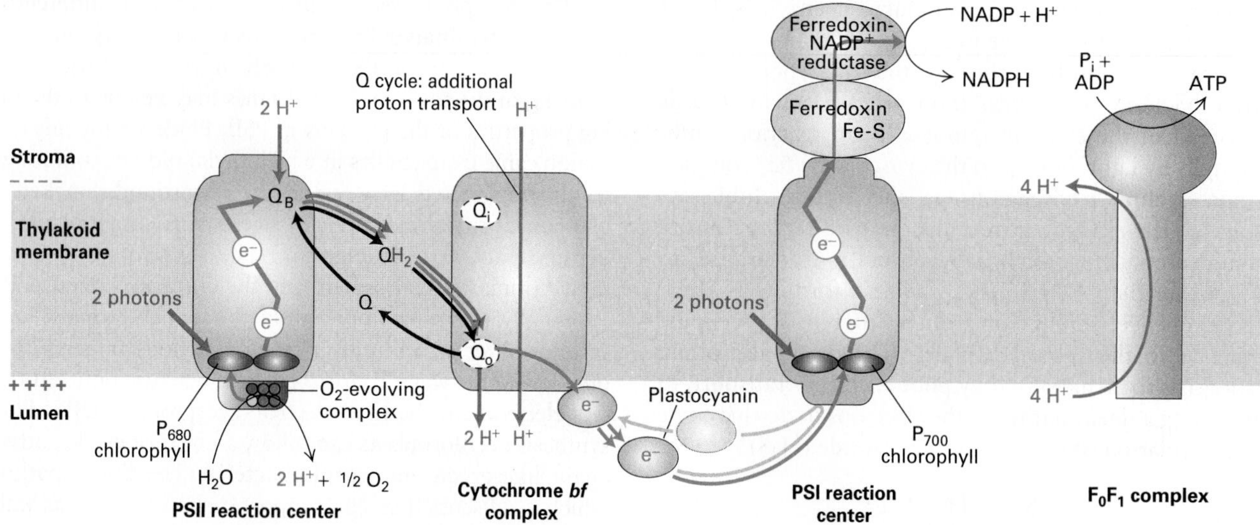

▲ **FIGURE 12-37 Linear electron flow in plants, which requires both chloroplast photosystems PSI and PSII.** Blue arrows indicate flow of electrons; red arrows indicate proton movement. LHCs are not shown. *(Left)* In the PSII reaction center, two sequential light-induced excitations of the same P_{680} chlorophylls result in reduction of the primary electron acceptor Q_B to QH_2. On the luminal side of PSII, electrons removed from H_2O in the thylakoid lumen are transferred to $P_{680}{}^+$, restoring the reaction-center chlorophylls to the ground state and generating O_2. *(Center)* The cytochrome *bf* complex then accepts electrons from QH_2, coupled to the release of two protons into the lumen. Operation of a Q cycle in the cytochrome *bf* complex translocates additional protons across the membrane to the thylakoid lumen, increasing the proton-motive force. *(Right)* In the PSI reaction center, each electron released from light-excited P_{700} chlorophylls moves via a series of carriers in the reaction center to the stromal surface, where soluble ferredoxin (an Fe-S protein) transfers the electron to ferredoxin-NADP$^+$ reductase (FNR). This enzyme uses the prosthetic group flavin adenine dinucleotide (FAD) and a proton to reduce NADP$^+$, forming NADPH. $P_{700}{}^+$ is restored to its ground state by addition of an electron carried from PSII via the cytochrome *bf* complex and plastocyanin, a soluble electron carrier.

where it is accepted by ferredoxin, an iron-sulfur (Fe-S) protein. Electrons excited in PSI can be transferred from ferredoxin via the enzyme ferredoxin-NADP$^+$ reductase (FNR). This enzyme uses the prosthetic group FAD as an electron carrier to reduce NADP$^+$, forming, together with one proton picked up from the stroma, the reduced molecule NADPH.

F_0F_1 complexes in the thylakoid membrane use the proton-motive force generated during linear electron flow to synthesize ATP on the stromal side of membrane. Thus this pathway exploits the energy from multiple photons absorbed by both PSII and PSI and their antennas to generate both NADPH and ATP in the stroma of the chloroplast, where they are utilized for CO_2 fixation.

An Oxygen-Evolving Complex Is Located on the Luminal Surface of the PSII Reaction Center

Somewhat surprisingly, the structure of the PSII reaction center, which removes electrons from H_2O to form O_2, resembles that of the reaction center of photosynthetic purple bacteria, which does not form O_2. Like the bacterial reaction center, the PSII reaction center contains two molecules of chlorophyll *a* (P_{680}), as well as two other accessory chlorophylls, two pheophytins, two quinones (Q_A and Q_B), and one nonheme iron atom. These small molecules are bound to two proteins in PSII, called D1 and D2, whose sequences are remarkably similar to the sequences of the L and M subunits of the bacterial reaction center, attesting to their common evolutionary origins (see Figure 12-35).

When PSII absorbs a photon with a wavelength of <680 nm, it triggers the loss of an electron from a P_{680} molecule, generating $P_{680}{}^+$. As in photosynthetic purple bacteria, the electron is transported rapidly, possibly via an accessory chlorophyll, to a pheophytin, then to a quinone (Q_A), and then to the primary electron acceptor, Q_B, on the outer (stromal) surface of the thylakoid membrane (Figures 12-37 and 12-38).

The photochemically oxidized reaction-center chlorophyll of PSII, $P_{680}{}^+$, is the *strongest* biological oxidant known. The reduction potential of $P_{680}{}^+$ is more positive than that of water, and thus it can oxidize water to generate O_2 and H^+ ions. Photosynthetic bacteria cannot oxidize water because the excited chlorophyll a^+ in the bacterial reaction center is not a sufficiently strong oxidant. (As noted earlier, purple bacteria use H_2S and H_2 as electron donors to reduce chlorophyll a^+ in linear electron flow.)

The splitting of H_2O, which provides the electrons for reduction of $P_{680}{}^+$ in PSII, is catalyzed by a three-protein complex, the *oxygen-evolving complex,* located on the luminal surface of the thylakoid membrane. The oxygen-evolving complex contains four manganese (Mn) ions as well as bound Cl^- and Ca^{2+} ions (Figure 12-38); this is one of the very few cases in which Mn plays a role in a biological system. These Mn ions together with the three extrinsic proteins can be removed from the reaction center by treatment with solutions of concentrated salts; this abolishes O_2 formation but does not affect light absorption or the initial stages of electron transport.

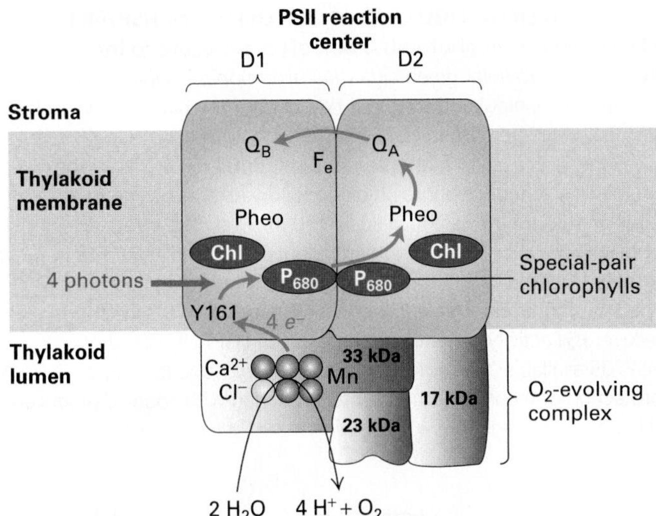

▲ **FIGURE 12-38 Electron flow and O_2 evolution in chloroplast PSII.** The PSII reaction center, comprising two integral proteins, D1 and D2, special-pair chlorophylls (P_{680}), and other electron carriers, is associated with an oxygen-evolving complex on the luminal surface. Bound to the three extrinsic proteins (33, 23, and 17 kDa) of the oxygen-evolving complex are four manganese ions (Mn, red), a Ca^{2+} ion (blue), and a Cl^- ion (yellow). These bound ions function in the splitting of H_2O and maintain the environment essential for high rates of O_2 evolution. Tyrosine-161 (Y161) of the D1 polypeptide conducts electrons from the Mn ions to the oxidized reaction-center chlorophyll ($P_{680}{}^+$), reducing it to the ground state P_{680}. [Adapted from C. Hoganson and G. Babcock, 1997, *Science* **277**:1953.]

The oxidation of two molecules of H_2O to form O_2 requires the removal of four electrons, but absorption of each photon by PSII results in the transfer of just one electron. A simple experiment, described in Figure 12-39, resolved whether the formation of O_2 depends on a single PSII or multiple ones acting in concert. The results indicated that a single PSII must lose an electron and then oxidize the oxygen-evolving complex four times in a row for an O_2 molecule to be formed.

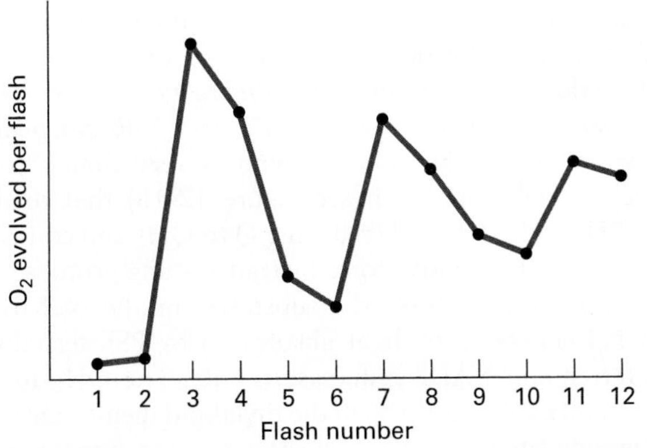

▲ EXPERIMENTAL FIGURE 12-39 **A single PSII absorbs a photon and transfers an electron four times to generate one O_2.** Dark-adapted chloroplasts were exposed to a series of closely spaced, short (5 μs) pulses of light that activated virtually all the PSIIs in the preparation. The peaks in O_2 evolution occurred after every fourth pulse, indicating that absorption of four photons by one PSII is required to generate each O_2 molecule. Because the dark-adapted chloroplasts were initially in a partially reduced state, the peaks in O_2 evolution occurred after flashes 3, 7, and 11. [From J. Berg et al., 2002, *Biochemistry*, 5th ed., W. H. Freeman and Company.]

Manganese is known to exist in multiple oxidation states with from two to five positive charges. Indeed, spectroscopic studies showed that the bound Mn ions in the oxygen-evolving complex cycle through five different oxidation states, S_0–S_4. In this S cycle, a total of two H_2O molecules are split into four protons, four electrons, and one O_2 molecule. The electrons released from H_2O are transferred, one at a time, via the Mn ions and a nearby tyrosine side chain on the D1 subunit to the reaction-center $P_{680}{}^+$, where they regenerate the reduced chlorophyll, P_{680} ground state. The protons released from H_2O remain in the thylakoid lumen.

Herbicides that inhibit photosynthesis not only are very important in agriculture but also have proved useful in dissecting the pathway of photoelectron transport in plants. One such class of herbicides, the *s*-triazines (e.g., atrazine), binds specifically to the D1 subunit in the PSII reaction center, thus inhibiting binding of oxidized Q_B to its site on the stromal surface of the thylakoid membrane. When added to illuminated chloroplasts, *s*-triazines cause all downstream electron carriers to accumulate in the oxidized form, since no electrons can be released from PSII. In atrazine-resistant mutants, a single amino acid change in D1 renders it unable to bind the herbicide, so photosynthesis proceeds at normal rates. Such resistant weeds are prevalent and present a major agricultural problem. ▮

Cells Use Multiple Mechanisms to Protect Against Damage from Reactive Oxygen Species During Photoelectron Transport

As we saw earlier in the case of ROS generation by the mitochondrion, ATP generation via the electron transport chain brings with it potential deleterious side effects. The same is true for the chloroplast. Even though the PSI and PSII photosystems with their associated light-harvesting complexes are remarkably efficient at converting radiant energy to useful chemical energy in the form of ATP and NADPH, they are not perfect. Depending on the intensity of the light and the physiologic conditions of the cells, a relatively small—but significant—amount of energy absorbed by chlorophylls in the light-harvesting antennas and reactions centers results in the chlorophyll being converted to an activated state called *"triplet"* chlorophyll. In this state, the chlorophyll can transfer some of its energy to molecular oxygen (O_2), converting it from its normal, *relatively* unreactive ground state, called triplet oxygen ($^3O_2$) to a very highly reactive (ROS) singlet state form, $^1O_2$. If the $^1O_2$ is not quickly quenched by reacting with specialized $^1O_2$ "scavenger molecules," it will react with and usually damage nearby molecules. This damage can suppress the efficiency of thylakoid activity and is called *photoinhibition*. Carotenoids (polymers of unsaturated isoprene groups, including beta-carotene, which gives carrots their

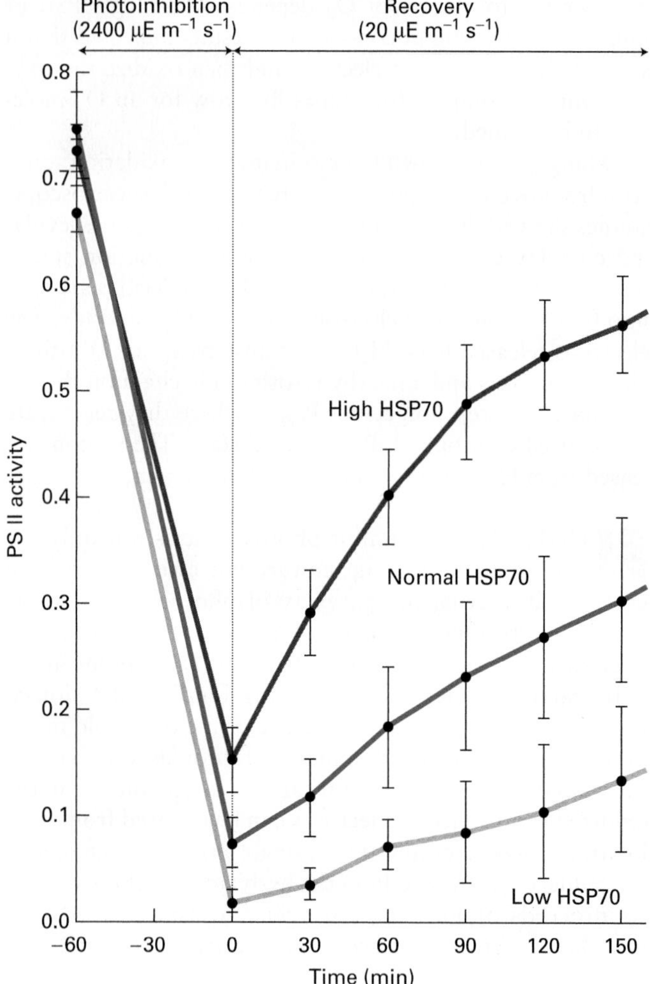

◀ **EXPERIMENTAL FIGURE 12-40 The chaperone HSP70B helps PSII recover from photoinhibition after exposure to intense light.** The unicellular green alga *Chlamydomonas reinhardtii* was genetically manipulated so that it had abnormally high or low levels of the chaperone protein HSP70B. The high, low, and normal strains were then exposed to high-intensity light (2400 μE m^{-2} s^{-1}) for 60 minutes to induce photoinhibition followed by exposure to low light (20 μE m^{-2} s^{-1}) for up to 150 minutes. The effects of photoinhibition by the high-intensity light and the ability of PSII to recover from the photoinhibition were measured using fluorescence spectroscopy to determine PSII activity. The ability of the cells to recover PSII activity depends on the levels of HSP70B—the more HSP70B available, the more rapid the recovery—due to HSP70B protection of the PSII reaction centers that had withstood $^1O_2$-induced D1 subunit damage. [From Schroda et al., 1999, *Plant Cell* **11:**1165.]

orange color) and α-tocopherol (a form of vitamin E) are hydrophobic small molecules that play important roles as $^1O_2$ quenchers to protect plants. For example, inhibition of tocopherol synthesis in the unicellular green alga *Chlamydomonas reinhardtii* by the herbicide pyrazolynate can result in greater light-induced photoinhibition. To help further limit the potential damage in light-harvesting antennas, carotenoid molecules siphon off energy from the dangerous triplet chlorophyll, thus preventing $^1O_2$ formation.

Under intense illumination, photosystem PSII is especially prone to generating $^1O_2$, whereas PSI will produce other ROS, including superoxide, hydrogen peroxide, and hydroxyl radicals. The D1 subunit in the PSII reaction center (see Figure 12-38) is, even under low light conditions, subjected to almost constant $^1O_2$-mediated damage. The damaged reaction center moves from the grana to the unstacked regions of the thylakoid, where the D1 subunit is degraded by a protease and replaced by newly synthesized D1 proteins in what is called the D1 protein damage-repair cycle. The rapid replacement of damaged D1, which requires a high rate of D1 synthesis, helps the PSII recover from photoinactivation and maintain sufficient activity. The experiment in Figure 12-40 shows that an important component in the damage-repair cycle is the chaperone protein HSP70B (see Chapter 3), which binds to the damaged PSII and helps prevent loss of the other components of the complex as the D1 subunit is replaced. The extent of photoinhibition can depend on the amount of HSP70B available to the chloroplasts.

Cyclic Electron Flow Through PSI Generates a Proton-Motive Force but No NADPH or O_2

As we've seen, electrons from reduced ferredoxin in PSI are transferred to $NADP^+$ during linear electron flow, resulting in production of NADPH (see Figure 12-37). In some circumstances cells must generate relative amounts of ATP and NADPH that differ from those produced by linear electron flow (e.g., greater ATP-to-NADPH ratio). To do this, they photosynthetically produce ATP from PSI without concomitant NADPH production. This is accomplished using a PSII-independent process called *cyclic photophosphorylation.* In this process electrons cycle between PSI, ferredoxin, plastoquinone (Q), and the cytochrome *bf* complex (Figure 12-41); thus no net NADPH is generated, and there is no need to oxidize water and produce O_2. There are two distinct cyclic electron flow pathways: the *NAD(P)H dehydrogenase (Ndh)-dependent* (shown in Figure 12-41) and *Ndh-independent* pathways. Ndh is an enzyme complex very similar to the mitochondrial complex I (see Figure 12-16) that oxidizes NADPH or NADH while reducing Q to QH_2 and contributing to the proton motive force by transporting protons. During cyclic electron flow, the substrate for the Ndh is the NADPH generated by light absorption by PSI, ferredoxin, and ferredoxin-NADP reductase (FNR). The QH_2 formed by Ndh then diffuses through the thylakoid membrane to the Q_o binding site on the luminal surface of the cytochrome *bf* complex. There it releases two electrons to the cytochrome *bf* complex and two protons to the thylakoid lumen, generating a proton-motive force. As in linear electron flow, these electrons return to PSI via plastocyanin. This cyclic electron flow is similar to the cyclic process that occurs in the single photosystem of purple bacteria (see Figure 12-36). A Q cycle operates in the cytochrome *bf* complex during cyclic electron flow, leading to transport of two additional protons into the lumen for each pair of electrons transported and a greater proton-motive force.

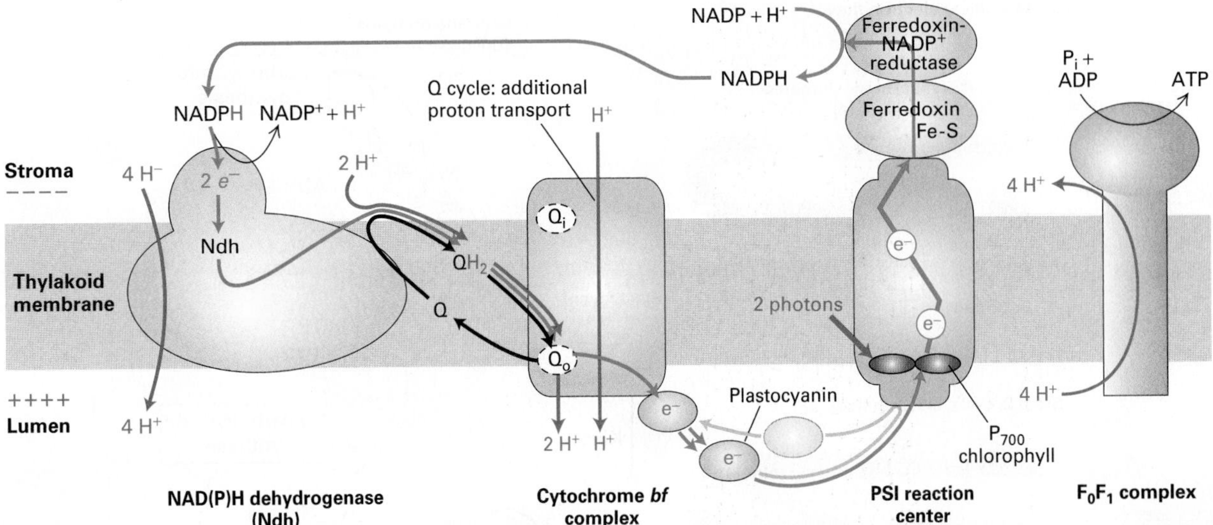

▲ **FIGURE 12-41 Cyclic electron flow in plants, which generates a proton-motive force and ATP but no oxygen or net NADPH.** In the NAD(P)H-dehydrogenase (Ndh)-dependent pathway for cyclic electron flow, light energy is used by PSI to transport electrons in a cycle to generate a proton-motive force and ATP without oxidizing water. The NADPH formed via the PSI/ferredoxin/FNR—instead of being used to fix carbon—is oxidized by Ndh. The released electrons are transferred to plastoquinone (Q) within the membrane to generate QH_2, which then transfers the electrons to the cytochrome *bf* complex, then to plastocyanin, and finally back to PSI, as is the case for the linear electron flow pathway (see Figure 12-37).

In Ndh-independent cyclic electron flow, the mechanism of which has not yet been completely defined, electrons from the ferredoxin are used to reduce Q, either via a hypothetical membrane-associated ferredoxin:plastoquinone oxidoreductase (FQR) or via the Q_i site, which is part of the Q cycle in the cytochrome *bf* complex.

Relative Activities of Photosystems I and II Are Regulated

In order for PSII, which is preferentially located in the stacked grana, and PSI, which is preferentially located in the unstacked thylakoid membranes, to act in sequence during linear electron flow, the amount of light energy delivered to the two reaction centers must be controlled so that each center activates the same number of electrons. This balanced condition is called state 1 (Figure 12-42). If the two photosystems are not equally excited, then cyclic electron flow occurs in PSI and PSII becomes less active (state 2). Variations in the wavelengths and intensities of ambient light (as a consequent of the time of day, clouds, etc.) can change the relative activation of the two photosystems, potentially upsetting the appropriate relative amounts of linear and cyclic electron flow necessary for production of optimal ratios of ATP and NADPH.

One mechanism for regulating the relative contributions of PSI and PSII, in response to varying lighting conditions and thus the relative amounts of linear and cyclic electron flow, entails redistributing the light-harvesting complex LHCII between the two photosystems. The more LHCII associated with a particular photosystem, the more efficiently that system will be activated by light and the greater its contribution to electron flow. The distribution of LHCII between PSI and PSII is mediated by its reversible phosphorylation and dephosphorylation by a regulated membrane-associated kinase and an apparently constitutively active phosphatase. LHCII's unphosphorylated form is preferentially associated with PSII, and the phosphorylated form diffuses in the thylakoid membrane from the grana to the unstacked region and associates with PSI more than the unphosphorylated form. Light conditions in which there is preferential absorption of light by PSII result in the production of high levels of QH_2 that bind to the cytochrome *bf* complex (see Figure 12-37). Consequent conformational changes in this complex are apparently responsible for activation of the LHCII kinase, increased LHCII phosphorylation, compensatory increased activation of PSI relative to PSII, and thus an increase in cyclic electron flow in state 2 (Figure 12-42).

Regulating the supramolecular organization of the photosystems in plants thus has the effect of directing them toward ATP production (state 2) or toward generation of reducing equivalents (NADPH) and ATP (state 1), depending on ambient light conditions and the metabolic needs of the plant. Both NADPH and ATP are required to convert CO_2 to sucrose or starch, the fourth stage in photosynthesis, which we cover in the last section of this chapter.

KEY CONCEPTS OF SECTION 12.5

Molecular Analysis of Photosystems

- In the single photosystem of purple bacteria, cyclic electron flow from light-excited, special-pair chlorophyll *a* molecules in the reaction center generates a proton-motive

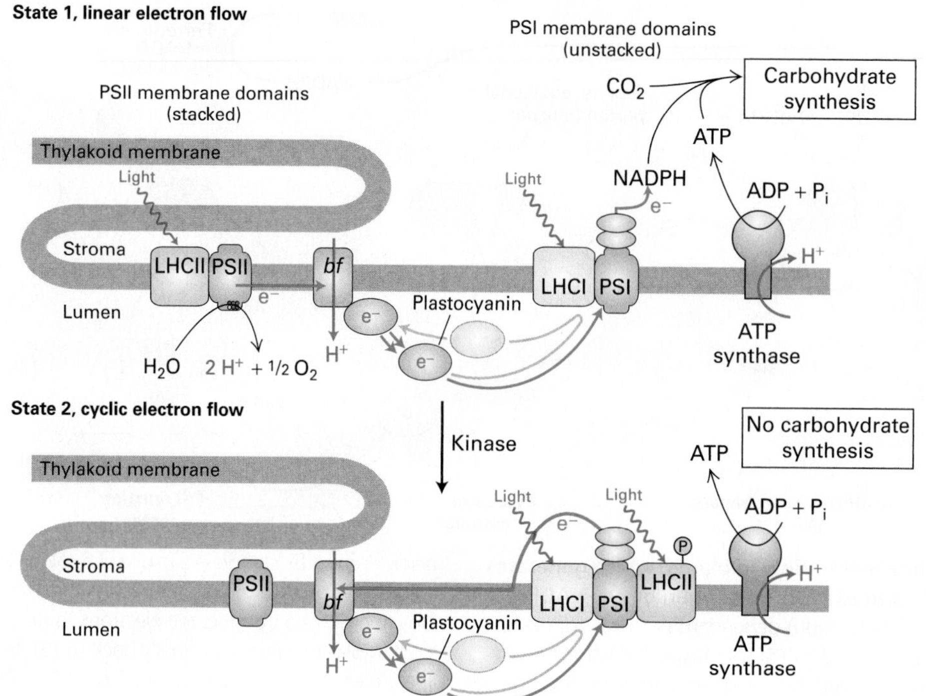

▲ **FIGURE 12-42 Phosphorylation of LHCII and the regulation of linear versus cyclic electron flow.** *(Top)* In normal sunlight, PSI and PSII are equally activated, and the photosystems are organized in state 1. In this arrangement, light-harvesting complex II (LHCII) is not phosphorylated and is tightly associated with the PSII reaction center in the grana. As a result, PSII and PSI can function in parallel in linear electron flow. *(Bottom)* When light excitation of the two photosystems is unbalanced (e.g., too much via PSII), LHCII becomes phosphorylated, dissociates from PSII, and diffuses into the unstacked membranes, where it associates with PSI and its permanently associated LHCI. In this alternative supramolecular organization (state 2), most of the absorbed light energy is transferred to PSI, supporting cyclic electron flow and ATP production but no formation of NADPH and thus no CO_2 fixation. [Adapted from F. A. Wollman, 2001, *EMBO J.* **20:**3623.]

force, which is used mainly to power ATP synthesis by the F_0F_1 complex in the plasma membrane (see Figure 12-36).

- Plants contain two photosystems, PSI and PSII, which have different functions and are physically separated in the thylakoid membrane. PSII splits H_2O into O_2, and PSI reduces $NADP^+$ to NADPH. Cyanobacteria have two analogous photosystems.

- In chloroplasts, light energy absorbed by light-harvesting complexes (LHCs) is transferred to chlorophyll *a* molecules in the reaction centers (P_{680} in PSII and P_{700} in PSI).

- Electrons flow through PSII via the same carriers that are present in the bacterial photosystem. In contrast to the bacterial system, photochemically oxidized $P_{680}{}^+$ in PSII is regenerated to P_{680} by electrons derived from the splitting of H_2O with evolution of O_2 (see Figure 12-37, *left*).

- In linear electron flow, photochemically oxidized $P_{700}{}^+$ in PSI is reduced, regenerating P_{700}, by electrons transferred from PSII via the cytochrome *bf* complex and soluble plastocyanin. Electrons released from P_{700} following excitation of PSI are transported via several carriers ultimately to $NADP^+$, generating NADPH (see Figure 12-37, *right*).

- The absorption of light by pigments in the chloroplast can generate toxic reactive oxygen species (ROS), including singlet oxygen, $^1O_2$, and hydrogen peroxide, H_2O_2. Small molecule scavengers and antioxidant enzymes help to protect against ROS-induced damage; however, singlet oxygen damage to the D1 subunit of PSII still occurs, causing photoinhibition. An HSP70 chaperone helps PSII recover from the damage.

- In contrast to linear electron flow, which requires both PSII and PSI, cyclic electron flow in plants involves only PSI. In this pathway, neither net NADPH nor O_2 is formed although a proton-motive force is generated.

- Reversible phosphorylation and dephosphorylation of the light-harvesting complex II control the functional organization of the photosynthetic apparatus in thylakoid membranes. State 1 favors linear electron flow, whereas state 2 favors cyclic electron flow (see Figure 12-42).

12.6 CO_2 Metabolism During Photosynthesis

Chloroplasts perform many metabolic reactions in green leaves. In addition to CO_2 fixation—incorporation of gaseous CO_2 into small organic molecules and then sugars—the synthesis of almost all amino acids, all fatty acids and carotenes, all pyrimidines, and probably all

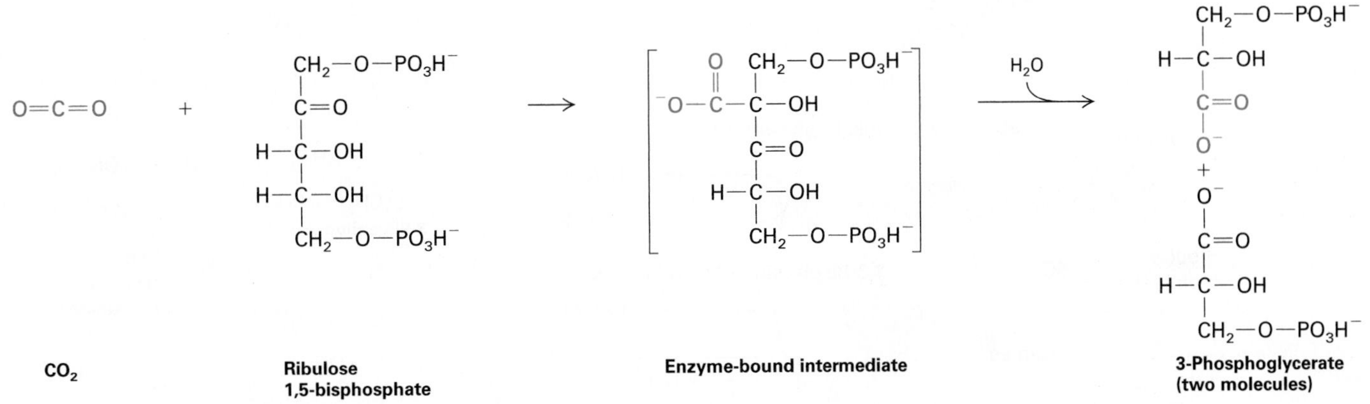

CO_2 + Ribulose 1,5-bisphosphate → [Enzyme-bound intermediate] $\xrightarrow{H_2O}$ 3-Phosphoglycerate (two molecules)

▲ **FIGURE 12-43 The initial reaction of rubisco that fixes CO_2 into organic compounds.** In this reaction, catalyzed by ribulose 1,5-bisphosphate carboxylase (rubisco), CO_2 condenses with the five-carbon sugar ribulose 1,5-bisphosphate. The products are two molecules of 3-phosphoglycerate.

purines occurs in chloroplasts. However, the synthesis of sugars from CO_2 is the most extensively studied biosynthetic pathway in plant cells. We first consider the unique pathway, known as the **Calvin cycle** (after discoverer Melvin Calvin), that fixes CO_2 into three-carbon compounds, powered by energy released during ATP hydrolysis and oxidation of NADPH.

Rubisco Fixes CO_2 in the Chloroplast Stroma

The enzyme **ribulose 1,5-bisphosphate carboxylase,** or **rubisco,** fixes CO_2 into precursor molecules that are subsequently converted into carbohydrates. Rubisco is located in the stromal space of the chloroplast. This enzyme adds CO_2 to the five-carbon sugar ribulose 1,5-bisphosphate to form two molecules of the three-carbon-containing 3-phosphoglycerate (Figure 12-43). Rubisco is a large enzyme ($\approx$500 kDa) composed of eight identical large and eight identical small subunits. One subunit is encoded in chloroplast DNA; the other, in nuclear DNA. Because the catalytic rate of rubisco is quite low, many copies of the enzyme are needed to fix sufficient CO_2. Indeed, this enzyme makes up almost 50 percent of the chloroplast protein and is believed to be the most abundant protein on earth.

When photosynthetic algae are exposed to a brief pulse of ^{14}C-labeled CO_2 and the cells are then quickly disrupted, 3-phosphoglycerate is radiolabeled most rapidly, and all the radioactivity is found in the carboxyl group. Because CO_2 is initially incorporated into a three-carbon compound, the Calvin cycle is also called the *C_3 pathway* of carbon fixation (Figure 12-44).

The fate of 3-phosphoglycerate formed by rubisco is complex: some is converted to hexoses incorporated into starch or sucrose, but some is used to regenerate ribulose 1,5-bisphosphate. At least nine enzymes are required to regenerate ribulose 1,5-bisphosphate from 3-phosphoglycerate. Quantitatively, for every 12 molecules of 3-phosphoglycerate generated by rubisco (a total of 36 C atoms), 2 of them (6 C atoms) are converted to 2 molecules of glyceraldehyde 3-phosphate (and later to 1 hexose), whereas 10 molecules (30 C atoms) are converted to 6 molecules of ribulose 1,5-bisphosphate (Figure 12-44, *top*). The fixation of six CO_2 molecules and the net formation of two glyceraldehyde 3-phosphate molecules require the consumption of 18 ATPs and 12 NADPHs, generated by the light-requiring processes of photosynthesis.

Synthesis of Sucrose Using Fixed CO_2 Is Completed in the Cytosol

After its formation in the chloroplast stroma, glyceraldehyde 3-phosphate is transported to the cytosol in exchange for phosphate. The final steps of sucrose synthesis (Figure 12-44, *bottom*) occur in the cytosol of leaf cells.

An antiporter transport protein in the chloroplast membrane brings fixed CO_2 (as glyceraldehyde 3-phosphate) into the cytosol when the cell is exporting sucrose vigorously. No fixed CO_2 leaves the chloroplast unless phosphate is fed into it to replace the phosphate carried out of the stroma in the form of glyceraldehyde 3-phosphate. During the synthesis of sucrose from glyceraldehyde 3-phosphate, inorganic phosphate groups are released (Figure 12-44, *bottom left*). Thus the synthesis of sucrose facilitates the transport of additional glyceraldehyde 3-phosphate from the chloroplast to the cytosol by providing phosphate for the antiporter.

Light and Rubisco Activase Stimulate CO_2 Fixation

The Calvin cycle enzymes that catalyze CO_2 fixation are rapidly inactivated in the dark, thereby conserving ATP that is generated in the dark for other synthetic reactions, such as lipid and amino acid biosynthesis. One mechanism that contributes to this control is the pH dependence of several Calvin cycle enzymes. Because protons are transported from the stroma into the thylakoid lumen during photoelectron transport (see Figure 12-37), the pH of the stroma increases from $\approx$7 in the dark to $\approx$8 in the light. The increased activity of several Calvin cycle enzymes at the higher pH promotes CO_2 fixation in the light.

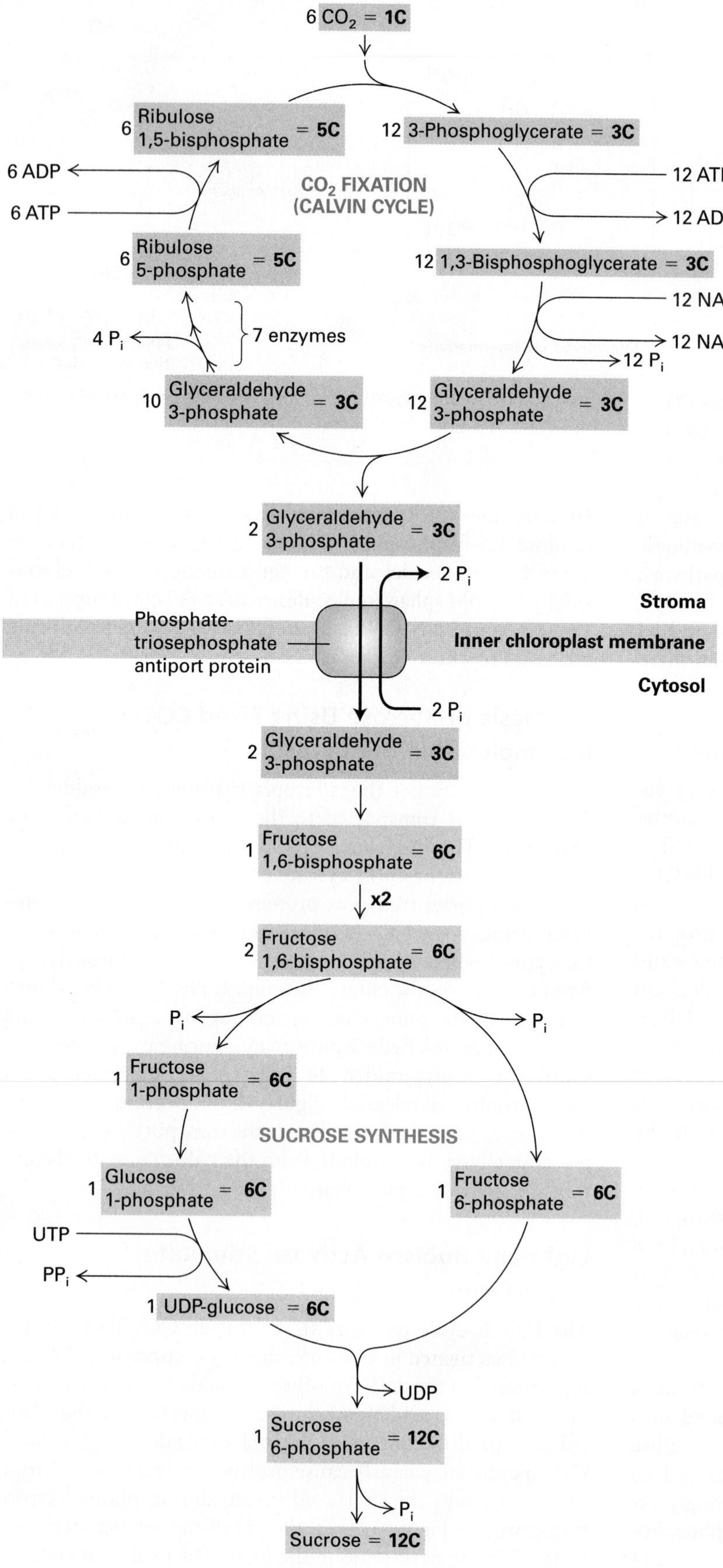

6 CO2 = 1C
6 Ribulose 1,5-bisphosphate = 5C
12 3-Phosphoglycerate = 3C
6 ADP
6 ATP
CO2 FIXATION (CALVIN CYCLE)
12 ATP
12 ADP
6 Ribulose 5-phosphate = 5C
12 1,3-Bisphosphoglycerate = 3C
12 NADPH
12 NADP+
12 Pi
4 Pi
7 enzymes
10 Glyceraldehyde 3-phosphate = 3C
12 Glyceraldehyde 3-phosphate = 3C
2 Glyceraldehyde 3-phosphate = 3C
2 Pi
Stroma
Phosphate-triosephosphate antiport protein
Inner chloroplast membrane
Cytosol
2 Pi
2 Glyceraldehyde 3-phosphate = 3C
1 Fructose 1,6-bisphosphate = 6C
x2
2 Fructose 1,6-bisphosphate = 6C
Pi
Pi
1 Fructose 1-phosphate = 6C
SUCROSE SYNTHESIS
1 Glucose 1-phosphate = 6C
1 Fructose 6-phosphate = 6C
UTP
PPi
1 UDP-glucose = 6C
UDP
1 Sucrose 6-phosphate = 12C
Pi
Sucrose = 12C

CO2 O=C=O
3-Phosphoglycerate
Ribulose 5-phosphate
1,3-Bisphosphoglycerate
Ribulose 1,5-bisphosphate
Glyceraldehyde 3-phosphate

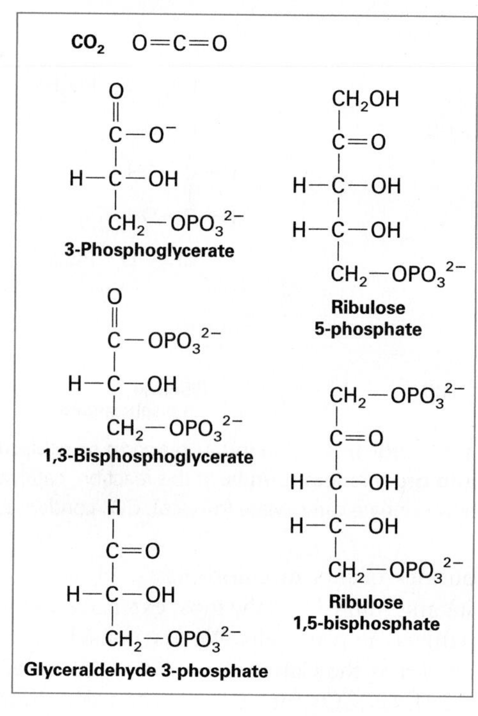

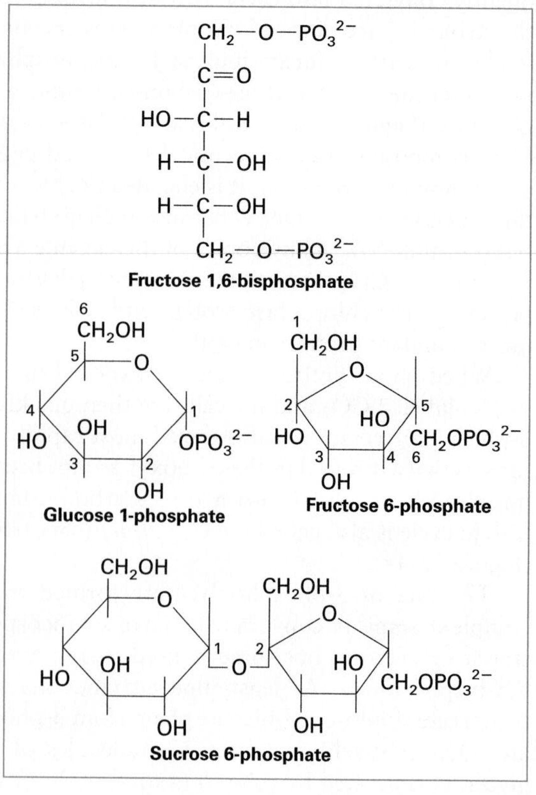

Fructose 1,6-bisphosphate
Glucose 1-phosphate
Fructose 6-phosphate
Sucrose 6-phosphate

◀ **FIGURE 12-44 The pathway of carbon during photosynthesis.** *(Top)* Six molecules of CO_2 are converted into two molecules of glyceraldehyde 3-phosphate. These reactions, which constitute the Calvin cycle, occur in the stroma of the chloroplast. Via the phosphate/triosephosphate antiporter, some glyceraldehyde 3-phosphate is transported to the cytosol in exchange for phosphate. *(Bottom)* In the cytosol, an exergonic series of reactions converts glyceraldehyde 3-phosphate to fructose 1,6-bisphosphate. Two molecules of fructose 1,6-bisphosphate are used to synthesize one of the disaccharide sucrose. Some glyceraldehyde 3-phosphate (not shown here) is also converted to amino acids and fats, compounds essential for plant growth.

A stromal protein called *thioredoxin (Tx)* also plays a role in controlling some Calvin cycle enzymes. In the dark, thioredoxin contains a disulfide bond; in the light, electrons are transferred from PSI, via ferredoxin, to thioredoxin, reducing its disulfide bond:

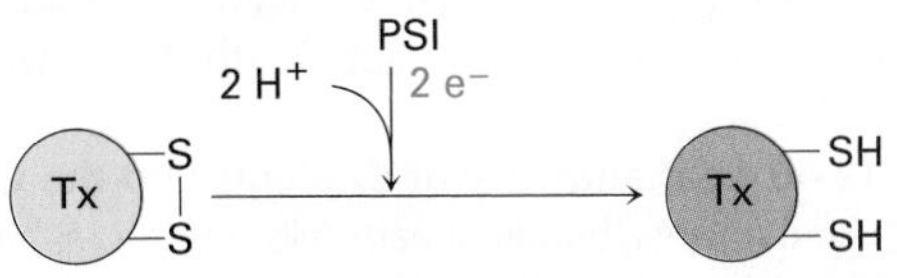

Reduced thioredoxin then activates several Calvin cycle enzymes by reducing disulfide bonds in them. In the dark, when thioredoxin becomes reoxidized, these enzymes are reoxidized and so inactivated. Thus these enzymes are sensitive to the redox state of the stroma, which in turn is light sensitive—an elegant mechanism for regulating enzymatic activity by light.

Rubisco is one such light/redox-sensitive enzyme, although its regulation is very complex and not yet fully understood. Rubisco is spontaneously activated in the presence of high CO_2 and Mg^{2+} concentrations. The activating reaction entails covalent addition of CO_2 to the side-chain amino group of lysine in the active site, forming a carbamate group that then binds a Mg^{2+} ion required for activity. Under normal conditions, however, with ambient levels of CO_2, the reaction is slow and usually requires catalysis by *rubisco activase,* an enzyme that simultaneously hydrolyzes ATP and uses the energy to attach a CO_2 to the lysine. Rubisco activase also accelerates an activating conformational change in rubisco (inactive-closed to active-opened state). The regulation of rubisco activase by thioredoxin is, at least in part in some species, responsible for rubisco's light/redox sensitivity. Furthermore, rubisco activase's activity is sensitive to the ratio of ATP:ADP. If that ratio is low (relatively high ADP), then the activase will not activate rubisco (and so the cell will expend less of its scarce ATP to fix carbon). Given the key role of rubisco in controlling energy utilization and carbon flux—both in an individual chloroplast and, in a sense, throughout the entire biosphere—it is not surprising that its activity is tightly regulated.

Photorespiration, Which Competes with Photosynthesis, Is Reduced in Plants That Fix CO_2 by the C_4 Pathway

As noted above, rubisco catalyzes the incorporation of CO_2 into ribulose 1,5-bisphosphate. It can catalyze a second, distinct, and *competing* reaction with the same substrate—ribulose 1,5-bisphosphate—but with O_2 in place of CO_2 as a second substrate (Figure 12-45). The products of the second reaction are one molecule of 3-phosphoglycerate and one molecule of the two-carbon compound phosphoglycolate. The first (carbon-fixing) reaction is favored when the ambient CO_2 concentration is relatively high, whereas the second is favored when

▲ **FIGURE 12-45 CO_2 fixation and photorespiration.** These competing pathways are both initiated by ribulose 1,5-bisphosphate carboxylase (rubisco), and both utilize ribulose 1,5-bisphosphate. CO_2 fixation, pathway **1**, is favored by high CO_2 and low O_2 pressures; photorespiration, pathway **2**, occurs at low CO_2 and high O_2 pressures (that is, under normal atmospheric conditions). Phosphoglycolate is recycled via a complex set of reactions that take place in peroxisomes and mitochondria, as well as chloroplasts. The net result: for every two molecules of phosphoglycolate formed by photorespiration (four C atoms), one molecule of 3-phosphoglycerate is ultimately formed and recycled and one molecule of CO_2 is lost.

CO_2 is low and O_2 relatively high. The pathway initiated by the second reaction with O_2 is called **photorespiration**—a process that takes place in light, consumes O_2, and converts ribulose 1,5-bisphosphate in part to CO_2. As Figure 12-45 shows, photorespiration is wasteful to the energy economy of the plant: it consumes ATP and O_2, and it generates CO_2 without fixing carbon. Indeed, when CO_2 is low and O_2 is high, much of the CO_2 fixed by the Calvin cycle is lost as the result of photorespiration. Recent studies have suggested that this surprising, wasteful alternative reaction catalyzed by rubisco may be a consequence of the inherent difficulty the enzyme has in specifically binding the relatively featureless CO_2 molecule and of the ability of both CO_2 and O_2 to react and form distinct products with the same initial enzyme/ribulose 1,5-bisphosphate intermediate.

Excessive photorespiration could become a problem for plants in a hot, dry environment, because they must keep the gas-exchange pores (stomata) in their leaves closed much of the time to prevent excessive loss of moisture. As a consequence, the CO_2 level inside the leaf can fall below the K_m of rubisco for CO_2. Under these conditions, the rate of photosynthesis is slowed, photorespiration is greatly favored, and the plant might be in danger of fixing inadequate amounts of CO_2. Corn, sugarcane, crabgrass, and other plants that can grow in hot, dry environments have evolved a way to avoid this problem by utilizing a two-step pathway of CO_2 fixation in which a CO_2-hoarding step precedes the Calvin cycle. The pathway has been named the *C_4 pathway* because $[^{14}C]CO_2$ labeling showed that the first radioactive molecules formed during photosynthesis in this pathway are four-carbon compounds, such as oxaloacetate and malate, rather than the three-carbon molecules that initiate the Calvin cycle (C_3 pathway).

The C_4 pathway involves two types of cells: *mesophyll cells,* which are adjacent to the air spaces in the leaf interior, and *bundle sheath cells,* which surround the vascular tissue and are sequestered away from the high oxygen levels to which mesophyll cells are exposed (Figure 12-46a). In the mesophyll cells of C_4 plants, phosphoenolpyruvate, a three-carbon molecule derived from pyruvate, reacts with CO_2 to generate

(a)

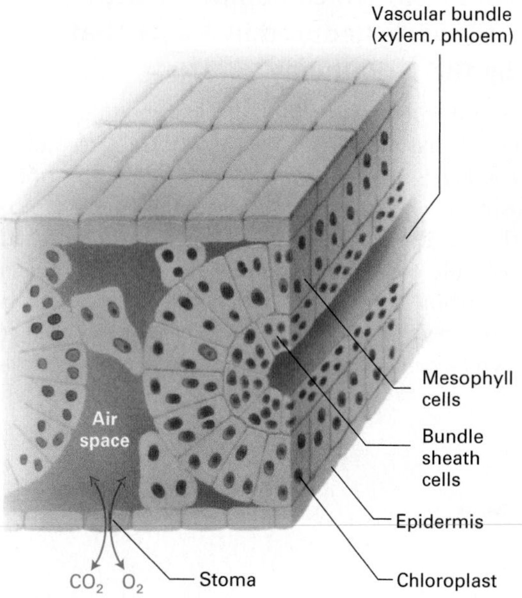

(b)

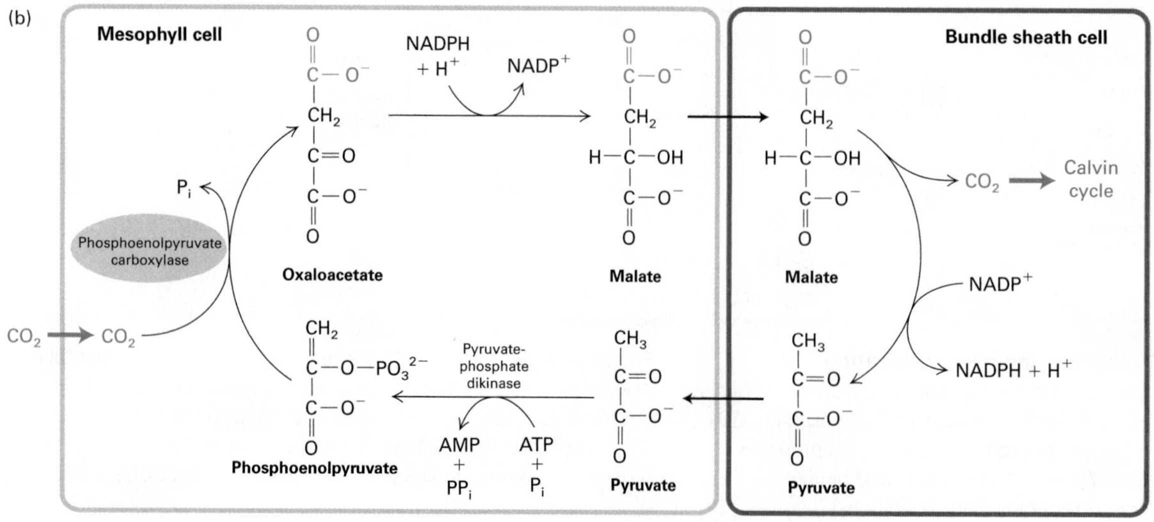

◄ **FIGURE 12-46 Leaf anatomy of C_4 plants and the C_4 pathway.** (a) In C_4 plants, bundle sheath cells line the vascular bundles containing the xylem and phloem. Mesophyll cells, which are adjacent to the substomal air spaces, can assimilate CO_2 into four-carbon molecules at low ambient CO_2 and deliver it to the interior bundle sheath cells. Bundle sheath cells contain abundant chloroplasts and are the sites of photosynthesis and sucrose synthesis. Sucrose is carried to the rest of the plant via the phloem. In C_3 plants, which lack bundle sheath cells, the Calvin cycle operates in the mesophyll cells to fix CO_2. (b) The key enzyme in the C_4 pathway is phosphoenolpyruvate carboxylase, which assimilates CO_2 to form oxaloacetate in mesophyll cells. Decarboxylation of malate or other C_4 intermediates in bundle sheath cells releases CO_2, which enters the standard Calvin cycle (see Figure 12-44, *top*).

oxaloacetate, a four-carbon compound (Figure 12-46b). The enzyme that catalyzes this reaction, *phosphoenolpyruvate carboxylase,* is found almost exclusively in C_4 plants and unlike rubisco is insensitive to O_2. The overall reaction from pyruvate to oxaloacetate involves the hydrolysis of one ATP and has a negative ΔG. Therefore, CO_2 fixation will proceed even when the CO_2 concentration is low. The oxaloacetate formed in mesophyll cells is reduced to malate, which is transferred by a special transporter to the bundle sheath cells, where the CO_2 released by decarboxylation enters the Calvin cycle (Figure 12-46b).

Because of the transport of CO_2 from mesophyll cells, the CO_2 concentration in the bundle sheath cells of C_4 plants is much higher than it is in the normal atmosphere. Bundle sheath cells are also unusual in that they lack PSII and carry out only cyclic electron flow catalyzed by PSI, so no O_2 is evolved. The high CO_2 and reduced O_2 concentrations in the bundle sheath cells favor the fixation of CO_2 by rubisco to form 3-phosphoglycerate and inhibit the utilization of ribulose 1,5-bisphosphate in photorespiration.

In contrast, the high O_2 concentration in the atmosphere favors photorespiration in the mesophyll cells of C_3 plants (pathway 2 in Figure 12-45); as a result, as much as 50 percent of the carbon fixed by rubisco may be reoxidized to CO_2 in C_3 plants. C_4 plants are superior to C_3 plants in utilizing the available CO_2, since the C_4 enzyme phosphoenolpyruvate carboxylase has a higher affinity for CO_2 than does rubisco in the Calvin cycle. However, one ATP is consumed in the cyclic C_4 process (to generate phosphoenolpyruvate from pyruvate); thus the overall efficiency of the photosynthetic production of sugars from NADPH and ATP is lower than it is in C_3 plants, which use only the Calvin cycle for CO_2 fixation. Nonetheless, the net rates of photosynthesis for C_4 grasses, such as corn or sugarcane, can be two to three times the rates for otherwise similar C_3 grasses, such as wheat, rice, or oats, owing to the elimination of losses from photorespiration.

Of the two carbohydrate products of photosynthesis, starch remains in the mesophyll cells of C_3 plants and the bundle sheaf cells in C_4 plants. In these cells, starch is subjected to glycolysis, mainly in the dark, forming ATP, NADH, and small molecules that are used as building blocks for the synthesis of amino acids, lipids, and other cellular constituents. Sucrose, in contrast, is exported from the photosynthetic cells and transported throughout the plant.

KEY CONCEPTS OF SECTION 12.6

CO_2 Metabolism During Photosynthesis

- In the Calvin cycle, CO_2 is fixed into organic molecules in a series of reactions that occur in the chloroplast stroma. The initial reaction, catalyzed by rubisco, forms a three-carbon intermediate. Some of the glyceraldehyde 3-phosphate generated in the cycle is transported to the cytosol and converted to sucrose (see Figure 12-44).
- The light-dependent activation of several Calvin cycle enzymes and other mechanisms increases fixation of CO_2 in the light. The redox state of the stroma plays a key role in this regulation as does the regulation of the activity of rubisco by rubisco activase.
- In C_3 plants, a substantial fraction of the CO_2 fixed by the Calvin cycle can be lost as the result of photorespiration, a wasteful reaction catalyzed by rubisco that is favored at low CO_2 and high O_2 levels (see Figure 12-45).
- In C_4 plants, CO_2 is fixed initially in the outer mesophyll cells by reaction with phosphoenolpyruvate. The four-carbon molecules so generated are shuttled to the interior bundle sheath cells, where the CO_2 is released and then used in the Calvin cycle. The rate of photorespiration in C_4 plants is much lower than in C_3 plants.

Perspectives for the Future

Although the overall processes of photosynthesis and mitochondrial oxidation are well understood, many important details remain to be uncovered by a new generation of scientists. For example, little is known about how complexes I and IV in mitochondria couple proton and electron movements to create a proton-motive force. Similarly, although the binding-change mechanism for ATP synthesis by the F_0F_1 complex is now generally accepted, we do not understand how conformational changes in each β subunit are coupled to the cyclical binding of ADP and P_i, formation of ATP, and then release of ATP. In addition, many questions remain about the precise mechanism of action of transport proteins in the inner mitochondrial and chloroplast membranes that play key roles in oxidative phosphorylation and photosynthesis.

We now know that release of cytochrome *c* and other proteins from the intermembrane space of mitochondria into the cytosol plays a major role in triggering apoptosis (Chapter 21). Certain members of the Bcl-2 family of apoptotic proteins and ion channels localized in part to the outer mitochondrial membrane participate in this process. The connections between energy metabolism and mechanisms underlying apoptosis remain to be clearly defined.

The generation of toxic ROS in the chloroplast induces compensatory activation of protective genes in the nucleus. Determining the mechanisms of these signaling pathways will provide much greater insight into the logic of such regulatory pathways. As we better understand the mechanisms underlying photosynthesis, particularly the action of rubisco and it regulation, it is possible that we will be able to exploit these insights to improve crop yields to provide abundant and inexpensive food to all who need it.

Key Terms

aerobic oxidation *479*
ATP synthase *504*
Calvin cycle *525*
carbon fixation *511*
cellular respiration *485*
C_4 pathway *528*
chemiosmosis *480*
chlorophylls *511*
chloroplasts *479*
citric acid cycle *479*
cytochromes *495*
electron carriers *487*

electron transport chain *481*
endosymbiont hypothesis *485*
fermentation *485*
F_0F_1 complex *505*
glycolysis *481*
mitochondria *479*
oxidative phosphorylation *481*
photoelectron transport *515*
photorespiration *528*
photosynthesis *479*
photosystems *514*
proton-motive force *480*
Q cycle *500*
reactive oxygen species *502*
respiratory control *510*
rubisco *525*
substrate-level phosphorylation *481*
thylakoids *511*
uncouplers *510*

Review the Concepts

1. The proton-motive force (pmf) is essential for both mitochondrial and chloroplast function. What produces the pmf, and what is its relationship to ATP?

2. The mitochondrial inner membrane exhibits all of the fundamental characteristics of a typical cell membrane, but it also has several unique characteristics that are closely associated with its role in oxidative phosphorylation. What are these unique characteristics? How does each contribute to the function of the inner membrane?

3. Maximal production of ATP from glucose involves the reactions of glycolysis, the citric acid cycle, and the electron transport chain. Which of these reactions requires O_2, and why? Which, in certain organisms or physiological conditions, can proceed in the absence of O_2?

4. Describe how the electrons produced by glycolysis are delivered to the electron transport chain. What would be the consequence for overall ATP yield per glucose molecule if a mutation inactivated this delivery system? What would be the longer-term consequence for the activity of the glycolytic pathway?

5. Mitochondrial oxidation of fatty acids is a major source of ATP, yet fatty acids can be oxidized elsewhere. What organelle, besides the mitochondrion, can oxidize fatty acids? What is the fundamental difference between oxidation occurring in this organelle and mitochondrial oxidation?

6. Each of the cytochromes in the mitochondria contains prosthetic groups. What is a prosthetic group? Which type of prosthetic group is associated with the cytochromes? What property of the various cytochromes ensures unidirectional electron flow along the electron transport chain?

7. It is estimated that each electron pair donated by NADH leads to the synthesis of approximately three ATP molecules, whereas each electron pair donated by $FADH_2$ leads to the synthesis of approximately two ATP molecules. What is the underlying reason for the difference in yield for electrons donated by $FADH_2$ versus NADH?

8. Much of our understanding of ATP synthase is derived from research on aerobic bacteria. What makes these organisms useful for this research? Where do the reactions of glycolysis, the citric acid cycle, and the electron transport chain occur in these organisms? Where is the pmf generated in aerobic bacteria? What other cellular processes depend on the pmf in these organisms?

9. An important function of the mitochondrial inner membrane is to provide a selectively permeable barrier to the movement of water-soluble molecules and thus generate different chemical environments on either side of the membrane. However, many of the substrates and products of oxidative phosphorylation are water soluble and must cross the inner membrane. How does this transport occur?

10. The Q cycle plays a major role in the electron transport chain of mitochondria, chloroplasts, and bacteria. What is the function of the Q cycle, and how does it carry out this function? What electron transport components participate in the Q cycle in mitochondria, in purple bacteria, and in chloroplasts?

11. Write the overall reaction of oxygen-generating photosynthesis. Explain the following statement: the O_2 generated by photosynthesis is simply a by-product of the pathway's generation of carbohydrates and ATP.

12. Photosynthesis can be divided into multiple stages. What are the stages of photosynthesis, and where does each occur within the chloroplast? Where is the sucrose produced by photosynthesis generated?

13. The photosystems responsible for absorption of light energy are composed of two linked components, the reaction center and an antenna complex. What is the pigment composition and role of each in the process of light absorption? What evidence exists that the pigments found in these components are involved in photosynthesis?

14. Photosynthesis in green and purple bacteria does not produce O_2. Why? How can these organisms still use photosynthesis to produce ATP? What molecules serve as electron donors in these organisms?

15. Chloroplasts contain two photosystems. What is the function of each? For linear electron flow, diagram the flow of electrons from photon absorption to NADPH formation. What does the energy stored in the form of NADPH synthesize?

16. The Calvin cycle reactions that fix CO_2 do not function in the dark. What are the likely reasons for this? How are these reactions regulated by light?

17. Rubisco, which may be the most abundant protein on earth, plays a key role in the synthesis of carbohydrates in organisms that use photosynthesis. What is rubisco, where is it located, and what function does it serve?

Analyze the Data

A proton gradient can be analyzed with fluorescent dyes whose emission-intensity profiles depend on pH. One of the most useful dyes for measuring the pH gradient across mitochondrial membranes is the membrane-impermeant, water-soluble fluorophore 2′,7′-bis-(2-carboxyethyl)-5(6)-carboxyfluorescein (BCECF). The effect of pH on the emission intensity of BCECF, excited at 505 nm, is shown in the accompanying figure. In one study, sealed vesicles containing this compound were prepared by mixing unsealed, isolated inner mitochondrial membranes with BCECF; after resealing of the membranes, the vesicles were collected by centrifugation and then resuspended in nonfluorescent medium.

a. When these vesicles were incubated in a physiological buffer containing NADH, ADP, P_i, and O_2, the fluorescence of BCECF trapped inside gradually decreased in intensity. What does this decrease in fluorescent intensity suggest about this vesicular preparation?

b. How would you expect the concentrations of ADP, P_i, and O_2 to change during the course of the experiment described in part a? Why?

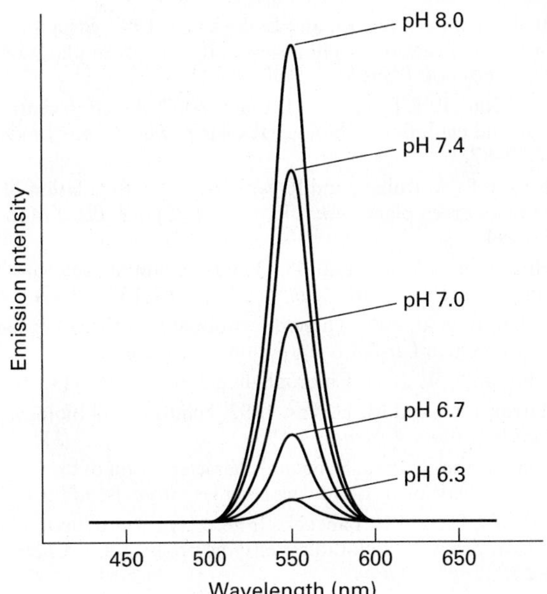

c. After the vesicles were incubated in buffer containing ADP, P_i, and O_2 for a period of time, addition of dinitrophenol caused an increase in BCECF fluorescence. In contrast, addition of valinomycin produced only a small transient effect. Explain these findings.

d. Predict the outcome of an experiment performed as described in part a if brown-fat tissue was used as a source of unsealed, isolated inner mitochondrial membranes. Explain your answer.

References

First Steps of Glucose and Fatty Acid Catabolism: Glycolysis and the Citric Acid Cycle

Berg, J., J. Tymoczko, and L. Stryer. 2002. *Biochemistry,* 5th ed. W. H. Freeman and Company, chaps. 16 and 17.

Canfield, D. E. 2005. The early history of atmospheric oxygen: homage to Robert M. Garrels. *Annu. Rev. Earth Planet. Sci.* **33**:1–36.

Chan, D.C. 2006. Mitochondria: dynamic organelles in disease, aging, and development. *Cell* **125**(7):1241–1252.

Depre, C., M. Rider, and L. Hue. 1998. Mechanisms of control of heart glycolysis. *Eur. J. Biochem.* **258**:277–290.

Eaton, S., K. Bartlett, and M. Pourfarzam. 1996. Mammalian mitochondrial beta-oxidation. *Biochem. J.* **320**(Part 2):345–557.

Fersht, A. 1999. *Structure and Mechanism in Protein Science: A Guide to Enzyme Catalysis and Protein Folding.* W. H. Freeman and Company.

Fothergill-Gilmore, L. A., and P. A. Michels. 1993. Evolution of glycolysis. *Prog. Biophys. Mol. Biol.* **59**:105–135.

Guest, J. R., and G. C. Russell. 1992. Complexes and complexities of the citric acid cycle in *Escherichia coli. Curr. Top. Cell Reg.* **33**:231–247.

Krebs, H. A. 1970. The history of the tricarboxylic acid cycle. *Perspect. Biol. Med.* **14**:154–170.

Nelson, D. L., and M. M. Cox. 2000. *Lehninger Principles of Biochemistry.* Worth, chaps. 14–17, 19.

Pilkis, S. J., T. H. Claus, I. J. Kurland, and A. J. Lange. 1995. 6-Phosphofructo-2-kinase/fructose-2,6-bisphosphatase: a metabolic signaling enzyme. *Ann. Rev. Biochem.* **64**:799–835.

Rasmussen, B., and R. Wolfe. 1999. Regulation of fatty acid oxidation in skeletal muscle. *Ann. Rev. Nutrition* **19**:463–484.

Velot, C., M. Mixon, M. Teige, and P. Srere. 1997. Model of a quinary structure between Krebs TCA cycle enzymes: a model for the metabolon. *Biochemistry* **36**:14271–14276.

Wanders, R. J., and H. R. Waterham. 2006. Biochemistry of mammalian peroxisomes revisited. *Annu. Rev. Biochem.* **75**:295–332.

The Electron Transport Chain and Generation of the Proton-Motive Force

Babcock, G. 1999. How oxygen is activated and reduced in respiration. *Proc. Nat'l. Acad. Sci. USA* **96**:12971–12973.

Beinert, H., R. Holm, and E. Münck. 1997. Iron-sulfur clusters: nature's modular, multipurpose structures. *Science* **277**:653–659.

Brandt, U. 2006. Energy Converting NADH:quinone oxidoreductase (complex I). *Annu. Rev. Biochem.* **75**:165–187.

Brandt, U., and B. Trumpower. 1994. The protonmotive Q cycle in mitochondria and bacteria. *Crit. Rev. Biochem. Mol. Biol.* **29**:165–197.

Darrouzet, E., C. Moser, P. L. Dutton, and F. Daldal. 2001. Large scale domain movement in cytochrome bc1: a new device for electron transfer in proteins. *Trends Biochem. Sci.* **26**:445–451.

Grigorieff, N. 1999. Structure of the respiratory NADH:ubiquinone oxidoreductase (complex I). *Curr. Opin. Struc. Biol.* **9**:476–483.

Hosler, J. P., S. Ferguson-Miller,, and D. A. Mills. 2006. Energy transduction: proton transfer through the respiratory complexes. *Annu. Rev. Biochem.* **75**:165–187.

Michel, H., J. Behr, A. Harrenga, and A. Kannt. 1998. Cytochrome *c* oxidase. *Ann. Rev. Biophys. Biomol. Struc.* **27**:329–356.

Mitchell, P. 1979. Keilin's respiratory chain concept and its chemiosmotic consequences. *Science* **206**:1148–1159. (Nobel Prize Lecture.)

Ramirez, B. E., B. Malmström, J. R. Winkler, and H. B. Gray. 1995. The currents of life: the terminal electron-transfer complex of respiration. *Proc. Nat'l. Acad. Sci. USA* **92**:11949–11951.

Ruitenberg, M., et al. 2002. Reduction of cytochrome *c* oxidase by a second electron leads to proton translocation. *Nature* **417**:99–102.

Saraste, M. 1999. Oxidative phosphorylation at the fin de siècle. *Science* **283**:1488–1492.

Schafer, E., et al. 2006. Architecture of active mammalian respiratory chain supercomplexes. *J. Biol. Chem.* **281**(22):15370–15375.

Schultz, B., and S. Chan. 2001. Structures and proton-pumping strategies of mitochondrial respiratory enzymes. *Ann. Rev. Biophys. Biomol. Struc.* **30**:23–65.

Sheeran, F. L., and S. Pepe. 2006. Energy deficiency in the failing heart: linking increased reactive oxygen species and disruption of oxidative phosphorylation rate. *Biochim. Biophys. Acta.* **1757**(5–6):543–552.

Tsukihara, T., et al. 1996. The whole structure of the 13-subunit oxidized cytochrome *c* oxidase at 2.8 Å. *Science* **272**:1136–1144.

Walker, J. E. 1995. Determination of the structures of respiratory enzyme complexes from mammalian mitochondria. *Biochim. Biophys. Acta* **1271**:221–227.

Wallace, D. C. 2005. A mitochondrial paradigm of metabolic and degenerative diseases, aging, and cancer: a dawn for evolutionary medicine. *Annu. Rev. Genet.* 359–407.

Xia, D., et al. 1997. Crystal structure of the cytochrome *bc*$_1$ complex from bovine heart mitochondria. *Science* **277**:60–66.

Zaslavsky, D., and R. Gennis. 2000. Proton pumping by cytochrome oxidase: progress and postulates. *Biochim. Biophys. Acta* **1458**:164–179.

M. Zhang, E. Mileykovskaya, and W. Dowhan. 2005. Cardiolipin is essential for organization of complexes III and IV into a supercomplex in intact yeast mitochondria. *J. Biol. Chem.* **280**(33):29403–29408.

Zhang, Z., et al. 1998. Electron transfer by domain movement in cytochrome *bc*$_1$. *Nature* **392**:677–684.

Harnessing the Proton-Motive Force for Energy-Requiring Processes

Aksimentiev, A., I. A. Balabin, R. H. Fillingame, and K. Schulten. 2004. Insights into the molecular mechanism of rotation in the F_0 sector of ATP synthase. *Biophys. J.* **86**(3):1332–1344.

Bianchet, M. A., J. Hullihen, P. Pedersen, and M. Amzel. 1998 The 2.8 Å structure of rat liver F1-ATPase: configuration of a critical intermediate in ATP synthesis/hydrolysis. *Proc. Nat'l. Acad. Sci. USA* **95**:11065–11070.

Boyer, P. D. 1997. The ATP synthase—a splendid molecular machine. *Ann. Rev. Biochem.* **66**:717–749.

Capaldi, R., and R. Aggeler. 2002. Mechanism of the F_0F_1-type ATP synthase—a biological rotary motor. *Trends Biochem. Sci.* **27**:154–160.

Elston, T., H. Wang, and G. Oster. 1998. Energy transduction in ATP synthase. *Nature* **391**:510–512.

Hinkle, P. C. 2005. P/O ratios of mitochondrial oxidative phosphorylation. *Biochim. Biophys. Acta.* **1706**(1–2):1–11.

Kinosita, K., et al. 1998. F_1-ATPase: a rotary motor made of a single molecule. *Cell* **93**:21–24.

Klingenberg, M., and S. Huang. 1999. Structure and function of the uncoupling protein from brown adipose tissue. *Biochim. Biophys. Acta* **1415**:271–296.

Nury, H., et al.. 2006. Relations between structure and function of the mitochondrial ADP/ATP carrier. *Annu. Rev. Biochem.* **75**:713–741.

Tsunoda, S., R. Aggeler, M. Yoshida, and R. Capaldi. 2001. Rotation of the **c** subunit oligomer in fully functional F_0F_1 ATP synthase. *Proc. Nat'l. Acad. Sci. USA* **98**:898–902.

Vercesi, A. E., et al. 2006. Plant uncoupling mitochondrial proteins. *Annu. Rev. Plant Biol.* **57**:383–404.

Yasuda, R., et al. 2001. Resolution of distinct rotational substeps by submillisecond kinetic analysis of F1-ATPase. *Nature* **410**:898–904.

Photosynthetic Stages and Light-Absorbing Pigments

Ben-Shem, A., F. Frolow, and N. Nelson. 2003. Crystal structure of plant photosystem I. *Nature* **426**(6967):630–635.

Blankenship, R. E. 2002. *Molecular Mechanisms of Photosynthesis.* Blackwell.

Deisenhofer, J., and J. R. Norris, eds. 1993. *The Photosynthetic Reaction Center,* vols. 1 and 2. Academic Press.

McDermott, G., et al. 1995. Crystal structure of an integral membrane light-harvesting complex from photosynthetic bacteria. *Nature* **364**:517.

Nelson, N., and C. F. Yocum. 2006. Structure and function of photosystems I and II. *Annu. Rev. Plant Biol.* **57**:521–565.

Prince, R. 1996. Photosynthesis: the Z-scheme revisited. *Trends Biochem. Sci.* **21**:121–122.

Wollman, F. A. 2001. State transitions reveal the dynamics and flexibility of the photosynthetic apparatus. *EMBO J.* **20**:3623–3630.

Molecular Analysis of Photosystems

Allen, J. F. 2002. Photosynthesis of ATP—electrons, proton pumps, rotors, and poise. *Cell* **110**:273–276.

Aro, E. M., I. Virgin, and B. Andersson. 1993. Photoinhibition of photosystem II: Inactivation, protein damage, and turnover. *Biochim. Biophys. Acta* **1143**:113–134.

Deisenhofer, J., and H. Michel. 1989. The photosynthetic reaction center from the purple bacterium *Rhodopseudomonas viridis. Science* **245**:1463–1473. (Nobel Prize Lecture.)

Deisenhofer, J., and H. Michel. 1991. Structures of bacterial photosynthetic reaction centers. *Ann. Rev. Cell Biol.* **7**:1–23.

Dekker, J. P., and E. J. Boekema. 2005. Supramolecular organization of thylakoid membrane proteins in green plants. *Biochim. Biophys. Acta.* **1706**(1–2):12–39.

Finazzi, G. 2005. The central role of the green alga *Chlamydomonas reinhardtii* in revealing the mechanism of state transitions *J. Exp. Bot.* **56**(411):383–388.

Haldrup, A., P. Jensen, C. Lunde, and H. Scheller. 2001. Balance of power: a view of the mechanism of photosynthetic state transitions. *Trends Plant Sci.* **6**:301–305.

Hankamer, B., J. Barber, and E. Boekema. 1997. Structure and membrane organization of photosystem II from green plants. *Ann. Rev. Plant Physiol. Plant Mol. Biol.* **48**:641–672.

Heathcote, P., P. Fyfe, and M. Jones. 2002. Reaction centres: the structure and evolution of biological solar power. *Trends Biochem. Sci.* **27**:79–87.

Horton, P., A. Ruban, and R. Walters. 1996. Regulation of light harvesting in green plants. *Ann. Rev. Plant Physiol. Plant Mol. Biol.* **47**:655–684.

Joliot, P., and A. Joliot. 2005. Quantification of cyclic and linear flows in plants. *Proc. Natl. Acad. Sci. USA* **102**(13):4913–4918.

Jordan, P., et al. 2001 Three-dimensional structure of cyanobacterial photosystem I at 2.5 Å resolution. *Nature* **411**:909–917.

Kühlbrandt, W. 2001. Chlorophylls galore. *Nature* **411**:896–898.

Martin, J. L., and M. H. Vos. 1992. Femtosecond biology. *Ann. Rev. Biophys. Biomol. Struc.* **21**:199–222.

Penner-Hahn, J. 1998. Structural characterization of the Mn site in the photosynthetic oxygen-evolving complex. *Struc. Bonding* **90**:1–36.

Tommos, C., and G. Babcock. 1998. Oxygen production in nature: a light-driven metalloradical enzyme process. *Acc. Chem. Res.* **31**:18–25.

CO_2 Metabolism During Photosynthesis

Buchanan, B. B. 1991. Regulation of CO_2 assimilation in oxygenic photosynthesis: the ferredoxin/thioredoxin system. Perspective on its discovery, present status, and future development. *Arch. Biochem. Biophys.* **288**:1–9.

Gutteridge, S., and J. Pierce. 2006. A unified theory for the basis of the limitations of the primary reaction of photosynthetic CO_2 fixation: was Dr. Pangloss right? *Proc. Natl. Acad. Sci. USA* **103**:7203–7204.

Portis, A. 1992. Regulation of ribulose 1,5-bisphosphate carboxylase/oxygenase activity. *Ann. Rev. Plant Physiol. Plant Mol. Biol.* **43**:415–437.

Rawsthorne, S. 1992. Towards an understanding of C_3-C_4 photosynthesis. *Essays Biochem.* **27**:135–146.

Rokka, A., I. Zhang, and E.-M. Aro. 2001. Rubisco activase: an enzyme with a temperature-dependent dual function? *Plant J.* **25**:463–472.

Sage, R., and J. Colemana. 2001. Effects of low atmospheric CO_2 on plants: more than a thing of the past. *Trends Plant Sci.* **6**:18–24.

Schneider, G., Y. Lindqvist, and C. I. Branden. 1992. Rubisco: structure and mechanism. *Ann. Rev. Biophys. Biomol. Struc.* **21**:119–153.

Tcherkez, G. G., G. D. Farquhar, and T. J. Andrews. 2006. Despite slow catalysis and confused substrate specificity, all ribulose bisphosphate carboxylases may be nearly perfectly optimized. *Proc. Natl. Acad. Sci. USA* **103**(19):7246–7251.

Wolosiuk, R. A., M. A. Ballicora, and K. Hagelin. 1993. The reductive pentose phosphate cycle for photosynthetic CO_2 assimilation: enzyme modulation. *FASEB J.* **7**:622–637.

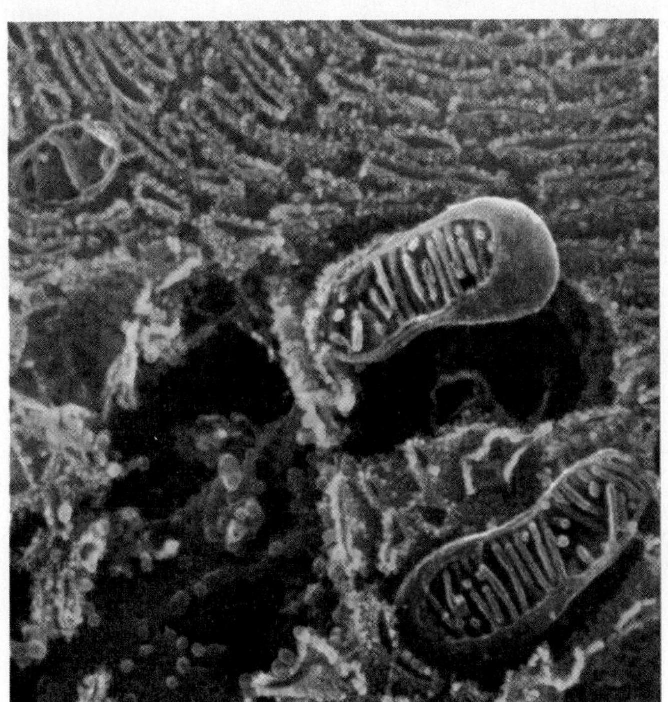

A scanning electron micrograph of a cell showing two of the intracellular compartments to which newly synthesized proteins are targeted. Membranes of the rough ER receive secretory proteins and membrane proteins destined for the cell surface. The cytosolic proteins involved in respiration are targeted to the various compartments of the mitochondria, including the matrix, inner membrane, and intermembrane space. [Professors Pietro M. Motta & Tomonori Naguro/Photo Researchers, Inc.]

CHAPTER

13

MOVING PROTEINS INTO MEMBRANES AND ORGANELLES

A typical mammalian cell contains up to 10,000 different kinds of proteins; a yeast cell, about 5000. The vast majority of these proteins are synthesized by cytosolic ribosomes, and many remain within the cytosol. However, as many as half of the different kinds of proteins produced in a typical cell are delivered to a particular organelle within the cell or to the cell surface. For example, many hormone receptor proteins and transporter proteins must be delivered to the plasma membrane, some water-soluble enzymes such as RNA and DNA polymerases must be targeted to the nucleus, and components of the extracellular matrix as well as digestive enzymes and polypeptide signaling molecules must be directed to the cell surface for secretion from the cell. These and all the other proteins produced by a cell must reach their correct locations for the cell to function properly.

The delivery of newly synthesized proteins to their proper cellular destinations, usually referred to as *protein targeting* or *protein sorting*, encompasses two very different kinds of processes. The first general process involves targeting of a protein to the membrane of an intracellular organelle and can occur either during translation or soon after synthesis of the protein is complete. For membrane proteins, targeting leads to insertion of the protein into the lipid bilayer of the membrane, whereas for water-soluble proteins, targeting leads to translocation of the entire protein across the membrane into the aqueous interior of the organelle. Proteins are sorted to the endoplasmic reticulum (ER), mitochondria, chloroplasts, peroxisomes, and nucleus by this general process (Figure 13-1).

A second general sorting process is known as the **secretory pathway.** This process begins in the ER; thus all proteins slated to enter the secretory pathway are initially targeted to the ER membrane. These proteins include not only soluble and membrane proteins that reside in the ER itself but also proteins that are secreted from the cell, resident proteins in the lumen of the Golgi complex and lysosomes, and integral proteins in the membranes of these organelles as

OUTLINE

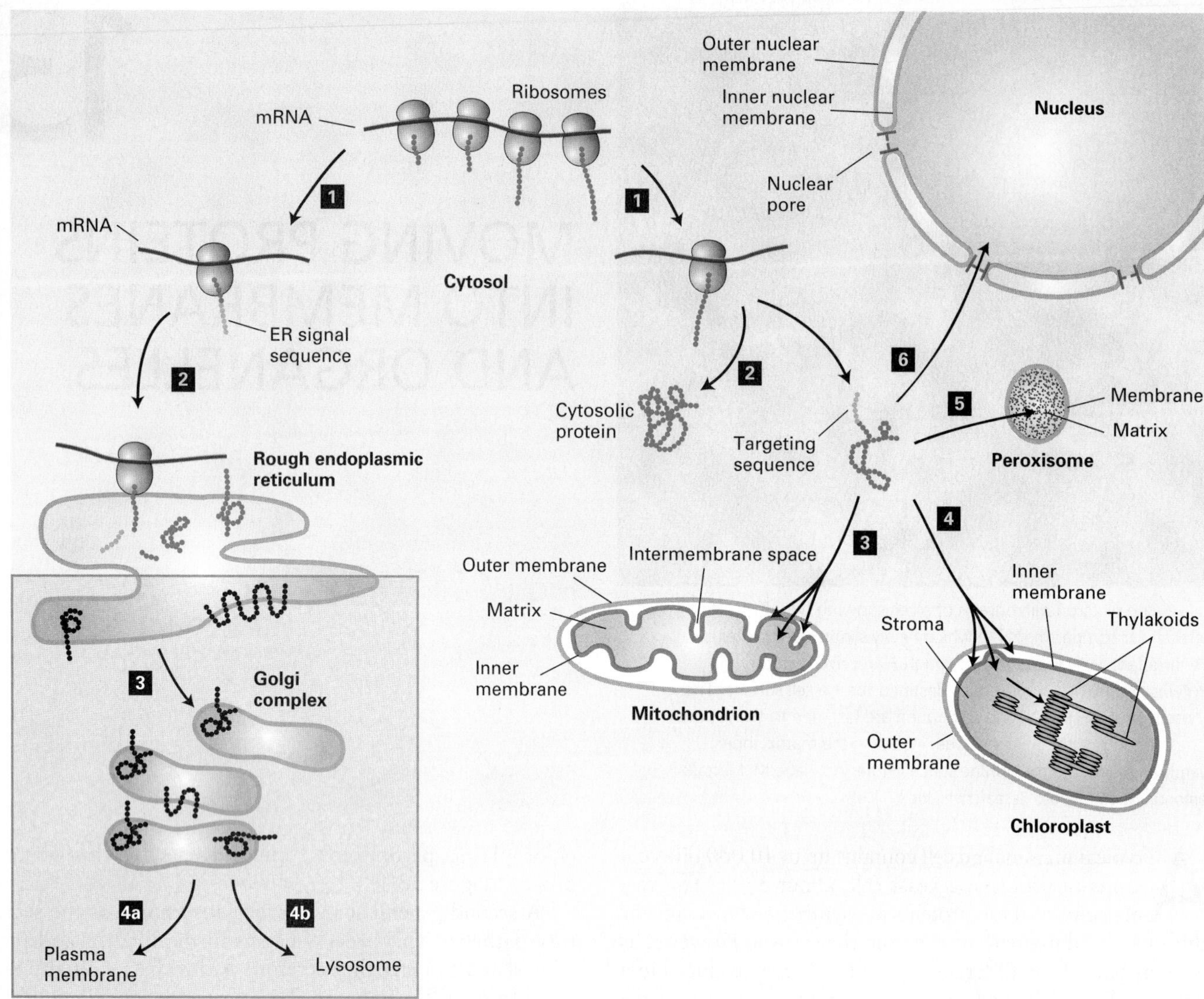

▲ **FIGURE 13-1 Overview of major protein-sorting pathways in eukaryotes.** All nuclear-encoded mRNAs are translated on cytosolic ribosomes. *Right (nonsecretory pathways):* Synthesis of proteins lacking an ER signal sequence is completed on free ribosomes (step **1**). Those proteins that contain no targeting sequence are released into the cytosol and remain there (step **2**). Proteins with an organelle-specific targeting sequence (pink) first are released into the cytosol (step **2**) but then are imported into mitochondria, chloroplasts, peroxisomes, or the nucleus (steps **3**–**6**). Mitochondrial and chloroplast proteins typically pass through the outer and inner membranes to enter the matrix or stromal space, respectively. Other proteins are sorted to other subcompartments of these organelles by additional sorting steps. Nuclear proteins enter and exit through visible pores in the nuclear envelope. *Left (secretory pathway):* Ribosomes synthesizing nascent proteins in the secretory pathway are directed to the rough endoplasmic reticulum (ER) by an ER signal sequence (pink; steps **1**, **2**). After translation is completed on the ER, these proteins can move via transport vesicles to the Golgi complex (step **3**). Further sorting delivers proteins either to the plasma membrane or to lysosomes (steps **4a**, **4b**). The processes underlying the secretory pathway (steps **3**, **4**, *shaded box*) are discussed in Chapter 14.

well as in the plasma membrane. Targeting to the ER generally involves *nascent* proteins still in the process of being synthesized. Once translocated across the ER membrane, proteins are assembled into their native conformation by protein-folding catalysts present in the lumen of the ER. This process is monitored carefully, and only after their folding and assembly is complete are proteins permitted to be transported out of the ER to other organelles. Proteins are also modified in various ways after translocation into the ER. These modifications can include addition of carbohydrate groups, stabilization of protein structure through disulfide bond formation, and specific proteolytic cleavages. Proteins whose final destination is the Golgi, lysosome, or cell surface are transported along the secretory pathway by the action of small vesicles that bud from the membrane of one organelle and then fuse with the membrane of another (see Figure 13-1, *shaded box*). We discuss vesicle-based protein sorting in the next chapter because mechanistically it differs significantly from protein targeting to the membranes of intracellular organelles.

In this chapter, we examine how proteins are targeted to the membranes of intracellular organelles and subsequently inserted into the organelle membrane or moved into its interior. Two features of this protein-sorting process initially were quite baffling: how a given protein could be directed to only one specific membrane and how relatively large protein molecules could be translocated across a membrane without disrupting the bilayer as a barrier to ions and small molecules. Using a combination of biochemical purification methods and genetic screens for identifying mutants unable to execute particular translocation steps, cell biologists have identified many of the cellular components required for translocation across each of the different intracellular membranes. In addition, many of the major translocation processes in the cell have been reconstituted using the purified protein components incorporated into artificial lipid bilayers. Such in vitro systems can be freely manipulated experimentally.

These studies have shown that despite some variations, the same basic mechanisms govern protein sorting to all the various intracellular organelles. We now know, for instance, that the information to target a protein to a particular organelle destination is encoded within the amino acid sequence of the protein itself, usually within sequences of 20–50 amino acids, known generically as **signal sequences** (see Figure 13-1); these are also called *uptake-targeting sequences* or *signal peptides*. Each organelle carries a set of receptor proteins that bind only to specific kinds of signal sequences, thus ensuring that the information encoded in a signal sequence governs the specificity of targeting. Once a protein containing a signal sequence has interacted with the corresponding receptor, the protein chain is transferred to some kind of *translocation channel* that allows the protein to pass through the membrane bilayer. The unidirectional transfer of a protein into an organelle, without sliding back out into the cytoplasm, is usually achieved by coupling translocation to an energetically favorable process such as hydrolysis of ATP. Some proteins are subsequently sorted further to reach a subcompartment within the target organelle; such sorting depends on yet other signal sequences and other receptor proteins. Finally, signal sequences often are removed from the mature protein by specific proteases once translocation across the membrane is completed.

For each of the protein-targeting events discussed in this chapter, we will seek to answer four fundamental questions:

1. What is the nature of the *signal sequence*, and what distinguishes it from other types of signal sequences?

2. What is the *receptor* for the signal sequence?

3. What is the structure of the *translocation channel* that allows transfer of proteins across the membrane bilayer? In particular, is the channel so narrow that proteins can pass through only in an unfolded state, or will it accommodate folded protein domains?

4. What is the source of *energy* that drives unidirectional transfer across the membrane?

In the first part of the chapter, we cover targeting of proteins to the ER, including the post-translational modifications that occur to proteins as they enter the secretory pathway. We then describe targeting of proteins to mitochondria, chloroplasts, and peroxisomes. Finally, we cover the transport of proteins into and out of the nucleus through nuclear pores.

13.1 Translocation of Secretory Proteins Across the ER Membrane

All eukaryotic cells use essentially the same secretory pathway for synthesizing and sorting secreted proteins and proteins that reside in the aqueous luminal space or the membrane of the ER, Golgi, and lysosomes (Figure 13-1, *left*). For simplicity, we refer to these proteins collectively as *secretory proteins*. The basic mechanism of the secretory pathway involves three steps: (1) protein synthesis and translocation across the ER membrane, (2) protein folding and modification inside the ER lumen, and (3) protein transport to the Golgi, lysosomes, or cell surface through budding and fusing of vesicles. In the first two sections of this chapter we discuss how secretory proteins are targeted to the ER and how some proteins can be inserted into the ER membrane, because these processes are mechanistically similar to targeting of proteins to the other organelles discussed in this chapter. We also consider here, in Section 13.3, how proteins fold and are modified as they enter the ER. The process of vesicle transport is addressed in the following chapter.

Although all cells secrete a variety of proteins (e.g., extracellular matrix proteins), certain types of cells are specialized for secretion of large amounts of specific proteins. Pancreatic acinar cells, for instance, synthesize large quantities of several digestive enzymes that are secreted into ductules that lead to the intestine. Because such secretory cells contain the organelles of the secretory pathway (e.g., ER and Golgi) in great abundance, they have been widely used in studying this pathway.

The sequence of events that occur immediately after the synthesis of a secretory protein were first elucidated by pulse-labeling experiments with pancreatic acinar cells. In such cells, radioactively labeled amino acids are incorporated into secretory proteins as they are synthesized on ribosomes that are bound to the surface of the ER. The portion of the ER that receives proteins entering the secretory pathway is known as the *rough ER* because it is so densely studded with ribosomes that its surface appears morphologically distinct from other ER membrane (Figure 13-2). When cells are homogenized, the rough ER breaks up into small closed vesicles, termed *rough microsomes*, with the same orientation

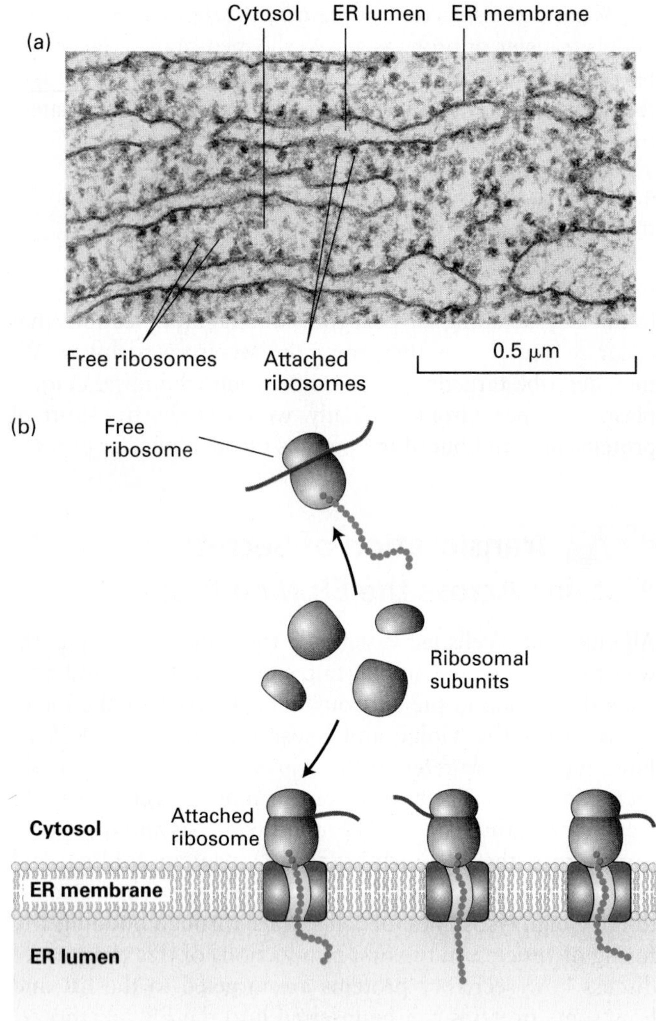

▲ **FIGURE 13-2 Structure of the rough ER.** (a) Electron micrograph of ribosomes attached to the rough ER in a pancreatic acinar cell. Most of the proteins synthesized by this type of cell are to be secreted and are formed on membrane-attached ribosomes. A few membrane-unattached (free) ribosomes are evident; presumably, these are synthesizing cytosolic or other nonsecretory proteins. (b) Schematic representation of protein synthesis on the ER. Note that membrane-bound and free cytosolic ribosomes are identical. Membrane-bound ribosomes get recruited to the endoplasmic reticulum during protein synthesis of a polypeptide containing an ER signal sequence. [Part (a) Courtesy of G. Palade.]

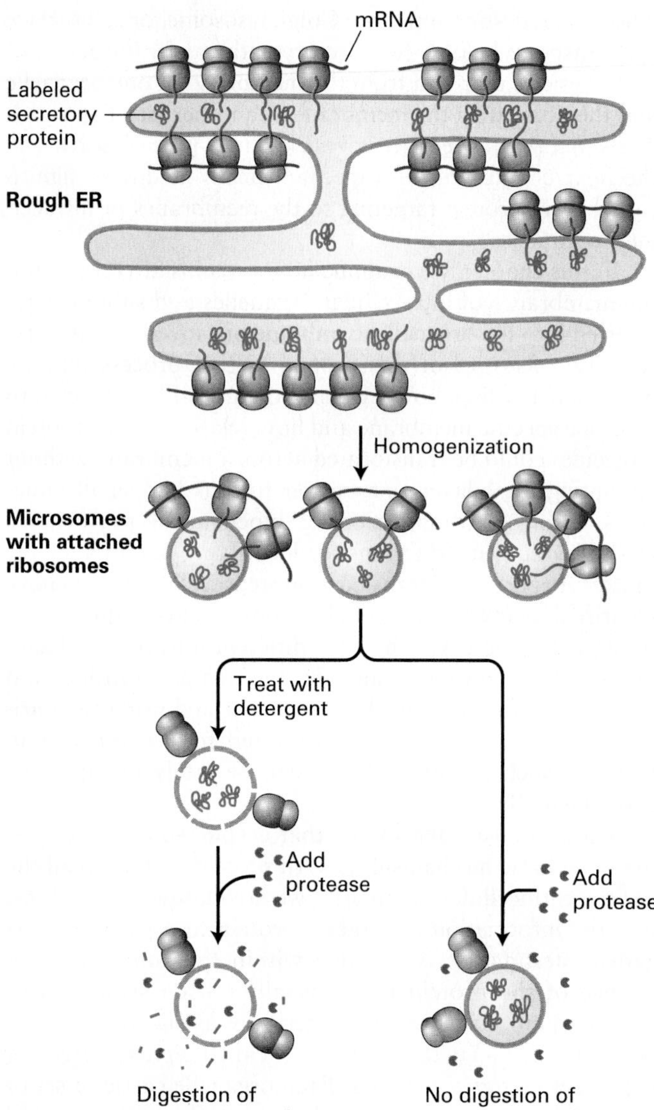

► **EXPERIMENTAL FIGURE 13-3 Secretory proteins enter the ER.** Labeling experiments demonstrate that secretory proteins are localized to the ER lumen shortly after synthesis. Cells are incubated for a brief time with radiolabeled amino acids so that only newly synthesized proteins become labeled. The cells then are homogenized, fracturing the plasma membrane and shearing the rough ER into small vesicles called *microsomes.* Because they have bound ribosomes, microsomes have a much greater buoyant density than other membranous organelles and can be separated from them by a combination of differential and sucrose density-gradient centrifugation (Chapter 9). The purified microsomes are treated with a protease in the presence or absence of a detergent. The labeled secretory proteins associated with the microsomes are digested by the protease only if the permeability barrier of the microsomal membrane is first destroyed by treatment with detergent. This finding indicates that the newly made proteins are inside the microsomes, equivalent to the lumen of the rough ER.

as that found in the intact cell (ribosomes on the outside). The experiments depicted in Figure 13-3, in which microsomes isolated from pulse-labeled cells are treated with a protease, demonstrate that although secretory proteins are synthesized on ribosomes bound to the cytosolic face of the ER membrane, the polypeptides produced by these ribosomes end up within the lumen of ER vesicles. Experiments such as this raised the question of how polypeptides are recognized as secretory proteins shortly after their synthesis begins and how the N-terminus of a nascent secretory protein is threaded across the ER membrane.

A Hydrophobic N-Terminal Signal Sequence Targets Nascent Secretory Proteins to the ER

After synthesis of a secretory protein begins on free ribosomes in the cytosol, a 16- to 30-residue ER signal sequence in the nascent protein directs the ribosome to the ER

▶ **EXPERIMENTAL FIGURE 13-4 Translocation and translation occur simultaneously.** Cell-free experiments demonstrate that translocation of secretory proteins into microsomes is coupled to translation. Treatment of microsomes with EDTA, which chelates Mg^{+2} ions, strips them of associated ribosomes, allowing isolation of ribosome-free microsomes, which are equivalent to ER membranes (see Figure 13-3). Synthesis is carried out in a cell-free system containing functional ribosomes, tRNAs, ATP, GTP, and cytosolic enzymes to which mRNA encoding a secretory protein is added. The secretory protein is synthesized in the absence of microsomes (a) but is translocated across the vesicle membrane and loses its signal sequence only if microsomes are present during protein synthesis (b).

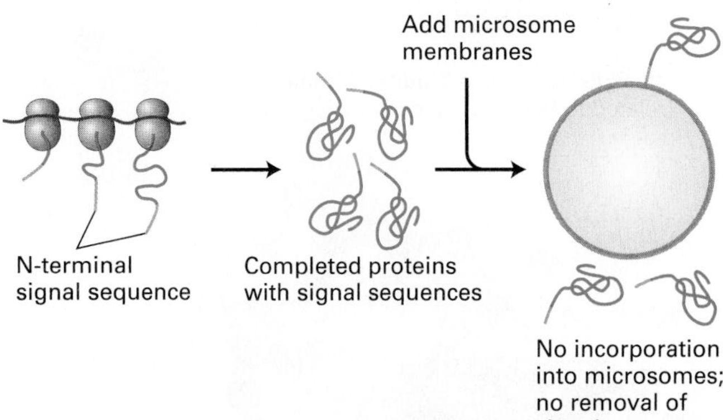

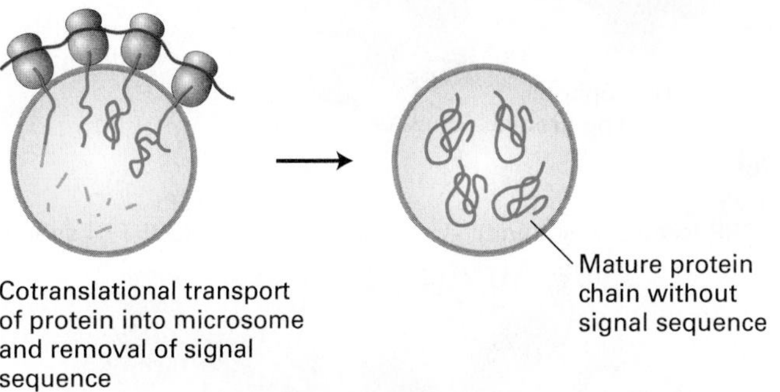

membrane and initiates translocation of the growing polypeptide across the ER membrane (see Figure 13-1, *left*). An ER signal sequence typically is located at the N-terminus of the protein, the first part of the protein to be synthesized. The signal sequences of different secretory proteins all contain one or more positively charged amino acids adjacent to a continuous stretch of 6–12 hydrophobic residues (the core), but otherwise they have little in common. For most secretory proteins, the signal sequence is cleaved from the protein while it is still elongating on the ribosome; thus signal sequences are usually not present in the "mature" proteins found in cells.

The hydrophobic core of ER signal sequences is essential for their function. For instance, the specific deletion of several of the hydrophobic amino acids from a signal sequence or the introduction of charged amino acids into the hydrophobic core by mutation can abolish the ability of the N-terminus of a protein to function as a signal sequence. As a consequence, the modified protein remains in the cytosol, unable to cross the ER membrane into the lumen. Using recombinant DNA techniques, researchers have produced cytosolic proteins with added N-terminal amino acid sequences. Provided the added sequence is sufficiently long and hydrophobic, such a modified cytosolic protein is translocated to the ER lumen. Thus the hydrophobic residues in the core of ER signal sequences form a binding site that is critical for the interaction of signal sequences with the machinery responsible for targeting the protein to the ER membrane.

Biochemical studies utilizing a cell-free protein-synthesizing system, mRNA encoding a secretory protein, and microsomes stripped of their own bound ribosomes have clarified the function and fate of ER signal sequences. Initial experiments with this system demonstrated that a typical secretory protein is incorporated into microsomes and has its signal sequence removed only if the microsomes are present during protein synthesis. If microsomes are added to the system after protein synthesis is completed, no protein transport into the microsomes occurs (Figure 13-4). Subsequent experiments were designed to determine the precise stage of protein synthesis at which microsomes must be present in order for translocation to occur. In these experiments, microsomes were added to the reaction mixtures at different times after protein synthesis had begun. These experiments showed that microsomes must be added before the first 70 or so amino acids are linked together in order for the completed secretory protein to be localized in the microsomal lumen. At this point, the first 40 amino acids or so protrude from the ribosome, including the signal sequence that later will be cleaved off, and the next 30 or so amino acids are still buried within a channel in the ribosome (see Figure 4-26). Thus the transport of most secretory proteins into the ER lumen begins while the incompletely synthesized (nascent) protein is still bound to the ribosome, a process referred to as **cotranslational translocation.**

Cotranslational Translocation Is Initiated by Two GTP-Hydrolyzing Proteins

Since secretory proteins are synthesized in association with the ER membrane but not with any other cellular membrane, a signal-sequence recognition mechanism must target them there. The two key components in this targeting are the **signal-recognition particle (SRP)** and its receptor, located in the ER membrane. The SRP is a cytosolic ribonucleoprotein particle that transiently binds to both the ER signal sequence in a nascent protein as well as the large ribosomal subunit, forming a large complex; SRP then targets the nascent protein-ribosome complex to the ER membrane by binding to the SRP receptor on the membrane.

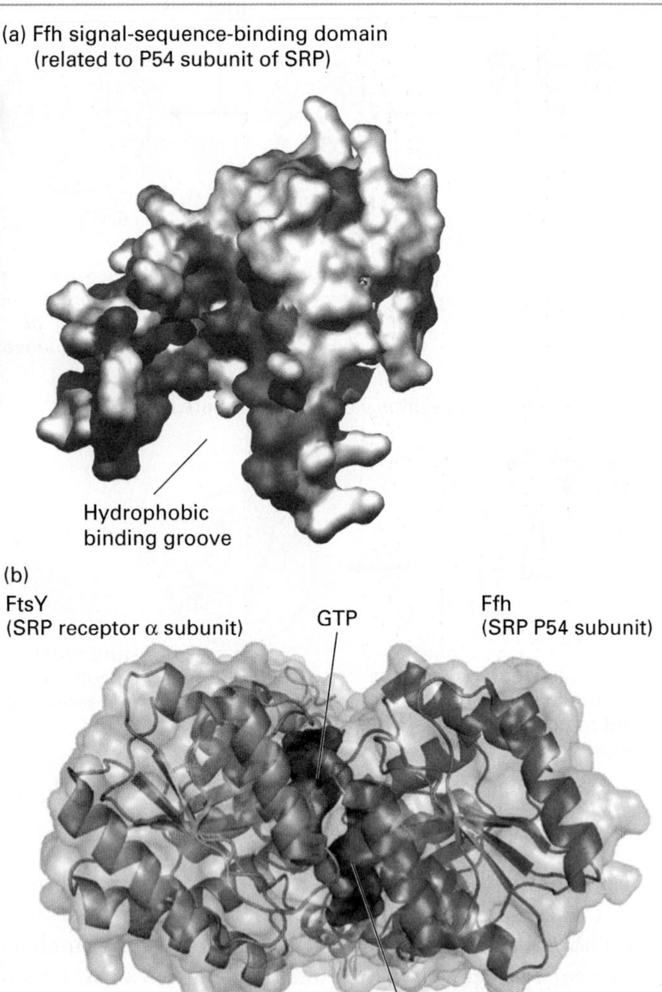

◄ **FIGURE 13-5 Structure of the signal-recognition particle (SRP).** (a) The bacterial Ffh protein is homologous to the portion of P54 that binds ER signal sequences. This surface model shows the binding domain in Ffh, which contains a large cleft lined with hydrophobic amino acids (purple) whose side chains interact with signal sequences. (b) The structure of GTP bound to FtsY (the bacterial homolog of the α subunit of SRP receptor) and Ffh proteins illustrates how the interaction between these proteins is controlled by GTP binding and hydrolysis. Ffh and FtsY each can bind to one molecule of GTP, and when Ffh and FtsY bind to each other, the two bound molecules of GTP fit in the interface between the protein subunits and stabilize the dimer. Assembly of the semi-symmetrical dimer allows formation of two active sites for the hydrolysis of both bound GTP molecules. Hydrolysis to GDP destabilizes the interface, causing disassembly of the dimer. [Part (a) adapted from R. J. Keenan et al., 1998, *Cell* **94:**181. Part (b) adapted from P. J. Focia et al., 2004, *Science* **303:**373.]

The SRP is made up of six discrete polypeptides bound to a 300-nucleotide RNA, which acts as a scaffold for the hexamer. One of the SRP proteins (P54) can be chemically cross-linked to ER signal sequences, evidence that this particular protein is the subunit that binds to the signal sequence in a nascent secretory protein. A region of P54 containing many amino acid residues with hydrophobic side chains is homologous to a bacterial protein known as Ffh, which performs an analogous function to P54 in the translocation of proteins across the cytoplasmic membrane of bacterial cells. The structure of Ffh contains a cleft whose inner surface is lined by hydrophobic side chains (Figure 13-5a). The hydrophobic region of P54 is thought to contain an analogous cleft that interacts with the hydrophobic N-terminal signals of nascent secretory proteins and selectively targets them to the ER membrane. Other polypeptides in the SRP interact with the ribosome or are required for protein translocation into the ER lumen.

The SRP brings the nascent chain-ribosome complex to the ER membrane by docking with the SRP receptor, an integral protein of the ER membrane made up of two subunits: an α subunit and a smaller β subunit. Interaction of the SRP/nascent chain/ribosome complex with the SRP receptor is strengthened when both the P54 subunit of SRP and the α subunit of the SRP receptor are bound to GTP. The structure of analogs of the P54 subunit of SRP (Ffh) and the SRP receptor α subunit (FtsY), from the archaebacteria *Thermus aquaticus*, provides insight into how a cycle of GTP binding and hydrolysis can drive the binding and dissociation of these proteins. Figure 13-5b shows that the Ffh and FtsY each bound to a single molecule of GTP come together to form a pseudo-symmetrical heterodimer. Neither subunit alone contains a complete active site for the hydrolysis of GTP, but when the two proteins come together, they form two complete active sites that are capable of hydrolyzing both bound GTP molecules.

Figure 13-6 summarizes our current understanding of secretory protein synthesis and the role of the SRP and its receptor in this process. Hydrolysis of the bound GTP accompanies disassembly of the SRP and SRP receptor and, in a manner that is not understood, initiates transfer of the nascent chain and ribosome to a site on the ER membrane, where translocation can take place. After dissociating from each other, SRP and its receptor each release their bound GDP, SRP recycles back to the cytosol, and both are ready to initiate another round of interaction between ribosomes synthesizing nascent secretory proteins and the ER membrane.

In the cell-free translation system described previously, the presence of SRP slows elongation of a secretory protein when microsomes are absent, thereby inhibiting synthesis of the complete protein (see Figure 13-4). This finding suggests that interaction of the SRP with both the nascent chain of a secretory protein and with the free ribosome prevents the nascent chain from becoming too long for translocation into the ER. Only after the SRP/nascent chain/ribosome complex has bound to the SRP receptor in the ER membrane does

Focus Animation: Synthesis of Secreted and Membrane-Bound Proteins

MEDIA CONNECTIONS

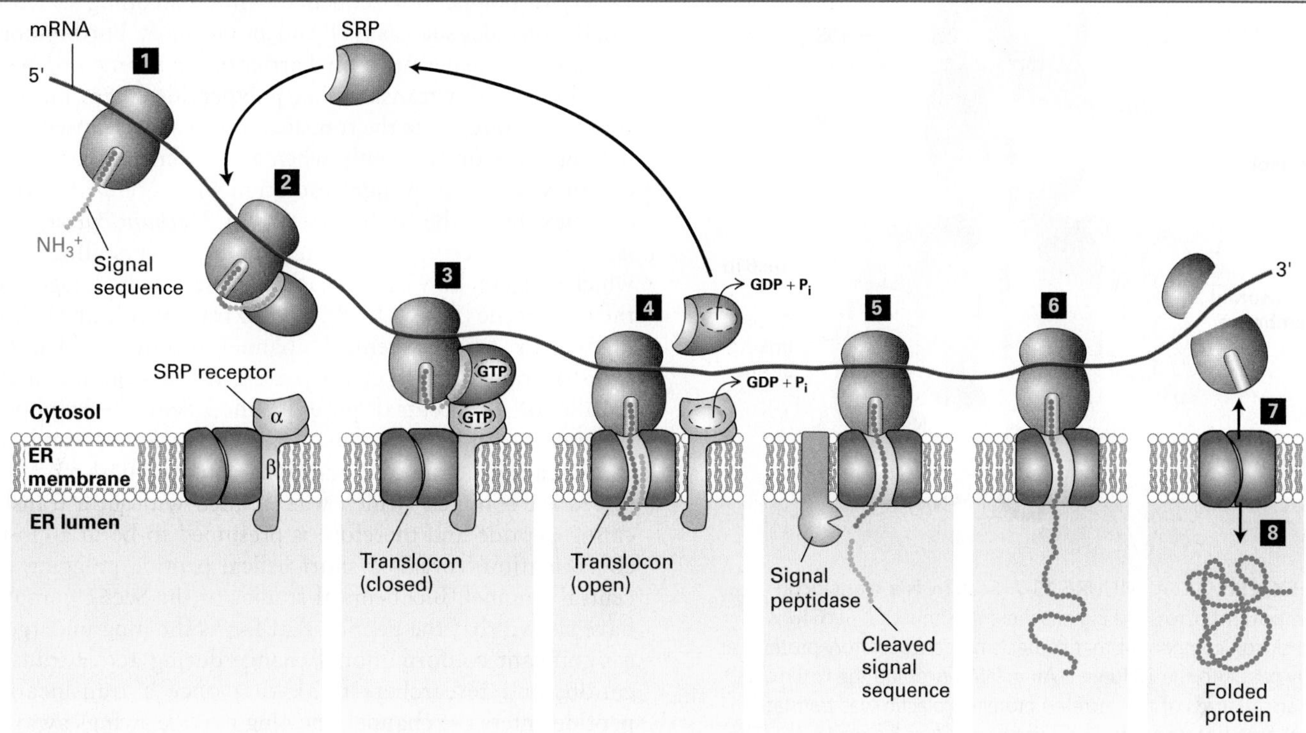

▲ **FIGURE 13-6 Cotranslational translocation.** Steps **1**, **2**: Once the ER signal sequence emerges from the ribosome, it is bound by a signal-recognition particle (SRP). Step **3**: The SRP delivers the ribosome/nascent polypeptide complex to the SRP receptor in the ER membrane. This interaction is strengthened by binding of GTP to both the SRP and its receptor. Step **4**: Transfer of the ribosome/nascent polypeptide to the translocon leads to opening of this translocation channel and insertion of the signal sequence and adjacent segment of the growing polypeptide into the central pore. Both the SRP and SRP receptor, once dissociated from the translocon, hydrolyze their bound GTP and then are ready to initiate the insertion of another polypeptide chain. Step **5**: As the polypeptide chain elongates, it passes through the translocon channel into the ER lumen, where the signal sequence is cleaved by signal peptidase and is rapidly degraded. Step **6**: The peptide chain continues to elongate as the mRNA is translated toward the 3′ end. Because the ribosome is attached to the translocon, the growing chain is extruded through the translocon into the ER lumen. Steps **7**, **8**: Once translation is complete, the ribosome is released, the remainder of the protein is drawn into the ER lumen, the translocon closes, and the protein assumes its native folded conformation.

SRP release the nascent chain, allowing elongation to continue at the normal rate. Thus the SRP and SRP receptor not only help mediate interaction of a nascent secretory protein with the ER membrane but also act together to permit elongation and synthesis of complete proteins only when ER membranes are present.

Passage of Growing Polypeptides Through the Translocon Is Driven by Energy Released During Translation

Once the SRP and its receptor have targeted a ribosome synthesizing a secretory protein to the ER membrane, the ribosome and nascent chain are rapidly transferred to the **translocon,** a protein-lined channel within the membrane. As translation continues, the elongating chain passes directly from the large ribosomal subunit into the central pore of the translocon. The 60S ribosomal subunit is aligned with the pore of the translocon in such a way that the growing chain is never exposed to the cytoplasm and is prevented from folding until it reaches the ER lumen (Figure 13-6).

The translocon was first identified by mutations in the yeast gene encoding Sec61α, which caused a block in the translocation of secretory proteins into the lumen of the ER. Subsequently, three proteins called the *Sec61 complex* were found to form the mammalian translocon: Sec61α, an integral membrane protein with 10 membrane-spanning α helices, and two smaller proteins, termed Sec61β and Sec61γ. Chemical cross-linking experiments demonstrated that the translocating polypeptide chain comes into contact with the Sec61α protein in both yeast and

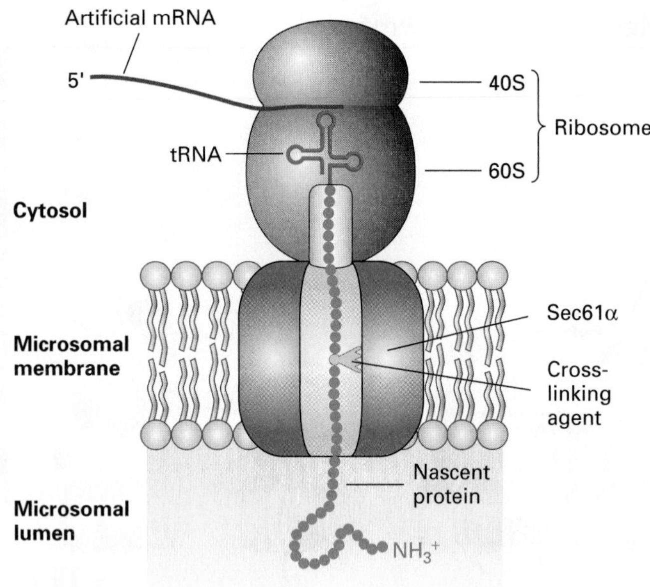

▲ **EXPERIMENTAL FIGURE 13-7 Sec61α is a translocon component.** Cross-linking experiments show that Sec61α is a translocon component that contacts nascent secretory proteins as they pass into the ER lumen. An mRNA encoding the N-terminal 70 amino acids of the secreted protein prolactin was translated in a cell-free system containing microsomes (see Figure 13-4b). The mRNA lacked a chain-termination codon and contained one lysine codon, near the middle of the sequence. The reactions contained a chemically modified lysyl-tRNA in which a light-activated cross-linking reagent was attached to the lysine side chain. Although the entire mRNA was translated, the completed polypeptide could not be released from the ribosome without a chain-termination codon and thus became "stuck" crossing the ER membrane. The reaction mixtures then were exposed to an intense light, causing the nascent chain to become covalently bound to whatever proteins were near it in the translocon. When the experiment was performed using microsomes from mammalian cells, the nascent chain became covalently linked to Sec61α. Different versions of the prolactin mRNA were created so that the modified lysine residue would be placed at different distances from the ribosome; cross-linking to Sec61α was observed only when the modified lysine was positioned within the translocation channel. [Adapted from T. A. Rapoport, 1992, *Science* **258:**931, and D. Görlich and T. A. Rapoport, 1993, *Cell* **75:**615.]

mammalian cells, confirming its identity as a translocon component (Figure 13-7).

When microsomes in the cell-free translocation system were replaced with reconstituted phospholipid vesicles containing only the SRP receptor and Sec61 complex, nascent secretory protein was translocated from its SRP/ribosome complex into the vesicles. This finding indicates that the SRP receptor and the Sec61 complex are the only ER-membrane proteins absolutely required for translocation. Because neither of these can hydrolyze ATP or otherwise provide energy to drive the translocation, the energy derived from chain elongation at the ribosome appears to be sufficient to push the polypeptide chain across the membrane in one direction.

The translocon must be able to allow passage of a wide variety of polypeptide sequences while remaining sealed to small molecules such as ATP and amino acids. Furthermore, to maintain the permeability barrier of the ER membrane in the absence of a translocating polypeptide, there must be some way to regulate the translocon so that it is closed in its default state, opening only when a ribosome-nascent chain complex is bound. A high-resolution structure of the Sec61 complex from the archaebacterium *Methanococcus jannaschii* was recently determined by x-ray crystallography, which suggests how the translocon preserves the integrity of the membrane (Figure 13-8). The 10 transmembrane helices of Sec61α form a central channel through which the translocating peptide chain passes. A constriction in the middle of the central pore is lined with hydrophobic isoleucine residues that may form a gasket around the translocating peptide. In addition, the structural model of the Sec61 complex (which was isolated without a translocating peptide and therefore is presumed to be in a closed conformation) reveals a short helical peptide plugging the central channel. Biochemical studies of the Sec61 complex have shown that the peptide that forms the plug undergoes a significant conformational change during active translocation, and researchers think that once a translocating peptide enters the channel, the plug peptide swings away to allow translocation to proceed.

As the growing polypeptide chain enters the lumen of the ER, the signal sequence is cleaved by *signal peptidase,* which is a transmembrane ER protein associated with the translocon (see Figure 13-6). Signal peptidase recognizes a sequence on the C-terminal side of the hydrophobic core of the signal peptide and cleaves the chain specifically at this sequence once it has emerged into the luminal space of the ER. After the signal sequence has been cleaved, the growing polypeptide moves through the translocon into the ER lumen. The translocon remains open until translation is completed and the entire polypeptide chain has moved into the ER lumen.

Electron microscopy of the Sec61 complex isolated from the ER of eukaryotic cells reveals that three or four copies of Sec61α coassemble in the plane of the membrane. The functional significance of this association between translocon channels is not understood at this time, but the oligomerization of translocon channels may facilitate the association between the translocon, signal peptidase, and other luminal protein complexes that participate in the translocation process.

ATP Hydrolysis Powers Post-translational Translocation of Some Secretory Proteins in Yeast

In most eukaryotes, secretory proteins enter the ER by cotranslational translocation. In yeast, however, some secretory proteins enter the ER lumen after translation has been completed. In such *post-translational translocation,* the translocating protein passes through the same Sec61 translocon that is used in cotranslational translocation. However, the SRP

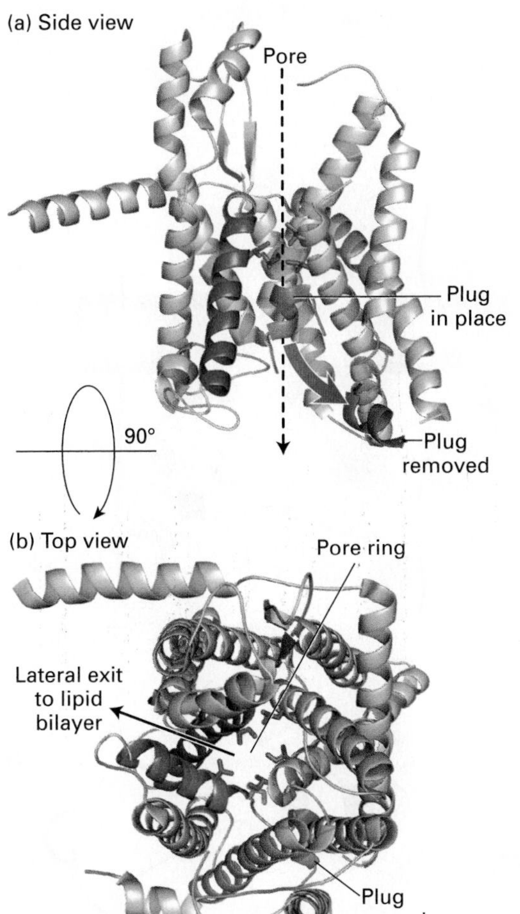

▲ **EXPERIMENTAL FIGURE 13-8 Structure of a bacterial Sec61 complex.** The structure of the detergent-solubilized Sec61 complex from the archaebacterium *M. jannaschii* (also known as the SecY complex) was determined by x-ray crystallography. (a) A side view shows the hourglass-shaped channel through the center of the pore. A ring of isoleucine residues at the constricted waist of the pore may form a gasket that keeps the channel sealed to small molecules even as a translocating polypeptide passes through the channel. When no translocating peptide is present, the channel is closed by a short helical plug. This plug is thought to move out of the channel during translocation. In this view the front half of protein has been removed to better show the pore. (b) A view looking through the center of the channel shows a region (on the left side) where helices may separate allowing lateral passage of a hydrophobic transmembrane domain into the lipid bilayer. [Adapted from A. R. Osborne et al., 2005, *Ann. Rev. Cell Dev. Biology* **21:**529.]

and SRP receptor are not involved in post-translational translocation, and in such cases a direct interaction between the translocon and the signal sequence of the completed protein appears to be sufficient for targeting to the ER membrane. In addition, the driving force for unidirectional translocation across the ER membrane is provided by an additional protein complex known as the *Sec63 complex* and a member of the Hsc70 family of **molecular chaperones** known as *BiP*. The tetrameric Sec63 complex is embedded in the ER membrane in the vicinity of the translocon, whereas BiP is within the ER lumen. Like other members of the Hsc70 family, BiP has a peptide-binding domain and an ATPase domain. These chaperones bind and stabilize unfolded or partially folded proteins (see Figure 3-16).

The current model for post-translational translocation of a protein into the ER is outlined in Figure 13-9. Once the N-terminal segment of the protein enters the ER lumen, signal peptidase cleaves the signal sequence just as in cotranslational translocation (step **1**). Interaction of BiP·ATP with the luminal portion of the Sec63 complex causes hydrolysis of the bound ATP, producing a conformational change in BiP that promotes its binding to an exposed polypeptide chain (step **2**). Since the Sec63 complex is located near the translocon, BiP is thus activated at sites where nascent polypeptides can enter the ER. Certain experiments suggest that in the absence of binding to BiP, an unfolded polypeptide slides back and forth within the translocon channel. Such random sliding motions rarely result in the entire polypeptide's crossing the ER membrane. Binding of a molecule of BiP·ADP to the luminal portion of the polypeptide prevents backsliding of the polypeptide out of the ER. As further inward random sliding exposes more of the polypeptide on the luminal side of the ER membrane, successive binding of BiP·ADP molecules to the polypeptide chain acts as a ratchet, ultimately drawing the entire polypeptide into the ER within a few seconds (steps **3** and **4**). On a slower time scale, the BiP molecules spontaneously exchange their bound ADP for ATP, leading to release of the polypeptide, which can then fold into its native conformation (steps **5** and **6**). The recycled BiP·ATP then is ready for another interaction with Sec63. BiP and the Sec63 complex are also required for cotranslational translocation. The details of their role in this process are not well understood, but they are thought to act at an early stage of the process such as threading the signal peptide into the pore of the translocon.

The overall reaction carried out by BiP is an important example of how the chemical energy released by the hydrolysis of ATP can power the mechanical movement of a protein across a membrane. Bacterial cells also use an ATP-driven process for translocating completed proteins across the plasma membrane. In bacteria the driving force for translocation comes from a cytosolic ATPase known as the SecA protein. SecA binds to the cytoplasmic side of the translocon and hydrolyzes cytosolic ATP. By a mechanism that is not well understood, the SecA protein pushes segments of the polypeptide through the membrane in a mechanical cycle coupled to the hydrolysis of ATP.

KEY CONCEPTS OF SECTION 13.1

Translocation of Secretory Proteins Across the ER Membrane

■ Synthesis of secreted proteins, integral plasma-membrane proteins, and proteins destined for the ER, Golgi complex, or lysosome begins on cytosolic ribosomes, which become attached to the membrane of the ER, forming the rough ER (see Figure 13-1, *left*).

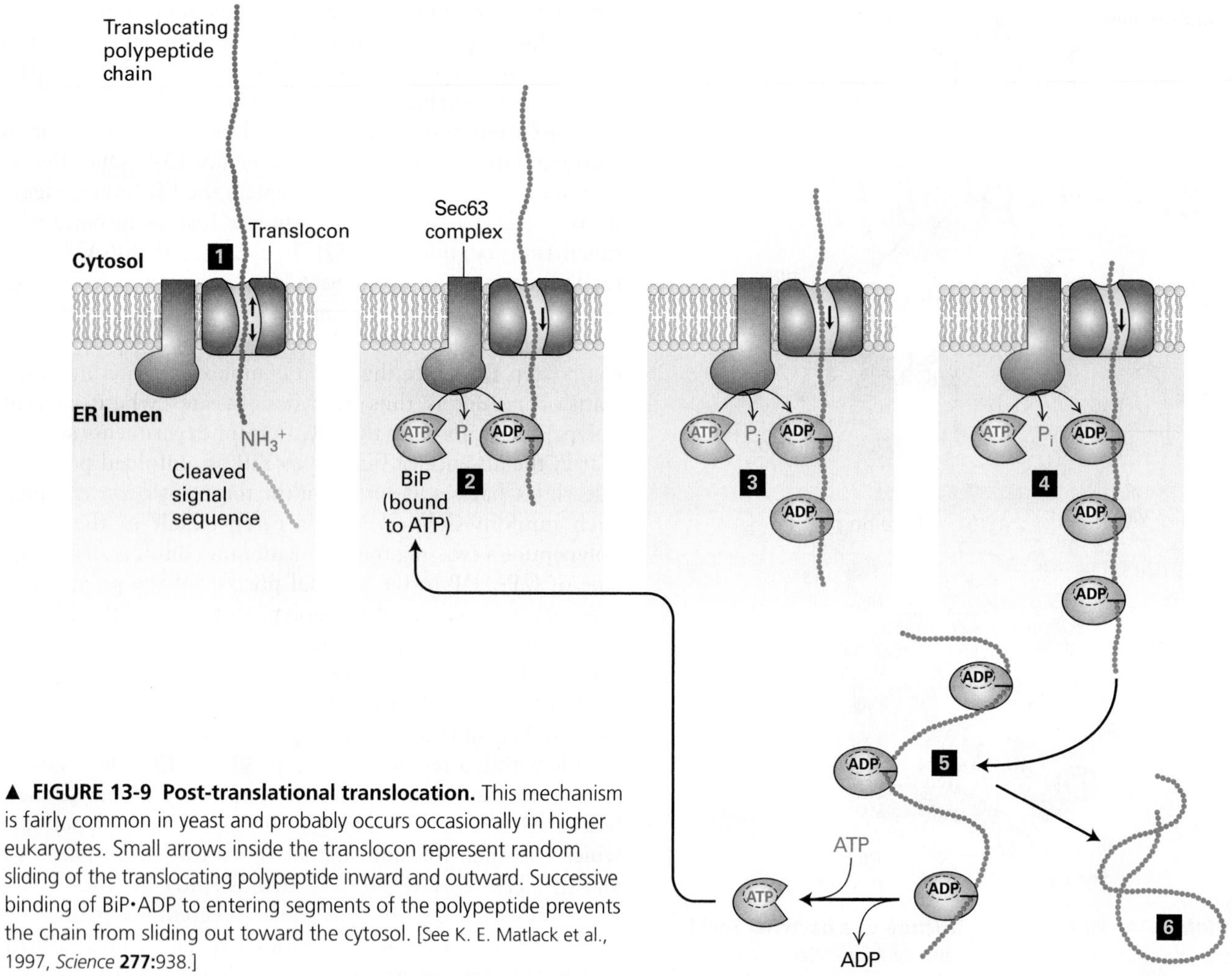

▲ **FIGURE 13-9 Post-translational translocation.** This mechanism is fairly common in yeast and probably occurs occasionally in higher eukaryotes. Small arrows inside the translocon represent random sliding of the translocating polypeptide inward and outward. Successive binding of BiP·ADP to entering segments of the polypeptide prevents the chain from sliding out toward the cytosol. [See K. E. Matlack et al., 1997, *Science* **277**:938.]

■ The ER signal sequence on a nascent secretory protein consists of a segment of hydrophobic amino acids, generally located at the N-terminus.

■ In cotranslational translocation, the signal-recognition particle (SRP) first recognizes and binds the ER signal sequence on a nascent secretory protein and in turn is bound by an SRP receptor on the ER membrane, thereby targeting the ribosome/nascent chain complex to the ER.

■ The SRP and SRP receptor then mediate insertion of the nascent secretory protein into the translocon (Sec61 complex). Hydrolysis of two molecules of GTP by the SRP and its receptor drive this docking process and cause the dissociation of SRP (see Figures 13-5 and 13-6). As the ribosome attached to the translocon continues translation, the unfolded protein chain is extruded into the ER lumen. No additional energy is required for translocation.

■ The translocon contains a central channel lined with hydrophobic residues that allows transit of an unfolded protein chain while remaining sealed to ions and small hydrophilic molecules. In addition, the channel is gated so that it only is open when a polypeptide is being translocated.

■ In post-translational translocation, a completed secretory protein is targeted to the ER membrane by interaction of the signal sequence with the translocon. The polypeptide chain is then pulled into the ER by a ratcheting mechanism that requires ATP hydrolysis by the chaperone BiP, which stabilizes the entering polypeptide (see Figure 13-9). In bacteria, the driving force for post-translational translocation comes from SecA, a cytosolic ATPase that pushes polypeptides through the translocon channel.

■ In both cotranslational and post-translational translocation, a signal peptidase in the ER membrane cleaves the ER signal sequence from a secretory protein soon after the N-terminus enters the lumen.

13.2 Insertion of Proteins into the ER Membrane

In previous chapters we have encountered many of the vast array of integral (transmembrane) proteins that are present throughout the cell. Each such protein has a unique orientation with respect to the membrane's phospholipid bilayer.

Integral membrane proteins located in the ER, Golgi, and lysosomes and also proteins in the plasma membrane, which are all synthesized on the rough ER, remain embedded in the membrane in their unique orientation as they move to their final destinations along the same pathway followed by soluble secretory proteins (see Figure 13-1, *left*). During this transport, the orientation of a membrane protein is preserved; that is, the same segments of the protein always face the cytosol, whereas other segments always face in the opposite direction. Thus the final orientation of these membrane proteins is established during their biosynthesis on the ER membrane. In this section, we first see how integral proteins can interact with membranes and then examine how several types of sequences, collectively known as **topogenic sequences,** direct the membrane insertion and orientation of various classes of integral proteins. These processes occur via modifications of the basic mechanism used to translocate soluble secretory proteins across the ER membrane.

Several Topological Classes of Integral Membrane Proteins Are Synthesized on the ER

The *topology of a membrane protein* refers to the number of times that its polypeptide chain spans the membrane and the orientation of these membrane-spanning segments within the membrane. The key elements of a protein that determine its topology are membrane-spanning segments themselves, which usually are α-helices containing 20–25 hydrophobic amino acids that contribute to energetically favorable interactions within the hydrophobic interior of the phospholipid bilayer.

Most integral membrane proteins fall into one of the four topological classes illustrated in Figure 13-10. Topological classes I, II, and III comprise *single-pass* proteins, which have only one membrane-spanning α-helical segment. Type I proteins have a cleaved N-terminal ER signal sequence and are anchored in the membrane with their hydrophilic N-terminal region on the luminal face (also known as the **exoplasmic face**) and their hydrophilic C-terminal region on the cytosolic face. Type II proteins do not contain a cleavable ER signal sequence and are oriented with their hydrophilic N-terminal region on the cytosolic face and their hydrophilic C-terminal region on the exoplasmic face (i.e., opposite to type I proteins). Type III proteins have the same orientation as type I proteins but do not contain a cleavable signal sequence. These different topologies reflect distinct mechanisms used by the cell to establish the membrane orientation of transmembrane segments, as discussed in the next section.

The proteins forming topological class IV contain two or more membrane-spanning segments and are sometimes called *multipass* proteins. For example, many of the membrane transport proteins discussed in Chapter 11 and the numerous G protein–coupled receptors covered in Chapter 15 belong to this class. A final type of membrane protein

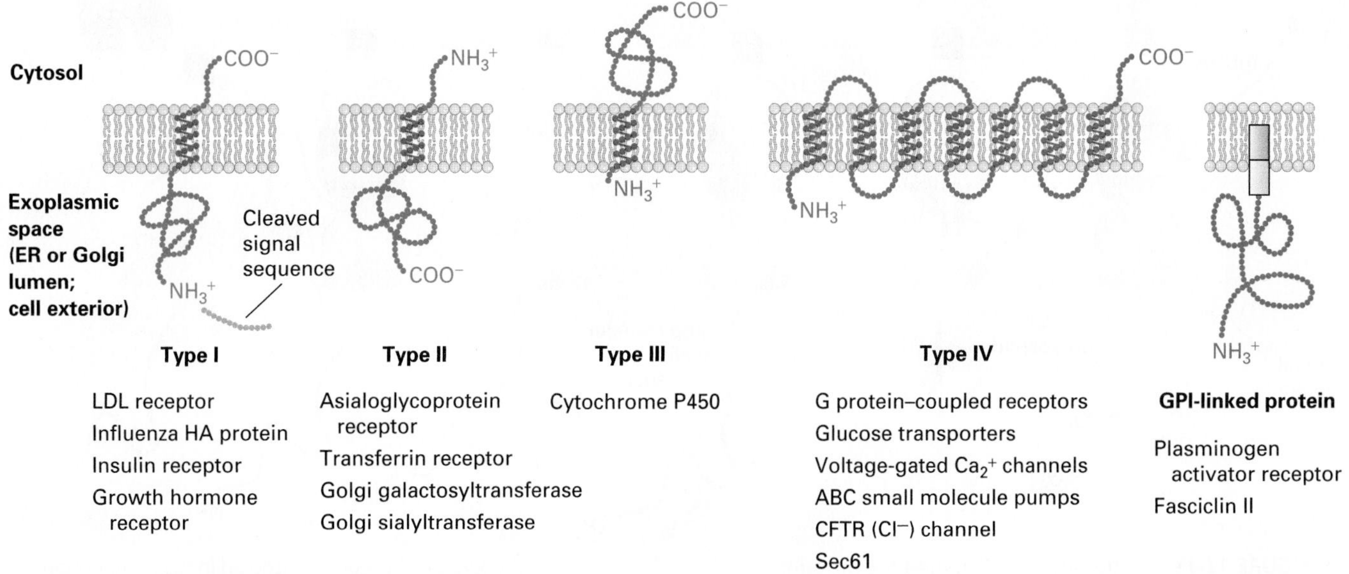

▲ **FIGURE 13-10 ER membrane proteins.** Four topological classes of integral membrane proteins are synthesized on the rough ER as well as a fifth type tethered to the membrane by a phospholipid anchor. Membrane proteins are classified by their orientation in the membrane and the types of signals they contain to direct them there. In classes I–IV, the hydrophobic segments of the protein chain form α helices embedded in the membrane bilayer; the regions outside the membrane are hydrophilic and fold into various conformations. All type IV proteins have multiple transmembrane α helices. The type IV topology depicted here corresponds to that of G protein–coupled receptors: seven α helices, the N-terminus on the exoplasmic side of the membrane, and the C-terminus on the cytosolic side. Other type IV proteins may have a different number of helices and various orientations of the N-terminus and C-terminus. [See E. Hartmann et al., 1989, *Proc. Nat'l. Acad. Sci. USA* **86:**5786, and C. A. Brown and S. D. Black, 1989, *J. Biol. Chem.* **264:**4442.]

lacks a hydrophobic membrane-spanning segment altogether; instead, these proteins are linked to an amphipathic phospholipid anchor that is embedded in the membrane (Figure 13-10, *right*).

Internal Stop-Transfer and Signal-Anchor Sequences Determine Topology of Single-Pass Proteins

We begin our discussion of how membrane protein topology is determined with the membrane insertion of integral proteins that contain a single, hydrophobic membrane-spanning segment. Two sequences are involved in targeting and orienting type I proteins in the ER membrane, whereas type II and type III proteins contain a single, internal topogenic sequence. As we will see, there are three main types of topogenic sequences that are used to direct proteins to the ER membrane and to orient them within it. We have already been introduced to one, the N-terminal ER signal sequence. The other two, introduced here, are internal sequences known as stop-transfer anchor sequences and signal-anchor sequences.

Type I Proteins All type I transmembrane proteins possess an N-terminal signal sequence that targets them to the ER as well as an internal hydrophobic sequence that becomes the membrane-spanning α helix. The N-terminal signal sequence on a nascent type I protein, like that of a secretory protein, initiates cotranslational translocation of the protein through the combined action of the SRP and SRP receptor. Once the N-terminus of the growing polypeptide enters the lumen of the ER, the signal sequence is cleaved, and the growing chain continues to be extruded across the ER membrane. However, unlike the case with secretory proteins, when the sequence of approximately 22 hydrophobic amino acids that will become a transmembrane domain of the nascent chain enters the translocon, it stops transfer of the protein through the channel (Figure 13-11). The structure of the Sec61 complex suggests that the channel may be able to open like a clamshell, allowing the hydrophobic transmembrane segment of the translocating peptide to move laterally between the protein domains constituting the translocon wall (see Figure 13-8). When the peptide exits the translocon in this manner, it becomes anchored in the phospholipid bilayer of the membrane. Because of the dual function of such a sequence to both stop passage of the polypeptide chain through the translocon and to become a hydrophobic transmembrane segment in the membrane bilayer, it is called a *stop-transfer anchor sequence.*

Once translocation is interrupted, translation continues at the ribosome, which is still anchored to the now unoccupied and closed translocon. As the C-terminus of the protein chain is synthesized, it loops out on the cytosolic side of the membrane. When translation is completed, the ribosome is

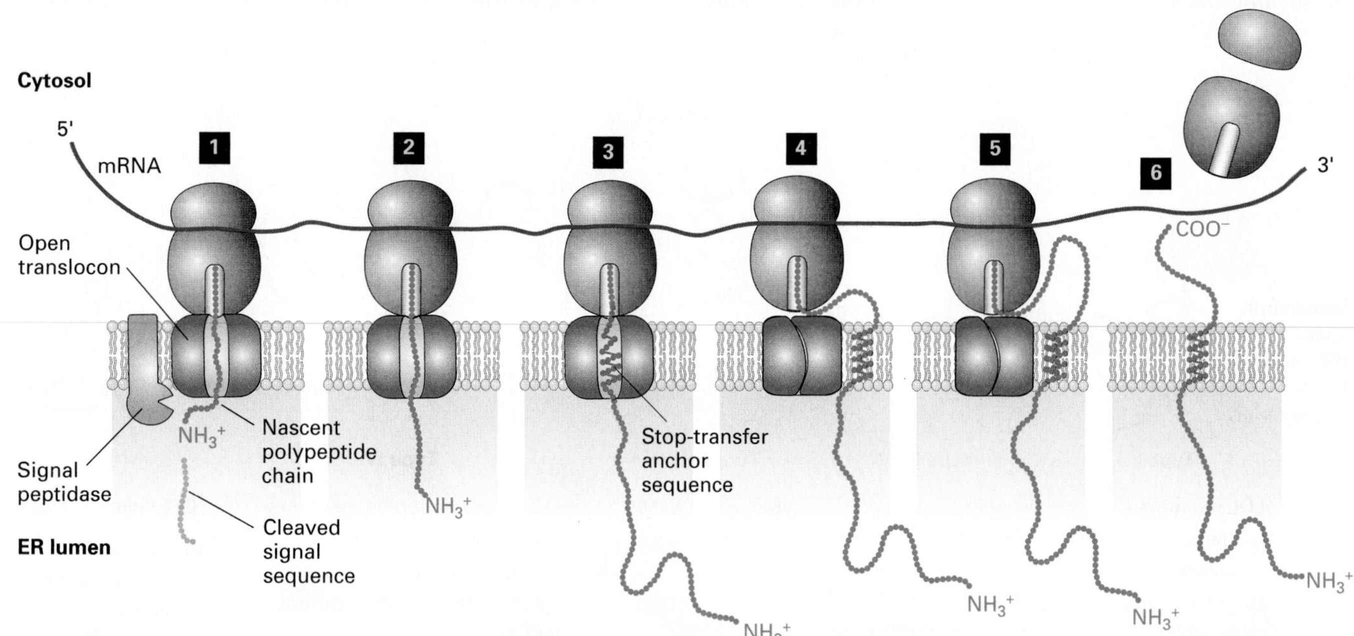

▲ **FIGURE 13-11 Positioning type I single-pass proteins.** Step **1**: After the ribosome/nascent chain complex becomes associated with a translocon in the ER membrane, the N-terminal signal sequence is cleaved. This process occurs by the same mechanism as the one for soluble secretory proteins (see Figure 13-6). Steps **2**, **3**: The chain is elongated until the hydrophobic stop-transfer anchor sequence is synthesized and enters the translocon, where it prevents the nascent chain from extruding farther into the ER lumen. Step **4**: The stop-transfer anchor sequence moves laterally between the translocon subunits and becomes anchored in the phospholipid bilayer. At this time, the translocon probably closes. Step **5**: As synthesis continues, the elongating chain may loop out into the cytosol through the small space between the ribosome and translocon. Step **6**: When synthesis is complete, the ribosomal subunits are released into the cytosol, leaving the protein free to diffuse in the membrane. [See H. Do et al., 1996, *Cell* **85:**369, and W. Mothes et al., 1997, *Cell* **89:**523.]

released from the translocon and the C-terminus of the newly synthesized type I protein remains in the cytosol.

Support for this mechanism has come from studies in which cDNAs encoding various mutant receptors for human growth hormone (HGH) are expressed in cultured mammalian cells. The wild-type HGH receptor, a typical type I protein, is transported normally to the plasma membrane. However, a mutant receptor that has charged residues inserted into the single α-helical membrane-spanning segment or that is missing most of this segment is translocated entirely into the ER lumen and is eventually secreted from the cell as a soluble protein. These kinds of experiments establish that the hydrophobic membrane-spanning α helix of the HGH receptor and of other type I proteins functions both as a stop-transfer sequence and a membrane anchor that prevents the C-terminus of the protein from crossing the ER membrane.

Type II and Type III Proteins Unlike type I proteins, type II and type III proteins lack a cleavable N-terminal ER signal sequence. Instead, both possess a single internal hydrophobic *signal-anchor sequence* that functions as both an ER signal sequence and membrane-anchor sequence. Recall that type II and type III proteins have opposite orientations in the membrane (see Figure 13-10); this difference depends on the orientation that their respective signal-anchor sequences assume within the translocon. The internal signal-anchor sequence in type II proteins directs insertion of the nascent chain into the ER membrane so that the N-terminus of the chain faces the cytosol, using the same SRP-dependent mechanism described for signal sequences (Figure 13-12a). However, the internal signal-anchor sequence is *not* cleaved and moves laterally between the protein domains of the translocon wall into the phospholipid bilayer, where it functions as a membrane anchor. As elongation continues, the C-terminal region of the growing chain is extruded through the translocon into the ER lumen by cotranslational translocation.

In the case of type III proteins, the signal-anchor sequence, which is located near the N-terminus, inserts the nascent chain into the ER membrane with its N-terminus facing the lumen, in the opposite orientation of the signal anchor in type II proteins. The signal-anchor sequence of type III proteins also functions like a stop-transfer sequence and prevents further extrusion of the nascent chain into the ER lumen (Figure 13-12b). Continued elongation of the

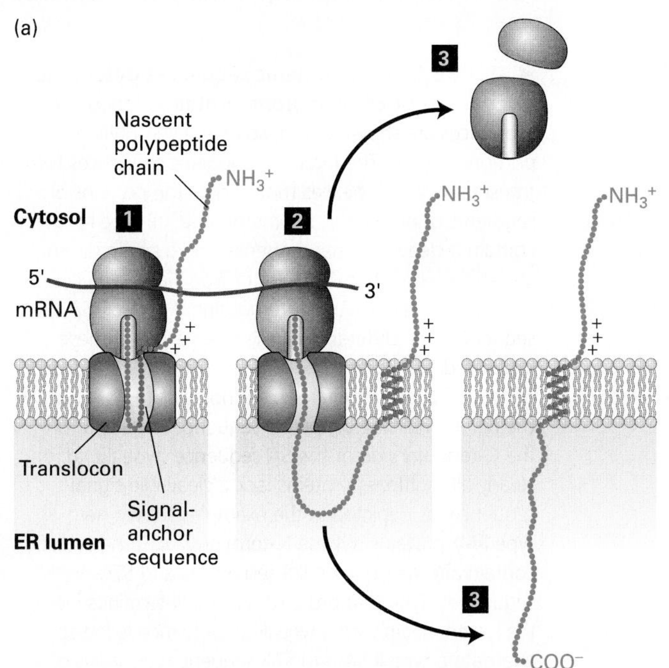

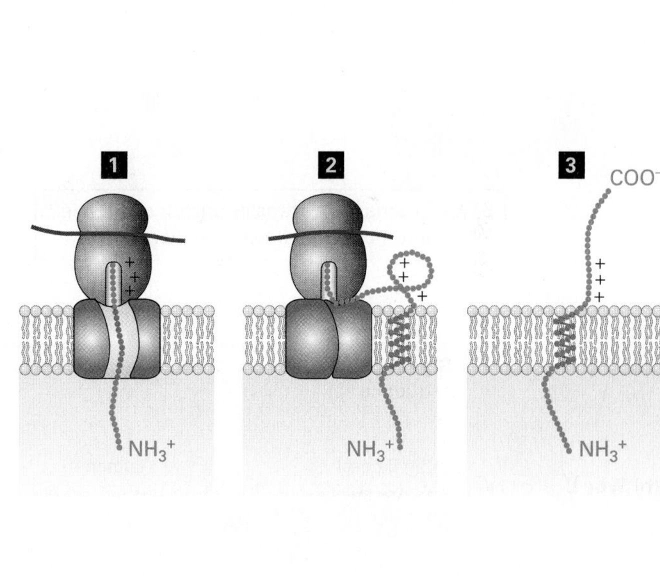

▲ **FIGURE 13-12 Positioning type II and type III single-pass proteins.** (a) Type II proteins. Step **1**: After the internal signal-anchor sequence is synthesized on a cytosolic ribosome, it is bound by an SRP (not shown), which directs the ribosome/nascent chain complex to the ER membrane. This is similar to targeting of soluble secretory proteins except that the hydrophobic signal sequence is not located at the N-terminus and is not subsequently cleaved. The nascent chain becomes oriented in the translocon with its N-terminal portion toward the cytosol. This orientation is believed to be mediated by the positively charged residues shown N-terminal to the signal-anchor sequence. Step **2**: As the chain is elongated and extruded into the lumen, the internal signal-anchor moves laterally out of the translocon and anchors the chain in the phospholipid bilayer. Step **3**: Once protein synthesis is completed, the C-terminus of the polypeptide is released into the lumen, and the ribosomal subunits are released into the cytosol. (b) Type III proteins. Step **1**: Assembly is by a similar pathway to that of type II proteins except that positively charged residues on the C-terminal side of the signal-anchor sequence cause the transmembrane segment to be oriented within the translocon with its C-terminal portion oriented to the cytosol and the N-terminal side of the protein in the ER lumen. Steps **2**, **3**: Chain elongation of the C-terminal portion of the protein is completed in the cytosol, and ribosomal subunits are released. [See M. Spiess and H. F. Lodish, 1986, *Cell* **44**:177, and H. Do et al., 1996, *Cell* **85**:369.]

chain C-terminal to the signal-anchor/stop-transfer sequence proceeds as it does for type I proteins, with the hydrophobic sequence moving laterally between the translocon subunits to anchor the polypeptide in the ER membrane (see Figure 13-11).

One of the features of signal-anchor sequences that appears to determine their insertion orientation is a high density of positively charged amino acids adjacent to one end of the hydrophobic segment. For reasons that are not well understood, these positively charged residues tend to remain on the cytosolic side of the membrane, not traversing the membrane into the ER lumen. Thus the position of the charged residues dictates the orientation of the signal-anchor sequence within the translocon as well as whether or not the rest of the polypeptide chain continues to pass into the ER lumen: type II proteins tend to have positively charged residues on the N-terminal side of their signal-anchor sequence, orienting the N-terminus in the cytosol and allowing passage of the C-terminal side into the ER (Figure 13-12a), whereas type III proteins tend to have positively charged residues on the C-terminal side of their signal-anchor sequence, inserting the N-terminus into the translocon and restricting the C-terminus to the cytosol (Figure 13-12b).

A striking experimental demonstration of the importance of the flanking charge in determining membrane orientation is provided by neuraminidase, a type II protein in the surface coat of influenza virus. Three arginine residues are located just N-terminal to the internal signal-anchor sequence in neuraminidase. Mutation of these three positively charged residues to negatively charged glutamate residues causes neuraminidase to acquire the reverse orientation. Similar experiments have shown that other proteins, with either type II or type III orientation, can be made to "flip" their orientation in the ER membrane by mutating charged residues that flank the internal signal-anchor segment.

Multipass Proteins Have Multiple Internal Topogenic Sequences

Figure 13-13 summarizes the arrangements of topogenic sequences in single-pass and multipass transmembrane proteins. In multipass (type IV) proteins, each of the membrane-spanning α helices acts as a topogenic sequence in the ways that we have already discussed: they can act to direct the protein to the ER, to anchor the protein in the ER membrane, or to stop transfer of the protein through the membrane. Multipass proteins fall into one of two types depending on whether the N-terminus extends into the cytosol or the exoplasmic space (e.g., the ER lumen, cell exterior). This N-terminal topology usually is determined

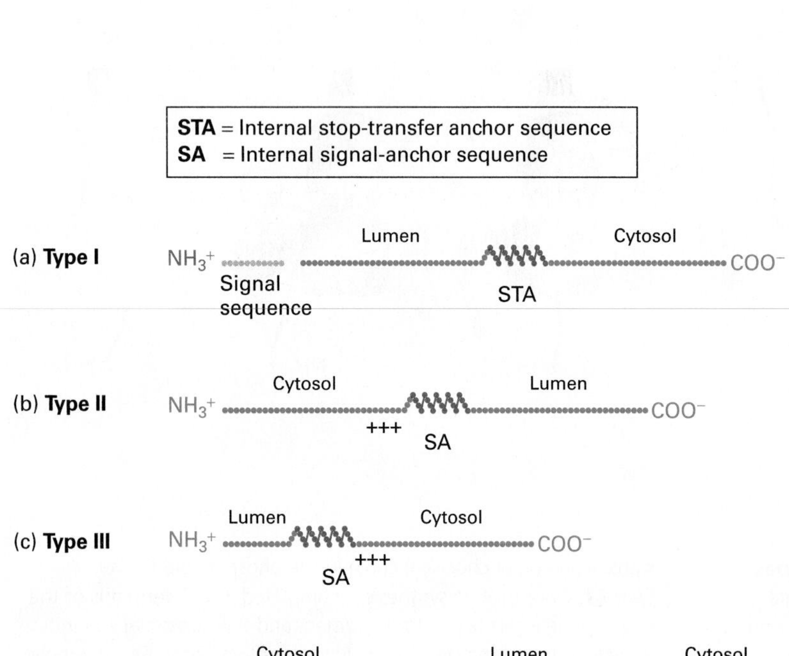

◄ **FIGURE 13-13 Topogenic sequences determine orientation of ER membrane proteins.** Topogenic sequences are shown in red; soluble, hydrophilic portions in blue. The internal topogenic sequences form transmembrane α helices that anchor the proteins or segments of proteins in the membrane. (a) Type I proteins contain a cleaved signal sequence and a single internal stop-transfer anchor (STA). (b, c) Type II and type III proteins contain a single internal signal-anchor (SA) sequence. The difference in the orientation of these proteins depends largely on whether there is a high density of positively charged amino acids (+++) on the N-terminal side of the SA sequence (type II) or on the C-terminal side of the SA sequence (type III). (d, e) Nearly all multipass proteins lack a cleavable signal sequence, as depicted in the examples shown here. Type IV-A proteins, whose N-terminus faces the cytosol, contain alternating type II S sequences and STA sequences. Type IV-B proteins, whose N-terminus faces the lumen, begin with a type III SA sequence followed by alternating type II SA and STA sequences. Proteins of each type with different numbers of α helices (odd or even) are known.

by the hydrophobic segment closest to the N-terminus and the charge of the sequences flanking it. If a type IV protein has an *even* number of transmembrane α helices, both its N-terminus and C-terminus will be oriented toward the same side of the membrane (Figure 13-13d). Conversely, if a type IV protein has an *odd* number of α helices, its two ends will have opposite orientations (Figure 13-13e).

Type IV Proteins with N-Terminus in Cytosol Among the multipass proteins whose N-terminus extends into the cytosol are the various glucose transporters (GLUTs) and most ion-channel proteins, discussed in Chapter 11. In these proteins, the hydrophobic segment closest to the N-terminus initiates insertion of the nascent chain into the ER membrane with the N-terminus oriented toward the cytosol; thus this α-helical segment functions like the internal signal-anchor sequence of a type II protein (see Figure 13-12a). As the nascent chain following the first α helix elongates, it moves through the translocon until the second hydrophobic α helix is formed. This helix prevents further extrusion of the nascent chain through the translocon; thus its function is similar to that of the stop-transfer anchor sequence in a type I protein (see Figure 13-11).

After synthesis of the first two transmembrane α helices, both ends of the nascent chain face the cytosol and the loop between them extends into the ER lumen. The C-terminus of the nascent chain then continues to grow into the cytosol, as it does in synthesis of type I and type III proteins. According to this mechanism, the third α helix acts as another type II signal-anchor sequence and the fourth as another stop-transfer anchor sequence (Figure 13-13d). Apparently, once the first topogenic sequence of a multipass polypeptide initiates association with the translocon, the ribosome remains attached to the translocon, and topogenic sequences that subsequently emerge from the ribosome are threaded into the translocon without the need for the SRP and the SRP receptor.

Experiments that use recombinant DNA techniques to exchange hydrophobic α helices have provided insight into the functioning of the topogenic sequences in type IV-A multipass proteins. These experiments indicate that the order of the hydrophobic α helices relative to each other in the growing chain largely determines whether a given helix functions as a signal-anchor sequence or stop-transfer anchor sequence. Other than its hydrophobicity, the specific amino acid sequence of a particular helix has little bearing on its function. Thus the first N-terminal α helix and the subsequent odd-numbered ones function as signal-anchor sequences, whereas the intervening even-numbered helices function as stop-transfer anchor sequences.

Type IV Proteins with N-Terminus in the Exoplasmic Space The large family of G protein–coupled receptors, all of which contain seven transmembrane α helices, constitute the most numerous type IV-B proteins, whose N-terminus extends into the exoplasmic space. In these proteins, the hydrophobic α helix closest to the N-terminus often is followed by a cluster of positively charged amino acids, similar to a type III signal-anchor sequence (see Figure 13-12b). As a result, the first α helix inserts the nascent chain into the translocon with the N-terminus extending into the lumen (see Figure 13-13e). As the chain is elongated, it is inserted into the ER membrane by alternating type II signal-anchor sequences and stop-transfer sequences, as just described for type IV-A proteins.

A Phospholipid Anchor Tethers Some Cell-Surface Proteins to the Membrane

Some cell-surface proteins are anchored to the phospholipid bilayer not by a sequence of hydrophobic amino acids but by a covalently attached amphipathic molecule, *glycosylphosphatidylinositol (GPI)* (Figure 13-14a and Chapter 10). These proteins are synthesized and initially anchored to the ER membrane exactly like type I transmembrane proteins, with a cleaved N-terminal signal sequence and internal stop-transfer anchor sequence directing the process (see Figure 13-11). However, a short sequence of amino acids in the luminal domain, adjacent to the membrane-spanning domain, is recognized by a transamidase located within the ER membrane. This enzyme simultaneously cleaves off the original stop-transfer anchor sequence and transfers the luminal portion of the protein to a preformed GPI anchor in the membrane (Figure 13-14b).

Why change one type of membrane anchor for another? Attachment of the GPI anchor, which results in removal of the cytosol-facing hydrophilic domain from the protein, can have several consequences. Proteins with GPI anchors, for example, can diffuse relatively rapidly in the plane of the phospholipid bilayer membrane. In contrast, many proteins anchored by membrane-spanning α helices are impeded from moving laterally in the membrane because their cytosol-facing segments interact with the cytoskeleton. In addition, the GPI anchor targets the attached protein to the apical domain of the plasma membrane in certain polarized epithelial cells, as we discuss in Chapter 14.

The Topology of a Membrane Protein Often Can Be Deduced from Its Sequence

As we have seen, various topogenic sequences in integral membrane proteins synthesized on the ER govern interaction of the nascent chain with the translocon. When scientists begin to study a protein of unknown function, the identification of potential topogenic sequences within the corresponding gene sequence can provide important clues about the protein's topological class and function. Suppose, for example, that the gene for a protein known to be required for a cell-to-cell signaling pathway contains nucleotide sequences that encode an apparent N-terminal signal sequence and an internal hydrophobic sequence. These findings suggest that the protein is a type I integral membrane protein

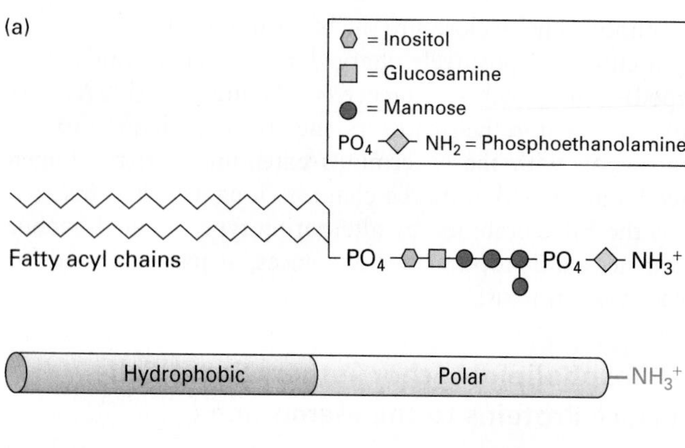

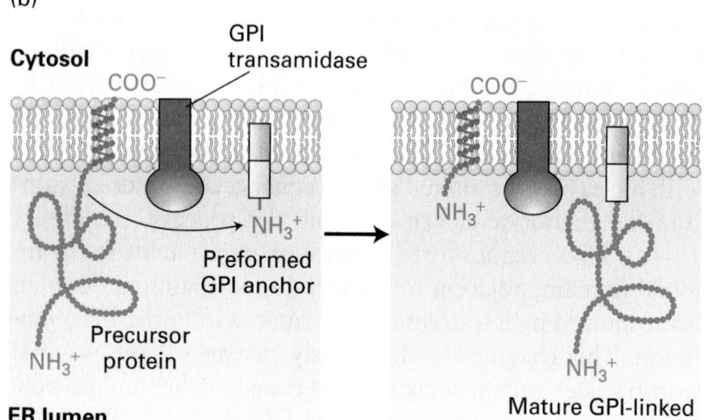

▲ **FIGURE 13-14 GPI-anchored proteins.** (a) Structure of a glycosylphosphatidylinositol (GPI) from yeast. The hydrophobic portion of the molecule is composed of fatty acyl chains, whereas the polar (hydrophilic) portion of the molecule is composed of carbohydrate residues and phosphate groups. In other organisms, both the length of the acyl chains and the carbohydrate moieties may vary somewhat from the structure shown. (b) Formation of GPI-anchored proteins in the ER membrane. The protein is synthesized and initially inserted into the ER membrane as shown in Figure 13-11. A specific transamidase simultaneously cleaves the precursor protein within the exoplasmic-facing domain, near the stop-transfer anchor sequence (red), and transfers the carboxyl group of the new C-terminus to the terminal amino group of a preformed GPI anchor. [See C. Abeijon and C. B. Hirschberg, 1992, *Trends Biochem. Sci.* **17**:32, and K. Kodukula et al., 1992, *Proc. Nat'l. Acad. Sci. USA* **89**:4982.]

and therefore may be a cell-surface receptor for an extracellular ligand.

Identification of topogenic sequences requires a way to scan sequence databases for segments that are sufficiently hydrophobic to be either a signal sequence or a transmembrane anchor sequence. Topogenic sequences can often be identified with the aid of computer programs that generate a *hydropathy profile* for the protein of interest. The first step is to assign a value known as the *hydropathic index* to each amino acid in the protein. By convention, hydrophobic amino acids are assigned positive values and hydrophilic amino acids negative values. Although different scales for the hydropathic index exist, all assign the most positive values to amino acids with side chains made up of mostly hydrocarbon residues (e.g., phenylalanine and methionine) and the most negative values to charged amino acids (e.g., arginine and aspartate). The second step is to identify longer segments of sufficient overall hydrophobicity to be N-terminal signal sequences or internal stop-transfer sequences and signal-anchor sequences. To accomplish this, the total hydropathic index for each successive segment of 20 consecutive amino acids is calculated along the entire length of the protein. Plots of these calculated values against position in the amino acid sequence yield a hydropathy profile.

Figure 13-15 shows the hydropathy profiles for three different membrane proteins. The prominent peaks in such plots identify probable topogenic sequences as well as their position and approximate length. For example, the hydropathy profile of the human growth hormone receptor reveals the presence of both a hydrophobic signal sequence at the extreme N-terminus of the protein and an internal hydrophobic stop-transfer sequence (Figure 13-15a). On the basis of this profile, we can deduce, correctly, that the human growth hormone receptor is a type I integral membrane protein. The hydropathy profile of the asialoglycoprotein receptor, a cell-surface protein that mediates removal of abnormal extracellular glycoproteins, reveals a prominent internal hydrophobic signal-anchor sequence but gives no indication of a hydrophobic N-terminal signal sequence (Figure 13-15b). Thus we can predict that the asialoglycoprotein receptor is a type II or type III membrane protein. The distribution of charged residues on either side of the signal-anchor sequence often can differentiate between these possibilities since positively charged amino acids flanking a membrane-spanning segment usually are oriented toward the cytosolic face of the membrane. For instance, in the case of the asialoglycoprotein receptor, examination of the residues flanking the signal-anchor sequence reveals that the residues on the N-terminal side carry a net positive charge, thus correctly predicting that this is a type II protein.

The hydropathy profile of the GLUT1 glucose transporter, a multipass membrane protein, shows the presence of many segments that are sufficiently hydrophobic to be membrane-spanning helices (Figure 13-15c). The complexity of this profile illustrates the difficulty both in unambiguously identifying all the membrane-spanning segments in a multipass protein and in predicting the topology of individual signal-anchor and stop-transfer sequences. More sophisticated computer algorithms have been developed that take into account the presence of positively charged amino acids adjacent to hydrophobic segments as well as the length of and spacing between segments. Using all this information, the best algorithms can predict the complex topology of multipass proteins with an accuracy of greater than 75 percent.

(a) Human growth hormone receptor (type I)

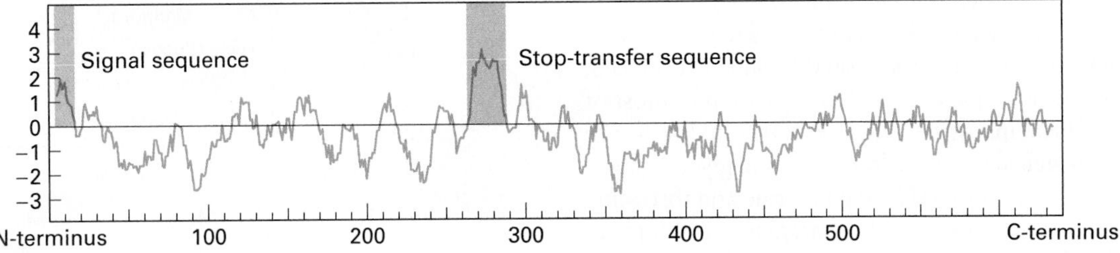

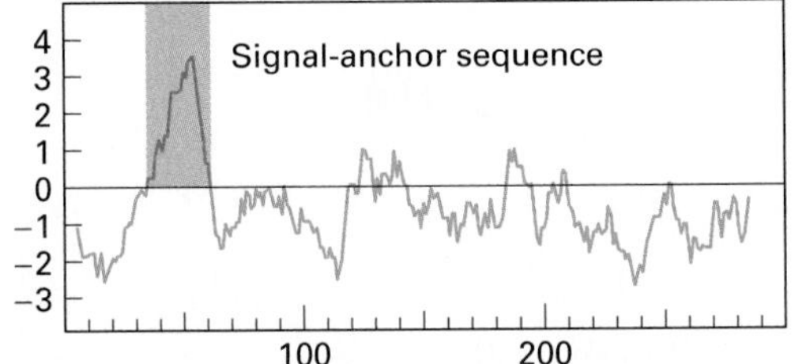

(c) GLUT1 (type IV)

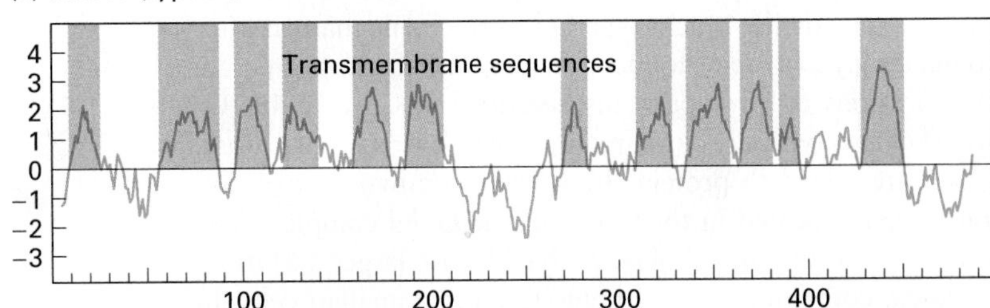

▲ EXPERIMENTAL FIGURE 13-15 **Hydropathy profiles.** Hydropathy profiles can identify likely topogenic sequences in integral membrane proteins. They are generated by plotting the total hydrophobicity of each segment of 20 contiguous amino acids along the length of a protein. Positive values indicate relatively hydrophobic portions; negative values, relatively polar portions of the protein. Probable topogenic sequences are marked. The complex profiles for multipass (type IV) proteins, such as GLUT1 in part (c), often must be supplemented with other analyses to determine the topology of these proteins.

Finally, sequence homology to a known protein may permit accurate prediction of the topology of a newly discovered multipass protein. For example, the genomes of multicellular organisms encode a very large number of multipass proteins with seven transmembrane α helices. The similarities between the sequences of these proteins strongly suggest that all have the same topology as the well-studied G protein–coupled receptors, which have the N-terminus oriented to the exoplasmic side and the C-terminus oriented to the cytosolic side of the membrane.

KEY CONCEPTS OF SECTION 13.2

Insertion of Proteins into the ER Membrane

- Integral membrane proteins synthesized on the rough ER fall into four topological classes as well as a lipid-linked type (see Figure 13-10).
- Topogenic sequences—N-terminal signal sequences, internal stop-transfer anchor sequences, and internal signal-anchor sequences—direct the insertion and orientation of nascent proteins within the ER membrane. This orientation is retained during transport of the completed membrane protein to its final destination.
- Single-pass membrane proteins contain one or two topogenic sequences. In multipass membrane proteins, each α-helical segment can function as an internal topogenic sequence, depending on its location in the polypeptide chain and the presence of adjacent positively charged residues (see Figure 13-13).
- Some cell-surface proteins are initially synthesized as type I proteins on the ER and then are cleaved with their luminal domain transferred to a GPI anchor (see Figure 13-14).
- The topology of membrane proteins can often be correctly predicted by computer programs that identify hydrophobic topogenic segments within the amino acid sequence and generate hydropathy profiles (see Figure 13-15).

13.3 Protein Modifications, Folding, and Quality Control in the ER

Membrane and soluble secretory proteins synthesized on the rough ER undergo four principal modifications before they reach their final destinations: (1) covalent addition and processing of carbohydrates *(glycosylation)* in the ER and Golgi, (2) formation of disulfide bonds in the ER, (3) proper folding of polypeptide chains and assembly of multisubunit proteins in the ER, and (4) specific proteolytic cleavages in the ER, Golgi, and secretory vesicles. Generally speaking, these modifications promote folding of secretory proteins into their native structures and add structural stability to proteins exposed to the extracellular environment. Modifications such as glycosylation also allow the cell to produce a vast array of chemically distinct molecules at the cell surface that are the basis of specific molecular interactions used in cell-to-cell adhesion and communication.

One or more carbohydrate chains are added to the vast majority of proteins that are synthesized on the rough ER; indeed, glycosylation is the principal chemical modification to most of these proteins. Proteins with attached carbohydrates are known as **glycoproteins.** Carbohydrate chains in glycoproteins may be attached to the hydroxyl group in serine and threonine residues or to the amide nitrogen of asparagine. These are referred to as **O-linked oligosaccharides** and **N-linked oligosaccharides,** respectively. The various types of O-linked oligosaccharides include the mucin-type O-linked chains (named after abundant glycoproteins found in mucus) and the carbohydrate modifications on proteoglycans described in Chapter 19. O-linked chains typically contain only one to four sugar residues, which are added to proteins by enzymes known as glycosyltransferases, located in the lumen of the Golgi complex. The more common *N*-linked oligosaccharides are larger and more complex, containing several branches in mammalian cells. In this section we focus on *N*-linked oligosaccharides, whose initial synthesis occurs in the ER. After the initial N-glycosylation of a protein in the ER, the oligosaccharide chain is modified in the ER and commonly in the Golgi as well.

Disulfide bond formation, protein folding, and assembly of multimeric proteins, which take place exclusively in the rough ER, also are discussed in this section. Only properly folded and assembled proteins are transported from the rough ER to the Golgi complex and ultimately to the cell surface or other final destination. Unfolded, misfolded, or partly folded and assembled proteins are selectively retained in the rough ER. We consider several features of such "quality control" in the latter part of this section.

As discussed previously, N-terminal ER signal sequences are cleaved from secretory proteins and type I membrane proteins in the ER. Some proteins also undergo other specific proteolytic cleavages in the Golgi complex or secretory vesicles. We cover these cleavages, as well as carbohydrate modifications that occur primarily or exclusively in the Golgi complex, in the next chapter.

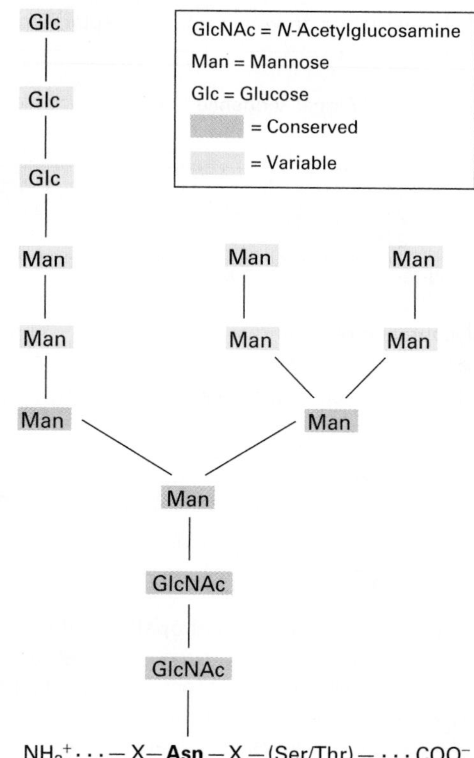

▲ **FIGURE 13-16 Common precursor of *N*-linked oligosaccharides.** This 14-residue precursor of *N*-linked oligosaccharides is added to nascent proteins in the rough ER. Subsequent removal and in some cases addition of specific sugar residues occur in the ER and Golgi complex. The core region, composed of five residues highlighted in purple, is retained in all *N*-linked oligosaccharides. The precursor can be linked only to asparagine (Asn) residues that are separated by one amino acid (X) from a serine (Ser) or threonine (Thr) on the carboxyl side.

A Preformed *N*-Linked Oligosaccharide Is Added to Many Proteins in the Rough ER

Biosynthesis of all *N*-linked oligosaccharides begins in the rough ER with addition of a preformed oligosaccharide precursor containing 14 residues (Figure 13-16). The structure of this precursor is the same in plants, animals, and single-celled eukaryotes—a branched oligosaccharide, containing three glucose (Glc), nine mannose (Man), and two *N*-acetylglucosamine (GlcNAc) molecules, which can be written as $Glc_3Man_9(GlcNAc)_2$. Once added to a protein, this branched carbohydrate structure is modified by addition or removal of monosaccharides in the ER and Golgi compartments. The modifications to N-linked chains differ from one glycoprotein to another and differ among different organisms, but a core of 5 of the 14 residues is conserved in the structures of all *N*-linked oligosaccharides on secretory and membrane proteins.

Prior to transfer to a nascent chain in the lumen of the ER, the precursor oligosaccharide is assembled on a membrane-attached anchor called *dolichol phosphate,* a long-chain polyisoprenoid lipid (Chapter 10). After the first sugar, GlcNAc, is attached to the dolichol phosphate by a pyrophosphate bond, the other sugars are added by glycosidic bonds in a complex set of reactions catalyzed by enzymes attached to the cytosolic or luminal faces of the rough ER membrane (Figure 13-17). The final dolichol pyrophosphoryl oligosaccharide is oriented so that the oligosaccharide portion faces the ER lumen.

The entire 14-residue precursor is transferred from the dolichol carrier to an asparagine residue on a nascent polypeptide as it emerges into the ER lumen (Figure 13-18, step **1**). Only asparagine residues in the tripeptide sequences Asn-X-Ser and Asn-X-Thr (where X is any amino acid except proline) are substrates for *oligosaccharyl transferase,* the enzyme that catalyzes this reaction. Two of the three subunits of this enzyme are ER membrane proteins whose cytosol-facing domains bind to the ribosome, localizing a third subunit of the transferase, the catalytic subunit, near the growing polypeptide chain in the ER lumen. Not all

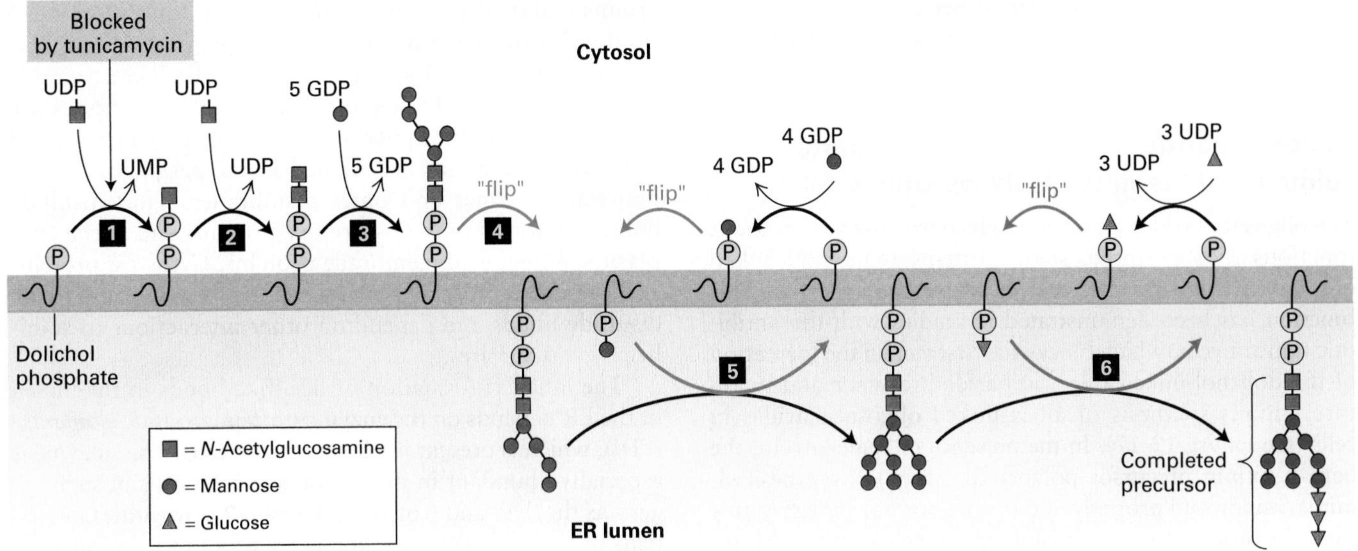

▲ **FIGURE 13-17 Biosynthesis of the oligosaccharide precursor.** Dolichol phosphate is a strongly hydrophobic lipid, containing 75–95 carbon atoms, that is embedded in the ER membrane. Two *N*-acetylglucosamine (GlcNAc) and five mannose residues are added one at a time to a dolichol phosphate on the cytosolic face of the ER membrane (steps **1**–**3**). The nucleotide-sugar donors in these and later reactions are synthesized in the cytosol. Note that the first sugar residue is attached to dolichol by a high-energy pyrophosphate linkage. Tunicamycin, which blocks the first enzyme in this pathway, inhibits the synthesis of all *N*-linked oligosaccharides in cells. After the seven-residue dolichol pyrophosphoryl intermediate is flipped to the luminal face (step **4**), the remaining four mannose and all three glucose residues are added one at a time (steps **5**, **6**). In the later reactions, the sugar to be added is first transferred from a nucleotide-sugar to a carrier dolichol phosphate on the cytosolic face of the ER; the carrier is then flipped to the luminal face, where the sugar is transferred to the growing oligosaccharide, after which the "empty" carrier is flipped back to the cytosolic face. [After C. Abeijon and C. B. Hirschberg, 1992, *Trends Biochem. Sci.* **17**:32.]

Asn-X-Ser/Thr sequences become glycosylated, and it is not possible to predict from the amino acid sequence alone which potential N-linked glycosylation sites will be modified; for instance, rapid folding of a segment of a protein containing an Asn-X-Ser/Thr sequence may prevent transfer of the oligosaccharide precursor to it.

Immediately after the entire precursor, $Glc_3Man_9(GlcNAc)_2$, is transferred to a nascent polypeptide, three different enzymes, called glycosidases, remove all three glucose residues and one particular mannose residue (Figure 13-18, steps **2**–**4**). The three glucose residues, which are the last residues added during synthesis of the precursor

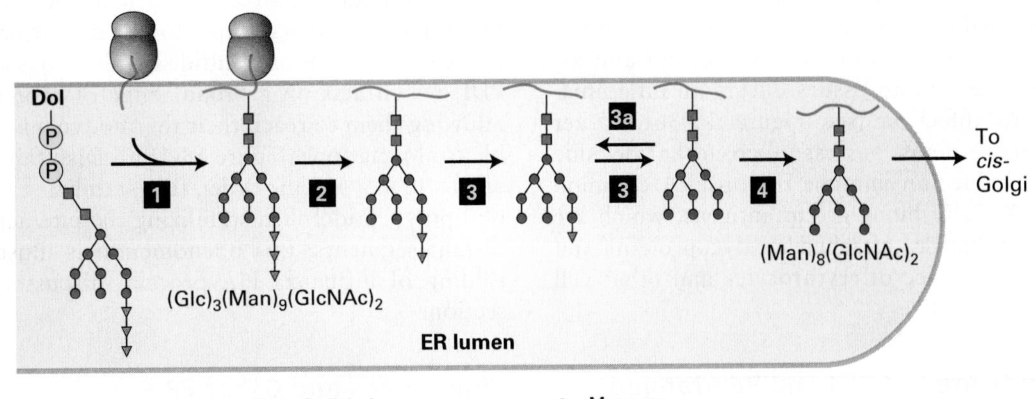

▲ **FIGURE 13-18 Addition and initial processing of *N*-linked oligosaccharides.** In the rough ER of vertebrate cells, the $Glc_3Man_9(GlcNAc)_2$ precursor is transferred from the dolichol carrier to a susceptible asparagine residue on a nascent protein as soon as the asparagine crosses to the luminal side of the ER (step **1**). In three separate reactions, first one glucose residue (step **2**), then two glucose residues (step **3**), and finally one mannose residue (step **4**) are removed. Re-addition of one glucose residue (step **3a**) plays a role in the correct folding of many proteins in the ER, as discussed later. [See R. Kornfeld and S. Kornfeld, 1985, *Ann. Rev. Biochem.* **45**:631, and M. Sousa and A. J. Parodi, 1995, *EMBO J.* **14**:4196.]

on the dolichol carrier, appear to act as a signal that the oligosaccharide is complete and ready to be transferred to a protein.

Oligosaccharide Side Chains May Promote Folding and Stability of Glycoproteins

The oligosaccharides attached to glycoproteins serve various functions. For example, some proteins require *N*-linked oligosaccharides in order to fold properly in the ER. This function has been demonstrated in studies with the antibiotic tunicamycin, which blocks the first step in the formation of the dolichol-linked oligosaccharide precursor and therefore inhibits synthesis of all *N*-linked oligosaccharides in cells (see Figure 13-17). In the presence of tunicamycin, the hemagglutinin precursor polypeptide (HA_0) is synthesized, but it cannot fold properly and form a normal trimer; in this case, the protein remains, misfolded, in the rough ER. Moreover, mutation of a particular asparagine in the HA sequence to a glutamine residue prevents addition of an *N*-linked oligosaccharide to that site and causes the protein to accumulate in the ER in an unfolded state.

In addition to promoting proper folding, *N*-linked oligosaccharides also confer stability on many secreted glycoproteins. Many secretory proteins fold properly and are transported to their final destination even if the addition of all *N*-linked oligosaccharides is blocked, for example, by tunicamycin. However, such nonglycosylated proteins have been shown to be less stable than their glycosylated forms. For instance, glycosylated fibronectin, a normal component of the extracellular matrix, is degraded much more slowly by tissue proteases than is nonglycosylated fibronectin.

Oligosaccharides on certain cell-surface glycoproteins also play a role in cell-cell adhesion. For example, the plasma membrane of white blood cells (leukocytes) contains cell-adhesion molecules (CAMs) that are extensively glycosylated. The oligosaccharides in these molecules interact with a sugar-binding domain in certain CAMs found on endothelial cells lining blood vessels. This interaction tethers the leukocytes to the endothelium and assists in their movement into tissues during an inflammatory response to infection (see Figure 19-36). Other cell-surface glycoproteins possess oligosaccharide side chains that can induce an immune response. A common example is the A, B, O blood-group antigens, which are *O*-linked oligosaccharides attached to glycoproteins and glycolipids on the surface of erythrocytes and other cell types (see Figure 10-20).

Disulfide Bonds Are Formed and Rearranged by Proteins in the ER Lumen

In Chapter 3 we learned that both intramolecular and intermolecular **disulfide bonds** (–S–S–) help stabilize the tertiary and quaternary structure of many proteins. These covalent bonds form by the oxidative linkage of **sulfhydryl groups** (–SH), also known as *thiol* groups, on two cysteine residues in the same or different polypeptide chains. This reaction can proceed spontaneously only when a suitable oxidant is present. In eukaryotic cells, disulfide bonds are formed only in the lumen of the rough ER; in bacterial cells, disulfide bonds are formed in the periplasmic space between the inner and outer membranes. Thus disulfide bonds are found only in secretory proteins and in the exoplasmic domains of membrane proteins. Cytosolic proteins and organelle proteins synthesized on free ribosomes lack disulfide bonds and depend on other interactions to stabilize their structures.

The efficient formation of disulfide bonds in the lumen of the ER depends on the enzyme *protein disulfide isomerase* (PDI), which is present in all eukaryotic cells. This enzyme is especially abundant in the ER of secretory cells in such organs as the liver and pancreas, where large quantities of proteins that contain disulfide bonds are produced. As shown in Figure 13-19a, the disulfide bond in the active site of PDI can be readily transferred to a protein by two sequential thiol-disulfide transfer reactions. The reduced PDI generated by this reaction is returned to an oxidized form by the action of an ER-resident protein, called *Ero1*, which carries a disulfide bond that can be transferred to PDI. Ero1 itself becomes oxidized by reaction with molecular oxygen that has diffused into the ER.

In proteins that contain more than one disulfide bond, the proper pairing of cysteine residues is essential for normal structure and activity. Disulfide bonds commonly are formed between cysteines that occur sequentially in the amino acid sequence while a polypeptide is still growing on the ribosome. Such sequential formation, however, sometimes yields disulfide bonds between the wrong cysteines. For example, proinsulin, a precursor to the peptide hormone insulin, has three disulfide bonds that link cysteines 1 and 4, 2 and 6, and 3 and 5. In this case, a disulfide bond that initially formed sequentially (e.g., between cysteines 1 and 2) would have to be rearranged for the protein to achieve its proper folded conformation. In cells, the rearrangement of disulfide bonds also is accelerated by PDI, which acts on a broad range of protein substrates, allowing them to reach their thermodynamically most stable conformations (Figure 13-19b). Disulfide bonds generally form in a specific order, first stabilizing small domains of a polypeptide, then stabilizing the interactions of more distant segments; this phenomenon is illustrated by the folding of influenza HA protein, discussed in the next section.

Chaperones and Other ER Proteins Facilitate Folding and Assembly of Proteins

Although many denatured proteins can spontaneously refold into their native state in vitro, such refolding usually requires hours to reach completion. Yet new soluble and membrane proteins produced in the ER generally fold into

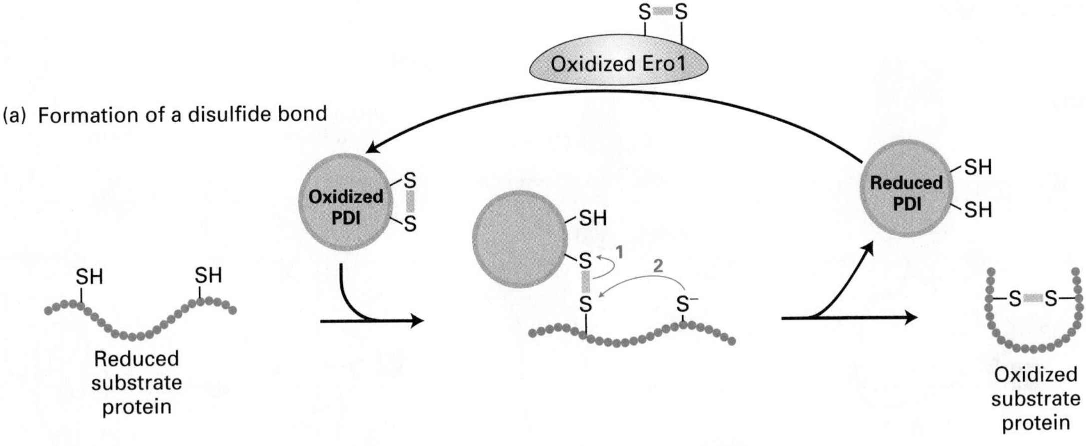

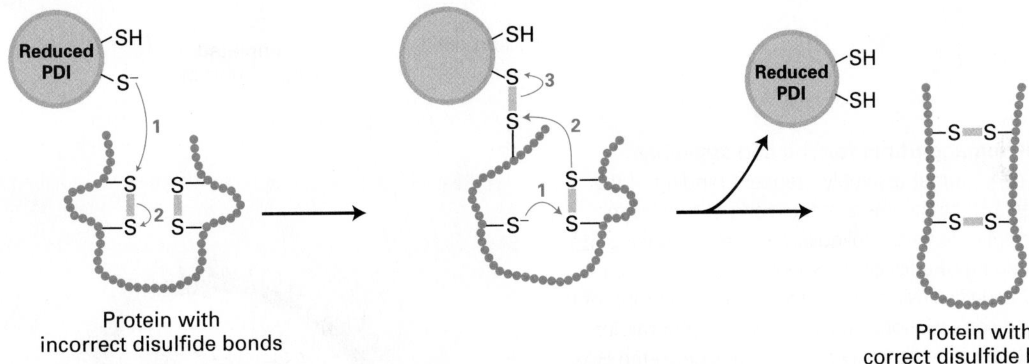

▲ **FIGURE 13-19 Action of protein disulfide isomerase (PDI).** PDI forms and rearranges disulfide bonds via an active site with two closely spaced cysteine residues that are easily interconverted between the reduced dithiol form and the oxidized disulfide form. Numbered red arrows indicate the sequence of electron transfers. Yellow bars represent disulfide bonds. (a) In the formation of disulfide bonds, the ionized ($-S^-$) form of a cysteine thiol in the substrate protein reacts with the disulfide (S—S) bond in oxidized PDI to form a disulfide-bonded PDI–substrate protein intermediate. A second ionized thiol in the substrate then reacts with this intermediate, forming a disulfide bond within the substrate protein and releasing reduced PDI. PDI, in turn, transfers electrons to a disulfide bond in the luminal protein Ero1, thereby regenerating the oxidized form of PDI. (b) Reduced PDI can catalyze rearrangement of improperly formed disulfide bonds by similar thiol-disulfide transfer reactions. In this case, reduced PDI both initiates and is regenerated in the reaction pathway. These reactions are repeated until the most stable conformation of the protein is achieved. [See M. M. Lyles and H. F. Gilbert, 1991, *Biochemistry* **30:**619.]

their proper conformation within minutes after their synthesis. The rapid folding of these newly synthesized proteins in cells depends on the sequential action of several proteins present within the ER lumen. We have already seen how the molecular chaperone BiP can drive post-translational translocation in yeast by binding fully synthesized polypeptides as they enter the ER (see Figure 13-9). BiP can also bind transiently to nascent chains as they enter the ER during cotranslational translocation. Bound BiP is thought to prevent segments of a nascent chain from misfolding or forming aggregates, thereby promoting folding of the entire polypeptide into the proper conformation. Protein disulfide isomerase (PDI) also contributes to proper folding, because correct 3-D conformation is stabilized by disulfide bonds in many proteins.

As illustrated in Figure 13-20, two other ER proteins, the homologous **lectins** (carbohydrate-binding proteins) *calnexin* and *calreticulin,* bind selectively to certain *N*-linked oligosaccharides on growing nascent chains. The ligand for these two lectins, which contains a single glucose residue, is generated by a specific glucosyltransferase in the ER lumen (see Figure 13-18, step **3a**). This enzyme acts only on polypeptide chains that are unfolded or misfolded, and in this respect the glucosyltransferase acts as one of the primary surveillance mechanisms to ensure quality control of protein folding in the ER. Binding of calnexin and calreticulin to unfolded nascent chains marked with glucosylated *N*-linked oligosaccharides prevents aggregation of adjacent segments of a protein as it is being made on the ER. Thus calnexin and calreticulin, like BiP, help prevent premature, incorrect folding of segments of a newly made protein.

Other important protein-folding catalysts in the ER lumen are *peptidyl-prolyl isomerases,* a family of enzymes that

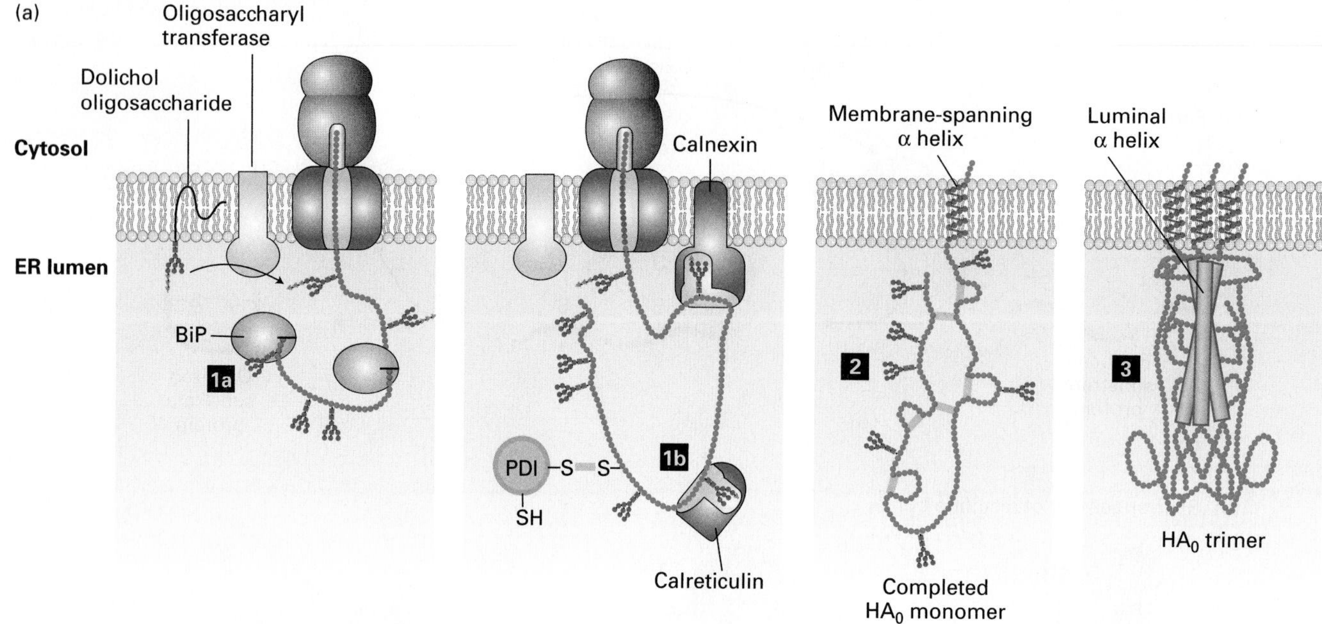

(b)

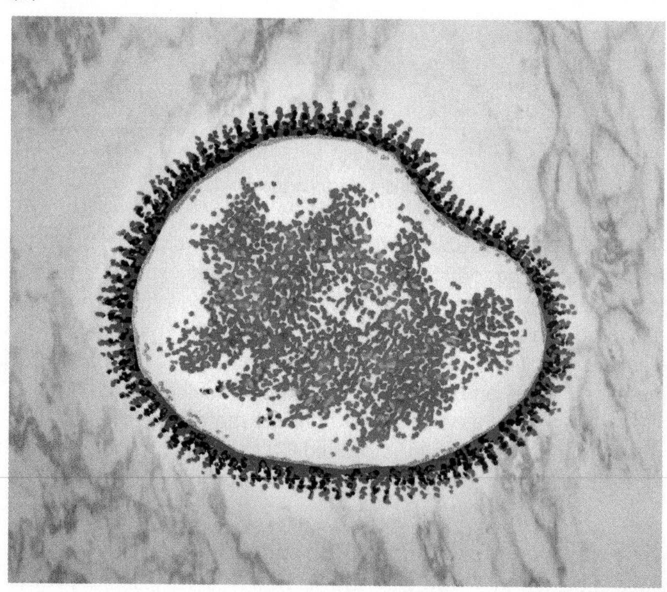

▶ **FIGURE 13-20 Hemagglutinin folding and assembly.** (a) Mechanism of (HA_0) trimer assembly. Transient binding of the chaperone BiP (step **1a**) to the nascent chain and of two lectins, calnexin and calreticulin, to certain oligosaccharide chains (step **1b**) promotes proper folding of adjacent segments. A total of seven *N*-linked oligosaccharide chains are added to the luminal portion of the nascent chain during cotranslational translocation, and PDI catalyzes the formation of six disulfide bonds per monomer. Completed HA_0 monomers are anchored in the membrane by a single membrane-spanning α helix with their N-terminus in the lumen (step **2**). Interaction of three HA_0 chains with one another, initially via their transmembrane α helices, apparently triggers formation of a long stem containing one α helix from the luminal part of each HA_0 polypeptide. Finally, interactions between the three globular heads occur, generating a stable HA_0 trimer (step **3**). (b) Electron micrograph of a complete influenza virion showing trimers of HA protein protruding as spikes from the surface of the viral membrane. [Part (a) See U. Tatu et al., 1995, *EMBO J.* **14:**1340, and D. Hebert et al., 1997, *J. Cell Biol.* **139:**613. Part (b) Chris Bjornberg/Photo Researchers, Inc.]

accelerate the rotation about peptidyl-prolyl bonds at proline residues in unfolded segments of a polypeptide:

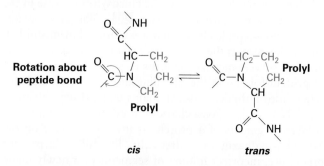

Rotation about peptide bond

Prolyl (*cis*) ⇌ **Prolyl** (*trans*)

Such isomerizations sometimes are the rate-limiting step in the folding of protein domains. Many peptidyl-prolyl isomerases can catalyze the rotation of exposed peptidyl-prolyl bonds indiscriminately in numerous proteins, but some have very specific protein substrates.

Many important secretory and membrane proteins synthesized on the ER are built of two or more polypeptide subunits. In all cases, the assembly of subunits constituting these multisubunit (**multimeric**) proteins occurs in the ER. An important class of multimeric secreted proteins is the immunoglobulins, which contain two heavy (H) and two light (L) chains, all linked by intrachain disulfide bonds. Hemagglutinin (HA) is another multimeric protein that provides a good illustration of folding and subunit assembly (Figure 13-20). This trimeric protein forms the spikes that protrude from the surface of an influenza virus particle. The HA trimer is formed within the ER of an infected host cell from

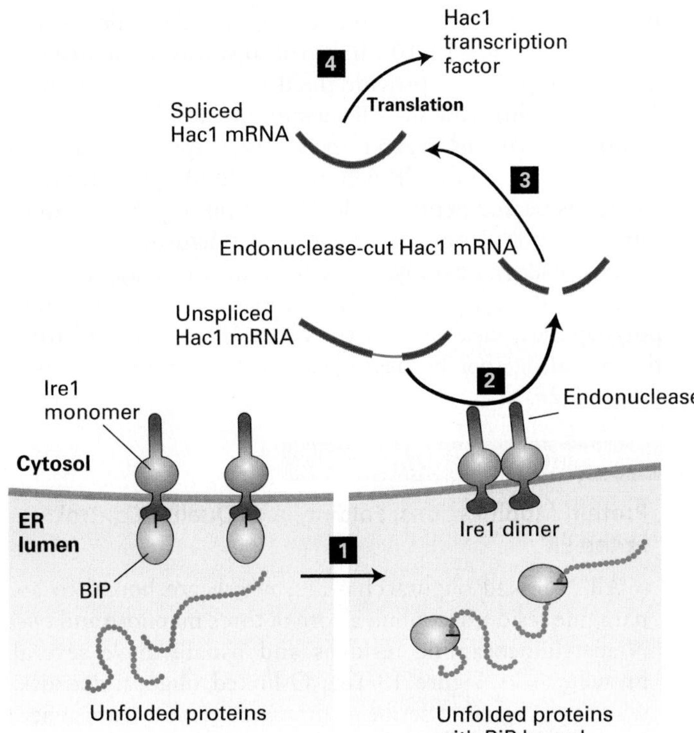

◀ **FIGURE 13-21 The unfolded-protein response.** Ire1, a transmembrane protein in the ER membrane, has a binding site for BiP on its luminal domain; the cytosolic domain contains a specific RNA endonuclease. Step **1**: Accumulating unfolded proteins in the ER lumen bind BiP molecules, releasing them from monomeric Ire1. Dimerization of Ire1 then activates its endonuclease activity. Steps **2**, **3**: The unspliced mRNA precursor encoding the transcription factor Hac1 is cleaved by dimeric Ire1, and the two exons are joined to form functional Hac1 mRNA. Current evidence indicates that this processing occurs in the cytosol, although pre-mRNA processing generally occurs in the nucleus. Step **4**: Hac1 is translated into Hac1 protein, which then moves back into the nucleus and activates transcription of genes encoding several protein-folding catalysts. [See U. Ruegsegger et al., 2001, *Cell* **107:**103; A. Bertolotti et al., 2000, *Nature Cell Biol.* **2:**326; and C. Sidrauski and P. Walter, 1997, *Cell* **90:**1031.]

three copies of a precursor protein termed HA_0, which has a single membrane-spanning α helix. In the Golgi complex, each of the three HA_0 proteins is cleaved to form two polypeptides, HA_1 and HA_2; thus each HA molecule that eventually resides on the viral surface contains three copies of HA_1 and three of HA_2 (see Figure 3-10). The trimer is stabilized by interactions between the large exoplasmic domains of the constituent polypeptides, which extend into the ER lumen; after HA is transported to the cell surface, these domains extend into the extracellular space. Interactions between the smaller cytosolic and membrane-spanning portions of the HA subunits also help stabilize the trimeric protein. Studies have shown that it takes just 10 minutes for the HA_0 polypeptides to fold and assemble into their proper trimeric conformation.

Improperly Folded Proteins in the ER Induce Expression of Protein-Folding Catalysts

Wild-type proteins that are synthesized on the rough ER cannot exit this compartment until they achieve their completely folded conformation. Likewise, almost any mutation that prevents proper folding of a protein in the ER also blocks movement of the polypeptide from the ER lumen or membrane to the Golgi complex. The mechanisms for retaining unfolded or incompletely folded proteins within the ER probably increase the overall efficiency of folding by keeping intermediate forms in proximity to folding catalysts, which are most abundant in the ER. Improperly folded proteins retained within the ER generally are seen bound to the ER chaperones BiP and calnexin. Thus these luminal folding catalysts perform two related functions: assisting in the folding of normal proteins by preventing their aggregation and binding to irreversibly misfolded proteins.

Both mammalian cells and yeasts respond to the presence of unfolded proteins in the rough ER by increasing transcription of several genes encoding ER chaperones and other folding catalysts. A key participant in this *unfolded-protein response* is Ire1, an ER membrane protein that exists as both a monomer and a dimer. The dimeric form, but not the monomeric form, promotes formation of Hac1, a transcription factor in yeast that activates expression of the genes induced in the unfolded-protein response. As depicted in Figure 13-21, binding of BiP to the luminal domain of monomeric Ire1 prevents formation of the Ire1 dimer. Thus the quantity of free BiP in the ER lumen probably determines the relative proportion of monomeric and dimeric Ire1. Accumulation of unfolded proteins within the ER lumen sequesters BiP molecules, making them unavailable for binding to Ire1. As a result the level of dimeric Ire1 increases, leading to an increase in the level of Hac1 and production of proteins that assist in protein folding.

Mammalian cells contain an additional regulatory pathway that operates in response to unfolded proteins in the ER. In this pathway, accumulation of unfolded proteins in the ER triggers proteolysis of ATF6, a transmembrane protein in the ER membrane, at a site within the membrane-spanning segment. The cytosolic domain of ATF6 released by proteolysis then moves to the nucleus, where it stimulates transcription of the genes encoding ER chaperones. Activation of a transcription factor by such *regulated intramembrane proteolysis* also occurs in the Notch signaling pathway and during activation of the cholesterol-responsive transcription factor SREBP (see Figures 16-36 and 16-38).

A hereditary form of emphysema illustrates the detrimental effects that can result from misfolding of proteins in the ER. This disease is caused by a point mutation in α_1-antitrypsin, which normally is secreted by hepatocytes and macrophages. The wild-type protein binds to and inhibits trypsin and also the blood protease elastase. In the absence of α_1-antitrypsin, elastase degrades the fine tissue in the lung that participates in the absorption of oxygen, eventually producing the symptoms of emphysema. Although the mutant α_1-antitrypsin is synthesized in the rough ER, it does not fold properly, forming an almost crystalline aggregate that is not exported from the ER. In hepatocytes, the secretion of other proteins also becomes impaired as the rough ER is filled with aggregated α_1-antitrypsin. ∎

Unassembled or Misfolded Proteins in the ER Are Often Transported to the Cytosol for Degradation

Misfolded secretory and membrane proteins, as well as the unassembled subunits of multimeric proteins, often are degraded within an hour or two after their synthesis in the rough ER. For many years, researchers thought that proteolytic enzymes within the ER lumen catalyzed degradation of misfolded or unassembled polypeptides, but such proteases were never found. More recent studies have shown that misfolded membrane and secretory proteins are recognized by specific ER membrane proteins and are targeted for transport from the ER lumen into the cytosol, by a process known as *dislocation* or *retrotranslocation*.

The dislocation of misfolded proteins out of the ER depends on a set of proteins located in the ER membrane and in the cytosol that perform three basic functions. The first function is recognition of misfolded proteins that will be substrates for the dislocation reaction. The recognition takes place in the lumen of the ER and in some cases may involve binding of BiP to the unfolded protein. However, the details of how misfolded proteins are recognized are not well understood. In particular, it is not known how proteins that cannot fold properly, and are thus legitimate substrates for the dislocation process, are distinguished from normal proteins that have transient partially folded states as they acquire their fully folded conformation.

Once an unfolded protein has been identified, it is targeted for dislocation across the ER membrane. Some kind of channel must exist for dislocation of misfolded proteins across the ER membrane. The Sec61 translocon was thought to participate in the dislocation reaction, but current evidence suggests that dislocated polypeptides do not actually traverse "backward" through the channel in the Sec61 complex.

As segments of the dislocated polypeptide are exposed to the cytosol, they encounter cytosolic enzymes that drive retrotranslocation. One of these enzymes is an ATPase called p97, a member of a protein family known as the **AAA ATPase family.** Other members of the AAA ATPase family are known to couple the energy of ATP hydrolysis to disassembly of protein complexes. For example, in Chapter 14 we will encounter an AAA ATPase family member called NSF that is an important component of the secretory pathway, using the energy of ATP hydrolysis to disassemble protein complexes generated during vesicle budding and fusion (see Figure 14-10). In retrotranslocation, hydrolysis of ATP by p97 may provide the driving force to pull misfolded proteins from the ER membrane into the cytosol. As misfolded proteins reenter the cytosol, specific ubiquitin ligase enzymes in the ER membrane add ubiquitin residues to the dislocated peptide. Like the action of p97, the ubiquitinylation reaction is coupled to ATP hydrolysis; this release of energy possibly also contributes to trapping proteins in the cytosol. The resulting polyubiquitinylated polypeptides, now fully in the cytosol, are removed from the cell altogether by degradation in the proteasome (see Figure 3-29).

KEY CONCEPTS OF SECTION 13.3

Protein Modifications, Folding, and Quality Control in the ER

- All *N*-linked oligosaccharides, which are bound to asparagine residues, contain a core of three mannose and two *N*-acetylglucosamine residues and usually have several branches (see Figure 13-16). *O*-linked oligosaccharides, which are bound to serine or threonine residues, are generally short, often containing only one to four sugar residues.
- Formation of all *N*-linked oligosaccharides begins with assembly of a conserved 14-residue high-mannose precursor on dolichol, a lipid in the membrane of the rough ER (see Figure 13-17). After this preformed oligosaccharide is transferred to specific asparagine residues of nascent polypeptide chains in the ER lumen, three glucose residues and one mannose residue are removed (see Figure 13-18).
- Oligosaccharide side chains may assist in the proper folding of glycoproteins, help protect the mature proteins from proteolysis, participate in cell-cell adhesion, and function as antigens.
- Disulfide bonds are added to many secretory proteins and the exoplasmic domain of membrane proteins in the ER. Protein disulfide isomerase (PDI), present in the ER lumen, catalyzes both the formation and the rearrangement of disulfide bonds (see Figure 13-19).
- The chaperone BiP, the lectins calnexin and calreticulin, and peptidyl-prolyl isomerases work together to ensure proper folding of newly made secretory and membrane proteins in the ER. The subunits of multimeric proteins also assemble in the ER.
- Only properly folded proteins and assembled subunits are transported from the rough ER to the Golgi complex in vesicles.
- The accumulation of abnormally folded proteins and unassembled subunits in the ER can induce increased expression of ER protein-folding catalysts via the unfolded-protein response (see Figure 13-21).
- Unassembled or misfolded proteins in the ER often are transported back to the cytosol, where they are degraded in the ubiquitin/proteasome pathway.

13.4 Sorting of Proteins to Mitochondria and Chloroplasts

In the remainder of this chapter, we examine how proteins synthesized on cytosolic ribosomes are sorted to mitochondria, chloroplasts, peroxisomes, and the nucleus (see Figure 13-1). In both mitochondria and chloroplasts an internal lumen called the *matrix* is surrounded by a double membrane, and internal subcompartments exist within the matrix. In contrast, peroxisomes are bounded by a single membrane and have a single luminal matrix compartment. Because of these and other differences, we consider peroxisomes separately in the next section. Likewise, the mechanism of protein transport into and out of the nucleus differs considerably from sorting to mitochondria and chloroplasts; this is discussed in the last section.

In addition to being bounded by two membranes, mitochondria and chloroplasts share similar types of electron transport proteins and use an F-class ATPase to synthesize ATP (see Figure 12-2). Remarkably, gram-negative bacteria also exhibit these characteristics. Also like bacterial cells, mitochondria and chloroplasts contain their own DNA, which encodes organelle rRNAs, tRNAs, and some proteins (Chapter 6). Moreover, growth and division of mitochondria and chloroplasts are not coupled to nuclear division. Rather, these organelles grow by the incorporation of cellular proteins and lipids, and new organelles form by division of preexisting organelles. The numerous similarities of free-living bacterial cells with mitochondria and chloroplasts have led scientists to hypothesize that these organelles arose by the incorporation of bacteria into ancestral eukaryotic cells, forming endosymbiontic organelles (see Figure 6-20). The sequence similarity of many membrane translocation proteins shared by mitochondria, chloroplasts, and bacteria provides the most striking evidence for this ancient evolutionary relationship. In this section we will examine these membrane translocation proteins in detail.

Proteins encoded by mitochondrial DNA or chloroplast DNA are synthesized on ribosomes within the organelles and directed to the correct subcompartment immediately after synthesis. The majority of proteins located in mitochondria and chloroplasts, however, are encoded by genes in the nucleus and are imported into the organelles after their synthesis in the cytosol. Apparently, over aeons of evolution, much of the genetic information from the ancestral bacterial DNA in these endosymbiotic organelles moved, by an unknown mechanism, to the nucleus. Precursor proteins synthesized in the cytosol that are destined for the matrix of mitochondria or the equivalent space in chloroplasts, the stroma, usually contain specific N-terminal uptake-targeting sequences that specify binding to receptor proteins on the organelle surface. Generally, this sequence is cleaved once it reaches the matrix or stroma. Clearly, these uptake-targeting sequences are similar in their location and general function to the signal sequences that direct nascent proteins to the ER lumen. Although the three types of signals share some common sequence features, their specific sequences differ considerably, as summarized in Table 13-1.

In both mitochondria and chloroplasts, protein import requires energy and occurs at points where the outer and inner organelle membranes are in close contact. Because mitochondria and chloroplasts contain multiple membranes and

TABLE 13-1 Uptake-Targeting Sequences That Direct Proteins from the Cytosol to Organelles*

TARGET ORGANELLE	LOCATION OF SEQUENCE WITHIN PROTEIN	REMOVAL OF SEQUENCE	NATURE OF SEQUENCE
Endoplasmic reticulum (lumen)	N-terminus	Yes	Core of 6–12 hydrophobic amino acids, often preceded by one or more basic amino acids (Arg, Lys)
Mitochondrion (matrix)	N-terminus	Yes	Amphipathic helix, 20–50 residues in length, with Arg and Lys residues on one side and hydrophobic residues on the other
Chloroplast (stroma)	N-terminus	Yes	No common motifs; generally rich in Ser, Thr, and small hydrophobic residues and poor in Glu and Asp
Peroxisome (matrix)	C-terminus (most proteins); N-terminus (few proteins)	No	PTS1 signal (Ser-Lys-Leu) at extreme C-terminus; PTS2 signal at N-terminus
Nucleus (nucleoplasm)	Varies	No	Multiple different kinds; a common motif includes a short segment rich in Lys and Arg residues

*Different or additional sequences target proteins to organelle membranes and subcompartments.

▶ **EXPERIMENTAL FIGURE 13-22 Import of mitochondrial precursor proteins is assayed in a cell-free system.** Inside mitochondria, proteins are protected from the action of proteases such as trypsin. When no mitochondria are present, mitochondrial proteins synthesized in the cytosol are degraded by added protease. Protein uptake occurs only with energized (respiring) mitochondria, which have a proton electrochemical gradient (proton-motive force) across the inner membrane. The imported protein must contain an appropriate uptake-targeting sequence. Uptake also requires ATP and a cytosolic extract containing chaperone proteins that maintain the precursor proteins in an unfolded conformation. This assay has been used to study targeting sequences and other features of the translocation process.

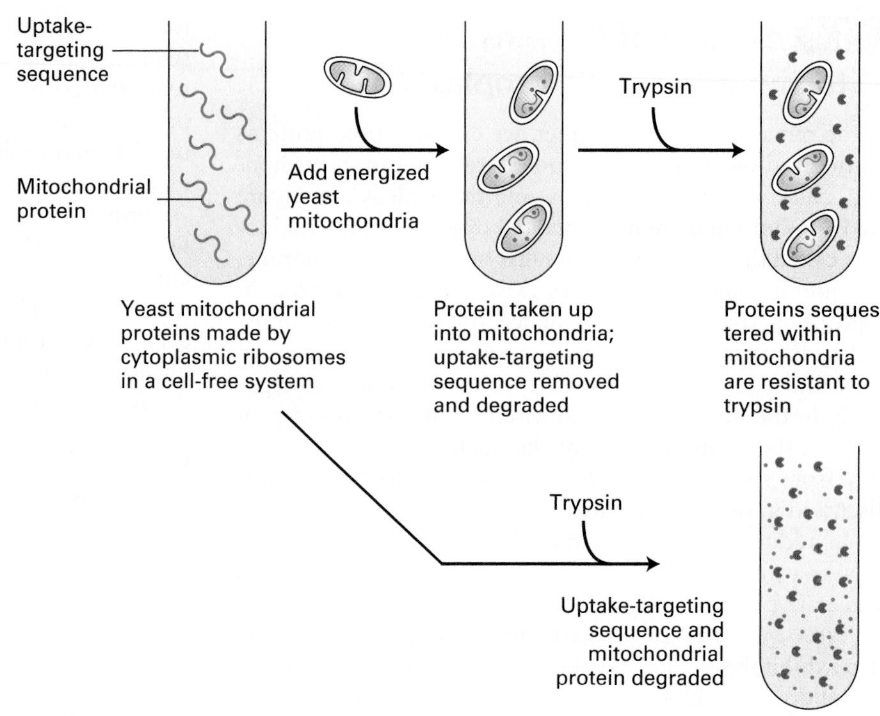
Uptake-targeting sequence
Mitochondrial protein
Add energized yeast mitochondria
Trypsin
Yeast mitochondrial proteins made by cytoplasmic ribosomes in a cell-free system
Protein taken up into mitochondria; uptake-targeting sequence removed and degraded
Proteins sequestered within mitochondria are resistant to trypsin
Trypsin
Uptake-targeting sequence and mitochondrial protein degraded

membrane-limited spaces, sorting of many proteins to their correct location often requires the sequential action of two targeting sequences and two membrane-bound translocation systems: one to direct the protein into the organelle and the other to direct it into the correct organellar compartment or membrane. As we will see, the mechanisms for sorting various proteins to mitochondria and chloroplasts are related to some of the mechanisms discussed previously.

Amphipathic N-Terminal Signal Sequences Direct Proteins to the Mitochondrial Matrix

All proteins that travel from the cytosol to the same mitochondrial destination have targeting signals that share common motifs, although the signal sequences are generally not identical. Thus the receptors that recognize such signals are able to bind to a number of different but related sequences. The most extensively studied sequences for localizing proteins to mitochondria are the *matrix-targeting sequences.* These sequences, located at the N-terminus, are usually 20–50 amino acids in length. They are rich in hydrophobic amino acids, positively charged basic amino acids (arginine and lysine), and hydroxylated ones (serine and threonine) but tend to lack negatively charged acidic residues (aspartate and glutamate).

Mitochondrial matrix-targeting sequences are thought to assume an α-helical conformation in which positively charged amino acids predominate on one side of the helix and hydrophobic amino acids predominate on the other side. Sequences such as these that contain both hydrophobic and hydrophilic regions are said to be **amphipathic.** Mutations that disrupt this amphipathic character usually disrupt targeting to the matrix, although many other amino acid substitutions do not. These findings indicate that the amphipathicity of matrix-targeting sequences is critical to their function.

The cell-free assay outlined in Figure 13-22 has been widely used in studies on the import of mitochondrial precursor proteins. In this system, respiring (energized) mitochondria extracted from cells can incorporate mitochondrial precursor proteins carrying appropriate uptake-targeting sequences that have been synthesized in the absence of mitochondria. Successful incorporation of the precursor into the organelle can be assayed either by resistance to digestion by an added protease such as trypsin or, in most cases, by cleavage of the N-terminal targeting sequences by specific mitochondrial proteases. The uptake of completely presynthesized mitochondrial precursor proteins by the organelle in this system contrasts with the cell-free cotranslational translocation of secretory proteins into the ER, which generally occurs only when microsomal (ER-derived) membranes are present during synthesis (see Figure 13-4).

Mitochondrial Protein Import Requires Outer-Membrane Receptors and Translocons in Both Membranes

Figure 13-23 presents an overview of protein import from the cytosol into the mitochondrial matrix, the route into the mitochondrion followed by most imported proteins. We will discuss in detail each step of protein transport into the matrix and then consider how some proteins subsequently are targeted to other compartments of the mitochondrion.

After synthesis in the cytosol, the soluble precursors of mitochondrial proteins (including hydrophobic integral membrane proteins) interact directly with the mitochondrial membrane. In general, only unfolded proteins can be imported into the mitochondrion. Chaperone proteins such as

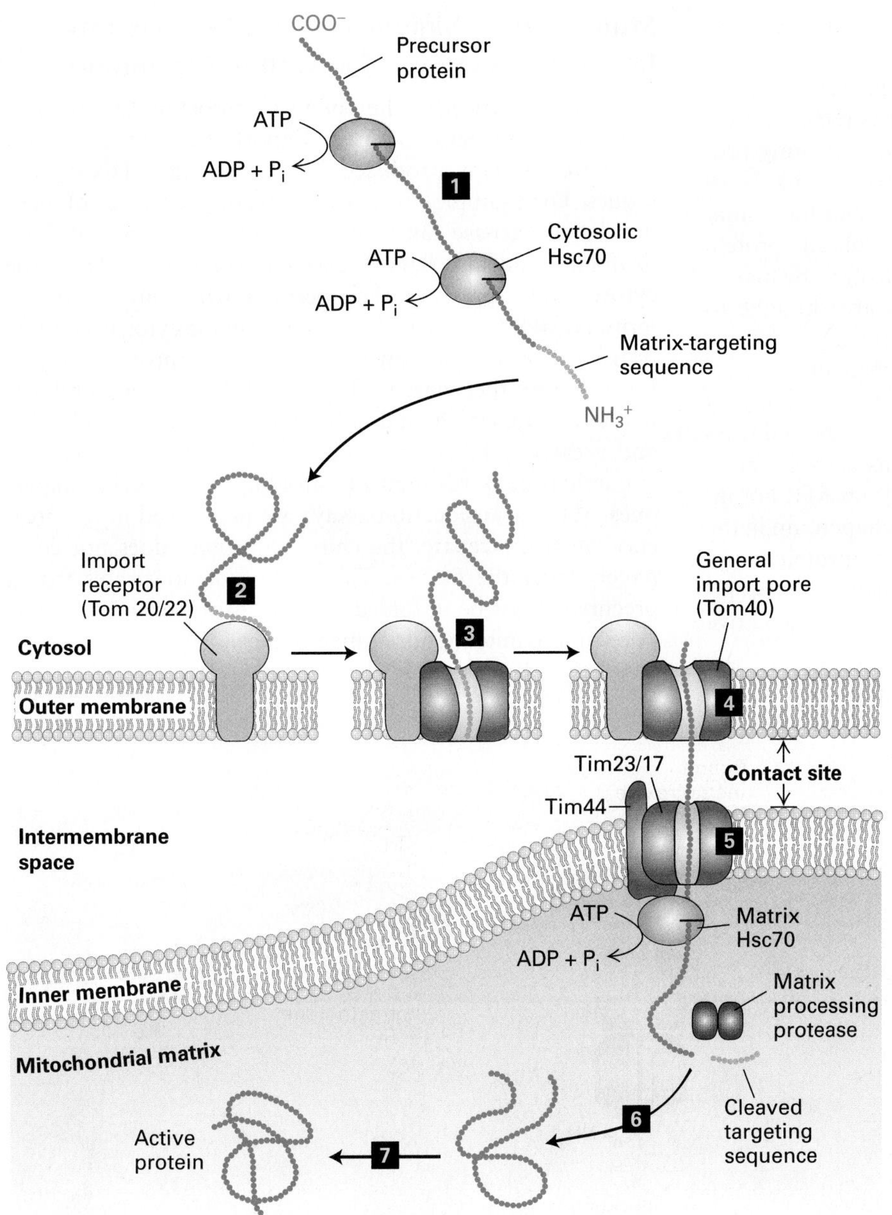

◀ **FIGURE 13-23 Protein import into the mitochondrial matrix.** Precursor proteins synthesized on cytosolic ribosomes are maintained in an unfolded or partially folded state by bound chaperones, such as Hsc70 (step **1**). After a precursor protein binds to an import receptor near a site of contact with the inner membrane (step **2**), it is transferred into the general import pore (step **3**). The translocating protein then moves through this channel and an adjacent channel in the inner membrane (steps **4**, **5**). Note that translocation occurs at rare "contact sites" at which the inner and outer membranes appear to touch. Binding of the translocating protein by the matrix chaperone Hsc70 and subsequent ATP hydrolysis by Hsc70 helps drive import into the matrix. Once the uptake-targeting sequence is removed by a matrix protease and Hsc70 is released from the newly imported protein (step **6**), it folds into the mature, active conformation within the matrix (step **7**). Folding of some proteins depends on matrix chaperonins. [See G. Schatz, 1996, *J. Biol. Chem.* **271:**31763, and N. Pfanner et al., 1997, *Ann. Rev. Cell Devel. Biol.* **13:**25.]

cytosolic Hsc70 keep nascent and newly made proteins in an unfolded state so that they can be taken up by mitochondria. This process requires ATP hydrolysis. Import of an unfolded mitochondrial precursor is initiated by the binding of a mitochondrial targeting sequence to an *import receptor* in the outer mitochondrial membrane. These receptors were first identified by experiments in which antibodies to specific proteins of the outer mitochondrial membrane were shown to inhibit protein import into isolated mitochondria. Subsequent genetic experiments, in which the genes for specific mitochondrial outer-membrane proteins were mutated, showed that specific receptor proteins were responsible for the import of different classes of mitochondrial proteins. For example, N-terminal matrix-targeting sequences are recognized by Tom20 and Tom22. (Proteins in the outer mitochondrial membrane involved in targeting and import are designated *Tom* proteins for *t*ranslocon of the *o*uter *m*embrane.)

The import receptors subsequently transfer the precursor proteins to an import channel in the outer membrane. This channel, composed mainly of the Tom40 protein, is known as the *general import pore* because all known mitochondrial precursor proteins gain access to the interior compartments of the mitochondrion through this channel. When purified and incorporated into liposomes, Tom40 forms a transmembrane channel with a pore wide enough to accommodate an unfolded polypeptide chain. The general import pore forms a largely passive channel through the outer mitochondrial membrane, and the driving force for unidirectional transport into mitochondria comes from within the mitochondrion. In the case of precursors destined for the mitochondrial matrix, transfer through the outer membrane occurs simultaneously with transfer through an inner-membrane channel composed of the Tim23 and Tim17 proteins. (*Tim* stands for *t*ranslocon of the *i*nner *m*embrane.) Translocation into the matrix thus occurs at "contact sites" where the outer and inner membranes are in close proximity.

Soon after the N-terminal matrix-targeting sequence of a protein enters the mitochondrial matrix, it is removed by a protease that resides within the matrix. The emerging protein also is bound by matrix Hsc70, a chaperone that is localized to the translocation channels in the inner mitochondrial membrane by interacting with transmembrane protein Tim44. This interaction stimulates ATP hydrolysis by matrix Hsc70, and together these two proteins are thought to power translocation of proteins into the matrix.

Some imported proteins can fold into their final, active conformation without further assistance. Final folding of many matrix proteins, however, requires a **chaperonin.** As discussed in Chapter 3, chaperonin proteins actively facilitate protein folding in a process that depends on ATP. For instance, yeast mutants defective in Hsc60, a chaperonin in the mitochondrial matrix, can import matrix proteins and cleave their uptake-targeting sequence normally, but the imported polypeptides fail to fold and assemble into the native tertiary and quaternary structures.

Studies with Chimeric Proteins Demonstrate Important Features of Mitochondrial Import

Dramatic evidence for the ability of mitochondrial matrix-targeting sequences to direct import was obtained with chimeric proteins produced by recombinant DNA techniques. For example, the matrix-targeting sequence of alcohol dehydrogenase can be fused to the N-terminus of dihydrofolate reductase (DHFR), which normally resides in the cytosol. In the presence of chaperones, which prevent the C-terminal DHFR segment from folding in the cytosol, cell-free translocation assays show that the chimeric protein is transported into the matrix (Figure 13-24a). The inhibitor methotrexate, which binds tightly to the active site of DHFR and greatly stabilizes its folded conformation, renders the chimeric protein resistant to unfolding by cytosolic chaperones. When translocation assays are performed in the presence of methotrexate, the chimeric protein does not completely enter the matrix. This finding demonstrates that a precursor must be unfolded in order to traverse the import pores in the mitochondrial membranes.

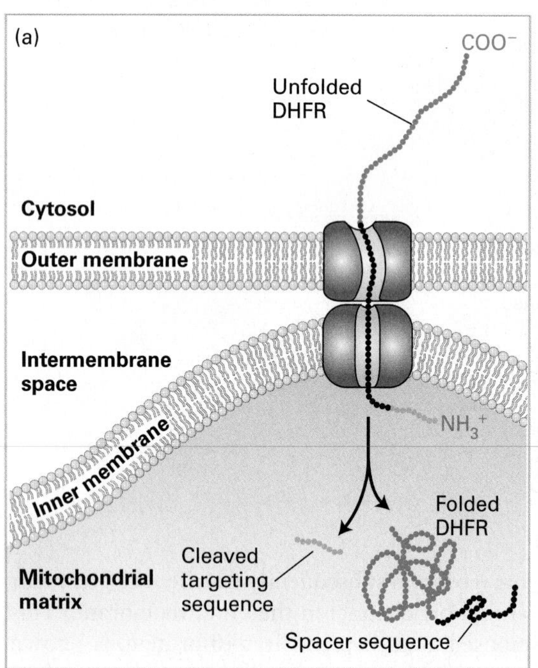

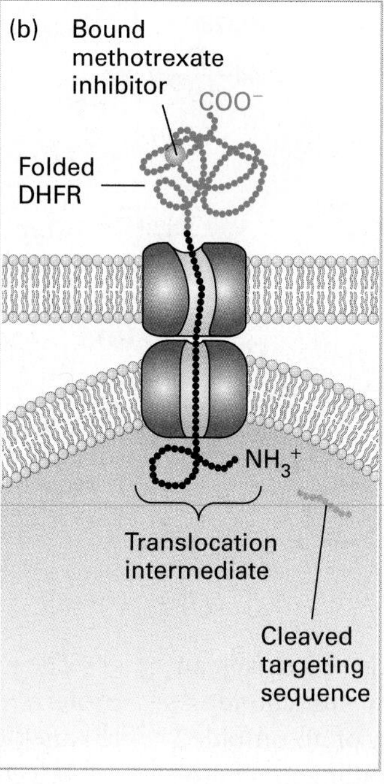

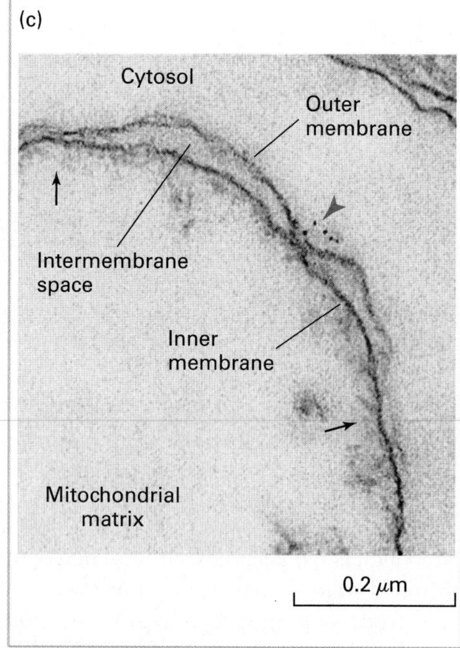

▲ **EXPERIMENTAL FIGURE 13-24 Experiments with chimeric proteins elucidate mitochondrial protein import.** These experiments show that a matrix-targeting sequence alone directs proteins to the mitochondrial matrix and that only unfolded proteins are translocated across both membranes. The chimeric protein in these experiments contained a matrix-targeting signal at its N-terminus (red), followed by a spacer sequence of no particular function (black) and then by dihydrofolate reductase (DHFR), an enzyme normally present only in the cytosol. (a) When the DHFR segment is unfolded, the chimeric protein moves across both membranes to the matrix of energized mitochondria and the matrix-targeting signal then is removed. (b) When the C-terminus of the chimeric protein is locked in the folded state by binding of methotrexate, translocation is blocked. If the spacer sequence is long enough to extend across both transport channels, a stable translocation intermediate, with the targeting sequence cleaved off, is generated in the presence of methotrexate, as shown here. (c) The C-terminus of the translocation intermediate in (b) can be detected by incubating the mitochondria with antibodies that bind to the DHFR segment, followed by gold particles coated with bacterial protein A, which binds nonspecifically to antibody molecules (see Figure 9-21). An electron micrograph of a sectioned sample reveals gold particles (red arrowhead) bound to the translocation intermediate at a contact site between the inner and outer membranes. Other contact sites (black arrows) also are evident. [Parts (a) and (b) adapted from J. Rassow et al., 1990, *FEBS Letters* **275:**190. Part (c) from M. Schweiger et al., 1987, *J. Cell Biol.* **105:**235, courtesy of W. Neupert.]

Additional studies revealed that if a sufficiently long spacer sequence separates the N-terminal matrix-targeting sequence and DHFR portion of the chimeric protein, then in the presence of methotrexate a translocation intermediate that spans both membranes can be trapped if enough of the polypeptide protrudes into the matrix to prevent the polypeptide chain from sliding back into the cytosol, possibly by stably associating with matrix Hsc70 (Figure 13-24b). In order for such a stable translocation intermediate to form, the spacer sequence must be long enough to span both membranes; a spacer of 50 amino acids extended to its maximum possible length is adequate to do so. If the chimera contains a shorter spacer—say, 35 amino acids—no stable translocation intermediate is obtained because the spacer cannot span both membranes. These observations provide further evidence that translocated proteins can span both inner and outer mitochondrial membranes and traverse these membranes in an unfolded state.

Microscopic studies of stable translocation intermediates show that they accumulate at sites where the inner and outer mitochondrial membranes are close together, evidence that precursor proteins enter only at such sites (Figure 13-24c). The distance from the cytosolic face of the outer membrane to the matrix face of the inner membrane at these *contact sites* is consistent with the length of an unfolded spacer sequence required for formation of a stable translocation intermediate. Moreover, stable translocation intermediates can be chemically cross-linked to the protein subunits that comprise the translocation channels of both the outer and inner membranes. This finding demonstrates that imported proteins can simultaneously engage channels in both the outer and inner mitochondrial membrane, as depicted in Figure 13-23. Since roughly 1000 stuck chimeric proteins can be observed in a typical yeast mitochondrion, it is thought that mitochondria have approximately 1000 general import pores for the uptake of mitochondrial proteins.

Three Energy Inputs Are Needed to Import Proteins into Mitochondria

As noted previously and indicated in Figure 13-23, ATP hydrolysis by Hsc70 chaperone proteins in both the cytosol and the mitochondrial matrix is required for import of mitochondrial proteins. Cytosolic Hsc70 expends energy to maintain bound precursor proteins in an unfolded state that is competent for translocation into the matrix. The importance of ATP to this function was demonstrated in studies in which a mitochondrial precursor protein was purified and then denatured (unfolded) by urea. When tested in the cell-free mitochondrial translocation system, the denatured protein was incorporated into the matrix in the absence of ATP. In contrast, import of the native, undenatured precursor required ATP for the normal unfolding function of cytosolic chaperones.

The sequential binding and ATP-driven release of multiple matrix Hsc70 molecules to a translocating protein may simply trap the unfolded protein in the matrix. Alternatively, the matrix Hsc70, anchored to the membrane by the Tim44 protein, may act as a molecular motor to pull the protein into the matrix (see Figure 13-23). In this case, the functions of matrix Hsc70 and Tim44 would be analogous to those of the chaperone BiP and Sec63 complex, respectively, in post-translational translocation into the ER lumen (see Figure 13-9).

The third energy input required for mitochondrial protein import is a H^+ electrochemical gradient, or **proton-motive force,** across the inner membrane. Recall from Chapter 12 that protons are pumped from the matrix into the intermembrane space during electron transport, creating transmembrane potential across the inner membrane. In general, only mitochondria that are actively undergoing respiration, and therefore have generated a proton-motive force across the inner membrane, are able to translocate precursor proteins from the cytosol into the mitochondrial matrix. Treatment of mitochondria with inhibitors or uncouplers of oxidative phosphorylation, such as cyanide or dinitrophenol, dissipates this proton-motive force. Although precursor proteins still can bind tightly to receptors on such poisoned mitochondria, the proteins cannot be imported, either in intact cells or in cell-free systems, even in the presence of ATP and chaperone proteins. Scientists do not fully understand how the proton-motive force is used to facilitate entry of a precursor protein into the matrix. Once a protein is partially inserted into the inner membrane, it is subjected to a transmembrane potential of 200 mV (matrix space negative). This seemingly small potential difference is established across the very narrow hydrophobic core of the lipid bilayer, which gives an enormous electric gradient, equivalent to about 400,000 V/cm. One hypothesis is that the positive charges in the amphipathic matrix-targeting sequence could simply be "electrophoresed," or pulled, into the matrix space by the inside-negative membrane electric potential.

Multiple Signals and Pathways Target Proteins to Submitochondrial Compartments

Unlike targeting to the matrix, targeting of proteins to the intermembrane space, inner membrane, and outer membrane of mitochondria generally requires more than one targeting sequence and occurs via one of several pathways. Figure 13-25 summarizes the organization of targeting sequences in proteins sorted to different mitochondrial locations.

Inner-Membrane Proteins Three separate pathways are known to target proteins to the inner mitochondrial membrane. One pathway makes use of the same machinery that is used for targeting of matrix proteins (Figure 13-26, path A). A cytochrome oxidase subunit called CoxVa is a protein transported by this pathway. The precursor form of CoxVa, which contains an N-terminal matrix-targeting sequence recognized by the Tom20/22 import receptor, is transferred through the general import pore of the outer membrane and the inner-membrane Tim23/17 translocation complex. In addition to the matrix-targeting sequence, which is cleaved during import, CoxVa contains a hydrophobic stop-transfer sequence. As the protein passes through the Tim23/17 channel, the stop-transfer sequence blocks translocation of the

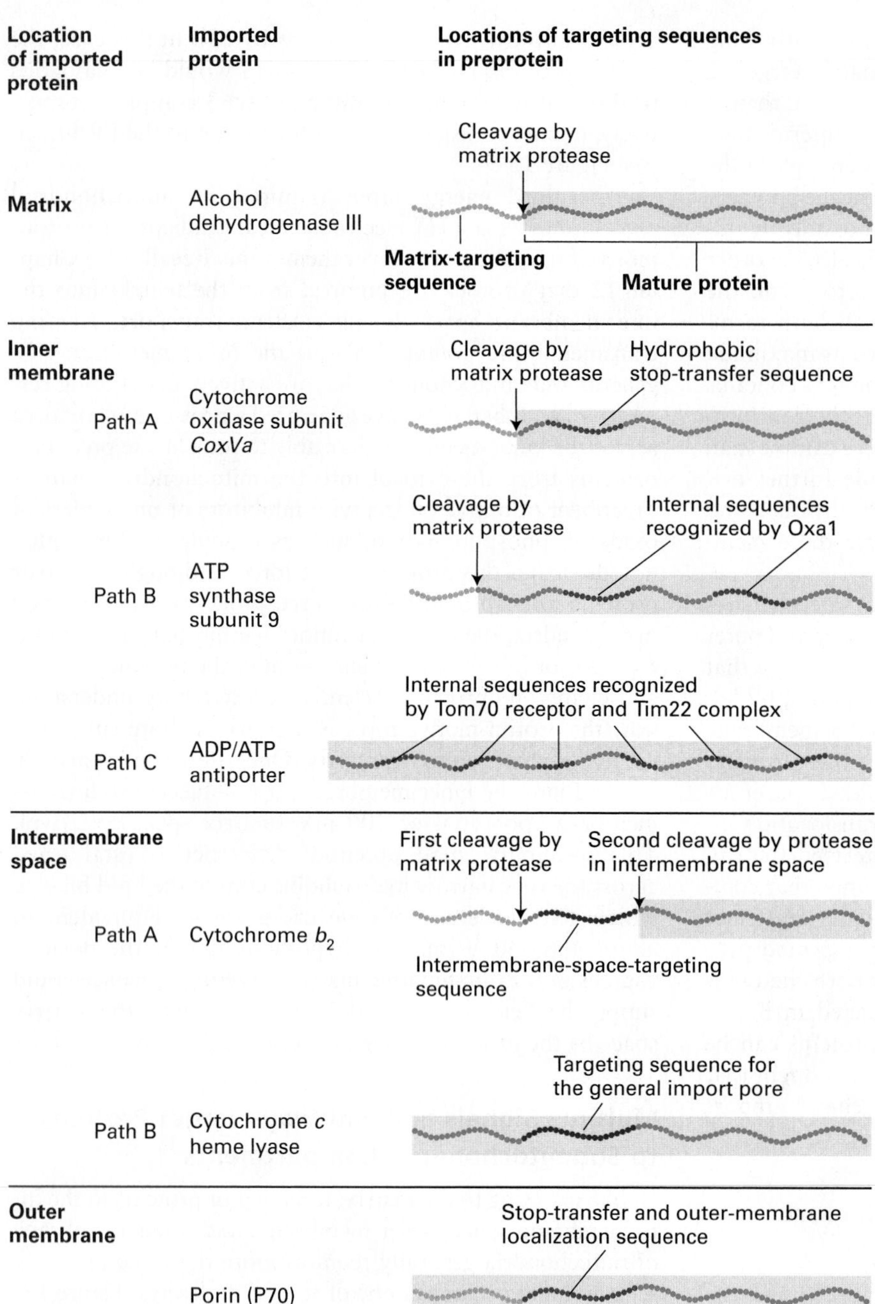

◀ **FIGURE 13-25 Targeting sequences in imported mitochondrial proteins.** Most mitochondrial proteins have an N-terminal matrix-targeting sequence (pink) that is similar but not identical in different proteins. Proteins destined for the inner membrane, the intermembrane space, or the outer membrane have one or more additional targeting sequences that function to direct the proteins to these locations by several different pathways. The lettered pathways correspond to those illustrated in Figures 13-26 and 13-27. [See W. Neupert, 1997, *Ann. Rev. Biochem.* **66**:863.]

C-terminus across the inner membrane. The membrane-anchored intermediate is then transferred laterally into the bilayer of the inner membrane much as type I integral membrane proteins are incorporated into the ER membrane (see Figure 13-11).

A second pathway to the inner membrane is followed by proteins (e.g., ATP synthase subunit 9) whose precursors contain both a matrix-targeting sequence and internal hydrophobic domains recognized by an inner-membrane protein termed *Oxa1*. This pathway is thought to involve translocation of at least a portion of the precursor into the matrix via the Tom40 and Tim23/17 channels. After cleavage of the matrix-targeting sequence, the protein is inserted into the inner membrane by a process that requires interaction with Oxa1 and perhaps other inner-membrane proteins (Figure 13-26, path B). Oxa1 is related to a bacterial protein involved in inserting some inner-membrane proteins in bacteria. This relatedness suggests that Oxa1 may have descended from the translocation machinery in the endosymbiotic bacterium that eventually became the mitochondrion. However, the proteins forming the inner-membrane channels in mitochondria are not related to the proteins in bacterial translocons. Oxa1 also participates in the inner-membrane insertion of certain proteins (e.g., subunit II of cytochrome oxidase) that are encoded by mitochondrial DNA and synthesized in the matrix by mitochondrial ribosomes.

The final pathway for insertion in the inner mitochondrial membrane is followed by multipass proteins that

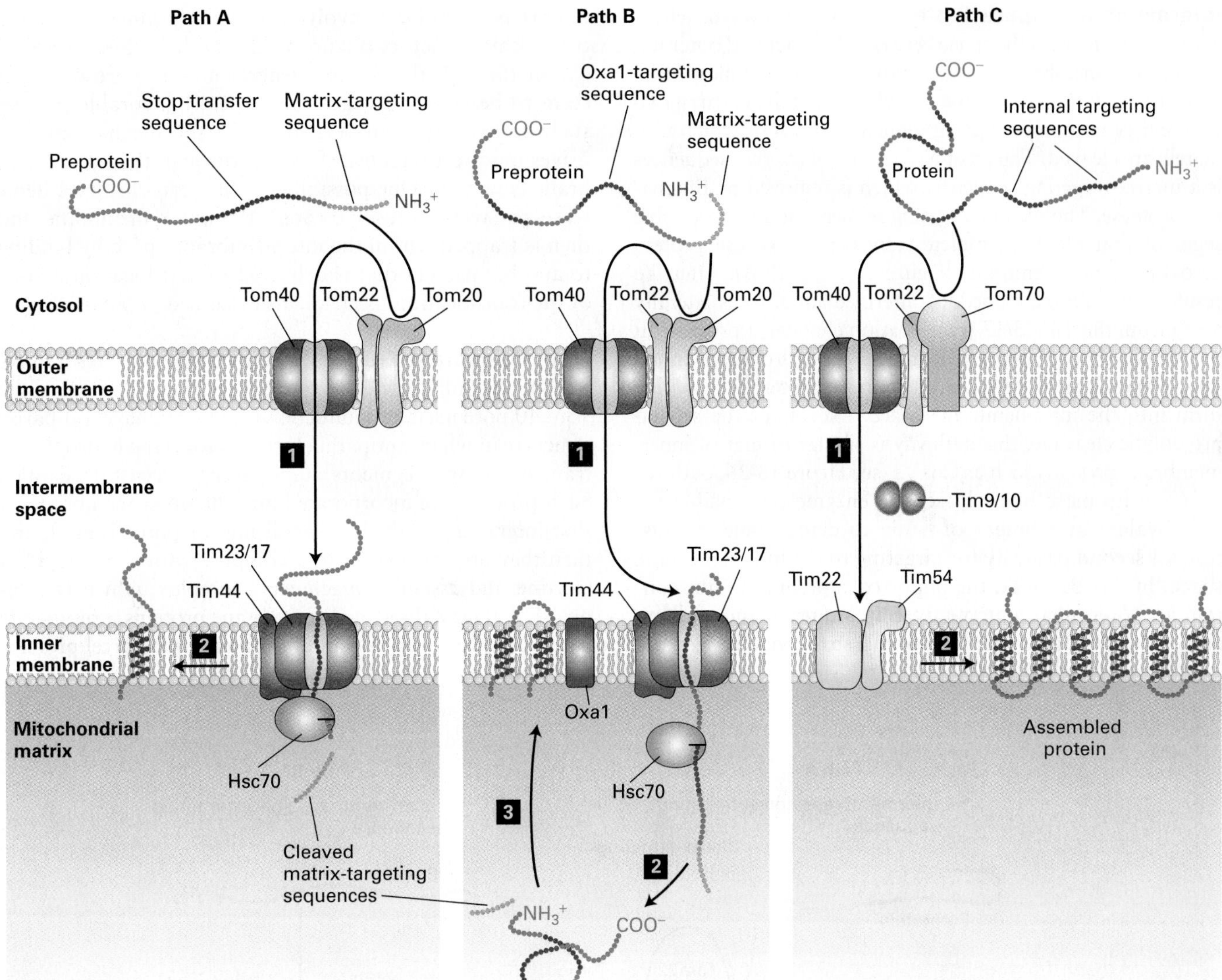

▲ **FIGURE 13-26 Three pathways to the inner mitochondrial membrane from the cytosol.** Proteins with different targeting sequences are directed to the inner membrane via different pathways. In all three pathways, proteins cross the outer membrane via the Tom40 general import pore. Proteins delivered by pathways A and B contain an N-terminal matrix-targeting sequence that is recognized by the Tom20/22 import receptor in the outer membrane. Although both these pathways use the Tim23/17 inner-membrane channel, they differ in that the entire precursor protein enters the matrix and then is redirected to the inner membrane in pathway B. Matrix Hsc70 plays a role similar to its role in the import of soluble matrix proteins (see Figure 13-23). Proteins delivered by pathway C contain internal sequences that are recognized by the Tom70/Tom22 import receptor; a different inner-membrane translocation channel (Tim22/54) is used in this pathway. Two intermembrane proteins (Tim9 and Tim10) facilitate transfer between the outer and inner channels. See the text for discussion. [See R. E. Dalbey and A. Kuhn, 2000, *Ann. Rev. Cell Devel. Biol.* **16:**51, and N. Pfanner and A. Geissler, 2001, *Nature Rev. Mol. Cell Biol.* **2:**339.]

contain six membrane-spanning domains, such as the ADP/ATP antiporter. These proteins, which lack the usual N-terminal matrix-targeting sequence, contain multiple internal mitochondrial targeting sequences. After the internal sequences are recognized by a second import receptor composed of outer-membrane proteins Tom70 and Tom22, the imported protein passes through the outer membrane via the general import pore (Figure 13-26, path C). The protein then is transferred to a second translocation complex in the inner membrane composed of the Tim22 and Tim54 proteins. Transfer to the Tim22/54 complex depends on a multimeric complex of two small proteins, Tim9 and Tim10, that reside in the intermembrane space. The small Tim proteins are thought to act as chaperones, guiding imported protein precursors from the general import pore to the Tim22/54 complex in the inner membrane by binding to their hydrophobic regions, preventing them from forming insoluble aggregates in the aqueous environment of the intermembrane space. Ultimately the Tim22/54 complex is responsible for incorporating the multiple hydrophobic segments of the imported protein into the inner membrane.

Intermembrane-Space Proteins Two pathways deliver cytosolic proteins to the space between the inner and outer mitochondrial membranes. The major pathway is followed by proteins, such as cytochrome b_2, whose precursors carry two different N-terminal targeting sequences, both of which ultimately are cleaved. The most N-terminal of the two sequences is a matrix-targeting sequence, which is removed by the matrix protease. The second targeting sequence is a hydrophobic segment that blocks complete translocation of the protein across the inner membrane (Figure 13-27, path A). After the resulting membrane-embedded intermediate diffuses laterally away from the Tim23/17 translocation channel, a protease in the membrane cleaves the protein near the hydrophobic transmembrane segment, releasing the mature protein in a soluble form into the intermembrane space. Except for the second proteolytic cleavage, this pathway is similar to that of inner-membrane proteins such as CoxVa (see Figure 13-26, path A).

Cytochrome *c* heme lyase, the enzyme responsible for the covalent attachment of heme to cytochrome *c*, illustrates a second pathway for targeting to the intermembrane space. In this pathway, the imported protein does not contain an N-terminal matrix-targeting sequence and is delivered directly to the intermembrane space via the general import pore without involvement of any inner-membrane translocation factors (Figure 13-27, path B). Since translocation through the Tom40 general import pore does not seem to be coupled to any energetically favorable process such as hydrolysis of ATP or GTP, the mechanism that drives unidirectional translocation through the outer membrane is unclear. One possibility is that cytochrome *c* heme lyase passively diffuses through the outer membrane and then is trapped within the intermembrane space by binding to another protein that is delivered to that location by one of the translocation mechanisms discussed previously.

Outer-Membrane Proteins Many of the proteins that reside in the mitochondrial outer membrane, including the Tom 40 pore itself and mitochondrial porin, have a β-barrel structure in which antiparallel strands form the hydrophobic transmembrane segments surrounding a central channel. Such proteins are incorporated into the outer membrane by first interacting with the general import pore, Tom40, and then they are transferred to a complex known as the SAM (*s*orting and *a*ssembly *m*achinery) complex, which is composed of at least three outer membrane proteins. Presumably it is the very stable hydrophobic nature of β-barrel proteins

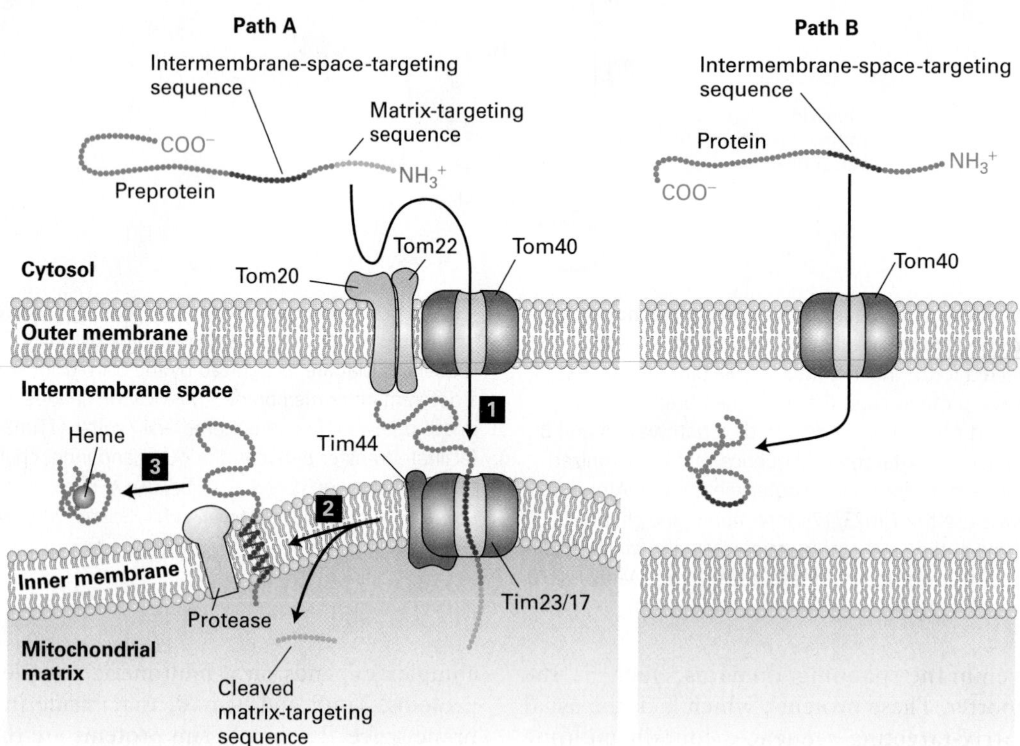

▲ **FIGURE 13-27 Two pathways to the mitochondrial intermembrane space.** Pathway A, the major one for delivery of proteins from the cytosol to the intermembrane space, is similar to pathway A for delivery to the inner membrane (see Figure 13-26). The major difference is that the internal targeting sequence in proteins such as cytochrome b_2 destined for the intermembrane space is recognized by an inner-membrane protease, which cleaves the protein on the intermembrane-space side of the membrane. The released protein then folds and binds to its heme cofactor within the intermembrane space. Pathway B involves direct delivery to the intermembrane space through the Tom40 general import pore in the outer membrane. [See R. E. Dalbey and A. Kuhn, 2000, *Ann. Rev. Cell Devel. Biol.* **16:**51; N. Pfanner and A. Geissler, 2001, *Nature Rev. Mol. Cell Biol.* **2:**339; and K. Diekert et al., 1999, *Proc. Nat'l. Acad. Sci. USA* **96:**11752.]

that ultimately causes them to be stably incorporated into the outer membrane, but precisely how the SAM complex facilitates this process is not known.

Targeting of Chloroplast Stromal Proteins Is Similar to Import of Mitochondrial Matrix Proteins

Among the proteins found in the chloroplast stroma are the enzymes of the Calvin cycle, which functions in fixing carbon dioxide into carbohydrates during photosynthesis (Chapter 12). The large (L) subunit of ribulose 1,5-bisphosphate carboxylase (rubisco) is encoded by chloroplast DNA and synthesized on chloroplast ribosomes in the stromal space. The small (S) subunit of rubisco and all the other Calvin cycle enzymes are encoded by nuclear genes and transported to chloroplasts after their synthesis in the cytosol. The precursor forms of these stromal proteins contain an N-terminal *stromal-import* sequence (see Table 13-1).

Experiments with isolated chloroplasts, similar to those with mitochondria illustrated in Figure 13-22, have shown that they can import the S-subunit precursor after its synthesis. After the unfolded precursor enters the stromal space, it binds transiently to a stromal Hsc70 chaperone and the N-terminal sequence is cleaved. In reactions facilitated by Hsc60 chaperonins that reside within the stromal space, eight S subunits combine with the eight L subunits to yield the active rubisco enzyme.

The general process of stromal import appears to be very similar to that for importing proteins into the mitochondrial matrix (see Figure 13-23). At least three chloroplast outer-membrane proteins, including a receptor that binds the stromal-import sequence and a translocation channel protein, and five inner-membrane proteins are known to be essential for directing proteins to the stroma. Although these proteins are functionally analogous to the receptor and channel proteins in the mitochondrial membrane, they are not structurally homologous. The lack of sequence similarity between these chloroplast and mitochondrial proteins suggests that they may have arisen independently during evolution.

The available evidence suggests that chloroplast stromal proteins, like mitochondrial matrix proteins, are imported in the unfolded state. Import into the stroma depends on ATP hydrolysis catalyzed by a stromal Hsc70 chaperone whose function is similar to that of Hsc70 in the mitochondrial matrix and BiP in the ER lumen. Unlike mitochondria, chloroplasts do not generate an electrochemical gradient (proton-motive force) across their inner membrane. Thus protein import into the chloroplast stroma appears to be powered solely by ATP hydrolysis.

Proteins Are Targeted to Thylakoids by Mechanisms Related to Translocation Across the Bacterial Inner Membrane

In addition to the double membrane that surrounds them, chloroplasts contain a series of internal interconnected membranous sacs, the **thylakoids** (see Figure 12-29). Proteins localized to the thylakoid membrane or lumen carry out photosynthesis. Many of these proteins are synthesized in the cytosol as precursors containing multiple targeting sequences. For example, plastocyanin and other proteins destined for the thylakoid lumen require the successive action of two uptake-targeting sequences. The first is an N-terminal stromal-import sequence that directs the protein to the stroma by the same pathway that imports the rubisco S subunit. The second sequence targets the protein from the stroma to the thylakoid lumen. The role of these targeting sequences has been shown in experiments measuring the uptake of mutant proteins generated by recombinant DNA techniques into isolated chloroplasts. For instance, mutant plastocyanin that lacks the thylakoid-targeting sequence but contains an intact stromal-import sequence accumulates in the stroma and is not transported into the thylakoid lumen.

Four separate pathways for transporting proteins from the stroma into the thylakoid have been identified. All four pathways have been found to be closely related to analogous transport mechanisms in bacteria, illustrating the close evolutionary relationship between the stromal membrane and the bacterial inner membrane. Transport of plastocyanin and related proteins into the thylakoid lumen from the stroma occurs by a chloroplast SRP-dependent pathway that utilizes a translocon similar to SecY, the bacterial version of the Sec61 complex (Figure 13-28, *left*). A second pathway for transporting proteins into the thylakoid lumen involves a protein related to bacterial protein SecA, which uses the energy from ATP hydrolysis to drive protein translocation through the SecY translocon. A third pathway, which targets proteins to the thylakoid membrane, depends on a protein related to the mitochondrial Oxa1 protein and the homologous bacterial protein (see Figure 13-26, path B). Some proteins encoded by chloroplast DNA and synthesized in the stroma or transported into the stroma from the cytosol are inserted into the thylakoid membrane via this pathway.

Finally, thylakoid proteins that bind metal-containing cofactors follow another pathway into the thylakoid lumen (Figure 13-28, *right*). The unfolded precursors of these proteins are first targeted to the stroma, where the N-terminal stromal-import sequence is cleaved off and the protein then folds and binds its cofactor. A set of thylakoid-membrane proteins assists in translocating the folded protein and bound cofactor into the thylakoid lumen, a process powered by the pH gradient normally maintained across the thylakoid membrane. The thylakoid-targeting sequence that directs a protein to this pH-dependent pathway includes two closely spaced arginine residues that are crucial for recognition. Bacterial cells also have a mechanism for translocating folded proteins with a similar arginine-containing sequence across the inner membrane. The molecular mechanism whereby these large folded globular proteins are transported across the thylakoid membrane is currently under intense study.

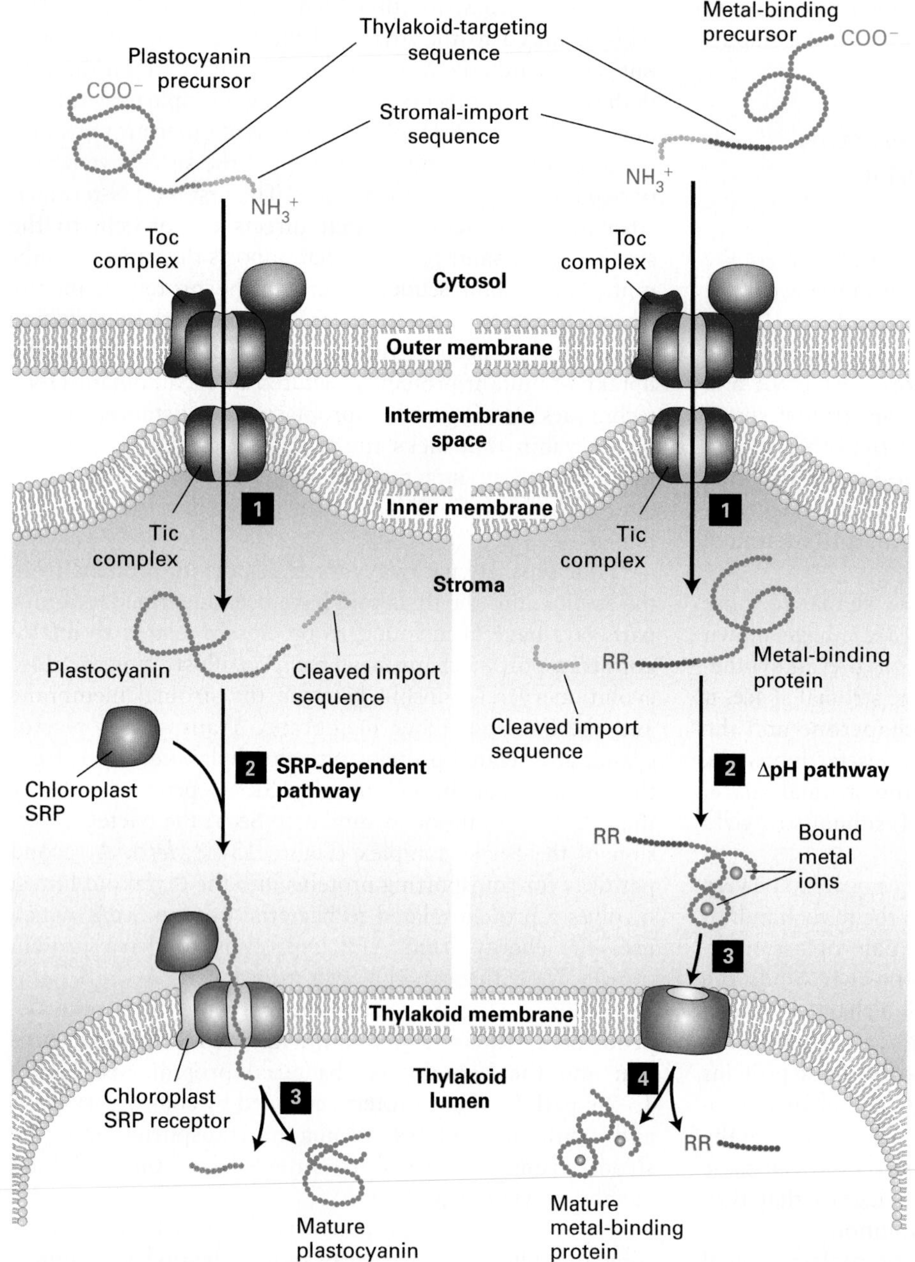

◀ **FIGURE 13-28 Transporting proteins to chloroplast thylakoids.** Two of the four pathways for transporting proteins from the cytosol to the thylakoid lumen are shown here. In these pathways, unfolded precursors are delivered to the stroma via the same outer-membrane proteins that import stromal-localized proteins. Cleavage of the N-terminal stromal-import sequence by a stromal protease then reveals the thylakoid-targeting sequence (step **1**). At this point the two pathways diverge. In the SRP-dependent pathway (*left*), plastocyanin and similar proteins are kept unfolded in the stromal space by a set of chaperones (not shown) and, directed by the thylakoid-targeting sequence, bind to proteins that are closely related to the bacterial SRP, SRP receptor, and SecY translocon, which mediate movement into the lumen (step **2**). After the thylakoid-targeting sequence is removed in the thylakoid lumen by a separate endoprotease, the protein folds into its mature conformation (step **3**). In the pH-dependent pathway (*right*), metal-binding proteins fold in the stroma, and complex redox cofactors are added (step **2**). Two arginine residues (RR) at the N-terminus of the thylakoid-targeting sequence and a pH gradient across the inner membrane are required for transport of the folded protein into the thylakoid lumen (step **3**). The translocon in the thylakoid membrane is composed of at least four proteins related to proteins in the bacterial inner membrane. The thylakoid targeting sequence containing the two arginine residues is cleaved in the thylakoid lumen (step **4**). [See R. Dalbey and C. Robinson, 1999, *Trends Biochem. Sci.* **24:**17; R. E. Dalbey and A. Kuhn, 2000, *Ann. Rev. Cell Devel. Biol.* **16:**51; and C. Robinson and A. Bolhuis, 2001, *Nature Rev. Mol. Cell Biol.* **2:**350.]

KEY CONCEPTS OF SECTION 13.4

Sorting of Proteins to Mitochondria and Chloroplasts

- Most mitochondrial and chloroplast proteins are encoded by nuclear genes, synthesized on cytosolic ribosomes, and imported post-translationally into the organelles.
- All the information required to target a precursor protein from the cytosol to the mitochondrial matrix or chloroplast stroma is contained within its N-terminal uptake-targeting sequence. After protein import, the uptake-targeting sequence is removed by proteases within the matrix or stroma.
- Cytosolic chaperones maintain the precursors of mitochondrial and chloroplast proteins in an unfolded state. Only unfolded proteins can be imported into the organelles. Translocation in mitochondria occurs at sites where the outer and inner membranes of the organelles are close together.
- Proteins destined for the mitochondrial matrix bind to receptors on the outer mitochondrial membrane and then are transferred to the general import pore (Tom40) in the outer membrane. Translocation occurs concurrently through the outer and inner membranes, driven by the proton-motive force across the inner membrane and ATP hydrolysis by the Hsc70 ATPase in the matrix (see Figure 13-23).
- Proteins sorted to mitochondrial destinations other than the matrix usually contain two or more targeting sequences, one of which may be an N-terminal matrix-targeting sequence (see Figure 13-25).

- Some mitochondrial proteins destined for the intermembrane space or inner membrane are first imported into the matrix and then redirected; others never enter the matrix but go directly to their final location.

- Protein import into the chloroplast stroma occurs through inner-membrane and outer-membrane translocation channels that are analogous in function to mitochondrial channels but composed of proteins unrelated in sequence to the corresponding mitochondrial proteins.

- Proteins destined for the thylakoid have secondary targeting sequences. After entry of these proteins into the stroma, cleavage of the stromal-targeting sequences reveals the thylakoid-targeting sequences.

- The four known pathways for moving proteins from the chloroplast stroma to the thylakoid closely resemble translocation across the bacterial inner membrane (see Figure 13-28). One of these systems can translocate folded proteins.

13.5 Sorting of Peroxisomal Proteins

Peroxisomes are small organelles bounded by a single membrane. Unlike mitochondria and chloroplasts, peroxisomes lack DNA and ribosomes. Thus all peroxisomal proteins are encoded by nuclear genes, synthesized on ribosomes free in the cytosol, and then incorporated into preexisting or newly generated peroxisomes. As peroxisomes are enlarged by addition of protein (and lipid), they eventually divide, forming new ones, as is the case with mitochondria and chloroplasts.

The size and enzyme composition of peroxisomes vary considerably in different kinds of cells. However, all peroxisomes contain enzymes that use molecular oxygen to oxidize various substrates such as amino acids and fatty acids, breaking them down into smaller components for use in biosynthetic pathways. The hydrogen peroxide (H_2O_2) generated by these oxidation reactions is extremely reactive and potentially harmful to cellular components; however, the peroxisome also contains enzymes, such catalase, that efficiently convert H_2O_2 into H_2O. In mammals, peroxisomes are most abundant in liver cells, where they constitute about 1 to 2 percent of the cell volume.

Cytosolic Receptor Targets Proteins with an SKL Sequence at the C-Terminus into the Peroxisomal Matrix

The import of catalase and other proteins into rat liver peroxisomes can be assayed in a cell-free system similar to that used for monitoring mitochondrial protein import (see Figure 13-22). By testing various mutant catalase proteins in this system, researchers discovered that the sequence Ser-Lys-Leu (SKL in one-letter code) or a related sequence at the C-terminus was necessary for peroxisomal targeting. Further, addition of the SKL sequence to the C-terminus of a normally cytosolic protein leads to uptake of the altered protein by peroxisomes in cultured cells. All but a few of the many different peroxisomal matrix proteins bear a sequence of this type, known as *peroxisomal-targeting sequence 1*, or simply *PTS1*.

The pathway for import of catalase and other PTS1-bearing proteins into the peroxisomal matrix is depicted in Figure 13-29. The PTS1 binds to a soluble carrier protein in the cytosol (Pex5), which in turn binds to a receptor in the peroxisome membrane (Pex14). The soluble and membrane-associated peroxisomal import receptors appear to have a function analogous to that of the SRP and SRP receptor in targeting proteins to the ER lumen, except that the soluble PTS1-binding protein functions post-translationally. The protein to be imported then moves through a multimeric

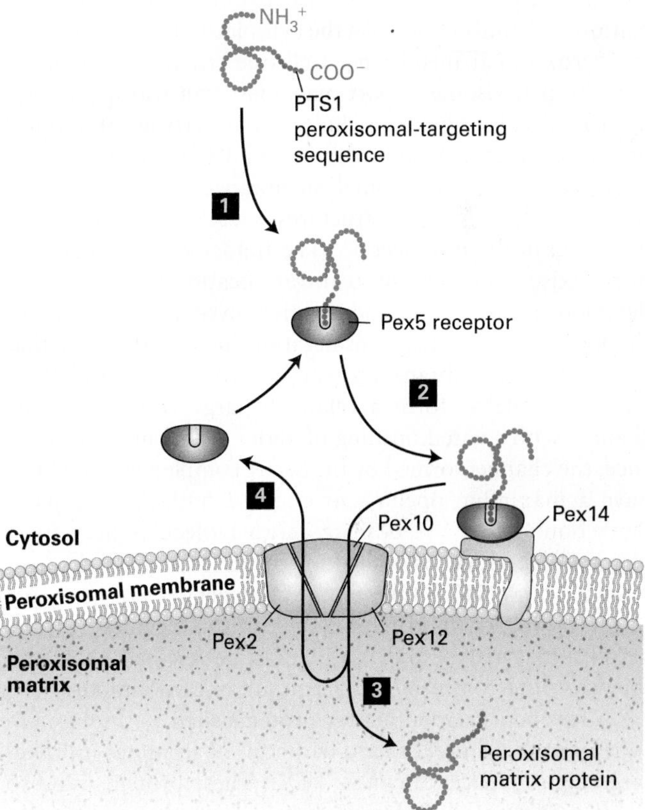

▲ **FIGURE 13-29 PTS1 directed import of peroxisomal matrix proteins.** Step **1**: Catalase and most other peroxisomal matrix proteins contain a C-terminal PTS1 uptake-targeting sequence (red) that binds to the cytosolic receptor Pex5. Step **2**: Pex5 with the bound matrix protein interacts with the Pex14 receptor located on the peroxisome membrane. Step **3**: The matrix protein–Pex5 complex is then transferred to a set of membrane proteins (Pex10, Pex12, and Pex2) that are necessary for translocation into the peroxisomal matrix by an unknown mechanism. Step **4**: At some point, either during translocation or in the lumen, Pex5 dissociates from the matrix protein and returns to the cytosol, a process that involves the Pex2/10/12 complex and additional membrane and cytosolic proteins not shown. Note that folded proteins can be imported into peroxisomes and that the targeting sequence is not removed in the matrix. [See P. E. Purdue and P. B. Lazarow, 2001, *Ann. Rev. Cell Devel. Biol.* **17**:701; S. Subramani et al., 2000, *Ann. Rev. Biochem.* **69**:399; and V. Dammai and S. Subramani, 2001, *Cell* **105**:187.]

translocation channel while still bound to Pex5, a feature that differs from protein import into the ER lumen. At some stage either during or after entry into the matrix, Pex5 dissociates from the peroxisomal matrix protein and is recycled back to the cytoplasm. In contrast to the N-terminal uptake-targeting sequences on proteins destined for the ER lumen, mitochondrial matrix, and chloroplast stroma, the PTS1 sequence is not cleaved from proteins after their entry into a peroxisome. Protein import into peroxisomes requires ATP hydrolysis, but it is not known how the energy released from ATP is used to power unidirectional translocation across the peroxisomal membrane.

The peroxisome import machinery, unlike most systems that mediate protein import into the ER, mitochondria, and chloroplasts, can translocate folded proteins across the membrane. For example, catalase assumes a folded conformation and binds to heme in the cytoplasm before traversing the peroxisomal membrane. Cell-free studies have shown that the peroxisome import machinery can transport large macromolecular objects, including gold particles of about 9 nm in diameter, as long as they have a PTS1 tag attached to them. However, peroxisomal membranes do not appear to contain large stable pore structures, such as the nuclear pore described in the next section. The fundamental mechanism of peroxisomal matrix protein translocation is not well understood but is a topic under active investigation. Some of the mechanisms under consideration include the idea that peroxisomal membrane proteins Pex10, Pex12, and Pex2 may assemble to form a relatively large transmembrane channel with a gated opening of about 10–15 nm (for reference, the channel formed by the Sec61 complex is thought to have a maximum opening of about 2 nm). Alternatively, Pex5 bound to PTS1-bearing cargo molecules may form oligomeric complexes embedded in the peroxisomal membrane. According to this idea, ATP-dependent disassembly of the complex would release the cargo molecules into the peroxisomal matrix and Pex5 would be released back into the cytosol to complete another round of cargo import.

A few peroxisomal matrix proteins such as thiolase are synthesized as precursors with an N-terminal uptake-targeting sequence known as *PTS2*. These proteins bind to a different cytosolic receptor protein, but otherwise import is thought to occur by the same mechanism as for PTS1-containing proteins.

Peroxisomal Membrane and Matrix Proteins Are Incorporated by Different Pathways

Autosomal recessive mutations that cause defective peroxisome assembly occur naturally in the human population. Such defects can lead to severe developmental defects often associated with craniofacial abnormalities. In *Zellweger syndrome* and related disorders, for example, the transport of many or all proteins into the peroxisomal matrix is impaired: newly synthesized peroxisomal enzymes remain in the cytosol and are eventually degraded. Genetic analyses of cultured cells from different Zellweger patients and of yeast cells carrying similar mutations have identified more than 20 genes that are required for peroxisome biogenesis. ■

Studies with peroxisome-assembly mutants have shown that different pathways are used for importing peroxisomal matrix proteins versus inserting proteins into the peroxisomal membrane. For example, analysis of cells from some Zellweger patients led to identification of genes encoding the putative translocation channel proteins Pex10, Pex12, and Pex2. Mutant cells defective in any one of these proteins cannot incorporate matrix proteins into peroxisomes; nonetheless, the cells contain empty peroxisomes that have a normal complement of peroxisomal membrane proteins (Figure 13-30b).

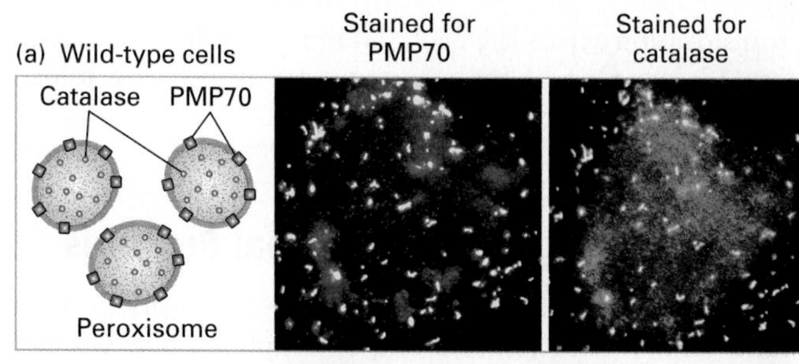

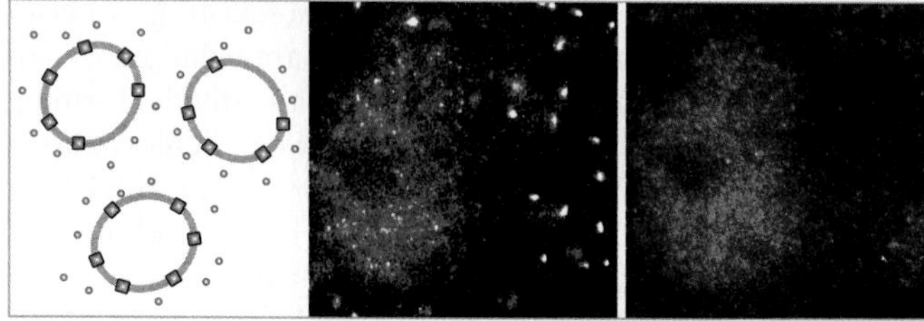

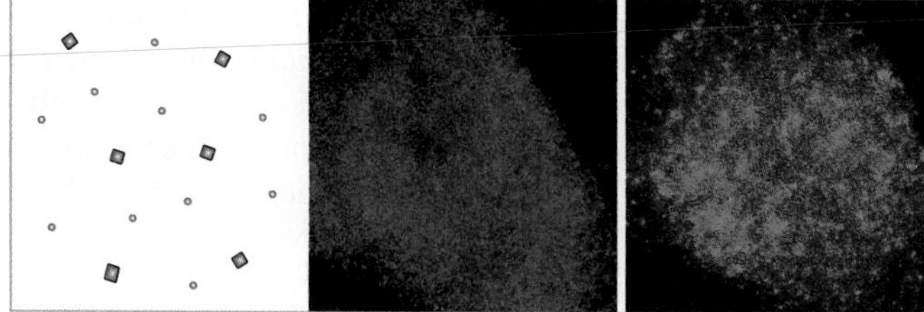

▲ **EXPERIMENTAL FIGURE 13-30 Studies reveal different pathways for incorporation of peroxisomal membrane and matrix proteins.** Cells were stained with fluorescent antibodies to PMP70, a peroxisomal membrane protein, or with fluorescent antibodies to catalase, a peroxisomal matrix protein, then viewed in a fluorescent microscope. (a) In wild-type cells, both peroxisomal membrane and matrix proteins are visible as bright foci in numerous peroxisomal bodies. (b) In cells from a Pex12-deficient patient, catalase is distributed uniformly throughout the cytosol, whereas PMP70 is localized normally to peroxisomal bodies. (c) In cells from a Pex3-deficient patient, peroxisomal membranes cannot assemble, and as a consequence peroxisomal bodies do not form. Thus both catalase and PMP70 are mis-localized to the cytosol. [Courtesy of Stephen Gould, Johns Hopkins University.]

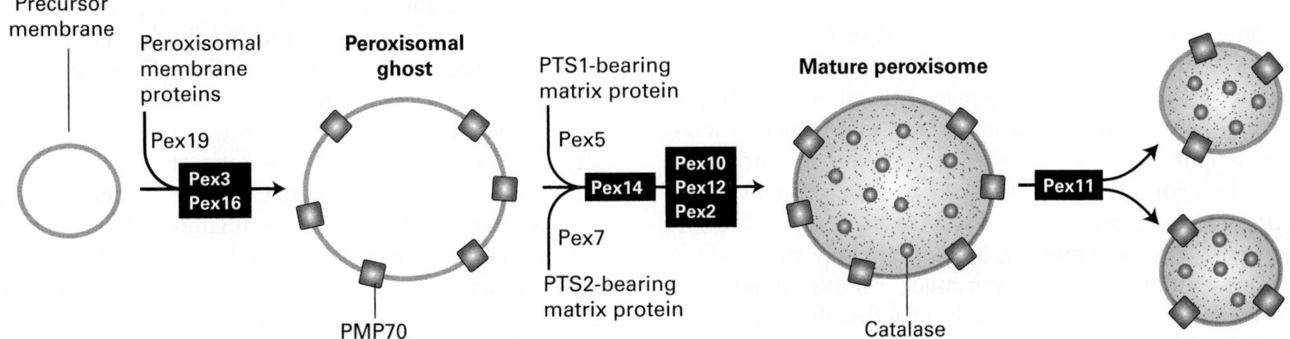

▲ **FIGURE 13-31 Model of peroxisomal biogenesis and division.** The first stage in the de novo formation of peroxisomes is the incorporation of peroxisomal membrane proteins into precursor membranes derived from the ER. Pex19 acts as the receptor for membrane-targeting sequences. Pex3 and Pex16 are required for the proper insertion of proteins into the forming peroxisomal membrane. Insertion of all peroxisomal membrane proteins produces a peroxisomal ghost, which is capable of importing proteins targeted to the matrix. The pathways for importing PTS1- and PTS2-bearing matrix proteins differ only in the identity of the cytosolic receptor (Pex5 and Pex7, respectively) that binds the targeting sequence (see Figure 13-29). Complete incorporation of matrix proteins yields a mature peroxisome. The proliferation of peroxisomes requires division of mature peroxisomes, a process that depends on the Pex11 protein.

Mutations in any one of three other genes were found to block insertion of peroxisomal membrane proteins as well as import of matrix proteins (Figure 13-30c). These findings demonstrate that one set of proteins translocates soluble proteins into the peroxisomal matrix but a different set is required for insertion of proteins into the peroxisomal membrane. This situation differs markedly from that of the ER, mitochondrion, and chloroplast, for which, as we have seen, membrane proteins and soluble proteins share many of the same components for their insertion into these organelles.

Although most peroxisomes are generated by division of preexisting organelles, these organelles can arise de novo by the three-stage process depicted in Figure 13-31. In this case, peroxisome assembly begins in the ER. At least two peroxisomal membrane proteins, Pex3 and Pex16, are inserted into the ER membrane by the mechanisms described in Section 13.2. Pex3 and Pex16 recruit additional peroxisomal proteins such as Pex19 forming a specialized region of the ER membrane that can bud off of the ER to form a peroxisomal precursor membrane. Analysis of mutant cells revealed that Pex19 is the receptor protein responsible for targeting of peroxisomal membrane proteins, whereas Pex3 and Pex16 are necessary for their proper insertion into the membrane. These three proteins are thought to be responsible for peroxisomal membrane protein assembly in mature peroxisomes as well as during the de novo formation of new peroxisomes. The insertion of peroxisomal membrane proteins generates membranes that have all the components necessary for import of matrix proteins, leading to the formation of mature, functional peroxisomes. Division of mature peroxisomes, which largely determines the number of peroxisomes within a cell, depends on still another protein, Pex11. Overexpression of the Pex11 protein causes a large increase in the number of peroxisomes, suggesting that this protein controls the extent of peroxisome division. The small peroxisomes generated by division can be enlarged by incorporation of additional matrix and membrane proteins via the same pathways described previously.

KEY CONCEPTS OF SECTION 13.5

Sorting of Peroxisomal Proteins

- All peroxisomal proteins are synthesized on cytosolic ribosomes and incorporated into the organelle posttranslationally.
- Most peroxisomal matrix proteins contain a C-terminal PTS1 targeting sequence; a few have an N-terminal PTS2 targeting sequence. Neither targeting sequence is cleaved after import.
- All proteins destined for the peroxisomal matrix bind to a cytosolic carrier protein, which differs for PTS1- and PTS2-bearing proteins, and then are directed to common import receptor and translocation machinery on the peroxisomal membrane (see Figure 13-29).
- Translocation of matrix proteins across the peroxisomal membrane depends on ATP hydrolysis. Many peroxisomal matrix proteins fold in the cytosol and traverse the membrane in a folded conformation, which is different than for protein import into organelles such as the ER, mitochondrion, and chloroplast.
- Proteins destined for the peroxisomal membrane contain different targeting sequences than peroxisomal matrix proteins and are imported by a different pathway.
- Unlike mitochondria and chloroplasts, peroxisomes can arise de novo from precursor membranes probably derived from the ER as well as by division of preexisting organelles (see Figure 13-31).

13.6 Transport into and out of the Nucleus

The nucleus is separated from the cytoplasm by two membranes, which form the **nuclear envelope** (see Figure 9-1). The nuclear envelope is continuous with the ER and forms a part of it. Transport of proteins from the cytoplasm into the nucleus and movement of macromolecules, including mRNPs, tRNAs, and ribosomal subunits, out of the nucleus occur through *nuclear pores,* which span both membranes of the nuclear envelope. Import of proteins into the nucleus shares some fundamental features with protein import into other organelles. For example, imported nuclear proteins carry specific targeting sequences known as nuclear localization sequences, or NLSs. However, proteins are imported into the nucleus in a folded state, and thus nuclear import differs fundamentally from protein translocation across the membranes of the ER, mitochondrion, and chloroplast, where proteins are unfolded during translocation. In this section we discuss the main mechanism by which proteins and some ribonuclear proteins such as ribosomes enter and exit the nucleus. We will also discuss how mRNAs and other ribonuclear protein complexes are exported from the nucleus by a process that differs mechanistically from nuclear protein import.

Large and Small Molecules Enter and Leave the Nucleus via Nuclear Pore Complexes

Numerous pores perforate the nuclear envelope in all eukaryotic cells. Each nuclear pore is formed from an elaborate structure termed the **nuclear pore complex (NPC),** which is one of the largest protein assemblages in the cell. The total mass of the pore structure is 60–80 million Da in vertebrates, which is about 16 times larger than a ribosome. An NPC is made up of multiple copies of some 30 different proteins called **nucleoporins.** Electron micrographs of nuclear pore complexes reveal a roughly octagonal, membrane-embedded structure from which eight approximately 100-nm-long filaments extend into the nucleoplasm (Figure 13-32). The distal ends of these filaments are joined by the terminal ring, forming a structure called the *nuclear basket.* The membrane-embedded portion of the NPC is also attached directly to the **nuclear lamina,** a network of lamin intermediate filaments that form a meshwork extending over the inner surface of the nuclear envelope (see Figure 20-16). Cytoplasmic filaments extend from the cytoplasmic side of the NPC into the cytosol.

Ions, small metabolites, and globular proteins up to 20–40 kDa can passively diffuse through the central aqueous region of the nuclear pore complex. However, large proteins and ribonucleoprotein complexes cannot diffuse in and out of the nucleus. Rather, these macromolecules are actively transported through the NPC with the assistance of soluble transporter proteins that bind macromolecules and also interact with nucleoporins.

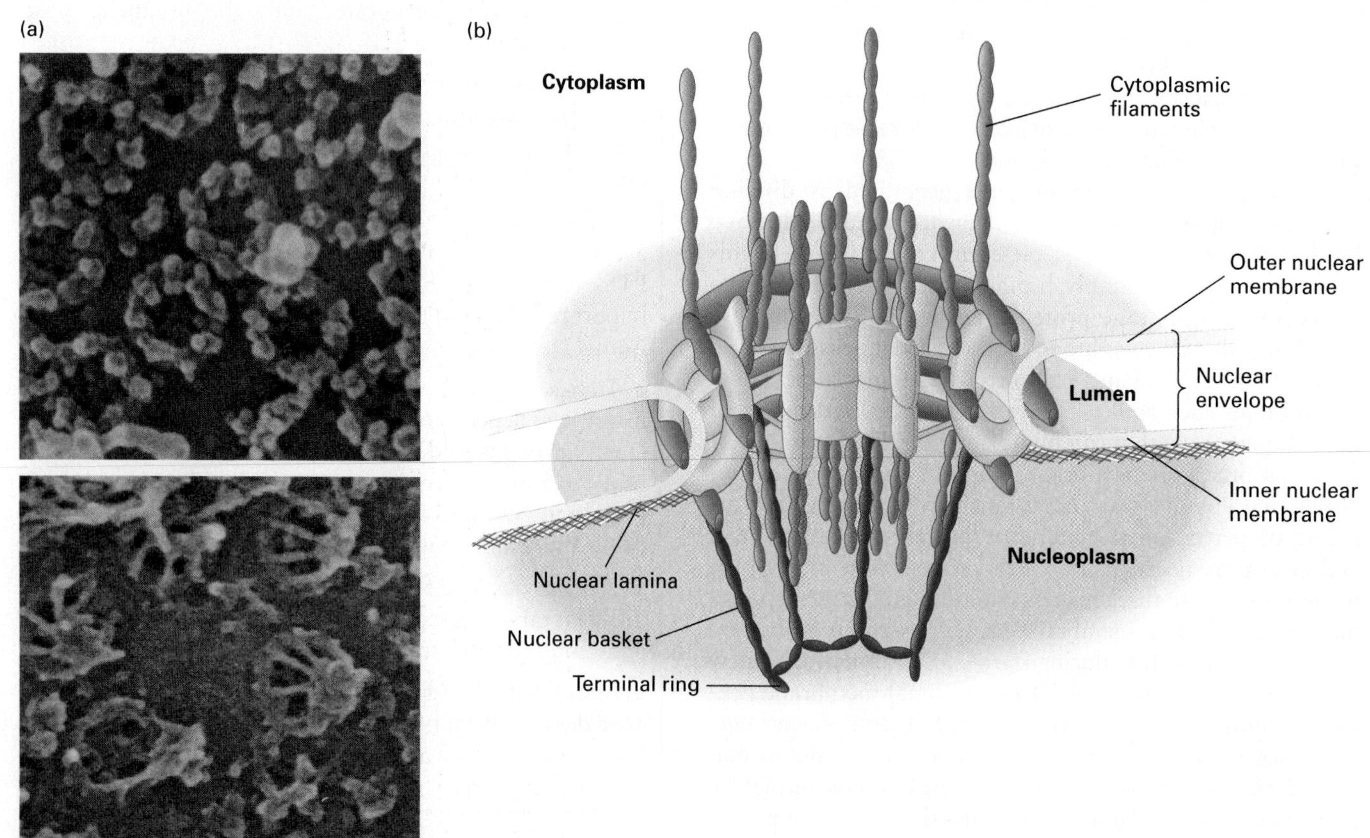

▲ **FIGURE 13-32 Nuclear pore complex.** (a) Nuclear envelopes microdissected from the large nuclei of *Xenopus* oocytes visualized by field emission in-lens scanning electron microscopy. *Top:* View of the cytoplasmic face reveals octagonal shape of membrane-embedded portion of nuclear pore complexes. *Bottom:* View of the nucleoplasmic face shows the nuclear basket that extends from the membrane portion. (b) Cutaway model of the pore complex. [Part (a) from V. Doye and E. Hurt, 1997, *Curr. Opin. Cell Biol.* **9:**401; courtesy of M. W. Goldberg and T. D. Allen. Part (b) adapted from M. P. Rout and J. D. Atchison, 2001, *J. Biol. Chem.* **276:**16593.]

Importins Transport Proteins Containing Nuclear-Localization Signals into the Nucleus

All proteins found in the nucleus—such as histones, transcription factors, and DNA and RNA polymerases—are synthesized in the cytoplasm and imported into the nucleus through nuclear pore complexes. Such proteins contain a *nuclear-localization signal (NLS)* that directs their selective transport into the nucleus. NLSs were first discovered through the analysis of mutants of a virus, simian virus 40 (SV40), that produced an abnormal form of the viral protein called large T-antigen. The wild-type form of this protein is localized to the nucleus in virus-infected cells, whereas some mutated forms of large T-antigen accumulate in the cytoplasm. The mutations responsible for this altered cellular localization all occur within a specific seven-residue sequence rich in basic amino acids near the C-terminus of the protein: Pro-Lys-Lys-Lys-Arg-Lys-Val. Experiments with engineered hybrid proteins in which this sequence was fused to a cytosolic protein demonstrated that it directs transport into the nucleus and consequently functions as an NLS (Figure 13-33). NLS sequences subsequently were identified in numerous other proteins imported into the nucleus. Many of these are similar to the basic NLS in SV40 large T-antigen, whereas other NLSs are chemically quite different. For instance, an NLS in the RNA-binding protein hnRNP A1 is hydrophobic. Thus there must be multiple mechanisms for the recognition of these very different sequences.

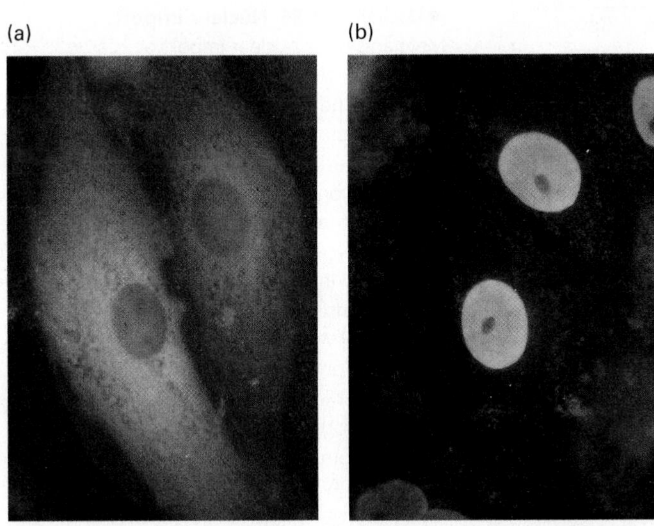

▲ EXPERIMENTAL FIGURE 13-33 **Nuclear-localization signal (NLS) directs proteins to the cell nucleus.** Cytoplasmic proteins can be localized to the nucleus when fused to a nuclear localization signal. (a) Normal pyruvate kinase, visualized by immunofluorescence after cultured cells were treated with a specific antibody (yellow), is localized to the cytoplasm. This very large cytosolic protein functions in carbohydrate metabolism. (b) When a chimeric pyruvate kinase protein containing the SV40 NLS at its N-terminus was expressed in cells, it was localized to the nucleus. The chimeric protein was expressed from a transfected engineered gene produced by fusing a viral gene fragment encoding the SV40 NLS to the pyruvate kinase gene. [From D. Kalderon et al., 1984, *Cell* **39**:499; courtesy of Dr. Alan Smith.]

Early work on the mechanism of nuclear import focused on proteins containing a basic NLS, similar to the one in SV40 large T-antigen. A digitonin-permeabilized cell system provided an in vitro assay for analyzing soluble cytosolic components required for nuclear import (Figure 13-34). Digitonin is a detergent that permeabilizes the plasma membrane, allowing cytoplasmic components to leak out of the cell while leaving the nuclear envelope and NPCs intact. Using this assay system, researchers purified three required proteins: Ran, importin α, and importin β. *Ran* is a monomeric **G protein** that exists in either GTP- or GDP-bound conformations (see Figure 3-32). The two **importins** can form a heterodimeric *nuclear-import receptor:* the α subunit binds to a basic NLS in a "cargo" protein to be transported into the nucleus, whereas the β subunit interacts with proteins in the nuclear pore to shuttle the cargo protein through it. The importin β subunit can also bind directly to certain NLS sequences and thus can act alone as a nuclear import receptor for these proteins.

(a) Effect of digitonin

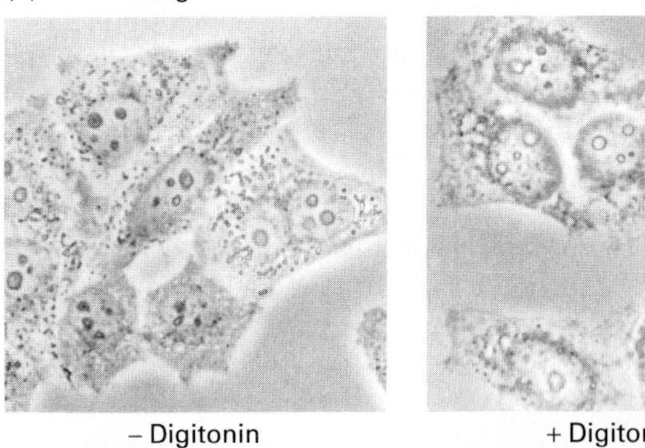

(b) Nuclear import by permeabilized cells

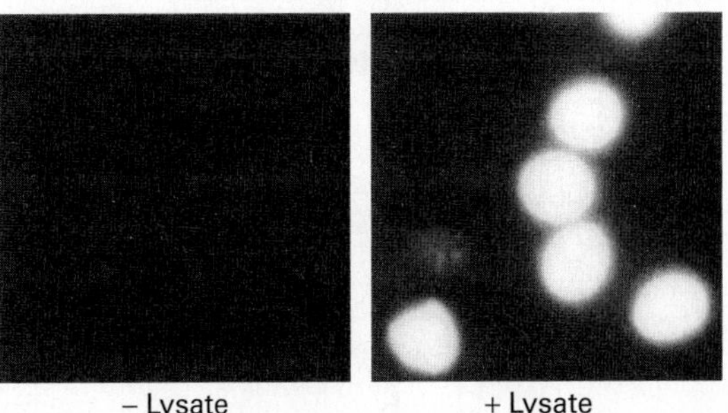

▲ EXPERIMENTAL FIGURE 13-34 **Cytosolic proteins are required for nuclear transport.** The failure of nuclear transport to occur in permeabilized cultured cells in the absence of lysate demonstrates the involvement of soluble cytosolic components in the process. (a) Phase-contrast micrographs of untreated and digitonin-permeabilized HeLa cells. Treatment of a monolayer of cultured cells with the mild, nonionic detergent digitonin permeabilizes the plasma membrane so that cytosolic constituents leak out but leaves the nuclear envelope and NPCs intact. (b) Fluorescence micrographs of digitonin-permeabilized HeLa cells incubated with a fluorescent protein chemically coupled to a synthetic SV40 T-antigen NLS peptide in the presence and absence of cytosol (lysate). Accumulation of this transport substrate in the nucleus occurred only when cytosol was included in the incubation (*right*). [From S. Adam et al., 1990, *J. Cell. Biol.* **111**:807; courtesy of Dr. Larry Gerace.]

The mechanism for import of cytoplasmic cargo proteins mediated by an importin is shown in Figure 13-35 (the general mechanism is the same for either a monomeric or dimeric importin). Free importin in the cytoplasm binds to its cognate NLS in a cargo protein, forming a *bimolecular cargo complex*. The cargo complex then translocates through the NPC channel as the importin β subunit interacts with a class of nucleoporins called *FG-nucleoporins*. These nucleoporins, which line the channel of the nuclear pore complex and also are found in the nuclear basket and the cytoplasmic filaments, contain multiple repeats of short hydrophobic sequences rich in phenylalanine (F) and glycine (G) residues (FG-repeats). The hydrophobic FG-repeats are thought to occur in regions of extended, otherwise hydrophilic polypeptide chains that fill the central transporter channel and in some way allow the relatively hydrophobic importin complexes to traverse the channel efficiently while excluding unchaperoned hydrophilic proteins larger than 20–40 kDa.

When the cargo complex reaches the nucleoplasm, the importin interacts with Ran·GTP, causing a conformational change in the importin that decreases its affinity for the NLS, releasing the cargo protein into the nucleoplasm. The importin-Ran·GTP complex then diffuses back through the NPC. Once the importin-Ran·GTP complex reaches the cytoplasmic side of the NPC, Ran interacts with a specific *GTPase-activating protein (Ran-GAP)* that is a component of the NPC cytoplasmic filaments. This stimulates Ran to hydrolyze its bound GTP to GDP, causing it to convert to a conformation that has low affinity for the importin, so that the free importin is released into the cytoplasm, where it can participate in another cycle of import. Ran·GDP travels back through the pore to the nucleoplasm, where it encounters a specific *guanine nucleotide-exchange factor (Ran-GEF)* that causes Ran to release its bound GDP in favor of GTP. The net result of this series of reactions is the coupling of the hydrolysis of GTP to the transfer of an NLS-bearing protein from the cytoplasm to the nuclear interior, thus providing a driving force for nuclear transport.

The import complex travels through the pore by diffusion, a random process. Yet transport is unidirectional. The direction of transport is a consequence of the rapid dissociation of the import complex when it reaches the nucleoplasm. As a result, there is a concentration gradient of the importin-cargo complex across the NPC: high in the cytoplasm, where the complex assembles and low in the nucleoplasm, where it dissociates. This concentration gradient is responsible for the unidirectional nature of nuclear import. A similar concentration gradient is responsible for driving the importin in the nucleus back into the cytoplasm. The concentration of the importin-Ran·GTP complex is higher in the nucleoplasm, where it assembles, than on the cytoplasmic side of the NPC, where it dissociates. Ultimately, the direction of the transport processes is dependent on the asymmetric distribution of the Ran-GEF and the Ran-GAP. Ran-GEF in the nucleoplasm maintains Ran in the Ran·GTP state, where it promotes dissociation of the cargo complex. Ran-GAP on the cytoplasmic side of the NPC

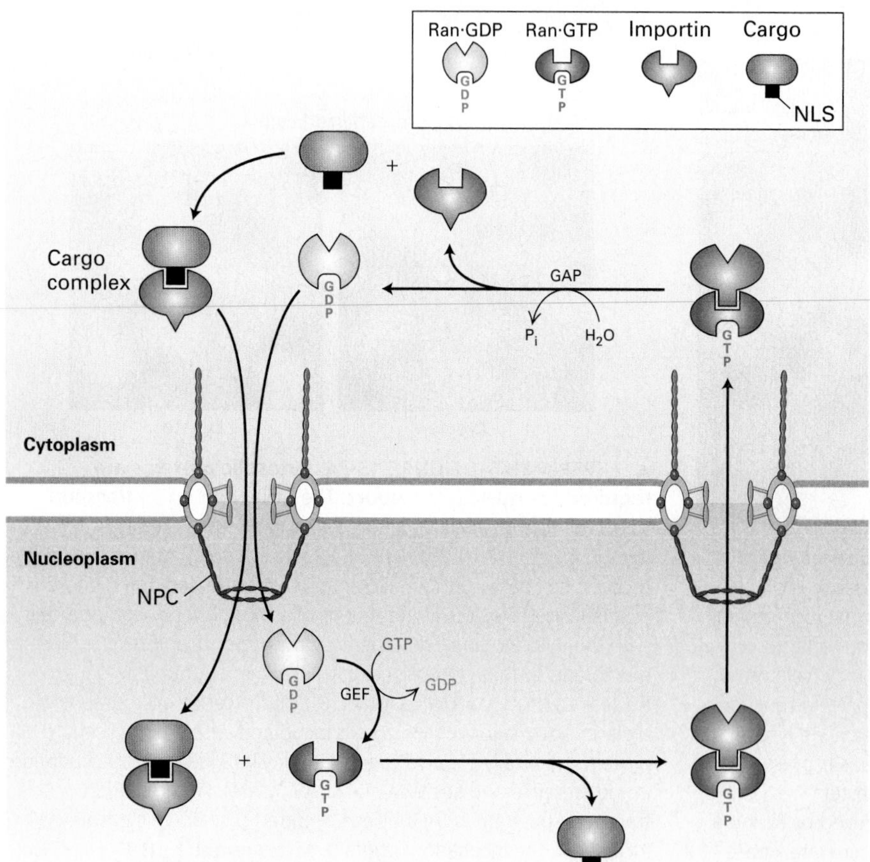

◀ **FIGURE 13-35 Nuclear import.** Mechanism for nuclear import of "cargo" proteins. In the cytoplasm (*bottom*), a free importin binds to the NLS of a cargo protein, forming a bimolecular cargo complex. In the case of a basic NLS, the adapter protein importin α bridges the NLS and importin β, forming a trimolecular cargo complex (not shown). The cargo complex diffuses through the NPC by interacting with successive FG-nucleoporins. In the nucleoplasm, interaction of Ran·GTP with the importin causes a conformational change that decreases its affinity for the NLS, releasing the cargo. To support another cycle of import, the importin-Ran·GTP complex is transported back to the cytoplasm. A GTPase-accelerating protein (GAP) associated with the cytoplasmic filaments of the NPC stimulates Ran to hydrolyze the bound GTP. This generates a conformational change causing dissociation from the importin, which can then initiate another round of import. Ran·GDP is returned to the nucleoplasm, where a guanine nucleotide-exchange factor (GEF) causes release of GDP and rebinding of GTP.

converts Ran·GTP to Ran·GDP, dissociating the importin-Ran·GTP complex and releasing free importin into the cytosol.

Exportins Transport Proteins Containing Nuclear-Export Signals out of the Nucleus

A very similar mechanism is used to export proteins, tRNAs, and ribosomal subunits from the nucleus to the cytoplasm. This mechanism initially was elucidated from studies of certain ribonuclear protein complexes that "shuttle" between the nucleus and cytoplasm. Such "shuttling" proteins contain a *nuclear-export signal (NES)* that stimulates their export from the nucleus to the cytoplasm through nuclear pores, in addition to an NLS that results in their uptake into the nucleus. Experiments with engineered hybrid genes encoding a nucleus-restricted protein fused to various segments of a protein that shuttles in and out of the nucleus have identified at least three different classes of NESs: a leucine-rich sequence found in PKI (an inhibitor of protein kinase A) and in the Rev protein of human immunodeficiency virus (HIV), as well as two sequences identified in two different heterogeneous ribonucleoprotein particles (hnRNPs). The functionally significant structural features that specify nuclear export remain poorly understood.

The mechanism whereby shuttling proteins are exported from the nucleus is best understood for those containing a leucine-rich NES. According to the current model, shown in Figure 13-36a, a specific **exportin,** or *nuclear-export receptor,* in the nucleus, called exportin 1, first forms a complex with Ran·GTP and then binds the NES in a cargo protein. Binding of exportin 1 to Ran·GTP causes a conformational change in exportin 1 that increases its affinity for the NES so that a *trimolecular cargo complex* is formed. Like importins, exportin 1 interacts transiently with FG-repeats in FG-nucleoporins and diffuses through the NPC. The cargo complex dissociates when it encounters the Ran-GAP in the NPC cytoplasmic filaments, which stimulates Ran to hydrolyze the bound GTP, shifting it into a conformation that has low affinity for exportin 1. The released exportin 1 changes conformation to a structure that has low affinity for the NES, releasing the cargo into the cytosol. The direction of the export process is driven by this dissociation of the cargo from exportin 1 in the cytoplasm, which causes a concentration gradient of the cargo complex across the NPC that is high in the nucleoplasm and low in the cytoplasm. Exportin 1 and the Ran·GDP are then transported back into the nucleus through an NPC.

By comparing this model for nuclear export with that in Figure 13-35 for nuclear import, we can see one obvious difference: Ran·GTP is part of the cargo complex during export but not during import. Apart from this difference, the two transport processes are remarkably similar. In both processes, association of a transport signal receptor with Ran·GTP in the nucleoplasm causes a conformational change that affects its affinity for the transport signal. During import, the interaction causes release of the cargo, whereas during export, the interaction promotes association with the cargo. In both export and import, stimulation of Ran·GTP hydrolysis in the cytoplasm by Ran-GAP produces a conformational change in Ran that releases the transport signal receptor. During nuclear export, the cargo is also released. Importins and exportins both are thought to diffuse through the NPC channel by successive interactions with FG-repeats in FG-nucleoporins. Localization of the Ran-GAP and -GEF to the cytoplasm and nucleus, respectively, is the basis for the unidirectional transport of cargo proteins across the NPC.

In keeping with the similarity in function of importins and exportins, the two types of transport proteins are highly homologous in sequence and structure. The entire family is called the importin β family, or **karyopherins.** There are 14 karyopherins in yeast and more than 20 in mammalian cells. The NESs or NLSs to which they bind have been determined for only a fraction of them. Remarkably, some individual karyopherins function as both an importin and an exportin.

A similar shuttling mechanism has been shown to export other cargoes from the nucleus. For example, exportin-t functions to export tRNAs. Exportin-t binds fully processed tRNAs in a complex with Ran·GTP that diffuses through NPCs and dissociates when it interacts with Ran-GAP in the NPC cytoplasmic filaments, releasing the tRNA into the cytosol. A Ran-dependent process is also required for the nuclear export of ribosomal subunits through NPCs once the protein and RNA components have been properly assembled in the nucleolus. Likewise, certain specific mRNAs that associate with particular hnRNP proteins can be exported by a Ran-dependent mechanism.

Most mRNAs Are Exported from the Nucleus by a Ran-Independent Mechanism

Once the processing of an mRNA is completed in the nucleus, it remains associated with specific hnRNP proteins in a *messenger ribonuclear protein complex,* or mRNP. The principle transporter of mRNPs out of the nucleus is the **mRNP exporter,** a heterodimeric protein composed of a large subunit called *nuclear export factor 1 (NXF1)* or *TAP* and a small subunit, *nuclear export transporter 1 (Nxt1).* The large subunit binds to nuclear mRNPs through cooperative interactions with the RNA and other mRNP adapter proteins that associate with nascent pre-mRNAs during transcription elongation and pre-mRNA processing. It seems likely that multiple TAP/Nxt1 mRNP exporters bound along the length of an mRNP assist in its export. TAP/Nxt1 acts like a karyopherin in the sense that both subunits interact with the FG-domains of FG-nucleoporins, allowing them to diffuse through the central channel of the NPC. Tap also binds reversibly to the protein Gle2, which in turn binds a nucleoporin in the nuclear basket, presumably positioning the mRNP for export through the nuclear pore. A nucleoporin in the cytoplasmic filaments of the NPC is also required for mRNP export. This nucleoporin binds an RNA helicase that is proposed to function in the dissociation of the mRNP exporter and other hnRNP proteins from the mRNP as it reaches the cytoplasm.

The TAP/Nxt1 mRNP exporters do not appear to interact with Ran, and thus the unidirectional transport of mRNA out of the nucleus requires a source of energy other than GTP hydrolysis by Ran. As the mRNP complex is transported through an NPC, the proteins associated with

(a)

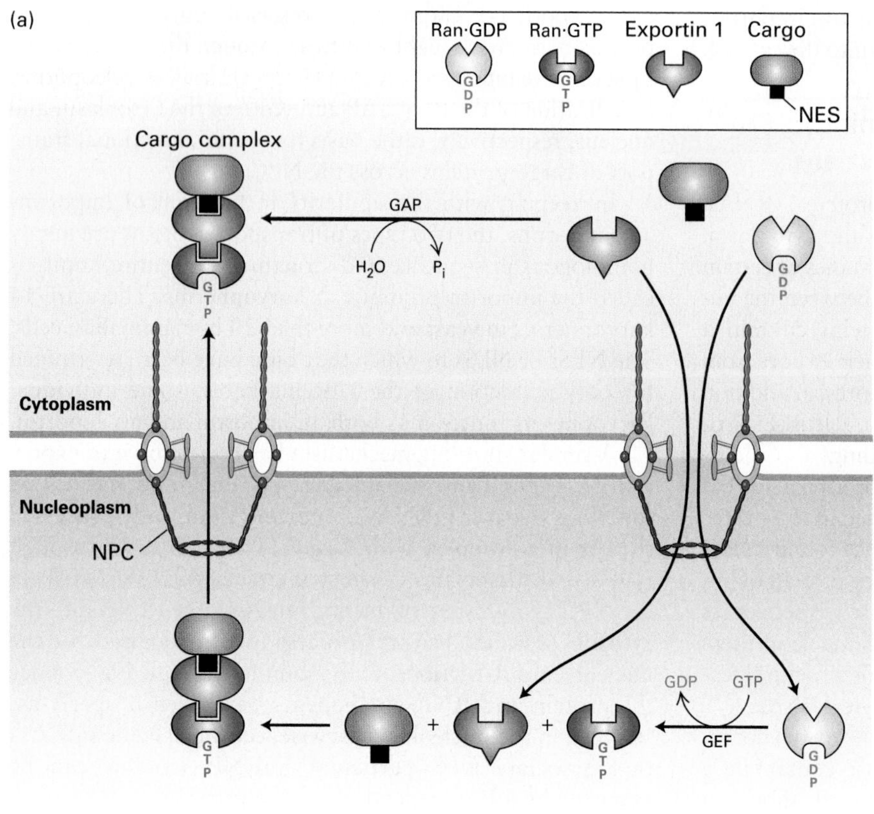

(b)

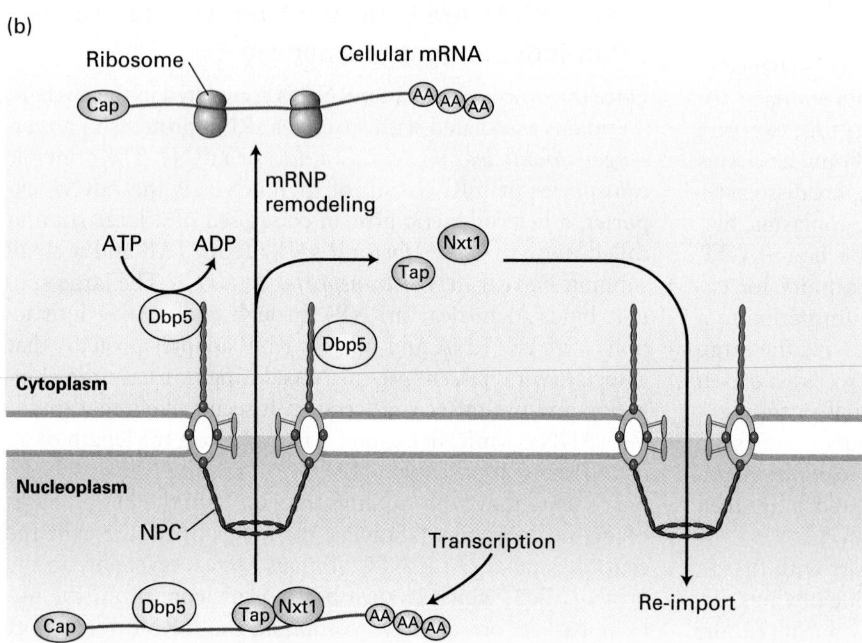

◀ **FIGURE 13-36 Ran-dependent and Ran-independent nuclear export.** (a) Ran-dependent mechanism for nuclear export of cargo proteins containing a leucine-rich nuclear-export signal (NES). In the nucleoplasm, the protein exportin 1 binds cooperatively to the NES of the cargo protein to be transported and to Ran·GTP. After the resulting cargo complex diffuses through an NPC via transient interactions with FG repeats in FG-nucleoporins, the Ran-GAP associated with the NPC cytoplasmic filaments stimulates conversion of Ran·GTP to Ran·GDP. The accompanying conformational change in Ran leads to dissociation of the complex. The NES-containing cargo protein is released into the cytosol, whereas exportin 1 and Ran·GDP are transported back into the nucleus through NPCs. Ran-GEF in the nucleoplasm then stimulates conversion of Ran·GDP to Ran·GTP. (b) Ran-independent nuclear export of mRNAs. The heterodimeric TAP/Nxt1 complex binds to mRNA-protein complexes (mRNPs) in the nucleus. Association of TAP/Nxt1 with components of the NPC direct the associated mRNP to the central channel of the pore. An RNA helicase (Dbp5) located on the cytoplasmic side of the NPC is thought to provide the driving force by hydrolysis for moving the mRNP through the pore. The helicase also frees the mRNA from TAP and Nxt1 proteins, which are recycled back into the nucleus by the Ran-dependent import process depicted in Figure 13-35.

it are exchanged for another set of proteins in the cytoplasm, a process called mRNP remodeling (Figure 13-36b). Several nuclear mRNP proteins dissociate from the mRNP before it reaches the cytoplasmic side of the NPC. These remain in the nucleus, where they bind to newly synthesized nascent pre-mRNA. Other nuclear mRNP proteins, including the TAP/Nxt1 mRNP exporter, are exported through NPCs into the cytoplasm. Once they reach the cytoplasmic side of the NPC, they dissociate from the mRNP with the help of the RNA helicase, Dbp5, which associates with cytoplasmic NPC filaments. Recall that RNA helicases use the energy derived from hydrolysis of ATP to move along RNA molecules, separating double-stranded RNA chains and dissociating RNA-protein complexes (Chapter 4). This leads to the simple idea that Dpb5, which associates with the cytoplasmic side of the nuclear pore complex, acts as an ATP-driven motor to move mRNP complexes through the nuclear pore.

After remodeling is completed, the TAP and Nxt1 proteins that have been stripped from the mRNA by Dbp5 helicase are imported back into the nucleus by an importin, where they can function in the export of another mRNP. Consequently, these nuclear mRNP proteins shuttle between the nucleus and cytoplasm, carrying mRNPs through NPCs (Figure 13-36b).

KEY CONCEPTS OF SECTION 13.6

Transport into and out of the Nucleus

- The nuclear envelope contains numerous nuclear pore complexes (NPCs), large, complicated structures composed of multiple copies of 30 proteins called *nucleoporins* (see Figure 13-32). FG-nucleoporins, which contain multiple repeats of a short hydrophobic sequence (FG-repeats), line the central transporter channel and play a role in transport of all macromolecules through nuclear pores.
- Transport of macromolecules larger than 20–40 kDa through nuclear pores requires the assistance of proteins that interact with both the transported molecule and FG-repeats of FG-nucleoporins.
- Proteins imported to or exported from the nucleus contain a specific amino acid sequence that functions as a nuclear-localization signal (NLS) or a nuclear-export signal (NES). Nucleus-restricted proteins contain an NLS but not an NES, whereas proteins that shuttle between the nucleus and cytoplasm contain both signals.
- Several different types of NES and NLS have been identified. Each type of nuclear-transport signal is thought to interact with a specific receptor protein (exportin or importin) belonging to a family of homologous proteins termed *karyopherins*.
- A "cargo" protein bearing an NES or NLS translocates through nuclear pores bound to its cognate receptor protein (karyopherin), which also interacts with FG-nucleoporins. Importins and exportins are thought to diffuse through the channel, filled with a hydrophobic matrix of FG-repeats. Both transport processes also require participation of Ran, a monomeric G protein that exists in different conformations when bound to GTP or GDP.
- After a cargo complex reaches its destination (the cytoplasm during export and the nucleus during import), it dissociates, freeing the cargo protein and other components. The latter then are transported through nuclear pores in the reverse direction to participate in transporting additional molecules of cargo protein (see Figures 13-35 and 13-36).
- The unidirectional nature of protein export and import through nuclear pores results from localization of the Ran guanine nucleotide-exchange factor (GEF) in the nucleus and of Ran GTPase-activating protein (GAP) in the cytoplasm. The interaction of import cargo complexes with the Ran-GTP in the nucleoplasm causes dissociation of the complex, releasing the cargo into the nucleoplasm (see Figure 13-35). Export cargo complexes dissociate in the cytoplasm when they interact with Ran·GAP localized to the NPC cytoplasmic filaments (see Figure 13-36).
- Most mRNPs are exported from the nucleus by a heterodimeric mRNP exporter that interacts with FG-repeats of FG-nucleoporins. The direction of transport (nucleus to cytoplasm) may result from the action of an RNA helicase associated with the cytoplasmic filaments of the nuclear pore complexes.

Perspectives for the Future

As we have seen in this chapter, we now understand many aspects of the basic processes responsible for selectively transporting proteins into the endoplasmic reticulum (ER), mitochondrion, chloroplast, peroxisome, and nucleus. Biochemical and genetic studies, for instance, have identified cis-acting signal sequences responsible for targeting proteins to the correct organelle membrane and the membrane receptors that recognize these signal sequences. We also have learned much about the underlying mechanisms that translocate proteins across organelle membranes and have determined whether energy is used to push or pull proteins across the membrane in one direction, the type of channel through which proteins pass, and whether proteins are translocated in a folded or an unfolded state. Nonetheless, many fundamental questions remain unanswered, including how fully folded proteins move across a membrane and how the topology of multipass membrane proteins is determined.

The peroxisomal import machinery provides one example of the translocation of folded proteins. It not only is capable of translocating fully folded proteins with bound cofactors into the peroxisomal matrix but can even direct the import of a large gold particle decorated with a (PTS1) peroxisomal targeting peptide. Some researchers have speculated that the mechanism of peroxisomal import may be related to that of nuclear import, the best-understood example of post-translational translocation of folded proteins. Both the peroxisomal and nuclear import machinery can transport folded molecules of very divergent sizes, and both appear to involve a component that cycles between the cytosol and the organelle interior—the Pex5 PTS1 receptor in the case of peroxisomal import and the Ran-importin complex in the case of nuclear import. However, there also appear to be crucial differences between the two translocation processes. For example, nuclear pores represent large, stable macromolecular assemblies readily observed by electron microscopy, whereas analogous porelike structures have not been observed in the peroxisomal membrane. Moreover, small molecules can readily pass through nuclear pores, whereas peroxisomal membranes maintain a permanent barrier to the diffusion of small hydrophilic molecules. Taken together, these observations suggest that peroxisomal import may require an entirely new type of translocation mechanism.

The evolutionarily conserved mechanisms for translocating folded proteins across the cytoplasmic membrane of bacterial cells and across the thylakoid membrane of chloroplasts also are poorly understood. A better understanding of all of these processes for translocating folded proteins across a membrane will likely hinge on future development of in vitro translocation systems that allow investigators to define the biochemical mechanisms driving translocation and to identify the structures of trapped translocation intermediates.

Compared with our understanding of how soluble proteins are translocated into the ER lumen and mitochondrial matrix, our understanding of how cis-acting sequences specify the topology of multipass membrane proteins is quite elementary. For instance, we do not know how the translocon channel accommodates polypeptides that are oriented differently with respect to the membrane, nor do we understand how local polypeptide sequences interact with the translocon channel both to set the orientation of transmembrane spans and to signal for lateral passage into the membrane bilayer. A better understanding of how the amino acid sequences of membrane proteins can specify membrane topology will be crucial for decoding the vast amount of structural information for membrane proteins contained within databases of genomic sequences.

A more detailed understanding of all translocation processes should continue to emerge from genetic and biochemical studies, both in yeasts and in mammals. These studies will undoubtedly reveal additional key proteins involved in the recognition of targeting sequences and in the translocation of proteins across lipid bilayers. Finally, the structural studies of translocon channels will likely be extended in the future to reveal at resolutions on the atomic scale the conformational states that are associated with each step of the translocation cycle.

Key Terms

biomolecular cargo complex *572*
cotranslational translocation *537*
dislocation *556*
dolichol phosphate *550*
exportin *573*
FG-nucleoporins *572*
general import pore *559*
hydropathy profile *548*
importins *571*
karyopherins *574*
molecular chaperones *541*
N-linked oligosaccharides *550*
nuclear pore complex (NPC) *570*
O-linked oligosaccharides *550*
post-translational translocation *540*
Ran protein *571*
signal-anchor sequence *544*
signal-recognition particle (SRP) *537*
signal (uptake-targeting) sequences *535*
stop-transfer anchor sequence *544*
topogenic sequences *543*
topology of a membrane protein *543*
translocon *539*
trimolecular cargo complex *573*
unfolded-protein response *555*

Review the Concepts

1. Describe the source or sources of energy needed for unidirectional translocation across the membrane in (a) cotranslational translocation into the endoplasmic reticulum (ER); (b) post-translational translocation into the ER; (c) translocation across the bacterial cytoplasmic membrane; and (d) translocation into the mitochondrial matrix.

2. Translocation into most organelles usually requires the activity of one or more cytosolic proteins. Describe the basic function of three different cytosolic factors required for translocation into the ER, mitochondria, and peroxisomes, respectively.

3. Describe the typical principles used to identify topogenic sequences within proteins and how these can be used to develop computer algorithms. How does the identification of topogenic sequences lead to prediction of the membrane arrangement of a multipass protein? What is the importance of the arrangement of positive charges relative to the membrane orientation of a signal-anchor sequence?

4. The endoplasmic reticulum (ER) is an important site of "quality control" for newly synthesized proteins. What is meant by "quality control" in this context? What accessory proteins are typically involved in the processing of newly synthesized proteins within the ER? Cells generally degrade ER-exit-incompetent proteins. Where within the cell does such degradation occur and what is the relationship of the Sec61 protein translocon and p97 to the degradation process?

5. Temperature-sensitive yeast mutants have been isolated that block each of the enzymatic steps in the synthesis of the dolichol-oligosaccharide precursor for *N*-linked glycosylation (see Figure 13-17). Propose an explanation for why mutations that block synthesis of the intermediate with the structure dolichol-PP-$(GlcNAc)_2Man_5$ completely prevent addition of *N*-linked oligosaccharide chains to secretory proteins, whereas mutations that block conversion of this intermediate into the completed precursor—dolichol-PP-$(GlcNAc)_2Man_9Glc_3$—allow the addition of *N*-linked oligosaccharide chains to secretory glycoproteins.

6. Name four different proteins that facilitate the modification and/or folding of secretory proteins within the lumen of the ER. Indicate which of these proteins covalently modifies substrate proteins and which brings about only conformational changes in substrate proteins.

7. Because you are interested in studying how a particular secretory protein folds within the ER, you wish to determine whether BiP binds to the newly synthesized protein in ER extracts. You find that you can isolate some of the newly synthesized secretory protein bound to BiP when ADP is added to the cell extract but not when ATP is added to the extract. Explain this result based on the mechanism for BiP binding to substrate proteins.

8. Describe what would happen to the precursor of a mitochondrial matrix protein in the following types of mitochondrial mutants: (a) a mutation in the Tom22 signal receptor, (b) a mutation in the Tom70 signal receptor, (c) a mutation in the matrix Hsc70, and (d) a mutation in the matrix signal peptidase.

9. Describe the similarities and differences between the mechanism of import into the mitochondrial matrix and the chloroplast stroma.

10. Design a set of experiments using chimeric proteins, composed of a mitochondrial precursor protein fused to dihydrofolate reductase (DHFR), that could be used to determine how much of the precursor protein must protrude into the mitochondrial matrix in order for the matrix-targeting sequence to be cleaved by the matrix-processing protease (see Figure 13-24).

11. Protein targeting to both mitochondria and chloroplasts involves the sorting of proteins to multiple sites within the respective organelle. Briefly list these sites. Taking the mitochondrion as an example and the proteins ADP/ATP anti-porter and cytochrome b_2 as the specific cases, compare and contrast the extent to which a common mechanism is used for the site-specific targeting of these two proteins.

12. Peroxisomes contain enzymes that use molecular oxygen to oxidize various substrates, but in the process hydrogen peroxide forms and must be degraded. What is the name of the enzyme responsible for the breakdown of hydrogen peroxide to water and what mechanism and associated proteins allow for its import into the peroxisome?

13. Suppose that you have identified a new mutant cell line that lacks functional peroxisomes. Describe how you could determine experimentally whether the mutant is primarily defective for insertion/assembly of peroxisomal membrane proteins or matrix proteins.

14. Evidence throughout Chapter 13 reveals that specific motifs within polypeptides are necessary to direct or target these proteins across membranes and into organelles. The nuclear import of proteins having a molecular mass more than approximately 40 kDa is no different, and they must be actively imported through nuclear pore complexes. What is the name given to the amino acid sequence that allows the selective transport of macromolecular cargo proteins into the nucleus? Name three proteins that are required for this import and briefly describe how they function.

15. Why is localization of Ran-GAP in the nucleus and Ran-GEF in the cytoplasm necessary for unidirectional transport of cargo proteins containing an NES?

Analyze the Data

Imagine that you are evaluating the early steps in translocation and processing of the secretory protein prolactin. By using an experimental approach similar to that shown in Figure 13-7, you can use truncated prolactin mRNAs to control the length of nascent prolactin polypeptides that are synthesized. When prolactin mRNA that lacks a chain-termination (stop) codon is translated in vitro, the newly synthesized polypeptide ending with the last codon included on the mRNA will remain attached to the ribosome, thus allowing a polypeptide of defined length to extend from the ribosome. You have generated a set of mRNAs that encode segments of the N-terminus of prolactin of increasing length, and each mRNA can be translated in vitro by a cytosolic translation extract containing ribosomes, tRNAs, aminoacyl-tRNA synthetases, GTP, and translation initiation and elongation factors. When radiolabeled amino acids are included in the translation mixture, only the polypeptide encoded by the added mRNA will be labeled. After completion of translation, each reaction mixture was resolved by SDS poly-acrylamide gel electrophoresis, and the labeled polypeptides were identified by autoradiography.

a. The autoradiogram depicted below shows the results of an experiment in which each translation reaction was carried out either in the presence (+) or the absence (−) of microsomal membranes. Based on the gel mobility of peptides synthesized in the presence or absence of microsomes, deduce how long the prolactin nascent chain must be in order for the prolactin signal peptide to enter the ER lumen and to be cleaved by signal peptidase. (Note that microsomes carry significant quantities of SRP weakly bound to the membranes.)

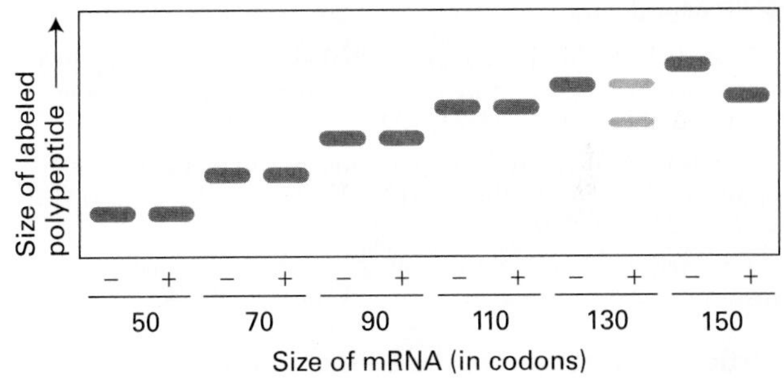

b. Given this length, what can you conclude about the conformational state(s) of the nascent prolactin polypeptide when it is cleaved by signal peptidase? The following lengths will be useful for your calculation: the prolactin signal sequence is cleaved after amino acid 31; the channel within the ribosome occupied by a nascent polypeptide is about 150 Å long; a membrane bilayer is about 50 Å thick; in polypeptides with an α-helical conformation, one residue extends 1.5 Å, whereas in fully extended polypeptides, one residue extends about 3.5 Å.

c. The experiment described in part (a) is carried out in an identical manner except that microsomal membranes are not present during translation but are added after translation is complete. In this case none of the samples shows a difference in mobility in the presence or absence of microsomes. What can you conclude about whether prolactin can be translocated into isolated microsomes post-translationally?

d. In another experiment, each translation reaction was carried out in the presence of microsomes, and then the microsomal membranes and bound ribosomes were separated from free ribosomes and soluble proteins by centrifugation. For each translation reaction, both the total reaction (T) and the membrane fraction (M) were resolved in neighboring gel lanes. Based on the amounts of labeled polypeptide in the membrane fractions in the autoradiogram depicted below, deduce how long the prolactin nascent chain must be in order for ribosomes engaged in translation to engage the SRP and thereby become bound to microsomal membranes.

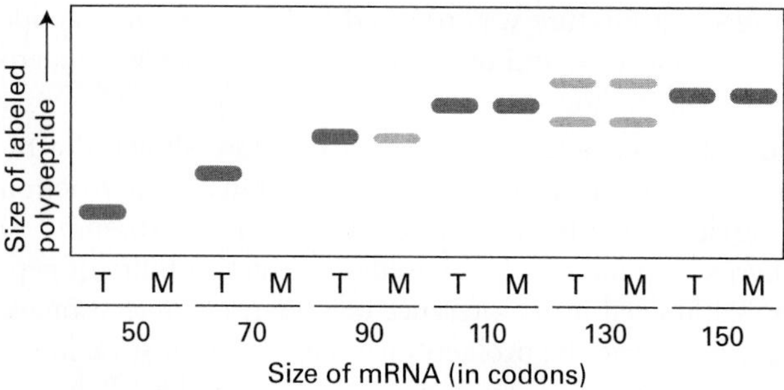

References

Translocation of Secretory Proteins Across the ER Membrane

Egea, P. F., R. M. Stroud, and P. Walter. 2005. Targeting proteins to membranes: structure of the signal recognition particle. *Curr. Opin. Struct. Biol.* **15**:213–220.

Osborne, A. R., T. A. Rapoport, and B. van den Berg. 2005. Protein translocation by the Sec61/SecY channel. *Annu. Rev. Cell Dev. Biol.* **21**:529–550.

Wickner, W., and R. Schekman. 2005. Protein translocation across biological membranes. *Science* **310**:1452–1456.

Insertion of Proteins into the ER Membrane

Englund, P. T. 1993. The structure and biosynthesis of glycosylphosphatidylinositol protein anchors. *Ann. Rev. Biochem.* **62**:121–138.

Goder, V., and M. Spiess. 2001. Topogenesis of membrane proteins: determinants and dynamics. *FEBS Lett.* **504**:87–93.

Mothes, W., et al. 1997. Molecular mechanism of membrane protein integration into the endoplasmic reticulum. *Cell* **89**:523–533.

von Heijne, G. 1999. Recent advances in the understanding of membrane protein assembly and structure. *Q. Rev. Biophys.* **32**:285–307.

Protein Modifications, Folding, and Quality Control in the ER

Helenius, A., and M. Aebi. 2004. Roles of N-linked glycans in the endoplasmic reticulum. *Annu. Rev. Biochem.* **73**:1019–1049.

Kornfeld, R., and S. Kornfeld. 1985. Assembly of asparagine-linked oligosaccharides. *Ann. Rev. Biochem.* **45**:631–664.

Patil, C., and P. Walter. 2001. Intracellular signaling from the endoplasmic reticulum to the nucleus: the unfolded protein response in yeast and mammals. *Curr. Opin. Cell Biol.* **13**:349–355.

Meusser, B., C. Hirsch, E. Jarosch, and T. Sommer. 2005. ERAD: the long road to destruction. *Nature Cell Biol.* **7**:766–772.

Sevier, C. S., and C. A. Kaiser. 2002. Formation and transfer of disulphide bonds in living cells. *Nature Rev. Mol. Cell Biol.* **3**:836–847.

Silberstein, S., and R. Gilmore. 1996. Biochemistry, molecular biology, and genetics of the oligosaccharyltransferase. *FASEB J.* **10**:849–858.

Trombetta, E. S., and A. J. Parod. 2003. Quality control and protein folding in the secretory pathway. *Annu. Rev. Cell Dev. Biol.* **19**:649–676.

Tsai, B., Y. Ye, and T. A. Rapoport. 2002. Retro-translocation of proteins from the endoplasmic reticulum into the cytosol. *Nature Rev. Mol. Cell Biol.* **3**:246–255.

Sorting of Proteins to Mitochondria and Chloroplasts

Koehler, C. M. 2004. New developments in mitochondrial assembly. *Ann. Rev. Cell Dev. Biol.* **20**:309–335.

Dolezal, P., V. Likic, J. Tachezy, and T. Lithgow. 2006. Evolution of the molecular machines for protein import into mitochondria. *Science* **313**:314–318.

Dalbey, R. E., and A. Kuhn. 2000. Evolutionarily related insertion pathways of bacterial, mitochondrial, and thylakoid membrane proteins. *Ann. Rev. Cell Devel. Biol.* **16**:51–87.

Matouschek, A., N. Pfanner, and W. Voos. 2000. Protein unfolding by mitochondria: the Hsp70 import motor. *EMBO Rept.* **1**:404–410.

Neupert, W., and M. Brunner. 2002. The protein import motor of mitochondria. *Nature Rev. Mol. Cell Biol.* **3**:555–565.

Rapaport, D. 2005. How does the TOM complex mediate insertion of precursor proteins into the mitochondrial outer membrane? *J. Cell Biol.* **171**:419–423.

Robinson, C., and A. Bolhuis. 2001. Protein targeting by the twin-arginine translocation pathway. *Nature Rev. Mol. Cell Biol.* **2**:350–356.

Soll, J., and E. Schleiff. 2003. Protein import into chloroplasts. *Nature Rev. Mol. Cell Biol.* **5**:198–208.

Truscott, K. N., K. Brandner, and N. Pfanner. 2003. Mechanisms of protein import into mitochondria. *Curr. Biol.* **13**:R326–R337.

Sorting of Peroxisomal Proteins

Dammai, V., and S. Subramani. 2001. The human peroxisomal targeting signal receptor, Pex5p, is translocated into the peroxisomal matrix and recycled to the cytosol. *Cell* **105**:187–196.

Gould, S. J., and C. S. Collins. 2002. Opinion: peroxisomal-protein import: is it really that complex? *Nature Rev. Mol. Cell Biol.* **3**:382–389.

Gould, S. J., and D. Valle. 2000. Peroxisome biogenesis disorders: genetics and cell biology. *Trends Genet.* **16**:340–345.

Hoepfner, D., D. Schildknegt, I. Braakman, P. Philippsen, and H. F. Tabak. 2005. Contribution of the endoplasmic reticulum to peroxisome formation. *Cell* **122**:85-95.

Purdue, P. E., and P. B. Lazarow. 2001. Peroxisome biogenesis. *Ann. Rev. Cell Devel. Biol.* **17**:701–752.

Subramani, S., A. Koller, and W. B. Snyder. 2000. Import of peroxisomal matrix and membrane proteins. *Ann. Rev. Biochem.* **69**:399–418.

Transport into and out of the Nucleus

Chook, Y. M., and G. Blobel. 2001. Karyopherins and nuclear import. *Curr. Opin. Struct. Biol.* **11**:703–715.

Cole, C. N., and J. J. Scarcelli. 2006. Transport of messenger RNA from the nucleus to the cytoplasm. *Curr. Opin. Cell Biol.* **18**:299–306.

Johnson, A. W., E. Lund, and J. Dahlberg. 2002. Nuclear export of ribosomal subunits. *Trends Biochem. Sci.* **27**:580–585.

Ribbeck, K., and D. Gorlich. 2001. Kinetic analysis of translocation through nuclear pore complexes. *EMBO J.* **20**:1320–1330.

Rout, M. P., and J. D. Aitchison. 2001. The nuclear pore complex as a transport machine. *J. Biol. Chem.* **276**:16593–16596.

Schwartz, T. U. 2005. Modularity within the architecture of the nuclear pore complex. *Curr. Opin. Struct. Biol.* **15**:221–226.

Suntharalingam, M., and S. R. Wente. 2003. Peering through the pore: nuclear pore complex structure, assembly, and function. *Dev. Cell.* **4**:775–789.

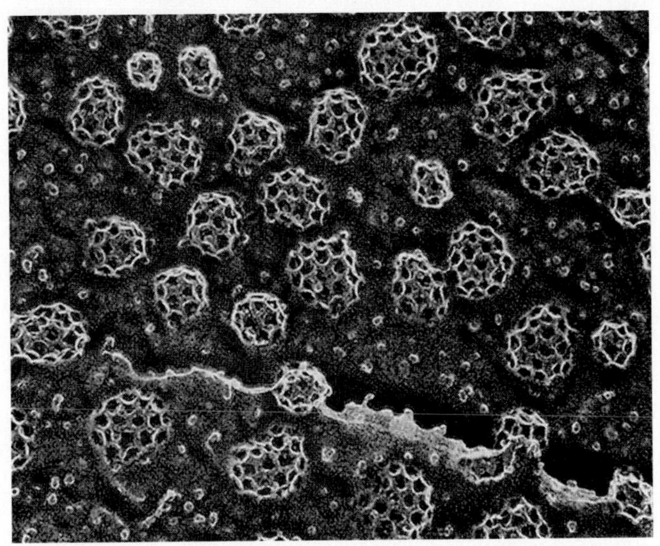

Scanning electron micrograph showing the formation of clathrin-coated vesicles on the cytosolic face of the plasma membrane. [John Heuser, Washington University School of Medicine.]

CHAPTER

14

VESICULAR TRAFFIC, SECRETION, AND ENDOCYTOSIS

In the previous chapter we explored how proteins are targeted to and translocated across the membranes of several different intracellular organelles, including the endoplasmic reticulum, mitochondria and chloroplasts, peroxisomes, and the nucleus. In this chapter we turn our attention to the **secretory pathway** and the mechanisms that allow soluble and membrane proteins to be delivered to the plasma membrane and the lysosome. We will also discuss the related processes of endocytosis and autophagy, which deliver proteins and small molecules from either outside the cell or from the cytoplasm to the interior of the lysosome for degradation.

Soluble and membrane proteins slated to function at the cell surface or in the lysosome are transported to their final destination via the secretory pathway. Proteins delivered to the plasma membrane include cell-surface receptors, transporters for nutrient uptake, and ion channels that maintain the proper ionic and electrochemical balance across the plasma membrane. Soluble secreted proteins follow the same pathway to the cell surface as plasma membrane proteins, but instead of remaining embedded in the membrane, secreted proteins are released into the aqueous extracellular environment in soluble form. Examples of secreted proteins are digestive enzymes, peptide hormones, serum proteins, and collagen. As described in Chapter 9, the lysosome is an organelle with an acidic interior that is generally used for degradation of unwanted proteins and the storage of small molecules such as amino acids. Accordingly, the types of proteins delivered to the lysosomal membrane are subunits of the V-class proton pump that pumps H^+ from the cytosol into the acidic lumen of the lysosome as well as transporters to release small molecules stored in the lysosome into the cytoplasm. Soluble proteins delivered by this pathway include lysosomal digestive enzymes such as proteases, glycosidases, phosphatases, and lipases.

In contrast to the secretory pathway, which is generally used to deliver newly synthesized membrane proteins to their correct address, the **endocytic pathway** is used to take up substances from the cell surface into the interior of the cell. The endocytic pathway is used to take up certain nutrients that are too large to be transported across the plasma membrane by one of the transport mechanisms discussed in Chapter 11. For example, the endocytic pathway is utilized in the uptake of cholesterol carried in LDL particles and iron atoms carried by the iron-binding protein transferrin. In addition, the endocytic pathway can be used remove receptor proteins from the cell surface as a way to down-regulate their activity.

OUTLINE

A single unifying principle governs all protein trafficking in the secretory and endocytic pathways: transport of membrane and soluble proteins from one membrane-bounded compartment to another is mediated by **transport vesicles** that collect "*cargo" proteins* in buds arising from the membrane of one compartment and then deliver these cargo proteins to the next compartment by fusing with the membrane of that compartment. Importantly, as transport vesicles bud from one membrane and fuse with the next, the same face of the membrane remains oriented toward the cytosol. Therefore once a protein has been inserted into the membrane or the lumen of the ER, the protein can be carried along the secretory pathway, moving from one organelle to the next without being translocated across another membrane or altering its orientation within the membrane. Similarly, the endocytic pathway uses vesicle traffic to transport proteins from the plasma membrane to the endosome and lysosome and thus preserves their orientation in the membrane of these organelles. Figure 14-1 outlines the major routes for protein trafficking in the cell.

Reduced to its simplest elements, the secretory pathway for delivery of newly synthesized proteins to the plasma membrane or the lysosome operates in two stages. The first stage takes place in the rough endoplasmic reticulum (ER), as described in Chapter 13. Newly synthesized soluble and membrane proteins are translocated into the ER, where they fold into their proper conformation and receive covalent modifications such as N-linked and O-linked carbohydrates and disulfide bonds. Once newly synthesized proteins are properly folded and have received their correct modifications in the ER lumen, they progress to the second stage of the secretory pathway, transport through the Golgi. In the ER, secretory proteins are packaged into anterograde (forward-moving) transport vesicles. These vesicles fuse with each other to form a flattened membrane-bounded compartment known as the *cis*-Golgi **cisterna.** Certain proteins, mainly ER-localized proteins, are retrieved from the *cis*-Golgi to the ER via a different set of retrograde (backward-moving) transport vesicles. A new *cis*-Golgi cisterna with its cargo of proteins physically moves from the *cis* position (nearest the ER) to the *trans* position (farthest from the ER), successively becoming first a *medial*-Golgi cisterna and then a *trans*-Golgi cisterna. This process, known as cisternal maturation, does not involve the budding off and fusion of anterograde transport vesicles. During cisternal maturation, enzymes and other Golgi-resident proteins are constantly being retrieved from later to earlier Golgi cisternae by retrograde transport vesicles, thereby remaining localized to the *cis*-, *medial*-, or *trans*-Golgi cisternae. As secretory proteins move through the Golgi, they can receive further modifications to linked carbohydrates by specific glycosyl transferases that are housed in the different Golgi compartments.

Proteins in the secretory pathway that are destined for the plasma membrane or lysosome eventually reach a complex network of membranes and vesicles termed the ***trans*-Golgi network** (TGN). The TGN is a major branch point in the secretory pathway, and through a process known as *protein sorting*, a protein can be loaded into one of at least three different kinds of vesicles that bud from the TGN. After budding from the *trans*-Golgi network, the first type of vesicle immediately moves to and fuses with the plasma membrane in a process known as **exocytosis,** thus releasing its contents to the exterior of the cell while the membrane proteins from the vesicle become incorporated into the plasma membrane. In all cell types, at least some proteins are loaded into such vesicles and secreted continuously in this manner. The second type of vesicle to bud from the *trans*-Golgi network, known as **secretory vesicles,** are stored inside the cell until a signal for exocytosis causes release of their contents at the plasma membrane. Among the proteins released by such regulated secretion are peptide hormones (e.g., insulin, glucagon, ACTH) from various endocrine cells, precursors of digestive enzymes from pancreatic acinar cells, milk proteins from the mammary gland, and neurotransmitters from neurons. The third type of vesicle that buds from the *trans*-Golgi network is directed to the **lysosome,** an organelle responsible for the intracellular degradation of macromolecules, and to lysosome-like storage organelles in certain cells. Secretory proteins destined for lysosomes are first transported by vesicles from the *trans*-Golgi network to a compartment usually called the **late endosome;** proteins then are transferred to the lysosome by direct fusion of the endosome with the lysosomal membrane.

Endocytosis is related mechanistically to the secretory pathway. In the endocytic pathway, vesicles bud from the plasma membrane, bringing membrane proteins and their bound **ligands** into the cell (see Figure 14-1). After being internalized by endocytosis, some proteins are transported to lysosomes via the late endosome, whereas others are recycled back to the cell surface.

In this chapter we first discuss how our knowledge of the secretory pathway and endocytosis has expanded through experimental techniques. Then we focus on the general mechanisms of membrane budding and fusion. We will see that although different kinds of transport vesicles utilize distinct sets of proteins for their formation and fusion, all vesicles use the same general mechanism for budding, selection of particular sets of cargo molecules, and fusion with the appropriate target membrane. The following two sections show how coordination between particular vesicle trafficking steps can maintain the identity (i.e., a stable set of resident proteins) of the different compartments along the secretory pathway and how cargo selection by vesicles is used to sort proteins to different intracellular locations. Next we will turn our attention to the endocytic pathway to examine how endocytosis is used to transport macromolecules from the extracellular environment into the cell interior. Finally, we will examine the variety of ways that membrane proteins and macromolecules from the cell interior are transported to the lysosome for degradation.

14.1 Techniques for Studying the Secretory Pathway

The key to understanding how proteins are transported through the organelles of the secretory pathway has been to develop a basic description of the function of transport vesicles. Many components required for the formation and

Overview Animation: Protein Secretion

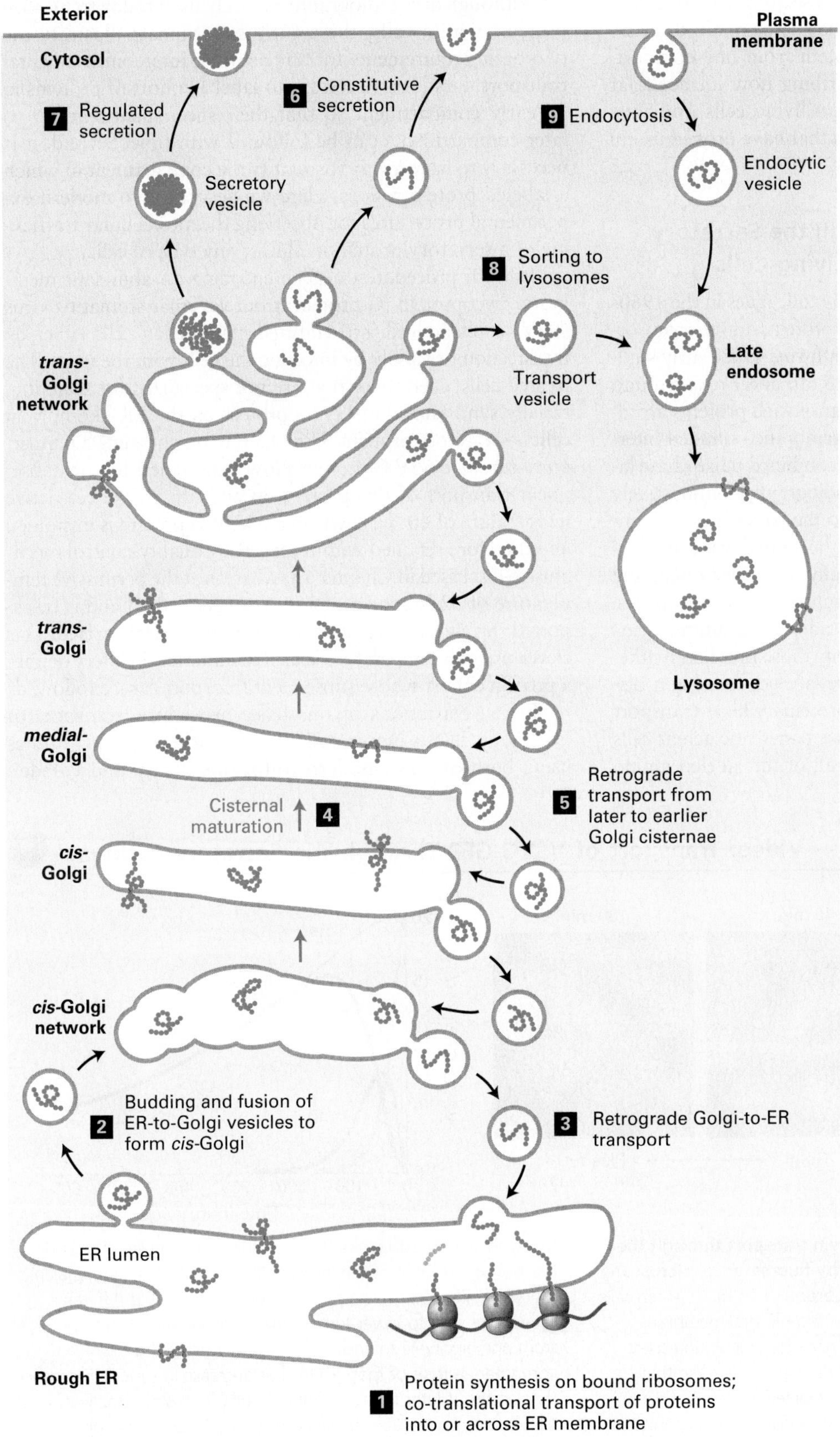

MEDIA CONNECTIONS

Exterior
Plasma membrane
Cytosol
7 Regulated secretion
6 Constitutive secretion
9 Endocytosis
Secretory vesicle
Endocytic vesicle
8 Sorting to lysosomes
Transport vesicle
Late endosome
trans-Golgi network
trans-Golgi
Lysosome
medial-Golgi
5 Retrograde transport from later to earlier Golgi cisternae
Cisternal maturation 4
cis-Golgi
cis-Golgi network
2 Budding and fusion of ER-to-Golgi vesicles to form *cis*-Golgi
3 Retrograde Golgi-to-ER transport
ER lumen
Rough ER
1 Protein synthesis on bound ribosomes; co-translational transport of proteins into or across ER membrane

◄ **FIGURE 14-1 Overview of the secretory and endocytic pathways of protein sorting.** *Secretory pathway:* Synthesis of proteins bearing an ER signal sequence is completed on the rough ER **1**, and the newly made polypeptide chains are inserted into the ER membrane or cross it into the lumen (Chapter 13). Some proteins (e.g., ER enzymes or structural proteins) remain within the ER. The remainder are packaged into transport vesicles **2** that bud from the ER and fuse to form new *cis*-Golgi cisternae. Missorted ER-resident proteins and vesicle membrane proteins that need to be reused are retrieved to the ER by vesicles **3** that bud from the *cis*-Golgi and fuse with the ER. Each *cis*-Golgi cisterna, with its protein content, physically moves from the *cis* to the *trans* face of the Golgi complex **4** by a nonvesicular process called cisternal maturation. Retrograde transport vesicles **5** move Golgi-resident proteins to the proper Golgi compartment. In all cells, certain soluble proteins move to the cell surface in transport vesicles **6** and are secreted continuously (constitutive secretion). In certain cell types, some soluble proteins are stored in secretory vesicles **7** and are released only after the cell receives an appropriate neural or hormonal signal (regulated secretion). Lysosome-destined membrane and soluble proteins, which are transported in vesicles that bud from the *trans*-Golgi **8**, first move to the late endosome and then to the lysosome. *Endocytic pathway:* Membrane and soluble extracellular proteins taken up in vesicles that bud from the plasma membrane **9** also can move to the lysosome via the endosome.

fusion of transport vesicles have been identified in the past decade by a remarkable convergence of the genetic and biochemical approaches described in this section. All studies of intracellular protein trafficking employ some method for assaying the transport of a given protein from one compartment to another. We begin by describing how intracellular protein transport can be followed in living cells and then consider genetic and in vitro systems that have proved useful in elucidating the secretory pathway.

Transport of a Protein Through the Secretory Pathway Can Be Assayed in Living Cells

The classic studies of G. Palade and his colleagues in the 1960s first established the order in which proteins move from organelle to organelle in the secretory pathway. These early studies also showed that secretory proteins are never released into the cytosol, the first indication that transported proteins are always associated with some type of membrane-bounded intermediate. In these experiments, which combined **pulse-chase** labeling (see Figure 3-39) and **autoradiography,** radioactively labeled amino acids were injected into the pancreas of a hamster. At different times after injection, the animal was sacrificed and the pancreatic cells were chemically fixed, sectioned, and subjected to autoradiography to visualize the location of the radiolabeled proteins. Because the radioactive amino acids were administered in a short pulse, only those proteins synthesized immediately after injection were labeled, forming a distinct group, or cohort, of labeled proteins whose transport could be followed. In addition, because pancreatic acinar cells are dedicated secretory cells, almost all of the labeled amino acids in these cells are incorporated into secretory proteins, facilitating the observation of transported proteins.

Although autoradiography is rarely used today to localize proteins within cells, these early experiments illustrate the two basic requirements for any assay of intercompartmental transport. First, it is necessary to label a cohort of proteins in an early compartment so that their subsequent transfer to later compartments can be followed with time. Second, it is necessary to have a way to identify the compartment in which a labeled protein resides. Here we describe two modern experimental procedures for observing the intracellular trafficking of a secretory protein in almost any type of cell.

In both procedures, a gene encoding an abundant membrane glycoprotein (G protein) from vesicular stomatitis virus (VSV) is introduced into cultured mammalian cells either by **transfection** or simply by infecting the cells with the virus. The treated cells, even those that are not specialized for secretion, rapidly synthesize the VSV G protein on the ER like normal cellular secretory proteins. Use of a mutant encoding a temperature-sensitive VSV G protein allows researchers to turn subsequent transport of this protein on and off. At the restrictive temperature of 40 °C, newly made VSV G protein is misfolded and therefore retained within the ER by quality-control mechanisms discussed in Chapter 13, whereas at the permissive temperature of 32 °C, the protein is correctly folded and is transported through the secretory pathway to the cell surface. This clever use of a temperature-sensitive mutation in effect defines a protein cohort whose subsequent transport can be followed.

In two variations of this basic procedure, transport of VSV G protein is monitored by different techniques. Studies using both of these modern trafficking assays and Palade's

MEDIA CONNECTIONS

Video: Transport of VSVG-GFP Through the Secretory Pathway 

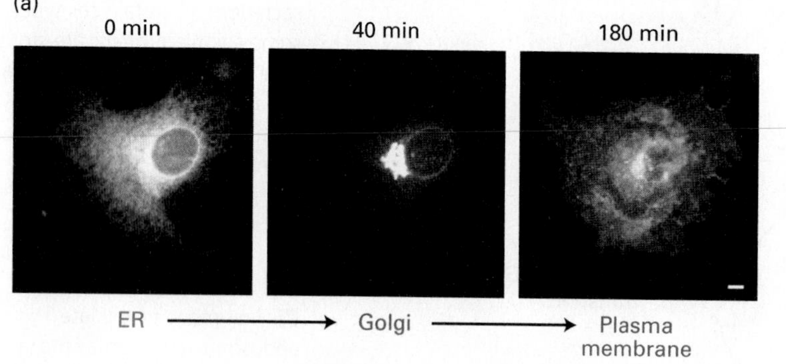

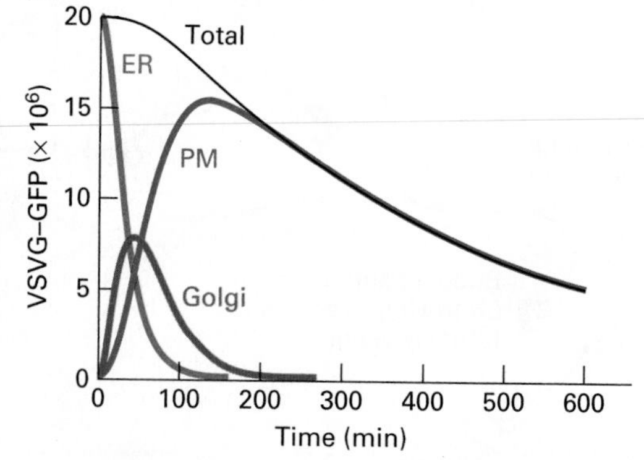

▲ EXPERIMENTAL FIGURE 14-2 **Protein transport through the secretory pathway can be visualized by fluorescence microscopy of cells producing a GFP-tagged membrane protein.** Cultured cells were transfected with a hybrid gene encoding the viral membrane glycoprotein VSV G protein linked to the gene for green fluorescent protein (GFP). A mutant version of the viral gene was used so that newly made hybrid protein (VSVG-GFP) is retained in the ER at 40 °C but is released for transport at 32 °C. (a) Fluorescence micrographs of cells just before and at two times after they were shifted to the lower temperature. Movement of VSVG-GFP from the ER to the Golgi and finally to the cell surface occurred within 180 minutes. The scale bar is 5 μm. (b) Plot of the levels of VSVG-GFP in the endoplasmic reticulum (ER), Golgi, and plasma membrane (PM) at different times after shift to lower temperature. The kinetics of transport from one organelle to another can be reconstructed from computer analysis of these data. The decrease in total fluorescence that occurs at later times probably results from slow inactivation of GFP fluorescence. [From Jennifer Lippincott-Schwartz and Koret Hirschberg, Metabolism Branch, National Institute of Child Health and Human Development.]

early experiments all came to the same conclusion: in mammalian cells vesicle-mediated transport of a protein molecule from its site of synthesis on the rough ER to its arrival at the plasma membrane takes from 30 to 60 minutes.

Microscopy of GFP-Labeled VSV G Protein One approach for observing transport of VSV G protein employs a hybrid gene in which the viral gene is fused to the gene encoding *green fluorescent protein (GFP),* a naturally fluorescent protein (Chapter 9). The hybrid gene is transfected into cultured cells by techniques described in Chapter 5. When cells expressing the temperature-sensitive form of the hybrid protein (VSVG-GFP) are grown at the restrictive temperature, VSVG-GFP accumulates in the ER, which appears as a lacy network of membranes when cells are observed in a fluorescent microscope. When the cells are subsequently shifted to a permissive temperature, the VSVG-GFP can be seen to move first to the membranes of the Golgi apparatus, which are densely concentrated at the edge of the nucleus, and then to the cell surface (Figure 14-2a). By analyzing the distribution of VSVG-GFP at different times after shifting cells to the permissive temperature, researchers have determined how long VSVG-GFP resides in each organelle of the secretory pathway (Figure 14-2b).

Detection of Compartment-Specific Oligosaccharide Modifications A second way to follow the transport of secretory proteins takes advantage of modifications to their carbohydrate side chains that occur at different stages of the secretory pathway. To understand this approach, recall that many secretory proteins leaving the ER contain one or more copies of the **N-linked oligosaccharide** $Man_8(GlcNAc)_2$, which are synthesized and attached to secretory proteins in the ER (see Figure 13-18). As a protein moves through the Golgi complex, different enzymes localized to the *cis-*, *medial-*, and *trans*-Golgi cisternae catalyze an ordered series of reactions to these core $Man_8(GlcNAc)_2$ chains, as discussed in a later section of this chapter. For instance, glycosidases that reside specifically in the *cis*-Golgi compartment sequentially trim mannose residues off the core oligosaccharide to yield a "trimmed" form $Man_5(GlcNAc)_2$. Scientists can use a specialized carbohydrate-cleaving enzyme known as endoglycosidase D to distinguish glycosylated proteins that remain in the ER from those that have entered the *cis*-Golgi: trimmed *cis*-Golgi-specific oligosaccharides are cleaved from proteins by endoglycosidase D, whereas the core (untrimmed) oligosaccharide chains on secretory proteins within the ER are resistant to cleavage by this enzyme (Figure 14-3a). Because a deglycosylated protein produced by endoglycosidase D digestion moves faster on an SDS gel

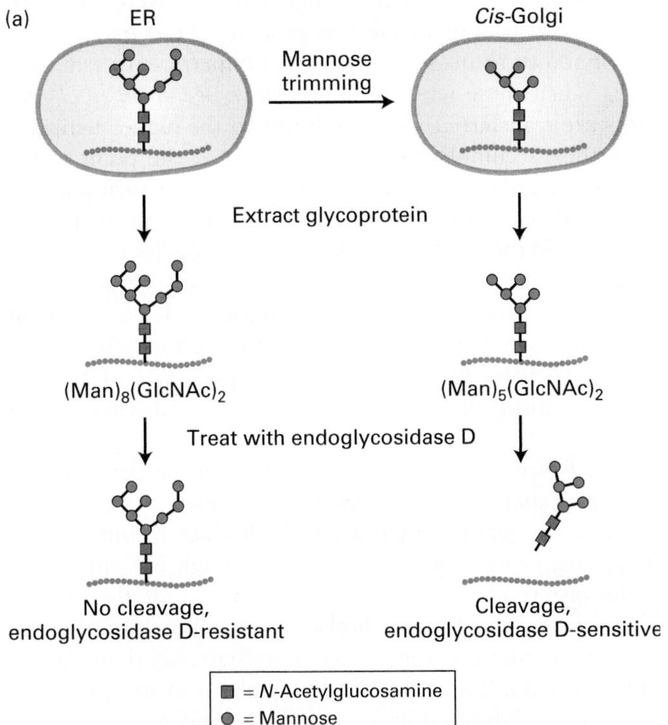

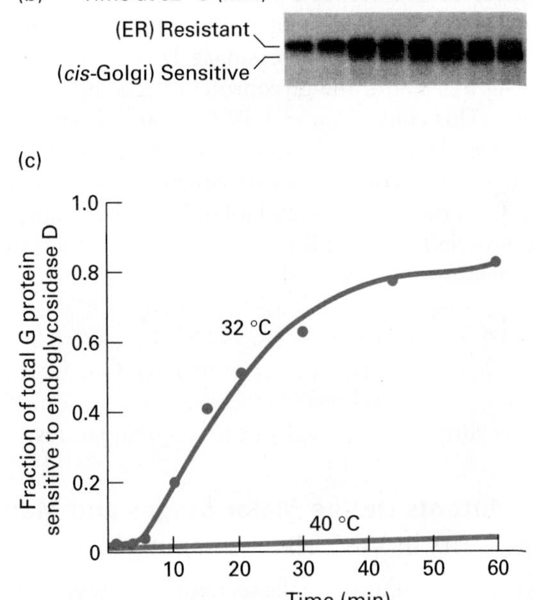

▲ EXPERIMENTAL FIGURE 14-3 **Transport of a membrane glycoprotein from the ER to the Golgi can be assayed based on sensitivity to cleavage by endoglycosidase D.** Cells expressing a temperature-sensitive VSV G protein (VSVG) were labeled with a pulse of radioactive amino acids at the nonpermissive temperature so that labeled protein was retained in the ER. At periodic times after a return to the permissive temperature of 32 °C, VSVG was extracted from cells and digested with endoglycosidase D. (a) As proteins move to the *cis*-Golgi from the ER, the core oligosaccharide $Man_8(GlcNAc)_2$ is trimmed to $Man_5(GlcNAc)_2$ by enzymes that reside in the *cis*-Golgi compartment. Endoglycosidase D cleaves the oligosaccharide chains from proteins processed in the *cis*-Golgi but not from proteins in the ER. (b) SDS gel electrophoresis of the digestion mixtures resolves the resistant, uncleaved (slower-migrating) and sensitive, cleaved (faster-migrating) forms of labeled VSVG. As this electrophoretogram shows, initially all of the VSVG was resistant to digestion, but with time an increasing fraction is sensitive to digestion, reflecting protein transported from the ER to the Golgi and processed there. In control cells kept at 40 °C, only slow-moving, digestion-resistant VSVG was detected after 60 minutes (not shown). (c) Plot of the proportion of VSVG that is sensitive to digestion, derived from electrophoretic data, reveals the time course of ER → Golgi transport. [From C. J. Beckers et al., 1987, *Cell* **50:**523.]

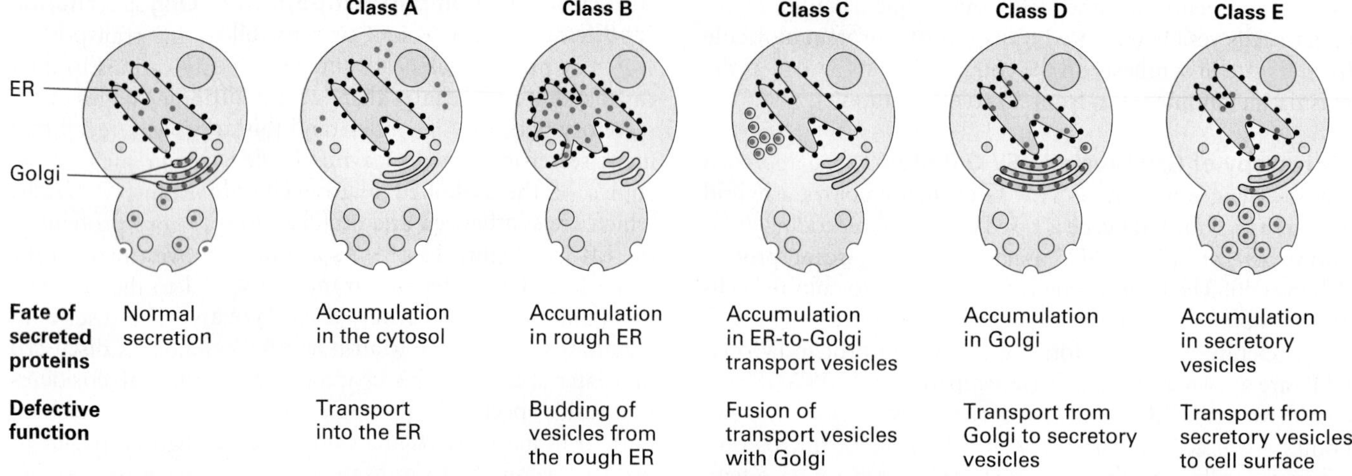

▲ EXPERIMENTAL FIGURE 14-4 **Phenotypes of yeast *sec* mutants identified stages in the secretory pathway.** These temperature-sensitive mutants can be grouped into five classes based on the site where newly made secreted proteins (red dots) accumulate when cells are shifted from the permissive temperature to the higher, nonpermissive one. Analysis of double mutants permitted the sequential order of the steps to be determined. [See P. Novick et al., 1981, *Cell* **25**:461, and C. A. Kaiser and R. Schekman, 1990, *Cell* **61**:723.]

than the corresponding glycosylated protein, these proteins can be readily distinguished (Figure 14-3b).

This type of assay can be used to track movement of VSV G protein in virus-infected cells pulse-labeled with radioactive amino acids. Immediately after labeling, all the extracted labeled VSV G protein is still in the ER and is resistant to digestion by endoglycosidase D, but with time an increasing fraction of the glycoprotein becomes sensitive to digestion. This conversion of VSV G protein from an endoglycosidase D–resistant form to an endoglycosidase D–sensitive form corresponds to vesicular transport of the protein from the ER to the *cis*-Golgi. Note that transport of VSV G protein from the ER to the Golgi takes about 30 minutes as measured by either the assay based on oligosaccharide processing or fluorescence microscopy of VSVG-GFP (Figure 14-3c). A variety of assays based on specific carbohydrate modifications that occur in later Golgi compartments have been developed to measure progression of VSV G protein through each stage of the Golgi apparatus.

Yeast Mutants Define Major Stages and Many Components in Vesicular Transport

The general organization of the secretory pathway and many of the molecular components required for vesicle trafficking are similar in all eukaryotic cells. Because of this conservation, genetic studies with yeast have been useful in confirming the sequence of steps in the secretory pathway and in identifying many of the proteins that participate in vesicular traffic. Although yeasts secrete few proteins into the growth medium, they continuously secrete a number of enzymes that remain localized in the narrow space between the plasma membrane and the cell wall. The best studied of these, invertase, hydrolyzes the disaccharide sucrose to glucose and fructose.

A large number of yeast mutants initially were identified based on their ability to secrete proteins at one temperature and inability to do so at a higher, nonpermissive temperature. When these temperature-sensitive *secretion (sec) mutants* are transferred from the lower to the higher temperature, they accumulate secreted proteins at the point in the pathway blocked by the mutation. Analysis of such mutants identified five classes (A–E) characterized by protein accumulation in the cytosol, rough ER, small vesicles taking proteins from the ER to the Golgi complex, Golgi cisternae, or constitutive secretory vesicles (Figure 14-4). Subsequent characterization of *sec* mutants in the various classes has helped elucidate the fundamental components and molecular mechanisms of vesicle trafficking that we discuss in later sections.

To determine the order of the steps in the pathway, researchers analyzed double *sec* mutants. For instance, when yeast cells contain mutations in both class B and class D functions, proteins accumulate in the rough ER, not in the Golgi cisternae. Since proteins accumulate at the earliest blocked step; this finding shows that class B mutations must act at an earlier point in the secretory pathway than class D mutations do. These studies confirmed that as a secreted protein is synthesized and processed, it moves sequentially from the cytosol → rough ER → ER-to-Golgi transport vesicles → Golgi cisternae → secretory vesicles and finally is exocytosed.

The three methods outlined in this section have delineated the major steps of the secretory pathway and have contributed to the identification of many of the proteins responsible for vesicle budding and fusion. Currently each of the individual steps in the secretory pathway is being studied in mechanistic detail, and increasingly, biochemical assays and molecular genetic studies are used to study each of these steps in terms of the function of individual protein molecules.

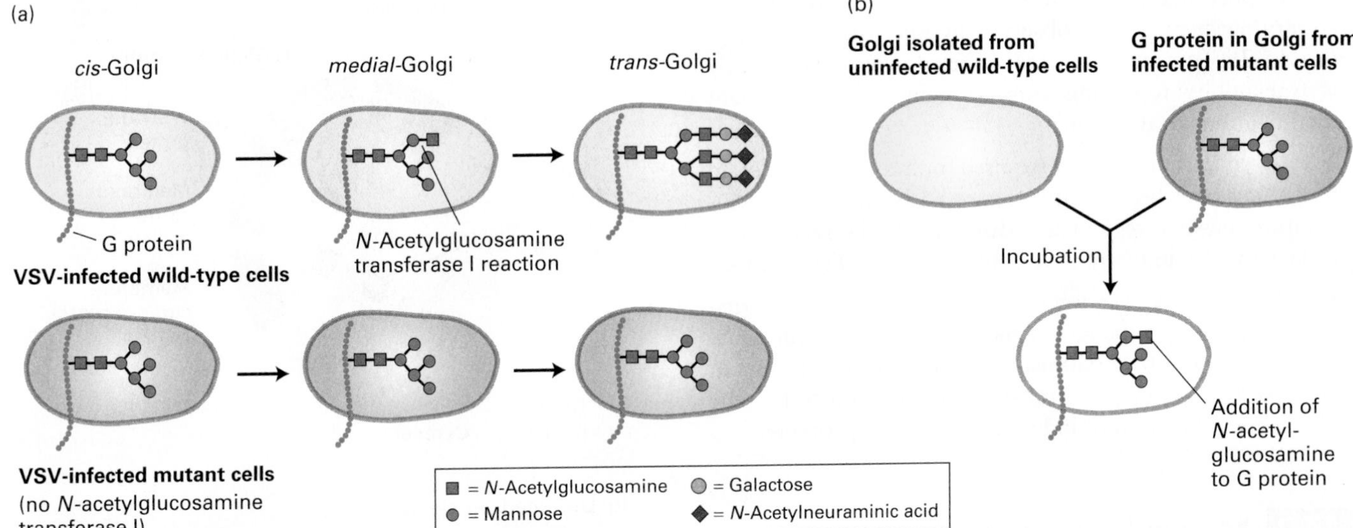

▲ EXPERIMENTAL FIGURE 14-5 **A cell-free assay demonstrates protein transport from one Golgi cisterna to another.** (a) A mutant line of cultured fibroblasts is essential in this type of assay. In this example, the cells lack the enzyme *N*-acetylglucosamine transferase I (step **2** in Figure 14-14). In wild-type cells, this enzyme is localized to the *medial*-Golgi and modifies N-linked oligosaccharides by the addition of one *N*-acetylglucosamine. In VSV-infected wild-type cells, the oligosaccharide on the viral G protein is modified to a typical complex oligosaccharide, as shown in the *trans*-Golgi panel. In infected mutant cells, however, the G protein reaches the cell surface with a simpler high-mannose oligosaccharide containing only two *N*-acetylglucosamine and five mannose residues. (b) When Golgi cisternae isolated from infected mutant cells are incubated with Golgi cisternae from normal, uninfected cells, the VSV G protein produced in vitro contains the additional *N*-acetylglucosamine. This modification is carried out by transferase enzyme that is moved by transport vesicles from the wild-type *medial*-Golgi cisternae to the mutant *cis*-Golgi cisternae in the reaction mixture. [See W. E. Balch et al., 1984, *Cell* **39:**405 and 525; W. A. Braell et al., 1984, *Cell* **39:**511; and J. E. Rothman and T. Söllner, 1997, *Science* **276:**1212.]

Cell-Free Transport Assays Allow Dissection of Individual Steps in Vesicular Transport

In vitro assays for intercompartmental transport are powerful complementary approaches to studies with yeast *sec* mutants for identifying and analyzing the cellular components responsible for vesicular trafficking. In one application of this approach, cultured mutant cells lacking one of the enzymes that modify N-linked oligosaccharide chains in the Golgi are infected with vesicular stomatitis virus (VSV). For example, if infected cells lack *N*-acetylglucosamine transferase I, they produce abundant amounts of VSV G protein but cannot add *N*-acetylglucosamine residues to the oligosaccharide chains in the *medial*-Golgi as wild-type cells do (Figure 14-5a). When Golgi membranes isolated from such mutant cells are mixed with Golgi membranes from wild-type, uninfected cells, the addition of *N*-acetylglucosamine to VSV G protein is restored (Figure 14-5b). This modification is the consequence of vesicular transport of *N*-acetylglucosamine transferase I from the wild-type *medial*-Golgi to the *cis*-Golgi compartment from virally infected mutant cells. Successful intercompartmental transport in this cell-free system depends on requirements that are typical of a normal physiological process, including a cytosolic extract, a source of chemical energy in the form of ATP and GTP, and incubation at physiological temperatures.

In addition, under appropriate conditions a uniform population of the transport vesicles that move *N*-acetylglucosamine transferase I from the *medial*- to *cis*-Golgi can be purified away from the donor wild-type Golgi membranes by centrifugation. By examining the proteins that are enriched in these vesicles, scientists have been able to identify many of the integral membrane proteins and peripheral vesicle coat proteins that are the structural components of this type of vesicle. Moreover, fractionation of the cytosolic extract required for transport in cell-free reaction mixtures has permitted isolation of the various proteins required for formation of transport vesicles and of proteins required for the targeting and fusion of vesicles with appropriate acceptor membranes. In vitro assays similar in general design to the one shown in Figure 14-5 have been used to study various transport steps in the secretory pathway.

KEY CONCEPTS OF SECTION 14.1

Techniques for Studying the Secretory Pathway

- All assays for following the trafficking of proteins through the secretory pathway in living cells require a way to label a cohort of secretory proteins and a way to identify the compartments where labeled proteins subsequently are located.
- Pulse labeling with radioactive amino acids can specifically label a cohort of newly made proteins in the ER. Alternatively, a temperature-sensitive mutant protein that is retained in the ER at the nonpermissive temperature will be released as a cohort for transport when cells are shifted to the permissive temperature.

- Transport of a fluorescently labeled protein along the secretory pathway can be observed by microscopy (see Figure 14-2). Transport of a radiolabeled protein commonly is tracked by following compartment-specific covalent modifications to the protein.
- Many of the components required for intracellular protein trafficking have been identified in yeast by analysis of temperature-sensitive *sec* mutants defective for the secretion of proteins at the nonpermissive temperature (see Figure 14-4).
- Cell-free assays for intercompartmental protein transport have allowed the biochemical dissection of individual steps of the secretory pathway. Such in vitro reactions can be used to produce pure transport vesicles and to test the biochemical function of individual transport proteins.

14.2 Molecular Mechanisms of Vesicular Traffic

Small membrane-bounded vesicles that transport proteins from one organelle to another are common elements in the secretory and endocytic pathways (see Figure 14-1). These vesicles bud from the membrane of a particular *"parent" (donor) organelle* and fuse with the membrane of a particular *"target" (destination) organelle*. Although each step in the secretory and endocytic pathways employs a different type of vesicle, studies employing genetic and biochemical techniques have revealed that each of the different vesicular transport steps is simply a variation on a common theme. In this section we explore the basic mechanisms underlying vesicle budding and fusion, that all vesicle types have in common.

The budding of vesicles from their parent membrane is driven by the polymerization of soluble protein complexes onto the membrane to form a proteinaceous vesicle coat (Figure 14-6a). Interactions between the cytosolic portions of integral membrane proteins and the vesicle coat gather the appropriate cargo proteins into the forming vesicle. Thus the coat not only gives curvature to the membrane to form a vesicle but also acts as the filter to determine which proteins are admitted into the vesicle.

The integral membrane proteins in a budding vesicle include **v-SNAREs**, which are crucial to eventual fusion of the vesicle with the correct target membrane. Shortly after formation of a vesicle is completed, the coat is shed, exposing the vesicle's v-SNARE proteins. The specific joining of v-SNAREs in the vesicle membrane with cognate **t-SNAREs** in the target membrane brings the membranes into close apposition, allowing the two bilayers to fuse (Figure 14-6b). We will examine these mechanisms of budding, docking, and fusion in greater detail now before moving on to the specifics of the secretory and endocytic pathways in subsequent sections.

Assembly of a Protein Coat Drives Vesicle Formation and Selection of Cargo Molecules

Three types of coated vesicles have been characterized, each with a different type of protein coat and each formed by

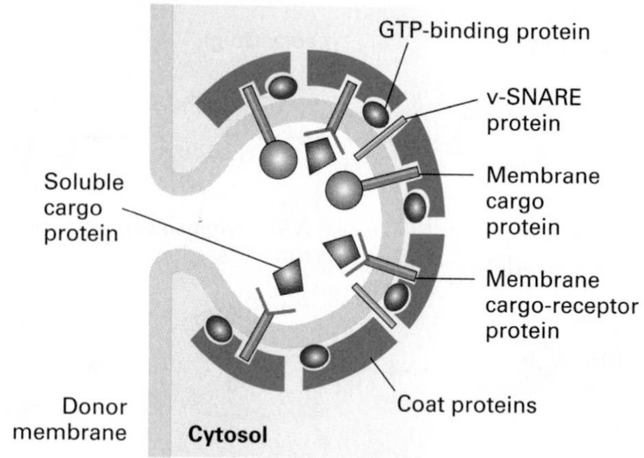

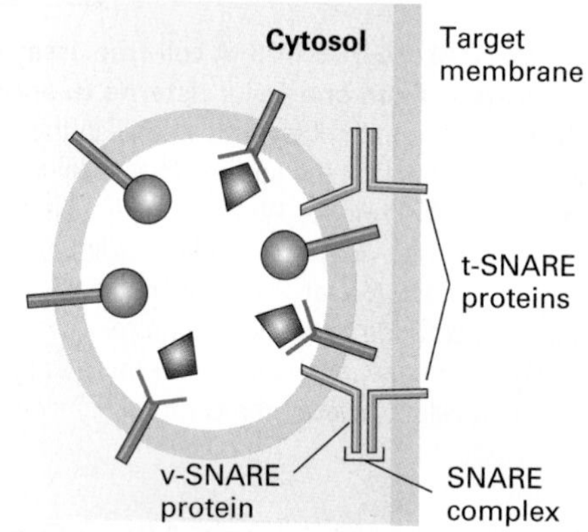

▲ **FIGURE 14-6 Overview of vesicle budding and fusion with a target membrane.** (a) Budding is initiated by recruitment of a small GTP-binding protein to a patch of donor membrane. Complexes of coat proteins in the cytosol then bind to the cytosolic domain of membrane cargo proteins, some of which also act as receptors that bind soluble proteins in the lumen, thereby recruiting luminal cargo proteins into the budding vesicle. (b) After being released and shedding its coat, a vesicle fuses with its target membrane in a process that involves interaction of cognate SNARE proteins.

reversible polymerization of a distinct set of protein subunits (Table 14-1). Each type of vesicle, named for its primary coat proteins, transports cargo proteins from particular parent organelles to particular destination organelles:

- **COPII** vesicles transport proteins from the rough ER to the Golgi.
- **COPI** vesicles mainly transport proteins in the retrograde direction between Golgi cisternae and from the *cis*-Golgi back to the rough ER.
- **Clathrin** vesicles transport proteins from the plasma membrane (cell surface) and the *trans*-Golgi network to late endosomes.

Every vesicle-mediated trafficking step is thought to utilize some kind of vesicle coat; however, a specific coat

TABLE 14-1 Coated Vesicles Involved in Protein Trafficking

VESICLE TYPE	TRANSPORT STEP MEDIATED	COAT PROTEINS	ASSOCIATED GTPase
COPII	ER to *cis*-Golgi	Sec23/Sec24 and Sec13/Sec31 complexes, Sec16	Sar1
COPI	*cis*-Golgi to ER Later to earlier Golgi cisternae	Coatomers containing seven different COP subunits	ARF
Clathrin and adapter proteins*	*trans*-Golgi to endosome	Clathrin + AP1 complexes	ARF
	trans-Golgi to endosome	Clathrin + GGA	ARF
	Plasma membrane to endosome	Clathrin + AP2 complexes	ARF
	Golgi to lysosome, melanosome, or platelet vesicles	AP3 complexes	ARF

*Each type of AP complex consists of four different subunits. It is not known whether the coat of AP3 vesicles contains clathrin.

protein complex has not been identified for every type of vesicle. For example, researchers have not yet identified the coat proteins surrounding the vesicles that move proteins from the *trans*-Golgi to the plasma membrane during either constitutive or regulated secretion.

The general scheme of vesicle budding shown in Figure 14-6a applies to all three known types of coated vesicles. Experiments with isolated or artificial membranes and purified coat proteins have shown that polymerization of the coat proteins onto the cytosolic face of the parent membrane is necessary to produce the high curvature of the membrane that is typical of a transport vesicle about 50 nm in diameter. Electron micrographs of in vitro budding reactions often reveal structures that exhibit discrete regions of the parent membrane bearing a dense coat accompanied by the curvature characteristic of a completed vesicle (Figure 14-7). Such structures, usually called *vesicle buds,* appear to be intermediates that are visible after the coat has begun to polymerize but before the completed vesicle pinches off from the parent membrane. The polymerized coat proteins are thought to form some type of curved lattice that drives the formation of a vesicle bud by adhering to the cytosolic face of the membrane.

A Conserved Set of GTPase Switch Proteins Controls Assembly of Different Vesicle Coats

Based on in vitro vesicle-budding reactions with isolated membranes and purified coat proteins, scientists have determined the minimum set of coat components required to form each of the three major types of vesicles. Although most of the coat proteins differ considerably from one type of vesicle to another, the coats of all three vesicles contain a small GTP-binding protein that acts as a regulatory subunit to control coat assembly (see Figure 14-6a). For both COPI and clathrin vesicles, this GTP-binding protein is known as *ARF protein.* A different but related GTP-binding protein known as *Sar1 protein* is present in the coat of COPII vesicles. Both ARF and Sar1 are monomeric proteins with an overall structure similar to that of Ras, a key intracellular signal-transducing protein (see Figure 16-24). ARF and Sar1 proteins, like Ras, belong to the **GTPase superfamily** of switch proteins that cycle between inactive GDP-bound and active GTP-bound forms (see Figure 3-32).

The cycle of GTP binding and hydrolysis by ARF and Sar1 are thought to control the initiation of coat assembly, as schematically depicted for the assembly of COPII vesicles

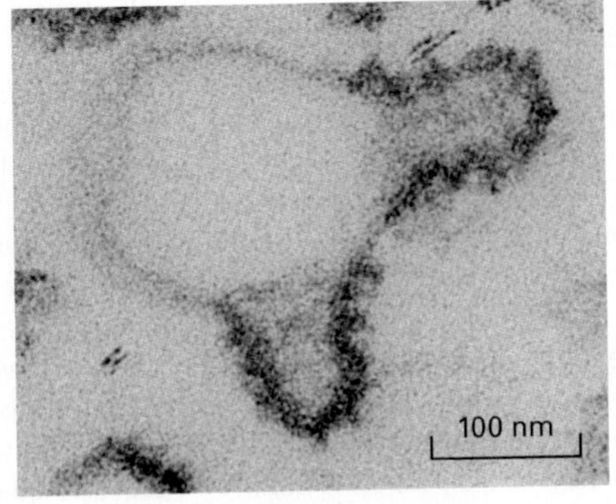

▲ EXPERIMENTAL FIGURE 14-7 **Vesicle buds can be visualized during in vitro budding reactions.** When purified COPII coat components are incubated with isolated ER vesicles or artificial phospholipid vesicles (liposomes), polymerization of the coat proteins on the vesicle surface induces emergence of highly curved buds. In this electron micrograph of an in vitro budding reaction, note the distinct membrane coat, visible as a dark protein layer, present on the vesicle buds. [From K. Matsuoka et al., 1988, *Cell* **93**(2):263.]

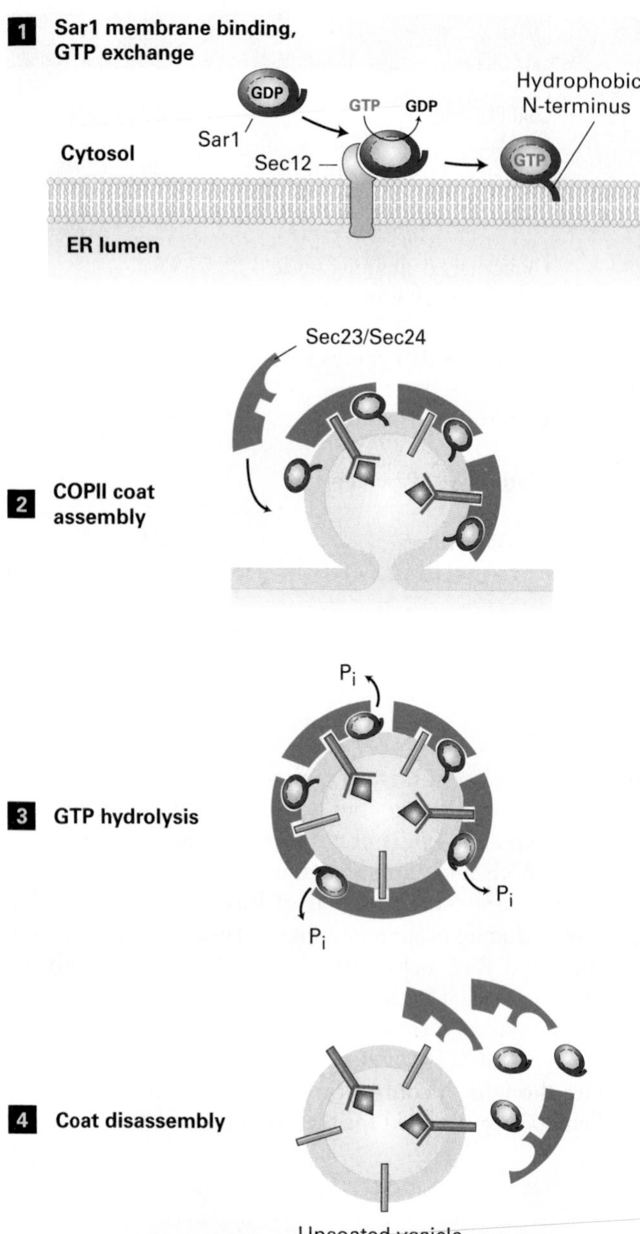

▲ **FIGURE 14-8 Model for the role of Sar1 in the assembly and disassembly of COPII coats.** Step **1**: Interaction of soluble GDP-bound Sar1 with the exchange factor Sec12, an ER integral membrane protein, catalyzes exchange of GTP for GDP on Sar1. In the GTP-bound form of Sar1, its hydrophobic N-terminus extends outward from the protein's surface and anchors Sar1 to the ER membrane. Step **2**: Sar1 attached to the membrane serves as a binding site for the Sec23/Sec24 coat protein complex. Membrane cargo proteins are recruited to the forming vesicle bud by binding of specific short sequences (sorting signals) in their cytosolic regions to sites on the Sec23/Sec24 complex. Some membrane cargo proteins also act as receptors that bind soluble proteins in the lumen. The coat is completed by assembly of a second type of coat complex composed of Sec13 and Sec31 (not shown). Step **3**: After the vesicle coat is complete, the Sec23 coat subunit promotes GTP hydrolysis by Sar1. Step **4**: Release of Sar1·GDP from the vesicle membrane causes disassembly of the coat. [See S. Springer et al., 1999, *Cell* **97:**145.]

in Figure 14-8. First, an ER membrane protein known as Sec12 catalyzes release of GDP from cytosolic Sar1·GDP and binding of GTP. The Sec12 guanine nucleotide–exchange factor apparently receives and integrates multiple as yet unknown signals, probably including the presence of cargo proteins in the ER membrane that are ready to be transported. Binding of GTP causes a conformational change in Sar1 that exposes its hydrophobic N-terminus, which then becomes embedded in the phospholipid bilayer and tethers Sar1·GTP to the ER membrane. The membrane-attached Sar1·GTP drives polymerization of cytosolic complexes of COPII subunits on the membrane, eventually leading to formation of vesicle buds. Once COPII vesicles are released from the donor membrane, the Sar1 GTPase activity hydrolyzes Sar1·GTP in the vesicle membrane to Sar1·GDP with the assistance of one of the coat subunits. This hydrolysis triggers disassembly of the COPII coat. Thus Sar1 couples a cycle of GTP binding and hydrolysis to the formation and then dissociation of the COPII coat.

ARF protein undergoes a similar cycle of nucleotide exchange and hydrolysis coupled to the assembly of vesicle coats composed either of COPI or of clathrin and other coat proteins (AP complexes), discussed later. A covalent protein modification known as a myristate anchor on the N-terminus of ARF protein weakly tethers ARF·GDP to the Golgi membrane. When GTP is exchanged for the bound GDP by a nucleotide-exchange factor attached to the Golgi membrane, the resulting conformational change in ARF allows hydrophobic residues in its N-terminal segment to insert into the membrane bilayer. The resulting tight association of ARF·GTP with the membrane serves as the foundation for further coat assembly.

Drawing on the structural similarities of Sar1 and ARF to other small GTPase switch proteins, researchers have constructed genes encoding mutant versions of the two proteins that have predictable effects on vesicular traffic when transfected into cultured cells. For example, in cells expressing mutant versions of Sar1 or ARF that cannot hydrolyze GTP, vesicle coats form and vesicle buds pinch off. However, because the mutant proteins cannot trigger disassembly of the coat, all available coat subunits eventually become permanently assembled into coated vesicles that are unable to fuse with target membranes. Addition of a nonhydrolyzable GTP analog to in vitro vesicle-budding reactions causes a similar blocking of coat disassembly. The vesicles that form in such reactions have coats that never dissociate, allowing their composition and structure to be more readily analyzed. The purified COPI vesicles shown in Figure 14-9 were produced in such a budding reaction.

Targeting Sequences on Cargo Proteins Make Specific Molecular Contacts with Coat Proteins

In order for transport vesicles to move specific proteins from one compartment to the next, vesicle buds must be able to discriminate among potential membrane and soluble cargo proteins, accepting only those cargo proteins that should advance to the next compartment and excluding those that

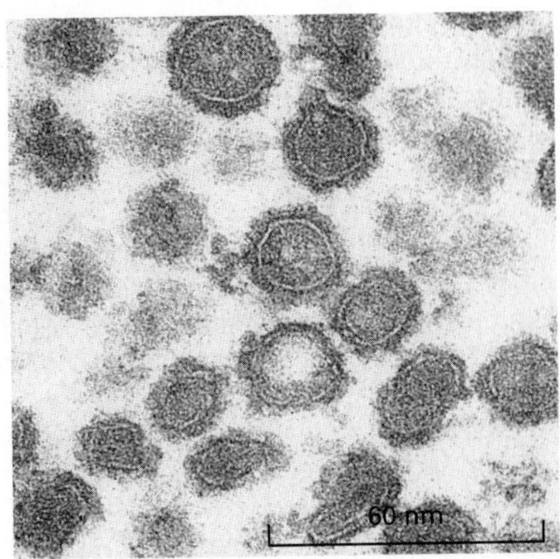

▲ EXPERIMENTAL FIGURE 14-9 **Coated vesicles accumulate during in vitro budding reactions in the presence of a nonhydrolyzable analog of GTP.** When isolated Golgi membranes are incubated with a cytosolic extract containing COPI coat proteins, vesicles form and bud off from the membranes. Inclusion of a nonhydrolyzable analog of GTP in the budding reaction prevents disassembly of the coat after vesicle release. This micrograph shows COPI vesicles generated in such a reaction and separated from membranes by centrifugation. Coated vesicles prepared in this way can be analyzed to determine their components and properties. [Courtesy of L. Orci.]

should remain as residents in the donor compartment. In addition to sculpting the curvature of a donor membrane, the vesicle coat functions in selecting specific proteins as cargo. The primary mechanism by which the vesicle coat selects cargo molecules is by directly binding to specific sequences, or **sorting signals,** in the cytosolic portion of membrane cargo proteins (see Figure 14-6a). The polymerized coat thus acts as an affinity matrix to cluster selected membrane cargo proteins into forming vesicle buds. Since soluble proteins within the lumen of parent organelles cannot contact the coat directly, they require a different kind of sorting signal. Soluble luminal proteins often contain what can be thought of as *luminal sorting signals,* which bind to the luminal domains of certain membrane cargo proteins that act as receptors for luminal cargo proteins. The properties of several known sorting signals in membrane and soluble proteins are summarized in Table 14-2. We describe the role of these signals in more detail in later sections.

Rab GTPases Control Docking of Vesicles on Target Membranes

A second set of small GTP-binding proteins, known as *Rab proteins,* participate in the targeting of vesicles to the appropriate target membrane. Like Sar1 and ARF, Rab proteins

TABLE 14-2 Known Sorting Signals That Direct Proteins to Specific Transport Vesicles

SIGNAL SEQUENCE*	PROTEINS WITH SIGNAL	SIGNAL RECEPTOR	VESICLES THAT INCORPORATE SIGNAL-BEARING PROTEIN
LUMENAL SORTING SIGNALS			
Lys-Asp-Glu-Leu (KDEL)	ER-resident soluble proteins	KDEL receptor in *cis*-Golgi membrane	COPI
Mannose 6-phosphate (M6P)	Soluble lysosomal enzymes after processing in *cis*-Golgi	M6P receptor in *trans*-Golgi membrane	Clathrin/AP1
	Secreted lysosomal enzymes	M6P receptor in plasma membrane	Clathrin/AP2
CYTOPLASMIC SORTING SIGNALS			
Lys-Lys-X-X (KKXX)	ER-resident membrane proteins	COPI α and β subunits	COPI
Di-acidic (e.g., Asp-X-Glu)	Cargo membrane proteins in ER	COPII Sec24 subunit	COPII
Asn-Pro-X-Tyr (NPXY)	LDL receptor in plasma membrane	AP2 complex	Clathrin/AP2
Tyr-X-X-Φ (YXXΦ)	Membrane proteins in *trans*-Golgi	AP1 (μ1 subunit)	Clathrin/AP1
	Plasma membrane proteins	AP2 (μ2 subunit)	Clathrin/AP2
Leu-Leu (LL)	Plasma membrane proteins	AP2 complexes	Clathrin/AP2

*X = any amino acid; Φ = hydrophobic amino acid. Single-letter amino acid abbreviations are in parentheses.

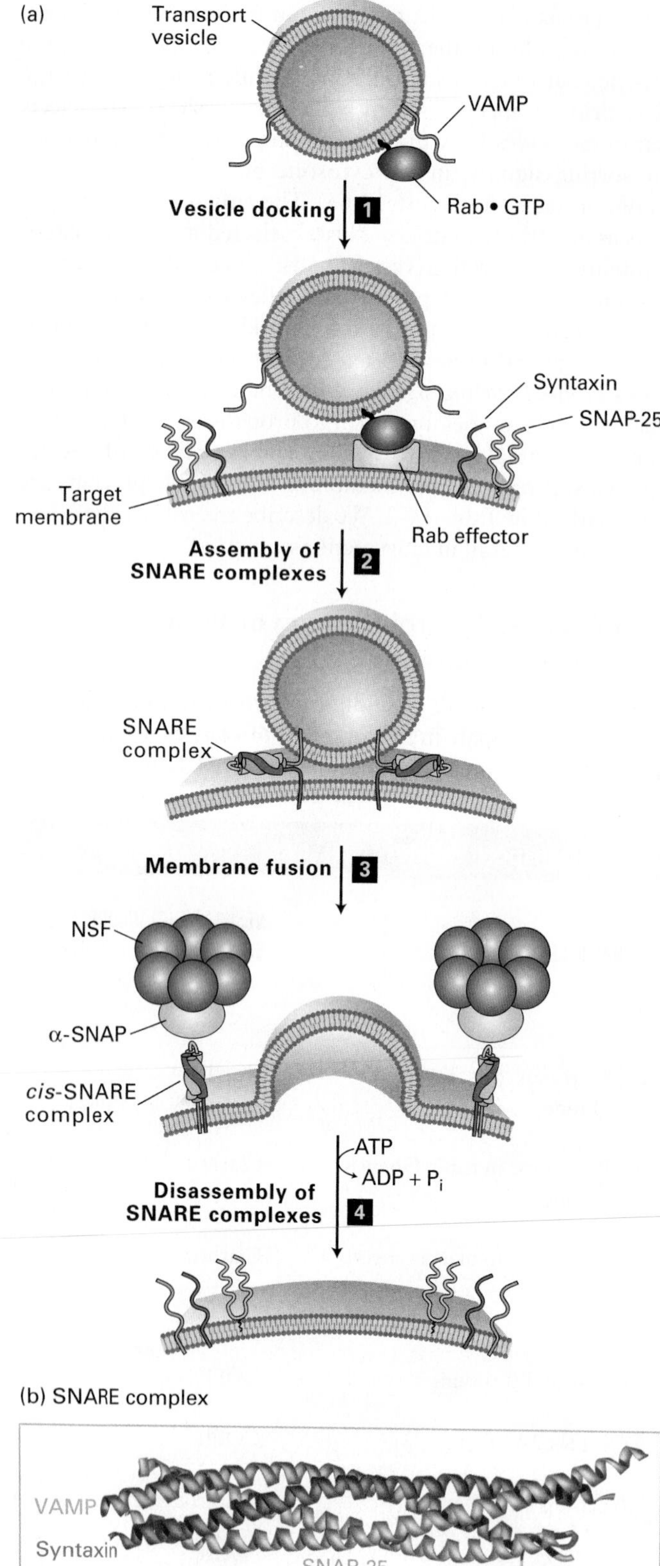

◀ **FIGURE 14-10 Model for docking and fusion of transport vesicles with their target membranes.** (a) The proteins shown in this example participate in fusion of secretory vesicles with the plasma membrane, but similar proteins mediate all vesicle-fusion events. Step **1**: A Rab protein tethered via a lipid anchor to a secretory vesicle binds to an effector protein complex on the plasma membrane, thereby docking the transport vesicle on the appropriate target membrane. Step **2**: A v-SNARE protein (in this case, VAMP) interacts with the cytosolic domains of the cognate t-SNAREs (in this case, syntaxin and SNAP-25). The very stable coiled-coil SNARE complexes that are formed hold the vesicle close to the target membrane. Step **3**: Fusion of the two membranes immediately follows formation of SNARE complexes, but precisely how this occurs is not known. Step **4**: Following membrane fusion, NSF in conjunction with α-SNAP protein binds to the SNARE complexes. The NSF-catalyzed hydrolysis of ATP then drives dissociation of the SNARE complexes, freeing the SNARE proteins for another round of vesicle fusion. Also at this time, Rab·GTP is hydrolyzed to Rab·GDP and dissociates from the Rab effector (not shown). (b) The SNARE complex. Numerous noncovalent interactions between four long α helices, two from SNAP-25 and one each from syntaxin and VAMP, stabilize the coiled-coil structure. [See J. E. Rothman and T. Söllner, 1997, *Science* **276:**1212, and W. Weis and R. Scheller, 1998, *Nature* **395:**328. Part (b) from Y. A. Chen and R. H. Scheller, 2001, *Nat. Rev. Mol. Cell Biol.* **2**(2):98.]

belong to the GTPase superfamily of switch proteins. Conversion of cytosolic Rab·GDP to Rab·GTP, catalyzed by a specific guanine nucleotide–exchange factor, induces a conformational change in Rab that enables it to interact with a surface protein on a particular transport vesicle and insert its isoprenoid anchor into the vesicle membrane. Once Rab·GTP is tethered to the vesicle surface, it is thought to interact with one of a number of different large proteins, known as Rab effectors, attached to the target membrane. Binding of Rab·GTP to a Rab effector docks the vesicle on an appropriate target membrane (Figure 14-10, step **1**). After vesicle fusion occurs, the GTP bound to the Rab protein is hydrolyzed to GDP, triggering the release of Rab·GDP, which then can undergo another cycle of GDP-GTP exchange, binding, and hydrolysis.

Several lines of evidence support the involvement of specific Rab proteins in vesicle-fusion events. For instance, the yeast *SEC4* gene encodes a Rab protein, and yeast cells expressing mutant Sec4 proteins accumulate secretory vesicles that are unable to fuse with the plasma membrane (class E mutants in Figure 14-4). In mammalian cells, Rab5 protein is localized to endocytic vesicles, also known as early endosomes. These uncoated vesicles form from clathrin-coated vesicles just after they bud from the plasma membrane during endocytosis (see Figure 14-1, step **9**). The fusion of early endosomes with each other in cell-free systems requires the presence of Rab5, and addition of Rab5 and GTP to cell-free extracts accelerates the rate at which these vesicles fuse with each other. A long coiled protein known as EEA1 (*e*arly *e*ndosome *a*ntigen *1*), which resides on the membrane of the early endosome, functions as the effector for Rab5. In this case, Rab5·GTP on one endocytic vesicle is thought to specifically bind to EEA1 on the membrane of another endocytic vesicle, setting the stage for fusion of the two vesicles.

A different type of Rab effector appears to function for each vesicle type and at each step of the secretory pathway. Many questions remain about how Rab proteins are targeted to the correct membrane and how specific complexes form between the different Rab proteins and their corresponding effector proteins.

Paired Sets of SNARE Proteins Mediate Fusion of Vesicles with Target Membranes

As noted previously, shortly after a vesicle buds off from the donor membrane, the vesicle coat disassembles to uncover a vesicle-specific membrane protein, a v-SNARE (see Figure 14-6b). Likewise, each type of target membrane in a cell contains t-SNARE membrane proteins. After Rab-mediated docking of a vesicle on its target (destination) membrane, the interaction of cognate SNAREs brings the two membranes close enough together that they can fuse.

One of the best-understood examples of SNARE-mediated fusion occurs during exocytosis of secreted proteins (Figure 14-10, steps **2** and **3**). In this case, the v-SNARE, known as *VAMP* (*v*esicle-*a*ssociated *m*embrane *p*rotein), is incorporated into secretory vesicles as they bud from the *trans*-Golgi network. The t-SNAREs are *syntaxin,* an integral membrane protein in the plasma membrane, and *SNAP-25,* which is attached to the plasma membrane by a hydrophobic lipid anchor in the middle of the protein. The cytosolic region in each of these three SNARE proteins contains a repeating heptad sequence that allows four α helices—one from VAMP, one from syntaxin, and two from SNAP-25—to coil around one another to form a four-helix bundle. The unusual stability of this bundled SNARE complex is conferred by the arrangement of hydrophobic and charged amino residues in the heptad repeats. The hydrophobic amino acids are buried in the central core of the bundle, and amino acids of opposite charge are aligned to form favorable electrostatic interactions between helices. As the four-helix bundles form, the vesicle and target membranes are drawn into close apposition by the embedded transmembrane domains of VAMP and syntaxin.

In vitro experiments have shown that when **liposomes** containing purified VAMP are incubated with other liposomes containing syntaxin and SNAP-25, the two classes of membranes fuse, albeit slowly. This finding is strong evidence that the close apposition of membranes resulting from formation of SNARE complexes is sufficient to bring about membrane fusion. Fusion of a vesicle and target membrane occurs more rapidly and efficiently in the cell than it does in liposome experiments in which fusion is catalyzed only by SNARE proteins. The likely explanation for this difference is that in the cell, other proteins such as Rab proteins and their effectors are involved in targeting vesicles to the correct membrane.

Yeast cells, like all eukaryotic cells, express more than 20 different related v-SNARE and t-SNARE proteins. Analyses of yeast mutants defective in each of the SNARE genes have identified specific membrane-fusion events in which each SNARE protein participates. For all fusion events that have been examined, the SNAREs form four-helix bundled complexes, similar to the VAMP/syntaxin/SNAP-25 complexes that mediate fusion of secretory vesicles with the plasma membrane. However, in other fusion events (e.g., fusion of COPII vesicles with the *cis*-Golgi network), each participating SNARE protein contributes only one α helix to the bundle (unlike SNAP-25, which contributes two helices); in these cases the SNARE complexes comprise one v-SNARE and three t-SNARE molecules.

Using the in vitro liposome fusion assay, researchers have tested the ability of various combinations of individual v-SNARE and t-SNARE proteins to mediate fusion of donor and target membranes. Of the very large number of different combinations tested, only a small number could efficiently mediate membrane fusion. To a remarkable degree, the functional combinations of v-SNAREs and t-SNAREs revealed in these in vitro experiments correspond to the actual SNARE protein interactions that mediate known membrane-fusion events in the yeast cell. Thus the specificity of the interaction between SNARE proteins can account for much of the specificity of fusion between a particular vesicle type and its target membrane.

Dissociation of SNARE Complexes After Membrane Fusion Is Driven by ATP Hydrolysis

After a vesicle and its target membrane have fused, the SNARE complexes must dissociate to make the individual SNARE proteins available for additional fusion events. Because of the stability of SNARE complexes, which are held together by numerous noncovalent intermolecular interactions, their dissociation depends on additional proteins and the input of energy.

The first clue that dissociation of SNARE complexes required the assistance of other proteins came from in vitro transport reactions depleted of certain cytosolic proteins. The observed accumulation of vesicles in these reactions indicated that vesicles could form but were unable to fuse with a target membrane. Eventually two proteins, designated *NSF* and *α-SNAP,* were found to be required for ongoing vesicle fusion in the in vitro transport reaction. The function of NSF in vivo can be blocked selectively by *N*-ethylmaleimide (NEM), a chemical that reacts with an essential –SH group on NSF (hence the name, *N*EM-*s*ensitive *f*actor).

Among the class C yeast *sec* mutants are strains that lack functional Sec18 or Sec17, the yeast counterparts of mammalian NSF and α-SNAP, respectively. When these class C mutants are placed at the nonpermissive temperature, they accumulate ER-to-Golgi transport vesicles; when the cells are shifted to the lower, permissive temperature, the accumulated vesicles are able to fuse with the *cis*-Golgi.

Subsequent to the initial biochemical and genetic studies identifying NSF and α-SNAP, more sophisticated in vitro transport assays were developed. Using these newer assays, researchers have shown that NSF and α-SNAP proteins are not necessary for actual membrane fusion but rather are required for regeneration of free SNARE proteins. NSF, a hexamer of identical subunits, associates with a SNARE complex with the aid of α-SNAP (*s*oluble *N*SF *a*ttachment *p*rotein). The bound NSF then hydrolyzes ATP, releasing sufficient energy to dissociate the SNARE complex (Figure 14-10, step **4**). Evidently, the defects in vesicle fusion observed in the earlier in vitro fusion assays and in the yeast mutants after a loss of Sec17 or Sec18 were a consequence of free SNARE proteins rapidly becoming sequestered in undissociated SNARE complexes and thus being unavailable to mediate membrane fusion.

KEY CONCEPTS OF SECTION 14.2

Molecular Mechanisms of Vesicular Traffic

- The three well-characterized transport vesicles—COPI, COPII, and clathrin vesicles—are distinguished by the proteins that form their coats and the transport routes they mediate (see Table 14-1).
- All types of coated vesicles are formed by polymerization of cytosolic coat proteins onto a donor (parent) membrane to form vesicle buds that eventually pinch off from the membrane to release a complete vesicle. Shortly after vesicle release, the coat is shed, exposing proteins required for fusion with the target membrane (see Figure 14-6).
- Small GTP-binding proteins (ARF or Sar1) belonging to the GTPase superfamily control polymerization of coat proteins, the initial step in vesicle budding (see Figure 14-8). After vesicles are released from the donor membrane, hydrolysis of GTP bound to ARF or Sar1 triggers disassembly of the vesicle coats.
- Specific sorting signals in membrane and luminal proteins of donor organelles interact with coat proteins during vesicle budding, thereby recruiting cargo proteins to vesicles (see Table 14-2).
- A second set of GTP-binding proteins, the Rab proteins, regulate docking of vesicles with the correct target membrane. Each Rab appears to bind to a specific Rab effector associated with the target membrane.
- Each v-SNARE in a vesicular membrane specifically binds to a complex of cognate t-SNARE proteins in the target membrane, inducing fusion of the two membranes. After fusion is completed, the SNARE complex is disassembled in an ATP-dependent reaction mediated by other cytosolic proteins (see Figure 14-10).

14.3 Early Stages of the Secretory Pathway

In this section we take a closer look at vesicular traffic through the ER and Golgi stages of the secretory pathway and some of the evidence supporting the general mechanisms discussed in the previous section. Recall that *anterograde transport* from the ER to Golgi, the first step in the secretory pathway, is mediated by COPII vesicles. These vesicles contain newly synthesized proteins destined for the Golgi, cell surface, or lysosomes as well as vesicle components such as v-SNAREs that are required to target vesicles to the *cis*-Golgi membrane. Proper sorting of proteins between the ER and Golgi also requires reverse *retrograde transport* from the *cis*-Golgi to the ER and is mediated by COPI vesicles (Figure 14-11). This retrograde vesicle transport serves to retrieve v-SNARE proteins and the membrane itself back to the ER to provide the necessary material for additional rounds of vesicle budding from the ER. COPI-mediated retrograde transport also retrieves missorted ER-resident proteins from the *cis*-Golgi to correct sorting mistakes. Proteins that have been correctly delivered to the Golgi advance through successive compartments of the Golgi by cisternal maturation.

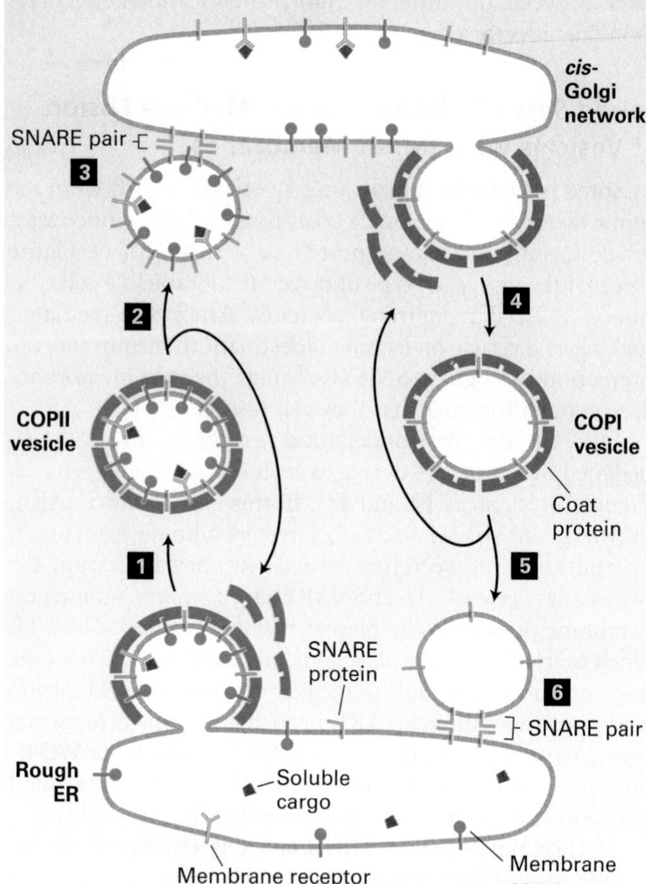

▲ **FIGURE 14-11 Vesicle-mediated protein trafficking between the ER and *cis*-Golgi.** Steps **1**–**3**: Forward (anterograde) transport is mediated by COPII vesicles, which are formed by polymerization of soluble COPII coat protein complexes (green) on the ER membrane. v-SNAREs (orange) and other cargo proteins (blue) in the ER membrane are incorporated into the vesicle by interacting with coat proteins. Soluble cargo proteins (magenta) are recruited by binding to appropriate receptors in the membrane of budding vesicles. Dissociation of the coat recycles free coat complexes and exposes v-SNARE proteins on the vesicle surface. After the uncoated vesicle becomes tethered to the *cis*-Golgi membrane in a Rab-mediated process, pairing between the exposed v-SNAREs and cognate t-SNAREs in the Golgi membrane allows vesicle fusion, releasing the contents into the *cis*-Golgi compartment (see Figure 14-10). Steps **4**–**6**: Reverse (retrograde) transport, mediated by vesicles coated with COPI proteins (purple), recycles the membrane bilayer and certain proteins, such as v-SNAREs and missorted ER-resident proteins (not shown), from the *cis*-Golgi to the ER. All SNARE proteins are shown in orange although v-SNAREs and t-SNAREs are distinct proteins.

COPII Vesicles Mediate Transport from the ER to the Golgi

COPII vesicles were first recognized when cell-free extracts of yeast rough ER membranes were incubated with cytosol and a nonhydrolyzable analog of GTP. The vesicles that formed from the ER membranes had a distinct coat similar to that on COPI vesicles but composed of different proteins, designated COPII proteins. Yeast cells with mutations in the genes for COPII proteins are class B *sec* mutants and accumulate proteins in the rough ER (see Figure 14-4). Analysis of such mutants has revealed several proteins required for formation of COPII vesicles.

As described previously, formation of COPII vesicles is triggered when Sec12, a guanine nucleotide–exchange factor in the ER membrane, catalyzes the exchange of bound GDP for GTP on cytosolic Sar1. This exchange induces binding of Sar1 to the ER membrane followed by binding of a complex of Sec23 and Sec24 proteins (see Figure 14-8). The resulting ternary complex formed between Sar1·GTP, Sec23, and Sec24 is shown in Figure 14-12. After this complex forms on the ER membrane, a second complex comprising Sec13 and Sec31 proteins binds to complete the coat structure. Pure Sec13 and Sec31 proteins can spontaneously assemble into cagelike lattices. It is thought that Sec13 and Sec31 form the structural scaffold for COPII vesicles. Finally, a large fibrous protein, called Sec16, which is bound to the cytosolic surface of the ER, interacts with Sar1·GTP and the Sec13/31 and Sec23/24 complexes and acts to organize the other coat proteins, increasing the efficiency of coat polymerization. Similar to Sec 13/31, clathrin also has this ability to self-assemble into a coatlike structure, as will be discussed in Section 14.4.

Certain integral ER membrane proteins are specifically recruited into COPII vesicles for transport to the Golgi. The cytosolic segments of many of these proteins contain a *di-acidic sorting signal* (the key residues in this sequence are Asp-X-Glu, or DXE in the one-letter code) (see Table 14-2). This sorting signal binds to the Sec24 subunit of the COPII coat and is essential for the selective export of certain membrane proteins from the ER (Figure 14-12). Biochemical and genetic studies currently are under way to identify additional signals that help direct membrane cargo proteins into COPII vesicles. Other ongoing studies seek to determine how soluble cargo proteins are selectively loaded into COPII vesicles. Although purified COPII vesicles from yeast cells have been found to contain a membrane protein that binds the soluble α mating factor, the receptors for other soluble cargo proteins such as invertase are not yet known.

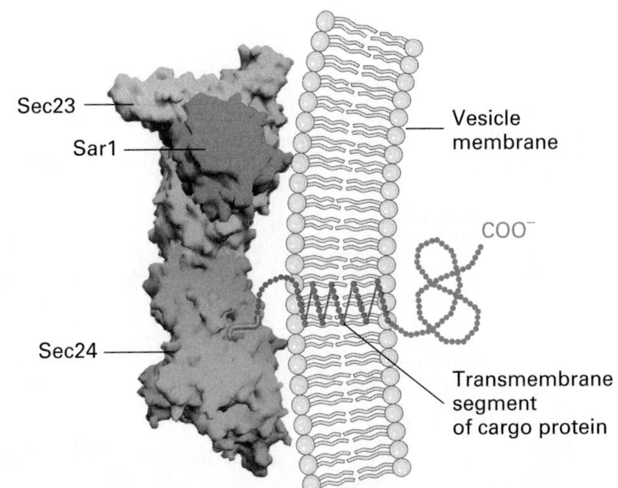

▲ **FIGURE 14-12 Three-dimensional structure of ternary complex comprising the COPII coat proteins Sec23 and Sec24 and Sar1·GTP.** Early in the formation of the COPII coat, Sec23 (orange)/Sec24 (green) complexes are recruited to the ER membrane by Sar1 (red) in its GTP-bound state. In order to form a stable ternary complex in solution for structural studies, the nonhydrolyzable GTP analog GppNHp was used. A cargo protein in the ER membrane can be recruited to COPII vesicles by interaction of a tripeptide di-acidic signal (purple) in the cargo's cytosolic domain with Sec24. The likely position of the COPII vesicle membrane and the transmembrane segment of the cargo protein are indicated. The N-terminal segment of Sar1 that tethers it to the membrane is not shown. [See X. Bi et al., 2002, *Nature* **419**:271; interaction with peptide courtesy of J. Goldberg.]

The inherited disease cystic fibrosis is caused by mutations in a protein known as CFTR, which is synthesized as an integral membrane protein in the ER and is transported to the Golgi before being transported to the plasma membranes of epithelial cells, where it functions as a chloride channel. Researchers have recently shown that the CFTR protein contains a di-acidic sorting signal that binds to the Sec24 subunit of the COPII coat and is necessary for transport of the CFTR protein out of the ER. The most common CFTR mutation is a deletion of a phenylalanine at position 508 in the protein sequence (known as ΔF508). This mutation prevents normal transport of CFTR to the plasma membrane by blocking its packaging into COPII vesicles budding from the ER. Although the ΔF508 mutation is not in the vicinity of the di-acidic sorting signal, this mutation may change the conformation of the cytosolic portion of CFTR so that the di-acidic signal is unable to bind to Sec24. ■

The experiments described previously in which the transit of VSVG-GFP in cultured mammalian cells is followed by fluorescence microscopy (see Figure 14-2) provided insight into the intermediates in ER-to-Golgi transport. In some cells, small fluorescent vesicles containing VSVG-GFP could be seen to form from the ER, move less than 1 μm, and then fuse directly with the *cis*-Golgi. In other cells, in which the ER was located several micrometers from the Golgi complex, several ER-derived vesicles were seen to fuse with each other shortly after their formation, forming what is termed the *ER-to-Golgi intermediate compartment.* These larger structures then were transported along microtubules to the *cis*-Golgi, much in the way vesicles in nerve cells are transported from the cell body, where they are formed, down the long axon to the axon terminus (Chapter 18). Microtubules function much as "railroad tracks" enabling these large aggregates of transport vesicles to move long distances to their *cis*-Golgi destination. At the time the ER-to-Golgi intermediate compartment is formed, some COPI vesicles bud off from it, recycling some proteins back to the ER.

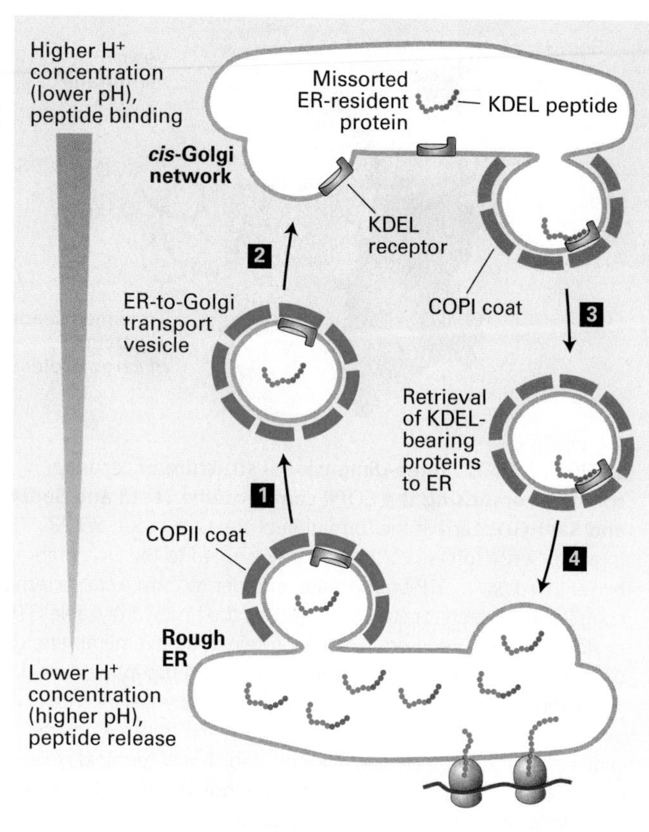

◀ **FIGURE 14-13 Role of the KDEL receptor in retrieval of ER-resident luminal proteins from the Golgi.** ER luminal proteins, especially those present at high levels, can be passively incorporated into COPII vesicles and transported to the Golgi (steps **1** and **2**). Many such proteins bear a C-terminal KDEL (Lys-Asp-Glu-Leu) sequence (red) that allows them to be retrieved. The KDEL receptor, located mainly in the *cis*-Golgi network and in both COPII and COPI vesicles, binds proteins bearing the KDEL sorting signal and returns them to the ER (steps **3** and **4**). This retrieval system prevents depletion of ER luminal proteins such as those needed for proper folding of newly made secretory proteins. The binding affinity of the KDEL receptor is very sensitive to pH. The small difference in the pH of the ER and Golgi favors binding of KDEL-bearing proteins to the receptor in Golgi-derived vesicles and their release in the ER. [Adapted from J. Semenza et al., 1990, *Cell* **61**:1349.]

COPI Vesicles Mediate Retrograde Transport within the Golgi and from the Golgi to the ER

COPI vesicles were first discovered when isolated Golgi fractions were incubated in a solution containing cytosol and a nonhydrolyzable analog of GTP (see Figure 14-9). Subsequent analysis of these vesicles showed that the coat is formed from large cytosolic complexes, called *coatomers*, composed of seven polypeptide subunits. Yeast cells containing temperature-sensitive mutations in COPI proteins accumulate proteins in the rough ER at the nonpermissive temperature and thus are categorized as class B *sec* mutants (see Figure 14-4). Although discovery of these mutants initially suggested that COPI vesicles mediate ER-to-Golgi transport, subsequent experiments showed that their main function is retrograde transport, both between Golgi cisternae and from the *cis*-Golgi to the rough ER (see Figure 14-11, *right*). Because COPI mutants cannot recycle key membrane proteins back to the rough ER, the ER gradually becomes depleted of ER proteins such as v-SNAREs necessary for COPII vesicle function. Eventually, vesicle formation from the rough ER grinds to a halt; secretory proteins continue to be synthesized but accumulate in the ER, the defining characteristic of class B *sec* mutants. The general ability of *sec* mutants involved in either COPI or COPII vesicle function to eventually block both anterograde and retrograde transport illustrates the fundamental interdependence of these two transport processes.

As discussed in Chapter 13, the ER contains several soluble proteins dedicated to the folding and modification of newly synthesized secretory proteins. These include the chaperone BiP and the enzyme protein disulfide isomerase, which are necessary for the ER to carry out its functions. Although such ER-resident luminal proteins are not specifically selected by COPII vesicles, their sheer abundance causes them to be continuously loaded passively into vesicles destined for the *cis*-Golgi. The transport of these soluble proteins back to the ER, mediated by COPI vesicles, prevents their eventual depletion.

Most soluble ER-resident proteins carry a Lys-Asp-Glu-Leu (KDEL in the one-letter code) sequence at their C-terminus (see Table 14-2). Several experiments demonstrated that this *KDEL sorting signal* is both necessary and sufficient to cause a protein bearing this sequence to be located in the ER. For instance, when a mutant protein disulfide isomerase lacking these four residues is synthesized in cultured fibroblasts, the protein is secreted. Moreover, if a protein that normally is secreted is altered so that it contains the KDEL signal at its C-terminus, the protein is located in the ER. The KDEL sorting signal is recognized and bound by the *KDEL receptor,* a transmembrane protein found primarily on small transport vesicles shuttling between the ER and the *cis*-Golgi and on the *cis*-Golgi reticulum. In addition, soluble ER-resident proteins that carry the KDEL signal have oligosaccharide chains with modifications that are catalyzed by enzymes found only in the *cis*-Golgi or *cis*-Golgi network; thus at some time these proteins must have left the ER and been transported at least as far as the *cis*-Golgi network. These findings indicate that the KDEL receptor acts mainly to retrieve soluble proteins containing the KDEL sorting signal that have escaped to the *cis*-Golgi network and return them to the ER (Figure 14-13). The KDEL receptor binds more tightly to its ligand at low pH, and it is thought that the receptor is able to bind KEDL peptides in the *cis*-Golgi but to release these peptides in the ER because the pH of the Golgi is slightly lower than that of the ER.

The KDEL receptor and other membrane proteins that are transported back to the ER from the Golgi contain a Lys-Lys-X-X sequence at the very end of their C-terminal segment, which faces the cytosol (see Table 14-2). This *KKXX sorting signal,* which binds to a complex of the COPI α and β subunits (two of the seven polypeptide subunits in the COPI coatomer), is both necessary and sufficient to incorporate membrane proteins into COPI vesicles for retrograde transport to the ER. Temperature-sensitive yeast mutants lacking COPIα or COPIβ not only are unable to bind the KKXX signal but also are unable to retrieve proteins bearing this signal back to the ER, indicating that COPI vesicles mediate retrograde Golgi-to-ER transport.

Clearly, the partitioning of proteins between the ER and Golgi complex is a highly selective and regulated process ultimately controlled by the specificity of cargo loading into both COPII (anterograde) and COPI (retrograde) vesicles. The selective entry of proteins into membrane-bounded transport vesicles, the recycling of membrane phospholipids and proteins, and the recycling of soluble luminal proteins between the two compartments are fundamental features of vesicular protein trafficking that also occur in later stages of the secretory pathway.

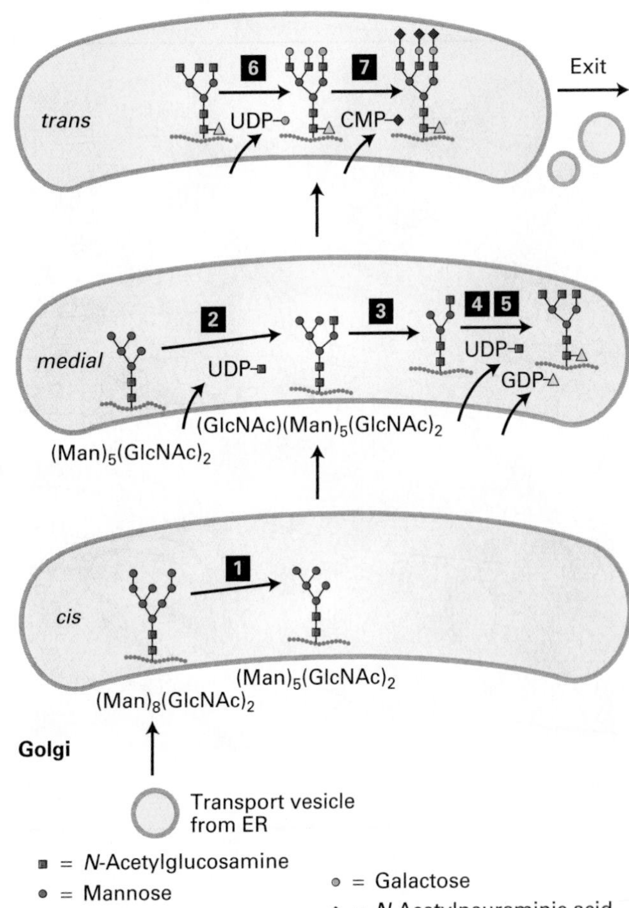

▲ **FIGURE 14-14 Processing of N-linked oligosaccharide chains on glycoproteins within *cis*-, *medial*-, and *trans*-Golgi cisternae in vertebrate cells.** The enzymes catalyzing each step are localized to the indicated compartments. After removal of three mannose residues in the *cis*-Golgi (step **1**), the protein moves by cisternal maturation to the *medial*-Golgi. Here, three *N*-acetylglucosamine (GlcNAc) residues are added (steps **2** and **4**), two more mannose residues are removed (step **3**), and a single fucose is added (step **5**). Processing is completed in the *trans*-Golgi by addition of three galactose residues (step **6**) and finally by linkage of an *N*-acetylneuraminic acid residue to each of the galactose residues (step **7**). Specific transferase enzymes add sugars to the oligosaccharide, one at a time, from sugar nucleotide precursors imported from the cytosol. This pathway represents the Golgi processing events for a typical mammalian glycoprotein. Variations in the structure of N-linked oligosaccharides can result from differences in processing steps in the Golgi. [See R. Kornfeld and S. Kornfeld, 1985, *Ann. Rev. Biochem.* **45**:631.]

Anterograde Transport Through the Golgi Occurs by Cisternal Maturation

The Golgi complex is organized into three or four subcompartments, which are often arranged in a stacked set of flattened sacs, or cisterna. The subcompartments of the Golgi differ from one another according to the enzymes that they contain. Many of the enzymes are glycosidases and glycosyltransferases that are involved in modifying N-linked or O-linked carbohydrates attached to secretory proteins as they transit the Golgi stack. On the whole, the Golgi complex operates much like an assembly line, with proteins moving in sequence through the Golgi stack, the modified carbohydrate chains in one compartment serving as the substrates for the modifying enzymes of the next compartment (see Figure 14-14 for a representative sequence of modification steps).

For many years it was thought that the Golgi complex was an essentially static set of compartments with small transport vesicles carrying secretory proteins forward, from the *cis*- to the *medial*-Golgi and from the *medial*- to the *trans*-Golgi. Indeed, electron microscopy reveals many small vesicles associated with the Golgi complex that move proteins from one Golgi compartment to another (Figure 14-15). However, these vesicles most likely mediate retrograde transport, retrieving ER or Golgi enzymes from a later compartment and transporting them to an earlier compartment in the secretory pathway. Thus

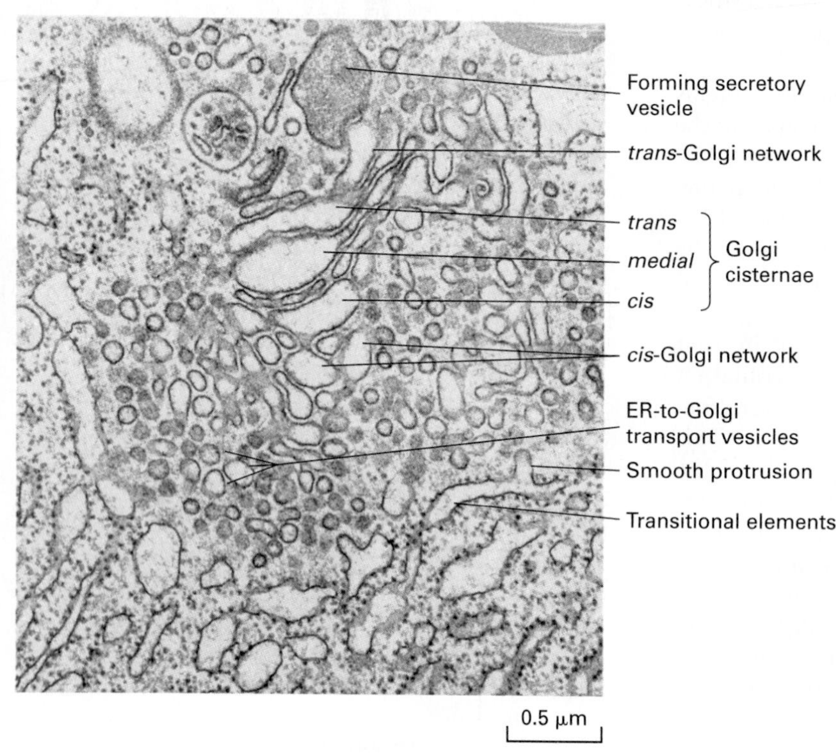

◀ EXPERIMENTAL FIGURE 14-15 **Electron micrograph of the Golgi complex in an exocrine pancreatic cell reveals both anterograde and retrograde transport vesicles.** A large secretory vesicle can be seen forming from the *trans*-Golgi network. Elements of the rough ER are on the bottom and left in this micrograph. Adjacent to the rough ER are transitional elements from which smooth protrusions appear to be budding. These buds form the small vesicles that transport secretory proteins from the rough ER to the Golgi complex. Interspersed among the Golgi cisternae are other small vesicles now known to function in retrograde, not anterograde, transport. [Courtesy G. Palade.]

the Golgi appears to have a highly dynamic organization. To see the effect this retrograde transport has on the organization of the Golgi, consider the net effect on the *medial*-Golgi compartment as enzymes from the *trans*-Golgi move to the *medial*-Golgi while enzymes from the *medial*-Golgi are transported to the *cis*-Golgi. As this process continues, the *medial*-Golgi acquires enzymes from the *trans*-Golgi while losing *medial*-Golgi enzymes and thus gradually becomes a new *trans*-Golgi compartment. In this way, secretory cargo proteins can acquire carbohydrate modification in the proper sequential order without being moved from one cisterna to another via anterograde vesicle transport.

The first evidence that the forward transport of cargo proteins from the *cis*- to the *trans*-Golgi occurs by a nonvesicular mechanism, called *cisternal maturation*, came from careful microscopic analysis of the synthesis of algal scales. These cell-wall glycoproteins are assembled in the *cis*-Golgi into large complexes visible in the electron microscope. Like other secretory proteins, newly made scales move from the *cis*- to the *trans*-Golgi, but they can be 20 times larger than the usual transport vesicles that bud from Golgi cisternae. Similarly, in the synthesis of collagen by fibroblasts, large aggregates of the procollagen precursor often form in the lumen of the *cis*-Golgi (see Figure 19-24). The procollagen aggregates are too large to be incorporated into small transport vesicles, and investigators could never find such aggregates in transport vesicles. These observations suggest that the forward movement of these and perhaps all secretory proteins from one Golgi compartment to another does *not* occur via small vesicles.

A particularly elegant demonstration of cisternal maturation in yeast takes advantage of different-colored versions of GFP to image two different Golgi proteins simultaneously. Figure 14-16 shows how a *cis*-Golgi resident protein labeled with a green fluorescent protein and a *trans*-Golgi protein labeled with a red fluorescent protein behave in the same yeast cell. At any given moment individual Golgi cisterna appear to have a distinct compartmental identity, in the sense that they contain either the *cis*-Golgi protein or the *trans*-Golgi protein but only rarely contain both proteins. However, over time an individual cisterna labeled with the *cis*-Golgi protein can be seen to progressively lose this protein and acquire the *trans*-Golgi protein. This behavior is exactly that predicted for the cisternal maturation model, in which the composition of an individual cisterna changes as Golgi resident proteins move from later to earlier Golgi compartments.

Numerous controversial questions concerning membrane flow within the Golgi stack remain unresolved. For example, although most protein traffic appears to move through the Golgi complex by a cisternal maturation mechanism, there is evidence that at least some of the COPI trans-

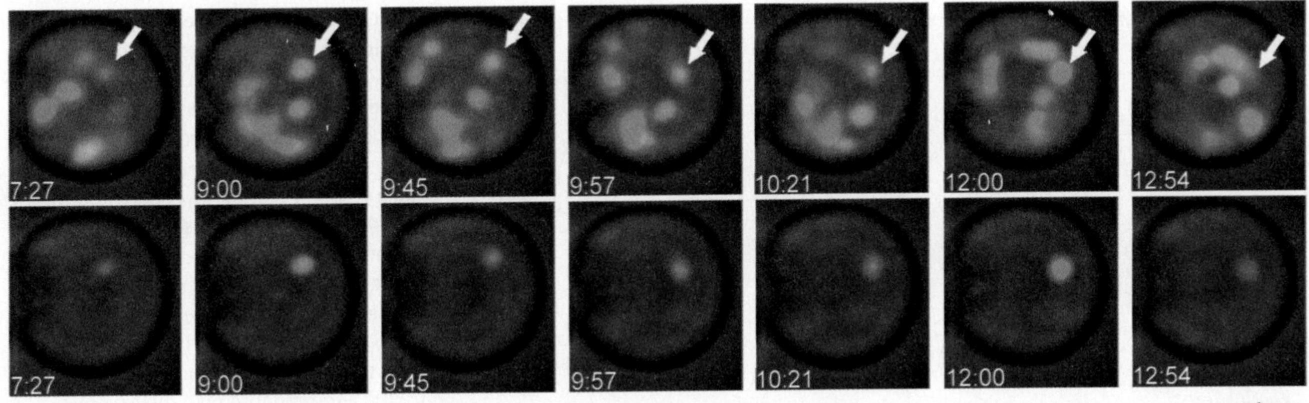

▲ EXPERIMENTAL FIGURE 14-16 **Fluorescence-tagged fusion proteins demonstrate Golgi cisternal maturation in a living yeast cell.** Yeast cells expressing the early Golgi protein Vrg4 fused to GFP (green fluorescence) and the late Golgi protein Sec7 fused to DsRed (red fluorescence) are imaged by time-lapse microscopy. The top series of images, taken approximately 1 minute apart, shows a collection of Golgi cisternae, which at any one time are labeled either with Vrg4 or Sec7. The bottom series of images show just one Golgi cisterna, isolated by digital processing of the image. First only Vrg4-GFP is located in the isolated cisterna and then Sec7-DsRed alone is located in the isolated cisterna, following a brief period in which both proteins are co-localized in this compartment. This experiment is a direct demonstration of the cisternal maturation hypothesis, showing that the composition of individual cisternae follow a process of maturation characterized by loss of early Golgi proteins and gain of late Golgi proteins [From Losev et al., 2006, *Nature* **441:**1002.]

port vesicles that bud from Golgi membranes contain cargo proteins (rather than Golgi enzymes) and move in an anterograde (rather than retrograde) direction.

KEY CONCEPTS OF SECTION 14.3

Early Stages of the Secretory Pathway

- COPII vesicles transport proteins from the rough ER to the *cis*-Golgi; COPI vesicles transport proteins in the reverse direction (see Figure 14-11).
- COPII coats comprise three components: the small GTP-binding protein Sar1, a Sec23/Sec24 complex, and a Sec13/Sec31 complex.
- Components of the COPII coat bind to membrane cargo proteins containing a di-acidic or other sorting signal in their cytosolic regions (see Figure 14-12). Soluble cargo proteins probably are targeted to COPII vesicles by binding to a membrane protein receptor.
- Many soluble ER-resident proteins contain a KDEL sorting signal. Binding of this retrieval sequence to a specific receptor protein in the *cis*-Golgi membrane recruits missorted ER proteins into retrograde COPI vesicles (see Figure 14-13).
- Membrane proteins needed to form COPII vesicles can be retrieved from the *cis*-Golgi by COPI vesicles. One of the sorting signals that directs membrane proteins into COPI vesicles is a KKXX sequence, which binds to subunits of the COPI coat.
- COPI vesicles also carry Golgi-resident proteins from later to earlier compartments in the Golgi stack.
- Soluble and membrane proteins advance through the Golgi complex by cisternal maturation, a process of anterograde transport that depends on resident Golgi enzymes moving by COPI vesicular transport in a retrograde direction.

14.4 Later Stages of the Secretory Pathway

As cargo proteins move from the *cis* face to the *trans* face of the Golgi complex by cisternal maturation, modifications to their oligosaccharide chains are carried out by Golgi-resident enzymes. The retrograde trafficking of COPI vesicles from later to earlier Golgi compartments maintains sufficient levels of these carbohydrate-modifying enzymes in their functional compartments. Eventually, properly processed cargo proteins reach the *trans*-Golgi network, the most distal Golgi compartment. Here, they are sorted into one of a number of different kinds of vesicles for delivery to their final destination. In this section we discuss the different kinds of vesicles that bud from the *trans*-Golgi network, the mechanisms that segregate cargo proteins among them, and key processing events that occur late in the secretory pathway. The various types of vesicles that bud from the trans-Golgi are summarized in Figure 14-17.

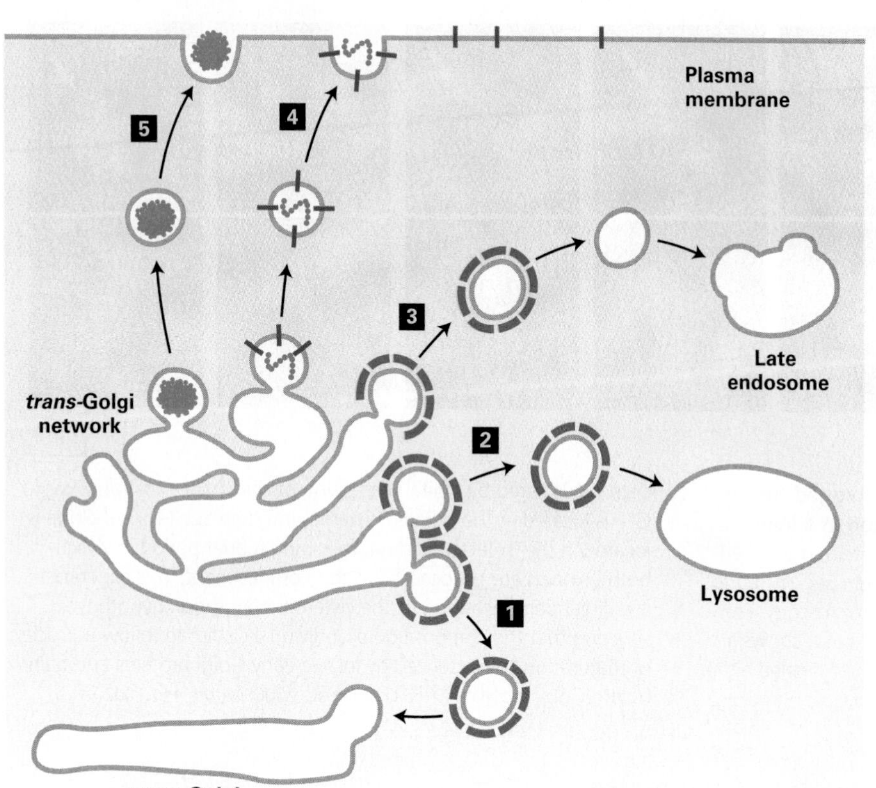

◀ **FIGURE 14-17 Vesicle-mediated protein trafficking from the *trans*-golgi network.** COPI (purple) vesicles (**1**) mediate retrograde transport within the Golgi. Proteins that function in the lumen or in the membrane of the lysosome are transported from the *trans*-Golgi network via clathrin coated (red) vesicles (**3**). After uncoating, these vesicles fuse with late endosomes, which deliver their contents to the lysosome. The coat on most clathrin vesicles contains additional proteins (AP complexes) not indicated here. Some vesicles from the *trans*-Golgi carrying cargo destined for the lysosome fuse with the lysosome directly (**2**), bypassing the endosome. These vesicles are coated with a type of AP complex (blue); it is unknown whether these vesicles also contain clathrin. The coat proteins surrounding constitutive (**4**) and regulated (**5**) secretory vesicles are not yet characterized; these vesicles carry secreted proteins and plasma-membrane proteins from the *trans*-Golgi network to the cell surface.

MEDIA CONNECTIONS

Video: Birth of a Clathrin Coat

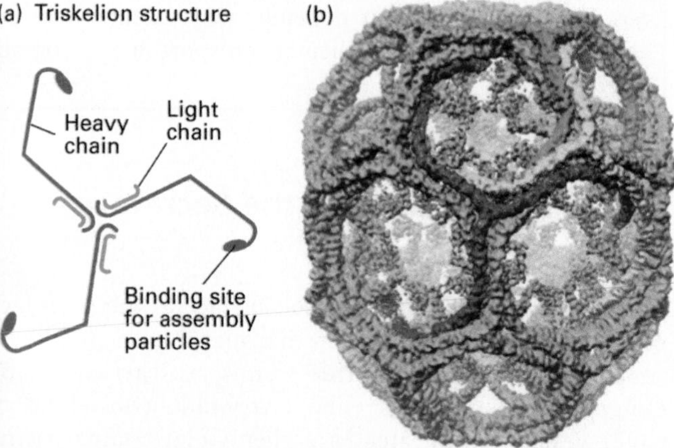

▲ **FIGURE 14-18 Structure of clathrin coats.** (a) A clathrin molecule, called a triskelion, is composed of three heavy and three light chains. It has an intrinsic curvature due to the bend in the heavy chains. (b) Clathrin coats were formed in vitro by mixing purified clathrin heavy and light chains with AP2 complexes in the absence of membranes. Cryoelectron micrographs of more than 1000 assembled hexagonal clathrin barrel particles were analyzed by digital image processing to generate an average structural representation. The processed image shows only the clathrin heavy chains in a structure composed of 36 triskelions. Three representative triskelions are highlighted in red, yellow, and green. Part of the AP2 complexes packed into the interior of the clathrin cage are also visible in this the processed representation. [See B. Pishvaee and G. Payne, 1998, *Cell* **95**:443. Part (b) from Fotin et al., 2004, *Nature* **432**:573.]

Vesicles Coated with Clathrin and/or Adapter Proteins Mediate Several Transport Steps

The best-characterized vesicles that bud from the *trans*-Golgi network (TGN) have a two-layered coat: an outer layer composed of the fibrous protein clathrin and an inner layer composed of *adapter protein (AP) complexes.* Purified clathrin molecules, which have a three-limbed shape, are called *triskelions,* from the Greek for "three-legged" (Figure 14-18a). Each limb contains one clathrin heavy chain (180,000 MW) and one clathrin light chain (≈35,000–40,000 MW). Triskelions polymerize to form a polygonal lattice with an intrinsic curvature (Figure 14-18b). When clathrin polymerizes on a donor membrane, it does so in association with AP complexes, which assemble between the clathrin lattice and the membrane. Each AP complex (340,000 MW) contains one copy each of four different adapter subunit proteins. A specific association between the globular domain at the end of each clathrin heavy chain in a triskelion and one subunit of the AP complex both promotes the co-assembly of clathrin triskelions with AP complexes and adds to the stability of the completed vesicle coat.

By binding to the cytosolic face of membrane proteins, adapter proteins determine which cargo proteins are specifically included in (or excluded from) a budding transport vesicle. Three different AP complexes are known (AP1, AP2, AP3), each with four subunits of different, though related, proteins. Recently, a second general type of adapter protein known as GGA has been shown to contain in a single 70,000 MW polypeptide both clathrin- and cargo-binding elements similar to those found in the much larger hetero-tetrameric AP complexes. Vesicles containing each type of adapter complex (AP or GGA) have been found to mediate specific transport steps (see Table 14-1). All vesicles whose coats contain one of these complexes utilize ARF to initiate coat assembly onto the donor membrane. As discussed previously, ARF also initiates assembly of COPI coats. The additional features of the membrane or protein factors that determine which type of coat will assemble after ARF attachment are not well understood at this time.

Vesicles that bud from the *trans*-Golgi network en route to the lysosome by way of the late endosome have clathrin coats associated with either AP1 or GGA. Both AP1 and GGA bind to the cytosolic domain of cargo proteins in the donor membrane. Recent studies have shown that membrane proteins containing a Tyr-X-X-Φ sequence, where X is any amino acid and Φ is a bulky hydrophobic amino acid, are recruited into clathrin/AP1 vesicles budding from the *trans*-Golgi network. This *YXXΦ sorting signal* interacts with one of the AP1 subunits in the vesicle coat. As we discuss in the next section, vesicles with clathrin/AP2 coats, which bud from the plasma membrane during endocytosis, also can recognize the YXXΦ sorting signal. Vesicles coated with GGA proteins and clathrin bind cargo molecules with a different kind of sorting sequence. Cytosolic sorting signals that specifically bind to GGA adapter proteins include Asp-X-Leu-Leu and Asp-Phe-Gly-X-Φ sequences (where X and Φ are defined as above).

Some vesicles that bud from the *trans*-Golgi network have coats composed of the AP3 complex. Although the AP3 complex does contain a binding site for clathrin similar to the AP1 and AP2 complexes, it is not clear whether clathrin is necessary for function of AP3-containing vesicles since mutants of AP3 that lack the clathrin binding site appear to be fully functional. AP3-coated vesicles mediate trafficking to the lysosome, but they appear to bypass the late endosome and fuse directly with the lysosomal membrane. In certain types of cells, such AP3 vesicles mediate protein transport to specialized storage compartments related to the lysosome. For example, AP3 is required for delivery of proteins to melanosomes, which contain the black pigment melanin in skin cells, and to platelet storage vesicles in megakaryocytes, a large cell that fragments into dozens of platelets. Mice with mutations in either of two different subunits of AP3 not only have abnormal skin pigmentation but also exhibit bleeding disorders. The latter occur because tears in blood vessels cannot be repaired without platelets that contain normal storage vesicles.

Dynamin Is Required for Pinching Off of Clathrin Vesicles

A fundamental step in the formation of a transport vesicle that we have not yet considered is how a vesicle bud is pinched off from the donor membrane. In the case of clathrin/AP-coated vesicles, a cytosolic protein called *dynamin* is essential for release of complete vesicles. At the later stages of bud formation, dynamin polymerizes around the neck portion and then hydrolyzes GTP. The energy derived from GTP hydrolysis is thought to drive a conformational change in dynamin that stretches the vesicle neck until the vesicle pinches off (Figure 14-19). Interestingly, COPI and COPII vesicles appear to pinch off from donor membranes without the aid of a GTPase such as dynamin. At present this fundamental difference in the process of pinching off among the different types of vesicles is not understood.

Incubation of cell extracts with a nonhydrolyzable derivative of GTP provides dramatic evidence for the importance of dynamin in pinching off of clathrin/AP vesicles during endocytosis. Such treatment leads to accumulation of clathrin-coated vesicle buds with excessively long necks that are surrounded by polymeric dynamin but do not pinch off (Figure 14-20). Likewise, cells expressing mutant forms of dynamin that cannot bind GTP do not form clathrin-coated vesicles and instead accumulate similar long-necked vesicle buds encased with polymerized dynamin.

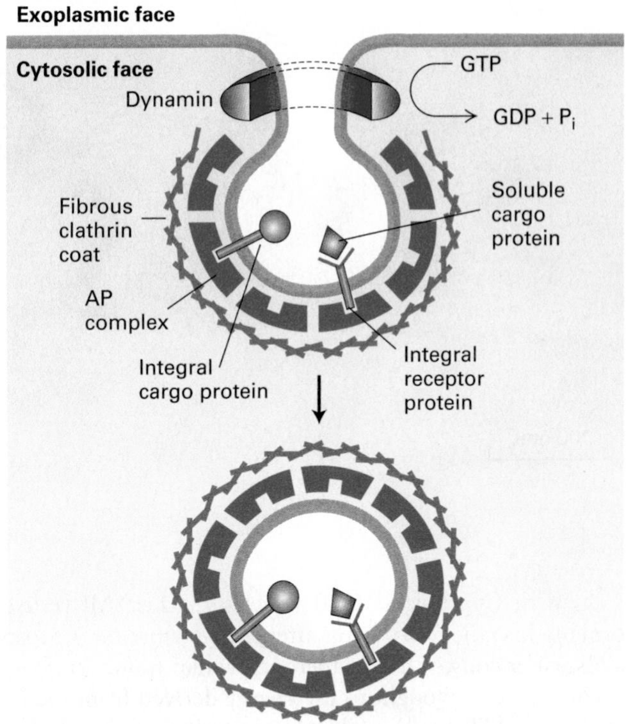

▲ **FIGURE 14-19 Model for dynamin-mediated pinching off of clathrin/AP-coated vesicles.** After a vesicle bud forms, dynamin polymerizes over the neck. By a mechanism that is not well understood, dynamin-catalyzed hydrolysis of GTP leads to release of the vesicle from the donor membrane. Note that membrane proteins in the donor membrane are incorporated into vesicles by interacting with AP complexes in the coat. [Adapted from K. Takel et al., 1995, *Nature* **374**:186.]

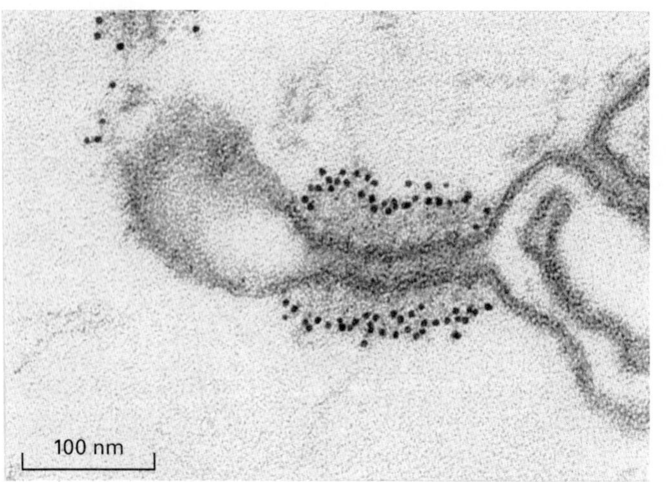

◀ **EXPERIMENTAL FIGURE 14-20 GTP hydrolysis by dynamin is required for pinching off of clathrin-coated vesicles in cell-free extracts.** A preparation of nerve terminals, which undergo extensive endocytosis, was lysed by treatment with distilled water and incubated with GTP-γ-S, a nonhydrolyzable derivative of GTP. After sectioning, the preparation was treated with gold-tagged anti-dynamin antibody and viewed in the electron microscope. This image, which shows a long-necked clathrin/AP-coated bud with polymerized dynamin lining the neck, reveals that buds can form in the absence of GTP hydrolysis, but vesicles cannot pinch off. The extensive polymerization of dynamin that occurs in the presence of GTP-γ-S probably does not occur during the normal budding process. [From K. Takel et al., 1995, *Nature* **374:**186; courtesy of Pietro De Camilli.]

As with COPI and COPII vesicles, clathrin/AP vesicles normally lose their coat soon after their formation. Cytosolic Hsc70, a constitutive chaperone protein found in all eukaryotic cells, is thought to use energy derived from the hydrolysis of ATP to drive depolymerization of the clathrin coat into triskelions. Uncoating not only releases triskelions for reuse in the formation of additional vesicles but also exposes v-SNAREs for use in fusion with target membranes. Conformational changes that occur when ARF switches from the GTP-bound to GDP-bound state are thought to regulate the timing of clathrin coat depolymerization. How the action of Hsc70 might be coupled to ARF switching is not well understood.

Mannose 6-Phosphate Residues Target Soluble Proteins to Lysosomes

Most of the sorting signals that function in vesicular trafficking are short amino acid sequences in the targeted protein. In contrast, the sorting signal that directs soluble lysosomal enzymes from the *trans*-Golgi network to the late endosome is a carbohydrate residue, *mannose 6-phosphate (M6P)*, which is formed in the *cis*-Golgi. The addition and initial processing of one or more preformed N-linked oligosaccharide precursors in the rough ER is the same for lysosomal enzymes as for membrane and secreted proteins, yielding core $Man_8(GlcNAc)_2$ chains (see Figure 13-18). In the *cis*-Golgi, the N-linked oligosaccharides present on most lysosomal enzymes undergo a two-step reaction sequence that generates M6P residues (Figure 14-21). The addition of M6P residues to the oligosaccharide chains of soluble lysosomal enzymes prevents these proteins from undergoing the further processing reactions characteristic of secreted and membrane proteins (see Figure 14-14).

As shown in Figure 14-22, the segregation of M6P-bearing lysosomal enzymes from secreted and membrane proteins occurs in the *trans*-Golgi network. Here, transmembrane *mannose 6-phosphate receptors* bind the M6P residues on lysosome-destined proteins very tightly and specifically. Clathrin/AP1 vesicles containing the M6P receptor and bound lysosomal enzymes then bud from the *trans*-Golgi network, lose their coats, and subsequently fuse with

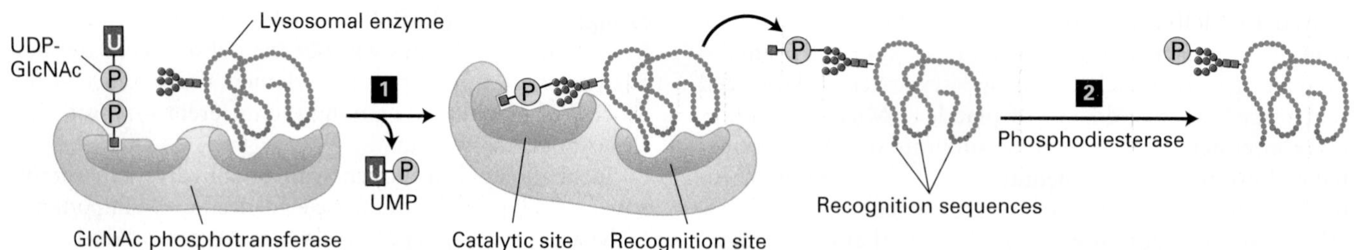

▲ **FIGURE 14-21 Formation of mannose 6-phosphate (M6P) residues that target soluble enzymes to lysosomes.** The M6P residues that direct proteins to lysosomes are generated in the *cis*-Golgi by two Golgi-resident enzymes. Step **1**: An *N*-acetylglucosamine (GlcNAc) phosphotransferase transfers a phosphorylated GlcNAc group to carbon atom 6 of one or more mannose residues. Because only lysosomal enzymes contain sequences (red) that are recognized and bound by this enzyme, phosphorylated GlcNAc groups are added specifically to lysosomal enzymes. Step **2**: After release of a modified protein from the phosphotransferase, a phosphodiesterase removes the GlcNAc group, leaving a phosphorylated mannose residue on the lysosomal enzyme. [See A. B. Cantor et al., 1992, *J. Biol. Chem.* **267:**23349, and S. Kornfeld, 1987, *FASEB J.* **1:**462.]

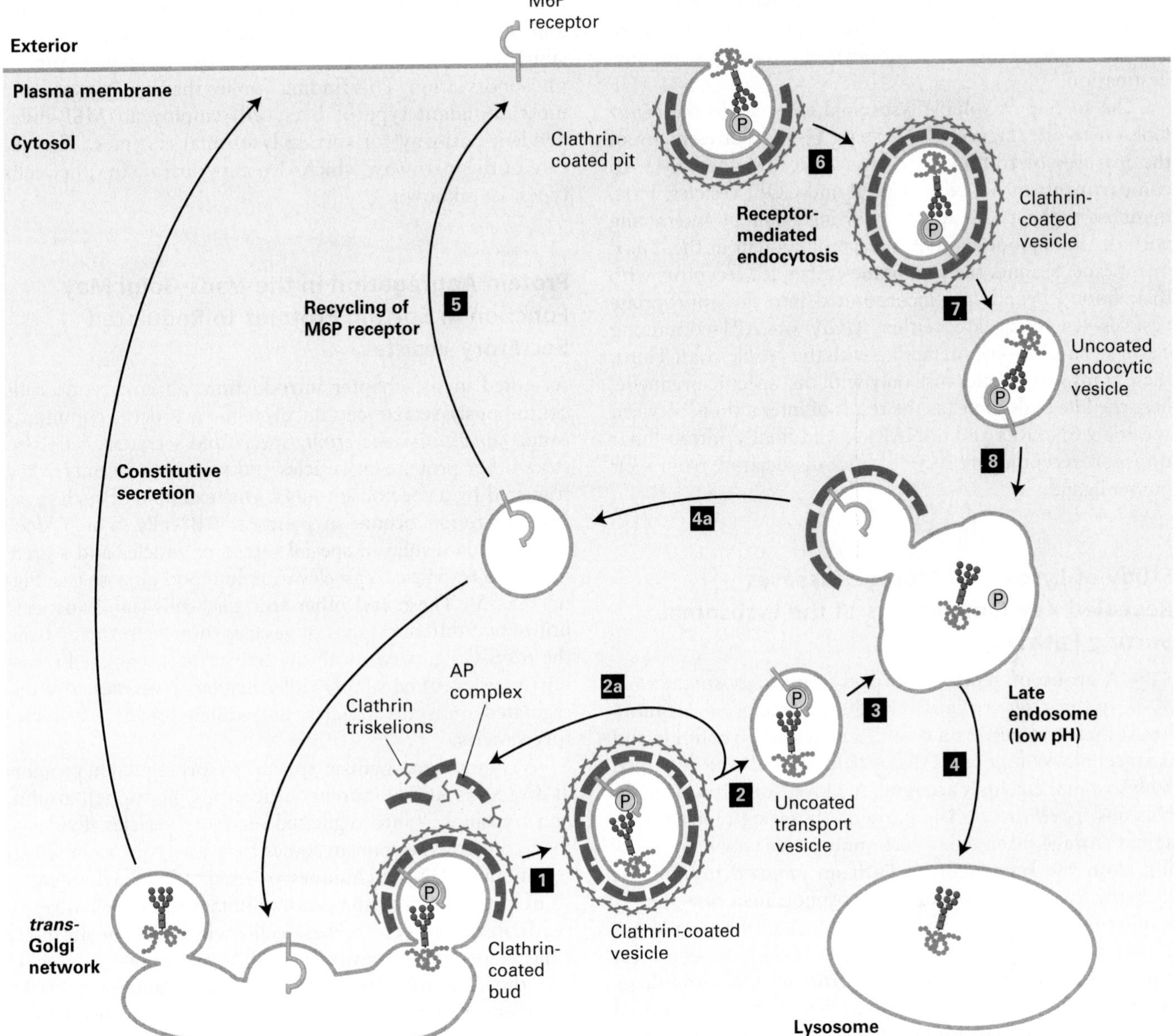

▲ **FIGURE 14-22 Trafficking of soluble lysosomal enzymes from the *trans*-Golgi network and cell surface to lysosomes.** Newly synthesized lysosomal enzymes, produced in the ER, acquire mannose 6-phosphate (M6P) residues in the *cis*-Golgi (see Figure 14-21). For simplicity, only one phosphorylated oligosaccharide chain is depicted, although lysosomal enzymes typically have many such chains. In the *trans*-Golgi network, proteins that bear the M6P sorting signal interact with M6P receptors in the membrane and thereby are directed into clathrin/AP1 vesicles (step **1**). The coat surrounding released vesicles is rapidly depolymerized (step **2**), and the uncoated transport vesicles fuse with late endosomes (step **3**). After the phosphorylated enzymes dissociate from the M6P receptors and are dephosphorylated, late endosomes subsequently fuse with a lysosome (step **4**). Note that coat proteins and M6P receptors are recycled (steps **2a** and **4a**), and some receptors are delivered to the cell surface (step **5**). Phosphorylated lysosomal enzymes occasionally are sorted from the *trans*-Golgi to the cell surface and secreted. These secreted enzymes can be retrieved by receptor-mediated endocytosis (steps **6**–**8**), a process that closely parallels trafficking of lysosomal enzymes from the *trans*-Golgi network to lysosomes. [See G. Griffiths et al., 1988, *Cell* **52**:329; S. Kornfeld, 1992, *Ann. Rev. Biochem.* **61**:307; and G. Griffiths and J. Gruenberg, 1991, *Trends Cell Biol.* **1**:5.]

the late endosome by mechanisms described previously. Because M6P receptors can bind M6P at the slightly acidic pH (≈6.5) of the *trans*-Golgi network but not at a pH less than 6, the bound lysosomal enzymes are released within late endosomes, which have an internal pH of 5.0–5.5. Furthermore, a phosphatase within late endosomes usually removes the phosphate from M6P residues on lysosomal enzymes, preventing any rebinding to the M6P receptor that might occur in spite of the low pH in endosomes. Vesicles budding from late endosomes recycle the M6P receptor back to the *trans*-Golgi network or, on occasion, to the cell surface. Eventually, mature late endosomes fuse with lysosomes, delivering the lysosomal enzymes to their final destination.

The sorting of soluble lysosomal enzymes in the *trans*-Golgi network (Figure 14-22, steps **1**–**4**) shares many of the features of trafficking between the ER and *cis*-Golgi compartments mediated by COPII and COPI vesicles. First, mannose 6-phosphate acts as a sorting signal by interacting with the luminal domain of a receptor protein in the donor membrane. Second, the membrane-embedded receptors with their bound ligands are incorporated into the appropriate vesicles—in this case, either GGA or AP1-containing clathrin vesicles—by interacting with the vesicle coat. Third, these transport vesicles fuse only with one specific organelle, here the late endosome, as the result of interactions between specific v-SNAREs and t-SNAREs. And finally, intracellular transport receptors are recycled after dissociating from their bound ligand.

Study of Lysosomal Storage Diseases Revealed Key Components of the Lysosomal Sorting Pathway

A group of genetic disorders termed *lysosomal storage diseases* are caused by the absence of one or more lysosomal enzymes. As a result, undigested glycolipids and extracellular components that would normally be degraded by lysosomal enzymes accumulate in lysosomes as large inclusions. *I-cell disease* is a particularly severe type of lysosomal storage disease in which multiple enzymes are missing from the lysosomes. Cells from affected individuals lack the *N*-acetylglucosamine phosphotransferase that is required for formation of M6P residues on lysosomal enzymes in the *cis*-Golgi (see Figure 14-21). Biochemical comparison of lysosomal enzymes from normal individuals with those from patients with I-cell disease led to the initial discovery of mannose 6-phosphate as the lysosomal sorting signal. Lacking the M6P sorting signal, the lysosomal enzymes in I-cell patients are secreted rather than being sorted to and sequestered in lysosomes.

When fibroblasts from patients with I-cell disease are grown in a medium containing lysosomal enzymes bearing M6P residues, the diseased cells acquire a nearly normal intracellular content of lysosomal enzymes. This finding indicates that the plasma membrane of these cells contains M6P receptors, which can internalize extracellular phosphorylated lysosomal enzymes by receptor-mediated endocytosis. This process, used by many cell-surface receptors to bring bound proteins or particles into the cell, is discussed in detail in the next section. It is now known that even in normal cells, some M6P receptors are transported to the plasma membrane and some phosphorylated lysosomal enzymes are secreted (see Figure 14-22). The secreted enzymes can be retrieved by receptor-mediated endocytosis and directed to lysosomes. This pathway thus scavenges any lysosomal enzymes that escape the usual M6P sorting pathway.

Hepatocytes from patients with I-cell disease contain a normal complement of lysosomal enzymes and no inclusions, even though these cells are defective in mannose phosphorylation. This finding implies that hepatocytes (the most abundant type of liver cell) employ an M6P-independent pathway for sorting lysosomal enzymes. The nature of this pathway, which also may operate in other cells types, is unknown. ▮

Protein Aggregation in the *trans*-Golgi May Function in Sorting Proteins to Regulated Secretory Vesicles

As noted in the chapter introduction, all eukaryotic cells continuously secrete certain proteins, a process commonly called *constitutive secretion.* Specialized secretory cells also store other proteins in vesicles and secrete them only when triggered by a specific stimulus. One example of such *regulated secretion* occurs in pancreatic β cells, which store newly made insulin in special secretory vesicles and secrete insulin in response to an elevation in blood glucose (see Figure 15-33). These and other secretory cells simultaneously utilize two different types of vesicles to move proteins from the *trans*-Golgi network to the cell surface: regulated transport vesicles, often simply called secretory vesicles, and unregulated transport vesicles, also called constitutive secretory vesicles.

A common mechanism appears to sort regulated proteins as diverse as ACTH (adrenocorticotropic hormone), insulin, and trypsinogen into regulated secretory vesicles. Evidence for a common mechanism comes from experiments in which recombinant DNA techniques are used to induce the synthesis of insulin and trypsinogen in pituitary tumor cells already synthesizing ACTH. In these cells, which do not normally express insulin or trypsinogen, all three proteins segregate into the same regulated secretory vesicles and are secreted together when a hormone binds to a receptor on the pituitary cells and causes a rise in cytosolic Ca^{2+}. Although these three proteins share no identical amino acid sequences that might serve as a sorting sequence, they obviously have some common feature that signals their incorporation into regulated secretory vesicles.

Morphologic evidence suggests that sorting into the regulated pathway is controlled by selective protein aggregation. For instance, immature vesicles in this pathway—those that have just budded from the *trans*-Golgi network—contain diffuse aggregates of secreted protein that are visible in the electron microscope. These aggregates also are found in vesicles that are in the process of budding, indicating that proteins destined for regulated secretory vesicles selectively aggregate together before their incorporation into the vesicles.

Other studies have shown that regulated secretory vesicles from mammalian secretory cells contain three proteins, *chromogranin A, chromogranin B,* and *secretogranin II,* that together form aggregates when incubated at the ionic conditions ($pH \approx 6.5$ and 1 mM Ca^{2+}) thought to occur in the *trans*-Golgi network; such aggregates do not form at the

(a) Proinsulin antibody

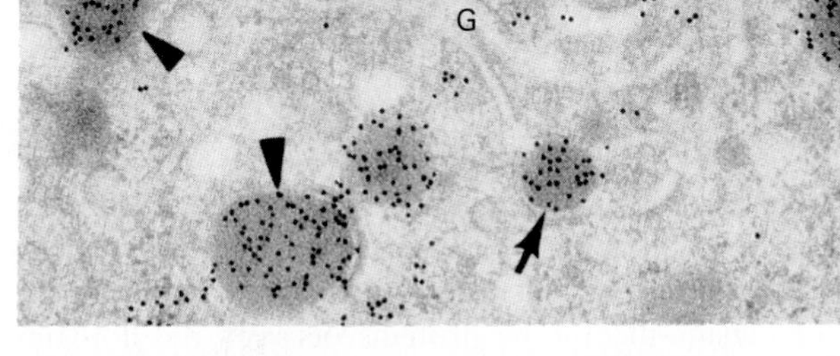

(b) Insulin antibody

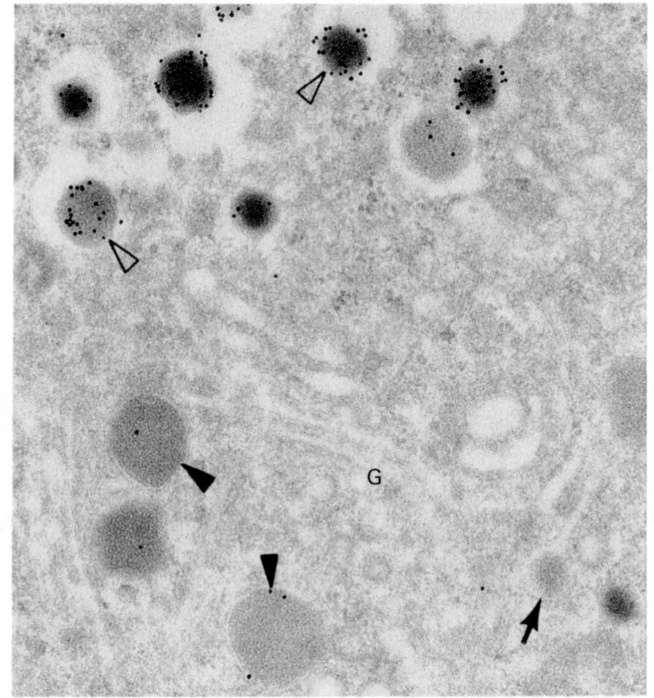

◀ **EXPERIMENTAL FIGURE 14-23 Proteolytic cleavage of proinsulin occurs in secretory vesicles after they have budded from the *trans*-Golgi network.** Serial sections of the Golgi region of an insulin-secreting cell were stained with (a) a monoclonal antibody that recognizes proinsulin but not insulin or (b) a different antibody that recognizes insulin but not proinsulin. The antibodies, which were bound to electron-opaque gold particles, appear as dark dots in these electron micrographs (see Figure 9-21). Immature secretory vesicles (closed arrowheads) and vesicles budding from the *trans*-Golgi (arrows) stain with the proinsulin antibody but not with insulin antibody. These vesicles contain diffuse protein aggregates that include proinsulin and other regulated secreted proteins. Mature vesicles (open arrowheads) stain with insulin antibody but not with proinsulin antibody and have a dense core of almost crystalline insulin. Since budding and immature secretory vesicles contain proinsulin (not insulin), the proteolytic conversion of proinsulin to insulin must take place in these vesicles after they bud from the *trans*-Golgi network. The inset in (a) shows a proinsulin-rich secretory vesicle surrounded by a protein coat (dashed line). [From L. Orci et al., 1987, *Cell* **49:**865; courtesy of L. Orci.]

neutral pH of the ER. The selective aggregation of regulated secreted proteins together with chromogranin A, chromogranin B, or secretogranin II could be the basis for sorting of these proteins into regulated secretory vesicles. Secreted proteins that do not associate with these proteins, and thus do not form aggregates, would be sorted into unregulated transport vesicles by default.

Some Proteins Undergo Proteolytic Processing After Leaving the *trans*-Golgi

For some secretory proteins (e.g., growth hormone) and certain viral membrane proteins (e.g., the VSV glycoprotein), removal of the N-terminal ER signal sequence from the nascent chain is the only known proteolytic cleavage required to convert the polypeptide to the mature, active species (see Figure 13-6). However, some membrane and many soluble secretory proteins initially are synthesized as relatively long-lived, inactive precursors, termed *proproteins,* that require further proteolytic processing to generate the mature, active proteins. Examples of proteins that undergo such processing are soluble lysosomal enzymes, many membrane proteins such as influenza hemagglutinin (HA), and secreted proteins such as serum albumin, insulin, glucagon, and the yeast α mating factor. In general, the proteolytic conversion of a proprotein to the corresponding mature protein occurs after the proprotein has been sorted in the *trans*-Golgi network to appropriate vesicles.

In the case of soluble lysosomal enzymes, the proproteins are called *proenzymes,* which are sorted by the M6P receptor as catalytically inactive enzymes. In the late endosome or lysosome a proenzyme undergoes a proteolytic cleavage that generates a smaller but enzymatically active polypeptide. Delaying the activation of lysosomal proenzymes until they reach the lysosome prevents them from digesting macromolecules in earlier compartments of the secretory pathway.

Normally, mature vesicles carrying secreted proteins to the cell surface are formed by fusion of several immature ones containing proprotein. Proteolytic cleavage of proproteins, such as proinsulin, occurs in vesicles after they move away from the *trans*-Golgi network (Figure 14-23). The proproteins of most constitutively secreted proteins (e.g., albumin) are cleaved only once at a site C-terminal to a

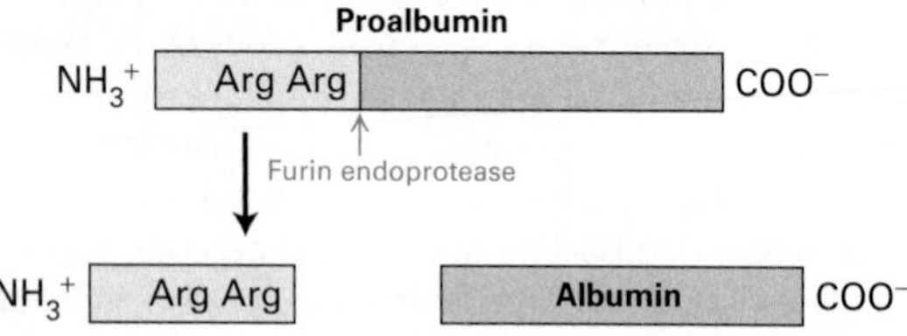

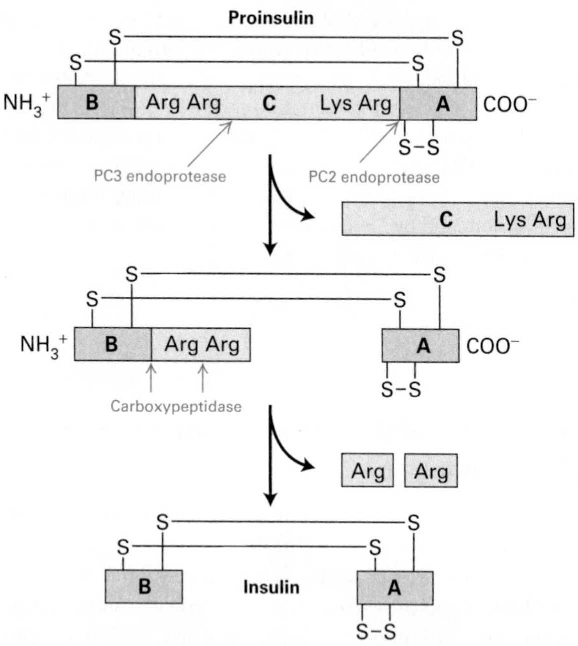

▲ **FIGURE 14-24 Proteolytic processing of proproteins in the constitutive and regulated secretion pathways.** The processing of proalbumin and proinsulin is typical of the constitutive and regulated pathways, respectively. The endoproteases that function in such processing cleave at the C-terminal side of two consecutive amino acids. (a) The endoprotease furin acts on the precursors of constitutive secreted proteins. (b) Two endoproteases, PC2 and PC3, act on the precursors of regulated secreted proteins. The final processing of many such proteins is catalyzed by a carboxypeptidase that sequentially removes two basic amino acid residues at the C-terminus of a polypeptide. [See D. Steiner et al., 1992, *J. Biol. Chem.* **267**:23435.]

dibasic recognition sequence such as Arg-Arg or Lys-Arg (Figure 14-24a). Proteolytic processing of proteins whose secretion is regulated generally entails additional cleavages. In the case of proinsulin, multiple cleavages of the single polypeptide chain yields the N-terminal B chain and the C-terminal A chain of mature insulin, which are linked by disulfide bonds, and the central C peptide, which is lost and subsequently degraded (Figure 14-24b).

The breakthrough in identifying the proteases responsible for such processing of secreted proteins came from analysis of yeast with a mutation in the *KEX2* gene. These mutant cells synthesized the precursor of the α mating factor but could not proteolytically process it to the functional form and thus were unable to mate with cells of the opposite mating type (see Figure 21-19). The wild-type *KEX2* gene encodes an endoprotease that cleaves the α-factor precursor at a site C-terminal to Arg-Arg and Lys-Arg residues. Using the *KEX2* gene as a DNA probe, researchers were able to clone a family of mammalian endoproteases, all of which cleave a protein chain on the C-terminal side of an Arg-Arg or Lys-Arg sequence. One, called *furin,* is found in all mammalian cells; it processes proteins such as albumin that are secreted by the continuous pathway. In contrast, the *PC2* and *PC3 endoproteases* are found only in cells that exhibit regulated secretion; these enzymes are localized to regulated secretory vesicles and proteolytically cleave the precursors of many hormones at specific sites.

Several Pathways Sort Membrane Proteins to the Apical or Basolateral Region of Polarized Cells

The plasma membrane of polarized epithelial cells is divided into two domains, **apical** and **basolateral;** tight junctions located between the two domains prevent the movement of plasma-membrane proteins between the domains (see Figure 19-9). Several sorting mechanisms direct newly synthesized membrane proteins to either the apical or basolateral domain of epithelial cells, and any one protein may be sorted by more than one mechanism. Although these sorting mechanisms are understood in general terms, the molecular signals underlying the vesicle-mediated transport of membrane proteins in polarized cells are not yet known. As a result of this sorting and the restriction on protein movement within the plasma membrane due to tight junctions, distinct sets of proteins are found in the apical or basolateral domain. This preferential localization of certain transport proteins is critical to a variety of important physiological functions, such as absorption of nutrients from the intestinal lumen and acidification of the stomach lumen (see Figures 11-29 and 11-30).

Microscopic and cell-fractionation studies indicate that proteins destined for either the apical or the basolateral membranes are initially located together within the membranes of the *trans*-Golgi network. In some cases, proteins destined for the apical membrane are sorted into their own transport vesicles that bud from the *trans*-Golgi network and then move to the apical region, whereas proteins destined for the basolateral membrane are sorted into other vesicles that move to the basolateral region. The different vesicle types can be distinguished by their protein constituents, including distinct Rab and v-SNARE proteins, which apparently target them to the appropriate plasma-membrane domain. In this mechanism, segregation of proteins destined for either the apical or basolateral membranes occurs as cargo proteins are incorporated into particular types of vesicles budding from the *trans*-Golgi network.

Such direct basolateral-apical sorting has been investigated in cultured Madin-Darby canine kidney (MDCK)

Video: Segregation of Apical and Basolateral Cargo in the Golgi of Live Cells

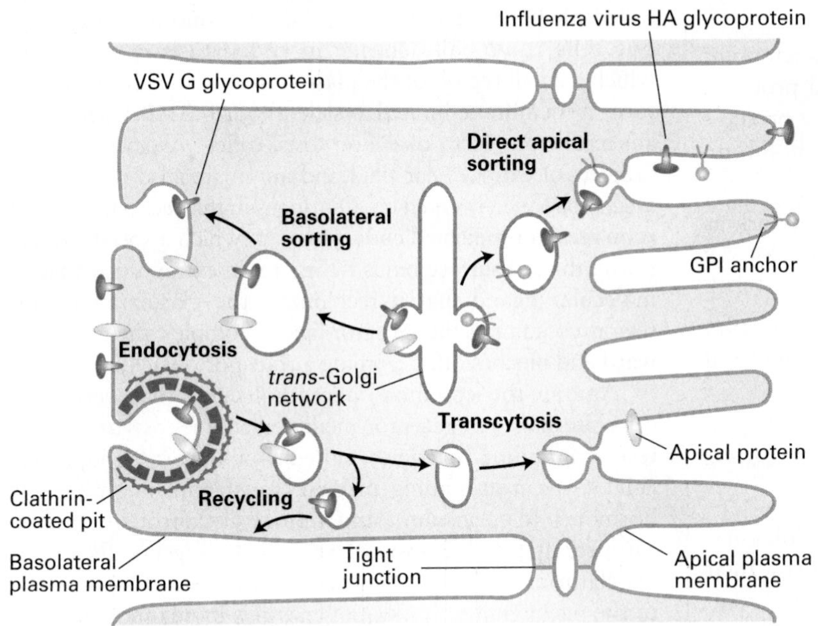

◀ **FIGURE 14-25 Sorting of proteins destined for the apical and basolateral plasma membranes of polarized cells.** When cultured MDCK cells are infected simultaneously with VSV and influenza virus, the VSV G glycoprotein (purple) is found only on the basolateral membrane, whereas the influenza HA glycoprotein (green) is found only on the apical membrane. Some cellular proteins (orange circle), especially those with a GPI anchor, are likewise sorted directly to the apical membrane and others to the basolateral membrane (not shown) via specific transport vesicles that bud from the *trans*-Golgi network. In certain polarized cells, some apical and basolateral proteins are transported together to the basolateral surface; the apical proteins (yellow oval) then move selectively, by endocytosis and transcytosis, to the apical membrane. [After K. Simons and A. Wandinger-Ness, 1990, *Cell* **62:**207, and K. Mostov et al., 1992, *J. Cell Biol.* **116:**577.]

cells, a line of cultured polarized epithelial cells (see Figure 9-34). In MDCK cells infected with the influenza virus, progeny viruses bud only from the apical membrane, whereas in cells infected with vesicular stomatitis virus (VSV), progeny viruses bud only from the basolateral membrane. This difference occurs because the HA glycoprotein of influenza virus is transported from the Golgi complex exclusively to the apical membrane and the VSV G protein is transported only to the basolateral membrane (Figure 14-25). Furthermore, when the gene encoding HA protein is introduced into uninfected cells by recombinant DNA techniques, all the expressed HA accumulates in the apical membrane, indicating that the sorting signal resides in the HA glycoprotein itself and not in other viral proteins produced during viral infection.

Among the cellular proteins that undergo similar apical-basolateral sorting in the Golgi are those with a *glycosylphosphatidylinositol (GPI) membrane anchor.* In MDCK cells and most other types of epithelial cells, GPI-anchored proteins are targeted to the apical membrane. In membranes GPI-anchored proteins are clustered into **lipid rafts,** which are rich in sphingolipids (see Chapter 10). This finding suggests that lipid rafts are localized to the apical membrane along with proteins that preferentially partition them in many cells. However, the GPI anchor is not an apical sorting signal in all polarized cells; in thyroid cells, for example, GPI-anchored proteins are targeted to the basolateral membrane. Other than GPI anchors, no unique sequences have been identified that are both necessary and sufficient to target proteins to either the apical or basolateral domain. Instead, each membrane protein may contain multiple sorting signals, any one of which can target it to the appropriate plasma-membrane domain. The identification of such complex signals and of the vesicle coat proteins that recognize them is currently being pursued for a number of different proteins that are sorted to specific plasma-membrane domains of polarized epithelial cells.

Another mechanism for sorting apical and basolateral proteins, also illustrated in Figure 14-25, operates in hepatocytes. The basolateral membranes of hepatocytes face the blood (as in intestinal epithelial cells), and the apical membranes line the small intercellular channels into which bile is secreted. In hepatocytes, newly made apical and basolateral proteins are first transported in vesicles from the *trans*-Golgi network to the basolateral region and incorporated into the plasma membrane by exocytosis (i.e., fusion of the vesicle membrane with the plasma membrane). From there, both basolateral and apical proteins are endocytosed in the same vesicles, but then their paths diverge. The endocytosed basolateral proteins are sorted into transport vesicles that recycle them to the basolateral membrane. In contrast, the apically destined endocytosed proteins are sorted into transport vesicles that move across the cell and fuse with the apical membrane, a process called **transcytosis.** This process also is used to move extracellular materials from one side of an epithelium to another. Even in epithelial cells, such as MDCK cells, in which apical-basolateral protein sorting occurs in the Golgi, transcytosis may provide a "fail-safe" sorting mechanism. That is, an apical protein sorted incorrectly to the basolateral membrane would be subjected to endocytosis and then correctly delivered to the apical membrane.

KEY CONCEPTS OF SECTION 14.4

Later Stages of the Secretory Pathway

- The *trans*-Golgi network (TGN) is a major branch point in the secretory pathway where soluble secreted proteins, lysosomal proteins, and in some cells membrane proteins destined for the basolateral or apical plasma membrane are segregated into different transport vesicles.
- Many vesicles that bud from the *trans*-Golgi network as well as endocytic vesicles bear a coat composed of AP (adapter protein) complexes and clathrin (see Figure 14-18).
- Pinching off of clathrin-coated vesicles requires dynamin, which forms a collar around the neck of the vesicle bud and hydrolyzes GTP (see Figure 14-19).
- Soluble enzymes destined for lysosomes are modified in the *cis*-Golgi, yielding multiple mannose 6-phosphate (M6P) residues on their oligosaccharide chains.
- M6P receptors in the membrane of the *trans*-Golgi network bind proteins bearing M6P residues and direct their transfer to late endosomes, where receptors and their ligand proteins dissociate. The receptors then are recycled to the Golgi or plasma membrane, and the lysosomal enzymes are delivered to lysosomes (see Figure 14-22).
- Regulated secreted proteins are concentrated and stored in secretory vesicles to await a neural or hormonal signal for exocytosis. Protein aggregation within the *trans*-Golgi network may play a role in sorting secreted proteins to the regulated pathway.
- Many proteins transported through the secretory pathway undergo post-Golgi proteolytic cleavages that yield the mature, active proteins. Generally, proteolytic maturation can occur in vesicles carrying proteins from the *trans*-Golgi network to the cell surface, in the late endosome, or in the lysosome.
- In polarized epithelial cells, membrane proteins destined for the apical or basolateral domains of the plasma membrane are sorted in the *trans*-Golgi network into different transport vesicles (see Figure 14-25). The GPI anchor is the only apical-basolateral sorting signal identified so far.
- In hepatocytes and some other polarized cells, all plasma-membrane proteins are directed first to the basolateral membrane. Apically destined proteins then are endocytosed and moved across the cell to the apical membrane (transcytosis).

14.5 Receptor-Mediated Endocytosis

In previous sections we have explored the main pathways whereby secretory and membrane proteins synthesized on the rough ER are delivered to the cell surface or other destinations. Cells also can internalize materials from their surroundings and sort these to particular destinations. A few cell types (e.g., macrophages) can take up whole bacteria and other large particles by **phagocytosis,** a nonselective actin-mediated process in which extensions of the plasma membrane envelop the ingested material, forming large vesicles called phagosomes (see Figure 9-2). In contrast, all eukaryotic cells continually engage in endocytosis, a process in which a small region of the plasma membrane invaginates to form a membrane-limited vesicle about 0.05–0.1 μm in diameter. In one form of endocytosis, called *pinocytosis,* small droplets of extracellular fluid and any material dissolved in it are nonspecifically taken up. Our focus in this section, however, is on **receptor-mediated endocytosis,** in which a specific receptor on the cell surface binds tightly to an extracellular macromolecular ligand that it recognizes; the plasma-membrane region containing the receptor-ligand complex then buds inward and pinches off, becoming a transport vesicle.

Among the common macromolecules that vertebrate cells internalize by receptor-mediated endocytosis are cholesterol-containing particles called low-density lipoprotein (LDL), the iron-binding protein transferrin, many protein hormones (e.g., insulin), and certain glycoproteins. Receptor-mediated endocytosis of such ligands generally occurs via clathrin/AP2-coated pits and vesicles in a process similar to the packaging of lysosomal enzymes by mannose 6-phosphate (M6P) in the *trans*-Golgi network (see Figure 14-22). As noted earlier, some M6P receptors are found on the cell surface, and these participate in the receptor-mediated endocytosis of lysosomal enzymes that are mistakenly secreted. In general, the transmembrane receptor proteins that function in the uptake of extracellular ligands are internalized from the cell surface during endocytosis and are then sorted and recycled back to the cell surface, much like the recycling of M6P receptors to the plasma membrane and *trans*-Golgi. The rate at which a ligand is internalized is limited by the amount of its corresponding receptor on the cell surface.

Clathrin/AP2 pits make up about 2 percent of the surface of cells such as hepatocytes and fibroblasts. Many internalized ligands have been observed in these pits and vesicles, which are thought to function as intermediates in the endocytosis of most (though not all) ligands bound to cell-surface receptors (Figure 14-26). Some receptors are clustered over clathrin-coated pits even in the absence of ligand. Other receptors diffuse freely in the plane of the plasma membrane but undergo a conformational change when binding to ligand, so that when the receptor-ligand complex diffuses into a clathrin-coated pit, it is retained there. Two or more types of receptor-bound ligands, such as LDL and transferrin, can be seen in the same coated pit or vesicle.

Cells Take Up Lipids from the Blood in the Form of Large, Well-Defined Lipoprotein Complexes

Lipids absorbed from the diet in the intestines or stored in adipose tissue can be distributed to cells throughout the body. To facilitate the mass transfer of lipids between cells, animals have evolved an efficient way to package from hundreds to thousands of lipid molecules into water-soluble, macromolecular carriers, called **lipoproteins,** that cells can take up from the circulation as an ensemble. A lipoprotein particle has a shell composed of proteins (*apolipoproteins*)

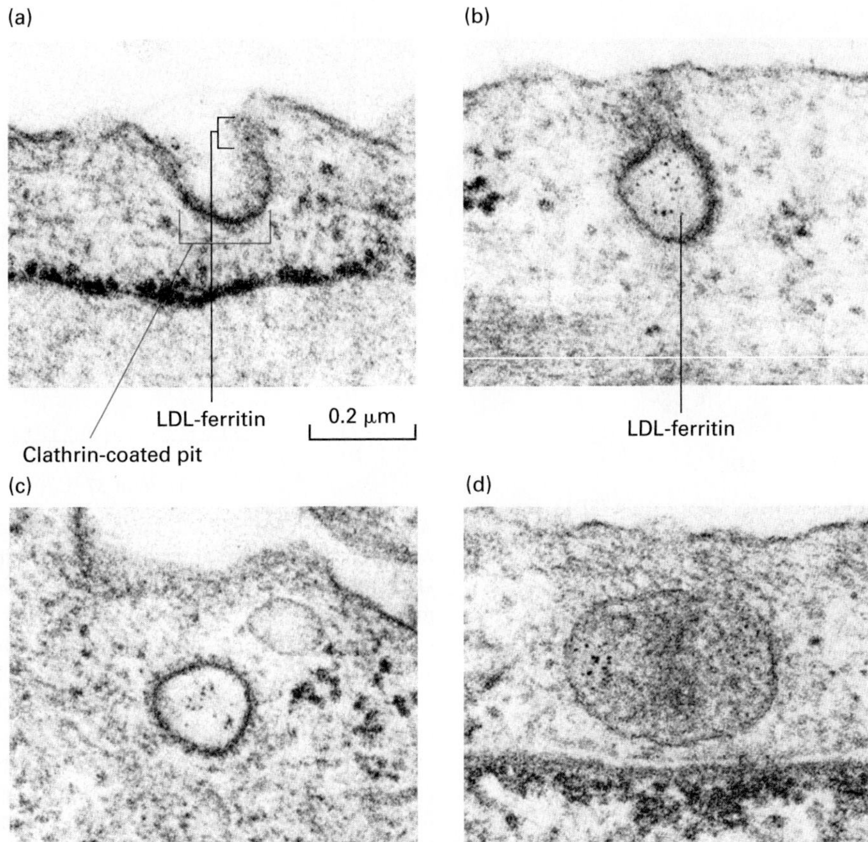

▲ **EXPERIMENTAL FIGURE 14-26 The initial stages of receptor-mediated endocytosis of low-density lipoprotein (LDL) particles are revealed by electron microscopy.** Cultured human fibroblasts were incubated in a medium containing LDL particles covalently linked to the electron-dense, iron-containing protein ferritin; each small iron particle in ferritin is visible as a small dot under the electron microscope. Cells initially were incubated at 4 °C; at this temperature LDL can bind to its receptor, but internalization does not occur. After excess LDL not bound to the cells was washed away, the cells were warmed to 37 °C and then prepared for microscopy at periodic intervals. (a) A coated pit, showing the clathrin coat on the inner (cytosolic) surface of the pit, soon after the temperature was raised. (b) A pit containing LDL apparently closing on itself to form a coated vesicle. (c) A coated vesicle containing ferritin-tagged LDL particles. (d) Ferritin-tagged LDL particles in a smooth-surfaced early endosome 6 minutes after internalization began. [Photographs courtesy of R. Anderson. Reprinted by permission from J. Goldstein et al., *Nature* **279:**679. Copyright 1979, Macmillan Journals Limited. See also M. S. Brown and J. Goldstein, 1986, *Science* **232:**34.]

and a cholesterol-containing phospholipid monolayer. The shell is **amphipathic** because its outer surface is hydrophilic, making these particles water soluble, and its inner surface is hydrophobic. Adjacent to the hydrophobic inner surface of the shell is a core of neutral lipids containing mostly cholesteryl esters, triglycerides, or both. Mammalian lipoproteins fall into different classes, defined by their differing buoyant densities. The class we will consider here is **low-density lipoprotein (LDL)**. A typical LDL particle, depicted in Figure 14-27, is a sphere 20–25 nm in diameter. The amphipathic outer shell is composed of a phospholipid monolayer and a single molecule of a large protein known as *apoB-100;* the core of a particle is packed with cholesterol in the form of cholesteryl esters.

Two general experimental approaches have been used to study how LDL particles enter cells. The first method makes use of LDL that has been labeled by the covalent attachment of radioactive ^{125}I to the side chains of tyrosine residues in apoB-100 on the surfaces of the LDL particles. After cultured cells are incubated for several hours with the labeled LDL, it is possible to determine how much LDL is bound to the surfaces of cells, how much is internalized, and how much of the apoB-100 component of the LDL is degraded by enzymatic hydrolysis to individual amino acids. The degradation of apoB-100 can be detected by the release of ^{125}I-tyrosine into the culture medium. Figure 14-28 shows the time course of events in receptor-mediated cellular LDL processing determined by pulse-chase experiments with a fixed concentration of ^{125}I-labeled LDL. These experiments clearly demonstrate the order of events: surface binding of LDL $\rightarrow$ internalization $\rightarrow$ degradation. The second approach involves tagging LDL particles with an electron-dense label that can be detected by electron microscopy. Such studies can reveal the details of how LDL particles first bind to the surface of cells at clathrin-coated endocytic pits and then remain associated with the coated pits as they invaginate and bud off to form coated vesicles and finally are transported to endosomes (see Figure 14-26).

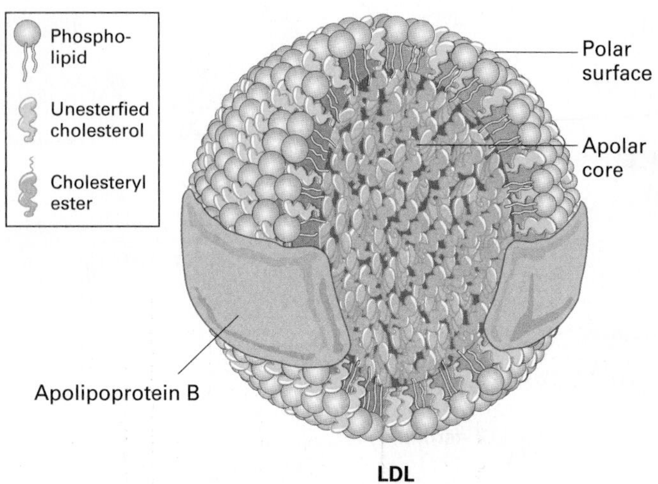

▲ **FIGURE 14-27 Model of low-density lipoprotein (LDL).** This class and the other classes of lipoproteins have the same general structure: an amphipathic shell, composed of a phospholipid monolayer (not bilayer), cholesterol, and protein, and a hydrophobic core, composed mostly of cholesteryl esters or triglycerides or both but with minor amounts of other neutral lipids (e.g., some vitamins). This model of LDL is based on electron microscopy and other low-resolution biophysical methods. LDL is unique in that it contains only a single molecule of one type of apolipoprotein (apoB), which appears to wrap around the outside of the particle as a band of protein. The other lipoproteins contain multiple apolipoprotein molecules, often of different types. [Adapted from M. Krieger, 1995, in E. Haber, ed., *Molecular Cardiovascular Medicine,* Scientific American Medicine, pp. 31–47.]

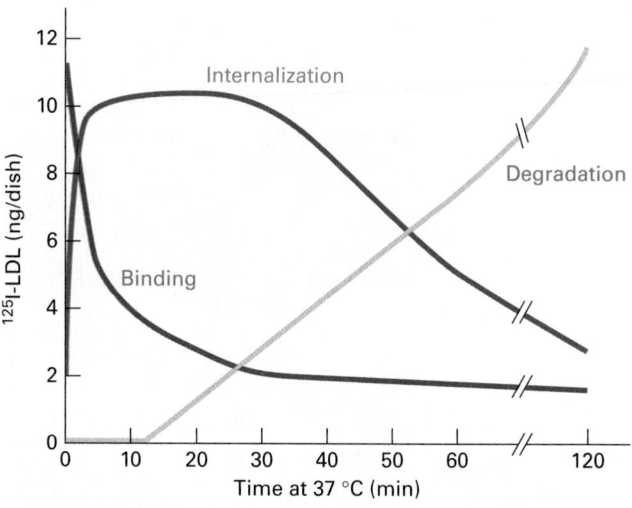

▲ EXPERIMENTAL FIGURE 14-28 **Pulse-chase experiment demonstrates precursor-product relations in cellular uptake of LDL.** Cultured normal human skin fibroblasts were incubated in a medium containing ^{125}I-LDL for 2 hours at 4 °C (the pulse). After excess ^{125}I-LDL not bound to the cells was washed away, the cells were incubated at 37 °C for the indicated amounts of time in the absence of external LDL (the chase). The amounts of surface-bound, internalized, and degraded (hydrolyzed) ^{125}I-LDL were measured. Binding but not internalization or hydrolysis of LDL apoB-100 occurs during the 4 °C pulse. The data show the very rapid disappearance of bound ^{125}I-LDL from the surface as it is internalized after the cells have been warmed to allow membrane movements. After a lag period of 15–20 minutes, lysosomal degradation of the internalized ^{125}I-LDL commences. [See M. S. Brown and J. L. Goldstein, 1976, *Cell* **9**:663.]

Receptors for Low-Density Lipoprotein and Other Ligands Contain Sorting Signals That Target Them for Endocytosis

The key to understanding how LDL particles bind to the cell surface and are then taken up into endocytic vesicles came from discovery of the *LDL receptor (LDLR)*. The LDL receptor is an 839-residue glycoprotein with a single transmembrane segment; it has a short C-terminal cytosolic segment and a long N-terminal exoplasmic segment that contains a β-propeller domain and a ligand-binding domain. Seven cysteine-rich imperfect repeats form the ligand-binding domain, which interacts with the apoB-100 molecule in a LDL particle. Figure 14-29 shows how LDL receptor proteins facilitate internalization of LDL particles by receptor-mediated endocytosis. After internalized LDL particles reach lysosomes, lysosomal proteases hydrolyze their surface apolipoproteins and lysosomal cholesteryl esterases hydrolyze their core cholesteryl esters. The unesterified cholesterol is then free to leave the lysosome and be used as necessary by the cell in the synthesis of membranes or various cholesterol derivatives.

The discovery of the LDL receptor and an understanding of how it functions came from studying cells from patients with *familial hypercholesterolemia (FH)*, a hereditary disease that is marked by elevated plasma LDL cholesterol and is now known to be caused by mutations in the *LDLR* gene. In patients who have one normal and one defective copy of the *LDLR* gene (heterozygotes), LDL cholesterol in the blood is increased about twofold. Those with two defective *LDLR* genes (homozygotes) have LDL cholesterol levels that are from fourfold to sixfold as high as normal. FH heterozygotes commonly develop cardiovascular disease about 10 years earlier than normal people do, and FH homozygotes usually die of heart attacks before reaching their late 20s.

A variety of mutations in the gene encoding the LDL receptor can cause familial hypercholesterolemia. Some mutations prevent the synthesis of the LDLR protein; others prevent proper folding of the receptor protein in the ER, leading to its premature degradation (Chapter 13); still other mutations reduce the ability of the LDL receptor to bind LDL tightly. A particularly informative group of mutant receptors are expressed on the cell surface and bind LDL normally but cannot mediate the internalization of bound LDL. In individuals with this type of defect, plasma-membrane receptors for other ligands are internalized normally, but the mutant LDL receptor is not recruited into coated pits. Analysis of this mutant receptor and other mutant LDL receptors generated experimentally and expressed in fibroblasts identified a four-residue motif in the cytosolic segment of the receptor that is crucial for its internalization: Asn-Pro-X-Tyr, where X can be any amino acid. This *NPXY sorting signal* binds to the AP2 complex,

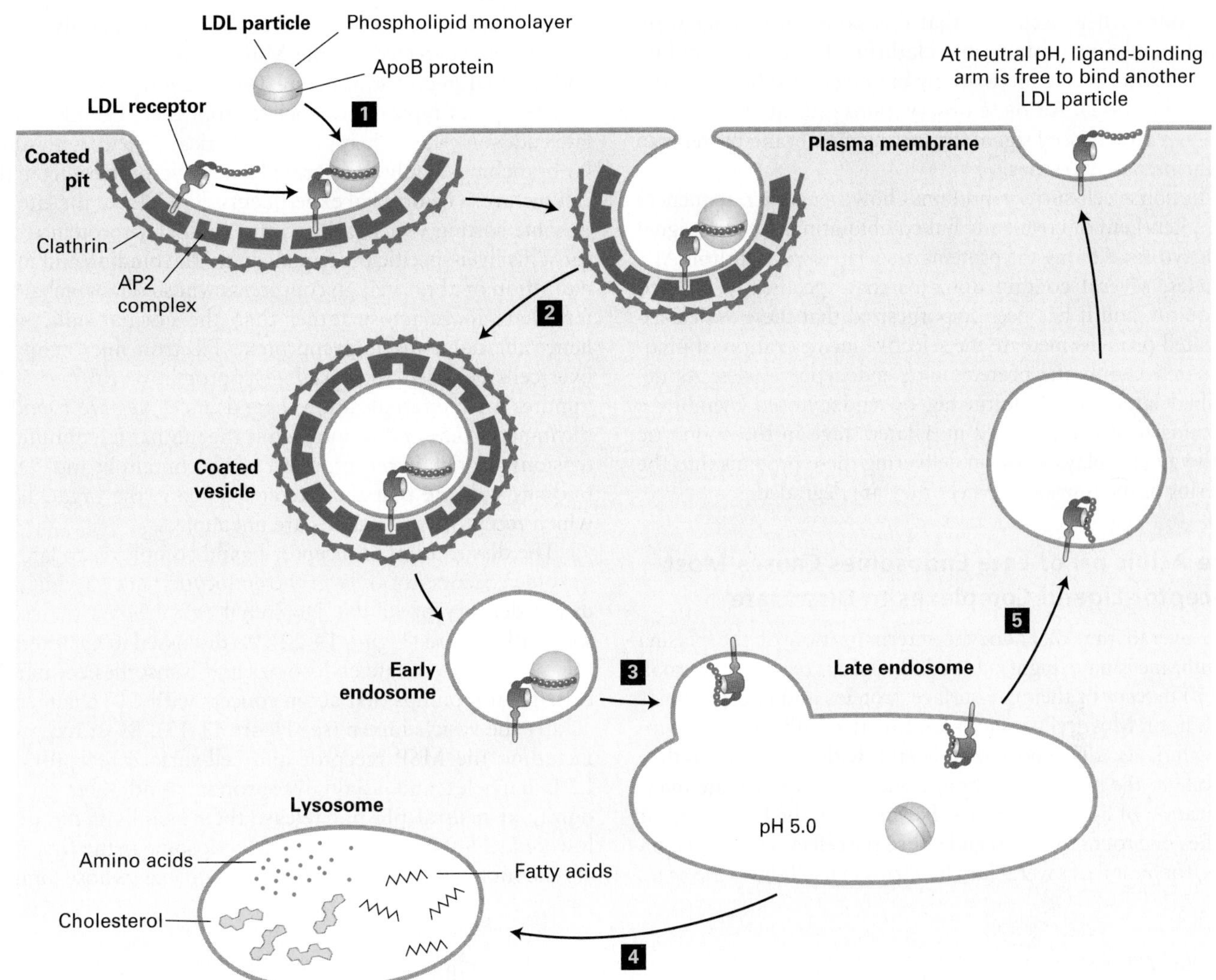

▲ **FIGURE 14-29 Endocytic pathway for internalizing low-density lipoprotein (LDL).** Step **1**: Cell-surface LDL receptors bind to an apoB protein embedded in the phospholipid outer layer of LDL particles. Interaction between the NPXY sorting signal in the cytosolic tail of the LDL receptor and the AP2 complex incorporates the receptor-ligand complex into forming endocytic vesicles. Step **2**: Clathrin-coated pits (or buds) containing receptor-LDL complexes are pinched off by the same dynamin-mediated mechanism used to form clathrin/AP1 vesicles on the *trans*-Golgi network (see Figure 14-19). Step **3**: After the vesicle coat is shed, the uncoated endocytic vesicle (early endosome) fuses with the late endosome. The acidic pH in this compartment causes a conformational change in the LDL receptor that leads to release of the bound LDL particle. Step **4**: The late endosome fuses with the lysosome, and the proteins and lipids of the free LDL particle are broken down to their constituent parts by enzymes in the lysosome. Step **5**: The LDL receptor recycles to the cell surface, where at the neutral pH of the exterior medium the receptor undergoes a conformational change so that it can bind another LDL particle. [See M. S. Brown and J. L. Goldstein, 1986, *Science* **232**:34, and G. Rudenko et al., 2002, *Science* **298**:2353.]

linking the clathrin/AP2 coat to the cytosolic segment of the LDL receptor in coated pits. A mutation in any of the conserved residues of the NPXY signal will abolish the ability of the LDL receptor to be incorporated into coated pits.

A small number of individuals who exhibit the usual symptoms associated with familial hypercholesterolemia produce normal LDL receptors. In these individuals, the gene encoding the AP2 subunit protein that binds the NPXY sorting signal is defective. As a result, LDL receptors are not incorporated into clathrin/AP2 vesicles and endocytosis of LDL particles is compromised. Analysis of patients with this genetic disorder highlights the importance of adapter proteins in protein trafficking mediated by clathrin vesicles. ▌

Mutational studies have shown that other cell-surface receptors can be directed into forming clathrin/AP2 pits by a different sorting signal: Tyr-X-X-Φ, where X can be any amino acid and Φ is a bulky hydrophobic amino acid. This Tyr-X-X-Φ sorting signal in the cytosolic segment of a receptor protein binds to a specific cleft in one of the protein subunits of the AP2 complex. Because the tyrosine and Φ residues mediate this binding, a mutation in either one reduces or abolishes the ability of the receptor to be incorporated into clathrin/AP2-coated pits. Moreover, if influenza HA protein, which is not normally endocytosed, is genetically engineered to contain this four-residue sequence in its cytosolic domain, the mutant HA is internalized. Recall from our earlier discussion that this same sorting signal recruits membrane proteins into clathrin/AP1 vesicles that bud from the *trans*-Golgi network by binding to a subunit of AP1 (see Table 14-2). All these observations indicate that Tyr-X-X-Φ is a widely used signal for sorting membrane proteins to clathrin-coated vesicles.

In some cell-surface proteins, however, other sequences (e.g., Leu-Leu) or covalently linked ubiquitin molecules signal endocytosis. Among the proteins associated with clathrin/AP2 vesicles, several contain domains that specifically bind to ubiquitin, and it has been hypothesized that these vesicle-associated proteins mediate the selective incorporation of ubiquitinated membrane proteins into endocytic vesicles. As described later, the ubiquitin tag on endocytosed membrane proteins is also recognized at a later stage in the endocytic pathway and plays a role in delivering these proteins into the interior of the lysosome, where they are degraded.

The Acidic pH of Late Endosomes Causes Most Receptor-Ligand Complexes to Dissociate

The overall rate of endocytic internalization of the plasma membrane is quite high: cultured fibroblasts regularly internalize 50 percent of their cell-surface proteins and phospholipids each hour. Most cell-surface receptors that undergo endocytosis will repeatedly deposit their ligands within the cell and then recycle to the plasma membrane, once again to mediate internalization of ligand molecules. For instance, the LDL receptor makes one round trip into and out of the cell every 10–20 minutes, for a total of several hundred trips in its 20-hour life span.

Internalized receptor-ligand complexes commonly follow the pathway depicted for the M6P receptor in Figure 14-22 and the LDL receptor in Figure 14-29. Endocytosed cell-surface receptors typically dissociate from their ligands within late endosomes, which appear as spherical vesicles with tubular branching membranes located a few micrometers from the cell surface. The original experiments that defined the late endosome sorting vesicle utilized the asialoglycoprotein receptor. This liver-specific protein mediates the binding and internalization of abnormal glycoproteins whose oligosaccharides terminate in galactose rather than the normal sialic acid; hence the name *asialo*glycoprotein. Electron microscopy of liver cells perfused with asialoglycoprotein reveal that 5–10 minutes after internalization, ligand molecules are found in the lumen of late endosomes, while the tubular membrane extensions are rich in receptor and rarely contain ligand. These findings indicate that the late endosome is the organelle in which receptors and ligands are uncoupled.

The dissociation of receptor-ligand complexes in late endosomes occurs not only in the endocytic pathway but also in the delivery of soluble lysosomal enzymes via the secretory pathway (see Figure 14-22). As discussed in Chapter 11, the membranes of late endosomes and lysosomes contain V-class proton pumps that act in concert with Cl^- channels to acidify the vesicle lumen (see Figure 11-13). Most receptors, including the M6P receptor and cell-surface receptors for LDL particles and asialoglycoprotein, bind their ligands tightly at neutral pH but release their ligands if the pH is lowered to 6.0 or below. The late endosome is the first vesicle encountered by receptor-ligand complexes whose luminal

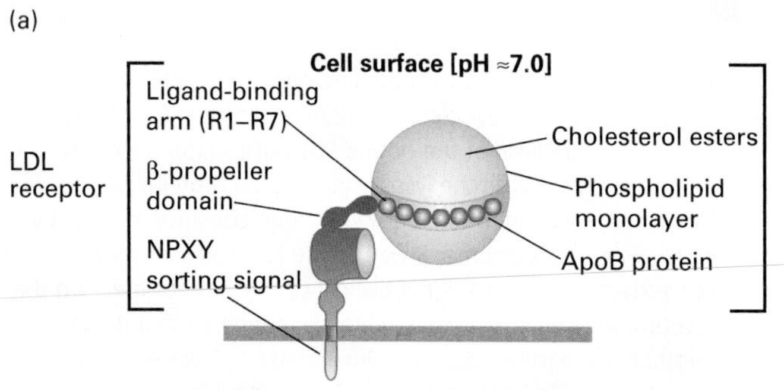

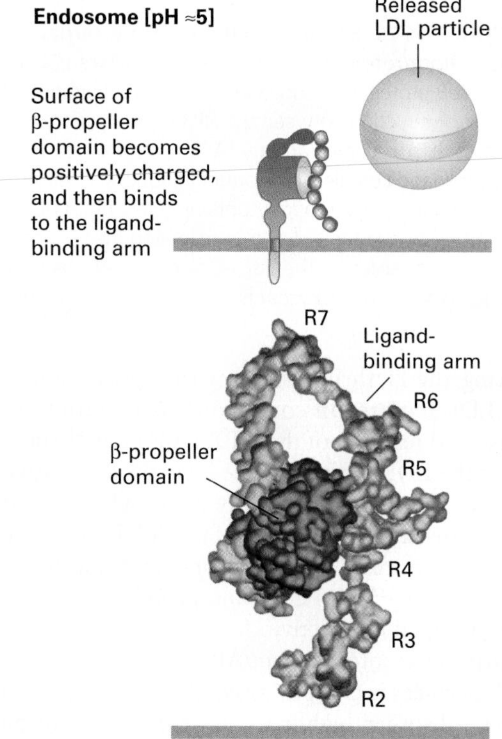

▲ **FIGURE 14-30 Model for pH-dependent binding of LDL particles by the LDL receptor.** Schematic depiction of LDL receptor at neutral pH found at the cell surface (a) and at the acidic pH found in the interior of the late endosome (b). (a) At the cell surface, apoB-100 on the surface of a LDL particle binds tightly to the receptor. Of the seven repeats (R1–R7) in the ligand-binding arm, R4 and R5 appear to be most critical for LDL binding. (b, *top*) Within the endosome, histidine residues in the β-propeller domain of the LDL receptor become protonated. The positively charged propeller can bind with high affinity to the ligand-binding arm, which contains negatively charged residues, causing release of the LDL particle. (b, *bottom*) Experimental electron density and C_α trace model of the extracellular region of the LDL receptor at pH 5.3 based on x-ray crystallographic analysis. In this conformation, extensive hydrophobic and ionic interactions occur between the β propeller and the R4 and R5 repeats. [Part (b) from G. Rudenko et al., 2002, *Science* **298:**2353.]

pH is sufficiently acidic to promote dissociation of most endocytosed receptors from their tightly bound ligands.

The mechanism by which the LDL receptor releases bound LDL particles is now understood in detail (Figure 14-30). At the endosomal pH of 5.0–5.5, histidine residues in the β-propeller domain of the receptor become protonated, forming a site that can bind with high affinity to the negatively charged repeats in the ligand-binding domain. This intramolecular interaction sequesters the repeats in a conformation that cannot simultaneously bind to apoB-100, thus causing release of the bound LDL particle.

The Endocytic Pathway Delivers Iron to Cells without Dissociation of the Receptor-Transferrin Complex in Endosomes

The endocytic pathway involving the transferrin receptor and its ligand differs from the LDL pathway in that the receptor-ligand complex does not dissociate in late endosomes. Nonetheless, changes in pH also mediate the sorting of receptors and ligands in the transferrin pathway, which functions to deliver iron to cells.

A major glycoprotein in the blood, transferrin transports iron to all tissue cells from the liver (the main site of iron storage in the body) and from the intestine (the site of iron absorption). The iron-free form, *apotransferrin,* binds two Fe^{3+} ions very tightly to form *ferrotransferrin.* All mammalian cells contain cell-surface transferrin receptors that avidly bind ferrotransferrin at neutral pH, after which the receptor-bound ferrotransferrin is subjected to endocytosis. Like the components of an LDL particle, the two bound Fe^{3+} atoms remain in the cell, but the apotransferrin part of the ligand does not dissociate from the receptor, and within minutes after being endocytosed, apotransferrin is secreted from the cell.

As depicted in Figure 14-31, the explanation for the behavior of the transferrin receptor–ligand complex lies in the

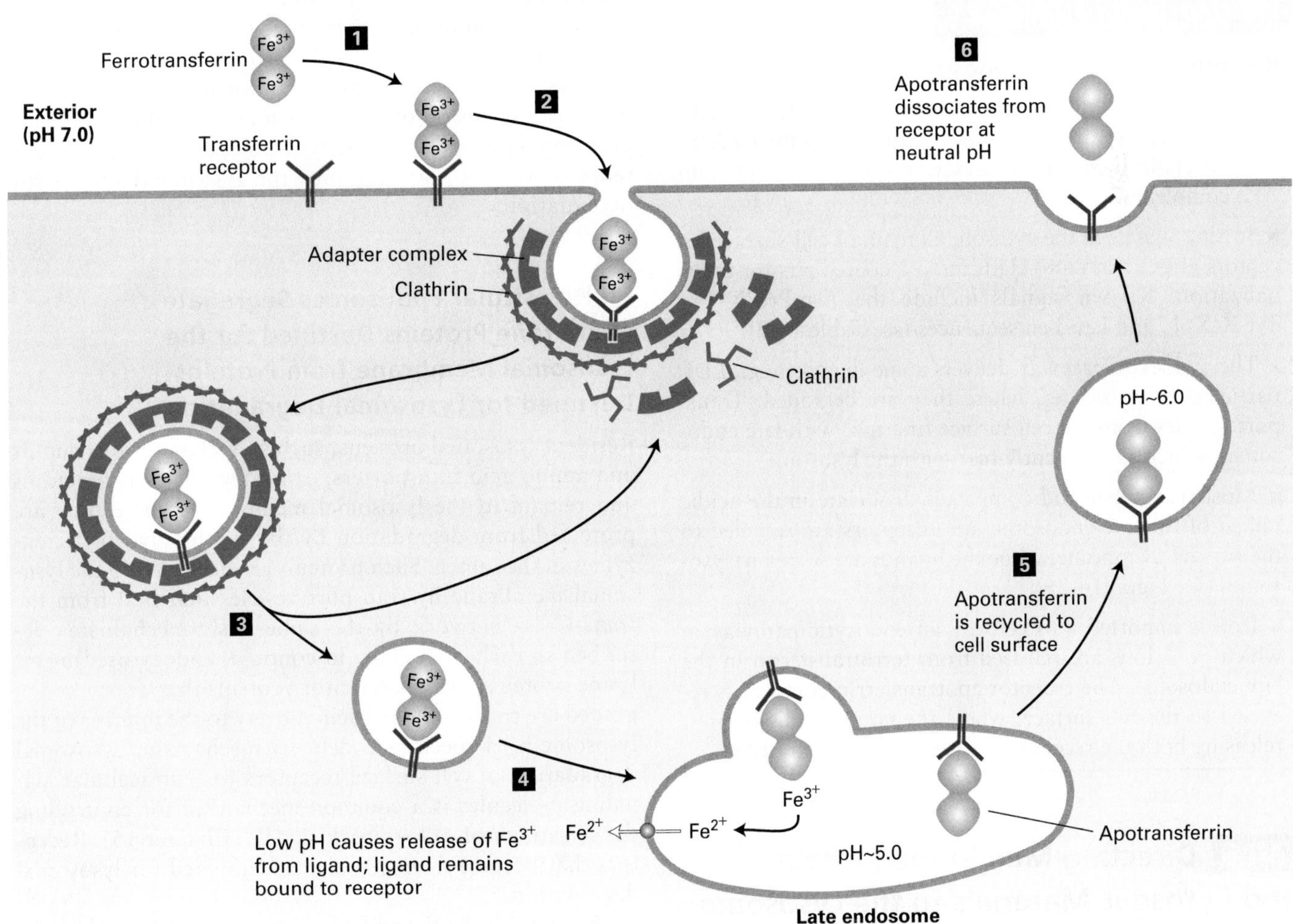

▲ **FIGURE 14-31 The transferrin cycle, which operates in all growing mammalian cells.** Step **1**: The transferrin dimer carrying two bound atoms of Fe^{+3}, called ferrotransferrin, binds to the transferrin receptor at the cell surface. Step **2**: Interaction between the tail of the transferrin receptor and the AP2 adapter complex incorporates the receptor-ligand complex into endocytic clathrin-coated vesicles. Steps **3** and **4**: The vesicle coat is shed and the endocytic vesicles fuse with the membrane of the endosome. Fe^{+3} is released from the receptor-ferrotransferrin complex in the acidic late endosome compartment. Step **5**: The apotransferrin protein remains bound to its receptor at this pH, and they recycle to the cell surface together. Step **6**: The neutral pH of the exterior medium causes release of the iron-free apotransferrin. [See A. Ciechanover et al., 1983, *J. Biol. Chem.* **258:**9681.]

unique ability of apotransferrin to remain bound to the transferrin receptor at the low pH (5.0–5.5) of late endosomes. At a pH of less than 6.0, the two bound Fe^{3+} atoms dissociate from ferrotransferrin, are reduced to Fe^{2+} by an unknown mechanism, and then are exported into the cytosol by an endosomal transporter specific for divalent metal ions. The receptor-apotransferrin complex remaining after dissociation of the iron atoms is recycled back to the cell surface. Although apotransferrin binds tightly to its receptor at a pH of 5.0 or 6.0, it does not bind at neutral pH. Hence the bound apotransferrin dissociates from the transferrin receptor when the recycling vesicles fuse with the plasma membrane and the receptor-ligand complex encounters the neutral pH of the extracellular interstitial fluid or growth medium. The recycled receptor is then free to bind another molecule of ferrotransferrin, and the released apotransferrin is carried in the bloodstream to the liver or intestine to be reloaded with iron.

KEY CONCEPTS OF SECTION 14.5

Receptor-Mediated Endocytosis

- Some extracellular ligands that bind to specific cell-surface receptors are internalized, along with their receptors, in clathrin-coated vesicles whose coats also contain AP2 complexes.
- Sorting signals in the cytosolic domain of cell-surface receptors target them into clathrin/AP2-coated pits for internalization. Known signals include the Asn-Pro-X-Tyr, Tyr-X-X-Φ, and Leu-Leu sequences (see Table 14-2).
- The endocytic pathway delivers some ligands (e.g., LDL particles) to lysosomes, where they are degraded. Transport vesicles from the cell surface first fuse with late endosomes, which subsequently fuse with the lysosome.
- Most receptor-ligand complexes dissociate in the acidic milieu of the late endosome; the receptors are recycled to the plasma membrane, while the ligands are sorted to lysosomes (see Figure14-29).
- Iron is imported into cells by an endocytic pathway in which Fe^{3+} ions are released from ferrotransferrin in the late endosome. The receptor-apotransferrin complex is recycled to the cell surface, where the complex dissociates, releasing both the receptor and apotransferrin for reuse.

14.6 Directing Membrane Proteins and Cytosolic Materials to the Lysosome

The major function of lysosomes is to degrade extracellular materials taken up by the cell and intracellular components under certain conditions. Materials to be degraded must be delivered to the lumen of the lysosome, where the various degradative enzymes reside. As just discussed, endocytosed ligands (e.g., LDL particles) that dissociate from their receptors in the late endosome subsequently enter the lysosomal lumen when the membrane of the late endosome fuses with the membrane of the lysosome (see Figure 14-29). Likewise, phagosomes carrying bacteria or other particulate matter can fuse with lysosomes, releasing their contents into the lumen for degradation.

It is apparent how the general vesicular trafficking mechanism discussed in this chapter can be used to deliver the luminal contents of an endosomal organelle to the lumen of the lysosome for degradation. However, vesicular trafficking cannot allow for delivery of membrane proteins or cytosolic materials to the lysosomal lumen. As we will see in this section, the cell has two different specialized pathways for delivery of these molecules to the lysosome interior for degradation. The first pathway is used to degrade endocytosed membrane proteins and utilizes an unusual type of vesicle that buds into the lumen of the endosome to produce a multivesicular endosome. The second pathway, known as autophagy, involves the de novo formation of a double membrane organelle known as an autophagosome that envelops cytosolic material, such as soluble cytosolic proteins or sometimes organelles such as peroxisomes or mitochondria. Both pathways lead to fusion of either the multivesicular endosome or autophagosome with the lysosome, depositing the contents of these organelles into the lysosomal lumen for degradation.

Multivesicular Endosomes Segregate Membrane Proteins Destined for the Lysosomal Membrane from Proteins Destined for Lysosomal Degradation

Resident lysosomal proteins, such as V-class proton pumps and amino acid transporters, can carry out their functions and remain in the lysosomal membrane, where they are protected from degradation by the soluble hydrolytic enzymes in the lumen. Such proteins are delivered to the lysosomal membrane by transport vesicles that bud from the *trans*-Golgi network by the same basic mechanisms described in earlier sections. In contrast, endocytosed membrane proteins such as receptor proteins that are to be degraded are transferred in their entirety to the interior of the lysosome by a specialized delivery mechanism. Lysosomal degradation of cell-surface receptors for extracellular signaling molecules is a common mechanism for controlling the sensitivity of cells to such signals (Chapter 15). Receptors that become damaged also are targeted for lysosomal degradation.

Early evidence that membranes can be delivered to the lumen of compartments came from electron micrographs showing membrane vesicles and fragments of membranes within endosomes and lysosomes (see Figure 9-2c). Parallel experiments in yeast revealed that endocytosed receptor proteins targeted to the vacuole (the yeast organelle equivalent to the lysosome) were primarily associated with membrane fragments and small vesicles within the interior of the vacuole rather than with the vacuole surface membrane.

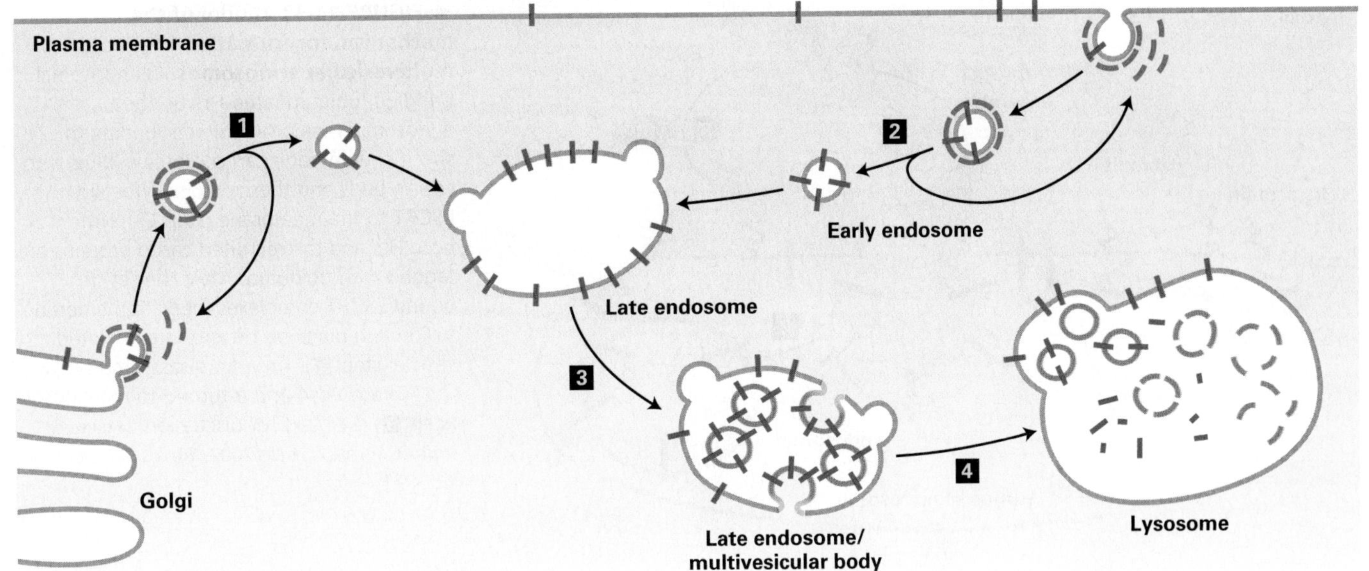

▲ **FIGURE 14-32 Delivery of plasma-membrane proteins to the lysosomal interior for degradation.** Early endosomes carrying endocytosed plasma-membrane proteins (blue) and vesicles carrying lysosomal membrane proteins (green) from the *trans*-Golgi network fuse with the late endosome, transferring their membrane proteins to the endosomal membrane (steps **1** and **2**). Proteins to be degraded, such as those from the early endosome, are incorporated into vesicles that bud *into* the interior of the late endosome, eventually forming a multivesicular endosome containing many such internal vesicles (step **3**). Fusion of a multivesicular endosome directly with a lysosome releases the internal vesicles into the lumen of the lysosome, where they can be degraded (step **4**). Because proton pumps and other lysosomal membrane proteins normally are not incorporated into internal endosomal vesicles, they are delivered to the lysosomal membrane and are protected from degradation. [See F. Reggiori and D. J. Klionsky, 2002, *Eukaryot. Cell* **1**:11, and D. J. Katzmann et al., 2002, *Nature Rev. Mol. Cell Biol.* **3**:893.]

These observations suggest that endocytosed membrane proteins can be incorporated into specialized vesicles that form at the endosomal membrane (Figure 14-32). Although these vesicles are similar in size and appearance to transport vesicles, they differ topologically. Transport vesicles bud *outward* from the surface of a donor organelle into the cytosol, whereas vesicles within the endosome bud *inward* from the surface into the lumen (away from the cytosol). Mature endosomes containing numerous vesicles in their interior are usually called *multivesicular endosomes* (or bodies). Eventually, the surface membrane of a multivesicular endosome fuses with the membrane of a lysosome, thereby delivering its internal vesicles and the membrane proteins they contain into the lysosome interior for degradation. Thus the sorting of proteins in the endosomal membrane determines which ones will remain on the lysosome surface (e.g., pumps and transporters) and which ones will be incorporated into internal vesicles and ultimately degraded in lysosomes.

Many of the proteins required for inward budding of the endosomal membrane were first identified by mutations in yeast that blocked delivery of membrane proteins to the interior of the vacuole. More than 10 such "budding" proteins have been identified in yeast, most with significant similarities to mammalian proteins that evidently perform the same function in mammalian cells. The current model of endosomal budding to form multivesicular endosomes in mammalian cells is based primarily on studies in yeast (Figure 14-33). Most cargo proteins that enter the multivesicular endosome are tagged with ubiquitin. Cargo proteins destined to enter the multivesicular endosome usually receive their ubiquitin tag at the plasma membrane, the TGN, or the endosomal membrane. We have already seen how ubiquitin tagging can serve as a signal for degradation of cytosolic or misfolded ER proteins by the proteasome (see Chapters 3 and 13). When used as a signal for proteasomal degradation, the ubiquitin tag usually consists of a chain of covalently linked ubiquitin molecules (polyubiquitin), whereas ubiquitin used to tag proteins for entry into the multivesicular endosome usually takes the form of a single (monoubiquitin) molecule. In the membrane of the endosome a ubiquitin-tagged peripheral membrane protein, known as Hrs, facilitates loading of specific monoubiquitinated membrane cargo proteins into vesicle buds directed into the interior of the endosome. The ubiquitinated Hrs protein then recruits a set of three different protein complexes to the membrane. These *ESCRT* (*e*ndo*s*omal sorting *c*omplexes *r*equired for *t*ransport) *proteins* include the ubiquitin-binding protein Tsg101. The membrane-associated ESCRT proteins act to complete vesicle budding, leading to release of a vesicle carrying specific membrane cargo into the interior of the endosome. Finally, an ATPase, known as Vps4, uses the energy from ATP hydrolysis to disassemble the ESCRT, releasing them into the

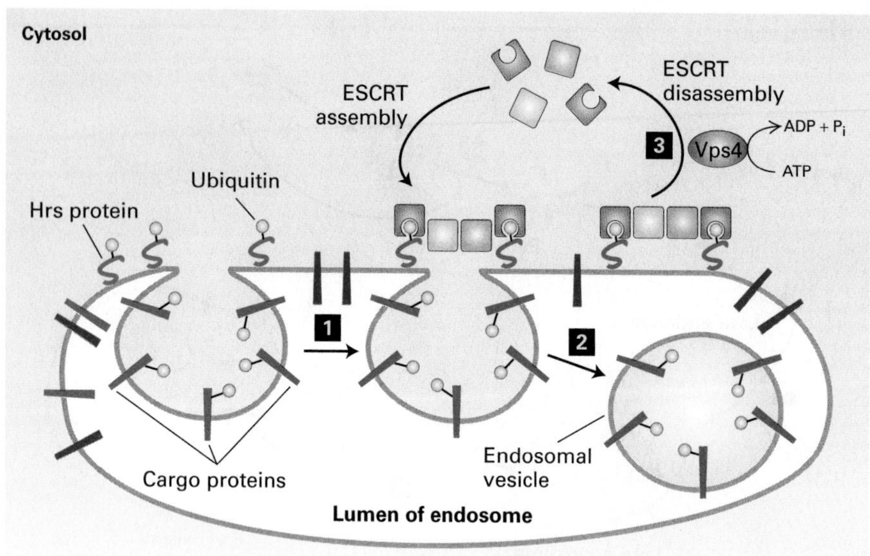

◄ **FIGURE 14-33 Model of the mechanism for formation of multivesicular endosomes.** In endosomal budding, ubiquitinated Hrs on the endosomal membrane directs loading of specific membrane cargo proteins (blue) into vesicle buds and then recruits cytosolic ESCRT to the membrane (step **1**). Note that both Hrs and the recruited cargo proteins are tagged with ubiquitin. After the set of bound ESCRT complexes mediate membrane fusion and pinching off of the completed vesicle (step **2**), they are disasssembled by the ATPase Vps4 and returned to the cytosol (step **3**). See text for discussion. [Adapted from O. Pornillos et al., 2002, *Trends Cell Biol.* **12**:569.]

cytosol for another round of budding. In the fusion event that pinches off a completed endosomal vesicle, the ESCRT proteins and Vps4 may function like SNAREs and NSF, respectively, in the typical membrane-fusion process discussed previously (see Figure 14-10).

Retroviruses Bud from the Plasma Membrane by a Process Similar to Formation of Multivesicular Endosomes

The vesicles that bud into the interior of endosomes have a topology similar to that of enveloped virus particles that bud from the plasma membrane of virus-infected cells. Moreover, recent experiments demonstrate that a common set of proteins is required for both types of membrane-budding events. In fact, the two processes so closely parallel each other in mechanistic detail as to suggest that enveloped viruses have evolved mechanisms to recruit the cellular proteins used in inward endosomal budding for their own purposes.

The human immunodeficiency virus (HIV) is an enveloped **retrovirus** that buds from the plasma membrane of infected cells in a process driven by viral Gag protein, the major structural component of completed virus particles. Gag protein binds to the plasma membrane of an infected cell and ≈4000 Gag molecules polymerize into a spherical shell, producing a structure that looks like a vesicle bud protruding outward from the plasma membrane. Mutational studies with HIV have revealed that the N-terminal segment of Gag protein is required for association with the plasma membrane, whereas the C-terminal segment is required for pinching off of complete HIV particles. For instance, if the portion of the viral genome encoding the C-terminus of Gag is removed, HIV buds will form in infected cells, but pinching off does not occur, and thus no free virus particles are released.

The first indication that HIV budding employs the same molecular machinery as vesicle budding into endosomes came from the observation that Tsg101, an ESCRT protein, binds to the C-terminus of Gag protein. Subsequent findings have clearly established the mechanistic parallels between the two processes. For example, Gag is ubiquitinated as part of the process of virus budding, and in cells with mutations in Tsg101 or Vps4, HIV virus buds accumulate but cannot pinch off from the membrane (Figure 14-34). Moreover, when a segment from the cellular Hrs protein is added to a truncated Gag protein, proper budding and release of virus particles is restored. Taken together, these results indicate that Gag protein mimics the function of Hrs, redirecting ESCRT to the plasma membrane, where they can function in the budding of virus particles.

Other enveloped retroviruses such as murine leukemia virus and Rous sarcoma virus also have been shown to require ESCRT complexes for their budding, although each virus appears to have evolved a somewhat different mechanism to recruit ESCRT complexes to the site of virus budding.

The Autophagic Pathway Delivers Cytosolic Proteins or Entire Organelles to Lysosomes When cells are placed under stress such as conditions of starvation, they have the capacity to recycle macromolecules for use as nutrients in a process of lysosomal degradation known as *autophagy* ("eating oneself"). The autophagic pathway involves the formation of a flattened double-membrane cup-shaped structure that envelops a region of the cytosol or an entire organelle (e.g., mitochondrion), forming an *autophagosome,* or *autophagic vesicle* (Figure 14-35). The outer membrane

Podcast: HIV Budding from the Plasma Membrane

MEDIA CONNECTIONS

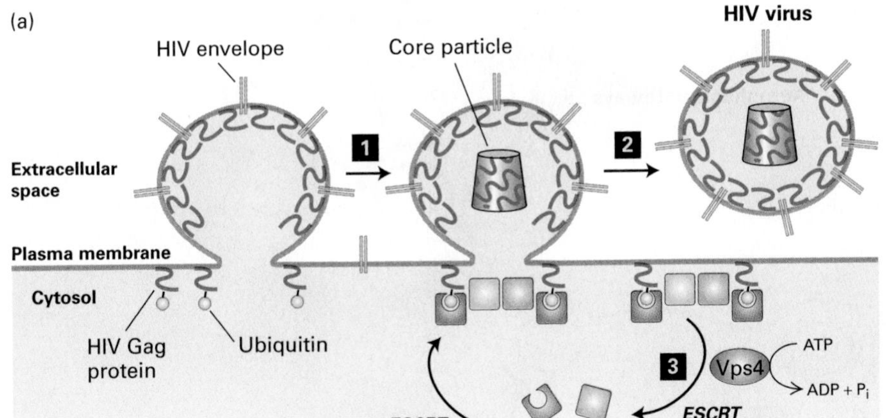

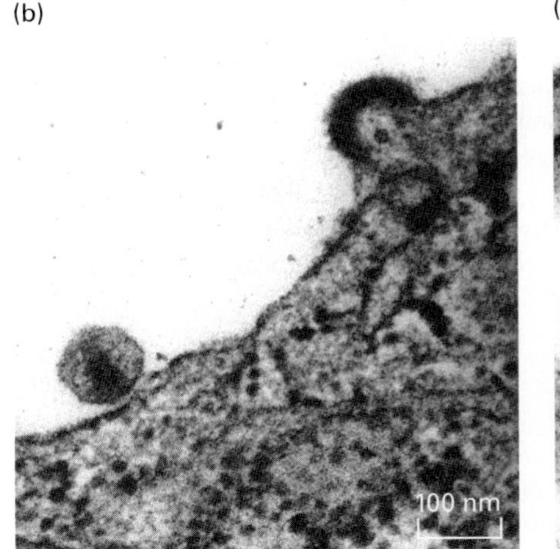

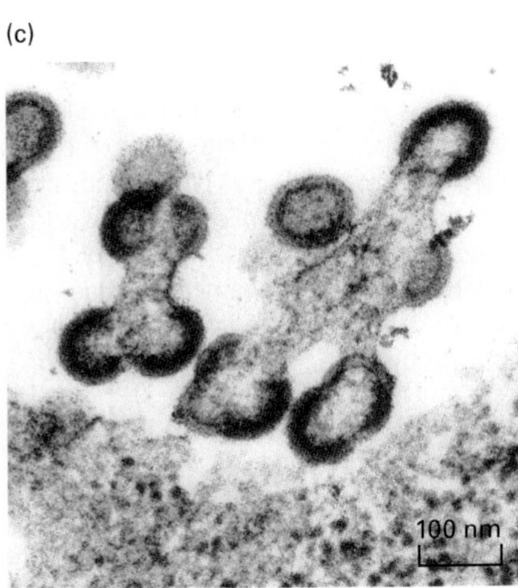

◀ **FIGURE 14-34 Mechanism for budding of HIV from the plasma membrane.** Proteins required for formation of multivesicular endosomes are exploited by HIV for virus budding from the plasma membrane. (a) Budding of HIV particles from HIV-infected cells occurs by a similar mechanism as in Figure 14-33, using the virally encoded Gag protein and cellular ESCRT and Vps4 (steps **1**–**3**). Ubiquitinated Gag near a budding particle functions like Hrs. See text for discussion. (b) In wild-type cells infected with HIV, virus particles bud from the plasma membrane and are rapidly released into the extracellular space. (c) In cells that lack the functional ESCRT protein Tsg101, the viral Gag protein forms dense viruslike structures, but budding of these structures from the plasma membrane cannot be completed and chains of incomplete viral buds still attached to the plasma membrane accumulate. [Wes Sundquist, University of Utah.]

of an autophagic vesicle can fuse with the lysosome, delivering a large vesicle, bounded by a single membrane bilayer, to the interior of the lysosome. Similar to the situation that occurs when multivesicular endosomes are delivered to the lysosome, lipases and proteases within the lysosome will degrade the autophagic vesicle and its contents into their molecular components. Amino acid permeases in the lysosomal membrane then allow for transport of free amino acids back into the cytosol for use in synthesis of new proteins.

The formation and fusion of autophagic vesicles are thought to take place in three basic steps. Although the underlying mechanisms for each of these steps remain poorly understood, they are thought to be related to the basic mechanisms for vesicular trafficking discussed in this chapter.

Autophagic Vesicle Nucleation The autophagic vesicle is thought to originate from a fragment of a membrane-bounded organelle. Although the origin of this membrane is not known, most studies suggest that the autophagic vesicle is initially derived from a fragment of the ER. Autophagy often appears to be a nonspecific process in which a random portion of the cytoplasm becomes enveloped by an autophagosome. In these cases, the site of nucleation is probably random. In other cases, the autophagosome surrounds particular organelles. For example, in a process known as *pexophagy,* peroxisomes are enveloped and degraded when they are not needed. In such cases of autophagy of specific organelles, some type of signal or binding site must be present on the surface of the organelle to target nucleation of the autophagic vesicle.

Autophagic Vesicle Growth and Completion New membrane must be delivered to the autophagosome membrane in order for this cup-shaped organelle to grow. This

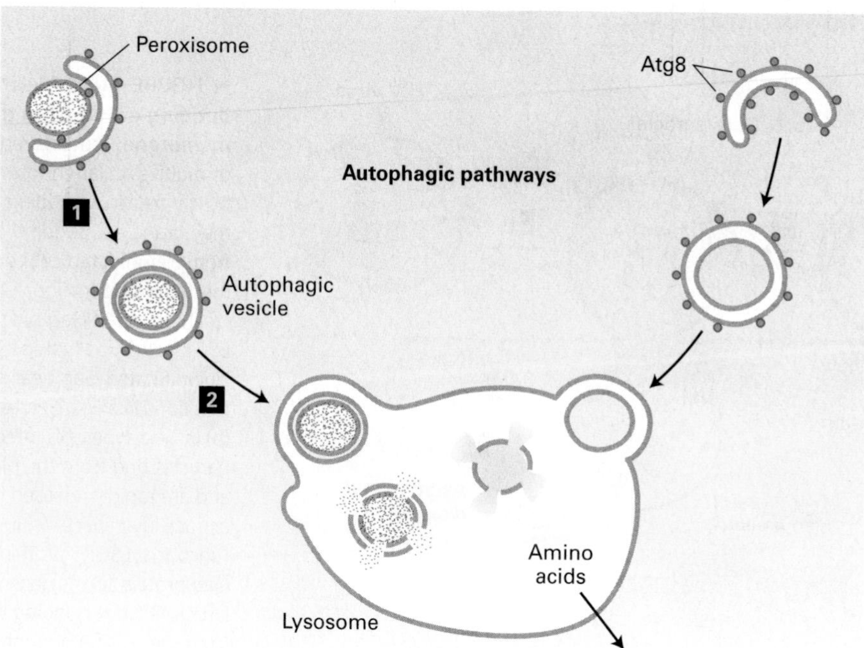

▲ **FIGURE 14-35 The autophagic pathway.** The autophagic pathway allows cytosolic proteins and organelles to be delivered to the lysosomal interior for degradation. In the autophagic pathway, a cup-shaped structure forms around portions of the cytosol or an organelle such as a peroxisome, as shown here. Continued addition of membrane eventually leads to the formation of an autophagosome vesicle that envelops its contents by two complete membranes (step **1**). Fusion of the outer membrane with the membrane of a lysosome releases a single-layer vesicle and its contents into the lysosome interior (step **2**). After degradation of the protein and lipid by hydrolases in the lysosome interior, the released amino acids are transported across the lysosomal membrane into the cytosol. Proteins known to participate in the autophagic pathway include Atg8, which forms a coat structure around the autophagosome.

growth is likely to occur by the fusion of some type of transport vesicle with the membrane of the autophagosome. Although the origin of such vesicles is not known, the endosome is a likely candidate. Some of the proteins that participate in the formation of autophagosomes have been identified in genetic screens in yeast for mutants that are defective in autophagy. A subset of these proteins appear to form a coat structure on the surface of the autophagosome. One of these proteins is Atg8, shown in Figure 14-35, which is covalently linked to the lipid phosphatidylethanolamine and thus becomes attached to the cytoplasmic leaflet of the autophagic vesicle. This coat may give the autophagosome its cup-shaped structure.

Autophagic Vesicle Targeting and Fusion The outer membrane of the completed autophagosome is thought to contain a set of proteins that target fusion with the membrane of the lysosome. Two vesicle-tethering proteins have been found to be required for autophagosome fusion with the lysosome, but the corresponding SNARE proteins have not been identified. Fusion of the autophagosome with the lysosome occurs after Atg8 has been released from the membrane by proteolytic cleavage, and this proteolysis step only occurs once the autophagic vesicle has completely formed a sealed double-membrane system. Thus Atg8 protein appears to mask fusion proteins and to prevent premature fusion of the autophagosome with the lysosome.

KEY CONCEPTS OF SECTION 14.6

Directing Membrane Proteins and Cytosolic Materials to the Lysosome

- Endocytosed membrane proteins destined for degradation in the lysosome are incorporated into vesicles that bud into the interior of the endosome. Multivesicular endosomes, which contain many of these internal vesicles, can fuse with the lysosome to deliver the vesicles to the interior of the lysosome (see Figure 14-32).

- Some of the cellular components (e.g., ESCRT) that mediate inward budding of endosomal membranes are used in the budding and pinching off of enveloped viruses such as HIV from the plasma membrane of virus-infected cells (see Figures 14-33 and 14-34).

- A portion of the cytoplasm or an entire organelle (e.g., peroxisome) can be enveloped in a flattened membrane and eventually incorporated into a double-membrane autophagic vesicle. Fusion of the outer vesicle membrane with the lysosome delivers the enveloped contents to the interior of the lysosome for degradation (see Figure 14-35).

Perspectives for the Future

The biochemical, genetic, and structural information presented in this chapter shows that we now have a basic understanding of how protein traffic flows from one membrane-bounded compartment to another. Our understanding of these processes has come largely from experiments on the function of various types of transport vesicles. These studies have led to the identification of many vesicle components and the discovery of how these components work together to drive vesicle budding, to incorporate the correct set of cargo molecules from the donor organelle, and then to mediate fusion of a completed vesicle with the membrane of a target organelle.

Despite these advances, important stages of the secretory and endocytic pathways remain about which we know relatively little. For example, we do not yet know what types of proteins form the coats of either the regulated or the constitutive secretory vesicles that bud from the *trans*-Golgi network. Moreover, the types of signals on cargo proteins that might target them for packaging into secretory vesicles have not yet been defined. Another baffling process is the formation of vesicles that bud away from the cytosol, such as the vesicles that enter multivesicular endosomes. Although some of the proteins that participate in formation of these "internal" endosome vesicles are known, we do not know what determines their shape or what type of process causes them to pinch off from the donor membrane. Similarly, the origin and growth of the membrane of the autophagic vesicle is also poorly understood. In the future, it should be possible for these and other poorly understood vesicle-trafficking steps to be dissected through the use of the same powerful combination of biochemical and genetic methods that have delineated the working parts of COPI, COPII, and clathrin/AP vesicles.

Questions still remain about vesicle trafficking between the ER and *cis*-Golgi, between Golgi stacks, and between the *trans*-Golgi and endosome, the best-characterized transport steps. In particular, our understanding of how proteins are actually sorted between these organelles is incomplete largely because of the highly dynamic nature of all the organelles along the secretory pathway. Although we know many of the details of how particular vesicle components function, we cannot account for why their functions are restricted to specific stages in the overall flow of anterograde and retrograde transport steps. For example, we cannot explain why COPII vesicles fuse with one another to form a new *cis*-Golgi stack, whereas COPI vesicles fuse with the membrane of the ER, since both vesicle types appear to contain similar sets of v-SNARE proteins. In the same vein, we do not know what feature of the Golgi membrane actually distinguishes a COPI-coated vesicle bud from a clathrin/AP-coated bud. In both cases, binding of ARF protein to the Golgi membrane appears to initiate vesicle budding. The solution to these problems will require a more integrated understanding of the flow of vesicular traffic in the context of the entire secretory pathway. Recent improvements in our ability to image vesicular transport of cargo proteins in live cells gives hope that some of these more subtle aspects of vesicle function may be clarified in the near future.

Key Terms

AP (adapter protein) complexes *598*
anterograde transport *592*
ARF protein *587*
autophagy *614*
cisternal maturation *680*
clathrin *586*
constitutive secretion *602*
COPI *586*
COPII *586*
dynamin *599*
endocytic pathway *579*
ESCRT proteins *613*
late endosome *580*
low-density lipoprotein (LDL) *607*
mannose 6-phosphate (M6P) *600*
multivesicular endosomes *613*
Rab proteins *589*
receptor-mediated endocytosis *606*
regulated secretion *24*
retrograde transport *592*
sec mutants *584*
secretory pathway *580*
sorting signals *589*
transcytosis *605*
trans-Golgi network (TGN) *580*
transport vesicles *580*
t-SNAREs *586*
v-SNAREs *586*

Review the Concepts

1. The studies of Palade and colleagues used pulse-chase labeling with radioactively labeled amino acids and autoradiography to visualize the location of newly synthesized proteins in pancreatic acinar cells. These early experiments provided invaluable information on protein synthesis and intercompartmental transport. New methods have evolved, but two basic requirements are still necessary for any assay to study this type of protein transport. What two requirements must be met? Briefly describe the experimental approaches needed to meet the criteria.

2. *Sec18* is a yeast gene that encodes NSF. It is a class C mutant in the yeast secretory pathway. What is the mechanistic role of NSF in membrane trafficking? As indicated by its class C phenotype, why does an NSF mutation produce accumulation of vesicles at what appears to be only one stage of the secretory pathway?

3. Vesicle budding is associated with coat proteins. What is the role of coat proteins in vesicle budding? How are coat proteins recruited to membranes? What kinds of molecules are likely to be included or excluded from newly formed vesicles? What is the best-known example of a protein likely to be involved in vesicle pinching off?

4. Treatment of cells with the drug brefeldin A (BFA) has the effect of decoating Golgi apparatus membranes, resulting in a cell in which the vast majority of Golgi proteins are found in the ER. What inferences can be made from this

observation regarding roles of coat proteins other than promoting vesicle formation? Predict what type of mutation in Arf1 might have the same effect as treating cells with BFA.

5. An antibody to an exposed "hinge" region of βCOPI known as EAGE blocks the function of βCOPI when microinjected into HeLa cells. Predict what the consequences of this functional block might be for anterograde transport from the ER to the plasma membrane. Propose an experiment to test whether the effect of EAGE microinjection is initially on anterograde or retrograde transport.

6. Specificity in fusion between vesicles involves two discrete and sequential processes. Describe the first of the two processes and its regulation by GTPase switch proteins. What effect on the size of early endosomes might result from overexpression of a mutant form of Rab5 that is stuck in the GTP-bound state?

7. Sorting signals that cause retrograde transport of a protein in the secretory pathway are sometimes known as retrieval sequences. List the two known examples of retrieval sequences for soluble and membrane proteins of the ER. How does the presence of a retrieval sequence on a soluble ER protein result in its retrieval from the *cis*-Golgi complex? Describe how the concept of a retrieval sequence is essential to the cisternal-progression model.

8. Clathrin adapter protein (AP) complexes bind directly to the cytosolic face of membrane proteins and also interact with clathrin. What are the four known adapter protein complexes? Why may clathrin be considered to be an accessory protein to a core coat composed of adapter proteins?

9. I-cell disease is a classic example of an inherited human defect in protein targeting that affects an entire class of proteins, soluble enzymes of the lysosome. What is the molecular defect in I-cell disease? Why does it affect the targeting of an entire class of proteins? What other types of mutations might produce the same phenotype?

10. The TGN, *trans*-Golgi network, is the site of multiple sorting processes as proteins and lipids exit the Golgi complex. Compare and contrast the sorting of proteins to lysosomes versus the packaging of proteins into regulated secretory granules such as those containing insulin. Compare and contrast the sorting of proteins to the basolateral versus apical cell surfaces in MDCK cells versus hepatocytes.

11. Describe how pH plays a key role in regulating the interaction between mannose 6-phosphate and the mannose 6-phosphate receptor. Why does elevating endosomal pH lead to the secretion of newly synthesized lysosomal enzymes into the extracellular medium?

12. What mechanistic features are shared by (a) the formation of multivesicular endosomes by budding into the interior of the endosome and (b) the outward budding of HIV virus at the cell surface? You wish to design a peptide inhibitor/competitor of HIV budding and decide to mimic in a synthetic peptide a portion of the HIV Gag protein. Which portion of the HIV Gag protein would be a logical choice? What normal cellular process might this inhibitor block?

13. The phagocytic and autophagic pathways serve two fundamental roles, but both deliver their vesicles to the lysosome. What are the fundamental differences between the two pathways? Describe the three basic steps in the formation and fusion of autophagic vesicles.

Analyze the Data

In order to examine the specificity of membrane fusion conferred by specific v-SNAREs and t-SNAREs (see MacNew et al., 2000, *Nature* **407**:153–159), liposomes (artificial lipid membranes) were reconstituted with specific t-SNARE complexes or with v-SNAREs. To measure fusion, the v-SNARE liposomes also contained a fluorescent lipid at a relatively high concentration such that its fluorescence is quenched. (Quenching is reduced fluorescence relative to that expected. In this case, quenching occurs because the fluorescent lipids are too concentrated and interfere with each other's ability to become excited.) On fusion of these liposomes with those lacking the fluorescent lipid, the fluorescent lipids are diluted, and quenching is alleviated. Three sets of liposomes were prepared using yeast t-SNARE complexes: those containing plasma membrane t-SNAREs, Golgi t-SNAREs, or vacuolar t-SNAREs. Each of these was mixed with fluorescent liposomes containing one of three different yeast v-SNAREs. The following data were obtained.

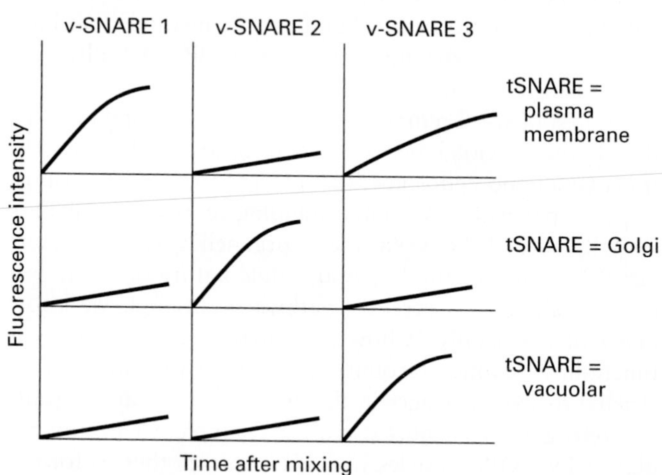

a. What can be deduced from these data about the specificity of membrane fusion events?

b. Where might you expect to find v-SNAREs 1, 2, and 3 in yeast?

c. What kind of experiment could be designed to determine where in the secretory pathway a given v-SNARE is required in vivo?

d. The cytoplasmic domain of v-SNARE 2 has been expressed and purified from *E. coli*. Various amounts of this domain are incubated either with the Golgi t-SNARE liposomes or with v-SNARE 2 liposomes. The liposomes are then washed free of unbound protein. The various liposomes are then mixed, as indicated below, and the fluorescence of each sample is measured 1 hour after mixing. How can the data be explained? What would you predict the outcome to be if yeast were to overexpress the cytoplasmic domain of v-SNARE 2?

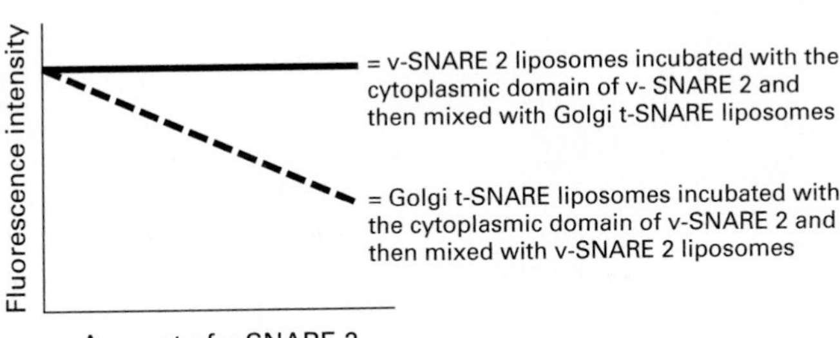

References

Techniques for Studying the Secretory Pathway

Beckers, C. J., et al. 1987. Semi-intact cells permeable to macromolecules: use in reconstitution of protein transport from the endoplasmic reticulum to the Golgi complex. *Cell* **50**:523–534.

Kaiser, C. A., and R. Schekman. 1990. Distinct sets of SEC genes govern transport vesicle formation and fusion early in the secretory pathway. *Cell* **61**:723–733.

Novick, P., et al. 1981. Order of events in the yeast secretory pathway. *Cell* **25**:461–469.

Lippincott-Schwartz, J., et al. 2001. Studying protein dynamics in living cells. *Nature Rev. Mol. Cell Biol.* **2**:444–456.

Orci, L., et al. 1989. Dissection of a single round of vesicular transport: sequential intermediates for intercisternal movement in the Golgi stack. *Cell* **56**:357–368.

Palade, G. 1975. Intracellular aspects of the process of protein synthesis. *Science* **189**:347–358.

Molecular Mechanisms of Vesicular Traffic

Bonifacino, J. S., and B. S. Glick. 2004. The mechanisms of vesicle budding and fusion. *Cell* **116**:153–66.

Grosshans, B. L., D. Ortiz, and P. Novick. 2006. Rabs and their effectors: achieving specificity in membrane traffic. *Proc. Natl. Acad. Sci. USA* **103**:11821–11827.

Jahn, R., et al. 2003. Membrane fusion. *Cell* **112**:519–533.

Kirchhausen, T. 2000. Three ways to make a vesicle. *Nature Rev. Mol. Cell Biol.* **1**:187–198.

McNew, J. A., et al. 2000. Compartmental specificity of cellular membrane fusion encoded in SNARE proteins. *Nature* **407**:153–159.

Ostermann, J., et al. 1993. Stepwise assembly of functionally active transport vesicles. *Cell* **75**:1015–1025.

Schimmöller, F., I. Simon, and S. Pfeffer. 1998. Rab GTPases, directors of vesicle docking. *J. Biol. Chem.* **273**:22161–22164.

Weber, T., et al. 1998. SNAREpins: minimal machinery for membrane fusion. *Cell* **92**:759–772.

Wickner, W., and A. Haas. 2000. Yeast homotypic vacuole fusion: a window on organelle trafficking mechanisms. *Ann. Rev. Biochem.* **69**:247–275.

Zerial, M., and H. McBride. 2001. Rab proteins as membrane organizers. *Nature Rev. Mol. Cell Biol.* **2**:107–117.

Early Stages of the Secretory Pathway

Barlowe, C. 2003. Signals for COPII-dependent export from the ER: what's the ticket out? *Trends Cell Biol.* **13**:295–300.

Behnia, R., and S. Munro. 2005. Organelle identity and the signposts for membrane traffic. *Nature* **438**:597–604.

Bi, X., et al. 2002. Structure of the Sec23/24-Sar1 pre-budding complex of the COPII vesicle coat. *Nature* **419**:271–277.

Gurkan, C., et al. 2006. The COPII cage: unifying principles of vesicle coat assembly. *Nat. Rev. Mol. Cell Biol.* **7**:727–738.

Lee, M. C., et al. 2004. Bi-directional protein transport between the ER and Golgi. *Ann. Rev. Cell Dev. Biol.* **20**:87–123.

Letourneur, F., et al. 1994. Coatomer is essential for retrieval of dilysine-tagged proteins to the endoplasmic reticulum. *Cell* **79**:1199–1207.

Losev, E., et al. 2006. Golgi maturation visualized in living yeast. *Nature* **441**:1002–1006.

Pelham, H. R. 1995. Sorting and retrieval between the endoplasmic reticulum and Golgi apparatus. *Curr. Opin. Cell Biol.* **7**:530–535.

Later Stages of the Secretory Pathway

Bonifacino, J. S. 2004. The GGA proteins: adaptors on the move. *Nat. Rev. Mol. Cell Biol.* **5**:23–32.

Bonifacino, J. S., and E. C. Dell'Angelica. 1999. Molecular bases for the recognition of tyrosine-based sorting signals. *J. Cell Biol.* **145**:923–926.

Edeling, M. A., C. Smith, and D. Owen. 2006. Life of a clathrin coat: insights from clathrin and AP structures. *Nat. Rev. Mol. Cell Biol.* **7**:32–44.

Fotin, A., et al. 2004. Molecular model for a complete clathrin lattice from electron cryomicroscopy. *Nature* **432**:573–579.

Ghosh, P., et al. 2003. Mannose 6-phosphate receptors: new twists in the tale. *Nature Rev. Mol. Cell Bio.* **4**:202–213.

Mostov, K. E., M. Verges, and Y. Altschuler. 2000. Membrane traffic in polarized epithelial cells. *Curr. Opin. Cell Biol.* **12**:483–490.

Schmid, S. 1997. Clathrin-coated vesicle formation and protein sorting: an integrated process. *Ann. Rev. Biochem.* **66**:511–548.

Simons, K., and E. Ikonen. 1997. Functional rafts in cell membranes. *Nature* **387**:569–572.

Song, B. D., and S. L. Schmid. 2003. A molecular motor or a regulator? Dynamin's in a class of its own. *Biochemistry* **42**:1369–1376.

Steiner, D. F., et al. 1996. The role of prohormone convertases in insulin biosynthesis: evidence for inherited defects in their action in man and experimental animals. *Diabetes Metab.* **22**:94–104.

Tooze, S. A., et al. 2001. Secretory granule biogenesis: rafting to the SNARE. *Trends Cell Biol.* **11**:116–122.

Receptor-Mediated Endocytosis

Brown, M. S., and J. L. Goldstein. 1986. Receptor-mediated pathway for cholesterol homeostasis. Nobel Prize Lecture. *Science* **232**:34–47.

Kaksonen, M., C. P. Toret, and D. G. Drubin. 2006. Harnessing actin dynamics for clathrin-mediated endocytosis. *Nat. Rev. Mol. Cell Biol.* **7**:404–414.

Rudenko, G., et al. 2002. Structure of the LDL receptor extracellular domain at endosomal pH. *Science* **298**:2353–2358.

Directing Membrane Proteins and Cytosolic Materials to the Lysosome

Hurley, J. H., and S. D. Emr. The ESCRT complexes: structure and mechanism of a membrane-trafficking network. *Ann. Rev. Biophys. Biomol. Struct.* **35**:277–298.

Katzmann, D. J., et al. 2002. Receptor downregulation and multivesicular-body sorting. *Nature Rev. Mol. Cell Biol.* **3**:893–905.

Khalfan, W. A., and D. J. Klionsky. 2002. Molecular machinery required for autophagy and the cytoplasm to vacuole targeting (Cvt) pathway in *S. cerevisiae*. *Curr. Opin. Cell Biol.* **14**:468–475.

Lemmon, S. K., and L. M. Traub. 2000. Sorting in the endosomal system in yeast and animal cells. *Curr. Opin. Cell Biol.* **12**:457–466.

Pornillos, O., et al. 2002. Mechanisms of enveloped RNA virus budding. *Trends Cell Biol.* **12**:569–579.

Shintani, T., and D. J. Klionsky. 2004. Autophagy in health and disease: a double-edged sword. *Science* **306**:990–995.

CLASSIC EXPERIMENT 14.1

FOLLOWING A PROTEIN OUT OF THE CELL

J. Jamieson and G. Palade, 1966, *Proc. Natl. Acad. Sci. USA* **55**(2):424–431

The advent of electron microscopy allowed researchers to see the cell and its structures at an unprecedented level of detail. George Palade utilized this tool not only to look at the fine details of the cell but also to analyze the process of secretion. By combining electron microscopy with pulse-chase experiments, Palade uncovered the path proteins follow to leave the cell.

Background

In addition to synthesizing proteins to carry out cellular functions, many cells must also produce and secrete additional proteins that perform their duties outside the cell. Cell biologists, including Palade, wondered how secreted proteins make their passage from the inside to the outside of the cell. Early experiments suggesting that proteins destined for secretion are synthesized in a particular intracellular location and then follow a pathway to the cell surface employed methods to disrupt cells synthesizing a particular secreted protein and to separate their various organelles by centrifugation. These cell-fractionation studies showed that secreted proteins can be found in membrane-bounded vesicles derived from the endoplasmic reticulum (ER), where they are synthesized, and with zymogen granules, from which they are eventually released from the cell. Unfortunately, results from these studies were hard to interpret due to difficulties in obtaining clean separation of all of the different organelles that contain secretory proteins. To further clarify the pathway, Palade turned to a newly developed technique, high-resolution autoradiography, that allowed him to detect the position of radioactively labeled proteins in thin cell sections that had been prepared for electron microscopy of intracellular organelles. His work led to the seminal finding that secreted proteins travel within vesicles from the ER to the Golgi complex and then to the plasma membrane.

The Experiment

Palade wanted to identify which cell structures and organelles participate in protein secretion. To study such a complex process, he carefully chose an appropriate model system for his studies, the pancreatic exocrine cell, which is responsible for producing and secreting large amounts of digestive enzymes. Because these cells have the unusual property of expressing only secretory proteins, a general label for newly synthesized protein, such as radioactively labeled leucine, will only be incorporated into protein molecules that are following the secretory pathway.

Palade first examined the protein secretion pathway in vivo by injecting live guinea pigs with [^{3}H]-leucine, which was incorporated into newly made proteins, thereby radioactively labeling them. At time points from 4 minutes to 15 hours, the animals were sacrificed, and the pancreatic tissue was fixed. By subjecting the specimens to autoradiography and viewing them in an electron microscope, Palade could trace where the labeled proteins were in cells at various times. As expected, the radioactivity localized in vesicles at the ER at time points immediately following the [^{3}H]-leucine injection and at the plasma membrane at the later time points. The surprise came in the middle time points. Rather than traveling straight from the ER to the plasma membrane, the radioactively labeled proteins appeared to stop off at the Golgi complex in the middle of their journey. In addition, there never was a time point where the radioactively labeled proteins were not confined to vesicles.

The observation that the Golgi complex was involved in protein secretion was both surprising and intriguing. To thoroughly address the role of this organelle in protein secretion, Palade turned to in vitro pulse-chase experiments, which permitted more precise monitoring of the fate of labeled proteins. In this labeling technique, cells are exposed to radiolabeled precursor, in this case [^{3}H]-leucine, for a short period known as the *pulse*. The radioactive precursor is then replaced with its nonlabeled form for a subsequent *chase* period. Proteins synthesized during the pulse period will be labeled and detected by autoradiography, whereas those synthesized during the chase period, which are nonlabeled, will not be detected. Palade began by cutting guinea pig pancreas into thick slices, which were then incubated for 3 minutes in media containing [^{3}H]-leucine. At the end of the pulse, he added excess unlabeled leucine. The tissue slices were then either fixed for autoradiography or used for cell fractionation. To ensure that his results were an accurate reflection of protein secretion in vivo, Palade meticulously characterized the system. Once convinced that his in vitro system accurately mimicked protein secretion in vivo, he proceeded to the critical experiment. He pulse-labeled tissue slices with [^{3}H]-leucine for 3 minutes, then chased the label for 7, 17, 37, 57, and 117 minutes with unlabeled leucine. Radioactivity, again confined in vesicles, began at the ER, then traveled in vesicles to the Golgi complex and remained in the vesicles

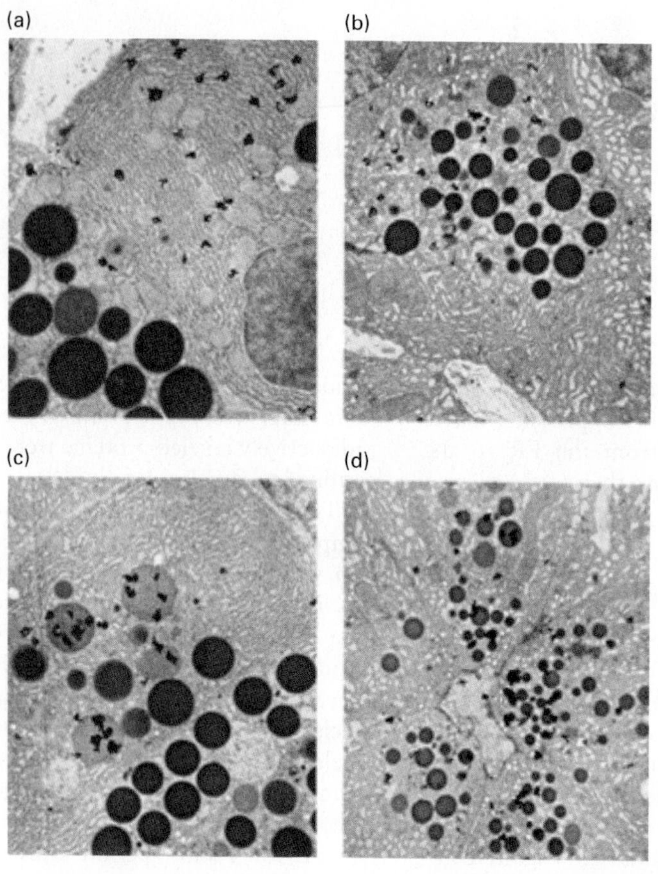

◄ **FIGURE 1 The synthesis and movement of guinea pig pancreatic secretory proteins as revealed by electron microscope autoradiography.** After a period of labeling with [^{3}H]-leucine, the tissue is fixed, sectioned for electron microscopy, and subjected to autoradiography. The radioactive decay of [^{3}H] in newly synthesized proteins produces autoradiographic grains in an emulsion placed over the cell section (which appear in the micrograph as dense, wormlike granules) that mark the position of newly synthesized proteins. (a) At the end of a 3-minute labeling period autoradiographic grains are over the rough ER. (b) Following a 7-minute chase period with unlabeled leucine, most of the labeled proteins have moved to the Golgi vesicles. (c) After a 37-minute chase, most of the proteins are over immature secretory vesicles. (d) After a 117-minute chase, the majority of the proteins are over mature zymogen granules. [Courtesy of J. Jamieson and G. Palade.]

as they passed through the Golgi and onto the plasma membrane (see Figure 1). As the vesicles traveled farther along the pathway, they became more densely packed with radioactive protein. From his remarkable series of autoradiograms at different chase times, Palade concluded that secreted proteins travel in vesicles from the ER to the Golgi and onto the plasma membrane and that throughout this process, they remain in vesicles and do not mix with the rest of the cell.

Discussion

Palade's experiments gave biologists the first clear look at the stages of the secretory pathway. His studies on pancreatic exocrine cells yielded two fundamental observations. First, that secreted proteins pass through the Golgi complex on their way out of the cell. This was the first function assigned to the Golgi complex. Second, secreted proteins never mix with cellular proteins in the cytosol; they are segregated into vesicles throughout the pathway. These findings were predicated from two important aspects of the experimental design. Palade's careful use of electron microscopy and autoradiography allowed him to look at the fine details of the pathway. Of equal importance was the choice of a cell type devoted to secretion, the pancreatic exocrine cell, as a model system. In a different cell type, significant amounts of nonsecreted proteins would have also been produced during the labeling, obscuring the fate of secretory proteins in particular.

Palade's work set the stage for more detailed studies. Once the secretory pathway was clearly described, entire fields of research were opened up to investigation in the synthesis and movement of both secreted and membrane proteins. For this groundbreaking work, Palade was awarded the Nobel Prize for Physiology and Medicine in 1974.

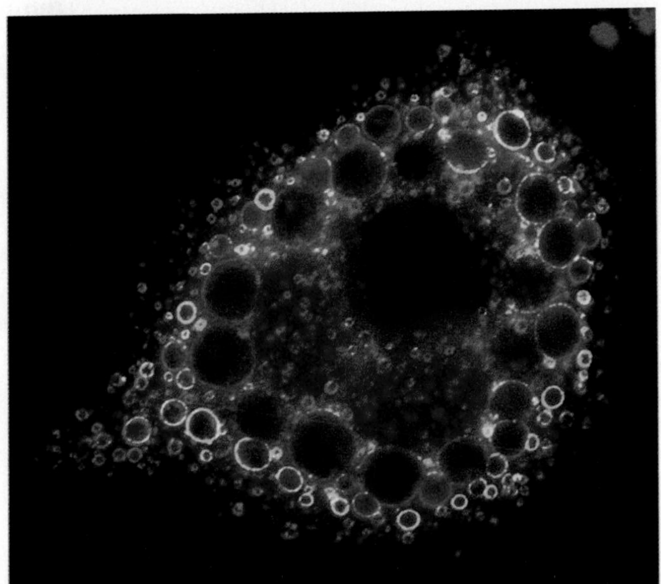

The storage and metabolism of triglycerides is regulated by many important hormones. This 3T3-L1 adipocyte is stained for three proteins that line triglyceride droplets at different stages of their maturation—perilipin in blue, adipophilin in green, and TIP47 in red. Courtesy Perry Bickel.

CHAPTER

15

CELL SIGNALING I: SIGNAL TRANSDUCTION AND SHORT-TERM CELLULAR RESPONSES

No cell lives in isolation; cellular communication is a fundamental property of all cells and shapes the function and abilities of every living organism. Even single-celled organisms have the ability to communicate with each other or other organisms. Eukaryotic microorganisms, such as yeasts, slime molds, and protozoans, use secreted molecules called **pheromones** to coordinate the aggregation of free-living cells for sexual mating or differentiation under certain environmental conditions. Yeast mating-type factors are a well-understood example of pheromone-mediated cell-to-cell signaling (Chapter 21). More important in plants and animals are extracellular **signaling molecules** that function *within* an organism to control metabolism of sugars, fats, and amino acids, the growth and differentiation of tissues, the synthesis and secretion of proteins, and the composition of intracellular and extracellular fluids. Animals also respond to many signals from their environment, including light, oxygen, odorants, and tastants in food.

Many extracellular signaling molecules are synthesized and released by *signaling cells* within the organism. In all cases signaling molecules produce a specific response only in *target cells* that have **receptors** for the signaling molecules. Many types of chemicals are used as signals: small molecules (e.g., amino acid or lipid derivatives, acetylcholine), peptides (e.g., ACTH and vasopressin), soluble proteins (e.g., insulin and growth hormone) and many proteins tethered to the surface of a cell or bound to the extracellular matrix. Most receptors bind a single molecule or a group of closely related molecules.

Some signaling molecules, especially hydrophobic molecules such as steroids, retinoids, and thyroxine, spontaneously diffuse through the plasma membrane and bind to intracellular receptors; signaling from such *intracellular* receptors is discussed in detail Chapter 7. Some small signaling molecules are hydrophilic and are transported by membrane proteins into the cell cytoplasm in order to influence cell behavior.

Most signaling molecules, however, are too large and too hydrophilic to penetrate through the plasma membrane. These bind to *cell-surface* receptors that are integral proteins

OUTLINE

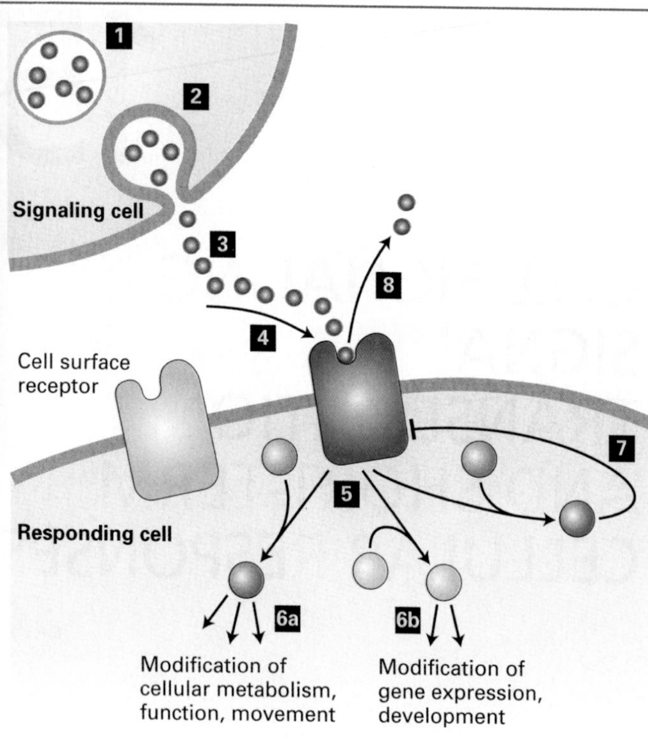

◄ **FIGURE 15-1 General principles of signaling by cell-surface receptors.** Communication by extracellular signals usually involves the following steps: Synthesis of the signaling molecule by the signaling cell and its incorporation into small intracellular vesicles (1), its release into the extracellular space by exocytosis (2), and transport of the signal to the target cell (3) where the signaling molecule binds to a specific cell-surface receptor protein leading to activation of the receptor (4). The activated receptor then initiates one or more intracellular signal-transduction pathways (5) leading to specific changes, usually short-term, in cellular function, metabolism, or movement (6a) or to long-term changes in gene expression or development (6b). Termination of the cellular response is caused by intracellular signaling molecules that inhibit receptor function (7) and by removal of the extracellular signal (8).

on the plasma membranes. Cell-surface receptors generally consist of three discrete segments: a segment on the extracellular surface, a segment that spans the plasma membrane, and a segment facing the cytosol. The signaling molecule acts as a **ligand,** which binds to a structurally complementary site on the extracellular or membrane-spanning domains of the receptor. Binding of the ligand induces a conformational change in the receptor that is transmitted through the membrane-spanning domain to the cytosolic domain, resulting in binding to and subsequent activation (or inhibition) of other proteins in the cytosol or attached to the plasma membrane. The overall process of converting extracellular signals into intracellular responses, as well as the individual steps in this process, is termed **signal transduction** (Figure 15-1).

In all eukaryotes there are only about a dozen classes of cell-surface receptors, and only a few dozen types of highly conserved intracellular signal-transduction pathways and proteins that are activated in the cytosol. Our knowledge of these common themes has advanced greatly in recent years, in large measure because these pathways are highly conserved and function in essentially the same way in organisms as diverse as worms, flies, and mice. Genetic studies combined with biochemical analyses have enabled researchers to trace many entire signaling pathways from binding of ligand to receptors to the final cellular responses.

Unfortunately, the terminology for designating pathways can be confusing. Pathways commonly are named based either on the general class of receptor involved (e.g., G protein–coupled receptors, receptor tyrosine kinases), the type of ligand (e.g., TGFβ, Wnt, Hedgehog), or a key intracellular signal-transduction component (e.g., NF- κB). In some cases, the same pathway may be referred to by different names. Fortunately, as researchers have discovered the molecular details of more and more receptors and pathways, some overarching principles and mechanisms are beginning to emerge. These shared features can help us make sense of the wealth of new information concerning cell-to-cell signaling.

Perhaps the most numerous class of receptors—found in organisms from yeast to human—are commonly called **G protein–coupled receptors (GPCRs).** The human genome encodes about 900 G protein–coupled receptors including receptors in the visual, olfactory (smell), and gustatory (taste) systems, many neurotransmitter receptors, and most of the receptors for hormones that control carbohydrate, amino acid, and fat metabolism. Essentially the same signaling pathway is used in yeast for signaling by mating factors (Chapter 21). This chapter focuses mainly on these receptors, which usually induce short-term changes in cell function. Activation of many cell-surface receptors alters the pattern of gene expression by the cell, leading to cell differentiation and other long-term consequences. These receptors and the intracellular signaling pathways they activate are explored in Chapter 16. Certain other cell-surface receptors are normally closed ion channels that open in response to ligand binding, allowing a particular type of ion to cross the plasma membrane. These receptors, called ligand-gated ion channels, are especially important in nerve cells; they are discussed in Chapter 23.

In this chapter we first review the general principles of cell signaling and describe how cell-surface receptors are identified and characterized. Next we discuss several features of many signal-transduction pathways and their regulation that have been conserved throughout evolution and are found in a wide variety of organisms. We then describe the common elements in G protein–coupled signaling pathways and examine in molecular detail the mechanisms by which diverse cellular functions are controlled by a relatively small set of G protein–coupled receptors and their associated signaling pathways. Finally, we will see that a signal-transduction pathway activated by one receptor can affect the signaling pathway downstream of a second kind of receptor—sometimes positively, at other times in a negative manner—in order to integrate cellular responses to multiple extracellular signals. This kind of signal integration plays a key role in the body's response to differing needs for glucose.

15.1 From Extracellular Signal to Cellular Response

Communication by extracellular signals within an organism usually involves the following steps (see Figure 15-1): **1** synthesis and **2** release of the signaling molecule by the signaling cell; **3** transport of the signal to the target cell; **4** binding of the signal by a specific receptor protein leading to a conformational change; **5** initiation of one or more intracellular signal-transduction pathways by the activated receptor; **6** specific changes in cellular function, metabolism, or development; and feedback regulation usually involving **7** deactivation of the receptor and **8** removal of the signaling molecule, which together terminate the cellular response.

Signaling Cells Produce and Release Signaling Molecules

In humans rapid responses to changes in the environment are primarily mediated by the nervous system and by **hormones** including small peptides (e.g., insulin and adrenocorticotropic hormone, ACTH) and small (nonpeptide) molecules such as the *catecholamines* (e.g., epinephrine, norepinephrine, and dopamine). The cells that make these signaling molecules are found in the pancreas (insulin), the pituitary gland (ACTH), the adrenal glands (epinephrine and norepinephrine), neurons (norepinephrine) and the part of the brain termed the hypothalamus (dopamine). Small signaling molecules, including catecholamine neurotransmitters, are synthesized in the cytosol and then transported into secretory vesicles (Chapter 23), whereas peptide and protein hormones are synthesized and processed by the secretory pathway detailed in Chapter 14. In both cases vesicles containing these signaling molecules accumulate just under the plasma membrane where they are held, awaiting a release signal (see Figure 15-1, step **1**).

Stimulation of signaling cells is almost always coupled to a rise in the local concentration of Ca^{2+} near the vesicles, causing immediate fusion of the membranes of the vesicles and the plasma membrane, leading to exocytosis of the stored peptide or protein hormone or small molecule into the surrounding medium or blood (see Figure 15-1, step **2**).

Released peptide hormones persist in the blood for only seconds to minutes before being degraded by blood and tissue proteases. Released small molecules such as catecholamines are rapidly inactivated by enzymes or taken up via transporters into specific cells. The initial actions of these signaling molecules on target cells (the activation or inhibition of specific enzymes) usually only lasts seconds or minutes. Thus the catecholamines and some peptide hormones can mediate short-term response that are terminated by their own degradation. For sustained or longer-term responses, such as cell division or cell differentiation, the cells must be exposed to signaling molecules for extended periods, usually hours.

Signaling Molecules Can Act Locally or at a Distance

Released signaling molecules travel to their target cells (see Figure 15-1, step **3**). Some are transported long distances by the blood; others have more local effects. In animals, signaling by extracellular molecules can be classified into three types—endocrine, paracrine, or autocrine—based on the distance over which the signal acts (Figure 15-2a–c). In addition, certain membrane-bound proteins on one cell can directly signal an adjacent cell.

(a) Endocrine signaling

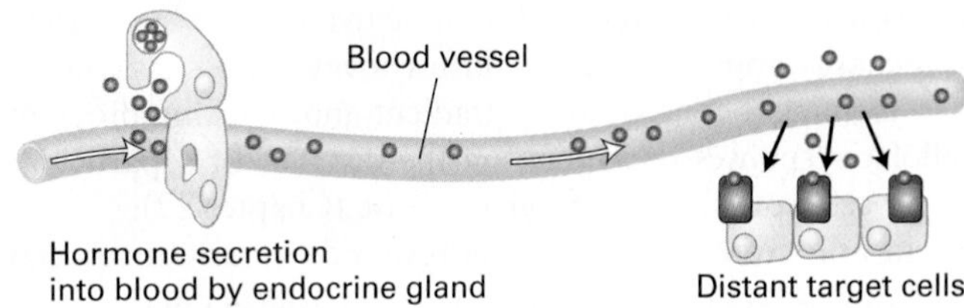

(b) Paracrine signaling

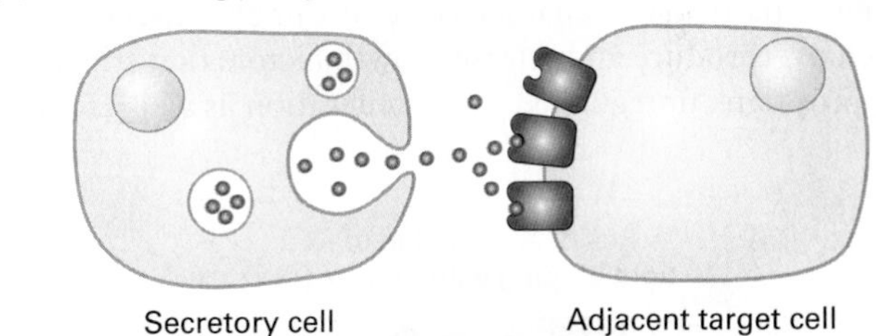

(c) Autocrine signaling

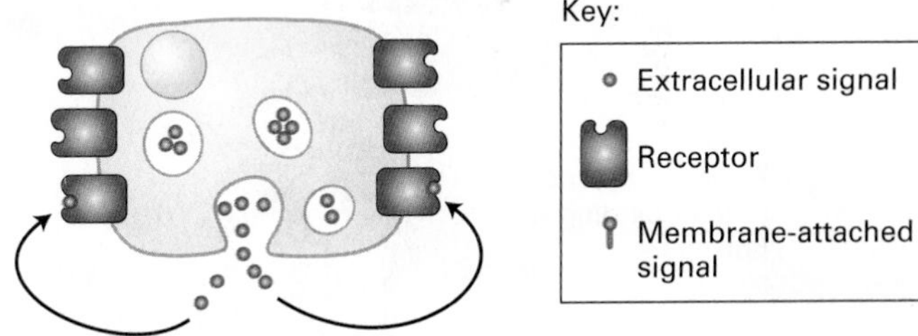

(d) Signaling by plasma membrane–attached proteins

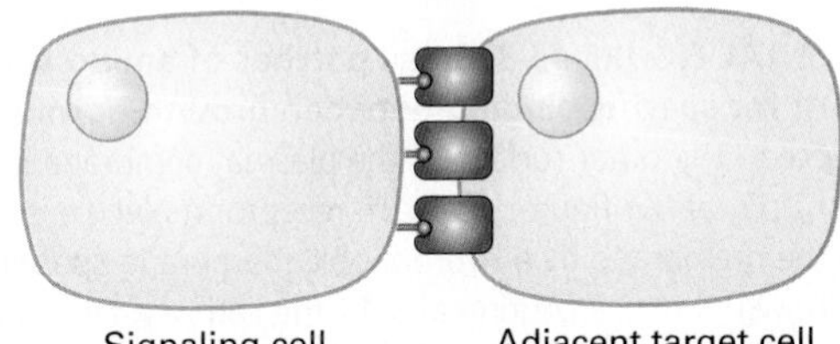

▲ **FIGURE 15-2 General schemes of intracellular signaling.** (a–c) Cell-to-cell signaling by extracellular chemicals occurs over distances from a few micrometers in autocrine and paracrine signaling to several meters in endocrine signaling. (d) Proteins attached to the plasma membrane of one cell can interact directly with cell-surface receptors on adjacent cells.

In **endocrine** signaling, the signaling molecules are synthesized and secreted by signaling cells (endocrine cells), transported through the circulatory system of the organism, and finally act on target cells distant from their site of synthesis. The term *hormone* generally refers to signaling molecules that mediate endocrine signaling.

In **paracrine** signaling, the signaling molecules released by a cell affect only those target cells in close proximity. The conduction by a neurotransmitter of a signal from one nerve cell to another or from a nerve cell to a muscle cell (inducing or inhibiting muscle contraction) occurs via paracrine signaling. Many **growth factors** regulating development in multicellular organisms also act at short range. Some of these molecules bind tightly to the extracellular matrix, unable to signal, but subsequently can be released in an active form. Many developmentally important signals diffuse away from the signaling cell, forming a concentration gradient and inducing different cellular responses depending on the distance of a particular target cell from the site of signal release (Chapter 22).

In **autocrine** signaling, cells respond to substances that they themselves release. Some growth factors act in this fashion, and cultured cells often secrete growth factors that stimulate their own growth and proliferation. This type of signaling is particularly characteristic of tumor cells, many of which overproduce and release growth factors that stimulate inappropriate, unregulated self-proliferation as well as influencing adjacent nontumor cells; this process may lead to formation of a tumor mass.

Signaling molecules that are integral membrane proteins located on the cell surface also play an important role in development (Figure 15-2d). In some cases, such membrane-bound signals on one cell bind receptors on the surface of an adjacent target cell to trigger its differentiation. In other cases, proteolytic cleavage of a membrane-bound signaling protein releases the extracellular segment, which functions as a soluble signaling molecule.

Some signaling molecules can act at both short and long ranges. Epinephrine, for example, functions as a neurotransmitter (paracrine signaling) and as a systemic hormone (endocrine signaling). Another example is epidermal growth factor (EGF), which is synthesized as an integral plasma membrane protein. Membrane-bound EGF can bind to and signal an adjacent cell by direct contact. Cleavage by an extracellular matrix protease releases a soluble form of EGF, which can signal in either an autocrine or a paracrine manner.

Binding of Signaling Molecules Activates Receptors on Target Cells

When a signaling molecule arrives at a target cell, it binds to a receptor (see Figure 15-1, step **4**). The vast majority of receptors are activated by binding of secreted or membrane-bound

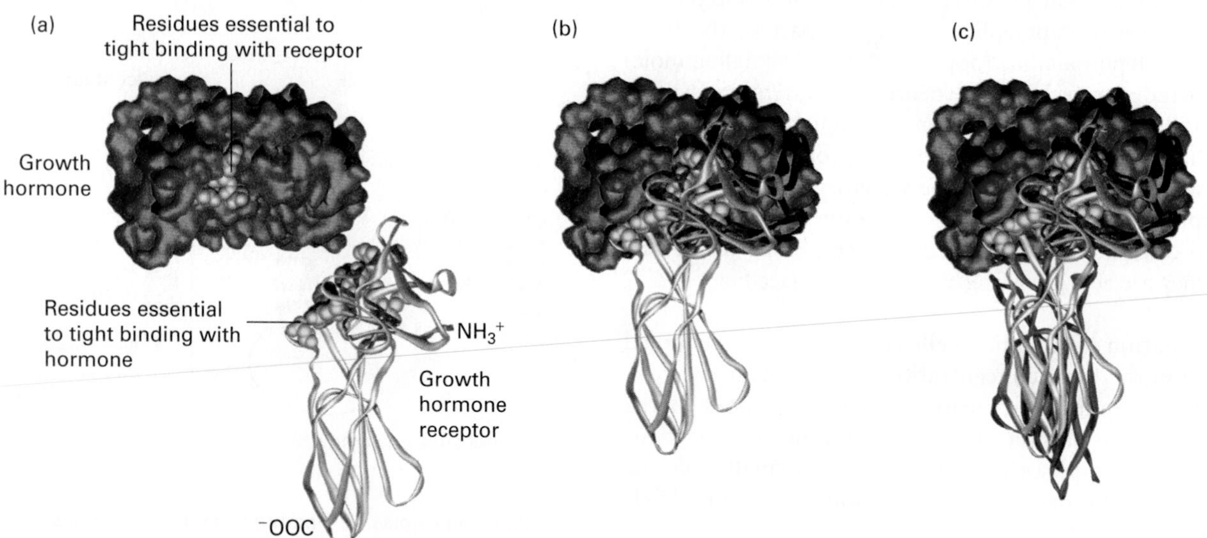

▲ EXPERIMENTAL FIGURE 15-3 **Small patches of amino acids are important for specific binding between growth hormone and its receptor.** The outer surface of the plasma membrane is toward the bottom of the figure, and each receptor molecule is anchored to the membrane by a hydrophobic membrane-spanning α helix (not shown) that is a continuation of the carboxyl terminus depicted in the figure. As determined from the three-dimensional structure of the growth hormone–growth hormone receptor complex, 28 amino acids in the hormone are at the binding interface with one receptor. Each of these amino acids was mutated, one at a time, to alanine, and the effect on receptor binding was determined. (a) From this study it was found that only eight amino acids on growth hormone (pink) contribute 85 percent of the binding energy; these amino acids are distant from each other in the primary sequence but adjacent in the folded protein. Similar studies showed that two tryptophan residues (blue) in the receptor contribute most of the energy for binding growth hormone, although other amino acids at the interface with the hormone (yellow) are also important. (b) Binding of growth hormone to one receptor molecule is followed by (c) binding of a second receptor (purple) to the opposing side of the hormone; this involves the same set of yellow and blue amino acids on the receptor but different residues on the hormone. As we see in the next chapter, such hormone-induced receptor dimerization is a common mechanism for activation of receptors for protein hormones. [After B. Cunningham and J. Wells, 1993, *J. Mol. Biol.* **234**:554, and T. Clackson and J. Wells, 1995, *Science* **267**:383.]

molecules (e.g., hormones, growth factors, neurotransmitters, and pheromones). Some receptors, however, are activated by changes in the concentration of a metabolite (e.g., oxygen or nutrients) or by physical stimuli (e.g., light, touch, heat).

As noted earlier, the specific receptor protein for any hydrophilic extracellular signaling molecule is almost always located on the surface of the target cell. The signaling molecule, or ligand, binds to a site on the extracellular domain of the receptor as a consequence of their molecular complementarity (Figure 15-3; see also Chapter 2). Binding of a ligand causes a conformational change in the membrane-bound receptor that initiates a sequence of reactions leading to a specific cellular response inside the cell. Different cell types may have different sets of receptors for the same ligand, each of which can induce a different response. Or the same receptor may be found on various cell types, and binding of a particular ligand to a particular receptor may trigger a different response in each type of cell. In these ways, one ligand can induce different cells to respond in a variety of ways.

Ligand binding to G protein–coupled receptors triggers an intracellular protein (a trimeric G protein) to exchange one bound GDP nucleotide for a GTP. The conformational change induced by GTP binding, in turn, affects interaction of the G protein with downstream signal-transduction proteins. The human receptors for the hormones epinephrine, serotonin, and glucagon are examples of G protein–coupled receptors. As we discuss in other chapters, the conformational change resulting from ligand binding to some other classes of receptors activates (or occasionally inhibits) the phosphorylating, or kinase, activity of the receptor's intracellular domain, or of an associated cytosolic kinase enzyme. Subsequent phosphorylation of substrate proteins in the cytosol modifies their activities. Activation of still other receptors involves proteolysis or disassembly on intracellular multiprotein complexes, leading to release of signal-transduction proteins.

KEY CONCEPTS OF SECTION 15.1

From Extracellular Signal to Cellular Response

■ Extracellular signaling molecules regulate interactions between unicellular organisms and are critical regulators of physiology and development in multicellular organisms.

■ Extracellular signaling molecules bind to receptors, inducing a conformational change in the receptor (activation) and consequent alteration in intracellular activities and functions.

■ Signals from one cell act on nearby cells in paracrine signaling, on distant cells in endocrine signaling, or on the signaling cell itself in autocrine signaling (see Figure 15-2).

■ External signals include membrane-anchored and secreted proteins and peptides (both dissolved in the extracellular fluid or embedded in extracellular matrix), small lipophilic molecules (e.g., steroid hormones, thyroxine), small hydrophilic molecules (e.g., epinephrine), gases (e.g., nitric oxide), and physical stimuli (e.g., light).

■ Binding of extracellular signaling molecules to cell-surface receptors triggers a conformational change in the receptor, which in turn leads to activation of intracellular signal-transduction pathways that ultimately modulate cellular metabolism, function, or gene expression (see Figure 15-1).

15.2 Studying Cell-Surface Receptors

The response of a cell or tissue to specific external signals is dictated (a) by the cell's complement of receptors that can recognize the signals, (b) the signal-transduction pathways activated by those receptors, and (c) the intracellular processes affected by those pathways. Recall that the interaction between ligand and receptor causes a conformational change in the receptor protein that enables it to interact with other proteins, thereby initiating a signaling cascade (see Figure 15-1, steps **4** and **5**). In this section we explore the biochemical basis for the specificity of receptor–ligand binding, as well as the ability of different concentrations of ligand to activate a pathway. We also examine experimental techniques used to purify and characterize receptor proteins. Many of these methods are applicable to receptors that mediate endocytosis (Chapter 14) or cell adhesion (Chapter 19) as well as those receptors that mediate signaling.

Typical cell-surface receptors are present in 1000 to 50,000 copies per cell. This may seem like a large number, but a "typical" mammalian cell contains $\approx 10^{10}$ total protein molecules and $\approx 10^6$ proteins on the plasma membrane. Thus the receptor for a particular signaling molecule commonly constitutes only 0.1 to 5 percent of plasma membrane proteins. This low abundance complicates the isolation and purification of cell-surface receptors. The purification of receptors is also difficult because these integral membrane proteins first must be solubilized from the membrane with a nonionic detergent (see Figure 10-23) and then separated from other cellular proteins.

Receptor Proteins Bind Ligands Specifically

Each receptor generally binds only a single signaling molecule (ligand) or a group of very closely structurally related molecules. The *binding specificity* of a receptor refers to its ability to distinguish closely related substances. Ligand binding depends on weak, multiple noncovalent forces (i.e., ionic, van der Waals, and hydrophobic interactions) and **molecular complementarity** between the interacting surfaces of a receptor and ligand (see Figure 2-12). The insulin receptor, for example, binds insulin and related hormones called insulin-like growth factors 1 and 2 (IGF-1 and IGF-2), but no other hormones. Similarly, the receptor that binds growth hormone does not bind other hormones of similar structure, and acetylcholine receptors bind only this small molecule and not others that differ only slightly in chemical structure. As these examples illustrate, all receptors are highly specific for their ligands.

Not only is each receptor protein characterized by its binding specificity for a particular ligand, the resulting receptor–ligand complex exhibits *effector specificity* because the complex mediates a specific cellular response. Many signaling molecules bind to multiple types of receptors, each of which can activate different intracellular signaling pathways and thus induce different cellular responses. For instance, the surfaces of skeletal muscle cells, heart muscle cells, and the pancreatic acinar cells that produce hydrolytic digestive enzymes each have different types of receptors for acetylcholine. In a skeletal muscle cell, release of acetylcholine from an adjacent neuron triggers contraction by activating an acetylcholine-gated ion channel. In the autoimmune paralytic disease myasthenia gravis, the body makes antibodies that block the activity of its own acetylcholine receptors. In heart muscle, the release of acetylcholine by different neurons activates a G protein–coupled receptor and slows the rate of contraction, and thus the heart rate. Release of acetylcholine near pancreatic acinar cells triggers a rise in cytosolic $[Ca^{2+}]$ that induces exocytosis of the digestive enzymes stored in secretory granules to facilitate digestion of a meal. Thus formation of different acetylcholine–receptor complexes in different cell types leads to different cellular responses.

On the other hand, different receptors of the same class that bind different ligands often induce the same cellular responses in a cell; thus different extracellular signals can cause a common change in the behavior of a cell. In liver cells, for instance, the hormones epinephrine, glucagon, and ACTH bind to different members of the G protein–coupled receptor family, but all these receptors activate the same signal-transduction pathway, one that promotes synthesis of a small intracellular signaling molecule (cyclic AMP) that in turn regulates various metabolic functions, including glycogen breakdown. As a result, all three hormones have the same effect on liver-cell metabolism, as further detailed in Sections 15.6 and 15.8.

The Dissociation Constant Is a Measure of the Affinity of a Receptor for Its Ligand

Ligand binding to a receptor usually can be viewed as a simple reversible reaction, where the receptor is represented as R, the ligand as L, and the receptor–ligand complex as RL:

$$\mathrm{R} + \mathrm{L} \underset{k_{\mathrm{on}}}{\overset{k_{\mathrm{off}}}{\rightleftarrows}} \mathrm{RL} \qquad (15\text{-}1)$$

k_{off} is the rate constant for dissociation of a ligand from its receptor, and k_{on} is the rate constant for formation of a receptor-ligand complex from free ligand and receptor.

At equilibrium the rate of formation of the receptor–ligand complex is equal to the rate of its dissociation, and can be described by the simple equilibrium-binding equation

$$K_{\mathrm{d}} = \frac{[\mathrm{R}][\mathrm{L}]}{[\mathrm{RL}]} \qquad (15\text{-}2)$$

where [R] and [L] are the concentrations of free receptor (that is, receptor without bound ligand) and ligand, respectively, at equilibrium, and [RL] is the concentration of the receptor–ligand complex. K_{d}, the **dissociation constant,** is a measure of the *affinity* of the receptor for its ligand. For a simple binding reaction, $K_{\mathrm{d}} = k_{\mathrm{off}}/k_{\mathrm{on}}$. The *lower* k_{off} is relative to k_{on}, the more *stable* the RL complex—the tighter the binding—and thus the *lower* the value of K_{d}. Another way of seeing this key point is that K_{d} equals the concentration of ligand at which half of the receptors have a ligand bound; at this ligand concentration [R] = [RL] and thus, from Equation 15-2, K_{d} = [L]. The lower the K_{d}, the lower the ligand concentration required to bind 50 percent of the cell-surface receptors. The K_{d} for a binding reaction here is essentially equivalent to the Michaelis constant K_{m}, which reflects the affinity of an enzyme for its substrate (Chapter 3).

Like all equilibrium constants, however, the value of K_{d} does not depend on the *absolute* values of k_{off} and k_{on}, only on their ratio. In the next section we learn how K_{d} values are experimentally determined.

Binding Assays Are Used to Detect Receptors and Determine Their Affinities for Ligands

Usually receptors are detected and measured by their ability to bind radioactive ligands to intact cells or to cell fragments. Figure 15-4 illustrates such a *binding assay* for interaction of insulin with insulin receptors on liver cells. The amounts of radioactive insulin bound to its receptor on cells growing in Petri dishes (vertical axis) were measured as a function of increasing insulin added to the extracellular fluid (horizontal axis). Both the number of ligand-binding sites per cell and the K_{d} value are easily determined from the specific binding curve (curve B), usually by applying readily available computer curve-fitting programs to the experimental values. Assuming each receptor generally binds just one ligand molecule, the number of ligand-binding sites equals the number of active receptors per cell. In the example shown in Figure 15-4, the value of K_{d} is 1.4×10^{-10} M. In other words, an insulin concentration of 1.4×10^{-10} M in the extracellular fluid is required for 50 percent of a cell's insulin receptors to have a bound insulin. A receptor usually has a different affinity for each of the ligands that it can bind. For instance, similar binding assays showed that the K_{d} for binding of insulin-like growth factor 1 (IGF-1) to the insulin receptor is 3×10^{-8} M. Thus an ≈200-fold higher concentration of IGF-1 than insulin is required to bind 50 percent of the insulin receptors.

Direct binding assays like the one in Figure 15-4 are feasible with receptors that have a strong affinity for their ligands, such as the erythropoietin receptor ($K_{\mathrm{d}} = 1 \times 10^{-10}$ M) and the insulin receptor on liver cells ($K_{\mathrm{d}} = 1.4 \times 10^{-10}$ M). However, many ligands such as epinephrine and other catecholamines bind to their receptors with much lower affinity. If the K_{d} for binding is greater than $\approx 1 \times 10^{-7}$ M, a case when the rate constant k_{off} is relatively large

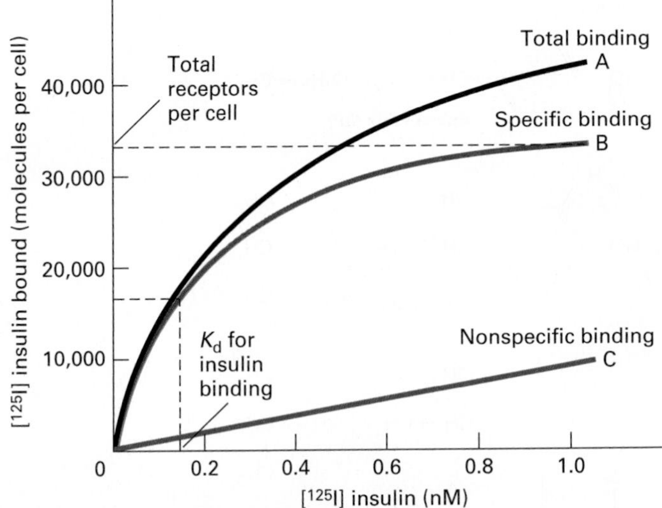

▲ EXPERIMENTAL FIGURE 15-4 **For high-affinity ligands, binding assays can determine the K_d and the number of receptors per cell.** Shown here are data for insulin-specific receptors on the surface of liver cells. A suspension of cells is incubated for 1 hour at 4° C with increasing concentrations of ^{125}I-labeled insulin; the low temperature is used to prevent endocytosis of the cell-surface receptors. The cells are separated from unbound insulin, usually by centrifugation, and the amount of radioactivity bound to them is measured. The total binding curve A represents insulin specifically bound to high-affinity receptors as well as insulin nonspecifically bound with low affinity to other molecules on the cell surface. The contribution of nonspecific binding to total binding is determined by repeating the binding assay in the presence of a 100-fold excess of unlabeled insulin, which saturates all the specific high-affinity sites. In this case, all the labeled insulin binds to nonspecific sites, yielding curve C. The specific binding curve B is calculated as the difference between curves A and C. As determined by the maximum of the specific binding curve B, the number of specific insulin-binding sites (surface receptors) per cell is 33,000. The K_d is the concentration of insulin required to bind to 50 percent of the surface insulin receptors (in this case about 16,500 receptors/cell). Thus, the K_d is about 1.4×10^{-10} M, or 0.14 nM. [Adapted from A. Ciechanover et al., 1983, *Cell* **32**:267.]

compared to k_{on}, then it is likely that during the seconds to minutes required to measure the amount of bound ligand, some of the receptor-bound ligand will dissociate and thus the observed values will be systematically too low.

One way to detect weak binding of a ligand to its receptor is in a *competition assay* with another ligand that binds to the same receptor with high affinity (low K_d value). In this type of assay, increasing amounts of an unlabeled, low-affinity ligand (the competitor) are added to a cell sample with a constant amount of the radiolabeled, high-affinity ligand (Figure 15-5). Binding of unlabeled competitor blocks binding of the radioactive ligand to the receptor; the concentration of competitor required to inhibit binding of half the radioactive ligand approximates the K_d value for binding of the competitor to the receptor. It is possible to accurately measure the amount of the high-affinity ligand bound in this assay because little dissociates during the experimental manipulations required for the measurement (relatively low k_{off}).

Competitive binding is often used to study synthetic analogs of natural hormones that activate or inhibit receptors. These analogs, which are widely used in research on cell-surface receptors and as drugs, fall into two classes: **agonists,** which mimic the function of a natural hormone by binding to its receptor and inducing the normal response, and **antagonists,** which bind to the receptor but induce no response. By occupying ligand-binding sites on a receptor, an antagonist can block binding of the natural hormone (or agonist) and thus reduce the usual physiological activity of the hormone. In other words, antagonists inhibit receptor signaling.

Consider for instance the drug isoproterenol used to treat asthma. Isoproterenol is made by the addition of two methyl groups to epinephrine (see Figure 15-5, *right*). Isoproterenol, an agonist of the epinephrine-responsive G protein–coupled receptors on bronchial smooth muscle cells, binds about tenfold more strongly (10-fold lower K_d) than does epinephrine (see Figure 15-5, *left*). Because activation of these receptors promotes relaxation of bronchial smooth muscle and thus opening of the air passages in the lungs, isoproterenol is used in treating bronchial asthma, chronic bronchitis, and emphysema. In contrast, activation of a different type of epinephrine-responsive G protein–coupled receptors on cardiac muscle cells increases the heart contraction rate. Antagonists of this receptor, such as alprenolol and related compounds, are referred to as *beta-blockers*; such antagonists are used to slow heart contractions in the treatment of cardiac arrhythmias and angina. ▮

Maximal Cellular Response to a Signaling Molecule Usually Does Not Require Activation of All Receptors

All signaling systems evolved such that a rise in the level of extracellular signaling molecules induces a proportional response in the responding cell. For this to happen the binding affinity (K_d value) of a cell-surface receptor for a circulating hormone must be greater than the normal (unstimulated) level of that hormone in the extracellular fluids or blood. We can see this principle in practice by comparing the levels of insulin present in the body and the K_d for binding of insulin to its receptor on liver cells, 1.4×10^{-10} M. Suppose, for instance, that the normal concentration of insulin in the blood is 5×10^{-12} M. By substituting this value and the insulin K_d into Equation 15-2, we can calculate the fraction of insulin receptors with bound insulin—[RL]/([RL] +[R])—at equilibrium as 0.0344; that is, about 3 percent of the total insulin receptors will be bound with insulin. If the insulin concentration rises fivefold to 2.5×10^{-11} M, the number of receptor–hormone complexes will rise proportionately, almost fivefold, so that about 15 percent of the total receptors will have bound insulin. If the extent of the induced cellular response

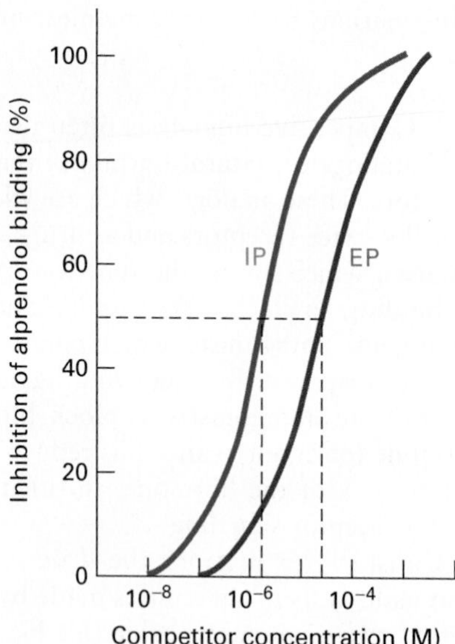

▲ EXPERIMENTAL FIGURE 15-5 **For low-affinity ligands, binding can be detected in competition assays.** In this example, the synthetic ligand alprenolol, which binds with high affinity to the epinephrine receptor on liver cells ($K_d \cong 3 \times 10^{-9}$ M), is used to detect the binding of two low-affinity ligands, the natural hormone epinephrine (EP) and a synthetic ligand called isoproterenol (IP). Assays are performed as described in Figure 15-4 but with a constant amount of [^{3}H]alprenolol to which increasing amounts of unlabeled epinephrine or isoproterenol are added. At each competitor concentration, the amount of bound labeled alprenolol is determined. In a plot of the inhibition of [^{3}H]alprenolol binding versus epinephrine or isoproterenol concentration, such as shown here, the concentration of the competitor that inhibits alprenolol binding by 50 percent approximates the K_d value for competitor binding. Note that the concentrations of competitors are plotted on a logarithmic scale. The K_d for binding of epinephrine to its receptor on liver cells is only $\approx 5 \times 10^{-5}$ M and would not be measurable by a direct binding assay with [^{3}H]epinephrine. The K_d for binding of isoproterenol, which induces the normal cellular response, is more than tenfold lower.

parallels the number of hormone–receptor complexes, [RL], as is often the case, then the cellular responses also will increase by about fivefold.

On the other hand, suppose that the normal concentration of insulin in the blood is the same as the K_d value of 1.4×10^{-10} M; in this case, 50 percent of the total receptors would have a bound insulin. A fivefold increase in the insulin concentration to 7×10^{-10} M would result only in a 66 percent increase in the fraction of receptors with bound insulin (to 83 percent bound). Thus, in order for a rise in hormone concentration to cause a proportional increase in the fraction of receptors with bound ligand, the normal concentration of the hormone must be well below the K_d value.

In general the maximal cellular response to a particular ligand is induced when much less than 100 percent of its receptors are bound to the ligand. This phenomenon can be revealed by determining the extent of the response and of receptor–ligand binding at different concentrations of ligand (Figure 15-6). For example, a typical red blood (erythroid) progenitor cell has $\approx$1000 surface receptors for erythropoietin, the protein hormone that induces these cells to proliferate and differentiate into red blood cells. Because only 100 of these receptors need to bind erythropoietin to induce division of a progenitor cell, the ligand concentration needed to induce 50 percent of the maximal cellular response is proportionally lower than the K_d value for binding. In such cases, a plot of the percentage of maximal binding versus ligand concentration differs from a plot of the percentage of maximal cellular response versus ligand concentration.

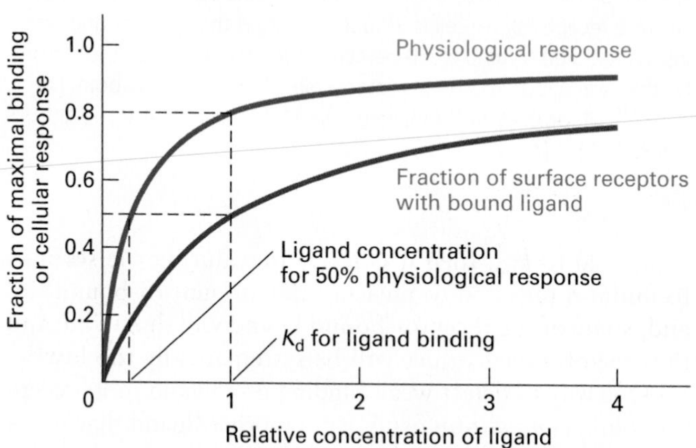

▲ EXPERIMENTAL FIGURE 15-6 **The maximal physiological response to an external signal occurs when only a fraction of the receptors are occupied by ligand.** For signaling pathways that exhibit this behavior, plots of the extent of ligand binding to the receptor and of physiological response at different ligand concentrations differ. In the example shown here, 50 percent of the maximal physiological response is induced at a ligand concentration at which only 18 percent of the receptors are occupied. Likewise, 80 percent of the maximal response is induced when the ligand concentration equals the K_d value, at which 50 percent of the receptors are occupied.

Sensitivity of a Cell to External Signals Is Determined by the Number of Surface Receptors and Their Affinity for Ligand

Because the cellular response to a particular signaling molecule depends on the number of receptor–ligand complexes, the fewer receptors present on the surface of a cell, the less *sensitive* the cell is to that ligand. As a consequence, a higher ligand concentration is necessary to induce the physiological response than would be the case if more receptors were present.

To illustrate the important relationship between receptor number and hormone sensitivity, let's extend our example of a typical erythroid progenitor cell. The K_d for binding of erythropoietin (Epo) to its receptor is about 10^{-10} M. As we noted above, only 10 percent of the ≈1000 erythropoietin receptors on the surface of a cell must be bound to ligand to induce the maximal cellular response. We can determine the ligand concentration, [L], needed to induce the maximal response by rewriting Equation 15-2 as follows:

$$[L] = \frac{K_d}{\dfrac{R_T}{[RL]} - 1} \qquad (15\text{-}3)$$

where R_T = [R] + [RL], the total number of receptors per cell. If the total number of Epo receptors per cell, R_T, is 1000, K_d is 10^{-10} M, and [RL] is 100 (the number of Epo-occupied receptors needed to induce the maximal response), then an Epo concentration ([L]) of 1.1×10^{-11} M will elicit the maximal response. If the total number of Epo receptors (R_T) is reduced to 200/cell, then a ninefold higher Epo concentration (10^{-10} M) is required to occupy 100 receptors and induce the maximal response. Clearly, therefore, a cell's sensitivity to a hormone is heavily influenced by the number of receptors for that hormone that are present as well as the K_d.

Epithelial growth factor (EGF), as its name implies, stimulates the proliferation of many types of epithelial cells, including those that line the ducts of the mammary gland. In about 25 percent of breast cancers the tumor cells produce elevated levels of one particular EGF receptor called HER2. The overproduction of HER2 makes the cells hypersensitive to ambient levels of EGF that normally are too low to stimulate cell proliferation; as a consequence growth of these tumor cells is inappropriately stimulated by EGF. Understanding of the role of HER2 in certain breast cancers led to development of monoclonal antibodies that bind HER2, and thereby block binding of EGF; these antibodies have proven useful in treatment of these breast cancer patients. ▮

The HER2–breast cancer connection vividly demonstrates that regulation of the number of receptors for a given signaling molecule expressed by a cell plays a key role in directing physiological and developmental events. Such regulation can occur at the levels of transcription, translation, and post-translational processing, or by controlling the rate of receptor degradation. Alternatively, endocytosis of receptors on the cell surface can sufficiently reduce the number present to terminate the usual cellular response at the prevailing signal concentration. As we discuss in later sections, other mechanisms can reduce a receptor's affinity for ligand, and thus reduce the cell's response to a given concentration of ligand. Thus reduction in a cell's sensitivity to a particular ligand, called *desensitization,* can result from various mechanisms and is critical to the ability of cells to respond appropriately to external signals.

Receptors Can Be Purified by Affinity Techniques

Because of their low abundance special techniques are necessary to isolate and purify receptors. Cell-surface receptors can be identified and followed through isolation procedures by *affinity labeling.* In this technique, cells are mixed with an excess of a radiolabeled ligand for the receptor of interest. After unbound ligand is washed away, the cells are treated with a chemical agent that covalently cross-links the bound labeled ligand molecules to receptors on the cell surface. Once a radiolabeled ligand is covalently cross-linked to its receptor, it remains bound even in the presence of detergents and other denaturing agents that are used to solubilize receptor proteins from the cell membrane. The labeled ligand provides a means for detecting the receptor during purification procedures.

Another technique often used in purifying cell-surface receptors that retain their ligand-binding ability when solubilized by detergents is similar to *affinity chromatography* using antibodies (see Figure 3-37c). To purify a receptor by this technique, a ligand for the receptor of interest, rather than an antibody, is chemically linked to the beads used to form a column. A crude, detergent-solubilized preparation of membrane proteins is passed through the column; only the receptor binds, while other proteins are washed away. Passage of an excess of the soluble ligand through the column causes the bound receptor to be displaced from the beads and eluted from the column. In some cases, a receptor can be purified as much as 100,000-fold in a single affinity chromatographic step.

Receptors Are Frequently Expressed from Cloned Genes

Once the amino acid sequence of a purified receptor has been determined its gene can be cloned. A *functional expression assay* of the cloned cDNA in a mammalian cell that normally lacks the encoded receptor can provide definitive proof that the proper protein indeed has been obtained (Figure 15-7). Such expression assays also permit investigators to study the effects of mutating specific amino acids on ligand binding or on "downstream" signal transduction, thereby pinpointing the receptor amino acids responsible for interacting with the ligand or with critical signal-transduction proteins.

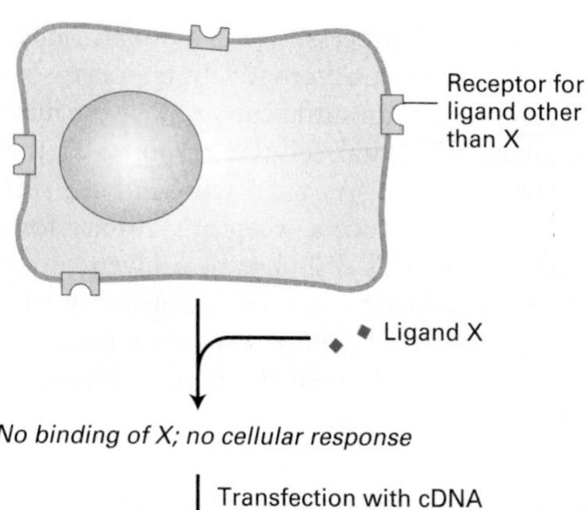

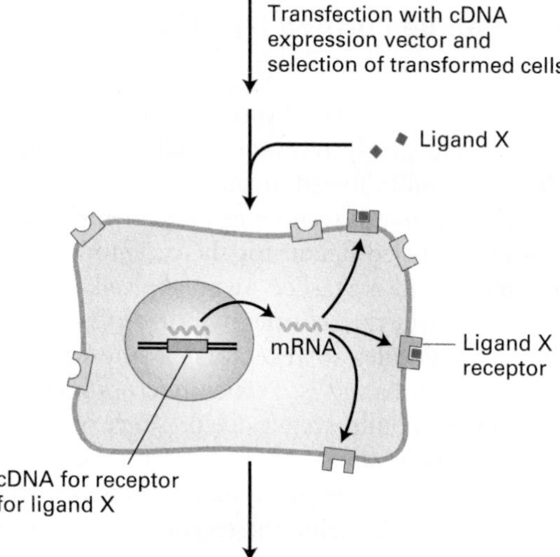

▲ **EXPERIMENTAL FIGURE 15-7 Functional expression assay can identify a cDNA encoding a cell-surface receptor.** Target cells lacking receptors for a particular ligand (X) are stably transfected with a cDNA expression vector encoding the receptor. The design of the expression vector permits selection of transformed cells from those that do not incorporate the vector into their genome (see Figure 5-32b). Providing that these cells already express all the relevant signal-transduction proteins, the transfected cells exhibit the normal cellular response to X if the cDNA in fact encodes the functional receptor.

Cell-surface receptors for many signaling molecules are present in such small amounts that they cannot be purified by affinity chromatography and other conventional biochemical techniques. These low-abundance receptor proteins can now be identified and cloned by various recombinant DNA techniques, eliminating the need to isolate and purify them from cell extracts. In one technique, a library of cloned cDNAs prepared from the entire mRNA extracted from cells that produce the receptor is inserted into expression vectors by techniques described in Chapter 5. The recombinant vectors then are transfected into cells that normally do not synthesize the receptor of interest (see Figure 15-7). Only the very few transfected cells that contain the cDNA encoding the desired receptor synthesize it; other transfected cells produce irrelevant proteins. The rare cells expressing the desired receptor can be detected and purified by various techniques such as fluorescence-activated cell sorting using a fluorescent-labeled ligand for the receptor of interest (see Figure 9-28). Once a cDNA clone encoding the receptor is identified, the sequence of the cDNA can be determined and that of the receptor protein deduced from the cDNA sequence. Cells overexpressing the receptor protein can be used to purify large amounts of the protein, which can be used to determine its three-dimensional structure. This structural information can provide additional insights into the mechanisms by which the receptor functions, as well as suggesting how new types of drugs might interact with the receptor to treat diseases.

Genomic studies coupled with functional expression assays are now being used to identify genes for previously unknown receptors. In this approach, stored DNA sequences are analyzed for similarities with sequences known to encode receptor proteins (Chapter 6). Any putative receptor genes that are identified in such a search then can be tested for their ability to bind a signaling molecule or induce a response in cultured cells by a functional expression assay.

KEY CONCEPTS OF SECTION 15.2

Characterizing Cell-Surface Receptors

- Receptors bind ligands with considerable specificity, which is determined by noncovalent interactions between a ligand and specific amino acids in the receptor protein (see Figure 15-3).
- The concentration of ligand at which half its receptors are occupied, the K_d, can be determined experimentally and is a measure of the affinity of the receptor for the ligand (see Figure 15-4).
- The maximal response of a cell to a particular ligand generally occurs at ligand concentrations at which most of its receptors are still not occupied (see Figure 15-6).
- Because the amount of a particular receptor expressed is generally quite low (ranging from $\approx$1000 to 50,000 molecules per cell), biochemical purification may not be feasible. Genes encoding low-abundance receptors for specific ligands often can be isolated from cDNA libraries transfected into cultured cells.
- Functional expression assays can determine if a cDNA encodes a particular receptor and are useful in studying the effects on receptor function of specific mutations in its sequence (see Figure 15-7).

15.3 Highly Conserved Components of Intracellular Signal-Transduction Pathways

External signals induce two major types of cellular responses: (1) changes in the activity or function of specific enzymes and other proteins that pre-exist in the cell, and (2) changes in the amounts of specific proteins produced by a cell, most commonly by modification of **transcription factors** that stimulate or repress gene expression (see Figure 15-1, step 6). In general, the first type of response occurs more rapidly than the second. Signaling from G protein–coupled receptors, described in detail later in this chapter, often results in changes in the activity of pre-existing proteins, although activation of these receptors on some cells also induces changes in gene expression. Other classes of receptors operate primarily but not exclusively to modulate gene expression. Transcription factors activated in the cytosol by these pathways move into the nucleus, where they stimulate (or occasionally repress) transcription of specific target genes. We consider these signaling pathways, which regulate transcription of many genes essential for cell division and for many cell differentiation processes, in the following chapter.

Several intracellular proteins or small molecules are employed in a variety of signal-transduction pathways. These include cytosolic enzymes that add or remove phosphate groups from specific target proteins. Ligand binding to a receptor activates or inhibits these enzymes, whose action in turn activates or inhibits the function of their target proteins. G proteins, another component of many signal-transduction pathways, shuttle between a state with a bound GTP that is capable of activating other proteins and a state with a bound GDP that is inactive. A number of small molecules (e.g., Ca^{2+} and cyclic AMP) are also frequently used in intracellular signal-transduction pathways: A rise in the concentration of one of these molecules results in its binding to an intracellular target protein, causing a conformational change in the protein that modulates its function. Here we review the basic properties of these intracellular signal-transducing molecules. The rules and exceptions that govern how they are used in particular signaling pathways are developed further in subsequent sections of this chapter.

GTP-Binding Proteins Are Frequently Used As On/Off Switches

We introduced the large group of intracellular switch proteins that form the **GTPase superfamily** in Chapter 3. These guanine nucleotide–binding proteins are turned "on" when they bind GTP and turned "off" when the GTP is hydrolyzed to GDP (see Figure 3-32). Signal-induced conversion of the inactive to active state is mediated by a *guanine nucleotide–exchange factor (GEF)*, which causes release of GDP from the switch protein. Subsequent binding of GTP, favored by its high intracellular concentration relative to its binding affinity, induces a conformational change in at least two highly conserved segments of the protein, termed switch I and switch II, allowing the protein to bind to and activate other downstream signaling proteins (Figure 15-8). The intrinsic GTPase activity of the protein then hydrolyzes the bound GTP to GDP and P_i, thus changing the conformation of switch I and switch II from the active form back to the

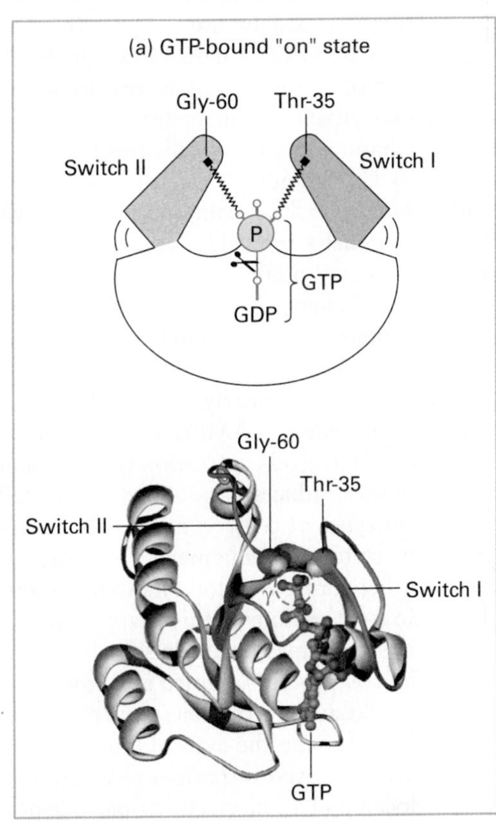

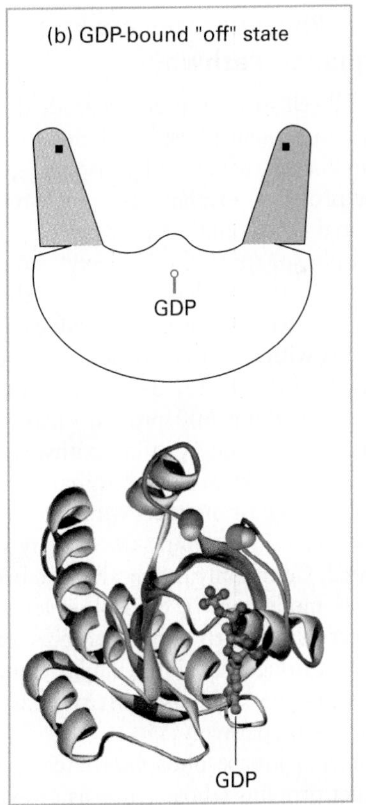

▲ **FIGURE 15-8 Switching mechanism for G proteins.** The ability of a G protein to interact with other proteins and thus transduce a signal differs in the GTP-bound "on" state and GDP-bound "off" state. (a) In the active "on" state, two domains, termed switch I (green) and switch II (blue), are bound to the terminal γ phosphate of GTP through interactions with the backbone amide groups of a conserved threonine and glycine residue. (b) Release of the γ phosphate by GTPase-catalyzed hydrolysis causes switch I and switch II to relax into a different conformation, the inactive "off" state. The ribbon models shown here represent both conformations of Ras, a monomeric G protein. A similar spring-loaded mechanism switches the α subunit in trimeric G proteins between the active and inactive conformations by movement of three switch segments. [Adapted from I. Vetter and A. Wittinghofer, 2001, *Science* **294:**1299.]

inactive form. The rate of GTP hydrolysis regulates the length of time the switch protein remains in the active conformation and able to signal downstream: The slower the rate of GTP hydrolysis, the longer the protein remains in the active state. The rate of GTP hydrolysis is often modulated by other proteins. For instance, both *GTPase-activating proteins (GAP)* and *regulator of G protein signaling (RGS) proteins* accelerate GTP hydrolysis. Many regulators of G protein activity are themselves controlled by extracellular signals.

Two large classes of GTPase switch proteins are used in signaling. **Trimeric (large) G proteins,** as noted already, directly bind to and are activated by certain cell-surface receptors. The activated receptor functions as a GEF and triggers release of GDP and binding of GTP. **Monomeric (small) G proteins,** such as Ras and various Ras-like proteins, play crucial roles in many pathways that regulate cell division and cell motility; these G proteins frequently undergo activating mutations in cancers. Ras is linked indirectly to receptors via adapter proteins and GEF proteins discussed in the next chapter. All G switch proteins contain regions like switch I and switch II that modulate the activity of specific effector proteins by direct protein–protein interactions when the G protein is bound to GTP. Despite these similarities, the two classes of GTP-binding proteins are regulated in very different ways.

Protein Kinases and Phosphatases are Employed in Virtually All Signaling Pathways

Activation of virtually all cell-surface receptors leads directly or indirectly to changes in protein phosphorylation through the activation of protein **kinases,** which add phosphate groups to specific residues, or protein **phosphatases,** which remove phosphate groups. Animal cells contain two types of protein kinases: those that add phosphate to the hydroxyl group on tyrosine residues and those that add phosphate to the hydroxyl group on serine or threonine (or both) residues. Phosphatases can act in concert with kinases to switch the function of various proteins on or off (see Figure 3-33). At last count the human genome encodes about 600 protein kinases and 100 different phosphatases. In some signaling pathways, the receptor itself possesses intrinsic kinase or phosphatase activity; in other pathways, the receptor interacts with cytosolic or membrane-associated kinases. Importantly the activity of all kinases is highly regulated. Commonly the catalytic activity of a protein kinase itself is modulated by phosphorylation by other kinases, by direct binding to other proteins or by changes in the levels of various small intracellular signaling molecules. The resulting cascades of kinase activity are a common feature of many signaling pathways.

In general, each protein kinase phosphorylates specific residues in a set of target proteins whose patterns of expression generally differ in different cell types. Many proteins are substrates for multiple kinases each of which phosphorylates different amino acids. Each phosphorylation event can modify the activity of a particular target protein in different ways, some activating its function, others inhibiting it. An example we encounter later is glycogen phosphorylase kinase, a key regulatory enzyme in glucose metabolism.

The activity of all protein kinases is opposed by the activity of protein phosphatases, some of which are themselves regulated by extracellular signals. Thus the activity of a protein in a cell can be a complex function of the activities of the usually multiple kinases and phosphatases that act on it. Several examples of this phenomenon occur in regulation of the cell cycle and are described in Chapter 20.

Second Messengers Carry and Amplify Signals from Many Receptors

The binding of ligands ("first messengers") to many cell-surface receptors leads to a short-lived increase (or decrease) in the concentration of certain low-molecular-weight intracellular signaling molecules termed **second messengers.** These, in turn, bind to other proteins modifying their activity.

One second messenger used in virtually all metazoan cells is Ca^{2+} ions. We noted in Chapter 11 that the concentration of Ca^{2+} free in the cytosol is kept very low ($\approx 10^{-7}$ M) by ATP-powered pumps that continually transport Ca^{2+} out of the cell or into the endoplasmic reticulum (ER). The cytosolic Ca^{2+} level can increase from 10- to 100-fold by a signal-induced release of Ca^{2+} from ER stores or by its import through calcium channels from the extracellular environment. In muscle, a signal-induced rise in cytosolic Ca^{2+} triggers contraction (see Figure 17-33). In endocrine cells, a similar increase in Ca^{2+} induces exocytosis of secretory vesicles containing hormones. In nerve cells, a Ca^{2+} increase leads to the exocytosis of neurotransmitter-containing vesicles (see Chapter 23). In all cells this rise in cytosolic Ca^{2+} is sensed by Ca^{2+}-binding proteins, particularly those of the *EF hand family,* such as calmodulin, all of which contain the helix-loop-helix motif (see Figure 3-9a). The binding of Ca^{2+} to calmodulin and other EF hand proteins causes a conformational change that permits the protein to bind various target proteins, thereby switching their activities on or off (see Figure 3-31).

Another nearly universally used second messenger is **cyclic AMP (cAMP).** In many eukaryotic cells, a rise in cAMP triggers activation of a particular protein kinase that in turn induces various changes in cell metabolism in different types of cells. In other cells, cAMP regulates the activity of certain ion channels. The structures of cAMP and three other common second messengers are shown in Figure 15-9. In later sections of this chapter, we examine the specific roles of second messengers in signaling pathways activated by various G protein–coupled receptors.

Because second messengers such as Ca^{2+} and cAMP diffuse through the cytosol much faster than do proteins, they are employed in pathways where the downstream target is located in an intracellular particle or organelle (such as a secretory vesicle) distant from the plasma membrane receptor. Another advantage of second messengers is that they facilitate amplification of an extracellular signal. Activation of a *single* cell-surface receptor molecule can result in an increase in perhaps thousands of cAMP molecules or Ca^{2+} ions in the cytosol. Each of these, in turn, by activating its target protein affects the activity of multiple downstream proteins.

Focus Animation: Second Messengers in Signaling Pathways

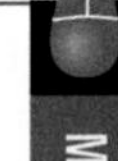

MEDIA CONNECTIONS

3′,5′-Cyclic AMP (cAMP)

Activates protein kinase A (PKA)

3′,5′-Cyclic GMP (cGMP)

Activates protein kinase G (PKG) and opens cation channels in rod cells

Fatty acyl groups

Glycerol

1,2-Diacylglycerol (DAG)

Activates protein kinase C (PKC)

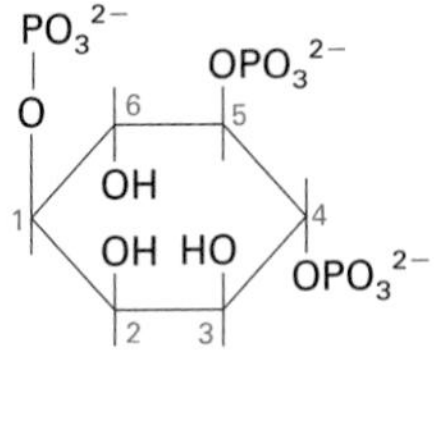

Inositol 1,4,5-trisphosphate (IP_3)

Opens Ca^{2+} channels in the endoplasmic reticulum

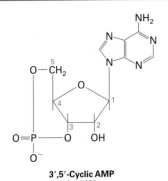

▲ **FIGURE 15-9 Four common intracellular second messengers.** The major direct effect or effects of each compound are indicated below its structural formula. Calcium ion (Ca^{2+}) and several membrane-bound phosphatidylinositol derivatives also act as second messengers.

KEY CONCEPTS OF SECTION 15.3

Highly Conserved Components of Intracellular Signal-Transduction Pathways

- Protein kinases and phosphatases are employed in virtually all signaling pathways; their activities are highly regulated by receptors.
- Other conserved proteins that act in many signal-transduction pathways include monomeric and trimeric G switch proteins (see Figure 15-8).
- The cytosolic concentrations of second messengers, such as Ca^{2+} and cAMP, increase or occasionally decrease in response to binding of ligand to cell-surface receptors (see Figure 15-9). These nonprotein, low-molecular-weight intracellular signaling molecules, in turn, regulate the activities of enzymes and nonenzymatic proteins in signaling pathways.

15.4 General Elements of G Protein–Coupled Receptor Systems

As noted above, perhaps the most numerous class of receptors—found in organisms from yeast to man—are the G protein–coupled receptors (GPCRs). Receptor activation by ligand binding triggers activation of the coupled trimeric G protein, which interacts with downstream signal-transduction proteins. All GPCR signaling pathways share the following common elements: (1) a receptor that contains seven membrane-spanning domains; (2) a coupled trimeric G protein, which functions as a switch by cycling between active and inactive forms; (3) a membrane-bound effector protein; and (4) feedback regulation and desensitization of the signaling pathway. A second messenger also occurs in many GPCR pathways. These components are modular and can be "mixed and matched" to achieve an astonishing number of different pathways. GPCR pathways usually have short-term effects in the cell by quickly modifying existing proteins, either enzymes or ion channels. Thus these pathways allow cells to respond rapidly to a variety of signals, whether they be environmental stimuli such as light or hormonal stimuli such as epinephrine.

In this section, we first consider the general features of GPCR signal transduction and then discuss each of the membrane-bound components in turn: the receptor, the trimeric G protein, and the effector proteins. In Section 15.5–15.7 we describe GPCR pathways that involve several different effector proteins. The short-term signaling described in these sections of this chapter often can be turned into long-term signaling involving changes in transcription and, as a consequence, cell differentiation, as described in Chapter 16.

G Protein–Coupled Receptors Are a Large and Diverse Family with a Common Structure and Function

All G protein–coupled receptors have the same orientation in the membrane and contain seven transmembrane α-helical regions (H1–H7), four extracellular segments, and four cytosolic segments (Figure 15-10). Invariably the N-terminus is on the exoplasmic face and the C-terminus is on the cytosolic face of the plasma membrane. The carboxyl-terminal segment (C4), the C3 loop, and, in some receptors, also the C2 loop are involved in interactions with a coupled trimeric G protein. There are many subfamilies of G protein–coupled receptors that are conserved through evolution; members of these subfamilies are especially similar in amino acid sequence and structure.

The exterior surface of all G protein–coupled receptors mainly consists of hydrophobic amino acids that allow the protein to be stably anchored in the hydrophobic core of the plasma membrane. One G protein–coupled receptors whose structure is known in molecular detail is *rhodopsin,* which is composed of the protein *opsin* covalently bound to the

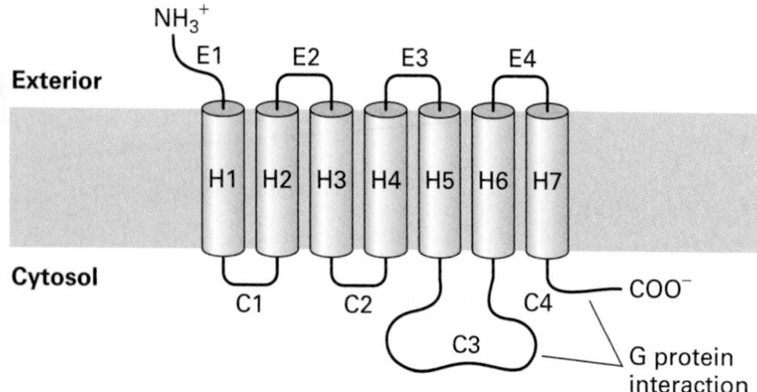

▲ **FIGURE 15-10 General structure of G protein–coupled receptors.** All receptors of this type have the same orientation in the membrane and contain seven transmembrane α-helical regions (H1–H7), four extracellular segments (E1–E4), and four cytosolic segments (C1–C4). The carboxyl-terminal segment (C4), the C3 loop, and, in some receptors, also the C2 loop are involved in interactions with a coupled trimeric G protein.

light-absorbing visual pigment 11-*cis*-retinal. In rhodopsin, the seven membrane-embedded α helices of opsin completely surround a central segment to which retinal is covalently bound. In this case, binding of the ligand, retinal, does not trigger a conformational change in the receptor; rather, absorption of a quantum of light by the bound retinal induces a change in opsin conformation that activates it, as we discuss in Section 15.5.

The amino acids that form the interior of different G protein–coupled receptors are diverse, allowing different receptors to bind very different small molecules, be they hydrophilic like epinephrine or hydrophobic like many odorants and retinal. Figure 15-11 depicts a model of the complex formed between the β_2-adrenergic receptor and the hormone epinephrine. As with retinal, epinephrine is thought to bind (in this case, noncovalently) in the middle of the plane of the membrane, interacting with amino acids in the interior-facing side of several of the membrane-spanning α helices.

As an example of the diversity and functionality of GPCR proteins, we will consider the different G protein–coupled receptors for epinephrine that are found in different types of mammalian cells. The hormone **epinephrine** is particularly important in mediating the body's response to stress (flight-or-flight response), such as fear or heavy exercise, when tissues may have an increased need to catabolize glucose and fatty acids to produce ATP. These principal metabolic fuels can be supplied to the blood in seconds by the rapid breakdown of glycogen to glucose in the liver and of triacylglycerols to fatty acids in adipose (fat) cells.

In mammals, the liberation of glucose and fatty acids can be triggered by binding of epinephrine (or its derivative norepinephrine) to *β-adrenergic receptors* on the surface of hepatic (liver) and adipose cells. Epinephrine bound to β-adrenergic receptors on heart muscle cells increases the contraction rate, which increases the blood supply to the tissues. In contrast, epinephrine stimulation of β-adrenergic receptors on smooth muscle cells of the intestine causes them to relax. Another type of epinephrine receptor, the *α-adrenergic receptor,* is found on smooth muscle cells lining the blood vessels in the intestinal tract, skin, and kidneys. Binding of epinephrine to these receptors causes the arteries to constrict, cutting off circulation to these peripheral organs.

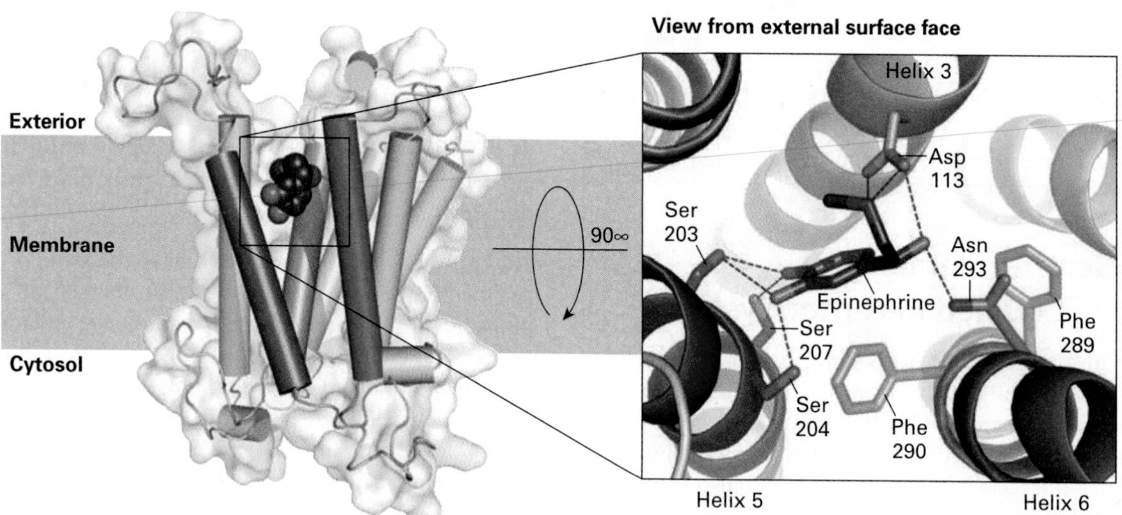

▲ **FIGURE 15-11 Structural model of complex formed between epinephrine and the β_2-adrenergic receptor.** (left) side view. The approximate location of the membrane phospholipid bilayer is indicated. The three α-helices that participate in epinephrine binding are colored red (Helix 5), green (Helix 3), and purple (Helix 6). (right) view from external face. Epinephrine atoms are colored grey (C), red (O) and purple (N). Epinephrine interacts with several residues in the receptor that are within the plane of the membrane. Its amino group forms an ionic bond with the carboxylate side chain of aspartate 113 (D^{113}) in H3; the catechol ring engages in hydrophobic interactions with phenylalanine 290 (F^{290}) in H6; and two hydroxyl groups on the catechol ring hydrogen-bond to the hydroxyl groups in three serine residues (S^{203}, S^{204} and S^{207}) in H5. [See P. L. Freddolino et al., 2004, *Proc. Nat'l. Acad. Sci. USA* **101:**2736; adapted from model provided by W. A. Goddard.]

Effect on adenylyl cyclase

Exterior — NH_3^+

Cytosol — COO^-

α_2-Adrenergic receptor (wild type) — Inhibits (binds $G_{\alpha i}$)

β_2-Adrenergic receptor (wild type) — Activates (binds $G_{\alpha s}$)

Chimeric receptor 1 — Activates (binds $G_{\alpha s}$)

Chimeric receptor 2 — Inhibits (binds $G_{\alpha i}$)

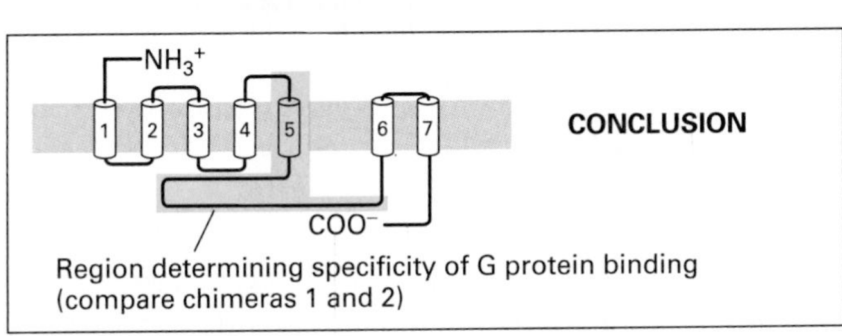

▲ EXPERIMENTAL FIGURE 15-12 **Studies with chimeric adrenergic receptors identify the long C3 loop as critical to interaction with G proteins.** *Xenopus* oocytes were microinjected with mRNA encoding a wild-type α_2-adrenergic, β_2-adrenergic, or chimeric α-β receptors. Although *Xenopus* oocytes do not normally express adrenergic receptors, they do express G proteins that can couple to the foreign receptors expressed on the surface of microinjected oocytes. The adenylyl cyclase activity of the injected cells in the presence of epinephrine agonists was determined and indicated whether the expressed adrenergic receptor bound to the stimulatory ($G_{\alpha s}$) or inhibitory ($G_{\alpha i}$) type of oocyte G protein. Comparison of chimeric receptor 1, which interacts with $G_{\alpha s}$, and chimeric receptor 2, which interacts with $G_{\alpha i}$, shows that the G protein specificity is determined primarily by the source of the cytosol-facing C3 loop (yellow) between α helices 5 and 6. [See B. Kobilka et al., 1988, *Science* **240:**1310.]

These diverse effects of epinephrine help orchestrate integrated responses throughout the body all directed to a common end: supplying energy that can be used for the rapid movement of major locomotor muscles in response to bodily stress.

Although all α- and β-adrenergic receptors bind epinephrine, different receptors are coupled to different G proteins that induce different downstream signaling pathways, leading to different cellular responses. Studies with chimeric adrenergic receptors, like those outlined in Figure 15-12, suggest that the long C3 loop between α helices 5 and 6 is important for interactions between a receptor and its coupled G protein. Presumably, ligand binding causes these helices to move relative to each other in a way that allows the loop to bind and activate the transducing G protein. Other evidence indicates that the C2 loop, joining helices 3 and 4, also contributes to the interaction of some receptors with a G protein.

G Protein–Coupled Receptors Activate Exchange of GTP for GDP on the α Subunit of a Trimeric G Protein

Trimeric G proteins contain three subunits designated α, β, and γ. Both the G_α and G_γ subunits are linked to the membrane by covalently attached lipids. During intracellular signaling, the β and γ subunits remain bound together and are usually referred to as the $G_{\beta\gamma}$ subunit. In the resting state, when no ligand is bound to the receptor, the G_α subunit has a bound GDP and is complexed with $G_{\beta\gamma}$. Binding of a normal hormonal ligand (e.g., epinephrine) or an agonist (e.g., isoproterenol) to a G protein–coupled receptor changes the conformation of its cytosol-facing loops and enables the receptor to bind to the G_α subunit (Figure 15-13, steps **1** and **2**). This binding releases the bound GDP; thus the activated ligand-bound receptor functions as a guanine nucleotide–exchange factor (GEF) for the G_α subunit (step **3**). Next GTP rapidly binds to the "empty" guanine nucleotide site in the G_α subunit, causing a change in the conformation of its switch segments (see Figure 15-8). These changes weaken the binding of G_α with both the receptor and the $G_{\beta\gamma}$ subunit (step **4**).

In most cases, G_α·GTP, which remains anchored in the membrane, then interacts with and activates an associated effector protein, as depicted in Figure 15-13 (step **5**). In some cases, G_α·GTP inhibits the effector. Moreover, depending on the type of cell and G protein, the $G_{\beta\gamma}$ subunit, freed from its α subunit, will sometimes transduce a signal by interacting with an effector protein.

In any case, the active G_α·GTP state is short-lived because the bound GTP is hydrolyzed to GDP in minutes, catalyzed by the intrinsic GTPase activity of the G_α subunit (see Figure 15-13, step **6**). The conformation of the G_α then switches back to the inactive G_α·GDP state, blocking any further activation of effector proteins. The rate of GTP hydrolysis is further enhanced by binding of the G_α·GTP complex to the effector; the effector thus functions as a GTPase-activating protein (GAP). This mechanism significantly reduces the duration of effector activation and avoids a cellular overreaction. In many cases, a noneffector RGS protein also accelerates GTP hydrolysis by the G_α subunit, further reducing the time during which the effector remains activated. The resulting G_α·GDP quickly reassociates with $G_{\beta\gamma}$ and the complex becomes ready to interact with an activated receptor and

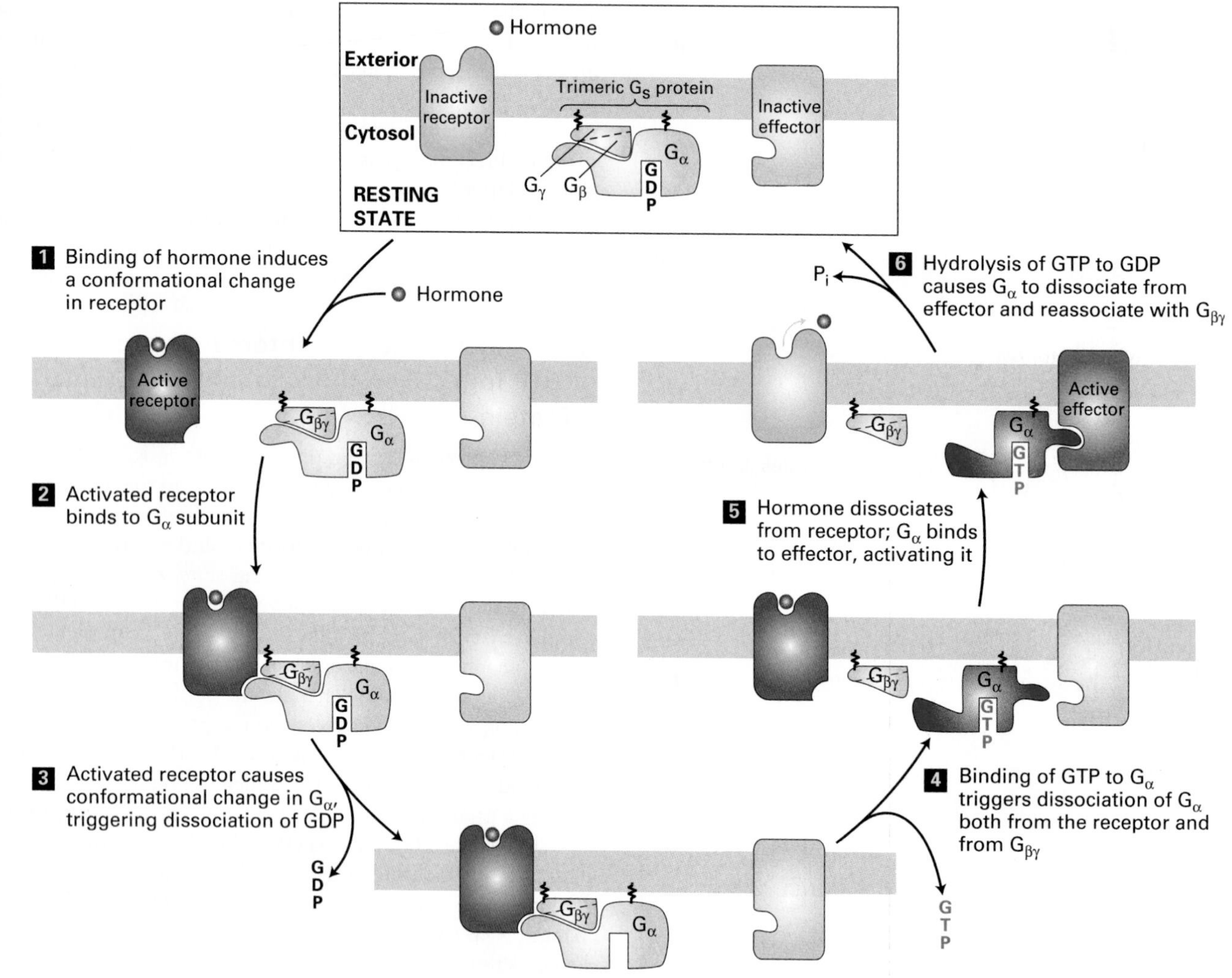

▲ **FIGURE 15-13 General mechanism of the activation of effector proteins associated with G protein–coupled receptors.** The G_α and $G_{\beta\gamma}$ subunits of trimeric G proteins are tethered to the membrane by covalently attached lipid molecules (wiggly black lines). Following ligand binding, exchange of GDP with GTP, and dissociation of the G protein subunits, (steps 1–4), the free G_α·GTP binds to and activates an effector protein (step 5). Hydrolysis of GTP terminates signaling and leads to reassembly of the trimeric G protein, returning the system to the resting state (step 6). Binding of another ligand molecule causes repetition of the cycle. In some pathways, the effector protein is activated by the free $G_{\beta\gamma}$ subunit. [After W. Oldham and H. Hamm, 2006, (*Quart. Rev. Biophys.* **40**: (In press)].

start the process all over again. Thus the GPCR signal-transduction system contains a built-in feedback mechanism that ensures the effector protein becomes activated only for a few seconds or minutes following receptor activation; continual activation of receptors via ligand binding is essential for prolonged activation of the effector.

Early evidence supporting the model shown in Figure 15-13 came from studies with compounds that can bind to G_α subunits as well as GTP does, but cannot be hydrolyzed by the intrinsic GTPase. In some of these compounds, the P–O–P phosphodiester linkage connecting the β and γ phosphates of GTP is replaced by a nonhydrolyzable P–CH_2–P or P–NH–P linkage. Addition of such a GTP analog to a plasma membrane preparation in the presence of the natural ligand or an agonist for a particular receptor results in a much longer-lived activation of the associated effector protein than occurs with GTP. In this experiment, once the nonhydrolyzable GTP analog is exchanged for GDP bound to G_α, it remains permanently bound to G_α. Because the G_α·analog complex is as functional as the normal G_α·GTP complex in activating the effector protein, the effector remains permanently active.

Podcast: Activation of G Proteins Measured by Fluorescence Resonance Energy Transfer (FRET)

MEDIA CONNECTIONS

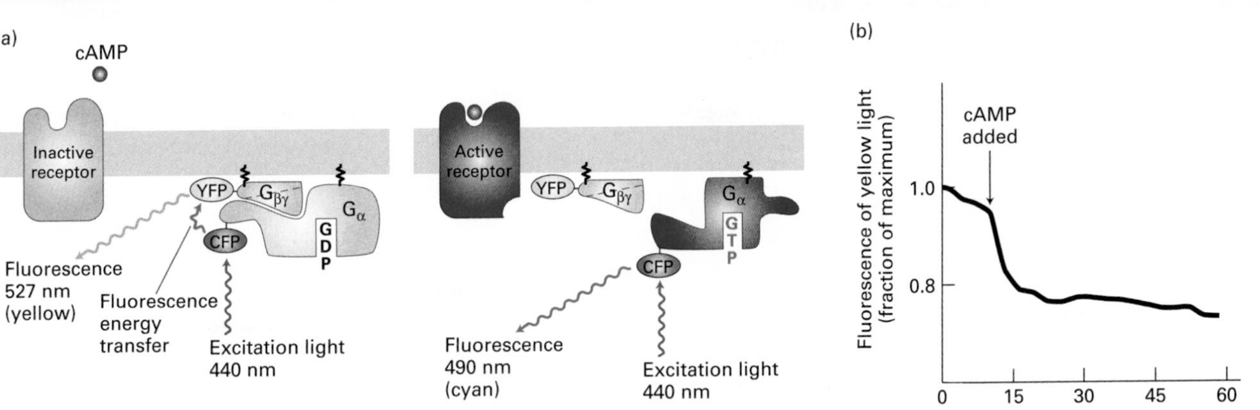

▲ **EXPERIMENTAL FIGURE 15-14 Activation of G proteins occurs within seconds of ligand binding in amoeba cells.** In the amoeba *Dictyostelium discoideum* cell, cAMP acts as an extracellular signaling molecule and binds to a G protein–coupled receptor; it is not a second messenger. Amoeba cells were transfected with genes encoding two fusion proteins: a G_α fused to cyan fluorescent protein (CFP), a mutant form of green fluorescent protein (GFP), and a G_β fused to another GFP variant, yellow fluorescent protein (YFP). CFP normally fluoresces 490-nm light; YFP, 527-nm light. (a) When CFP and YFP are nearby, as in the resting $G_\alpha \cdot G_{\beta\gamma}$ complex, fluorescence energy transfer can occur between CFP and YFP *(left)*. As a result, irradiation of resting cells with 440-nm light (which directly excites CFP but not YFP) causes emission of 527-nm (yellow) light, characteristic of YFP. However, if ligand binding leads to dissociation of the G_α and $G_{\beta\gamma}$ subunits, then fluorescence energy transfer cannot occur. In this case, irradiation of cells at 440 nm causes emission of 490-nm light (cyan) characteristic of CFP *(right)*. (b) Plot of the emission of yellow light (527 nm) from a single transfected amoeba cell before and after addition of extracellular cyclic AMP (arrows), the ligand for the G protein–coupled receptor in these cells. The drop in yellow fluorescence, which results from the dissociation of the G_α-CFP fusion protein from the $G_{\beta\gamma}$-YFP fusion protein, occurs within seconds of cAMP addition. [Adapted from C. Janetopoulos et al., 2001, *Science* **291:**2408.]

GPCR-mediated dissociation of trimeric G proteins recently has been detected in living cells. These studies have exploited the phenomenon of *fluorescence energy transfer,* which changes the wavelength of emitted fluorescence when two fluorescent proteins interact. Figure 15-14 shows how this experimental approach has demonstrated the dissociation of the $G_\alpha \cdot G_{\beta\gamma}$ complex within a few seconds of ligand addition, providing further evidence for the model of G protein cycling. This general experimental protocol can be used to follow the formation and dissociation of other protein–protein complexes in living cells.

Different G Proteins Are Activated by Different GPCRs and In Turn Regulate Different Effector Proteins

All effector proteins in GPCR pathways are either membrane-bound ion channels or enzymes that catalyze formation of the second messengers shown in Figure 15-9. The variations on the theme of GPCR signaling that we examine in Sections 15.5–15.7 arise because multiple G proteins are encoded in eukaryotic genomes. At last count humans have 21 different G_α subunits encoded by 16 genes, several of which undergo alternative splicing; 6 G_β subunits; and 12 G_γ subunits. So far as is known, the different $G_{\beta\gamma}$ subunits function similarly.

Table 15-1 summarizes the functions of the major classes of G proteins with different G_α subunits. For example, the different types of epinephrine receptors mentioned previously are coupled to different G proteins that influence effectors differently, and thus have distinct effects on cell behavior in a target cell. Both subtypes of β-adrenergic receptors, termed β_1 and β_2, are coupled to a *stimulatory* G protein (G_s) whose alpha subunit ($G_{\alpha s}$) activates a membrane-bound effector enzyme called **adenylyl cyclase.** Once activated, this enzyme catalyzes synthesis of the second messenger cAMP. In contrast, the α_2-adrenergic receptor is coupled to a $G_{\alpha i}$ protein that inhibits adenylyl cyclase, the same effector enzyme associated with β-adrenergic receptors. The $G_{\alpha q}$ protein, which is coupled to the α_1-adrenergic receptor, activates a different effector enzyme, **phospholipase C,** that generates two other second messengers (DAG and IP_3). Examples of signaling pathways that use each of the G_α proteins listed in Table 15-1 are described in the following three sections.

Some bacterial toxins contain a subunit that penetrates the plasma membrane of target mammalian cells and in the cytosol catalyzes a chemical modification of G_α proteins that prevents hydrolysis of bound GTP to GDP. For example, toxins produced by the bacterium *Vibrio cholera,* which causes cholera, or certain strains of *E. coli* modify the $G_{\alpha s}$ protein in intestinal epithelial cells.

TABLE 15-1 Major Classes of Mammalian Trimeric G Proteins and Their Effectors*

G_α CLASS	ASSOCIATED EFFECTOR	2ND MESSENGER	RECEPTOR EXAMPLES
$G_{\alpha s}$	Adenylyl cyclase	cAMP (increased)	β-Adrenergic (epinephrine) receptor; receptors for glucagon, serotonin, vasopressin
$G_{\alpha i}$	Adenylyl cyclase K^+ channel ($G_{\beta\gamma}$ activates effector)	cAMP (decreased) Change in membrane potential	α_2-Adrenergic receptor Muscarinic acetylcholine receptor
$G_{\alpha olf}$	Adenylyl cyclase	cAMP (increased)	Odorant receptors in nose
$G_{\alpha q}$	Phospholipase C	IP_3, DAG (increased)	α_1-Adrenergic receptor
$G_{\alpha o}$	Phospholipase C	IP_3, DAG (increased)	Acetylcholine receptor in endothelial cells
$G_{\alpha t}$	cGMP phosphodiesterase	cGMP (decreased)	Rhodopsin (light receptor) in rod cells

*A given G_α subclass may be associated with more than one effector protein. To date, only one major $G_{\alpha s}$ has been identified, but multiple $G_{\alpha q}$ and $G_{\alpha i}$ proteins have been described. Effector proteins commonly are regulated by G_α but in some cases by $G_{\beta\gamma}$ or the combined action of G_α and $G_{\beta\gamma}$. IP_3 = inositol 1,4,5-trisphosphate; DAG = 1,2-diacylglycerol.

SOURCES: See L. Birnbaumer, 1992, *Cell* **71**:1069; Z. Farfel et al., 1999, *New Eng. J. Med.* **340**:1012; and K. Pierce et al., 2002, *Nature Rev. Mol. Cell Biol.* **3**:639.

As a result, $G_{\alpha s}$ remains in the active state, continuously activating the effector adenylyl cyclase in the absence of hormonal stimulation. The resulting excessive rise in intracellular cAMP leads to the loss of electrolytes and water into the intestinal lumen, producing the watery diarrhea characteristic of infection by these bacteria. The toxin produced by *Bordetella pertussis*, a bacterium that commonly infects the respiratory tract and causes whooping cough, catalyzes a modification of $G_{\alpha i}$ that prevents release of bound GDP. As a result, $G_{\alpha i}$ is locked in the inactive state, reducing the inhibition of adenylyl cyclase. The resulting increase in cAMP in epithelial cells of the airways promotes loss of fluids and electrolytes and mucus secretion ▮

KEY CONCEPTS OF SECTION 15.4

General Elements of G Protein-Coupled Receptor Systems

- G protein–coupled receptors are a large and diverse family with a common structure of seven membrane-spanning α helices.
- Trimeric G proteins transduce signals from coupled cell-surface receptors to associated membrane-bound effector proteins, which are either enzymes that form second messengers (e.g., adenylyl cyclase) or ion channel proteins (see Table 15-1).
- Signals most commonly are transduced by G_α, a GTPase switch protein that alternates between an active ("on") state with bound GTP and inactive ("off") state with GDP. The β and γ subunits, which remain bound together, occasionally transduce signals.
- Hormone-occupied receptors act as guanine nucleotide–exchange factors (GEFs) for G_α proteins, catalyzing dissociation of GDP and enabling GTP to bind. The resulting change in conformation of switch regions in G_α causes it to dissociate from the $G_{\beta\gamma}$ subunit and the receptor and interact with an effector protein (see Figure 15-13).
- Fluorescence energy transfer experiments demonstrate receptor-mediated dissociation of coupled G_α and $G_{\beta\gamma}$ subunits in living cells (see Figure 15-14).

15.5 G Protein–Coupled Receptors That Regulate Ion Channels

One of the simplest cellular responses to a signal is the opening of ion channels essential for transmission of nerve impulses. Nerve impulses are essential to the sensory perception of environmental stimuli such as light and odors, to transmission of information to and from the brain, and to the stimulation of muscle movement. During transmission of nerve impulses, the rapid opening and closing of ion channels causes changes in the membrane potential. Many neurotransmitter receptors are simply ligand-gated ion channels, which open in response to binding of a ligand. Such receptors include some types of glutamate, serotonin, and acetylcholine receptors, including the acetylcholine receptor found at nerve–muscle synapses. Ligand-gated ion channels that function as neurotransmitter receptors are covered in Chapter 23.

Some neurotransmitter receptors, however, are G protein–coupled receptors whose effector proteins are a Na^+ or K^+ channel. Neurotransmitter binding to these receptors causes the associated ion channel to open or close, leading to changes in the membrane potential. Still other neurotransmitter receptors, as well as odorant receptors in the nose and photoreceptors in the eye, are G protein–coupled receptors that indirectly modulate the activity of ion channels via the action of second messengers. In this section, we consider two G protein–coupled receptors that illustrate the direct and indirect mechanisms for regulating ion channels: the muscarinic acetylcholine receptor of the heart and the light-activated rhodopsin of the eye.

Acetylcholine Receptors in the Heart Muscle Activate a G Protein That Opens K^+ Channels

Activation of *muscarinic acetylcholine receptors* in cardiac muscle *slows* the rate of heart muscle contraction. (Because muscarine, an acetylcholine analog, also activates these receptors, they are termed "muscarinic.") This type of acetylcholine receptor is coupled to a $G_{\alpha i}$ protein, and ligand binding leads to opening of associated K^+ channels (the effector protein) in the plasma membrane (see Table 15-1). The subsequent efflux of K^+ ions from the cytosol causes an increase in the magnitude of the usual inside-negative potential across the plasma membrane that lasts for several seconds. This state of the membrane, called **hyperpolarization,** reduces the frequency of muscle contraction. This effect can be determined experimentally by direct addition of acetylcholine to isolated heart muscle cells and measurement of the potential using a microelectrode inserted into the cell (see Figure 11-18).

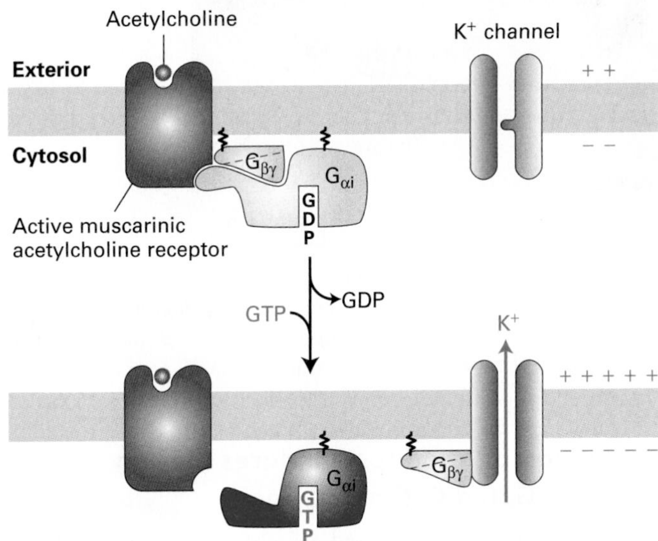

▲ **FIGURE 15-15 Activation of the muscarinic acetylcholine receptor and its effector K^+ channel in heart muscle.** Binding of acetylcholine triggers activation of the $G_{\alpha i}$ subunit and its dissociation from the $G_{\beta\gamma}$ subunit in the usual way (see Figure 15-13). In this case, the released $G_{\beta\gamma}$ subunit (rather than $G_{\alpha i}$·GTP) binds to and opens the associated effector protein, a K^+ channel. The increase in K^+ permeability hyperpolarizes the membrane, which reduces the frequency of heart muscle contraction. Though not shown here, activation is terminated when the GTP bound to $G_{\alpha i}$ is hydrolyzed to GDP and $G_{\alpha i}$·GDP recombines with $G_{\beta\gamma}$. [See K. Ho et al., 1993, *Nature* **362:**31, and Y. Kubo et al., 1993, *Nature* **362:**127.]

As depicted in Figure 15-15, the signal from activated muscarinic acetylcholine receptors is transduced to the effector protein by the released $G_{\beta\gamma}$ subunit rather than by G_α·GTP. That $G_{\beta\gamma}$ directly activates the K^+ channel was demonstrated by patch-clamping experiments, which can measure ion flow through a single ion channel in a small patch of membrane (see Figure 11-21). When purified $G_{\beta\gamma}$ protein was added to the cytosolic face of a patch of heart muscle plasma membrane, K^+ channels opened immediately, even in the absence of acetylcholine or other neurotransmitters—clearly indicating that it is the $G_{\beta\gamma}$ protein that is responsible for opening the effector K^+ channels and not G_α·GTP.

Light Activates $G_{\alpha t}$-Coupled Rhodopsins

The human retina contains two types of photoreceptor cells, *rods* and *cones,* which are the primary recipients of visual stimulation. Cones are involved in color vision, while rods are stimulated by weak light like moonlight over a range of wavelengths. The photoreceptors synapse on layer upon layer of interneurons that are innervated by different combinations of photoreceptor cells. All these signals are processed and interpreted by the part of the brain called the *visual cortex.*

As noted already, rhodopsin consists of the protein opsin, which has the usual GPCR structure, covalently linked to the light-absorbing pigment 11-*cis*-retinal. Rhodopsin is localized to the thousand or so flattened membrane disks that make up the outer segment of rod cells (Figure 15-16). A human rod cell contains about 4×10^7 molecules of rhodopsin. The trimeric G protein coupled to rhodopsin, called *transducin* (G_t), contains the $G_{\alpha t}$ subunit (see Table 15-1); like rhodopsin, $G_{\alpha t}$ is found only in rod cells.

Upon absorption of a photon, the retinal moiety of rhodopsin is immediately converted from the cis to the all-trans isomer, causing a conformational change in the opsin portion that activates it (Figure 15-17). This is equivalent to the conformational change that occurs upon ligand binding by other G protein–coupled receptors. Analogous to other G protein–coupled receptors, the light-activated form of rhodopsin interacts with and activates a G_α protein, in this case $G_{\alpha t}$. Activated opsin is unstable and spontaneously dissociates into its component parts, releasing opsin and all-*trans*-retinal, thereby terminating visual signaling. In the dark, free all-*trans*-retinal is converted back to 11-*cis*-retinal, which can then rebind to opsin, re-forming rhodopsin.

In the dark, the membrane potential of a rod cell is about −30 mV, considerably less than the resting potential (−60 to −90 mV) typical of neurons and other electrically active cells. This state of the membrane, called **depolarization,** causes rod cells in the dark to constantly secrete neurotransmitters, and thus the neurons with which they synapse are continually being stimulated. The depolarized state of the plasma membrane of resting rod cells is due to the presence of a large number of open *nonselective* ion channels that admit Na^+ and Ca^{2+}, as well as K^+. Absorption

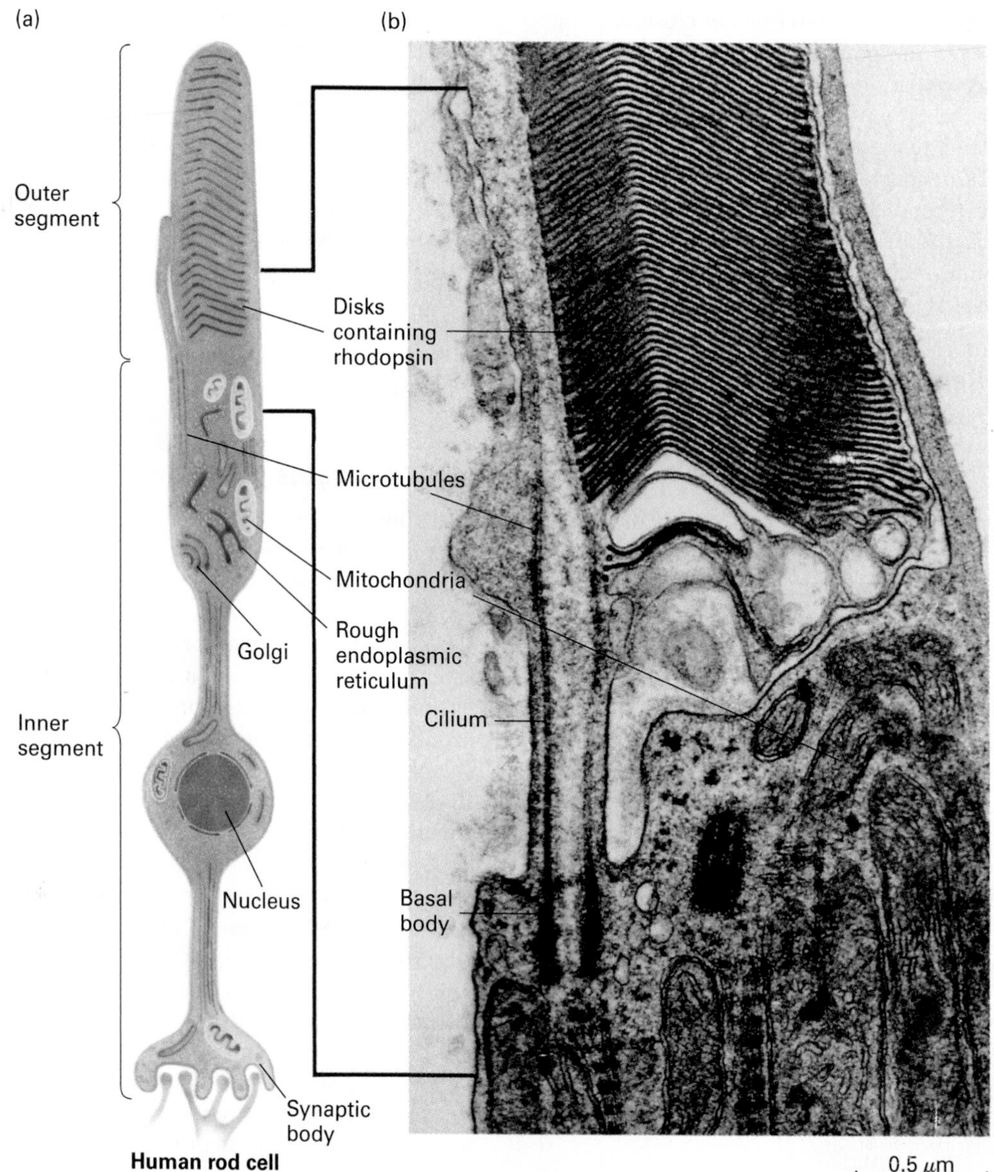

▲ **FIGURE 15-16 Human rod cell.** (a) Schematic diagram of an entire rod cell. At the synaptic body, the rod cell forms synapses with one or more bipolar interneurons. Rhodopsin, a light-sensitive G protein–coupled receptor, is located in the flattened membrane disks of the outer segment. (b) Electron micrograph of the region of the rod cell indicated by the bracket in (a). This region includes the junction of the inner and outer segments. [Part (b) from R. G. Kessel and R. H. Kardon, 1979, *Tissues and Organs: A Text-Atlas of Scanning Electron Microscopy,* W. H. Freeman and Company, p. 91.]

of light by rhodopsin leads to closing of these channels, causing the membrane potential to become *more* inside negative.

The more photons absorbed by rhodopsin, the more channels are closed, the fewer Na^+ and Ca^{2+} ions cross the membrane from the outside, the more negative the membrane potential becomes, and the less neurotransmitter is released. This change is transmitted to the brain where it is perceived as light. Remarkably, a single photon absorbed by a resting rod cell produces a measurable response, a more inside negative change in the membrane potential of about 1 mV, which in amphibians lasts a second or two. Humans are able to detect a flash of as few as five photons.

Activation of Rhodopsin Induces Closing of cGMP-Gated Cation Channels

Opening of GPCR-stimulated K^+ channels in the heart requires only an activated G protein (see Figure 15-15). In contrast, the closing of cation channels in the rod-cell plasma membrane requires changes in the concentration of the second messenger **cyclic GMP,** or **cGMP** (see Figure 15-9). Rod outer segments contain an unusually high concentration ($\approx$0.07 mM) of cGMP, which is continuously formed from GTP in a reaction catalyzed by guanylyl cyclase, which appears to be unaffected by light. However, light absorption by rhodopsin induces activation of a *cGMP phosphodiesterase,* which hydrolyzes cGMP to 5′-GMP. As a result, the cGMP

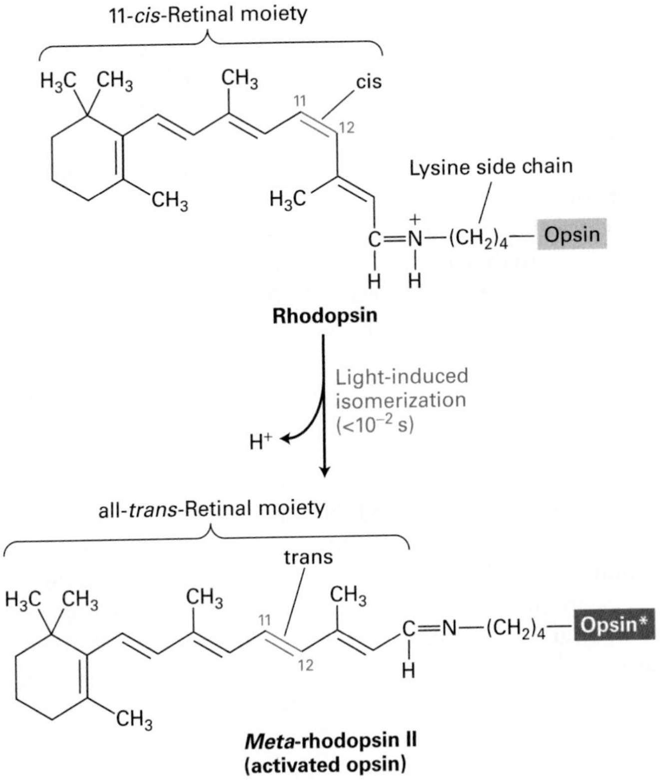

11-*cis*-Retinal moiety

Lysine side chain

Opsin

Rhodopsin

Light-induced isomerization ($<10^{-2}$ s)

H^+

all-*trans*-Retinal moiety

Opsin*

***Meta*-rhodopsin II (activated opsin)**

◀ **FIGURE 15-17 The light-triggered step in vision.** The light-absorbing pigment 11-*cis*-retinal is covalently bound to the amino group of a lysine residue in opsin, the protein portion of rhodopsin. Absorption of light causes rapid photoisomerization of the bound *cis*-retinal to the all-*trans* isomer, forming the unstable intermediate *meta*-rhodopsin II, or activated opsin, which activates G_t proteins. Within seconds, all-*trans*-retinal dissociates from opsin and is converted by an enzyme back to the *cis* isomer, which then rebinds to another opsin molecule. [See J. Nathans, 1992, *Biochemistry* **31**:4923.]

concentration decreases upon illumination. The high level of cGMP present in the dark acts to keep *cGMP-gated cation channels* open; the light-induced decrease in cGMP leads to channel closing, membrane hyperpolarization, and reduced neurotransmitter release.

As depicted in Figure 15-18, cGMP phosphodiesterase is the effector protein for $G_{\alpha t}$. The free $G_{\alpha t}$·GTP complex that is generated after light absorption by rhodopsin binds to the two inhibitory γ subunits of cGMP phosphodiesterase, releasing the active catalytic α and β subunits, which then convert cGMP to GMP. This is a clear example of how signal-induced removal of an inhibitor can quickly activate an enzyme, a common mechanism in signaling pathways. A single molecule of activated opsin in the disk membrane can activate 500 $G_{\alpha t}$ molecules, each of which in turn activates one cGMP phosphodiesterase, thereby amplifying the original light signal early in this pathway.

Direct support for the role of cGMP in rod-cell activity has been obtained in patch-clamping studies using isolated patches of rod outer-segment plasma membrane, which contains abundant cGMP-gated cation channels. When cGMP is added to the cytosolic surface of these patches, there is a rapid increase in the number of open ion channels; cGMP binds directly to a site on the channel protein to keep them open, indicating that these are nucleotide-gated channels.

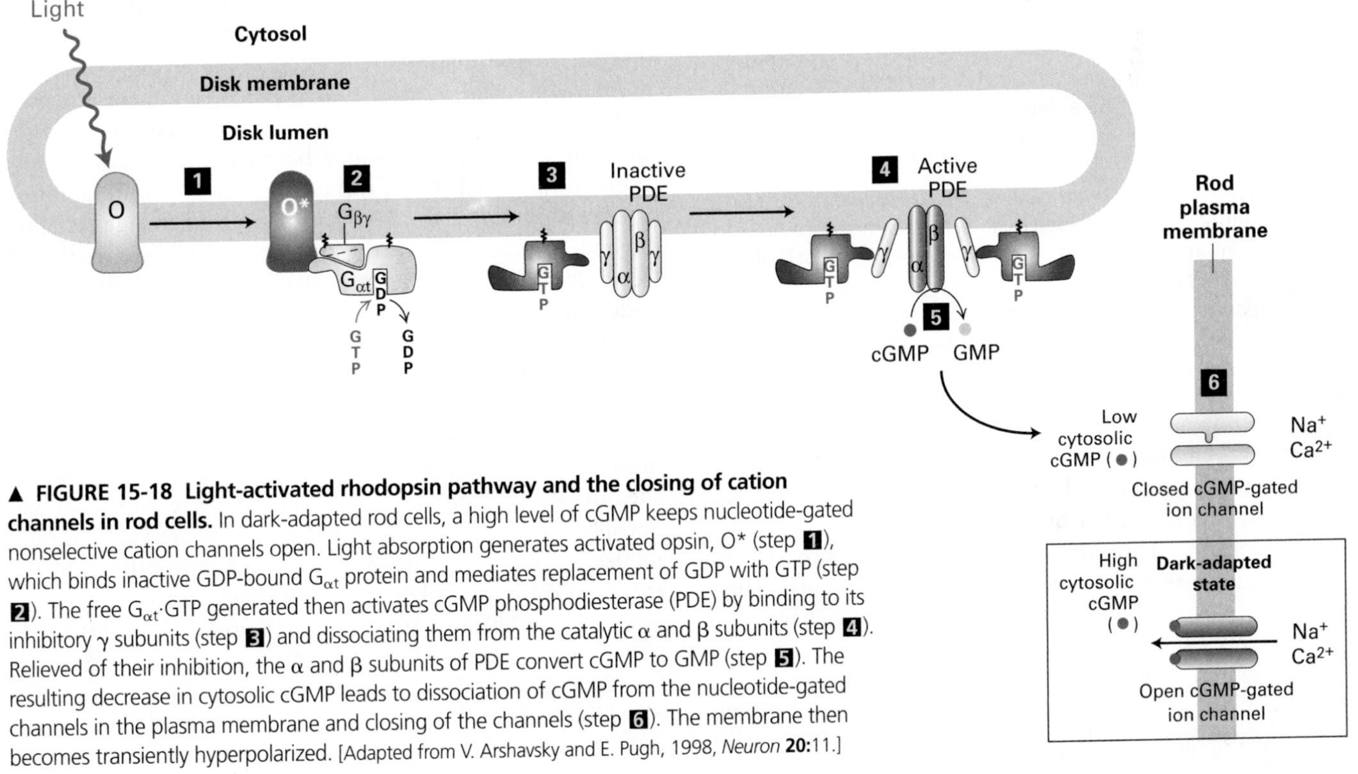

▲ **FIGURE 15-18 Light-activated rhodopsin pathway and the closing of cation channels in rod cells.** In dark-adapted rod cells, a high level of cGMP keeps nucleotide-gated nonselective cation channels open. Light absorption generates activated opsin, O* (step **1**), which binds inactive GDP-bound $G_{\alpha t}$ protein and mediates replacement of GDP with GTP (step **2**). The free $G_{\alpha t}$·GTP generated then activates cGMP phosphodiesterase (PDE) by binding to its inhibitory γ subunits (step **3**) and dissociating them from the catalytic α and β subunits (step **4**). Relieved of their inhibition, the α and β subunits of PDE convert cGMP to GMP (step **5**). The resulting decrease in cytosolic cGMP leads to dissociation of cGMP from the nucleotide-gated channels in the plasma membrane and closing of the channels (step **6**). The membrane then becomes transiently hyperpolarized. [Adapted from V. Arshavsky and E. Pugh, 1998, *Neuron* **20**:11.]

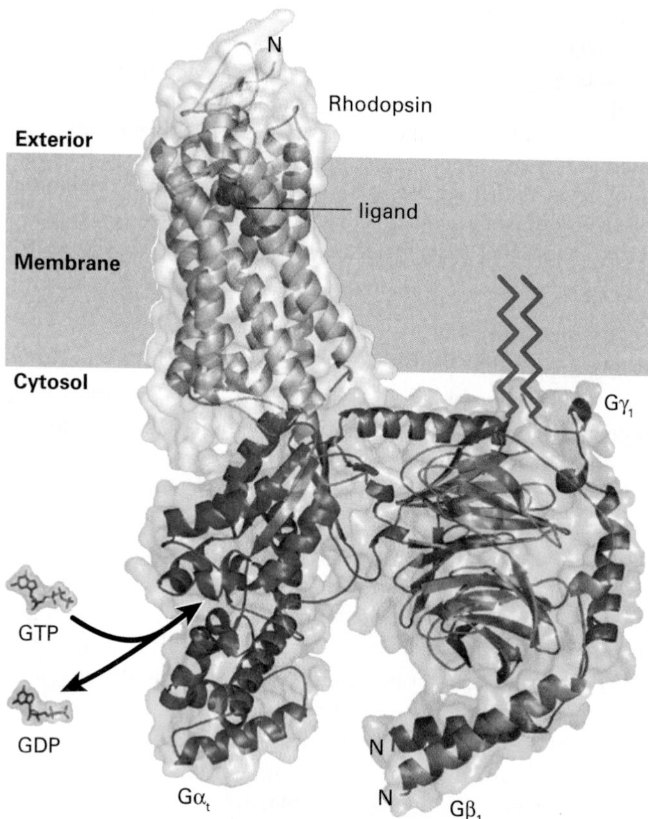

▲ **FIGURE 15-19 Structural model of rhodopsin and its associated trimeric G protein, transducin (G_t).** The structures of rhodopsin and the $G_{\alpha t}$ and $G_{\beta\gamma}$ subunits were obtained by x-ray crystallography. The C-terminal segment of rhodopsin that follows transmembrane helix H7 and extends into the cytosol is not shown in this model. The orientation of $G_{\alpha t}$ with respect to rhodopsin and the membrane is hypothetical; it is based on the charge and hydrophobicity of the protein surfaces and the known rhodopsin-binding sites on $G_{\alpha t}$. As in other trimeric G proteins, the $G_{\alpha t}$ and G_γ subunits contain covalently attached lipids (red and blue jagged lines) that are thought to be inserted into the membrane. In the GDP-bound form shown here, the α subunit (blue) and the β subunit (green) interact with each other, as do the β subunit and γ subunit (red), but the small γ subunit, which contains just two α helices, does not contact the α subunit. Several segments of the α subunit are thought to interact with an activated rhodopsin, causing a conformational change that promotes release of GDP and binding of GTP. Binding of GTP, in turn, induces large conformational changes in the switch regions of $G_{\alpha t}$ leading to its dissociation from $G_{\beta\gamma}$. [Adapted from H. Hamm, 2001, *Proc. Nat'l. Acad. Sci. USA* **98:**4819, and W. Oldham and H. Hamm 2006 *Quart. Rev. Biophys.* **40:** (In press).]

Like the K^+ channels discussed in Chapter 11, the cGMP-gated channel protein contains four subunits [see Figure 11-19]. In this case each of the subunits is able to bind a cGMP molecule. Three or four cGMP molecules must bind per channel in order to open it; this allosteric interaction makes channel opening very sensitive to small changes in cGMP levels.

Conversion of active $G_{\alpha t}$·GTP back to inactive $G_{\alpha t}$·GDP is accelerated by a specific GTPase-activating protein (GAP). In mammals $G_{\alpha t}$ normally remains in the active GTP-bound state for only a fraction of a second. Thus cGMP phosphodiesterase rapidly becomes inactivated, and the cGMP level gradually rises to its original level when the light stimulus is removed. This allows rapid responses of the eye toward moving or changing objects.

Recent x-ray crystallographic studies reveal how the subunits of G_t protein interact with each other and with light-activated rhodopsin and provide clues about how binding of GTP leads to dissociation of G_α from $G_{\beta\gamma}$. As revealed in the structural model in Figure 15-19, two surfaces of $G_{\alpha t}$ interact with G_β: an N-terminal region near the membrane surface and the two adjacent switch I and switch II regions, which are found in all G_α proteins. G_γ directly contacts G_β but not $G_{t\alpha}$. These models also suggest that the nucleotide-binding domain of $G_{\alpha t}$, together with the lipid anchors at the C-terminus of G_γ and the N-terminus of $G_{\alpha t}$, form a surface that binds to light-activated rhodopsin (O* in Figure 15-18), promoting the release of GDP from $G_{\alpha t}$ and the subsequent binding of GTP. The subsequent conformational changes in $G_{t\alpha}$, particularly those within switches I and II, disrupt the molecular interactions between $G_{\alpha t}$ and $G_{\beta\gamma}$, leading to their dissociation. The structural studies with rhodopsin and $G_{\alpha t}$ are consistent with data concerning other G protein–coupled receptors and are thought to be generally applicable to all receptors of this type.

Rod Cells Adapt to Varying Levels of Ambient Light Because of Opsin Phosphorylation and Binding of Arrestin

Cone cells are insensitive to low levels of illumination, and the activity of rod cells is inhibited at high light levels. Thus when we move from bright daylight into a dimly lighted room, we are initially blinded. As the rod cells slowly become sensitive to the dim light, we gradually are able to see and distinguish objects. This process of *visual adaptation* permits a rod cell to perceive contrast over a 100,000-fold range of ambient light levels. This wide range of sensitivity is possible because differences in light levels in the visual field, rather than the absolute amount of absorbed light, are used to form visual images. Light-dependent regulation of the rhodopsin signaling pathway (see Figure 15-18) is responsible for this extraordinarily wide sensitivity range.

One process contributing to visual adaptation involves phosphorylation of opsin in its active conformation (O*) but not in its inactive, or dark form (O) by *rhodopsin kinase* (Figure 15-20), a member of a class of GPCR kinases. Each opsin molecule has three principal serine phosphorylation sites on its cytosol-facing surface; the more sites that are phosphorylated, the less able O* is to activate $G_{\alpha t}$ and thus induce closing of cGMP-gated cation channels. Because the extent of opsin phosphorylation by rhodopsin kinase is proportional to the amount of time each opsin molecule spends in the light-activated form, it is a measure of the background (ambient) level of light. Under bright-light conditions, opsin phosphorylation is increased, and consequently its ability to activate $G_{\alpha t}$ is reduced. In other words, rhodopsin is desensitized

Overview Animation: Extracellular Signaling

MEDIA CONNECTIONS

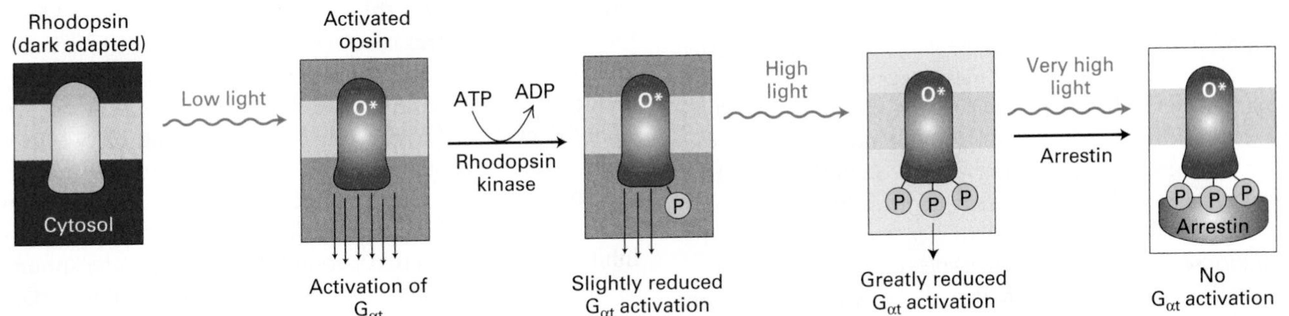

▲ **FIGURE 15-20 Rod-cell adaptation to ambient light level changes and opsin phosphorylation.** Light-activated opsin (O*), but not dark-adapted rhodopsin, is a substrate for rhodopsin kinase. The extent of opsin phosphorylation is directly proportional to the amount of time each opsin molecule spends in the light-activated form and thus to the average ambient light level over the previous few minutes. The ability of O* to activate $G_{\alpha t}$ is inversely proportional to the number of phosphorylated residues on O*. Thus the higher the ambient light level, the greater the extent of opsin phosphorylation and the larger the increase in light level needed to activate the same number of $G_{\alpha t}$ molecules. At very high light levels, arrestin binds to the completely phosphorylated opsin, forming a complex that cannot activate $G_{\alpha t}$ at all. [See L. Lagnado and D. Baylor, 1992, *Neuron* **8**:995, and A. Mendez et al., 2000, *Neuron* **28**:153.]

by bright light, and thus a greater increase in light intensity is necessary to generate a change in cGMP levels and a visual signal. When the level of ambient light is reduced, the opsins become dephosphorylated and the ability to activate $G_{\alpha t}$ increases; in this case, relatively fewer additional photons are necessary to generate a visual signal. The importance of opsin phosphorylation in visual adaptation is supported by studies with rod cells from mice with mutant rhodopsins bearing zero or only one of the target serine residues. These rod cells show a much slower than normal rate of deactivation in bright light.

The light-dependent desensitization of rod cells is further increased by binding of the cytosolic protein *β-arrestin.* At high ambient light (such as noontime outdoors), β-arrestin binds to the phosphorylated serine residues on the C-terminal opsin segment. Bound β-arrestin completely prevents interaction of $G_{\alpha t}$ with phosphorylated O*, totally blocking formation of the active $G_{\alpha t}$·GTP complex and causing a shutdown of all rod-cell activity.

The negative-feedback regulation of rod-cell activity by rhodopsin kinase and arrestin is similar to adaptation (or desensitization) of other G protein–coupled receptors to high ligand levels. Another mechanism, which appears unique to rod cells, also contributes to visual adaptation. In dark-adapted cells virtually all the $G_{\alpha t}$ and $G_{\beta\gamma}$ subunits are in the outer segments. But exposure for 10 minutes to moderate daytime intensities of light causes over 80 percent of the $G_{\alpha t}$ and $G_{\beta\gamma}$ subunits to move out of the outer segments into other cellular compartments. Although the mechanism by which these proteins move is not yet known, the result is that $G_{\alpha t}$ proteins are physically unable to bind activated opsin. As occurs in other signaling pathways, multiple mechanisms are thus used to inactivate signaling during visual adaptation, presumably to allow strict control of activation of the signaling pathway over broad ranges of illumination.

KEY CONCEPTS OF SECTION 15.5

G Protein–Coupled Receptors That Regulate Ion Channels

- The cardiac muscarinic acetylcholine receptor is a GPCR whose effector protein is a K^+ channel. Receptor activation causes release of the $G_{\beta\gamma}$ subunit, which opens K^+ channels (see Figure 15-15). The resulting hyperpolarization of the cell membrane slows the rate of heart muscle contraction.
- Rhodopsin, the photosensitive GPCR in rod cells, comprises the protein opsin linked to 11-*cis*-retinal. Light-induced isomerization of the 11-*cis*-retinal moiety produces activated opsin, which then activates the coupled trimeric G protein transducin (G_t) by catalyzing exchange of free GTP for bound GDP on the $G_{\alpha t}$ subunit.
- The effector protein in the rhodopsin pathway is cGMP phosphodiesterase, which is activated by the $G_{\alpha t}$·GTP–mediated release of inhibitory subunits. Reduction in the cGMP level by this enzyme leads to closing of cGMP-gated Na^+/Ca^{2+} channels, hyperpolarization of the membrane, and decreased release of neurotransmitter (see Figure 15-18).
- As with other G_α proteins, binding of GTP to $G_{\alpha t}$ causes conformational changes in the protein that disrupt its molecular interactions with $G_{\beta\gamma}$ and enable $G_{\alpha t}$·GTP to bind to its downstream effector.
- Phosphorylation of light-activated opsin by rhodopsin kinase interferes with its ability to activate $G_{\alpha t}$; subsequent binding of arrestin to phosphorylated opsin further inhibits its ability to activate $G_{\alpha t}$ (see Figure 15-20). This general mechanism of adaptation, or desensitization, is utilized by other GPCRs at high ligand levels.

15.6 G Protein–Coupled Receptors That Activate or Inhibit Adenylyl Cyclase

GPCR pathways that utilize adenylyl cyclase as an effector protein and cAMP as the second messenger are found in most mammalian cells. These pathways follow the general GPCR mechanism outlined in Figure 15-13: Ligand binding to the receptor activates a coupled trimeric G protein that activates adenylyl cyclase, which synthesizes the diffusible second messenger cAMP. cAMP, in turn, activates a cAMP-dependent protein kinase that phosphorylates specific target proteins.

To explore this GPCR/cAMP pathway we focus on the first such pathway discovered—the hormone-stimulated generation of glucose from glycogen, a storage polymer of glucose. The breakdown of glycogen *(glycogenolysis)* occurs in muscle and liver cells in response to epinephrine, glucagon, and other hormones whose receptors are coupled to the $G_{\alpha s}$ protein (see Table 15-1). This system exemplifies how activation of a pathway can coordinate the activity of a group of intracellular enzymes toward a common purpose, the release of glucose from its stored form.

Adenylyl Cyclase Is Stimulated and Inhibited by Different Receptor–Ligand Complexes

Under conditions where demand for glucose is high because of low blood sugar, glucagon is released by the pancreatic islets; in case of sudden stress, epinephrine is released by the adrenal glands. Both glucagon and epinephrine signal cells to break down glycogen, releasing individual glucose molecules. In the liver, glucagon and epinephrine bind to different G protein–coupled receptors, but both receptors interact with and activate the same $G_{\alpha s}$ that activates adenylyl cyclase. Hence, both hormones induce the same metabolic responses. Activation of adenylyl cyclase, and thus the cAMP level, is proportional to the total concentration of $G_{\alpha s}$·GTP resulting from binding of both hormones to their respective receptors.

Positive (activation) and negative (inhibition) regulation of adenylyl cyclase activity occurs in many cell types, providing fine-tuned control of the cAMP level (Figure 15-21). For example, the breakdown of triacylglycerols to fatty acids in adipose cells (*lipolysis*) is stimulated by binding of epinephrine, glucagon, or ACTH to receptors that activate adenylyl cyclase. On the other hand, binding of two other hormones, prostaglandin PGE_1 or adenosine, to their respective G protein–coupled receptors inhibits adenylyl cyclase. The prostaglandin and adenosine receptors activate an inhibitory G_i protein that contains the same β and γ subunits as the stimulatory G_s protein but a different α subunit ($G_{\alpha i}$). After the active $G_{\alpha i}$·GTP complex dissociates from $G_{\beta\gamma}$, it binds to but inhibits (rather than stimulates) adenylyl cyclase, resulting in lower cAMP levels.

Structural Studies Established How $G_{\alpha s}$·GTP Binds to and Activates Adenylyl Cyclase

X-ray crystallographic analysis has pinpointed the regions in $G_{\alpha s}$·GTP that interact with adenylyl cyclase. This enzyme is a multipass transmembrane protein with two large cytosolic segments containing the catalytic domains (Figure 15-22a). Because such transmembrane proteins are notoriously difficult to crystallize, scientists prepared two protein fragments encompassing the two catalytic domains of adenylyl cyclase that tightly associate with one another in a heterodimer. When these catalytic fragments are allowed to associate in the presence of $G_{\alpha s}$·GTP and forskolin, they are stabilized in their active conformations.

The resulting water-soluble complex (two adenylyl cyclase domain fragments/$G_{\alpha s}$·GTP /forskolin) was catalytically active and showed pharmacological and biochemical properties similar to those of intact full-length adenylyl cyclase. In this complex, two regions of $G_{\alpha s}$·GTP, the switch II helix and the α3-β5 loop, contact the adenylyl cyclase fragments (Figure 15-22b). These contacts are thought to be responsible for the activation of the enzyme by $G_{\alpha s}$·GTP. Recall that switch II

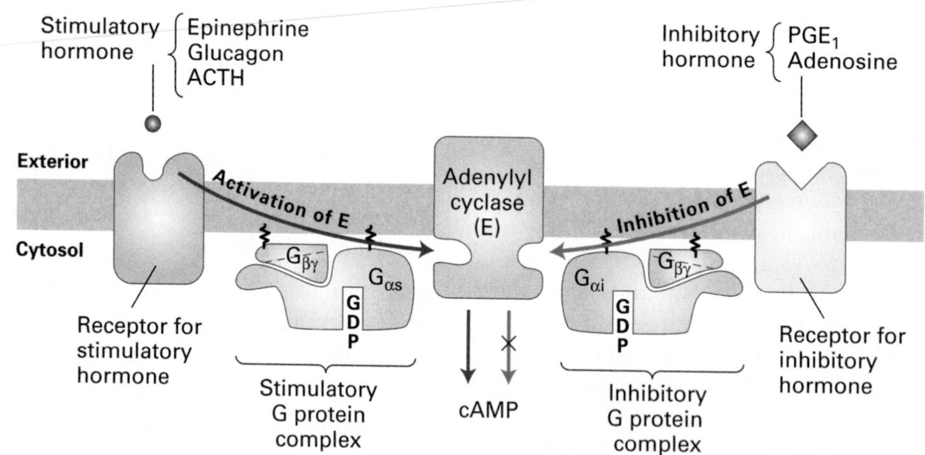

▲ **FIGURE 15-21 Hormone-induced activation and inhibition of adenylyl cyclase in adipose cells.** Ligand binding to $G_{\alpha s}$-coupled receptors causes activation of adenylyl cyclase, whereas ligand binding to $G_{\alpha i}$-coupled receptors causes inhibition of the enzyme. The $G_{\beta\gamma}$ subunit in both stimulatory and inhibitory G proteins is identical; the G_{α} subunits and their corresponding receptors differ. Ligand-stimulated formation of active G_{α}·GTP complexes occurs by the same mechanism in both $G_{\alpha s}$ and $G_{\alpha i}$ proteins (see Figure 15-13). However, $G_{\alpha s}$·GTP and $G_{\alpha i}$·GTP interact differently with adenylyl cyclase, so that one stimulates and the other inhibits its catalytic activity. [See A. G. Gilman, 1984, *Cell* **36**:577.]

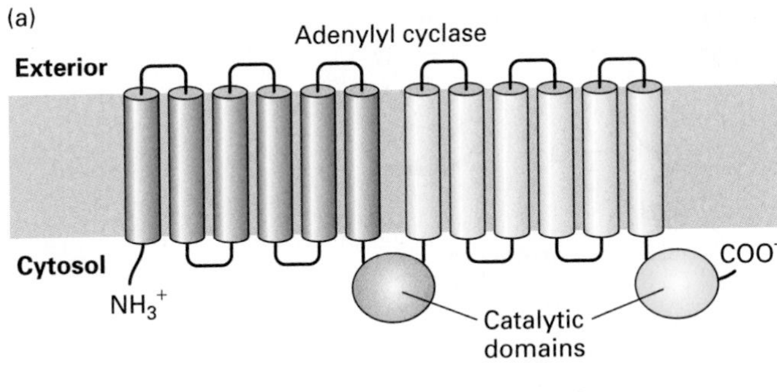

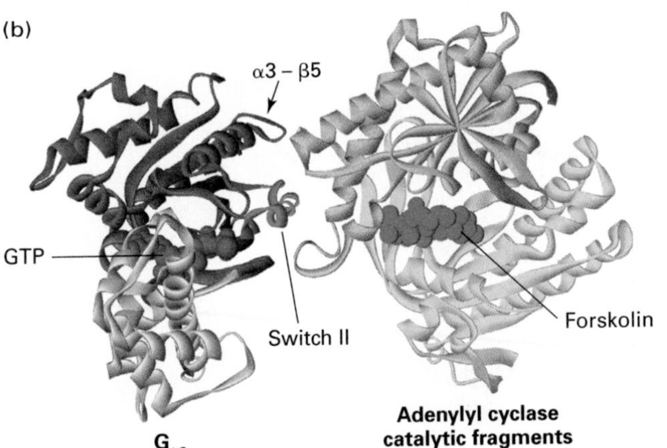

▲ **FIGURE 15-22 Structure of mammalian adenylyl cyclases and their interaction with $G_{\alpha s}$·GTP.** (a) Schematic diagram of mammalian adenylyl cyclases. The membrane-bound enzyme contains two similar catalytic domains on the cytosolic face of the membrane and two integral membrane domains, each of which is thought to contain six transmembrane α helices. (b) Model of the three-dimensional structure of $G_{\alpha s}$·GTP complexed with two fragments encompassing the catalytic domain of adenylyl cyclase determined by x-ray crystallography. The α3-β5 loop and the helix in the switch II region (blue) of $G_{\alpha s}$·GTP interact simultaneously with a specific region of adenylyl cyclase. The darker-colored portion of $G_{\alpha s}$ is the GTPase domain, which is similar in structure to Ras (see Figure 15-8); the lighter portion is a helical domain. The two adenylyl cyclase fragments are shown in orange and yellow. Forskolin (green) locks the cyclase fragments in their active conformations. [Part (a) see W.-J. Tang and A. G. Gilman, 1992, *Cell* **70**:869. Part (b) adapted from J. J. G. Tesmer et al., 1997, *Science* **278**:1907.]

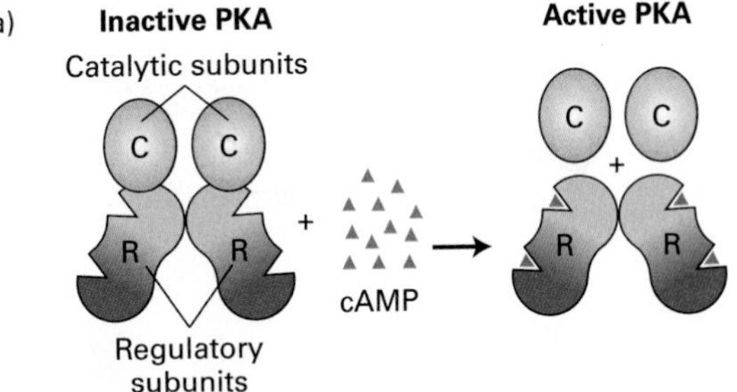

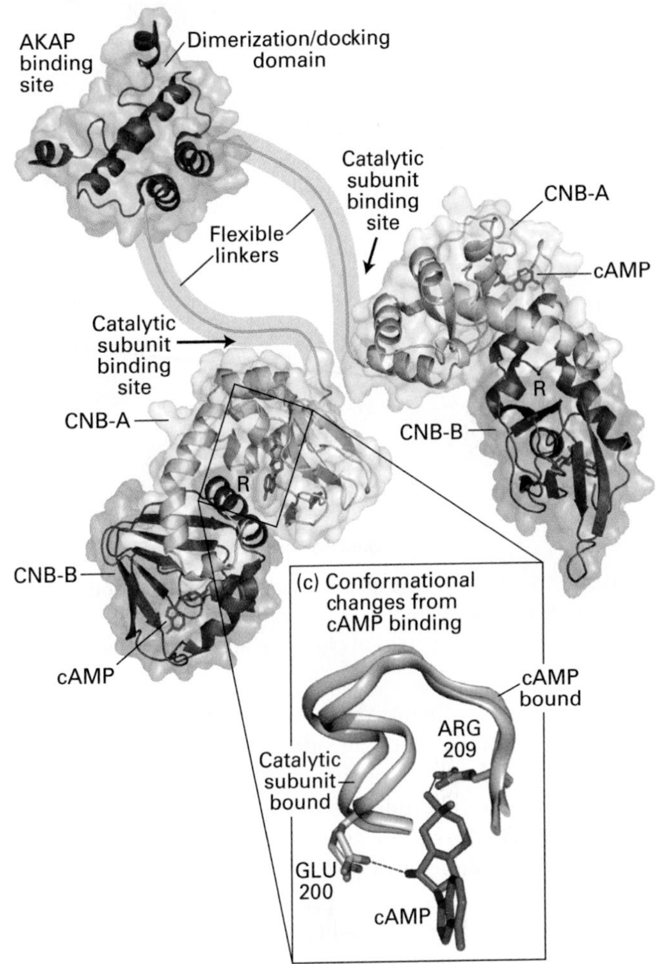

▲ **FIGURE 15-23 Structure of the regulatory (R) subunits of protein kinase A and its activation by cAMP.** (a) Protein kinase A (PKA) consists of two regulatory (R) subunits (green) and two catalytic (C) subunits. When cAMP (red triangle) binds to the regulatory subunit, the catalytic subunit is released, thus activating PKA. (b) The two regulatory subunits are joined by a flexible linker and a dimerization/docking domain where A-kinase activating protein (AKAP, Figure 15-28) can bind. Each R subunit has two cAMP-binding domains, CNB-A and CNB-B, and a binding site for a catalytic subunit (arrow). (c) Binding of cAMP to the CNB-A domain displaces the catalytic subunit leading to its activation. Without bound cAMP, one loop of the CNB-A domain (purple) is in a conformation that can bind the catalytic (C) subunit. A glutamate (E200) and arginine (R209) residue participate in binding of cAMP (red), which causes a conformational change (green) in the loop that prevents binding of the loop to the C subunit. [Part (b) after S. S. Taylor et al., 2005, *Biochim. Biophys. Acta* **1754**:25. Part (c) after C. Kim, N-H Xuong, and S. S. Taylor, 2005, *Science* **307**:690.]

is one of the segments of a G_α protein whose conformation is different in the GTP-bound and GDP-bound states (see Figure 15-8). The GTP-induced conformation of $G_{\alpha s}$ that favors its dissociation from $G_{\beta\gamma}$ is precisely the conformation essential for binding of $G_{\alpha s}$ to adenylyl cyclase. Other studies indicate that $G_{\alpha i}$ binds to a different region of adenylyl cyclase, accounting for its different effect.

cAMP Activates Protein Kinase A by Releasing Catalytic Subunits

In multicellular animals, virtually all the diverse effects of cAMP are mediated through **protein kinase A (PKA),** also called *cAMP-dependent protein kinase*. Inactive PKA is a tetramer consisting of two regulatory (R) subunits and two catalytic (C) subunits (Figure 15-23a). Each R subunit contains a

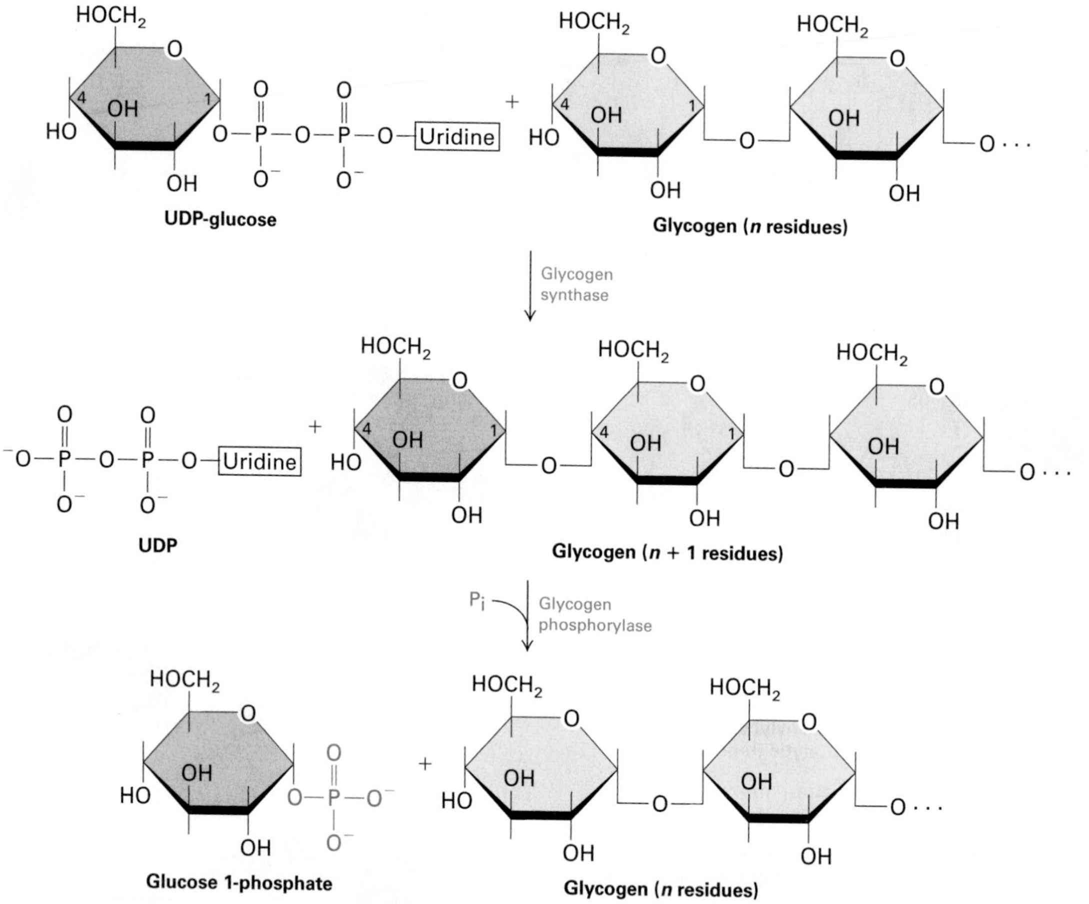

▲ **FIGURE 15-24 Synthesis and degradation of glycogen.** Incorporation of glucose from UDP-glucose into glycogen is catalyzed by glycogen synthase. Removal of glucose units from glycogen is catalyzed by glycogen phosphorylase. Because two different enzymes catalyze the formation and degradation of glycogen, the two reactions can be independently regulated.

pseudosubstrate sequence that binds to the active site in a catalytic domain. By blocking substrate binding, the R subunits inhibit the activity of the catalytic subunits. Inactive PKA is turned on by binding of cAMP. Each R subunit has two distinct cAMP-binding sites, called CNB-A and CNB-B (Figure 15-23b). Binding of cAMP to an R subunit causes a conformational change in the pseudosubstrate domain that leads to release of the associated C subunit, unmasking its catalytic site and activating its kinase activity (Figure 15-23c).

Binding of cAMP by an R subunit of protein kinase A occurs in a cooperative fashion; that is, binding of the first cAMP molecule to CNB-B lowers the K_d for binding of the second cAMP to CNB-A. Thus small changes in the level of cytosolic cAMP can cause proportionately large changes in the amount of dissociated C subunits and, hence, in kinase activity. Rapid activation of enzymes by hormone-triggered dissociation of an inhibitor is a common feature of many signaling pathways.

Glycogen Metabolism Is Regulated by Hormone-Induced Activation of Protein Kinase A

Glycogen, a large glucose polymer, is the major storage form of glucose in animals. Like all biopolymers, glycogen is synthesized by one set of enzymes and degraded by another (Figure 15-24). Degradation of glycogen, or glycogenolysis, involves the stepwise removal of glucose residues from one end of the polymer by a phosphorolysis reaction, catalyzed by *glycogen phosphorylase,* yielding glucose 1-phosphate.

In both muscle and liver cells, glucose 1-phosphate produced from glycogen is converted to glucose 6-phosphate. In muscle cells, this metabolite enters the glycolytic pathway and is metabolized to generate ATP for use in powering muscle contraction (Chapter 12). Unlike muscle cells, liver cells contain a phosphatase that hydrolyzes glucose 6-phosphate to glucose, which is exported from these cells in part by a glucose transporter (GLUT2) in the plasma membrane (Chapter 11). Thus glycogen stores in the liver are primarily broken down to glucose, which is immediately released into the blood and transported to other tissues, particularly the muscles and brain.

The epinephrine-stimulated activation of adenylyl cyclase, resulting increase in cAMP, and subsequent activation of protein kinase A (PKA), enhances the conversion of glycogen to glucose 1-phosphate in two ways: by *inhibiting* glycogen synthesis and by *stimulating* glycogen degradation (Figure 15-25a). PKA phosphorylates and in so doing inactivates glycogen synthase, the enzyme that synthesizes glycogen. PKA promotes glycogen degradation indirectly by phosphorylating and thus activating an intermediate kinase,

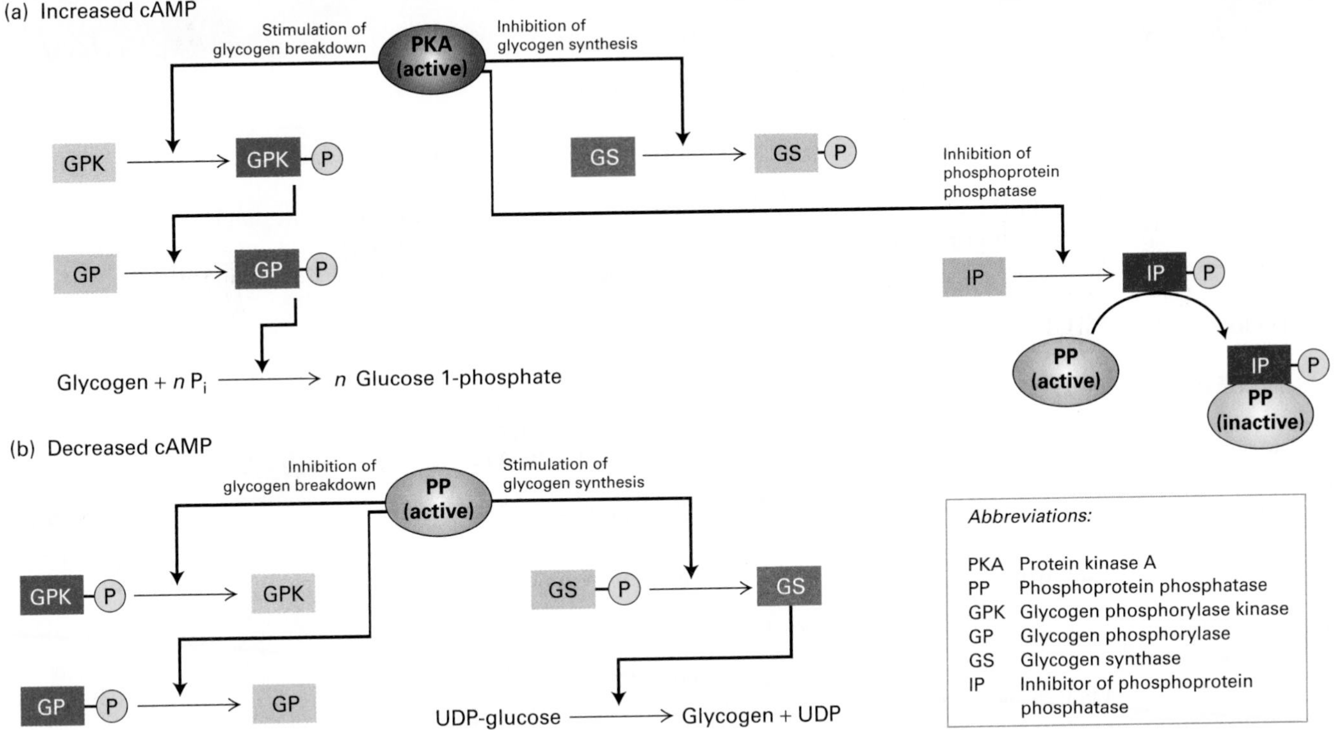

▲ **FIGURE 15-25 Regulation of glycogen metabolism by cAMP in liver and muscle cells.** Active enzymes are highlighted in darker shades; inactive forms, in lighter shades. (a) An increase in cytosolic cAMP activates protein kinase A (PKA), which inhibits glycogen synthesis directly and promotes glycogen degradation via a protein kinase cascade. At high cAMP, PKA also phosphorylates an inhibitor of phosphoprotein phosphatase (PP). Binding of the phosphorylated inhibitor to PP prevents this phosphatase from dephosphorylating the activated enzymes in the kinase cascade or the inactive glycogen synthase. (b) A decrease in cAMP inactivates PKA, leading to release of the active form of PP. The action of this enzyme promotes glycogen synthesis and inhibits glycogen degradation.

glycogen phosphorylase kinase (GPK), that in turn phosphorylates and activates glycogen phosphorylase, the enzyme that degrades glycogen.

The entire process is reversed when epinephrine is removed and the level of cAMP drops, inactivating protein kinase A (PKA). This reversal is mediated by *phosphoprotein phosphatase,* which removes the phosphate residues from the inactive form of glycogen synthase, thereby activating it, and from the active forms of glycogen phosphorylase kinase and glycogen phosphorylase, thereby inactivating them (Figure 15-25b). Phosphoprotein phosphatase itself is regulated by PKA. An inhibitor of phosphoprotein phosphatase is normally inactive. When activated PKA phosphorylates this inhibitory protein, it can bind to phosphoprotein phosphatase, inhibiting its activity (see Figure 15-25a). At low cAMP levels, when PKA is inactive, the inhibitor is not phosphorylated and phosphoprotein phosphatase is active. As a result, in the absence of cAMP the synthesis of glycogen by glycogen synthase is enhanced and the degradation of glycogen by glycogen phosphorylase is inhibited.

Epinephrine-induced glycogenolysis thus exhibits dual regulation: activation of the enzymes catalyzing glycogen degradation and inhibition of enzymes promoting glycogen synthesis. Such coordinate regulation of synthetic and degradative pathways provides an efficient mechanism for achieving a particular cellular response and is a common phenomenon in regulatory biology.

cAMP-Mediated Activation of Protein Kinase A Produces Diverse Responses in Different Cell Types

In adipose cells, epinephrine-induced activation of protein kinase A (PKA) promotes phosphorylation and activation of the phospholipase that hydrolyzes stored triglycerides to yield free fatty acids and glycerol. These fatty acids are released into the blood and taken up as an energy source by cells in other tissues such as the kidney, heart, and muscles. Therefore, activation of PKA by epinephrine in two different cell types, liver and adipose, has different effects. Indeed cAMP and PKA mediate a large array of hormone-induced cellular responses in multiple body cells (Table 15-2).

Although protein kinase A acts on different substrates in different types of cells, it always phosphorylates a serine or threonine residue that occurs within the same sequence motif: X-Arg-(Arg/Lys)-X-(Ser/Thr)-Φ, where X denotes any amino acid and Φ denotes a hydrophobic amino acid. Other serine/threonine kinases phosphorylate target residues within other sequence motifs.

TABLE 15-2 Cellular Responses to Hormone-Induced Rise in cAMP in Various Tissues*

TISSUE	HORMONE INDUCING RISE IN cAMP	CELLULAR RESPONSE
Adipose	Epinephrine; ACTH; glucagon	Increase in hydrolysis of triglyceride; decrease in amino acid uptake
Liver	Epinephrine; norepinephrine; glucagon	Increase in conversion of glycogen to glucose; inhibition of glycogen synthesis; increase in amino acid uptake; increase in gluconeogenesis (synthesis of glucose from amino acids)
Ovarian follicle	FSH; LH	Increase in synthesis of estrogen, progesterone
Adrenal cortex	ACTH	Increase in synthesis of aldosterone, cortisol
Cardiac muscle	Epinephrine	Increase in contraction rate
Thyroid gland	TSH	Secretion of thyroxine
Bone	Parathyroid hormone	Increase in resorption of calcium from bone
Skeletal muscle	Epinephrine	Conversion of glycogen to glucose
Intestine	Epinephrine	Fluid secretion
Kidney	Vasopressin	Resorption of water
Blood platelets	Prostaglandin I	Inhibition of aggregation and secretion

*Nearly all the effects of cAMP are mediated through protein kinase A (PKA), which is activated by binding of cAMP.
SOURCE: E. W. Sutherland, 1972, *Science* **177**:401.

Signal Amplification Commonly Occurs in Many Signaling Pathways

Receptors are low-abundance proteins, typically present in only a few thousand copies per cell. Yet the cellular responses induced by binding of a relatively small number of hormone molecules to the available receptors may require production of tens of thousands or even millions of second messenger or activated enzyme molecules per cell. Thus substantial *signal amplification* often must occur in order for a hormone signal to induce a significant cellular response.

In the case of G protein–coupled receptors, signal amplification is possible in part because both receptors and G proteins can diffuse rapidly in the plasma membrane. A single epinephrine–GPCR complex causes conversion of up to 100 inactive $G_{\alpha s}$ molecules to the active form before epinephrine dissociates from the receptor. Each active $G_{\alpha s}$·GTP, in turn, activates a single adenylyl cyclase molecule, which then catalyzes synthesis of many cAMP molecules during the time $G_{\alpha s}$·GTP is bound to it.

The amplification that occurs in such an amplification cascade depends on the number of steps in it and the relative concentrations of the various components. In the epinephrine-induced cascade shown in Figure 15-26, blood levels of epinephrine as low as 10^{-10} M can stimulate liver glycogenolysis and release of glucose. An epinephrine stimulus of this magnitude generates an intracellular cAMP concentration of 10^{-6} M,

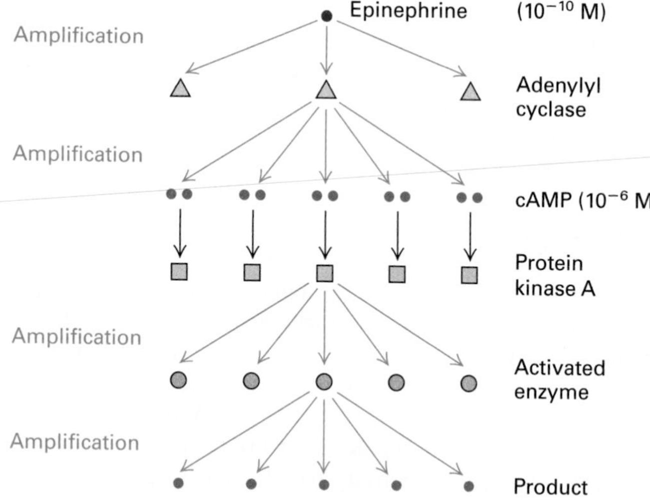

▲ **FIGURE 15-26 Amplification of an external signal downstream from a cell-surface receptor.** In this example, binding of a single epinephrine molecule to one $G_{\alpha s}$ protein–coupled receptor molecule induces synthesis of a large number of cAMP molecules, the first level of amplification. Four molecules of cAMP activate two molecules of protein kinase A (PKA), but each activated PKA phosphorylates and activates multiple product molecules. This second level of amplification may involve several sequential reactions in which the product of one reaction activates the enzyme catalyzing the next reaction. The more steps in such a cascade, the greater the signal amplification possible.

an amplification of 10^4 fold. Because three more catalytic steps precede the release of glucose, another 10^4 amplification can occur, resulting in a 10^8 amplification of the epinephrine signal. In striated muscle the amplification is less dramatic, because the concentrations of the three successive enzymes in the glycogenolytic cascade—protein kinase A, glycogen phosphorylase kinase, and glycogen phosphorylase—are in a 1:10:240 ratio (a potential 240-fold maximal amplification). The epinephrine-induced GPCR pathway leading to glycogenolysis, whether in liver or striated muscle cells, dramatically illustrates how the effects of an external signal can be amplified.

Several Mechanisms Down-Regulate Signaling from G Protein–Coupled Receptors

For cells to respond effectively to changes in their environment, mechanisms must exist to terminate the activation of signaling pathways. Several mechanisms contribute to termination of cellular responses to hormones mediated by β-adrenergic receptors and other G protein–coupled receptors coupled to $G_{\alpha s}$. First, the affinity of the receptor for its ligand decreases when the GDP bound to $G_{\alpha s}$ is replaced with GTP. This increase in the K_d of the receptor–hormone complex enhances dissociation of the ligand from the receptor and thereby limits the number of $G_{\alpha s}$ proteins that are activated. Second, the intrinsic GTPase activity of $G_{\alpha s}$ converts the bound GTP to GDP, resulting in inactivation of the protein and decreased adenylyl cyclase activity. Importantly, the rate of hydrolysis of GTP bound to $G_{\alpha s}$ is enhanced when $G_{\alpha s}$ binds to adenylyl cyclase, lessening the duration of cAMP production; thus adenylyl cyclase functions as a GAP for $G_{\alpha s}$. More generally, binding of most if not all G_α·GTP complexes to their respective effector proteins accelerates the rate of GTP hydrolysis. Finally, *cAMP phosphodiesterase* acts to hydrolyze cAMP to 5′-AMP, terminating the cellular response. Thus the continuous presence of hormone at a high enough concentration is required for continuous activation of adenylyl cyclase and maintenance of an elevated cAMP level. Once the hormone concentration falls sufficiently, the cellular response quickly terminates.

Receptors can also be down-regulated by *feedback repression,* in which the end product of a pathway blocks an early step in the pathway. For instance, when a $G_{\alpha s}$ protein–coupled receptor is exposed to hormonal stimulation for several hours, several serine and threonine residues in the cytosolic domain of the receptor become phosphorylated by protein kinase A (PKA), the end product of the $G_{\alpha s}$ pathway. The phosphorylated receptor can bind its ligand but cannot efficiently activate $G_{\alpha s}$; thus ligand binding to the phosphorylated receptor leads to reduced activation of adenylyl cyclase compared with ligand binding to a nonphosphorylated receptor. Because the activity of PKA is enhanced by the high cAMP level induced by any hormone that activates $G_{\alpha s}$, prolonged exposure to one such hormone, say, epinephrine, desensitizes not only β-adrenergic receptors but also other $G_{\alpha s}$ protein–coupled receptors that bind different ligands (e.g., glucagon receptor in liver). This cross-regulation is called *heterologous desensitization.*

Exposure of cells to epinephrine also leads to other forms of desensitization. Particular residues in the cytosolic domain of the β-adrenergic receptor, not those phosphorylated by PKA, can be phosphorylated by the enzyme *β-adrenergic receptor kinase (BARK),* but *only* when epinephrine or an agonist is bound to the receptor and the receptor is in its active conformation. This process is called *homologous desensitization,* because only those receptors that are in their active conformations are subject to deactivation by phosphorylation. Another example of this regulatory mechanism is the desensitization of rhodopsin by rhodopsin kinase.

Recall from our discussion of the rhopodsin pathway that binding of β-arrestin to extensively phosphorylated opsin completely inhibits activation of coupled G proteins by activated opsin (see Figure 15-20). In fact, β-arrestin plays a similar role in desensitizing other G protein–coupled receptors, including β-adrenergic receptors. An additional function of β-arrestin in regulating cell-surface receptors initially was suggested by the observation that disappearance of β-adrenergic receptors from the cell surface in response to ligand binding is stimulated by overexpression of BARK and β-arrestin. Subsequent studies revealed that β-arrestin binds not only to phosphorylated receptors but also to clathrin and an associated protein termed AP2, two key components of the coated vesicles that are involved in one type of endocytosis. These interactions promote the formation of coated pits and endocytosis of the associated receptors, thereby decreasing the number of receptors exposed on the cell surface (Figure 15-27). Eventually some of the internalized receptors are

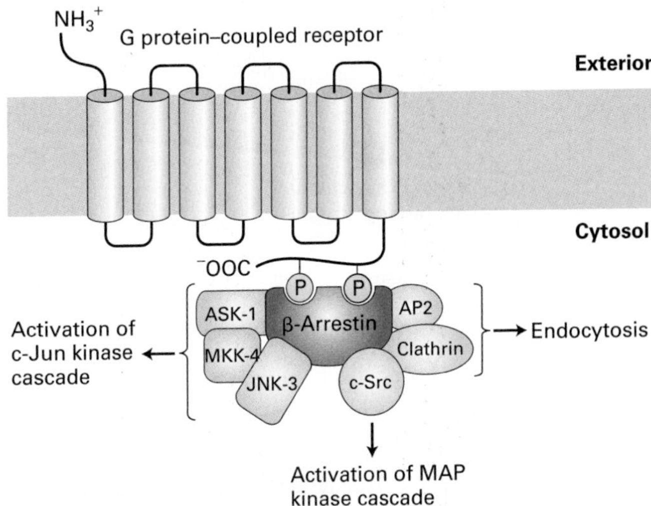

▲ **FIGURE 15-27 Role of β-arrestin in GPCR desensitization and signal transduction.** β-Arrestin binds to phosphorylated serine and threonine residues in the C-terminal segment of G protein–coupled receptors (GPCRs). Clathrin and AP2, two other proteins bound by β-arrestin, promote endocytosis of the receptor. β-Arrestin also functions in transducing signals from activated receptors by binding to and activating several cytosolic protein kinases. c-Src activates the MAP kinase pathway, leading to phosphorylation of key transcription factors (Chapter 16). Interaction of β-arrestin with three other proteins, including JNK-3 (a Jun N-terminal kinase), results in phosphorylation and activation of another transcription factor, c-Jun. [Adapted from W. Miller and R. J. Lefkowitz, 2001, *Curr. Opin. Cell Biol.* **13:**139, and K. Pierce et al., 2002, *Nature Rev. Mol. Cell Biol.* **3:**639]

degraded intracellularly, and some are dephosphorylated in endosomes. Following dissociation of β-arrestin, the resensitized (dephosphorylated) receptors recycle to the cell surface, similar to recycling of the LDL receptor (Chapter 14).

Desensitization of many GPCRs and other classes of receptors occurs by receptor phosphorylation, arrestin binding, and endocytosis of ligand-occupied receptors, leading to their sequestration inside the cell. In addition to its role in regulating receptor activity, β-arrestin also functions as an adapter protein in transducing signals from G protein–coupled receptors to the nucleus (Chapter 16). The multiple functions of β-arrestin illustrate the importance of adapter proteins in both regulating signaling and transducing signals from cell-surface receptors.

Anchoring Proteins Localize Effects of cAMP to Specific Regions of the Cell

In many cell types, a rise in the cAMP level may produce a response that is required in one part of the cell but is unneeded, perhaps deleterious, in another. A family of anchoring proteins localizes isoforms of protein kinase A (PKA) to specific subcellular locations, thereby restricting cAMP-dependent responses to these locations. These proteins, referred to as *A kinase–associated proteins (AKAPs)*, have a two-domain structure with one domain conferring a specific subcellular location and another that binds to a regulatory subunit of protein kinase A (see Figure 15-23b).

One such anchoring protein (AKAP15) is tethered to the cytosolic face of the plasma membrane near a particular type of gated Ca^{2+} channel in certain heart muscle cells. In the heart, activation of β-adrenergic receptors by epinephrine (as part of the fight-or-flight response) leads to PKA-catalyzed phosphorylation of these Ca^{2+} channels, causing them to open; the resulting influx of Ca^{2+} increases the rate of heart muscle contraction. The binding of AKAP15 to protein kinase A localizes the kinase next to these channels, thereby reducing the time that otherwise would be required for diffusion of PKA catalytic subunits from their sites of generation to their Ca^{2+}-channel substrates.

A different AKAP in heart muscle anchors both protein kinase A and cAMP phosphodiesterase (PDE) to the outer nuclear membrane. Because of the close proximity of PDE to protein kinase A, negative feedback provides tight local control of the cAMP concentration and hence local PKA activity (Figure 15-28). The localization of protein kinase A near the nuclear membrane also facilitates entry of its catalytic subunits into the nucleus, where they phosphorylate and activate certain transcription factors (see Chapter 16).

KEY CONCEPTS OF SECTION 15.6

G Protein–Coupled Receptors That Activate or Inhibit Adenylyl Cyclase

■ Ligand activation of G protein–coupled receptors that activate $G_{\alpha s}$ results in the activation of the membrane-

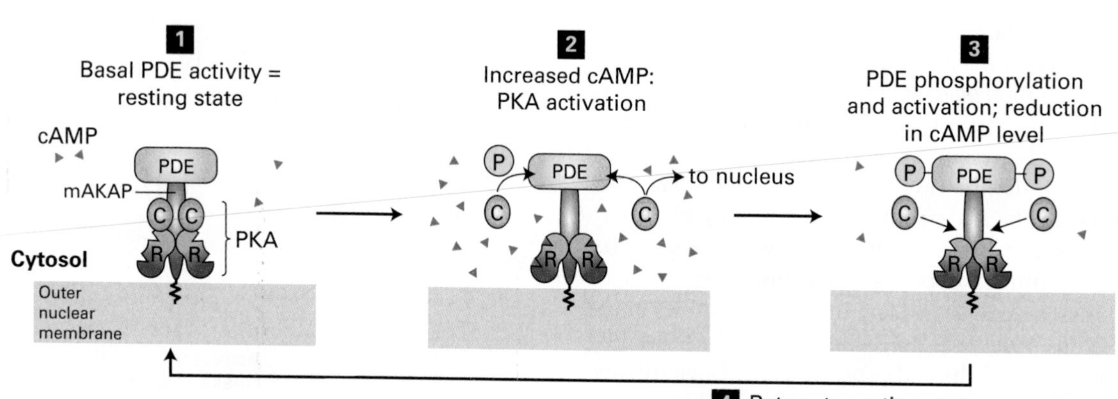

▲ **FIGURE 15-28 Localization of protein kinase A (PKA) to the nuclear membrane in heart muscle by an A kinase-associated protein.** This member of the AKAP family, designated mAKAP, anchors both cAMP phosphodiesterase (PDE) and the regulatory subunit of PKA to the nuclear membrane, maintaining them in a negative feedback loop that provides close local control of the cAMP level and PKA activity. Step **1**: The basal level of PDE activity in the absence of hormone (resting state) keeps cAMP levels below those necessary for PKA activation. Steps **2** and **3**: Activation of β-adrenergic receptors causes an increase in cAMP level in excess of that which can be degraded by PDE. The resulting binding of cAMP to the regulatory (R) subunits of PKA releases the active catalytic (C) subunits into the cytosol. Some C subunits enter into the nucleus, where they phosphorylate and thus activate certain transcription factors (Chapter 16). Concomitant phosphorylation of PDE by active PKA catalytic subunits stimulates its catalytic activity, thereby hydrolyzing cAMP and driving cAMP levels back to basal and causing reformation of the inactive PKA. Step **4**: Subsequent dephosphorylation of PDE returns the complex to the resting state. [Adapted from K. L. Dodge et al., 2001, *EMBO J.* **20:**1921.]

bound enzyme adenylyl cyclase, which converts ATP to the second messenger cyclic AMP (cAMP).

- Ligand activation of G protein–coupled receptors that activate $G_{\alpha i}$ results in the inhibition of adenylyl cyclase and lower levels of cAMP.
- The switch regions in the activated forms of $G_{\alpha s}$·GTP and $G_{\alpha i}$·GTP bind to the heterodimeric active site domains in adenylyl cyclase to activate or inhibit the enzyme, respectively.
- cAMP binds cooperatively to a regulatory subunit of protein kinase A (PKA) releasing the active kinase catalytic subunit (see Figure 15-23).
- PKA mediates the diverse effects of cAMP in most cells (see Table 15-2). The substrates for PKA and thus the cellular response to hormone-induced activation of PKA vary among cell types.
- In liver and muscle cells, activation of PKA induced by epinephrine and other hormones exerts a dual effect, inhibiting glycogen synthesis and stimulating glycogen breakdown via a kinase cascade (see Figure 15-25).
- Signaling pathways involving second messengers and kinase cascades amplify an external signal tremendously (see Figure 15-26).
- BARK phosphorylates ligand-bound β-adrenergic receptors, leading to the binding of β-arrestin and endocytosis of the receptors. The consequent reduction in the number of cell-surface receptors renders the cell less sensitive to additional hormone.
- Localization of PKA to specific regions of the cell by anchoring proteins restricts the effects of cAMP to particular subcellular locations.

15.7 G Protein–Coupled Receptors That Activate Phospholipase C

Calcium ions play an essential role in regulating cellular responses to external signals and internal metabolic changes. As we saw in Chapter 11 the level of Ca^{2+} in the cytosol is maintained at a submicromolar level (<0.2 μM) by the continuous action of ATP-powered Ca^{2+} pumps, which transport Ca^{2+} ions across the plasma membrane to the cell exterior or into the lumens of the endoplasmic reticulum and other vesicles. Much intracellular Ca^{2+} is also sequestered in the mitochondria.

A small rise in cytosolic Ca^{2+} induces a variety of cellular responses including hormone secretion by endocrine cells, secretion of digestive enzymes by pancreatic exocrine cells, and contraction of muscle (Table 15-3). For example, acetylcholine stimulation of G protein–coupled receptors in secretory cells of

TABLE 15-3 Cellular Responses to Hormone-Induced Rise in Cytosolic Ca^{2+} in Various Tissues*

TISSUE	HORMONE INDUCING RISE IN CA^{2+}	CELLULAR RESPONSE
Pancreas (acinar cells)	Acetylcholine	Secretion of digestive enzymes, such as amylase and trypsinogen
Parotid (salivary) gland	Acetylcholine	Secretion of amylase
Vascular or stomach smooth muscle	Acetylcholine	Contraction
Liver	Vasopressin	Conversion of glycogen to glucose
Blood platelets	Thrombin	Aggregation, shape change, secretion of hormones
Mast cells	Antigen	Histamine secretion
Fibroblasts	Peptide growth factors (e.g., bombesin and PDGF)	DNA synthesis, cell division

*Hormone stimulation leads to production of inositol 1,4,5-trisphosphate (IP_3), a second messenger that promotes release of Ca^{2+} stored in the endoplasmic reticulum.
SOURCE: M. J. Berridge, 1987, *Ann. Rev. Biochem.* **56**:159; M. J. Berridge and R. F. Irvine, 1984, *Nature* **312**:315.

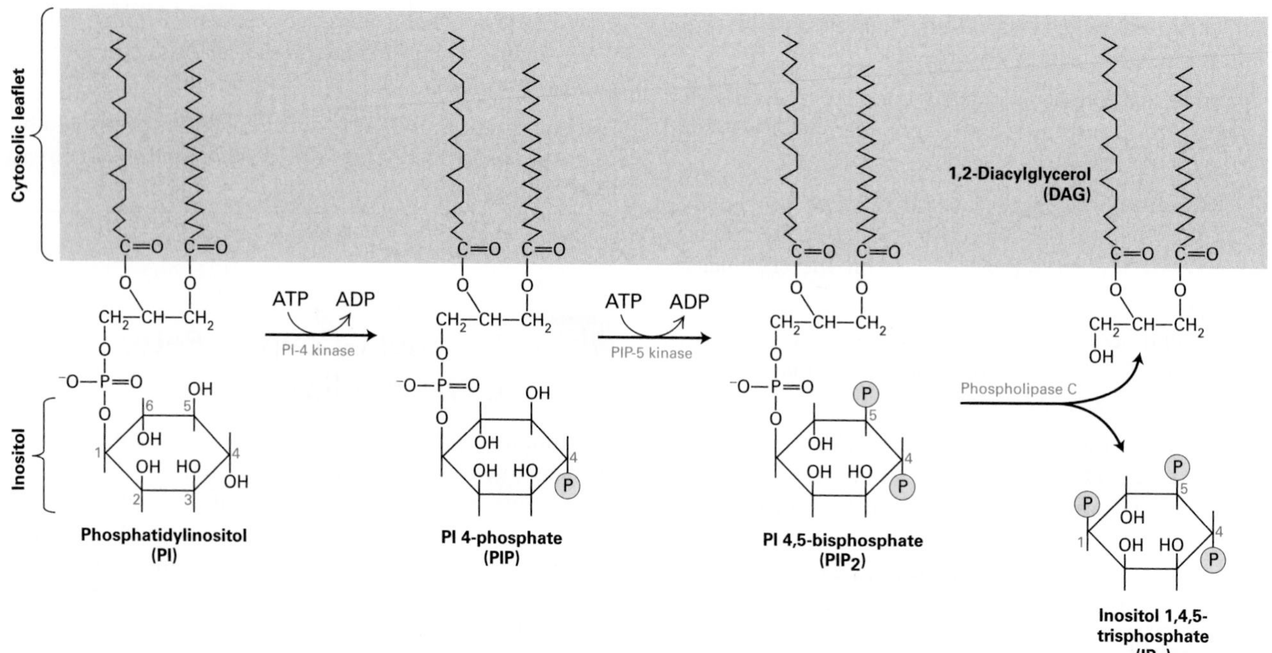

▲ **FIGURE 15-29 Synthesis of second messengers DAG and IP_3 from phosphatidylinositol (PI).** Each membrane-bound PI kinase places a phosphate (yellow circles) on a specific hydroxyl group on the inositol ring, producing the phosphorylated derivatives PIP and PIP_2. Cleavage of PIP_2 by phospholipase C yields the two important second messengers DAG and IP_3. [See A. Toker and L. C. Cantley, 1997, *Nature* **387**:673, and C. L. Carpenter and L. C. Cantley, 1996, *Curr. Opin. Cell Biol.* **8**:153.]

the pancreas and parotid (salivary) gland induces a rise in Ca^{2+} that triggers the fusion of secretory vesicles with the plasma membrane and release of their protein contents into the extracellular space. In blood platelets, the rise in Ca^{2+} induced by thrombin stimulation triggers a conformational change in these cell fragments leading to their aggregation, an important step in blood clotting to prevent leakage out of blood vessels.

In this section, we discuss an important GPCR-triggered signal-transduction pathway that results in an elevation of cytosolic Ca^{2+} ions. Binding of many hormones to their G protein–coupled receptors on liver, fat, and other cells activates G proteins containing either $G_{\alpha o}$ or $G_{\alpha q}$. The effector protein activated by GTP-bound $G_{\alpha o}$ or $G_{\alpha q}$ is phospholipase C (PLC), which hydrolyzes a phosphoester bond in certain phospholipids yielding two second messengers that function in elevating the cytosolic Ca^{2+} level and activating protein kinase C (PKC). This kinase in turn affects many important cellular processes such as growth and differentiation. These pathways also produce second messengers that are important for remodeling the actin cytoskeleton (Chapter 17) and for binding of proteins important for endocytosis and vesicle fusions (Chapter 14).

Phosphorylated Derivatives of Inositol Are Important Second Messengers

A number of important second messengers, used in several signal-transduction pathways, are derived from the membrane lipid *phosphatidylinositol (PI)*. The inositol group in this phospholipid, which always faces the cytosol, can be reversibly phosphorylated at one or more positions by the combined actions of various kinases and phosphatases discussed in Chapter 16. One derivative of PI, the lipid phosphatidyl inositol 4,5-bisphosphate (PIP_2), is cleaved by activated phospholipase C into two important second messengers: **1,2-diacylglycerol (DAG)**, a lipophilic molecule that remains associated with the membrane, and **inositol 1,4,5-trisphosphate (IP_3)**, which can freely diffuse in the cytosol (Figure 15-29). We refer to downstream events involving these two second messengers collectively as the *IP_3/DAG pathway.*

Calcium Ion Release from the Endoplasmic Reticulum Is Triggered by IP_3

G protein–coupled receptors that activate phospholipase C induce an elevation in cytosolic Ca^{2+} even when Ca^{2+} ions are absent from the surrounding extracellular fluid. In this situation, Ca^{2+} is released into the cytosol from the ER lumen through operation of the *IP_3-gated Ca^{2+} channel* in the ER membrane, as depicted in Figure 15-30 (step **4**). This large channel protein is composed of four identical subunits, each of which contains an IP_3-binding site in the N-terminal cytosolic domain. IP_3 binding induces opening of the channel, allowing Ca^{2+} to flow down its concentration gradient from the ER into the cytosol. When various phosphorylated inositols found in cells are added to preparations of ER vesicles, only IP_3 causes release of Ca^{2+} ions from the vesicles. This simple experiment demonstrates the specificity of the IP_3 effect.

The IP_3-mediated rise in the cytosolic Ca^{2+} level is transient because Ca^{2+} pumps located in the plasma membrane and ER membrane actively transport Ca^{2+} from the cytosol to the cell exterior and ER lumen, respectively. Furthermore, within a second of its generation, the phosphate linked to the

Focus Animation: Second Messengers in Signaling Pathways

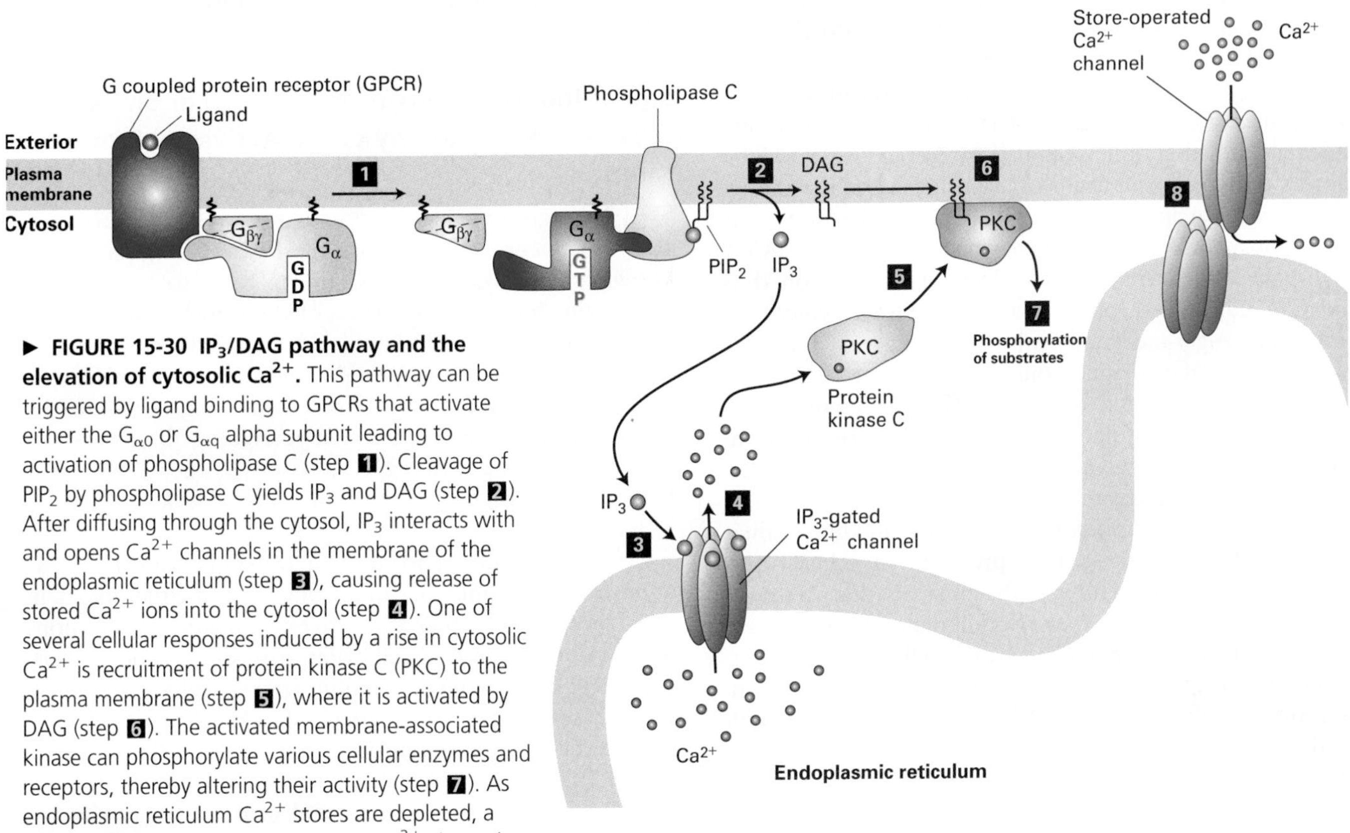

MEDIA CONNECTIONS

► **FIGURE 15-30 IP_3/DAG pathway and the elevation of cytosolic Ca^{2+}.** This pathway can be triggered by ligand binding to GPCRs that activate either the $G_{\alpha o}$ or $G_{\alpha q}$ alpha subunit leading to activation of phospholipase C (step **1**). Cleavage of PIP_2 by phospholipase C yields IP_3 and DAG (step **2**). After diffusing through the cytosol, IP_3 interacts with and opens Ca^{2+} channels in the membrane of the endoplasmic reticulum (step **3**), causing release of stored Ca^{2+} ions into the cytosol (step **4**). One of several cellular responses induced by a rise in cytosolic Ca^{2+} is recruitment of protein kinase C (PKC) to the plasma membrane (step **5**), where it is activated by DAG (step **6**). The activated membrane-associated kinase can phosphorylate various cellular enzymes and receptors, thereby altering their activity (step **7**). As endoplasmic reticulum Ca^{2+} stores are depleted, a protein associated with the IP_3-gated Ca^{2+} channels binds to and opens store-operated Ca^{2+} channels in the plasma membrane, allowing influx of extracellular Ca^{2+} (step **8**). [Adapted from J. W. Putney, 1999, *Proc. Nat'l. Acad. Sci. USA* **96:**14669.]

carbon-5 of IP_3 (see Figure 15-29) is hydrolyzed, yielding inositol 1,4-bisphosphate. This compound cannot bind to the IP_3-gated Ca^{2+} channel protein and thus does not stimulate Ca^{2+} release from the ER.

Without some means for replenishing depleted stores of intracellular Ca^{2+}, a cell would soon be unable to increase the cytosolic Ca^{2+} level in response to hormone-induced IP_3. Patch-clamping studies (see Figure 11-21) have revealed that a plasma membrane Ca^{2+} channel, called the *store-operated channel,* opens in response to depletion of ER Ca^{2+} stores. In a way that is not fully understood, depletion of Ca^{2+} in the ER lumen leads to a conformational change in a protein associated with the IP_3-gated Ca^{2+} channel that allows it to bind to the store-operated Ca^{2+} channel in the plasma membrane, causing the latter to open (see Figure 15-30, step **8**).

Continuous activation of certain G protein–coupled receptors induces rapid, repeated spikes in the level of cytosolic Ca^{2+}. These bursts in cytosolic Ca^{2+} levels are caused by a complex interaction between the cytosolic Ca^{2+} concentration and the IP_3-gated Ca^{2+} channel protein. The submicromolar level of cytosolic Ca^{2+} in the resting state potentiates opening of these channels by IP_3, thus facilitating the rapid rise in cytosolic Ca^{2+} following hormone stimulation of the cell-surface G protein–coupled receptor. However, the higher cytosolic Ca^{2+} levels reached at the peak of the spike inhibit IP_3-induced release of Ca^{2+} from intracellular stores by decreasing the affinity of the Ca^{2+} channels for IP_3. As a result, the channels close, and the cytosolic Ca^{2+} level drops rapidly. Calcium ion spikes occur in the pituitary gland cells that secrete luteinizing hormone (LH), which plays an important role in controlling ovulation and thus female fertility. LH secretion is induced by binding of luteinizing hormone–releasing hormone (LHRH) to its G protein–coupled receptors on these cells; LHRH binding induces repeated Ca^{2+} spikes. Each Ca^{2+} spike induces exocytosis of a few LH-containing secretory vesicles, presumably those close to the plasma membrane. The advantage of hormone-induced fluctuation in the cytosolic Ca^{2+} level and protein secretion, rather than a sustained rise in cytosolic Ca^{2+}, is not understood.

The Ca^{2+}/Calmodulin Complex Mediates Many Cellular Responses to External Signals

The small ubiquitous cytosolic protein **calmodulin** functions as a multipurpose switch protein that mediates many cellular effects of Ca^{2+} ions. Binding of Ca^{2+} to four sites on calmodulin yields a complex that interacts with and modulates the activity of many enzymes and other proteins (see Figure 3-31). Because four Ca^{2+} bind to calmodulin in a cooperative fashion, a small change in the level of cytosolic Ca^{2+} leads to a large change in the level of active calmodulin. One well-studied enzyme activated by the Ca^{2+}/calmodulin complex is myosin light-chain kinase, which regulates the activity of myosin in muscle cells (Chapter 17). Another is cAMP phosphodiesterase, the enzyme that degrades cAMP to 5′-AMP and terminates its effects. This reaction thus links Ca^{2+} and cAMP, one of many examples in which two second messenger-mediated pathways interact to fine-tune certain aspects of cell regulation.

In many cells, the rise in cytosolic Ca^{2+} following receptor signaling via phospholipase C–generated IP_3 leads to activation of specific transcription factors. In some cases, Ca^{2+}/calmodulin activates protein kinases that, in turn, phosphorylate transcription factors, thereby modifying their activity and regulating gene expression. In other cases, Ca^{2+}/calmodulin activates a phosphatase that removes phosphate groups from a transcription factor. An important example of this mechanism involves T cells of the immune system in which Ca^{2+} ions enhance the activity of an essential transcription factor called NFAT (nuclear factor of activated T cells). In unstimulated cells, phosphorylated NFAT is located in the cytosol. Following receptor stimulation and elevation of cytosolic Ca^{2+}, the Ca^{2+}/calmodulin complex binds to and activates calcineurin, a protein-serine phosphatase. Activated calcineurin then dephosphorylates key residues on cytosolic NFAT, exposing a nuclear localization sequence that allows NFAT to move into the nucleus and stimulate expression of genes essential for the function of T cells.

Diacylglycerol (DAG) Activates Protein Kinase C, Which Regulates Many Other Proteins

After its formation by phospholipase C–catalyzed hydrolysis of PIP_2, the secondary messenger DAG remains associated with the plasma membrane (see Figure 15-29). The principal function of DAG is to activate a family of protein kinases collectively termed **protein kinase C (PKC)**. In the absence of hormone stimulation, protein kinase C is present as a soluble cytosolic protein that is catalytically inactive. A rise in the cytosolic Ca^{2+} level causes protein kinase C to translocate to the cytosolic leaflet of the plasma membrane, where it can interact with membrane-associated DAG (see Figure 15-30, steps **5** and **6**). Activation of protein kinase C thus depends on an increase of both Ca^{2+} ions and DAG, suggesting an interaction between the two branches of the IP_3/DAG pathway.

The activation of protein kinase C in different cells results in a varied array of cellular responses, indicating that it plays a key role in many aspects of cellular growth and metabolism. In liver cells, for instance, protein kinase C helps regulate glycogen metabolism by phosphorylating and thus inhibiting glycogen synthase. Protein kinase C also phosphorylates various transcription factors; depending on the cell type, these induce synthesis of mRNAs that trigger cell division.

Signal-Induced Relaxation of Vascular Smooth Muscle Is Mediated by cGMP-Activated Protein Kinase G

Nitroglycerin has been used for over a century as a treatment for the intense chest pain of angina. It was known to slowly decompose in the body to *nitric oxide (NO)*, which causes relaxation of the smooth muscle cells surrounding the blood vessels that "feed" the heart muscle itself, thereby increasing the diameter of the blood vessels and increasing the flow of oxygen-bearing blood to the heart muscle. One of the most intriguing discoveries in modern medicine is that NO, a toxic gas found in car exhaust, is in fact a natural signaling molecule. ■

Definitive evidence for the role of NO in inducing relaxation of smooth muscle came from a set of experiments in which acetylcholine was added to experimental preparations of the smooth muscle cells that surround blood vessels. Direct application of acetylcholine to these cells caused them to contract, the expected effect of acetylcholine on these muscle cells. But addition of acetylcholine to the lumen of small isolated blood vessels caused the underlying smooth muscles to relax, not contract. Subsequent studies showed that in response to acetylcholine the endothelial cells that line the lumen of blood vessels were releasing some substance that in turn triggered muscle cell relaxation. That substance turned out to be NO.

We now know that endothelial cells contain a G_o protein–coupled receptor that binds acetylcholine and activates phospholipase C, leading to an increase in the level of cytosolic Ca^{2+}. After Ca^{2+} binds to calmodulin, the resulting complex stimulates the activity of NO synthase, an enzyme that catalyzes formation of NO from O_2 and the amino acid arginine. Because NO has a short half-life (2–30 seconds), it can diffuse only locally in tissues from its site of synthesis. In particular NO diffuses from the endothelial cell into neighboring smooth muscle cells, where it triggers muscle relaxation (Figure 15-31).

The effect of NO on smooth muscle is mediated by the second messenger cGMP, which is formed by an intracellular NO receptor expressed by smooth muscle cells. Binding of NO to the heme group in this receptor leads to a conformational change that increases its intrinsic guanylyl cyclase activity, leading to a rise in the cytosolic cGMP level. Most of the effects of cGMP are mediated by a cGMP-dependent protein kinase, also known as *protein kinase G (PKG)*. In vascular smooth muscle, protein kinase G activates a signaling pathway that results in inhibition of the actin–myosin complex, relaxation of the cell, and thus dilation of the blood vessel. In this case, cGMP acts indirectly via protein kinase G, whereas in rod cells cGMP acts directly by binding to and thus opening cation channels in the plasma membrane (see Figure 15-18).

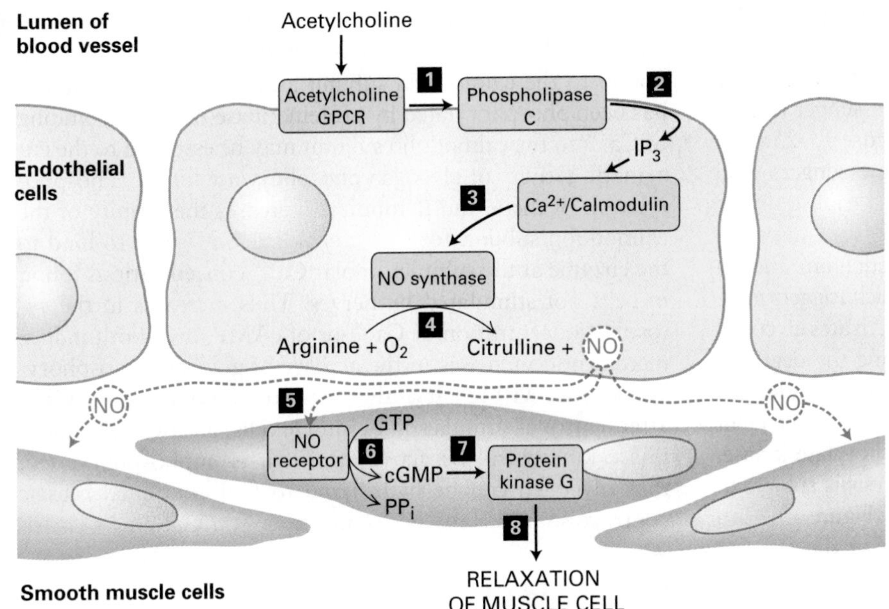

◄ **FIGURE 15-31 The nitric oxide (NO)/cGMP pathway and the relaxation of arterial smooth muscle.** Nitric oxide is synthesized in endothelial cells in response to acetylcholine and the subsequent elevation in cytosolic Ca^{2+} (**1**–**4**). NO diffuses locally through tissues and activates an intracellular NO receptor with guanylyl cyclase activity in nearby smooth muscle cells (**5**). The resulting rise in cGMP activates protein kinase G (**6** and **7**), leading to relaxation of the muscle and thus vasodilation (**8**). The cell-surface receptor for atrial natriuretic factor (ANF) also has intrinsic guanylyl cyclase activity (not shown); stimulation of this receptor on smooth muscle cells also leads to increased cGMP and subsequent muscle relaxation. PP_i = pyrophosphate. [See C. S. Lowenstein et al., 1994, *Ann. Intern. Med.* **120:**227.]

Relaxation of vascular smooth muscle also is triggered by binding of atrial natriuretic factor (ANF) and some other peptide hormones to their receptors on smooth muscle cells. The cytosolic domain of these cell-surface receptors, like the intracellular NO receptor, possesses intrinsic guanylyl cyclase activity. When an increased blood volume stretches cardiac muscle cells in the heart atrium, they release ANF. Circulating ANF binds to ANF receptors on the surface of smooth muscle cells surrounding blood vessels, inducing activation of their guanylyl cyclase activity and formation of cGMP. Subsequent activation of protein kinase G causes dilation of the vessel by the mechanism described above. This vasodilation reduces blood pressure and counters the stimulus that provoked the initial release of ANF.

KEY CONCEPTS OF SECTION 15.7

G Protein–Coupled Receptors That Activate Phospholipase C

- Simulation of some G protein–coupled receptors leads to activation of G proteins containing the $G_{\alpha o}$ or $G_{\alpha q}$ alpha subunit.
- These G proteins activate phospholipase C, which generates two second messengers: diffusible IP_3 and membrane-bound DAG (see Figure 15-29).
- IP_3 triggers opening of IP_3-gated Ca^{2+} channels in the endoplasmic reticulum and elevation of cytosolic free Ca^{2+}. In response to elevated cytosolic Ca^{2+}, protein kinase C is recruited to the plasma membrane, where it is activated by DAG (see Figure 15-30).
- A small rise in cytosolic Ca^{2+} induces a variety of cellular responses including hormone secretion, contraction of muscle, and platelet aggregation (see Table 15-3).
- The Ca^{2+}/calmodulin complex regulates the activity of many different proteins, including cAMP phosphodiesterase, nitric oxide synthase, and protein kinases or phosphatases that control the activity of various transcription factors.
- Stimulation of acetylcholine G protein–coupled receptors on endothelial cells induces an increase in cytosolic Ca^{2+} and subsequent synthesis of NO. After diffusing into surrounding smooth muscle cells, NO activates an intracellular guanylate cyclase to synthesize cGMP. The resulting increase in cGMP leads to activation of protein kinase G, which triggers a pathway resulting to muscle relaxation and vasodilation (see Figure 15-31).
- cGMP is also produced in vascular smooth muscle cells by stimulation of cell-surface receptors that have intrinsic guanylate cyclase activity. These include receptors for atrial natriuretic factor (ANF).

15.8 Integrating Responses of Cells to Environmental Influences

Just as no cell lives in isolation from other cells, no intracellular signaling pathway functions alone. All cells constantly receive multiple signals from their environment including changes in hormone levels, metabolites, and gases such as oxygen. All body cells constantly respond to behavioral demands as well as to injury or infection. In this section, we consider the cellular responses to variations in the demand for the key metabolite glucose. Cellular responses to changes in other nutrients and to oxygen, which are largely reflected in alterations in gene expression, are covered in Chapter 7.

Integration of Multiple Second Messengers Regulates Glycogenolysis

One way for cells to respond appropriately to a complex environment is to sense and integrate its responses to more than one signal. Again, the breakdown of glycogen to glucose (glycogenolysis) provides an excellent example. As described in Section 15.6, epinephrine stimulation of muscle and liver cells leads to a rise in the second messenger cAMP, which promotes glycogen breakdown (see Figure 15-25a). In both muscle and liver cells, other second messengers also produce the same cellular response.

In muscle cells, stimulation by nerve impulses causes the release of Ca^{2+} ions from the sarcoplasmic reticulum and an increase in the cytosolic Ca^{2+} concentration, which triggers muscle contraction. The rise in cytosolic Ca^{2+} also activates glycogen phosphorylase kinase (GPK), thereby stimulating the degradation of glycogen to glucose 1-phosphate, which fuels prolonged contraction. Recall that phosphorylation by cAMP-dependent protein kinase A also activates glycogen phosphorylase kinase. Thus this key regulatory enzyme in glycogenolysis is subject to both neural and hormonal regulation in muscle (Figure 15-32a).

In liver cells, hormone-induced activation of the effector protein phospholipase C also regulates glycogen breakdown by generating two second messengers, DAG and IP_3. As we saw Section 15.7, IP_3 induces an increase in cytosolic Ca^{2+}, which activates glycogen phosphorylase kinase as in muscle cells, leading to glycogen degradation. Moreover, the combined effect of DAG and increased Ca^{2+} activates protein kinase C (see Figure 15-30). This kinase can phosphorylate glycogen synthase, thereby inhibiting the enzyme and reducing the rate of glycogen synthesis. In this case, multiple intracellular signal-transduction pathways are activated by the same signal (Figure 15-32b).

The dual regulation of glycogen phosphorylase kinase by Ca^{2+} and protein kinase A in both muscle and liver results from its multimeric subunit structure $(\alpha\beta\gamma\delta)_4$. The γ subunit is the catalytic enzyme; the regulatory α and β subunits, which are similar in structure, are phosphorylated by protein kinase A; and the δ subunit is calmodulin. Glycogen phosphorylase kinase is maximally active when Ca^{2+} ions are bound to the calmodulin subunit and at least the α subunit has been phosphorylated by protein kinase A. In fact, binding of Ca^{2+} to the calmodulin subunit may be essential to the enzymatic activity of glycogen phosphorylase kinase. Phosphorylation of the α and β subunits increases the affinity of the calmodulin subunit for Ca^{2+}, enabling Ca^{2+} ions to bind to the enzyme at the submicromolar Ca^{2+} concentrations found in cells not stimulated by nerves. Thus increases in the cytosolic concentration of Ca^{2+} or of cAMP or of both induce incremental increases in the activity of glycogen phosphorylase kinase. As a result of the elevated level of cytosolic Ca^{2+} after neuronal stimulation of muscle cells, glycogen phosphorylase kinase will be active even if it is unphosphorylated; thus glycogen can be hydrolyzed to fuel continued muscle contraction in the absence of hormone stimulation.

Insulin and Glucagon Work Together to Maintain a Stable Blood Glucose Level

During normal daily living the maintenance of normal blood glucose concentrations depends on the balance between two peptide hormones, **insulin** and **glucagon**, which are made in distinct pancreatic islet cells and elicit different cellular responses. Insulin, which contains two polypeptide chains linked by disulfide bonds, is synthesized by the β cells in the islets; glucagon, a monomeric peptide, is produced by the α islet cells. Insulin *reduces* the level of blood glucose, whereas glucagon *increases* blood glucose. The availability of blood glucose is regulated during periods of abundance (following a meal) or scarcity (following fasting) by the adjustment of insulin and glucagon concentrations in the blood.

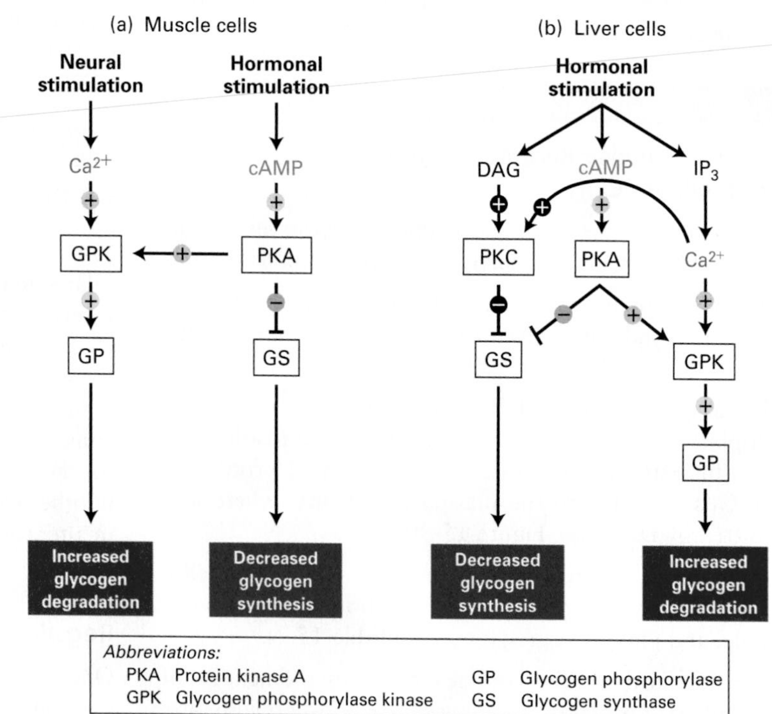

▶ **FIGURE 15-32 Integrated regulation of glycogenolysis.** (a) Neuronal stimulation of striated muscle cells or epinephrine binding to β-adrenergic receptors on their surfaces leads to increased cytosolic concentration of the second messengers Ca^{2+} or cAMP, respectively. The key regulatory enzyme glycogen phosphorylase kinase (GPK) is activated by Ca^{2+} ions and by phosphorylation by cAMP-dependent protein kinase A (PKA). (b) In liver cells, hormonal stimulation of β-adrenergic receptors leads to increased cytosolic concentrations of cAMP and two other second messengers, diacylglycerol (DAG) and inositol 1,4,5-trisphosphate (IP_3). Enzymes are marked by white boxes. (+) = activation of enzyme activity; (−) = inhibition.

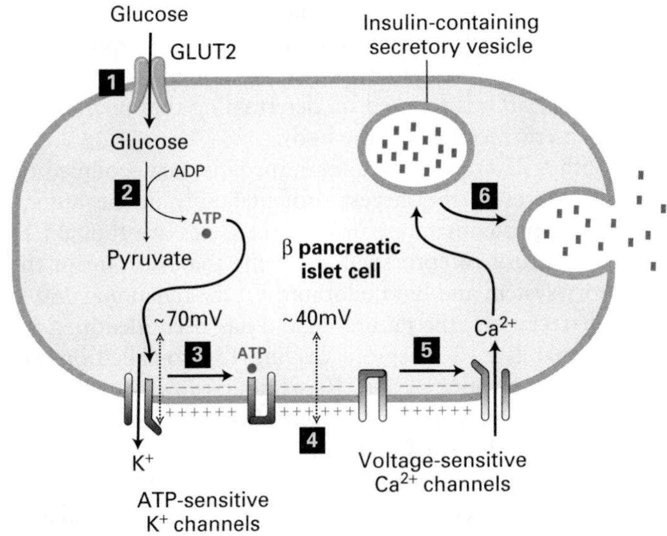

▲ **FIGURE 15-33 Secretion of insulin in response to a rise in blood glucose.** The entry of glucose into pancreatic β cells is mediated by the GLUT2 glucose transporter (1). Because the K_m for glucose of GLUT2 is ≈20 mM, a rise in extracellular glucose from 5 mM, characteristic of the fasting state, causes a proportionate increase in the rate of glucose entry (see Figure 11-4). The conversion of glucose into pyruvate is thus accelerated, resulting in an increase in the concentration of ATP in the cytosol (2). The binding of ATP to ATP-sensitive K^+ channels closes these channels (3), thus reducing the efflux of K^+ ions from the cell. The resulting small depolarization of the plasma membrane (4) triggers the opening of voltage-sensitive Ca^{2+} channels (5). The influx of Ca^{2+} ions raises the cytosolic Ca^{2+} concentration, triggering the fusion of insulin-containing secretory vesicles with the plasma membrane and the secretion of insulin (6). [Adapted from J. Q. Henquin, 2000, *Diabetes* **49:**1751.]

After a meal, when blood glucose rises above its normal level of 5 mM, the pancreatic β cells respond to the rise in glucose (and amino acids) by releasing insulin into the blood (Figure 15-33). The released insulin circulates in the blood and binds to insulin receptors present on many different kinds of cells, including muscle and adipocytes (fat-storing cells). The insulin receptor belongs to the class of receptors termed **receptor tyrosine kinases (RTKs),** which we describe in Chapter 16. It can transduce signals through an intracellular pathway leading to the activation of **protein kinase B.** By an unknown mechanism, protein kinase B triggers the fusion of intracellular vesicles containing the glucose transporter GLUT4 with the plasma membrane (Figure 15-34). The resulting tenfold increase in the number of GLUT4 molecules on the cell surface increases glucose influx proportionally, thus lowering blood glucose.

As the blood glucose level drops, insulin secretion and blood levels drop, and insulin receptors are no longer being activated as strongly. In response, cell-surface GLUT4 is internalized by endocytosis, lowering the level of cell-surface GLUT4 and thus glucose import. Insulin stimulation of muscle cells also promotes the conversion of glucose into glycogen, and it enhances the degradation of glucose to pyruvate. Insulin also acts on hepatocytes (liver cells) to inhibit glucose synthesis from smaller molecules, such as lactate and acetate, and to enhance glycogen synthesis from glucose. The net effect of all these actions is to lower blood glucose back to the fasting concentration of about 5 mM while storing the excess glucose intracellularly as glycogen for future use.

If the blood glucose level falls below about 5 mM, for example due to sudden muscular activity, reduced insulin secretion from pancreatic β cells induces pancreatic α cells to increase their secretion of glucagon into the blood. Like the

Technique Animation: Reporter Constructs

MEDIA CONNECTIONS

(a) Resting cell

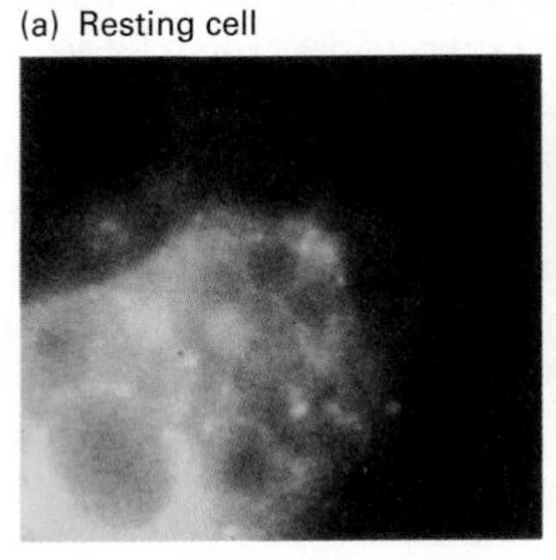

(b) 2.5 min

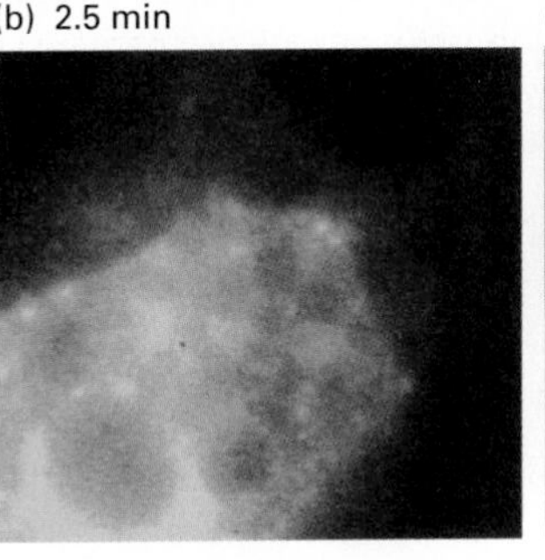

(c) 5 min

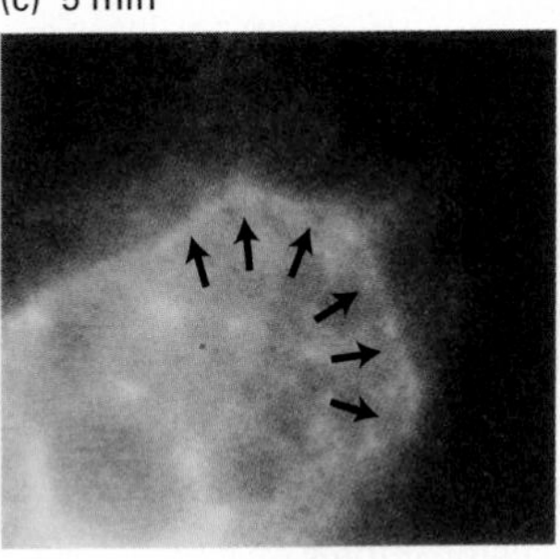

(d) 10 min

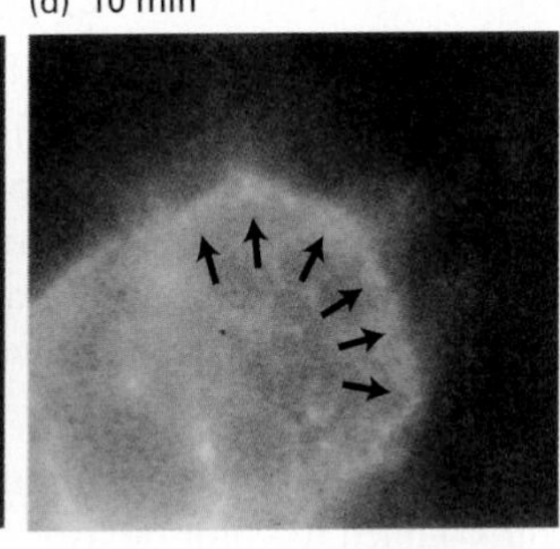

▲ EXPERIMENTAL FIGURE 15-34 **Insulin stimulation of fat cells induces translocation of GLUT4 from intracellular vesicles to the plasma membrane.** In this experiment, fat cells were engineered to express a chimeric protein whose N-terminal end corresponded to the GLUT4 sequence, followed by the entirety of the GFP sequence. When a cell is exposed to light of the exciting wavelength, GFP fluoresces yellow-green, indicating the position of GLUT4 within cells. In resting cells (a), most GLUT4 is in internal membranes that are not connected to the plasma membrane. Successive images of the same cell after treatment with insulin for 2.5, 5, and 10 minutes show that with time, increasing numbers of these GLUT4-containing membranes fuse with the plasma membrane, thereby moving GLUT4 to the cell surface (arrows) and enabling it to transport glucose from the blood into the cell. Muscle cells also contain insulin-responsive GLUT4 transporters. [Courtesy of J. Bogan; see J. Bogan et al., 2001, *Mol. Cell Biol.* **21:**4785.]

epinephrine receptor, the glucagon receptor, found primarily on liver cells, is coupled to the $G_{\alpha s}$ protein, whose effector protein is adenylyl cyclase. Glucagon stimulation of liver cells induces a rise in cAMP, leading to activation of protein kinase A, which inhibits glycogen synthesis and promotes glycogenolysis, yielding glucose 1-phosphate (see Figures 15-25a and 15-32b). Liver cells convert glucose 1-phosphate into glucose, which is released into the blood, thus raising blood glucose back toward its normal fasting level.

Unfortunately, these intricate and powerful control systems sometimes fail, causing serious, even life-threatening disease. *Diabetes mellitus* results from a deficiency in the amount of insulin released from the pancreas in response to rising blood glucose (type I) or from a decrease in the ability of muscle and fat cells to respond to insulin (type II). In both types, the regulation of blood glucose is impaired, leading to persistent elevated blood glucose concentrations (hyperglycemia) and other possible complications if left untreated. Type I diabetes is caused by an autoimmune process that destroys the insulin-producing β cells in the pancreas. Also called insulin-dependent diabetes, this form of the disease is generally responsive to insulin therapy. Most Americans with diabetes mellitus have type II, or insulin-independent diabetes, but the underlying cause of this form of the disease is not well understood. Further identification of the signaling pathways that control energy metabolism is expected to provide insight into the pathophysiology of diabetes, hopefully leading to new methods for its prevention and treatment. ∎

KEY CONCEPTS OF SECTION 15.8

Integrating Responses of Cells to Environmental Influences

- Glycogen breakdown and synthesis is regulated by multiple second messengers induced by neural or hormonal stimulation (see Figure 15-32).
- A rise in blood glucose stimulates the release of insulin from pancreatic β cells (see Figure 15-33). Subsequent binding of insulin to its receptor on muscle cells and adipocytes leads to the activation of protein kinase B, which promotes glucose uptake and glycogen synthesis, resulting in a decrease in blood glucose.
- A lowering of blood glucose stimulates glucagon release from pancreatic α cells. Binding of glucagon to its G protein–coupled receptor on liver cells promotes glycogenolysis by the cAMP-triggered kinase cascade (similar to epinephrine stimulation under stress conditions) and an increase in blood glucose.

Perspectives for the Future

In this chapter we focused primarily on signal-transduction pathways activated by individual G protein–coupled receptors. However, even these relatively simple pathways presage the more complex situation within living cells. Many G protein–coupled receptors form homodimers or heterodimers with other G protein–coupled receptors that bind ligands with different specificities and affinities. Much current research is focused on determining the functions of these dimeric receptors in the body.

With ≈720 members in total, the G protein–coupled receptors represent the largest protein family in the human genome. Approximately half of these genes are thought to encode sensory receptors; of these the majority are in the olfactory system and bind odorants. Of the remaining 360 G protein–receptors, the natural ligand has been identified for approximately 210 receptors, leaving 150 so-called *orphan GPCRs*, that is putative GPCRs without known cognate ligands. Many of these orphan receptors are likely to bind heretofore unidentified signaling molecules, including new peptide hormones. G protein–coupled receptors already represent the largest class of target molecules for drugs available in the clinic, and therefore orphan GPCRs represent a fruitful resource for drug discovery by the pharmaceutical industry.

One approach that has proven fruitful in identifying ligands of orphan GPCRs involves expressing the receptor genes in transfected cells and using them as a reporter system to detect substances in tissue extracts that activate signal-transduction pathways in these cells. This approach has already led to stunning insights into human behavior. One example is two novel peptides, termed orexin-A and orexin-B (from the Greek *orexis*, meaning appetite), that were identified as the ligands for two orphan GPCRs. Further research showed that the *orexin* gene is expressed only in the hypothalamus, the part of the brain that regulates feeding. Injection of orexin into the brain ventricles caused animals to eat more, and expression of the *orexin* gene increased markedly during fasting. Both of these findings are consistent with orexin's role in increasing appetite. Strikingly, mice deficient for orexins suffer from narcolepsy, a disorder characterized in humans by excessive daytime sleepiness (for mice, nighttime sleepiness). Moreover, very recent reports suggest that the orexin system is dysfunctional in a majority of human narcolepsy patients: Orexin peptides cannot be detected in their cerebrospinal fluid (although there is no evidence of mutation in their *orexin* genes). These findings firmly link orexin neuropeptides and their receptors to both feeding behavior and sleep in both animals and humans.

One can only wonder about what other peptides and small-molecule hormones remain to be discovered, and the insights that study of these will provide for our understanding of human metabolism, growth, and behavior.

Key Terms

adenylyl cyclase *639*
adrenergic receptors *636*
agonist *629*
arrestin *645*
autocrine *626*
calmodulin *655*
cAMP *634*
competition assay *629*
desensitization *631*
endocrine *626*
functional expression assay *631*
glucagon *659*

G protein–coupled receptors *624*
insulin *659*
IP_3/DAG pathway *654*
muscarinic acetylcholine receptors *641*
paracrine *626*
phospholipase C *639*
protein kinase A *647*
protein kinase C *657*
rhodopsin *635*
second messengers *634*
signal amplification *649*
signal transduction *624*
stimulatory G protein *639*
transducin *641*
trimeric G proteins *634*
visual adaptation *644*

Review the Concepts

1. What common features are shared by most of the different cell signaling systems?

2. Signaling by soluble extracellular molecules can be classified as endocrine, paracrine, or autocrine. Describe how these three types of cellular signaling differ. Growth hormone is secreted from the pituitary, which is located at the base of the brain, and acts through growth hormone receptors located on the liver. Is this an example of endocrine, paracrine, or autocrine signaling? Why?

3. A ligand binds two different receptors with a K_d value of 10^{-7} M for receptor 1 and a K_d value of 10^{-9} M for receptor 2. For which receptor does the ligand show the greater affinity? Calculate the fraction of receptors that have a bound ligand ($[RL]/R_T$) in the case of receptor 1 and receptor 2, if the concentration of free ligand is 10^{-8} M.

4. A study of the properties of cell-surface receptors can be greatly enhanced by isolation or cloning of the cell-surface receptor. Describe how a cell-surface receptor can be isolated by affinity chromatography. How can you clone a cell-surface receptor using a functional expression assay?

5. Signal-transducing trimeric G proteins consist of three subunits designated α, β, γ. The G_α subunit is a GTPase switch protein that cycles between active and inactive states depending upon whether it is bound to GTP or to GDP. Review the steps for ligand-induced activation of effector proteins mediated by the trimeric G proteins. Suppose that you have isolated a mutant G_α subunit that has an increased GTPase activity. What effect would this mutation have on the G protein and the effector protein?

6. Explain how second messengers such as Ca^{2+} and cAMP can transmit and amplify an extracellular signal.

7. The cholera toxin, produced by the bacterium *Vibrio cholera,* causes a watery diarrhea in infected individuals. What is the molecular basis for this effect of cholera toxin?

8. Epinephrine binds to both β-adrenergic and α-adrenergic receptors. Describe the opposite actions on the effector protein, adenylyl cyclase, elicited by the binding of epinephrine to these two types of receptors. Describe the effect of adding an agonist or antagonist to a β-adrenergic receptor on the activity of adenylyl cyclase.

9. Both rhodopsin in vision and the muscarinic acetylcholine receptor system in cardiac muscle are coupled to ion channels via G proteins. Describe the similarities and differences between these two systems.

10. In liver and muscle cells, epinephrine stimulates the release of glucose from glycogen by inhibiting glycogen synthesis and stimulating glycogen breakdown. Outline the molecular events that occur after epinephrine binds to its receptor and the resultant increase in the concentration of intracellular cAMP. How are the cAMP levels returned to normal? Describe the events that occur after cAMP levels decline.

11. Continuous exposure of a G_s protein–coupled receptor to its ligand leads to a phenomenon known as desensitization. Describe several molecular mechanisms for receptor desensitization. How can a receptor be reset to its original sensitized state? What effect would a mutant receptor lacking serine or threonine phosphorylation sites have on a cell?

12. Visual adaptation and receptor desensitization involve similar phosphorylation mechanisms. Describe how the β-adrenergic receptor kinase (BARK) and rhodopsin kinase play important roles in these processes. What role does dephosphorylation play in these reactions?

13. Sometimes cells need to localize the effects of signaling systems to specific subcellular regions. One example is localization of cAMP signals in heart muscle. What proteins are involved? How does this system work?

14. Inositol 1,4,5-trisphosphate (IP_3) and diacylglycerol (DAG) are second messenger molecules derived from the cleavage of the phosphatidylinositol 4,5-bisphosphate (PIP_2) by activated phospholipase C. Describe the role of IP_3 in the release of Ca^{2+} from the endoplasmic reticulum. How do cells replenish the endoplasmic reticulum stores of Ca^{2+}? What is the principal function of DAG?

15. Recent research has identified a surprising molecular link between feeding behavior and sleep. Describe the signaling factors that may be shared by both systems.

Analyze the Data

Mutations in trimeric G proteins can cause many diseases in humans. Patients with acromegaly often have pituitary tumors that oversecrete the pituitary hormone called growth hormone (GH). A subset of these growth hormone (GH)– secreting pituitary tumors result from mutations in G proteins. GH-releasing hormone (GHRH) stimulates GH release from the pituitary by binding to GHRH receptors and stimulating adenylyl cyclase. Cloning and sequencing of the wild-type and mutant $G_{\alpha s}$ gene

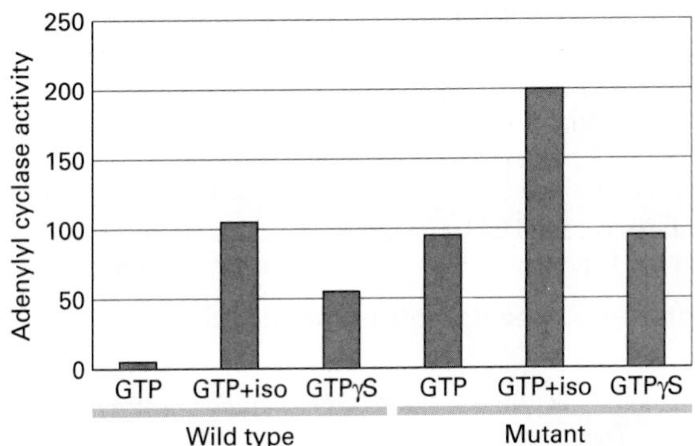

from normal individuals and patients with the pituitary tumors revealed a missense mutation in the $G_{\alpha s}$ gene sequence.

a. To investigate the effect of the mutation on $G_{\alpha s}$ activity, wild-type and mutant $G_{\alpha s}$ cDNAs were transfected into cells that lack the $G_{\alpha s}$ gene. These cells express a β_2-adrenergic receptor, which can be activated by isoproterenol, a β_2-adrenergic receptor agonist. Membranes were isolated from transfected cells and assayed for adenylyl cyclase activity in the presence of GTP or the hydrolysis-resistant GTP analog, GTP-γS. From the figure above, what do you conclude about the effect of the mutation on $G_{\alpha s}$ activity in the presence of GTP alone compared with GTP-γS alone or GTP plus isoproterenol (iso)?

b. In the transfected cells described in part a, what would you predict would be the cAMP levels in cells transfected with the wild-type $G_{\alpha s}$ and the mutant $G_{\alpha s}$? What effect might this have on the cells?

c. To further characterize the molecular defect caused by this mutation, the intrinsic GTPase activity present in both wild-type and mutant $G_{\alpha s}$ was assayed. Assays for GTPase activity showed that the mutation reduced the $k_{cat\text{-}GTP}$ (catalysis rate constant for GTP hydrolysis) from a wild-type value of 4.1 min^{-1} to the mutant value of 0.1 min^{-1}. What do you conclude about the effect of the mutation on the GTPase activity present in the mutant $G_{\alpha s}$ subunit? How do these GTPase results explain the adenylyl cyclase results shown in part a?

References

15.2 Studying Cell-Surface Receptors

Coughlin, S. R. 2005. Protease-activated receptors in hemostasis, thrombosis and vascular biology. *J Thromb Haemost.* **3**:1800–1814.

Gross, A., and H. F. Lodish. 2006. Cellular trafficking and degradation of erythropoietin and NESP. *J. Biol. Chem.* **281**:2024–2032.

Simonsen, H., and H. F. Lodish. 1994. Cloning by function: expression cloning in mammalian cells. *Trends Pharmacol. Sci.* **15**:437–441.

15.3 Highly Conserved Components of Intracellular Signal-Transduction Pathways

Cabrera-Vera, T. M. et al. 2003. Insights into G protein structure, function, and regulation. *Endocr. Rev.* **24**:765–781.

Vetter, I. R., and A. Wittinghofer. 2001. The guanine nucleotide-binding switch in three dimensions. *Science* **294**:1299–1304.

15.4 General Elements of G Protein–Coupled Receptor Systems

Bourne, H. R. 1997. How receptors talk to trimeric G proteins. *Curr. Opin. Cell Biol.* **9**:134–142.

Bourne, H. R. 2001. Receptor activation: what does the rhodopsin structure tell us? *Trends Pharmacol. Sci.* **22**:587–593.

Farfel, Z., H. Bourne, and T. Iiri. 1999. The expanding spectrum of G protein diseases. *New Eng. J. Med.* **340**: 1012–1020.

Pierce, K. L., R. T. Premont, and R. J. Lefkowitz. 2002. Seven-transmembrane receptors. *Nature Rev. Mol. Cell Biol.* **3**:639–652.

Oldham, W. M., and H. Hamm. 2006. Structural basis of function in heterotrimeric G proteins. *Quart Rev. Biophys.* **40**: (In press).

15.5 G Protein–Coupled Receptors That Regulate Ion Channels

Chin, D., and A. R. Means. 2000. Calmodulin: a prototypical calcium sensor. *Trends Cell Biol.* **10**:322–328.

Filipek, S., et al. 2003. G protein-coupled receptor rhodopsin: a prospectus. *Ann. Rev. Physiol.* **65**:851–879.

Filipek, S., et al. 2003. The crystallographic model of rhodopsin and its use in studies of other G protein–coupled receptors. *Ann. Rev. Biophys Biomol. Struc.* **32**:375–397.

Hurley, J. H., and J. A. Grobler. 1997. Protein kinase C and phospholipase C: bilayer interactions and regulation. *Curr. Opin. Struc. Biol.* 7:557–565.

Nathans, J. 1999. The evolution and physiology of human color vision: insights from molecular genetic studies of visual pigments. *Neuron* **24**:299–312.

Palczewski, K. 2006. G protein–coupled receptor rhodopsin. Ann. Rev. Biochem. **75**:743–767.

Ramsey, I. S., M. Delling, and D. Clapham. 2006. An introduction to TRP channels. *Ann. Rev. Physiol.* **68**:619–647.

Singer, W. D., H. A. Brown, and P. C. Sternweis. 1997. Regulation of eukaryotic phosphatidylinositol-specific phospholipase C and phospholipase D. *Ann. Rev. Biochem.* **66**:475–509.

15.6 G Protein–Coupled Receptors That Activate or Inhibit Adenylyl Cyclase

Browner, M., and R. Fletterick. 1992. Phosphorylase: a biological transducer. *Trends Biochem. Sci.* **17**:66–71.

Hurley, J. H. 1999. Structure, mechanism, and regulation of mammalian adenylyl cyclase. *J. Biol. Chem.* **274**:7599–7602.

Johnson, L. N. 1992. Glycogen phosphorylase: control by phosphorylation and allosteric effectors. *FASEB J.* **6**: 2274–2282.

Taylor, S. S., et al. 2005. Dynamics of signaling by PKA. *Biochim. Biophys. Acta* **1754**:25–37.

Lefkowitz, R. J., and S. K. Shenoy. 2005. Transduction of receptor signals by β- arrestins. *Science* **308**:512–517.

Shenoy, S., and R. J. Lefkowitz. 2003. Multifaceted roles of β-arrestins in the regulation of seven-membrane spanning receptor trafficking and signaling. *Biochem. J.* **375**:503–515.

Smith, F. D., L. K. Langeberg, and J. D. Scott. 2006. The where's and when's of kinase anchoring. *Trends Biochem. Sci.* **31**:316–323.

Witters L. A., B. E. Kemp, and A. R. Means. 2006. Chutes and ladders: the search for protein kinases that act on AMPK. *Trends Biochem. Sci.* **31**:13–16.

Wong, W., and J. D. Scott. 2004. AKAP signalling complexes: focal points in space and time. *Nature Rev. Mol. Cell Biol.* **5**:959–970.

15.7 G Protein–Coupled Receptors That Activate Phospholipase C

Carlton, J. G., and P. J. Cullen. 2005. Coincidence detection in phosphoinositide signaling. *Trends Cell Biol.* **15**:540–547.

Kahl, C. R., and A. R. Means. 2003. Regulation of cell cycle progression by calcium/calmodulin-dependent pathways. *Endocr. Rev.* **24**:719–736.

Patterson, R., D. Boehning, and S. Snyder. 2004. Inositol 1, 4, 5, trisphosphate receptors as signal integrators. *Ann. Rev. Biochem.* 73:437–465.

15.8 Integrating Responses of Cells to Environmental Influences

Papin, J. A., et al. 2005. Reconstruction of cellular signalling networks and analysis of their properties. *Nature Rev. Mol. Cell Biol.* **6**:99–111.

Rothman, D., M. Shults, and B. Imperiali. 2005. Chemical approaches for investigating phosphorylation in signal transduction networks. *Trends Cell Biol.* **15**:502–510.

Taniguchi, C., B. Emanuelli, and C. R. Kahn. 2006. Critical nodes in signalling pathways: insights into insulin action. *Nature Rev. Mol. Cell Biol.* **7**:85–96.

Watson, R. T., and J. Pessin. 2006. Bridging the GAP between insulin signaling and GLUT4 translocation. *Trends Biochem. Sci.* **31**:215–222.

CLASSIC EXPERIMENT 15.1

THE INFANCY OF SIGNAL TRANSDUCTION—GTP STIMULATION OF cAMP SYNTHESIS

M. Rodbell et al., 1971, *J. Biol. Chem.* **246**:1877

In the late 1960s the study of hormone action blossomed following the discovery that cyclic adenosine monophosphate (cAMP) functioned as a second messenger, coupling the hormone-mediated activation of a receptor to a cellular response. In setting up an experimental system to investigate the hormone-induced synthesis of cAMP, Martin Rodbell discovered an important new player in intracellular signaling—guanosine triphosphate (GTP).

Background

The discovery of GTP's role in regulating signal transduction began with studies on how glucagon and other hormones send a signal across the plasma membrane that eventually evokes a cellular response. At the outset of Rodbell's studies, it was known that binding of glucagon to specific receptor proteins embedded in the membrane stimulates production of cAMP. The formation of cAMP from ATP is catalyzed by a membrane-bound enzyme called adenyl cyclase. It had been proposed that the action of glucagon, and other cAMP-stimulating hormones, relied on additional molecular components that couple receptor activation to the production of cAMP. However, in studies with isolated fat-cell membranes known as "ghosts," Rodbell and his coworkers were unable to provide any further insight into how glucagon binding leads to an increase in production of cAMP. Rodbell then began a series of studies with a newly developed cell-free system, purified rat liver membranes, which retained both membrane-bound and membrane-associated proteins. These experiments eventually led to the finding that GTP is required for the glucagon-induced stimulation of adenyl cyclase.

The Experiment

One of Rodbell's first goals was to characterize the binding of glucagon to the glucagon receptor in the cell-free rat liver membrane system. First, purified rat liver membranes were incubated with glucagon labeled with the radioactive isotope of iodine (^{125}I). Membranes were then separated from the unbound [^{125}I] glucagon by centrifugation. Once it was established that labeled glucagon would indeed bind to the purified rat liver cell membranes, the study went on to determine if this binding led directly to activation of adenyl cyclase and production of cAMP in the purified rat liver cell membranes.

The production of cAMP in the cell-free system required the addition of ATP; the substrate for adenyl cyclase, Mg^{2+}; and an ATP-regenerating system consisting of creatine kinase and phosphocreatine. Surprisingly, when the glucagon-binding experiment was repeated in the presence of these additional factors, Rodbell observed a 50 percent decrease in glucagon binding. Full binding could be restored only when ATP was omitted from the reaction. This observation inspired an investigation of the effect of nucleoside triphosphates on the binding of glucagon to its receptor. It was shown that relatively high (i.e., millimolar) concentrations of not only ATP but also uridine triphosphate (UTP) and cytidine triphosphate (CTP) reduced the binding of labeled glucagon. In contrast, the reduction of glucagon binding in the presence of GTP occurred at far lower (micromolar) concentrations. Moreover, low concentrations of GTP were found to stimulate the dissociation of bound glucagon from the receptor. Taken together, these studies suggested that GTP alters the glucagon receptor in a manner that lowers its affinity for glucagon. This decreased affinity both affects the ability of glucagon to bind to the receptor and encourages the dissociation of bound glucagon.

The observation that GTP was involved in the action of glucagon led to a second key question: Can GTP also exert an affect on adenyl cyclase? Addressing this question experimentally required the addition of both ATP, as a substrate for adenyl cyclase, and GTP, as the factor being examined, to the purified rat liver membranes. However, the previous study had shown that the concentration of ATP required as a substrate for adenyl cyclase could affect glucagon binding. Might it also stimulate adenyl cyclase? The concentration of ATP used in the experiment could not be reduced because ATP was readily hydrolyzed by ATPases present in the rat liver membrane. To get around this dilemma, Rodbell replaced ATP with an AMP analog, 5′-adenyl-imidodiphosphate (AMP-PNP), which can be converted to cAMP by adenyl cyclase, yet is resistant to hydrolysis by membrane ATPases. The critical experiment now could be performed. Purified rat liver membranes were treated with glucagon both in the presence and absence of GTP, and the production of cAMP from AMP-PNP was measured. The addition of GTP clearly stimulated the production of cAMP when compared to glucagon alone (Figure 1) indicating that GTP affects not only the binding of glucagon to its receptor but also stimulates the activation of adenylyl cyclase.

Discussion

Two key factors led Rodbell and his colleagues to detect the role of GTP in signal transduction, whereas previous

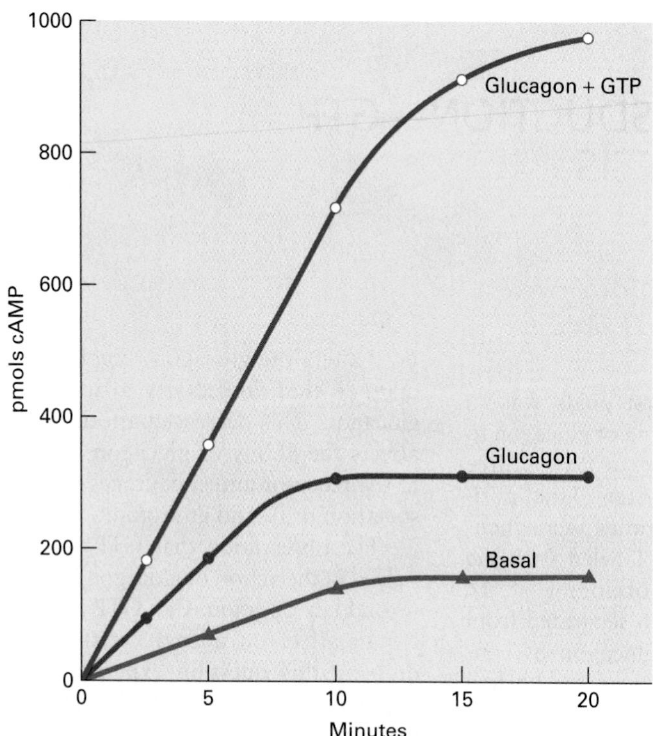

◀ **FIGURE 1 Effect of GTP on glucagon-stimulated cAMP production from AMP-PNP by purified rat liver membranes.** In the absence of GTP, glucagon stimulates cAMP formation about twofold over the basal level in the absence of added hormone. When GTP also is added, cAMP production increases another fivefold. [Adapted from M. Rodbell et al., 1971, *J. Biol. Chem.* **246:**1877.]

studies had failed to do so. First, by switching from fat-cell ghosts to the rat liver membrane system, the Rodbell researchers avoided contamination of their cell-free system with GTP, a problem associated with the procedure for isolating ghosts. Such contamination would mask the effects of GTP on glucagon binding and activation of adenyl cyclase. Second, when ATP was first shown to influence glucagon binding, Rodbell did not simply accept the plausible explanation that ATP, the substrate for adenyl cyclase, also affects binding of glucagon. Instead, he chose to test the effects on binding of the other common nucleoside triphosphates. Rodbell later noted that he knew commercial preparations of ATP often are contaminated with low concentrations of other nucleoside triphosphates. The possibility of contamination suggested to him that small concentrations of GTP might exert large effects on glucagon binding and the stimulation of adenyl cyclase.

This critical series of experiments stimulated a large number of studies on the role of GTP in hormone action, eventually leading to the discovery of G proteins, the GTP-binding proteins that couple certain receptors to the adenyl cyclase. Subsequently, an enormous family of receptors that require G proteins to transduce their signals were identified in eukaryotes from yeast to humans. These G protein-coupled receptors are involved in the action of many hormones as well as in a number of other biological activities, including neurotransmission and the immune response. It is now known that binding of ligands to their cognate G protein-coupled receptors stimulates the associated G proteins to bind GTP. This binding causes transduction of a signal that stimulates adenyl cyclase to produce cAMP and also desensitization of the receptor, which then releases its ligand. Both of these affects were observed in Rodbell's experiments on glucagon action. For these seminal observations, Rodbell was awarded the Nobel Prize for Physiology and Medicine in 1994.

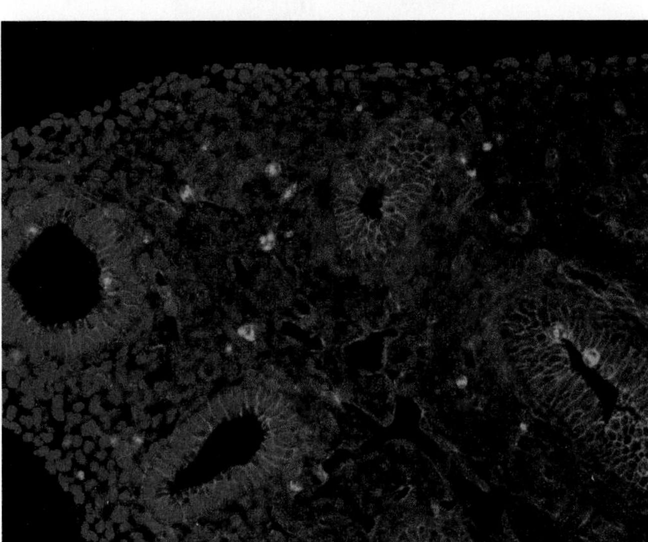

MAP kinase signaling in embryonic Day 13.5 mouse lung. Activated ERK is detected by a primary antibody that detects phosphorylated ERK followed by a secondary antibody conjugated to green fluorescing FITC. © Stijn Delanghe, Saverio Bellusci, Denise Tefft and David Warburton.

CHAPTER

16

CELL SIGNALING II: SIGNALING PATHWAYS THAT CONTROL GENE ACTIVITY

A cell's ability to respond to its environment is essential to its survival. Short-term responses to environmental stimuli, which can occur rapidly and are usually reversible, most often result from modification of existing proteins, as detailed in Chapter 15. Longer term responses, which are discussed in this chapter, are usually the result of changes in transcription of genes. Transcription is influenced by chromatin structure and the cell's complement of transcription factors and other proteins (Chapter 7). These determine which genes the cell can potentially transcribe at any given time. We think of these properties as the cell's "memory" determined by its history and response to previous signals. But many key regulatory transcription factors are held in an inactive state in the cytosol or nucleus and become activated in response to external signals. In this chapter we focus on how ligands that bind to cell-surface receptors trigger activation of specific transcription factors that, in turn, determine the precise pattern of cellular gene expression.

Extracellular signals that induce long-term responses affect many aspects of cell function: division, differentiation, and even communication with other cells. Alterations in these signaling pathways cause many human diseases, including cancer, diabetes, and immune defects. In addition to the crucial roles external signals play in development, signals are essential in enabling differentiated cells to respond to their environment by changing their shape, metabolism, or movement. For example, one type of transcription factor (NF-κB) ultimately impacts expression of more than 150 genes involved in the immune response to infection; NF-κB is activated by many protein hormones that act on immune system cells. Another family of extracellular signaling molecules, the cytokines, is involved in maintaining appropriate levels of blood cells such as erythrocytes (red blood cells), leukocytes (white blood cells), and platelets.

In order to illustrate the variety of mechanisms used to activate key transcription factors, we focus in this chapter on eight classes of cell-surface receptors and the intracellular signaling pathways that they activate. Ligand binding to many receptors causes activation by inducing two (or more) receptor molecules to form a complex on the cell surface. Most signaling pathways involve one or more protein kinases.

OUTLINE

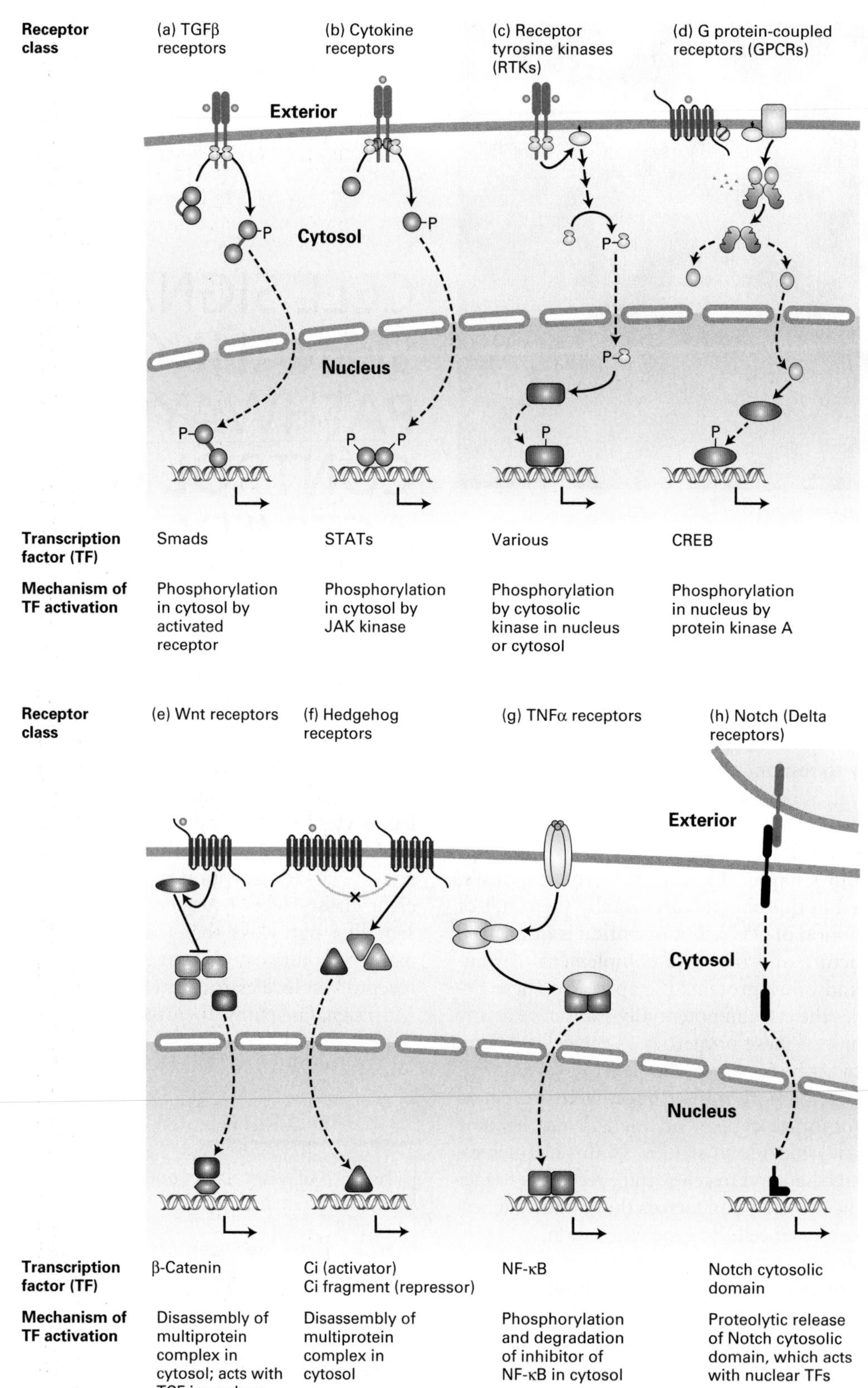

▲ **FIGURE 16-1 Overview of eight major classes of cell-surface receptors.** In many signaling pathways, ligand binding to a receptor leads to activation of transcription factors (TFs) in the cytosol, permitting them to translocate into the nucleus and stimulate (or occasionally repress) transcription of their target genes (a, b). Alternatively, receptor stimulation may lead to activation of cytosolic protein kinases that then translocate into the nucleus and regulate the activity of nuclear TFs (c, d). In other pathways, active TFs are released from cytosolic multiprotein complexes (e, f) or by proteolysis (g, h). Some receptor classes can trigger more than one intracellular pathway, as shown in Figure 16-2. [After A. H. Brivanlou and J. Darnell, 2002, *Science* **295:**813.]

The kinase may be an intrinsic part of the receptor protein or be tightly bound to the receptor. In either case, kinase activity is activated by ligand binding, resulting—directly or indirectly—in activation of specific transcription factors located in the cytosol (Figure 16-1a,b). In other pathways, receptor stimulation leads to activation of cytosolic protein kinases that translocate into the nucleus and phosphorylate specific nuclear transcription factors (Figure 16-1c,d). Binding of ligand to receptors for other types of signaling proteins causes disassembly of multiprotein complexes in the cytosol, releasing transcription factors that then translocate into the nucleus (Figure 16-1e,f). In still other signaling pathways, proteolytic cleavage of an inhibitor or the receptor itself releases an active transcription factor (Figure 16-1g,h). Importantly, all of these pathways are highly regulated, often by negative feedback, in order to control the level and duration of the signal's effects on cellular gene expression.

A typical mammalian cell expresses ≈100 different types of cell-surface receptors, many of which activate the same or similar signal-transduction pathways. As shown in Figure 16-2, several classes of receptors can transduce signals by more than one pathway, and some pathways are activated to a greater extent in certain cells than others. Moreover, many genes are regulated by multiple transcription factors, each of which can be activated by one or more extracellular signals. Especially during early development, such "cross talk" between signaling pathways and the resultant sequential alterations in the pattern of gene expression eventually can become so extensive that the cell assumes a different developmental fate. The receiving cell's prior history and regulatory state can alter the effect of a signal; the same signal applied to different cells will elicit distinct responses.

The pathways we discuss in this chapter have been conserved throughout evolution and operate in much the same manner in flies, worms, and man. The substantial homology exhibited among many proteins in these signaling pathways has enabled researchers to employ a variety of experimental approaches and systems to identify and study the function of extracellular signaling molecules, receptors, and intracellular signal-transduction proteins. For instance, the secreted signaling protein Hedgehog (Hh) and its receptor were first identified in *Drosophila* mutants with developmental defects. Subsequently, the human and mouse homologs of these proteins were cloned and shown to participate in a number of important signaling events during differentiation. Abnormal activation of the Hh pathway occurs in several human tumors. The examples explored in this chapter illustrate the importance of studying signaling pathways both genetically—in flies, mice, worms, yeasts, and other organisms—and biochemically.

Most of the discussion in this chapter is organized in terms of individual signaling pathways. That is, we consider the signaling molecules, their receptors, the intracellular

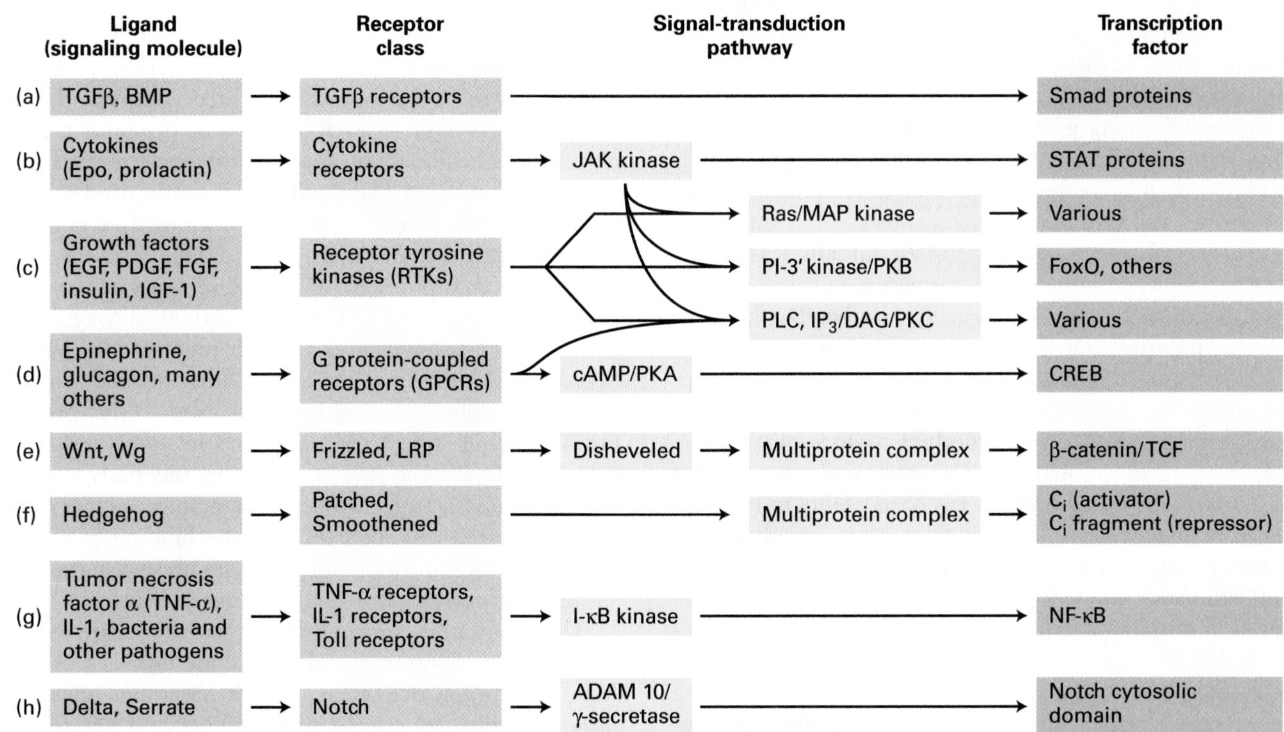

▲ **FIGURE 16-2 Components and modularity of major signaling pathways.** Signaling pathways are initiated by the binding of a signaling molecule (the ligand), which activates a receptor. Activated receptors then trigger various intracellular signal-transduction pathways that result in generation of active transcription factors either in the cytosol or nucleus. Transcription factors that are activated in the cytosol translocate to the nucleus (see Figure 16-1). Many receptor classes, including cytokine receptors, receptor tyrosine kinases (RTKs), and G protein–coupled receptors, can transduce signals by more than one pathway. PKA = protein kinase A; PKB = protein kinase B; PKC = protein kinase C; PLC = phospholipase C.

signal-transduction pathway(s), the regulated transcription factors, and regulation of the pathway itself for each of the receptor classes shown in Figure 16-1. In Chapters 21 and 22, we examine how extracellular signals affect gene regulation during several crucial developmental stages, and in particular how cells integrate the responses to multiple signals. In Chapter 25 we illustrate how abnormalities in several signal-transduction pathways described in this chapter can lead to cancer.

16.1 TGFβ Receptors and the Direct Activation of Smads

We begin our survey of signaling systems that control gene activity with one of the simplest: one family of signaling molecules (the TGFβ superfamily) binds to its receptors (the TGFβ receptors) and activates one class of transcription factors (the Smads), which are located in the cytosol; activated Smads then move into the nucleus to regulate transcription (see Figure 16-1a). Unlike many of the other signaling systems presented in this chapter, the TGFβ receptor activates only one type of transcription factor, and the transcription factor is activated by only one type of receptor. However, in spite of its simplicity, the TGFβ pathway can have widely diverse effects in different types of cells because different members of the TGFβ superfamily activate different members of the TGFβ receptor family that activate different members of the Smad class of transcription factors. Additionally, the same activated Smad protein will partner with different transcription factors in different cell types and thus activate different sets of genes in these cells.

The **transforming growth factor β (TGFβ)** superfamily includes a number of related extracellular signaling molecules that play widespread roles in regulating development in both invertebrates and vertebrates. One member of this superfamily, *bone morphogenetic protein (BMP),* initially was identified by its ability to induce bone formation in cultured cells. Now called BMP7, it is used clinically to strengthen bone after severe fractures. Of the numerous BMP proteins subsequently recognized, many help induce key steps in development, including formation of mesoderm and the earliest blood-forming cells. Most have nothing to do with bones.

Another member of the TGFβ superfamily, now called TGFβ-1, was identified on the basis of its ability to induce a malignant phenotype in certain cultured mammalian cells ("transforming growth factor"). However, the three human TGFβ isoforms that are known all potently *prevent* proliferation of most mammalian cells by inducing synthesis of proteins that inhibit the cell cycle. TGFβ is produced by many cells in the body and inhibits growth both of the secreting cell (autocrine signaling) and neighboring (paracrine signaling) cells. Loss of TGFβ receptors or any of several intracellular signal-transduction proteins in the TGFβ pathway, thereby releasing cells from this growth inhibition, occurs frequently in human tumors. TGFβ proteins also promote expression of cell-adhesion molecules and extracellular-matrix molecules, which play important roles in tissue organization (Chapter 19). A *Drosophila* homolog of TGFβ, called Dpp protein, participates in dorsal-ventral patterning in fly embryos. Other mammalian members of the TGFβ superfamily, the activins and inhibins, affect early development of the genital tract. We consider such developmentally important TGFb proteins in Chapter 22.

Despite the complexity of cellular effects induced by various members of the TGFβ superfamily, the signaling pathway is basically a simple one (see Figure 16-2a). Once activated, receptors for these ligands directly phosphorylate and activate a particular type of transcription factor. The response of a given cell to this activated transcription factor depends on the constellation of other transcription factors it already contains. In this section, we will progress sequentially through the TGFβ pathway, considering first the family of signal molecules, then the TGFβ receptors and their discovery. Next we present information about how these receptors activate Smad transcription factors and the feedback loops that regulate signaling by this pathway. The role that TGFβ plays in cancer completes our examination of TGFβ-Smad signaling.

A TGFβ Signaling Molecule Is Formed by Cleavage of an Inactive Precursor

Most animal cell types produce and secrete members of the TGFβ superfamily in an inactive form that is stored nearby in the extracellular matrix. Release of the active form from the matrix by protease digestion or inactivation of an inhibitor leads to quick mobilization of the signal already in place—an important feature of many signaling pathways.

In humans TGFβ consists of three isoforms, TGFβ-1, TGFβ-2, and TGFβ-3, each encoded by a unique gene and expressed in both a tissue-specific and developmentally regulated fashion. Each TGFβ isoform is synthesized as part of a larger dimeric precursor, linked by a disulfide bond, that contains a pro-domain (often called LAP). After the precursor is secreted, LAP is cleaved off but remains noncovalently bound to the mature TGFβ via interactions between specific four amino acid sequences in each polypeptide. Most secreted TGFβ is stored in the extracellular matrix as a latent, inactive complex containing the cleaved TGFβ precursor and a disulfide-linked protein called *latent TGFβ-binding protein (LTBP)*. Binding of LAP by the matrix protein thrombospondin triggers release of mature, active dimeric TGFβ. Alternatively, digestion of the binding proteins by serum proteases or by metalloproteases present in the matrix can result in activation of TGFβ (Figure 16-3a).

The monomeric form of TGFβ growth factors contains three conserved intramolecular disulfide linkages. An additional cysteine in the center of each monomer links TGFβ monomers into functional homodimers and heterodimers (Figure 16-3b). Much of the sequence variation among different TGFβ proteins is observed in the N-terminal regions, the loops joining the β strands, and the α helices. Different heterodimeric combinations may increase the functional diversity of these proteins beyond that generated by differences in the primary sequence of the monomer.

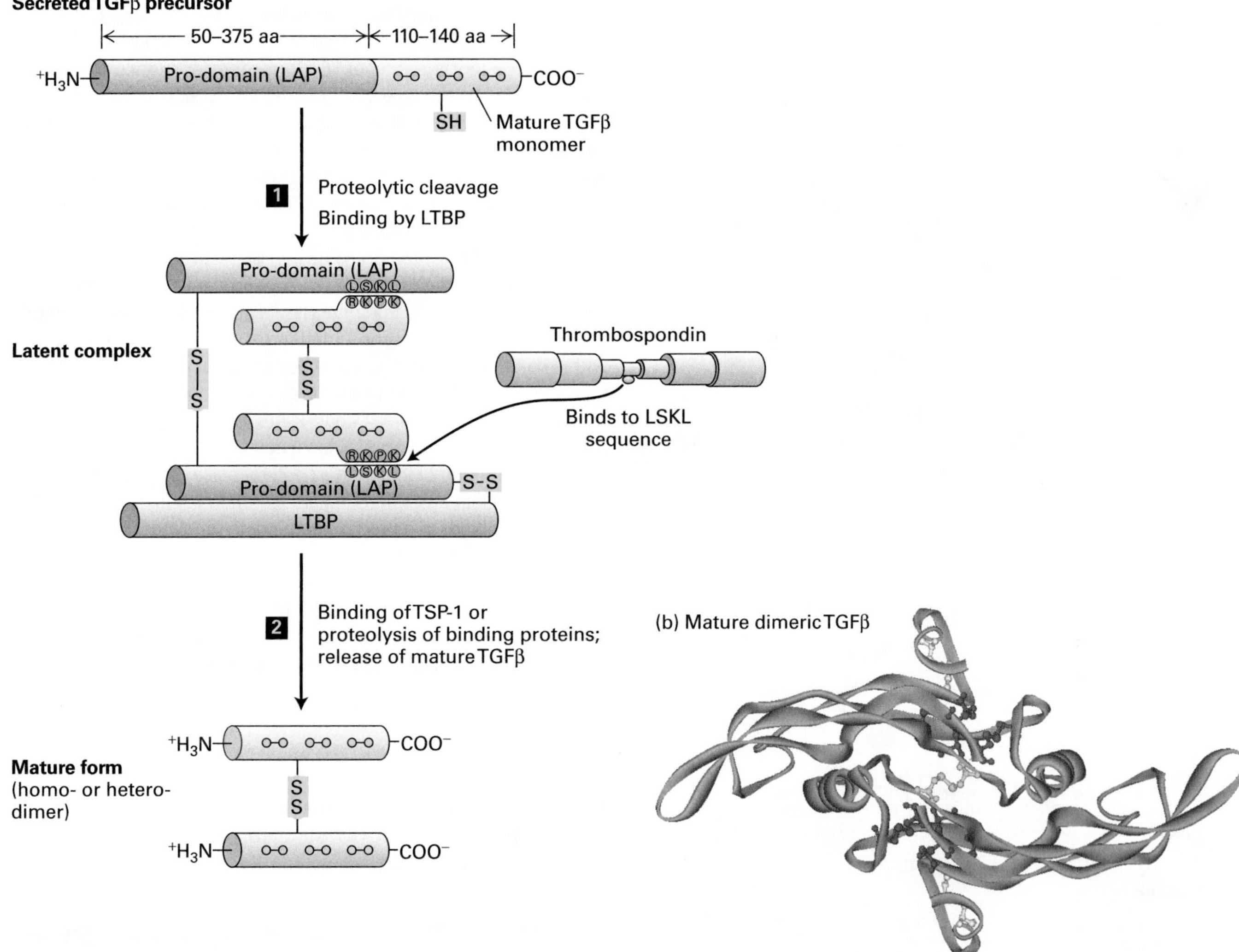

▲ **FIGURE 16-3 Formation and structure of TGFβ superfamily of signaling molecules.** (a) Step **1**: Dimeric TGFβ precursors form inside the cell, although only the monomeric form is shown in the top diagram. Soon after being secreted, a precursor molecule is cleaved, but the pro-domain, often called LAP, and the mature TGFβ remain noncovalently bound by specific interactions between LSKL and RKPK amino acid sequences, respectively, in a complex that also contains latent TGFβ-binding protein (LTBP, orange). Mature monomeric TGFβ (blue and green) contains six conserved cysteine residues (yellow circles), which form three intrachain disulfide bonds; a single disulfide bond connects two monomers (shown in second diagram). The entire complex resulting from cleavage is stored in the intracellular matrix. Step **2**: Mature homo- or heterodimeric TGFβ can be released from this complex by binding of the extracellular matrix protein thrombospondin-1 (TSP-1) to the LSKL sequence in the LAP protein. Alternatively, serum proteases can digest the binding proteins, releasing active TGFβ. (b) In this ribbon diagram of mature TGFβ dimer, the two subunits are shown in green and blue. Disulfide-linked cysteine residues (yellow and red) are shown in ball-and-stick form. The three intrachain disulfide linkages (red) in each monomer form a cystine-knot domain, which is resistant to degradation. [Part (a) see J. Massagué and Y.-G. Chen, 2000, *Genes and Devel.* **14**:627, and J. E. Murphy-Ullrich and M. Poczatek, 2000, *Cytokine Growth Factor Rev.* **11**:59. Part (b) from S. Daopin et al., 1992, *Science* **257**:369.]

Radioactive Tagging Was Used to Identify TGFβ Receptors

Researchers first identified the TGFβ signaling molecule as a growth inhibitory factor, but to understand the way it worked, they had to find the receptors to which it bound. The logic of how they went about their search is representative of typical biochemical approaches to identifying receptors. Investigators first reacted the purified growth factor with the radioisotope iodine-125 (^{125}I) under conditions such that the iodine covalently bonds to exposed tyrosine residues—effectively tagging them with a radioactive label. The ^{125}I-labeled TGFβ protein was incubated with cultured cells, and the incubation mixture then was treated with a chemical agent that covalently cross-linked the labeled TGFβ to its receptors on the cell surface. Purification of the ^{125}I-labeled TGFβ–receptor complexes revealed three different polypeptides with apparent molecular weights of 55, 85, and 280 kDa, referred to as types RI, RII, and RIII TGFβ receptors, respectively.

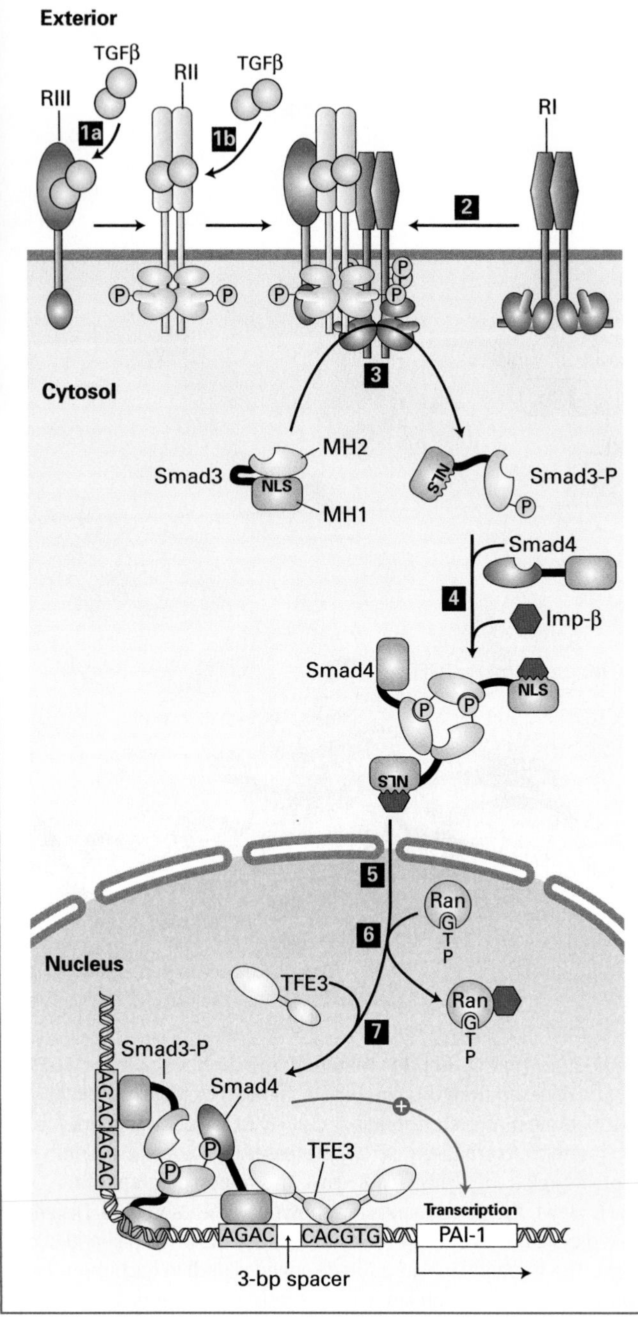

◀ **FIGURE 16-4 TGFβ/Smad signaling pathway.** Step **1a**: In some cells, TGFβ binds to the type III TGFβ receptor (RIII), which increases the concentration of TGFβ near the cell surface and also presents TGFβ to the type II receptor (RII). Step **1b**: In other cells, TGFβ binds directly to RII, a constitutively phosphorylated and active kinase. Step **2**: Ligand-bound RII recruits and phosphorylates the juxtamembrane segment of the type I receptor (RI), which does not directly bind TGFβ. This releases the inhibition of RI kinase activity that otherwise is imposed by the segment of RI between the membrane and its kinase domain. Step **3**: Activated RI then phosphorylates Smad3 (shown here) or another R-Smad, causing a conformational change that unmasks its nuclear-localization signal (NLS). Step **4**: Two phosphorylated molecules of Smad3 interact with a co-Smad (Smad4), which is not phosphorylated, and with importin β (Imp-β), forming a large cytosolic complex. Steps **5** and **6**: After the entire complex translocates into the nucleus, Ran·GTP causes dissociation of Imp-β as discussed in Chapter 13. Step **7**: A nuclear transcription factor (e.g., TFE3) then associates with the Smad3/Smad4 complex, forming an activation complex that cooperatively binds in a precise geometry to regulatory sequences of a target gene. Shown at the bottom is the activation complex for the gene encoding plasminogen activator inhibitor (PAI-1). [See Z. Xiao et al., 2000, *J. Biol. Chem.* **275:**23425; J. Massagué and D. Wotton, 2000, *EMBO J.* **19:**1745; X. Hua et al., 1999, *Proc. Nat'l. Acad. Sci. USA* **96:**13130; and A. Moustakas and C.-H. Heldin, 2002, *Genes Devel.* **16:**1867.]

Figure 16-4 (steps **1** and **2**) depicts the relationship and function of the three TGFβ receptor proteins. The most abundant, RIII, is a cell-surface proteoglycan, also called β-glycan. RIII, a monomeric transmembrane protein, binds and concentrates mature TGFβ molecules near the cell surface, facilitating their binding to RII receptors. The type I and type II receptors are dimeric transmembrane proteins with serine/threonine kinases as part of their cytosolic domains. RII exhibits **constitutive** kinase activity; that is, it is active even when not bound to TGFβ. Binding of TGFβ induces the formation of complexes containing two copies each of RI and RII. An RII subunit then phosphorylates serine and threonine residues in a highly conserved sequence of the RI subunit adjacent to the cytosolic face of the plasma membrane, thereby activating the RI kinase activity.

Activated TGFβ Receptors Phosphorylate Smad Transcription Factors

Researchers identified the transcription factors downstream from TGFβ receptors in *Drosophila* from genetic studies using mutant fruit flies. These transcription factors in *Drosophila* and the related vertebrate proteins are now called **Smads.** Three types of Smad proteins function in the TGFβ signaling pathway: *R-Smads* (receptor-regulated Smads), *co-Smads,* and *I-Smads* (inhibitory Smads).

As illustrated in Figure 16-4, an R-Smad (Smad2 or Smad3) contains two domains, MH1 and MH2, separated by a flexible linker region. The N-terminal MH1 domain contains the specific DNA-binding segment and also a sequence called the *nuclear-localization signal (NLS).* NLSs are present in virtually all transcription factors found in the cytosol and are required for their transport into the nucleus (Chapter 13). When R-Smads are in their inactive, nonphosphorylated state, however, the NLS is masked and the MH1 and MH2 domains associate in such a way that they cannot bind to DNA or to a co-Smad. Phosphorylation of three serine residues near the C-terminus of an R-Smad by activated type I TGFβ receptors separates the domains, permitting binding of importin β (see Figure 13-35) to the NLS, which enables entrance of the Smad into the nucleus.

Simultaneously a complex containing two molecules of Smad3 (or Smad2) and one molecule of a co-Smad (Smad4) forms in the cytosol. This complex is stabilized by binding of two phosphorylated serines in each Smad3 to phosphoserine-binding sites in both the Smad3 and the Smad4 MH2 domains. The bound importin β then mediates translocation of the heteromeric R-Smad/co-Smad complexes into the nucleus. After importin β dissociates inside the nucleus, the Smad2/Smad4 or Smad3/Smad4 complexes bind to other transcription factors to activate transcription of specific target genes.

Within the nucleus R-Smads are continuously being dephosphorylated, which results in the dissociation of the R-Smad/co-Smad complex and export of these Smads from the nucleus. Because of this continuous nucleocytoplasmic shuttling of the Smads, the concentration of active Smads within the nucleus closely reflects the levels of activated TGFβ receptors on the cell surface.

Virtually all mammalian cells secrete at least one TGFβ isoform, and most have TGFβ receptors on their surface. However, because different types of cells contain different sets of transcription factors with which the activated Smads can bind, the cellular responses induced by TGFβ vary among cell types. In epithelial cells and fibroblasts, for example, TGFβ induces expression not only of extracellular-matrix proteins (e.g., collagens) but also of proteins that inhibit serum proteases, which otherwise would degrade these extracellular-matrix proteins. This inhibition stabilizes the matrix, allowing cells to form stable tissues. The inhibitory proteins include plasminogen activator inhibitor 1 (PAI-1). Transcription of the *PAI-1* gene requires formation of a complex of the transcription factor TFE3 with the Smad3/Smad4 complex and binding of all these proteins to specific sequences within the regulatory region of the *PAI-1* gene (see Figure 16-4, *bottom*). By partnering with other transcription factors, Smad2/Smad4 and Smad3/Smad4 complexes promote expression of proteins such as p15, which arrests the cell cycle at the G_1 stage and thus blocks cell proliferation (Chapter 20).

As just discussed, binding of any one TGFβ isoform to its specific receptors leads to phosphorylation of Smad2 or Smad3 followed by formation of Smad2/Smad4 or Smad3/Smad4 complexes, and eventually transcriptional activation of specific target genes (e.g., the *PAI-1* gene). On the other hand, BMP proteins, which also belong to the TGFβ superfamily, bind to and activate a different set of receptors that are similar to the TGFβ RI and RII proteins but phosphorylate Smad1 (rather than Smad2 or Smad3). Smad1 then dimerizes with Smad4, and the Smad1/Smad4 complex activates different transcriptional responses than those induced by Smad2/Smad4 or Smad3/Smad4.

Negative Feedback Loops Regulate TGFβ/Smad Signaling

Most signaling pathways are modulated in such a way that the response to a growth factor or other signaling molecule is decreased (or occasionally increased) with time; this enables the fine-tuned control of cellular responses. TGFβ/Smad pathways are regulated by several intracellular proteins, including two cytosolic proteins called *SnoN* and *Ski* (Ski stands for "*S*loan-*K*ettering Cancer *I*nstitute"). These proteins were originally identified as cancer-causing **oncoproteins** because they cause abnormal cell proliferation when overexpressed in cultured primary fibroblast cells. How they accomplish this was not understood until years later when SnoN and Ski were found to bind to both Smad4 and phosphorylated Smad3 after TGFβ stimulation. SnoN and Ski do not prevent formation of Smad3/Smad4 complexes or affect the ability of the Smad complexes to bind to DNA control regions. Rather, they block transcription activation by the bound Smad complexes, thereby rendering cells resistant to the growth-inhibitory effects of TGFβ (Figure 16-5). Interestingly, stimulation by TGFβ causes the rapid degradation of Ski and SnoN, but after a few hours, expression of both Ski and SnoN becomes strongly induced via mechanisms not yet understood. The increased levels of these proteins are thought to dampen long-term signaling effects due to continued exposure to TGFβ.

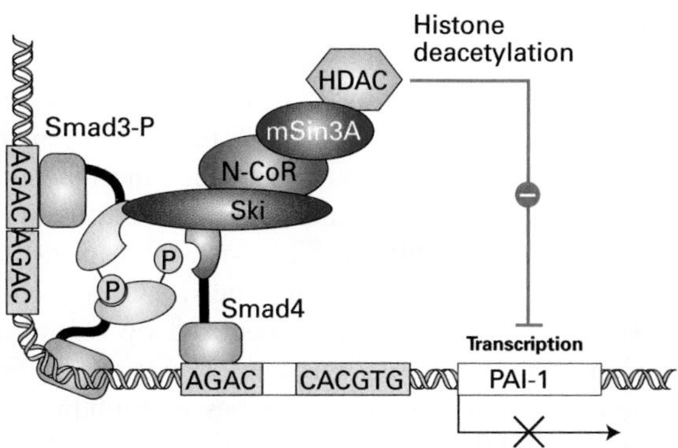

▲ **FIGURE 16-5 Model of Ski-mediated down-regulation of Smad transcription-activating function.** Ski represses Smad function by binding directly to Smad4. Since the Ski-binding domain on Smad4 significantly overlaps with the domain required for binding the phosphorylated tail of Smad3, binding of Ski disrupts the normal interactions between Smad3 and Smad4. Additionally, Ski also recruits the protein N-CoR, which binds directly to mSin3A; in turn mSin3A interacts with histone deacetylase (HDAC), an enzyme that promotes histone deacetylation (Chapter 7). As a result, transcription activation induced by TGFβ and mediated by Smad complexes is shut down. [See K. Luo, 2004, *Curr. Opin. Genet. and Dev.* **14:**65.]

Among the proteins induced after TGFβ stimulation are the I-Smads, especially Smad7. Smad7 blocks the ability of activated type I receptors (RI) to phosphorylate R-Smads proteins, and it may also target TGFβ receptors for degradation. In these ways Smad7, like Ski and SnoN, participates in negative feedback loops: Its induction inhibits intracellular signaling by long-term exposure to the stimulating hormone. In later sections we see how signaling by other cell-surface receptors is also restrained by negative feedback loops.

Loss of TGFβ Signaling Plays a Key Role in Cancer

Many human tumors contain inactivating mutations in either TGFβ receptors or Smad proteins, and thus are resistant to growth inhibition by TGFβ (see Figure 25-24).

Most human pancreatic cancers, for instance, contain a deletion in the gene encoding Smad4 and thus cannot induce p15 and other cell-cycle inhibitors in response to TGFβ. In fact, Smad4 was originally called *DPC* (*d*eleted in *p*ancreatic *c*ancer). Retinoblastoma, colon and gastric cancer, hepatoma, and some T- and B-cell malignancies are also unresponsive to TGFβ growth inhibition. This loss of responsiveness correlates with loss of type I or type II TGFβ receptors; responsiveness to TGFβ can be restored by recombinant expression of the "missing" protein. Mutations in Smad2 also commonly occur in several types of human tumors. Not only is TGFβ signaling essential for controlling cell proliferation, as these examples show, but it also causes some cells to differentiate along specific pathways, as discussed in Chapter 22. ∎

KEY CONCEPTS OF SECTION 16.1

TGFβ Receptors and the Direct Activation of Smads

- TGFβ is produced as an inactive precursor that is stored in the extracellular matrix. Several mechanisms can release the active, mature dimeric growth factor (see Figure 16-3).
- Stimulation by TGFβ leads to activation of the intrinsic serine/threonine kinase activity in the cytosolic domain of the type I (RI) receptor, which then phosphorylates an R-Smad, exposing a nuclear-localization signal.
- After phosphorylated R-Smad binds a co-Smad, the resulting complex translocates into the nucleus, where it interacts with various transcription factors to induce expression of target genes (see Figure 16-4).
- Oncoproteins (e.g., Ski and SnoN) and I-Smads (e.g., Smad7) act as negative regulators of TGFβ signaling.
- TGFβ signaling generally inhibits cell proliferation. Loss of various components of the signaling pathway contributes to abnormal cell proliferation and malignancy.

16.2 Cytokine Receptors and the JAK/STAT Pathway

We turn now to another type of signaling pathway that leads to long-term genetic effects by activation of transcription factors. The signal molecules in this pathway, the **cytokines,** play many important roles in growth and differentiation of cells, especially blood and immune system cells. Like the TGFβ pathway just described, the cytokine pathway involves only a few steps. The cytosolic domain of all **cytokine receptors** tightly binds a member of a family of cytosolic protein tyrosine kinases, the *JAK kinases*. Activated JAK kinases, in turn, directly phosphorylate and activate transcription factors that are members of the *STAT* (*S*ignal *T*ransduction and *A*ctivation of *T*ranscription) family. Activated cytokine receptors activate additional pathways (see Figure 16-2b) that are also activated by other receptor classes. However, the *JAK/STAT pathway* described in this section is initiated mainly by activation of cytokine receptors, although it can be activated by other receptors.

We begin by discussing the cytokine family of signaling molecules and the cytokine receptors. Next, we use an experimental approach to explore how the JAK/STAT pathway was discovered. Then we consider the details of how a JAK protein activates a STAT transcription factor, followed by a discussion of how cytokine signaling paths are regulated. This section concludes with a biological application exploring the regulation of red blood cells in the human body.

Cytokines Influence Development of Many Cell Types

The cytokines form a family of relatively small, secreted signaling molecules (generally containing about 160 amino acids) that control many aspects of growth and differentiation of specific types of cells. During pregnancy, for example, the cytokine *prolactin* induces epithelial cells lining the immature ductules of the mammary gland to differentiate into the acinar cells that produce milk proteins and secrete them into the ducts. Other cytokines, the **interleukins,** are essential for proliferation and functioning of T cells and antibody-producing B cells of the immune system. Another family of cytokines, the **interferons,** are produced and secreted by certain cell types following virus infection. The secreted interferons act on nearby cells to induce enzymes that render these cells more resistant to virus infection. The role of interleukins and interferons in immune responses are covered in Chapter 24.

Many cytokines induce formation of important types of blood cells. For instance, granulocyte colony stimulating factor (G-CSF) induces a particular type of progenitor cell in the bone marrow to divide several times and then differentiate into granulocytes, the type of white blood cell that inactivates bacteria and other pathogens. Because many cancer therapies reduce granulocyte formation by the body, G-CSF often is administered to patients to stimulate proliferation and differentiation of granulocyte progenitor cells, thus restoring the normal level of granulocytes in the blood. Thrombopoietin, a "cousin" of G-CSF, similarly acts on megakaryocyte progenitors to divide and differentiate into megakaryocytes. These then fragment into the cell pieces called platelets, which are critical for blood clotting. ∎

Another cytokine, **erythropoietin (Epo),** triggers production of erythrocytes (red blood cells) by inducing the proliferation and differentiation of erythroid progenitor cells in the bone marrow (Figure 16-6). Erythropoietin is synthesized by kidney cells that monitor the concentration of oxygen in the blood. A drop in blood oxygen signifies a lower than optimal level of erythrocytes, whose major function is to transport oxygen complexed to hemoglobin. By means of the oxygen-sensitive transcription factor HIF-1α, the kidney cells respond to low oxygen by synthesizing more erythropoietin and secreting it into the blood. As the level of erythropoietin rises, more and more erythroid progenitors are

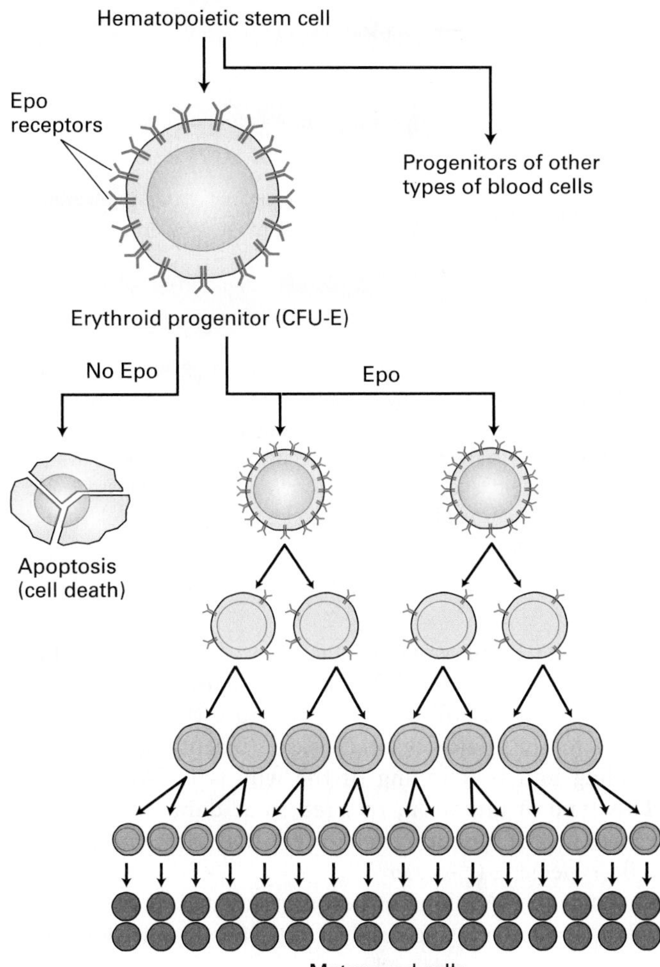

▲ **FIGURE 16-6 Erythropoietin and formation of red blood cells (erythrocytes).** Erythroid progenitor cells, called colony-forming units erythroid (CFU-E), are derived from hematopoietic stem cells, which also give rise to progenitors of other blood cell types. In the absence of erythropoietin (Epo), CFU-E cells undergo apoptosis. Binding of Epo to its receptors on a CFU-E induces transcription of several genes whose encoded proteins prevent programmed cell death (apoptosis), allowing the cell to survive. Other Epo-induced proteins trigger the developmental program of three to five terminal cell divisions. The Epo receptor and other membrane proteins are lost from these cells as they undergo differentiation. If CFU-E cells are cultured with Epo in a semisolid medium (e.g., containing methylcellulose), daughter cells cannot move away, and thus each CFU-E produces a colony of 30–100 erythroid cells, hence its name. [See M. Socolovsky et al., 2001, *Blood* **98:**3261.]

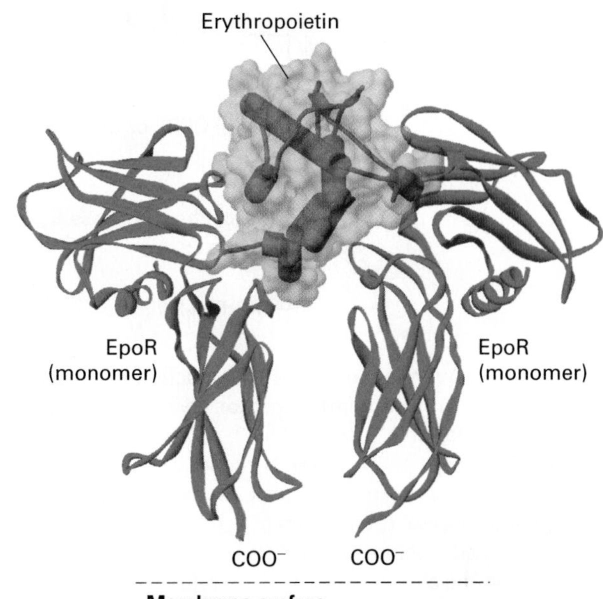

▲ **FIGURE 16-7 Structure of erythropoietin bound to an erythropoietin receptor.** Erythropoietin (Epo) contains four conserved long α helices that are folded in a particular arrangement. The activated erythropoietin receptor (EpoR) is a dimer of identical subunits; the extracellular domain of each monomer is constructed of two subdomains each containing seven conserved β strands folded in a characteristic fashion. Side chains of residues on two of the helices in Epo contact loops on one EpoR monomer, while residues on the two other Epo helices bind to the same loop segments in a second receptor monomer, thereby stabilizing the dimeric receptor in a specific conformation. The structures of other cytokines and their receptors are similar to Epo and EpoR. [Adapted from R. S. Syed et al., 1998, *Nature* **395:**511.]

saved from death, allowing each to produce ≈50 or so erythrocytes in a period of only a few days. In this way, the body can respond to the loss of blood by accelerating the production of erythrocytes.

Cytokine Receptors Have Similar Structures and Activate Similar Signaling Pathways

Strikingly, all cytokines have a similar tertiary structure, consisting of four long conserved α helices folded together in a specific orientation. The structural homology among cytokines is evidence that they all evolved from a common ancestral protein. Likewise, the various cytokine receptors undoubtedly evolved from a single common ancestor as all cytokine receptors have similar structures. Their extracellular domains are constructed of two subdomains, each of which contains seven conserved β strands folded together in a characteristic fashion. The interaction of one erythropoietin molecule with two identical erythropoietin receptor (EpoR) proteins, depicted in Figure 16-7, exemplifies the binding of a cytokine to its receptor.

Whether or not a cell responds to a particular cytokine depends simply on whether or not it expresses the corresponding (cognate) receptor. Although all cytokine receptors activate similar intracellular signal-transduction pathways, the response of any particular cell to a cytokine signal depends on the cell's constellation of transcription factors, chromatin structures, and other proteins relating to the developmental history of the cell. If receptors for prolactin or thrombopoietin, for example, are expressed experimentally in an erythroid progenitor cell, the cell will respond to these cytokines by dividing and differentiating into erythrocytes, not into mammary cells or megakaryocytes.

Figure 16-8 summarizes the intracellular signaling pathways activated when the EpoR binds erythropoietin. Stimulation of other cytokine receptors by their specific ligands

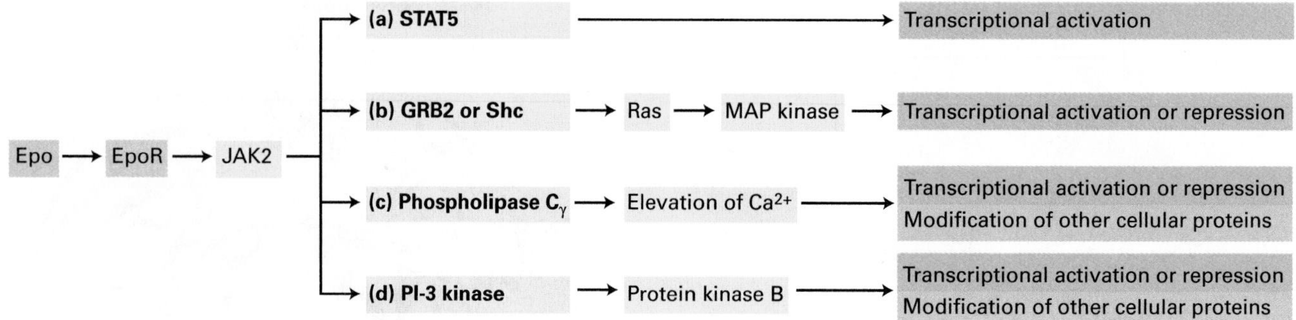

▲ **FIGURE 16-8 Overview of signal-transduction pathways triggered by the erythropoietin receptor.** Erythropoietin (Epo) binds to the erythropoietin receptor (EpoR), activating the associated JAK kinase. Four major pathways can transduce a signal from the activated, phosphorylated EpoR–JAK complex. Each pathway ultimately regulates transcription of different sets of genes. (a) In the most direct pathway, discussed in this section, the transcription factor STAT5 is phosphorylated and activated directly in the cytosol. (b) Binding of adapter proteins (GRB2 or Shc) to an activated EpoR leads to activation of the Ras/MAP kinase pathway (Section 16.4). (c, d) Two phosphoinositide pathways are triggered by recruitment of phospholipase C_γ and PI-3 kinase to the membrane (Section 16.5) following activation of EpoR. Elevated levels of Ca^{2+} and activated protein kinase B also modulate the activity of cytosolic proteins that are not involved in control of transcription.

activates similar pathways. All these pathways eventually lead to activation of transcription factors, causing an increase or decrease in expression of particular target genes. Here we focus on the JAK/STAT pathway; the other pathways are discussed in later sections.

JAK Kinases Activate STAT Transcription Factors

To understand how JAK and STAT proteins function, we examine the pathway downstream of the erythropoietin receptor (EpoR), one of the best-understood cytokine signaling pathways. A JAK2 kinase is tightly bound to the cytosolic domain of all cytokine receptors (see Figure 16-1b). Like the three other members of the JAK family of kinases, JAK2 contains an N-terminal receptor-binding domain, a C-terminal kinase domain that is normally poorly active catalytically, and a middle domain that regulates kinase activity by an unknown mechanism. JAK2, erythropoietin, and the EpoR are all required for formation of adult-type erythrocytes, which normally begins at day 12 of embryonic development in mice. As Figure 16-9 shows, embryonic mice lacking functional genes encoding the EpoR cannot form adult-type erythrocytes and eventually die owing to the inability to transport oxygen to the fetal organs. Mice lacking functional genes encoding Epo or JAK2 show similar blocks in fetal development.

As already noted, erythropoietin binds simultaneously to the extracellular domains of two EpoR monomers on the cell surface (see Figure 16-7). As a result, the associated JAKs are brought close enough together so that one can phosphorylate the other on a critical tyrosine in a region of the protein called the *activation lip* (Figure 16-10). As with other kinases, phosphorylation of the activation lip leads to a conformational change that reduces the K_m for ATP or the substrate to be phosphorylated, thus increasing kinase activity. One piece of evidence for this activation mechanism comes from study of a mutant JAK2 in which the critical tyrosine is mutated to phenylalanine. The mutant JAK2 binds normally to the EpoR but cannot be phosphorylated. In erythroid cells, expression of this mutant JAK2 in greater than normal amounts totally blocks EpoR signaling, because the mutant JAK2 binds to the majority of cytokine receptors, preventing binding and functioning of the wild-type JAK2 protein. This type of mutation, referred to as **dominant-negative,** causes loss of function even in cells that carry copies of the wild-type gene (Chapter 5).

Once the JAK kinases become activated, they phosphorylate several tyrosine residues on the cytosolic domain of

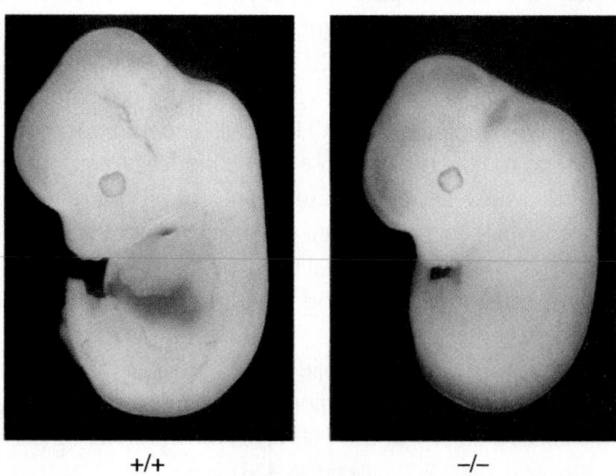

▲ EXPERIMENTAL FIGURE 16-9 **Studies with mutant mice reveal that the erythropoietin receptor (EpoR) is essential for development of erythrocytes.** Mice in which both alleles of the gene encoding EpoR are "knocked out" develop normally until embryonic day 13, when they begin to die of anemia due to the lack of erythrocyte-mediated transport of oxygen to the fetal organs. The red organ in the wild-type embryos (+/+) is the fetal liver, the major site of erythrocyte production at this developmental stage. The absence of color in the mutant embryos (−/−) indicates the absence of erythrocytes containing hemoglobin. Otherwise the mutant embryos appear normal, indicating that the main function of the EpoR in early mouse development is to support production of erythrocytes. [From H. Wu et al., 1995, *Cell* **83:**59].

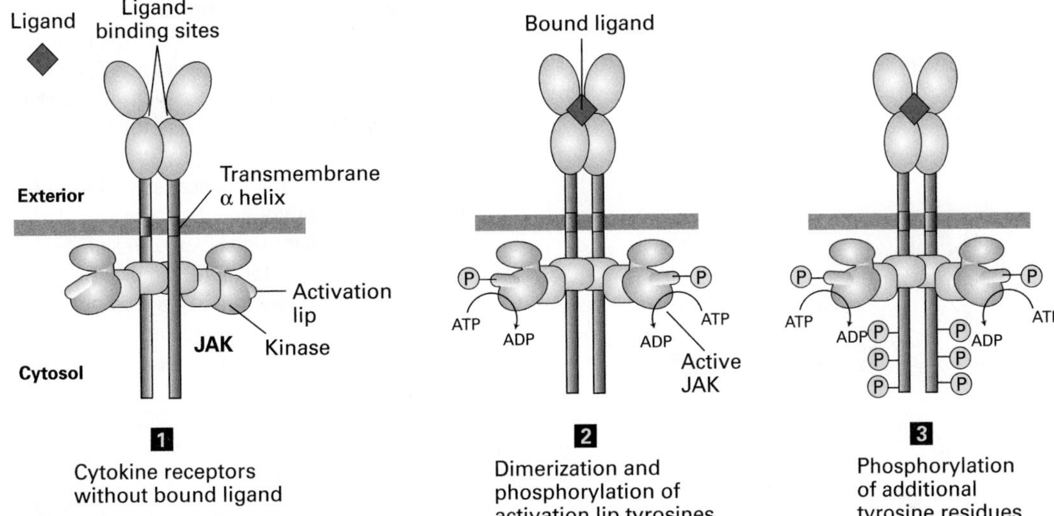

▲ **FIGURE 16-10 General structure and activation of cytokine receptors.** The cytosolic domain of cytokine receptors tightly and irreversibly associates with a separate JAK kinase. In the absence of ligand (**1**), the receptors form a homodimer but the JAK kinases are poorly active. Ligand binding causes a conformational change that brings together the associated JAK kinase domains, which then phosphorylate each other on a tyrosine residue in the activation lip (**2**). Phosphorylation causes the lip to move out of the kinase catalytic site, thus increasing the ability of ATP and the protein substrate to bind. The activated kinase then phosphorylates several tyrosine residues in the receptor's cytosolic domain (**3**). The resulting phosphotyrosines function as docking sites for inactive STAT transcription factors and other signal-transduction proteins that contain SH2 or PTB domains.

the receptor (see Figure 16-10, **3**). Several of these phosphotyrosine residues then serve as binding sites for *SH2 domains*, which are part of many signal-transduction proteins, including the STAT group of transcription factors. The SH2 domain derived its full name, the Src *h*omology 2 domain, from its homology with a region in the prototypical Src cytosolic tyrosine kinase encoded by the *src* gene. (Src is an acronym for sarcoma, and a mutant form of the cellular *src* gene was found in chickens with sarcomas, as Chapter 25 details.) The three-dimensional structures of SH2 domains in different proteins are very similar, but each binds to a distinct sequence of amino acids surrounding a phosphotyrosine residue. The unique amino acid sequence of each SH2 domain determines the specific phosphotyrosine residues it binds (Figure 16-11). Variations in the hydrophobic socket in the SH2 domains of different STATs and other signal-transduction proteins allow them to bind to phosphotyrosines adjacent to different sequences, accounting for differences in their binding partners.

All STAT proteins contain an N-terminal DNA-binding domain, an SH2 domain that binds to a specific phosphotyrosine in a cytokine receptor's cytosolic domain, and a C-terminal domain with a critical tyrosine residue. Once a STAT is bound to the receptor, the C-terminal tyrosine is phosphorylated by an associated JAK kinase (Figure 16-12a). This arrangement ensures that in a particular cell only those STAT proteins with an SH2 domain that can bind to a particular receptor protein will be activated. The erythropoietin receptor, for example, activates STAT5 but not STATs 1, 2, 3, or 4. A phosphorylated STAT dissociates spontaneously from the receptor, and two phosphorylated

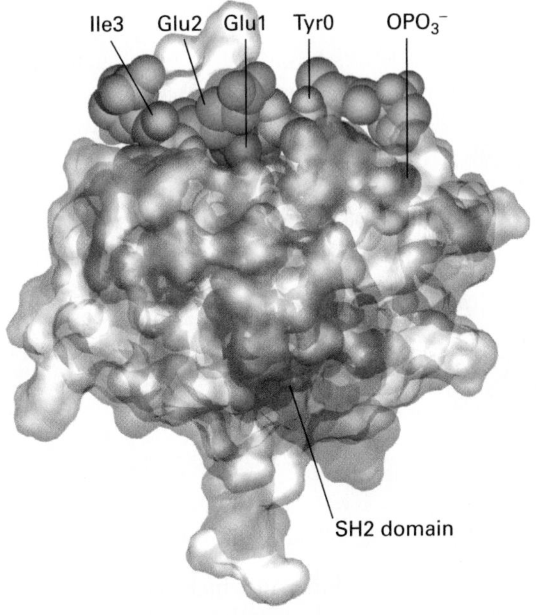

▲ **FIGURE 16-11 Surface model of a SH2 domain bound to a phosphotyrosine-containing peptide.** The peptide bound by this SH2 domain from Src tyrosine kinase (gray) is shown in spacefill. The SH2 domain binds strongly to short target peptides containing a critical four-residue core sequence: phosphotyrosine (Tyr0 and OPO_3^-)–glutamic acid (Glu1)–glutamic acid (Glu2)–isoleucine (Ile3). Binding resembles the insertion of a two-pronged "plug"—the phosphotyrosine and isoleucine side chains of the peptide—into a two-pronged "socket" in the SH2 domain. The two glutamate residues are bound to sites on the surface of the SH2 domain between the two sockets. Nonbinding residues on the target peptide are colored green. [See G. Waksman et al., 1993, *Cell* **72**:779.]

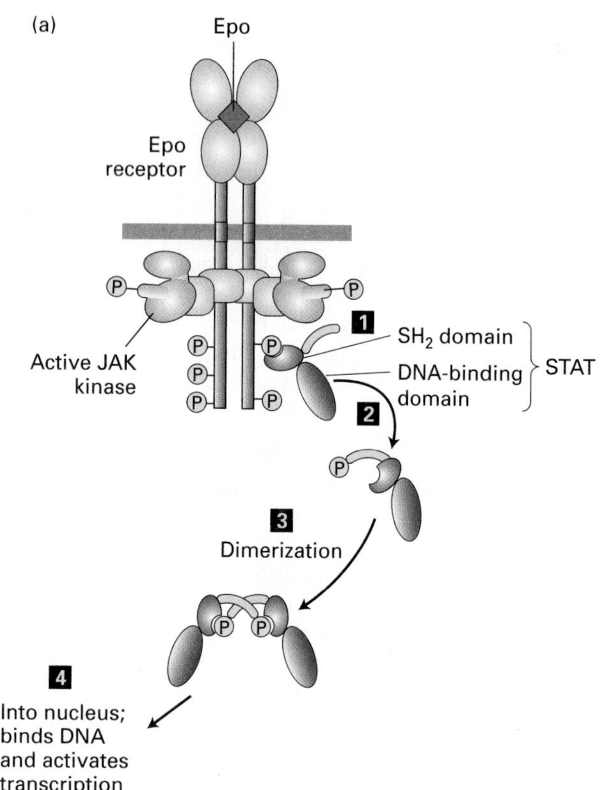

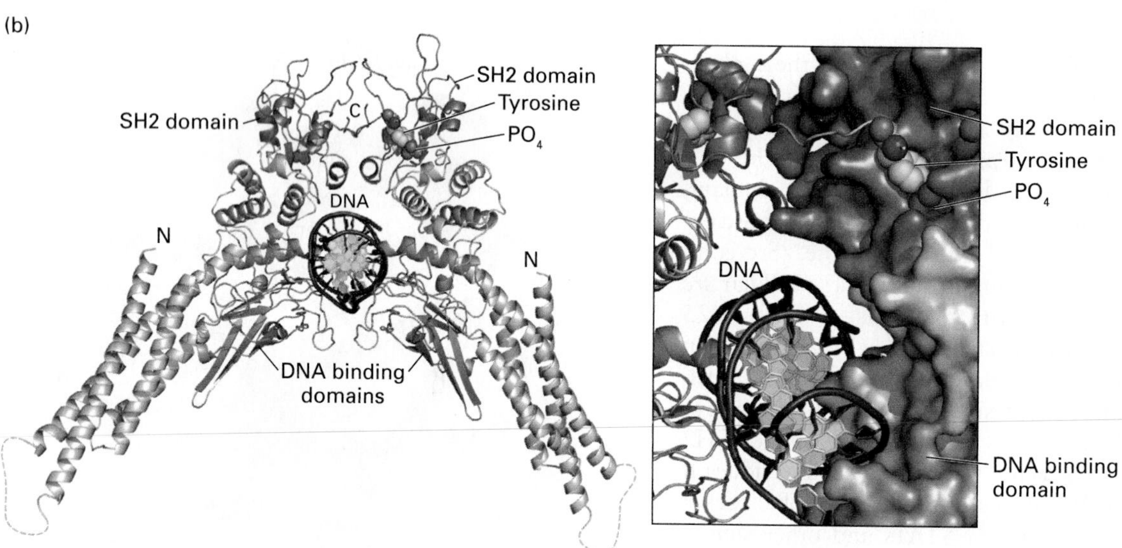

◄ **FIGURE 16-12 Activation and structure of STAT proteins.** (a) Phosphorylation and dimerization of STAT proteins. Step **1**: Following activation of a cytokine receptor (see Figure 16-10), an inactive monomeric STAT transcription factor binds to a phosphotyrosine in the receptor, bringing the STAT close to the active JAK associated with the receptor. The JAK then phosphorylates the C-terminal tyrosine in the STAT. Steps **2** and **3**: Phosphorylated STATs spontaneously dissociate from the receptor and spontaneously dimerize. Because the STAT homodimer has two phosphotyrosine–SH2 domain interactions, whereas the receptor–STAT complex is stabilized by only one such interaction, phosphorylated STATs tend not to rebind to the receptor. Step **4**: The STAT dimer moves into the nucleus, where it can bind to promoter sequences and activate transcription of target genes. (b) Ribbon diagram of the STAT1 dimer bound to DNA (black). The STAT1 dimer forms a C-shaped clamp around DNA that is stabilized by reciprocal and highly specific interactions between the SH2 domain (purple) of one monomer and the phosphorylated tyrosine residue (yellow) on the C-terminal segment of the other. The phosphotyrosine-binding site of the SH2 domain in each monomer is coupled structurally to the DNA-binding domain (magenta), suggesting a potential role for the SH2-phosphotyrosine interaction in the stabilization of DNA interacting elements. [Part (b) after X. Chen et al., 1998, *Cell* **93:**827.]

STAT proteins form a dimer in which the SH2 domain on each binds to the phosphotyrosine in the other. Because dimerization also exposes the nuclear-localization signal (NLS), STAT dimers move into the nucleus, where they bind to specific **enhancers** (DNA regulatory sequences) controlling target genes (Figure 16-12b).

Different STATs activate different genes in different cells. As just noted, stimulation of erythroid progenitor cells by erythropoietin (Epo) leads to activation of STAT5. The major protein induced by active STAT5 is Bcl-x_L, which prevents the programmed cell death, or **apoptosis,** of these progenitors, allowing them to proliferate and differentiate into erythroid cells (see Figure 16-6). In the normal state, when Epo levels are low, bone marrow stem cells continuously create progenitor erythroid cells that are quickly destroyed. This energetically expensive process allows the body to respond to the need for more erythrocytes very quickly in response to a rise in Epo levels. Indeed, mice lacking STAT5 are highly anemic because many of the erythroid progenitors undergo apoptosis even in the presence of high erythropoietin levels. Such mutant mice produce some erythrocytes and thus survive, because the erythropoietin receptor is linked to other anti-apoptotic pathways that do not involve STAT proteins (see Figure 16-8).

Complementation Genetics Revealed That JAK and STAT Proteins Transduce Cytokine Signals

Researchers have long known about cytokine signaling molecules, and once radioactive tagging (or expression cloning) was applied to cytokine-responsive cells, it did not take long to find the cytokine receptors, as described in the previous section for TGFβ receptors. However, without new techniques researchers had no way to explore the intracellular signaling pathways that transduce cytokine signals for many years. A new type of functional genetic screening (functional complementation) led researchers to both JAK proteins and STAT transcription factors.

In the first part of these studies, a bacterial **reporter gene** encoding guanine phosphoribosyl transferase (GPRT) was linked to an upstream promoter that is activated by the antiviral cytokine interferon. The resulting construct was introduced into cultured mammalian cells that were genetically deficient in the human homolog HGPRT. Either GPRT or HGPRT must be expressed for a cell to incorporate purines in the culture medium into ribonucleotides and then into DNA or RNA. As shown in Figure 16-13a, HGPRT-negative cells carrying the reporter gene responded to interferon treatment by expressing GPRT, thus acquiring the ability to grow in HAT medium. Cells lacking GPRT or HGPRT cannot grow in HAT medium since synthesis of purines by the cells is blocked by aminopterin (the A in HAT); thus DNA synthesis by the cells is dependent on incorporation of purines from the culture medium (see Figure 9-36). Simultaneously the cells acquired sensitivity to killing by the nonnatural purine analog 6-thioguanine, which is converted into the corresponding ribonucleotide by GPRT; incorporation of this purine into DNA in place of guanosine eventually causes cell death.

The reporter cells were then heavily treated with mutagens in an attempt to inactivate both alleles of the genes encoding critical signal-transduction proteins in the interferon signaling pathway. Researchers looked for mutant cells that expressed the interferon receptor (as evidenced by the cell's ability to bind radioactive interferon) but did not express GPRT in response to interferon and thus survived killing by 6-thioguanine when cells were cultured in the presence of interferon (Figure 16-13b). After many such interferon-nonresponding mutant cell lines were obtained, they were used to screen a genomic or cDNA library for the wild-type genes that complemented the mutated genes in nonresponding cells, a technique called **functional complementation** (see Figure 5-18). In this case, mutant cells expressing the corresponding recombinant wild-type gene grew on HAT medium and were sensitive to killing by 6-thioguanine in the presence of interferon. That is, they acted like wild-type cells.

Cloning of the genes identified by this procedure led to recognition of two key signal-transduction proteins that are activated by the cytokine interferon: a JAK tyrosine kinase and a STAT transcription factor. Subsequent work showed that one (sometimes two) of the four human JAK proteins and at least one of several STAT proteins are involved in signaling downstream from all cytokine receptors.

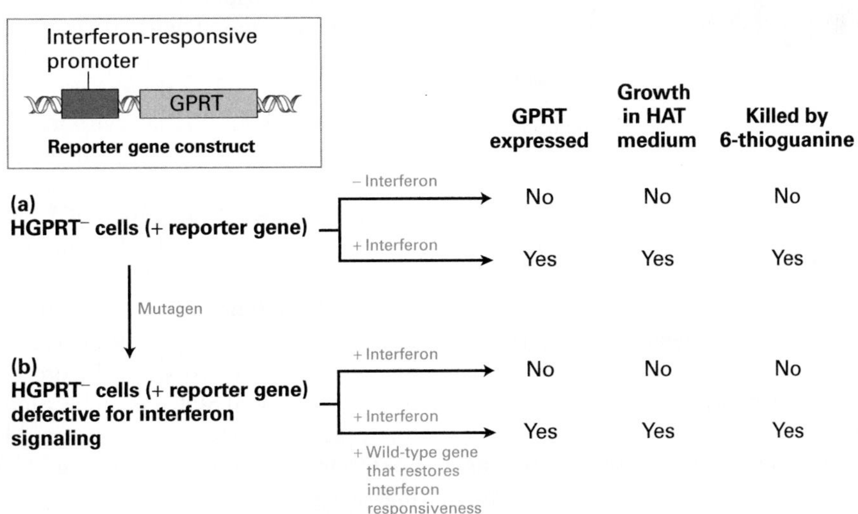

▲ **EXPERIMENTAL FIGURE 16-13 Mutagenized cells carrying a reporter gene were used to identify JAKs and STATs as essential signal-transduction proteins.** A reporter gene was constructed consisting of an interferon-responsive promoter upstream of the bacterial gene encoding GPRT, a key enzyme in the purine salvage pathway (see Figure 9-36). (a) Introduction of this construct into mammalian cells lacking the mammalian homolog HGPRT yielded reporter cells that grew in HAT medium and were killed by 6-thioguanine in the presence but not the absence of interferon. (b) Following treatment of reporter cells with a mutagen, cells with defects in the signaling pathway initiated by interferon do not induce HGPRT in response to interferon and thus cannot incorporate the toxic purine 6-thioguanine. Restoration of interferon responsiveness by functional complementation with wild-type DNA clones identified genes encoding JAKs and STATs. See the text for details. [See R. McKendry et al., 1991, *Proc. Nat'l. Acad. Sci. USA* **88:**11455; D. Watling et al., 1993, *Nature* **366:**166; and G. Stark and A. Gudkov, 1999, *Human Mol. Genet.* **8:**1925.]

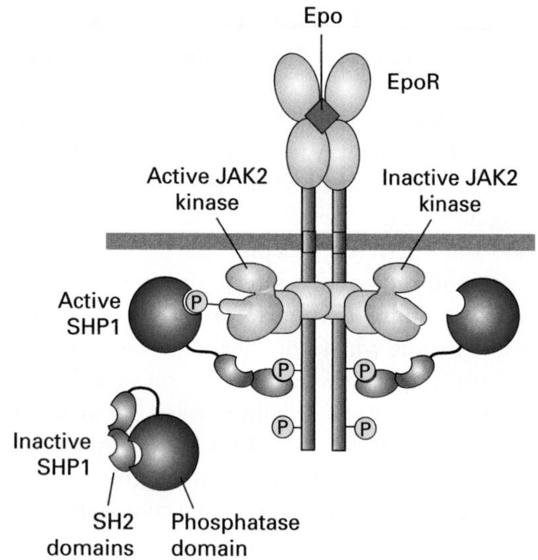

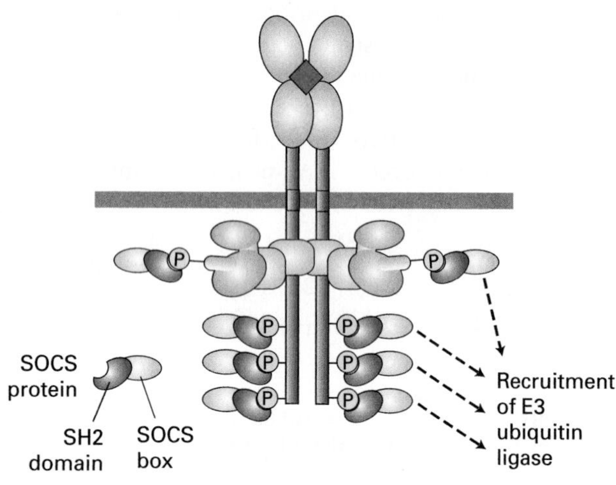

◀ **FIGURE 16-14 Two mechanisms for terminating signal transduction from the erythropoietin receptor (EpoR).** (a) Short-term regulation: SHP1, a phosphotyrosine phosphatase, is present in an inactive form in unstimulated cells. Binding of an SH2 domain in SHP1 to a particular phosphotyrosine in the activated receptor unmasks its phosphatase catalytic site and positions it near the phosphorylated tyrosine in the lip region of JAK2. Removal of the phosphate from this tyrosine inactivates the JAK kinase. (b) Long-term regulation: SOCS proteins, whose expression is induced by STAT proteins in erythropoietin-stimulated erythroid cells, inhibit or permanently terminate signaling over longer time periods. Binding of SOCS to phosphotyrosine residues on EpoR or JAK2 blocks binding of other signaling proteins *(left).* The SOCS box can also target proteins such as JAK2 for degradation by the ubiquitin-proteasome pathway *(right).* Similar mechanisms regulate signaling from other cytokine receptors. [Part (a) adapted from S. Constantinescu et al., 1999, *Trends Endocrin. Metabol.* **10:**18. Part (b) adapted from B. T. Kile and W. S. Alexander, 2001, *Cell. Mol. Life Sci.* **58:**1.]

Signaling from Cytokine Receptors Is Regulated by Negative Signals

Signal-induced transcription of target genes for too long a period can be as dangerous for the cell as too little induction. Thus cells must be able to turn off a signaling pathway quickly unless the extracellular signal remains continuously present. In various progenitor cells, two classes of proteins serve to dampen signaling from cytokine receptors, one over the short term (minutes) and the other over longer periods of time (hours to days).

Short-Term Regulation by SHP1, a Phosphotyrosine Phosphatase Mutant mice lacking a protein called *SHP1* die because of excess production of erythrocytes and several other types of blood cells. Analysis of these mutant mice offered the first suggestion that SHP1, a phosphotyrosine **phosphatase,** negatively regulates signaling from several types of cytokine receptors in several types of progenitor cells.

SHP1 dampens cytokine signaling by binding to a cytokine receptor and inactivating the associated JAK protein, as is depicted in Figure 16-14a. In addition to a phosphatase catalytic domain, SHP1 has two SH2 domains. When cells are in the resting state, unstimulated by a cytokine, one of the SH2 domains in SHP1 physically binds to and inactivates the catalytic site in the phosphatase domain. In the stimulated state, however, this blocking SH2 domain binds to a specific phosphotyrosine residue in the activated receptor. The conformational change that accompanies this binding unmasks the SHP1 catalytic site and also brings it adjacent to the phosphotyrosine residue in the activation lip of the JAK associated with the receptor. By removing this phosphate, SHP1 inactivates the JAK, so that it can no longer phosphorylate the receptor or other substrates (e.g., STATs) unless additional cytokine molecules bind to cell-surface receptors, initiating a new round of signaling.

Long-Term Regulation by SOCS Proteins In a classic example of negative feedback, among the genes whose transcription is induced by STAT proteins are those encoding a class of small proteins, termed *SOCS proteins,* that terminate signaling from cytokine receptors. These negative regulators act in two ways (Figure 16-14b). First, the SH2 domain in several SOCS proteins binds to phosphotyrosines on an activated receptor, preventing binding of other SH2-containing signaling proteins (e.g., STATs) and thus inhibiting receptor signaling. One SOCS protein, SOCS-1, also binds to the critical phosphotyrosine in the activation lip of activated JAK2 kinase, thereby inhibiting its catalytic activity. Second, all SOCS proteins contain a domain, called the SOCS box, that recruits components of E3 ubiquitin ligases (see Figure 3-29). As a result of binding SOCS-1, for instance, JAK2 becomes polyubiquitinated and then degraded in **proteasomes,** thus permanently turning off all JAK2-mediated signaling pathways. The observation that proteasome inhibitors prolong JAK2 signal transduction supports this mechanism.

Studies with cultured mammalian cells have shown that the receptor for growth hormone, which belongs to the cytokine receptor superfamily, is down regulated by another SOCS protein, SOCS-2. Strikingly, mice deficient in SOCS-2 grow significantly larger than their wild-type counterparts and have long bone lengths and proportionate enlargement of most organs. Thus SOCS proteins play an essential negative role in regulating intracellular signaling from the receptors for erythropoietin, growth hormone, and other cytokines.

Mutant Erythropoietin Receptor That Cannot Be Turned Off Leads to Increased Numbers of Erythrocytes

In normal adult men and women the percentage of erythrocytes in the blood (the *hematocrit*) is maintained very close to 45–47 percent. A drop in the hematocrit results in increased production of erythropoietin by the kidney, and the elevated erythropoietin level causes more erythroid progenitors to undergo terminal proliferation and differentiation into mature erythrocytes, soon restoring the hematocrit to its normal level. In endurance sports, such as cross-country skiing, where oxygen transport to the muscles may become limiting, an excess of erythrocytes may confer a competitive advantage. For this reason, use of supplemental erythropoietin to increase the hematocrit above the normal level is banned in all international athletic competitions, and athletes are regularly tested for elevated hematocrit and for the presence of commercial recombinant erythropoietin in their urine.

Supplemental erythropoietin not only confers a possible competitive advantage but also can be dangerous. Too many erythrocytes can cause the blood to become sluggish and clot in small blood vessels, especially in the brain. Several athletes who doped themselves with erythropoietin have died of a stroke while exercising.

Discovery of a mutant, unregulated erythropoietin receptor (EpoR) explained a suspicious situation in which a winner of three gold medals in Olympic cross-country skiing was found to have a hematocrit above 60 percent. Testing for erythropoietin in his blood and urine, however, revealed lower-than-normal amounts. Subsequent DNA analysis showed that the athlete was heterozygous for a mutation in the gene encoding the erythropoietin receptor. The mutant allele encoded a truncated receptor missing several of the tyrosines that normally become phosphorylated after stimulation by erythropoietin. As a consequence, the mutant receptor was able to activate STAT5 and other signaling proteins normally, but was unable to bind the negatively acting SHP1 phosphatase, which usually terminates signaling (see Figure 16-14a). Thus the very low level of erythropoietin produced by this athlete induced prolonged intracellular signaling in his erythroid progenitor cells, resulting in production of higher-than-normal numbers of erythrocytes. This example vividly illustrates the fine level of control over signaling from the erythropoietin receptor in the human body. ▮

KEY CONCEPTS OF SECTION 16.2

Cytokine Receptors and the JAK/STAT Pathway

- All cytokines are constructed of four α helices that are folded in a characteristic arrangement.
- Erythropoietin, a cytokine secreted by kidney cells, prevents apoptosis and promotes proliferation and differentiation of erythroid progenitor cells in the bone marrow. An excess of erythropoietin or mutations in its receptor that prevent down-regulation result in production of elevated numbers of erythrocytes.
- All cytokine receptors are closely associated with a JAK protein tyrosine kinase, which can activate several downstream signaling pathways leading to changes in transcription of target genes or in the activity of proteins that do not regulate transcription (see Figure 16-8).
- The JAK/STAT pathway operates downstream of all cytokine receptors. STAT monomers bound to receptors are phosphorylated by receptor-associated JAKs, then dimerize and move to the nucleus, where they activate transcription (see Figure 16-12). Signaling from cytokine receptors is terminated by the phosphotyrosine phosphatase SHP1 and several SOCS proteins (see Figure 16-14).

16.3 Receptor Tyrosine Kinases

We now turn our attention to another large and important class of cell-surface receptors—the **receptor tyrosine kinases (RTKs)**—that regulate many aspects of cell proliferation and differentiation, cell survival, and cellular metabolism. The signaling molecules that activate RTKs are soluble or membrane-bound peptide or protein hormones including many that were initially identified as growth factors for specific types of cells. These RTK ligands include nerve growth factor (NGF), platelet-derived growth factor (PDGF), fibroblast growth factor (FGF), and epidermal growth factor (EGF). Many RTKs and their ligands were identified in studies on human cancers associated with mutant forms of growth-factor receptors that stimulate proliferation even in the absence of growth factor. Other RTKs have been uncovered during analysis of developmental mutations that lead to blocks in differentiation of certain cell types in *C. elegans, Drosophila*, and the mouse.

Like cytokine receptors, RTKs signal through a protein tyrosine kinase. However, unlike cytokine receptors, which associate with a separate cytosolic kinase protein (a JAK), RTKs have an intrinsic kinase as part of their cytosolic domain. Ligand-induced dimerization and activation of an RTK stimulates its tyrosine kinase activity, which triggers several intracellular signal-transduction pathways (see Figure 16-2). In this section we discuss how ligand binding leads to activation of the RTKs; subsequently we consider how the numbers of cell-surface RTKs are controlled. In the following section we examine one pathway that is triggered by virtually every RTK, the Ras/MAP kinase pathway.

Ligand Binding Leads to Phosphorylation and Activation of Intrinsic Kinase in RTKs

All RTKs have three essential components: an extracellular domain containing a ligand-binding site, a single hydrophobic transmembrane α helix, and a cytosolic domain that includes a region with protein tyrosine kinase activity. Most RTKs are monomeric, and ligand binding to the extracellular domain induces formation of receptor dimers. Some monomeric ligands for RTKs, including fibroblast growth factor (FGF), bind tightly to heparan sulfate, a negatively charged polysaccharide component of the extracellular matrix (Chapter 19); this association enhances ligand binding to the monomeric receptor and formation of a dimeric receptor–ligand complex (Figure 16-15). The ligands for some RTKs are dimeric; their binding brings two receptor monomers together directly. Yet other RTKs, such as the insulin receptor, form disulfide-linked dimers even in the absence of hormone; binding of ligand to this type of RTK alters its conformation in such a way that the receptor becomes activated.

Regardless of the mechanism by which a ligand binds and locks an RTK into a functional dimeric state, the next step is universal. In the resting, unstimulated state, the intrinsic kinase activity of an RTK is very low (Figure 16-16, step **1**). In the ligand-bound dimeric receptor, however, the kinase in one subunit phosphorylates one tyrosine residue in the activation lip of the catalytic site in the other subunit (step **2**). This leads to a conformational change that facilitates binding of ATP in some receptors (e.g., insulin receptor) and binding of protein substrates in other receptors (e.g., FGF receptor). The resulting enhanced kinase activity then phosphorylates additional tyrosine residues in the cytosolic domain of the receptor (step **3**). This ligand-induced activation of RTK kinase activity is analogous to the activation of the JAK kinases associated with cytokine receptors (see Figure 16-10). The difference resides in the location of the kinase catalytic site, which is within the cytosolic domain of RTKs, but within a separate, associated, cytosolic JAK kinase in the case of cytokine receptors.

Overexpression of HER2, a Receptor Tyrosine Kinase, Occurs in Some Breast Cancers

Four receptor tyrosine kinases (RTKs) participate in signaling by the many members of the **epidermal growth factor (EGF)** family of signaling molecules. In humans, the four members of the **HER** (*h*uman *e*pidermal growth factor *r*eceptor) **family** are denoted HER1, 2, 3, and 4. HER1 directly binds three EGF family members: EGF, heparin-binding EGF (HB-EGF), and tumor-derived growth factor alpha (TGF-α). Binding of any of these ligands to the extracellular domain of a HER1 monomer leads to homodimerization of the HER1 extracellular domain (Figure 16-17). Binding of EGF triggers a conformational change in the receptor extracellular domain so that it "clamps" down on the ligand. This pushes out a loop located between the two EGF-binding domains, and interactions between the two extended ("activated") loop segments allow

(a) Side view

Heparan sulfate

Heparan sulfate

FGF

FGF

FGFR

FGFR

Membrane surface

(b) Top-down view

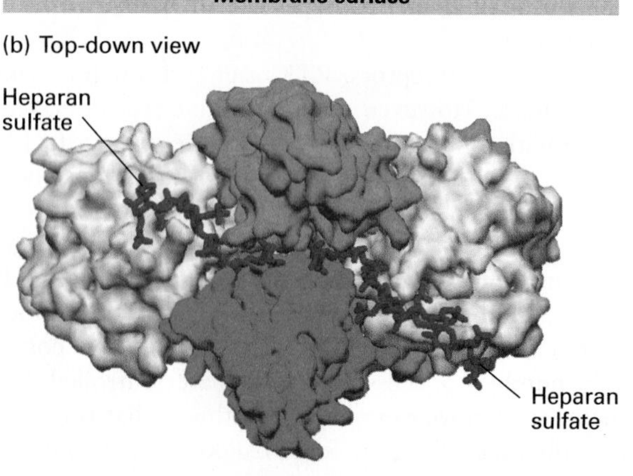

◄ **FIGURE 16-15 Structure of the fibroblast growth factor (FGF) receptor, stabilized by heparan sulfate.** Shown here are side and top-down views of the complex comprising the extracellular domains of two FGF receptor (FGFR) monomers (green and blue), two bound FGF molecules (white), and two short heparan sulfate chains (purple), which bind tightly to FGF. (a) In the side view, the upper domain of one receptor monomer (blue) is seen situated behind that of the other (green); the plane of the plasma membrane is at the bottom. A small segment of the extracellular domain whose structure is not known connects to the membrane-spanning α-helical segment of each of the two receptor monomers (not shown) that protrude downwards into the membrane. (b) In the top view, the heparan sulfate chains are seen threading between and making numerous contacts with the upper domains of both receptor monomers. These interactions promote binding of the ligand to the receptor and receptor dimerization. [Adapted from J. Schlessinger et al., 2000, *Mol. Cell* **6**:743.]

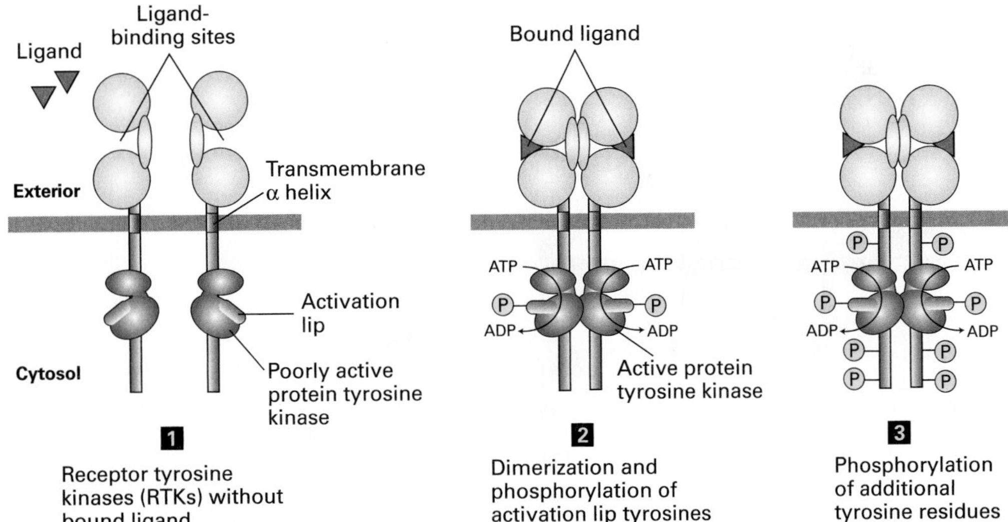

▲ **FIGURE 16-16 General structure and activation of receptor tyrosine kinases (RTKs).** The cytosolic domain of RTKs contains a protein tyrosine kinase catalytic site. In the absence of ligand (1), RTKs exist as monomers with poorly active kinases. Ligand binding causes a conformational change that promotes formation of a functional dimeric receptor, bringing together two poorly active kinases that then phosphorylate each other on a tyrosine residue in the activation lip (2). Phosphorylation causes the lip to move out of the kinase catalytic site, thus allowing ATP or a protein substrate to bind. The activated kinase then phosphorylates other tyrosine residues in the receptor's cytosolic domain (3). The resulting phosphotyrosines function as docking sites for various signal-transduction proteins.

formation of the receptor dimer. Dimerization of HER1 leads to activation of the receptor's kinase activity through phosphorylation of the activation lip in the kinase cytosolic domain (see Figure 16-16).

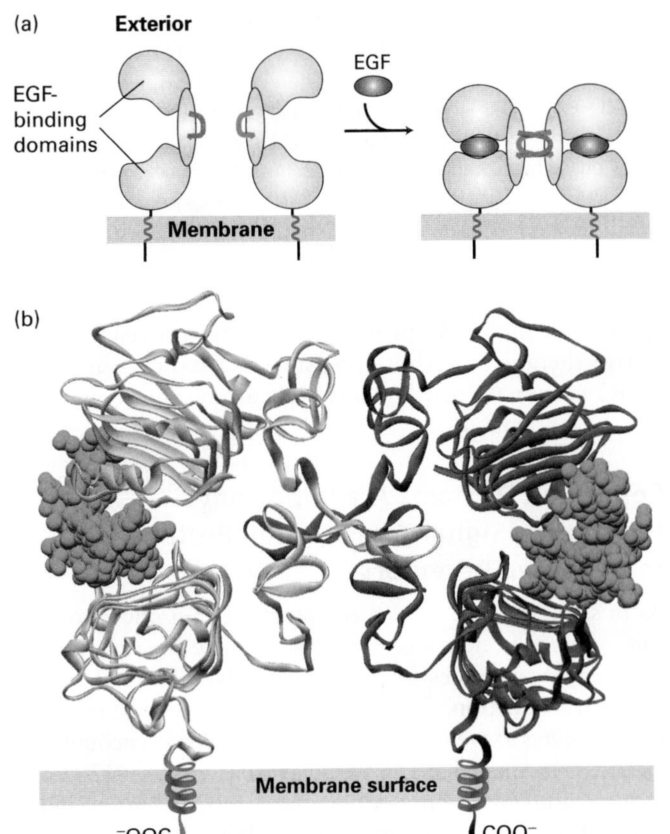

Two other members of the EGF family, neuregulins 1 and 2 (NRG1 and NRG2) bind to both HER3 and HER4; HB-EGF also binds to HER4. Importantly, HER2 does not directly bind a ligand, but exists on the membrane in a preactivated conformation with the loop segment protruding outward and the ligand-binding domains in close proximity (Figure 16-18a). HER2 signals by forming heterocomplexes with ligand-bound HER1, HER3, or HER4 and facilitates signaling by all EGF family members (Figure 16-18b). HER3 lacks a functional kinase domain; after binding a ligand it dimerizes with HER2 and becomes phosphorylated by the HER2 kinase. This activates downstream signal-transduction pathways.

◀ **FIGURE 16-17 Ligand-induced dimerization of HER1, a human receptor for epidermal growth factor (EGF).** (a) Schematic depiction of the extracellular and transmembrane domains of HER1, which is a receptor tyrosine kinase. Binding of one EGF molecule to a monomeric receptor causes an alteration in the structure of a loop located between the two EGF-binding domains. Dimerization of two identical ligand-bound receptor monomers in the plane of the membrane occurs primarily through interactions between the two "activated" loop segments. (b) Structure of the dimeric HER1 protein bound to transforming growth factor α (TGFα), a member of the EGF family. The receptor's extracellular domains are shown in white *(left)* and blue *(right)*; the transmembrane domain is shown in red as an alpha helix but its structure is not known in detail. The two smaller TGFα molecules are colored green. Note the interaction between the "activated" loop segments in the two receptor monomers. [Part (a) adapted from J. Schlessinger, 2002, *Cell* **110:**669. Part (b) from T. Garrett et al., 2002, *Cell* **110:**763.]

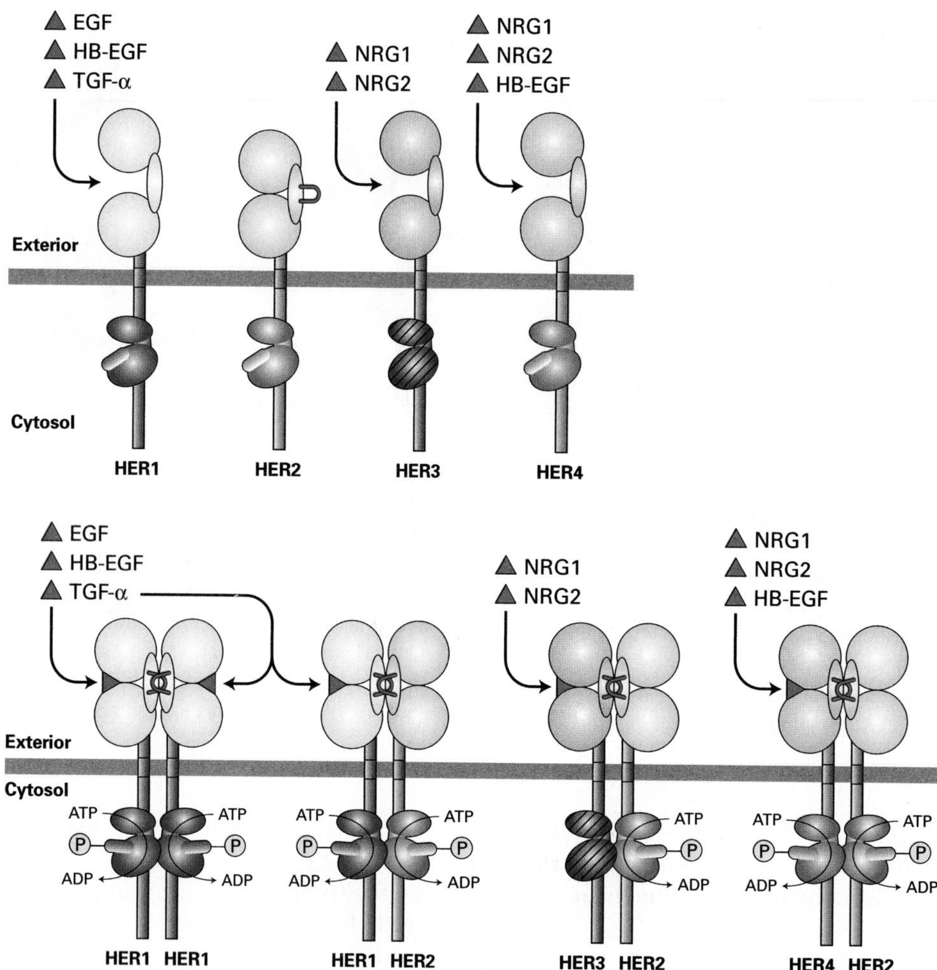

▲ **FIGURE 16-18 The HER family of receptors and their ligands.** Humans express four receptor tyrosine kinases—denoted HER1, 2, 3, and 4—that bind epidermal growth factor (EGF) and other EGF family members. (a) As shown, the HER proteins differentially bind EGF, heparin-binding EGF (HB-EGF), tumor-derived growth factor alpha (TGFα), and neuregulins 1 and 2 (NRG1 and NRG2). Note that HER2, which does not directly bind a ligand, exists in the plasma surface membrane in a pre-activated state indicated by red hook. (b) Ligand-bound HER1 can form activated homodimers bound together by loop segments (red hooks), as detailed in Figure 16-17. HER2 forms heterodimers with ligand-bound HER1, HER3, and HER4 and facilitates signaling by all EGF family members. HER3 lacks a functional kinase domain and can signal only when complexed with HER2. [After N. E. Hynes and H. A. Lane, 2005, *Nature Rev. Cancer* **5:**341(erratum in *Nature Rev. Cancer* **5:**580), and A. B. Singh and R. C. Harris, 2005, *Cell Signal* **17(Oct.):**1183.]

Normal epithelial cells express a small amount of HER2 protein on their plasma membranes in a tissue-specific pattern. In tumor cells, errors in DNA replication often result in multiple copies of a gene on a single chromosome, an alteration known as gene amplification (Chapter 25). Amplification of the *HER2* gene occurs in approximately 25 percent of breast cancer patients, resulting in overexpression of HER2 protein in the tumor cells. Breast cancer patients with HER2 overexpression have a worse prognosis, including shortened survival, than do patients without this abnormality. As Figure 16-18 emphasizes, overexpression of HER2 makes the tumor cells sensitive to growth stimulation by low levels of any member of the EGF family of growth factors, levels that would not stimulate proliferation of cells with normal HER2 levels. Discovery of the role of HER2 overexpression in certain breast cancers led researchers to develop monoclonal antibodies specific for the HER2 protein. These have proven to be effective therapies for those breast cancer patients in which HER2 is overexpressed, reducing recurrence by about 50% in these patients. ▮

Conserved Domains Are Important for Binding Signal-Transduction Proteins to Activated Receptors

As in signaling by cytokine receptors, certain phosphotyrosine residues in activated receptor tyrosine kinases (RTKs) serve as docking sites for proteins involved in downstream signal transduction. These phosphotyrosine residues bind *PTB domains* as well as SH2 domains, which we encountered in discussing the JAK/STAT pathway (see Figure 16-12). Both of these domains are present in a large array of intracellular signal-transduction proteins that couple activated RTKs and cytokine receptors to downstream components of signal-transduction pathways.

Once bound to an activated receptor, some signal-transduction proteins are phosphorylated by the receptor-associated kinase to achieve their active form. Many enzymes that function in signal-transduction pathways are present in the cytosol in unstimulated cells. Binding to an activated receptor positions these enzymes near their substrates, which are localized in the plasma membrane; the modified substrate then triggers a downstream signal-transduction pathway. Several cytokine receptors (e.g., the IL-4 receptor) and RTKs (e.g., the insulin receptor) bind *multidocking proteins,* like IRS-1, via a PTB domain in the docking protein (Figure 16-19). The activated receptor then phosphorylates the bound docking protein, forming many phosphotyrosines that in turn serve as docking sites for SH2-containing signaling proteins. Some of these proteins in turn may also be phosphorylated by the activated receptor.

As noted earlier, each SH2 domain binds to a distinct sequence of amino acids surrounding a phosphotyrosine residue. The unique amino acid sequence of each SH2 domain determines which amino acid sequence containing a phosphotyrosine it will bind. This specificity plays an important role in determining which signal-transduction proteins bind to which receptors. The SH2 domain of the Src tyrosine kinase for example, binds strongly to any peptide containing a critical four-residue core sequence: phosphotyrosine–glutamic acid–glutamic acid–isoleucine (see Figure 16-11). These four amino acids make intimate contact with the peptide-binding site in the Src SH2 domain. Binding resembles the insertion of a two-pronged "plug"—the phosphotyrosine and isoleucine side chains of the peptide—into a two-pronged "socket" in the SH2 domain. The two glutamic acids fit snugly onto the surface of the SH2 domain between the phosphotyrosine socket and the hydrophobic socket that accepts the isoleucine residue. The binding specificity of SH2 domains is largely determined by residues C-terminal to the phosphotyrosine in a target peptide. In contrast, the binding specificity of PTB domains is determined by the specific residues on the N-terminal side of a phosphotyrosine residue. Sometimes a PTB domain binds to a target peptide even if the tyrosine is not phosphorylated.

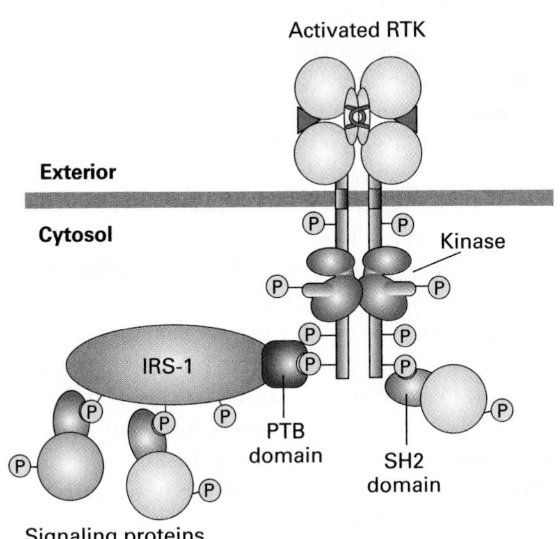

▲ **FIGURE 16-19 Recruitment of intracellular signal-transduction proteins to the cell membrane by binding to phosphotyrosine residues in receptors or receptor-associated proteins.** Cytosolic proteins with SH2 (purple) or PTB (maroon) domains can bind to specific phosphotyrosine residues in activated RTKs (shown here) or cytokine receptors. In some cases, these signal-transduction proteins then are phosphorylated by the receptor's intrinsic or associated protein tyrosine kinase, enhancing their activity. Certain RTKs and cytokine receptors utilize multidocking proteins such as IRS-1 to increase the number of signaling proteins that are recruited and activated. Subsequent phosphorylation of a receptor-bound IRS-1 by the receptor kinase creates additional docking sites for SH2-containing signaling proteins.

Down-regulation of RTK Signaling Occurs by Endocytosis and Lysosomal Degradation

We have already seen several ways that signal-transduction pathways are controlled. Intracellular proteins such as Ski and SOCS negatively regulate their respective signal-transduction pathways after their expression is induced by TGFβ or cytokines. Phosphorylation of receptors and downstream signaling proteins is reversed by the carefully controlled action of phosphatases. Here we discuss two related mechanisms by which RTK signaling is restrained: ligand-induced endocytosis of surface receptor–ligand complexes, and sorting of the internalized receptor or ligand to the lysosome for degradation. Thus treatment of cells with ligand for several hours reduces the number of available cell-surface receptors such that they no longer respond to that concentration of hormone. This prevents inappropriate prolonged receptor activity, but under these conditions cells usually will respond if the hormone level is increased further.

In the absence of epidermal growth factor (EGF), for instance, cell-surface HER1 receptors for this ligand are relatively long-lived, with an average half-life of 10 to 15 hours. To a large measure this is because the unbound receptors are internalized via clathrin-coated pits into endosomes at a relatively slow rate, on average once every 30 minutes, and are returned rapidly to the plasma membrane. Following binding of an EGF ligand, the rate of endocytosis of HER1 is increased ≈10 fold, and only about half of the internalized receptors return to the plasma membrane, depending on the cell type; the rest are degraded in lysosomes. Thus each time a HER1–EGF complex is internalized, via the process termed **receptor-mediated endocytosis** (see Figure 14-29), the receptor has about a 50 percent chance of being degraded. Exposure of a fibroblast cell to high levels of EGF for several hours induces several rounds of endocytosis, resulting in degradation of most cell-surface receptor molecules and thus a reduction in the cell's sensitivity to EGF. In this way, prolonged treatment with EGF desensitizes the cell to that level of hormone, though the cell may respond if the level of EGF is increased further.

HER1 mutants that lack kinase activity do not undergo accelerated endocytosis in the presence of ligand. It is likely that ligand-induced activation of the kinase activity in normal HER1 induces a conformational change in the cytosolic tail, exposing a sorting motif that facilitates receptor recruitment into clathrin-coated pits and subsequent internalization of the receptor–ligand complex. Despite extensive study of mutant HER1 cytosolic domains, the identity of these "sorting motifs" is controversial, and most likely multiple motifs function to enhance endocytosis. Interestingly, internalized receptors can continue to signal from endosomes or other intracellular compartments before their degradation as evidenced by their binding to signaling proteins such as Grb-2, and SOS, which are discussed in the next section.

After internalization, some cell-surface receptors (e.g., the LDL cholesterol receptor) are efficiently recycled to the surface (see Figure 14-29). In contrast, the fraction of activated HER1 receptors that are sorted to lysosomes can vary from 20 to 80 percent in different cell types. There is a strong correlation between monoubiquitination of the HER1 cytosolic domain by c-Cbl, an E3 ubiquitin ligase (see Figure 3-29), and HER1 degradation. c-Cbl contains an EGFR-binding domain, which binds directly to phosphorylated EGF receptors, and a RING finger domain, which recruits ubiquitin-conjugating enzymes and mediates transfer of ubiquitin to the receptor. The ubiquitin functions as a "tag" on the receptor that stimulates its incorporation from endosomes into multivesicular bodies (see Figure 14-33) that ultimately are degraded inside lysosomes. A role for c-Cbl in EGF receptor trafficking emerged from genetic studies in *C. elegans,* which established that c-Cbl negatively regulates the nematode EGF receptor (Let-23), probably by inducing its degradation. Similarly, knockout mice lacking c-Cbl show hyperproliferation of mammary gland epithelia, consistent with a role of c-Cbl as a negative regulator of EGF signaling.

Experiments with mutant cell lines demonstrate that internalization of RTKs plays an important role in regulating cellular responses to EGF and other growth factors. For instance, a mutation in the EGF receptor (HER1) that prevents it from being incorporated into coated pits makes it resistant to receptor-mediated (ligand-induced) endocytosis. As a result, this mutation leads to substantially above-normal numbers of EGF receptors on cells and thus increased sensitivity of cells to EGF as a mitogenic signal. Such mutant cells are prone to EGF-induced **transformation** into tumor cells. Interestingly, the other EGF family receptors—HER2, HER3, and HER4—do not undergo ligand-induced internalization, an observation that emphasizes how each receptor evolved to be regulated in its own appropriate manner.

KEY CONCEPTS OF SECTION 16.3

Receptor Tyrosine Kinases

- Receptor tyrosine kinases (RTKs), which bind to peptide and protein hormones, may exist as preformed dimers or dimerize during binding to ligands. Ligand binding triggers formation of functional dimeric receptors and phosphorylation of the activation lip in the intrinsic protein tyrosine kinases, enhancing their catalytic activity (see Figure 16-16). The activated receptor also phosphorylates tyrosine residues in the receptor cytosolic domain and in other protein substrates.
- Humans express four RTKs that bind different members of the epidermal growth factor family of signaling molecules (see Figure 16-18). One of these receptors, HER2, does not bind ligand; it forms active heterodimers with ligand-bound monomers of the other three HER proteins. Overexpression of HER2 is implicated in about 25 percent of breast cancers.
- Short peptide sequences containing phosphotyrosine residues are bound by SH2 and PTB domains, which are found in many signal-transducing proteins. Such protein–protein interactions are important in many signaling pathways.
- Endocytosis of receptor–hormone complexes and their degradation in lysosomes is a principal way of reducing the number of receptor tyrosine kinases and cytokine receptors on the cell surface, thus decreasing the sensitivity of cells to many peptide hormones.

16.4 Activation of Ras and MAP Kinase Pathways

Almost all receptor tyrosine kinases can activate the *Ras/MAP kinase pathway* (see Figure 16-2c). The **Ras protein,** a monomeric (small) G protein, belongs to the **GTPase superfamily** of intracellular switch proteins (see Figure 15-8). Activated Ras promotes formation, at the membrane, of signal-transduction complexes containing three sequentially acting protein kinases. This *kinase cascade* culminates in activation of certain members of the **MAP kinase** family, which can translocate into the nucleus and phosphorylate many different proteins. Among the target proteins for MAP kinase are transcription factors that regulate expression of proteins with important roles in the cell cycle and in differentiation. Many cytokine receptors also can activate the Ras/MAP kinase pathway (see Figure 16-2b). Moreover, different types of extracellular signals often activate different signaling pathways that result in activation of different members of the MAP kinase family.

Because an activating mutation in a RTK, Ras, or a protein in the MAP kinase cascade is found in almost all types of human tumors, the RTK/Ras/MAP kinase pathway has been subjected to extensive study and a great deal is known about the components of this pathway. We begin our discussion by reviewing how Ras cycles between the active and inactive state. We then describe how Ras is activated and passes a signal to the MAP kinase pathway. Finally we examine recent studies indicating that both yeasts and cells of higher eukaryotes contain multiple MAP kinase pathways, and consider the ways in which cells keep different MAP kinase pathways separate from one another through the use of scaffold proteins.

Ras, a GTPase Switch Protein, Cycles Between Active and Inactive States

Like the G_α subunits in trimeric G proteins, Ras alternates between an active on state with a bound GTP and an inactive off state with a bound GDP. As discussed in Chapter 15, trimeric G proteins are directly linked to G protein–coupled receptors (GPCRs) on the cell surface and transduce signals, usually via the G_α subunit, to various effectors such as adenylyl cyclase. In contrast, Ras is not directly linked to cell-surface receptors.

The activity of the Ras protein is regulated by several factors. Ras activation is accelerated by a *guanine nucleotide–exchange factor (GEF),* which binds to the Ras·GDP complex, causing dissociation of the bound GDP (see Figure 3-32). Because GTP is present in cells at a higher concentration than GDP, GTP binds spontaneously to "empty" Ras molecules, with release of GEF and formation of the active Ras·GTP. Subsequent hydrolysis of the bound GTP to GDP deactivates Ras. Unlike the deactivation of G_α·GTP, deactivation of Ras·GTP requires the assistance of another protein, a *GTPase-activating protein (GAP).* Binding of GAP to Ras·GTP accelerates the intrinsic GTPase activity of Ras by more than a hundredfold. Thus the average lifetime of a GTP bound to Ras is about 1 minute, which is much longer than the average lifetime of a G_α·GTP complex. In cells, GAP binds to specific phosphotyrosines in activated RTKs, bringing it close enough to membrane-bound Ras·GTP to exert its accelerating effect on GTP hydrolysis. The actual hydrolysis of GTP is catalyzed by amino acids from both Ras and GAP. In particular, insertion of an arginine side chain on GAP into the Ras active site stabilizes an intermediate in the hydrolysis reaction.

Ras ($\approx$170 amino acids) is smaller than G_α proteins ($\approx$300 amino acids), but the GTP-binding domains of the two proteins have a similar structure (see Figure 15-8). Structural and biochemical studies show that G_α also contains a GAP domain that increases the rate of GTP hydrolysis by G_α. Because this domain is not present in Ras, it has an intrinsically slower rate of GTP hydrolysis.

Mammalian Ras proteins have been studied in great detail because mutant Ras proteins are associated with many types of human cancer. These mutant proteins, which bind but cannot hydrolyze GTP, are permanently in the "on" state and contribute to oncogenic transformation (Chapter 25). Determination of the three-dimensional structure of the Ras–GAP complex and tests of mutant forms of Ras explained the puzzling observation that most oncogenic, constitutively active Ras proteins (Ras^D) contain a mutation at position 12. Replacement of the normal glycine-12 with any other amino acid (except proline) blocks the functional binding of GAP and in essence "locks" Ras in the active GTP-bound state. ▮

Receptor Tyrosine Kinases Are Linked to Ras by Adapter Proteins

The first indication that Ras functions downstream from RTKs in a common signaling pathway came from experiments in which cultured fibroblast cells were induced to proliferate by treatment with a mixture of two protein hormones: platelet-derived growth factor (PDGF) and epidermal growth factor (EGF). Microinjection of anti-Ras antibodies into these cells blocked cell proliferation. Conversely, injection of Ras^D, a constitutively active mutant Ras protein that hydrolyzes GTP very inefficiently and thus persists in the active state, caused the cells to proliferate in the absence of the growth factors. These findings are consistent with studies showing that addition of FGF to fibroblasts leads to a rapid increase in the proportion of Ras present in the GTP-bound active form.

However, an activated RTK (e.g., a ligand-bound EGF receptor) cannot directly activate Ras. Rather, two cytosolic proteins—GRB2 and Sos—must first be recruited to provide a link between the receptor and Ras (Figure 16-20). An SH2 domain in GRB2 binds to a specific phosphotyrosine residue in the activated receptor. GRB2 also contains two SH3 domains, which bind to and activate Sos. GRB2 thus functions as an *adapter protein* for many receptor tyrosine kinases. Sos is a guanine nucleotide–exchange protein (GEF), which catalyzes conversion of inactive GDP-bound Ras to the active GTP-bound form.

Genetic Studies in *Drosophila* Identified Key Signal-Transducing Proteins in the Ras/MAP Kinase Pathway

Our knowledge of the proteins involved in the Ras/MAP kinase pathway came principally from genetic analyses of mutant fruit flies (*Drosophila*) and worms (*C. elegans*) which were blocked at particular stages of differentiation. To illustrate the power of this experimental approach, we consider development of a particular type of cell in the compound eye of *Drosophila.*

The compound eye of the fly is composed of some 800 individual eyes called *ommatidia* (Figure 16-21a). Each ommatidium consists of 22 cells, eight of which are photosensitive neurons called *retinula*, or R cells, designated R1–R8 (Figure 16-21b). An RTK called *Sevenless (Sev)* specifically regulates development of the R7 cell and is not essential for any other known function. In flies with a mutant *sevenless (sev)* gene, the R7 cell in each ommatidium does not form (Figure 16-21c). Since the R7 photoreceptor is necessary for flies to see in ultraviolet light, mutants that lack functional R7 cells but are otherwise normal are easily isolated. Therefore fly R7 cells are an ideal genetic system to study cell development.

During development of each ommatidium, a protein called *Boss (Bride of Sevenless)* is expressed on the surface of the R8 cell. This membrane-tethered protein is the ligand for the Sev RTK on the surface of the neighboring R7 precursor cell, signaling it to develop into a photosensitive

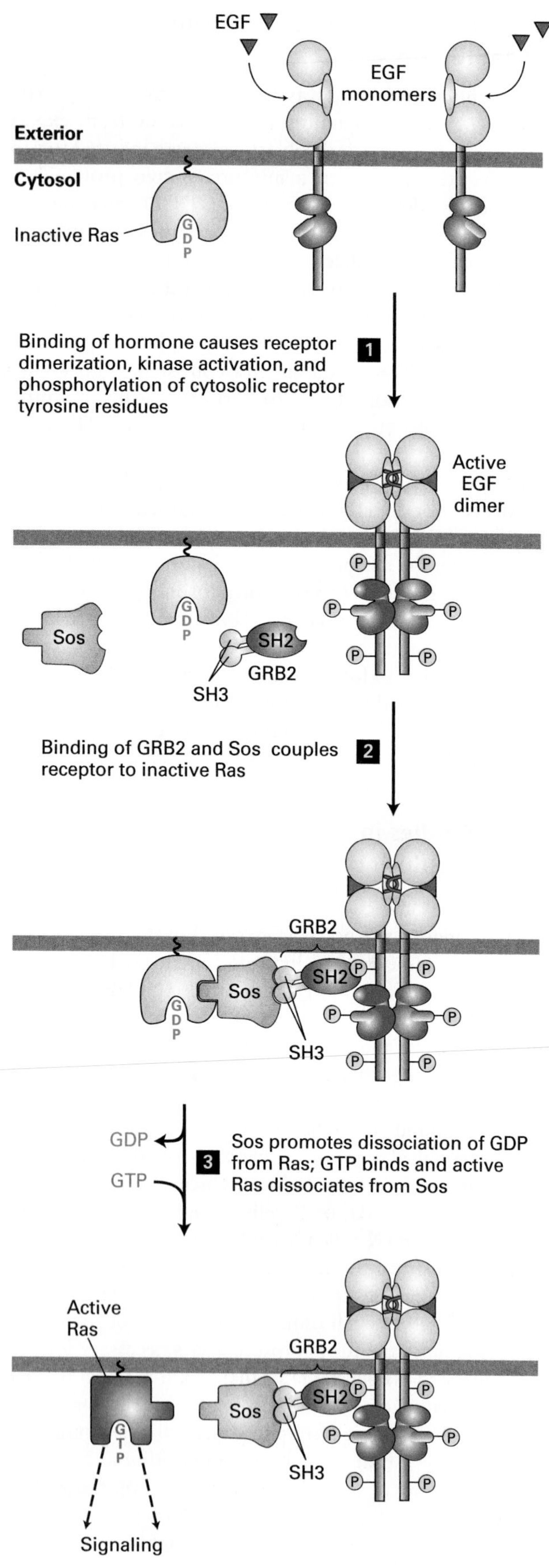

◀ **FIGURE 16-20 Activation of Ras following ligand binding to receptor tyrosine kinases (RTKs).** The receptors for epidermal growth factor (EGF) and many other growth factors are RTKs. The cytosolic adapter protein GRB2 binds to a specific phosphotyrosine on an activated, ligand-bound receptor and to the cytosolic Sos protein, bringing it near its substrate, the inactive Ras·GDP. The guanine nucleotide–exchange factor (GEF) activity of Sos then promotes formation of active Ras·GTP. Note that Ras is tethered to the membrane by a hydrophobic farnesyl anchor (see Figure 10-19). [See J. Schlessinger, 2000, *Cell* **103:**211, and M. A. Simon, 2000, *Cell* **103:**13.]

neuron (Figure 16-22a). In mutant flies that do not express a functional Boss protein or Sev RTK, interaction between the Boss and Sev proteins cannot occur, and no R7 cells develop (Figure 16-22b); this is the origin of the name "Sevenless" for the RTK in the R7 cells.

To identify intracellular signal-transducing proteins in the Sev RTK pathway, investigators produced mutant flies expressing a temperature-sensitive Sev protein. When these flies were maintained at a permissive temperature, all their ommatidia contained R7 cells; when they were maintained at a nonpermissive temperature, no R7 cells developed. At a particular intermediate temperature, however, just enough of the Sev RTK was functional to mediate normal R7 development. The investigators reasoned that at this intermediate temperature, the signaling pathway would become defective (and thus no R7 cells would develop) if the level of another protein involved in the pathway was reduced, thus reducing the activity of the overall pathway below the level required to form an R7 cell. A recessive mutation affecting such a protein would have this effect because, in diploid organisms like *Drosophila,* a heterozygote containing one wild-type and one mutant allele of a gene will produce half the normal amount of the gene product; hence, even if such a recessive mutation is in an essential gene, the organism will usually be viable. However, a fly carrying a temperature-sensitive mutation in the *sev* gene and a second mutation affecting another protein in the signaling pathway would be expected to lack R7 cells at the intermediate temperature.

By use of this screen, researchers identified three genes encoding important proteins in the Sev pathway: an SH2-containing adapter protein exhibiting 64 percent amino acid sequence identity to human GRB2; a guanine nucleotide–exchange factor called Sos (Son of Sevenless) exhibiting 45 percent identity with its mouse counterpart; and a Ras protein exhibiting 80 percent identity with its mammalian counterparts (see Figure 16-20). These three proteins later were found to function in other signaling pathways initiated by ligand binding to different RTKs and used at different times and places in the developing fly.

In subsequent studies, researchers introduced a mutant ras^D gene into fly embryos carrying the sevenless mutation. As noted earlier, the ras^D gene encodes a constitutive Ras protein that is present in the active GTP-bound form even in the absence of a hormone signal. Although no functional Sev RTK was expressed in these double-mutants (sev^-; ras^D), R7 cells formed normally, indicating that presence of an activated

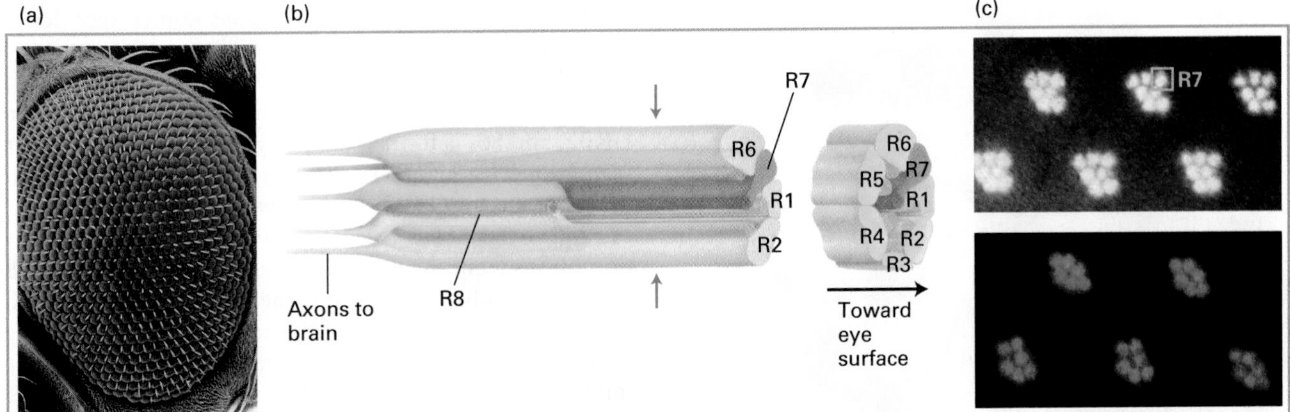

▲ **FIGURE 16-21 The compound eye of *Drosophila melanogaster.*** (a) Scanning electron micrograph showing individual ommatidia that comprise the fruit fly eye. (b) Longitudinal and cutaway views of a single ommatidium. Each of these tubular structures contains eight photoreceptors, designated R1–R8, which are long, cylindrically shaped light-sensitive cells. R1–R6 (yellow) extend throughout the depth of the retina, whereas R7 (brown) is located toward the surface of the eye, and R8 (blue) toward the backside where the axons exit. (c) Comparison of eyes from wild-type and *sevenless* mutant flies viewed by a special technique that can distinguish the photoreceptors in an ommatidium. The plane of sectioning is indicated by the blue arrows in (b), and the R8 cell is out of the plane of these images. The seven photoreceptors in this plane are easily seen in the wild-type ommatidia *(top)*, whereas only six are visible in the mutant ommatidia *(bottom)*. Flies with the *sevenless* mutation lack the R7 cell in their eyes. [Part (a) from E. Hafen and K. Basler, 1991, *Development* **1**(suppl.):123. Part (b) adapted from R. Reinke and S. L. Zipursky, 1988, *Cell* **55**:321. Part (c) courtesy of U. Banerjee.]

Ras protein is sufficient for induction of R7-cell development (Figure 16-22c). This finding, which is consistent with the results with cultured fibroblasts described earlier, supports the conclusion that activation of Ras is a principal step in intracellular signaling by most if not all RTKs.

Binding of Sos Protein to Inactive Ras Causes a Conformational Change That Activates Ras

In addition to its SH2 domain that binds to activated RTKs, the GRB2 adapter protein contains two *SH3 domains*, which bind to Sos, the Ras guanine nucleotide–exchange factor (see Figure 16-20). Like phosphotyrosine-binding SH2 and PTB domains, SH3 domains are present in a large number of proteins involved in intracellular signaling. Although the three-dimensional structures of various SH3 domains are similar, their specific amino acid sequences differ. The SH3 domains in GRB2 selectively bind to proline-rich sequences in Sos; different SH3 domains in other proteins bind to proline-rich sequences distinct from those in Sos.

Proline residues play two roles in the interaction between an SH3 domain in an adapter protein (e.g., GRB2) and a proline-rich sequence in another protein (e.g., Sos). First, the proline-rich sequence assumes an extended conformation

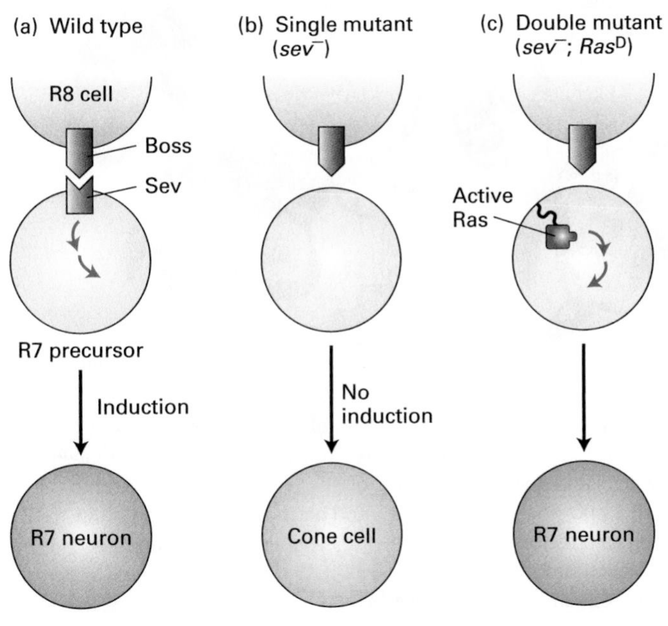

◀ **EXPERIMENTAL FIGURE 16-22 Genetic studies reveal that activation of Ras induces development of R7 photoreceptors in the *Drosophila* eye.** (a) During larval development of wild-type flies, the R8 cell in each developing ommatidium expresses a cell-surface protein, called Boss, that binds to the Sev RTK on the surface of its neighboring R7 precursor cell. This interaction induces changes in gene expression that result in differentiation of the precursor cell into a functional R7 neuron. (b) In fly embryos with a mutation in the *sevenless (sev)* gene, R7 precursor cells cannot bind Boss and therefore do not differentiate normally into R7 cells. Rather the precursor cell enters an alternative developmental pathway and eventually becomes a cone cell. (c) Double-mutant larvae (sev^-; Ras^D) express a constitutively active Ras (Ras^D) in the R7 precursor cell, which induces differentiation of R7 precursor cells in the absence of the Boss-mediated signal. This finding shows that activated Ras is sufficient to mediate induction of an R7 cell. [See M. A. Simon et al., 1991, *Cell* **67**:701, and M. E. Fortini et al., 1992, *Nature* **355**:559.]

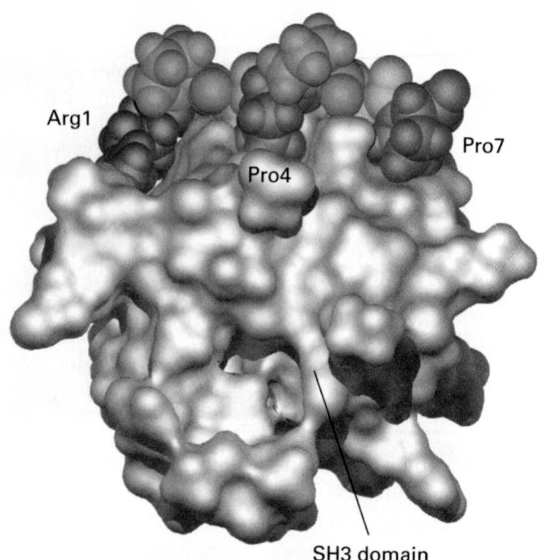

▲ **FIGURE 16-23 Surface model of an SH3 domain bound to a target peptide.** The short, proline-rich target peptide is shown as a space-filling model. In this target peptide, two prolines (Pro4 and Pro7, dark blue) fit into binding pockets on the surface of the SH3 domain. Interactions involving an arginine (Arg1, red), two other prolines (light blue), and other residues in the target peptide (green) determine the specificity of binding. [After H. Yu et al., 1994, *Cell* **76:**933.]

that permits extensive contacts with the SH3 domain, thereby facilitating interaction. Second, a subset of these prolines fit into binding "pockets" on the surface of the SH3 domain (Figure 16-23). Several nonproline residues also interact with the SH3 domain and are responsible for determining the binding specificity. Hence the binding of proteins to SH3 and to SH2 domains follows a similar strategy: certain residues provide the overall structural motif necessary for binding, and neighboring residues confer specificity to the binding.

Following activation of an RTK (e.g., Sevenless or the EGF receptor), a complex containing the activated receptor, GRB2, and Sos is formed on the cytosolic face of the plasma membrane (see Figure 16-20). Complex formation depends on the ability of GRB2 to bind *simultaneously* to the receptor and to Sos. Thus receptor activation leads to relocalization of Sos from the cytosol to the membrane, bringing Sos near to its substrate, namely, membrane-bound Ras·GDP. Binding of Sos to Ras·GDP leads to conformational changes in the Switch I and Switch II segments of Ras, thereby opening the binding pocket for GDP so it can diffuse out (Figure 16-24). GTP then binds to and activates Ras. Binding of GTP to Ras, in turn, induces a specific conformation of Switch I and Switch II that allows Ras·GTP to activate the first downstream protein kinase of the MAP kinase pathway.

Signals Pass from Activated Ras to a Cascade of Protein Kinases

Biochemical and genetic studies in yeast, *C. elegans*, *Drosophila*, and mammals have revealed a highly conserved cascade of protein kinases, culminating in MAP kinase, that operates downstream from activated Ras. Although activation of the kinase cascade does not yield the same biological results in all cells, a common set of sequentially acting kinases defines the MAP kinase pathway, as outlined in Figure 16-25. Active Ras·GTP binds to the N-terminal regulatory domain of *Raf*, a serine/threonine kinase, thereby activating it (step **2**). Hydrolysis of Ras·GTP to Ras·GDP releases active Raf (step **3**), which phosphorylates and thereby activates *MEK* (step **4**). (A dual-specificity protein kinase, MEK phosphorylates its target proteins on both tyrosine and serine/threonine residues.) Active MEK then phosphorylates

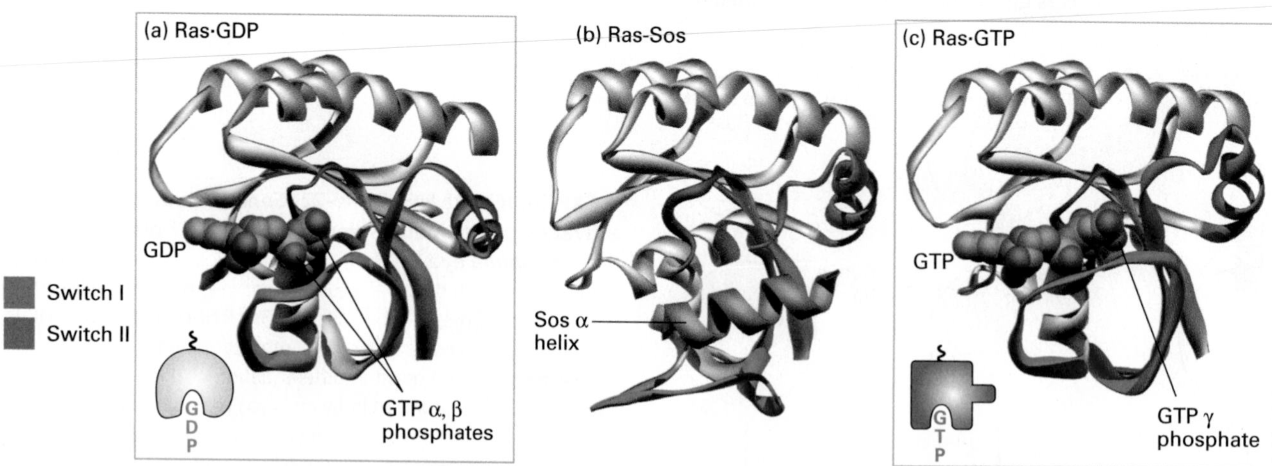

▲ **FIGURE 16-24 Structures of Ras bound to GDP, Sos protein, and GTP.** (a) In Ras·GDP, the Switch I (green) and Switch II (blue) segments do not directly interact with GDP. (b) One α helix (orange) in Sos binds to both switch regions of Ras·GDP, leading to a massive conformational change in Ras. In effect, Sos pries Ras open by displacing the Switch I region, thereby allowing GDP to diffuse out. (c) GTP is thought to bind to the Ras–Sos complex first through its base (guanine); subsequent binding of the GTP phosphates completes the interaction. The resulting conformational change in Switch I and Switch II segments of Ras, allowing both to bind to the GTP γ phosphate, displaces Sos and promotes interaction of Ras·GTP with its effectors (discussed later). See Figure 15-8 for another depiction of Ras·GDP and Ras·GTP. [Adapted from P. A. Boriack-Sjodin and J. Kuriyan, 1998, *Nature* **394:**341.]

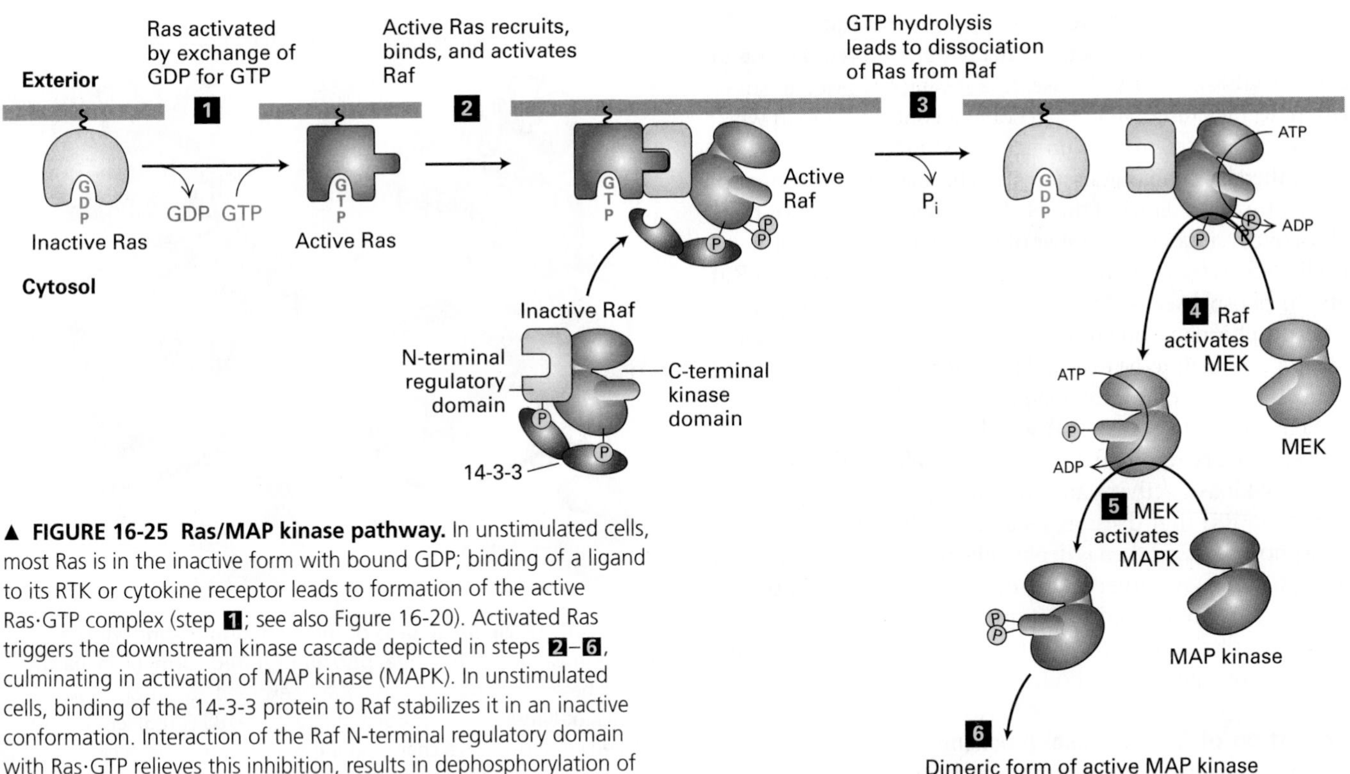

▲ **FIGURE 16-25 Ras/MAP kinase pathway.** In unstimulated cells, most Ras is in the inactive form with bound GDP; binding of a ligand to its RTK or cytokine receptor leads to formation of the active Ras·GTP complex (step **1**; see also Figure 16-20). Activated Ras triggers the downstream kinase cascade depicted in steps **2**–**6**, culminating in activation of MAP kinase (MAPK). In unstimulated cells, binding of the 14-3-3 protein to Raf stabilizes it in an inactive conformation. Interaction of the Raf N-terminal regulatory domain with Ras·GTP relieves this inhibition, results in dephosphorylation of one of the serines that bind Raf to 14-3-3, and leads to activation of Raf kinase activity. Note that in contrast to many other protein kinases, activation of Raf does not depend on phosphorylation of the activation lip. After inactive Ras·GDP dissociates from Raf, it presumably can be reactivated by signals from activated receptors, thereby recruiting additional Raf molecules to the membrane. See the text for details. [See E. Kerkhoff and U. Rapp, 2001, *Adv. Enzyme Regul.* **41:**261; J. Avruch et al., 2001, *Recent Prog. Hormone Res.* **56:**127; and M. Yip-Schneider et al., 2000, *Biochem. J.* **351:**151.]

and activates MAP kinase, another serine/threonine kinase also known as ERK (step **5**). MAP kinase phosphorylates many different proteins, including nuclear transcription factors, that mediate cellular responses (step **6**).

Several types of experiments have demonstrated that Raf, MEK, and MAP kinase lie downstream from Ras and have revealed the sequential order of these proteins in the pathway. For example, mutant Raf proteins missing the N-terminal regulatory domain are constitutively active and induce quiescent cultured cells to proliferate in the absence of stimulation by growth factors. These mutant Raf proteins were initially identified in tumor cells; like the constitutively active Ras^D protein, such mutant Raf proteins are said to be encoded by **oncogenes,** whose encoded proteins promote transformation of the cells in which they are expressed (Chapter 25). Conversely, cultured mammalian cells that express a mutant, nonfunctional Raf protein cannot be stimulated to proliferate uncontrollably by a constitutively active Ras^D protein. This finding established a link between the Raf and Ras proteins. In vitro binding studies further showed that the purified Ras·GTP complex binds directly to the N-terminal regulatory domain of Raf and activates its catalytic activity.

That MAP kinase is activated in response to Ras activation was demonstrated in quiescent cultured cells expressing a constitutively active Ras^D protein. In these cells activated MAP kinase is generated in the absence of stimulation by growth-promoting hormones. More importantly, R7 photoreceptors develop normally in the developing eye of *Drosophila* mutants that lack a functional Ras or Raf protein but express a constitutively active MAP kinase. This finding indicates that activation of MAP kinase is sufficient to transmit a proliferation or differentiation signal normally initiated by ligand binding to a receptor tyrosine kinase such as Sevenless (see Figure 16-22). Biochemical studies showed, however, that Raf cannot directly phosphorylate MAP kinase or otherwise activate its activity.

The final link in the kinase cascade activated by Ras·GTP emerged from studies in which scientists fractionated extracts of cultured cells searching for a kinase activity that could phosphorylate MAP kinase and that was present only in cells stimulated with growth factors, not quiescent cells. This work led to identification of MEK, a kinase that specifically phosphorylates one threonine and one tyrosine residue on the activation lip of MAP kinase, thereby activating its catalytic activity. (The acronym MEK comes from MAP and ERK kinase.) Later studies showed that MEK binds to the C-terminal catalytic domain of Raf and is phosphorylated by the Raf serine/threonine kinase; this phosphorylation activates the catalytic activity of MEK. Hence, activation of Ras induces a kinase cascade that includes Raf, MEK, and MAP kinase: activated RTK → Ras → Raf → MEK → MAP kinase.

Activation of Raf Kinase The mechanism for activating Raf differs from that used to activate many other protein kinases including MEK and MAP kinase. In a resting cell prior to stimulation, Raf is present in the cytosol in a conformation in which the N-terminal regulatory domain is bound to the kinase domain, thereby inhibiting its activity. This inactive conformation is stabilized by a dimer of the *14-3-3 protein*, which binds phosphoserine residues in a number of important signaling proteins. Each 14-3-3 monomer binds to a phosphoserine residue in Raf, one to phosphoserine-259 in the N-terminal domain and the other to phosphoserine-621 (see Figure 16-25). These interactions are thought to be essential for Raf to achieve a conformational state such that it can bind to activated Ras.

The binding of Ras·GTP, which is anchored to the membrane, to the N-terminal domain of Raf relieves the inhibition of Raf's kinase activity and also induces a conformational change in Raf that disrupts its association with 14-3-3. Raf phosphoserine-259 then is dephosphorylated (by an unknown phosphatase) and other serine or threonine residues on Raf become phosphorylated by yet other kinases. These reactions incrementally increase the Raf kinase activity by mechanisms that are not fully understood.

Activation of MAP Kinase Biochemical and x-ray crystallographic studies have provided a detailed picture of how phosphorylation activates MAP kinase. As in JAK kinases and receptor tyrosine kinases, the catalytic site in the inactive, unphosphorylated form of MAP kinase is blocked by a stretch of amino acids, the activation lip (Figure 16-26a). Binding of MEK to MAP kinase destabilizes the lip structure, resulting in exposure of tyrosine-185, which is buried in the inactive conformation. Following phosphorylation of this critical tyrosine, MEK phosphorylates the neighboring threonine-183 (Figure 16-26b).

Both the phosphorylated tyrosine and the phosphorylated threonine residue in MAP kinase interact with additional amino acids, thereby conferring an altered conformation to the lip region, which in turn permits binding of ATP to the catalytic site. The phosphotyrosine residue (pY185) also plays a key role in binding specific substrate proteins to the surface of MAP kinase. Phosphorylation promotes not only the catalytic activity of MAP kinase but also its dimerization. The dimeric form of MAP kinase (but not the monomeric form) can be translocated to the nucleus, where it regulates the activity of many nuclear transcription factors.

MAP Kinase Regulates the Activity of Many Transcription Factors Controlling Early-Response Genes

Addition of a growth factor (e.g., EGF or PDGF) to quiescent cultured mammalian cells causes a rapid increase in the expression of as many as 100 different genes. These are called *early-response genes* because they are induced well before cells enter the S phase and replicate their DNA (see Chapter 20). One important early-response gene encodes the transcription factor c-Fos. Together with other transcription

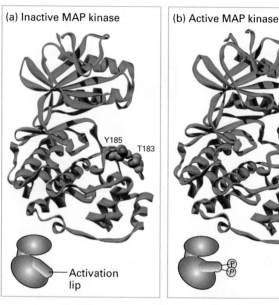

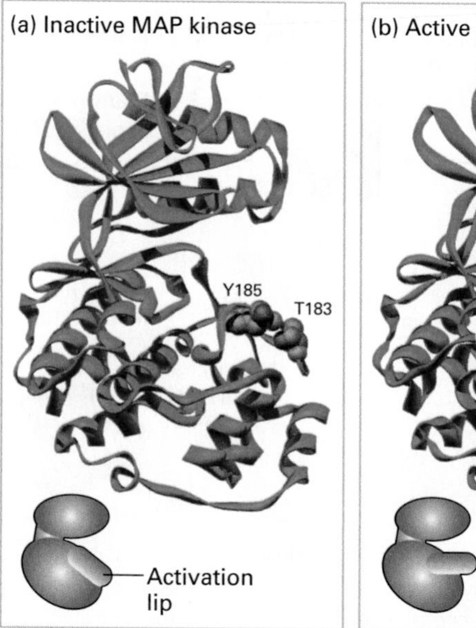

▲ **FIGURE 16-26 Structures of inactive, unphosphorylated MAP kinase and the active, phosphorylated form.** (a) In inactive MAP kinase the activation lip is not fully exposed. (b) Phosphorylation by MEK at tyrosine-185 (Y185) and threonine-183 (T183) leads to a marked conformational change in the activation lip. This activating change promotes dimerization of MAP kinase and binding of its substrates—ATP and its target proteins. A similar phosphorylation-dependent mechanism activates JAK kinases, the intrinsic kinase activity of RTKs, and MEK. [After B. J. Canagarajah et al., 1997, *Cell* **90:**859.]

factors, such as c-Jun, c-Fos induces expression of many genes encoding proteins necessary for cells to progress through the cell cycle. Most RTKs that bind growth factors utilize the MAP kinase pathway to activate genes encoding proteins like c-Fos that in turn propel the cell through the cell cycle.

The enhancer that regulates the c-*fos* gene contains a *serum response element (SRE)*, so named because it is activated by many growth factors in serum. This complex enhancer contains DNA sequences that bind multiple transcription factors. Some of these are activated by MAP kinase; others by different protein kinases that function in other signaling pathways. As depicted in Figure 16-27, activated (phosphorylated) dimeric MAP kinase induces transcription of the c-*fos* gene by direct activation of one transcription factor, *ternary complex factor (TCF)*, and indirect activation of another, *serum response factor (SRF)*. In the cytosol, MAP kinase phosphorylates and activates a kinase called $p90^{RSK}$, which translocates to the nucleus where it phosphorylates a specific serine in SRF. After translocating to the nucleus, MAP kinase directly phosphorylates specific serines in TCF. Association of phosphorylated TCF with two molecules of phosphorylated SRF forms an active trimeric factor that binds strongly to the SRE DNA segment.

As evidence for this model, abundant expression in cultured mammalian cells of a mutant dominant-negative TCF that lacks the serine residues phosphorylated by MAP kinase

Overview Animation: Extracellular Signaling

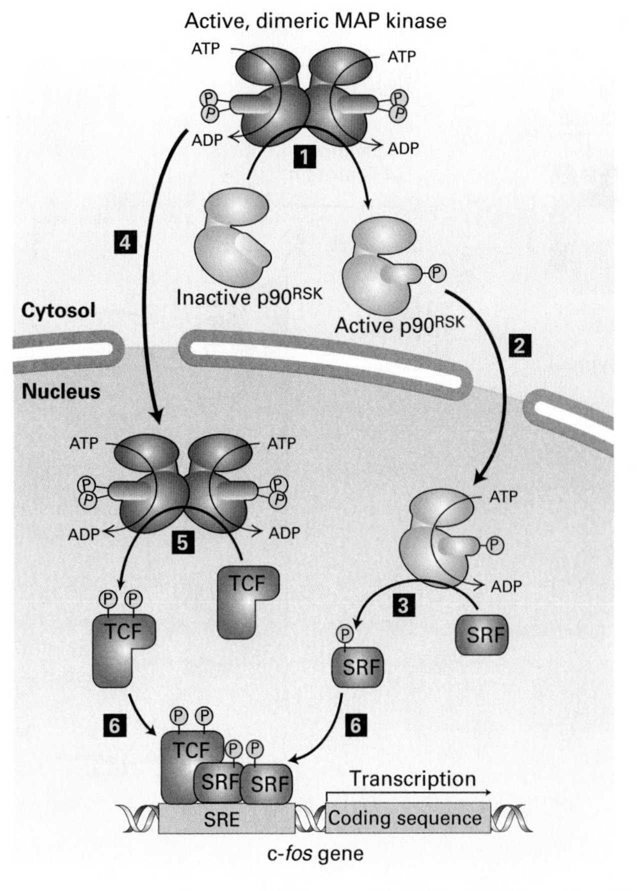

▶ **FIGURE 16-27 Induction of gene transcription by MAP kinase.** Steps **1**–**3**: In the cytosol, MAP kinase phosphorylates and activates the kinase $p90^{RSK}$, which then moves into the nucleus and phosphorylates the SRF transcription factor. Steps **4** and **5**: After translocating into the nucleus, MAP kinase directly phosphorylates the transcription factor TCF. Step **6**: Phosphorylated TCF and SRF act together to stimulate transcription of genes (e.g., *c-fos*) that contain an SRE sequence in their promoter. See the text for details. [See R. Marais et al., 1993, *Cell* **73**:381, and V. M. Rivera et al., 1993, *Mol. Cell Biol.* **13**:6260.]

blocks the ability of MAP kinase to activate gene expression driven by the SRE enhancer. Moreover, biochemical studies showed directly that phosphorylation of SRF by active $p90^{RSK}$ increases the rate and affinity of its binding to SRE sequences in DNA, accounting for the increase in the frequency of transcription initiation. Thus both transcription factors are required for maximal growth factor–induced stimulation of gene expression via the MAP kinase pathway, although only TCF is directly activated by MAP kinase.

G Protein–Coupled Receptors Transmit Signals to MAP Kinase in Yeast Mating Pathways

Although many MAP kinase pathways are initiated by RTKs or cytokine receptors, signaling from other receptors can activate MAP kinase in different cell types of higher eukaryotes. Moreover, yeasts and other single-celled eukaryotes, which lack cytokine receptors or RTKs, do possess several MAP kinase pathways. To illustrate, we consider the mating pathway in *S. cerevisiae,* a well-studied example of a MAP kinase cascade linked to G protein–coupled receptors (GPCRs), in this case for two secreted peptide pheromones, the **a** and α factors.

As discussed in Chapter 21, these pheromones control mating between haploid yeast cells of the opposite mating type, **a** or α. An **a** haploid cell secretes the **a** mating factor and has cell-surface receptors for the α factor; an α cell secretes the α factor and has cell-surface receptors for the **a** factor (see Figure 21-19). Thus each type of cell recognizes the mating factor produced by the opposite type. Activation of the MAP kinase pathway by either the **a** or α receptors induces transcription of genes that inhibit progression of the cell cycle and others that enable cells of opposite mating type to fuse together and ultimately form a diploid cell.

Ligand binding to either of the two yeast pheromone GPCRs triggers the exchange of GTP for GDP on the G_α subunit and dissociation of G_α·GTP from the $G_{\beta\gamma}$ complex. This activation process is identical to that for the GPCRs discussed in the previous chapter (see Figure 15-13). In many mammalian GPCR-initiated pathways, the active G_α transduces the signal. In contrast, mutant studies have shown that the dissociated $G_{\beta\gamma}$ complex mediates all the physiological responses induced by activation of the yeast pheromone receptors. For instance, in yeast cells that lack G_α, the $G_{\beta\gamma}$ subunit is always free. Such cells can mate in the absence of mating factors; that is, the mating response is constitutively on. However, in cells defective for the G_β or G_γ subunit, the mating pathway cannot be induced at all. If dissociated G_α were the transducer, in these mutant cells the pathway would be expected to be constitutively active.

In yeast mating pathways, $G_{\beta\gamma}$ functions by triggering a kinase cascade that is analogous to the one downstream from Ras. The components of this cascade were uncovered mainly through analyses of mutants that possess functional **a** and α receptors and G proteins but are sterile (*Ste*), or defective in mating responses. The physical interactions between the components were assessed through immunoprecipitation experiments with extracts of yeast cells and other types of studies. Based on these studies, scientists have proposed the kinase cascade shown in Figure 16-28a. Free $G_{\beta\gamma}$, which is tethered to the membrane via the lipid bound to the γ subunit, binds the Ste5 protein, thus recruiting it and its bound kinases to the plasma membrane. Ste5 has no obvious catalytic function and acts as a scaffold for assembling other components in the cascade (Ste11, Ste7, and Fus3). Next, Ste20 protein kinase, a protein localized to the plasma membrane, phosphorylates and activates Ste11, a serine/threonine kinase analogous to Raf and other mammalian MEKK proteins. Activated Ste11 then phosphorylates Ste7, a dual-specificity MEK that then phosphorylates and activates

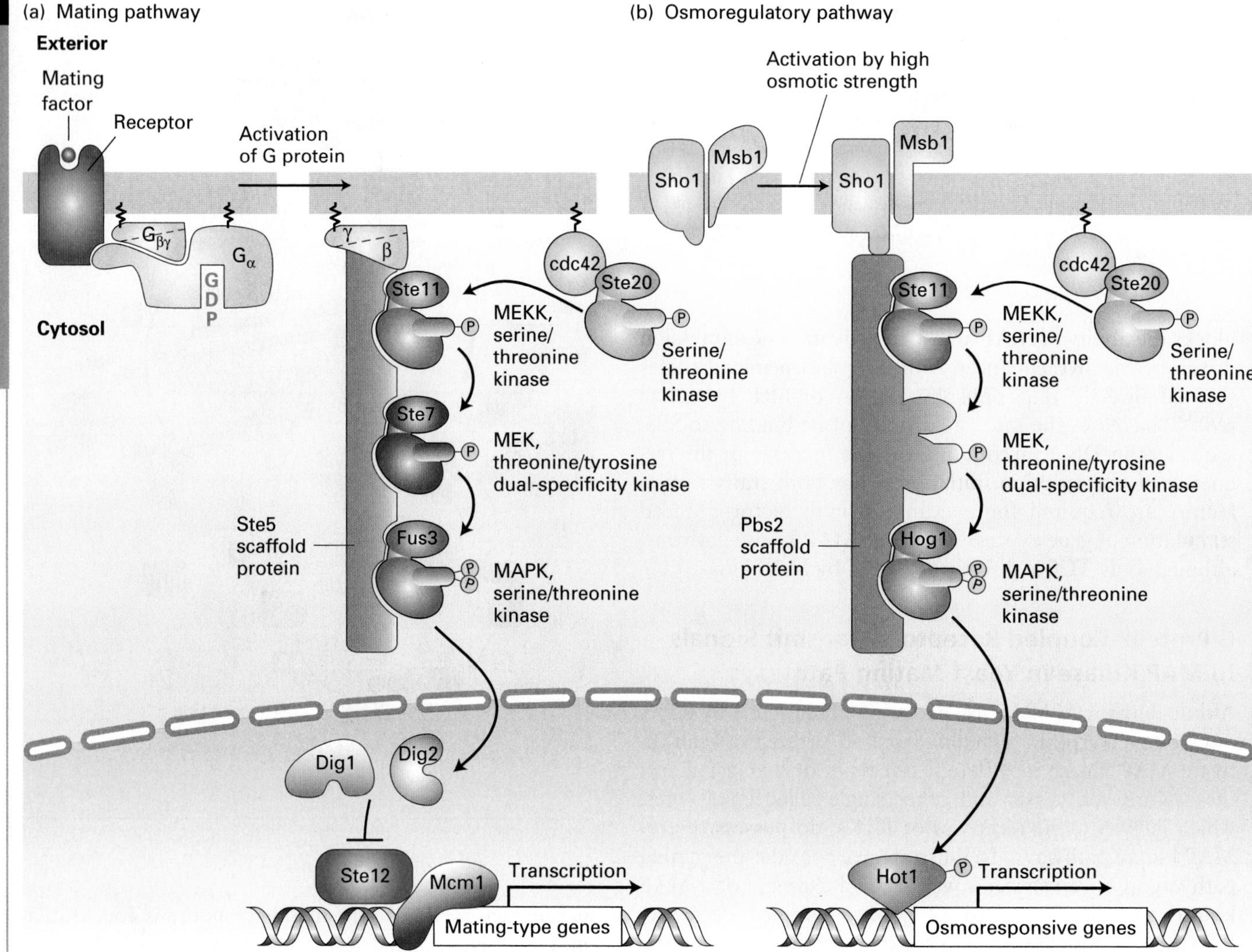

▲ **FIGURE 16-28 Yeast MAP kinase cascades in the mating and osmoregulatory pathways.** In yeast, different receptors activate multiple MAP kinase pathways, two of which are outlined here. (a) Mating pathway: The receptors for yeast **a** and α mating factors are coupled to the same trimeric G protein. Following ligand binding and dissociation of the G protein, the membrane-tethered $G_{\beta\gamma}$ subunit binds the Ste5 scaffold to the plasma membrane. The resident Ste20 kinase then phosphorylates and activates Ste11, which is analogous to Raf and other mammalian MEK kinase (MEKK) proteins. This initiates a kinase cascade in which the final component, Fus3, is functionally equivalent to MAP kinase (MAPK) in higher eukaryotes. Like other MAP kinases, activated Fus3 then translocates into the nucleus. There it phosphorylates two proteins, Dig1 and Dig2, relieving their inhibition of the Ste12 transcription factor, allowing it to bind to DNA and initiate transcription of mating-type genes. (b) Osmoregulatory pathway: Two plasma membrane proteins, Sho1 and Msb1, are activated in an unknown manner by exposure of yeast cells to media of high osmotic strength. Activated Sho1 recruits the Pbs2 scaffold protein, which contains a MEK domain, to the plasma membrane. At the plasma membrane, Ste20 phosphorylates and activates Ste11, initiating a kinase cascade that activates Hog1, a MAP kinase. After translocating to the nucleus, Hog1 phosphorylates the Hot1 transcription factor, allowing it to promote transcription of genes encoding proteins that catalyze synthesis of proteins required for survival in high osmotic strength media. [After M. A. Schwartz and H. D. Madhani, 2004, *Ann. Rev. Genet.* **38:**725, and N. Dard and M. Peter, 2006, *BioEssays* **28:**146.]

Fus3, a serine/threonine kinase equivalent to MAP kinase. After translocation to the nucleus, Fus3 phosphorylates two proteins, Dig1 and Dig2, relieving their inhibition of the Ste12 transcription factor. Activated Ste12 in turn induces expression of proteins involved in mating-specific cellular responses.

Scaffold Proteins Separate Multiple MAP Kinase Pathways in Eukaryotic Cells

In addition to the MAP kinases discussed above, both yeasts and higher eukaryotic cells contain other members of the MAP kinase family. Mammalian MAP kinases include *Jun*

N-terminal kinases (JNKs) and *p38 kinases,* which become activated by various types of stresses; several members of the MAP kinase family in yeasts are described below. All members of the MAP kinase family are serine/threonine kinases that are activated in the cytosol in response to specific extracellular signals and then translocate to the nucleus. Activation of all known MAP kinases requires phosphorylation of both a tyrosine and a threonine residue in the lip region by a member of the MEK family of dual-specificity kinases (see Figure 16-26). Thus in all eukaryotic cells, binding of a wide variety of extracellular signaling molecules triggers highly conserved kinase cascades culminating in activation of a particular MAP kinase.

Current genetic and biochemical studies in the mouse and *Drosophila* are aimed at determining which MAP kinases are required for mediating the response to which signals in higher eukaryotes. This has already been accomplished in large part for the simpler organism *S. cerevisiae.* Each of the six MAP kinases encoded in the *S. cerevisiae* genome has been assigned by genetic analyses to specific signaling pathways triggered by various extracellular signals, such as pheromones, high osmolarity, starvation, hypotonic shock, and carbon/nitrogen deprivation. Each of these MAP kinases mediates very specific cellular responses, as exemplified by Fus3 in the mating pathway and Hog1 in the osmoregulatory pathway (see Figure 16-28).

In both yeasts and higher eukaryotic cells, different MAP kinase cascades share some common components. For instance, Ste11 functions in the three yeast signaling pathways that regulate mating, the response to high osmotic strength conditions, and filamentous growth, which is induced by starvation. Nevertheless, each pathway activates its own MAP kinase: Fus3 in the mating pathway, Hog1 in the osmoregulatory pathway, and Kss1 in the filamentation pathway. Similarly, in mammalian cells, common upstream signal-transducing proteins participate in activating multiple JNK kinases.

Once the sharing of components among different MAP kinase pathways was recognized, researchers wondered how the specificity of the cellular responses to particular signals could be achieved. Studies with yeast provided the initial evidence that pathway-specific *scaffold proteins* enable the signal-transducing kinases in a particular pathway to interact with one another but not with kinases in other pathways. For example, the scaffold protein Ste5 stabilizes a large complex that includes Ste11 and other kinases in the mating pathway; similarly Pbs2 binds Ste11 and other kinases in the osmoregulatory pathway (see Figure 16-28). In each pathway in which Ste11 participates, it is constrained within a large complex that forms in response to a specific extracellular signal, and signaling downstream from Ste11 is restricted to the complex in which it is localized. As a result, exposure of yeast cells to mating factors induces activation of a single MAP kinase, Fus3, whereas exposure to a high osmolarity induces activation of a different MAP kinase, Hog1.

Scaffolds for MAP kinase pathways are well documented in yeast, fly, and worm cells, but their presence in mammalian cells has been difficult to demonstrate. Perhaps the best-documented scaffold protein is *Ksr* (*k*inase *s*uppressor of *R*as), which binds both MEK and MAP kinase. Loss of the *Drosophila* Ksr homolog blocks signaling by a constitutively active Ras protein, suggesting a positive role for Ksr in the Ras/MAP kinase pathway in fly cells. Although knockout mice that lack Ksr are phenotypically normal, activation of MAP kinase by growth factors or cytokines is lower than normal in several types of cells in these animals. This finding suggests that Ksr functions as a scaffold that enhances but is not essential for Ras/MAP kinase signaling in mammalian cells. Other proteins also have been found to bind to specific mammalian MAP kinases. Thus the signal specificity of different MAP kinases in animal cells may arise from their association with various scaffold-like proteins, but much additional research is needed to test this possibility.

The Ras/MAP Kinase Pathway Can Induce Diverse Cellular Responses

The Ras/MAP kinase pathway can be activated in many if not all vertebrate cells by a wide variety of receptor tyrosine kinases (RTKs). In particular, signaling through this pathway is used repeatedly in the course of development, yet the outcome in regard to cell-fate specification varies in different tissues. Why does one cell respond by dividing, another by differentiating, and still another by dying? If there is no specificity beyond the ligand and receptor, an activated Ras might substitute for any signal. In fact, activated Ras can do so in many cell types. In one DNA microarray study of fibroblasts, for instance, the same set of genes was transcriptionally induced by platelet-derived growth factor (PDGF) and by fibroblast growth factor (FGF), suggesting that exposure to either signaling molecule had similar effects. The PDGF receptor and the FGF receptor are both receptor tyrosine kinases, and the binding of ligand to either receptor can activate Ras.

Although several mechanisms for producing diverse cellular responses to a particular signaling molecule have been uncovered, here we focus on two: (1) the strength or duration of the signal governs the nature of the response; and (2) different intracellular pathways are activated by the same receptor in different cell types.

Differences in Signal Strength or Duration Evidence supporting the use of the first mechanism comes from studies with PC12 cells, a cultured cell line capable of differentiating into adipocytes or neurons. Nerve growth factor (NGF) promotes the formation of neurons, whereas epidermal growth factor (EGF) promotes the formation of adipocytes. Strengthening the EGF signal by prolonging exposure to it causes neuronal differentiation. Although both NGF and EGF are RTK ligands, NGF is a much stronger activator of the Ras/MAP kinase pathway than is EGF. The EGF receptor can apparently activate this pathway only after prolonged stimulation.

Differences in Downstream Pathways Signaling through cell type–specific pathways downstream of an RTK has been demonstrated in *C. elegans*. In worms, EGF signals induce at least five distinct responses, each one in a different type of cell. Four of the five responses are mediated by the common Ras/MAP kinase pathway; the fifth, hermaphrodite ovulation, employs a different downstream pathway in which the second messenger inositol 1,4,5-trisphosphate (IP_3) is generated. Binding of IP_3 to its receptor, an IP_3-gated Ca^{2+} channel, in the endoplasmic reticulum (ER) membrane leads to the release of stored Ca^{2+} from the ER (see Figure 15-30). The rise in cytosolic Ca^{2+} then triggers ovulation. This alternative pathway was discovered with a genetic screen that implicated the IP_3 receptor in EGF signaling—a good example of how a mutation in an unexpected gene can lead to a discovery.

KEY CONCEPTS OF SECTION 16.4

Activation of Ras and MAP Kinase Pathways

- Ras is an intracellular GTPase switch protein that acts downstream from most RTKs. Like G_α, Ras cycles between an inactive GDP-bound form and an active GTP-bound form. Ras cycling requires the assistance of two proteins: a guanine nucleotide–exchange factor (GEF) and a GTPase-activating protein (GAP).
- RTKs are linked indirectly to Ras via two proteins: GRB2, an adapter protein, and Sos, which has GEF activity (see Figure 16-20).
- The SH2 domain in GRB2 binds to a phosphotyrosine in activated RTKs, while its two SH3 domains bind Sos, thereby bringing Sos close to membrane-bound Ras·GDP and activating its nucleotide-exchange activity.
- Binding of Sos to inactive Ras causes a large conformational change that permits release of GDP and binding of GTP, forming active Ras (see Figure 16-24). GAP, which accelerates GTP hydrolysis, is localized near Ras·GTP by binding to activated RTKs.
- Activated Ras triggers a kinase cascade in which Raf, MEK, and MAP kinase are sequentially phosphorylated and thus activated. Activated MAP kinase dimerizes and translocates to the nucleus (see Figure 16-25).
- Activation of MAP kinase following stimulation of a growth factor receptor leads to phosphorylation and activation of two transcription factors, which associate into a trimeric complex that promotes transcription of various early-response genes (see Figure 16-27).
- Different extracellular signals induce activation of different MAP kinase pathways, which regulate diverse cellular processes.
- The upstream components of MAP kinase cascades assemble into large pathway-specific complexes stabilized by scaffold proteins (see Figure 16-28). This assures that activation of one pathway by a particular extracellular signal does not lead to activation of other pathways containing shared components.

16.5 Phosphoinositides as Signal Transducers

In previous sections, we have seen how signal transduction from cytokine receptors and receptor tyrosine kinases (RTKs) begins with formation of multiprotein complexes associated with the plasma membrane (see Figures 16-12 and 16-20). Here we discuss how these same receptors initiate signaling pathways that involve membrane-bound phosphorylated inositol lipids, collectively referred to as **phosphoinositides.** Many of these pathways have short-term effects on cell metabolism and all have long-term effects on the pattern of gene expression. We begin with the branch of the phosphoinositide pathway that also is mediated by G protein–coupled receptors and then consider another branch that is not shared with these receptors.

Phospholipase C_γ Is Activated by Some RTKs and Cytokine Receptors

As discussed in Chapter 15, hormonal stimulation of some G protein–coupled receptors leads to activation of phospholipase C (PLC), specifically the β isoform (PLC_β). This membrane-associated enzyme then cleaves phosphatidylinositol 4,5-bisphosphate (PIP_2) to generate two important **second messengers,** 1,2-diacylglycerol (DAG) and inositol 1,4,5-trisphosphate (IP_3). Signaling via the *IP_3/DAG pathway* described in Chapter 15 leads to an increase in cytosolic Ca^{2+} and to activation of **protein kinase C** (see Figure 15-30). This pathway has both short-term effects on cell metabolism and movement and long-term effects on gene expression.

Many RTKs and cytokine receptors also can initiate the IP_3/DAG pathway by activating another isoform of phospholipase C, the γ isoform (PLC_γ). The SH2 domains of PLC_γ bind to specific phosphotyrosines on the activated receptors, thus positioning the enzyme close to its membrane-bound substrate, phosphatidyl inositol 4,5-bisphosphate (PIP_2). In addition, the kinase activity associated with receptor activation phosphorylates tyrosine residues on the bound PLC_γ, enhancing its hydrolase activity. Thus activated RTKs and cytokine receptors promote PLC_γ activity in two ways: by localizing the enzyme to the membrane and by phosphorylating it.

Recruitment of PI-3 Kinase to Hormone-Stimulated Receptors Leads to Synthesis of Phosphorylated Phosphatidylinositols

Many activated RTKs and cytokine receptors initiate another phosphoinositide pathway by recruiting the enzyme phosphatidylinositol-3 (PI-3) kinase to the membrane. In some cells, this *PI-3 kinase pathway* can trigger cell division and prevent programmed cell death (apoptosis), thus assuring cell survival. In other cells, this pathway induces specific changes in cell metabolism.

PI-3 kinase was first identified in studies of the polyoma virus, a DNA virus that transforms certain mammalian cells to uncontrolled growth. Transformation requires several viral-encoded oncoproteins, including one termed "middle T." In an attempt to discover how middle T functions, investigators uncovered PI-3 kinase protein in partially purified preparations of middle T, suggesting a specific interaction between the two. Then they set out to determine how PI-3 kinase might affect cell behavior.

When an inactive, dominant-negative version of PI-3 kinase was expressed in polyoma virus–transformed cells, it inhibited the uncontrolled cell proliferation characteristic of virus-transformed cells. This finding suggested that the normal kinase is important in certain signaling pathways essential for cell proliferation or for the prevention of apoptosis. Subsequent work showed that PI-3 kinases participate in many signaling pathways related to cell growth and apoptosis. Of the nine PI-3 kinase homologs encoded by the human genome, the best characterized contains a p110 subunit with catalytic activity and a p85 subunit with an SH2 phosphotyrosine-binding domain.

PI-3 kinase is recruited to the plasma membrane by binding of its SH2 domain to phosphotyrosines on the cytosolic domain of many activated RTKs and cytokine receptors. This recruitment of PI-3 kinase to the plasma membrane positions its catalytic domain near its phosphoinositide substrates on the cytosolic face of the plasma membrane, leading to formation of PI 3,4-bisphosphate or PI 3,4,5-trisphosphate (Figure 16-29). By acting as docking sites for various signal-transducing proteins, these membrane-bound PI 3-phosphates in turn transduce signals downstream in several important pathways.

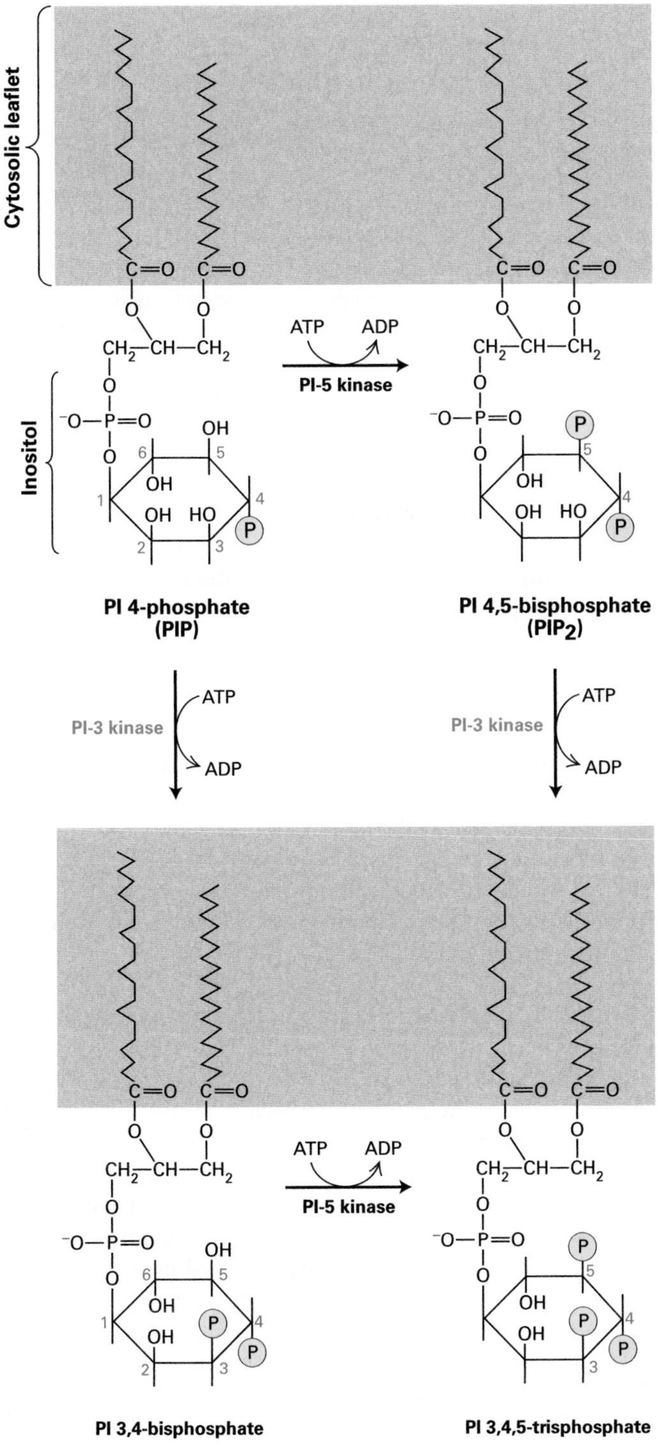

◄ **FIGURE 16-29 Generation of phosphatidylinositol 3-phosphates.** The enzyme phosphatidylinositol-3 kinase (PI-3 kinase) is recruited to the membrane by many activated receptor tyrosine kinases (RTKs) and cytokine receptors. The 3-phosphate added by this enzyme, to yield PI 3,4-bisphosphate or PI 3,4,5-trisphosphate, is a binding site for various signal-transduction proteins, such as the PH domain of protein kinase B. PI 4,5-bisphosphate also is the substrate of phospholipase C (see Figure 15-29). [See L. Rameh and L. C. Cantley, 1999, *J. Biol. Chem.* **274:**8347.]

Accumulation of PI 3-Phosphates in the Plasma Membrane Leads to Activation of Several Kinases

Many protein kinases become activated by binding to phosphatidyl inositol 3-phosphates in the plasma membrane. In turn, these kinases affect the activity of many cellular proteins. One important kinase that binds to PI 3-phosphates is **protein kinase B (PKB)**, a serine/threonine kinase that is also called **Akt.** Besides its kinase domain, protein kinase B also contains a PH domain that can tightly bind the 3-phosphate in both PI 3,4-bisphosphate and PI 3,4,5-trisphosphate. In unstimulated, resting cells, the level of both these compounds is low, and protein kinase B is present in the cytosol in an inactive form (Figure 16-30). Following hormone stimulation and the resulting rise in PI 3-phosphates, protein kinase B binds to them and becomes localized at the plasma membrane. Binding of protein kinase B to PI 3-phosphates not only recruits the enzyme to the plasma membrane but also releases inhibition of the catalytic site by the PH domain. Maximal activation of protein kinase B, however, depends on recruitment of two other kinases: PDK1 and PDK2.

PDK1 is recruited to the plasma membrane via binding of its own PH domain to PI 3-phosphates. Both membrane-associated protein kinase B and PDK1 can diffuse in the plane of the membrane, bringing them close enough that PDK1 can phosphorylate protein kinase B on a critical

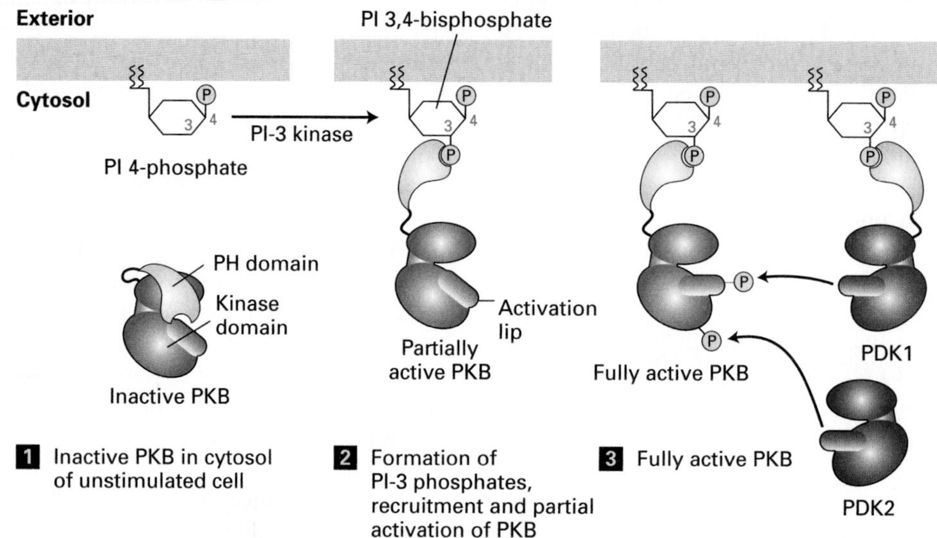

▲ **FIGURE 16-30 Recruitment and activation of protein kinase B (PKB) in PI-3 kinase pathways.** In unstimulated cells (**1**), PKB is in the cytosol with its PH domain bound to the catalytic kinase domain, inhibiting its activity. Hormone stimulation leads to activation of PI-3 kinase and subsequent formation of phosphatidylinositol (PI) 3-phosphates (see Figure 16-29). The 3-phosphate groups serve as docking sites on the plasma membrane for the PH domain of PKB (**2**) and another kinase, PDK1. Full activation of PKB requires phosphorylation both in the activation lip by PDK1 and at the C-terminus by a second kinase, PDK2 (**3**). [Adapted from A. Toker and A. Newton, 2000, *Cell* **103:**185, and S. Sarbassov et al., 2005, *Curr. Opin. Cell Biol.* **17:**596.]

threonine residue in its activation lip—yet another example of kinase activation by phosphorylation. Phosphorylation of a second serine, not in the lip segment, by PDK2 is necessary for maximal protein kinase B activity (see Figure 16-30). Similar to the regulation of Raf activity (see Figure 16-25), release of an inhibitory domain and phosphorylation by other kinases regulate the activity of protein kinase B.

Activated Protein Kinase B Induces Many Cellular Responses

Once fully activated, protein kinase B can dissociate from the plasma membrane and phosphorylate its many target proteins, which have a wide range of effects on cell behavior. Although activation of PKB takes only 5–10 minutes, its effects can last as long as several hours.

Promotion of Cell Survival In many cells activated protein kinase B directly phosphorylates and inactivates pro-apoptotic proteins such as Bad, a short-term effect that prevents activation of an apoptotic pathway leading to cell death (Chapter 21). Activated protein kinase B also promotes survival of many cultured cells by phosphorylating the Forkhead transcription factor FOXO 3A on multiple serine/threonine residues, thereby reducing its pro-apoptotic effect and contributing to cell survival.

In the absence of growth factors, FOXO 3A is unphosphorylated and mainly localizes to the nucleus, where it activates transcription of several genes encoding pro-apoptotic proteins. When growth factors are added to the cells, protein kinase B becomes active and phosphorylates FOXO 3A. This allows the cytosolic phosphoserine-binding protein 14-3-3 to bind FOXO 3A and thus sequester it in the cytosol. (Recall that 14-3-3 also retains phosphorylated Raf protein in an inactive state in the cytosol; see Figure 16-25.) Withdrawal of growth factor leads to inactivation of protein kinase B and dephosphorylation of FOXO 3A, thus favoring its accumulation in the nucleus and transcription of apoptosis-inducing genes. A FOXO 3A mutant in which the three serine target residues for protein kinase B are mutated to alanines is "constitutively active" and initiates apoptosis even in the presence of activated protein kinase B. This finding demonstrates the importance of FOXO 3A and protein kinase B in controlling apoptosis of cultured cells. Deregulation of protein kinase B is implicated in the pathogenesis both of cancer and diabetes.

Promotion of Glucose Uptake and Storage by Insulin As we learned in Chapter 15, insulin acts on muscle, liver, and fat cells to lower the level of blood glucose by increasing its uptake from the blood. In muscle and liver, insulin also promotes storage of glucose as glycogen. The insulin receptor is a dimeric receptor tyrosine kinase that triggers the Ras/MAP kinase pathway, leading to changes in gene expression. Insulin stimulation also can initiate the PI-3 kinase/protein kinase B pathway. The resultant, activated protein kinase B exerts several short-term effects that lower blood glucose and promote glycogen synthesis. The principal short-term effect is increased import of glucose by fat and muscle cells. The GLUT4 glucose transporter is normally retained in intracellular membrane vesicles by a protein called AS160.

Activated protein kinase B phosphorylates AS160; through mechanisms that are not fully understood this causes movement of GLUT4 to the cell surface (see Figure 15-34). The resulting increased influx of glucose into these cells lowers blood glucose levels.

In both liver and muscle, insulin stimulation also leads to short-term activation of glycogen synthase (GS), which synthesizes glycogen from UDP-glucose (see Figure 15-24). In resting cells (i.e., in the absence of insulin), glycogen synthase kinase 3 (GSK3) is active and phosphorylates glycogen synthase, thereby blocking its activity. In insulin-stimulated cells, activated protein kinase B phosphorylates and thereby inactivates GSK3, relieving the GSK3-mediated inhibition of glycogen synthase and promoting glycogen synthesis. As a result, the intracellular concentration of glucose and its metabolites is reduced, stimulating glucose uptake from the blood. This insulin-dependent effect represents another mechanism for reducing the blood glucose level.

The PI-3 Kinase Pathway Is Negatively Regulated by PTEN Phosphatase

Like virtually all intracellular signaling events, phosphorylation by PI-3 kinase is reversible. The relevant phosphatase, termed *PTEN phosphatase,* has an unusually broad specificity. Although PTEN can remove phosphate groups attached to serine, threonine, and tyrosine residues in proteins, its ability to remove the 3-phosphate from PI 3,4,5-trisphosphate is thought to be its major function in cells. Overexpression of PTEN in cultured mammalian cells promotes apoptosis by reducing the level of PI 3,4,5-trisphosphate and hence the activation and anti-apoptotic effect of protein kinase B.

In multiple types of advanced human cancers, the *PTEN* gene is deleted. The resulting loss of PTEN protein contributes to the uncontrolled growth of cells. Indeed, cells lacking PTEN have elevated levels of PI 3,4,5-trisphosphate and PKB activity. Since protein kinase B exerts an anti-apoptotic effect, loss of PTEN indirectly reduces the programmed cell death that is the normal fate of many cells. In certain cells, such as neuronal stem cells, absence of PTEN not only prevents apoptosis but also leads to stimulation of cell-cycle progression and an enhanced rate of proliferation. Knockout mice lacking PTEN have big brains with excess numbers of neurons, attesting to PTEN's importance in control of normal development. ▮

KEY CONCEPTS OF SECTION 16.5

Phosphoinositides as Signal Transducers

- Many RTKs and cytokine receptors can initiate the IP_3/DAG signaling pathway by activating phospholipase C_γ (PLC_γ), a different PLC isoform than the one activated by G protein–coupled receptors.
- Activated RTKs and cytokine receptors also can initiate another phosphoinositide pathway by binding PI-3 kinases, thereby allowing the enzymes access to their membrane-bound phosphoinositide substrates, which then become phosphorylated at the 3 position (see Figure 16-29).
- The PH domain in various proteins binds to PI 3-phosphates, forming signaling complexes associated with the plasma membrane.
- Protein kinase B (PKB) becomes partially activated by binding to PI 3-phosphates. Its full activation requires phosphorylation by another kinase, PDK1, which also is recruited to the membrane by binding to PI 3-phosphates and by a second kinase, PDK2, (see Figure 16-30).
- Activated protein kinase B promotes survival of many cells by directly phosphorylating and inactivating several pro-apoptotic proteins and by phosphorylating and inactivating a transcription factor that otherwise induces synthesis of pro-apoptotic proteins.
- Activation of the insulin receptor, a receptor tyrosine kinase, on fat and muscle cells initiates the PI-3 kinase pathway. The resulting activated protein kinase B promotes glucose uptake and glycogen synthesis.
- Signaling via the PI-3 kinase pathway is terminated by the PTEN phosphatase, which hydrolyzes the 3-phosphate in PI 3-phosphates. Loss of PTEN, a common occurrence in human tumors, promotes cell survival and proliferation.

16.6 Activation of Gene Transcription by Seven-Spanning Cell-Surface Receptors

Chapter 15 focused on intracellular signal-transduction pathways initiated by ligand binding to G protein–coupled receptors (GPCRs). These seven-spanning receptors often have short-term (seconds to minutes) effects on cell metabolism, primarily by modulating the activity of preexisting enzymes or other proteins. However, GPCR signaling pathways also can have long-term effects (hours to days) owing to activation or repression of gene transcription. We have already seen how yeast G protein–coupled receptors for mating factors activate a MAP kinase pathway leading to long-term changes in gene expression (see Figure 16-28a). In the first part of this section, we discuss two other ways that G protein–coupled receptors affect gene expression. The first mechanism operates through phosphorylation of transcription factors by protein kinase A, which is activated downstream of G_s-coupled receptors (see Figure 16-2d). The second mechanism acts through binding of arrestin to many ligand-occupied G protein–coupled receptors and subsequent binding of enzymes in the MAP kinase and other pathways.

In the remainder of this section, we consider two other classes of seven-spanning receptors—those that bind **Wnt** and **Hedgehog,** two protein signals that play key roles in development (see Figure 16-2e,f). Although similar in structure to G protein–coupled receptors, the receptors for Wnt and

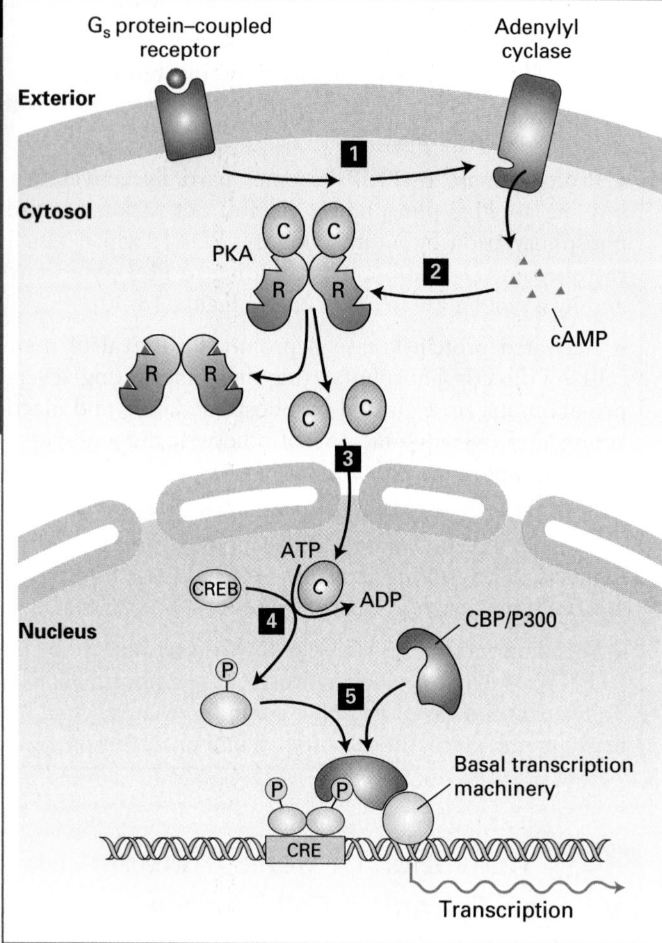

◀ **FIGURE 16-31 Activation of CREB transcription factor following ligand binding to G_s protein–coupled receptors.** Receptor stimulation (**1**) leads to activation of protein kinase A, PKA (**2**). Catalytic subunits of PKA translocate to the nucleus (**3**) and there phosphorylate and activate the CREB transcription factor (**4**). Phosphorylated CREB associates with the co-activator CBP/P300 (**5**) to stimulate various target genes controlled by the CRE regulatory element. See the text for details. [See K. A. Lee and N. Masson, 1993, *Biochim. Biophys. Acta* **1174:**221, and D. Parker et al., 1996, *Mol. Cell Biol.* **16(2):**694.]

Hedgehog do not activate G proteins. These pathways have been elucidated mainly through genetic analysis of developmental mutants in *Drosophila* but are operative in humans as well. Activation of these receptors leads to expression of key genes required for a cell to acquire a new identify or fate. In Chapter 22, we explore the roles of these receptors in several key developmental pathways and also illustrate how these signaling pathways interact with others, activated by different receptors, to specify the precise fate of many cells during development.

CREB Links cAMP and Protein Kinase A to Activation of Gene Transcription

In mammalian cells, an elevation in the cytosolic cAMP level results in activation of **protein kinase A (PKA)**, leading to many different types of short-term responses in different cell types (see Table 15-2). One of the most important short-term PKA-mediated effects is activation of glycogenolysis in liver and muscle, increasing the level of glucose in the blood (see Figure 15-25a). Activation of protein kinase A also stimulates the expression of many genes, leading to long-term effects on the cells that often enhance the short-term effects of activated protein kinase A. For instance, in liver cells, protein kinase A induces expression of several enzymes involved in converting three-carbon compounds such as pyruvate (Figure 12-3) to glucose, thus increasing the level of glucose in the blood.

All genes regulated by protein kinase A contain a cis-acting DNA sequence, the *cAMP-response element (CRE),* that binds the phosphorylated form of a transcription factor called *CRE-binding (CREB) protein,* which is found only in the nucleus. As detailed in Chapter 15, binding of neurotransmitters and hormones to G_s protein–coupled receptors results in the release of the active catalytic subunit of protein kinase A. Some of the catalytic subunits then translocate to the nucleus and phosphorylate serine-133 on CREB protein. Phosphorylated CREB protein binds to CRE-containing target genes and also binds to a *co-activator* termed *CBP/300,* which links CREB to the basal transcriptional machinery, thereby permitting CREB to stimulate transcription (Figure 16-31). As discussed in Chapter 7, other signal-regulated transcription factors rely on CBP/P300 to exert their activating effect. Thus this co-activator plays an important role in integrating signals from multiple signaling pathways that regulate gene transcription.

GPCR-Bound Arrestin Activates Several Kinase Cascades

In higher organisms, activation of the MAP kinase pathway is often triggered by G protein-coupled receptors (GPCRs). As we discussed in Chapter 15, β-arrestin binds to phosphorylated serines in the cytosolic domain of activated G protein–coupled receptors and desensitizes cells to further hormone stimulation in two ways: by inhibiting activation of a G_α protein and by promoting endocytosis of the GPCR–arrestin complex. The GPCR–arrestin complex also acts as a scaffold for binding and activating several cytosolic kinases (see Figure 15-27). These include c-Src, a cytosolic protein tyrosine kinase that activates the MAP kinase pathway and other pathways leading to transcription of genes needed for cell division. A complex of three arrestin-bound proteins, including a Jun N-terminal kinase (JNK-1), initiates a kinase cascade that ultimately activates the c-Jun transcription factor. Activated c-Jun promotes expression of certain growth-promoting enzymes and other proteins that help cells respond to stresses.

Wnt Signals Trigger Release of a Transcription Factor from Cytosolic Protein Complex

Like G protein–coupled receptors, the receptors for Wnt proteins span the plasma membrane seven times, but there the similarity ends. The first vertebrate *Wnt* gene to be discovered, the mouse *Wnt-1* gene, attracted notice because it was overexpressed in certain mammary cancers. Subsequent work showed that overexpression was caused by insertion of a mouse mammary tumor virus (MMTV) provirus near the *Wnt-1* gene. Hence Wnt-1 is a **proto-oncogene,** a normal cellular gene whose inappropriate expression promotes the onset of cancer (Chapter 25). The word Wnt is an amalgamation of *wingless,* the corresponding fly gene, with *int* for the retrovirus integration site in mouse.

Activation of the *Wnt pathway* controls numerous critical developmental events, such as brain development, limb patterning, and organogenesis. A major role for Wnt signaling in bone formation was revealed by the finding that mutations in Wnt pathway components affect bone density in humans. Wnt signaling is now known to control formation of osteoblasts (bone-forming cells). Additionally, Wnt signals are important in controlling stem cells (Chapter 21) and in many other aspects of development (Chapter 22). Disturbances in signaling through the Wnt pathway are associated with various human cancers, particularly colon cancer (Chapter 25).

Because of the conservation of the Wnt signaling pathway in metazoan evolution, genetic studies in *Drosophila* and *C. elegans,* studies of mouse proto-oncogenes and tumor-suppressor genes, and studies of cell-junction components have all contributed to identifying various pathway components. Wnt proteins, the extracellular signaling molecules in the pathway, are modified by addition of a palmitate group near their N termini. This hydrophobic group is thought to tether Wnt proteins to the plasma membrane of Wnt secreting cells, thus limiting their range of action to adjacent cells. Wnt proteins act through two cell-surface receptor proteins: *Frizzled (Fz),* which contains seven transmembrane α helices and directly binds Wnt; and a co-receptor designated *LRP,* which appears to associate with Frizzled in a Wnt signal–dependent manner (see Figure 16-2e). Mutations in the genes encoding Wnt proteins, Frizzled, or LRP (called Arrow in *Drosophila*) all have similar effects on the development of embryos.

According to a current model of the *Wnt pathway,* the central player in intracellular Wnt signal transduction is called *β-catenin* in vertebrates and Armadillo in *Drosophila.* This multi-talented protein functions both as a transcriptional activator and as a membrane–cytoskeleton linker protein (see Figure 19-12). In the absence of a Wnt signal, β-catenin is phosphorylated by a complex containing glycogen synthase kinase 3 (GSK3), the same protein kinase that functions in regulation of blood glucose (Section 16.5); the adenomatosis polyposis coli (APC) protein, an important human tumor suppressor; and Axin, a scaffold protein. Phosphorylated β-catenin is then ubiquitinated and degraded in proteasomes (Figure 16-32a).

In the presence of Wnt, Axin binds to the cytosolic domain of the LRP co-receptor. This binding disrupts the complex containing GSK3 and β-catenin, prevents phosphorylation of β-catenin by GSK3, and stabilizes β-catenin in the cytosol (Figure 16-22b). Wnt-induced stabilization of β-catenin also requires the Dishevelled (Dsh) protein, which is bound to the cytosolic domain of the receptor Frizzled (Fz). The freed β-catenin translocates to the nucleus where it associates with the TCF transcription factor to control

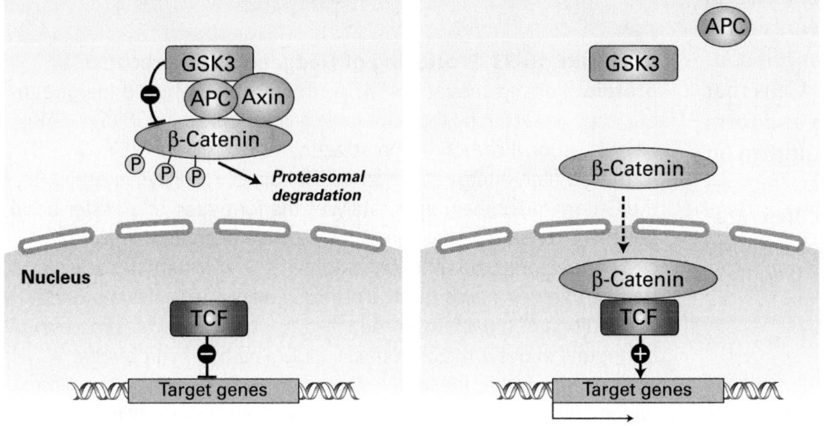

◄ FIGURE 16-32 Wnt signaling pathway. (a) In the absence of Wnt, β-catenin is found in a complex with Axin (a scaffold protein), APC, and the kinase GSK3, which phosphorylates β-catenin, leading to its degradation. The Axin-mediated formation of this complex facilitates phosphorylation of β-catenin by GSK3 by an estimated factor of >20,000. The TCF transcription factor in the nucleus acts as a repressor of target genes unless altered by Wnt signaling. (b) Binding of Wnt to its receptor Frizzled (Fz), triggers phosphorylation of the LRP co-receptor by GSK3 and another kinase, and thus allows subsequent binding of Axin. This disrupts the Axin–APC–GSK3–β-catenin complex, preventing phosphorylation of β-catenin by GSK3 and leads to accumulation of β-catenin in the cell. After translocation to the nucleus, β-catenin may act with TCF to activate target genes or, alternatively, cause the export of TCF from the nucleus and possibly its activation in the cytosol. [After R. Nusse, 2005, *Nature* **438:**747; see also The Wnt Gene Homepage, www.stanford.edu/~rnusse/wntwindow.html.]

expression of particular target genes. (Recall that TCF also functions in the MAP kinase pathway; see Figure 16-27.) Among the Wnt target genes are many that also control Wnt signaling, indicating a high degree of feedback regulation. The importance of β-catenin stability and location means that Wnt signals affect a critical balance between the three pools of β-catenin in the cell: the cytoskeleton, cytosol, and nucleus.

Wnt signaling also requires binding to cell-surface proteoglycans. A **proteoglycan** consists of a core protein bound to **glycosaminoglycan** (GAG) chains such as heparin sulfate and chondroitin sulfate (see Figure 19-29). Evidence for the participation of proteoglycans in Wnt signaling comes from *Drosophila sugarless (sgl)* mutants, which lack a key enzyme needed to synthesize heparin and chondroitin sulfate. These mutants have greatly depressed levels of Wingless, the fly Wnt protein, and exhibit other phenotypes associated with defects in Wnt signaling. Mutations in two other fly genes, *dally* and *dally-like,* both of which encode core proteins of cell-surface proteoglycans, also are associated with defective Wnt signaling in *Drosophila.* How proteoglycans facilitate Wnt signaling is unknown, but perhaps binding of Wnt to specific glycosaminoglycan chains is required for it to bind to its receptor Fz or co-receptor LRP. This mechanism would be analogous to the binding of fibroblast growth factor (FGF) to heparan sulfate, which enhances binding of FGF to its receptor tyrosine kinase (see Figure 16-15).

Hedgehog Signaling Relieves Repression of Target Genes

The *Hedgehog (Hh) pathway* is similar to the Wnt pathway in that two membrane proteins, one with seven membrane-spanning segments, are required to receive and transduce a signal (see Figure 16-2f). The Hh pathway also involves disassembly of an intracellular complex containing a transcription factor, like the Wnt pathway. However, in contrast to Wnt signaling, the Hh protein, the extracellular signal in the pathway, is synthesized as a precursor that is cleaved, and the two membrane proteins involved in Hh signaling are thought to move between the plasma membrane and intracellular vesicles.

Although Hedgehog is a secreted protein, it moves only a short distance from a signaling cell, on the order of 1–20 cells, and is bound by receptors on receiving cells. Thus Hh signals, like Wnt signals, have quite localized effects. As Hh diffuses away from secreting cells, however, its concentration decreases. As we learn in Chapter 22, different Hh concentrations induce different fates in receiving cells: Cells that receive a large amount of Hh turn on certain genes and form certain structures; cells that receive a smaller amount turn on different genes and thus form different structures. Signals that induce different cell fates depending on their concentration are referred to as **morphogens.** During development, the production of Hedgehog and other morphogens is tightly regulated in time and space.

Hedgehog signaling, which is conserved throughout the animal kingdom, functions in the formation of many tissues and organs. Mutations in components of the Hedgehog signaling pathway have been implicated in human birth defects such as cyclopia, a single eye resulting from union of the right and left brain primordia, and in multiple forms of cancer. ■

Processing of Hh Precursor Protein Hedgehog is formed from a precursor protein with autoproteolytic activity that enables the protein to cut itself in half. The cleavage produces an N-terminal fragment, which is subsequently secreted to signal to other cells, and a C-terminal fragment, which is degraded. As shown in Figure 16-33, cleavage of the precursor is accompanied by covalent addition of the lipid cholesterol to the new carboxyl terminus of the N-terminal fragment. The C-terminal domain of the precursor, which catalyzes this reaction, is found in

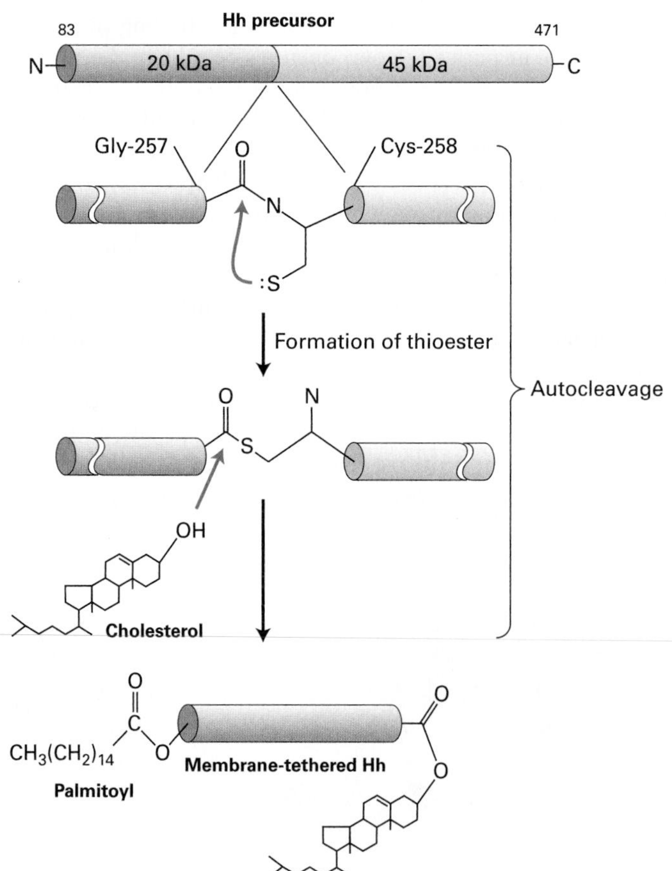

▲ **FIGURE 16-33 Processing of Hedgehog (Hh) precursor protein.** Cells synthesize a 45-kDa Hh precursor, which undergoes a nucleophilic attack by the thiol side chain of cysteine 258 (Cys-258) on the carbonyl carbon of the adjacent residue glycine 257 (Gly-257), forming a high-energy thioester intermediate. An enzymic activity in the C-terminal domain then catalyzes the formation of an ester bond between the β-3 hydroxyl group of cholesterol and glycine 257, cleaving the precursor into two fragments. The N-terminal signaling fragment (blue) retains the cholesterol moiety and is also modified by the addition of a palmitoyl group to the N-terminus. This processing is thought to occur mostly intracellularly. The two hydrophobic anchors may tether the secreted, processed Hh protein to the plasma membrane. [Adapted from J. A. Porter et al., 1996, *Science* **274:**255.]

other proteins and may promote the linkage of these proteins to membranes by a similar autoproteolytic mechanism.

A second modification to Hedgehog, the addition of a palmitoyl group to the N-terminus, makes the protein even more hydrophobic. Together, the two attached hydrophobic groups may cause secreted Hedgehog to bind nonspecifically and reversibly to cell plasma membranes, thereby limiting its diffusion and thus its range of action in tissues. Spatial restriction plays a crucial role in constraining the effects of powerful inductive signals like Hh. Recall that the palmitoyl group also is added to Wnt proteins and likely also causes Wnt to bind reversibly to cells, thereby restricting Wnt signaling to cells adjacent to the signaling cell.

Hh Pathway in *Drosophila* Genetic studies in *Drosophila* indicate that two membrane proteins, *Smoothened (Smo)* and *Patched (Ptc)*, are required to receive and transduce a Hedgehog signal to the cell interior. Smoothened has seven membrane-spanning a helices and is related in sequence to the Wnt receptor Fz. Patched is predicted to contain 12 transmembrane a helices and is most similar structurally to the Niemann-Pick C1 (NPC1) protein, a member of the **ABC superfamily** of membrane proteins (see Table 11-3). The NPC1 protein, which probably functions as an ATP-powered pump, is necessary for normal intracellular movement of sterols and other substances through vesicle-trafficking pathways. In humans, mutations in the *NPC1* gene cause a rare, autosomal recessive disorder marked by defects in movements of late endosomes and in the handling of cholesterol in endosomes and lysosomes. A related protein, NPC1L1, is the major cholesterol uptake transporter in the mammalian intestine. Patched likely evolved from a NPC1-like ancestor, since NPC1 but not Patched is clearly present in yeast. This may be an example of how a cell component needed for fundamental cell metabolism was adapted as a component of a developmental signaling pathway. Duplication of the *npc1* gene would have been followed by divergent evolution of one of the copies.

Figure 16-34 depicts a current model of the *Hedgehog (Hh) pathway*, based largely on work in *Drosophila*. Evidence supporting this model comes from study of fly embryos with loss-of-function mutations in the *hedgehog (hh)* or *smoothened (smo)* genes. Both types of mutant embryos have very similar developmental phenotypes. Moreover, both the *hh* and *smo* genes are required to activate transcription

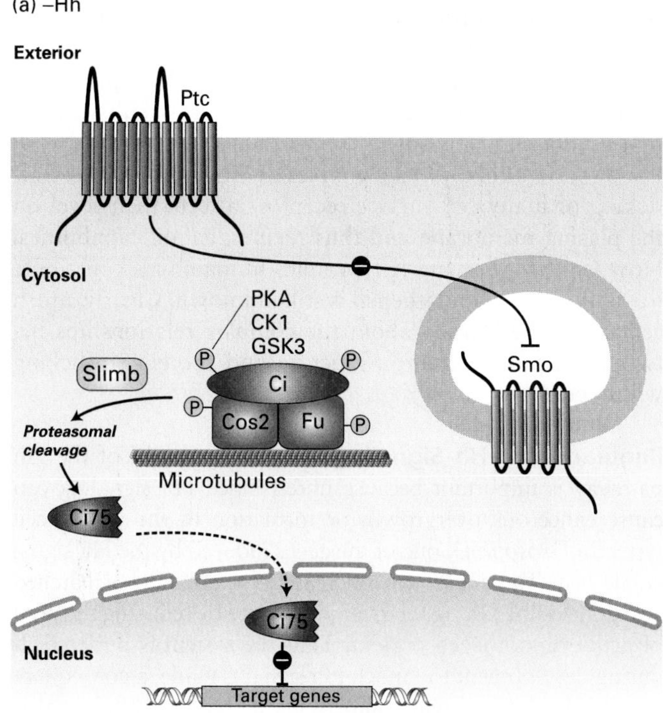

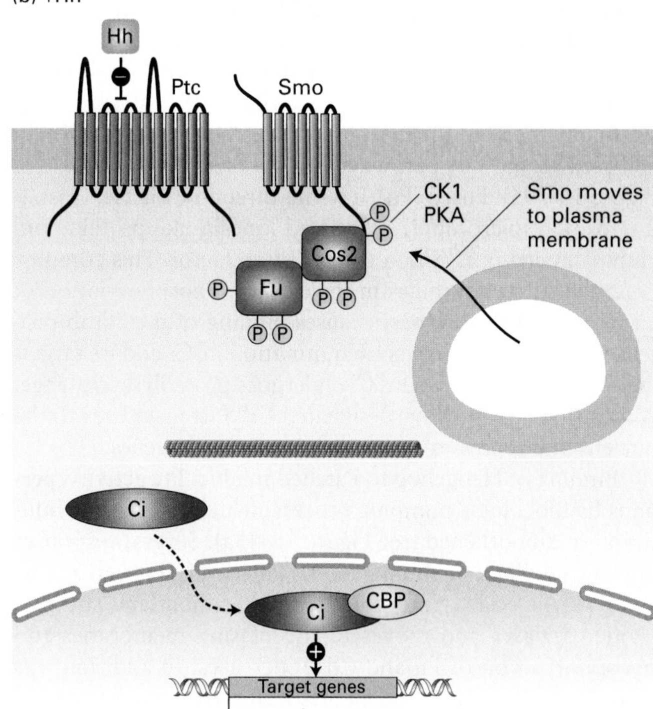

▲ **FIGURE 16-34 Hedgehog (Hh) signaling pathway.** (a) In the absence of Hh, Patched (Ptc) protein inhibits Smoothened (Smo), which is present largely in the membrane of internal vesicles. A complex containing Fused (Fu), a kinase, Costal-2 (Cos2), a kinesin-related motor protein, and Cubitis interruptus (Ci), a zinc-finger transcription factor, binds to microtubules. Ci is phosphorylated in a series of steps involving protein kinase A (PKA), glycogen synthase kinase 3 (GSK3), and casein kinase 1 (CK1). The phosphorylated Ci is then proteolytically cleaved in a process requiring the protein Slimb and the ubiquitin/proteasome pathway, generating the fragment Ci75, which functions as a transcriptional repressor of Hh-responsive genes. (b) In the presence of Hh, Hh binds to Ptc, causing some Ptc to move to internal compartments (not shown) and relieving the inhibition of Smo. Smo then moves to the plasma membrane, is phosphorylated, binds Cos2, and is stabilized from degradation. Both Fu and Cos2 become extensively phosphorylated, and most importantly the Fu/Cos2/Ci complex is dissociated from microtubules. This leads to the stabilization of a full-length, alternately modified Ci, which functions as a transcriptional activator in conjunction with CREB-binding protein (CBP). The exact membrane compartments in which Ptc and Smo respond to Hh and function are unknown. [After M.A. Price, 2006, *Genes Dev.* **20:**399.]

of the same target genes (e.g., *patched* and *wingless*) during embryonic development. In contrast, loss-of-function mutations in the *patched (ptc)* gene produce a quite different phenotype, one similar to the effect of flooding the embryo with Hedgehog protein. Thus Patched appears to antagonize the actions of Hedgehog and vice versa. These findings suggest that, in the absence of Hedgehog, Patched represses target genes by inhibiting a signaling pathway needed for gene activation. The additional observation that Smoothened is required for the transcription of target genes in mutants lacking *patched* function, places Smoothened downstream of Patched in the Hh pathway. The evidence indicates that Hedgehog binds directly to Patched and prevents Patched from blocking Smoothened action, thus activating the transcription of target genes.

Immunostaining of appropriate *Drosophila* embryonic cells with antibodies to Hh, Ptc, and Smo has shown that in the absence of Hedgehog, Patched is enriched in the plasma membrane, but Smoothened is in internal vesicle membranes. Following binding of Hedgehog to Patched, both proteins move from the cell surface into internal vesicles, while Smoothened moves from internal vesicles to the surface. The similarity of Patched to transporter proteins suggests that in the absence of Hh binding, it either pumps a small-molecule inhibitor toward Smoothened or pumps an activator away from it. This mechanism is supported by the finding that a number of natural and synthetic small molecules bind to and regulate Smo activity.

In the absence of the Hh signal, the cytosolic protein complex in the Hh pathway consists of three proteins (see Figure 16-34a): Fused (Fu), a serine-threonine kinase; Costal-2 (Cos2), a microtubule-associated kinesin-like protein; and Cubitis interruptus (Ci), a transcription factor. This complex is bound to microtubules in the cytosol. Phosphorylation of Ci by at least three kinases causes binding of the Slimb protein. Slimb in turn directs ubiquitination of Ci and its targeting to proteasomes, where Ci undergoes proteolytic cleavage. A resulting Ci fragment, designated Ci75, translocates to the nucleus and represses expression of Hh target genes.

Binding of Hedgehog to Patched inhibits Ptc activity, perhaps by blocking a pumping process, which relieves the inhibition of Smoothened (see Figure 16-43a). Several responses are triggered by Hh binding: Some Patched is internalized by the receiving cells; Smoothened is phosphorylated by two protein kinases and moves to the plasma membrane; and phosphorylation of Fu and Cos2 increases. In addition, the complex of Fu, Cos2, and Ci dissociates from microtubules, and Cos2 becomes associated with the C-terminal tail of Smoothened. The resulting disruption of the Fus/Cos2/Ci complex causes a reduction in both phosphorylation and cleavage of Ci. As a result, a modified form of full-length Ci is generated and translocates to the nucleus where it binds to the transcriptional co-activator CREB-binding protein (CBP), promoting the expression of target genes.

Hh Pathway in Mammals The Hh signaling pathway in mammals shares many features with the *Drosophila* pathway, but there are also some striking differences. First, mammalian genomes contain three *hh* genes and two *ptc* genes, which are expressed differentially among various tissues. Second, mammals express three Gli transcription factors that divide up the roles of the single Ci protein in *Drosophila*. Third, there appears to be no mammalian Cos2 ortholog, and the role of possible Fu orthologs is unclear.

The most fascinating aspect of the mammalian Hh pathway is the newly recognized involvement of *intraflagellar transport (IFT) proteins*. IFT proteins are required to move materials within **flagella** and **cilia**, long plasma membrane–enveloped structures that protrude from the cell surface. The roles of the abundant cilia in the trachea in moving materials along the tracheal surface and of flagella in sperm locomotion are well known (Chapter 18). Most cells, however, have a single immotile cilium called the *primary cilium*. The function of a primary cilium has been rather mysterious, but there is increasing evidence for its involvement in signal transduction, especially in the mammalian Hh signaling pathway. For example, mutations that eliminate IFT function cause induction of Hh pathway target genes, similar to the effect of inactivating mutations in Patched. In addition, several components of the Hh pathway, including Smoothened, are located, in part, in primary cilia. The primary cilium may be a signal-transduction center, and its functions may substitute for the apparently missing Cos2 kinesin-like protein that is found in the *Drosophila* Hh pathway. The absence of IFT proteins in Hh signaling in *Drosophila* provides some support for this substitution hypothesis.

We have already seen examples of how intracellular trafficking of many cell-surface receptors affects their level on the plasma membrane and thus their signaling capabilities. How Cos2 in flies and IFT proteins in mammals contribute to trafficking of Smoothened is still unknown. Clearly, much remains to be learned about the complex relationships between receptor signal transduction and protein trafficking within cells.

Regulation of Hh Signaling Feedback control of the Hh pathway is important because unrestrained Hh signaling can cause cancerous overgrowth or formation of the wrong cell types. In *Drosophila*, one of the genes induced by the Hh signal is *patched*. The subsequent increase in expression of Patched antagonizes the Hh signal in large measure by reducing the pool of active Smoothened protein. Thus the system is buffered: If during development too much Hh signal is made, a consequent increase in Patched will compensate; if too little Hh signal is made, the amount of Patched is decreased.

KEY CONCEPTS OF SECTION 16.6

Activation of Gene Transcription by Seven-Spanning Cell-Surface Receptors

- Downstream of activated G protein–coupled receptors, signal-induced activation of protein kinase A (PKA) often leads to phosphorylation of nuclear CREB protein, which together with the CBP/300 co-activator stimulates transcription of many target genes (see Figure 16-31).

- The GPCR–arrestin complex activates several cytosolic kinases, initiating cascades that lead to transcriptional activation of many genes controlling cell growth (see Figure 15-27).
- Both Hedgehog and Wnt signal proteins contain lipid anchors that can tether them to cell membranes, thereby reducing their signaling range.
- Wnt signals act through two cell-surface proteins, the receptor Frizzled and co-receptor Lrp, and an intracellular complex containing β-catenin (see Figure 16-32). Binding of Wnt promotes the stability and nuclear localization of β-catenin, which either directly or indirectly promotes activation of the TCF transcription factor.
- The Hedgehog signal also acts through two cell-surface proteins, Smoothened and Patched, and an intracellular complex containing the Cubitis interruptus (Ci) transcription factor (see Figure 16-34). An activating form of Ci is generated in the presence of Hedgehog; a repressing Ci fragment is generated in the absence of Hedgehog. Both Patched and Smoothened change their subcellular location in response to Hedgehog binding to Patched.

16.7 Pathways That Involve Signal-Induced Protein Cleavage

All the signaling pathways discussed so far are reversible and thus can be turned off relatively quickly if the signal is removed. In this section, we discuss several essentially irreversible pathways in which a component is proteolytically cleaved. First we examine the *NF-κB pathway* in which an inactive transcription factor is sequestered in the cytosol bound to an inhibitor (see Figure 16-1g); several stress-inducing conditions cause immediate degradation of the inhibitor enabling cells to respond immediately and vigorously by activating gene transcription. Next we consider signaling pathways involving protein cleavage outside of the cell by members of the *matrix metalloprotease (MMP) family.* In the *Notch/Delta pathway* (see Figure 16-1h), for instance, extracellular MMP cleavage of the receptor is followed by its cleavage within the plasma membrane by a different protease. This pathway determines the fates of many types of cells during development. Many growth factors, including members of the epidermal growth factor (EGF) family, are made as membrane-spanning precursors; cleavage of these proteins by matrix metalloproteases releases the active growth factor into the extracellular medium. This process goes awry in many cancers and may lead to an often fatal enlargement of the heart. Inappropriate MMP cleavage of yet another membrane-spanning protein has been implicated in the pathology of Alzheimer's disease. We conclude our discussion by describing the intramembrane cleavage of a transcription factor precursor within the Golgi membrane in response to low cholesterol levels. This pathway is essential for maintaining the proper balance of cholesterol and phospholipids for constructing cell membranes (Chapter 10).

Degradation of an Inhibitor Protein Activates the NF-κB Transcription Factors

The examples in previous sections, like TGFβ receptors and MAP kinases, have demonstrated the importance of signal-induced phosphorylation in modulating the activity of many transcription factors. Another mechanism for regulating transcription factor activity in response to extracellular signals was revealed in studies with both mammalian cells and *Drosophila.* This mechanism, which involves phosphorylation and subsequent ubiquitin-mediated degradation of an inhibitor protein, is exemplified by the NF-κB transcription factor.

NF-κB is rapidly activated in mammalian immune-system cells in response to bacterial and viral infection, inflammation, and a number of other stressful situations, such as ionizing radiation. As we learn in Chapter 24, the NF-κB pathway is activated in some cells of the immune system when components of bacterial or fungal cell walls bind to certain *Toll-like receptors* on the cell surface. This pathway is also activated by so-called inflammatory cytokines, such as *tumor necrosis factor alpha (TNFα)* and *interleukin 1 (IL-1),* that are released by nearby cells in response to infection. In all cases, binding of ligand to its receptor induces assembly of a multiprotein complex in the cytosol. Formation of this complex triggers a signaling pathway that results in activation of the NF-κB transcription factor.

NF-κB was originally discovered on the basis of its transcriptional activation of the gene encoding the light chain of antibodies (immunoglobulins) in B cells. It is now thought to be the master transcriptional regulator of the immune system in mammals. Although flies do not make antibodies, NF-κB homologs in *Drosophila* induce synthesis of a large number of secreted antimicrobial peptides in response to bacterial and viral infection (Chapter 24). This phenomenon indicates that the NF-κB regulatory system has been conserved during evolution and is more than half a billion years old.

Biochemical studies in mammalian cells and genetic studies in flies have provided important insights into the operation of the NF-κB pathway. The two subunits of heterodimeric NF-κB (p65 and p50) share a region of homology at their N-termini that is required for their dimerization and binding to DNA. In cells that are not undergoing a stress or responding to signs of an infection, NF-κB is sequestered in an inactive state in the cytosol by direct binding to an inhibitor called *I-κB.* A single molecule of I-κB binds to the paired N-terminal domains of the p50/p65 heterodimer, thereby masking their nuclear-localization signals. The protein kinase complex termed *I-κB kinase* is the point of convergence of all of the extracellular signals that activate NF-κB. Within minutes of stimulation of the cell by an infectious agent or inflammatory cytokine, I-κB kinase becomes activated and phosphorylates two N-terminal serine residues on I-κB (Figure 16-35, steps **1** and **2**). An E3 ubiquitin ligase then binds to these phosphoserines and polyubiquitinates I-κB, triggering its immediate degradation by a proteasome (steps **3** and **4**). In cells expressing mutant forms of I-κB in

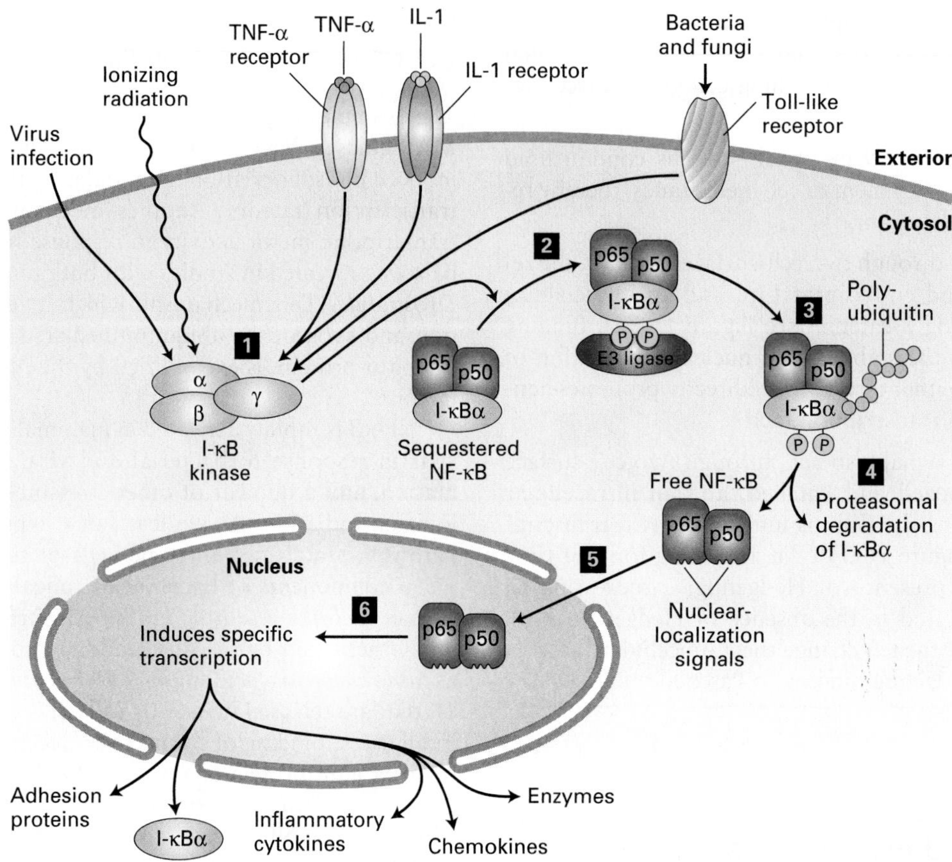

▲ **FIGURE 16-35 NF-κB signaling pathway.** In resting cells, the dimeric transcription factor NF-κB, composed of p50 and p65 subunits, is sequestered in the cytosol, bound to the inhibitor I-κBα. Step **1**: Activation of the trimeric I-κB kinase is stimulated by many agents including virus infection, ionizing radiation, binding of the proinflammatory cytokines TNFα or IL-1 to their respective receptors, or activation of any of several Toll-like receptors by components of invading bacteria or fungi. Step **2**: I-κB kinase then phosphorylates the inhibitor I-κBα, which then binds an E3 ubiquitin ligase. Steps **3** and **4**: Subsequent polyubiquitination of I-κB targets it for degradation by proteasomes. Step **5**: The removal of I-κB unmasks the nuclear-localization signals (NLS) in both subunits of NF-κB, allowing their translocation to the nucleus. Step **6**: In the nucleus, NF-κB activates transcription of numerous target genes, including the gene encoding I-κBα, which acts to terminate signaling, and genes encoding various inflammatory cytokines, which promote signaling. [See R. Khush et al., 2001, *Trends Immunol.* **22:**260, and J-L Luo et al., 2005, *J. Clin Invest.* **115:**2625.]

which these two serines have been changed to alanine, and thus cannot be phosphorylated, NF-κB is permanently inactive, demonstrating that phosphorylation of I-κB is essential for pathway activation.

The degradation of I-κB exposes the nuclear-localization signals on NF-κB, which then translocates into the nucleus and activates transcription of a multitude of target genes (see Figure 16-35, steps **5** and **6**). Despite its activation by proteolysis, NF-κB signaling eventually is turned off by a negative feedback loop, since one of the genes whose transcription is immediately induced by NF-κB encodes I-κB. The resulting increased levels of the I-κB protein bind active NF-κB in the nucleus and return it to the cytosol.

In many immune-system cells, NF-κB stimulates transcription of more than 150 genes, including those encoding cytokines and chemokines; the latter attract other immune-system cells and fibroblasts to sites of infection. NF-κB also promotes expression of receptor proteins that enable neutrophils (a type of white blood cell) to migrate from the blood into the underlying tissue (see Figure 19-36). In addition, NF-κB stimulates expression of iNOS, the inducible isoform of the enzyme that produces nitric oxide, which is toxic to bacterial cells, as well as expression of several anti-apoptotic proteins, which prevent cell death. Thus this single transcription factor coordinates and activates the body's defense either directly by responding to pathogens and stress or indirectly by responding to signaling molecules released from other infected or wounded tissues and cells.

Besides its roles in inflammation and immunity, NF-κB plays a key role during mammalian development. For instance, NF-κB is essential for survival of developing liver cells. Mouse embryos that cannot express one of the I-κB kinase subunits die at mid-gestation due to liver degeneration caused by excessive apoptosis of cells that would normally survive. As we will see in Chapter 20, phosphorylation-dependent degradation of a cyclin kinase–dependent inhibitor plays a central role in regulating progression through the cell cycle in *S. cerevisiae.* It seems likely that phosphorylation-dependent protein degradation may emerge as a common regulatory mechanism in many different cellular processes.

Ligand-Activated Notch Is Cleaved Twice, Releasing a Transcription Factor

Both the receptor called Notch and its ligand Delta are single-pass transmembrane proteins found on the cell surface. Notch also has other ligands, including Serrate, but the molecular mechanisms of activation are the same with each ligand. Delta binds to Notch, activating Notch so that it undergoes two cleavage events; these result in release of the Notch cytosolic domain, which functions as a transcription factor. For activation to occur, Notch and Delta must be located in the membranes of adjacent cells. Their location in different cells is essential, because they participate in a highly conserved and important differentiation process in both invertebrates and vertebrates, called **lateral inhibition.** In this process, adjacent and developmentally equivalent cells assume completely different fates. In effect, one cell in a group of equivalent cells instructs the others around it to choose a different fate. This process, discussed in detail in Chapter 22, is particularly important in preventing too many nerve precursor cells forming from an undifferentiated layer of epithelial cells.

Notch protein is synthesized as a monomeric membrane protein in the endoplasmic reticulum. In the Golgi complex, it undergoes a proteolytic cleavage that generates an extracellular subunit and a transmembrane-cytosolic subunit; the two subunits remain noncovalently associated with each other in the absence of interaction with Delta residing on another cell. Following binding of Delta, the Notch protein on the responding cell undergoes two proteolytic cleavages in a proscribed order (Figure 16-36). The first is catalyzed by ADAM 10, a matrix metalloprotease. (The name ADAM stands for *A D*isintegrin *a*nd *M*etalloprotease. Disintegrin refers to an ADAM domain that binds integrins (Chapter 19) and disrupts cell-matrix interactions.) The second cleavage occurs within the hydrophobic membrane-spanning region of Notch, and is catalyzed by a four-protein transmembrane complex termed *γ-secretase.* This cleavage releases the Notch cytosolic segment, which immediately translocates to the nucleus where it affects transcription of various target genes. Such signal-induced *regulated intramembrane proteolysis (RIP)* also occurs in the response of cells to low cholesterol (see below) and to the presence of unfolded proteins in the endoplasmic reticulum (Chapter 13).

The γ-secretase complex contains a protein termed *presenilin 1* and three other essential subunits, aph-1, pen-2, and nicastrin. Presenilin 1 (PS1) was first identified as the

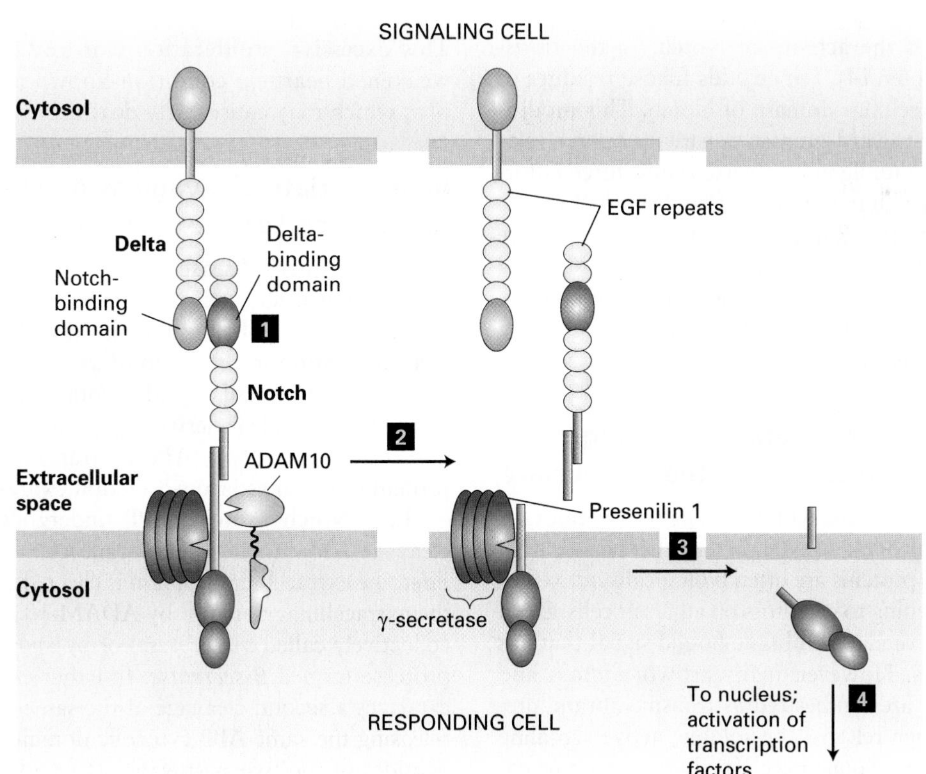

▲ **FIGURE 16-36 Notch/Delta signaling pathway.** In the absence of Delta, the extracellular subunit of Notch on a responding cell is noncovalently associated with its transmembrane-cytosolic subunit. When Notch binds to its ligand Delta on an adjacent signaling cell (step **1**), Notch is first cleaved by the matrix metalloprotease ADAM 10, which is bound to the membrane, releasing the extracellular Notch segment (step **2**). γ-Secretase, a complex of four membrane proteins including the presumed protease presenilin 1, then associates with the remaining portion of Notch and catalyzes an intramembrane cleavage that releases the cytosolic segment of Notch (step **3**). Following translocation to the nucleus, this Notch segment interacts with several transcription factors to affect expression of genes that in turn influence the determination of cell fate during development (step **4**). [See M. S. Brown et al., 2000, *Cell* **100:**391, and D. Seals and S. Courtneidge, 2003, *Genes Dev.* **17:**7.]

product of a gene that commonly is mutated in patients with an early-onset autosomal dominant form of Alzheimer's disease. Studies on cells lacking nicastrin revealed why γ-secretase can only cleave proteins that have first been cleaved by an ADAM or other matrix metalloprotease. Nicastrin binds to the N-terminal extracellular stump of the membrane protein that is generated by the first protease (see Figure 16-36). Without this stump, nicastrin and thus the entire γ-secretase complex cannot interact with its target protein. We examine the role of ADAM proteins and γ-secretase in the development of Alzheimer's disease later.

In *Drosophila,* the released intracellular segment of Notch forms a complex with a DNA-binding protein called Suppressor of Hairless, or Su(H). This complex stimulates transcription of many genes whose net effect is to influence the determination of cell fate during development. One of the proteins increased in this manner is Notch itself, and Delta production is correspondingly reduced (see Figure 22-42). As we see in Chapter 22, reciprocal regulation of the receptor and ligand in this fashion is an essential feature of the interaction between initially equivalent cells that causes them to assume different cell fates.

Further studies have revealed several lines of evidence that the Notch pathway is carefully tuned with many built-in checks and balances. Genetic studies in *Drosophila* led to the discovery of the Fringe (Fng) protein, a glycosyl transferase that influences the activity of Notch. In the *trans*-Golgi network (Chapter 14), Fringe adds fucose residues to a region in the extracellular domain of Notch. This modification biases Notch toward greater sensitivity to its Delta ligand than to its Serrate ligand. In vertebrates, three colorfully named proteins that are related to Fringe (Lunatic fringe, Manic fringe, and Radical fringe) alter the relative sensitivity of Notch to its three ligands (Delta, Jagged1, and Jagged2). In flies and in mammals, the biases imposed by Fng proteins influence developmental outcomes because Fng modifies Notch in some cells but not others.

Matrix Metalloproteases Catalyze Cleavage of Many Signaling Proteins from the Cell Surface

Many signaling molecules are synthesized as transmembrane proteins whose signal domain extends into the extracellular space. Such signaling proteins are often biologically active but can signal only by binding to receptors on adjacent cells. Delta is a good example of such a membrane-bound signal that has very localized effects. However, many growth factors and other protein signals are synthesized as transmembrane precursors whose cleavage releases the soluble, active signaling molecule into the extracellular space. The human genome encodes 19 metalloproteases in the ADAM family and many are involved in cleaving the precursors of signaling proteins just outside their transmembrane segment. This ADAM-mediated proteolysis of such precursors is similar to the cleavage of Notch by ADAM 10 (see Figure 16-36) except that the released extracellular segment has signaling activity. ADAM activity and hence the release of active signaling proteins must be tightly regulated by the cell, but how this happens is not yet clear. A breakdown in the mechanisms for regulating ADAM proteases can lead to abnormal cell proliferation.

Medically important examples of the regulated cleavage of signal protein precursors are members of the EGF family, including EGF, HB-EGF, TGF-α, NRG1, and NRG2 (see Figure 16-18). The increased activity of one or more ADAMs that is seen in many cancers can promote cancer development in two ways. First, heightened ADAM activity can lead to high levels of extracellular EGF family growth factors that stimulate secreting cells (autocrine signaling) or adjacent cells (paracrine signaling) to proliferate inappropriately. Second, by destroying components of the extracellular matrix, increased ADAM activity is thought to facilitate metastasis, the movement of tumor cells to other sites in the body.

ADAM proteases also are an important factor in heart disease. As we learned in the last chapter, activation by adrenaline of β-adrenergic receptors in heart muscle causes glycogenolysis and an increase in the rate of muscle contraction. Prolonged treatment of heart muscle cells with epinephrine, however, leads to activation of ADAM 9 by an unknown mechanism. This matrix metalloprotease cleaves the transmembrane precursor of HB-EGF. The released HB-EGF then binds to EGF receptors on the signaling heart muscle cells and stimulates their inappropriate growth. This excessive proliferation can lead to an enlarged but weakened heart—a condition known as cardiac hypertrophy, which may cause early death. ■

Inappropriate Cleavage of Amyloid Precursor Protein Can Lead to Alzheimer's Disease

Alzheimer's disease is another disorder marked by the inappropriate activity of matrix metalloproteases. A major pathologic change associated with Alzheimer's disease is accumulation in the brain of *amyloid plaques* containing aggregates of a small peptide containing 42 residues termed $A\beta_{42}$. This peptide is derived by proteolytic cleavage of *amyloid precursor protein (APP),* a transmembrane cell-surface protein of still mysterious function expressed by neurons.

Like Notch protein, APP undergoes one extracellular cleavage and one intramembrane cleavage (Figure 16-37). First, the extracellular domain is cleaved at one of two sites in the extracellular domain: by ADAM 10 or ADAM 17 (often collectively called *α-secretase*) or by another matrix metalloprotease termed *β-secretase.* In either case, γ-secretase then catalyzes a second cleavage at the same intramembrane site, releasing the same APP cytosolic domain but different small peptides in the two pathways. The pathway initiated by α-secretase generates a 26-residue peptide that apparently does no harm. In contrast, the pathway initiated by β-secretase generates the pathologic $A\beta_{42}$ peptide, which spontaneously forms oligomers and then the larger amyloid plaques found in the brain of patients with Alzheimer's disease.

APP was recognized as a major player in Alzheimer's disease through a genetic analysis of the small percentage of patients with a family history of the disease. Many had

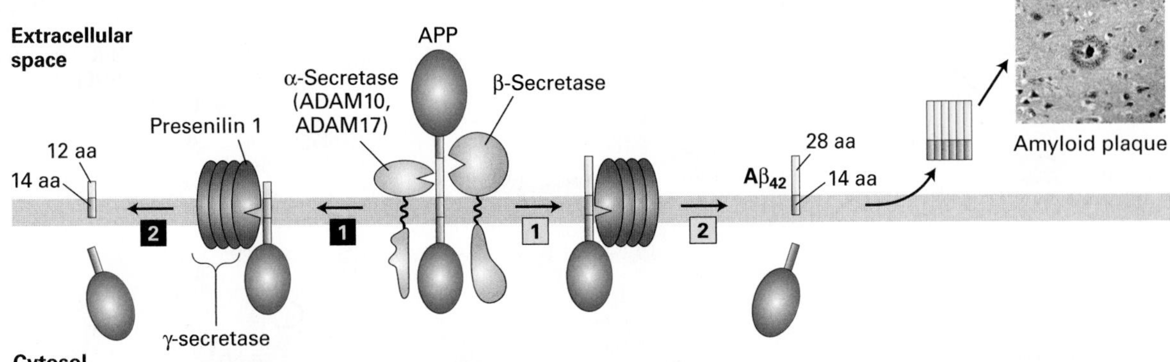

▲ **FIGURE 16-37 Proteolytic cleavage of APP and Alzheimer's disease.** (*Left*) Sequential proteolytic cleavage by α-secretase (ADAM 10 or ADAM 17) (**1**) and γ-secretase (**2**) produces an innocuous membrane-embedded peptide of 26 amino acids. (*Right*) Cleavage in the extracellular domain by β-secretase (1) followed by cleavage within the membrane by γ-secretase (2) generates the 42-residue $A\beta_{42}$ peptide that spontaneously forms oligomers and then the larger amyloid plaques found in the brain of patients with Alzheimer's disease. In both pathways the cytosolic segment of APP is released into the cytosol, but its function is not known. [See S. Lichtenthaler and C. Haass, 2004, *J. Clin. Invest.* **113:**1384, and V. Wilquet and B. De Strooper, 2004, *Curr. Opin. Neurobiol.* **14:**582. Inset © ISM/Phototake]

mutations in the APP protein, and intriguingly these mutations are clustered around the cleavage sites of α-, β-, or γ-secretase depicted in Figure 16-37. Other cases of familial Alzheimer's disease involve missense mutations in presenilin 1, a subunit of γ-secretase, that enhances the formation of the $A\beta_{42}$ peptide, leading to plaque formation and eventually to the death of neurons. ▮

γ-Secretase catalyzes the regulated intramembrane cleavage of over 100 cell-surface proteins, including Notch (see Figure 16-36). Evidence supporting the involvement of the presenilin 1 subunit of γ-secretase in Notch signaling came from genetic studies in the roundworm *C. elegans.* Mutations in the worm homolog of presenilin 1 caused developmental defects similar to those caused by Notch mutations. Later work showed that mammalian Notch does not undergo signal-induced intramembrane proteolysis in mouse neuronal cells genetically missing presenilin 1. But whether presenilin 1 is the actual γ-secretase protease or an essential cofactor of the "real" protease is not yet certain. Within its membrane-spanning segments, presenilin 1 has two aspartate residues in a configuration that resembles that of the two aspartates in the active site of water-soluble "aspartyl proteases," and mutation of either of these aspartate residues in presenilin 1 abolishes its ability to stimulate cleavage of Notch. Current data are thus consistent with the notion that presenilin 1 is the protease that cleaves Notch, APP, and many other proteins within their transmembrane segments.

Regulated Intramembrane Proteolysis of SREBP Releases a Transcription Factor That Acts to Maintain Phospholipid and Cholesterol Levels

A cell would soon face a crisis if it did not have enough lipids to make adequate amounts of membranes or had so much cholesterol that large crystals formed and damaged cellular structures. To prevent such disastrous events, cells normally maintain appropriate lipid levels by regulating their supply and utilization of lipids. Coordinate regulation of the metabolism of phospholipids and cholesterol is necessary to maintain the proper composition of membranes. Regulated intramembrane proteolysis, which occurs in the Notch pathway, also plays an important role in the cellular response to low cholesterol levels.

As we learned in Chapter 14, **low-density lipoprotein (LDL)** is rich in cholesterol and functions in transporting cholesterol within the circulation (see Figure 14-27). Both the cholesterol biosynthetic pathway (see Figure 10-26) and cellular levels of LDL receptors are down-regulated when cholesterol is in excess. Since LDL is imported into cells via receptor-mediated endocytosis (see Figure 14-29), a decrease in the number of LDL receptors leads to reduced cellular import of cholesterol. Both cholesterol biosynthesis and import are regulated at the level of gene transcription. For example, when cultured cells are incubated with increasing concentrations of LDL, the level and the activity of HMG-CoA reductase, the rate-controlling enzyme in cholesterol biosynthesis is suppressed, whereas the activity of acyl:cholesterol acyl transferase (ACAT), which converts cholesterol into the esterified storage form, is increased.

Cholesterol-dependent transcriptional regulation often depends on 10-base-pair *sterol regulatory elements (SREs),* or SRE half-sites, in the promoters of regulated target genes. (These SREs differ from the *s*erum *r*esponse *e*lements that control many early-response genes, as discussed in Section 16.4.) The interaction of cholesterol-dependent transcription factors called **SRE-binding proteins (SREBPs)** with these response elements modulates the expression of the target genes. The SREBP-mediated pathway begins in the membranes of the endoplasmic reticulum (ER) and includes at least two other proteins besides SREBP.

When cells have adequate concentrations of cholesterol, SREBP is found in the ER membrane complexed with SCAP

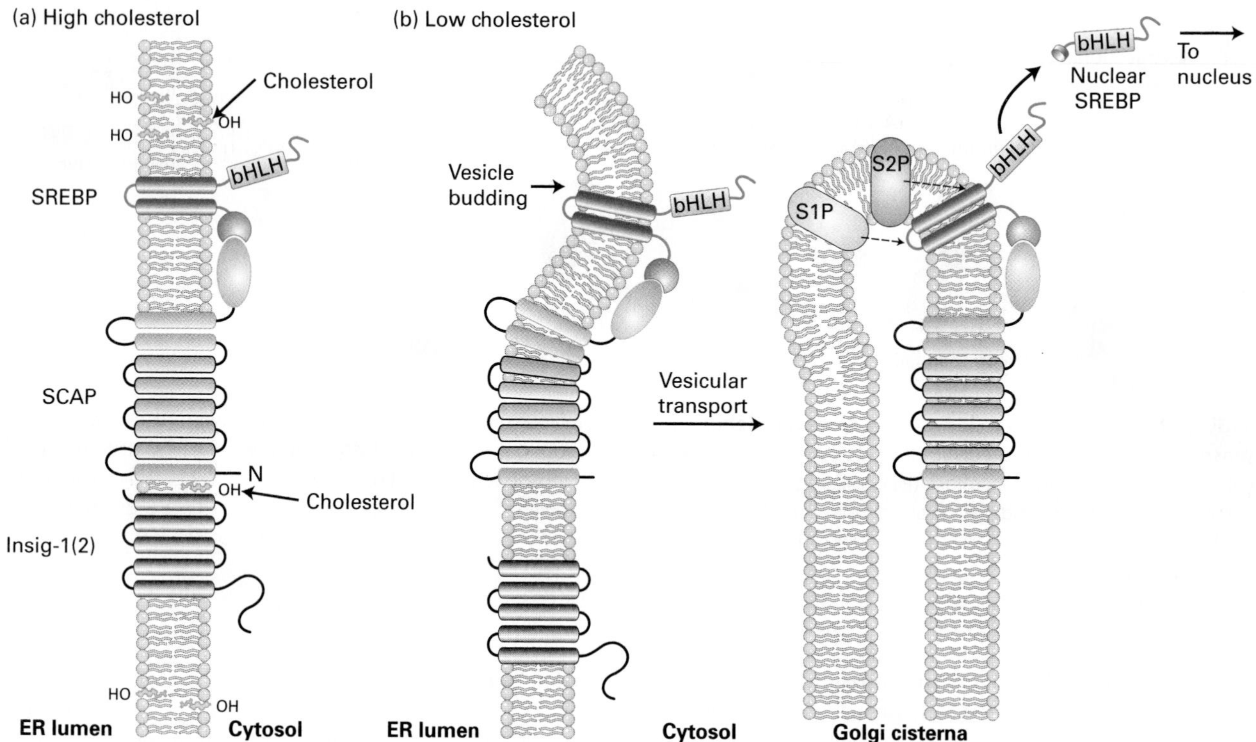

▲ **FIGURE 16-38 Cholesterol-sensitive control of SREBP activation.** The cellular pool of cholesterol is monitored by the combined action of insig-1(2) and SCAP, both transmembrane proteins located in the ER membrane. (a) When cholesterol levels are high, insig-1(2) binds to the sterol-sensing domain in SCAP, anchoring the SCAP–SREBP complex in the ER membrane. (b) The dissociation of insig-1(2) from SCAP at low cholesterol levels allows the SCAP–SREBP complex to move to the Golgi complex by vesicular transport. In the Golgi, the sequential cleavage of SREBP by the site 1 and site 2 proteases (S1P, S2P) releases the N-terminal bHLH domain of SREBP. After this released domain, called nuclear SREBP (nSREBP), translocates into the nucleus, it controls the transcription of genes containing sterol regulatory elements (SREs) in their promoters. [Adapted from T. F. Osborne, 2001, *Genes Devel.* **15:**1873; see T. Yang et al., 2002, *Cell* **110:**489.]

(SREBP cleavage-activating protein), insig-1 (or its close homolog insig-2), and perhaps other proteins (Figure 16-38a). SREBP has three distinct domains: an N-terminal cytosolic domain, containing a basic helix-loop-helix (bHLH) DNA-binding motif (see Figure 7-26), that functions as a transcription factor when cleaved from the rest of SREBP; a central membrane-anchoring domain containing two transmembrane α helices; and a C-terminal cytosolic regulatory domain. SCAP has eight transmembrane α helices and a large C-terminal cytosolic domain that interacts with the regulatory domain of SREBP. Five of the transmembrane helices in SCAP form a *sterol-sensing domain* similar to that in HMG CoA reductase (see Section 10.3). When the sterol-sensing domain in SCAP is bound to cholesterol, the protein also binds to insig-1(2). When insig-1(2) is tightly bound to the SCAP–cholesterol complex, it blocks the binding of SCAP to coat proteins on COPII vesicles, thereby preventing incorporation of the SCAP/SREBP complex into ER-to-Golgi transport vesicles (see Chapter 14). Thus the cholesterol-dependent binding of insig to the SCAP–cholesterol–SREBP complex traps that complex in the ER.

When cellular cholesterol levels drop, some of the cholesterol bound to SCAP is released. Consequently, insig-1(2) no longer binds to the cholesterol-free SCAP, and the SCAP–SREBP complex moves from the ER to the Golgi apparatus via COPII vesicles (Figure 16-38b). In the Golgi, SREBP is cleaved sequentially at two sites by two membrane-bound proteases, S1P and S2P, additional examples of regulated intramembrane proteolysis. The second cleavage at site 2 releases the N-terminal bHLH-containing domain into the cytosol. This fragment, called *nSREBP (nuclear SREBP)*, is rapidly translocated into the nucleus. There it activates transcription of genes containing *sterol regulatory elements* (SREs) in their promoters, such as those encoding the LDL receptor and HMG-CoA reductase. Thus a reduction in cellular cholesterol, by activating the *insig1(2)/SCAP/SREBP pathway*, triggers expression of genes encoding proteins that both import cholesterol into the cell (LDL receptor) and synthesize cholesterol from small precursor molecules (HMG-CoA reductase).

After cleavage of SREBP in the Golgi, SCAP apparently recycles back to the ER where it can interact with insig-1(2) and another intact SREBP molecule. High-level transcription of SRE-controlled genes requires the ongoing generation of new nSREBP because it is degraded fairly rapidly by the ubiquitin-mediated proteasomal pathway (Chapter 3). The rapid generation and degradation of nSREBP help cells respond quickly to changes in levels of intracellular cholesterol.

Under some circumstances (e.g., during cell growth), cells need an increased supply of all the essential membrane lipids and their fatty acid precursors (coordinate regulation). But cells sometimes need greater amounts of some lipids, such as cholesterol to make steroid hormones, than others, such as phospholipids (differential regulation). The complex regulation of lipid metabolism characteristic of higher eukaryotes depends in large part on abundant transcription factors, including multiple SREBPs, that control the expression of proteins involved in lipid metabolism. For example, various SREBPs regulate the transcription of genes encoding many proteins participating in the cellular uptake of lipids (e.g., the LDL receptor) and most enzymes in the pathways for synthesizing cholesterol, fatty acids, triglycerides, and phospholipids. Mammals express three known isoforms of SREBP: SREBP-1a and SREBP-1c, which are generated from alternatively spliced RNAs produced from the same gene, and SREBP-2, which is encoded by a different gene. Together these protease-regulated transcription factors control the availability not only of cholesterol but also of fatty acids and the triglycerides and phospholipids made from fatty acids. In mammalian cells, SREBP-1a and SREBP-1c exert a greater influence on fatty acid metabolism than on cholesterol metabolism, whereas the reverse is the case for SREBP-2.

Because the risk for atherosclerotic disease is directly proportional to the plasma levels of LDL cholesterol (the so-called "bad" cholesterol) and inversely proportional to those of HDL cholesterol, a major public health goal has been to lower LDL and raise HDL cholesterol levels. The most successful drugs for controlling the LDL:HDL ratio are the statins, which cause reductions in plasma LDL. As discussed in Chapter 10, these drugs bind to HMG-CoA reductase and directly inhibit its activity, thereby lowering cholesterol biosynthesis and the pool of cholesterol in the liver. Activation of SREBP in response to this cholesterol depletion promotes increased synthesis of HMG-CoA reductase and the LDL receptor. Of greatest importance here is the resulting increased numbers of hepatic LDL receptors, which mediate increased import of LDL cholesterol from the plasma and thus lower the level of LDL cholesterol in the circulation. Statins also appear to inhibit atherosclerosis by suppressing the inflammation that triggers the process. Although the mechanism of this inhibition is not well understood, it apparently contributes to the atheroprotective effect of statins. ▮

KEY CONCEPTS OF SECTION 16.7

Pathways That Involve Signal-Induced Protein Cleavage

- The NF-κB transcription factor regulates many genes that permit cells to respond to infection and inflammation.
- In unstimulated cells, NF-κB is localized to the cytosol, bound to the inhibitor protein I-κB. In response to many types of extracellular signals, phosphorylation-dependent ubiquitination and degradation of I-κB in proteasomes releases active NF-κB, which translocates to the nucleus (see Figure 16-35).
- Upon binding to its ligand Delta on the surface of an adjacent cell, the receptor Notch protein undergoes two proteolytic cleavages (see Figure 16-36). The released Notch cytosolic segment then translocates into the nucleus and modulates transcription of target genes critical in determining cell fate during development.
- Cleavage of membrane-bound precursors of members of the EGF family of signaling molecules is catalyzed by ADAM proteases. Inappropriate cleavage of these precursors can result in abnormal cell proliferation, potentially leading to cancer, cardiac hypertrophy, and other diseases.
- γ-Secretase, which catalyzes the regulated intramembrane proteolysis of Notch, also participates in the cleavage of amyloid precursor protein (APP) into a peptide that forms plaques characteristic of Alzheimer's disease (see Figure 16-37).
- In the insig-1(2)/SCAP/SREBP pathway, the active nSREBP transcription factor is released from the Golgi membrane by intramembrane proteolysis when cellular cholesterol is low (see Figure 16-38). It then stimulates the expression of genes encoding proteins that function in cholesterol biosynthesis (e.g., HMG-CoA reductase) and cellular import of cholesterol (e.g., LDL receptor). When cholesterol is high, SREBP is retained in the ER membrane complexed with insig-1(2) and SCAP.

Perspectives for the Future

The confluence of genetics, biochemistry, and structural biology has given us an increasingly detailed view of how signals are transmitted from the cell surface and transduced into changes in cellular behavior. The multitude of different extracellular signals, receptors for them, and intracellular signal-transduction pathways fall into a relatively small number of classes, and one major goal is to understand how similar signaling pathways often regulate very different cellular processes. For instance, STAT5 activates very different sets of genes in erythroid precursor cells, following stimulation of the erythropoietin receptor, than in mammary epithelial cells, following stimulation of the prolactin receptor. Presumably STAT5 binds to different groups of transcription factors in these and other cell types, but the nature of these proteins and how they collaborate to induce cell-specific patterns of gene expression remain to be uncovered.

Conversely, activation of the same signal-transduction component in the same cell through different receptors often elicits different cellular responses. One commonly held view is that the duration of activation of the MAP kinase and other signaling pathways affects the pattern of gene expression. But how this specificity is determined remains an outstanding question in signal transduction. Genetic and molecular studies in flies, worms, and mice will contribute to our understanding of the interplay between different pathway components and the underlying regulatory principles controlling specificity in multicellular organisms.

Researchers have determined the three-dimensional structures of various signaling proteins during the past several years, permitting more detailed analysis of several signal-transduction pathways. The molecular structures of different kinases, for example, exhibit striking similarities and important variations that impart to them novel regulatory features. The activity of several kinases, such as Raf and protein kinase B (PKB), is controlled by inhibitory domains as well as by multiple phosphorylations catalyzed by several other kinases. Our understanding of how the activity of these and other kinases is precisely regulated to meet the cell's needs will require additional structural and cell biological studies.

Abnormalities in signal transduction underlie many different diseases, including the majority of cancers and many inflammatory conditions. Detailed knowledge of the signaling pathways involved and the structure of their protein components will continue to provide important molecular clues for the design of specific therapies. Despite the close structural relationship between different signaling molecules (e.g., kinases), recent studies suggest that inhibitors selective for specific subclasses can be designed. In many tumors of epithelial origin, the EGF receptor has undergone a mutation that increases its activity. Remarkably a small-molecule drug (Iressa™) inhibits the kinase activity of the mutant EGF receptor but has no effect on the normal EGF receptor or other receptors. Thus the drug slows cancer growth only in patients with this particular mutation. Similarly, monoclonal antibodies or decoy receptors (soluble proteins that contain the ligand-binding domain of a receptor and thus sequester the ligand) that prevent pro-inflammatory cytokines like IL-1 and TNFα from binding to their cognate receptors are now being used in treatment of several inflammatory diseases such as arthritis.

Key Terms

activation lip *674*
adapter protein *685*
constitutive *670*
cytokines *672*
erythropoietin *672*
Hedgehog (Hh) pathway *700*
HER family *680*
insig-1(2)/SCAP/SREBP pathway *708*
JAK/STAT pathway *672*
kinase cascade *684*
MAP kinase *684*
matrix metalloprotease (MMP) family *703*
NF-κB pathway *703*
Notch/Delta pathway *703*
nuclear-localization signal (NLS) *670*
phosphoinositides *694*
PI-3 kinase pathway *694*
presenilin 1 *705*
protein kinase B (PKB) *695*
PTEN phosphatase *697*
Ras protein *684*
receptor tyrosine kinases (RTKs) *679*
regulated intramembrane proteolysis *705*
scaffold proteins *693*
SH2 domains *675*
Smads *670*
SRE-binding protein (SREBP) *707*
transforming growth factor β (TGFβ) *668*
Wnt pathway *699*

Review the Concepts

1. Binding of TGFβ to its receptors can elicit a variety of responses in different cell types. For example, TGFβ induces plasminogen activator inhibitor in epithelial cells and specific immunoglobulins in B cells. In both cell types, Smad3 is activated. Given the conservation of the signaling pathway, what accounts for the diversity of the response to TGFβ in various cell types?

2. How is the signal generated by binding of TGFβ to cell-surface receptors transmitted to the nucleus where changes in target gene expression occur?

3. Name three features common to the activation of cytokine receptors and receptor tyrosine kinases. Name one difference with respect to the enzymatic activity of these receptors.

4. The intracellular events that proceed when erythropoietin binds to its cell-surface receptor are well-characterized examples of cell-signaling pathways that activate gene expression. What molecule translocates from the cytosol to the nucleus after (a) JAK2 activates STAT5 and (b) GRB2 binds to the Epo receptor?

5. Once an activated signaling pathway has elicited the proper changes in target gene expression, the pathway must be inactivated. Otherwise, pathologic consequences may result, as exemplified by persistent growth factor–initiated signaling in many cancers. Many signaling pathways possess intrinsic negative feedback by which a downstream event in a pathway turns off an upstream event. Describe the negative feedback that down-regulates signals induced by (a) TGFβ and (b) erythropoietin.

6. Even though GRB2 lacks intrinsic enzymatic activity, it is an essential component of the epidermal growth factor (EGF) signaling pathway that activates MAP kinase. What is the function of GRB2? What role do the SH2 and SH3 domains play in the function of GRB2? Many other signaling proteins possess SH2 domains. What determines the specificity of SH2 interactions with other molecules?

7. A mutation in the Ras protein renders Ras constitutively active (Ras^D). What is constitutive activation? How is constitutively active Ras cancer-promoting? What type of mutation might render the following proteins constitutively active: (a) Smad3; (b) MAP kinase; and (c) NF-κB?

8. The enzyme Ste11 participates in several distinct MAP kinase signaling pathways in the budding yeast *S. cerevisiae*. What is the substrate for Ste11 in the mating factor signaling pathway? When a yeast cell is stimulated by mating factor, what prevents induction of osmolytes required for survival in high osmotic strength media since Ste11 also participates in the MAP kinase pathway initiated by high osmolarity?

9. Describe the events required for full activation of protein kinase B. Name two effects of insulin mediated by protein kinase B in muscle cells.

10. Describe the function of the PTEN phosphatase in the PI-3 kinase signaling pathway. Why is a loss-of-function mutation in PTEN cancer-promoting? Predict the effect of constitutively active PTEN on cell growth and survival.

11. Activation of protein kinase A (PKA) can have both short- and long-term effects in cells. What are some of the long-term effects of activated PKA in liver cells? What pathway mediates these effects?

12. Similar to Wnt proteins, the extracellular signaling protein Hedgehog can remain anchored to cell membranes. What modifications to Hedgehog enable it to be membrane-bound? Why is this property useful?

13. Intraflagellar transport (IFT) proteins are required to move materials within cilia and flagella. Most cells have a single immobile cilium called the primary cilium. What evidence shows the primary cilium may be involved in signal transduction?

14. Why is the signaling pathway that activates NF-κB considered to be relatively irreversible compared with cytokine or RTK signaling pathways? Nonetheless, NF-κB signaling must be down-regulated eventually. How is the NF-κB signaling pathway turned off?

15. What biochemical reaction is catalyzed by γ-secretase? What is the role of γ-secretase in transducing the signal induced by the binding of Delta to its receptor, Notch? How are mutations in presenilin 1, one of the subunits of γ-secretase, thought to contribute to Alzheimer's disease?

Analyze the Data

G. Johnson and colleagues have analyzed the MAP kinase cascade in which MEKK2 participates in mammalian cells. By a yeast two-hybrid screen (see Chapter 7), MEKK2 was found to bind MEK5, which can phosphorylate a MAP kinase. To elucidate the signaling pathway transduced by MEKK2 in vivo, the following studies were performed in human embryonic kidney (HEK293) cells in culture.

a. HEK293 cells were transfected with a plasmid encoding recombinant, tagged MEKK2, along with a plasmid encoding MEK5 or a control vector that did not encode a protein (mock). Recombinant MEK5 was precipitated from the cell extract by absorption to a specific antibody. The immunoprecipitated material was then resolved by polyacrylamide gel electrophoresis, transferred to a membrane, and examined by Western blotting with an antibody that recognized tagged MEKK2. The results are shown in part (a) of the figure below. What information about this MAP kinase cascade do we learn from this experiment? Do the data in part (a) of the figure prove that MEKK2 activates MEK5, or vice versa?

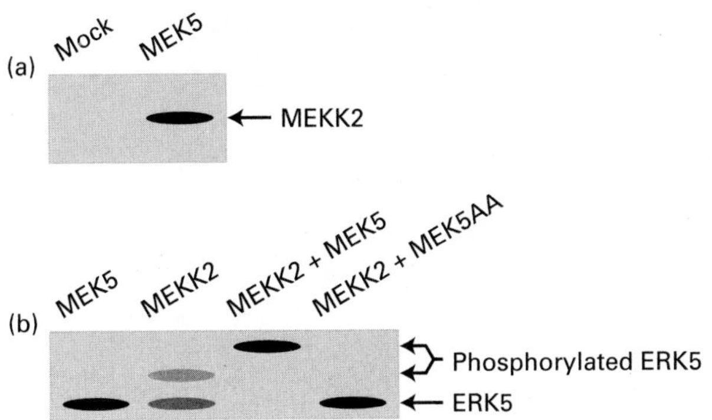

b. ERK5 is a MAP kinase previously shown to be activated when phosphorylated by MEK5. When ERK5 is phosphorylated by MEK5, its migration on a polyacrylamide gel is retarded. In another experiment, HEK293 cells were transfected with a plasmid encoding ERK5 along with plasmids encoding MEK5, MEKK2, MEKK2 and MEK5, or MEKK2 and MEK5AA. MEK5AA is a mutant, inactive version of MEK5 that functions as a dominant-negative. Expression of MEK5AA in HEK293 cells prevents signaling through active, endogenous MEK5. Lysates of transfected cells were analyzed by Western blotting with an antibody against recombinant ERK5. From the data in part (b) of the figure, what can we conclude about the role of MEKK2 in the activation of ERK5? How do the data obtained when cells are cotransfected with ERK5, MEKK2, and MEK5AA help to elucidate the order of participants in this kinase cascade?

References

16.1 TGFβ Receptors and the Direct Activation of Smads

Luo, K. 2004. Ski and SnoN: negative regulators of TGF-β signaling. *Curr. Opin. Genet. Devel.* **14**:65–70.

Massagué, J., J. Seoane, and D. Wotton. 2005. Smad transcription factors. *Genes & Devel.* **19**:2783–2810.

Moustakas, A., S. Souchelnytskyi, and C. Heldin. 2001. Smad regulation in TGF-beta signal transduction. *J. Cell Sci.* **114**:4359–4369.

Xu, L. and J. Massagué. 2004. Nucleocytoplasmic shuttling of signal transducers. *Nature Rev. Mol. Cell Biol.* **5**:209–219.

16.2 Cytokine Receptors and the JAK/STAT Pathway

Levy, D. E., and J. E. Darnell, Jr. 2002. Stats: transcriptional control and biological impact. *Nature Rev. Mol. Cell Biol.* **3**:651–62.

Naka, T., et al. 2005. Negative regulation of cytokine and TLR signalings by SOCS and others. *Adv. Immunol.* **87**:61–122.

Ozaki, K., and W. Leonard. 2002. Cytokine and cytokine receptor pleiotropy and redundancy. *J. Biol. Chem.* **277**:29355–29358.

Wornold, S., and D. Hilton. 2004. Inhibitors of cytokine signal transduction. *J. Biol. Chem.* **279**:821–824.

Yoshimura, A. 2006. Signal transduction of inflammatory cytokines and tumor development. *Cancer Sci.* 97:439–447.

16.3 Receptor Tyrosine Kinases

Eswarakumar, V. P., I. Lax, and J. Schlessinger. 2005. Cellular signaling by fibroblast growth factor receptors. *Cytokine & Growth Factor Rev.* **16**:139–149.

Hubbard, S. R. 2004. Juxtamembrane autoinhibition in receptor tyrosine kinases. *Nature Rev. Molec. Cell Biol.* **5**:464–471.

Hynes, N. E., and H. A. Lane. 2005. ERB receptors and cancer: the complexity of targeted inhibitors. *Nature Rev. Cancer* **5**:341–354.

Mohammadi, M., S. K. Olsen, and O. A. Ibrahimi. 2005. Structural basis for fibroblast growth factor receptor activation. *Cytokine and Growth Factor Rev.* **16**:107–137.

Schwartz, M. A. and H. Madhani. 2004. Principles of MAP kinase signaling specificity in *Saccharomyces cerevisiae*. *Ann. Rev. Genet.* **38**:725–748.

Simon, M. 2000. Receptor tyrosine kinases: specific outcomes from general signals. *Cell* **103**:13–15.

Singh, A. B., and R. C. Harris. 2005. Autocrine, paracrine and juxtacrine signaling by EGFR ligands. *Cell Signaling* **17**:1183–1193.

Wiley, H. S. 2003. Trafficking of the ErbB receptors and its influence on signaling. *Exp. Cell Res.* **284:**78–88.

Wiley, H. S., S. Y. Shvartsman, and D. A. Lauffenburger. 2003. Computational modeling of the EGF-receptor system: a paradigm for systems biology. *Trends Cell Biol.* **13:** 43–50.

16.4 Activation of Ras and MAP Kinase Pathways

Avruch, J., et al. 2001. Ras activation of the Raf kinase: tyrosine kinase recruitment of the MAP kinase cascade. *Recent Prog. Horm. Res.* **56:**127–155.

Bhattacharyya, R. P., et al. 2006. The Ste5 scaffold allosterically modulates signaling output of the yeast mating pathway. *Science* **311:**822–826.

Dard, N., and M. Peter. 2006. Scaffold proteins in MAP kinase signaling: more than simple passive activating platforms. *BioEssays* **28:**146–156.

Elion, E., M. Qi, and W. Chen. 2005. Signaling specificity in yeast. *Science* **307:**687–688.

Gastel, M. 2006. MAPKAP kinases—MKs—two's company, three's a crowd. *Nature Rev. Molec. Cell Biol.* **7:**211–224.

Huse, M., and J. Kuriyan. 2002. The conformational plasticity of protein kinases. *Cell* **109:**275–282.

Kerkhoff, E., and U. Rapp. 2001. The Ras-Raf relationship: an unfinished puzzle. *Adv. Enz. Regul.* **41:**261–267.

Kolch, W. 2005. Coordinating ERK/MAPK signalling through scaffolds and inhibitors *Nature Rev. Molec. Cell Biol.* **6:**827–837.

Murphy, L. O., and J. Blenis. 2006. MAPK signal specificity: the right place at the right time. *Trends Biochem. Sci.* **31:**268–275.

Tzivion, G., and J. Avruch. 2002. 14-3-3 proteins: active cofactors in cellular regulation by serine/threonine phosphorylation. *J. Biol. Chem.* **277:**3061–3064.

16.5 Phosphoinositides as Signal Transducers

Brazil, D., Z.-Z. Yang, and B. A. Hemmings. 2004. Advances in protein kinase B signalling: AKTion on multiple fronts. *Trends Biochem. Sci.* **29:**233–242.

Cantley, L. 2002. The phosphoinositide 3-kinase pathway. *Science* **296:**1655–1657.

Carlton, J. G., and P. J. Cullen. 2005. Coincidence detection in phosphoinositide signaling. *Trends Cell Biol.* **15:**540–547.

Kahl, C. R., and A. R. Means. 2003. Regulation of cell cycle progression by calcium/calmodulin-dependent pathways. *Endocr. Rev.* **24:**719–736.

Michell, R. H., et al. 2006. Phosphatidylinositol 3,5-bisphosphate: metabolism and cellular functions. *Trends Biochem. Sci.* **31:**52–63.

Niggli, V. 2005. Regulation of protein activities by phosphoinositide phosphates. *Ann. Rev. Cell Devel. Biol.* **21:**57–79.

Patterson, R., D. Boehning, and S. Snyder. 2004. Inositol 1, 4, 5, trisphosphate receptors as signal integrators. *Ann. Rev. Biochem.* **73:**437–465.

Toker, A., and A. Newton. 2000. Cellular signaling: pivoting around PDK-1. *Cell* **103:**185–188.

16.6 Activation of Gene Transcription by Seven-Spanning Cell-Surface Receptors

Gordon, M. D., and R. Nusse. 2006. Wnt signaling: multiple pathways, multiple receptors, and multiple transcription factors. *J. Biol. Chem.* **281:**22429–22433.

Logan, C. Y., and R. Nusse. 2004. The Wnt signaling pathway in development and disease. *Ann. Rev. Cell Devel. Biol.* **20:** 781–810.

Lum, L., and P. A. Beachy. 2004. The Hedgehog response network: sensors, switches, and routers. *Science* **304:**1755–1759.

Mann, R. K., and P. A. Beachy. 2004. Novel lipid modifications of secreted protein signals. *Ann. Rev. Biochem.* **73:**891–923.

Moon, R. T., et al. 2002. The promise and perils of Wnt signaling through beta-catenin. *Science* **296:**1644–1646.

Nusse, R. 2005. Relays at the membrane: the Wnt signaling pathway. *Nature* **438:**747–749.

Price, M. A. 2006. CKI, there's more than one: casein kinase I family members in Wnt and Hedgehog signaling. *Genes & Devel.* **20:**399–410.

Seto, E. S. and Bellen, H. 2004. The ins and outs of Wingless signaling. *Trends Cell Biol.* **14:**45–53.

Singla, V., and J. F. Reiter. 2006. The primary cilium as the cell's antenna: signaling at a sensory organelle. *Science* **313:**629–633.

16.7 Pathways That Involve Signal-Induced Protein Cleavage

Brown, M., J. Ye, R. Rawson, and J. Goldstein. 2000. Regulated intramembrane proteolysis: a control mechanism conserved from bacteria to humans. *Cell* **100:**391–398.

De Strooper, B. 2005. Nicastrin: gatekeeper of the γ-secretase complex. *Cell* **122:**318–320.

Goldstein, J. L., R. A. DeBose-Boyd, and M. S. Brown. 2006. Protein sensors for membrane sterols. *Cell* **124:**35–46.

Lazarov, V. K., et al. 2006. Electron microscopic structure of purified, active γ-secretase reveals an aqueous intramembrane chamber and two pores. *Proc. Nat'l. Acad. Sci. USA* **103:**6889–6894.

Lichtenthaler, S. F., and C. Haass. 2004. Amyloid at the cutting edge: activation of α-secretase prevents amyloidogenesis in an Alzheimer disease mouse model. *J. Clin. Invest.* **113:**1384–1387.

Luo, J. L., H. Kamata, and M. Karin. 2005. IKK/NF-κB signaling: balancing life and death—a new approach to cancer therapy. *J. Clin. Invest.* **115:**2625–2632.

Seals, D., and S. A. Courtneidge. 2003. The ADAMs family of metalloproteases: multidomain proteins with multiple functions. *Genes & Devel.* **17:**7–30.

Weihofen, A., and B. Martoglio. 2003. Intramembrane-cleaving proteases. *Trends Cell Biol.* **13:**71–78.

Wilquet, V., and B. De Strooper. 2004. Amyloid-beta precursor protein processing in neurodegeneration. *Curr. Opin. Neurobiol.* **14:**582–588.

Yamamoto, Y., and R. B. Gaynor. 2004. IκB kinases: key regulators of the NF-κB pathway. *Trends Biochem. Sci.* **29:**72–79.

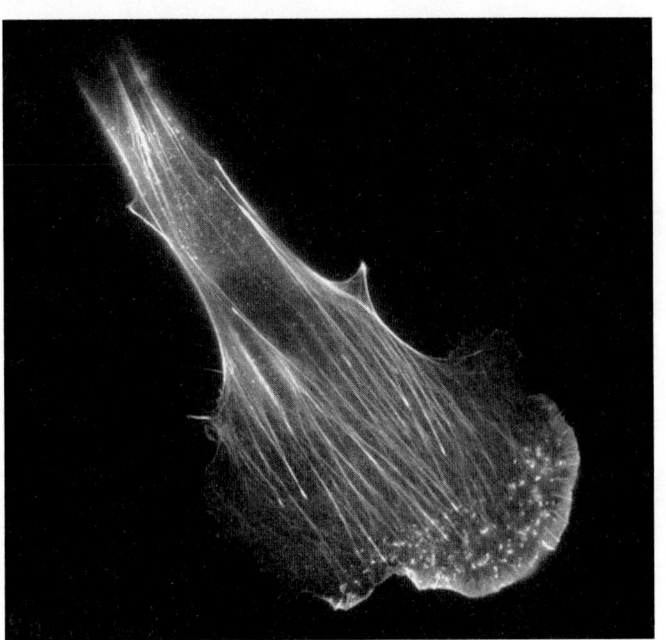

Migrating cell stained with fluorescent phalloidin, a drug that specifically binds F-actin. [Courtesy of K. Rottner and J. V. Small].

CHAPTER

17

CELL ORGANIZATION AND MOVEMENT I: MICROFILAMENTS

When we look through a microscope at the wonderful diversity of cells in nature, the variety of cell shapes and movements we can discern is astonishing. At first we may notice that some cells, such as vertebrate sperm, ciliates such as *Tetrahymena,* or flagellates such as *Chlamydomonas* swim rapidly, propelled by cilia and flagella. Other cells, such as amebas and human macrophages, move more sedately, propelled not by external appendages but by coordinated movement of the cell itself. We also might notice that some cells in tissues attach to one another, forming a pavementlike sheet, whereas other cells—neurons, for example—have long processes, up to 3 ft in length, and make selective contacts between cells. Looking more closely at the internal organization of cells, we see that organelles have characteristic locations; for example, the Golgi apparatus is generally near the central nucleus. How is this diversity of shape, cellular organization, and motility achieved? Why is it important for cells to have a distinct shape and clear internal organization? Let us first consider two examples of cells with very different functions and organizations.

The epithelial cells that line the intestine form a tight, pavementlike layer of brick-shaped cells, known as an **epithelium** (Figure 17-1a, b). Their function is to import nutrients (such as glucose) across the apical (top) plasma membrane and export them across the basolateral (bottom-side) plasma membrane into the bloodstream. To perform this directional transport, the apical and basolateral plasma membranes of epithelial cells must have different protein compositions. Epithelial cells are attached and sealed together by cellular junctions (Chapter 19), which also separate the apical and basolateral domains. This separation allows the cell to place the correct transport proteins in the plasma membranes of the two surfaces. In addition, the apical membrane has a unique morphology, with numerous fingerlike projections called **microvilli** that increase the area of the plasma membrane available for absorption. To achieve this organization, epithelial cells must have some kind of internal structure to give them shape and to deliver the appropriate proteins to the correct membrane surface.

Also consider macrophages, a type of white blood cell whose job it is to seek out infectious agents and destroy them by phagocytosis. Bacteria release chemicals that attract the macrophage and guide it to the infection. As the macrophage crawls along the chemical gradient, twisting and turning to

OUTLINE

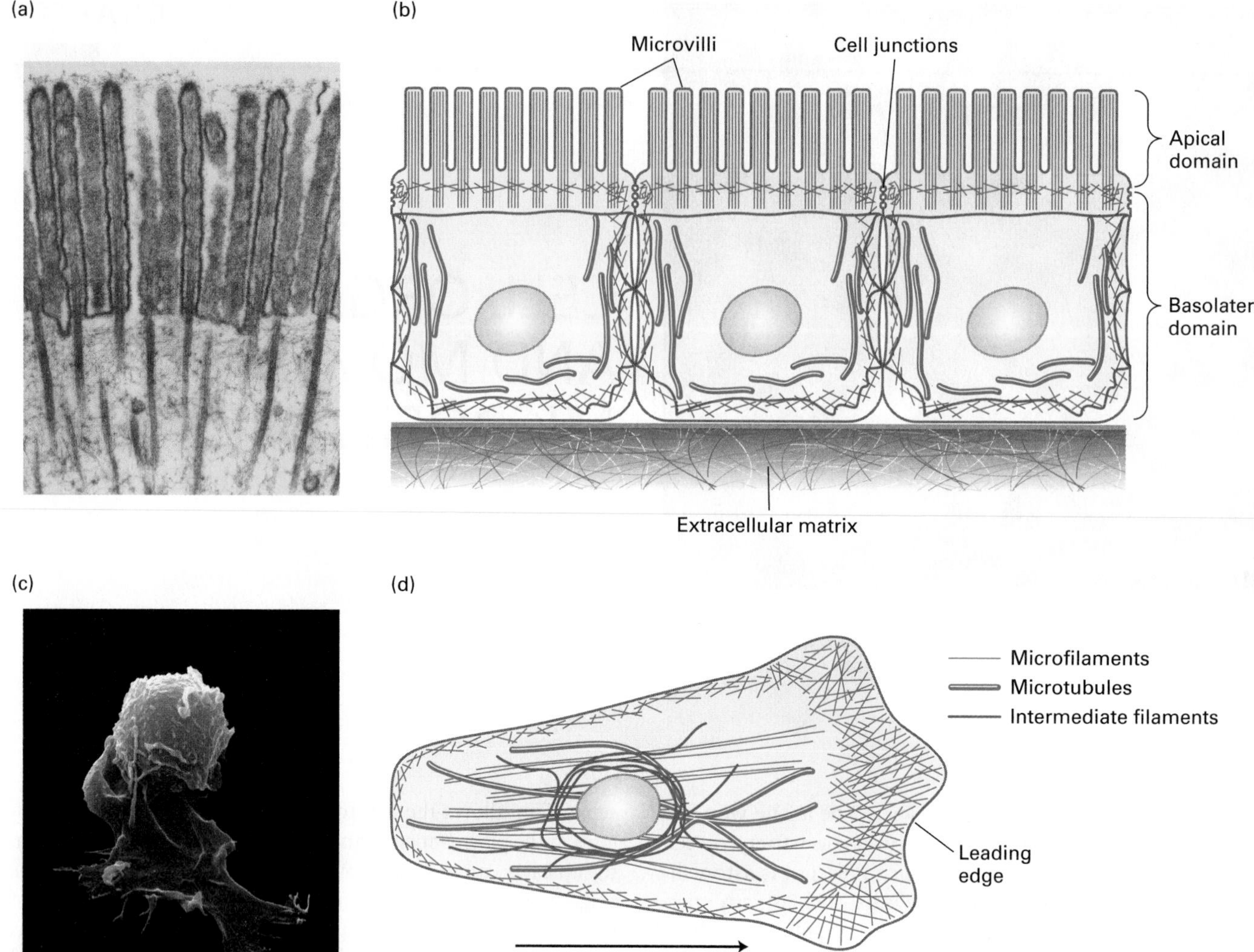

▲ **FIGURE 17-1 Overview of the cytoskeletons of an epithelial cell and a migrating cell.** (a) Transmission electron micrograph of a thin section of an epithelial cell from the small intestine. (b) Epithelial cells are highly polarized, with distinct apical and basolateral domains. An intestinal epithelial cell transport nutrients into the cell through the apical domain and out across the basolateral domain. (c) Scanning electron micrograph of a migrating cell. The thin leading edge (also known as a lamellipodium) at the front can be seen, with the large cell body behind it. (d) A migrating cell, such as a fibroblast or a macrophage, has morphologically distinct domains, with a leading edge at the front. Microfilaments are indicated in red, microtubules in green, and intermediate filaments in dark blue. The position of the nucleus (light blue oval) is also shown. [Part (a), Courtesy of Mark Mooseker; Part (b), Courtesy of Science Photo Library.]

get to the bacteria and phagocytose them, it has to constantly reorganize its cell locomotion machinery. As we will see, its internal motile machinery must be oriented in one direction as it crawls (Figure 17-1c, d).

These are just two examples of **cell polarity**—the ability of cells to generate functionally distinct regions. In fact, as you think about all types of cells, you will realize that most of them have some form of cell polarity. An additional and fundamental example of cell polarity is the ability of cells to divide: they must first select an axis for cell division and then set up the machinery to segregate their organelles along that axis.

A cell's shape and its functional polarity are provided by a three-dimensional filamentous protein network called the **cytoskeleton.** The cytoskeleton extends throughout the cell and is attached to the plasma membrane and internal organelles, so providing a framework for cellular organization. The term *cytoskeleton* does not imply a fixed structure like a skeleton. In fact, the cytoskeleton can be very dynamic, with components capable of reorganization in less than a minute, or it can be quite stable for several hours. As a result, the lengths and dynamics of filaments can vary greatly, filaments can be assembled into diverse types of structures, and they can be regulated locally in the cell.

The cytoskeleton is composed of three major filament systems (Figure 17-2), all of which are organized and regulated in time and space. Each filament system is composed of a polymer

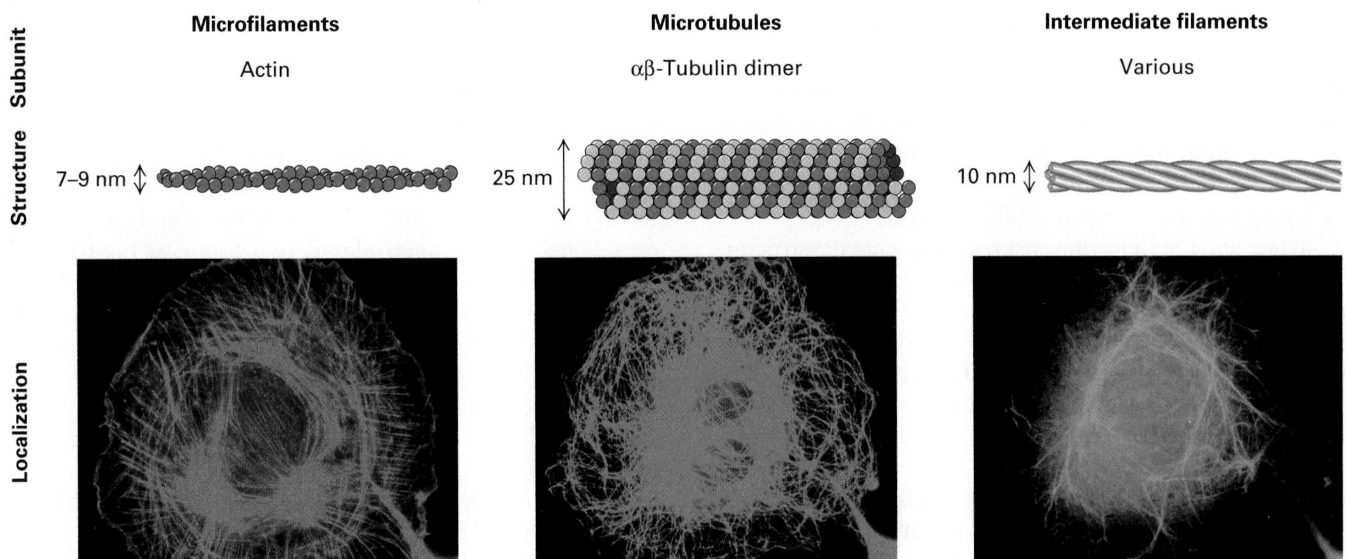

▲ **FIGURE 17-2 The components of the cytoskeleton.** Each filament type is assembled from specific subunits in a reversible process so that cells can assemble and disassemble filaments as needed. Bottom panels show localization of the three filament systems in the same cultured cell as seen by immunofluorescence microscopy of actin, tubulin, and an intermediate filament protein, respectively. [Courtesy of J. V. Small.]

of assembled subunits. The subunits that make up the filaments undergo regulated assembly and disassembly, giving the cell the flexibility to lay down or disassemble different types of structures as needed. **Microfilaments** are polymers of the protein actin organized into functional bundles and networks by actin-binding proteins. Microfilaments are especially important in the organization of the plasma membrane, including surface structures such as microvilli. Microfilaments can function on their own or serve as tracks for ATP-powered myosin **motor proteins,** which provide a contractile function (as in muscle) or ferry cargo along microfilaments. **Microtubules** are long tubes formed by the protein tubulin and organized by microtubule-associated proteins. They often extend throughout the cell, providing an organizational framework for associated organelles and structural support to cilia and flagella. They also make up the structure of the mitotic spindle, the machine for separating duplicated chromosomes at mitosis. Molecular motors called kinesins and dyneins transport cargo along microtubules and are also powered by ATP hydrolysis. **Intermediate filaments** are tissue-specific filamentous structures providing a number of different functions, including structural support to the nuclear membrane, structural integrity to cells in tissues, and structural and barrier functions in skin, hair, and nails. No known motors use intermediate filaments as tracks.

As we can see in Figure 17-1, cells can construct very different arrangements of their cytoskeletons. To establish these arrangements, cells must sense signals—either from soluble factors bathing the cell, adjacent cells, or the extracellular matrix—and interpret them (Figure 17-3). These signals are detected by cell-surface receptors that activate signal transduction pathways that converge on factors that regulate cytoskeletal organization.

The importance of the cytoskeleton for normal cell function and motility is evident when a defect in a cytoskeletal component—or in cytoskeletal regulation—causes a disease. For example, about 1 in 500 people has a defect that affects

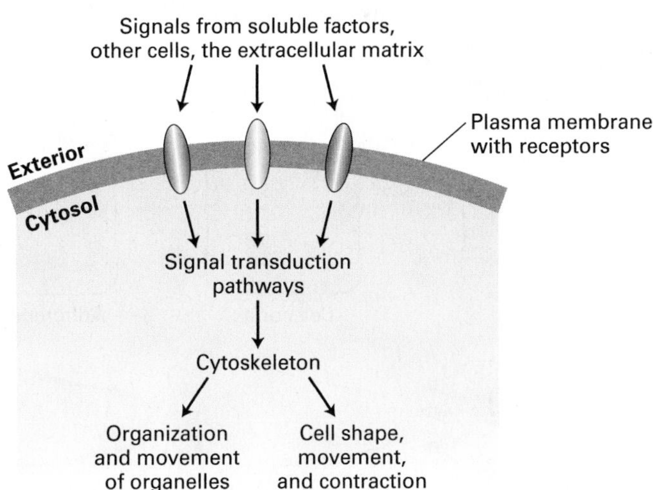

▲ **FIGURE 17-3 Cell signaling regulates cytoskeleton function.** Cells use cell-surface receptors to sense external signals from the extracellular matrix, other cells, or soluble factors. These signals are transmitted across the plasma membrane and activate specific cytosolic signaling pathways. Signals—often integrated from more than one receptor—lead to the organization of the cytoskeleton to provide cells with their shape, as well as to determine organelle distribution and movement. In the absence of external signals, cells still organize their internal structure, but not in a polarized manner.

the contractile apparatus of the heart, which results in cardiomyopathies varying in degree of severity. Many diseases of the red blood cell affect the cytoskeletal components that support these cells' plasma membrane. Metastatic cancer cells exhibit unregulated motility, breaking away from their tissue of origin and migrating to new locations to form new colonies of uncontrolled growth.

In this and the following chapter, we discuss the structure, function, and regulation of the cytoskeleton. We will see how a cell arranges its cytoskeleton to determine cell shape and polarity, to provide organization and motility to its organelles, and to be the structural framework for such processes as cell swimming and cell crawling. We will discuss how cells assemble the three different filament systems and how signal-transduction pathways regulate these structures both locally and globally. Our focus in this chapter is on microfilaments and actin-based structures. Although we initially examine the cytoskeletal systems separately, in the next chapter we will see that microfilaments cooperate with microtubules and intermediate filaments in the normal functioning of cells.

17.1 Microfilaments and Actin Structures

Microfilaments can assemble into many different types of structures within a cell (Figure 17-4). Each of these structures underlies particular cellular functions. Microfilaments can exist as a tight bundle of filaments making up the core of slender, fingerlike cell-surface projections called **microvilli** but can also be in a less-ordered network, known as the *cell cortex,* beneath the plasma membrane, where they provide support and organization. In epithelial cells, microfilaments form a contractile band around the cell, the *adherens belt,* that is intimately associated with adherens junctions (Chapter 19) to provide strength to the epithelium. In migrating cells, a network of microfilaments is found at the front of the cell in the *leading edge,* or *lamellipodium,* which can also have protruding bundles of filaments called *filopodia.* Many cells have contractile microfilaments called *stress fibers,* which attach to the substratum through specialized regions called *focal adhesions* or *focal contacts.* At a late stage of cell division, after all the organelles have been duplicated and segregated, a *contractile ring* forms and constricts to generate two daughter cells in a process known as **cytokinesis.** The electron micrograph in Figure 17-4b shows microfilaments in microvilli. Different arrangements of microfilaments can coexist within a single cell, as shown in Figure 17-4c, in this case a migrating fibroblast.

The basic building block of microfilaments is **actin,** a protein that has the remarkable property of being able to reversibly assemble into a polarized filament with functionally distinct ends. These filaments are then molded into various structures by actin-binding proteins. Cells use actin filaments in many ways: in a structural role, by harnessing the power of actin polymerization to do work, or as tracks for myosin motors. In this section, we look at the actin protein itself and the filaments into which it assembles.

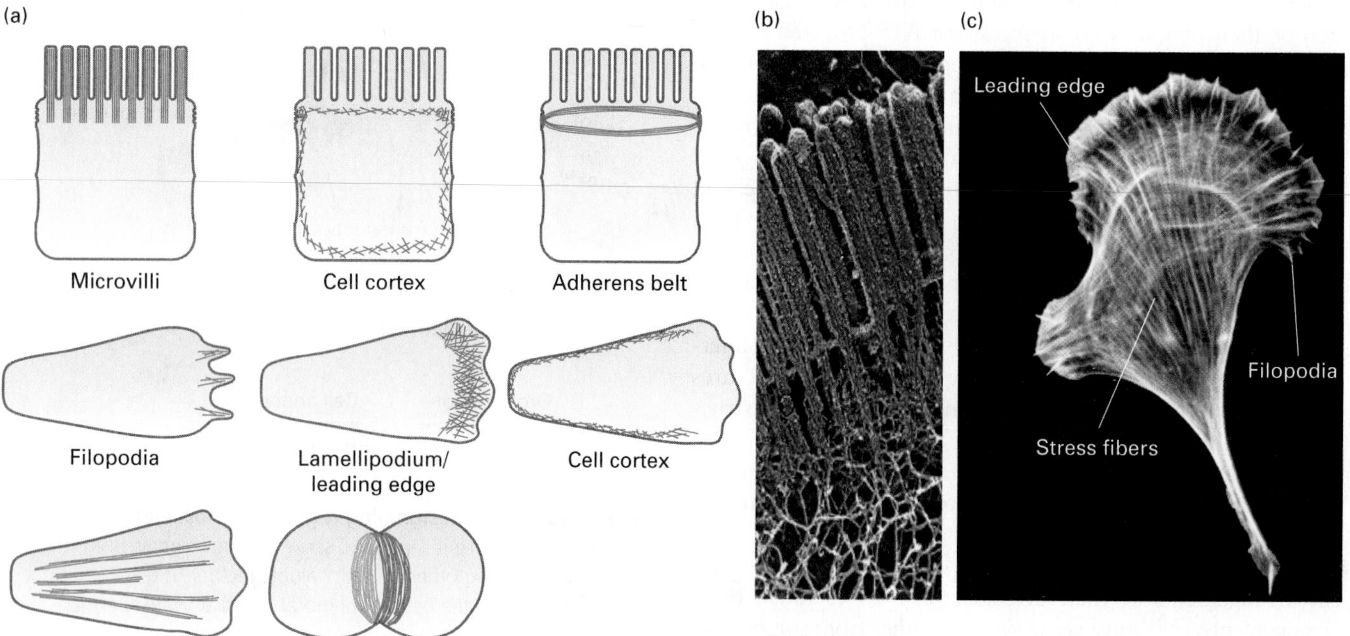

▲ **FIGURE 17-4 Examples of microfilament-based structures.** (a) In each panel, microfilaments are depicted in red. (b) Electron micrograph of the apical region of a polarized epithelial cell, showing the bundles of actin filaments that make up the cores of the microvilli (c) A cell moving to the top, stained for actin with fluorescent phalloidin, a drug that specifically binds F-actin. Note how many different organizations can exist in one cell. [Part (b), courtesy of N. Hirokawa; Part (c) courtesy of J. V. Small.]

Actin Is Ancient, Abundant, and Highly Conserved

Actin is an abundant intracellular protein in eukaryotic cells. In muscle cells, for example, actin comprises 10 percent by weight of the total cell protein; even in nonmuscle cells, actin makes up 1–5 percent of the cellular protein. The cytosolic concentration of actin in nonmuscle cells ranges from 0.1 to 0.5 mM; in special structures such as microvilli, however, the local actin concentration can be as high as 5 mM. To grasp how much actin cells contain, consider a typical liver cell, which has 2×10^4 insulin receptor molecules but approximately 5×10^8, or half a billion, actin molecules. Because they form structures that extend across large parts of the cell interior, cytoskeletal proteins are among the most abundant proteins in a cell.

A moderate-sized protein with a molecular weight of 43,000, actin is encoded by a large, highly conserved gene family. Actin arose from an ancestral bacterial gene, which then evolved as eukaryotic cells became specialized. Some single-celled organisms, such as rod-shaped bacteria, yeasts, and amebas, have one or two ancestral actin genes, whereas multicellular organisms often contain multiple actin genes. For instance, humans have six actin genes, which encode **isoforms** of the protein, and some plants have more than 60 actin genes, although most are **pseudogenes,** which do not encode functional actin proteins. In vertebrates, four actin isoforms are present in various muscle cells, and two isoforms, β-actin and γ-actin, are found in nonmuscle cells. These six isoforms differ at only about 25 of the 375 residues in the complete protein. Although these differences among isoforms seem minor, the isoforms have different functions: α-actin is associated with contractile structures, γ-actin accounts for filaments in stress fibers, and β-actin is enriched in the cell cortex and leading edge of motile cells. Sequencing of actins from different sources has revealed that they are among the most conserved proteins in a cell, comparable with histones, the structural proteins of chromatin (Chapter 6). The protein sequences of actins from amebas and from animals are identical at 80 percent of the amino acid positions despite about a billion years of evolution.

G-Actin Monomers Assemble into Long, Helical F-Actin Polymers

Actin exists as a globular monomer called *G-actin* and as a filamentous polymer called *F-actin,* which is a linear chain of G-actin subunits. (The microfilaments visualized in a cell by electron microscopy are F-actin filaments plus any bound proteins.) Each actin molecule contains a Mg^{2+} ion complexed with either ATP or ADP. The importance of the interconversion between the ATP and the ADP forms of actin in the assembly of the cytoskeleton is discussed later.

Although G-actin appears globular in the electron microscope, x-ray crystallographic analysis reveals that it is separated into two lobes by a deep cleft (Figure 17-5a). The cleft is oriented in the same direction (*top*) in all actin subunits in the filament. The end of a filament with an exposed binding cleft is the (−) end; the opposite end is the (+) end.

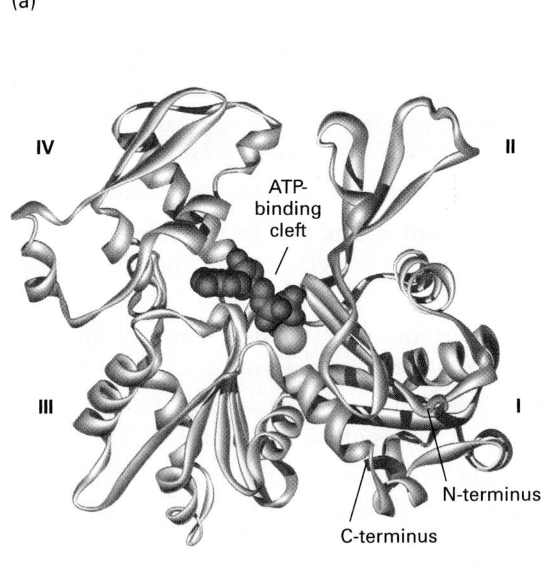

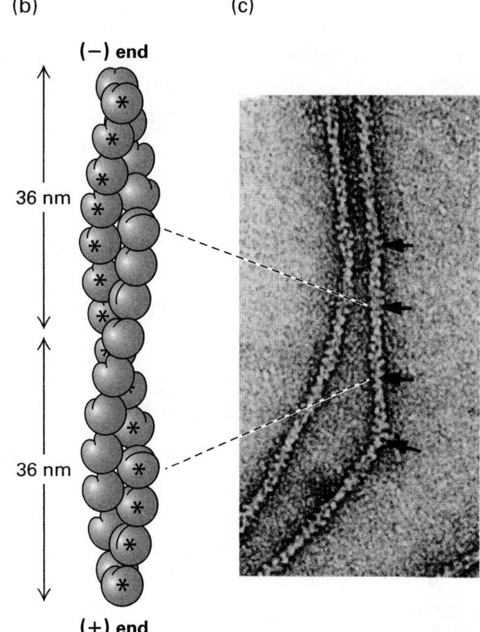

▲ **FIGURE 17-5 Structures of monomeric G-actin and F-actin filaments.** (a) Model of actin monomer (measuring 5.5 × 5.5 × 3.5 nm) shows it is divided by a central cleft into two approximately equally sized lobes and four subdomains, numbered I–IV. ATP (red) binds at the bottom of the cleft and contacts both lobes (the yellow ball represents Mg^{2+}). The N- and C-termini lie in subdomain I. (b) An actin filament appears as two strands of subunits. One repeating unit consists of 28 subunits (14 in each strand, indicated by * for one strand), covering a distance of 72 nm. The ATP-binding cleft is oriented in the same direction (*top*) in all actin subunits in the filament. The end of a filament with an exposed binding cleft is the (−) end; the opposite end is the (+) end. (c) In the electron microscope, negatively stained actin filaments appear as long, flexible, and twisted strands of beaded subunits. Because of the twist, the filament appears alternately thinner (7-nm diameter) and thicker (9-nm diameter) (arrows). [Part (a) adapted from C. E. Schutt et al., 1993, *Nature* **365**:810; courtesy of M. Rozycki. Part (c) courtesy of R. Craig.]

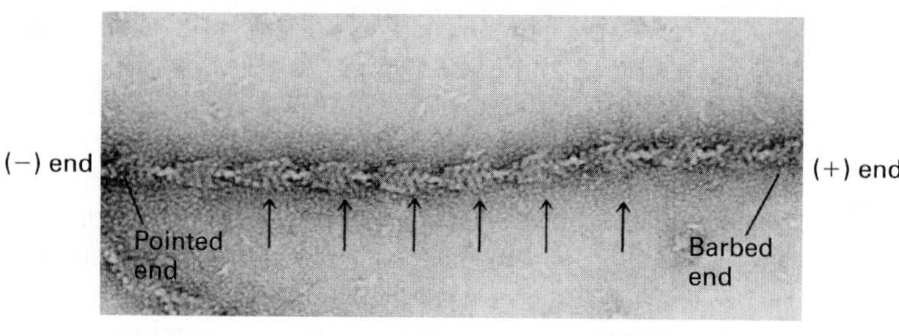

◄ **EXPERIMENTAL FIGURE 17-6 Decoration demonstrates the polarity of an actin filament.** Myosin S1 head domains bind to actin subunits in a particular orientation. When bound to all the subunits in a filament, S1 appears to spiral around the filament. This coating of myosin heads produces a series of arrowheadlike decorations (arrows), most easily seen at the wide views of the filament. The polarity in decoration defines a pointed (−) end and a barbed (+) end; the former corresponds to the top of the model in Figure 17-5. [Courtesy of R. Craig.]

lobes and the cleft compose the *ATPase fold,* the site where ATP and Mg^{2+} are bound. In actin, the floor of the cleft acts as a hinge that allows the lobes to flex relative to each other. When ATP or ADP is bound to G-actin, the nucleotide affects the conformation of the molecule; in fact, without a bound nucleotide, G-actin denatures very quickly. The addition of cations—Mg^{2+}, K^{+}, or Na^{+}—to a solution of G-actin will induce the polymerization of G-actin into F-actin filaments. The process is reversible: F-actin depolymerizes into G-actin when the ionic strength of the solution is lowered. The F-actin filaments that form in vitro are indistinguishable from microfilaments isolated from cells, indicating that actin alone makes up the filamentous structure of microfilaments.

When negatively stained with uranyl acetate for electron microscopy, F-actin appears as a twisted string whose diameter varies between 7 and 9 nm (Figure 17-5c). From the results of x-ray diffraction studies of actin filaments and from the actin monomer structure shown in Figure 17-5a, scientists have produced a model of an actin filament in which the subunits are organized in a helical structure (Figure 17-5b). In this arrangement, the filament can be considered as two helical strands wound around each other. Each subunit in the structure contacts one subunit above, one below, and two on the side in the other strand. The subunits in a single strand wind around the back of the other strand and repeat after 72 nm or 14 actin subunits. Since there are two strands, the actin filament appears to repeat every 36 nm (Figure 17-5b).

F-Actin Has Structural and Functional Polarity

All subunits in an actin filament point toward the same end of the filament. Consequently, a filament exhibits polarity; that is, one end differs from the other. As we will see, one end of the filament is favored for the addition of actin subunits and is designated the (+) end, whereas the other end is favored for dissociation, designated the (−) end. At the (+) end, the ATP-binding cleft of the terminal actin subunit contacts the neighboring subunit, whereas on the (−) end, the cleft is exposed to the surrounding solution (Figure 17-5b).

Without the atomic resolution afforded by x-ray crystallography, the cleft in an actin subunit and therefore the polarity of a filament are not detectable. However, the polarity of actin filaments can be demonstrated by electron microscopy in "decoration" experiments, which exploit the ability of the protein myosin to bind specifically to actin filaments. In this type of experiment, an excess of myosin S1, the actin-binding globular head domain of myosin, is mixed with actin filaments and binding is permitted to take place. Myosin attaches to the sides of a filament with a slight tilt. When all the actin subunits are bound by myosin, the filament appears coated ("decorated") with arrowheads that all point toward one end of the filament (Figure 17-6).

The ability of the myosin S1 head to bind and decorate F-actin is very useful experimentally—it has allowed researchers to identify the polarity of actin filaments, both in vitro and in cells. The arrowhead points to the (−) end, and so the (−) end is often called the "pointed" end of an actin filament; the (+) end is known as the "barbed" end. Because myosin binds to actin filaments and not to microtubules or intermediate filaments, arrowhead decoration is one criterion by which actin filaments can be definitively identified among the other cytoskeletal fibers in electron micrographs of thin-sectioned cells.

KEY CONCEPTS OF SECTION 17.1

Microfilaments and Actin Structures

- Microfilaments can be assembled into diverse structures, many associated with the plasma membrane (see Figure 17-4).
- Actin, the basic building block of microfilaments, is a major protein of eukaryotic cells and is highly conserved.
- Actin can reversibly assemble into filaments that consist of two helices of actin subunits.
- The actin subunits in a filament are all oriented in the same direction, with the nucleotide-binding site exposed on the (−) end (see Figure 17-5).

17.2 Dynamics of Actin Filaments

The actin cytoskeleton is not a static, unchanging structure consisting of bundles and networks of filaments. Although microfilaments may be relatively static in some structures, in others they are shrinking or growing in length. These changes in the organization of actin filaments can generate forces that cause large changes in the shape of a cell or drive intracellular movements. In this section, we consider the mechanism and regulation of actin polymerization, which is largely responsible for the dynamic nature of the cytoskeleton.

Actin Polymerization in Vitro Proceeds in Three Steps

The in vitro polymerization of G-actin (for Globular, or monomeric actin) to form F-actin filaments can be monitored by viscometry, sedimentation, fluorescence spectroscopy, or fluorescence microscopy (Chapter 9). When actin filaments become long enough to become entangled, the viscosity of the solution increases, which is measured as a decrease in its flow rate in a viscometer. The basis of the sedimentation assay is the ability of ultracentrifugation (100,000*g* for 30 minutes) to sediment F-actin but not G-actin. The third assay makes use of G-actin covalently labeled with a fluorescent dye; the fluorescence spectrum of the labeled G-actin monomer changes when it is polymerized into F-actin. Finally, growth of the fluorescently labeled filaments can be imaged with fluorescence video microscopy. These assays are useful for kinetic studies of actin polymerization and for characterization of actin-binding proteins to determine how they affect actin dynamics or how they cross-link actin filaments.

The mechanism of actin assembly has been studied extensively. Remarkably, one can purify G-actin at a high protein concentration without it forming filaments—provided it is maintained in a buffer with ATP and low levels of cations. However, as we saw above, if the cation level is increased (e.g., to 100 mM K^+ and 2 mM Mg^{2+}), G-actin will polymerize, with the kinetics of the reaction depending on the starting concentration of G-actin. The polymerization of pure G-actin in vitro proceeds in three sequential phases (Figure 17-7a). The first, *nucleation phase* is marked by a lag period in which G-actin subunits combine into short, unstable oligomers. When the oligomer reaches three subunits in length, it can act as a stable seed, or nucleus, which in the second, *elongation phase* rapidly increases in filament length by

Animation: Actin Polymerization

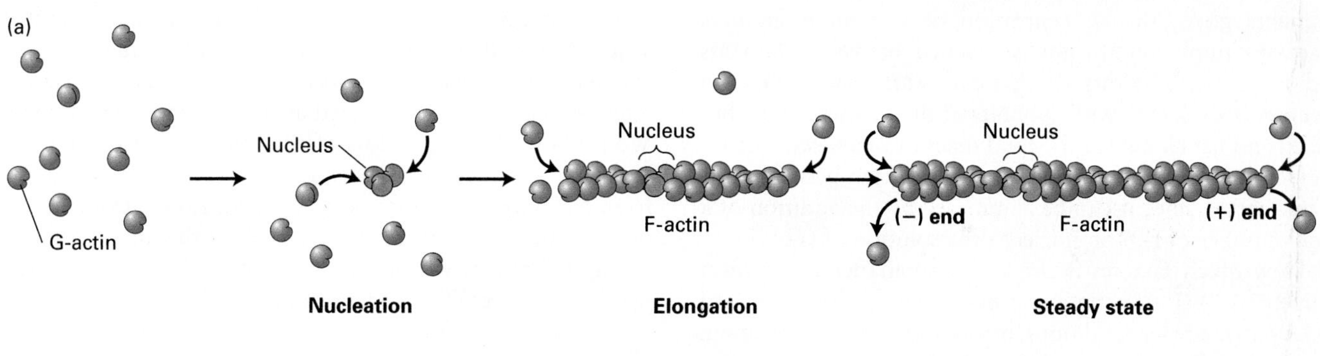

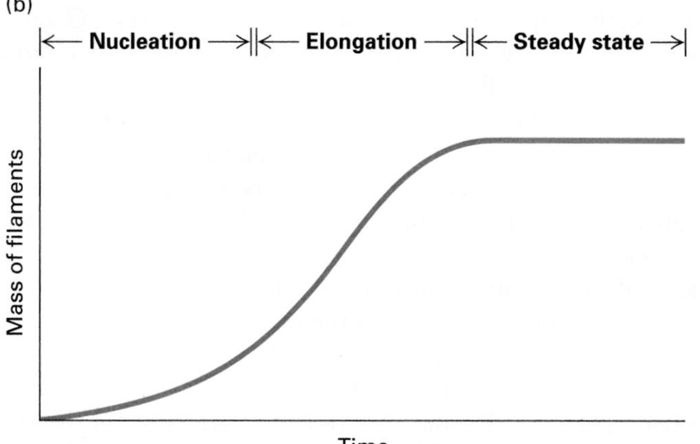

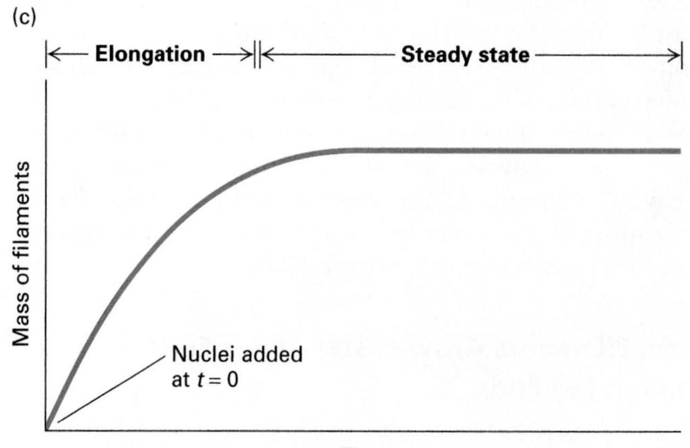

▲ **FIGURE 17-7 Polymerization of G-actin in vitro occurs in three phases.** (a) In the initial nucleation phase, ATP–G-actin monomers (red) slowly form stable complexes of actin (purple). These nuclei are rapidly elongated in the second phase by the addition of subunits to both ends of the filament. In the third phase, the ends of actin filaments are in a steady state with monomeric G-actin. (b) Time course of the in vitro polymerization reaction reveals the initial lag period associated with nucleation, the elongation phase, and steady state. (c) If some short stable actin filament fragments are added at the start of the reaction to act as nuclei, elongation proceeds immediately without any lag period.

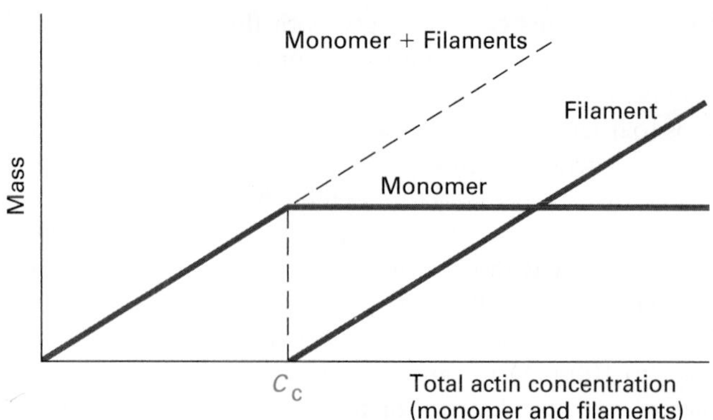

▲ **FIGURE 17-8 Concentration of actin determines filament formation.** The critical concentration (C_c) is the concentration of G-actin monomers in equilibrium with actin filaments. At monomer concentrations below the C_c, no polymerization takes place. When polymerization is induced at monomer concentrations above the C_c, filaments assemble until steady state is reached and the monomer concentration falls to C_c.

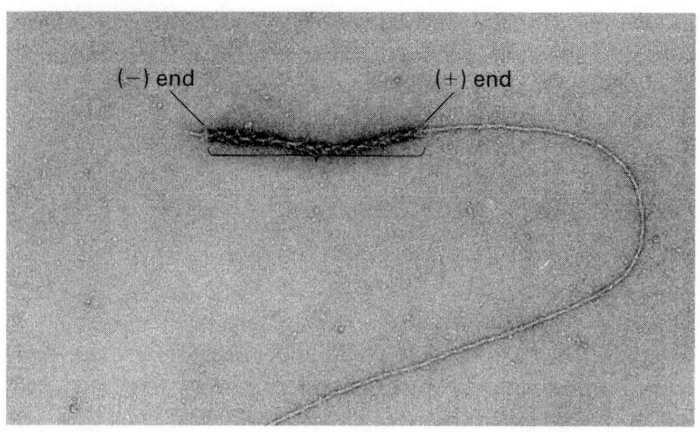

▲ EXPERIMENTAL FIGURE 17-9 **The two ends of a myosin-decorated actin filament grow unequally.** When short actin filaments are decorated with a myosin S1 head and then used to nucleate actin polymerization, the resulting actin subunits add much more efficiently to the (+) end than the (−) end of the nucleating filament. This result indicates that G-actin monomers are added much faster at the (+) end than at the (−) end. [Courtesy of T. Pollard.]

the addition of actin monomers to both of its ends. As F-actin filaments grow, the concentration of G-actin monomers decreases until equilibrium is reached between filaments and monomers. In this third, *steady-state phase,* G-actin monomers exchange with subunits at the filament ends, but there is no net change in the total mass of filaments. The kinetic curves in Figure 17-7c show that the lag period is due to nucleation since it can be eliminated by the addition of a small number of F-actin nuclei to the solution of G-actin.

How much G-actin is needed to spontaneously form a filament? If ATP–G-actin at various concentrations is placed under polymerizing conditions, below a certain concentration, filaments do not assemble (Figure 17-8). Above this concentration, filaments are formed, and when steady state is reached, the addition of free subunits is balanced by the dissociation of subunits from filament ends to yield a mixture of filaments and monomers. The concentration at which filaments are formed is known as the overall *critical concentration*, C_c. Below C_c, filaments will not form; above C_c, filaments form. At steady state, the concentration of monomeric actin remains at the critical concentration (Figure 17-8).

Actin Filaments Grow Faster at (+) Ends Than at (−) Ends

We saw earlier that myosin S1 head decoration experiments reveal an inherent structural polarity of F-actin (see Figure 17-6). If free ATP−G-actin is added to a preexisting myosin-decorated filament, the two ends grow at very different rates (Figure 17-9). In fact, for a given free ATP−G-actin concentration, the rate of addition of ATP−G-actin is nearly 10 times faster at one end, the (+) end, than at the (−) end. The rate of addition is, of course, determined by the concentration of free ATP−G-actin. Kinetic experiments have shown that the rate of addition at the (+) end is about 12 $\mu M^{-1}\ s^{-1}$ and at the (−) end about 1.3 $\mu M^{-1}\ s^{-1}$ (Figure 17-10a). This means that if 1 μM free ATP−G-actin is added to preformed filaments, on average 12 subunits will be added to the (+) end every second, whereas only 1.3 will be added at the (−) end every second. What about the rate of loss of subunits from each end? By contrast, the rates of dissociation of ATP−G-actin subunits from the two ends are quite similar, about 1.4 s^{-1} from the (+) end and 0.8 s^{-1} from the (−) end. Since this dissociation is simply the rate at which subunits leave ends, it does not depend on the free ATP−G-actin concentration.

What implications does this have for actin dynamics? First let's consider just one end, the (+) end. As we noted above, the rate of addition depends on the free ATP−G-actin concentration, whereas the rate of loss of subunits does not. Thus subunits will be added at high free ATP−G-actin concentrations, but as the concentration is lowered, a point will be reached at which the rate of addition is balanced by the rate of loss and no net growth occurs at that end. This is called the critical concentration C^{+}_{c} for the (+) end, and we can calculate it by setting the rate of assembly equal to the rate of disassembly. Thus at the critical concentration, the rate of assembly is C^{+}_{c} times the measured rate of addition of 12 $\mu M^{-1}\ s^{-1}$, whereas the rate of disassembly is independent of the free actin concentration, namely 1.4 s^{-1}. Setting these equal to each other yields a C^{+}_{c} of 0.12 μM for the (+) end. Above this free ATP−G-actin concentration, subunits add to the (+) end and net growth occurs, whereas below it, there is a net loss of subunits and shrinkage occurs.

Now let's consider just the (−) end. Because the rate of addition is much lower, 1.3 $\mu M^{-1}\ s^{-1}$, yet the rate of dissociation is about the same, 0.8 s^{-1}, the critical concentration C^{-}_{c} at the (−) end is calculated to be about 0.8 s^{-1}/1.3 $\mu M^{-1}\ s^{-1}$, that is, 0.6 μM. Thus at less than 0.6 μM free ATP−G-actin,

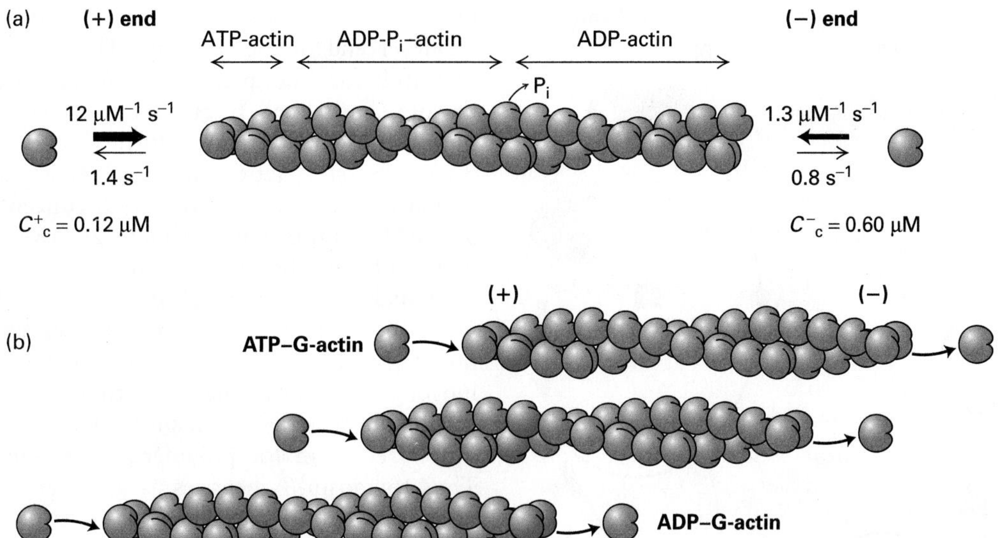

▲ **FIGURE 17-10 ATP-actin subunits add faster at the (+) end than the (−) end, resulting in a lower critical concentration and treadmilling at steady state.** (a) The rate of addition of ATP−G-actin is much faster at the (+) end than at the (−) end, whereas the rate of dissociation of G-actin is similar. This difference results in a lower critical concentration at the (+) end. At steady state, ATP-actin is added preferentially at the (+) end, giving rise to a short region of the filament containing ATP-actin and regions containing ADP-P_i-actin and ADP-actin toward the (−) end. (b) At steady state, ATP−G-actin subunits add preferentially to the (+) end, while ADP-actin subunits disassemble from the (−) end, giving rise to treadmilling of subunits.

say, 0.3μM, the (−) end will lose subunits. But at this concentration, the (+) end will grow, since 0.3 μM is above C^+_c. Because the critical concentrations are different, at steady state the free ATP−G-actin will be intermediate between C^+_c and C^-_c, so the (+) end will grow and the (−) end will lose subunits. This phenomenon is known as *treadmilling* (see Figure 17-10b).

The ability of actin filaments to treadmill is powered by hydrolysis of ATP. When ATP−G-actin binds to a (+) end, ATP is hydrolyzed to ADP and P_i. The P_i is slowly released from the subunits in the filament, so that the filament becomes asymmetric, with ATP-actin subunits at the (+) end of the filament followed by a region with ADP-P_i–actin and ADP-actin subunits toward the (−) end (Figure 17-10a). During hydrolysis of ATP and subsequent release of P_i from subunits in a filament, actin undergoes a conformational change that is responsible for the different association and dissociation rates at the two ends. Here we have considered only the kinetics of ATP−G-actin, but in reality it is ADP−G-actin that dissociates from the (−) end. Our analysis also relies on a plentiful supply of ATP−G-actin, which, as we will see, turns out to be the case in vivo. Thus actin can use the power generated by hydrolysis of ATP to treadmill and treadmilling filaments can do work in vivo, as we will see later.

Actin Filament Treadmilling Is Accelerated by Profilin and Cofilin

Measurements of the rate of actin treadmilling in vivo show that it can be several times higher than can be achieved with pure actin in vitro under physiological conditions. Consistent with a treadmilling model, growth of actin filaments in vivo only ever occurs at the (+) end. How is enhanced treadmilling achieved, and how does the cell recharge the ADP-actin dissociating from the (−) end to ATP-actin for assembly at the (+) end? Two different actin-binding proteins make important contributions to these processes.

The first is *profilin,* a small protein that binds G-actin on the side opposite to the nucleotide-binding cleft. When profilin binds ADP-actin, it opens the cleft and greatly enhances the loss of ADP, which is replaced by the more abundant cellular ATP, yielding a profilin-ATP-actin complex. This complex cannot bind to the (−) end because profilin blocks the sites on G-actin for (−) end assembly. However, the profilin-ATP-actin complex can bind efficiently to the (+) end, and then profilin dissociates after the new actin subunit is bound (Figure 17-11). This does not enhance treadmilling but provides a supply of ATP-actin from released ADP-actin; as a consequence, essentially all the G-actin in a cell has bound ATP.

Profilin has another important property: it can bind proteins with sequences rich in proline residues at the same time as binding actin. We will see later (Figure 17-14) how this property is important in actin filament assembly.

Cofilin is also a small protein, but it binds specifically to F-actin in which the subunits contain ADP, which are the older subunits in the filament toward the (−) end. Cofilin binds by bridging two actin monomers and inducing a small change in the twist of the filament. This small twist destabilizes the filament, breaking it into short pieces. By breaking the filament in this way, cofilin generates many more free (−) ends and therefore greatly enhances the disassembly of

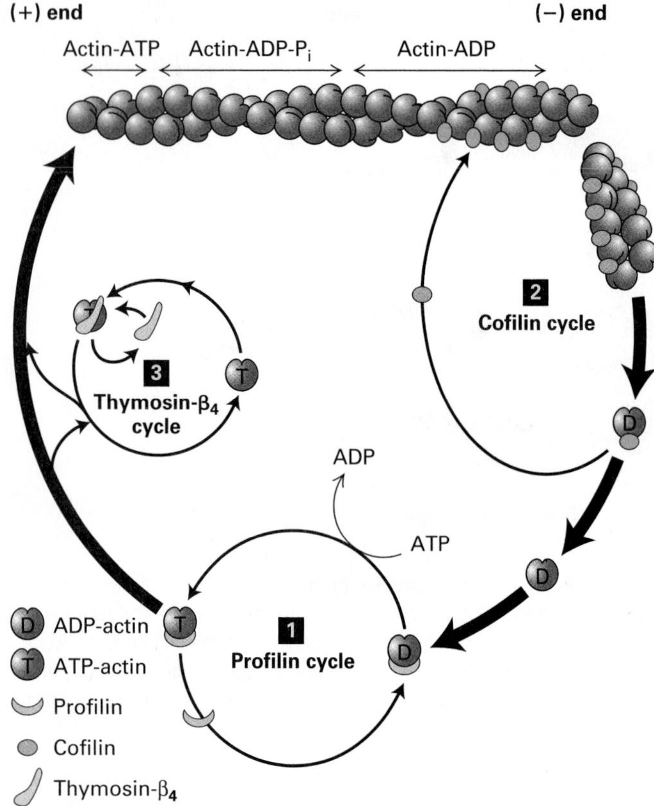

▲ **FIGURE 17-11 Actin-binding proteins regulate the rate of assembly and disassembly as well as the availability of G-actin for polymerization.** In the profilin cycle (**1**), profilin binds ADP–G-actin and catalyzes the exchange of ADP for ATP. The ATP–G-actin–profilin complex can either deliver actin to the (+) end of a filament or dissociate. In the cofilin cycle (**2**), cofilin binds preferentially to filaments containing ADP-actin, inducing them to fragment and thus enhancing depolymerization by making more filament ends. In the thymosin-β$_4$ cycle (**3**), G-actin is bound by thymosin-β$_4$, sequestering it from polymerization. As the free G-actin concentration is lowered by polymerization, G-actin–thymosin-β$_4$ dissociates to make available free G-actin for polymerization.

the (−) end of the filament (see Figure 17-11). The released ADP-actin subunits are then recharged by profilin and added to the (+) end as described above. In this way, profilin and cofilin can enhance treadmilling in vitro more than tenfold, up to levels seen in vivo.

Thymosin-β_4 Provides a Reservoir of Actin for Polymerization

It has long been known that cells often have a very large pool of unpolymerized actin, sometimes as much as half the actin in the cell. Since cellular actin levels can be as high as 100–400 μM, this means that there can be 50–200 μM unpolymerized actin in cells. Since the critical concentration in vitro is about 0.2 μM, why doesn't all this actin polymerize? The answer lies, at least in part, in the presence of actin monomer sequestering proteins. One of these is thymosin-β_4, a small protein that binds to ATP–G-actin in such a way that it inhibits addition of the actin subunit to either end of the filament. Thymosin-β_4 can be very plentiful; for example, in human blood platelets. These discoid-shaped cell fragments are very abundant in the blood, and when activated during blood clotting, they undergo a burst of actin assembly. Platelets are rich in actin: they are estimated to have a total concentration of about 550 μM actin, of which about 220 μM is in the unpolymerized form. They also contain about 500 μM thymosin-β_4, which sequesters much of the free actin. However, as in any protein-protein interaction, free actin and free thymosin-β_4 are in a dynamic equilibrium with the actin–thymosin-β_4. If some of the free actin is used up for polymerization, more actin–thymosin-β_4 will dissociate, providing more free actin for polymerization (see Figure 17-11). Thus thymosin-β_4 behaves as a buffer of unpolymerized actin for when it is needed.

Capping Proteins Block Assembly and Disassembly at Actin Filament Ends

The treadmilling and dynamics of actin filaments are regulated in cells by *capping proteins* that specifically bind to the ends of the filaments. If this were not the case, actin filaments would continue to grow and disassemble in an uncontrolled manner. As one might expect, two classes of proteins have been discovered: ones that bind the (+) end and ones that bind the (−) end (Figure 17-12).

A protein known as *CapZ,* consisting of two closely related subunits, binds with a very high affinity (~0.1 nM) to the (+) end of actin filaments, thereby inhibiting subunit addition or loss. The concentration of CapZ in cells is generally sufficient to rapidly cap any newly formed (+) ends. So how can filaments grow at their (+) ends? At least two mechanisms regulate the activity of CapZ. First, the capping activity of CapZ is inhibited by the regulatory lipid

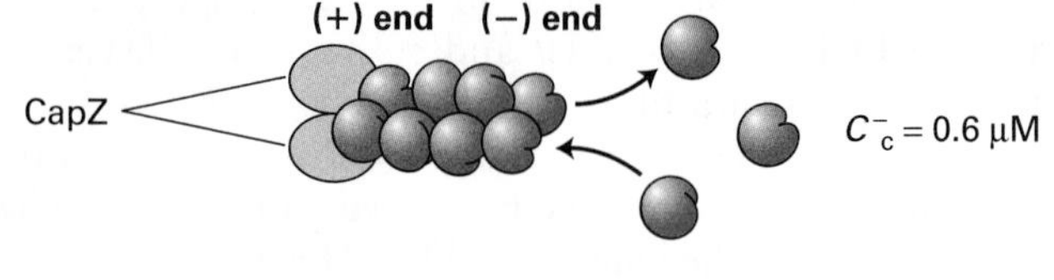

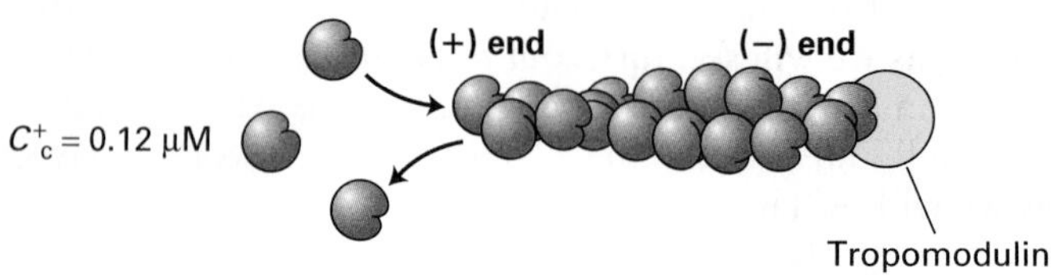

▲ **FIGURE 17-12 Capping proteins block assembly and disassembly at filament ends.** CapZ blocks the (+) end, which is where filaments normally grow, so its function is to limit growth at the (+) end. A capping protein such as tropomodulin blocks (–) ends, where filament disassembly normally occurs; thus the major function of tropomodulin is to stabilize filaments.

PI(4,5,)P2, found in the plasma membrane (Chapter 16). Second, recent work has shown that certain regulatory proteins are able to bind the (+) end and simultaneously protect it from CapZ while still allowing assembly there. Thus cells have evolved an elaborate mechanism to block assembly of actin filaments at their (+) ends except when and where assembly is needed.

Another protein, called *tropomodulin,* binds to the (−) end of actin filaments, also inhibiting assembly and disassembly. This protein is found predominantly in cells in which actin filaments need to be highly stabilized. Two examples we will encounter later in this chapter are the short actin filaments in the cortex of the red blood cell and the actin filaments in muscle. As we will see, in both cases tropomodulin works with another protein, tropomyosin, that lies along the filament to stabilize it. Tropomodulin binds to both tropomyosin and actin at the (−) end to greatly stabilize the filament.

In addition to CapZ, another class of proteins can cap the (+) ends of actin filaments. These proteins also can sever actin filaments. One member of this family, *gelsolin,* is regulated by increased levels of Ca^{2+}. On binding Ca^{2+}, gelsolin undergoes a conformational change that allows it to bind to the side of an actin filament and then insert itself between subunits of the helix, thereby breaking the filament. It then remains bound to the (+) end, generating a new (−) end that can disassemble. Actin cross-linking proteins can turn a solution of filaments into a gel, and under conditions of elevated Ca^{2+}, gelsolin can break the filaments and turn it into a sol; hence its name.

KEY CONCEPTS OF SECTION 17.2

Dynamics of Actin Filaments

- The rate-limiting step in actin assembly is the formation of a short actin oligomer (nucleus) that can then be elongated into filaments.
- The critical concentration (C_c) is the concentration of free G-actin at which the assembly onto a filament end is balanced by loss from that end.
- When the concentration of G-actin is above the C_c, the filament end will grow; when it is less than the C_c, the filament will shrink (see Figure 17-8).
- ATP-G-actin adds much faster at the (+) end than the (−) end, resulting in a lower critical concentration at the (+) end than at the (−) end.
- At steady state, actin subunits treadmill through a filament. ATP-actin is added at the (+) end, ATP is then hydrolyzed to ADP and P_i, P_i is lost, and ADP-actin dissociates from the (−) end.
- The length and rate of turnover of actin filaments is regulated by specialized actin-binding proteins (see Figure 17-11). Profilin enhances the exchange of ADP for ATP on G-actin; cofilin enhances the rate of loss of ADP-actin from the filament (−) end, and thymosin-β_4 binds G-actin to provide reserve actin when it is needed. Capping proteins bind to filament ends, blocking assembly and disassembly.

17.3 Mechanisms of Actin Filament Assembly

The rate-limiting step of actin polymerization in vitro is the formation of an initial actin nucleus from which a filament can grow (see Figure 17-7a). In cells, this inherent property of actin is used as a control point to determine where actin filaments are assembled—this is how the different actin assemblies within a single cell are generated (see Figures 17-1 and 17-4). Two classes of *actin nucleating proteins*, the *formin* protein family and the *Arp2/3 complex*, nucleate actin assembly under the control of signal-transduction pathways. Moreover, they nucleate the assembly of different actin organizations: formins lead to the assembly of long actin filaments, whereas the Arp2/3 complex leads to branched networks. We will discuss each separately and see how the power of actin polymerization can power motile processes in a cell.

Formins Assemble Unbranched Filaments

Formins are found in essentially all eukaryotic cells as quite a diverse family of proteins: seven different classes are present in vertebrates. Although diverse, all formin family members have two adjacent domains in common, the so-called FH1 and FH2 domains (formin-homology domains 1 and 2). Two FH2 domains, which provide the basic nucleating function of formins, associate to form a doughnut-shaped complex (Figure 17-13a). This complex has the ability to nucleate actin assembly by binding two actin subunits, holding them so that the (+) end is toward the FH2 domains. The nascent filament can now grow at the (+) end while the FH2 domain dimer remains attached. How can it do this? As we saw earlier, an actin filament can be thought of as two intertwined strands of subunits. The FH2 dimer can bind to the two terminal subunits. It then probably rocks between the two end subunits, letting go of one to allow addition of a new subunit and then binding the newly added subunit and freeing up space for the addition of another subunit to the other strand. In this way, rocking between the two subunits on the end, it can remain attached while simultaneously allowing growth at the (+) end (see Figure 17-13a).

The FH1 domain adjacent to the FH2 domain also makes an important contribution to actin filament growth (Figure 17-14). This domain is rich in proline residues that are sites for the binding of several profilin molecules. We discussed earlier how profilin can exchange the ADP nucleotide on G-actin to generate profilin-ATP-actin. The FH1 domain behaves as a landing site to increase the local concentration of profilin–G-actin–ATP complexes. In a way that is not yet completely understood, these complexes are provided to the FH2 domain to add actin to the (+) end of the filament, thereby allowing rapid assembly to occur between the FH2 domain dimer on the growing filament (Figure 17-14). Since the formin allows addition of actin subunits to the (+) end, long filaments with a formin at their (+) end are generated (see Figure 17-13b). In this manner, formins nucleate actin

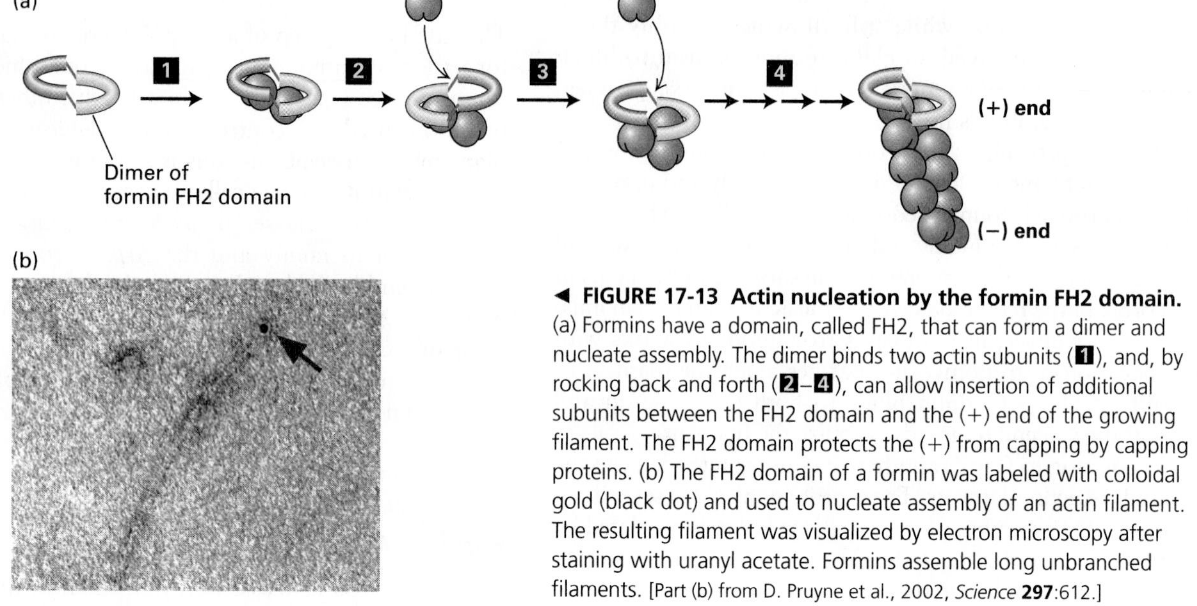

◀ **FIGURE 17-13 Actin nucleation by the formin FH2 domain.** (a) Formins have a domain, called FH2, that can form a dimer and nucleate assembly. The dimer binds two actin subunits (1), and, by rocking back and forth (2–4), can allow insertion of additional subunits between the FH2 domain and the (+) end of the growing filament. The FH2 domain protects the (+) from capping by capping proteins. (b) The FH2 domain of a formin was labeled with colloidal gold (black dot) and used to nucleate assembly of an actin filament. The resulting filament was visualized by electron microscopy after staining with uranyl acetate. Formins assemble long unbranched filaments. [Part (b) from D. Pruyne et al., 2002, *Science* **297**:612.]

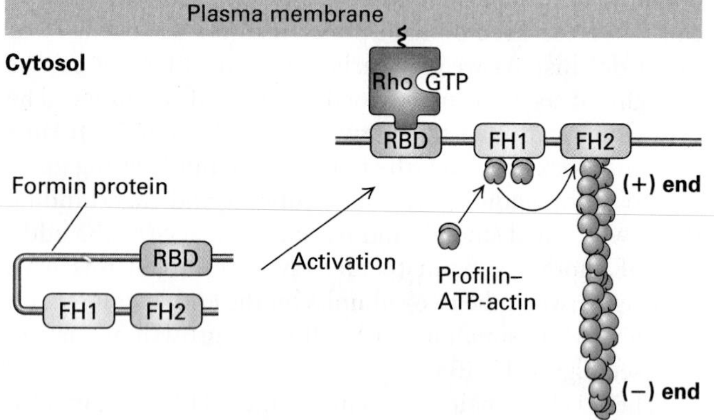

▲ **FIGURE 17-14 Regulation of formins by an intramolecular interaction.** At least one of the seven formin classes found in vertebrates is regulated by an intramolecular interaction. The inactive formin is activated by binding membrane-bound active Rho-GTP to its Rho binding domain (RBD), resulting in exposure of the FH2 domain, which can then nucleate the assembly of a new filament. All formins have an FH1 domain adjacent to the FH2 domain: the proline-rich FH1 domain is a site for recruitment of profilin–ATP–G-actin complexes that can then be "fed" into the growing (+) end. For simplicity of representation, a single formin protein is shown, but as shown in Figure 17-13, the protein functions as a dimer to nucleate actin assembly.

assembly and have the remarkable ability to remain bound to the (+) end while also allowing rapid assembly there. To ensure the continued growth of the filaent, formins bind to the (+) end in such a way that precludes binding of a (+) end capping protein such as CapZ, which would normally terminate assembly.

Formin activity has to be regulated. At least some formins exist in a folded inactive conformation as a result of an interaction between the first half of the protein and the C-terminal tail. These formins are activated by membrane-bound Rho-GTP, a Ras-related small GTPase (Figure 17-14). Thus when Rho is switched from the inactive Rho-GDP form into its activated Rho-GTP state, it can bind and activate the formin.

Recent studies have shown that formins are responsible for the assembly of long actin filaments such as those found in stress fibers and in the contractile ring during cytokinesis (see Figure 17-4). The actin-nucleating role of formins was only discovered recently, so the roles performed by this diverse protein family are only now being uncovered. Since there are many different formin classes in animals, it is likely that formins will be found to assemble additional actin-based structures.

The Arp2/3 Complex Nucleates Branched Filament Assembly

The Arp2/3 complex is a protein machine consisting of seven subunits, two of which are actin-related proteins

("Arp"), explaining its name. It is found in essentially all eukaryotes, including plants, yeasts, and animal cells. The Arp2/3 complex alone is a very poor nucleator when added into an actin assembly assay. To be active, Arp2/3 needs to bind both a regulatory protein, an example of which is WASp (Wiskott-Aldrich syndrome protein), and a preformed actin filament (Figure 17-15). Thus if you add into an actin assembly assay activated WASp and preformed actin filaments, Arp2/3 is a potent nucleator. How does the Arp2/3 complex nucleate filaments? When it binds to the side of F-actin in the presence of an activator, it changes conformation so that the two actin-related proteins, Arp2 and Arp3, resemble the (+) end of an actin filament (Figure 17-15a). This provides a template for the assembly of a new filament with a free (+) end. This new (+) end then grows as long as ATP–G-actin is available or until it is capped by a (+) end-capping protein such as CapZ. The angle between the old filament and the new one is 70° (Figure 17-15b). This is also the angle observed experimentally in branched filaments at the leading edge of motile cells, which is believed to be formed by the action of the activated Arp2/3 complex (Figure 17-15c). As we discuss in the next section, the Arp2/3 complex can be used to drive actin polymerization to power intracellular motility.

Actin nucleation by the Arp2/3 complex is exquisitely controlled, and the WASp protein is part of that regulatory process. WASp exists in a folded inactive conformation, so that the Arp2/3 activation domain at the carboxy-terminal end of the protein is not available (Figure 17-16). One mechanism to activate the protein involves the small Ras-related GTP-binding protein Cdc42 (Figure 17-16; Section 17.7), which in the GTP state binds to and opens WASp, making the acidic activation domain accessible to Arp2/3. WASp also has an actin-monomer-binding site adjacent to the C-terminal

Video: Direct Observation of Actin Filament Branching Mediated by Activated Arp2/3

MEDIA CONNECTIONS

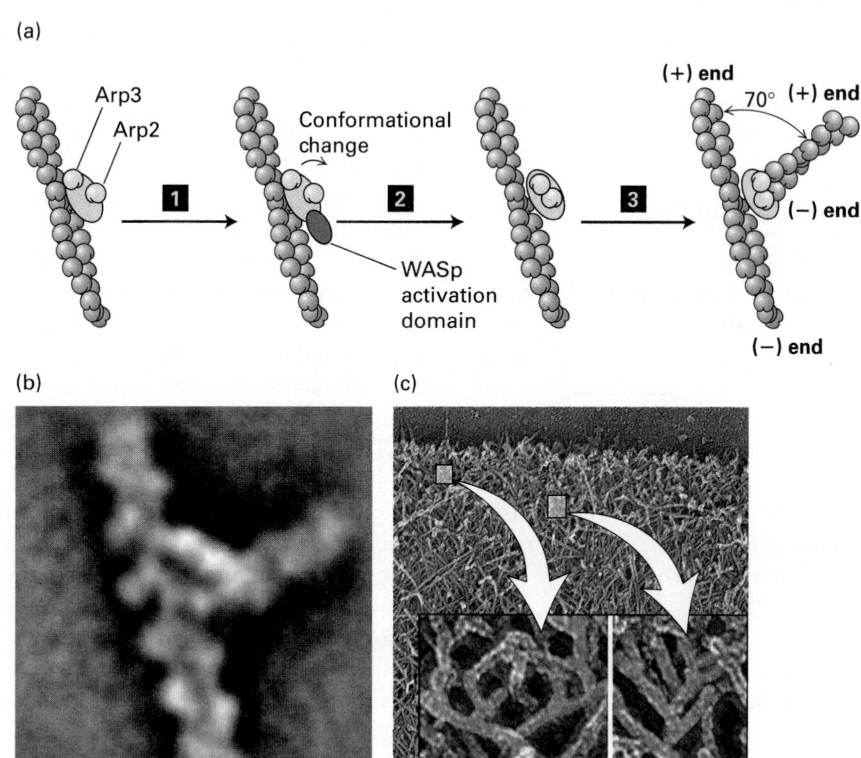

▲ **FIGURE 17-15 Actin nucleation by the Arp2/3 complex.** (a) To nucleate actin assembly efficiently, the Arp2/3 complex binds to the side of an actin filament (1) and to an activator such as the C-terminal domain of the WASp protein (2). This induces a conformational change in Arp2/3 so that the complex can now bind two actin subunits in a conformation corresponding to a free (+) end (3), and assembly occurs. The Arp2/3 branch makes a characteristic 70° angle between the filaments. (b) Averaged image compiled from several electron micrographs of Arp2/3 at an actin branch. (c) Image of actin filaments in the leading edge, with a magnification and coloring of individual branched filaments. [Part (a) adapted from A. E. Kelley et al., 2006, *J. Biol. Chem.* **281**:10589. Part (b) from C. Egile et al., 2006, *PLoS Biol.* **3**:e383. Part (c) from T. M. Svitkina and G. G. Borisy, 1999, *J. Cell Biol.* **145**:1009.]

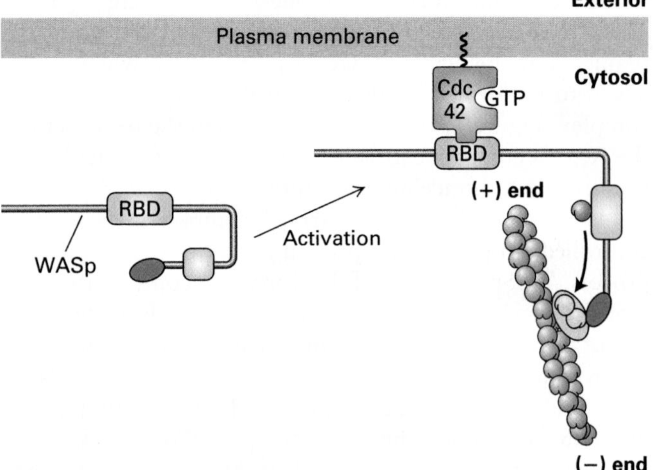

▲ **FIGURE 17-16 Regulation of the Arp2/3 complex by WASp.** The WASp is inactive due to an intramolecular interaction, thereby masking the activation domain. On binding the membrane-bound active small G-protein Cdc42-GTP (a member of the Rho family) through the Rho binding domain (RBD), the intermolecular interaction is relieved, exposing the acidic domain (purple) for activation of the Arp2/3 complex. WASp also has a G-actin binding region (brown) that aids in actin nucleation of the activated Arp2/3 complex.

Arp2/3 activation domain that facilitates the first steps in nucleation of a new filament (Figure 17-16).

Intracellular Movements Can Be Powered by Actin Polymerization

How can actin polymerization be harnessed to do work? As we have seen, actin polymerization involves the hydrolysis of actin-ATP to actin-ADP, which allows actin to grow preferentially at the (+) end and disassemble at the (−) end. If an actin filament were to become fixed in the meshwork of the cytoskeleton and you could bind and ride on the assembling (+) end, you would be transported across the cell. This is just what the intracellular bacterial parasite *Listeria monocytogenes* does. The study of *Listeria* motility was, in fact, the way the function of the Arp2/3 protein was discovered.

Listeria is a food-borne pathogen that causes mild gastrointestinal symptoms in most adults but can be fatal to the elderly or immunocompromised individuals. It can enter animal cells and divide in the cytoplasm. To move from one cell to another in the animal, it moves around the cell by polymerizing actin into a comet tail like the plume behind a rocket (Figure 17-17a, b), and when it runs into the plasma membrane, it pushes its way into the adjacent cell to infect it. How does it recruit the host cell actin to move around? *Listeria* has on its surface a protein called ActA, which can bind and activate the Arp2/3 complex (Figure 17-17c). The ActA protein also binds a protein known as VASP, which has three important properties. First, VASP has a proline-rich region that can bind profilin-ATP-actin for enhancing ATP-actin assembly into the newly formed barbed end generated by the Arp2/3 complex. Second, it can hold on to the end of the newly formed filament. Third, it can protect the (+) end of the growing filament from capping by CapZ. The assembling filament then pushes on the bacterium. Since the filament is embedded in the cytoskeletal matrix of the cell, it is stationary, and so the bacterial cell moves forward (ahead of the polymerizing actin). Researchers have reconstituted *Listeria* motility in the test tube using purified proteins to see what the minimal requirements for *Listeria* motility are. Remarkably, the bacteria will move when just four proteins are added: ATP–G-actin, the Arp2/3 complex, CapZ, and cofilin (Figure 17-17b, c). We have discussed the role of actin and Arp2/3, but for what are CapZ and cofilin needed? As we have seen earlier, CapZ rapidly caps the free (+) end of actin filaments, so when a growing filament no longer contributes to bacterial movement, it is rapidly capped and inhibited from further elongation. In this way, assembly occurs only adjacent to the bacterium where ActA is stimulating the Arp2/3 complex. Cofilin is necessary to accelerate the disassembly of the (–) end of the actin filament to regenerate actin to keep the polymerization cycle going (see Figure 17-11). This minimal rate of motility can be increased by the presence of other proteins, such as VASP and profilin, as mentioned above.

To move inside cells, the *Listeria* bacterium, as well as other opportunistic pathogens such as the *Shigella* species that cause dysentery, has hijacked a normal, regulated cellular process involved in cell locomotion. As we discuss in more detail later (Section 17.7), moving cells have a thin sheet of cytoplasm that protrudes from the front of the cell called the leading edge (see Figure 17-4 and 17-15c). This thin sheet of cytoplasm consists of a dense meshwork of actin filaments that are continually elongating at the front of the cell to push the membrane forward. Factors in the leading edge membrane activate the Arp2/3 complex to nucleate these filaments. Thus the power of actin assembly pushes the membrane forward to contribute to cell locomotion.

The power of actin assembly is also used during endocytosis. As we discussed in Chapter 14, **endocytosis** involves the pinching in of the plasma membrane to make an endocytic vesicle that is transported into the cell. Recent studies have shown that after clathrin-coated vesicles pinch off from the membrane, they are driven into the cytoplasm, powered by a rapid and very short-lived burst (a few seconds in duration) of actin polymerization, in a manner very similar to *Listeria* motility.

Toxins That Perturb the Pool of Actin Monomers Are Useful for Studying Actin Dynamics

Certain fungi and sponges have developed toxins that target the polymerization cycle of actin and are therefore toxic to animal cells. Two types of toxins have been characterized. The first class is represented by two unrelated toxins, cytochalasin D and latrunculin, which promote the depolymerization of filaments, though by different mechanisms. Cytochalasin D, a fungal alkaloid, depolymerizes actin filaments by binding to the (+) end of F-actin, where it blocks

Video: In Vivo Assembly of Actin Tails in *Listeria*-Infected Bacteria

(a)

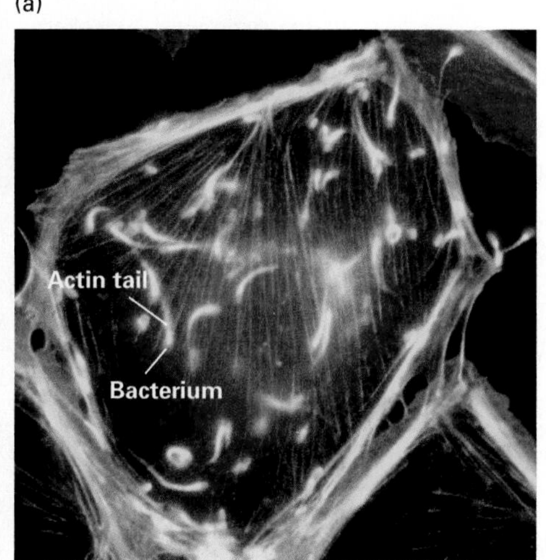

(b)

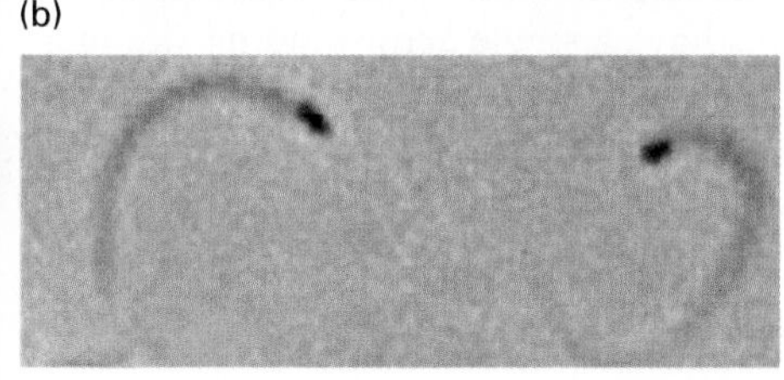

(c)

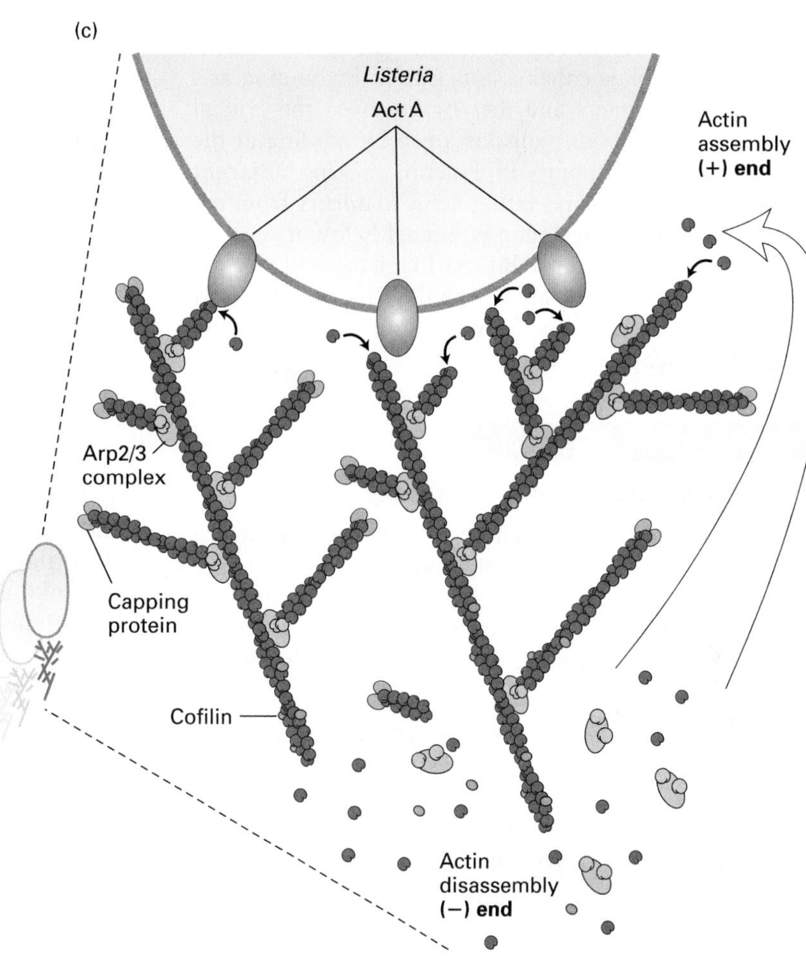

▲ **EXPERIMENTAL FIGURE 17-17** ***Listeria*** **utilizes the power of actin polymerization for intracellular movement.** (a) Fluorescence microscopy of a cultured cell stained with an antibody to the bacterial surface protein (red) and fluorescent phalloidin to localize F-actin (green). Behind each bacterium is an actin "comet tail" that propels the bacterium forward by actin polymerization. When the bacterium runs into the plasma membrane, it pushes the membrane out into a structure like a filopodium, which protrudes into a neighboring cell. (b) Listeria motility can be reconstituted in vitro with bacteria and just four proteins: ATP–G-actin, Arp2/3 complex, CapZ, and cofilin. The phase micrograph shows bacteria (black), behind which are the phase-dense actin tails. (c) A model of how *Listeria* moves using just four proteins. The ActA protein on the cell surface activates the Arp2/3 complex to nucleate new filament assembly from preexisting filaments. Filaments grow at their (+) end until capped by CapZ. Actin is recycled through the action of cofilin, which enhances depolymerization at the (−) end of the filaments. In this way, polymerization is confined to the back of the bacterium, which propels it forward. [Part (a) courtesy of J. Theriot and T. Michison. Part (b) from T. P. Loisel et al., 1999, *Nature* **401**:613.]

further addition of subunits. Latrunculin, a toxin secreted by sponges, binds and sequesters G-actin, inhibiting it from adding to a filament end. Exposure to either toxin thus increases the monomer pool. When cytochalasin or latrunculin is added to live cells, the actin cytoskeleton disassembles and cell movements such as locomotion and cytokinesis are inhibited. These observations were among the first to implicate actin filaments in cell motility. Latrunculin is especially useful because it binds actin monomers and prevents any new actin assembly. Thus if you add latrunculin to a cell, the rate at which actin-based structures disappear reflects their normal rate of turnover. This has revealed that some structures have half-times of much less than a minute, whereas others are much more stable. For example, it has been found that the leading edge of motile cells turns over every 30–180 seconds, whereas stress fibers turn over every 5–10 minutes.

In contrast, the monomer-polymer equilibrium is shifted in the direction of filaments by jasplakinolide, another sponge toxin, and by phalloidin, which is isolated from *Amanita phalloides* (the "angel of death" mushroom). Jasplakinolide enhances nucleation by binding and stabilizing actin dimers and thereby lowering the critical concentration. Phalloidin poisons a cell by binding at the interface between subunits in F-actin, locking adjacent subunits together and preventing actin filaments from depolymerizing. Even when actin is diluted below its critical concentration, phalloidin-stabilized filaments will not depolymerize. Fluorescent-labeled phalloidin, which binds only to F-actin, is commonly used to stain actin filaments for light microscopy (see Figure 17-4).

KEY CONCEPTS OF SECTION 17.3

Mechanisms of Actin Filament Assembly

- Actin assembly is nucleated by two classes of proteins: formins nucleate the assembly of unbranched filaments (see Figure 17-13), whereas the Arp2/3 complex nucleates the assembly of branched actin networks (see Figure 17-15). The activities of formins and Arp2/3 are regulated by signal-transduction pathways.
- Functionally different actin-based structures are assembled by formins and Arp2/3 nucleators. Formins drive the assembly of stress fibers and the contractile ring, whereas Arp2/3 nucleates the assembly of branched actin filaments found in the leading edge of motile cells.
- The power of actin polymerization can be harnessed to do work, as is seen in the Arp2/3-dependent intracellular movement of pathogenic bacteria (see Figure 17-17) and inward movement of endocytic vesicles.
- Several toxins affect the dynamics of actin polymerization; some, such as latrunculin, bind and sequester actin monomers, whereas others, such as phalloidin, stabilize filamentous actin. Fluorescently labeled phalloidin is useful for staining actin filaments.

17.4 Organization of Actin-Based Cellular Structures

So far in this chapter, we have seen that actin filaments can be assembled into a wide variety of different arrangements and how many associated proteins nucleate actin assembly and regulate filament turnover. Dozens of proteins in a vertebrate cell organize these filaments into diverse functional structures. Here we discuss just a few of these proteins, giving examples of typical types of actin cross-linking proteins found in cells, and also discuss the proteins involved in making functional links between actin and membrane proteins. One fascinating problem, about which very little is known, is how cells assemble different actin-based structures within the same cytoplasm of a cell. Some of this organization must be due to local regulation, a topic we come to at the end of the chapter.

Cross-Linking Proteins Organize Actin Filaments into Bundles or Networks

When one assembles actin filaments in a test tube, they form a tangled network. However, in cells, actin filaments are found in a variety of structures, such as the highly ordered filament bundles in microvilli or the meshwork characteristic of the leading edge (see Figure 17-4a). These different organizations are determined by the presence of actin *cross-linking proteins*. To be able to organize actin, an actin cross-linking protein must have two F-actin-binding sites (Figure 17-18).

Cross-linking of F-actin can be achieved by having two actin-binding sites within a single polypeptide, as with *fimbrin*, a protein found in microvilli to build bundles of filaments all having the same polarity. Other actin cross-linking proteins have a single actin-binding site in a polypeptide chain and then associate to form dimers to bring together two actin-binding sites. These dimeric cross-linking proteins can assemble to have a rigid rod connecting the two binding sites, as happens with *α-actinin*, which also holds parallel actin filaments but farther apart than fimbrin. Another protein, called *spectrin*, is a tetramer with two actin-binding sites; spectrin spans an even greater distance between actin filaments and makes networks under the plasma membrane (see Figure 17-19). Other types of cross-linking proteins, such as *filamin*, have a highly flexible region between the two binding sites, functioning like a molecular leaf spring, so they can make stabilizing cross-links between filaments in a meshwork (Figure 17-18), as is found in the leading edge of motile cells. The Arp2/3 complex, which we discussed in terms of its ability to nucleate actin filament assembly, is also an important cross-linking protein, attaching the (−) end of one filament to the side of another filament (see Figure 17-15).

Adaptor Proteins Link Actin Filaments to Membranes

To contribute to the structure of cells and also harness the power of actin polymerization, actin filaments are attached to membranes or are associated with intracellular structures. Actin filaments are especially abundant in the cell cortex underlying the plasma membrane, to which they give support. Actin filaments can interact with membranes either laterally or at their end.

Our first example of actin filaments attached to membranes is the human erythrocyte—the red blood cell. The erythrocyte consists essentially of plasma membrane enclosing a high concentration of the protein hemoglobin to

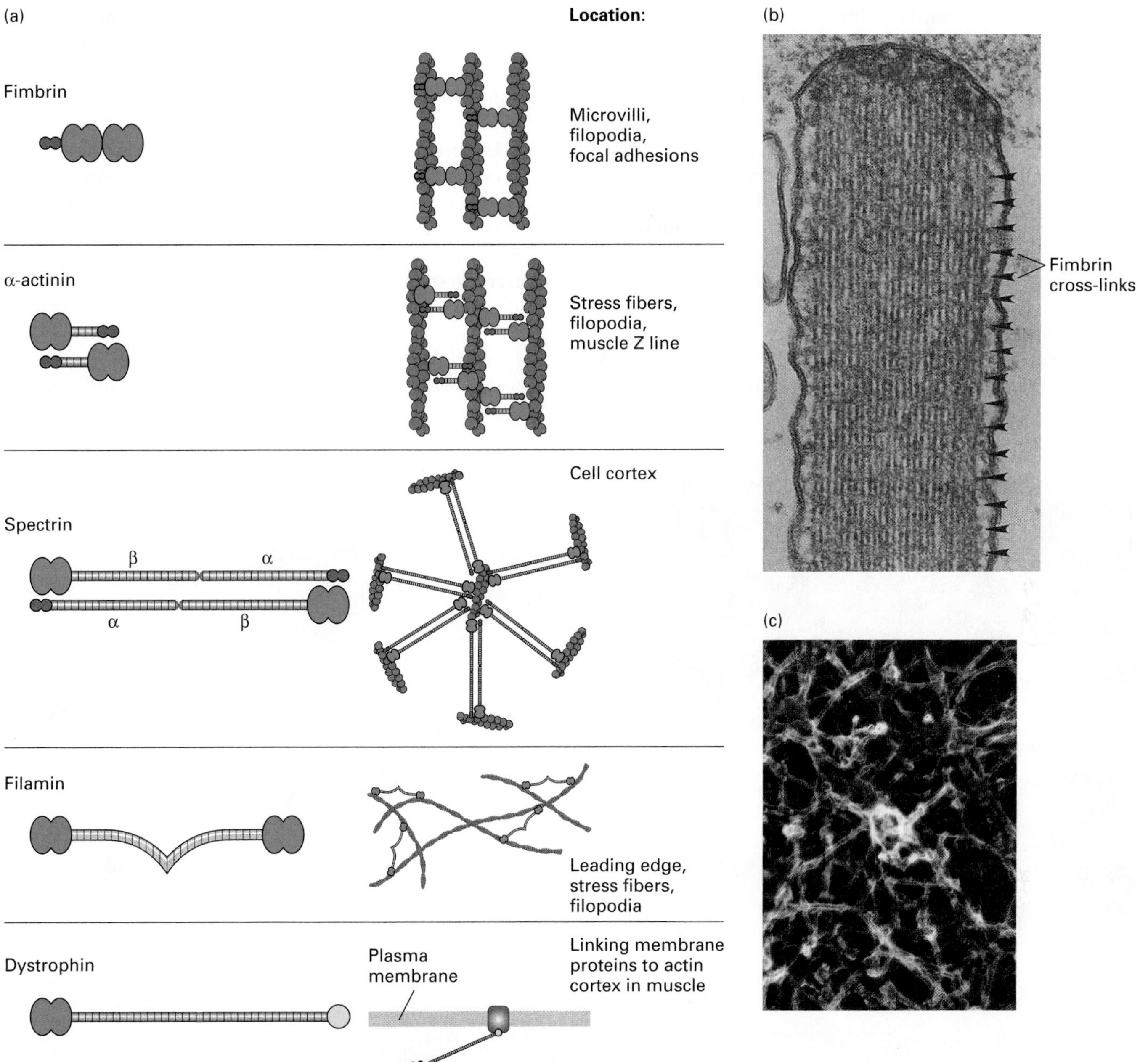

▲ **FIGURE 17-18 Actin cross-linking proteins mold F-actin filaments into diverse structures.** (a) Examples of four F-actin cross-linking proteins, all of which have two domains (blue) that bind F-actin. Some have a Ca^{2+}-binding site (purple) that inhibits their activity at high levels of free Ca^{2+}. Also shown is dystrophin, which has an actin-binding site on its N-terminal end and a C-terminal domain that binds the membrane protein dystroglycan. (b) Transmission electron micrograph of a thin section of a stereocilium (an unfortunate name since it is really a giant microvillus) on a sensory hair cell in the inner ear. This structure contains a bundle of actin filaments cross-linked by fimbrin. (c) Long cross-linking proteins such as filamin are flexible and can thus cross-link actin filaments into a network. [Part (b) from L. G. Tilney, 1983, *J. Cell Biol.* **96**:822. Part (c) courtesy of J. Hartwig.]

transport oxygen from the lungs to tissues and carbon dioxide from tissues back to the lung—all powered by the magnificent muscle known as the heart. Erythrocytes must be able to survive the raging torrents of blood flow in the heart, then flow down arteries and survive squeezing through narrow capillaries before being cycled through the lungs via the heart. To survive this grueling process for thousands of cycles, erythrocytes have a microfilament-based network underlying their plasma membrane that gives them both the tensile strength and the flexibility necessary for their journey. This network is based on short actin filaments of about 14 subunits in length, stabilized on their sides by tropomyosin (discussed in more detail in Section 17.6) and by the capping protein tropomodulin on the (−) end. These short filaments serve as hubs for binding about six flexible spectrin molecules, generating a fishing-net type of structure (Figure 17-19a). This network gives the erythrocyte both its strength and flexibility. Spectrin is attached to membrane proteins through two mechanisms—through a protein called *ankyrin* to the bicarbonate transporter (a transmembrane protein also known as *band 3*) and through a spectrin and F-actin-binding protein called *band 4.1* to another transmembrane protein called *glycophorin* C (Figure 17-19b). Although this linkage is highly developed in the erythrocyte, similar types of linkages occur in many cell types. For example, a related type of ankyrin-spectrin attachment links the Na^+/K^+ ATPase to the actin cytoskeleton on the basolateral membrane of epithelial cells.

Genetic defects in proteins of the red blood cell cytoskeleton can result in cells that rupture easily, giving rise to diseases known as hereditary spherocytic anemias (*spherocytic* because the cells are rounder, *anemias* because there is a shortage of red blood cells) and hence a shorter life span. In human patients, mutations in spectrin, band 4.1, and ankyrin can cause this disease. ▮

In addition to the spectrin-based type of support in the cell cortex, microfilaments provide the support for

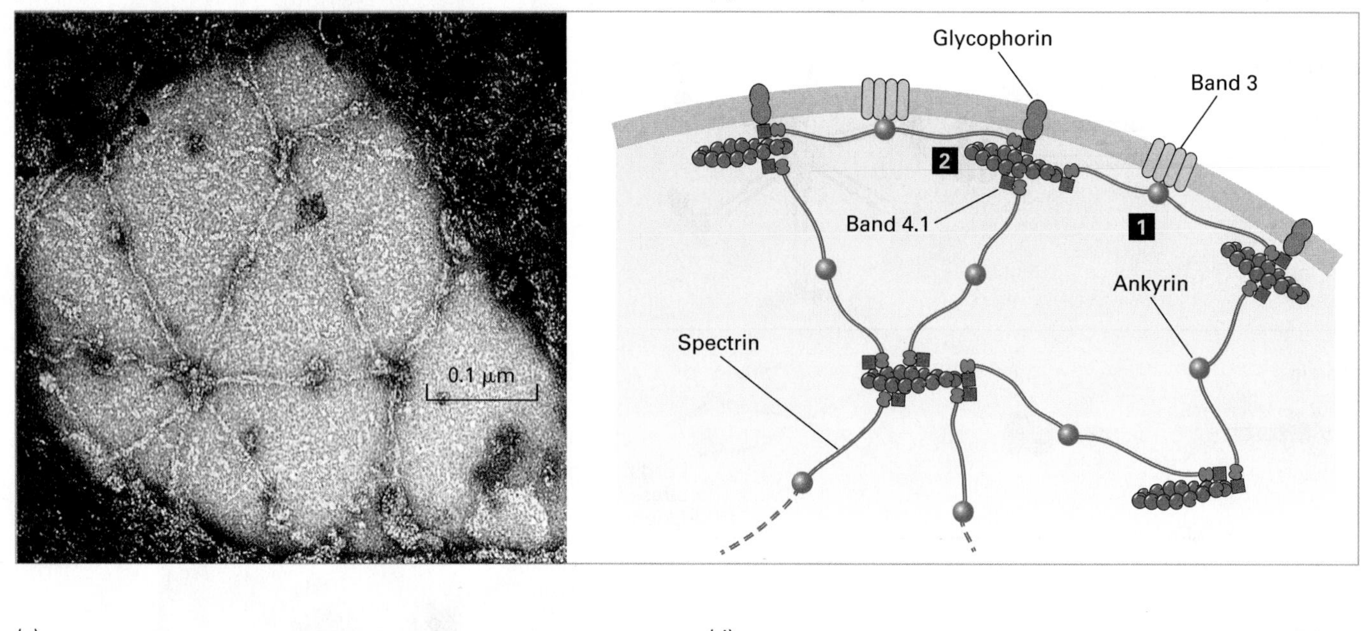

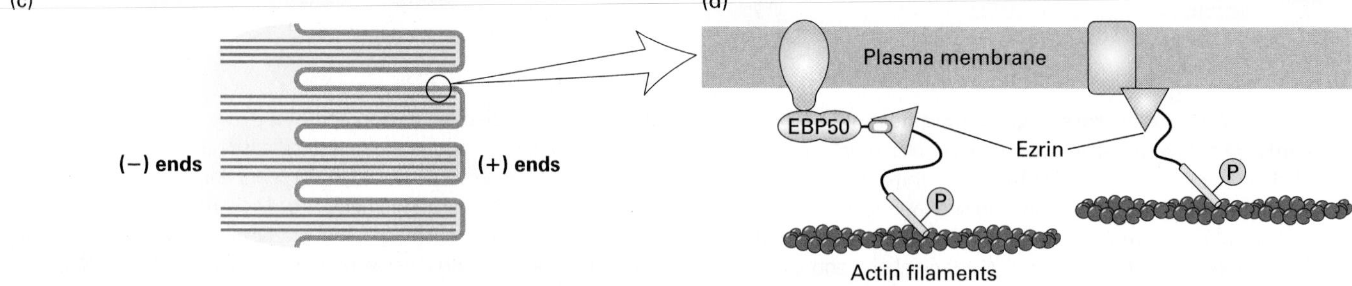

▲ **FIGURE 17-19 Lateral attachment of microfilaments to membranes.** (a) Electron micrograph of the erythrocyte membrane showing the spoke-and-hub organization of the cortical cytoskeleton supporting the plasma membrane in human erythrocytes. The long spokes are composed mainly of spectrin and can be seen to intersect at the hubs, or membrane-attachment sites. The darker spots along the spokes are ankyrin molecules, which cross-link spectrin to integral membrane proteins. (b) Diagram of the erythrocyte cytoskeleton, showing the two main types of attachment: (1) ankyrin and (2) band 4.1. (c) Microvilli on the apical aspect of an epithelial cell, showing the polarity of the actin filaments. (d) Ezrin, a member of the ezrin-radixin-moesin (ERM family), links actin filaments laterally to the plasma membrane in surface structures such as microvilli; attachment can be direct or indirect. Ezrin, activated by phosphorylation (P), links directly to the cytoplasmic region of transmembrane proteins (*right*) or indirectly through a scaffolding protein such as EBP50 (*left*). [Part (a) from T. J. Byers and D. Branton, 1985, *Proc. Nat'l. Acad. Sci. USA* **82**:6153. Courtesy of D. Branton. Part (b) adapted from S. E. Lux, 1979, *Nature* **281**:426, and E. J. Luna and A. L. Hitt, 1992, *Science* **258**:955. Part (d) adapted from A. Bretscher et al., 2002, *Nature Rev. Mol. Cell Biol.* **3**:586.]

cell-surface structures such as microvilli and membrane ruffles. If we think of a microvillus, it is clear that it must have an end-on attachment at the tip and lateral attachments down its length. What is the orientation of actin filaments in microvilli? Decoration of microvillar filaments by the S1 fragment of myosin show that it is the (+) end at the tip (Figure 17-19c). Moreover, when fluorescent actin is added to a cell, it is incorporated at the tip of a microvillus, showing that not only is the (+) end there but that actin filament assembly occurs there. At present it is not known how actin filaments are attached at the microvillus tip, but a likely candidate is a formin protein. This (+) end orientation of actin filaments with respect to the plasma membrane is almost universally found—not just in microvilli but also, for example, in the leading edge of motile cells. The lateral attachments to the plasma membrane are believed to be provided, at least in part, by the ERM (ezrin-radixin-moesin) family of proteins. These are regulated proteins that exist in a folded, inactive form. When activated, often by the membrane-bound regulatory lipid PIP_2 (phosphatidylinositol $(4,5,)P_2$) and subsequent phosphorylation, F-actin and membrane-protein-binding sites of the ERM protein are exposed to provide a lateral linkage to actin filaments (Figure 17-19d). At the plasma membrane, ERM proteins can link the actin filaments directly or indirectly through scaffolding proteins to the cytoplasmic domain of membrane proteins.

The two types of actin membrane linkages we have discussed do not involve areas of the plasma membrane directly attached to other cells or the extracellular matrix. In contrast, highly specialized regions of the plasma membrane of epithelial cells, called adherens junctions, make contacts between cells (see Figure 17-1b). Other specialized regions of association, called focal adhesions, mediate attachment of cells to the extracellular matrix. These specialized types of attachments in turn connect to the cytoskeleton and are described in more detail when we discuss cell migration (Section 17.7) and cells in the context of tissues (Chapter 19).

Muscular dystrophies are genetic diseases often characterized by the progressive weakening of skeletal muscle. One of these genetic diseases, Duchenne muscular dystrophy, affects the protein dystrophin, whose gene is located on the X-chromosome, and so the disease is much more prevalent in males. *Dystrophin* is a modular protein whose function is to link the cortical actin network of muscle cells to a complex of membrane proteins that link to the extracellular matrix. Thus dystrophin has an N-terminal actin-binding domain, followed by a series of spectrinlike repeats and terminating in a domain that binds the transmembrane dystroglycan complex to the extracellular matrix protein laminin (see Figure 17-18a). In the absence of dystrophin, the plasma membrane of muscle cells becomes weakened by cycles of muscle contraction and eventually ruptures, resulting in death of the muscle myofibril. ▮

KEY CONCEPTS OF SECTION 17.4

Organization of Actin-Based Cellular Structures

- Actin filaments are organized by cross-linking proteins that have two F-actin-binding sites. Actin cross-linking proteins can be long or short, rigid or flexible, depending on the type of structure involved (see Figure 17-18).
- Actin filaments are attached laterally to the plasma membrane by specific classes of proteins, such as are seen in the red blood cell or in cell-surface structures such as microvilli (see Figure 17-19).
- The (+) end of actin filaments can also be attached to membranes, with assembly mediated between the filament end and the membrane.
- Several diseases have been traced to defects in the microfilament-based cortical cytoskeleton that underlies the plasma membrane.

17.5 Myosins: Actin-Based Motor Proteins

We have discussed how actin polymerization nucleated by the Arp2/3 complex can be harnessed to do work. In addition to actin-polymerization-based motility, cells have a large family of motor proteins, called **myosins,** that can move along actin filaments powered by ATP hydrolysis. The first myosin discovered, myosin II, was isolated from skeletal muscle. For a long time, biologists thought that this was the only type of myosin found in nature. However, they then discovered other types of myosins and began to ask how many different functional classes might exist. Today we know that there are several different classes of myosins, in addition to the myosin II of skeletal muscle, that provide a motor function. The other classes of myosin provide a myriad of functions, such as moving organelles and other structures around cells as well as contributing to cell migration. Indeed, with the discovery and analysis of all these actin-based motors and the corresponding microtubule-based motors described in the next chapter, the former relatively static view of a cell has been replaced with the realization that it is incredibly dynamic—more like an organized but busy freeway system with motors busily ferrying components around.

To begin to understand myosins, we first discuss their general domain organization. Armed with this information, we explore the diversity of myosins in different organisms and describe in more detail some of those that are common in eukaryotes. Myosins have the amazing ability to convert the energy released by ATP hydrolysis into mechanical work. All myosins convert ATP hydrolysis into work, yet different myosins can perform very different types of functions. For example, many molecules of myosin II pull together on actin filaments to bring about muscle contraction, whereas myosin V binds to vesicular cargo to transport it along actin filaments. To understand how such diverse functions can be accommodated by one type of motor mechanism, we will investigate the basic mechanism of how the energy released by ATP hydrolysis is converted into work and then see how this mechanism is modified to tailor the properties of specific myosin classes for their specific functions.

Myosins Have Head, Neck, and Tail Domains with Distinct Functions

Much of what we know about myosins comes from studies of myosin II from skeletal muscle. In skeletal muscle, myosin II is assembled into so-called bipolar thick filaments containing hundreds of individual myosin II proteins (Figure 17-20a) with opposite orientations in each half of the bipolar filament. These myosin filaments interdigitate with actin thin filaments to bring about muscle contraction. We will discuss how this system works in a later section, but here we investigate the properties of the myosin itself.

It is possible to dissolve the myosin thick filament in a solution of ATP and high salt. The resulting soluble myosin II protein consists of six polypeptides—two associated heavy chains and four light chains. Two light chains associate with the "neck" region of each heavy chain (Figure 17-20b). The soluble myosin has ATPase activity, reflecting its ability to power movements by hydrolysis of ATP. But which domain of myosin is responsible for this activity? To identify functional domains in a protein, a standard approach is to cleave it into fragments with specific proteases and ask which fragments have the activity. Soluble myosin II can be cleaved into two pieces by gentle treatment with the protease chymotrypsin to yield two fragments, one called heavy mero-myosin (HMM; *mero* means "part of") and the other

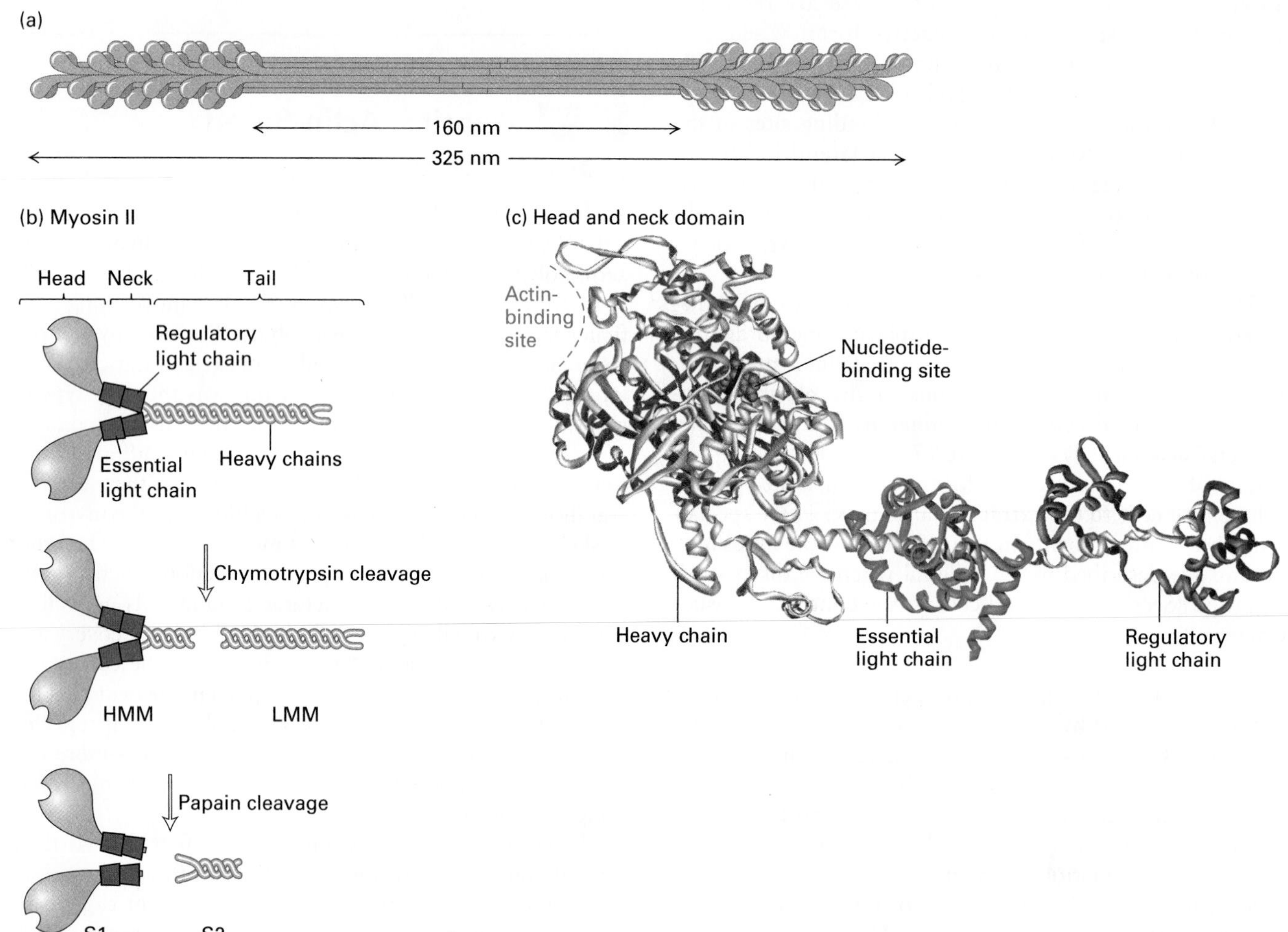

▲ **FIGURE 17-20 Structure of myosin II.** (a) Organization of myosin II in filaments isolated from skeletal muscle. Myosin II assembles into bipolar filaments in which the tails form the shaft of the filament with heads exposed. Extraction of bipolar filaments with high salt and ATP disassembles the filament into individual myosin II molecules. (b) Myosin II molecules consist of two identical heavy chains (light blue) and four light chains (green and blue). The tail of the heavy chains forms a coiled-coil to dimerize; the neck region of each heavy chain has two light chains associated with it. Limited proteolytic cleavage of myosin II generates tail fragments—LMM and S2—and the S1 motor domain. (c) Three-dimensional model of a single S1 head domain shows that it has a curved, elongated shape and is bisected by a cleft. The nucleotide-binding pocket lies on one side of this cleft, and the actin-binding site lies on the other side near the tip of the head. Wrapped around the shaft of the α-helical neck are two light chains. These chains stiffen the neck so that it can act as a lever arm for the head. Shown here is the ADP-bound conformation.

Animation: In Vitro Motility Myosin Assay

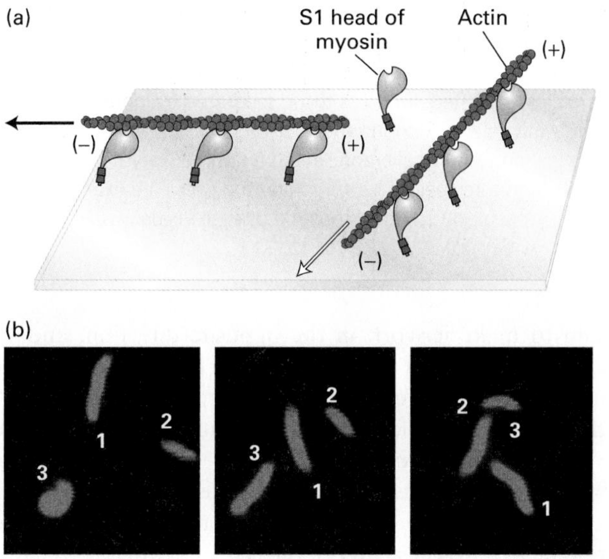

◄ **EXPERIMENTAL FIGURE 17-21 Sliding-filament assay is used to detect myosin-powered movement.** (a) After myosin molecules are adsorbed onto the surface of a glass coverslip, excess unbound myosin is removed; the coverslip then is placed myosin-side down on a glass slide to form a chamber through which solutions can flow. A solution of actin filaments, made visible and stable by staining with rhodamine-labeled phalloidin, is allowed to flow into the chamber. (The coverslip in the diagram is shown inverted from its orientation on the flow chamber to make it easier to see the positions of the molecules.) In the presence of ATP, the myosin heads walk toward the (+) end of filaments by the mechanism illustrated in Figure 17-24. Because myosin tails are immobilized, walking of the heads causes sliding of the filaments. Movement of individual filaments can be observed in a fluorescence light microscope. (b) These photographs show the positions of three actin filaments (numbered 1, 2, 3) at 30-second intervals recorded by video microscopy. The rate of filament movement can be determined from such recordings. [Part (b) courtesy of M. Footer and S. Kron.]

light meromyosin (LMM) (Figure 17-20b). The heavy meromyosin can be further cleaved with the protease papain to yield subfragment 1 (S1) and subfragment 2 (S2). By analyzing the properties of the various fragments—S1, S2, and LMM—it was found that the intrinsic ATPase activity of myosin resides in the S1 fragment, as does an F-actin-binding site. Moreover, it was found that the ATPase activity of the S1 fragment was greatly enhanced by the presence of filamentous actin, so it is said to have an *actin-activated ATPase activity,* which is a hallmark of all myosins. The S1 fragment of myosin II consists of the *head* and *neck* domains, whereas the S2 and LMM regions make up the *tail* domain (Figure 17-20b). X-ray crystallographic analysis of the head and neck domains revealed its shape, the positions of the light chains, and the locations of the ATP-binding and actin-binding sites. The elongated myosin head is attached at one end to the α-helical neck (Figure 17-20c). Two light-chain molecules lie at the base of the head, wrapped around the neck like C-clamps. In this position, the light chains stiffen the neck region.

How much of myosin II is necessary and sufficient for "motor" activity? To answer this question, one needs a simple in vitro motility assay. In one such assay, the *sliding-filament assay,* myosin molecules are tethered to a coverslip to which is added stabilized fluorescence-labeled actin filaments. Because the myosin molecules are tethered, they cannot slide; thus any force generated by interaction of myosin heads with actin filaments forces the filaments to move relative to the myosin (Figure 17-21a). If ATP is present, added actin filaments can be seen to glide along the surface of the coverslip; if ATP is absent, no filament movement is observed. Using this assay, one can show that the S1 head of myosin II is sufficient to bring about movement of actin filaments. This movement is caused by the tethered myosin S1 fragments (bound to the coverslip) trying to "move" toward the (+) end of a filament; thus the filaments move with the (−) end leading. The rate at which myosin moves an actin filament can be determined from video camera recordings of sliding-filament assays (Figure 17-21b).

All myosins have a domain related to the S1 domain of myosin II, which is responsible for their motor activity. The tail domain does not contribute to motility but rather defines what is moved by the S1-related domain. Thus, as might be expected, the tail domains can be very different and are tailored to bind specific cargoes.

Myosins Make Up a Large Family of Mechanochemical Motor Proteins

Since all myosins have related S1-motor domains with considerable similarity in primary amino acid sequence, it is possible to determine how many myosin genes, and how many different classes of myosins, exist in a sequenced genome. There are about 40 myosin genes in the human genome (Figure 17-22), nine in *Drosophila,* and five in the budding yeast. Computer analysis of the sequence relationships between the myosin head domains suggests that about 20 distinct classes of myosins have evolved in eukaryotes, with greater sequence similarity within a class than between. As indicated in Figure 17-22, the genetic basis for some diseases has been traced to genes encoding myosins. All myosin head domains convert ATP hydrolysis into mechanical work using the same general mechanism. However, as we will see, subtle differences in this mechanism can have profound effects on the functional properties of different classes of myosin. How do these different classes relate to tail domains? Amazingly, if one takes just the protein sequences of the tail domains of the myosins and uses this information to place them in classes, they fall into the same groupings as the motor domains. This implies that motor domains with specific properties have co-evolved with specific classes of tail

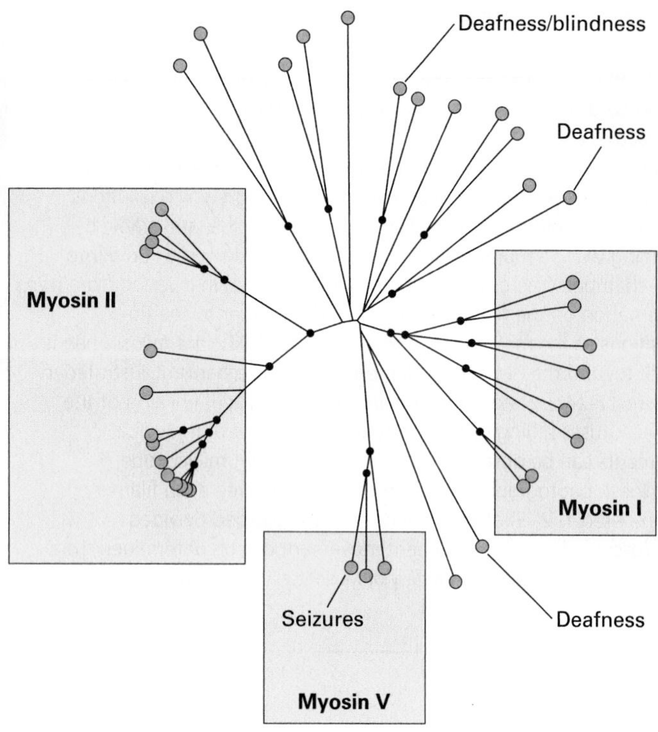

◀ **FIGURE 17-22 The myosin superfamily in humans.** Computer analysis of the relatedness of S1 head domains of all of the approximately 40 myosins encoded by the human genome. Each myosin is indicated by a line, with the length of the line indicating phylogenetic distance relationships. Thus myosins connected by short lines are closely related, whereas those separated by longer lines are more distantly related. Among these myosins are three classes—myosins I, II, and V—widely represented among eukaryotes, with others having more specialized functions. Indicated are examples in which loss of a specific myosin causes a disease. [Redrawn and modified from R. E. Cheney, 2001, *Mol. Biol. Cell* **12**:780.]

domains, which makes a lot of sense, suggesting that each class of myosin has evolved to carry out a specific function.

In every case that has been tested so far, myosins move toward the (+) end of an actin filament—with one exception, *myosin VI.* This remarkable myosin has an insert in its head domain to make it work in the opposite direction, and so motility is toward the (−) end of an actin filament. Myosin VI in animal cells is believed to contribute to endocytosis by moving the endocytic vesicles along actin filaments away from the plasma membrane. Recall that membrane-associated actin filaments have their (+) ends toward the membrane, so a motor directed toward the (−) end would take them away from the membrane toward the center of the cell.

Among all these different classes of myosins are three especially well-studied ones, which are commonly found in animals and fungi: the so-called *myosin I, myosin II,* and *myosin V* families (Figure 17-23). In humans, eight genes encode heavy chains for the myosin I family, 14 for the myosin II family, and three for the myosin V family.

The myosin II class assembles into bipolar filaments, which is important for its involvement in contractile functions; indeed, this is the only class of myosins involved in contractile functions. The large number of members in this

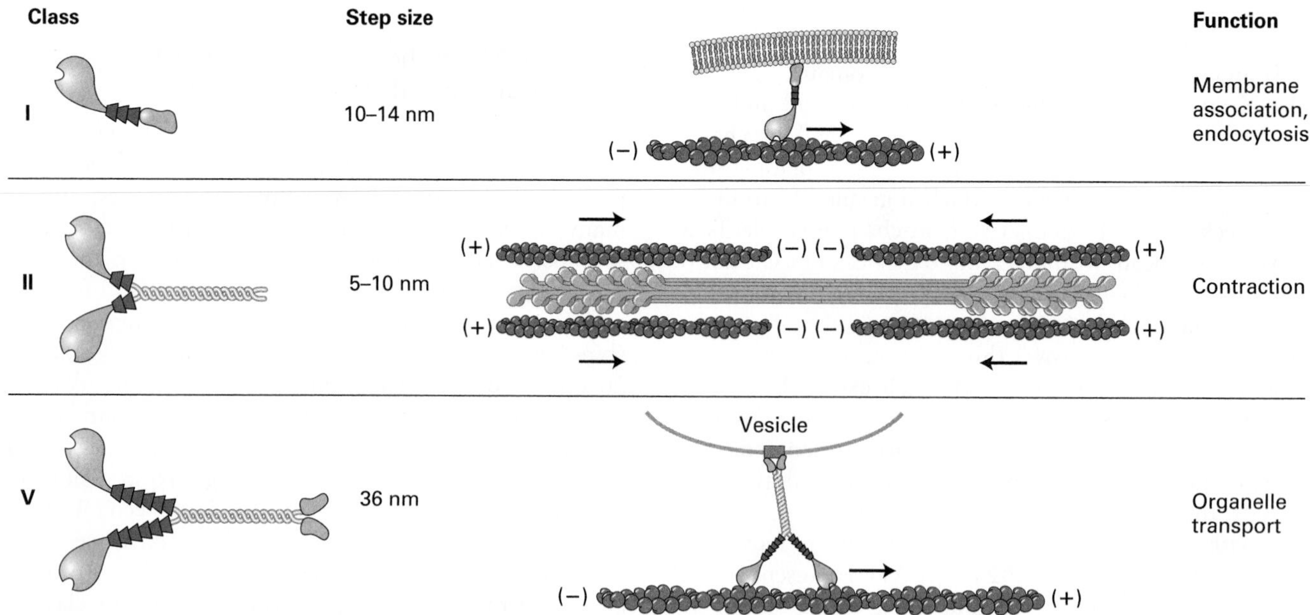

▲ **FIGURE 17-23 Three common classes of myosin.** Myosin I consist of a head domain with a variable number of light chains associated with the neck domain. Members of the myosin I class are the only myosins to have a single head domain. Some of these myosins are believed to associate directly with membranes through lipid interactions. Myosin IIs have two head domains and two light chains per neck and are the only class that can assemble into bipolar filaments. Myosin Vs have two head domains and six light chains per neck. They bind specific receptors (brown box) on organelles, which they transport. All myosins in these three classes move toward the (+) end of actin filaments.

Podcast: Myosin Movement Against Actin Filaments
Animation: Myosin-Actin Cross-Bridge Cycle

MEDIA CONNECTIONS

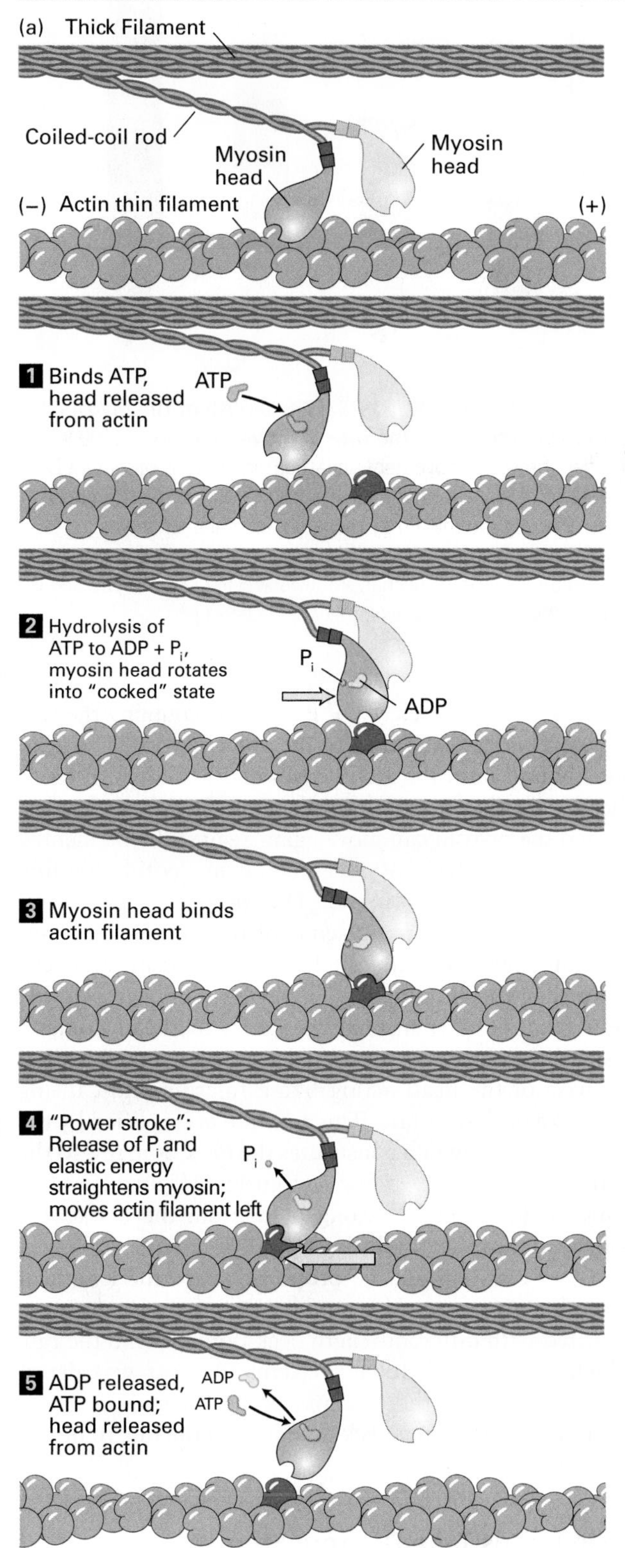

(a) Thick Filament
Coiled-coil rod
Myosin head
Myosin head
(−) Actin thin filament
(+)
1 Binds ATP, head released from actin
ATP
2 Hydrolysis of ATP to ADP + P_i, myosin head rotates into "cocked" state
P_i
ADP
3 Myosin head binds actin filament
4 "Power stroke": Release of P_i and elastic energy straightens myosin; moves actin filament left
P_i
5 ADP released, ATP bound; head released from actin
ADP
ATP

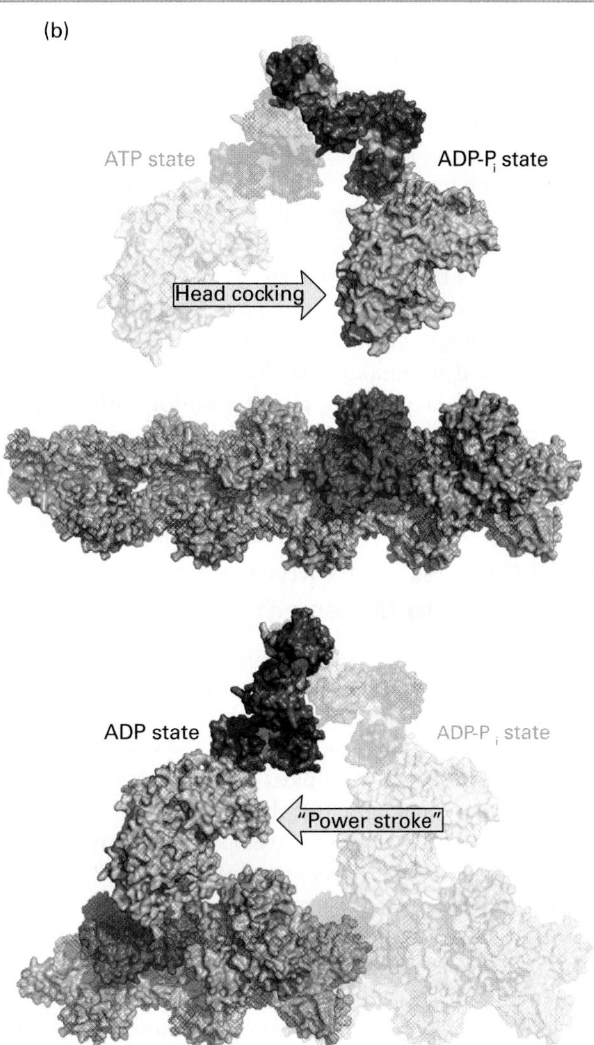

(b)
ATP state
ADP-P_i state
Head cocking
ADP state
ADP-P_i state
"Power stroke"

◄ **FIGURE 17-24 The myosin head uses ATP to pull on an actin filament.** (a) In the absence of ATP, the myosin head is firmly attached to the actin filament. Although this state is very short-lived in living muscle, it is the state responsible for muscle stiffness in death (rigor mortis). Step (**1**): On binding ATP, the myosin head releases from the actin filament. Step (**2**): The head hydrolyzes the ATP to ADP and P_i which induces a rotation in the head with respect to the neck. This "cocked state" stores the energy released by ATP hydrolysis as elastic energy, like a stretched spring. Step (**3**): Myosin in the "cocked" state binds actin. Step (**4**): When bound to actin the myosin head couples release of P_i with release of the elastic energy to move the actin filament. This is known as the "power stroke" as it involves moving the actin filament with respect to the end of the myosin neck domain. Step (**5**): The head remains tightly bound to the filament as ADP is released and before fresh ATP is bound by the head. (b) Molecular models of the conformational changes in the myosin head involved in "cocking" the head (upper panel) and during the power stroke (lower panel). The myosin light chains are shown in dark blue and green; the rest of the myosin head and neck are colored in light blue, and actin is red [Part (a) adapted from R. D. Vale and R. A. Milligan, 2002, *Science* **288**:88. Part (b) based on Geeves and Holmes 2005 (unpublished)].

class reflects the need for myosin IIs with the slightly different contractile properties seen in different muscles (e.g., skeletal, cardiac, and various types of smooth muscle) as well as in nonmuscle cells.

The myosin II class is the only one that assembles into bipolar filaments. All myosin II members have a relatively short neck domain, with two light chains per heavy chain. The myosin I class is quite large, has a variable number of light chains associated with the neck region, and is the only one in which heavy chains are not associated through their tail domains and so are single-headed. The large size and diversity of the myosin I class suggests that these myosins perform many functions, most of which remain to be determined, but some members of this family connect actin filaments to membranes, and others are implicated in endocytosis. Members of the myosin V class have two heavy chains, giving a motor with two heads, long neck regions with six light chains each, and tail regions that dimerize and terminate in domains that bind to specific organelles to be transported.

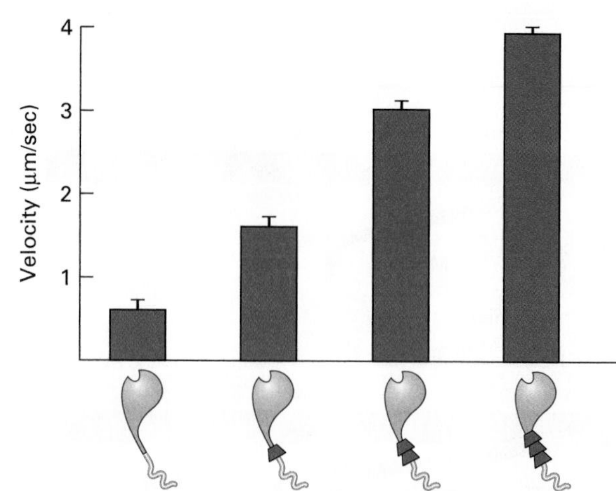

▲ **EXPERIMENTAL FIGURE 17-25 The length of the myosin II neck domain determines the rate of movement.** To test the lever-arm model of myosin movement, investigators used recombinant DNA techniques to make myosin heads attached to different-length neck domains. The rate at which they move on actin filaments was determined. The longer the lever arm, the faster the myosin moves, supporting the proposed mechanism. [Redrawn from K. A. Ruppel and J. A. Spudich, 1996, *Annu. Rev. Cell Mol. Biol.* **12**:543–573.]

Conformational Changes in the Myosin Head Couple ATP Hydrolysis to Movement

The results of studies of muscle contraction provided the first evidence that myosin heads slide or walk along actin filaments. Unraveling the mechanism of muscle contraction was greatly aided by the development of in vitro motility assays and single-molecule force measurements. On the basis of information obtained with these techniques and the three-dimensional structure of the myosin head, researchers developed a general model for how myosin harnesses the energy released by ATP hydrolysis to move along an actin filament (Figure 17-24 on page 23). Because all myosins are thought to use the same basic mechanism to generate movement, we will ignore whether the myosin tail is bound to a vesicle or is part of a thick filament, as it is in muscle. One important aspect of this model is that the hydrolysis of a single ATP molecule is coupled to each step taken by a myosin molecule along an actin filament.

A question that has intrigued biologists is how myosin can convert the chemical energy released by ATP hydrolysis into mechanical work. It has been known for a long time that the S1 head of myosin is an ATPase, having the ability to hydrolyze ATP into ADP and P_i. Biochemical analysis reveals how this works (Figure 17-24a). In the absence of ATP, the head of myosin binds very tightly to F-actin. When ATP binds, the affinity of the head for F-actin is greatly reduced and releases from actin. The myosin head then hydrolyzes the ATP, and the hydrolysis products, ADP and P_i, remain bound. The energy provided by the hydrolysis of ATP induces a conformation change in the head that results in the head domain rotating with respect to the neck. This is known as the "cocked" position of the head (Figure 17-24b). In the absence of F-actin, release of P_i is exceptionally slow, the slowest part of the ATPase cycle. However, in the presence of actin, the head binds F-actin tightly, inducing both release of P_i and rotation of the head back to its original position, thus moving the actin filament relative to the neck domain (Figure 17-24b). In this way, binding to F-actin induces the movement of the head and release of P_i, thereby coupling the two processes. This step is known as the *power stroke*. The head remains bound until the ADP leaves and a fresh ATP binds the head, releasing it from the filament. The cycle then repeats, and the myosin can move again against the filament.

How is hydrolysis of ATP in the nucleotide-binding pocket converted into force? The results of structural studies of myosin in the presence of nucleotides, and nucleotide analogs that mimic the various steps in the cycle, indicate that the binding and hydrolysis of a nucleotide cause a small conformational change in the head domain. This small movement is amplified by a "converter" region at the base of the head acting like a fulcrum and causing the leverlike neck to rotate. This rotation is amplified by the rodlike lever arm, which constitutes the neck domain, so the actin filament moves by a few nanometers (Figure 17-24b).

This model makes a strong prediction: the distance a myosin head moves actin during hydrolysis of one ATP—the myosin *step size*—should be proportional to the length of the neck domain. To test this, mutant myosin molecules were constructed with different-length neck domains and the rate at which they moved down an actin filament was determined. Remarkably, there is an excellent correspondence between the length of the neck domain and the rate of movement (Figure 17-25).

Myosin Heads Take Discrete Steps Along Actin Filaments

The most critical feature of myosin is its ability to generate a force that powers movements. Researchers have used *optical traps* to measure the forces generated by single myosin molecules (Figure 17-26). In this approach, myosin is immobilized on beads at a low density. An actin filament, held between two optical traps, is lowered toward the bead until it contacts the myosin molecule. When ATP is added, the myosin pulls on the actin filament. Using a mechanical feedback mechanism controlled by a computer, one can measure the distance pulled and the forces and duration of the movement (Figure 17-26). The results of optical trap studies show that myosin II does not interact with the actin filament continuously but rather binds, moves, and releases it. In fact, myosin II spends on average only about 10 percent of each ATPase cycle in contact with F-actin—it is said to have a *duty ratio* of 10 percent. This will be important later when we consider that in muscle, hundreds of myosin heads pull on actin filaments, so that at any one time, 10 percent of the heads are engaged to provide a smooth contraction.

When myosin II does contact F-actin, it takes discrete steps, approximately 5–15 nm long (Figure 17-27), and generates 3–5 piconewtons (pN) of force, approximately the same force as that exerted by gravity on a single bacterium.

If we now look at a similar optical trap experiment with myosin V, the curves look completely different (Figure 17-27). Now we can easily discern clear steps of about 36 nm in length. This larger step size reflects the longer neck domain—the lever arm—of myosin V. Moreover, we see that the motor takes many sequential steps without releasing from the actin—it is said to move *processively*. This is because its ATPase cycle is modified to have a much higher duty ratio (>70 percent) by slowing the rate of ADP release; thus the head remains in contact with the actin filament for a much larger percentage of the cycle. Since a single myosin V molecule has two heads, a duty ratio of >50 percent ensures that one head is in contact at all times as it moves down an actin filament so that it does not fall off.

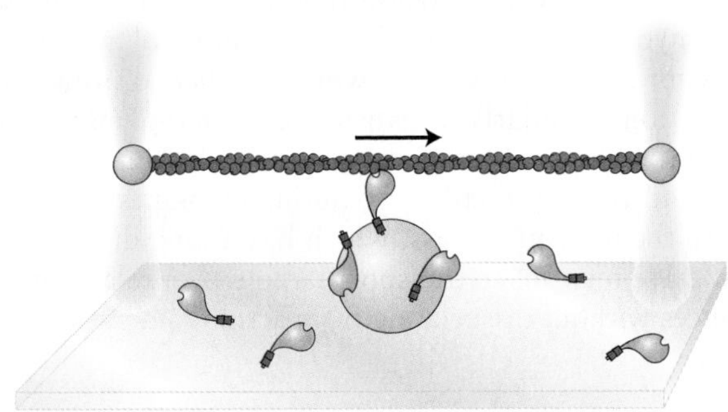

▲ **FIGURE 17-26 Optical trap determines the step size and force generated by a single myosin molecule.** In an optical trap, the beam of an infrared laser is focused by a light microscope on a latex bead (or any other object that does not absorb infrared light), which captures and holds the bead in the center of the beam. The strength of the force holding the bead is adjusted by increasing or decreasing the intensity of the laser beam. In this experiment, an actin filament is held between two optical traps. The actin filament is then lowered onto another bead with a dilute concentration of myosin molecules attached to it. If the actin filament encounters a myosin molecule in the presence of ATP, the myosin will pull on the actin filament, which allows the investigators to measure both the force generated and the step size the myosin takes.

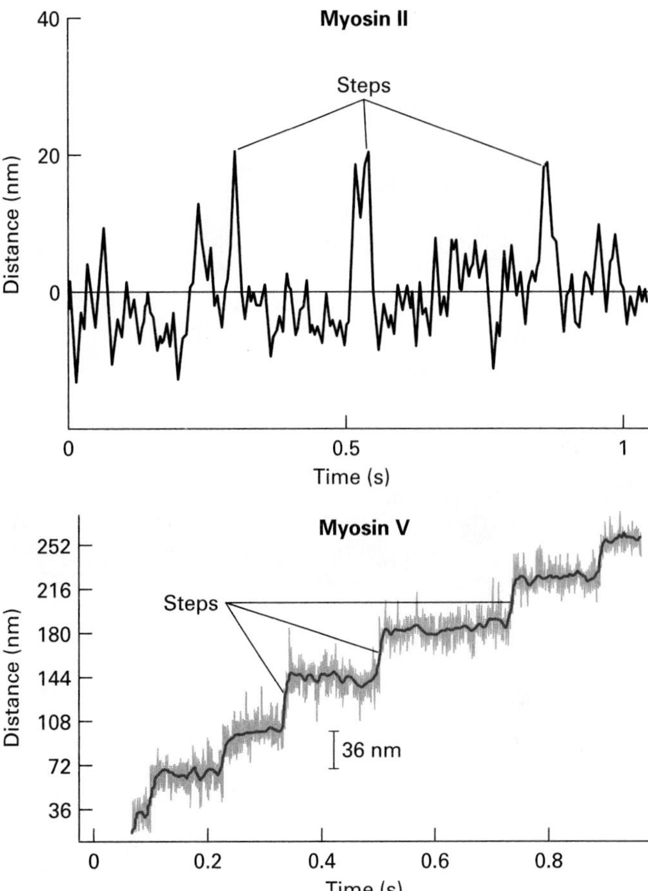

▲ EXPERIMENTAL FIGURE 17-27 **Measuring the step size and processivity of myosins.** Using an optical trap setup similar to the one described in Figure 17-26, investigators have analyzed the behavior of myosin II (top trace) and myosin V (bottom trace). As shown by the peaks in the trace, myosin II takes erratic small steps (5–15 nm), which means it binds the actin filament, moves, and then lets go. It is therefore a nonprocessive motor. By contrast, myosin V takes clear 36-nm steps one after the other, so it has a step size of 36 nm and is highly processive—that is, it does not let go of the actin filament. [Part (a) from Finer et al., 1994, *Nature* **368**:113. Part (b) from M. Rief et al., 2000, *Proc. Nat'l. Acad. Sci. USA* **97**:9482.]

Myosin V Walks Hand Over Hand Down an Actin Filament

The next question is, how do the two heads of myosin V work together to move down a filament? One model proposes that the two heads walk down a filament hand over hand with alternately leading heads (Figure 17-28a). An alternative possibility is an inchworm model, in which the leading head takes a step, the second head is pulled up behind it, and then the leading head takes another step (Figure 17-28b). How can one distinguish between these models? In the inchworm model, each individual head takes 36-nm steps, whereas in the walking model, each takes 72-nm steps. Scientists have managed to attach a fluorescent probe to just one neck region of myosin V and watch it walk down an actin filament: it takes 72-nm steps (Figure 17-28c), and so it walks hand over hand down a filament (72 nm is the

(a) Hand over hand

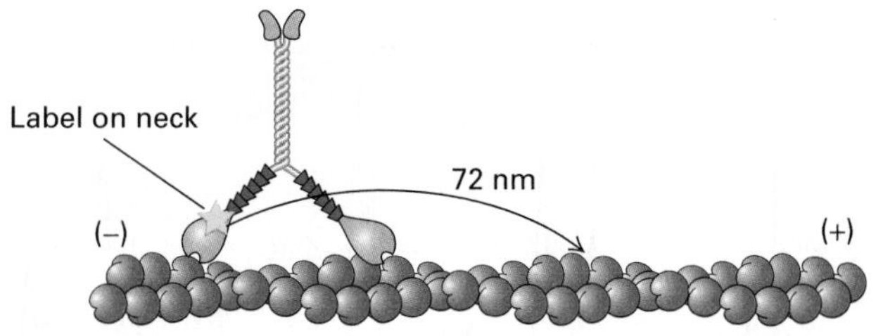

(b) Inchworm

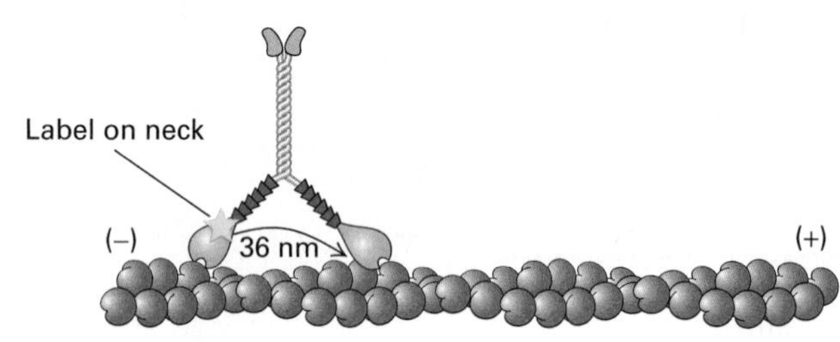

(c)

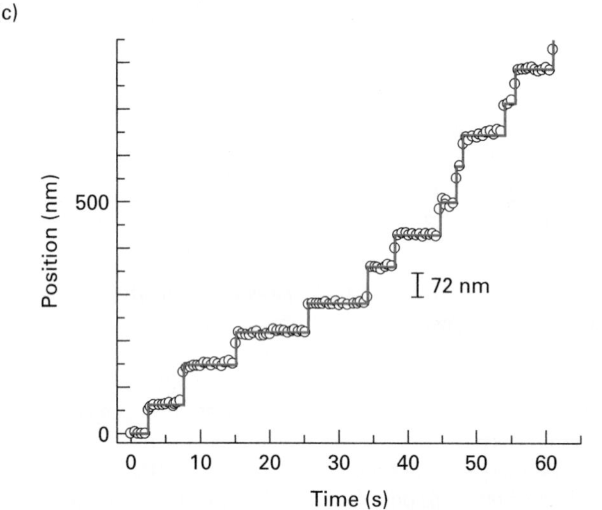

▲ EXPERIMENTAL FIGURE 17-28 **Myosin V has a step size of 36 nm, yet each head moves in 72-nm steps, so it moves hand over hand.** Two models for myosin V movement down a filament have been suggested. (a) In the hand-over-hand model, one head binds an actin filament, and the other then swings around and binds a site 72 nm ahead. (b) In the inchworm model, the leading head moves 36 nm, then the lagging head moves up behind it, allowing the leading head to take another 36-nm step. (c) Myosin V labeled with a fluorescent tag on just one head can be seen to have a step size of 72 nm. Thus myosin V walks hand over hand. [Adapted from A. Yildiz et al., 2003, *Science* **300**:2061.]

step size for each head; the step size for the double-headed motor is 36 nm). Why is the step size of myosin V so large? If we compare its step size of 36 nm to the structure of the actin filament, we see that it is the same as the length between helical repeats in the filament (see Figures 17-5 and 17-28a), so myosin V steps between equivalent binding sites as it walks down one side of an actin filament. Myosin V has presumably evolved to take large steps the size of the helical repeat of actin and to do this very processively so it rarely dissociates from an actin filament. These are exactly the properties one would expect for a motor designed to transport cargo along an actin filament.

KEY CONCEPTS FOR SECTION 17.5

Myosins: Actin-Based Motor Proteins

- Myosins are actin-based motors powered by ATP hydrolysis.
- Myosins have a motor head domain, a lever-arm neck domain, and a cargo-binding tail domain (see Figure 17-20).
- There are many classes of myosin, with three classes present in many eukaryotes: myosin I has a single head domain, myosin II has two heads and assembles into bipolar filaments, and myosin V has two heads and does not assemble in filaments (see Figure 17-23).
- Myosins convert ATP hydrolysis to mechanical work by amplifying a small conformational change in their head through their neck domain when the head is bound to F-actin (see Figure 17-24).
- Myosin heads take discrete steps along an actin filament, which can be small (5–15 nm) and nonprocessive in the case of myosin II or large (36 nm) and processive for myosin V.

17.6 Myosin-Powered Movements

We have already discussed how myosins have head and neck domains responsible for their motor properties. We now come to the tail regions, which define the cargoes that myosins move. The function of many of the newly discovered classes of myosins found in metazoans is not yet known. Below, we give just two examples where we have a good idea of what myosins do. Our first example is skeletal muscle, which is where myosin II was discovered. In muscle, many myosin II heads joined together in bipolar filaments, each with a short duty cycle, work together to bring about contraction. Similarly organized contractile machineries function in the contraction of smooth muscle and in stress fibers and the contractile ring during cytokinesis. We then turn to the myosin V class, which has a long duty cycle to allow these myosins to transport cargoes over relatively long distances without dissociating from actin filaments.

Myosin Thick Filaments and Actin Thin Filaments in Skeletal Muscle Slide Past One Another During Contraction

Muscle cells have evolved to carry out one highly specialized function—contraction. Muscle contractions must occur quickly and repetitively, and they must occur through long distances and with enough force to move large loads. A typical skeletal muscle cell is cylindrical, large (1–40 mm in length and 10–50 μm in width), and multinucleated (containing as many as 100 nuclei) (Figure 17-29a). Within the muscle cell are many **myofibrils** consisting of a regular repeating array of a specialized structure called a **sarcomere** (Figure 17-29b). A sarcomere, which is about 2 μm long in resting muscle, shortens by about 70 percent of its length during contraction. Electron microscopy and biochemical analysis have shown that each sarcomere contains two major types of filaments: *thick filaments*, composed of myosin II, and *thin filaments*, containing actin and associated proteins (Figure 17-29c).

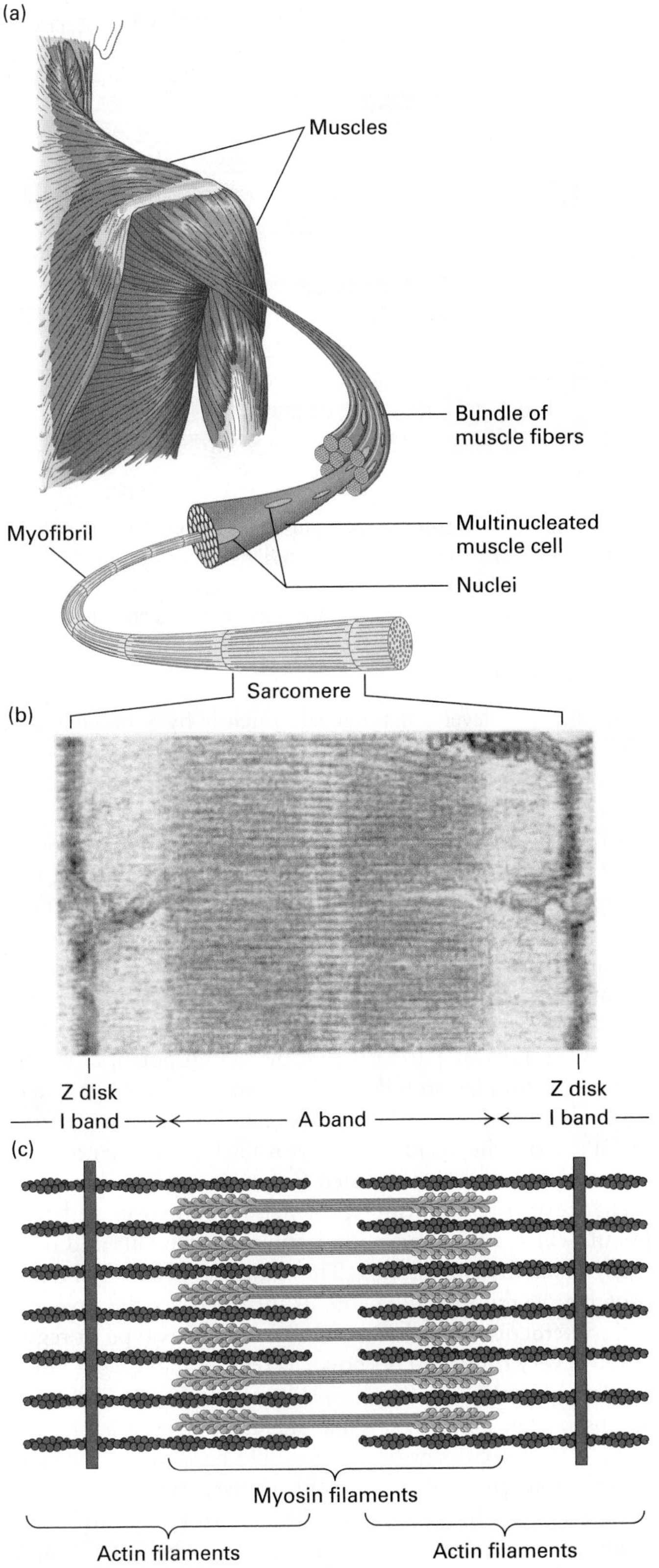

◀ **FIGURE 17-29 Structure of the skeletal muscle sarcomere.** (a) Skeletal muscles consist of muscle fibers made of bundles of multinucleated cells. Each cell contains a bundle of myofibrils, which consist of thousands of repeating contractile structures called sarcomeres. (b) Electron micrograph of mouse striated muscle in longitudinal section, showing one sarcomere. On either side of the Z disks are the lightly stained I bands, composed entirely of actin thin filaments. These thin filaments extend from both sides of the Z disk to interdigitate with the dark-stained myosin thick filaments in the A band. (c) Diagram of the arrangement of myosin and actin filaments in a sarcomere. [Part (b) courtesy of S. P. Dadoune.]

The thick filaments are composed of myosin II bipolar filaments, in which the heads on each half of the filament have opposite orientations (see Figure 17-20a). The thin actin filaments are assembled with their (+) ends embedded in a densely staining structure known as the *Z disk* (Figure 17-29b), so that the two sets of actin filaments in a sarcomere have opposite orientations (Figure 17-30). To understand how a muscle contracts, consider the interactions between one myosin head (among the hundreds in a thick filament) and a thin (actin) filament, as diagrammed in Figure 17-24. During these cyclical interactions, also called the *cross-bridge cycle*, the hydrolysis of ATP is coupled to the movement of a myosin head toward the Z disk, which corresponds to the (+) end of the thin filament. Because the thick filament is bipolar, the action of the myosin heads at opposite ends of the thick filament draws the thin filaments toward the center of the thick filament and therefore toward the center of the sarcomere (Figure 17-30). This movement shortens the sarcomere until the ends of the thick filaments abut the Z disk. Contraction of an intact muscle results from the activity of hundreds of myosin heads on a single

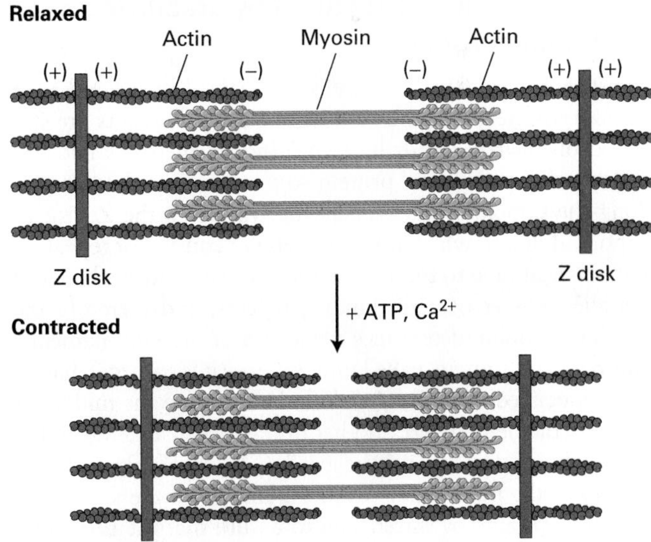

▲ **FIGURE 17-30 The sliding-filament model of contraction in striated muscle.** The arrangement of thick myosin and thin actin filaments in the relaxed state is shown in the top diagram. In the presence of ATP and Ca^{2+}, the myosin heads extending from the thick filaments walk toward the (+) ends of the thin filaments. Because the thin filaments are anchored at the Z disks (purple), movement of myosin pulls the actin filaments toward the center of the sarcomere, shortening its length in the contracted state, as shown in the bottom diagram.

thick filament, amplified by the hundreds of thick and thin filaments in a sarcomere and thousands of sarcomeres in a muscle fiber. We can now see why myosin II is both nonprocessive and needs to have a short duty cycle (see Figure 17-27): each head pulls a short distance on the actin filament and then lets go to allow other heads to pull, and so many heads working together allow the smooth contraction of the sarcomere.

The heart is an amazing contractile organ—it contracts without interruption about 3 million times a year, or a fifth of a billion times in a lifetime. The muscle cells of the heart contain contractile machinery very similar to that of skeletal muscle except that they are mono- and bi-nucleated cells. In each cell, the end sarcomeres insert into structures at the plasma membrane called intercalated disks, which link the cells into a contractile chain. Since heart muscle cells are only generated early in human life, they cannot be replaced in response to damage, such as occurs during a heart attack. Many different mutations in proteins of the heart contractile machinery give rise to *hypertrophic cardiomyopathies*—thickening of the heart wall muscle, which compromises its function. For example, many mutations have been documented in the cardiac myosin heavy chain that compromise its contractile function even in heterozygous individuals. In such individuals, the heart tries to compensate by hypertrophy, often resulting in fatal heart arrhythmia (irregular beating). In addition to myosin heavy-chain defects, defects that result in cardiomyopathies have been traced to mutations in other components of the contractile machinery, including actin, myosin light chains, tropomyosin and troponin, and structural components such as titin (discussed below). ▮

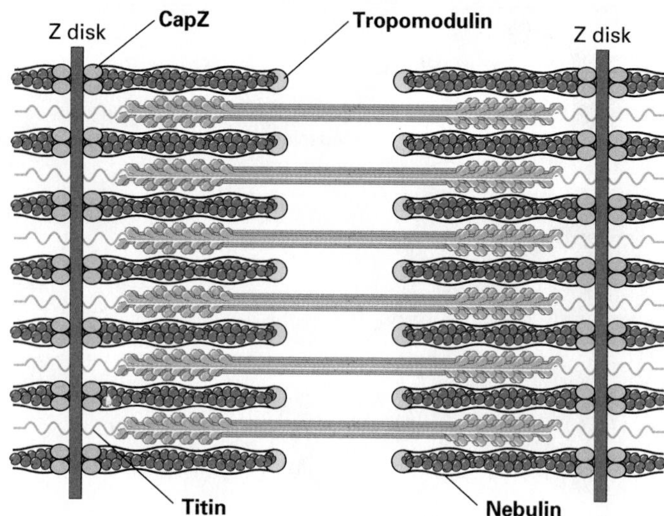

▲ **FIGURE 17-31 Accessory proteins found in skeletal muscle.** To stabilize the actin filaments, CapZ caps the (+) end of the thin filaments at the Z disk, whereas tropomodulin caps the (–) end. The giant protein titin extends through the thick filaments and attaches to the Z disk. Nebulin binds actin subunits and determines the length of the thin filament.

Skeletal Muscle Is Structured by Stabilizing and Scaffolding Proteins

The structure of the sarcomere is maintained by a number of accessory proteins (Figure 17-31). The actin filaments are stabilized on their (+) ends by *CapZ* and on their (−) ends by *tropomodulin.* A giant protein known as *nebulin* extends along the thin actin filament all the way from the Z disk to tropomodulin, to which it binds. Nebulin consists of repeating domains that bind to the actin in the filament, and it is believed that the number of actin binding repeats, and therefore the length of nebulin, determines the length of the thin filaments. Another giant protein, called *titin* (because it is so large), has its head associated with the Z disk and extends to the middle of the thick filament, where another titin molecule extends to the subsequent Z disk. Titin is believed to be an elastic molecule that holds the thick filaments in the middle of the sarcomere and also prevents overstretching to ensure that the thick filaments remain interdigitated between the thin filaments.

Contraction of Skeletal Muscle Is Regulated by Ca^{2+} and Actin-Binding Proteins

Like many cellular processes, skeletal muscle contraction is initiated by an increase in the cytosolic Ca^{2+} concentration. As described in Chapter 11, the Ca^{2+} concentration of the cytosol is normally kept low, below 0.1 μM. In skeletal muscle cells, a low cytosolic Ca^{2+} level is maintained primarily by a unique Ca^{2+} ATPase that continually pumps Ca^{2+} ions from the cytosol containing the myofibrils into the **sarcoplasmic reticulum (SR),** a specialized endoplasmic reticulum of the muscle cells (Figure 17-32). This activity establishes a reservoir of Ca^{2+} in the SR.

The arrival of a nerve impulse (or **action potential;** see Chapter 23) at a neuromuscular junction triggers an action potential in the muscle-cell plasma membrane (also known as the *sarcolemma*). The action potential travels down invaginations of the plasma membrane known as *transverse tubules,* which penetrate the cell to lie around each myofibril (Figure 17-32). The arrival of the action potential in the transverse tubules stimulates the opening of voltage-gated Ca^{2+} channels in the SR membrane, and the ensuing release of Ca^{2+} from the SR raises the cytosolic Ca^{2+} concentration in the myofibrils. This elevated Ca^{2+} concentration induces a change in two accessory proteins, tropomyosin and troponin, which are bound to the actin thin filaments and normally block myosin binding. The change in position of these proteins on the thin filaments in turn permits the myosin-actin interactions and hence contraction. This type of regulation is very rapid and is known as *thin filament regulation.*

Tropomyosin (TM) is a ropelike molecule, about 40 nm in length, that binds to seven actin subunits in an actin filament. TM molecules are strung together head to tail, forming a continuous chain along each side of the actin thin filament (Figure 17-33a, b). Associated with each tropomyosin is *troponin* (TN), a complex of three subunits, TN-T, TN-I, and TN-C. Troponin-C is the calcium-binding subunit of troponin. TN-C controls the position of TM on the surface of an actin filament through the TN-I and TN-T subunits.

Under the control of Ca^{2+} and TN, TM can occupy two positions on a thin filament—switching from a state of muscle relaxation to contraction. In the absence of Ca^{2+} (the relaxed state), TM blocks myosin's interaction with F-actin and the

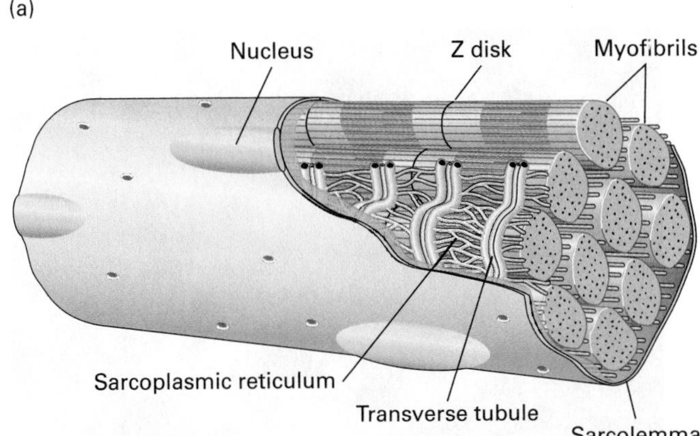

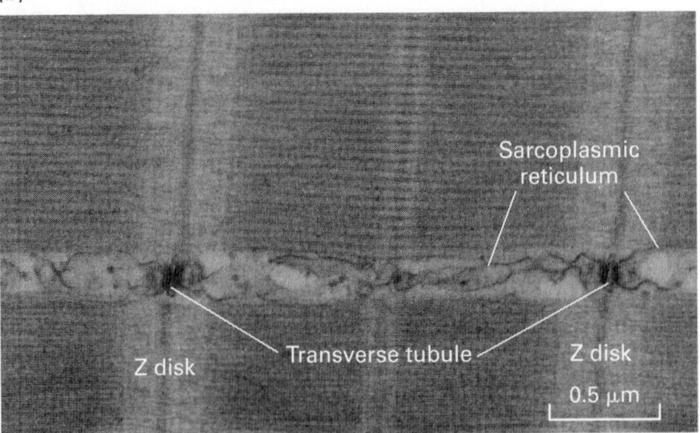

▲ **FIGURE 17-32 The sarcoplasmic reticulum regulates the level of free Ca^{2+} in myofibrils.** (a) When a nerve impulse stimulates a muscle cell, the action potential is transmitted down a transverse tubule (yellow), which is continuous with the plasma membrane (sarcolemma), leading to release of Ca^{2+} from the adjacent sarcoplasmic reticulum into the myofibrils. (b) Thin-section electron micrograph of skeletal muscle, showing the intimate relationship of the sarcoplasmic reticulum to the muscle fibers. [Part (b) from Porter KR, Franzini-Armstrong C. ASCB Image & Video Library. August 2006:FND-14. Available at: http://cellimages.ascb.org/u?/p4041coll1, 83:]

muscle is relaxed. Binding of Ca^{2+} ions to TN-C triggers movement of TM to a new site on the filament, thereby exposing the myosin-binding sites on actin (Figure 17-33b). Thus, at Ca^{2+} concentrations greater than 1 μM, the inhibition exerted by the TM-TN complex is relieved and contraction occurs. The Ca^{2+}-dependent cycling between relaxation and contraction states in skeletal muscle is summarized in Figure 17-33c.

Actin and Myosin II Form Contractile Bundles in Nonmuscle Cells

In skeletal muscle, actin thin filaments and myosin II thick filaments can assemble into contractile structures. Nonmuscle cells contain several types of related **contractile bundles** composed of actin and myosin II filaments, which are similar to skeletal muscle fibers but much less well organized. Moreover, they lack the troponin regulatory system and are instead regulated by myosin phosphorylation, as we will discuss later.

In epithelial cells, contractile bundles are most commonly found as an *adherens belt,* which encircles the inner surface of the cell at the level of the adherens junction (see Figure 17-4a), and are important in maintaining the integrity of the epithelium. Stress fibers, which are seen along the lower surfaces of cells cultured on artificial (glass or plastic) surfaces or in extracellular matrices, are a second type of contractile bundle (see Figure 17-4a, c) important in cell adhesion, especially on

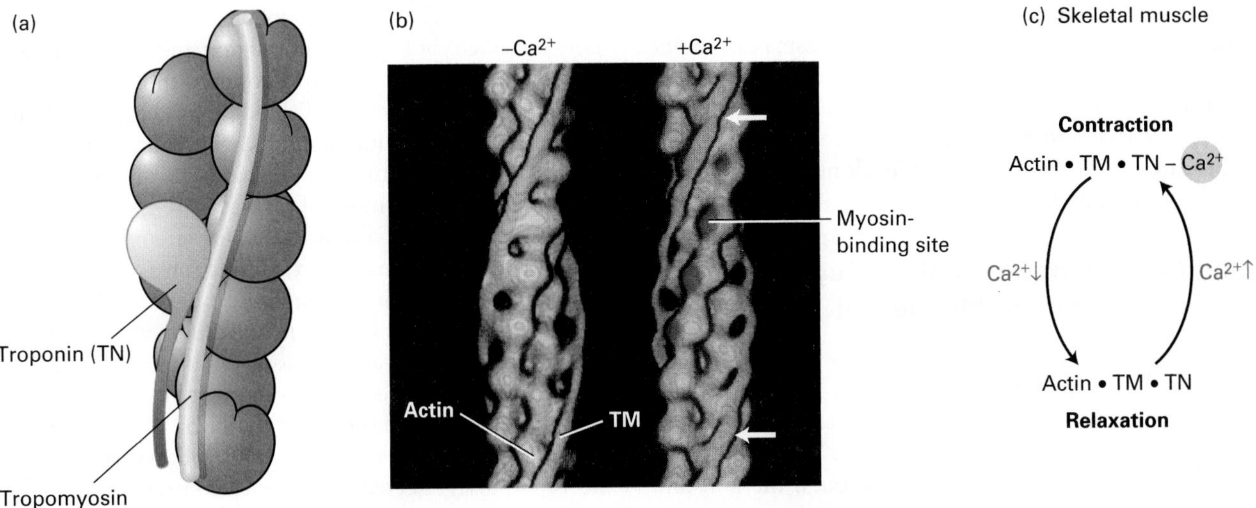

▲ **FIGURE 17-33 Ca^{2+}-dependent thin-filament regulation of skeletal muscle contraction.** (a) Model of the tropomyosin-troponin regulatory complex on a thin filament. Troponin is a clublike protein complex that is bound to the long α-helical tropomyosin molecule. (b) Three-dimensional electron-microscopic reconstructions of the tropomyosin helix (yellow) on a thin filament from scallop muscle. Tropomyosin in the relaxed state (*left*) shifts to a new position (arrow) in the state inducing contraction (*right*) when the Ca^{2+} concentration increases. This movement exposes myosin-binding sites (red) on actin. (Troponin is not shown in this representation but remains bound to tropomyosin in both states.) (c) Summary of the regulation of skeletal muscle contraction by Ca^{2+} binding to troponin. [Part (b) adapted from W. Lehman, R. Craig, and P. Vibert, 1993, *Nature* **123**:313; courtesy of P. Vibert.]

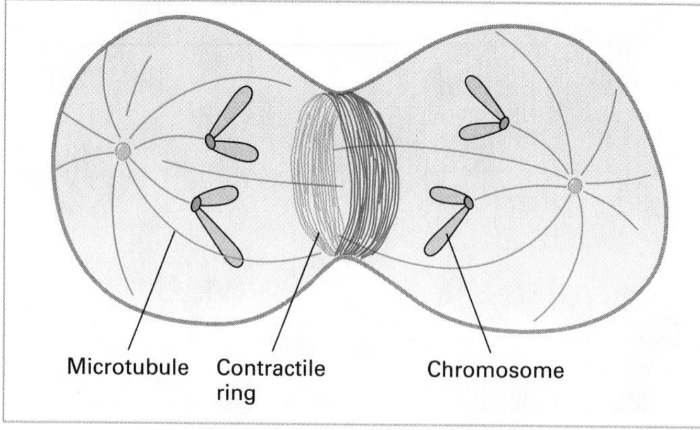

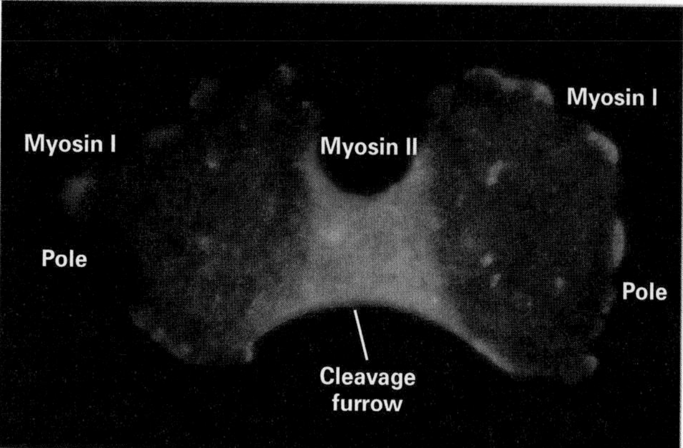

▲ EXPERIMENTAL FIGURE 17-34 **Fluorescent antibodies reveal the localization of myosin I and myosin II during cytokinesis.** (a) Diagram of a cell going through cytokinesis, showing the mitotic spindle (microtubules green, chromosomes blue) and the contractile ring with actin filaments (red). (b) Fluorescence micrograph of a *Dictyostelium* ameba during cytokinesis reveals that myosin II (red) is concentrated in the cleavage furrow, whereas myosin I (green) is localized at the poles of the cell. The cell was stained with antibodies specific for myosin I and myosin II, with each antibody preparation linked to a different fluorescent dye. [Courtesy of Y. Fukui.]

deformable substrates. The ends of stress fibers terminate at integrin-containing focal adhesions, special structures that attach a cell to the underlying substratum (see Figure 17-39). Circumferential belts and stress fibers contain several proteins found in the contractile apparatus of smooth muscle and exhibit some organizational features resembling those of muscle sarcomeres. A third type of contractile bundle, referred to as a contractile ring, is a transient structure that assembles at the equator of a dividing cell, encircling the cell midway between the poles of the mitotic spindle (Figure 17-34a). As division of the cytoplasm (cytokinesis) proceeds, the diameter of the contractile ring decreases, and so the cell is pinched into two parts by a deepening cleavage furrow. Dividing cells stained with antibodies against myosin I and myosin II show that myosin II is localized to the contractile ring, whereas myosin I is at the distal regions, where it links the actin cortex to the plasma membrane (Figure 17-34b). This localization indicates that myosin II, but not myosin I, takes part in cytokinesis. Cells in which the myosin II gene has been deleted are unable to undergo cytokinesis. Instead, these cells form a multinucleated syncytium because cytokinesis, but not nuclear division, is inhibited.

Myosin-Dependent Mechanisms Regulate Contraction in Smooth Muscle and Nonmuscle Cells

Smooth muscle is a specialized tissue composed of contractile cells found in many internal organs. For example, smooth muscle surrounds blood vessels to regulate blood pressure, surrounds the intestine to move food through the gut, and restricts airway passages in the lung. Smooth muscle cells contains large, loosely aligned contractile bundles that resemble the contractile bundles in epithelial cells. The contractile apparatus of smooth muscle and its regulation constitute a valuable model for understanding how myosin activity is regulated in a nonmuscle cell. As we have just seen, skeletal muscle contraction is regulated by the tropomyosin-troponin complex bound to the actin thin filament switching between the contraction-inducing state in the presence of Ca^{2+} and the relaxed state in its absence. In contrast, smooth muscle contraction is regulated by the cycling of myosin II between on and off states. Myosin II cycling, and thus contraction of smooth muscle and nonmuscle cells, is regulated in response to many extracellular signaling molecules.

Contraction of vertebrate smooth muscle is regulated primarily by a pathway in which the *myosin regulatory light chain (LC)* associated with the myosin II neck domain (see Figure 17-20b) undergoes phosphorylation and dephosphorylation. When the regulatory light chain is not phosphorylated, the myosin II ATPase cycle is inactive. The smooth muscle contracts when the regulatory LC is phosphorylated by the enzyme *myosin LC kinase* (Figure 17-35). Because this enzyme is activated by Ca^{2+}, the cytosolic Ca^{2+} level indirectly regulates the extent of LC phosphorylation and hence contraction. The Ca^{2+}-dependent regulation of myosin LC kinase activity is mediated through the Ca^{2+}-binding protein calmodulin (see Figure 3-31). Calcium first binds to calmodulin, and the Ca^{2+}/calmodulin complex then binds to myosin LC kinase and activates it. This mode of regulation relies on the diffusion of Ca^{2+} over greater distances than in sarcomeres and on the action of protein kinases, so contraction is much slower in smooth muscle than in skeletal muscle. Because this regulation involves myosin, it is known as *thick filament regulation.*

The role of activated myosin LC kinase can be demonstrated by microinjecting a kinase inhibitor into smooth muscle cells. Even though the inhibitor does not block the rise in the cytosolic Ca^{2+} level that follows the arrival of a nerve impulse, injected cells cannot contract.

Unlike skeletal muscle, which is stimulated to contract solely by nerve impulses, smooth muscle cells and nonmuscle cells are regulated by many types of external signals in addition to nervous stimuli. For example, norepinephrine, angiotensin, endothelin, histamine, and other signaling molecules can modulate or induce the contraction of smooth muscle or elicit changes in the shape and adhesion of nonmuscle cells by triggering various signal-transduction

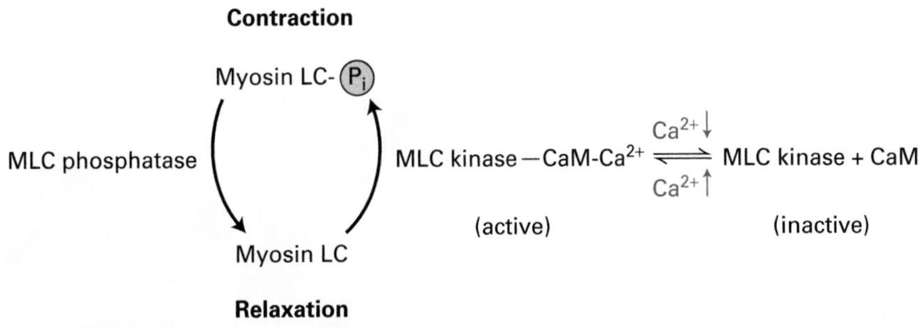

Contraction

Myosin LC-P_i

MLC phosphatase

MLC kinase—CaM-Ca^{2+} $\underset{Ca^{2+}\uparrow}{\overset{Ca^{2+}\downarrow}{\rightleftharpoons}}$ MLC kinase + CaM

(active) (inactive)

Myosin LC

Relaxation

▲ **FIGURE 17-35 Myosin phosphorylation mechanism for regulating smooth muscle contraction.** In vertebrate smooth muscle, phosphorylation of the myosin regulatory light chain (LC) by Ca^{2+}-dependent myosin LC kinase activates contraction. At Ca^{2+} concentrations $<10^{-6}$ M, the myosin LC kinase is inactive, and a myosin LC phosphatase, which is not dependent on Ca^{2+} for activity, dephosphorylates the myosin LC, causing muscle relaxation.

pathways. Some of these pathways lead to an increase in the cytosolic Ca^{2+} level; as previously described, this increase can stimulate myosin activity by activating myosin LC kinase (see Figure 17-35). As we will discuss below, other pathways activate *Rho kinase,* which is also able to activate myosin activity by phosphorylating the regulatory light chain, although in a Ca^{2+}-independent manner.

Myosin-V-Bound Vesicles Are Carried Along Actin Filaments

In contrast to the contractile functions of myosin II filaments, the myosin V family of proteins are the most processive myosin motors known and transport cargo down actin filaments. In the next chapter we discuss how they can work together with microtubule motors to bring about transport of organelles. Although not a lot is known about their functions in mammalian cells, myosin V motors are not unimportant: defects in a specific myosin V protein can cause severe diseases, such as seizures (see Figure 17-22).

Much more is known about myosin V motors in more experimentally accessible and simpler systems such as the budding yeast. This well-studied organism grows by budding, which requires its secretory machinery to target newly synthesized material to the growing bud (Figure 17-36a). Myosin V transports secretory vesicles along actin filaments at 3 μm/sec into the bud. However, this is not the only function of myosin Vs in yeast. At a later stage of the cell cycle,

Animation: Movement of Multiple Cargoes by Myosin V in Yeast

MEDIA CONNECTIONS

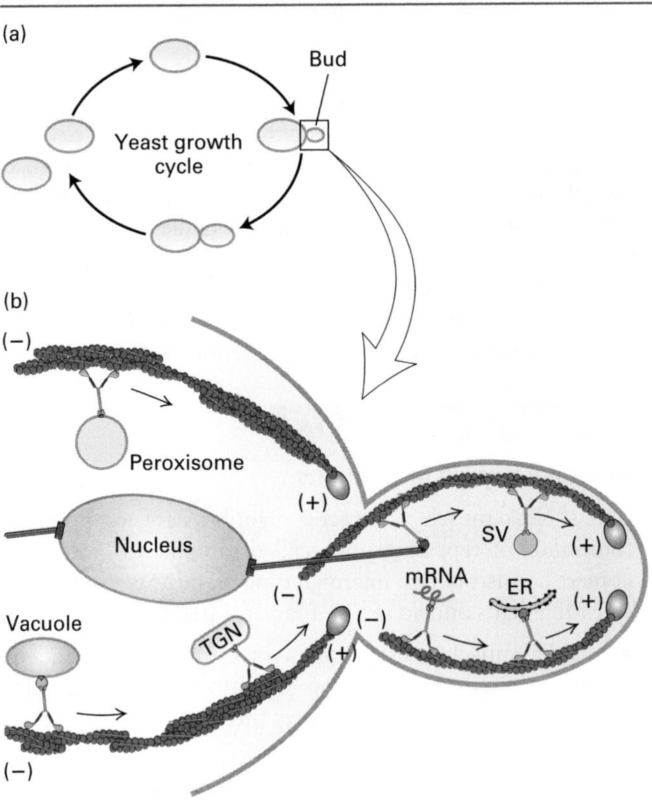

◀ **FIGURE 17-36 Myosin Vs carry many different cargoes in budding yeast.** (a) The yeast *Saccharomyces cerevisiae* (used in making bread, beer, and wine) grows by budding. Secretory vesicles are transported into the bud, which swells to about the size of the mother cell. The cells then go through cytokinesis and each divide again. (b) Diagram of a medium-sized bud showing how myosin Vs transport secretory vesicles (SV) down actin cables nucleated by formins (purple) located at the bud tip and bud neck. Myosins Vs are also used to segregate organelles, such as the vacuole (the yeast equivalent of a lysosome), peroxisomes, endoplasmic reticulum (ER), *trans*-Golgi network (TGN), and even selected mRNAs into the bud. Myosin V also binds the end of cytoplasmic microtubules to orient the nucleus in preparation for mitosis. [Adapted from D. Pruyne et al., 2004, *Ann. Rev. Cell Biol.* **20:**559.]

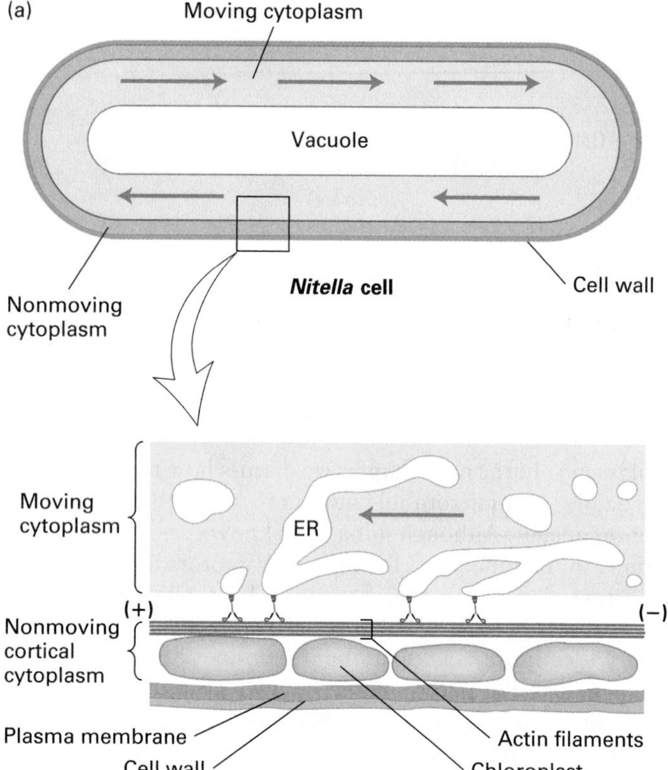

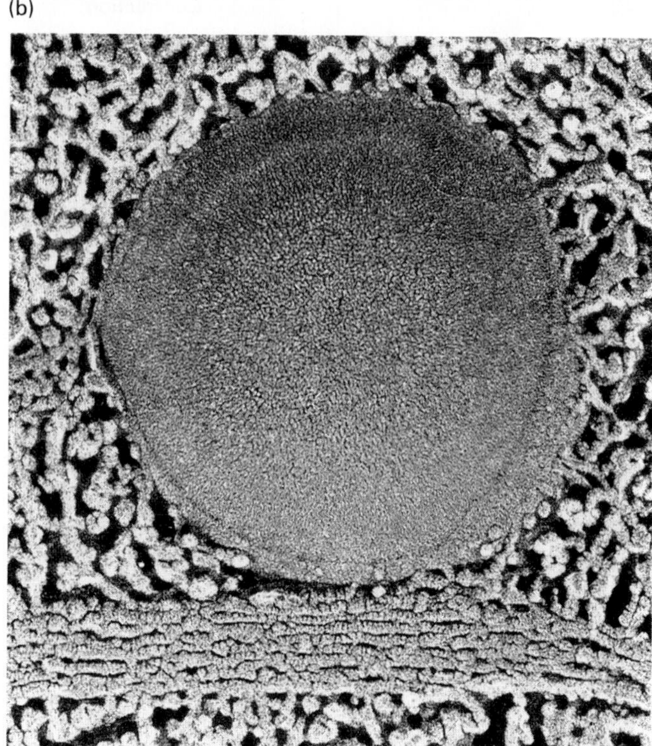

▲ **FIGURE 17-37 Cytoplasmic streaming in cylindrical giant algae.** (a) The center of a *Nitella* cell is filled with a single large water-filled vacuole, which is surrounded by a layer of moving cytoplasm (blue arrows). A nonmoving layer of cortical cytoplasm filled with chloroplasts lies just under the plasma membrane (enlarged in bottom figure). On the inner side of this layer are bundles of stationary actin filaments (red), all oriented with the same polarity. A motor protein (blue), a plant myosin V, carries parts of the endoplasmic reticulum (ER) along the actin filaments. The movement of the ER network propels the entire viscous cytoplasm, including organelles that are enmeshed in the ER network. (b) Electron micrograph of the cortical cytoplasm showing a large vesicle connected to an underlying bundle of actin filaments. This vesicle, which is part of the ER network, contacts the stationary actin filaments and moves along them by a myosin motor protein. [Part (b) from B. Kachar.]

all the organelles have to be distributed between the mother and daughter cells. Remarkably, myosin Vs in yeast provide the transport system for segregation of many organelles, including peroxisomes, lysosomes (also known as vacuoles), endoplasmic reticulum, and the *trans*-Golgi network and even transport the ends of microtubules and some specific messenger RNAs into the bud (Figure 17-36b). Whereas budding yeast uses myosin V and polarized actin filaments in the transport of many organelles, animal cells, which are much larger, employ microtubules and their motors to transport many of these organelles over relatively long distances. We discuss these transport mechanisms in the next chapter.

Perhaps the most dramatic use of myosin Vs is seen in the giant green algae, such as *Nitella* and *Chara*. In these large cells, which can be 2 cm in length, cytosol flows rapidly, at a rate approaching 4.5 mm/min, in an endless loop around the inner circumference of the cell (Figure 17-37). This *cytoplasmic streaming* is a principal mechanism for distributing cellular metabolites, especially in large cells such as plant cells and amebas.

Close inspection of objects caught in the flowing cytosol, such as the endoplasmic reticulum (ER) and other membrane-bounded vesicles, shows that the velocity of streaming increases from the cell center (zero velocity) to the cell periphery. This gradient in the rate of flow is most easily explained if the motor generating the flow lies at the membrane. In electron micrographs, bundles of actin filaments can be seen aligned along the length of the cell, lying across chloroplasts embedded at the membrane. Attached to the actin bundles are vesicles of the ER network. The bulk cytosol is propelled by myosin attached to parts of the ER adjacent to the actin filaments. The flow rate of the cytosol in *Nitella* is at least 15 times as fast as the movement produced by any other known myosin.

KEY CONCEPTS OF SECTION 17.6

Myosin-Powered Movements

- In skeletal muscle, contractile myofibrils are composed of thousands of repeating units called sarcomeres. Each sarcomere consists of two interdigitating filament types: myosin thick filaments and actin thin filaments (see Figure 17-29).
- Skeletal muscle contraction involves the ATP-dependent sliding of myosin thick filaments along actin thin filaments to shorten the sarcomere and hence the myofibril (see Figure 17-30).
- The ends of the actin thin filaments in skeletal muscle are stabilized by CapZ at the (+) end and by tropomodulin at

Video: Mechanisms of Fish Keratinocyte Migration
Video: Actin Filaments in the Lamellipodium of a Fish Keratinocyte

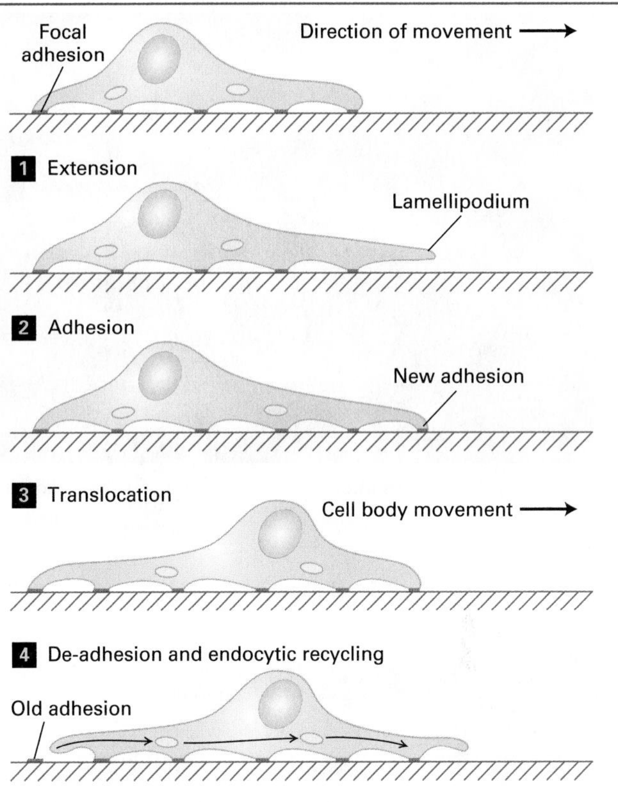

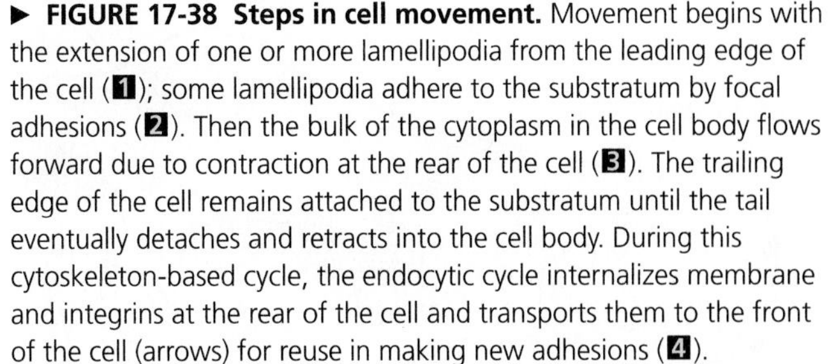

▶ **FIGURE 17-38 Steps in cell movement.** Movement begins with the extension of one or more lamellipodia from the leading edge of the cell (**1**); some lamellipodia adhere to the substratum by focal adhesions (**2**). Then the bulk of the cytoplasm in the cell body flows forward due to contraction at the rear of the cell (**3**). The trailing edge of the cell remains attached to the substratum until the tail eventually detaches and retracts into the cell body. During this cytoskeleton-based cycle, the endocytic cycle internalizes membrane and integrins at the rear of the cell and transports them to the front of the cell (arrows) for reuse in making new adhesions (**4**).

the (−) end. Two large proteins, nebulin associated with the thin filaments and titin with the thick filaments, also contribute to skeletal muscle organization.

- Skeletal muscle contraction is subject to thin filament regulation. At low levels of free Ca^{2+}, the muscle is relaxed and tropomyosin blocks the interaction of myosin and F-actin. At elevated levels of free Ca^{2+}, the troponin complex associated with tropomyosin binds Ca^{2+} and moves the tropomyosin to uncover the myosin-binding sites on actin, allowing contraction (see Figure 17-33).
- Smooth and nonmuscle cells have contractile bundles of actin and myosin filaments, with a similar organization to skeletal muscle but less well ordered.
- Contractile bundles are subject to thick filament regulation. A myosin light chain is phosphorylated by myosin light-chain kinase, which activates myosin and hence induces contraction. The myosin light-chain kinase is activated by binding Ca^{2+}-calmodulin when the free Ca^{2+} concentration rises (see Figure 17-35).
- Myosin V transports cargo by walking processively along actin filaments.

17.7 Cell Migration: Signaling and Chemotaxis

We have now examined the different mechanisms used by cells to create movement—from the assembly of actin filaments and the formation of actin-filament bundles and networks to the contraction of bundles of actin and myosin and the transport of organelles by myosin molecules along actin filaments. Some of these mechanisms constitute the major processes whereby cells generate the forces needed to migrate. *Cell migration* results from the coordination of motions generated in different parts of a cell, integrated with a directed endocytic cycle.

The study of cell migration is important to many fields of biology and medicine. For example, an essential feature of animal development is the migration of specific cells along predetermined paths. Epithelial cells in an adult animal migrate to heal a wound, and white blood cells migrate to sites of infection. Less obvious are the continual slow migration of intestinal epithelial cells along the villi in the intestine and the slow but constant migration of endothelial cells that line the blood vessels. The inappropriate migration of cancer cells away from their normal tissue results in metastasis.

Cell migration is initiated by the formation of a large, broad membrane protrusion at the leading edge of a cell. Video microscopy reveals that a major feature of this movement is the polymerization of actin at the membrane. Actin filaments at the leading edge are rapidly cross-linked into bundles and networks in a protruding region, called a *lamellipodium* in vertebrate cells. In some cases, slender, fingerlike membrane projections, called *filopodia,* also extend from the leading edge. These structures form stable contacts with the underlying surface and prevent the leading edge membrane from retracting. In this section, we take a closer look at how cells employ the various force-generating processes to move across a surface. We also consider the role of signaling pathways in coordinating and integrating the actions of the cytoskeleton, a major focus of current research.

Cell Migration Coordinates Force Generation with Cell Adhesion and Membrane Recycling

A moving fibroblast (connective tissue cell) displays a characteristic sequence of events—initial extension of a membrane protrusion, attachment to the substratum, forward flow of cytosol, and retraction of the rear of the cell (Figure 17-38). Although we describe each of these events separately, all of them are occurring simultaneously in a coordinated way.

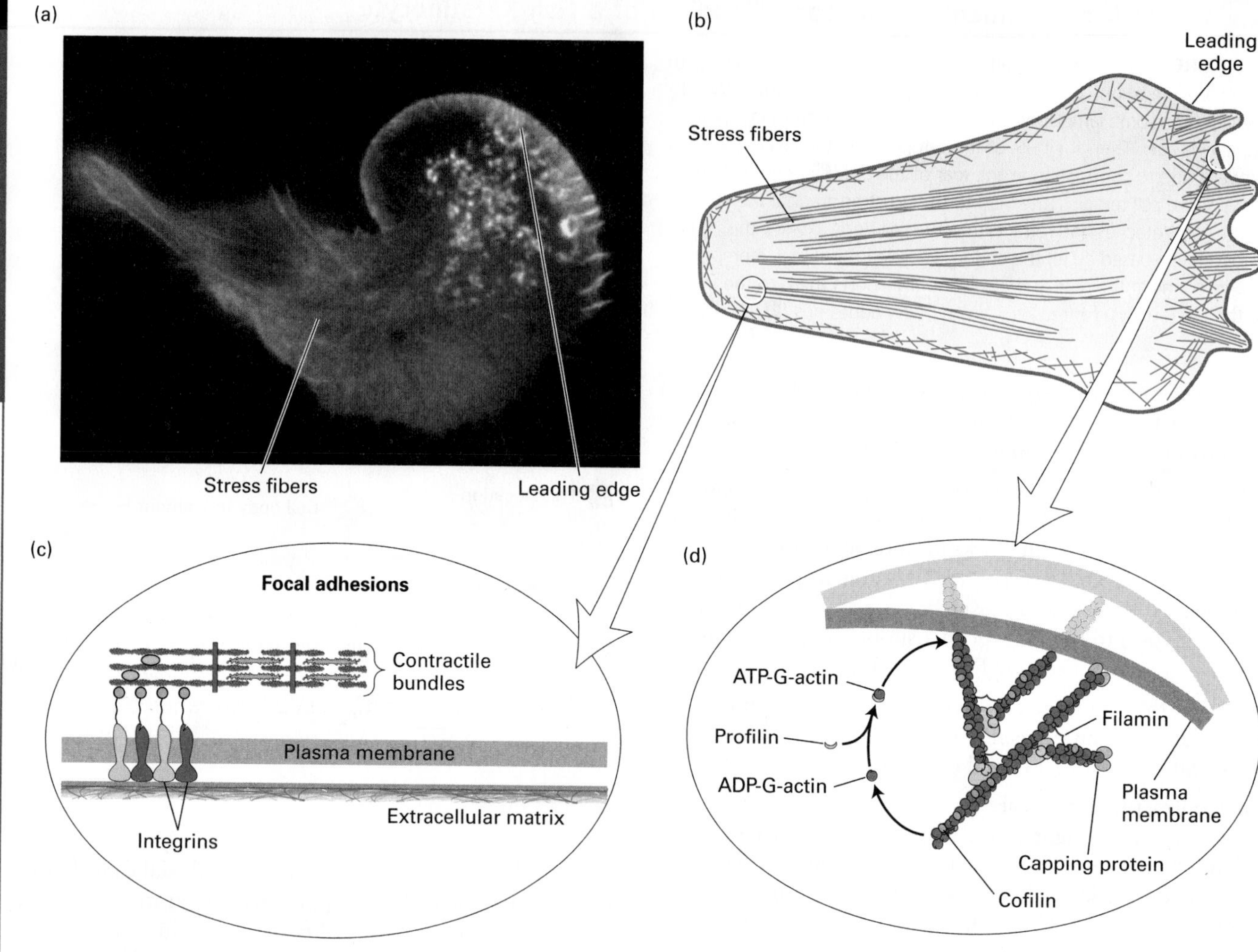

▲ **FIGURE 17-39 Actin-based structures involved in cell locomotion.** (a) Localization of actin in a fibroblast expressing GFP-actin. (b) Diagram of the classes of microfilaments involved in cell migration. The meshwork of actin filaments in the leading edge advances the cell forward. Contractile fibers in the cell cortex squeeze the cell body forward, and stress fibers terminating in focal adhesions also pull the bulk of the cell body up as the rear adhesions are released. (c) The structure of focal adhesions involves the attachment of the ends of stress fibers through integrins to the underlying extracellular matrix. Focal adhesions also contain many signaling molecules important for cell locomotion. (d) The dynamic actin meshwork in the leading edge is nucleated by the Arp2/3 complex and employs the same set of factors that control assembly and disassembly of actin filaments in the *Listeria* tail (see Figure 17-17).

Membrane Extension The network of actin filaments at the leading edge is a type of cellular engine that pushes the membrane forward by an actin-polymerization-based mechanism in a manner very similar to the propulsion of *Listeria* by actin polymerization (Figure 17-39d; for *Listeria* see Figure 17-17c). Thus at the membrane of the leading edge, actin is nucleated by the activated Arp2/3 complex and filaments are elongated by assembly onto (+) ends adjacent to the plasma membrane. As the actin meshwork is fixed with respect to the substratum, the front membrane is pushed out as the filaments elongate. Similarly, the *Listeria* "ride" on the polymerizing actin tail, which is also fixed within the cytoplasm, by nucleating actin assembly. Actin turnover, and thus treadmilling, is mediated, as it is in the comet tails of *Listeria,* by the action of profilin and cofilin (Figure 17-39d). New membrane at the leading edge is probably supplied by exocytosis of membrane internalized by endocytosis at the rear, as diagrammed in Figure 17-38, step **4**.

Cell-Substrate Adhesions When the membrane has been extended and the cytoskeleton has been assembled, the plasma membrane becomes firmly attached to the substratum. Time-lapse microscopy shows that actin bundles in the leading edge become anchored to structures known as focal adhesions (Figure 17-39c). The attachment serves two purposes: it prevents the leading lamella from retracting, and it attaches the cell to the substratum, allowing the cell to move forward. Given the importance of focal adhesions and their regulation during cell locomotion, it is not surprising that they have been found to be very rich in molecules involved in signal-transduction pathways. Focal adhesions are discussed in more detail in Chapter 19, when we discuss cell-matrix interactions.

The membrane proteins involved in attachments are called **integrins,** the cell-adhesion molecules that mediate most cell-matrix interactions. These proteins have an external domain that binds to specific components of the extracellular matrix, such as fibronectin and collagen, and a cytoplasmic domain that links them to the actin cytoskeleton (Figure 17-39 and Chapter 19). The cell makes attachments at the front, and as the cell migrates forward, the adhesions eventually assume positions toward the back. When they reach the back of the cell, the integrins are internalized by endocytosis and transported, using both the microfilament and microtubule cytoskeletons (Chapter 18) to the front of the cell to make new adhesions (see Figure 17-38, step **4**). This cycle of adhesion molecules in a migrating cell resembles the way a tank uses its treads to move forward.

Cell-Body Translocation After the forward attachments have been made, the bulk contents of the cell body are translocated forward (see Figure 17-38). It is believed that the nucleus and the other organelles embedded in the cytoskeleton are moved forward by myosin-dependent cortical contraction in the rear part of the cell, like squeezing the lower half of a tube of toothpaste. Consistent with this model, myosin II is localized to the rear cell cortex.

Breaking Cell Attachments Finally, in the last step of movement (de-adhesion), the focal adhesions at the rear of the cell are broken, the integrins recycled, and the freed tail brought forward. In the light microscope, the tail is seen to "snap" loose from its connections—perhaps by the contraction of stress fibers in the tail or by elastic tension—and it sometimes leaves a little bit of its membrane behind, still firmly attached to the substratum.

The ability of a cell to move corresponds to a balance between the mechanical forces generated by the cytoskeleton and the resisting forces generated by cell adhesions. Cells cannot move if they are either too strongly attached or not attached to a surface. This relation can be demonstrated by measuring the rate of movement in cells that express varying levels of integrins. Such measurements show that the fastest migration occurs at an intermediate level of adhesion, with the rate of movement falling off at high and low levels of adhesion. Cell locomotion thus results from traction forces exerted by the cell on the underlying substratum.

The Small GTP-Binding Proteins Cdc42, Rac, and Rho Control Actin Organization

A striking feature of a moving cell is its polarity: a cell has a front and a back. When a cell makes a turn, a new leading edge forms in the new direction. If these extensions formed in all directions at once, the cell would be unable to pick a new direction of movement. To sustain movement in a particular direction, a cell requires signals to coordinate events at the front of the cell with events at the back and, indeed, signals to tell the cell where its front is. Understanding how this coordination occurs emerged from studies with growth factors.

Growth factors, such as epidermal growth factor (EGF) and platelet-derived growth factor (PDGF), bind to specific cell-surface receptors (Chapter 16) and stimulate cells to move and then to divide. For example, in a wound, blood platelets become activated by being exposed to collagen in the extracellular matrix at the wound edge, which helps the blood to clot. Activated platelets also secrete PDGF to attract fibroblasts and epithelial cells to enter the wound and repair it. It is possible to watch part of this process in vitro. If you grow cells in a culture dish and, after starving them of growth factors, you add some fresh growth factor, within a minute or two the cells respond by forming membrane ruffles. Membrane ruffles are very similar to the lamellipodia of migrating cells: they are a result of activation of the machinery that controls actin assembly. Scientists knew that growth factors bind to very specific receptors on the cell surface and induce a signal-transduction pathway on the inner surface of the plasma membrane (Chapter 15), but how that linked up to the actin machinery was mysterious. The scientists then found that the signal-transduction pathway activates *Rac,* a member of the small **GTPase superfamily** of Ras-related proteins (Chapter 15). Rac is one member of a family of proteins that regulate microfilament organizations; two others are *Cdc42* and *Rho.* Unfortunately, due to the history of their discovery, this family of proteins also has been collectively named "Rho proteins," of which Cdc42, Rac, and Rho are members. To understand how these proteins work, we have to first recall the way small GTP-binding proteins function.

Like all small GTPases of the Ras superfamily, Cdc42, Rac, and Rho act as molecular switches, inactive in the GDP-bound state and active in the GTP-bound state (Figure 17-40). In their GDP-bound state, they exist free in the cytoplasm in an inactive

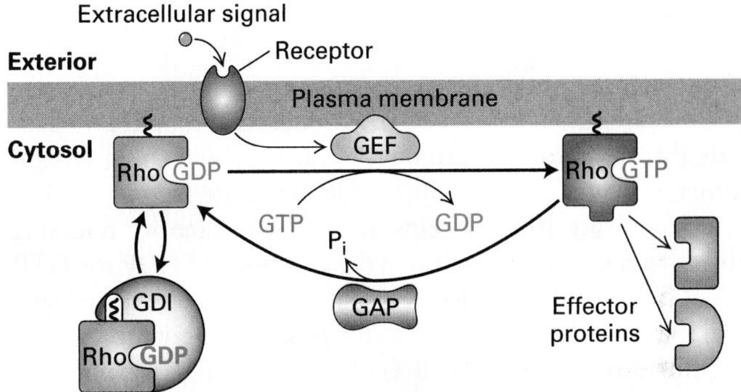

▲ **FIGURE 17-40 The Rho family of small GTPases are molecular switches regulated by accessory proteins.** Rho proteins exist in the Rho-GDP bound form complexed with a protein known as GDI (guanine nucleotide displacement inhibitor), which retains them in an inactive state in the cytosol. Membrane-bound signaling pathways bring Rho proteins to the membrane and through the action of a GEF (guanine nucleotide exchange factor) exchange the GDP for GTP, thus activating them. Membrane-bound activated Rho-GTP can then bind effector proteins that cause changes in the actin cytoskeleton. The Rho protein remains in the active Rho-GTP state until acted on by a GAP (GTPase-activating protein), which returns it to the cytoplasm. [Adapted from S. Etienne-Manneville and A. Hall, 2002, *Nature* **420**:629.]

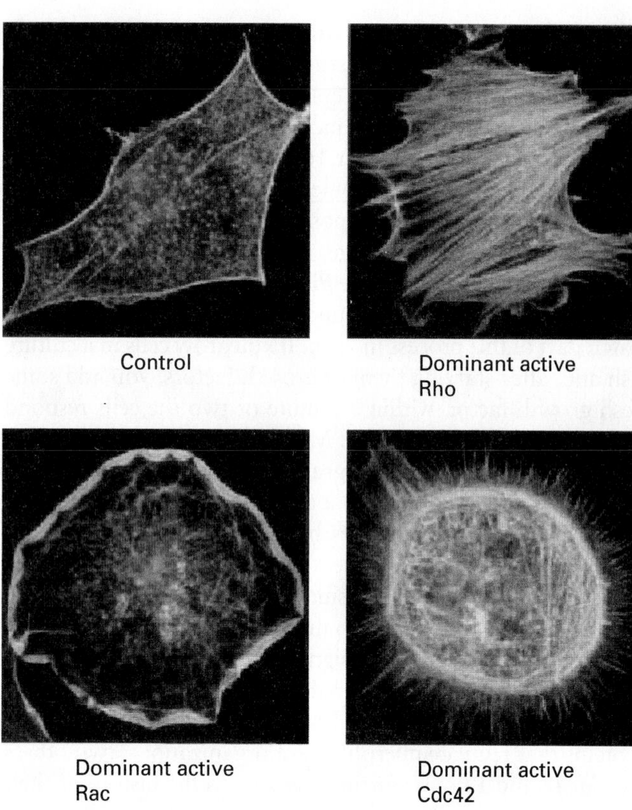

▲ EXPERIMENTAL FIGURE 17-41 **Dominant-active Rac, Rho, and Cdc42 induce different actin-containing structures.** To look at the effects of constitutively active Rac, Rho, and Cdc42, growth-factor-starved fibroblasts were microinjected with plasmids to express dominant-active versions of the three proteins. The cells were then treated with fluorescent phalloidin, which stains filamentous actin. Dominant-active Rac induces the formation of peripheral membrane ruffles, whereas dominant-active Rho induces abundant contractile stress fibers and dominant active Cdc42 induces filopodia. [From A. Hall, 1998, *Science* **279**:509–514.]

form bound to a protein known as guanine nucleotide displacement inhibitor (GDI). Growth factors can bind and activate their receptors to turn on specific membrane-bound regulatory proteins, guanine nucleotide exchange factors (GEFs), which activate Rho proteins at the membrane by releasing them from GDI and catalyzing the exchange of GDP for GTP. The GTP-bound active Rho protein associates with the plasma membrane, where it binds *effector proteins* to initiate the biological response. The small GTPase remains active until the GTP is hydrolyzed to GDP, which is stimulated by specific GTPase-activating proteins (GAPs). An important approach to unraveling the functions of Rho proteins has been to introduce into cells mutant proteins that are either locked in the active—Rho-GTP—state or in the inactive—Rho-GDP—state. A mutant small GTPase that is locked in the active state is said to be a *dominant-active* protein. Such a dominant-active protein binds the effector molecules constitutively, and one can then assess the biological outcomes. Alternatively, one can introduce a different mutant that is *dominant negative,* which permanently binds GDP and cannot be activated by a GEF. Thus introduction of a dominant-negative protein interferes with the signal-transduction pathway, so one can now assess what processes are blocked.

Cdc42, Rac, and Rho were implicated in regulation of microfilament organizations because introduction of dominant-active mutants had dramatic effects on the actin cytoskeleton, even in the absence of growth factors. It was discovered that dominant-active Cdc42 resulted in the appearance of filopodia, dominant-active Rac resulted in the appearance of membrane ruffles, and dominant-active Rho resulted in the formation of stress fibers that then contracted (Figure 17-41). How can one tell if dominant-active Rac and growth factor stimulation, both of which stimulate membrane ruffle formation, operate in the same signal-transduction pathway? If growth factor stimulation leads to Rac activation, introduction of a dominant-negative Rac protein into a cell should block the ability of a growth factor to induce membrane ruffling. This is precisely what is found. Using this and many other biochemical strategies, scientists have identified signaling pathways involving Cdc42, Rac, and Rho (Figure 17-42).

Some of the pathways that these proteins regulate contain proteins we are familiar with. Thus activation of Cdc42 stimulates actin assembly by Arp2/3 through activation of WASp, resulting in the formation of filopodia. Activation of Rac also induces Arp2/3 but through the WAVE complex, leading to the assembly of branched actin filaments in the leading edge (Figure 17-42). Activation of Rho has at least two effects: it can activate assembly of unbranched F-actin through a formin pathway or induce activation of nonmuscle myosin II by catalyzing, through a Rho-activated protein kinase, the phosphorylation of the myosin light chain and the phosphorylation and inhibition of the myosin light-chain phosphatase. Both actions of Rho kinase lead to a higher level of myosin light-chain phosphorylation and therefore higher myosin activity and contraction. The three Rho proteins, Cdc42, Rac, and Rho, are also linked by activation and inhibition pathways, as shown in Figure 17-42.

Cell Migration Involves the Coordinate Regulation of Cdc42, Rac, and Rho

How do each of these small GTP-binding proteins contribute to the regulation of cell migration? To answer this question, researchers developed an in vitro wound-healing assay (Figure 17-43a). Cells in culture are grown in a petri dish with growth factors, allowing them to grow until they are confluent and form a tight monolayer, at which point they stop dividing. The cell monolayer is then scratched with a needle to remove a swath of cells to generate a "wound" containing a free edge of cells. The cells on the edge sense the loss of their neighbors and, in response to components of the extracellular matrix now exposed on the dish surface, move to fill up the empty wound area. To do this, they orient themselves toward the free area, first putting out a lamellipodium and then moving in that direction. In this way one can study the induction of directed cell migration in vitro.

Using this system, researchers have introduced dominant-negative Rac into cells on the wound edge and asked how it

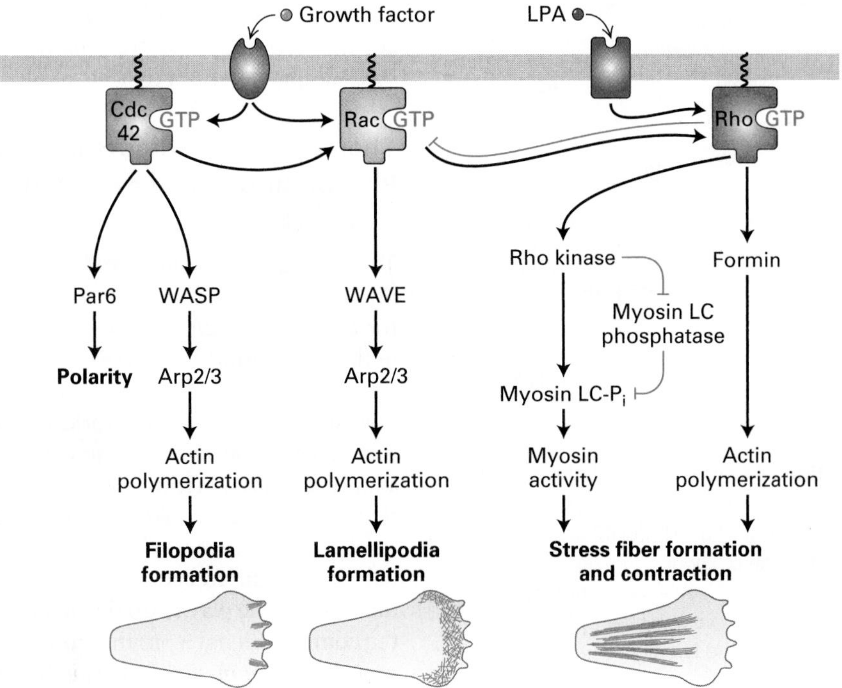

▲ **FIGURE 17-42 Summary of signal-induced changes in the actin cytoskeleton.** Specific signals are detected by cell-surface receptors. Detection leads to the activation of the small GTP-binding proteins, which then interact with effectors to bring about cytoskeletal changes as indicated.

affects the ability of the cells to migrate and fill the wound. Since Rac is needed for activation of the Arp2/3 complex to form the lamellipodium, it is not surprising that the cells failed to form this structure and did not migrate, and so the wound did not close (Figure 17-43c). A very interesting result is obtained when dominant-negative Cdc42 is introduced into the cells at the wound edge: they can form a leading edge but do not orient in the correct direction—in fact, they try to migrate in random directions. This suggests that Cdc42 is critical for regulating the overall polarity of the cell. Studies from yeast (where Cdc42 was first described), wounded-cell monolayers, epithelial cells, and neurons reveal that Cdc42 is a master regulator of polarity in many different systems. Part of this regulation in animals involves its binding to its effector, Par-6, a polarity protein that functions in nematodes (where it was first discovered), neurons, and epithelial cells.

Studies such as these suggest a general model of how cell migration is controlled (Figure 17-44). Signals from the environment are transmitted to Cdc42, which orients the cell. The oriented cell has high Rac activity at the front to induce the formation of the leading edge; Rho activity is high in the rear to assemble contractile structures and activate the myosin-II-based contractile machinery. It is important to notice that different regions of the cell can have different levels of active Cdc42, Rac, or Rho, so these regulators are controlled locally within the cell. Part of this spatial regulation occurs because some small G proteins can work antagonistically. For example, active Rho can stimulate pathways that

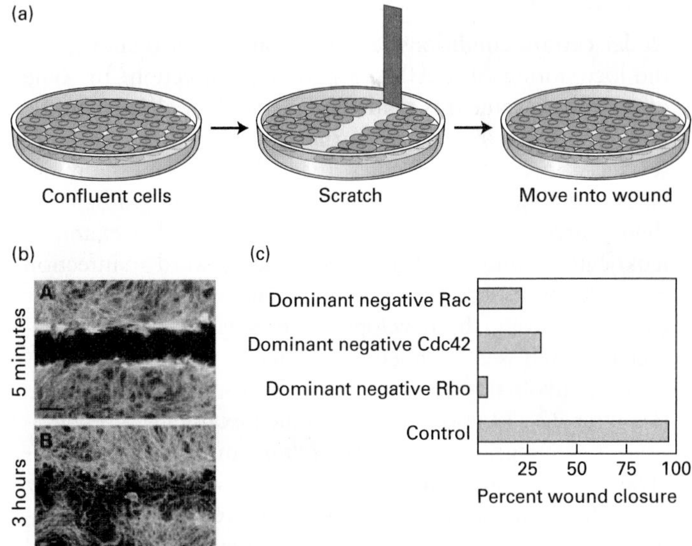

▲ EXPERIMENTAL FIGURE 17-43 **The wounded-cell monolayer assay can be used to dissect signaling pathways in directed cell movement.** (a) A confluent layer of cells is scratched to remove a swath of about three cells wide to generate a free cell border. The cells detect the free space and newly exposed extracellular matrix and over a period of hours fill the area. (b) Localization of actin in a wounded monolayer 5 minutes and 3 hours after scratching; the cells have migrated into the wounded area. (c) Effect of introducing dominant-negative Cdc42, Rac, and Rho into cells at the wound edge; all affect wound closure. [Parts (b) and (c) from C. D. Nobes and A. Hall, 1999, *J. Cell Biol.* **144**:1235–1244.]

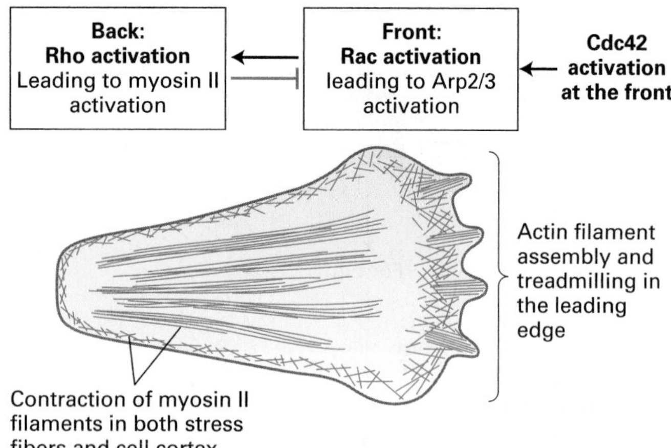

▲ **FIGURE 17-44 Contribution of Cdc42, Rac, and Rho to cell movement.** The overall polarity of a migrating cell is controlled by Cdc42, which is activated at the front of a cell. Cdc42 activation leads to active Rac in the front of the cell, which generates the leading edge, and active Rho at the back of the cell, which causes myosin II activation and contraction. Active Rho inhibits Rac activation, ensuring the asymmetry of the two active G-proteins.

lead to the inactivation of Rac. This might help ensure no leading edge structures form at the rear of the cell.

Migrating Cells Are Steered by Chemotactic Molecules

Under certain conditions, extracellular chemical cues guide the locomotion of a cell in a particular direction. In some cases, the movement is guided by insoluble molecules in the underlying substratum, as in the wound-healing assay described above. In other cases, the cell senses soluble molecules and follows them, along a concentration gradient, to their source—a process known as **chemotaxis.** For example, leukocytes (white blood cells) are guided toward an infection by a tripeptide secreted by many bacterial cells. In another example, during the development of skeletal muscle, a secreted protein signal called scatter factor guides the migration of myoblasts to the proper locations in limb buds (Chapter 22). One of the best-studied examples of chemotaxis is the migration of *Dictyostelium* amebas along an increasing concentration of cAMP, which is an extracellular chemotactic agent in this organism (Figure 17-45a). Following cAMP to its source, the amebas aggregate into a slug and then differentiate into a fruiting body.

Despite the variety of different chemotactic molecules—sugars, peptides, cell metabolites, cell-wall or membrane lipids—they all work through a common and familiar mechanism: binding to cell-surface receptors, activation of intracellular signaling pathways, and remodeling of the cytoskeleton through the activation or inhibition of various actin-binding proteins. What is quite amazing is that just a 2 percent difference in the concentration of chemotactic molecules between the front and back of the cell is sufficient to induce directed cell migration. Equally amazing is the finding that the internal signal-transduction pathways used in chemotaxis have been conserved between *Dictyostelium* amebas and human leukocytes despite almost a billion years of evolution.

Chemotactic Gradients Induce Altered Phosphoinositide Levels Between the Front and Back of a Cell

To investigate how *Dictyostelium* amebas sense a chemotactic gradient, investigators have studied the cell-surface receptors for extracellular cAMP and downstream signaling pathways in the expectation that these must somehow sense the concentration gradient. Before we discuss the details, let's consider how such a system might work. If a cell can sense a 2 percent difference in concentration across its length, it is unlikely that simply activating actin assembly 2 percent more at the front than at the back could lead to directed movement. Rather, there must be some mechanism that amplifies this small external signal difference into a large internal biochemical difference. One way to do this would be for the cell to subtract the common signal from the front and back and only respond to a *difference in signal.* This is believed to be how it works. To try to understand this, investigators have looked at the concentration of active components of the signaling pathway to see where the amplification occurs.

Micrographs of cAMP receptors tagged with green fluorescent protein (GFP) show that the receptors are distributed uniformly on the surface of an ameba cell (Figure 17-45b); therefore an internal gradient must be established by another component of the signaling pathway. Because cAMP receptors signal through trimeric G proteins (Chapter 16), a subunit of the trimeric G protein and other downstream signaling proteins were tagged with GFP to look at their distributions. Fluorescence micrographs show that the concentration of trimeric G proteins is also rather uniform. Downstream of the trimeric G proteins is PI-3 kinase, an enzyme that phosphorylates membrane-bound inositol phospholipids (**phosphoinositides**), such as PI4,5-biphosphate [PI(4,5)P_2] to the signaling lipid PI3,4,5-triphosphate [PI(3,4,5)P_3] (see Figure 16-29). Remarkably, the enzyme PI-3 kinase is highly enriched at the front of a migrating cell, as are its products. The phosphatase, PTEN, that dephosphorylates the signaling lipid PI(3,4,5)P3 back to PI(4,5)P2 is enriched in the tail of the migrating cell (Figure 17-45b, c). This asymmetry is believed to be set up in the following way. Prior to seeing a gradient, the phosphatase PTEN is associated uniformly with the plasma membrane. When a cell "sees" a gradient, PI 3-kinase is activated a bit more at the front than the back. This results in slightly higher levels of the signaling phospholipid at the front. The association of the phosphatase PTEN with the membrane is very sensitive to the level of the newly formed PI(3,4,5)P_3—so it is preferentially depleted from the front. Since it is less effective at dephosphorylating the PI(3,4,5)P_3 at the front and more effective at dephosphorylating PI(3,4,5)P_3 at the rear, a strong asymmetry of PI(3,4,5)P_3 results. Thus the phosphatase PTEN contributes to the background subtraction necessary for a cell to sense a shallow gradient of chemoattractant.

 Video: Chemotaxis of a Single *Dictyostelium* Cell to cAMP

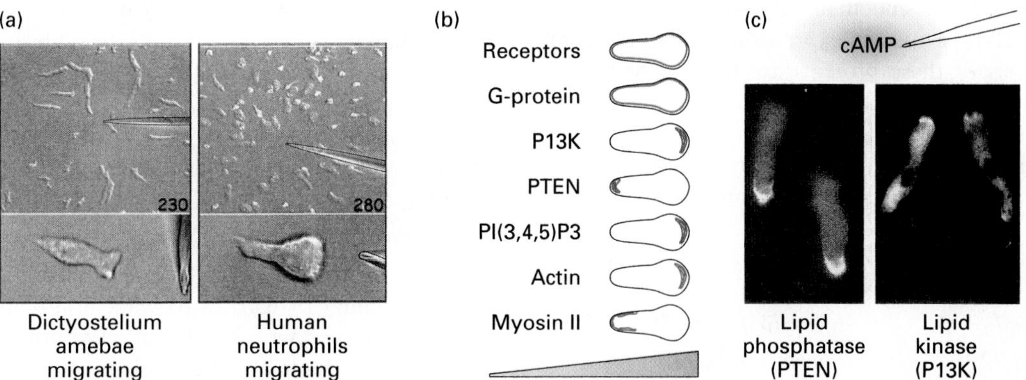

▲ EXPERIMENTAL FIGURE 17-45 **Chemotaxis involves elevated levels of signaling phosphoinositides, which signal to the actin cytoskeleton.** (a) *Dictyostelium* cells migrate toward a pipet of cAMP (*left*), and human neutrophils (a type of leukocyte) migrate toward a pipet of fMLP (formylated Met-Leu-Phe), a chemotactic peptide produced by bacteria (*right*). In the lower two panels are individual chemotaxing *Dictyostelium* and neutrophil cells that look remarkably similar, despite about 800 million years of evolution separating them. (b) Summary of results of studies exploring the concentration of components of signaling pathways (green) in *Dictyostelium* cells undergoing chemotaxis toward cAMP. Also shown are the concentration of actin and myosin (red). (c) The enzyme PI3 kinase, which generates PI (3,4,5)P3, is enriched at the front of chemotaxing cells, whereas PTEN, the phosphatase that hydrolyses PI (3,4,5)P2, is enriched at the back. These distributions result in elevated PI (3,4,5)P3 at the front of the cells, which signals the polarity for movement. [Part (a) from C. Parent, 2004, *Curr. Opin. Cell Biol.* **16**:4–13. Part (c) from M. Iijima et al., 2002, *Dev. Cell* **3**:469–478.]

The difference in local PI(3,4,5)P_3 concentration now signals to the actin cytoskeleton to assemble a leading edge at the front and contraction at the rear (Figure 17-45b), and the cell is on its way to the source of chemoattractant. This cell polarization is not stable in the absence of the chemotactic gradient, so if the gradient changes, as might happen with a moving bacterium, the cell will also change its direction and follow the gradient to its source.

KEY CONCEPTS OF SECTION 17.7

Cell Migration: Signaling and Chemotaxis

- Cell migration involves the extension of an actin-rich leading edge at the front of the cell, the formation of adhesive contacts that move backward with respect to the cell, and their subsequent release, combined with rear contraction to push the cell forward (see Figure 17-38).
- The assembly and function of actin filaments is controlled by signaling pathways through small GTP-binding proteins of the Rho family. Cdc42 regulates overall polarity and the formation of filopodia, Rac regulates actin meshwork formation through the Arp2/3 complex, and Rho regulates both actin filament formation by formins as well as contraction through regulation of myosin II (see Figure 17-42).
- Chemotaxis, the directed movement toward an attractant, involves signaling pathways that establish differences in phosphoinositides between the front and rear of the cell, which in turn regulate the actin cytoskeleton and direction of cell migration (see Figure 17-45).

Perspectives for the Future

In this chapter, we have seen that cells have intricate mechanisms for the spatial and temporal assembly, turnover, and attachment of microfilaments to membranes. Biochemical analyses of actin-binding proteins coupled with protein inventories provided by sequences of whole genomes have allowed the cataloging of many different classes of actin-binding proteins. To understand how this large group of proteins can assemble specific structures in a cell, it will be important to know the concentration of all the components, how they interact, and their regulation by signaling pathways. Although this might seem like a daunting task, new microscopic methods to detect the location of specific protein-protein interactions and the location of many of the key signaling pathways suggest that rapid progress will be made in this area.

The protein inventories provided by genomic sequences have also documented the large number of myosin families, yet the biochemical properties of many of these motors, or their biological functions, remain to be elucidated. Again, recent technical developments, including the ability to tag motors with fluorescent tracers such as GFP, or knocking down their expression with RNAi technologies are providing very powerful avenues to help elucidate motor functions. However, some important aspects of motors remain largely unexplored. For example, a motor that transports an organelle down a filament has to first bind the organelle, then transport it, and then release it at the destination—little is known about how these different events are coordinated or how these types of myosin-based motors are returned to pick up new cargo.

Finally, although we have generally discussed microfilaments without regard to tissue type—except for the specializations found in skeletal and smooth muscles—many actin-binding proteins show cell-type specific expression, and so the array and relative levels of these proteins are tailored to specific functions of different cell types. This is clearly the case, as revealed by proteomic analysis of cell-specific protein expression and by the fact that many diseases are a consequence of tissue-specific expression of actin-binding proteins or myosins.

Key Terms

Arp2/3 complex *723*
capping protein *722*
chemotaxis *750*
cofilin *721*
contractile bundles *741*
critical concentration, C_c *720*
cross-linking protein *728*
duty ratio *737*
F-actin *717*
filopodia *716*
focal adhesion/focal contact *716*
formin *723*
G-actin *717*
GTPase superfamily *747*
integrin *747*
intermediate filaments *715*
lamellipodium *716*
leading edge *716*
microfilaments *715*
microtubules *715*
microvilli *716*
motor protein *715*
myofibril *738*
myosin head domain *733*
myosin LC kinase *742*
myosin tail domain *733*
nucleation *719*
polarity *718*
profilin *721*
sarcomere *730*
sliding-filament assay *733*
stress fibers *716*
thick filaments *739*
thin filaments *739*
thymosin β_4 *722*
treadmilling *721*

Review the Concepts

1. Three systems of cytoskeletal filaments exist in most eukaryotic cells. Compare them in terms of composition, function, and structure.

2. Actin filaments have a defined polarity. What is filament polarity? How is it generated at the subunit level? How is filament polarity detectable?

3. In cells, actin filaments form bundles and/or networks. How do cells form such structures, and what specifically determines whether actin filaments will form a bundle or a network?

4. Much of our understanding of actin assembly in the cell is derived from experiments using purified actin in vitro. What techniques can be used to study actin assembly in vitro? Explain how each of these techniques works.

5. The predominant forms of actin inside a cell are ATP–G-actin and ADP–F-actin. Explain how the interconversion of the nucleotide state is coupled to the assembly and disassembly of actin subunits. What would be the consequence for actin filament assembly/disassembly if a mutation prevented actin's ability to bind ATP? What would be the consequence if a mutation prevented actin's ability to hydrolyze ATP?

6. Actin filaments at the leading edge of a crawling cell are believed to undergo treadmilling. What is treadmilling, and what accounts for this assembly behavior?

7. Although purified actin can reversibly assemble in vitro, various actin-binding proteins regulate the assembly of actin filaments in the cell. Predict the effect on a cell's actin cytoskeleton if function-blocking antibodies against each of the following were independently microinjected into cells: profilin, thymosin-β_4, CapZ, and the Arp2/3 complex.

8. There are at least 20 different types of myosin. What properties do all types share, and what makes them different?

9. The ability of myosin to walk along an actin filament may be observed with the aid of an appropriately equipped microscope. Describe how such assays are typically performed. Why is ATP required in these assays? How can such assays be used to determine the direction of myosin movement or the force produced by myosin?

10. Contractile bundles occur in nonmuscle cells, although the structures are less organized than the sarcomeres of muscle cells. What is the purpose of nonmuscle contractile bundles?

11. How does myosin convert the chemical energy released by ATP hydrolysis into mechanical work?

12. Myosin II has a duty ratio of 10 percent, and its step size is about 5–15 nm. In contrast, myosin V has a much higher duty ratio (about 70 percent) and takes 36-nm steps as it walks down an actin filament. What differences between myosin II and myosin V account for their different properties? How do the different structures and properties of myosin II and myosin V reflect their different functions in cells?

13. Contraction of both skeletal and smooth muscle is triggered by an increase in cytosolic Ca^{2+}. Compare the mechanisms by which each type of muscle converts a rise in Ca^{2+} into contraction.

14. Several types of cells utilize the actin cytoskeleton to power locomotion across surfaces. How are different assemblies of actin filaments involved in locomotion?

15. To move in a specific direction, migrating cells must utilize extracellular cues to establish which portion of the cell will act as the front and which will act as the back. Describe how G-proteins appear to be involved in the signaling pathways used by migrating cells to determine direction of movement.

16. Cell motility has been described as like the motion of tank treads. At the leading edge, actin filaments form rapidly into bundles and networks that make protrusions and move the cell forward. At the rear, cell attachments are broken and the tail end of the cell is brought forward. What provides the traction for moving cells? How does cell-body translocation happen? How are cell attachments released as cells move forward?

Analyze the Data

Myosin V is an abundant nonmuscle myosin that is responsible for the transport of cargo such as organelles in many cell types. Structurally, it consists of two identical polypeptide chains that dimerize to form a homodimer. The motor domains reside at the N-terminus of each chain and contain both ATP- and actin-binding sites. The motor domain is followed by a neck region containing six "IQ" motifs, each of which binds calmodulin, a Ca^{2+}-binding protein. The neck domain is followed by a region capable of forming coiled coils, via which the two chains dimerize. The final 400 amino acid residues form a globular tail domain (GTD), to which cargo binds. Myosin V would consume large amounts of ATP if its motor domain were always active, and a number of studies have been conducted to understand how this motor is regulated.

a. The rate of ATP hydrolysis (i.e., ATP molecules hydrolyzed per second per myosin V) was measured in the presence of increasing amounts of free Ca^{2+}. The concentration of cytosolic free Ca^{2+} is normally less than 10^{-6} M but can be elevated in localized areas of the cell and is often elevated in response to a signaling event. What do these data suggest about myosin V regulation?

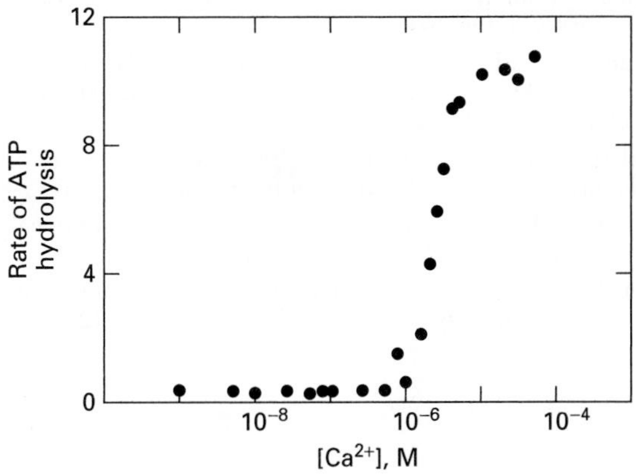

b. In additional studies, the ATPase activity of myosin V was measured in the presence of increasing amounts of F actin in the presence or absence of 10^{-6} M free Ca^{2+}. What additional information about myosin V regulation do these data provide?

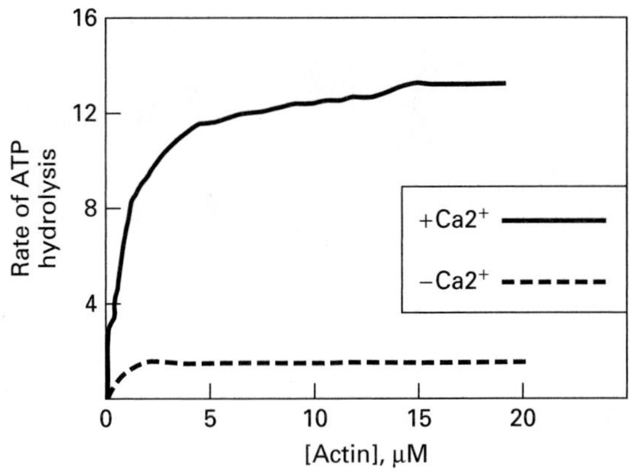

c. Next, the behavior of truncated myosin V, which lacks just its C-terminal globular tail, was examined and compared to the behavior of intact myosin V. From this experiment, what can you deduce about the mechanism by which myosin V is regulated?

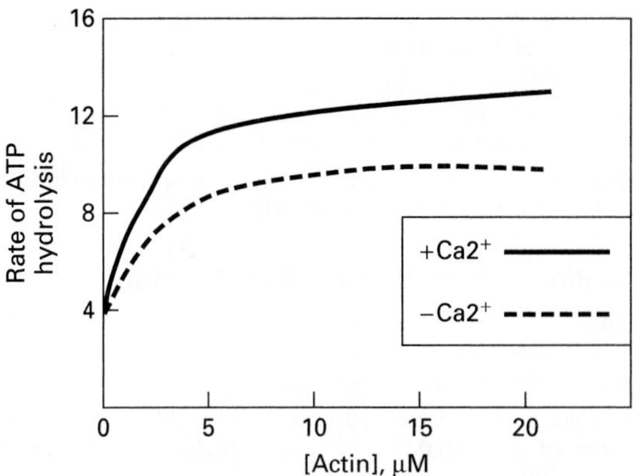

References

General References

Bray, D. 2001. *Cell Movements.* Garland.

Howard, J. 2001. *The Mechanics of Motor Proteins and the Cytoskeleton.* Sinauer.

Kreis, T., and R. Vale. 1999. *Guidebook to the Cytoskeletal and Motor Proteins.* Oxford University Press.

Web Sites

The myosin home page:
http://www.mrc-lmb.cam.ac.uk/myosin/myosin.html

The cytokinetic mafia home page—all things cytokinetic:
http://www.unc.edu/depts/salmlab/mafia/mafia.html

Microfilaments and Actin Structures

Cooper, J. A., and D. A. Schafer. 2000. Control of actin assembly and disassembly at filament ends. *Curr. Opin. Cell Biol.* **12**:97–103.

Holmes, K. C., D. Popp, W. Gebhard, and W. Kabsch. 1990. Atomic model of the actin filament. *Nature* **347**:44–49.

Kabsch, W., H. G. Mannherz, D. Suck, E. F. Pai, and K. C. Holmes. 1990. Atomic structure of the actin:DNase I complex. *Nature* **347**:37–44.

Pollard, T. D., L. Blanchoin, and R. D. Mullins. 2000. Molecular mechanisms controlling actin filament dynamics in nonmuscle cells. *Ann. Rev. Biophys. Biomol. Struc.* **29**:545–576.

Dynamics of Actin Filaments

Paavilainen, V. O., E. Bertling, S. Falck, and P. Lappalainen. 2004. Regulation of cytoskeletal dynamics by actin-monomer-binding proteins. *Trends Cell Biol.* 14:386–94.

Theriot, J. A. 1997. Accelerating on a treadmill: ADF/cofilin promotes rapid actin filament turnover in the dynamic cytoskeleton. *J. Cell Biol.* **136**:1165–1168.

Mechanisms of Actin Filament Assembly

Gouin, E., M. D. Welch, and P. Cossart. 2005. Actin-based motility of intracellular pathogens. *Curr. Opin. Microbiol.* **8**:35–45.

Higgs, H. N. 2005. Formin proteins: a domain-based approach. *Trends Biochem. Sci.* **30**:342–353.

Higgs, H. N., and T. D. Pollard. 2000. Regulation of actin filament network formation through ARP2/3 complex: activation by a diverse array of proteins. *Ann. Rev. Biochem.* **70**:619–676.

Pruyne, D., et al. 2002 Role of formins in actin assembly: nucleation and barbed end association. *Science* **297**:612–615.

Volkmann, N., et al. 2001. Structure of Arp2/3 complex in its activated state and in actin filament branch junctions. *Science* **293**:2456–2459.

Welch, M. D., and R. D. Mullins. 2002. Cellular control of actin nucleation. *Annu. Rev. Cell Dev. Biol.* **18**:247–288.

Zigmond, S. H. 2004. Formin-induced nucleation of actin filaments. *Curr. Opin. Cell Biol.* **16**:99–105.

Organization of Actin-Based Cellular Structures

Bennett, V., and A. J. Baines. 2001. Spectrin and ankyrin-based pathways: metazoan inventions for integrating cells into tissues. *Physiological Reviews* **81**:1353–1392.

Bretscher, A., K. Edwards, and R. Fehon. 2002. ERM proteins and merlin: integrators at the cell cortex. *Nature Rev. Mol. Cell Biol.* 3:586–599.

McGough, A. 1998. F-actin-binding proteins. *Curr. Opin. Struc. Biol.* 8:166–176.

Stossel, T. P., et al. 2001. Filamins as integrators of cell mechanics and signalling. *Nature Rev. Mol. Cell Biol.* 2(2):138–145.

Myosins: Actin-based Motor Proteins

Berg, J. S., B. C. Powell, and R. E. Cheney. 2001. A millennial myosin census. *Mol. Biol. Cell* **12**:780–794.

Mermall, V., P. L. Post, and M. S. Mooseker. 1998. Unconventional myosin in cell movement, membrane traffic, and signal transduction. *Science* **279**:527–533.

Rayment, I. 1996. The structural basis of the myosin ATPase activity. *J. Biol. Chem.* **271**:15850–15853.

Tyska, M. J., and M. S. Mooseker. 2003. Myosin-V motility: these levers were made for walking. *Trends Cell Biol.* **13**:447–451.

Vale, R. D. 2003. The molecular motor toolbox for intracellular transport. *Cell* **112**:467–480.

Vale, R. D., and R. A. Milligan. 2000. The way things move: looking under the hood of molecular motor proteins. *Science* **288**:88–95.

Myosin-Powered Movements

Bretscher A. 2003. Polarized growth and organelle segregation in yeast—the tracks, motors, and receptors. *J. Cell Biol.* **160**:811–816.

Clark, K. A., A. S. McElhinny, M. C. Beckerle, and C. C. Gregorio. 2002. Striated muscle cytoarchitecture: an intricate web of form and function. *Annu. Rev. Cell Dev. Biol.* **18**:637–706.

Grazier, H. L., and S. Labeit. 2004. The giant protein titin: a major player in myocardial mechanics, signaling, and disease. *Circ. Res.* **94**:284–295.

Cell Migration: Signaling and Chemotaxis

Borisy, G. G., and T. M. Svitkina. 2000. Actin machinery: pushing the envelope. *Curr. Opin. Cell Biol.* **12**:104–112.

Burridge, K., and K. Wennerberg. 2004. Rho and Rac take center stage. *Cell* **116**:167–179.

Etienne-Manneville, S. 2004. Cdc42—the centre of polarity. *J. Cell Sci.* **117**:1291–1300.

Etienne-Manneville, S., and A. Hall. 2002. Rho GTPases in cell biology. *Nature* **420**:629–635.

Manahan, C. L., P. A. Iglesias, Y. Long, and P. N. Devreotes. 2004. Chemoattractant signaling in *Dictyostelium discoideum*. *Ann. Rev. Cell Dev. Biol.* **20**:223–253.

Pollard, T. D., and G. G. Borisy. 2003. Cellular motility driven by assembly and disassembly of actin filaments. *Cell* **112**:453–465.

Ridley, A. J., et al. 2003. Cell migration: integrating signals from the front to back. *Science* **302**:1704–1709.

Small, J. V., T. Strada, E. Vignal, and K. Rottner. 2002. The lamellipodium: where motility begins. *Trends Cell Biol.* **12**:112–120.

CLASSIC EXPERIMENT 17.1

LOOKING AT MUSCLE CONTRACTION

H. Huxley and J. Hanson, 1954, *Nature* **173:**973–976

The contraction and relaxation of striated muscles allow us to perform all of our daily tasks. How does this happen? Scientist have long looked to see how fused muscles cells, called myofibrils, differ from other cells that cannot perform powerful movement. In 1954, Jean Hanson and Hugh Huxley published their microscopy studies on muscle contraction, which demonstrated the mechanism by which it occurs.

Background

The ability of muscles to perform work has long been a fascinating process. Voluntary muscle contraction is performed by striated muscles, which are named for their appearance when viewed under the microscope. By the 1950s, biologists studying myofibrils, the cells that make up muscles, had named many of the structures they had observed under the microscope. One contracting unit, called a sarcomere, is made up of two main regions called the A band, and the I band. The A band contains two darkly colored thick striations and one thin striation. The I band is made up primarily of light-colored striations, which are divided by a darkly colored line known as the Z disk. Although these structures had been characterized, their role in muscle contraction remained unclear. At the same time, biochemists also tried to tackle this problem by looking for proteins that are more abundant in myofibrils than in nonmuscle cells. They found muscles to contain large amounts of the structural proteins actin and myosin in a complex with each other. Actin and myosin form polymers that can shorten when treated with adenosine triphosphate (ATP).

With these observations in mind, Hanson and Huxley began their study of cross striations in muscle. In a few short years, they united the biochemical data with the microscopy observations and developed a model for muscle contraction that holds true today.

The Experiment

Hanson and Huxley primarily used phase-contrast microscopy in their studies of striated muscles that they isolated from rabbits. The technique allowed them to obtain clear pictures of the sarcomere and to take careful measurements of the A and the I bands. By treating the muscles with a variety of chemicals, then studying them under the phase-contrast microscope, they were able to successfully combine biochemistry with microscopy to describe muscle structure as well as the mechanism of contraction.

In their first set of studies, Hanson and Huxley employed chemicals that are known to specifically extract either myosin or actin from myofibrils. First, they treated myofibrils with a chemical that specifically removes myosin from muscle. They used phase-contrast microscopy to compare untreated myofibrils to myosin-extracted myofibrils. In the untreated muscle, they observed the previously identified sarcometic structure, including the darkly colored A band. When they looked at the myosin-extracted cells, however, the darkly colored A band was not observed. Next, they extracted actin from the myosin-extracted muscle cells. When they extracted both myosin and actin from the myofibril, they could see no identifiable structure to the cell under phase-contrast microscopy. From these experiments, they concluded that myosin was located primarily in the A band, whereas actin is found throughout the myofibril.

With a better understanding of the biochemical nature of muscle structures, Huxley and Hanson went on to study the mechanism of muscle contraction. They isolated individual myofibrils from muscle tissue and treated them with ATP, causing them to contract at a slow rate. Using this technique, they could take pictures of various stages of muscle contraction observed using phase-contrast microscopy. They could also mechanically induce stretching by manipulating the coverslip, which allowed them to also observe the relaxation process. With these techniques in hand, they examined how the structure of the myofibril changes during contraction and stretch.

First, Huxley and Hanson treated myofibrils with ATP, then photographed the images they observed under phase-contrast microscopy. These pictures allowed them to measure the lengths of both the A band and the I band at various stages of contraction. When they looked at myofibrils freely contracting, they noticed a consistent shortening of the lightly colored I band, whereas the length of the A band remained constant (Figure 1). Within the A band, they observed the formation of an increasingly dense area throughout the contraction.

Next, the two scientists examined how the myofibril structure changes during a simulated muscle stretch. They stretched isolated myofibrils mounted on glass slides by manipulating the coverslip. They again photographed phase-contrast microscopy images and measured the lengths of the A and the I bands. During stretch the length of the I band increased, rather than shortened, as it had in contraction.

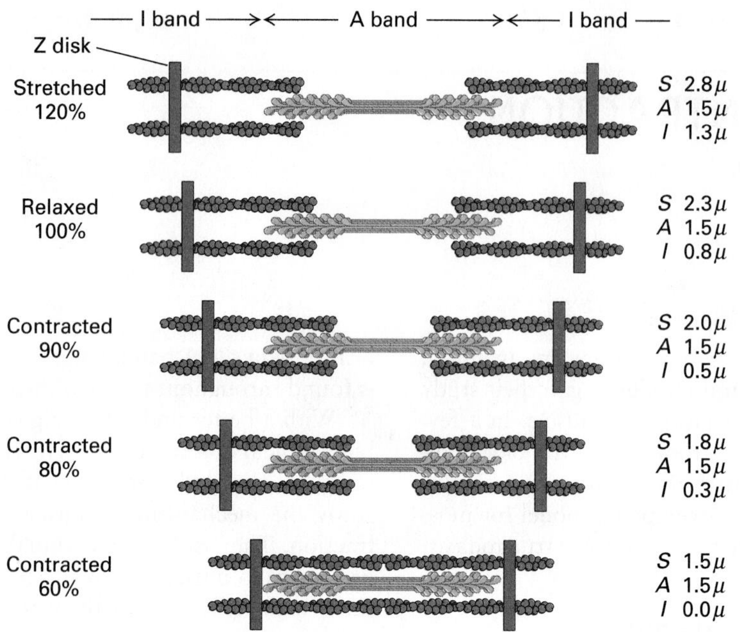

◀ **FIGURE 1 Schematic diagram of muscle contraction and stretch observed by Hanson and Huxley.** The lengths of the sarcomere (S), the A band (A), and the I band (I) were measured in muscle samples contracted 60 percent in length relative to the relaxed muscle (bottom) or stretched to 120 percent (top). The lengths of the sarcomere, the I band, and the A band are noted on the right. Notice that from 120 percent stretch to 60 percent contraction the A band does not change in length. However, the length of the I band can stretch to 1.3 microns, and at 60 percent contraction, it disappears as the sarcomere shortens to the overall length of the A band. [Adapted from J. Hanson and H. E. Huxley, 1955, *Symp. Soc. Exp. Biol. Fibrous Proteins and Their Biological Significance* **9:**249.]

Once again, the length of the A band remained unchanged. The dense zone that formed in the A band during contraction became less dense during stretch.

From their observations, Hanson and Huxley developed a model for muscle contraction and stretch (Figure 1). In their model, the actin filaments in the I band are drawn up into the A during contraction, and thus the I band becomes shorter. This allows for increased interaction between the myosin located in the A band and the actin filaments. As the muscle stretches, the actin filaments withdraw from the A band. From these data, Hanson and Huxley proposed that muscle contraction is driven by actin filaments moving in and out of a mass of stationary myosin filaments.

Discussion

By combining microscopic observations with known biochemical treatments of muscle fibers, Hanson and Huxley were able to describe the biochemical nature of muscle structur and outline a mechanism for musc contraction. A large body of resear continues to focus on understandi the process of muscle contraction. Sc entists now know that muscles co tract by ATP hydrolysis, driving a co formational change in myosin th allows it to pull on actin. Researche are continuing to uncover the molec lar details of this process, whereas th mechanism contraction proposed b Hanson and Huxley remains in place

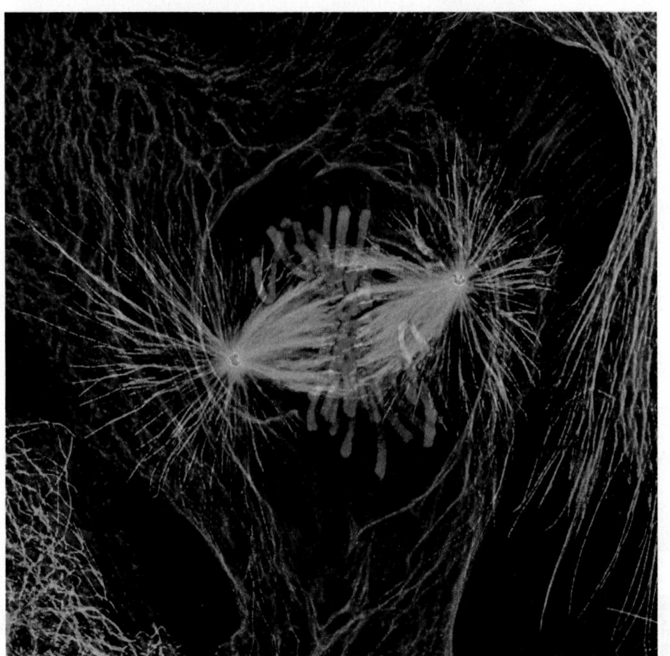

Newt lung cell in mitosis stained for centrosomes (magenta), microtubules (green), chromosomes (blue), keratin intermediate filaments (red). [Courtesy of A. Khodjakor, from *Nature* **408**:423–424 (2000).]

CHAPTER

18

CELL ORGANIZATION AND MOVEMENT II: MICROTUBULES AND INTERMEDIATE FILAMENTS

Three types of filaments make up the animal-cell cytoskeleton: microfilaments, microtubules, and intermediate filaments. Why have these three distinct types of filaments evolved? It seems likely that their physical properties are suited to different functions. In Chapter 17 we described how actin filaments are often cross-linked into networks of bundles to form flexible and dynamic structures and to serve as tracks for the many different classes of myosin motors. Microtubules are stiff tubes that can exist as a single structure extending up to 20 μm in cells or as the bundled structures as seen in cilia and flagella. A consequence of their tubular design is the ability of microtubules to generate pulling and pushing forces without buckling, a property that allows single tubules to extend large distances within a cell and bundles to slide past each other, as occurs in flagella and in the mitotic spindle. Microtubules' ability to extend long distances in the cell, together with their intrinsic polarity, is exploited by microtubule-dependent motors, which use microtubules as tracks for long-range transport of organelles. Microtubules can be highly dynamic—being assembled and disassembled from their ends—providing the cell with the flexibility to reorganize microtubule organization as needed.

In contrast to microfilaments and microtubules, intermediate filaments have great tensile strength and have evolved to withstand much larger stresses and strains. With properties akin to strong molecular ropes, they are ideally suited to endow both cells and tissues with structural integrity and contribute to cellular organization. Intermediate filaments do not have an intrinsic polarity like microfilaments and microtubules, so it is not surprising that there are no known motor proteins that use intermediate filaments as tracks. Although we discuss microtubules and intermediate filaments together in this chapter—and their localization in the cytoplasm can look superficially quite similar—we will see their dynamics and functions are very different. A summary of the similarities and differences among the three cytoskeletal systems is presented in Figure 18-1.

OUTLINE

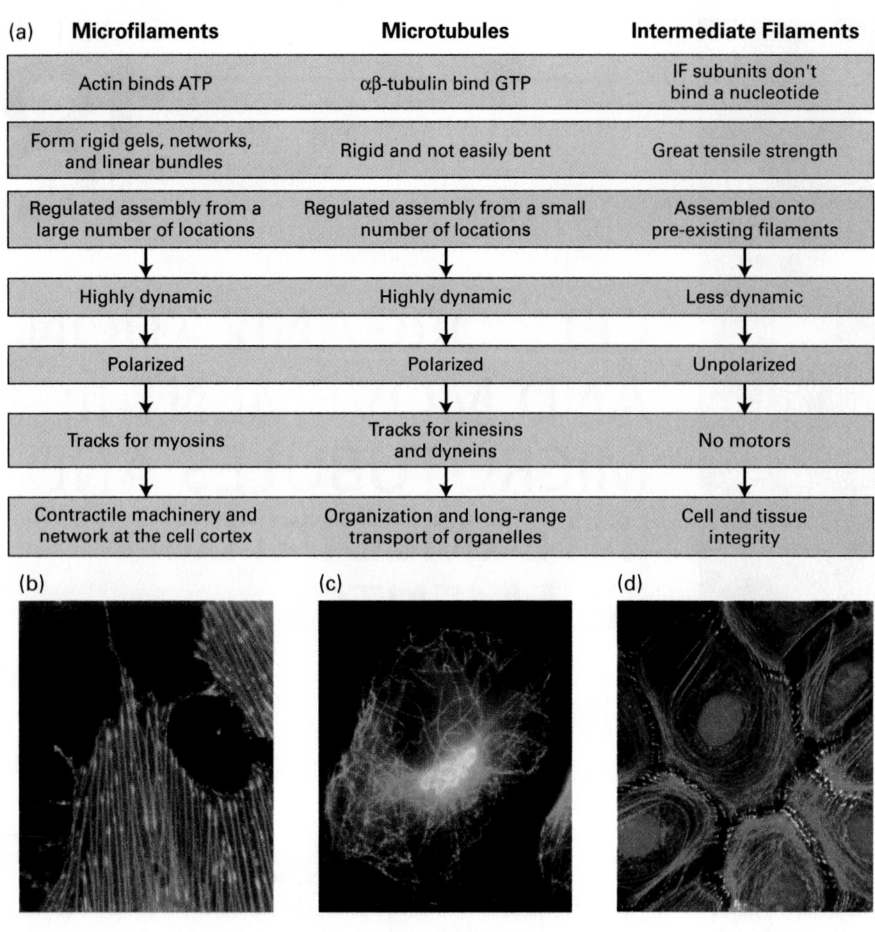

(a)
Microfilaments
Microtubules
Intermediate Filaments
Actin binds ATP
αβ-tubulin bind GTP
IF subunits don't bind a nucleotide
Form rigid gels, networks, and linear bundles
Rigid and not easily bent
Great tensile strength
Regulated assembly from a large number of locations
Regulated assembly from a small number of locations
Assembled onto pre-existing filaments
Highly dynamic
Highly dynamic
Less dynamic
Polarized
Polarized
Unpolarized
Tracks for myosins
Tracks for kinesins and dyneins
No motors
Contractile machinery and network at the cell cortex
Organization and long-range transport of organelles
Cell and tissue integrity
(b)
(c)
(d)

◄ **FIGURE 18-1 Overview of the physical properties and functions of the three cytoskeletal systems in animal cells.** (a) Biophysical and biochemical properties (orange) and biological properties (green) are shown for each filament type. The micrographs show examples of each filament type in a particular cellular context, but note that microtubules also make up other structures, and intermediate filaments also line the inner surface of the nucleus. (b) Well-spread cultured cells stained for actin (green) and sites of actin attachment to the substratum (red). (c) Localization of microtubules (green) and the Golgi apparatus (yellow). Notice the central location of the Golgi apparatus, which is collected there by transport along microtubules. (d) Localization of cytokeratins (red), a type of intermediate filament, and a component of desmosomes (green) in epithelial cells. Cytokeratins from individual cells are attached to each other through the desmosomes. [Part (b) courtesy of K. Burridge. Part (c) courtesy of W. Brown. Part (d) courtesy of E. Fuchs.]

This chapter covers four main topics. First, we discuss the structure and dynamics of microtubules and their motor proteins. Second, we examine how microtubules and their motors contribute to the movement of the specialized structures cilia and flagella. Third, we discuss the role of microtubules in the mitotic spindle—a molecular machine that accurately segregates duplicated chromosomes. Finally, we explore the roles of the different classes of intermediate filaments that provide structure to the nuclear envelope as well as strength and organization to cells and tissues. Although we consider microtubules, microfilaments, and intermediate filaments individually, the three cytoskeletal systems do not act independently of one another, and we consider some examples of this interdependence in the last section of the chapter.

18.1 Microtubule Structure and Organization

In the early days of electron microscopy, cells biologists noted long tubules in the cytoplasm that they called **microtubules.** Morphologically similar microtubules were seen making up the fibers of the mitotic spindle, as components of axons, and as the structural elements in cilia and flagella (Figure 18-2a, b). A careful examination of all the microtubules seen in transverse section indicated that they were made up of 13 longitudinal repeating units (Figure 18-2c), now called *protofilaments,* suggesting they all had a common structure. Microtubules purified from brain were then found to consist of a major protein, **tubulin,** and associated proteins, **microtubule-associated proteins (MAPs).** Purified tubulin alone can assemble into a microtubule under favorable conditions, proving that it is the structural component of the microtubule wall. MAPs modify the assembly and dynamics of the microtubules assembled from tubulin.

Microtubule Walls Are Polarized Structures Built from αβ-Tubulin Dimers

Purified soluble tubulin is a dimer consisting of two closely related subunits of about 55,000 daltons, α- and β-tubulin. Genomic analyses reveal that genes encoding both α- and β-tubulins are present in all eukaryotes, with considerable expansion in the number of genes in multicellular organisms. For example, budding yeast has two genes specifying α-tubulin and one for β-tubulin, whereas the soil nematode *Caenorhabditis elegans* has nine genes

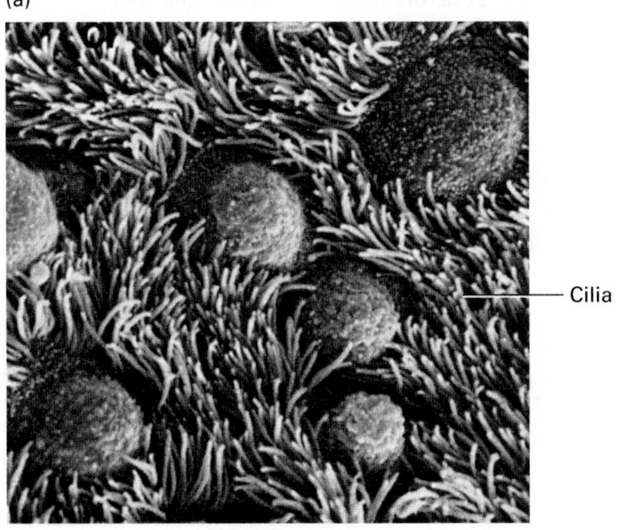

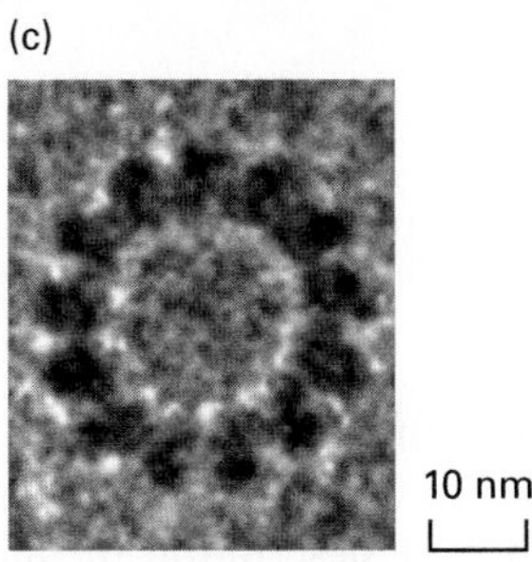

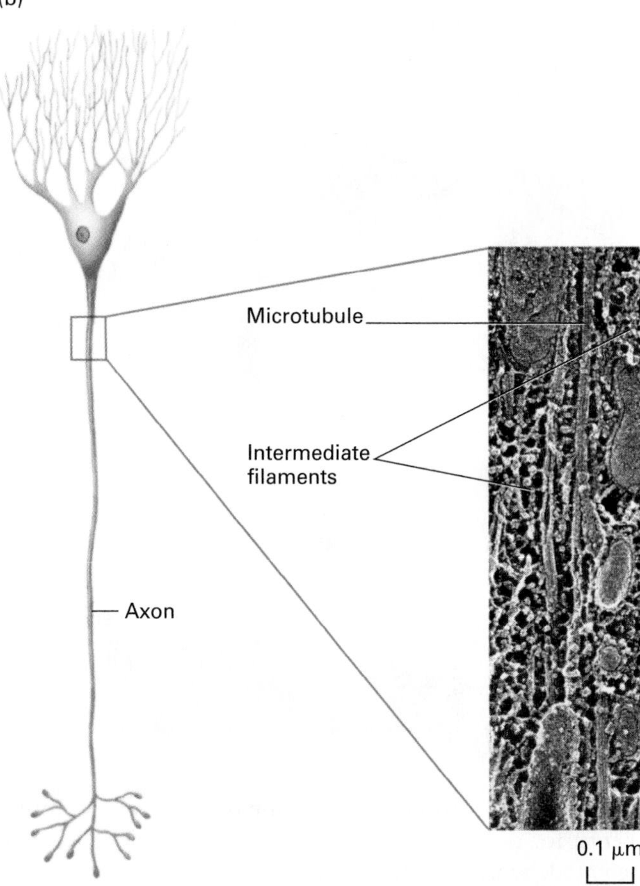

▲ **FIGURE 18-2 Microtubules are found in many different locations, and all have similar structures.** (a) Surface of the ciliated epithelium lining a rabbit oviduct viewed in a scanning electron microscope. Beating cilia, which have a core of microtubules, propel an egg down the oviduct. (b) Microtubules and intermediate filaments in a quick-frozen and deep-etched frog axon visualized in a transmission electron microscope. (c) High-magnification view of a single microtubule showing 13 repeating units known as protofilaments. [Part (a) from R. G. Kessels and R. H. Kardon, 1975, *Tissues and Organs,* W. H. Freeman and Company. Part (b) from N. Hirokawa, 1982, *J. Cell Biol.* **94:**129; courtesy of N. Hirokawa. Part (c) courtesy C. Bouchet-Marquis, 2007, *Biology of the Cell* **99:**45.]

encoding α-tubulin and six for β-tubulin. In addition to α- and β-tubulin, all these organisms also have genes specifying a third tubulin, γ-tubulin, which serves a regulatory function. Additional isoforms of tubulin have also been discovered that are present only in organisms that have centrioles and basal bodies, suggesting they are important for those structures.

Each subunit of the tubulin dimer can bind one molecule of GTP (Figure 18-3a). The GTP in the α-tubulin subunit is never hydrolyzed and is trapped by the interface between the α- and β-subunits. By contrast, the β-subunit has a GTP-binding site in the dimer that is exchangeable with free GTP, and this bound GTP can be hydrolyzed.

Microtubules consist of 13 laterally associated protofilaments forming a tubule whose external diameter is about 25 nm (Figure 18-3b). Each protofilament is made up of αβ-tubulin dimers, so that the subunits alternate down a protofilament, with each subunit type repeating each 8 nm. Because the protofilament is made up of tubulin dimers, each protofilament has an α-subunit at one end and a β-subunit at the other—so the protofilaments have an intrinsic **polarity.** In a microtubule, all the laterally associated protofilaments have the same polarity, so the microtubule also has an overall polarity. The (+) end of the microtubule favored for polymerization is the end with exposed β-subunits, whereas the (−) end has exposed α-subunits. In microtubules, the heterodimers in adjacent protofilaments are staggered slightly, forming tilted rows of α- and β-tubulin monomers in the microtubule wall. If you follow a row of β-subunits, for example, spiraling around a microtubule for one full turn, you will end up precisely three subunits up the protofilament, abutting an α-subunit. Thus all microtubules have a single longitudinal *seam,* where an α-subunit in one protofilament meets a β-subunit in the next protofilament.

Most microtubules in a cell consist of a simple tube, a *singlet* microtubule, built from 13 protofilaments. In rare

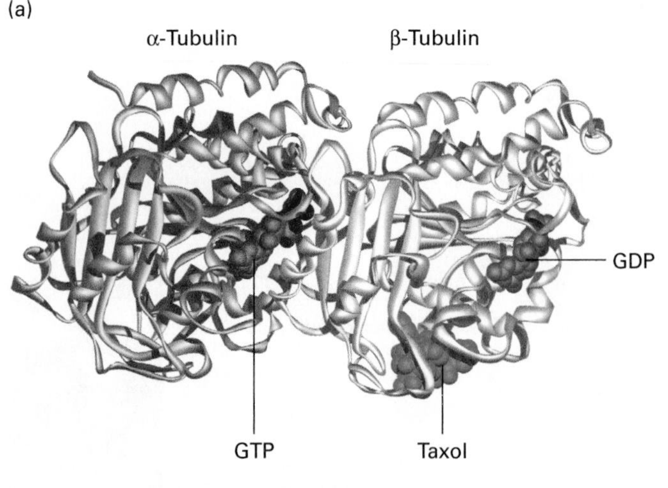

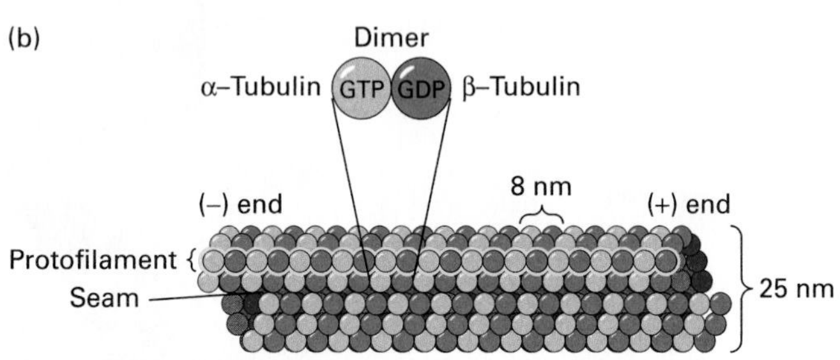

◄ **FIGURE 18-3 Structure of tubulin dimers and their organization into microtubules.** (a) Ribbon diagram of the tubulin dimer. The GTP (red) bound to the α-tubulin monomer is nonexchangeable, whereas the GDP (blue) bound to the β-tubulin monomer is exchangeable with free GTP. Taxol, a drug that stabilizes microtubules and is used to treat some cancers, binds to another part of the β-subunit. (b) The organization of tubulin subunits in a microtubule. The dimers are aligned end to end into protofilaments, which pack side by side to form the wall of the microtubule. The protofilaments are slightly staggered so that α-tubulin in one protofilament is in contact with α-tubulin in the neighboring protofilaments, except at the seam, where an α-subunit contacts a β-subunit. The microtubule displays a structural polarity in that subunits are added preferentially at the end, designated the (+) end, at which β-tubulin monomers are exposed. [Part (a) modified from E. Nogales et al., 1998, *Nature* **391:**199; courtesy of E. Nogales.]

cases, singlet microtubules contain more or fewer protofilaments; for example, certain microtubules in the neurons of nematode worms contain 11 or 15 protofilaments. In addition to the simple singlet structure, *doublet* or *triplet* microtubules are found in specialized structures such as cilia and flagella (doublet microtubules) and centrioles and basal bodies (triplet microtubules), structures we discuss in more detail below. Each doublet or triplet contains one complete 13-protofilament microtubule (A tubule) and one or two additional tubules (B and C) consisting of 10 protofilaments each (Figure 18-4).

Microtubules Are Assembled from MTOCs to Generate Diverse Organizations

With the identification of tubulin as the major structural component of microtubules, antibodies to tubulin could be generated and used in immunofluorescence microscopy to localize microtubules in cells (Figure 18-5a, b). This approach, coupled with the description of microtubules seen by electron microscopy, showed that microtubules are assembled from specific sites to generate many different types of organization (Figure 18-5).

The nucleation phase of microtubule assembly is such an unfavorable reaction that spontaneous nucleation does not play a significant role in microtubule assembly in vivo. Rather, all microtubules are nucleated from structures known as **microtubule-organizing centers,** or **MTOCs.** In most cases the (−) end of the microtubule stays anchored in the MTOC. In non-mitotic cells, also known as *interphase* cells, the MTOC is known as the **centrosome** and is generally located near the nucleus, producing a radial array of microtubules with their (+) ends toward the cell periphery (Figure 18-5c). This radial display provides tracks for microtubule-based

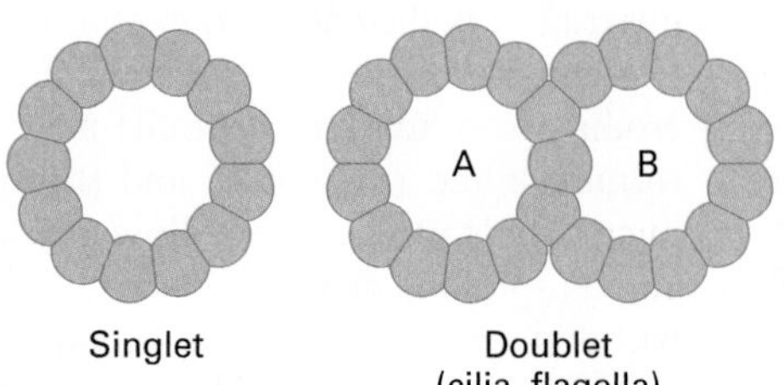

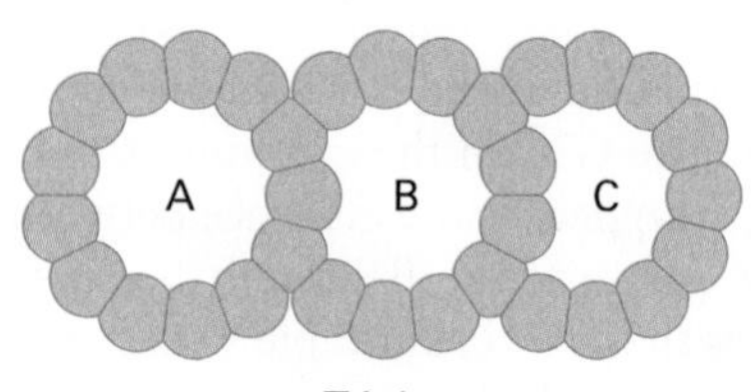

▲ **FIGURE 18-4 Singlet, doublet, and triplet microtubules.** In cross section, a typical microtubule, a singlet, is a simple tube built from 13 protofilaments. In a doublet microtubule, an additional set of 10 protofilaments forms a second tubule (B) by fusing to the wall of a singlet (A) microtubule. Attachment of another 10 protofilaments to the (B) tubule of a doublet microtubule creates a (C) tubule and a triplet structure.

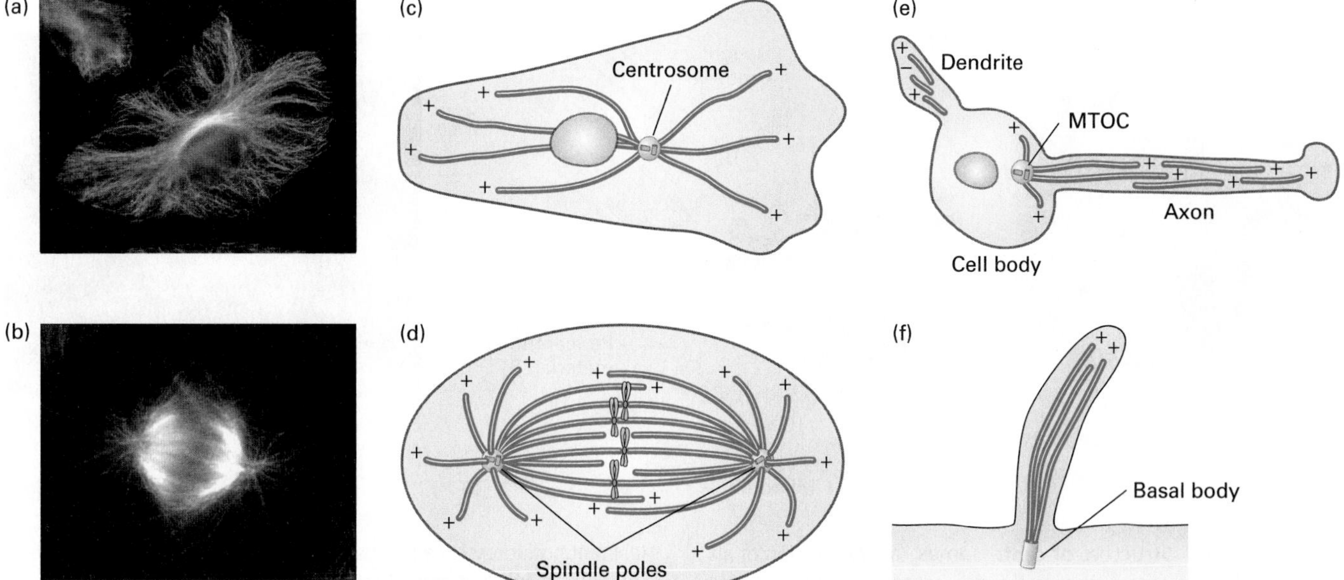

▲ **FIGURE 18-5 Microtubules are assembled from microtubule organizing centers (MTOCs).** The distribution of microtubules in cultured cells as seen by immunofluorescence microscopy using antibodies to tubulin in an interphase cell (a) and a cell in mitosis (b). (c–f) Diagrams of the distribution of microtubules in cells and structures, all of which are assembled from distinct MTOCs. In an interphase cell (c), the MTOC is called a centrosome (the nucleus is indicated by a blue oval); in a mitotic cell (d), the two MTOCs are called spindle poles (the chromosomes are shown in blue); in a neuron (e), microtubules in both axons and dendrites are assembled from an MTOC in the cell body and then released from it; the microtubules that make up the shaft of a cilium or flagellum (f) are assembled from an MTOC known as a basal body. The polarity of microtubules is indicted by (+) and (−). [Part (a) courtesy of A. Bretscher. Part (b) courtesy of T. Wittmann.]

motor proteins to organize and transport membrane-bound compartments, such as those comprising the secretory and endocytic pathways. During mitosis, cells completely reorganize their microtubules to form a bipolar spindle, assembled from two MTOCs called the *spindle poles,* to accurately segregate copies of the duplicated chromosomes (Figure 18-5d). In another example, neurons have long processes called axons, in which organelles are transported in both directions along microtubules (Figure 18-5e). The microtubules in axons, which can be as long as 1 meter in length, are not continuous and have been released from the MTOC but nevertheless are all of the same polarity. In the same cells, the microtubules in the dendrites have mixed polarity, although the functional significance of this is not clear. In cilia and flagella (Figure 18-5f), microtubules are assembled from an MTOC called a basal body.

In interphase cells microtubules are assembled from the centrosome. Electron microscopy shows that centrosomes in animal cells consist of a pair of orthogonally arranged cylindrical *centrioles* surrounded by apparently amorphous material called **pericentriolar material** (Figure 18-6a, arrowheads). Centrioles, which are about 0.5 μm long and 0.2 μm in diameter, are highly organized and stable structures that consist of nine sets of triplet microtubules and are closely related in structure to basal bodies, found at the base of cilia and flagella. Centrosomes are critical for nucleating microtubule assembly in the cytoplasm. It is not the centrioles themselves that nucleate the cytoplasmic microtubule array but factors in the pericentriolar material. An important component is the *γ-tubulin ring complex (γ-TURC)* (Figures 18-6b and 18-7). This protein complex is located in the pericentriolar material and consists of many copies of γ-tubulin, a form of tubulin distinct from α- and β-tubulin, associated with several other proteins. It is currently believed that γ-TURC acts like a split-washer template to bind αβ-tubulin dimers for the formation of a new microtubule (Figure 18-7b). In addition to nucleating the assembly of microtubules, centrosomes anchor and regulate the dynamics of the (−) end of the microtubules, which are located there.

Basal bodies have a structure similar to each centriole and are the MTOC found at the base of cilia and flagella. The A and B tubules of their triplet microtubules provide a template for the assembly of the microtubules that make up the core structure of cilia and flagella.

KEY CONCEPTS OF SECTION 18.1

Microtubule Structure and Organization

- Tubulin is the major structural component of microtubules and, together with microtubule-associated proteins (MAPs), makes up microtubules (see Figure 18-3).

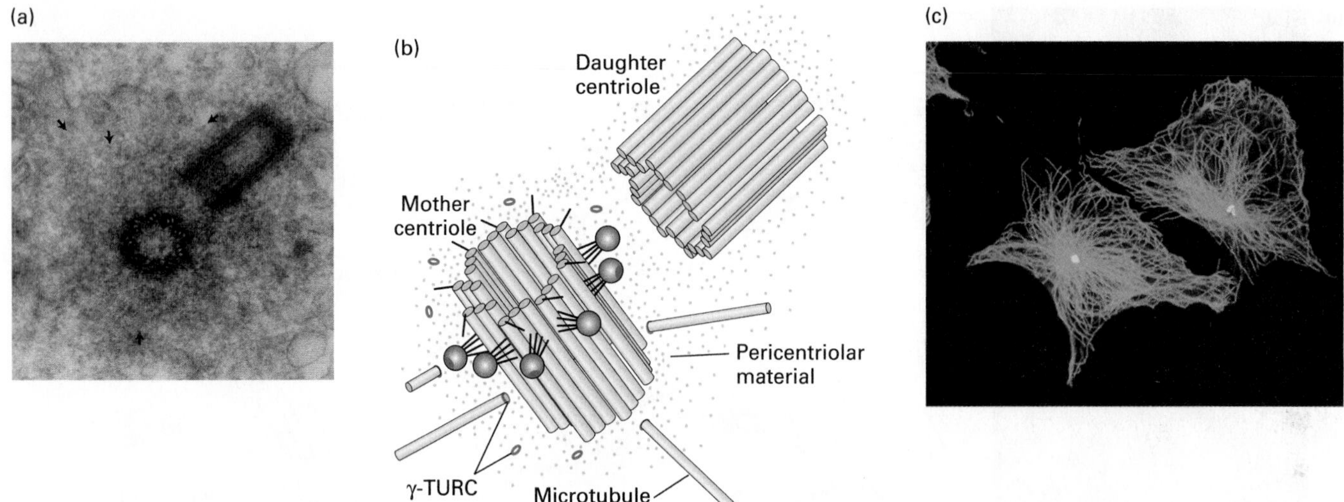

▲ **FIGURE 18-6 Structure of centrosomes.** (a) Thin section of an animal-cell centrosome showing the two centrioles at right angles to each other surrounded by pericentiolar material (arrows). (b) Diagram of a centrosome showing the two centrioles, each of which consists of nine linked outer triplet microtubules embedded in pericentriolar material that contains γ–TURC nucleating structures. (c) Immunofluorescence microscopy showing the microtubule display (green) in a cultured animal cell and the location of the MTOC, using an antibody to a centrosomal protein (yellow). [Parts (a) and (b) from G. Sluder, 2005, *Nature Rev. Mol. Cell Biol.* **6:**743. Part (c) courtesy of R. Kuriyama.]

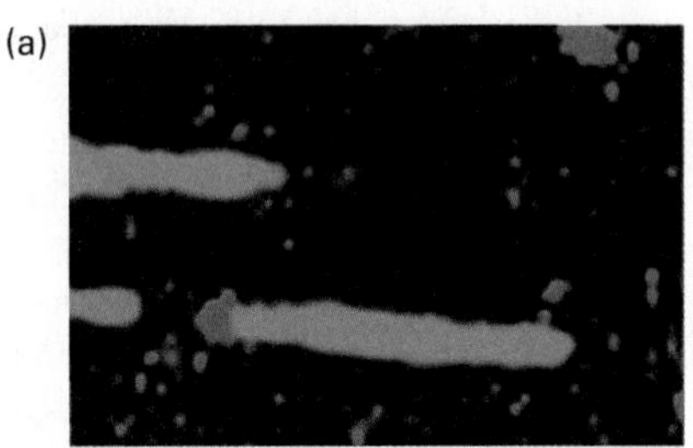

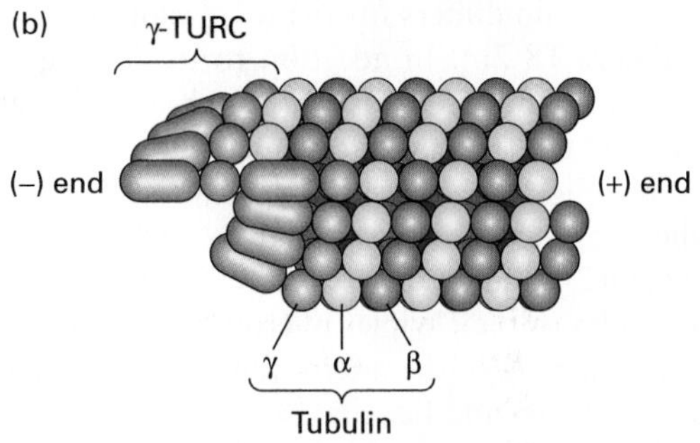

▲ **FIGURE 18-7 The γ-tubulin ring complex (γ-TURC), which nucleates microtubule assembly.** (a) An immunofluorescence micrograph of in vitro assembled microtubules is labeled green and a component of the γ-TURC is labeled red, showing that it is located specifically at one end of the microtubule. (b) Model of how γ-TURC may nucleate assembly of a microtubule by forming a template corresponding to the (−) end of a microtubule. [Part (a) Modified from T. J. Keating and G. G. Borisy, 2000, *Nature Cell Biol.* **2:**352; courtesy of T. J. Keating and G. G. Borisy.]

- Free tubulin exists as an αβ-dimer, with the α-subunit binding a trapped and nonhydrolyzable GTP and the β-subunit binding an exchangeable and hydrolyzable GTP.
- αβ-tubulin assembles into microtubules having 13 laterally associated protofilaments, with an α-subunit exposed at the (−) end and a β-subunit at the (+) end of each protofilament.
- In cilia and flagella, centrioles and basal bodies, doublet or triplet microtubules exist in which the additional microtubules have 10 protofilaments (see Figure 18-4).
- All microtubules are nucleated from microtubule-organizing centers (MTOCs), and many remain anchored with their (−) end there. Thus the end away from the MTOC is always the (+) end.
- The centrosome is the MTOC that nucleates the radial array of microtubules in nonmitotic cells, spindle poles are the MTOCs that nucleate the microtubules of the mitotic spindle, and basal bodies are the MTOCs that assemble microtubules of cilia and flagella (see Figure 18-5).
- Centrosomes consist of two centrioles and the pericentriolar material that contains the γ–TURC microtubule nucleating complex (see Figure 18-6).

18.2 Microtubule Dynamics

Microtubules are dynamic structures due to assembly and disassembly at their ends. The degree of dynamics can vary enormously, with an average microtubule lifetime of less than 1 minute for cells in mitosis and about 5–10 minutes for the microtubules that make up the radial array seen in nonmitotic cells. Microtubule lifetime is longer in axons and much longer in cilia and flagella. To elucidate how these differences occur, we first discuss polymerization properties of microtubule protein and the dynamic behavior of the ends of assembled microtubules.

Microtubules Are Dynamic Structures Due to Kinetic Differences at Their Ends

Microtubules assemble by the polymerization of dimeric αβ-tubulin in a process that is greatly catalyzed by the presence of microtubule-associated proteins (MAPs). Microtubules assembled from a mixture of αβ-tubulin and MAPs, collectively called *microtubular protein,* disassemble when chilled to 4 °C and then reassemble into microtubules again on warming to 37 °C. This property allowed investigators to purify microtubular protein and to explore the mechanism of microtubule assembly and dynamics when a solution of microtubule protein is warmed from 4 °C to 37 °C.

Analysis of the bulk polymerization properties of a solution of microtubule protein reveals that it has several features in common with the polymerization of actin. However, since the assembly properties of microtubules are heavily influenced by the specific MAPs in the reaction mixture with αβ-tubulin, only a few general features will be highlighted here. First, a time course of polymerization reveals a slow nucleation phase, followed by a rapid elongation phase, and then a steady state phase in which assembly is balanced by disassembly, just like the assembly of actin into filaments (see Figure 17-7). Second, for assembly to occur, the αβ-tubulin concentration must be above the *critical concentration* (C_c). Above this concentration, dimers polymerize into microtubules, whereas at concentrations below the C_c, microtubules depolymerize, similar to the behavior of G-actin and F-actin (Figure 18-8a).

Third, at αβ-tubulin concentrations higher than the C_c for polymerization, dimers add faster to one end (Figure 18-9). By analogy with F-actin assembly, the preferred end for assembly is designated the (+) end, which is the end with β-tubulin exposed (see Figure 18-8b). Fourth, the critical concentration is lower at the (+) end than at the (−) end. As a result of this property, at steady state the free αβ–tubulin concentration is higher than the critical concentration at the (+) end but lower than at the (−) end, so subunits are added to the (+) end and subunits dissociate from the (−) end. This results in net addition to the (+) end and loss from the (−) end, so subunits appear to move down a microtubule in a process known as *treadmilling* (see Figure 18-8c), similar to that seen for actin (see Figure 17-10b). Because the intracellular concentration of tubulin (10–20 μM) is much higher than the critical concentration (C_c) for assembly (0.03 μM), polymerization is highly favored in a cell. However, mechanisms exist in cells to regulate where and when microtubules polymerize.

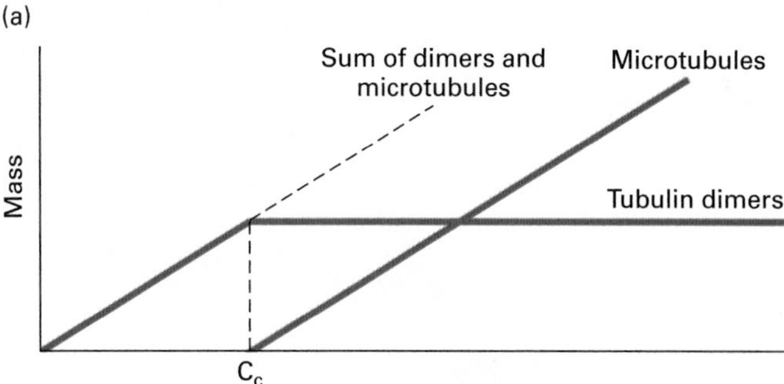

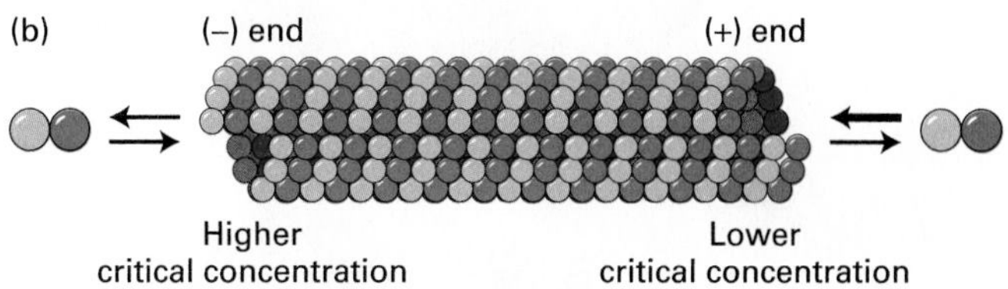

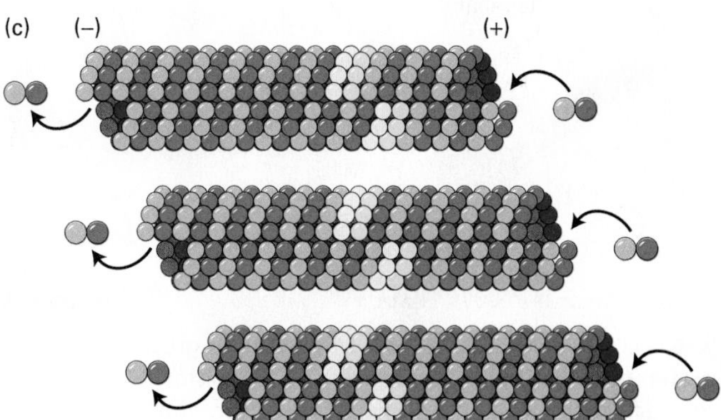

▲ **FIGURE 18-8 Microtubule treadmilling.** (a) αβ-tubulin dimers assemble into microtubules only when present above the critical concentration (C_c). Above C_c, microtubules at steady state are in equilibrium with free αβ-tubulin dimers. (b) The critical concentrations for αβ-tubulin dimers (with bound GTP) at the two ends is different since the rate of addition is higher at the (+) end. (c) At steady state, αβ-tubulin dimers add preferentially to the (+) end and are lost from the (−) end, so that subunits in the microtubule (yellow) appear to move down the microtubule, or "treadmill."

Individual Microtubules Exhibit Dynamic Instability

The properties of microtubule assembly are similar to those of the assembly of actin into filaments when one considers a population of growing microtubules. However, researchers found an additional phenomenon when they examined the behavior of individual microtubules within a population. They did a very simple experiment. Microtubules were assembled in vitro and then sheared to break them into shorter pieces whose length could be analyzed. Under these conditions, one would expect the short microtubules to

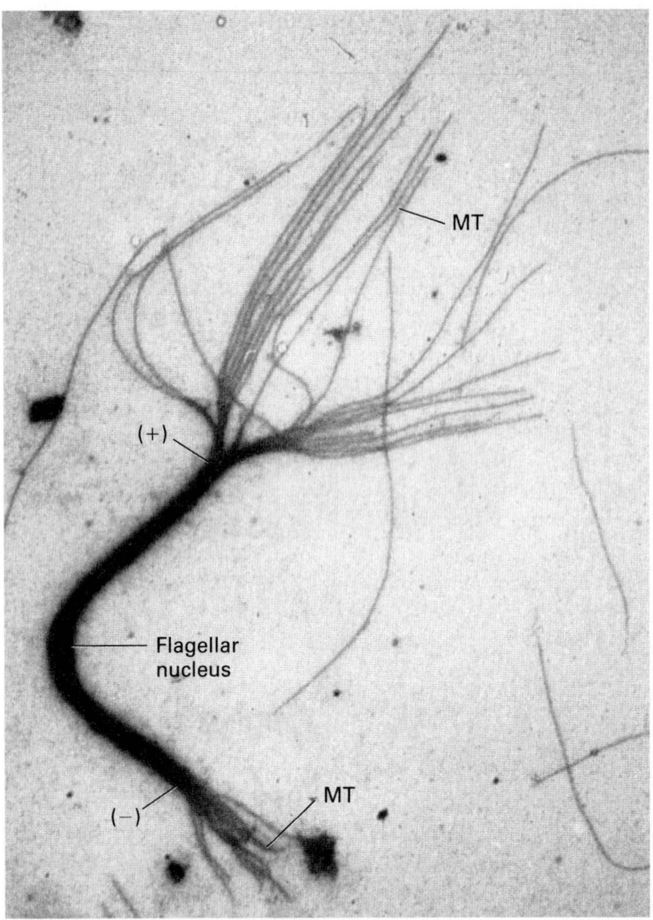

▲ **EXPERIMENTAL FIGURE 18-9 Microtubules grow preferentially at the (+) end.** Fragment of a microtubule bundle from a flagellum was used as a nucleus for the in vitro addition of αβ-tubulin. The nucleating flagellar fragment is the thick bundle seen in this electron micrograph, with the newly formed microtubules (MT) radiating from its ends. The greater length of the microtubules at one end, the (+) end, indicates that tubulin subunits are added preferentially to this end. [Courtesy of G. Borisy.]

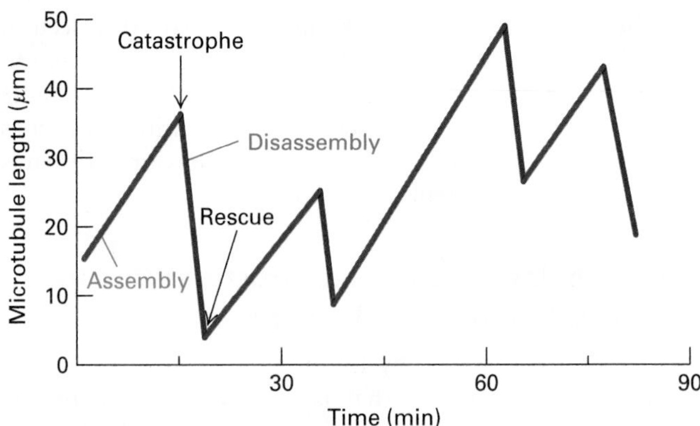

▲ **FIGURE 18-10 Dynamic instability of microtubules in vitro.** Individual microtubules can be observed in the light microscope and their lengths plotted at different times during assembly and disassembly. Assembly and disassembly each proceed at uniform rates, but there is a big difference between the rate of assembly and that of disassembly, as seen in the different slopes of the lines. Shortening of a microtubule is much more rapid (7 μm/min) than growth (1 μm/min). Notice the abrupt transitions to the shrinkage stage (catastrophe) and to the elongation stage (rescue). [Adapted from P. M. Bayley, K. K. Sharma, and S. R. Martin, 1994, in *Microtubules*, Wiley-Liss, p. 118.]

treadmill. However, the researchers found that some of the microtubules grew in length, whereas others shortened very rapidly—thus indicating two distinct populations of microtubules. Further studies showed that individual microtubules could grow and then suddenly undergo a *catastrophe* to a shrinking phase. Moreover, sometimes a depolymerizing microtubule end could go through a *rescue* and begin growing again (Figure 18-10). Although this phenomenon was first seen in vitro, analysis of fluorescently labeled tubulin microinjected into live cells showed that microtubules in cells also undergo periods of growth and shrinkage (Figure 18-11). This process of alternating between growing and shrinking states is known as *dynamic instability*. Thus dynamic instability is a feature of individual microtubule ends distinct from the ability of microtubules to treadmill by dimer addition at the (+) end and loss at the (−) end.

What is the molecular basis of dynamic instability? If you look carefully by electron microscopy at the ends of growing and shrinking microtubules, you can see they are quite different. A growing microtubule has a blunt end, whereas a depolymerizing end has protofilaments peeling off like ram's horns (Figure 18-12).

Recent studies have provided a simple structural explanation for the two classes of microtubule ends. As we noted above, the β-subunit of the αβ-tubulin dimer is exposed on the (+) end of each protofilament. Using GTP and GDP analogs, researchers have found that artificially made *single* protofilaments—where there are no lateral interactions—made up of repeating αβ-tubulin dimers containing GDP-β-tubulin are curved, like a ram's horn. However, artificially made single protofilaments made up of αβ-tubulin dimers with GTP-β-tubulin are straight. Thus growing microtubules with blunt ends terminate in GTP-β-tubulin, whereas shrinking ones with curled ends terminate in GDP-β-tubulin. Therefore if the GTP in the terminal β-tubulins of a microtubule become hydrolyzed, as will inevitably happen in a random manner, a formerly blunt-end growing microtubule will curl and a catastrophe ensue. These relationships are summarized in Figure 18-12.

These results have an additional implication. For this we have to consider in more detail a growing microtubule. The addition of another dimer to the protofilament (+) end of a growing microtubule involves an interaction between the new α-subunit and the terminal β-subunit. This interaction stimulates the hydrolysis of the GTP to GDP in the former terminal β-subunit. However, the β-tubulin in the newly added dimer contains GTP. Thus a

Video: Cytoplasmic Assembly of Microtubules in Cultured Cells

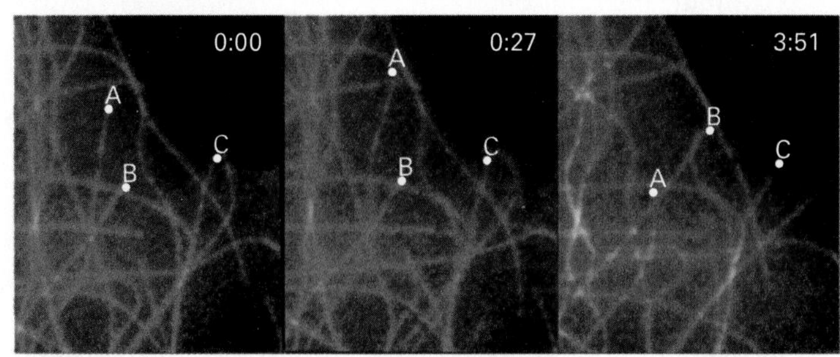

◄ **EXPERIMENTAL FIGURE 18-11 Fluorescence microscopy reveals growth and shrinkage of individual microtubules in vivo.** Fluorescently labeled tubulin was microinjected into cultured human fibroblasts. The cells were chilled to depolymerize preexisting microtubules into tubulin dimers and were then incubated at 37 °C to allow repolymerization, thus incorporating the fluorescent tubulin into all the cell's microtubules. A region of the cell periphery was viewed in the fluorescence microscope at 0 second, 27 seconds later, and 3 minutes, 51 seconds later (left to right panels). In this period, several microtubules elongate and shorten. The letters and white dots mark the position of ends of three microtubules. [Modified from P. J. Sammak and G. Borisy, 1988, *Nature* **332:**724.]

growing microtubule protofilament has GDP-β-tubulin down its length and is capped by GTP-β-tubulin. As we mentioned above, an *isolated* protofilament containing GDP-β-tubulin is curved along its length, so when it is present in a microtubule, why doesn't it break out and peel away? The lateral protofilament-protofilament interactions in the β-tubulin-GTP cap are sufficiently tight that they do not allow the microtubule to unpeel at its end—and so the protofilaments behind the GTP-β-tubulin cap are constrained from unpeeling (see Figure 18-12). The energy released by GTP hydrolysis of the subunits behind the cap is stored within the lattice as structural strain waiting to be released when the GTP-β-tubulin is lost. If the GTP-β-tubulin is lost, the stored energy can do work if some structure, such as a chromosome, is attached to the disassembling microtubule end.

Thus we see that the ability of β-tubulin to bind and hydrolyze GTP has three important consequences for microtubule biology:

- Distinct critical concentrations at the two ends give rise to treadmilling
- The (+) end is subject to dynamic instability
- Microtubules can store torsional energy to do work

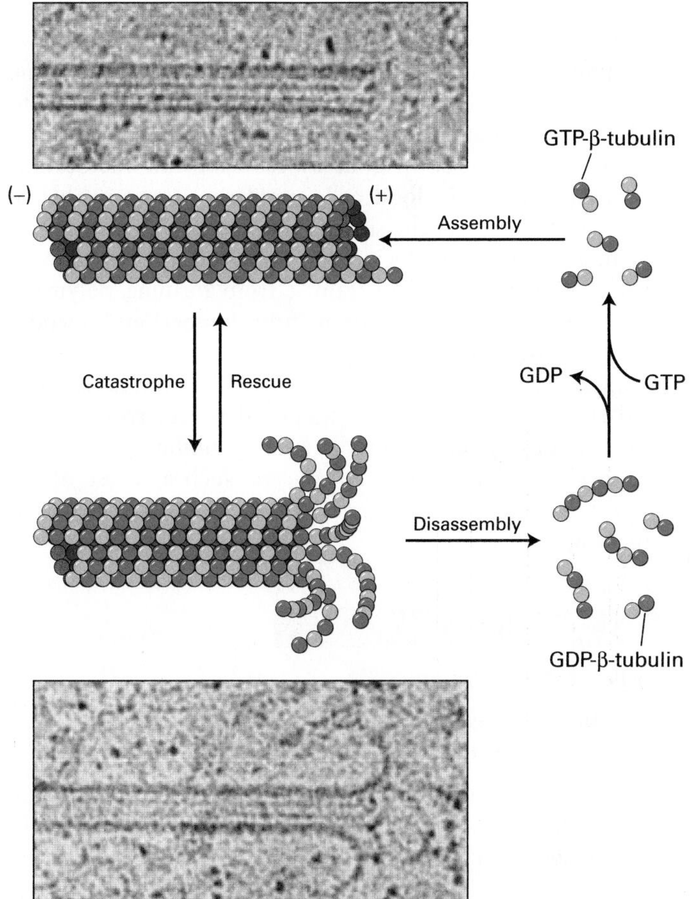

◄ **FIGURE 18-12 Dynamic instability depends on the presence or absence of a GTP-β-tubulin cap.** Images taken in the electron microscope of frozen samples of a growing microtubule (upper) and a shrinking microtubule (lower). Notice that the end of the growing microtubule has a blunt end, whereas the shrinking one has curls like a ram's horns. The diagram shows that a microtubule with GTP-β-tubulin on the end of each protofilament is strongly favored to grow. However, a microtubule with GDP-β-tubulin at the end of the protofilaments forms a curved structure and will undergo rapid disassembly. Switching between growing and shrinking, called rescues and catastrophes, can occur, and the rate of switching is regulated by associated proteins. [Images from E-M Mandelkow et al., 1991, *J. Cell Biol.* **114:**977. Diagram modified from A. Desai and T. J. Mitchison, 1997, *Annu. Rev. Cell Dev. Biol.* **13:**83–117.]

Localized Assembly and "Search-and-Capture" Help Organize Microtubules

We have now presented two concepts relating to microtubule organization and (+) end dynamics: microtubules are assembled from localized sites known as MTOCs, and individual microtubules can undergo dynamic instability. Together these two processes contribute to the distribution of microtubules in cells.

In an interphase cell growing in culture, microtubules are constantly being nucleated from the centrosome and spreading out randomly "searching" the cytoplasmic space. The frequency of catastrophes and rescues, together with growth and shrinkage rates, determines the length of each microtubule—if the microtubule is subject to a high catastrophe frequency and low rescue, it will shrink back to the cell center and disappear, whereas if it has few catastrophes and is readily rescued, it will continue to grow. If the searching microtubule encounters a structure or organelle that stabilizes its (+) end to protect it from catastrophes—thereby "capturing" it—the microtubule end will remain attached to the structure, whereas unattached microtubules will have a greater frequency of being disassembled. So the dynamics of the microtubule end is a very important determinant of microtubule life cycle and function. "Search and capture" is part of the mechanism determining the overall organization of microtubules in a cell. Moreover, by changing the rate of nucleation or local microtubule dynamics and capture sites, a cell can rapidly change its overall microtubule distribution.

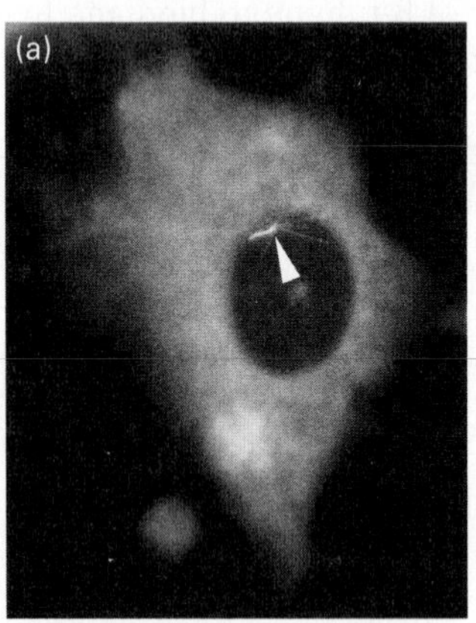

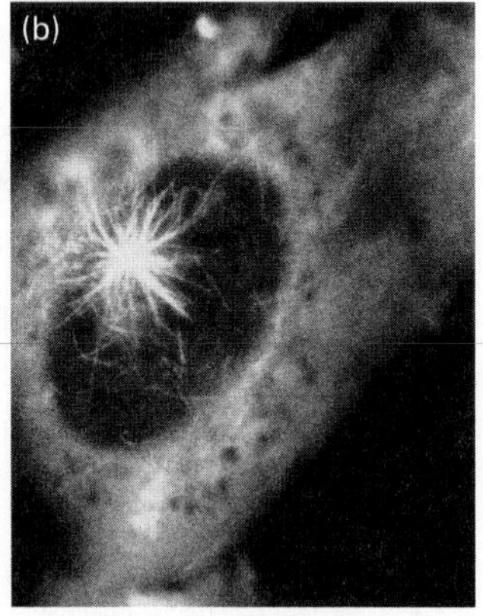

▲ **EXPERIMENTAL FIGURE 18-13 Microtubules grow from the MTOC.** To investigate from where microtubules assemble in vivo, a cultured fibroblast was treated with colchicine until almost all the cytoplasmic microtubules were disassembled. The cell was then stained with antibodies to tubulin and viewed by immunofluorescence microscopy (panel [a]). The colchicine was then washed out to allow the reassembly of microtubules. Panel (b) shows the first stages of reassembly, revealing microtubules growing from a central region above the nucleus (dark areas). Note in panel (a) the remaining primary cilium (arrowhead) associated with the centrosome that is not depolymerized by colchicine treatment under these conditions. Note also the fluorescence from the cytoplasm, which is from unpolymerized αβ-tubulin dimers. [From M. Osborn and K. Weber, 1976, *Proc. Natl. Acad. Sci. USA* **73:**867–871.]

Drugs Affecting Tubulin Polymerization Are Useful Experimentally and to Treat Diseases

The conserved nature of tubulins and their essential involvement in critical processes such as mitosis make them prime targets for both naturally occurring and synthetic drugs that affect polymerization or depolymerization. Historically, the first known such drug was colchicine, present in extracts of the meadow saffron, which binds tubulin dimers so that they cannot polymerize into a microtubule. Since most microtubules are in a dynamic state between dimers and polymers, the addition of colchicine sequesters all free dimers in the cytoplasm, resulting in loss of microtubules due to their natural turnover. Treatment of cultured cells with colchicine for a short time results in the depolymerization of all the cytoplasmic microtubules, leaving the more stable tubulin-containing centrosome (Figure 18-13a). Also seen after colchicine treatment is the primary cilium, a solitary cilium on the surface of the cell that is assembled from one of the centrioles acting as its basal body (discussed below). When the colchicine is washed out to allow regrowth of the microtubules, they can be seen to grow from the centrosome, revealing its ability to nucleate new microtubule assembly (Figure 18-13b).

Colchicine has been used for hundreds of years to relieve the joint pain of acute gout—a famous patient was King Henry VIII of England, who was treated with colchicine to relieve this ailment. A low level of colchicine relieves the inflammation caused in gout by reducing the microtubule dynamics of white blood cells, rendering them unable to migrate efficiently to the site of inflammation.

In addition to colchicine, a number of other drugs bind the tubulin dimer and restrain it from forming polymers. These include podophyllotoxin (from juniper) and nocodazole (a synthetic drug).

Taxol, a plant alkaloid from the Pacific yew tree, binds and stabilizes microtubules against depolymerization. Because taxol stops cells from dividing by inhibiting mitosis, it has been used to treat some cancers, such as those of the breast and ovary, where the cells are especially sensitive to the drug. ▮

KEY CONCEPTS OF SECTION 18.2

Microtubule Dynamics

- Microtubules can treadmill using the energy of GTP hydrolysis (see Figure 18-8).
- Individual microtubule (+) ends can undergo dynamic instability, with alternating periods of growth or rapid shrinkage (see Figure 18-10).
- The β-tubulin in all microtubules binds primarily GDP. However, growing microtubules (+) ends are capped by

GTP-β-tubulin and have blunt ends, whereas shrinking microtubules have lost the GTP-β-tubulin cap and the protofilaments peel outward and disassemble (see Figure 18-12).

- Growing microtubules store the energy from GTP hydrolysis in the microtubule lattice, so they have the potential to do work when disassembling.
- Microtubules assembled from the centrosome and exhibiting dynamic instability can "search" the cytoplasm and become "captured" by structures or organelles that stabilize their (+) end. In this way, assembly coupled with "search and capture" can contribute to the overall distribution of microtubules in a cell.

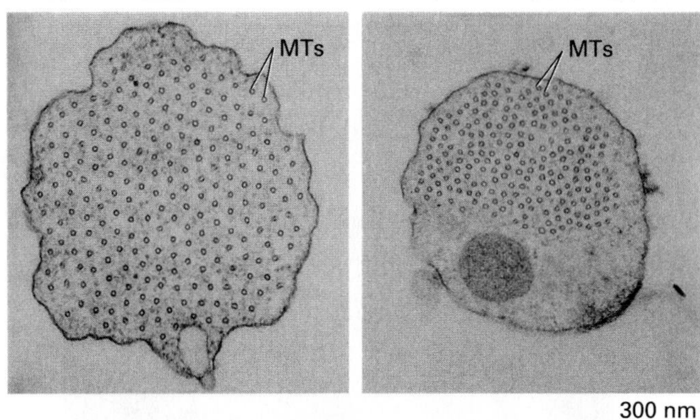

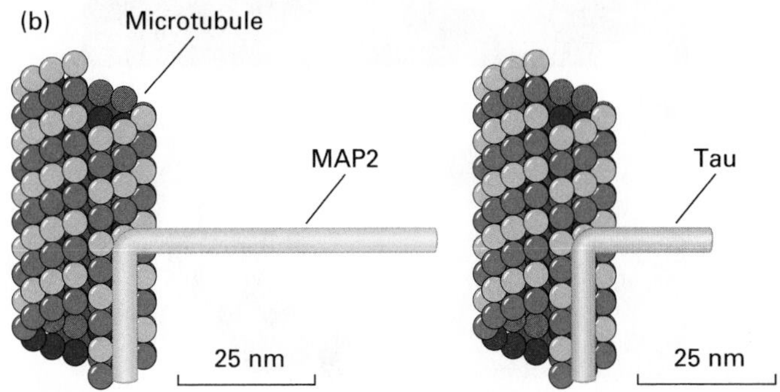

▲ EXPERIMENTAL FIGURE 18-14 **Spacing of microtubules depends on the length of the projection domain of microtubule-associated proteins.** Insect cells transfected to express MAP2, which has a long arm, or to express Tau protein, which has a short arm, grow long axonlike processes. (a) Electron micrographs of cross sections through the processes induced by the expression of MAP2 (*left*) or Tau (*right*) in transfected cells. Note that the spacing between microtubules (MTs) in MAP2-containing cells is larger than in Tau-containing cells. Both cell types contain approximately the same number of microtubules, but the effect of MAP2 is to enlarge the caliber of the axonlike process. (b) Diagrams of association between microtubules and MAPs. Note the difference in the lengths of the projection arms in MAP2 and Tau. [Part (a) from J. Chen et al., 1992, *Nature* **360:**674.]

18.3 Regulation of Microtubule Structure and Dynamics

Although the wall of microtubules is built from αβ-tubulin dimers, highly purified αβ-tubulin will assemble in vitro into microtubules only under special unphysiological conditions. Assembly of microtubules in vitro under physiological conditions requires the presence of stabilizing microtubule associated proteins, or MAPs. Stabilizing MAPs represent just one class of protein that interacts with tubulin in microtubules; other classes modify the growth properties of microtubules or destabilize them. We discuss the various classes separately.

Microtubules Are Stabilized by Side- and End-Binding Proteins

Several different classes of proteins stabilize microtubules, many of them showing cell-type-specific expression. Among the best studied are the *tau* family of proteins, which includes tau itself, MAP2, and MAP4. These proteins have a modular design, with 18-residue positively charged sequences, repeated three to four times, that binds to the negatively charged tubulin surface and a domain that projects from the microtubule wall (Figure 18-14). Tau proteins are believed to stabilize microtubules and also to act as spacers between them (Figure 18-14). MAP2 is found only in dendrites, where it forms fibrous cross-bridges between microtubules and links microtubules to intermediate filaments. Tau, which is much smaller than most other MAPs, is present in both axons and dendrites. The basis for this selectivity is still a mystery.

When stabilizing MAPs coat the outer wall of a microtubule, they can increase the growth rate of microtubules or suppress the catastrophe frequency. In many cases, the activity of the MAPs is regulated by the reversible phosphorylation of their projection domain. Phosphorylated MAPs are unable to bind to microtubules; thus phosphorylation promotes microtubule disassembly. For example, microtubule-affinity-regulating kinase (MARK/Par-1) is a key modulator of tau proteins. Some MAPs are also phosphorylated by a **cyclin-dependent kinase** (**CDK**) that plays a major role in controlling the activities of proteins in the course of the cell cycle (Chapter 20).

Some MAPs have recently been found to have the surprising ability to associate with the (+) ends of microtubules and in some cases only the (+) ends of growing, not shrinking ones (Figure 18-15). This class of proteins is known as +TIPs, and they perform varied functions when present at the microtubule tip. Some +TIPs selectively stabilize the (+) end against a catastrophe or enhance the frequency of rescues, thus promoting the continued growth of the microtubule. Other +TIPs are attachment proteins, so that when the growing microtubule encounters a structure or organelle, it can become attached to it. For example, microtubules extending into the leading edge of a migrating

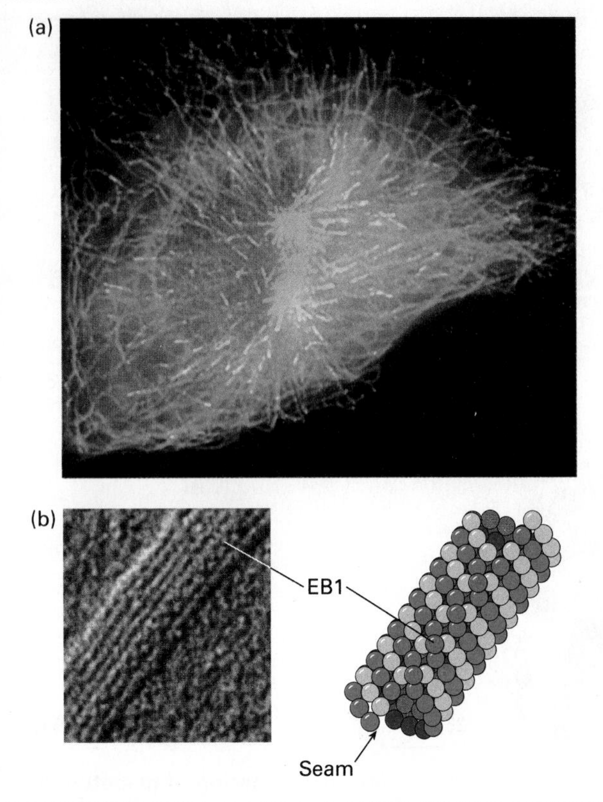

◀ **EXPERIMENTAL FIGURE 18-15 The +TIP protein EB1 associates with the (+) ends of microtubules.** (a) A cultured cell stained with antibodies to tubulin (green) and the + TIP protein EB1 (red). EB1 is enriched in the region of the microtubule (+) end. (b) In vitro studies have shown that EB1 binds specifically to the microtubule seam, as seen in the enhanced electron micrograph and model, but how this binding leads to enrichment at the (+) end is not yet known. [Part (a) courtesy of T. Wittmann; part (b) modified from Sandblad et al. 2006 *Cell* **127:**1415.]

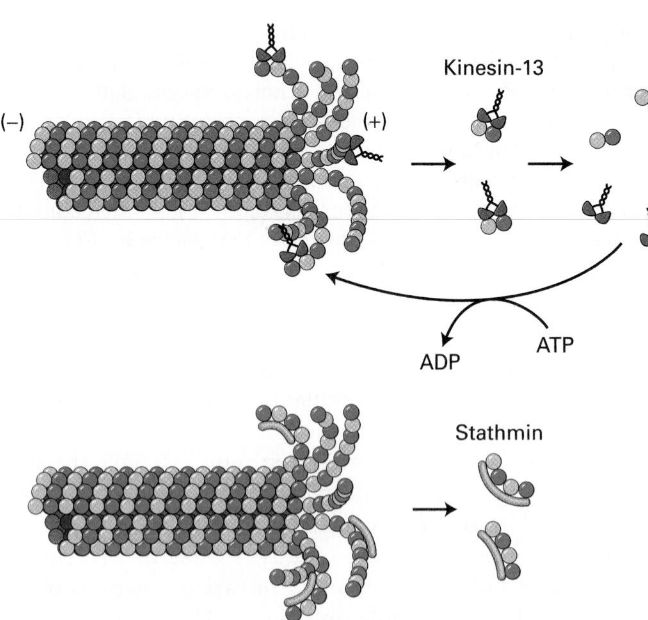

▲ **FIGURE 18-16 Proteins that destabilize the ends of microtubules.** (a) A member of the kinesin-13 family enriched at microtubule ends can enhance the disassembly of that end. These proteins are ATPases, and ATP enhances their activity by dissociating them from the αβ-tubulin dimer. (b) Stathmin binds selectively to curved protofilaments and enhances their dissociation from a microtubule end. Stathmin's activity is inhibited by phosphorylation.

cell can become attached and stabilized by binding components there.

Microtubules Are Disassembled by End Binding and Severing Proteins

Just as for microfilaments, mechanisms exist for enhancing the disassembly of microtubules. Although most of the regulation of microtubule dynamics appears to happen at the (+) end, in some situations, such as in mitosis, it can occur at both ends.

Three mechanisms for microtubule destabilization are known. One of these involves the kinesin-13 family of proteins (Figure 18-16). The kinesin-13 proteins bind and curve the end of the tubulin protofilaments into the GDP-β-tubulin conformation. They then facilitate the removal of terminal tubulin dimers, thereby greatly enhancing the frequency of catastrophes. They act catalytically in the sense that they need to hydrolyze ATP to sequentially remove terminal tubulin dimers (Figure 18-16).

Another protein, known as Op18/stathmin, was originally identified as a protein highly overexpressed in certain cancers; hence part of its name (Oncoprotein 18). It is a small protein that binds two tubulin dimers in a curved, GDP-β-tubulin-like conformation, which also enhances the rate of catastrophes. It may work by enhancing the hydrolysis of the GTP in the terminal tubulin dimer and aiding in its dissociation from the end of the microtubule. As might be expected for a regulator of microtubule ends, it is subject to negative regulation by phosphorylation by a large variety of kinases. In fact, it has been found that Op18/stathmin is inactivated by phosphorylation near the leading edge of motile cells, which contributes to preferential growth of microtubules toward the front of the cell.

A third mechanism for destabilizing microtubules is through the action of a protein known as katanin (from the Japanese for "sword"). Katanin plays a role in MTOCs, where it severs and releases anchored microtubules.

KEY CONCEPTS OF SECTION 18.3

Regulation of Microtubule Structure and Dynamics

- Microtubules can be stabilized by side-binding microtubule-associated proteins (MAPs) (see Figure 18-14).

■ Some MAPs, called +TIPs, bind selectively to the (+) end of microtubules and can alter the dynamic properties of the microtubule or localize components to the searching (+) end of the microtubule (see Figure 18-15).

■ Microtubule ends can be destabilized by some proteins, such as the kinesin-13 family of proteins and stathmin, to enhance the frequency of catastrophes (see Figure 18-16).

18.4 Kinesins and Dyneins: Microtubule-Based Motor Proteins

Organelles in cells are frequently transported distances of many micrometers along well-defined routes in the cytosol and delivered to particular intracellular locations. Diffusion alone cannot account for the rate, directionality, and destinations of such transport processes. Findings from early experiments with fish-scale pigment cells and nerve cells first demonstrated that microtubules function as tracks in the intracellular transport of various types of "cargo."

As already discussed, polymerization and depolymerization of microtubules can do work using the energy provided by GTP hydrolysis. In addition, **motor proteins** move along microtubules powered by ATP hydrolysis. Two main families of motor proteins—**kinesins** and **dyneins**—are known to mediate transport along microtubules. In this section we discuss how these motor proteins work and the functions they perform in interphase cells. In subsequent sections, we discuss their functions in cilia and flagella and in mitosis.

Organelles in Axons Are Transported Along Microtubules in Both Directions

A neuron must constantly supply new materials—proteins and membranes—to an axon terminal to replenish those lost in the exocytosis of neurotransmitters at the junction (synapse) with another cell (Chapter 23). Because proteins and membranes are primarily synthesized in the cell body, these materials must be transported down the axon, which can be as much as a meter in length, to the synaptic region. This movement of materials is accomplished on microtubules, which are all oriented with their (+) ends toward the axon terminal (see Figure 18-5e).

The results of classic pulse-chase experiments in which radioactive precursors were microinjected into the dorsal-root ganglia near the spinal cord and then tracked along their nerve axons showed that **axonal transport** occurs in both directions. *Anterograde* transport proceeds from the cell body to the synaptic terminals and is associated with axonal growth and the delivery of synaptic vesicles. In the opposite, *retrograde,* direction, "old" membranes from the synaptic terminals move along the axon rapidly toward the cell body, where they will be degraded in lysosomes. Findings from such experiments also revealed that different

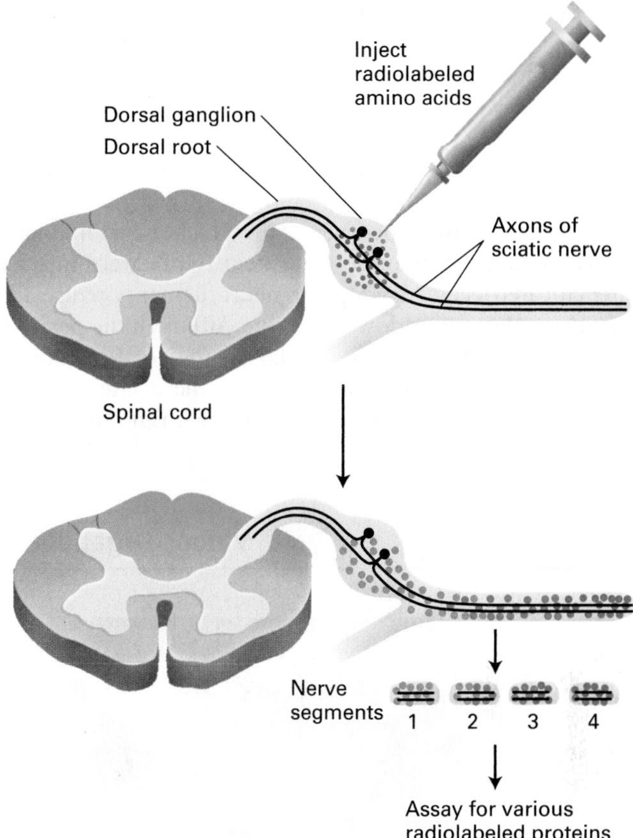

▲ EXPERIMENTAL FIGURE 18-17 **The rate of axonal transport in vivo can be determined by radiolabeling and gel electrophoresis.** The cell bodies of neurons in the sciatic nerve are located in dorsal-root ganglia (near the spinal cord). Radioactive amino acids injected into these ganglia in experimental animals are incorporated into newly synthesized proteins, which are then transported down the axon to the synapse. Animals are sacrificed at various times after injection and the dissected sciatic nerve is cut into small segments to see how far radioactively labeled proteins have been transported; these proteins can be identified after gel electrophoresis and autoradiography. The red, blue, and purple dots represent groups of proteins that are transported down the axon at different rates, red most rapidly, purple least rapidly.

materials move at different speeds (Figure 18-17). The fastest-moving material, consisting of membrane-limited vesicles, has a velocity of about 3 μm/s, or 250 mm/day—requiring about four days to travel from a cell body in your back down an axon that terminates in your big toe. The slowest-moving material, comprising tubulin subunits and neurofilaments (the intermediate filaments found in neurons), moves only a fraction of a millimeter per day. Organelles such as mitochondria move down the axon at an intermediate rate.

Axonal transport can be directly observed by video microscopy of cytoplasm extruded from a squid giant axon. The movement of vesicles along microtubules in this cell-free system requires ATP, its rate is similar to that of fast axonal transport in intact cells, and it can proceed in both the anterograde and the retrograde directions (Figure 18-18a). Electron microscopy of the same region of the axon cytoplasm reveals organelles attached to individual microtubules (Figure 18-18b). These pioneering in vitro experiments established definitively that organelles move along individual microtubules and that their movement requires ATP.

Findings from experiments in which neurofilaments tagged with green fluorescent protein (GFP) were injected into cultured cells suggest that neurofilaments pause frequently as they move down an axon. Although the peak velocity of neurofilaments is similar to that of fast-moving vesicles, their numerous pauses lower the average rate of transport. These findings suggest that there is no fundamental difference between fast and slow axonal transport, although why neurofilament transport stops periodically is unknown.

Kinesin-1 Powers Anterograde Transport of Vesicles Down Axons Toward the (+) End of Microtubules

The protein responsible for anterograde organelle transport was first purified from axonal extracts. Researchers found that by mixing three components—purified organelles from squid axons, an organelle-free cytoplasmic axonal extract, and taxol-stabilized microtubules—organelles could be seen moving on the microtubules in an ATP-dependent manner.

Video: Organelle Movement Along Microtubules in a Squid Axon 

MEDIA CONNECTIONS

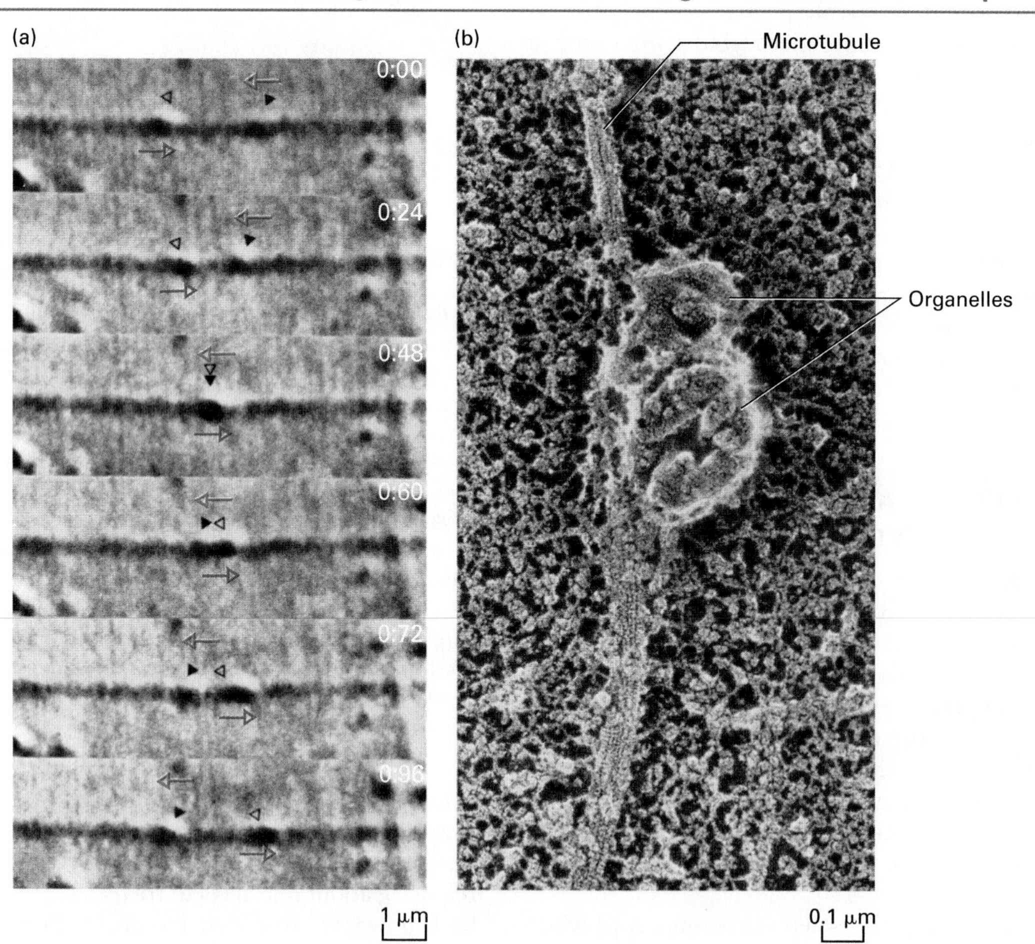

▲ **EXPERIMENTAL FIGURE 18-18 DIC microscopy demonstrates microtubule-based vesicle transport in vitro.** (a) The cytoplasm was squeezed from a squid giant axon with a roller onto a glass coverslip. After buffer containing ATP was added to the preparation, it was viewed by differential interference contrast (DIC) microscopy, and the images were recorded on videotape. In the sequential images shown, the two organelles indicated by open and solid triangles move in opposite directions (indicated by colored arrows) along the same filament, pass each other, and continue in their original directions. Elapsed time in seconds appears at the upper-right corner of each video frame. (b) A region of cytoplasm similar to that shown in part (a) was freeze-dried, rotary shadowed with platinum, and viewed in the electron microscope. Two large structures attached to one microtubule are visible; these structures presumably are small vesicles that were moving along the microtubule when the preparation was frozen. [See B. J. Schnapp et al., 1985, *Cell* **40:**455; courtesy of B. J. Schnapp, R. D. Vale, M. P. Sheetz, and T. S. Reese.]

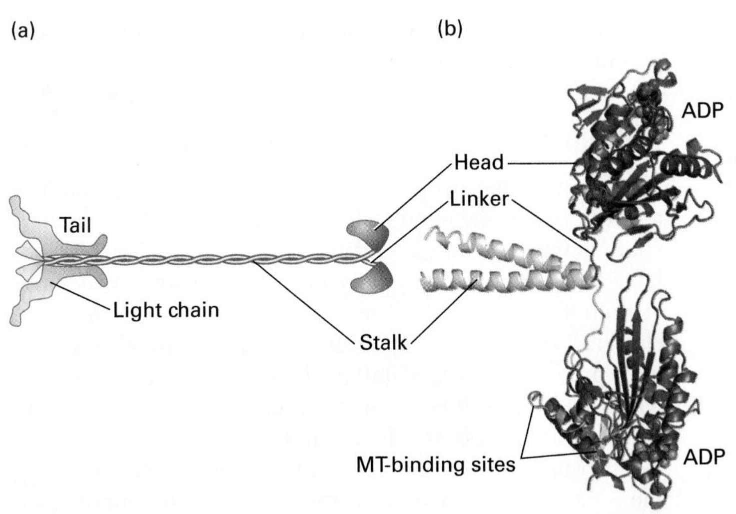

◀ **FIGURE 18-19 Structure of kinesin-1.** (a) Representation of kinesin-1, with its two heavy chains each with a motor domain and associated light chains. Each head is attached to the coiled-coil stalk by a flexible linker domain. (b) X-ray structure of the kinesin heads with the microtubule and nucleotide binding sites indicated. [Part (a) modified from R. D. Vale, 2003, *Cell* **112**:467. Part (b) courtesy of E. Mandelkow and E. M. Mandelkow, adapted from M. Thormahlen et al., 1998, *J. Struc. Biol.* **122**:30.]

However, if they omitted the axonal extract, the organelles neither bound nor moved along the microtubules, suggesting that the extract contributes a motor protein. The strategy for purifying the motor protein was based on additional observations of organelles moving on microtubules. It was known that if ATP was hydrolyzed to ADP, the organelles fell off the microtubules. However, if the nonhydrolyzable ATP analog AMPPNP was added, organelles remained associated with microtubules but did not move. This suggested that the motor linked the organelles to the microtubules very tightly in AMPPNP but then was released from the microtubule when the AMPPNP was replaced by ATP and its subsequent hydrolysis to ADP. Researchers used this clue to purify the motor.

Kinesin-1 isolated from squid giant axons is a dimer of two heavy chains, each associated with a light chain, with a total molecular weight of about 380,000. The molecule comprises a pair of globular *head domains* connected by a short flexible *linker domain* to a long *central stalk* and terminating in a pair of small globular *tail domains,* which associate with the light chains (Figure 18-19). Each domain carries out a particular function: the head domain, which binds microtubules and ATP, is responsible for the motor activity of kinesin; the linker domain is critical for forward motility; the stalk domain is involved in dimerization of the two heavy chains; and the tail domain is responsible for binding to receptors on the membrane of cargoes.

Kinesin-1-dependent movement of vesicles can be tracked by in vitro motility assays similar to those used to study myosin-dependent movements (see Figure 17-21). In one type of assay, a vesicle or a plastic bead coated with kinesin-1 is added to a glass slide along with a preparation of stabilized microtubules. In the presence of ATP, the beads can be observed microscopically to move along a microtubule in one direction. Researchers found that the beads coated with kinesin-1 always moved from the (−) to the (+) end of a microtubule (Figure 18-20). Thus kinesin-1 is a (+) end–directed microtubule motor protein, and additional evidence shows that it mediates anterograde axonal transport.

Kinesins Form a Large Protein Family with Diverse Functions

Following the discovery of kinesin-1, a number of proteins with kinesin-related motor domains were identified both in genetic screens and using molecular biology approaches. There are now 14 known classes of kinesins in animals, defined as sharing amino acid sequence homology with the motor domain of kinesin-1. The family of kinesin-related proteins is encoded by about 45 genes in the human genome. Although the function of all of these proteins has not yet

Video: Kinesin-1-Driven Transport of Vesicles Along Microtubules in Vitro

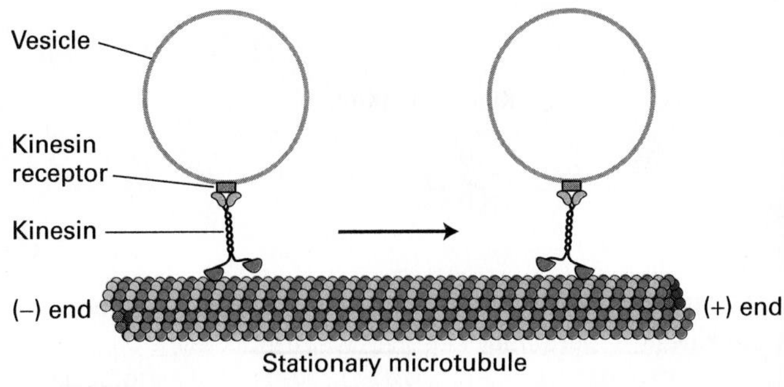

▲ **FIGURE 18-20 Model of kinesin-1-catalyzed vesicle transport.** Kinesin-1 molecules, attached to receptors on the vesicle surface, transport the vesicles from the (−) end to the (+) end of a stationary microtubule. ATP is required for movement. [Adapted from R. D. Vale et al., 1985, *Cell* **40**:559, and T. Schroer et al., 1988, *J. Cell Biol.* **107**:1785.]

MEDIA CONNECTIONS

been elucidated, some of the best-studied kinesins are involved in processes such as organelle, mRNA, and chromosome transport and microtubule sliding and microtubule depolymerization.

As with the different classes of myosin motors, the kinesin-related motor domain is fused to a variety of class-specific nonmotor domains (Figure 18-21). Whereas kinesin-1 has two heavy and two light chains, members of the kinesin-2 family, also involved in organelle transport, have two different related heavy-chain motor domains and a third polypeptide that associates with the tail and binds cargo. Members of the bipolar kinesin-5 family have four heavy chains, forming bipolar motors that can pull on antiparallel microtubules toward the (+) ends. The kinesin-14 motors are the only known class to move toward the (−) end of a microtubule; this class functions in mitosis. Yet another type, the kinesin-13 family, have two subunits but with the kinesin-related domain in the middle of the polypeptide. Kinesin-13 proteins do not have motor activity, but recall that these are special ATP-hydrolyzing proteins that can enhance the depolymerization of microtubule ends (see Figure 18-16).

Kinesin-1 Is a Highly Processive Motor

How does kinesin-1 move down a microtubule? Optical trap and fluorescent-labeling techniques similar to those used to characterize myosin (see Figures 17-26, 17-27, and 17-28) have been used to study how kinesin-1 moves down a microtubule and how ATP hydrolysis is converted into mechanical work. Such experiments demonstrate that it is a very processive motor—taking hundreds of steps walking hand over hand down a microtubule without dissociating. During this process, the double-headed molecule takes 8-nm steps from one β-tubulin subunit to the next, tracking down the same protofilament in the microtubule. This entails each *individual* head taking 16-nm steps. The two heads work in a highly coordinated manner so that one is always attached to the microtubule.

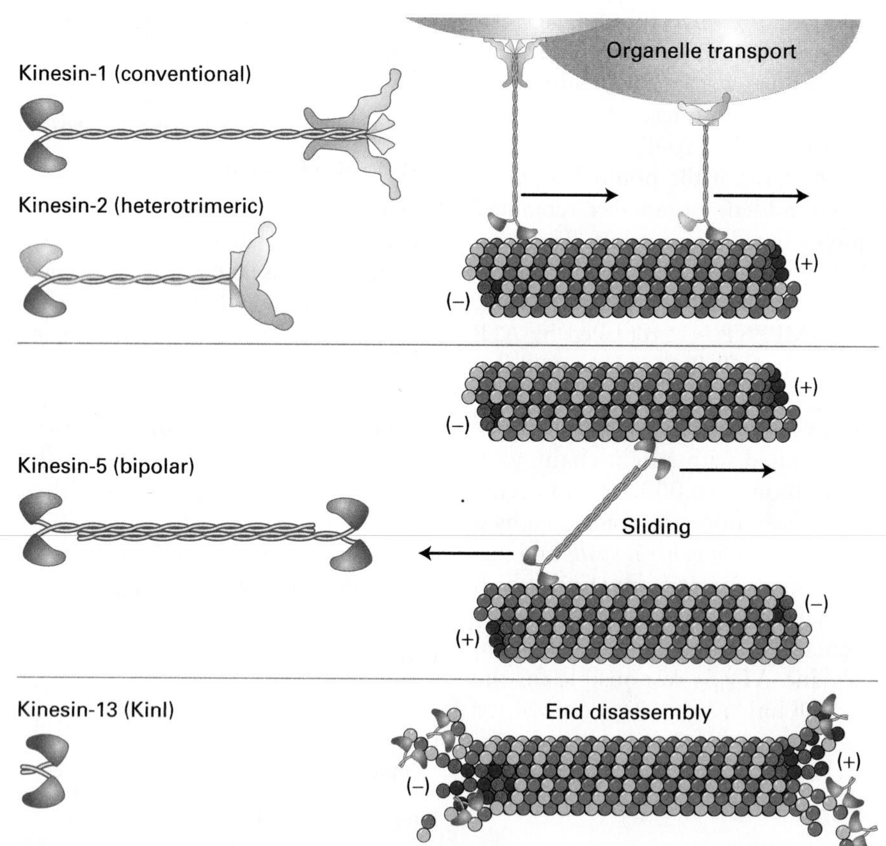

▲ **FIGURE 18-21 Structure and function of selected members of the kinesin superfamily.** Kinesin-1, which includes the original kinesin isolated from squid axons, is a (+) end–directed microtubule motor involved in organelle transport. The kinesin-2 family have two different, but closely related, heavy chains, and a third cargo binding subunit; this class also transports organelles in a (+) end–directed manner. The kinesin-5 family have four heavy chains assembled in a bipolar manner to interact with two antiparallel microtubules and also move toward the (+) end. Kinesin-13 family members have the motor domain in the middle of their heavy chains and do not have motor activity but destabilize microtubule ends. Additional kinesin family members are mentioned in the text. Different kinesins have been given many different names, some of which are shown in parentheses. We use here the unified nomenclature described in C. J. Lawrence et al., 2004, *J. Cell Biol.* 167:19–22. [Diagrams modified from R. D. Vale, 2003, *Cell* **112:**467.]

Podcast: Kinesin Movement Along a Microtubule
Animation: Kinesin-Microtubule Stepping Cycle

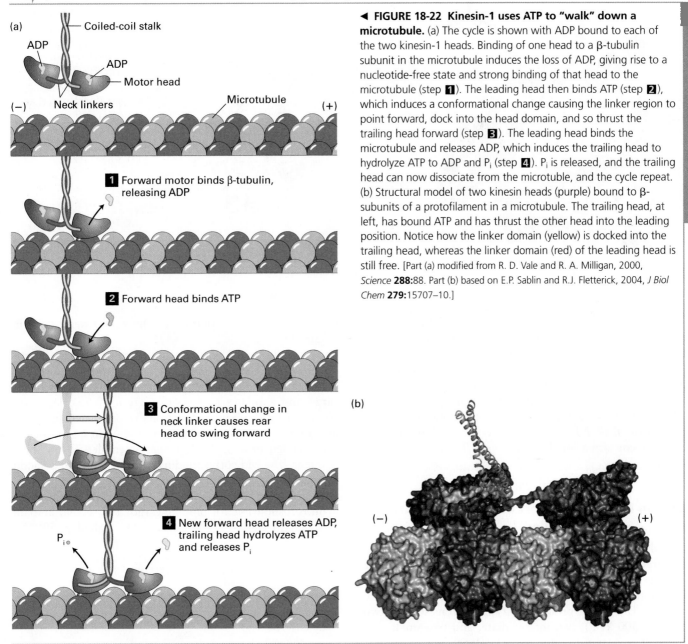

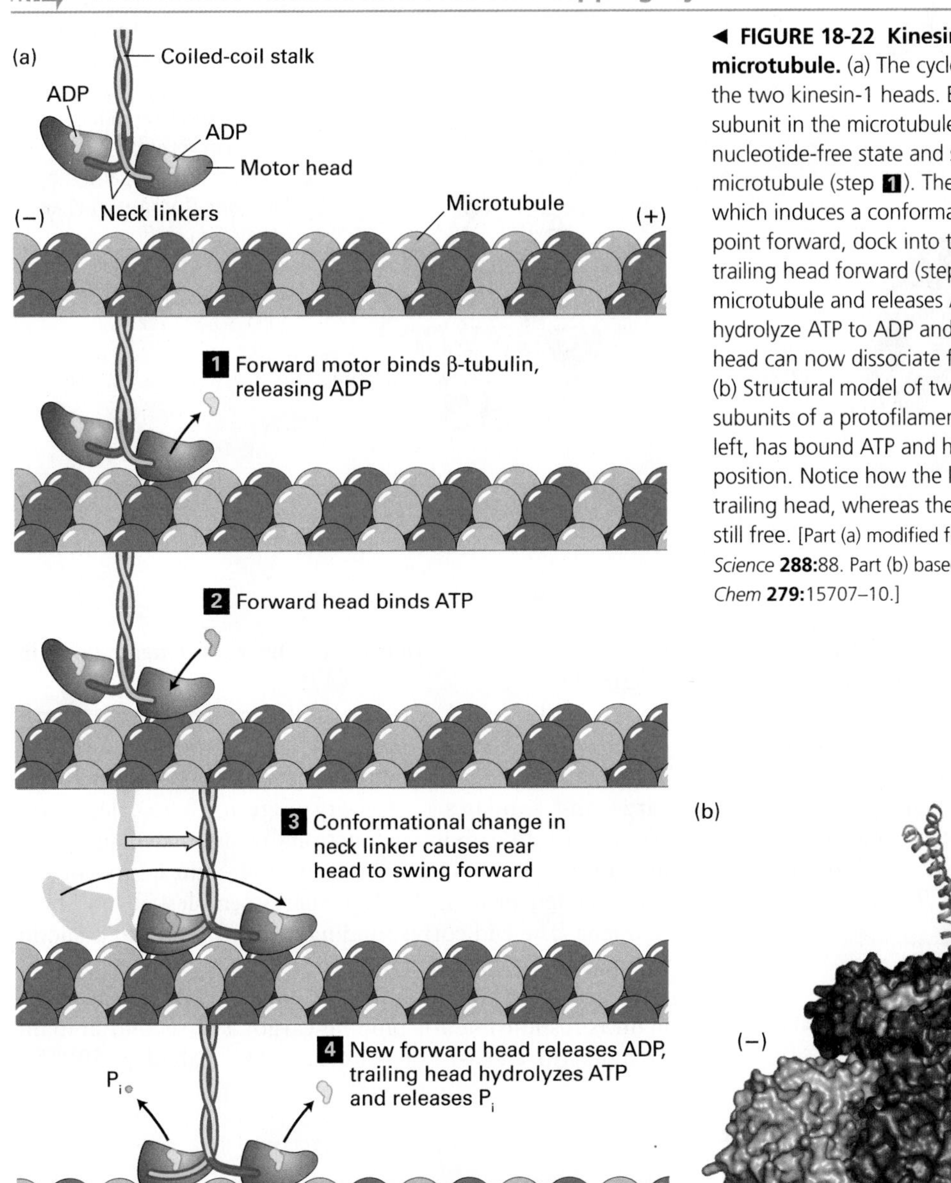

(a)
Coiled-coil stalk
ADP
ADP
Motor head
Neck linkers
Microtubule
(−)
(+)
1 Forward motor binds β-tubulin, releasing ADP
2 Forward head binds ATP
3 Conformational change in neck linker causes rear head to swing forward
4 New forward head releases ADP, trailing head hydrolyzes ATP and releases P_i
P_i
(b)
(−)
(+)

◀ **FIGURE 18-22 Kinesin-1 uses ATP to "walk" down a microtubule.** (a) The cycle is shown with ADP bound to each of the two kinesin-1 heads. Binding of one head to a β-tubulin subunit in the microtubule induces the loss of ADP, giving rise to a nucleotide-free state and strong binding of that head to the microtubule (step **1**). The leading head then binds ATP (step **2**), which induces a conformational change causing the linker region to point forward, dock into the head domain, and so thrust the trailing head forward (step **3**). The leading head binds the microtubule and releases ADP, which induces the trailing head to hydrolyze ATP to ADP and P_i (step **4**). P_i is released, and the trailing head can now dissociate from the microtuble, and the cycle repeat. (b) Structural model of two kinesin heads (purple) bound to β-subunits of a protofilament in a microtubule. The trailing head, at left, has bound ATP and has thrust the other head into the leading position. Notice how the linker domain (yellow) is docked into the trailing head, whereas the linker domain (red) of the leading head is still free. [Part (a) modified from R. D. Vale and R. A. Milligan, 2000, *Science* **288**:88. Part (b) based on E.P. Sablin and R.J. Fletterick, 2004, *J Biol Chem* **279:**15707–10.]

MEDIA CONNECTIONS

The ATP cycle of kinesin-1 movement is most easily viewed by starting with the leading head in a nucleotide-free state, under which conditions it is strongly bound to a β-subunit of a protofilament, and with the trailing head in a weakly bound state containing ADP (Figure 18-22). The energy associated with binding ATP to the leading head induces a forward motion of the linker domain of that head that then physically docks into the core head domain. This movement results in the linker domain pointing forward and physically swinging the trailing head—like throwing a ballet dancer—into a position where it becomes the leading head. The new leading head finds the next β-tubulin subunit, which induces dissociation of the ADP and tight binding. Importantly, this step also induces the now trailing head to hydrolyze ATP to ADP and release P_i and be converted into a weakly bound state. The leading head is now ready to bind ATP and swing the trailing head in front of it to repeat the cycle. Because this cycle requires one head to always be firmly attached to a β-tubulin subunit in a protofilament, kinesin-1 is exceptionally processive as it moves down a microtubule.

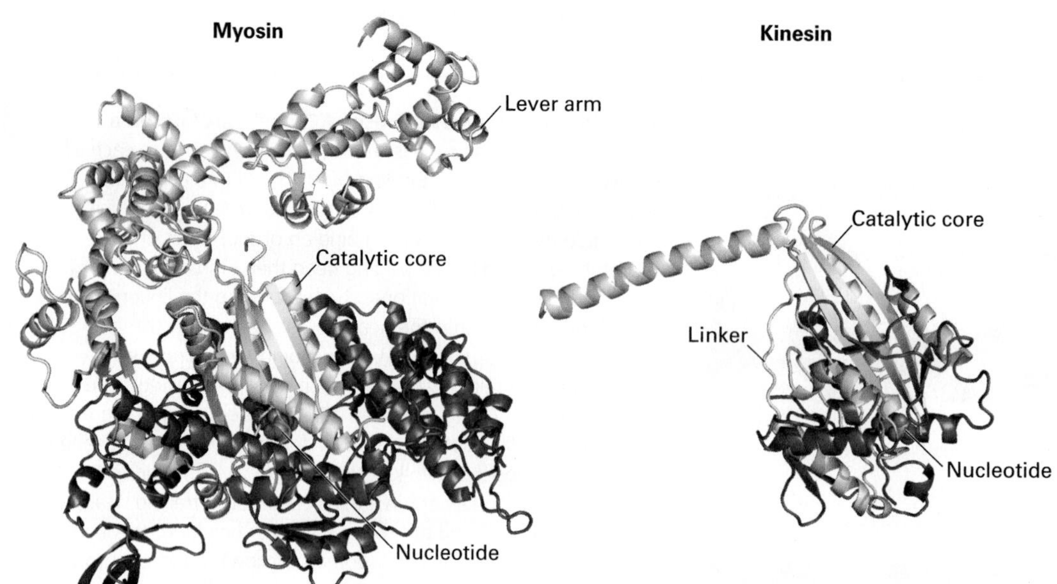

◄ **FIGURE 18-23 Convergent structural evolution of the ATP-binding core of myosin head and kinesin.** The common catalytic cores of myosin and kinesin are shown in yellow, the nucleotide in red, and the lever arm (for myosin-II) and linker domain (for kinesin-1) in light purple. [Modified from R. D. Vale and R. A. Milligan, 2000, *Science* **288:**88.]

When the x-ray structure of the kinesin head was determined, it revealed a major surprise—the catalytic core has the same overall structure as myosin's (Figure 18-23)! This occurs despite no amino acid sequence conservation, arguing strongly for convergent evolution to make a fold that can utilize the hydrolysis of ATP to generate work. Moreover, the same type of three-dimensional structure is seen in small GTP-binding proteins, such as Ras, that undergo a conformational change on GTP hydrolysis (see Figure 15-8).

Dynein Motors Transport Organelles Toward the (−) End of Microtubules

In addition to kinesin motors, which primarily mediate anterograde (+) end directed transport of organelles, cells use another motor, *cytoplasmic dynein,* to transport organelles in a retrograde fashion toward the (−) end of microtubules. This motor protein is very large, consisting of two large (>500 kDa), two intermediate, and two small subunits. It is responsible for the ATP-dependent retrograde transport of organelles toward the (−) ends of microtubules in axons, as well as many other functions we consider in the next sections. Compared to myosins and kinesins, the family of dynein-related proteins is not very diverse.

Like kinesin-1, cytoplasmic dynein is a two-headed molecule, built around two identical or nearly identical heavy chains. However, because of the enormous size of the motor domain, dynein has been less well characterized in terms of its mechanochemical activity. A single dynein heavy chain consists of a stem and a round head domain containing the ATPase activity, from which protrudes a stalk (Figure 18-24). At the end of the stalk is a microtubule-binding site. Electron microscopy suggests that the power stroke of dynein involves the rotation of the round head domain (Figure 18-25).

Unlike kinesin-1, dynein cannot mediate cargo transport by itself. Rather, dynein-related transport requires *dynactin,* a large protein complex that both links dynein to its cargo and regulates its activity (Figure 18-26). Dynactin consists of 11 subunits, functionally organized into two domains. One domain is built around eight copies of the actin-related protein Arp1, which assembles into a short filament. The end corresponding to the (+) end of an actin filament is capped by capping protein; a number of subunits are associated with the (−) end. This Arp-related domain is responsible for binding cargo. The second domain of dynactin consists of a long protein called p150^{Glued},

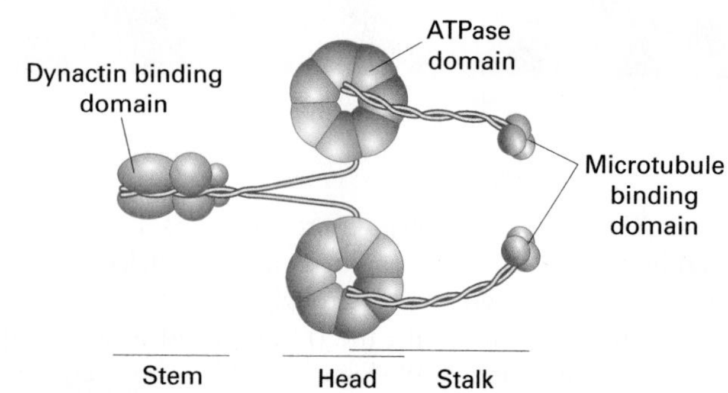

▲ **FIGURE 18-24 The domain structure of cytoplasmic dynein.** Each ATPase head domain of dynein consists of seven repeated motifs like petals on a flower. Emerging from this is a coiled-coil domain with a microtubule binding site at the end. At left is shown a number of additional subunits that are associated with the heavy chains and link dynein to cargo through dynactin. [Modified from R. D. Vale, 2003, *Cell* **112:**467.]

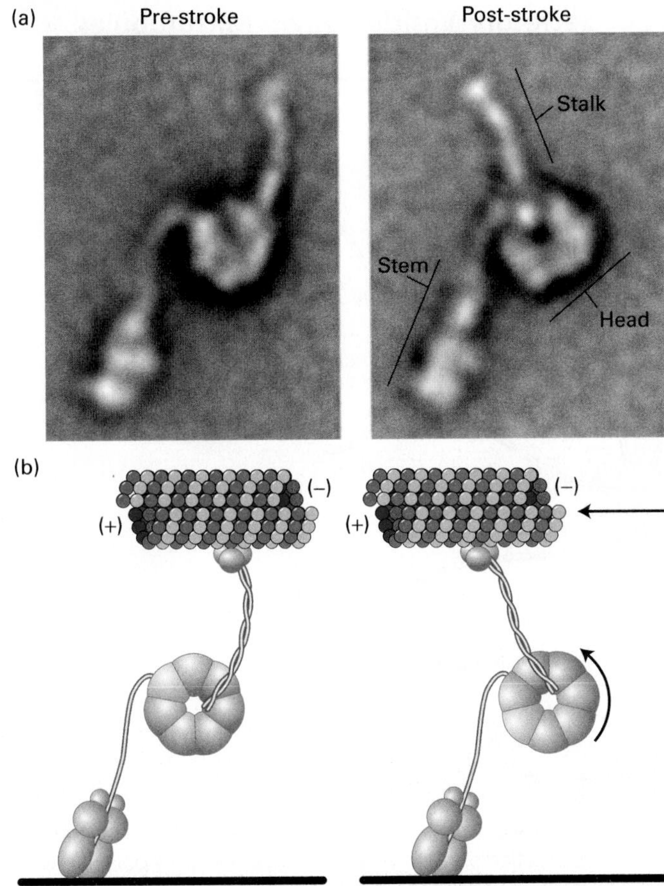

▲ **FIGURE 18-25 The power stroke of dynein.** (a) Multiple images of purified single-headed dynein molecules in their prestroke and poststroke states were recorded in an electron microscope and then averaged. The left image shows dynein in the ADP-P_i state, which represents the prestroke state, and the right image in a nucleotide-free poststroke state. (b) A comparison of the images shows that the force-generation mechanism involves a change in orientation of the head relative to the stem, causing a movement of the microtubule-binding stalk. [Part (a) modified from S. A. Burgess et al., 2003, *Nature* **421**:715; courtesy of S. A. Burgess.]

which contains the dynein binding site and has a microtubule binding site at one end. Holding the two dynactin domains together is a protein called *dynamitin*—so named because when it is overexpressed, it dissociates (or "blows apart") the two domains, making a nonfunctional complex. This feature has been very useful experimentally because it has allowed researchers to identify processes that are dependent on dynein-dynactin, which do not occur in dynamitin-overexpressing cells.

Kinesins and Dyneins Cooperate in the Transport of Organelles Throughout the Cell

Both dynein and kinesin family members play important roles in the microtubule-dependent organization of organelles in cells (Figure 18-27). Because the orientation of microtubules is fixed by the MTOC, the direction of transport—toward or away from the cell center—depends on the motor protein. For example, the Golgi apparatus collects in the vicinity of the centrosome, where the (−) ends of microtubules lie, and is driven there by dynein-dynactin. In addition, secretory cargo emerging from the endoplasmic reticulum is transported to the Golgi by dynein-dynactin. Conversely, the endoplasmic reticulum is spread throughout the cytoplasm and is transported there by kinesin-1, which moves toward the peripheral (+) ends of microtubules. Some organelles of the endocytic pathway are also associated with dynein-dynactin, including late endosomes and lysosomes. Kinesins have also been shown to transport mitochondria, as well as nonmembranous cargo such as specific mRNAs encoding proteins that need to be localized during development.

We have seen how kinesin-1 transports organelles in an anterograde fashion down axons. What happens to the motor when it gets to the end of the axon? The answer is that it is carried back in a retrograde fashion on organelles

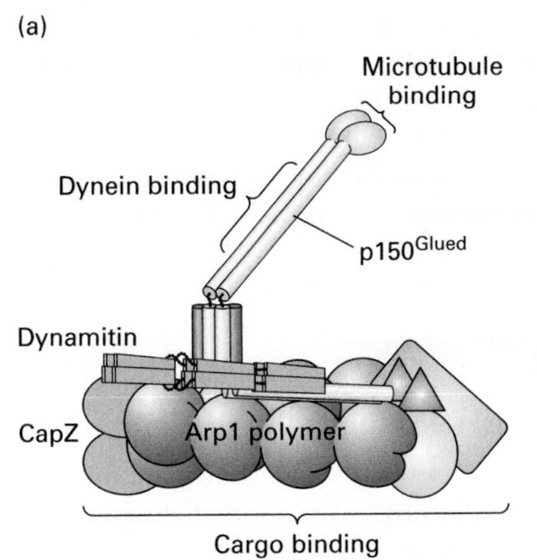

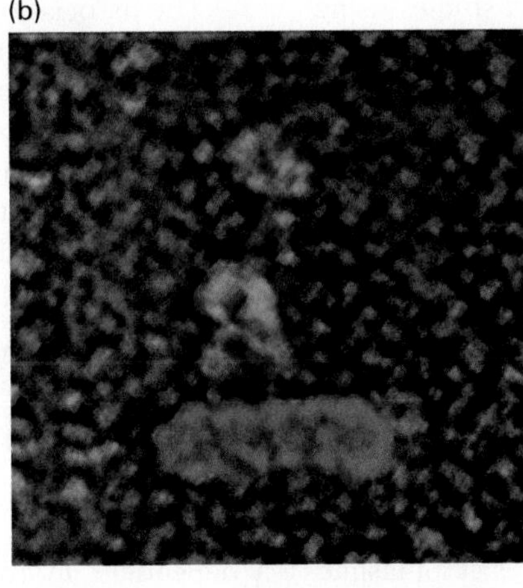

◀ **FIGURE 18-26 The dynactin complex linking dynein to cargo.** (a) One domain of the complex binds cargo and is built around a short filament made up of about eight subunits of the actin-related protein Arp1 capped by CapZ. Another domain consists of the protein p150^{glued}, which has a microtubule binding site on its distal end and is involved in attaching cytoplasmic dynein to the complex. Dynamitin holds the two parts of the dynactin complex together. (b) Electron micrograph of a metal replica of the dynactin complex isolated from brain. The Arp1 minifilament (purple) and the dynamitin/p150^{glued} side arm (blue) are highlighted. [Part (a) modified from T. A. Schroer, 2004, *Annu. Rev. Cell Dev. Biol.* **20**:759. Part (b) from D. M. Eckley et al., 1999, *J. Cell Biol.* **147**:307.]

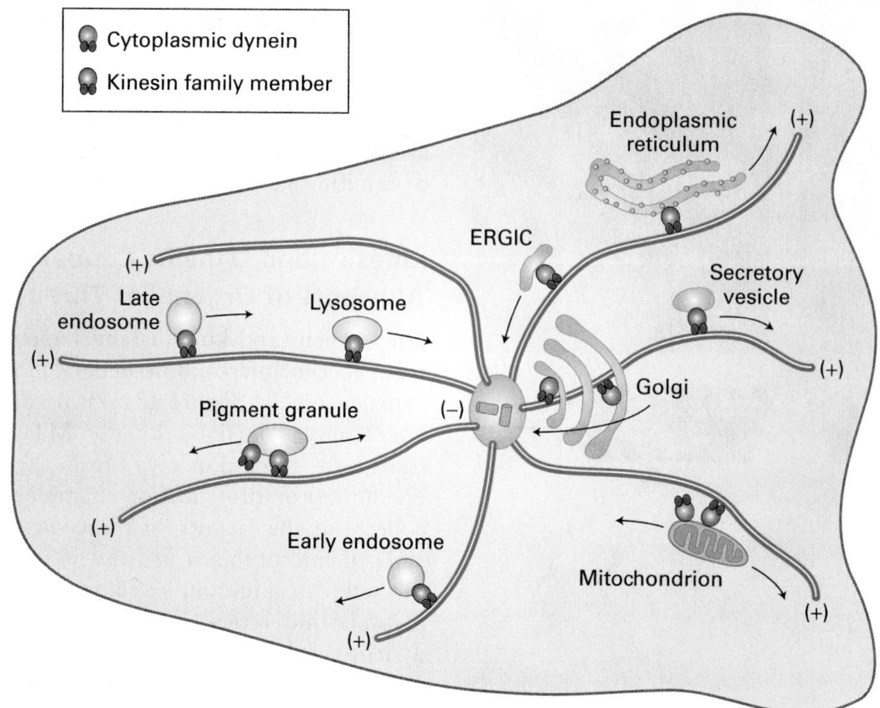

▲ **FIGURE 18-27 Organelle transport by microtubule motors.** Cytoplasmic dyneins (red) mediate retrograde transport of organelles toward the (−) end of microtubules (cell center); kinesins (purple) mediate anterograde transport toward the (+) end (cell periphery). Most organelles have one or more microtubule-based motors associated with them. It should be noted that the association of motors with organelles varies by cell type, so some of these associations may not exist in all cells, whereas others not shown here also exist. ERGIC = ER-to-Golgi intermediate compartment.

transported by cytoplasmic dynein. Thus kinesin-1 and dynein can associate with the same organelle and a mechanism must exist that turns one motor off while activating the other, although such mechanisms are not yet fully understood.

Much of what we know about the regulation of microtubule-based organelle transport comes from studies using fish (e.g., angelfish) or frog melanophores. Melanophores are cells of the vertebrate skin that contain hundreds of dark melanin-filled pigment granules called melanosomes. Melanophores either have their pigment granules dispersed, in which case they make the skin darker, or aggregated at the center, which makes the skin paler (Figure 18-28). These changes in skin color, mediated by neurotransmitters in the fish and regulated by hormones in the frog, serve in either camouflage in the case of the fish or enhance social interactions in the frog. The movement of the granules is mediated by changes in intracellular cAMP and is dependent on microtubules. Studies investigating which motors are involved have shown that pigment granule dispersion requires kinesin-2, whereas aggregation requires cytoplasmic dynein/dynactin. The first hints of how these activities might be coordinated came from the finding that overexpression of dynamitin inhibited granule transport in both directions. This surprising result was explained when it was found that dynactin binds not only to cytoplasmic dynein but also to kinesin-2—and may coordinate the activity of the two motors.

The association of dynein and kinesin-2 with the same organelle is not limited to melanosomes; it has recently been suggested that these motors may cooperate to localize late endosomes/lysosomes and mitochondria appropriately in some cells. Thus the concept that organelles can have a number of distinct motors associated with them is not the exception but an emerging theme.

KEY CONCEPTS OF SECTION 18.4

Kinesins and Dyneins: Microtubule-Based Motor Proteins

- Kinesin-1 is a microtubule (+) end–directed ATP-dependent motor that transports membrane-bound organelles (see Figure 18-20).

Video: Aggregation and Dispersion of Pigment Granules in Fish Melanophores

(a)

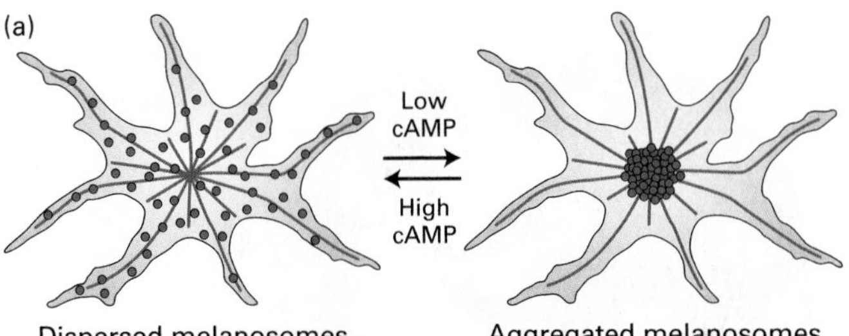

(b)

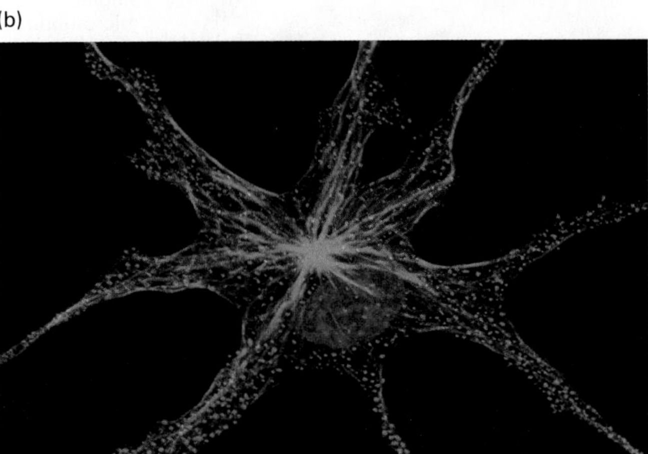

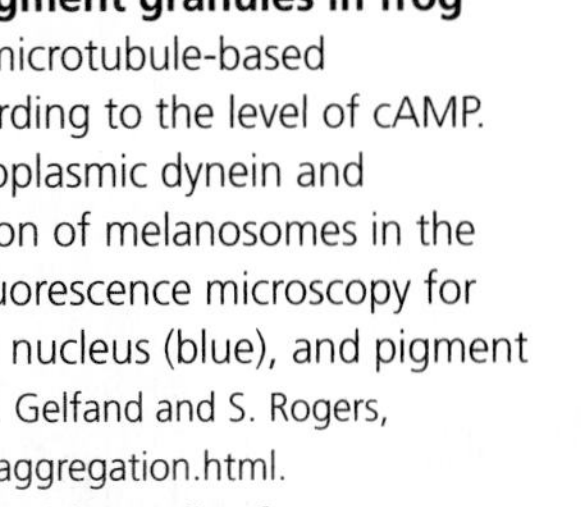

◀ **FIGURE 18-28 Movement of pigment granules in frog melanophores.** (a) Diagram of the microtubule-based reorganization of melanosomes according to the level of cAMP. Melanosomes are aggregated by cytoplasmic dynein and dispersed by kinesin-2. (b) Visualization of melanosomes in the dispersed state as seen by immunofluorescence microscopy for microtubules (green), the DNA in the nucleus (blue), and pigment granules (red). [Part (a) modified from V. Gelfand and S. Rogers, http://www.proweb.org/kinesin/Pigment_aggregation.html. Part (b) from S. Rogers, www.itg.uiuc.edu/exhibits/gallery/]

MEDIA CONNECTIONS

- Kinesin-1 consists of two heavy chains, each with an N-terminal motor domain and two light chains that associate with cargo (see Figure 18-19).
- The kinesin superfamily includes motors that function in interphase and mitotic cells, transporting organelles, and sliding antiparallel microtubules and even includes one class that is not motile but destabilizes microtubule ends (see Figure 18-21).
- Kinesin-1 is a highly processive motor because it coordinates ATP hydrolysis between its two heads so that one head is always firmly bound to a microtubule (see Figure 18-22).
- Cytoplasmic dynein is a microtubule (−) end–directed ATP-dependent motor that associates with the dynactin complex to transport cargo (see Figure 18-25).
- Kinesins and dyneins associate with many different organelles to organize their location in cells (see Figure 18-27).

18.5 Cilia and Flagella: Microtubule-Based Surface Structures

Cilia and **flagella** are related microtubule-based and membrane-bound extensions that project from many protozoa and most animal cells. Abundant motile cilia are found on the surface of specific epithelia, such as those that line the trachea, where they beat in an orchestrated wavelike fashion to move fluids. Animal-cell flagella, which are longer but have a very similar structure, can propel a cell, such as sperm, through liquid. Cilia and flagella contain many different microtubule-based motors: axonemal dyneins are responsible for the beating of flagella and cilia, whereas kinesin-2 and cytoplasmic dynein are responsible for flagella and cilia assembly and turnover.

Eukaryotic Cilia and Flagella Contain Long Doublet Microtubules Bridged by Dynein Motors

Cilia and flagella range in length from a few micrometers to more than 2 mm for some insect sperm flagella. They possess a central bundle of microtubules, called the **axoneme,** which consists of a so-called 9 + 2 arrangement of nine doublet microtubules surrounding a central pair of singlet, yet ultrastructurally distinct, microtubules (Figure 18-29a, b). Each of the nine outer doublets consists of an A microtubule with 13 protofilaments and a B microtubule with 10 protofilaments. All the microtubules in cilia and flagella have the same polarity: the (+) ends are located at the tip. At its point of attachment in the cell, the axoneme connects with the **basal body,** a complicated structure containing nine triplet microtubules (see Figure 18-29a).

The structure of the axoneme is held together by three sets of protein cross-links (see Figure 18-29b). The central pair of singlet microtubules is connected by periodic bridges, like rungs on a ladder. A second set of linkers, composed of the protein *nexin,* joins adjacent outer doublet microtubules. *Radial spokes* project from each A tubule of the outer doublets toward the central pair.

The major motor protein present in cilia and flagella is *axonemal dynein,* a large, multisubunit protein related to cytoplasmic dynein. Two rows of dynein motors are attached periodically down the length of each A tubule of the outer doublet microtubules; these are called the *inner-arm* and *outer-arm* dyneins (see Figure 18-29b). It is these dynein motors interacting with the adjacent B tubule that bring about cilia and flagella bending.

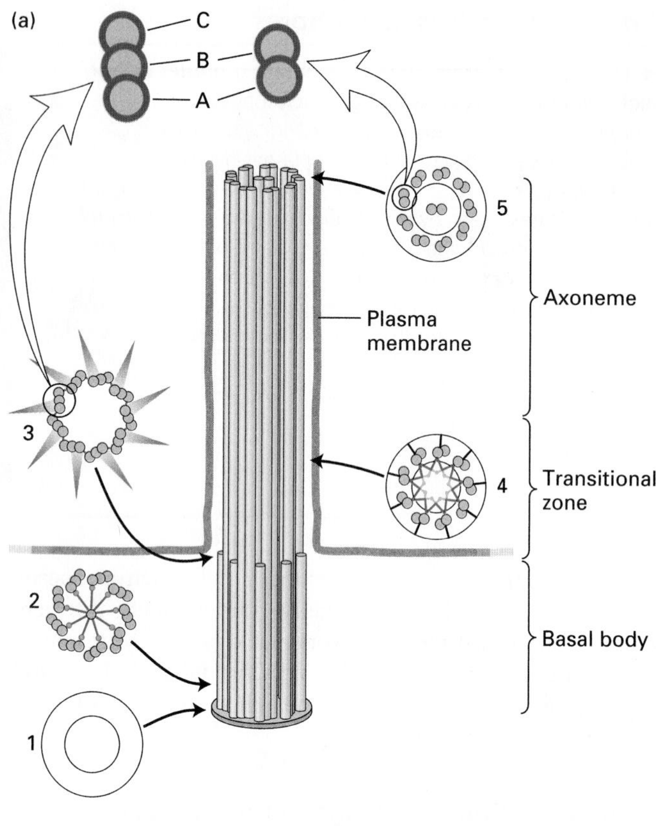

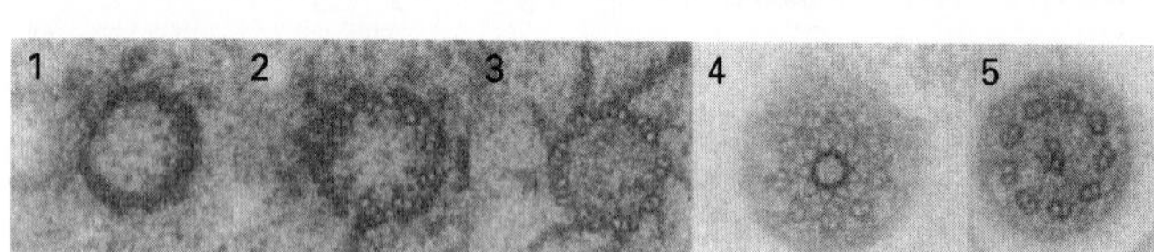

(b)

4
5
3
B tubule
A tubule
Dynein arms
Radial spoke
6
2
Central
singlet
microtubule
1
7
9
8
50 nm

Plasma
membrane
Outer-arm
dynein
Nexin
Central pair
of singlet
microtubules
Inner-arm
dynein
4
3
5
2
6
1
7
9
8
Radial spoke head
A tubule
B tubule
Doublet microtubule

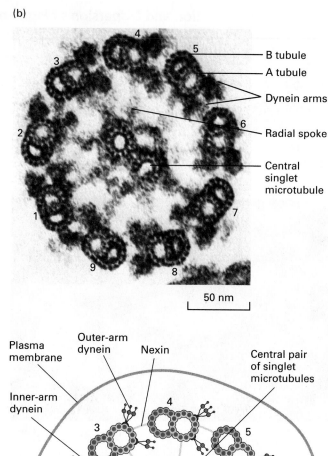

▲ **FIGURE 18-29 Structural organization of cilia and flagella.** (a) Cilia and flagella are assembled from a basal body, a structure built around nine linked triplet microtubules. Continuous with the A and B microtubules of the basal body are the A and B tubules of the axoneme—the membrane-bound core of the cilium or flagellum. Between the basal body and axoneme is the transitional zone. The diagram and accompanying transverse sections of the basal body, transitional zone, and axoneme show their intricate structures. (b) Thin section of a transverse section of a cilium (with plasma membrane removed) with a diagram to show the identity of the structures. [Part (a) modified from S. K. Dutcher, 2001, *Curr. Opin. Cell Biol.* **13:**49–54; courtesy of S. Dutcher. Part (b) courtesy of L. Tilney.]

Ciliary and Flagellar Beating Are Produced by Controlled Sliding of Outer Doublet Microtubules

Cilia and flagella are motile structures because activation of the axonemal dynein motors induces bending in them. A close examination of this motility using video microscopy reveals that a bend starts at the base of a cilium or flagellum and then propagates along the structure (Figure 18-30). A clue to how this occurs came from studies of isolated axonemes. In classic experiments, axonemes were treated with a protease that cleaves the nexin links. When ATP was added to the treated axonemes, the doublet microtubules slid past one another as dynein, attached to the A tubule of one doublet, "walked"

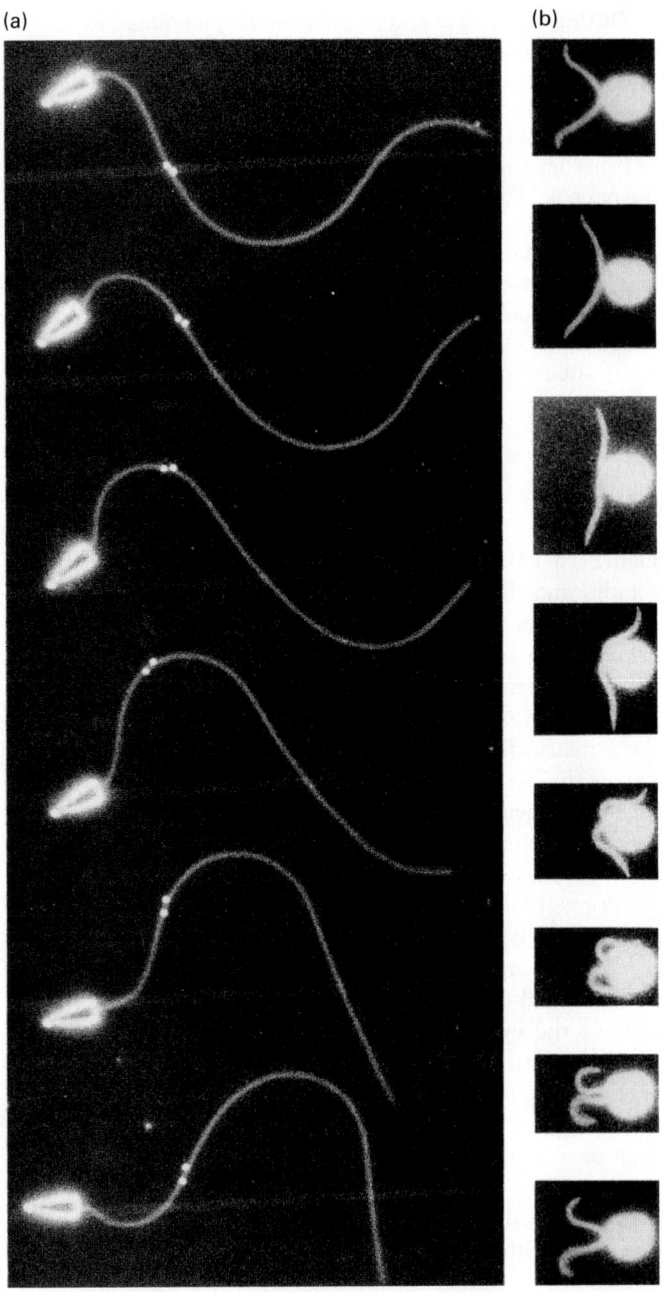

◀ **EXPERIMENTAL FIGURE 18-30 Video microscopy shows flagellar movements that propel sperm and *Chlamydomonas* forward.** In both cases, the cells are moving to the left. (a) In the typical sperm flagellum, successive waves of bending originate at the base and are propagated out toward the tip; these waves push against the water and propel the cell forward. Captured in this multiple-exposure sequence, a bend at the base of the sperm in the first (*top*) frame has moved distally halfway along the flagellum by the last frame. A pair of gold beads on the flagellum are seen to slide apart as the bend moves through their region. (b) Beating of the two flagella on *Chlamydomonas* occurs in two stages, called the *effective stroke* (top three frames) and the *recovery stroke* (remaining frames). The effective stroke pulls the organism through the water. During the recovery stroke, a different wave of bending moves outward from the bases of the flagella, pushing the flagella along the surface of the cell until they reach the position to initiate another effective stroke. Beating commonly occurs 5–10 times per second. [Part (a) from C. Brokaw, 1991, *J. Cell Biol.* **114**(6): cover photograph; courtesy of C. Brokaw. Part (b) courtesy of S. Goldstein.]

down the B tubule of the adjacent doublet (Figure 18-31b, c). In an axoneme with intact nexin links, the action of dynein induces flagellar bending as the microtubule doublets are connected to one another (Figure 18-31a). How specific subsets of dynein are activated and how a wave of activation is propagated down the axoneme are not yet understood.

Intraflagellar Transport Moves Material Up and Down Cilia and Flagella

Although axonemal dynein is involved in bending flagella, another type of motility has been more recently observed. Careful examination of flagella on the biflagellate green alga *Chlamydomonas reinhardtii* revealed particles moving at a constant speed of ≈2.5 μm/s toward the tip (anterograde movement) and other particles moving at ≈4 μm/s from the tip to the base (retrograde movement). This transport is known as *intraflagellar transport* (IFT) and occurs on both cilia and flagella. Light and electron microscopy revealed that the particles moved between the outer doublet microtubules and the plasma membrane (Figure 18-32). Analysis of algal mutants demonstrated that the anterograde movement is powered by kinesin-2 and that retrograde movement is powered by cytoplasmic dynein.

What is the function of IFT? Because all the microtubules have their growing (+) ends at the flagella tip, this is the site at which new tubulin subunits are added. In cells defective for kinesin-2, flagella shrink, suggesting that IFT transports new material to the tip for growth. Since IFT is a continually occurring process, what happens to the kinesin-2 molecules when they get to the tip, and where do the dynein motors come from to transport the particles retrogradely? Remarkably, dynein is carried as a cargo to the tip on the anterograde-moving particles powered by kinesin-2, and then kinesin-2 becomes cargo as the particles are transported back to the base by dynein.

The anterograde and retrograde particles transported on *Chlamydomonas* flagella have been isolated and their composition determined. All have homologs present in organisms containing cilia, such as nematodes, fruit flies, mice, and humans, but these particles are absent from the genomes of yeasts and plants that lack cilia, suggesting they are specific to IFT.

(a)

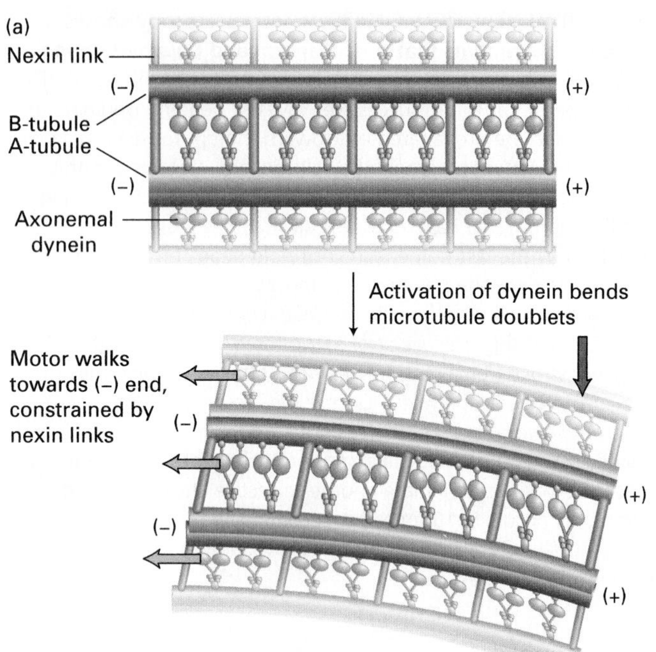

(b) Nexin links removed by protease

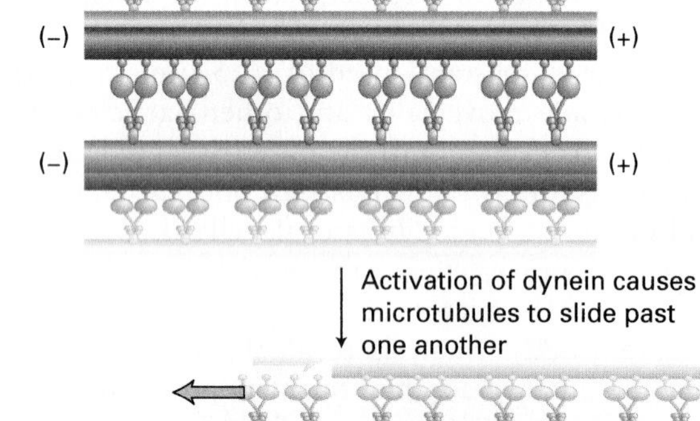

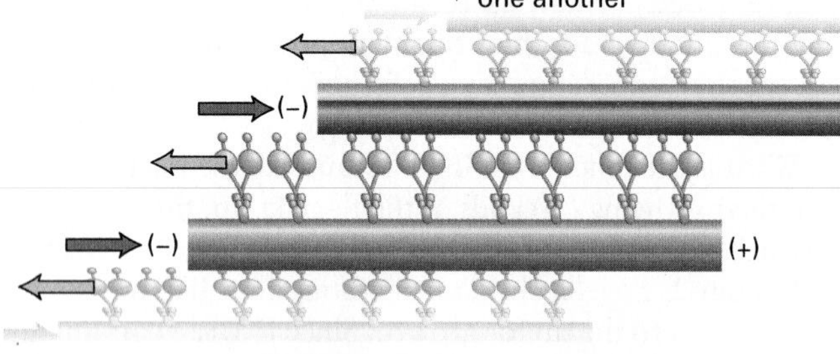

(c)

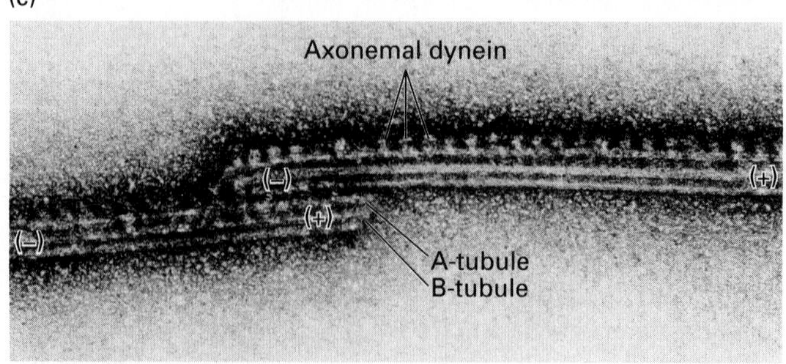

◀ **FIGURE 18-31 Cilia and flagella bending mediated by axonemal dynein.** (a) Axonemal dynein attached to an A tubule of an outer doublet pulls on the B tubule of the adjacent tubule trying to move to the (−) end. Because the adjacent tubules are tethered by nexin, the force generated by dynein bends the cilium or flagellum. (b) Experimental evidence for the model in (a). When the nexin links are cleaved with a protease and ATP added to induce dynein activity, the microtubule doublets slide past one another. (c) Electron micrograph of two doublet microtubules in a protease-treated axoneme incubated with ATP. In the absence of cross-linking proteins, doublet microtubules slide excessively. The dynein arms can be seen projecting from A tubules and interacting with B tubules of the left microtubule doublet. [Part (c) courtesy of P. Satir.]

Defects in Intraflagellar Transport Cause Disease by Affecting Sensory Primary Cilia

Most vertebrate cells contain a solitary nonmotile cilium, known as the *primary cilium* (Figure 18-33; see Figure 18-13), which lacks both the central pair of microtubules and the dynein side arms. The primary cilium is assembled from the centrosome, with one of the centrioles functioning as its basal body. Clues to the function of the primary cilium came from the discovery that loss of a mammalian homolog of a *Chlamydomonas* IFT protein results in defects in the primary cilium and causes autosomal recessive polycystic kidney disease (ADPKD). It is believed that the primary cilia on the epithelial cells of the kidney collecting tubule act as mechanochemical sensors to measure the rate of fluid flow by the degree to which they are bent.

Primary cilia are also involved in other sensory roles. The sense of smell is due to reception of odorants by the odorant receptors located in the primary cilium of olfactory sensory neurons in the nose. In another example, the rod and cone cells of the eye have a primary cilium with a greatly expanded tip to accommodate proteins involved in photoreception. The retinal protein opsin moves through the cilium at about 2000 molecules per minute, transported by kinesin-2 as part of the intraflagellar transport system. Defects in this transport cause retinal degeneration. Given these roles for primary cilia in sensory detection, it is not surprising that defects in primary cilia can have wide-ranging consequences. For example, patients with Bardet-Biedl syndrome, which affects basal bodies and cilia, have retinal degeneration and cannot smell as well as several other disorders, suggesting that primary cilia are involved in many processes yet to be uncovered. ■

KEY CONCEPTS OF SECTION 18.5

Cilia and Flagella: Microtubule-Based Surface Structures

- Motile cilia and flagella are microtubule-based cell-surface structures with a characteristic central pair of singlet microtubules and nine sets of outer doublet microtubules (see Figure 18-29).
- All cilia and flagella grow from basal bodies, structures with nine sets of outer triplet microtubules and closely related to centrioles.

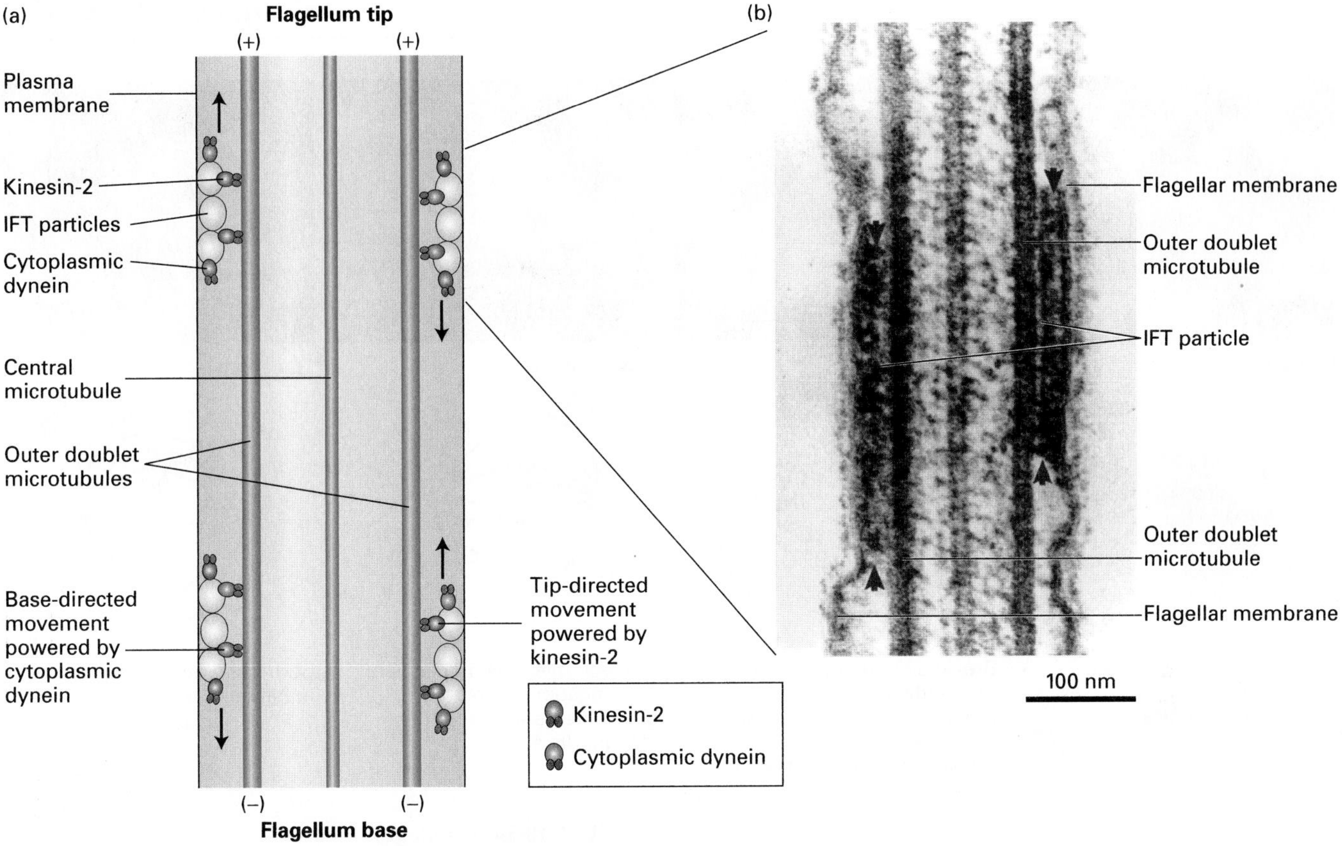

▲ **FIGURE 18-32 Intraflagellar transport.** (a) Particles are transported between the plasma membrane and the outer doublet microtubules. Transport of the particles to the tip is dependent on kinesin-2, whereas transport toward the base is mediated by cytoplasmic dynein. (b) Thin-section electron micrograph shows IFT particles in a section of a *Chlamydomonas* flagellum. [Part (b) from J. L. Rosenbaum and G. B Witman, 2003, *Nature Reviews Molecular Cell Biology* **3:**813–825.]

- Axonemal dyneins attached to the A tubule on one doublet microtubule interact with the B tubule of another to bend cilia and flagella.

- Cilia and flagella have a mechanism, intraflagellar transport, to transport organelles to the tip by kinesin-2 and from the tip back to the base by cytoplasmic dynein. This transport regulates the function and length of cilia and flagella.

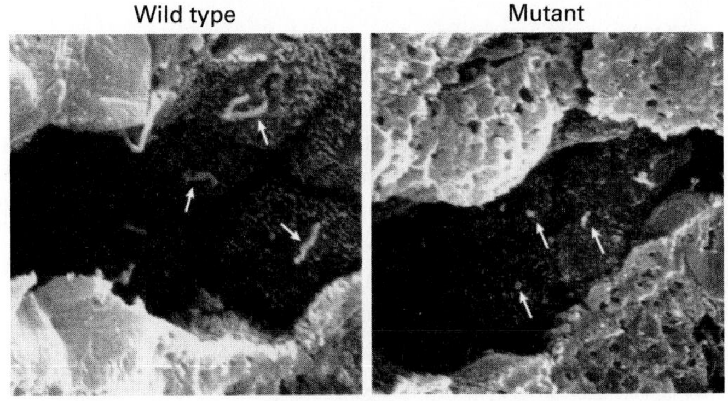

▲ **FIGURE 18-33 Defective primary cilium in mouse mutants lacking components of the IFT particle.** Scanning electron micrographs of epithelial cells of a collecting tubule from a wild-type mouse (*left*) and a mutant mouse defective in a component of the IFT particles. Arrows point to the primary cilia, which are short stubs in the mutant mouse. [From G. Pazour et al., 2000, *J. Cell Biol.* **151:**709–718.]

18.6 Mitosis

Of all the events that permit the existence and perpetuation of life, perhaps the most critical is the ability of cells to accurately duplicate and then faithfully segregate their chromosomes at each cell division. During the **cell cycle,** a highly regulated process discussed in Chapter 20, cells duplicate their chromosomes precisely once during a period known as S phase (for synthesis phase). Once the individual chromosomes are duplicated, they are held together along their length by proteins called *cohesins.* The cells then pass through a period called G_2 (for gap 2) and enter **mitosis,** the process by which the duplicated chromosomes are segregated to the daughter cells. This process has to be very precise—loss or gain of a chromosome can either be lethal (in which case it is often not detected) or cause severe

MEDIA CONNECTIONS

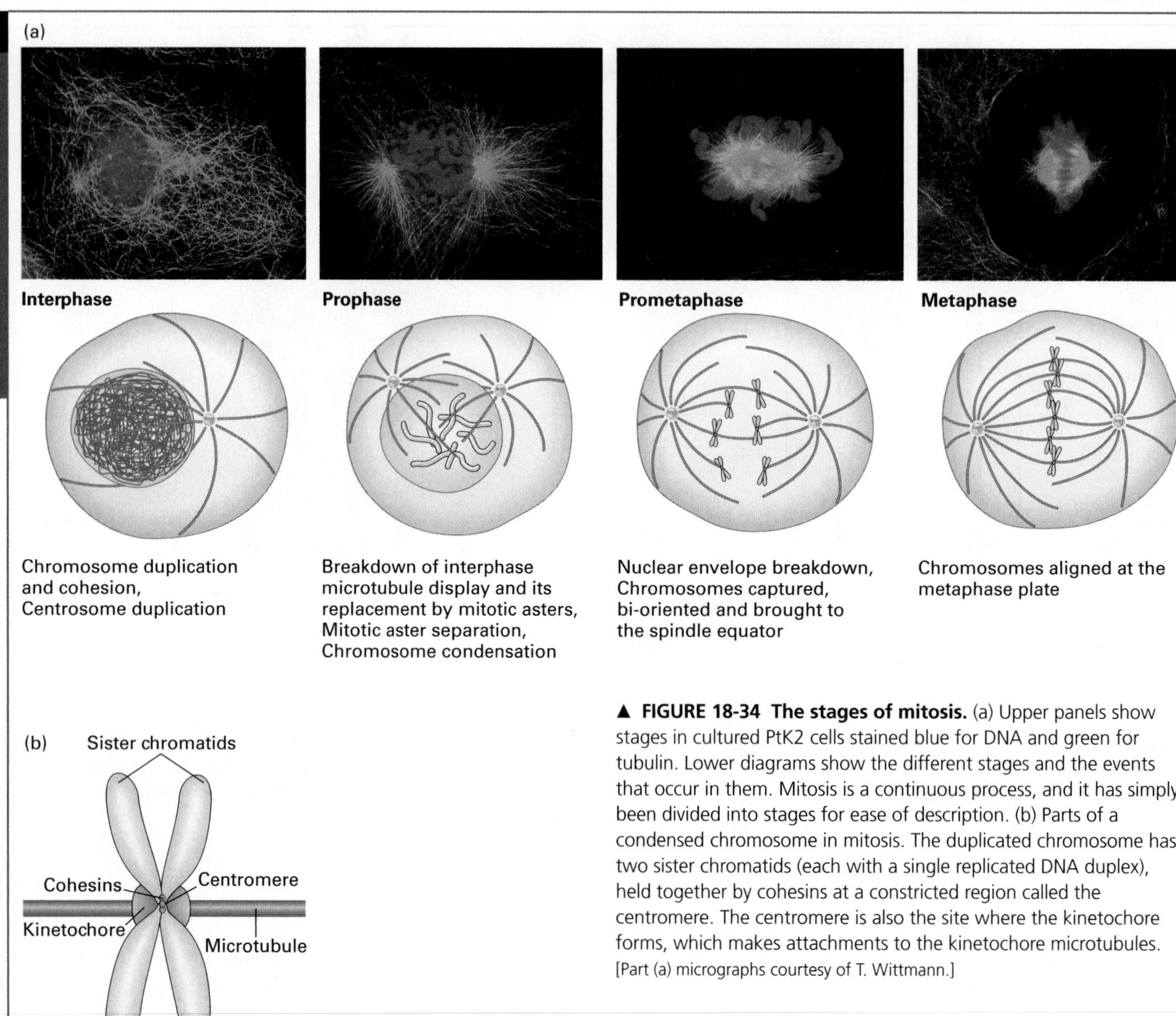

▲ **FIGURE 18-34 The stages of mitosis.** (a) Upper panels show stages in cultured PtK2 cells stained blue for DNA and green for tubulin. Lower diagrams show the different stages and the events that occur in them. Mitosis is a continuous process, and it has simply been divided into stages for ease of description. (b) Parts of a condensed chromosome in mitosis. The duplicated chromosome has two sister chromatids (each with a single replicated DNA duplex), held together by cohesins at a constricted region called the centromere. The centromere is also the site where the kinetochore forms, which makes attachments to the kinetochore microtubules. [Part (a) micrographs courtesy of T. Wittmann.]

complications. It is estimated that yeast only missegregates one of its 16 chromosomes every 100,000 cell divisions, which makes it one of the most accurate processes in biology.

Mitosis Can Be Divided into Six Phases

Mitosis has been divided up into several stages for ease of description (Figure 18-34a), but in reality it is a continuous process. The first stage of mitosis, called **prophase,** is signaled by a number of coordinated and dramatic events. First, the interphase array of microtubules is replaced as the duplicated centrosomes (see below) become more active in microtubule nucleation. This provides two sites of assembly of dynamic microtubules known as **asters.** The two asters then move to opposite sides of the nucleus and will become the two poles of the **mitotic spindle,** the microtubule-based structure that separates chromosomes. Second, protein synthesis stops, and the internal order of membranes systems, normally dependent on the interphase array of microtubules, is disassembled. In addition, endocytosis and exocytosis are halted, and the microfilament organization is generally rearranged to give rise to a rounded cell. In the nucleus, the nucleolus breaks down and chromosomes begin to condense. Cohesins holding the duplicated chromosome together, or sister **chromatids** as they are called at this stage, are degraded except at the centromeric region, where the two sister chromatids remain

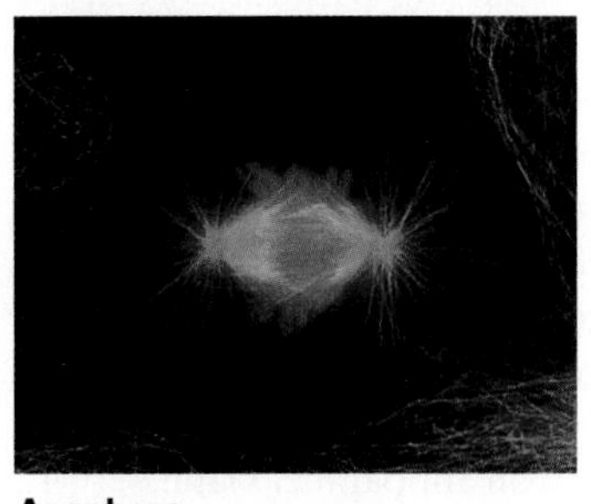

Anaphase

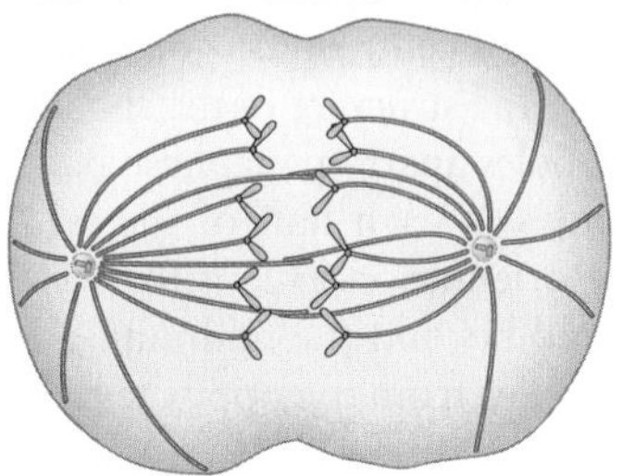

APC/C activated and cohesins degraded
Anaphase A: Chromosome movement to poles
Anaphase B: Spindle pole separation

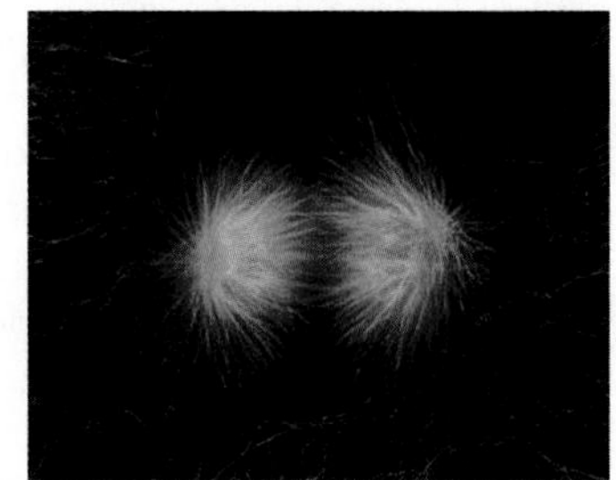

Telophase

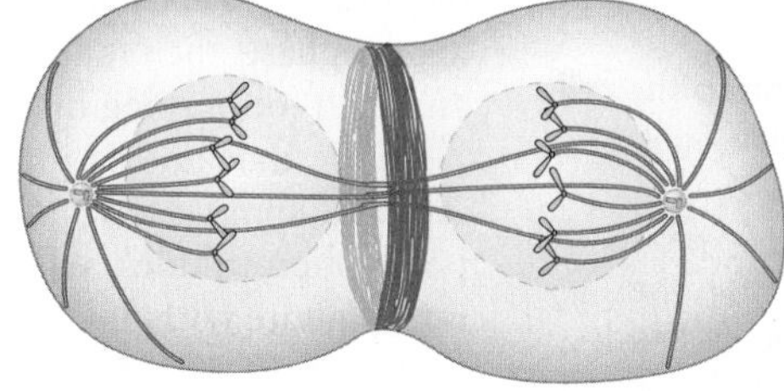

Nuclear envelope reassembly, Assembly of contractile ring

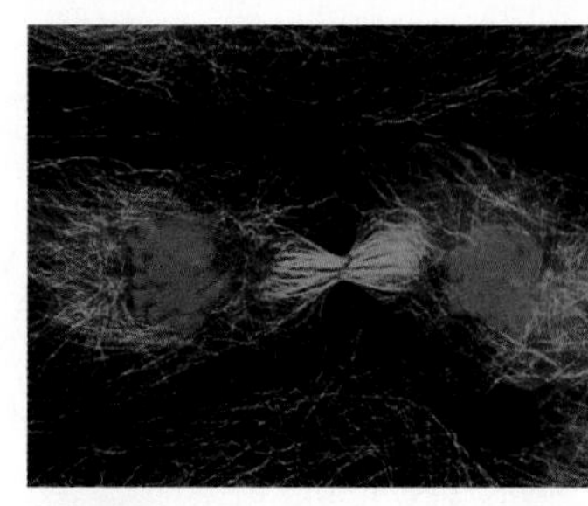

Cytokinesis

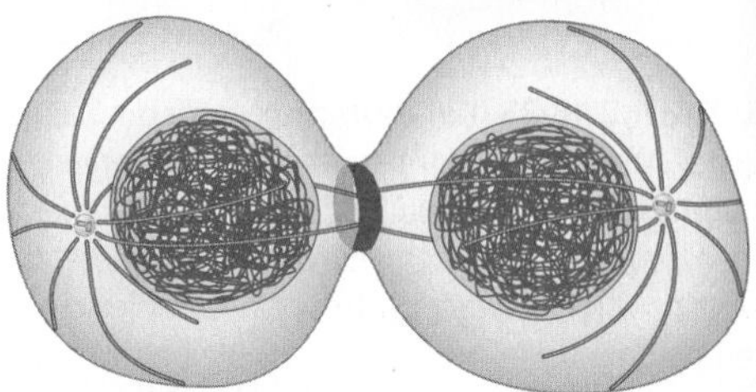

Reformation of interphase microtubule array, Contractile ring forms cleavage furrow

attached by cohesins (Figure 18-34b). As discussed in more detail in Chapter 20, all these events are coordinated by a rapid increase in the activity of the mitotic cyclin-CDK complex, which is a kinase that phosphorylates multiple proteins.

The next stage of mitosis, **prometaphase,** is initiated by the breakdown of the nuclear envelope and the nuclear pores and disassembly of the lamin-based nuclear lamina. Microtubules assembled from the spindle poles search and "capture" chromosome pairs at specialized structures called **kinetochores.** Each chromatid has a kinetochore, so sister chromatids have two kinetochores, each of which becomes attached to opposite spindle poles during prometaphase. When they are attached to both spindle poles, sister chromatids align equidistant from the two spindle poles in a process known as *congression.* Prometaphase continues until all chromosomes have congressed, at which point the cell enters the next stage, **metaphase,** defined as the stage when all the chromosomes are aligned at the *metaphase plate.*

The next stage, **anaphase,** is induced by activation of the anaphase-promoting complex/cyclosome (APC/C). The activated APC/C leads to the destruction of the cohesins, proteins holding the sister chromatids together, so that now each separated chromosome can be pulled to its respective pole by the microtubules attached to its kinetochore. This movement is known as *anaphase A.* A separate and distinct movement also occurs: the movement of the spindle poles farther apart in a process known as *anaphase B.* Now that the chromosomes have separated, the cell enters **telophase,** when the nuclear envelope reforms, the chromosomes decondense, and the cell is pinched into two daughter cells by the contractile ring during **cytokinesis.**

Centrosomes Duplicate Early in the Cell Cycle in Preparation for Mitosis

In order to separate the chromosomes at mitosis, cells duplicate their MTOCs—their centrosomes—coordinately with the duplication of their chromosomes in S phase (Figure 18-35). The duplicated centrosomes separate and become the two MTOCs—the spindle poles—of the mitotic spindle. The number of centrosomes in animal cells has to be very carefully controlled. In fact, many tumor cells have more than two centrosomes, which contribute to genetic instability resulting in missegregation of chromosomes and hence **aneuploidy** (unequal numbers of chromosomes).

As cells enter mitosis, the activity of the two MTOCs—their ability to nucleate microtubules—increases greatly as they accumulate more pericentriolar material. Because the microtubules radiating from these two MTOCs now resemble stars, they are often called mitotic asters.

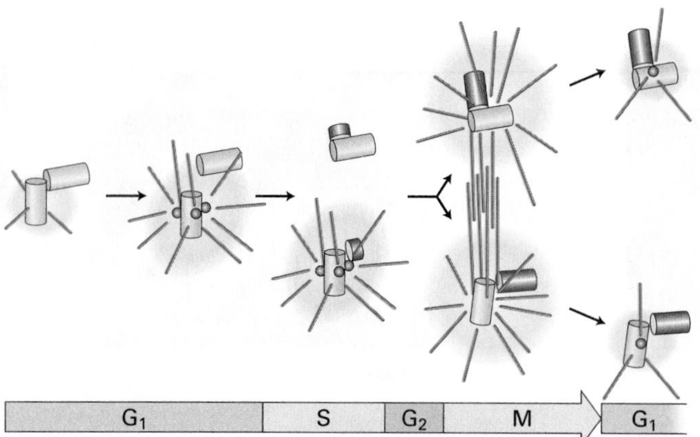

▲ **FIGURE 18-35 Relation of centrosome duplication to the cell cycle.** After the pair of parent centrioles (green) separates slightly, a daughter centriole (blue) buds from each and elongates. By G_2, growth of the daughter centrioles is complete, but the two pairs remain within a single centrosomal complex. Early in mitosis, the centrosome splits, and each centriole pair migrates to opposite sides of the nucleus. The amount of pericentriolar material and the activity to nucleate microtubule assembly increases greatly in mitosis. In mitosis, these MTOCs are called spindle poles.

The Mitotic Spindle Contains Three Classes of Microtubules

Before we discuss the mechanisms involved in this remarkable process, it is important to understand the three distinct classes of microtubules that emanate from the spindle poles, which is where all their (−) ends are embedded. The first class is the *astral microtubules,* which extend from the spindle poles to the cell cortex (Figure 18-36). By interacting with the cortex, the astral microtubules perform the critical function of orienting the spindle with the axis of cell division. The second set link the spindle poles to the kinetochores on the chromosomes and are therefore called *kinetochore microtubules.* This set of microtubules first finds the chromosomes, then attaches them through the two kinetochores to both spindle poles and at anaphase A transports them to the poles. The third set of microtubules extend from each spindle pole body toward the opposite one and interact together in an antiparallel manner; these are called *polar microtubules.* These microtubules are responsible initially for pushing the duplicated centrosomes apart during prophase, then for maintaining the structure of the spindle, and then for pushing the spindle poles apart in anaphase B.

Note that all the microtubules in each half of the symmetrical spindle have the same orientation except for some polar microtubules, which extend beyond the midpoint and interdigitate with polar microtubules from the opposite pole.

Microtubule Dynamics Increases Dramatically in Mitosis

Although we have drawn static images of the stages of mitosis, microtubules in all stages of mitosis are highly dynamic. As we saw above, as cells enter mitosis, the ability of their centrosomes to nucleate assembly of microtubules increases significantly (see Figure 18-35). In addition, microtubules become much more dynamic. How was this determined? In principle, you could watch microtubules and

(a)

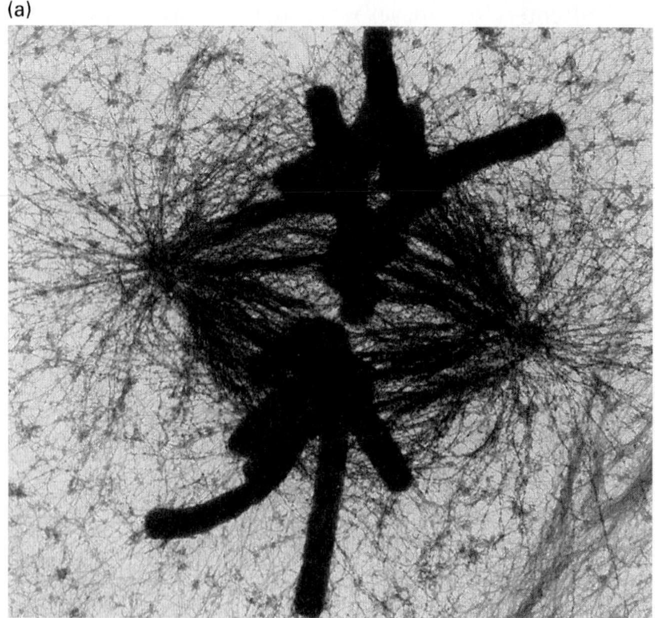

▲ **FIGURE 18-36 Mitotic spindles have three distinct classes of microtubules.** (a) In this high-voltage electron micrograph, microtubules were stained with biotin-tagged anti-tubulin antibodies to increase their size. The large cylindrical objects are chromosomes. (b) Schematic diagram corresponding to the metaphase cell in (a). Three sets of microtubules (MTs) make up the mitotic apparatus. All the microtubules have their (−) ends at the poles. Astral microtubules project toward the cortex and are linked to it. Kinetochore microtubules are connected to chromosomes. Polar microtubules project toward the cell center with their distal (+) ends overlapping. The spindle pole and associated microtubules is also known as a mitotic aster. [Part (a) courtesy of J. R. McIntosh.]

follow their individual behaviors. However, in general, there are too many microtubules in a mitotic spindle to do this. To get an average value for how dynamic microtubules are, researchers have introduced fluorescent labeled tubulin into cells, which becomes incorporated randomly into all microtubules. They then bleach the fluorescent label in a small region of the mitotic spindle and measure the rate at which fluorescence comes back in a technique known as *fluorescence recovery after photobleaching* (FRAP) (see Figure 10-12). Since the recovery of fluorescence is due to assembly of new microtubules from soluble fluorescent tubulin dimers, this represents the average rate at which microtubules turn over. In a mitotic spindle, their half-life is about 15 seconds, whereas in an interphase cell, it is about 5 minutes. It should be noted that these are bulk measurements and individual populations of microtubules can be more stable, as we will see.

What makes microtubules more dynamic in mitosis? Dynamic instability is a measure of relative contributions of growth rates, shrinkage rates, catastrophes, and rescues. Analysis of microtubule dynamics in vivo shows that the enhanced dynamics of individual microtubules in mitosis is mostly generated by increased catastrophes and fewer rescues, with little change in rates of growth (i.e., lengthening) or shrinkage (i.e., shortening). Studies with extracts from frog oocytes have suggested that the main factor enhancing catastrophes in both interphase and mitotic extracts is depolymerizing by kinesin-13 proteins. This can be seen in an in vitro assay where microtubule assembly from pure tubulin is nucleated from purified centrosomes (Figure 18-37a). If kinesin-13 is added into the assay, many fewer microtubules are formed. However, if the stabilizing microtubule-associated protein called XMAP215 is added with the kinesin-13, many microtubules are formed due to a dramatic reduction in catastrophe frequency. It turns out that the activity of kinesin-13 does not change significantly during the cell cycle, whereas the activity of XMAP215 is inhibited by its phosphorylation during mitosis (Figure 18-37b). This results in much more unstable microtubules as the cell enters mitosis (Figure 18-37c).

Microtubules Treadmill During Mitosis

In addition to being highly dynamic in terms of assembly and disassembly, microtubules in the mitotic spindle are treadmilling—that is, constantly adding dimers at the microtubule (+) end and losing them at the (−) end. Treadmilling can be revealed by expressing small amounts of GFP-tubulin in mitotic cells, which is incorporated randomly into microtubules, giving rise to fluorescent speckles where the concentration of incorporated GFP-tubulin happens to be higher. These speckles provide a marker on the microtubules, so by following them in a living cell, one can determine if the microtubules are stationary with respect to spindle poles or moving (Figure 18-38). Experiments such as these show that microtubules are constantly treadmilling in prometaphase, metaphase, and anaphase.

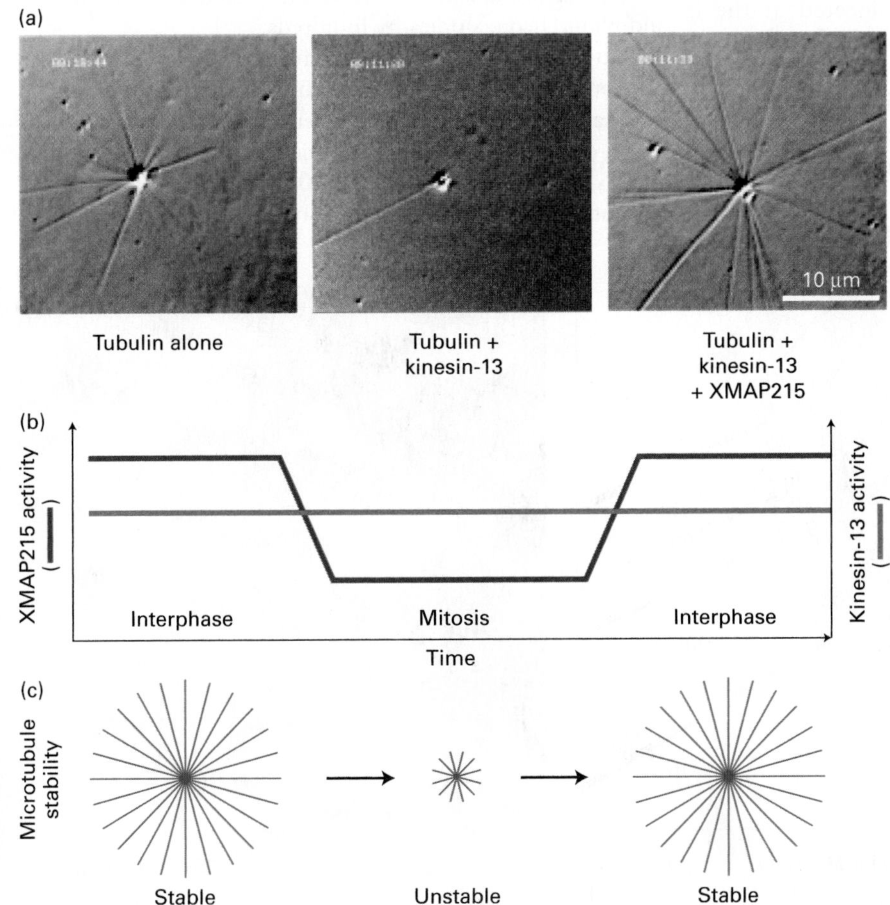

◀ **EXPERIMENTAL FIGURE 18-37** **Microtubule dynamics increases in mitosis due to loss of a stabilizing MAP.** (a) These three panels reveal the ability of centrosomes to assemble microtubules from pure tubulin (*left*); tubulin and the destabilizing protein kinesin-13 (*middle*); or tubulin, kinesin-13, and the stabilizing protein XMAP215 (*Xenopus* MAP of 215kD) (*right*). Further analysis shows that the major effect of XMAP215 is to suppress catastrophes induced by kinesin-13. (b) The increased dynamics of microtubules in mitosis is due to the inactivation of XMAP215 by phosphorylation.(c) Diagram relating the dfferent stabilities of microtubules in interphase and mitosis. Note that in addition to differential *stability* between interphase and mitosis, the ability of MTOCs to *nucleate* microtubules also increases dramatically in mitosis (see Figure 18-35). [Part (a) from Kinoshita et al., 2001, *Science* **294**:1340–1343. Part (b) from Kinoshita et al., 2002, *Trends Cell Biol.* **12**:267–273.]

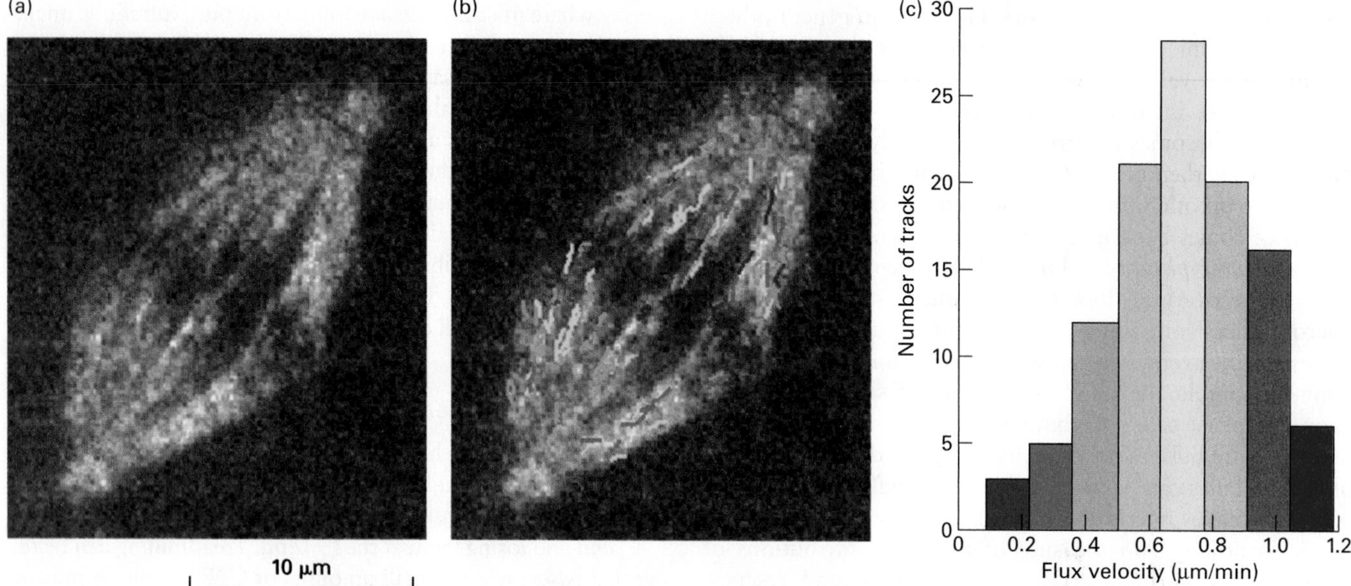

▲ **EXPERIMENTAL FIGURE 18-38 Microtubules in mitosis treadmill toward the spindle poles.** (a) A small amount of GFP-tubulin was expressed in cultured PtK1 cells to visualize microtubules and generate "speckles" along them due to random uneven incorporation of GFP-tubulin. (b) By following speckles in time, the direction and rate of movement of microtubules can be determined, color-coded according to the rate shown in (c). This analysis shows that microtubules treadmill (or "flux") in these cells at about 0.7 μm/min toward the poles. [From L. A. Cameron, 2006, *J. Cell Biol.* **173:** 173–179.]

The Kinetochore Captures and Helps Transport Chromosomes

To attach to microtubules, each chromosome has a specialized structure called a kinetochore. It is located at the **centromere,** a constricted region of the condensed chromosome defined by centromeric DNA. Centromeric DNA can vary enormously in size; in budding yeast it is about 125 bp, whereas in humans it is on the order of 1 Mbp. Kinetochores contain many protein complexes to link the centromeric DNA eventually to microtubules. In animal cells, the kinetochore consists of an inner centromere and inner and outer kinetochore layers, with the (+) ends of the kinetochore microtubules terminating in the outer layer (Figure 18-39). Yeast kinetochores are attached by a single microtubule to their pole, human kinetochores are attached by about 30, and plant chromosomes by hundreds.

How does a kinetochore become attached to microtubules in prometaphase? Microtubules nucleated from the spindle poles are very dynamic, and when they contact the kinetochore, either laterally or at their end, this can lead to chromosomal attachment (Figure 18-40a, steps **1a** and **1b**). Microtubules "captured" by kinetochores are selectively

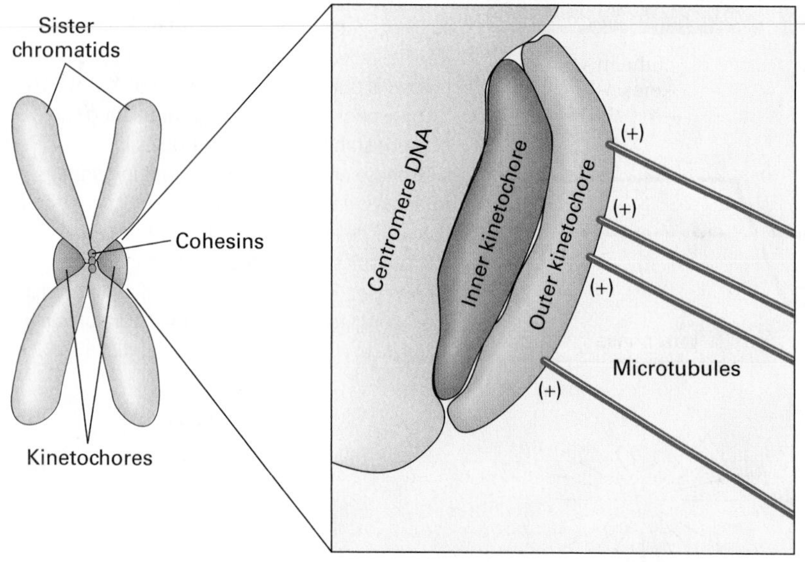

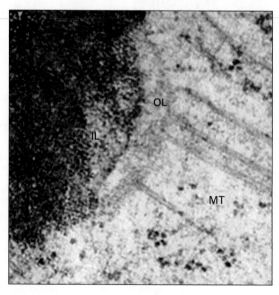

▲ **FIGURE 18-39 The structure of a mammalian kinetochore.** Diagram and electron micrograph of a mammalian kinetochore. [Modified from B. McEwen et al., 1998, *Chromosoma* **107:**366; courtesy of B. McEwen.]

stabilized by reducing the level of catastrophes, thereby promoting the chance that the attachment will persist.

Recent studies have uncovered a mechanism involving the small Ran GTPase that enhances the chance that microtubules will encounter kinetochores. Recall that the Ran GTPase cycle is involved in transport of proteins in and out of the nucleus through nuclear pores (Chapter 13; see Figure 13-36). During mitosis, when the nuclear membrane and pores have disassembled, an exchange factor for the Ran GTPase is bound to chromosomes, thereby generating a higher local concentration of Ran-GTP. Because the enzyme that stimulates GTP hydrolysis on Ran—Ran GAP—is evenly distributed in the cytosol, this generates a gradient of Ran-GTP centered on the chromosomes. Ran-GTP induces the release of factors that promote the growth of microtubules, in this way biasing growth of microtubules nucleated from spindle poles toward chromosomes.

Once attached to microtubules, the motor protein dynein/dynactin located at the kinetochore moves the chromosome pair down the microtubule toward the spindle pole (Figure 18-40a, step **2**). In this orientation, the unoccupied kinetochore on the opposite side is pointing toward the distal spindle pole, and eventually a microtubule from the distal pole will capture the free kinetochore. The chromosome pair is now said to be *bi-oriented* (Figure 18-40a, step **3**). With the two kinetochores attached to opposite poles, the duplicated chromosome is now under tension, being pulled in both directions.

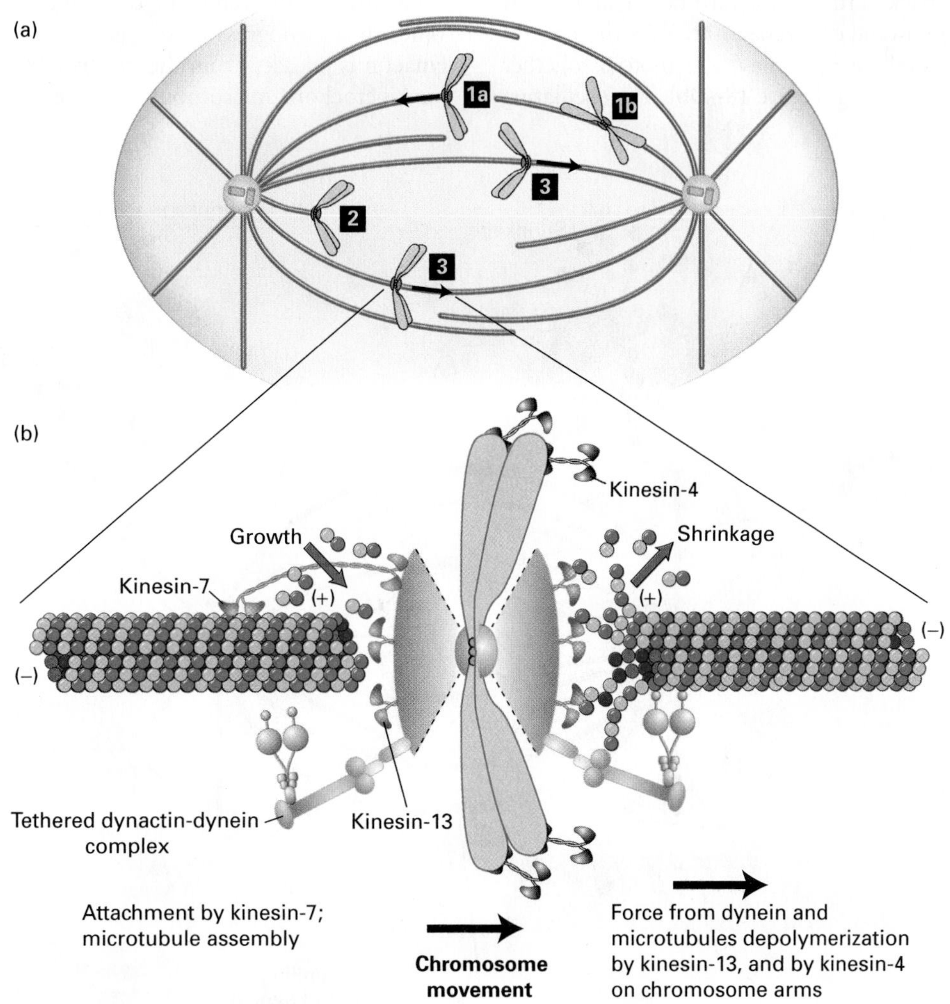

▲ **FIGURE 18-40 Chromosome capture and congression in prometaphase.** (a) In the first stage of prometaphase, chromosomes become attached, either to the end of a microtubule (**1a**) or to the side of a microtubule (**1b**). The duplicated chromosome is then drawn toward the spindle pole by kinetochore-associated dynein/dynactin as this motor moves toward the (−) end of a microtubule (**2**). Eventually, a microtubule from the opposite pole finds and becomes attached to the free kinetochore, and the chromosome is now said to be bi-oriented (**3**). The bi-oriented chromosomes then move to a central point between the spindle poles in a process known as chromosome congression. Note that during these steps, chromosome arms point away from the closest spindle pole: this is due to chromokinesin/kinesin-4 motors on the chromosome arms moving toward the (+) ends of the polar microtubules. (b) Congression involves bidirectional oscillations of chromosomes, with one set of kinetochore microtubules shortening on one side and the other set lengthening on the other. On the shortening side, a kinesin-13 protein stimulates microtubule disassembly and a dynein/dynactin complex moves the chromosome toward the pole. On the side with lengthening microtubules, a CENP-E/kinesin-7 protein holds on to the growing microtubule. The kinetochore also contains many additional protein complexes not shown here. [Modified from Cleveland et al., *Cell* **112**:407–421.]

This tension is a key part of the mechanism whereby a cell properly segregates its chromosomes. When the chromosome is correctly bi-oriented, the cell senses the tension produced, and the attachments are stabilized. What happens if both kinetochores accidentally become attached to the same pole? In this situation, there is very little tension on the kinetochore microtubules, and turnover of kinetochore microtubules is enhanced. Exactly how the cell senses tension is unclear.

Duplicated Chromosomes Are Aligned by Motors and Treadmilling Microtubules

During prometaphase, the chromosomes come to lie at the midpoint between the two spindle poles, in a process known as chromosome *congression.* During this process, chromosome pairs often oscillate backward and forward before arriving at the midpoint. Chromosome congression involves the coordinated activity of several microtubule-based motors together with microtubule treadmilling (Figure 18-40b). The oscillating behavior involves lengthening of microtubules attached to one kinetochore and shortening microtubules attached to the other kinetochore, all without losing their attachments. In metazoans, several microtubule-based motors contribute to this process. First, dynein/dynactin provides the strongest force pulling the chromosome pair toward the more *distant* pole. This movement requires simultaneous shortening of the microtubule, which is enhanced by kinetochore-localized kinesin-13. The microtubules associated with the other kinetochore have to grow as the chromosome moves. Anchored at this kinetochore is a kinesin-related motor, kinesin-7, that holds on to the growing (+) end of the lengthening microtubule. Contributing to congression is also another kinesin, kinesin-4, associated with the chromosome arms. Kinesin-4, a (+) end–directed motor, interacts with the polar microtubules to pull the chromosomes toward the center of the spindle. When the chromosomes have congressed to the metaphase plate, dynein/dynactin is released from the kinetochores and streams down the kinetochore microtubules to the poles. These different

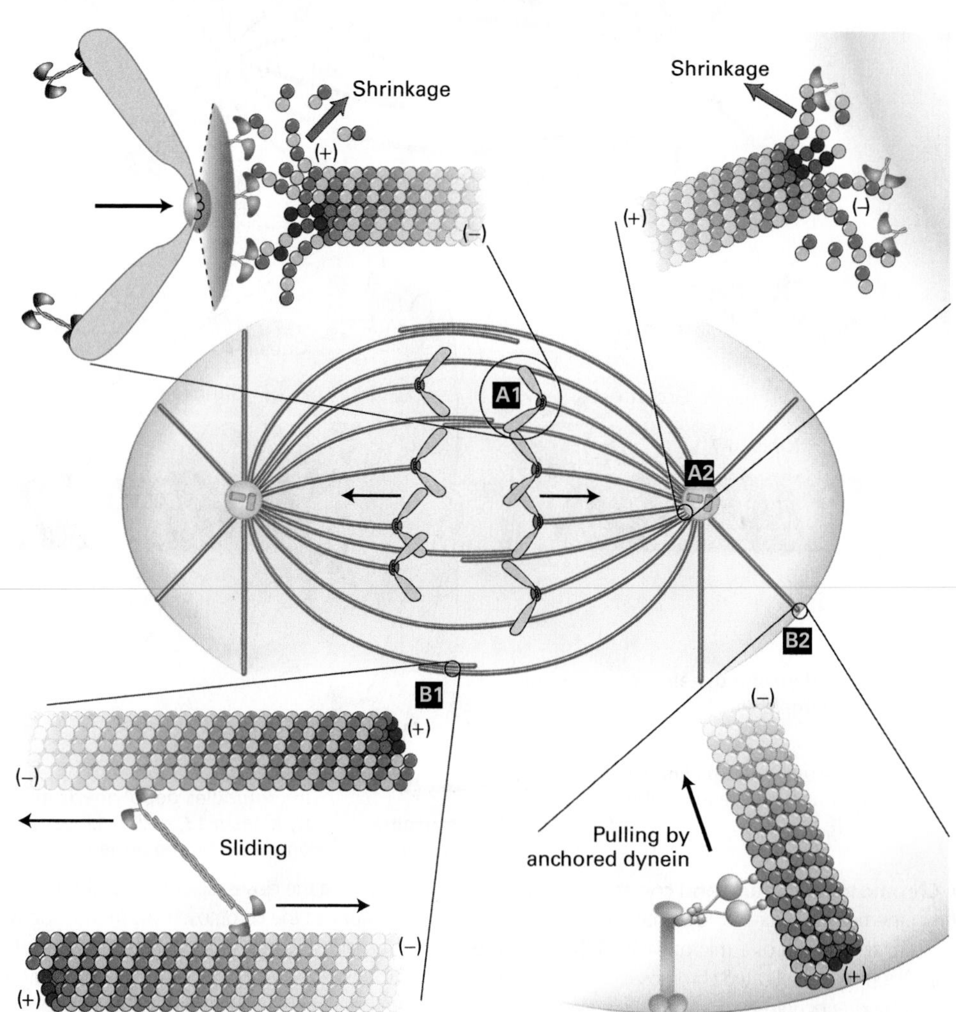

▲ **FIGURE 18-41 Chromosome movement and spindle pole separation in anaphase.** Anaphase A movement is powered by microtubule-shortening kinesin-13 proteins at the kinetochore (A1) and at the spindle pole (A2). Note that the chromosome arms still point away from the spindle poles due to associated chromokinesin/kinesin-4 members, so the depolymerization force has to be able to overcome the force pulling the arms toward the center of the spindle. Anaphase B also has two components: sliding of antiparallel polar microtubules powered by a kinesin-5 (+) end–directed motor (B1) and by dynein/dynactin located at the cell cortex pulling on astral microtubules (B2). [Modified from Cleveland et al., *Cell* **112**:407–421.]

activities and opposing forces work together to bring all the chromosomes to the metaphase plate, and when successfully there, the cell is ready for anaphase. As we discuss in Chapter 20, the cell has a mechanism—the spindle checkpoint—to ensure that anaphase does not proceed until all the chromosomes have arrived at the metaphase plate.

Anaphase A Moves Chromosomes to Poles by Microtubule Shortening

The onset of anaphase A is one of the most dramatic movements that can be observed in the light microscope. Suddenly, the two sister chromatids separate from each other and are drawn to their respective poles. This appears to work so fast because the kinetochore microtubules are under tension, and as soon as the cohesin attachments between chromosomes are released, the chromosomes are free to move.

Experiments with isolated metaphase chromosomes have shown that anaphase A movement can be powered by microtubule shortening, utilizing the stored structural strain released by removing the GTP cap. This can be nicely demonstrated in vitro. When metaphase chromosomes are added to purified microtubules, they bind preferentially to the (+) ends of the microtubules. Dilution of the mixture to reduce the concentration of free tubulin dimers results in the movement of the chromosomes toward the (−) ends by microtubule depolymerization at the chromosome-bound (+) end. Recent experiments have shown that in *Drosophila* two kinesin-13 proteins, a class of microtubule depolymerizing proteins (see Figure 18-16), contribute to chromosome movement in anaphase A. One of the kinesin-13 proteins is localized at the kinetochore and enhances disassembly there (Figure 18-41, **A1**), and the other is localized at the spindle pole, enhancing depolymerization there (Figure 18-41, **A2**). Thus, at least in the fly, anaphase A is powered in part by kinesin-13 proteins specifically localized at the kinetochore and spindle pole to shorten the kinetochore microtubules at both their (+) and (−) ends to draw the chromosomes to the poles.

Anaphase B Separates Poles by the Combined Action of Kinesins and Dynein

The second part of anaphase involves separation of the spindle poles in a process known as anaphase B. A major contributor to this movement is the involvement of the bipolar kinesin-5 proteins (Figure 18-41, **B1**). These motors associate with the overlapping polar microtubules, and since they are (+) end–directed motors, they push the poles apart. While this is happening, the polar microtubules have to grow to accommodate the increased distance between the spindle poles—at the same time as the kinetochore microtubules are shortening for anaphase A! Another motor—the microtubule (−) end–directed motor cytoplasmic dynein, localized and anchored on the cell cortex—pulls on the astral microtubules and thus helps separate the spindle poles (Figure 18-41, **B2**).

Additional Mechanisms Contribute to Spindle Formation

There are a number of cases in vivo where spindles form in the absence of centrosomes. This implies that nucleation of microtubules from centrosomes is not the only way a spindle can form. Studies exploiting mitotic extracts from frog eggs—extracts that do not contain centrosomes—show that the addition of beads covered with DNA is sufficient to assemble a relatively normal mitotic spindle (Figure 18-42). In this system, the beads recruit preformed microtubules, and factors in the extract cooperate to make a spindle. One of the factors necessary for this reaction is cytoplasmic dynein, which is proposed to bind to two microtubules and migrate to their (−) ends and thereby draw them together.

Cytokinesis Splits the Duplicated Cell in Two

During late anaphase and telophase in animal cells, the cell assembles a microfilament-based *contractile ring* attached to

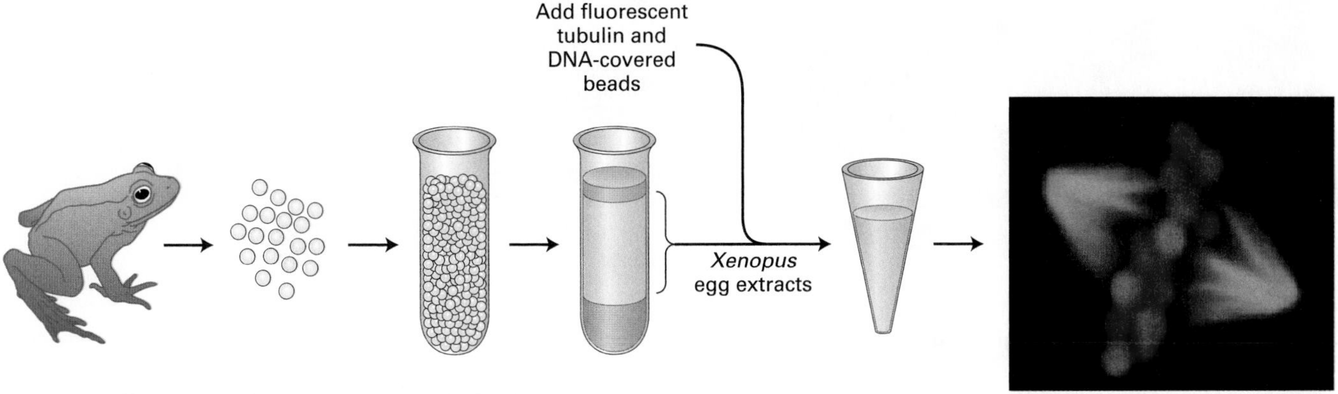

▲ **EXPERIMENTAL FIGURE 18-42 Mitotic spindles can form in the absence of centrosomes.** Centrosome-free extracts can be isolated from frog oocytes arrested in mitosis by centrifuging eggs to separate a soluble material from the organelles and yolk. When fluorescently labeled tubulin (green) is added to extracts together with beads covered with DNA (red), mitotic spindles spontaneously form around the beads from randomly nucleated microtubules. [Modified from Kinoshita et al., 2002, *Trends Cell Biol.* **12**:267–273, and Antonio et al., 2000, *Cell* **102**:425.]

the plasma membrane that will eventually contract and pinch the cell into two, a process known as cytokinesis (see Figure 18-34). The contractile ring is a thin band of actin filaments of mixed polarity interspersed with myosin-II bipolar filaments (see Figure 17-34). On receiving a signal, the ring contracts first to generate a *cleavage furrow* and then to pinch the cell into two. Two aspects of the contractile ring are essential to its function. First, it has to be appropriately placed in the cell. It is known that this placement is determined by signals provided by the spindle, so that the ring forms equidistant between the two spindle poles. Recent work with *Drosophila* mutants has suggested that components that accumulate at the central point of the spindle during anaphase direct the formation of the contractile ring. However, the molecular signals involved in coordinating spindle position and the position of the contractile ring are still being unraveled.

The second important aspect of the contractile ring is the timing of its contraction—if it contracted before all chromosomes have moved to their respective poles, disastrous genetic consequences would ensue. Again, the mechanism whereby this timing is determined is being unraveled.

Plant Cells Reorganize Their Microtubules and Build a New Cell Wall in Mitosis

Interphase plant cells lack a central MTOC that organizes microtubules into the radiating interphase array typical of animal cells. Instead, numerous MTOCs line the cortex of plant cells and nucleate the assembly of transverse bands of microtubules below the cell wall (Figure 18-43, *left*). The microtubules, which are of mixed polarity, are released from the cortical MTOCs by the action of katanin, a microtubule-severing protein; loss of katanin gives rise to very long microtubules and misshapen cells. The reason for this is that these cortical microtubules, which are cross-linked by plant-specific MAPs, aid in laying down extracellular cellulose microfibrils, the main component of the rigid cell wall (see Figure 19-37).

Although mitotic events in plant cells are generally similar to those in animal cells, formation of the spindle and cytokinesis have unique features in plants (Figure 18-43). Plant cells bundle their cortical microtubules into the *preprophase band* and reorganize them into a spindle at prophase without the aid of centrosomes. At metaphase,

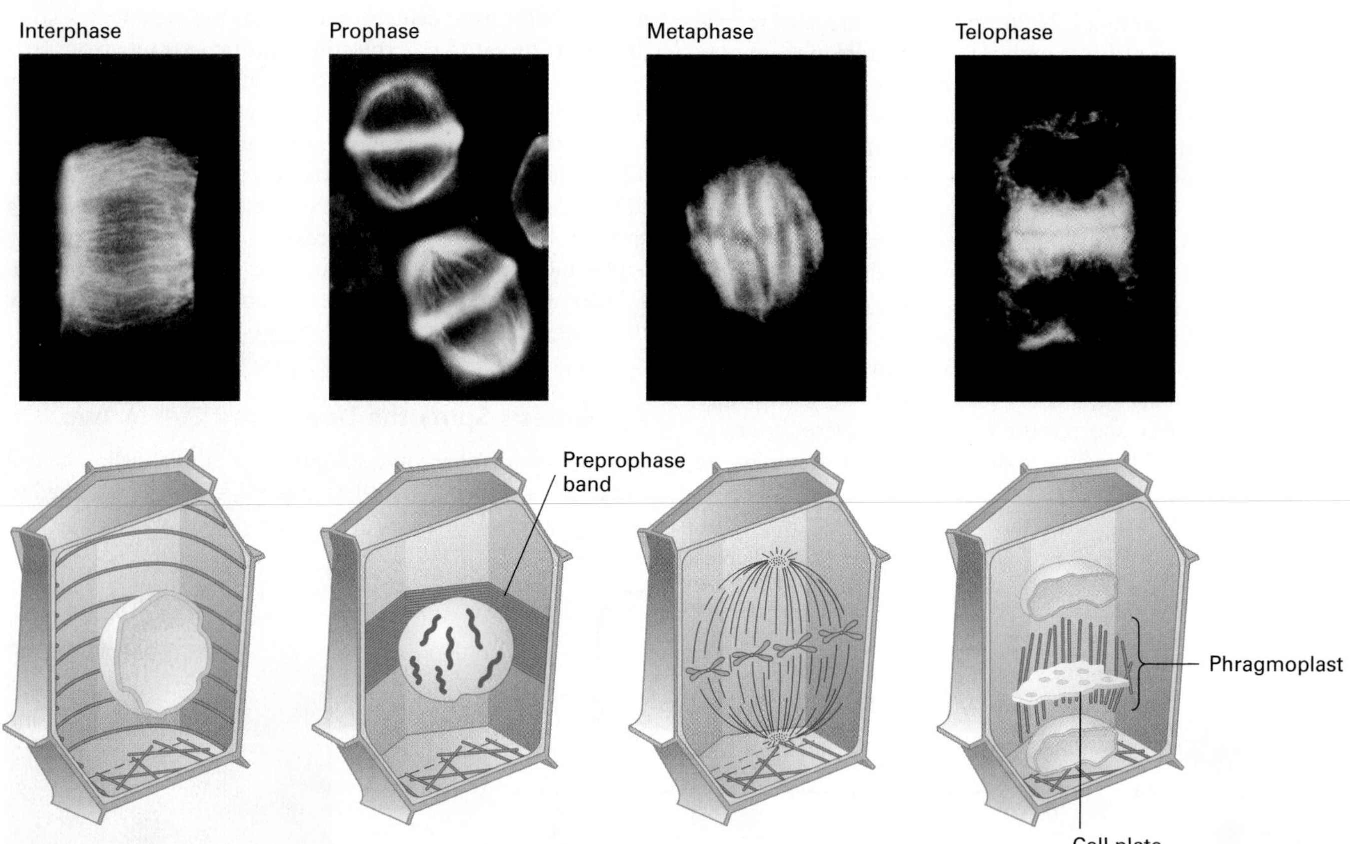

▲ FIGURE 18-43 Mitosis in a higher plant cell. Immunofluorescence micrographs (*top*) and corresponding diagrams (*bottom*) showing arrangement of microtubules in interphase and mitotic plant cells. A cortical array of microtubules girdles a cell during interphase. Webs of microtubules cap the growing ends of plant cells and remain intact during cell division. As a cell enters prophase, the microtubules are bundled around the nucleus and reorganized into a spindle that appears similar to that in metaphase animal cells. By late telophase, the nuclear membrane has re-formed around the daughter nuclei and the Golgi-derived phragmoplast has assembled at the equatorial plate. Additional small vesicles derived from the Golgi complex accumulate at the equatorial plate and fuse with the phragmoplast to form the new cell plate. [Adapted from R. H. Goddard et al., 1994, *Plant Physiol.* **104:**1; micrographs courtesy of Susan M. Wick.]

the mitotic apparatus appears much the same in plant and animal cells. However, the division of the cell into two is quite different from animal cells. Golgi-derived vesicles, which appear at metaphase, are transported into the mitotic apparatus along microtubules that radiate from each end of the spindle. At telophase, these vesicles line up near the center of the dividing cell and then fuse to form the **phragmoplast,** a membrane structure that replaces the animal-cell contractile ring. The membranes of the vesicles forming the phragmoplast become the plasma membranes of the daughter cells. The contents of these vesicles, such as polysaccharide precursors of cellulose and pectin, form the early cell piate, which develops into the new cell wall between the daughter cells.

KEY CONCEPTS OF SECTION 18.6

Mitosis

- Mitosis—the accurate separation of duplicated chromosomes—involves a molecular machine comprising dynamic treadmilling microtubules and molecular motors.
- The mitotic spindle has three classes of microtubules, all emanating from the spindle poles—kinetochore microtubules, which attach to chromosomes; polar microtubules from each spindle pole, which overlap in the middle of the spindle; and astral microtubules, which extend to the cell cortex (see Figure 18-36).
- In the first stage of mitosis, prophase, the nuclear chromosomes condense and the spindle poles move to either side of the nucleus (see Figure 18-34).
- At prometaphase, the nuclear envelope breaks down and microtubules emanating from the spindle poles capture chromosome pairs at their kinetochores. Each of the two kinetochores on the duplicated chromosomes becomes attached to the two spindle poles, which allows the chromosomes to congress to the middle of the spindle.
- At metaphase, chromosomes are aligned on the metaphase plate. The spindle checkpoint system monitors unattached kinetochores and delays anaphase until all chromosomes are attached.
- At anaphase, duplicated chromosomes separate and move toward the spindle poles by shortening of the kinetochore microtubules at both the kinetochore and spindle pole (anaphase A). The spindle poles also move apart, pushed by bipolar kinesin-5 moving toward the (+) ends of the polar microtubules (anaphase B). Spindle separation is also facilitated by cortically located dynein pulling on astral microtubules (see Figure 18-40, 18-41).
- Redundant mechanisms contribute to the fidelity of mitosis since the mitotic spindle has the ability to self-assemble in the absence of MTOCs.
- The position of the actin-myosin based cleavage furrow is determined by the position of the spindle and at cytokinesis contracts to pinch the cell in two.

18.7 Intermediate Filaments

The third major filament system of eukaryotes is collectively called **intermediate filaments.** This name reflects their diameter of about 10 nm, which is intermediate between the 6–8 nm of microfilaments and myosin thick filaments of skeletal muscle. Intermediate filaments extend throughout the cytoplasm as well as line the inner nuclear envelope of interphase animal cells (Figure 18-44). Intermediate filaments have several unique properties that distinguish them from microfilaments and microtubules. First, they are biochemically much more heterogeneous—that is, many different, but evolutionarily related, intermediate filament subunits exist that are often expressed in a tissue-dependent manner. Second, they have great tensile strength, which is most clearly demonstrated by hair and nail that consist primarily of the intermediate filaments of dead cells. Third, they do not have an intrinsic polarity like microfilaments and microtubules, and their constituent subunits do not bind a nucleotide. Fourth, because they have no intrinsic polarity, it is not surprising that there are no known motors that use them as tracks. Fifth, although they are dynamic in terms of subunit exchange, they are much more stable than microfilaments and microtubules because the exchange rate

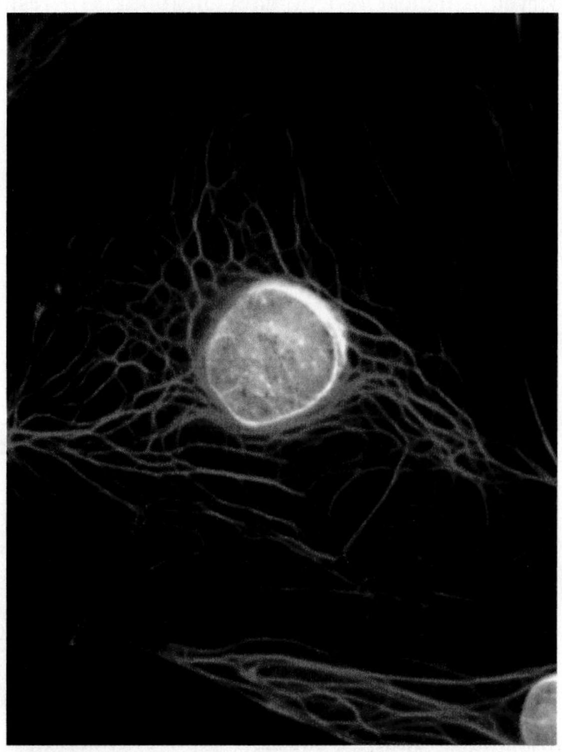

▲ EXPERIMENTAL FIGURE 18-44 **Localization of two types of intermediate filaments in an epithelial cell.** Immunofluorescence micrograph of a PtK2 cell doubly stained with keratin (red) and lamin (blue) antibodies. A meshwork of lamin intermediate filaments can be seen underlying the nuclear membrane, whereas the keratin filaments extend from the nucleus to the plasma membrane. [Courtesy of R. D. Goldman.]

is much slower. Indeed, a standard way to purify intermediate filaments is to subject cells to harsh extraction conditions in a detergent so that all membranes, microfilaments, and microtubules are solubilized, leaving a residue that is almost exclusively intermediate filaments. Finally, intermediate filaments are not found in all eukaryotes. Fungi and plants do not have intermediate filaments, and insects only have one class, represented by two genes that express lamin A/C and B.

These properties make intermediate filaments unique and important structures of metazoans. The importance of intermediate filaments is underscored by the identification of more than 40 clinical disorders, some of which are discussed below, associated with defects in genes encoding intermediate filament proteins. To understand their contributions to cell and tissue structure, we first examine the structure of intermediate filament proteins and how they assemble into a filament. We then describe the different classes of intermediate filaments and the functions they perform.

Intermediate Filaments Are Assembled from Subunit Dimers

Intermediate filaments (IFs) are encoded in the human genome by 70 different genes in at least five subfamilies. The defining feature of IF proteins is the presence of a conserved α-helical rod domain of about 310 residues that has the sequence features of a coiled-coil motif (see Figure 3-9a). Flanking the rod domain are nonhelical N- and C-terminal domains of different sizes, characteristic of each IF class.

The primary building block of intermediate filaments is a dimer (Figure 18-45a) held together through the rod domains that associate as a coiled-coil. These dimers then associate in an offset fashion to make tetramers, where the two dimers are in opposite orientations (Figure 18-45b). Tetramers are assembled end to end and interlocked into long *protofilaments.* Four protofilaments associate into a *protofibril,* and four protofibrils associate side to side to generate the 10-nm filament. Thus an intermediate filament has 16 protofilaments in it. Because the tetramer is symmetric, intermediate filaments have no polarity. This description of the filament is based on its structure rather than its mechanism of assembly: at present it is not yet clear how intermediate filaments are assembled in vivo. Unlike microfilaments and microtubules, there are no known intermediate filament nucleating, sequestering, capping, or filament-severing proteins.

Intermediate Filaments Proteins Are Expressed in a Tissue-Specific Manner

Sequence analysis of IF proteins reveals that they fall into at least five distinct homology classes, with four classes showing a strong correspondence between the sequence class and

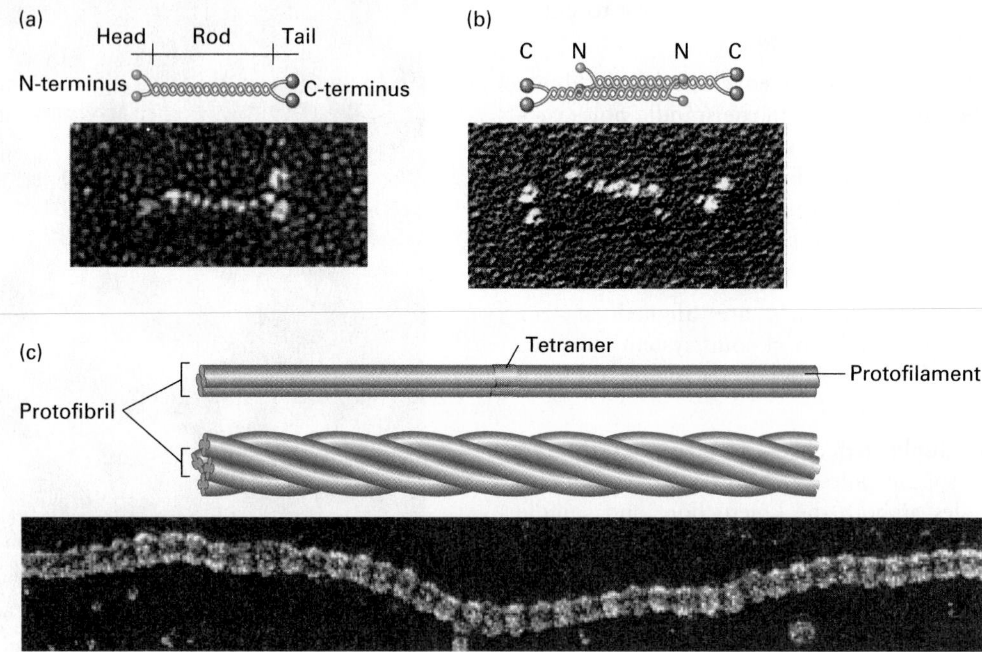

▲ **FIGURE 18-45 Structure and assembly of intermediate filaments.** Electron micrographs and drawings of IF protein dimers, tetramers, and mature intermediate filaments from *Ascaris,* an intestinal parasitic worm. (a) IF proteins form parallel dimers through a highly conserved coiled-coil core domain. The globular heads and tails are quite variable in length and sequence between intermediate filament classes. (b) A tetramer is formed by antiparallel, staggered side-by-side aggregation of two identical dimers. (c) Tetramers aggregate end to end and laterally into a protofibril. In a mature filament, consisting of four protofibrils, the globular domains form beaded clusters on the surface. [Adapted from N. Geisler et al., 1998, *J. Mol. Biol.* **282:**601; courtesy of Ueli Aebi.]

TABLE 18-1 The Major Classes of Intermediate Filaments in Mammals			
CLASS	**PROTEIN**	**DISTRIBUTION**	**PROPOSED FUNCTION**
I	Acidic keratins	Epithelial cells	Tissue strength and integrity
II	Basic keratins	Epithelial cells	Tissue strength and integrity
III	Desmin, GFAP, vimentin	Muscle, glial cells, mesenchymal cells	Sarcomere organization, integrity
IV	Neurofilaments (NFL, NFM, and NFH)	Neurons	Axon organization
V	Lamins	Nucleus	Nuclear structure and organization

the developmental origin of the cell type in which the IF protein is expressed (Table 18-1).

Keratins that make up classes I and II are found in epithelia, class III intermediate filaments are generally found in cells of mesodermal origin, and class IV make up the **neurofilaments** found in neurons. The **lamins** make up class V, which are found lining the nucleus of all animal tissues. We briefly summarize the five different homology classes and discuss their roles in specific tissues.

Keratins Keratins provide strength to epithelial cells. The first two homology classes are the so-called *acidic* and *basic keratins*. There are about 50 genes in the human genome encoding keratins, about evenly split between the acidic and basic classes. These keratin subunits assemble into an obligate dimer, so that each dimer consists of one basic chain and one acidic chain; these are then assembled into a filament as described above.

The keratins are by far the most diverse of the IF protein families, with keratin pairs showing different expression patterns between distinct epithelia and also showing differentiation-specific regulation. Among these are the so-called hard keratins that make up hair and nails. These keratins are rich in cysteine residues that become oxidized to form disulfide bridges, thereby strengthening them. This property is exploited by hair stylists: if you do not like the shape of your hair, the disulfide bonds in your hair keratin can be reduced, the hair reshaped, and the disulfide bonds reformed by oxidation—known as a "permanent."

The so-called soft, or *cyto-keratins*, are found in epithelial cells. The epidermal-cell layers that make up the skin provide a good example of the function of keratins. The lowest layer of cells, the *basal layer*, which is in contact with the basal lamina, proliferates constantly, giving rise to cells called *keratinocytes*. After they leave the basal layer, the keratinocytes differentiate and express abundant cytokeratins. The cytokeratins associate with desmosomes, specialized attachment sites between cells, thereby providing sheets of cells that can withstand abrasion. The cells eventually die, leaving dead cells from which all cell organelles have disappeared. This dead cell layer provides an essential barrier to water evaporation, without which we could not survive. The life of a skin cell, from birth to its loss from the animal as a skin flake, is about one month.

In all epithelia, keratin filaments associate with desmosomes, which link adjacent cells together, and hemidesmosomes, which link cells to the extracellular matrix, thereby giving cells and tissues their strength. These structures are described in more detail in Chapter 19.

In addition to simply providing structural support, there is increasing evidence that keratin filaments provide some organization to organelles and participate in signal-transduction pathways. For example, in response to tissue injury, rapid cell growth is induced. In epithelial cells it has been shown that the growth signal requires an interaction between a cell-growth-signaling molecule and a specific keratin.

Desmin The class III intermediate filament proteins include vimentin, found in mesenchymal cells; GFAP (glial fibrillary acidic protein), found in glial cells; and *desmin,* found in muscle cells. Desmin provides strength and organization to muscle cells.

In smooth muscle, desmin filaments link cytoplasmic *dense bodies,* into which the contractile myofibrils are also attached, to the plasma membrane to ensure that cells resist overstretching. In skeletal muscle, a lattice composed of a band of desmin filaments surrounds the sarcomere. The desmin filaments encircle the Z disk and are cross-linked to the plasma membrane. Longitudinal desmin filaments cross to neighboring Z disks within the myofibril, and connections between desmin filaments around Z disks in adjacent myofibrils serve to cross-link myofibrils into bundles within a muscle cell. The lattice is also attached to the sarcomere through interactions with myosin thick filaments. Because the desmin filaments lie outside the sarcomere, they do not actively participate in generating contractile forces. Rather, desmin plays an essential structural role in maintaining muscle integrity. In transgenic mice lacking desmin, for example, this supporting architecture is disrupted and Z disks are misaligned. The location and morphology of mitochondria in these mice are also abnormal, suggesting that these intermediate filaments may also contribute to organization of organelles.

Neurofilaments Type IV intermediate filaments consist of the three related subunits—NF-L, NF-M, NF-H (for NF light, medium, and heavy)—that make up the neurofilaments found in the axons of nerve cells (see Figure 18-2). The three subunits differ mainly in the size of their C-terminal domains, and all form obligate heterodimers. Experiments with transgenic mice reveal that neurofilaments are necessary to establish the correct diameter of axons, which determines the rate at which nerve impulses are propagated down axons.

Lamins The most widespread intermediate filaments are the class V lamins, which provide strength and support to the inner surface of the nuclear membrane (see Figure 18-44). Lamins are the progenitor of all IF proteins, with the cytoplasmic IFs arising by gene duplication and mutation. The lamins provide a two-dimensional network lying between the nuclear envelope and the chromatin in the nucleus. In humans, three genes encode lamins: one encodes A-type lamin and two encode B type. The B-type lamin appears to be the primordial gene and is expressed in all cells, whereas A-type lamins are developmentally regulated. B-lamins are post-translationally isoprenylated, which presumably helps them associate with the inner nuclear envelope membrane. In addition, they bind inner nuclear membrane proteins such as emerin and lamin-associated polypeptides (LAP2). Lamins bind multiple proteins and have been proposed to play roles in large-scale chromatin organization and the spacing of nuclear pores. As cells en-

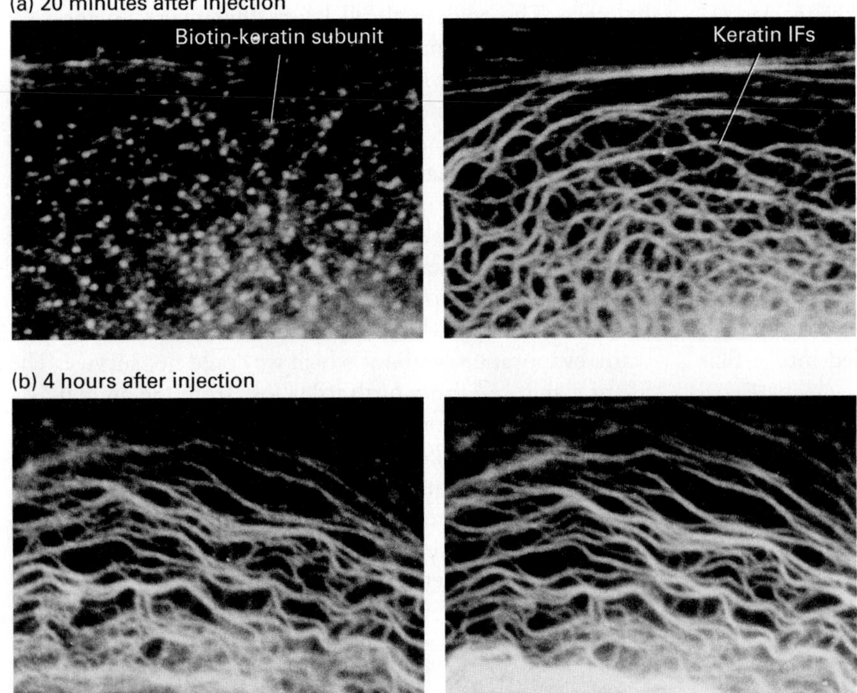

◀ **EXPERIMENTAL FIGURE 18-46 Keratin intermediate filaments are dynamic as soluble keratin is incorporated into filaments.** Monomeric type I keratin was purified, chemically labeled with biotin, and microinjected into living epithelial cells. The cells were then fixed at different times after injection and stained with an antibody to biotin and with antibodies to keratin. (a) At 20 minutes after injection, the injected biotin-labeled keratin is concentrated in small foci scattered through the cytoplasm (*left*) and has not been integrated into the endogenous keratin cytoskeleton (*right*). (b) By 4 hours, the biotin-labeled (*left*) and the keratin filaments (*right*) display identical patterns, indicating that the microinjected protein has become incorporated into the existing cytoskeleton. [From R. K. Miller, K. Vistrom, and R. D. Goldman, 1991, *J. Cell Biol.* **113**:843; courtesy of R. D. Goldman.]

ter mitosis, lamins become hyperphosphorylated and disassemble; in telophase they reassemble with the reassembling nuclear membrane.

Intermediate Filaments Are Dynamic

Although intermediate filaments are much more stable than microtubules and microfilaments, IF protein subunits have been shown to be in dynamic equilibrium with the existing IF cytoskeleton. In one experiment, a biotin-labeled type I keratin was injected into fibroblasts; within 2 hours, the labeled protein had been incorporated into the already existing keratin cytoskeleton (Figure 18-46). The results of this experiment and others demonstrate that IF subunits in a soluble pool are able to add themselves to preexisting filaments and that subunits are able to dissociate from intact filaments.

The relative stability of intermediate filaments presents special challenges in mitotic cells, which must reorganize all three cytoskeletal networks in the course of the cell cycle. In particular, breakdown of the nuclear envelope early in mitosis depends on the disassembly of the lamin filaments that form a meshwork supporting the membrane. As discussed in Chapter 20, the phosphorylation of nuclear lamins by a cyclin-dependent kinase that becomes active early in mitosis (prophase) induces the disassembly of intact filaments and prevents their reassembly. Later in mitosis (telophase), removal of these phosphates by specific phosphatases promotes lamin reassembly, which is critical to re-formation of a nuclear envelope around the daughter chromosomes. The opposing actions of kinases and phosphatases thus provide a rapid mechanism for controlling the assembly state of lamin intermediate filaments. Other intermediate filaments undergo similar disassembly and reassembly in the cell cycle.

Defects in Lamins and Keratins Cause Many Diseases

There are about 50 known mutations in the human gene for type-A lamin that are known to cause disease, many of which cause forms of Emery-Dreifuss muscular dystrophy (EDMD). Other mutations in the lamin-A gene cause dilated cardiomyopathy. It is not yet clear why these type-A lamin mutations cause EDMD, but perhaps in muscle tissues the fragile nuclei cannot stand the stress and strains of the tissue, so they are the first to show symptoms. Interestingly, other forms of EDMD have been traced to mutations in emerin, the lamin-binding membrane protein of the inner nuclear envelope. Yet other mutations in type-A lamin cause progeria—accelerated aging. Thus the Hutchison-Gilford progeria ("prematurely old") syndrome is caused by a splicing error that results in a lamin A with a defective C-terminal domain.

The structural integrity of the skin is essential in order to withstand abrasion. In humans and mice, the K4 and K14 keratin isoforms form heterodimers that assemble into protofilaments. A mutant K14 with deletions in either the N- or the C-terminal domain can form heterodimers in vitro but does not assemble into protofilaments. The expression of such mutant keratin proteins in cells causes IF networks to break down into aggregates. Transgenic mice that express a mutant K14 protein in the basal stem cells of the epidermis display gross skin abnormalities, primarily blistering of the epidermis, that resemble the human skin disease *epidermolysis bullosa simplex* (EBS). Histological examination of the blistered area reveals a high incidence of dead basal cells. Death of these cells appears to be caused by mechanical trauma from rubbing of the skin during movement of the limbs. Without their normal bundles of keratin filaments, the mutant basal cells become fragile and easily damaged, causing the overlying epidermal layers to delaminate and blister (Figure 18-47).

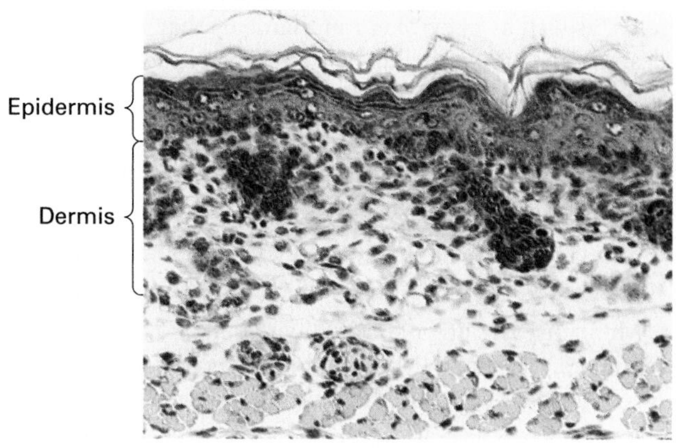

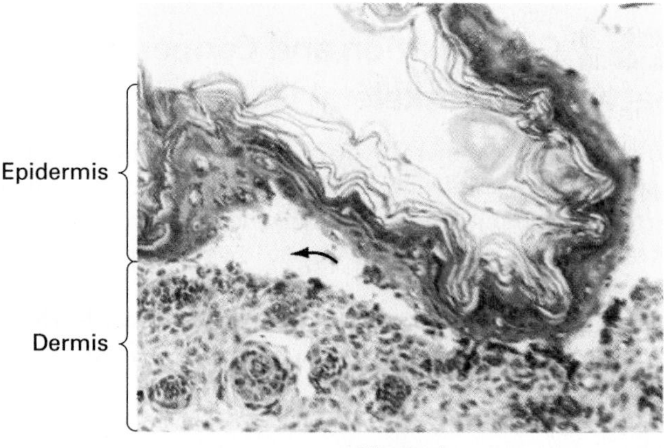

▲ EXPERIMENTAL FIGURE 18-47 **Transgenic mice carrying a mutant keratin gene exhibit blistering similar to that in the human disease epidermolysis bullosa simplex.** Histological sections through the skin of a normal mouse and a transgenic mouse carrying a mutant K14 keratin gene are shown. In the normal mouse, the skin consists of a hard outer epidermal layer covering in contact with the soft inner dermal layer. In the skin from the transgenic mouse, the two layers are separated (arrow) due to weakening of the cells at the base of the epidermis. [From P. Coulombe et al., 1991, *Cell* **66:**1301; courtesy of E. Fuchs.]

Like the role of desmin filaments in supporting muscle tissue, the general role of keratin filaments appears to be to maintain the structural integrity of epithelial tissues by mechanically reinforcing the connections between cells. ∎

KEY CONCEPTS OF SECTION 18.7

Intermediate Filaments

- Intermediate filaments are the only nonpolarized fibrous component of the cytoskeleton and do not have associated motor proteins. Intermediate filaments are built from coiled-coil dimers that associate in an antiparallel fashion into tetramers and then into protofilaments, 16 of which make up the filament (see Figure 18-45).
- There are five major classes of intermediate proteins, with the nuclear lamins (class V) being the most ancient and ubiquitous in animal cells. The other four classes show tissue-specific expression (see Table 18-1).
- Keratins (classes I and II) are found in animal hair and nails, as well as in cytokeratin filaments that associate with desmosomes in epithelial cells to provide the cells and tissue with strength.
- The class III filaments include vimentin, GFAP, and desmin, which provide structure and order to muscle Z disks and restrain smooth muscle from overextension.
- The neurofilaments make up class IV and are important for the structure of axons.
- Many diseases are associated with defects in intermediate filaments, especially laminopathies, which include a variety of conditions, and mutations in keratin genes, which can cause severe defects in skin (see Figure 18-47).

18.8 Coordination and Cooperation between Cytoskeletal Elements

So far, we have generally discussed the three cytoskeletal filaments classes—microfilaments, microtubules, and intermediate filaments—as though they function independently of one another. However, the example that the microtubule-based mitotic spindle determines the site of formation of the microfilament-based contractile ring is just one example of how these two cytoskeletal systems are coordinated. Here we mention some other examples of linkages, physical and regulatory, between cytoskeletal elements and their integration into other aspects of cellular organization.

Intermediate Filament–Associated Proteins Contribute to Cellular Organization

A group of proteins collectively called *intermediate filament–associated proteins* (IFAPs) have been identified that co-purify with intermediate filaments. Among these are the family of **plakins,** which are involved in attaching intermediate filaments to other structures. Some of these associate with keratin filaments to link them to desmosomes, which are junctions between epithelial cells that provides stability to a tissue, and hemidesmosomes, which are located at regions of the plasma membrane where intermediate filaments are linked to the extra-cellular matrix (these topics are covered in detail in Chapter 19). Other plakins are found along intermediate filaments and have binding sites for microfilaments and microtubules. One of these proteins, called plectin, can be seen by immunoelectron microscopy to provide connections between microtubules and intermediate filaments (Figure 18-48).

Microfilaments and Microtubules Cooperate to Transport Melanosomes

Studies of mutant mice with light-colored coats has uncovered a pathway in which microtubules and microfilaments cooperate to transport pigment granules. The color pigment in the hair is produced in cells called melanocytes, cells very similar to the fish and frog melanophores discussed earlier (see Figure 18-28). Melanocytes are found in the hair follicle at the base of the hair shaft and contain pigment-laden granules called melanosomes. Melanosomes are transported to the dendritic extensions of melanocytes for subsequent exocytosis to the surrounding epithelial cells. Transport to the cell periphery is mediated, just as in frog melanophores, by a kinesin family member. At the periphery, they are then handed off to myosin V and delivered for exocytosis. If the myosin V system is defective, the melanosomes are not captured and stay in the cell body. Thus microtubules are responsible for the long-range transport of melanosomes, whereas microfilament-based myosin V is responsible for capture and delivery at the cell

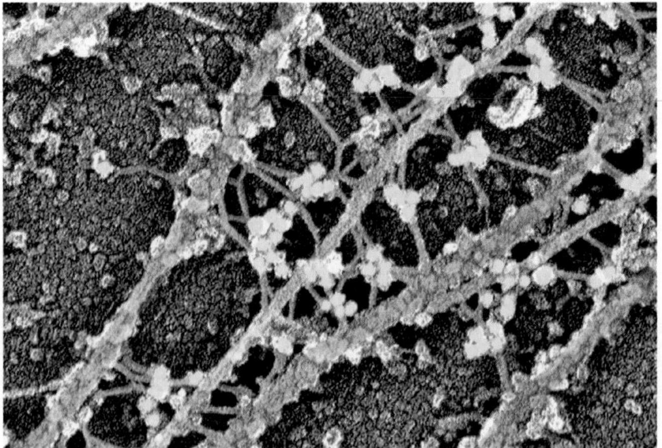

▲ EXPERIMENTAL FIGURE 18-48 **Gold-labeled antibody identifies plectin cross-links between intermediate filaments and microtubules.** In this immunoelectron micrograph of a fibroblast cell, microtubules are highlighted in red; intermediate filaments, in blue; and the short connecting fibers between them, in green. Staining with gold-labeled antibodies to plectin (yellow) reveals that these fibers contain plectin. [From T. M. Svitkina, A. B. Verkhovsky, and G. G. Borisy, 1996, *J. Cell Biol.* **135:**991; courtesy of T. M. Svitkina.]

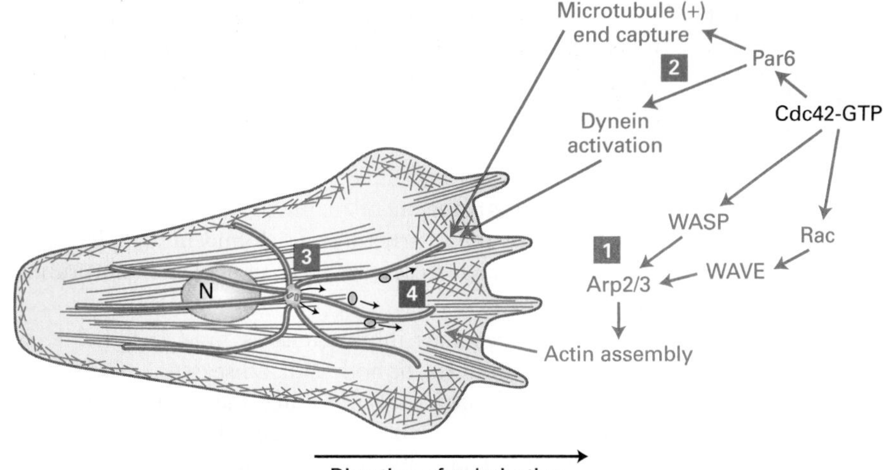

◀ **FIGURE 18-49 Cdc42 regulates microfilaments and microtubules independently to polarize a migrating cell.** Active Cdc42-GTP at the front of the cell leads to Rac activation, which results in the assembly of a microfilament-based leading edge (step **1**). Independently, Cdc42-GTP also leads to the capture of microtubule (+) ends and the activation of dynein (step **2**). Together these pull on microtubules to orient the centrosome (step **3**) toward the front of the cell. This reorientation polarizes the secretory pathway for the delivery along microtubules of adhesion molecules carried in secretory vesicles (step **4**). [Based on studies in S. Etienne-Manneville et al., 2005, *J. Cell Biol.* **170:**895–901.]

cortex. This type of division of labor—long-range transport by microtubules and short range by microfilaments—has been found in many different systems, from transport in filamentous fungi to transport along axons.

Cdc42 Coordinates Microtubules and Microfilaments During Cell Migration

In Chapter 17, we discussed how the polarity of a migrating cell is regulated by Cdc42, which results in the formation of an actin-based leading edge at the front of the cell and contraction at the back (see Figures 17-44 and 18-49, step **1**). It turns out that Cdc42 activation at the cell front also leads to polarization of the microtubule cytoskeleton. This was originally studied in wound-healing assays (see Figure 17-43) where it was noticed that when the cells at the edge are induced to polarize and move to fill up the scratch, the Golgi complex is moved to the front of the nucleus toward the cell front. Since Golgi localization is dependent on the location of the MTOC (see Figures 18-1c, 18-27), this was because the centrosome came to lie in front of the nucleus. Recent studies have suggested how this happens. Cdc42 activation at the front of the cell binds the polarity factor Par6, which results in the recruitment of the dynein/dynactin complex (Figure 18-49, step **2**). Cortically localized dynein/dynactin then interacts with microtubules pulling on them to orient the centrosome and hence the whole radial array of microtubules (Figure 18-49, step **3**). This reorientation of the microtubule system leads to the reorganization of the secretory pathway to deliver secretory products, especially integrins to bind the extracellular matrix, to the front of the cell for attachment to the substratum for cell migration (Figure 18-49, step **4**).

KEY CONCEPTS OF SECTION 18.8

Coordination and Cooperation between Cytoskeletal Elements

- Intermediate filaments are linked both to specific attachment sites, called desmosomes and hemidesmosomes, on the plasma membrane, as well to microfilaments and microtubules (see Figure 18-48).
- In animal cells, microtubules are generally utilized for the long-range delivery of organelles, whereas microfilaments handle their local delivery.
- The signaling molecule Cdc42 coordinately regulates microfilaments and microtubules during cell migration.

Perspectives for the Future

In Chapters 17 and 18 we have seen how microfilaments, microtubules, and intermediate filaments provide structure and organization to cells. Without this elaborate system, cells would lack all order and hence all possibility of function or division. The name "cytoskeleton" suggests a relatively static structure on which the cell organization is hung. However, the cytoskeleton is actually a dynamic framework responding to signal-transduction pathways and operating both locally and globally to provide cells with order to undertake their functions.

In outline, we have elucidated many of the distinct and common functions of the three filament systems. We know many of the components and probably all the motors. However, in many ways this is just an exciting beginning. With the available sequenced genomes and at least in principle a complete inventory of the cytoskeletal components, we have a parts list. However, a parts list is just that; what we need is to understand how the parts come together in specific processes.

A very active area of research today is to use the parts list to systematically identify the localization (through GFP fusions), functions (through RNAi knockdown), and associated partners (through isolation of protein complexes) of all cytoskeletal components. Consider that there are about 45 genes in animals that encode members of the kinesin family, yet we only know what a small subset of them do or what cargo they carry and for what purpose. In each case it is reasonable to assume the motors are regulated, about which very little is currently known. As we begin to put all the pieces in place, it will be increasingly possible to reconstitute specific processes in vitro. Some aspects of the mitotic spindle have already been reconstituted, which is an encouraging beginning, but it will be some time before it is possible to reconstitute the whole process.

Structural biology is going play a major role because it will allow us to see in detail how different components work. Consider the large number of proteins that associate with the microtubule (+) end, the so-called +TIPs. We do not know structurally how they maintain their association at the tip, and recent work has suggested that associations can change in different parts of the cell—again, we are only just beginning to see how these processes are regulated.

Perhaps the biggest—and most exciting—challenge is to uncover how signal-transduction pathways coordinate functions between all the different cytoskeletal elements. We are beginning to see glimpses of what is in store from the signal-transduction pathways that regulate cell polarity and allow cell migration.

Although all these studies are likely to be aimed at basic cell biology, as we can see from the studies of intraflagellar transport, such studies often open a window into the underlying basis of disease, from which strategies for treatments can be developed. The interplay between basic cell biology and medicine contributes immensely to the excitement and social value of working in this area.

Key Terms

anaphase *783*
asters *782*
axonal transport *769*
axoneme *777*
basal body *761*
centromere *786*
centrosome *760*
cytokinesis *783*
desmin *794*
dynamic instability *764*
dyneins *769*
intermediate filament–associated proteins (IFAPs) *796*
keratins *793*
kinesins *769*
kinetochores *783*
lamins *794*
metaphase *783*
microtubule-associated proteins (MAPs) *758*
microtubule-organizing center (MTOC) *760*
mitosis *781*
mitotic spindle *782*
plakins *796*
prophase *782*
protofilament *758*
telophase *783*
tubulin *758*

Review the Concepts

1. Microtubules are polar filaments; that is, one end is different from the other. What is the basis for this polarity, how is polarity related to microtubule organization within the cell, and how is polarity related to the intracellular movements powered by microtubule-dependent motors?

2. Microtubules both in vitro and in vivo undergo dynamic instability, and this type of assembly is thought to be intrinsic to the microtubule. What is the current model that accounts for dynamic instability?

3. In cells, microtubule assembly depends on other proteins as well as tubulin concentration and temperature. What types of proteins influence microtubule assembly in vivo, and how does each type affect assembly?

4. Microtubules within a cell appear to be arranged in specific arrays. What cellular structure is responsible for determining the arrangement of microtubules within a cell? How many of these structures are found in a typical cell? Describe how such structures serve to nucleate microtubule assembly.

5. Many drugs that inhibit mitosis bind specifically to tubulin, microtubules, or both. What diseases are such drugs used to treat? Functionally speaking, these drugs can be divided into two groups based on their effect on microtubule assembly. What are the two mechanisms by which such drugs alter microtubule structure?

6. Kinesin-1 was the first member of the kinesin motor family to be identified and therefore is perhaps the best-characterized family member. What fundamental property of kinesin was used to purify it?

7. Certain cellular components appear to move bidirectionally on microtubules. Describe how this is possible given that microtubule orientation is fixed by the MTOC.

8. The motile properties of kinesin motor proteins involve both the motor domain and the linker domain. Describe the role of each domain in kinesin movement, direction of movement, or both.

9. Cell swimming depends on appendages containing microtubules. What is the underlying structure of these appendages, and how do these structures generate the force required to produce swimming?

10. The mitotic spindle is often described as a microtubule-based cellular machine. The microtubules that constitute the mitotic spindle can be classified into three distinct types. What are the three types of spindle microtubules, and what is the function of each?

11. Mitotic spindle function relies heavily on microtubule motors. For each of the following motor proteins, predict the effect on spindle formation, function, or both of adding a drug that specifically inhibits only that motor: kinesin-5, kinesin-13, and kinesin-4.

12. The poleward movement of kinetochores, and hence chromatids, during anaphase A requires that kinetochores maintain a hold on the shortening microtubules. How does a kinetochore hold on to shortening microtubules?

13. Anaphase B involves the separation of spindle poles. What forces have been proposed to drive this separation? What underlying molecular mechanisms are thought to provide these forces?

14. Cytokinesis, the process of cytoplasmic division, occurs shortly after the separated sister chromatids have neared the opposite spindle poles. How is the plane of cytokinesis determined? What are the respective roles of microtubules and actin filaments in cytokinesis?

15. Explain why there are no known motors that use intermediate filaments as tracks.

Analyze the Data

a. Kinesin-1 contains two identical heavy chains and therefore has two identical motor domains. In contrast, kinesin-5 contains four identical heavy chains. Electron microscopic analysis of metal-shadowed kinesins results in the images shown in the top panel. Pretreatment of these kinesin with an antibody that binds specifically to the kinesin motor domain results in the images shown in the lower panel. All four images are at the same approximate magnification. What can you deduce about the structure of kinesin-5 from these data?

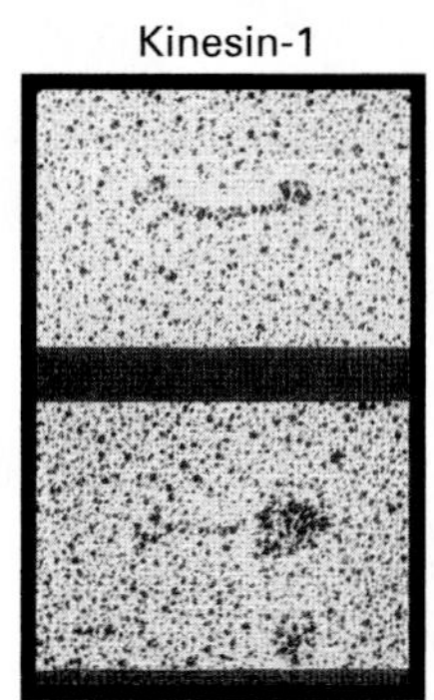

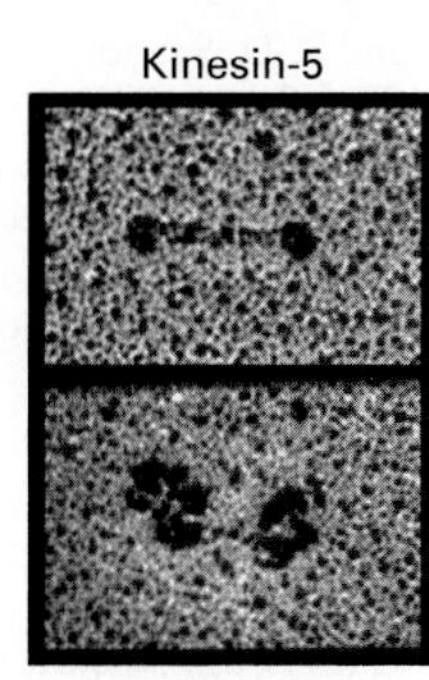

b. To determine if kinesin-5 is a (+) or (−) end microtubule motor, polarity-marked microtubules are generated by assembling short microtubules from brightly fluorescent tubulin and then elongating those short, bright microtubules using less fluorescent tubulin. As a result, the microtubules are very fluorescent at one end and less fluorescent along most of their length. A perfusion chamber is then coated with purified kinesin-5, which becomes immobilized on the glass surface. The chamber is then perfused with the polarity labeled microtubules and ATP, and microtubule gliding with respect to the immobilized kinesin-5 is observed. The following sequence of images is obtained. Which end of these microtubules, the bright or the less bright end, is the (+) end? Do these microtubules glide on kinesin-5 with their (+) or (−) end leading? Based on these data, is kinesin-5 a (+) or (−) end microtubule motor?

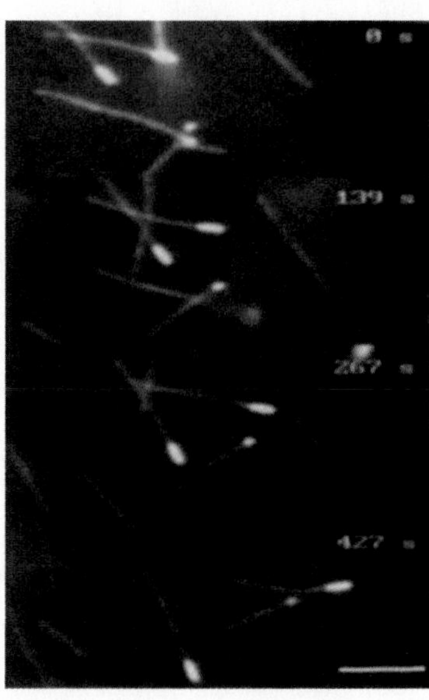

c. Kinesin-5 can cross-bridge adjacent microtubules. Polarity marked microtubules are assembled in which tubulin attached to a red fluorescent dye is assembled to form short red microtubules, which are then elongated with tubulin attached to a green fluorescent dye. The microtubules are mixed with kinesin-5 and observed by fluorescence microscopy as ATP is added. The following images show a time sequence of two overlapping and cross-bridged microtubules as ATP is added. The arrowhead is in a fixed position. Can you explain what happens when ATP is added to microtubules cross-bridged by kinesin-5?

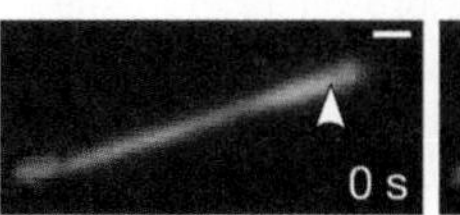

d. Eg5 is a kinesin-5 family member in *Xenopus*. To understand Eg5 function in vivo, cells are transfected with RNAi directed against this motor. The following images are obtained of mitotic cells. What function might Eg5 play in cells?

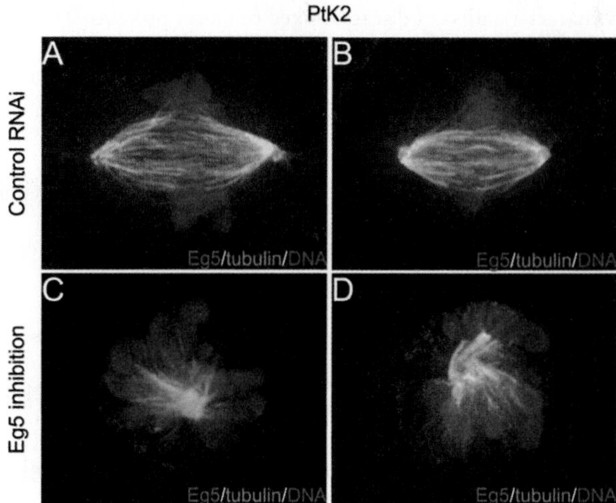

References

Microtubule Structure and Organization

Badano, J. L., T. M. Teslovich, and N. Katsanis. 2005. The centrosome in human genetic disease. *Nature Rev. Mol. Cell Biol.* **6**:194–205.

Doxsey, S. 2001. Re-evaluating centrosome function. *Nature Rev. Mol. Cell Biol.* **2**:688–698.

Dutcher, S. K. 2001. The tubulin fraternity: alpha to eta. *Curr. Opin. Cell Biol.* **13**:49–54.

Nogales, E., and H-W Wang. 2006. Structural intermediates in microtubule assembly and disassembly: how and why? *Curr. Opin. Cell Biol.* **18**:179–184.

Microtubule Dynamics

Cassimeris, L. 2002. The oncoprotein 18/Stathmin family of microtubule destabilizers. *Curr. Opin. Cell Biol.* **14**:18–24.

Desai, A., and T. J. Mitchison. 1997. Microtubule polymerization Dyanamics. *Annu. Rev.Cell Dev. Biol.* **13**:83–117.

Howard, J., and A. A. Hyman. 2003. Dynamics and mechanics of the microtubule plus end. *Nature* **422**:753–758.

Regulation of Microtubule Structure and Dynamics

Akhmanova, A., and C. C. Hoogenraad. 2005. Microtubule plus-end-tracking proteins: mechanisms and functions. *Curr. Opin. Cell Biol.* **17**:47–54.

Galjart, N. 2005. CLIPs and CLASPs and cellular dynamics. *Nature Rev. Mol. Cell Biol.* **6**:487–498.

Kinesins and Dyneins—Microtubule-Based Motor Proteins

Web site: kinesin home page, http://www.proweb.org/kinesin/

Burgess, S. A., et al. 2003. Dynein structure and power stroke. *Nature* **421**:715–718.

Dell, K. R. 2003. Dynactin polices two-way organelle traffic. *J. Cell Biol.* **160**:291–293.

Dujardin, D. L., and R. B. Vallee. 2002. Dynein at the cortex. *Curr. Opin. Cell Biol.* **14**:44–49.

Goldstein, L. S. 2001. Kinesin molecular motors: transport pathways, receptors, and human disease. *Proc. Nat'l. Acad. Sci. USA* **98**:6999–7003.

Hirokawa, N. 1998. Kinesin and dynein superfamily proteins and the mechanism of organelle transport. *Science* **279**:519–526.

Hirokawa, N., and R. Takemure. 2003. Biochemical and molecular characterization of diseases linked to motor proteins. *Trends Cell Biol.* **28**:558–565.

Lawrence, C. J., et al. 2004. A standardized kinesin nomenclature. *J. Cell Biol.* **167**:19–22.

Schroer, T. A. 2004. Dynactin. *Ann. Rev. Cell Dev. Biol.* **20**:759–779.

Vale, R. D. 2003. The molecular motor toolbox for intracellular transport. *Cell* **112**:467–480.

Vale, R. D., and R. A. Milligan. 2000. The way things move: looking under the hood of molecular motor proteins. *Science* **288**:88–95.

Wordeman, L. 2005. Microtubule-depolymerizing kinesins. *Curr. Opin. Cell Biol.* **17**:82–88.

Yildiz, A., M. Tomishige, R. D. Vale, and P. R. Selvin. 2004. Kinesin walks hand-over-hand. *Science* **303**:676–678.

Cilia and Flagella: Microtubule-Based Surface Structures

Rosenbaum, J. L., and G. B. Witman. 2002. Intraflagellar transport. *Nature Rev. Mol. Cell Biol.***3**: 813–825.

Singla, V., and J. F. Reiter. 2006. The primary cilium as the cells' antenna: signaling at a sensory organelle. *Science* **313**:629–633.

Mitosis

Web site: http://www.proweb.org/kinesin/FxnSpindleMotility.html

Cleveland, D. W., Y. Mao, Y., and K. F. Sullivan. 2003. Centromeres and kinetochores: from epigenetics to mitotic checkpoint signaling. *Cell* **112**:407–421.

Gadde, S., and R. Heald. 2004. Mechanisms and molecules of the mitotic spindle. *Curr. Biol.* **14**:R797–R805.

Heald, R., et al. 1997. Spindle assembly in *Xenopus* egg extracts: respective roles of centrosomes and microtubule self-organization. *J. Cell Biol.* **138**:615–628.

Kinoshita, K., B. Habermann, and A. A. Hyman. 2002. XMAP215: a key component of the dynamic microtubule cytoskeleton. *Trends Cell Biol.* **12**:267–273.

Mitchison, T. J., and E. D. Salmon. 2001. Mitosis: a history of division. *Nature Cell Biol.* **3**:E17–E21.

Rogers, G. C., et al. 2004. Two mitotic kinesins cooperate to drive sister chromatid separation during anaphase. *Nature* **427**:364–370.

Wittmann, T., A. Hyman, and A. Desai. 2001. The spindle: a dynamic assembly of microtubules and motors. *Nature Cell Biol.* **3**:E28–E34.

Intermediate Filaments

Intermediate Filaments Database: http://www.interfil.org/index.php

Goldman, R. D., et al. 2002. Nuclear lamins: building blocks of nuclear architecture. *Genes Dev.* **16**:533–547.

Herrmann, H., and U. Aebi. 2000. Intermediate filaments and their associates: multi-talented structural elements specifying cytoarchitecture and cytodynamics. *Curr. Opin. Cell Biol.* **12**: 79–90.

Mattout, A., et al. 2006. Nuclear lamins, disease and aging. *Curr. Opin. Cell Biol.* **18**:335–341.

Coordination and Cooperation between Cytoskeletal Elements

Web site: Melanophores, http://www.proweb.org/kinesin/Melanophore.html

Chang, L., and R. D. Goldman. 2004. Intermediate filaments mediate cytoskeletal crosstalk. *Nature Rev. Mol. Cell Biol.* **5**: 601–613.

Etienne-Manneville, S., et al. 2005. Cdc42 and Par6-PKCζ regulate the spatially localized association of Dlg1 and APC to control cell polarization. *J. Cell Biol.* **170**:895–901.

Kodama, A., T. Lechler, and E. Fuchs. 2004. Coordinating cytoskeletal tracks to polarize cellular movements. *J. Cell Biol.* **167**: 203–207.

Wu, X., X. Xiang, and J. A. Hammer III. 2006. Motor proteins at the microtubule plus-end. *Trends Cell Biol.* **16**:135–143.

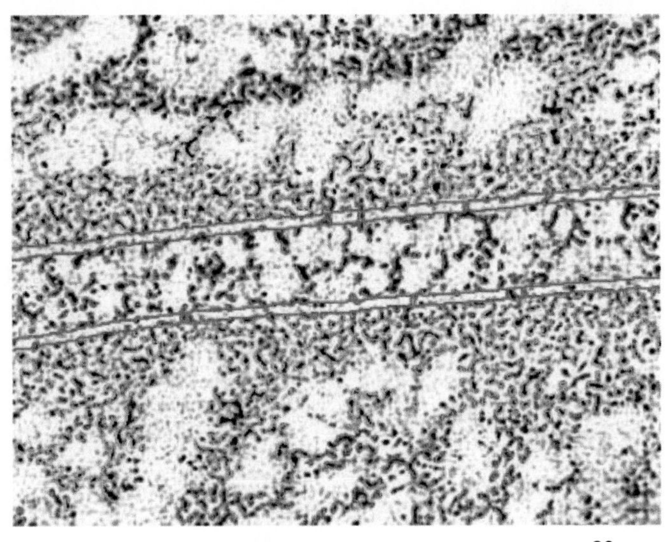

False-color image of a section through a desmosome from neonatal mouse epidermis. Electron microscope tomography was used to generate an image of a specialized cellular junction, called a desmosome, that helps hold cells in the skin together. Cell-adhesion molecules called cadherins are blue, membranes from adjacent cells are each outlined in red, and related intracellular molecules, cytoplasmic plaque and intermediate filaments, are orange and light green, respectively. Scale bar is 30 nm. [From W. He, P. Cowin, and D. L. Stokes, 2003, *Science* **302**(5642):109–113.]

CHAPTER

19

INTEGRATING CELLS INTO TISSUES

In the development of complex multicellular organisms such as plants and animals, progenitor cells differentiate into distinct "types" that have characteristic compositions, structures, and functions. Cells of a given type often aggregate into a *tissue* to cooperatively perform a common function: muscle contracts; nervous tissues conduct electrical impulses; xylem tissue in plants transports water. Different tissues can be organized into an *organ,* again to perform one or more specific functions. For instance, the muscles, valves, and blood vessels of a heart work together to pump blood. The coordinated functioning of many types of cells and tissues permits the organism to move, metabolize, reproduce, and carry out other essential activities.

Even simple animals exhibit complex tissue organization. The adult form of the roundworm *Caenorhabditis elegans* contains a mere 959 cells, yet these cells fall into 12 different general cell types and many distinct subtypes. Vertebrates have hundreds of different cell types, including leukocytes (white blood cells) and erythrocytes (red blood cells); photoreceptors in the retina; adipocytes, which store fat; fibroblasts in connective tissue; and hundreds of different subtypes of neurons in the human brain. Despite their diverse forms and functions, all animal cells can be classified as being components of just five main classes of tissue: *epithelial tissue, connective tissue, muscular tissue, nervous tissue,* and *blood.* Various cell types are arranged in precise patterns of staggering complexity to generate tissues and organs. The costs of such complexity include increased requirements for information, material, energy, and time during the development of an individual organism. Although the physiological costs of complex tissues and organs are high, they confer the ability to thrive in varied and variable environments, a major evolutionary advantage.

OUTLINE

The complex and diverse morphologies of plants and animals are examples of the whole being greater than the sum of the individual parts, more technically described as the emergent properties of a complex system. For example, the distinct mechanical properties of rigid bones, flexible joints, and contracting muscles permit vertebrates to move efficiently and achieve substantial size. One of the defining characteristics of animals with complex tissues and organs (metazoans) such as ourselves is that the external and internal surfaces of most of their tissues and organs, and indeed the exterior of the entire organism, are built from tightly packed sheetlike layers of cells known as **epithelia.** The formation of an epithelium and its subsequent remodeling into more complex collections of epithelial and nonepithelial tissues is a hallmark of the development of metazoans. Sheets of tightly attached epithelial cells act as regulatable, selective permeability barriers, which permit the generation of chemically and functionally distinct compartments in an organism (e.g., stomach and bloodstream). As a result, distinct and sometimes opposite functions (e.g., digestion and synthesis) can efficiently proceed simultaneously within an organism. Such compartmentalization also permits more sophisticated regulation of diverse biological functions. In many ways, the roles of complex tissues and organs in an organism are analogous to those of organelles and membranes in individual cells.

The assembly of distinct tissues and their organization into organs are determined by molecular interactions at the

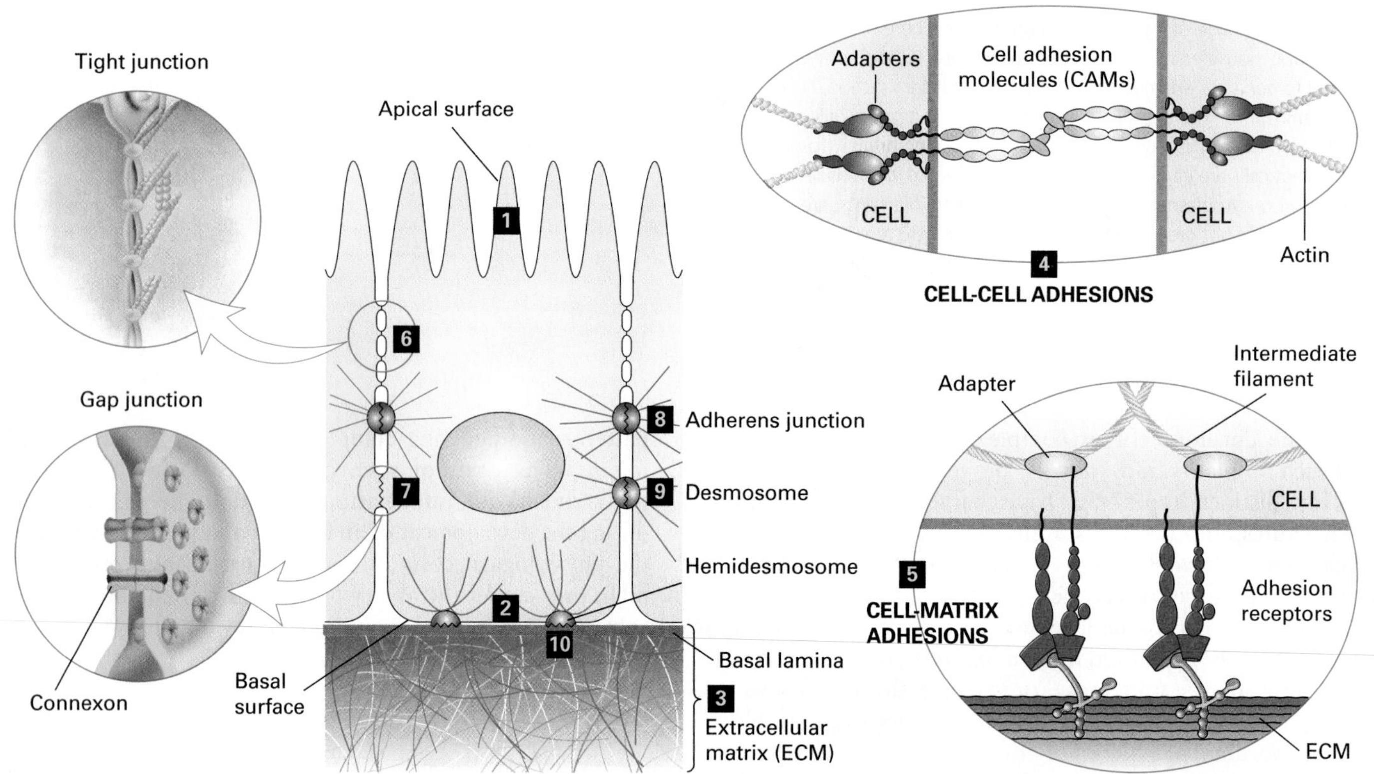

▲ **FIGURE 19-1 Overview of major cell-cell and cell-matrix adhesive interactions.** Schematic cutaway drawing of a typical epithelial tissue, such as in the intestines. The apical (upper) surface of these cells is packed with fingerlike microvilli **1** that project into the intestinal lumen, and the basal (bottom) surface **2** rests on extracellular matrix (ECM). The ECM associated with epithelial cells is usually organized into various interconnected layers (e.g., the basal lamina, connecting fibers, connective tissue), in which large, interdigitating ECM macromolecules bind to one another and to the cells **3**. Cell-adhesion molecules (CAMs) bind to CAMs on other cells, mediating cell-cell adhesions **4**, and adhesion receptors bind to various components of the ECM, mediating cell-matrix adhesions **5**. Both types of cell-surface adhesion molecules are usually integral membrane proteins whose cytosolic domains often bind to multiple intracellular adapter proteins. These adapters, directly or indirectly, link the CAM to the cytoskeleton (actin or intermediate filaments) and to intracellular signaling pathways. As a consequence, information can be transferred by CAMs and the macromolecules to which they bind from the cell exterior into the intracellular environment and vice versa. In some cases, a complex aggregate of CAMs, adapters, and associated proteins is assembled. Specific localized aggregates of CAMs or adhesion receptors form various types of cell junctions that play important roles in holding tissues together and facilitating communication between cells and their environment. Tight junctions **6**, lying just under the microvilli, prevent the diffusion of many substances through the extracellular spaces between the cells. Gap junctions **7** allow the movement through connexon channels of small molecules and ions between the cytosols of adjacent cells. The remaining three types of junctions, adherens junctions **8**, spot desmosomes **9**, and hemidesmosomes **10**, link the cytoskeleton of a cell to other cells or the ECM. [See V. Vasioukhin and E. Fuchs, 2001, *Curr. Opin. Cell Biol.* **13**:76.]

cellular level and would not be possible without the temporally and spatially regulated expression of a wide array of adhesive molecules. Cells in tissues can adhere directly to one another (*cell-cell adhesion*) through specialized membrane proteins called **cell-adhesion molecules (CAMs)** that often cluster into specialized cell junctions (Figure 19-1). In the fruit fly *Drosophila melanogaster,* at least 500 genes (~4% of the total) are estimated to be involved in cell adhesion. Cells in animal tissues also adhere indirectly (*cell-matrix adhesion*) through the binding of **adhesion receptors** in the plasma membrane to components of the surrounding **extracellular matrix (ECM)**, a complex interdigitating meshwork of proteins and polysaccharides secreted by cells into the spaces between them. These two basic types of interactions not only allow cells to aggregate into distinct tissues but also provide a means for the bidirectional transfer of information between the exterior and the interior of cells. Such information transfer is important to many biological processes, including cell survival, proliferation, differentiation, and migration. Therefore it is not surprising that defects that interfere with the adhesive interactions and the associated flow of information can cause or contribute to diseases, including a wide variety of neuromuscular and skeletal disorders and cancer.

In this chapter, we examine various types of adhesive molecules and how they interact. Because of the particularly well-understood nature of the adhesive molecules in tissues that form tight epithelia, as well as their very early evolutionary development, we will initially focus on epithelial tissues, such as the walls of the intestinal tract and those that form skin. Epithelial cells are normally nonmotile (sessile); however, during development, wound healing, and in certain pathologic states (e.g., cancer), epithelial cells can transform into more motile cells. Changes in expression and function of adhesive molecules play a key role in this transformation, as they do in normal biological processes involving cell movement, such as the crawling of white blood cells into sites of infection. We therefore follow the discussion of epithelial tissues with a discussion of adhesion in nonepithelial, developing, and motile tissues.

The evolution of plants and animals diverged before multicellular organisms arose. Thus multicellularity and the molecular means for assembling tissues and organs must have arisen independently in animal and plant lineages. Not surprisingly, then, animals and plants exhibit many differences in the organization and development of tissues. For this reason, we first consider the organization of tissues in animals and then deal separately with plants.

19.1 Cell-Cell and Cell-Matrix Adhesion: An Overview

We begin with a brief orientation to the various types of adhesive molecules, their major functions in organisms, and their evolutionary origin. In subsequent sections, we examine in detail the unique structures and properties of the various participants in cell-cell and cell-matrix interactions.

Cell-Adhesion Molecules Bind to One Another and to Intracellular Proteins

A large number of CAMs fall into four major families: the cadherins, immunoglobulin (Ig) superfamily, integrins, and selectins. As the schematic structures in Figure 19-2 illustrate, many CAMs and other adhesion molecules are mosaics of multiple distinct domains, many of which can be found in more than one kind of protein. Some of these domains confer the binding specificity that characterizes a particular protein. Other membrane proteins, whose structures do not belong to any of the major classes of CAMs, also participate in cell-cell adhesion in various tissues.

CAMs mediate, through their extracellular domains, adhesive interactions between cells of the same type (*homotypic* adhesion) or between cells of different types (*heterotypic* adhesion). A CAM on one cell can directly bind to the same kind of CAM on an adjacent cell (*homophilic* binding) or to a different class of CAM (*heterophilic* binding). CAMs can be broadly distributed along the regions of plasma membranes that contact other cells or clustered in discrete patches or spots called **cell junctions.** Cell-cell adhesions can be tight and long lasting or relatively weak and transient. The associations between nerve cells in the spinal cord or the metabolic cells in the liver exhibit tight adhesion. In contrast, immune-system cells in the blood often exhibit only weak, short-lasting interactions, allowing them to roll along and pass through a blood vessel wall on their way to fight an infection within a tissue.

The cytosol-facing domains of CAMs recruit sets of multifunctional adapter proteins (see Figure 19-1). These adapters act as linkers that directly or indirectly connect CAMs to elements of the cytoskeleton (Chapters 17 and 18); they can also recruit intracellular molecules that function in signaling pathways to control protein activity—both intracellular proteins and the CAMs themselves—and gene expression (Chapters 15 and 16). In many cases, a complex aggregate of CAMs, adapter proteins, and other associated proteins is assembled at the inner surface of the plasma membrane. Because cell-cell adhesions are intrinsically associated with the cytoskeleton and signaling pathways, a cell's surroundings influence its shape and functional properties ("outside-in" effects); likewise, cellular shape and function influence a cell's surroundings ("inside-out" effects). Thus *connectivity* and *communication* are intimately related properties of cells in tissues.

The formation of many cell-cell adhesions entails two types of molecular interactions (Figure 19-3). First, CAMs on one cell associate laterally through their extracellular domains, cytosolic domains, or both into homodimers or higher-order oligomers in the plane of the cell's plasma membrane; these interactions are called *intracellular, lateral,* or *cis* interactions. Second, CAM oligomers on one cell bind to the same or different CAMs on an adjacent cell;

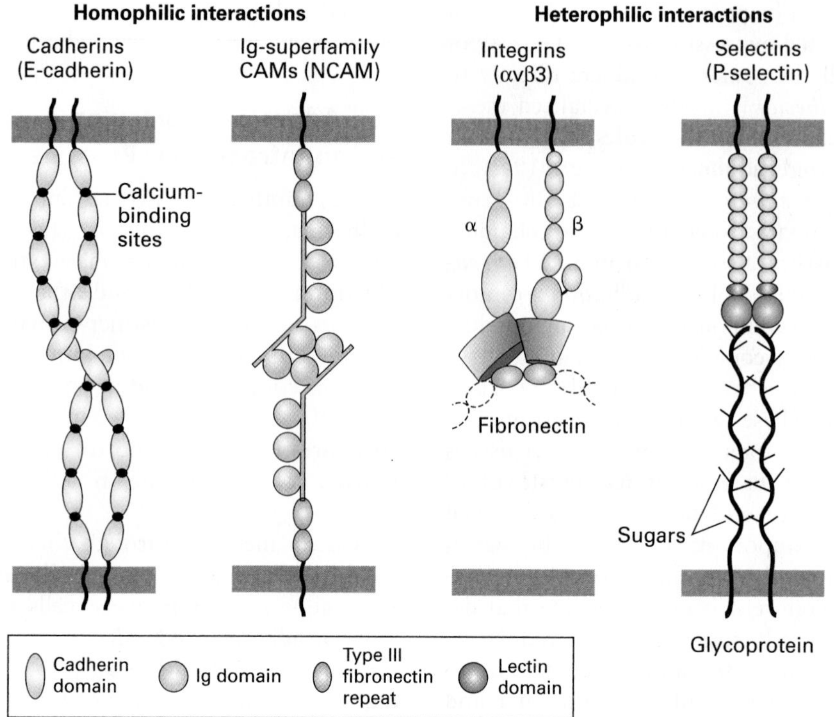

▲ **FIGURE 19-2 Major families of cell-adhesion molecules (CAMs) and adhesion receptors.** Dimeric E-cadherins most commonly form homophilic (self) cross-bridges with E-cadherins on adjacent cells. Members of the immunoglobulin (Ig) superfamily of CAMs can form both homophilic linkages (shown here) and heterophilic (nonself) linkages. Heterodimeric integrins (for example, αv and β3 chains) function as CAMs or as adhesion receptors (shown here) that bind to very large, multiadhesive matrix proteins such as fibronectin, only a small part of which is shown here. Selectins, shown as dimers, contain a carbohydrate-binding lectin domain that recognizes specialized sugar structures on glycoproteins (shown here) and glycolipids on adjacent cells. Note that CAMs often form higher-order oligomers within the plane of the plasma membrane. Many adhesive molecules contain multiple distinct domains, some of which are found in more than one kind of CAM. The cytoplasmic domains of these proteins are often associated with adapter proteins that link them to the cytoskeleton or to signaling pathways. [See R. O. Hynes, 1999, *Trends Cell Biol.* **9**(12):M33, and R. O. Hynes, 2002, *Cell* **110:**673–687.]

these interactions are called *intercellular* or *trans* interactions. Trans interactions sometimes induce additional cis interactions and, as a consequence, yet even more trans interactions.

Adhesive interactions between cells vary considerably, depending on the particular CAMs participating and the tissue. Just like Velcro, very tight adhesion can be generated when many weak interactions are combined, and this is especially the case when CAMs are concentrated in small, well-defined areas, such as cellular junctions. Some CAMs require calcium ions to form effective adhesions; others do not. Furthermore, the association of intracellular molecules with the cytosolic domains of CAMs can dramatically influence the intermolecular interactions of

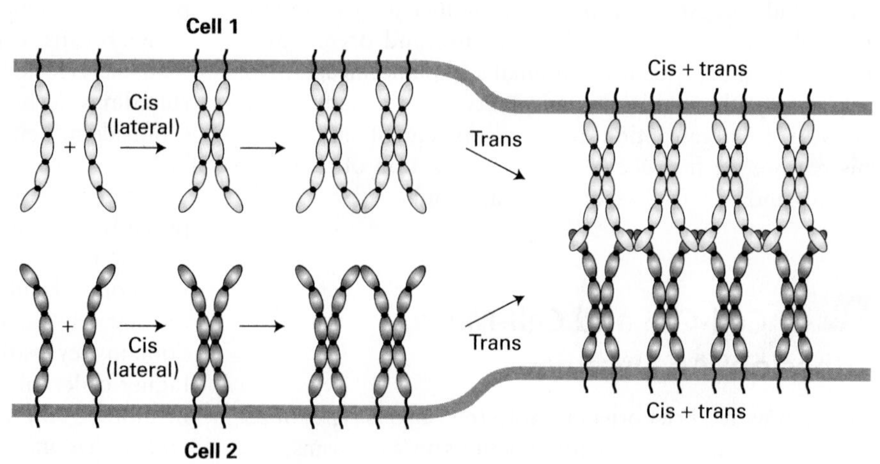

► **FIGURE 19-3 Model for the generation of cell-cell adhesions.** Lateral interactions between cell-adhesion molecules (CAMs) within the plasma membrane of a cell form dimers and larger oligomers. The parts of the molecules that participate in these cis interactions vary among the different CAMs. Subsequent trans interactions between distal domains of CAMs on adjacent cells generate a Velcro-like strong adhesion between the cells. [Adapted from M. S. Steinberg and P. M. McNutt, 1999, *Curr. Opin. Cell Biol.* **11:**554.]

CAMs by promoting their cis association (clustering) or by altering their conformation. Among the many variables that determine the nature of adhesion between two cells are the binding affinity of the interacting molecules (thermodynamic properties), the overall "on" and "off" rates of association and dissociation for each interacting molecule (kinetic properties), the spatial distribution or density of adhesion molecules (ensemble properties), the active versus inactive states of CAMs with respect to adhesion (biochemical properties), and external forces such as stretching and pulling in muscle or the laminar and turbulent flow of cells and surrounding fluids in the circulatory system (mechanical properties).

The Extracellular Matrix Participates in Adhesion, Signaling, and Other Functions

Components of the extracellular matrix bind both to certain adhesion receptors, such as integrins, located on the cell surface (see Figure 19-2) and to one another. As a consequence of ECM interactions with receptors on adjacent cells, the ECM mediates indirect cell adhesion. ECM components include proteoglycans, a unique type of glycoprotein; collagens, proteins that often form fibers; soluble multiadhesive matrix proteins; and others (Table 19-1). Multiadhesive matrix proteins, such as the proteins fibronectin and laminin, are long, flexible molecules that contain multiple domains responsible for binding various types of collagen, other matrix proteins, polysaccharides, cell-surface adhesion receptors, and extracellular signaling molecules. These proteins are important for organizing the other components of the extracellular matrix. They also regulate cell-matrix adhesion—and thus cell migration and cell shape. The relative volumes of cells versus matrix vary greatly among different animal tissues. Some connective tissue, for instance, is mostly matrix, whereas many organs are composed of very densely packed cells with relatively little matrix. The density of packing of the molecules within the ECM itself can also vary greatly.

H. V. Wilson's classic studies of adhesion in marine sponge cells showed conclusively that one primary function of ECM is literally holding tissue together. Figure 19-4a, b, which re-creates Wilson's classic work, shows that when sponges are mechanically dissociated and individual cells from two species of sponge are mixed, the cells of one species will adhere to each other but not to those from the other species. This specificity is due, in part, to different adhesive proteins in the ECM that bind to the cells via surface receptors. These adhesive proteins can be purified and used to coat colored beads, which, when mixed, aggregate with each other with a specificity similar to intact sponge cells (Figure 19-4c, d).

Different combinations of ECM components tailor the extracellular matrix for specific purposes: strength in a tendon, tooth, or bone; cushioning in cartilage; and adhesion in most tissues. In addition, the composition of the matrix, which can vary depending on the anatomical site and physiological status of a tissue, can provide environmental cues, letting a cell know where it is and what it should do. Changes in ECM components, which are constantly being remodeled, degraded, and resynthesized locally, can modulate the interactions of a cell with its environment. The matrix also serves as a reservoir for many extracellular signaling molecules that control cell growth and differentiation. In addition, the matrix provides a lattice through or on which cells can move, particularly in the early stages of tissue assembly. Morphogenesis—the stage of embryonic development in which tissues, organs,

TABLE 19-1 Extracellular Matrix Proteins

Proteoglycans	Perlecan	
Collagens	Sheet forming (e.g., type IV)	
	Fibular collagens (e.g., types I, II, and III)	
Multiadhesive matrix proteins	Laminin	
	Fibronectin	
	Nidogen/entactin	

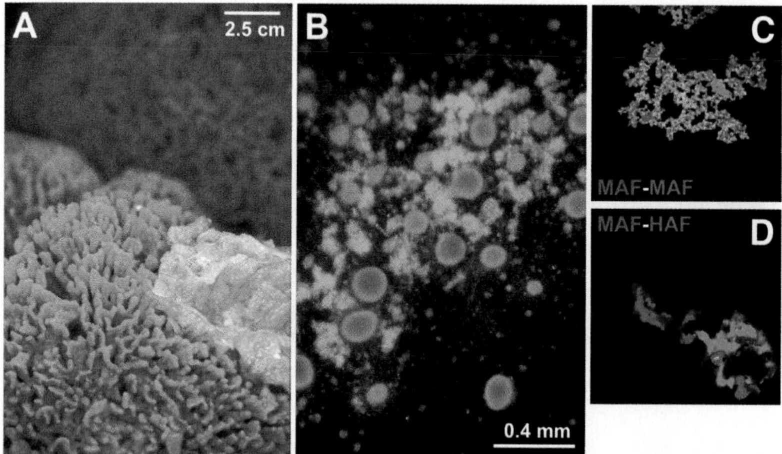

▲ EXPERIMENTAL FIGURE 19-4 **Mechanically separated marine sponges reassemble through species-specific homotypic cell adhesion.** (a) Two sponges *Microciona prolifera* (orange) and *Halichondria panicea* (yellow) growing in the wild. (b) After mechanical disruption and mixing of both types of intact sponges, the individual cells were allowed to reassociate for ~30 minutes with gentle stirring. The cells aggregate with species-specific homotypic adhesion, forming micro-clumps of *Microciona prolifera* cells (orange) and *Halichondria panicea* cells (yellow). (c) and (d) Red or green fluorescently labeled beads were coated with the proteoglycan aggregation factor (AF) from the ECM of either *Microciona prolifera* (MAF) or *Halichondria panicea* (HAF). Panel C shows that when both colored beads were coated only with MAF, they all aggregated together, forming yellow aggregates (combination of red and green). Panel D shows that MAF (red) and HAF (green) coated beads do not readily form mixed aggregates but rather assemble into distinct clumps held together by homotypic adhesion. (magnification, 40×) [Adapted from X. Fernandez-Busquets and M. M. Burger, 2003, *Cell Mol. Life Sci.* **60:**88–112, and J. Jarchow and M. M. Burger, 1998, *Cell Adhes. Commun.* **6:**405–414.]

and body parts are formed by cell movements and rearrangements—also is critically dependent on cell-matrix adhesion as well as cell-cell adhesion. For example, branching morphogenesis (formation of branching structures) to form the air sacs in the lung, blood vessels, mammary and salivary glands, and other structures requires cell-matrix interactions (Figure 19-5). Disruptions in cell-matrix and cell-cell interactions can have devastating consequences for the development of tissues, as is seen in the dramatic changes in the skeletal system of embryonic mice when the genes for either of two key ECM molecules, collagen II or perlecan, are inactivated (Figure 19-6).

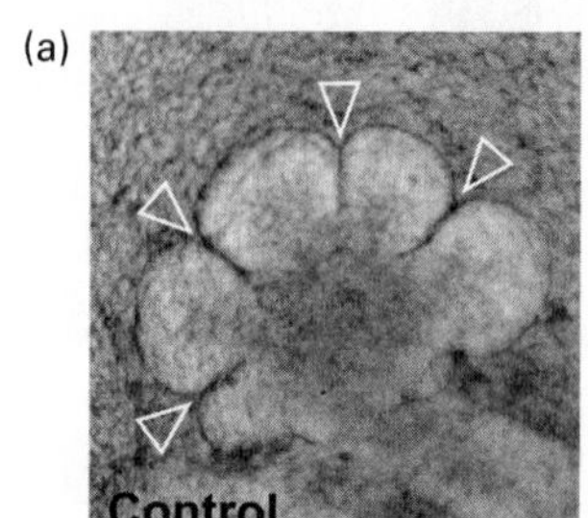

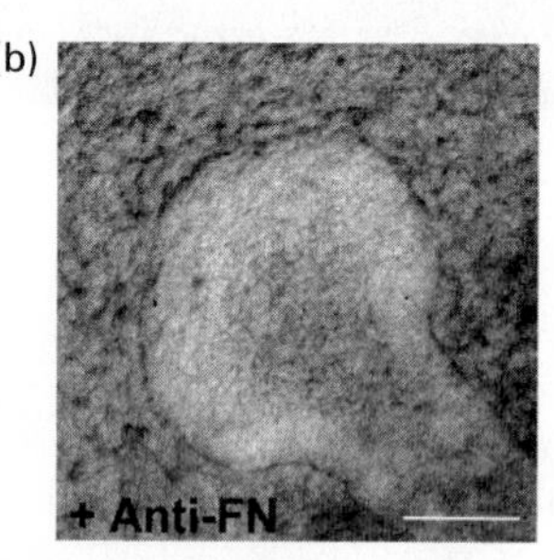

▲ EXPERIMENTAL FIGURE 19-5 **Antibodies to fibronectin block branching morphogenesis in developing mouse tissues.** Immature salivary glands were isolated from murine embryos and allowed to undergo branching morphogenesis in vitro for 10 hours in the absence (a) or presence (b) of an antibody that binds to and blocks the activity of the ECM molecule fibronectin. Anti-fibronectin antibody (Anti-FN) treatment blocked branch formation (arrowheads). Inhibition of fibronectin's adhesion receptor (an integrin) also blocks branch formation (not shown). [Takayoshi et al., 2003, *Nature* **423:**876–881.]

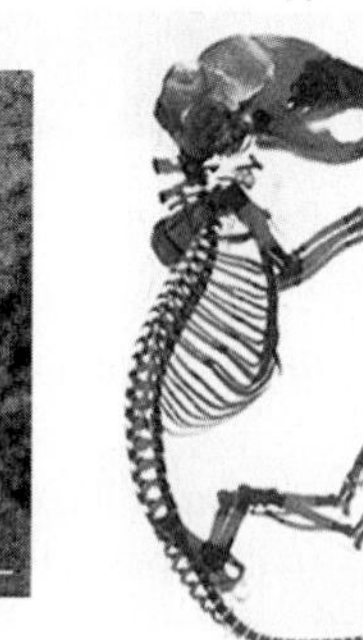

▲ EXPERIMENTAL FIGURE 19-6 **Inactivating the genes for some ECM proteins results in defective skeletal development in mice.** These photographs show skeletons of normal (*left*), collagen II–deficient (*center*), or perlecan-deficient (*right*) murine embryos that were isolated and stained to visualize the cartilage (blue) and bone (red). Absence of these key ECM components leads to dwarfism, with many skeletal elements shortened and disfigured. [From E. Gustafsson et al., 2003, *Ann. NY Acad. Sci.* **995:**140–150.]

Disruptions in adhesion also are characteristic of various diseases, such as metastatic cancer, in which cancerous cells leave their normal locations and spread throughout the body.

Although many CAMs and adhesion receptors were initially identified and characterized because of their adhesive properties, they also play a major role in signaling, using many of the pathways discussed in Chapters 15 and 16. Figure 19-7 illustrates how one adhesion receptor, integrin, physically and functionally interacts via adaptors and signaling kinases with a broad array of intracellular signaling pathways, including those initiated by receptor tyrosine kinases, to influence cell survival, gene transcription, cytoskeletal organization, cell motility, and cell proliferation. Conversely, changes in the activities of signaling pathways inside of cells can influence the structures of CAMs and adhesion receptors and thus modulate their ability to interact with other cells and ECM.

The Evolution of Multifaceted Adhesion Molecules Enabled the Evolution of Diverse Animal Tissues

Cell-cell and cell-matrix adhesions are responsible for the formation, composition, architecture, and function of animal tissues. Not surprisingly, some adhesion molecules are evolutionarily ancient and are among the most highly conserved proteins in multicellular organisms. Sponges, the most primitive multicellular organisms, express certain CAMs and multiadhesive ECM molecules whose structures are strikingly similar to those of the corresponding human proteins. The evolution of organisms with complex tissues and organs (metazoans) has depended on the evolution of diverse adhesion molecules with novel properties and functions, whose levels of expression differ in different types of cells. Some CAMs (e.g., cadherins), adhesion receptors (e.g., integrins and immunoglobulin

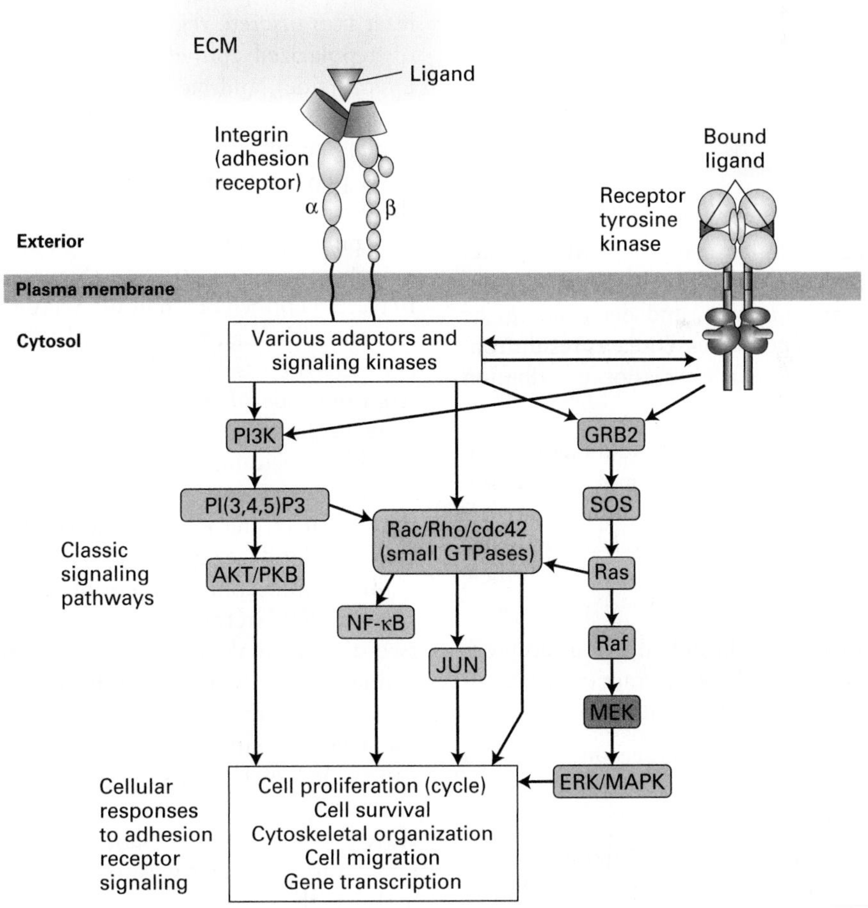

▲ **FIGURE 19-7 Integrin adhesion receptor-mediated signaling pathways that control diverse cell functions.** Binding of integrins to their ligands induces conformational changes in their cytoplasmic domains, alternating their interactions with cytoplasmic proteins. These include signaling kinases (src-family kinases, focal adhesion kinase [FAK], integrin-linked kinase [ILK]) and adaptor proteins (e.g., talin, paxillin, vinculin) that transmit signals via diverse signaling pathways, thereby influencing cell proliferation, cell survival, cytoskeletal organization, cell migration, and gene transcription. Many of the components of the pathways shown here are shared with other cell-surface-activated signaling pathways, discussed in Chapters 15 and 16. [Modified from W. Guo and F. G. Giancotti, 2004, *Nat. Rev. Mol. Cell Biol.* **5**(10):816–826, and R. O. Hynes, 2002, *Cell* **110:**673–687.]

superfamily CAMs), and ECM components (type IV collagen, laminin, nidogen/entactin, and perlecan-like proteoglycans) are highly conserved, whereas others are not. For example, fruit flies do not have certain types of collagen or the ECM protein fibronectin that play crucial roles in mammals. A common feature of adhesive proteins is repeating domains forming very large proteins. The overall length of these molecules, combined with their ability to bind numerous ligands via distinct functional domains, likely played a role their evolution.

The diversity of adhesive molecules arises in large part from two phenomena that can generate numerous closely related proteins, called **isoforms**, that constitute a protein family. In some cases, the different members of a protein family are encoded by multiple genes that arose from a common ancestor by gene duplication and divergent evolution (Chapter 6). In other cases, a single gene produces an RNA transcript that can undergo alternative splicing to yield multiple mRNAs, each encoding a distinct protein isoform (Chapter 8). Both phenomena contribute to the diversity of some protein families such as the cadherins. Particular isoforms of an adhesive protein are often expressed in some cell types and tissues but not others.

KEY CONCEPTS OF SECTION 19.1

Cell-Cell and Cell-Matrix Adhesion: An Overview

- Cell-cell and cell-extracellular matrix (ECM) interactions are critical for assembling cells into tissues, controlling cell shape and function, and determining the developmental fate of cells and tissues. Diseases result from abnormalities in the structures or expression of adhesion molecules.
- Cell-adhesion molecules (CAMs) mediate direct cell-cell adhesions (homotypic and heterotypic), and cell-surface adhesion receptors mediate cell-matrix adhesions (see Figure 19-1). These interactions bind cells into tissues and facilitate communication between cells and their environments.
- The cytosolic domains of CAMs and adhesion receptors bind adapter proteins that mediate interaction with cytoskeletal fibers and intracellular signaling proteins.
- The major families of cell-surface adhesion molecules are the cadherins, selectins, Ig-superfamily CAMs, and integrins (see Figure 19-2).
- Tight cell-cell adhesions entail both cis (lateral or intracellular) oligomerization of CAMs and trans (intercellular) interaction of like (homophilic) or different (heterophilic) CAMs (see Figure 19-3). The combination of cis and trans interactions produces a Velcro-like adhesion between cells.
- The extracellular matrix (ECM) is a complex meshwork of proteins and polysaccharides that contributes to the structure and function of a tissue. The major classes of ECM molecules are proteoglycans, collagens, and multiadhesive matrix proteins (fibronectin, laminin).
- The evolution of adhesion molecules with specialized structures and functions permits cells to assemble into diverse classes of tissues with varying functions.

19.2 Cell-Cell and Cell-ECM Junctions and Their Adhesion Molecules

Cells in epithelial and nonepithelial tissues use many, but not all, of the same cell-cell and cell-matrix adhesion molecules. Because of the relatively simple organization of epithelia, as well as their fundamental role in evolution and development, we begin our detailed discussion of adhesion with the epithelium.

Epithelial Cells Have Distinct Apical, Lateral, and Basal Surfaces

Cells that form epithelial tissues are said to be **polarized** because their plasma membranes are organized into at least two discrete regions. Typically, the distinct surfaces of a polarized epithelial cell are called the **apical** (top), **lateral** (side), and **basal** (base or bottom) surfaces (Figure 19-8). The area of the apical surface is often greatly expanded by the formation of microvilli. Adhesion molecules play essential roles in generating and maintaining these structures.

Epithelia in different body locations have characteristic morphologies and functions (Figure 19-8). Stratified (multilayered) epithelia commonly serve as barriers and protective surfaces (e.g., the skin), whereas simple (single-layer) epithelia often selectively move ions and small molecules from one side of the layer to the other. For instance, the simple columnar epithelium lining the stomach secretes hydrochloric acid into the lumen; a similar epithelium lining the small intestine transports products of digestion from the lumen of the intestine across the basolateral surface into the blood (see Figure 11-19).

In simple columnar epithelia, adhesive interactions between the lateral surfaces hold the cells together into a two-dimensional sheet, whereas those at the basal surface connect the cells to a specialized underlying extracellular matrix called the **basal lamina.** Often the basal and lateral surfaces are similar in composition and together are called the **basolateral** surface. The basolateral surfaces of most simple epithelia are usually on the side of the cell closest to the blood vessels, whereas the apical surface is not in direct contact with other cells or the ECM. In animals with closed circulatory systems, blood flows through vessels whose inner lining is composed of flattened epithelial cells called endothelial cells. The apical side of endothelial cells, which faces the blood, is usually called the *luminal* surface and the opposite basal side, the *abluminal* surface.

In general, epithelial cells are sessile, immobile cells, in that adhesion molecules firmly and stably attach them to one another and their associated ECM. One especially important

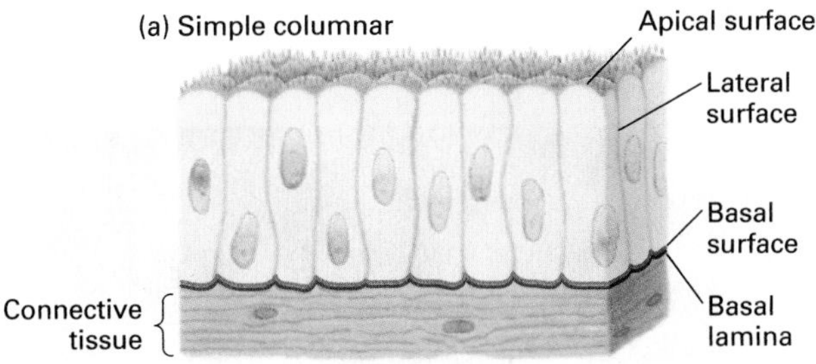

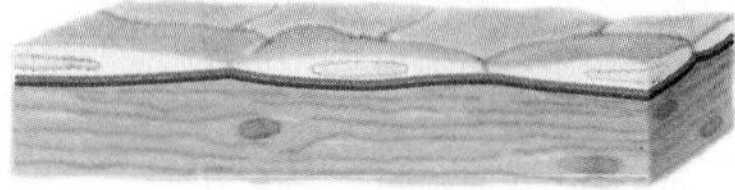

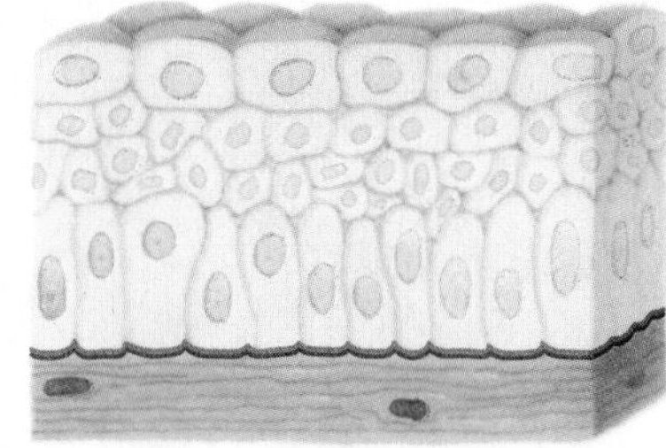

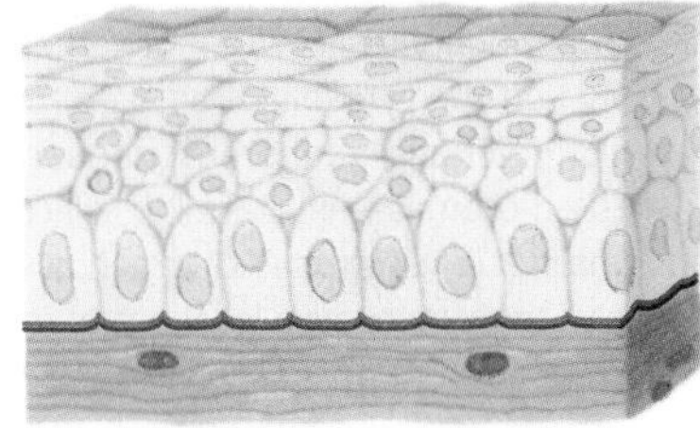

▲ **FIGURE 19-8 Principal types of epithelia.** The apical and basolateral surfaces of epithelial cells exhibit distinctive characteristics. (a) Simple columnar epithelia consist of elongated cells, including mucus-secreting cells (in the lining of the stomach and cervical tract) and absorptive cells (in the lining of the small intestine). (b) Simple squamous epithelia, composed of thin cells, line the blood vessels (endothelial cells/endothelium) and many body cavities. (c) Transitional epithelia, composed of several layers of cells with different shapes, line certain cavities subject to expansion and contraction (e.g., the urinary bladder). (d) Stratified squamous (nonkeratinized) epithelia line surfaces such as the mouth and vagina; these linings resist abrasion and generally do not participate in the absorption or secretion of materials into or out of the cavity. The basal lamina, a thin fibrous network of collagen and other ECM components, supports all epithelia and connects them to the underlying connective tissue.

mechanism used to generate strong, stable adhesions is to concentrate subsets of these molecules into clusters called junctions.

Three Types of Junctions Mediate Many Cell-Cell and Cell-ECM Interactions

All epithelial cells in a sheet are connected to one another and the extracellular matrix by specialized junctions. Although hundreds of individual adhesion-molecule-mediated interactions are sufficient to cause cells to adhere, junctions play special roles in imparting strength and rigidity to a tissue, transmitting information between the extracellular and the intracellular space, controlling the passage of ions and molecules across cell layers, and serving as conduits for the movement of ions and molecules from the cytoplasm of one cell to that of its immediate neighbor. Particularly important to epithelial sheets is the formation of junctions that help form tight seals between the cells and thus allow the sheet to serve as a barrier to the flow of molecules from one side of the sheet to the other.

Three major classes of animal cell junctions are prominent features of simple columnar epithelia (Figure 19-9 and Table 19-2). **Anchoring junctions** and **tight junctions** perform the key task of holding the tissue together. These junctions are organized into three parts: (1) adhesive proteins in the plasma membrane that connect one cell to another cell on the lateral surfaces (CAMs) or to the extracellular matrix on the basal surfaces (adhesion receptors); (2) adapter proteins, which connect the CAMs or adhesion receptors to cytoskeletal filaments and signaling molecules; and (3) the cytoskeletal filaments themselves. Tight junctions also control the flow of solutes through the extracellular spaces between the cells, forming an epithelial sheet. Tight junctions are found primarily in epithelial cells, whereas anchoring junctions can be seen in both epithelial and nonepithelial cells. The third class of junctions, **gap junctions,** permit the rapid diffusion of small, water-soluble molecules between the cytoplasm of adjacent cells. They share with anchoring and tight junctions the role of helping a cell communicate with its environments but are structurally very different from anchoring junctions and tight junctions and do not play a key role in strengthening cell-cell and cell-ECM adhesions. Gap junctions, found in both epithelial and nonepithelial cells, resemble cell-cell junctions in plants called plasmodesmata, which we discuss in Section 19.6.

Three types of anchoring junctions are present in cells. Two participate in cell-cell adhesion, whereas the third participates in cell-matrix adhesion. *Adherens junctions* connect the lateral membranes of adjacent epithelial cells and are usually located near the apical surface, just below the tight junctions (Figure 19-9). A circumferential belt of actin and myosin filaments in a complex with the adherens junction functions as a tension cable that can internally brace the cell and thereby control its shape. Epithelial and some other types of cells, such as smooth muscle and heart cells, are also bound tightly together by *desmosomes,* snaplike points of contact sometimes called spot desmosomes. *Hemidesmosomes,* found mainly on the basal surface of epithelial cells, anchor an epithelium to components of the underlying extracellular matrix, much like nails holding down a carpet. Bundles of intermediate filaments running parallel to the cell surface or through the cell interconnect spot desmosomes and hemidesmosomes, imparting shape and rigidity to the cell. Adherens junctions and desmosomes are found in many different types of cells; hemidesmosomes appear to be restricted to epithelial cells.

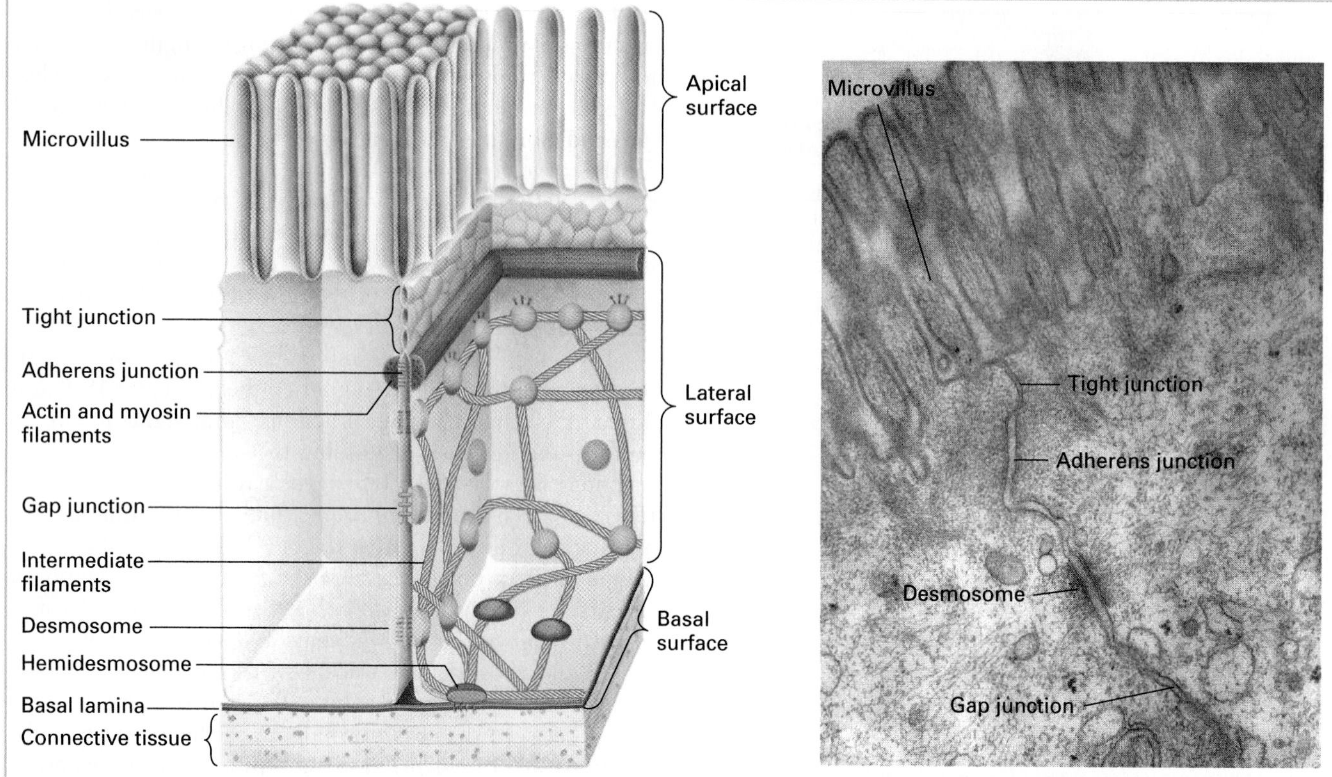

▲ **FIGURE 19-9 Principal types of cell junctions connecting the columnar epithelial cells lining the small intestine.** (a) Schematic cutaway drawing of intestinal epithelial cells. The basal surface of the cells rests on a basal lamina, and the apical surface is packed with fingerlike microvilli that project into the intestinal lumen. Tight junctions, lying just under the microvilli, prevent the diffusion of many substances between the intestinal lumen and the blood through the extracellular space between cells. Gap junctions allow the movement of small molecules and ions between the cytosols of adjacent cells. The remaining three types of junctions—adherens junctions, spot desmosomes, and hemidesmosomes—are critical to cell-cell and cell-matrix adhesion and signaling. (b) Electron micrograph of a thin section of intestinal epithelial cells, showing relative locations of the different junctions. [Part (b) C. Jacobson et al., 2001, *Journal Cell Biol.* **152:**435–450.]

Desmosomes and hemidesmosomes help transmit shear forces from one region of a cell layer to the epithelium as a whole, providing strength and rigidity to the entire epithelial cell layer. They are especially important in maintaining the integrity of skin epithelia. For instance, mutations that interfere with hemidesmosomal anchoring in the skin can lead to blistering in which the epithelium becomes detached from its matrix foundation and extracellular fluid accumulates at the basolateral surface, forcing the skin to balloon outward.

Cadherins Mediate Cell-Cell Adhesions in Adherens Junctions and Desmosomes

The primary CAMs in adherens junctions and desmosomes belong to the **cadherin** family. In vertebrates, this protein family of more than 100 members can be grouped into at least six subfamilies, including *classical cadherins* and *desmosomal cadherins,* which we will describe below, as well as protocadherins and others. The diversity of cadherins arises from the presence of multiple cadherin genes and alternative RNA splicing. It is not surprising that there are many different types of cadherins in vertebrates, because many different types of cells in widely diverse tissues use these CAMs to mediate adhesion and communication. The brain expresses the largest number of different cadherins, presumably owing to the necessity of forming many very specific cell-cell contacts to help establish its complex wiring diagram. Invertebrates, however, are able to function with fewer than 20 cadherins.

Classical Cadherins The "classical" cadherins include E-, N-, and P-cadherins. E- and N-cadherins are the most widely expressed, particularly during early differentiation. Sheets of polarized epithelial cells, such as those that line the small intestine or kidney tubules, contain abundant E-cadherin along their lateral surfaces. Although E-cadherin is concentrated in adherens junctions, it is present throughout the lateral surfaces, where it is thought to link adjacent cell membranes. The results of experiments with L cells, a line of cultured mouse fibroblasts, demonstrated that E-cadherins preferentially mediate homophilic interactions. L cells express no cadherins and adhere poorly to themselves or to other cells. When the

TABLE 19-2 Cell Junctions

JUNCTION	ADHESION TYPE	PRINCIPAL CAMs OR ADHESION RECEPTORS	CYTOSKELETAL ATTACHMENT	FUNCTION
Anchoring junctions				
1. Adherens junctions	Cell-cell	Cadherins	Actin filaments	Shape, tension, signaling
2. Desmosomes	Cell-cell	Desmosomal cadherins	Intermediate filaments	Strength, durability, signaling
3. Hemidesmosomes	Cell-matrix	Integrin ($\alpha6\beta4$)	Intermediate filaments	Shape, rigidity, signaling
Tight junctions	Cell-cell	Occludin, claudin, JAMs	Actin filaments	Controlling solute flow, signaling
Gap junctions	Cell-cell	Connexins, innexins, pannexins	Possible indirect connections to cytoskeleton through adapters to other junctions	Communication; small-molecule transport between cells
Plasmodesmata (plants only)	Cell-cell	Undefined	Actin filaments	Communication; molecule transport between cells

E-cadherin gene was introduced into L cells, the engineered cadherin-expressing L cells were found to adhere preferentially to other cells expressing E-cadherin (Figure 19-10). These L cells expressing E-cadherin formed epithelial-like aggregates with one another and with epithelial cells isolated from lungs. Although most E-cadherins exhibit primarily homophilic binding, some mediate heterophilic interactions.

The adhesiveness of cadherins depends on the presence of extracellular Ca^{2+}, the property that gave rise to their name (*c*alcium *adhering*). For example, the adhesion of L cells expressing E-cadherin is prevented when the cells are bathed in a solution that is low in Ca^{2+} (Figure 19-10). Some adhesion molecules require some minimal amount of Ca^{2+} in the extracellular fluid to function properly, whereas others (e.g., IgCAMs) are Ca^{2+}-independent.

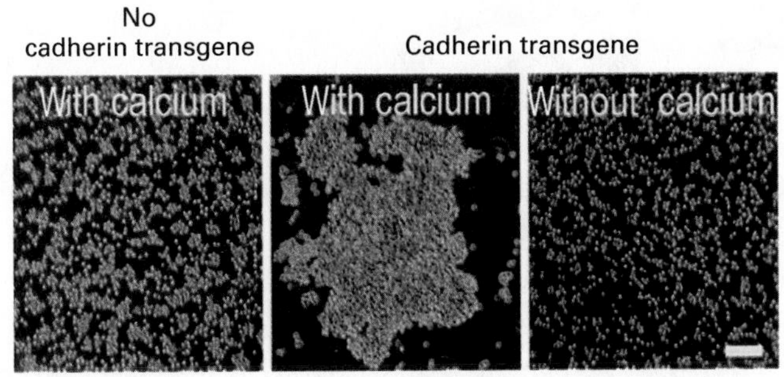

▲ **EXPERIMENTAL FIGURE 19-10 E-cadherin mediates Ca^{2+}-dependent adhesion of L cells.** Under standard cell culture conditions in the presence of calcium in the extracellular fluid, L cells do not aggregate into sheets (*left*). Introduction of a gene that causes the expression of E-cadherin in these cells results in their aggregation into epithelial-like clumps in the presence of calcium (*center*) but not in its absence (*right*). Bar, 60 μm. [From Cynthia L. Adams et al., 1998, *J. Cell Biol.* **142**(4):1105–1119.]

The role of E-cadherin in adhesion can also be demonstrated in experiments with cultured epithelial cells called *Madin-Darby canine kidney (MDCK)* cells (Figure 9-34). A green fluorescent-protein-labeled form of E-cadherin has been used in these cells to show that clusters of E-cadherin mediate the initial attachment and subsequent zippering up of the cells into sheets (Figure 19-11). In this experimental system, the addition of an antibody that binds to E-cadherin, preventing its homophilic interactions, blocks the Ca^{2+}-dependent attachment of suspended MDCK cells to each other and the subsequent formation of intercellular adherens junctions.

Each classical cadherin contains a single transmembrane domain, a relatively short C-terminal cytosolic domain, and five extracellular "cadherin" domains (see Figure 19-2). The extracellular domains are necessary for Ca^{2+} binding and cadherin-mediated cell-cell adhesion. Cadherin-mediated adhesion entails both lateral (intracellular) and trans (intercellular) molecular interactions (see Figure 19-3). The Ca^{2+}-binding sites, located between the cadherin repeats, serve to rigidify the cadherin oligomers. The cadherin oligomers subsequently form intercellular complexes to generate cell-cell adhesion and then additional lateral contacts, resulting in a "zippering up" of cadherins into clusters. In this way, multiple low-affinity interactions sum to produce a very tight intercellular adhesion.

MEDIA CONNECTIONS

Podcast: E-cadherin Zipper

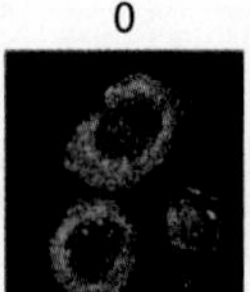

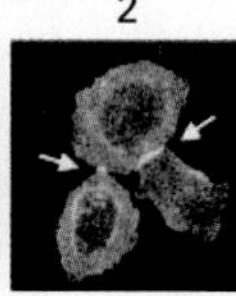

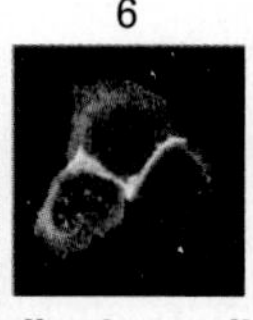

▲ **EXPERIMENTAL FIGURE 19-11 E-cadherin mediates adhesive connections in cultured MDCK epithelial cells.** An E-cadherin gene fused to green fluorescent protein (GFP) was introduced into cultured MDCK cells. The cells were then mixed together in a calcium-containing medium and the distribution of fluorescent E-cadherin was visualized over time (shown in hours). Clusters of E-cadherin mediate the initial attachment and subsequent zippering up of the epithelial cells. [From Cynthia L. Adams et al., 1998, *J. Cell Biol.* **142**(4):1105–1119.]

The results of domain swap experiments, in which an extracellular domain of one kind of cadherin is replaced with the corresponding domain of a different cadherin, have indicated that the specificity of binding resides, at least in part, in the most distal (farthest from the membrane) extracellular domain, the N-terminal domain. Cadherin-mediated adhesion was commonly thought to require only head-to-head interactions between the N-terminal domains of cadherin oligomers on adjacent cells, as depicted in Figure 19-12. However, some experiments suggest that under some conditions, at least three cadherin domains from each molecule, not just the N-terminal domains, participate by interdigitation in trans associations.

The C-terminal cytosolic domain of classical cadherins is linked to the actin cytoskeleton by adapter proteins (Figure 19-12). These linkages are essential for strong adhesion, apparently owing primarily to their contributing to increased lateral associations. For example, disruption of the interactions between classical cadherins and α- or β-catenin—two common adapter proteins that link these cadherins to actin filaments—dramatically reduces cadherin-mediated cell-cell adhesion. This disruption occurs spontaneously in tumor cells, which sometimes fail to express α-catenin, and can be induced experimentally by depleting the cytosolic pool of accessible β-catenin. The cytosolic domains of cadherins also interact with intracellular signaling molecules such as β-catenin and p120-catenin. Interestingly, β-catenin not only mediates cytoskeletal attachment but can also translocate to the nucleus and alter gene transcription in the Wnt signaling pathway (see Figure 16-32).

Cadherins play a critical role during tissue differentiation. Each classical cadherin has a characteristic tissue distribution. In the course of differentiation, the amount or nature of the cell-surface cadherins changes, affecting many aspects of cell-cell adhesion and cell migration. For instance, the reorganization of tissues during morphogenesis is often accompanied by the conversion of nonmotile epithelial cells into motile precursor cells for other tissues (mesenchymal cells). Such *epithelial-mesenchymal transitions* are associated with a reduction in the expression of E-cadherin (Figure 19-13a, b). The conversion of epithelial cells into malignant carcinoma cells, such as in certain

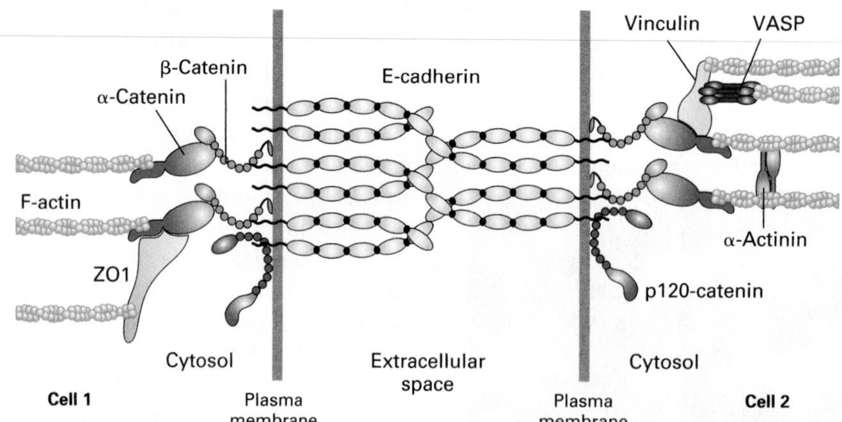

▲ **FIGURE 19-12 Protein constituents of typical adherens junctions.** The exoplasmic domains of E-cadherin dimers clustered at adherens junctions on adjacent cells form Ca^{+2}-dependent homophilic interactions. The cytosolic domains of the E-cadherins bind directly or indirectly to multiple adapter proteins (e.g., β-catenin) that connect the junctions to actin filaments (F-actin) of the cytoskeleton and participate in intracellular signaling pathways. Somewhat different sets of adapter proteins are illustrated in the two cells to emphasize that a variety of adapters can interact with adherens junctions. Some of these adapters, such as ZO1, can interact with several different CAMs. [Adapted from V. Vasioukhin and E. Fuchs, 2001, *Curr. Opin. Cell Biol.* **13**:76.]

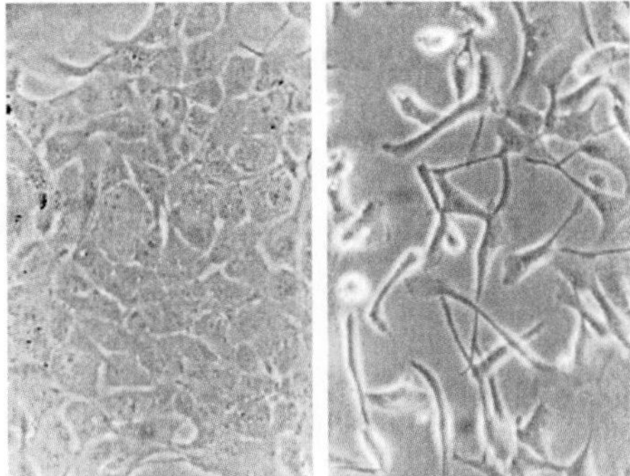

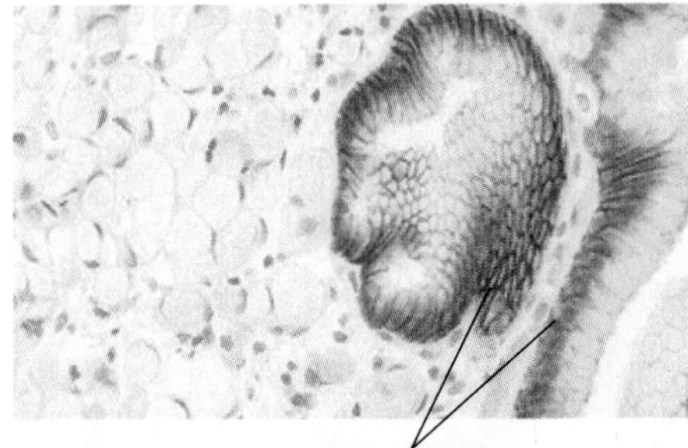

▲ **EXPERIMENTAL FIGURE 19-13 E-cadherin activity is lost during the epithelial-mesenchymal transition and cancer progression.** A protein called Snail that suppresses the expression of E-cadherin is associated with epithelial-mesenchymal transitions. (a) Normal epithelial MDCK cells grown in culture. (b) Expression of the *snail* gene in MDCK cells causes them to undergo an epithelial-mesenchymal transition. (c) Distribution of E-cadherin detected by immunohistochemical staining (dark brown) in thin sections of tissue from a patient with hereditary diffuse gastric cancer. E-cadherin is seen at the intercellular boarders of normal stomach gastric gland epithelial cells (*upper right*); no E-cadherin is seen at the boarders of underlying invasive carcinoma cells. [Panels (a) and (b) from Alfonso Martinez Arias, 2001, *Cell* **105:**425–431; Images courtesy of M.A. Nieto.). Panel (c) from F. Carneiro et al., 2004, *J. Pathol.* **203**(2):681–687.]

ductal breast tumors or hereditary diffuse gastric cancer (Figure 19-13c), is also marked by a loss of E-cadherin activity.

Desmosomal Cadherins Desmosomes (Figure 19-14) contain two specialized cadherin proteins, *desmoglein* and *desmocollin,* whose cytosolic domains are distinct from those in the classical cadherins. The cytosolic domains of desmosomal cadherins interact with adapters proteins such as plakoglobin (similar in structure to β-catenin), plakophilins, and a member of the plakin family of adapters called desmoplakin. These adapters, which form the thick cytoplasmic plaques characteristic of desmosomes, in turn interact with intermediate filaments.

The cadherin desmoglein was identified through studies of an unusual but revealing skin disease called *pemphigus vulgaris,* an autoimmune disease. Patients with autoimmune disorders synthesize antibodies that bind to a normal body protein. In pemphigus vulgaris the autoantibodies disrupt adhesion between epithelial cells, causing blisters of the skin and mucous membranes. The predominant autoantibody was shown to be specific for

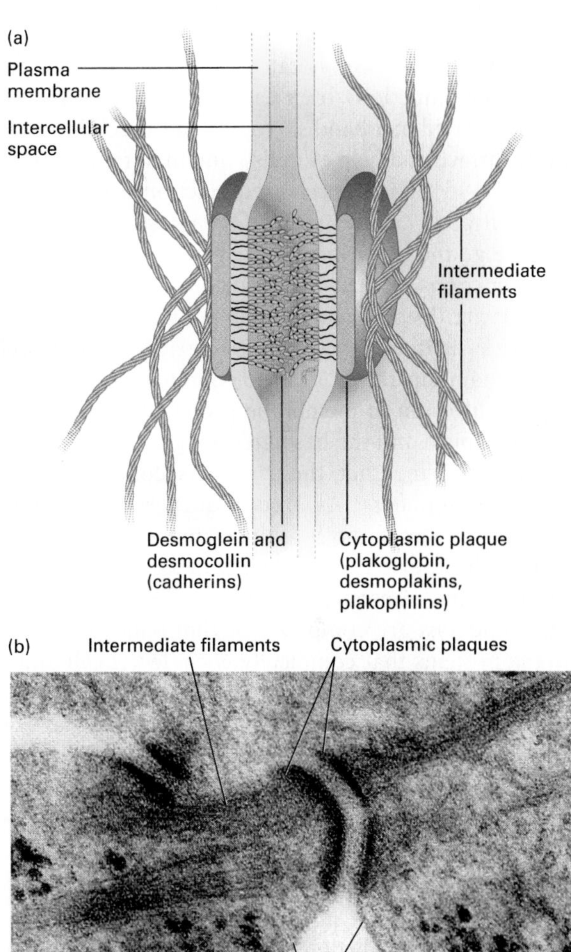

▲ **FIGURE 19-14 Desmosomes.** (a) Model of a desmosome between epithelial cells with attachments to the sides of intermediate filaments. The transmembrane CAMs desmoglein and desmocollin belong to the cadherin family. Adapter proteins bound to the cytoplasmic domains of the CAMs include plakoglobin, desmoplakins, and plakophilins. (b) Electron micrograph of a thin section of a desmosome connecting two cultured differentiated human keratinocytes. Bundles of intermediate filaments radiate from the two darkly staining cytoplasmic plaques that line the inner surface of the adjacent plasma membranes. [Part (a), see B. M. Gumbiner, 1993, *Neuron* **11:**551, and D. R. Garrod, 1993, *Curr. Opin. Cell Biol.* **5:**30. Part (b) courtesy of R. van Buskirk.]

desmoglein; indeed, the addition of such antibodies to normal skin induces the formation of blisters and disruption of cell adhesion. ∎

The firm epithelial cell-cell adhesions mediated by cadherins in adherens junctions permits the formation of a second class of intercellular junctions in epithelia—tight junctions.

Tight Junctions Seal Off Body Cavities and Restrict Diffusion of Membrane Components

For polarized epithelial cells to function as barriers and mediators of selective transport, extracellular fluids surrounding their apical and basolateral membranes must be kept separate. Tight junctions between adjacent epithelial cells are usually located in a band surrounding the cell just below the apical surface and help establish and maintain cell polarity (Figure 19-15). These specialized junctions form a barrier that seals off body cavities such as the intestinal lumen and the blood (e.g., the blood-brain barrier).

Tight junctions prevent the diffusion of macromolecules and, to varying degrees, small water-soluble molecules and ions across an epithelial sheet via the spaces between cells. They also maintain the polarity of epithelial cells by preventing the diffusion of membrane proteins and glycolipids between the apical and the basolateral regions of the plasma membrane, ensuring that these regions contain different membrane components. As a consequence, movement of many nutrients across the intestinal epithelium is in large part through the *transcellular pathway* via specific membrane-bound transport proteins (see Figure 11-29).

Tight junctions are composed of thin bands of plasma-membrane proteins that completely encircle the cell and are in contact with similar thin bands on adjacent cells. When thin sections of cells are viewed in an electron microscope, the lateral surfaces of adjacent cells appear to touch each other at intervals and even to fuse in the zone just below the apical surface (see Figure 19-9b). In freeze-fracture preparations, tight junctions appear as an interlocking network of ridges and grooves in the plasma membrane (Figure 19-15a).

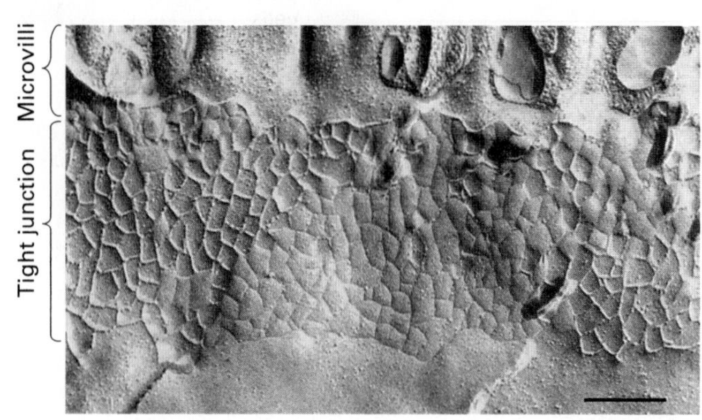

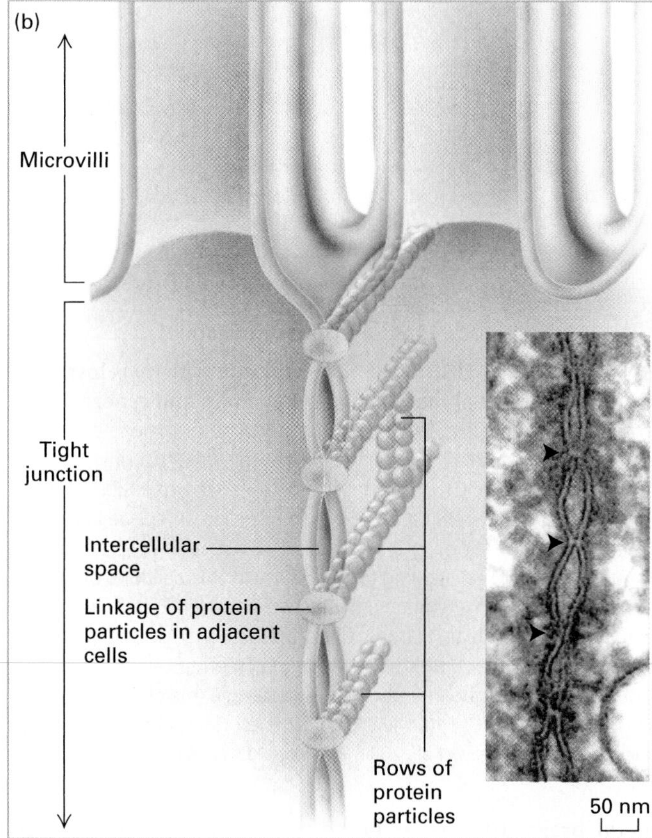

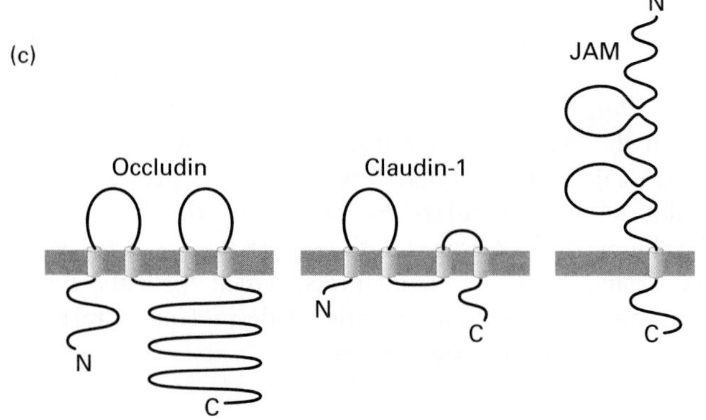

▶ **FIGURE 19-15 Tight junctions.** (a) Freeze-fracture preparation of tight junction zone between two intestinal epithelial cells. The fracture plane passes through the plasma membrane of one of the two adjacent cells. A honeycomb-like network of ridges and grooves below the microvilli constitutes the tight junction zone. (b) Schematic drawing shows how a tight junction might be formed by the linkage of rows of protein particles in adjacent cells. In the inset micrograph of an ultrathin sectional view of a tight junction, the adjacent cells can be seen in close contact where the rows of proteins interact. (c) As shown in these schematic drawings of the major proteins in tight junctions, both occludin and claudin-1 contain four transmembrane helices, whereas the junction adhesion molecule (JAM) has a single transmembrane domain and a large extracellular region. [Part (a) courtesy of L. A. Staehelin. Drawing in part (b) adapted from L. A. Staehelin and B. E. Hull, 1978, *Sci. Am.* **238**(5):140, and D. Goodenough, 1999, *Proc. Nat'l. Acad. Sci. USA* **96**:319. Photograph in part (b) courtesy of S. Tsukita et al., 2001, *Nature Rev. Mol. Cell Biol.* **2**:285. Drawing in part (c) adapted from S. Tsukita et al., 2001, *Nature Rev. Mol. Cell Biol.* **2**:285.]

Very high magnification reveals that rows of protein particles 3–4 nm in diameter form the ridges seen in freeze-fracture micrographs of tight junctions. In the model shown in Figure 19-15b, the tight junction is formed by a double row of these particles, one row donated by each cell. Treatment of an epithelium with the protease trypsin destroys the tight junctions, supporting the proposal that proteins are essential structural components of these junctions. The two principal integral-membrane proteins found in tight junctions are *occludin* and *claudin.* When investigators engineered mice with mutations inactivating the occludin gene, which was thought to be essential for tight junction formation, the mice still had morphologically distinct tight junctions. Further analysis led to the discovery of claudin. Each of these proteins has four membrane-spanning α helices (Figure 19-15c). The claudin multigene family encodes numerous homologous proteins that exhibit distinct tissue-specific patterns of expression. A group of *junction adhesion molecules (JAMs)* have been found to contribute to homophilic adhesion and other functions of tight junctions. These molecules, which contain a single transmembrane α helix, belong to the Ig superfamily of CAMs. The extracellular domains of rows of occludin, claudin, and JAM proteins in the plasma membrane of one cell apparently form extremely tight links with similar rows of the same proteins in an adjacent cell, creating a tight seal. Ca^{2+}-dependent cadherin-mediated adhesion also plays an important role in tight junction formation, stability, and function.

The long C-terminal cytosolic segment of occludin binds to PDZ domains in certain large cytosolic adapter proteins. These domains are found in various cytosolic proteins and mediate binding to the C-termini of particular plasma-membrane proteins or to each other. PDZ-containing adapter proteins associated with occludin are bound, in turn, to other cytoskeletal and signaling proteins and to actin fibers. These interactions appear to stabilize the linkage between occludin and claudin molecules that is essential for maintaining the integrity of tight junctions. The C-termini of claudins also bind to the intracellular, multiple-PDZ-domain-containing adaptor protein ZO-1, which is also found in adherens junctions (see Figure 19-12). Thus, as is the case for adherens junctions and desmosomes, cytosolic adaptor proteins and their connections to the cytoskeleton are critical components of tight junctions.

Plasma-membrane proteins cannot diffuse in the plane of the membrane past tight junctions. These junctions also restrict the lateral movement of lipids in the exoplasmic leaflet of the plasma membrane in the apical and basolateral regions of epithelial cells. Indeed, the lipid compositions of the exoplasmic leaflet in these two regions are distinct. Essentially all glycolipids are present in the exoplasmic face of the apical membrane, as are all proteins linked to the membrane by a glycosylphosphatidylinositol (GPI) anchor (see Figure 10-19). In contrast, lipids in the cytosolic leaflet in the apical and basolateral regions of epithelial cells have the same composition and can apparently diffuse laterally from one region of the membrane to the other.

A simple experiment demonstrates the impermeability of certain tight junctions to many water-soluble substances. In this experiment, lanthanum hydroxide (an electron-dense colloid of high molecular weight) is injected into the pancreatic blood vessel of an experimental animal; a few minutes later, the pancreatic epithelial acinar cells are fixed and prepared for microscopy. As shown in Figure 19-16, the lanthanum hydroxide diffuses from the blood into the space that separates the lateral surfaces of adjacent acinar cells but cannot penetrate past the tight junction.

The barrier to diffusion provided by tight junctions is not absolute. Owing at least in part to the varying properties of the different types of claudin molecules located in different tight junctions, their permeability to ions, small molecules, and water varies enormously among different epithelial tissues. In epithelia with "leaky" tight junctions, small molecules can move from one side of the cell layer to the other through the *paracellular pathway* in addition to the transcellular pathway (Figure 19-17).

The leakiness of tight junctions can be altered by intracellular signaling pathways, especially G protein and cyclic AMP–coupled pathways (Chapter 15). The regulation of tight junction permeability is often studied by measuring ion flux (electrical resistance) or the movement of radioactive or fluorescent molecules across monolayers of MDCK cells.

The importance of paracellular transport is illustrated in several human diseases. In hereditary hypomagnesemia, defects in the *claudin16* gene prevent the normal paracellular flow of magnesium in the kidney. This results in an abnormally low blood level of magnesium, which can lead to convulsions. Furthermore, a mutation in

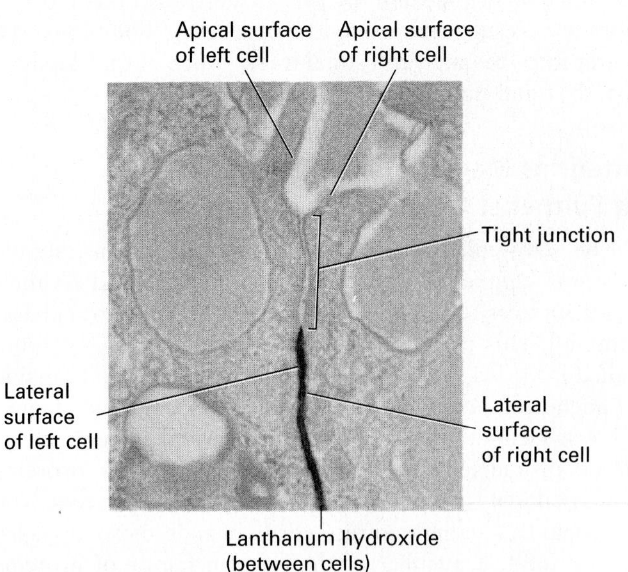

▲ EXPERIMENTAL FIGURE 19-16 **Tight junctions prevent passage of large molecules through extracellular space between epithelial cells.** Tight junctions in the pancreas are impermeable to the large water-soluble colloid lanthanum hydroxide (dark stain) administered from the basolateral side of the epithelium. [Courtesy of D. Friend.]

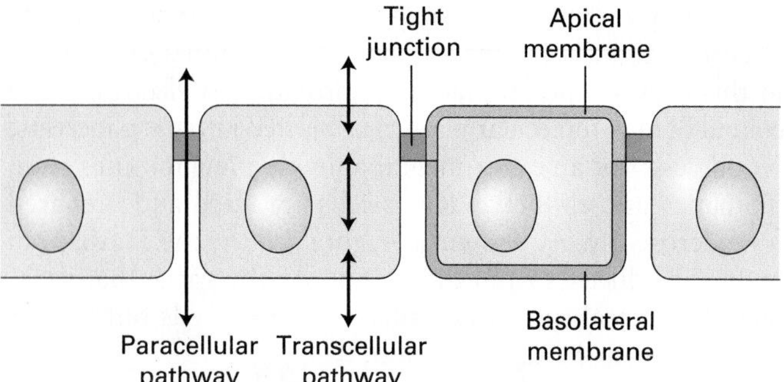

▲ **FIGURE 19-17 Transcellular and paracellular pathways of transepithelial transport.** Transcellular transport requires the cellular uptake of molecules on one side and subsequent release on the opposite side by mechanisms discussed in Chapter 11. In paracellular transport, molecules move extracellularly through parts of tight junctions, whose permeability to small molecules and ions depends on the composition of the junctional components and the physiologic state of the epithelial cells. [Adapted from S. Tsukita et al., 2001, *Nature Rev. Mol. Cell Biol.* **2**:285.]

the *claudin14* gene causes hereditary deafness, apparently by altering transport around hair-cell epithelia in the cochlea of the inner ear.

Toxins produced by *Vibrio cholerae,* which causes cholera, and several other enteric (gastrointestinal tract) bacteria alter the permeability barrier of the intestinal epithelium by altering the composition or activity of tight junctions. Other bacterial toxins can affect the ion-pumping activity of membrane transport proteins in intestinal epithelial cells. Toxin-induced changes in tight junction permeability (increased paracellular transport) and in protein-mediated ion pumping (increased transcellular transport) can result in massive loss of internal body ions and water into the gastrointestinal tract, which in turn leads to diarrhea and potentially lethal dehydration. ▮

Integrins Mediate Cell-ECM Adhesions in Epithelial Cells

To be stably anchored to solid tissue and organs, simple columnar epithelial sheets must be firmly attached via their basal surfaces to the underlying extracellular matrix (basal lamina). This attachment occurs via adhesion receptors called integrins, which are located both within and outside of anchoring junctions called *hemidesmosomes* (see Figure 19-9a). Hemidesmosomes comprise several integral membrane proteins linked via cytoplasmic adaptor proteins (e.g., plakins) to keratin-based intermediate filaments. The principal ECM adhesion receptor in hemidesmosomes is integrin α6β4, a member of the **integrin** family of proteins (see Figure 19-2).

Integrins function as adhesion receptors and CAMs in a wide variety of epithelial and nonepithelial cells, mediating many cell-matrix and cell-cell interactions (Table 19-3). In vertebrates, at least 24 integrin heterodimers, composed of 18 types of α subunits and 8 types of β subunits in various combinations, are known. A single β chain can interact with any one of multiple α chains, forming integrins that bind different ligands. This phenomenon of *combinatorial diversity* allows a relatively small number of components to serve a large number of distinct functions. Although most cells express several distinct integrins that bind the same or different ligands, many integrins are expressed predominantly in certain types of cells. Not only do many integrins bind more than one ligand but several of their ligands bind to multiple integrins.

All integrins appear to have evolved from two ancient general subgroups: those that bind proteins containing the tripeptide sequence Arg-Gly-Asp, usually called the *RGD sequence* (e.g., fibronectin) and those that bind laminin. Several integrin α subunits contain a distinctive inserted domain, the *I-domain,* which can mediate binding of certain integrins (e.g., α1β1 and α2β1) to various collagens in the ECM. Some integrins with I-domains are expressed exclusively on leukocytes and red and white blood cell precursor (hematopoietic) cells. These domains recognize cell-adhesion molecules on other cells, including members of the Ig superfamily (e.g., ICAMs, VCAMs), and thus participate in cell-cell adhesion.

Integrins typically exhibit low affinities for their ligands, with dissociation constants K_D between 10^{-6} and 10^{-7} mol/L. However, the multiple weak interactions generated by the binding of hundreds or thousands of integrin molecules to their ligands on cells or in the extracellular matrix allow a cell to remain firmly anchored to its ligand-expressing target.

Parts of both the α and the β subunits of an integrin molecule contribute to the primary extracellular ligand-binding site (see Figure 19-2). Ligand binding to integrins also requires the simultaneous binding of divalent cations. Like other cell-surface adhesive molecules, the cytosolic region of integrins interacts with adapter proteins that in turn bind to the cytoskeleton and intracellular signaling molecules. Most integrins are linked to the actin cytoskeleton, such as the α6β1 and α3β1 integrins that connect the basal surface of epithelial cells to the basal lamina via laminin. However, the cytosolic domain of the β4 chain in the α6β4 integrin in hemidesmosomes, which is much longer than those of other β integrins, binds to specialized adapter proteins that in turn interact with keratin-based intermediate filaments.

As we will see, the diversity of integrins and their ECM ligands enables integrins to participate in a wide array of key biological processes, including the migration of cells to their correct locations in the formation of the body plan of an embryo (morphogenesis) and in the inflammatory response. The importance of integrins in diverse processes is highlighted by the defects exhibited by knockout mice engineered to have mutations in each of almost all of the integrin subunit genes. These defects include major abnormalities in development, blood vessel formation, leukocyte function, inflammation, bone remodeling, and hemostasis. Despite their differences, all these processes depend on integrin-mediated regulated interactions between the cytoskeleton and either the ECM or CAMs on other cells.

TABLE 19-3 Selected Vertebrate Integrins*

SUBUNIT COMPOSITION	PRIMARY CELLULAR DISTRIBUTION	LIGANDS
α1β1	Many types	Mainly collagens
α2β1	Many types	Mainly collagens; also laminins
α3β1	Many types	Laminins
α4β1	Hematopoietic cells	Fibronectin; VCAM-1
α5β1	Fibroblasts	Fibronectin
α6β1	Many types	Laminins
αLβ2	T lymphocytes	ICAM-1, ICAM-2
αMβ2	Monocytes	Serum proteins (e.g., C3b, fibrinogen, factor X); ICAM-1
αIIbβ3	Platelets	Serum proteins (e.g., fibrinogen, von Willebrand factor, vitronectin); fibronectin
α6β4	Epithelial cells	Laminin

*The integrins are grouped into subfamilies having a common β subunit. Ligands shown in red are CAMs; all others are ECM or serum proteins. Some subunits can have multiple spliced isoforms with different cytosolic domains.
SOURCE: R. O. Hynes, 1992, *Cell* **69**:11.

In addition to their adhesion function, integrins can mediate outside-in and inside-out signaling (see Figure 19-7). The engagement of integrins by their extracellular ligands can, through adapter proteins bound to the integrin cytosolic region, influence the cytoskeleton and intracellular signaling pathways (outside-in signaling). Conversely, intracellular signaling pathways can alter, from the cytoplasm, the structure of integrins and consequently their abilities to adhere to their extracellular ligands and mediate cell-cell and cell-matrix interactions (inside-out signaling). Integrin-mediated signaling pathways influence processes as diverse as cell survival, cell proliferation, and programmed cell death (Chapter 21).

Gap Junctions Composed of Connexins Allow Small Molecules to Pass Directly Between Adjacent Cells

Early electron micrographs of virtually all animal cells that were in contact revealed sites of cell-cell contact with a characteristic intercellular gap (Figure 19-18a). This feature prompted early morphologists to call these regions gap junctions. In retrospect, the most important feature of these junctions is not the gap itself but a well-defined set of cylindrical particles that cross the gap and compose pores connecting the cytoplasms of adjacent cells.

In many tissues, large numbers of gap junctional particles cluster together in patches (e.g., along the lateral surfaces of epithelial cells; see Figure 19-9). When the plasma membrane is purified and then sheared into small fragments, some pieces mainly containing patches of gap junctions are generated. Owing to their relatively high protein content, these fragments have a higher density than that of the bulk of the plasma membrane and can be purified by equilibrium density-gradient centrifugation (see Figure 9-26). When these preparations are viewed perpendicular to the membrane, the gap junctions appear as arrays of hexagonal particles that enclose water-filled channels (Figure 19-18b).

The effective pore size of gap junctions can be measured by injecting a cell with a fluorescent dye covalently linked to molecules of various sizes and observing with a fluorescence microscope whether the dye passes into neighboring cells. Gap junctions between mammalian cells permit the passage of molecules as large as 1.2 nm in diameter. In insects, these junctions are permeable to molecules as large as 2 nm in diameter. Generally speaking, molecules smaller than 1200 Da pass freely and those larger than 2000 Da do not pass; the passage of intermediate-sized molecules is variable and limited. Thus ions, many low-molecular-weight precursors of cellular macromolecules, products of intermediary metabolism, and small intracellular signaling molecules can pass from cell to cell through gap junctions.

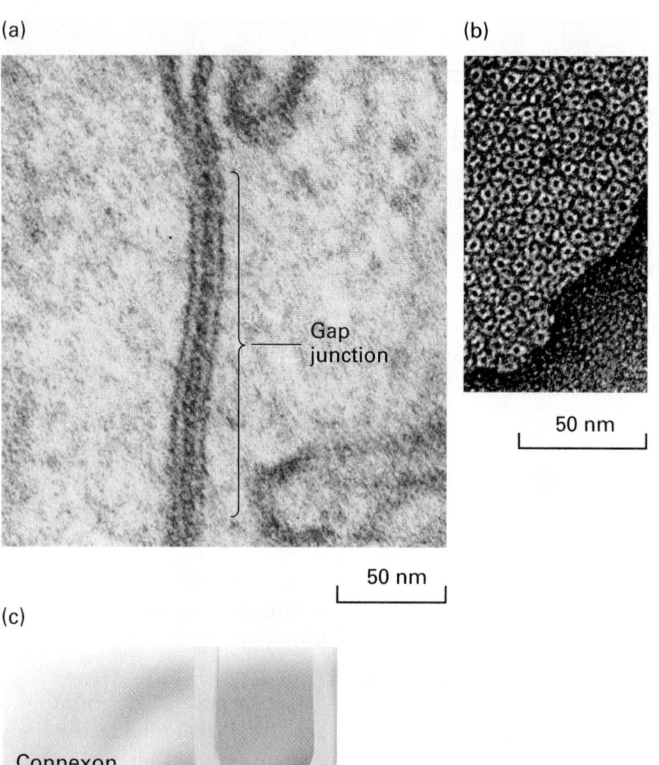

◄ **FIGURE 19-18 Gap junctions.** (a) In this thin section through a gap junction connecting two mouse liver cells, the two plasma membranes are closely associated for a distance of several hundred nanometers, separated by a "gap" of 2–3 nm. (b) Numerous roughly hexagonal particles are visible in this perpendicular, or en face, view of the cytosolic face of a region of plasma membrane enriched in gap junctions. Each particle aligns with a similar particle on an adjacent cell, forming a channel connecting the two cells. (c) Schematic model of a gap junction connecting two plasma membranes. Both membranes contain connexon hemichannels, cylinders of six dumbbell-shaped connexin molecules. Two connexons join in the gap between the cells to form a gap-junction channel, 1.5–2.0 nm in diameter, that connects the cytosols of the two cells. (d) Electron density of a recombinant gap-junction channel determined by electron crystallography. (*Left*) Side view of the complete structure oriented as in part (c). M = membrane bilayer; E = extracellular gap; C = cytosol. (*Right*) View looking down on the connexon from the cytosol perpendicular to the membrane bilayers. Superimposed on the electron density map are models of the transmembrane α helices (gold), four per subunit, 24 per connexon hemichannel. [Part (a) courtesy of D. Goodenough. Part (b) courtesy of N. Gilula. Part (d) adapted from S. J. Fleishman et al., 2004, *Mol. Cell.* **15**(6):879–888.]

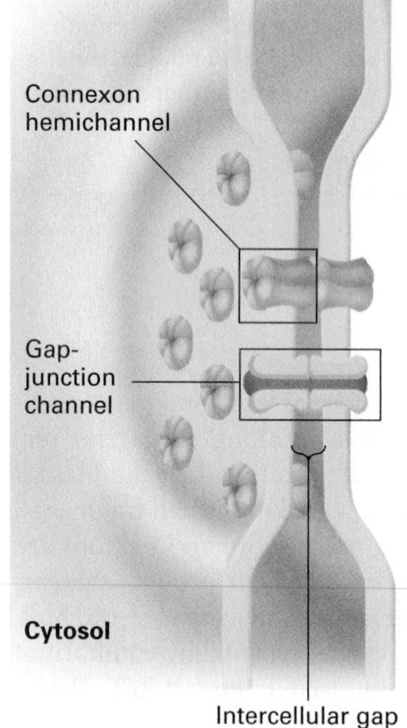

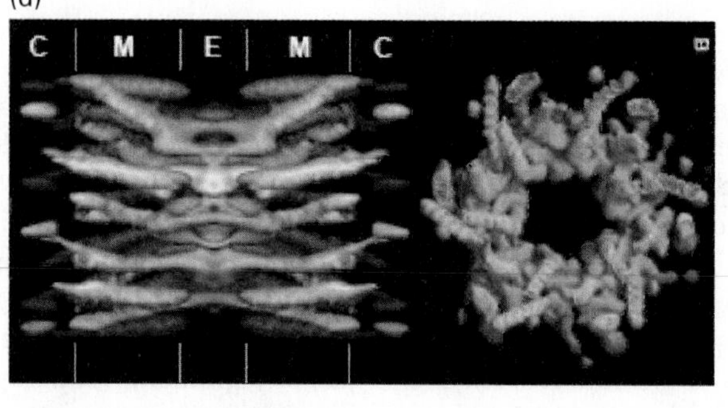

In nervous tissue, some neurons are connected by gap junctions through which ions pass rapidly, thereby allowing very rapid transmission of electric signals. Impulse transmission through these connections, called electrical synapses, is almost a thousandfold as rapid as at chemical synapses (Chapter 23). Gap junctions are also present in many non-neuronal tissues, where they help to integrate the electrical and metabolic activities of many cells. In the heart, for instance, gap junctions rapidly pass ionic signals among muscle cells, which are tightly interconnected via desmosomes, and thus contribute to the electrically stimulated coordinate contraction of cardiac muscle cells during a beat. As discussed in Chapter 15, some extracellular hormonal signals induce the production or release of small intracellular signaling molecules called **second messengers** (e.g., cyclic AMP, IP_3, and Ca^{2+}) that regulate cellular metabolism. Because second messengers can be transferred between cells through gap junctions, hormonal stimulation of one cell can trigger a coordinated response by that same cell and many of its neighbors. Such gap junction–mediated signaling plays an important role, for example, in the secretion of digestive enzymes by the pancreas and in the coordinated muscular contractile waves (peristalsis) in the intestine. Another vivid example of gap junction–mediated transport is the phenomenon of *metabolic coupling*, or *metabolic cooperation*, in which a cell transfers nutrients or intermediary metabolites to a neighboring cell that is itself unable to synthesize them. Gap junctions play critical roles in the development of egg cells in the ovary by mediating the movement of both metabolites and signaling molecules between an oocyte and its surrounding granulosa cells as well as between neighboring granulosa cells.

A current model of the structure of the gap junction is shown in Figure 19-18c, d. Vertebrate gap junctions are composed of **connexins,** a family of structurally related transmembrane proteins with molecular weights between 26,000 and 60,000. A completely different family of proteins, the innexins, forms the gap junctions in invertebrates. A third family of innexin-like proteins, called pannexins, was recently discovered in both vertebrates and invertebrates. Each vertebrate hexagonal particle consists of 12 noncovalently associated connexin molecules: 6 form a cylindrical connexin hemichannel in one plasma membrane that is joined to a connexin hemichannel in the adjacent cell membrane, forming the continuous aqueous channel between the cells. Each individual connexin molecule spans the plasma membrane four times with a topology similar to that of occludin (see Figure 19-15). Pannexins are capable of forming intercellular channels as well; however, pannexin hemichannels may also function to permit direct exchange between the intracellular and extracellular spaces.

There are 21 different connexin genes in humans, with different sets of connexins expressed in different cell types. This diversity, together with the generation of mutant mice with inactivating mutations in connexin genes, has highlighted the importance of connexins in a wide variety of cellular systems. Some cells express a single connexin that forms homotypic channels. Most cells, however, express at least two connexins; these different proteins assemble into heteromeric connexins, which in turn form heterotypic gap-junction channels. Diversity in channel composition leads to differences in channel permeability. For example, channels made from a 43-kDa connexin isoform, Cx43—the most ubiquitously expressed connexin—are more than 100-fold as permeable to ADP and ATP as those made from Cx32 (32 kDa).

The permeability of gap junctions can be regulated by changes in the intracellular pH and Ca^{2+} concentration and phosphorylation of connexin. One example of the physiological regulation of gap junctions is mammalian childbirth. The muscle cells in the mammalian uterus must contract strongly and synchronously during labor to expel the fetus. To facilitate this coordinate activity, immediately prior to and during labor there is an approximately five- to tenfold increase in the amount of the major myometrial connexin, Cx43, and an increase in the number and size of gap junctions, which decrease rapidly postpartum.

Assembly of connexins, their trafficking within cells, and formation of functional gap junctions apparently depend on N-cadherin and its associated junctional proteins (e.g., α- and β-catenins ZO-1, ZO-2) as well as desmosomal proteins (plakoglobin, desmoplakin, and plakophilin-2). PDZ domains in ZO-1 and ZO-2 bind to the C-terminus of Cx43 and apparently mediate its interaction with catenins and N-cadherin. The relevance of these relationships is particularly evident in the heart, which depends on adjacent gap junctions (for rapid coordinated electrical coupling) and adherens junctions and desmosomes (for mechanical coupling between cardiomyocytes) for the intercellular integration of electrical activity and movement required for normal cardiac function. It is noteworthy that ZO-1 serves as an adaptor for adherens (see Figure 19-12), tight, and gap junctions, suggesting this and other adapters can help integrate the formation and functions of these diverse junctions.

Mutations in connexin genes cause at least eight human diseases, including neurosensory deafness (Cx26 and Cx31), cataract or heart malformations (Cx43, Cx46, and Cx50), and the X-linked form of Charcot-Marie-Tooth disease (Cx32), which is marked by progressive degeneration of peripheral nerves. ▮

KEY CONCEPTS OF SECTION 19.2

Cell-Cell and Cell-ECM Junctions and Their Adhesion Molecules

- Polarized epithelial cells have distinct apical, basal, and lateral surfaces. Microvilli projecting from the apical surfaces of many epithelial cells considerably expand the cells' surface areas.

- Three major classes of cell junctions—anchoring junctions, tight junctions, and gap junctions—assemble epithelial cells into sheets and mediate communication between them (see Figures 19-1 and 19-9). Anchoring junctions can be further subdivided into adherens junctions, desmosomes, and hemidesmosomes.

- Adherens junctions and desmosomes are cadherin-containing anchoring junctions that bind the membranes of adjacent cells, giving strength and rigidity to the entire tissue.

- Cadherins are cell-adhesion molecules (CAMs) responsible for Ca^{2+}-dependent interactions between cells in epithelial and other tissues. They promote strong cell-cell adhesion by mediating both lateral intracellular and intercellular interactions.

- Adapter proteins that bind to the cytosolic domain of cadherins and other CAMs and adhesion receptors mediate the association of cytoskeletal and signaling molecules with the plasma membrane (see Figure 19-12). Strong cell-cell adhesion depends on the linkage of the interacting CAMs to the cytoskeleton.

- Tight junctions block the diffusion of proteins and some lipids in the plane of the plasma membrane, contributing to the polarity of epithelial cells. They also limit and regulate the extracellular (paracellular) flow of water and solutes from one side of the epithelium to the other (see Figure 19-17).

- Hemidesmosomes are integrin-containing anchoring junctions that attach cells to elements of the underlying extracellular matrix.

- Integrins are a large family of αβ heterodimeric cell-surface proteins that mediate both cell-cell and cell-matrix adhesions and inside-out and outside-in signaling in numerous tissues.

- Gap junctions are constructed of multiple copies of connexin proteins, assembled into a transmembrane channel that interconnects the cytoplasms of two adjacent cells (see Figure 19-18). Small molecules and ions can pass through gap junctions, permitting metabolic and electrical coupling of adjacent cells.

19.3 The Extracellular Matrix I: The Basal Lamina

In animals, the extracellular matrix helps organize cells into tissues and coordinates their cellular functions by activating intracellular signaling pathways that control cell growth, proliferation, and gene expression. Many functions of the matrix require transmembrane adhesion receptors that bind directly to ECM components and that also interact, through adapter proteins, with the cytoskeleton. A principal class of adhesion receptors that mediate cell-matrix adhesion is integrins (Section 19.2). However, other types of molecules also function as important adhesion receptors.

Adhesion receptors bind to three types of molecules abundant in the extracellular matrix of all tissues:

- **Proteoglycans,** a group of glycoproteins that cushion cells and bind a wide variety of extracellular molecules

- **Collagen** fibers, which provide structural integrity and mechanical strength and resilience

- Soluble **multiadhesive matrix proteins,** such as laminin and fibronectin, which bind to and cross-link cell-surface adhesion receptors and other ECM components.

We begin our description of the structures and functions of these major ECM components in the context of the basal lamina—the specialized extracellular matrix sheet that plays a particularly important role in determining the overall architecture and function of epithelial tissue. In the next section, we discuss the ECM molecules commonly found in nonepithelial tissues, including connective tissue.

The Basal Lamina Provides a Foundation for Assembly of Cells into Tissues

In animals, most organized groups of cells in epithelial and nonepithelial tissues are underlain or surrounded by the basal lamina, a sheetlike meshwork of ECM components usually no more than 60–120 nm thick (Figure 19-19). The basal lamina is structured differently in different tissues. In columnar and other epithelia (e.g., intestinal lining, skin), it is a foundation on which only one surface of the cells rests. In other tissues, such as muscle or fat, the basal lamina surrounds each cell. Basal laminae play important roles in regeneration after tissue damage and in embryonic development. For instance, the basal lamina helps four- and eight-celled embryos adhere together in a ball. In the development of the nervous system, neurons migrate along ECM pathways that contain basal lamina components. In higher animals, two distinct basal laminae are employed to form a tight barrier that limits diffusion of molecules between the blood and the brain (blood-brain barrier), and in the kidney a specialized basal lamina serves as a selective permeability blood filter. Thus the basal lamina is important for organizing cells into tissues and distinct compartments, tissue repair, and guiding

(a)

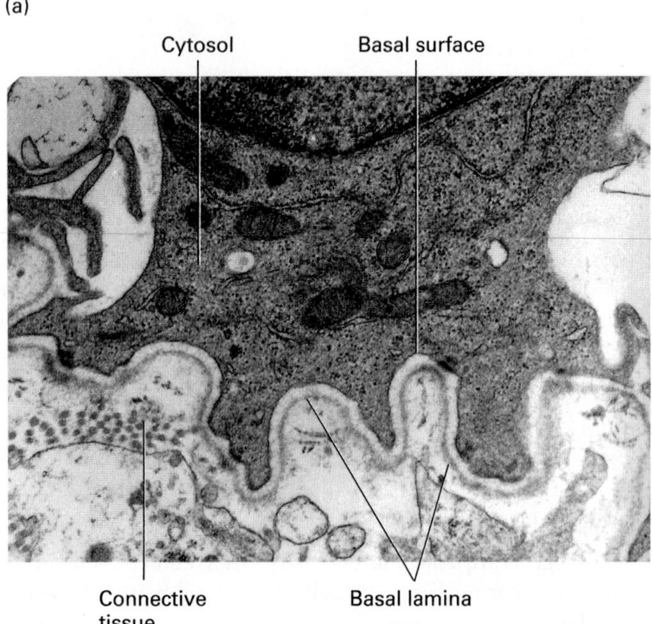

(b)

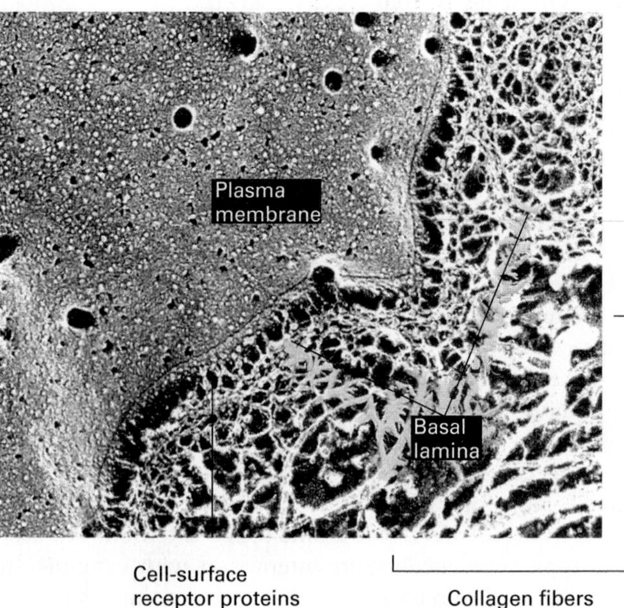

▲ **FIGURE 19-19 The basal lamina separating epithelial cells and some other cells from connective tissue.** (a) Transmission electron micrograph of a thin section of cells (*top*) and underlying connective tissue (*bottom*). The electron-dense layer of the basal lamina can be seen to follow the undulation of the basal surface of the cells. (b) Electron micrograph of a quick-freeze deep-etch preparation of skeletal muscle showing the relation of the plasma membrane, basal lamina, and surrounding connective tissue. In this preparation, the basal lamina is revealed as a meshwork of filamentous proteins that associates with the plasma membrane and the thicker collagen fibers of the connective tissue. [Part (a) courtesy of P. FitzGerald. Part (b) from D. W. Fawcett, 1981, *The Cell,* 2d ed., Saunders/Photo Researchers; courtesy of John Heuser.]

migrating cells during development. It is therefore not surprising that basal lamina components have been highly conserved throughout evolution.

Most of the ECM components in the basal lamina are synthesized by the cells that rest on it. Four ubiquitous protein components are found in basal laminae (Figure 19-20):

- *Type IV collagen,* trimeric molecules with both rodlike and globular domains that form a two-dimensional network
- *Laminins,* a family of multiadhesive, cross-shaped proteins that form a fibrous two-dimensional network with type IV collagen and that also bind to integrins and other adhesion receptors
- *Perlecan,* a large multidomain proteoglycan that binds to and cross-links many ECM components and cell-surface molecules
- *Nidogen* (also called *entactin*), a rodlike molecule that cross-links type IV collagen, perlecan, and laminin and helps incorporate other components into the ECM

Other ECM molecules are incorporated into various basal laminae, depending on the tissue and particular functional requirements of the basal lamina.

As depicted in Figure 19-1, one side of the basal lamina is linked to cells by adhesion receptors, including $\alpha6\beta4$ integrin in hemidesmosomes, which binds to laminin in the basal lamina. The other side of the basal lamina is anchored to the adjacent connective tissue by a layer of fibers of collagen embedded in a proteoglycan-rich matrix. In stratified squamous epithelia (e.g., skin), this linkage is mediated by anchoring fibrils of type VII collagen. Together, the basal lamina and collagen-anchoring fibrils form the structure called the *basement membrane.*

Laminin, a Multiadhesive Matrix Protein, Helps Cross-link Components of the Basal Lamina

Laminin, the principal multiadhesive matrix protein in basal laminae, is a heterotrimeric, cross-shaped protein with a total molecular weight of 820,000 (Figure 19-21). At least 15 laminin isoforms, each containing slightly different polypeptide chains, have been identified. Globular *LG domains* at the C-terminus of the laminin α subunit mediate Ca^{2+}-dependent binding to specific carbohydrates on certain cell-surface molecules such as syndecan and dystroglycan, which will be described further in Section 19.4. LG domains are found in a wide variety of proteins and can mediate binding to steroids and proteins as well as carbohydrates. For example, LG domains in the α chain of laminin can mediate binding to certain integrins, including $\alpha6\beta4$ integrin in hemidesmosomes on the basal surfaces of epithelial cells. Laminin is the principal basal laminal ligand of $\alpha6\beta4$ and other integrins (Table 19-3).

Sheet-Forming Type IV Collagen Is a Major Structural Component of the Basal Lamina

Type IV collagen is a principal structural component of all basal laminae and can bind to certain integrin adhesion receptors. Collagen IV is one of more than 20 types of collagen that participate in the formation of distinct extracellular matrices in various tissues (Table 19-4). Although they differ in certain structural features and tissue distribution, all collagens are trimeric proteins made from three polypeptides, usually called collagen α chains. All three α chains can be identical (homotrimeric) or different (heterotrimeric). A trimeric collagen molecule contains one or more three-stranded segments, each with a similar triple-helical structure (Figure 19-22a). Each strand contributed by one of the α chains is twisted into a left-handed helix, and three such strands from the three α chains wrap around each other to form a right-handed triple helix.

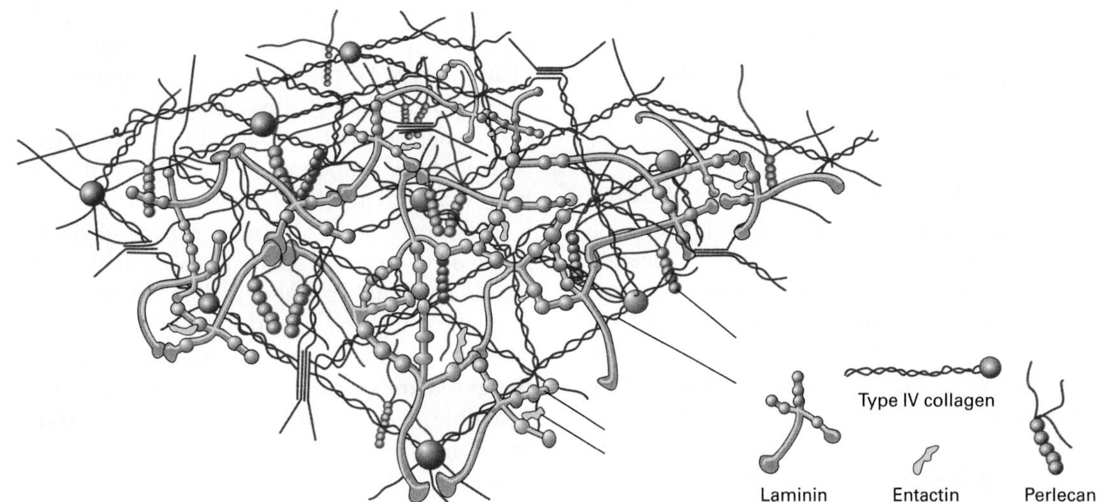

▲ **FIGURE 19-20 Major protein components of the basal lamina.** Type IV collagen and laminin each form two-dimensional networks, which are cross-linked by entactin and perlecan molecules. [Adapted from B. Alberts et al., 1994, *Molecular Biology of the Cell,* 3d ed., Garland, p. 991.]

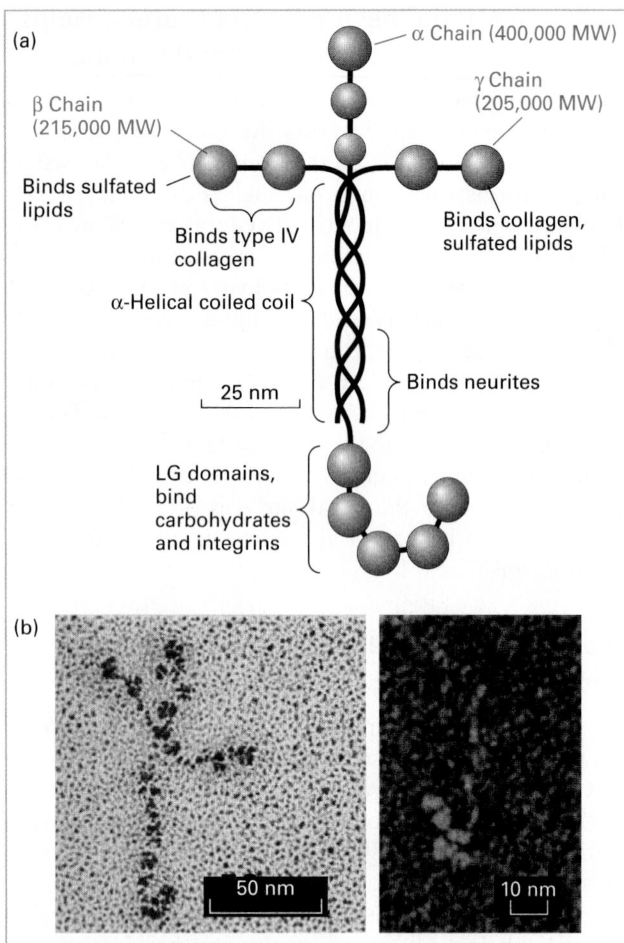

▲ **FIGURE 19-21 Laminin, a heterotrimeric multiadhesive matrix protein found in all basal laminae.** (a) Schematic model showing the general shape, location of globular domains, and coiled-coil region in which laminin's three chains are covalently linked by several disulfide bonds. Different regions of laminin bind to cell-surface receptors and various matrix components. (b) Electron micrographs of intact laminin molecule, showing its characteristic cross appearance (*left*) and the carbohydrate-binding LG domains near the C-terminus (*right*). [Part (a) adapted from G. R. Martin and R. Timpl, 1987, *Ann. Rev. Cell Biol.* **3:**57, and K. Yamada, 1991, *J. Biol. Chem.* **266:**12809. Part (b) from R. Timpl et al., 2000, *Matrix Biol.* **19:**309; photograph at right courtesy of Jürgen Engel.]

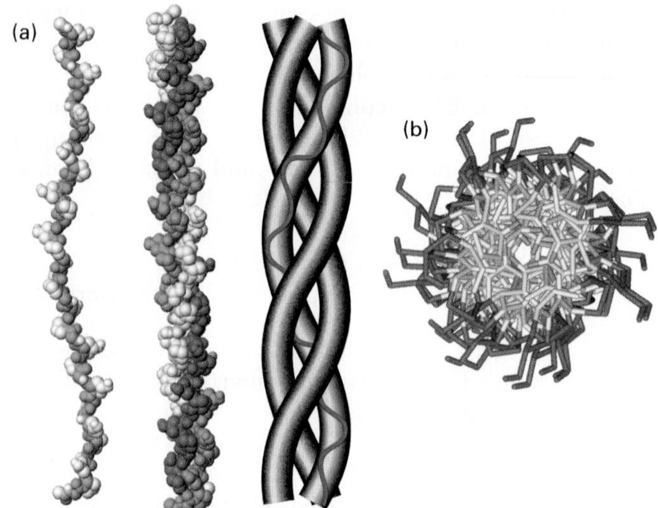

▲ **FIGURE 19-22 The collagen triple helix.** (a) (*Left*) Side view of the crystal structure of a polypeptide fragment whose sequence is based on repeating sets of three amino acids, Gly-X-Y, characteristic of collagen α chains. (*Center*) Each chain is twisted into a left-handed helix, and three chains wrap around each other to form a right-handed triple helix. The schematic model (*right*) clearly illustrates the triple helical nature of the structure. (b) View down the axis of the triple helix. The proton side chains of the glycine residues (orange) point into the very narrow space between the polypeptide chains in the center of the triple helix. In mutations in collagen in which other amino acids replace glycine, the proton in glycine is replaced by larger groups that disrupt the packing of the chains and destabilize the triple-helical structure. [Adapted from R. Z. Kramer et al., 2001, *J. Mol. Biol.* **311**(1):131.]

The collagen triple helix can form because of an unusual abundance of three amino acids: glycine, proline, and a modified form of proline called hydroxyproline (see Figure 2-15). They make up the characteristic repeating motif Gly-X-Y, where X and Y can be any amino acid but are often proline and hydroxyproline and less often lysine and hydroxylysine. Glycine is essential because its small side chain, a hydrogen atom, is the only one that can fit into the crowded center of the three-stranded helix (Figure 19-22b). Hydrogen bonds help hold the three chains together. Although the rigid peptidyl-proline and peptidyl-hydroxyproline linkages are not compatible with formation of a classic single-stranded α helix, they stabilize the distinctive three-stranded collagen helix. The hydroxyl group in hydroxyproline helps hold its ring in a conformation that stabilizes the three-stranded helix.

The unique properties of each type of collagen are due mainly to differences in (1) the number and lengths of the collagenous, triple-helical segments; (2) the segments that flank or interrupt the triple-helical segments and that fold into other kinds of three-dimensional structures; and (3) the covalent modification of the α chains (e.g., hydroxylation, glycosylation, oxidation, cross-linking). For example, the chains in type IV collagen, which is unique to basal laminae, are designated IVα chains. Mammals express six homologous IVα chains, which assemble into a series of type IV collagens with distinct properties. All subtypes of type IV collagen, however, form a 400-nm-long triple helix (Figure 19-23) that is interrupted about 24 times with nonhelical segments and flanked by large globular domains at the C-termini of the chains and smaller globular domains at the N-termini. The nonhelical regions introduce flexibility into the molecule. Through both lateral associations and interactions entailing the globular N- and C-termini, type IV collagen molecules assemble into a branching, irregular two-dimensional fibrous network that forms the lattice on which the basal lamina is built (Figure 19-23).

In the kidney, a double basal lamina, the glomerular basement membrane, separates the epithelium that lines the urinary space from the endothelium that lines the

TABLE 19-4 Selected Collagens

TYPE	MOLECULE COMPOSITION	STRUCTURAL FEATURES	REPRESENTATIVE TISSUES
FIBRILLAR COLLAGENS			
I	$[\alpha1(I)]_2[\alpha2(I)]$	300-nm-long fibrils	Skin, tendon, bone, ligaments, dentin, interstitial tissues
II	$[\alpha1(II)]_3$	300-nm-long fibrils	Cartilage, vitreous humor
III	$[\alpha1(III)]_3$	300-nm-long fibrils; often with type I	Skin, muscle, blood vessels
V	$[\alpha1(V)_2\ \alpha2(V)]$, $[\alpha1(V)_3]$	390-nm-long fibrils with globular N-terminal extension; often with type I	Cornea, teeth, bone, placenta, skin, smooth muscle
FIBRIL-ASSOCIATED COLLAGENS			
VI	$[\alpha1(VI)][\alpha2(VI)]$	Lateral association with type I; periodic globular domains	Most interstitial tissues
IX	$[\alpha1(IX)][\alpha2(IX)][\alpha3(IX)]$	Lateral association with type II; N-terminal globular domain; bound GAG	Cartilage, vitreous humor
SHEET-FORMING AND ANCHORING COLLAGENS			
IV	$[\alpha1(IV)]_2[\alpha2(IV)]$	Two-dimensional network	All basal laminae
VII	$[\alpha1(VII)]_3$	Long fibrils	Below basal lamina of the skin
XV	$[\alpha1(XV)]_3$	Core protein of chondroitin sulfate proteoglycan	Widespread; near basal lamina in muscle
TRANSMEMBRANE COLLAGENS			
XIII	$[\alpha1(XIII)]_3$	Integral membrane protein	Hemidesmosomes in skin
XVII	$[\alpha1(XVII)]_3$	Integral membrane protein	Hemidesmosomes in skin
HOST DEFENSE COLLAGENS			
Collectins		Oligomers of triple helix; lectin domains	Blood, alveolar space
C1q		Oligomers of triple helix	Blood (complement)
Class A scavenger receptors		Homotrimeric membrane proteins	Macrophages

SOURCES: K. Kuhn, 1987, in R. Mayne and R. Burgeson, eds., *Structure and Function of Collagen Types,* Academic Press, p. 2, and M. van der Rest and R. Garrone, 1991, *FASEB J.* 5:2814.

surrounding blood-filled capillaries. Defects in this structure, which is responsible for ultrafiltration of the blood and initial urine formation, can lead to renal failure. For instance, mutations that alter the C-terminal globular domain of certain IVα chains are associated with progressive renal failure as well as sensorineural hearing loss and ocular abnormalities, a condition known as *Alport's syndrome.* In *Goodpasture's syndrome,* a relatively rare autoimmune disease, self-attacking, or "auto," antibodies bind to the α3 chains of type IV collagen found in the glomerular basement membrane and lungs. This binding sets off an immune response that causes cellular damage resulting in progressive renal failure and pulmonary hemorrhage. ▮

Perlecan, a Proteoglycan, Cross-links Components of the Basal Lamina and Cell-Surface Receptors

Perlecan, the major secreted proteoglycan in basal laminae, consists of a large multidomain core protein (≈470 kDa) made up of repeats of five distinct domains, including laminin-like LG domains and Ig domains. The many globular repeats give it the appearance of a string of pearls when visualized by electron microscopy; hence the name perlecan. Perlecan is a glycoprotein (a protein with covalently attached sugars) that contains three types of covalent polysaccharide chains: *N*-linked (Chapter 14), *O*-linked (Section 19.4), and glycosaminoglycans (GAGs), which are long polymers of repeating disaccharides (see below). Glycoproteins containing covalently attached GAG chains are called proteoglycans. The protein constituent of a proteoglycan is usually called the *core* protein, onto which the GAG chains are attached. Both the protein and the GAG components of perlecan contribute to its ability to incorporate into and define the structure and function of basal laminae. Because of its multiple domains and polysaccharide chains that have distinctive binding properties, perlecan binds to dozens of other molecules, including other ECM components (e.g., laminin, nidogen), cell-surface receptors, and polypeptide growth factors. Simultaneous binding to these molecules results in perlecan-mediated cross-linking. Perlecan can be found in basal laminae and non-basal-laminae ECM. The adhesion receptor dystroglycan can bind perlecan directly, via its LG domains, and indirectly, via its binding to laminin. In humans, mutations in the perlecan gene can lead either to dwarfism or to muscle abnormalities, apparently due to dysfunction of the neuromuscular junction that controls muscle firing.

KEY CONCEPTS OF SECTION 19.3

The Extracellular Matrix I: The Basal Lamina

- The basal lamina, a thin meshwork of extracellular matrix (ECM) molecules, separates most epithelia and other organized groups of cells from adjacent connective tissue. Together, the basal lamina and the immediately adjacent collagen network form a structure called the basement membrane.

- Four ECM proteins are found in all basal laminae (see Figure 19-20): laminin (a multiadhesive matrix protein), type IV collagen, perlecan (a proteoglycan), and entactin/nidogen.

- Cell-surface adhesion receptors (e.g., α6β4 integrin in hemidesmosomes) anchor cells to the basal lamina, which in turn is connected to other ECM components (see Figure 19-1). Laminin in the basal lamina is the principal ligand of α6β4 integrin.

- Laminin and other multiadhesive matrix proteins are multidomain molecules that bind multiple adhesion receptors and ECM components.

- The large, flexible molecules of type IV collagen interact end to end and laterally to form a meshlike scaffold to which other ECM components and adhesion receptors can bind (see Figure 19-23).

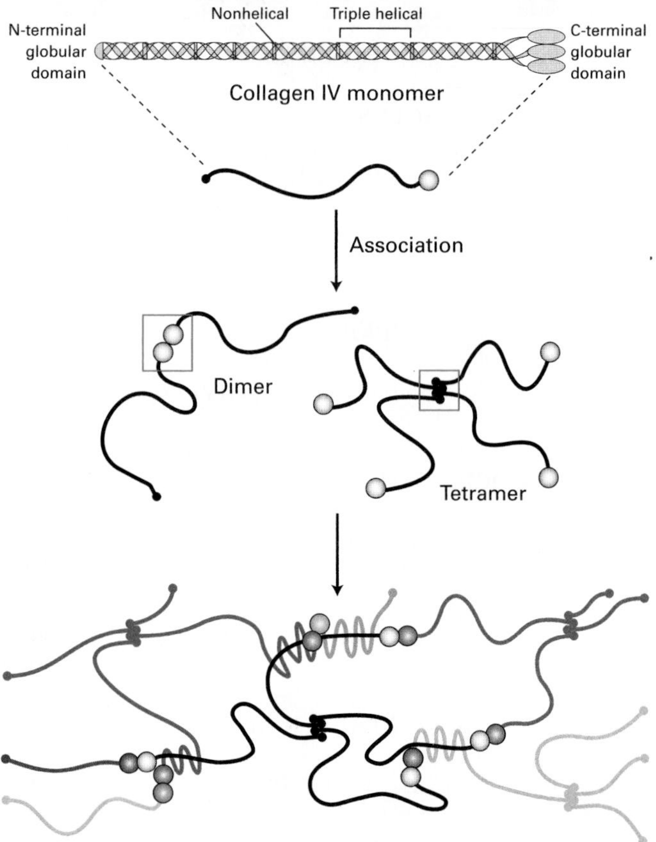

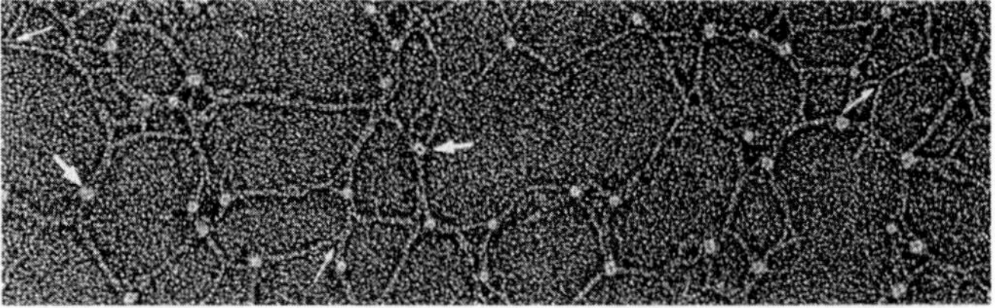

▲ **FIGURE 19-23 Structure and assembly of type IV collagen.** (a) Schematic representation of type IV collagen. This 400-nm-long molecule has a small noncollagenous globular domain at the N-terminus and a large globular domain at the C-terminus. The triple helix is interrupted by nonhelical segments that introduce flexible kinks in the molecule. Lateral interactions between triple helical segments, as well as head-to-head and tail-to-tail interactions between the globular domains, form dimers, tetramers, and higher-order complexes, yielding a sheetlike network. (b) Electron micrograph of type IV collagen network formed in vitro. The lacy appearance results from the flexibility of the molecule, the side-to-side binding between triple-helical segments (thin arrow), and the interactions between C-terminal globular domains (thick arrow). [Part (a) adapted from A. Boutaud, 2000, *J. Biol. Chem.* **275:**30716. Part (b) courtesy of P. Yurchenco; see P. Yurchenco and G. C. Ruben, 1987, *J. Cell Biol.* **105:**2559.]

- Type IV collagen is a member of the collagen family of proteins distinguished by the presence of repeating tripeptide sequences of Gly-X-Y that give rise to the collagen triple-helical structure (see Figure 19-22). Different collagens are distinguished by the length and chemical modifications of their α chains and by the presence of segments that interrupt or flank their triple-helical regions.

- Perlecan, a large secreted proteoglycan present primarily in the basal lamina, binds many ECM components and adhesion receptors. Proteoglycans consist of membrane-associated or secreted core proteins covalently linked to one or more specialized polysaccharide chains called glycosaminoglycans (GAGs).

19.4 The Extracellular Matrix II: Connective and Other Tissues

Connective tissue, such as tendon and cartilage, differs from other solid tissues in that most of its volume is made up of extracellular matrix rather than cells. This matrix is packed with insoluble protein fibers and contains proteoglycans, various multiadhesive proteins, and a specialized glycosaminoglycan called hyaluronan. The most abundant fibrous protein in connective tissue is collagen. Rubberlike elastin fibers, which can be stretched and relaxed, also are present in deformable sites (e.g., skin, tendons, heart). The fibronectins, a family of multiadhesive matrix proteins, form their own distinct fibrils in the matrix of most connective tissues. Although several types of cells are found in connective tissues, the various ECM components are produced largely by cells called **fibroblasts.**

Fibrillar Collagens Are the Major Fibrous Proteins in the ECM of Connective Tissues

About 80–90 percent of the collagen in the body consists of *fibrillar collagens* (types I, II, and III), located primarily in connective tissues (see Table 19-4). Because of its abundance in tendon-rich tissue such as rat tail, type I collagen is easy to isolate and was the first collagen to be characterized. Its fundamental structural unit is a long (300-nm), thin (1.5-nm-diameter) triple helix (see Figure 19-22) consisting of two α1(I) chains and one α2(I) chain, each 1050 amino acids in length. The triple-stranded molecules associate into higher-order polymers called collagen *fibrils,* which in turn often aggregate into larger bundles called collagen *fibers* (Figure 19-24).

Quantitatively minor classes of collagen include *fibril-associated collagens,* which link the fibrillar collagens to one another or to other ECM components; *sheet-forming and anchoring collagens,* which form two-dimensional networks in basal laminae (type IV) and connect the basal lamina in skin to the underlying connective tissue (type VII); *transmembrane collagens,* which function as adhesion receptors (e.g., BP180 in hemidesmosomes); and *host defense*

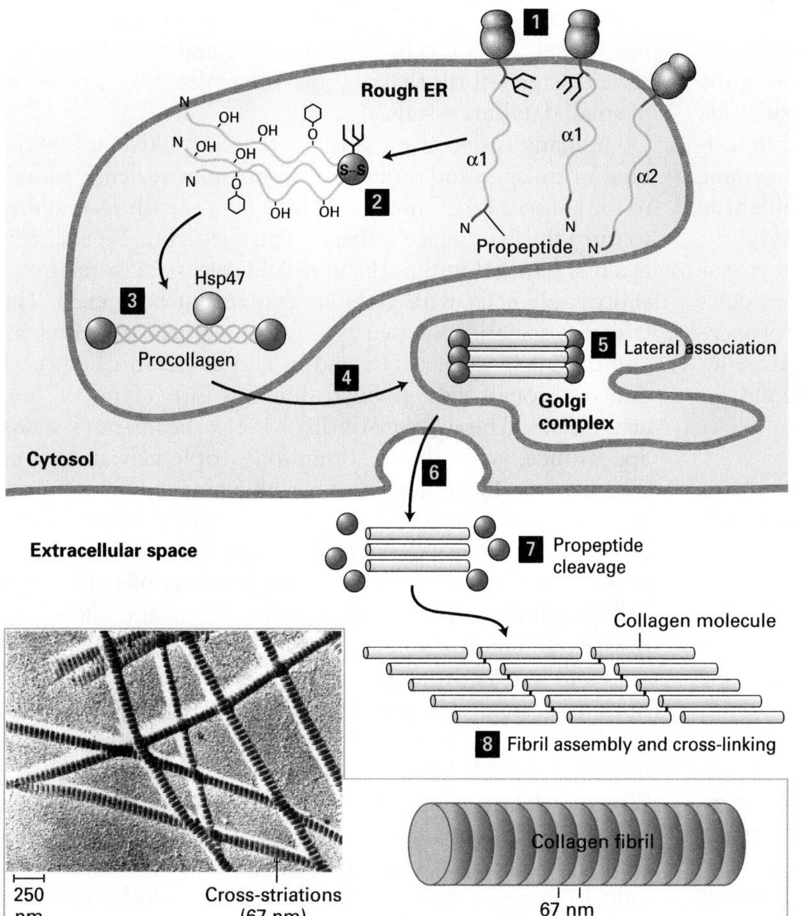

◄ **FIGURE 19-24 Biosynthesis of fibrillar collagens.** Step **1**: Procollagen α chains are synthesized on ribosomes associated with the endoplasmic reticulum (ER) membrane, and asparagine-linked oligosaccharides are added to the C-terminal propeptide. Step **2**: Propeptides associate to form trimers and are covalently linked by disulfide bonds, and selected residues in the Gly-X-Y triplet repeats are covalently modified (certain prolines and lysines are hydroxylated, galactose [Gal] or galactose-glucose [hexagons] is attached to some hydroxylysines, prolines are cis → trans isomerized). Step **3**: The modifications facilitate zipperlike formation, stabilization of triple helices, and binding by the chaperone protein Hsp47 (Chapter 13), which may stabilize the helices or prevent premature aggregation of the trimers or both. Steps **4** and **5**: The folded procollagens are transported to and through the Golgi apparatus, where some lateral association into small bundles takes place. The chains are then secreted (step **6**), the N- and C-terminal propeptides are removed (step **7**), and the trimers assemble into fibrils and are covalently cross-linked (step **8**). The 67-nm staggering of the trimers gives the fibrils a striated appearance in electron micrographs (*inset*). [Adapted from A. V. Persikov and B. Brodsky, 2002, *Proc. Nat'l. Acad. Sci. USA* **99**(3):1101–1103.]

collagens, which help the body recognize and eliminate pathogens. Interestingly, several collagens (e.g., types XVIII and XV) are also proteoglycans with covalently attached GAGs (see Table 19-4).

Fibrillar Collagen Is Secreted and Assembled into Fibrils Outside of the Cell

Fibrillar collagens are secreted proteins, produced primarily by fibroblasts in the ECM. Collagen biosynthesis and secretion follow the normal pathway for a secreted protein, described in detail in Chapters 13 and 14. The collagen α chains are synthesized as longer precursors, called pro-α chains, by ribosomes attached to the endoplasmic reticulum (ER). The pro-α chains undergo a series of covalent modifications and fold into triple-helical *procollagen* molecules before their release from cells (see Figure 19-24).

After the secretion of procollagen from the cell, extracellular peptidases remove the N-terminal and C-terminal propeptides. In fibrillar collagens, the resulting molecules, which consist almost entirely of a triple-stranded helix, associate laterally to generate fibrils with a diameter of 50–200 nm. In fibrils, adjacent collagen molecules are displaced from one another by 67 nm, about one-quarter of their length. This staggered array produces a striated effect that can be seen in both light and electron microscopic images of collagen fibrils (Figure 19-24, *inset*). The unique properties of the fibrous collagens (e.g., types I, II, III) are mainly due to the formation of fibrils.

Short non-triple-helical segments at either end of the fibrillar collagen α chains are of particular importance in the formation of collagen fibrils. Lysine and hydroxylysine side chains in these segments are covalently modified by extracellular lysyl oxidases to form aldehydes in place of the amine group at the end of the side chain. These reactive aldehyde groups form covalent cross-links with lysine, hydroxylysine, and histidine residues in adjacent molecules. These cross-links stabilize the side-by-side packing of collagen molecules and generate a very strong fibril. The removal of the propeptides and covalent cross-linking take place in the extracellular space to prevent the potentially catastrophic assembly of fibrils within the cell.

The post-translational modifications of pro-α chains are crucial for the formation of mature collagen molecules and their assembly into fibrils. Defects in these modifications have serious consequences, as ancient mariners frequently experienced. For example, ascorbic acid (vitamin C) is an essential cofactor for the hydroxylases responsible for adding hydroxyl groups to proline and lysine residues in pro-α chains. In cells deprived of ascorbate, as in the disease *scurvy,* the pro-α chains are not hydroxylated sufficiently to form stable triple-helical procollagen at normal body temperature, and the procollagen that forms cannot assemble into normal fibrils. Without the structural support of collagen, blood vessels, tendons, and skin become fragile. Fresh fruit in the diet can supply sufficient vitamin C to support the formation of normal collagen. Historically, British sailors were provided with limes to prevent scurvy, leading to their being called "limeys." Mutations in lysyl hydroxylase genes also can cause connective-tissue defects. ▮

Type I and II Collagens Associate with Nonfibrillar Collagens to Form Diverse Structures

Collagens differ in the structures of the fibers they form and how these fibers are organized into networks. Of the predominant types of collagen found in connective tissues, type I collagen forms long fibers, whereas networks of type II collagen are more meshlike. In tendons, for instance, the long type I collagen fibers connect muscles to bones and must withstand enormous forces. Because type I collagen fibers have great tensile strength, tendons can be stretched without being broken. Indeed, gram for gram, type I collagen is stronger than steel. Two quantitatively minor fibrillar collagens, type V and type XI, coassemble into fibers with type I collagen, thereby regulating the structures and properties of the fibers. Incorporation of type V collagen, for example, results in smaller-diameter fibers.

Type I collagen fibrils are also used as the reinforcing rods in the construction of bone. Bones and teeth are hard and strong because they contain large amounts of dahllite, a crystalline calcium- and phosphate-containing mineral. Most bones are about 70 percent mineral and 30 percent protein, the vast majority of which is type I collagen. Bones form when certain cells (chondrocytes and osteoblasts) secrete collagen fibrils that are then mineralized by deposition of small dahllite crystals.

In many connective tissues, particularly skeletal muscle, type VI collagen and proteoglycans are noncovalently bound to the sides of type I fibrils and may bind the fibrils together to form thicker collagen fibers (Figure 19-25a). Type VI collagen is unusual in that the molecule consists of a relatively short triple helix with globular domains at both ends. The lateral association of two type VI monomers generates an "antiparallel" dimer. The end-to-end association of these dimers through their globular domains forms type VI "microfibrils." These microfibrils have a beads-on-a-string appearance, with about 60-nm-long triple-helical regions separated by 40-nm-long globular domains.

The fibrils of type II collagen, the major collagen in cartilage, are smaller in diameter than type I fibrils and are oriented randomly in a viscous proteoglycan matrix. The rigid collagen fibrils impart strength to the matrix and allow it to resist large deformations in shape. Type II fibrils are cross-linked to matrix proteoglycans by type IX collagen, another fibril-associated collagen. Type IX collagen and several related types have two or three triple-helical segments connected by flexible kinks and an N-terminal globular segment (Figure 19-25b). The globular N-terminal segment of type IX collagen extends from the fibrils at the end of one of its helical segments, as does a GAG chain that is sometimes linked to one of the type IX chains. These protruding nonhelical structures are thought to anchor the type II fibril to

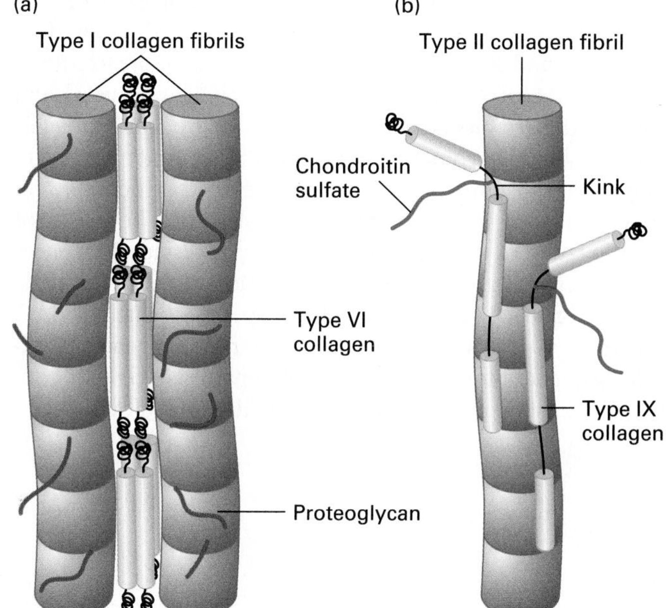

▲ **FIGURE 19-25 Interactions of fibrous collagens with nonfibrous fibril-associated collagens.** (a) In tendons, type I fibrils are all oriented in the direction of the stress applied to the tendon. Proteoglycans and type VI collagen bind noncovalently to fibrils, coating the surface. The microfibrils of type VI collagen, which contain globular and triple-helical segments, bind to type I fibrils and link them together into thicker fibers. (b) In cartilage, type IX collagen molecules are covalently bound at regular intervals along type II fibrils. A chondroitin sulfate chain, covalently linked to the α2(IX) chain at the flexible kink, projects outward from the fibril, as does the globular N-terminal region. [Part (a), see R. R. Bruns et al., 1986, *J. Cell Biol.* **103**:393. Part (b), see L. M. Shaw and B. Olson, 1991, *Trends Biochem. Sci.* **18**:191.]

proteoglycans and other components of the matrix. The interrupted triple-helical structure of type IX and related collagens prevents them from assembling into fibrils, although they can associate with fibrils formed from other collagen types and form covalent cross-links to them.

Mutations affecting type I collagen and its associated proteins cause a variety of human diseases. Certain mutations in the genes encoding the type I collagen α1(I) or α2(I) chains lead to *osteogenesis imperfecta*, or brittle-bone disease. Because every third position in a collagen α chain must be a glycine for the triple helix to form (see Figure 19-22), mutations of glycine to almost any other amino acid are deleterious, resulting in poorly formed and unstable helices. Only one defective α chain of the three in a collagen molecule can disrupt the whole molecule's triple-helical structure and function. A mutation in a single copy (allele) of either the α1(I) gene or the α2(I) gene, which are located on nonsex chromosomes (autosomes), can cause this disorder. Thus it normally shows autosomal dominant inheritance.

Absence or malfunctioning of collagen-fibril-associated microfibrils in muscle tissue due to mutations in the type VI collagen genes cause dominant or recessive congenital muscular dystrophies with generalized muscle weakness, respiratory insufficiency, muscle wasting, and muscle-related joint abnormalities. Skin abnormalities have also been reported with type VI collagen disease. ▮

Proteoglycans and Their Constituent GAGs Play Diverse Roles in the ECM

As we saw with perlecan in the basal lamina, proteoglycans play an important role in cell-ECM adhesion. Proteoglycans are a subset of secreted or cell-surface-attached glycoproteins containing covalently linked specialized polysaccharide chains called **glycosaminoglycans (GAGs)**. GAGs are long linear polymers of specific repeating disaccharides. Usually one sugar is either a uronic acid (D-glucuronic acid or L-iduronic acid) or D-galactose; the other sugar is *N*-acetylglucosamine or *N*-acetylgalactosamine (Figure 19-26). One or both of the sugars contain at least one anionic group (carboxylate or sulfate). Thus each GAG chain bears many negative charges. GAGs are classified into several major types based on the nature of the repeating disaccharide unit: heparan sulfate, chondroitin sulfate, dermatan sulfate, keratan sulfate, and hyaluronan. A hypersulfated form of heparan sulfate called *heparin*, produced mostly by mast cells, plays a key role in allergic reactions. It is also used medically as an anticlotting drug because of its ability to activate a natural clotting inhibitor called antithrombin III.

With the exception of hyaluronan, which is discussed below, all the major GAGs occur naturally as components of proteoglycans. Like other secreted and transmembrane glycoproteins, proteoglycan core proteins are synthesized on the endoplasmic reticulum (Chapter 13). The GAG chains are assembled on these cores in the Golgi complex. To generate heparan or chondroitin sulfate chains, a three-sugar "linker" is first attached to the hydroxyl side chains of certain serine residues in a core protein; thus the linker is an **O-linked oligosaccharide** (Figure 19-27a). In contrast, the linkers for the addition of keratan sulfate chains are oligosaccharide chains attached to asparagine residues; such **N-linked oligosaccharides** are present in many glycoproteins (Chapter 14), although only a subset carry GAG chains. All GAG chains are elongated by the alternating addition of sugar monomers to form the disaccharide repeats characteristic of a particular GAG; the chains are often modified subsequently by the covalent linkage of small molecules such as sulfate. The mechanisms responsible for determining which proteins are modified with GAGs, the sequence of disaccharides to be added, the sites to be sulfated, and the lengths of the GAG chains are unknown. The ratio of polysaccharide to protein in all proteoglycans is much higher than that in most other glycoproteins.

Function of GAG Chain Modifications As is the case with the sequence of amino acids in proteins, the arrangement of the sugar residues in GAG chains and the modification of specific sugars (e.g., addition of sulfate) in the chains can determine their function and that of the proteoglycans containing them. For example, groupings of certain modified sugars in the GAG chains of heparin sulfate proteoglycans can control the binding of growth factors to certain receptors or the activities of proteins in the blood-clotting cascade.

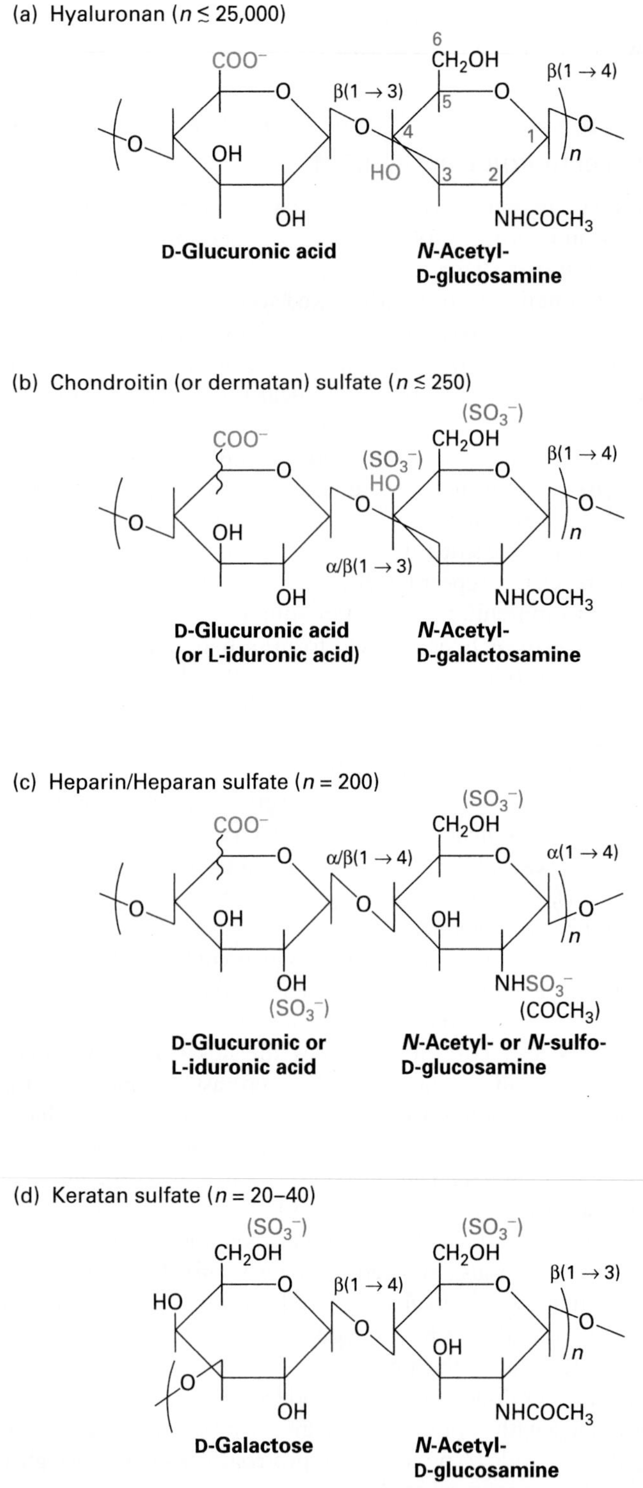

(a) Hyaluronan ($n \leq 25{,}000$)

D-Glucuronic acid — $\beta(1 \rightarrow 3)$ — N-Acetyl-D-glucosamine — $\beta(1 \rightarrow 4)$

(b) Chondroitin (or dermatan) sulfate ($n \leq 250$)

D-Glucuronic acid (or L-iduronic acid) — $\alpha/\beta(1 \rightarrow 3)$ — N-Acetyl-D-galactosamine — $\beta(1 \rightarrow 4)$

(c) Heparin/Heparan sulfate ($n = 200$)

D-Glucuronic or L-iduronic acid — $\alpha/\beta(1 \rightarrow 4)$ — N-Acetyl- or N-sulfo-D-glucosamine — $\alpha(1 \rightarrow 4)$

(d) Keratan sulfate ($n = 20–40$)

D-Galactose — $\beta(1 \rightarrow 4)$ — N-Acetyl-D-glucosamine — $\beta(1 \rightarrow 3)$

▲ **FIGURE 19-26 The repeating disaccharides of glycosaminoglycans (GAGs), the polysaccharide components of proteoglycans.** Each of the four classes of GAGs is formed by polymerization of monomer units into repeats of a particular disaccharide and subsequent modifications, including addition of sulfate groups and inversion (epimerization) of the carboxyl group on carbon 5 of D-glucuronic acid to yield L-iduronic acid. Heparin is generated by hypersulfation of heparan sulfate, whereas hyaluronan is unsulfated. The number (n) of disaccharides typically found in each glycosaminoglycan chain is given. The squiggly lines represent covalent bonds that are oriented either above (D-glucuronic acid) or below (L-iduronic acid) the ring.

For years, the chemical and structural complexity of proteoglycans posed a daunting barrier to an analysis of their structures and an understanding of their many diverse functions. In recent years, investigators employing classical and state-of-the-art biochemical techniques (e.g., capillary high-pressure liquid chromatography), mass spectrometry, and genetics have begun to elucidate the detailed structures and functions of these ubiquitous ECM molecules. The results of ongoing studies suggest that sets of sugar-residue sequences containing some modifications in common, rather than single unique sequences, are responsible for specifying distinct GAG functions. A case in point is a set of five-residue (pentasaccharide) sequences found in a subset of heparin GAGs that control the activity of antithrombin III (ATIII), an inhibitor of the

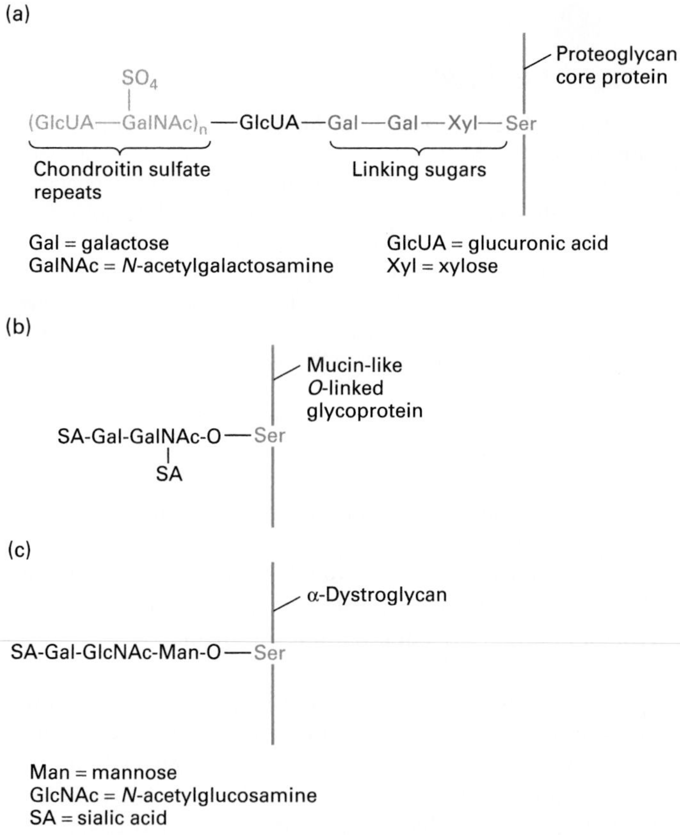

(a)

$(GlcUA—GalNAc)_n$—GlcUA—Gal—Gal—Xyl—Ser (Proteoglycan core protein); SO_4 on GalNAc

Chondroitin sulfate repeats; Linking sugars

Gal = galactose
GalNAc = N-acetylgalactosamine
GlcUA = glucuronic acid
Xyl = xylose

(b)

SA-Gal-GalNAc-O—Ser (Mucin-like O-linked glycoprotein), with SA on GalNAc

(c)

SA-Gal-GlcNAc-Man-O—Ser (α-Dystroglycan)

Man = mannose
GlcNAc = N-acetylglucosamine
SA = sialic acid

▲ **FIGURE 19-27 Hydroxyl (O-) linked polysaccharides.** (a) Synthesis of a glycosaminoglycan (GAG), in this case chondroitin sulfate, is initiated by transfer of a xylose residue to a serine residue in the core protein, most likely in the Golgi complex, followed by sequential addition of two galactose residues. Glucuronic acid and N-acetylgalactosamine residues are then added sequentially to these linking sugars, forming the chondroitin sulfate chain. Heparan sulfate chains are connected to core proteins by the same three-sugar linker. (b) Mucin-type O-linked chains are covalently bound to glycoproteins via an N-acetylgalactosamine (GalNAc) monosaccharide to which are covalently attached a variety of other sugars. (c) Certain specialized O-linked oligosaccharides, such as those found in the protein dystroglycan, are bound to proteins via mannose (Man) monosaccharides.

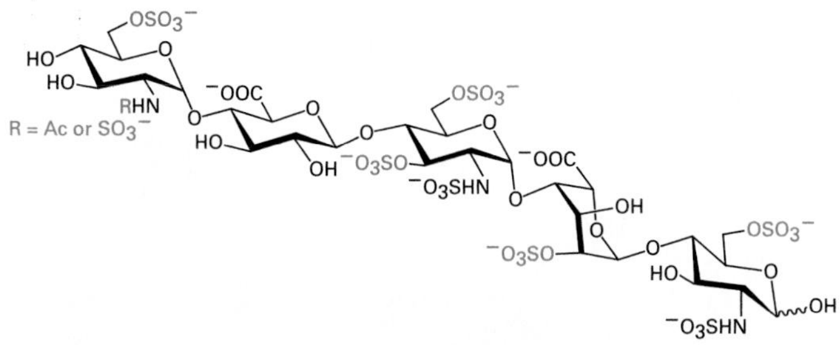

▲ **FIGURE 19-28 Pentasaccharide GAG sequence that regulates the activity of antithrombin III (ATIII).** Sets of modified five-residue sequences in the much longer GAG called heparin with the composition shown here bind to ATIII and activate it, thereby inhibiting blood clotting. The sulfate groups in red type are essential for this heparin function; the modifications in blue type may be present but are not essential. Other sets of modified GAG sequences are thought to regulate the activity of other target proteins.

key blood-clotting protease thrombin. When these pentasaccharide sequences in heparin are sulfated at two specific positions, heparin can activate ATIII, thereby inhibiting clot formation (Figure 19-28). Several other sulfates can be present in the active pentasaccharide in various combinations, but they are not essential for the anticlotting activity of heparin. The rationale for generating sets of similar active sequences rather than a single unique sequence and the mechanisms that control GAG biosynthetic pathways, permitting the generation of such active sequences, are not well understood.

Diversity of Proteoglycans The proteoglycans constitute a remarkably diverse group of molecules that are abundant in the extracellular matrix of all animal tissues and are also expressed on the cell surface. For example, of the five major classes of heparan sulfate proteoglycans, three are located in the extracellular matrix (perlecan, agrin, and type XVIII collagen) and two are cell-surface proteins. The latter include integral membrane proteins (syndecans) and GPI-anchored proteins (glypicans); the GAG chains in both types of cell-surface proteoglycans extend into the extracellular space. The sequences and lengths of proteoglycan core proteins vary considerably, and the number of attached GAG chains ranges from just a few to more than 100. Moreover, a core protein is often linked to two different types of GAG chains, generating a "hybrid" proteoglycan. The basal laminal proteoglycan perlecan is primarily a heparan sulfate proteoglycan (HSPG) with three to four GAG chains, although it sometimes can have a bound chondroitin sulfate chain. Additional diversity in proteoglycans occurs because the numbers of chains, compositions, and sequences of the GAGs attached to otherwise identical core proteins can differ considerably. Laboratory generation and analysis of mutants with defects in proteoglycan production in *Drosophila melanogaster* (fruit fly), *C. elegans* (roundworm), and mice have clearly shown that proteoglycans play critical roles in development, most likely as modulators of various signaling pathways.

Syndecans are cell-surface proteoglycans expressed by epithelial and nonepithelial cells that bind to collagens and multiadhesive matrix proteins (e.g., fibronectin), anchoring cells to the extracellular matrix. Like that of many integral membrane proteins, the cytosolic domain of syndecan interacts with the actin cytoskeleton and in some cases with intracellular regulatory molecules. In addition, cell-surface proteoglycans like syndecan bind many protein growth factors and other external signaling molecules, thereby helping to regulate cellular metabolism and function. For instance, syndecans in the hypothalamic region of the brain modulate feeding behavior in response to food deprivation. They do so by participating in the binding to cell-surface receptors of antisatiety peptides that help control feeding behavior. In the fed state, the syndecan extracellular domain decorated with heparan sulfate chains is released from the surface by proteolysis, thus suppressing the activity of the antisatiety peptides and feeding behavior. In mice engineered to overexpress the syndecan-1 gene in the hypothalamic region of the brain and other tissues, normal control of feeding by antisatiety peptides is disrupted and the animals overeat and become obese.

Hyaluronan Resists Compression, Facilitates Cell Migration, and Gives Cartilage Its Gel-like Properties

Hyaluronan, also called hyaluronic acid (HA) or hyaluronate, is a nonsulfated GAG (see Figure 19-26a) made by a plasma-membrane-bound enzyme (HA synthase) that is directly secreted into the extracellular space. (A similar approach is used by plant cells to make their ECM component cellulose.) HA is a major component of the extracellular matrix that surrounds migrating and proliferating cells, particularly in embryonic tissues. In addition, hyaluronan forms the backbone of complex proteoglycan aggregates found in many extracellular matrices, particularly cartilage. Because of its remarkable physical properties, hyaluronan imparts stiffness and resilience as well as a lubricating quality to many types of connective tissue such as joints.

Hyaluronan molecules range in length from a few disaccharide repeats to ≈25,000. The typical hyaluronan in joints such as the elbow has 10,000 repeats for a total mass of 4×10^6 Da and length of 10 μm (about the diameter of a small cell). Individual segments of a hyaluronan molecule fold into a rodlike conformation because of the β glycosidic linkages between the sugars and extensive intrachain hydrogen bonding. Mutual repulsion between negatively charged carboxylate groups that protrude outward at regular intervals also contributes to these local rigid structures. Overall, however, hyaluronan is not a long, rigid rod as is fibrillar collagen; rather, in solution it is very flexible, bending and twisting into many conformations, forming a random coil.

Because of the large number of anionic residues on its surface, the typical hyaluronan molecule binds a large amount of water and behaves as if it were a large hydrated sphere with a diameter of ≈500 nm. As the concentration of hyaluronan increases, the long chains begin to entangle, forming a viscous gel. Even at low concentrations, hyaluronan forms a hydrated gel; when placed in a confining space, such as in a matrix between two cells, the long hyaluronan molecules will tend to push outward. This outward pushing creates a swelling, or *turgor pressure,* within the extracellular space. In addition, the binding of cations by COO^- groups on the surface of hyaluronan increases the concentration of ions and thus the osmotic pressure in the gel. As a result, large amounts of water are taken up into the matrix, contributing to the turgor pressure. These swelling forces give connective tissues their ability to resist compression forces, in contrast with collagen fibers, which are able to resist stretching forces.

Hyaluronan is bound to the surface of many migrating cells by a number of adhesion receptors (e.g., one called CD44) containing HA-binding domains, each with a similar three-dimensional conformation. Because of its loose, hydrated, porous nature, the hyaluronan "coat" bound to cells appears to keep cells apart from one another, giving them the freedom to move about and proliferate. The cessation of cell movement and the initiation of cell-cell attachments are frequently correlated with a decrease in hyaluronan, a decrease in HA-binding cell-surface molecules, and an increase in the extracellular enzyme hyaluronidase, which degrades hyaluronan in the matrix. These functions of hyaluronan are particularly important during the many cell migrations that facilitate differentiation and in the release of a mammalian egg cell (oocyte) from its surrounding cells after ovulation.

The predominant proteoglycan in cartilage, called *aggrecan,* assembles with hyaluronan into very large aggregates, illustrative of the complex structures that proteoglycans sometimes form. The backbone of the cartilage proteoglycan aggregate is a long molecule of hyaluronan to which multiple aggrecan molecules are bound tightly but noncovalently (Figure 19-29a). A single aggrecan aggregate, one of the largest macromolecular complexes known, can be more than 4 mm long and have a volume larger than that of a bacterial cell. These aggregates give cartilage its unique gel-like properties and its resistance to deformation, essential for distributing the load in weight-bearing joints.

The aggrecan core protein (≈250,000 MW) has one N-terminal globular domain that binds with high affinity to a specific decasaccharide sequence within hyaluronan. This specific sequence is generated by covalent modification of some of the repeating disaccharides in the hyaluronan chain. The interaction between aggrecan and hyaluronan is facilitated by a link protein that binds to both the aggrecan core protein and hyaluronan (Figure 19-29b). Aggrecan and the link protein have in common a "link" domain, ≈100 amino acids long, that is found in numerous matrix and cell-surface hyaluronan-binding proteins in both cartilaginous and noncartilaginous tissues. Almost certainly these proteins arose in the course of evolution from a single ancestral gene that encoded just this domain.

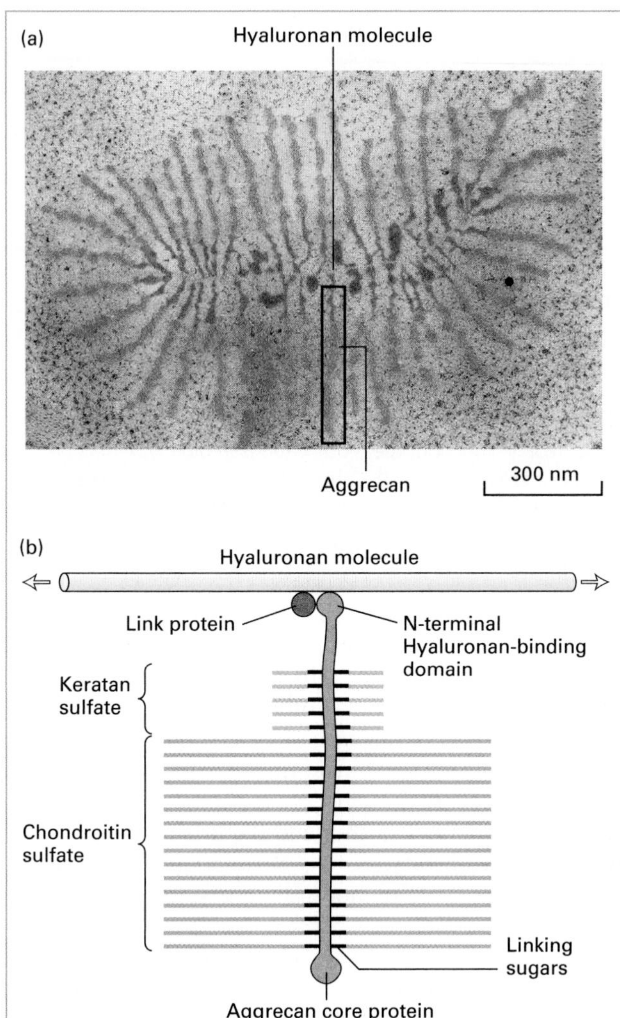

▲ **FIGURE 19-29 Structure of proteoglycan aggregate from cartilage.** (a) Electron micrograph of an aggrecan aggregate from fetal bovine epiphyseal cartilage. Aggrecan core proteins are bound at ≈40-nm intervals to a molecule of hyaluronan. (b) Schematic representation of an aggrecan monomer bound to hyaluronan. In aggrecan, both keratan sulfate and chondroitin sulfate chains are attached to the core protein. The N-terminal domain of the core protein binds noncovalently to a hyaluronan molecule. Binding is facilitated by a link protein, which binds to both the hyaluronan molecule and the aggrecan core protein. Each aggrecan core protein has 127 Ser-Gly sequences at which GAG chains can be added. The molecular weight of an aggrecan monomer averages 2×10^6. The entire aggregate, which may contain upward of 100 aggrecan monomers, has a molecular weight in excess of 2×10^8 and is about as large as the bacterium *E. coli.* [Part (a) from J. A. Buckwalter and L. Rosenberg, 1983, *Coll. Rel. Res.* **3:**489; courtesy of L. Rosenberg.]

Fibronectins Interconnect Cells and Matrix, Influencing Cell Shape, Differentiation, and Movement

Many different cell types synthesize **fibronectin,** an abundant multiadhesive matrix protein found in all vertebrates. The discovery that fibronectin functions as an adhesive molecule stemmed from observations that it is present on the surfaces of normal fibroblastic cells, which adhere tightly to petri dishes in laboratory experiments, but is absent from the surfaces of tumorigenic (i.e., cancerous) cells, which adhere weakly. The 20 or so isoforms of fibronectin are generated by alternative splicing of the RNA transcript produced from a single gene (see Figure 4-16). Fibronectins are essential for the migration and differentiation of many cell types in embryogenesis. These proteins are also important for wound healing because they promote blood clotting and facilitate the migration of macrophages and other immune cells into the affected area.

Fibronectins help attach cells to the extracellular matrix by binding to other ECM components, particularly fibrous collagens and heparan sulfate proteoglycans, and to cell-surface adhesion receptors such as integrins (see Figure 19-2). Through their interactions with adhesion receptors (e.g., α5β1 integrin), fibronectins influence the shape and movement of cells and the organization of the cytoskeleton. Conversely, by regulating their receptor-mediated attachments to fibronectin and other ECM components, cells can sculpt the immediate ECM environment to suit their needs.

Fibronectins are dimers of two similar polypeptides linked at their C-termini by two disulfide bonds; each chain is about 60–70 nm long and 2–3 nm thick. Partial digestion of fibronectin with low amounts of proteases and analysis of the fragments showed that each chain comprises several functional regions with different ligand-binding specificities (Figure 19-30a). Each region, in turn, contains multiple copies of certain sequences that can be classified into one of three types. These classifications are designated fibronectin type I, II, and III repeats, on the basis of similarities in amino acid sequence, although the sequences of any two repeats of a given type are not identical. These linked repeats give the molecule the appearance of beads on a string. The combination of different repeats composing the regions confers on fibronectin its ability to bind multiple ligands.

One of the type III repeats in the cell-binding region of fibronectin mediates binding to certain integrins. The results of studies with synthetic peptides corresponding to parts of this repeat identified the tripeptide sequence Arg-Gly-Asp, usually called the *RGD sequence,* as the minimal sequence within this repeat required for recognition by those integrins. In one study, heptapeptides containing the RGD sequence or a variation of this sequence were tested for their ability to mediate the adhesion of rat kidney cells to a culture dish. The results showed that heptapeptides containing the RGD sequence mimicked intact fibronectin's ability to stimulate integrin-mediated adhesion, whereas variant heptapeptides lacking this sequence were ineffective (Figure 19-31).

A three-dimensional model of fibronectin binding to integrin based on structures of parts of both fibronectin and integrin has been assembled. In a high-resolution structure of the integrin-binding fibronectin type III repeat and its neighboring type III domain, the RGD sequence is at the apex of a loop that protrudes outward from the molecule, in a position facilitating binding to integrins (see Figure 19-30b). Although the RGD sequence is required for binding to several integrins, its affinity for integrins is substantially less than that of intact fibronectin or of the entire cell-binding region in fibronectin. Thus structural features near the RGD sequence in fibronectins (e.g., parts of adjacent repeats, such as the synergy region; see Figure 19-30b) and in other RGD-containing proteins enhance their binding to certain integrins. Moreover, the simple soluble dimeric forms of fibronectin produced by the liver or fibroblasts are initially in a nonfunctional conformation that binds poorly to integrins because the RGD sequence is not readily accessible. The adsorption of fibronectin to a collagen matrix or the basal lamina or, experimentally, to a plastic tissue-culture dish results in a conformational change that enhances its ability to bind to cells. Possibly, this conformational change increases the accessibility of the RGD sequence for integrin binding.

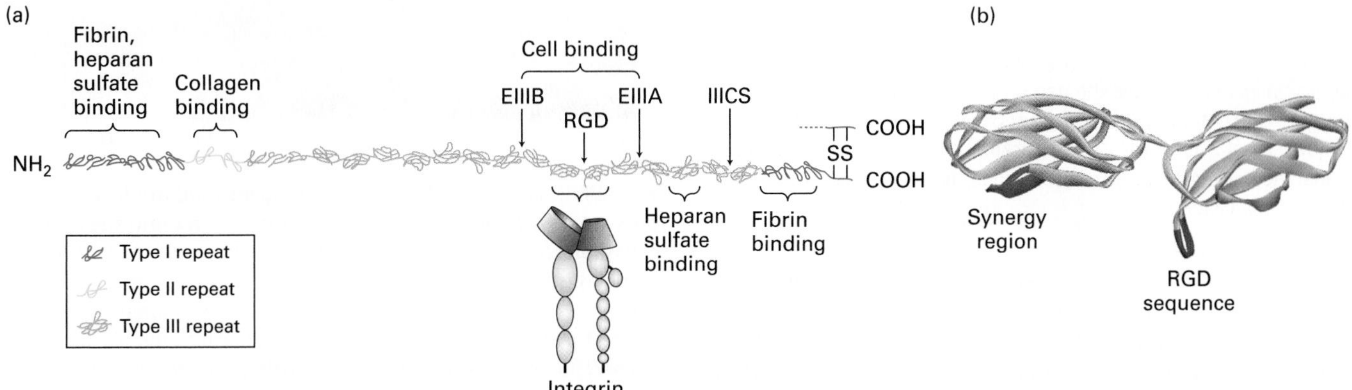

▲ **FIGURE 19-30 Organization of fibronectin and its binding to integrin.** (a) Scale model of fibronectin is shown docked by two type III repeats to the extracellular domains of integrin. Only one of the two similar chains, which are linked by disulfided bonds near their C-termini, in the dimeric fibronectin molecule is shown. Each chain contains about 2446 amino acids and is composed of three types of repeating amino acid sequences (type I, II, or III repeats). The EIIIA, EIIIB—both type III repeats—and IIICS domain are variably spliced into the structure at locations indicated by arrows. Circulating fibronectin lacks one or both of EIIIA and EIIIB. At least five different sequences may be present in the IIICS region as a result of alternative splicing (see Figure 4-16). Each chain contains several multirepeat-containing regions, some of which contain specific binding sites (made up of multiple-binding repeats) for heparan sulfate, fibrin (a major constituent of blood clots), collagen, and cell-surface integrins. The integrin-binding domain is also known as the cell-binding domain. Structures of fibronectin's domains were determined from fragments of the molecule. (b) A high-resolution structure shows that the RGD binding sequence (red) extends outward in a loop from its compact type III domain on the same side of fibronectin as the synergy region (blue), which also contributes to high-affinity binding to integrins. [Adapted from D. J. Leahy et al., 1996, *Cell* **84**:161.]

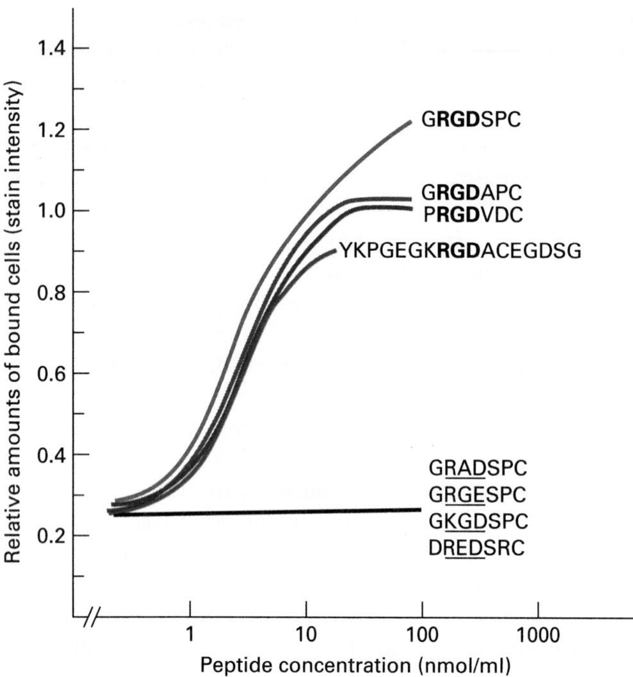

▲ EXPERIMENTAL FIGURE 19-31 **A specific tripeptide sequence (RGD) in the cell-binding region of fibronectin is required for adhesion of cells.** The cell-binding region of fibronectin contains an integrin-binding hexapeptide sequence, GRGDSP in the single-letter amino acid code (see Figure 2-14). Together with an additional C-terminal cysteine (C) residue this heptapeptide and several variants were synthesized chemically. Different concentrations of each synthetic peptide were added to polystyrene dishes that had the protein immunoglobulin G (IgG) firmly attached to their surfaces; the peptides were then chemically cross-linked to the IgG. Subsequently, cultured normal rat kidney cells were added to the dishes and incubated for 30 minutes to allow adhesion. After the nonbound cells were washed away, the relative amounts of cells that had adhered firmly were determined by staining the bound cells with a dye and measuring the intensity of the staining with a spectrophotometer. The results shown here indicate that cell adhesion increased above the background level with increasing peptide concentration for those peptides containing the RGD sequence but not for the variants lacking this sequence (modification underlined). [From M. D. Pierschbacher and E. Ruoslahti, 1984, *Proc. Nat'l. Acad. Sci. USA* **81:**5985.]

Microscopy and other experimental approaches (e.g., biochemical binding experiments) have demonstrated the role of integrins in cross-linking fibronectin and other ECM components to the cytoskeleton. For example, the colocalization of cytoskeletal actin filaments and integrins within cells can be visualized by fluorescence microscopy (Figure 19-32a). The binding of cell-surface integrins to fibronectin in the matrix induces the actin cytoskeleton–dependent movement of some integrin molecules in the plane of the membrane. The ensuing mechanical tension due to the relative movement of different integrins bound to a single fibronectin dimer stretches the fibronectin. This stretching promotes self-association of fibronectins into multimeric fibrils.

The force needed to unfold and expose functional self-association sites in fibronectin is much less than that needed to disrupt fibronectin-integrin binding. Thus fibronectin molecules remain bound to integrin while cell-generated mechanical forces induce fibril formation. In effect, the integrins through adapter proteins transmit the intracellular forces generated by the actin cytoskeleton to extracellular fibronectin. Gradually, the initially formed fibronectin fibrils mature into highly stable matrix components by covalent cross-linking. In some electron micrographic images, exterior fibronectin fibrils appear to be aligned in a seemingly continuous line with bundles of actin fibers within the cell (Figure 19-32b). These observations and the results from other studies provided the first example of a molecularly

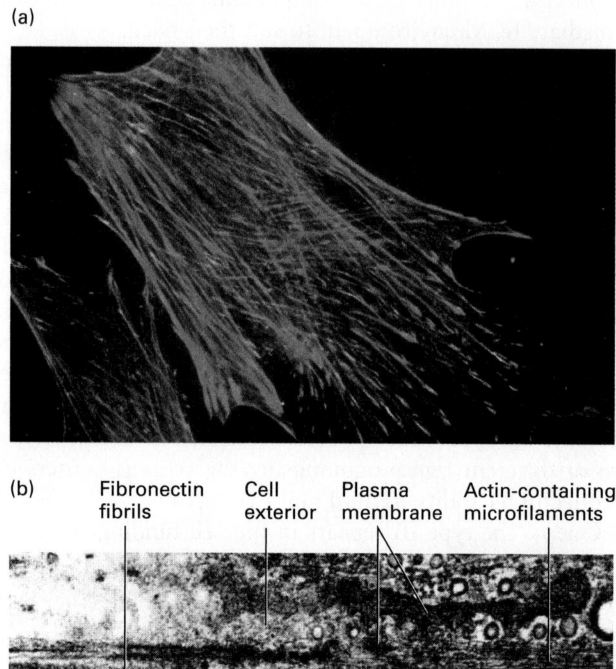

▲ EXPERIMENTAL FIGURE 19-32 **Integrins mediate linkage between fibronectin in the extracellular matrix and the cytoskeleton.** (a) Immunofluorescent micrograph of a fixed cultured fibroblast showing colocalization of the $\alpha5\beta1$ integrin (green) and actin-containing stress fibers (red). The cell was incubated with two types of monoclonal antibody: an integrin-specific antibody linked to a green fluorescing dye and an actin-specific antibody linked to a red fluorescing dye. Stress fibers are long bundles of actin microfilaments that radiate inward from points where the cell contacts a substratum. At the distal end of these fibers, near the plasma membrane, the coincidence of actin (red) and fibronectin-binding integrin (green) produces a yellow fluorescence. (b) Electron micrograph of the junction of fibronectin and actin fibers in a cultured fibroblast. Individual actin-containing 7-nm microfilaments, components of a stress fiber, end at the obliquely sectioned cell membrane. The microfilaments appear aligned in close proximity to the thicker, densely stained fibronectin fibrils on the outside of the cell. [Part (a) from J. Duband et al., 1988, *J. Cell Biol.* 107:1385. Part (b) from I. J. Singer, 1979, *Cell* 16:675; courtesy of I. J. Singer; copyright 1979, MIT.]

well defined adhesion receptor (i.e., an integrin) forming a bridge between the intracellular cytoskeleton and the extracellular matrix components—a phenomenon now known to be widespread.

KEY CONCEPTS OF SECTION 19.4

The Extracellular Matrix II: Connective and Other Tissues

- Connective tissue, such as tendon and cartilage, differs from other solid tissues in that most of its volume is made up of extracellular matrix (ECM) rather than cells.
- The synthesis of fibrillar collagen (e.g., types I, II, and III) begins inside the cell with the chemical modification of newly made α chains and their assembly into triple-helical procollagen within the endoplasmic reticulum. After secretion, procollagen molecules are cleaved, associate laterally, and are covalently cross-linked into bundles called fibrils, which can form larger assemblies called fibers (see Figure 19-24).
- The various collagens are distinguished by the ability of their helical and nonhelical regions to associate into fibrils, to form sheets, or to cross-link other collagen types (see Table 19-4).
- Proteoglycans consist of membrane-associated or secreted core proteins covalently linked to one or more glycosaminoglycan (GAG) chains, which are linear polymers of disaccharides that are often modified by sulfation.
- Cell-surface proteoglycans such as the syndecans facilitate cell-matrix interactions and help present certain external signaling molecules to their cell-surface receptors.
- Hyaluronan, a highly hydrated GAG, is a major component of the ECM of migrating and proliferating cells. Certain cell-surface adhesion receptors bind hyaluronan to cells.
- Large proteoglycan aggregates containing a central hyaluronan molecule noncovalently bound to the core protein of multiple proteoglycan molecules (e.g., aggrecan) contribute to the distinctive mechanical properties of the matrix (see Figure 19-29).
- Fibronectins are abundant multiadhesive matrix proteins that play a key role in migration and cellular differentiation. They contain binding sites for integrins and ECM components (collagens, proteoglycans) and thus can attach cells to the matrix (see Figure 19-30).
- The tripeptide RGD sequence Arg-Gly-Asp, found in fibronectins and some other matrix proteins, is recognized by several integrins.

19.5 Adhesive Interactions in Motile and Nonmotile Cells

After adhesive interactions in epithelia form during differentiation, they often are very stable and can last throughout the life span of the cells or until the epithelium undergoes further differentiation. Although such long-lasting (nonmotile) adhesion also exists in nonepithelial tissues, some nonepithelial cells must be able to crawl across or through a layer of extracellular matrix or other cells. Moreover, during development or wound healing and in certain pathologic states (e.g., cancer), epithelial cells can transform into more motile cells (the epithelial-mesenchymal transition). Changes in expression of adhesive molecules play a key role in this transformation, as they do in other biological processes involving cell movement, such as the crawling of white blood cells into tissue sites of infection. In this section, we describe various cell-surface structures that mediate transient adhesive interactions that are especially adapted for the movement of cells as well as those that mediate long-lasting adhesion. The detailed intracellular mechanisms used to generate the mechanical forces that propel cells and modify their shapes are covered in Chapters 17 and 18.

Integrins Relay Signals Between Cells and Their Three-Dimensional Environment

As already discussed, integrins connect epithelial cells to the basal lamina and, through adapter proteins, to intermediate filaments of the cytoskeleton (see Figure 19-1). That is, integrins form a bridge between the ECM and cytoskeleton; they do the same in nonepithelial cells. In nonepithelial cells, integrins in the plasma membrane also are clustered with other molecules in various adhesive structures called focal adhesions, focal contacts, focal complexes, 3-D adhesions, and fibrillar adhesions, as well as in circular adhesions called podosomes. These structures are readily observed by fluorescence microscopy with the use of antibodies that recognize integrins or other coclustered molecules (Figure 19-33). Like cell-matrix anchoring junctions in epithelial cells, these various adhesive structures attach nonepithelial cells to the extracellular matrix through, for example, binding to fibronectin (see Figure 19-32). They also contain dozens of intracellular adapter and associated proteins that mediate attachment to cytoskeletal actin filaments and activate adhesion-dependent signals for cell growth and cell motility (see Figure 19-7).

Although found in many nonepithelial cells, integrin-containing adhesive structures have been studied most frequently in fibroblasts grown in cell culture on flat glass or plastic substrata. These conditions only poorly approximate the three-dimensional ECM environment that normally surrounds such cells in vivo. When fibroblasts are cultured in three-dimensional ECM matrices derived from cells or tissues, they form adhesions to the three-dimensional ECM substratum, called 3-D adhesions. These structures differ somewhat in composition, shape, distribution, and activity from the focal or fibrillar adhesions seen in cells growing on the flat substratum typically used in cell-culture experiments (Figure 19-33). Cultured fibroblasts with these "more natural" anchoring junctions display greater adhesion and mobility, increased rates of cell proliferation, and spindle-shaped morphologies more like those of fibroblasts in tissues than do cells cultured on flat surfaces. These observations indicate that the topological, compositional, and mechanical

(a) Focal adhesion

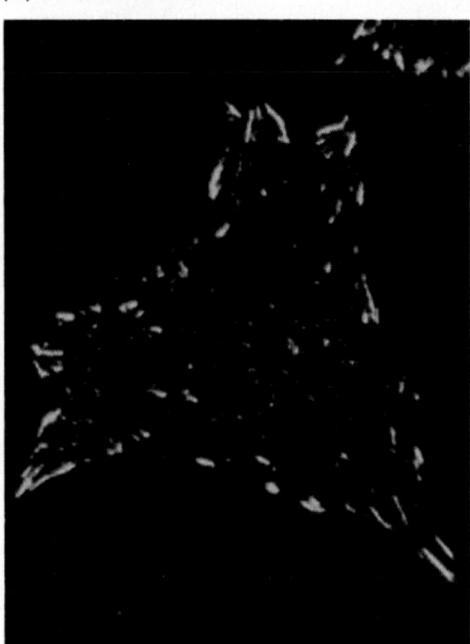

(b) 3-D adhesion

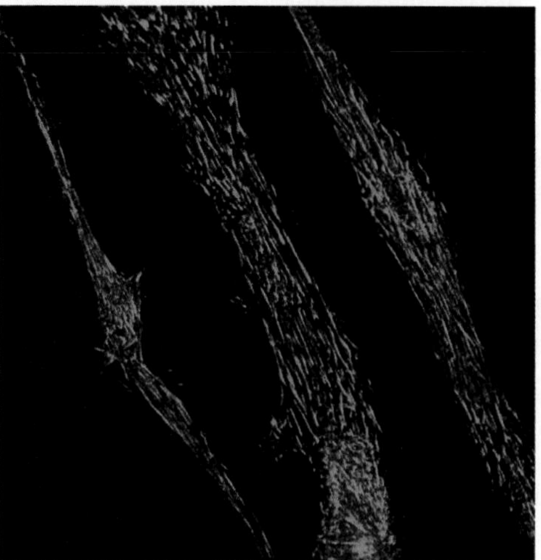

◀ **EXPERIMENTAL FIGURE 19-33 Integrins cluster into adhesive structures with various morphologies in nonepithelial cells.** Immunofluorescence methods were used to detect adhesive structures (green) on cultured cells. Shown here are focal adhesions (a) and 3-D adhesions (b) on the surfaces of human fibroblasts. Cells were grown directly on the flat surface of a culture dish (a) or on a three-dimensional matrix of ECM components (b). The shape, distribution, and composition of the integrin-based adhesions formed by cells vary, depending on the cells' environment. [Part (a) from B. Geiger et al., 2001, *Nature Rev. Mol. Cell Biol.* **2:**793. Part (b) courtesy of K. Yamada and E. Cukierman; see E. Cukierman et al., 2001, *Science* **294:**1708–1712.]

(e.g., flexibility) properties of the extracellular matrix all play a role in controlling the shape and activity of a cell. Tissue-specific differences in these matrix characteristics probably contribute to the tissue-specific properties of cells.

The importance of the 3-D environment of cells has been highlighted by cell culture studies of the morphogenesis, functioning, and stability of specialized milk-producing mammary epithelial cells and their cancerously transformed counterparts. For example, the 3-D matrix-dependent outside-in signaling mediated by integrins influences the epidermal growth factor tyrosine kinase receptor signaling system and vice versa. The 3-D matrix also permits the mammary epithelial cells to generate in vivo-like circular epithelial structures, called acini, that secrete the major protein constituents of milk, and permits comparisons of the responses of normal and cancerous cells to potential chemotherapeutic agents. Analogous systems employing both natural and synthetic 3-D ECMs are being developed to provide more in vivo–like conditions to study other complex tissues and organs, such as the liver.

Regulation of Integrin-Mediated Adhesion and Signaling Controls Cell Movement

Cells can exquisitely control the strength of integrin-mediated cell-matrix interactions by regulating the number of integrins (expression levels), their ligand-binding activities, or both. Such regulation is critical to the role of these interactions in cell migration and other functions involving cell movement.

Integrin Binding Many, if not all, integrins can exist in two conformations: a low-affinity (inactive) form and a high-affinity (active) form (Figure 19-34). The results of structural studies and experiments investigating the binding of ligands by integrins have provided a model of the changes that take place when integrins are activated. In the inactive state, the αβ heterodimer is bent, the conformation of the ligand-binding site at the tip of the molecule allows only low-affinity ligand binding, and the cytoplasmic C-terminal tails of the two subunits are closely bound together. In the "straight," active state, alterations in the conformation of the domains that form the binding site permit tighter (high-affinity) ligand binding, and the cytoplasmic tails separate.

These structural models provide an attractive explanation for the ability of integrins to mediate outside-in and inside-out signaling. The binding of certain ECM molecules or CAMs on other cells to the bent, low-affinity structure would force the molecule to straighten and consequently separate the cytoplasmic tails. Intracellular adapters could "sense" the separation of the tails and, as a result, either bind to or dissociate from the tails. The changes in these adapters could then alter the cytoskeleton and activate or inhibit intracellular signaling pathways. Conversely, changes in the metabolic state of the cells could cause intracellular adapters to bind to the tails or to dissociate from them and thus force the tails to either separate or associate. As a consequence, the integrin would either bend (inactivate) or straighten (activate), thereby altering its interaction with the ECM or other cells.

Platelet function, discussed in more detail below, provides a good example of how cell-matrix interactions are modulated by controlling integrin binding activity. Platelets are cell fragments that circulate in the blood and clump together with ECM molecules in forming a blood clot. In its basal state, the αIIbβ3 integrin present on the plasma membranes of platelets normally cannot bind tightly to its protein ligands (e.g., fibrinogen, fibronectin), all of which participate in the formation of a blood clot, because it is in the inactive (bent) conformation. In a forming clot, the binding via other signaling receptors of a platelet to collagen or a large ECM protein called von Willabrand factor or the platelet's activation by ADP or the clotting enzyme thrombin generates intracellular signals. These signals induce changes in signaling pathways in the cytoplasm that result in an activating conformational change in αIIbβ3 integrin. As a consequence, this integrin can then tightly bind to extracellular clotting proteins and participate in clot formation. People with genetic defects in the β3 integrin subunit are prone to excessive bleeding, attesting to the role of this integrin in the formation of blood clots.

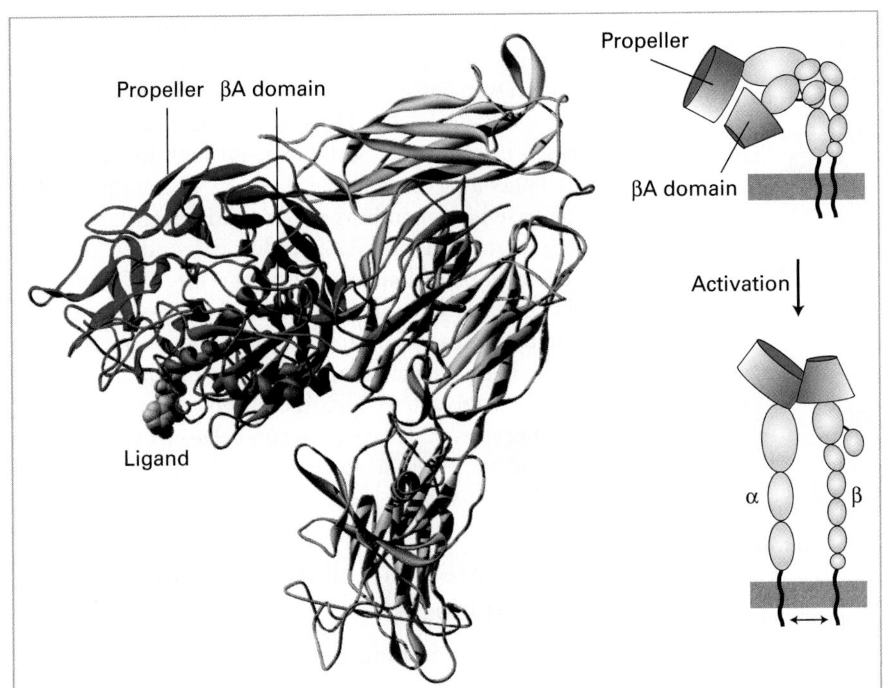

◀ **FIGURE 19-34 Model for integrin activation.** (*Left*) The molecular model is based on the x-ray crystal structure of the extracellular region of αvβ3 integrin in its inactive, low-affinity ("bent") form, with the α subunit in shades of blue and the β subunit in shades of red. The major ligand-binding sites are at the tip of the molecule, where the propeller domain of the α subunit (dark blue) and βA domain (dark red) interact. An RGD peptide ligand is shown in yellow. (*Right*) Activation of integrins is thought to be due to conformational changes that include straightening of the molecule; key movements near the propeller and βA domains, which increases the affinity for ligands; and separation of the cytoplasmic domains, resulting in altered interactions with adapter proteins. [Adapted from M. Arnaout et al., 2002, *Curr. Opin. Cell Biol.* **14:**641, and R. O. Hynes, 2002, *Cell* **110:**673.]

Integrin Expression The attachment of cells to ECM components can also be modulated by altering the number of integrin molecules exposed on the cell surface. The α4β1 integrin, which is found on many hematopoietic cells, offers an example of this regulatory mechanism. For these hematopoietic cells to proliferate and differentiate, they must be attached to fibronectin synthesized by supportive ("stromal") cells in the bone marrow. The α4β1 integrin on hematopoietic cells binds to a Glu-Ile-Leu-Asp-Val (EILDV) sequence in fibronectin, thereby anchoring the cells to the matrix. This integrin also binds to a sequence in a CAM called vascular CAM-1 (VCAM-1), which is present on stromal cells of the bone marrow. Thus hematopoietic cells directly contact the stromal cells as well as attach to the matrix. Late in their differentiation, hematopoietic cells decrease their expression of α4β1 integrin; the resulting reduction in the number of α4β1 integrin molecules on the cell surface is thought to allow mature blood cells to detach from the matrix and stromal cells in the bone marrow and subsequently enter the circulation.

Connections Between the ECM and Cytoskeleton Are Defective in Muscular Dystrophy

The importance of the adhesion-receptor-mediated linkage between ECM components and the cytoskeleton is highlighted by a set of hereditary muscle-wasting diseases, collectively called muscular dystrophies. Duchenne muscular dystrophy (DMD), the most common type, is a sex-linked disorder, affecting 1 in 3300 boys, that results in cardiac or respiratory failure, usually in the late teens or early twenties. The first clue to understanding the molecular basis of this disease came from the discovery that people with DMD carry mutations in the gene encoding a protein named *dystrophin.* This very large protein was found to be a cytosolic adapter protein, binding to actin filaments and to an adhesion receptor called *dystroglycan.*

Dystroglycan is synthesized as a large glycoprotein precursor that is proteolytically cleaved into two subunits. The α subunit is a peripheral membrane protein, and the β subunit is a transmembrane protein whose extracellular domain associates with the α subunit (Figure 19-35). Multiple *O*-linked oligosaccharides are attached covalently to side-chain hydroxyl groups of serine and threonine residues in the α subunit. Unlike the most abundant *O*-linked (also called mucin-like) oligosaccharides in which an *N*-acetylgalactosamine (GalNAc) is the first sugar in the chain linked directly to the hydroxy group of the side chain of serine or theonine or the linkage in proteoglycans, many of the *O*-linked chains in dystroglycan are directly linked to the hydroxyl group via a mannose sugar (see Figure 19-27).

These specialized *O*-linked oligosaccharides bind to various basal lamina components, including the LG domains of multiadhesive matrix protein laminin and the proteoglycans perlecan and agrin. The neurexins, a family of adhesion molecules expressed by neurons, also are bound via these oligosaccharides, whose detailed heterogeneous structures and mechanisms of synthesis have not been fully elucidated.

The transmembrane segment of the dystroglycan β subunit associates with a complex of integral membrane proteins; its cytosolic domain binds dystrophin and other adapter proteins, as well as various intracellular signaling proteins (Figure 19-35). The resulting large, heteromeric assemblage, the *dystrophin glycoprotein complex (DGC),* links the extracellular matrix to the cytoskeleton and signaling pathways within

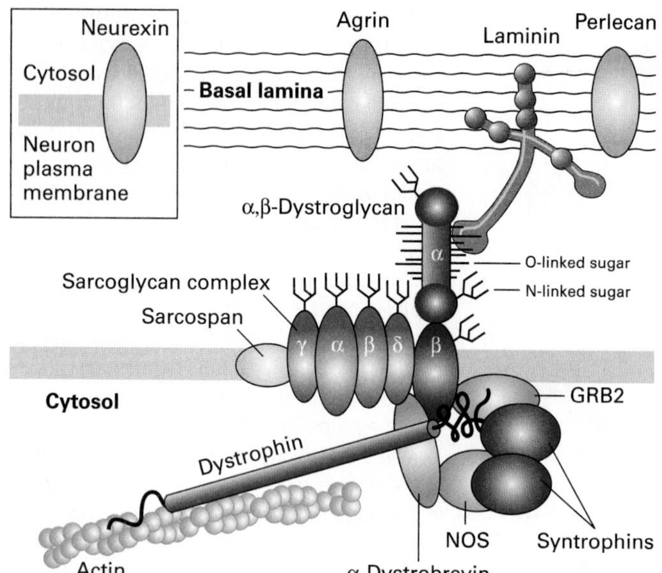

▲ **FIGURE 19-35 Dystrophin glycoprotein complex (DGC) in skeletal muscle cells.** This schematic model shows that the DGC comprises three subcomplexes: the α,β dystroglycan subcomplex; the sarcoglycan/sarcospan subcomplex of integral membrane proteins; and the cytosolic adapter subcomplex comprising dystrophin, other adapter proteins, and signaling molecules. Through its *O*-linked sugars, β-dystroglycan binds to components of the basal lamina, such as laminin and perlecan, and cell surface proteins, such as neurexin in neurons. Dystrophin—the protein defective in Duchenne muscular dystrophy—links β-dystroglycan to the actin cytoskeleton, and α-dystrobrevin links dystrophin to the sarcoglycan/sarcospan subcomplex. Nitric oxide synthase (NOS) produces nitric oxide, a gaseous signaling molecule, and GRB2 is a component of signaling pathways activated by certain cell-surface receptors (Chapter 15). [Adapted from S. J. Winder, 2001, *Trends Biochem. Sci.* **26:**118, and D. E. Michele and K. P. Campbell, 2003, *J. Biol. Chem.* **278(18):**15457–15460.]

muscle and other types of cells. For instance, the signaling enzyme nitric oxide synthase (NOS) is associated through syntrophin with the cytosolic dystrophin subcomplex in skeletal muscle. The rise in intracellular Ca^{2+} during muscle contraction activates NOS to produce nitric oxide (NO), a signaling molecule that diffuses into smooth muscle cells surrounding nearby blood vessels. NO promotes smooth muscle relaxation, leading to a local rise in the flow of blood supplying nutrients and oxygen to the skeletal muscle.

Mutations in dystrophin, other DGC components, laminin, or enzymes that add the O-linked sugars to dystroglycan can all disrupt the DGC-mediated link between the exterior and the interior of muscle cells and cause muscular dystrophies. In addition, dystroglycan mutations have been shown to greatly reduce the clustering of acetylcholine receptors on muscle cells at the neuromuscular junctions, which also is dependent on the basal lamina proteins laminin and agrin. These and possibly other effects of DGC defects apparently lead to a cumulative weakening of the mechanical stability of muscle cells as they undergo contraction and relaxation, resulting in deterioration of the cells and muscular dystrophy.

Dystroglycan provides an elegant—and medically relevant—example of the intricate networks of connectivity in cell biology. Dystroglycan was originally discovered in the context of studying DMD. However, it was later shown to be expressed in nonmuscle cells and, through its binding to laminin, to play a key role in the assembly and stability of at least some basement membranes. Thus it is essential for normal development (Chapter 22). Additional studies led to its identification as a cell-surface receptor for the virus that causes the frequently fatal human disease Lassa fever and other related viruses, all of which bind via the same specialized O-linked sugars that mediate binding to laminin. Furthermore, dystroglycan is also the receptor on specialized cells in the nervous system—Schwann cells—to which binds the pathogenic bacterium *Mycobacterium leprae*, the causative organism of leprosy. ■

IgCAMs Mediate Cell-Cell Adhesion in Neuronal and Other Tissues

Numerous transmembrane proteins characterized by the presence of multiple immunoglobulin domains (repeats) in their extracellular regions constitute the immunoglobulin (Ig) superfamily of CAMs, or **IgCAMs.** The Ig domain is a common protein motif, containing 70–110 residues, that was first identified in antibodies, the antigen-binding immunoglobulins, but has a much older evolutionary origin in CAMs. The human, *D. melanogaster,* and *C. elegans* genomes include about 765, 150, and 64 genes, respectively, that encode proteins containing Ig domains. Immunoglobulin domains are found in a wide variety of cell-surface proteins, including T-cell receptors produced by lymphocytes and many proteins that take part in adhesive interactions. Among the IgCAMs are neural CAMs; intercellular CAMs (ICAMs), which function in the movement of leukocytes into tissues; and junction adhesion molecules (JAMs), which are present in tight junctions.

As their name implies, neural CAMs are of particular importance in neural tissues. One type, the NCAMs, primarily mediate homophilic interactions. First expressed during morphogenesis, NCAMs play an important role in the differentiation of muscle, glial, and nerve cells. Their role in cell adhesion has been directly demonstrated by the inhibition of adhesion with anti-NCAM antibodies. Numerous NCAM isoforms, encoded by a single gene, are generated by alternative mRNA splicing and by differences in glycosylation. Other neural CAMs (e.g., L1-CAM) are encoded by different genes. In humans, mutations in different parts of the L1-CAM gene cause various neuropathologies (e.g., mental retardation, congenital hydrocephalus, and spasticity).

An NCAM comprises an extracellular region with five Ig repeats and two fibronectin type III repeats, a single membrane-spanning segment, and a cytosolic segment that interacts with the cytoskeleton (see Figure 19-2). In contrast, the extracellular region of L1-CAM has six Ig repeats and four fibronectin type III repeats. As with cadherins, cis (intracellular) interactions and trans (intercellular) interactions probably play key roles in IgCAM-mediated adhesion (see Figure 19-3); however, adhesion mediated by IgCAMs is Ca^{2+}-independent.

The covalent attachment of multiple chains of sialic acid, a negatively charged sugar derivative, to NCAMs alters their adhesive properties. In embryonic tissues such as brain, polysialic acid constitutes as much as 25 percent of the mass of NCAMs. Possibly because of repulsion between the many negatively charged sugars in these NCAMs, cell-cell contacts are fairly transient, being made and then broken, a property necessary for the development of the nervous system. In contrast, NCAMs from adult tissues contain only one-third as much sialic acid, permitting more stable adhesions.

Leukocyte Movement into Tissues Is Orchestrated by a Precisely Timed Sequence of Adhesive Interactions

In adult organisms, several types of white blood cells (leukocytes) participate in the defense against infection caused by foreign invaders (e.g., bacteria and viruses) and tissue damage due to trauma or inflammation. To fight infection and clear away damaged tissue, these cells must move rapidly from the blood, where they circulate as unattached, relatively quiescent cells, into the underlying tissue at sites of infection, inflammation, or damage. We know a great deal about the movement into tissue, termed *extravasation,* of four types of leukocytes: neutrophils, which release several antibacterial proteins; monocytes, the precursors of macrophages, which can engulf and destroy foreign particles; and T and B lymphocytes, the antigen-recognizing cells of the immune system (Chapter 24).

Extravasation requires the successive formation and breakage of cell-cell contacts between leukocytes in the blood and endothelial cells lining the vessels. Some of these contacts are mediated by **selectins,** a family of CAMs that mediate leukocyte–vascular cell interactions. A key player in these interactions is *P-selectin,* which is localized to the blood-facing surface of endothelial cells. All selectins contain a Ca^{2+}-dependent *lectin domain,* which is located at the distal end of the extracellular region of the molecule and recognizes oligosaccharides in glycoproteins or glycolipids (see Figure 19-2). For example, the primary ligand for P- and E-selectins is an oligosaccharide called the *sialyl Lewis-x antigen,* a part of longer oligosaccharides present in abundance on leukocyte glycoproteins and glycolipids.

Figure 19-36 illustrates the basic sequence of cell-cell interactions leading to the extravasation of leukocytes. Various inflammatory signals released in areas of infection or inflammation first cause activation of the endothelium. P-selectin exposed on the surface of activated endothelial cells mediates the weak adhesion of passing leukocytes. Because of the force of the blood flow and the rapid "on" and "off" rates of P-selectin binding to its ligands, these "trapped" leukocytes are slowed but not stopped and literally roll along the surface of the endothelium. Among the signals that promote activation of the endothelium are **chemokines,** a group of small secreted proteins (8–12 kDa) produced by a wide variety of cells, including endothelial cells and leukocytes.

For tight adhesion to occur between activated endothelial cells and leukocytes, β2-containing integrins on the surfaces of leukocytes also must be activated by chemokines or other local activation signals such as *platelet-activating factor (PAF).* Platelet-activating factor is unusual in that it is a phospholipid, rather than a protein; it is exposed on the surface of activated endothelial cells at the same time that P-selectin is exposed. The binding of PAF or other activators to their receptors on leukocytes leads to activation of the leukocyte integrins to their high-affinity form (see Figure 19-34). (Most of the receptors for chemokines and PAF are members of the G protein–coupled receptor superfamily discussed in Chapter 15.) Activated integrins on leukocytes then bind to distinct IgCAMs on the surface of endothelial cells. These include ICAM-2, which is expressed constitutively, and ICAM-1. ICAM-1, whose synthesis along with that of E-selectin and P-selectin is induced by activation, does not usually contribute substantially to leukocyte endothelial cell adhesion immediately after activation but rather participates at later times in cases of chronic inflammation. The resulting tight adhesion mediated by the Ca^{2+}-independent-integrin–ICAM interactions leads to the cessation of rolling and to the spreading of leukocytes on the surface of the endothelium; soon the adhered cells move between adjacent endothelial cells and into the underlying tissue.

The selective adhesion of leukocytes to the endothelium near sites of infection or inflammation thus depends on the sequential appearance and activation of several different CAMs on the surfaces of the interacting cells. Different types of leukocytes express specific integrins containing the β2 subunit: for example, αLβ2 by T lymphocytes and αMβ2 by monocytes, the circulating precursors of tissue macrophages. Nonetheless, all leukocytes move into tissues by the same general mechanism depicted in Figure 19-36.

Many of the CAMs used to direct leukocyte adhesion are shared among different types of leukocytes and target tissues. Yet often only a particular type of leukocyte is directed to a particular tissue. How is this specificity achieved? A three-step model has been proposed to account for the cell-type specificity of such leukocyte–endothelial-cell interactions. First, endothelial activation promotes initial relatively weak, transient, and reversible binding (e.g., the interaction of selectins and their carbohydrate ligands). Without additional local activation signals, the leukocyte will quickly move on. Second, cells in the immediate vicinity of the site of infection or inflammation release or express on their surfaces chemical signals (e.g., chemokines, PAF) that activate only special subsets (depending on their complement of chemokine receptors) of the transiently attached leukocytes. Third, additional activation-dependent CAMs (e.g., integrins) engage their binding partners, leading to strong sustained adhesion. Only if the proper combination of CAMs, binding partners, and activation signals are engaged together with the appropriate

Animation: Cell-Cell Adhesion in Leukocyte Extravasation

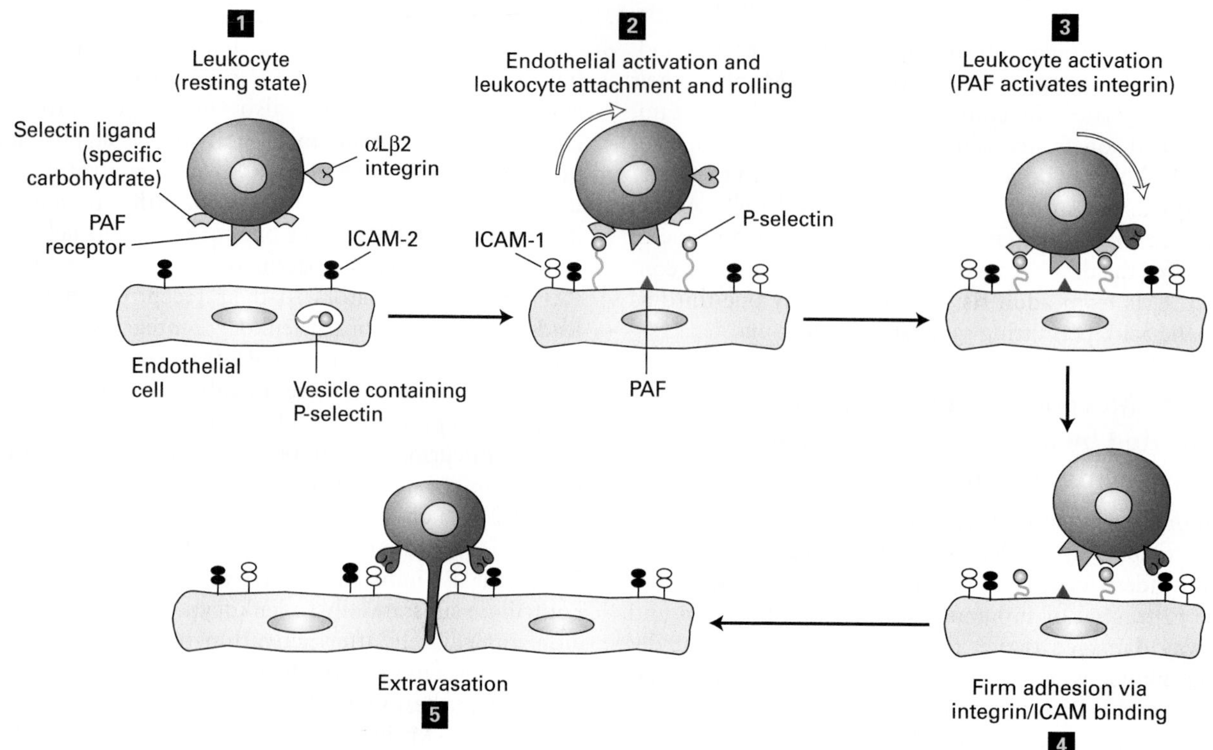

▲ **FIGURE 19-36 Sequence of cell-cell interactions leading to tight binding of leukocytes to activated endothelial cells and subsequent extravasation.** Step **1**: In the absence of inflammation or infection, leukocytes and endothelial cells lining blood vessels are in a resting state. Step **2**: Inflammatory signals released only in areas of inflammation, infection, or both activate resting endothelial cells to move vesicle-sequestered selectins to the cell surface. The exposed selectins mediate loose binding of leukocytes by interacting with carbohydrate ligands on leukocytes. Activation of the endothelium also causes synthesis of platelet-activating factor (PAF) and ICAM-1, both expressed on the cell surface. PAF and other usually secreted activators, including chemokines, then induce changes in the shapes of the leukocytes and activation of leukocyte integrins such as αLβ2, which is expressed by T lymphocytes **3**. The subsequent tight binding between activated integrins on leukocytes and CAMs on the endothelium (e.g., ICAM-2 and ICAM-1) results in firm adhesion **4** and subsequent movement (extravasation) into the underlying tissue **5**. [Adapted from R. O. Hynes and A. Lander, 1992, *Cell* **68:**303.]

timing at a specific site will a given leukocyte adhere strongly. Such combinatorial diversity and cross talk allows a small set of CAMs to serve diverse functions throughout the body—a good example of biological parsimony.

Leukocyte-adhesion deficiency is caused by a genetic defect in the synthesis of the integrin β2 subunit. People with this disorder are susceptible to repeated bacterial infections because their leukocytes cannot extravasate properly and thus fight the infection within the tissue.

Some pathogenic viruses have evolved mechanisms to exploit for their own purposes cell-surface proteins that participate in the normal response to inflammation. For example, many of the RNA viruses that cause the common cold (rhinoviruses) bind to and enter cells through ICAM-1, and chemokine receptors can be important entry sites for human immunodeficiency virus (HIV), the cause of AIDS. Integrins appear to participate in the binding and/or internalization of a wide variety of viruses, including reoviruses (causing fever and gastroenteritis, especially in infants), adenoviruses (causing conjunctivitis, acute respiratory disease), and foot-and-mouth disease virus (causing fever in cattle and pigs). ▮

KEY CONCEPTS OF SECTION 19.5

Adhesive Interactions in Diverse Motile and Nonmotile Cells

- Many cells have integrin-containing aggregates (e.g., focal adhesions, 3-D adhesions, podosomes) that physically and functionally connect cells to the extracellular matrix and facilitate inside-out and outside-in signaling.
- Via interaction with integrins, the three-dimensional structure of the ECM surrounding a cell can profoundly influence the behavior of the cell.

■ Integrins exist in two conformations that differ in the affinity for ligands and interactions with cytosolic adapter proteins (see Figure 19-34); switching between these two conformations allows regulation of integrin activity, which is important for control of cell adhesion and movements.

■ Dystroglycan, an adhesion receptor, forms a large complex with dystrophin, other adapter proteins, and signaling molecules (see Figure 19-35). This complex links the actin cytoskeleton to the surrounding matrix, providing mechanical stability to muscle. Mutations in various components of this complex cause different types of muscular dystrophy.

■ Neural cell-adhesion molecules, which belong to the immunoglobulin (Ig) family of CAMs, mediate Ca^{2+}-independent cell-cell adhesion in neural and other tissues.

■ The combinatorial and sequential interaction of several types of CAMs (e.g., selectins, integrins, and ICAMs) is critical for the specific and tight adhesion of different types of leukocytes to endothelial cells in response to local signals induced by infection or inflammation (see Figure 19-36).

19.6 Plant Tissues

We turn now to the assembly of plant cells into tissues. The overall structural organization of plants is generally simpler than that of animals. For instance, plants have only four broad types of cells, which in mature plants form four basic classes of tissue: *dermal tissue* interacts with the environment; *vascular tissue* transports water and dissolved substances (e.g., sugars, ions); space-filling *ground tissue* constitutes the major sites of metabolism; and *sporogenous tissue* forms the reproductive organs. Plant tissues are organized into just four main organ systems: *stems* have support and transport functions, *roots* provide anchorage and absorb and store nutrients, *leaves* are the sites of photosynthesis, and *flowers* enclose the reproductive structures. Thus at the cell, tissue, and organ levels, plants are generally less complex than most animals.

Moreover, unlike animals, plants do not replace or repair old or damaged cells or tissues; they simply grow new organs. Indeed, the developmental fate of any given plant cell is primarily based on its position in the organism rather than its lineage (Chapter 21), whereas both are important in animals. Thus in both plants and animals a cell's direct communication with it neighbors is important. Most importantly for this chapter and in contrast with animals, few cells in plants directly contact one another through molecules incorporated into their plasma membranes. Instead, plant cells are typically surrounded by a rigid **cell wall** that contacts the cell walls of adjacent cells (Figure 19-37a). Also in contrast with animal cells, a plant cell rarely changes its position in the organism relative to other cells. These features of plants and their organization have determined the distinctive molecular mechanisms by

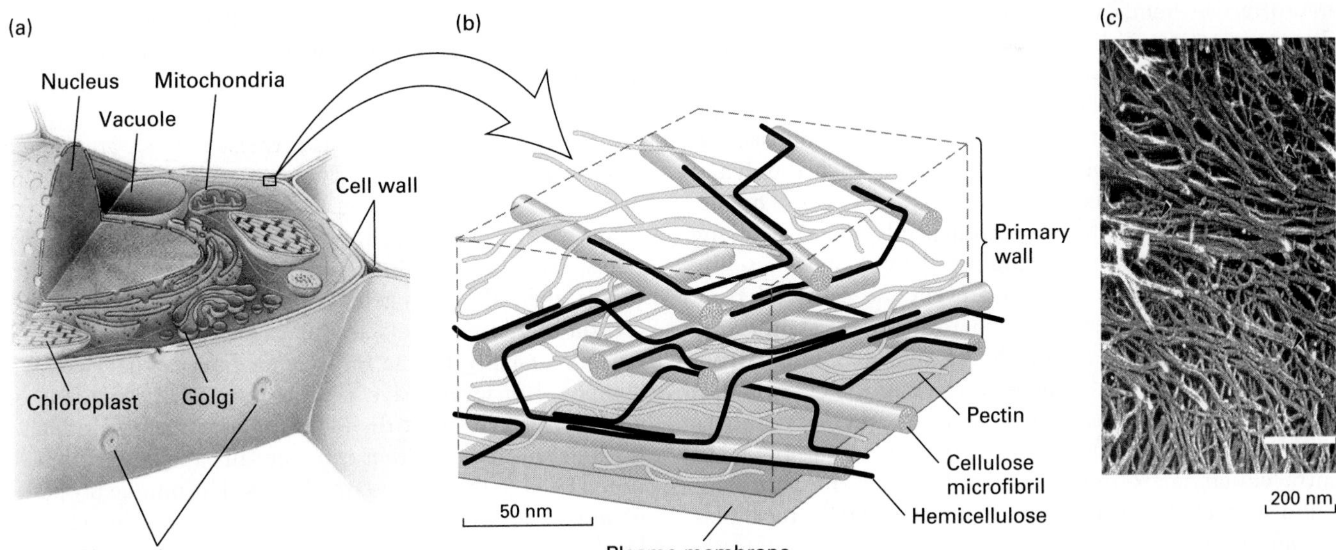

▲ **FIGURE 19-37 Structure of the plant cell wall.** (a) Overview of the organization of a typical plant cell, in which the organelle-filled cell with its plasma membrane is surrounded by a well-defined extracellular matrix called the cell wall. (b) Schematic representation of the cell wall of an onion. Cellulose and hemicellulose are arranged into at least three layers in a matrix of pectin polymers. The size of the polymers and their separations are drawn to scale. To simplify the diagram, most of the hemicellulose cross-links and other matrix constituents (e.g., extensin, lignin) are not shown. (c) Fast-freeze, deep-etch electron micrograph of the cell wall of the garden pea in which some of the pectin polysaccharides were removed by chemical treatment. The abundant thicker fibers are cellulose microfibrils, and the thinner fibers are hemicellulose cross-links (arrowheads). [Part (b) adapted from M. McCann and K. R. Roberts, 1991, in C. Lloyd, ed., *The Cytoskeletal Basis of Plant Growth and Form,* Academic Press, p. 126 as modified in Somerville C., Bauer S., Brininstool G., Facette M., Hamann T., Milne J., Osborne E., Paredez A., Persson S., Raab T., Vorwerk S., Youngs H. Part (c) from T. Fujino and T. Itoh, 1998, *Plant Cell Physiol.*, **39**(12):1315–1323.]

which their cells are incorporated into tissues and communicate with one another.

The Plant Cell Wall Is a Laminate of Cellulose Fibrils in a Matrix of Glycoproteins

The plant extracellular matrix, or cell wall, which is mainly composed of polysaccharides and is ≈0.2 μm thick, completely coats the outside of the plant cell's plasma membrane. This structure serves some of the same functions as those of the ECM produced by animal cells, even though the two structures are composed of entirely different macromolecules and have a different organization. About 1000 genes in the plant *Arabidopsis* are devoted to the synthesis and functioning of its cell wall, including approximately 414 glycosyltransferase and more than 316 glycosyl hydrolase genes. Like animal cell ECM, the plant cell wall connects cells into tissues, signals a plant cell to grow and divide, and controls the shape of plant organs. It is a dynamic structure that plays important roles in controlling the differentiation of plant cells during embryogenesis and growth and provides a barrier to protect against pathogen infection. Just as the extracellular matrix helps define the shapes of animal cells, the cell wall defines the shapes of plant cells. When the cell wall is digested away from plant cells by hydrolytic enzymes, spherical cells enclosed by a plasma membrane are left.

Because a major function of a plant cell wall is to withstand the osmotic turgor pressure of the cell (between 14.5 and 435 pounds per square inch!), the cell wall is built for lateral strength. It is arranged into layers of **cellulose** microfibrils—bundles of 30–36 chains of long (as much as 7 μm or greater), linear, extensively hydrogen-bonded polymers of glucose in β glycosidic linkages. The cellulose microfibrils are embedded in a matrix composed of *pectin,* a polymer of D-galacturonic acid and other monosaccharides, and *hemicellulose,* a short, highly branched polymer of several five- and six-carbon monosaccharides. The mechanical strength of the cell wall depends on cross-linking of the microfibrils by hemicellulose chains (Figure 19-37b, c). The layers of microfibrils prevent the cell wall from stretching laterally. Cellulose microfibrils are synthesized on the exoplasmic face of the plasma membrane from UDP-glucose and ADP-glucose formed in the cytosol. The polymerizing enzyme, called *cellulose synthase,* moves within the plane of the plasma membrane along tracks of intracellular microtubules as cellulose is formed, providing a distinctive mechanism for intracellular/extracellular communication.

Unlike cellulose, pectin and hemicellulose are synthesized in the Golgi apparatus and transported to the cell surface, where they form an interlinked network that helps bind the walls of adjacent cells to one another and cushions them. When purified, pectin binds water and forms a gel in the presence of Ca^{2+} and borate ions—hence the use of pectins in many processed foods. As much as 15 percent of the cell wall may be composed of *extensin,* a glycoprotein that contains abundant hydroxyproline and serine. Most of the hydroxyproline residues are linked to short chains of arabinose (a five-carbon monosaccharide), and the serine residues are linked to galactose. Carbohydrate accounts for about 65 percent of extensin by weight, and its protein backbone forms an extended rodlike helix with the hydroxyl or O-linked carbohydrates protruding outward. *Lignin*—a complex, insoluble polymer of phenolic residues—associates with cellulose and is a strengthening material. Like cartilage proteoglycans, lignin resists compression forces on the matrix.

The cell wall is a selective filter whose permeability is controlled largely by pectins in the wall matrix. Whereas water and ions diffuse freely across cell walls, the diffusion of large molecules, including proteins larger than 20 kDa, is limited. This limitation may account for why many plant hormones are small, water-soluble molecules, which can diffuse across the cell wall and interact with receptors in the plasma membrane of plant cells.

Loosening of the Cell Wall Permits Plant Cell Growth

Because the cell wall surrounding a plant cell prevents it from expanding, the wall's structure must be loosened when the cell grows. The amount, type, and direction of plant-cell growth are regulated by small-molecule hormones (e.g., indoleacetic acid) called *auxins.* The auxin-induced weakening of the cell wall permits the expansion of the intracellular vacuole by uptake of water, leading to elongation of the cell. We can grasp the magnitude of this phenomenon by considering that, if all cells in a redwood tree were reduced to the size of a typical liver cell, the tree would have a maximum height of only 1 meter.

The cell wall undergoes its greatest changes at the **meristem** of a root or shoot tip. These sites are where cells divide and grow. Young meristematic cells are connected by thin primary cell walls, which can be loosened and stretched to allow subsequent cell elongation. After cell elongation ceases, the cell wall is generally thickened, either by the secretion of additional macromolecules into the primary wall or, more usually, by the formation of a secondary cell wall composed of several layers. Most of the cell eventually degenerates, leaving only the cell wall in mature tissues such as the xylem—the tubes that conduct salts and water from the roots through the stems to the leaves. The unique properties of wood and of plant fibers such as cotton are due to the molecular properties of the cell walls in the tissues of origin.

Plasmodesmata Directly Connect the Cytosols of Adjacent Cells in Higher Plants

The presence of a cell wall separating cells in plants imposes barriers to cell-cell communication—and thus cell-type differentiation—not faced by animals. One distinctive mechanism used by plant cells to communicate directly is through specialized cell-cell junctions called **plasmodesmata,** which extend through the cell wall. Like gap junctions, plasmodesmata are channels that connect the cytosol of a cell with that of an adjacent cell. The diameter of the channel is about 30–60 nm, and its length can vary and be greater than 1 μm. The density of plasmodesmata varies depending on the plant and cell type, and even the smallest meristematic cells have more than 1000 interconnections with their neighbors. Although a variety of proteins that are physically or functionally associated with plasmodesmata have been identified, key structural protein components of plasmodesmata remain to be identified.

Molecules smaller than about 1000 Da, including a variety of metabolic and signaling compounds (ions, sugars, amino acids), generally can diffuse through plasmodesmata. However, the size of the channel through which molecules pass is highly regulated. In some circumstances, the channel is clamped shut; in others, it is dilated sufficiently to permit the passage of molecules larger than 10,000 Da. Among the factors that affect the permeability of plasmodesmata is the cytosolic Ca^{2+} concentration, with an increase in cytosolic Ca^{2+} reversibly inhibiting movement of molecules through these structures.

Although plasmodesmata and gap junctions resemble each other functionally with respect to forming channels for small molecule diffusion, their structures differ dramatically in two significant ways (Figure 19-38). The plasma membranes of the adjacent plant cells merge to form a continuous channel, the *annulus,* at each plasmodesmata, whereas the membranes of cells at a gap junction are not continuous with each other. In addition, plasmodesmata exhibit many additional complex structural and functional characteristics. For example, they contain within the channel an extension of the endoplasmic reticulum called a *desmotubule* that passes through the annulus, which connects the cytosols of adjacent plant cells. They also have a variety of specialized proteins at the entrance of the channel and running throughout the length of the channel, including special cytoskeletal, motor, and docking proteins that regulate the sizes and types of molecules that can pass through the channel. Many types of molecules spread from cell to cell through plasmodesmata, including proteins called non-cell-autonomous proteins (NCAPs, including some transcription factors), nucleic acid/protein complexes, metabolic products, and plant viruses. It appears that some of these require special chaperones to facilitate transport. Specialized kinases may also phosphorylate plasmodesmal components to regulate their activities (e.g., opening of the channels). Soluble molecules pass through the cytosolic annulus, or sleeve (about 3–4 nm in diameter), that lies between the plasma membrane and desmotubule, whereas membrane-bound molecules or certain proteins within the ER lumen can pass from cell to cell via the desmotubule. Plasmodesmata appear to play an especially important role in regulating the development of plant cells and tissues, as is suggested by their ability to mediate intracellular movement of transcription factors and ribonuclear protein complexes.

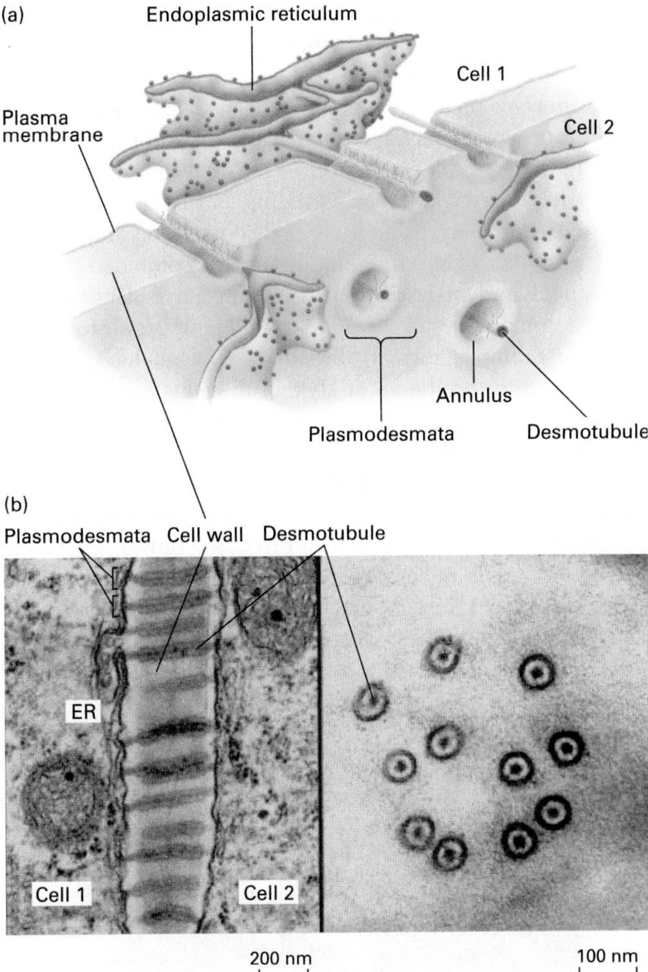

▲ **FIGURE 19-38 Plasmodesmata.** (a) Schematic model of a plasmodesma showing the desmotubule, an extension of the endoplasmic reticulum (ER), and the annulus, a plasma-membrane-lined channel filled with cytosol that interconnects the cytosols of adjacent cells. (b) Electron micrographs of thin sections of a sugarcane leaf (brackets indicate individual plasmodesmata). (*Left*) Longitudinal view showing ER and desmotubule running through each annulus. (*Right*) Perpendicular cross-sectional views of plasmodesmata, in some of which spoke structures connecting the plasma membrane to the desmotubule can be seen. [Part (b) from K. Robinson-Beers and R. F. Evert, 1991, *Planta* **184:**307–318.]

Only a Few Adhesive Molecules Have Been Identified in Plants

Systematic analysis of the *Arabidopsis* genome and biochemical analysis of other plant species provide no evidence for the existence of plant homologs of most animal CAMs, adhesion receptors, and ECM components. This finding is not surprising, given the dramatically different nature of cell-cell and cell-matrix/wall interactions in animals and plants.

Among the adhesive-type proteins apparently unique to plants are five wall-associated kinases (WAKs) and WAK-like proteins expressed in the plasma membrane of *Arabidopsis* cells. The extracellular regions in all these proteins contain multiple epidermal growth factor (EGF) repeats, frequently found in animal cell-surface receptors, which may directly participate in binding to other molecules. Some WAKs have been shown to bind to glycine-rich proteins in the cell wall, thereby mediating membrane-wall contacts. These *Arabidopsis* proteins have a single transmembrane domain and an intracellular cytosolic tyrosine kinase domain, which may participate in signaling pathways somewhat like the receptor tyrosine kinases discussed in Chapter 16.

The results of in vitro binding assays combined with in vivo studies and analyses of plant mutants have identified several macromolecules in the ECM that are important for adhesion. For example, normal adhesion of pollen, which contains sperm cells, to the stigma or style in the female reproductive organ of the Easter lily requires a cysteine-rich protein called stigma/stylar cysteine-rich adhesin (SCA) and a specialized pectin that can bind to SCA (Figure 19-39). A small, probably ECM embedded, ~10 kDa protein called chymocyanin works in conjunction with SCA to help direct the movement of the sperm-containing pollen tube (chemotaxis) to the ovary.

Disruption of the gene encoding glucuronyltransferase 1, a key enzyme in pectin biosynthesis, has provided a striking illustration of the importance of pectins in intercellular adhesion in plant meristems. Normally, specialized pectin molecules help hold the cells in meristems tightly together. When grown in culture as a cluster of relatively undifferentiated cells, called a callus, normal meristematic cells adhere tightly and can differentiate into chlorophyll-producing cells, giving the callus a green color. Eventually the callus will generate shoots. In contrast, mutant cells with an inactivated glucuronyltransferase 1 gene are large, associate loosely with each other, and do not differentiate normally, forming a yellow callus. The introduction of a normal glucuronyltransferase 1 gene into the mutant cells restores their ability to adhere and differentiate normally.

The paucity of plant adhesive molecules identified to date, in contrast with the many well-defined animal adhesive molecules, may be due to the technical difficulties in working with the ECM/cell wall of plants. Adhesive interactions are often likely to play different roles in plant and animal biology, at least in part because of their differences in development and physiology.

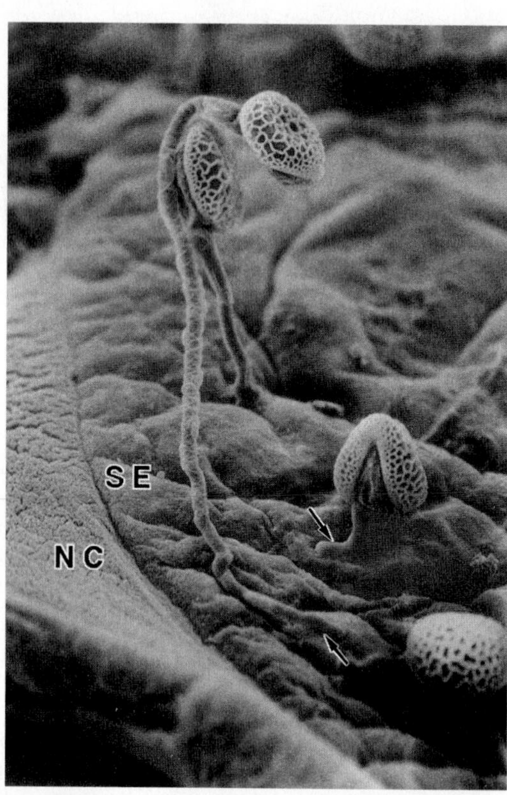

▲ EXPERIMENTAL FIGURE 19-39 **An in vitro assay was used to identify molecules required for adherence of pollen tubes to the stylar matrix.** In this assay, extracellular stylar matrix collected from lily styles (SE) or an artificial matrix is dried onto nitrocellulose membranes (NC). Pollen tubes containing sperm are then added and their binding to the dried matrix is assessed. In this scanning electron micrograph, the tips of pollen tubes (arrows) can be seen binding to dried stylar matrix. This type of assay has shown that pollen adherence depends on stigma/stylar cysteine-rich adhesin (SCA) and a pectin that binds to SCA. [From G. Y. Jauh et al., 1997, *Sex Plant Reprod.* **10:**173.]

KEY CONCEPTS OF SECTION 19.6

Plant Tissues

- The integration of cells into tissues in plants is fundamentally different from the assembly of animal tissues, primarily because each plant cell is surrounded by a relatively rigid cell wall.
- The plant cell wall comprises layers of cellulose microfibrils embedded within a matrix of hemicellulose, pectin, extensin, and other less abundant molecules.
- Cellulose, a large, linear glucose polymer, assembles spontaneously into microfibrils stabilized by hydrogen bonding.
- The cell wall defines the shapes of plant cells and restricts their elongation. Auxin-induced loosening of the cell wall permits elongation.
- Adjacent plant cells can communicate through plasmodesmata, junctions that allow molecules to pass through complex channels connecting the cytosols of adjacent cells (see Figure 19-35).
- Plants do not produce homologs of the common adhesive molecules found in animals. Only a few adhesive molecules unique to plants have been well documented to date.

Perspectives for the Future

A deeper understanding of the integration of cells into tissues in complex organisms will draw on insights and techniques from virtually all subdisciplines of molecular cell biology—biochemistry, biophysics, microscopy, genetics, genomics, proteomics, and developmental biology—together with bioengineering and computer science. This area of cell biology is undergoing explosive growth.

An important set of questions for the future deals with the mechanisms by which cells detect and respond to mechanical forces on them and the extracellular matrix, as well as the influence of their three-dimensional arrangements and interactions. A related question is how this information is used to control cell and tissue structure and function. These issues involve the fields of biomechanics and mechanotransduction. Shear or other stresses can induce distinct patterns of gene expression and cell growth and can greatly alter cell metabolism and responses to extracellular stimuli. Mechanosensitive nonselective cation channels (NSC_{MS}), a least some of which appear to be members of the transient receptor potential (TRP) cation channel family, are activated by the stretch of plasma membrane and are important players in mechanotransduction, such as that involved in sensing sound in the ear, which is mediated, in part, by specialized cadherins. Most of the classes of molecules discussed in this chapter—ECM, adhesion receptors, CAMs, intracellular adapters, and the cytoskeleton—are thought to play crucial roles in mechanosensing and mechanotransduction. Future research should give us a far more sophisticated understanding of the roles of the three-dimensional organization of cells and ECM components and the forces acting on them under normal and pathological conditions in controlling the structures and activities of tissues. Applications of such understanding will provide new methods to explore basic cell/tissue biology and provide improved technologies for the search for novel therapies for disease.

Although junctions help play a key role in forming stable epithelial tissues and defining the shapes and functional properties of epithelia, they are not static. Remodeling in terms of replacement of older molecules with more recently synthesized molecules is ongoing, and the dynamic properties of junctions open the door to more substantial changes when necessary (the epithelial-mesenchymal transition during development, wound healing, etc.). Understanding the molecular mechanisms underlying the relationship between stability and dynamic change will provide new insights into morphogenesis, maintaining tissue integrity and function, and response to (or induction of) pathology.

Numerous questions relate to intracellular signaling from CAMs and adhesion receptors. Such signaling must be integrated with other cellular signaling pathways that are activated by various external signals (e.g., growth factors) so that the cell responds appropriately and in a coordinated fashion to many different simultaneous internal and external stimuli. It appears that small GTPase proteins participate in at least some of the integrated pathways associated with signaling between cellular junctions. How are the logic circuits constructed that allow cross talk between diverse signaling pathways? How do these circuits integrate the information from these pathways? How is the combination of outside-in and inside-out signaling mediated by CAMs and adhesion receptors merged into such circuits?

We can expect ever-increasing progress in the exploration of the influence of glycobiology (the study of biology of oligo- and polysaccharides) on cell biology. The importance of specialized GAG sequences in controlling cellular activities, especially interactions between some growth factors and their receptors, is now clear. With the identification of the biosynthetic mechanisms by which these complex structures are generated and the development of tools to manipulate GAG structures and test their functions in cultured systems and intact animals, we can expect a dramatic increase in our understanding of the cell biology of GAGs in the next several years. There is still much to learn about the biosynthesis, structures, and functions of many other glycoconjugates, such as the O-linked sugars on dystroglycan that are essential for its binding to its ECM ligands (laminin, etc.).

A structural hallmark of CAMs, adhesion receptors, and ECM proteins is the presence of multiple domains that impart diverse functions to a single polypeptide chain. It is generally agreed that such multidomain proteins arose evolutionarily by the assembly of distinct DNA sequences encoding the distinct domains. Genes encoding multiple domains provide opportunities to generate enormous sequence and functional diversity by alternative splicing and the use of alternate promoters within a gene. Thus even though the number of independent genes in the human genome seems surprisingly small in comparison with other organisms, far more distinct protein molecules can be produced than predicted from the number of genes. Such diversity seems very well suited to the generation of proteins that take part in specifying adhesive connections in the nervous system, especially the brain. Indeed, several groups of proteins expressed by neurons appear to have just such combinatorial diversity of structure. They include the protocadherins, a family of cadherins with many proteins encoded per gene (14–19 for the three genes in mammals); the neurexins, which comprise more than 1000 proteins encoded by three genes; and the Dscams, members of the IgCAM superfamily encoded by a *Drosophila* gene that has the potential to express 38,016 distinct proteins owing to alternative splicing. A continuing goal for future work will be to describe and understand the molecular basis of functional cell-cell and cell-matrix attachments—the "wiring"—in the nervous system and how that wiring ultimately permits complex neuronal control and, indeed, the intellect required to understand molecular cell biology.

Key Terms

adhesion receptor *803*
adherens junction *809*
anchoring junction *809*
basal lamina *808*
cadherin *803*
cell-adhesion molecule (CAM) *803*
cell wall *839*
collagen *805*
connexin *819*
desmosome *809*
epithelium *802*
epithelial-mesenchymal transition *833*
extracellular matrix (ECM) *833*
fibrillar collagen *825*
fibronectin *805*
gap junction *809*
glycosaminoglycan (GAG) *822*
hyaluronan *825*
immunoglobulin cell-adhesion molecule (IgCAM) *836*
integrin *803*
laminin *821*
multiadhesive matrix protein *805*
paracellular pathway *815*
plasmodesmata *841*
proteoglycan *805*
RGD sequence *816*
selectin *803*
syndecan *829*
tight junction *809*

Review the Concepts

1. Using specific examples, describe the two phenomena that give rise to the diversity of adhesive molecules.

2. Cadherins are known to mediate homophilic interactions between cells. What is a homophilic interaction, and how can it be demonstrated experimentally for E-cadherins?

3. Together with their role in connecting the lateral membranes of adjacent epithelial cells, adherins junctions play a role in controlling cell shape. What proteins and structures are involved in this role?

4. What is the normal function of tight junctions? What can happen to tissues when tight junctions do not function properly?

5. What is collagen, and how is it synthesized? How do we know that collagen is required for tissue integrity?

6. Using structural models, explain how integrins mediate outside-in and inside-out signaling.

7. Compare the functions and properties of each of three types of macromolecules that are abundant in the extracellular matrix of all tissues.

8. Many proteoglycans have cell-signaling roles. Regulation of feeding behavior by sydecans in the hypothalamic region of the brain is one example. How is this regulation accomplished?

9. You have synthesized an oligopeptide containing an RGD sequence surrounded by other amino acids. What is the effect of this peptide when added to a fibroblast cell culture grown on a layer of fibronectin absorbed to the tissue culture dish? Why does this happen?

10. Blood clotting is a crucial function for mammalian survival. How do the multiadhesive properties of fibronectin lead to the recruitment of platelets to blood clots?

11. How do changes in molecular connections between the extracellular matrix (ECM) and cytoskeleton give rise to Duchenne muscular dystrophy?

12. To fight infection, leukocytes move rapidly from the blood into the tissue sites of infection. What is this process called? How are adhesion molecules involved in this process?

13. The structure of a plant cell wall needs to loosen to accommodate cell growth. What signaling molecule controls this process?

14. Compare plasmodesmata in plant cells to gap junctions in animal cells.

Analyze the Data

Researchers have isolated two E-cadherin mutant isoforms that are hypothesized to function differently from the isoform of the wild-type E-cadherin. An E-cadherin negative mammary carcinoma cell line was transfected with the mutant E-cadherin genes A (part a in the figure, triangles) or B (part b) (triangles) or the wild-type E-cadherin gene (black circles) and

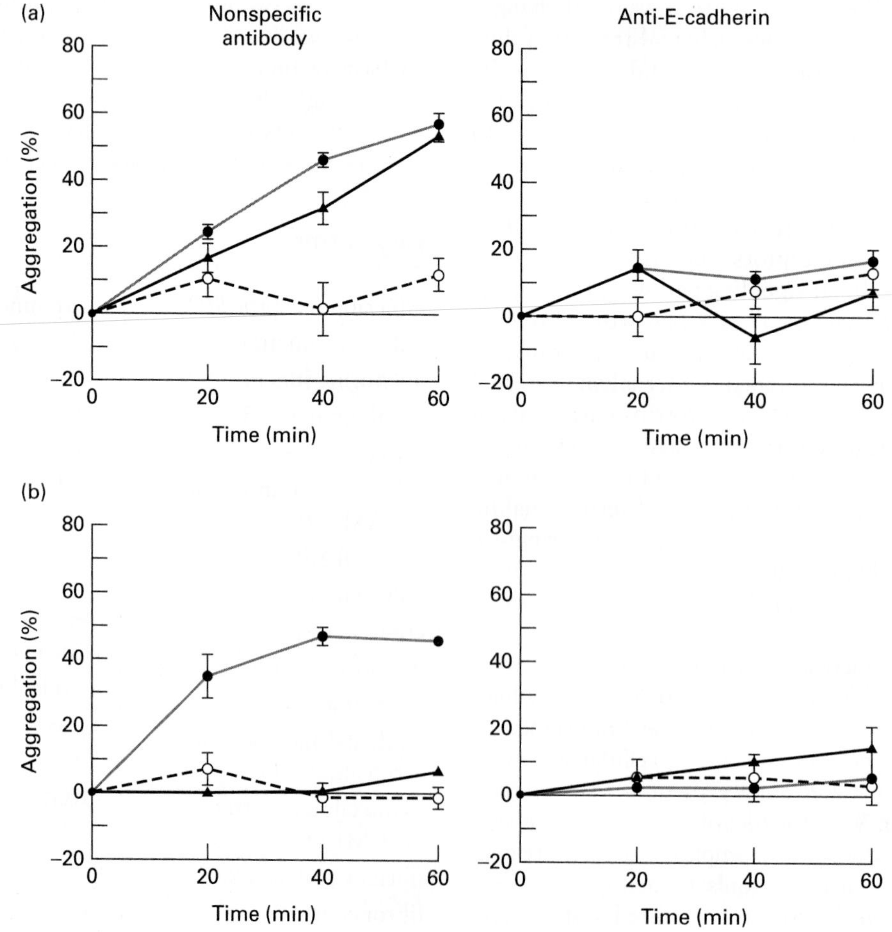

compared to untransfected cells (open circles) in an aggregation assay. In this assay, cells are first dissociated by trypsin treatment and then allowed to aggregate in solution over a period of minutes. Aggregating cells from mutants A and B are presented in panels a and b respectively. To demonstrate that the observed adhesion was cadherin mediated, the cells were pretreated with a nonspecific antibody (left panel) or a function-blocking anti-E-cadherin monoclonal antibody (right panel).

a. Why do cells transfected with the wild-type E-cadherin gene have greater aggregation than control, untransfected cells?

b. From these data, what can be said about the function of mutants A and B?

c. Why does the addition of the anti-E-cadherin monoclonal antibody, but not the nonspecific antibody, block aggregation?

d. What would happen to the aggregation ability of the cells transfected with the wild-type E-cadherin gene if the assay were performed in media low in Ca^{2+}?

References

Cell-Cell and Cell-Matrix Adhesion: An Overview

Carthew, R. W. 2005. Adhesion proteins and the control of cell shape. *Curr. Opin. Genet. Dev.* **15**(4):358–363.

M. Cereijido, R. G. Contreras, and L. Shoshani. 2004. Cell adhesion, polarity, and epithelia in the dawn of metazoans. *Physiol. Rev.* **84**:1229–1262.

Gumbiner, B. M. 1996. Cell adhesion: the molecular basis of tissue architecture and morphogenesis. *Cell* **84**:345–357.

Hynes, R. O. 1999. Cell adhesion: old and new questions. *Trends Cell Biol.* **9**(12):M33–M337. Millennium issue.

Hynes, R. O. 2002. Integrins: bidirectional, allosteric signaling machines. *Cell* **110**:673–687.

Hynes, R. O., and Q. Zhao. 2000. The evolution of cell adhesion. *J. Cell Biol.* **150**(2):F89–F96.

Ingber, D. E. 2006. Cellular mechanotransduction: putting all the pieces together again. *FASEB J.* **20**(7):811–827.

Jamora, C., and E. Fuchs. 2002. Intercellular adhesion, signalling and the cytoskeleton. *Nature Cell Biol.* **4**(4):E101–E108.

Juliano, R. L. 2002. Signal transduction by cell adhesion receptors and the cytoskeleton: functions of integrins, cadherins, selectins, and immunoglobulin-superfamily members. *Ann. Rev. Pharmacol. Toxicol.* **42**:283–323.

Leahy, D. J. 1997. Implications of atomic resolution structures for cell adhesion. *Ann. Rev. Cell Devel. Biol.* **13**:363–393.

Thiery, Jean Paul, and Jonathan P. Sleeman. 2006. Complex networks orchestrate epithelial-mesenchymal transitions. *Nature Rev. Mol. Cell Biol.* **7**:131–142.

Vogel, V. 2006. Mechanotransduction involving multimodular proteins: converting force into biochemical signals. *Annu. Rev. Biophys. Biomol. Struct.* 35:459–488.

Vogel, V., and M. Sheetz. 2006. Local force and geometry sensing regulate cell functions. *Nat. Rev. Mol. Cell Biol.* 7(4):265–275.

Cell-Cell and Cell-ECM Junctions and Their Adhesion Molecules

The Cadherin Resource:

http://calcium.uhnres.utoronto.ca/cadherin/pub_pages/classify/index.htm

Chu, Y. S., et al. 2004. Force measurements in E-cadherin-mediated cell doublets reveal rapid adhesion strengthened by actin cytoskeleton remodeling through Rac and Cdc42. *J. Cell Biol.* **167**:1183–1194.

Clandinin, T. R., and S. L. Zipursky. 2002. Making connections in the fly visual system. *Neuron* **35**:827–841.

Conacci-Sorrell, M., J. Zhurinsky, and A. Ben-Ze'ev. 2002. The cadherin-catenin adhesion system in signaling and cancer. *J. Clin. Invest.* **109**:987–991.

Fuchs, E., and S. Raghavan. 2002. Getting under the skin of epidermal morphogenesis. *Nature Rev. Genet.* **3**(3):199–209.

Gates, J., and M. Peifer. 2005. Can 1000 reviews be wrong? Actin, alpha-catenin, and adherens junctions. *Cell* **123**(5):769–772.

Gumbiner, Barry M. 2005. Regulation of cadherin-mediated adhesion in morphogenesis. *Nature Rev. Mol. Cell Biol.* **6**:622–634.

Hatzfeld, M. 2007. Plakophilins: multifunctional proteins or just regulators of desmosomal adhesion? *Biochim. Biophys. Acta* **1773**:69–77.

Hobbie, L., et al. 1987. Restoration of LDL receptor activity in mutant cells by intercellular junctional communication. *Science* **235**:69–73.

Hollande, F., A. Shulkes, and G. S. Baldwin. 2005. Signaling the junctions in gut epithelium. *Science* **2005**(277):pe13.

Hunter, W. R. J. Barker, C. Zhu, and R. G. Gourdie. 2005. Zonula occludens-1 alters connexin43 gap junction size and organization by influencing channel accretion. *Mol. Biol. Cell* **16**(12):5686–5698.

Jefferson, Julius J., Conrad L. Leung, and Ronald K. H. Liem. 2004. Plakins: goliaths that link cell junctions and the cytoskeleton. *Nature Rev. Mol. Cell Biol.* **5**:542–553.

Laird, D. W. 2006. Life cycle of connexins in health and disease. *Biochem. J.* **394**(pt. 3):527–543.

Lee, J. M., S. Dedhar, R. Kalluri, and E. W. Thompson. 2006. The epithelial-mesenchymal transition: new insights in signaling, development, and disease *J. Cell Biol.* **172**(7):973–981.

Li, J., V. V. Patel, and G. L. Radice. 2006. Dysregulation of cell adhesion proteins and cardiac arrhythmogenesis. *Clin. Med. Res.* **4**(1):42–52.

Litjens, S. H., J. M. de Pereda, and A. Sonnenberg. 2006. Current insights into the formation and breakdown of hemidesmosomes. *Trends Cell Biol.* **16**(7):376–383.

Pierschbacher, M. D., and E. Ruoslahti. 1984. Cell attachment activity of fibronectin can be duplicated by small synthetic fragments of the molecule. *Nature* **309**(5963):30–33.

Schöck, F., and N. Perrimon. 2002. Molecular mechanisms of epithelial morphogenesis. *Ann. Rev. Cell Devel. Biol.* **18**:463–493.

Tanoue, T., and M. Takeichi. 2005. New insights into fat cadherins. *J. Cell Sci.* **118**(pt. 11):2347–2353.

Tsukita, S., M. Furuse, and M. Itoh. 2001. Multifunctional strands in tight junctions. *Nature Rev. Mol. Cell Biol.* **2**:285–293.

Vogelmann, R., M. R. Amieva, S. Falkow, and W. J. Nelson. 2004. Breaking into the epithelial apical-junctional complex—news from pathogen hackers. *Curr. Opin. Cell Biol.* **16**(1):86–93.

The Extracellular Matrix I: The Basal Lamina

Boutaud, A., et al. 2000. Type IV collagen of the glomerular basement membrane: evidence that the chain specificity of network assembly is encoded by the noncollagenous NC1 domains. *J. Biol. Chem.* **275**:30716–30724.

Esko, J. D., and U. Lindahl. 2001. Molecular diversity of heparan sulfate. *J. Clin. Invest.* **108**:169–173.

Hallmann, Rupert, et al. 2005. Expression and function of laminins in the embryonic and mature vasculature. *Physiol. Rev.* **85**:979–1000.

Hohenester, E., and J. Engel. 2002. Domain structure and organisation in extracellular matrix proteins. *Matrix Biol.* **21**(2):115–128.

Iozzo, R. V. 2005. Basement membrane proteoglycans: from cellar to ceiling. *Nature Rev. Mol. Cell Biol.* **6**(8):646–656.

Kanagawa, M., et al. 2005. Disruption of perlecan binding and matrix assembly by post-translational or genetic disruption of dystroglycan function. *FEBS Lett.* **579**(21):4792–4796.

Nakato, H., and K. Kimata. 2002. Heparan sulfate fine structure and specificity of proteoglycan functions. *Biochim. Biophys. Acta* **1573**:312–318.

Perrimon, N., and M. Bernfield. 2001. Cellular functions of proteoglycans: an overview. *Semin. Cell Devel. Biol.* **12**(2):65–67.

Rosenberg, R. D., et al. 1997. Heparan sulfate proteoglycans of the cardiovascular system: specific structures emerge but how is synthesis regulated? *J. Clin. Invest.* **99**:2062–2070.

Sasaki, T., R. Fassler, and E. Hohenester. 2004. Laminin: the crux of basement membrane assembly. *J Cell Biol.* **164**(7):959–963.

The Extracellular Matrix II: Connective and Other Tissues

Canty, E. G., and K. E. Kadler. 2005. Procollagen trafficking, processing and fibrillogenesis. *J. Cell Sci.* **118**:1341–1353.

Comelli, E. M., et al. 2006. A focused microarray approach to functional glycomics: transcriptional regulation of the glycome. *Glycobiology* **16**(2):117–131.

Couchman, J. R. 2003. Syndecans: proteoglycan regulators of cell-surface microdomains? *Nature Rev. Mol. Cell Biol.* **4**:926–938.

Kramer, R. Z., J. Bella, B. Brodsky, and H. M. Berman. 2001. The crystal and molecular structure of a collagen-like peptide with a biologically relevant sequence. *J. Mol. Biol.* **311**:131–147.

Mao, J. R., and J. Bristow. 2001. The Ehlers-Danlos syndrome: on beyond collagens. *J. Clin. Invest.* **107**:1063–1069.

Sakai, T., M. Larsen, and K. Yamada. 2003. Fibronectin requirement in branching morphogenesis. *Nature* **423**:876–881.

Shaw, L. M., and B. R. Olsen. 1991. FACIT collagens: diverse molecular bridges in extracellular matrices. *Trends Biochem. Sci.* **16**(5):191–194.

Weiner, S., W. Traub, and H. D. Wagner. 1999. Lamellar bone: structure-function relations. *J. Struc. Biol.* **126**:241–255.

Adhesive Interactions in Diverse Motile and Nonmotile Cells

Barresi, R., and K. P. Campbell. 2006. Dystroglycan: from biosynthesis to pathogenesis of human disease. *J. Cell Sci.* **119**(pt. 2):199–207.

Bartsch, U. 2003. Neural CAMs and their role in the development and organization of myelin sheaths. *Front. Biosci.* **8**:D477–D490.

Brummendorf, T., and V. Lemmon. 2001. Immunoglobulin superfamily receptors: cis-interactions, intracellular adapters and alternative splicing regulate adhesion. *Curr. Opin. Cell Biol.* **13**:611–618.

Cukierman, E., R. Pankov, and K. M. Yamada. 2002. Cell interactions with three-dimensional matrices. *Curr. Opin. Cell Biol.* **14**:633–639.

Even-Ram, S., and K. M. Yamada. 2005. Cell migration in 3D matrix. *Curr. Opin. Cell Biol.* **17**(5):524–532.

Geiger, B., A. Bershadsky, R. Pankov, and K. M. Yamada. 2001. Transmembrane crosstalk between the extracellular matrix and the cytoskeleton. *Nature Rev. Mol. Cell Biol.* **2**:793–805.

Griffith, L. G., and M. A. Swartz. 2006. Capturing complex 3D tissue physiology in vitro. *Nature Rev. Mol. Cell Biol.* **7**(3):211–224.

Lawrence, M. B., and T. A. Springer. 1991. Leukocytes roll on a selectin at physiologic flow rates: distinction from and prerequisite for adhesion through integrins. *Cell* **65**:859–873.

Nelson, C. M., and M. J. Bissell. 2005. Modeling dynamic reciprocity: engineering three-dimensional culture models of breast architecture, function, and neoplastic transformation. *Sem. Cancer Biol.* **15**(5):342–352.

Reizes, O., et al. 2001. Transgenic expression of syndecan-1 uncovers a physiological control of feeding behavior by syndecan-3. *Cell* **106**:105–116.

Rougon, G., and O. Hobert. 2003. New insights into the diversity and function of neuronal immunoglobulin superfamily molecules. *Annu. Rev. Neurosci.* **26**:207–238.

Shimaoka, M., J. Takagi, and T. A. Springer. 2002. Conformational regulation of integrin structure and function. *Ann. Rev. Biophys. Biomol. Struc.* **31**:485–516.

Somers, W. S., J. Tang, G. D. Shaw, and R. T. Camphausen. 2000. Insights into the molecular basis of leukocyte tethering and rolling revealed by structures of P- and E-selectin bound to SLe(X) and PSGL-1. *Cell* **103**:467–479.

Stein, E., and M. Tessier-Lavigne. 2001. Hierarchical organization of guidance receptors: silencing of netrin attraction by Slit through a Robo/DCC receptor complex. *Science* **291**:1928–1938.

Xiong, J. P., et al. 2001. Crystal structure of the extracellular segment of integrin αVβ3. *Science* **294**:339–345.

Plant Tissues

Bacic, A. 2006. Breaking an impasse in pectin biosynthesis. *Proc. Nat'l. Acad. Sci. USA* **103**(15):5639–5640.

Delmer, D. P., and C. H. Haigler. 2002. The regulation of metabolic flux to cellulose, a major sink for carbon in plants. *Metab. Eng.* **4**:22–28.

Iwai, H., N. Masaoka, T. Ishii, and S. Satoh. 2000. A pectin glucuronyltransferase gene is essential for intercellular attachment in the plant meristem. *Proc. Nat'l. Acad. Sci. USA* **99**:16319–16324.

Jorgensen, R. A., and W. J. Lucas. 2006. Teaching resources: movement of macromolecules in plant cells through plasmodesmata. *Science* **2006**(323):tr2.

Kim, S., J. Dong, and E. M. Lord. 2004. Pollen tube guidance: the role of adhesion and chemotropic molecules. *Curr. Top. Dev. Biol.* **61**:61–79.

Lord, E. M., and J. C. Mollet. 2002. Plant cell adhesion: a bioassay facilitates discovery of the first pectin biosynthetic gene. *Proc. Nat'l. Acad. Sci. USA* **99**:15843–15845.

Lord, E. M., and S. D. Russell. 2002. The mechanisms of pollination and fertilization in plants. *Ann. Rev. Cell Devel. Biol.* **18**:81–105.

Lough, T. J., and W. J. Lucas. 2006. Integrative plant biology: role of phloem long-distance macromolecular trafficking. *Annu. Rev. Plant Biol.* **57**:203–232.

Pennell, R. 1998. Cell walls: structures and signals. *Curr. Opin. Plant Biol.* **1**:504–510.

Roberts, A. G., and K. J. Oparka. 2003. Plasmodesmata and the control of symplastic transport. *Plant Cell Environ.* **26**:103–124.

Ross, W. Whetten, J. J. MacKay, and R. R. Sederoff. 1998. Recent advances in understanding lignin biosynthesis. *Ann. Rev. Plant Physiol. Plant Mol. Biol.* **49**:585–609.

Somerville, C., et al. 2004. Toward a systems approach to understanding plant cell walls. *Science* **306**(5705):2206–2211.

Zambryski, P., and K. Crawford. 2000. Plasmodesmata: gatekeepers for cell-to-cell transport of developmental signals in plants. *Ann. Rev. Cell Devel. Biol.* **16**:393–421.

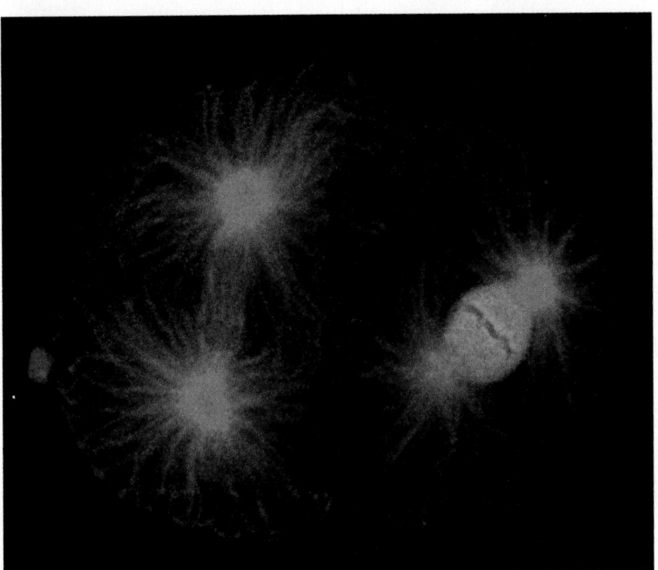

A two-cell *C. elegans* embryo stained with antibodies against tubulin (red) and CeBUB-1, a spindle checkpoint protein (green). DNA is stained with DAPI (blue). CeBUB-1 is localized on the chromosomes and kinetochore-attached spindle microtubules during metaphase in the smaller, posterior cell (*right*). It is presumed to monitor chromosome attachment and tension. The larger, anterior cell (*left*) has already entered anaphase, and CeBUB-1 is no longer detectable on the chromosomes or spindle microtubules. Thus asynchrony of this second cell cycle in the *C. elegans* embryo allows the observation of both the presence of a functional spindle checkpoint protein during metaphase, and its absence after initiation of anaphase. [Encanada et al., 2005, *Mol. Biol. Cell* **16:**1056.]

CHAPTER

20

REGULATING THE EUKARYOTIC CELL CYCLE

Proper control of **cell division** is vital to all organisms. In unicellular organisms, cell division must be balanced with cell growth so that cell size is properly maintained. If several divisions occur before parental cells have reached the proper size, daughter cells eventually become too small to be viable. If cells grow too large before cell division, the cells function improperly and the number of cells increases slowly. In developing multicellular organisms, the replication of each cell must be precisely controlled and timed to faithfully and reproducibly complete the developmental program in every individual. Each type of cell in every tissue must control its replication precisely for normal development of complex organs such as the brain or the kidney. In a normal adult, cells divide only when and where they are needed. However, loss of normal controls on cell replication is the fundamental defect in cancer, an all-too-familiar disease that kills one in every six people in the developed world (Chapter 25). The molecular mechanisms regulating eukaryotic cell division discussed in this chapter have gone a long way in explaining the loss of replication control in cancer cells. Appropriately, the initial experiments that elucidated the master regulators of cell division in all eukaryotes were awarded the Nobel prize in 2001.

The term **cell cycle** refers to the ordered series of macromolecular events that lead to cell division and the production of two daughter cells, each containing chromosomes identical with those of the parental cell. Two main molecular processes take place during the cell cycle, with resting intervals in between: during the S phase of the cycle, the parental

OUTLINE

chromosomes are duplicated; in mitosis (M phase), the resulting daughter chromosomes are distributed to each daughter cell (Figure 20-1). High accuracy and fidelity are required to assure that each daughter cell inherits the correct number of each chromosome. Further, chromosome replication and cell division must occur in the proper order in every cell division. If a cell undergoes the events of mitosis before the replication of all chromosomes has been completed, at least one daughter cell will lose genetic information. If a second round of replication occurs in one region of a chromosome before cell division occurs, the genes encoded in that region are increased in number out of proportion to other genes, a phenomenon that often leads to an imbalance of gene expression that is incompatible with viability.

To achieve the required level of accuracy and fidelity in chromosome replication and chromosome **segregation** to daughter cells during mitosis, and to coordinate these with cell growth and developmental programs, cell division is controlled by checkpoint surveillance mechanisms that prevent initiation of each step in cell division until earlier steps on which it depends have been completed. Mutations that inactivate or alter the normal operation of these checkpoints contribute to the generation of cancer cells because they result in chromosomal rearrangements and abnormal numbers of chromosomes, which lead to mutations and changes in gene expression level that cause uncontrolled cell growth (Chapter 25). Normally, such chromosomal abnormalities are prevented by multiple layers of control mechanisms that regulate the eukaryotic cell cycle.

In the late 1980s, it became clear that the molecular processes regulating the two key events in the cell cycle—chromosome replication and segregation—are fundamentally similar in all eukaryotic cells. Initially, it was surprising to many researchers that cells as diverse as baker's yeast and developing human neurons use nearly identical proteins to regulate their division. However, like transcription and protein synthesis, control of cell division appears to be a fundamental cellular process that evolved and was largely optimized early in eukaryotic evolution. Because of this similarity, research with diverse organisms, each with its own particular experimental advantages, has contributed to a growing understanding of how these events are coordinated and controlled. Biochemical, genetic, imaging, and micromanipulation techniques all have been employed in studying various aspects of the eukaryotic cell cycle. These studies have revealed that cell division is controlled primarily by regulating the timing of nuclear DNA replication and mitosis.

The master controllers of these events are a small number of *heterodimeric protein kinases* that contain a regulatory subunit (cyclin) and catalytic subunit (cyclin-dependent kinase, or CDK). These kinases regulate the activities of multiple proteins involved in DNA replication and mitosis by phosphorylating them at specific regulatory sites, activating some and inhibiting others to coordinate their activities. Regulated degradation of proteins also plays a prominent role in important cell-cycle transitions. Since protein degradation is irreversible, this ensures that the processes move in only one direction.

In this chapter, we first present an overview of the cell cycle and introduce the many different ways in which it is regulated. We then examine each cell-cycle transition in greater detail, and the work in various experimental systems that has led to our current understanding of these crucial regulatory mechanisms. Next we discuss the components of the mammalian cell cycle, and the system of checkpoints that further ensures proper progression through the cell cycle. Finally, we conclude with a discussion of meiosis, a special type of cell division that generates haploid germ cells (egg and sperm), and the molecular mechanisms that distinguish it from mitosis.

Overview Animation: Cell-Cycle Control

MEDIA CONNECTIONS

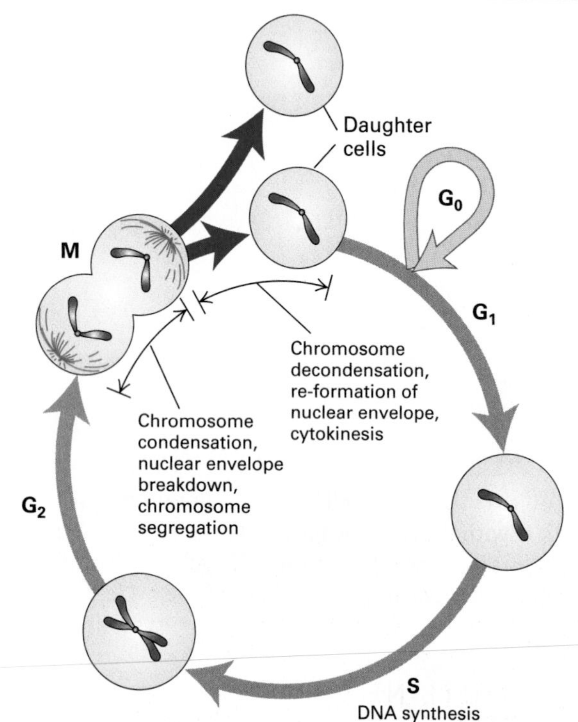

▲ **FIGURE 20-1 The fate of a single parental chromosome throughout the eukaryotic cell cycle.** Following mitosis (M), daughter cells contain $2n$ chromosomes in diploid organisms and $1n$ chromosomes in haploid organisms. In proliferating cells, G_1 is the period between the "birth" of a cell following mitosis and the initiation of DNA synthesis, which marks the beginning of the S phase. At the end of the S phase, cells enter G_2 containing twice the number of chromosomes as G_1 cells ($4n$ in diploid organisms, $2n$ in haploid organisms). The end of G_2 is marked by the onset of mitosis, during which numerous events leading to cell division occur. The G_1, S, and G_2 phases are collectively referred to as *interphase*, the period between one mitosis and the next. Most nonproliferating cells in vertebrates leave the cell cycle in G_1, entering the G_0 state. Although chromosomes condense only during mitosis, here they are shown in condensed form throughout the cell cycle to emphasize the number of chromosomes at each stage. For simplicity, the nuclear envelope is not depicted.

20.1 Overview of the Cell Cycle and Its Control

We begin our discussion by reviewing the stages of the eukaryotic cell cycle, presenting a summary of the current model of how the cycle is regulated, and briefly describing key experimental systems that have provided revealing information about cell-cycle regulation. As mentioned earlier, since the fundamental molecules involved in cell-cycle control are highly homologous in all eukaryotes, virtually everything learned about control of the cell cycle, whether it is from studies of yeast, sea urchins, or frogs, is relevant to control of the cell cycle in human cells.

The Cell Cycle Is an Ordered Series of Events Leading to Cell Replication

As illustrated in Figure 20-1, the cell cycle is divided into four major phases. Cycling (replicating) mammalian somatic cells grow in size and synthesize RNAs and proteins required for DNA synthesis during the **G_1 (first gap) phase.** When cells have reached the appropriate size and have synthesized the required proteins, they enter the **S (synthesis) phase**, the period in which they are actively replicating their chromosomes. After progressing through a second gap phase, the **G_2 phase**, cells begin the complicated process of mitosis, also called the **M (mitotic) phase**, which is divided into several stages (see Figure 20-2, *top*).

In discussing mitosis, we commonly use the term **chromosome** for the *replicated* structures that condense and become visible in the light microscope during the **prophase** period of mitosis. Thus each chromosome is composed of the two daughter DNA molecules resulting from DNA replication, plus the histones and other chromosomal proteins associated with them (see Figure 6-40). The two identical daughter DNA molecules and associated chromosomal proteins that form one chromosome are called sister **chromatids.** Sister chromatids are attached to each other by protein cross-links along their lengths. In vertebrates, these cross-links become confined to a single region of association called the **centromere** as chromosome condensation progresses.

During **interphase**, the portion of the cell cycle between the end of one M phase and the beginning of the next, the outer nuclear membrane is continuous with the endoplasmic reticulum (see Figure 9-1, **4**). With the onset of mitosis in prophase, the nuclear envelope retracts into the endoplasmic reticulum in most cells from higher eukaryotes, and Golgi membranes break down into vesicles. As described in Chapter 18, cellular microtubules form the **mitotic apparatus,** consisting of a football-shaped bundle of microtubules (the spindle) with a star-shaped cluster of microtubules radiating from each end, or spindle pole. During the **metaphase** period of mitosis, a multiprotein complex, the **kinetochore**, assembles at each centromere. The kinetochores of sister chromatids then associate with microtubules coming from opposite spindle poles (see Figure 18-36), and chromosomes align in a plane in the center of the cell. During the **anaphase** period of mitosis, sister chromatids separate. They initially are pulled by motor proteins along the spindle microtubules toward the opposite poles and then are further separated as the mitotic spindle elongates (see Figure 18-41).

Once chromosome separation is complete, the mitotic spindle disassembles and chromosomes decondense during **telophase.** The nuclear envelope re-forms around the segregated chromosomes as they decondense. The physical division of the cytoplasm, called **cytokinesis,** then yields two daughter cells as the Golgi complex re-forms in each daughter cell. Following mitosis, cycling cells enter the G_1 phase, embarking on another turn of the cycle. In yeasts and other fungi, the nuclear envelope does not break down during mitosis. In these organisms, the mitotic spindle forms within the nuclear envelope, which then pinches off, forming two nuclei at the time of cytokinesis.

In vertebrates and diploid yeasts, cells in G_1 have a **diploid** number of chromosomes ($2n$), one inherited from each parent. In haploid yeasts, cells in G_1 have one of each chromosome ($1n$), the **haploid** number. Rapidly replicating human cells progress through the full cell cycle in about 24 hours: mitosis takes ≈30 minutes; G_1, 9 hours; the S phase, 10 hours; and G_2, 4.5 hours. In contrast, the full cycle takes only ≈90 minutes in rapidly growing yeast cells.

In multicellular organisms, most differentiated cells "exit" the cell cycle and survive for days, weeks, or in some cases (e.g., nerve cells and cells of the eye lens) even the lifetime of the organism without dividing again. Such *postmitotic* cells generally exit the cell cycle in G_1, entering a phase called **G_0** (see Figure 20-1). Some G_0 cells can return to the cell cycle and resume replicating; this reentry is regulated, thereby providing control of cell proliferation.

Regulated Protein Phosphorylation and Degradation Control Passage Through the Cell Cycle

As mentioned in the chapter introduction, passage through the cell cycle is controlled by heterodimeric protein kinases. The concentrations of the **cyclins**, the regulatory subunits of the heterodimers, increase and decrease as cells progress through the cell cycle. The concentrations of the catalytic subunits of these kinases, called **cyclin-dependent kinases (CDKs)**, do not fluctuate in such a characteristic manner in yeast cells, but they have no kinase activity unless they are associated with a cyclin. Each CDK can associate with a small number of different cyclins that determine the substrate specificity of the complex, i.e., which proteins it phosphorylates. Cyclin-CDK complexes activate or inhibit hundreds of proteins involved in cell-cycle progression by phosphorylating them at specific regulatory sites. Thus proper progression through the cell cycle is governed by activation of the appropriate cyclin-CDK complex at the appropriate time. As we shall see, cells employ a variety of mechanisms for regulating the activities of each cyclin-CDK heterodimer.

Cyclin-CDKs are divided into three classes: G_1 cyclin-CDKs, S-phase cyclin-CDKs, and mitotic cyclin-CDKs (Figure 20-2). Two ubiquitin-protein ligases, *SCF* and the *anaphase-promoting complex*, are also key regulators of cell-cycle transitions. The anaphase-promoting complex is also sometimes called the *cyclosome*, and in this chapter is abbreviated as *APC/C*. Recall that ubiquitin-protein ligases polyubiquitinylate substrate proteins, marking them for degradation by proteasomes (see Figure 3-29).

G_1 Cyclin-CDKs G_1 cyclin-CDKs control entry into S phase by phosphorylating and thereby regulating transcription factors controlling genes required for chromosome replication. These include the enzymes that synthesize deoxynucleoside triphosphates, DNA polymerases and other replication proteins, and the S-phase cyclin-CDKs required for initiating DNA replication. In higher organisms, control of cell division is achieved primarily by regulating the synthesis and activity of G_1 cyclin-CDK complexes.

Focus Animation: Mitosis 

MEDIA CONNECTIONS

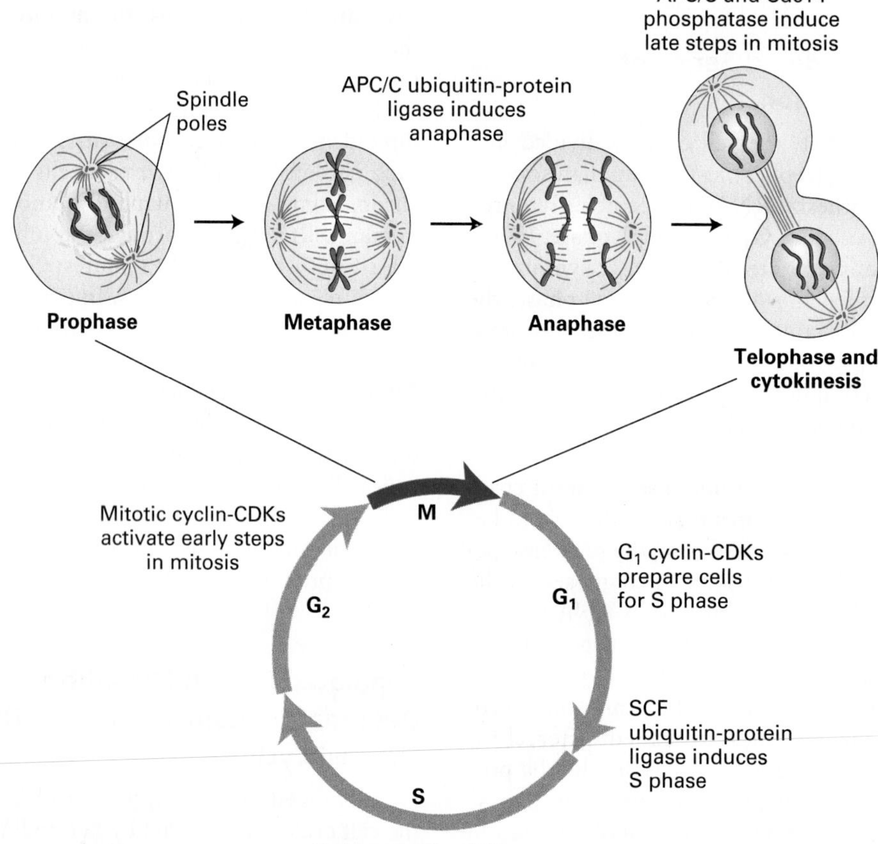

▲ **FIGURE 20-2 Regulation of cell-cycle transitions.** Cell-cycle transitions are regulated by cyclin-CDK protein kinases, protein phosphatases, and ubiquitin-protein ligases. The cell cycle is diagrammed, with the major stages of mitosis shown at the top. In early G_1, no cyclin-CDKs are active. In mid-G_1, G_1 cyclin-CDKs activate transcription of genes required for DNA replication. S phase is initiated by the SCF ubiquitin-protein ligase that polyubiquitinylates inhibitors of S-phase cyclin-CDKs, marking them for degradation by proteasomes. The S-phase cyclin-CDKs then activate DNA replication origins, and DNA synthesis commences. Once DNA replication is complete, cells enter G_2. In late G_2, mitotic cyclin-CDKs activate early steps of mitosis: nuclear envelope breakdown, remodeling of microtubules to form the mitotic spindle apparatus, chromosome condensation during prophase, and attachment of chromosome kinetochores to microtubules of the spindle apparatus in metaphase, as discussed in Chapter 18. Microtubule motors and shortening of spindle microtubules pull the daughter chromatids toward opposite spindle poles, but they cannot separate until the anaphase-promoting complex (APC/C), a ubiquitin-protein ligase, polyubiquitinylates securin, marking it for degradation by proteasomes. This results in degradation of protein complexes linking the sister chromatids at the centromere, and the onset of anaphase as sister chromatids separate. After chromosome movement to the spindle poles, the APC/C polyubiquitinylates mitotic cyclins, causing their degradation by proteasomes. The resulting drop in mitotic cyclin-CDK activity, along with the action of Cdc14 phosphatase, results in chromosome decondensation, reassembly of nuclear membranes around the daughter-cell nuclei, remodeling of microtubules into daughter-cell cytoskeletons, and cytokinesis.

S-phase Cyclin-CDKs S-phase cyclin-CDKs are synthesized in late G_1. However, they are immediately bound by inhibitors that prevent them from functioning. When G_1 cyclin-CDKs reach maximal activity, they phosphorylate these S-phase cyclin-CDK inhibitors, allowing the inhibitory proteins to be bound and polyubiquitinylated by the SCF ubiquitin-protein ligase. This results in a sudden increase in S-phase cyclin-CDK activity. The active S-phase cyclin-CDKs phosphorylate and activate protein components of prereplication complexes, which are assembled on DNA replication origins during G_1. DNA replication initiates, marking the onset of S phase.

Mitotic Cyclin-CDKs Mitotic cyclin-CDK complexes are synthesized during S phase and G_2, but their activities are held in check by phosphorylation at inhibitory sites until DNA synthesis is completed. Once activated, mitotic cyclin-CDKs phosphorylate and activate hundreds of proteins, including chromatin-associated proteins that promote chromosome condensation, nuclear envelope and nuclear pore complex proteins that cause retraction of the nuclear envelope into the ER, microtubule-associated proteins that remodel the microtubule cytoskeleton into the mitotic apparatus, kinetochore proteins that cause assembly of the kinetochore and its attachment to microtubules, as well as other protein kinases. These additional protein kinases, once activated by mitotic cyclin-CDKs, assist in the regulation of proteins involved in the events of mitosis. Once every kinetochore properly attaches to spindle microtubules and chromosomes are aligned at the metaphase plate, the APC/C ubiquitin-protein ligase polyubiquitinylates a key regulatory protein, securin, leading to its degradation by proteasomes. Securin normally acts to prevent degradation of the proteins that link sister chromatids at their centromeres; destruction of securin allows sister chromatids to separate, resulting in the onset of anaphase. The APC/C also directs degradation of mitotic cyclins, late in anaphase, once daughter chromosomes have segregated. The resulting fall in mitotic cyclin-CDK activity allows dephosphorylation of its target proteins. This leads to the onset of telophase and the decondensation of chromosomes, reassembly of the nuclear envelope and nuclear pore complexes around the segregated chromosomes to generate two daughter-cell nuclei, and cytokinesis, which completes cell division.

Ubiquitin-Protein Ligases SCF and APC/C ubiquitin-protein ligases control passage through three critical cell-cycle transitions—$G_1 \rightarrow$ S phase, metaphase $\rightarrow$ anaphase, and anaphase $\rightarrow$ telophase and cytokinesis—by regulating both cyclin-CDK activity and centromeric cohesion between sister chromatids. Because these transitions are triggered by the regulated degradation of proteins, an irreversible process, cells are forced to traverse the cell cycle in one direction only.

Checkpoint Mechanisms Surveillance mechanisms known as **checkpoints** operate to ensure that major cell-cycle transitions are not initiated until the previous process on which they depend has been completed. Cell-cycle transitions monitored by checkpoints include initiation of S phase, initiation of mitosis, separation of daughter chromosomes at anaphase, and onset of telophase and cytokinesis. Checkpoint surveillance mechanisms are responsible for the extraordinarily high fidelity of cell division, ensuring that each daughter cell receives the correct number of accurately replicated chromosomes. These checkpoint mechanisms function by controlling the protein kinase activities of the cyclin-CDKs through a variety of mechanisms: regulation of the synthesis and degradation of cyclins required for CDK protein kinase activity, phosphorylation of CDKs at inhibitory and activating sites, regulation of the synthesis and stability of CDK inhibitors (CKIs) that bind and inactivate CDKs or cyclin-CDK complexes, and regulation of the APC/C ubiquitin-protein ligase.

Diverse Experimental Systems Have Been Used to Identify and Isolate Cell-Cycle Control Proteins

The first evidence that diffusible factors regulate the cell cycle came from cell-fusion experiments with cultured mammalian cells. When interphase cells in the G_1, S, or G_2 phase of the cell cycle were fused to cells in mitosis, their nuclear envelopes retracted and their chromosomes condensed (Figure 20-3). This finding indicated that some diffusible component or components in the cytoplasm of the mitotic cells forced interphase nuclei to undergo many of the processes associated with early mitosis. We now know that these factors are the mitotic cyclin-CDK complexes.

Similarly, when cells in G_1 were fused to cells in S phase and the fused cells exposed to radiolabeled thymidine, the label was incorporated into the DNA of the G_1 nucleus as well as the S-phase nucleus, indicating that DNA synthesis began in the G_1 nucleus shortly after fusion. However, when cells in G_2 were fused to S-phase cells, no incorporation of labeled thymidine occurred in the G_2 nuclei. Thus diffusible factors in an S-phase cell can enter the nucleus of a G_1 cell and stimulate DNA synthesis, but these factors cannot induce DNA synthesis in a G_2 nucleus. We now know that these factors are S-phase cyclin-CDK complexes, which can activate the prereplication complexes assembled on DNA replication origins in early G_1 nuclei. DNA synthesis does not occur in G_2 nuclei because there are no prereplication complexes assembled on the DNA. Although these cell-fusion experiments demonstrated that diffusible factors control entry into the S and M phases of the cell cycle, genetic and biochemical experiments were needed to identify these factors.

The budding yeast *Saccharomyces cerevisiae* and the distantly related fission yeast *Schizosaccharomyces pombe* have been especially useful for isolation of mutants that are blocked at specific steps in the cell cycle or that exhibit altered regulation of the cycle. In both of these yeasts, organisms with **temperature-sensitive mutations** causing defects in specific proteins required to progress through the cell cycle are readily recognized microscopically and therefore easily isolated (see Figure 5-6). Such cells are called *cdc* (cell-*d*ivision *c*ycle) mutants. The wild-type alleles of recessive temperature-sensitive *cdc* mutant alleles can be isolated readily by transforming haploid mutant cells with

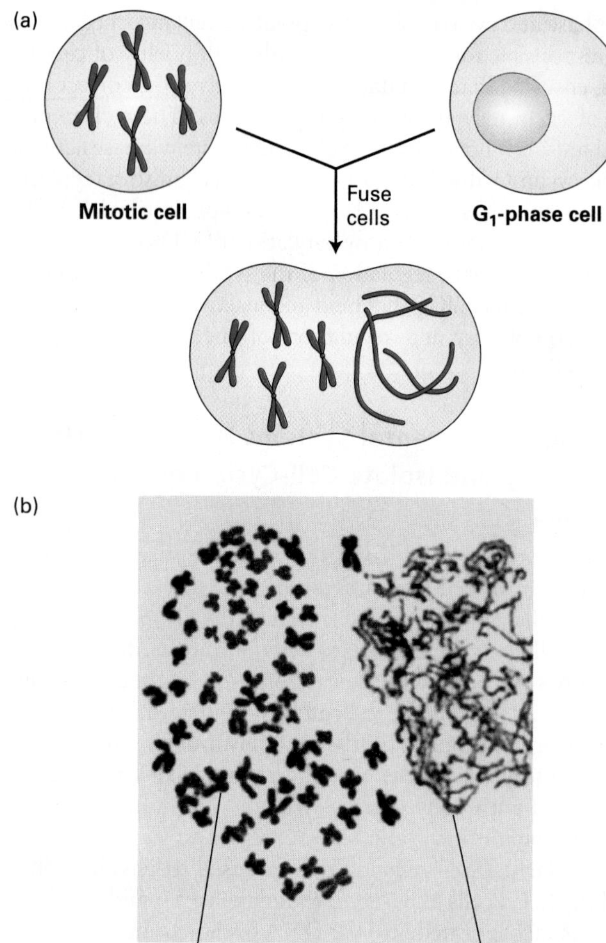

▲ **EXPERIMENTAL FIGURE 20-3 A diffusible factor in mitotic cells induces mitosis in interphase cells.** (a) Diagram of cell-fusion experiment. In unfused interphase cells (*right*), the nuclear envelope is intact and the chromosomes are not condensed, so individual chromosomes cannot be distinguished (see Figures 1-2b and 6-33a). In mitotic cells (*left*), the nuclear envelope is absent and the individual replicated chromosomes are highly condensed. When a cell in G_1 is fused to a cell in mitosis, the nuclear envelope of the G_1 cell retracts into the ER and the chromosomes condense, although not to the same extent as chromosomes in the mitotic cell. (b) A micrograph of chromosomes in a G_1 cell fused to a cell in mitosis as diagrammed in (a). [Part (b) from R. T. Johnson and P. N. Rao, 1970, *Biol. Rev.* **46**:97.]

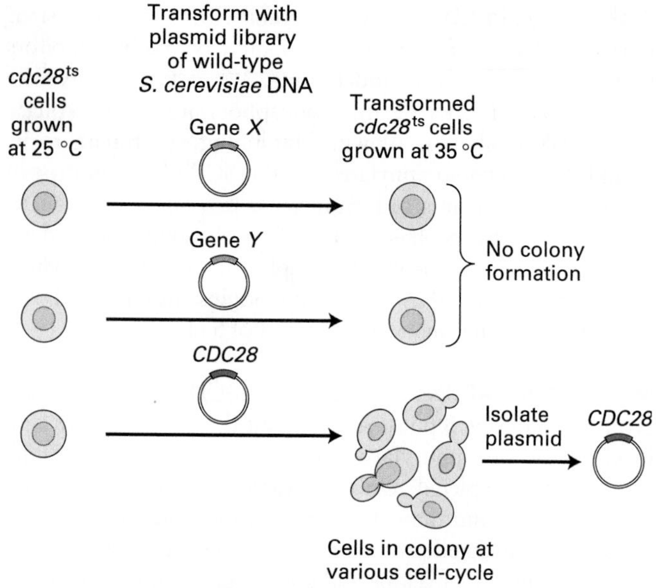

▲ **EXPERIMENTAL FIGURE 20-4 Wild-type cell-division cycle (*CDC*) genes can be isolated from a *S. cerevisiae* genomic library by functional complementation of *cdc* mutants.** Mutant cells with a temperature-sensitive mutation in a *CDC* gene are transformed with a genomic library prepared from wild-type cells and plated on nutrient agar at the nonpermissive temperature (35 °C). Each transformed cell takes up a single plasmid containing one genomic DNA fragment. Most such fragments include genes (e.g., *X* and *Y*) that do not encode the defective Cdc protein; transformed cells that take up such fragments do not form colonies at the nonpermissive temperature. The rare cell that takes up a plasmid containing the wild-type version of the mutant gene (in this case *CDC28*, a cyclin-dependent kinase) is complemented, allowing the cell to replicate and form a colony at the nonpermissive temperature. Plasmid DNA isolated from this colony carries the wild-type *CDC* gene corresponding to the gene that is defective in the mutant cells. The same procedure is used to isolate wild-type *cdc*$^+$ genes in *S. pombe*. See Figures 5-17 and 5-18 for more detailed illustrations of the construction and screening of a yeast genomic library.

a plasmid library prepared from wild-type cells and then plating the transformed cells at the nonpermissive temperature (Figure 20-4). When plated out, the haploid mutant cells cannot form colonies at the nonpermissive temperature. However, a transformed mutant cell can grow into a colony if it also contains a plasmid that carries the wild-type allele that complements the recessive mutation; the plasmids bearing the wild-type allele can then be recovered from those cells. Because many of the proteins that regulate the cell cycle are highly conserved, human cDNAs cloned into yeast expression vectors often can complement yeast cell-cycle mutants, leading to the rapid isolation of human genes encoding cell-cycle control proteins.

Biochemical studies require the preparation of cell extracts from many cells. For biochemical studies of the cell cycle, the eggs and early embryos of amphibians and marine invertebrates are particularly suitable. In these organisms, multiple synchronous cell cycles follow fertilization of a large egg. By isolating large numbers of eggs from females and fertilizing them simultaneously by addition of sperm (or treating them in ways that mimic fertilization), researchers can obtain extracts from cells at specific points in the cell cycle for analysis of proteins and enzymatic activities.

In the following sections we describe critical experiments that led to the current model of eukaryotic cell-cycle regulation summarized in Figure 20-2, and present further details of the various regulatory events. As we will see, results obtained with different experimental systems and approaches have provided insights about each of the key transition points in the cell cycle. For historical reasons, the names of various cyclins and cyclin-dependent kinases from yeasts and vertebrates differ. Table 20-1 lists the names of those that we discuss in this chapter and indicates when in the cell cycle they are active. Whenever possible, we will use general terms to describe cell-cycle regulators, instead of species-specific terminology.

TABLE 20-1 Selected Cyclins and Cyclin-Dependent Kinases (CDKs)

ORGANISM /PROTEIN	NAME
S. POMBE	
CDK (one only)	Cdc2
Mitotic cyclin (one only)	Cdc13
S. CEREVISIAE	
CDK (one only)	Cdc28
Mid-G_1 cyclin	Cln3
Late-G_1 cyclins	Cln1, Cln2
Early S-phase cyclins	Clb5, Clb6
Late S-phase and early mitotic cyclins	Clb3, Clb4
Late mitotic cyclins	Clb1, Clb2
VERTEBRATES	
Mid-G_1 CDKs	CDK4, CDK6
Late-G_1 and S-phase CDK	CDK2
Mitotic CDKs	CDK1, CDK2
Mid-G_1 cyclins	D-type cyclins
Late-G_1 and S-phase cyclin	Cyclin E
S-phase and mitotic cyclin	Cyclin A
Mitotic cyclins	Cyclin A, Cyclin B

NOTE: Those cyclins and CDKs discussed in this chapter are listed and classified by the period in the cell cycle in which they function. A heterodimer composed of a mitotic cyclin and CDK is commonly referred to as a *mitosis-promoting factor (MPF)*.

KEY CONCEPTS OF SECTION 20.1

Overview of the Cell Cycle and Its Control

- The eukaryotic cell cycle is divided into four phases: M (mitosis), G_1 (the period between mitosis and the initiation of nuclear DNA replication), S (the period of nuclear DNA replication), and G_2 (the period between the completion of nuclear DNA replication and mitosis) (see Figure 20-1).
- Cyclin-CDK complexes, composed of a regulatory cyclin subunit and a catalytic cyclin-dependent kinase subunit, regulate progress of a cell through the cell cycle (see Figure 20-2). Large multisubunit ubiquitin-protein ligases also polyubiquitinylate key cell-cycle regulators, marking them for degradation by proteasomes.
- Cell-cycle transitions are monitored by checkpoint surveillance mechanisms to ensure that each process is completed correctly before the next step is initiated.
- Cell-fusion experiments with cultured mammalian cells first demonstrated that diffusible factors regulate the cell cycle. Mitotic cyclin-CDK complexes cause chromosome condensation and disassembly of the nuclear envelope in G_1 and G_2 cells when they are fused to mitotic cells. Similarly, S-phase cyclin-CDK complexes stimulate DNA replication in the nuclei of G_1 cells when they are fused to S-phase cells.
- Genetic studies in yeast allowed the isolation of cell-division cycle *(cdc)* mutants, which led to the identification of genes that regulate the cell cycle (see Figure 20-4).
- Amphibian and invertebrate eggs and early embryos from synchronously fertilized eggs provide sources of extracts for biochemical studies of cell-cycle events.

20.2 Control of Mitosis by Cyclins and MPF Activity

The initial discovery of a protein that controls a critical cell-cycle transition came from research on mitosis. These studies led to two remarkable discoveries about control of the cell cycle. First, complex molecular processes in the early stages of mitosis such as chromosome condensation, nuclear envelope breakdown, mitotic spindle formation, and attachment of chromosome kinetochores to spindle microtubules can all be regulated and coordinated by a small number of master cell-cycle regulatory proteins. Second, these master regulators and the proteins that control them are highly conserved, so that cell-cycle studies in fungi, sea urchins, insects, frogs, and other species are directly applicable to all eukaryotic cells, including human cells.

A breakthrough in identification of the factor that induces mitosis came from studies of oocyte maturation in the frog *Xenopus laevis*. To understand these experiments, we must first lay out the events of oocyte maturation, which can be duplicated in vitro. As oocytes develop in the frog ovary, they replicate their DNA and become arrested in G_2 for 8 months, during which time they grow in size to a diameter of 1 mm, stockpiling all the materials needed for the multiple cell divisions required to generate a swimming, feeding tadpole (see Figure 5-3, *right,* for an overview of meiosis). When stimulated by a male, an adult female's ovarian cells secrete the steroid hormone progesterone, which induces the G_2-arrested oocytes to enter meiosis I and progress to the second meiotic metaphase (Figure 20-5). At this stage the cells are called *eggs*. When fertilized by sperm, the egg nucleus is released from its metaphase II arrest and completes meiosis. The resulting haploid egg pronucleus then fuses with the haploid sperm pronucleus, producing a diploid **zygote** nucleus. DNA replication follows and the first mitotic division of early embryogenesis begins. The resulting embryonic cells then proceed through 11 more rapid, synchronous cell cycles, generating a hollow sphere, the blastula. Cell division then slows, and subsequent divisions are nonsynchronous, with cells at different positions in the blastula dividing at different times. The advantage of using this system to study factors involved in mitosis is that large numbers of oocytes and eggs can be prepared that are all proceeding synchronously through the events that follow progesterone treatment and fertilization. This makes it possible to prepare sufficient amounts of cell extract for biochemical experiments.

In this section, we explore how experiments with *Xenopus* allowed the discovery and characterization of the factor that controls entry into and exit from mitosis. We'll also discuss the identification and characterization of some of its individual protein components, and how they regulate mitotic events.

Maturation-Promoting Factor (MPF) Stimulates Meiotic Maturation in Oocytes and Mitosis in Somatic Cells

When G_2-arrested *Xenopus* oocytes are removed from the ovary of an adult female frog and treated with progesterone in vitro, they undergo *meiotic maturation,* the process of oocyte maturation from a G_2-arrested oocyte to the egg arrested in metaphase of meiosis II (see Figure 20-5). Alternatively, microinjection of cytoplasm from eggs arrested in metaphase of meiosis II into G_2-arrested oocytes stimulates the oocytes to mature into eggs in the absence of proges-

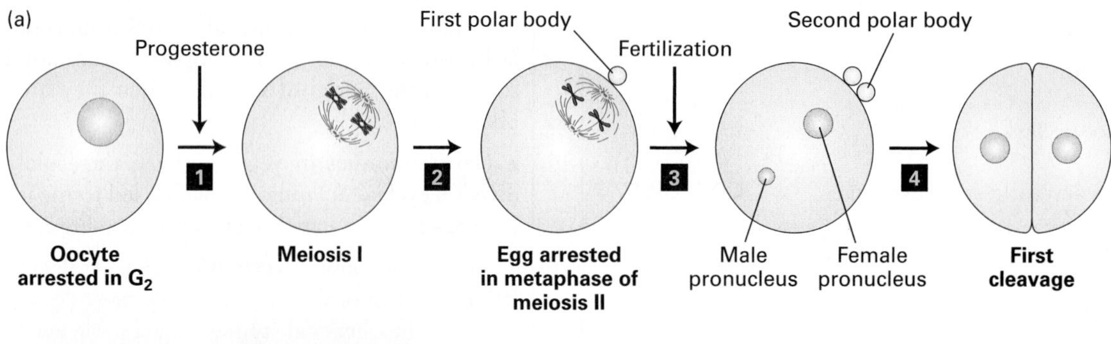

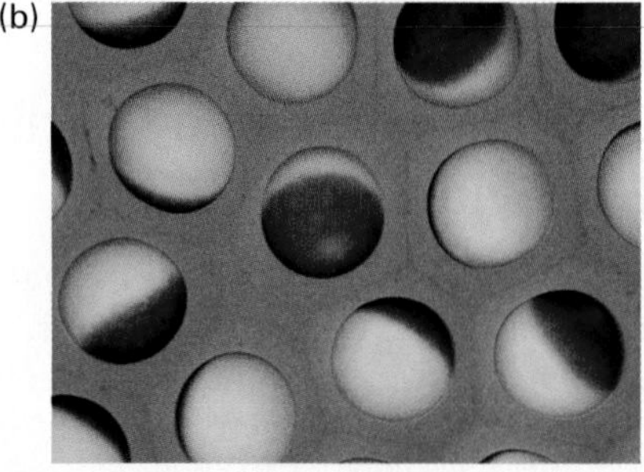

▲ EXPERIMENTAL FIGURE 20-5 **Progesterone stimulates meiotic maturation of *Xenopus* oocytes.** (a) Step 1: Treatment of G_2-arrested *Xenopus* oocytes surgically removed from the ovary of an adult female with progesterone causes the oocytes to enter meiosis I. Two pairs of synapsed homologous chromosomes (blue) connected to mitotic spindle microtubules (green) are shown schematically to represent cells in metaphase of meiosis I. Step 2: Segregation of homologous chromosomes and a highly asymmetrical cell division expels half the chromosomes into a small cell called the *first polar body*. The oocyte immediately commences meiosis II and arrests in metaphase to yield an egg. Two chromosomes connected to spindle microtubules are shown schematically to represent egg cells arrested in metaphase of meiosis II. Step 3: Fertilization by sperm releases eggs from their metaphase arrest, allowing them to proceed through anaphase of meiosis II and undergo a second highly asymmetrical cell division that eliminates one chromatid of each chromosome in a second polar body. Step 4: The resulting haploid female pronucleus fuses with the haploid sperm pronucleus to produce a diploid zygote, which undergoes DNA replication and the first mitosis of 12 synchronous early embryonic cleavages. (b) Micrograph of Xenopus oocytes. [Part (b): Copyright © ISM/Phototake]

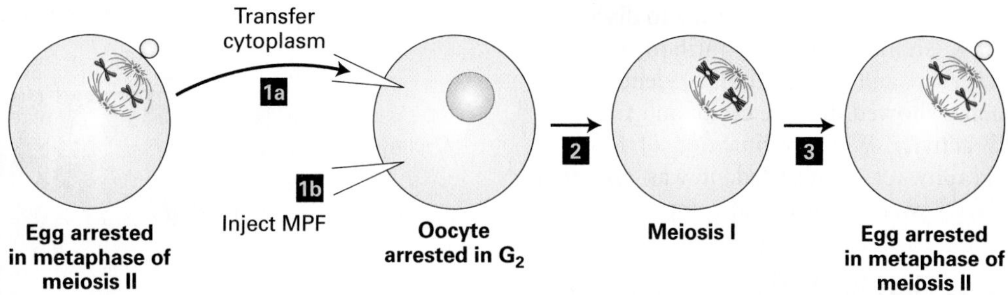

▲ **EXPERIMENTAL FIGURE 20-6 A diffusible factor in arrested *Xenopus* eggs promotes meiotic maturation.** When ≈5 percent of the cytoplasm from an unfertilized *Xenopus* egg arrested in metaphase of meiosis II is microinjected into a G_2-arrested oocyte (step **1a**), the oocyte enters meiosis I (step **2**) and proceeds to metaphase of meiosis II (step **3**), generating a mature egg in the absence of progesterone. This process can be repeated multiple times without further addition of progesterone, showing that egg cytoplasm contains an oocyte maturation-promoting factor (MPF). Microinjection of G_2-arrested oocytes provided the first assay for MPF activity (step **1b**) at different stages of the cell cycle and from different organisms. [See Y. Masui and C. L. Markert, 1971, *J. Exp. Zool.* **177:**129.]

terone (Figure 20-6). This system not only led to the initial identification of a factor in egg cytoplasm that stimulates maturation of oocytes in vitro but also provided an assay for this factor, called *maturation-promoting factor (MPF).* As we will see shortly, MPF turned out to be a mitotic cyclin-CDK heterodimer, the key factor that regulates the initiation of mitosis in all eukaryotic cells.

Using the microinjection system to assay MPF activity at different times during oocyte maturation in vitro, researchers found that untreated G_2-arrested oocytes have low levels of MPF activity, but treatment with progesterone induces MPF activity as the cells enter meiosis I (Figure 20-7). MPF activity falls as the cells enter the interphase between meiosis I and II, but then rises again as the cells enter meiosis II and remains high in the egg cells arrested in metaphase II. Following fertilization, MPF activity falls again until the zygote (fertilized egg) enters the first mitosis of embryonic development. Throughout the following 11 synchronous cycles of mitosis in the early embryo, MPF activity is low in the interphase periods between mitoses and then rises as the cells enter mitosis.

Although initially discovered in frogs, MPF activity has been found in mitotic cells from all species assayed. For example, cultured mammalian cells can be arrested in mitotic metaphase by treatment with compounds (e.g., colchicine) that inhibit assembly of microtubules. This is similar to the natural arrest of *Xenopus* eggs in metaphase of meiosis II until they are fertilized (Figure 20-5). When cytoplasm from such mitotically arrested mammalian cells was injected into G_2-arrested *Xenopus* oocytes, the oocytes matured into eggs; that is, the mammalian somatic mitotic cells contained a cytosolic factor that exhibited frog MPF activity. This finding suggested that MPF controls the entry of mammalian somatic cells into mitosis as well as the entry of frog oocytes into meiosis. When cytoplasm from mitotically arrested mammalian somatic cells was injected into interphase cells, the interphase cells entered mitosis; that is, their nuclear membranes retracted into the ER and their chromosomes condensed. Thus MPF is the diffusible factor, first revealed in cell-fusion experiments (see Figure 20-3), that promotes entry of cells into mitosis. Conveniently, the acronym MPF also can stand for **mitosis-promoting factor,** a name that denotes the more general activity of this factor.

Because the oocyte injection assay initially used to measure MPF activity is cumbersome, several years passed before MPF was purified by column chromatography and directly characterized. Before that was completed, the individual components of MPF were recognized through different experimental approaches. First we discuss here how the regulatory cyclin subunit was identified and then describe, in

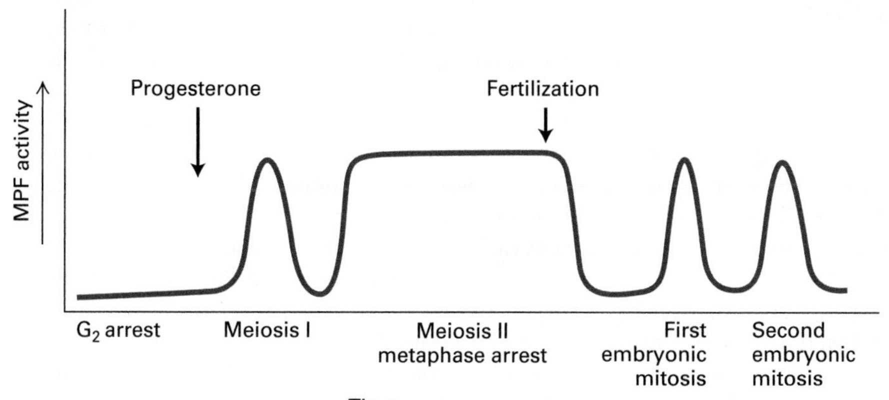

◀ **EXPERIMENTAL FIGURE 20-7 MPF activity in *Xenopus* oocytes, eggs, and early embryos peaks as cells enter meiosis and mitosis.** Diagrams of the cell structures corresponding to each stage are shown in Figure 20-5. MPF activity was determined by the microinjection assay shown in Figure 20-6 and quantitated by making dilutions of cell extracts. See text for discussion. [See J. Gerhart et al., 1984, *J. Cell Biol.* **98:**1247; adapted from A. Murray and M. W. Kirschner, 1989, *Nature* **339:**275.]

Section 20.3, how yeast genetic experiments led to discovery of the catalytic CDK subunit. Specific antibodies raised against the mitotic cyclin and CDK subunits identified by these other approaches showed that the cyclin and CDK co-purified with MPF activity. When purification of MPF by column chromatography was completed, it was apparent that the cyclin and CDK were its principal components.

Mitotic Cyclin Was First Identified in Early Sea Urchin Embryos

Experiments with protein synthesis inhibitors showed that new protein synthesis is required for the increase in MPF activity during the mitotic phase of each cell cycle in early frog embryos. As in early frog embryos, the initial cell cycles in the early sea urchin embryo occur synchronously, with all the embryonic cells entering mitosis simultaneously. The following biochemical experiment with sea urchin eggs and embryos identified the cyclin component of MPF. Large numbers of sea urchin eggs were collected, fertilized synchronously, and incubated in media containing radioactive amino acids. Small samples of synchronously dividing embryos were removed at multiple intervals after fertilization, and radiolabeled proteins were analyzed by SDS-gel electrophoresis and autoradiography of the resulting gel (Figure 20-8). One abundantly expressed protein peaked in concentration early in mitosis, fell abruptly just before cell cleavage, and then accumulated again during the following interphase to peak early in the next mitosis, falling abruptly just before the second cleavage (Figure 20-8). Careful analysis showed that this protein, named *cyclin B,* is synthesized continuously during the embryonic cell cycles but is abruptly destroyed following each anaphase. Since its concentration peaks in mitosis, cyclin B functions as a *mitotic cyclin.*

In subsequent experiments, a cDNA clone encoding sea urchin cyclin B was used as a probe to isolate a homologous cyclin B cDNA from *Xenopus laevis.* Expression of this cDNA generated pure *Xenopus* cyclin B protein, which was used to produce antibody specific for cyclin B. In Western blots, this antibody recognizes a polypeptide that co-purifies with MPF activity from *Xenopus* eggs, demonstrating that cyclin B is indeed one component of MPF.

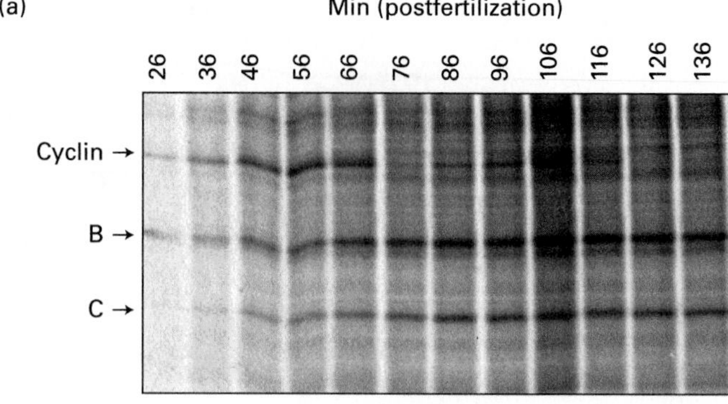

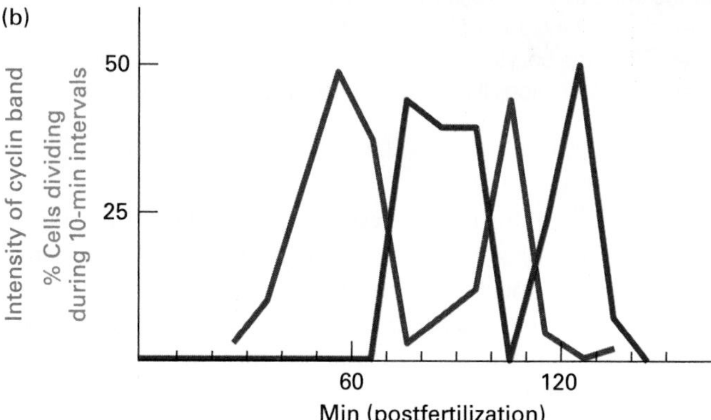

▲ EXPERIMENTAL FIGURE 20-8 **Autoradiography permits the detection of cyclical synthesis and destruction of mitotic cyclin in sea urchin embryos.** A suspension of sea urchin eggs was synchronously fertilized by the addition of sea urchin sperm, and ^{35}S-methionine was added. At 10-minute intervals beginning 26 minutes after fertilization, samples were taken for protein analysis on an SDS-polyacrylamide gel and for detection of cell cleavage by microscopy. (a) Autoradiogram of the SDS gel showing samples removed at each time point. Most proteins, such as B and C, continuously increased in intensity. In contrast, cyclin suddenly decreased in intensity at 76 minutes after fertilization and then began increasing again at 86 minutes. The cyclin band peaked again at 106 min and decreased again at 126 min. (b) Plot of the intensity of the cyclin band (red line) and the fraction of cells that had undergone cleavage during the previous 10-minute interval (cyan line). Note that the amount of cyclin fell precipitously just before cell cleavage. [From T. Evans et al., 1983, *Cell* **33**:389; courtesy of R. Timothy Hunt, Imperial Cancer Research Fund.]

Cyclin B Levels and Kinase Activity of Mitosis-Promoting Factor (MPF) Change Together in Cycling *Xenopus* Egg Extracts

Two unusual aspects of the synchronous cell cycles in early *Xenopus* embryos provided a way to study the role of mitotic cyclin in controlling MPF activity. First, following fertilization of *Xenopus* eggs, the G_1 and G_2 periods are minimized during the initial 12 synchronous cell cycles. That is, once mitosis is complete, the early embryonic cells proceed immediately into the next S phase, and once DNA replication is complete, the cells progress almost immediately into the next mitosis. Second, the oscillation in MPF activity that occurs as early frog embryos enter and exit mitosis is observed in the cytoplasm of a fertilized frog egg even when the nucleus is removed and no transcription can occur. This phenomenon was discovered by removing the nucleus of a fertilized *Xenopus* egg and then removing cytoplasm after varying periods and assaying MPF activity by the oocyte maturation assay (see Figure 20-6). This independence of a "cell-cycle clock" occurs only in the early divisions of animal embryos where many of the cell cycle checkpoints we will discuss later do not operate. These observations indicate that all the cellular components required for progress through the truncated cell cycles in early *Xenopus* embryos are stored in the unfertilized egg. In somatic cells generated later in development and in yeasts considered in later sections, specific mRNAs must be produced at specific points in the cell cycle; however, in early *Xenopus* embryos, all the mRNAs necessary for the early cell divisions are present in the unfertilized egg.

Extracts prepared from unfertilized *Xenopus* eggs contain all the materials required for multiple cell cycles, including the enzymes and precursors needed for DNA replication, the histones and other chromatin proteins involved in assembling the replicated DNA into chromosomes, and the proteins and phospholipids required in formation of the nuclear envelope. These egg extracts also synthesize proteins encoded by mRNAs in the extract, including cyclin B.

When nuclei prepared from *Xenopus* sperm are added to such a *Xenopus* egg extract, the nucleus and DNA inside it are induced to behave as if progressing through the cell cycle. Sperm nuclei are used in this experiment because they are readily isolated in large numbers and contain highly condensed chromatin. The dramatic change in morphology that occurs when the sperm chromatin decondenses is obvious. When the highly condensed sperm nuclei are added to an egg extract, the sperm chromatin decondenses, and the DNA replicates one time. The replicated sperm chromosomes then condense and the nuclear envelope disassembles, just as it does in intact cells entering mitosis. About 10 minutes after the nuclear envelope disassembles, all the cyclin B in the extract suddenly is degraded, as it is in intact cells following anaphase. Following cyclin B degradation, the sperm chromosomes decondense and a nuclear envelope reforms around them, as in an intact cell during telophase. After about 20 minutes, the cycle begins again. These remarkable *Xenopus* egg extracts can mediate several of these cycles, which mimic the rapid synchronous cycles of an early frog embryo.

Studies with this egg extract experimental system were aided by development of a new assay for MPF activity. Using MPF purified with the help of the oocyte injection assay (see Figure 20-6), researchers found that MPF phosphorylates histone H1. This H1 kinase activity provided a much simpler and more easily quantitated assay for MPF activity than the oocyte injection assay. Armed with a convenient assay, researchers tracked the kinase activity of MPF and concentration of cyclin B in cycling *Xenopus* egg extracts. These studies showed that MPF activity rises and falls in synchrony with the concentration of cyclin B (Figure 20-9a). The early

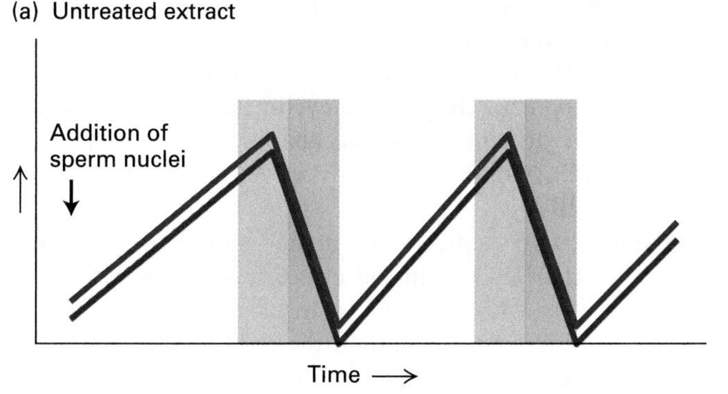

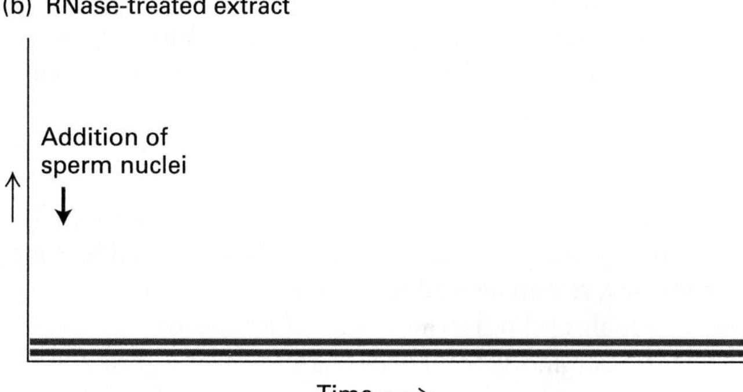

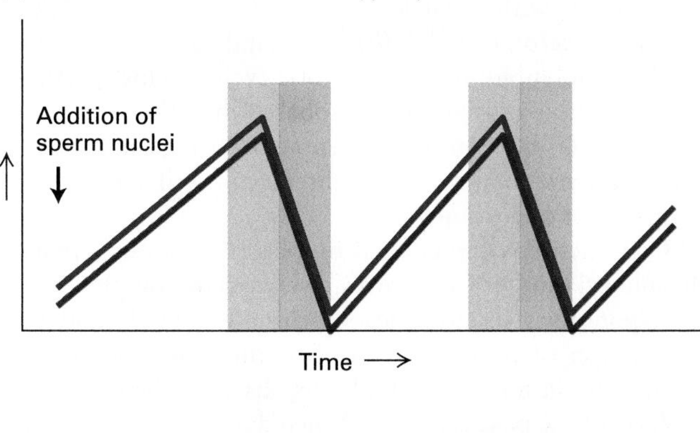

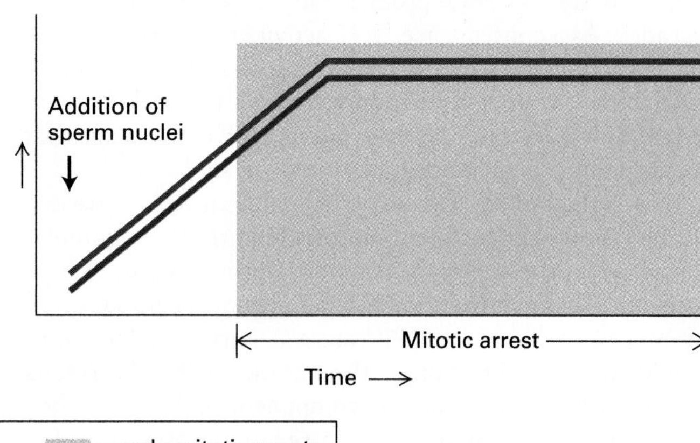

= MPF activity
= cyclin B concentration
= early mitotic events
= late mitotic events

▲ EXPERIMENTAL FIGURE 20-9 **Cycling of MPF activity and mitotic events in *Xenopus* egg extracts depend on synthesis and degradation of cyclin B.** In all cases, MPF kinase activity and cyclin B concentration were determined at various times after addition of sperm nuclei to a *Xenopus* egg extract treated as indicated in each panel. Microscopic observations determined the occurrence of early mitotic events (blue shading), including chromosome condensation and nuclear envelope disassembly, and of late events (orange shading), including chromosome decondensation and nuclear envelope reassembly. See text for discussion. [See A. W. Murray et al., 1989, *Nature* **339**:275; adapted from A. Murray and T. Hunt, 1993, *The Cell Cycle: An Introduction,* W. H. Freeman and Company.]

events of mitosis—chromosome condensation and nuclear envelope disassembly—occurred when MPF activity reached its highest levels in parallel with the rise in cyclin B concentration. The late events of mitosis—chromosome decondensation and reformation of the nuclear envelope—occurred when MPF activity fell in parallel with the cyclin B concentration.

To test the functions of cyclin B in these cell-cycle events, all mRNAs in the egg extract were degraded by digestion with a low concentration of RNase, which then was inactivated by addition of a specific inhibitor. This treatment destroys mRNAs without affecting the tRNAs and rRNAs required for protein synthesis. When sperm nuclei were added to the RNase-treated extracts, the nuclei replicated their DNA, but the increase in MPF activity and the early mitotic events, which the untreated extract supports, did not occur (Figure 20-9b). Addition of *cyclin B* mRNA, produced in vitro from cloned *cyclin B* cDNA, to the RNase-treated egg extract and sperm nuclei restored the parallel oscillations in the cyclin B concentration and MPF activity and the characteristic early and late mitotic events as observed with the untreated egg extract (Figure 20-9c). Since cyclin B is the only protein synthesized under these conditions, these results demonstrate that it is the crucial protein whose synthesis is required to regulate MPF activity and the cycles of chromosome condensation and nuclear envelope disassembly mediated by *Xenopus* egg extracts.

In these experiments, the late mitotic events of chromosome decondensation and nuclear envelope assembly coincided with decreases in the cyclin B level and MPF activity. To determine whether degradation of cyclin B is required for exit from mitosis, researchers added a mutant mRNA encoding a nondegradable cyclin B to a mixture of RNase-treated *Xenopus* egg extract and sperm nuclei. As shown in Figure 20-9d, MPF activity increased in parallel with the level of the mutant cyclin B, triggering condensation of the sperm chromatin and nuclear envelope disassembly (early mitotic events). However, the mutant cyclin B produced in this reaction was never degraded. As a consequence, MPF activity remained elevated, and the late mitotic events of chromosome decondensation and nuclear envelope formation were both blocked. This experiment demonstrates that the fall in MPF activity and exit from mitosis depend on degradation of cyclin B.

The results of the two experiments with RNase-treated extracts show that entry into mitosis requires the accumulation of cyclin B protein, the *Xenopus* mitotic cyclin, to high levels, and that exit from mitosis requires the degradation of this mitotic cyclin. Since MPF kinase activity varied in parallel with the concentration of the mitotic cyclin, the results implied that cyclin B was one component of MPF. Further, high MPF kinase activity causes entry into mitosis, and a fall in MPF kinase activity is required to exit mitosis.

Anaphase-Promoting Complex (APC/C) Controls Degradation of Mitotic Cyclins and Exit from Mitosis

Further studies revealed that vertebrate cells contain three proteins that can function like cyclin B to stimulate *Xenopus* oocyte maturation: two closely related cyclin Bs, and cyclin A. Collectively they are known as *B-type cyclins,* grouped because they are each degraded rapidly late in mitosis. In intact cells, degradation of all the B-type cyclins begins after the onset of anaphase, the period of mitosis when sister chromatids are separated and pulled toward opposite spindle poles.

Biochemical studies with *Xenopus* egg extracts showed that at the time of their degradation, wild-type B-type cyclins are modified by covalent addition of multiple **ubiquitin** molecules. As mentioned earlier, this process of polyubiquitination marks proteins for rapid degradation in eukaryotic cells by **proteasomes**, multiprotein cylindrical structures containing numerous proteases. *Ubiquitin-protein ligases* mediate addition of ubiquitin chains to B-type cyclins or other target proteins (see Figure 3-29). Many ubiquitin-protein ligases are multisubunit protein complexes.

Sequencing of cDNAs encoding several B-type cyclins from various eukaryotes showed that all contain a homologous nine-residue sequence near the N-terminus called the *destruction box*. Deletion of this destruction box, as in the mutant mRNA used in the experiment depicted in Figure 20-9d, prevents the polyubiquitination of B-type cyclins and thus makes them nondegradable. The ubiquitin-protein ligase that recognizes the mitotic cyclin destruction box is a multisubunit protein called the anaphase-promoting complex (APC/C), which we introduced earlier in the chapter (see Figure 20-2).

Figure 20-10 depicts the current model that best explains the changes in mitotic cyclin levels seen in cycling *Xenopus* early embryonic cells. In early embryos, mitotic cyclin is synthesized throughout the cell cycle from a stable mRNA. The observed fall in its concentration in late anaphase results from its APC/C-stimulated degradation at this point in the cell cycle. As we discuss in Section 20.5, genetic studies with yeast led to identification of an *APC/C specificity factor,* called *Cdh1,* that binds to APC/C and directs it to polyubiquitinylate mitotic cyclins. This specificity factor is active only in late anaphase when the segregating chromosomes have moved far enough apart in the dividing cell to assure that both daughter cells will contain one complete set of chromosomes.

Cdh1 activity is regulated by phosphorylation events—phosphorylation of Cdh1 by G_1 cyclin-CDK complexes during G_1 inhibits its association with the APC/C and thus degradation of mitotic cyclin. This inhibition permits the gradual rise in mitotic cyclin levels observed throughout interphase of the next cell cycle. When daughter chromosomes have segregated properly in late anaphase, the *Cdc14 phosphatase* is activated and dephosphorylates Cdh1, allowing it to bind to the APC/C. This interaction quickly leads to APC/C-mediated polyubiquitination and proteasomal degradation of B-type cyclins, and hence MPF inactivation (see Figure 20-10). In Section 20.7, we will see how the activity of Cdc14 itself is controlled to assure that a cell exits mitosis only when its chromosomes have segregated properly.

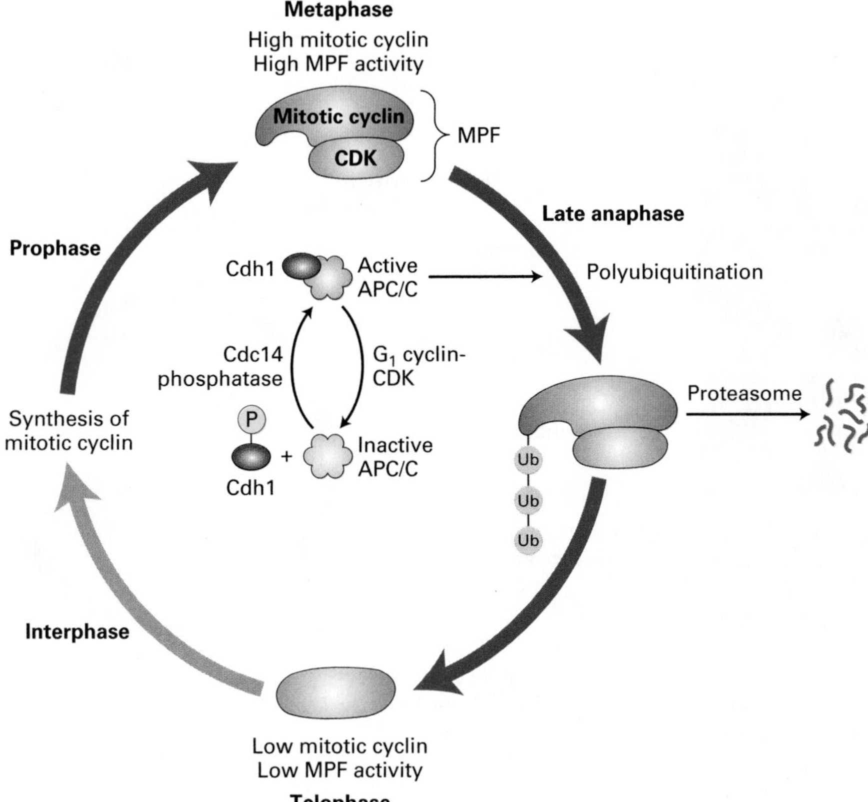

▲ **FIGURE 20-10 Regulation of mitotic cyclin levels in cycling *Xenopus* early embryonic cells.** In late anaphase, the anaphase-promoting complex (APC/C) polyubiquitinylates mitotic cyclins. As the cyclins are degraded by proteasomes, MPF kinase activity declines precipitously, triggering the onset of telophase. APC/C activity is directed toward mitotic cyclins by a specificity factor, called *Cdh1*, which is phosphorylated and thereby inactivated by G_1 cyclin-CDK complexes. A specific phosphatase called *Cdc14* removes the regulatory phosphate from the specificity factor late in anaphase. Once the specificity factor is inhibited in G_1, the concentration of mitotic cyclin increases, eventually reaching a high enough level to stimulate entry into the subsequent mitosis.

KEY CONCEPTS OF SECTION 20.2

Control of Mitosis by Cyclins and MPF Activity

- MPF is a protein kinase that requires a mitotic cyclin for activity. The protein kinase activity of MPF stimulates the onset of mitosis by phosphorylating multiple specific protein substrates, most of which remain to be identified.
- In the synchronously dividing cells of early *Xenopus* and sea urchin embryos, the concentration of mitotic cyclins (e.g., cyclin B) and MPF activity increase as cells enter mitosis and then fall as cells exit mitosis (see Figures 20-7 and 20-8).
- The rise and fall in MPF activity during the cell cycle result from concomitant synthesis and degradation of mitotic cyclin protein (see Figure 20-9).
- The multisubunit anaphase-promoting complex (APC/C) is a ubiquitin ligase that recognizes a conserved destruction box sequence in mitotic cyclins and promotes their polyubiquitination, marking the proteins for rapid degradation by proteasomes. The resulting decrease in MPF activity leads to completion of mitosis.
- The ubiquitin ligase activity of APC/C is controlled so that mitotic cyclins are polyubiquitinylated only during late anaphase (see Figure 20-10). Deactivation of APC/C in G_1 permits accumulation of mitotic cyclins during the next cell cycle. This results in the cyclical increases and decreases in MPF activity that cause the entry into and exit from mitosis.

20.3 Cyclin-Dependent Kinase Regulation During Mitosis

The studies with *Xenopus* egg extracts described in the previous section showed that continuous synthesis of a mitotic cyclin followed by its periodic degradation at late anaphase is required for the rapid cycles of mitosis observed in early *Xenopus* embryos. Identification of the catalytic protein kinase subunit of MPF and insight into its regulation initially came from genetic analysis of the cell cycle in the fission yeast *S. pombe*. An advantage of genetic studies is that genes involved in a process can be identified (and readily cloned from yeasts) without any prior knowledge of the biochemical activities of the proteins they encode.

S. pombe grows as a rod-shaped cell that increases in length as it grows and then divides in the middle during mitosis to

(a)

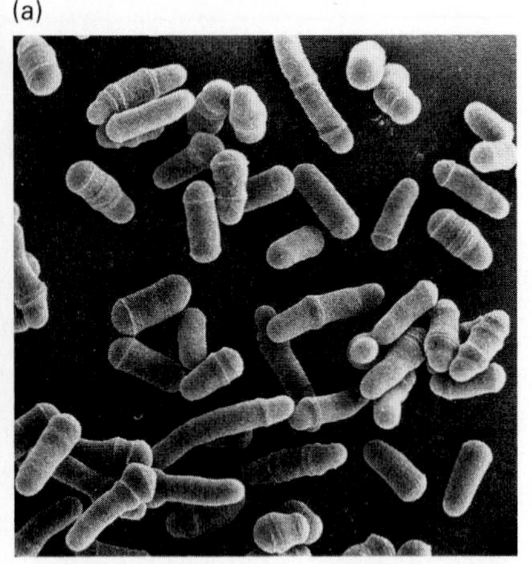

(b)

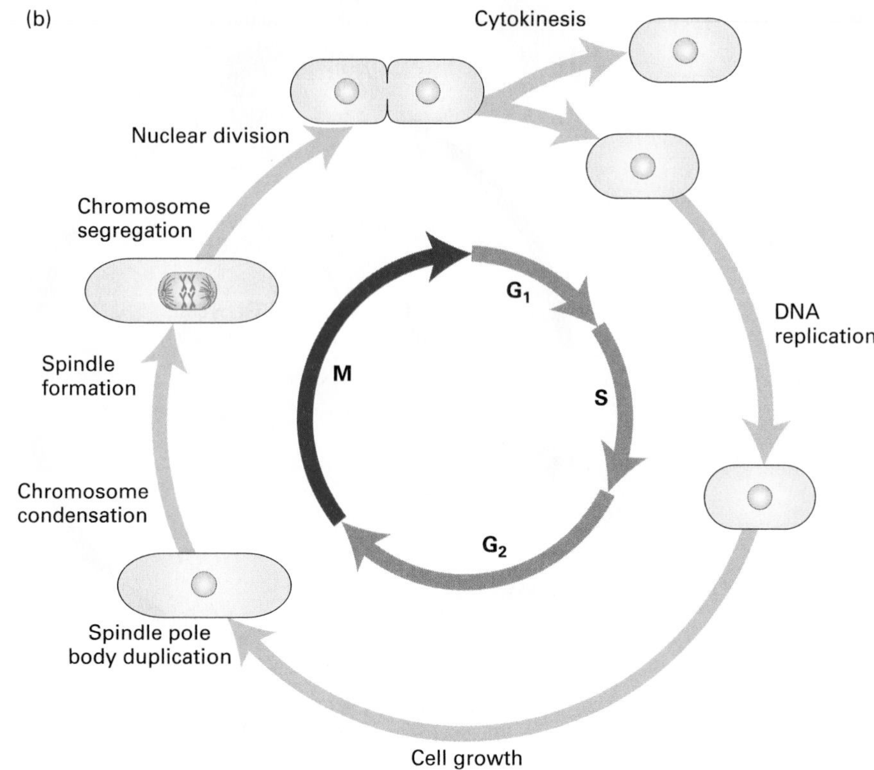

▲ **FIGURE 20-11 The fission yeast *S. pombe*.** (a) Scanning electron micrograph of *S. pombe* cells at various stages of the cell cycle. Long cells are about to enter mitosis; short cells have just passed through cytokinesis. (b) Main events in the *S. pombe* cell cycle. Note that the nuclear envelope does not disassemble during mitosis in *S. pombe* and other yeasts. [Part (a) courtesy of N. Hajibagheri.]

produce two daughter cells of equal size (Figure 20-11). Unlike most mammalian cells that grow primarily during G_1, this yeast does most of its growing during the G_2 phase of the cell cycle. Entry into mitosis is carefully regulated in response to cell size in order to properly coordinate cell division with cell growth. Consequently, this organism is ideal for isolating mutants in genes that regulate entry into mitosis since mutations that alter the timing of mitosis yield cells of abnormal size, a readily observed phenotype.

Temperature-sensitive mutants of *S. pombe* with conditional defects in the ability to progress through the cell cycle are easily recognized because they cause characteristic changes in cell length at the nonpermissive temperature. Many such mutants have been isolated, and fall into two groups. In the first group are *cdc* mutants, which fail to progress through one of the phases of the cell cycle at the nonpermissive temperature; they form extremely long cells because they continue to grow in length, but fail to divide. In contrast, *wee* mutants form smaller-than-normal cells because they are defective in the proteins that normally prevent cells from dividing when they are too small.

In *S. pombe* wild-type genes are indicated in italics with a superscript plus sign (e.g., *cdc2*$^+$); genes with a recessive mutation, in italics with a superscript minus sign (e.g., *cdc2*$^-$). The protein encoded by a particular gene is designated by the gene symbol in roman type with an initial capital letter (e.g., Cdc2).

In this section we see how genetic analyses as well as structural studies of the proteins involved allowed elucidation of the basic mechanisms controlling entry into mitosis. First we discuss how mitotic regulatory genes were identified in *S. pombe*, and how they were shown to be analogous to *Xenopus* MPF. Next we explore the mechanisms used by *S. pombe* to regulate mitotic cyclin-CDK activity. Mammalian cells regulate mitotic cyclin-CDK in a similar manner, and we end the section with an analysis of the structure of a human CDK and how its activity depends on phosphorylation-induced conformational changes.

MPF Components Are Conserved Between Lower and Higher Eukaryotes

Mutations in *cdc2,* one of several different *cdc* genes in *S. pombe,* produce opposite phenotypes depending on whether the mutation is recessive or dominant (Figure 20-12). Recessive mutations (*cdc2*$^-$) give rise to abnormally long cells, whereas dominant mutations (*cdc2*^{D}) give rise to abnormally small cells, the wee phenotype. As discussed in Chapter 5, **recessive** mutations generally cause a *loss* of the wild-type protein function; in diploid cells, both alleles must be mutant in order for the mutant phenotype to be observed. In contrast, **dominant** mutations generally result in a *gain* in protein function, either because of overproduction or lack of regulation; in this case, the presence of only one mutant allele confers the mutant phenotype in diploid cells. The finding that a loss of Cdc2 activity (*cdc2*$^-$ mutants) prevents entry into mitosis and a gain of Cdc2 activity (*cdc2*^{D} mutants) brings on mitosis earlier than normal identified Cdc2 as a key regulator of entry into mitosis in *S. pombe*.

The wild-type *cdc2*$^+$ gene contained in a *S. pombe* plasmid library was identified and isolated by its ability to

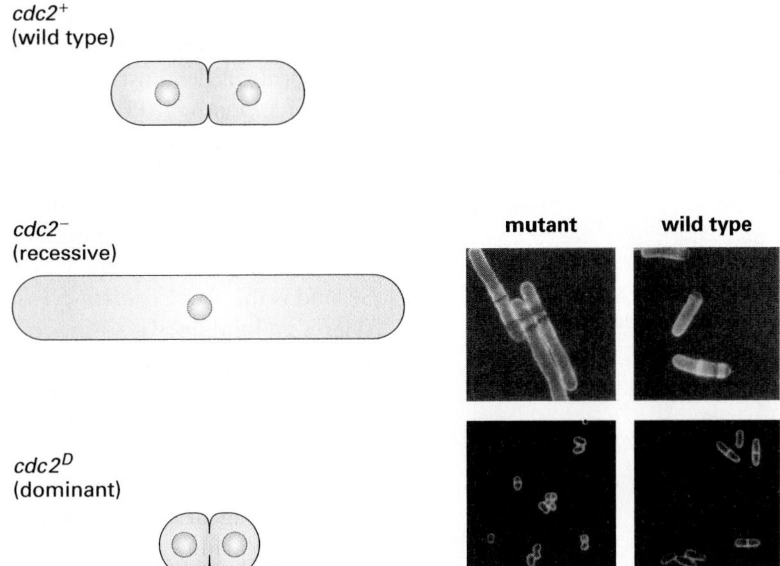

▲ EXPERIMENTAL FIGURE 20-12 **Recessive and dominant *S. pombe cdc2* mutants have opposite phenotypes.** The wild-type cell (*cdc2*$^+$) is depicted just before cytokinesis with two normal-size daughter cells. A recessive *cdc2*$^-$ mutant cannot enter mitosis at the nonpermissive temperature and appears as an elongated cell with a single nucleus, which contains duplicated chromosomes. A dominant *cdc2*^{D} mutant enters mitosis prematurely before reaching normal size in G_2; thus, the two daughter cells resulting from cytokinesis are smaller than normal—they have the wee phenotype. Upper micrographs on the right compare a *cdc2*$^-$ ts mutant and wt cells 5 h after shift to the non-permissive temperature. Lower micrographs compare a *cdc2*^{D} mutant to wt using an alternative cell fixation method. [Top photos: P. Nurse, P. Thuriaux, and K. Nasmyth, 1976, *Molec. Gen. Genet.* **146:**167; Bottom photos: P. Nurse, 2002, *Chem. Bio. Chem.* **3:**596.]

complement *cdc2*$^-$ mutants (see Figure 20-4). Sequencing showed that *cdc2*$^+$ encodes a 34-kDa protein with homology to eukaryotic protein kinases. In subsequent studies, researchers identified cDNA clones from other organisms that could complement *S. pombe cdc2*$^-$ mutants. Remarkably, they isolated a human cDNA encoding a protein identical with *S. pombe* Cdc2 in 63 percent of its residues. At the time of this experiment it was a tremendous surprise to scientists that a human protein could perform the essential functions of a protein from so distantly related an organism as *S. pombe*. The complementation of *S. pombe cdc2*$^-$ mutants by human Cdc2 was one of the first demonstrations that proteins performing fundamental cellular processes are highly conserved between all eukaryotic organisms.

Isolation and sequencing of another *S. pombe cdc* gene (*cdc13*$^+$), which also is required for entry into mitosis, revealed that it encodes a protein with homology to sea urchin and *Xenopus* cyclin B. Further studies showed that a heterodimer of Cdc13 and Cdc2 forms the *S. pombe* MPF. Like *Xenopus* MPF, this heterodimer has protein kinase activity that phosphorylates histone H1. Moreover, the H1 protein kinase activity rises as *S. pombe* cells enter mitosis and falls as they exit mitosis in parallel with the rise and fall in the Cdc13 protein level. These findings, which are completely analogous to the results obtained with *Xenopus* egg extracts (see Figure 20-9a), identified Cdc13 as the mitotic cyclin in *S. pombe*. Further studies showed that the isolated Cdc2 protein and its homologs in other eukaryotes have little protein kinase activity until they are bound by a cyclin. Hence, this family of protein kinases became known as cyclin-dependent kinases, or CDKs.

Researchers soon found that antibodies raised against a highly conserved region of Cdc2 recognize a polypeptide that co-purifies with MPF purified from *Xenopus* eggs. Thus *Xenopus* MPF is also composed of a CDK (called *CDK1*) plus a mitotic cyclin, cyclin B. This convergence of findings from biochemical studies in an invertebrate (sea urchin) and a vertebrate (*Xenopus*) and from genetic studies in a yeast indicated that entry into mitosis is controlled by analogous mitotic cyclin-CDK complexes in all eukaryotes (Figure 20-2).

Phosphorylation of the CDK Subunit Regulates the Kinase Activity of MPF

As we saw from the studies in *Xenopus* egg extracts and comparable biochemical studies in *S. pombe*, one way of regulating MPF activity is to control the stability of mitotic cyclins. Mitotic cyclins are suddenly degraded in late anaphase because they are polyubiquitinylated by the activated APC/C. A similar APC/C complex operates in *S. pombe* and all eukaryotes. Since a cyclin must be bound to a CDK for it to have significant kinase activity, the degradation of mitotic cyclin causes a drop in MPF activity. However, in *S. pombe* and all other eukaryotes, additional layers of regulation are used by the cell to ensure that cyclin-CDK complexes are active only at the appropriate time in the cell cycle. These additional layers of control were first revealed by studying mutations in *S. pombe* genes other than *cdc2*$^+$ (encoding the *S. pombe* CDK) or *cdc13*$^+$ (encoding the *S. pombe* mitotic cyclin) that also affect cell size at the nonpermissive temperature. For example, temperature-sensitive *cdc25*$^-$ mutants cannot enter mitosis at the nonpermissive temperature, producing elongated cells. On the other hand, overexpression of

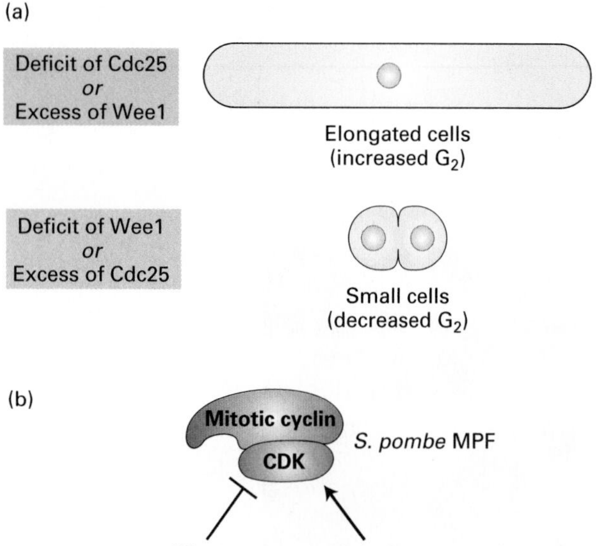

▲ EXPERIMENTAL FIGURE 20-13 **Cdc25 and Wee1 have opposing effects on *S. pombe* MPF activity.** (a) Cells that lack Cdc25 or Wee1 activity, as a result of recessive temperature-sensitive mutations in the corresponding genes, have the opposite phenotype. Likewise, cells with multiple copies of plasmids containing wild-type *cdc25*$^+$ or *wee1*$^+$, and which thus produce an excess of the encoded proteins, have opposite phenotypes. (b) These phenotypes imply that the mitotic cyclin-CDK complex is activated (→) by Cdc25 and inhibited (—|) by Wee1. See text for further discussion.

Cdc25 from a plasmid present in multiple copies per cell decreases the length of G_2, causing premature entry into mitosis and small (wee) cells (Figure 20-13a). Conversely, loss-of-function mutations in the *wee1*$^+$ gene causes premature entry into mitosis resulting in small cells, whereas overproduction of Wee1 protein increases the length of G_2 and results in elongated cells. A logical interpretation of these findings is that Cdc25 protein stimulates the kinase activity of *S. pombe* MPF, whereas Wee1 protein inhibits MPF activity (Figure 20-13b).

In subsequent studies, the wild-type *cdc25*$^+$ and *wee1*$^+$ genes were isolated, sequenced, and used to produce the encoded proteins with suitable expression vectors. The deduced sequences of Cdc25 and Wee1 and biochemical studies of the proteins demonstrated that they regulate the kinase activity of *S. pombe* MPF by phosphorylating and dephosphorylating specific regulatory sites in the CDK subunit of MPF.

Figure 20-14 illustrates the functions of four proteins that regulate the protein kinase activity of the *S. pombe* CDK. First is the mitotic cyclin of *S. pombe,* which associates with the CDK to form MPF with extremely low activity. Second is the *Wee1 protein-tyrosine kinase,* which phosphorylates an inhibitory tyrosine residue (Y15) in the CDK subunit. Third is another kinase, designated *CDK-activating kinase (CAK),* which phosphorylates an activating threonine residue (T161). When both residues are phosphorylated, MPF is inactive. Finally, the *Cdc25 phosphatase* removes the phosphate from Y15, yielding highly active MPF. Site-specific mutagenesis that changed the Y15 in *S. pombe* CDK to a phenylalanine, which cannot be phosphorylated, produced mutants with the wee phenotype, similar to that of *wee1*$^-$ mutants. Both mutations prevent the inhibitory phosphorylation at Y15, resulting in the inability to properly regulate MPF activity, leading to premature entry into mitosis.

As discussed further in Section 20.7, the checkpoint surveillance systems that ensure that chromosome replication is complete and that there is no unrepaired damage to chromosomes or DNA before initiating mitosis function by regulating the activities of the inhibitory Wee1 kinase and the activating Cdc25 phosphatase. Wee1 and Cdc25 homologs exist in higher eukaryotes, and very similar checkpoint control systems operate in human cells.

Conformational Changes Induced by Cyclin Binding and Phosphorylation Increase MPF Activity

Unlike both fission and budding yeasts, each of which produce just one CDK, vertebrates produce several CDKs (see Table 20-1). The three-dimensional structure of one human cyclin-dependent kinase (CDK2) has been determined and

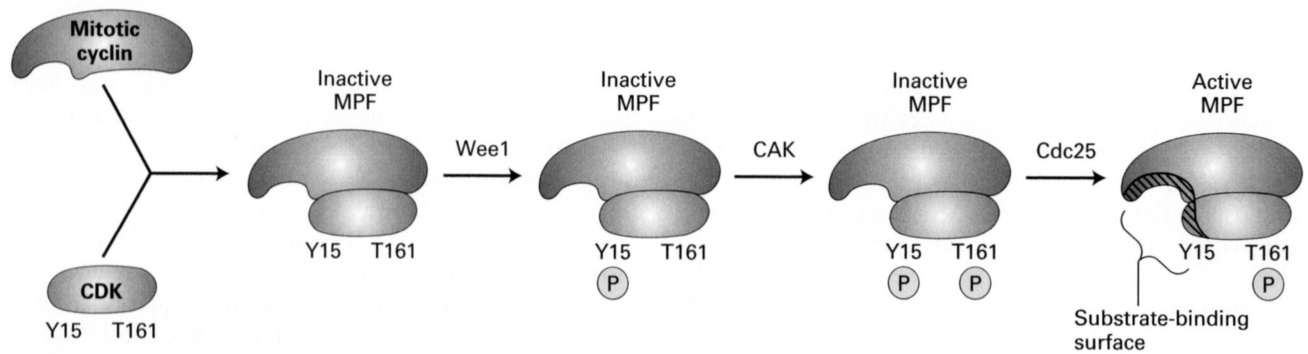

▲ FIGURE 20-14 **Regulation of the kinase activity of *S. pombe* mitosis-promoting factor (MPF).** Interaction of mitotic cyclin (Cdc13) with cyclin-dependent kinase (Cdc2) forms MPF. The CDK subunit can be phosphorylated at two regulatory sites: by Wee1 at tyrosine 15 (Y15) and by CDK-activating kinase (CAK) at threonine 161 (T161). Removal of the phosphate on Y15 by Cdc25 phosphatase yields active MPF in which the CDK subunit is phosphorylated at T161, and is unphosphorylated at Y15. The mitotic cyclin subunit contributes to the specificity of substrate binding by MPF, probably by forming part of the substrate-binding surface (crosshatch), which also includes the inhibitory Y15 residue.

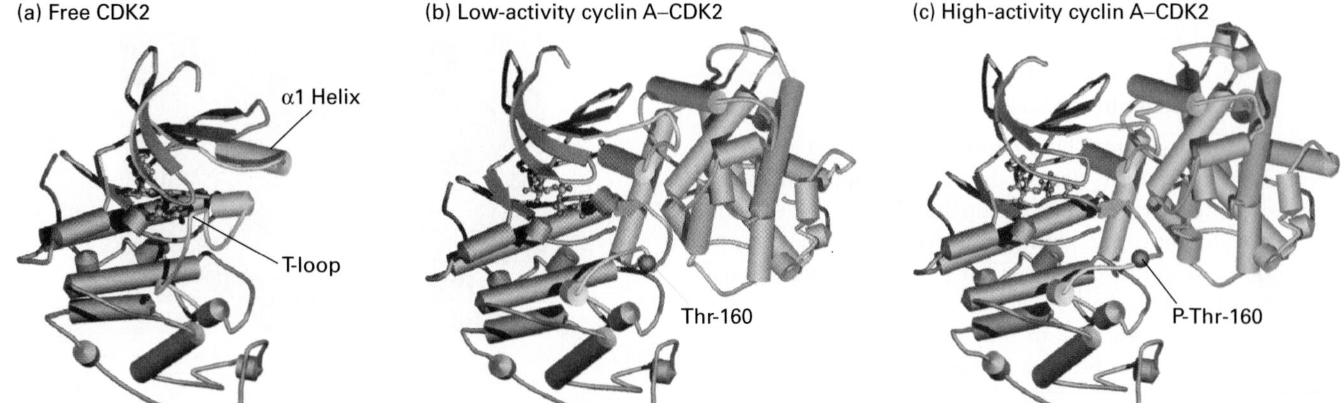

▲ **FIGURE 20-15 Structural models of human CDK2, which is homologous to the *S. pombe* cyclin-dependent kinase (CDK).** (a) Free, inactive CDK2 unbound to cyclin A. In free CDK2, the T-loop blocks access of protein substrates to the γ-phosphate of the bound ATP, shown as a ball-and-stick model. The conformations of the regions highlighted in yellow are altered when CDK is bound to cyclin A. (b) Unphosphorylated, low-activity cyclin A–CDK2 complex. Conformational changes induced by binding of a domain of cyclin A (green) cause the T-loop to pull away from the active site of CDK2, so that substrate proteins can bind. The α1 helix in CDK2, which interacts extensively with cyclin A, moves several angstroms into the catalytic cleft, repositioning key catalytic side chains required for the phosphotransfer reaction. The red ball marks the position equivalent to threonine 161 in *S. pombe* Cdc2. (c) Phosphorylated, high-activity cyclin A–CDK2 complex. The conformational changes induced by phosphorylation of the activating threonine (red ball) alter the shape of the substrate-binding surface, greatly increasing the affinity for protein substrates. [Courtesy of P. D. Jeffrey. See A. A. Russo et al., 1996, *Nature Struct. Biol.* **3**:696.]

provides insight into how cyclin binding and phosphorylation of CDKs regulate their protein kinase activity. Although the three-dimensional structures of the *S. pombe* CDK and most other CDKs have not been determined, their extensive sequence homology with human CDK2 suggests that all these CDKs have a similar structure and are regulated by a similar mechanism.

Unphosphorylated, inactive CDK2 contains a flexible region, called the *T-loop*, that blocks access of protein substrates to the active site where ATP is bound (Figure 20-15a). Steric blocking by the T-loop largely explains why free CDK2, unbound to cyclin, has little protein kinase activity. Unphosphorylated CDK2 bound to one of its cyclin partners has minimal but detectable protein kinase activity in vitro, although it may be essentially inactive in vivo. Extensive interactions between the cyclin and the T-loop cause a dramatic shift in the position of the T-loop, thereby exposing the CDK2 active site (Figure 20-15b). Binding of the cyclin also shifts the position of the α1 helix in CDK2, modifying its substrate-binding surface. High activity of the cyclin-CDK complex requires phosphorylation of the activating threonine, located in the T-loop, causing additional conformational changes in the cyclin-CDK2 complex that greatly increase its affinity for protein substrates (Figure 20-15c). As a result, the kinase activity of the phosphorylated complex is a hundredfold greater than that of the unphosphorylated complex.

The inhibitory tyrosine residue (Y15) in the *S. pombe* CDK is in the region of the protein that binds the ATP phosphates. Vertebrate CDK2 proteins contain an additional inhibitory residue, threonine-14 (T14), that is located in the same region of the protein. Phosphorylation of Y15 and T14 in these proteins prevents binding of ATP because of electrostatic repulsion between the phosphates linked to the protein and the phosphates of ATP. Thus these phosphorylations inhibit protein kinase activity even when the CDK protein is bound by a cyclin and the activating residue is phosphorylated.

So far we have discussed two mechanisms for controlling cyclin-CDK activity: (1) regulation of the concentration of mitotic cyclins as outlined in Figure 20-10 and (2) regulation of the kinase activity of MPF as outlined in Figure 20-14. In Section 20.5 we shall see that the protein kinase activities of cyclin-CDK complexes can also be regulated by CDK inhibitory proteins that bind to CDKs or cyclin-CDK complexes, blocking their ability to interact with substrates.

KEY CONCEPTS OF SECTION 20.3

Cyclin-Dependent Kinase Regulation During Mitosis

- In the fission yeast *S. pombe,* the *cdc2*$^+$ gene encodes a cyclin-dependent protein kinase (CDK) that associates with a mitotic cyclin encoded by the *cdc13*$^+$ gene. The resulting mitotic cyclin-CDK heterodimer is equivalent to *Xenopus* MPF. Mutants that lack either the mitotic cyclin or the CDK fail to enter mitosis and, therefore, form elongated cells.
- The protein kinase activity of the mitotic cyclin-CDK complex (MPF) depends on the phosphorylation state of two residues in the catalytic CDK subunit (see Figure 20-14). The activity is greatest when threonine 161 is

phosphorylated and is inhibited by Wee1-catalyzed phosphorylation of tyrosine 15, which interferes with correct binding of ATP. This inhibitory phosphate is removed by the Cdc25 protein phosphatase.

- The human cyclin-CDK2 complex is similar to MPF from *Xenopus* and *S. pombe*. Structural studies with the human proteins reveal that cyclin binding to CDK2 and phosphorylation of the activating threonine (equivalent to threonine 161 in the *S. pombe* CDK) cause conformational changes that expose the active site and modify the substrate-binding surface so that it has high activity and affinity for protein substrates (see Figure 20-15).

20.4 Molecular Mechanisms for Regulating Mitotic Events

In the previous sections, we have seen that a regulated increase in MPF activity induces entry into mitosis. Presumably, the entry into mitosis is a consequence of the phosphorylation of specific proteins by the protein kinase activity of MPF. However, until recently, the vast majority of proteins phosphorylated by MPF were not determined; consequently, precisely how MPF induces entry into mitosis is not well understood. Although many of the recently identified substrates of MPF remain to be studied, analysis of a small number of substrates has provided examples that show how their phosphorylation by MPF mediates many of the early events of mitosis: chromosome condensation, formation of the mitotic spindle, and disassembly of the nuclear envelope (see Figure 18-34).

Recall that a decrease in mitotic cyclins and the associated inactivation of MPF coincides with the later stages of mitosis (see Figure 20-9a). Just before this, in early anaphase, sister chromatids separate and move to opposite spindle poles. During telophase, microtubule dynamics return to interphase conditions, the chromosomes decondense, the nuclear envelope re-forms, the Golgi complex is remodeled, and cytokinesis occurs. Some of these processes are triggered by dephosphorylation; others, by protein degradation.

In this section, we discuss the molecular mechanisms and specific proteins associated with some of the events that characterize early and late mitosis. These mechanisms illustrate how cyclin-CDK complexes together with ubiquitin-protein ligases control passage through the mitotic phase of the cell cycle.

Phosphorylation of Nuclear Lamins and Other Proteins Promotes Early Mitotic Events

The nuclear envelope is a double-membrane extension of the rough endoplasmic reticulum containing many nuclear pore complexes (see Figure 9-1 and Figure 13-32). The lipid bilayer of the inner nuclear membrane is supported by the

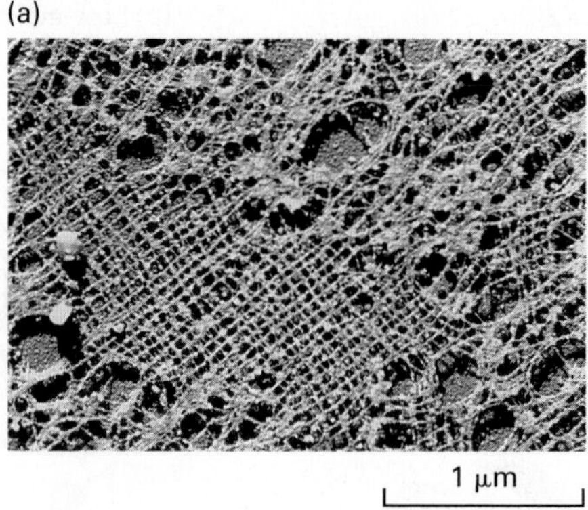

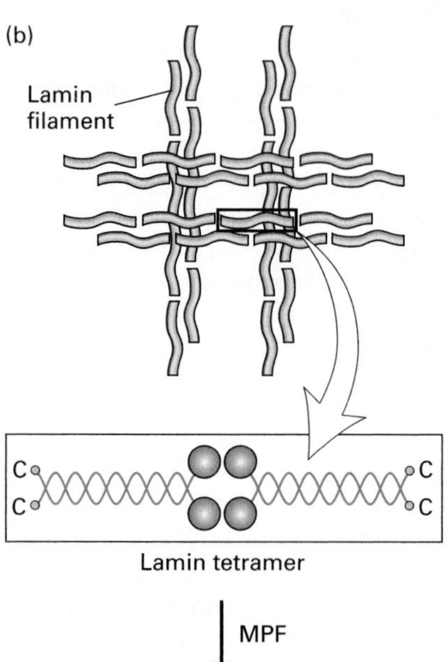

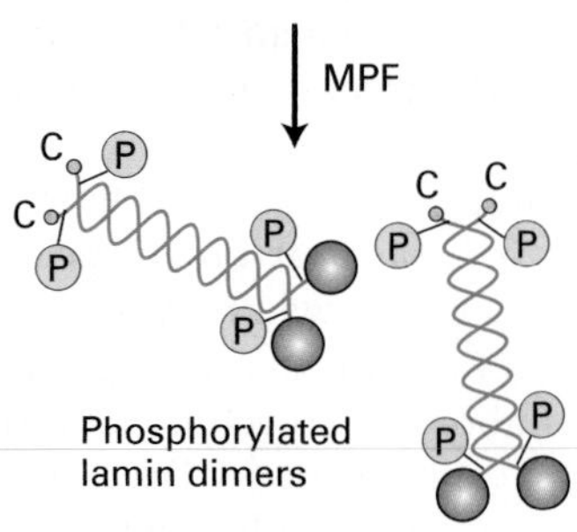

▲ **FIGURE 20-16 The nuclear lamina and its depolymerization.** (a) Electron micrograph of the nuclear lamina from a *Xenopus* oocyte. Note the regular meshlike network of lamin intermediate filaments. This structure lies adjacent to the inner nuclear membrane (see Figure 18-44). (b) Schematic diagrams of the structure of the nuclear lamina. Two perpendicular sets of 10-nm-diameter filaments built of lamins A, B, and C form the nuclear lamina (*top*). Individual lamin filaments are formed by end-to-end polymerization of lamin tetramers, which consist of two lamin coiled-coil dimers (*middle*). The red and green circles represent the globular N-terminal and C-terminal domains, respectively. Phosphorylation of specific serine residues near the ends of the rodlike central section of lamin dimers causes the tetramers to depolymerize (*bottom*). As a result, the nuclear lamina disintegrates. [Part (a) from U. Aebi et al., 1986, *Nature* **323:**560; courtesy of U. Aebi; part (b) adapted from A. Murray and T. Hunt, 1993, *The Cell Cycle: An Introduction*, W. H. Freeman and Company.]

nuclear lamina, a meshwork of lamin filaments located adjacent to the inside face of the nuclear envelope (Figure 20-16a). The three nuclear **lamins** (A, B, and C) present in vertebrate cells belong to the class of cytoskeletal proteins, the intermediate filaments, that are critical in supporting cellular membranes (Chapter 18).

Lamins A and C, which are encoded by the same transcription unit and produced by alternative splicing of a single pre-mRNA, are identical except for a 133-residue region at the C-terminus of lamin A, which is absent in lamin C. Lamin B, encoded by a different transcription unit, is modified post-transcriptionally by the addition of a hydrophobic isoprenyl group near its carboxyl terminus. This fatty acid becomes embedded in the inner nuclear membrane, thereby anchoring the nuclear lamina to the membrane (see Figure 10-19). All three nuclear lamins form dimers containing a rodlike α-helical coiled-coil central section and globular head and tail domains; polymerization of these dimers through head-to-head and tail-to-tail associations generates the intermediate filaments that compose the nuclear lamina (see Figure 18-45).

Once MPF is activated at the end of G_2 through events described in the last section, MPF phosphorylates specific serine residues in all three nuclear lamins. This causes depolymerization of the lamin intermediate filaments (Figure 20-16b). The phosphorylated lamin A and C dimers are released into solution, whereas the phosphorylated lamin B dimers remain associated with the nuclear membrane via their isoprenyl anchor. Depolymerization of the nuclear lamins leads to disintegration of the nuclear lamina meshwork and contributes to disassembly of the nuclear envelope. The experiment summarized in Figure 20-17 shows that disassembly of the nuclear envelope, which normally occurs early in mitosis, depends on phosphorylation of lamin A.

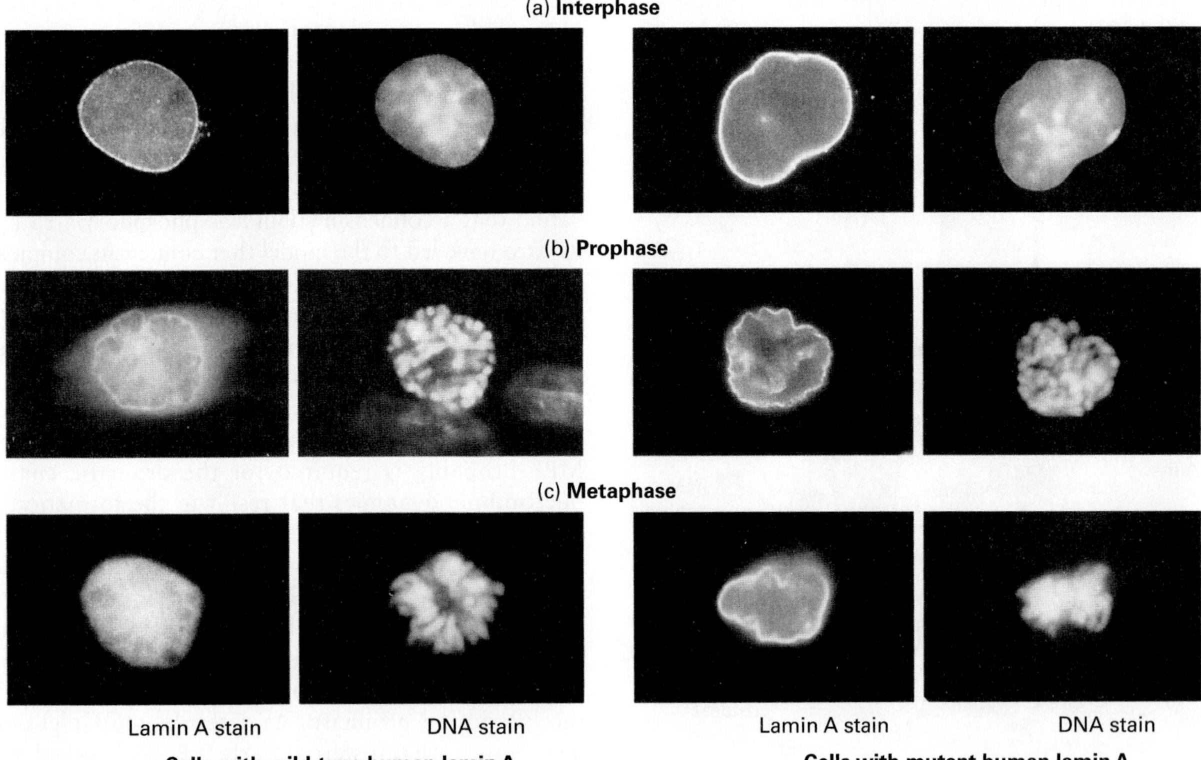

▲ **EXPERIMENTAL FIGURE 20-17 Phosphorylation of human lamin A causes lamin depolymerization.** Site-directed mutagenesis was used to prepare a mutant human *lamin A* gene encoding a protein in which alanines replace the serines that normally are phosphorylated in wild-type lamin A (see Figure 20-16b). As a result, the mutant lamin A cannot be phosphorylated. Expression vectors carrying the wild-type or mutant human gene were separately transfected into cultured hamster cells. Because the transfected *lamin* genes are expressed at much higher levels than the endogenous hamster *lamin* gene, most of the lamin A produced in transfected cells is human lamin A. Transfected cells at various stages in the cell cycle then were stained with a fluorescent-labeled monoclonal antibody specific for human lamin A and with a fluorescent dye that binds to DNA. The bright band of fluorescence around the perimeter of the nucleus in interphase cells stained for human lamin A represents polymerized (unphosphorylated) lamin A (a). In cells expressing the wild-type human lamin A, the diffuse lamin staining throughout the cytoplasm in prophase and metaphase (b and c) and the absence of the bright peripheral band in metaphase (c) indicate depolymerization of lamin A. In contrast, little lamin depolymerization occurred in cells expressing the mutant lamin A. DNA staining showed that the chromosomes were fully condensed by metaphase in cells expressing either wild-type or mutant lamin A. [From R. Heald and F. McKeon, 1990, *Cell* **61**:579.]

Curiously, spontaneous dominant mutations in the lamin A/C gene *(LMNA)* cause the rare syndrome Hutchinson-Guilford progeria. Patients expressing one of these mutant lamins A and C undergo a greatly accelerated rate of aging. Other LMNA mutations cause striated muscular diseases, abnormal fat cell function, and peripheral nerve cell diseases. While the molecular mechanisms underlying these symptoms are not understood, the observation that different mutations in the *LMNA* gene produce distinct syndromes suggests that lamin A and C perform several different functions in normal cells. If that were the case, mutations that affect one or another of these functions might produce the distinct group of symptoms that constitute the different syndromes associated with lamin A/C mutations. ∎

MPF-catalyzed phosphorylation of specific **nucleoporins** (Figure 20-18 1) causes nuclear pore complexes to dissociate into subcomplexes during prophase. Phosphorylation of integral membrane proteins of the inner nuclear membrane (Figure 20-18 2) is thought to decrease their affinity for chromatin and further contribute to disassembly of the nuclear envelope. The weakening of the associations between the inner nuclear membrane proteins and the nuclear lamina (Figure 20-18 3) and chromatin (Figure 20-18 4) allows sheets of inner nuclear membrane to retract into the endoplasmic reticulum, which is continuous with the outer nuclear membrane.

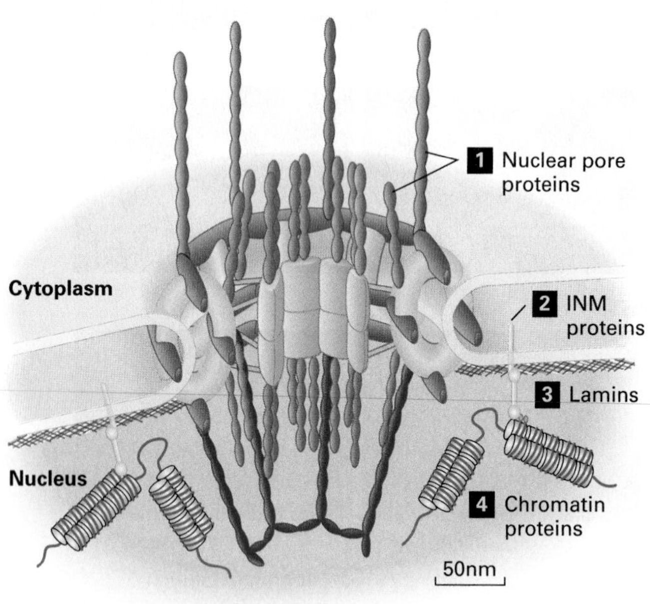

▲ **FIGURE 20-18 Nuclear envelope proteins phosphorylated by MPF.** (1) Components of the nuclear pore complex (NPC) are phosphorylated by MPF in prophase, causing NPCs to dissociate into soluble and membrane-associated NPC subcomplexes. (2) MPF phosphorylation of inner nuclear membrane (INM) proteins inhibits their interactions with the nuclear lamina and chromatin. (3) MPF phosphorylation of nuclear lamins causes their depolymerization and dissolution of the nuclear lamina. (4) MPF phosphorylation of chromatin proteins induces chromatin condensation and inhibits interactions between chromatin and the nuclear envelope. [Adapted from B. Burke and J. Ellenberg, 2002, *Nat. Rev. Mol. Cell Biol.* **3**:487.]

Several lines of evidence indicate that MPF-catalyzed phosphorylation also plays a role in chromosome condensation and formation of the mitotic spindle apparatus. For instance, genetic experiments in the budding yeast *S. cerevisiae* identified a family of *s*tructural *m*aintenance of *c*hromosomes proteins, or **SMC proteins,** that are required for normal chromosome segregation. These large proteins (≈1200 amino acids) contain characteristic ATPase domains at their N- and C-termini and long regions that participate in coiled-coil structures (see Figure 6-38a).

Immunoprecipitation studies with antibodies specific for *Xenopus* SMC proteins revealed that in cycling egg extracts some SMC proteins are part of a multiprotein complex called *condensin,* which becomes phosphorylated as cells enter mitosis. When the anti-SMC antibodies were used to deplete condensin from an egg extract, the extract lost its ability to condense added sperm chromatin following the initial decondensation phase. Other in vitro experiments showed that phosphorylated purified condensin binds to DNA and winds it into supercoils (see Figure 4-8), whereas unphosphorylated condensin does not. These results and the observation that a condensin subunit is phosphorylated by MPF in vitro have led to the model that condensin complexes are activated by phosphorylation catalyzed by MPF. Once activated, condensin complexes are proposed to bind to DNA at intervals along the chromosome, forming successively smaller loops that result in chromosome condensation (see Figure 6-38c).

Phosphorylation of microtubule-associated proteins by MPF probably is required for the dramatic changes in microtubule dynamics that result in the formation of the mitotic spindle and asters (Chapter 18). In addition, phosphorylation of proteins associated with the endoplasmic reticulum (ER) and Golgi complex, by MPF or other protein kinases activated by MPF-catalyzed phosphorylation, is thought to alter the trafficking of vesicles between the ER and Golgi to favor trafficking in the direction of the ER during prophase. As a result, the Golgi complex membranes are transferred to the ER, and vesicular traffic from the ER through the Golgi to the cell surface (Chapter 14), seen in interphase cells, does not occur during mitosis.

Many of the direct substrates of MPF have been identified in *S. cerevisiae* by engineering a CDK mutant that can utilize an analog of ATP that is not bound by other kinases (Figure 20-19). This ATP analog has a bulky benzyl group attached to N_6 of the adenine. This makes the analog too large to fit into the ATP-binding pocket of wild-type protein kinases. However, the ATP-binding pocket of the mutant CDK was modified to accommodate this large ATP analog. Consequently, only the mutant CDK can utilize this

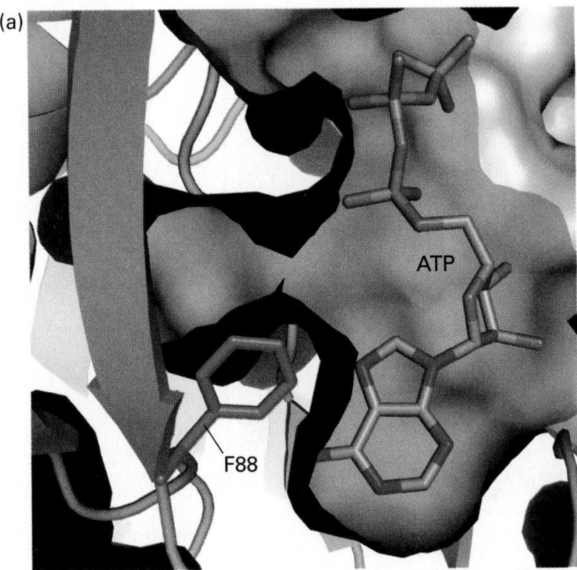

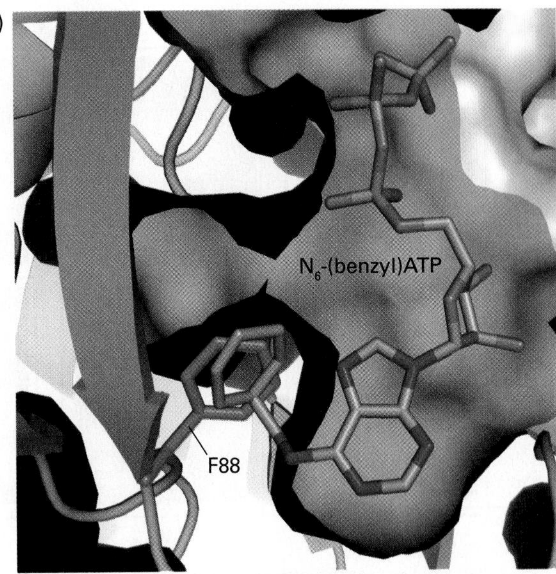

▲ **FIGURE 20-19 ATP analog–dependent CDK mutant.** (a) Representation of the ATP-binding and catalytic sites of wild-type *S. cerevisiae* CDK (Cdc28). Bound ATP and a phenylalanine side chain (pink) in the vicinity of the binding pocket are shown in stick format. (b) Bulky ATP analogs such as those containing a benzyl group bound to the N_6 amino nitrogen are too large to fit into the ATP-binding pocket of wild-type protein kinases and thus cannot be utilized by them. In the *S. cerevisiae* CDK mutant, the phenylalanine at position 88 is changed to glycine, which lacks a large side chain. The mutant exhibits high protein kinase activity using N_6-(benzyl)ATP. These models of *S. cerevisiae* CDK are based on crystal structures of the PKA kinase domain, which shares extensive homology with the kinase domain of *S. cerevisiae* CDK. [See J.A. Ubersax et al., 2003, *Nature* **425:**859; K. Shah et al., 1997, *Proc. Nat'l Acad. Sci. USA* **94:**3565.]

ATP analog as a substrate for transferring its γ-phosphate to a protein side chain. When the N_6-benzyl ATP analog with a labeled γ-phosphate was incubated with yeast cell extracts and recombinant yeast MPF containing the mutant CDK, multiple proteins were labeled. True yeast MPF in vivo substrates could be verified among these potential substrates by treatment of cells expressing the mutant CDK in place of the wild-type protein with a similar derivative of another ATP analog that inhibits protein kinases. This derivative of the kinase inhibitor also contains a bulky substitution at the adenine N_6 position so that it can bind to and inhibit only the mutant CDK. It is sterically blocked from binding to all other kinases and consequently inhibits only the mutant CDK engineered in these cells. Treatment of cells with this specific mutant CDK inhibitor resulted in the dephosphorylation of most of the putative MPF targets identified initially, indicating that these proteins are indeed phosphorylated by the CDK in vivo as well as in vitro. This procedure identified most of the known CDK substrates plus more than 150 additional yeast proteins. These are currently being analyzed for their functions in cell cycle processes.

Unlinking of Sister Chromatids Initiates Anaphase

We saw earlier that in late anaphase, polyubiquitination of mitotic cyclins by the anaphase-promoting complex (APC/C) leads to the proteasomal destruction of these cyclins (see Figure 20-10). Additional experiments with *Xenopus* egg extracts provided evidence that degradation of cyclin B, the *Xenopus* mitotic cyclin, and the resulting decrease in MPF activity are required for chromosome decondensation but not for chromosome segregation (Figure 20-20a, b).

To determine if ubiquitin-dependent degradation of another protein is required for chromosome segregation, researchers prepared a peptide containing the cyclin destruction-box sequence and the site of polyubiquitination. When this peptide was added to a reaction mixture containing untreated egg extract and sperm nuclei, decondensation of the chromosomes and, more interestingly, movement of chromosomes toward the spindle poles were greatly delayed at peptide concentrations of 20–40 μg/ml and blocked altogether at higher concentrations (Figure 20-20c). The added excess destruction-box peptide is thought to act as a substrate for the APC/C-directed polyubiquitination system, competing with the normal endogenous target proteins and thereby delaying or preventing their degradation by proteasomes. Competition with cyclin B delays cyclin B degradation, accounting for the observed inhibition of chromosome decondensation. The observation that chromosome segregation also was inhibited in this experiment but not in the experiment with mutant nondegradable cyclin B (see Figure 20-20b) indicated that segregation depends on polyubiquitination of a

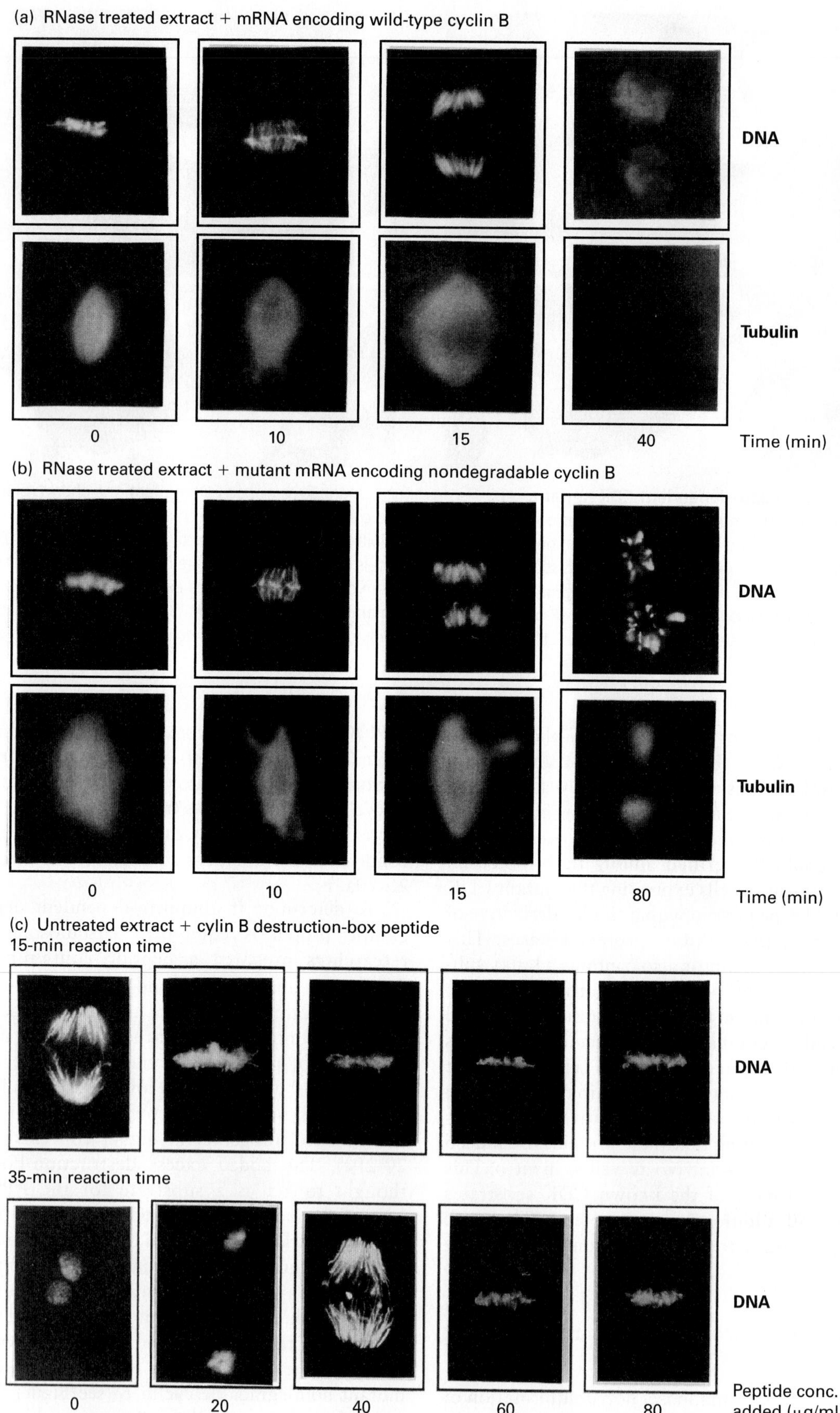
(a) RNase treated extract + mRNA encoding wild-type cyclin B
DNA
Tubulin
0
10
15
40
Time (min)
(b) RNase treated extract + mutant mRNA encoding nondegradable cyclin B
DNA
Tubulin
0
10
15
80
Time (min)
(c) Untreated extract + cylin B destruction-box peptide
15-min reaction time
DNA
35-min reaction time
DNA
0
20
40
60
80
Peptide conc.
added (μg/ml)

◀ **EXPERIMENTAL FIGURE 20-20 Onset of anaphase depends on polyubiquitination of proteins other than cyclin B.** The reaction mixtures contained an untreated or RNase-treated *Xenopus* egg extract and isolated *Xenopus* sperm nuclei, plus other components indicated below. Chromosomes were visualized with a fluorescent DNA-binding dye. Fluorescent rhodamine-labeled tubulin in the reactions was incorporated into microtubules, permitting observation of the mitotic spindle apparatus. (a, b) After the egg extract was treated with RNase to destroy endogenous mRNAs, an RNase inhibitor was added. Then mRNA encoding either wild-type cyclin B or a mutant nondegradable cyclin B was added. The time at which the condensed chromosomes and assembled spindle apparatus became visible after addition of sperm nuclei is designated 0 minutes. In the presence of wild-type cyclin B (a), condensed chromosomes attached to the spindle microtubules and segregated toward the poles of the spindle. By 40 minutes, the spindle had depolymerized (thus is not visible), and the chromosomes had decondensed (diffuse DNA staining) as cyclin B was degraded. In the presence of nondegradable cyclin B (b), chromosomes segregated to the spindle poles by 15 minutes, as in (a), but the spindle microtubules did not depolymerize and the chromosomes did not decondense even after 80 minutes. These observations indicate that degradation of cyclin B is not required for chromosome segregation during anaphase, although it is required for depolymerization of spindle microtubules and chromosome decondensation during telophase. (c) Various concentrations of a short peptide containing the cyclin B destruction box were added to extracts that had not been treated with RNase; the samples were stained for DNA at 15 or 35 minutes after formation of the spindle apparatus. The two lowest peptide concentrations delayed chromosome segregation, and the higher concentrations completely inhibited chromosome segregation. In this experiment, the added destruction box peptide is thought to competitively inhibit APC/C-mediated polyubiquitination of cyclin B as well as another target protein whose degradation is required for chromosome segregation. [From S. L. Holloway et al., 1993, *Cell* **73:**1393; courtesy of A. W. Murray.]

different target protein by the same ubiquitin-protein ligase that binds the cyclin B destruction box and the isolated destruction-box peptide.

As mentioned earlier, each sister chromatid of a metaphase chromosome is attached to microtubules via its kinetochore, a complex of proteins assembled at the centromere. The opposite ends of these kinetochore microtubules associate with one of the spindle poles (see Figure 18-36). At metaphase, the spindle is in a state of tension, with forces pulling the two kinetochores toward the opposite spindle poles balanced by forces pushing the spindle poles apart. Sister chromatids do not separate, because they are held together at their centromeres by multiprotein complexes called *cohesins*. Among the proteins composing the cohesin complexes are members of the SMC protein family discussed in the previous section (see Figure 6-38). When *Xenopus* egg extracts were depleted of cohesin by treatment with antibodies specific for the cohesin SMC proteins, the depleted extracts were able to replicate the DNA in added sperm nuclei, but the resulting sister chromatids did not associate properly with each other. Furthermore, in *S. cerevisiae* with temperature-sensitive mutations in cohesin subunits, incubation at the nonpermissive temperature causes errors in chromosome segregation during mitosis. Since attachment of sister chromatids to spindle fibers from opposite spindle poles requires linkage between them, this is the expected result if sister chromatids of these mutant cells are not associated during mitosis. These findings demonstrate that cohesin is necessary for the cohesion between sister chromatids.

Cohesin molecules associate with chromosomes in late G_1. Figure 20-21 presents one model for how the circular cohesin complexes link daughter chromosomes as they replicate in S phase. According to this model, either the DNA replication fork passes through the cohesin circles, or cohesin circles open to let the replication fork pass and then close around both daughter chromatids. This leaves cohesin links along the full length of the daughter chromatids. In some organisms, such as *C. elegans*, protein links persist in the chromosome arms until they are broken in anaphase. However, in *S. cerevisiae* and vertebrates, phosphorylation of cohesins by protein kinases that are activated by MPF causes cohesin complexes to dissociate from the chromatid arms in late prophase. As opposed to cohesin complexes in the chromosome arms, the same type of cohesin molecules in the vicinity of the centromere do not dissociate, and continue to hold sister chromatids in the region of the centromere. Analysis of *S. cerevisiae* mutants defective in mitotic chromosomal segregation revealed that a specific isoform of a protein phosphatase called *PP2A* normally associates with centromeres (observed in human chromosomes in Figure 20-22). This phosphatase rapidly dephosphorylates cohesin complexes phosphorylated in late prophase by the kinase mentioned above. This occurs only in the vicinity of the centromere where the phosphatase is bound, so that centromere-associated cohesin complexes do not dissociate during late prophase like cohesin complexes in the chromosome arms, but rather continue to link chromatids at the centromere.

Further studies of yeast mutants have led to the model depicted in Figure 20-23 for how the APC/C regulates sister chromatid separation to initiate anaphase. Cohesin SMC proteins link sister chromatids at the centromere. The cross-linking activity of cohesin depends on *securin*, which is found in all eukaryotes. Prior to anaphase, securin binds to and inhibits separase, a ubiquitous protease. Once all chromosome kinetochores have attached to spindle microtubules, the APC/C is directed by a specificity factor called *Cdc20* to polyubiquitinylate securin (note that this specificity factor is distinct from Cdh1, which directs the APC/C to polyubiquitinylate B-type cyclins). Polyubiquitinylated securin is rapidly degraded by proteasomes, thereby releasing separase. Free from its inhibitor, separase cleaves a small subunit of cohesin called *kleisin*, breaking the protein circles linking sister chromatids. Once this link is broken, anaphase begins, as the poleward

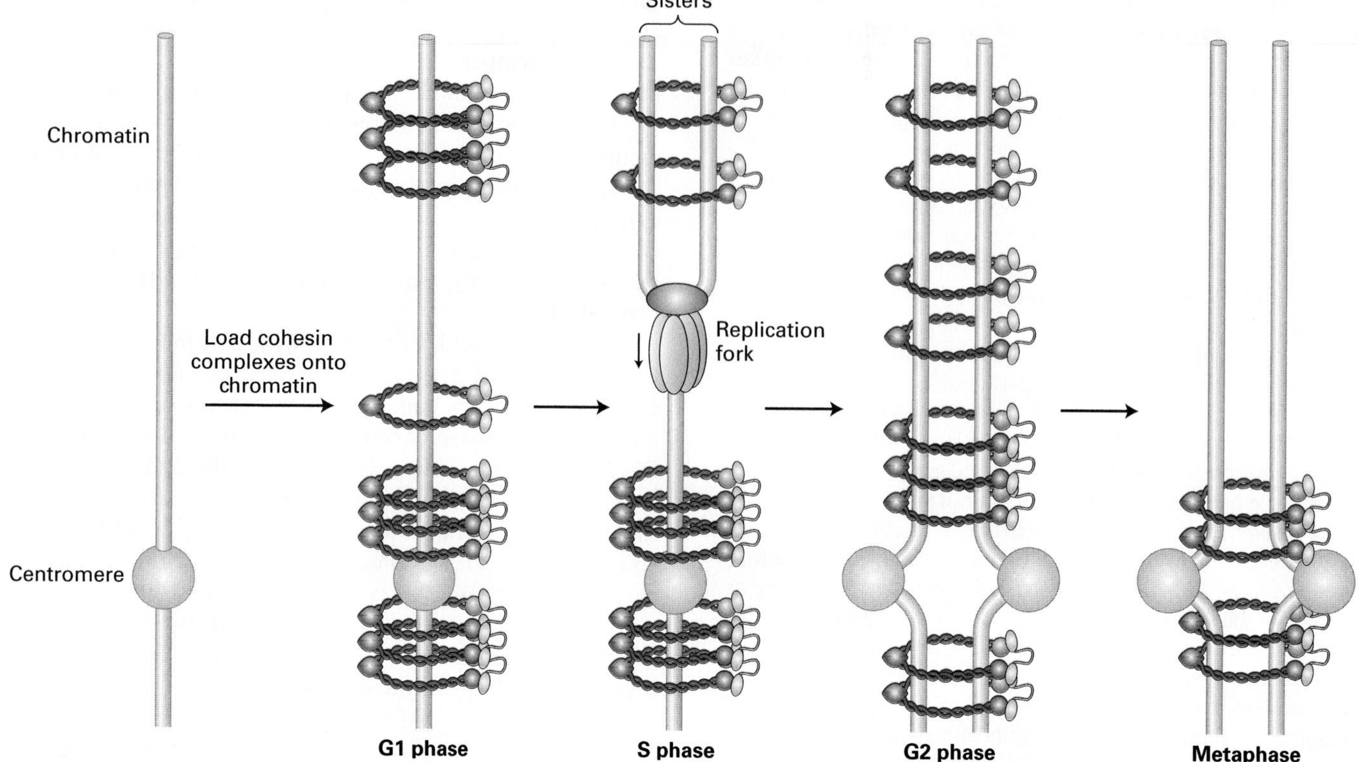

▲ **FIGURE 20-21 Model for cohesin linkage of daughter chromosomes.** There is strong evidence that the cohesin complex is circular, like other SMC protein complexes (see Figure 6-38), but it is not known whether a single cohesin ring links daughter chromatids, or whether two rings, one each around the separate sister chromatids, are linked to each other like links in a chain, possibly involving several linked cohesin circles between sister chromatids. Passage of a replication fork through a cohesin ring results in linking of sister chromatids. In vertebrate cells, cohesins are released from chromosome arms during prophase and early metaphase, and by the end of metaphase are retained only in the region of the centromere. [From K. Nasmyth and C. H. Haering, 2005, *Ann. Rev. Biochem.* **74**:595.]

force exerted on kinetochores moves sister chromatids toward the opposite spindle poles.

Because Cdc20—the specificity factor that directs APC/C to securin—is activated before Cdh1—the specificity factor that directs APC/C to mitotic cyclins—MPF activity does not decrease until after the chromosomes have segregated. As a result of this temporal order in the activation of the two APC/C specificity factors, the chromosomes remain in the condensed state and reassembly of the nuclear envelope does not occur until chromosomes are moved to the proper position. As we shall see in Section 20.7, Cdc20 and Cdh1 are regulated by checkpoint surveillance mechanisms. Cdc20 is inhibited until every kinetochore has attached to a spindle fiber and tension is applied to the kinetochores of all sister chromatids, pulling them toward opposite spindle poles. Cdh1, on the other hand, is inhibited until daughter chromosomes have been separated by a sufficient distance in anaphase to ensure that the separated chromosomes are included in separate nuclei as the nuclear envelopes re-form and the cell divides.

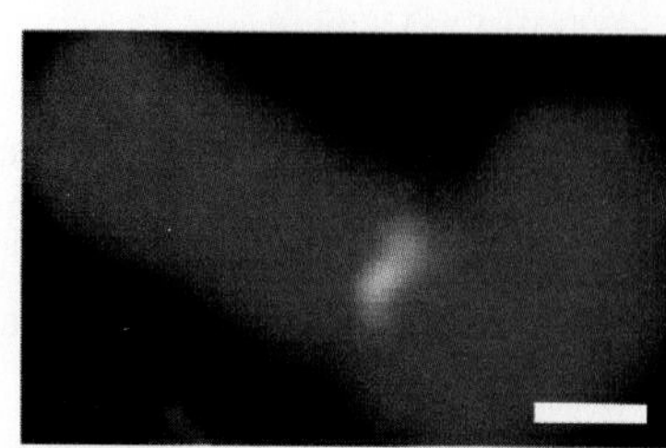

▲ **FIGURE 20-22 Localization of PP2A subtype at the centromere of a human metaphase chromosome.** DNA is stained blue. A marker protein for centromeres was detected with a specific antibody (red), as was PP2A subtype B56α (green). Bar = 1 μm. [From S. Tomoya et al., 2006, *Nature* **441**:46.]

Chromosome Decondensation and Reassembly of the Nuclear Envelope Depend on Dephosphorylation of MPF Substrates

Earlier we discussed how MPF-mediated phosphorylation of nuclear lamins, nucleoporins, and proteins in the inner nuclear membrane contributes to the dissociation of nuclear pore complexes and retraction of the nuclear membrane into the ER. When chromosomes have separated sufficiently during anaphase, the chromosome segregation checkpoint surveillance mechanism activates the

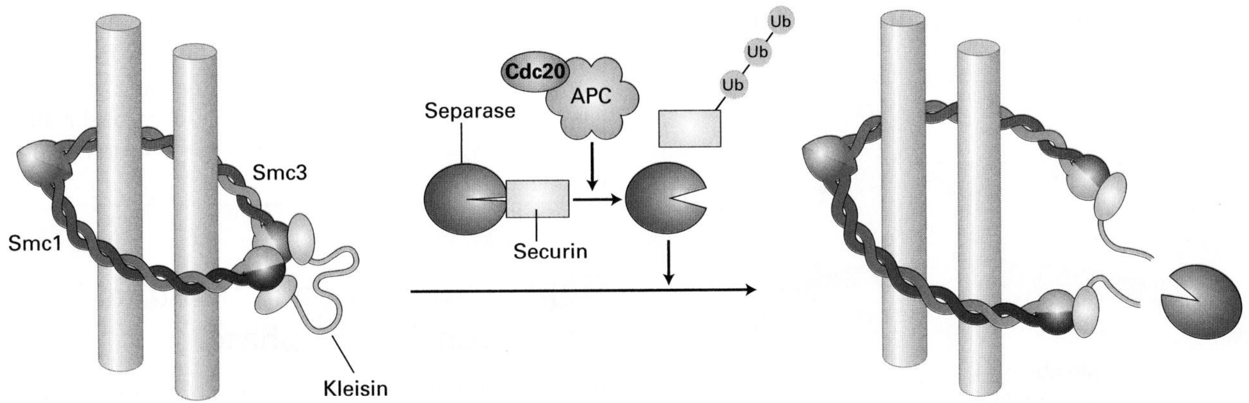

▲ **FIGURE 20-23 Regulation of cohesin cleavage.** Separase, a protease that can cleave the small kleisin subunit of cohesin complexes, is inhibited before anaphase by the binding of securin. When all the kinetochores have attached to spindle microtubules and the spindle apparatus is properly assembled and oriented, the Cdc20 specificity factor associates with the APC/C and directs it to polyubiquitinylate securin. Following securin degradation by proteasomes, the released separase cleaves the kleisin subunit, breaking the cohesin circles, allowing sister chromatids to be pulled apart by the spindle apparatus that is pulling them toward opposite spindle poles. [Adapted from K. Nasmyth and C. H. Haering, 2005, *Ann. Rev. Biochem.* **74:**595.]

protein phosphatase Cdc14. Cdc14 removes phosphate groups that were added to proteins by MPF and, consequently, is an active antagonist of MPF function. Importantly, Cdc14 also dephosphorylates and consequently activates the Cdh1 specificity factor. This allows Cdh1 to bind to the APC/C complex, directing it to polyubiquitinylate mitotic cyclins, inducing their degradation (see Figure 20-10).

Reversal of MPF phosphorylation changes the activities of many proteins back to their usual state in interphase cells. Dephosphorylation of condensins, histone H1 and other chromatin-associated proteins leads to the decondensation of mitotic chromosomes in telophase. Dephosphorylated inner nuclear membrane proteins are thought to bind to chromatin once again. As a result, multiple projections of regions of the ER membrane containing these proteins are thought to associate with the surface of the decondensing chromosomes and then fuse with one another to form a continuous double membrane around each chromosome (Figure 20-24). Dephosphorylation of nuclear pore subcomplexes allows them to reassemble into complete NPCs traversing the inner and outer membranes soon after fusion of the ER projections. Ran·GTP, required for driving most nuclear import and export (Chapter 13), stimulates both fusion of the ER projections to form daughter nuclear envelopes and assembly of NPCs from the nuclear pore subcomplexes that were generated by MPF phosphorylation of nucleoporins in prophase (Figure 20-24). The Ran·GTP concentration is highest in the microvicinity of the decondensing chromosomes because the Ran-guanine nucleotide–exchange factor (Ran-GEF) is bound to chromatin. Consequently, membrane fusion is stimulated at the surfaces of decondensing chromosomes, forming sheets of nuclear membrane with inserted NPCs.

The reassembly of nuclear envelopes containing NPCs around each chromosome forms individual mini-nuclei called *karyomeres*. Subsequent fusion of the karyomeres associated with each spindle pole generates the two daughter-cell nuclei, each containing a full set of chromosomes. Dephosphorylated lamins A and C appear to be imported through the reassembled NPCs during this period and reassemble into a new nuclear lamina. Reassembly of the nuclear lamina in the daughter nuclei probably is initiated on lamin B molecules, which remain associated with the ER membrane via their isoprenyl anchors throughout mitosis and become localized to the inner membrane of the reassembled nuclear envelopes of karyomeres.

KEY CONCEPTS OF SECTION 20.4

Molecular Mechanisms for Regulating Mitotic Events

- Early in mitosis, MPF-catalyzed phosphorylation of lamins A, B, and C, and of nucleoporins and inner nuclear envelope proteins causes depolymerization of lamin filaments (see Figure 20-16) and dissociation of nuclear pores into pore subcomplexes, leading to disassembly of the nuclear envelope and its retraction into the ER.
- Phosphorylation of condensin complexes by MPF or a kinase regulated by MPF promotes chromosome condensation early in mitosis.
- Sister chromatids formed by DNA replication in S phase are linked at the centromere by cohesin complexes that contain DNA-binding SMC proteins and other proteins.
- At the onset of anaphase, the APC/C is directed by Cdc20 to polyubiquitinylate securin, which subsequently is degraded by proteasomes. This activates separase, which cleaves kleisin, a subunit of cohesin, thereby unlinking sister chromatids (see Figure 20-23).

MEDIA CONNECTIONS

Video: Nuclear Envelope Dynamics During Mitosis

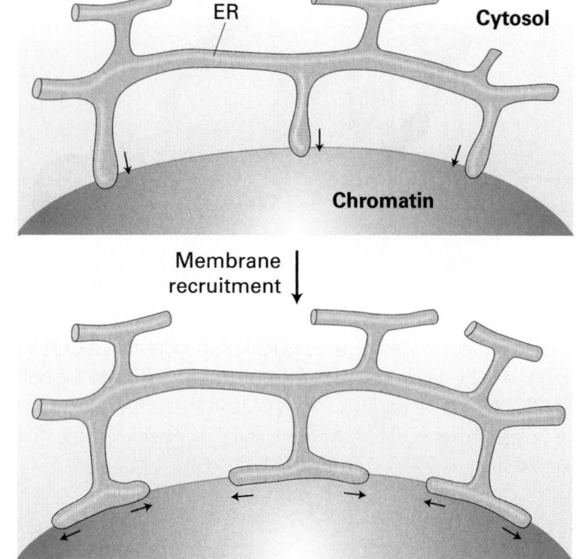

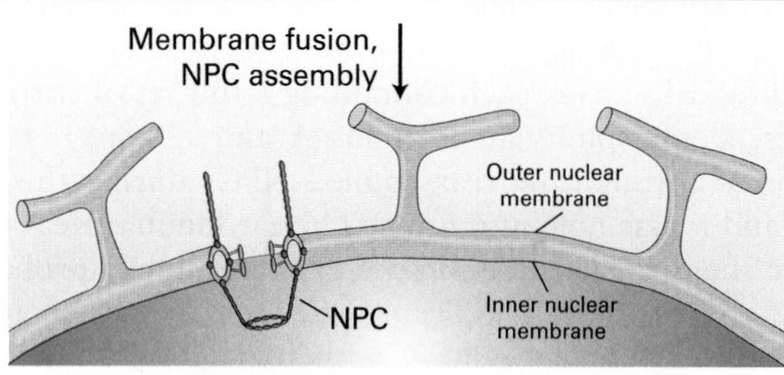

▲ **FIGURE 20-24 Model for reassembly of the nuclear envelope during telophase.** Extensions of the endoplasmic reticulum (ER) associate with each decondensing chromosome and then fuse with one another, forming a double membrane around the chromosome. Dephosphorylated nuclear pore subcomplexes reassemble into nuclear pores, forming individual mini-nuclei called *karyomeres*. The enclosed chromosome further decondenses, and subsequent fusion of the nuclear envelopes of all the karyomeres at each spindle pole forms a single nucleus containing a full set of chromosomes. NPC = nuclear pore complex. [Adapted from B. Burke and J. Ellenberg, 2002, *Nature Rev. Mol. Cell Biol.* **3:**487.]

- After sister chromatids have moved to the spindle poles, the APC/C is directed by Cdh1 to polyubiquitinylate mitotic cyclins, leading to their destruction and causing the decrease in MPF activity that marks the onset of telophase.
- The fall in MPF activity in telophase allows phosphatases such as Cdc14 to remove the regulatory phosphates from condensin, lamins, nucleoporins, and other nuclear membrane proteins, permitting the decondensation of chromosomes and the reassembly of the nuclear membrane, nuclear lamina, and nuclear pore complexes.
- The association of Ran-GEF with chromatin results in a high local concentration of Ran·GTP near the decondensing chromosomes, promoting the fusion of nuclear envelope extensions from the ER around each chromosome. This forms karyomeres that then fuse to form daughter cell nuclei (see Figure 20-24).

20.5 Cyclin-CDK and Ubiquitin-Protein Ligase Control of S phase

In most vertebrate cells, the key decision determining whether or not a cell will divide is the decision to enter the S phase. In most cases, once a vertebrate cell has become committed to entering the S phase, it does so a few hours later and progresses through the remainder of the cell cycle until it completes mitosis. The budding yeast *Saccharomyces cerevisiae* regulates its proliferation similarly, and much of our current understanding of the molecular mechanisms controlling entry into the S phase and the control of DNA replication originated with genetic studies of *S. cerevisiae.*

S. cerevisiae cells replicate by budding (Figure 20-25). Both mother and daughter cells remain in the G_1 period of the cell cycle while growing, although it takes the initially larger mother cells a shorter time to reach a size compatible with cell division. When *S. cerevisiae* cells in G_1 have grown sufficiently, they begin a program of gene expression that leads to entry into the S phase. If G_1 cells are shifted from a rich medium to a medium low in nutrients before they reach a critical size, they remain in G_1 and grow slowly until they are large enough to enter the S phase. However, once G_1 cells reach the critical size, they become committed to completing the cell cycle, entering the S phase and proceeding through G_2 and mitosis, even if they are shifted to a medium low in nutrients. The point in late G_1 of growing *S. cerevisiae* cells when they become irrevocably committed to entering the S phase and traversing the entire cell cycle is called *START*. As we shall see in Section 20.6, a comparable phenomenon occurs in replicating mammalian cells.

In this section, our focus is on the $G_1 \rightarrow S$ transition as we explore the molecular events that constitute START. Just as in mitosis, entry into S phase is controlled by the activity of cyclin-CDKs. However, the regulatory mechanisms governing the activity of these cyclin-CDKs differ from those used by mitotic cyclin-CDK complexes. We discuss the roles of G_1 cyclin-CDKs and S-phase cyclin-CDKs in initiating DNA synthesis and ensuring that DNA replication occurs only once per cell cycle, as well as how the cell cycle "resets" after mitosis, in preparation for the next cell division.

A Cyclin-Dependent Kinase (CDK) Is Critical for S-Phase Entry in *S. cerevisiae*

All *S. cerevisiae* cells carrying a mutation in a particular *cdc* gene arrest with the same size bud at the nonpermissive

Video: Mitosis and Budding in *S. cerevisiae*

(a)

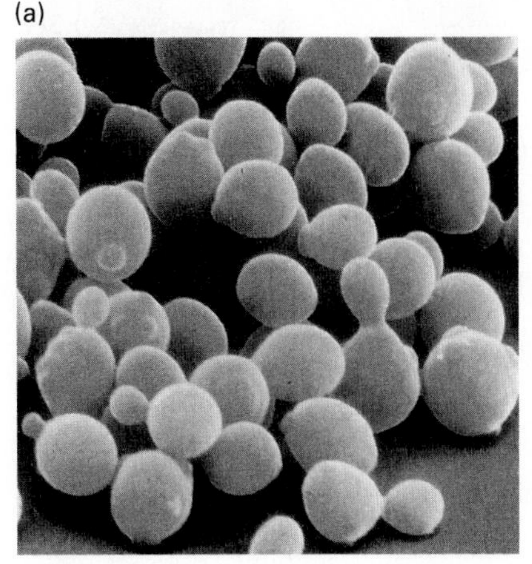

(b)

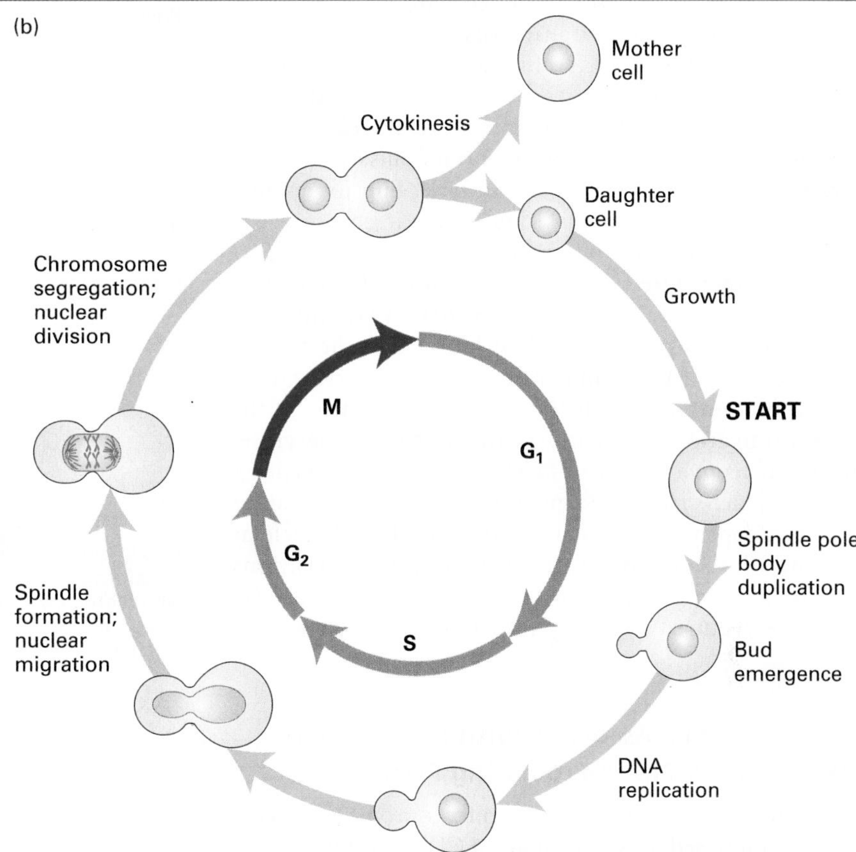

▲ **FIGURE 20-25 The budding yeast *S. cerevisiae*.** (a) Scanning electron micrograph of *S. cerevisiae* cells at various stages of the cell cycle. The larger the bud, which emerges at the end of the G_1 phase, the further along in the cycle the cell is. (b) Main events in the *S. cerevisiae* cell cycle. Daughter cells are born smaller than mother cells and must grow to a greater extent in G_1 before they are large enough to enter the S phase. As in *S. pombe,* the nuclear envelope does not break down during mitosis. Unlike *S. pombe* chromosomes, the small *S. cerevisiae* chromosomes do not condense sufficiently to be visible by light microscopy. [Part (a) courtesy of E. Schachtbach and I. Herskowitz.]

temperature (see Figure 5-6b). Each type of mutant has a terminal phenotype with a particular bud size: no bud, intermediate-sized buds, or large buds. Note that in *S. cerevisiae* wild-type genes are indicated in italic capital letters (e.g., *CDC28*) and recessive mutant genes in italic lowercase letters (e.g., *cdc28*); the corresponding wild-type protein is written in roman letters with an initial capital (e.g., Cdc28), similar to *S. pombe* proteins.

Temperature-sensitive mutants in the *cdc28* gene, now known to encode the *S. cerevisiae* CDK, do not form buds at the nonpermissive temperature. This phenotype indicates that Cdc28 function is required for entry into the S phase. When these mutants are shifted to the nonpermissive temperature, they behave like wild-type cells suddenly deprived of nutrients; that is, *cdc28* mutant cells that have grown large enough to pass START at the time of the temperature shift continue through the cell cycle normally and undergo mitosis, whereas those that are too small to have passed START when shifted to the nonpermissive temperature do not enter the S phase even though nutrients are plentiful. Even though *cdc28* cells blocked in G_1 continue to grow in size at the nonpermissive temperature, they cannot pass START and enter the S phase. Thus they appear as large cells with no bud.

The wild-type *CDC28* gene was isolated by its ability to complement mutant *cdc28* cells at the nonpermissive temperature (see Figure 20-4). Sequencing of *CDC28* showed that the encoded protein is homologous to known protein kinases, and when Cdc28 protein was expressed in *E. coli,* it exhibited low protein kinase activity. Like *S. pombe, S. cerevisiae* contains only a single cyclin-dependent protein kinase (CDK) that functions directly in cell-cycle control. Sequence comparisons have shown that the CDKs in the two species are highly homologous.

The difference in the phenotypes of *S. pombe* and *S. cerevisiae* cells with temperature-sensitive mutations in their CDK genes can be explained by the physiology of the two yeasts. In *S. pombe* cells growing in rich media, cell-cycle control is exerted primarily at the $G_2 \rightarrow M$ transition (i.e., entry to mitosis). In many *S. pombe* CDK recessive mutants, including those initially isolated which gave the phenotype depicted in Figure 20-12, enough CDK activity is maintained at the nonpermissive temperature to permit cells to enter the S phase, but not enough to permit entry into mitosis. Such mutant cells are observed to be elongated cells arrested in G_2. At the nonpermissive temperature, cultures of completely defective CDK mutants include some cells arrested in G_1 and some arrested in G_2, depending on their location in the cell cycle at the time of the temperature shift. Conversely, cell-cycle regulation in *S. cerevisiae* is exerted primarily at the $G_1 \rightarrow S$ transition (i.e., entry to the S phase). Therefore, partially defective mutants of CDK are arrested in G_1, but completely defective CDK mutants are arrested in either G_1 or G_2, depending on their location in the cell cycle at the time of the temperature shift. These observations demonstrate that both the *S. pombe* and the *S. cerevisiae* CDKs are required for entry into both the S phase and mitosis.

Three G_1 Cyclins Associate with *S. cerevisiae* CDK to Form S-Phase-Promoting Factors

By the late 1980s, it was clear that mitosis-promoting factor (MPF) is composed of two subunits: a CDK and a mitotic B-type cyclin required to activate the catalytic subunit. By analogy, it seemed likely that *S. cerevisiae* contains an **S-phase-promoting factor (SPF)** that phosphorylates and regulates proteins required for DNA synthesis. Similar to MPF, SPF was proposed to be a heterodimer composed of the *S. cerevisiae* CDK and a cyclin, in this case one that acts in G_1 (see Figure 20-2).

To identify this putative *G_1 cyclin,* researchers looked for genes that, when expressed at high concentration, could suppress certain temperature-sensitive mutations in the *S. cerevisiae* CDK. The rationale of this approach is illustrated in Figure 20-26. Researchers isolated two such genes, designated *CLN1* and *CLN2*. Using a different approach, researchers identified a dominant mutation in a third gene called *CLN3*.

Sequencing of the three *CLN* genes showed that they encoded related proteins, each of which includes an ≈100-residue region exhibiting significant homology with B-type cyclins from sea urchin, *Xenopus*, human, and *S. pombe*. This region encodes the cyclin domain that interacts with CDKs and is included in the domain of the human cyclin shown in Figure 20-15b, c. The finding that the three Cln proteins contain this region of homology with mitotic cyclins suggested that they were the sought-after *S. cerevisiae* G_1 cyclins. (Note that the homologous CDK-binding domain found in various cyclins differs from the destruction box mentioned earlier, which is found only in B-type cyclins.)

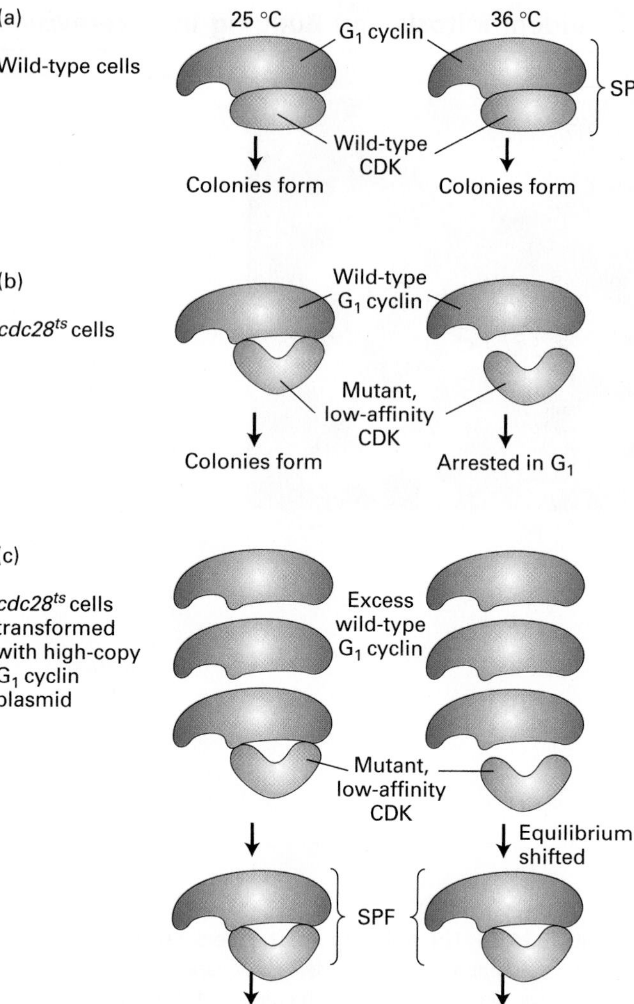

▲ EXPERIMENTAL FIGURE 20-26 **Genes encoding two *S. cerevisiae* G_1 cyclins were identified by their ability to suppress a temperature-sensitive mutant CDK.** This genetic screen is based on differences in the interactions between G_1 cyclins and wild-type and temperature-sensitive (ts) *S. cerevisiae* CDKs. (a) Wild-type cells produce a normal CDK that associates with G_1 cyclins, forming the active S-phase-promoting factor (SPF), resulting in colony formation at both the permissive and the nonpermissive temperature (i.e., 25 ° and 36 °C). (b) Some *cdc28^ts* mutants express a mutant CDK with low affinity for G_1 cyclin at 36 °C. These mutants produce enough G_1 cyclin-CDK (SPF) to support growth and colony development at 25 °C, but not at 36 °C. (c) When *cdc28^ts* cells were transformed with a *S. cerevisiae* genomic library cloned in a high-copy plasmid, three types of colonies formed at 36 °C: one contained a plasmid carrying the wild-type *CDC28* gene; the other two contained plasmids carrying either the *CLN1* or *CLN2* gene. In transformed cells carrying the *CLN1* or *CLN2* gene, the concentration of the encoded G_1 cyclin is high enough to offset the low affinity of the mutant CDK for a G_1 cyclin at 36 °C, so that enough SPF forms to support entry into the S phase and subsequent mitosis. Untransformed *cdc28^ts* cells and cells transformed with plasmids carrying other genes are arrested in G_1 and do not form colonies. [See J. A. Hadwiger et al., 1989, *Proc. Nat'l Acad. Sci. USA* **86:**6255.]

Gene-knockout experiments showed that *S. cerevisiae* cells can grow in rich medium if they carry any one of the three G_1 cyclin genes. As the data presented in Figure 20-27 indicate, overproduction of one G_1 cyclin decreases the fraction of cells in G_1, demonstrating that high levels of the G_1 cyclin-CDK complex drive cells through START prematurely. Moreover, in the absence of all three of the G_1 cyclins, cells become arrested in G_1, indicating that a G_1 cyclin-CDK heterodimer, or SPF, is required for *S. cerevisiae* cells to enter the S phase. These findings are reminiscent of the results for the *S. pombe* mitotic cyclin with regard to passage through G_2 and entry into mitosis. Overproduction of the mitotic cyclin caused a shortened G_2 and premature entry into mitosis, whereas inhibition of the mitotic cyclin by mutation resulted in a lengthened G_2 (see Figure 20-12). Thus these results confirmed that the *S. cerevisiae* Cln proteins are *G_1 cyclins* that regulate passage through the G_1 phase of the cell cycle.

In wild-type yeast cells, Cln3 functions as a *mid-G_1 cyclin,* because it is maximally expressed at about the midpoint of G_1 in continuously cycling cells. Its mRNA is produced at a nearly constant level throughout the cell cycle, but its translation is regulated in response to nutrient levels. The *CLN3* mRNA contains a short upstream

Podcast: G_1-cyclin Control of Entry into S-phase

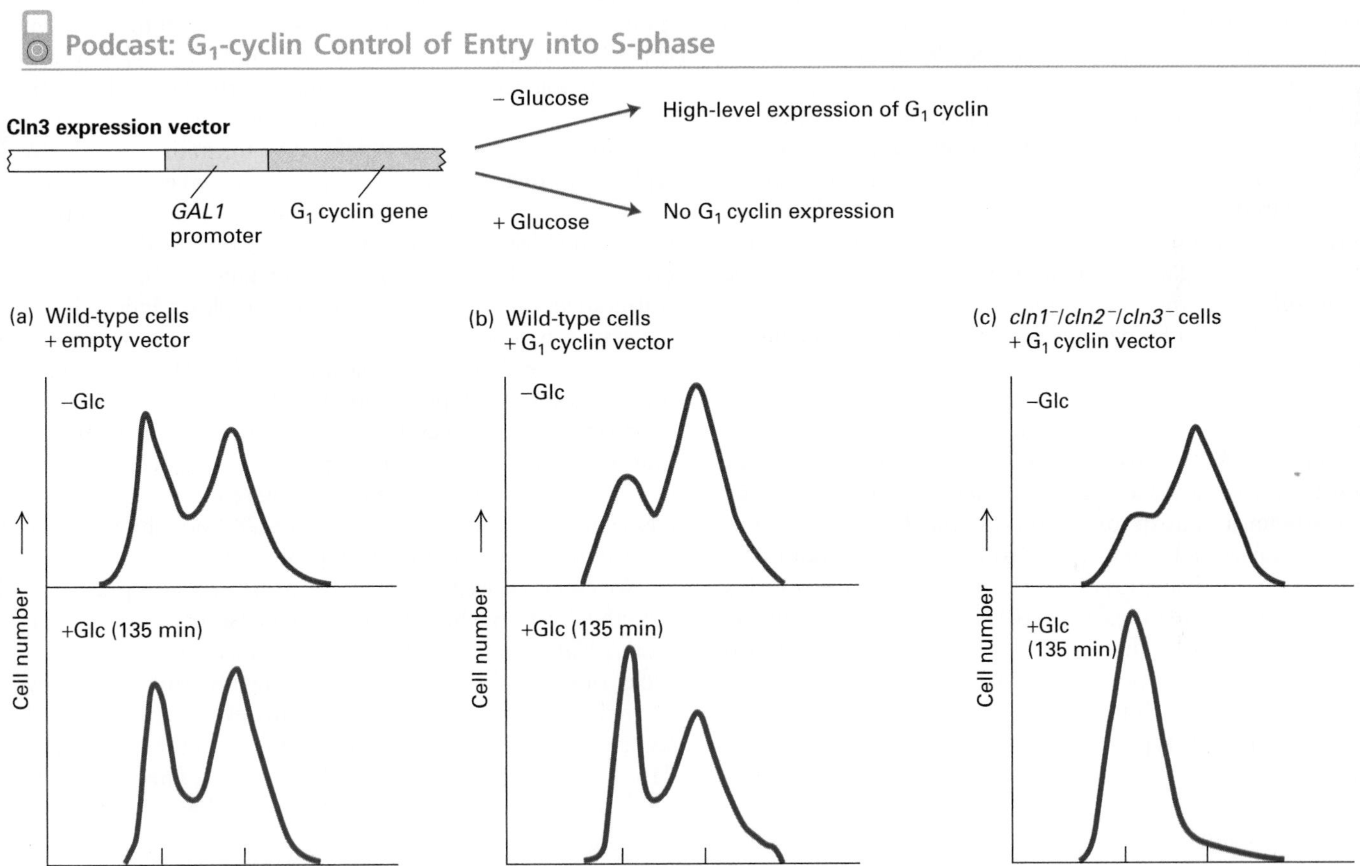

▲ **EXPERIMENTAL FIGURE 20-27 G_1 cyclin is required for *S. cerevisiae* cells to enter S phase, and overexpression of G_1 cyclin prematurely drives them into the S phase.** The yeast expression vector used in these experiments (*top*) carried one of the three *S. cerevisiae* G_1 cyclin genes linked to the strong *GAL1* promoter, which is turned off when glucose is present in the medium. To determine the proportion of cells in G_1 and G_2, cells were exposed to a fluorescent dye that binds to DNA and then were passed through a fluorescence-activated cell sorter (see Figure 9-28). Since the DNA content of G_2 cells is twice that of G_1 cells, this procedure can distinguish cells in the two cell-cycle phases. (a) Wild-type cells transformed with an empty expression vector displayed the normal distribution of cells in G_1 and G_2 in the absence of glucose (Glc) and after addition of glucose. (b) In the absence of glucose, wild-type cells transformed with the G_1 cyclin expression vector displayed a higher-than-normal percentage of cells in the S phase and G_2 because overexpression of the G_1 cyclin decreased the G_1 period (*top curve*). When expression of the G_1 cyclin from the vector was shut off by addition of glucose, the cell distribution returned to normal (*bottom curve*). (c) Cells with mutations in all three G_1 cyclin genes and transformed with the G_1 cyclin expression vector also showed a high percentage of cells in S and G_2 in the absence of glucose (*top curve*). Moreover, when expression of G_1 cyclin from the vector was shut off by addition of glucose, the cells completed the cell cycle and arrested in G_1 (*bottom curve*), indicating that a G_1 cyclin is required for entry into the S phase. [Adapted from H. E. Richardson et al., 1989, *Cell* **59**:1127.]

open-reading frame that inhibits translation initiation at the Cln3 open reading frame. This inhibition is diminished when nutrients are in abundance, leading to activation of the TOR pathway and the subsequent increase in translation initiation factor activity (see Figure 8-30). Since Cln3 is a highly unstable protein, its concentration fluctuates with the translation rate of its mRNA. Consequently, the amount and activity of mid-G_1 cyclin-CDK complexes, which depend on the concentration of the mid-G_1 cyclin protein, are largely regulated by the nutrient level.

Once sufficient mid-G_1 cyclin is synthesized from its mRNA, the mid-G_1 cyclin-CDK complex phosphorylates and activates two related transcription factors, SBF and MBF. These induce transcription of the late-G_1 cyclin genes, *CLN1* and *CLN2*, whose encoded proteins accelerate entry into the S phase. Thus regulation of *CLN3* mRNA translation in response to the concentration of nutrients in the medium is thought to be primarily responsible for controlling the length of G_1 in *S. cerevisiae*. In addition to the late-G_1 cyclins, SBF and MBF also stimulate transcription of several other genes required for DNA replication, including genes encoding DNA polymerase subunits, RPA subunits (the eukaryotic ssDNA-binding protein), DNA ligase, and enzymes required for deoxyribonucleotide synthesis.

One of the important substrates of the late-G_1 cyclin–CDK complexes is Cdh1. Recall that this specificity factor directs the APC/C to polyubiquitinylate B-type cyclins during late anaphase, marking them for proteolysis by proteasomes. Phosphorylation of Cdh1 by G_1 cyclin-CDKs causes it to dissociate from the APC/C complex, inhibiting further polyubiquitination of B-type cyclins during late G_1 (see Figure 20-10). The MBF transcription factor activated by the mid-G_1 cyclin-CDK complex also stimulates transcription of two B-type cyclins in addition to the late-G_1 cyclins. Because the complexes formed between these B-type cyclins and the *S. cerevisiae* CDK are required for initiation of DNA synthesis, they are called *early S-phase cyclins*. Inactivation of Cdh1 allows the S-phase cyclin-CDK complexes to accumulate in late G_1. The specificity factor Cdh1 is phosphorylated and inactivated by both late-G_1 and B-type cyclin-CDK complexes, and thus remains inhibited throughout S, G_2, and M phase until late anaphase when the Cdc14 phosphatase is activated and removes the inhibitory phosphate from Cdh1.

Degradation of the S-Phase Inhibitor Triggers DNA Replication

As the S-phase cyclin-CDK heterodimers accumulate in late G_1, they are immediately inactivated by binding of an inhibitor, called *Sic1*, that is expressed late in mitosis and in early G_1. Because Sic1 specifically inhibits B-type cyclin-CDK complexes, but has no effect on the G_1 cyclin-CDK complexes, it functions as an *S-phase inhibitor*.

Entry into the S phase is defined by the initiation of DNA replication. In *S. cerevisiae* cells this occurs when the Sic1 inhibitor is precipitously degraded following its polyubiquitination by the distinct ubiquitin-protein ligase called SCF mentioned earlier (Figure 20-28; see also Figure 20-2). Once Sic1 is degraded, the S-phase cyclin-CDK complexes induce DNA replication by phosphorylating several proteins in prereplication complexes bound to replication origins. This mechanism for activating the S-phase cyclin-CDK complexes—that is, inhibiting them as the cyclins are synthesized and then precipitously degrading the inhibitor—permits the sudden activation of large numbers of complexes, as opposed to the gradual increase in kinase activity that would result if no inhibitor were present during synthesis of the S-phase cyclins.

We can now see that regulated proteasomal degradation directed by two ubiquitin-protein ligase complexes, SCF and APC/C, controls three major transitions in the cell cycle: onset of the S phase through degradation of Sic1 by SCF, the beginning of anaphase through degradation of securin by the APC/C, and exit from mitosis through degradation of B-type cyclins by the APC/C. The APC/C is

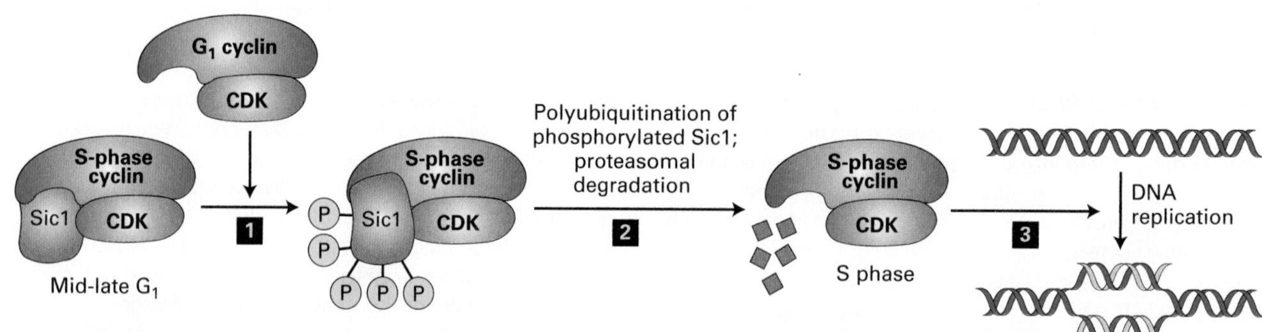

▲ **FIGURE 20-28 Control of S phase onset in *S. cerevisiae* by regulated proteolysis of the S-phase inhibitor, Sic1.** The S-phase cyclin-CDK complexes (Clb5-CDK and Clb6-CDK) begin to accumulate in G_1, but are inhibited by Sic1. This inhibition prevents initiation of DNA replication until the cell is fully prepared. G_1 cyclin-CDK complexes assembled in late G_1 (Cln1-CDK and Cln2-CDK) phosphorylate Sic1 at multiple sites (step **1**), marking it for polyubiquitination by the SCF ubiquitin ligase, and subsequent proteasomal degradation (step **2**). The active S-phase cyclin-CDK complexes then trigger initiation of DNA synthesis (step **3**) by phosphorylating components of pre-initiation complexes assembled on DNA replication origins in early G_1. [Adapted from R. W. King et al., 1996, *Science* **274:**1652.]

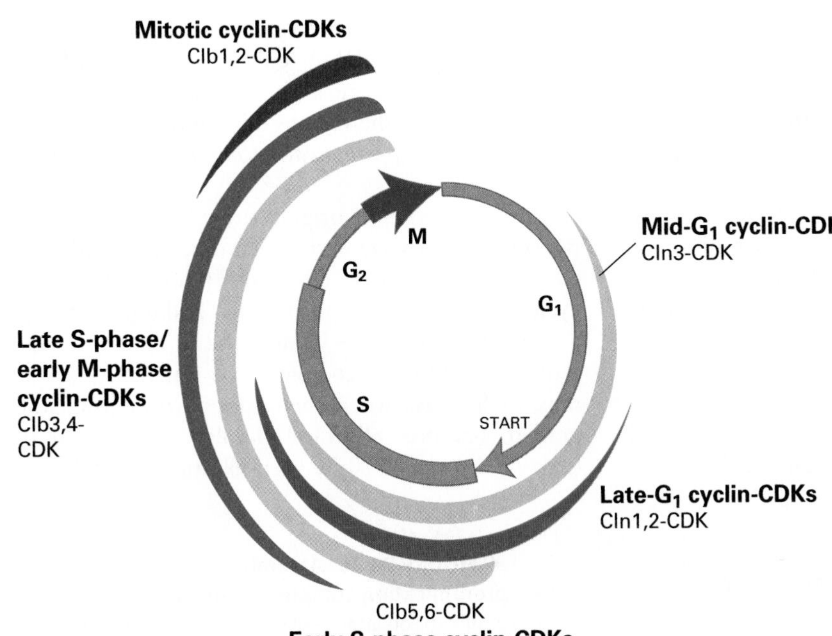

◀ **FIGURE 20-29 Activity of *S. cerevisiae* cyclin-CDK complexes through the course of the cell cycle.** The width of the colored bands is approximately proportional to the demonstrated or proposed protein kinase activity of the indicated cyclin-CDK complexes. *S. cerevisiae* produces a single cyclin-dependent kinase (CDK) whose activity is controlled by the various cyclins, which are expressed during different portions of the cell cycle.

directed to polyubiquitinylate the anaphase inhibitor securin by the Cdc20 specificity factor (see Figure 20-23). The APC/C-Cdc20 complex also directs the degradation of S-phase cyclins and much of the mitotic cyclin, but sufficient mitotic cyclin remains to maintain chromosome condensation until late anaphase. Then, another specificity factor, Cdh1, targets the APC/C to the remaining B-type cyclins (see Figure 20-10).

In contrast to the APC/C, the SCF ubiquitin-protein ligase is not regulated by phosphorylation of specificity factors, but rather by phosphorylation of its substrate, Sic1. Sic1 is phosphorylated by G_1 cyclin-CDKs (see Figure 20-28). It must be phosphorylated at at least six sites, which are relatively poor substrates for the G_1 cyclin-CDKs, before it is bound sufficiently well by SCF to be polyubiquitinylated. This difference in strategy for regulating the ubiquitin-protein ligase activities of SCF and APC/C probably occurs because the APC/C has several substrates, including securin and B-type cyclins, which must be degraded at different times in the cycle. In contrast, entry into the S phase requires the degradation of only a single protein, the Sic1 inhibitor. Also, the requirement for phosphorylating multiple weak sites in Sic1 delays the onset of S phase until G_1 cyclin-CDK activity has reached its peak and virtually all other G_1 cyclin-CDK substrates have been phosphorylated. An obvious advantage of proteolysis for controlling passage through these critical points in the cell cycle is that protein degradation is an irreversible process, ensuring that cells proceed irreversibly in one direction through the cycle.

Multiple Cyclins Regulate the Kinase Activity of *S. cerevisiae* CDK During Different Cell-Cycle Phases

As budding yeast cells progress through the S phase, they begin transcribing genes encoding two additional B-type cyclins. These form heterodimeric late S-phase/early M-phase cyclin-CDK complexes that, together with the early S-phase cyclin-CDK complexes, activate DNA replication origins throughout S phase. The late S-phase/early M-phase cyclin-CDK complexes are named so because they also initiate formation of the mitotic spindle at the beginning of mitosis, with the help of two other mitotic cyclins. These additional mitotic cyclins are expressed when *S. cerevisiae* cells complete chromosome replication and enter G_2. They function as *late mitotic cyclins*, associating with the CDK to form complexes that mediate the events of mitosis.

Each group of cyclins thus directs the *S. cerevisiae* CDK to specific functions associated with various cell-cycle phases, as outlined in Figure 20-29. The mid-G_1 cyclin-CDK induces expression of late G_1 cyclins and other proteins in mid-G_1 by phosphorylating and activating the SBF and MBF transcription factors. The G_1 cyclin-CDKs inhibit the APC/C, allowing B-type cyclins (S-phase and M-phase cyclins) to accumulate; the G_1 cyclin-CDKs also activate degradation of the S-phase inhibitor Sic1. The early S-phase cyclin-CDK complexes then trigger DNA synthesis. The late S-phase cyclins also function as mitotic cyclins by triggering formation of mitotic spindles. The remaining two *S. cerevisiae* cyclins are mitotic cyclins whose concentrations peak midway through mitosis. They are also involved in mitotic spindle formation, but they mainly function as mitotic cyclins, forming complexes with CDK that trigger chromosome segregation and nuclear division (see Table 20-1).

Replication at Each Origin Is Initiated Only Once During the Cell Cycle

As discussed in Chapter 4, eukaryotic chromosomes are replicated from multiple **replication origins.** Initiation of replication from these origins occurs throughout S phase.

However, no eukaryotic origin initiates more than once per S phase. Moreover, the S phase continues until replication from multiple origins along the length of each chromosome results in complete replication of the entire chromosome. These two factors ensure that the correct gene copy number is maintained as cells proliferate.

Yeast replication origins contain an 11-base-pair conserved core sequence to which is bound a hexameric protein, the *origin-recognition complex (ORC),* required for initiation of DNA synthesis. DNase I footprinting analysis (Figure 7-17) and immunoprecipitation of chromatin proteins cross-linked to specific DNA sequences (Figure 7-37) during various phases of the cell cycle indicate that the ORC remains associated with origins during all phases of the cycle. Several additional replication initiation factors required to initiate DNA synthesis at origins were identified in genetic studies in *S. cerevisiae.* These DNA replication initiation factors associate with the ORC at origins during G_1, but not during G_2 or M. During G_1 the various initiation factors assemble with the ORC into a *prereplication complex* at each origin (Figure 20-30).

The restriction of origin "firing" to once and only once per cell cycle in *S. cerevisiae* is enforced by the alternating cycle of B-type cyclin-CDK activity levels through the cell cycle: low in telophase through G_1 and high in S, G_2, and M through anaphase (see Figure 20-29). As we just discussed, S-phase cyclin-CDK complexes become active at the beginning of S phase when their specific inhibitor, Sic1, is degraded. The prereplication complexes assembled at origins early in G_1 (Figure 20-30, step **1**) initiate DNA synthesis in

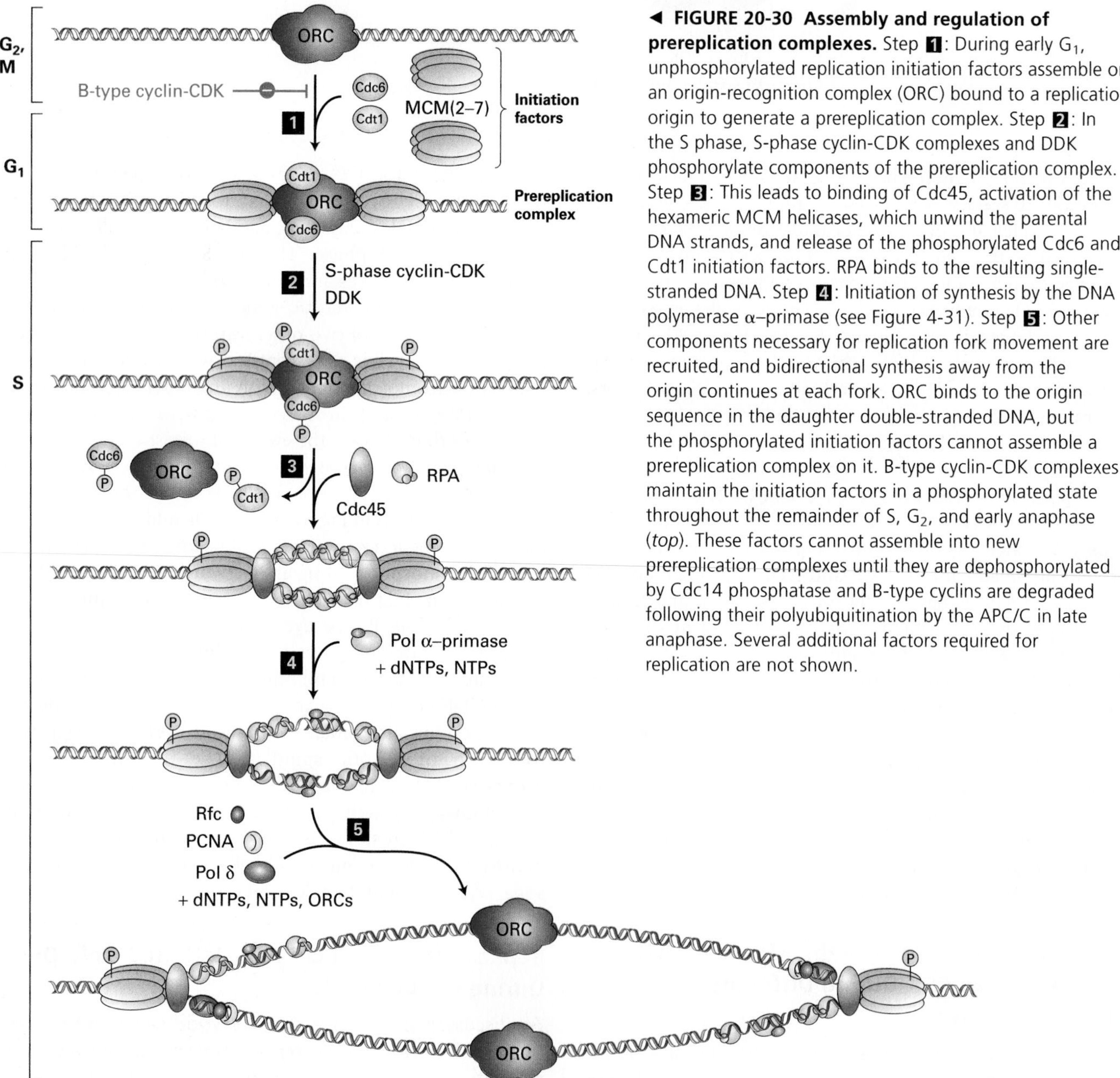

◀ **FIGURE 20-30 Assembly and regulation of prereplication complexes.** Step **1**: During early G_1, unphosphorylated replication initiation factors assemble on an origin-recognition complex (ORC) bound to a replication origin to generate a prereplication complex. Step **2**: In the S phase, S-phase cyclin-CDK complexes and DDK phosphorylate components of the prereplication complex. Step **3**: This leads to binding of Cdc45, activation of the hexameric MCM helicases, which unwind the parental DNA strands, and release of the phosphorylated Cdc6 and Cdt1 initiation factors. RPA binds to the resulting single-stranded DNA. Step **4**: Initiation of synthesis by the DNA polymerase α–primase (see Figure 4-31). Step **5**: Other components necessary for replication fork movement are recruited, and bidirectional synthesis away from the origin continues at each fork. ORC binds to the origin sequence in the daughter double-stranded DNA, but the phosphorylated initiation factors cannot assemble a prereplication complex on it. B-type cyclin-CDK complexes maintain the initiation factors in a phosphorylated state throughout the remainder of S, G_2, and early anaphase (*top*). These factors cannot assemble into new prereplication complexes until they are dephosphorylated by Cdc14 phosphatase and B-type cyclins are degraded following their polyubiquitination by the APC/C in late anaphase. Several additional factors required for replication are not shown.

S phase when they are phosphorylated by the S-phase cyclin-CDKs and a second heterodimeric protein kinase, DDK, expressed in G_1 along with other proteins involved in DNA replication (step **2**). Although the complete set of proteins that must be phosphorylated to activate initiation of DNA synthesis has not yet been determined, there is evidence that phosphorylation of at least one subunit of the hexameric *MCM helicase* and of another initiation factor called *Cdc6* is required. Following their phosphorylation, the helicase unwinds the DNA, and the resulting single-stranded DNA is bound by the single-stranded binding protein RPA and other replication factors (Figure 20-30 steps **3**, **4**, and **5**; see also Figure 4-31).

As the replication forks progress away from each origin, the phosphorylated initiation factors are displaced from the chromatin. However, ORC complexes immediately bind to the origin sequence in the replicated daughter duplex DNAs and remain bound throughout the cell cycle (see Figure 20-30, step **5**). Origins can fire only once during the S phase because the phosphorylated initiation factors cannot reassemble into a prereplication complex. Consequently, phosphorylation of components of the prereplication complex by S-phase cyclin-CDK complexes and the DDK complex simultaneously activates initiation of DNA replication at an origin and inhibits re-initiation of replication at that origin. As we have noted, B-type cyclin-CDK complexes remain active throughout the S phase, G_2, and early anaphase, maintaining the phosphorylated state of the replication initiation factors that prevents the assembly of new prereplication complexes (step **1**).

When the Cdc14 phosphatase is activated in late anaphase and the APC/C-Cdh1 complex triggers degradation of all B-type cyclins in telophase, phosphates on the initiation factors are removed by the unopposed Cdc14 phosphatase. This allows the reassembly of prereplication complexes during early G_1. As discussed previously, the inhibition of APC/C activity in G_1 sets the stage for accumulation of the S-phase cyclins needed for onset of the next S phase. This regulatory mechanism has two consequences: (1) prereplication complexes are assembled only during G_1, when the activity of B-type cyclin-CDK complexes is low, and (2) each origin initiates replication one time only during the S phase, when S phase cyclin-CDK complex activity is high. As a result, chromosomal DNA is replicated only one time each cell cycle.

KEY CONCEPTS OF SECTION 20.5

Cyclin-CDK and Ubiquitin-Protein Ligase Control of S phase

- *S. cerevisiae* expresses a single cyclin-dependent protein kinase (CDK), which interacts with several different cyclins during different phases of the cell cycle (see Figure 20-29).
- Three G_1 cyclins are active in G_1. The concentration of the mid-G_1 cyclin mRNA does not vary significantly through the cell cycle, but its translation is regulated by the availability of nutrients.
- Once active mid-G_1 cyclin-CDK complexes accumulate in mid-late G_1, they phosphorylate and activate two transcription factors that stimulate expression of the late-G_1 cyclins, as well as enzymes and other proteins required for DNA replication, and the early S-phase B-type cyclins.
- The late-G_1 cyclin-CDK complexes phosphorylate and inhibit Cdh1, the specificity factor that directs the anaphase-promoting complex (APC/C) to B-type cyclins, thus permitting accumulation of S-phase B-type cyclins in late G_1.
- S-phase cyclin-CDK complexes initially are inhibited by Sic1. Polyubiquitination of Sic1 by the SCF ubiquitin-protein ligase marks Sic1 for proteasomal degradation, releasing activated S-phase cyclin-CDK complexes that trigger onset of the S phase (see Figure 20-28).
- Late S-phase/early M-phase B-type cyclins, expressed later in the S phase, form heterodimers with the CDK that also promote DNA replication and initiate spindle formation early in mitosis.
- Late M-phase B-type cyclins, expressed in G_2, form heterodimers with the CDK that stimulate mitotic events.
- In late anaphase, the specificity factor Cdh1 is activated by dephosphorylation and then directs APC/C to polyubiquitinylate all the B-type cyclins. Their subsequent proteasomal degradation inactivates MPF activity, permitting exit from mitosis (see Figure 20-10).
- DNA replication is initiated from prereplication complexes assembled at origins during early G_1. S-phase cyclin-CDK complexes simultaneously trigger initiation from prereplication complexes and inhibit assembly of new prereplication complexes by phosphorylating components of the prereplication complex (see Figure 20-30).
- Initiation of DNA replication occurs at each origin, but only once, until a cell proceeds through anaphase, when activation of APC/C leads to the degradation of B-type cyclins. The block on re-initiation of DNA replication until replicated chromosomes have segregated assures that daughter cells contain the proper number of chromosomes per cell.

20.6 Cell-Cycle Control in Mammalian Cells

In multicellular organisms, precise control of the cell cycle during development and growth is critical for determining the size and shape of each tissue. Cell replication is controlled by a complex network of signaling pathways that integrate extracellular signals about the identity and numbers of neighboring cells and intracellular cues about cell size and developmental program (Chapters 21 and 22). Most differentiated cells withdraw from the cell cycle during G_1, entering the G_0 state (see Figure 20-1). Some differentiated cells (e.g., fibroblasts and lymphocytes) can be stimulated to reenter the cycle and replicate. Many postmitotic

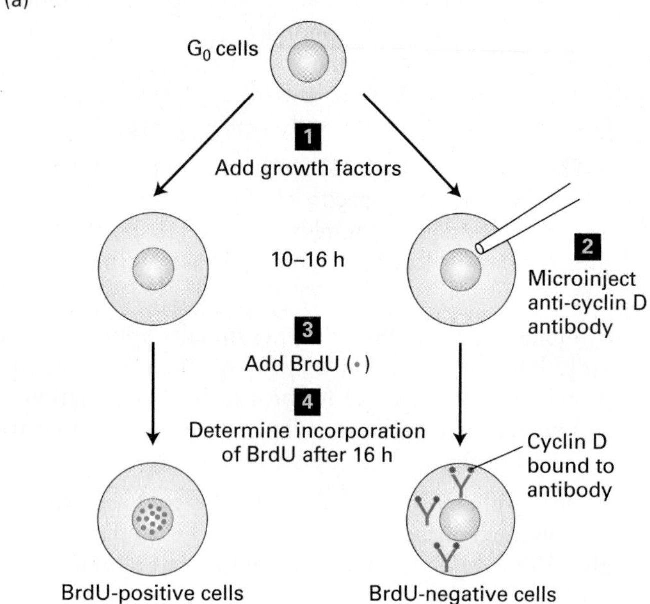

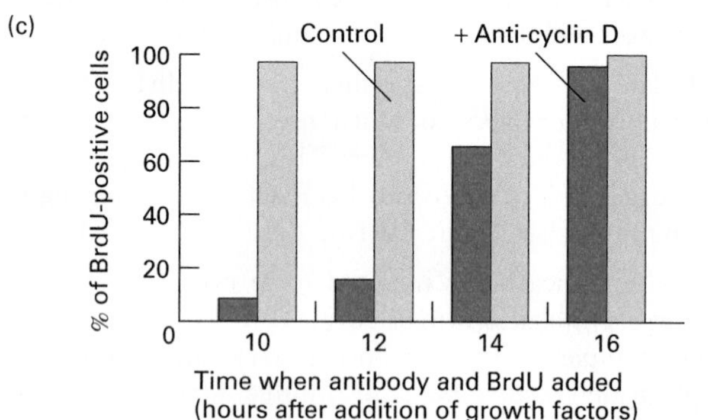

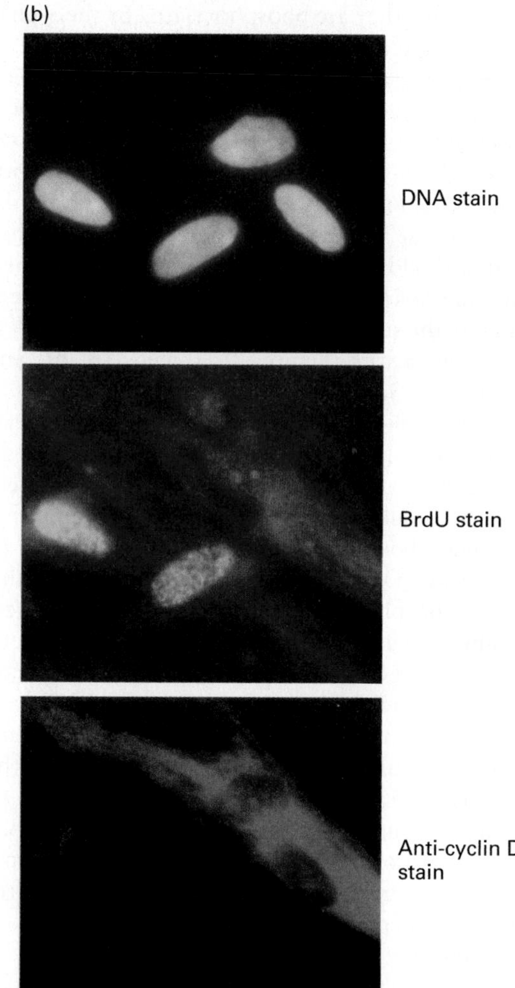

▲ EXPERIMENTAL FIGURE 20-31 **Microinjection experiments with anti-cyclin D antibody demonstrate that cyclin D is required for passage through the restriction point.** The G_0-arrested mammalian cells used in these experiments pass the restriction point 14–16 hours after addition of growth factors and enter the S phase 6–8 hours later. (a) Outline of experimental protocol. At various times 10–16 hours after addition of growth factors (**1**), some cells were microinjected with rabbit antibodies against cyclin D (**2**). Bromodeoxyuridine (BrdU), a thymidine analog, was then added to the medium (**3**), and the uninjected control cells (*left*) and microinjected experimental cells (*right*) were incubated for an additional 16 hours. Each sample was then analyzed to determine the percentage of cells that had incorporated BrdU into newly synthesized DNA (**4**), indicating that they had entered the S phase. (b) Analysis of control cells and experimental cells injected with anti-cyclin D antibody 8 hours after addition of growth factors. The three micrographs show the same field of cells stained 16 hours after addition of BrdU to the medium. Cells were stained with different fluorescent agents to visualize DNA (*top*), BrdU (*middle*), and anti-cyclin D antibody (*bottom*). Note that the two cells in this field injected with anti-cyclin D antibody (the red cells in the bottom micrograph) did not incorporate BrdU into nuclear DNA, as indicated by their lack of staining in the middle micrograph. (c) Percentage of control cells (blue bars) and experimental cells (red bars) that incorporated BrdU. Most cells injected with anti-cyclin D antibodies 10 or 12 hours after addition of growth factors failed to enter the S phase, indicated by the low level of BrdU incorporation. In contrast, anti-cyclin D antibodies had little effect on entry into the S phase and DNA synthesis when injected at 14 or 16 hours, that is, after cells had passed the restriction point. These results indicate that cyclin D is required to pass the restriction point, but once cells have passed the restriction point, they do not require cyclin D to enter the S phase 6–8 hours later. [Parts (b) and (c) adapted from V. Baldin et al., 1993, *Genes & Devel.* **7**:812.]

differentiated cells, however, never reenter the cell cycle to replicate again. As we discuss in this section, the cell-cycle regulatory mechanisms uncovered in yeasts and *Xenopus* eggs and early embryos also operate in the somatic cells of higher eukaryotes, including humans and other mammals.

Mammalian Restriction Point Is Analogous to START in Yeast Cells

Most studies of mammalian cell-cycle control have been done with cultured cells that require certain polypeptide growth factors (**mitogens**) to stimulate cell proliferation.

Binding of these growth factors to specific receptor proteins that span the plasma membrane initiates a cascade of signal transduction that ultimately influences transcription and cell-cycle control (Chapters 15 and 16).

Mammalian cells cultured in the absence of growth factors are arrested with a diploid complement of chromosomes in the G_0 period of the cell cycle. If growth factors are added to the culture medium, these *quiescent cells* pass through the **restriction point** 14–16 hours later, enter the S phase 6–8 hours after that, and traverse the remainder of the cell cycle (see Figure 20-2). The restriction point is the time after addition of growth factors when cells no longer require the presence of growth factors to enter S phase. Like START in yeast cells, the restriction point is the point in the cell cycle at which mammalian cells become committed to entering the S phase and completing the cell cycle, which takes about 24 hours for most cultured mammalian cells.

Multiple CDKs and Cyclins Regulate Passage of Mammalian Cells Through the Cell Cycle

Unlike *S. pombe* and *S. cerevisiae,* which each produce a single cyclin-dependent kinase (CDK) to regulate the cell cycle, mammalian cells use a small family of related CDKs to regulate progression through the cell cycle. Four CDKs are expressed at significant levels in most mammalian cells and play a role in regulating the cell cycle. Named CDK1, 2, 4, and 6, these proteins were identified by the ability of their cDNA clones to complement certain *cdc* yeast mutants or by their homology to other CDKs.

Like *S. cerevisiae,* mammalian cells express multiple cyclins. Cyclin A and cyclin B, which function in the S phase, G_2, and early mitosis, initially were detected as proteins whose concentration oscillates in experiments with synchronously cycling early sea urchin and clam embryos (see Figure 20-8). Homologous cyclin A and cyclin B proteins have been found in all multicellular animals examined. The cDNAs encoding three related human D-type cyclins and cyclin E were isolated based on their ability to complement *S. cerevisiae* cells mutant in all three genes encoding G_1 cyclins. The relative amounts of the three D-type cyclins expressed in various cell types differ. Here we refer to them collectively as cyclin D.

Cyclins D and E are the mammalian mid- and late-G_1 cyclins, respectively. Experiments in which cultured mammalian cells were microinjected with anti-cyclin D antibody at various times after addition of growth factors demonstrated that cyclin D is essential for passage through the restriction point (Figure 20-31).

Figure 20-32 presents a current model for the periods of the cell cycle in which different cyclin-CDK complexes act in G_0-arrested mammalian cells stimulated to divide by the addition of growth factors. In the absence of growth factors, cultured G_0 cells express neither cyclins nor CDKs; the absence of these critical proteins explains why G_0 cells do not progress through the cell cycle and replicate.

Table 20-1, presented early in this chapter, summarizes the various cyclins and CDKs that we have mentioned and the portions of the cell cycle in which they are active. The cyclins fall into two major groups, G_1 cyclins and B-type cyclins, which function in S, G_2, and M. Although it is not possible to draw a simple one-to-one correspondence between the functions of the several cyclins and CDKs in *S. pombe, S. cerevisiae,* and vertebrates, the various cyclin-CDK complexes they form can be broadly considered in terms of their functions in mid-G_1, late-G_1, S, and M phases. All B-type cyclins contain a conserved destruction box sequence that is recognized by the APC/C-Cdh1 ubiquitin-protein ligase, whereas G_1 cyclins lack this sequence. Thus the APC/C regulates only the activity of those cyclin-CDK complexes that include B-type cyclins.

Regulated Expression of Two Classes of Genes Returns G_0 Mammalian Cells to the Cell Cycle

Addition of growth factors to G_0-arrested mammalian cells induces transcription of multiple genes, most of which fall into one of two classes—*early-response* or *delayed-response* genes—depending on how soon their encoded mRNAs appear. Transcription of early-response genes is induced within a few minutes after addition of growth factors by signal-transduction cascades that activate preexisting transcription factors in the cytosol or nucleus (Chapter 16). Many of the early-response genes encode transcription factors, such as c-Fos and c-Jun, that stimulate transcription of the delayed-response genes. Mutant, unregulated forms of both c-Fos and c-Jun are expressed by oncogenic retroviruses (Chapter 25); the discovery that the activated viral forms of these proteins (v-Fos and v-Jun) can transform normal cells into cancer cells led to identification of the normal, regulated cellular forms of these transcription factors.

After peaking at about 30 minutes following addition of growth factors, the concentrations of the early-response mRNAs fall to a lower level that is maintained as long as growth factors are present in the medium. This decrease in early-response mRNA levels is mediated by early-response proteins.

Expression of delayed-response genes depends on proteins encoded by early-response genes. Some delayed-response genes encode additional transcription factors (see below); others encode mid- and late-G_1 cyclins and CDKs. The mid-G_1 cyclins and their associating CDKs are expressed first, followed by the late-G_1 cyclin and its CDK (see Figure 20-32). If growth factors are withdrawn before passage through the restriction point, transcription of the genes encoding these G_1 cyclins and CDKs ceases. Since these proteins and the mRNAs encoding them are unstable, their concentrations fall precipitously. As a consequence, the cells do not pass the restriction point and do not replicate.

In addition to being controlled by transcription of the gene encoding the mid-G_1 cyclin (cyclin D), the concentration of this cyclin is also regulated by controlling *translation* of cyclin D mRNA. In this regard, the mammalian mid-G_1 cyclin is similar to *S. cerevisiae* mid-G_1 cyclin. Addition of growth factors to cultured mammalian cells triggers signal transduction via the PI-3 kinase pathway discussed in Chapter 16, leading to activation of the mTOR pathway and the resulting activation of translation initiation factors (see

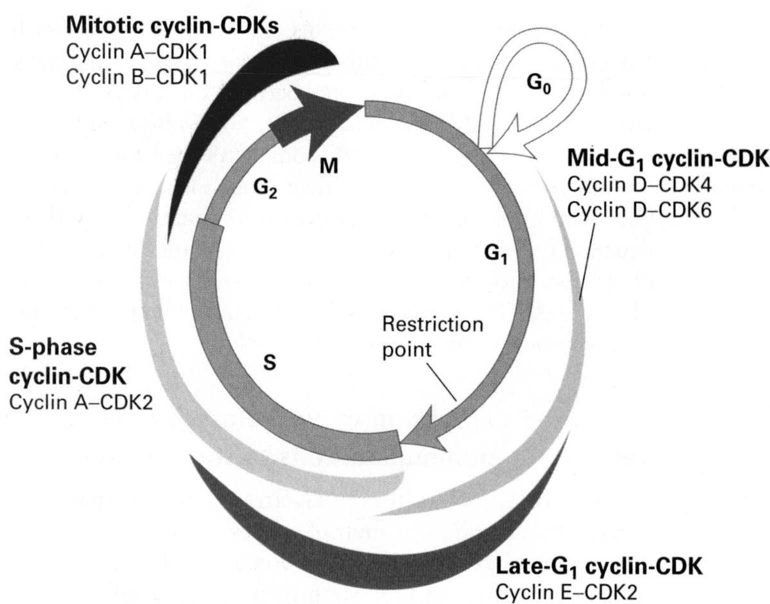

◄ **FIGURE 20-32 Activity of mammalian cyclin-CDK complexes through the course of the cell cycle.** Cultured G_0 cells are induced to divide by treatment with growth factors. The width of the colored bands is approximately proportional to the protein kinase activity of the indicated complexes. "Cyclin D" refers to all three D-type cyclins.

Figure 8-30). As a result, translation of cyclin D mRNA and other mRNAs is stimulated. Agents that inhibit activation of translation initiation factors, such as TGF-β, inhibit translation of cyclin D mRNA and thus inhibit cell proliferation.

Passage Through the Restriction Point Depends on Phosphorylation of the Tumor-Suppressor Rb Protein

Some members of a small family of related transcription factors, referred to collectively as *E2F factors,* are encoded by delayed-response genes. These transcription factors activate genes encoding many of the proteins involved in DNA synthesis. They also stimulate transcription of genes encoding the late-G_1 cyclin, the S-phase cyclin, and the S-phase CDK. Thus the E2Fs function in late G_1 similarly to the *S. cerevisiae* transcription factors SBF and MBF. In addition, E2Fs autostimulate transcription of their own genes. E2Fs function as transcriptional repressors when bound to *Rb protein,* which in turn binds histone deacetylase and methylase complexes. As discussed in Chapter 7, histone deacetylation and methylation of specific histone lysines causes chromatin to assume a condensed, transcriptionally inactive form.

Rb protein was initially identified as the product of the prototype **tumor-suppressor gene,** *RB*. The products of tumor-suppressor genes function in various ways to inhibit progression through the cell cycle (Chapter 25). Loss-of-function mutations in *RB* are associated with the disease *hereditary retinoblastoma.* A child with this disease inherits one normal RB^+ allele from one parent and one mutant RB^- allele from the other. If the RB^+ allele in any of the trillions of cells that make up the human body becomes mutated to a RB^- allele, then no functional Rb protein is expressed and the cell or one of its descendants is likely to become cancerous. For reasons that are not understood, this generally happens in a retinal cell leading to the retinal tumors that characterize this disease. Subsequently it was discovered that Rb function is inactivated in almost all cancer cells, either by mutations in both alleles of *RB*, or by abnormal regulation of Rb phosphorylation. ■

Rb protein is one of the most significant substrates of mammalian G_1 cyclin-CDK complexes. Phosphorylation of Rb protein at multiple sites prevents its association with E2Fs, thereby permitting E2Fs to activate transcription of genes required for entry into S phase. As shown in Figure 20-33, phosphorylation of Rb protein is initiated by the mid-G_1 cyclin-CDK complexes in mid G_1. Once the late-G_1 cyclin and CDK are induced by phosphorylation of some Rb, the resulting late-G_1 cyclin-CDK complex further phosphorylates Rb in late G_1. When late-G_1 cyclin-CDK accumulates to a

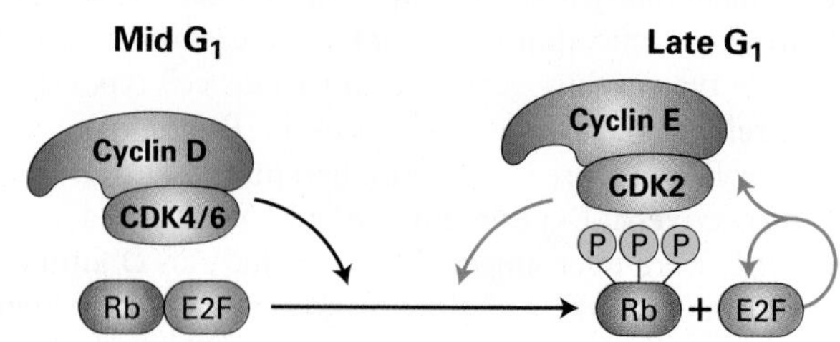

▲ **FIGURE 20-33 Regulation of Rb and E2F activities in mid-late G_1.** Stimulation of G_0 cells with mitogens induces expression of CDK4, CDK6, D-type cyclins, and the E2F transcription factors, all encoded by delayed-response genes. Rb protein initially inhibits E2F activity. When signaling from mitogens is sustained, the resulting cyclin D–CDK4/6 complexes begin phosphorylating Rb, releasing some E2F, which stimulates transcription of the genes encoding cyclin E, CDK2, and E2F itself (autostimulation). The cyclin E–CDK2 complexes further phosphorylate Rb, resulting in positive feedback loops (blue arrows) that lead to a rapid rise in the expression and activity of both E2F and cyclin E–CDK2 as the cell approaches the $G_1 \rightarrow S$ transition.

critical threshold level, further phosphorylation of Rb by the late-G_1 complex continues even when mid-G_1 cyclin-CDK activity is removed. This is one of the principal biochemical events responsible for passage through the restriction point. At this point, further phosphorylation of Rb by the late-G_1 cyclin-CDK occurs even when mitogens are withdrawn and mid-G_1 cyclin-CDK levels fall. Since E2F stimulates its own expression and that of the late-G_1 cyclin and CDK, positive cross-regulation of E2F and late-G_1 cyclin-CDK produces a rapid rise of both activities in late G_1.

As they accumulate, S-phase cyclin-CDK and mitotic cyclin-CDK complexes maintain Rb protein in the phosphorylated state throughout the S, G_2, and early M phases. After cells complete anaphase and enter early G_1 or G_0, the fall in cyclin-CDK levels leads to dephosphorylation of Rb. As a consequence, hypophosphorylated Rb is available to inhibit E2F activity during early G_1 of the next cycle and in G_0-arrested cells.

Cyclin A Is Required for DNA Synthesis and CDK1 for Entry into Mitosis

High levels of E2Fs activate transcription of the *cyclin A* gene as mammalian cells approach the $G_1 \rightarrow S$ transition. (Despite its name, cyclin A is a B-type cyclin, see Table 20-1.) Disruption of cyclin A function inhibits DNA synthesis in mammalian cells, suggesting that cyclin A is the S-phase cyclin and that, along with CDK2, it may function like *S. cerevisiae* S-phase cyclin-CDK complexes to trigger initiation of DNA synthesis. There is also evidence that the mammalian late-G_1 cyclin-CDK complexes also contribute to activation of prereplication complexes. Note that CDK2 complexes with both the late-G_1 and the S-phase cyclins (see Figure 20-32).

Three related *CDK inhibitory* proteins, or *CKIs* ($p27^{KIP1}$, $p57^{KIP2}$, and $p21^{CIP}$), appear to share the function of the *S. cerevisiae* S-phase inhibitor Sic1 (see Figure 20-28). Phosphorylation of $p27^{KIP1}$ by late-G_1 cyclin-CDK targets it for polyubiquitination by the mammalian SCF complex (see Figure 20-28). The mechanisms for degrading $p21^{CIP}$ and $p57^{KIP2}$ are less well understood.

The activity of mammalian cyclin-CDK2 complexes is also regulated by phosphorylation and dephosphorylation mechanisms similar to those controlling the *S. pombe* mitosis-promoting factor, MPF (see Figure 20-14). The *Cdc25A phosphatase,* which removes the inhibitory phosphate from CDK2, is a mammalian equivalent of *S. pombe* Cdc25 except that it functions at the $G_1 \rightarrow S$ transition rather than the $G_2 \rightarrow M$ transition. The mammalian phosphatase normally is activated late in G_1, but is degraded in the response of mammalian cells to DNA damage to prevent the cells from entering S phase (see Section 20.7).

Once late-G_1 cyclin-CDK and S-phase cyclin-CDK are activated by Cdc25A and the S-phase inhibitors have been degraded, DNA replication is initiated at prereplication complexes. The general mechanism is thought to parallel that in *S. cerevisiae* (see Figure 20-30), although small differences are found in vertebrates. Phosphorylation of DNA replication preinitiation complexes at replication origins by late-G_1 cyclin-CDK and S-phase cyclin-CDK likely promotes initiation of DNA replication. As in yeast, phosphorylation of these initiation factors likely prevents reassembly of prereplication complexes until the cell passes through mitosis, thereby assuring that replication from each origin occurs only once during each cell cycle. In metazoans, a second small protein, geminin, contributes to the inhibition of re-initiation at origins until cells complete a full cell cycle. *Geminin* is expressed in late G_1; it binds and inhibits replication initiation factors as they are released from preinitiation complexes once DNA replication is initiated during S phase (Figure 20-30, step **3**), contributing to the inhibition of re-initiation at an origin. Geminin contains a destruction box at its N-terminus that is recognized by the APC/C-Cdh1, causing it to be polyubiquitinylated in late anaphase and degraded by proteasomes. This frees the replication initiation factors, which are dephosphorylated by Cdc14 phosphatase, to bind to ORC on replication origins forming preinitiation complexes during the following G_1 phase.

The principal mammalian CDK in G_2 and mitosis is CDK1 (see Figure 20-32). This CDK, which is highly homologous with *S. pombe* CDK, associates with cyclins A and B. The mRNAs encoding either of these mammalian cyclins can promote meiotic maturation when injected into *Xenopus* oocytes arrested in G_2 (see Figure 20-6), demonstrating that they function as mitotic cyclins. In somatic vertebrate cells, cyclin A–CDK1 and cyclin B–CDK1 function together as the equivalent of the *S. pombe* MPF (mitotic cyclin-CDK). The kinase activity of these mammalian complexes also is regulated by proteins analogous to those that control the activity of the *S. pombe* MPF (see Figure 20-14). The inhibitory phosphate on CDK1 is removed by *Cdc25C phosphatase,* which is analogous to *S. pombe* Cdc25 phosphatase.

In cycling mammalian cells, cyclin B is first synthesized late in the S phase and increases in concentration as cells proceed through G_2, peaking during metaphase and dropping after late anaphase. This parallels the time course of cyclin B expression in *Xenopus* cycling egg extracts (see Figure 20-9). In human cells, cyclin B first accumulates in the cytosol and then enters the nucleus just before the nuclear envelope retracts into the ER early in mitosis. Thus MPF activity is controlled not only by phosphorylation and dephosphorylation but also by regulation of its import into the nucleus. In fact, cyclin B shuttles between the nucleus and cytosol, and the change in its localization during the cell cycle results from a change in the relative rates of import and export. As in *Xenopus* eggs and *S. cerevisiae,* cyclins A and B are polyubiquitinylated by the APC/C-Cdh1 complex during late anaphase and then are degraded by proteasomes (see Figure 20-10).

Two Types of Cyclin-CDK Inhibitors Contribute to Cell-Cycle Control in Mammals

As noted above, three related CKIs—$p21^{CIP}$, $p27^{KIP1}$, and $p57^{KIP2}$—inhibit late-G_1 cyclin-CDK and S-phase cyclin-CDK activity and must be degraded before DNA replication can begin. These same CDK inhibitory proteins also can bind to and inhibit the other mammalian cyclin-CDK complexes involved in cell-cycle control. As we discuss later, $p21^{CIP}$ plays a role in the response of mammalian cells to DNA damage. Experiments with knockout mice lacking $p27^{KIP1}$ have shown that this CKI is particularly important in controlling generalized cell proliferation soon after birth. Although $p27^{KIP1}$ knockouts are larger than normal, most develop normally otherwise. In contrast, $p57^{KIP2}$ knockouts exhibit defects in cell differentiation, and most die shortly after birth owing to defective development of various organs.

A second class of cyclin-CDK inhibitors called *INK4s* (*in*hibitors of *k*inase *4*) includes several small, closely related proteins that interact only with the mid-G_1 CDKs, CDK4 and CDK6, and thus function specifically in controlling the mid-G_1 phase. Binding of INK4s to CDK4 and CDK6 blocks their interaction with cyclin D and hence their protein kinase activity. The resulting decreased phosphorylation of Rb protein prevents transcriptional activation by E2Fs and entry into the S phase. One INK4 called *p16* is a tumor suppressor, like Rb protein discussed earlier. The presence of two mutant p16 alleles in a large fraction of human cancers is evidence for the important role of p16 in controlling the cell cycle (Chapter 25).

KEY CONCEPTS OF SECTION 20.6

Cell-Cycle Control in Mammalian Cells

- Various polypeptide growth factors called *mitogens* stimulate cultured mammalian cells to proliferate by inducing expression of early-response genes. Many of these encode transcription factors that stimulate expression of delayed-response genes encoding the G_1 CDKs, G_1 cyclins, and E2F transcription factors.

- Once cells pass the restriction point, they can enter the S phase and complete S, G_2, and mitosis in the absence of growth factors.

- Mammalian cells use several CDKs and cyclins to regulate passage through the cell cycle. Cyclin D–CDK4 and cyclin D–CDK6 function in mid to late G_1; cyclin E–CDK2, in late G_1 and early S; cyclin A–CDK2, in S; and cyclin A–CDK1 and cyclin B–CDK1 in G_2 and M through anaphase (see Figure 20-32).

- Unphosphorylated Rb protein binds to E2Fs, converting them into transcriptional repressors. Phosphorylation of Rb by the mid-G_1 cyclin-CDK liberates E2Fs to activate transcription of genes encoding the late-G_1 cyclin and CDK, as well as other proteins required for the S phase. E2Fs also autostimulate transcription of their own genes.

- The late-G_1 cyclin-CDK further phosphorylates Rb, further activating E2Fs. Once a critical level of late-G_1 cyclin-CDK has been expressed, a positive feedback loop with E2F results in a rapid rise of both activities that drives passage through the restriction point (see Figure 20-33).

- The activity of S-phase cyclin-CDK, induced by high E2F activity, initially is held in check by CKIs, which function like an S-phase inhibitor, and by the presence of an inhibitory phosphate on CDK2, the S-phase CDK. Proteasomal degradation of the inhibitors and activation of the Cdc25A phosphatase, as cells approach the $G_1 \rightarrow S$ transition, generate active S-phase cyclin-CDK. Along with the late-G_1 cyclin-CDK, this complex activates prereplication complexes to initiate DNA synthesis by a mechanism similar to that in *S. cerevisiae* (see Figure 20-30).

- Cyclin A–CDK1 and cyclin B–CDK1 induce the events of mitosis through early anaphase. Cyclins A and B are polyubiquitinylated by the anaphase-promoting complex (APC/C) during late anaphase and then are degraded by proteasomes.

- The activity of mammalian mitotic cyclin-CDK complexes is regulated by phosphorylation and dephosphorylation similarly to the mechanism in *S. pombe,* with the Cdc25C phosphatase removing inhibitory phosphates (see Figure 20-14).

- The activities of mammalian cyclin-CDK complexes also are regulated by CDK inhibitors (CKIs), which bind to and inhibit each of the mammalian cyclin-CDK complexes, and INK4 proteins, which block passage through G_1 by specifically inhibiting G_1 CDKs (CDK4 and CDK6).

20.7 Checkpoints in Cell-Cycle Regulation

Before proceeding, let's review the major steps in the eukaryotic cell cycle summarized in Figure 20-34. In continuously cycling cells, cyclin-CDK complexes are absent in early G_1. Hypophosphorylated DNA replication initiation factors are free to bind to ORC complexes at DNA replication origins, generating prereplication complexes that are inactive until they are phosphorylated by an S-phase cyclin-CDK (step **1**). In mid-G_1, mid-G_1 cyclin-CDKs are expressed and phosphorylate the APC/C specificity factor Cdh1, inactivating it and allowing newly synthesized B-type cyclins (and geminin in vertebrates) to accumulate when they are expressed (step **2**). The mid-G_1 cyclin-CDKs also phosphorylate specific transcription factors, activating expression of late-G_1 and S-phase cyclins (and CDK in vertebrate somatic cells) (step **3**). However, as B-type cyclins are expressed, they are immediately bound by inhibitors. When G_1 cyclin-CDK activities reach peak levels, they phosphorylate these inhibitors at multiple sites (step **4**), marking them for polyubiquitination by the SCF ubiquitin-protein ligase, and subsequent degradation by proteasomes (step **5**).

This rapid degradation of S-phase cyclin-CDK inhibitors releases S-phase cyclin-CDK activities, which phosphorylate key regulatory sites in prereplication complexes, stimulating initiation of DNA replication at multiple origins (step **6**).

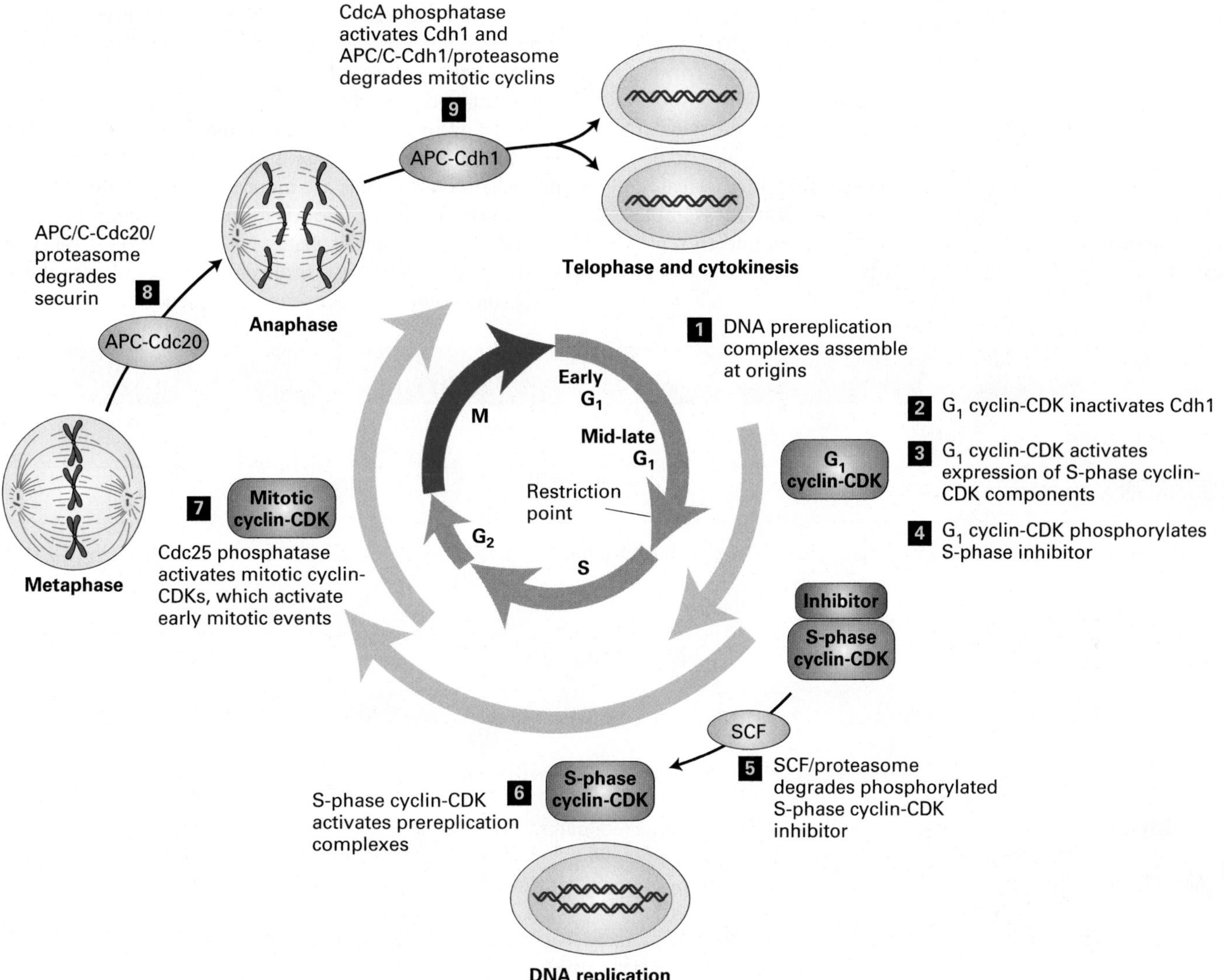
CdcA phosphatase
activates Cdh1 and
APC/C-Cdh1/proteasome
degrades mitotic cyclins
9
APC-Cdh1
Telophase and cytokinesis
APC/C-Cdc20/
proteasome
degrades
securin
8
APC-Cdc20
Anaphase
1 DNA prereplication
complexes assemble
at origins
Early
G1
M
Mid-late
G1
2 G1 cyclin-CDK inactivates Cdh1
G1
cyclin-CDK
3 G1 cyclin-CDK activates
expression of S-phase cyclin-
CDK components
Restriction
point
4 G1 cyclin-CDK phosphorylates
S-phase inhibitor
7
Mitotic
cyclin-CDK
G2
S
Metaphase
Cdc25 phosphatase
activates mitotic cyclin-
CDKs, which activate
early mitotic events
Inhibitor
S-phase
cyclin-CDK
SCF
5 SCF/proteasome
degrades phosphorylated
S-phase cyclin-CDK
inhibitor
S-phase cyclin-CDK
activates prereplication
complexes
6
S-phase
cyclin-CDK
DNA replication

▲ **FIGURE 20-34 Fundamental processes in the eukaryotic cell cycle.** See the text for discussion.

Mitotic cyclin-CDKs are expressed in late S phase and G_2. When DNA replication has been completed, they are activated by Cdc25 phosphatase, and either they, or other protein kinases that they activate, phosphorylate specific regulatory sites in more than a hundred proteins including histone H1, condensins and cohesins, additional chromatin-associated proteins, microtubule-associated proteins, nuclear lamins, inner nuclear membrane proteins, and nuclear pore complex proteins. These multiple, specific phosphorylations induce the early events of mitosis including chromosome condensation, remodeling of microtubules into the mitotic spindle apparatus, and, in animals and plants, retraction of the nuclear envelope into the ER (step **7**).

Once every kinetochore of each sister chromatid has attached to spindle microtubule fibers during metaphase, inhibition of the Cdc20 specificity factor is lifted. This results in active APC/C-Cdc20 and polyubiquitination and proteasomal degradation of securin (step **8**). Securin degradation releases the proteolytic activity of separase, which then cleaves the cohesin rings at centromeres that hold sister chromatids together. The forces exerted by the mitotic spindle apparatus then pull the released sister chromatids toward opposite spindle poles. The resulting sudden separation of all sister chromatids marks the beginning of *anaphase*.

Once the daughter chromosomes have separated sufficiently to ensure equal segregation of all chromosomes to daughter cells during cytokinesis, the Cdc14 phosphatase is activated. Cdc14 dephosphorylates and activates the Cdh1 APC/C specificity factor, resulting in the polyubiquitination and proteasomal degradation of all B-type cyclins (and geminin in vertebrates), and consequently, the loss of MPF activity (step **9**). Sites on the multiple proteins that were phosphorylated by cyclin-CDKs are dephosphorylated by Cdc14. This returns the proteins to their interphase functions, resulting in decondensation of chromosomes, formation of an interphase microtubule cytoskeleton with a single microtubule organizing center, and reassembly of the nuclear envelope during telophase, followed by cytokinesis. The dephosphorylated DNA replication-initiation factors (released by geminin degradation in vertebrates) then reassemble preinitiation complexes on ORC complexes bound to replication origins in daughter cells, in preparation for the next cell cycle (step **1**).

Successful completion of the cell cycle has several general requirements. Each process summarized in Figure 20-34 must go to completion before subsequent steps are undertaken, and the steps must occur in the correct order. Catastrophic genetic damage can occur if cells progress to the next phase of the cell cycle before the previous phase is properly completed. For example, when S-phase cells are induced to enter mitosis by fusion to a cell in mitosis, the MPF present in the mitotic cell forces the chromosomes of the S-phase cell to condense. This premature entry into mitosis results in fragmentation of the S-phase chromosomes, a disastrous consequence for a cell.

TABLE 20-2 Regulators of Cyclin-CDK Activity

TYPE OF REGULATOR	FUNCTION
KINASES AND PHOSPHATASES	
CAK kinase	Activates cyclin-CDKs
Wee1 kinase	Inhibits cyclin-CDKs
Cdc25 phosphatase	Activates cyclin-CDKs
Cdc14 phosphatase	Activates Cdh1 to inhibit mitotic cyclin-CDK
Cdc25A phosphatase	Activates vertebrate S-phase cyclin-CDK
Cdc25C phosphatase	Activates vertebrate mitotic cyclin-CDK
ATM/ATR kinases	Checkpoint controls, activate Chk1/Chk2 kinases
Chk1/Chk2 kinases	Checkpoint controls, inactivate Cdc25C and Cdc25A phosphatases to induce cell-cycle arrest
INHIBITORY PROTEINS	
Sic1	Binds and inhibits S-phase cyclin-CDKs
CKIs $p27^{KIP1}$, $p57^{KIP2}$, and $p21^{CIP}$	Bind and inhibit cyclin-CDKs
INK4	Binds and inhibits mid-G_1 CDKs
Mad2	Spindle-assembly checkpoint control, binds Cdc20 and prevents onset of anaphase and inactivation of B-type cyclin-CDKs
Rb	Binds E2Fs, preventing transcription of multiple cell cycle genes
UBIQUITIN-PROTEIN LIGASES	
SCF	Degradation of phosphorylated Sic1 or $p27^{KIP1}$ to activate S-phase cyclin-CDKs
APC/C + Cdc20	Induces degradation of Securin, initiating anaphase. Induces partial degradation of B-type cyclins
APC/C + Cdh1	Induces complete degradation of B-type cyclins to initiate telophase, and geminin in metazoans to allow formation of prereplication complexes on DNA replication origins

Another example of the importance of order of events in the cell cycle concerns attachment of kinetochores to microtubules of the mitotic spindle during metaphase. If anaphase is initiated before both kinetochores of a replicated chromosome become attached to microtubules from opposite spindle poles, daughter cells are produced that have missing or extra chromosomes, an outcome called *nondisjunction*. When nondisjunction occurs in mitotic cells, it can lead to the misregulation of genes, and contribute to the development of cancer. When nondisjunction occurs during the meiotic division that generates a human egg, Down syndrome can occur from trisomy of chromosome 21, resulting in developmental abnormalities and mental retardation. Other mechanisms can also generate trisomy. (Trisomy of any of the human chromosomes can occur, but for every other chromosome except chromosome 21, trisomy results in embryonic lethality or death shortly after birth.) ▮

We have seen how progression through the cell cycle is governed by precise regulation of the activities of multiple cyclin-CDK complexes. Table 20-2 summarizes the various types of regulators of cyclin-CDK activity. The key cell-cycle events of DNA replication and chromosome segregation must be accomplished with extraordinary accuracy and fidelity. To ensure that these processes occur correctly and in the proper order, cells have evolved multiple additional levels of regulation controlling these fundamental cell-cycle events. Collectively, these additional regulatory mechanisms are known as checkpoints (Figure 20-35). Several examples of cell-cycle checkpoints have been discussed earlier in the chapter. In this section, we consider these and additional checkpoints in terms of the major cell-cycle processes summarized above. Control mechanisms that operate at these checkpoints ensure that chromosomes are intact and that each stage of the cell cycle is completed before

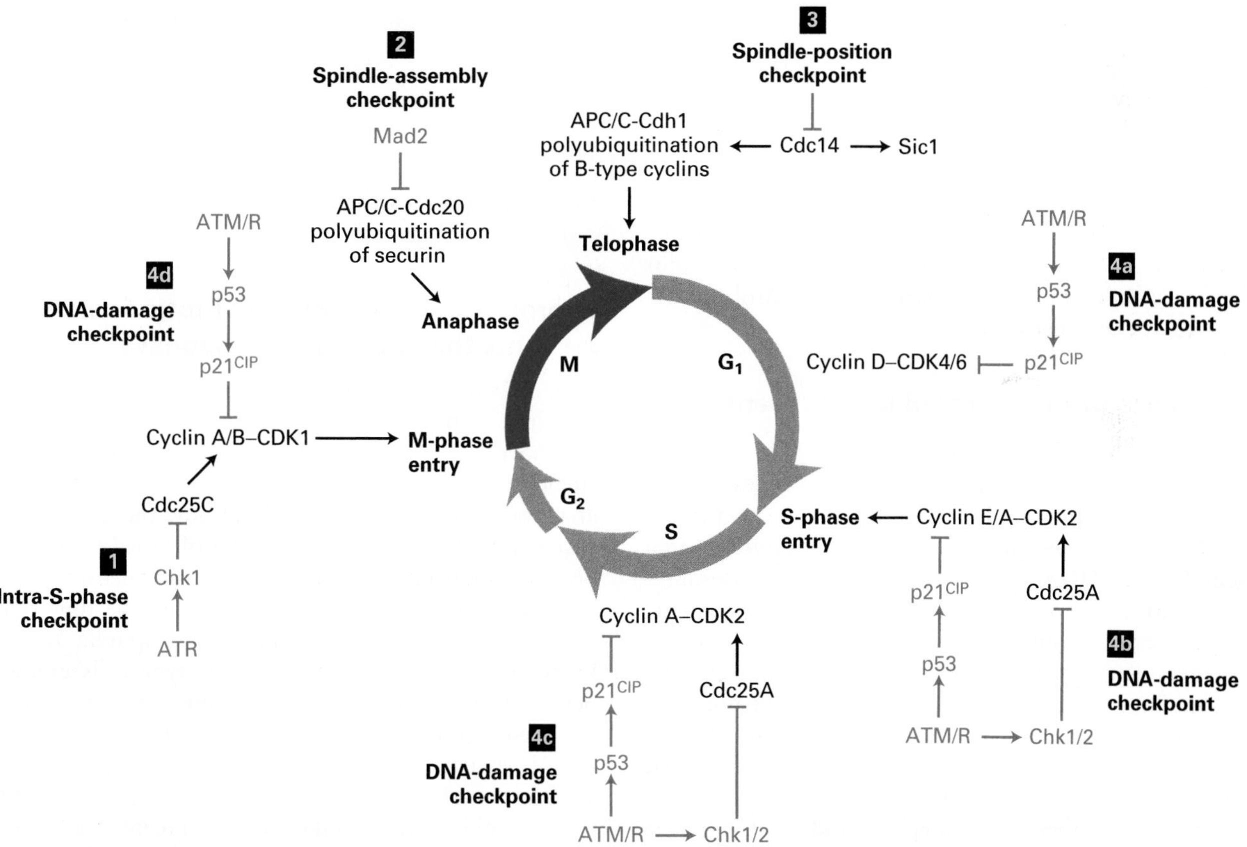

▲ **FIGURE 20-35 Overview of checkpoint controls in the cell cycle.** The intra-S-phase checkpoint (**1**) prevents activation of cyclin A–CDK1 and cyclin B–CDK1 (i.e., mitosis-promoting factor, MPF) by activation of an ATR-Chk1 protein kinase cascade that phosphorylates and inactivates Cdc25C, thereby inhibiting entry into mitosis. In the spindle-assembly checkpoint (**2**), Mad2 and other proteins inhibit activation of the APC/C specificity factor Cdc20 required for polyubiquitination of securin, thereby preventing entry into anaphase. The spindle-position checkpoint (**3**) prevents release of the Cdc14 phosphatase from nucleoli, thereby blocking activation of the APC/C specificity factor (Cdh1) required for APC/C polyubiquitination of B-type cyclins as well as induction of Sic1. As a result, the decrease in MPF activity required for the events of telophase does not occur. In the initial phase of the DNA-damage checkpoint (**4**), the ATM or ATR protein kinase (ATM/R) is activated. The active kinases then trigger two pathways: the Chk-Cdc25A pathway (**4b** and **4c**), blocking entry into or passage through S phase, and the p53-p21^{CIP} pathway, leading to arrest in G_1, S, and G_2 (**4a**–**4d**). See the text for further discussion. Red symbols indicate pathways that inhibit progression through the cell cycle.

TABLE 20-3 Checkpoint Proteins

CHECKPOINT	PURPOSE	SENSOR	ACTION
Intra-S phase checkpoint	Ensures all DNA replication is complete before entering M-phase	ATR detects replication forks	Inhibition of Cdc25C to prevent activation of mitotic cyclin-CDKs, blocking early mitotic events
Spindle-assembly checkpoint	Ensures all chromosome kinetochores are attached to spindle microtubules before anaphase	Mad2 detects kinetochores unattached to microtubules	Inhibition of Cdc20 to prevent activation of separase and onset of anaphase
Spindle-position checkpoint	Ensures all chromosomes are properly segregated to daughter cells before telophase and cytokinesis	*(S. cerevisiae)* Tem-1 detects proper position of spindle pole body in bud	Prevention of Cdc14 activation and degradation of mitotic cyclins, blocking late mitotic events
DNA-damage checkpoint	Detects damage to DNA throughout the cell cycle	ATM, ATR detect DNA damage	Inhibition of Cdc25A to prevent entry into S phase; $p21^{CIP}$ inhibition of all cyclin-CDK complexes to induce cell cycle arrest

the following stage is initiated. Our understanding of these control mechanisms at the molecular level has advanced considerably in recent years. Table 20-3 lists four major cell cycle checkpoints and summarizes the control mechanisms used at each checkpoint.

The Presence of Unreplicated DNA Prevents Entry into Mitosis

Cells that fail to replicate all their chromosomes do not normally enter mitosis. Operation of the *intra-S-phase checkpoint* control involves the recognition of unreplicated DNA and stalled DNA replication forks, which causes inhibition of MPF activation (see Figure 20-35, **1**). Genetic studies in the yeasts and biochemical studies with *Xenopus* egg extracts demonstrated that the ATR and Chk1 protein kinases inhibit entry into mitosis by cells that have not completed DNA synthesis. The association of ATR with replication forks is thought to activate its protein kinase activity, leading to phosphorylation and activation of the Chk1 kinase. Active Chk1 then phosphorylates and inactivates the Cdc25 phosphatase (Cdc25C in vertebrates), which otherwise removes the inhibitory phosphate from mitotic CDKs. As a result, the mitotic cyclin-CDK complexes remain inhibited and cannot phosphorylate targets required to initiate mitosis. ATR continues to initiate this protein kinase cascade until all replication forks complete DNA replication and disassemble. This mechanism makes the initiation of mitosis *dependent* on the completion of chromosome replication. This dependency or requirement that a cell-cycle phase must be completed before the next phase can be initiated is a critical aspect of checkpoint function required for orderly progression of the fundamental processes of the cell cycle (see Figure 20-34).

Improper Assembly of the Mitotic Spindle Prevents the Initiation of Anaphase

The *spindle-assembly checkpoint* prevents entry into anaphase until every single kinetochore of every chromatid is properly associated with spindle microtubules. If even a single kinetochore is unattached to a spindle microtubule, anaphase is inhibited. Clues about how this checkpoint operates initially came from isolation of yeast mutants in the presence of benomyl, a microtubule-depolymerizing drug. Low concentrations of benomyl increase the time required for yeast cells to assemble the mitotic spindle and attach kinetochores to microtubules. Wild-type cells exposed to benomyl do not begin anaphase until these processes are completed and then proceed on through mitosis, producing normal daughter cells. In contrast, mutants defective in the spindle-assembly checkpoint proceed through anaphase before assembly of the spindle and attachment of kinetochores is complete; consequently, they mis-segregate their chromosomes, producing abnormal daughter cells that die.

Analysis of these mutants identified a protein called *Mad2* (*m*itotic *a*rrest *d*efective 2) and other proteins that regulate Cdc20, the specificity factor required to target the APC/C to securin (see Figure 20-35, **2**). Recall that APC/C-Cdc20-mediated polyubiquitination of securin and its subsequent degradation is required for activation of separase and entry into anaphase (see Figure 20-23). Mad2 has been shown to associate with kinetochores that are unattached to microtubules. Kinetochore-bound Mad2 rapidly exchanges with a soluble form of Mad2 that inhibits all the Cdc20 in the cell. When microtubules attach to kinetochores, the kinetochores release the bound Mad2 and cease the process by which the inhibitory, soluble form of Mad2 is produced. However, when even a single kinetochore is unattached to microtubules from the opposite spindle pole of its sister, sufficient soluble inhibitory Mad2 is produced at the unattached kinetochore to inhibit all the Cdc20 in the cell. The current model for how this regulatory mechanism functions (Figure 20-36) was suggested by X-ray crystallography and NMR data revealing the structures of interacting proteins involved in the process. The model is supported by directed mutagenesis studies guided by these structures, biochemical studies of the protein-protein interactions, and microscopic studies in living cells using GFP-labeled proteins. This elegant model for the spindle-assembly checkpoint can account for the ability of a single unattached kinetochore to inhibit all the cellular Cdc20 until the kinetochore becomes properly associated with spindle microtubules.

Proper Segregation of Daughter Chromosomes Is Monitored by the Mitotic Exit Network

Once chromosomes have segregated properly, telophase commences. The various events of telophase and subsequent cytokinesis, collectively referred to as the *exit from*

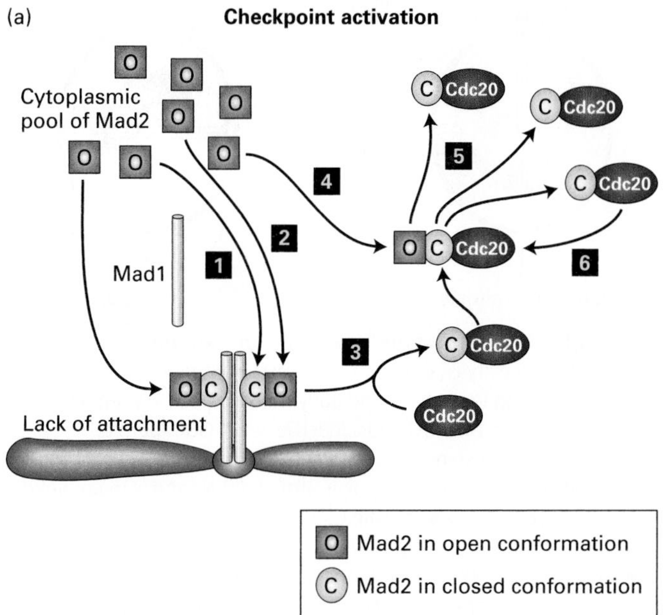

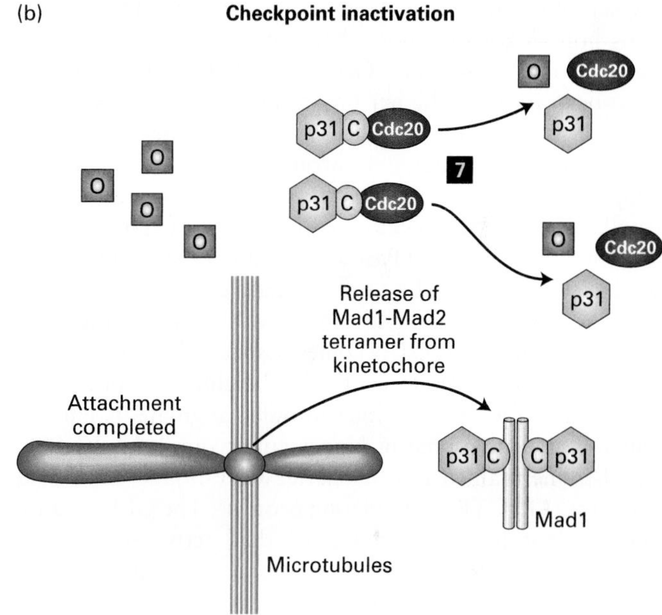

▲ **FIGURE 20-36 Model for Cdc20 Regulation.** The spindle-assembly checkpoint is active until every single kinetochore has attached properly to spindle microtubules. (a) The Mad2 protein exists in two conformations, one "open" (red squares) and the other "closed" (orange circles). According to the current model, Mad2 in the open conformation can bind either Mad1 or Cdc20. Binding to Mad1 or Cdc20 converts Mad2 to the closed conformation, which is stably bound to these proteins. Cdc20 bound by the closed conformation of Mad2 is inactive. Two Mad2 proteins in the same conformation do not interact, but closed Mad2 and open Mad2 can bind to each other transiently through a site on Mad2 distinct from the one that associates with either Mad1 or Cdc20. Mad1 and the closed conformation of Mad2 form a tetramer that binds to unattached kinetochores via the Mad1 subunit (**1**). Mad2 in the open conformation can bind transiently to the Mad2 in the closed conformation bound to Mad1 at the kinetochore (**2**). This interaction with the closed Mad2 stimulates open Mad2 to bind a Cdc20. Open Mad2 can bind Cdc20 only while it is interacting with a closed Mad2. This converts the open Mad2 protein to the closed conformation, causing it to dissociate from the Mad2 in the closed conformation at the kinetochore (**3**). The stable interaction of closed Mad2 with Cdc20 prevents Cdc20 from binding to the APC/C. Further, the closed Mad2 bound to Cdc20 can interact transiently with another Mad2 in the open conformation (**4**), causing it to bind another Cdc20 molecule. This converts this Mad2 to the closed conformation bound to Cdc20. This newly formed closed Mad2-Cdc20 complex dissociates from the first Mad2-Cdc20 pair, generating two Mad2-Cdc20 complexes (**5**). Thus free Mad2 in the open conformation is quickly converted to closed Mad2 bound to Cdc20 as this cycle repeats (**6**). The source of closed Mad2 that initiates this chain reaction is the closed Mad2 bound to Mad1 associated with a kinetochore, explaining how a single unattached kinetochore can cause inactivation of all the Cdc20 in the cell through the formation of closed Mad2-Cdc20 complexes. (b) Attachment of microtubules (green) to kinetochores causes the displacement of the Mad1/Mad2 tetramer. Mad2 in the displaced tetramer cannot interact with open Mad2, but rather, binds and activates another protein, $p31^{comet}$, which then binds Mad2 in Mad2-Cdc20 complexes, releasing active Cdc20 (**7**). However, a small number of Mad1-Mad2 tetramers bound to kinetochores can generate enough Mad2-Cdc20 complexes by the mechanism shown in (a) to overcome the activity of p31. Once all kinetochores have attached to microtubules causing the release of all Mad1-Mad2 tetramers, p31 activity predominates, releasing active Cdc20, which binds to the APC/C, resulting in polyubiquitination and proteasomal degradation of securin and the onset of anaphase. [Modified from A. De Antoni et al., 2005, *Curr. Biol.* **15:**214; see also K. Nasmyth, 2005, *Cell* **120:**739.]

mitosis, require inactivation of MPF. As discussed earlier, dephosphorylation of the APC/C specificity factor Cdh1 by the Cdc14 phosphatase leads to degradation of mitotic cyclins and loss of MPF activity late in anaphase (see Figure 20-10). During interphase and early mitosis, Cdc14 is sequestered in the nucleolus and inactivated. The *spindle-position checkpoint,* which monitors the location of the segregating daughter chromosomes at the end of anaphase, determines whether active Cdc14 is released from the nucleolus to promote exit from mitosis (see Figure 20-35, **3**).

Operation of this checkpoint in *S. cerevisiae* depends, in part, on a set of proteins referred to as the *mitotic exit network*. Regulation of Cdc14 activation operates similarly in most eukaryotes. In the fission yeast *S. pombe,* formation of the septum that divides daughter cells is regulated by proteins homologous to those that constitute the mitotic exit network in *S. cerevisiae*. Genes encoding similar proteins have been found in higher organisms where the homologs function in an analogous checkpoint that leads to arrest in late mitosis when daughter chromosomes do not segregate properly. A key component of the mitotic exit network is a small (monomeric) GTPase, called *Tem1* (Figure 20-37). This member of the **GTPase superfamily** of switch proteins controls the activity of a protein kinase cascade similarly to the way Ras controls MAP kinase pathways (Chapter 16). During anaphase, Tem1 becomes associated with the spindle pole body (SPB) closest to the daughter cell bud. (The SPB, from which spindle microtubules originate, is equivalent to the centrosome in higher eukaryotes.) At the SPB, Tem1 is maintained in the inactive GDP-bound state by a specific GAP (GTPase-activating protein). The GEF (guanosine nucleotide–exchange factor) that activates Tem1 is localized to the cortex of the bud and is absent from the mother cell. Another protein, Kin4 protein kinase, is localized to the mother cell cortex and is absent from the bud. When spindle microtubule elongation at the end of anaphase has correctly positioned segregating daughter chromosomes into the bud, Tem1 comes into contact with its GEF and the Tem1-GAP becomes phosphorylated and inhibited. As a consequence, Tem1 is converted into its active GTP-bound state. The terminal kinase in the cascade triggered by Tem1·GTP then phosphorylates the nucleolar anchor that binds and inhibits Cdc14, releasing the Cdc14 phosphatase into the cytoplasm and nucleoplasm in both the bud and mother cell (Figure 20-37, **1**). Once active Cdc14 is available, a cell can proceed through telophase and cytokinesis.

The mitotic exit network is a good example of the dependency of one cycle phase on completion of the previous phase. Telophase and cytokinesis cannot initiate until the chromosome segregation mechanism carries daughter chromosomes into the bud. This is because initiation of telophase depends on the activation of Cdc14, and the activation of Cdc14 requires that Tem1 be pushed all the way to the bud cortex.

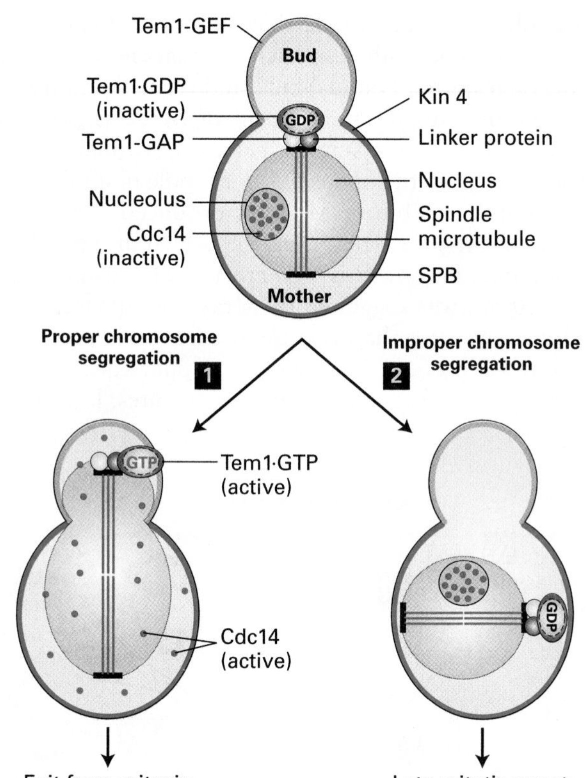

▲ **FIGURE 20-37 The spindle-position checkpoint.** Cdc14 phosphatase activity is required for the exit from mitosis. *Top:* In *S. cerevisiae,* during interphase and early mitosis, Cdc14 is sequestered and inactivated in the nucleolus. Inactive Tem1·GDP (purple) associates with the spindle pole body (SPB) nearest to the bud early in anaphase with the aid of a linker protein (blue) and is maintained in the inactive state by a specific GAP (GTPase-accelerating protein, yellow). If chromosome segregation occurs properly (**1**), extension of the spindle microtubules inserts the daughter SPB into the bud, causing Tem1 to come in contact with a specific GEF (guanine nucleotide–exchange factor) localized to the cortex of the bud (orange), and the inactivation of the Tem1-GAP. This converts inactive Tem1·GDP to active Tem1·GTP, which triggers a protein kinase cascade leading to release of active Cdc14 and exit from mitosis. If the spindle apparatus fails to place the daughter SPB in the bud (**2**), the SPB encounters Kin4 (cyan) localized to the mother cell cortex. This activates the Tem1-GAP, maintaining Tem1 in the inactive GDP-bound state and Cdc14 remains associated with nucleoli. Arrest in late mitosis results. [Adapted from G. Pereira and E. Schiebel, 2001, *Curr. Opin. Cell Biol.* **13:**762.]

If daughter chromosomes fail to segregate into the bud, Tem1 does not encounter the Tem1-GEF. Instead, the Kin4 kinase associated with the mother cell cortex maintains the Tem1-GAP in an activated state. Tem1, consequently, remains in its inactive, GDP-bound state, Cdc14 is not released from the nucleolus, and mitotic exit is blocked (Figure 20-37, **2**). Kin4 is not required in cells that segregate their chromosomes properly, but only in the small fraction of cells that fail to do so, giving them more time to push the daughter chromosomes into the bud. In *S. cerevisiae*, the error rate for mis-segregation of chromosomes is <1 in 10^5 cell divisions.

Cell-Cycle Arrest of Cells with Damaged DNA Depends on Tumor Suppressors

The proteins of the *DNA-damage checkpoint* sense DNA damage and block progression through the cell cycle until the damage is repaired. Damage to DNA can result from chemical agents and from irradiation with ultraviolet (UV) light or γ-rays. Arrest in G_1 and S prevents copying of damaged bases, which would fix mutations in the genome. Replication of damaged DNA also promotes chromosomal rearrangements that can contribute to the onset of cancer. Arrest in G_2 allows DNA double-stranded breaks to be repaired before mitosis. If a double-stranded break is not repaired, the broken distal portion of the damaged chromosome is not properly segregated because it is not physically linked to a centromere, which is pulled toward a spindle pole during anaphase.

As we discuss in detail in Chapter 25, inactivation of tumor-suppressor genes contributes to the development of cancer. The proteins encoded by several tumor-suppressor genes, including *ATM* and *Chk2,* normally function in the DNA-damage checkpoint. Patients with mutations in both copies of *ATM* or *Chk2* develop cancers far more frequently than normal. Both of these genes encode protein kinases.

DNA damage due to UV light is sensed by proteins that signal the presence of UV-damaged DNA to the ATM kinase, activating it. Activated ATM then phosphorylates and activates Chk2, which then phosphorylates the Cdc25A phosphatase, marking it for polyubiquitination by an ubiquitin-protein ligase and subsequent proteasomal degradation. Recall that removal of the inhibitory phosphate from mammalian CDK2 by Cdc25A is required for onset of and passage through the S phase, mediated by cyclin E–CDK2 and cyclin A–CDK2. Degradation of Cdc25A resulting from activation of the ATM-Chk2 pathway in G_1 or S-phase cells thus leads to G_1 or S arrest (see Figure 20-35, **4b** and **4c**). A similar pathway consisting of the protein kinases ATR and Chk1 leads to phosphorylation and polyubiquitination of Cdc25A in response to γ-irradiation. As discussed earlier for the intra-S-phase checkpoint, Chk1 also inactivates Cdc25C, preventing the activation of CDK1 and entry into mitosis.

Another tumor suppressor, **p53 protein**, contributes to arrest of cells with damaged DNA. Cells with functional p53 arrest in G_1 and G_2 when exposed to γ-irradiation, whereas cells lacking functional p53 do not arrest in G_1. Although the p53 protein is a transcription factor, under normal conditions it is extremely unstable and generally does not accumulate to high enough levels to stimulate transcription. The instability of p53 results from its polyubiquitination by a ubiquitin-protein ligase called *Mdm2* and subsequent proteasomal degradation. The rapid degradation of p53 is inhibited by ATM and ATR, which phosphorylate p53 at a site that interferes with Mdm2 binding. This and other modifications of p53 in response to DNA damage greatly increase its ability to activate transcription of specific genes that help the cell cope with DNA damage. One of these genes encodes $p21^{CIP}$, a generalized CKI that binds and inhibits all mammalian cyclin-CDK complexes. As a result, cells are arrested in G_1 and G_2 until the DNA damage is repaired and p53, and subsequently $p21^{CIP}$, levels fall (see Figure 20-35, **4a**–**4d**).

Under some circumstances, such as when DNA damage is extensive, p53 also activates expression of genes that lead to apoptosis, the process of programmed cell death that normally occurs in specific cells during the development of multicellular animals. In vertebrates, the p53 response evolved to induce apoptosis in the face of extensive DNA damage, presumably to prevent the accumulation of multiple mutations that might convert a normal cell into a cancer cell. The dual role of p53 in both cell-cycle arrest and the induction of apoptosis may account for the observation that nearly all cancer cells have mutations in both alleles of the *p53* gene or in the pathways that stabilize p53 in response to DNA damage (Chapter 25). The consequences of mutations in *p53, ATM,* and *Chk2* provide dramatic examples of the significance of cell-cycle checkpoints to the health of a multicellular organism. ▮

KEY CONCEPTS OF SECTION 20.7

Checkpoints in Cell-Cycle Regulation

- Checkpoint controls function to ensure that chromosomes are intact and that critical stages of the cell cycle are completed before the following stage is initiated.
- The intra-S-phase checkpoint operates during S and G_2 to prevent the activation of MPF before DNA synthesis is complete by inhibiting the activation of CDK1 by Cdc25C (see Figure 20-35, **1**).
- The spindle-assembly checkpoint, which prevents premature initiation of anaphase, utilizes Mad2 and other proteins to regulate the APC/C specificity factor Cdc20 that targets securin for polyubiquitination (see Figure 20-35, **2**, and Figure 20-36).
- The spindle-position checkpoint prevents telophase and cytokinesis until daughter chromosomes have been properly segregated, so that the daughter cell has a full set of chromosomes (see Figure 20-35, **3**).
- In the spindle-position checkpoint, the small GTPase Tem1 controls the availability of Cdc14 phosphatase, which in turn activates the APC/C specificity factor Cdh1 that targets B-type cyclins for degradation, causing inactivation of MPF (see Figure 20-10).
- The DNA-damage checkpoint arrests the cell cycle in response to DNA damage until the damage is repaired. Three

types of tumor-suppressor proteins (ATM/ATR, Chk1/2, and p53) are critical to this checkpoint.

- Activation of the ATM or ATR protein kinases in response to DNA damage due to UV light or γ-irradiation leads to arrest in G_1 and the S phase via a pathway that leads to loss of Cdc25A phosphatase activity. A second pathway from activated ATM/R stabilizes p53, which stimulates expression of $p21^{CIP}$. Subsequent inhibition of multiple CDK-cyclin complexes by $p21^{CIP}$ causes prolonged arrest in G_1 and G_2 (see Figure 20-35, **4a**–**4d**).
- In response to extensive DNA damage, p53 also activates genes that induce apoptosis.

20.8 Meiosis: A Special Type of Cell Division

In nearly all diploid eukaryotes, **meiosis** generates haploid germ cells (eggs and sperm), which can then fuse with a germ cell from another individual to generate a diploid zygote that develops into a new individual. Meiosis is a fundamental aspect of the biology and evolution of all eukaryotes because it results in the reassortment of the chromosome sets received from an individual's two parents. Both chromosome reassortment and **recombination** between parental DNA molecules during meiosis guarantees that each haploid germ cell generated will receive a unique combination of gene alleles that is distinct from each parent as well as from every other haploid germ cell formed.

The mechanisms of meiosis are analogous to those of mitosis. However, several key differences in meiosis allow this process to generate haploid cells with incredible genetic diversity (see Figure 5-3). In this section, we will discuss the parallels between molecular mechanisms of mitosis and meiosis, as well as the mechanistic differences responsible for the significant distinctions between these two fundamental processes of cell division.

Key Features Distinguish Meiosis from Mitosis

During meiosis (Figure 20-38), a single round of DNA replication is followed by two cycles of cell division, termed *meiosis I* and *meiosis II,* each distinct from the mitotic divisions of somatic cells. Figure 20-39 summarizes the distinctions between mitosis and meiosis. In G_2 and prophase of meiosis I, the two replicated chromatids of each chromosome (Figure 20-38, step **1**) are associated with each other by cohesin complexes along the full length of the chromosome arms, just as they are following DNA replication in a mitotic cell cycle (see Figure 20-21, G_2 phase). A major difference between meiosis and mitosis is that in prophase of meiosis I, homologous chromosomes (i.e., the maternal and paternal chromosome 1, the maternal and paternal chromosome 2, etc.) pair with each other, a process known as *synapsis* (Figure 20-39, row 4). This forms a *bivalent chromosome*, or *tetrad*, composed of four homologous chromatids, two maternal and two paternal. Significantly, at least one recombination event occurs between a maternal and a paternal chromatid in every tetrad (Figure 20-39, row 5). The **crossing over** of chromatids produced by recombination can be observed microscopically in the first meiotic prophase and metaphase as structures called *chiasmata* (singular, *chiasma*). In contrast, no pairing between homologous chromosomes occurs during mitosis, and recombination between nonsister chromatids is rare.

▶ **FIGURE 20-38 Meiosis.** Premeiotic cells have two copies of each chromosome (2*n*), one derived from the paternal parent and one from the maternal parent. For simplicity, the paternal and maternal homologs of only one chromosome are diagrammed. Step **1**: All chromosomes are replicated during the S phase before the first meiotic division, giving a 4*n* chromosomal complement. Cohesin complexes (not shown) link the sister chromatids composing each replicated chromosome along their full lengths. Step **2**: As chromosomes condense during the first meiotic prophase, replicated homologs become paired as the result of at least one crossover event between a paternal and a maternal chromatid. This pairing of replicated homologous chromosomes is called *synapsis*. At metaphase, shown here, both chromatids of one chromosome associate with microtubules emanating from one spindle pole, but each member of a homologous chromosome pair associates with microtubules emanating from opposite poles. Step **3**: During anaphase of meiosis I, the homologous chromosomes, each consisting of two chromatids, are pulled to opposite spindle poles. Step **4**: Cytokinesis yields the two daughter cells (now 2*n*), which enter meiosis II without undergoing DNA replication. At metaphase of meiosis II, shown here, the chromatids composing each replicated chromosome associate with spindle microtubules from opposite spindle poles, as they do in mitosis. Steps **5** and **6**: Segregation of chromatids to opposite spindle poles during the second meiotic anaphase followed by cytokinesis generates haploid germ cells (1*n*) containing one copy of each chromosome. Micrographs on the left show meiotic metaphase I and metaphase II in developing gametes from *Lilium* (Lily) ovules. Chromosomes are aligned at the metaphase plate. [Photos courtesy of Ed Reschke/Peter Arnold, Inc.]

Another key difference between mitosis and meiosis is that in the metaphase of meiosis I, the kinetochores at the centromeres of sister chromatids attach to spindle fibers emanating from the *same* spindle pole, rather than from opposite spindle poles as in mitosis. However, the kinetochores of the maternal and paternal chromosomes of each tetrad attach to spindle microtubules from opposite spindle poles (Figure 20-38, step **2**; Figure 20-39, row 6). Also, cohesion between the full length of the sister chromatid arms is maintained throughout metaphase of meiosis I. This is in contrast to mitosis, where cohesion between sister chromatid arms is lost during prophase in mitotic cells from most organisms, so that cohesion is maintained only in the region of the centromere during mitotic metaphase (Figure 20-21; Figure 20-39, row 7). Because nonsister chromatids have recombined at least once by metaphase of meiosis I, and because cohesion is maintained between the arms of sister chromatids, as the maternal and paternal chromosomes are pulled toward opposite spindle poles in metaphase, the

Focus Animation: Meiosis

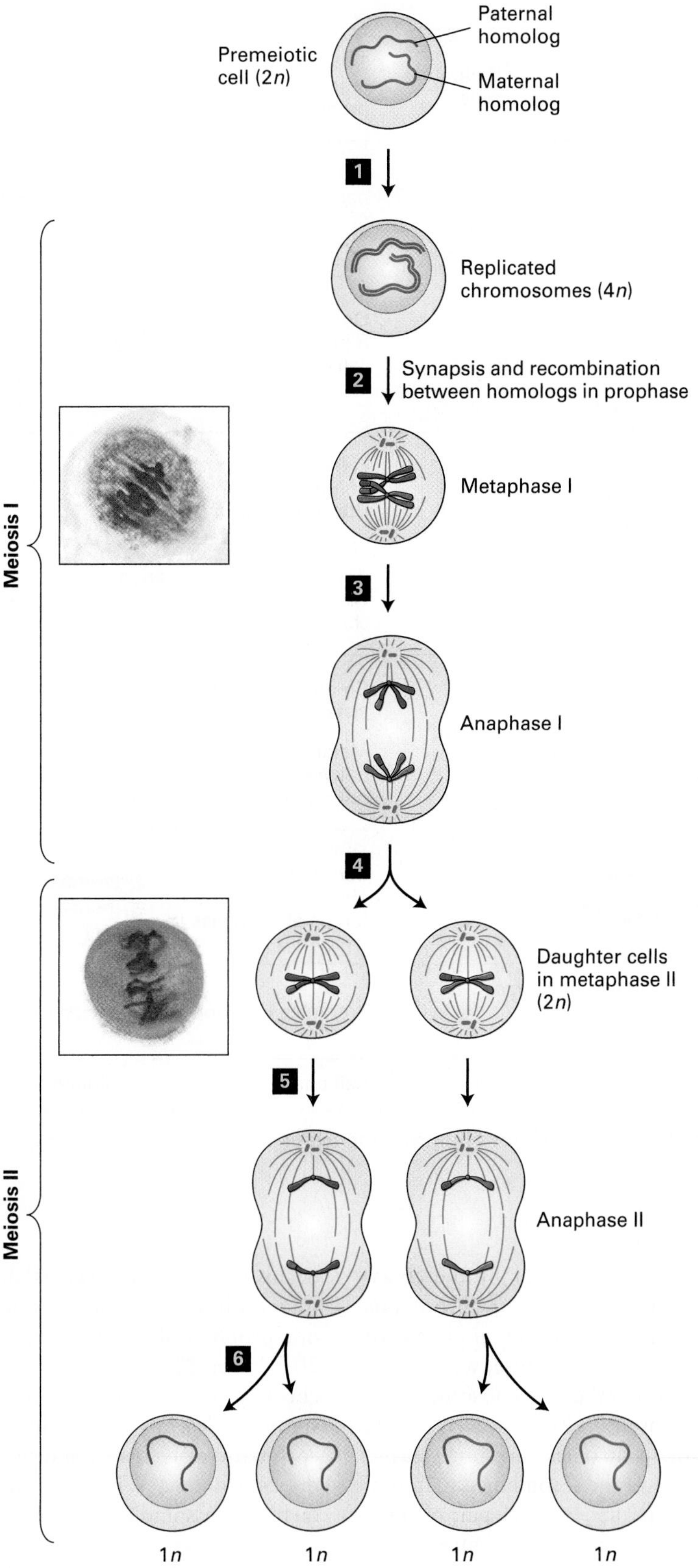

MEDIA CONNECTIONS

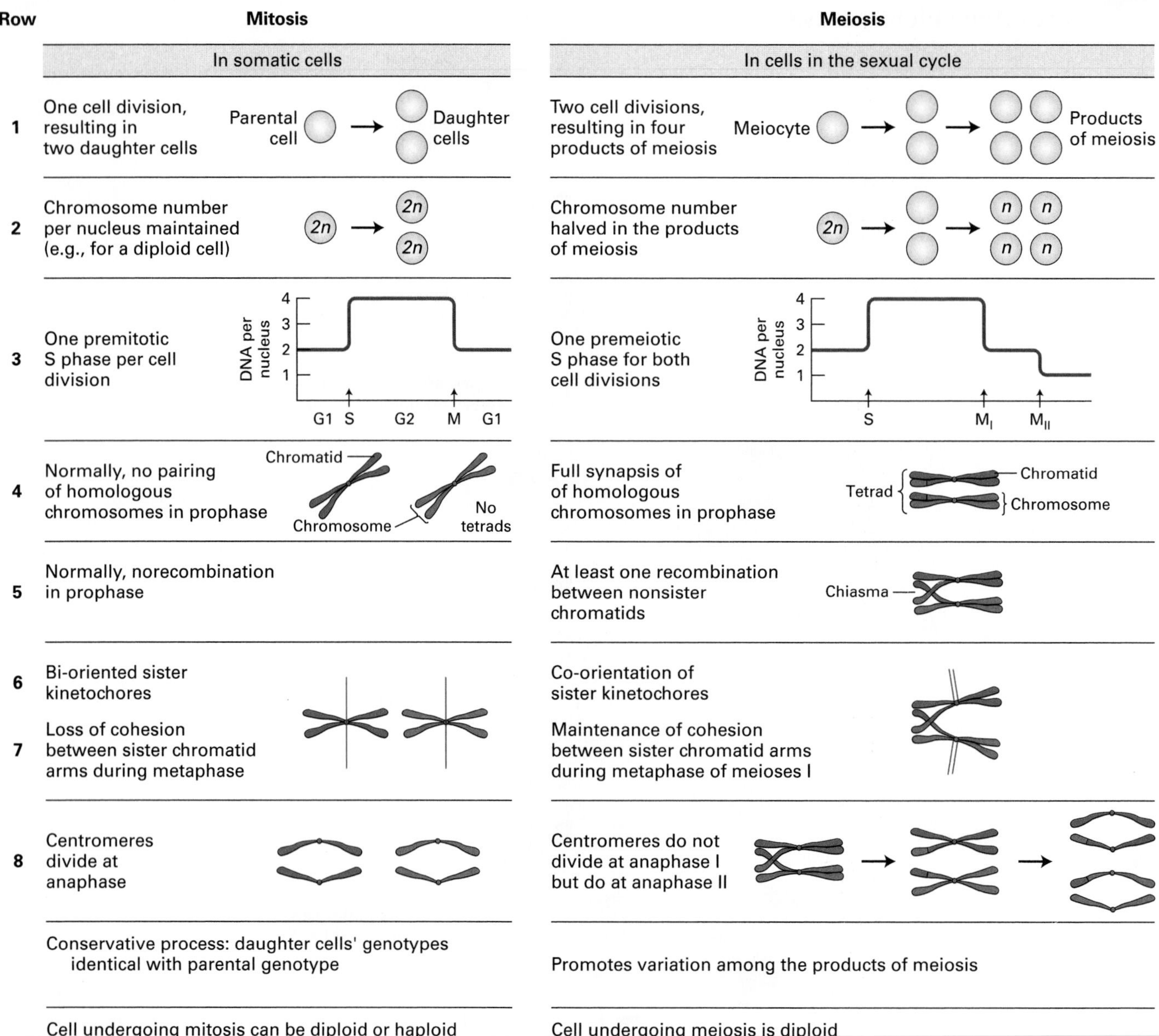

Row
Mitosis
Meiosis
In somatic cells
In cells in the sexual cycle
1
One cell division, resulting in two daughter cells
Parental cell
Daughter cells
Two cell divisions, resulting in four products of meiosis
Meiocyte
Products of meiosis
2
Chromosome number per nucleus maintained (e.g., for a diploid cell)
2n
Chromosome number halved in the products of meiosis
n
3
One premitotic S phase per cell division
DNA per nucleus
G1 S G2 M G1
One premeiotic S phase for both cell divisions
S
M$_I$
M$_{II}$
4
Normally, no pairing of homologous chromosomes in prophase
Chromatid
Chromosome
No tetrads
Full synapsis of of homologous chromosomes in prophase
Tetrad
Chromatid
Chromosome
5
Normally, norecombination in prophase
At least one recombination between nonsister chromatids
Chiasma
6
Bi-oriented sister kinetochores
Co-orientation of sister kinetochores
7
Loss of cohesion between sister chromatid arms during metaphase
Maintenance of cohesion between sister chromatid arms during metaphase of meioses I
8
Centromeres divide at anaphase
Centromeres do not divide at anaphase I but do at anaphase II
Conservative process: daughter cells' genotypes identical with parental genotype
Promotes variation among the products of meiosis
Cell undergoing mitosis can be diploid or haploid
Cell undergoing meiosis is diploid

▲ **FIGURE 20-39 Comparison of the main features of mitosis and meiosis.** [Adapted from A. J. F. Griffiths et al., 1999, *Modern Genetic Analysis*, W.H. Freeman and Company.]

homologous chromosomes are held together by the chiasmata between them and cohesion of the sister chromatids distal to the crossover (Figure 20-40). During anaphase of meiosis I, securin degradation releases separase, which then cleaves the cohesin rings holding the chromosome arms together as during mitosis (Figure 20-23). However, during meiosis I, cohesin rings at the centromere are not cleaved (Figure 20-39, row 8). This allows the recombined maternal and paternal chromosomes to separate, but each pair of chromatids remains associated at the centromere (Figure 20-38, step **3**; Figure 20-39, row 8).

In some organisms, meiosis II proceeds without decondensation of the chromosomes and assembly of a nuclear envelope. In other organisms, these typical interphase events occur, but the interphase is short and nuclear envelope retraction into the ER and chromosome condensation of meiotic prophase II follow rapidly. During metaphase II (Figure 20-38, step **4**), as in mitosis, the kinetochores of each sister chromatid attach to spindle microtubules from opposite spindle poles. Also, as during mitosis, cohesion between the chromatid arms is lost and maintained only in the region of the centromere. When the final kinetochore is properly attached to a spindle microtubule, anaphase II occurs (Figure 20-38, step **5**), followed by telophase II and cytokinesis to generate four haploid germ cells (Figure 20-38, step **6**). For each chromosome, at least two of the haploid germ cells have recombinant chromosomes generated from recombination between maternal and paternal chromosomes during

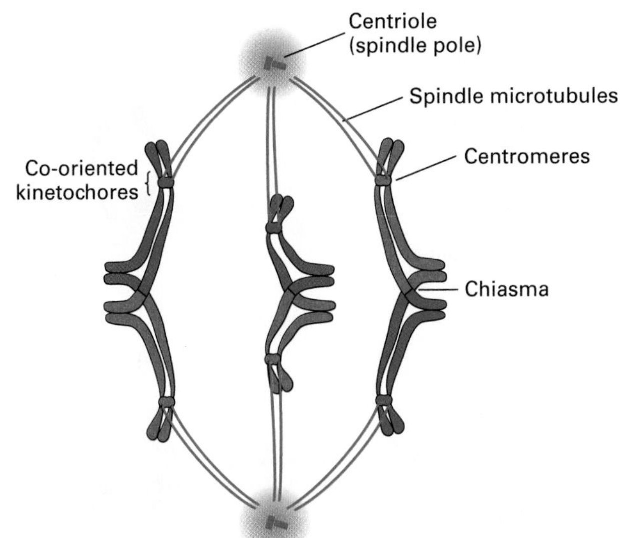

▲ **FIGURE 20-40 Cohesion between homologous chromosomes in meiosis I metaphase.** Connections between chromosomes during meiosis I are most easily visualized in organisms with acrocentric centromeres, such as the grasshopper. The kinetochores at the centromeres of sister chromatids attach to spindle microtubules emanating from the same spindle pole, with the kinetochores of the maternal (red) and paternal (blue) chromosomes attaching to spindle microtubules from opposite spindle poles. The maternal and paternal chromosomes are attached at the chiasmata formed by recombination between them and the cohesion between sister chromatid arms that persists throughout meiosis I metaphase. Note that elimination of cohesion between sister chromatid arms is all that is required for the homologous chromosomes to separate at anaphase. [Adapted from L. V. Paliulis and R. B. Nicklas, 2000, *J. Cell Biol.* **150:**1223.]

prophase of meiosis I (Figure 20-38, step **2**). Thus the recombination between nonsister chromatids that occurs in prophase of meiosis I has at least two functional consequences: First, it holds homologous chromosomes together during meiosis I metaphase (Figure 20-40). Second, it contributes to genetic diversity among individuals of a species by ensuring new combinations of gene alleles in different individuals. Genetic diversity also arises from the independent reassortment of maternal and paternal homologs during the meiotic divisions.

Repression of G_1 Cyclins and a Meiosis-Specific Protein Kinase Promote Premeiotic S Phase

In *S. cerevisiae* and *S. pombe,* depletion of nitrogen and carbon sources induces diploid cells to undergo meiosis, yielding haploid spores (see Figure 1-6). This process is analogous to the formation of germ cells in higher eukaryotes. Multiple yeast mutants that cannot form spores have been isolated, and the wild-type proteins encoded by these genes have been analyzed. These studies have identified specialized cell-cycle proteins required for meiosis.

Under starvation conditions, expression of G_1 cyclins in *S. cerevisiae* is repressed, blocking the normal progression of G_1 in cells as they complete mitosis. Instead, a set of early meiotic proteins is induced. Among these is *Ime2,* a protein kinase closely homologous to CDKs that performs the essential G_1 cyclin-CDK functions required to enter S phase: (1) phosphorylation of the APC/C specificity factor Cdh1, inactivating it so that B-type cyclins can accumulate, (2) phosphorylation of transcription factors to induce genes required for S phase including DNA polymerases and S-phase cyclins and CDKs, and (3) phosphorylation of the S-phase inhibitor Sic1, leading to release of active S-phase cyclin-CDK complexes and the onset of DNA replication in meiosis I.

The cell uses Ime2 during meiosis rather than the standard G_1 cyclin-CDKs so that its protein kinase activity can be regulated differently. While the transcription and translation of G_1 cyclins required for activity of G_1 cyclin-CDKs is repressed in nutrient-starved cells, the transcription and translation of Ime2 is activated. Also, the Ime2 kinase activity is regulated differently than for the G_1 cyclins. It does not require a cyclin partner for kinase activity and is not regulated by the same protein kinases and phosphatases that regulate the activities of G_1 cyclin-CDKs.

The mechanism by which DNA replication is suppressed between meiosis I and II is currently an active area of investigation. Following meiosis I anaphase, MPF activity does not fall as low as it does following mitotic anaphase. This intermediate level of MPF activity is required for normal meiosis II. It appears that MPF activity falls low enough to allow partial or complete cytokinesis, but not low enough to allow dephosphorylation of DNA replication initiation factors. Presumably, DNA replication does not occur between meiosis I and II in part because DNA replication initiation factors are maintained in a hyperphosphorylated form that cannot assemble prereplication complexes on DNA (see Figure 20-30). A second rise in MPF activity occurs that is required for formation of the meiosis II spindle. After all sister kinetochores have attached to microtubules from opposite spindle poles, the activity of Cdc20 is derepressed, separase is activated and cells proceed into meiosis II anaphase (Figure 20-38, step **5**), telophase, and cytokinesis, to generate haploid germ cells.

Recombination and a Meiosis-Specific Cohesin Subunit Are Necessary for the Specialized Chromosome Segregation in Meiosis I

As discussed earlier, in metaphase of meiosis I, both sister chromatids in one (replicated) chromosome associate with microtubules emanating from the *same* spindle pole, rather than from opposite poles as they do in mitosis. Two physical links between homologous chromosomes are thought to resist the pulling force of the spindle until anaphase: (a) crossing over between chromatids, one from each pair of homologous chromosomes, and (b) cohesin cross-links between sister chromatids distal to the crossover point (see Figure 20-40).

Evidence for the function of recombination during meiosis in *S. cerevisiae* comes from the observation that when recombination is blocked by mutations in proteins essential for the process, chromosomes segregate randomly during meiosis I; that is, homologous chromosomes do not necessarily segregate to opposite spindle poles. Such segregation to opposite spindle poles normally occurs because both chromatids of the maternal and paternal homologous chromosome pairs associate with spindle fibers emanating from opposite spindle poles (see Figure 20-38, step **3**, and Figure 20-40). As discussed above, this in turn requires that homologous chromosomes pair during prophase of meiosis I. Consequently, the finding that mutations that block recombination also block proper segregation in meiosis I implies that recombination is required for synapsis of homologous chromosomes in *S. cerevisiae.*

Unlike the mitotic anaphase (Figure 20-41a), at the onset of meiotic anaphase I, the cohesin cross-links between chromosome arms are cleaved by separase, allowing the homologous chromosomes to separate, but cohesin complexes at the centromere remain linked (Figure 20-41b, *top*). The maintenance of centromeric cohesion during meiosis I is necessary for the proper segregation of chromatids during meiosis II. Studies with a *S. pombe* mutant have shown that a specialized cohesin kleisin subunit (see Figure 20-23), *Rec8,* maintains centromeric cohesion between sister chromatids during meiosis I. Expressed only during meiosis, Rec8 is homologous to the cohesin subunit that closes the cohesin ring in the cohesin complex of mitotic cells. Immunolocalization experiments in *S. pombe* revealed that during early anaphase of meiosis I, Rec8 is lost from chromosome arms but is retained at centromeres. However, during early anaphase of meiosis II, centromeric Rec8 is degraded by separase, so the chromatids can segregate, as they do in mitosis (Figure 20-41b, *bottom*). *S. cerevisiae* Rec8 has been shown to localize and function similarly to *S. pombe* Rec8, and homologs of Rec8 also have been identified in higher organisms. Consequently, understanding the regulation of Rec8-cohesin complex cleavage is central to understanding chromosome segregation in meiosis I.

Micromanipulation experiments during grasshopper spermatogenesis indicated that chromosome-associated factors protect centromeric Rec8 from cleavage during meiosis I but not during meiosis II. These experiments also demonstrated that the attachment of sister kinetochores to microtubules emanating from the same spindle pole during meiosis I, as opposed to attachment to spindle fibers from opposite spindle poles in meiosis II and mitosis, also results from factors associated with the chromosomes (Figure 20-42). Thus crossing over, Rec8, and special kinetochore-associated proteins appear to function in meiosis in all eukaryotes.

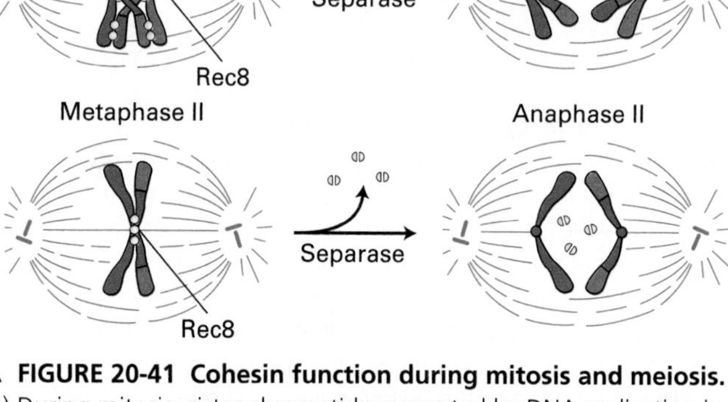

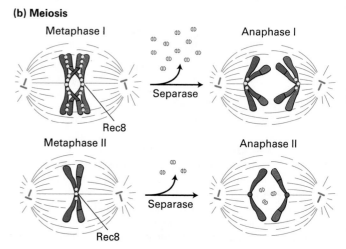

▲ **FIGURE 20-41 Cohesin function during mitosis and meiosis.** (a) During mitosis, sister chromatids generated by DNA replication in the S phase are initially associated by cohesin complexes along the full length of the chromatids. During chromosome condensation, cohesin complexes (yellow) become restricted to the region of the centromere at metaphase, as depicted here. Once separase cleaves the kleisin cohesin subunit (see Figure 20-23), sister chromatids can separate, marking the onset of anaphase. (b) In prophase of meiosis I, recombination between maternal and paternal chromatids produces synapsis of homologous parental chromosomes. By metaphase the chromatids of each replicated chromosome are cross-linked by cohesin complexes along their full length. Rec8, a meiosis-specific homolog of the mitotic kleisin, is cleaved in chromosome arms but not in the centromere, allowing homologous chromosome pairs to segregate to daughter cells. Centromeric Rec8 is cleaved during meiosis II, allowing individual chromatids to segregate to daughter cells. [Modified from F. Uhlmann, 2001, *Curr. Opin. Cell Biol.* **13:**754.]

Special Properties of Rec8 Regulate Its Cleavage in Meiosis I and II

The mechanism that protects *S. cerevisiae* Rec8 from degradation at centromeres during meiosis I is similar to the mechanism that protects kleisin cohesin subunits at centromeres during mitosis. Recall that during mitotic prophase, protein kinases activated by the mitotic cyclin-CDKs phosphorylate cohesins in the chromatid arms, causing them to dissociate, eliminating cohesion in chromatid arms by metaphase in most organisms. However, cohesion at the centromeres is maintained because a specific isoform of protein phosphatase 2A (PP2A) is localized to centromeric chromatin and keeps cohesin in a hypophosphorylated state that does not dissociate from chromatin (see Figure 20-22). Then, when the last kinetochore is properly associated with spindle microtubules, Cdc20 is derepressed and associates with the APC/C, causing polyubiquitination of securin. This releases separase activity, which cleaves the kleisin whether it is

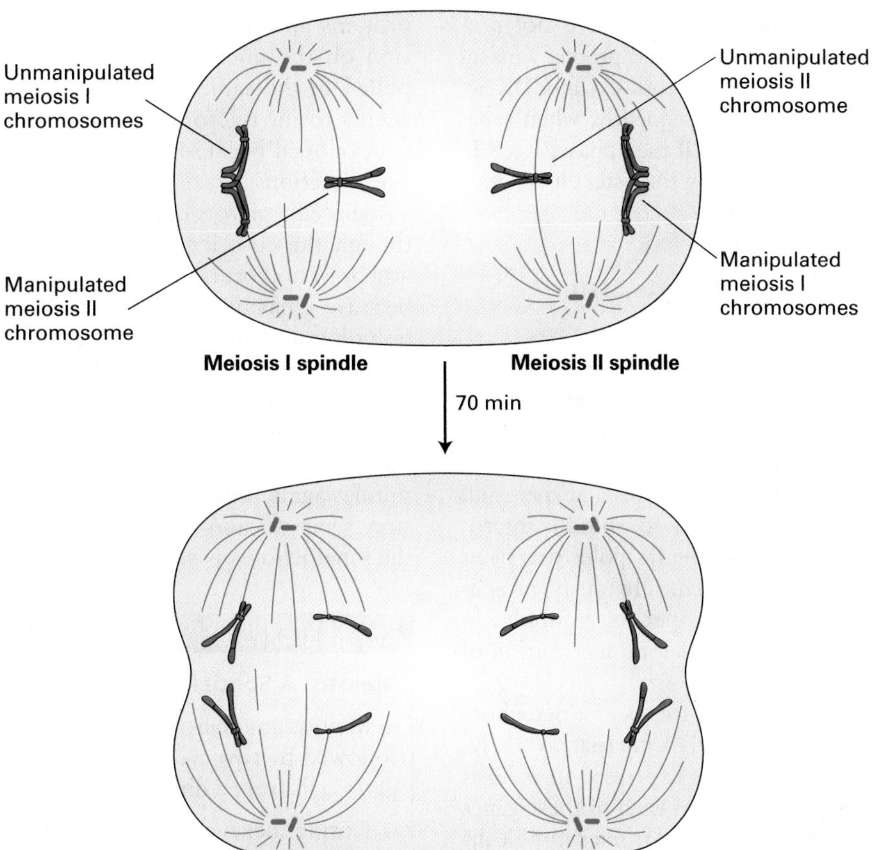

▲ **EXPERIMENTAL FIGURE 20-42 Anaphase movements and cohesion of meiotic chromosomes are determined by proteins associated with the chromosomes.** Grasshopper spermatocytes in meiosis I and II were fused so that both types of spindles with their associated chromosomes were present in a single fused cell. Then a micromanipulation needle was used to move some meiosis I chromosomes and meiosis II chromosomes from one spindle to another; other chromosomes were left attached to their normal spindles. After 70 minutes, both spindles with their attached chromosomes had completed anaphase movements. The paired meiosis I chromosomes separated normally (i.e., one homologous pair toward one spindle pole and the other homologous pair toward the other) whether attached to the meiosis I spindle (*left*) or meiosis II spindle (*right*). Similarly, the meiosis II chromosomes separated normally (i.e., one chromatid toward one spindle pole and the other chromatid toward the opposite pole) independent of which spindle they were attached to. These results indicate that the association of kinetochores with spindle microtubules and the stability of cohesins linking chromosomes are determined by factors that are associated with the chromosomes and not by the spindles or soluble components of cells in meiosis I and II. [See L. V. Paliulis and R. B. Nicklas, 2000, *J. Cell Biol.* **150:**1223.]

phosphorylated or not, eliminating cohesion at the centromere and allowing chromatid separation in anaphase (see Figure 20-41a).

This mechanism differs for meiosis I because when Rec8 replaces the mitotic kleisin in the cohesin complex, the complex does not dissociate in prophase when it is phosphorylated by mitotic protein kinases. Rec8 also differs from the mitotic kleisin in that it must be phosphorylated by these mitotic protein kinases to be cleaved by separase. During meiosis I the centromere-specific isoform of PP2A is bound to centromeric chromatin by a linker protein called *shugoshin* (Japanese for "guardian spirit"). However, shugoshin is not expressed during meiosis II.

These properties of Rec8 and shugoshin are required for the specialized chromosome segregation of meiosis I. Cohesin linkages between daughter chromatids are maintained in the chromosome arms because Rec8-containing cohesin complexes are resistant to dissociation when they are phosphorylated by the mitotic protein kinases. They are cleaved, however, when Cdc20 is a derepressed and separase is activated because the required phosphorylation by these mitotic protein kinases has occurred. This results in loss of cohesion at sister chromatid arms, which is all that is necessary for homologous chromosomes to dissociate at meiosis I anaphase (Figure 20-40 and Figure 20-41b, *top*). However, cohesion between sister chromatid centromeres is maintained because shugoshin is expressed in meiosis I and binds PP2A to the centromere, where it dephosphorylates nearby Rec8, making it resistant to cleavage by separase. In meiosis II, there is no cohesion between chromatid arms because cohesin complexes were removed from the arms during meiosis I anaphase. Since shugoshin is not expressed in meiosis II, PP2A is not localized to centromeric chromatin and centromeric cohesin is not protected from phosphorylation by the mitotic protein kinases that are activated by the second rise in mitotic cyclin-CDK activity during meiosis II prophase. Consequently, when separase is activated at the end of meiosis II metaphase, Rec8 in centromeric cohesin is cleaved, allowing the sister chromatids to be pulled to opposite spindle poles.

The Monopolin Complex Co-orients Sister Kinetochores in Meiosis I

As discussed earlier, in mitosis and meiosis II, sister kinetochores attach to spindle microtubules emanating from *opposite* spindle poles; the kinetochores are said to be *bi-oriented*. This is essential for segregation of sister chromatids to different daughter cells. In contrast, at meiosis I metaphase, sister kinetochores attach to spindle microtubules emanating from the *same* spindle pole; the sister kinetochores are said to be *co-oriented*. Obviously, attachment of sister kinetochores to the proper microtubules in meiosis I and II is critical for correct meiotic segregation of chromosomes.

Identification of a protein required for sister kinetochore cohesion in *S. cerevisiae* began with DNA microarray analysis of *S. cerevisiae* that revealed genes expressed in meiosis and not mitosis. Systematic knockout of each of these genes revealed that when one, *MAM1* (*m*onopolar microtubule *a*ttachment during *m*eiosis *I*) encoding a protein called *monopolin*, was inactivated, sister chromatids in metaphase of meiosis I associated with spindle microtubules emanating from opposite spindle poles, as though they were mitotic chromatids or chromatids in meiosis II.

Identification of proteins that interact with monopolin and in vivo fluorescence microscopy studies with GFP-labeled proteins revealed two proteins that form a *monopolin complex* with monopolin that are also found at kinetochores in mitotic cells and cells in meiosis II. These results have led to the model that the monopolin complex clamps microtubule binding sites in the sister kinetochores in the same orientation so that they interact with the + end of microtubules coming from the same spindle pole. In the absence of monopolin in meiosis II and mitotic cells, the other subunits of the monopolin complex that are expressed are proposed to clamp the spindle microtubule binding sites of sister chromatids in the opposite orientation so they can bind only the + ends of microtubules coming from opposite directions.

Tension on Spindle Microtubules Contributes to Proper Spindle Attachment

Additional micromanipulation experiments and genetic experiments in yeast have provided considerable evidence that stable attachment of spindle microtubules to kinetochores, and the stability of the microtubules themselves, requires that the microtubules be under tension during metaphase. If microtubules from the wrong spindle pole attach to sister chromatids early in metaphase, motor proteins and microtubule shortening do not produce tension on the microtubules because both kinetochores are pulled in the same direction. However, when attachment occurs to the microtubules from the correct poles, tension is developed because the kinetochores are pulled in opposite directions. During meiotic metaphase I, kinetochore-associated microtubules are also under tension (even though the co-oriented kinetochores of sister chromatids attach to microtubules coming from the same spindle pole) because chiasmata generated by recombination between homologous chromosomes prevent them from being pulled to the poles (see Figure 20-40). Since kinetochore binding to microtubules is unstable in the absence of tension, kinetochores that attach to the wrong spindle fibers release the incorrect microtubules, enabling them to bind microtubules again until attachments are made that generate tension. Once tension is generated, microtubule attachment to the kinetochores is stabilized.

KEY CONCEPTS OF SECTION 20.8

Meiosis: A Special Type of Cell Division

- Meiosis involves one cycle of chromosome replication followed by two cycles of cell division to produce haploid germ cells from a diploid premeiotic cell (see Figure 20-38).
- During meiosis I, replicated homologous chromosomes pair along their lengths in a process called synapsis. At least one recombination event between chromatids of homologous chromosomes almost invariably occurs.
- Most of the cell-cycle proteins that function in mitotically dividing cells also function in cells undergoing meiosis, but some proteins are unique to meiosis.
- In *S. cerevisiae,* expression of G_1 cyclins is repressed throughout meiosis. Meiosis-specific Ime2 performs the function of G_1 cyclin-CDK complexes in promoting initiation of DNA replication during meiosis I.
- In *S. cerevisiae,* recombination (crossing over) between chromatids of homologous parental chromatids as well as cohesin cross-links between chromatids distal to the crossover are responsible for holding the homologous chromosomes together during prophase and metaphase of meiosis I. A specialized cohesin subunit, Rec8, replaces the mitotic kleisin cohesin subunit during meiosis.
- At the onset of early anaphase of meiosis I, Rec8 in the chromosome arms is cleaved, but a meiosis-specific protein associated with the kinetochore protects Rec8 from cleavage in the region of the centromere. As a result, the chromatids of homologous chromosomes remain associated during segregation in meiosis I. Cleavage of centromeric Rec8 during anaphase of meiosis II allows individual chromatids to segregate into germ cells (see Figure 20-41b).
- Monopolin, another meiosis-specific protein complex, is required for both chromatids of homologous chromosomes to associate with microtubules emanating from the same spindle poles during meiosis I.

Perspectives for the Future

The remarkable pace of cell-cycle research over the last 25 years has led to the model of eukaryotic cell-cycle control outlined in Figure 20-34. A beautiful logic underlies these molecular controls. Each regulatory event has two important functions: to activate a step of the cell cycle and to prepare the cell for the next event of the cycle. This strategy ensures that the phases of the cycle occur in the proper order.

Although the general logic of cell-cycle regulation now seems well established, many critical details remain to be discovered. For instance, although researchers have identified some components of the prereplication complex that must be phosphorylated by S-phase cyclin-CDK complexes to initiate DNA replication, other components remain to be determined. As discussed earlier, substantial progress has been made recently in identifying substrates phosphorylated by mitotic cyclin-CDK complexes. Much work remains to be done to understand how the modification of these proteins leads to chromosome condensation and the remarkable reorganization of microtubules that results in assembly of the beautiful mitotic spindle. Much remains to be learned about how the activities of Wee1 kinase and Cdc25 phosphatase are controlled; these proteins in turn regulate the kinase activity of the CDK subunit in most cyclin-CDK complexes.

Much has been discovered recently about the operation of cell-cycle checkpoints, but the mechanisms that activate ATM and ATR in the DNA-damage checkpoint are poorly understood. Likewise, much remains to be learned about the control and mechanism of regulation of Mad2 in the spindle-assembly checkpoint, and of Cdc14 in the chromosome-segregation checkpoint in higher cells. As we learn in the next chapter, asymmetric cell division plays a critical role in the normal development of multicellular organisms. Many questions remain about how the plane of cytokinesis and the localization of daughter chromosomes are determined in cells that divide asymmetrically. Similarly, the mechanisms that underlie the unique segregation of chromatids during meiosis I have not been elucidated yet.

Understanding these detailed aspects of cell-cycle control will have significant consequences, particularly for the treatment of cancers. Cancer cells often have defects in cell-cycle checkpoints that led to the accumulation of multiple mutations and DNA rearrangements that result in the cancer phenotype. However, the absence of these checkpoints can make specific types of cancers particularly vulnerable to extensive DNA damage induced with radiation therapy or chemotherapy. Normal cells activate cell-cycle checkpoints that arrest the cell cycle until the DNA damage is repaired. But these cancer cells do not, and as a consequence suffer sufficient genetic damage to induce apoptosis. If more were understood about cell-cycle controls and checkpoints, it might be possible to design ever more effective therapeutic strategies, especially against types of cancer that are largely resistant to today's conventional therapies. It seems very likely that better understanding of the molecular processes involved will allow the design of more effective treatments in the future. ▮

Key Terms

anaphase-promoting complex (APC/C) *850*
APC/C specificity factor *858*
Cdc14 phosphatase *858*
Cdc25 phosphatase *862*
checkpoints *848*
CKIs *851*
cohesins *869*
condensin *866*
crossing over *892*
cyclin-dependent kinases (CDKs) *848*
destruction box *858*
E2F factors *882*
G_1 cyclins *874*
meiosis *892*
mitogens *880*
mitosis-promoting factor (MPF) *855*
mitotic cyclin *856*
p53 protein *891*
quiescent cells *881*
Rb protein *882*
restriction point *881*
SCF *876*
S-phase cyclins *876*
S-phase inhibitor *876*
S-phase-promoting factor (SPF) *874*
securin *851*
synapsis *892*
Wee1 protein-tyrosine kinase *862*

Review the Concepts

1. What strategy ensures that passage through the cell cycle is unidirectional and irreversible? What is the molecular machinery that underlies this strategy?

2. When fused with an S-phase cell, cells in which of the following phases of the cell cycle will initiate DNA replication prematurely—G_1? G_2? M? Predict the effect of fusing a cell in G_1 and a cell in G_2 with respect to the timing of S phase in each cell.

3. Tim Hunt shared the 2001 Nobel prize for his work in the discovery and characterization of cyclin proteins in eggs and embryos. What experimental evidence indicates that cyclin B is required for a cell to enter mitosis? What evidence indicates that cyclin B must be destroyed for a cell to exit mitosis?

4. In *Xenopus,* one of the substrates of MPF is the Cdc25 phosphatase. When phosphorylated by MPF, Cdc25 is activated. What is the substrate of Cdc25? How does this information help to explain the rapid rise in MPF activity as cells enter mitosis?

5. In 2001, the Nobel prize in physiology or medicine was awarded to three cell-cycle scientists. Sir Paul Nurse was recognized for his studies with the fission yeast *S. pombe,* in particular for the discovery and characterization of the *wee1* gene. What is the wee phenotype? What did the characterization of the *wee1* gene tell us about cell-cycle control?

6. Three known substrates of MPF or kinases regulated by MPF are the nuclear lamins, nucleoporins, and subunits of

condensin. Describe how the phosphorylation of each of these proteins affects its function and progression through mitosis.

7. Describe the series of events by which the APC/C promotes the separation of sister chromatids at anaphase.

8. At the end of telophase, nuclear envelope extensions from the ER fuse around each chromosome to form karyomeres that then fuse to form daughter-cell nuclei. What are the roles of Ran-GEF and Ran·GTP in this process?

9. Explain how CDK activity is modulated by the following proteins: (a) cyclin, (b) CAK, (c) Wee1, (d) p21.

10. A common feature of cell-cycle regulation is that the events of one phase ensure progression into a subsequent phase. In *S. cerevisiae*, mid- and late-G_1 cyclin-CDKs catalyze progression through G_1, and multiple B-type cyclin-CDK complexes catalyze progression through S, G_2, and M. Name three ways in which the activities of mid- and late-G_1 cyclin-CDKs promote the activation of S-phase and mitotic cyclin-CDK complexes.

11. For S phase to be completed in a timely manner, DNA replication initiates from multiple origins in eukaryotes. In *S. cerevisiae*, what role do S-phase cyclin-CDK complexes play to ensure that the entire genome is replicated once and only once per cell cycle?

12. What is the functional definition of the restriction point? Cancer cells typically lose restriction-point controls. Explain how the following mutations, which are found in some cancer cells, lead to a bypass of restriction-point controls: (a) overexpression of cyclin D, (b) loss of Rb function, (c) infection by retroviruses encoding v-Fos and v-Jun.

13. The Rb protein has been called the "master brake of the cell cycle." Describe how the Rb protein acts as a cell-cycle brake. How is the brake released in mid to late G_1 to allow the cell to proceed to the S phase?

14. Leeland Hartwell, the third recipient of the 2001 Nobel prize, was acknowledged for his characterization of cell-cycle checkpoints in the budding yeast *S. cerevisiae*. What is a cell-cycle checkpoint? Where do checkpoints occur in the cell cycle? How do cell-cycle checkpoints help to preserve the fidelity of the genome?

15. What role does p53 play in mediating cell-cycle arrest for cells with DNA damage?

16. Individuals with the hereditary disorder ataxia telangiectasia suffer from neurodegeneration, immunodeficiency, and increased incidence of cancer. The genetic basis for ataxia telangiectasia is a loss-of-function mutation in the ATM gene (ATM = *a*taxia *t*elangiectasia-*m*utated). Besides p53, what other substrate is phosphorylated by ATM? How does the phosphorylation of this substrate lead to inactivation of CDKs to enforce cell-cycle arrest at a checkpoint?

17. Meiosis and mitosis are overall analogous processes involving many of the same proteins. However, some proteins function uniquely in each of these cell-division events. Explain the meiosis-specific function of the following: (a) Ime2, (b) Rec8, (c) monopolin.

Analyze the Data

Many of the proteins that regulate transit through the cell cycle have been characterized. Recently, a new protein, Xnf7, has been identified in extracts of *Xenopus* eggs. This protein binds to the anaphase-promoting complex/cyclosome (APC/C). To elucidate the function of this protein, studies have been undertaken in which Xnf7 either has been depleted from extracts using an antibody raised against it, or has been augmented in the extracts through addition of extra Xnf7. The consequences on transit through mitosis were then assessed (see J. B. Casaletto et al., 2005, *J. Cell Biol.* **169**: 61–71).

a. *Xenopus* egg extracts, arrested in metaphase, were either depleted of Xnf7 or were mock-depleted (subjected to the same treatment as the first sample but without Xnf7 antibody), then released from metaphase arrest by addition of Ca^{2+}. Aliquots of the extract were then sampled at various times after Ca^{2+} addition and the amounts of cyclin B determined, as shown on the Western blot below. What information do these data provide about a possible function for Xnf7?

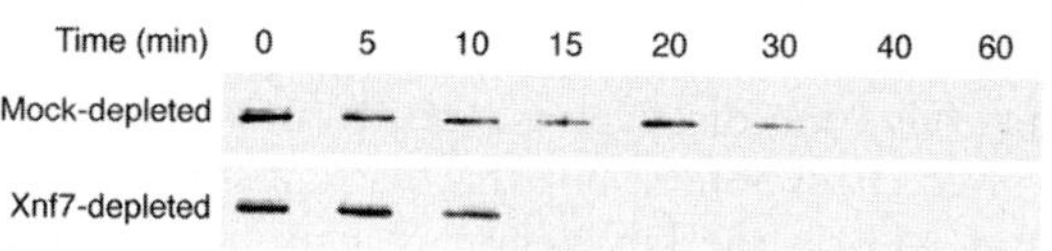

b. In additional studies, exogenous Xnf7 was added to *Xenopus* egg extracts, arrested at metaphase, so that the total amount of this protein in the extracts was higher than normal. The extracts, released from arrest by Ca^{2+} addition, were then assessed at various times after release for cyclin B ubiquitination (cyclin-Ub conjugates). What is the rationale for examining ubiquitination? Using the following figure, determine what information these studies add beyond that obtained from part (a)?

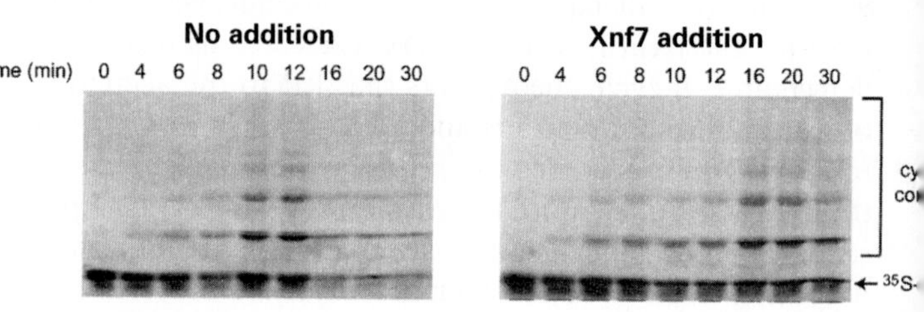

c. The spindle checkpoint prevents cells with unattached kinetochores from proceeding into anaphase. Thus, cells in which this checkpoint has been activated do not enter anaphase and do not degrade cyclin B. Nocodazole, a drug that prevents microtubule assembly, can be used to activate the spindle checkpoint. Cells in nocodazole become arrested in early mitosis because they cannot form a spindle, and thus all kinetochores remain unattached. To determine if Xnf7 is required for a functional spindle checkpoint, *Xenopus* egg extracts, arrested in metaphase, were subjected to various protocols (see the following figure): untreated (−nocodazole), or treated with nocodazole and either mock-depleted (preimmune) or immuno-depleted of Xnf7 (α-Xnf7). The extracts were then treated with Ca^{2+} to overcome arrest, and aliquots of the extracts were assessed at various times for cyclin B, as shown on the Western blot below. What can you conclude about Xnf7 from these data?

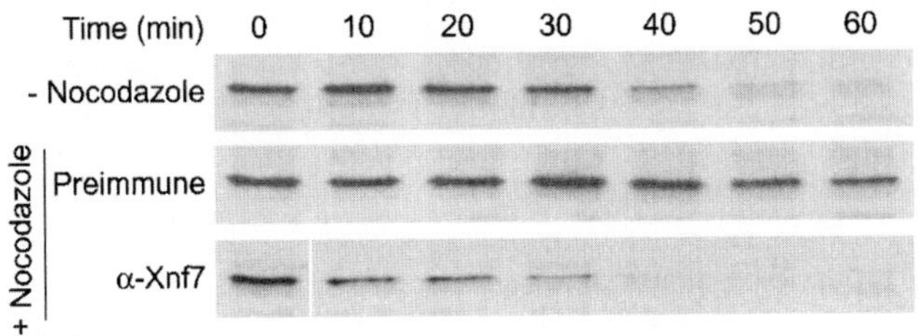

References

Overview of the Cell Cycle and Its Control

Morgan, D. O. 2006. *The Cell Cycle: Principles of Control.* New Science Press.

Nasmyth, K. 2001. A prize for proliferation. *Cell* **107**:689–701.

Control of Mitosis by Cyclins and MPF Activity

Doree, M., and T. Hunt. 2002. From Cdc2 to Cdk1: when did the cell cycle kinase join its cyclin partner? *J. Cell Sci.* **115**:2461–2464.

Masui, Y. 2001. From oocyte maturation to the in vitro cell cycle: the history of discoveries of Maturation-Promoting Factor (MPF) and Cytostatic Factor (CSF). *Differentiation* **69**:1–17.

Cyclin-Dependent Kinase Regulation During Mitosis

Nurse, P. 2002. Cyclin dependent kinases and cell cycle control (Nobel lecture). *Chembiochem.* **3**:596–603.

Molecular Mechanisms for Regulating Mitotic Events

Hirano, T. 2005. Condensins: organizing and segregating the genome. *Curr. Biol.* **15**:R265–R275.

Kline-Smith, S. L., S. Sandall, and A. Desai. 2005. Kinetochore-spindle microtubule interactions during mitosis. *Curr. Opin. Cell Biol.* **17**:35–46.

Meyer, H. H. 2005. Golgi reassembly after mitosis: the AAA family meets the ubiquitin family. *Biochim. Biophys. Acta* **1744**:108–119.

Nasmyth, K., and C. H. Haering. 2005. The structure and function of SMC and kleisin complexes. *Ann. Rev. Biochem.* **74**:595–648.

Nigg, E. A. 2001. Mitotic kinases as regulators of cell division and its checkpoints. *Nature Rev. Mol. Cell Biol.* **2**:21–32.

Peters, J. M. 2006. The anaphase promoting complex/cyclosome: a machine designed to destroy. *Nature Rev. Mol. Cell Biol.* **7**:644–656.

Roux, K. J., and B. Burke. 2006. From pore to kinetochore and back: regulating envelope assembly. *Dev. Cell* **11**:276–278.

Wirth, K. G., et al. 2006. Separase: a universal trigger for sister chromatid disjunction but not chromosome cycle progression. *J. Cell Biol.* **172**:847–860.

Yanagida, M. 2005. Basic mechanism of eukaryotic chromosome segregation. *Phil. Trans. R. Soc. Lond. B Biol. Sci.* **360**:609–621.

Cyclin-CDK and Ubiquitin-Protein Ligase Control of S-phase

Bell, S. P., and A. Dutta. 2002. DNA replication in eukaryotic cells. *Ann. Rev. Biochem.* **71**:333–374.

Deshaies, R. J. 1999. SCF and Cullin/Ring H2-based ubiquitin ligases. *Ann. Rev. Cell Devel. Biol.* **15**:435–467.

Diffley, J. F. 2004. Regulation of early events in chromosome replication. *Curr. Biol* **14**:R778–R786.

Nakayama, K. I., and K. Nakayama. 2005. Regulation of the cell cycle by SCF-type ubiquitin ligases. *Semin. Cell Dev. Biol.* **16**:323–333.

Reed, S. I. 2006. The ubiquitin-proteasome pathway in cell cycle control. *Results Probl. Cell Differ.* **42**:147–181.

Cell-Cycle Control in Mammalian Cells

Barr, F. A. 2004. Golgi inheritance: shaken but not stirred. *J. Cell Biol.* **164**:955–958.

DePamphilis, M. L., et al. 2006. Regulating the licensing of DNA replication origins in metazoa. *Curr. Opin. Cell Biol.* **18**:231–239.

Ekholm, S. V., and S. I. Reed. 2000. Regulation of G(1) cyclin-dependent kinases in the mammalian cell cycle. *Curr. Opin. Cell Biol.* **12**:676–684.

Machida, Y. J., J. L. Hamlin, and A. Dutta. 2005. Right place, right time, and only once: replication initiation in metazoans. *Cell* **123**:13–24.

Porter, L. A., and D. J. Donoghue. 2003. Cyclin B1 and CDK1: nuclear localization and upstream regulators. *Prog. Cell Cycle Res.* **5**:335–347.

Sears, R. C., and J. R. Nevins. 2002. Signaling networks that link cell proliferation and cell fate. *J. Biol. Chem.* **277**:11617–11620.

Sherr, C. J. 2001. The INK4a/ARF network in tumour suppression. *Nature Rev. Mol. Cell Biol.* **2**:731–737.

Checkpoints in Cell-Cycle Regulation

Bartek, J., C. Lukas, and J. Lukas. 2004. Checking on DNA damage in S phase. *Nature Rev. Mol. Cell Biol.* **5**:792–804.

Cheeseman, I. M., and A. Desai. 2004. Cell division: feeling tense enough? *Nature* **428**:32–33.

Gottifredi, V., and C. Prives. 2005. The S phase checkpoint: when the crowd meets at the fork. *Semin. Cell Dev. Biol.* **16**:355–368.

Hartwell, L. H. 2002. Yeast and cancer (Nobel lecture). *Biosci. Rep.* **22**:373–394.

Kastan, M. B., and J. Bartek. 2004. Cell-cycle checkpoints and cancer. *Nature* **432**:316–323.

Kitagawa, R., and M. B. Kastan. 2005. The ATM-dependent DNA damage signaling pathway. *Cold Spring Harbor Symp. Quant. Biol.* **70**:99–109.

Lambert, S., and A. M. Carr. 2005. Checkpoint responses to replication fork barriers. *Biochimie* **87**:591–602.

Lew, D. J., and D. J. Burke. 2003. The spindle assembly and spindle position checkpoints. *Ann. Rev. Genet.* **37**:251–282.

McGowan, C. H., and P. Russell. 2004. The DNA damage response: sensing and signaling. *Curr. Opin. Cell Biol.* **16**:629–633.

Nasmyth, K. 2005. How do so few control so many? *Cell* **120**:739–746.

Pereira, G., and E. Schiebel. 2001. The role of the yeast spindle pole body and the mammalian centrosome in regulating late mitotic events. *Curr. Opin. Cell Biol.* **13**:762–769.

Seshan, A., and A. Amon. 2004. Linked for life: temporal and spatial coordination of late mitotic events. *Curr. Opin. Cell Biol.* **16**:41–48.

Stark, G. R., and W. R. Taylor. 2006. Control of the G2/M transition. *Mol. Biotechnol.* **32**:227–248.

Stegmeier, F., and A. Amon. 2004. Closing mitosis: the functions of the Cdc14 phosphatase and its regulation. *Ann. Rev. Genet.* 38:203–232.

Takeda, D. Y., and A. Dutta. 2005. DNA replication and progression through S phase. *Oncogene* **24**:2827–2843.

Uhlmann, F. 2003. Separase regulation during mitosis. *Biochem. Soc. Symp.* **70**:243–251.

Meiosis: A Special Type of Cell Division

Marston, A. L., and A. Amon. 2004. Meiosis: cell-cycle controls shuffle and deal. *Nature Rev. Mol. Cell Biol.* **5**:983–997.

Petronczki, M., M. F. Siomos, and K. Nasmyth. 2003. Un ménage a quatre: the molecular biology of chromosome segregation in meiosis. *Cell* **112**:423–440.

Watanabe, Y. 2004. Modifying sister chromatid cohesion for meiosis. *J. Cell Sci.* **117**:4017–4023.

CLASSIC EXPERIMENT 20.1

CELL BIOLOGY EMERGING FROM THE SEA: THE DISCOVERY OF CYCLINS

T. Evans et al., 1983, *Cell* **33**:391.

From the first cell divisions after fertilization to aberrant divisions that occur in cancers, biologists have long been interested in how cells control when they divide. The processes of cell division have been separated into stages known collectively as the *cell cycle*. While studying early development in marine invertebrates in the early 1980s, Joan Ruderman and Tim Hunt discovered the cyclins, key regulators of the cell cycle.

Background

The question of how an organism develops from a fertilized egg continues to drive a large body of scientific research. Whereas such research was classically the concern of embryologists, the developing understanding of gene expression in the 1980s brought new approaches to answer this question. One such approach was to examine the pattern of gene expression in both the oocyte and the newly fertilized egg. Ruderman and Hunt were among the biologists who took this approach to the study of early development.

Biologists had well characterized the early development of a number of marine invertebrate systems. During the early stages of development, the embryonic cells grow synchronously, which allows an entire population of cells to be studied at the same stage of the cell cycle. Researchers had established that a large portion of the mRNA in the unfertilized oocyte is not translated. Upon fertilization, these maternal mRNA are rapidly translated. Previous studies had shown that when fertilized eggs are treated with drugs that inhibit protein synthesis, cell division could not take place. This suggested that the initial burst of protein synthesis from the maternal mRNA is required at the earliest stages of development. Ruderman and Hunt, while teaching a physiology course at the Marine Biological Lab in Woods Hole, Massachusetts, began a set of experiments designed to uncover the genes that were expressed at this point as well as the mechanism by which this burst of protein synthesis was controlled.

The Experiment

In a collaborative project, Ruderman and Hunt looked at regulation of gene expression in the fertilized egg of the surf clam *Spisula solidissima*. Whereas it was known that overall protein synthesis rapidly increased upon fertilization, they wanted to find out whether the proteins expressed in the earliest stage of development, the two-cell embryo, were different from those expressed in the unfertilized egg. When either oocytes or two-cell clam embryos are treated with radioactively labeled amino acids, the cell takes up the amino acids, which are subsequently incorporated into newly synthesized proteins. Using this technique, Ruderman and Hunt monitored the pattern of protein synthesis by breaking open the cells, separating the proteins using SDS-polyacrylamide gel electrophoresis (SDS-PAGE), and then visualizing the radioactively labeled proteins by autoradiography. When they compared the pattern of protein synthesis in the oocyte with that in the two-cell embryo, they saw that three different proteins that were either not expressed or expressed at an extremely low level in the oocyte were highly expressed in the embryo. In a subsequent study, Ruderman examined the pattern of protein expression in the oocytes of the starfish *Asterias forbesi* as they mature. She again observed the increased expression of three proteins of similar size to those that she and Hunt had seen in surf clam embryos.

Soon afterward, in a third study, Hunt examined the changes in protein expression during the maturation and fertilization of sea urchin oocytes. This time he performed the experiment in a slightly different manner. Rather than treating the oocytes and embryo with radioactively labeled amino acids for a set time period, he labeled the cells continuously for more than 2 hours, removing samples for analysis at 10-minute intervals. Now, he could monitor the changes in protein expression throughout the early stages of development. As had been shown in other organisms, the pattern of protein synthesis was altered when the sea urchin oocyte was fertilized. Three proteins—represented by three prominent bands on an autoradiograph—were expressed in the embryos, but not in the oocytes. Interestingly, the intensity of one of these bands changed over time; the band was intense at the early time points, then barely visible after 85 minutes. It increased in intensity again between 95 and 105 minutes. The intensity of the band, representing the amount of the protein in the cell, appeared to be oscillating over time. This suggested that the protein had been quickly degraded and then synthesized again.

Because the time frame of the experiment coincided with early embryonic cell divisions, Hunt next asked whether the synthesis and destruction of the protein was correlated with progression of the cell cycle. He examined a portion of cells from each time point under a microscope, counting the number of cells dividing at each time

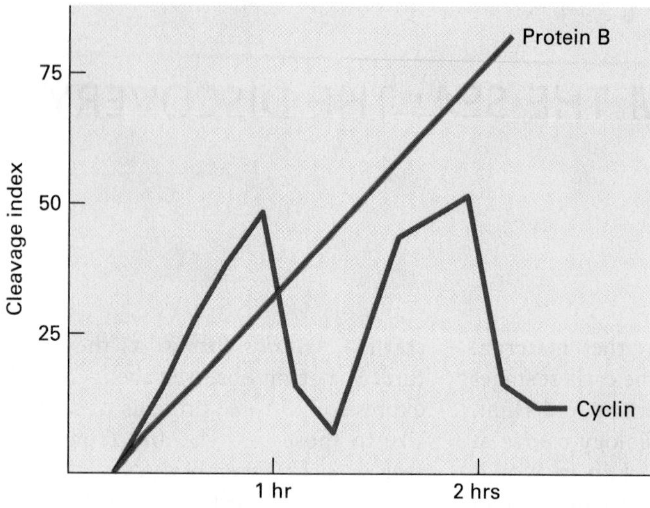

◀ **FIGURE 1 This figure compares the changing levels of sea urchin cyclin (drawn in blue) with a control protein (drawn in purple) as early embryonic cells progress through the cell cycle.** The overall level of cyclin increases over time, and then it is rapidly destroyed as the cells approach division. This pattern appears to repeat through each cell division. Meanwhile, the overall level of the control protein continues to increase throughout the time period of the experiment. [Adapted from T. Evans et al., 1983, *Cell* **33:**391.]

point where samples had been taken for protein analysis. Hunt then correlated the amount of the protein present in the cell with the proportion of cells dividing at each time point. He noticed that the level of expression of one of the proteins was highest before the cell divided and lowest upon cell division (see Figure 1), suggesting a correlation with the stage of the cell cycle. When the same experiment was performed in the surf clam, Hunt saw that two of the proteins that he and Ruderman had described previously displayed the same pattern of synthesis and destruction. Hunt called these proteins *cyclins* to reflect their changing expression through the cell cycle.

Discussion

The discovery of the cyclins heralded an explosion of investigation into the cell cycle. It is now known that these proteins regulate the cell cycle by associating with cyclin-dependent kinases, which in turn regulate the activities of a variety of transcription and replication factors, as well as other proteins involved in the complex alterations in cell architecture and chromosome structure that occur during mitosis. In brief, cyclin-CDK complexes direct and regulate through the cell cycle. As with so many key regulators of cellular functions, it was soon shown that the cyclins discovered in sea urchins and surf clams are conserved in eukaryotes from yeast to man. Since the identification of the first cyclins, scientists have identified at least 15 other cyclins that regulate all phases of the cell cycle.

In addition to the basic research interest in these proteins, the cyclins' central role in cell division has made them a focal point in cancer research. Cyclins are involved in the regulation of several genes that are known to play prominent roles in tumor development. Scientists have shown that at least one cyclin, cyclin D1, is overexpressed in a number of tumors. The role of these proteins in both normal and aberrant cell division continues to be an active and exciting area of research today.

CHAPTER

21

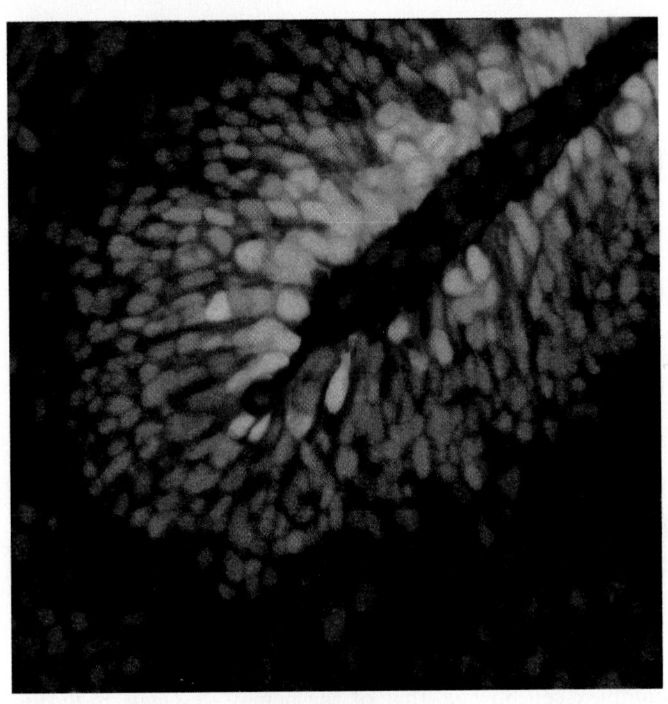

Cells being born in the developing cerebellum. All nuclei are labeled in red; the green cells are dividing and migrating into internal layers of the neural tissue. [Courtesy of Tal Raveh, Matthew Scott, and Jane Johnson.]

CELL BIRTH, LINEAGE, AND DEATH

During the evolution of multicellular organisms, new mechanisms arose to diversify cell types, to coordinate their production, to regulate their size and number, to organize them into functioning tissues, and to eliminate extraneous or aged cells. Signaling between cells became even more important than it was for single-celled organisms. The mode of reproduction also changed, with some cells becoming specialized as **germ cells** (e.g., eggs, sperm), which give rise to new organisms, as distinct from all other body cells, called **somatic cells.** Under normal conditions somatic cells will never be part of a new individual.

The formation of working tissues and organs during **development** of multicellular organisms depends in part on specific patterns of mitotic cell division. A series of such cell divisions akin to a family tree is called a *cell lineage.* A cell lineage traces the birth order of cells, the progressive restriction of their developmental potential, and their **differentiation** into specialized cell types (Figure 21-1). Cell lineages are controlled by cell-intrinsic (internal) factors—cells acting according to their history and internal regulators—as well as by cell-extrinsic (external) factors such as cell–cell signals and environmental inputs. A cell lineage begins with **stem cells,** unspecialized cells that can potentially reproduce themselves and generate more-specialized cells indefinitely. Their name comes from the image of a plant stem, which grows upward, continuing to form more stem, while sending off leaves and branches to the side. A cell lineage ultimately culminates in formation of terminally differentiated cells such as skin cells, neurons, or muscle cells. Terminal differentiation generally is irreversible, and the resulting highly specialized cells often cannot divide; they survive, carry out their functions for varying lengths of time, and then die.

Many cell lineages contain intermediate cells, referred to as *precursor cells* or *progenitor cells* or, if they are rapidly dividing, *transient amplifying (TA) cells.* The potential of such intermediate cells to form different kinds of differentiated cells is more limited than that of the stem cells from which they arise. (Although some researchers distinguish between precursor and progenitor cells, we will use these terms interchangeably.) Once a new precursor cell type is created, it often produces **transcription factors** characteristic of its fate. These transcription factors coordinately activate, or repress,

OUTLINE

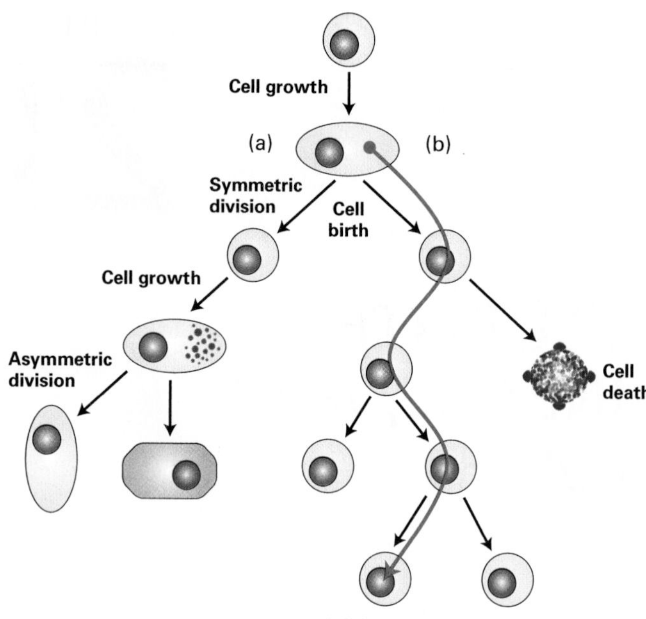

▲ **FIGURE 21-1 Overview of the birth, lineage, and death of cells.** Following growth, cells are "born" as the result of symmetric or asymmetric cell division. (a) The two daughter cells resulting from symmetric division are essentially identical to each other and to the parental cell. Such daughter cells subsequently can have different fates if they are exposed to different signals. The two daughter cells resulting from asymmetric division differ from birth and consequently have different fates. Asymmetric division commonly is preceded by the localization of regulatory molecules (green dots) in one part of the parent cell. (b) A series of symmetric and/or asymmetric cell divisions, called a cell lineage, gives birth to each of the specialized cell types found in a multicellular organism. The pattern of cell lineage can be under tight genetic control. Programmed cell death occurs during normal development (e.g., in the webbing that initially develops when fingers grow) and also in response to infection or poison. A series of specific programmed events, called apoptosis, is activated in these situations.

batteries of genes that direct the differentiation process. For instance, a few key regulatory transcription factors create the different mating types of budding yeast and similarly, a small number of such factors produced in sequence trigger the steps in forming differentiated muscle cells from precursors. We discuss both these examples in this chapter.

Typically we think of cell fates in terms of the differentiated cell types that are formed. A quite different cell fate, **programmed cell death,** also is absolutely crucial in the formation and maintenance of many tissues. A precise genetic regulatory system, with checks and balances, controls cell death just as other genetic programs control cell differentiation. In this chapter, then, we consider the life cycle of cells—their birth, their patterns of division (lineage), and their death. These aspects of cell biology converge with developmental biology and are among the most important processes regulated by the signaling pathways discussed in earlier chapters.

21.1 The Birth of Cells: Stem Cells, Niches, and Lineage

Many descriptions of cell division imply that the parental cell gives rise to two daughter cells that look and behave exactly like the parental cell: that is, cell division is *symmetric,* and the progeny retain the same properties as the parental cell. But if this were always the case, none of the hundreds of differentiated cell types present in complex organisms would ever be formed. Differences among cells can arise when two initially identical daughter cells diverge upon receiving distinct developmental or environmental signals. Alternatively, the two daughter cells may differ from "birth," with each inheriting different parts of the parental cell (see Figure 21-1). Daughter cells produced by such **asymmetric cell division** may differ in size, shape, and/or composition, or their genes may be in different states of activity or potential activity. The differences in these internal signals confer different fates on the two cells.

Here we discuss some general features of how different cell types are generated, culminating with the best-understood complex cell lineage, that of the nematode *Caenorhabditis elegans.* In later sections, we focus on examples of the molecular mechanisms that determine particular cell types in yeast, *Drosophila,* and mammals.

Stem Cells Give Rise to Both Stem Cells and Differentiating Cells

Stem cells, which give rise to the specialized cells composing the tissues of the body, exhibit several patterns of cell division (Figure 21-2). A stem cell may divide symmetrically

(a) Maintain stem-cell population

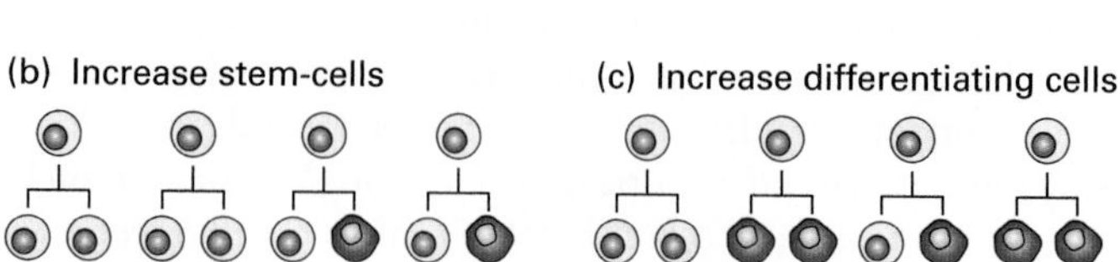

(b) Increase stem-cells

(c) Increase differentiating cells

▲ **FIGURE 21-2 Patterns of stem-cell division.** Divisions of stem cells (yellow) must maintain the stem-cell population, sometimes increase the number of stem cells, and at the right time produce differentiating cells (green). (a) Stem cells that undergo asymmetric divisions produce one stem cell and one differentiating cell. This does not increase the population of stem cells. (b) Some stem cells in a population may divide symmetrically to increase their population, which may be useful in normal development or during recovery from injury, while at the same time others are dividing asymmetrically as in (a). (c) In a third pattern, some stem cells may divide as in (b), while at the same time others produce two differentiating progeny. [Adapted from S. Morrison and J. Kimble, 2006, *Nature* **441:**1068.]

to yield two daughter stem cells identical to itself. Alternatively, a stem cell may divide asymmetrically to generate a copy of itself and a derivative stem cell that has more-restricted capabilities, such as dividing for a limited period of time or giving rise to fewer types of progeny compared with the parental stem cell. A *pluripotent* (or multipotent) stem cell has the capability of generating a number of different cell types, but not all. For instance, a pluripotent blood stem cell will form more of itself plus multiple types of blood cells, but never a skin cell. In contrast, a *unipotent* stem cell divides to form a copy of itself plus a cell that can form only one cell type. For example stem cells in the intestine continuously reproduce themselves, while the other daughter cell differentiates into an intestinal epithelial cell, as we discuss in greater detail below. In many cases, asymmetric division of a stem cell generates a progenitor cell, which embarks on a path of differentiation, or even a terminally differentiating cell.

The two critical properties of stem cells that together distinguish them from all other cells are the ability to reproduce themselves indefinitely, often called *self-renewal,* and the ability to divide asymmetrically to form one daughter stem cell identical to itself and one daughter cell of more restricted potential. Many stem-cell divisions are symmetric, producing two stem cells, but at some point some progeny need to differentiate. In this way, mitotic division of stem cells can either enlarge a population of undifferentiated cells or maintain a stem-cell population while steadily producing a stream of differentiating cells. Although some types of precursor cells can divide symmetrically to form more of themselves, they do so only for limited periods of time. Moreover, in contrast to stem cells, if a precursor cell divides asymmetrically, it generates two distinct daughter cells, neither of which is identical to the parental precursor cell.

The fertilized egg, or **zygote,** is the ultimate *totipotent* cell because it has the capability to generate all the cell types of the body. Although not technically a stem cell because it is not self-renewing, the zygote does give rise to cells with stem-cell properties. For example, the early mouse embryo passes through an eight-cell stage in which each cell can form every tissue; that is, they are totipotent. Thus the subdivision of body parts and tissue fates among the early embryonic cells has not irreversibly occurred at the eight-cell stage. At the 16-cell stage, this is no longer true; some of the cells are committed to particular differentiation paths.

Cell Fates Are Progressively Restricted During Development

The eight cells resulting from the first three divisions of a mammalian zygote (fertilized egg) all look the same. As demonstrated experimentally in sheep, each of the cells has the potential to give rise to a complete animal. Additional divisions produce a mass, composed of ≈64 cells, that separates into two cell types: trophectoderm, which will form extra-embryonic tissues like the placenta, and the *inner cell mass,* which gives rise to the embryo proper. The inner cell mass eventually forms three germ layers, each with distinct fates. One layer, the **ectoderm,** will make neural and epidermal cells; another, the **mesoderm,** will make muscle and connective tissue; the third layer, the **endoderm,** will make gut epithelia (Figure 21-3).

ECTODERM	MESODERM	ENDODERM
Central nervous system	Skull	Stomach
Retina and lens	Head, skeletal muscle	Colon
Cranial and sensory	Skeleton	Liver
Ganglia and nerves	Dermis of skin	Pancreas
Pigment cells	Connective tissue	Urinary bladder
Head connective tissue	Urogenital system	Epithelial parts of
Epidermis	Heart	trachea
Hair	Blood, lymph cells	lungs
Mammary glands	Spleen	pharynx
		thyroid
		intestine

▲ **FIGURE 21-3 Fates of the germ layers in animals.** Some of the tissue derivatives of the three germ layers are listed.

Once the three germ layers are established, they subsequently divide into cell populations with different fates. For instance, the ectoderm becomes divided into those cells that are precursors to the skin epithelium and those that are precursors to the nervous system. There appears to be a progressive restriction in the range of cell types that can be formed from stem cells and precursor cells as development proceeds. An early embryonic stem cell, as we've seen, can form every type of cell, an ectodermal cell has a choice between neural and epidermal fates, while a keratinocyte precursor can form skin but not neurons.

Another restriction that occurs early in animal development is the setting aside of cells that will form the **germ line**—the stem cells and precursor cells that eventually will give rise to eggs in a female and sperm in a male. Only the genome of the germ line will ever be passed on to progeny. The setting aside of germ-line cells early in development has been hypothesized to protect chromosomes from damage by reducing the number of rounds of replication they undergo. Whatever the reason, the early segregation of the germ-line is widespread (though not universal) among animals. In contrast, plants do nothing of the sort; meristems, growing tips of roots and shoots, can often give rise to germ-line cells and there is no germ-line lineage set aside early.

One consequence of the early segregation of germ-line cells is that the loss or rearrangement of genes in somatic cells will not affect the inherited genome of a future zygote. Nonetheless, although segments of the genome are rearranged and lost during development of lymphocytes from hematopoietic precursors, most somatic cells seem to have an intact genome, equivalent to that in the germ line (Chapter 24). Evidence that at least some somatic cells have a complete and functional genome comes from the successful production of cloned animals by *nuclear-transfer cloning*. In this procedure, the nucleus of an adult (somatic) cell is introduced into an egg that lacks its nucleus; the manipulated egg, which contains the diploid number of chromosomes and is equivalent to a zygote, then is implanted into a foster mother. The only source of genetic information to guide development of the embryo is the nuclear genome of the donor somatic cell. The frequent failure of such cloning experiments, however, raises questions about how many adult somatic cells do in fact have a complete functional genome. Even the successes, like the famous cloned sheep "Dolly," often have medical problems. The extent to which differentiated cells harbor fully functional genomes is still not fully understood. A cell could, for example, have an intact genome, but be unable to properly reactivate certain genes due to inherited chromatin states.

These observations raise two important questions: How are cell fates progressively restricted during development? Are these restrictions irreversible? In addressing these questions, it is important to remember that a cell's capabilities in its normal in vivo location may differ from what it is capable of doing if manipulated experimentally. Thus the observed limits to what a cell can do may result from natural regulatory mechanisms or may reflect a failure to find conditions that reveal the cell's full potential.

Although our focus in this chapter is on how cells become different, their ability to remain the same also is critical to the functioning of tissues and the whole organism. Non-dividing differentiated cells with particular characteristics often retain these features for many decades. Stem cells that divide regularly, such as a skin stem cell, must produce one daughter cell with the properties of the parental cell, retaining its characteristic composition, shape, behavior, and responses to specific external signals. Meanwhile, the other daughter cell, with its own distinct inheritance as the result of asymmetric cell division, embarks on a particular differentiation pathway, which may be determined both by the signals the cell receives and by intrinsic bias in the cell's potential, such as the previous activation of certain genes.

The Complete Cell Lineage of *C. elegans* Is Known

In the development of some organisms, cell lineages are under tight genetic control and thus are identical in all individuals of a species. In other organisms the exact number and arrangement of cells vary substantially among different individuals. The best-documented example of a reproducible pattern of cell divisions comes from the nematode *C. elegans*. Scientists have traced the lineage of all the somatic cells in *C. elegans* from the fertilized egg to the mature worm by following the development of live worms using Nomarski differential interference contrast (DIC) microscopy (Figure 21-4).

About 10 rounds of cell division, or fewer, create the adult worm, which is about 1 mm long and 70 μm in diameter. The adult worm has 959 somatic cell nuclei (hermaphrodite form) or 1031 (male). The number of somatic cells is somewhat fewer than the number of nuclei because some cells contain multiple nuclei (i.e., they are syncytia). Remarkably, the pattern of cell divisions starting from a *C. elegans* fertilized egg is nearly always the same. As we discuss later in the chapter, many cells that are generated during development undergo programmed cell death and are missing in the adult worm. The consistency of the *C. elegans* cell lineage does not result entirely from each newly born cell inheriting specific information about its destiny. That is, their birth cells are not necessarily "hard wired" by their own internal inherited instructions to follow a particular path of differentiation. In some cases, various signals direct initially identical cells to different fates, and the outcomes of these signals are consistent from one animal to the next.

The first few cell divisions in *C. elegans* produce six different *founder cells*, each with a separate fate as shown in

MEDIA CONNECTIONS

Video: *C. elegans* Crawling 

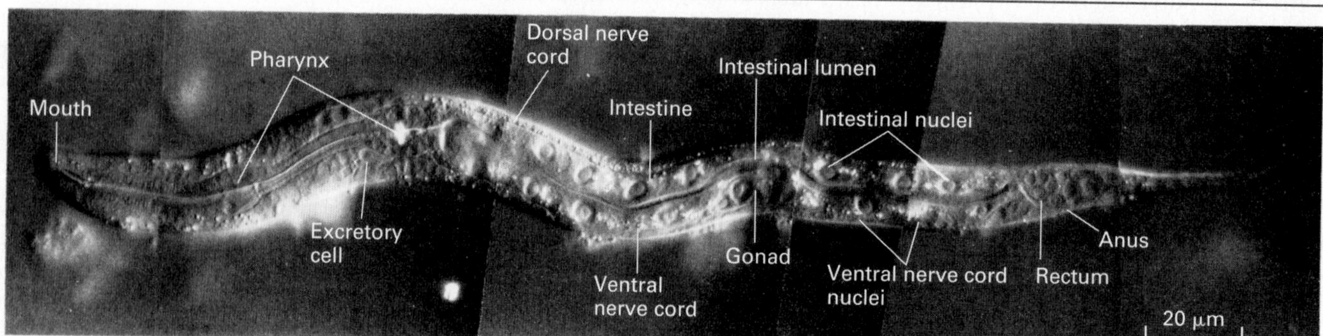

▲ **FIGURE 21-4 Newly hatched larva of *C. elegans*.** Some of the 959 somatic-cell nuclei in the hermaphrodite form are visualized in this micrograph obtained by differential interference contrast (DIC) microscopy, sometimes called Nomarski microscopy. The most easily seen are the intestinal nuclei, which appear as round discs. [From J. E. Sulston and H. R. Horvitz, 1977, *Devel. Biol.* **56**:110.]

Figure 21-5a, b. The initial division is asymmetric, giving rise to P1 and the AB founder cell. Further divisions in the P lineage form the other five founder cells. Some of the signals controlling division and fate asymmetry are known. For example, Wnt signals from the P2 precursor control the asymmetric division of the EMS cell into E and MS founder cells. Wnt signaling (see Figure 16-32) is also used in other asymmetric divisions in worms. Some of the embryonic cells function as stem cells, dividing repeatedly to form more of themselves or another type of precursor cell, while also generating differentiated cells that give rise to a particular tissue. The complete lineage of *C. elegans* is shown in Figure 21-5c. This organism has been a powerful model system for genetic studies to identify the regulators that control cell lineages in time and space.

Heterochronic Mutants Provide Clues About Control of Cell Lineage

Intriguing evidence for the genetic control of cell lineage has come from isolation and analysis of *heterochronic mutants*. In these mutants, a developmental event typical of one stage of development occurs too early (precocious development) or too late (retarded development). An example of the

Video: Time-Lapse Imaging of *C. elegans* Embryogenesis

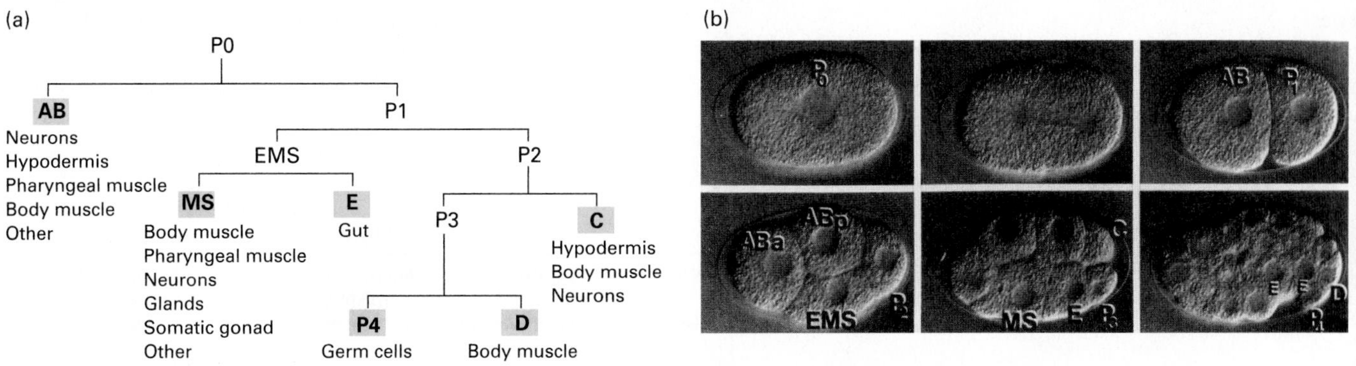

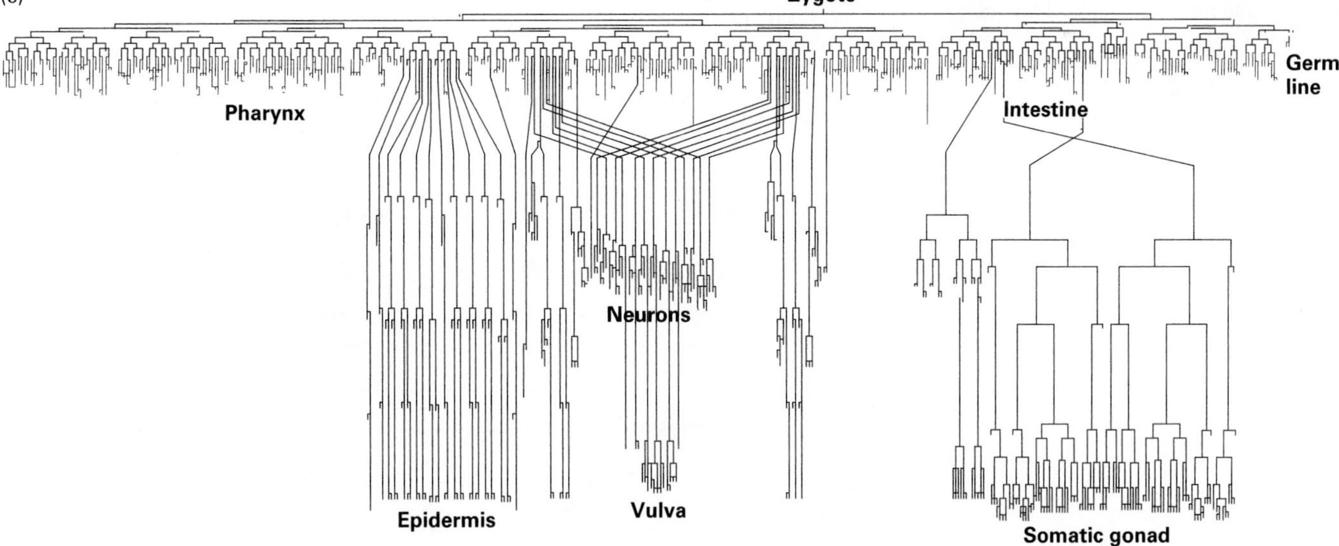

▲ **FIGURE 21-5 *C. elegans* lineage.** (a) Pattern of the first few divisions starting with P0 (the zygote) and leading to formation of the six founder cells (yellow highlight). The first division is asymmetric, producing P1 and AB, a founder cell. Further divisions in the P lineage generate the other five founder cells. Note that more than one lineage can lead to the same tissue type (e.g., muscle or neurons). The EMS cell is so named because it is the precursor to most of the endoderm and mesoderm. The lineage beginning with the P4 cell gives rise to all of the germ-line cells, which are set aside very early, as in most animals. All the other lineages give rise to somatic cells. (b) Light micrographs of the first few divisions of the embryo that generate the founder cells with cells labeled as in part (a). The texture of the cells shows the presence of organelles. (c) Full lineage of the entire body of the worm, showing some of the tissues formed. Note that any particular cell undergoes relatively few divisions, typically fewer than 15. [Part (b) from Einhard Schierenberg Zoologisches Institut, Universität Köln.]

former is premature occurrence of a cell division that yields a cell that differentiates and a cell that dies; as a result, the lineage that should have followed from the dead cell never happens. An example of the latter is the delayed occurrence of a lineage, causing juvenile structures to be produced, incorrectly, in more mature animals. In both cases, the character of a parental cell is, in essence, changed to the character of a cell at a different stage of development.

The study of heterochronic genes has been important to understanding the mechanisms of development and gene regulation. One example of precocious development in *C. elegans* comes from loss-of-function mutations in the *lin-14* gene, which cause premature formation of a certain neural precursor, the PDNB neuroblast (Figure 21-6a). The *lin-14* gene and several others found to be defective in heterochronic worm mutants encode RNA-binding or DNA-binding proteins, which presumably coordinate expression of other genes.

Two other genes (*lin-4* and *let-7*) discovered in heterochronic *C. elegans* mutants were initially puzzling because they appeared to encode small RNAs that do not encode any protein. To discover the products of these genes, scientists first determined which pieces of genomic DNA could restore gene function, and therefore proper cell lineage, to mutants defective in each gene. They then did the same thing with genomic DNA from the corresponding genomic regions of different species of worm. Comparison of the "rescuing" fragments from the different species revealed that they shared common short sequences with little protein-coding potential.

The short RNA molecules encoded by *lin-4* and *let-7* were subsequently shown to inhibit translation of the mRNAs encoded by *lin-14* and other heterochronic genes (Figure 21-6b). These small RNAs, called **micro RNAs (miRNAs)**, are produced by RNA polymerase II and are complementary to sequences in the 3′ untranslated parts of target mRNAs. The miRNAs direct post-transcriptional silencing of mRNAs by hybridizing to them and blocking translation or stimulating degradation (see Figure 8-25). Temporal changes in the production of *lin-4*, *let-7*, and other miRNAs during the life cycle of *C. elegans* serve as a regulatory clock for cell lineage.

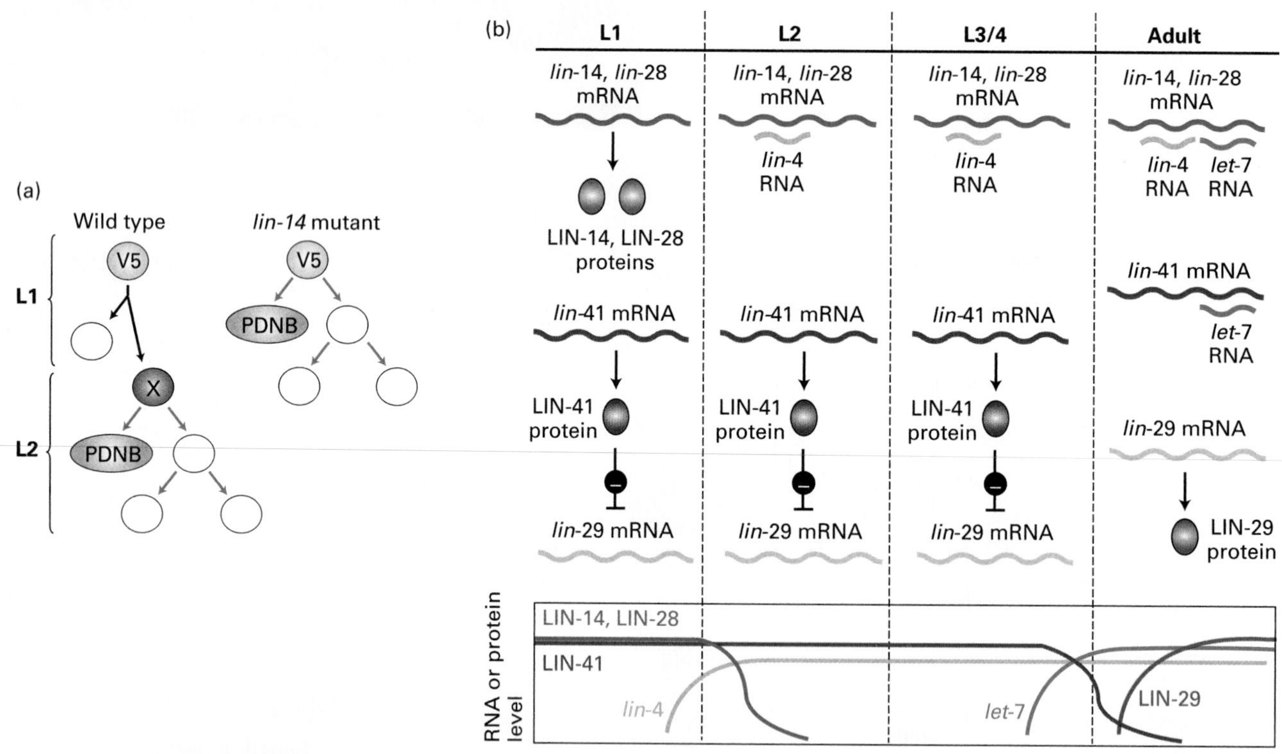

▲ **FIGURE 21-6 Timing of cell divisions during development of *C. elegans*.** (a) The pattern of cell division for the V5 cell of *C. elegans* is shown for normal (wild-type) worms and for a heterochronic mutant called *lin-14*. In the *lin-14* mutant, the pattern of cell division (red arrows) that normally occurs only in the second larval stage (L2) occurs in the first larval stage (L1), causing the PDNB neuroblast to be generated prematurely. In the mutant, the V5 cell behaves during L1 like cell "X" (purple) normally does in L2. The inference is that the LIN-14 protein prevents L2-type cell divisions, although precisely how it does so is unknown. (b) Two small regulatory RNAs, *lin-4* and *let-7*, serve as coordinating timers of gene expression. Binding of the *lin-4* RNA to the 3′ untranslated regions (UTRs) of *lin-14* and *lin-28* mRNAs prevents translation of these mRNAs into protein. This occurs following the first larval (L1) stage, permitting development to proceed to the later larval stages. Starting in the fourth larval stage (L4), production of *let-7* RNA begins. It hybridizes to *lin-14, lin-28*, and *lin-41* mRNAs, preventing their translation. LIN-41 protein is an inhibitor of translation of the *lin-29* mRNA, so the appearance of *let-7* RNA allows production of LIN-29 protein, which is needed for generation of adult cell lineages. LIN-4 may also bind to *lin-41* RNA at later stages. Only the 3′ UTRs of the mRNAs are depicted. [Adapted from B. J. Reinhart et al., 2000, *Nature* **403**:901.]

miRNAs have been identified in many other animals including vertebrates and insects. More than 300 are encoded in the human genome, perhaps as many as a thousand. Since production of miRNAs is temporally and spatially regulated, they are likely to control a broad range of events, perhaps including timed events as in *C. elegans*. How production of these regulatory miRNAs is temporally controlled is not yet known, but they have turned out to play many roles in regulating gene expression (Chapter 8).

Cultured Embryonic Stem Cells Can Differentiate into Various Cell Types

Embryonic stem (ES) cells can be isolated from early mammalian embryos and grown in culture (Figure 21-7a). Cultured ES cells can differentiate into a wide range of cell types, either in vitro or after reinsertion into a host embryo. When grown in suspension culture, human ES cells first differentiate into multicellular aggregates, called *embryoid bodies,* which resemble early embryos in the variety of tissues they form. When these are subsequently transferred to a solid medium, they grow into confluent cell sheets containing a variety of differentiated cell types including neural cells and pigmented and non-pigmented epithelial cells (Figure 21-7b). Under other conditions, ES cells have been induced to differentiate into precursors for various types of blood cells. What properties give ES cells their remarkable plasticity? A variety of actors play a role: signaling proteins, DNA methylation, micro RNAs, transcription factors, and chromatin regulators can all affect which genes become active (Chapters 7 and 8).

During the earliest stages of embryogenesis, as the fertilized egg begins to divide, both the paternal and maternal DNA becomes demethylated (see the discussion of DNA methylation in Chapter 7). This happens because a key maintenance methyltransferase (Dnmt1), which normally is present in the nucleus, is transiently excluded from the nucleus. During the first few cell divisions the pattern of methylation is reset, erasing earlier epigenetic marking of the DNA and creating a condition where cells have greater potential for diverse pathways of development. Mice engineered to lack Dnmt1 die as early embryos with drastically undermethylated DNA. ES cells prepared from such embryos are able to divide in culture, but in contrast to normal ES cells cannot undergo in vitro differentiation.

The properties of mouse ES cell are critically dependent on the action of three transcription factors produced shortly after fertilization: Nanog, Sox2, and Oct4. The genes that are bound by these factors have been identified using chromatin

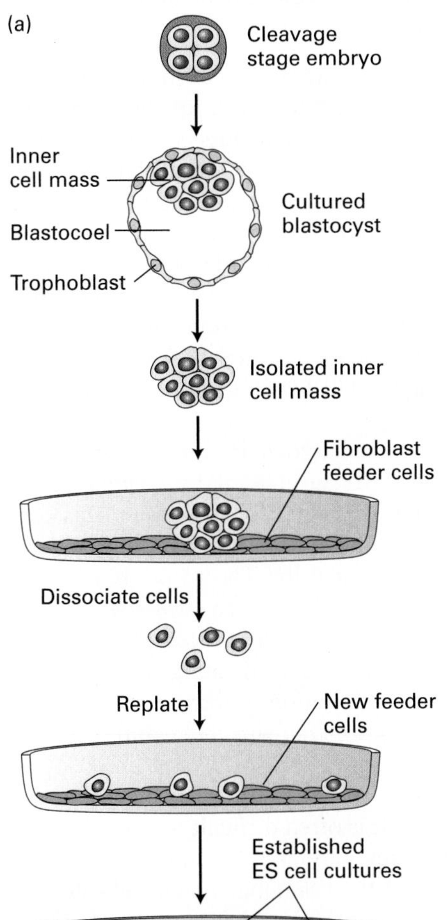

(b)

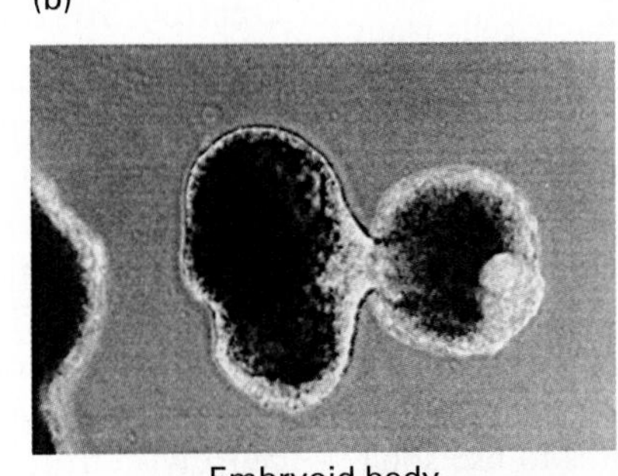
Embryoid body

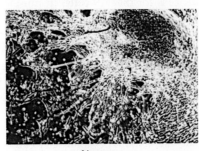

Neurons

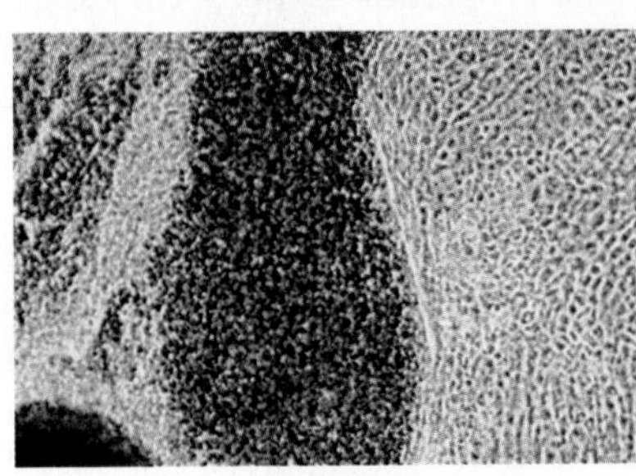
Epithelial cells

◄ **EXPERIMENTAL FIGURE 21-7** **Embryonic stem (ES) cells can be maintained in culture and can form differentiated cell types.** (a) Human blastocysts are grown from cleavage-stage embryos produced by in vitro fertilization. The inner cell mass is separated from the surrounding extra-embryonic tissues and plated onto a layer of fibroblast cells that help to nourish the embryonic cells. Individual cells are replated and form colonies of ES cells, which can be maintained for many generations and can be stored frozen. (b) In suspension culture, human ES cells differentiate into multicellular aggregates called embryoid bodies (*top*). After embryoid bodies are transferred to a gelatinized solid medium, they differentiate further into confluent cell sheets containing a variety of differentiated cell types including neural cells (*middle*), and pigmented and nonpigmented epithelial cells (*bottom*). [Parts (a) and (b) adapted from J. S. Odorico et al., 2001, *Stem Cells* **19:**193.]

immunoprecipitation experiments (see Figure 7-37). Each protein is found at more than a thousand chromosome locations. At about 350 locations, all three proteins are found. DNA microarrays have revealed which genes are active in ES cells. About half of the 350 loci where all three transcription factors accumulate are at or near genes that are transcribed in ES cells. The target genes regulated by these transcription factors encode a wide variety of proteins, including the Oct4, Nanog, and Sox2 proteins themselves, signaling components of the BMP and JAK/STAT pathways, and chromatin factors.

Chromatin regulators that control gene transcription (Chapter 7) are also important in ES cells. In *Drosophila*, polycomb group proteins form complexes that maintain gene repression states that have been previously established by DNA-binding transcription factors. Two mammalian protein complexes related to the fly Polycomb proteins, PRC1 and PRC2, are produced in ES cells. Early mouse embryos lacking components of PRC2 have abnormal development of the inner cell mass (the embryo proper), and ES cells cannot be made from embryos lacking PRC2 functions. The PRC2 complex of proteins acts by adding methyl groups to lysine 27 of histone H3, thus altering chromatin structure to repress genes. Remember that this type of regulation is distinct from methylation of DNA.

The possibility of using stem cells therapeutically to restore or replace damaged tissue is fueling much research on how to recognize and culture these remarkable cells from embryos and from various tissues in postnatal (adult) animals. For example, if neurons that produce the neurotransmitter dopamine could be generated from stem cells grown in culture, it might be possible to treat people with Parkinson's disease who have lost such neurons. For such an approach to succeed, a way must be found to direct a population of embryonic or other stem cells to form the right type of dopamine-producing neurons, and rejection by the immune system must be prevented. One way to prevent immune rejection is to use adult stem cells from a patient to produce therapeutic cells for that same patient. This is exactly what is done at present in some bone marrow transplants, as we shall see below. However it is not yet possible to isolate adult stem cells with similar capabilities for most other tissues. In animal experiments, embryonic stem cells have proven considerably more adept than adult stem cells at forming a variety of tissues. One approach for exploiting the advantages of ES cells while reducing immunological rejection may be to insert a nucleus from a patient into the environment of an embryonic cell, replacing the endogenous nucleus with one that will confer patient-specific properties upon the cells. Lines of cells that would be accepted by particular patients could then be established. Stem cells grown from blastocysts in this way may become an option for treating Parkinson's disease and perhaps other neurodegenerative conditions such as Alzheimer's disease.

Recent work has been directed at exploring whether embryonic or adult stem cells can be induced to differentiate into cell types that would be useful therapeutically. For example, mouse ES cells have been treated with inhibitors of phosphatidylinositol-3 kinase, a regulator in one of the phosphoinositide signaling pathways (Chapter 16). The treated ES cells differentiate into cells that resemble pancreatic β cells in their production of insulin, their sensitivity to glucose levels, and their aggregation into structures reminiscent of pancreas structures. Implantation of these differentiated cells into diabetic mice restored their growth, weight, glucose levels, and survival rates to normal. Many important questions must be answered before the feasibility of using human stem cells for such purposes can be assessed adequately. ▮

Apart from their possible benefit in treating disease, ES cells have already proven invaluable for producing mouse mutants useful in studying a wide range of diseases, developmental mechanisms, behavior, and physiology. By techniques described in Chapter 5, it is possible to eliminate or modify the function of a specific gene in ES cells (see Figure 5-40). Then the mutated ES cells can be employed to produce mice with a **gene knockout** (see Figure 5-41). Analysis of the effects caused by deleting or modifying a gene in this way often provides clues about the normal function of the gene and its encoded protein.

We will now examine the properties and regulation of some postnatal (adult) stem cells, descendants of ES cells, that build the various organs and tissues in animals.

Adult Stem Cells for Different Animal Tissues Occupy Sustaining Niches

Many differentiated cell types are sloughed from the body or have life spans that are shorter than that of the organism. Disease and trauma also can lead to loss of differentiated cells. Since differentiated cells generally do not divide, they must be replenished from nearby stem-cell populations. Postnatal (adult) animals contain stem cells for many tissues including the blood, intestine, skin, ovaries and testes, muscle, and liver. Even some parts of the adult brain, where little cell division normally occurs, have a population of stem cells. In muscle and liver, stem cells are most important in healing, as relatively little cell division occurs in the adult tissues at other times.

Stem cells need the right microenvironment to maintain themselves. In addition to intrinsic regulatory signals—like the presence of certain regulatory proteins—stem cells rely on extrinsic regulatory signals from surrounding cells to maintain their status as stem cells. The location where a stem-cell fate can be maintained is called a *stem-cell niche* by analogy to an ecological niche, which is a location that supports the existence and competitive advantage of a particular organism. The right combination of intrinsic and extrinsic regulation, imparted by a niche, will create and sustain a population of stem cells.

In order to investigate or use stem cells, they must be found and characterized. It is often difficult to identify stem cells precisely because they may lack distinctive shapes or gene expression. Much of the time, many stem cells do not divide particularly rapidly, being held in reserve, dividing slowly if at all, until stimulated by signals that convey the need for new cells. For example, an inadequate oxygen supply can stimulate blood stem cells to divide, and injury to the skin can stimulate regenerative cell division starting with the activation of stem cells. Some stem cells, however, including those that form the continuously shed epithelium of the intestine, are continuously dividing, usually at a slow rate.

One approach for identifying stem cells in a mixed cell population depends on their relatively slow rate of division. In this approach, a type of **pulse-chase** experiment, cells are provided with a brief pulse of labeled DNA precursors, such as bromodeoxyuridine (BrdU), then later examined to see which cells are labeled. After the BrdU pulse, cells that are not dividing will not be labeled at all, and rapidly dividing cells will dilute the BrdU label with normal unlabeled nucleotides (the chase). Stem cells, in contrast, will incorporate BrdU during their slow division process. Since they divide relatively rarely, stem cells will retain the BrdU label longer than most other cells, marking them as *label-retaining cells.* This sort of label retention is often a useful way to identify stem cells.

Germ-line Stem Cells The germ line is the line of cells that produces oocytes and sperm. It is distinct from the somatic cells that make all the other tissues but are not passed on to progeny. The germ line, like somatic-cell lineages, has stem cells. Stem-cell niches have been especially well defined in studies of germ-line stem cells in *Drosophila* and *C. elegans*. Germ-line stem cells are present in adult flies and worms and the location of the stem cells is well known. The stem cells were identified by BrdU label retention.

In the fly ovary, the niche where oocyte precursors form and begin to differentiate is located next to the tip of the *germarium* (Figure 21-8a). There are two or three germ-line stem cells in this location next to a few cap cells, which create the niche by secreting two **transforming growth factor β (TGFβ)** proteins (Dpp and Gbb) and **Hedgehog (Hh)** protein (Figure 21-8b). These secreted protein signals were introduced in Chapter 16. When the stem cells divide, they produce two daughters, one of which remains adjacent to the cap cells and is therefore a stem cell like the mother cell. The other daughter divides to produce two cystoblast cells that will differentiate into germ-line cells. The cystoblast cells embark on a path of differentiation because they are too far from the cap cells to receive the cap cell–derived signals, including Hh, Dpp, and Gbb, and direct cell-cell

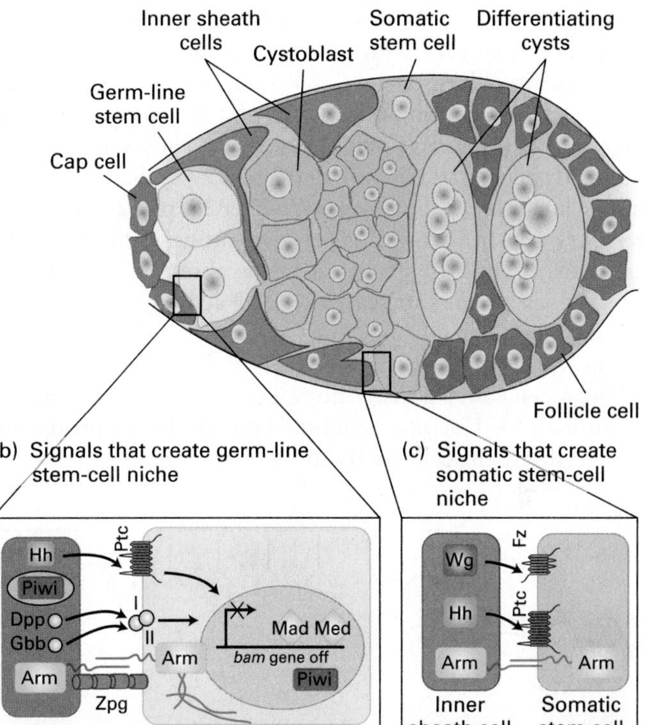

Cap cell	Germ-line stem cell	Inner sheath cell	Somatic stem cell
• Secretes Hh signal. • Secretes two TGFβ signals, Dpp & Gbb. • Produces Arm and Zpg surface proteins. • Has PIWI protein in the nucleus.	• Receives Hh through Ptc receptor, promoting self-renewal. • Receives Dpp and Gbb through TGFβ receptor subunits I and II, promoting self-renewal. • TGFβ protein signals cause activation of Mad and Med transcription factors to repress *bam* gene and allow self-renewal. • Produces Arm and Zpg surface proteins, which interact with themselves on the cap cell. • Has PIWI protein in the nucleus to promote stem cell fate.	• Secretes two signals, Wg and Hh. • Produces Arm protein on its surface.	• Receives Wg signals through the Fz receptors, promoting self-renewal. • Receives Hh signal through the Ptc receptor, promoting self-renewal. • Producs Arm, which interacts with Arm on inner sheath cell.

▶ **FIGURE 21-8 *Drosophila* germarium and the signals that create its stem-cell niches.** (a) Cross-section of the germarium showing female germ-line stem cells (yellow) and some somatic stem cells (gold) in their niches and the progeny cells derived from them. The germ-line stem cells produce cytoblasts (green), which differentiate into oocytes; the somatic stem cells produce follicle cells (brown) that will make the eggshell. The cap cells (dark green) create and maintain the niche for germ-line stem cells, while the inner sheath cells (blue) produce the niche for somatic stem cells. (b) Signaling pathways that control the properties of germ-line stem cells. The signaling molecules—the TGFβ proteins Dpp and Gpp as well as Hh— are produced by cap cells. Binding of these ligands to receptors on the surface of a stem cell results in repression of the *bam* gene by two transcription factors, Mad and Med. Repression of *bam* allows germ-line stem cells to renew, whereas activation of *bam* promotes differentiation. Two surface proteins, Arm and Zpg, which physically link cap cells and stem cells, are also important in maintaining the stem-cell niche. Cells out of reach of Arm and Zpg differentiate rather than renew. Micro RNAs are increasingly recognized as critical regulators of cell differentiation, including germ-line cells. Some of them bind to the Piwi protein, an essential germ-line regulator in both cap and stem cells. (c) Signaling pathways that control the properties of somatic stem cells. The Wnt signal Wingless (Wg) is produced by the inner sheath cells and is received by the Frizzled receptor (Fz) on a somatic stem cell. Hh is similarly produced, and is received by the Ptc receptor. Both receptors signal to control transcription resulting in self-renewal of somatic stem cells. [Adapted from L. Li and T. Xie, 2005, *Ann. Rev. Cell Devel. Biol.* **21**:605.]

interactions mediated by the cell-surface proteins Arm and Zpg, which together direct a cell to remain a stem cell. Both cap cells and germ-line stem cells produce Piwi proteins, which bind micro RNAs. Piwi proteins and their bound miRNAs regulate gene expression and control germ-line cell development in a wide variety of animals as well as stem-cell development in plants. Thus they constitute an ancient mechanism of developmental regulation. Separate somatic stem cells in the germarium produce follicle cells that will make the eggshell. The somatic stem cells have a niche too, created by the inner sheath cells, which produce Wingless (Wg) protein—a fly **Wnt** signal—and Hh protein (Figure 21-8c). Thus two different populations of stem cells can work in close coordination to produce different parts of an egg.

Micro RNAs control the division properties of *Drosophila* female germ-line stem cells. The Dicer protein, a double-stranded RNase, produces micro RNAs (see Figure 8-25). Germ-line stem cells with *dicer* mutations fail to pass successfully through the G_1 to S transition of the cell cycle; as a result, the population of stem cells and therefore oocytes is depleted. The absence of miRNA function causes, directly or indirectly, increased function of the p21/27 cyclin-dependent kinase inhibitor. As discussed in Chapter 20, p21/27 normally restricts G_1 to S transitions by regulating cyclin E–CDK complexes. Thus the net effect of the absent miRNA function, which permits increased p21/27 activity, is to restrict cell division.

In worms the long tubelike arms of the gonads have tips where a cell called a distal tip cell creates a stem-cell niche (Figure 21-9). The transmembrane protein Delta, produced by the distal tip cell, binds to the Notch receptor on the germ-line stem cells. The Delta/Notch signaling pathway (see Figure 16-36) promotes mitotic division of worm germ-line stem cells, thus creating more stem cells. Meiosis (i.e., germ-line differentiation) is blocked by the Delta signal until the stem cells move beyond the range of the signal from the distal tip cell. Mutations that activate Notch in the germ-line stem cells, even in the absence of Delta signal, cause a gonadal tumor with massive numbers of extra germ-line stem cells, due to excessive mitosis and little meiosis.

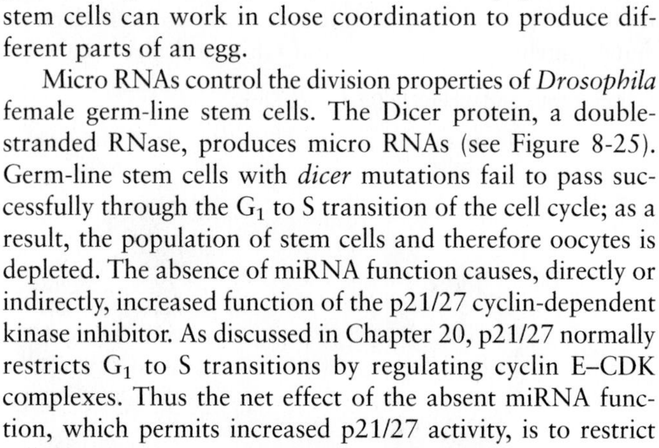

▲ **FIGURE 21-9** ***C. elegans*** **germ-line stem-cell niche.** A cross-section of the tip of a gonad arm shows stem cells in their niche and progeny derived from them. The single distal tip cell (green) in each gonad arm creates and maintains the niche. Self-renewing mitotic stem cells produced by the germ-line stem cells convert to meiosis when they move beyond the range of the Delta signal from the distal tip cell. During these stages, the cells are only partially separated by membranes ("Y" shapes) and are therefore a syncytium. [Adapted from L. Li and T. Xie, 2005, *Ann. Rev. Cell Devel. Biol.* **21:**605.]

The identification and characterization of *Drosophila* and *C. elegans* germ-line stem cells were important because they showed convincingly the existence of stem-cell niches and permitted experiments to identify the niche-made signals that cause cells to become and remain stem cells. Thus a stem-cell niche is a set of cells and the signals they produce, not just a location. Identification of specific molecules that maintain the stem-cell state in *Drosophila* and *C. elegans* brought an unexpected bonus: Some of these molecules are also used to form a stem-cell niche and control stem-cell fates in mammals. For example, germ-line stem cells in the mouse testis are dependent on a TGFβ signaling protein (GDNF) derived from somatic cells. Each seminiferous tubule contains exactly one germ-line stem cell, which divides asymmetrically to re-create itself and to produce a spermatogonial cell. This cell proliferates and its progeny become spermatocytes, which go through the extraordinary differentiation process that builds a sperm. The niche is created by a specialized region of the Sertoli cell along with a myoid cell and a basement membrane produced by the myoid cell, though many details of the molecular signals remain to be explored.

Skin/Hair Stem Cells in Mammals Epithelial stem cells that give rise to skin and hair in mammals are located in hair follicles and in the basal layer of the epithelium between follicles. In the hair follicle, the stem cells occupy a niche called the *bulge* (Figure 21-10a). The stem cells divide asymmetrically to produce more stem cells and to make precursor cells of at least two kinds. One type of precursor will rise toward the surface of the skin and form *keratinocytes,* the major cell type of skin, which is a multilayered epithelium (the epidermis). Other cells emerge from the stem cells to become hair-matrix progenitors that move down deeper in the hair follicle and form a complex set of structures including the hair itself.

The molecular regulators in the skin stem-cell niche are incompletely known. However, as in flies, TGFβ signals, arising from mesenchymal cells that surround the bulge cells, and a Wnt signal, arising from the dermal papilla, are important in controlling stem-cell renewal and differentiation into either skin or hair (Figure 21-10b). Evidence for the importance of Wnt signaling came from manipulations of the expression of *β-catenin,* a protein that helps link certain cell–cell junctions to the cytoskeleton (see Figure 19-12) and also functions as a signal transducer in the

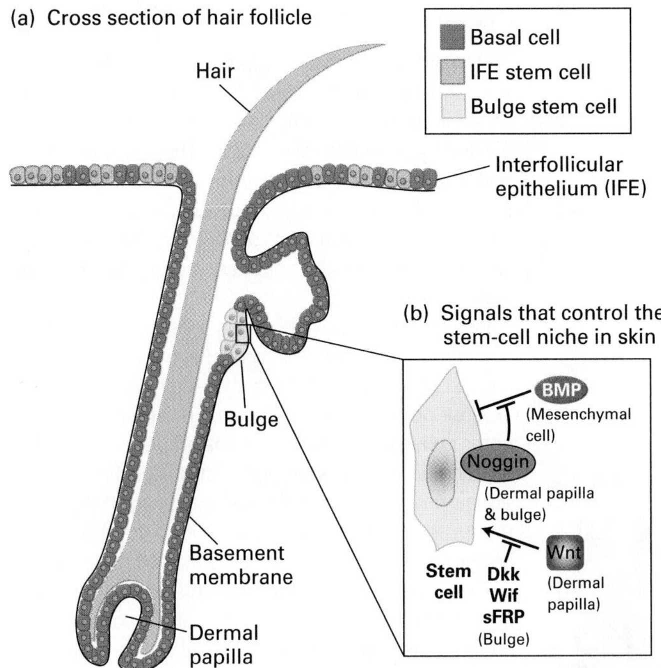

▲ **FIGURE 21-10 Skin/hair stem-cell niche in mammals and the signals that control it.** (a) A hair follicle showing stem cells (yellow) in the bulge and interfollicular stem cells (light green) among the basal epithelial cells (green) outside the hair follicle. The progeny of bulge stem cells migrate down to contribute to hair formation near the dermal papilla. The dark green cells have yet to begin terminal differentiation; they are transient amplifying cells that are still dividing but cannot generate stem cells. Note that only the basal layer of epithelial cells adjacent to the basement membrane are shown. Overlaying these basal cells, outside of the follicle, are several layers of differentiating keratinocytes. (b) Signaling events in the stem-cell niche. The source of each signal is indicated in parentheses. A Wnt signal promotes formation of new hair cells, but in the bulge—home of the stem cells—at least three Wnt inhibitors (Dkk, Wif, and sFRP) block differentiation and preserve the stem-cell state. Away from those inhibitors, the lack of Wnt signaling allows differentiation of stem cells into skin cells (keratinocytes). BMP, which belongs to the TGFβ family of signaling proteins, is produced by mesenchymal cells adjacent to the bulge. During development, when new hair is to be grown, the dermal papilla makes Noggin, which blocks the BMP signal, and more Wnt, which overcomes the inhibitors, allowing stem cells to differentiate into hair cells. BMPs have complex and incompletely understood roles in skin development, and their functions change with the stage of development. [Adapted from F. M. Watt et al., 2006, *Curr. Opin. Genet. Devel.* **16:**518, and L. Li and T. Xie, 2005, *Ann. Rev. Cell Devel. Biol.* **21:**605.]

Wnt pathway (see Figure 16-32). Activation of β-catenin changes the fate of cells from epidermis (skin) to hair. In contrast, removal of β-catenin from the skin of engineered mice eliminates formation of hair cells. Epithelial stem cells then form only epidermis, not hair cells. Thus β-catenin acts as a switch that controls which type of precursor arises from epithelial stem cells. Wnt signals also have stimulatory effects on cell division, which can be restrained by Wnt pathway inhibitors, such as DKK and sFRP, that are present in the bulge—a Battle of the Bulge of sorts.

Newly formed keratinocytes move toward the outer surface, becoming increasingly flattened and filled with keratin intermediate filaments (Chapter 18). It normally takes about 15–30 days for a newly "born" keratinocyte in the lowest layer of the skin to differentiate and move to the topmost layer. The "cells" forming the topmost layer are actually dead and are continually shed from the surface.

In addition to keratinocytes, skin contains dendritic epidermal T cells, an immune-system cell that produces a certain form of the T-cell receptor (Chapter 24). When dendritic epidermal T cells are genetically modified so they do not produce T-cell receptors, wound healing is slow and less complete than in normal skin. Normal healing is restored by addition of keratinocyte growth factor. The current hypothesis is that when dendritic epidermal T cells recognize antigens on cells in damaged tissue, they respond by producing stimulating proteins, such as keratinocyte growth factor, that promote the production of more keratinocytes and wound healing. Many other signals—including Hedgehog, calcium, and transforming growth factor α (TGFα)—control the production of skin cells from stem cells. Discovering how all these signals work together to control growth and stimulate healing will advance our understanding of diseases such as psoriasis and skin cancer and perhaps pave the way for effective treatments. ■

Intestinal Stem Cells In contrast to epidermis, the epithelium lining the small intestine is a single cell thick (see Figure 19-8). This thin layer is enormously important for keeping toxins and pathogens from entering our bodies; it also transports nutrients essential for survival from the intestinal lumen into the body (see Figure 11-29). The cells of the intestinal epithelium continuously regenerate from a stem-cell population located deep in the intestinal wall in pits called *crypts* (Figure 21-11a). By identifying the label-retaining cells in the intestinal epithelium, researchers determined that the stem cells are located precisely four or five cells above the bottom of a crypt. The niche is created by mesenchymal cells that abut the crypts at the level of the stem cells. These cells produce a Wnt signal, a BMP (TGFβ) signal, and possibly a ligand for the Notch receptor on stem cells (Figure 21-11b). Overproduction of β-catenin in intestinal cells leads to excess proliferation of those cells, as though they were receiving too much Wnt signal (which stabilizes β-catenin). Blocking the function of β-catenin by interfering with the TCF transcription factor that it activates abolishes the stem cells in the intestine. Thus Wnt signaling, acting through β-catenin, plays a critical role in maintaining the intestinal stem-cell population. BMP has the opposite effect, promoting differentiation and restraining the effect of Wnt.

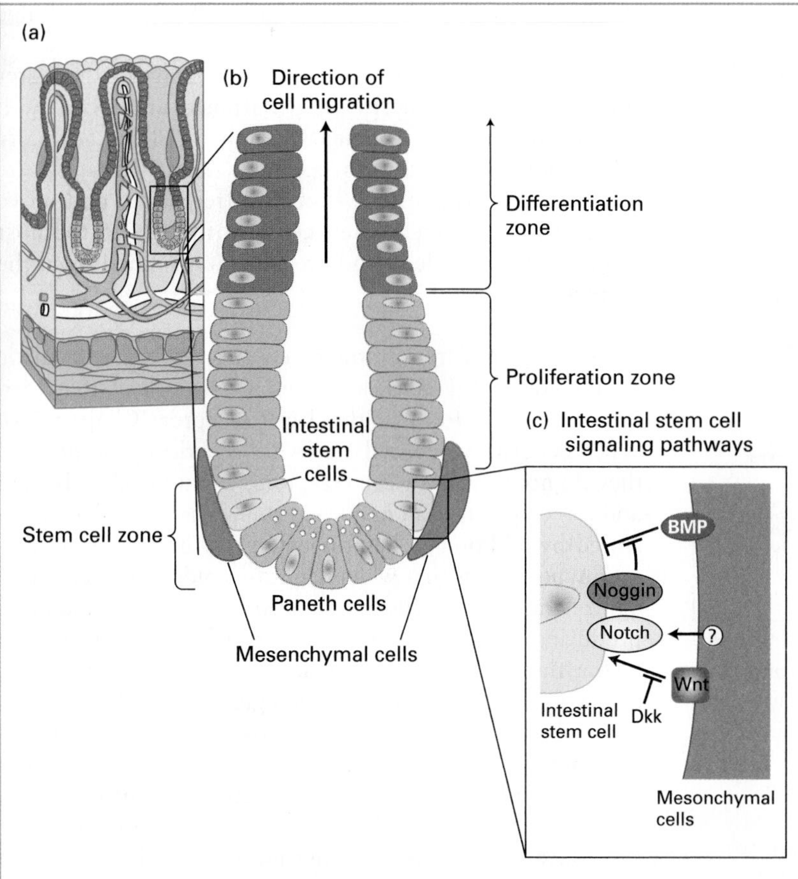

◀ **FIGURE 21-11 Intestinal stem-cell niche and the signals that control it.** (a) Finger-like projections of the inner surface of the intestine called villi are divided by deep pits called crypts. The single-cell thick epithelium protects us from infection and allows selective transport of useful nutrients into the blood stream. (b) An intestinal crypt showing the intestinal stem cells (yellow), their proliferating mitotic progeny (light green), and precursors (dark green) in the final stages of differentiation. Mesenchymal cells, shown in blue, create this stem-cell niche. Paneth cells (orange), located at the base of the crypt, secrete antimicrobial defense proteins called defensins. These proteins form pores in bacterial membranes, leading to death of the bacteria. Signaling events in the stem-cell niche. Wnt signals promote stem-cell fates, offsetting BMP signals that push toward differentiation. Noggin restrains the BMP signals and thus promotes stem-cell proliferation, whereas Dkk restrains Wnt signaling when growth is not needed. The Notch receptor is also involved, although its ligand on mesenchymal cells is unknown. [Adapted from L. Li and T. Xie, 2005, *Ann. Rev. Cell Devel. Biol.* **21:**605.].

Intestinal stem cells produce precursor cells that proliferate and differentiate as they ascend the sides of crypts to form the surface layer of the finger-like gut projections called *villi,* across which intestinal absorption occurs. Pulse-chase labeling experiments with BrdU have shown that the time from cell birth in the crypts to the loss of cells at the tip of the villi is only about 2 to 3 days. Thus enormous numbers of cells must be produced continually to keep the epithelium intact. The production of new cells is precisely controlled: too little division would eliminate villi and lead to breakdown of the intestinal surface; too much division would create an excessively large epithelium and also might be a step toward cancer. Indeed, mutations that inappropriately activate Wnt signal transduction are a major contributor to the progression of colon cancer, as we shall see in Chapter 25.

Neural Stem Cells The great interest in the formation of the nervous system and in finding better ways to prevent or treat neurodegenerative diseases has made the characterization of neural stem cells an important goal. The earliest stages of vertebrate neural development involve the rolling up of ectoderm to form the *neural tube,* which extends the length of the embryo from head to tail (Figure 21-12a). Initially the neural tube is composed of a single layer of cells, the neural stem cells (NSCs); these will give rise to the entire central nervous system (brain and spinal cord). Labeling and tracing experiments have shown where neural cells are born and where they go after birth. The most active region of cell division is the *subventricular zone,* which has the properties of a stem-cell niche and is named for its proximity to the central fluid-filled *ventricle.*

The embryonic neural stem cells that line the ventricle divide can symmetrically, produce two daughter stem cells side-by-side (Figure 21-12b), or asymmetrically, produce a cell that remains a stem cell and another that migrates radially outward. The migrating cells are often transient amplifying (TA) cells, which in turn divide to form neural precursors called neuroblasts. Once formed, TA cells and neuroblasts migrate radially outward and form successive layers of the neural tissue in an inside-out order, whereas the stem cells remain in contact with the ventricle (see Figure 21-12b). Newly formed cells therefore traverse the layers of preexisting cells before taking up residence on the outside.

Tracing experiments with viruses have shown that a neuroblast can produce two daughters, one a neuron and one a glial cell. In these experiments, a library of defective

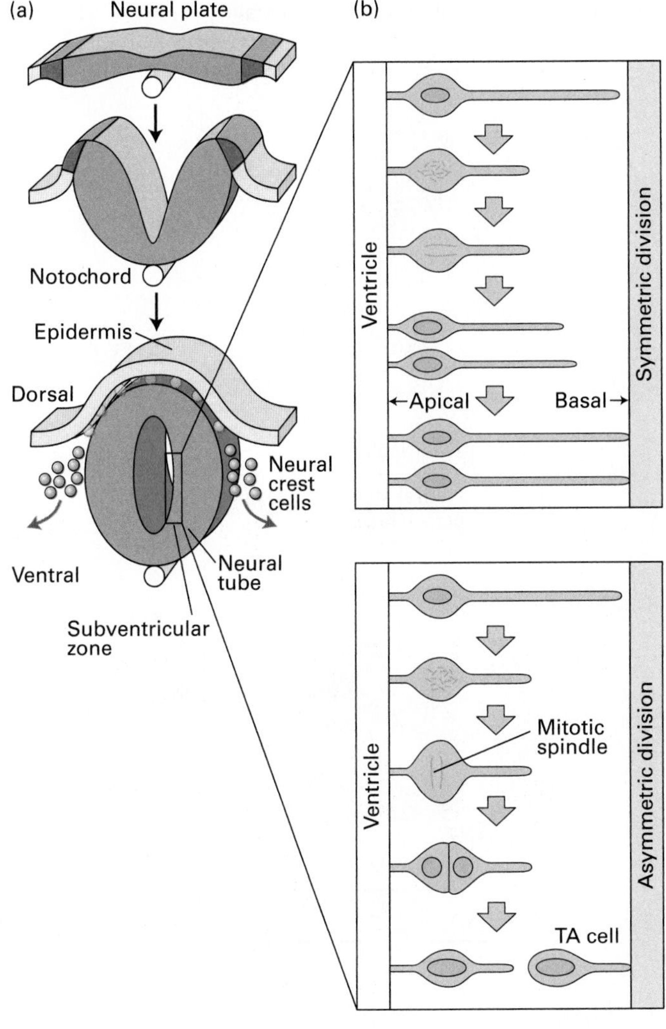

◀ **FIGURE 21-12 Formation of the neural tube and division of neural stem cells.** (a) Early in development, a portion of the ectoderm rolls up and separates from the rest of the embryonic cells. This forms the epidermis (gray) and the neural tube (blue). Near the interface between the two, neural crest cells form and then migrate to contribute to skin pigmentation, nerve formation, craniofacial skeleton, heart valves, peripheral neurons, and other structures. The notochord, a rod of mesoderm for which we are named (chordates), provides signals that affect cell fates in the neural tube (Chapter 22). The interior of the neural tube will become a series of fluid-filled chambers called ventricles. Neural stem cells located adjacent to the ventricles in the subventricular zone will divide to form neurons that migrate radially outwards to form the layers of the nervous system. (b) Neural stem cells in the subventricular zone can divide symmetrically *(top),* along their apical-basal axis, to give rise to side-by-side daughter stem cells, both in contact with the ventricle. Alternatively, stem cells can divide, along their other axis, to produce one daughter that is a stem cell able to self-renew and a daughter cell, called a transient amplifying (TA) cell, that begins to migrate and differentiate *(bottom).* A key difference between the two patterns of division is the orientation of the mitotic spindle. [Adapted from L. R. Wolpert et al. 2002. *Principles of Development*, 2d ed., Oxford University Press.]

retroviruses, each able to infect only once and containing a unique DNA sequence, was prepared (Figure 21-13a). Any cell infected by a single virion would give rise to a clone of cells that all carry that particular virus' DNA sequence. In this way, all the cells that derived from a single neural stem cell or TA cell can be identified as a clone (Figure 21-13b). The results of these tracing experiments were surprising. First, some neurons were found to have migrated considerable distances laterally, abandoning their radial migration to the outer cortical layer. Second, in some cases, a single neuron and single glial cell, shared the same viral DNA sequence. A neural precursor had evidently been infected and had then divided once to give rise to two quite different cell types.

Most mammalian brain cells stop dividing by adulthood but some cells in the subventricular zone and at least one other part of the brain continue to act as stem cells and generate new neurons (Figure 21-14). In the subventricular zone of adults, the neural stem cells are astrocytes, a somewhat confusing nomenclature since more traditionally "astrocyte" meant a type of glia cell. Evidently neural stem cells are a subset of astrocytes not previously recognized for their special stem-cell qualities. Neural stem cells have some properties of astrocytes, such as producing glial fibrillary acidic protein (GFAP), but they also can divide asymmetrically to renew themselves and to produce TA cells.

The subventricular stem-cell niche is created by mostly unknown signals from the ependymal cells that form a layer just inside the neural tube (lining the ventricle) and by endothelial cells that form blood vessels in the vicinity (Figure 21-14b). The endothelial cells, and the basal lamina they form, are in direct contact with neural stem cells and are believed to be essential in forming the niche. Each neural stem cell extends a single cilium through the ependymal cell layer to directly contact the ventricle. Though the function of the cilium is unknown, it may function as an antenna for receiving signals that would otherwise be inaccessible to the neural stem cell. The signals that create the niche are not completely characterized, but there is evidence for a blend of factors including FGFs, BMPs, IGF, VEGF, TGFα, and BDNF. The BMPs appear to favor astrocyte differentiation over neural differentiation, and over-expression of IGF (insulin-like growth factor) causes mice to develop with abnormally large brains. Later stages of neural development are discussed in Chapter 23.

Hematopoietic Stem Cells Like the intestinal epithelium the blood is a continuously replenished tissue. The stem cells

(a)

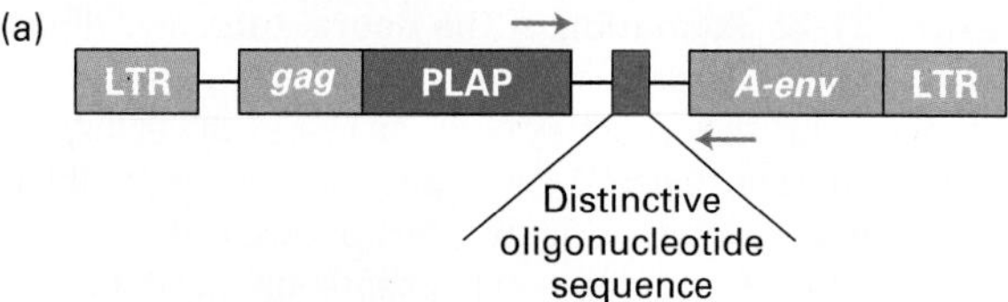

(b)

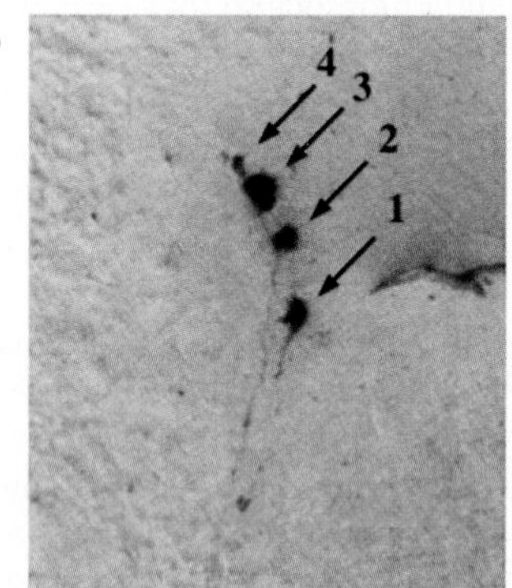

▲ EXPERIMENTAL FIGURE 21-13 **Retrovirus infection can be used to trace cell lineage.** (a) Engineered viral genome. The long terminal repeats (LTRs) are standard retroviral repeats. Viral proteins required for infection are encoded in the *gag* and in *A-env* genes. PLAP is an introduced gene for an alkaline phosphatase. Detection of this enzyme by histochemical staining is used to determine which cells carry a virus. The oligonucleotide sequence, synthesized by providing random nucleotides, is different in each virus and can be amplified by PCR, using primers for sequences that are in all viruses (purple arrows), and then sequenced. A library of more than 107 distinct viruses was made. Because these viruses lack the genes required for production of new virions in infected cells, each defective virus can infect only once. (b) Tissue section showing cells infected with defective viruses. The DNA from each stained clone of cells can be extracted and amplified by PCR to determine the sequence of the infecting virus. Cells descended from the same initially infected cell will have the same oligonucleotide sequence, whereas separate infection events will give different sequences. [From J. A. Golden et al., 1995, *Proc. Nat'l. Acad. Sci. USA* **92:**5704.].

(a) Cross section of whole developing brain

Lateral ventricle

(b) Subventricular zone

Neural stem cells (NSC)

Ependymal cells

Neuroblast

Transit-amplifying cells

Lateral ventricle

Blood vessel

NSC — Expansion — Neuroblast

▲ **FIGURE 21-14 Neural stem-cell niche.** (a) Cross-section of the developing nervous system showing the lateral ventricle, a fluid-filled space inside the neural tube. The area just surrounding the ventricle, called the subventricular zone, is the site of stem cells from which neural precursors arise. (b) The neural stem cells (yellow), a subset of astrocytes, are in contact with blood vessels (blue) and adjacent ependymal cells (pink), both of which provide signals or direct contacts that maintain the stem-cell population. Neural stem cells (NSC) divide to renew themselves and to form a dividing population of transit amplifying (TA) cells (light green; see Figure 21-12b). TA cells in turn divide to form neuroblasts (dark green), which give rise to neurons. [Adapted from L. Li and T. Xie, 2005, *Ann. Rev. Cell Devel. Biol.* **21:**605.]

that give rise to the different types of blood cells are located in the bone marrow in adult animals. All types of blood cells derive from a single type of pluripotent *hematopoietic stem cell,* which gives rise to the more-restricted *myeloid* and *lymphoid stem cells* (Figure 21-15). Although myeloid and lymphoid stem cells are capable of self-renewal, each type is committed to one of the two major hematopoietic lineages. Thus these cells function as both stem cells and precursor cells.

After hematopoietic stem cells form, numerous extracellular growth factors called **cytokines** regulate proliferation and differentiation of the precursor cells for various blood-cell lineages. Each branch of the blood-cell lineage tree has different cytokine regulators, allowing exquisite control of the production of specific cells types. For example, after a bleeding injury, when all blood cells are needed, multiple cytokines are produced. But when a person is traveling at high altitude and needs additional erythrocytes, **erythropoietin**—a cytokine that acts only on erythrocyte precursors—is produced. Erythropoietin activates several different intracellular signal-transduction pathways, leading to changes in gene expression that promote formation of erythrocytes (see Figure 16-6). In contrast GM-CSF, a different cytokine, stimulates production of granulocytes, macrophages, eosinophils, and megakaryocytes.

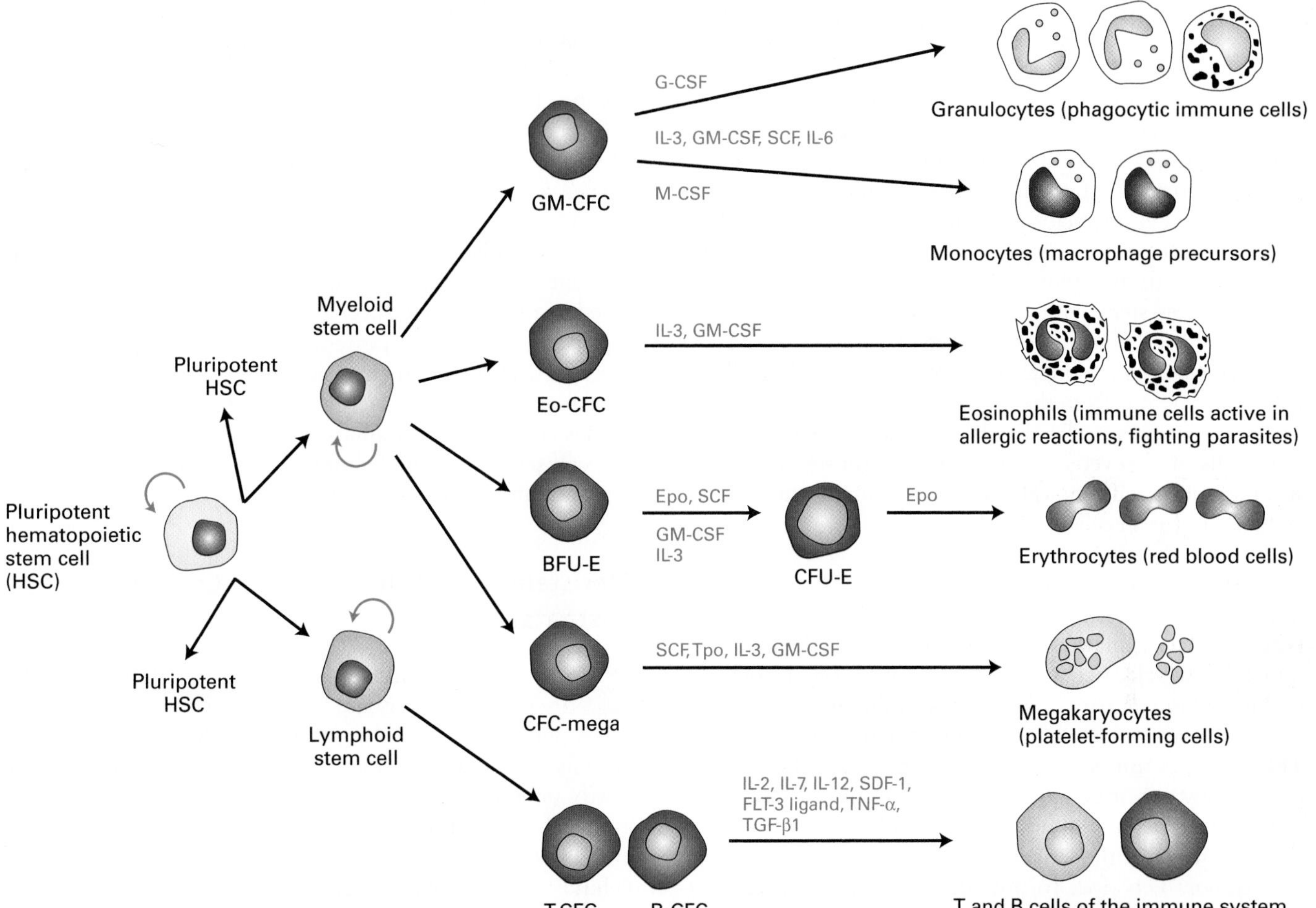

▲ **FIGURE 21-15 Formation of blood cells from hematopoietic stem cells in the bone marrow.** Pluripotent stem cells (yellow) may divide symmetrically to self-renew (blue curved arrow) or divide asymmetrically to form a myeloid or lymphoid stem cell (light green) and a daughter cell that is pluripotent like the parental cell. Although myeloid and lymphoid stem cells are capable of self-renewal, each type is committed to one of the two major hematopoietic lineages. Depending on the types and amounts of cytokines present, the myeloid and lymphoid stem cells generate different precursor cells (dark green), which are incapable of self-renewal. Precursor cells are called colony-forming cells (CFCs) because of their ability to form colonies containing the differentiated cell types shown at right. The colonies are detected on the spleen of animals that have had their own hematopoietic cells eliminated and the precursor cells introduced. Further cytokine-induced proliferation, commitment, and differentiation of the precursor cells give rise to the various types of blood cells. Some of the cytokines that support this process are indicated (red labels). GM = granulocyte-macrophage; Eo = eosinophil; E = erythrocyte; mega = megakaryocyte; T = T-cell; B = B-cell; CFU = colony-forming unit; CSF = colony-stimulating factor; IL = interleukin; SCF = stem cell factor; Epo = erythropoietin; Tpo = thrombopoietin; TNF = tumor necrosis factor; TGF = transforming growth factor; SDF = stromal cell–derived factor; FLT-3 ligand = ligand for fms-like tyrosine kinase receptor 3. [Adapted from M. Socolovsky et al., 1998, *Proc. Nat'l. Acad. Sci. USA* **95:**6573.]

The hematopoietic lineage originally was worked out by injecting the various types of precursor cells into mice whose precursor cells had been wiped out by irradiation. By observing which blood cells were restored in these transplant experiments, researchers could infer which precursors (e.g., GM-CFC, BFU-E) or terminally differentiated cells (e.g., erythrocytes, monocytes) arise from a particular type of precursor. Remarkably, a single hematopoietic stem cell is sufficient to restore the entire blood system of an irradiated mouse. The first step in these experiments was to separate the different types of hematopoietic precursors. This separation is possible because each type of precursor produces unique combinations of cell-surface proteins that can serve as type-specific markers. If bone marrow extracts are treated with fluorochrome-labeled antibodies for these markers, cells with different surface markers can be separated in a fluorescence-activated cell sorter (see Figure 9-28).

The frequency of hematopoietic stem cells is about 1 cell per 10^4 bone marrow cells. Activation of the *Hoxb4* gene in embryonic stem cells drives the formation of hematopoietic stem cells. (As described in Chapter 22, *Hoxb4* also plays a role in pattern formation along the head-to-tail body axis.) The *Bmi* gene is also required for self-renewal of hematopoietic stem cells, and of neural stem cells as well. This gene encodes a polycomb-type chromatin regulator protein that represses certain genes including some **Hox genes**, a group of developmentally important genes discussed in Chapter 22. Bmi1 is a component of the PRC1 protein complex that we discussed above in relation to embryonic stem cells. Thus members of the polycomb group of proteins are important in both embryonic and adult stem cells.

Like other stem cells, hematopoietic stem cells are residents of a niche. The niche is formed by spindle-shaped cells on the surface of the bone in the bone marrow. N-cadherin fastens the stem cells to these niche cells. A Delta-like ligand produced by niche cells signals to Notch receptors on the stem cells, and several other growth factor–receptor pairs stimulate their self-renewal and differentiation into myeloid or lymphoid stem cells.

Stem cells can become cancerous. For example, leukemia is a cancer of the white blood cells. This type of cancer is marked by two types of cells: leukemic tumor cells, which arise from differentiated white blood cells and have limited growth abilities, and the more dangerous leukemic tumor stem cells with unlimited growth abilities. The leukemia tumor stem cells, which are capable of initiating a new tumor on their own, are present in a human tumor about once for every million dividing leukemic cells. Thus the bulk of the leukemic cells are not able to grow a new tumor. Therefore to provide an enduring cure, treatments must ensure the death or limited mitosis of the tumor stem cells. This is particularly challenging because many cancer stem cells divide slowly or not at all for substantial periods, making them selectively *resistant* to chemotherapy drugs and irradiation—both treatments that target rapidly dividing cells.

To date, bone marrow transplants, a treatment for leukemia and other blood disorders, represent the most successful and widespread use of stem cells in medicine. In 1959 a patient with end-stage (fatal) leukemia was irradiated to destroy her cancer cells. She was transfused with bone marrow cells from her identical twin sister, thus avoiding an immune response, and was in remission for three months. This promising beginning led to present-day treatments that can often lead to a complete cure for leukemia.

The stem cells in transplanted bone marrow can generate new, functional blood cells in patients with certain hereditary blood diseases and in cancer patients who have received irradiation and/or chemotherapy. Both chemotherapy and irradiation destroy the bone marrow cells as well as cancer cells. Bone marrow transplants go beyond eliminating cancer cells as done with irradiation. Instead, an immune attack on the leukemic cells can be mounted by the injected donor cells. More than two dozen diseases are now commonly treated with bone marrow transplantation. These include leukemias and anemias of various types, lymphomas, severe combined immunodeficiency, and certain autoimmune disorders. The effectiveness of bone marrow transplants varies among diseases and patients from minor improvement to full cures. Much research is under way to employ other types of stem cells in the treatment of diseases involving nonblood tissues. ▮

You have probably noticed that all the molecular regulators of stem cells are familiar signal proteins rather than dedicated regulators that specialize in stem-cell control. Each type of signal is used repeatedly to control cell fates and proliferation. These are ancient signaling systems, at least a half billion years old, for which new uses have emerged as cells, tissues, organs, and animals have evolved new variations. We will see many more examples of such evolutionary conservation in the next chapter on development.

Meristems Are Niches for Stem Cells in Postnatal Plants

Stem cells in plants are located in **meristems,** populations of undifferentiated cells found at the tips of growing shoots. Shoot apical meristems (SAMs) produce leaves and shoots as well as more stem cells that constitute the nearly immortal meristems. Meristems can persist for thousands of years in long-lived species such as redwood trees and bristlecone pines. As a plant grows, the cells "left behind" the meristems are encased in rigid cell walls and can no longer grow. SAMs can split to form branches, each branch with its own SAM, or be converted into floral meristems (Figure 21-16). Floral meristems give rise to the four floral organs—sepals, stamens, carpels, and petals—that form flowers. Unlike SAMs, floral meristems are gradually depleted as they give rise to the floral organs.

A meristem is a stem-cell niche, but much remains to be learned about how the niche is created and maintained. Numerous genes have been found to regulate the formation, maintenance, and properties of meristems. Many of these genes encode transcription factors that direct progeny of stem cells down different paths of differentiation. For instance, a hierarchy of regulators, particularly transcription factors, controls the separation of differentiating cells from SAMs as leaves form; similarly, three types of regulators control formation of the floral organs from floral meristems (see Figure 22-36). In both cases, a cascade of gene interactions occurs, with earlier transcription factors causing production of later ones. At the same time, cells are dividing and the differentiating ones are spreading away from their original birth sites. One signal that creates the plant niche is Zwille/Pinhead, which encodes a protein related to the Piwi protein that supports stem-cell niches in animals (see Figure 21-8). These are "argonaute" family proteins, which repress genes in response to certain small RNA molecules. ▮

(a) Regions of shoot apical and floral meristems

(b) Fates of cells in L2 layer

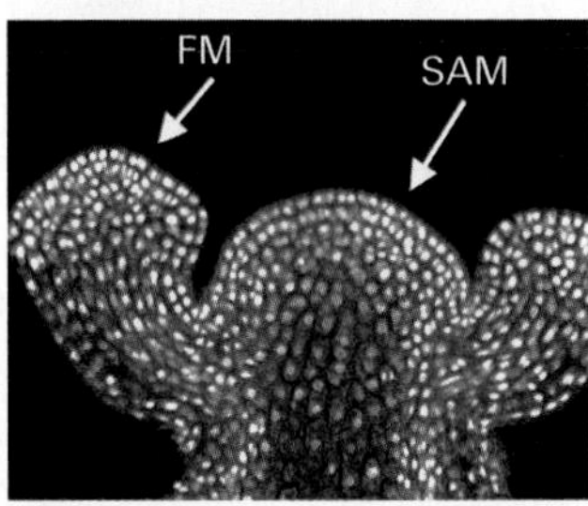

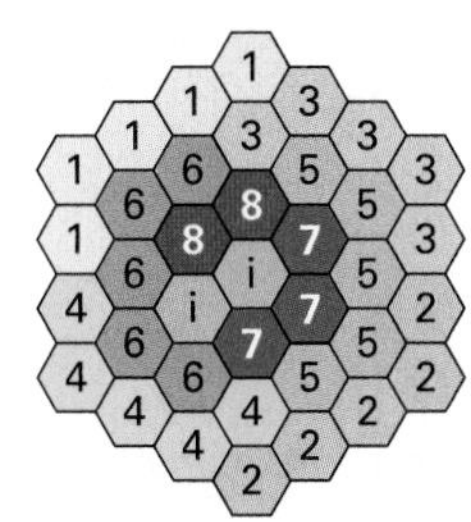

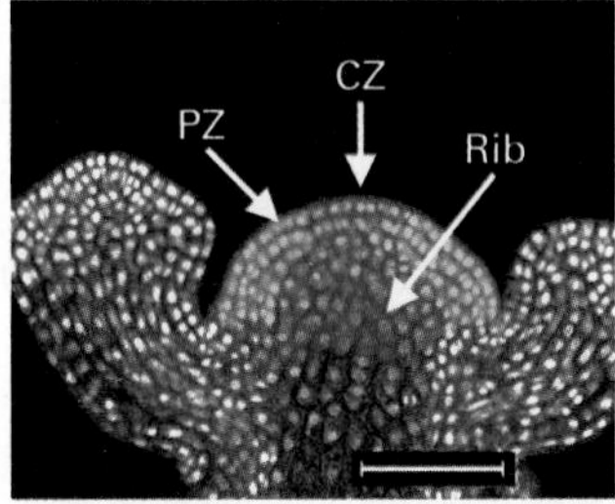

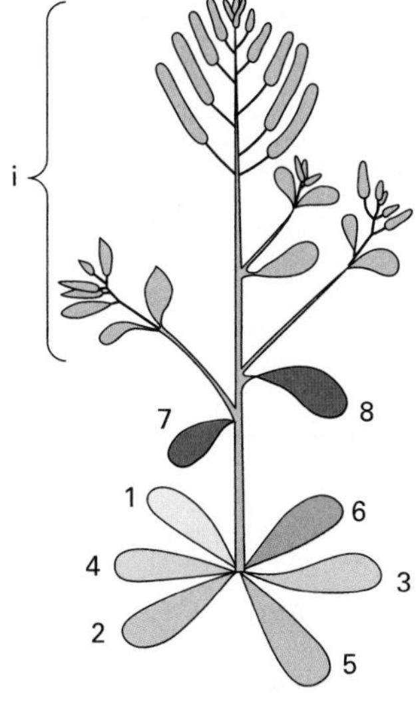

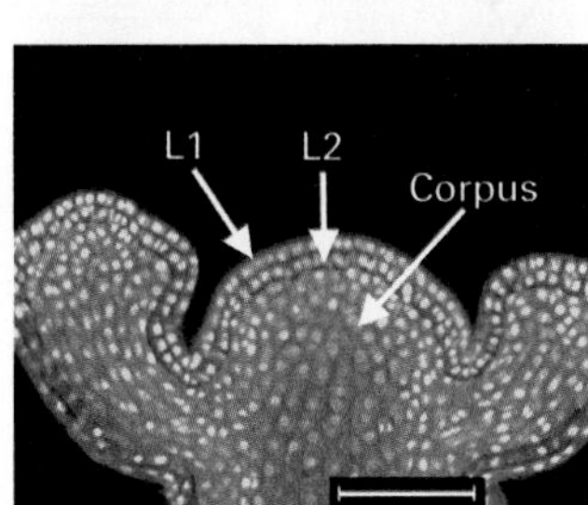

▲ **FIGURE 21-16 Cell fates in meristems of *Arabidopsis*.** (a) In these longitudinal sections, cell nuclei are revealed by staining with propidium iodide, which binds to DNA. *Top:* The shoot apical meristem (SAM) produces shoots, leaves, and more meristem. Flower production occurs when the meristem switches from leaf/shoot production to flower production, concomitant with an increase in the number of meristem cells to form floral meristems (FMs), as shown here. *Middle:* Cells in a SAM exhibit different fates and behaviors. Cells divide rapidly in the peripheral zone (PZ, green) to produce leaves and in the rib zone (Rib, blue) to produce central shoot structures. Cells in the central zone (CZ, red) divide more slowly, producing an ongoing source of meristem and contributing cells to the PZ and Rib. *Bottom:* The layers of the meristem, colored blue, pink and yellow here, are each derived (cloned) from the same precursor cell. Scale bars, 50 μm. (b) Color coding shows the fates of cells in different positions in the L2 layer. The color code does not correspond to that in part (a). [Part (a) from E. Meyerowitz, 1997, *Cell* **88**:299; micrographs courtesy of Elliot Meyerowitz. Part (b) after C. Wolpert et al., 2002, *Principles of Development,* 2d ed., Oxford University Press.]

KEY CONCEPTS OF SECTION 21.1

The Birth of Cells: Stem Cells, Niches, and Lineage

- In asymmetric cell division, two different types of daughter cells are formed from one mother cell. In contrast, both daughter cells formed in symmetric divisions are identical but may have different fates if they are exposed to different external signals (see Figure 21-1).
- Pluripotent stem cells can produce more than one type of descendant cell, including in some cases a stem cell with a more-restricted potential to produce differentiated cell types.
- Embryonic development of *C. elegans* begins with asymmetric division of the fertilized egg (zygote). The lineage of all the cells in adult worms is known and is highly reproducible (see Figure 21-5).
- Short regulatory RNAs (micro RNAs) control the timing of developmental cell divisions by preventing translation of mRNAs whose encoded proteins control cell lineages (see Figure 21-6).
- Cultured embryonic stem cells (ES cells) are capable of giving rise to many kinds of differentiated cell types. They are useful in production of genetically altered mice and offer potential for therapeutic uses. Specific transcription factors and chromatin regulators are important in giving ES cells their unusual properties.
- Stem cells are formed in niches that provide signals to maintain a population of nondifferentiating stem cells. The niche must maintain stem cells without allowing their excess proliferation and must block differentiation.
- Germ-line cells give rise to eggs or sperm. By definition, all other cells are somatic cells.
- Populations of stem cells associated with the gonads, skin, intestinal epithelium and most other tissues regenerate differentiated tissue cells that are damaged or sloughed or become aged (see Figures 21-8–21-11, 21-14).
- In the blood cell lineage, different precursor types form and proliferate under the control of distinct cytokines (see Figure 21-15). This allows the body to specifically induce the replenishment of some, or all, of the necessary types.
- Stem cells are prevented from differentiating by specific controls that operate in the niche. A high level of β-catenin, a component of the Wnt signaling pathway, has been implicated in preserving stem cells in the skin and intestine by directing cells toward division rather than differentiation states.
- Plant stem cells persist for the life of the plant in the meristem. Meristem cells can give rise to a broad spectrum of cell types and structures (see Figure 21-16).

21.2 Cell-Type Specification in Yeast

In the previous section, we saw that stem cells and precursor/progenitor cells produce progeny that embark upon specific differentiation paths. The elegant regulatory mechanisms of differentiation are referred to as *cell-type specification.* Specification usually involves a combination of external signals with internal signal-transduction mechanisms like those described in Chapters 15 and 16. The transition from an undifferentiated cell to a differentiating one often involves the production of one or a small number of transcription factors. The newly produced transcription factors are powerful switches that trigger the activation (and sometimes repression) of large batteries of subservient genes. Thus an initially modest change can cause massive changes in gene expression that confer a new character on the cell.

Our first example of cell-type specification comes from the budding yeast, *S. cerevisiae.* We introduced this useful unicellular eukaryote way back in Chapter 1 and have encountered it in several other chapters. *S. cerevisiae* forms three cell types: haploid **a** and α cells, and diploid **a**/α cells. Each type has its own distinctive set of active genes; many other genes are active in all three cell types. In a pattern common to many organisms and tissues, cell-type specification in yeast is controlled by a small number of transcription factors that coordinate the activities of many other genes. Similar regulatory features are found in the responses of higher eukaryotic cells to environmental signals and in the specification and patterning of cells and tissues during development (Chapter 22).

DNA microarray studies have provided a genome-wide picture of the fluctuations of gene expression in the different cell types and different stages of the *S. cerevisiae* life cycle (see Figure 5-29 for an explanation of the DNA microarray technique). Among other things, these studies identified 32 genes that are transcribed at more than twofold higher levels in α cells than in **a** cells. Another 50 genes are transcribed at more than twofold higher levels in **a** cells than in α cells. The products of these 82 genes, which initially are activated by cell-type specification transcription regulators, convey many of the critical differences between the two cell types. The results clearly demonstrate that changes in expression of only a small fraction of the genome, in this case <2 percent of the ≈6000 yeast genes, can significantly alter the behavior and properties of cells. Transcription of a much larger number of genes, about 25 percent of the total assayed, differed substantially in diploid cells compared with haploid cells. These differences in expression patterns make sense since **a** and α cells are very similar (hence, expression of relatively few genes differ between them), whereas haploid and diploid cells are quite different.

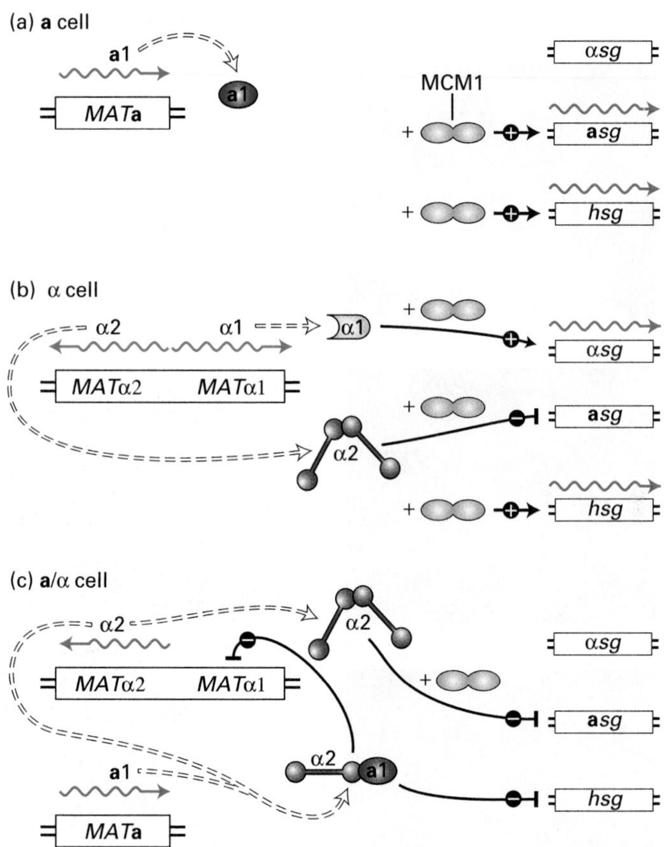

▲ **FIGURE 21-17 Transcriptional control of cell type–specific genes in *S. cerevisiae*.** The coding sequences carried at the *MAT* locus differ in haploid α and **a** cells and in diploid cells. Three type-specific transcription factors (α1, α2, and **a**1) encoded at the *MAT* locus act with MCM1, a constitutive transcription factor produced by all three cell types, to produce a distinctive pattern of gene expression in each of the three cell types. **a***sg* = **a**-specific genes/mRNAs; α*sg* = α-specific genes/mRNAs; *hsg* = haploid-specific genes/mRNAs.

Mating-Type Transcription Factors Specify Cell Types

Each of the three *S. cerevisiae* cell types expresses a unique set of regulatory genes that is responsible for all the differences among the three cell types. All haploid cells express certain haploid-specific genes; in addition, **a** cells express **a**-specific genes, and α cells express α-specific genes. In **a**/α diploid cells, diploid-specific genes are expressed, whereas haploid-specific, **a**-specific, and α-specific genes are not. As illustrated in Figure 21-17, three cell type–specific transcription factors (α1, α2, and **a**1) encoded at the *MAT locus* act in combination with a general transcription factor called *MCM1,* which is expressed in all three cell types, to mediate cell type–specific gene expression in *S. cerevisiae.* Thus the actions of just three transcription factors can set the yeast cell on a specific differentiation pathway culminating in a particular cell type. From the DNA microarray experiments we know one effect of these key players: the activation or repression of many dozens of genes that control cell characteristics.

MCM1 was the first member of the *MADS family* of transcription factors to be discovered. (MADS is an acronym for the initial four factors identified in this family.) The DNA-binding proteins composing this family dimerize and contain a similar N-terminal MADS domain. In Section 21.3 we will encounter other MADS transcription factors that participate in development of skeletal muscle. MADS transcription factors also specify cell types in floral organs (see Figure 22-36). Acting alone, MCM1 activates transcription of **a**-specific genes in **a** cells and of haploid-specific genes in both α and **a** cells (see Figure 21-17a, b). In haploid α cells,

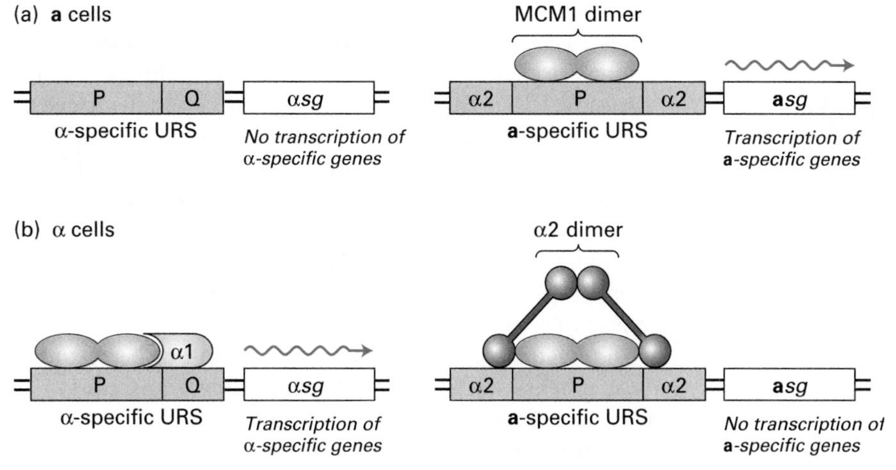

◀ **FIGURE 21-18 Activity of MCM1 in a and α yeast cells.** MCM1 binds as a dimer to the P site in α-specific and **a**-specific upstream regulatory sequences (URSs), which control transcription of α-specific genes and **a**-specific genes, respectively. (a) In **a** cells, MCM1 stimulates transcription of **a**-specific genes. MCM1 does not bind efficiently to the P site in α-specific URSs in the absence of α1 protein. (b) In α cells, the activity of MCM1 is modified by its association with α1 or α2. The α1-MCM1 complex stimulates transcription of α-specific genes, whereas the α2-MCM1 complex blocks transcription of **a**-specific genes. The α2-MCM1 complex also is produced in diploid cells, where it has the same blocking effect on transcription of **a**-specific genes (see Figure 21-17c).

however, the activity of MCM1 also is determined by its association with the α1 or α2 transcription factor. As a result of this combinatorial action, MCM1 promotes transcription of α-specific genes and represses transcription of **a**-specific genes in α cells. Now let's take a closer look at how MCM1 and the *MAT*-encoded proteins exert their effects.

MCM1 and α1–MCM1 Complexes Activate Gene Transcription

In **a** cells, homodimeric MCM1 binds to the so-called P box sequence in the upstream regulatory sequences (URSs) of **a**-specific genes, stimulating their transcription (Figure 21-18a). Transcription of α-specific genes is controlled by two adjacent sequences—the P box and the Q box—located in the URSs associated with these genes. Although MCM1 alone binds to the P box in **a**-specific URSs, it does not bind to the P box in α-specific URSs. Thus **a** cells do not transcribe α-specific genes.

In α cells, which produce the α1 transcription factor encoded by *MAT*α, the simultaneous binding of MCM1 and α1 to PQ sites occurs with high affinity (Figure 21-18b). This binding turns on transcription of α-specific genes. Therefore, **a**-specific transcription is a simple matter of a single transcription factor binding to its target genes, while α-specific transcription requires a combination of two factors—neither of which can activate target genes alone.

α2–MCM1 and α2–a1 Complexes Repress Transcription

Highly specific binding occurs as a consequence of the interaction of α2 with other transcription factors at different sites in DNA. Flanking the P box in each **a**-specific URS are two α2-binding sites. Both MCM1 and α2 can bind independently to an **a**-specific URS with relatively low affinity. In α cells, however, highly cooperative, simultaneous binding of both α2 and MCM1 proteins to these sites occurs with high affinity. This high-affinity binding represses transcription of **a**-specific genes, ensuring that they are not expressed in α cells and diploid cells (see Figure 21-18b, *right*). MCM1 promotes binding of α2 to an **a**-specific URS by orienting the two DNA-binding domains of the α2 dimer to the α2-binding sequences in this URS. Since a dimeric α2 molecule binds to both sites in an α-specific URS, each DNA site is referred to as a half-site. The relative positions of both half-sites and their orientation are highly conserved among different **a**-specific URSs.

Combinations of transcription factors create additional specificity in gene regulation. The presence of numerous α2-binding sites in the genome and the "relaxed" specificity of α2 protein may expand the range of genes that it can regulate. For instance, in **a**/α diploid cells, α2 forms a heterodimer with **a**1 that represses both haploid-specific genes and the gene encoding α1 (see Figure 21-17c). The example of α2 suggests that relaxed specificity may be a general strategy for increasing the regulatory range of a single transcription factor.

Pheromones Induce Mating of α and a Cells to Generate a Third Cell Type

An important feature of the yeast life cycle is the ability of haploid **a** and α cells to mate, that is, attach and fuse giving rise to a diploid **a**/α cell (see Figure 1-6). Each haploid cell type secretes a different *mating factor,* a small polypeptide **pheromone,** and expresses a cell-surface G protein–coupled receptor that recognizes the pheromone secreted by cells of the other type. Thus **a** and α cells both secrete and respond to pheromones (Figure 21-19). Binding of the mating factors to their receptors induces expression of a set of genes encoding proteins that direct arrest of the cell cycle in G_1 and promote attachment/fusion of haploid cells to form diploid cells. In the presence of sufficient nutrients, the diploid cells will continue to grow. Starvation, however, induces diploid cells to progress through meiosis, each yielding four haploid spores. If the environmental conditions become conducive to vegetative growth, the spores will germinate and undergo mitotic division.

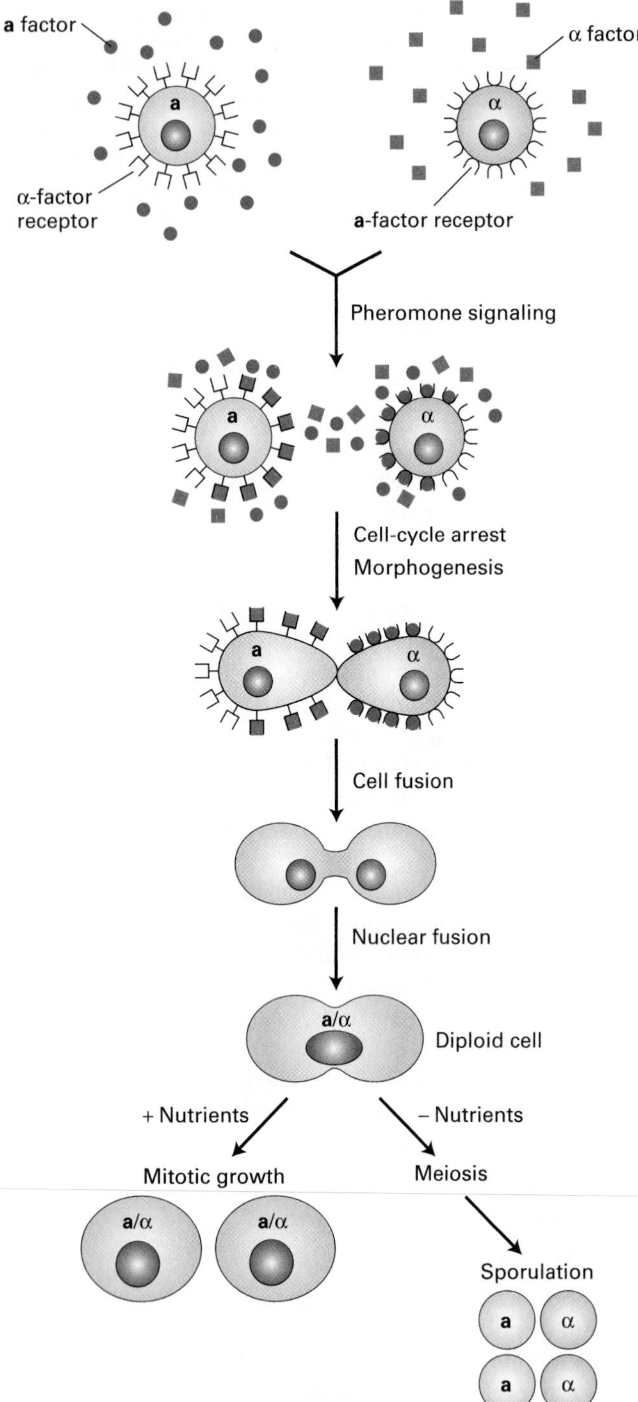

▲ **FIGURE 21-19 Pheromone-induced mating of haploid yeast cells.** The α cells produce α mating factor and **a**-factor receptor; the **a** cells produce **a** factor and α-factor receptor. Binding of the mating factors to their cognate receptors on cells of the opposite type leads to gene activation, resulting in mating and production of diploid cells. In the presence of sufficient nutrients, these cells will grow as diploids. Without sufficient nutrients, cells will undergo meiosis and form four haploid spores.

Studies with yeast mutants have provided insights into how the **a** and α pheromones induce mating. For instance, haploid yeast cells carrying mutations in the *sterile 12 (STE12)* locus cannot respond to pheromones and do not mate. The *STE12* gene encodes a transcription factor that binds to a DNA sequence referred to as the pheromone-responsive element, which is present in many different **a**- and α-specific URSs. Binding of mating factors to cell-surface receptors induces a cascade of signaling events, resulting in phosphorylation of various proteins including the Ste12 protein (see Figure 16-28a). This rapid phosphorylation is correlated with an increase in the ability of Ste12 to stimulate transcription. It is not yet known, however, whether Ste12 must be phosphorylated to stimulate transcription in response to pheromone.

Interaction of Ste12 protein with DNA has been studied most extensively at the URS controlling transcription of *STE2,* an **a**-specific gene encoding the receptor for the α pheromone. Pheromone-induced production of the α receptor encoded by *STE2* increases the efficiency of the mating process. Adjacent to the **a**-specific URS in the *STE2* gene is a pheromone-responsive element that binds Ste12. When **a** cells are treated with α pheromone, transcription of the *STE2* gene increases in a process that requires Ste12 protein. Ste12 protein has been found to bind most efficiently to the pheromone-responsive element in the *STE2* URS when MCM1 is simultaneously bound to the adjacent P site. We saw previously that MCM1 can act as an activator or a repressor at different URSs depending on whether it complexes with α1 or α2. In this case, the function of MCM1 as an activator is stimulated by the binding of yet another transcription factor, Ste12, whose activity is modified by extracellular signals.

KEY CONCEPTS OF SECTION 21.2

Cell-Type Specification in Yeast

- Specification of each of the three yeast cell types—the **a** and α haploid cells and the diploid **a**/α cells—is determined by a unique set of transcription factors acting in different combinations at specific regulatory sites in the yeast genome (see Figure 21-17).
- Some transcription factors can act as repressors or activators depending on the specific regulatory sites they bind and the presence or absence of other transcription factors bound to neighboring sites.
- Binding of mating-type pheromones by haploid yeast cells activates expression of genes encoding proteins that mediate mating, thereby generating the third yeast cell type (see Figure 21-19).

21.3 Specification and Differentiation of Muscle

An impressive array of molecular strategies, some analogous to those found in yeast cell-type specification, have evolved to carry out the complex developmental pathways that characterize multicellular organisms. Muscle cells have been the focus of many such studies because their development can be studied in cultured cells as well as in intact animals. Early advances in understanding the formation of muscle cells (myogenesis) came from discovering regulatory genes that could convert cultured cells into muscle cells. Then mouse mutations affecting those genes were created and studied to learn the functions of the proteins encoded by these genes, following which scientists have investigated how the muscle regulatory genes control other genes.

Recent microarray studies have looked for genes whose transcription differs in various subtypes of muscle in mice. These studies have identified 49 genes out of the 3000 genes examined that are transcribed at substantially different levels in red (endurance) muscle and white (fast response) muscle. Clues to the molecular basis of the functional differences between red and white muscle are likely to come from studying those 49 genes and their products.

Here we examine the role of certain transcription factors in creating skeletal muscle in vertebrates. These muscle regulators illustrate how coordinated transcription of sets of target genes can produce differentiated cell types and how a cascade of transcriptional events and signals is necessary to coordinate cell behaviors and functions.

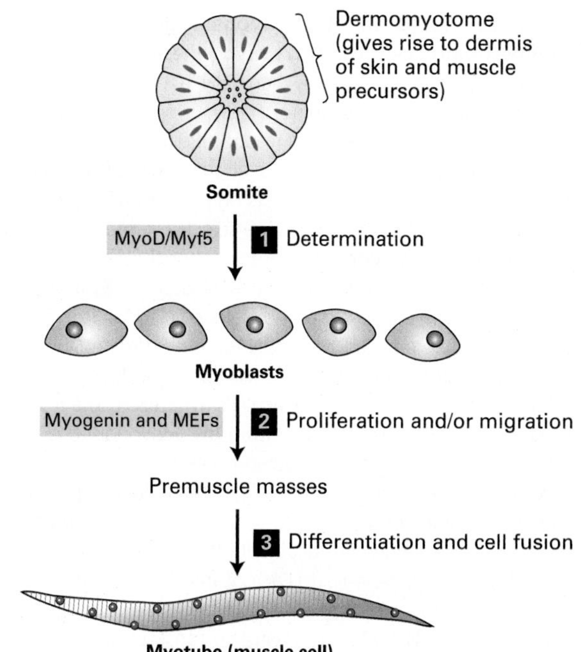

▲ **FIGURE 21-20 Three stages in development of vertebrate skeletal muscle.** Somites are epithelial spheres of embryonic mesoderm cells, some of which (the myotome) become determined as myoblasts after receiving signals from other tissues (**1**). After the myoblasts proliferate and migrate to the limb buds and elsewhere (**2**), they undergo terminal differentiation into multinucleate skeletal muscle cells, called myotubes (**3**). Key transcription factors that help drive the myogenic program are highlighted in yellow. See also Figure 21-23.

Embryonic Somites Give Rise to Myoblasts

Vertebrate skeletal myogenesis proceeds through three stages: **determination** of the precursor muscle cells, called *myoblasts,* which commits them to a muscle cell fate; proliferation and in some cases migration of myoblasts; and their terminal differentiation into mature muscle (Figure 21-20). In the first stage, myoblasts arise from blocks of mesoderm cells, called *somites,* that are located next to the neural tube in the embryo. Specific signals from surrounding tissue play an important role in determining where myoblasts will form in the developing somite. At the molecular level, the "decision" of a mesoderm cell to adopt a muscle cell fate reflects the activation of genes encoding particular transcription factors.

As myoblasts proliferate and migrate, say, to a developing limb bud, they become aligned, stop dividing, and fuse to form a **syncytium** (a cell containing many nuclei but sharing a common cytoplasm). We refer to this multinucleate cell as a *myotube.* Concomitant with cell fusion is a dramatic rise in the expression of genes necessary for further muscle development and function.

The specific extracellular signals that induce determination of each group of myoblasts are expressed only transiently. These signals trigger production of intracellular factors that maintain the myogenic program after the inducing signals are gone. We discuss the identification and functions of these myogenic proteins, and their interactions, in the next several sections.

Myogenic Genes Were First Identified in Studies with Cultured Fibroblasts

Myogenic genes are a fine example of how transcription factors control the progressive differentiation that occurs in a cell lineage. In vitro studies with the fibroblast cell line designated C3H 10T½ have played a central role in dissecting the transcription control mechanisms regulating skeletal myogenesis. When incubated in the presence of 5-azacytidine (a cytidine derivative that cannot be methylated), C3H 10T½ cells differentiate into myotubes. Upon entry into cells, 5-azacytidine is converted to 5-azadeoxycytidine triphosphate and then is incorporated into DNA in place of deoxycytidine. Because methylated deoxycytidine residues commonly are present in transcriptionally inactive DNA regions, replacement of cytidine residues with a derivative that cannot be methylated may permit activation of genes previously repressed by methylation.

The high frequency at which azacytidine-treated C3H 10T½ cells are converted into myotubes suggested to early workers that reactivation of one or a small number of closely linked genes is sufficient to drive a myogenic program. To test this hypothesis, researchers isolated DNA from C3H 10T½ cells grown in the presence of 5-azacytidine and transfected it into untreated cells. The observation that 1 in 10^4 cells transfected with this DNA was converted into a myotube is consistent with the hypothesis that one or a small set of closely linked genes is responsible for converting fibroblasts into myotubes.

Subsequent studies led to the isolation and characterization of four different but related genes that can convert

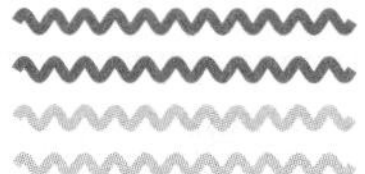

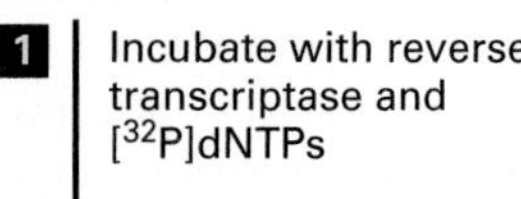

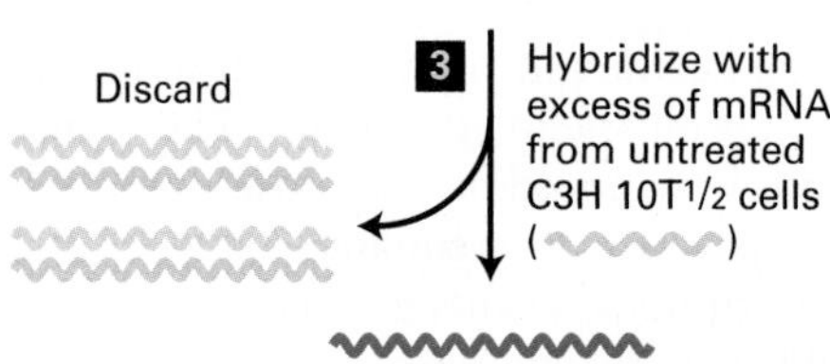

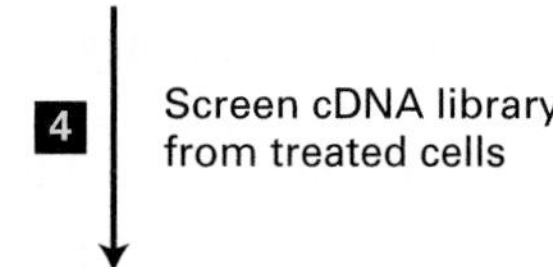

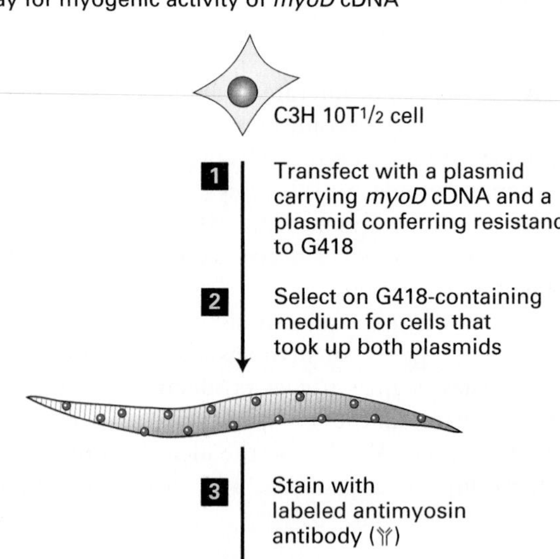

2 Remove mRNAs

2 Select on G418-containing medium for cells that took up both plasmids
3 Stain with labeled antimyosin antibody

▶ EXPERIMENTAL FIGURE 21-21 **Myogenic genes isolated from azacytidine-treated cells can drive myogenesis when transfected into other cells.** (a) When C3H 10T½ cells (a fibroblast cell line) are treated with azacytidine, they develop into myotubes at high frequency. To isolate the genes responsible for converting azacytidine-treated cells into myotubes, all the mRNAs from treated cells first were isolated from cell extracts on an oligo-dT column. Because of their poly(A) tails, mRNAs are selectively retained on this column. The red mRNAs represent molecules enriched in azacytidine-treated cells; the pink ones are all other mRNAs. Steps **1** and **2**: The isolated mRNAs were converted to radiolabeled cDNAs. Step **3**: When the cDNAs were mixed with mRNAs from *untreated* C3H 10T½ cells, only cDNAs derived from mRNAs (light red) produced by both azacytidine-treated cells and untreated cells hybridized. The resulting double-stranded DNA was separated from the unhybridized cDNAs (dark blue) produced only by azacytidine-treated cells. Step **4**: The cDNAs specific for azacytidine-treated cells then were used as probes to screen a cDNA library from azacytidine-treated cells (see Figure 5-16). At least some of the clones identified with these probes correspond to genes required for myogenesis. (b) Each of the cDNA clones identified in part (a) was incorporated into a plasmid carrying a strong promoter. Steps **1** and **2**: C3H 10T½ cells were cotransfected with each recombinant plasmid plus a second plasmid carrying a gene conferring resistance to an antibiotic called G418; only cells that have incorporated the plasmids will grow on a medium containing G418. One of the selected clones, designated *myoD,* was shown to drive conversion of C3H 10T½ cells into muscle cells, identified by their binding of labeled antibodies against myosin, a muscle-specific protein (step **3**). [See R. L. Davis et al., 1987, *Cell* **51:**987.]

C3H 10T½ cells into muscle. Figure 21-21 outlines the experimental protocol for identifying and assaying one of these genes, called the *myogenic determination (myoD)* gene. C3H 10T½ cells transfected with *myoD* cDNA and those treated with 5-azacytidine both formed myotubes. The *myoD* cDNA also was able to convert a number of other cultured cell lines into muscle. Based on these findings, the *myoD* gene was proposed to play a key role in muscle development. A similar approach identified three other genes—*myogenin, myf5,* and *mrf4*—that also function in muscle development.

Two Classes of Regulatory Factors Act in Concert to Guide Production of Muscle Cells

The four myogenic proteins—MyoD, Myf5, myogenin, and MRF4—are all members of the **basic helix-loop-helix** (bHLH) family of DNA-binding transcription factors (see Figure 7-26). Near the center of these proteins is a basic DNA-binding (B) region adjacent to the HLH domain, which mediates dimer formation. Flanking this central DNA- binding/dimerization region are two activation domains. We refer to the four myogenic bHLH proteins collectively as *muscle regulatory factors,* or *MRFs* (Figure 21-22a).

bHLH proteins form homo- and heterodimers that bind to a 6-bp DNA site with the consensus sequence CANNTG (N = any nucleotide). Referred to as the E box, this sequence is present in many different locations within the genome (on a purely random basis the E box will be found every 256 nucleotides). Thus some mechanism(s) must ensure that MRFs

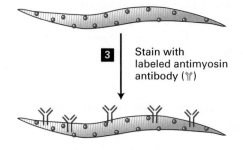

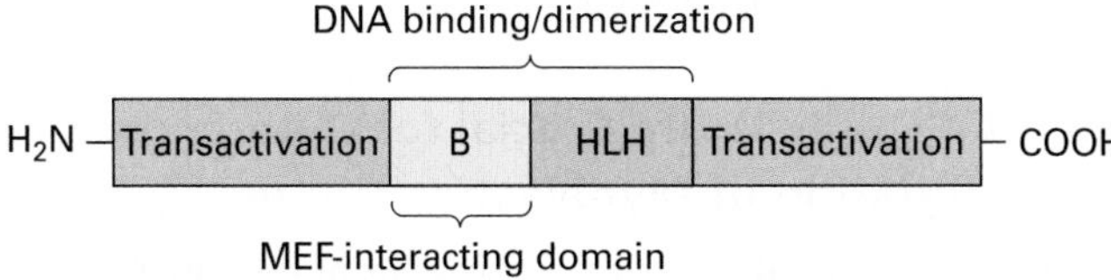

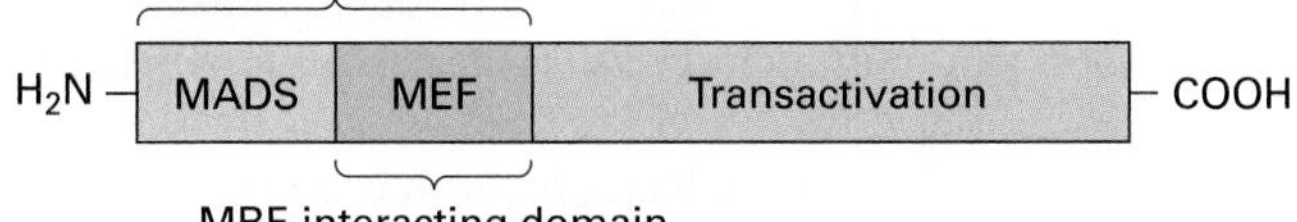

▲ **FIGURE 21-22 General structures of two classes of transcription factors that participate in myogenesis.** MRFs (muscle regulatory factors), such as myogenin and MyoD, are bHLH (basic helix-loop-helix) proteins produced only in developing muscle. MEFs (myocyte-enhancing factors), which are produced in several tissues in addition to developing muscle, belong to the MADS family. The myogenic activity of MRFs is enhanced by their interaction with MEFs. The domain structures of the proteins are shown, including transactivation, basic (B), helix-loop-helix (HLH), MADS and MEF domains.

specifically regulate muscle-specific genes and not other genes containing E boxes in their transcription-control regions. One clue to how this myogenic specificity is achieved was the finding that the DNA-binding affinity of MyoD is tenfold greater when it binds as a heterodimer complexed with E2A, another bHLH protein, than when it binds as a homodimer. Moreover, in azacytidine-treated C3H 10T½ cells, MyoD is found as a heterodimer complexed with E2A, and both proteins are required for myogenesis in these cells. The DNA-binding domains of E2A and MyoD have similar but not identical amino acid sequences, and both proteins recognize E box sequences. The other MRFs also form heterodimers with E2A that have properties similar to MyoD–E2A complexes. This heterodimerization restricts activity of the myogenic transcription factors to genes that contain at least two E boxes located close to each other.

Since E2A is expressed in many tissues, the requirement for E2A is not sufficient to confer myogenic specificity. Subsequent studies suggested that specific amino acids in the bHLH domain of all the MRFs confer myogenic specificity by allowing MRF–E2A complexes to bind specifically to another family of DNA-binding proteins called *myocyte enhancing factors,* or *MEFs.* MEFs were considered excellent candidates for interaction with MRFs for two reasons. First, many muscle-specific genes contain recognition sites for both MEFs and MRFs. Second, although MEFs cannot induce myogenic conversion of azacytidine-treated C3H 10T½ cells by themselves, they enhance the ability of MRFs to do so. This enhancement requires physical interaction between a MEF and MRF–E2A heterodimer. MEFs belong to the MADS family of transcription factors and contain a MEF domain, adjacent to the MADS domain, that mediates interaction with myogenin (Figure 21-22b). The synergistic action of the MEF homodimer and MRF–E2A heterodimer is thought to drive high-level expression of muscle-specific genes.

Knockout mice and *Drosophila* mutants have been used to explore the roles of MRF and MEF proteins in conferring myogenic specificity in intact animals, extending the work in cell culture. These experiments demonstrated the importance of three of the MRF proteins (MyoD, Myf5, and myogenin) and of MEF proteins for distinct steps in muscle development (see Figure 21-20). The function of the fourth myogenic protein, Mrf4, is not entirely clear.

Differentiation of Myoblasts Is Under Positive and Negative Control

Powerful developmental regulators like the MRFs cannot be allowed to run rampant. In fact, their actions are circumscribed at several levels. First, production of the muscle regulators is activated only in mesoderm cells in response to locally acting signals, such as Hedgehog, Wnt, and BMP, that are produced at the right time and place in the embryo. Other proteins mediate additional mechanisms for assuring tight control over myogenesis: chromatin-remodeling proteins are needed to make target genes accessible to MRFs; inhibitory proteins can restrict when MRFs act; and antagonistic relations between cell-cycle regulators and differentiation factors like MRFs ensure that differentiating cells will not divide, and vice versa. All these factors control when and where muscles form.

Activating Chromatin-Remodeling Proteins MRF proteins control batteries of muscle-specific genes, but can do so only if chromatin factors allow access. Remodeling of chromatin, which usually is necessary for gene activation, is carried out by large protein complexes (e.g., SWI/SNF complex) that have ATPase and perhaps helicase activity. These complexes recruit histone acetylases that modify chromatin to make genes accessible to transcription factors (Chapter 7). The hypothesis that remodeling complexes help myogenic factors was tested using dominant-negative versions of the ATPase proteins that form the cores of these complexes. (Recall from Chapter 5 that a dominant-negative mutation produces a mutant phenotype even when a normal allele of the gene also is present.) When genes carrying these dominant-negative mutations were transfected into C3H 10T½ cells, the subsequent introduction of myogenic genes no longer converted the cells into myotubes. In addition, a muscle-specific gene that is normally activated did not exhibit its usual pattern of chromatin changes in the doubly transfected C3H 10T½ cells. These results indicate that transcription activation by myogenic proteins depends on a suitable chromatin structure in the regions of muscle-specific genes. MEF2 recruits histone acetylases such as p300/CBP, through another protein that serves as a mediator, thus activating transcription of target genes. Chromatin immunoprecipitation experiments with antibodies against acetylated histone H4 show that the acetylated histone level associated with MEF2-regulated genes is higher in differentiated myotubes than in myoblasts (see Figure 7-37).

Inhibitory Proteins Screening for genes related to *myoD* led to identification of a related protein that retains the HLH dimerization region but lacks the basic DNA-binding region and hence is unable to bind to E box sequences in DNA. By binding to MyoD or E2A, this protein inhibits formation of MyoD–E2A heterodimers and hence their high-affinity binding to DNA. Accordingly, this protein is referred to as Id, for inhibitor of DNA binding. Id prevents cells that produce MyoD and E2A from activating transcription of the muscle-specific gene encoding creatine kinase. As a result, the cells remain in a proliferative growth state. When these cells are induced to differentiate into muscle (for instance, by the removal of serum, which contains the growth factors required for proliferative growth), the Id concentration falls. MyoD–E2A dimers now can form and bind to the regulatory regions of target genes, driving differentiation of C3H 10T½ cells into myoblast-like cells.

The role of histone deacetylases, which inhibit transcription, in muscle development was revealed in experiments in which scientists first introduced extra *myoD* genes into cultured C3H 10T½ cells to raise the level of MyoD. This resulted in increased activation of target genes and more rapid differentiation of the cells into myotubes. However, when genes encoding histone deacetylases also were introduced into the C3H 10T½ cells, the muscle-inducing effect of MyoD was blocked and the cells did not differentiate into myotubes. The explanation for how histone deacetylases inhibit MyoD-induced muscle differentiation came from the surprising finding that the muscle gene activator MEF2 can bind, through its MADS domain, to a histone deacetylase. This interaction, which can prevent MEF2 function and muscle differentiation, is normally blocked during differentiation because the histone deacetylase is phosphorylated by a Ca^{2+}/calmodulin-dependent protein kinase; the phosphorylated deacetylase then is moved from the nucleus to the cytoplasm. Taken together, these results indicate that activation of muscle genes by MyoD and MEF2 is in competition with inactivation of muscle genes by repressive chromatin structures.

Cell-Cycle Proteins The onset of terminal differentiation in many cell types is associated with arrest of the cell cycle, most commonly in G_1, suggesting that the transition from the determined to differentiated state may be influenced by cell-cycle proteins including **cyclins** and **cyclin-dependent kinases** (Chapter 20). For instance, certain inhibitors of cyclin-dependent kinases can induce muscle differentiation in cell culture, and the amounts of these inhibitors are markedly higher in differentiating muscle cells than in nondifferentiating ones in vivo. Conversely, differentiation of cultured myoblasts can be inhibited by transfecting the cells with DNA encoding cyclin D1 under the control of a constitutively active promoter. Expression of cyclin D1, which normally occurs only during G_1, is induced by mitogenic factors in many cell types and drives the cell cycle (see Figure 20-32). The ability of cyclin D1 to prevent myoblast differentiation in vitro may mimic aspects of the in vivo signals that antagonize the differentiation pathway. The antagonism between negative and positive regulators of G_1 progression is likely to play an important role in controlling myogenesis in vivo.

Cell–Cell Signals Are Crucial for Determination and Migration of Myoblasts

As noted already, after myoblasts arise from somites, they must move to their proper locations and form the correct attachments as they differentiate into muscle cells (Figure 21-23). Expression of myogenic genes often occurs after elaborate events that tell certain somite cells to delaminate from the somite epithelium and guide their subsequent movements to muscle assembly sites.

A transcription factor, Pax3, is produced in the subset of somite cells that will form muscle. Pax3 appears to be at the top of the regulatory hierarchy controlling muscle formation in the body wall and limbs. Myoblasts that will migrate, but not cells that remain behind, also produce a transcription factor called Lbx1. If Pax3 is not functional, Lbx1 transcripts are not produced and myoblasts do not migrate. Both Pax3 and Lbx1 can affect expression of *myoD*. The departure of myoblasts from somites also depends upon reception of a secreted protein signal appropriately called *scatter factor,* or *hepatocyte growth factor (SF/HGF)*. This signal, which is produced by embryonic connective tissue cells (mesenchyme) in the limb buds, attracts migrating myoblasts, thus directing them to their proper destination.

Production of SF/HGF is previously induced by still other secreted signals. If the SF/HGF signal or its receptor on myoblasts is not functional, somite cells will produce Lbx1 but not go on to migrate; thus no muscles will form in the limbs. Expression of the *myogenin* and *mrf4* genes does not begin until migrating myoblasts approach their limb-bud destinations (see Figure 21-20).

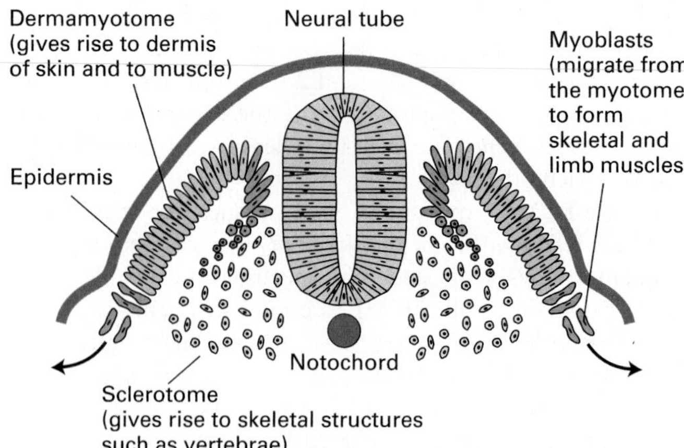

▲ **FIGURE 21-23 Embryonic determination and migration of myoblasts in mammals.** After formation of the neural tube, each somite forms sclerotome, which develops into skeletal structures, and dermomyotome, which gives rise to the dermis of the skin and to the muscles. Lateral myoblasts migrate from the dermomyotome to the limb bud; medial myoblasts develop into the trunk muscles. The remainder of a dermomyotome gives rise to the connective tissue of the skin. [Adapted from M. Buckingham, 1992, *Trends Genet.* **8**:144.]

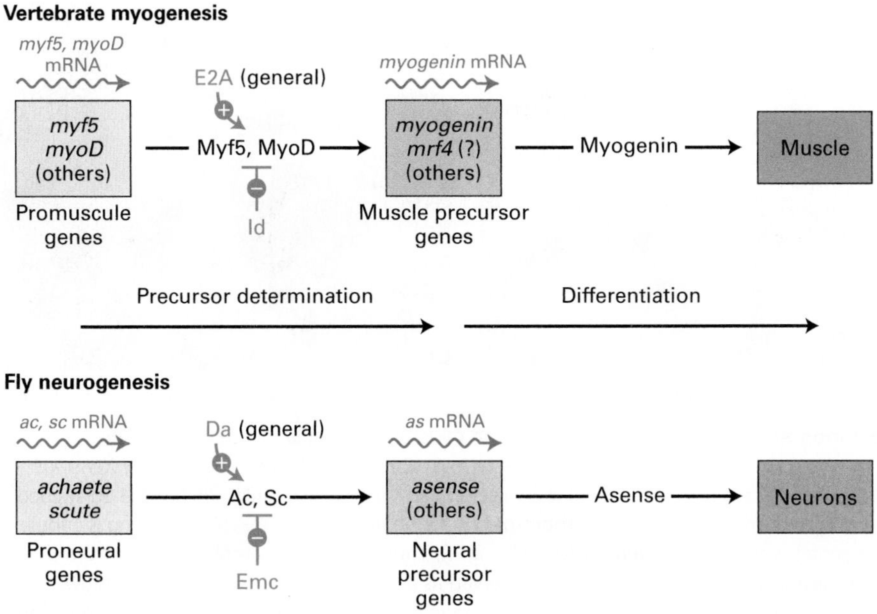

Vertebrate myogenesis

myf5, myoD mRNA — E2A (general) — myogenin mRNA

myf5 myoD (others)	— Myf5, MyoD →	myogenin mrf4 (?) (others)	— Myogenin →	Muscle
Promuscule genes	Id	Muscle precursor genes		

Precursor determination → Differentiation →

Fly neurogenesis

ac, sc mRNA — Da (general) — as mRNA

achaete scute	— Ac, Sc →	asense (others)	— Asense →	Neurons
Proneural genes	Emc	Neural precursor genes		

◀ **FIGURE 21-24 Comparison of genes that regulate vertebrate myogenesis and fly neurogenesis.** bHLH transcription factors have analogous functions in determination of neural and muscle precursor cells and their subsequent differentiation into mature neurons and muscle cells. In both cases, the proteins encoded by the earliest-acting genes (*left*) are under both positive and negative control by other related proteins (blue type). See text for details. [Adapted from Y. N. Jan and L. Y. Jan, 1993, *Cell* **75:**827.]

We have touched on just a few of the many external signals and transcription factors that participate in development of a properly patterned muscle. The function of all these regulatory molecules must be coordinated both in space and in time during myogenesis.

bHLH Regulatory Proteins Function in Creation of Other Tissues

Four bHLH transcription factors that are remarkably similar to the myogenic bHLH proteins control neurogenesis in *Drosophila*. Similar proteins appear to function in neurogenesis in vertebrates and perhaps in the determination and differentiation of hematopoietic cells.

The neurogenic *Drosophila* bHLH proteins are encoded by an ~100-kb stretch of genomic DNA, termed the *achaete-scute complex* (AS-C), containing four genes designated *achaete (ac), scute (sc), lethal of scute (l'sc),* and *asense (a)*. Analysis of the effects of loss-of-function mutations indicate that the Achaete (Ac) and Scute (Sc) proteins participate in determination of neuronal stem cells, called neuroblasts in flies, while the Asense (As) protein is required for differentiation of the progeny of these cells into neurons. (Note that the term *neuroblasts* refers to stem cells in flies but to precursor cells in mammals.) These functions are analogous to the roles of MyoD and Myf5 in muscle determination and of myogenin in differentiation. Two other *Drosophila* proteins, designated Da and Emc, are analogous in structure and function to vertebrate E2A and Id, respectively. For example, heterodimeric complexes of Da with Ac or Sc bind to DNA better than the homodimeric forms of Ac and Sc. Emc, like Id, lacks any DNA-binding domain; it binds to Ac and Sc proteins, thus inhibiting their association with Da and binding to DNA. The similar functions of these myogenic and neurogenic proteins are depicted in Figure 21-24.

A family of bHLH proteins related to the *Drosophila* Achaete and Scute proteins has been identified in vertebrates. One of these, called neurogenin, controls the formation of neuroblasts. In situ hybridization experiments showed that neurogenin is produced at an early stage in the developing nervous system and induces production of NeuroD, another bHLH protein that acts later (Figure 21-25a). Injection of large amounts of *neurogenin* mRNA into *Xenopus* embryos further demonstrated the ability of neurogenin to induce neurogenesis (Figure 21-25b). The function of neurogenin is analogous to that of the Achaete and Scute in *Drosophila;* likewise, NeuroD and Asense may have analogous functions in vertebrates and *Drosophila.*

KEY CONCEPTS OF SECTION 21.3

Specification and Differentiation of Muscle

- Development of skeletal muscle begins with the signal-induced determination of certain mesoderm cells in somites as myoblasts. Following their proliferation and migration, myoblasts stop dividing and differentiate into multinucleate muscle cells (myotubes) that express muscle-specific proteins (see Figure 21-20).
- Four myogenic bHLH transcription factors—MyoD, myogenin, Myf5, and MRF4, collectively called muscle-regulatory factors (MRFs)—associate with E2A and MEFs to form large transcriptional complexes that drive myogenesis and expression of muscle-specific genes.
- Dimerization of bHLH transcription factors with different partners modulates the specificity or affinity of their binding to specific DNA regulatory sites, and also may prevent their binding entirely.
- The myogenic program driven by MRFs depends on the SWI/SNF chromatin-remodeling complex, which makes target genes accessible.

(a)

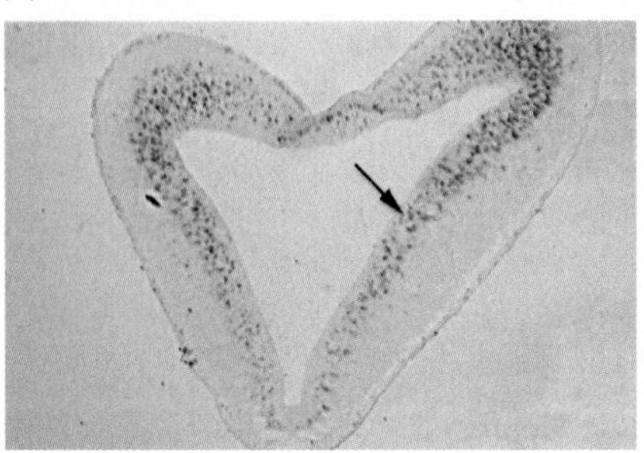

neurogenin mRNA

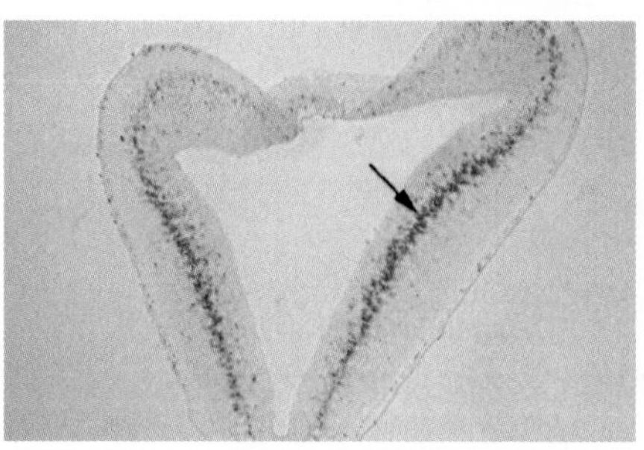

neuroD mRNA

(b)

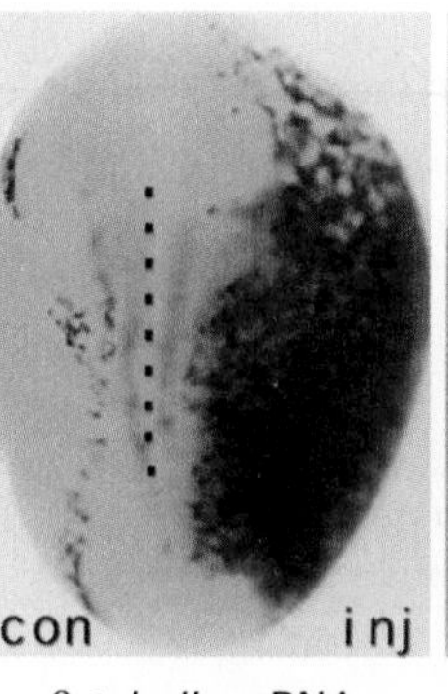

β-tubulin mRNA

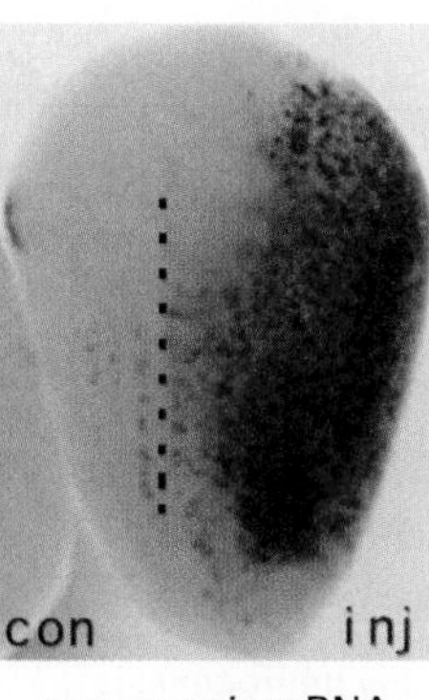

neurogenin mRNA

▲ EXPERIMENTAL FIGURE 21-25 **In situ hybridization and injection experiments demonstrate that neurogenin acts before NeuroD in vertebrate neurogenesis.** (a) Sections of rat neural tube were treated with a probe specific for *neurogenin* mRNA (*left*) or *neuroD* mRNA (*right*). The open space in the center is the ventricle, and the cells lining this cavity constitute the subventricular layer. All the neural cells are born in the subventricular layer and then migrate outward (see Figure 21-12). As illustrated in these micrographs, *neurogenin* mRNA is produced in proliferating neuroblasts in the subventricular layer (arrow), whereas *neuroD* mRNA is present in migrating neuroblasts that have left the ventricular zone (arrow). (b) One of the two cells in early *Xenopus* embryos was injected with *neurogenin* mRNA (inj) and then stained with a probe specific for neuron-specific mRNAs encoding β-tubulin (*left*) or NeuroD (*right*). The region of the embryo derived from the uninjected cell served as a control (con). The *neurogenin* mRNA induced a massive increase in the number of neuroblasts expressing *neuroD* mRNA and neurons expressing *β-tubulin* mRNA in the region of the neural tube derived from the injected cell. [From Q. Ma et al., 1996, *Cell* **87:**43; courtesy of D. J. Anderson.]

- The myogenic program is inhibited by binding of Id protein to MyoD, which blocks binding of MyoD to DNA, and by histone deacetylases, which repress activation of target genes by MRFs.
- Migration of myoblasts to the limb buds is induced by scatter factor/hepatocyte growth factor (SF/HGF), a protein signal secreted by mesenchymal cells (see Figure 21-23). Myoblasts must express both the Pax3 and Lbx1 transcription factors to migrate.
- Terminal differentiation of myoblasts and induction of muscle-specific proteins do not occur until myoblasts stop dividing and migrating.
- Neurogenesis in *Drosophila* depends upon a set of four neurogenic bHLH proteins that are conceptually and structurally similar to the vertebrate myogenic proteins (see Figure 21-24).
- A related vertebrate protein, neurogenin, is required for formation of neural precursors and also determines their fate as neurons or glial cells.

21.4 Regulation of Asymmetric Cell Division

During **embryogenesis,** the earliest stage in animal development, asymmetric cell division often creates the initial diversity that ultimately culminates in formation of specific differentiated cell types. Both stem cells and precursor cells can divide asymmetrically via similar mechanisms, although the details vary with the tissue. Even in bacteria, cell division may yield unequal daughter cells, for example, one that remains attached to a stalk and one that develops flagella used for swimming.

Essential to asymmetric cell division is *polarization* of the parental cell and then differential incorporation of parts of the parental cell into the two daughters (Figure 21-26). A variety of molecular mechanisms are employed to create and propagate the initial asymmetry that polarizes the parental cell. In addition to being different, the daughter cells must often be placed in a specific orientation with respect to surrounding structures. When stem cells divide asymmetrically, the cell that remains in contact with niche signals will persist as a stem cell. Therefore, the mitotic spindles and cell polarity must be aligned with the overall tissue so that differentiating cells move off in the right direction, and so that at least one daughter remains in the stem-cell niche to perpetuate the stem-cell population. This phenomenon is exemplified in the division of neural stem cells during embryonic development (see Figure 21-12b).

We begin with an especially well-understood example of asymmetric cell division, the budding of yeast cells, and move on to recently discovered protein complexes important for asymmetric cell divisions in multicellular organisms. We see in the yeast example an elegant system that links asymmetric division to the process of controlling cell type.

Yeast Mating-Type Switching Depends upon Asymmetric Cell Division

S. cerevisiae cells use a remarkable mechanism to control the differentiation of the cells as the cell lineage progresses. Whether a haploid yeast cell exhibits the α or **a** mating type

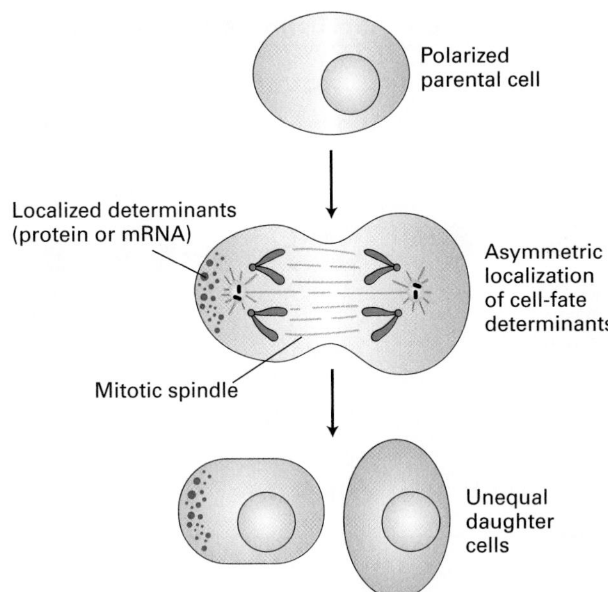

▲ **FIGURE 21-26 General features of asymmetric cell division.** Various mechanisms can lead to asymmetric distribution of cytoplasmic components, such as particular proteins or mRNAs (red dots), thereby forming a polarized parental cell. Division of a polarized cell will be asymmetric if the mitotic spindle is oriented so that the localized cytoplasmic components are distributed unequally to the two daughter cells, as shown here. However, if the spindle is positioned differently relative to the localized cytoplasmic components, division of a polarized cell may produce equivalent daughter cells.

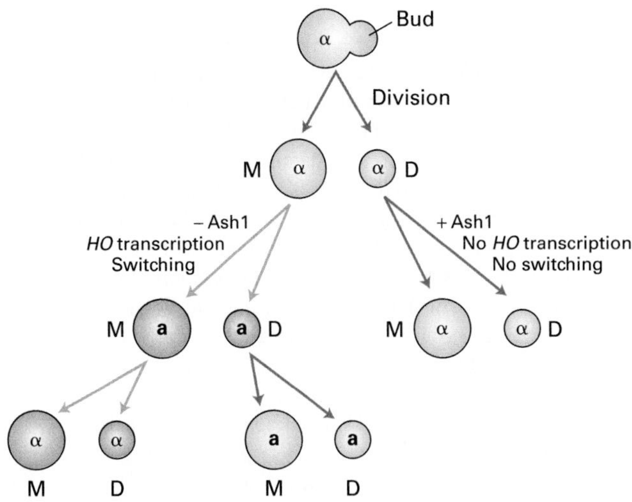

▲ **FIGURE 21-27 Switching of mating type in haploid yeast cells.** Division by budding forms a larger mother cell (M) and smaller daughter cell (D), both of which have the same mating type as the original cell (α in this example). The mother cell can switch mating type during G_1 of the next cell cycle and then divide again, producing two cells of the opposite **a** type. Switching depends on transcription of the *HO* gene, which occurs only in the absence of Ash1 protein. The smaller daughter cells, which produce Ash1 protein, cannot switch; after growing in size through interphase, they divide to form a mother cell and daughter cell. Orange cells and arrows indicate switch events.

is determined by which genes are present at the *MAT* locus (see Figure 21-17). As described in Chapter 7, the *MAT* locus in the *S. cerevisiae* genome is flanked by two "silent," transcriptionally inactive loci containing the alternative α or **a** sequences (see Figure 7-33). A specific DNA rearrangement brings the genes that encode the α-specific or **a**-specific transcription factors from these silent loci to the active *MAT* locus where they can be transcribed.

Interestingly, some haploid yeast cells can switch repeatedly between the α and **a** types. *Mating-type switching* occurs when the α allele occupying the *MAT* locus is replaced by the **a** allele, or vice versa. The first step in this process is catalyzed by HO endonuclease, which is expressed in mother cells but not in daughter cells. Thus mating-type switching occurs only in mother cells (Figure 21-27). Transcription of the *HO* gene is dependent on the SWI/SNF chromatin-remodeling complex (see Figure 7-43), the same complex that we encountered earlier in our discussion of myogenesis. Daughter yeast cells arising by budding from mother cells contain a protein called Ash1p (for *A*symmetric *s*ynthesis of *HO*) that prevents recruitment of the SWI/SNF complex to the *HO* gene, thereby preventing its transcription. The absence of Ash1 from mother cells allows them to transcribe the *HO* gene.

Recent experiments have revealed how the asymmetry in the distribution of Ash1 between mother and daughter cells is established. *ASH1* mRNA accumulates in the growing bud that will form a daughter cell due to the action of a myosin motor protein (Chapter 17). This motor protein, called Myo4p, moves the *ASH1* mRNA, as a ribonucleoprotein complex, along actin filaments in one direction only, toward the bud (Figure 21-28). Two connector proteins, termed She2p and She3p (for *S*WI5p-dependent *HO* *e*xpression) tether the *ASH1* mRNA to the Myo4p motor protein. By the time the bud separates from the mother cell, the mother cell is largely depleted of *ASH1* mRNA and thus can switch mating type in the following G_1 before additional *ASH1* mRNA is produced and before DNA replication in the S phase.

Budding yeasts use a relatively simple mechanism to create molecular differences between the two cells formed by division. In higher organisms, as in yeast, the mitotic spindle must be oriented in such a way that each daughter cell receives its own set of cytoplasmic components. Genetic studies in *C. elegans* and *Drosophila* have revealed the key participants, a first step in understanding at the molecular level how asymmetric cell division is regulated in multicellular organisms. To illustrate these complexities, we focus on asymmetric division of neuroblasts in *Drosophila*.

Proteins That Regulate Asymmetry Are Localized at Opposite Ends of Dividing Neuroblasts in *Drosophila*

Fly neuroblasts, which are stem cells, arise from a sheet of ectoderm cells that is one cell thick. As in vertebrates, the *Drosophila* ectoderm forms both epidermis and the nervous system, and many ectoderm cells have the potential to

MEDIA CONNECTIONS

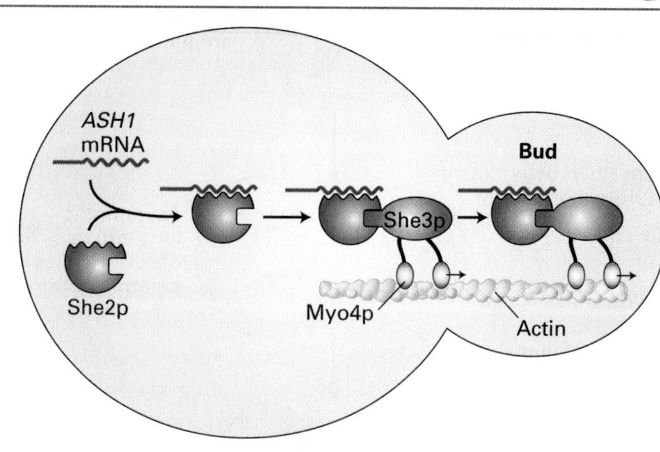

▶ **FIGURE 21-28 Model for restriction of mating-type switching to mother cells in *S. cerevisiae*.** Ash1 protein prevents a cell from transcribing the *HO* gene whose encoded protein initiates the DNA rearrangement that results in mating-type switching from **a** to α or α to **a**. Switching occurs only in the mother cell, after it separates from a newly budded daughter cell, because of the presence of Ash1 protein only in the daughter cell. The molecular basis for this differential localization of Ash1 is the one-way transport of *ASH1* mRNA into the bud. A linking protein, She2p, binds to specific 3′ untranslated sequences in the *ASH1* mRNA and also binds to She3p protein. This protein in turn binds to a myosin motor, Myo4p, which moves along actin filaments into the bud. [See S. Koon and B. J. Schnapp, 2001, *Curr. Biology* **11**:R166.]

assume either a neural or epidermal fate. Under the control of genes that become active only in certain cells, some of the cells increase in size and begin to loosen from the ectodermal layer. At this point, the delaminating cells use the Delta/Notch signaling pathway to mediate **lateral inhibition** of their neighbors, causing them to retain the epidermal fate (see Figures 16-36 and 22-42). The delaminating cells move inside and become spherical neuroblasts, while the prospective epidermal cells remain behind and close up to form a tight sheet. This process generates 60 neuroblasts in each body segment, which will give rise to about 700 neurons per segment.

Once formed, the neuroblasts undergo asymmetric divisions, at each division recreating themselves and producing a *ganglion mother cell (GMC)* on the basal side of the neuroblast (Figure 21-29). A single neuroblast will produce several GMCs; each GMC in turn forms two neurons. Depending on where they form in the embryo and consequent regulatory events, neuroblasts may form more or fewer GMCs. Neuroblasts and GMCs in different locations exhibit different

Video: Ectodermal Cell Divisions in the *Drosophila* Embryo

MEDIA CONNECTIONS

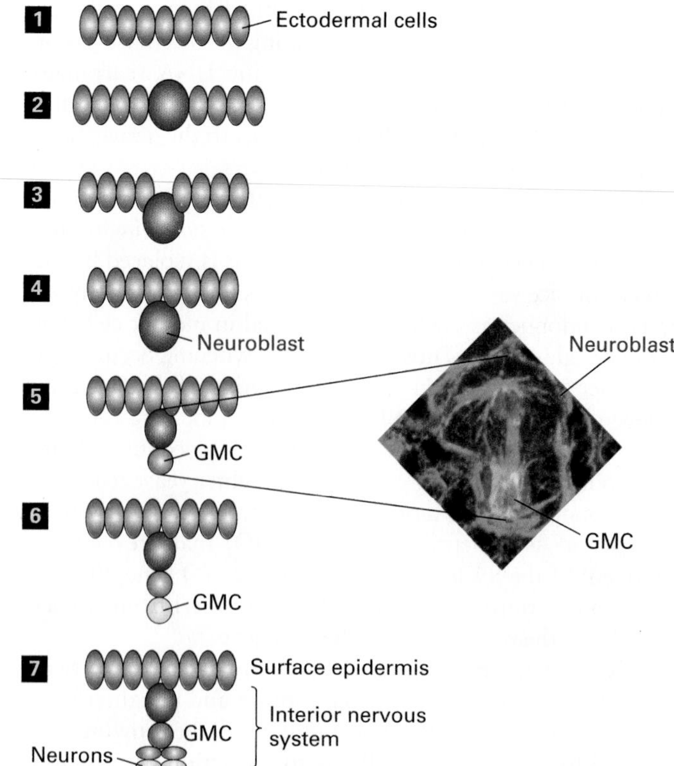

▶ **FIGURE 21-29 Asymmetric cell division during *Drosophila* neurogenesis.** The ectodermal sheet (1) of the early embryo gives rise to both epidermal cells and neural cells. Neuroblasts, the stem cells for the fly nervous system, are formed when ectoderm cells enlarge, separate from the ectodermal epithelium, and move into the interior of the embryo (2–4). Each neuroblast that arises divides asymmetrically to recreate itself and produce a ganglion mother cell, or GMC (5). Subsequent divisions of a neuroblast produce more GMCs, creating a stack of these precursor cells (6). Each GMC divides once to give rise to two neurons (7). Neuroblasts and their neuronal descendants can have different fates depending on their location. The micrograph shows an asymmetrically dividing *Drosophila* neuroblast. The apical end (red) will form a new neuroblast and the basal end (blue and red) will form a GMC. The microtubules are labeled in green. [Photograph courtesy of Dr. C. Q. Doe, University of Oregon.]

patterns of gene expression, an indicator of their fates. Analysis of fly mutants led to the discovery of key proteins that (1) establish apical-basal polarity in neuroblasts, (2) align the mitotic spindle of dividing neuroblasts with their polarity, and (3) direct formation of daughter cells whose fate and size differ from that of neuroblasts. Genetic studies of asymmetric cell divisions in the *C. elegans* early embryo independently led to the discovery of important cell division asymmetry proteins. The machinery controlling asymmetric cell division is highly conserved and readily recognized from worms to flies to mammals, indicating conservation of protein functions for more than half a billion years.

Basal and apical protein complexes congregate during each neuroblast division, disperse, and then localize again for the next round of division. Four protein complexes, which we denote as MPSB, BPP, DSL, and IPLG, govern the entire process (Figure 21-30):

- *BPP,* an apical complex also known as the PAR complex, is responsible for defining the end of the cell that will remain a neuroblast. It comprises *B*azooka and *P*ar6, both of which contain PDZ domains, and a*P*KC, an atypical isoform of protein kinase C.

- *IPLG,* a second apical complex, is composed of *I*nscuteable (Insc), *P*artner of inscuteable (Pins), *L*ocomotion defects (Loco), and *G*$_i$, a heterotrimeric G protein (Chapter 15). This complex is critical for orienting the spindle during asymmetric divisions.

- *DSL,* a complex that is fairly evenly distributed around the cell, is composed of *D*iscs-large (Dlg), *S*cribble (Scrib), and *L*ethal giant larvae (Lgl). Lgl reversibly associates with the cytoskeleton. The DSL complex is mostly employed in localizing basal proteins.

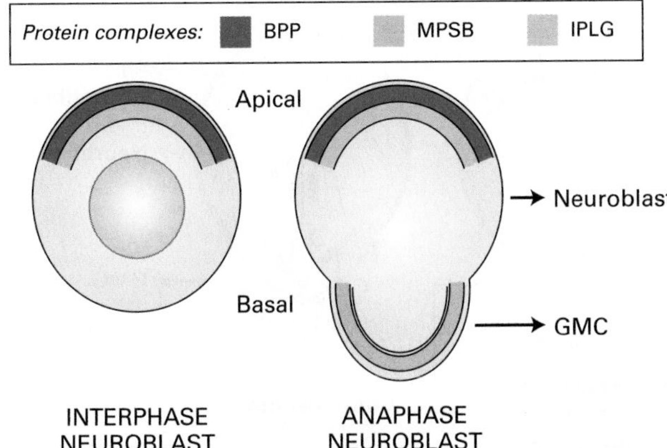

▲ **FIGURE 21-30 Localized protein complexes that control asymmetric cell division.** (a) In the *Drosophila* neuroblast, the BPP complex is localized apically in ectoderm cells and in delaminating neuroblasts (steps 1–3 in Figure 21-29). The IPLG complex is also apically localized. The DSL complex (not shown) is distributed fairly evenly around the cells. In response to regulation by BPP, the MPSB complex localizes to the basal side, where it will be incorporated into the ganglion mother cell (GMC). Mutations in genes that encode polarized proteins disrupt asymmetric cell division and are therefore fatal. Motor protein–mediated transport along cytoskeletal filaments localizes the MPSB basal complex. [See C. Q. Doe and B. Bowerman, 2001, *Curr. Opin. Cell Biol.* **13**:68, and A. Wodarz, 2005, *Curr. Opin. Cell Biol.* **17**:475.]

- *MPSB,* a basal complex, confers GMC cell fate. It includes the coiled-coil scaffold protein *M*iranda, the homeodomain-class transcription factor *P*rospero, the RNA-binding protein *S*taufen, and a translational repressor protein called *B*rain tumor (Brat).

With the key players in neuroblast asymmetric division introduced, let's examine their functions more closely.

Apical Complexes and Spindle Orientation For localized protein complexes to be differentially incorporated into two daughter cells, the plane of cell division must be appropriately oriented. In dividing fly neuroblasts, the **mitotic spindle** first aligns perpendicular to the apical-basal axis and then turns 90 degrees to align with it at the same time that the basal complexes become localized to the basal side (Figure 21-31). The apical IPLG and BPP complexes, which are already in place before spindle rotation, control the final orientation of the spindle. This is supported by the finding that mutations in any of the components of these complexes eliminate the coordination of the spindle with apical-basal polarity, causing the spindle orientation to become random.

The two apical protein complexes have different roles in spindle orientation. First, the BPP complex responds to extrinsic cues from the overlying ectoderm to form an apical crescent at late interphase. In this way, the BPP complex aligns neuroblast polarity with the surrounding tissue so that the apical side of the neuroblast is always next to the ectoderm. Second, the BPP complex recruits the IPLG complex to the apical cortex, and the BPP complex anchors one spindle pole to the apical cortex, thereby aligning the spindle along the apical-basal axis. The direct link with the spindle is mediated by the NuMA protein, which joins the Pins protein of the IPLG complex to microtubules. The IPLG complex is sufficient for anchoring the spindle and promoting asymmetric division.

Basal Complex and the Determination of GMC Fate During fly neuroblast divisions, the MPSB complex becomes localized to the basal cortex prior to each division and remains there while the basal part of the neuroblast becomes a new GMC (see Figure 21-30). Miranda provides a scaffold for the other three proteins in the complex (Prospero, Staufen, and Brat) and is needed to muster them near the basal plasma membrane. After each division, MPSB proteins in the basally located daughter cell inhibit neuroblast properties and confer GMC properties. Prospero negatively regulates transcription of cell-cycle genes, which remain active in the dividing neuroblast. Brat post-transcriptionally inhibits the transcription factor Myc, a positive regulator of cell division and negative regulator of cell size. In this way Brat helps keep the GMC small and restrains its division. Genetic studies provide support for such involvement of Prospero and Brat in determination of GMCs. For example, *brat* mutations

Video: Mitotic Spindle Asymmetry in *Drosophila* Neuroblast Cell Division

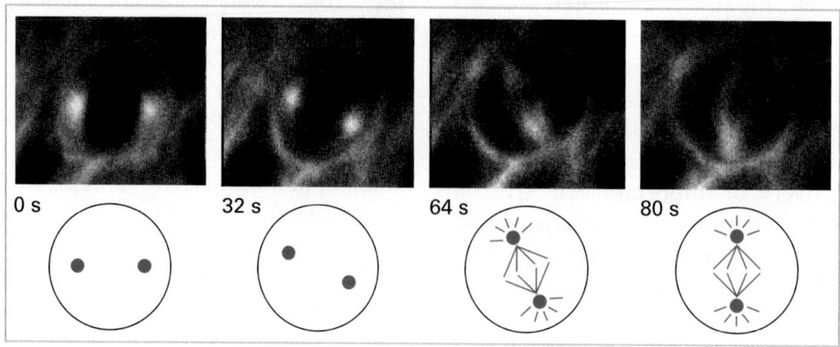

▲ **EXPERIMENTAL FIGURE 21-31 Time-lapse fluorescence imaging reveals rotation of the mitotic spindle in asymmetrically dividing neuroblasts.** Early *Drosophila* embryos were injected with a hybrid gene composed of the gene encoding green fluorescent protein (GFP) fused to the gene encoding Tau, a protein that binds to microtubules. At the top are time-lapse images of a single dividing neuroblast in a live embryo. The basal side is at the top, and the apical side at the bottom. At time 0, equivalent to prophase, the two centrosomes are visible on opposite sides of the cell. These function as the spindle poles; as mitosis proceeds, the microtubules forming the mitotic spindle are assembled from the poles (see Figure 18-34). In successive images (at 32, 64, and 80 seconds), the bipolar spindle can be seem to form and rotate 90 degrees to align with the apical-basal axis, as schematically depicted at the bottom. [From J. A. Kaltschmidt et al., 2000, *Nature Cell Biol.* **2:**7; courtesy of J. Kaltschmidt and A. H. Brand, Wellcome/CRC Institute, Cambridge University.]

cause GMCs to enlarge into neuroblasts and keep dividing, whereas loss of Prospero causes GMCs to remain small but maintain neuroblast-style gene expression and proliferation.

How does the MPSB complex become located basally prior to each neuroblast division? The answer is more complex than Ash1p localization in yeast. Both apical BPP and uniform cortical DSL complexes are involved. The BPP complex controls MPSB localization. The atypical protein kinase C (aPKC), a component of the BPP complex, phosphorylates and thus inactivates the Lgl protein, a component of the DSL complex. Lgl is required to bring MPSB proteins to the basal cortex. Since aPKC is located at the apical end of the cells, Lgl is active only in basal regions. The restriction of active Lgl to basal cortex explains how MPSB proteins are brought to the basal cortex where they cause one daughter cell to become a GMC.

How does active, basal Lgl control localization of MPSB? Although the full story is not yet known, genetic and biochemical studies show that actin, myosin II, and myosin VI are involved. For instance, drug-induced disruption of actin microfilaments blocks MPSB targeting to the neuroblast cortex. Myosin VI binds Miranda (the "M" of MPSB) directly and is also required for basal targeting of the MPSB complex.

Asymmetry of Daughter Cell Size A notable feature of neuroblast asymmetric division is the pronounced difference in the size of neuroblasts and GMCs. This cell-size difference is regulated by the Pins and $G_{\alpha i}$ components of the IPLG complex. The IPLG complex is brought to the apical cortex by virtue of the association of its Insc component with the Baz component of the BPP complex. When a neuroblast divides to produce a GMC and a neuroblast, the GMC is usually considerably smaller. The spindle is oriented, as we have described, in the apical-basal direction, and at metaphase of the neuroblast division, the two halves of the spindle are about equal in size. However the two centrosomes, one at each pole of the spindle, behave differently. The basal centrosome, marking the pole where the GMC will form, has few astral microtubules, while the apical centrosome enlarges and grows a bushy mass of astral microtubules that make that pole of the dividing neuroblast considerably larger (Figure 21-32).

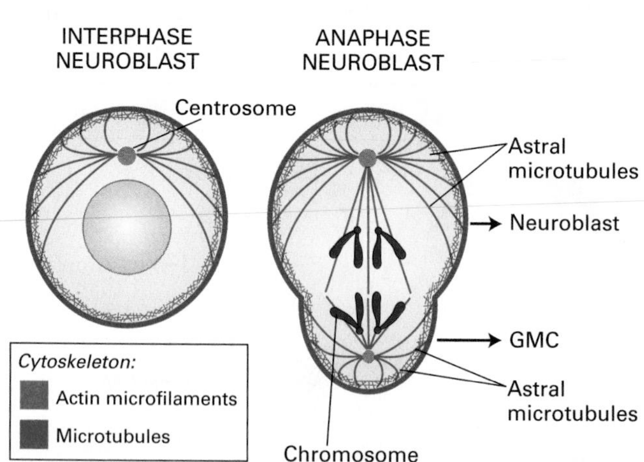

▲ **FIGURE 21-32 Orientation of mitotic spindle and difference in daughter cell size in asymmetric division of neuroblasts.** Interactions between spindle microtubules and the IPLG apical complex orient the spindle. Actin microfilaments (red) lie just under the cell surface at all times. Microtubules (blue) radiate from the centrosome during interphase and then assemble into the mitotic spindle, attached to the duplicated centrosomes, during cell division. Note the biased location of the centrosome during interphase, at the apical end of the cell. The asymmetry in daughter cell size begins with differential assembly of astral microtubules, which are considerably shorter and less abundant in the basal end of the dividing cell, which forms the daughter ganglion mother cell (GMC). [Adapted from C. Q. Doe and B. Bowerman, 2001, *Curr. Opin. Cell Biol.* **13:**68.]

Genetic tests have shown redundant control of the cell-size asymmetry between the two daughter cells. If either the BPP apical complex or the IPLG apical complex is functional, the cells formed will be the usual small GMC and large neuroblast. In contrast, a double mutant with defects in both the BPP and IPLG complexes (e.g., a double *pins* and *baz* mutant) produces two daughter cells that are equal in size. Double mutations that inactivate the G_β and G_γ subunits, but not the $G_{\alpha i}$ subunit, of the G_i protein component of the IPLG complex, cause numerous astral microtubules to form on both centrosomes. Over-production of the G_β protein has the opposite effect—no astral tubules on either centrosome. From these genetic analyses we may conclude that a normal function of the G_β protein is to selectively prevent assembly of astral microtubules at the basal centrosome. This regulation would involve the action of G_β/G_γ subunits that are *not* part of the apical IPLG complexes. Indeed, G_β is uniformly distributed all around the neuroblast cortex.

Heterotrimeric G proteins like the one in the IPLG complex often are controlled by a G protein–coupled receptor that, when activated, dissociates the trimer by binding G_α and releasing active G_β/G_γ subunits (Chapter 15). No sign of a G protein–coupled receptor has been found in the search for proteins controlling neuroblast asymmetry. Instead Pins and Loco, components of the IPLG complex, substitute for a receptor in triggering dissociation of the inactive heterotrimeric G protein. Pins and Loco are partially redundant; mutating both causes defects equivalent to mutation of either the G_β or G_γ subunit. Pins and Loco bind to GDP·$G_{\alpha i}$ and act like guanine nucleotide dissociation inhibitors, thereby keeping $G_{\alpha i}$ associated with GDP and allowing $G_{\alpha i}$·GDP and G_β/G_γ to act upon their (as yet unknown) targets. As would be expected for a typical G protein cycle, a GTPase-activating protein (GAP) and a GDP exchange factor (GEF) have also been found to regulate neuroblast division asymmetry. The GAP reaction inactivates the G protein by breaking down GTP to GDP, while the GEF reaction recharges the G protein for activity by bringing in a new GTP. The mechanism by which the G protein component of the IPLG complex regulates the activity of the centrosome remains unknown.

Summary of Asymmetry-Determining Protein Complexes The initial course of events in polarizing and organizing asymmetric cell division can be summarized as having three phases: (1) establishment of cell polarity, (2) alignment of the mitotic spindle with cell polarity, and (3) specification of distinct sibling fates.

For phase 1, the BPP complex is already apically located in all ectoderm cells. When some of those cells sink beneath the surface to become neuroblasts, the apical localization of BPP persists. The IPLG complex becomes apically located after the BPP complex. Acting through the evenly distributed DSL complex, the two apical complexes direct the basal localization of MPSB.

For phase 2, the orientation of the spindle relative to the ectoderm is an indirect outcome of the apically located BPP complex, which links the already oriented ectoderm to the IPLG complex. Mitotic spindle microtubules are joined to the apical IPLG complex by the fly NuMA protein, thus orienting the spindle.

For phase 3, as asymmetric cell divisions commence, each neuroblast renews itself while budding off a smaller daughter GMC in the interior (basal) direction. Different sibling fates are specified by proteins incorporated into the daughter cells. Neuroblasts have stem-cell properties and are determined to retain that stem-cell fate by aPKC, a part of the BPP complex that promotes neuroblast self-renewal. Brat and Prospero, components of the MPSB complex, are located in the smaller interior (basal) daughter cell and promote GMC differentiation. The G protein component of the IPLG complex activates astral microtubule assembly in a dividing neuroblast, determining the size of the two poles and hence of the daughter cells. Since IPLG is apically localized, the apical daughter cell (a neuroblast) is larger than the basal daughter cell (a GMC).

From this summary, we can see how a set of protein complexes coordinates multiple critical events during asymmetric cell division: localization and activation of asymmetry regulators, differential segregation of cell fate–determining regulatory proteins, orientation of the spindle, and generation of different sized daughter cells.

KEY CONCEPTS OF SECTION 21.4

Regulation of Asymmetric Cell Division

- Asymmetric cell division requires polarization of the dividing cell, which usually entails localization of some cytoplasmic components, and then the unequal distribution of these components to the daughter cells (see Figure 21-26).
- In the asymmetric division of budding yeasts, a myosin-dependent transport system carries *ASH1* mRNA into the bud (see Figure 21-28).
- Ash1 protein is produced in the daughter cell soon after division and prevents expression of HO endonuclease, which is necessary for mating-type switching. Thus a daughter cell cannot switch mating type, whereas the mother cell from which it arises can (see Figure 21-27).
- Asymmetric cell division in *Drosophila* neuroblasts depends on two apical protein complexes (BPP, IPLG), a basal complex (MPSB), and an evenly distributed complex (DSL). The basal proteins are incorporated into the ganglion mother cell (GMC) and contain proteins that determine cell fate (see Figure 21-30).
- Asymmetry factors exert their influence at least in part by controlling the orientation of the mitotic spindle, so that asymmetrically localized proteins and structures are differentially incorporated into the two daughter cells (see Figure 21-32).
- The atypical protein kinase C (aPKC) in the BPP apical complex phosphorylates the LGL protein, a component of the DSL complex, but can do so only in the apical region, since that is where BPP is located. Nonphosphorylated LGL, which therefore exists only in basal regions, is active in bringing MPSB to the cortex where it requires the actin cytoskeleton to be anchored.

■ The general process of asymmetric cell division and the protein complexes controlling it are highly conserved through evolutionary time.

21.5 Cell Death and Its Regulation

Programmed cell death is a counter-intuitive but essential cell fate. Cell death keeps our hands from being webbed, our embryonic tails from persisting, our immune system from responding to our own proteins, and our brain from being filled with useless electrical connections. In fact, the majority of cells generated during brain development subsequently die.

Cellular interactions regulate cell death in two fundamentally different ways. First, most if not all cells in multicellular organisms require signals to stay alive. In the absence of such survival signals, frequently referred to as **trophic factors,** cells activate a "suicide" program. Second, in some developmental contexts, including the immune system, specific signals induce a "murder" program that kills cells. Whether cells commit suicide for lack of survival signals or are murdered by killing signals from other cells, death is mediated by a common molecular pathway. In this section, we first distinguish

Video: Cells Undergoing Apoptosis

MEDIA CONNECTIONS

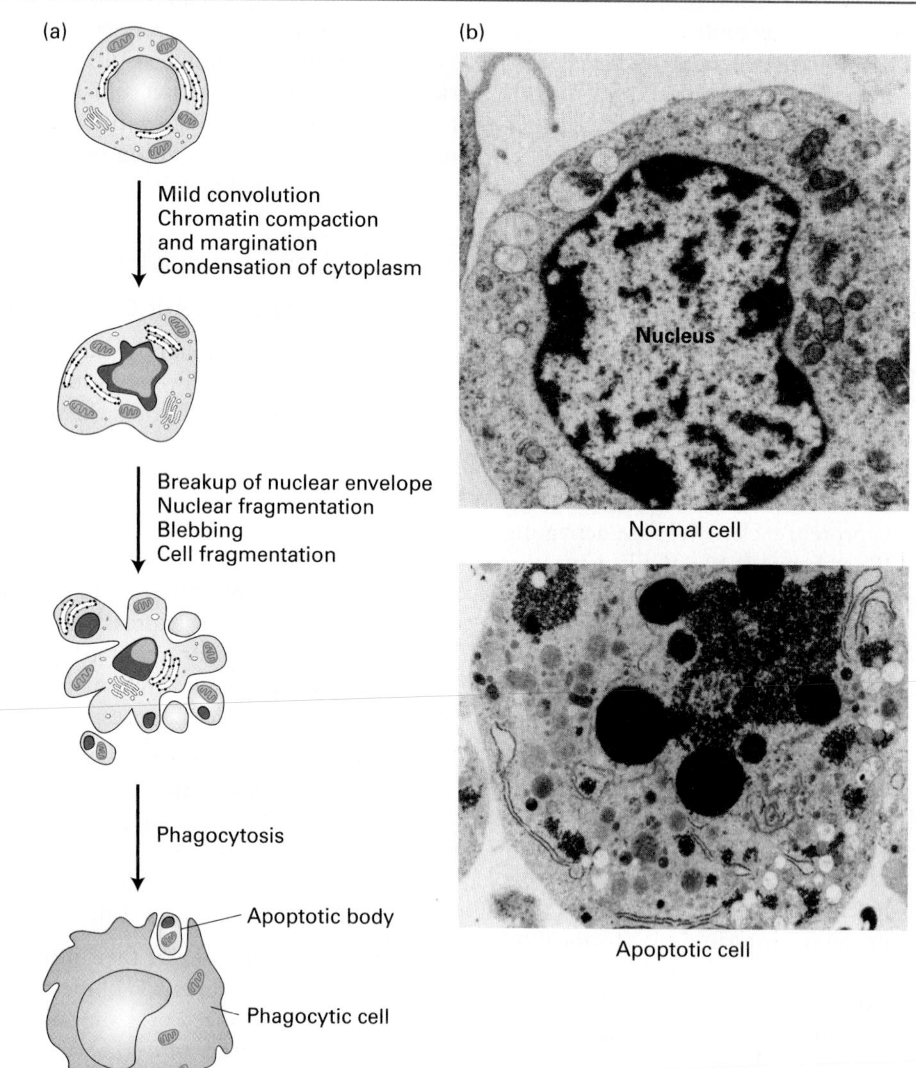

▲ **FIGURE 21-33 Ultrastructural features of cell death by apoptosis.** (a) Schematic drawings illustrating the progression of morphologic changes observed in apoptotic cells. Early in apoptosis, dense chromosome condensation occurs along the nuclear periphery. The cell body also shrinks, although most organelles remain intact. Later both the nucleus and cytoplasm fragment, forming apoptotic bodies, which are phagocytosed by surrounding cells. (b) Photomicrographs comparing a normal cell (*top*) and apoptotic cell (*bottom*). Clearly visible in the latter are dense spheres of compacted chromatin as the nucleus begins to fragment. [Part (a) adapted from J. Kuby, 1997, *Immunology,* 3d ed., W. H. Freeman & Co., p. 53. Part (b) from M. J. Arends and A. H. Wyllie, 1991, *Int'l. Rev. Exp. Pathol.* **32:**223.]

programmed cell death from death due to tissue injury, then consider the role of trophic factors in neuronal development, and finally describe the evolutionarily conserved effector pathway that leads to cell suicide or murder.

Programmed Cell Death Occurs Through Apoptosis

The demise of cells by programmed cell death is marked by a well-defined sequence of morphological changes, collectively referred to as **apoptosis,** a Greek word that means "dropping off" or "falling off," as leaves from a tree. Dying cells shrink and condense and then fragment, releasing small membrane-bound apoptotic bodies, which generally are engulfed by other cells (Figure 21-33; see also Figure 1-19). The nuclei condense and the DNA is fragmented. Importantly, the intracellular constituents are not released into the extracellular milieu where they might have deleterious effects on neighboring cells. The stereotypical changes in cells during apoptosis, like condensation of the nucleus and engulfment by surrounding cells, suggested to early workers that this type of cell death was under the control of a strict program. This program is critical during both embryonic and adult life to maintain normal cell number and composition.

The genes involved in controlling cell death encode proteins with three distinct functions:

- "Killer" proteins are required for a cell to begin the apoptotic process.
- "Destruction" proteins do things like digest DNA in a dying cell.
- "Engulfment" proteins are required for phagocytosis of the dying cell by another cell.

At first glance, engulfment seems to be simply an after-death cleanup process, but some evidence suggests that it is part of the final death decision. For example, mutations in killer genes always prevent cells from initiating apoptosis, whereas mutations that block engulfment sometimes allow cells to survive that would normally die. That is, cells with engulfment-gene mutations can initiate apoptosis but sometimes recover. Engulfment involves the assembly of a halo of actin around the dying cell, triggered by apoptosis proteins that activate Rac, a monomeric G protein that helps regulate actin polymerization (see Figure 17-42). A signal on the surface of the dying cell also stimulates a receptor on neighboring cells, which initiates membrane changes leading to engulfment.

In contrast to apoptosis, cells that die in response to tissue damage exhibit very different morphological changes, referred to as **necrosis.** Typically, cells that undergo this process swell and burst, releasing their intracellular contents, which can damage surrounding cells and frequently cause inflammation.

Neurotrophins Promote Survival of Neurons

The earliest studies demonstrating the importance of trophic factors in cellular development came from analyses of the developing nervous system. When neurons grow to make connections to other neurons or to muscles, sometimes over considerable distances, more cells grow than will eventually survive. The neurons' cell bodies are located in the spinal cord and adjacent ganglia, while their processes extend far outside these regions. Those that make connections prevail and survive; those that fail to connect die.

In the early 1900s the number of neurons innervating the periphery was shown to depend upon the size of the tissue to which they would connect, the so-called "target field." For instance, removal of limb buds from the developing chick embryo leads to a reduction in the number of sensory neurons and motoneurons innervating the bud (Figure 21-34).

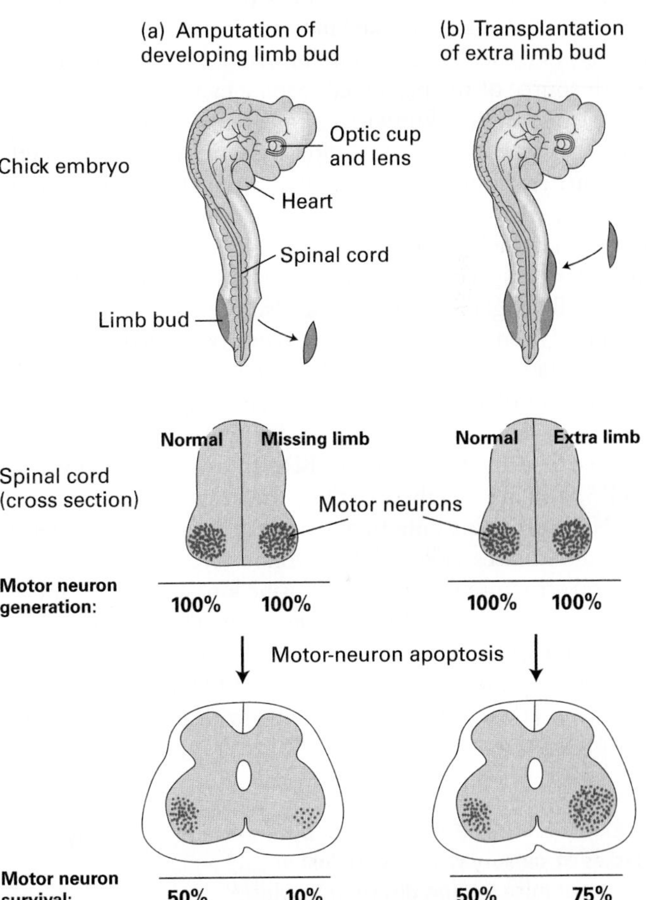

▲ EXPERIMENTAL FIGURE 21-34 **The survival of motor neurons depends on the size of the muscle target field they innervate.** (a) Removal of a limb bud from one side of a chick embryo at about 2.5 days results in a marked decrease in the number of motor neurons on the affected side. In an amputated embryo, normal numbers of motor neurons are generated on both sides (*middle*). Later in development, many fewer motor neurons remain on the side of the spinal cord with the missing limb than on the normal side (*bottom*). Note that only about 50 percent of the motor neurons that originally are generated normally survive. (b) Transplantation of an extra limb bud into an early chick embryo produces the opposite effect, more motor neurons on the side with additional target tissue than on the normal side. [Adapted from D. Purves, 1988, *Body and Brain: A Trophic Theory of Neural Connections,* Harvard University Press, and E. R. Kandel, J. H. Schwartz, and T. M. Jessell, 2000, *Principles of Neural Science,* 4th ed., McGraw-Hill,, p. 1054, Figure 53-11.]

Conversely, grafting additional limb tissue to a limb bud leads to an increase in the number of neurons in corresponding regions of the spinal cord and sensory ganglia. Indeed, incremental increases in the target-field size are accompanied by commensurate incremental increases in the number of neurons innervating the target field. This relation was found to result from the selective survival of neurons rather than changes in their differentiation or proliferation. The observation that many sensory and motor neurons die after reaching their peripheral target field suggested that these neurons compete for survival factors produced by the target tissue.

Subsequent to these early observations, scientists discovered that transplantation of a mouse sarcoma tumor into a chick led to a marked increase in the numbers of certain types of neurons. This finding implicated the tumor as a rich source of the presumed trophic factor. To isolate and purify this factor, known simply as nerve growth factor (NGF), scientists used an in vitro assay in which outgrowth of neurites from sensory ganglia (nerves) was measured. Neurites are extensions of the cell cytoplasm that can grow to become the long wires of the nervous system, the **axons** and **dendrites** (see Figure 23-2). The later discovery that the submaxillary gland in the mouse also produces large quantities of NGF enabled biochemists to purify and to sequence it. A homodimer of two 118-residue polypeptides, NGF belongs to a family of structurally and functionally related trophic factors collectively referred to as **neurotrophins.** Brain-derived neurotrophic factor (BDNF) and neurotrophin-3 (NT-3) also are members of this protein family.

Neurotrophins bind to and activate a family of receptor tyrosine kinases called *Trks* (pronounced "tracks"). (The general structure of receptor tyrosine kinases and the intracellular signaling pathways they activate are covered in Chapter 16.) Each neurotrophin binds with high affinity to one Trk receptor: NGF binds to TrkA; BDNF, to TrkB; and NT-3, to TrkC. NT-3 can also bind with lower affinity to both TrkA and TrkB. Binding of these factors to their receptors provides a survival signal for different classes of neurons. As neurons grow from the spinal cord to the periphery, neurotrophins produced by target tissues bind to Trk receptors on the growth cones of the extending axons, promoting survival of neurons that successfully reach targets. In addition, neurotrophins bind to a distinct type of receptor called $p75^{NTR}$ (NTR = neurotrophin receptor) but with lower affinity. However, $p75^{NTR}$ forms heteromultimeric complexes with the different Trk receptors; this association increases the affinity of Trks for their ligands. Depending on the cell type, binding of NGF and BDNF to $p75^{NTR}$ in the absence of TrkA may promote cell death rather than prevent it. (The phenomenon of multiple neurotrophins interacting with multiple similar receptors is comparable to EGF-like ligands and their HER receptors, illustrated in Figure 16-18).

To critically address the role of the neurotrophins in development, scientists produced mice with knockout mutations in each of the neurotrophins and their receptors. These studies revealed that different neurotrophins and their corresponding receptors are required for the survival of different classes of sensory neurons (Figure 21-35). For instance, pain-sensitive (nociceptive) neurons, which express TrkA, are selectively lost from the dorsal root ganglion of knockout mice lacking NGF or TrkA, whereas TrkB- and TrkC-expressing neurons are unaffected in such knockouts. In contrast, TrkC-expressing proprioceptive neurons, which detect the position of the limbs, are missing from the dorsal root ganglion in *TrkC* and *NT-3* mutants.

A Cascade of Caspase Proteins Functions in One Apoptotic Pathway

Neurotrophins and other signals that keep cells alive act upon an evolutionarily conserved cell-death control system. Key insights into the molecular mechanisms regulating cell

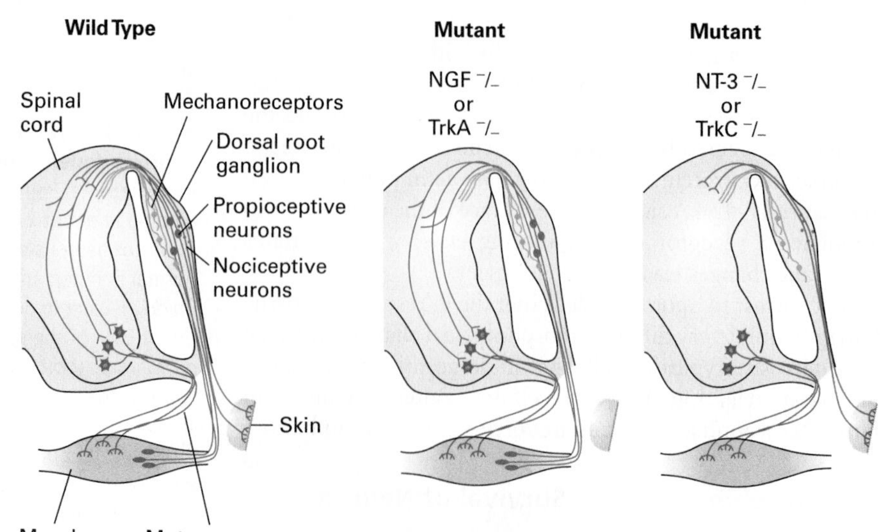

▶ **EXPERIMENTAL FIGURE 21-35 Different classes of sensory neurons are lost in knockout mice lacking different trophic factors or their receptors.** In animals lacking nerve growth factor (NGF) or its receptor TrkA, small nociceptive (pain-sensing) neurons (blue) that innervate the skin are missing. These neurons express TrkA receptor and innervate NGF-producing targets. In animals lacking either neurotrophin-3 (NT-3) or its receptor TrkC, large propioceptive neurons (red) innervating muscle spindles are missing. Muscle produces NT-3 and the propioceptive neurons express TrkC. Mechanoreceptors (orange), another class of sensory neurons in the dorsal root ganglion, are unaffected in these mutants. [Adapted from W. D. Snider, 1994, *Cell* **77:**627.]

Video: Programmed Cell Death in *C. elegans* Embryonic Development

(a)

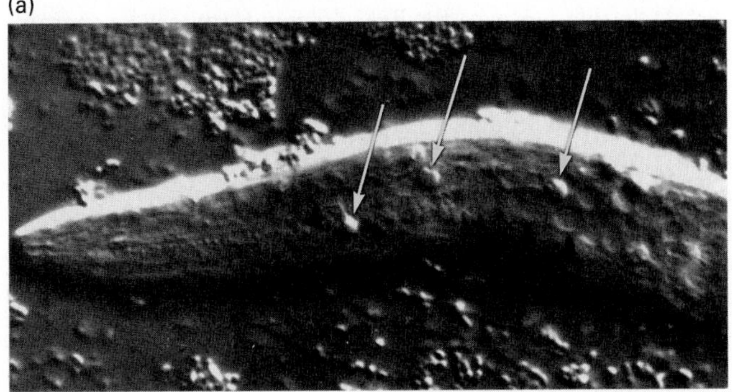

(b)

◀ **EXPERIMENTAL FIGURE 21-36 Mutations in the *ced-3* gene block programmed cell death in *C. elegans*.** (a) Newly hatched mutant larva carry a mutation in the *ced-1* gene. Because mutations in this gene prevent engulfment of dead cells, highly refractile dead cells accumulate (arrows), facilitating their visualization. (b) Newly hatched larva with mutations in both the *ced-1* and *ced-3* genes. The absence of refractile dead cells in these double mutants indicates that no cell deaths occurred. Thus CED-3 protein is required for programmed cell death. [From H. M. Ellis and H. R. Horvitz, 1986, *Cell* **91:**818; courtesy of Hilary Ellis.]

death came from genetic studies using *C. elegans*. Of the 947 nongonadal cells generated during development of the adult hermaphrodite form, 131 cells undergo programmed cell death. Specific mutations have identified four genes whose encoded proteins play an essential role in controlling programmed cell death during *C. elegans* development: *ced-3, ced-4, ced-9,* and *egl-1*. In *ced-3* or *ced-4* mutants, for example, the 131 "doomed" cells survive (Figure 21-36). The mammalian proteins that correspond most closely to the worm CED-3, CED-4, CED-9, and EGL-1 proteins are indicated in Figure 21-37. In discussing the worm proteins we will include the mammalian names in parentheses to make it easier to keep the relationships clear.

The confluence of genetic studies in worms and studies on human cancer cells first suggested that an evolutionarily conserved pathway mediates apoptosis. The first mammalian apoptotic gene to be cloned, *bcl-2,* was isolated from human follicular lymphomas. A mutant form of this gene, created in lymphoma cells by a chromosomal rearrangement, was shown to act as an **oncogene** that promoted cell survival rather than cell death (Chapter 25). The chromosome rearrangement joins the coding region of the *bcl-2* gene to an immunoglobulin gene enhancer. The combination results in over-production of Bcl-2 protein that keeps cancer cells alive when, otherwise, they would become programmed to die. The human Bcl-2 protein and worm CED-9 protein are homologous, and a *bcl-2* transgene can block the extensive cell death found in *ced-9* mutant worms even though the two proteins are only 23 percent homologous. Thus both proteins act as regulators that suppress the apoptotic pathway (see Figure 21-37). In addition, both proteins contain a single transmembrane domain and are localized to the outer mitochondrial, nuclear, and endoplasmic reticulum membranes, where they serve as sensors that control the apoptotic pathway in response to external stimuli. As we discuss below, other regulators promote apoptosis.

In the worm apoptotic pathway, CED-3 (caspase 9) is required to destroy cell components during apoptosis. CED-4 (Apaf-1) is a protease-activating factor that causes autocleavage of (and by) the CED-3 precursor protein, creating an active CED-3 protease that initiates cell death (see Figure 21-37). Cell death does not occur in *ced-3* and *ced-4* single mutants or in *ced-9/ced-3* double mutants, whereas all cells die during embryonic life in *ced-9* mutants, so the adult form never develops. These genetic studies indicate that the CED-3 and CED-4 are "killer" proteins required for cell death, that CED-9 (Bcl-2) suppresses apoptosis, and that the apoptotic pathway can be activated in all cells. Moreover, the absence of cell death in *ced-9/ced-3* double mutants suggests that CED-9 acts "upstream" of CED-3 to suppress the apoptotic pathway.

The mechanism by which CED-9 (Bcl-2) controls CED-3 (caspase 9) is now known. CED-9 protein, which is normally tethered to the outside of mitochondria, forms a complex with CED-4 (Apaf-1), thereby preventing activation of CED-3 by CED-4. As a result, the cell survives. This mechanism fits with the genetics, which shows that the absence of CED-9 has no effect if CED-3 is also missing (*ced-3/ced-9* double mutants have no cell death). The crystal structure of the trimeric CED-4/CED-9 complex reveals a huge contact surface between each of the two CED-4 molecules and the single CED-9 molecule (Figure 21-38). The large contact

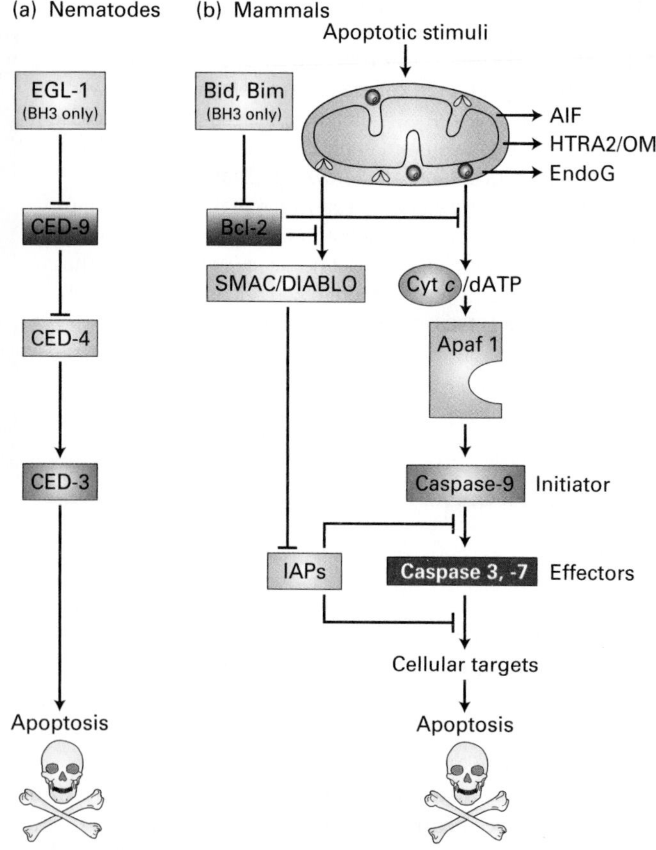

▲ **FIGURE 21-37 Evolutionary conservation of apoptosis pathways.** Similar proteins, shown in identical colors, play corresponding roles in both nematodes and mammals. (a) In nematodes, the protein called EGL-1 binds to CED-9 on the surface of mitochondria; this interaction releases CED-4 from the CED-9/CED-4 complex. Free CED-4 then activates autoproteolysis of the caspase CED-3, which destroys cell proteins to drive apoptosis. These relationships are shown as a genetic pathway, with EGL-1 inhibiting CED-9, which in turn inhibits CED-4. Active CED-4 activates CED-3. (b) In mammals, homologs of the nematode proteins and other proteins regulate apoptosis. The Bcl-2 protein is similar to CED-9 in promoting cell survival by preventing activation Apaf-1, which is similar to CED-4. Two BH3-only proteins, Bid and Bim, inhibit Bcl-2 to allow apoptosis. Apoptotic stimuli damage mitochondria, leading to release of several proteins that stimulate cell death. In particular, cytochrome c released from mitochondria activates Apaf-1, which in turn activates caspase-9. This initiator caspase then activates effector caspases-3 and -7, eventually leading to cell death. See text for discussion of other mammalian proteins (SMAC/DIABLO and IAPs) that have no nematode homologs. [Adapted from S. J. Riedl and Y. Shi, 2004, *Nature Rev. Mol. Cell Biol.* **5**(11):897]

surface makes the association highly specific, but in such a way that the dissociation of the complex can be regulated.

Transcription of *egl-1*, the fourth genetically defined apoptosis regulator gene, is stimulated in response to death signals. Newly produced EGL-1 protein catalyzes the release of CED-4 from CED-9. Both EGL-1 and CED-9 contain a 12-amino acid BH3 domain. Since EGL-1 lacks most of the other domains of CED-9, EGL-1 is called a BH3-only protein. The closest mammalian BH3-only proteins are Bim and Bid. Insight into how EGL-1 disrupts the CED-4/CED-9 complex comes from the crystal structure of EGL-1 (Bid/Bim) complexed with CED-9 (Bcl-2). In this complex, the BH3 domain forms the key part of the contact surface between the two proteins. CED-9 has a different conformation when bound by EGL-1 than when bound by CED-4. This finding suggests that EGL-1 binding distorts CED-9, making its interaction with CED-4 less probable and less stable. Once EGL-1 causes dissociation of the CED-4/CED-9 complex, the released CED-4 dimer dimerizes again to make a tetramer, which then activates CED-3. Cell death soon follows (Figure 21-39). Evidence for similar events has been found in human cultured cells.

Evidence that the steps described here are sufficient for caspase activation comes from experiments in which the events were reconstituted in vitro (i.e., in solution) with purified components. CED-3, CED-4, a truncated CED-9 that lacked its transmembrane mitochondrial membrane anchor, and EGL-1 were purified, as was a CED-4/CED-9 complex. Purified CED-4 (Apaf-1) was able to accelerate the autocatalysis of purified CED-3 (caspase 9), but addition of the truncated CED-9 (Bcl-2) to the reaction mixture inhibited the autocleavage. When the CED-4/CED-9 complex was mixed with CED-3, autocleavage did *not* occur, but addition of EGL-1 to the reaction restored CED-3 autocleavage.

The effector proteins in the apoptotic pathway, the **caspases**, are named because they contain a key cysteine residue in the catalytic site and selectively cleave proteins at sites just C-terminal to aspartate residues. Caspases work as homodimers,

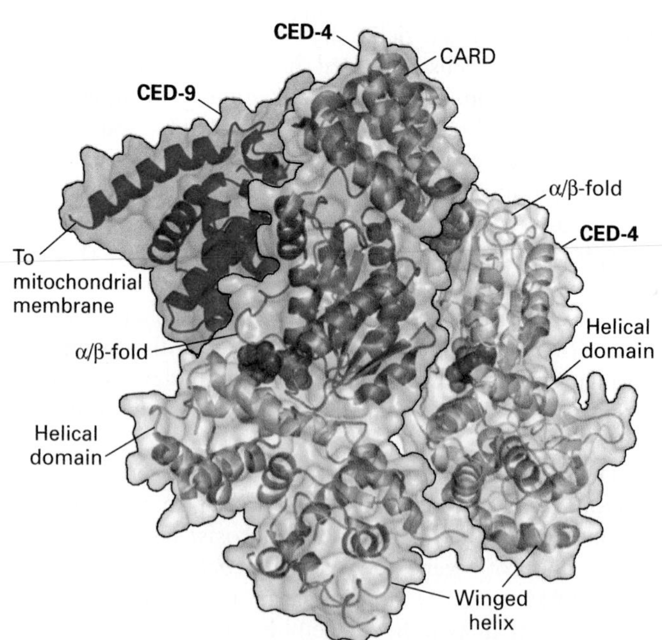

▲ **FIGURE 21-38 Structure of the CED-4/CED-9 protein complex.** The crystal structure has two CED-4 molecules associated with one CED-9 molecule. The C-terminus of CED-9 (dark blue) serves to tether the complex to the mitochondrial membrane. CED-4 is composed of four domains (CARD, α/β folds, winged helix domain, and another helical domain). Each CED-4 molecule has a bound ATP and a Mg^{2+} ion which are visible as a cluster of orange and blue atoms within each subunit. [Based on N. Yan et al., 2005, Nature **437:**831.]

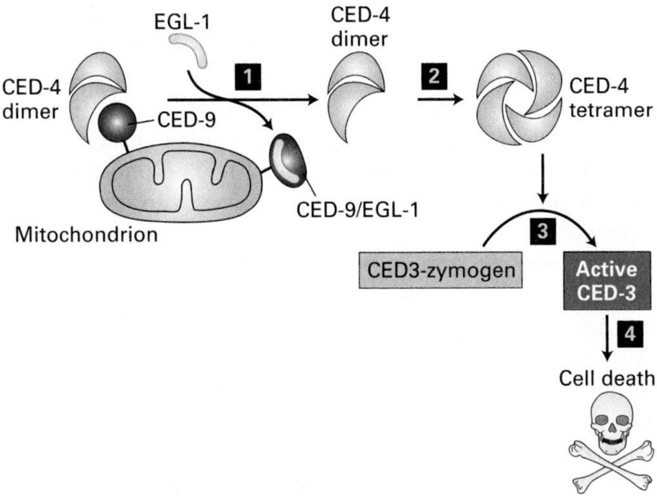

▲ **FIGURE 21-39 Activation of CED-3 protease in *C. elegans*.** EGL-1 protein, which is produced in response to signals that trigger cell death, displaces CED-4 dimer from its association with CED-9 on the surface of mitochondria (**1**). The free CED-4 dimer combines with another to form a tetramer (**2**), which catalyzes the conversion of the CED-3 zymogen (an enzymatically inactive precursor of a protease) into active CED-3 protease (**3**). This effector caspase then begins to destroy cell components and thus initiate apoptosis, leading to cell death (**4**). [Adapted from N. Yan et al., 2005, *Nature* **437**:831.]

with one domain of each stabilizing the active site of the other. The principal effector caspase in *C. elegans* is CED-3, while humans have 15 different caspases. All caspases are initially made as procaspases that must be cleaved to become active. Such proteolytic processing of proproteins is used repeatedly in blood clotting, generation of digestive enzymes, and generation of hormones. In vertebrates, initiator caspases (e.g., caspase-9) are activated by autoproteolysis induced by other types of proteins (e.g., Apaf-1), which help the initiators to aggregate. Activated initiator caspases cleave effector caspases (e.g., caspase-3) and thus quickly amplify the total caspase activity level in the dying cell. The various effector caspases recognize and cleave short amino acid sequences in many different target proteins. They differ in their preferred target sequences. Their specific intracellular targets include proteins of the nuclear lamina and cytoskeleton whose cleavage leads to the demise of a cell.

In mammals and flies but not worms, apoptosis is regulated by several other proteins (see Figure 21-37, *right*). For instance, a family of inhibitor of apoptosis proteins (IAPs), provides another way to restrain both initiator and effector caspases. IAPs have one or more zinc-binding domains that can bind directly to caspases and inhibit their protease activity. (Baculovirus, a type of insect virus, produces a protein that similarly binds to and inhibits caspases, thus preventing an infected cell from committing suicide to stop a viral infection before new viruses can be made.) The inhibition of caspases by IAPs, however, creates a problem when a cell needs to undergo apoptosis. Mitochondria enter the picture once again, since they are the source of a family of proteins, called SMAC/DIABLOs, that inhibit IAPs. After cell injury, SMAC/DIABLOs released from mitochondria bind to IAPs in the cytosol, thereby blocking the IAPs from binding to caspases. By relieving IAP-mediated inhibition, SMAC/DIABLOs promote caspase activity and cell death. Three other mitochondria-associated proteins—Htra2/Omi serine protease, apoptosis-inducing factor (AIF), and endonuclease G—also help to kill cells upon their release from mitochondria following cell injury. Htra2/Omi cleaves IAPs, thus relieving their restraint of apoptosis. Since this regulation is catalytic, Htra2/Omi is a more powerful antagonist of IAPs than is SMAC/DIABLO. AIF, a flavoprotein that normally acts as a NADH oxidase, is cleaved by proteases and moves to the nucleus where it causes chromosome condensation and DNA fragmentation. These effects are caspase-independent, so not all apoptosis involves caspases.

Pro-Apoptotic Regulators Permit Caspase Activation in the Absence of Trophic Factors

Having introduced the major participants in the apoptotic pathway, we now take a closer look at the workings of the mitochondrial membrane proteins that regulate apoptosis. Although the normal function of CED-9 and Bcl-2 is to suppress the cell-death pathway, other intracellular regulatory proteins promote apoptosis. The first pro-apoptotic regulator to be identified, named Bax, is related in sequence to CED-9 and Bcl-2. Overproduction of Bax induces cell death rather than protecting cells from apoptosis, as CED-9 and Bcl-2 do. Thus this family of regulatory proteins comprises both *anti-apoptotic* members (e.g., CED-9, Bcl-2) and *pro-apoptotic* members (e.g., Bax). All members of this family, which we refer to as the **Bcl-2 family,** are single-pass transmembrane proteins and can participate in oligomeric interactions. In mammals, six Bcl-2 family members prevent apoptosis and nine promote it. Thus the fate of a given cell—survival or death—may reflect the particular spectrum of Bcl-2 family members made by that cell and the intracellular signaling pathways regulating them.

Some Bcl-2 family members preserve or disrupt the integrity of the outer mitochondrial membrane, thereby controlling release of mitochondrial proteins such as cytochrome *c*. In normal healthy cells, cytochrome *c* is localized between the inner and outer mitochondrial membrane, but in cells undergoing apoptosis, cytochrome *c* is released into the cytosol. Overproduction of Bcl-2 blocks release of cytochrome *c* and blocks apoptosis; conversely, overproduction of Bax promotes release of cytochrome *c* into the cytosol and promotes apoptosis. Moreover, injection of cytochrome *c* into the cytosol of cells induces apoptosis. A variety of death-inducing stimuli cause Bax monomers to move from the cytosol to the outer mitochondrial membrane where they oligomerize. Bax homodimers, but not Bcl-2 homodimers or Bcl-2/Bax heterodimers, permit influx of ions through the mitochondrial membrane. It remains unclear how this ion influx triggers the release of cytochrome *c*.

The effect of Bcl-2 family members on the permeability of the mitochondrial outer membrane has been mimicked in vitro using vesicles composed of outer mitochondrial membrane. Addition of purified Bax in the presence of a BH3-only protein (e.g., worm EGL-1, vertebrate Bid/Bim) caused permeabilization of the membrane. This biochemical function of Bcl-2 family members appears to reflect their general ability to alter mitochondrial membranes. In addition to increased permeability, mitochondria normally undergo dramatic changes in their network morphology by fusion and fission during the cell-death process. Bcl-xL, a vertebrate member of the Bcl-2 family, and CED-9 introduced into mammalian cells, can induce mitochondrial fusion. Thus these proteins appear to have profound abilities to engineer the properties of outer mitochondrial membranes.

Once cytochrome *c* is released into the cytosol, it binds to Apaf-1 (the mammalian homolog of CED-4) and promotes activation of a caspase cascade leading to cell death (see Figure 21-37, *right*). In the absence of cytochrome *c*, monomeric Apaf-1 is bound to dATP. After binding cytochrome *c*, Apaf-1 cleaves its bound dATP into dADP and undergoes a dramatic assembly process into a disc-shaped heptamer, a 1.4 megadalton wheel of death called the **apoptosome** (Figure 21-40). The apoptosome serves as an activation machine for initiator and effector caspases.

Some Trophic Factors Induce Inactivation of a Pro-Apoptotic Regulator

We saw earlier that neurotrophins such as nerve growth factor (NGF) protect neurons from cell death. In the absence of trophic factors, however, the nonphosphorylated form of a pro-apoptotic protein called Bad is associated with Bcl-2/Bcl-xl at the mitochondrial membrane (Figure 21-41a). Binding of Bad inhibits the anti-apoptotic function of Bcl-2/Bcl-xl, thereby promoting cell death. Phosphorylated Bad, however, cannot bind to Bcl-2/Bcl-xl and is found in the cytosol complexed to the phosphoserine-binding protein 14-3-3. Hence, signaling pathways leading to Bad phosphorylation would be particularly attractive candidates for transmitting survival signals.

A number of trophic factors including NGF have been shown to trigger the PI-3 kinase signaling pathway, leading to activation of protein kinase B (see Figure 16-30). Activated protein kinase B phosphorylates Bad at sites known to inhibit its pro-apoptotic activity. Moreover, a constitutively active form of protein kinase B can rescue cultured neurotrophin-deprived neurons, which otherwise would undergo apoptosis and die. These findings support the mechanism for the survival action of trophic factors depicted in Figure 21-41b. In other cell types, different trophic factors may promote cell survival through post-translational modification of other components of the cell-death machinery.

Another mechanism by which neurotrophins can affect apoptosis, this time positively, involves $p75^{NTR}$, the low-affinity neurotrophin receptor mentioned above. This protein can either promote or inhibit apoptosis depending on the cellular context. In certain neurons, neurotrophin signals such as BDNF stimulate apoptosis by acting through $p75^{NTR}$. In these neurons, cleavage of $p75^{NTR}$ by a membrane-bound

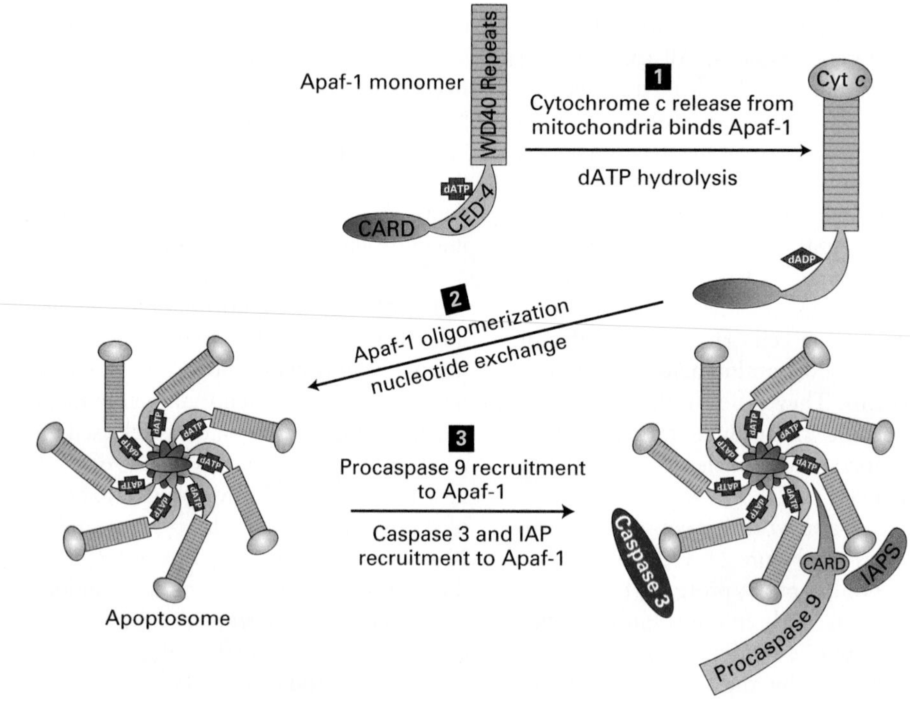

▲ **FIGURE 21-40 Assembly of the mammalian apoptosome.** In the absence of apoptosis triggers, Apaf-1 exists in the cytosol as an inactive monomer bound to dATP. Apaf-1 contains multiple WD40 repeats, a dATP-binding CED-4 domain, and a CARD domain. Step **1**: When apoptosis is triggered, damage to mitochondria allows release of cytochrome c, which binds to Apaf-1. This interaction leads to hydrolysis of the bound dATP to dADP and a change in the conformation of Apaf-1. Step **2**: In its extended conformation, Apaf-1 assembles into a seven-subunit complex, the apoptosome. Step **3**: Interaction of the apoptosome with the initiator procaspase-9 stimulates the autocleavage and dimerization of the procaspase, which is necessary for its activity. Active caspase-9 then acts upon effector caspases such as caspase-3. Inhibitor of apoptosis proteins (IAPs) are also bound by the apoptosome, although their exact actions there are not understood. [Adapted from Z. T. Schafer and S. Kornbluth, 2006, *Devel. Cell* **10**:549.]

Animation: Apoptosis

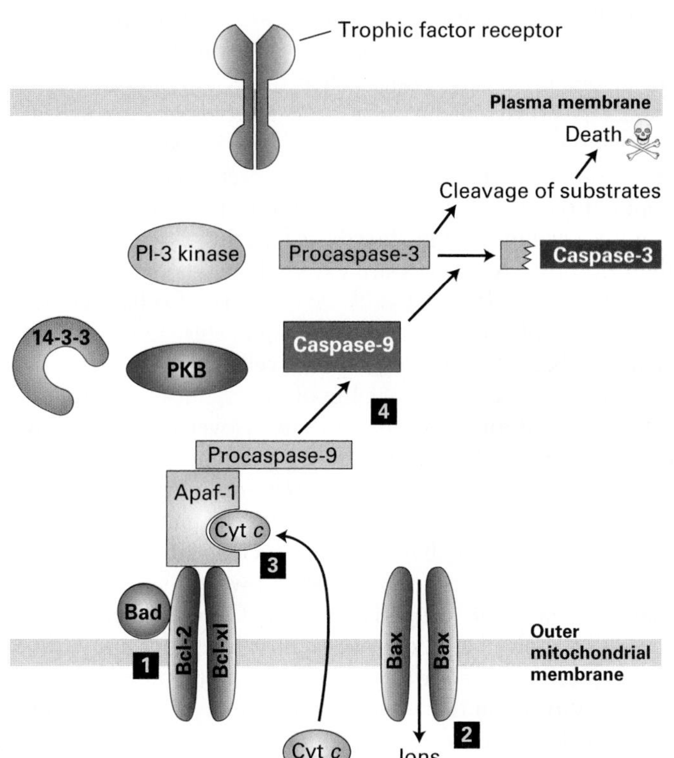

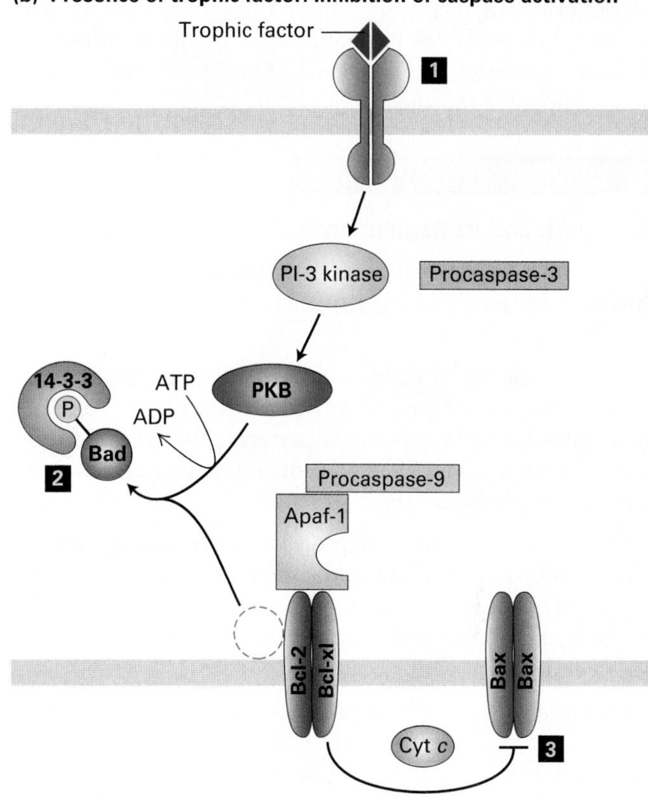

▲ **FIGURE 21-41 Proposed intracellular pathways leading to cell death by apoptosis or to trophic factor–mediated cell survival in mammalian cells.** (a) In the absence of a trophic factor, the soluble pro-apoptotic protein Bad binds to the anti-apoptotic proteins Bcl-2 and Bcl-xl, which are inserted into the mitochondrial membrane (**1**). Bad binding prevents the anti-apoptotic proteins from interacting with Bax, a membrane-bound pro-apoptotic protein. As a consequence, Bax forms homo-oligomeric channels in the membrane that mediate ion flux [**2**]. Through an as-yet-unknown mechanism, this flux leads to the release of cytochrome *c* into the cytosol, where it binds to the adapter protein Apaf-1 (**3**), promoting a caspase cascade that leads to cell death (**4**). (b) In some cells, binding of a trophic factor, such as NGF (**1**) stimulates PI-3 kinase activity, leading to the downstream activation of protein kinase B (PKB), which phosphorylates Bad. Phosphorylated Bad then forms a complex with the 14-3-3 protein (**2**). With Bad sequestered in the cytosol, the anti-apoptotic Bcl-2/Bcl-xl proteins can inhibit the activity of Bax (**3**), thereby preventing the release of cytochrome *c* and activation of the caspase cascade. [Adapted from B. Pettman and C. E. Henderson, 1998, *Neuron* **20**:633.]

protease called γ-secretase, releases the receptor's intracellular domain, which is associated with a DNA-binding protein called NRIF. The cleavage of $p75^{NTR}$ leads to the ubiquitination of NRIF and its movement to the nucleus where it stimulates apoptosis, perhaps by regulating transcription. γ-Secretase is the same protease that catalyzes the intramembrane cleavage of the receptor Notch, thus activating it, and also of amyloid precursor protein (APP) in the genesis of Alzheimer's disease (see Figures 16-36 and 16-37).

Tumor Necrosis Factor and Related Death Signals Promote Cell Murder by Activating Caspases

Although cell death can arise as a default in the absence of survival factors, apoptosis can also be stimulated by positively acting *death signals*. For instance, tumor necrosis factor (TNFα), which is released by macrophages, triggers the cell death and tissue destruction seen in certain chronic inflammatory diseases (Chapter 24). Another important death-inducing signal, the Fas ligand, is a cell-surface protein produced by activated natural killer cells and cytotoxic T lymphocytes. This signal can trigger death of virus-infected cells, some tumor cells, and foreign graft cells.

Both TNF and Fas ligand act through cell-surface "death" receptors that have a single transmembrane domain and are activated when ligand binding brings three receptor molecules into close proximity. The trimeric receptor complex attracts a protein called FADD (Fas-associated death domain), which serves as an adapter to recruit and in some way activate caspase-8, an initiator caspase, in cells receiving a death signal. The death domain found in FADD is a sequence that is present in a number of proteins involved in apoptosis. Once activated, caspase-8 activates other caspases and the amplification cascade begins. To test the ability of the Fas receptor to induce cell death, researchers incubated cells with antibodies against the receptor. These antibodies, which bind and cross-link Fas receptors, were found to stimulate cell death, indicating that activation of the Fas receptor is sufficient to trigger apoptosis.

KEY CONCEPTS OF SECTION 21.5

Cell Death and Its Regulation

- All cells require trophic factors to prevent apoptosis and thus survive. In the absence of these factors, cells commit suicide.
- Genetic studies in *C. elegans* defined an evolutionarily conserved apoptotic pathway with three major components: membrane-bound regulatory proteins, cytosolic regulatory proteins, and effector proteases called caspases in vertebrates (see Figure 21-37).
- Once activated, apoptotic proteases cleave specific intracellular substrates leading to the demise of a cell. Proteins (e.g., CED-4, Apaf-1), which bind regulatory proteins and caspases, are required for caspase activation (see Figures 21-39 and 21-40).
- Pro-apoptotic regulator proteins (e.g., Bax, Bad) promote caspase activation, and anti-apoptotic regulators (e.g., Bcl-2) suppress activation. Direct interactions between pro-apoptotic and anti-apoptotic proteins lead to cell death in the absence of trophic factors. Binding of extracellular trophic factors can trigger changes in these interactions, resulting in cell survival (see Figure 21-41).
- The Bcl-2 family contains both pro-apoptotic and anti-apoptotic proteins; all are single-pass transmembrane proteins and engage in protein–protein interactions. Bcl-2 molecules can restrain the release of cytochrome *c* from mitochondria, inhibiting cell death, while pro-apoptotic factors stimulate membrane breakdown that allows cytochrome C to escape, bind to Apaf-1, and thus activate caspases.
- Binding of extracellular death signals, such as tumor necrosis factor and Fas ligand, to their receptors activates an associated protein (FADD) that in turn triggers the caspase cascade leading to cell murder.

Perspectives for the Future

Cell birth, lineage, and death, which lie at the heart of the development, growth, and healing of an organism, are also central to disease processes, most notably cancer. Few transformations seem more remarkable than the blooming of cell types during development. A lineage beginning with a fertilized egg, a "plain vanilla" 200-μm sphere, produces neurons a yard long, pulsating multinucleate muscle cells, exquisitely light-sensitive retina cells, ravenous macrophages that recognize and engulf germs, and all the hundreds of other cell types. Regulators of cell lineage produce this rich variety by controlling two critical decisions: (1) when and where to activate the cell cycle (Chapter 20) and (2) whether the two daughter cells will be the same or different. A cell may be just like its parent, or it may embark on a new path.

Cell birth is normally carefully restricted to specific locales and times, such as the basal layer of the skin or the root meristem. Liver regenerates when there is injury, but liver cancer is prevented by restricting unnecessary growth at other times. Cell lineage is patterned by the asymmetric distribution of key regulators to the daughter cells of a division. Some of these regulators are intrinsic to the parent cell, becoming asymmetrically distributed during polarization of the cell; other regulators are external signals that differentially reach the daughter cells. Asymmetry of cells becomes asymmetry of tissues and whole organisms. Our left and right hands differ only as a result of cell asymmetry.

Some cells persist for the life of the organism, but others such as blood and intestinal cells turn over rapidly. Many cells live for awhile and are then programmed to die and be replaced by others arising from a stem-cell population. Programmed cell death is also the basis for the meticulous elimination of potentially harmful cells, such as autoreactive immune cells, which attack the body's own cells, or neurons that have failed to properly connect. Cell-death programs have also evolved as a defense against infection, and virus-infected cells are selectively murdered in response to death signals. Viruses, in turn, devote much of their effort to evading host defenses. For example, p53, a transcription factor that senses cell stresses and damage and activates transcription of pro-apoptotic members of the *bcl-2* gene family, is inhibited by the adenovirus E1B protein. It has been estimated that about a third of the adenovirus genome is directed at evading host defenses. Cell death is relevant to toxic chemicals as well as viral infections; malformations due to poisons often originate from excess apoptosis.

Failures of programmed cell death can lead to uncontrolled cancerous growth (Chapter 25). The proteins that prevent the death of cancer cells therefore become possible targets for drugs. A tumor may contain a mixture of cells, some capable of seeding new tumors or continued uncontrolled growth, and some capable only of growing in place or for a limited time. In this sense the tumor has its own stem cells, and they must be found and studied, so they become vulnerable to medical intervention. One option is to manipulate the cell-death pathway by sending signals that will make cancer cells destroy themselves.

Much attention is now being given to the regulation of stem cells in an effort to understand how dividing populations of cells are created and maintained. This has clear implications for repair of tissue: for example, to restore damaged eyes, torn cartilage, degenerating brain tissue, or failing organs. One interesting possibility is that some populations of stem cells with the potential to generate or regenerate tissue are normally eliminated by cell death during later development. If so, finding ways to selectively block the death of these cells could make regeneration more likely. Could the elimination of such cells during mammalian development be the difference between an amphibian capable of limb regeneration and a mammal that is not?

Key Terms

apoptosis *937*
apoptosome *942*
asymmetric cell division *906*
Bcl-2 family *941*
BPP complex *933*
caspases *940*
cell lineage *905*
death signals *943*
determination *925*
differentiation *905*
ectoderm *907*
embryonic stem (ES) cells *911*
endoderm *907*
ganglion mother cell (GMC) *932*
germ line *907*
heterochronic mutants *909*
MAT locus *922*
mating factor *923*
mating-type switching *931*
meristems *920*
mesoderm *907*
micro RNAs (miRNAs) *910*
MPSB complex *933*
muscle regulatory factors (MRFs) *926*
neurotrophins *938*
nuclear-transfer cloning *908*
pluripotent *907*
precursor (progenitor) cells *905*
somatic cells *905*
stem cells *905*
stem-cell niche *912*
transient amplifying (TA) cells *905*
totipotent *907*
trophic factors *936*

Review the Concepts

1. What two properties define a stem cell? Distinguish between a totipotent stem cell, a pluripotent stem cell, and a precursor (progenitor) cell.

2. Where are stem cells located in plants? Where are stem cells located in adult animals? How does the concept of stem cell differ between animal and plant systems?

3. In 1997, Dolly the sheep was cloned by a technique called somatic cell nuclear transfer (or nuclear-transfer cloning). A nucleus from an adult mammary cell was transferred into an egg from which the nucleus had been removed. The egg was allowed to divide several times in culture, then the embryo was transferred to a surrogate mother who gave birth to Dolly. Dolly died in 2003 after mating and giving birth herself to viable offspring. What does the creation of Dolly tell us about the potential of nuclear material derived from a fully differentiated adult cell? Does the creation of Dolly tell us anything about the potential of an intact, fully differentiated adult cell? Name three types of information that function to preserve cell type. Which of these types of information was shown to be reversible by the Dolly experiment?

4. The roundworm *C. elegans* has proven to be a valuable model organism for studies of cell birth, cell lineage, and cell death. What properties of *C. elegans* render it so well suited for these studies? Why is so much information from *C. elegans* experiments of use to investigators interested in mammalian development?

5. How are retroviruses used in tracing experiments that map cell lineages?

6. In the budding yeast *S. cerevisiae,* what is the role of the MCM1 protein in the following?

   a. transcription of **a**-specific genes in **a** cells
   b. blocking transcription of α-specific genes in **a** cells
   c. transcription of α-specific genes in α cells
   d. blocking transcription of **a**-specific genes in α cells

7. In *S. cerevisiae,* what ensures that **a** and α cells mate with one another rather than with cells of the same mating type (i.e., **a** with **a** or α with α)?

8. Exposure of C3H 10T½ cells to 5-azacytidine, a nucleotide analog, is a model system for muscle differentiation. How was 5-azacytidine treatment used to isolate the genes involved in muscle differentiation?

9. Through the experiments on C3H 10T½ cells treated with 5-azacytidine, MyoD was identified as a key transcription factor in regulating the differentiation of muscle. To what general class of DNA-binding proteins does MyoD belong? How do the interactions of MyoD with the following proteins affect its function? (a) E2A, (b) MEFs, (c) Id.

10. The mechanisms that regulate muscle differentiation in mammals and neural differentiation in *Drosophila* (and probably mammals as well) bear remarkable similarities. What proteins function analogous to MyoD, myogenin, Id, and E2A in neural cell differentiation in *Drosophila*? Based on these analogies, predict the effect of microinjection of *myoD* mRNA on the development of *Xenopus* embryos.

11. Predict the effect of the following mutations on the ability of mother and daughter cells of *S. cerevisiae* to undergo mating-type switching following cell division:

   a. loss-of-function mutation in the HO endonuclease
   b. gain-of-function mutation that renders HO endonuclease gene constitutively expressed independent of SWI/SNF
   c. gain-of-function mutation in SWI/SNF that renders it insensitive to Ash1

12. Asymmetric cell division often relies on cytoskeletal elements to generate or maintain the asymmetric distribution of cellular factors. In *S. cerevisiae,* what factor is localized to the bud by myosin motors? In *Drosophila* neuroblasts, what factors are localized apically by microtubules?

13. How do studies of brain development in knockout mice support the statement that apoptosis is a default pathway in neuronal cells?

14. What morphologic features distinguish programmed cell death and necrotic cell death? TNF and Fas ligand bind cell-surface receptors to trigger cell death. Although the death signal is generated external to the cell, why do we consider the death induced by these molecules to be apoptotic rather than necrotic?

15. Predict the effects of the following mutations on the ability of a cell to undergo apoptosis:

   a. mutation in Bad such that it cannot be phosphorylated by protein kinase B (PKB)
   b. overexpression of Bcl-2
   c. mutation in Bax such that it cannot form homodimers

One common characteristic of cancer cells is a loss of function in the apoptotic pathway. Which of the mutations listed above might you expect to find in some cancer cells?

16. How do IAPs (inhibitor of apoptosis proteins) interact with caspases to prevent apoptosis? How do mitochondrial proteins interact with IAPs to prevent inhibition of apoptosis?

Analyze the Data

The immortal-strand hypothesis postulates that when a stem cell divides asymetrically to produce one "new" stem cell and one progenitor cell, the new stem cell receives the sister chromatids containing the oldest strand of DNA (the immortal strand). The other daughter cell, a progenitor cell that eventually gives rise to differentiated cells, receives the sister chromatids containing more recent DNA strands (see diagram below). If the immortal-strand mechanism actually operates, it would prevent the accumulation of mutations in adult stem cells that otherwise would occur during each round of DNA replication.

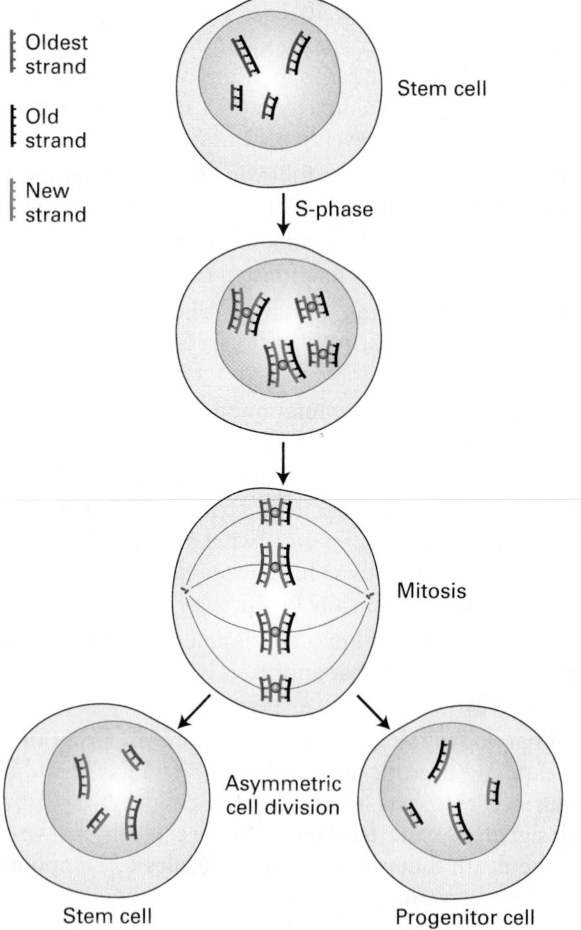

Muscle satellite cells are progenitors for myoblasts and are the source of cells that result in muscle growth after birth and muscle repair after injury. The satellite cells can replenish themselves, suggesting that they also have properties of stem cells. To test the immortal-strand hypothesis, researchers recently conducted the following studies:

a. Satellite cells were isolated from mouse muscle fibers and cultured in vitro in the presence of BrdU, a nucleotide analog that is incorporated into DNA during replication. After 4 days in BrdU, all satellite cells in the culture were extensively labeled with BrdU (pulse), as expected if these cells underwent symmetric divisions. The cells were then incubated for 18 hours in the absence of BrdU (chase), a period of time that corresponds to approximately two cell divisions in these cells. The images below show two examples of muscle satellite cells undergoing division after this 18-hr incubation in the absence of BrdU. The blue labeling (Hoechst) shows total DNA, the red labeling shows where BrdU-containing DNA is located.

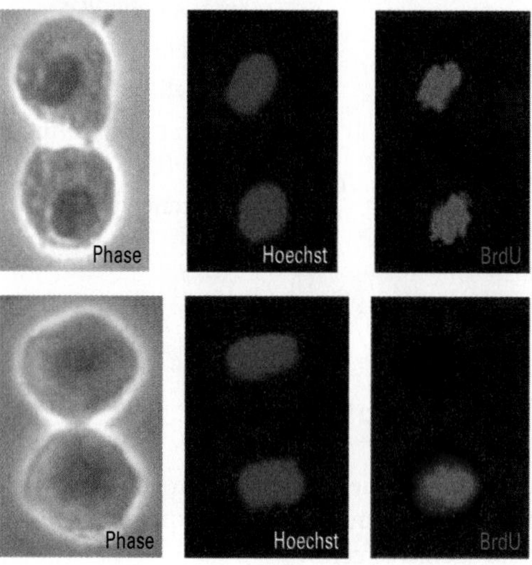

The majority of the cells appear like the dividing cell in the top panel, but about 1.5 percent of dividing cells appear like that in the lower panel. Can you explain these observations? Given that mice have 40 chromosomes, could the segregation of the BrdU-labeled sister chromatids, observed in the lower panel, have occurred by chance?

b. Satellite cells were subjected to a BrdU pulse-chase experiment similar to that described in part (a) above and then were assessed for the production of Numb, a protein whose presence or absence allows two daughter cells to adopt different developmental fates. The micrographs below show a dividing cell stained for Numb (green) and for BrdU-containing DNA (red). What do you expect to be the outcome of the daughter cell that acquires Numb? How might you determine if Numb is involved in generating co-segregation of the older DNA strands?

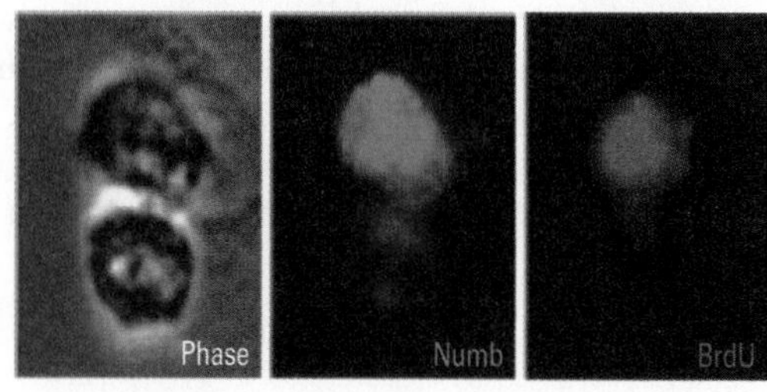

c. Suppose you conducted a pulse-chase experiment using an established cultured cell line, but asymmetric divisions like that observed in the lower panel in part (a) above were not observed. Explain this result.

References

The Birth of Cells: Stem Cells, Niches, and Lineage

Aurelio, O., T. Boulin, and O. Hobert. 2003. Identification of spatial and temporal cues that regulate postembryonic expression of axon maintenance factors in the *C. elegans* ventral nerve cord. *Development* **130**:599–610.

Buszczak, M., and A. C. Spradling. 2006. Searching chromatin for stem cell identity. *Cell* **125**:233–236.

Chopra, V.S., and R. K. Mishra. 2005. To SIR with Polycomb: linking silencing mechanisms. *Bioessays* **27**:119–121.

Copelan, E.A. 2006. Hematopoietic stem-cell transplantation. *N. Engl. J. Med.* **354**:1813–1826.

Edenfeld, G., J. Pielage, and C. Klambt. 2002. Cell lineage specification in the nervous system. *Curr. Opin. Genet. Devel.* **12**:473–477.

Feinberg, A. P., R. Ohlsson, and S. Henikoff. 2006. The epigenetic progenitor origin of human cancer. *Nature Rev. Genet.* **7**:21–33.

Golden, J. A., S. C. Fields-Berry, and C. L. Cepko. 1995. Construction and characterization of a highly complex retroviral library for lineage analysis. *Proc. Nat'l. Acad. Sci. U SA* **92**:5704–5708.

Hatfield, S.D., et al. 2005. Stem cell division is regulated by the micro RNA pathway. *Nature* **435**:974–978.

Hochedlinger, K., and R. Jaenisch. 2006. Nuclear reprogramming and pluripotency. *Nature* **441**:1061–1067.

Hori, Y., et al. 2002. Growth inhibitors promote differentiation of insulin-producing tissue from embryonic stem cells. *Proc. Nat'l. Acad. Sci. USA* **99**:16105–16110.

Huelsken, J., et al. 2001. β-Catenin controls hair follicle morphogenesis and stem cell differentiation in the skin. *Cell* **105**:533–545.

Li, L., and T. Xie. 2005. Stem cell niche: structure and function. *Ann. Rev. Cell Devel. Biol.* **21**:605–631.

Marshman, E., C. Booth, and C. S. Potten. 2002. The intestinal epithelial stem cell. *Bioessays* **24**:91–98.

Morrison, S. J., and J. Kimble. 2006. Asymmetric and symmetric stem-cell divisions in development and cancer. *Nature* **441**:1068–1074.

Orkin, S. H. 2000. Diversification of haematopoietic stem cells to specific lineages. *Nature Rev. Genet.* **1**:57–64.

Reinhart, B. J., et al. 2000. The 21-nucleotide *let-7* RNA regulates developmental timing in *Caenorhabditis elegans. Nature* **403**:901–906.

Sanchez Alvardo, A. 2006. Planarian regeneration: its end is its beginning. *Cell* **124**:241–245.

Shafritz, D.A., et al. 2006. Liver stem cells and prospects for liver reconstitution by transplanted cells. *Hepatology* **43**(2 Suppl 1):S89–98.

Watt, F. M., C. Lo Selso, and V. Silva-Vargas. 2006. Epidermal stem cells: an update. *Curr. Opin. Genet. Devel.* **16**:518–524.

Wu, H., and Y. E. Sun. 2006. Epigenetic regulation of stem cell differentiation. *Pediatr. Res.* **59**(4 Pt 2):21R–25R.

Cell-Type Specification in Yeast

Bagnat, M., and K. Simons. 2002. Cell surface polarization during yeast mating. *Proc. Nat'l. Acad. Sci. USA* **99**:14183–14188.

Coic, E., G-F. Richard, and J. E. Haber. 2006. Cell cycle-dependent regulation of *Saccharomyces cerevisiae* donor preference during mating-type switching by SBF (Swi4/Swi6) and Fkh1. *Mol. Cell Biol.* **26**:5470–5480.

Cosma, M. P. 2004. Daughter-specific repression of *Saccharomyces cerevisiae* HO: Ash1 is the commander. *EMBO Rep.* **5**:953–957.

Dittmar, G. A., et al. 2002. Role of a ubiquitin-like modification in polarized morphogenesis. *Science* **295**:2442–2446.

Dohlman, H. G., and J. W. Thorner. 2001. Regulation of G protein-initiated signal transduction in yeast: paradigms and principles. *Ann. Rev. Biochem.* **70**:703–754.

Hall, I. M., et al. 2002. Establishment and maintenance of a heterochromatin domain. *Science* **297**:2232–2237.

Kirchmaier, A. L., and J. Rine. 2006. Cell cycle requirements in assembling silent chromatin in *Saccharomyces cerevisiae. Mol. Cell Biol.* **26**:852–862.

Lau, A., H. Blitzblau, and S. P. Bell. 2002. Cell-cycle control of the establishment of mating-type silencing in *S. cerevisiae. Genes Devel.* **16**:2935–2945.

Miller, M. G., and A. D. Johnson. 2002. White-opaque switching in *Candida albicans* is controlled by mating-type locus homeodomain proteins and allows efficient mating. *Cell* **110**:293–302.

Takizawa, P. A., and R. D. Vale. 2000. The myosin motor, Myo4p, binds Ash1 mRNA via the adapter protein, She3p. *Proc. Nat'l. Acad. Sci. USA* **97**:5273–5278.

Specification and Differentiation of Muscle

Bailey, P., T. Holowacz, and A. B. Lassar. 2001. The origin of skeletal muscle stem cells in the embryo and the adult. *Curr. Opin. Cell Biol.* **13**:679–689.

Berkes, C.A, and S. J. Tapscott. 2005. MyoD and the transcriptional control of myogenesis. *Semin. Cell. Devel. Biol.* **16**:585–595.

Buckingham, M., S. Meilhac, and S. Zaffran. 2005. Building the mammalian heart from two sources of myocardial cells. *Nature Rev. Genet.* **6**:826–835.

Dhawan, J., and T. A. Rando. 2005. Stem cells in postnatal myogenesis: molecular mechanisms of satellite cell quiescence, activation and replenishment. *Trends Cell Biol.* **15**:666–673.

Gustafsson, M. K., et al. 2002. Myf5 is a direct target of long-range Shh signaling and Gli regulation for muscle specification. *Genes Devel.* **16**:114–126.

McKinsey, T. A., C. L. Zhang, and E. N. Olson. 2002. Signaling chromatin to make muscle. *Curr. Opin. Cell Biol.* **14**:763–772.

Pipes, G. C., E. E. Creemers, and E. N. Olson. 2006. The myocardin family of transcriptional coactivators: versatile regulators of cell growth, migration, and myogenesis. *Genes Devel.* **20**:1545–1556.

Yan, Z., et al. 2003. Highly coordinated gene regulation in mouse skeletal muscle regeneration. *J. Biol. Chem.* **278**:8826–8836.

Regulation of Asymmetric Cell Division

Bellaiche, Y., and M. Gotta. 2005. Heterotrimeric G proteins and regulation of size asymmetry during cell division. *Curr. Opin. Cell Biol.* **17**:658–663.

Betschinger, J., and J. A. Knoblich. 2004. Dare to be different: asymmetric cell division in *Drosophila, C. elegans* and vertebrates. *Curr. Biol.* **14**:R674–685.

Betschinger, J., K. Mechtler, and J. A. Knoblich. 2003. The Par complex directs asymmetric cell division by phosphorylating the cytoskeletal protein Lgl. *Nature* **422**:326–330.

Betschinger, J., K. Mechtler, and J. A. Knoblich. 2006. Asymmetric segregation of the tumor suppressor brat regulates self-renewal in *Drosophila* neural stem cells. *Cell* **124**:1241–1253.

Bhalerao, S., et al. 2005. Localization-dependent and -independent roles of numb contribute to cell-fate specification in *Drosophila. Curr. Biol.* **15**:1583–1590.

Bowman, S. K., et al. 2006. The *Drosophila* NuMA Homolog Mud regulates spindle orientation in asymmetric cell division. *Devel. Cell* **10**:731–742.

Cleary, M. D., and C. Q. Doe. 2006. Regulation of neuroblast competence: multiple temporal identity factors specify distinct neuronal fates within a single early competence window. *Genes Devel.* **20**:429–434.

Cowan, C.R., and A. A. Hyman. 2004. Asymmetric cell division in *C. elegans:* cortical polarity and spindle positioning. *Ann. Rev. Cell Devel. Biol.* **20**:427–453.

Fichelson, P., et al. 2005. Cell cycle and cell-fate determination in *Drosophila* neural cell lineages. *Trends Genet.* **21**:413–420.

Heidstra, R., D. Welch, and B. Scheres. 2004. Mosaic analyses using marked activation and deletion clones dissect *Arabidopsis* SCARECROW action in asymmetric cell division. *Genes Devel.* **18**:1964–1969.

Helariutta, Y., et al. 2000. The SHORT-ROOT gene controls radial patterning of the *Arabidopsis* root through radial signaling. *Cell* **101**:555–567.

Hutterer, A., et al. 2004. Sequential roles of Cdc42, Par-6, aPKC, and Lgl in the establishment of epithelial polarity during *Drosophila* embryogenesis. *Devel. Cell* **6**:845–854.

Kipreos, E. T. 2005. *C. elegans* cell cycles: invariance and stem cell divisions. *Nature Rev. Mol. Cell Biol.* **6**:766–776.

Lee, C-Y., K. J. Robinson, and C. Q. Doe. 2006. Lgl, Pins and aPKC regulate neuroblast self-renewal versus differentiation. *Nature* **439**:594–598.

Lee, C-Y., et al. 2006. Brat is a Miranda cargo protein that promotes neuronal differentiation and inhibits neuroblast self-renewal. *Devel. Cell* **10**:441–449.

Lu, H., and D. Bilder. 2005. Endocytic control of epithelial polarity and proliferation in *Drosophila. Nature Cell Biol.* **7**:1232–1239.

Nance, J. 2005. PAR proteins and the establishment of cell polarity during *C. elegans* development. *Bioessays* **27**:126–135.

O'Donnell, K.A., et al. 2005. c-Myc-regulated micro RNAs modulate E2F1 expression. *Nature* 435:839–843.

Petritsch, C., et al. 2003. The *Drosophila* myosin VI Jaguar is required for basal protein targeting and correct spindle orientation in mitotic neuroblasts. *Devel. Cell* **4**:273–281.

Plant, P. J., et al. 2003. A polarity complex of mPar-6 and atypical PKC binds, phosphorylates and regulates mammalian Lgl. *Nature Cell Biol.* 5:301–308.

Shapiro, L., H. H. McAdams, and R. Losick. 2002. Generating and exploiting polarity in bacteria. *Science* **298**:1942–1946.

Siegrest, S. E., and C. Q. Doe. 2005. Microtubule-induced Pins/$G_{\alpha i}$ cortical polarity in *Drosophila* neuroblasts. *Cell* **123**:1323–1335.

Siegrest, S. E., and C. Q. Doe. 2006. Extrinsic cues orient the cell division axis in Drosophila embryonic neuroblasts. *Development* 133:529–536.

Siller, K. H., C. Cabernard, and C. Q. Doe. 2006. The NuMA-related Mud protein binds Pins and regulates spindle orientation in *Drosophila* neuroblasts. *Nature Cell Biol.* **8**:594–600.

Wang, H., and W. Chia. 2005. *Drosophila* neural progenitor polarity and asymmetric division. *Biol. Cell* **97**:63–74.

Wodarz, A. 2005. Molecular control of cell polarity and asymmetric cell division in *Drosophila* neuroblasts. *Curr. Opin. Cell Biol.* **17**:475–481.

Zarnescu, D. C., et al. 2005. Fragile X protein functions with lgl and the par complex in flies and mice. *Devel. Cell* **8**:43–52.

Zgurski, J. M., et al. Asymmetric auxin response precedes asymmetric growth and differentiation of asymmetric leaf1 and asymmetric leaf2 *Arabidopsis* leaves. *Plant Cell* **17**:77–91.

Cell Death and Its Regulation

Aderem, A. 2002. How to eat something bigger than your head. *Cell* **110**:5–8.

Alvarez-Garcia, I., and E. A. Miska. 2005. Micro RNA functions in animal development and human disease. *Development* **132**:4653–4662.

Ambrose, V. 2003. Micro RNA pathways in flies and worms: growth, death, fat, stress, and timing. *Cell* **113**:673–676.

Baehrecke, E. H. 2002. How death shapes life during development. *Nature Rev. Molec. Cell Biol.* **3**:779–787.

Bao, Q., S. T. Riedl, and Y. Shi. 2005. Structure of Apaf-1 in the auto-inhibited form: a critical role for ADP. *Cell Cycle* **8**: 1001-1003.

Brennecke, J., et al. 2003. bantam encodes a developmentally regulated micro RNA that controls cell proliferation and regulates the proapoptotic gene hid in *Drosophila. Cell* **113**:25–36.

Cory, S., and J. M. Adams. 2002. The Bcl2 family: regulators of the cellular life-or-death switch. *Nature Rev. Cancer* **2**: 647–656.

Estaquier, J., and D. Arnoult. 2006. CED-9 and EGL-1: a duo also regulating mitochondrial network morphology. *Molec. Cell* **21**:730–732.

Hay, B. A., and M. Guo. 2006. Caspase-dependent cell death in *Drosophila. Ann. Rev. Cell Devel. Biol.* **22**:623–650.

Green, D. R., and G. Kroemer. 2004. The pathophysiology of mitochondrial cell death. *Science* **305**:626–629.

Jacks, T., and R. A. Weinberg. 2002. Taking the study of cancer cell survival to a new dimension. *Cell* **111**:923–925.

Kinchen, J. M., et al. 2005. Two pathways converge at CED-10 to mediate actin rearrangement and corpse removal in *C. elegans. Nature* **434**:93–99.

Kinchen, J. M., and M. O. Hengartner. 2005. Tales of cannibalism, suicide, and murder: programmed cell death in *C. elegans. Curr. Top. Devel. Biol.* **65**:1–45.

Lakhani, S. A., et al. 2006. Caspases 3 and 7: key mediators of mitochondrial events of apoptosis. *Science* **10**:847–851.

Marsden, V. S., and A. Strasser. 2003. Control of apoptosis in the immune system: Bcl-2, BH3-only proteins and more. *Ann. Rev. Immunol.* **21**:71–105.

Penninger, J. M., and Kroemer, G. 2003. Mitochondria, AIF, and caspases—rivaling for cell death execution. *Nature Cell Biol.* **5**:97–99.

Riedl, S. J., and Y. Shi. 2004. Molecular mechanisms of caspase regulation during apoptosis. *Nature Rev. Molec. Cell Biol.* **5**:897–907.

Schafer, Z. T., and S. Kornbluth. 2006. The apoptosome: physiological, developmental, and pathological modes of regulation. *Devel. Cell* **10**:549–561.

Vaccari, T., and D. Bilder. 2005. The *Drosophila* tumor suppressor vps25 prevents nonautonomous overproliferation by regulating notch trafficking. *Devel. Cell* **9**:687–698.

Xu, P., M. Guo, and B. A. Hay. 2004. Micro RNAs and the regulation of cell death. *Trends Genet.* **20**:617–624.

Yan, N., et al. 2004. Structural, biochemical, and functional analyses of CED-9 recognition by the proapoptotic proteins EGL-1 and CED-4. *Molec. Cell* **24**:999–1006.

Yan, N., et al. 2005. Structure of the CED-4–CED-9 complex provides insights into programmed cell death in *Caenorhabditis elegans. Nature* **437**:831–837.

Yan, N., and Y. Shi. 2005. Mechanisms of apoptosis through structural biology. *Ann. Rev. Cell Devel. Biol.* **21**: 35–56.

Zuzarte-Luis, V., and J. M. Hurle. 2002. Programmed cell death in the developing limb. *Int'l. J. Devel. Biol.* **46**: 871–876.

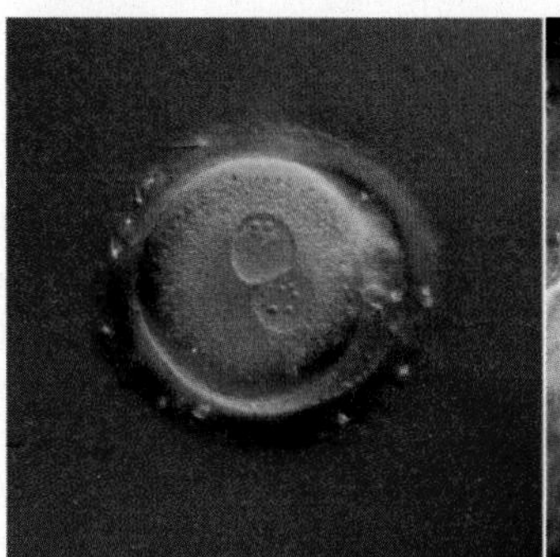
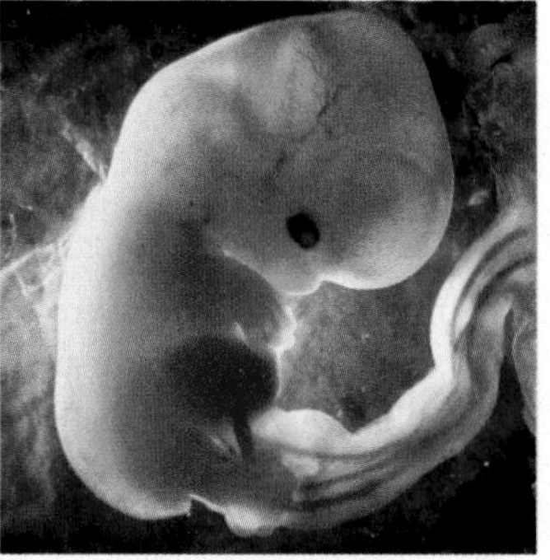

From a single cell to a human embryo. Twenty hours after the fertilization of a human egg, the male and female pronuclei are about to fuse and combine the genetic information from father and mother. Forty-six days later the embryo, 2 cm long, is beginning to develop organs and tissues, nourished by blood entering through the umbilical cord. [Courtesy of The Lennart Nilsson Award Board.]

CHAPTER

22

THE MOLECULAR CELL BIOLOGY OF DEVELOPMENT

Just as notes and chords blend into a symphony, genes, proteins, and cells act as an integrated system during **embryogenesis**, the development of an embryo. Signals flow within and between cells, and massive waves of gene expression allow thousands of types and shapes of cells to form (Chapters 7, 15, 16). For normal embryogenesis, the cell cycle must be regulated, as described in Chapter 20, so that cell growth and division occur at the right times and places; cell lineages like those described in Chapter 21 must be organized in time and space; mechanisms described in Chapters 17–19 must organize cells into tissues, organs, and whole bodies; and cell death (Chapter 21) must be programmed so that the webbing—not the fingers—is removed.

In this chapter we explore the regulation of early-stage animal embryos to observe developmental mechanisms in context. We concentrate on insects and mammals, with some examples from other animal species, as well as plants. After a brief summary of early development, we describe how eggs and sperm are made, how fertilization occurs, and the special genetic properties of early mammalian cells. Then we look at the earliest cell divisions in mammalian development and the creation of different layers of tissues. The formation of repeating segments in animal embryos and the genes that eventually cause those segments to differ are discussed next. We also examine several particularly informative aspects of later animal development, including formation of the left-right asymmetry of the body, control of cell fates in the early nervous system, and the patterning of limbs. As we cover various topics, we will see how different experimental approaches—lineage tracing, genetic screens, mosaic animals, manipulations of signaling proteins, and transplantation—have been used to discover and analyze key molecular and cellular events that build animals.

Let's start by thinking about a simple situation in which a sheet of cells has formed through cell division, but all the cells are identical. To form a working tissue, each cell has to do its job. Some may divide, some may bend, some may send out a signal. Each cell must somehow learn its location and fate, and start to differentiate appropriately. **Differentiation** may entail activation of certain genes, production of particular proteins, an increase or decrease in cell division, changed shape, changed surface properties and adhesion to other cells, the release of secreted signals, the acquisition of electrical activity, polarization along one or more axes,

OUTLINE

migration, or a combination of any of these. A mistake in any aspect of cell differentiation early in development can be fatal to the organism.

The fascination of developmental cell biology lies in discovering how the integrated system of development works and why it is so successful despite variations in environment, inherited genes, cell numbers, and nutrition. At the same time, this field offers a new way to explore evolution and human origins: how animal forms and species arose, are maintained, and change. In addition, many diseases are most readily understood in the context of normal developmental processes gone awry. And it all begins with a single cell!

22.1 Highlights of Development

Single-celled organisms can go through complex developmental life cycles during which their shapes and behaviors may change dramatically. An example is the life cycle of the malaria plasmodium discussed in Chapter 1 (see Figure 1-4). In this chapter, however, we concentrate on development of multicellular animals. The earliest stages of animal development accomplish several crucial goals: combination of maternal and paternal genomes in a new organism; an increase in the number of cells; formation of three main layers of cells, the first step in creation of different cell and tissue types; and the laying out of the main organization of the embryo—front to back, head to tail, and left to right.

Development Progresses from Egg and Sperm to an Early Embryo

The development of a new organism begins with the fusion of male and female **gametes:** an egg (*oocyte*), carrying a set of chromosomes from the mother, and a *sperm*, carrying a set of chromosomes from the father. The gametes, or sex cells, are **haploid** because they have gone through **meiosis** and thus contain only one set of chromosomes (see Figure 5-3). They combine in a process called fertilization, creating the initial single cell, the **zygote**, that has two sets of chromosomes (one maternal and one paternal) and is therefore **diploid.** The zygote begins to divide in a process called **cleavage,** which produces a mass of cells that often look rather similar to each other (Figure 22-1). The progeny of these initial cells will gradually become different, forming all the organs and tissues.

Early in embryogenesis the cells divide into two distinct sets: **germ-line cells,** which will give rise to gametes, and **somatic cells,** which will form most of the body but are not passed on to future progeny. Germ-line cells are carriers of genetic alterations and, in some cases, inherited disease states, but anything that happens to the DNA of somatic cells cannot be passed on to future generations.

In order for a functioning organism to develop, the embryo must become **polarized.** That is, cells on one side (left-right or head-tail or front-back) behave differently than cells on another side. The information that sets the polarities of

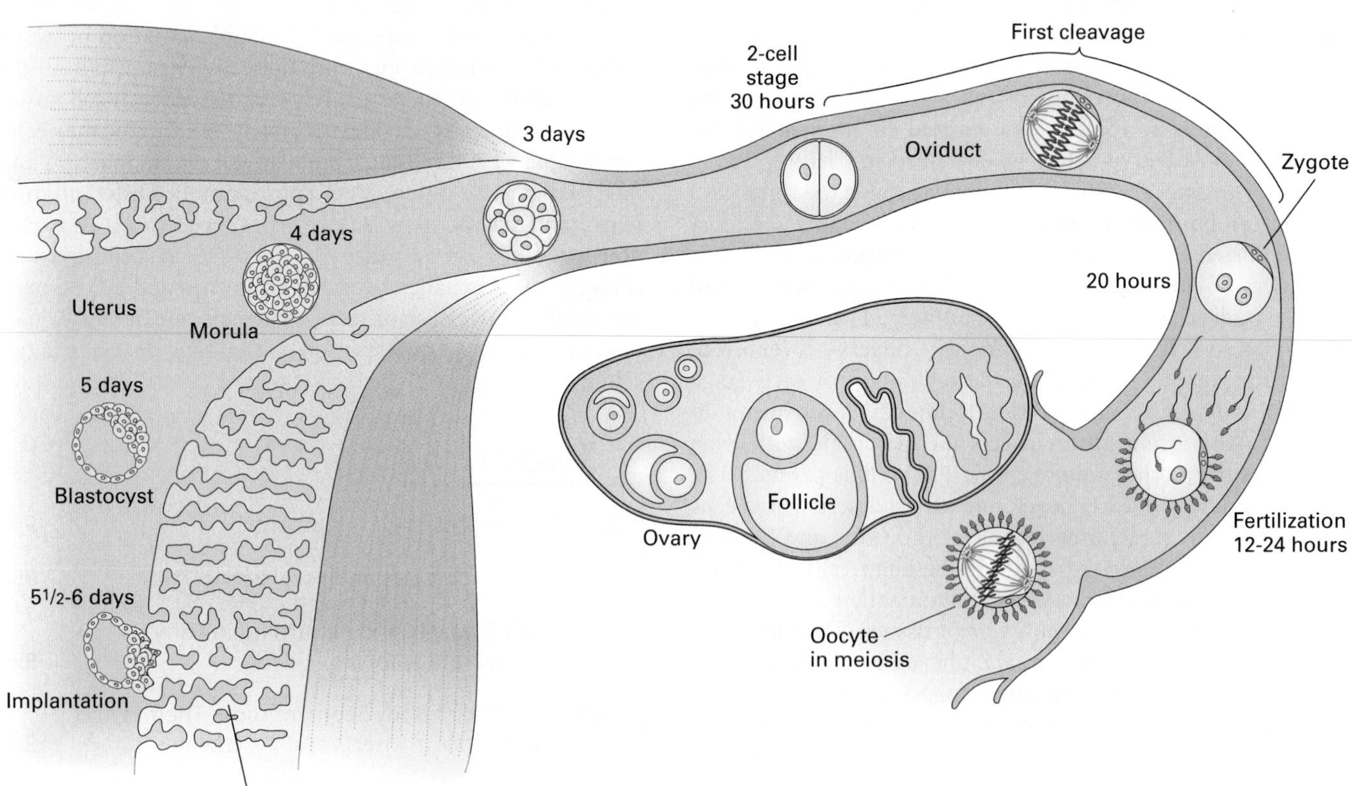

▲ **FIGURE 22-1 Early stages in human development.** This drawing depicts fertilization, cleavage (cell divisions), and implantation in the context of the oviduct and uterus where these events occur. The timing is indicated in terms of hours or days after release of an egg from the ovary. [Adapted from T.W. Sadler, *Langman's Medical Embryology*, 9th ed., Lippencott Williams and Wilkins, Figure 2-11.]

the embryo may be provided in the form of localized proteins or RNA molecules in the egg or the location where the sperm enters the egg, which can cause local changes that begin the process of polarization. Alternatively, a random asymmetry of the early embryo may become fixed as the embryo's final polarity.

To create the initial tissue layers, sheets of cells fold and some cells move inward, a process called **gastrulation.** The embryo becomes more three-dimensional and consists of three **germ layers:** the ectoderm ("outer skin"), the mesoderm ("middle skin"), and the endoderm ("inner skin"). Although the concept of three germ layers is really an oversimplification of the more complex truth, each layer gives rise to particular tissues and structures, even in rather diverse animals, and thus the terms have had enough value to persist (see Figure 21-3). The cell movements that generate the germ layers require that the surface properties of cells change to release or cement connections between cells. Cells in the early embryo can form either loose-knit masses classified as **mesenchyme** or sheets called **epithelia.** In the course of development, cells can go back and forth between these two states. Epithelia are usually polarized, one side of the sheet being described as basal and the other apical.

Early in development, cells begin to signal using the various types of secreted proteins discussed in Chapters 15 and 16. For signaling to be useful in creating new cell types, some cells must make the signal, while others make the receptor. Cells that are able to sense and respond to a signal are described as *competent*. The polarization of the early embryo that begins to make cells different allows some cells to become signal producers and other cells to become signal receivers; some cells both produce and receive signals. **Induction** is the process whereby a signal sent from one cell or set of cells influences the fates of other cells. For example, a signal may induce precursor cells to form neural tissue.

Embryonic cells are constantly sampling their local environment to see how it conforms to their genetically programmed expectations. The absence of an expected signal or the presence of an inappropriate one can cause a cell to die, divide, move, change identity, or change shape. One signal can inform a cell about its location along the dorsal-ventral (back-front) body axis, while another tells the cell its anterior-posterior (head-tail) position. Yet a third signal could tell the cell to commence mitosis. A cell can *integrate* multiple signals and thus respond simultaneously to different signals. In some cases, a cell's final response will not be a simple summation of the multiple signals. Instead, one signal may modify the cell's response to another signal. In addition to local signals, long-distance signals such as hormones, which can reach much or all of a developing or mature animal, coordinate the timing of events, stimulating growth or triggering other changes.

Some developmental signals act in a concentration-dependent manner. That is, the receiving cells respond in one way to high signal levels and in a different way to low signal levels. Generally, cells close to the source of a signal will be exposed to high levels, and cells farther away to lower levels. Thus the location of the source, the location of cells that have the relevant receptor, and the ability of the signal to move through tissue will govern where particular responses to a concentration-dependent signal occur.

As the Embryo Develops, Cell Layers Become Tissues and Organs

Molecular cell biology underlies the ways individual cells differentiate into different cell types, say those forming a kidney or the skin. Critical to cell differentiation is activation of specific genes, such as the genes that encode various keratins, the fibrous proteins of the skin. In addition to the differential gene expression that creates and characterizes a *type* of cell, the formation of tissues and organs depends on the proper *arrangements* of cells. For example, the light-sensitive cells of the eye must be arrayed on the retina surface, and muscle cells must be arranged so that they can move the lens and focus the image on the retina. Thus formation of organs and tissues depends on communication between cells as well as movements and rearrangements of cells. The branching patterns of blood vessels, and the proper connections between nerves and muscles, are possible because developing cells communicate.

The process of forming all the working parts of the body—the heart, liver, kidney, gonads, lungs, etc.—is known as *organogenesis*. Organogenesis often involves bending and folding layers of cells. An epithelial sheet of cells can bend if cells along a stripe down the middle all constrict their apical end, while the other end is relaxed. Such coordinated changes in cytoskeletal shape and organization are crucial in organogenesis, as are signals between different tissue layers. Circulatory systems and hearts typically undergo organogenesis early, since abundant oxygen must be delivered to every cell.

Distinct cell types from distinct lineages collaborate to form an organ. The skin has epidermis from the ectoderm and dermis from the mesoderm; the lens of the eye derives from one cell lineage and retina from another; limbs are made of muscle and bone and nerves and skin. Signals between tissues of different origins allow formation of proper structures at the right times and places.

The organization of whole tissues, known as **pattern formation,** underlies much of the beauty of the natural world: flowers, butterfly wings, faces, coral reef fish; warning colors, camouflage, and mimicry. Symmetric patterns are common in animals: Our left hand mirrors our right hand, for instance. Both hands are made of the same stuff; only the pattern differs. Patterns often include repeating units like body segments or vertebrae or fingers. Variations on the repeats confer different functions, like opposable thumbs. Symmetry breaking, or asymmetric patterning, is crucial to development of many animals, including humans. For example, our growing heart twists in a particular helical way to begin forming heart chambers.

Although this chapter concentrates on development of the embryo, development continues after birth of an animal. Many tissues continue to grow and develop until the adult stage is reached. And, as discussed in Chapter 21, some tissues such as skin, blood, hippocampus (part of the brain), intestinal lining, and the eyes of fish continuously regenerate even in adults, or regenerate after injury (e.g., liver). In many animals, ecological complexities such as changing conditions or migrations trigger *metamorphosis*. Insect embryos, for instance, metamorphose to become larvae (juveniles), which subsequently metamorphose to become adults; salmon metamorphose to adapt from fresh to salt water.

Aging also can be viewed as a developmental process. As an animal ages, many cell types undergo physiological changes that underlie changes in organ and tissue functions. That aging is a genetically programmed aspect of development is apparent from observations of seemingly similar tissues aging at different rates in different animal species. Aging can also be strongly influenced by environmental factors, so as in many developmental processes there is an interplay between genes, nutrition, and experience.

Genes That Regulate Development Are at the Heart of Evolution

Evolution is a process of change in the forms and abilities of living organisms over generations. Just as no two humans are identical (even "identical" twins are not), differences are found within any group of organisms. Occasionally the differences confer an advantage in reproduction. In fact the process of evolution has been employed by farmers and others for thousands of years, selecting plants and animals with useful properties and selectively breeding those. The enormous variation among strains of dogs, for example, has resulted from selection by humans over perhaps 10,000–20,000 years. The fossil record shows that natural selection, which has operated for hundreds of millions of years, has led to changes far greater than the differences among dogs. Changes in climate, geology, location and the presence of other creatures created new ecological niches and animals evolved in ways that allowed life in many of them.

Since the form and functions of any organism are controlled by an increasingly well-understood battery of development-regulating genes, changes in such genes must account for the heritable changes that underlie the evolution of different animals and plants. However, the mutations underlying inherited evolutionary changes, whether the selection was by humans or by natural causes, generally have been unknown in the past. This is beginning to change with the current surge of knowledge about which genes matter to the development of tissues and organs, along with the increasing availability of DNA sequences from many organisms. These advances are revealing the molecular genetic differences that distinguish varieties of plants and animal.

Many changes in genes are harmful, and changes in genes that control growth and/or development might be expected to have major deleterious effects. Indeed, many inherited human genetic syndromes (a syndrome is a group of disease features that tend to occur together) involving birth defects or cancer have been linked to developmental genes.

The conservation of most types of developmentally important proteins across diverse animal species indicate that these proteins existed in animal ancestors half a billion years ago. In many cases, particular proteins have remained engaged in generating a particular organ or tissue type over all that time, despite the dramatic morphological distinctions between, say, insect and mammalian hearts. The similarities in gene functions among animal species have allowed inferences to be drawn about human gene functions and disease processes based on information from many different species.

Evolution of animal form could in principle depend upon the emergence of new genes and new proteins, but relatively few such cases are known. Instead changes in body form over generations appear to be due primarily to alterations in hundreds of thousands of short regulatory DNA sequences that influence the transcription of genes and the processing of RNA. Most of the working protein "hardware" is similar in animals of vastly different morphology, but the "software" can change readily and rapidly.

KEY CONCEPTS OF SECTION 22.1

Highlights of Development

- The paternal and maternal genomes come together in the fertilized egg.
- Germ-line cells make eggs and sperm and some are inherited by progeny; somatic cells make all the rest of the body but are not passed on to progeny.
- During embryogenesis, a largely symmetric fertilized human egg becomes an early embryo that is asymmetric along the anterior-posterior (head-tail), left-right, and dorsal-ventral (back-front) axes (see Figure 22-1).
- An early embryo forms three initial layers of cells—endoderm, mesoderm, and ectoderm—that give rise to different tissues and organs.
- Embryonic cells communicate with each other via secreted protein signals that bind to receptors on receiving cells, triggering changes in the receiving cells that lead to their differentiation into specific cell types. This process, in which one type of cell, or group of cells, sends a signal that affects the fates of other cells, is called induction.
- Organogenesis is the process of forming all the working tissues and organs of the body.
- Pattern formation is the process of organizing the shapes, sizes, and colors of tissues and organs during development.
- Many genes have been identified that control development, and damage to such genes can lead to birth defects, cancer, or tissue degeneration.
- Many types of development-regulating genes are evolutionarily conserved. Not only can the genes be identified in a wide spectrum of animal types, but in many cases a gene plays a similar role in seemingly very different animals. This reflects the evolution of animals from common ancestors.

22.2 Gametogenesis and Fertilization

Whether in the comfort of a womb or the shelter of a shell, embryos confront similar challenges: starting with the right chromosomes, obtaining nutrition, organizing growth, making diverse cell types, and creating pattern.

Germ-line Cells Are All That We Inherit

The setting aside of germ-line cells early in development has been hypothesized to protect chromosomes from damage by reducing the number of rounds of replication they undergo or by allowing special protection of the cells that are critical to heredity. Whatever the reason, early segregation of the germ line is widespread (though not universal) among animals. In contrast, plants do nothing of the sort: Most meristems, the groups of dividing cells at the tips of growing shoots and roots, can give rise to germ-line cells. One consequence of the early segregation of germ-line cells is that the loss or rearrangement of genes in somatic cells cannot affect the inherited genome of a future zygote.

Fruit flies (*Drosophila*) were critical in working out many fundamental aspects of chromosome behavior beginning almost a century ago. Now they are used to investigate development, genomics, and molecular cell biology, and a variety of mutant flies serve as models of human diseases. These model organisms also are useful for examining gametogenesis, the creation of eggs (*oogenesis*) and sperm (*spermatogenesis*).

Drosophila oogenesis begins with a stem cell that divides asymmetrically to generate a single germ-line cell (or simply **germ cell**), which divides four times to generate 16 cells. One of these cells will complete meiosis (see Figure 5-3), becoming an oocyte; the other 15 cells become *nurse cells,* which synthesize proteins and mRNAs that are transported through cytoplasmic bridges to the oocyte (Figure 22-2). The proteins and mRNAs provided by the nurse cells are necessary for maturation of the oocyte and for the early stages of embryogenesis; they also play a key role in organizing the main body of a fly. At least one-third of the *Drosophila* genome is represented in the mRNA contributed by the mother to the oocyte, a substantial dowry. Each group of 16 cells is surrounded by a single layer of somatic cells called the *follicle,* which deposits the eggshell. The mature oocyte, or egg, is released into the oviduct, where it is fertilized; the fertilized egg (the zygote) is then laid.

In mammals, about 2,500 germ-line cell precursors, the *primordial germ cells* (PGCs) form during gastrulation and migrate to a part of the abdominal cavity's mesoderm where the gonad (ovary or testis) will eventually form. These primordial germ cells produce Kit protein, a cell-surface receptor for Steel, a mitogenic protein signal secreted from cells along the path of PGC migration. Thus, as the primordial germ cells migrate, they divide under the influence of Steel. After arriving at their destination in the gonads, primordial germ cells prepare for either oogenesis or spermatogenesis.

As shown in Figure 22-3a, primordial germ cells in the developing human ovary continue dividing during months 2–7 of gestation to produce about six million *primary oocytes.* Of these, about 400,000 survive to puberty and about 500 will be ovulated during a lifetime. The primary oocytes begin meiosis but are arrested at the first meiotic prophase. Because of this prophase arrest, primary oocytes are tetraploid, which provides extra copies of the genome to help provision the unusually large egg cell. Some oocytes remain in the first meiotic prophase for nearly 50 years. Puberty triggers rapid growth of primary oocytes, which will eventually be about 200 μm in diameter. Meiosis continues at ovulation, but is completed only after fertilization. Each meiosis yields a single mature oocyte, or egg cell.

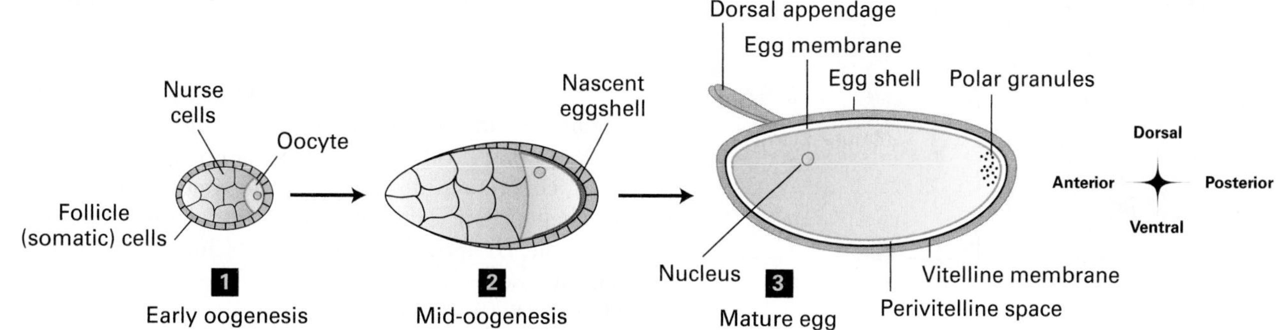

▲ **FIGURE 22-2 *Drosophila* oogenesis.** A single germ-cell precursor gives rise to fifteen nurse cells (green) and a single oocyte (yellow) early in oogenesis (**1**). The early oocyte is about the same size as the neighboring nurse cells; the follicle, a layer of somatic cells, surrounds the oocyte and nurse cells. The nurse cells begin to synthesize mRNAs and proteins necessary for oocyte maturation, and the follicle cells begin to form the eggshell. Midway through oogenesis (**2**), the oocyte has increased in size considerably. The mature egg (**3**) is surrounded by the completed eggshell (gray). The nurse cells have been discarded, but mRNAs synthesized and translocated to the oocyte by the nurse cells function in the early embryo. Polar granules located in the posterior region of the egg cytoplasm mark the region in which germ-line cells will arise. The asymmetry of the mature oocyte (e.g., the off-center position of the nucleus) sets the stage for the initial cell-fate determination in the embryo. After its release into the oviduct, fertilization of the egg triggers embryogenesis. [Adapted from A. J. F. Griffiths et al., 1993, *An Introduction to Genetic Analysis,* 5th ed., W. H. Freeman and Company, p. 643.]

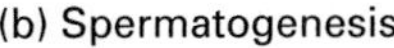

(a) Oogenesis

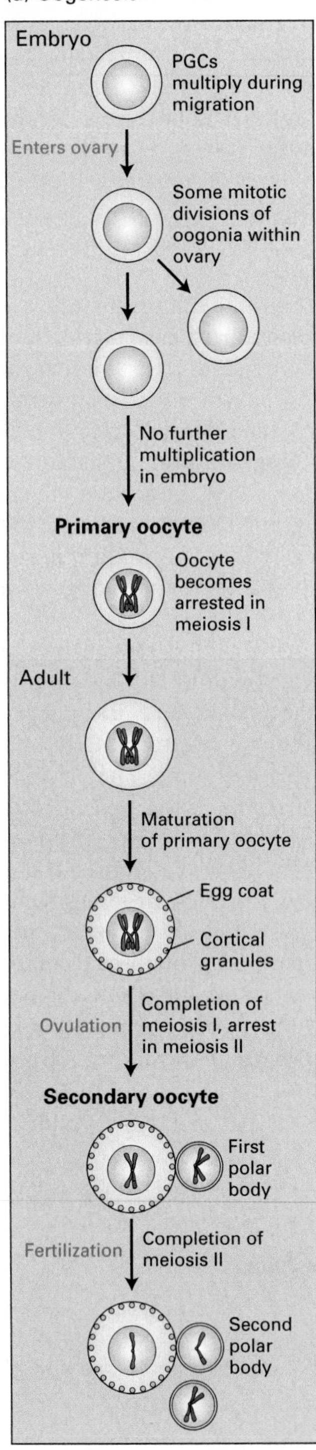

(b) Spermatogenesis

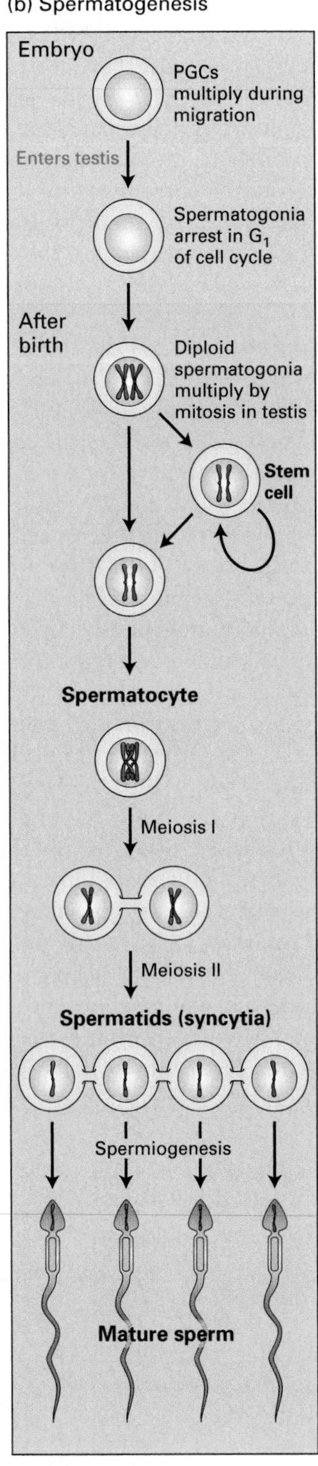

◀ **FIGURE 22-3 Mammalian gametogenesis.** In both males and females, the process of gamete formation begins in the embryo and is completed in the adult. (a) Oogenesis begins with proliferation of primordial germ cells (PGCs) in the early embryo and their accumulation at the site where the ovary will form. Primary oocytes become arrested in meiosis I until puberty. Upon ovulation, an oocyte completes meiosis I and arrests in meiosis II. Meiosis II is completed if the oocyte is fertilized. Each meiosis yields a single haploid oocyte; the other products of meiosis become polar bodies, which are nonfunctional. (b) Spermatogenesis also begins with PCGs multiplying in the embryo and accumulating in the developing testis. Unlike oocyte precursors, sperm precursors (spermatogonia) arrest in G_1 of the cell cycle and do not commence meiosis until after birth. Subsequent mitotic and meiotic cell divisions are incomplete, yielding haploid spermatids with bridged cytoplasm (syncytia). Spermatids differentiate into mature sperm in a process called spermiogenesis (see Figure 22-4). [Adapted from L. Wolpert et al., 2001, *Principles of Development*, 2nd ed., Oxford University Press, Figure 12-18.]

Primary germ cells that migrate to the developing male testis have a completely different fate (Figure 22-3b). In the testis, they are arrested in G_1 of the cell cycle and no meiosis occurs. After birth these arrested cells, called *spermatogonia*, resume mitosis. Their mitosis is a bit peculiar, in that incomplete cytokinesis causes cells with bridged cytoplasm (syncytia, "same cells") to form. From the spermatogonia come *primary spermatocytes*, the cells that enter meiosis, still with cytoplasmic bridges. Each meiosis produces four haploid gametes, or *spermatids*, connected to each other by cytoplasmic bridges. These bridges allow synchronization of germ-cell maturation and sharing of substances between cells; thus the haploid cells resulting from meiosis can share products of the unique X or Y chromosome each haploid cell carries.

Spermatids undergo a dramatic differentiation process (*spermiogenesis*) that generates mature sperm cells (Figure 22-4). The Golgi apparatus moves to one end of the cell to form the *acrosomal cap* over the nucleus, while the flagellum begins to form at the other end of the cell. Mitochondria coalesce near the base of the flagellum, ready to provide ATP for swimming. Much of the cytoplasm is extruded, the cytoplasmic bridges are lost, and the nucleus condenses to a mere shadow of its former self, driven by the arginine-rich protein protamine, which displaces the normal histones from the chromatin. The sperm is ready to go.

In humans, formation of a sperm cell from a spermatocyte takes about two months. Human males produce about 10^8 sperm/day and more than 10^{12} in a lifetime. Human sperm are about 50 μm long, but *Drosophila* sperm are considerably larger. Amazingly, the sperm in one fly species are about 5.8 cm long, which is twenty times longer than the male fly that produces them! A mature sperm consists of a head, containing the acrosome and condensed nucleus, and a flagellum tail. The tail contains a complex structure, the **axoneme,** composed of microtubules and dynein motor proteins whose movements propel the sperm toward the egg (see Figures 18-29–18-31). Mitochondria at the base of the tail provide energy in the form of ATP to power axoneme movement.

Mutations that damage motor proteins in sperm flagella can cause sterility. In the inherited disease Kartagener's syndrome, for instance, defects in the dynein arms of the axoneme (sometimes caused by mutation of the dynein gene itself) cause immotile cilia and flagella. The consequences include male infertility and situs inversus.

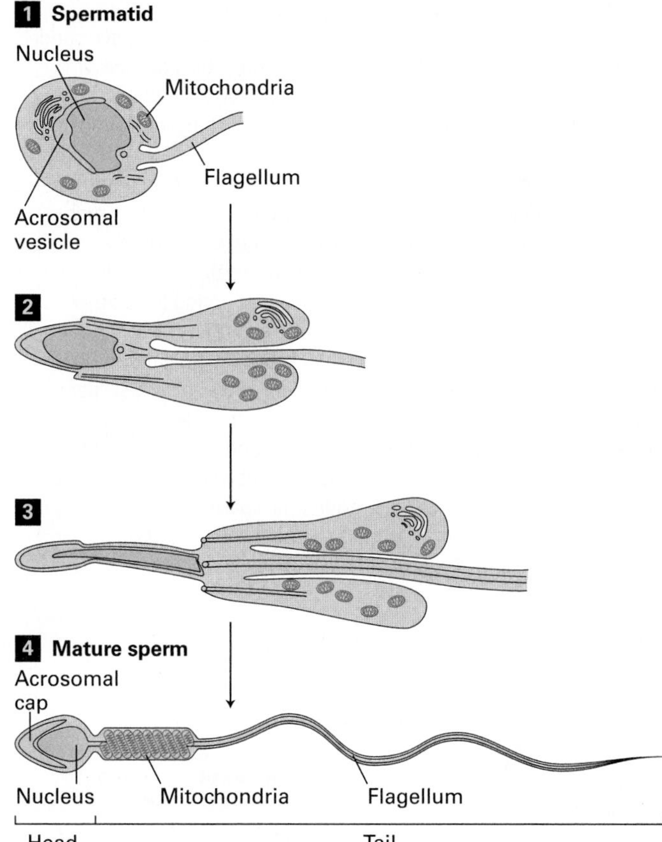

▲ **FIGURE 22-4 Spermiogenesis.** The differentiation of a spermatid (1) into a mature sperm cell (4) involves a series of dramatic morphological changes. At one end, the acrosomal cap forms over the nucleus, which becomes highly condensed. At the other end, the flagellum elongates, and much of the cytoplasm around its base is lost, leaving a sheath of mitochondria. Although not depicted, each spermatid is connected by cytoplasmic bridges to adjacent spermatids (see Figure 22-3b). These are severed during spermiogenesis, so each mature sperm cell can move independently. [Adapted from K. Kalthoff, 1996, *Analysis of Biological Development,* McGraw-Hill, Figure 3.9, and L. Wolpert et al., 1998, *Principles of Development*, 1st ed., Oxford University Press, Figure 12.21.]

Situs inversus is a birth defect in which the heart and other organs are located on the wrong side of the body. This abnormality results from the immotility of cilia that would normally polarize the left-right axis of the body during gastrulation, as we shall see in Section 22.3. ▮

Several critically important events happen during gametogenesis. In both sexes meiosis reduces the number of chromosomes to a haploid set. The sorting of chromosomes that goes on is itself an extraordinary generator of variation. Since there are 23 pairs of chromosomes in humans and there is an equal chance of retaining either one of a pair in a particular gamete, 2^{23} different possible combinations of chromosomes could emerge from meiosis. Among human populations, there is single nucleotide sequence heterogeneity about once in a thousand base pairs, so an average size human chromosome of about 130×10^6 bp will have 130,000 differences from its homolog. Add to this diversity the process of recombination (about one crossover per chromosome) that gives rise to new combinations of sequences, and the diversity is striking. Male gametogenesis also governs the sex of the next generation: Sperm carrying an X chromosome produce female offspring, while those carrying a Y chromosome produce male offspring.

Fertilization Unifies the Genome

Fertilization can be a spectacular event, as when all the coral animals of the Great Barrier Reef of Australia release eggs and sperm on the same night once each year at a full moon. The process of *fertilization,* which results in union of an egg and sperm, comprises several challenging events: penetration of an egg by a single sperm cell, assembly of a diploid (no more and no less) genome, completion of the egg's meiotic division, and initiation of a proper program of gene activation. Sperm were first described in the late 1600s, and eggs, being much larger, were recognized much earlier. Fertilization itself, however, was not directly observed and documented until the late 1800s. Pioneering studies on fertilization were done with sea urchin eggs and sperm because both eggs and sperm are easy to isolate, grow in culture, and watch. Observation of the fusion of the haploid pronuclei of the egg and sperm helped to dispel the earlier idea that sperm were contaminating animals living in semen.

It is remarkable that a sperm is ever able to reach and penetrate an egg. In humans, for instance, each sperm is competing with more than 100 million other sperm for a single egg, and it must swim a long distance to approach the egg. Moreover, the egg has multiple surrounding layers that restrict sperm entry. Although sperm are streamlined for speed and swimming ability, only a few dozen reach the egg in the oviduct (see Figure 22-1). The flagellum of human sperm contains about 9000 dynein motors that flex microtubules in the axoneme, causing successive bending that propels the sperm forward (see Figure 18-31). Estimates of the force produced by a single dynein motor protein range from 1–6 picoNewtons (pN). Since 1 pN is enough to lift a red blood cell, a sperm clearly packs a lot of motor power.

The acrosomal cap (or simply, acrosome), found at the tip of the sperm head, is a membrane-bound compartment specialized for interaction with the oocyte. The acrosome's membrane is just under the plasma membrane at the sperm head; on the other side of the acrosome, its membrane is juxtaposed to the nuclear membrane (see Figure 22-4). Inside the acrosome are soluble enzymes including hydrolases and proteases. Once a sperm approaches an egg, it must first penetrate a layer of cumulus cells that are derived from the ovarian follicle where the oocyte matured. The sperm then encounters the *zona pellucida,* a gelatinous extracellular matrix, ≈6 μm thick, that surrounds the egg (Figure 22-5a). The zona pellucida is composed largely of three glycoproteins

(a)

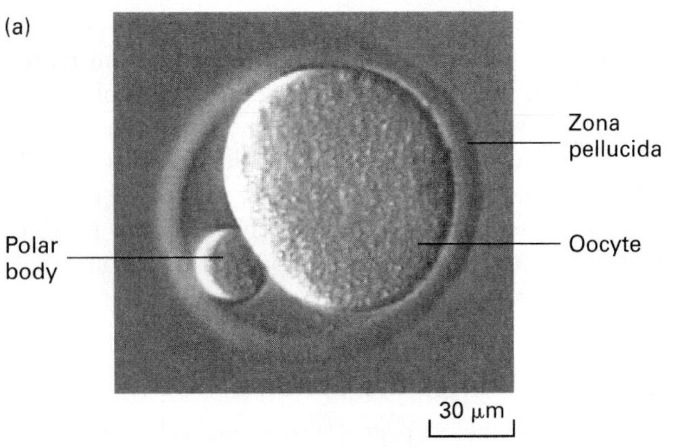

◀ **FIGURE 22-5 Gamete fusion during fertilization.** (a) Mammalian eggs, such as the mouse oocyte shown here, are surrounded by a ring of translucent material, the zona pellucida, which provides a binding matrix for sperm. The diameter of a mouse egg is ≈70 μm, and the zona pellucida is ≈6 μm thick. The polar body is a nonfunctional product of meiosis. Scale bar = 30 μm. (b) In the initial stage of fertilization, the sperm penetrates a layer of cumulus cells surrounding the egg (1) to reach the zona pellucida. Interactions between GalT, a protein on the sperm surface, and ZP3, a glycoprotein in the zona pellucida, trigger the acrosomal reaction (2), which releases enzymes from the acrosomal vesicle. Degradation of the zona pellucida by hydrolases and proteases released during the acrosomal reaction allows the sperm to begin entering the egg (3). Specific recognition proteins on the surfaces of egg and sperm facilitate fusion of their plasma membranes. Fusion and subsequent entry of the first sperm nucleus into the egg cytoplasm (4 and 5) trigger the release of Ca^{2+} within the oocyte. Cortical granules (orange) respond to the Ca^{2+} surge by fusing with the oocyte membrane and releasing enzymes that act on the zona pellucida to prevent binding of additional sperm. [Part (a) courtesy of Doug Kline. Part (b) adapted from L. Wolpert et al., 2001, *Principles of Development*, 2nd ed., Oxford Press, Figure 12-22.]

1
Binding of sperm to zona pellucida
Sperm recognizes ZP3, a zona pellucida protein

Cumulus cell layer
Acrosomal vesicle
Zona pellucida
Follicle cell

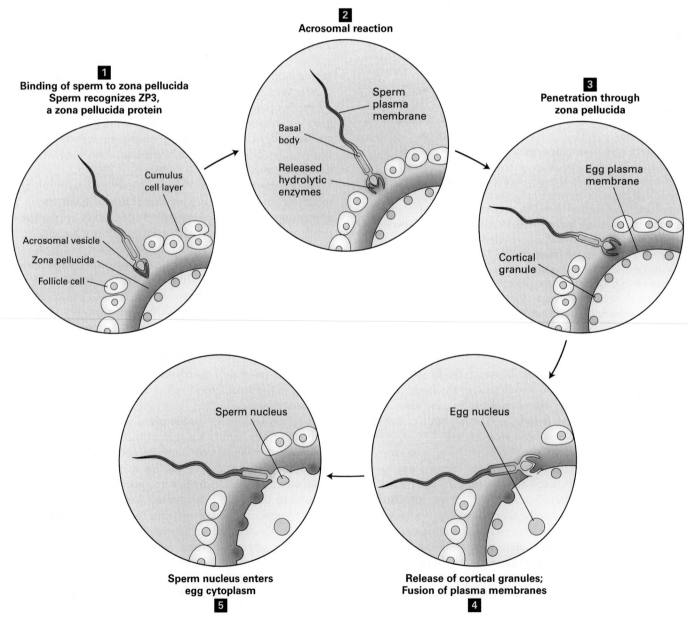

2
Acrosomal reaction

Sperm plasma membrane
Basal body
Released hydrolytic enzymes

3
Penetration through zona pellucida

Egg plasma membrane
Cortical granule

Egg nucleus

Release of cortical granules; Fusion of plasma membranes
4

Sperm nucleus

Sperm nucleus enters egg cytoplasm
5

called ZP1, 2, and 3, the most interesting of which is ZP3 because it binds sperm. The sugar residues on ZP3 are needed for the sperm to recognize the oocyte; their removal prevents fertilization. The sugars on ZP3 are bound by beta-1,4-galactosyltransferase-I (GalT) on the surface of the sperm. When ZP3 on the egg induces aggregation of multiple copies of GalT on the head of one sperm, a G protein cascade in the sperm cell is triggered, resulting in exocytosis of the acrosome and release of its contents onto the surface of the egg (Figure 22-5b, steps **1** and **2**). This process commonly is called the *acrosomal reaction*. Sperm lacking GalT are unable to bind ZP3 or undergo the acrosomal reaction, although other proteins of the sperm surface allow GalT-deficient sperm to still bind to the zona pellucida. The recognition of ZP3 is species specific, thus preventing the formation of an inviable embryo with a sperm from one species and oocyte from another.

As the released acrosomal enzymes digest the zona pellucida, the sperm can penetrate this barrier to its entry. The plasma membranes of the sperm and egg then touch and fuse, allowing the sperm nucleus to enter the egg (Figure 22-5, steps **3**–**5**). Again, specific recognition proteins have been discovered that mediate membrane fusion, including CD9, an integrin in the egg plasma membrane, and Izumo, an immunoglobulin-domain protein in the sperm plasma membrane (Figure 22-6). Proteins such as Izumo and CD9 could become targets for controlling fertility.

Following mating of animals, sperm are in a race to approach and fuse with an egg. Under normal circumstances, many sperm cells are likely to reach each available egg. The first sperm to successfully penetrate the zona pellucida and fuse with an egg triggers a dramatic response by the egg that prevents *polyspermy*, the entry of other sperm that would bring in excess chromosomes. After the first sperm succeeds in fusing with the surface of the oocyte, a flux of Ca^{2+} flows through the oocyte at about 5–10 μm/sec, starting from the site of sperm entry. One effect of this Ca^{2+} flux is to cause vesicles located just under the plasma membrane of the egg, called cortical granules, to release their contents through the plasma membrane and form a shielding fertilization membrane that blocks other sperm from entering. In this *cortical reaction*, the movement of the granules is controlled by a rich assembly of actin microfilaments that forms from globular actin in response to sperm entry. The Ca^{2+} flux induced by sperm–egg fusion also acts as an effective signal for beginning development of the zygote. Among the earliest events to occur is completion of meiosis by the egg (recall that it is blocked in meiosis I, see Figure 22-3a). The haploid egg and sperm nuclei can then fuse, yielding the diploid nucleus of the zygote.

An oocyte contributes a considerable dowry to the newly formed zygote. In mammals and many other animal species, all the mitochondrial DNA in the zygote comes from the egg; no sperm mitochondrial DNA survives fertilization. Female-specific mitochondrial DNA inheritance has been used to trace maternal heritage in human history, for example following early humans from their origins in Africa. Little or no transcription occurs during oocyte meiosis and the first embryo cleavages, so during this time the oocyte's RNA is crucial. Among the most abundant are histone mRNAs that have short poly-A tails and as a result are translated inefficiently. Translation of these histone mRNAs is regulated by a stem-loop binding protein (SLBP) that binds to nucleotides in the 3′ untranslated sequence. SLBP, regulated

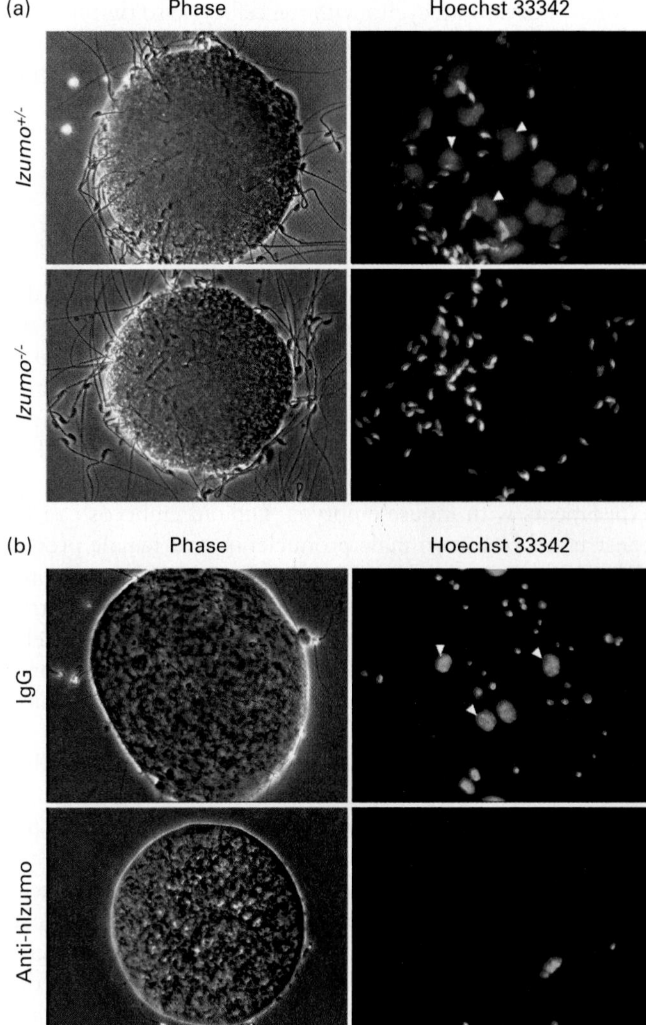

▲ **EXPERIMENTAL FIGURE 22-6 A sperm membrane protein mediates sperm–egg membrane fusion.** Izumo, a protein in the plasma membrane of both mouse and human sperm, facilitates fusion between sperm and oocyte membranes. (a) Hamster eggs stripped of their zona pellucida were inseminated with heterozygous *Izumo*$^{+/-}$ and homozygous *Izumo*$^{-/-}$ mouse sperm. Phase-contrast micrographs show a single egg surrounded by multiple sperm. Six hours after insemination, Hoechst 33342, which stains sperm heads blue, was added to the medium. In the *Izumo*$^{+/-}$ sperm, the white arrowheads indicate swelling sperm heads reacting with the egg after staining with Hoechst 33342. No fusion was observed with *Izumo*$^{-/-}$ sperm. (b) In a second experiment, zona-free hamster eggs were inseminated with wild-type human sperm in the presence of anti-human Izumo (anti-hIzumo) or control IgG antibody. After 6 hours, the samples were stained with Hoechst 33342. Fusion was observed in the control sample (white arrowheads) but not in the presence of anti-hIzumo antibody. The inactivated sperm do not trigger a polyspermy block. [From N. Inoue et al., 2005, *Nature* **434**:234].

by phosphorylation, cycles with the cell cycle so that histone mRNA is stabilized and translated preferentially during S phase. Other oocyte mRNAs are differentially activated during early embryonic development.

Genomic Imprinting Controls Gene Activation According to Maternal or Paternal Chromosome Origin

One might expect that the two pronuclei contributed by the sperm and oocyte would have equal capacity to activate genes, except for the difference between the X and Y chromosomes. Yet even after fertilization, when the paternally and maternally derived chromosomes are in the same nucleus, their male or female origin continues to have a lingering influence. This phenomenon has been revealed by experiments with mouse embryos. Diploid embryos can be constructed from two male pronuclei or two female pronuclei, which in theory should be adequate. However, zygotes with two male-derived haploid genomes form good extra-embryonic tissues but highly defective embryos. Conversely, zygotes with two female-derived haploid genomes form quite good embryos but poor yolk sacs and placentas, which are extra-embryonic tissues.

The explanation for these results is a process called **genomic imprinting,** which occurs during spermatogenesis and oogenesis. Imprinting is accomplished by modifications of the chromatin, but not the DNA sequence, in the developing gametes so that only certain genes are accessible for subsequent activation and transcription. In humans, for example, the gene for insulin-like growth factor 2 (*Igf2*) is present on both copies of chromosome 11 in the embryo, but is inactivated on the chromosome derived from the mother. Conversely, in some humans the *Igf-2r* gene on the male-derived chromosome 6 is inactive, whereas the female-derived allele is active. *Igf-2r* encodes a receptor for Igf-2, which transports Igf-2 to the lysosome for degradation. Imprinting does not alter the nucleotide sequence of DNA. For that reason either the maternal or paternal copy of every chromosome, if it ends up in a fertilized gamete, can function in development of progeny.

Abnormalities in gene imprinting that cause defects in growth, a variety of inherited diseases, and cancer, highlight the importance of this process. The mechanism of imprinting involves differential methylation of DNA during germ-line differentiation. The most common and important type of methylation alters CpG dinucleotides, which occur about 30 million times in the mammalian genome. Commonly 60–80 percent of the C residues in CpG dinucleotides are modified by several DNA methyltransferases. Most CpG methylations are erased in primordial germ-line cells during embryonic development, thus allowing reactivation of imprinted genes. Subsequent re-methylation as germ-line cells develop differentially affects some genes depending on whether a chromosome is going through oogenesis or spermatogenesis (i.e., whether the animal is male or female). Thus eggs and sperm end up with different, distinct imprints. After fertilization, a second wave of demethylation occurs in cleavage and blastocyst-stage mouse embryos, though this does not affect imprinted genes.

The reason imprinting has evolved is unclear. Relatively few genes, about eighty in mammals, are known to be imprinted. However, the genes regulated in this manner play a role in controlling embryonic growth and development. This observation suggests a possible relation between growth control and the maternal or paternal origin of the genes.

Too Much of a Good Thing: The X Chromosome Is Regulated by Dosage Compensation

Males have one X chromosome and females have two. Because of this difference in the number of X chromosomes in the two sexes, a female embryo potentially will make twice as much of each of the products encoded by X-chromosome genes as a male does. That double-dose turns out to be toxic, and a variety of mechanisms have evolved in many species, including humans, to restore a viable balance of sex-chromosome gene expression.

About four days after fertilization, mammalian embryos consist of an outer sheath of cells, which will become extra-embryonic tissue such as the placenta, and an inner mass of cells, which will form the embryo proper. At this stage, one of the two X chromosomes (either X_m from the maternal parent or X_p from the paternal parent) in each of the inner cells of a female embryo becomes highly condensed and transcriptionally inactive. Half the cells are left with an active X_p chromosome; the other half, with an active X_m chromosome. X inactivation is a mechanism of *dosage compensation* that ensures that males and females produce roughly similar levels of X-chromosome gene products. Once an embryonic cell undergoes X inactivation, the same X chromosome (X_m or X_p) remains active and the other remains inactive in all the descendants of that cell. Thus all women are, in this sense, genetic mosaics, since half their cells have an active X_m and half have an active X_p. Because the two X chromosomes differ on average at 1 in 1000 base pairs, the cells in females will no longer be identical with respect to active X-linked genes. Males, in contrast, have the same active X chromosome in every cell.

Mammalian dosage compensation requires a region of the X chromosome called the X-inactivation center (Figure 22-7a). Within this region are two genes, *Xist* and *Tsix,* that both encode long RNAs. Because the genes overlap and are transcribed in opposite directions, their RNA products are antisense with respect to each other (hence the reversed names). *Xist* RNA, which is made from only one of the two X chromosomes in a female cell, actually coats the chromosome from which it is made. Since *Xist* RNA does not move to the other X chromosome, that X is not coated and as a result remains active.

The mechanism for triggering expression of the *Xist* gene from only one chromosome is incompletely understood. Cis-acting control elements in the X-inactivation center appear to detect the presence of the two X chromosomes in female cells (Figure 22-7b). A clue about the sensing process comes from the observation that the two X-inactivation centers in a female

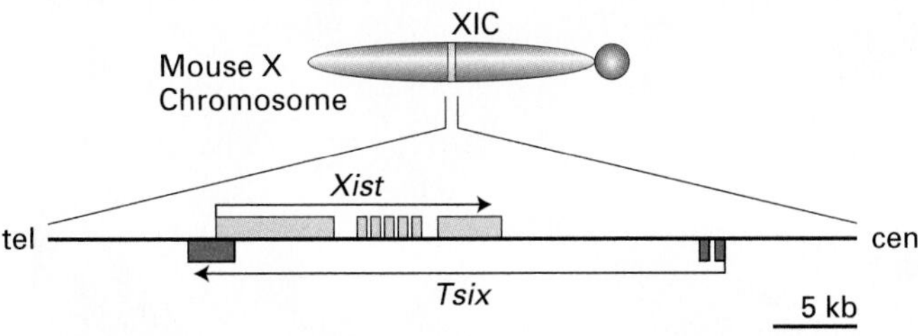
(a) X-inactivation center (XIC)
XIC
Mouse X Chromosome
Xist
tel
cen
Tsix
5 kb

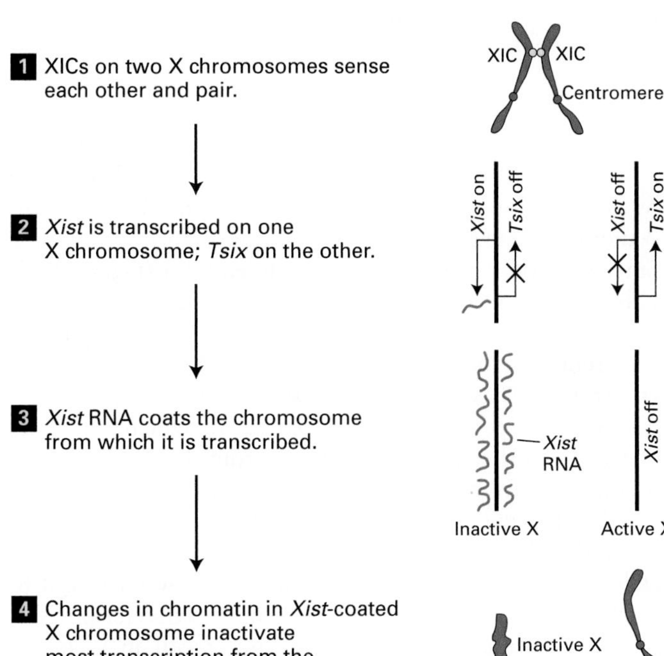
(b) Mechanism of dosage compensation in females (2X)
1 XICs on two X chromosomes sense each other and pair.
XIC
XIC
Centromere
2 Xist is transcribed on one X chromosome; Tsix on the other.
Xist on
Tsix off
Xist off
Tsix on
3 Xist RNA coats the chromosome from which it is transcribed.
Xist RNA
Xist off
Inactive X
Active X
4 Changes in chromatin in Xist-coated X chromosome inactivate most transcription from the chromosome.
Inactive X

▲ **FIGURE 22-7 Dosage compensation in mammalian females.** Early in embryogenesis, one of the female X chromosomes becomes highly compacted in a special chromatin structure; this compacted, transcriptionally inactive chromosome is called a Barr body. The mechanism that detects the presence of two X chromosomes in the female embryo involves a region on the X chromosome, the X-inactivation center (XIC), diagrammed in (a). The transient pairing of the XICs on the maternal and paternal X chromosomes in females initiates the process of dosage compensation, outlined in (b). Since pairing and subsequent events occur randomly in each cell, half the cells will have the maternally derived X inactivated and the other half will have the paternally derived X inactivated. Since male embryos have only one X chromosome no XIC pairing can occur and no *Xist* RNA is produced, hence there is no X-chromosome inactivation.

cell transiently co-localize prior to one of the X chromosomes becoming inactive. This co-localization may result in (1) low expression of *Tsix* from the chromosome that will be inactivated, which leads to (2) a temporary chromatin-regulated silencing of *Xist* transcription, followed by (3) a surge of high *Xist* expression leading to inactivation of the chromosome.

After production of *Xist* RNA, Polycomb-group proteins modify lysine 27 residues on histone H3 tails of the *Xist*-coated X chromosome. This and other early changes trigger a series of chromatin modifications that lead to compaction and transcriptional inactivity.

KEY CONCEPTS OF SECTION 22.2

Gametogenesis and Fertilization

- Early in mammalian embryogenesis, the primordial germ cells migrate to the gonads and begin, but do not complete, the production of gametes (oocytes or sperm) in utero.
- In mammalian oogenesis, primary oocytes arrested in meiosis form in the embryo and are stored in the ovary; they complete meiosis, forming haploid mature oocytes (eggs), following fertilization (see Figure 22-3a). Oocytes contain a pool of mitochondria, mRNAs, and other materials necessary for early development.
- In mammalian spermatogenesis, germ-line precursors form in the embryo and are stored in the testis; these diploid cells undergo meiosis after birth, forming haploid spermatids (see Figure 22-3b). Differentiation of spermatids (spermiogenesis) yields mature sperm, elongated cells with a long flagella and compressed nucleus.
- During fertilization, sperm initially recognize and bind to the gelatinous layer (zona pellucida) on the outside of the oocyte. This interaction, which involves glycoproteins (e.g., ZP3) in the zona pellucida and proteins on the sperm surface, triggers the acrosomal reaction in the sperm. Enzymes released from the sperm acrosome degrade the zona pellucida, leading to fusion of the sperm and egg plasma membranes (see Figure 22-5).
- Fertilization must be restricted to a single sperm to prevent chromosome imbalances. Sperm–egg fusion induces changes in the oocyte that exclude entry of other sperm.
- Although the sperm and oocyte each contribute a haploid genome, genomic imprinting restricts which alleles are active in each cell. In the case of some imprinted genes, the paternal copy is active in the embryo and the corresponding maternal copy is inactive, or vice versa.
- Males survive with one X chromosome, but the two X chromosomes in females could lead to a double dose of X-gene products that is harmful. This situation is avoided in human females by inactivation of one of the two X chromosomes in every cell of an early embryo, a process called dosage compensation (see Figure 22-7).

22.3 Cell Diversity and Patterning in Early Vertebrate Embryos

Fertilization produces a single cell, the zygote, which divides rapidly. Within a few days, the newly formed cells start sending and receiving various signals that largely determine their future fates. The first distinctions to appear are two different

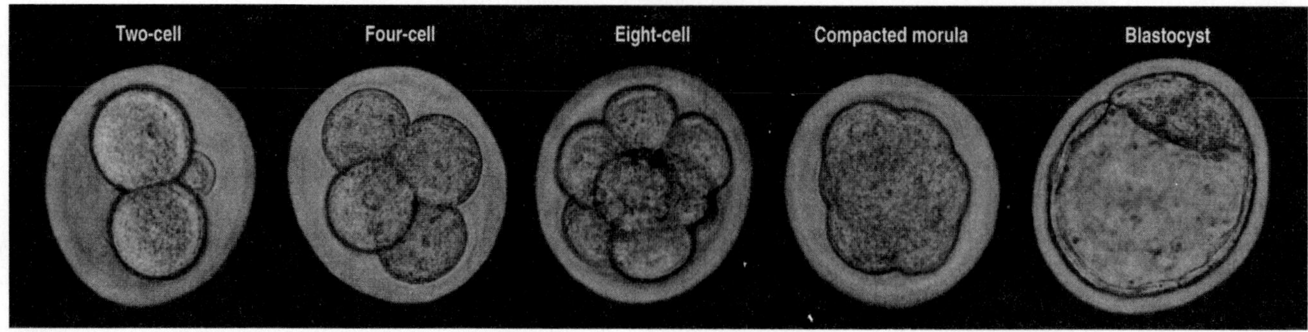

▲ **FIGURE 22-8 Cleavage divisions in the mouse embryo.** There is little cell growth during these divisions, so that the cells become progressively smaller. See text for discussion. [Copyright Oxford University Press, T. Fleming.]

cell types and then the initial polarization of the embryo along its various axes. In time, further distinctions lead to formation of multiple cell layers from which tissues and organs begin to form. Much of this early fate determination requires signaling systems that we discussed in Chapters 15 and 16. The general theme is that a few cells will begin to produce a signal, and other cells will produce the receptors that make them responsive to that signal; signal reception then induces the production of certain transcription factors that regulate batteries of genes to control the fate of the receiving cell.

Cleavage Leads to the First Differentiation Events

The zygote is the ultimate *totipotent* cell because it has the capability to generate all the cell types of the body. Fertilization is quickly followed by cleavage, cell divisions that take about one day each in the mouse. In mammals, cleavage divisions occur before the embryo is implanted in the uterus wall (see Figure 22-1). The condensed, transcriptionally inactive chromatin from the sperm returns to a more normal state by replacement of the sperm's special histones with normal histones provided by the oocyte.

Initially the cells are fairly spherical and loosely attached to each other (Figure 22-8). As demonstrated experimentally in sheep, each cell at the eight-cell stage has the potential to give rise to a complete animal. Three days after fertilization the 8-cell embryo divides again to form the 16-cell *morula* (from the Greek for "raspberry"), after which the cell affinities increase substantially and the embryo undergoes *compaction,* a process that depends in part upon the surface molecule E-cadherin (Chapter 19). Compaction is driven by increased cell–cell adhesion and initially results in a more solid mass of cells, the compacted morula. Next, some cell–cell adhesions diminish, and fluid begins to flow into an internal cavity, the *blastocoel.* Additional divisions produce a **blastocyst**-stage embryo, composed of ≈64 cells that have separated into two cell types: *trophectoderm (TE),* which will form extra-embryonic tissues like the placenta, and the *inner cell mass (ICM),* which gives rise to the embryo proper.

Whether a cell assumes a TE or ICM fate is determined by its location within very early embryos. This can be demonstrated experimentally by placing a labeled cell on the outside or the inside of a very early embryo (Figure 22-9) Labeled cells placed on the outside form extra-embryonic tissues (TE fate) almost exclusively, and those inserted inside preferentially form embryo tissues (ICM fate). DNA microarray analysis of gene expression at each stage of early development reveal dramatic changes in which genes are expressed as the embryo progresses from the 2-cell to blastocyst stage. Even these very early embryos use Wnt, Notch, and TGFβ signaling pathways (Chapter 16).

Both ICM and TE cells have the properties of stem cells: That is, each type can divide to renew itself and to start its own distinct lineage that produces diverse populations of differentiated cells. The inner cell mass is the source of **embryonic stem (ES) cells,** which can contribute to any part of the embryo (see Figure 21-7). The isolation of these cells as cell lines has enabled dramatic advances in mouse genetic manipulations. The earliest known sign of the ICM fate is expression of the *Oct4* gene, which is a critical regulator for maintaining cells in a plastic *pluripotent* state.

As shown in Figure 22-9a, ICM cells are located on one side of the blastocoel, while TE cells form a hollow ball around the inner cell mass and blastocoel. At this blastocyst stage, the TE cells are in an organized epithelial sheet, with cell–cell junctions between adjacent cells. In contrast, the ICM cells are a loose mass that can be described as *mesenchyme,* a term commonly applied to loosely organized and loosely attached cells. During development, cells often undergo an *epithelial-mesenchymal transition,* or the reverse. Epithelial cells form sheets that act as a barrier, move in harmony, and have a clear polar character from one side of the sheet to the other. Mesenchymal cells are bold individuals, less responsive to peer pressure. With their ability to cut loose from each other, they can migrate as individual cells, seed new organs, form circulating blood cells, and adhere in a three-dimensional mass, such as the inner cell mass.

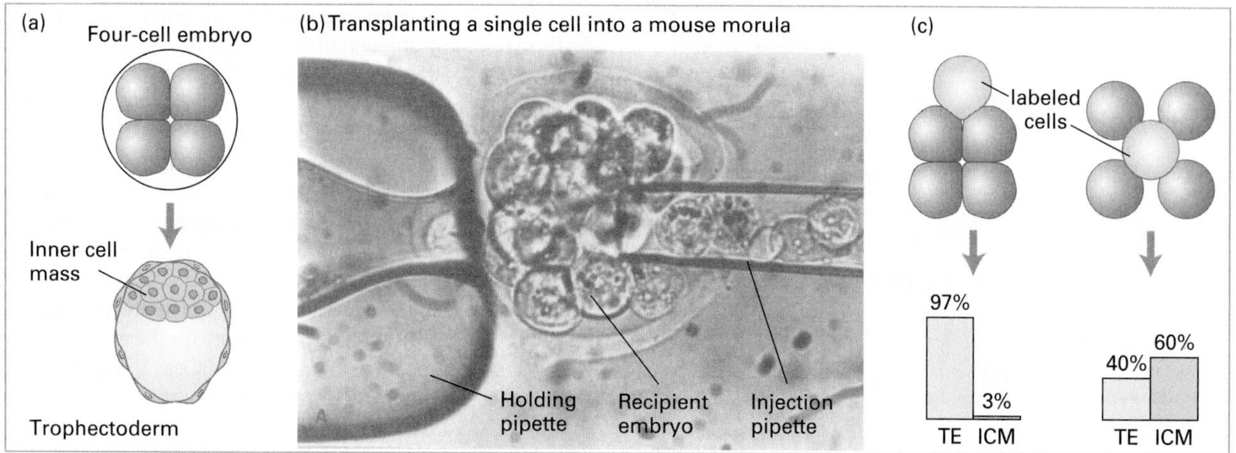

▲ **EXPERIMENTAL FIGURE 22-9 Cell location determines cell fate in the early embryo.** (a) A 4-cell embryo normally develops into a blastocyst consisting of trophectoderm (TE) cells on the outside and inner cell mass (ICM) cells inside. (b) In order to discover whether position affects the fates of cells, transplantation experiments were done with mouse embryos. First, recipient morula-stage embryos had cells removed to make room for implanted cells. Then donor morula-stage embryos were soaked in a dye that does not transfer between cells. Finally, labeled cells from the donor embryos were injected into inner or outer regions of the recipient embryos, as shown in the micrograph. The recipient embryo is held in place by a slight vacuum applied to the holding pipette. (c) The subsequent fates of the descendants of the transplanted labeled cells were monitored. For simplicity, 4-cell recipient cells are depicted, although morula-stage (16-cell) embryos were used as both donors and recipients. The results, summarized in the graphs, show that outer cells overwhelmingly form trophectoderm and inner cells tend to become inner cell mass but also form considerable trophectoderm. [Parts (a) and (c) adapted from L. Wolpert et al., 2001, *Principles of Development*, 2nd ed., Oxford Press, Figure 3-12. Part (b) from R. L. Gardner and J. Nichols, 1991, *Human Reprod.* **6**:25–35.]

The Genomes of Most Somatic Cells Are Complete

What is happening to the genome during early embryonic stages? Although different parts of the genome are transcribed in different cells, the genome itself is thought to be identical in nearly all cells. One well-documented exception occurs during development of lymphocytes from hematopoietic precursors. Segments of the genome are rearranged or lost during lymphocyte development, generating clones of lymphocytes whose genomes are unique (Chapter 24). Also, mature erythrocytes (red blood cells) lack a nucleus and thus have no nuclear genome. Most somatic cells, however, appear to have an intact genome, equivalent to that in the germ line.

Evidence that at least some somatic cells have a complete and functional genome comes from the successful production of cloned animals by nuclear-transfer cloning. In this procedure, the nucleus of an adult (somatic) cell is introduced into an egg that lacks its nucleus; the manipulated egg, which contains the diploid number of chromosomes and is equivalent to a zygote, then is implanted into a foster mother. The only source of genetic information to guide development of the embryo is the nuclear genome from the donor somatic cell. The frequent failure of such cloning experiments raises questions about how many adult somatic cells do in fact have a complete functional genome. Even the successes, like the famous cloned sheep "Dolly," have some medical problems. Even if differentiated cells have a physically complete genome, clearly only portions of it are transcriptionally active (Chapters 6 and 7). A cell could, for example, have an intact genome, but be unable to properly reactivate it due to inherited chromatin states.

Further evidence that the genome of a differentiated cell can revert to having the full developmental potential characteristic of an embryonic stem cell comes from experiments in which differentiated olfactory sensory neurons were genetically marked with green fluorescence protein (GFP) and then used as donors of nuclei. When the nuclei from differentiated olfactory cells were implanted into enucleated mouse oocytes, 14 percent of the oocytes developed into blastocysts that produced GFP. These GFP-marked blastocysts were used to derive embryonic stem (ES) cell lines, which were then used to generate mouse embryos. After implantation into female mice, these embryos, derived entirely from olfactory neuron genomes, formed healthy mice. Thus the genome of a differentiated cell can be reprogrammed completely to form all tissues of a mouse.

Gastrulation Creates Multiple Tissue Layers, Which Become Polarized

The embryonic cells composing the inner cell mass of the blastocyst have impressive abilities, but they hardly look like an embryo. Soon after the blastocyst implants into the uterine wall, the loose-knit conglomeration of ICM cells is turned into a multi-layered structure with head-tail, front-back, and

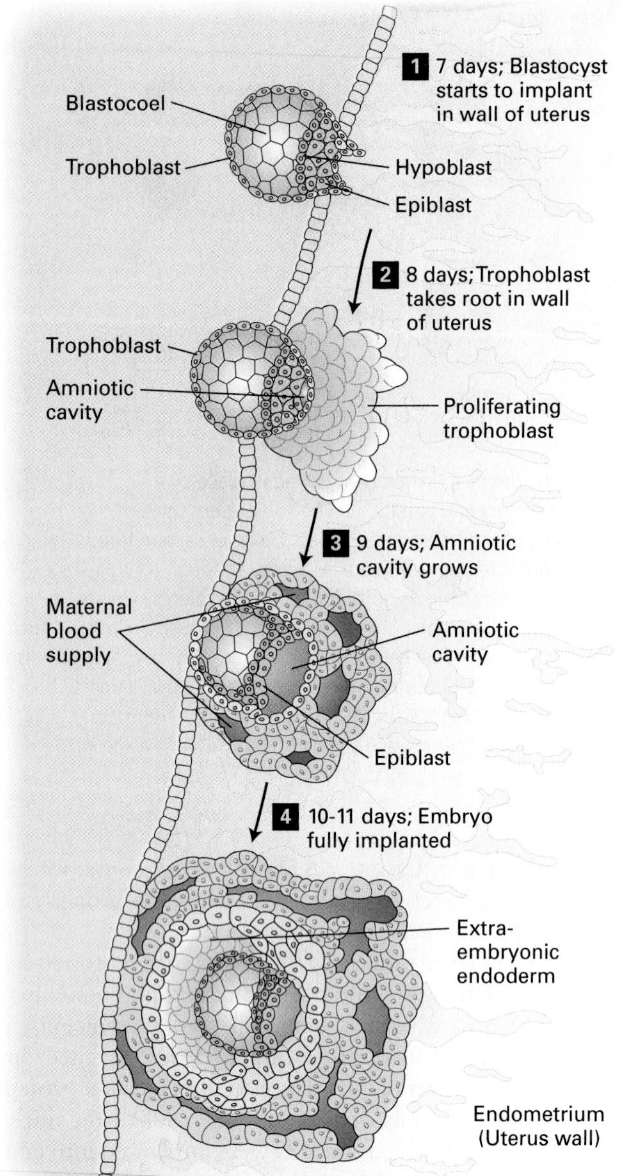

◄ **FIGURE 22-10 Human development from days 7 to 11 after fertilization.** Implantation of the blastocyst into the wall of the uterus occurs at about 7 days (see Figure 22-1 for earlier stages). The blastocyst is polarized in that the inner cell mass (ICM) is at one end and the blastocoel (fluid-filled cavity) at the other, both encased in trophectoderm cells, which are now known as the trophoblast. By day 7 (1), the ICM cells of the blastocyst have already begun to separate into two layers—the hypoblast (green) and the epiblast (blue). The epiblast will form the embryo proper and the hypoblast will form extra-embryonic structures (in addition to those formed by trophoblast). Next, the trophoblast cells proliferate and invade the wall of the uterus (2), which is necessary for the embryo to start receiving nourishment from the mother. By day 9 (3), the amniotic cavity (gray), which forms between the epiblast and the trophoblast, is enlarging. By 10–11 days (4), the embryo (now the size of a period on this page) is fully implanted and is being supplied by the maternal circulatory system. The extra-embryonic endoderm (white) is derived from the hypoblast and will form the yolk sac, a remnant of our evolutionary history when our ancestors required significant amounts of yolk to nourish the embryo. The two cell layers of the implanted embryo are transformed into the three germ layers during the next stage, gastrulation. [Adapted from S. F. Gilbert, 2006, *Developmental Biology*, 8th ed., Sinauer Press, Figure 11-33.]

(a)

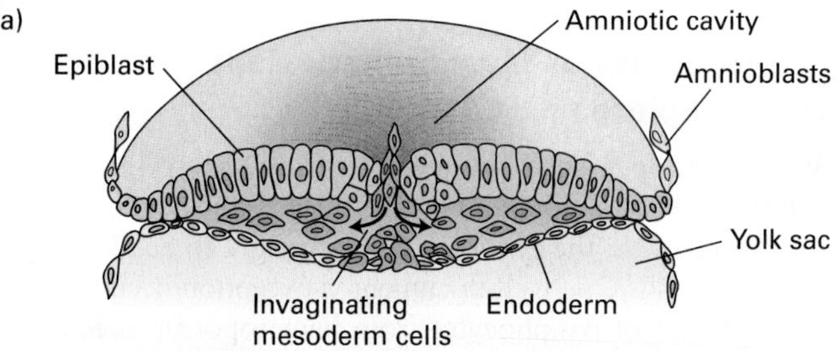

(b)

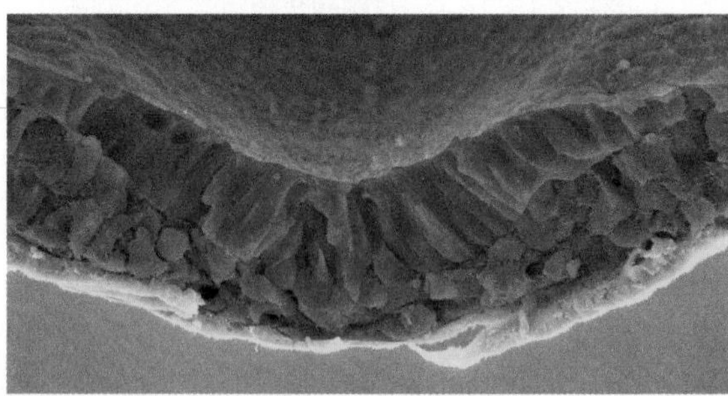

▲ **FIGURE 22-11 Gastrulation in animals.** During gastrulation, epiblast cells migrate to create the three germ layers: ectoderm, mesoderm, and endoderm. Cells in the different layers have largely distinct fates and therefore represent distinct cell lineages. (a) Shown here is a sketch of a human embryo about 16 days after fertilization of an egg. The first cells to move from the epiblast into the interior form endoderm. They displace the hypoblast (not shown) and appear as a layer along the bottom of the interior space. The next cells to invaginate become mesoderm. The remaining epiblast cells become ectoderm. (b) Scanning electron micrograph of a cross section of a similar-stage embryo. [Part (a) adapted from N. A. Campbell and J. B. Reece, 2005, *Biology,* 7th ed., Pearson/Benjamin Cummings, Figure 47.13. Part (b) courtesy of Kathy Sulik, University of North Carolina–Chapel Hill.]

left-right polarity. As depicted in Figure 22-10, the inner cell mass initially separates into two layers called the *hypoblast* and the *epiblast*. The epiblast will form the embryo proper. The hypoblast will form extra-embryonic structures that connect to the mother's circulation. Other extra-embryonic structures will form from the trophectoderm, called the *trophoblast* after implantation.

In the mouse, in which axis formation has been studied extensively, the first sign of the anterior-posterior axis becomes apparent at 6.5 days (2 weeks in humans). At this time a visible groove, the *primitive streak,* forms on the surface of the epiblast in the region that will become the posterior of the embryo. Along the primitive streak, cells leave the epiblast layer and move into the space between the epiblast and hypoblast in the process of gastrulation (Figure 22-11). The first cells to invaginate form embryonic **endoderm;**

later arrivals become **mesoderm.** Cells that do not invaginate to become endoderm or mesoderm remain behind in the epiblast to become **ectoderm.** The post-gastrulation embryo proper thus has three germ layers and is polarized along the dorsal-ventral axis and the anterior-posterior axis. The ectoderm will make neural and epidermal cells; the mesoderm will make muscle, connective tissue, and blood; the endoderm will make gut epithelia.

Once the three germ layers are established, they subsequently divide into cell populations with different fates. There appears to be a progressive restriction in the range of cell types that can be formed from stem cells and precursor cells as development proceeds. An early embryonic stem cell, as we've seen, can form every type of cell; an ectodermal cell has a choice between neural and epidermal fates; a keratinocyte precursor can form skin but not neurons. These observations raise two important questions: How are cell fates progressively restricted during development? Are these restrictions irreversible? Such questions are often addressed with transplant experiments, asking what a cell will do when moved to an abnormal position in an embryo. Will it retain the fate of the old location, or adopt a fate appropriate to the new location? In addressing such questions, it is important to remember that what a cell in its normal in vivo location will do may differ from what a cell is capable of doing if it is manipulated experimentally. Thus the observed limits to what a cell can do may result from natural regulatory mechanisms or may reflect a failure to find conditions that reveal the cell's full potential.

The fates of different parts of the early embryo were determined in the early days of embryology by marking amphibian or chick embryo cells with ink and tracking them. More recently, chimeric animals composed of chicken and quail cells have been used to study cell-fate determination during embryonic development. Embryos composed of cells from both bird species develop fairly normally, yet the cells derived from each donor are distinguishable under the microscope. Thus the contributions of the different donor cells to the final bird can be ascertained. In addition to testing which cells can form which types of tissue, the rigidity of cell-fate determination can be tested. When cells from one germ layer are transplanted into one of the other layers, they do *not* give rise to cells appropriate to their new location. Thus the endoderm, mesoderm, and ectoderm are not only morphologically distinct; they are firmly determined as different cell types with different fates.

In vertebrates, secreted protein signals are involved in directing not only the initial formation of the germ layers but also their polarization along the body axis. In the remainder of Section 22.3, we examine the role of secreted signals and their antagonists in early development.

Signal Gradients May Induce Different Cell Fates

The most intensive and revealing studies of gastrulation and formation of the initial tissues have been done in the clawed toad *Xenopus laevis* (more commonly described as a frog). By dissecting and transplanting *Xenopus* tissues, developmental biologists have identified powerful signals that direct cell fates in the early embryo. Some of these signals have been found to function in similar ways in early mammalian embryos.

The *Xenopus* egg is huge, about 1 mm in diameter. After fertilization, this giant cell divides, initially without growth, to make successively smaller cells that vary in size. Once the embryo is a ball of cells, a blastocoel cavity opens within the ball. The cells at one end of the embryo are large; this side of the embryo is called the *vegetal pole.* Cells at the opposite end are smaller; this side is known as the *animal pole.* Morphological and molecular markers clearly reveal the polarization of the embryo by the midblastula stage. For example, the proteins VegT and Vg-1 are molecular markers for the vegetal pole. On what will become the dorsal side ("back") of the embryo, the protein β-catenin accumulates to high levels. The position of β-catenin accumulation is controlled by the site of sperm entry. As we have seen in Chapter 16, β-catenin is activated as a transcription factor by Wnt signals (see Figure 16-32). Even in the absence of a Wnt signal, accumulated β-catenin can trigger the induction of specific genes.

The accumulation of β-catenin is the earliest known indicator of the frog dorsal-ventral axis. More importantly, β-catenin is essential for the formation of two signal-emitting centers on the dorsal side of the late blastula: the *Nieuwkoop center* and the *BCNE (blastula chordin and noggin expression) center* (Figure 22-12). The Nieuwkoop center forms where VegT, Vg-1, and high levels of β-catenin overlap, that is, in the vegetal part of the dorsal side. Some of the future endoderm is located in this same region. The Nieuwkoop center releases Nodal proteins, which are members of the transforming growth factor β (TGFβ) family of secreted protein signals (Chapter 16). The BCNE center, which forms where future neural-ectoderm cells arise, is detectable through its expression of genes encoding some transcription factors as well as Chordin and Noggin. These two secreted proteins are antagonists of bone morphogenetic proteins (BMPs), other members of the TGFβ family. Thus the BCNE center seems to be mainly involved in preventing BMP action. At the equator of the slightly later, gastrula-stage embryo, a central belt of cells that forms between the ectoderm at the animal pole and the endoderm at the vegetal pole becomes the mesoderm. During *Xenopus* gastrulation, the cells forming the future endoderm and mesoderm invaginate by the same sort of process that occurs in mammalian embryos (see Figure 22-11).

We know now that the mesoderm is induced by signals, particularly TGFβ signaling proteins. In the search for mesoderm-inducing factors, researchers added everything but the kitchen sink to early frog embryos. A remarkable range of molecules was found to induce mesoderm cells, including many that could not possibly be natural inducers. For a molecule to be considered a genuine, natural inducer, it has to meet three criteria: (1) the molecule has to be necessary for mesoderm induction to occur (based on interference or mutants); (2) it has to be sufficient to induce mesoderm from cells not normally fated to be mesoderm (based on ectopic expression tests, that is, expression in abnormal sites); and (3) it has to be produced at the right time and place in a normal embryo (based on antibody staining or in situ hybridization). Few

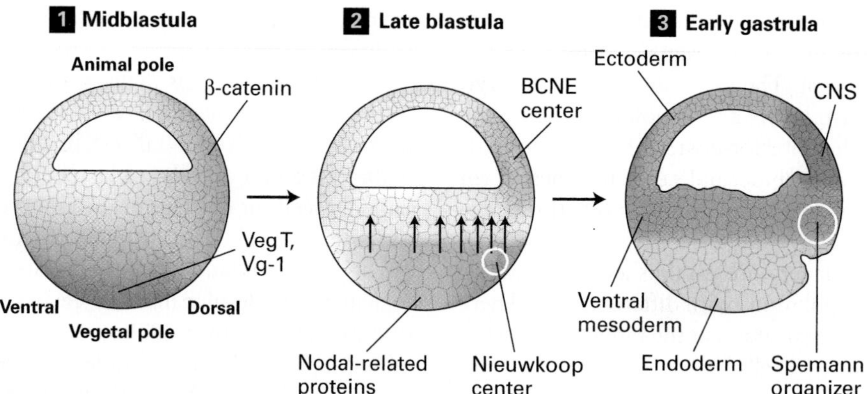

▲ **FIGURE 22-12 Signaling centers in the early *Xenopus* embryo.** Cross-sections of the embryo at three stages are shown. The three germ layers—endoderm (yellow), mesoderm (red), and ectoderm (blue)—are determined by the gastrula stage (3). They derive from earlier embryos (1 and 2), which are polarized by VegT (a transcription factor) and Vg-1 (a TGFβ signal) to form the animal pole and by Nodal signals from the Nieuwkoop center to form ventral-to-dorsal cells. Mesoderm is induced by signals coming from the vegetal region to the central embryo. Dorsal-ventral cell fates are also controlled by signals from the Spemann organizer in the gastrula. See text for further discussion. [Adapted from L. Wolpert et al., 2001, *Principles of Development*, 2nd ed., Oxford Press, Figure 3-35.]

molecules fit all these requirements, but some TGFβ proteins do.

In some cases, the induction of cell fates involves a binary choice: In the presence of a signal, the cell is directed down one developmental pathway; in the absence of the signal, the cell assumes a different developmental fate or fails to develop at all. Such signals can work in a *relay mode.* That is, an initial signal induces a cascade of induction in which cells close to the signal source are induced to assume specific fates; they, in turn, produce other signals to organize their neighbors (Figure 22-13a). Alternatively, a signal may induce different cell fates, depending on its concentration. In this *gradient mode,* the fate of a receiving cell is determined by the amount of the signal that reaches it, which is related to its distance from the signal source (Figure 22-13b). Any substance that can induce different responses depending on its concentration is often referred to as a **morphogen.**

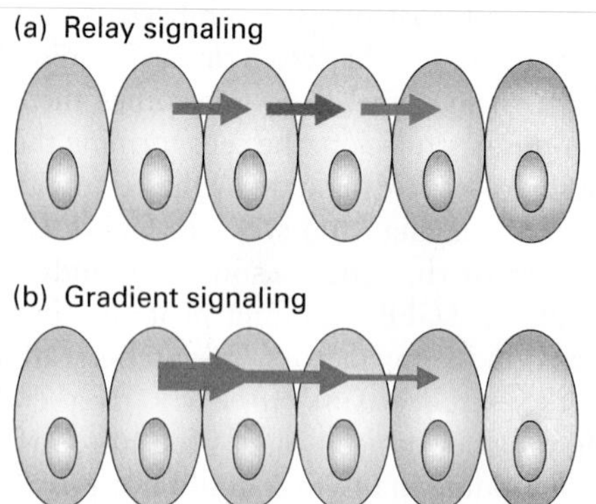

▲ **FIGURE 22-13 Two modes of inductive signaling.** In the relay mode (a), a short-range signal (red arrow) stimulates the receiving cell to send another signal (purple), and so on for one or more rounds. In the gradient mode (b), a signal produced in localized source cells (pink cells) reaches nearby cells in larger amounts than the amounts reaching distant cells. If the receiving cells respond differently to different concentrations of the signal (indicated by width of the arrows), then a single signal may create multiple cell types.

The concentration at which a signal induces a specific cellular response is called a *threshold.* A graded signal, or morphogen, exhibits several thresholds, each one corresponding to a specific response in receiving cells. For instance, a low concentration of an inductive signal causes a cell to assume fate A, but a higher signal concentration causes the cell to assume fate B. In the gradient mode of signaling, the signal is newly created, and so it has not built up to equal levels everywhere. Alternatively, the signal could be produced at one end of a field of cells and destroyed or inactivated at the other (the "source and sink" idea), so a graded distribution is maintained.

Studies with activin, a TGFβ-type signaling protein that can alter cell fate in early *Xenopus* embryos, have provided insight into how cells determine the concentration of a graded inductive signal. Activin helps organize the mesoderm along the dorsal-ventral axis of an animal. Specific genes are used as indicators of the tissue-creating effects of signals such as activin. For instance, a low concentration of activin induces expression of the *Xenopus brachyury (Xbra)* gene throughout the early mesoderm. Xbra is a transcription factor necessary for mesoderm development. Higher concentrations of activin induce expression of the *Xenopus goosecoid (Xgsc)* gene. Xgsc protein is able to transform ventral into dorsal mesoderm; so the local induction of *Xgsc* by activin causes the formation of dorsal, rather than ventral, mesodermal cells near the activin source.

Using ^{35}S-labeled activin, scientists demonstrated that *Xenopus* blastula cells each produce some 5000 type II TGFβ-like receptors that bind activin. Findings from additional experiments showed that maximal *Xbra* expression was achieved when about 100 receptors were occupied. At a concentration of activin at which 300 receptors were occupied, cells began expressing higher levels of *Xgsc*. Similar results were obtained with blastula cells experimentally manipulated to produce sevenfold higher levels of the activin type II receptor. These findings indicate that blastula cells measure the absolute number of ligand-bound receptors rather than the ratio of bound to unbound receptors, and confirm the importance of signal concentration.

Signal Antagonists Influence Cell Fates and Tissue Induction

The most dorsal of the mesoderm cells in the early frog gastrula become a famous signaling center called the **Spemann organizer** (Figure 22-12, **3**). This organizer, reported by Spemann and Mangold in 1924, can be transplanted into a host embryo, where it takes over part of the host embryo and causes the formation of a nearly complete new axis (Figure 22-14). The Spemann organizer is formed under the influence of Nodal and perhaps other signals from the Nieuwkoop center.

The Spemann organizer is the source of a remarkable number of secreted signal antagonists. These include Chordin and Noggin (BMP antagonists); Frzb-1, Crescent, sFRP2, and Dkk-1 (Wnt antagonists); and the multi-signal antagonist Cerberus, which dampens the effects of Wnt *and* BMP *and* Nodal signals. When Noggin is applied to embryos that lack a functional Spemann organizer, the embryos develop normally. This finding shows that Noggin all by itself can mimic the effect of the Spemann organizer. A ventral signaling center, at the opposite end of the mesoderm from the Spemann organizer, makes several signals including BMP4 (a TFGβ family signal). Thus the two signaling centers, one dorsal and one ventral, appear to battle each other for control of cell fates across the mesoderm.

▲ EXPERIMENTAL FIGURE 22-14 **A transplanted Spemann organizer directs formation of a new body axis in host embryo.** *(Top)* Normal frog embryo at swimming tadpole stage. *(Bottom)* When the Spemann organizer, a powerful signaling center, is transplanted into an early host embryo so that it now has two organizers (right, the implant is the white patch at the bottom), much of the main neural and mesodermal axis is duplicated (left). See text for discussion. [From E. de Robertis and H. Kuroda, 2004, *Ann. Rev. Cell Devel. Biol.* **20**:285].

You have undoubtedly noticed that many of the molecules involved in tissue formation in early embryos are signal antagonists. One effect of such an antagonist can be to sharpen or move a boundary between cell types. A signal coming from source cells is progressively less potent with distance; at some point, it falls below a threshold amount and is without effect. If a secreted antagonist comes from the opposite direction, it will block the action of the signal even in cells receiving above-threshold amounts.

We see an example of this phenomenon in the formation of neural cells in the anterior of frog embryos. Normally, secreted TGFβ proteins *prevent* the formation of neural cells near the animal pole of *Xenopus* embryo. When the animal pole is removed and placed into culture, it is called the "animal cap." Production of BMP4 (a TGFβ family signal; see Chapter 16) by an animal cap prevents formation of neural tissue in the culture. The effect of signals and other regulators on neural induction can be tested by exposing parts of the animal cap from *Xenopus* embryos to individual proteins and seeing whether neural cells form. This type of in vitro experiment first revealed the ability of Chordin to antagonize BMP4 and induce neural-cell identity. Only when BMP signaling is successful can non-neural cell types form. Together, these data led to a simple model in which Chordin prevents BMP4 from binding to its receptor. In principle, inhibition could occur by the direct binding of Chordin to BMP receptors or to BMP molecules themselves. Biochemical studies demonstrated that Chordin binds BMP2 and BMP4 homodimers or BMP4/BMP7 heterodimers with high affinity ($K_d = 3 \times 10^{-10}$ M) and prevents them from binding to their receptors (Figure 22-15). Chordin-mediated inhibition of BMP signaling is relieved by Xolloid protein, a protease that specifically cleaves Chordin in Chordin–BMP complexes, releasing active BMP.

Cerberus, another antagonist secreted from cells in the Spemann organizer, is named after the mythological guardian dog with three heads because it has binding sites for three different types of powerful signals—Wnt, Nodal, and BMP. The binding of these signals by Cerberus prevents activation of their respective receptors. By inactivating Wnt, Nodal, and BMP signals having roles in the development of the trunk and tail of the body, Cerberus promotes head development.

In *Xenopus*, neural induction seems to be the default state that must be actively blocked in order for other cell types to develop. In chicks and mammals, however, fibroblast growth factor (FGF), BMPs, and Wnt appear to be necessary signals for induction of neural tissue in the posterior embryo. In the anterior, the actions of BMP and Wnt are prevented by antagonists (Figure 22-16). These antagonists are produced in a region called the *node*, which is functionally equivalent to the Spemann organizer in frogs. The node is a depression at the anterior end of the primitive streak; during gastrulation, the streak with the node leading it gradually expands forward toward the head.

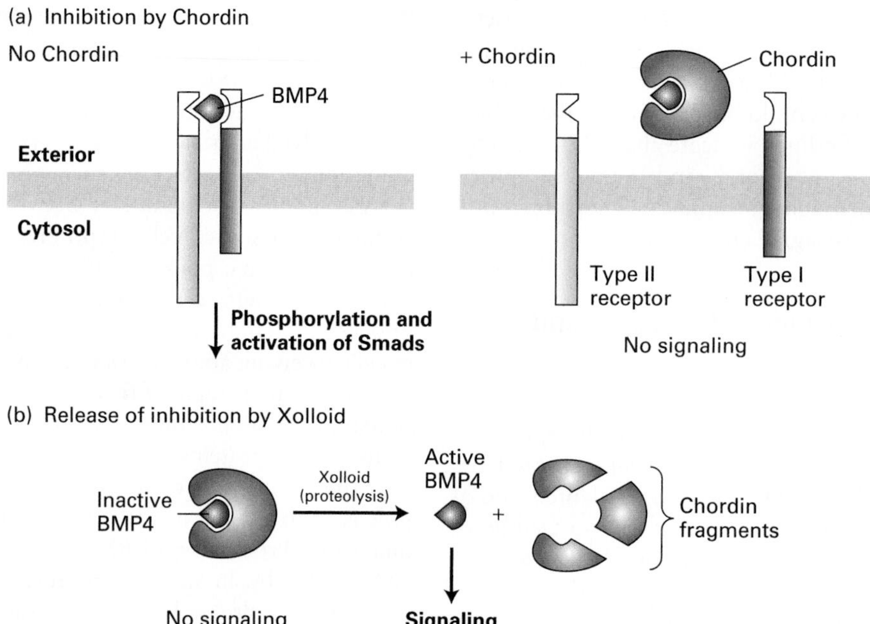

▲ **FIGURE 22-15 Modulation of BMP4 signaling in *Xenopus* by Chordin and Xolloid.** (a) Chordin binds BMP4, a TGFβ-family secreted protein signal, and prevents it from binding to its receptor. See Figure 16-4 to review the TGFβ signaling pathway. (b) Xolloid specifically cleaves Chordin in the Chordin–BMP4 complex, releasing BMP4 in a form that can bind to its receptor and trigger signaling. [See S. Piccolo et al., 1997, *Cell* **91:**407].

Sorting out all the signals and antagonists involved in determining cell fates along the three body axes will require much additional research. Nonetheless, biologists now have identified most of the powerful regulators that cause the formation of different tissue types in the early embryo, including those that control differences between the left and right sides of the body.

A Cascade of Signals Distinguishes Left from Right

Having discussed dorsal-ventral and anterior-posterior axis formation, we turn next to the genetic controls that organize the left-right axis. Antibody staining and in situ hybridization have been used to look at gene expression in late gastrula embryos. One of the most striking and interesting findings has been that some genes are active on only the left or only the right side of the embryo. These include genes encoding three secreted protein signals: Sonic hedgehog, FGF, and Nodal (Figure 22-17). Experiments in which genes with this type of expression pattern were disrupted have shown that such genes are not merely responding to left-right information; they are part of the system for controlling the subsequent pattern of the embryo. An animal tissue or organ may exhibit left-right asymmetry in two ways: by forming primarily on one side of the body or by having an asymmetric

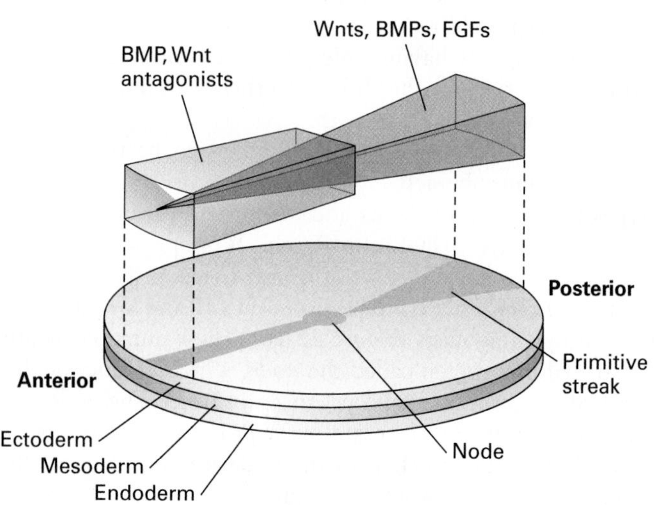

◄ **FIGURE 22-16 Regulators of anterior-posterior cell fates in the mouse embryo.** The late gastrula embryo, which is actually curled, is depicted here as a flattened oval with three layers, ectoderm, mesoderm, and endoderm. The primitive streak, where future mesoderm cells moved inside during gastrulation, is indicated as a purple area on the surface of the oval. The node (green circle) is a signaling center that forms earlier and eventually becomes the notochord (see Figure 22-39). Regulatory signals that control anterior-posterior patterning are indicated above the oval. Wnt, BMP, and FGF secreted signaling proteins are made by posterior tissues and direct cells to assume posterior-cell fates such as neural tissue. The actions of BMP and Wnt proteins are blocked in the anterior embryo by antagonistic proteins that are made by cells in the node and accumulate there. [Adapted from S.F. Gilbert, 2006, *Developmental Biology*, 8th ed., Sinauer Press, Figure 11-41.]

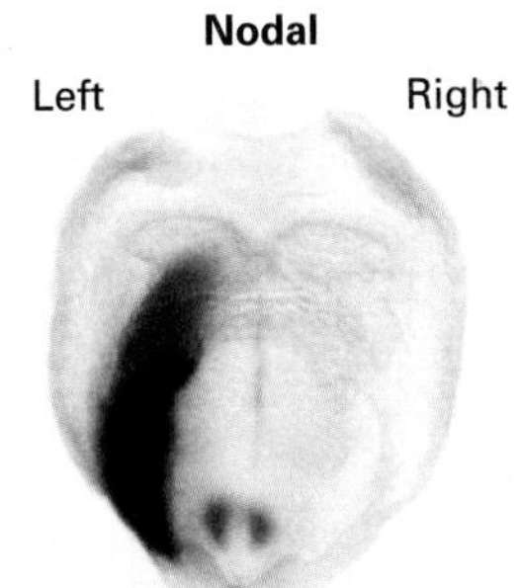

▲ **EXPERIMENTAL FIGURE 22-17 Nodal protein is produced only on side of a 8-day mouse embryo.** In this in situ hybridization experiment, the embryo was permeabilized and then incubated with a probe specific for *nodal* mRNA. The probe was made with nucleotide analogs that can be bound by antibodies. After hybridization, the specimen was exposed to an anti-probe antibody that is covalently linked to a reporter enzyme. After addition of the enzyme substrate, a colored precipitate formed where the probe hybridized to the nodal mRNA, obviously on one side only. This type of hybridization experiment is widely used in studying gene expression in embryos. Other experiments showed that the Nodal-producing side eventually forms left-side structures. [From H. Hamada et al., 2002, *Nature Genet.* **3**:103].

morphology. To see how left-right asymmetry can be created during development, we consider the heart, which is asymmetric in both ways.

The normal process of heart development in the mouse is shown in Figure 22-18. The most amazing feature of the heart is that a tiny organ formed early in embryogenesis begins beating at 23 days (human), when the entire embryo is about 2.5 mm long, and continues its reliable pumping as it grows to about $6 \times 8 \times 12$ cm and weighs about a third of a kilogram. This is like growing an outboard motor into the huge engine of an ocean liner, while all the time keeping the motor running!

At the earliest stages of heart development, the *Nkx2.5* gene, a member of a family of *Nkx* genes that encode homeodomain transcription factors, is expressed in heart precursor cells in the lateral plate mesoderm. The first *Nkx2.5*-type gene was discovered as a *Drosophila* mutant by virtue of its phenotype: no heart. The fly mutant was named *tinman* after the Wizard of Oz character, who thought he had no heart. Humans with only one copy of *Nkx2.5* have a variety of heart defects. The identification of *Nkx2.5* allowed researchers to detect early heart precursor cells that express

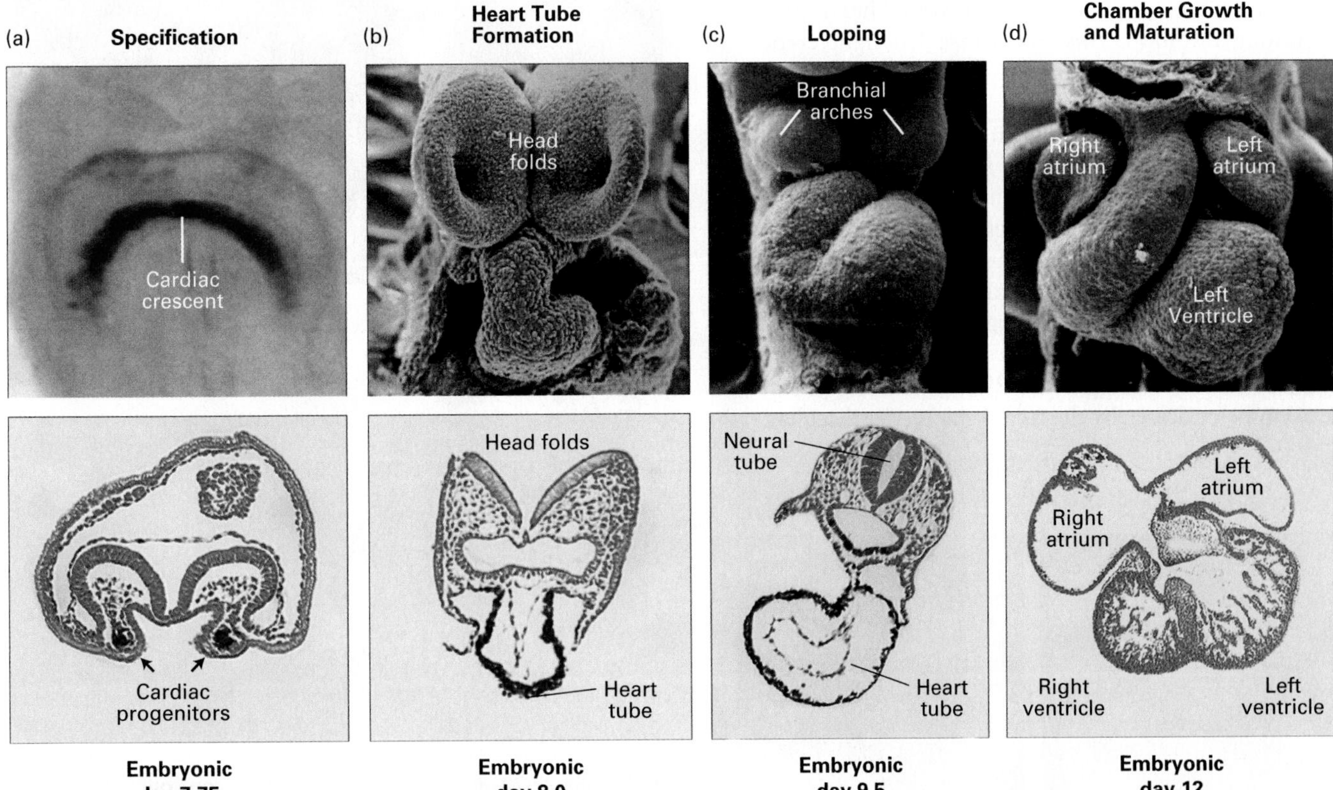

▲ **FIGURE 22-18 Heart development in the mouse embryo from day 7.75 to day 12.** Whole mount and scanning electron micrographs are shown at top; transverse histological sections, at the bottom. In situ hybridization was used to detect the heart primordial structures (blue) in whole mount and sections. (a) The cardiac crescent, the earliest heart primordium formed, is detectable by day 7.75. (b) The heart tube develops from the cardiac crescent underneath the head folds. (c) The heart tube begins looping beneath the branchial arches. (d) Chamber growth and maturation generates the four chambers of the heart by day 12. The blue stains in the tissue sections reveal proteins made under the control of key heart regulatory genes, including *Nkx2.5* (see text). Such regulatory genes are important for understanding cell-fate determination mechanisms and also serve, as shown here, as superb markers of early precursor cells that could not be recognized otherwise. [From E. N. Olson and M. D. Schneider, 2003, *Genes & Devel.* **17**:1937.]

the gene, cells that had not been distinguishable from other embryonic cells prior to having this powerful new marker gene (see Figure 22-18). Subsequent experiments using this marker gene have been designed to identify factors that control the earliest steps in heart formation. Similar discoveries of early regulators for other organs and tissues are providing tools for discovering what signals control their formation. The *Nkx2.5* story provides a dramatic example of the dedication of a particular transcription factor to a particular organ during half a billion years or so of evolutionary divergence. It also illustrates the value of *Drosophila* and other model organisms for discovering human genes involved in disease. Finally it shows that organs that serve the same purpose but look really different, like insect and mammalian hearts, may nonetheless be built using some of the same molecular regulators.

In order to form a functioning pump, the mammalian heart tube bends and twists in an asymmetric way. This looping brings together chamber primordia in such a way that openings can form between some of them (see Figure 22-18c, d). These openings will become valves. The bending and looping of the heart can happen in a left- or right-handed way depending, as we shall see, on information inherited from earlier asymmetric gene expression. Figure 22-19 shows a mouse mutant with reversed heart direction.

Among the mutations that cause left-right errors are some that affect the formation and function of cilia. We have already mentioned Kartagener's syndrome, in which mutations in the motor protein dynein can cause left-right reversal of the heart as well as male infertility due to immotile sperm flagella. These and other findings suggested a role of cilia or flagella in controlling left-right asymmetry. In the mouse, the relevant cilia turn out to be in the node (Figure 22-20a), the region in mammalian embryos that is more or less equivalent to the Spemann organizer in amphibians. The biased motion of cilia in the node causes fluids and signals to flow toward the left (Figure 22-20b). It is not yet clear that the asymmetry driven by the cilia is the very earliest left-right difference in all animals, since evidence exists for asymmetries that precede node formation. The cilia hypothesis is strongly supported by genetic analyses that in an unbiased way screened for left-right developmental defects.

The mystery of what is moved by the nodal cilia may have been resolved with the recent discovery of nodal vesicle particles (NVPs), which are 0.35- to 0.5-μm in diameter. NVPs contain Sonic hedgehog protein and retinoic acid, both known to be powerful regulators of cell fates. In response to FGF signals, NVPs are released from cells and are moved to the left side of the node where they stimulate heightened Ca^{2+} levels, one aspect of "leftness." The movement of the NVPs and/or other substances results in asymmetric expression of genes encoding TGFβ signals and consequent left-only

(a)

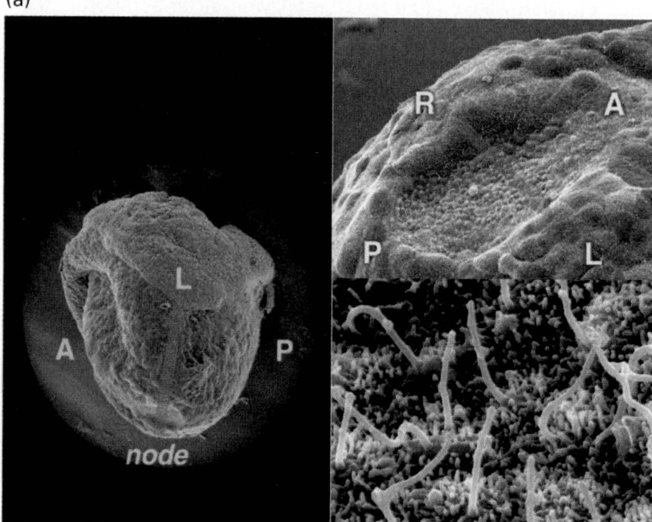

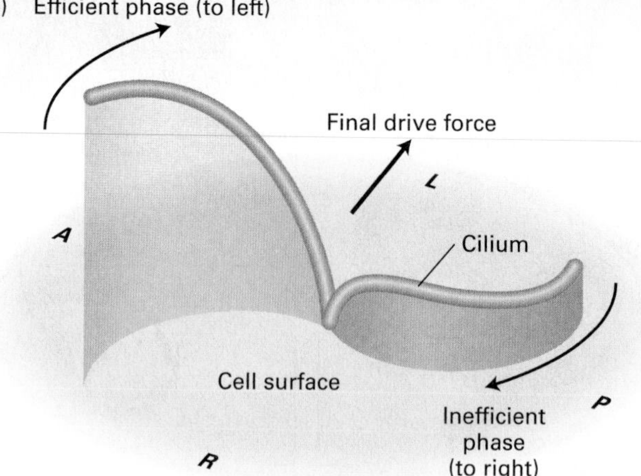

▲ **FIGURE 22-20 Role of cilia at the node in controlling left-right polarity in mammalian embryos.** (a) Mouse embryo at 8 days *(left)* and high magnification view of the node *(right)* showing cilia. A = anterior; P = posterior; L = left; R = right. (b) Model of cilia function. The cilia more effectively move fluid, and the signaling proteins carried by it, in the "power stroke" toward the left. The "recovery stroke" toward the right is inefficient in moving fluid. Thus the inherent asymmetry of the motor apparatus inside tiny cilia leads to the asymmetry of our entire body. [Part (a) from H. Hamada et al., 2002, *Nature Genetics* **3:**103.Part (b) adapted from S. Nonaka et al., 2005, *PLoS Biology* **3:**268.]

Normal

iv mutant

▲ **FIGURE 22-19 Normal and "reversed" hearts.** Hearts are genetically programmed to loop in a particular direction. In the normal mouse heart shown here, the looping is opposite to that of the heart from an *iv* mutant mouse, as indicated by the arrows. The *iv* gene encodes a dynein motor protein that is necessary for proper left-right asymmetry of the heart. [Copyright Oxford University Press, N. Brown.]

expression of the *Pitx2* gene, which encodes a transcription factor. These regulatory events set in motion the regulatory processes that give rise to different cell arrangements, and eventually organ structures, on the left and right sides.

KEY CONCEPTS OF SECTION 22.3

Cell Diversity and Patterning in Early Vertebrate Embryos

- The fertilized egg is a totipotent cell, which can give rise to all the kinds of cells in the body.
- Cleavage of the fertilized egg is a series of rapid cell divisions, initially giving rise to a clump of similar-looking cells (see Figure 22-8). By the time 64 cells have formed in mammalian embryos, some cells have moved to the outside, constituting the trophectoderm, while others remain inside as the inner cell mass.
- The trophectoderm will form extra-embryonic tissues like the placenta, and the inner cell mass (ICM) will form the embryo itself (see Figure 22-10). The ICM is the source of embryonic stem cells.
- Since most somatic cells have a complete (diploid) genome, cell differentiation usually does not depend on the physical loss of some genes. One exception is lymphocytes, in which portions of the genome are rearranged or lost during development.
- The inner cell mass, a clump of undifferentiated cells, begins to form multiple cell layers during gastrulation. Initially the cells form two layers and then cells in one layer move into the middle to form a third layer. This gives rise to the three germ layers: endoderm, mesoderm, and ectoderm (see Figure 22-11).
- Several signal-emitting centers form in early vertebrate embryos (see Figures 22-12 and 22-16). Cells that receive the signals, usually integrating more than one, activate certain genes in response. Secreted protein signals and signal antagonists coming from these centers thus determine cell fates along the anterior-posterior and dorsal-ventral axes.
- Signals that control cell fates in a dose-dependent way are called morphogens. In the case of such signals, receiving cells take on one fate at high levels; different fates at moderate levels; and a third, default fate in the absence of the signal.
- Antagonists of signals are important in restraining and balancing the effects of powerful inductive signals (see Figure 22-15). For example, mesoderm induction by TGFβ is prevented in regions of the embryo where the nervous system will form by secreted protein antagonists that compete with TGFβ.
- During gastrulation, differences in gene expression from left to right become evident (see Figure 22-17). This left-right polarized gene expression depends in part on cilia at the node, a depression at the anterior end of the primitive streak. These cilia appear to move signals to one side more than to the other (see Figure 22-20), setting in motion distinct cell fate decisions on the two sides of the embryo.

22.4 Control of Body Segmentation: Themes and Variations in Insects and Vertebrates

Forming structure and pattern in animals involves more than creating tissue types and determining head to tail, side to side, and front to back polarities. Insects, vertebrates, and many other creatures have repeating body parts. Indeed, vertebrates are named for the repeating bones (vertebrae) that make up the spinal column. Such repeating body parts are usually not perfectly identical; rather, they exhibit themes and variations that are functional for the animal. For example, some of our vertebrae have ribs attached, and others do not. When a wasp visits your picnic, the striped segments of its body are clearly visible. Some of the segments of an insect body have legs and others do not. In this section, we examine the genetic regulation of embryo segmentation in an insect and in a vertebrate to see two different ways to create a repeating pattern. Then we will look at genes that control the differences among the repeats. Remarkably, segment-specific differences are controlled by the same genes in insects and vertebrates, despite the half billion years or so since their common ancestor was alive.

The study of the molecular cell biology of body segmentation began with a powerful genetic screen designed to systematically identify every gene needed to form the segments in *Drosophila*. This sort of systematic screening had been applied to identifying components of more specific processes such as mating-type regulation in yeast, but had never before been applied to the development of a large complicated animal. The screen was enormously successful in identifying segmentation genes, opening up a new era of molecular analysis of pattern formation, one aspect of the generation of form (*morphogenesis*). Startlingly, though, the screen succeeded far beyond expectations in two ways: it identified many genes involved in development of internal tissues, not only the segmented outer skeleton, and the genes identified in flies turned out to be representative of genes used in all animals to control embryonic development. Since then systematic genetic screens have been applied to many other developmental and cell biological processes, such as neural and heart development, using a variety of model organisms including worms, flies, mice, and zebrafish (see Figure 1-25). As a consequence, the genes being identified in various genome projects can increasingly be described in terms of their functions, as well as the tissues, times, and processes in which they participate. That is a crucial step in understanding the whole network of regulators that drives development.

To understand how segmentation arises, we first describe early *Drosophila* development and then discuss the types and functions of genes that were found in the original genetic screen. Next we examine the segmentation process in vertebrates and finally conclude this section by considering pattern formation in plants.

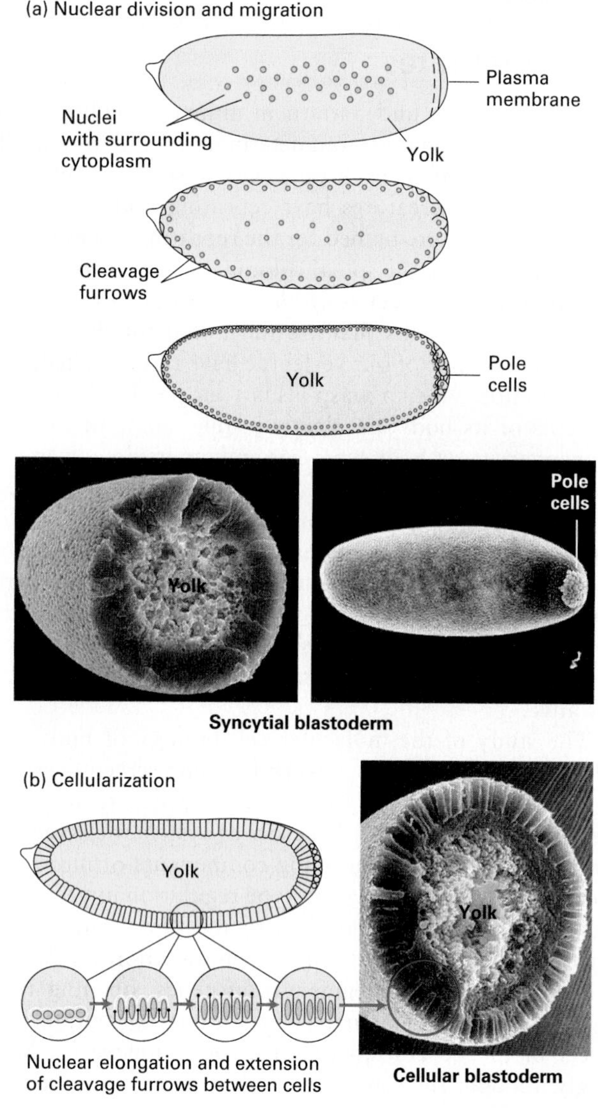

◀ **FIGURE 22-21 Formation of the cellular blastoderm during early *Drosophila* embryogenesis.** Stages from the syncytium (a) to cellular blastoderm (b) are illustrated in diagrams and electron micrographs. Nuclear division is not accompanied by cell division until about 6000 nuclei have formed and migrated outward to the plasma membrane. Before cellularization, the embryo displays surface bulges overlying individual nuclei, which remain within a common cytoplasm. No membranes other than that surrounding the entire embryo are present. After cellularization, cell membranes are evident around individual nuclei. Note the segregation of the nuclei of so-called pole cells, which give rise to germ-line cells, at the posterior end of the syncytial blastoderm. [Photographs courtesy of A. P. Mahowald; diagrams after P. A. Lawrence, *The Making of a Fly*, 1992, Blackwell Scientific, Oxford.]

Early *Drosophila* Development Is an Exercise in Speed

After fertilization and fusion of male and female pronuclei in *Drosophila*, the first 13 nuclear divisions of the zygote are synchronous and rapid, each division occurring about every 10 minutes. This DNA replication is the most rapid known for a eukaryote, with the entire 160 Mb of *Drosophila* chromosomal DNA copied in a cell-cycle S phase that lasts only 3 minutes. Because these nuclear divisions are *not* accompanied by cell divisions, they generate a multinucleated egg cell, a **syncytium,** with a common cytoplasm and plasma membrane (Figure 22-21a). As the nuclei divide, they begin to migrate outward toward the plasma membrane. The nuclei reach the surface about 2–3 hours after fertilization, forming the *syncytial blastoderm;* during the next hour or so, cell membranes form around the nuclei, generating the cellular blastoderm, or blastula (Figure 22-21b). All future tissues are derived from the 6000 or so epithelial cells on the surface of the blastula, whose interior is filled with yolk. Soon some of the epithelial cells move inside, the insect version of gastrulation, and eventually develop into the internal tissues.

In all animals, many of the important pattern-controlling regulators act when the embryo has relatively few cells. The fly blastula, for instance, is only about 100 cells long and 60 cells in circumference, but even at this early stage, the major parts of the body plan are laid out. Because the fly embryo is a syncytium for the first few hours of development, regulatory molecules can move in the common cytoplasm without having to cross plasma membranes. Some molecules form gradients, which are used in the earliest stages of cell-fate determination in *Drosophila* before subdivision of the syncytium into individual cells. Thus transcription factors, as well as secreted molecules, can function as morphogens in the syncytial fly embryo. Within 1 day of fertilization, an amazingly short time to form a whole organism, the embryo develops into a segmented larva (juvenile stage) that lacks wings and legs. Development continues through three larval stages (≈4 days total) and the ≈5-day pupal stage during which metamorphosis takes place and adult structures are created (Figure 22-22a). At the end of pupation, about 10 days after fertilization, the pupal case splits and an adult fly emerges.

The initially equivalent cells of the syncytial embryo rapidly begin to assume different fates, leading to a well-ordered pattern of distinct cell identities. These early patterning events set the stage for the later development and proper placement of different tissues (e.g., muscle, nerve, epidermis) and body parts, as well as the shapes of the appendages and the organization of cell types within them. The early embryo is symmetric side to side, and the initial differences among cells are created along the dorsal-ventral (back-front) and anterior-posterior (head-tail) axes. During the larval stages, groups of ectodermal cells, called *imaginal discs,* form at specific sites and stay largely quiescent during early larval stages (Figure 22-22b). Each disc will grow and give rise to specific adult structures during pupation, one for each wing, one for each leg, and so forth.

Different sets of genes are active in establishing cell fates along the dorsal-ventral (back-front) and anterior-posterior (head-tail) axes. The initial fate of every cell is governed by both dorsal-ventral-acting and anterior-posterior-acting regulators in a kind of two-dimensional grid. Both regulatory systems begin with information and molecules contributed

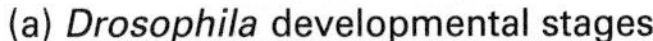

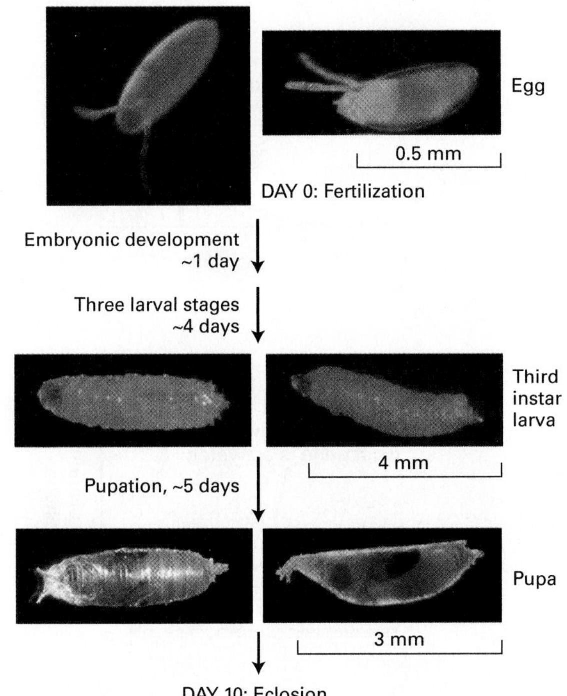

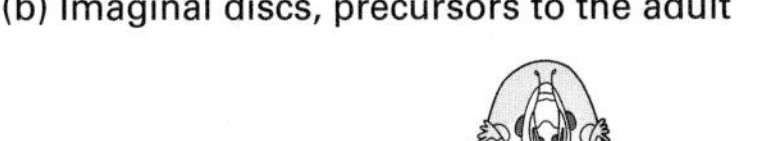

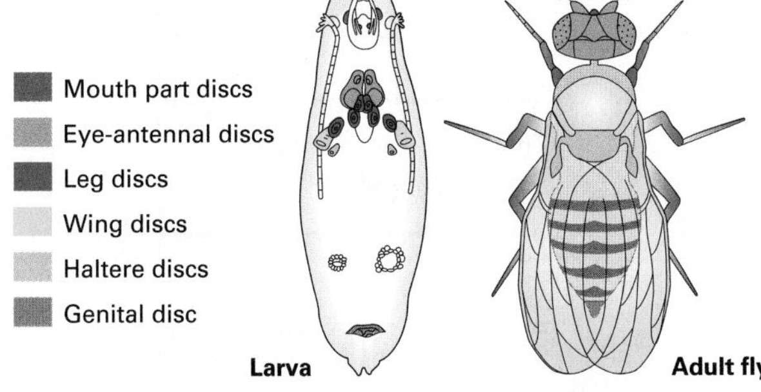

▲ **FIGURE 22-22 Major stages in the development of *Drosophila* and location of imaginal discs.** (a) The fertilized egg develops into a blastoderm and undergoes cellularization in a few hours. The larva, a segmented form, appears in about 1 day and passes through three stages (instars) over a 4-day period, developing into a prepupa. Pupation takes ≈4–5 days, ending with the emergence of the adult fly from the pupal case. (b) Groups of ectodermal cells called imaginal discs are set aside at specific sites in the larval body cavity. During pupation, these give rise to the various body parts indicated. Other precursor cells give rise to adult muscle, the nervous system, and other internal structures. [Part (a) courtesy Kaye Suyama. Part (b) adapted from J. W. Fristrom et al., 1969, in E. W. Hanly, ed., *Park City Symposium on Problems in Biology*, University of Utah Press, p. 381.]

to the oocyte as a dowry from the mother. The dorsal-ventral control system involves differential transport of a transcription factor called Dorsal into the nuclei of the syncytial embryo. This transcription factor, related to the vertebrate NF-κB protein, is present in its inactive form throughout the cytoplasm of the syncytial embryo. A signal protein concentrated on the ventral side of the embryo triggers the NF-κB signaling pathway (see Figure 16-35), leading to activation of Dorsal and its translocation into nuclei. As a result, Dorsal enters ventral nuclei at a high level, lateral nuclei at a modest level, and dorsal nuclei not at all (Figure 22-23). Thus Dorsal acts in a graded fashion and has the property of a morphogen, even though this protein is not secreted. The differential entry of the Dorsal transcription factor into nuclei, which controls subsequent cell fates, is governed by a complex signaling process that starts during oogenesis in the follicle cells and is transmitted to the developing embryo. We will not pursue the details of dorsal-ventral patterning further and will instead concentrate on anterior-posterior patterning.

To decipher the molecular basis of cell-fate determination and patterning along the three body axes, investigators have cloned genes identified in screens for mutations that affect the body plan; determined the spatial and temporal patterns of mRNA production for each gene and the distribution of the encoded proteins in the embryo; and assessed the effects of mutations on cell differentiation, tissue patterning, and the expression of other regulatory genes. The principles of cell-fate determination and tissue patterning learned from *Drosophila* have proved to have broad applicability to animal development.

Transcriptional Control Specifies the Embryo's Anterior and Posterior

We turn now to determination of the anterior-posterior axis in the early fly embryo while it is still a syncytium. The process begins during oogenesis when *maternal mRNAs* produced by nurse cells are transported into the oocyte and become localized in discrete spatial domains (see Figure 22-2). For example, *bicoid* mRNA is trapped at the most anterior region, or anterior pole, of the early fly embryo (Figure 22-24). The anterior localization of *bicoid* mRNA depends on its 3′-untranslated end and three maternally derived proteins. Embryos produced by female flies that are homozygous for *bicoid* mutations lack anterior body parts, attesting to the importance of Bicoid protein in specifying anterior cell fates.

Bicoid protein is a homeodomain-type transcription factor that activates expression of certain anterior-specific genes discussed later. In the syncytial fly embryo, Bicoid protein spreads through the common cytoplasm away from the anterior end where it is produced from the localized mRNA. As a result, a Bicoid protein gradient is established along the anterior-posterior axis of the syncytial embryo. Since the effects of Bicoid are concentration-dependent, it is acting as a morphogen. Evidence that the Bicoid protein gradient determines anterior structures was obtained through injection of synthetic *bicoid* mRNA at different locations in the embryo. This treatment led to the formation of anterior structures at the site of injection, with progressively more posterior structures forming at increasing distances from the injection site. Another test was to make flies that produced extra anterior Bicoid protein; in these flies, the anterior structures expanded to occupy a greater proportion of the embryo.

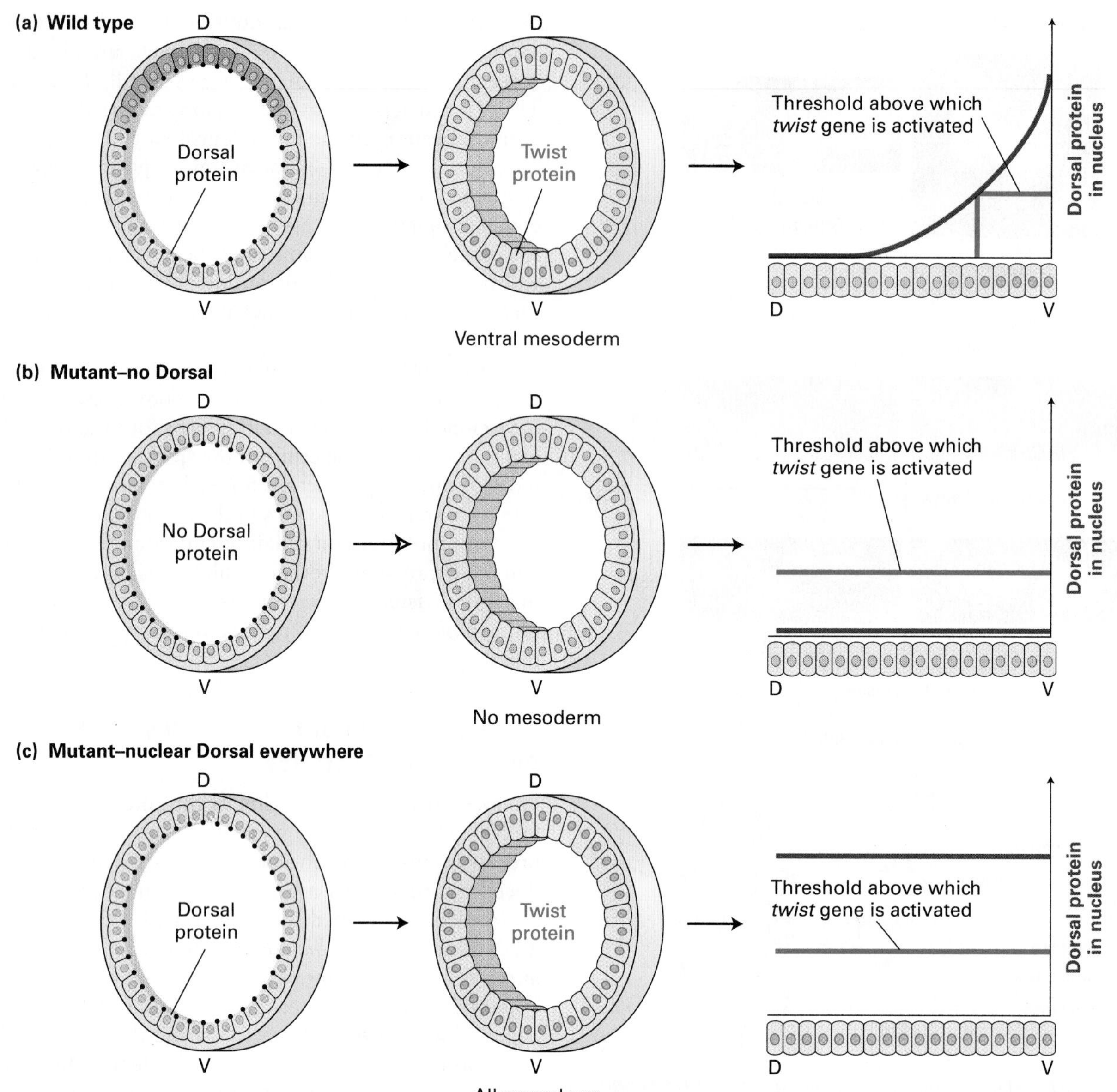

▲ EXPERIMENTAL FIGURE 22-23 **A gradient of the transcription factor Dorsal directs dorsal-ventral cell fates in the early embryo.** Dorsal is the *Drosophila* homolog of vertebrate NF-κB (See Figure 16-35). (a) In wild-type embryos, more Dorsal protein enters the nuclei on the ventral side. Once inside a nucleus, Dorsal activates target genes such as *twist,* which encodes a transcription factor that directs mesoderm formation. The differential entry of Dorsal thus creates a dorsal-ventral polarity in *twist* expression and mesoderm induction. Inactive cytoplasmic NF-κB on the dorsal side fails to activate *twist.* (b) A mutant lacking Dorsal makes no mesoderm cells. (c) Conversely, in a mutant that has Dorsal in the nuclei of all cells, all cells differentiate as mesoderm. [Adapted from L. Wolpert et al., *Principles of Development*, 2nd edition, Oxford Press, Figure 5-14.]

Bicoid protein promotes transcription of the *hunchback (hb)* gene from the embryo's genome. Transcription of *hunchback* is greatest in the anterior of the embryo where the Bicoid concentration is highest. Mutations in *hunchback* and several other genes in the embryo's genome lead to large gaps in the anterior-posterior pattern of the early embryo; hence these genes are collectively called **gap genes.** Several types of evidence indicate that Bicoid protein directly regulates transcription of *hunchback*. For example, increasing the number of copies of the *bicoid* gene expands the Bicoid and Hunchback (Hb) protein gradients posteriorly in parallel (Figure 22-25a–c). Analysis of the *hunchback* gene revealed that it contains three low-affinity and three high-affinity binding sites for Bicoid protein. Experiments with synthetic genes containing either all high-affinity or all low-affinity Bicoid-binding sites demonstrated that the affinity of the site determines the threshold concentration of Bicoid at which gene transcription is activated (Figure 22-25d, e). In addition, the number of Bicoid-binding sites occupied at a given concentration has been shown to determine the amplitude, or level, of the transcription response.

Overview Animation: Gene Control in Embryonic Development

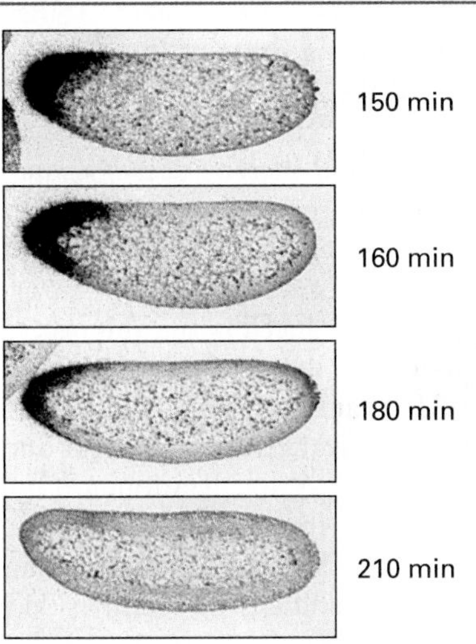

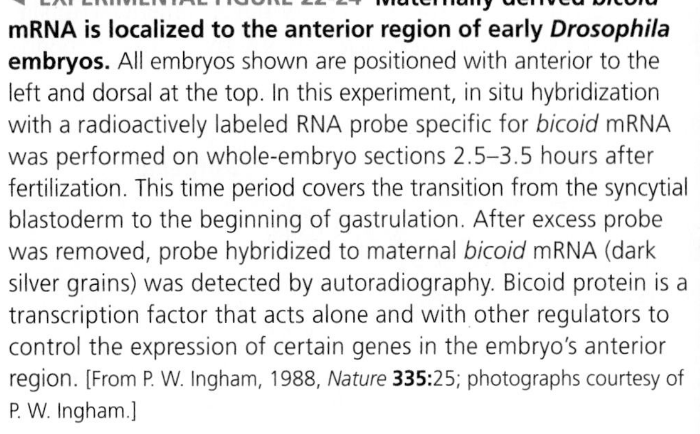
◀ EXPERIMENTAL FIGURE 22-24 **Maternally derived *bicoid* mRNA is localized to the anterior region of early *Drosophila* embryos.** All embryos shown are positioned with anterior to the left and dorsal at the top. In this experiment, in situ hybridization with a radioactively labeled RNA probe specific for *bicoid* mRNA was performed on whole-embryo sections 2.5–3.5 hours after fertilization. This time period covers the transition from the syncytial blastoderm to the beginning of gastrulation. After excess probe was removed, probe hybridized to maternal *bicoid* mRNA (dark silver grains) was detected by autoradiography. Bicoid protein is a transcription factor that acts alone and with other regulators to control the expression of certain genes in the embryo's anterior region. [From P. W. Ingham, 1988, *Nature* **335**:25; photographs courtesy of P. W. Ingham.]

MEDIA CONNECTIONS

Findings from studies of Bicoid's ability to regulate transcription of the *hunchback* gene show that variations in the levels of transcription factors, as well as in the number or affinity of specific regulatory sequences controlling different target genes, or both, contribute to generating diverse patterns of gene expression during *Drosophila* development. Similar mechanisms are employed in other developing organisms.

Translation Inhibitors Reinforce Anterior-Posterior Patterning

Cell fates at the posterior end of the fly embryo are specified by a different mechanism—one in which control is at the translational level rather than the transcriptional level. As just discussed, transcription of the embryo's *hunchback* gene, which promotes anterior cell fates, produces an anteriorly located band of *hunchback* mRNA and Hunchback protein because of the anterior → posterior gradient of maternally derived Bicoid protein. In addition, however, the early embryo contains *hunchback* mRNA synthesized by nurse cells. Even though this maternal *hunchback* mRNA is uniformly distributed throughout the embryo, its translation is prevented in the posterior region by another maternally derived protein called Nanos, which is localized to the posterior end of the embryo. The set of genes required for posterior localization of Nanos protein is also required for germ-line cells to form at the posterior end of the embryo. One of these genes (*staufen*) is necessary for development of primordial germ cells (PGCs) in zebrafish. Thus at least some germ-line regulators have existed since fish and flies had a common ancestor.

Copies of *bicoid* gene in mother | Bicoid protein gradient in embryo | Promoter | Expression pattern
(a) 0 [Bicoid] Anterior → Posterior A P
(b) 1 *hunchback* High-affinity Bicoid-binding site Low-affinity Bicoid-binding site Hunchback protein
(c) 2
(d) 2 Synthetic High-affinity Bicoid-binding sites Reporter-gene product
(e) 2 Synthetic Low-affinity Bicoid-binding sites

◀ EXPERIMENTAL FIGURE 22-25 **Maternally derived Bicoid controls expression of the embryonic *hunchback (hb)* gene along the anterior-posterior axis.** (a–c) Increasing the number of *bicoid* genes in mother flies changed the Bicoid gradient in the early embryo, leading to a corresponding change in the gradient of Hunchback protein produced from the *hunchback* gene in the embryo's genome. The *hunchback* promoter contains three high-affinity and three low-affinity Bicoid-binding sites. Transgenic flies carrying a reporter gene linked to a synthetic promoter containing either four high-affinity sites (d) or four low-affinity sites (e) were prepared. In response to the same Bicoid protein gradient in the embryo, expression of the reporter gene controlled by a promoter carrying high-affinity Bicoid-binding sites extended more posteriorly than did transcription of a reporter gene carrying low-affinity sites. This result indicates that the threshold concentration of Bicoid that activates *hunchback* transcription depends on the affinity of the Bicoid-binding site. Bicoid regulates other target genes in a similar fashion. [Adapted from D. St. Johnston and C. Nüsslein-Volhard, 1992, *Cell* **68**:201.]

Figure 22-26 illustrates how translational regulation by Nanos helps to establish the anterior → posterior Hunchback gradient needed for normal fly development. Translational repression of *hunchback* mRNA by Nanos depends on a specific sequence in the 3′-untranslated region of the mRNA, the Nanos-response elements (NRE). Along with two other RNA-binding proteins, Nanos binds to the NRE in *hunchback* mRNA. The results of genetic and molecular studies suggest that Nanos promotes deadenylation of *hunchback* mRNA and thereby decreases its translation. In the absence of Nanos, the accumulation of maternal Hb protein in the posterior region leads to failure of the posterior structures to form normally, and the embryo dies. Conversely, if Nanos is produced in the anterior, thereby inhibiting the production of Hb from both maternal and embryonic *hunchback* mRNA, anterior body parts fail to form, again a lethal consequence. Translational control due to the action of an inhibitor, mRNA localization, or both, may be a widely used strategy for regulating development. For instance, specific mRNAs are localized during the development of muscle cells, and during cell division in the budding yeast *Saccharomyces cerevisiae* (see Figure 21-28).

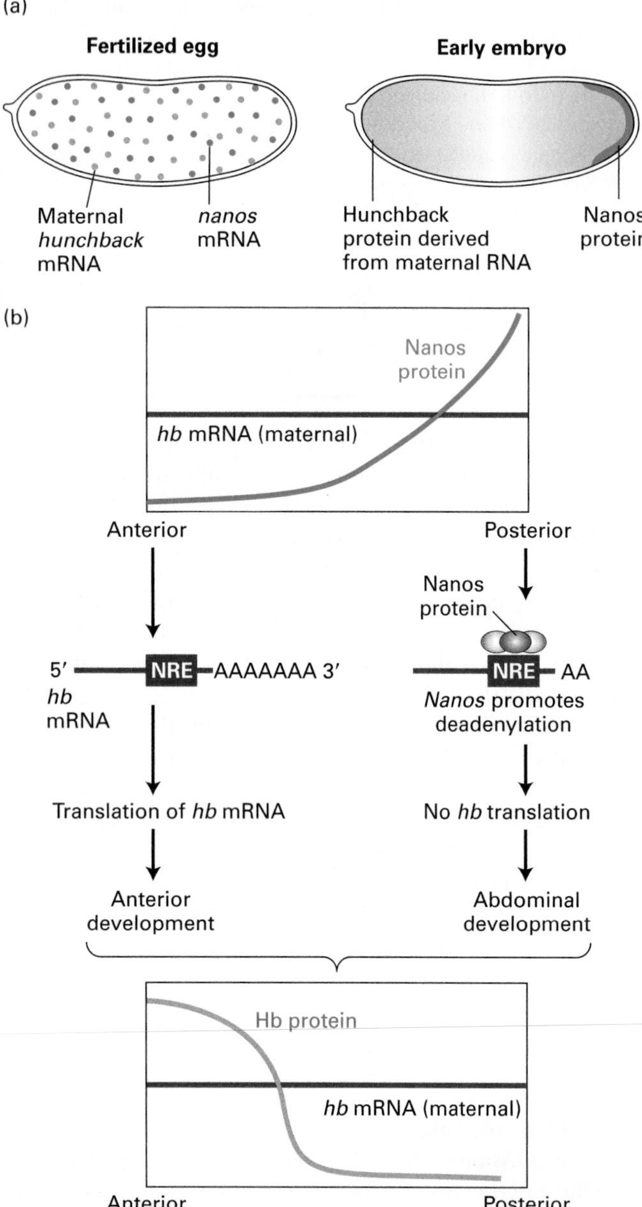

▲ **FIGURE 22-26 Role of Nanos protein in excluding maternally derived Hunchback (Hb) protein from the posterior region of *Drosophila* embryos.** (a) Both *nanos* (blue) and *hunchback* (red) mRNAs derived from the mother are distributed uniformly in the fertilized egg and early embryo. Nanos protein, which is produced only in the posterior region, subsequently inhibits translation of maternal *hb* mRNA posteriorly. (b) Diffusion of Nanos protein from its site of synthesis in the posterior region establishes a posterior → anterior Nanos gradient. A complex of Nanos and two other proteins inhibits translation of maternal *hb* mRNA. As a consequence, maternally derived Hb protein is expressed in a graded fashion that parallels and reinforces the Hb protein gradient resulting from Bicoid-controlled transcription of the embryo's *hb* gene (see Figure 22-25). [See C. Wreden et al., 1997, *Development* **124:**3015.]

Insect Segmentation Is Controlled by a Cascade of Transcription Factors

In both insects and vertebrates, the anterior-posterior (head-tail) axis is divided into a set of repeats, or more precisely repeats with variations: vertebrae and associated ganglia in vertebrates, body segments in insects. Specific genes control subdivision of the embryo into repeats, while other genes control the differences between repeats. As noted already, not all vertebrae have attached ribs, and only some of an insect's body segments have legs growing from them. We discuss the genes controlling segmentation of insects in this section and the rather different type of regulation underlying vertebrate segmentation in the next section. Then we delve into the genes that control differences between segments.

Once the gap genes have been properly activated in the *Drosophila* embryo, the next steps on the road to body segmentation are controlled by a *transcription-factor (TF) cascade* in which one TF controls a gene encoding another TF, which in turn controls expression of a third TF. At each step, more than one gene may be regulated. Such a TF cascade can generate a population of cells that may all look alike but differ at the transcriptional level. TF cascades have both a temporal and a spatial dimension. At each step in a cascade, for instance, RNA polymerase and ribosomes can take more than an hour to produce a transcription factor from its corresponding mRNA. Spatial factors come into play when cells at different positions within an embryo synthesize different transcription factors.

The rough outline of cell fates that is laid down in the syncytial fly embryo is refined into a system for precisely controlling the fates of individual cells. Discovery of the relevant regulators came from a genetic screen for mutants with altered embryo body segments. In addition to *hunchback*, four other gap genes—*Krüppel*, *knirps*, *giant*, and *tailless*—are transcribed in specific spatial domains beginning about 2 hours after fertilization (Figure 22-27a). Expression of these genes, like that of *hunchback*, is regulated first by maternal factors and then by cross-interactions among the gap genes.

Video: Expression of Segmentation Genes in a *Drosophila* Embryo

MEDIA CONNECTIONS

(a) Gap-gene proteins

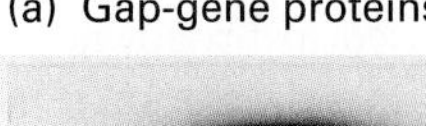

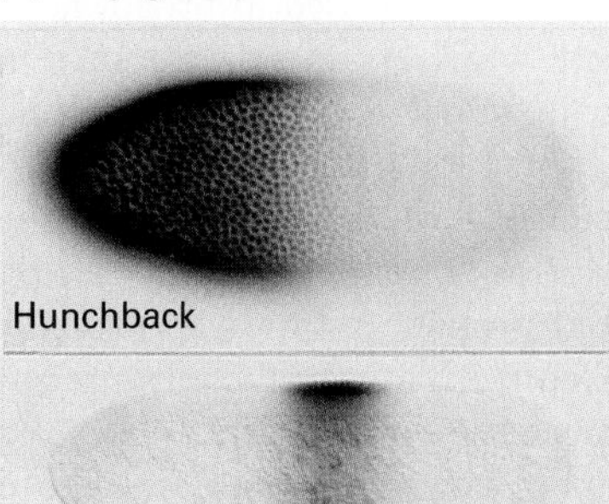

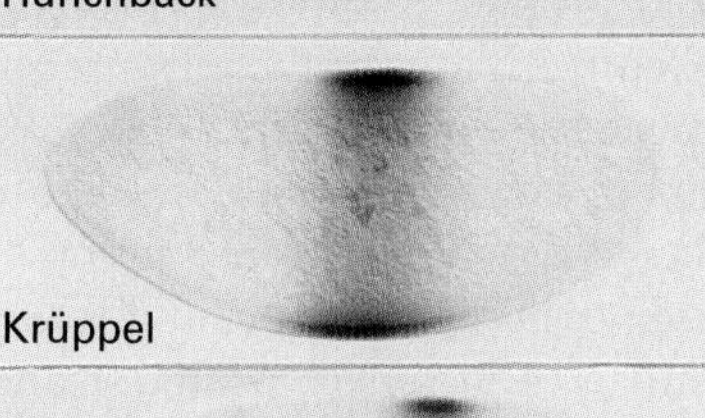

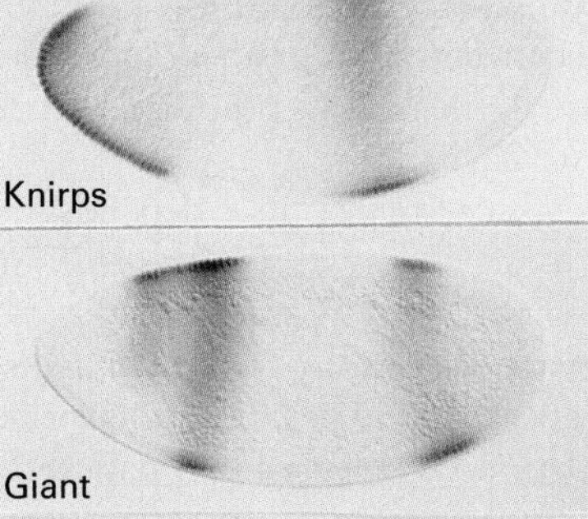

(b) Hunchback and Krüppel

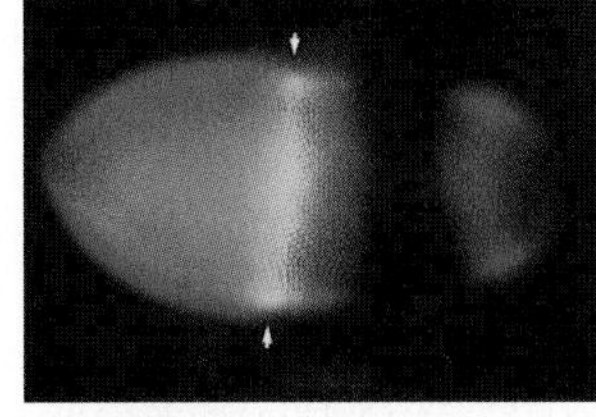

(c)

(d)

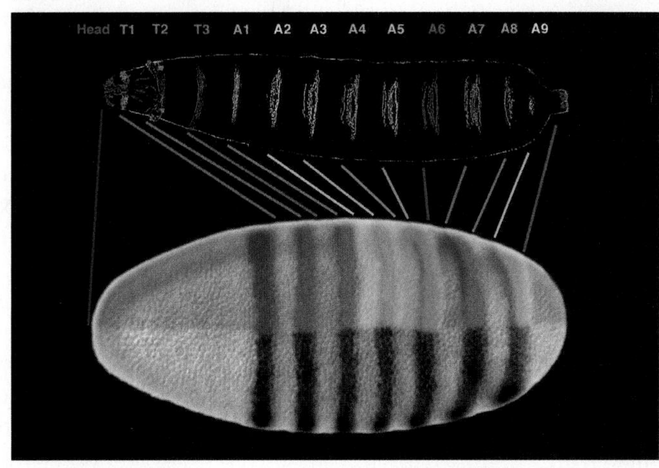

▲ EXPERIMENTAL FIGURE 22-27 **Gap genes and pair-rule genes are expressed in characteristic spatial patterns in early *Drosophila* embryos.** Fixed, permeabilized embryos were stained with fluorescence-labeled antibodies specific for a particular protein. All embryos shown are positioned with anterior to the left and dorsal at the top. (a) These syncytial embryos were stained individually for the proteins encoded by four of the five gap genes. Transcription of the *Krüppel, knirps,* and *giant* gap genes is regulated by Hunchback, Bicoid, and Caudal. (b) This syncytial embryo was doubly stained to visualize Hunchback protein (red) and Krüppel protein (green). The posterior Hunchback protein visible here is only weakly visible in part (a) due to the plane of focus. The yellow band identifies the region in which production of these two gap proteins overlaps. (c) In this blastoderm-stage embryo, Fushi tarazu (blue) and Even-skipped (brown) proteins, encoded by the pair-rule genes *ftz* and *eve,* respectively, are expressed in stripes. Each stripe corresponds to the primordial cells of one body segment. Altogether about 14 segments are formed. No morphological evidence of segmentation can be seen at this stage, but staining for the RNA or protein products of pair-rule genes reveals the beginnings of a segmented body plan. (d) The relationship between the early segment primordia (lower embryo), expression stripes of one pair-rule gene (dark gray), and the eventual larval segments that are formed (upper larva) is depicted. The colors indicate the different segments, from head to tail, and how they correspond from embryo to larva. The larval head is barely visible externally. Note that each segment develops from a primordium that is about four cells wide (in the head-to-tail direction) and about 60 cells around. Half of the segments develop from cells that express the pair-rule gene and half from the "interstripes" that do not; the interstripes express a different pair-rule gene. T = thoracic segments; A = abdominal segments. [Part (a) adapted from G. Struhl et al., 1992, *Cell* **69**:237. Part (b) courtesy of M. Levine. Part (c) courtesy of Peter Lawrence. Part (d) courtesy of Nipam Patel.]

All the gap-gene proteins are transcription factors. Because these proteins are distributed in broad overlapping peaks (Figure 22-27b), each cell along the anterior-posterior axis contains a particular combination of gap-gene proteins that activates or represses specific genes within that cell. Indeed, something like a battle ensues, because some gap proteins repress the transcription of genes encoding other gap proteins. Although they have no known extracellular ligands, some gap proteins resemble **nuclear receptors,** which are intracellular proteins that bind lipophilic ligands (e.g., steroid hormones) capable of crossing the plasma membrane. Most ligand–nuclear receptor complexes function as transcription factors (see Figure 7-50). The sequence similarity between gap proteins and nuclear receptors suggests that gap genes may have evolved from genes whose transcription was controlled by signals that could cross membranes, such as the steroid hormones. The use of such signal-controlled genes, rather than TF cascades, could explain how early cell-fate specification operates in animals that do not have a syncytial stage.

The fates of cells distributed along the anterior-posterior axis are specified early in fly development. At the same time, cells are responding to the dorsal-ventral control system.

Each cell is thus uniquely specified along both axes. If each of the five gap genes were expressed in its own section of the embryo, at just one concentration, only five cell types could be formed. The actual situation permits far greater diversity among the cells. The amount of each gap protein varies from low to high to low along the anterior-posterior axis, and the expression domains of different gap genes overlap. This complexity creates combinations of transcription factors that lead to the creation of many more than five cell types. Remarkably, the next step in *Drosophila* development generates a repeating pattern of cell types from the rather chaotic non-repeating pattern of gap-gene expression domains.

The first sign of segmentation in fly embryos is a pattern of repeating stripes of transcription of eight genes collectively called **pair-rule genes** (Figure 22-27c). The body of a larval fly consists of 14 segments, and each pair-rule gene is transcribed in half of the segment primordia, or seven stripes, separated by "interstripes" where that pair-rule gene is not transcribed (Figure 22-27d). Mutant embryos that lack the function of a pair-rule gene have their body segments fused together in pair-wise fashion—hence the name of this class of genes. The expression stripes for each pair-rule gene partly overlap with those of other pair-rule genes; so each gene must be responding in a unique way to gap-gene and other earlier regulators.

The transcription of pair-rule genes is controlled by transcription factors encoded by gap and maternal genes. Because gap and maternal genes are expressed in broad, non-repeating bands, the question arises: How can such a non-repeating pattern of gene activities confer a repeating pattern such as the striped expression of pair-rule genes? To answer this question, we consider the transcription of the *even-skipped (eve)* gene in stripe 2, which is controlled by the maternally derived Bicoid protein and the gap proteins Hunchback, Krüppel, and Giant. All four of these transcription factors bind to a clustered set of regulatory sites, or **enhancer,** located upstream of the *eve* promoter (Figure 22-28a). Hunchback and Bicoid activate the transcription of *eve* in a broad spatial domain, whereas Krüppel and Giant repress *eve* transcription, thus creating sharp posterior and anterior boundaries. The combined effects of these proteins, each of which has a unique concentration gradient along the anterior-posterior axis, initially demarcates the boundaries of stripe 2 expression (Figure 22-28b).

Expression of the other *eve* stripes also depends on specific enhancers. Each stripe of *eve* expression is formed in response to a different combination of transcriptional regulators acting on a specific enhancer, so the *non-repeating* distributions of regulators create *repeating* patterns of pair-rule gene repression and activation. If even one enhancer is bound by an activating combination of transcriptional

MEDIA CONNECTIONS

Video: Establishing Eve Expression in Drosophila Embryogenesis 

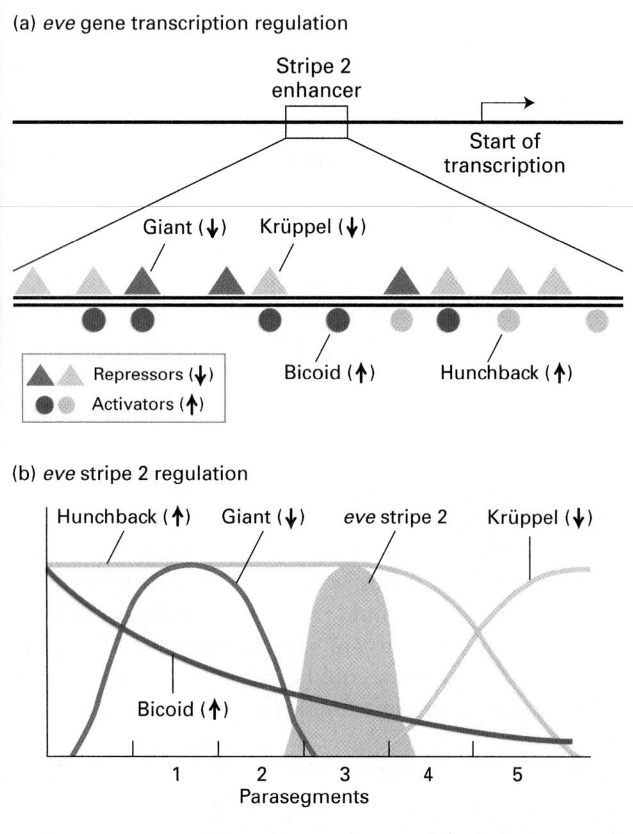

◀ **FIGURE 22-28 Control of *even-skipped (eve)* stripe 2 in the *Drosophila* embryo.** Only one of the *eve* gene stripes is represented. Within each *eve* stripe, a segment boundary will later form. Thus *eve* function gives rise to half the segment boundaries in the embryo. (a) Diagram of the 815-bp enhancer controlling transcription of the pair-rule gene *eve* in stripe 2. This regulatory region contains binding sites for Bicoid and Hunchback proteins, which activate the transcription of *eve,* and for Giant and Krüppel proteins, which repress its transcription. The enhancer is shown with all binding sites occupied, but in an embryo occupation of sites will vary with position along the anterior-posterior axis. (b) Concentration gradients of the four transcription factors that regulate *eve* stripe 2. The coordinated effect of the two repressors (↓) and two activators (↑) determine the precise boundaries of the second anterior *eve* stripe. Only in the orange region is the combination of regulators correct for the *eve* gene to be transcribed in response to the stripe 2 control element. Further anterior, Giant turns *eve* off; further posterior, the level of Bicoid activator is too low to overcome repression by Krüppel. Expression of other stripes is regulated independently by other combinations of transcription factors that bind to enhancers not depicted in part (a). [See S. Small et al., 1991, *Genes & Devel.* **8:**827.]

regulators, the presence of other enhancers in an inactive, "off" state (not bound to a regulator) will not prevent transcription. For instance, in Eve stripe 2, the right combination and amounts of Hunchback and Bicoid create an "on" state that activates transcription even though other enhancers are present in the inactive state. In each stripe, at least one enhancer is bound by an activating combination of regulators. Note that this system of gene control is flexible and could be used to produce non-repeating patterns of transcription if that was useful to an animal.

Similar responses to gap and maternal proteins govern the striped patterns of transcription of the two other pair-rule genes, *runt* and *hairy*. Because the enhancers of *runt* and *hairy* respond to different combinations of regulators, the *eve, runt,* and *hairy* expression stripes partly overlap one another, with each stripe for any one gene offset from a stripe for another gene. Subsequently, other pair-rule genes, including *fushi tarazu (ftz)* and *paired,* become active in response to the Eve, Runt, and Hairy proteins, which are transcription factors, as well as to maternal and gap proteins. The outcome of this transcription-factor cascade is a pattern of overlapping stripes.

The initial pattern of pair-rule stripes, which is not very sharp or precise, is sharpened by autoregulation. The Eve protein, for instance, binds to its own gene and increases transcription in the stripes, a positive autoregulatory loop. This enhancement does not occur at the edges of stripes where the initial Eve protein concentration is low; so the boundary between stripe and interstripe is fine-tuned.

The pair-rule genes direct formation of the embryo's segment boundaries. Since each pair-rule gene is expressed in stripes and each stripe overlaps one segment boundary, each pair-rule gene contributes to half the segment boundaries. Acting together, all the pair-rule genes form all the segment boundaries and also control other pattern elements within each segment. In early embryos each segment primordium is about four cells wide along the anterior-posterior axis, which corresponds to the approximate width of pair-rule expression stripes. With pair-rule genes active in alternating four-on four-off patterns, the repeat unit is about eight cells. Each cell expresses a combination of transcription factors that can distinguish it from any of the other seven cells in the repeat unit. Under the control of pair-rule proteins and the later-acting **segment-polarity genes**, the repeating morphology of segments begins to emerge; it is completed about 10 hours after fertilization. As cell-fate determination progresses in the fly embryo, a variety of signaling proteins begin to play a role. These include Hedgehog and Wnt, which are encoded by segment-polarity genes and are produced in stripes, one stripe within each segment, under the control of pair-rule gene products. The broader and earlier-formed stripes of pair-rule gene expression overlap in certain regions, and that's where particular combinations of pair-rule transcription factors give rise to the fine pattern of segment-polarity gene stripes. Note that the onset of signal-based controls allows cells to respond to what their neighbors have done and make adjustments. Otherwise parts of the pattern might be missing or duplicated.

From the broad maternal gradient of Bicoid to the single-cell precision of the segment-polarity genes, the fly embryo is progressively subdivided into repeating units. One can readily imagine how changes in stripe-specific enhancers, amounts of particular transcription factors, and the range of signals during evolution could modify the pattern of segments in different organisms.

Vertebrate Segmentation Is Controlled by Cyclical Expression of Regulatory Genes

Now we return to vertebrates to examine how segmentation in these animals compares with that in insects. After the three body axes have been established in a vertebrate embryo, dramatic changes take place along all of them. One of the most visible changes is the initiation of a repeating pattern that later gives rise to vertebrae and ribs. This is patterning along the anterior (head) to posterior (tail) axis. In mice and humans, the first sign of the vertebrae appears in the mesoderm that accumulates under the primitive streak. The mesoderm forms, you will recall, by an epithelial-mesenchymal transition in which cells along the primitive streak cut loose and migrate inside (see Figure 22-11). On each side of the midline axis, mesoderm composed of loose mesenchymal cells begins to round up to form pairs of spherical epithelia called *somites*. Somites initially form at the anterior and successively appear pair by pair in the posterior direction (Figure 22-29a), giving them value as a way to stage embryos. This striking case of a mesenchymal-epithelial transition has huge consequences for the embryo. From somites come the vertebrae and ribs, the muscles of the body wall and limbs, and the dermis (inner skin) of the back. Without somites, we would be blobs.

The mesoderm that has not yet formed somites is called presomitic mesoderm. As gastrulation spreads toward the posterior, somites arise from presomitic mesoderm, and more presomitic mesoderm forms posteriorly. The interplay of four signaling systems at the anterior limit of the presomitic mesoderm controls somite formation: an FGF signal from the tail, a retinoic acid signal from the head, and Wnt and Notch signals within the presomitic mesoderm. The *fgf8* gene is expressed in the presomitic mesoderm where it is forming near the posterior tip of the embryo. Because *fgf8* mRNA is unstable, the highest level of FGF8 protein builds up in the posterior and forms a gradient toward the anterior. High levels of FGF8 prevent maturation of cells into somites in the posterior. Conversely, retinoic acid from the head of the embryo acts in a graded fashion, highest in the anterior presomitic mesoderm, to stimulate formation of somites (Figure 22-29b).

The most remarkable aspect of somite-formation regulation in vertebrates was first discovered in studies of the chicken gene *hairy1,* which is related to one of the *Drosophila* pair-rule segmentation genes. In situ hybridization of developing somites in chick embryos showed that *hairy1* transcripts are produced in cycles, with the duration of one cycle corresponding to the time it takes to form one somite (90 minutes in a chick, longer in mammals). A wave

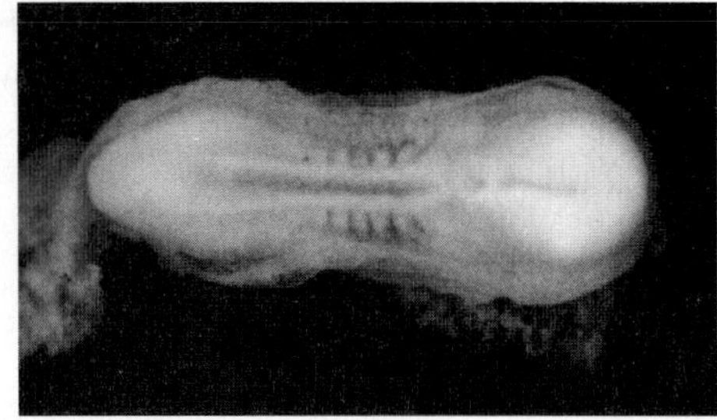

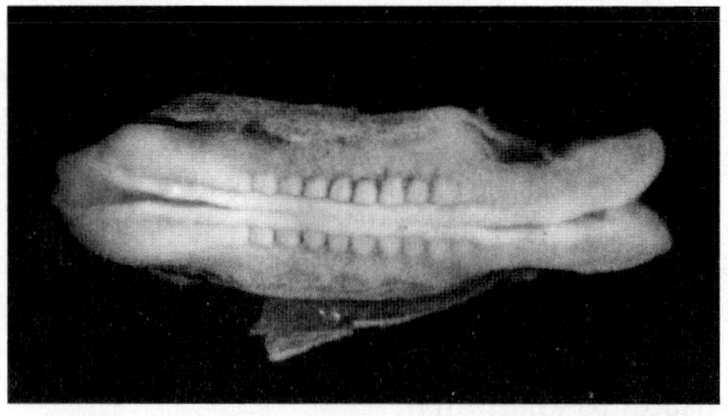

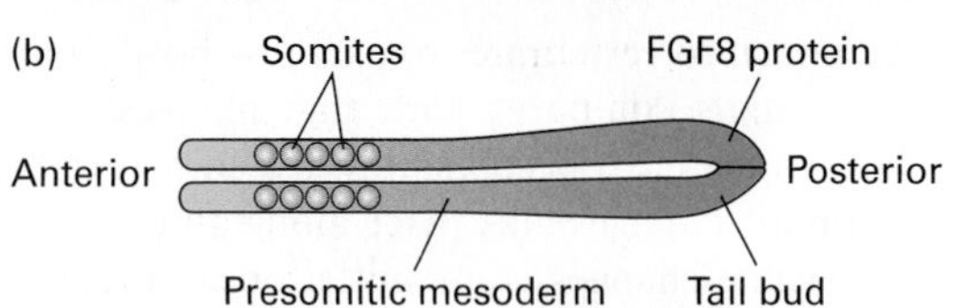

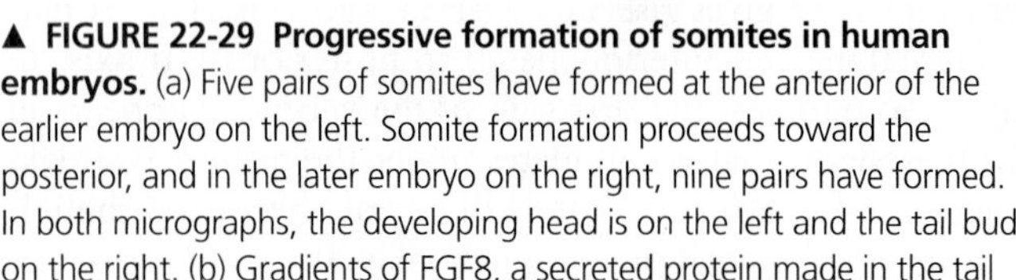

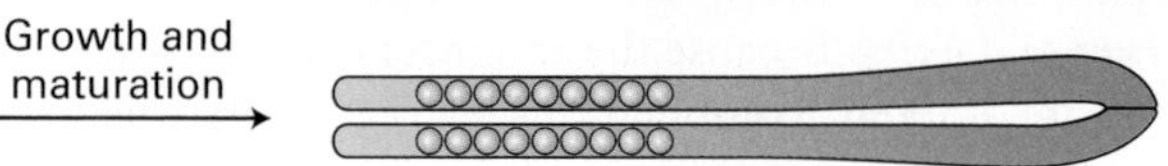

▲ **FIGURE 22-29 Progressive formation of somites in human embryos.** (a) Five pairs of somites have formed at the anterior of the earlier embryo on the left. Somite formation proceeds toward the posterior, and in the later embryo on the right, nine pairs have formed. In both micrographs, the developing head is on the left and the tail bud on the right. (b) Gradients of FGF8, a secreted protein made in the tail bud, and retinoic acid from the head, control somite formation from presomitic mesoderm, which first arises in the tail bud. High levels of FGF8 prevent maturation of presomitic cells into somites in the posterior, whereas high levels of retinoic acid acts to stimulate formation of somites. [Part (a) Kohei Shiota/Congenital Anomaly Research Center, Kyoto University. Part (b) adapted from A. F. Schier, 2004, *Nature* **427**:403.]

of *hairy1* expression moves from posterior to anterior in the presomitic (unsegmented) mesoderm. Subsequent investigations revealed a rather large number of genes that undergo cycles of expression, and all turned out to be related to the Notch or Wnt signaling pathways. Mutations in either pathway cause drastic defects in somite formation. In humans, for instance, mutations affecting Notch pathway components cause Alagille syndrome and Jarcho-Levin syndrome, both of which are associated with malformed vertebrae.

For both the Notch and Wnt pathways, feedback loops are established that cause temporal cycling of expression. For example, the *hes7* gene encodes a transcription factor involved in Notch signal transduction. When transcription of *hes7* is stimulated by a FGF signal from the posterior presomitic mesoderm, a burst of Hes7 protein production occurs (Figure 22-30a). Hes7 protein, in turn, controls the expression of target genes that contribute to somite formation. Because Hes7 also acts as a repressor of its own gene, binding to its gene and turning it off, the Hes7 protein accumulates only until it reaches a high enough level to represses *hes7* transcription. This negative autoregulation thus limits the duration of *hes7* expression. Each part of the presomitic mesoderm does the same thing in turn: *hes7* transcription on, Hes7 protein accumulates, *hes7* transcription off (Figure 22-30b). As the source of the FGF signal starts near the head and retreats posteriorly, *hes7* is transcribed first in the most anterior region, then in progressively more posterior regions. A similar feedback system operates for Wnt signaling, again causing a burst of gene expression just as a somite forms, another burst in the cells that will form the subsequent somite, and so on. If Notch or Wnt feedback loops are blocked, somites are highly abnormal, but the details of how both pathways control shaping of the somites are not known.

We can now understand the two different strategies for controlling the formation of repeating body parts in insects and vertebrates. In *Drosophila,* the regulation differs for each body segment: different combinations of gap transcription factors, activated in specific regions along the anterior-posterior axis by maternal influences, regulate pair-rule gene stripes, and pair-rule transcription factors in turn combine to regulate the still finer stripes of segment-polarity gene transcription. Each stripe has a distinct regulatory history involving different gap or pair-rule proteins. Thus repeat formation is controlled by *spatial* differences. In contrast, repeating vertebrate somites are formed by the same regulatory process occurring again and again. A remarkable feedback system creates a cyclical clock that causes the genes responsible for building somites to be expressed in bursts. Thus repeat formation is controlled by *temporal* (time) differences.

Differences Between Segments Are Controlled by Hox Genes

Despite the differences in how repeating body parts are formed in insects and vertebrates, the two groups of animals are reunified in employing the same family of genes to create variation among the repeats. These are the **Hox genes,** which control differences in cell identities and indeed the identities of whole parts of an animal along the anterior-posterior axis. These differences are superimposed upon the underlying repetitive nature of some of the tissues. Hox genes encode highly related transcription factors containing the **homeodomain** motif (Chapter 7). Indeed, what unifies the whole group of Hox proteins is a similar DNA-binding homeodomain sequence; the proteins have little else in common. The homeodomain sequences are also the basis for classifying all the Hox genes.

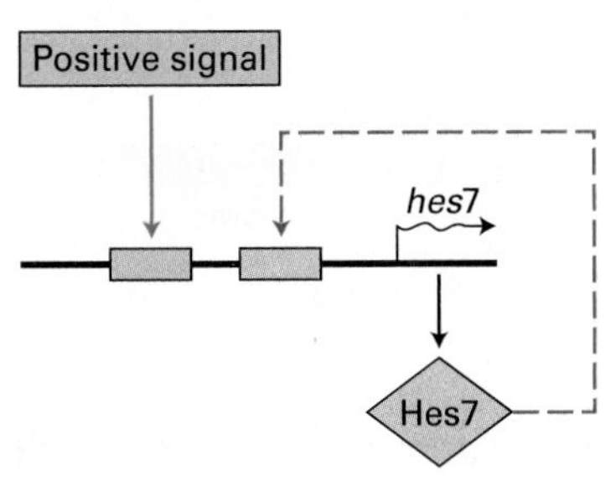

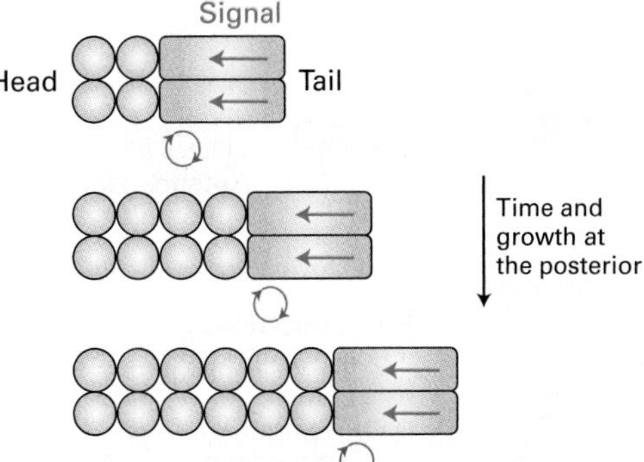

▲ **FIGURE 22-30 Control of cyclical gene expression in the developing somites.** (a) An initial burst of *hes7* transcription is triggered by a positive signal, probably FGF8. As Hes7 protein (a Notch pathway component) accumulates, it eventually binds to its own gene and turns off transcription. This process is repeated, once per forming somite, in increasingly posterior regions of the presomitic mesoderm. (b) The FGF8 signal continues to be at the posterior end, so progressively more posterior presomitic mesoderm is triggered to activate cyclical Notch and Wnt expression programs, leading to somite formation. The red circles represent bursts of transcription of genes encoding Notch and Wnt signaling components and the negative feedback loops that shut them down.

Mutations in Hox genes often cause **homeosis**—that is, the formation of a body part having the characteristics normally found in another part at a different site. For example, some mutant flies develop legs on their heads instead of antennae. Loss of function of a particular Hox gene in a location where it is normally active leads to homeosis if a different Hox gene becomes derepressed there; the result is the formation of cells and structures characteristic of the derepressed gene. A Hox gene that is abnormally expressed where it is normally inactive can take over and impose its own favorite developmental pathway on its new location (Figure 22-31).

Organization of Hox Genes Classical genetic studies in *Drosophila* led to discovery of the first Hox genes (e.g., *Antennapedia* and *Ultrabithorax*). Corresponding genes with similar functions (orthologs) have since been identified in most animal species. Each Hox gene is transcribed in a particular region along the anterior-posterior axis in a remarkable arrangement where the order of genes along the chromosomes is colinear with the order in which they are expressed along the anterior-posterior axis (Figure 22-32a). The fly Hox genes are located at two locations on the same chromosome but are effectively one cluster of eight Hox genes. At one end of the cluster are "head" genes, which are transcribed specifically in the head and are necessary for formation of head structures. Next to them are genes active and functional in the thorax, and at the other end of the cluster are abdomen genes. The arrangement reflects evolutionary gene duplications and is retained because the genes share regulatory sequences such as enhancers (Chapter 7). The expression domains of Hox genes can overlap, so the development of a particular body structure can depend upon more than one gene.

In *Drosophila*, the spatial pattern of Hox-gene transcription is regulated by maternal, gap, and pair-rule transcription factors. The protein encoded by a particular Hox gene controls the organization of cells within the region in which that Hox gene is expressed. For example, a Hox protein can direct

Normal

***Ubx* mutant**

▲ **FIGURE 22-31 Hox-gene phenotypes.** Like other Hox genes, the *Ultrabithorax (Ubx)* gene controls the organization of cells within the region in which it is expressed. It normally acts in preventing wing formation, so that normal flies have a single pair of wings. Mutations in Hox genes often lead to the formation of a body part where it does not normally exist. In the mutant shown here, the loss of *Ubx* function from the third thoracic segment allows wings to form where normally there are only balancer organs called halteres. [From E. B. Lewis, 1978, *Nature* **276:**565. Reprinted by permission from *Nature*, copyright 1978, Macmillan Journals Limited.]

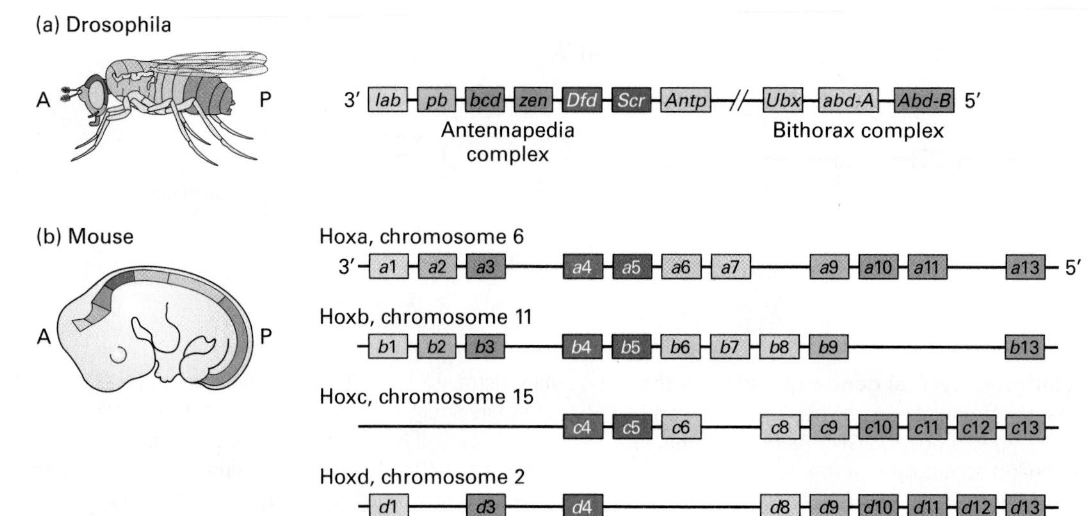

▲ **FIGURE 22-32 Relation between Hox gene clusters in *Drosophila* and mammals.** (a) The single *Drosophila* Hox cluster is split into two chromosomal locations; one gene group is the Antennapedia complex and the other is the Bithorax complex. The genes that control head formation are at the 3′ end of the cluster (yellow/red shades), those that control formation of the abdomen are at the 5′ end (blue/green shades), and the ones in-between control thoracic structures (purple), as illustrated in the fly drawing, which shows where the different genes are expressed. (b) The arrangement of genes in the mouse and human Hox gene clusters is similar to that in *Drosophila*, but there are four clusters and each of them is missing some of the set of genes. For example, the class 1 Hox genes of mice and humans are similar to the *lab* gene of *Drosophila,* based on encoded protein sequences. There is a class 1 gene in Hox a, b, and d clusters but not in the c cluster. Evidently three is enough. In contrast, the class 4 gene is represented in all of the mammalian clusters. Another difference from insect to mammal is that mammals have "extra" versions of the posterior genes (class 9 and up) that correspond to the *abd*-type genes in flies. The drawings of the fly and mouse embryo indicate where Hox genes are transcribed, from which it can be seen that the order has been preserved during the roughly half billion years since they had a common ancestor. The drawings are a simplification, since in many cases a gene is expressed with a sharp anterior boundary as well as a graded pattern going toward the tail, and expression patterns vary between different tissues. [Adapted from L. Wolpert et al., 2001, *Principles of Development*, 2nd ed., Oxford University Press, Box 4-4a].

or prevent the local production of a secreted signaling protein, cell-surface receptor, or transcription factor that is needed to build an appendage on a particular body segment. *Drosophila* Hox proteins control the transcription of target genes whose encoded proteins determine the diverse morphologies of body segments. Much remains to be learned about how morphology is controlled by these target genes, but some targets encode powerful Wnt and TGFβ signals. The association of Hox proteins with their binding sites on DNA is assisted by cofactors that bind to both Hox proteins and DNA, adding specificity and affinity to these interactions.

How about vertebrates? In contrast to flies, which have a single Hox gene cluster, mammals have four copies of the Hox cluster, a–d, located on different chromosomes (Figure 22-32b). Within each cluster, the genes are numbered 1, at the head end, to 13, at the tail end. The different copies of a particular gene (e.g., *Hox4*) in the four clusters are closer to each other in sequence than to Hox genes of another numerical class. Although the mammalian Hox genes are clearly related to the fly Hox genes, there has been an expansion of fly *abd*-type genes in vertebrates (Hox classes 8–13). Within each cluster some genes have been lost, evidently because the other two or three copies are sufficient. Experiments with mouse "knockouts" missing one or more Hox genes from one numerical class show considerable redundancy among the 39 total genes.

Evolution of Hox Gene Clusters Hox clusters are the most dramatic example of the conservation of gene groupings across a wide range of animals. Their organization is so striking that they serve as useful tools for studying evolution. The single Hox cluster in *Drosophila* is represented four times in mammals (see Figure 22-32). Comparisons of the genome sequences of a variety of vertebrates have revealed that the copies of the Hox clusters are far from perfect. During evolution, some species have lost one or more Hox genes; in other species, Hox genes have become duplicated within a cluster. The transitions between different clusters and losses and gains of individual Hox genes among present-day organisms allow ancestral forms to be deduced (Figure 22-33). For example, in the time since frogs and humans had a common ancestor, about 370 million years ago, frogs have lost Hox genes *b13* and *d12*. As more studies are done of how Hox genes control body morphology, it will be increasingly possible to relate evolutionary changes in Hox genes to pattern formation in the embryos

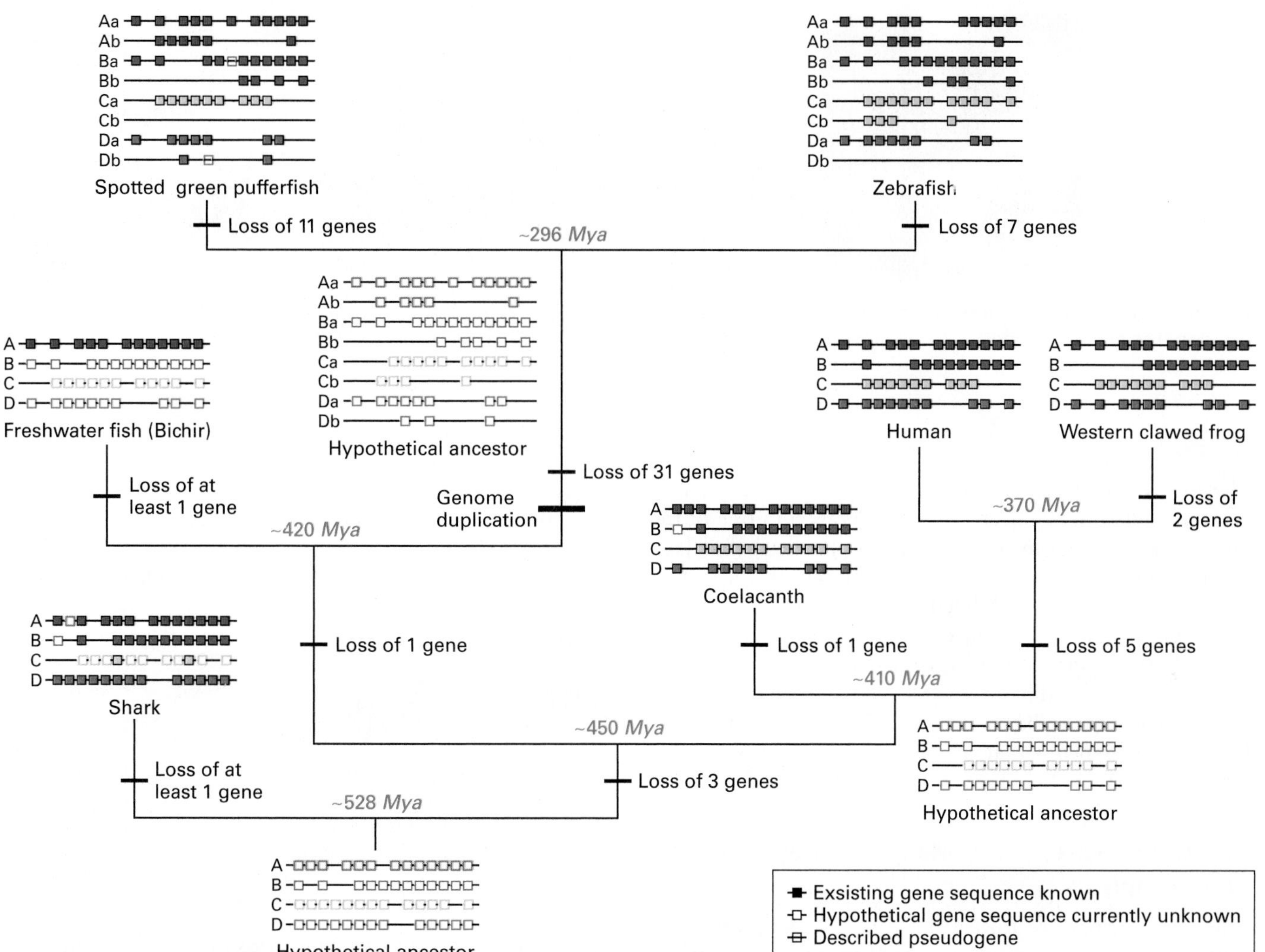

▲ **FIGURE 22-33 Evolution of vertebrate Hox clusters.** Genome projects in which all genes are sequenced from different organisms, as well as focused studies of Hox genes, are providing insight into the evolution of the Hox clusters. The Hox clusters shown here in dark green, blue, yellow, and red are based on sequencing of the genomes of the indicated present-day organisms. Comparing sequences from a variety of vertebrates reveals that the copies of the Hox clusters are far from perfect: In some species, one or more genes have been lost, while in others, genes have been duplicated within a cluster. Since the fossil record and studies of DNA sequence conservation have established the relationships between the organisms shown here, it is possible to deduce the arrangement of Hox genes in hypothetical ancestors (genes in light colors) and reconstruct an outline of the events that happened in the evolution of the Hox clusters. The approximate distances in time (Mya = million years ago) since any pair of species had a common ancestor are indicated in red type. Of course, not all ancestors are represented here, so not every step can be deduced. As more studies are done of how Hox genes control body morphology, it will be increasingly possible to relate evolutionary changes in Hox genes to pattern formation in the embryos they control. [Adapted from S. Hoegg and A. Meyer, 2005, *Trends Genet.* **21**:441.]

they control. This genomics approach to exploring evolution will yield more detailed biological histories as more genomes are sequenced.

Functions of Vertebrate Hox Proteins Vertebrate Hox proteins control the different morphologies of vertebrae, of repeated segments of the hindbrain, and of the digits of the limbs. Mutations affecting some of the most posterior Hox genes in humans cause inherited syndromes that involve polydactyly (extra fingers or toes) and syndactyly (fused fingers or toes). Particular Hox genes are active in many other tissues too. As in flies, mammalian Hox proteins often act in combination to control target genes, and they use cofactors of the same types used by flies. Mutations affecting some of these cofactors have been implicated in human cancer.

As in flies, Hox-gene expression domains in early vertebrate embryos respond to seemingly invisible boundaries that correspond later to transitions between morphologically distinct repeating body units. Each Hox gene is expressed in some somites but not others (Figure 22-34). They are expressed first in presomitic mesoderm, where they control the morphology of the vertebrae that will form later.

Vertebrae are classified into groups according to their morphology and position along the anterior-posterior axis.

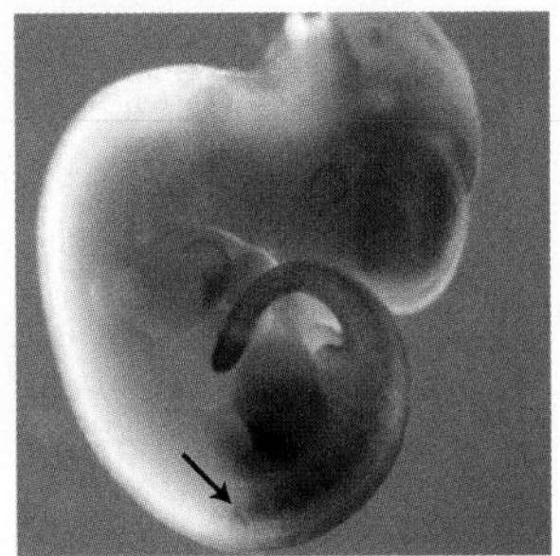

▲ **FIGURE 22-34 Expression of the mouse *Hoxc10* gene in somites.** The anterior boundary of *Hoxc10* expression (arrow) is clearly visible in this 9-day-old embryo. [From M. Carapuco et al., 2005, *Genes & Devel.* **19:**2116.]

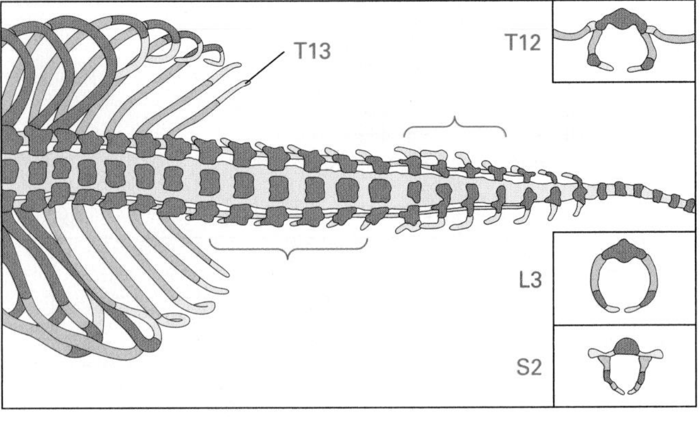

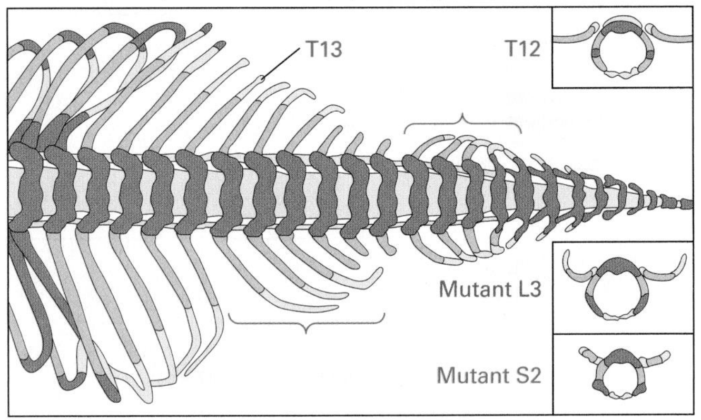

▲ EXPERIMENTAL FIGURE 22-35 ***Hox10* genes regulate vertebra shape.** Separate knockout mice were constructed, each lacking a portion of the *Hox10 a, c,* or *d* gene. Each of these genes is on a different chromosome. In each case, heterozygous mice survived because they have a second, wild-type, copy of the gene. The mice were crossed to construct a strain heterozygous for mutations at all three genetic loci. Crossing these mice together generated some homozygous mutant embryos lacking both copies of all three *Hox10* genes. Skeletons from 18.5 day mutant embryos and wild-type embryos were isolated and stained to reveal the details of the skeletons. These top-view and cross-sectional diagrams, based on the stained skeletons, illustrate the results. In wild-type mice, the thoracic (T) vertebrae have ribs, with the most posterior rib on T13. Lumbar (L) vertebrae (blue bracket) and sacral (S) vertebrate (red bracket) do not have ribs. Mutants that are homozygous for all three *Hox10* genes have ribs on lumbar vertebrae (e.g., L3) that would normally have been ribless, and even on some sacral ribs (e.g., S2). The insets show cross sections of different vertebrae and ribs to reveal their shapes in that dimension. From these results we can infer that Hox10 proteins are normally required to suppress rib formation on the developing lumbar and sacral vertebrae. Since the *Hox10* genes are normally expressed in this region of the embryo, this conclusion makes sense. If any one *Hox10* gene is functioning, the extra ribs are not seen, or only rarely seen, so the *Hox10* genes have at least partially redundant functions. [Adapted from D. M. Wellik and M. R. Capecchi, 2003, *Science* **301:**363.]

The effects of Hox mutations on vertebral development have been studied with engineered mouse mutants. In one of the most striking experiments, mice were produced that lack functional *Hoxa10, Hoxc10,* and *Hoxd10* genes (there is no *Hoxb10*). These mice exhibited a dramatic transformation of vertebral patterns. Lumbar and even sacral vertebrae, which normally have no ribs, developed with varying degrees of partial ribs (Figure 22-35). Probably even more Hox genes would have to be changed for a complete transformation. The inference is that the normal role of Hox10 transcription factors is to prevent rib development.

Hox-Gene Expression Is Maintained by a Variety of Mechanisms

When Hox genes are turned on, their transcription must continue to maintain cell properties in specific locations. As in the case of the pair-rule gene *even-skipped,* the regulatory regions of some Hox genes contain binding sites for their encoded proteins. Thus Hox proteins can help to maintain their own expression through many cell generations using an autoregulatory loop.

Another mechanism for maintaining normal patterns of Hox-gene expression requires proteins that modulate chromatin structure. These proteins are encoded by two classes of genes referred to as the Trithorax group and Polycomb group. The pattern of Hox-gene expression is initially normal in Polycomb-group mutants, but eventually Hox-gene transcription is derepressed in places where the genes should be inactive. The result is multiple homeotic transformations, indicating that the normal function of Polycomb proteins is to keep Hox genes in a transcriptionally inactive state. The results of immunohistological and biochemical studies have shown that Polycomb proteins bind to multiple chromosomal locations and form large complexes containing different proteins of the Polycomb group. The current view is that the transient repression of genes set up by patterning proteins earlier in development is "locked in" by Polycomb proteins. This stable Polycomb-dependent repression may result from the ability of these proteins to assemble inactive chromatin structures (Chapter 7). Polycomb complexes contain many proteins, including histone deacetylases, and appear to inactivate transcription by modifying histones to promote gene silencing.

Whereas Polycomb proteins repress expression of certain Hox genes, proteins encoded by the Trithorax group of genes are necessary for maintaining expression of Hox genes. Like Polycomb proteins, Trithorax proteins bind to multiple chromosomal sites and form large multiprotein complexes, some with a mass of $\approx 2 \times 10^6$ Da, about half the size of a ribosome. Some Trithorax-group proteins are homologous to the yeast SWI/SNF proteins, which are crucial for transcriptional activation of many yeast genes. Trithorax proteins stimulate gene expression by selectively remodeling the chromatin structure of certain loci to a transcriptionally active form (see Figure 7-43). The core of each complex is an ATPase, often of the Brm class of proteins. There is evidence that many or most genes require such complexes for transcription to take place.

Many regulators of Hox-gene expression have been implicated in leukemia. Chromosomal translocations that fuse the genes encoding these regulators to novel sequences, sometimes causing a gene encoding a chimeric protein to form, are frequently found in leukemia patients. Such fusions, for instance, can create oncogenes that cause white blood cells to grow uncontrollably (see Figure 25-20). Hox genes are active in blood cells, though Hox functions in those cells are incompletely understood.

Homeotic genes—that is, genes like the Hox genes that control development of whole parts of the body—are also important in plant development, as we shall see now. Again the homeotic genes superimpose a set of variations on an underlying repeat pattern.

Flower Development Requires Spatially Regulated Production of Transcription Factors

It may seem a long jump from animal segmentation to plants, but in terms of the molecules that control pattern formation, many principles are similar. Like vertebrae or insect segments, flowers have repeating parts. The basic mechanisms controlling development in plants are much like those in *Drosophila:* differential production of transcription factors, controlled in space and time, specifies cell identities. Our understanding of cell-identity control in plants benefited greatly from the choice of *Arabidopsis thaliana* as a model organism. This plant has many of the same advantages as flies and worms for use as a model system: It is easy to grow, mutants can be obtained, and transgenic organisms can be made. We will focus on certain transcription-control mechanisms regulating the formation of cell identity in flowers. These mechanisms are strikingly similar to those controlling cell-type and anterior-posterior regional specification in yeast and animals. ▮

Floral Organs A flower comprises four different organs called sepals, petals, stamens, and carpels, which are arranged in concentric circles called whorls. Whorl 1 is the outermost; whorl 4, the innermost. *Arabidopsis* has a complete set of floral organs, including four sepals in whorl 1, four petals in whorl 2, six stamens in whorl 3, and two carpels containing ovaries in whorl 4 (Figure 22-36a). These organs grow from a collection of undifferentiated, morphologically indistinguishable cells called the floral **meristem.** As cells within the center of the floral meristem divide, four concentric rings of primordia form sequentially. The outer-ring primordium, which gives rise to the sepals, forms first, followed by the primordium giving rise to the petals, then the stamen and carpel primordia.

Floral Organ–Identity Genes Genetic studies have shown that normal flower development requires three classes of *floral organ–identity genes*, designated A, B, and C genes. Mutations in these genes produce phenotypes equivalent to those associated with homeotic mutations in flies and mammals; that is, one part of the body is replaced by another. In plants lacking all A, B, and C function, the floral organs develop as leaves (Figure 22-36b, *left*).

The loss-of-function mutations that led to the identification of the A, B, and C gene classes are summarized in Figure 22-36b (*right*). On the basis of the various homeotic phenotypes observed, scientists proposed a model to explain how the three classes of genes control floral-organ identity. According to this ABC model for specifying floral organs, class A genes specify sepal identity in whorl 1 and do not require either class B or class C genes to do so. Similarly, class C genes specify carpel identity in whorl 4 and, again, do so independently of class A and B genes. In contrast with these structures, which are specified by only a single class of genes, the petals in whorl 2 are specified by class A and B genes, and the stamens in whorl 3 are specified by class B and C genes. To account for the observed effects of removing A genes or C genes, the model also postulates that A genes repress C genes in whorls 1 and 2 and, conversely, C genes repress A genes in whorls 3 and 4.

To determine if the actual expression patterns of class A, B, and C genes are consistent with this model, researchers cloned these genes and assessed the expression patterns of their mRNAs in the four whorls in wild-type *Arabidopsis* plants and in loss-of-function mutants (Figure 22-37a, b). Consistent with the ABC model, A genes are expressed in whorls 1 and 2, B genes in whorls 2 and 3, and C genes in whorls 3 and 4. Furthermore, in class A mutants, class C genes are also expressed in organ primordia of whorls 1 and 2; similarly, in class C mutants, class A genes are also expressed in whorls 3 and 4. These findings are consistent with the homeotic transformations observed in these mutants.

To test whether these patterns of expression are functionally important, scientists produced transgenic *Arabidopsis* plants in which floral organ–identity genes were expressed in inappropriate whorls. For instance, the introduction of a transgene carrying class B genes linked to an A-class promoter leads to the ubiquitous expression of class B genes in all whorls (Figure 22-37c). In such transgenics, whorl 1, now under the control of class A and B genes, develops into petals instead of sepals; likewise, whorl 4, under the control of both class B and class C genes, gives rise to stamens instead of carpels. These results support the functional importance of the ABC model for specifying floral identity.

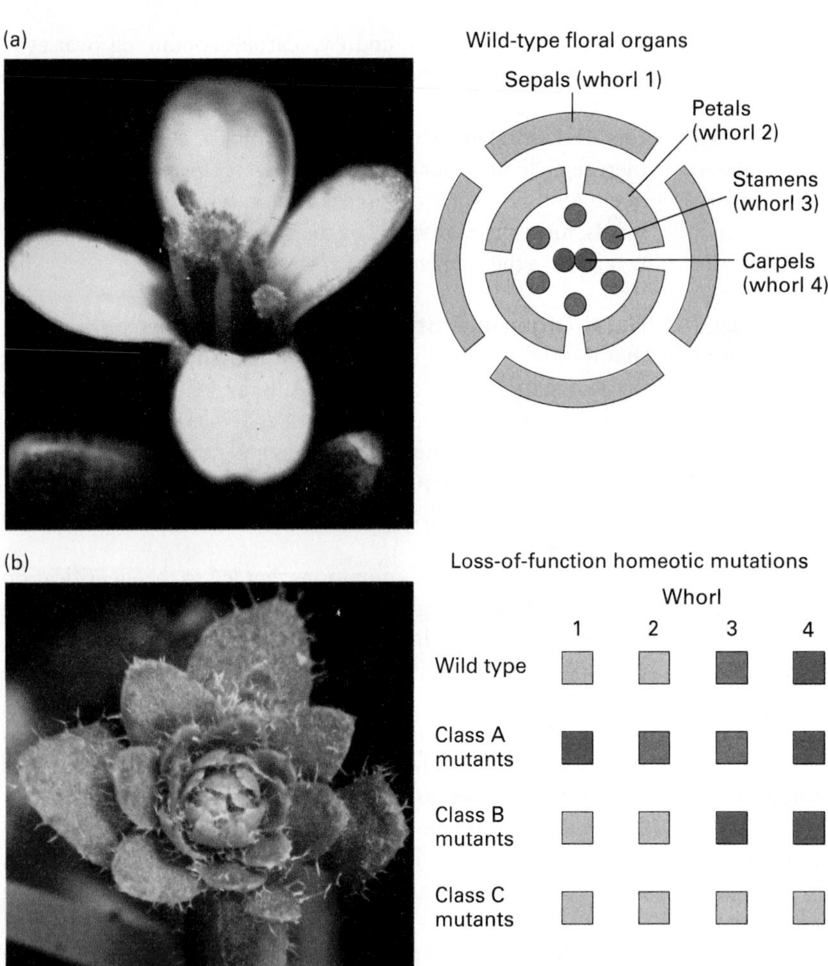

◀ **FIGURE 22-36 Floral organs and the effects of mutations in organ-identity genes.** (a) Flowers of wild-type *Arabidopsis thaliana* have four sepals in whorl 1, four petals in whorl 2, six stamens in whorl 3, and two carpels in whorl 4. The floral organs are found in concentric whorls as diagrammed at right. (b) In *Arabidopsis* with mutations in all three classes of floral organ–identity genes, the four floral organs are transformed into leaflike structures (*left*). Phenotypic analysis of mutants identified three classes of genes that control specification of floral organs in *Arabidopsis* (*right*). Class A mutations affect organ identity in whorls 1 and 2: sepals (green) become carpels (purple) and petals (orange) become stamens (red). Class B mutations cause transformation of whorls 2 and 3: petals become sepals and stamens become carpels. In class C mutations, whorls 3 and 4 are transformed: stamens become petals and carpels become sepals. [See D. Wiegel and E. M. Meyerowitz, 1994, *Cell* **78:**203.]

Sequencing of floral organ–identity genes revealed that many encode proteins belonging to the *MADS family* of transcription factors, which form homo- and hetero-dimers. Thus floral-organ identity may be specified by a combinatorial mechanism in which differences in the activities of different homo- and heterodimeric forms of various A, B, and C proteins regulate the expression of subordinate downstream genes necessary for the formation of the different cell types in each organ. Other MADS transcription factors function in cell-type specification in yeast and muscle (Chapter 21).

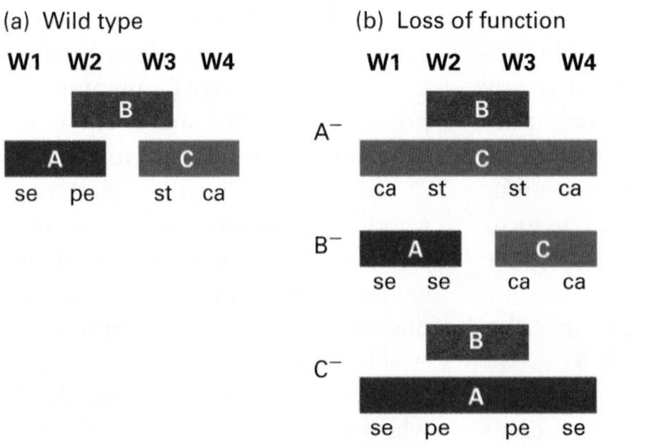

▲ **EXPERIMENTAL FIGURE 22-37 Expression patterns of class A, B, and C genes support the ABC model of floral organ specification.** Depicted here are the observed expression patterns of the floral organ–identity genes in wild-type, mutant, and transgenic *Arabidopsis*. Colored bars represent the A, B, and C mRNAs in each whorl (W1, W2, W3, W4). The observed floral organ in each whorl is indicated as follows: sepal = se; petals = pe; stamens = st; and carpels = ca. See text for discussion. [See D. Wiegel and E. M. Meyerowitz, 1994, *Cell* **78:**203, and B. A. Krizek and E. M. Meyerowitz, 1996, *Development* **122:**11.]

KEY CONCEPTS OF SECTION 22.4

Control of Body Segmentation: Themes and Variations in Insects and Vertebrates

■ Segmentation is a common feature of pattern formation in many animals, as indicated by our vertebrae and insect body segments.

■ In *Drosophila*, early development gives rise to an embryo with about 6000 cells in a sheet on the surface (see 22-21). Because the early embryo is a syncytium, both transcription factors and secreted proteins can function as graded regulators, or morphogens.

■ One system of genes and proteins controls anterior-posterior organization of the fly embryo; another controls dorsal-ventral patterning. Thus a cell in any particular position in the embryo integrates two types of informative signals. RNA and protein placed asymmetrically in the oocyte by the mother gives rise to the initial asymmetries in the embryo (see Figures 22-23 and 22-24).

■ Three classes of genes act sequentially to subdivide the fly embryo into segments along the anterior-posterior axis. Gap genes are expressed first, then pair-rule genes, and finally segment-polarity genes (see Figure 22-27). The gap and pair-rule genes encode transcription factors; segment-polarity genes encode components of the Hedgehog and Wnt signaling pathways.

■ The striped patterns of expression of pair-rule genes such as *even-skipped* are controlled by enhancers that activate transcription when bound by specific combinations of gap-gene proteins (see Figure 22-28). Each stripe forms where the proper combination of gap proteins occurs, in a certain position along the anterior-posterior axis. Similarly pair-rule proteins combine to regulate segment-polarity genes.

■ Vertebrate segmentation begins with the formation of repeating blocks of mesoderm called somites, first near the head and then sequentially in more posterior regions (see Figure 22-29).

■ In vertebrates, somite formation is controlled by the cyclical expression of genes such as *hes7* along the anterior-posterior axis. This pattern of gene expression depends on steady FGF signals from the developing tail and negative autoregulation of the genes involved in somite formation (see Figure 22-30).

■ In flies, humans, and many other animals, the different patterns of segments along the anterior-posterior axis are controlled by Hox genes (see Figure 22-32). For example, the differences in the shapes of vertebrae are regulated by Hox-gene products.

■ Hox genes encode transcription factors that regulate batteries of genes that in turn control cell and tissue morphology. Hox genes act in many tissue types; their expression patterns are maintained by the action of negative- and positive-acting chromatin proteins. Mutations in Hox genes cause body parts to form in the wrong places.

■ The shapes and patterns of flower petals, stamens, carpals, and sepals are controlled by the combined actions of MADS transcription factors encoded by three classes of *floral-organ identity genes*. In plants with mutations in all three classes, leaves rather than flowers develop (see Figure 22-36b), a transformation reminiscent of the Hox mutations in animals.

22.5 Cell-Type Specification in Early Neural Development

By the end of gastrulation, the animal embryo has three germ layers, and its three axes—head to tail, front to back, and left to right—are easily detectable. The ectoderm and endoderm are largely epithelial sheets, sandwiching a mass of more loose-knit mesenchymal mesoderm cells. Now the embryo has to make a nervous system. To do so, the embryo starts folding along the central axis of posterior ectoderm, a process called **neurulation** (Figure 22-38).

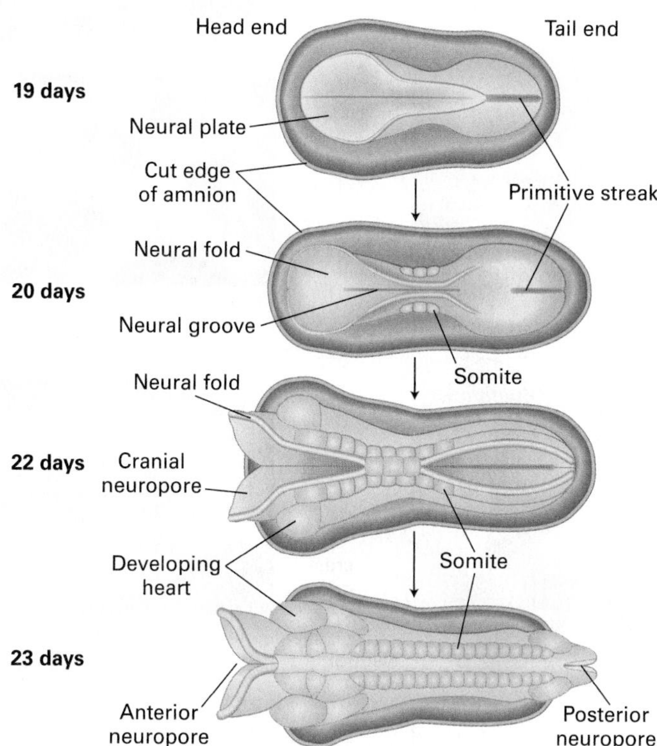

▲ **FIGURE 22-38 Neurulation in the human embryo.** In these drawings, the amniotic membrane has been "removed" to reveal the embryo, called the neurula at this stage. Folding of the ectoderm destined to become neural tissue begins in the center of the embryo at about 21 days of fetal development and proceeds toward both ends, forming the neural tube. At the same time, somites are progressively forming in the mesoderm. The cranial neuropore at the head end of the neural tube closes at about day 24 of human fetal development; the neural tube is completed at the tail end (caudal neuropore) on day 26. Defects in neural tube closure lead to a variety of serious and all-too-frequent birth defects, including spina bifida. [Adapted from W. J. Larsen, 1997, *Human Embryology*, 2nd ed., Churchill-Livingstone, Figure 4-9.]

Neurulation Begins Formation of the Brain and Spinal Cord

Signals that are still poorly understood designate cells along the midline as prospective neural tissue. The best evidence to date, from amphibians and mice, shows that BMP4 prevents neural induction, whereas antagonists of BMP4 such as Chordin, Noggin, and Cerberus can induce neural cell fates by opposing BMP4. Wnt signaling also may induce neural fates by acting earlier and preventing BMP4 expression in certain cells. Whatever the exact series of regulatory events, the prospective neural cells are located in the sheet of ectodermal cells along with flanking cells that will eventually form epidermis. Along what will become the spine, a portion of the ectodermal sheet (the neural plate) folds inward to make a U-shaped canyon (Figure 22-39a). This is the beginning of the *neural tube*. Folding of the neural plate results from highly organized changes in the cytoskeleton of participating cells. Rearrangement of microfilaments in some neural plate cells make the cells wedge-shaped, causing the ectodermal sheet to fold (Figure 22-39b). This is a clear example of how changes in individual cells can contribute to the organized shape changes of large embryonic structures. Each cell has to have enough information to contribute usefully to forming the tissue as a whole.

Neural tube formation also depends on **micro RNAs (miRNAs)**, short RNAs that can influence gene expression by controlling mRNA function or inhibiting translation (see Figures 8-26 and 8-27). Evidence for the involvement of miRNAs in neurulation came from studies with zebrafish in which Dicer, the enzyme that processes micro RNAs, was mutated, thus effectively eliminating all miRNA activity. The resulting embryos

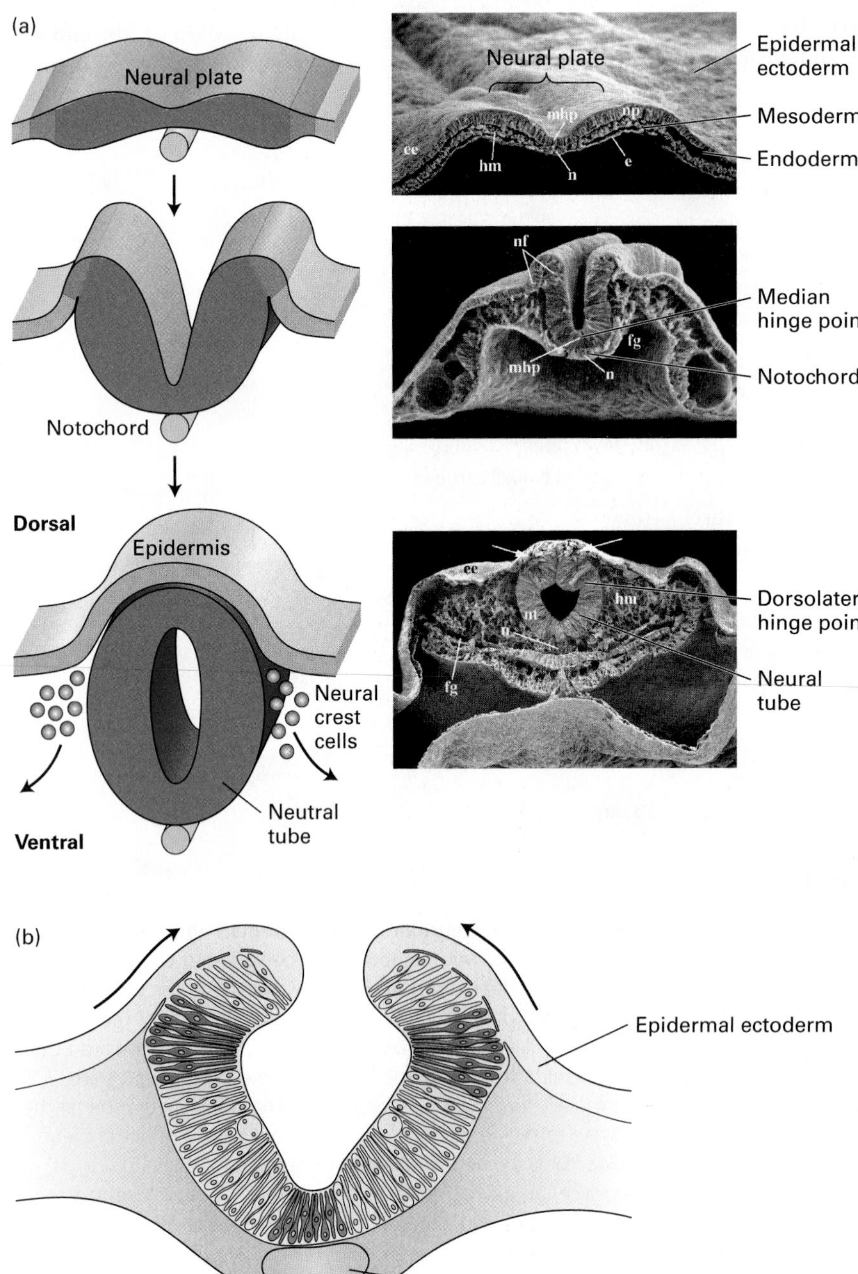

◄ **FIGURE 22-39 Neurulation in the chick.** (a) Drawings on the left show the ectoderm initially divided into prospective neural (blue and red) and epidermal (tan) cells. The neural primordium (neural plate) begins to fold, and eventually the neural folds fuse to make the neural tube. Part of the underlying mesoderm undergoes a mesenchymal-epithelial transition to form the rod-like notochord (from which the term *chordates* is derived). A population of cells (red) near the site of fusion undergo an epithelial-mesenchymal transition and migrate away from the neural tube. These neural crest cells will contribute to heart valves, most of the facial bones, peripheral nervous system, and melanocytes that form the pigment in skin. During neurulation, the epidermis loses its continuity with the neural tissue and fuses above it. Scanning electron micrographs show chick embryos at the same stages of neurulation. (b) In the cooperative (hinge-point) model of neurulation, microfilament rearrangements cause wedging of neuroepithelial cells within two dorsolateral hinge points (blue) and single median hinge point (red). Arrows indicate mediolateral expansion of the epidermal ectoderm. [Part (a) drawings adapted from L. Wolpert et al., 2001, *Principles of Development*, 2nd ed., Oxford Press, Figure 11-10; micrographs courtesy of J. F. Colas and G. C. Schoenwolf. Part (b) adapted from J. F. Colas and G. C. Schoenwolf, 2001, *Devel. Dyn.* **221**:117.]

had proper axis formation but failed to undergo neurulation. To demonstrate that miRNAs, and not some other molecule affected by Dicer, are responsible for the problem, scientists attempted to rescue the mutant fish embryos by injecting them with a prominent class of preformed miRNA duplexes. This succeeded to a remarkable degree, restoring much of the defective neural development. It will be fascinating to learn in the future how miRNAs control neural tube morphogenesis.

Signal Gradients and Transcription Factors Specify Cell Types in the Neural Tube and Somites

As we've seen already, developmental signals can act in relay fashion, a sort of bucket brigade in which each cell receives one signal and passes on another, or in graded fashion, in which different concentrations of one signal induce different cell fates (see Figure 22-13). Graded signals, or morphogens, are important in specifying cell fates in different parts of the neural tube: motor neurons in the ventral part, a variety of interneurons in the lateral parts, and sensory neurons in the dorsal part. The different cell types can be distinguished, prior to morphological differentiation, by the proteins that they produce.

Graded concentrations of Sonic hedgehog (Shh), a vertebrate equivalent of *Drosophila* Hedgehog, determine the fates of at least four cell types in the chick ventral neural tube. These cells are found at different positions along the dorsoventral axis in the following order from ventral to dorsal: floor-plate cells, motor neurons, V2 interneurons, and V1 interneurons. During development, Shh is initially produced at high levels in the notochord, which directly contacts the ventral-most region of the neural tube (Figure 22-40a). The Shh from the notochord induces the most ventral neural-tube cells to form floor-plate cells, a type of non-neuronal glial cell (Chapter 23). Floor-plate cells also produce Shh, forming a Shh-signaling center in the ventral-most region of the neural tube. Antibodies to Shh protein block the formation of the different ventral neural-tube cells in the chick, and these cell types fail to form in mice homozygous for mutations in the *Sonic hedgehog (Shh)* gene.

To determine whether Shh-triggered induction of ventral neural-tube cells is through a graded or a relay mechanism, scientists added different concentrations of Shh to chick neural-tube explants. In the absence of Shh, no ventral cells formed. In the presence of very high concentrations of Shh, floor-plate cells formed; whereas, at a slightly lower concentration, motor neurons formed. When the level of Shh was decreased another twofold, only V2 neurons formed. And, finally, only V1 neurons developed when the Shh concentration was decreased another twofold. These data strongly suggest that in the developing neural tube different cell types are formed in response to a ventral → dorsal gradient of Shh, though exactly when in development the signal has its impact is unknown. The accumulating evidence for gradients does not rule out additional relay signals that may yet be discovered.

Cell fates in the dorsal region of the neural tube are determined by BMP proteins (e.g., BMP4 and BMP7), which belong to the TGFβ family. Recall that TGFβ-type signals

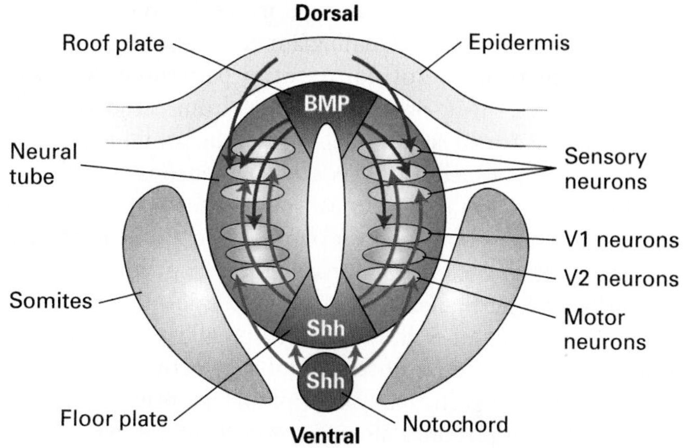

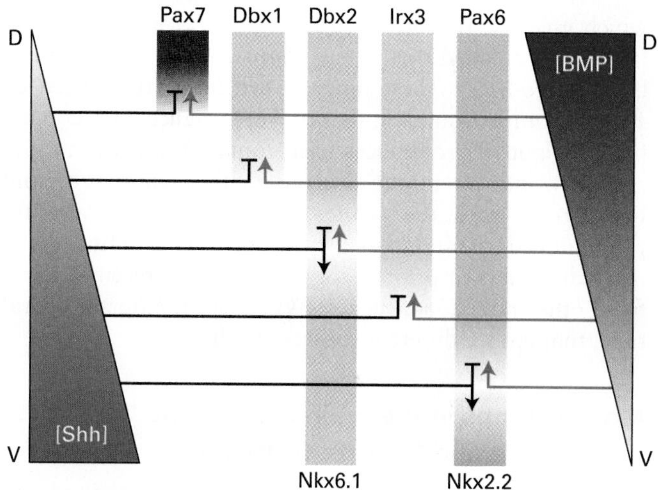

▲ **FIGURE 22-40 Regulation of neural-cell fate in vertebrates.** (a) Sonic hedgehog (Shh) secreted by cells in the notochord induces floor-plate development. The floor plate, in turn, produces Shh, which forms a ventral → dorsal gradient that induces additional cell fates. In the dorsal region, BMP proteins (TGFβ-type signals) secreted from overlying ectoderm cells and subsequently from roof plate cells. (b) The relative concentrations of Shh and BMP are indicated by the colored gradients. Cell fates in the neural tube can be detected by the differential production of all the transcription factors shown between the gradients. High to moderate levels of Shh induce expression of the *Nkx2.2* and *Nkx6.1* genes (↓) but block production of the five transcription factors indicated at the top (⊤). Even low levels of Shh can block expression of *Dbx1* and *Pax7*, both genes whose expression is promoted by BMP signals coming from cells in the dorsal neural tube. The border between Pax6 and Nkx2.2, and between Dbx2 and Nkx6.1, is further sharpened by mutually repressive interactions; each protein turns off the gene encoding the other protein. The outcome of all this is a set of different cell fates (e.g., motor neurons or sensory neurons) along the dorsal-ventral axis. Each cell type contains a unique blend of transcription factors, which presumably control expression of many other genes that confer distinctive properties on cells that type. [See T. M. Jessell, 2000, *Nature Rev. Genet.* **1**:20.]

and their antagonists are critical in determining dorsal cell fates in early frog embryos. A *Drosophila* TGFβ signal called Dpp also functions in determining dorsal cell fates in early fly embryos. Indeed, TGFβ signaling appears to be an evolutionarily ancient regulator of dorsoventral patterning. In vertebrate embryos, BMP proteins secreted from ectoderm cells overlying the dorsal side of the neural tube promote the formation of dorsal cells such as sensory neurons (see Figure 22-40a). Thus cells in the neural tube sense multiple signals that originate at opposite positions on the dorsoventral axis. By integrating the signals from both origins, each cell embarks on a particular course of differentiation.

The gradients of Shh and TGFβ spreading out from the ventral and dorsal sides of the neural tube activate or repress the production of particular transcription factors in neural-tube cells at different positions along the dorsal-ventral axis (Figure 22-40b). When these transcription factors are first made, the boundaries between them are fuzzy, but some of the boundaries sharpen due to mutually repressive cross-regulation.

Similar mechanisms determine cell fates in the somites, which give rise to body wall muscle (*myotome*), the inner layer of skin called the dermis (*dermatome*), and the vertebrae (*sclerotome*). These different cell fates are induced by signals from surrounding tissues. For instance, Shh coming from the notochord induces sclerotome. Thus the same signal, Shh, induces motor neurons in the neural tube and bone primordia in the somites. The receiving cells are preprogrammed to respond in distinct ways to the same inducer depending on their prior history. Somites also receive signals from other directions, such as Wnt from the dorsal neural tube, that induce different subsets of cells.

Most Neurons in the Brain Arise in the Innermost Neural Tube and Migrate Outward

The cortex of the brain is a thin sheet of cells organized into half a dozen layers. This portion of our anatomy—required for the most-advanced thinking abilities—is the source of our greatest pride and feelings of superiority, for better or worse, over other living things. Development of the neural tube, which is initially a single cell layer thick, into a brain requires both the generation of vast numbers of new, progenitor cells from stem cells and the organization of those new cells into layers.

The new cells form mostly in the *subventricular zone,* the inner lining of the neural tube lying closest to the ventricles, which are the fluid-filled cavities inside the brain that arise from the interior of the neural tube. Cells take up their final positions in a simple inside-out order, forming the innermost layer first and then, progressively, the outer ones (see Figure 21-12). Thus neurons born late have to migrate past the older cells that have already taken up their stations. The migration of some cells involves interactions with radial glia, support cells that elongate to span the entire distance from the subventricular zone to the outer layer of the cortex. In addition some cells undergo remarkable tangential migration at right angles to the ventricle-to-surface plane.

Scientists have used time-lapse microscopy to observe the behaviors of migrating neurons. Their movement is not smooth or continuous. Instead, the cells move vigorously for a time, pause and extend processes in what appears to be a testing of the waters, and then resume motion in either the original direction or another. In some cases neurons seem to move over each other; in other cases they follow the processes of glia cells. Migrating cells are responding to a wide variety of guidance cues in the form of surface molecules and secreted signals, while internally each cell undergoes dramatic shape changes.

Once neurons reach their final destinations, and often while they are in transit, they form long processes to communicate with other cells: the **dendrites,** which receive signals, and the **axons,** which transmit signals, sometimes over distances greater than a meter. We will discuss how the correct wiring pattern is built, to the extent that it is understood, in Chapter 23.

Lateral Inhibition Mediated by Notch Signaling Causes Early Neural Cells to Become Different

The development of the nervous system provides examples of an important general mechanism for ensuring that all necessary structures are built, but not in excessive numbers. This mechanism called **lateral inhibition,** consists in essence of a cell communicating to surrounding cells "I am doing this, so you should make something else." Adjacent cells with equivalent or near-equivalent potential are in this way directed toward distinct fates. Genetic analyses of *Drosophila* neural development first revealed the role of the highly conserved Notch/Delta pathway in lateral inhibition. The *Drosophila* proteins Notch and Delta are the prototype receptor and ligand, respectively, in this signaling pathway. Both proteins are large transmembrane proteins whose extracellular domains contain multiple EGF-like repeats and binding sites for the other protein. Although Delta is cleaved to make an apparently soluble version of its extracellular domain, findings from studies with genetically mosaic *Drosophila* have shown that the Delta signal reaches only adjacent cells.

Interaction between Delta and Notch triggers the proteolytic cleavage of Notch, releasing its cytosolic segment, which translocates to the nucleus and regulates the transcription of specific target genes (see Figure 16-36). In particular, Notch signaling activates the transcription of *Notch* itself and represses the transcription of *Delta,* thereby intensifying the difference between the interacting cells (Figure 22-41a). Notch-mediated signaling can give rise to a sharp boundary between two cell populations or can single out one cell from a cluster of cells (Figure 22-41b). Notch signaling controls cell fates in most tissues and has consequences for differentiation, proliferation, the creation of cell asymmetry, and apoptosis. Here we describe two examples of Notch signaling in cell-fate determination.

Loss-of-function mutations in the *Notch* or *Delta* genes produce a wide spectrum of phenotypes in *Drosophila.* One consequence of such mutations in either gene is an increase in the number of neuroblasts in the central nervous system. In *Drosophila* embryogenesis, a sheet of ectoderm cells becomes divided into two populations of cells: those that move inside the embryo develop into neuroblasts, which give rise to neurons; those that remain external form the epidermis and cuticle (see

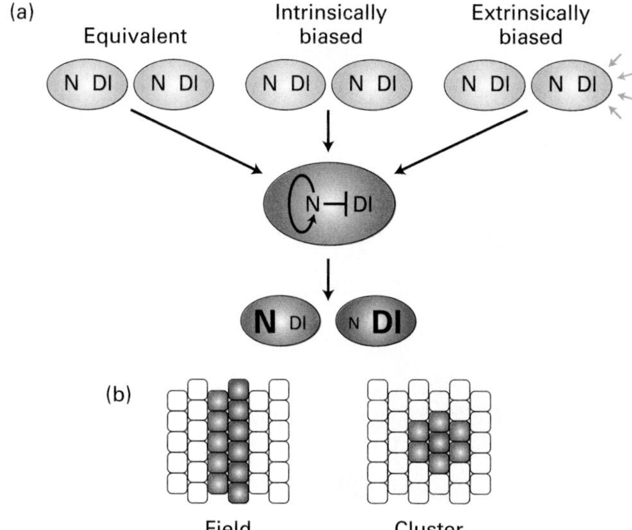

◀ **FIGURE 22-41 Amplification of an initial bias to create different cell types by Notch-mediated lateral inhibition.** (a) A difference between two initially equivalent cells may arise randomly (*left*). Alternatively, interacting cells may have an intrinsic bias (*center*) or an extrinsic bias (*right*). For instance, cells that have received different proteins in an asymmetric cell division will be intrinsically biased; those that have received different signals (orange) will be extrinsically biased. Regardless of how the small initial bias arises, Notch becomes predominant in one of the two cells, promoting its own expression and repressing production of its ligand Delta in that cell. In the other cell, production of Delta predominates. The outcome is reinforcement of the small initial difference. (b) Notch-mediated lateral inhibition may create a sharp boundary in an initial field of cells, such as along the edge of the developing *Drosophila* wing, or distinguish a central cell from a surrounding cluster of cells, as in neural precursor establishment. [Adapted from S. Artavanis-Tsakonas et al., 1999, *Science* **284:**770.]

Figure 21-29). As some of the cells enlarge and then loosen from the ectodermal sheet to become neuroblasts, they signal to surrounding cells to prevent their neighbors from becoming neuroblasts—a case of lateral inhibition. Notch/Delta signaling is used for this inhibition; in embryos lacking the Notch receptor or its ligand, all the ectoderm precursor cells become neural.

The role of Notch signaling in specifying neural cell fates has been studied extensively in the developing *Drosophila* peripheral nervous system. In flies, various sensory organs arise from proneural cell clusters, which produce bHLH transcription factors, such as Achaete and Scute, that promote neural cell fates. In normal development, one cell within a proneural cluster is somehow anointed to become a *sensory organ precursor (SOP)*. In the other cells of a cluster, Notch signaling leads to the repression of proneural genes, and so the neural fate is inhibited; these nonselected cells give rise to epidermis (Figure 22-42). Temperature-sensitive mutations that cause functional loss of either Notch or Delta lead to the development of additional SOPs from a proneural cluster. In contrast, in developing flies that produce a constitutively active form of Notch (i.e., active in the absence of a ligand), all the cells in a proneural cluster develop into epidermal cells.

To assess the role of the Notch pathway during primary neurogenesis in vertebrates, scientists injected mRNA encoding different forms of Notch and Delta into *Xenopus* embryos. Injection of mRNA encoding the constitutively active cytosolic segment of Notch inhibited the formation of neurons. In contrast, injection of mRNA encoding an altered form of Delta that prevents Notch activation led to

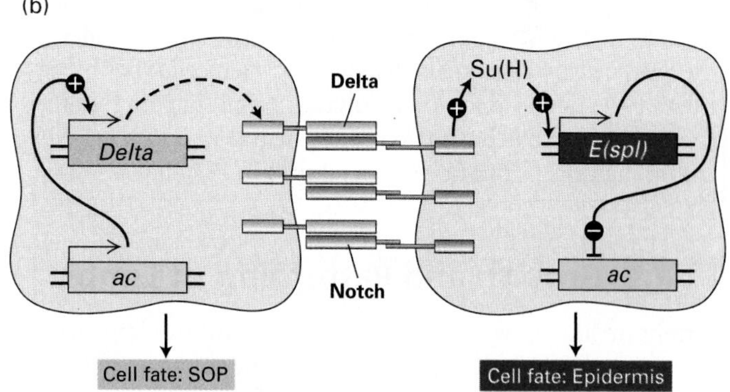

▲ **FIGURE 22-42 Role of Notch-mediated lateral inhibition in formation of sensory organ precursors (SOPs) in *Drosophila*.** (a) Extracellular signaling molecules and transcription factors, encoded by early-patterning genes, control the precise spatiotemporal pattern of proneural bHLH proteins such as Achaete and Scute (yellow). Most cells within the field express Emc (orange), a related protein that antagonizes Achaete and Scute. A small group of cells, a proneural cluster, produce proneural bHLH proteins. The region of a proneural cluster from which an SOP will form expresses lower levels of Emc, giving these cells a bias toward SOP formation. Interactions among these cells, mediated by Notch signaling, leads to accumulation of Notch-regulated E(spl) repressor proteins in neighboring cells (blue), restricting SOP formation to a single cell (green). (b) Initially, *achaete (ac)* and other proneural genes are transcribed in all the cells within a proneural cluster, as are *Notch* and *Delta*. Achaete and other proneural bHLH proteins promote expression of *Delta*. When one cell at random begins to produce slightly more Achaete (*left*), its production of Delta increases, leading to stronger Notch signaling in all its neighboring cells (*right*). In the receiving cells, the Notch signaling pathway activates a transcription factor designated Su(H), which in turn stimulates expression of *E(spl)* genes. *E(spl)* expression stays low in the high-Delta (left) cell, allowing a neural, i.e. SOP, fate. In the right cell, the E(spl) proteins specifically repress transcription of *ac* and other proneural genes. The resulting decrease in Achaete leads to a decrease in Delta, thus amplifying the initial random difference among the cells. As a consequence of these interactions and others, one cell of a proneural cluster is selected as a SOP; all the others lose their neural potential and develop into epidermal cells.

the formation of too many neurons. These findings indicate that in vertebrates, as in *Drosophila*, lateral inhibition mediated by Notch signaling controls neural precursor cell fates. Similarly in the nematode *Caenorhabditis elegans*, Notch signaling is used during vulva development to form distinct adjacent cell types by lateral inhibition. The Notch signaling pathway is used during the formation of many organs and tissues, but not always for lateral inhibition.

KEY CONCEPTS OF SECTION 22.5

Cell-Type Specification in Early Neural Development

- The vertebrate nervous system develops from ectoderm cells, which are initially arranged in a sheet formed during gastrulation. Folding of this ectodermal sheet (the neural plate) first forms the neural tube in the process of neurulation (see Figures 22-38 and 22-39).
- The notochord, a rod of mesoderm lying underneath (ventral to) the neural tube, is a source of signaling proteins that induce cell fates in surrounding tissues. For example, Sonic hedgehog (Shh) from the notochord influences the fates of nearby cells in the neural tube, somites, and endoderm.
- Shh is a morphogen that directs certain cell fates at high doses and other cell fates at lower doses. It tends to promote ventral fates, like floorplate. Its influence is tempered by other signals such as BMP, a TGFβ-type signal coming from overlying ectoderm that promotes dorsal neural fates (see Figure 22-40).
- The cells of the neural tube, which is initially one cell thick, proliferate and form neural precursor cells that migrate radially outward to form the layers of the brain and spinal cord.
- During normal *Drosophila* neurogenesis, only some ectodermal cells form neurons; others form other ectoderm-derived cells like skin. The balance between cell types is regulated by lateral inhibition involving Notch signaling. In this process, cells that have taken on a neural fate instruct surrounding cells not to do so (see Figures 22-41 and 22-42).

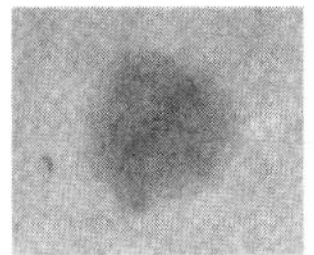

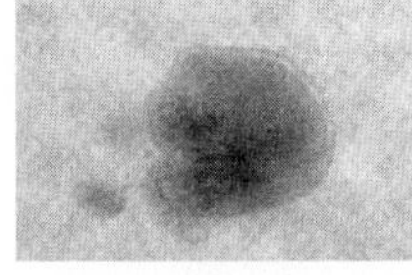

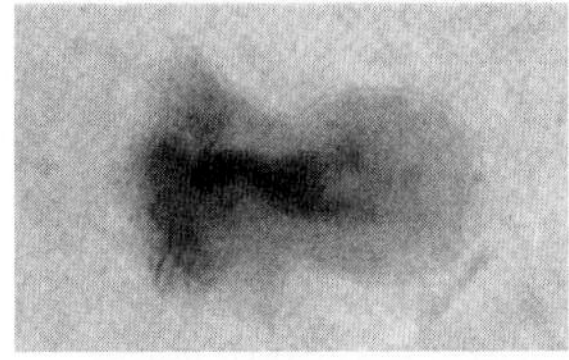

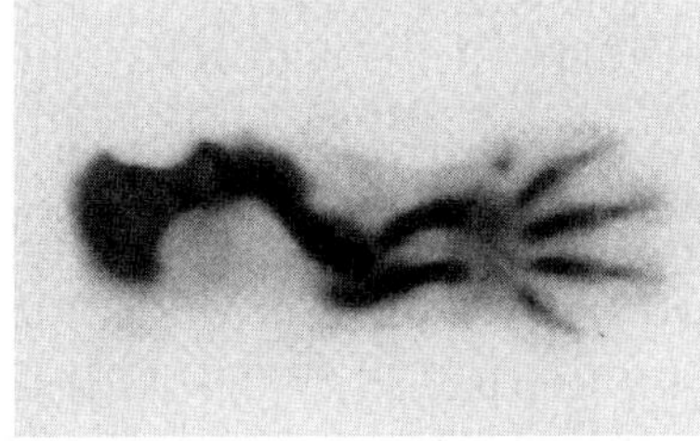

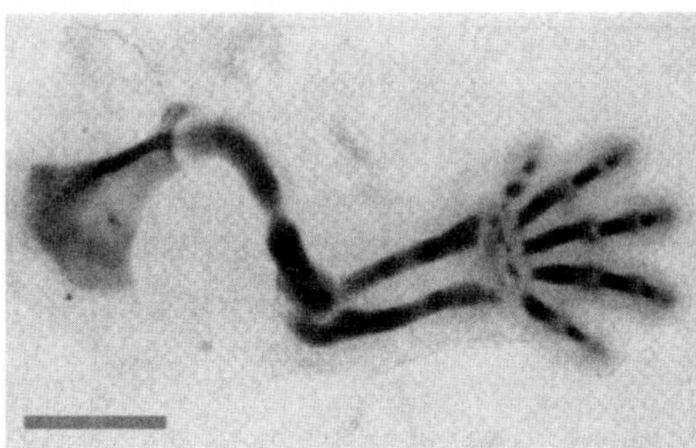

▲ **FIGURE 22-43 Limb development in the mouse.** Limb buds, which consist of an inner mass of mesoderm cells and outer layer of ectoderm, form at specific locations on the embryo's flank. Outgrowth and patterning along the three axes leads to formation of the all the limb structures. [From L. Wolpert et al., 2001, *Principles of Development,* 2nd ed., Oxford University Press, Figure 10-5.]

22.6 Growth and Patterning of Limbs

Vertebrate limb development provides a beautiful example of how the actions of individual cells combine to create pattern. Vertebrate limbs grow from small "buds" composed of an inner mass of mesoderm cells surrounded by a sheath of ectoderm (Figure 22-43). Hindlimbs and forelimbs are obviously related, as are left and right limbs. If the limbs were broken down into their constituent molecules or cells, the composition of all four would be nearly identical, yet their shapes differ in ways that are absolutely crucial to successful life. Pattern formation—that is, organizing those molecules and cells into a coherent whole—is a central goal of studying the molecular cell biology of limb development. Where would birds be with four legs instead of a pair each of legs and wings? Birds have in fact been a prime experimental animal for studies on limb development, since chick embryos can be readily accessed in the egg during the period of limb formation.

In testing various theories of limb development in the chick, researchers have subjected embryos to transplantation, injected them with signaling proteins or signal-producing cells, and introduced retroviruses that direct gene expression in abnormal sites. The continued growth of the animal in the egg then is observed to see the effect of any particular treatment. Because it is relatively easy to reduce or eliminate a gene's function in mice, genetic experiments have been useful in studying limb development in these animals. The results from the two experimental systems are largely in agreement.

Hox Genes Determine the Right Places for Limbs to Grow

The first event in vertebrate limb development is determination of where, along the head-tail axis, limb growth will begin. We are not millipedes; decisions must be made. In both insects

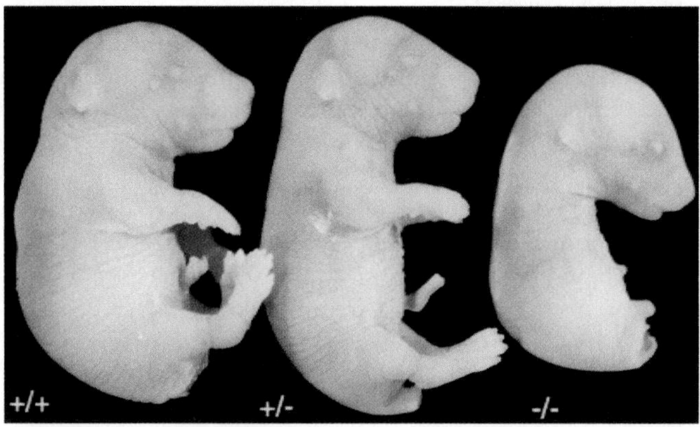

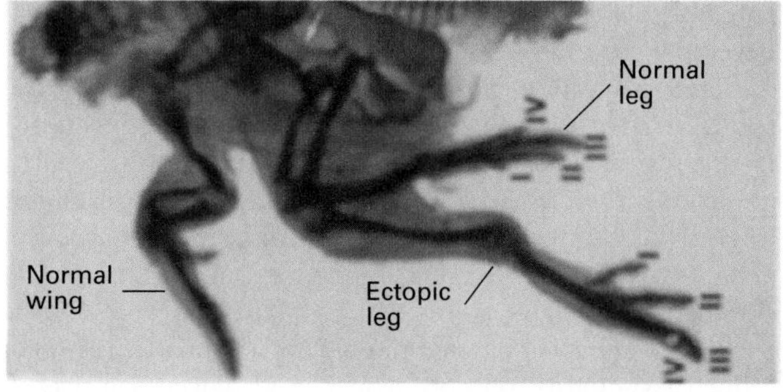

◄ EXPERIMENTAL FIGURE 22-44 **The effect of altering FGF10 function on limb development in the mouse.** (a) Gene knockout mice, lacking a functional part of the *fgf10* gene, were constructed by recombination. These photographs of mouse embryos a few days before birth show that mice heterozygous for the *fgf10* gene (+/−) are fairly normal, but homozygous mice (−/−) have severely impaired limb development compared with that in wild-type mice (+/+). These genetic data prove the importance of fibroblast growth factor 10 (FGF10), a secreted signal protein, for limb development. (b) To test whether FGF10 is capable of triggering limb development, experiments with chick embryos were done. These embryos are advantageous because the embryo can continue to grow after a surgical operation. Beads soaked in FGF10 protein were surgically implanted in the flank of a mid-stage chick embryo prior to limb development, in a region where no limb would normally develop. FGF10, but not control substances, was able to induce the formation of a fifth limb. The young chick shown here, from a FGF10-treated embryo, has an extra (ectopic) leg. Only half of the chick's skeleton is shown. [Part (a) from Min et al., 1998, *Genes & Devel.* **12:**3156. Part (b) from M. J. Cohn et al., 1995, *Cell* **80:**739.]

and vertebrates, the Hox genes control where limbs are made. The mesoderm that gives rise to limb buds, called *intermediate mesoderm,* is located adjacent to the somite mesoderm; mesoderm farther from somites is called *lateral plate mesoderm.* Expression of Hox genes in the intermediate mesoderm influences the position of limb-bud formation by controlling expression of genes (e.g., *Tbx* and *Pitx1*) that encode other transcription factors in the lateral plate mesoderm. The Tbx and Pitx1 proteins, in turn, control production of secreted signals that are required for limb development.

Among these signals are fibroblast growth factor 10 (FGF10), which is secreted from cells in the lateral plate mesoderm and initiates outgrowth of a limb from specific regions of the embryo's flank. Mouse mutants lacking FGF10 develop without limbs (Figure 22-44a). Conversely, implantation of a bead soaked in FGF10 at sites in the flank of a chick embryo where a limb does not normally form causes an extra limb to grow (Figure 22-44b). These results demonstrate the remarkable inductive capabilities of FGF. Wnt signaling also plays a role in the initial outgrowth of limb buds.

Limb Development Depends on Integration of Multiple Extracellular Signal Gradients

The three axes of a limb bud and developed limb—anterior-posterior (thumb to little finger), dorsal-ventral (back of hand versus palm), and proximal-distal (shoulder to fingers)—are shown in Figure 22-45. Early limb development is marked by formation of two important signaling centers: the *apical ectodermal ridge (AER),* a region of surface ectoderm at the distal tip of the emerging limb bud and the *zone of polarizing activity (ZPA)* in mesoderm at the posterior end of the bud.

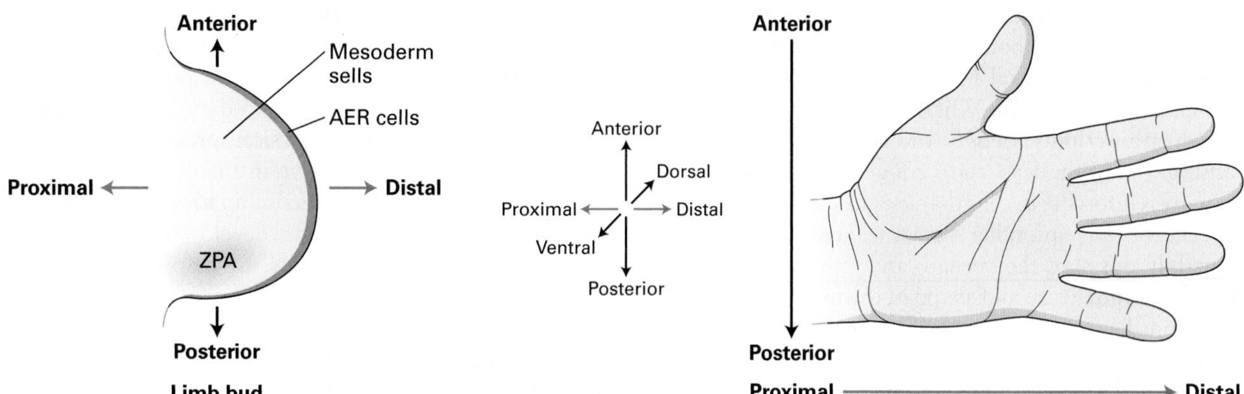

▲ **FIGURE 22-45 The axes of the limb bud and the hand.** A limb bud *(left)* and fully developed limb, a hand in this example *(right),* has three axes: anterior-posterior (thumb to little finger), proximal-distal (shoulder to fingers), and dorsal-ventral (back of hand versus palm). In the developing limb bud, the apical ectodermal ridge (AER) is a source of fibroblast growth factor (FGF) signals, and the zone of polarizing activity (ZPA) is a source of Sonic hedgehog (Shh).

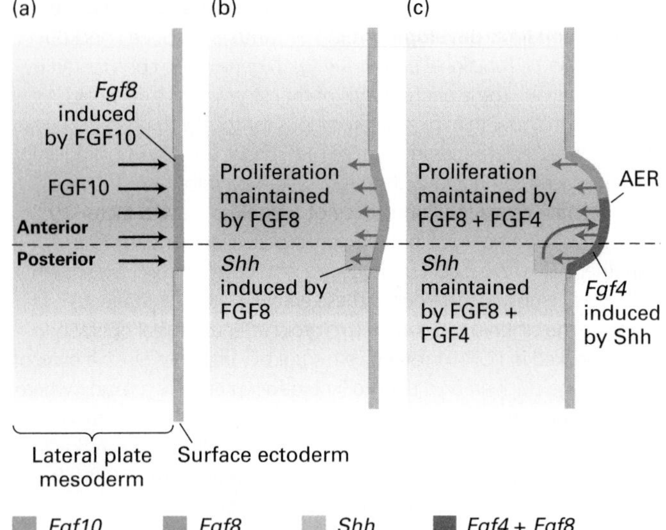

▲ **FIGURE 22-46 Integration of three signals in vertebrate limb development along the proximal-distal and anterior-posterior axes.** Each limb bud grows out of the flank of the embryo. (a) A fibroblast growth factor (FGF) signal, probably FGF10, comes from the mesoderm in specific regions of the embryo's flank, one region for each limb. FGF10 acts on a local region of surface ectoderm called the apical ectodermal ridge (AER) because it will form a prominent ridge. (b) The ectoderm that receives a FGF10 signal is induced to produce FGF8, another secreted signal. At the posterior end of the limb bud, FGF8 induces transcription of the *Sonic hedgehog (Shh)* gene. (c) Shh signaling induces transcription of the gene encoding FGF4 in the AER. FGF8 and FGF4 promote continued proliferation of the mesoderm cells, causing outgrowth of the limb bud. Shh also stimulates this outgrowth and confers posterior characteristics on the posterior part of the limb. Development along the dorsal-ventral axis depends on a Wnt signal that is not shown here.

In response to FGF10, the AER begins to secrete FGF8 and later FGF4. Both of these signals drive persistent division of mesoderm cells and therefore continued limb outgrowth. Sonic hedgehog (Shh) is produced in the ZPA of the posterior limb bud (Figure 22-46); indeed it is Shh secretion that defines the ZPA. The FGF signals confer a distal fate on cells in the limb bud, and Shh confers a posterior fate. If Shh is added to the anterior part of the bud, the limb that eventually forms will have two posterior patterns of bones in mirror-image, and no anterior. Along the dorsal-ventral axis, a Wnt7a signal directs cells to form ventral cell types. The Wnt7a, FGF4, and FGF8 signals promote the transcription of *Shh,* and Shh signaling promotes the transcription of the *Fgf4* and *Fgf8* genes. Thus the signals are mutually reinforcing in cells that are close enough; cells too far from one of the reinforcing signals will cease making their own signal. In this way, the strength and movement of signals is tied to the eventual size and shape of the limb.

The main developmental task in the formation of limbs and organs is to organize a few cell types (e.g., mesenchyme, vascular, epithelial) into complex multicellular structures. As we've just seen, cells in the midst of the limb bud are exposed to gradients of multiple signals (FGFs, Shh, and Wnt) that provide the initial guidelines for this building process. By integrating these signals, each cell is directed to express specific transcription factors depending on its location relative to the three limb axes. The action of these transcription factors in turn control formation of the various parts of the limb.

Since each of the limb-bud signals must be produced in exactly the right cells, there is much interest in learning how their genes are controlled. The *Shh* gene, for example, is transcribed only in the ZPA cells of the posterior limb bud in response to FGF and other factors. Initial efforts to describe the transcriptional enhancer that controls *Shh* transcription in the limb bud failed. The mystery was solved when four families were found with a high frequency of polydactyly, an inherited abnormality marked by the presence of extra digits on the hands and feet. The mutation involved in each family mapped to a region of the genome in the vicinity of the human *Shh* gene, although the protein-coding parts of *Shh* were normal. These findings strongly suggested that damage to the enhancer controlling *Shh* transcription caused the abnormal development resulting in polydactyly in these families.

The *Shh* limb enhancer is quite remarkable in two respects. First, this regulatory sequence is located inside the transcription unit of another gene. Second, it is located about one million base pairs away from the *Shh* promoter! This is the Guinness record for long-distance regulation of gene transcription. Recent sequence comparisons have shown a high degree of sequence similarity throughout the ≈1-kb *Shh* enhancer in mice and humans. A smaller region of sequence similarity, the core sequence, also occurs in ricefish (Figure 22-47). All four polydactylous families were found to have mutations that changed the sequence of the *Shh* enhancer, as did two polydactylous mouse strains. These findings demonstrate the power of using the evolutionary conservation of DNA sequences to understand gene regulation and account for human disease.

Hox Genes Also Control Fine Patterning of Limb Structures

Once limb-bud growth and Shh production have gotten underway, Shh induces Hox-gene expression, starting at the posterior of the bud. Subsequently, expression of the Hox genes undergoes a complex series of changes and spreads more anteriorly along the bud. The transcription factors produced from *Hoxd11–Hoxd13* in posterior cells in turn stimulate expression of *Shh* in the posterior. As the bud grows during the period 9.0–10.5 days after fertilization, the domains of *Hoxd*-gene expression spread out (Figure 22-48). The Hox genes remain active in the distal limb bud, where digits will be formed. The combinations of Hox transcription factors in the cells of the developing limb bud, in their elaborate overlapping patterns, are responsible for fine-scale control of limb pattern. Humans carrying mutations in *Hoxd13* have an inherited limb malformation syndrome called *synpolydactyly,* marked by duplications of fingers and toes. This defect demonstrates the crucial nature of *Hoxd13* for normal limb development. In mice, deleting all *Hoxa* and *Hoxd* genes causes limbs to develop as short stubs without any distal structures, confirming the importance of Hox genes in mouse limb development.

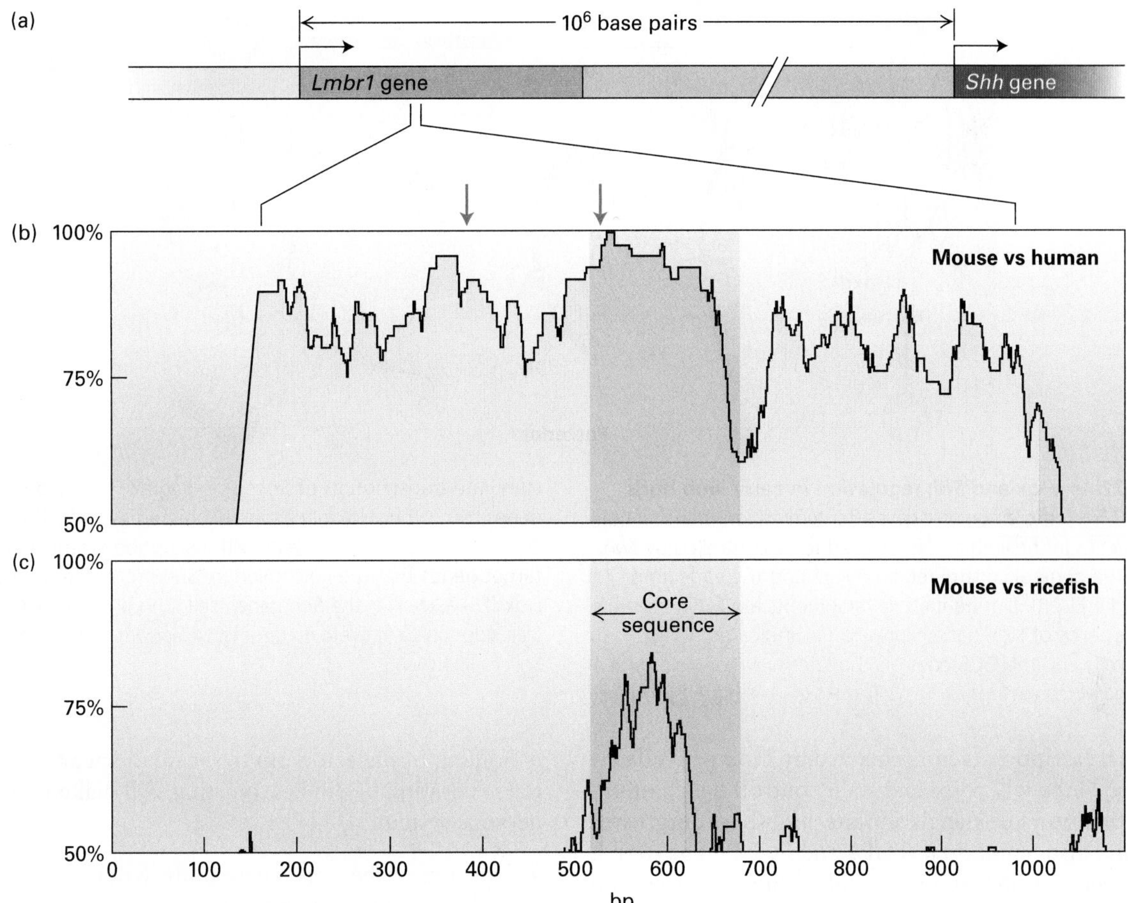

▲ EXPERIMENTAL FIGURE 22-47 **Mutations that cause polydactyly (extra digits) and evolutionary sequence conservation were used to identify the *Shh* enhancer.** Members of four human families with a particular inherited form of polydactyly (extra digits) and two mouse strains with a similar defect were found to have mutations in a region located about 1 megabase (Mb) from the *Sonic hedgehog (Shh)* promoter. This region is located within a 31kb intron of the *Lmbr1* gene *(top)*. To find out if the mutations in this region are in fact responsible for polydactyly, scientists compared the nucleotide sequences of DNA from mice, humans, and rice fish. The top graph, plotting the percentage sequence identity between mouse and human DNA, indicates that the similarity is quite high in the ≈1-kb region containing the human mutations (black arrows) and mouse mutations (red arrows). Outside this region, the sequence similarity drops off dramatically. The bottom graph shows a far more limited region of similarity, referred to as the core, between mouse and rice fish DNA. These results suggested that the 1-kb region contains an enhancer controlling expression of *Shh* in the limb bud. In agreement with this idea, the mutant mice were found to express *Shh* in areas of the anterior limb bud where the gene should be off. Apparently, damage to the *Shh* limb-bud enhancer that changes spatial expression of *Shh* can result in defective limb patterning. [See L. A. Lettice et al., 2003, *Human Molec. Genet.* **12:**1725. Graphs adapted from T. M. Sagai et al., 2005, *Development* **132:**797.]

The *Shh* and the *Hoxd* genes regulate each other, forming a positive feedback loop in the posterior bud that keeps both genes turned on. In the anterior bud, a transcription factor called Gli3 represses the *Shh* and *Hoxd* genes. Shh signaling represses transcription of the *Gli3* gene in posterior cells (see Figure 22-48). The production of Hoxd transcription factors in distinct regions of the limb bud begs the question of which genes they control and how those target genes affect the pattern of the limbs. Little is known about either the identity or function of Hoxd target genes. Among the direct targets of Hoxd proteins is the gene encoding one of the ephrins, a family of membrane-bound signaling proteins (see Figure 23-42). Thus much work remains to be done to understand how the production of Hox transcription factors in precise spatial patterns is translated into the exquisite and (within a species) highly reproducible shapes of limbs.

The final pattern in the limbs seems to be attained using repeated rounds of signaling and transcriptional control, with the early events roughing out the pattern and the later events refining it. At the same time that the limb is growing vigorously, the new cells are exposed to a mixture of pre-existing signals. Eventually apoptosis (Chapter 21) is triggered in specific cells, thus removing the webbing between the digits.

Our discussion of limb development has revealed an important feature of this and other developmental processes: the repeated use of the same kind of signals and regulators. Wnt signals initiate limb development and also control patterning along the dorsal-ventral axis of the limb bud.

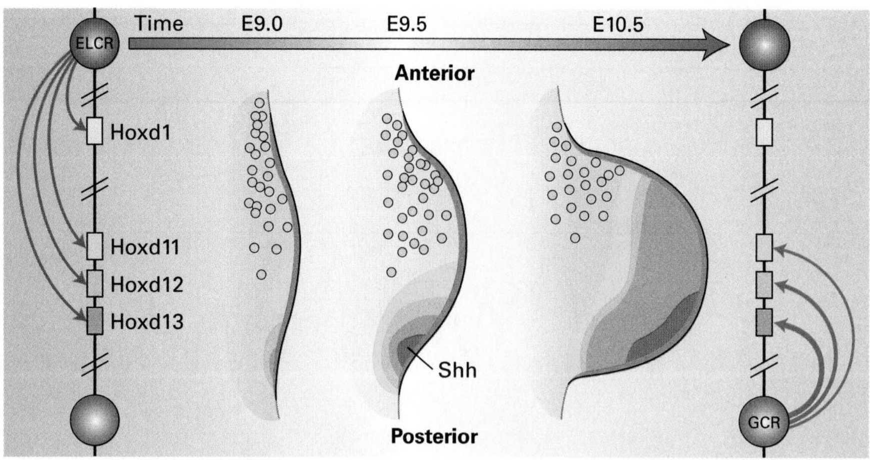

▲ **FIGURE 22-48 *Hox* and *Shh* regulation in early limb buds.** E9, E9.5, E10.5 refer to embryonic days after fertilization. Transcription of the Hox genes in the limb bud is believed to be controlled by two enhancers: the general control region (GCR) and the early limb control region (ELCR). During limb development, the ELCR controls the initial patterns of Hox-gene expression as indicated, which includes *hoxd1*, and the GCR controls the later wave of Hox-gene transcription, which excludes *hoxd1*. The Hoxd11–Hoxd13 proteins stimulate transcription of *Shh*. The diagram shows only *Hoxd* genes, but other Hox genes are expressed as well. The light blue circles represent Gli3 protein, a transcription factor that represses target genes that are controlled by Shh protein (e.g., the *Hoxd11–Hoxd13*) as well as the *Shh* gene itself. Shh protein turns off *Gli3* transcription to activate its targets. [Adapted from J. Deschamps, 2004, *Science* **304:**1610.]

Similarly, transcription factors encoded by Hox genes determine where limbs will form and then control digit patterning. And finally, FGF signals initiate limb-bud outgrowth (FGF10) and also maintain it (FGF4 and FGF8). The toolkit of developmentally important molecules is not too large, and the tools are used repeatedly to control cell fates.

So Far, So Good

In the preceding sections, we have witnessed many key events in the genesis of animal embryos:

- Construction of two quite different types of gametes and their fusion to restore a diploid genome with different patterns of gene expression by maternally and paternally derived chromosomes.

- Transformation of the fertilized egg (zygote) during cleavage and gastrulation into three layers of stem cells, the endoderm, mesoderm, and ectoderm, and the generation of a large number of inductive interactions between cells in different layers.

- Establishment of a dorsal-ventral axis, an anterior-posterior axis, and a left-right axis.

- Formation of the heart, the left-right asymmetric organ that will pump blood for a lifetime.

- Division of the early embryo into repeating body segments driven by different combinations of transcription factors distributed in space (insects) or cycles of gene expression that differ in time (vertebrates).

- Generation of distinctive features on each of the body-segment repeats, in both insects and vertebrates, as the result of Hox-gene activity.

- Folding of the ectoderm to create the neural tube and initial generation of the neurons that will make up the central nervous system.

- Selection of the sites where limbs will be located and their development along three axes.

But the embryo is not quite done. There is the matter of forming many other organs and tissues: skin, bone, muscle, eyes, vasculature, liver, kidney, gonads, lungs, gut and others . . . a lot of fascinating molecular cell biology. There is the sex-determination control system that makes half of us different from the other half, and the control of organ and body size. There is a need to wire the brain. There are regenerative processes and aging processes, all under the influence of genetic and environmental controls. We cannot discuss all these topics in this volume. But the principles of embryonic development described in this chapter are broadly applicable and allow scientists to use information about the development of one type of tissue or organ to guide studies of others.

KEY CONCEPTS OF SECTION 22.6

Growth and Patterning of Limbs

- The initiation of limb development at the right locations along the anterior-posterior body axis is initially controlled by Hox genes and other genes, leading to expression of signals that induce bud growth from specific regions of the embryo's flank.

- Limbs start as buds with an ectodermal sheath over a mesoderm filling (see Figure 22-45). Patterning within the limb bud involves cell-fate determination under the influence of three signaling systems that work simultaneously and influence the expression of each other in space- and time-dependent ways.

■ Outgrowth of the limb bud along the proximal-distal axis is driven by a FGF signal that emerges from the apical ectodermal ridge (AER) at the outermost part of the bud. Patterning of the bud along the anterior-posterior axis depends upon Sonic hedgehog (Shh) signal produced in the zone of polarizing activity (ZPA) in the posterior bud (see Figure 22-46). A Wnt signal directs cells to form ventral cell types.

■ Information from FGF, Shh, and Wnt signals is integrated by cells, so that each assumes its proper fate and role in the developing limb.

■ Shh regulates expression of Hox genes in complex overlapping patterns during limb-bud development (see Figure 22-48). The Hox genes are important for regulating detailed pattern in the limb, for example, of the different digits and the bone, muscle, and nerve they contain.

Perspectives for the Future

You have undoubtedly noticed that the same classes of signals and transcription factors are used again and again during development of a polarized embryo. There are three reasons for this. First, much to the amazement and relief of biologists, the number of different classes of signals and of transcription factors is not too large, perhaps 20 different types of signals for example. For each type of signal, the standard signal-transduction pathway is quite well known (Chapter 16). Variations on the standard are interesting, but it is often fairly safe to predict what will happen when a particular signal (e.g., BMP) reaches a cell, and how to detect and measure the cell response to the signal. Second, the protein classes and gene systems involved in development are, to a high degree, conserved in most animals. This conservation has created a common "molecular genetic" language among biologists working on the development of many organisms, allowing them to share in discovery and understanding. Third, the patterns of evolution of development are becoming clear. Understanding how development is controlled allows us to understand much better how distinct animal forms can arise from mutations that change the actions of developmental regulators. What use is all this work, you might ask. Two answers seem obvious: understanding our origins and using our knowledge of animal development to prevent and treat human diseases.

Darwin understood the importance of embryology to his theory and wrote extensively on the subject. But the genetic and molecular cell biology mechanisms underlying evolution were completely unknown in his time. Now we have rich detail about how genes control animal form, and a flood of new information is pouring in. New fossils are found regularly of intermediate ancestral forms that link present-day animals. For example, fossils that appear to represent common ancesters to whales and hippos have been found, as well as the creature *Tiktaalik* that looks like a fish with scales, but has four legs. Such ancient creatures provide clues about how changes in developmental processes—pattern forming processes—underlie evolution. As the networks of interactions that control development of each particular tissue are traced, we will be able to understand how alterations of some components can change animal form and function. Among all the changes that led to animal diversity—that made bats able to navigate by sonar and moths able to "sniff" single molecules and cheetahs that can run at 60 mph—were changes that made us what we are.

Genetic damage to any of the developmental regulators are likely to lead to birth defects, cancer, degenerative disease, and altered resistance to infection. Indeed, all of these conditions have been observed, and their relationship to faulty development established. Thus developmental biology is a rich source of new information about the causes of human disease. Since many proteins work in pathways, linking one developmental regulator to a disease has often led to identification of additional human genes tied to the same disease. Apart from *finding* human disease genes, there is the very real prospect of applying our knowledge of development to spur tissue regeneration and promote faster, better healing. Regeneration of blood cells from bone marrow transplants that contain hematopoietic stem cells is now a well-established procedure. There is every likelihood that a flood of new therapies will emerge as manipulation of signals and other developmental proteins becomes more precise and sophisticated. The ability to transfer knowledge from a wide variety of animal embryos to human biology, a triumph of evolutionary theory and practice, is a foundation for the next stages of medical advances.

Key Terms

acrosomal reaction *957*
apical ectodermal ridge (AER) *991*
blastocyst *960*
cleavage *950*
cortical reaction *957*
dosage compensation system *958*
epithelial-mesenchymal transitions *960*
floral organ–identity genes *983*
gametes *950*
gap genes *972*
gastrulation *951*
germ layers *951*
genomic imprinting *958*
homeosis *979*
Hox genes *978*
induction *951*
lateral inhibition *988*
maternal mRNAs *971*
morphogen *964*
morphogenesis *969*
neurulation *985*
notochord *987*
organogenesis *951*
pair-rule genes *976*
pattern formation *951*
segment-polarity genes *977*
sensory organ precursor (SOP) *989*
somites *977*
Spemann organizer *965*
syncytium *970*
transcription cascade *974*
zone of polarizing activity (ZPA) 991

Review the Concepts

1. In differentiated cells, only selected genes are transcribed. Yet, except for lymphocytes and erythrocytes, most somatic cells contain the same nuclear genome. What evidence shows this is true?

2. Compare and contrast the action of morphogens involved in dorsal-ventral specification in *Xenopus laevis* and *Drosophila melanogaster*.

3. Using in situ hybridization with a probe specific for *dorsal* mRNA, where in the syncytial *Drosophila* embryo would you expect *dorsal* to be expressed? Using immunohistochemistry with an antibody specific for Dorsal protein, where would you expect to detect Dorsal protein?

4. A microarray analysis of wild-type fly embryos and *dorsal* mutant embryos would be expected to yield information on all genes regulated by Dorsal protein. Why? Other than new genes regulated by Dorsal, one would expect to see changes in regulation of previously identified genes. Expression of which genes would be increased or decreased in *dorsal* mutants?

5. Deleting the 3′ UTR of the *bicoid* gene would yield what phenotype in a mutant fly? Why?

6. How can the group of five gap genes specify more than five types of cells in *Drosophila* embryos?

7. At which stage in embryological development do somites form? What factors affect their development?

8. What is homeosis? Give an example of a floral homeotic mutation and describe the phenotype of the mutant and the normal function of the wild-type gene product.

9. Hox-gene expression can be maintained through many cell generations. What molecular mechanisms are involved in this process?

10. Describe an experiment with *Xenopus* embryos that demonstrates the role of the Notch pathway in regulating the formation of neurons in vertebrates.

11. What is the evidence that a gradient of Sonic hedgehog leads to development of different cell types within the chick neural tube?

12. What evidence implicates fibroblast growth factor 10 (FGF10) in limb development?

13. Synpolydactyly is an inherited human abnormality characterized by duplicated fingers and toes. What type of mutation could cause this abnormality? What stage of development would be affected?

Analyze the Data

In *Drosophila,* maternally produced mRNAs and proteins that determine the body axes are transported from nurse cells into the developing oocyte. During early oogenesis, one of these maternal mRNAs, *oskar,* is dispersed throughout the oocyte and is not translated. By mid-oogenesis, *oskar* mRNA is transported to the posterior pole of the oocyte where it is translated; the Oskar protein then initiates formation of most abdominal segments and the germ-line cells. This developmental process is defective in oskar nonsense (*oskNS*) mutants. To further understand the role of *oskar* mRNA and its protein in *Drosophila* development, a new mutant (*oskX)* was generated (see A. Jenny et al., 2006, *Development* 133:2827-2833). In this fly mutant, the *oskar* gene contains a transposable element (TE) in its first exon (E1), as diagrammed here. For simplicity, introns are represented as thin black lines separating the exons:

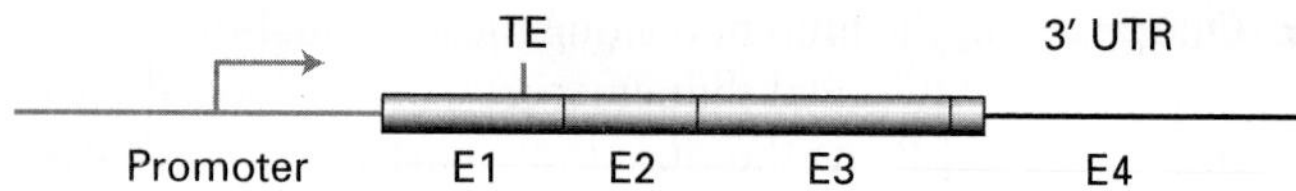

a. Northern blot analysis using a probe complementary to *oskar* mRNA or to *rp49* mRNA was performed on mRNA isolated from egg chambers, which contain nurse cells and the developing oocyte, in wild-type and mutant flies. The results are shown in part (a) of the figure below. Micrographs of two egg chambers that were subjected to in situ hybridization with a probe directed against *oskar* mRNA are shown in part (b). The egg chamber on the left is from a wild-type female; the egg chamber on the right is from an *oskX* female. Hybridization appears as white staining in the micrographs.

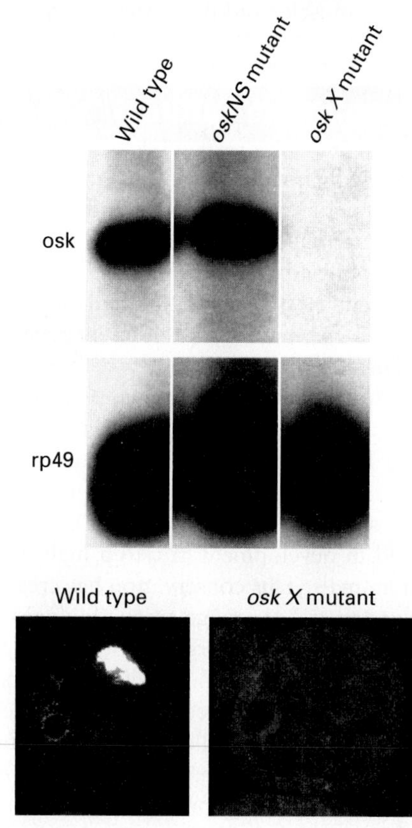

What do these data suggest about the *oskX* mutation relative to the *oskNS* mutation? What is the purpose of probing for *rp49* mRNA?

b. Further study revealed that females hemizygous (only one allele present in the animal) for *oskNS* lay eggs that, when fertilized, develop into embryos lacking posterior structures. In contrast, females hemizygous for *oskX* produce ooyctes that begin to form but then degenerate. What information do these observations provide about the function of Oskar protein?

c. The following rescue experiments were undertaken. Several transgenes carrying different domains of the *oskar* gene were constructed and introduced individually into females hemizygous for the *oskX* mutation. The ability of females expressing each transgene to lay eggs was then monitored. In all transgene constructs, the *oskar* promoter was replaced with a yeast inducible promoter (UAS). As diagrammed in part (a) of the figure below, the first transgene construct includes the entire wild-type *oskar* gene with its three introns (UAS oskWT). For simplicity, introns are not depicted in the drawings. In the second construct, the wild-type *oskar* gene lacks its three introns (UAS oskΔi(1,2,3)). In the third construct, 3′ untranslated region (UTR) of the wild-type *oskar* gene is replaced with the 3′ UTR from an irrelevant gene (UAS oskK10). The final construct contains only the 3′ UTR of the *oskar* gene (UAS osk3′ UTR). The relative egg-laying ability of the transgenic females is shown in part (b). The white bar (w^{1118}) is a measure of the eggs laid on average by nontransgenic wild-type females.

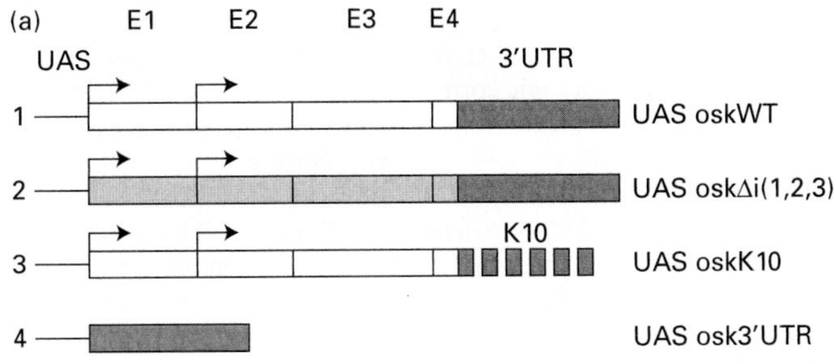

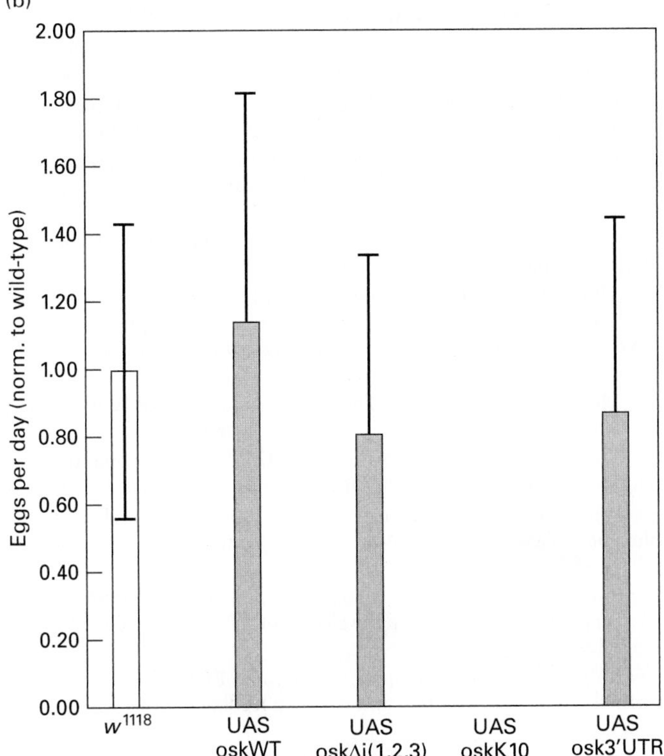

Offspring from females expressing transgenes 1 or 2 were normal, whereas those arising from females expressing transgene 4 lacked abdominal segments. What other information can be deduced from these observations about the role of *oskar* in *Drosophila* development?

References

Highlights of Development

Anderson, K. V., and P. W. Ingham. 2003. The transformation of the model organism: a decade of developmental genetics. *Nature Genet.* 33(Suppl):285–293.

Ashe, H. L., and J. Briscoe. 2006. The interpretation of morphogen gradients. *Development* **133**:385–394.

Blitz, I. L., G. Andelfinger, and M. E. Horb. 2006. Germ layers to organs: using *Xenopus* to study "later" development. *Semin. Cell Devel. Biol.* **17**:133–145.

Eggan, K., et al. 2004. Mice cloned from olfactory sensory neurons. *Nature* **428**:44–49.

Grimm, O. H., and J. B. Gurdon. 2002. Nuclear exclusion of Smad2 is a mechanism leading to loss of competence. *Nature Cell Biol.* **4**:519–522.

Gurdon, J. B. 2006. From nuclear transfer to nuclear reprogramming: the reversal of cell differentiation. *Ann. Rev. Cell Devel. Biol.* **22**:1–22.

Hoegg, S., and A. Meyer. 2005. Hox clusters as models for vertebrate genome evolution. *Trends Genet.* **21**:421–424.

Ingham, P. W., and M. Placzek. 2006. Orchestrating ontogenesis: variations on a theme by sonic hedgehog. *Nature Rev. Genet.* **7**:841–850.

Kamath, R. S., et al. 2003. Systematic functional analysis of the *Caenorhabditis elegans* genome using RNAi. *Nature* **421**:231–237.

Kim, S. K., et al. 2001. A gene expression map for *Caenorhabditis elegans. Science* **293**:2087–2092.

Kimmel, A. R., and R. A. Firtel. 2004. Breaking symmetries: regulation of *Dictyostelium* development through chemoattractant and morphogen signal-response. *Curr. Opin. Genet. Devel.* **14**(5):540–549.

Leptin, M. 2005. Gastrulation movements: the logic and the nuts and bolts. *Devel. Cell* **8**:305–320.

O'Connor, M. B., et al. 2006. Shaping BMP morphogen gradients in the *Drosophila* embryo and pupal wing. *Development* **133**(2):183–193.

Schier A. F., and W. S. Talbot. 2005. Molecular genetics of axis formation in zebrafish. *Ann. Rev. Genet.* **39**:561-613.

Tomancak, P., et al. 2002. Systematic determination of patterns of gene expression during *Drosophila* embryogenesis. *Genome Biol.* 3(12):research0088.1—0088.14.

Gametogenesis and Fertilization

Chow, J.C., et al. 2005. Silencing of the mammalian X chromosome. *Ann. Rev. Genomics Human Genet.* **6**:69–92.

Hoodbhoy, T., and J. Dean. 2004. Insights into the molecular basis of sperm–egg recognition in mammals. *Reproduction* **127**:417–422.

Inoue, N., et al. 2005. The immunoglobulin superfamily protein Izumo is required for sperm to fuse with eggs. *Nature* **434**:234–238.

Lee, J. T. 2005. Regulation of X-chromosome counting by *Tsix* and *Xite* sequences. *Science* **309**:768–771.

Navarro, P., et al. 2005. *Tsix* transcription across the *Xist* gene alters chromatin confirmation without affecting *Xist* transcription: implications for X-chromosome inactivation. *Genes & Devel* **19**:1474–1484.

Sado, T., and A. C. Ferguson-Smith. 2005. Imprinted X inactivation and reprogramming in the preimplantation mouse embryo. *Human Molec. Genet.* **1**:R59–R64.

Trasler, J. M. 2006. Gamete imprinting: setting epigenetic patterns for the next generation. *Reprod. Fertil. Devel.* **18**:63–69.

Wasserman, P. M., et al. 2004. Egg–sperm interactions at fertilization in mammals. *Eur. J. Obstet. Gynecol. Reprod. Biol.* **115S**:S57–S60.

Cell Diversity and Patterning in Early Vertebrate Embryos

Freeman, M., and J. B. Gurdon. 2002. Regulatory principles of developmental signaling. *Ann. Rev. Cell Devel. Biol.* **18**:515–539.

Hirokawa, N., et al. 2006. Nodal flow and the generation of left-right asymmetry. *Cell* **125**:33–45.

Olson, E. N. 2006. Gene regulatory networks in the evolution and development of the heart. *Science* **313**:1922–1927.

Shiratori, H., and H. Hamada. 2006. The left-right axis in the mouse: from origin to morphology. *Development* **133**:2095–2104.

Stern, C. D. Evolution of the mechanisms that establish the embryonic axes. 2006. *Curr. Opin. Genet. Devel.* **16**:413–418.

Tam, P. P., D. A. Loebel, and S. S. Tanaka. 2006. Building the mouse gastrula: signals, asymmetry and lineages. *Curr. Opin. Genet. Devel.* **16**:419–425.

Control of Body Segmentation: Themes and Variations in Insects and Vertebrates

Akam, M. 1987. The molecular basis for metameric pattern in the *Drosophila* embryo. *Development* **101**:1–22.

Andrioli, L. P., et al. 2002. Anterior repression of a *Drosophila* stripe enhancer requires three position-specific mechanisms. *Development* **129**:4931–4940.

Aulehla, A., and B. G. Herrmann. 2004. Segmentation in vertebrates: clock and gradient finally joined. *Genes Devel.* **18**:2060–2067.

Carapuco, M., et al. 2005. Hox genes specify vertebral types in the presomitic mesoderm. *Genes & Devel.* **19**:2116–2121.

Chang, A. J., and D. Morisato. 2002. Regulation of Easter activity is required for shaping the Dorsal gradient in the *Drosophila* embryo. *Development* **129**:5635–5645.

Chopra,V. S., and R. K. Mishra. 2006. "Mir"acles in hox gene regulation. *Bioessays* **28**:445–448.

Dale, J. K., et al. 2006. Oscillations of the *snail* genes in the presomitic mesoderm coordinate segmental patterning and morphogenesis in vertebrate somitogenesis. *Devel. Cell* **10**:355–366.

Davis, G. K., and N. H. Patel. 2002. Short, long, and beyond: molecular and embryological approaches to insect segmentation. *Ann. Rev. Entomol.* **47**:669–699.

Dubrulle, J., and O. Pourquie. 2004. Coupling segmentation to axis formation. *Development* **131**:5783–5793.

Ephrussi, A., and D. St. Johnston. 2004. Seeing is believing: the bicoid morphogen gradient matures. *Cell* **116**:143–152.

Freeman, M., and J. B. Gurdon. 2002. Regulatory principles of developmental signaling. *Ann. Rev. Cell Devel. Biol.* **18**:515–539.

Fujioka, M., et al. 1999. Analysis of an even-skipped rescue transgene reveals both composite and discrete neuronal and early blastoderm enhancers, and multi-stripe positioning by gap gene repressor gradients. *Development* **126**:2527–2538.

Gridley, T. The long and short of it: somite formation in mice. 2006. *Devel. Dyn.* **235**:2330–2336.

Houchmandzadeh, B., E. Wieschaus, and S. Leibler. 2002. Establishment of developmental precision and proportions in the early *Drosophila* embryo. *Nature* **415**:798–802.

Johnstone, O., and P. Lasko. 2001. Translational regulation and RNA localization in *Drosophila* oocytes and embryos. *Ann. Rev. Genet.* **35**:365–406.

Lemons, D., and W. McGinnis. 2006. Genomic evolution of Hox gene clusters. *Science* **313**:1918–1922.

Morgan, R. 2006. Hox genes: a continuation of embryonic patterning? *Trends Genet.* **22**:67–69.

Sanson, B. 2001. Generating patterns from fields of cells: examples from *Drosophila* segmentation. *EMBO Rep.* **2**:1083–1088.

Stathopoulos, A., et al. 2002. Whole-genome analysis of dorsal-ventral patterning in the *Drosophila* embryo. *Cell* **111**:687–701.

Swalla, B. J. 2006. Building divergent body plans with similar genetic pathways. *Heredity* **97**(3):235–243.

Takeda, K., T. Kaisho, and S. Akira. 2003. Toll-like receptors. *Ann. Rev. Immunol.* **21**:335–376.

Wang, M., and P. W. Sternberg. 2001. Pattern formation during *C. elegans* vulval induction. *Curr. Topics Devel. Bio.* **51**:189–220.

Wellik, D. M., and M. R. Capecchi. 2003. Hox10 and Hox11 genes are required to globally pattern the mammalian skeleton. *Science* **301**:363–367.

Cell-Type Specification in Early Neural Development

Colas, J-F., and G. C. Schoenwolf. 2003. Differential expression of two cell adhesion molecules, Ephrin-A5 and Integrin alpha6, during cranial neurulation in the chick embryo. *Devel. Neurosci.* **25**:357–365.

Cooke, J. E., and C. B. Moens. 2002. Boundary formation in the hindbrain: Eph only it were simple. *Trends Neurosci.* **25**:260–267.

Grandbarbe, L., et al. 2003. Delta-Notch signaling controls the generation of neurons/glia from neural stem cells in a stepwise process. *Development* **130**:1391–1402.

Jessell, T. M. 2000. Neuronal specification in the spinal cord: inductive signals and transcriptional codes. *Nature Rev. Genet.* **1**:20–29.

Kriegstein, A. R., and S. C. Noctor. 2004. Patterns of neuronal migration in the embryonic cortex. *Trends Neurosci.* **27**:392–399.

Miao, H., et al. 2001. Activation of EphA receptor tyrosine kinase inhibits the Ras/MAPK pathway. *Nature Cell Biol.* **3**:527–530.

Santiago, A., and C. A. Erickson. 2002. Ephrin-B ligands play a dual role in the control of neural crest cell migration. *Development* **129**:3621–3632.

Growth and Patterning of Limbs

Boulet, A. M., et al. 2004. The roles of *Fgf4* and *Fgf8* in limb bud initiation and outgrowth. *Devel. Biol.* **273**:361–372.

Colas, J-F., and G. C. Schoenwolf. 2001. Towards a cellular and molecular understanding of neurulation. *Devel. Dyn.* **221**:117–145.

Han, M., et al. 2005. Limb regeneration in higher vertebrates: developing a roadmap. *Anat. Rec. B New Anat.* **287**:14–24.

Li, C., et al. 2005. FGFR1 function at the earliest stages of mouse limb development plays an indispensable role in subsequent autopod morphogenesis. *Development* **132**:4755–4764.

Martin, G. 2001. Making a vertebrate limb: new players enter from the wings. *Bioessays* **23**:865–868.

Minguillon, C., J. Del Buono, and M. P. Logan. 2005. *Tbx5* and *Tbx4* are not sufficient to determine limb-specific morphologies but have common roles in initiating limb outgrowth. *Devel. Cell* **8**:75–84.

Moon, R. T., et al. 2002. The promise and perils of Wnt signaling through beta-catenin. *Science* **296**:1644–1646.

Nybakken, K., and N. Perrimon. 2002. Hedgehog signal transduction: recent findings. *Curr. Opin. Genet. Devel.* **12**:503–511.

Pandur, P., D. Maurus, and M. Kuhl. 2002. Increasingly complex: new players enter the Wnt signaling network. *Bioessays* **24**:881–884.

Pires-daSilva, A., and R. J. Sommer. 2003. The evolution of signaling pathways in animal development. *Nature Rev. Genet.* **4**:39–49.

Rubin, C., et al. 2003. Sprouty fine-tunes EGF signaling through interlinked positive and negative feedback loops. *Curr. Biol.* **13**:297–307.

Salsi, V., and V. Zappavigna. 2006. *Hoxd13* and *Hoxa13* directly control the expression of the EphA7 Ephrin tyrosine kinase receptor in developing limbs. *J. Biol. Chem.* **281**:1992–1999.

Tickle, C. 2006. Making digit patterns in the vertebrate limb. *Nature Rev. Molec. Cell Biol.* **7**:45–53.

Tickle, C. 2003. Patterning systems—from one end of the Lim to the other. *Devel. Cell* **4**:449–458.

Tickle, C., and A. Munsterberg. 2001. Vertebrate limb development: the early stages in chick and mouse. *Curr. Opin. Genet. Devel.* **11**:476–481.

Zuniga, A. 2005. Globalization reaches gene regulation: the case for vertebrate limb development. *Curr. Opin. Genet. Devel.* **15**:403–409.

C L A S S I C E X P E R I M E N T 2 2 . 1

USING LETHAL MUTATIONS TO STUDY DEVELOPMENT

C. Nüsslein-Volhard and E. Wieschaus, 1980, Nature **287**:795

One of the most fascinating questions in developmental biology concerns the proper formation of an embryo. How does a fertilized egg "know" how to form a complex organism? Scientists have puzzled over how to address this question for a long time. In 1980, Christiane Nüsslein-Volhard and Eric Wieschaus first reported studies with the fruit fly *Drosophila melanogaster* in which a genetic approach was used to address this question.

Background

To examine complex processes such as development of an embryo from a fertilized egg, biologists often collect mutant organisms that differ from the normal (wild-type) organism. To apply this genetic approach to developmental biology, geneticists first look for mutant organisms that display an obvious defect in overall formation. Early work uncovered a number of genes involved in the development of the fruit fly *Drosophila melanogaster.* In the first genes examined, the mutations resulted in the birth of flies with obvious physical defects, such as the presence of an extra set of wings. Because this approach relied on examining viable flies with physical malformations, it missed many developmentally important genes that, when mutated, result in the death of the fly embryo.

In the late 1970s, Nüsslein-Volhard and Wieschaus began their pioneering work on the development of *Drosophila* embryos. They sought to identify as many genes in the developmental process as possible by looking for genes that, when mutated, resulted in the death of the embryonic fly. Their work unveiled several key genes active in the early development of not only *Drosophila,* but higher organisms as well.

The Experiment

Geneticists develop systematic methods, known as genetic screens, to search for mutations that affect biological processes. Nüsslein-Volhard and Wieschaus had to consider several previous observations on *Drosophila* development when they designed their screen. First, they knew that genes expressed in the egg, called *maternal-effect genes,* as well as genes expressed after fertilization in the developing embryo, called *zygotically active genes,* controlled the early development of an embryo. They chose to focus on isolating mutations that affect the zygotically active genes. Second, they had to consider that the *Drosophila* genome is diploid, which means that the progeny receive a copy of each gene from both parents. Scientists had previously demonstrated that *Drosophila* required only a single wild-type copy of most genes in order to develop into a viable fly. This made it likely that recessive mutations in developmentally active genes would *not* result in embryonic death, the phenotypic screen used by Nüsslein-Volhard and Wieschaus. Therefore, they had to breed mutant *Drosophila* to obtain flies that were homozygous for the mutations of interest.

The overall mutation rate in a naturally occurring population is quite low. If a geneticist were to search for mutants in a natural population, he or she would have to examine a large number of individuals. To circumvent this difficulty, Nüsslein-Volhard and Wieschaus induced mutations in a *Drosophila* population at the onset of the screen by feeding a mutagenic chemical to male flies and mating them to a genetically defined population of wild-type female flies in a process known as a genetic cross. The resulting progeny would be heterozygous for any mutations on the chromosomes they received from the father. The heterozygotes were then bred as separate lines, in essence isolating each new mutation in a separate fly stock. Flies within each stock then were crossed with each other to generate homozygous embryos. If the mutation affected a gene needed for embryogenesis, the homozygous embryos would die but could be examined for phenotype. The mutant gene, however, would not be lost as it could be recovered in the heterozygous flies of that same stock.

Using this screen, Nüsslein-Volhard and Wieschaus amassed a large collection of mutant flies. The next step was to assign the various mutants to specific classes, based on their phenotype. They focused on the segmentation of the larvae. Whereas all mutants in their screen necessarily displayed the phenotype of embryonic lethality, they differed greatly in their segmentation defects. To classify these defects, Nüsslein-Volhard and Wieschaus examined the larvae under the microscope. They compared the body pattern of a wild-type viable larva to those of the embryonic lethal mutants. By comparing these patterns, they uncovered three classes of genes that affect segmentation, which they called *gap, pair-rule,* and *segment-polarity.*

Gap mutants are missing up to eight segments from the overall body, which results in a smaller body due to the death of some of the embryo cells. Three mutants—*knirps, hunchback,* and the previously characterized *Krüppel*—fell into this class, as shown in the accompanying photographs. The next class of mutants, the pair-rule mutants, had deletions of alternating segments of the body, which caused roughly half-normal-size larval bodies to form. Six

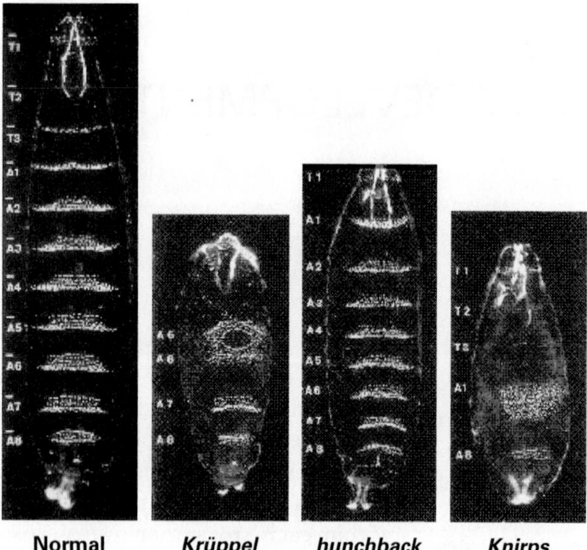

◀ **FIGURE 1** A viable, wild-type larva (Normal) is compared with three larvae that have mutations in the gap genes *Krüppel, hunchback,* or *knirps.* Thoracic segments are labeled T1–T3, whereas abdominal segments are designated A1–A8. Gap mutants are missing entire segments from the body plan, as illustrated by the labeled segments on the left. [From C. Nüsslein-Volhard and E. Wieschaus, 1980, *Nature* **287:**795.]

previously uncharacterized mutants—*paired, even-skipped, odd-skipped, odd-paired, fushi tarazu,* and *runt*—were placed in this class. The final set of mutants, the segment-polarity mutants, had the same overall number of segments as wild-type larvae. But a part of the body pattern within each segment was deleted in each mutant. The deleted portion of the pattern within each segment was replaced by a mirror image of the portion that remained. Nüsslein-Volhard and Wieschaus's initial screen uncovered six mutants of this class, three of which—*gooseberry, hedgehog,* and *patched*—had not been previously observed.

Discussion

In the first published report of their screen, Nüsslein-Volhard and Wieschaus described 15 mutations that affected segmentation. Of these, only five were in previously identified genes. When they completed the study—often referred to as the Heidelberg screens—they had identified 139 different genes that, when mutated, resulted in embryonic death. These mutants fell into 17 different classes including the ones described here. Together these genes act in a field of equivalent cells and direct each cell to take on a fate appropriate to its position within the field. In this way some cells form segment anterior, some make bristles, some make sensory organs, and so forth. Genes that control these sorts of organizing cell-fate decisions are called *pattern-formation genes.* As molecular techniques evolved, scientists cloned many of these genes and characterized their gene products. These mutants formed the base for the next quarter century of research into the development of *Drosophila.* Moreover, many vertebrate pattern-formation genes are close homologs of the corresponding fly genes, and use similar molecular mechanisms to organize cells, tissues, and organs. Thus the work of Nüsslein-Volhard and Wieschaus greatly advanced the study of development in all vertebrates, including ourselves.

The majority of genes identified by Nüsslein-Volhard and Wieschaus encode transcription factors, but their screen also uncovered genes encoding signaling molecules, receptors, enzymes, adhesion molecules, cytoskeleton proteins, and some proteins whose functions remain unknown. Mammalian genes related to some of the *Drosophila* genes uncovered by Nüsslein-Volhard and Wieschaus subsequently were found to be important in human diseases such as cancer and birth defects. In 1995, the Nobel Foundation awarded its prize for Physiology and Medicine to Nüsslein-Volhard and Wieschaus for their pioneering work.

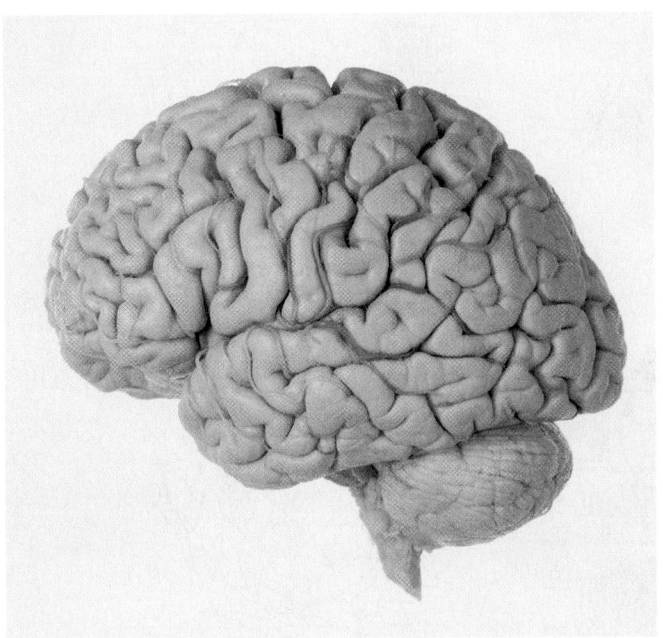

The convoluted surface of the cerebrum and (*lower right*) the cerebellum. [© Ralph Hutchings/Visuals Unlimited]

CHAPTER

23

NERVE CELLS

When humans wish to view themselves as the pinnacle of evolution, the assertion is usually related to the brain, since many of our other abilities fare poorly in comparison with other animals. The complexity of the human brain is staggering and seems adequate to account for its amazing abilities. In the 1.3-kg adult human brain (78 percent of which is water!), there are about 10^{11} nerve cells, called neurons. The number of human brain neurons is comparable to the estimated 10^{11} stars in our galaxy, the Milky Way. The neurons in one human brain are connected by some 10^{14} synapses, the junction points where two or more neurons communicate. With 6.5×10^9 people on the earth, there are about 6.5×10^{23} human synapses in existence, which is also about the total number of stars in the universe . . . so far as we know. Using our neurons, we can keep searching for more.

Right this moment you are vigorously employing neurons to detect and interpret visual information. Neurons gobble ATP, made exclusively from glucose, at a tremendous rate. Although the brain is only about 2 percent of the body's mass, it uses about 20 percent of the body's resting energy. This extensive brain energy is used to drive electrical signaling along neurons, which are often elongated cells, and chemical signaling between them. The electrical pulses that travel along neurons are called action potentials, and information is encoded as the frequency at which action potentials are fired. Owing to the speed of electrical transmission, neurons are champion signal transducers, much faster than cells that secrete hormones or developmental signaling proteins. The rapidity of neural signaling makes it possible to handle large amounts of information quickly. The amazing complexity of our neural network makes possible sophisticated perception, analysis, and response, and forms the cellular machinery underlying instinct, learning, memory, and emotion.

In this chapter we will focus on neurobiology at the cell and molecular level. We will start by looking at the general architecture of neurons and at how they carry signals (Figure 23-1). Next, we will look at ion flow, channel proteins, and membrane properties: how electrical pulses move rapidly along neurons. Third, we will examine communication between neurons: electrical signals traveling along a cell must be translated into a chemical pulse between cells and then back into an electrical signal in the receiving cells. In the

OUTLINE

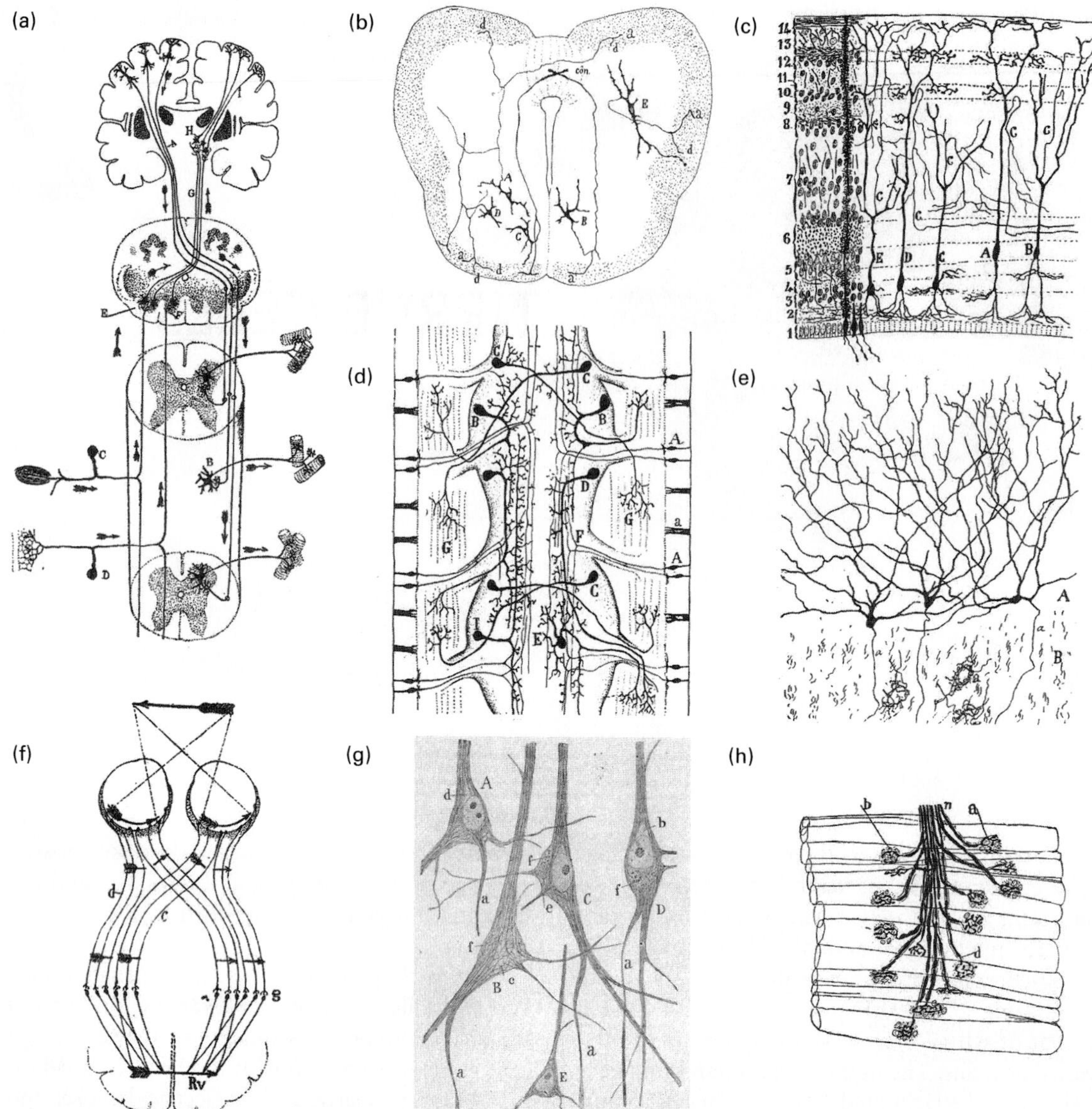

▲ **FIGURE 23-1 Illustrations of the nervous system and nerve cells by Ramon y Cajal (1852–1934).** (a) Sensory and motor nervous systems of a worm (*A* = sensory cells of the skin, *C* = motor cells with crossed processes, *G* = terminal ramifications of a motor neuron on a muscle). (b) Cross section of a spinal cord. (c) Section through the optic lobe of a chameleon (numbers indicate layers from deep to superficial; *A, C, D* = Shepherd's crook cells). (d, e) Cells in the nuclear layer of the kitten superior colliculus. (f) Chiasm (crossing place) and central projection of human visual pathways (*c* = crossed bundle of the optic nerve, *d* = large uncrossed bundle, *Rv* = projection of the mental image—the arrow—onto visual areas of the cerebral cortex). (g) Neurons in the fusiform layer of the human motor cortex (*A, E* = pyramidal cells, *a* = axon). (h) Motor neurons terminating on rabbit muscles (*a* = terminal arborization of an axon, *d* = point where myelin sheath ends, *n* = branch of a nerve).

fourth section we will examine neurons in sensory tissues: how visual, touch, auditory, taste, and olfactory information is collected, processed, and interpreted. Finally we will explore the systems that guide nerves as they grow: how signaling systems initially organize the "wiring diagram."

The speed, precision, and integrative power of neural signaling makes possible accurate and timely sensory perception of a swiftly changing environment, and is the basis of high-throughput signal processing and analysis. We refer to the processing as "thinking," and molecular cell biology is at the heart of it.

23.1 Neurons and Glia: Building Blocks of the Nervous System

In this section we take an initial look at the structure of neurons and how they propagate electrical and chemical signals. **Neurons** are distinguished by their elongated, asymmetric shape, by their highly localized proteins and organelles, and most of all by a set of proteins that controls the flow of ions across the plasma membrane. Because one neuron can respond to the inputs from multiple neurons, generate electrical signals, and transmit the signals to multiple neurons, a nervous system has considerable powers of signal analysis. For example, a neuron might pass on a signal only if it receives five simultaneous activating signals from input neurons. The receiving neuron measures both the total *amount* of incoming signal and whether the five signals are roughly *synchronous*. Input from one neuron to another can be either excitatory—combining with other signals to trigger electrical transduction in the receiving cell—or inhibitory, discouraging such transmission. Thus the properties and connections of individual neurons set the stage for integration and refinement of information. We will begin by looking at how signals are received and sent, and in subsequent parts of the chapter we will look at the molecular details of the machinery involved.

Information Flows Through Neurons from Dendrites to Axons

Neurons arise from roughly spherical *neuroblast* precursors. Newly born neurons can migrate long distances while still in the form of simple round cells before growing into dramatically elongated cells. Fully differentiated neurons take many forms, but generally share certain key features (see Figure 23-1). The nucleus is found in a rounded part of the cell called the *cell body* (Figure 23-2). Branching cell processes called **dendrites** (from the Greek for "treelike") are found at one end, and are the main structure where signals are received from other neurons via synapses. Incoming signals are also received at synapses that form on neuronal cell bodies. Neurons often have extremely long dendrites with complex branches, particularly in the central nervous system. This allows them to form synapses with, and receive signals from, a large number of other neurons—up to tens of thousands. Thus the converging dendritic branches allow signals from many cells to be received and integrated by a single neuron.

When a neuron is first differentiating, the end of the cell opposite the dendrites undergoes dramatic outgrowth to form a long extended arm called the **axon,** which is essentially a transmission wire. The growth of axons must be controlled so that proper connections are formed, a process called axon guidance that is discussed in Section 23.5. The diameters of axons vary from just a micrometer in certain nerves of the human brain to a millimeter in the giant fiber of the squid. Axons can be meters in length (in giraffe necks, for example), and are often partly covered with electrical insulation called the **myelin sheath** (see Figure 23-2b), which is made up of cells called **glia.** The insulation speeds electrical transmission and prevents short circuits. The short, branched ends of the axon at the opposite end of the neuron from the dendrites are called the *axon termini.* This is where signals are passed along to the next neuron. The asymmetry of the neuron, with dendrites at one end and axon termini at the other, is indicative of the unidirectional flow of information from dendrites to axons.

For all the impressiveness of neurons, they are a minority population of cells in the human brain. Glia, cells that play many roles in the brain but do not themselves conduct electrical impulses, outnumber neurons by about 10 to 1 in humans. One of their roles is to produce myelin sheaths, but they have other important roles. Much current work is devoted to learning how glia build the myelin insulators that control neuronal electrical transmission, provide growth factors and other signals to neurons, receive signals from neurons, and influence the formation of synapses.

Information Moves as Pulses of Ion Flow Called Action Potentials

Nerve cells are members of a class of *excitable cells,* which also includes muscle cells, cells in the pancreas, and some others. The term indicates that the cells can build up a voltage across their plasma membranes, the **membrane potential** and

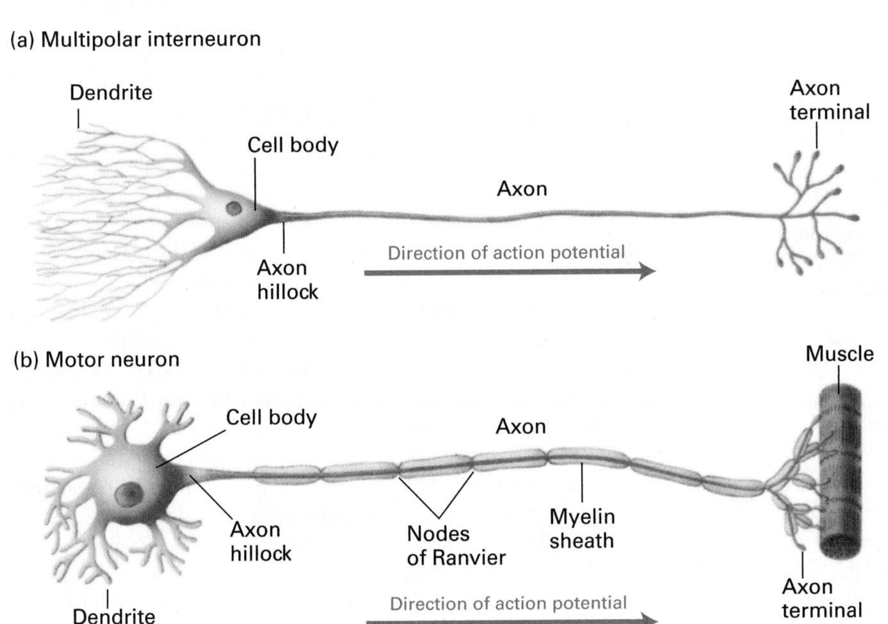

◀ **FIGURE 23-2 Typical morphology of two types of mammalian neurons.** Action potentials arise in the axon hillock and are conducted toward the axon terminus. (a) A multipolar interneuron has profusely branched dendrites, which *receive* signals at synapses with several hundred other neurons. Small voltage changes imparted by inputs in the dendrites can sum to give rise to the more massive action potential, which starts in the hillock. A single long axon that branches laterally at its terminus *transmits* signals to other neurons. (b) A motor neuron innervating a muscle cell typically has a single long axon extending from the cell body to the effector cell. In mammalian motor neurons, an insulating sheath of myelin usually covers all parts of the axon except at the nodes of Ranvier and the axon terminals. The myelin sheath is composed of cells called *glia*.

that this voltage can be discharged (allowed to come back to zero voltage or even swing to positive) in various ways and for various purposes (Chapter 11). The voltage in a typical neuron, called the *resting potential* because it is the state when no signal is in transit, is established by ion pumps in the plasma membrane. The pumps use energy, in the form of ATP, to move positively charged ions out of the cell. The result is a net negative charge inside the cell compared with the outside environment. A typical resting potential is −60 mV.

Neurons have a language all their own. The signals take the form of brief local voltage changes, from negative inside to positive, an event designated **depolarization.** A powerful surge of depolarizing voltage change, moving from one end of the neuron to the other, is called an **action potential.** "Depolarization" is somewhat of a misnomer, since the neuron goes from negative inside to neutral to positive inside, which could be accurately described as depolarization followed by the opposite polarization (Figure 23-3). At the peak of an action potential, the membrane potential can be as much as +50 mV (inside positive), a net change of ≈110 mV. As we shall see in greater detail in Section 23.2, the voltage change (which is eventually added to other voltage changes to create the action potential) begins at the dendrite end of the cell in response to inputs from other cells and moves along the axon to the axon terminus. Action potentials move at speeds up to 100 meters per second. In humans, for instance, axons may be more than a meter long, yet it takes only a few milliseconds for an action potential to move along their length. Neurons can fire repeatedly after a brief recovery period, for example, every 4 ms, as in Figure 23-3. After the action potential passes a sector of a neuron, channel proteins and pumps restore the negative-inside resting potential (*repolarization*). The restoration process chases the action potential down the axon to the terminus, leaving the neuron ready to signal again. Action potentials are all or none. Once the threshold to start one is reached, a full firing occurs. The signal information is therefore carried primarily not by the intensity of the action potentials, but by the timing and frequency of them.

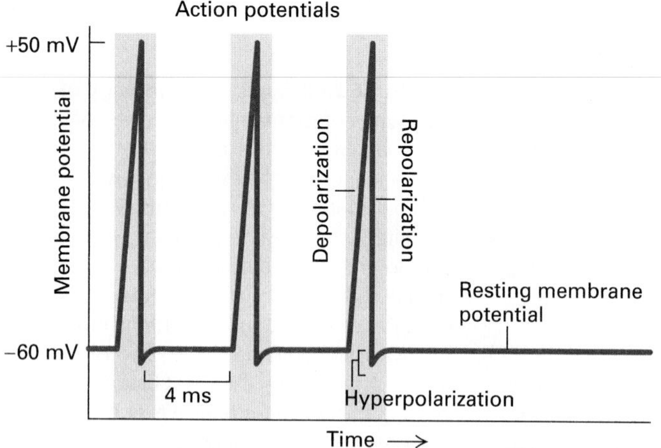

▲ EXPERIMENTAL FIGURE 23-3 **Recording of an axonal membrane potential over time reveals the amplitude and frequency of action poentials.** An action potential is a sudden, transient depolarization of the membrane, followed by repolarization to the resting potential of about −60 mV. The axonal membrane potential can be measured with a small electrode placed into it (see Figure 11-18). This recording of the axonal membrane potential in this neuron shows that it is generating one action potential about every 4 milliseconds.

Some excitable cells are not neurons. Muscle contraction is triggered by motor neurons that synapse directly with excitable muscle cells (see Figure 23-2b). Insulin secretion from the beta cells of the pancreas is triggered by neurons. In both cases the activating event involves an opening of plasma

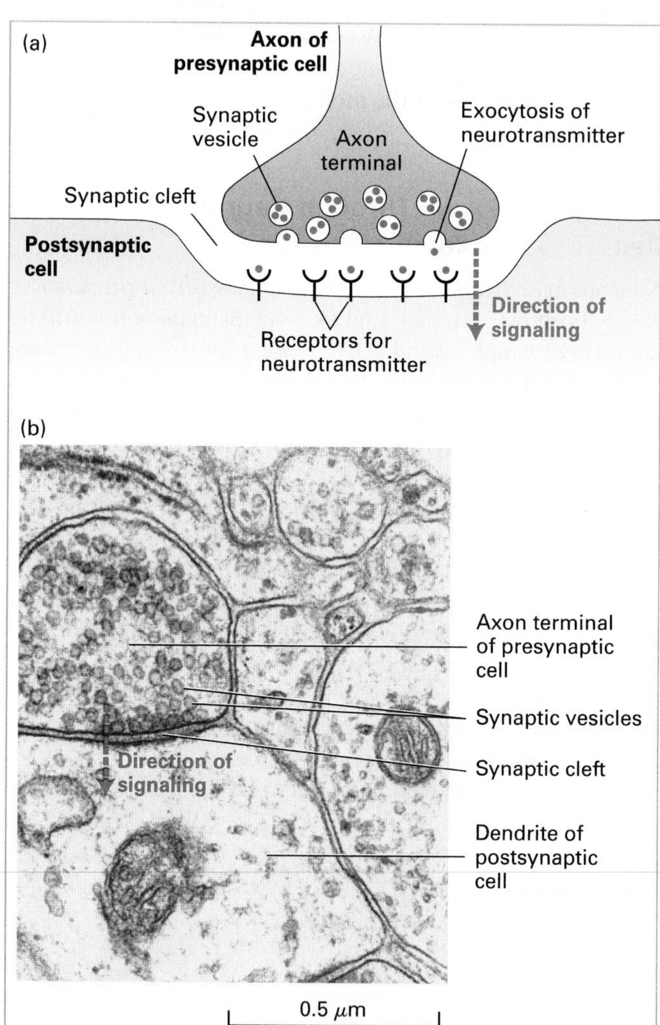

▲ **FIGURE 23-4 A chemical synapse.** (a) A narrow region—the synaptic cleft—separates the plasma membranes of the presynaptic and postsynaptic cells. Arrival of action potentials at a synapse causes release of neurotransmitters (red circles) by the presynaptic cell, their diffusion across the synaptic cleft, and their binding by specific receptors on the plasma membrane of the postsynaptic cell. Generally these signals depolarize the postsynaptic membrane (making the potential inside less negative), tending to induce an action potential in it. (b) Electron micrograph shows a dendrite synapsing with an axon terminal filled with synaptic vesicles. In the synaptic region, the plasma membrane of the presynaptic cell is specialized for vesicle exocytosis; synaptic vesicles containing a neurotransmitter are clustered in these regions. The opposing membrane of the postsynaptic cell (in this case, a neuron) contains receptors for the neurotransmitter. [Part (b) from C. Raine et al., eds., 1981, *Basic Neurochemistry,* 3d ed., Little, Brown, p. 32.]

membrane channels that causes changes in the transmembrane flow of ions and in the electrical properties of the regulated cells.

Information Flows Between Neurons via Synapses

What starts an action potential? Axon termini from one neuron are closely apposed to dendrites of another, at the junction called a **synapse** (Figure 23-4). The axon termini of the *presynaptic cell* use exocytosis to release small molecules called **neurotransmitters.** Neurotransmitters, glutamate or acetylcholine, for example, diffuse across the synapse in about 0.5 ms and bind to receptors on the dendrite of the adjacent neuron. Binding of neurotransmitter triggers opening or closing of specific ion channels in the plasma membrane of *postsynaptic cell* dendrites, leading to changes in the membrane potential at this point. The ensuing local depolarization, if large enough, triggers an action potential. Transmission is unidirectional, from the axon termini of the presynaptic cell to dendrites of the postsynaptic cell. In some synapses, the effect of the neurotransmitters is to hyperpolarize and therefore lower the likelihood of an action potential in the postsynaptic cell. A single axon in the central nervous system can synapse with many neurons and induce responses in all of them simultaneously. Conversely, sometimes multiple neurons must act on the postsynaptic cell roughly synchronously to have a strong enough impact to trigger an action potential. Neuronal integration of depolarizing and hyperpolarizing signals determines the likelihood of an action potential.

Thus neurons employ a combination of extremely fast electrical transmission *along* the axon with rapid chemical communication *between* cells. Now we will look at how a chain of neurons, a circuit, can achieve a useful function.

The Nervous System Uses Signaling Circuits Composed of Multiple Neurons

In complex multicellular animals, such as insects and mammals, various types of neurons form signaling circuits. A *sensory neuron* reports an event that has happened, like the arrival of a flash of light or the movement of a muscle. A *motor neuron* carries a signal to a muscle to stimulate its contraction (Figure 23-5, and see Figure 23-2b). An *interneuron* bridges other neurons, sometimes allowing integration or divergence of signals, sometimes extending the reach of a signal. In a simple type of circuit called a *reflex arc* interneurons connect multiple sensory and motor neurons, allowing one sensory neuron to affect multiple motor neurons and one motor neuron to be affected by multiple sensory neurons; in this way interneurons integrate and enhance reflexes. For example, the knee-jerk reflex in humans (see Figure 23-5) involves a complex reflex arc in which one muscle is stimulated to contract while another is inhibited from contracting. The reflex also sends information to the brain to announce what happened. Such circuits allow an organism to respond to a sensory input by the coordinated action of sets of muscles that together achieve a single purpose.

These simple signaling circuits, however, do not directly explain higher-order brain functions such as reasoning, computation, and memory development. Typical neurons in the brain receive signals from up to a thousand other neurons and, in turn, can direct chemical signals to many other neurons. The output of the nervous system depends on its circuit

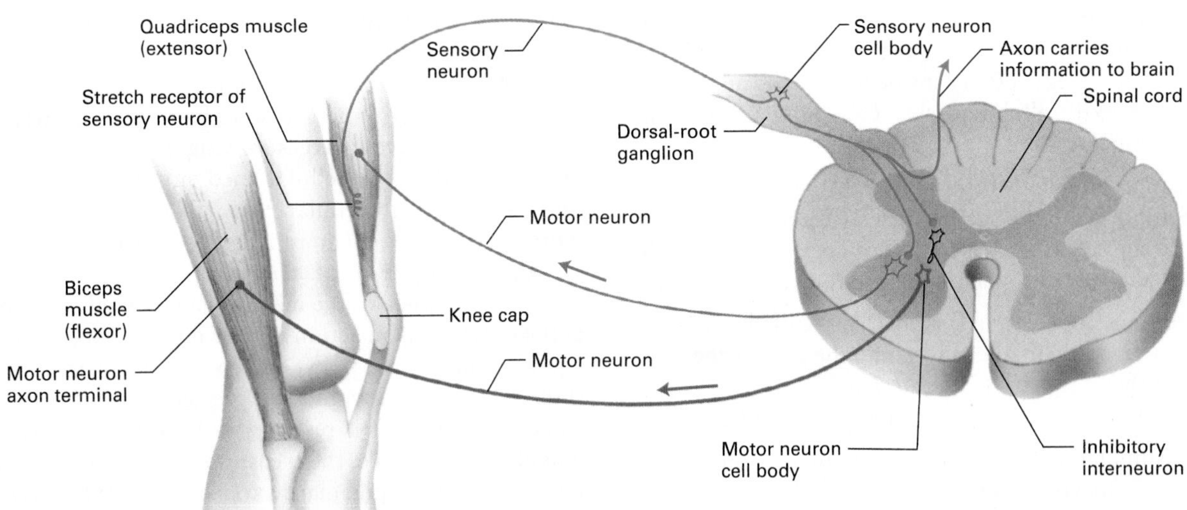

▲ **FIGURE 23-5 The knee-jerk reflex.** A tap of the hammer stretches the quadriceps muscle, thus triggering electrical activity in the stretch receptor sensory neuron. The action potential, traveling in the direction of the top blue arrow, sends signals to the brain so we are aware of what is happening, and also to two kinds of cells in the dorsal-root ganglion that is located in the spinal cord. One cell, a motor neuron that connects back to the quadriceps (red), stimulates muscle contraction so that you kick the person who hammered your knee. The second connection activates, or "excites," an inhibitory interneuron (black). The interneuron has a damping effect, blocking activity by a flexor motor neuron (green) that would, in other circumstances, activate the hamstring muscle that opposes the quadriceps. In this way, relaxation of the hamstring is coupled to contraction of the quadriceps. This is a reflex because movement requires no conscious decision.

properties—the wiring, or interconnections, between neurons and the strength of these interconnections. Complex aspects of the nervous system, such as vision and consciousness, cannot be understood at the single-cell level, but only at the level of networks of nerve cells that can be studied by techniques of systems analysis. The nervous system is constantly changing; alterations in the number and nature of the interconnections between individual neurons occur, for example, in the formation of new memories.

KEY CONCEPTS OF SECTION 23.1

Neurons and Glia: Building Blocks of the Nervous System

- Neuron are highly asymmetric cells composed of dendrites at one end, a cell body containing the nucleus, a long axon, and axon termini.
- Neurons carry information from one end to the other using pulses of ion flow across the plasma membrane. Branched cell processes, dendrites, at one end of the cell receive chemical signals from other neurons, triggering ion flow. The electrical signal moves rapidly to axon termini at the other end of the cell (see Figure 23-2).
- Glial cells are ten times as abundant as neurons and serve many purposes such as building the insulation that coats neurons and supporting the formation of new synapses.
- A resting neuron carrying no signal has protein pumps that move ions across the plasma membrane. The movement of ions such as K^+ and Na^+ and Cl^- creates a net negative charge inside the cell. This voltage is called the resting potential and usually is about −60 mV (see Figure 23-3).
- If a stimulus causes channel proteins to open so that ions can flow more freely, a strong pulse of voltage change may pass down the neuron from dendrites to axon termini. The cell goes from being −60 mV inside to +50 mV inside, relative to the extracellular world. This pulse is called an action potential.
- The action potential travels down the neuron because a change in voltage near the dendrites triggers a change in voltage in the cell body, which in turn does the same to the proximal and then distal axon, and so forth.
- Neurons connect across small gaps called synapses. Since an action potential cannot jump the gap, at the axon termini of the presynaptic cell the signal is converted from electrical to chemical to stimulate the postsynaptic cell.
- Upon stimulation by an action potential, axon termini release, by exocytosis, small packets of chemicals called neurotransmitters. Neurotransmitters diffuse across the synapse and bind to receptors on the dendrites on the other side of the synapse. These receptors initiate a new axon potential in the postsynaptic cell (see Figure 23-4).
- Neurons form circuits. They may consist of sensory neurons, interneurons, and motor neurons, as in the knee-jerk response (see Figure 23-5).

23.2 Voltage-Gated Ion Channels and the Propagation of Action Potentials in Nerve Cells

Action potentials are propagated because a change in voltage in one part of the cell triggers the opening of channels in the next section of the cell. *Voltage-gated channels* therefore lie at the heart of neural transmission (Chapter 11). In this section, we first introduce some of the key properties of action potentials, which move rapidly along the axon. We then describe how the voltage-gated channels responsible for propagating action potentials in neurons operate. Thus electric signals carry information within a nerve cell, while chemical signals, discussed in the next section, carry information from one neuron to another, or from a neuron to a muscle or other target cell.

The Magnitude of the Action Potential Is Close to E_{Na}

Operation of the Na^+/K^+ pump generates a high concentration of K^+ and a low concentration of Na^+ in the cytosol, relative to those in the extracellular medium (Chapter 11). The subsequent outward movement of K^+ ions through nongated K^+ channels is driven by the K^+ concentration gradient (cytosol > medium), generating the resting membrane potential. The entry of Na^+ ions into the cytosol from the medium also is thermodynamically favored, driven by the Na^+ concentration gradient (medium > cytosol) and the inside-negative membrane potential (see Figure 11-24). However, most Na^+ channels in the plasma membrane are closed in resting cells, so little inward movement of Na^+ ions can occur (Figure 23-6a).

If enough Na^+ channels open, the resulting influx of Na^+ ions will more than compensate for the efflux of K^+ ions through open resting K^+ channels. The result would be a *net* inward movement of cations, generating an excess of positive charges on the cytosolic face and a corresponding excess of negative charges (owing to the Cl^- ions "left behind" in the extracellular medium after influx of Na^+ ions) on the extracellular face (Figure 23-6b). In other words, the plasma membrane is depolarized to such an extent that the inside face becomes positive.

The magnitude of the membrane potential at the peak of depolarization in an action potential is very close to the Na^+ equilibrium potential E_{Na} given by the Nernst equation (Equation 11-2), as would be expected if opening of voltage-gated Na^+ channels is responsible for generating action potentials. For example, the measured peak value of the action potential for the squid giant axon is 35 mV, which is close to the calculated value of E_{Na} (55 mV) based on Na^+ concentrations of 440 mM outside and 50 mM inside. The relationship between the magnitude of the action potential and the concentration of Na^+ ions inside and outside the cell has been confirmed experimentally. For instance, if the concentration of Na^+ ions in the solution bathing the squid axon is reduced to one-third of normal, the magnitude of the depolarization is reduced by 40 mV, nearly as predicted.

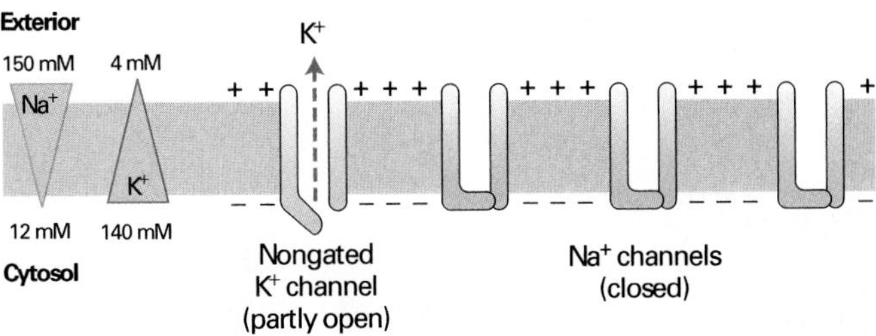

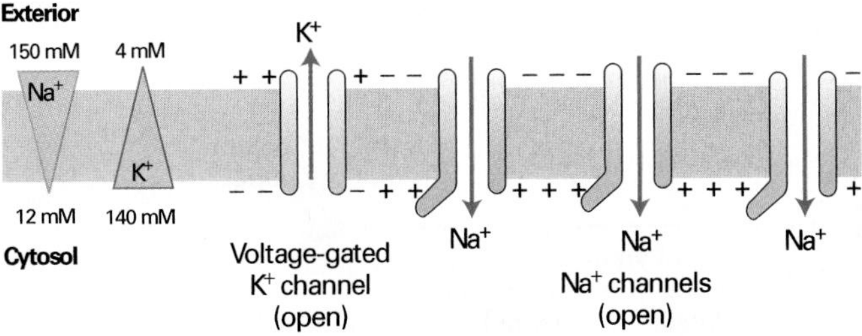

▲ **FIGURE 23-6 Depolarization of the plasma membrane due to opening of gated Na^+ channels.** (a) In resting neurons, a type of nongated K^+ channel is partially open, but the more numerous gated Na^+ channels are closed. The movement of K^+ ions outward establishes the inside-negative membrane potential characteristic of most cells. (b) Opening of gated Na^+ channels permits an influx of sufficient Na^+ ions to cause a reversal of the membrane potential. In the depolarized state, voltage-gated K^+ channels open and subsequently repolarize the membrane. Note that the flows of ions are too small to have much effect on the overall concentration of either Na^+ or K^+ in the cytosol or exterior fluid.

Sequential Opening and Closing of Voltage-Gated Na^+ and K^+ Channels Generate Action Potentials

The cycle of changes in membrane potential and return to the resting value that constitutes an action potential lasts 1–2 milliseconds and can occur hundreds of times a second in a typical neuron (see Figure 23-3). These cyclical changes in the membrane potential result from the sequential opening and closing first of *voltage-gated Na^+ channels* and then of *voltage-gated K^+ channels*. The role of these channels in the generation of action potentials was elucidated in classic studies done on the giant axon of the squid, in which multiple microelectrodes can be inserted without causing damage to the integrity of the plasma membrane. However, the same basic mechanism is used by all neurons.

Voltage-Gated Na^+ Channels As just discussed, voltage-gated Na^+ channels are closed in resting neurons. A small depolarization of the membrane causes a conformational change in these channel proteins that opens a gate on the cytosolic surface of the pore, permitting Na^+ ions to pass through the pore into the cell. The greater the initial membrane depolarization, the more voltage-gated Na^+ channels that open and the more Na^+ ions that enter.

As Na^+ ions flow inward through opened channels, the excess positive charges on the cytosolic face and negative charges on the exoplasmic face diffuse a short distance away from the initial site of depolarization. This *passive spread* of positive and negative charges depolarizes (makes the inside less negative) adjacent segments of the plasma membrane, causing opening of additional voltage-gated Na^+ channels in these segments and an increase in Na^+ influx. As more Na^+ ions enter the cell, the inside of the cell membrane becomes more depolarized, causing the opening of yet more voltage-gated Na^+ channels and even more membrane depolarization, setting into motion an explosive entry of Na^+ ions. For a fraction of a millisecond, the permeability of this region of the membrane to Na^+ becomes vastly greater than that for K^+, and the membrane potential approaches E_{Na}, the equilibrium potential for a membrane permeable only to Na^+ ions. As the membrane potential approaches E_{Na}, however, further net inward movement of Na^+ ions ceases, since the concentration gradient of Na^+ ions (outside > inside) is now offset by the inside-positive membrane potential E_{Na}. The action potential is, at its peak, close to the value of E_{Na}.

Figure 23-7 schematically depicts the critical structural features of voltage-gated Na^+ channels and the conformational changes that cause their opening and closing. In the resting state, a segment of the protein on the cytosolic face—the *gate*—obstructs the central pore, preventing passage of ions. A small depolarization of the membrane triggers movement of positively charged *voltage-sensing* α helices toward the exoplasmic surface, causing a conformational change in the gate that opens the channel and allows ion flow. After about 1 ms, further Na^+ influx is prevented by movement of the cytosol-facing *channel-inactivating segment* into the open channel. As long as the membrane remains depolarized, the channel-inactivating segment remains in the channel opening; during this *refractory period,* the channel is inactivated and cannot be reopened. A few milliseconds after the inside-negative resting potential is reestablished, the channel-inactivating segment swings away from the pore and the channel returns to the closed resting state, once again able to be opened by depolarization.

Voltage-Gated K^+ Channels The repolarization of the membrane that occurs during the refractory period is due largely to opening of voltage-gated K^+ channels. The subsequent increased efflux of K^+ from the cytosol removes the excess positive charges from the cytosolic face of the plasma membrane (i.e., makes it more negative), thereby restoring the inside-negative resting potential. Actually, for a brief instant the membrane becomes hyperpolarized. At the peak of this **hyperpolarization,** the potential approaches E_K, which is more negative than the resting potential (see Figure 23-3).

Opening of the voltage-gated K^+ channels is induced by the large depolarization of the action potential. Unlike voltage-gated Na^+ channels, most types of voltage-gated K^+ channels remain open as long as the membrane is depolarized, and close only when the membrane potential has returned to an inside-negative value. Because the voltage-gated K^+ channels open slightly after the initial depolarization, at the height of the action potential, they sometimes are called *delayed K^+ channels*. Eventually all the voltage-gated K^+ and Na^+ channels return to their closed resting state. The only

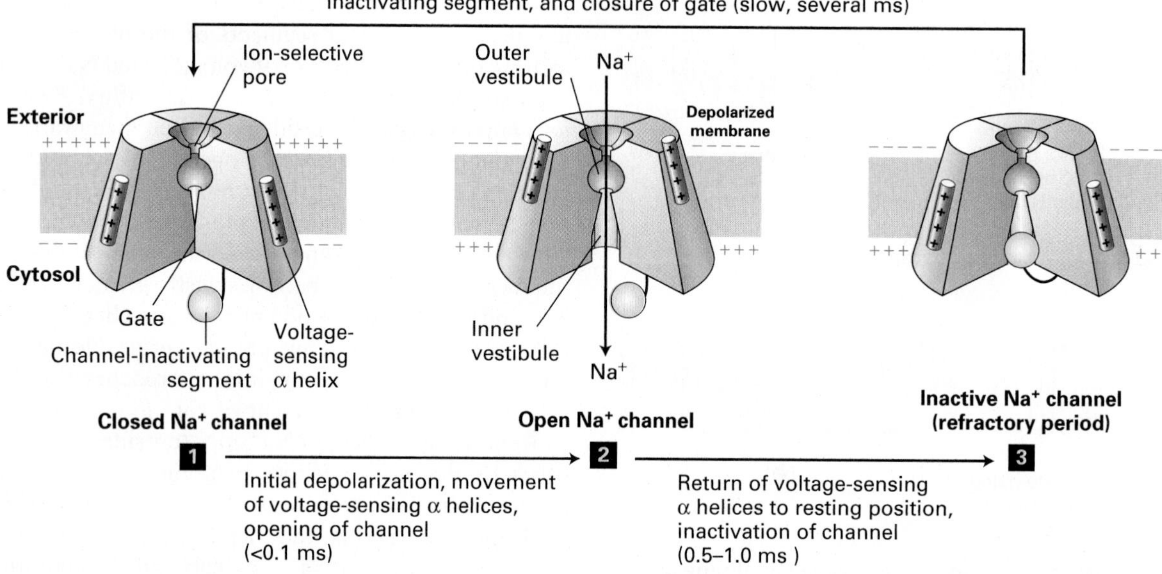

▲ **FIGURE 23-7 Operational model of the voltage-gated Na^+ channel.** Four transmembrane domains in the protein contribute to the central pore through which ions move. The critical components that control movement of Na^+ ions are shown here in the cutaway views depicting three of the four transmembrane domains. **1** In the closed, resting state, the voltage-sensing α helices, which have positively charged side chains every third residue, are attracted to the negative charges on the cytosolic side of the resting membrane. This keeps the gate segment in a position that blocks the channel, preventing entry of Na^+ ions. **2** In response to a small depolarization, the voltage-sensing helices rotate in a screwlike manner toward the outer membrane surface, causing an immediate conformational change in the gate segment that opens the channel. **3** The voltage-sensing helices rapidly return to the resting position, and the channel-inactivating segment moves into the open channel, preventing passage of further ions. **4** Once the membrane is repolarized, the channel-inactivating segment is displaced from the channel opening and the gate closes; the protein reverts to the closed, resting state and can be opened again by depolarization. [See W. A. Catterall, 2001, *Nature* **409:**988; M. Zhou et al., 2001, *Nature* **411:**657; and B. A. Yi and L. Y. Jan, 2000, *Neuron* **27:**423.]

open channels in this baseline condition are the non-gated K^+ channels that generate the resting membrane potential, which soon returns to its usual value (see Figure 23-6a).

The patch-clamp tracings in Figure 23-8 reveal the essential properties of voltage-gated K^+ channels. In this experiment, small segments of a neuronal plasma membrane were held clamped at different voltages, and the flux of electric charges through the patch due to flow of K^+ ions through open K^+ channels was measured. At the modest depolarizing voltage of −10 mV, the channels in the membrane patch open infrequently and remain open for only a few milliseconds, as judged, respectively, by the number and width of

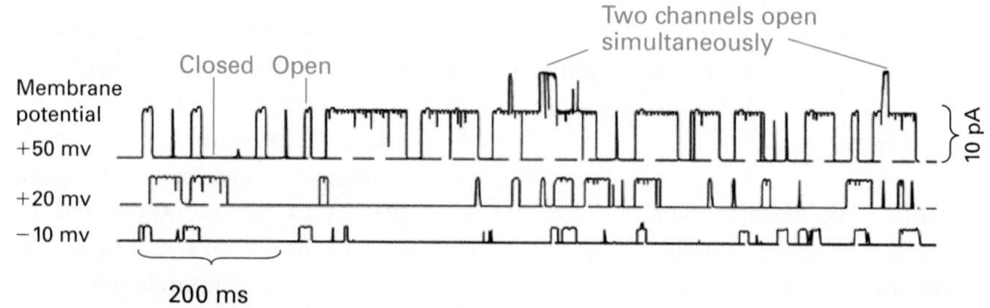

▲ EXPERIMENTAL FIGURE 23-8 **Probability of channel opening and current flux through individual voltage-gated K^+ channels increases with the extent of membrane depolarization.** These patch-clamp tracings were obtained from patches of neuronal plasma membrane clamped at three different potentials, +50, +20, and −10 mV. The upward deviations in the current indicate the opening of K^+ channels and movement of K^+ ions outward (cytosolic to exoplasmic face) across the membrane. Increasing the membrane depolarization (i.e., the clamping voltage) from −10 mV to +50 mV increases the probability a channel will open, the time it stays open, and the amount of electric current (numbers of ions) that pass through it. [From B. Pallota et al., 1981, *Nature* **293:**471, as modified by B. Hille, 1992, *Ion Channels of Excitable Membranes*, 2d ed., Sinauer, p. 122.]

the upward blips on the tracings. Further, the ion flux through them is rather small, as measured by the electric current passing through each open channel (the height of the blips). Depolarizing the membrane further to +20 mV causes these channels to open about twice as frequently. Also, more K^+ ions move through each open channel (the height of the blips is greater) because the force driving cytosolic K^+ ions outward is greater at a membrane potential of +20 mV than at −10 mV. Depolarizing the membrane further to +50 mV, the value at the peak of an action potential, causes opening of more K^+ channels and also increases the flux of K^+ through them. Thus, by opening during the peak of the action potential, these K^+ channels permit the outward movement of K^+ ions and repolarization of the membrane potential while the voltage-gated Na^+ channels are closed and inactivated.

More than 100 voltage-gated K^+ channel proteins have been identified in humans and other vertebrates. As we discuss later, all these channel proteins have a similar overall structure, but they exhibit different voltage dependencies, conductivities, channel kinetics, and other functional properties. Many open only at strongly depolarizing voltages, a property required for generation of the maximal depolarization characteristic of the action potential before repolarization of the membrane begins.

Action Potentials Are Propagated Unidirectionally Without Diminution

The generation of an action potential relates to the changes that occur in a small patch of the neuronal plasma membrane. At the peak of the action potential, passive spread of the membrane depolarization is sufficient to depolarize a neighboring segment of membrane. This causes a few voltage-gated Na^+ channels in this region to open, thereby increasing the extent of depolarization in this region and causing an explosive opening of more Na^+ channels and generation of an action potential. This depolarization soon triggers opening of voltage-gated K^+ channels and restoration of the resting potential. The action potential thus spreads as a traveling wave away from its initial site without diminution.

As noted earlier, during the refractory period voltage-gated Na^+ channels are inactivated for several milliseconds. Such previously opened channels cannot open during this period even if the membrane is depolarized owing to passive spread. As illustrated in Figure 23-9, the inability of Na^+ channels to reopen during the refractory period ensures that action potentials are propagated only in one direction, from the axon initial segment where they originate to the axon terminus. This property of the Na^+ channels also limits the number of action potentials per second that a neuron can conduct. This is important, since it is the frequency of action potentials that carries the information. Reopening of Na^+ channels upstream of an action potential (i.e., closer to the cell body) also is delayed by the membrane hyperpolarization that results from opening of voltage-gated K^+ channels.

Nerve Cells Can Conduct Many Action Potentials in the Absence of ATP

The depolarization of the membrane during an action potential results from movement of just a small number of Na^+ ions into a neuron and does not significantly affect the intracellular Na^+ concentration. A typical nerve cell has about 10 voltage-gated Na^+ channels per square micrometer (μm^2) of plasma membrane. Since each channel passes ≈5000–10,000 ions during the millisecond it is open (see Figure 11-22), a maximum of 10^5 ions per μm^2 of plasma membrane will move inward during each action potential.

To assess the effect of this ion flux on the cytosolic Na^+ concentration of 10 mM (0.01 mol/L), typical of a resting axon, we focus on a segment of axon 1 micrometer (μm) long and 10 μm in diameter. The volume of this segment is 78 μm^3, or 7.8×10^{-14} liters, and it contains 4.7×10^8 Na^+ ions: (10^{-2} mol/L) (7.8×10^{-14} L) (6×10^{23} Na^+/mol). The surface area of this segment of the axon is 31 μm^2, and during passage of one action potential, 10^5 Na^+ ions will enter per μm^2 of membrane. Thus this Na^+ influx increases the number of Na^+ ions in this segment by only one part in about 150: $(4.7 \times 10^8) \div (3.1 \times 10^6)$. Likewise, the repolarization of the membrane due to the efflux of K^+ ions through voltage-gated K^+ channels does not significantly change the intracellular K^+ concentration.

Because so few Na^+ and K^+ ions move across the plasma membrane during each action potential, the ATP-driven Na^+/K^+ pump that maintains the usual ion gradients plays no direct role in impulse conduction. Since the ion movements during each action potential involve only a minute fraction of the cell's K^+ and Na^+ ions, a nerve cell can fire hundreds or even thousands of times in the absence of ATP.

All Voltage-Gated Ion Channels Have Similar Structures

Having explained how the action potential is dependent on regulated opening and closing of voltage-gated channels, we turn to a molecular dissection of these remarkable proteins. After describing the basic structure of these channels, we focus on three questions:

- How do these proteins sense changes in membrane potential?
- How is this change transduced into opening of the channel?
- What causes these channels to become inactivated shortly after opening?

The initial breakthrough in understanding voltage-gated ion channels came from analysis of fruit flies *(Drosophila melanogaster)* carrying the *shaker* mutation. These flies shake vigorously under ether anesthesia, reflecting a loss of motor control and a defect in certain motor neurons that have an abnormally prolonged action potential. Researchers suspected that the *shaker* mutation caused a defect in channel function. Cloning of the gene involved confirmed that

► **FIGURE 23-9 Unidirectional conduction of an action potential due to transient inactivation of voltage-gated Na^+ channels.** At time 0, an action potential (red) is at the 2-mm position on the axon; the Na^+ channels at this position are open and Na^+ ions are flowing inward. The excess Na^+ ions diffuse in both directions along the inside of the membrane, passively spreading the depolarization. Because the Na^+ channels at the 1-mm position are still inactivated (green), they cannot yet be reopened by the small depolarization caused by passive spread; the Na^+ channels at the 3-mm position, in contrast, begin to open. Each region of the membrane is refractory (inactive) for a few milliseconds after an action potential has passed. Thus, the depolarization at the 2-mm site at time 0 triggers action potentials downstream only; at 1 ms an action potential is passing the 3-mm position, and at 2 ms an action potential is passing the 4-mm position.

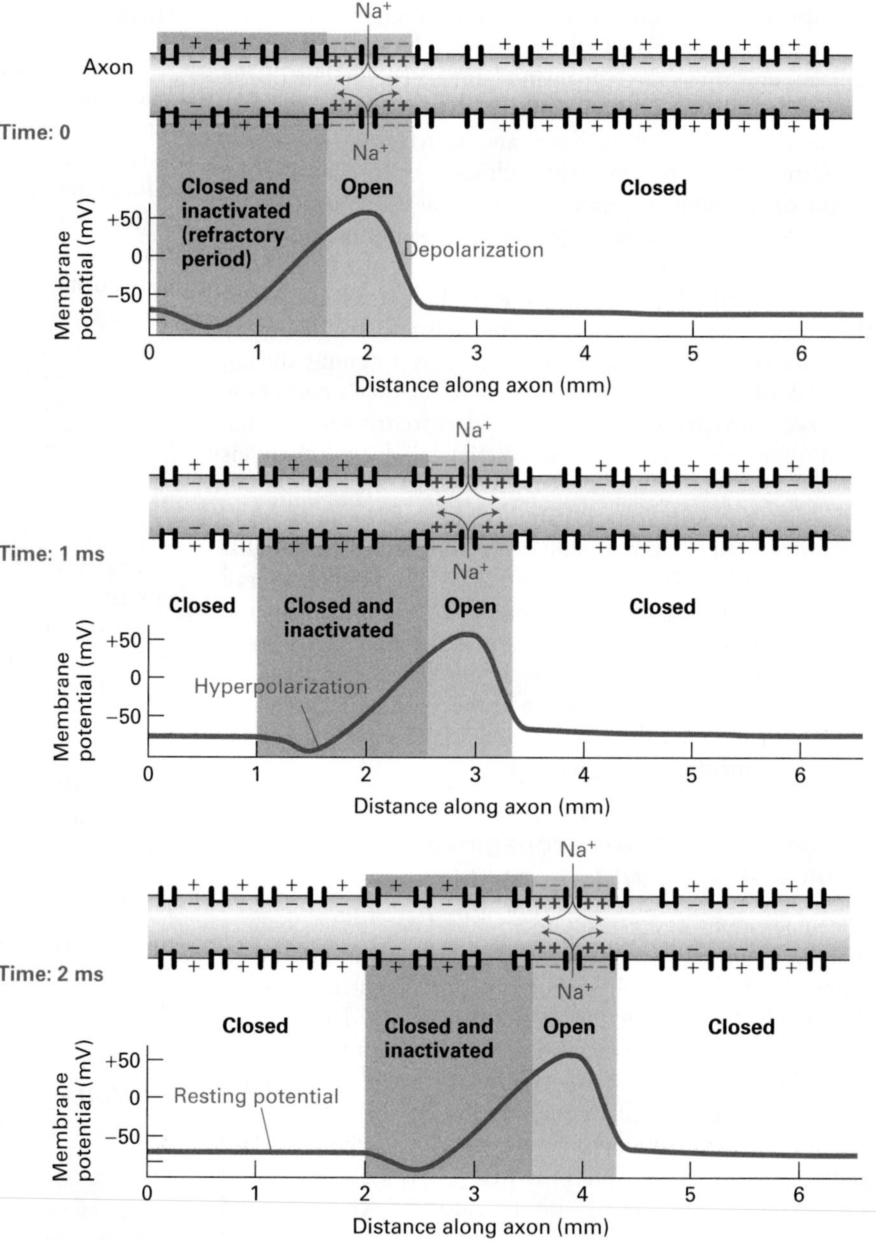

the defective protein was a channel. The *shaker* mutation prevents the mutant channel from opening normally immediately upon depolarization. To test whether the wild-type *shaker* gene encoded a K^+ channel, cloned wild-type *shaker* cDNA was used as a template to produce *shaker* mRNA in a cell-free system. Expression of this mRNA in frog oocytes and patch-clamp measurements on the newly synthesized channel protein showed that its functional properties were identical with those of the voltage-gated K^+ channel in the neuronal membrane, demonstrating conclusively that the *shaker* gene encodes this K^+-channel protein.

The Shaker K^+ channel and most other voltage-gated K^+ channels that have been identified are tetrameric proteins composed of four identical subunits arranged in the membrane around a central pore. Each subunit is constructed of six membrane-spanning α helixes, designated S1–S6, and a P segment (Figure 23-10a). The S5 and S6 helices and the P segment are structurally and functionally homologous to those in the nongated resting K^+ channel discussed earlier (see Figure 11-19). The S5 and S6 helices form the lining of the channel through which the ion travels. The S1–S4 helices act as a voltage sensor (with S4 acting as the primary sensor) and are described as paddles owing to the way they protrude from the central complex. The N-terminal "ball" extending into the cytosol from S1 is the channel-inactivating segment.

Voltage-gated Na^+ channels and Ca^{2+} channels are monomeric proteins organized into four homologous domains, I–IV (Figure 23-10b). Each of these domains is similar to a subunit of a voltage-gated K^+ channel. However, in contrast to voltage-gated K^+ channels, which have four channel-inactivating segments, the monomeric voltage-gated channels have a single channel-inactivating segment. Except for this

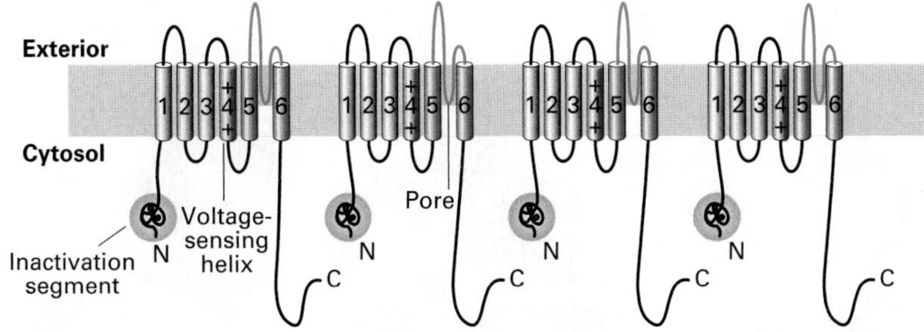

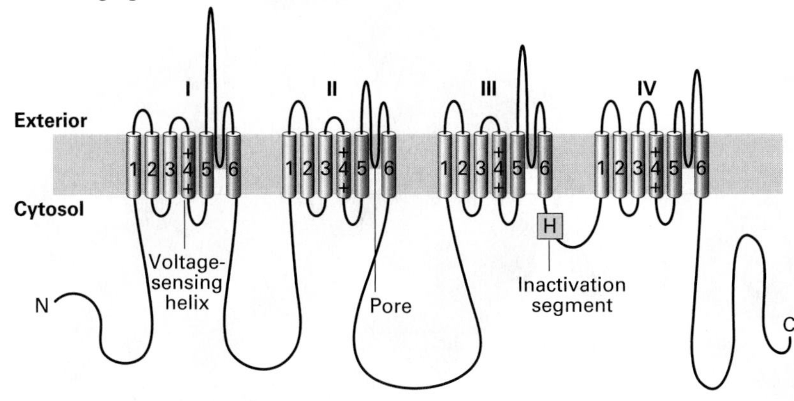

◀ **FIGURE 23-10 Schematic depictions of the secondary structures of voltage-gated K$^+$ and Na$^+$ channels.** (a) Voltage-gated K$^+$ channels are composed of four identical subunits, each containing 600–700 amino acids, and six membrane-spanning α helices, S1–S6. The N-terminus of each subunit, located in the cytosol and labeled N, forms a globular domain (orange ball) essential for inactivation of the open channel. The S5 and S6 helices (green) and the P segment (blue) are homologous to those in nongated resting K$^+$ channels, but each subunit contains four additional transmembrane α helices. One of these, S4 (red), is the primary voltage-sensing α helix and is assisted in this role by helices S1–3. (b) Voltage-gated Na$^+$ channels are monomers containing 1800–2000 amino acids organized into four transmembrane domains (I–IV) that are similar to the subunits in voltage-gated K$^+$ channels. The single channel-inactivating segment, located in the cytosol between domains III and IV, contains a conserved hydrophobic motif (H; yellow). Voltage-gated Ca^{2+} channels have a similar overall structure. Most voltage-gated ion channels also contain regulatory (β) subunits that are not depicted here. [Part (a) adapted from C. Miller, 1992, *Curr. Biol.* 2:573, and H. Larsson et al., 1996, *Neuron* **16:**387; part (b) adapted from W. A. Catterall, 2001, *Nature* **409:**988.]

minor structural difference and their varying ion permeabilities, all voltage-gated ion channels are thought to function in a similar manner and to have evolved from a monomeric ancestral channel protein that contained six transmembrane α helices.

Voltage-Sensing S4 α Helices Move in Response to Membrane Depolarization

The understanding of channel-protein biochemistry is advancing rapidly owing to new crystal structures for bacterial and *shaker* potassium channels and other channels. One method used to obtain crystals of these difficult membrane proteins was to surround them with bound fragments of monoclonal antibodies (Fab's; Chapter 24); in other cases they were crystallized in complexes with normal protein-binding partners.

The structures of the channels reveal remarkable arrangements of the voltage-sensing domains, and suggest how parts of the protein move in order to open the channel. The tetramer has a pore whose walls are formed by helices S5 and S6 (Figure 23-11a). Outside that core structure four arms, or "paddles," protrude into the surrounding membrane; these are the voltage sensors, and they are in minimal contact with the pore. Sensitive electric measurements suggested that the opening of a voltage-gated Na$^+$ or K$^+$ channel is accompanied by the movement of 12–14 protein-bound positive charges from the cytosolic to the exoplasmic surface of the membrane. The moving part of the protein is composed of helices S1–S4; S4 accounts for much of the positive charge and is therefore the primary voltage sensor, with a positively charged lysine or arginine every third or fourth residue. Arginines in S4 have been measured moving as much as 1.5 nm as the channel opens, which can be compared with the ≈5-nm thickness of the membrane or the 1.2-nm diameter of the α helix itself. The movement of these gating charges (or voltage sensors) under the force of the electric field triggers a conformational change in the protein that opens the channel. Thus the S4 helix is the key part of the voltage sensor, which then moves the S1–S4 helices across much of the membrane. The most unusual aspect of the voltage-sensitive channel structures is the presence of charged groups, e.g., arginines, in contact with lipid. The location of the voltage sensor helps to explain earlier experiments where a non-voltage-sensitive channel was converted into a voltage-sensing channel by adding to it voltage-sensing domains. Such an experiment would seem unlikely to work if the voltage sensors had to be deeply embedded in the core structure.

Studies with mutant Shaker K$^+$ channels support the importance of the S4 helix in voltage sensing. When one or more arginine or lysine residues in the S4 helix of the Shaker K$^+$ channel were replaced with neutral or acidic residues, fewer positive charges than normal moved across the membrane in response to a membrane depolarization, indicating that arginine and lysine residues in the S4 helix do indeed move across the membrane. In other studies, mutant Shaker proteins in which various S4 residues were converted to cysteine were tested for their reactivity with a water-soluble cysteine-modifying chemical agent that cannot cross the membrane. On the basis of whether the cysteines reacted with the agent added to one side or the other of the membrane, the results indicated that in the resting state amino acids near the C-terminus of the S4 helix face the cytosol; after the membrane is depolarized, some of these same amino

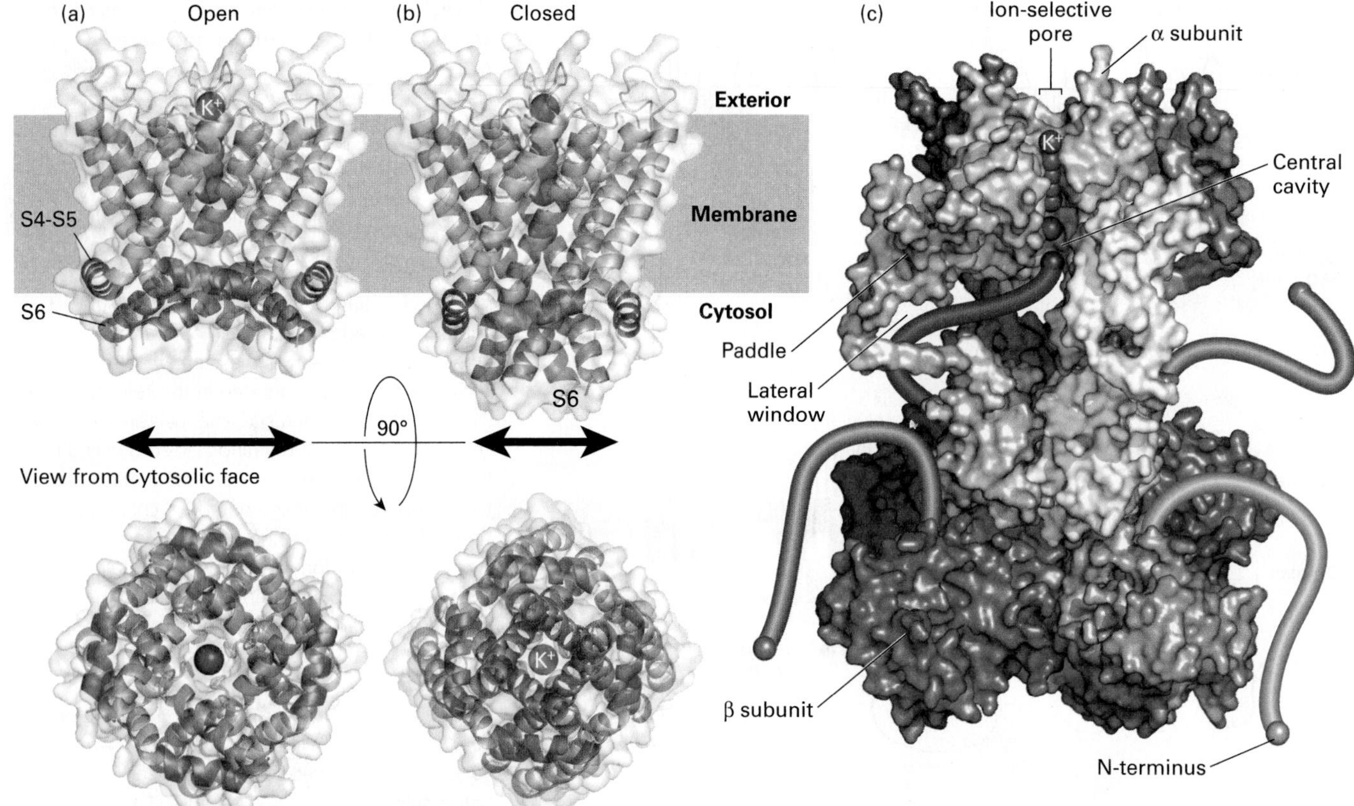

▲ **FIGURE 23-11 Molecular structure of a voltage-sensitive potassium channel.** The two ribbon diagrams show models of the potassium channel in (a) open and (b) closed states. Since the molecule is a tetramer of the same subunit, four copies of each helix are visible; the ones farthest away are seen only dimly. The brown (S5) and green (S6) alpha helices span the membrane, with the interior of the cell at the bottom and exterior at the top. The purple spheres represent K^+ ions, which pass through the open channel and occupy part of the closed channel without passing through. The S6 (green) helices line the pore. Note how the helices are tightly packed at the bottom in (b), closing the channel so that the K^+ ion cannot pass through. [compare the distances between S5 helices as shown by the arrows below (a) and (b).] The S4–S5 linker, located in the cytoplasm, connects the S4 helix (not shown) to the S5 helix (brown). For clarity, helices S1 through S4 have been omitted from the model; they would normally be attached to the end of the S4–S5 linker and protrude from the molecule as the voltage-sensing "paddles." These paddles move from near the interior to the exterior of the membrane in response to depolarization. Since each one is attached to an S4–S5 linker, each linker and its attached S5 helix is moved, in turn moving S6 helices, which opens the pore. The structure of the open mammalian channel (a) has been determined. The closed-channel structure in (b) is hypothetical, but is based on observations of a closed bacterial potassium-channel structure. (c) The ball-and-chain model for inactivation of voltage-gated K^+ channels in three-dimensional cutaway view of the inactive state. In addition to the four α subunits (tan and gray) that form the channel, these channel proteins have four regulatory β subunits (purple). At the N terminus of each of the β subunit proteins is a small domain (purple "ball" on the end of the purple "chain") that controls the opening of the central pore. In this illustration the N terminus of one subunit has moved through a lateral window to block the pore. [Parts (a) and (b) from S. B. Long, E. B. Campbell, and R. MacKinnon, 2005, *Science* **309:**903–908; part (c) adapted from R. Aldrich, 2001, *Nature* **411:**643, and M. Zhou et al., 2001, *Nature* **411:**657.]

acids become exposed to the exoplasmic surface of the channel. These experiments directly demonstrated movement of the S4 helix across the membrane, as schematically depicted in Figure 23-7 for voltage-gated Na^+ channels.

The structure of the open form of a mammalian Shaker K^+ channel has been contrasted with the closed structure of a crystallized bacterial K^+ channel. The results suggest a model for the closing of the channel in response to movements of the voltage sensors across the membrane (Figure 23-11a, b). In the model, the voltage sensors, composed of helices S1–S4, move in response to voltage and exert a torque on a linker helix that connects S4 to S5. In the open-channel conformation, the position of the S4–S5 linkers allows the S5 helices to lie at about a 45° angle to the plane of the membrane (Figure 23-11a, brown helices), and the pore inside has a 1.2-nm opening. When the cell is repolarized and the voltage sensor moves toward the intracellular membrane surface, the S4–S5 linkers are twisted down, toward the inside of the cell. The S5 helices are consequently moved more orthogonal to the plane of the membrane (Figure 23-11b, brown helices). This position leaves the S5 and S6 helices in closer proximity, squeezing the channel closed (Figure 23-11a, b; the double headed arrows indicate spacing between adjacent S5 helices). Thus the gate probably is composed of the cytosol-facing ends of the S5 and S6 helixes, where the pore is narrowest.

Movement of the Channel-Inactivating Segment into the Open Pore Blocks Ion Flow

An important characteristic of most voltage-gated channels is inactivation; that is, soon after opening they close spontaneously, forming an inactive channel that will not reopen until the membrane is repolarized. In the resting state, the positively charged globular balls at the N-termini of the four subunits in a voltage-gated K^+ channel are free in the cytosol. Several milliseconds after the channel is opened by depolarization, one ball moves through an opening (*lateral window*) between two of the subunits and binds in a hydrophobic pocket in the pore's central cavity, blocking the flow of K^+ ions (Figure 23-11c). After a few milliseconds, the ball is displaced from the pore, and the protein reverts to the closed, resting state. The ball-and-chain domains in K^+ channels are functionally equivalent to the channel-inactivating segment in Na^+ channels.

The experimental results shown in Figure 23-12 demonstrate that inactivation of K^+ channels depends on the ball domains, occurs after channel opening, and does not require the ball domains to be covalently linked to the channel protein. In other experiments, mutant K^+ channels lacking portions of the ≈40-residue chain connecting the ball to the S1 helix were expressed in frog oocytes. Patch-clamp measurements of channel activity showed that the shorter the chain, the more rapid the inactivation, as if a ball attached to a shorter chain can move into the open channel more readily. Conversely, addition of random amino acids to lengthen the normal chain slows channel inactivation.

The single channel-inactivating segment in voltage-gated Na^+ channels contains a conserved hydrophobic motif composed of isoleucine, phenylalanine, methionine, and threonine (see Figure 23-10b). Like the longer ball-and-chain domain in K^+ channels, this segment folds into and blocks the Na^+-conducting pore until the membrane is repolarized (see Figure 23-7).

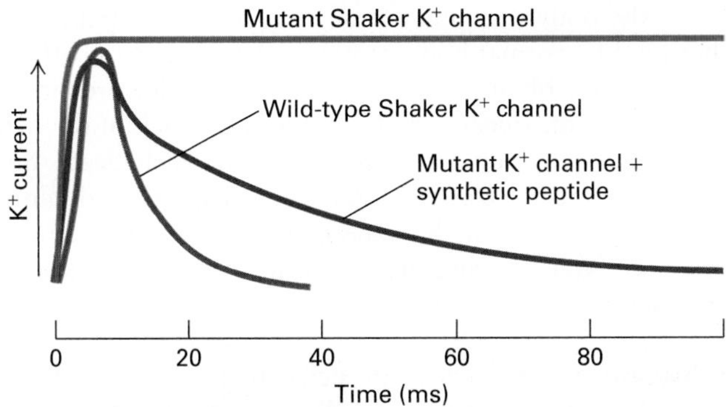

▲ EXPERIMENTAL FIGURE 23-12 **Experiments with a mutant K^+ channel lacking the N-terminal globular domains support the ball-and-chain inactivation model.** The wild-type Shaker K^+ channel and a mutant form lacking the amino acids composing the N-terminal ball were expressed in *Xenopus* oocytes. The activity of the channels then was monitored by the patch-clamp technique. When patches were depolarized from −0 to +30 mV, the wild-type channel opened for ≈5 ms and then closed (red curve). The mutant channel opened normally, but could not close (green curve). When a chemically synthesized ball peptide was added to the cytosolic face of the patch, the mutant channel opened normally and then closed (blue curve). This demonstrated that the added peptide inactivated the channel after it opened and that the ball does not have to be tethered to the protein in order to function. [From W. N. Zagotta et al., 1990, *Science* **250**:568.]

Myelination Increases the Velocity of Impulse Conduction

As we have seen, action potentials can move down an axon without diminution at speeds up to 1 meter per second. But even such fast speeds are insufficient to permit the complex movements typical of animals. In humans, for instance, the cell bodies of motor neurons innervating leg muscles are located in the spinal cord, and the axons are about a meter in length. The coordinated muscle contractions required for walking, running, and similar movements would be impossible if it took 1 second for an action potential to move from the spinal cord down the axon of a motor neuron to a leg muscle. The solution is to wrap cells in insulation that increases the rate of movement of an action potential. The insulation is called a **myelin sheath** (see Figure 23-2b). The presence of a myelin sheath around an axon increases the velocity of impulse conduction to 10–100 meters per second. As a result, in a typical human motor neuron, an action potential can travel the length of a 1-meter-long axon and stimulate a muscle to contract within 0.01 seconds.

In nonmyelinated neurons, the conduction velocity of an action potential is roughly proportional to the diameter of the axon, because a thicker axon will have a greater number of ions that can diffuse. The human brain is packed with relatively small, myelinated neurons. If the neurons in the human brain were not myelinated, their axonal diameters would have to increase about 10,000-fold to achieve the same conduction velocities as myelinated neurons. Thus vertebrate brains, with their densely packed neurons, never could have evolved without myelin.

Action Potentials "Jump" from Node to Node in Myelinated Axons

The myelin sheath surrounding an axon is formed from many glial cells. Each region of myelin formed by an individual glial cell is separated from the next region by an unmyelinated area of axonal membrane about 1 μm in length called the *node of Ranvier* (or simply, *node;* see Figure 23-2). The axonal membrane is in direct contact with the extracellular

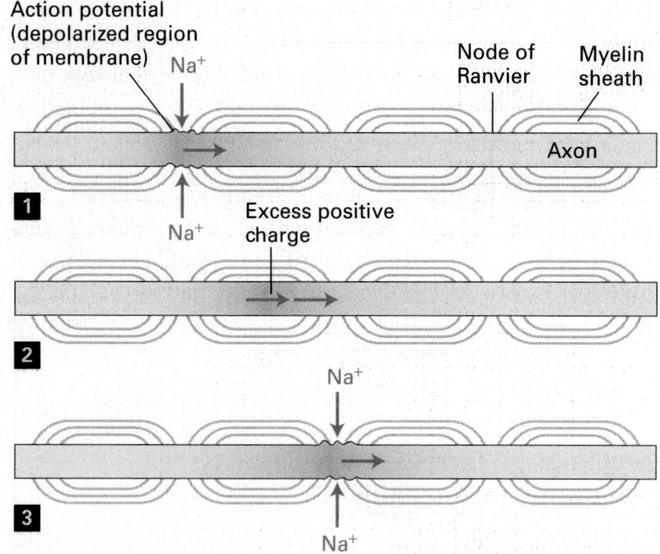

▲ **FIGURE 23-13 Conduction of action potentials in myelinated axons.** Because voltage-gated Na^+ channels are localized to the axonal membrane at the nodes of Ranvier, the influx of Na^+ ions associated with an action potential can occur only at nodes. When an action potential is generated at one node (step **1**), the excess positive ions in the cytosol, which cannot move outward across the sheath, diffuse rapidly down the axon, causing sufficient depolarization at the next node (step **2**) to induce an action potential at that node (step **3**). By this mechanism the action potential jumps from node to node along the axon.

fluid only at the nodes. Moreover, all the voltage-gated Na^+ channels and all the Na^+/K^+ pumps, which maintain the ionic gradients in the axon, are located in the nodes.

As a consequence of this localization, the inward movement of Na^+ ions that generates the action potential can occur only at the myelin-free nodes (Figure 23-13). The excess cytosolic positive ions generated at a node during the membrane depolarization associated with an action potential spread passively through the axonal cytosol to the next node with very little loss or attenuation, since they cannot cross the myelinated axonal membrane. This causes a depolarization at one node to spread rapidly to the next node, permitting the action potential, in effect, to jump from node to node. The transmission is called *saltatory conduction.* This phenomenon explains why the conduction velocity of myelinated neurons is about the same as that of much larger diameter unmyelinated neurons. For instance, a 12-μm-diameter myelinated vertebrate axon and a 600-μm-diameter unmyelinated squid axon both conduct impulses at 12 m/s.

Glia Produce Myelin Sheaths and Synapses

Of the four types of glia (three of which are shown in Figure 23-14), two produce myelin sheaths: *oligodendrocytes* make sheaths for the central nervous system (CNS), and *Schwann cells* make them for the peripheral nervous system. *Astrocytes,* a third type, are necessary for neurons to produce synapses and use them to communicate with other neurons. The fourth type, *microglia,* produce survival factors for cells and carry out immune functions. These cells participate in inflammatory responses and constitute a part of the CNS immune system. They can differentiate to form phagocytic cells with characteristics of macrophages (Chapter 24). Microglia form in the bone marrow, are not related by lineage to neurons or to other glia, and will not be discussed further.

Oligodendrocytes Oligodendrocytes form the spiral myelin sheath around axons of the central nervous system (Figure 23-14c). Each oligodendrocyte provides myelin sheaths to multiple neurons. The major protein constituents are myelin basic protein (MBP) and proteolipid protein (PLP). MBP, a peripheral membrane protein found in both the CNS and the PNS, has seven RNA splicing variants that encode different forms of the protein. It is synthesized by ribosomes located in the growing myelin sheath (Figure 23-14c), an example of specific transport of mRNAs to a peripheral cell region. The localization of MBP mRNA depends on microtubules.

Damage to proteins produced by oligodendrocytes underlies a prevalent human neurological disease, multiple sclerosis (MS). MS is usually characterized by spasms and weakness in one or more limbs, bladder dysfunction, local sensory losses, and visual disturbances. This disorder—the prototype *demyelinating disease*—is caused by patchy loss of myelin in areas of the brain and spinal cord. In MS patients, conduction of action potentials by the demyelinated neurons is slowed, and the Na^+ channels spread outward from the nodes, lowering their nodal concentration. The cause of the disease is not known but appears to involve either the body's production of auto-antibodies (antibodies that bind to normal body proteins) that react with MBP or the secretion of proteases that destroy myelin proteins. A mouse mutant, *shiverer,* has a deletion of much of the MBP gene, leading to tremors, convulsions, and early death. Similarly human (Pelizaeus-Merzbacher disease) and mouse (*jimpy*) mutations in the gene coding for the other major protein of CNS myelin, PLP, cause loss of oligodendrocytes and inadequate myelination. ▌

Schwann Cells Schwann cells form myelin sheaths around peripheral nerves. A Schwann cell myelin sheath is a remarkable spiral wrap (see Figure 23-14b). A long axon can have as many as several hundred Schwann cells along its length, each contributing insulation to an *internode* stretch of about 1–1.5 μm of axon. Not all axons are myelinated, for reasons that are not known. Mutations in mice that eliminate Schwann cells cause the death of most neurons.

In contrast to oligodendrocytes, Schwann cells each dedicate themselves to one axon. The sheaths are composed of about 70 percent lipid (rich in cholesterol) and 30 percent protein. In the peripheral nervous system the principle protein constituent (~80 percent) of myelin is called protein 0 (P_0), an integral membrane protein that has immunoglobulin (Ig) domains. MBP is also an abundant component. The extracellular Ig domains bind together the surfaces of

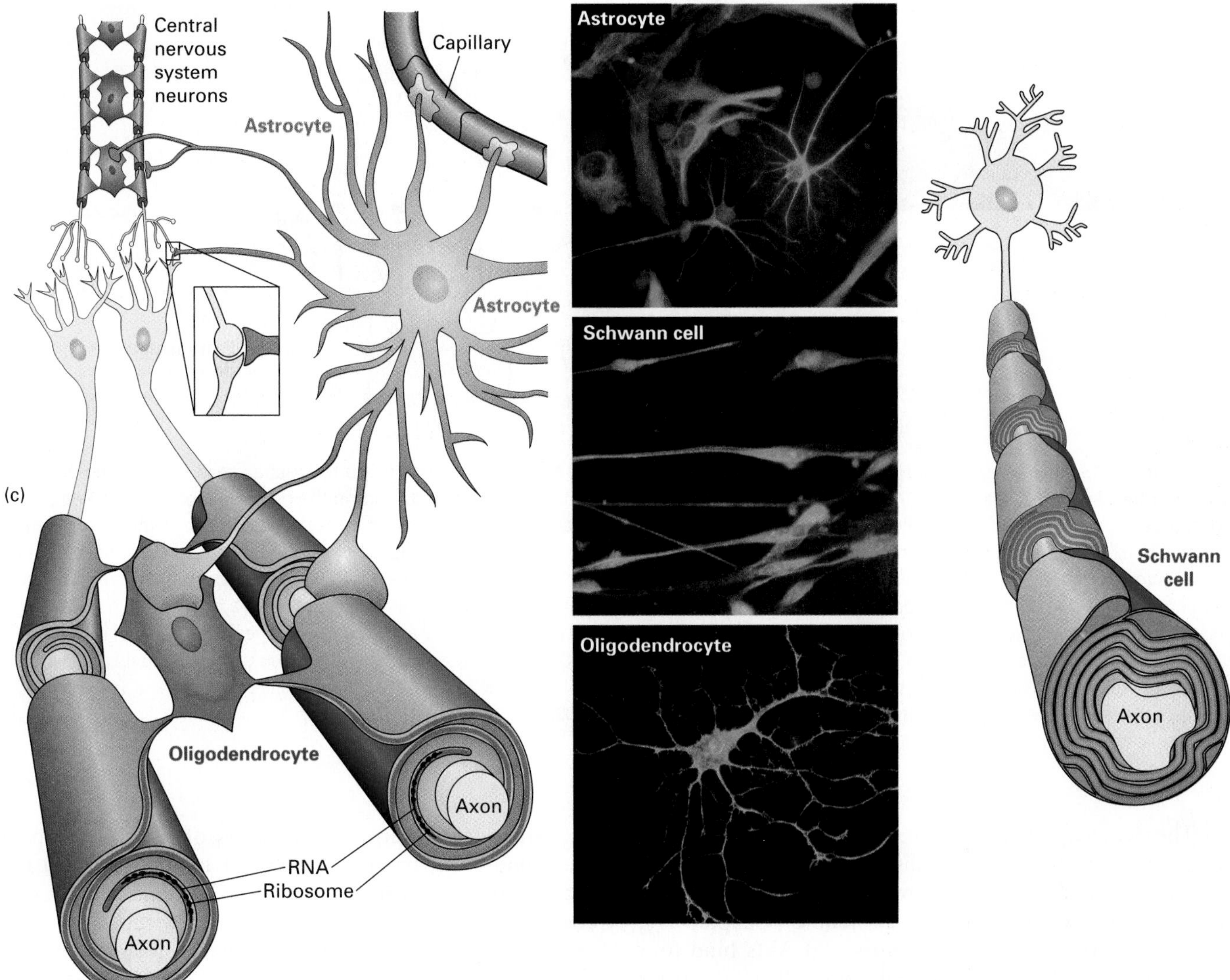

▲ **FIGURE 23-14 Three types of glia cells.** (a) Astrocytes interact with neurons but do not insulate them. (b) Each Schwann cell insulates a section of a single peripheral nervous system axon. (c) A single oligodendrocyte can myelinate multiple CNS axons. [From B. Stevens, 2003, *Curr. Biol.* **13**:469, and adapted from D. L. Sherman and P. Brophy, 2005, *Nature Rev. Neurosci.* **6**:683–690, Photos: Courtesy of Varsha Shukla and Doug Field from NIH.]

sequential wraps around the axon to compact the spiral of myelin sheath (Figure 23-15b). Other proteins play this kind of role in the CNS.

In humans, peripheral myelin, like CNS myelin, is a target of autoimmune disease, antibodies forming against P_0. The Guillain-Barre syndrome (GBS), also known as *acute inflammatory demyelinating polyneuropathy,* is one such disease. GBS is the most common cause of rapid-onset paralysis, occurring at a frequency of 10^{-5}. The cause is unknown. The common inherited neurological disorder called *Charcot-Marie-Tooth disease,* which damages peripheral motor and sensory nerve function, is due to overexpression of the gene that encodes PMP22 protein, another constituent of peripheral nerve myelin. ▮

Interactions between glia and neurons control the placement and spacing of myelin sheaths, and the assembly of nerve-transmission machinery at the nodes of Ranvier. Voltage-gated Na^+ channels and Na^+/K^+ pumps, for example, congregate at the nodes of Ranvier through interactions with cytoskeletal proteins. While all the details of the node assembly process are not understood, a number of key players have been identified. In the PNS, where the process has been most studied, surface adhesion molecules in the Schwann cell membrane interact with neuronal adhesion molecules. The glial membrane *i*mmuno*g*lobulin *c*ell-*a*dhesion *m*olecule (IgCAM) called *neurofascin155* contacts two axonal proteins, contactin and contactin-associated protein at the edge of the node. These cell-cell contact events create boundaries at each side of the node.

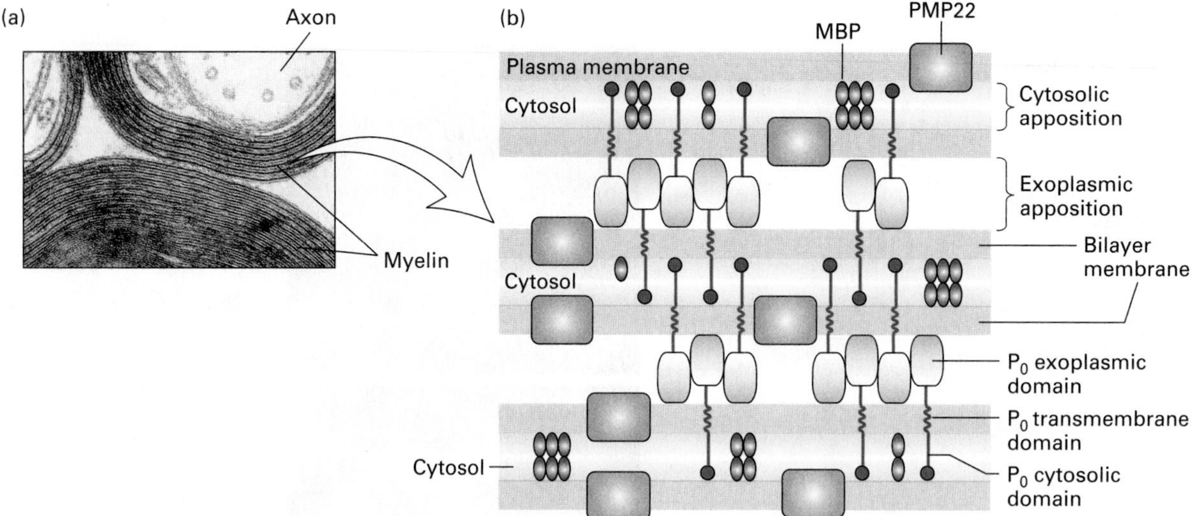

▲ **FIGURE 23-15 Formation and structure of a myelin sheath in the peripheral nervous system.** (a). At high magnification the specialized spiral myelin (My) membrane appears as a series of layers, or lamellae, of phospholipid bilayers wrapped around the axon (Ax); Mit = mitochondrion. (b) Close up view of three layers of the myelin membrane spiral. The two most abundant membrane peripheral myelin proteins, P_0 and PMP22, are produced only by Schwann cells. The exoplasmic domain of a P_0 protein, which has an immunoglobulin fold, associates with similar domains emanating from P_0 proteins in the opposite membrane surface, thereby "zippering" together the exoplasmic membrane surfaces in close apposition. These interactions are stabilized by binding of a tryptophan residue on the tip of the exoplasmic domain to lipids in the opposite membrane. Close apposition of the cytosolic faces of the membrane may result from binding of the cytosolic tail of each P_0 protein to phospholipids in the opposite membrane. PMP22 may also contribute to membrane compaction. Myelin basic protein (MBP), a cytosolic protein, remains between the closely apposed membranes as the cytosol is squeezed out. [Part (a) © Science VU/C. Raine/Visuals Unlimited; part (b) adapted from L. Shapiro et al., 1996, *Neuron* **17:**435, and E. J. Arroyo and S. S. Scherer, 2000, *Histochem. Cell Biol.* **113:**1.]

The channel proteins and other molecules that will accumulate at the node are initially dispersed through the axons. Then axonal proteins, including two IgCAMs called *NrCAM* and *neurofascin186,* as well as ankyrin G (Chapter 17), accumulate within the node. The two IgCAMs bind to a single transmembrane domain protein called *gliomedin* that is expressed in the glial cell. Experiments that eliminated gliomedin production showed that without it nodes do not form, so it is a key regulator. Ankyrin in the node contacts βIV spectrin, a major constituent of the cytoskeleton, thus tethering the node's protein complex to the cytoskeleton. Na^+ channels become associated with neurofascin186, NrCAM, and ankyrin G, firmly trapping the channel where it is needed. As a result of these multiple protein-protein interactions, the concentration of Na^+ channels is roughly a hundredfold higher in the nodal membrane of myelinated axons than in the axonal membrane of nonmyelinated neurons.

Astrocytes The third type of glial cell is the astrocyte, named for its starlike shape (see Figure 23-14a). These can constitute more than a third of brain mass and half of the brain's cells. There are two kinds. *Protoplasmic astrocytes* are in the gray matter (the areas rich in cell bodies); *fibrous astrocytes* are in the white matter (the areas composed mainly of axons; it is the myelin that makes it look white). The astrocytes make long thin processes that envelop all the brain's blood vessels.

Astrocytes are critical regulators of the formation of the blood-brain barrier. A mass of blood vessels in the brain supplies oxygen and removes CO_2, and delivers glucose and amino acids, with capillaries within a few micrometers of every cell. These capillaries form the blood-brain barrier, which prevents, for example, blood-borne circulating neurotransmitters and some drugs from entering the brain. The barrier consists of a set of tight junctions (Chapter 19) made by the endothelial cells that form the walls of capillaries. Astrocytes promote specialization of these endothelial cells, making them less permeable (Figure 23-16).

Many synapses and dendrites are also surrounded by astrocyte processes. Astrocytes produce abundant extracellular matrix proteins, some of which are used as guidance cues by migrating neurons, and a host of growth factors that carry a variety of types of information to neurons. The Ca^{2+}, K^+, Na^+, and Cl^- channels, among others, found in the plasma membranes of astrocytes influence the concentration of free ions in the extracellular space, thus affecting the membrane potentials of neurons and of the astrocytes themselves. Astrocytes also take up glutamate, a neurotransmitter, from extracellular spaces and turn it into glutamine. When nearby neurons have fired, glutamate binds to glutamate receptors on astrocytes, enabling the astrocytes to sense the event. Astrocytes are joined to each other by gap junctions, so changes in ionic composition in any of them are communicated to others, over distances of hundreds of microns.

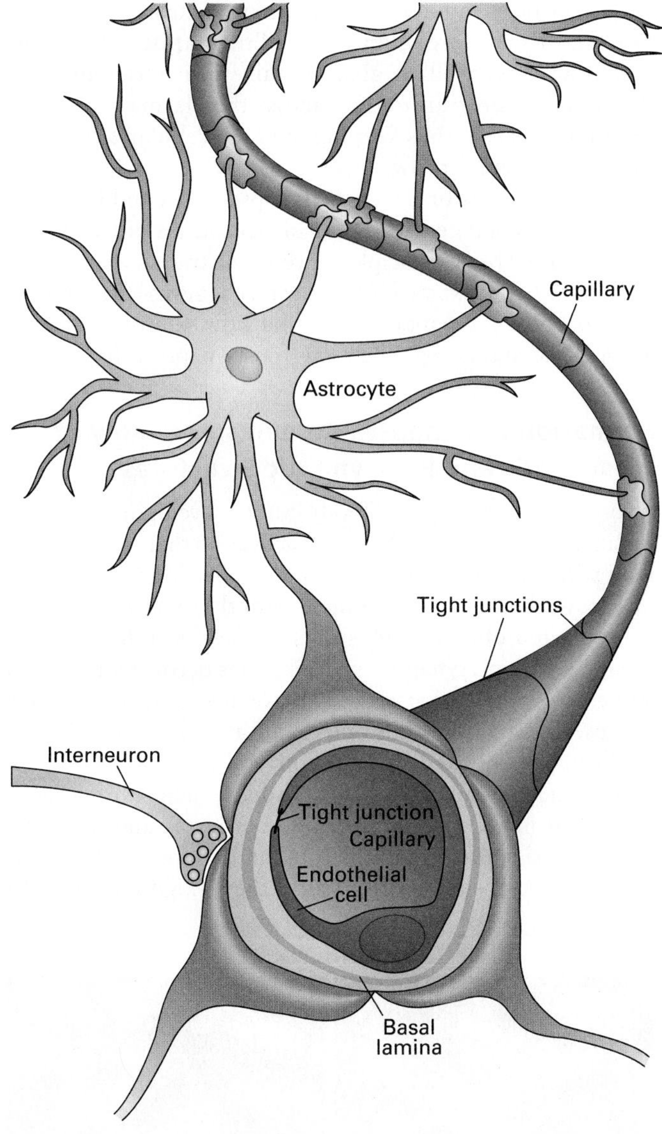

◀ **FIGURE 23-16 Astrocytes interact with endothelial cells at the blood-brain barrier.** The purpose of the blood-brain barrier is to control what types of molecules can travel out of the bloodstream into the brain and vice versa. The formation of the barrier by the endothelial cells that make up blood vessel walls entering the brain is directed by surrounding astrocytes. Capillaries in the brain are formed by endothelial cells with tight junctions that are impermeable to most molecules. Transport between cells is blocked, so only small molecules that can diffuse or substances specifically transported through cells can cross the barrier. The endothelial cells have transporters and permeability characteristics that allow oxygen and CO_2 across but selectively prevent other substances from crossing. Astrocytes surround the blood vessels, in contact with the endothelial cells, and send secreted protein signals to induce the endothelial cells to produce a selective barrier. Unless there is a specific carrier involved, lipid-soluble molecules stand the best chance of getting across, though they may travel poorly in the blood. Electrolytes, like Na^+ and Cl^-, are moved across by specific channel and transport proteins. The brain's endothelial cells have less vesicle transport than most endothelial cells, presumably since transport is more selective. The endothelial cells (burgandy) are ensheathed by a layer of basal lamina (orange) and contacted on the outside by astrocyte processes (tan). Pericytes are mesenchymal cells that provide support to the capillaries. [From N. J. Abbott, L. Rönnbäck, and E. Hansson, 2006, *Nature Rev. Neurosci.* **7**:41–53.]

KEY CONCEPTS OF SECTION 23.2

Voltage-Gated Ion Channels and the Propagation of Action Potentials in Nerve Cells

- Action potentials are sudden membrane depolarizations followed by a rapid repolarization. They originate at the axon initial segment and move down the axon toward the axon terminals, where the electric impulse is transmitted to other cells via a synapse (see Figures 23-3 and 23-6).

- An action potential results from the sequential opening and closing of voltage-gated Na^+ and K^+ channels in the plasma membrane of neurons and muscle cells (see Figure 23-9).

- Opening of voltage-gated Na^+ channels permits influx of Na^+ ions for about 1 ms, causing a sudden large depolarization of a segment of the membrane. The channels then close and become unable to open (refractory) for several milliseconds, preventing further Na^+ flow (see Figure 23-7).

- As the action potential reaches its peak, opening of voltage-gated K^+ channels permits efflux of K^+ ions, which repolarizes and then hyperpolarizes the membrane. As these channels close, the membrane returns to its resting potential (see Figure 23-3).

- The excess cytosolic cations associated with an action potential generated at one point on an axon spread passively to the adjacent segment, triggering opening of voltage-gated Na^+ channels and movement of the action potential along the axon.

- Because of the absolute refractory period of the voltage-gated Na^+ channels and the brief hyperpolarization resulting from K^+ efflux, the action potential is propagated in one direction only, toward the axon terminus.

- Voltage-gated Na^+ and Ca^{2+} channels are monomeric proteins containing four domains that are structurally and functionally similar to each of the subunits in the tetrameric voltage-gated K^+ channels.

- Each domain or subunit in voltage-gated cation channels contains six transmembrane α helices and a nonhelical P segment that forms the ion-selectivity pore (see Figure 23-10).

- Opening of voltage-gated channels results from movement of the positively charged S4 α helices toward the extracellular side of the membrane in response to a depolarization of sufficient magnitude.

- Closing and inactivation of voltage-gated cation channels result from movement of a cytosolic segment into the open pore (see Figure 23-11c).

- Myelination, which increases the rate of impulse conduction up to a hundredfold, permits the close packing of neurons characteristic of vertebrate brains.

- In myelinated neurons, voltage-gated Na^+ channels are concentrated at the nodes of Ranvier. Depolarization at one node spreads rapidly with little attenuation to the next node, so that the action potential jumps from node to node (see Figure 23-13).
- Myelin sheaths are produced by glial cells that wrap themselves in spirals around neurons. Oligodendrocytes produce myelin for the CNS; Schwann cells, for the PNS (see Figure 23-14).
- Astrocytes, a third type of glial cell, wrap their processes around synapses and blood vessels. Astrocytes secrete proteins that stimulate synapse formation, and also induce the endothelial cells of blood vessels to produce a blood-brain barrier that limits transepithelial flow of substances (see Figure 23-16).

23.3 Communication at Synapses

As we have discussed, electrical pulses transmit signals along neurons, but signals are transmitted between neurons and other excitable cells by chemical signals. Synapses are the junctions where *presynaptic* neurons release these chemical signals, or neurotransmitters, that act on a *postsynaptic* target cell (Figure 23-4), which can be another neuron or a muscle or gland cell. Neurotransmitters are small, water-soluble molecules (e.g., acetylcholine, dopamine). The cell-cell communication at chemical synapses goes in one direction: pre- to postsynaptic cell. Arrival of an action potential at an axon terminal leads to opening of voltage-sensitive Ca^{2+} channels and an influx of Ca^{2+}, causing a localized rise in the cytosolic Ca^{2+} concentration in the axon terminus. The rise in Ca^{2+} in turn triggers fusion of small (40–50-nm) synaptic vesicles containing neurotransmitters with the plasma membrane, releasing neurotransmitters from this presynaptic cell into the synaptic cleft, the narrow space separating it from postsynaptic cells. The membrane of the postsynaptic cell is located within approximately 50 nm of the presynaptic membrane.

Neurotransmitter receptors fall into two broad classes: ligand-gated ion channels, which open immediately upon neurotransmitter binding, and G protein–coupled receptors. Neurotransmitter binding to a G protein–coupled receptor induces the opening or closing of a *separate* ion-channel protein over a period of seconds to minutes. These "slow" neurotransmitter receptors are discussed in Chapter 15 along with G protein–coupled receptors that bind different types of ligands and modulate the activity of cytosolic proteins other than ion channels. Here we examine the structure and operation of the *nicotinic acetylcholine receptor* found at many nerve-muscle synapses. It was the first ligand-gated ion channel to be purified, cloned, and characterized at the molecular level, and provides a paradigm for other neurotransmitter-gated ion channels.

The duration of the neurotransmitter signal depends on the amount of transmitter released by the presynaptic cell, which in turn depends on the amount of transmitter that had been stored and upon the frequency of action potentials arriving at the synapse. The duration of signal also depends on how rapidly any remaining neurotransmitter is retrieved by the presynaptic cell. Presynaptic cell plasma membranes, as well as glia, contain transporter proteins that pump neurotransmitters across the plasma membrane back into the cell, thus keeping the extracellular concentrations of transmitter low.

Here we focus first on how synapses form and how they control the regulated secretion of neurotransmitters in the context of the basic principles of vesicular trafficking outlined in Chapter 14. Next we look at the mechanisms that limit the duration of the synaptic signal, and how neurotransmitters are received and interpreted by the postsynaptic cell.

Formation of Synapses Requires Assembly of Presynaptic and Postsynaptic Structures

Axons extend from the cell body during development, guided by signals from other cells along the way so that the axon termini will reach the correct location (see Section 23.5). As axons grow, they come into contact with the dendrites of other neurons, and often at such sites synapses form. In the CNS, synapses with presynaptic specializations occur frequently all along an axon, in contrast to motor neurons, which form synapses with muscle cells only at the axon termini.

Neurons cultured in isolation will not form synapses very efficiently, but when glia are added, the rate of synapse formation increases substantially. Astrocytes and Schwann cells send signals to neurons to stimulate the formation of synapses and then help to preserve them (Figure 23-17). To

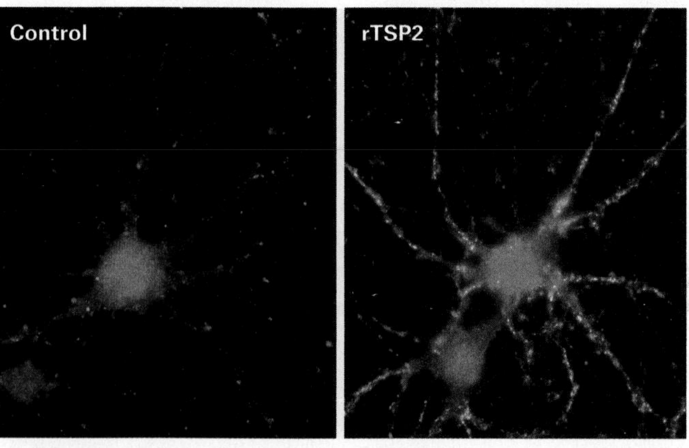

▲ EXPERIMENTAL FIGURE 23-17 **Signals from astrocytes have been shown to induce synapse formation.** Immunostaining for the presynaptic protein synaptotagmin (red) and the postsynaptic protein PSD-95 (yellow) yields few measured puncta (dots of stain) in control retinal ganglion cells cultured in the absence of astrocytes (*left*). However, a 3–5-fold increase in puncta occurs when these cells are cultured in the presence of astrocytes or the astrocyte protein product thrombospondin (*right*). (rTSP2 = the recombinant thrombospondin 2 that was used). Astrocytes secrete thrombospondin, which by itself has much the same effect on synapse formation as astrocytes themselves. Scale bar is 30 μm. [Reproduced with permission from K. S. Christopherson et al., 2005, *Cell* **120:**421–433.]

discover the signals involved, culture medium in which glia had been incubated was added to neuron cultures, and synapse formation was stimulated. By purifying different substances from that medium it was possible to identify the signal. Thrombospondin (TSP) protein, a component of extracellular matrix, was found to be the active agent. Confirmation came from mice lacking two *thrombospondin* genes: the mice had only 70 percent of the normal number of synapses in their brains. TSP probably does not work alone, as it is not as potent in inducing synapses as are whole glia. Another molecule that appears to account for some of the synapse-inducing activity of glia is cholesterol, and direct contact between glia and neurons may contribute as well.

Mutual communication between neurons and the glia that surround them is frequent and complex. The signals and information they carry is an area of active research. There is even evidence that neurons form synapses on glia. While glia do not have action potentials, they do have complex arrays of channels and ion fluxes.

At the site of a synapse, the presynaptic cell has hundreds to thousands of synaptic vesicles, some docked at the membrane and others waiting in reserve. The release into the synaptic cleft occurs in the *active zone,* a specialized region of the plasma membrane containing a remarkable assemblage of proteins whose functions include modifying the properties of the synaptic vesicles and bringing them into position for docking and fusing with the plasma membrane. Viewed by electron microscopy, the active zone has electron-dense material and fine cytoskeletal filaments (Figure 23-18). The active zone is assembled gradually, with synaptic vesicles accumulating first, then cytoskeletal elements, and then other proteins. A similarly dense region of specialized structures is seen across the synapse in the postsynaptic cell, the *postsynaptic density (PSD).* Cell-adhesion molecules that connect pre- and postsynaptic cells keep the active zone and PSD aligned. After release of synaptic vesicles in response to an action potential, the presynaptic neuron retrieves synaptic vesicle membrane proteins by **endocytosis** both within and outside the active zone.

The induction of PSD assembly has been extensively studied at the *neuromuscular junction* (NMJ). At these synapses **acetylcholine** is the neurotransmitter produced by motor neurons, and its receptor, AChR, is produced by the postsynaptic cell, which is a muscle cell. Muscle cell precursors, myoblasts, put into culture will spontaneously fuse into multinucleate myotubes that look similar to normal muscle cells (Chapter 21). As myotubes form, AChR is produced and inserted into the plasma membrane of the myotubes, reaching a density of about 1000 receptors/μm^2. The AChR is dispersed through the membrane, but if neurons are added to the culture, the AChR starts to concentrate at points of contact with the neurons. The neurons cause movement of preexisting AChR and also induce the myotubes to produce additional AChR. The density of receptors in a mature synapse reaches about 10,000–20,000/μm^2, while elsewhere in the plasma membrane the density is $<10/\mu m^2$.

These observations led to an investigation of the mutual signaling by neurons and myotubes. The conclusion from this work is that muscle cells begin to organize their postsynaptic structures before there is any discernible influence of motor axons. The arrival of an axon stabilizes and further modifies the structures that have formed.

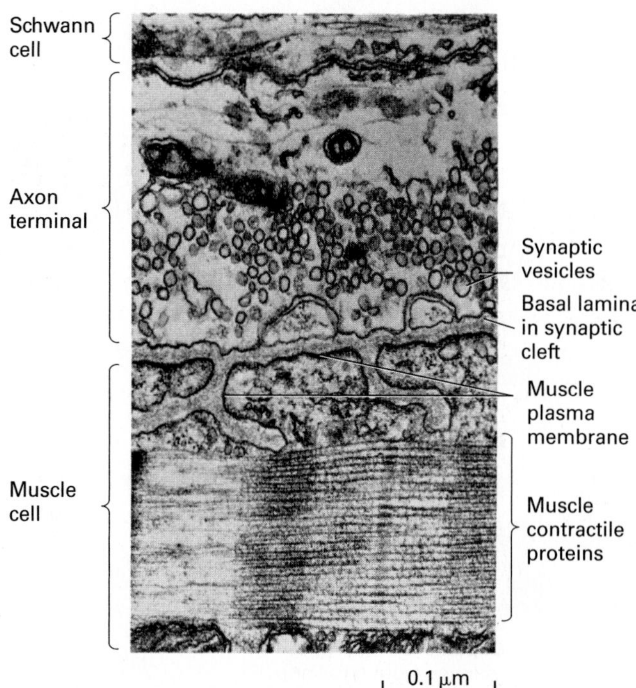

▲ **FIGURE 23-18 Synaptic vesicles in the axon terminal near the region where neurotransmitter is released.** In this longitudinal section through a neuromuscular junction, the basal lamina lies in the synaptic cleft separating the neuron from the muscle membrane, which is extensively folded. Acetylcholine receptors are concentrated in the postsynaptic muscle membrane at the top and part way down the sides of the folds in the membrane. A Schwann cell surrounds the axon terminal. [From J. E. Heuser and T. Reese, 1977, in E. R. Kandel, ed., *The Nervous System,* vol. 1, *Handbook of Physiology,* Williams and Wilkins, p. 266.]

Neurotransmitters Are Transported into Synaptic Vesicles by H^+-Linked Antiport Proteins

In this section, we focus on how neurotransmitters are packaged in membrane-bound *synaptic vesicles* in the axon terminus. Numerous small molecules function as neurotransmitters at various synapses. With the exception of acetylcholine, the neurotransmitters shown in Figure 23-19 are amino acids or derivatives of amino acids. Nucleotides such as ATP and the corresponding nucleosides, which lack phosphate groups, also function as neurotransmitters. Each neuron generally produces just one type of neurotransmitter.

All the "classic" neurotransmitters are synthesized in the cytosol and imported into membrane-bound synaptic vesicles within axon terminals, where they are stored. These vesicles are 40–50 nm in diameter, and their lumen has a low pH, generated by operation of a V-class proton pump in the vesicle membrane. Similar to the accumulation of metabolites

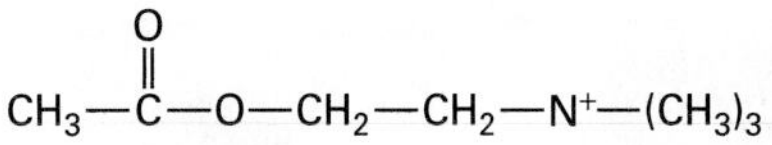

$CH_3-C(=O)-O-CH_2-CH_2-N^+-(CH_3)_3$

Acetylcholine

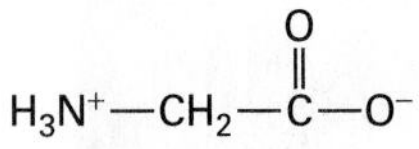

$H_3N^+-CH_2-C(=O)-O^-$

Glycine

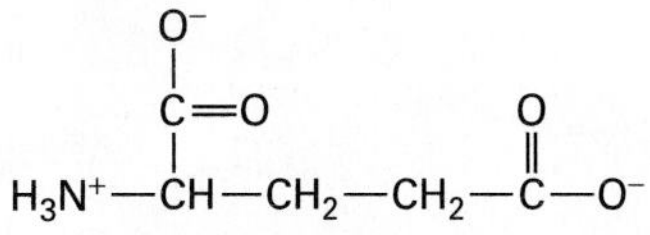

$H_3N^+-CH(COO^-)-CH_2-CH_2-C(=O)-O^-$

Glutamate

HO, HO — benzene ring — $CH_2-CH_2-NH_3^+$

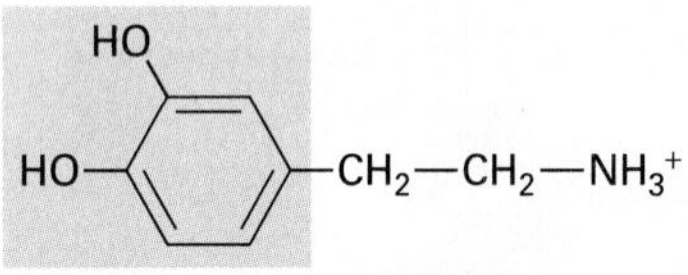

Dopamine
(derived from tyrosine)

HO, HO — benzene ring — $CH(OH)-CH_2-NH_3^+$

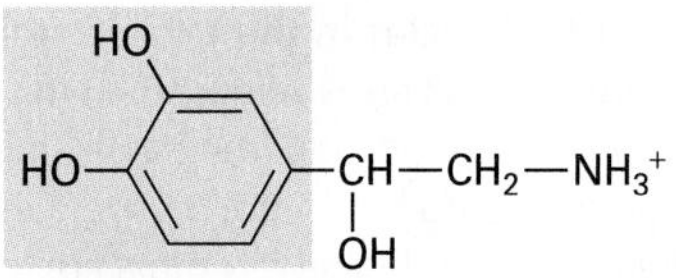

Norepinephrine
(derived from tyrosine)

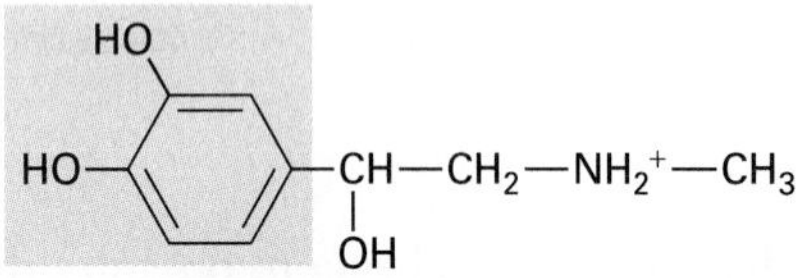

HO, HO — benzene ring — $CH(OH)-CH_2-NH_2^+-CH_3$

Epinephrine
(derived from tyrosine)

HO — indole ring (N—H) — $CH_2-CH_2-NH_3^+$

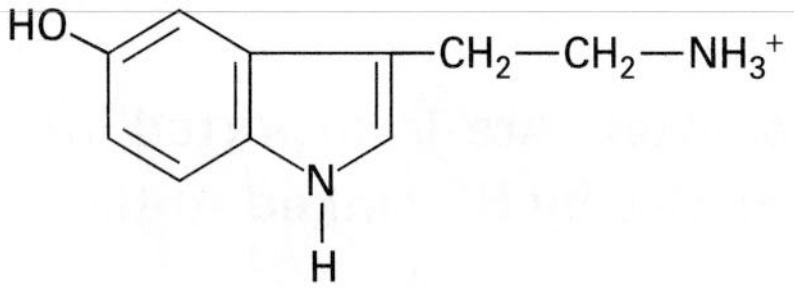

Serotonin, or **5-hydroxytryptamine**
(derived from tryptophan)

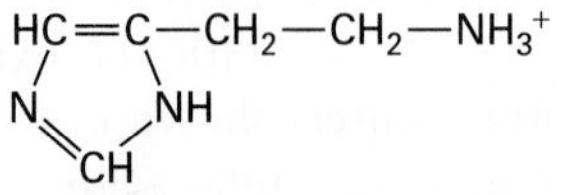

$HC=C-CH_2-CH_2-NH_3^+$ (imidazole ring: N, NH, CH)

Histamine
(derived from histidine)

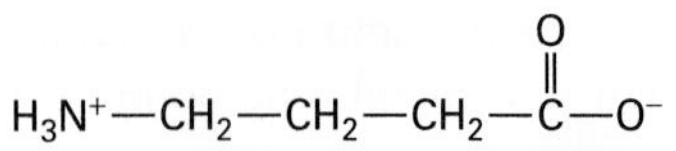

$H_3N^+-CH_2-CH_2-CH_2-C(=O)-O^-$

γ-Aminobutyric acid, or **GABA**
(derived from glutamate)

◀ **FIGURE 23-19 Structures of several small molecules that function as neurotransmitters.** Except for acetylcholine, all these are amino acids (glycine and glutamate) or derived from the indicated amino acids. The three transmitters synthesized from tyrosine, which contain the catechol moiety (blue highlight), are referred to as *catecholamines*.

in plant vacuoles (see Figure 11-28), this proton concentration gradient (vesicle lumen > cytosol) powers neurotransmitter import by ligand-specific H^+-linked antiporters in the vesicle membrane.

For example, acetylcholine is synthesized from acetyl coenzyme A (acetyl CoA), an intermediate in the degradation of glucose and fatty acids, and choline in a reaction catalyzed by choline acetyltransferase:

$$\underset{\text{Acetyl CoA}}{CH_3-\overset{O}{\overset{\|}{C}}-S-CoA} + \underset{\text{Choline}}{HO-CH_2-CH_2-N^+(CH_3)_3} \xrightarrow{\text{Choline acetyltransferase}}$$

$$\underset{\text{Acetylcholine}}{CH_3-\overset{O}{\overset{\|}{C}}-O-CH_2-CH_2-N^+(CH_3)_3} + CoA-SH$$

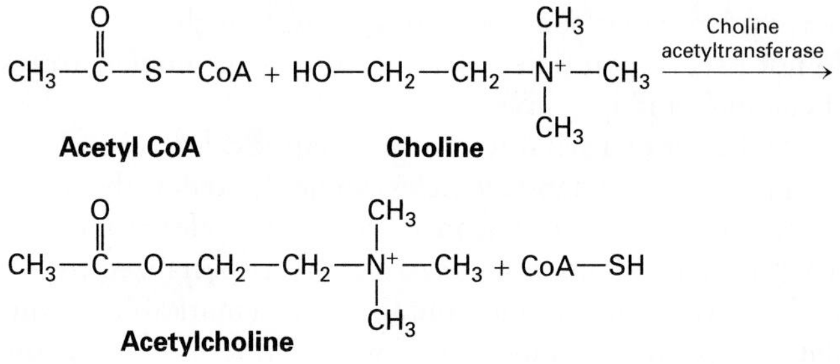

Synaptic vesicles take up and concentrate acetylcholine from the cytosol against a steep concentration gradient, using an H^+/acetylcholine antiporter in the vesicle membrane. Curiously, the gene encoding this antiporter is contained entirely within the first intron of the gene encoding choline acetyltransferase, a mechanism conserved throughout evolution for ensuring coordinate expression of these two proteins. Different H^+/neurotransmitter antiport proteins are used for import of other neurotransmitters into synaptic vesicles.

Synaptic Vesicles Loaded with Neurotransmitter Are Localized near the Plasma Membrane

Neurotransmitters are synthesized by enzymes in the cytosol and then transported into synaptic vesicles by transporter proteins dedicated to the task. For example glutamate is imported into synaptic vesicles by proteins called *vesicular glutamate transporters (VGLUTs)*. VGLUTs are highly specific for glutamate but have rather low substrate affinity (K_m = 1–3mM). The transporters are antiporters, moving glutamate into synaptic vesicles while protons move in the other direction. The membrane-potential gradient that drives the transport process is established by a vacuolar-type ATPase (Chapter 11).

The **exocytosis** of neurotransmitters from synaptic vesicles involves targeting and fusion events similar to those that lead to release of secreted proteins in the secretory pathway (Figure 23-20). However, several unique features permit the very rapid release of neurotransmitters in response to arrival of an action potential at the presynaptic axon terminal. For example, in resting neurons some neurotransmitter-filled synaptic vesicles are "docked" at the plasma membrane; others are in reserve in the active zone near the plasma membrane at the synaptic cleft.

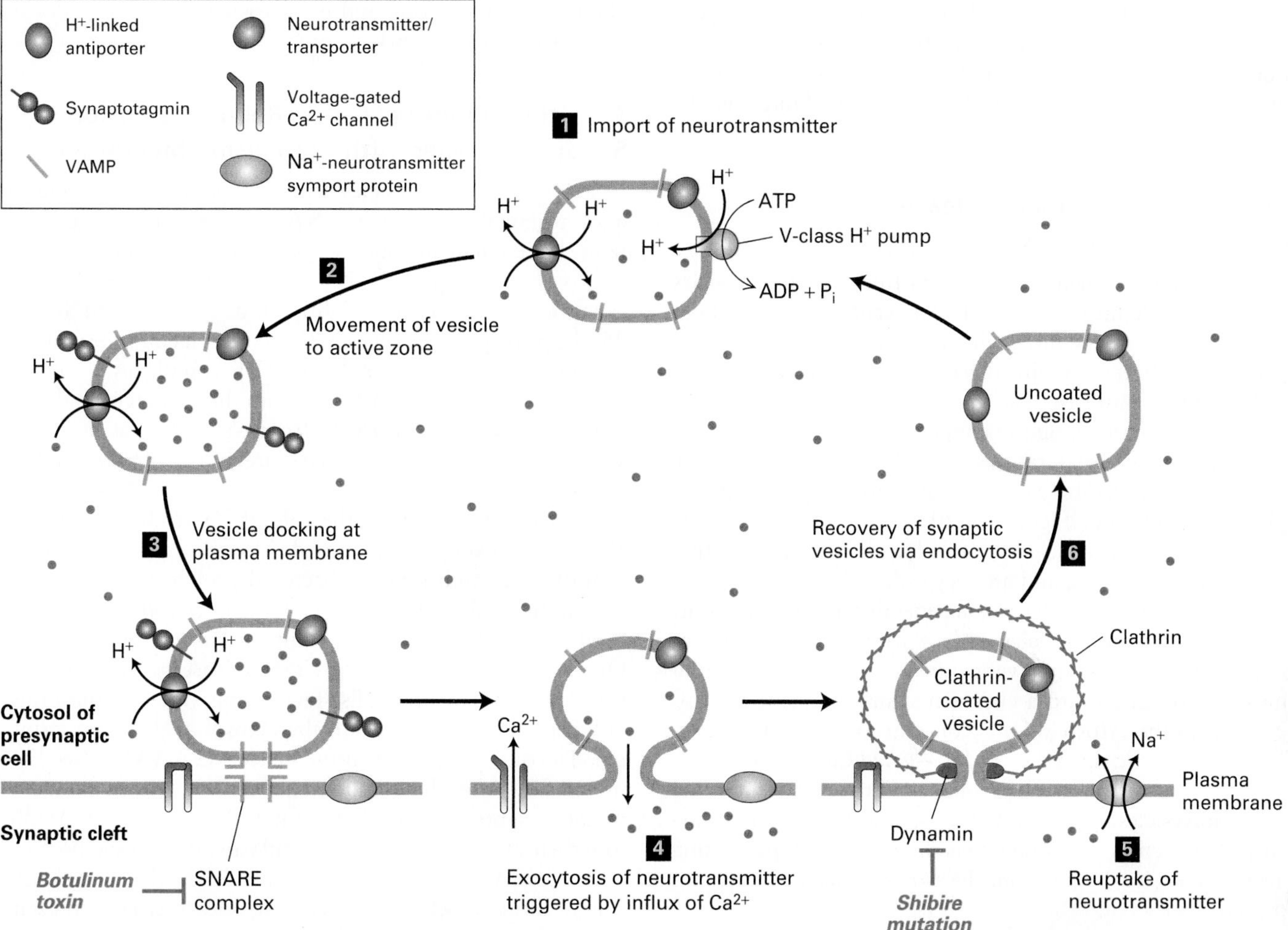

▲ **FIGURE 23-20 Cycling of neurotransmitters and of synaptic vesicles in axon terminals.** Most synaptic vesicles are formed by endocytic recycling as depicted here. The entire cycle typically takes about 60 seconds. Step **1**: The uncoated vesicles employ a variety of antiporters (blue) and other transport proteins (green) to import neurotransmitters (red dots) from the cytosol. Step **2**: Synaptic vesicles loaded with neurotransmitter move to the active zone. Step **3**: Vesicles dock at defined sites on the plasma membrane of a presynaptic cell. Synaptotagmin prevents membrane fusion and release of neurotransmitter. Botulinum toxin prevents exocytosis by proteolytically cleaving VAMP, the v-SNARE on vesicles. Synaptotagmin does not participate in steps **4**–**6** or **1**, though it is still present. For simplicity, it is not shown. Step **4**: In response to a nerve impulse (action potential), voltage-gated Ca^{2+} channels in the plasma membrane open, allowing an influx of Ca^{2+} from the extracellular medium. The resulting Ca^{2+}-induced conformational change in synaptotagmin leads to fusion of docked vesicles with the plasma membrane and release of neurotransmitters into the synaptic cleft. Step **5**: After clathrin/AP vesicles containing v-SNARE and neurotransmitter transporter proteins bud inward and are pinched off in a dynamin-mediated process, they lose their coat proteins. Dynamin mutations such as *shibire* in *Drosophila* block the re-formation of synaptic vesicles, leading to paralysis. At the same time, Na^+ symporter proteins take up neurotransmitter from the synaptic cleft, which limits the duration of the action potential and partially recharges the cell with transmitter. Step **6**: Vesicles are recovered by endocytosis, creating uncoated vesicles, ready to be refilled and begin the cycle anew. Unlike most neurotransmitters, acetylcholine is not recycled. See the text for details. [See K. Takei et al., 1996, *J. Cell Biol.* **133:**1237; V. Murthy and C. Stevens, 1998, *Nature* **392:**497; and R. Jahn et al., 2003, *Cell* **112:**519.]

In addition, the membrane of synaptic vesicles contains a specialized Ca^{2+}-binding protein that senses the rise in cytosolic Ca^{2+} after arrival of an action potential, triggering rapid fusion of docked vesicles with the presynaptic membrane.

A highly organized arrangement of cytoskeletal fibers in the axon terminal helps localize synaptic vesicles in the active zone. The vesicles themselves are linked together by *synapsin,* a fibrous phosphoprotein associated with the cytosolic surface of all synaptic-vesicle membranes. Filaments of synapsin also radiate from the plasma membrane and bind to vesicle-associated synapsin. These interactions probably keep synaptic vesicles close to the part of the plasma membrane facing the synapse. Indeed, synapsin knockout mice, although viable, are prone to seizures; during repetitive stimulation of many neurons in such mice, the number of synaptic vesicles that fuse with the plasma membrane is greatly reduced. Thus synapsins are thought to recruit synaptic vesicles to the active zone.

Rab3A, a GTP-binding protein located in the membrane of synaptic vesicles, is also required for targeting of neurotransmitter-filled vesicles to the active zone of presynaptic cells facing the synaptic cleft. Rab3A knockout mice, like synapsin-deficient mice, exhibit a reduced number of synaptic vesicles able to fuse with the plasma membrane after repetitive stimulation. The neuron-specific Rab3 is similar in sequence and function to other Rab proteins that participate in docking vesicles on particular target membranes in the secretory pathway.

Influx of Ca^{2+} Triggers Release of Neurotransmitters

The exocytosis of neurotransmitters from synaptic vesicles involves vesicle-targeting and fusion events similar to those that occur during the intracellular transport of secreted and plasma-membrane proteins (Chapter 13). Two features critical to synapse function differ from other secretory pathways: (1) secretion is tightly coupled to arrival of an action potential at the axon terminus, and (2) synaptic vesicles are recycled locally to the axon terminus after fusion with the plasma membrane. Figure 23-20 shows the entire cycle whereby synaptic vesicles are filled with neurotransmitter, release their contents, and are recycled.

Depolarization of the plasma membrane cannot, by itself, cause synaptic vesicles to fuse with the plasma membrane. In order to trigger vesicle fusion, an action potential must be converted into a chemical signal—namely, a localized rise in the cytosolic Ca^{2+} concentration. The transducers of the electric signals are *voltage-gated Ca^{2+} channels* localized to the region of the plasma membrane adjacent to the synaptic vesicles. The membrane depolarization due to arrival of an action potential opens these channels, permitting an influx of Ca^{2+} ions from the extracellular medium into the axon terminal.

A simple experiment demonstrates the importance of voltage-gated Ca^{2+} channels in release of neurotransmitters. A preparation of neurons in a Ca^{2+}-containing medium is treated with tetrodotoxin, a drug that blocks voltage-gated Na^+ channels and thus prevents conduction of action potentials. As expected, no neurotransmitters are secreted into the culture medium. If the axonal membrane then is artificially depolarized by making the medium ≈100 mM KCl in the presence of extracellular Ca^{2+}, neurotransmitters are released from the cells because of the influx of Ca^{2+} through open voltage-gated Ca^{2+} channels. Indeed, patch-clamping experiments show that voltage-gated Ca^{2+}channels, like voltage-gated Na^+ channels, open transiently upon depolarization of the membrane.

Two pools of neurotransmitter-filled synaptic vesicles are present in axon terminals: those docked at the plasma membrane, which can be readily exocytosed, and those in reserve in the active zone near the plasma membrane. Each rise in Ca^{2+} triggers exocytosis of about 10 percent of the docked vesicles. Membrane proteins unique to synaptic vesicles then are specifically internalized by endocytosis, usually via the same types of clathrin-coated vesicles used to recover other plasma-membrane proteins by other types of cells. After the endocytosed vesicles lose their clathrin coat, they are rapidly refilled with neurotransmitter. The ability of many neurons to fire 50 times a second is clear evidence that the recycling of vesicle membrane proteins occurs quite rapidly. The machinery of endocytosis and exocytosis is highly conserved, and is explained extensively in Chapter 14.

A Calcium-Binding Protein Regulates Fusion of Synaptic Vesicles with the Plasma Membrane

Fusion of synaptic vesicles with the plasma membrane of axon terminals depends on **SNAREs,** the same type of proteins that mediate membrane fusion of other regulated secretory vesicles (Figure 23-20). The principal v-SNARE in synaptic vesicles (VAMP) tightly binds syntaxin and SNAP-25, the principal t-SNAREs in the plasma membrane of axon terminals, to form four-helix SNARE complexes. After fusion, SNAP proteins and NSF within the axon terminal promote disassociation of VAMP from t-SNAREs, as in the fusion of secretory vesicles depicted previously (Figure 14-10).

Strong evidence for the role of VAMP in neurotransmitter exocytosis is provided by the mechanism of action of botulinum toxin, a bacterial protein that can cause the paralysis and death characteristic of *botulism,* a type of food poisoning. The toxin is composed of two polypeptides: One binds to motor neurons that release acetylcholine at synapses with muscle cells, facilitating entry of the other polypeptide, a protease, into the cytosol of the axon terminal. The only protein this protease cleaves is VAMP (see Figure 23-20). After the botulinum protease enters an axon terminal, synaptic vesicles that are not already docked rapidly lose their ability to fuse with the plasma membrane because cleavage of VAMP prevents assembly of SNARE complexes. The resulting block in acetylcholine release at neuromuscular synapses causes paralysis. However, vesicles that are already docked exhibit remarkable resistance to the toxin, indicating that SNARE complexes may already be in a partially assembled, protease-resistant state when vesicles are docked on the presynaptic membrane. ▮

The signal that triggers exocytosis of docked synaptic vesicles is a rise in the Ca^{2+} concentration in the cytosol near vesicles from <0.1 μM, characteristic of resting cells, to 1–100 μM following arrival of an action potential in stimulated cells. The speed with which synaptic vesicles fuse with the presynaptic membrane after a rise in cytosolic Ca^{2+} (less than 1 ms) indicates that the fusion machinery is entirely assembled in the resting state and can rapidly undergo a conformational change leading to exocytosis of neurotransmitter. A Ca^{2+}-binding protein called *synaptotagmin,* located in the membrane of synaptic vesicles, is thought to be a key component of the vesicle-fusion machinery that triggers exocytosis in response to Ca^{2+} (see Figure 23-20).

Several lines of evidence support a role for synaptotagmin as the Ca^{2+} sensor for exocytosis of neurotransmitters. Mutant embryos of *Drosophila* and *C. elegans* that completely lack synaptotagmin fail to hatch and exhibit very reduced, uncoordinated muscle contractions. Larvae with partial loss-of-function mutations of synaptotagmin survive, but their neurons are defective in Ca^{2+}-stimulated vesicle exocytosis. Moreover, in mice, mutations in synaptotagmin that decrease its affinity for Ca^{2+} cause a corresponding increase in the amount of cytosolic Ca^{2+} needed to trigger rapid exocytosis. The precise mechanism of synaptotagmin function is still unresolved.

Signaling at Synapses Is Terminated by Degradation or Reuptake of Neurotransmitters

Following their release from a presynaptic cell, neurotransmitters must be removed or destroyed to prevent continued stimulation of the postsynaptic cell. Signaling can be terminated by diffusion of a transmitter away from the synaptic cleft, but this is a slow process. Instead, one of two more rapid mechanisms terminates the action of neurotransmitters at most synapses.

Signaling by acetylcholine is terminated when it is hydrolyzed to acetate and choline by *acetylcholinesterase,* an enzyme localized to the synaptic cleft. Choline released in this reaction is transported back into the presynaptic axon terminal by a Na^+/choline symporter and used in synthesis of more acetylcholine. The operation of this transporter is similar to that of the Na^+-linked symporters used to transport glucose into cells against a concentration gradient (see Figure 11-25).

With the exception of acetylcholine, all the neurotransmitters shown in Figure 23-19 are removed from the synaptic cleft by transport into the axon terminals that released them. Thus these transmitters are recycled intact, as depicted in Figure 23-20 (step **5**). Transporters for GABA, norepinephrine, dopamine, and serotonin were the first to be cloned and studied. These four transport proteins are all Na^+-linked symporters. They are 60–70 percent identical in their amino acid sequences, and each is thought to contain 12 transmembrane α helices. As with other Na^+ symporters, the movement of Na^+ into the cell down its electrochemical gradient provides the energy for uptake of the neurotransmitter. To maintain electroneutrality, Cl^- often is transported via an ion channel along with the Na^+ and neurotransmitter.

Neurotransmitters and their transporters are targets of a variety of powerful and sometimes devastating drugs. Cocaine inhibits the transporters for norepinephrine, serotonin, and dopamine. Binding of cocaine to the dopamine transporter inhibits reuptake of dopamine, thus prolonging signaling at key brain synapses; indeed, the dopamine transporter is the principal brain "cocaine receptor." Therapeutic agents such as the antidepressant drugs fluoxetine (Prozac) and imipramine block serotonin uptake, and the tricyclic antidepressant desipramine blocks norepinephrine uptake. ■

Fly Mutants Lacking Dynamin Cannot Recycle Synaptic Vesicles

Synaptic vesicles are formed primarily by endocytic budding from the plasma membrane of axon terminals. Endocytosis usually involves clathrin-coated pits and is quite specific, in that several membrane proteins unique to the synaptic vesicles (e.g., neurotransmitter transporters) are specifically incorporated into the endocytosed vesicles. In this way, synaptic-vesicle membrane proteins can be reused and the recycled vesicles refilled with neurotransmitter (see Figure 23-20).

As in the formation of other clathrin/AP-coated vesicles, pinching off of endocytosed synaptic vesicles requires the GTP-binding protein *dynamin* (see Figure 14-19). Indeed, analysis of a temperature-sensitive *Drosophila* mutant called *shibire (shi),* which encodes the fly dynamin protein, provided early evidence for the role of dynamin in endocytosis. At the permissive temperature of 20 °C, the mutant flies are normal, but at the nonpermissive temperature of 30 °C, they are paralyzed (*shibire,* "paralyzed," in Japanese) because pinching off of clathrin-coated pits in neurons and other cells is blocked. When viewed in the electron microscope, the *shi* neurons at 30 °C show abundant clathrin-coated pits with long necks but few clathrin-coated vesicles. The appearance of nerve terminals in *shi* mutants at the nonpermissive temperature is similar to that of terminals from normal neurons incubated in the presence of a nonhydrolyzable analog of GTP (see Figure 14-20). Because of their inability to pinch off new synaptic vesicles, the neurons in *shi* mutants eventually become depleted of synaptic vesicles when flies are shifted to the nonpermissive temperature, leading to a cessation of synaptic signaling and to paralysis.

Opening of Acetylcholine-Gated Cation Channels Leads to Muscle Contraction

In this section we look at how binding of neurotransmitters by receptors on postsynaptic cells leads to changes in their membrane potential, using the communication between motor neurons and muscles as an example. At these synapses, often called neuromuscular junctions, acetylcholine is the neurotransmitter. A single axon terminus of a frog motor neuron may contain a million or more synaptic vesicles, each containing 1000–10,000 molecules of acetylcholine; these vesicles often accumulate in rows in the active zone (see Figure 23-18). Such a neuron can form synapses with a single skeletal muscle cell at several hundred points.

The nicotinic acetylcholine receptor, which is expressed in muscle cells, is a *ligand-gated channel* that admits both K^+ and Na^+. These receptors are also produced in the brain and are important in learning and memory; acetylcholine receptor loss is observed in schizophrenia, epilepsy, drug addiction, and Alzheimer's disease. Antibodies against acetylcholine receptors constitute a major part of the autoimmune reactivity in the disease myasthenia gravis. The receptor is so named because it is bound by nicotine; it has been implicated in addiction to nicotine by tobacco smokers. The receptor is also the target of potent neurotoxins, such as the conotoxins produced by certain Pacific Ocean snails. There are at least 14 different isoforms of the receptor, which assemble into homo- and heteropentamers with varied properties.

The effect of acetylcholine on this receptor can be determined by patch-clamping studies on isolated outside-out patches of muscle plasma membranes (see Figure 11-21c). Such measurements have shown that acetylcholine causes opening of a cation channel in the receptor capable of transmitting 15,000–30,000 Na^+ or K^+ ions per millisecond.

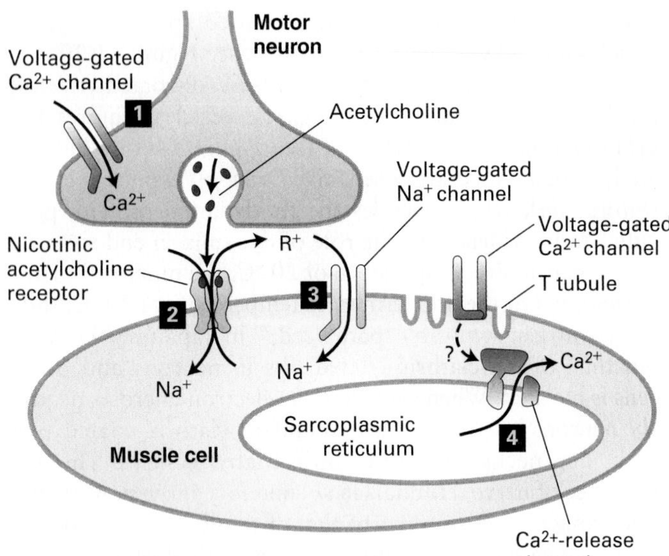

▲ **FIGURE 23-21 Sequential activation of gated ion channels at a neuromuscular junction.** Arrival of an action potential at the terminus of a presynaptic motor neuron induces opening of voltage-gated Ca^{2+} channels (step **1**) and subsequent release of acetylcholine, which triggers opening of the ligand-gated acetylcholine receptors in the muscle plasma membrane (step **2**). The open channel allows an influx of Na^+ and an efflux of K^+. The Na^+ influx produces a localized depolarization of the membrane, leading to opening of voltage-gated Na^+ channels and generation of an action potential (step **3**). When the spreading depolarization reaches T tubules, it is sensed by voltage-gated Ca^{2+} channels in the plasma membrane. Through an unknown mechanism (indicated as ?) these channels remain closed but influence Ca^{2+} channels in the sarcoplasmic reticulum membrane (a network of membrane-bound compartments in muscle), releasing stored Ca^{2+} into the cytosol (step **4**). The resulting rise in cytosolic Ca^{2+} causes muscle contraction by mechanisms discussed in Chapter 17.

However, since the resting potential of the muscle plasma membrane is near E_K, the potassium equilibrium potential, opening of acetylcholine receptor channels causes little increase in the efflux of K^+ ions; Na^+ ions, on the other hand, flow into the muscle cell, driven by the Na^+ electrochemical gradient.

The simultaneous increase in permeability to Na^+ and K^+ ions following binding of acetylcholine produces a net depolarization to about −15 mV from the muscle resting potential of −85 to −90 mV. As shown in Figure 23-21, this localized depolarization of the muscle plasma membrane triggers opening of voltage-gated Na^+ channels, leading to generation and conduction of an action potential in the muscle cell surface membrane by the same mechanisms described previously for neurons. When the membrane depolarization reaches T tubules, specialized invaginations of the plasma membrane, it affects Ca^{2+} channels in the plasma membrane apparently without causing them to open. Somehow this causes opening of adjacent Ca^{2+}-release channels in the sarcoplasmic reticulum membrane. The subsequent flow of stored Ca^{2+} ions from the sarcoplasmic reticulum into the cytosol raises the cytosolic Ca^{2+} concentration sufficiently to induce muscle contraction.

Careful monitoring of the membrane potential of the muscle membrane at a synapse with a cholinergic motor neuron has demonstrated spontaneous, intermittent, and random ≈2-ms depolarizations of about 0.5–1.0 mV in the absence of stimulation of the motor neuron. Each of these depolarizations is caused by the spontaneous release of acetylcholine from a single synaptic vesicle. Indeed, demonstration of such spontaneous small depolarizations led to the notion of the quantal release of acetylcholine (later applied to other neurotransmitters) and thereby led to the hypothesis of vesicle exocytosis at synapses. The release of one acetylcholine-containing synaptic vesicle results in the opening of about 3000 ion channels in the postsynaptic membrane, far short of the number needed to reach the threshold depolarization that induces an action potential. Clearly, stimulation of muscle contraction by a motor neuron requires the nearly simultaneous release of acetylcholine from numerous synaptic vesicles.

All Five Subunits in the Nicotinic Acetylcholine Receptor Contribute to the Ion Channel

The acetylcholine receptor from skeletal muscle is a pentameric protein with a subunit composition of $\alpha_2\beta\gamma\delta$. The α, β, γ, and δ subunits have considerable sequence homology; on average, about 35–40 percent of the residues in any two subunits are similar. The complete receptor has fivefold symmetry, and the actual cation channel is a tapered central pore lined by homologous segments from each of the five subunits (Figure 23-22).

The channel opens when the receptor cooperatively binds two acetylcholine molecules to sites located at the interfaces of the αδ and αγ subunits. Once acetylcholine is bound to a receptor, the channel is opened within a few microseconds. Studies measuring the permeability of different small cations suggest that the open ion channel is, at its narrowest, about 0.65–0.80 nm in diameter, in agreement with estimates from electron micrographs. This would be sufficient to allow passage of both Na^+ and K^+ ions with their shell of bound water molecules. Thus the acetylcholine receptor probably transports hydrated ions, unlike Na^+ and K^+ channels, both of which allow passage only of nonhydrated ions (see Figure 11-20).

The central ion channel is lined by five homologous transmembrane M2 α helices, one from each of the five subunits (see Figure 23-22a). The M2 helices are composed largely of hydrophobic or uncharged polar amino acids, but negatively charged aspartate or glutamate residues are located at each end, near the membrane faces, and several serine or threonine residues are near the middle. Mutant acetylcholine receptors in which a single negatively charged glutamate or aspartate in one M2 helix is replaced by a positively charged lysine have been expressed in frog oocytes. Patch-clamping measurements indicate that such altered proteins can function as channels, but the number of ions that pass through during the open state is reduced. The greater the number of glutamate or aspartate residues mutated (in one or multiple M2 helices), the greater the reduction in ion conductivity. These findings suggest that aspartate and glutamate residues form a ring of negative charges on the external surface of the pore that help to screen out anions and attract Na^+ or K^+ ions as they enter the channel. A similar ring of negative charges lining the cytosolic pore surface also helps select cations for passage (see Figure 23-22).

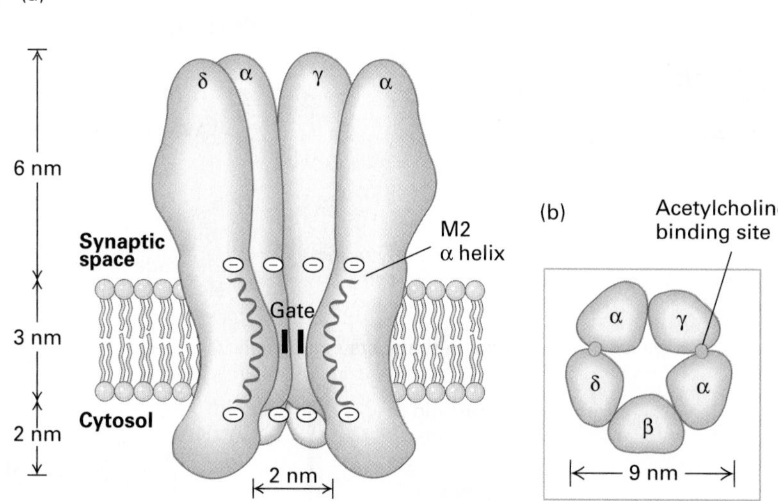

◀ **FIGURE 23-22 Three-dimensional structure of the nicotinic acetylcholine receptor.** (a) Schematic cutaway model of the pentameric receptor in the membrane; for clarity, the β subunit is not shown. Each subunit contains an M2 α helix (red) that faces the central pore. Aspartate and glutamate side chains at both ends of the M2 helices form two rings of negative charges that help exclude anions from and attract cations to the channel. The gate, which is opened by binding of acetylcholine, lies within the pore. (b) Cross section of the exoplasmic face of the receptor showing the arrangement of subunits around the central pore. The two acetylcholine binding sites are located about 3 nm from the membrane surface.

The two acetylcholine binding sites in the extracellular domain of the receptor lie ≈4 to 5 nm from the center of the pore. Binding of acetylcholine thus must trigger conformational changes in the receptor subunits that can cause channel opening at some distance from the binding sites. Receptors in isolated postsynaptic membranes can be trapped in the open or closed state by rapid freezing in liquid nitrogen. Images of such preparations suggest that the five M2 helices rotate relative to the vertical axis of the channel during opening and closing.

We have discussed the neuromuscular junction as an excellent example of how neurotransmitters and their receptors work. Similar ideas apply to glutamate and GABA, the two principal neurotransmitters in vertebrate brain. They use ligand-gated channels that work along the same principles as AchR.

Nerve Cells Make an All-or-None Decision to Generate an Action Potential

At the neuromuscular junction, virtually every action potential in the presynaptic motor neuron triggers an action potential in the postsynaptic muscle cell that propagates along the muscle fiber. The situation at synapses between neurons, especially those in the brain, is much more complex because the postsynaptic neuron commonly receives signals from many presynaptic neurons. The neurotransmitters released from presynaptic neurons may bind to an *excitatory receptor* on the postsynaptic neuron, thereby opening a channel that admits Na^+ ions or both Na^+ and K^+ ions. The acetylcholine receptor just discussed is one of many excitatory receptors, and opening of such ion channels leads to depolarization of the postsynaptic plasma membrane, promoting generation of an action potential. In contrast, binding of a neurotransmitter to an *inhibitory receptor* on the postsynaptic cell causes opening of K^+ or Cl^- channels, leading to an efflux of additional K^+ ions from the cytosol or an influx of Cl^- ions. In either case, the ion flow tends to hyperpolarize the plasma membrane, which inhibits generation of an action potential in the postsynaptic cell.

A single neuron can be affected simultaneously by signals received at multiple excitatory and inhibitory synapses. The neuron continuously integrates these signals and determines whether or not to generate an action potential. In this process, the various small depolarizations and hyperpolarizations generated at synapses move along the plasma membrane from the dendrites to the cell body and then to the axon initial segment, where they are summed together. An action potential is generated whenever the membrane at the axon initial segment becomes depolarized to a certain voltage called the *threshold potential* (Figure 23-23). Thus an action potential is generated in an all-or-nothing fashion: depolarization to the threshold always leads to an action potential, whereas any depolarization that does not reach the threshold potential never induces it.

Whether a neuron generates an action potential in the axon initial segment depends on the balance of the timing, amplitude, and localization of all the various inputs it receives; this signal computation differs for each type of neuron. In a sense, each neuron is a tiny computer that averages all the receptor activations and electric disturbances on its membrane and makes a decision whether to trigger an action potential and conduct it down the axon. An action potential will always have the same *magnitude* in any particular neuron. As we have noted, the *frequency* with which action potentials are generated in a particular neuron is the important parameter in its ability to signal other cells.

Gap Junctions Also Allow Neurons to Communicate

Chemical synapses employing neurotransmitters allow one-way communication at reasonably high speed. However sometimes signals go from cell to cell electrically, without the intervention of chemical synapses. *Electrical synapses* depend on **gap junction** channels that link two or more cells (Chapter 19). The effect of gap junction connections is to perfectly coordinate the activities of joined cells. An electrical synapse also is *bidirectional;* either neuron can excite the other. In the neocortex and thalamus and some other parts of the brain, electrical synapses are common. The key feature of electrical synapses is their speed. While it takes about 0.5–5 ms for a signal to cross a chemical synapse, transmission across

▶ **EXPERIMENTAL FIGURE 23-23 Incoming signals must reach the threshold potential to trigger an action potential in a postsynaptic cell.** In this example, the presynaptic neuron is generating about one action potential every 4 milliseconds. Arrival of each action potential at the synapse causes a small change in the membrane potential at the axon hillock of the postsynaptic cell, in this example a depolarization of ≈5 mV. When multiple stimuli cause the membrane of this postsynaptic cell to become depolarized to the threshold potential, here approximately −40 mV, an action potential is induced in it.

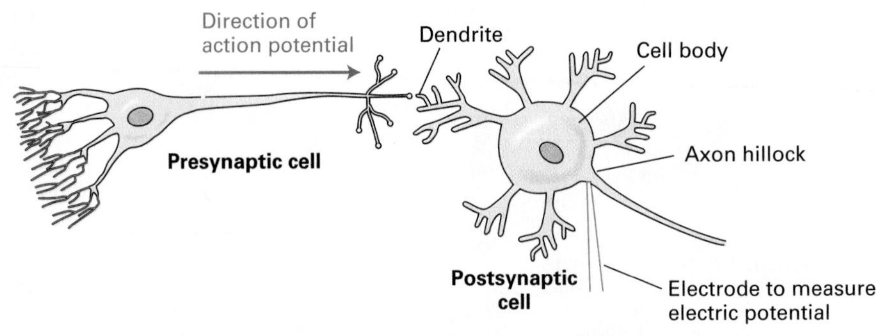

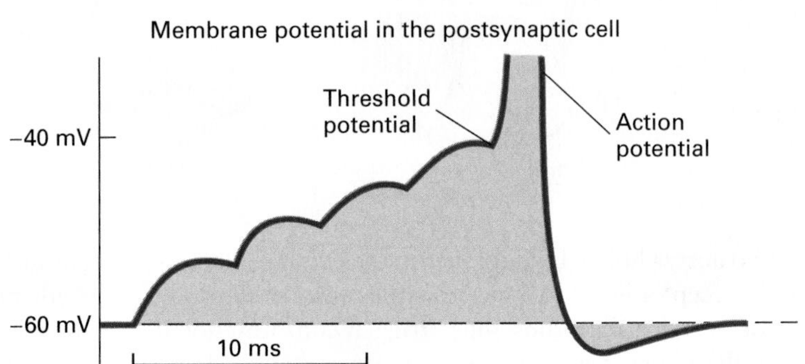

an electrical synapse is almost instantaneous, on the order of a fraction of a millisecond. The cytoplasm is continuous between the cells. In addition, the presynaptic cell (the one sending the signal) does not have to reach a threshold at which it can cause an action potential in the postsynaptic cell. Instead, any electrical current continues into the next cell and causes depolarization in proportion to the current.

An electrical synapse may contain thousands of gap channels, each composed of two hemichannels, one in each apposed cell. Gap junction channels have a structure similar to conventional gap junctions (Chapter 19). Each hemichannel is an assembly of six copies of the connexin protein. Since there are about 20 mammalian **connexin** genes, diversity in channel structure and function can arise from the different protein components. The 1.6–2.0-nm channel itself allows the diffusion of molecules up to about 1000 Da in size and has no trouble at all accommodating ions.

KEY CONCEPTS OF SECTION 23.3

Communication at Synapses

- Synapses are the junctions between a presynaptic cell and a postsynaptic cell, and consist of small gaps.
- Neurotransmitters are released by the presynaptic cell using exocytosis. They diffuse across the synapse and bind to receptors on the postsynaptic cell, which can be a neuron or a muscle.
- Chemical synapses of this sort are unidirectional (see Figure 23-4).
- Neurotransmitters (see Figure 23-19) are stored in hundreds to thousands of synaptic vesicles in the axon termini of the presynaptic cell (see Figure 23-18). When an action potential arrives there, voltage-sensitive Ca^{2+} channels open and the calcium causes synaptic vesicles to fuse with the plasma membrane, releasing neurotransmitter molecules into the synapse (see Figure 23-20).
- Communication between presynaptic and postsynaptic cells is abundant as a synapse is being formed. Cell-adhesion molecules keep the cells aligned. Neurons induce the accumulation of acetylcholine receptor, for example, in the postsynaptic muscle plasma membrane in the vicinity of a synapse.
- Synaptic vesicles fuse to the plasma membrane using cellular machinery that is standard issue for exocytosis, including SNAREs, syntaxin, and SNAP proteins. Synaptotagmin protein is the calcium sensor that detects the action potential–stimulated rise in calcium that leads to synaptic vesicle membrane fusion (see Figure 23-20).
- Dynamin, an endocytosis protein, is critical for the formation of new synaptic vesicles, probably to "pinch off" inbound vesicles.
- Neurotransmitter receptors fall into two classes: ligand-gated ion channels, which permit ion passage when open, and G protein–coupled receptors, which are linked to a separate ion channel.
- At synapses impulses are transmitted by neurotransmitters released from the axon terminal of the presynaptic cell and subsequently bound to specific receptors on the postsynaptic cell (see Figure 23-4).
- Low-molecular-weight neurotransmitters (e.g., acetylcholine, dopamine, epinephrine) are imported from the cytosol into synaptic vesicles by H^+-linked antiporters. V-class proton pumps maintain the low intravesicular pH that drives neurotransmitter import against a concentration gradient.
- Arrival of an action potential at a presynaptic axon terminal opens voltage-gated Ca^{2+} channels, leading to a localized rise in the cytosolic Ca^{2+} level that triggers exocytosis of

synaptic vesicles. Following neurotransmitter release, vesicles are formed by endocytosis and recycled (see Figure 23-20).

■ Coordinated operation of four gated ion channels at the synapse of a motor neuron and striated muscle cell leads to release of acetylcholine from the axon terminal, depolarization of the muscle membrane, generation of an action potential, and then contraction (see Figure 23-21).

■ The nicotinic acetylcholine receptor, a ligand-gated cation channel, contains five subunits, each of which has a transmembrane α helix (M2) that lines the channel (see Figure 23-22).

■ A postsynaptic neuron generates an action potential only when the plasma membrane at the axon hillock is depolarized to the threshold potential by the summation of small depolarizations and hyperpolarizations caused by activation of multiple neuronal receptors (see Figure 23-23).

■ Electrical synapses are direct, gap junction connections between neurons. Electrical synapses, unlike chemical synapses that employ neurotransmitter systems, are extremely fast in signal transmission and are bidirectional.

23.4 Sensational Cells: Seeing, Feeling, Hearing, Tasting, and Smelling

Dramatic progress has been made in understanding how our senses record impressions of the outside world, and how that information is processed by the brain. In this section we discuss cellular and molecular mechanisms and specialized nerve cells underlying vision, touch, hearing, taste, and olfaction.

The Eye Features Light-Sensitive Nerve Cells

While owls would favor hearing, and dogs smell, most humans would choose vision as the sense that provides the most effective window on the world. Light is fast, about 300,000 km/s, and moves in straight lines, so it is excellent for information transfer. The human eye is a complex structure that gathers light from the environment and focuses it on specialized light-sensitive nerve cells, which send signals to the brain, where they are translated into an image (Figure 23-24a).

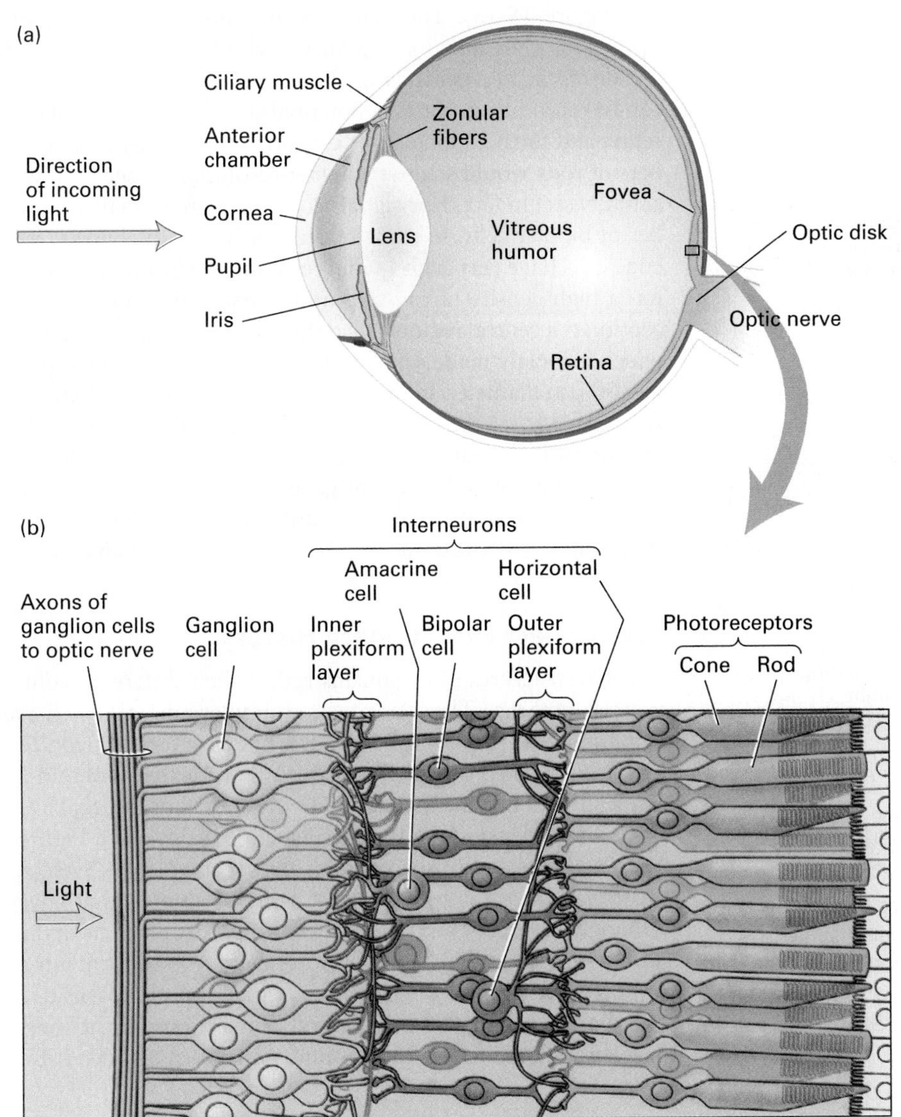

◄ **FIGURE 23-24 Structure of the human eye and three classes of neurons in the retina.** (a) The main tissues of the eye. Incoming light passes through the cornea, is focused by the lens, and activates light-sensing cells located in the retina. The iris restricts the amount of light entering the lens. The lens is supported by the zonular fibers, and moved by the ciliary muscle. The eye is filled with transparent, cushioning, vitreous fluid. The fovea is the location of the highest density of cells, and consequently senses the highest-resolution image. There is a blind spot (optic disc) where the optic nerve leaves the eye. (b) Detailed organization of cells in the retina. Note that incoming light has to pass through multiple layers of neurons before reaching the photoreceptor cells, the rods and cones. The interneurons include horizontal cells; bipolar cells, of which there are about a dozen types; and amacrine cells, of which there are more than 20 types. Retinal ganglion cells carry the signal information to the optic nerve. The two plexiform layers are where most connections are made. [Part (a) adapted from D. Randall, W. Burggren, and K. French, 2002, *Eckert Animal Physiology,* 5th ed., W. H. Freeman and Company, p. 259; part (b) from B. Kolb and I. Q. Whishaw, 2006, *An Introduction to Brain and Behavior,* 2d ed., Worth, p. 278.]

The human retina (Figure 23-24b) is about 200 μm thick. As we learned in Chapter 15, it contains two types of photoreceptors, *rods* and *cones,* which are the primary recipients of visual stimulation. Cones are involved in color vision, while rods are stimulated by weak light, like moonlight, over a range of wavelengths. In contrast to many other sensory neurons, stimulated photoreceptor cells become *hyperpolarized,* not depolarized. The photoreceptors synapse on layer upon layer of interneurons that are innervated by different combinations of photoreceptor cells. All these signals are processed and interpreted by the part of the brain called the *visual cortex*. In each eye we have about 6 million cones and 120 million rods, connecting to 540 million visual cortex cells, so a substantial part of our nervous system is devoted to detecting and interpreting light.

Rods detect faint light, as low as a single photon, owing to their sensitive visual pigment and their ability to amplify a weak signal. Rods have a single pigment that allows an equal response to a broad spectrum of wavelengths. Cones (Figure 23-25) come in the three types: red, green, and blue. The brain deduces color information by comparing the signals from a trio of cone cells, one of each kind, that share the same *receptive field*. The receptive field of each cell is measured as an angle with its vertex at the cell. If cells are densely packed and each has a small receptive field, highly detailed visual information is collected. If cells are less dense, have large receptive fields, or both, the image will have lower resolution.

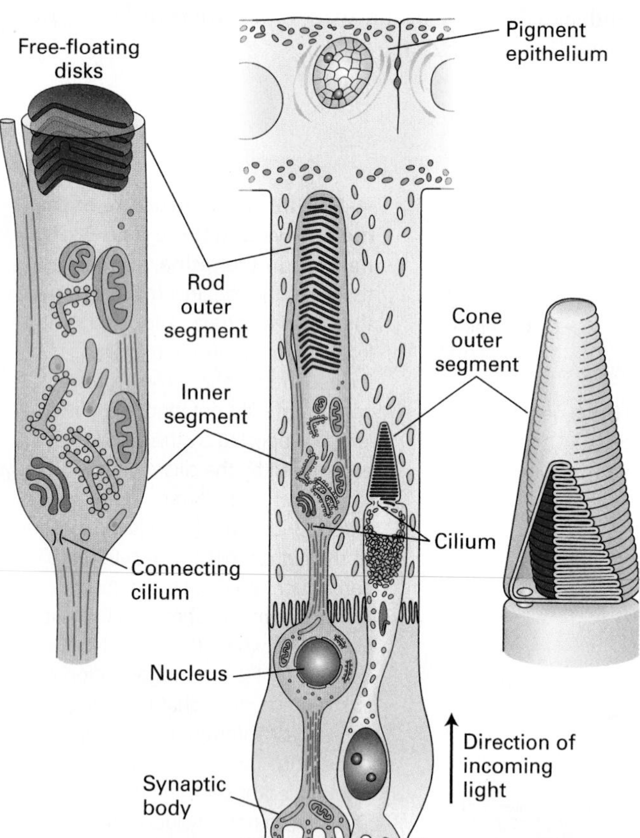

▲ **FIGURE 23-25 Rods and Cones.** Vertebrates have two types of photoreceptors, rods and cones, that differ morphologically and functionally. Cones detect color; rods detect light intensity but not color and are more sensitive than cones to low light levels. The pigments that absorb light are packed into flattened disks in the outer segments of rods and cones. Note that the outer segments containing the pigments are located on the innermost side of the retina, so light must traverse layers of cells before reaching the sensory organelles. [Adapted from from D. Randall, W. Burggren, and K. French, 2002, *Eckert Animal Physiology,* 5th ed., W. H. Freeman and Company, p. 261.]

Rods and cones are packed with light-absorbing pigments consisting of an *opsin* protein covalently bound to a small light-sensitive molecule called *11-cis-retinal.* These pigments are arranged in flattened membrane disks in the outer segments of rods and cones (Figure 23-25; also see Figure 15-16). Opsin-containing disks are continuously replaced, completely turning over about every 12 days. Opsins are G protein–coupled receptors that are activated when their bound retinal absorbs light. The pigment in rods is called *rhodopsin.* Light-induced isomerization of the retinal portion of rhodopsin alters the protein's conformation, triggering a signaling pathway that closes Na^+ and Ca^{2+} channels in the rod-cell membrane (see Figure 15-18). The pigments in cones contain different opsins, but they function similarly to rhodopsin.

Visual acuity depends on how finely the incoming light can be reproduced as action potential signals in the map of the retina and further along in the circuitry. The density and numbers of rods would suggest a higher-resolution image than we actually see. In fact their signals converge onto a smaller number of bipolar cells, so some resolution is lost. In contrast the cone cells have very high spatial resolution. Most of the retina has a high density of rod cells and sparse cone cells. The exception is a central region called the *fovea* (see Figure 23-24a), which is mostly made up of cone cells, about 150 cells across its 300-μm diameter. In the fovea the cone cells have small receptive fields, on the order of 0.01 degree, which allows high-resolution representation of incoming light patterns. The receptive fields of rod cells are generally larger, up to several degrees, lowering the sharpness of the image. In dim light our sight is blurry because we depend on rods rather than cones.

Eyes Reflect Evolutionary History

The eye structures of organisms reflect their different evolutionary histories. The eyes of some simpler animals, such as planarians, consist of a light-sensitive optic nerve accompanied by pigment cells, without lenses or other means to obtain a sharp image. In contrast, the compound eyes of insects have hundreds of lenses, one for each facet of the eye (see Figure 16-21). Yet, some of the proteins that regulate eye development play the same role in a vast array of animals that have strikingly different eyes. One of the most peculiar aspects of the human retina, caused by the way our eyes evolved from more primitive light-sensing structures, is that the photosensitive cells lie *behind* the mass of neural connections. Light must pass through the lens, vitreous fluid, and axons and dendrites and interneurons before reaching the photoreceptors (see Figure 23-24b; note direction of incoming light). Compared with optimal design of a light detector, the eye is backwards. The arrangement is also the reason we have a blind spot. Where the optic nerve connects, no light can be sensed.

Integrated Information from Multiple Ganglion Cells Forms Images of the World

If each photoreceptor cell responds to a tiny point of light, how is a larger image of the world assembled? The mass of neurons in the visual system makes this problem seem insurmountable. Fortunately the cells are organized in a rather simple hierarchical fashion that has allowed considerable progress to be made. The first stage of information processing is done in the retina, immediately after light is received by interneurons (see Figure 23-24b). In fact, processing of visual information begins at the very first synapses where the photoreceptor cell connects to interneurons. Interneurons allow signals from multiple photoreceptor cells to be combined and compared. By the time signals leave the eye via the axons of retinal ganglion cells that constitute the optic nerve, each signal conveys not a point of light but a *pattern* of light. Let us start by looking at what sorts of pattern information emerges from the retina.

The experimental approach to the problem uses electrical recordings done with electrodes inserted into individual ganglion cells. At the same time the eye of an anaesthetized animal is exposed to small spots of light shining on the retina. The first step is to identify the part of the retina that stimulates the particular cell into which the electrode is inserted. Next, variations in the size, position, and shape of the light spot are tested.

Remarkably, **bipolar cells** (Figure 23-24b, shown in purple) and the retinal ganglion cells to which they connect (Figure 23-24b, shown in yellow) are sensitive to particular patterns of retinal illumination. This is possible because each photoreceptor (rod or cone) cell simultaneously receives light *and*, a signal from nearby photoreceptor cells. These lateral signals are carried by *horizontal cells*, a type of interneuron (Figure 23-24b, shown in green). In essence, each photoreceptor cell compares what it sees to what it learns its neighbors are seeing.

Let's look at the consequences of such an arrangement. The receptive field of a single bipolar cell is approximately circular, and its information is passed on to a retinal ganglion cell that therefore has the same field. A single bipolar cell receives inputs from a group of photoreceptor cells covering the area of a circle. Each bipolar cell receives signals from a different circle of cells. This type of receptive field pattern is called *center-surround* (Figure 23-26). There is a critical added feature: a bipolar cell responds to light shining in the center of its receptive field—the group of photoreceptor cells near the center of the circle—by producing a set of action potentials. However, action potentials are *inhibited* if light strikes photoreceptor cells surrounding the central part of the circle (Figure 23-26a). This type of cell is an *on-center* cell because light directed at the middle of the field turns the cell on. Some bipolar cells, and therefore their corresponding retinal ganglion cells, have just

Podcast: Vision—Detecting and Recognizing Patterns

MEDIA CONNECTIONS

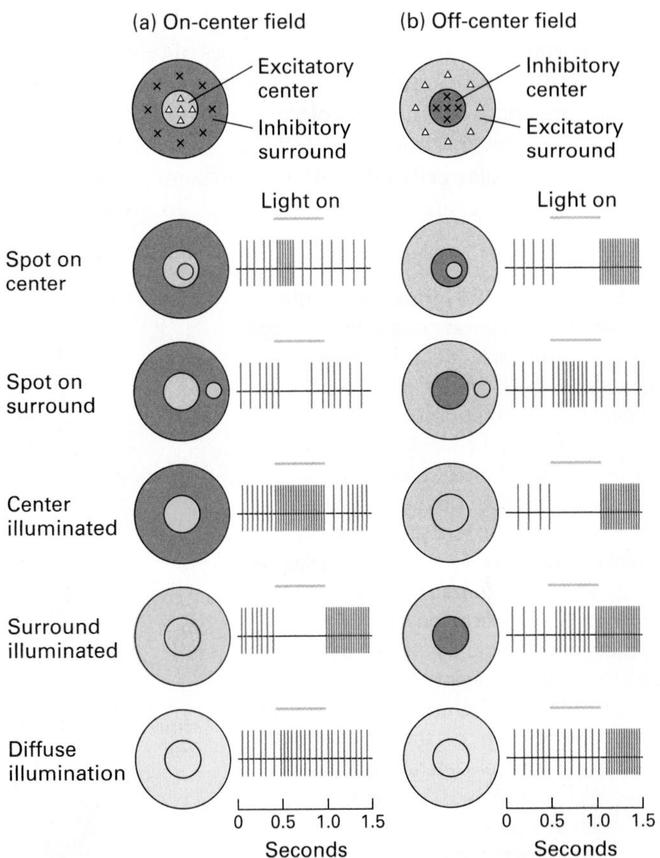

▲ **FIGURE 23-26 Center-surround receptive fields of retinal neurons.** Each retinal ganglion cell responds to a roughly circular receptive field that corresponds to a specific part of the retina. By exposing eyes to different illumination patterns and simultaneously recording individual retinal ganglion neurons, researchers discovered that each neuron responds to an annulus (doughnut)-shaped field, a pattern described as *center-surround*. Some cells fire trains of action potentials when the center is dark and the periphery is light ("off-center field"); some respond to the opposite pattern ("on-center field"). (a) An on-center field. *Spot on center:* a light spot focused on the center of the cell's field triggers a series, or "train," of action potentials. *Spot on surround:* a light spot focused on the peripheral region of the same cell's field inhibits action potentials. *Center illuminated:* illumination of the entire center spot of the field triggers a rapid burst of action potentials. *Surround illuminated:* illumination of the entire surround region has a strong damping effect on action potentials. *Diffuse illumination:* diffuse illumination of the cell's entire receptive field causes a weak response, i.e., a damping of the response seen if only the center is illuminated. (b) An "off-center" field of a cell that has the opposite property of the cell in (a). Illumination of a spot in the center of the off-center cell's field inhibits action potentials, while a spot of light in the periphery stimulates action potentials. [Adapted from F. Delcomyn, 1998, *Foundations of Neurobiology*, W. H. Freeman and Company, p. 265.]

the opposite response: light in the center reduces the frequency of action potentials, while light in the surround stimulates more frequent action potentials (Figure 23-26b). This is an *off-center* cell. The two types of cells, on-center and off-center, are present in roughly equal numbers. Both types of cells sense the relative light intensity in the center versus the surround, not the absolute amount of light in either place. *Bipolar and retinal ganglion cells are therefore contrast detectors.*

How do bipolar cells integrate photoreceptor cell information to detect center-surround patterns? To answer this we will look at the connections of cone photoreceptor cells to bipolar and horizontal interneurons (see Figure 23-24b). On-center and off-center bipolar cells differ in the types of channel proteins they use, giving opposite responses to the same glutamate neurotransmitter. For simplicity, here we will focus on bipolar on-center cells.

Bipolar and horizontal interneuron cells, like photoreceptor cells, lack voltage-gated Na^+ channels, so none of them generate action potentials. Instead, the secretion of neurotransmitters from the cells' synaptic termini is controlled by the

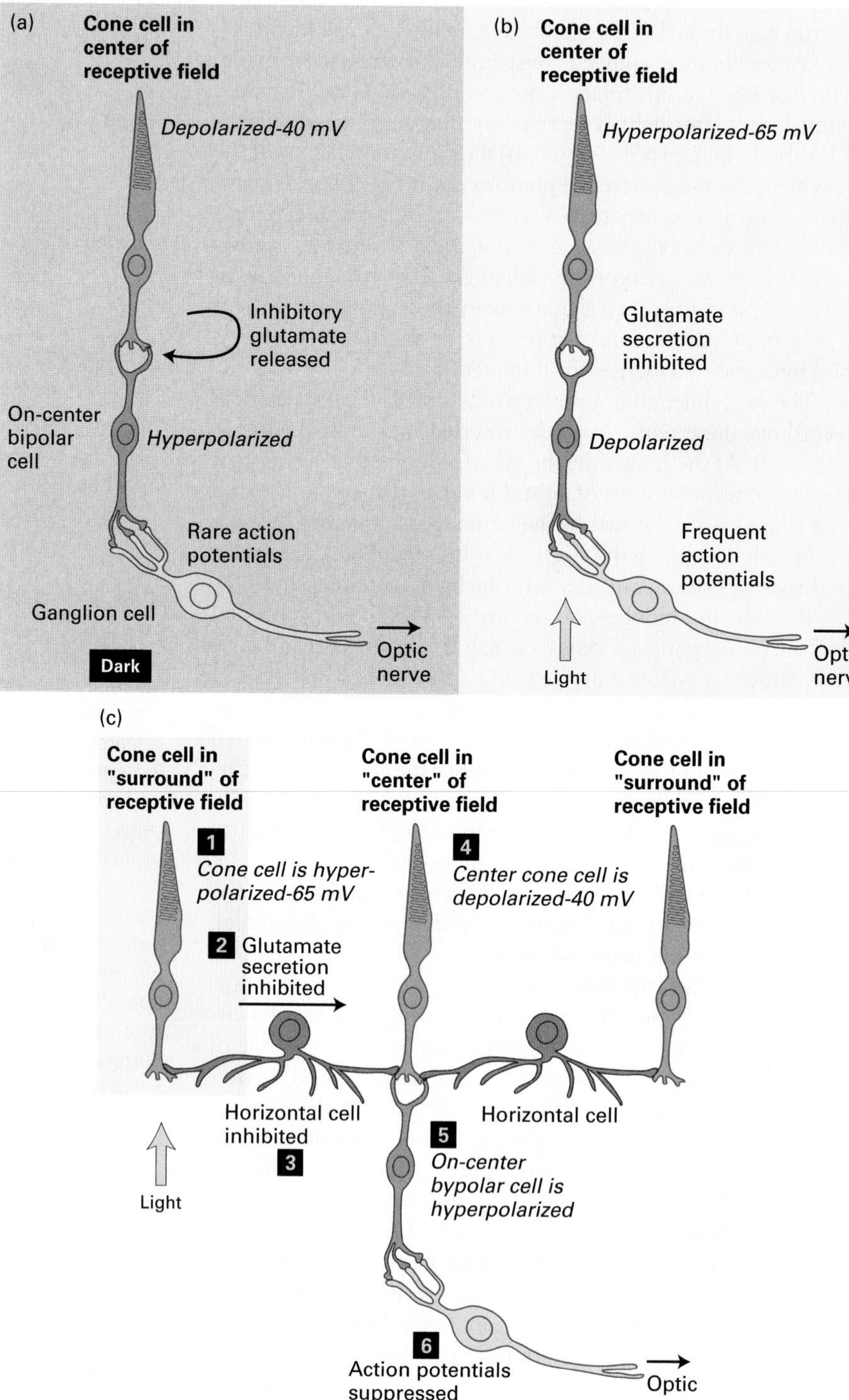

▶ **FIGURE 23-27 The influence of light and dark on on-center cone cells in image interpretation.** (a) In the dark, a cone cell in the center of the receptive field is depolarized, resulting in the release of glutamate. Glutamate inhibits the bipolar cell, causing it to hyperpolarize and thus preventing all but rare action potentials. Note that bipolar off-center cells have the opposite response to glutamate. (b) When light strikes the cone cell, it becomes hyperpolarized, with a resultant drop in glutamate secretion. Free of the inhibitory effect of glutamate, the bipolar cell depolarizes, and frequent action potentials result. (c) The activity of a cone cell is affected by surrounding cone cells through horizontal cells that connect cells laterally. If only the center cell is illuminated, the cell will stimulate the bipolar and consequently the ganglion cells. If both center *and* surround cells are illuminated, the center cell's signal to the bipolar cell will be inhibited. The system is therefore a contrast detector, looking for light patterns that illuminate a small center spot but not the surrounding retina. Here are the steps involved: Light striking the cone cell in the surround of a receptive field hyperpolarizes it **1**, which results in a reduction in the release of glutamate **2**. This, in turn, results in hyperpolarization of the horizontal cell, causing it to release less inhibitory transmitter to the center cone cell **3**. The center cone cell, in the absence of inhibition from surround cells, is depolarized **4**, desensitizing it to light and inducing an increase of glutamate release to the on-center bipolar cell, as in (a). The bipolar cell is therefore hyperpolarized **5**, resulting in suppression of action potentials **6**. The schemata shown are highly simplified, since all the cells can be connected to more than one cell at each stage of signal transmission.

degree of membrane polarization. In the dark, cone cells have a membrane potential of about −40 mV, which opens voltage-gated Ca^{2+} channels and causes the continuous release of the neurotransmitter glutamate (Figure 23-27a). This glutamate, coming from cone cells in the center of a receptive field, hyperpolarizes the bipolar neuron that covers that field, suppressing action potentials. Light striking the center field cone cells hyperpolarizes them to about −65 mV by suppressing an inward flow of Na^+ and other ions, closing the Ca^{2+} channels and reducing emission of glutamate (Figure 23-27b). This depolarizes the bipolar neuron, which in turn depolarizes ganglion cells and triggers action potentials to be sent to the brain.

On-center bipolar cells are most stimulated if the cone cells in the center of the receptive field are illuminated and the surround cone cells are in the dark. How do bipolar cells detect the light condition in the surround part of the receptive field? The surround input is mediated by horizontal cell interneurons (see Figure 23-24b, green cells). If there *is* light on a cone cell in the surround region of the field, the horizontal cell that is connected to that cone cell becomes hyperpolarized which means it reduces inhibitory transmitter release onto the cone cell in the center of the receptive field. The central cone cell becomes depolarized, as though it was in the dark, and consequently the bipolar cell in the center of the receptor field is hyperpolarized (see Figure 23-27c). Ganglion cell action potentials in the center of the field are suppressed despite the light falling on the central cone cells. Thus light in the surround inhibits sensing of the light in the center, a contrast detector.

The retina's processing of visual information is just the beginning of a hierarchical chain of pattern representation and interpretation events. The refined processing of visual information occurs, it must be remembered, by reading trains of action potentials coming from the eye. In the visual cortex, cells are found that are specifically sensitive to bars of light and dark, and each cell prefers bars at a certain angle. It is easy to see how the information that passes through retinal ganglion cells, the center-surround information, can be used to detect a bar of light or dark. If a visual cortex cell is stimulated by ganglion cells whose visual fields are arranged in a line, the integrated pattern would be a bar that passes through the centers of the individual ganglion cells' center-surround patterns (Figure 23-28). Further combinations can lead to recognition of more complex patterns by single cells. Some cells respond to a change of light—on to off or off to on. Others respond to edges, moving spots, or moving bars. Some cells in the higher levels of the visual cortex have even been found to recognize a certain face.

The discovery of cells that integrate spatial information from cells that have simple receptive fields was described by Nobel prizewinner David Hubel, who made the discovery with his collaborator Torsten Wiesel:

> Our first real discovery came about as a surprise. For three or four hours, we got absolutely nowhere. Then gradually we began to elicit some vague and inconsistent responses by stimulating somewhere in the midperiphery of the retina. We were inserting the glass slide with its black spot into the slot of the ophthalmoscope when suddenly, over the audio monitor, the cell went off like a machine gun.

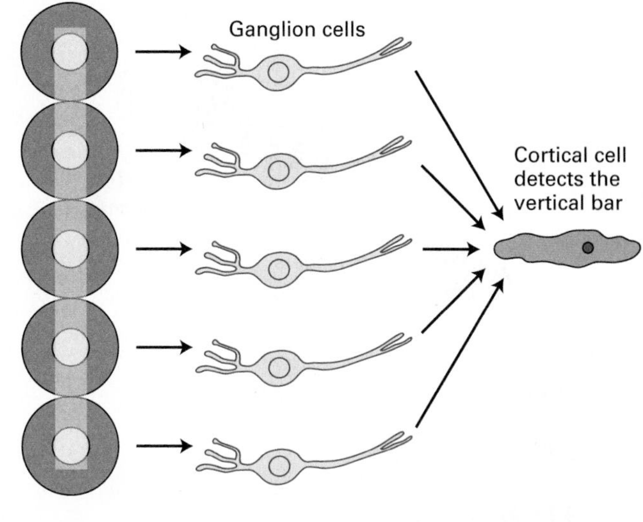

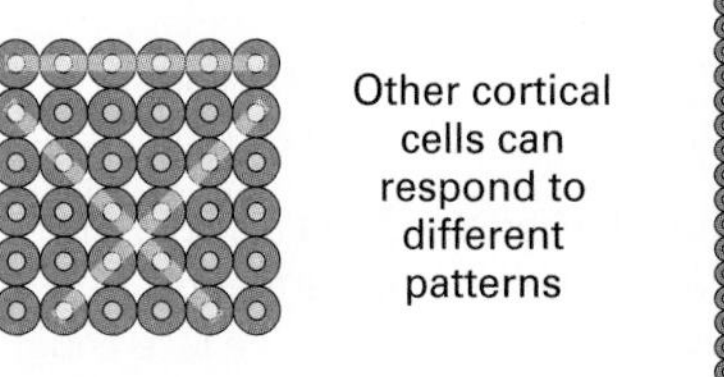

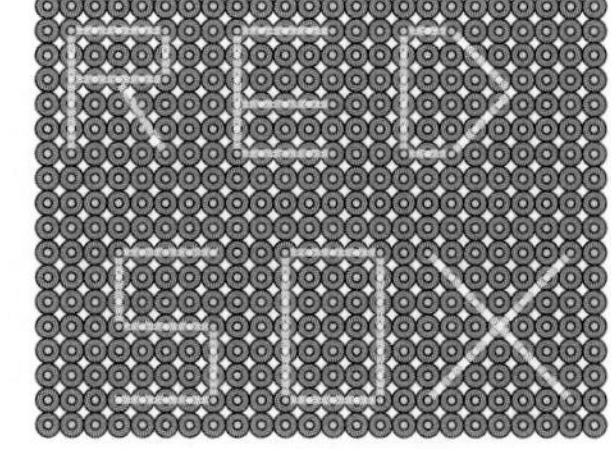

▲ **FIGURE 23-28 Complex pattern recognition.** A cortical cell responds to the sum of multiple ganglion cell "centers," thus detecting the vertical bar shown. Other cortical cells respond to different combinations of ganglion receptor fields or combinations of cortical neuron fields to recognize more complex patterns.

> After some fussing and fiddling, we found out what was happening. The response had nothing to do with the black dot. As the glass slide was inserted, its edge was casting onto the retina a faint but sharp shadow, a straight dark line on a light background. That was what the cell wanted, and it wanted, moreover, in just one narrow range of orientations. This was unheard of. It is hard now to think back and realize just how free we were from any idea of what cortical cells might be doing in an animal's daily life.

Mechanosensory Cells Detect Pain, Heat, Cold, Touch, and Pressure

Our skin, especially the skin of our fingers, is expert at collecting sensory information. Our whole body, in fact, has numerous **mechanosensors** embedded in its various tissues. These sensors frequently make us aware of touch, the positions and movements of our limbs or head (proprioception), pain, and temperature, though we often go through periods where we ignore the inputs. Mammals use one set of receptor cells to report on touch, and other sets of receptors for temperature, heat, and pain. Pain receptors, called **nociceptors,** respond to mechanical change, heat, and certain toxic chemicals (e.g., hot pepper). Genetic insensitivity to pain is

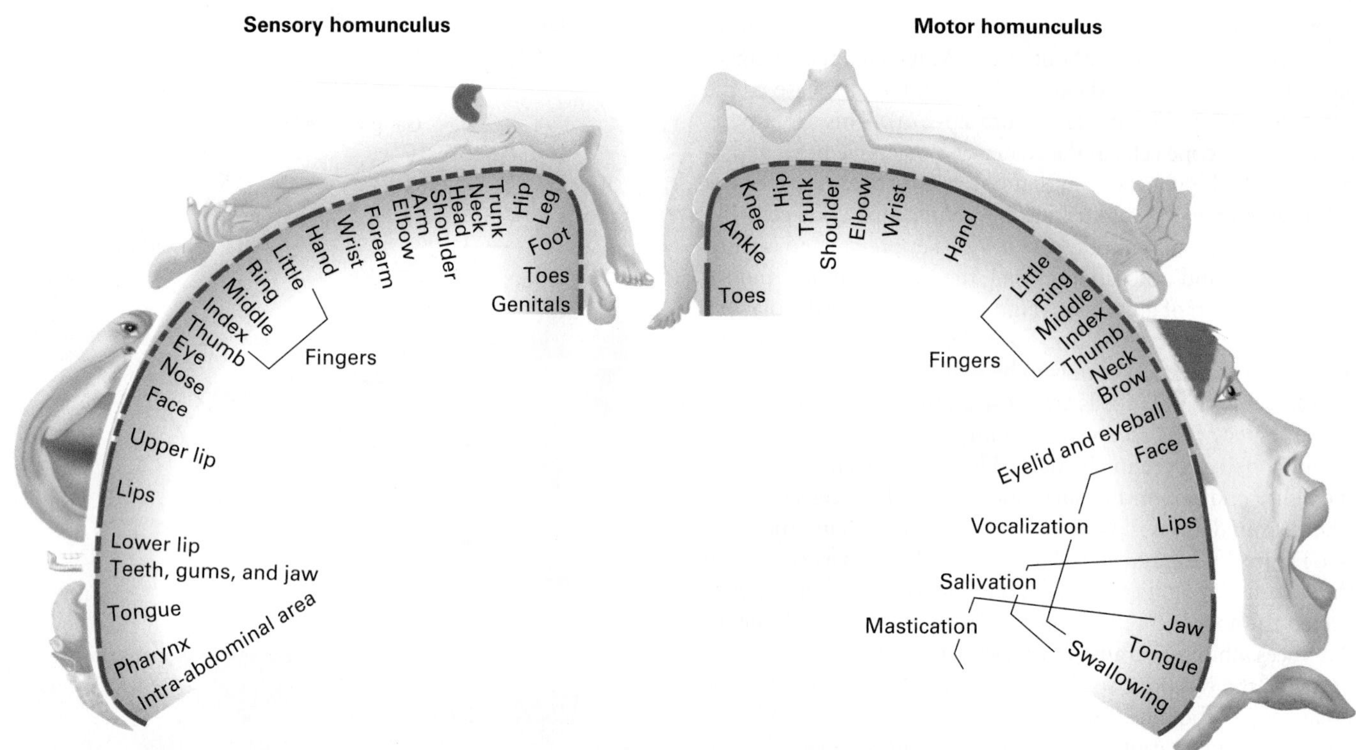

▲ **FIGURE 23-29 Homunculus.** The homunculus is a map of the regions of the brain's cortex that are dedicated to particular functions. Sensory and motor homunculi are shown. The face and hands take up large portions of the brain's sensory and motor capacities. [Adapted from D. Randall, W. Burggren, and K. French, 2002, *Eckert Animal Physiology,* 5th ed., W. H. Freeman and Company, p. 292.]

often due to mutations in a gene, *trkA*, that encodes a receptor for nerve growth factor (NGF), a protein mostly studied in different contexts. NGF and other neurotrophins have now been implicated as signals of pain. Thermal receptors detect temperature changes. These cells steadily send action potentials (2–5/s) that indicate the current temperature. Each temperature range has receptors tuned to it, so *which* cells are firing conveys the temperature.

Connecting from the skin's sensory cells to the brain does not take many synapses. Mechanosensors in the skin, for example, connect to the medulla, where they pass the signal along to neurons that go to the thalamus. A third neuron goes from there to the sensory cortex. A mere trio of neurons connects the periphery to the brain centers. In the cortex the sensory inputs are combined, through interneurons, with proprioceptive inputs that report the positions of muscles and joints. Presumably this makes it possible to know *what* you are feeling and *where* it must be if *that* arm in *that* position is feeling it. Proprioception receptors take multiple forms. Some of the most studied are muscle spindles, sensory assemblages that are buried in muscles to report on how much that muscle is extended. Such stretch receptors are crucial to smooth movement and well-timed responses.

Remarkably, the organization of the body is reflected in a map, or more accurately several maps, in the brain. The organization of the cortical neurons that respond to sensory signals is physically related to the spatial origins of the signals. In the brain the sensory neurons are laid out in a distorted map of the body. The motor neurons are also arranged in a map that can be aligned with the muscles they control. The maps are called the *sensory homunculus* and the *motor homunculus* (Figure 23-29). A *homunculus* is a "little human," an image of us. The map dimensions are not proportional to the body's dimensions, because the homunculi reflect the number of sensory or motor cells rather than the area of the body. The hands and feet are highly represented and take up much more space in the sensory map.

Inner Ear Cells Detect Sound and Motion

The outer ear captures sound, which moves three tiny bones (*ossicles*) in the middle ear, which in turn transmit sound-induced motions to the inner ear, or *cochlea* (Figure 23-30a). The cochlea is shaped like a snail, with nearly three turns, and indeed the name derives from the Greek word for "snail" (cochlos). The cochlea houses the **organ of Corti,** the sensory part of the inner ear, which transduces mechanical movement into electrical impulses. The human organ has about 16,000 hair cells arranged in four rows (Figure 23-30b, c), attached to about 30,000 afferent neurons that carry any signals to the brain. Hair cells produce **stereocilia,** which are moved by vibrations induced by sound. The oscillating vibrations alternately bend the stereocilia one way or the other, triggering depolarization events called receptor potentials in the 10 or so axons associated with each hair cell. These receptor potentials, which are milder than full action potentials, range up to 25 mV.

Hair cells and the neurons they influence are responsive to different sound frequencies. There is a gradient across the

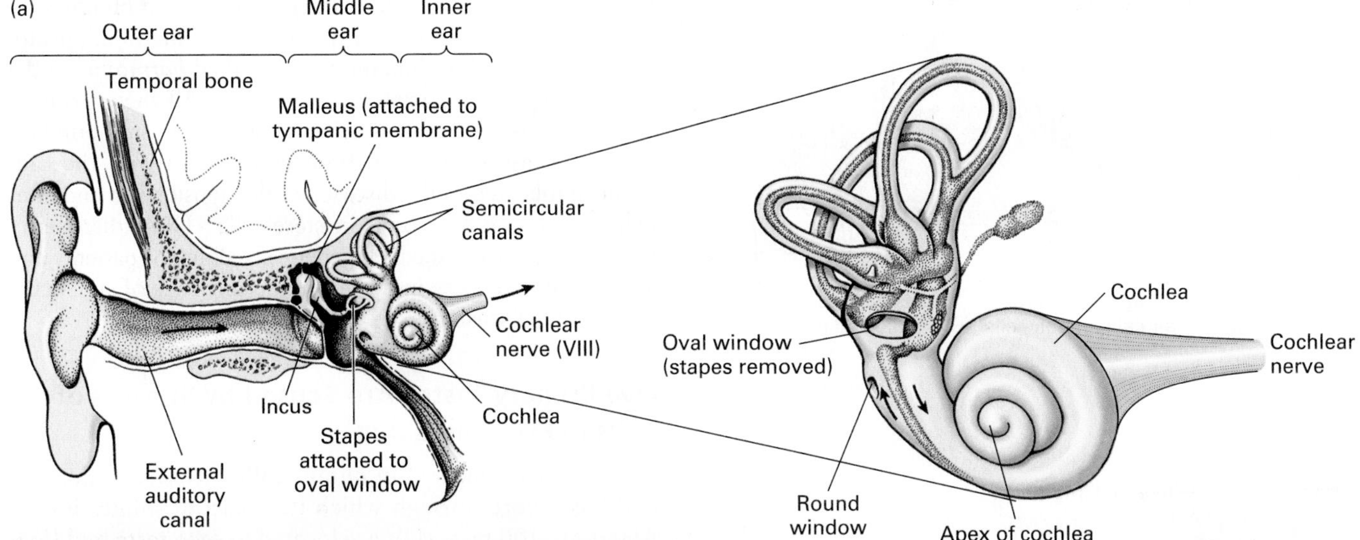

(b)

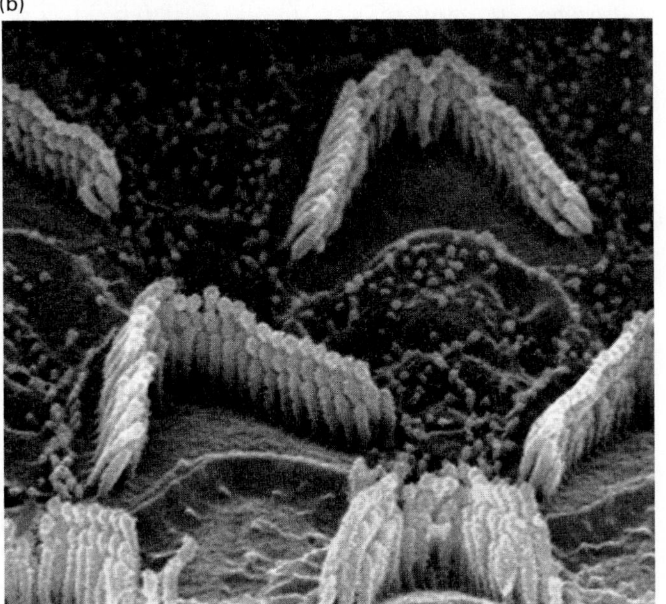

(c)

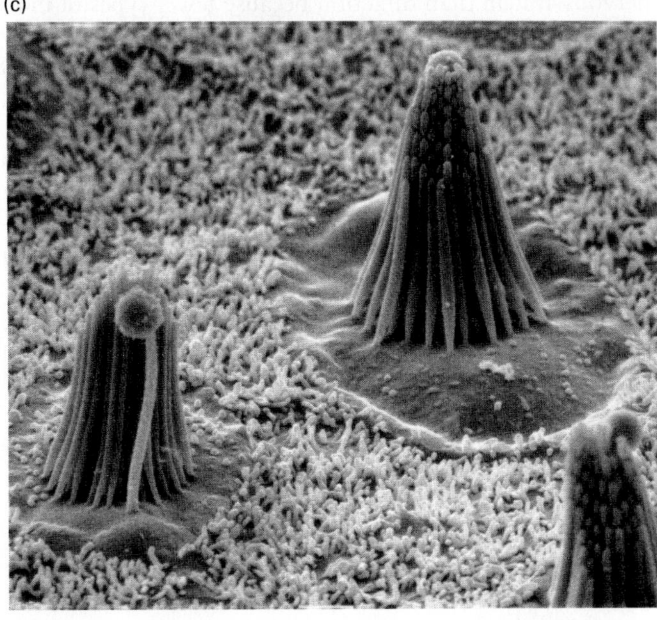

▲ **FIGURE 23-30 Structures of the ear.** (a) Sound enters the outer ear and travels to the middle ear, where three tiny bones (the malleus, incus, and stapes) transfer sound-induced vibration in the tympanic membrane across the middle ear to the cochlea in the inner ear, where mechanical vibration is transduced to electrical signals. The cochlea, unwound, would be 33 mm long. (b) The inner surface of the organ of Corti, found in the cochlea, as viewed by scanning electron microscopy. The tops of all the stereocilia (white) would be in contact with the tectorial membrane in the intact ear (see Figure 23-31). The stereocilia of the inner hair cells (*left row*) are arranged in a line, while the three rows of outer hair cells have stereocilia in V shapes. (c). Higher magnification of the stereocilia of outer hair cells. The hair cells are smooth except for the cilia, while surrounding support cells are covered by microvilli. [Part (a) adapted from D. Randall, W. Burggren, K. and French, 2002, *Eckert Animal Physiology,* 5th ed., W. H. Freeman and Company, p. 243. Parts (b) and (c): Courtesy of Bechara Kachar/National Institute of Health.]

cochlea of frequency sensitivity, so that the spatial locations of the cells that are stimulated reflect the frequency composition of a sound. Hair cells and neurons at one end of the cochlea hear low-frequency sounds and at the other end high frequencies. This is not because of a difference in either hairs or neurons. The graded sensitivity is due to a tapering tissue called the *basilar membrane* (Figure 23-31) that responds to low frequencies at one end and high frequencies at the other. Each frequency excites motion in a particular region of the 33-mm-long basilar membrane, which is then locally transferred to nearby hair cells. The orientation of the hair cells, and in particular of their bundles of stereocilia, with respect to the basilar membrane allows sensitive detection of deflections caused by sound. The polarity of the hair cells and their cytoskeletons are key to proper transduction of sound into electrical signals.

Some of the proteins that control the structure of hair cells and stereocilia have been identified through human genetics, tracking genes responsible for deafness. Five genes have been implicated in Usher type 1 syndrome, the

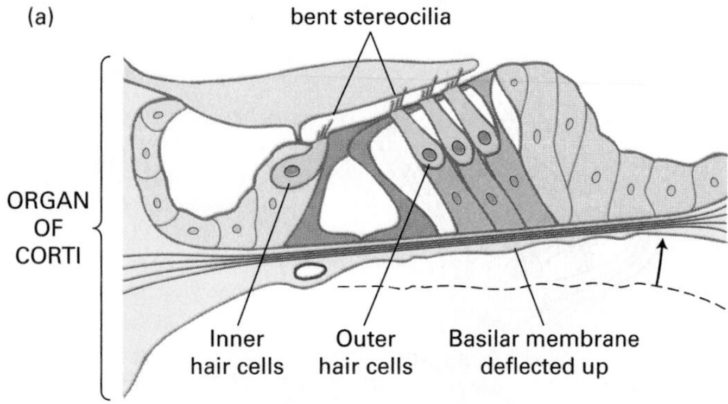

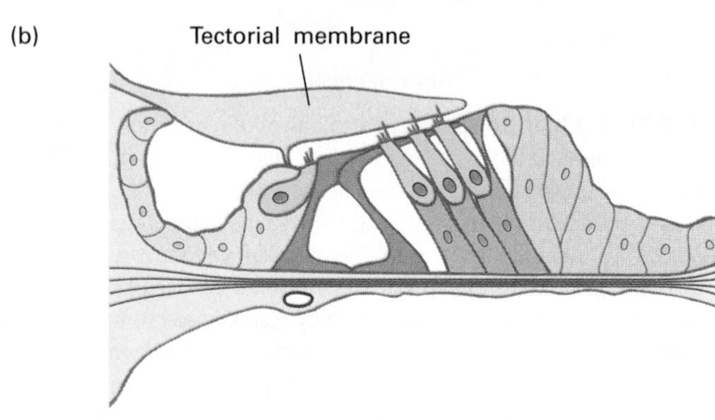

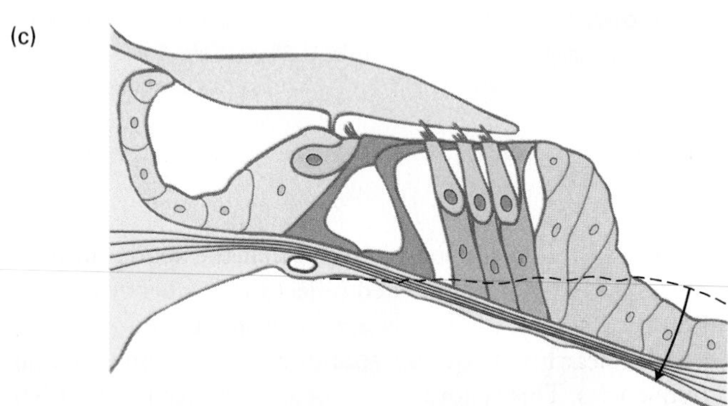

▲ **FIGURE 23-31 Sterocilia movement.** The stereocilia of the inner and outer hair cells (purple) are stimulated by a sideways movement with respect to the overhanging tectorial membrane, which in turn is influenced by oscillating fluid pressure changes in the organ of Corti. The fluid pressure in the organ oscillates at the frequency of the incoming sound. (a) As vibration begins, the basilar membrane (pink) is forced upward (indicated by the arrow) by fluid pressure changes, which translates into a leftward movement with respect to the tectorial membrane, thus bending the stereocilia to the right. (b) At the midpoint of the oscillation, the stereocilia relax. (c) When the oscillation goes the other way and the basilar membrane moves down (indicated by the arrow), the hair bundles are moved in the opposite direction by the shearing effect of the tectorial membrane. The motions are repeated with each wave. [Adapted from D. Randall, W. Burggren, and K. French, 2002, *Eckert Animal Physiology,* 5th ed., W. H. Freeman and Company, p. 248.]

most frequent cause of hereditary deafness and blindness in humans. The genes encode myosin VIIa, cadherin 23, protocadherin 15, a PDZ domain protein called *harmonin,* and a putative scaffolding protein called *Sans*. All these proteins are localized within stereocilia in auditory hair bundles. Harmonin associates with both F-actin and with the cadherins implicated in the disease, while myosin VIIa and Sans help to localize harmonin in stereocilia. These discoveries have emerged from medical genetic screening of patients and are revealing the key molecular underpinnings of stereocilia and auditory sensing.

Five Primary Tastes Are Sensed by Subsets of Cells in Each Taste Bud

Taste buds are located in bumps called *papillae,* and each bud has a pore through which fluid carries solutes inside. About 50–100 taste cells are located in each taste bud (Figure 23-32a, b). Cells in the tongue and other parts of the mouth are subjected to a lot of wear and tear, and taste bud cells are continuously replaced by cell divisions in the underlying epithelium. (A taste bud cell in rats has a lifetime of 10 days.)

Taste cells are epithelial cells that show some of the functions of neurons. Reception of a taste signal causes cell depolarization and receptor potentials that trigger action potentials; these, in turn cause Ca^{2+} uptake through voltage-dependent Ca^{2+} channels and release of neurotransmitters at synapses. Taste cells do not grow axons, instead signaling over short distances to other neurons. In contrast to most other sensory systems, there is, as yet, no known topographic representation at any level of the brain that corresponds to the different tastes.

We taste certain chemicals, all hydrophilic, nonvolatile molecules floating in saliva. Although all tastes are sensed on all areas of the tongue and there is no topographical taste map of the tongue, selective cells do respond preferentially to certain tastes. Taste is less demanding of the nervous system than olfaction because fewer types of molecules are monitored. What is impressive is the sensitivity of taste; bitter molecules can be detected at concentrations as low as 10^{-12} M. There are receptors for salt, sweet, sour, umami (e.g., monosodium glutamate and other amino acids), and bitter (Figure 23-32c, d, e, f) in all parts of the tongue. The receptors are of two different "flavors": channel proteins for salt and sour tastes and seven-transmembrane-domain proteins (G protein–coupled receptors) for sweetness, umami, and bitterness.

Salt is probably sensed by members of a family of Na^+ channels called *ENaC channels,* though definitive proof is lacking; other members of the family have diverse functions including neural memory. The influx of Na^+ through a channel depolarizes the cell. The role of ENaC channels as salt sensors is old, since ENaC proteins clearly detect salt in insects. In *Drosophila,* taste sensors are located in multiple places including the legs, so when the fly steps on something tasty, the proboscis extends to explore it further. However, the ENaC studies were done using fly larvae, which can respond

▼ **FIGURE 23-32 A mammalian taste bud and its receptors.** (a) The pink cells are the taste cells. These epithelial receptor cells contact the nerve cells (yellow). The chemical signals arrive at the microvilli seen at the top. (b) Photograph of a pair of taste buds, showing the receptor cells. The microvilli are clearly visible in the taste bud on the *left*. (c–f) Types of taste receptors. [Part (a) adapted from B. Kolb and I. Q. Whishaw, 2006, *An Introduction to Brain and Behavior,* 2d ed., Worth, p. 400: Part (b) from Ed Reschke/Peter Arnold.]

(b)

(a)

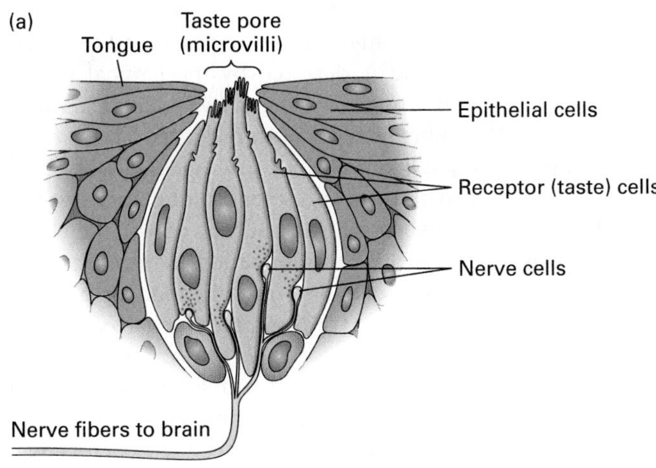

(c) Salt and acid

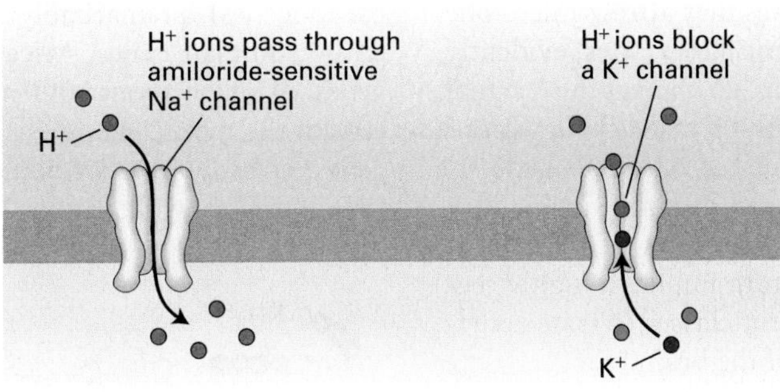

(d) Amino acid

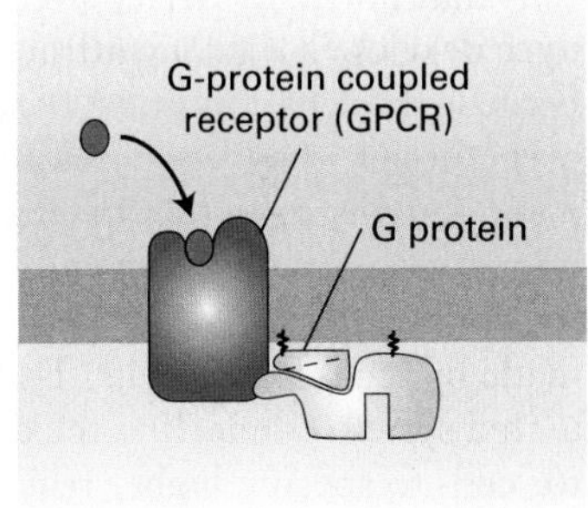

(e) Bitter compounds

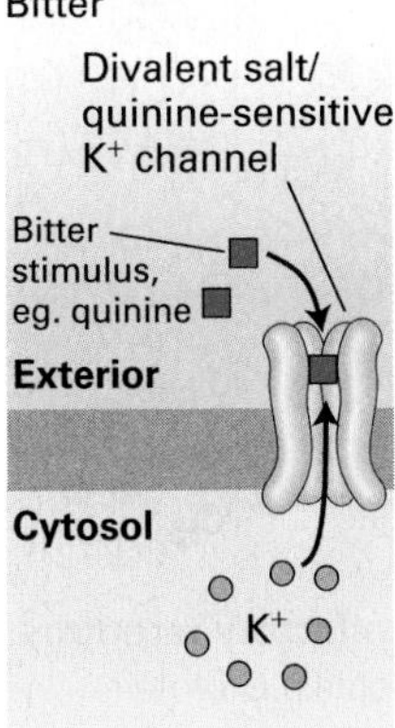

(f) Sugars and sweeteners

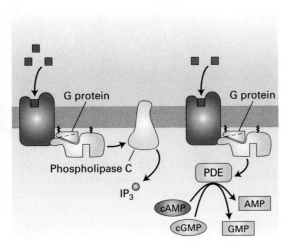

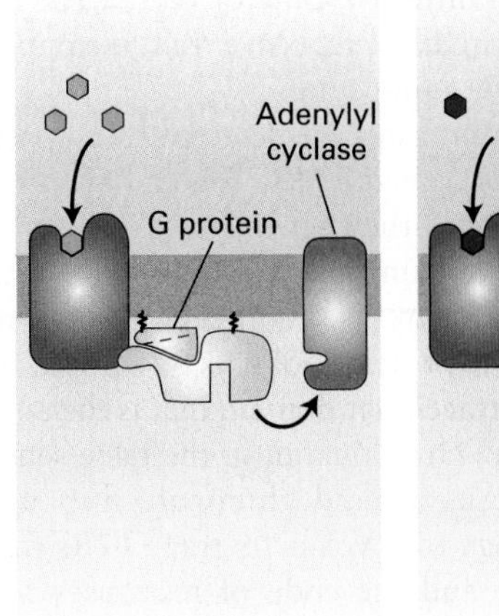

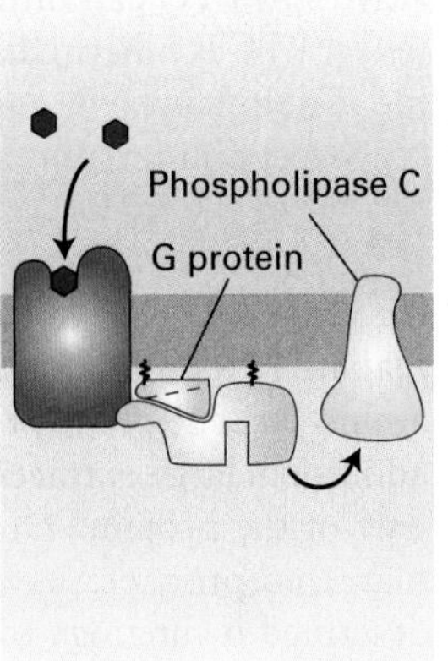

to salt if they have either of their two ENaC proteins. Sour reception is the detection of H^+ ions, which can move through the same channels as Na^+. H^+ may also be sensed due to its interference with K^+ channels and consequent increase in intracellular positive charge (i.e., a depolarizing effect).

Bitter tastes are more diverse than salt and have been found to depend on a diverse family of about 25 genes encoding various T2Rs, taste-receptor proteins with seven transmembrane α-helical domains; they are members of the class of proteins known as **G protein–coupled receptors (GPCRs)**. These GPCRs depolarize the cell by triggering, through G proteins, the activity of phospholipase C (PLC), which increases the concentration of IP_3 (see Figure 15-30). This causes Ca^{2+} release from intracellular compartments and consequent depolarization. One G protein involved in bitter taste transduction is called *gustducin*.

The first member of the T2R family to be identified came from human genetics studies that showed an important bitterness-detection gene on chromosome 5. Multiple T2R types can be expressed in the same taste cell, and about 15 percent of all taste cells express T2Rs. Bitter taste molecules are quite distinct in structure, which probably accounts for the need for the diverse family of T2Rs. Mice that have five amino acid changes in the receptor T2R5 are unable to taste the bitter taste of cycloheximide (a protein synthesis inhibitor, Chapter 8).

A dramatic gene regulation swap experiment was done to demonstrate the role of T2R proteins. Mice were engineered to express a bitter taste receptor, a T2R protein, in cells that normally detect sweet tastes that attract mice. The mice developed a strong attraction for bitter tastes, evidently because the cells continued to send a "go and eat this" signal even though they were detecting bitter taste. This experiment demonstrates that the specificity of taste cells is determined within the cells themselves, and that the signals they send are interpreted according to the neural connections made by that class of cells. This in turn implies a highly regulated system connecting the different classes of taste receptor cells to specific higher regions of the brain.

One bitter taste is especially famous because it is often used in genetics classes to teach about human variation. The chemical phenylthiocarbamide (PTC) tastes exceedingly bitter to many people but is tasteless to others. Human sensitivity to PTC can differ by a factor of 16. The inability to detect PTC is inherited as a recessive trait, meaning that tasting is dominant over nontasting.

Sweet and umami tastes are detected by a protein family related to the T2Rs, called *T1Rs*. When T1R proteins bind an appropriate *tastant,* they act through G proteins to trigger the release of calcium inside the cell. The three mammalian T1Rs differ from one another in a small number of amino acids. A T1R protein looks like a GPCR, but has an additional large extracellular domain that is the taste-binding part of the protein. This domain, in the taste-sensing glutamate receptor, closes around glutamate in a way that is described by analogy to a Venus fly trap. T1Rs form dimers and heterodimers, and the code of responses to different molecules is still under investigation. Mice lacking T1R2 or T1R3 fail to detect sugar; it is thought that the actual receptor is a heterodimer of the two. T1R3 appears to be a receptor for both sweet tastes and umami, and that is because it detects sweets when combined with T1R2 and umami when it combines with T1R1. Accordingly, taste cells express T1R1 or T1R2 but not both, as otherwise they would send an ambiguous message to the brain.

A Plethora of Receptors Detect Odors

The perception of volatile airborne chemicals imposes different demands than the perception of light, sound, touch, or taste. Light is sensed by only four molecules, tuned to different wavelengths. Sound is detected by mechanical effects through hairs that are tuned to different wavelengths. Touch requires a small number of types of transducers. The sense of taste measures a small number of substances dissolved in water. In contrast to all these other senses, olfactory systems can discriminate between many hundreds of volatile molecules moving through air. Discrimination between a large number of chemicals is useful in finding food or a mate, sensing pheromones, and avoiding predators, toxins, and fires. *Olfactory receptors* can work with enormous sensitivity. Male moths, for example, can detect single molecules of the signals sent drifting through the air by females.

In order to cope with so many signals, the olfactory system employs a large family of olfactory receptor proteins. Humans have about a thousand olfactory receptor genes, of which about a third are functional (the rest are unproductive pseudogenes), a remarkably large proportion of the estimated 25,000 human genes. Mice are more efficient, with 1300 genes, of which about 1000 are functional. That means 3 percent of the mouse genome is composed of olfactory receptor genes. *Drosophila* has about 60 olfactory receptor genes. In this section we will examine how olfactory receptor genes are

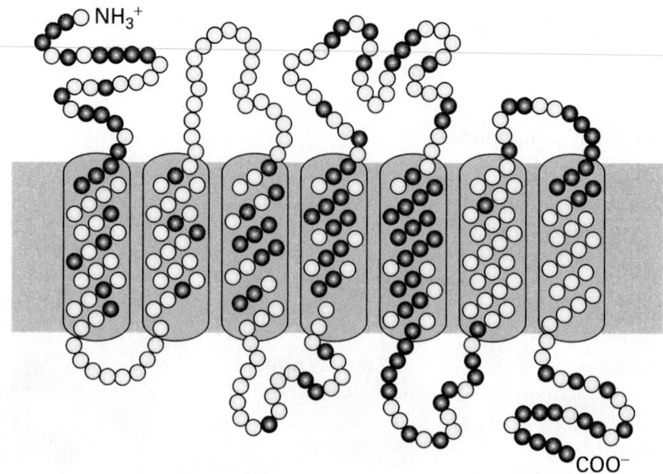

▲ **FIGURE 23-33 Sequence organization in olfactory receptors.** Olfactory receptors are seven-transmembrane-domain G protein–coupled receptor proteins. The cylinders indicate the extent of alpha helices that cross the membrane. Residues in black are highly variable, and some of these differences account for specific interactions with odorants. [From L. Buck and R. Axel, 1991, *Cell* **65:**175.]

employed, and how the brain can recognize which odor has been sensed—the initial stages of interpretation of our chemical world. Odor molecules are called *odorants*. They have diverse chemical structures, so olfactory receptors face some of the same challenges faced by antibodies—the need to bind and distinguish many variants of relatively small molecules.

Olfactory receptors are seven-transmembrane-domain proteins (Figure 23-33). In mammals, olfactory receptors are produced by cells of the nasal epithelium. These cells, called *olfactory receptor neurons (ORNs)*, transduce the chemical signal into action potentials (Figure 23-34). In *Drosophila*, ORNs are located in the antennae. The ORNs project their axons to the next higher level of the nervous system, which in mammals is located in the olfactory bulb of the brain. The ORN axons synapse with dendrites from *projection neurons* in insects (called *mitral neurons* in mammals); these synapses occur in the clusters of synaptic structures called *glomeruli*. The projection neurons connect to higher olfactory centers in the brain (Figure 23-35).

Each ORN produces only a single type of odorant receptor. Any electrical signal from that cell will convey to the brain a simple message: "my odor is binding to my receptors." Receptors are not always completely monospecific for odorants. Some receptors can bind more than one kind of molecule, but the molecules detected usually are closely related in structure. Conversely, some odorants bind to multiple receptors.

There are about a million ORNs in the mouse; so on average each of the thousand or so olfactory receptor genes is active in a thousand cells. There are about 2000 glomeruli (2 for each gene), so on average the axons from 500 ORNs converge on each glomerulus. From there the axons of about 50,000 mitral neurons, about 25 per glomerulus, connect to higher brain centers. Note that in contrast to the visual system, very little signal interpretation and refinement occurs in the sensory epithelium or even the projection neurons. The initial sensory information is carried to higher parts of the brain without processing, a simple report of what has been detected with no further analysis or commentary.

The one neuron–one receptor rule extends to *Drosophila*. Detailed studies have been done in larvae, where a simple olfactory system with only 21 ORNs uses about 10–20 olfactory receptor genes. It appears that a unique receptor is expressed in one ORN, which sends its projections to one glomerulus. ORNs can send either excitatory or inhibitory signals from their axon termini, probably in order to distinguish attractive versus repulsive odors. The ORNs project to glomeruli in the antennal lobe of the larval brain. The research began with tests of which odorants bind to which receptors (Figure 23-36a). Some odorants are detected by a single receptor, some by several, so the combinatorial pattern allows many odorants to be distinguished. The small total number of neurons has allowed a map to be constructed showing which odorants are detected by every glomerulus (Figure 23-36b). One striking finding was that glomeruli located near each other respond to odorants with related chemical structures, e.g., linear aliphatic compounds or aromatic compounds. The arrangement may reflect evolution of new receptors concomitant with a process of subdivision of the olfactory part of the brain.

The simple system of having each cell make only one receptor type also has some impressive difficulties: (1) Each receptor must be able to distinguish a type of odorant molecule or a set of molecules with specificity adequate to the needs of the organism. A receptor stimulated too frequently will probably not be too useful. (2) Each cell must produce one and only one receptor. All the other genes must be turned off. At the same time the collective efforts of all the cells in the nasal epithelium must allow the production of enough different receptors to give the animal adequate sensory versatility. It does little good to have hundreds of receptors if most of them are never used, but it is a regulatory challenge to turn on one and only one gene in each cell and yet use all the genes in the complete population of cells. (3) The wiring of the olfactory system must make discrimination among odorants possible. This could involve either a relatively random wiring that is then interpreted based on experience, or a programmed wiring that is reproducible from individual to individual. The brain must learn which cell is receiving which odorant so that electrical signals from the nose can be interpreted. It is often the case that a response to a particular odor is programmed in the genes, like a behavioral response to a pheromone. In such cases the brain must know which cells are detecting that pheromone. Otherwise the animal might be feeling romantic when it should be running away as fast as possible.

The solution to the first problem is the great variability of the olfactory receptor proteins, both within and between

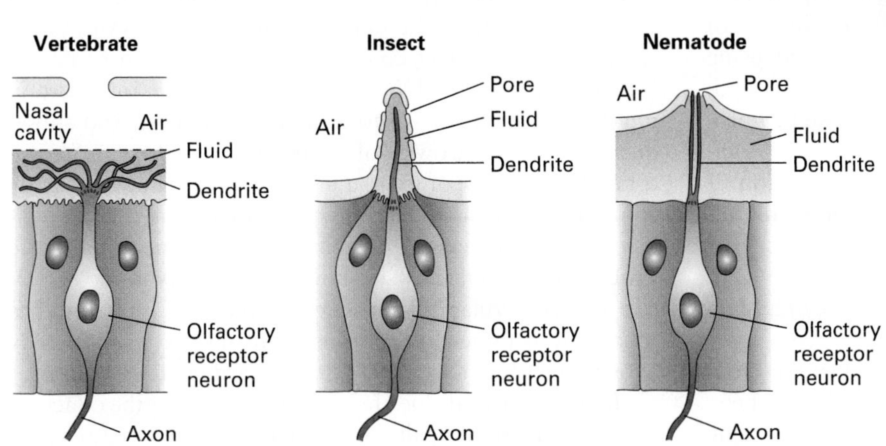

◀ **FIGURE 23-34 Structures of olfactory receptor neurons.** Across a vast span of evolutionary distance—vertebrate, insect, nematode—olfactory receptor neurons have similar forms. Each has fine processes exposed to volatile odorants dissolved in fluid. Highly specific olfactory receptors (not shown) in the cells sense the odorants. The cells shown are not drawn to the same scale. [Adapted from F. Delcomyn, 1998, *Foundations of Neurobiology*, W. H. Freeman and Company, p. 327.]

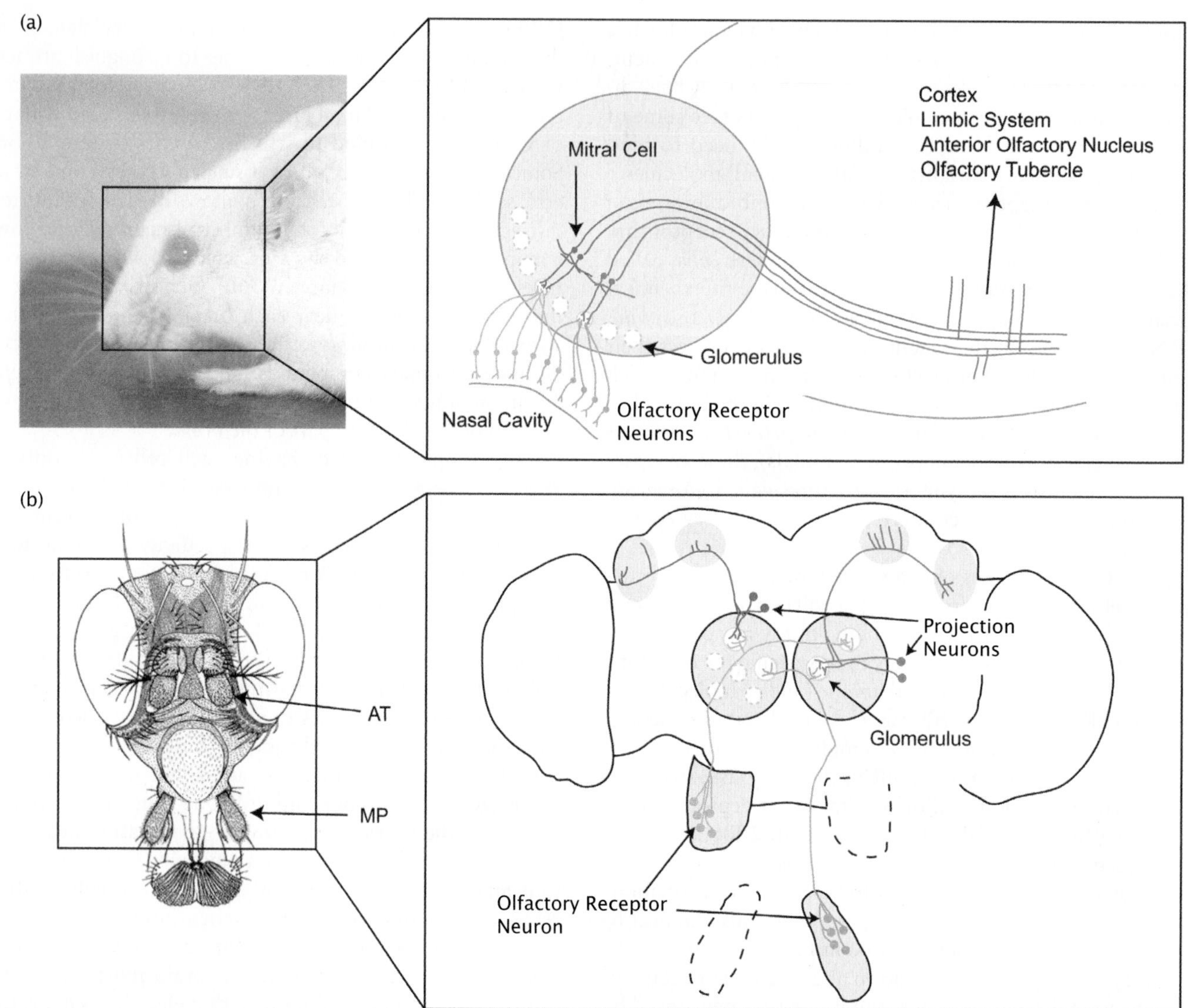

▲ **FIGURE 23-35 The anatomy of olfaction in mouse and fly.** In both the mouse (a) and the fly (b), olfactory receptor neurons (ORNs) that express a single type of receptor send their axons to the same glomerulus. In this figure the red and blue colors represent the neural connections for two distinct expressed receptors. In the mouse the glomeruli are located in the olfactory bulb; in the fly they are in the brain. In the glomeruli, the ORNs synapse with *projection neurons* in the fly, or *mitral neurons* in mammals. Each projection neuron (or mitral neuron) has its dendrites in a single glomerulus, thus carrying to higher centers of the brain information about a particular odorant. [From T. Komiyama and L. Luo, 2005, *Curr. Opin. Neurobiol.* **16:**67–73.]

species (see Figure 23-33; the black residues are highly variable). The solution to the second problem, the expression of a single olfactory receptor gene, has been explored using transgenic mice, but the mechanism is still not understood. When an engineered olfactory receptor gene is used to produce an olfactory receptor, other genes are turned off transcriptionally owing to some type of feedback regulation. If an engineered olfactory receptor gene is expressed that produces a reporter protein—not an olfactory receptor protein—then other genes can still be expressed. The feedback system must involve detection of the presence of a functional olfactory receptor protein.

The third problem, how the system is wired so the brain can understand which odor has been detected, has been partly answered. First, ORNs that have expressed the same receptor send their axons to the same glomerulus. Thus all cells that respond to the same odorant send processes to the same destination. This convergence process could be due to (1) an attractive signal that somehow is specific for a certain olfactory receptor or (2) to mutual recognition, and subsequent coordinate growth, of axons that have the same receptor on their surface or (3) to a pruning process in which many connections are made but only those that share the same olfactory receptor persist, possibly regulated by neuronal activity. Developmental analyses show that ORN axons do *not* arrive at a glomerular "blank slate." Rather the glomerulus has organized its projection neurons prior to the arrival of ORN axons. The system is to some degree hardwired.

In mice a crucial clue about the patterning of the olfactory system came from the discovery that olfactory receptors play

(a)

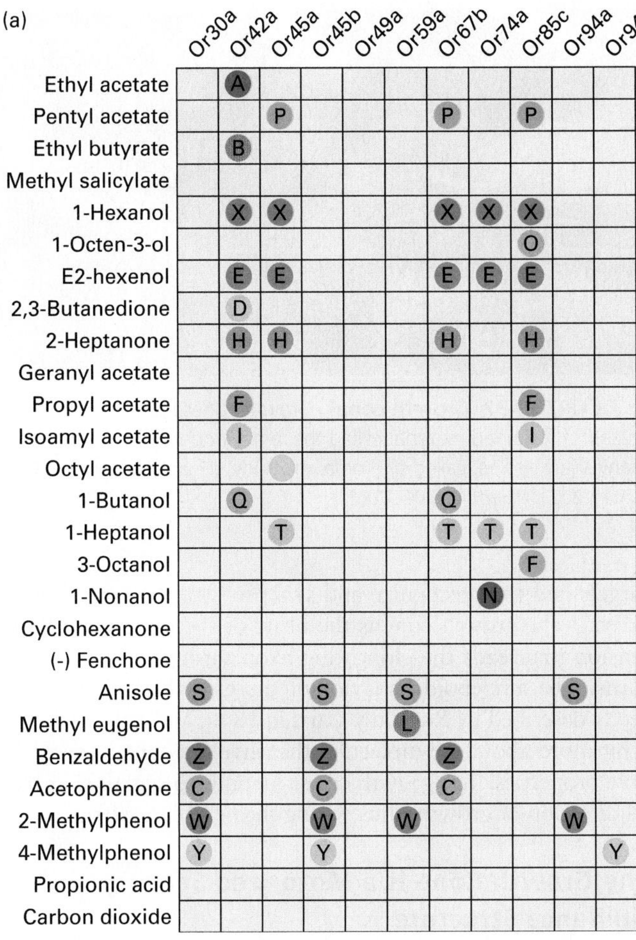

	Or30a	Or42a	Or45a	Or45b	Or49a	Or59a	Or67b	Or74a	Or85c	Or94a	Or94b
Ethyl acetate		A									
Pentyl acetate			P				P		P		
Ethyl butyrate		B									
Methyl salicylate											
1-Hexanol		X	X				X	X	X		
1-Octen-3-ol									O		
E2-hexenol		E	E				E	E	E		
2,3-Butanedione		D									
2-Heptanone		H	H				H		H		
Geranyl acetate											
Propyl acetate		F							F		
Isoamyl acetate		I							I		
Octyl acetate			●								
1-Butanol		Q					Q				
1-Heptanol			T				T	T	T		
3-Octanol									F		
1-Nonanol								N			
Cyclohexanone											
(-) Fenchone											
Anisole	S			S		S				S	
Methyl eugenol						L					
Benzaldehyde	Z			Z			Z				
Acetophenone	C			C			C				
2-Methylphenol	W			W		W				W	
4-Methylphenol	Y			Y							Y
Propionic acid											
Carbon dioxide											

(b)

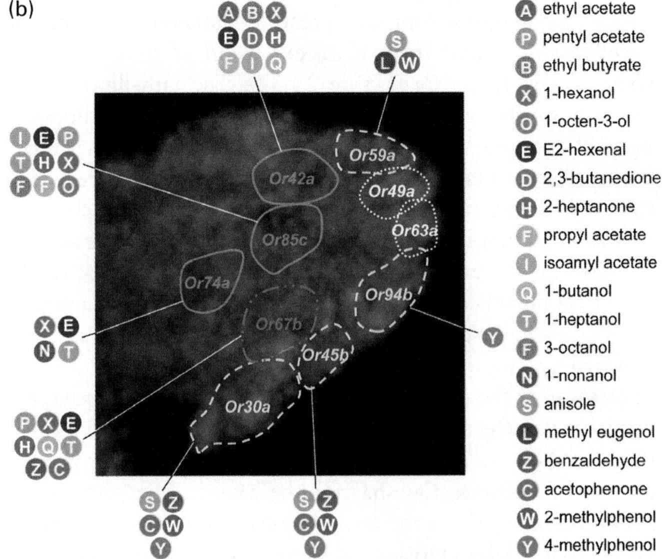

◀ EXPERIMENTAL FIGURE 23-36 **Individual olfactory receptor types can be experimentally linked to various odorants and traced to specific glomeruli in the *Drosophila* larval olfactory system.** (a) The different olfactory receptor proteins are listed across the top, and the 27 odorants tested are shown down the *left side*. Colored dots indicate strong odor responses. Note that some odorants stimulate multiple receptors (e.g., pentyl acetate), while others (e.g., ethyl butyrate) act on only a single receptor. Note that many receptors, such as Or42a or Or67b, respond primarily to aliphatic compounds, whereas others, such as Or30a and Or59a, respond to aromatic compounds. (b) Spatial map of olfactory information in glomeruli of the *Drosophila* larval brain. The mapping was done by expressing a reporter gene under the control of each of the relevant olfactory receptor neurons. The photograph indicates the glomeruli that receive projections from ORNs producing each of the ten indicated receptor protein types (Or42a, etc.). Also indicated are the odorants to which each receptor responds strongly. Note that with one exception (Or30a and Or45b) each glomerulus has unique sensory capacities. The exception might not be an exception if more olfactory gene expression patterns were tested. Glomeruli sensing odorants that are chemically similar tend to be situated next to one another. For example, the three glomeruli indicated by a blue solid line sense linear aliphatic compounds; those with yellow dashed lines, aromatic compounds. [Part (a) From S. A. Kreher, J. Y. Kwon, and J. R. Carlson, 2005, *Neuron* **46:**445–456. Part (b): Courtesy of Jae Young Kwon, Scott Kreher, and John Carlson.]

two roles in ORNs: odorant binding and, during development, axon guidance. Multiple ORN axons expressing the same receptor are guided to the same glomerulus destination. The full mechanism is not known, but it is clear that ORN axons respond both to their own olfactory receptor and to standard axon guidance molecules used in other parts of the nervous system. We will discuss those in the next section.

KEY CONCEPTS OF SECTION 23.4

Sensational Cells: Seeing, Feeling, Hearing, Tasting, and Smelling

- Eyes focus light on a light-sensitive surface, the retina, where two types of photoreceptor cells are located. Highly sensitive rod cells (for black, gray, and white) and less sensitive cone cells (for color) are stimulated to produce electrical signals in response to light (see Figure 23-24).
- Stimulated photoreceptor cells hyperpolarize, rather than depolarize like most neurons.
- Light-sensitive pigments called *opsins* and *rhodopsins* give photoreceptor cells the machinery to perceive light (see Figure 23-25).
- Combined signals from multiple photoreceptor cells are interpreted together to discern patterns of light and dark. Retinal ganglion cells, which carry signals from the retina to higher parts of the brain, are responsive to sets of photoreceptor cells in patterns of dark center and light surround or vice versa (see Figure 23-26).
- The initial simple center-surround patterns are combined by higher-level cells to give increasingly complex images of the world (see Figure 23-28).
- Mechanosensory cells in mammals detect body position (proprioception), pain, heat, cold, touch, and pressure. The signals generated by these receptors give rise to maps of the body on the surface of the brain, called *homunculi* (see Figure 23-29).
- The inner ear senses motion and allows balance to be maintained, and detects sound (see Figure 23-30). The sound detector is the cochlea, with the organ of Corti inside. Vibrations

carried by small bones from the outer to inner ear move fluid, which in turn moves fine hairs called *stereocilia* (see Figure 23-31). The movements of these cilia generate receptor potentials in the hair cells, which are perceived as sound.

- A small number of taste receptors detect chemicals dissolved in saliva. Some are seven-transmembrane-domain proteins, and these work in different homo- and heterodimeric combinations to detect different tastes (see Figure 23-32).
- Odorant receptors, which are seven-transmembrane G protein–coupled receptors, are encoded by a very large set of genes. Any one olfactory receptor neuron expresses one and only one olfactory receptor gene, so a signal from that cell to the brain unambiguously conveys the nature of the chemical sensed. ORNs that express the same receptor gene send their axons to the same destination in the brain (see Figures 23-33 to 23-36).

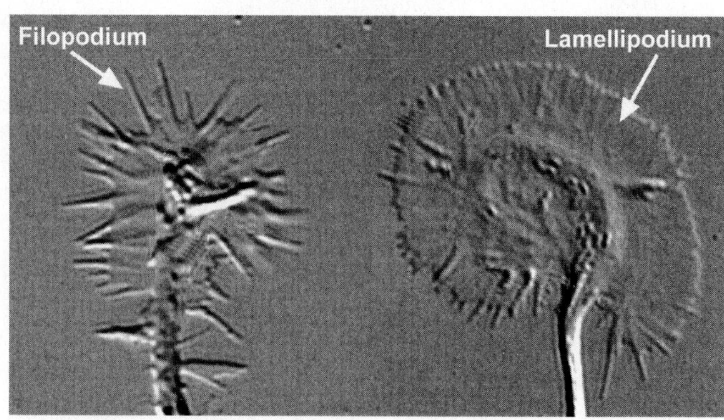

▲ **FIGURE 23-37 Growth cones.** Growth cones have a broad spread of flattened material called the *lamellipodium*, from which many sharp spikes, called *filopodia*, protrude. [From E. W. Bent and F. B. Gertler, 2003, *Neuron* **40**:209–227.]

23.5 The Path to Success: Controlling Axon Growth and Targeting

At the heart of the functions of the nervous system, especially the advanced functions, is complex circuitry. Having described the properties of nerve cells, electrical signal transduction, chemical transduction at synapses, and sensory cells, we now turn to the problem of how the connections between neurons are formed. Neural development can be viewed as having several steps: the formation of germinal zones where new neurons are born from stem cells, the determination of newly formed cells into neurons or glia, cell migration, axon and dendrite formation, growth of axons following guidance cues (sometimes over long distances), formation of synapses, pruning of exuberant connections, and the programmed death of some neurons.

A neural precursor cell often is unspectacular in appearance, a simple round cell. The growth of dendrites and axons is transforming. Axons grow in response to guidance systems, using signals from other cells. The leading, growing, end of an axon is called the **growth cone.** Signaling molecules bind to receptors at the growth cone, influence the cytoskeleton inside, and cause the growth cone to turn toward or away from the signal. The growth cone is a sensor and explorer.

> From the functional point of view, the growth cone may be regarded as a sort of club or battering ram, endowed with exquisite chemical sensitiveness, with rapid amoeboid movements, and with a certain impulsive force, thanks to which it is able to press forward and overcome obstacles met in its way, forcing cellular interstices until it arrives at its destination.
>
> —Ramón y Cajal, 1890, ref. in M. J. Zigmond et al., eds., 1999, *Fundamental Neuroscience,* Academic Press, chap. 18, p.519 [Cajal, S. R. y, 1890, Á quelle époque apparaissent les expansions des cellules nerveuses de la moëlle épinière du poulet? *Anatomomischer Anzeiger* **21–22:**609–639.]

In this section we focus on the formation and growth of axons, and the guidance systems that allow axons to extend toward targets and then recognize and synapse with the targets upon arrival. The growth cone is the heart of the matter, a cell extension that leads the elongating axon through its twists and turns. The amoeboid behavior of the growth cone was correctly described by Ramón y Cajal in 1890, and now we know a bit more about the molecules that give the cone its remarkable properties. The growth cone responds to signals that bias its direction of growth, thus wiring the nervous system.

The Growth Cone Is a Motorized Sensory Guidance Structure

The shapes of growth cones vary widely, but generally there are two major features: a broad spread of flattened material, the *lamellipodium,* and from it an extension of multiple sharp spikes called *filopodia* (Figure 23-37). The lamellipodium is typically on the order of 10–100 μm in breadth, the filopodia up to 20 μm long. As the growth cone moves away from the cell body, it leaves a tapering region behind that will be the shaft of the axon. Time-lapse studies show that as the growth cone advances, the filopodia that are located at the front move around the sides of the lamellipodium and retract at the back of the growth cone, where it is in the process of becoming an axon.

Three phases of growth-cone extension have been defined: protrusion, engorgement, and consolidation (Figure 23-38). *Protrusion* is the extension of lamellipodia and filopodia, *engorgement* is the swelling of both as the growth cone engulfs them, and *consolidation* is the active narrowing of the growth cone as it becomes the shaft of the axon. These three steps occur continuously as the growth-cone leading edge advances. Drugs that inhibit cytoskeletal elements have been used to investigate contributions of fiber systems to the three steps. Actin polymerization is required for protrusion and maintenance of the growth cone; cytochalasin B, which causes actin depolymerization (Chapter 17), led to retraction or misdirection of growth cones. Both lamellipodia and filopodia are formed by polymerization of actin. Microtubules form the backbone of the axon; colchicine, which causes microtubule depolymerization (Chapter 18), led to axon retraction but did not immediately perturb growth cones. Colchicine also caused the

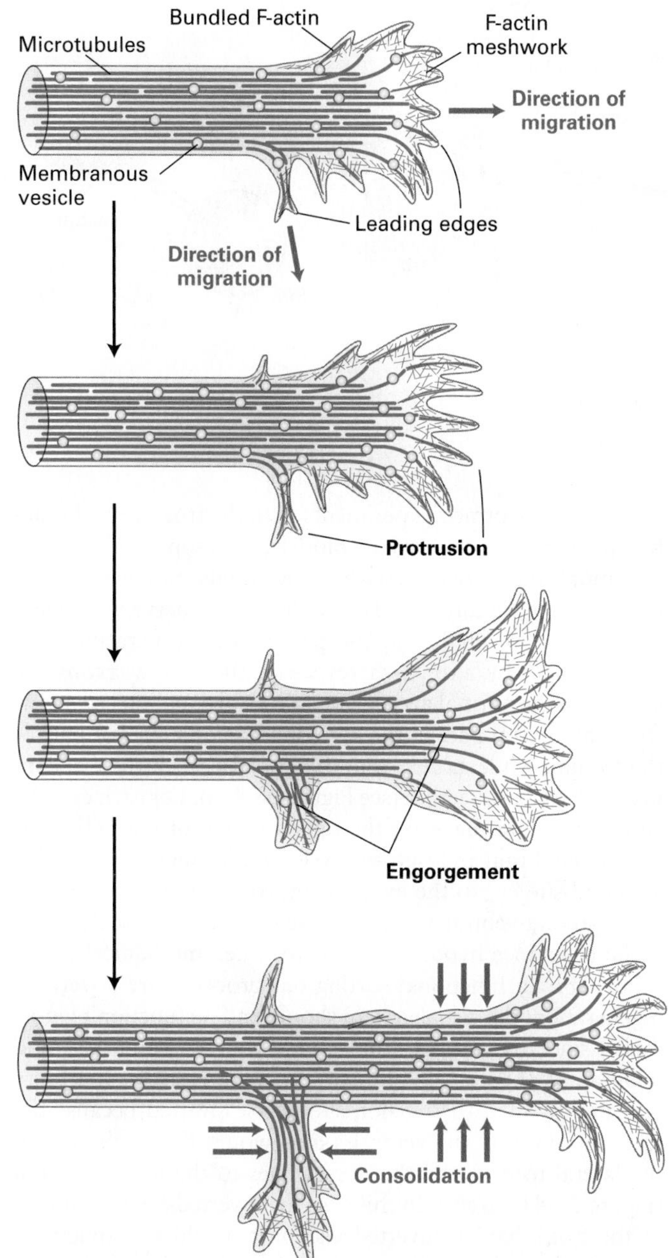

▲ **FIGURE 23-38 The advance of the growth cone.** During protrusion, filopodia and lamellipodia extend under pressure from intracellular F-actin meshworks and elongated bundles (ribs). During engorgement, microtubules are elongated into the protrusions and carry membrane-bound organelles into them. During consolidation, depolymerization of actin in the growth-cone neck is followed by narrowing of the cell around the microtubule bundle to form the axon shaft. A new protrusion can form by branching off the side of a cylindrical axon shaft. [From E. W. Bent and F. B. Gertler, 2003, *Neuron* **40:**209–227.]

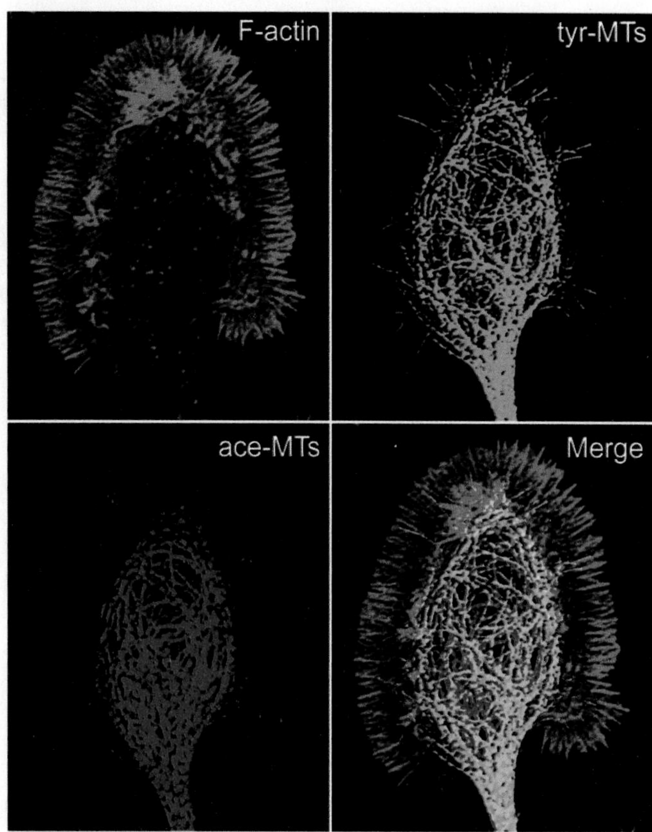

▲ EXPERIMENTAL FIGURE 23-39 **Antibody labeling reveals cytoskeleton components inside cultured hippocampal growth cones.** A single growth cone is shown three times, in each case labeled with an antibody for a different structure: F-actin, tyrosinylated microtubules (tyr-MTs), and acetylated microtubules (ace-MTs). The fourth panel shows a composite of the other three. Note the relative lack of microtubules at the leading edges and periphery, and the concentration of actin there. The microtubules are concentrated in the central region (although some tyr-MTs colocalize with actin bundle spikes in the peripheral region. The growth cone is paused, so the microtubules loop. [From E. W. Bent and F. B. Gertler, 2003, *Neuron* **40:**209–227.]

appearance of extra outgrowths of lamellipodia and filopodia along the axon shaft, implying that microtubules may normally prevent the assembly of microfilaments in the axon.

The concentration of actin in the periphery and leading edge of the growth cone, and of microtubules in the more central and lagging regions, reflects the different roles played by the two (Figure 23-39; see also Chapter 17). Actin is assembled into filaments in the leading cone, the filamentous mesh of actin flows backward as the cone advances, and actin is disassembled as the cone transforms into an axon. The transport of actin toward the rear occurs in the filopodia and lamellipodia and is driven by a myosin motor. Note that this is completely different from treadmilling (Chapter 17). Movement of the actin mesh with respect to the growth cone involves movement of the entire filament at 1–7 μm/min, attached to myosin motors. For a filopodium to advance, the rate of actin polymerization at the leading edge must exceed the rate of retrograde flow.

Actin filaments have been viewed as major determinants in turning growth cones, a process that is crucial for the neuron to respond to guiding signals. Microtubules are also involved, since microtubule-inhibiting drugs prevent turning as well. Microtubules are assembled at the neuron's centrosome and transported by dynein motors toward the advancing growth cone, with the plus end leading. While traveling, the microtubules

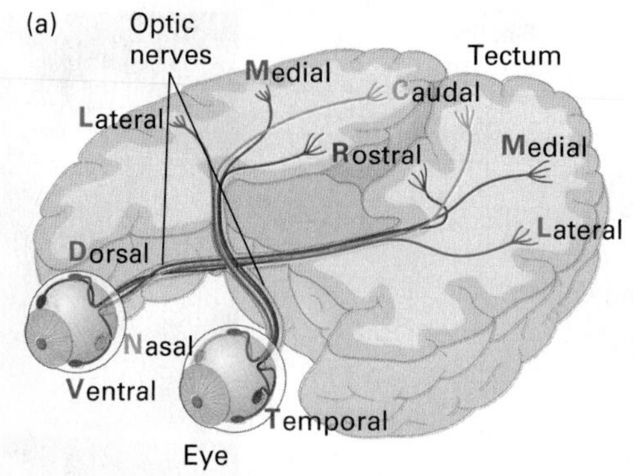

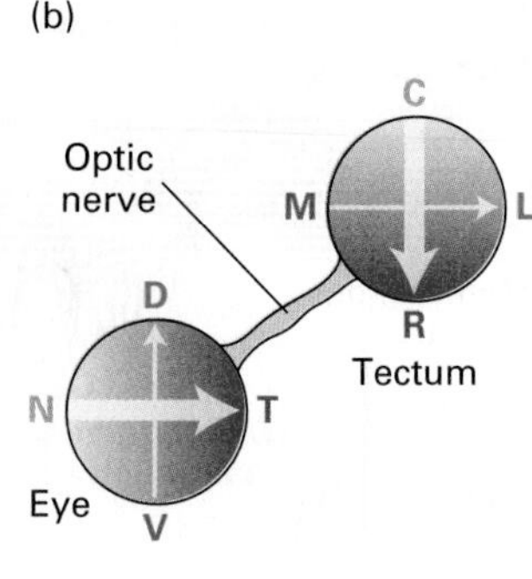

▶ **FIGURE 23-40 The retinotectal maps.** (a) The dorsal retina is connected to the lateral tectum on the opposite side of the brain, and the ventral retina is connected to medial tectum on the opposite side. Similarly the temporal-nasal (T-N) axis of the eye is reflected in the rostral-caudal (R-C) map of the tectum. (b) The maps of the visual world on the retina and the corresponding tectum are turned 90 degrees but otherwise are in register. The arrow shows how a pattern of light on the retina is reproduced as a set of retinal ganglion cell connections in the tectum. The tectum in mammals is referred to as the *superior colliculus (SC)* [G. Lemke and M. Reber, 2005, *Ann. Rev. Cell Dev. Biol.* **21**:551–580.]

can undergo polymerization and depolymerization. A tyrosinylated form of tubulin is preferentially present in more advanced parts of the growth cone, while acetylated tubulin is enriched in central and lagging parts and in the axon itself (see Figure 23-39). The roles of such post-translational modifications are described in Chapter 2. An ordered process of tubule assembly and modification underlies growth-cone advancement.

As we have seen (Chapter 17), actin polymerization is controlled by a startlingly large set of regulatory proteins. More than 20 actin-binding proteins have been found in growth cones, most of which control nucleation or polymerization of actin filaments, or tether the filaments to the membrane. Many of these actin-binding proteins are targets of signal transduction events triggered by axon guidance signals, as we shall see in the next section.

The Retinotectal Map Revealed an Ordered System of Axon Connections

Researchers long debated two general ideas about how neurons might get wired. One, the *"resonance hypothesis,"* proposed that cells extend axons along pathways that are defined by mechanical forces. After many paths are taken and connections formed, the ones that work are preserved while others are removed. The second idea, a *"specific pathways hypothesis,"* suggested that axons choose their path by chemical affinity, molecules on the growing axons contacting molecules along the way that provide guideposts or signals. In 1963 Roger Sperry proposed a version of the specific pathways idea called the *"chemoaffinity hypothesis."* He suggested that growth cones would find their way following molecular cues that form a gradient from start to destination, a seminal proposal that could not be properly tested for decades. Sperry's proposal was based on his studies of how the axons of the retinal ganglion cells, which form the optic nerve, are arranged when they arrive at the *optic tectum.* The optic tectum is located in the roof of the midbrain and is the destination of retinal ganglion cell axons that grow from the retina. The incoming retinal neurons form a map on the tectum (the **retinotectal map**) that reflects the arrangement of rods and cones in the retina, and indeed the visual world outside (Figure 23-40). The spatial map on the retina is in essence copied into the brain.

Sperry performed experiments with the frog eye and brain to distinguish the two models, "resonance" versus "chemoaffinity," that describe how axons may accomplish this mapping (Figure 23-41). The frog optic nerve will regenerate if it is severed, and the pattern of regeneration—the route-finding by axons—is revealing about how axons are guided. In the normal arrangement, retinal ganglion cell axons from the ventral part of the eye connect to the medial part of the tectum, while axons from the dorsal part of the eye connect to the lateral tectum (see Figure 23-41a). For each eye, the connections are made on the opposite side of the brain, left eye to right brain and right eye to left brain. Sperry next added a second surgery to the experiment, rotating one eye 180 degrees, so that ventral and dorsal are reversed (Figure 23-41b). If the resonance hypothesis is correct, i.e., mechanical forces followed by a functional sorting-out process were governing regeneration, the visual system should end up functioning normally, since the proper connections will be established and maintained (Figure 23-41c, *left*). If there is a chemoaffinity guidance system, then vision should be inverted because despite the rotation the ventral axons would find their way to the lateral tectum and the dorsal ones to the medial tectum (Figure 23-41c, *right*). In this case the inverted eye would lead to the frog's having inverted vision: it would see something above and would think it was below (Figure 23-41d). The results were clear: after regeneration the frog responded to a fly passing above by shooting its tongue *down.* The axons were originating from abnormal locations yet finding their way to the right connections, so the inverted eye was tricking the frog's brain. The chemoaffinity hypothesis was affirmed.

> It seems a necessary conclusion...that the cells and fibers of the brain and cord must carry some kind of individual identification tags, presumably cytochemical in nature, by which they are distinguished one from another almost, in many regions, to the level of the single neuron.
> —Roger Sperry, 1963

There is now substantial evidence that Sperry's general ideas were correct. Nonetheless much remains to be learned about how the enormous complexity of the neural circuitry is successfully assembled. The relative importance of local versus

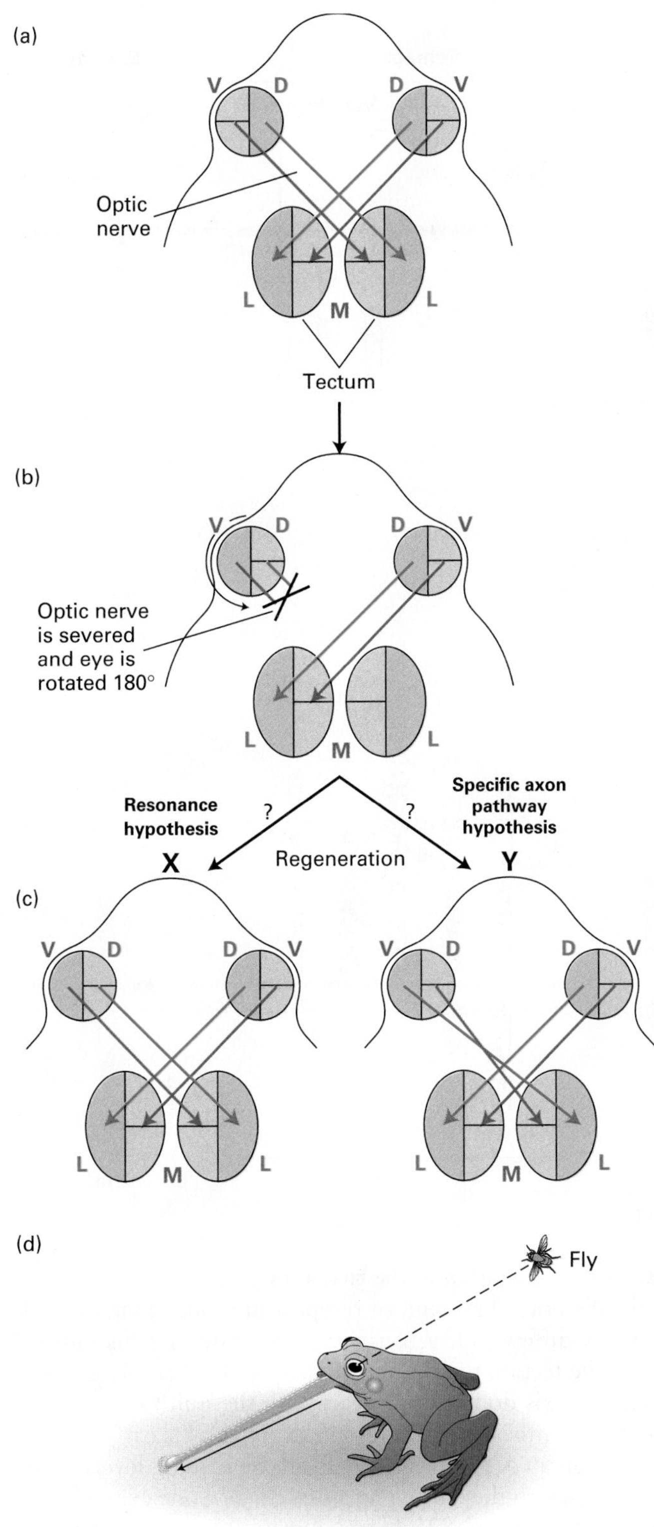

◀ EXPERIMENTAL FIGURE 23-41 **Eye rotation experiments test the properties of axon pathfinding.** (a) Normal projection of nerves, from dorsal (**D**) eye to lateral (**L**) tectum and ventral (**V**) eye to medial (**M**) tectum. (b) Schematic representation of two sequential operations, (1) a severing of the optic nerve and (2) a 180° rotation of the eye (or no rotation in the control experiment). Regeneration of the nerves was then allowed to occur. Note that in the rotated left eye, the dark-shaded half, now oriented dorsally (**D**), nevertheless still contains the ventral retina, as seen in (a). Likewise the light half of the eye, which encompasses the dorsal retina, is now oriented ventrally (**V**). (c) Two possible outcomes, depending on which hypothesis is correct. The resonance hypothesis predicts **X**: vision is restored because function selects proper connections. The specific axon pathway hypothesis predicts **Y**: vision is inverted because dorsal retinal axons still go to lateral tectum because of chemical affinity specified, even though the dorsal retina is now located in the ventral position. (d) The results support hypothesis **Y**; the frog's vision is inverted.

long-range signaling, the roles of glia, the influences of neural electrical activity, and the signal transduction and cytoskeletal changes that form the response to signals are important areas of current research. The lure of this field is substantial, since understanding how neurons are wired to one another underlies the working of the brain and at the same time is important for learning how to stimulate repair of damaged neural circuits.

There Are Four Families of Axon Guidance Molecules

For many years, attempts were made to identify key axon guidance molecules. Approaches included making panels of antibodies against surface molecules, culturing neurons and testing extracts for their ability to make growth cones turn, and using genetics to identify mutants that fail to properly wire the nervous system. All these attempts worked to a degree, but genetics was the most powerful approach, since it identified previously unknown molecules while at the same time convincingly demonstrating their importance in vivo.

We can set a high standard for what constitutes a proper guidance molecule: it must be produced by cells that actually guide neurons in vivo, it must be necessary for guidance, the guided cells must have sensors and signal-transduction machinery for responding, and the mislocalization of the signal must cause cells to turn the wrong way.

We will discuss here four families of proteins (Figure 23-42) that fulfill our criteria and, with their receptors, provide crucial information to growing axons: **Ephrins, Semaphorins, Netrins,** and two related proteins called **Robo and Slit.** These proteins have both attractive and repulsive effects on growing axons. Since the growth cone is an active *sender* of signals, as well, the communication is mutual. After the initial connections form, the wiring is refined by preserving connections that work and discarding those that do not contribute to neural function. Many cells that fail to make useful connections die by apoptosis.

Ephrins The retinotectal map described above is a striking example of the simplicity of the nervous system, which belies the seemingly chaotic mass of neurons in the visual part of the brain. This rather amazing phenomenon is due in part to a remarkable signaling system involving the **ephrins,** a family of cell-surface signaling proteins, and their receptors, the **Ephs.** (The word "Eph" comes from the *e*rythropoietin-*p*roducing *h*epatocellular carcinoma cell line where the proteins were originally found.) Ephs constitute the largest family of receptor tyrosine kinases (RTKs; Chapter 16), with 14 Ephs and 8 ephrins in mice. Although Ephs usually serve as receptors for ephrins,

▶ **FIGURE 23-42 Families of guidance molecules.** Four major families of signaling proteins provide crucial information to direct growing axons. The ligands (a) are Netrins, Robo/Slit, Sempaphorin, and Ephrin proteins. The receptors (b) are as follows: Netrin interacts with its receptor DCC in vertebrates; the corresponding but different receptor in nematodes is called Unc5. Both contain immunoglobulin (Ig) domains (blue crescents) and a variety of other domains as shown. The Slit ligand interacts with the Robo receptor, which also has Ig domains. The Semaphorins interact with diverse receptors, generically called Plexins. Some of the interactions are through "sema" domains (red bars), present in both ligand and receptor, others require Ig domains. Ephrin ligands interact with Eph receptors, although signaling appears to go in both directions and in some cases the Ephrins act more like receptors. All of the receptors have single transmembrane domains. The Netrin and Slit ligands are secreted and not membrane associated. The Semphorin ligands can associate with membranes to varying degrees—some not at all. The Ephrins are membrane-associated. See the text for a detailed discussion of these four families. In addition to these four families of proteins a few others, including the developmental regulators Wnt and Shh, contribute additional guidance information. [From R.P. Kruger, J. Aurandt, and K.-L. Guan, 2005, *Nature Rev. Mol. Cell Biol.* **6**:789–800, and B.J. Dickson, 2002, *Science* **298**:1959.]

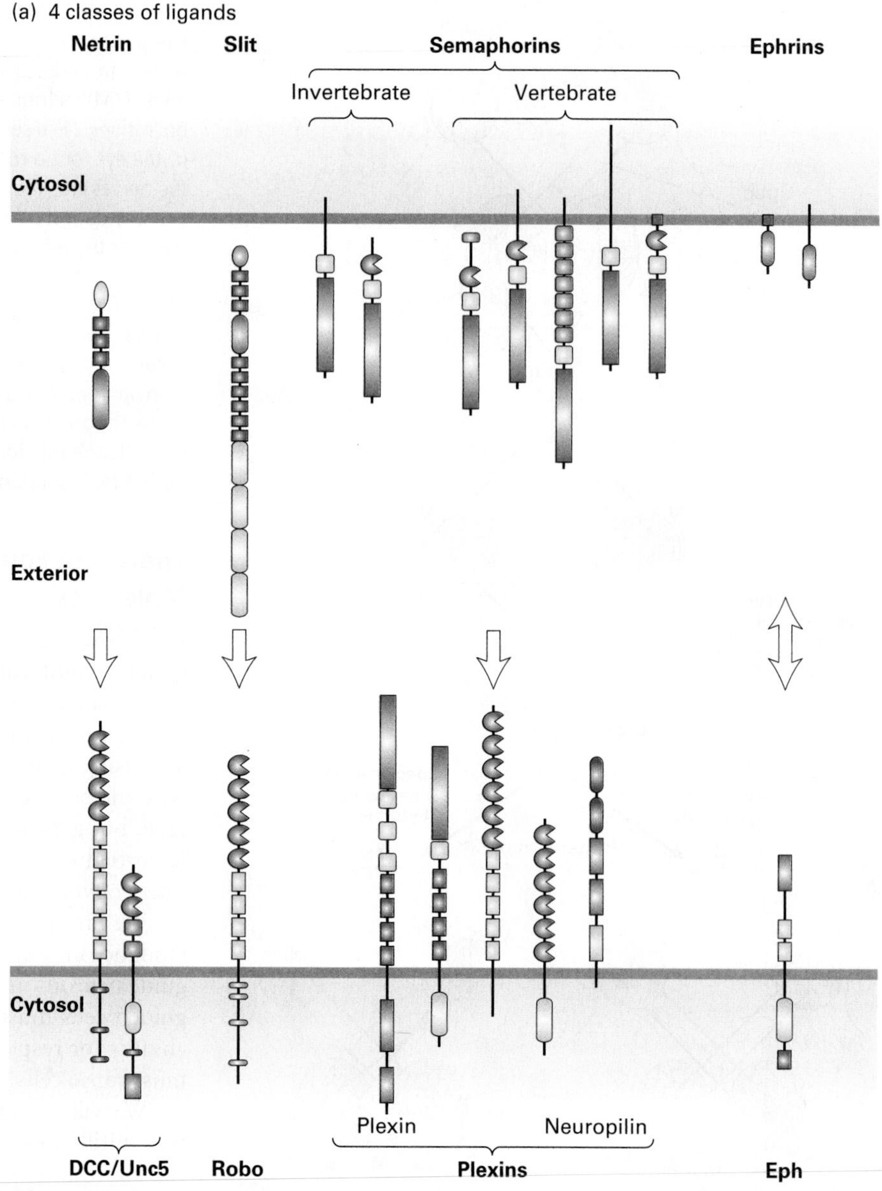

in some cell types signaling can go from an Eph-bearing in the other direction to an ephrin-bearing cell. In retinotectal map development, most signaling is from ephrin to Eph receptor. Each Eph receptor specifically binds one or a few ephrins. Ephs and one class of ephrins, the A class, are joined to cell-surface membranes by GPI links. Ephrin Bs, the other class of ephrins, are transmembrane proteins (see Figure 23-42). The ephrin proteins affect cell migration, axon guidance, synapse development, and vascular development, but we will focus on their roles in guidance and formation of the retinotectal map.

Ephs and ephrins are distributed in gradients so that advancing axons can recognize and grow toward appropriate targets. Antibodies against ephrin A5 show a gradient of protein with the highest levels in the anterior. Mice lacking ephrin A5 have guidance defects; axons that should have been targeting anterior tectum grow into posterior regions. Further investigations showed that cells express Eph receptor proteins in two orthogonal gradients, to control axon guidance along each of the two axes (Figure 23-43). In each axis, the graded amount of receptor in retinal ganglion cells confers differential sensitivity to specific ephrins emitted from the tectum targets. The EphA's and ephrin A's controlling one axis do not cross-react with the EphB's and ephrin B's for the other axis. Therefore, axons can "learn" their positions on an XY coordinate system by reading levels of ligand for which they have receptors. To give one example, axons that have EphB2, 3, and 4 receptors are attracted to places where there are high levels of ephrin B1 ligands, so axons originating in ventral retina tend to go to medial tectum. The full situation with all these gradients is more complex and not yet fully understood. Eph tyrosine kinase signal transduction influences the small GTPases Rho, Cdc42, and Rac, thus regulating assembly of the actin cytoskeleton and controlling guidance of the growth cone. The activation of an Eph receptor may cause attraction or repulsion of a growth cone, depending on the cell.

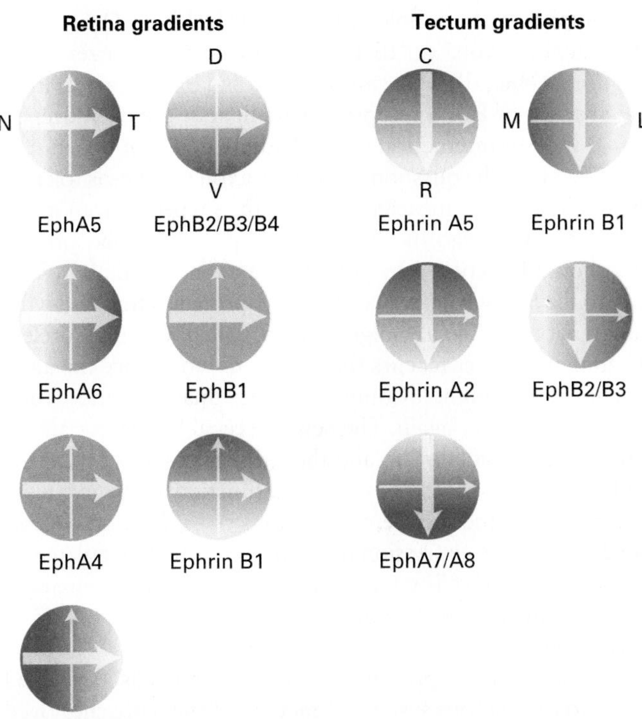

▲ **FIGURE 23-43 Gradients of Eph proteins form two orthogonal signaling systems.** Eph receptor gradients in the retina are shown on the *left*; ephrin gradients in the superior colliculus (tectum), on the *right*. Gradients along the nasal-temporal (retina) and rostral-caudal (tectum) axes are shown in blue; gradients along the dorsal-ventral (retina) and lateral-medial (tectum) axes are shown in red. The more color, the more protein present. The white arrows and arrow outlines indicate how the maps are related (see Figure 23-40). Ephrin A's are chemorepellent for EphA-expressing axons, so retinal ganglion cells carrying Eph A receptors tend to go from their origins at the temporal side of the retina to destinations in the rostral tectum, where the concentration of ephrin A's is lower.

Semaphorins Semaphorins are a diverse family (see Figure 23-42), and much remains to be learned about all their effects. They were named for the alphabetic signaling system of flags that was used to communicate over long distances. Semaphore signals can spell out any message, but in the nervous system semaphorins largely carry a single message: go away. They are potent repellents. The family of two invertebrate and five vertebrate semaphorin glycoproteins (Figure 23-42) includes some that are secreted and some that are membrane-bound. This implies that some of them act on adjacent cells, while others have a longer reach. Motor, sensory, olfactory, and hippocampal neurons can be repulsed by semaphorin signals. Semaphorins bind to receptors called **plexins** that are single-pass transmembrane proteins. The overall picture is of receptor proteins capable of acting as scaffolds both inside the cell and outside, assembling protein complexes and modifying their activities. How exactly this modifies growth-cone advancement is an area of current research; at least part of the answer is differential adhesion to cells on one side of the growth cone versus the other.

Netrins Netrins are secreted proteins related to laminin (see Figure 23-42). They were discovered in genetic screens looking for misrouting of neurons in *C. elegans,* the nematode worm for which every cell and every neuron has been identified. About 30 genes were found, three of which affected dorsal-ventral routing of the sensory and muscle neurons (Figure 23-44): *unc-6,* which encodes a netrin protein; *unc-40,* which encodes the netrin receptor (called *DCC* in mammals); and *unc-5,* which encodes a second type of netrin receptor. *Unc-6/Netrin* mutations affected both dorsally directed and ventrally directed axon extensions. Vertebrate netrins were found in studies of how commissural (crossing) neurons find their way through the spinal cord (see Figure 23-44). These neurons emerge from dorsal regions of the spinal cord and extend around the periphery of the cord toward the ventral midline.

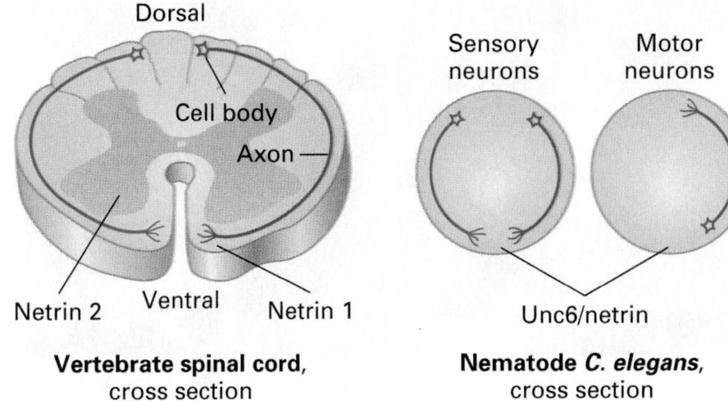

▲ **FIGURE 23-44 Evolutionary conservation of netrin signaling.** In the vertebrate spinal cord, commissural neurons with cell bodies in dorsal regions grow axons toward the ventral midline, attracted by netrin signals. In the worm *C. elegans*, sensory neurons follow a similar path toward ventral Unc6/Netrin, while motor neurons follow the opposite path away from Unc6/Netrin. Certain vertebrate axons are also repelled by netrins. Netrins were discovered in worms via mutations that altered the pathfinding by sensory neurons, and in vertebrates as secreted molecules that could influence pathfinding of commissural neurons in spinal cord explants.

To test for the presence of guidance molecules, parts of the spinal cord were cultured either separately or together. When the most dorsal part was cultured near a ventral part, axons grew out toward the ventral tissues. No axon outgrowth was observed when the two parts were cultured separately. Extracts from the ventral part had the same activity, stimulating axon outgrowth from dorsal tissue. A heroic protein purification, starting with about 20,000 embryonic chick brains, succeeded in identifying two proteins that were potent chemoattractant signals. Both were netrins. Netrin 1 is highly expressed in the spinal cord floor plate, the most ventral part. Further proof of the role of netrins came from mouse gene knockouts, in which the commissural neurons were unable to properly find their way (Figure 23-45). Netrins guide axons to the ventral midline in nematodes, flies, and vertebrates, an example of evolutionary conservation of protein function over more than half a billion years.

A puzzle remained. The worm version of the protein appeared to have two functions: attracting the axons of sensory neurons to grow toward the ventral midline and driving the axons of motor neurons *away* from the ventral midline. The simplest possibility was that netrin is attractive to some axons and repellent to others. Indeed evidence quickly emerged

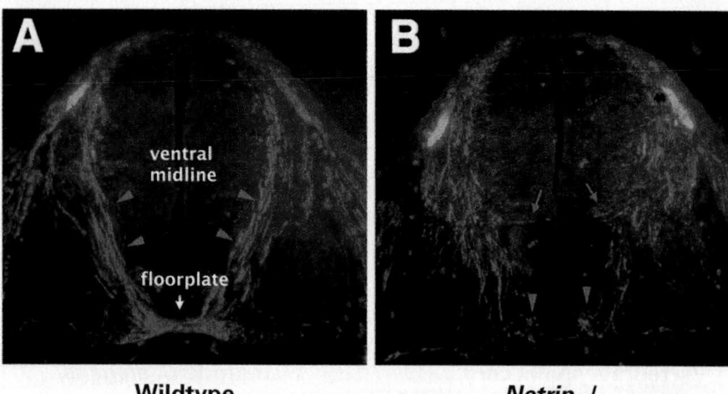

▲ EXPERIMENTAL FIGURE 23-45 **Mouse *netrin -/-* mutants have commissural neuron guidance defects.** (a) Wild-type tracing of commissural neurons (red) that originate dorsally (*top*) and grow toward and across the ventral midline (green arrowheads) under the influence of netrin produced by ventral midline (floor plate) cells. (b) Homozygous *netrin -/-* mouse mutant. Many commissural neurons wander off track (green arrows) before reaching the ventral regions, and others (green arrowheads) turn instead of crossing the ventral midline. [Courtesy of Marc Tessier-Lavigne, Genentech Inc.]

that certain vertebrate neurons found netrin repulsive. Genetic analyses in worms showed that Unc-40 receptor is required for attraction by netrin, while Unc-5 receptor in combination with Unc-40 is required for repulsion by netrin.

Another puzzle remained. If the axons of commissural neurons are attracted by netrin coming from the ventral midline, how do the axons continue to grow after they have crossed the midline? One would expect them to turn around and go right back. The solution to this puzzle awaited the discovery of still other key players in axon guidance, Slit, Robo, and Comm.

Robo and Slit Guidance Molecules The path of growing axons in the insect nerve cord is reminiscent of a subway map (Figure 23-46a). Genetics allowed the discovery of guidance genes and proteins that affect the pathfinding process, changing the map. A large set of random mutations was introduced into the *Drosophila* genome to generate lethal mutations. The mutations could be carried in heterozygotes, and when the heterozygotes of each line were crossed, a quarter of their progeny were homozygous for the newly induced mutation. These progeny were stained to show the embryonic nerve cord, the equivalent of our spinal cord, which in the wild type looks like a ladder (Figure 23-46b). In lines of flies where the mutation was in a gene necessary for axon guidance, defects in the nerve cord could be seen. Among the genes identified in this manner were three, *slit, roundabout (robo),* and *commissureless (comm)*. They defined yet another set of critically important and evolutionarily conserved axon guidance molecules.

Slit is a secreted protein (see Figure 23-42) made by midline glia. The Slit receptor is Robo, a single-pass transmembrane protein with only a short sequence in the cytoplasm and fibronectin and immunoglobulin domains on the outside of the plasma membrane (see Figure 23-42). The Robo/Slit complex is a chemorepellent interaction. The presence of Slit in the midline serves to repel axons containing Robo receptors, thereby ensuring that ipsilateral (same-side) neurons do not cross to the opposite side. In loss-of-function *slit* mutants, or in double mutants that lack two *Drosophila robo* genes, axons go to the midline but can never leave (Figure 23-46c, d). This can be explained by the fact that, in the absence of Robo receptor or its ligand, the chemorepellent Robo/Slit interaction at the midline does not occur.

This raises the question of how an axon that needs to cross the midline is first attracted to it and then repelled from it. An axon that will cross the midline does produce Robo protein and should be repelled by Slit, but the axon is initially refractory to the Slit signal because the Robo protein is trapped in the Golgi network by a Golgi protein called Commissureless (Comm) and never reaches the cell membrane. Once the axon reaches the midline, Comm becomes inactive and Robo reaches the surface again. The newly accessible receptor allows a response to midline Slit, and the axon grows away from the midline on the far side. Loss of *comm* function allows excessive Robo to reach the surface, so no axons cross the midline (Figure 23-46e). The expression of *comm* is normally regulated so that it is "off" in cells whose axons are supposed to remain in only the left or right longitudinal axon tract, and is "on" in cells whose axons must cross to the other side.

None of these protein guidance systems is dedicated solely to the nervous system; in fact, all of them are employed in other tissues for various purposes. Indeed most of the special attributes of neural cells appear to be more or less exaggerated versions of processes common to many or all cells. That is apparent (1) in the polarization of neurons from dendrite to axon, which employs cell-asymmetry proteins, (2) in the neuronal intracellular organelle-transport systems, which depend on variations of endo- and exocytosis, (3) in outgrowths of axons and dendrites, which have features of chemotaxis, and (4) in the use of channel proteins to control ion flow. Neurons are a variation on familiar cell biology themes, a variation with enormous functional power.

Developmental Regulators Also Guide Axons

In Chapter 22 we became familiar with a set of secreted signaling proteins that control cell fates and, in some cases, cell division. Since these were discovered for their roles in development and differentiation, they were not initially suspects in the hunt for axon guidance molecules. Yet it turns out that at least three cell fate regulators, Hedgehog (specifically Sonic hedgehog, Shh), BMP, and Wnt proteins, can also be axon guidance molecules. This is in a sense reassuring, as the sum total of complexity of the other guidance molecules still seems rather low compared with the challenge of routing millions of axons on complex paths.

In the developing spinal cord (see Figure 23-44), even in the absence of Netrin 1 some commissural axons still extend toward the ventral midline. Another protein, Shh, made in the floor plate, accounts for that remaining guidance activity. Proof of its role came from the discovery that (1) cultured cells that secrete Shh reorient commissural axons in tissue explants, (2) isolated neurons in culture turn toward a source of pure Shh protein, and (3) mutants affecting Shh signal transduction interfere with axon pathfinding. Thus commissural axons are guided toward the ventral midline by both Netrin

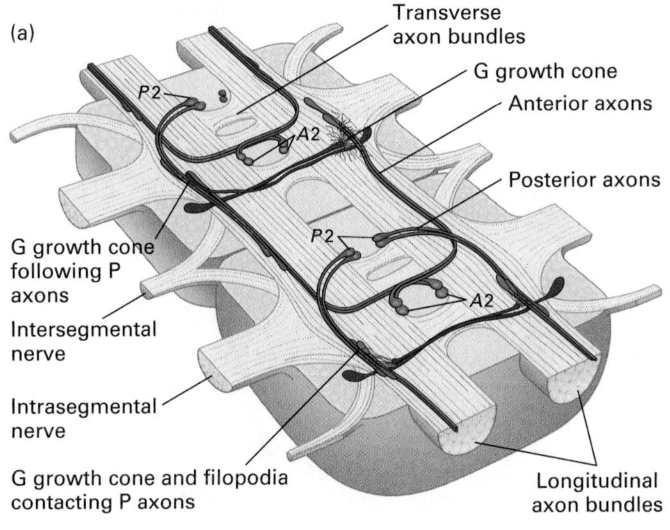

◄ **FIGURE 23-46 Phenotypes of mutations in axon guidance mutants in the insect nerve cord.** (a) This diagram shows examples of the paths taken by neurons that cross and form the commissures. Commissures are the "rungs of the ladder" which are connected by longitudinal axon bundles. The diagram was constructed from studies of axon growth in the grasshopper, where individual neurons can readily be identified and followed due to the large size of the embryo. For example the specific neurons A2 and P2, can be recognized and studied. The growth cones marked "G" (purple) can be recognized as following the axons of the "P" cells (blue). The circuitry is highly similar to that observed in the Drosophila embryo where genetic analyses have been used to identify molecules required for proper axon pathfinding (b-e). The brown-stained pattern of commissures in a Drosophila embryo can be used to detect mutants that do not properly carry out axon pathfinding. In (b) the wild-type pattern is seen. A few nuclei are stained blue in this panel to mark the midline. In (c-e) the nerve cords of embryos homozygous for three different mutations are shown. In each case there are dramatic failures of axon pathfinding. [(a) From C.S. Goodman and M. Bastiani. 1984. *Scientific American* **251:** 58. (b-e) From M. Seeger et al., 1993, *Neuron* **10:** 409–426]

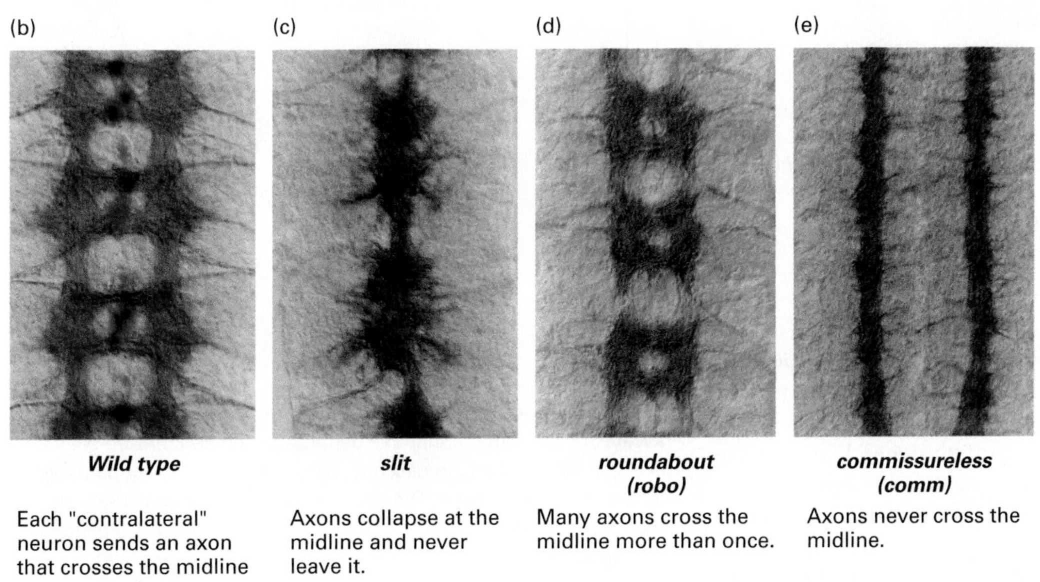

1 and Shh. Conversely, BMP proteins originating from dorsal regions repel commissural neurons. Mice mutant for certain BMP genes were found to have commissural axon guidance defects. Thus commissural neurons respond to "pulling" by Netrin 1 and Shh and "pushing" by BMPs.

In the developing spinal cord, after the axons of commissural neurons reach the ventral midline and cross it, they turn sharply toward the head (Figure 23-47). In fact, many axons must find their way along the anterior-posterior (A-P) axis (head to tail), but less is known about guidance in this dimension. Discoveries in *C. elegans, Drosophila,* and mice show that Wnt secreted-protein signals play an important role in A-P patterning. The use of a distinct signaling system for A-P versus D-V growth may prevent cells from getting confused about which axis is in play, when cells have to make decisions about both axes at the same time. *Wnt4* in particular is involved. It is expressed in a gradient in the neural tube floor plate, with the highest amount near the head. Neurons in culture can be attracted to Wnt signals, and interference with Wnt signal transduction causes defects in the anterior migration of commissural neurons. Wnt signals are also involved in guiding the growth of retinal ganglion neurons toward the optic tectum (see Figure 23-40).

Axon Guidance Molecules Cause the Growth Cone to Turn

The growth cone is in essence an explorer, probing its neighborhood to make decisions about which way to extend the axon. In addition to the protein signals we have discussed, growth cones in culture can be attracted by neurotransmitters such as acetylcholine. Guidance molecules affect the growth cone by directly regulating the actin and microtubule cytoskeleton. Repulsive forces cause the growth cone to stop advancing, or cause it to turn away. Attractants stimulate the growth cone to keep moving in their direction.

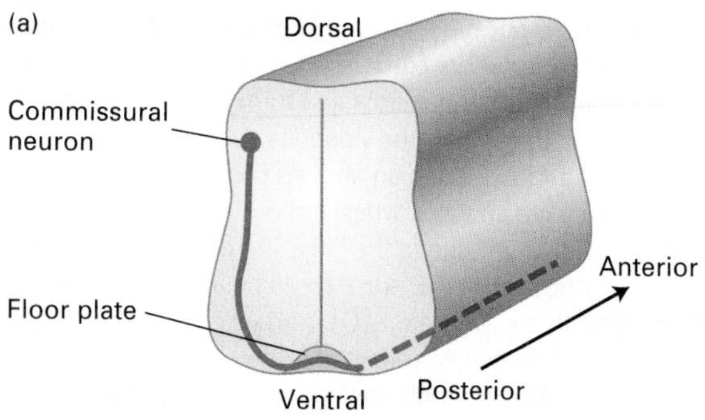

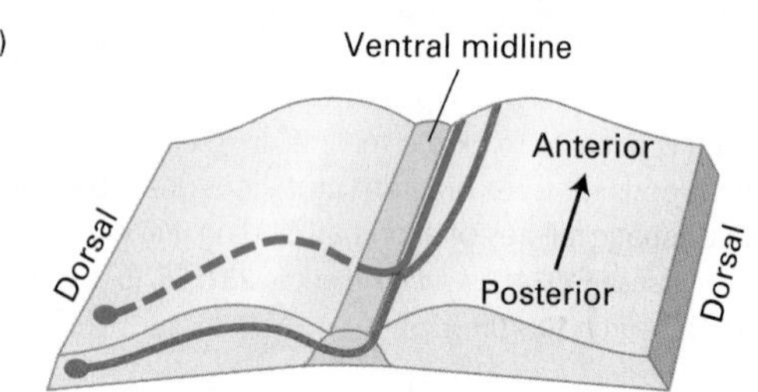

▲ **FIGURE 23-47 Wnt signals control the anterior-posterior elongation of commissural neurons in the spinal cord.** (a) The spinal cord of a 13-day-old rat embryo is shown with the path of a commissural neuron indicated in red. (b) "Open-book" view of the same path, but showing the different paths taken by (green) neurons that extend close to the midline and (red) neurons that take a more lateral path. Both respond to Wnt signals that are more anterior. [From A. Lyuksyutova et al., 2003, *Science* **302:**1984–1988.]

Turning of a growth cone is accomplished by causing filopodia or lamellipodia (or both) on one side of the cone to be more stable or to extend more vigorously, or conversely causing their collapse on the other side. How do all the guidance factors affect the growth cone? It is particularly important to understand how the growth cone responds to more than one guidance system at once, a fascinating problem in integrative molecular cell biology.

Ca^{2+} releases can strongly affect growth-cone behaviors. Dramatic changes in Ca^{2+} concentration occur over periods of less than a second (Figure 23-48), owing to the actions of

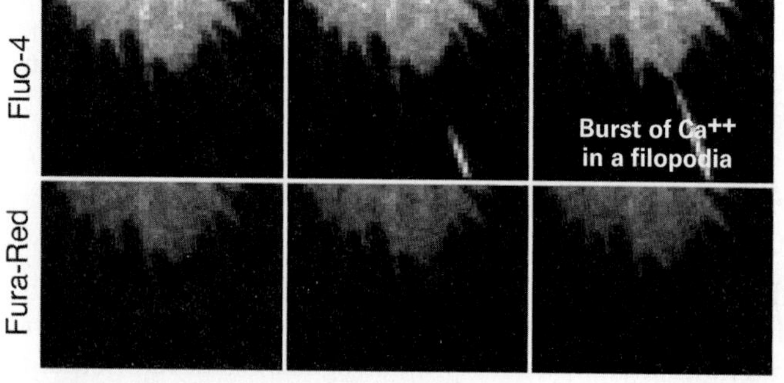

▲ EXPERIMENTAL FIGURE 23-48 **Ca^{2+} fluxes can be measured in a growth cone over time.** (a) Ca-sensitive dye (Fluo-4) is used to observe changes in intracellular Ca^{2+}; the Fura-Red dye is not Ca^{2+}-sensitive and serves as an internal control. Note the burst of Ca^{2+} in one of the filopodia. [From T. M. Gomez and J. Q. Zheng, 2006, *Nature Rev. Neurosci.* **7:**115–125.]

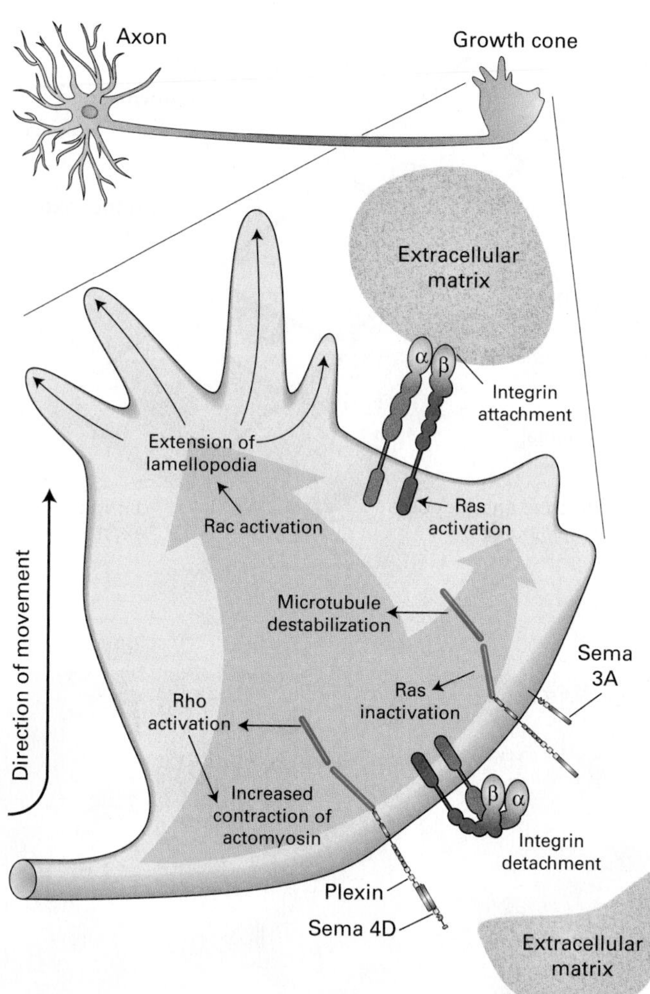

▲ **FIGURE 23-49 Semaphorins cause growth cones to turn away.** Soluble semaphorins like Sema3A or semaphorins on glia cells like Sema4D act through plexin receptors and intracellular pathways to turn the growth cone. The asymmetry of the signals causes, on the "right" of this cell, detachment of the growth cone from integrin (due to Ras inactivation), destabilization of microtubules, and increased contraction of actomyosin (through Rho activation). Meanwhile on the "left" the lamellipodia extend under the influence of active Rac, and integrin attachments to the extracellular matrix form under the influence of activated Ras. [From R. P. Kruger, J. Aurandt, and K.-L. Guan, 2005, *Nature Rev. Cell Mol. Biol.* **6:**789–800.]

pumps, channels, and Ca-sequestering proteins. Ca^{2+} inhibits neurite outgrowth in general, so reducing global $[Ca^{2+}]$, especially for prolonged times, can stimulate outgrowth. More local and more subtle changes in calcium can influence directional growth. Local $[Ca^{2+}]$ increases generally occur on the side of the growth cone closest to an attractant.

Semaphorins and other repellents stimulate the collapse of growth cones by causing the depolymerization of actin (Figure 23-49). The disassembly of the cytoskeleton is triggered preferentially on the side of the cone nearest the signal. Concomitantly, interference with integrins on that side of the cell selectively allows part of the growth cone to break free from extracellular matrix. Loss of actin appears to lead to a loss of microtubules where the growth cone abandons extension. Ephrin-dependent growth-cone repulsion depends on a series of steps that inactivate Rho and reduce actin assembly, therefore causing the growth cone to collapse.

KEY CONCEPTS OF SECTION 23.5

The Path to Success: Controlling Axon Growth and Targeting

- Growth cones are flattened fans called *lamellipodia* at the growing end of an axon (See Figure 23-37). They have small fingerlike protrusions called *filopodia*.

- Growth cones guide the growing axon and are therefore a major determinant of the initial pattern of wiring in the nervous system.

- The cytoskeleton that shapes the growth cone consists of a central bundle of microtubules immersed in a sea of shorter actin filaments (see Figure 23-38). Actin assembles into filaments at the growing tip of the growth cone and disassembles at the axon that forms after the growth cone has passed. Microtubules continue to extend during this process, to keep up with the advancing growth cone (see Figure 23-39).

- The spatial layout of light striking the retina is carried to the tectum, a visual part of the brain, and is represented as a map (see Figure 23-40). The retinal ganglion cells that connect retina to tectum maintain spatial order so that the map of the retina is aligned with the map in the tectum.

- Axons growing from retina to tectum follow guidance cues that inform as to medial-lateral and dorsal-ventral orientation (see Figure 23-41). The cues are provided by a variety of secreted and surface protein signals. Different strengths of binding between the surfaces of advancing growth cones and other neurons and the extracellular matrix influence the path chosen by growth cones.

- Four systems of guidance molecules have been identified (see Figure 23-42): ephrins and Ephs; semaphorins and plexins; netrins and DCC receptors; and Slits, Robos, and Comms. The multiple members of these protein families often make it difficult to precisely identify their functions. All the families have been highly conserved during evolution, and even some of their specific functions in development, like ventral midline axon growth guidance, have been conserved.

- Ephrins and Ephs are important in guiding retinal ganglion axons as they construct the retinotectal map (see Figure 23-43). Different classes of Eph proteins work in each of the two orthogonal axes, forming a two-dimensional system.

- Semaphorin signals, acting on their plexin receptors, have a repulsive effect on growth cones (see Figure 23-49).

- Netrin signals are usually attractive to growth cones that carry netrin receptors such as DCC, but netrins can also be repulsive signals (see Figure 23-44). Netrins guide commissural axons toward the ventral midline in the vertebrate spinal cord.

- Genetic screens led to finding Robo, Slit, and Comm proteins. Slit is a secreted signal that acts on Robo receptor, while Comm is a Golgi protein that controls whether Robo reaches the cell surface. Together they form a switch that allows axons to grow toward the midline, and subsequently grow away from it (see Figure 23-46).

- Three kinds of secreted proteins first identified as developmental regulators provide guidance functions: BMPs, Wnts, and Hedgehogs (see Figure 23-47).

- Axon guidance molecules, directly or indirectly, cause the growth cone to turn. Turning is the result of local effects of guidance factors on the membrane of the growth cone. The binding of a guidance signal can cause fluxes in calcium, changes in kinase and phosphatase activities, and differential activation of small GTPases like Rac that control cytoskeletal assembly (see Figure 23-49).

Perspectives for the Future

In this chapter we have provided an introduction to the remarkable properties of the nerve cells that serve as our interface with the world. We have seen how organelles and proteins—the universal systems of cell function—have become adapted to the special needs of neurons. The cytoskeleton shapes the cells, so that some are a meter long and others have enormously complex branched patterns. Transport processes and polarity control move organelles and macromolecules to their proper sites of action. Motors allow cells to migrate and extend processes. Signaling systems guide extending axons to their targets, including some signals that are also used in the development of many other tissues and organs. Cell-cell adhesion molecules provide selective and reversible contacts between cells. Most distinctively, channel proteins that probably evolved for import or export of metabolic molecules and for homeostatic control of ion concentrations have become adapted to the regulation of the electric properties of cells. Ligand- and voltage-gated channels create resting potentials and then allow unidirectional transmission of incredibly fast action potentials. The universal system of endocytosis and exocytosis has been adapted to the needs of chemical transduction across synapses. Other proteins have made sensory cells sensitive to light, touch, pain, sound, smell, and taste. Beyond the neurons themselves, glia provide insulation, stimulate synapse formation, and provide immune protection. Stem cells form neurons and glia initially, and some persist in the adult to allow continuing regeneration.

The advances in the cell biology of the nervous system have been paralleled by extraordinary advances in exploring how neural circuitry carries out the interpretation of sensory information, analytical thought, feedback mechanisms for motor control, establishment and retrieval of memories, inheritance of instincts, regulated hormonal control, and emotional responses. Some experiments are done with noninvasive imaging technologies, observing thousands to millions of neurons and detecting global electrical activity. Others are done observing a few cells at a time using inserted electrodes. It is difficult to bring the global views together with the single-cell views, but there have been notable successes, as we have seen here for the center-surround processing of visual information. From the standpoint of molecular cell biology, some of the greatest excitement has been in the area of research into mechanisms of memory. In most cases memory does not depend on forming new neurons. Instead, existing neurons are modified. Changes in the number and strength of synapses often underlie the establishment and endurance of memories. Current studies are directed at the molecular changes that alter synapses, both in the pre- and postsynaptic cells. Thus a fascinating problem that would seem to involve intractable numbers of neurons has become approachable through the techniques of molecular cell biology. A full understanding of memory will require finding ways to connect the molecular cell biology with more precise information about global circuits. This is likely to be accomplished by improvements in imaging methods, invasive and noninvasive, combined with the development of better ways to manipulate the activities of single neurons, or of large numbers of neurons simultaneously. There is every reason to expect these advances to happen, an exciting prospect for understanding the brain and doing a better job of treating diseases that affect the nervous system.

Key Terms

action potential *1004*
astrocytes *1014*
axon *1003*
axon guidance *1003*
center-surround *1029*
chemoaffinity hypothesis *1042*
cochlea *1032*
cones *1028*
dendrites *1003*
depolarization *1004*
endocytosis *1019*
excitatory receptor *1025*
glomeruli *1037*
growth cone *1040*
homunculus *1032*
inhibitory receptor *1025*
interneuron *1005*
ligand-gated channel *1023*
motor neuron *1005*
myelin sheath *1003*
neuroblast *1003*
neuromuscular junction *1019*
neuron *1002*
neurotransmitters *1005*
nociceptors *1031*
odorants *1037*
olfactory receptors *1036*
oligodendrocytes *1014*
ossicles *1032*
refractory period *1007*
repolarization *1004*
retinotectal map *1042*
rods *1028*
saltatory conduction *1014*
Schwann cells *1014*
sensory neuron *1005*
synapse *1005*
synaptic vesicles *1018*
voltage-gated channel *1006*

Review the Concepts

1. What is the role of glial cells in the brain and other parts of the nervous system?

2. The resting potential of a neuron is −60 mV inside compared with outside the cell. How is the resting potential maintained in animal cells?

3. Name the three phases of an action potential. Describe for each the underlying molecular basis and the ion involved. Why is the term *voltage-gated channel* applied to Na^+ channels involved in the generation of an action potential?

4. Explain how the crystal structures of potassium ion channels suggest the way in which the voltage-sensing domains interact with other parts of the proteins to open and close the ion channels. How does this structure-function relationship apply to other voltage-gated ion channels?

5. Explain why the strength of an action potential doesn't decrease as it travels down an axon.

6. What prevents a nerve signal from traveling "backwards" toward the cell body?

7. Myelination increases the velocity of action potential propagation along an axon. What is myelination? Myelination causes clustering of voltage-gated Na^+ channels and Na^+/K^+ pumps at nodes of Ranvier along the axon. Predict the consequences to action potential propagation of increasing the spacing between nodes of Ranvier by a factor of 10.

8. Acetylcholine is a common neurotransmitter released at the synapse. Predict the consequences for muscle activation of decreased acetylcholine esterase activity at nerve-muscle synapses.

9. Following the arrival of an action potential in stimulated cells, synaptic vesicles rapidly fuse with the presynaptic membrane. This happens in less than 1 ms. What mechanisms allow this process to take place at such great speed?

10. Neurons, particularly those in the brain, receive multiple excitatory and inhibitory signals. What is the name of the extension of the neuron at which such signals are received? How does the neuron integrate these signals to determine whether or not to generate an action potential?

11. What is the role of dynamin in recycling synaptic vesicles? What evidence supports this?

12. When a rod or cone cell is stimulated by light, a cascade of events is set in motion, which eventually leads to the transmission of visual information to the brain. Outline the molecular steps in this cascade from light absorption by rhodopsin to hyperpolarization of a rod or cone cell membrane.

13. Which cells are involved in visual contrast detection? How do these cells integrate photoreceptor cell information to detect contrast?

14. Compare the structures and functions of the receptor molecules for salt and sour taste; the taste receptor molecules

for sweetness, bitterness, and umami; and odor receptor molecules.

15. How did Roger Sperry's experiments in frogs distinguish between the resonance hypothesis and the chemoaffinity hypothesis for how neurons become wired?

Analyze the Data

Olfaction occurs when volatile compounds bind to specific odorant receptors. In mammals, each olfactory receptor neuron in the olfactory nasal epithelium expresses a single type of odorant receptor. These odorant receptors constitute a large multigene family (>1000 members) of related proteins. Binding of odorant induces a signaling cascade that is mediated via a G protein, $G_{\alpha olf}$. Recent studies suggest that there are a small number of olfactory sensory neurons in the nasal epithelium that express members of the trace-amine associated receptor (TAAR) family, chemoreceptors that are G protein–coupled receptors (GPCRs) but are unrelated to classical odorant receptors (see Liberles and Buck, 2006, *Nature* **442**:645–650). The mouse genome encodes 15 TAAR genes while the human genome encodes 6.

a. In order to examine the expression pattern of different TAARs in the olfactory nasal epithelium, researchers localized TAAR RNA by in situ hybridization in pairwise combinations. All possible pairwise combinations of the 15 mouse TAARs were examined. A typical example of the results obtained is shown in the top set of panels in the figure below in which TAAR6 and TAAR7 have been localized with fluorescent probes in the nasal epithelium of mouse. The Taar6 probe was labeled with a green fluor, the TAAR7 probe with a red fluor. The lower set of panels shows localization of mouse odorant receptor 28 (MOR28; green), a classical odorant receptor, and TAAR6 (red). Each stained patch in the images is the staining pattern of an individual olfactory neuron. The "merge" panels show the two other images superimposed. What do these data suggest about expression patterns of the TAARs?

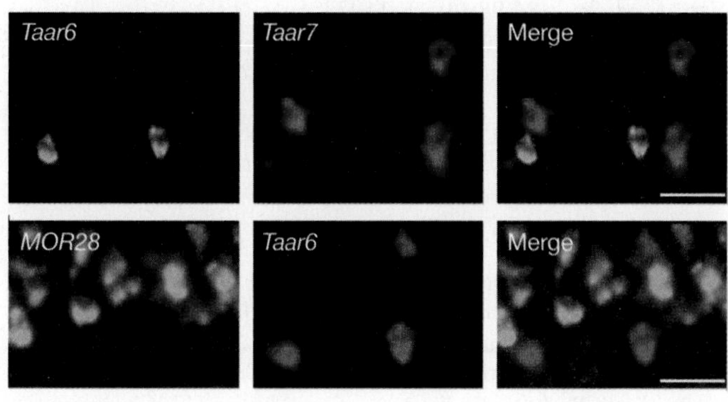

b. A number of cell lines that produce neither classical odorant receptors nor TAARs have each been transfected with the gene encoding a different TAAR. The cells have also been cotransfected with a gene encoding secreted alkaline phosphatase (SEAP) under control of a cAMP-responsive element. The cells are then exposed to various amines, as shown in the following figure, and SEAP activity in the medium is determined. The figure shows data for some representative TAARs (m = mouse, h = human). What do these data reveal about TAARs? What does the SEAP activity assay reveal about the signaling pathway utilized by chemoreception involving TAARs?

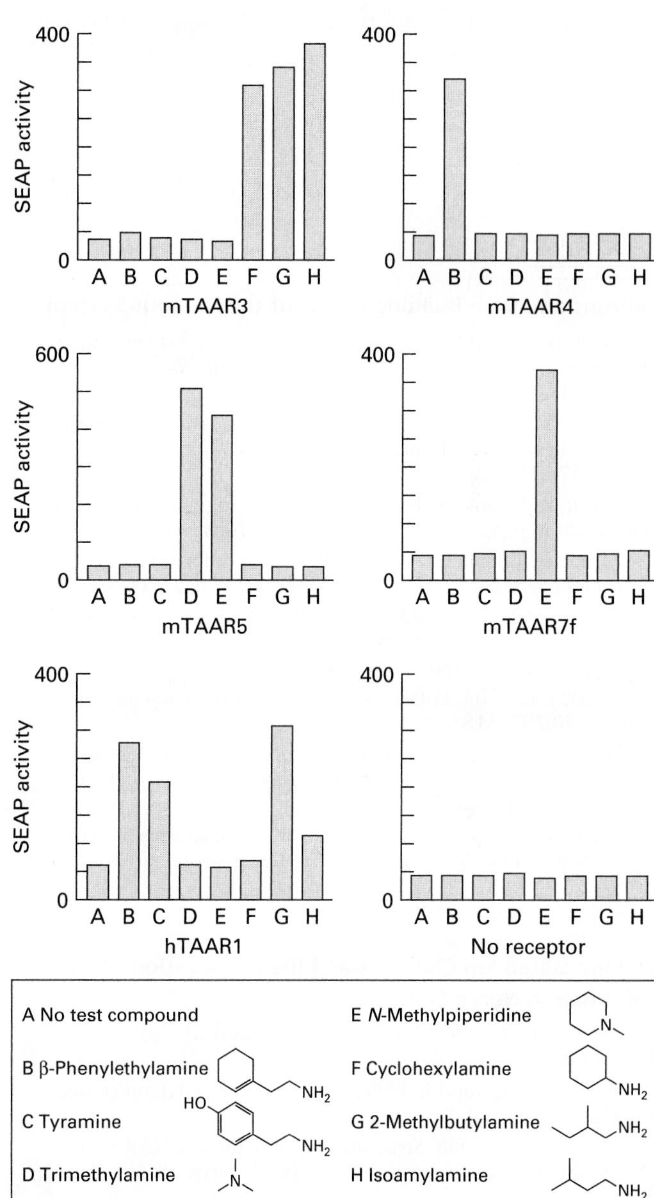

c. In a third set of studies, SEAP activity was measured in cells expressing mouse TAAR5 (mTAAR5) following exposure of the cells to diluted urine derived from two strains of mice or from humans, as indicated on the graphs below. Mice reach puberty at about one month of age. What do these data suggest may be a biological function for the TAAR5 neurons in mice? What additional studies would you conduct to support your hypothesis?

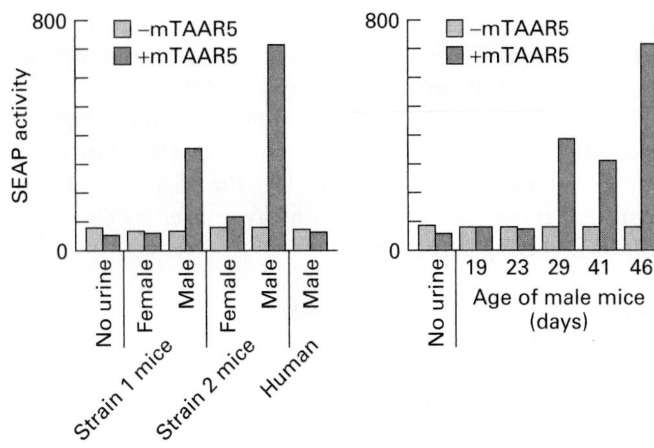

References

Neurons and Glia: Building Blocks of the Nervous System

Allen, N. J., and B. A. Barres. 2005. Signaling between glia and neurons: focus on synaptic plasticity. *Curr. Opin. Neurobiol.* **15**:542–548.

Eshed, Y., et al. 2005. Gliomedin mediates Schwann cell-asn interaction and the molecular assembly of the nodes of Ranvier. *Neuron* **47**:215–229.

Jessen, K. R., and R. Mirsky. 2005. The origin and development of glial cells in peripheral nerves. *Nature Rev. Neurosci.* **6**(9):671–682.

Parker, R. J., and V. J. Auld. 2006. Roles of glia in the *Drosophila* nervous system. *Semin. Cell Dev. Biol.* **17**(1):66–77.

Ramón y Cajal, S. 1911. *Histology of the Nervous System of Man and Vertebrates* (trans. N. Swanson and L. W. Swanson, 1995, Oxford University Press).

Salzer, J. L. 2003. Polarized domains of myelinated axons. *Neuron.* **40**:297–318.

Seifert, G., K. Schilling, and C. Steinhauser. 2006. Astrocyte dysfunction in neurological disorders: a molecular perspective. *Nature Rev. Neurosci.* **7**(3):194–206.

Sherman, D. L., and P. J. Brophy. 2005. Mechanisms of axon ensheathment and myelin growth. *Nature Rev. Neurosci.* **6**(9):683–690.

Stevens, B. 2003. Glia: much more than the neuron's side-kick. *Curr. Biol.* **13**:R469–R472.

Voltage-Gated Ion Channels and the Propagation of Action Potentials in Nerve Cells

Aldrich, R. W. 2001. Fifty years of inactivation. *Nature* **411**:643–644.

Armstrong, C., and B. Hille. 1998. Voltage-gated ion channels and electrical excitability. *Neuron* **20**:371–380.

Brunger, A. T. 2005. Structure and function of SNARE and SNARE-interacting proteins. *Quart. Rev. Biophys.* **38**(1):1–47.

Cannon, S. C. 2006. Pathomechanisms in channelopathies of skeletal muscle and brain. *Ann. Rev. Neurosci.* **29**:387–415.

Catterall, W. A. 2000. From ionic currents to molecular mechanisms: the structure and function of voltage-gated sodium channels. *Neuron* **26**:13–25.

Catterall, W. A. 2000. Structure and regulation of voltage-gated Ca^{2+} channels. *Ann. Rev. Cell Dev. Biol.* **16**:521–555.

Catterall, W. A. 2001. A 3D view of sodium channels. *Nature* **409**:988–989.

Clapham, D. 1999. Unlocking family secrets: K^+ channel transmembrane domains. *Cell* **97**:547–550.

Cooper, E. C., and L. Y. Jan. 1999. Ion channel genes and human neurological disease: recent progress, prospects, and challenges. *Proc. Nat'l Acad. Sci. USA* **96**:4759–4766.

del Camino, D., and G. Yellen. 2001. Tight steric closure at the intracellular activation gate of a voltage-gated K(+) channel. *Neuron* **32**:649–656.

Doyle, D. A., et al. 1998. The structure of the potassium channel: molecular basis of K^+ conduction and selectivity. *Science* **280**:69–77.

Dutzler, R., et al. 2002. X-ray structure of a ClC chloride channel at 3.0 Å reveals the molecular basis of anion selectivity. *Nature* **415**:287–294.

Gulbis, J. M., et al. 2000. Structure of the cytoplasmic beta subunit-T1 assembly of voltage-dependent K^+ channels. *Science* **289**:123–127.

Hanaoka, K., et al. 2000. Co-assembly of polycystin-1 and -2 produces unique cation-permeable currents. *Nature* **408**:990–994.

Hille, B. 2001. *Ion Channels of Excitable Membranes*, 3d ed. Sinauer Associates.

Jan, Y. N., and L. Y. Jan. 2001. Dendrites. *Genes Dev.* **15**:2627–2641.

Long, S. B., E. B. Campbell, and R. MacKinnon. 2005. Crystal structure of a mammalian voltage-dependent *Shaker* family K^+ channel. *Science.* **309**(5736):897–903.

Long, S. B., E. B. Campbell, and R. MacKinnon. 2005. Voltage sensor of Kv1.2: structural basis of electromechanical coupling. *Science.* **309**(5736): 903–908.

Miller, C. 2000. Ion channel surprises: prokaryotes do it again! *Neuron* **25**:7–9.

Montell, C., L. Birnbaumer, and V. Flockerzi. 2002. The TRP channels, a remarkably functional family. *Cell* **108**:595–598.

Neher, E. 1992. Ion channels for communication between and within cells. Nobel Lecture reprinted in *Neuron* **8**:605–612 and *Science* **256**:498–502.

Neher, E., and B. Sakmann. 1992. The patch clamp technique. *Sci. Am.* **266**(3):28–35.

Nichols, C., and A. Lopatin. 1997. Inward rectifier potassium channels. *Ann. Rev. Physiol.* **59**:171–192.

Sato, C., et al. 2001. The voltage-sensitive sodium channel is a bell-shaped molecule with several cavities. *Nature* **409**:1047–1051.

Shi, N., S. Ye, A. Alam, L. Chen, and Y. Jiang. 2006. Atomic structure of a Na^+- and K^+-conducting channel. *Nature* **440**(7083):570–574.

Yi, B. A., and L. Y. Jan. 2000. Taking apart the gating of voltage-gated K^+ channels. *Neuron* **27**:423–425.

Yi, B. A., et al. 2001. Controlling potassium channel activities: interplay between the membrane and intracellular factors. *Proc. Nat'l Acad. Sci. USA* **98**:11016–11023.

Zhou, M., et al. 2001. Potassium channel receptor site for the inactivation gate and quaternary amine inhibitors. *Nature* **411**:657–661.

Zhou, Y., et al. 2001. Chemistry of ion coordination and hydration revealed by a K^+ channel–Fab complex at 2 Å resolution. *Nature* **414**:43–48.

Communication at Synapses

Amara, S. G., and M. J. Kuhar. 1993. Neurotransmitter transporters: recent progress. *Ann. Rev. Neurosci.* **16**:73–93.

Bajjalieh, S. M., and R. H. Scheller. 1995. The biochemistry of neurotransmitter secretion. *J. Biol. Chem.* **270**:1971–1974.

Bamji, S. X. 2005. Cadherins: actin with the cytoskeleton to form synapses. *Neuron* **47**:175–178.

Betz, W., and J. Angleson. 1998. The synaptic vesicle cycle. *Ann. Rev. Physiol.* **60**:347–364.

Brejc, K., et al. 2001. Crystal structure of an ACh-binding protein reveals the ligand-binding domain of nicotinic receptors. *Nature* **411**:269–276.

Fatt, P., and B. Katz 1952. Spontaneous subthreshold activity at motor nerve endings. *J Physiol.* **117**:109–128.

Fernandez, J. M. 1997. Cellular and molecular mechanics by atomic force microscopy: capturing the exocytotic fusion pore in vivo? *Proc. Nat'l Acad. Sci. USA* **94**:9–10.

Ikeda, S. R. 2001. Signal transduction. Calcium channels—link locally, act globally. *Science* **294**:318–319.

Jan, L. Y., and C. F. Stevens. 2000. Signaling mechanisms: a decade of signaling. *Curr. Opin. Neurobiol.* **10**:625–630.

Karlin, A. 2002. Emerging structure of the nicotinic acetylcholine receptors. *Nature Rev. Neurosci.* **3**:102–114.

Kavanaugh, M. P. 1998. Neurotransmitter transport: models in flux. *Proc. Nat'l Acad. Sci. USA* **95**:12737–12738.

Klann, E., and T. E. Dever. 2004. Biochemical mechanisms for translational regulation in synaptic plasticity. *Nature Rev. Neurosci.* 5(12):931–942.

Kummer, T. T., T. Misgeld, and J. R. Sanes. 2006. Assembly of the postsynaptic membrane at the neuromuscular junction: paradigm lost. *Curr. Opin. Neurobiol.* **16**(1):74–82.

Lin, R. C., and R. H. Scheller. 2000. Mechanisms of synaptic vesicle exocytosis. *Ann. Rev. Cell Dev. Biol.* **16**:19–49.

Neher, E. 1998. Vesicle pools and Ca^{2+} microdomains: new tools for understanding their roles in neurotransmitter release. *Neuron* **20**:389–399.

Reith, M., ed. 1997. *Neurotransmitter Transporters: Structure, Function, and Regulation.* Humana Press.

Sakmann, B. 1992. Elementary steps in synaptic transmission revealed by currents through single ion channels. Nobel Lecture reprinted in *EMBO J.* **11**:2002–2016 and *Science* **256**:503–512.

Sosinsky, G. E., and B. J. Nicholson. 2005. Structural organization of gap junction channels. *Biochim. Biophys. Acta* **1711**(2):99–125.

Sudhof, T. C. 1995. The synaptic vesicle cycle: a cascade of protein-protein interactions. *Nature* **375**:645–653.

Ule, J., and R. B. Darnell. 2006. RNA binding proteins and the regulation of neuronal synaptic plasticity. *Curr. Opin. Neurobiol.* **16**(1):102–10.

Usdin, T. B., et al. 1995. Molecular biology of the vesicular ACh transporter. *Trends Neurosci.* **18**:218–224.

White-Grindley, E., and K. Si. 2006. RISC-y memories. *Cell* 124(1):23–60.

Ziv, N. E., and C. C. Garner. 2004. Cellular and molecular mechanisms of presynaptic assembly. *Nature Rev. Neurosci.* **5**(5):385–399.

Sensational Cells: Seeing, Feeling, Hearing, Tasting, and Smelling

Buck, L., and R. Axel. 1991. A novel multigene family may encode odorant receptors: a molecular basis for odor recognition. *Cell* **65**(1):175–187.

Cohen-Cory, S., and B. Lom. 2004. Neurotrophic regulation of retinal ganglion cell synaptic connectivity: from axons and dendrites to synapses. *Int'l J. Dev. Biol.* **48**(8–9):947–956.

Eatock, R. A., and K. M. Hurley. 2003. Functional development of hair cells. *Curr. Top. Dev. Biol.* **57**:389–448.

Esteve, P., and P. Bovolenta. 2006. Secreted inducers in vertebrate eye development: more functions for old morphogens. *Curr. Opin. Neurobiol.* **16**(1):13–19.

Gillespie, P. G., and R. G. Walker. 2001. Molecular basis of mechanosensory transduction. *Nature* **413**(6852):194–202.

Hoon, M. A., et al. 1999. Putative mammalian taste receptors: a class of taste-specific GPCRs with distinct topographic selectivity. *Cell* **96**(4):541–551.

Hudspeth, A. J. 2005. How the ear's works work: mechanoelectrical transduction and amplification by hair cells. *Comptes Rendus Biol.* **328**(2):155–162.

Lin, S. Y., and D. P. Corey. 2005. TRP channels in mechanosensation. *Curr. Opin. Neurobiol.* **15**(3):350–357.

Marin, E. C., et al. 2002. Representation of the glomerular olfactory map in the *Drosophila* brain. *Cell* **109**(2):243–255.

McKemy, D. D, W. M. Neuhausser, and D. Julius. 2002. Identification of a cold receptor reveals a general role for TRP channels in thermosensation. *Nature* **416**:52–58.

Mombaerts, P. 1999. Molecular biology of odorant receptors in vertebrates. *Ann. Rev. Neurosci.* **22**:487–509.

Nelson, G., et al. 2001. Mammalian sweet taste receptors. *Cell* **106**(3):381–390.

Zhang, X, and S. Firestein. 2002. The olfactory receptor gene superfamily of the mouse. *Nature Neurosci.* **5**(2):124–133.

The Path to Success: Controlling Axon Growth and Targeting

Chilton, J. K. 2006. Molecular mechanisms of axon guidance. *Dev. Biol.* **292**(1):13–24.

Dickson, B.J. and G.F. Gilestro. 2006. Regulation of commissural axon pathfinding by slit and its robo receptors. *Ann. Rev. Cell Dev. Biol.* **22**:651-75.

Gallo, G., and P. C. Letourneau. 2004. Regulation of growth cone actin filaments by guidance cues. *J. Neurobiol.* **58**(1):92–102.

Gomez, T.,M. and J. Q. Zheng. 2006. The molecular basis for calcium-dependent axon pathfinding. *Nature Rev. Neurosci.* 7(2):115–125.

Hedgecock, E. M., J. G. Culotti, and D. H. Hall. 1990. The unc-5, unc-6, and unc-40 genes guide circumferential migrations of pioneer axons and mesodermal cells on the epidermis in *C. elegans. Neuron* **4**(1):61–85.

Hilliard, M. A., and C.I. Bargmann. 2006. Wnt signals and frizzled activity orient anterior-posterior axon outgrowth in *C. elegans. Dev. Cell* **10**(3):379–390.

Hindges, R, et al. 2002. EphB forward signaling controls directional branch extension and arborization required for dorsal-ventral retinotopic mapping. *Neuron* **35**(3):475–487.

Jin, M., et al. 2005. Ca^{2+}-dependent regulation of rho GTPases triggers turning of nerve growth cones. *J. Neurosci.* **25**(9):2338–2347.

Kidd, T., K. S. Bland, and C. S. Goodman. 1999. Slit is the midline repellent for the robo receptor in *Drosophila. Cell* **96**(6):785–794.

Kruger, R. P., J. Aurandt, and K.-L. Guan. 2005. Semaphorins command cells to move. *Nature Rev. Mol. Cell Biol.* **6**(10):789–800.

Lemke, G., and M. Reber. 2005. Retinotectal mapping: new insights from molecular genetics. *Ann. Rev. Cell Dev. Biol.* **21**:551–580.

Li, W., et al. 2004. Activation of FAK and Src are receptor-proximal events required for netrin signaling. *Nature Neurosci.* 7(11):1213–1221.

Marquardt, T., et al. 2005. Coexpressed EphA receptors and ephrin-A ligands mediate opposing actions on growth cone navigation from distinct membrane domains. *Cell* **121**(1):127–139.

McLaughlin, T., and D. D. O'Leary. 2005. Molecular gradients and development of retinotopic maps. *Ann. Rev. Neurosci.* **28**:327–355.

Pan, C. L., et al. 2006. Multiple Wnts and frizzled receptors regulate anteriorly directed cell and growth cone migrations in *Caenorhabditis elegans. Dev. Cell.* **10**(3):367–377.

Schmitt, A.M., et al. 2006. Wnt-Ryk signalling mediates medial-lateral retinotectal topographic mapping. *Nature* **439**(7072):31–37.

Serafini T., et al. 1996. Netrin-1 is required for commissural axon guidance in the developing vertebrate nervous system. *Cell* **87**(6):1001–1014.

Sperry, R. W. 1963. Chemoaffinity in the orderly growth of nerve fiber patterns and connections. *Proc. Nat'l Acad. Sci. USA* **50**:703–710.

Tessier-Lavigne, M., et al. 1988. Chemotropic guidance of developing axons in the mammalian central nervous system. *Nature.* **336**(6201):775–778.

Walter, J., et al. 1987. Recognition of position-specific properties of tectal cell membranes by retinal axons in vitro. *Devel.* **101**(4):685–696.

Wen, Z., and J. Q. Zheng. 2006. Directional guidance of nerve growth cones. *Curr. Opin. Neurobiol.* **16**(1):52–58.

Xie, Y., et al. 2005. Phosphatidylinositol transfer protein-alpha in netrin-1-induced PLC signalling and neurite outgrowth. *Nature Cell Biol.* 7(11):1124–1132.

Zou, Y. 2006. Navigating the anterior-posterior axis with Wnts. *Neuron* **49**:787–789.

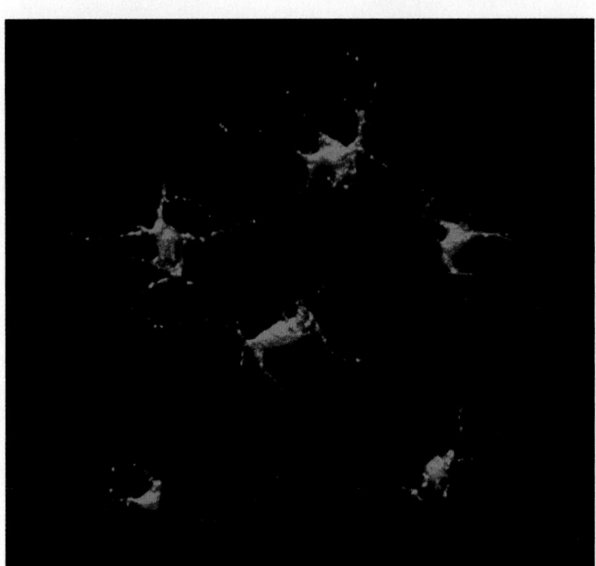

Dendritic cells in the skin have class II MHC molecules on their surface. Those shown here were engineered to express a class II MHC–GFP fusion protein, which fluoresces green. [Courtesy M. Boes and H. L. Ploegh.]

CHAPTER

24

IMMUNOLOGY

Immunity is a state of protection against the harmful effects of exposure to pathogens. Host defense can take many different forms, and all successful pathogens have found ways to disarm the immune system or manipulate it to their own advantage. Host–pathogen interactions are therefore an evolutionary work in progress. This explains why we continue to be assaulted by pathogenic viruses, bacteria, and parasites. The prevalence of infectious diseases illustrates the imperfections of host defense. But killing its host is not necessarily advantageous to a pathogen because complete elimination of the host would immediately remove the reservoir in which the pathogen replicates or survives. An immune system that could produce perfect sterilizing immunity would yield a world without pathogens, an outcome clearly at variance with life as we know it. Rather, the co-evolution of pathogens and their hosts allows pathogens, which have relatively short generation times, to continue to evolve sophisticated countermeasures, against which the host must respond by adjusting, if not improving, its defenses. Sophisticated defense comes at a price: An immune system capable of dealing with a massively diverse collection of rapidly evolving pathogens may mount an attack on the host organism's own cells and tissues, a phenomenon called *autoimmunity*.

In this chapter we deal mostly with the vertebrate immune system, with particular emphasis on those molecules, cell types, and pathways that uniquely distinguish the immune system from other types of cell and tissues. Host defense comprises three layers: (1) mechanical/chemical defenses, (2) innate immunity, and (3) adaptive immunity (Figure 24-1). Mechanical and chemical defenses operate continuously. Innate immune responses, which involve cells and molecules present at all times, are rapidly activated (minutes to hours), but their ability to distinguish among many different pathogens is somewhat limited. In contrast, adaptive immune responses take several days to develop fully and are highly specific; that is, they can distinguish between closely related pathogens based on very small molecular differences in structure.

The manner in which **antigens**—any material that can evoke an immune response—are recognized and how these foreign materials are eliminated involve molecular and cell biological principles unique to the immune system. We begin this chapter with a brief sketch of the organization of the

OUTLINE

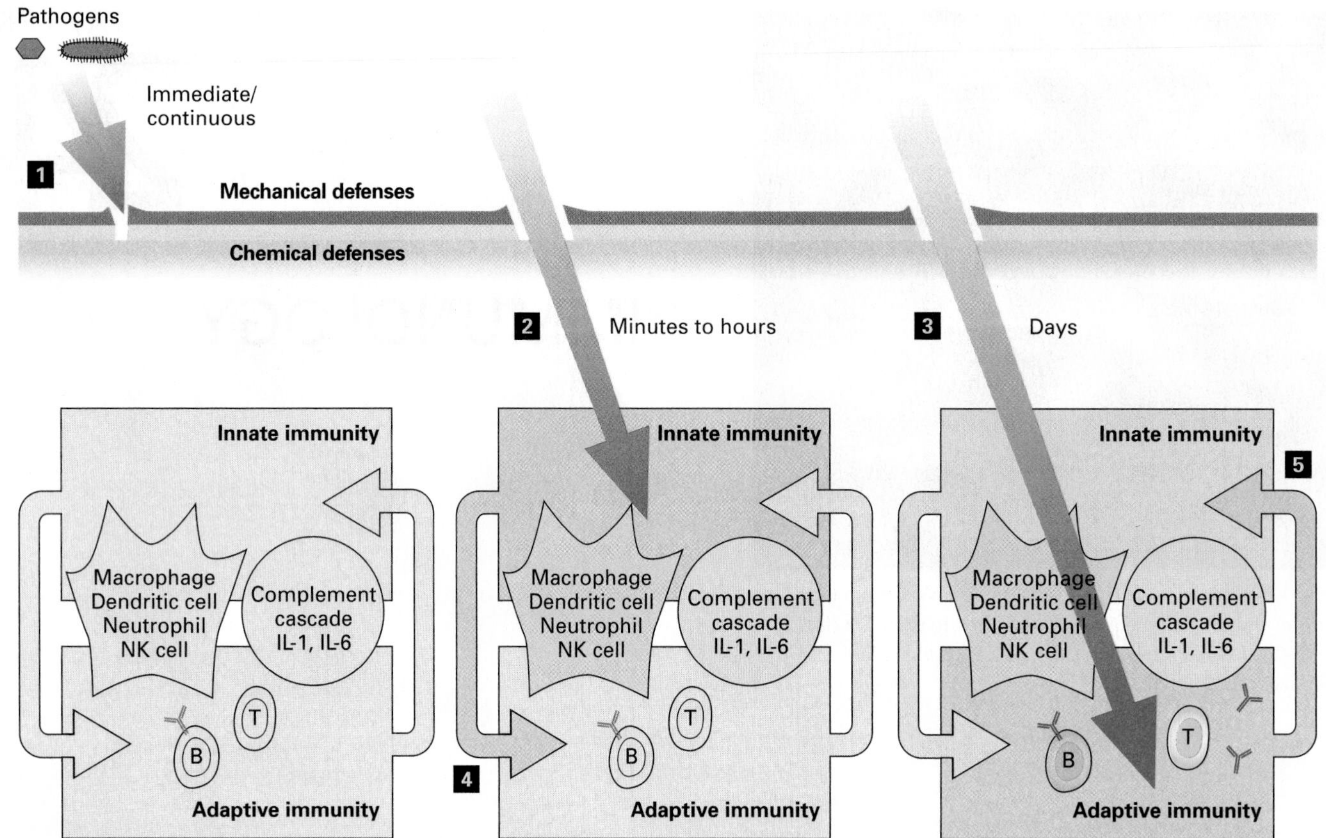

▲ **FIGURE 24-1 The three layers of vertebrate immune defenses.** *Left:* Mechanical defenses consist of epithelia and skin. Chemical defenses include the low pH of the gastric environment and antibacterial enzymes in tear fluid. These barriers provide continuous protection against invaders. Pathogens must physically breach these defenses (**1**) to infect the host. *Middle:* Pathogens that have breached the mechanical and chemical defenses (**2**) are handled by cells and molecules of the innate immune system (blue), which includes phagocytic cells (neutrophils, dendritic cells, macrophages), natural killer (NK) cells, complement proteins, and certain interleukins (IL-1, IL-6). Innate defenses are activated within minutes to hours of infection. *Right:* Pathogens that are not cleared by the innate immune system are dealt with by the adaptive immune system (**3**), in particular B and T lymphocytes. Full activation of adaptive immunity requires days. The products of an innate response may potentiate an ensuing adaptive response (**4**). Likewise, the products of an adaptive immune response, including antibodies (Y-shaped icons), may facilitate functioning of the innate immune system (**5**). Several cell types and secreted products straddle the fence between the innate and adaptive immune systems, and serve to connect these two layers of host defense.

mammalian immune system, introducing the essential players of innate and adaptive immunity and describing inflammation, a localized response to injury or infection that leads to the activation of immune-system cells and their recruitment to the affected site. In the next two sections, we discuss the structure and function of **antibody** (or immunoglobulin) molecules, which bind to specific molecular features on antigens, and how variability in antibody structure contributes to specific recognition of antigens. The enormous diversity of antigens that can be recognized by the immune system finds its explanation in unique rearrangements of the genetic material in B and T lymphocytes, commonly called **B cells** and **T cells,** which are the white blood cells that carry out antigen-specific recognition. These gene rearrangements not only control the specificity of antigen receptors on lymphocytes but also determine cell-fate decisions in the course of lymphocyte development.

Although the mechanisms that give rise to antigen-specific receptors on B and T cells are very similar, the manner in which these receptors recognize antigen is very different. The receptors on B cells can interact with intact antigens directly, but the receptors on T cells cannot. Instead, as described in Section 24.4, the receptors on T cells recognize cleaved (processed) forms of antigen, presented on the surface of target cells by glycoproteins encoded by the major histocompatibility complex (MHC). How MHC-encoded glycoproteins display these processed antigens is important for our understanding of how immune responses are initiated. MHC-encoded glycoproteins also help determine the developmental fate of T cells so that an organism's own cells and tissues (self antigens) normally do not evoke an immune response, whereas foreign antigens do. We conclude the chapter with an integrated view of the immune response to a pathogen, highlighting the collaboration between different immune-system cells that is required for an effective response.

24.1 Overview of Host Defenses

Because the immune system evolved to deal with pathogens, we begin our overview of host defenses by examining where typical pathogens are found and where they replicate. Then we introduce basic concepts of innate and adaptive immunity, including some of the key cellular and molecular players.

Pathogens Enter the Body Through Different Routes and Replicate at Different Sites

Pathogens affect all life forms capable of independent replication. Two general classes of pathogen, viruses and bacteria, have fundamentally different modes of propagation. With the exception of polymerases involved in copying genetic material, viruses generally lack the necessary machinery to synthesize their component parts and therefore are completely dependent on host cells for their propagation. In contrast, most bacteria are metabolically autonomous and do not rely on their host for replication, which allows them to be grown in the laboratory in the appropriate culture media. (An exception is those bacteria that can replicate only within a mammalian host cell.) Bacteria can cause disease because they possess virulence factors that act on the host's metabolism and physiology. Parasitic organisms also can cause disease. With increasingly complex life styles such as those of the protozoa that cause sleeping sickness (trypanosomes) or malaria (*Plasmodium* species) (see Figure 1-4), the pathogen's countermeasures also become increasingly complex. Bacteria, protozoa, and fungi—especially those that can cause disease in animals—often are called microbes.

Exposure to pathogens occurs via different routes. The skin itself has a surface area of ≈20 sq. ft.; the epithelial surfaces that line the airways, gastrointestinal tract, and genital tract present an even more formidable surface area of ≈4000 sq. ft. All these surfaces are continuously exposed to viruses and bacteria in the environment. Foodborne pathogens and sexually transmitted agents target the epithelia to which they are exposed. The sneeze of a flu-infected individual releases millions of virus particles in aerosolized form, ready for inhalation by the next person to be infected. Rupture of the skin, even if only by minor abrasions, or of the epithelial barriers that protect the underlying tissues, provides an easy route of entry for pathogens, which then gain access to a rich source of nutrients (for bacteria) and to the cells required for their replication (viruses).

Replication of viruses is strictly confined to the cytoplasm or nucleus of host cells, where protein synthesis and replication of the viral genetic material occur. Viruses spread to other cells either as free virus particles (virions) or by cell-to-cell spread. Many bacteria can replicate in the intercellular space, but some are specialized to invade host cells and survive there. Such intracellular bacteria reside either in membrane-delimited vesicles through which they enter cells by endocytosis or phagocytosis or in the cytoplasm if they escape from these vesicles. An effective host defense system, therefore, needs to be capable of eliminating not only cell-free viruses and free-living bacteria but also cells that harbor these pathogens.

Leukocytes Circulate Throughout the Body and Take Up Residence in Tissues and Lymph Nodes

With the exception of erythrocytes, few cells in the course of their assigned function cover such distances as do the cells that provide immunity. The mammalian circulation serves as the necessary transport vehicle for erythrocytes, leukocytes, and platelets. Although erythrocytes never leave the circulation (their oxygen-carrying function does not require it), leukocytes (white blood cells) use the circulation exclusively for transport and may leave and re-enter the circulation in the course of their tasks.

The immune system is an interconnected system of vessels, organs, and cells, divided into primary and secondary lymphoid structures (Figure 24-2). *Primary lymphoid organs* —the sites at which **lymphocytes** (the subset of leukocytes that includes B and T cells) are generated and acquire their

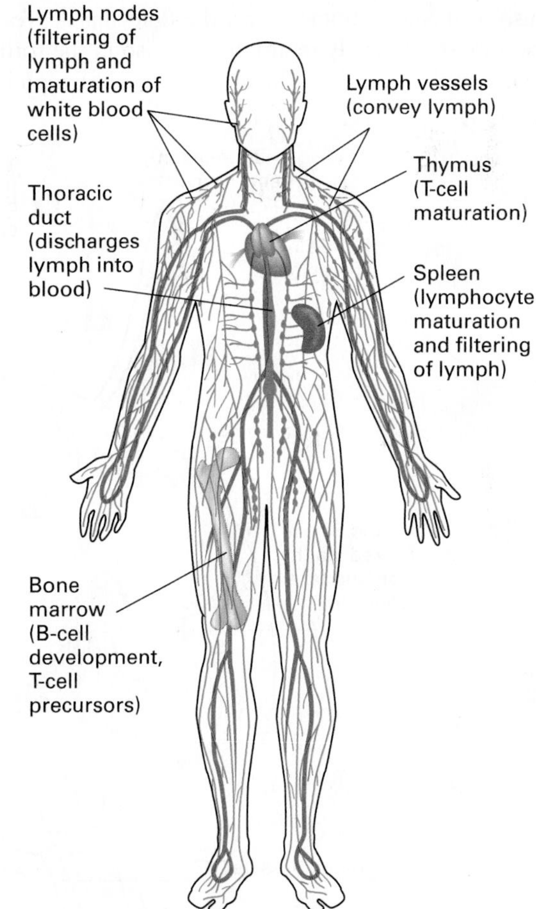

▲ **FIGURE 24-2 The circulatory and lymphatic systems.** Positive arterial pressure exerted by the pumping heart is responsible for loss of liquid from the circulation (red) into the interstitial spaces of the tissues, so that all cells of the body have access to nutrients and can dispose of waste. This interstitial fluid, whose volume is roughly three times that of all blood in the circulation, is returned to the circulation in the form of lymph, which passes through specialized anatomical structures called lymph nodes. The primary lymphoid organs, where lymphocytes are generated, are the bone marrow (B cells, T-cell precursors) and the thymus (T cells). The initiation of an immune response involves the secondary lymphoid organs (lymph nodes, spleen).

functional properties—include the thymus, where T cells are generated, and the bone marrow, where B cells are generated. Adaptive immune responses, which require functionally competent lymphocytes are initiated in *secondary lymphoid organs* including lymph nodes and the spleen. All of the lymphoid organs are populated by cells of hematopoietic origin (see Figure 21-15), generated in the fetal liver and throughout life in the bone marrow. The total number of lymphocytes in a young adult male is estimated to be 500×10^9, roughly 15 percent of which are found in the spleen, 40 percent in the other secondary lymphoid organs (tonsils, lymph nodes), 10 percent in the thymus, and 10 percent in the bone marrow; the remainder are circulating in the bloodstream.

Leukocytes must leave the bloodstream and enter tissues to perform their functions. Vertebrate blood vessels allow the escape of fluid from the circulation, driven by the positive arterial pressure exerted by the pumping heart. This fluid contains not only nutrients but also proteins that carry out defensive functions. To maintain homeostasis, the fluid that leaves the circulation must ultimately return and does so in the form of *lymph*, via lymphatic vessels. The total volume of lymph is up to three times the total blood volume. At their most distal ends, lymphatic vessels are open to collect the interstitial fluid that bathes the cells in tissues. The lymphatic vessels merge into larger collecting vessels, which deliver lymph to *lymph nodes*. A lymph node consists of a capsule, organized into areas that are defined by the cell types that inhabit them. Blood vessels entering a lymph node deliver B and T cells to it. The lymph that arrives in a lymph node carries cells that have encountered ("sampled") antigen, as well as soluble antigens, from the tissue drained by that particular afferent lymphatic vessel. In the lymph node, the cells and molecules required for the adaptive immune response interact, respond to the newly acquired antigenic information, and then execute the necessary effector functions to rid the body of the pathogen (Figure 24-3).

Lymph nodes can be thought of as filters in which antigenic information gathered from distal sites throughout the body is collected and displayed to the immune system in a form suitable to evoke an appropriate response. All the relevant steps that lead to lymphocyte activation take place in lymphoid organs. Cells that have received proper instructions to become functionally active leave the lymph node via efferent

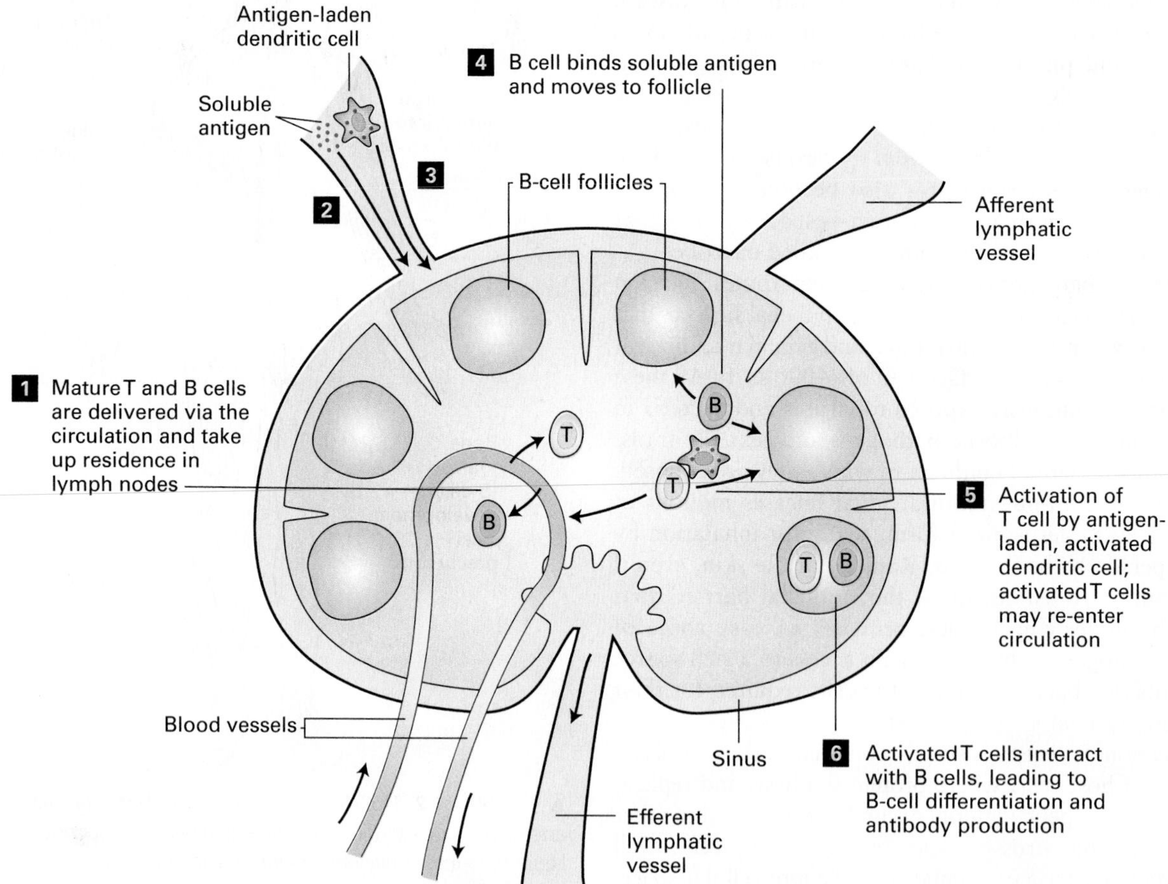

▲ **FIGURE 24-3 Initiation of the adaptive immune response in lymph nodes.** Recognition of antigen by B and T cells (lymphocytes) located in lymph nodes initiates an adaptive immune response. Lymphocytes leave the circulation and take up residence in lymph nodes (1). Lymph carries antigen in two forms—soluble antigen and antigen-laden dendritic cells; both are delivered to lymph nodes via afferent lymphatics (2, 3). Soluble antigen is recognized by B cells (4), and antigen-laden dendritic cells present antigen to T cells (5). Productive interactions between T and B cells (6) allow B cells to move into follicles and differentiate into plasma cells, which produce large amounts of secreted immunoglobulins (antibodies). Efferent lymphatic vessels return lymph from the lymph node to the circulation.

lymphatic vessels that ultimately drain into the circulation. Such activated cells recirculate through the bloodstream and—now ready for action—may reach a location where they again leave the circulation, move into tissues, and seek out pathogenic invaders or destroy virus-infected cells.

The exit of lymphocytes and other leukocytes from the circulation, recruitment of these cells to sites of infection, processing of antigenic information, and return of immune-system cells to the circulation are all carefully regulated processes that involve specific cell-adhesion events, chemotactic cues, and the traversal of endothelial barriers, as we discuss later.

Mechanical and Chemical Boundaries Form a First Layer of Defense Against Pathogens

As noted already, mechanical and chemical defenses form the first line of host defense against pathogens (see Figure 24-1). Mechanical defenses include the skin, epithelia, and arthropod exoskeleton, which are barriers that can be breached only by mechanical damage or through specific chemo-enzymatic attack. Chemical defenses include not only the low pH found in gastric secretions but also enzymes such as *lysozyme,* found in tear fluid, which can attack microbes directly.

The importance of mechanical defenses, which operate continuously, are immediately obvious in the case of burn victims: When the integrity of the epidermis and dermis is compromised, the rich source of nutrients in the underlying tissues is exposed, and airborne bacteria or otherwise harmless bacteria found on the skin can multiply unchecked, ultimately overwhelming the host. Viruses and bacteria have also evolved strategies to breach the integrity of these physical barriers. Enveloped viruses such as HIV, rabies virus, and influenza virus possess membrane proteins endowed with fusogenic properties. Following adhesion of a virion to the surface of the cell to be infected, direct fusion of the viral envelope with the host cell's membrane results in delivery of the viral genetic material into the host cytoplasm, where it is now available for transcription, translation, and replication (see Figures 4-47 and 4-49). Certain pathogenic bacteria (e.g., *S. aureus*) secrete collagenases that compromise the integrity of connective tissue and so facilitate entry of the bacteria.

Innate Immunity Provides a Second Line of Defense After Mechanical and Chemical Barriers Are Crossed

The innate immune system is activated once the mechanical and chemical defenses have failed, and the presence of an invader is sensed (see Figure 24-1). The innate immune system comprises cells and molecules that are immediately available for responding to pathogens. **Phagocytes,** cells that ingest and destroy pathogens, are widespread throughout tissues and epithelia and can be recruited to sites of infection. Several soluble proteins present constitutively in the blood, or produced in response to infection or inflammation, also contribute to innate defense. Animals that lack an adaptive immune system, such as insects, rely exclusively on innate defenses to combat infections.

Phagocytes and Antigen-Presenting Cells The innate immune system includes macrophages, neutrophils, and dendritic cells. All of these cells are phagocytic and come equipped with **Toll-like receptors (TLRs).** Members of this family of cell-surface proteins detect broad patterns of pathogen-specific markers and thus are key sensors for detecting the presence of viral or bacterial invaders. Engagement of Toll-like receptors is important in eliciting effector molecules, including antimicrobial peptides. **Dendritic cells** and **macrophages** whose Toll-like receptors have detected pathogens also function as **antigen-presenting cells (APCs)** by displaying processed foreign materials to antigen-specific T cells. The structure and function of Toll-like receptors and their role in activating dendritic cells are described in detail in Section 24.6.

Complement System Another important component of the innate immune system is **complement,** a collection of constitutive serum proteins that can bind directly to microbial or fungal surfaces. This binding activates a proteolytic cascade that culminates in formation of pore-forming proteins constituting the *membrane attack complex,* which is capable of permeabilizing the pathogen's protective membrane (Figure 24-4). The complement cascade is conceptually similar to the blood-clotting cascade, with amplification of the reaction at each successive stage of activation. At least three distinct pathways can activate complement. The *classical pathway* requires the presence of antibodies produced in the course of an adaptive response and bound to the surface of the microbe. Many microbial surfaces directly activate complement via the *alternative pathway*. Finally, pathogens that contain mannose-rich cell walls activate complement through the *mannose-binding lectin pathway.* The bound lectin then triggers activation of two mannose-binding lectin-associated proteases, MASP-1 and MASP-2, which allow activation of the downstream components of the complement cascade.

In the course of complement activation, the C3 and C4 complement proteins occupy a special role. These abundant serum proteins are synthesized as precursors that contain an internal, strained thioester linkage between a cysteine and a glutamate residue in close proximity. This thioester linkage becomes highly reactive upon proteolytic activation of C3 and C4 by their respective upstream partners. The activated thioester bond can react with primary amines or hydroxyls in close proximity, yielding a covalent bond linking C3 or C4 with a protein or carbohydrate close-by. If no such reactants are available, the thioester bond is simply hydrolyzed. This mode of action ensures that C3 and C4 fragments will be covalently deposited only on antigen–antibody complexes in close proximity.

Regardless of the pathway of complement activation engaged, activated C3 unleashes the terminal components of the complement cascade, C5 through C9, culminating in formation of the membrane attack complex, which inserts itself into most biological membranes and renders them permeable. The resulting loss of electrolytes and small solutes leads to lysis and death of the target cell. Whenever complement is activated, the membrane attack complex is formed and results in death of the cell onto which the complex is deposited. The direct microbicidal effect of a fully activated complement cascade is an important protective function.

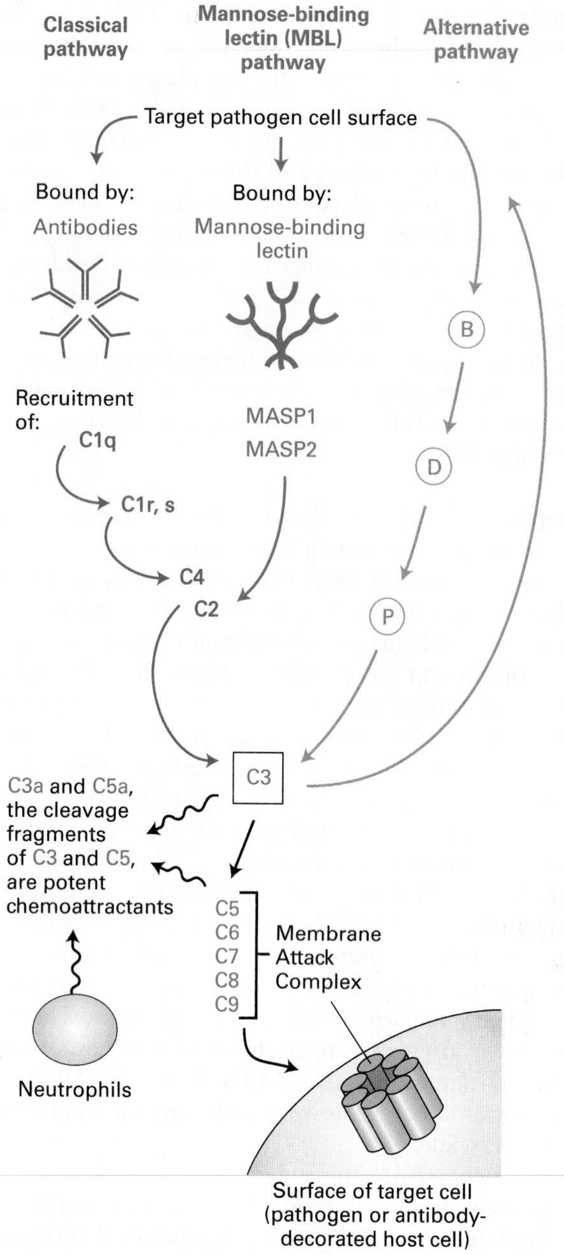

All three complement activation pathways also generate the C3a and 5a cleavage fragments, which bind to G protein–coupled receptors and function as chemoattractants for neutrophils and other cells involved in inflammation (see below). All three pathways also result in the covalent decoration of the structures targeted by complement activation with fragments of C3. Phagocytic cells make use of these C3-derived tags to recognize, ingest, and destroy the decorated particles, a process termed *opsonization.*

Natural Killer (NK) Cells In addition to bacterial invaders, the innate immune system also defends against viruses. When the presence of a virus-infected cell is detected, yet other cell types of the innate immune system become active, seek out the virus-infected targets, and kill them. For instance, many virus-infected cells produce type I **interferons,** which are good at activating **natural killer (NK) cells.** Activated NK cells not only afford direct protection by eliminating the factory of new virus particles, but they also secrete interferon γ (IFN-γ), which is essential for orchestrating many aspects of anti-viral defenses (Figure 24-5). Recognition by NK cells involves several classes of receptor, capable of delivering either stimulatory (promoting cell killing) or inhibitory signals. The interferons are classified as **cytokines,** small, secreted proteins that help regulate immune responses in a variety of ways. We will encounter

Classical pathway
Mannose-binding lectin (MBL) pathway
Alternative pathway
Target pathogen cell surface
Bound by: Antibodies
Bound by: Mannose-binding lectin
Recruitment of: C1q
C1r, s
C4
C2
MASP1
MASP2
B
D
P
C3
C3a and C5a, the cleavage fragments of C3 and C5, are potent chemoattractants
Neutrophils
C5 C6 C7 C8 C9 Membrane Attack Complex
Surface of target cell (pathogen or antibody-decorated host cell)

▲ **FIGURE 24-4 Three pathways of complement activation.** The classical pathway involves the formation of antibody–antigen complexes, which recruit the complement component C1q, leading to activation of C1r and C1s. This complex, in turn activates C4 and C2, which then convert C3 to its active form. In the mannose-binding lectin pathway, mannose-rich structures found on the surface of many pathogens are recognized by mannose-binding lectin, an interaction that results in activation of two serine proteases, MASP-1 and MASP-2. The alternative pathway requires deposition of a special form of the serum protein C3, a major complement component, onto a microbial surface. Subsequent activation of C3 involves factors B, D and P, found in serum. Each of the activation pathways is organized as a cascade of proteases in which the downstream component is itself a protease. Amplification of activity occurs with each successive step. All three pathways converge on C3, which triggers formation of the membrane attack complex, leading to destruction of target cells. The small fragments of C3 and C5 generated in the course of complement activation attract neutrophils, phagocytic cells that can kill bacteria at short range or upon ingestion.

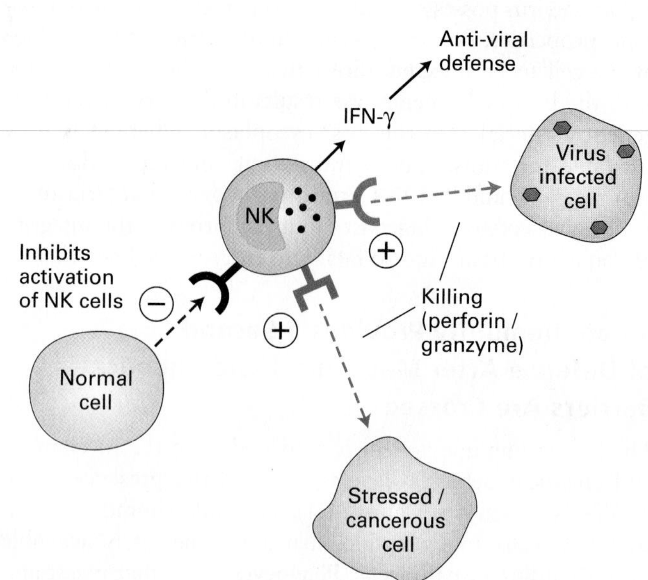

▲ **FIGURE 24-5 Natural killer cells.** Natural killer (NK) cells are an important source of the cytokine interferon γ (IFN-γ) and can kill virus-infected and cancerous cells by means of perforins. These pore-forming proteins allow access of serine proteases called granzymes to the cytoplasm of the cell about to be killed. Granzymes also can initiate apoptosis through activation of caspases (Chapter 21).

other cytokines and discuss some of their receptors as the chapter progresses.

Inflammation Is a Complex Response to Injury That Encompasses Both Innate and Adaptive Immunity

When a vascularized tissue is injured, the stereotypical response that follows is **inflammation.** Damage may be a simple paper cut or result from infection with a pathogen. Inflammation, or the inflammatory response, is characterized by four classical signs: *redness, swelling, heat,* and *pain.* These signs are caused by increased leakiness of blood vessels (vasodilation), the attraction of cells to the site of damage, and the production of soluble mediators responsible for the sensation of heat and pain. Inflammation has immediate protective value through the activation of the cell types and soluble products that together mount the innate immune response. Further, inflammation creates a local environment conducive to the initiation of the adaptive immune response. However, if not properly controlled, inflammation can also be a major cause of tissue damage.

Figure 24-6 depicts the key players in the inflammatory response to bacterial pathogens and the subsequent initiation of an adaptive immune response. Tissue-resident dendritic cells sense the presence of pathogens via their Toll-like receptors (TLRs) and respond to them by releasing soluble mediators such as cytokines and **chemokines;** the latter act as chemoattractants for immune-system cells. Neutrophils, a second important cell type in the inflammatory response, leave the circulation and migrate to wherever tissue injury or infection has occurred in response to various soluble mediators produced upon tissue damage. **Neutrophils,** which constitute almost half of all circulating leukocytes, are phagocytic, directly ingesting and destroying pathogenic bacteria. They also can interact with a wide variety of pathogen-derived macromolecules via their Toll-like receptors. Activation of these receptors allows neutrophils to produce cytokines and chemokines; the latter can attract more leukocytes—neutrophils, macrophages, and ultimately lymphocytes (T and B cells)—to the area. Activated neutrophils can release bacteria-destroying enzymes (e.g., lysozyme and proteases) as well as small peptides with microbicidal activity, collectively called *defensins.* Activated neutrophils also turn on the enzymes that generate superoxide anion and other reactive oxygen species (see Chapter 12, p. 502), which can kill microbes at short range. Another cell type contributing to the inflammatory response is tissue-resident *mast cells.* When activated by a variety of physical or chemical stimuli, mast cells release histamine, a mediator that increases vascular permeability and thereby facilitates access to the site of plasma proteins (e.g., complement) that can act against the invading pathogen.

A very important early response to infection or injury is activation of a variety of plasma proteases, including the proteins of the complement cascade discussed above (see Figure 24-4). The peptides produced during activation of these proteases possess chemoattractant activity, responsible for attracting neutrophils to the site of tissue damage. They further induce production of proinflammatory cytokines such as

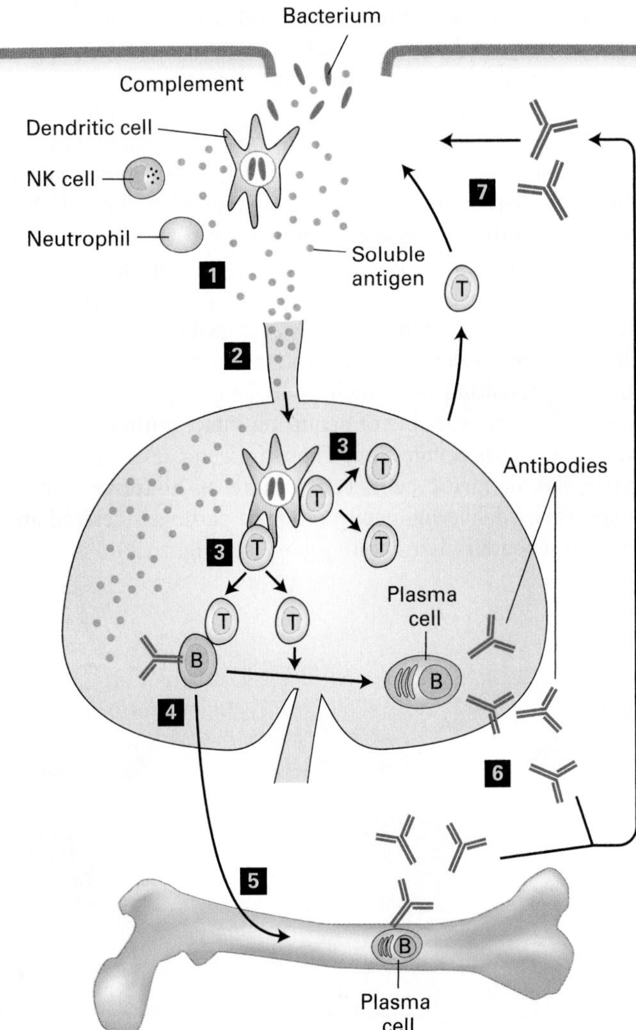

▲ **FIGURE 24-6 Interplay of innate and adaptive immune responses against a bacterial pathogen.** Once a bacterium breaches the host's mechanical and chemical defenses, the bacterium is exposed to components of the complement cascade, as well as to cells that confer immediate protection, such as neutrophils (**1**). Various inflammatory mediators induced by tissue damage contribute to a localized inflammatory response. Local destruction of the bacterium results in the release of bacterial antigens, which are delivered via the afferent lymphatics to the draining lymph node (**2**). Dendritic cells acquire antigen at the site of infection, become migratory in response to microbial products, and move to the lymph node, where they activate T cells (**3**). In the lymph node, antigen-stimulated T cells proliferate and acquire effector functions, including the ability to help B cells (**4**), some of which may move to the bone marrow and complete their differentiation into plasma cells there (**5**). In later stages of the immune response, activated T cells provide additional assistance to antigen-experienced B cells to yield plasma cells that secrete antigen-specific antibodies at a high rate (step **6**). Antibodies produced as a consequence of the initial exposure to bacteria act in synergy with complement to eliminate the infection (**7**), should it persist, or afford rapid protection in the case of re-exposure to the same pathogen.

interleukin 1 and 6 (IL-1 and IL-6). The recruitment of neutrophils also depends on an increase in vascular permeability, controlled in part by lipid mediators (e.g., prostaglandins and leukotrienes) that are derived from phospholipids and fatty acids. All of these events occur rapidly, starting within minutes of injury. A failure to resolve the cause of this immediate response may result in chronic inflammation, in which cells of the adaptive immune system play an important role.

When the pathogen burden at the site of tissue damage is high, it may exceed the capacity of innate defense mechanisms to deal with them. Moreover, some pathogens have acquired, in the course of evolution, tools to disable or bypass innate immune defenses. In such situations, the adaptive immune response is required to control the infection. This adaptive response depends on specialized cells that straddle the interface between adaptive and innate immunity, including antigen-presenting cells such as macrophages and dendritic cells, which are capable of acquiring intact pathogens and of killing them upon ingestion. These antigen-presenting cells, in particular dendritic cells, can initiate an adaptive immune response by delivering newly acquired pathogen-derived antigens to secondary lymphoid organs (see Figure 24-6).

Adaptive Immunity, the Third Line of Defense, Exhibits Specificity

Lymphocytes bearing antigen-specific receptors are the key cells responsible for adaptive immunity. An early indication of the specific nature of adaptive responses came with the discovery of antibodies, key effector molecules of adaptive immunity, by Emil von Behring and Shibasaburo Kitasato in 1905. They observed that when serum (the straw colored liquid that separates from cellular debris upon completion of the blood clotting process) from guinea pigs immunized with a sublethal dose of the deadly diphtheria toxin was transferred to animals never before exposed to the bacterium, the recipient animals were protected against a lethal dose of the same bacterium, which kills its host by production of a toxin (Figure 24-7). Transfer of serum from animals never exposed to diphtheria toxin failed to protect, and protection was limited to the microbe used as the source of toxin with which the animal that

Podcast: The Discovery of Antibodies 

MEDIA CONNECTIONS

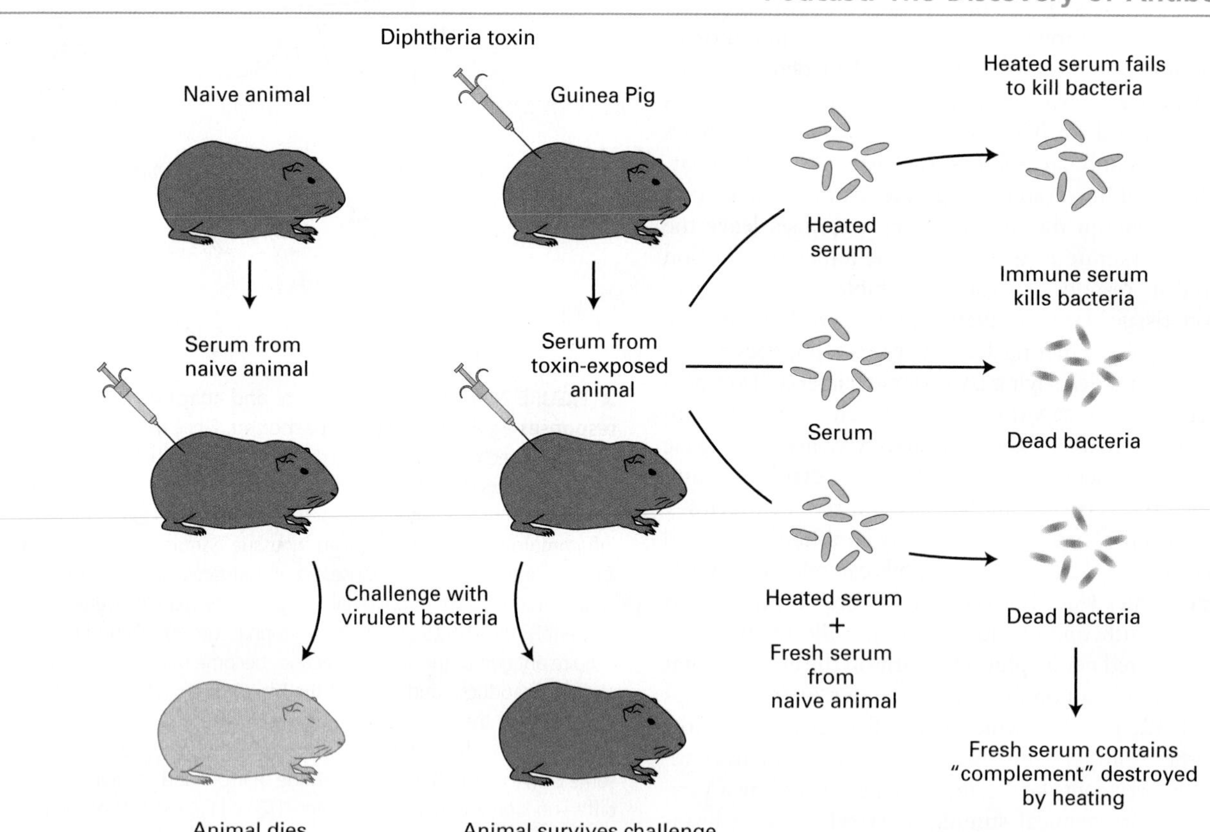

▲ EXPERIMENTAL FIGURE 24-7 **The existence of antibody in serum from infected animals was demonstrated by von Behring and Kitasato.** Exposure of animals to a sublethal dose of diphtheria toxin (or the bacteria that produce it) elicits in their serum a substance that protects against a subsequent challenge with a lethal dose of the toxin (or the bacteria that produce it). The protective effect of this serum substance can be transferred from an animal that has been exposed to the pathogen to a naive (nonexposed) animal. When the serum recipient is subsequently exposed to a lethal dose of the bacteria, the animal survives. This effect is specific for the pathogen used to elicit the response. Serum thus contains a transferable substance (antibody) that protects against the harmful effects of a virulent pathogen. Serum harvested from these animals, said to be immune, displays bactericidal activity in vitro. Heating of immune serum destroys its bactericidal activity. Addition of fresh nonheated serum from a naive animal restores the bactericidal activity of heated immune serum. Serum thus contains another substance that complements the activity of antibodies.

served as the serum donor was immunized. This experiment demonstrates *specificity*—that is, the ability to distinguish between two closely related substances of the same class. Such specificity is a hallmark of the adaptive immune system. Even proteins that differ by a single amino acid may be distinguished by immunological means.

From these experiments, von Behring inferred the existence of corpuscles ("Antikörper"), or antibodies, as the factor responsible for protection. The antibody-containing (immune) sera not only afforded protection in vivo, they also killed microbes in the test tube. Heating the immune sera to 56° C destroyed this killing activity, but it was restored by the addition of unheated fresh serum from naive animals (i.e., animals never exposed to the microbe). This finding suggested that a second factor, now called complement, acts in synergy with antibodies to kill bacteria. We now know that von Behring's antibodies are serum proteins referred to as **immunoglobulins** and that complement is actually a series of proteases (see Figure 24-4). Immunoglobulins can neutralize not only bacterial toxins, but also harmful agents such as viruses, by directly binding to them in a manner that prevents the virus from attaching itself to host cells. In the same vein, antibodies raised against snake venoms can be administered to the victims of snake bites to protect them from intoxication: The anti-snake venom antibodies bind to the venom, keep it from binding to its targets in the host, and in so doing neutralize it. Antibodies can thus have immediate protective effects.

KEY CONCEPTS OF SECTION 24.1

Overview of Host Defenses

- Mechanical and chemical defenses provide protection against most pathogens. This protection is immediate and continuous, yet possesses little specificity. Innate and adaptive immunity provide defenses against pathogens that breach the body's mechanical/chemical boundaries (see Figure 24-1).
- The circulatory and lymphatic systems distribute the molecular and cellular players in innate and adaptive immunity throughout the body (see Figure 24-2).
- Innate immunity is mediated by the complement system (see Figure 24-4) and several types of leukocytes, the most important of which are neutrophils and other phagocytic cells such as macrophages and dendritic cells. The cells and molecules of innate immunity are deployed rapidly (minutes to hours). Molecular patterns diagnostic of the presence of pathogens can be recognized by Toll-like receptors, but the specificity of recognition is modest.
- Adaptive immunity is mediated by T and B lymphocytes. These cells require days for full activation and deployment, but they can distinguish between closely related antigens. This specificity of antigen recognition is the key distinguishing feature of adaptive immunity.
- Innate and adaptive immunity act in a mutually synergistic fashion. Inflammation, an early response to tissue injury or infection, involves a series of events that combines elements of innate and adaptive immunity (see Figure 24-6).

24.2 Immunoglobulins: Structure and Function

Immunoglobulins, produced by B cells, are the best-understood molecules that confer adaptive immunity. In this section we describe the overall structural organization of immunoglobulins, their structural diversity, and how they bind to antigens.

Immunoglobulins Have a Conserved Structure Consisting of Heavy and Light Chains

Like complement, immunoglobulins are abundant serum proteins that can be classified in terms of their structural and functional properties. Fractionation of antisera, based on their functional activity (e.g., killing of microbes, binding of antigen), led to the identification of the immunoglobulins as the class of serum proteins responsible for antibody activity. Immunoglobulins are composed of two identical *heavy (H) chains*, covalently attached to two identical *light (L) chains* (Figure 24-8).

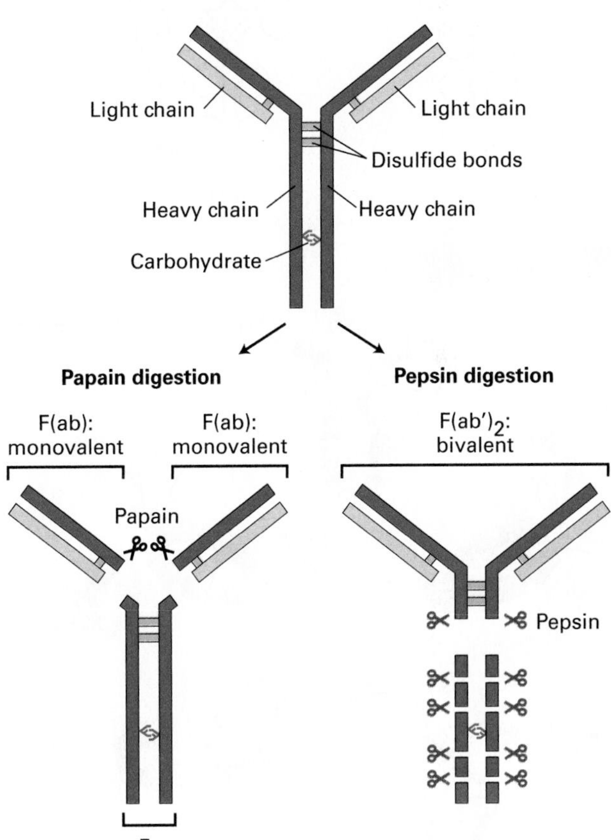

▲ **FIGURE 24-8 The basic structure of an immunoglobulin molecule.** Antibodies are serum proteins also known as immunoglobulins. They are two-fold symmetrical structures composed of two identical heavy chains and two identical light chains. Fragmentation of antibodies with proteases yields fragments that retain antigen-binding capacity. The protease papain yields monovalent F(ab) fragments, and the protease pepsin yields bivalent $F(ab')_2$ fragments. The Fc fragment is unable to bind antigen, but this portion of the intact molecule has other functional properties.

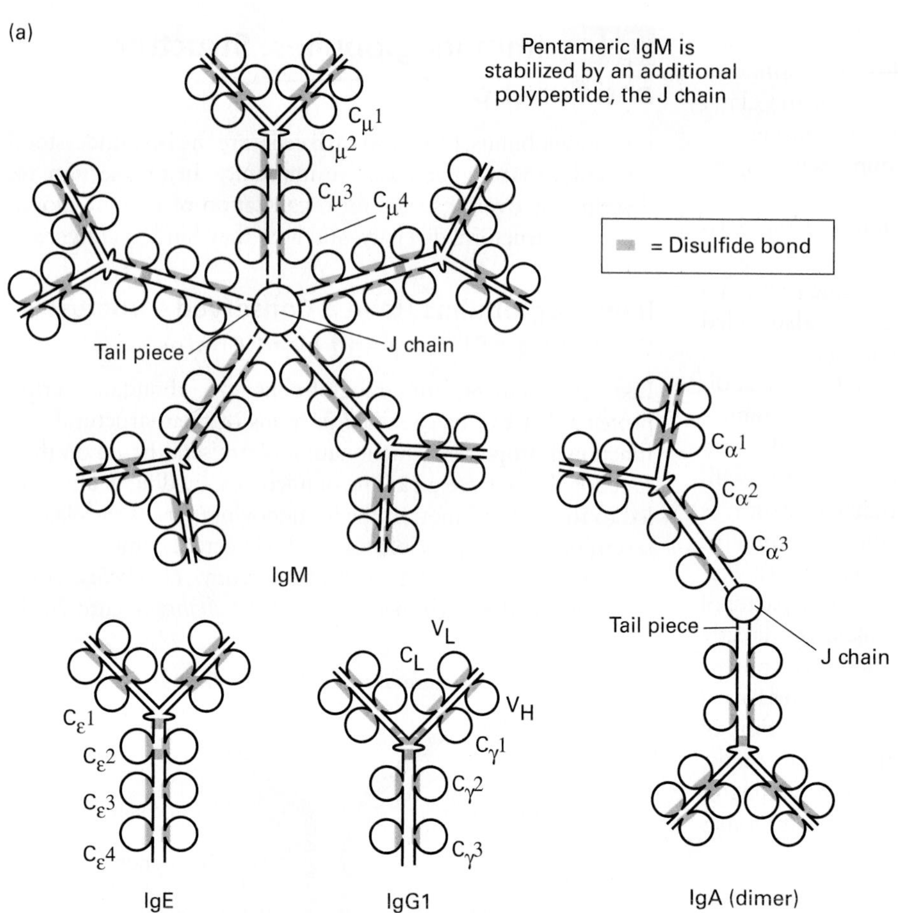

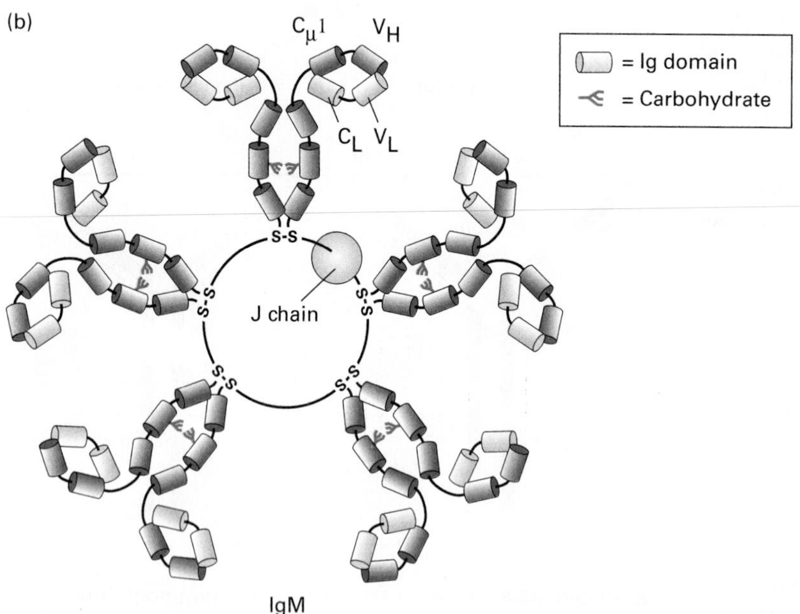

◀ **FIGURE 24-9 Immunoglobulin isotypes.** The different classes of immunoglobulins, called isotypes, may be distinguished biochemically and by immunological techniques. In mouse and humans there are two light-chain isotypes (κ and λ) and five heavy-chain isotypes (μ, δ, γ, ε, α). (a) Based on the identity of the heavy chain, each isotype defines a class of immunoglobulin. IgG, IgE, and IgD (not shown) are monomers with generally similar overall structures. IgM and IgA are unusual because they can occur in serum as pentamers and dimers, respectively, accompanied by an accessory subunit, the J chain, in covalent disulfide linkage. (b) This volume-rendered depiction of the IgM pentamer highlights the modular design of a typical immunoglobulin with each barrel representing an individual Ig domain. Each loop in the diagrams shown in part (a) also represent an Ig domain. Different isotypes have different functions. See Figure 24-12 for definitions of abbreviations.

The typical immunoglobulin therefore has a two-fold symmetrical structure, described as H_2L_2. An exception to this basic H_2L_2 architecture occurs in the camelids (camels, llamas, vicunas). These animals can make some immunoglobulins that are heavy-chain dimers (H_2) and lack light chains.

A biochemical approach was used to answer the question of how antibodies manage to distinguish between related antigens. Proteolytic enzymes were used to fragment immunoglobulins, which are rather large proteins, to identify the regions directly involved in antigen binding (see Figure 24-8). The protease papain yields monovalent fragments, called *F(ab)*, that can bind a single antigen molecule, whereas the protease pepsin yields bivalent fragments, referred to as *F(ab')₂* (F = fragment; ab = antibody). These enzymes are commonly used to convert intact immunoglobulin molecules into monovalent or bivalent reagents. Although F(ab) fragments are incapable of cross-linking, $F(ab')_2$ fragments can do so, a property frequently used to cross-link and so activate surface receptors. The portion released upon papain digestion and incapable of antigen binding is called *Fc*, because of its ease of crystallization (F = fragment; c = crystallizable). This biochemical approach using proteases was followed by peptide mapping and sequencing strategies to determine the primary structure of the immunoglobulins.

Multiple Immunoglobulin Isotypes Exist, Each with Different Functions

Based on their distinct biochemical properties, immunoglobulins are divided into different classes, or *isotypes*. There are two light-chain isotypes, κ and λ. The heavy chains show more variation: In mammals, the major heavy-chain isotypes are μ, δ, γ, α, and ε. These heavy chains can associate with either κ or λ light chains. Depending on the vertebrate species, further subdivisions occur for the α and γ chains, and fish possess an isotype not found in mammals. The fully assembled immunoglobulin (Ig) derives its name from the heavy chain: μ chains yield IgM; α chains, IgA; γ chains, IgG, δ chains, IgD; and ε chains, IgE. The general structures of the major Ig isotypes are depicted in Figure 24-9. By means of their unique structural features, each of the different Ig isotypes carries out specialized functions.

The IgM molecule is secreted as a pentamer, stabilized by disulfide bonds and an additional chain, the J chain. In its pentameric form, IgM possesses 10 identical antigen-binding sites, which allow high-avidity interactions with surfaces that display the corresponding (cognate) antigen. Upon deposition of IgM onto a surface that carries the antigen, the pentameric IgM molecule assumes a conformation highly conducive to activation of the complement cascade, an effective means of damaging the membrane onto which IgM is adsorbed and onto which complement proteins are deposited as a consequence.

The IgA molecule also interacts with the J chain, forming a dimeric structure. Dimeric IgA can bind to the polymeric IgA receptor on the basolateral side of epithelial cells, where its engagement results in receptor-mediated endocytosis. Subsequently, the IgA receptor is cleaved and dimeric IgA with the proteolytic receptor fragment (secretory piece) still attached is released from the apical side of the epithelial cell. This process, called **transcytosis,** is an effective means of delivering immunoglobulins from the basolateral side of an epithelium to the apical side (Figure 24-10a). Tear fluid and other secretions are rich in IgA and so provide protection against environmental pathogens.

The IgG isotype is important for neutralization of virus particles This isotype also helps prepare particulate antigens for acquisition by cells equipped with receptors specific for the Fc portion of IgG molecules (see below).

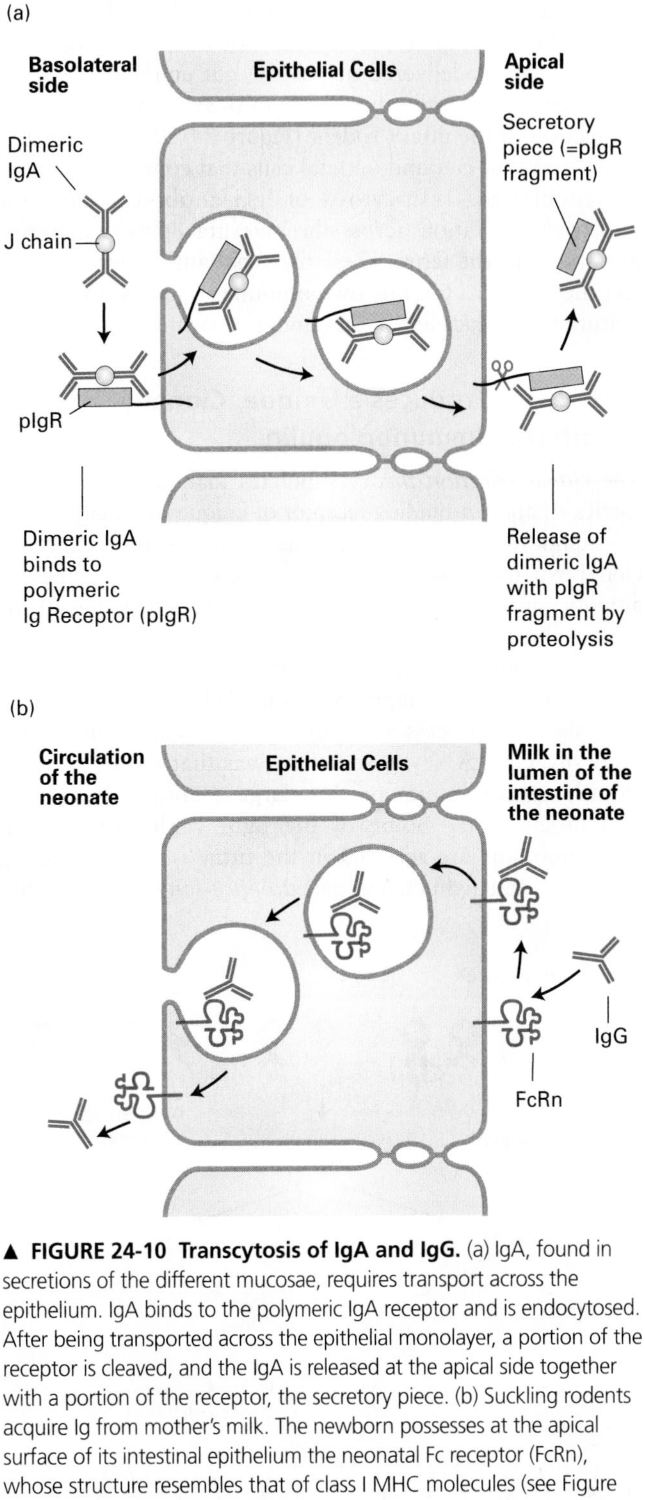

▲ **FIGURE 24-10 Transcytosis of IgA and IgG.** (a) IgA, found in secretions of the different mucosae, requires transport across the epithelium. IgA binds to the polymeric IgA receptor and is endocytosed. After being transported across the epithelial monolayer, a portion of the receptor is cleaved, and the IgA is released at the apical side together with a portion of the receptor, the secretory piece. (b) Suckling rodents acquire Ig from mother's milk. The newborn possesses at the apical surface of its intestinal epithelium the neonatal Fc receptor (FcRn), whose structure resembles that of class I MHC molecules (see Figure 24-21). After this receptor binds to the Fc portion of IgG, transcytosis moves the acquired IgG to the basolateral side of the epithelium. In humans, the syncytial trophoblast in the placenta expresses FcRn and so mediates acquisition of IgG from the maternal circulation and delivery to the fetus (transplacental transport).

The immune system of the newborn is immature, and in rodents protective antibodies are transferred from the mother to the fetus via the mother's milk. The receptor responsible for capturing maternal IgG is the neonatal Fc receptor, which is present on intestinal epithelial cells in rodents. By transcytosis, IgG captured on the luminal side of the newborn's intestinal tract is delivered across the gut epithelium and so makes maternal antibodies in the milk available for passive protection of the infant rodent (Figure 24-10b). In humans, the Fc receptor is found on fetal cells that contact the maternal circulation. Transcytosis of IgG antibodies from the maternal circulation across the placenta delivers maternal antibodies to the fetus. These maternal antibodies will protect the newborn until its own immune system is sufficiently mature to produce antibodies under its own steam.

Each B Cell Produces a Unique, Clonally Distributed Immunoglobulin

The *clonal selection theory* stipulates that each lymphocyte carries an antigen-binding receptor of unique specificity. When a lymphocyte encounters the antigen for which it is specific, clonal expansion (or multiplication) occurs and so allows an amplification of the response, culminating in clearing of the antigen (Figure 24-11).

B-cell tumors, which represent malignant clonal expansions of individual lymphocytes, enabled the first molecular analysis of the processes that underlie the generation of antibody diversity. A key observation was that tumors derived from lymphocytes may produce large quantities of secreted immunoglobulins. Some of the light chains of the immunoglobulins are secreted in the urine of tumor-bearing patients. These light chains, called *Bence-Jones proteins* after their discoverers, are readily purified and afforded the first target for a protein chemical analysis.

Two key observations emerged form this work: (1) no two tumors produced light chains of the identical biochemical properties, suggesting that they were all unique in sequence and (2) the differences in amino acid sequence that distinguish one light chain from another are not randomly distributed but occur clustered in a domain referred to as the *variable region of the light chain*, or V_L. This domain comprises the N-terminal

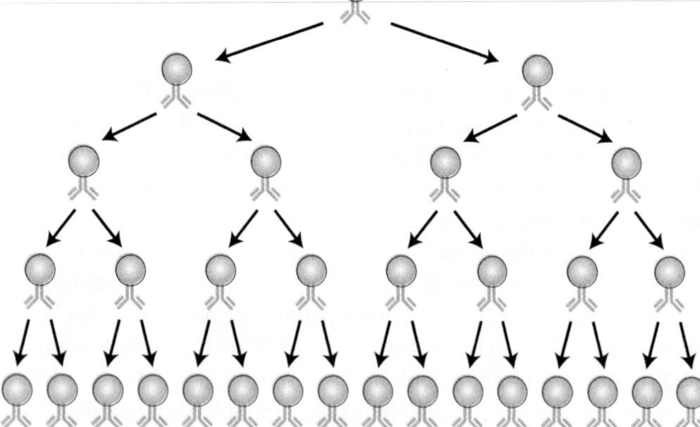

▲ **FIGURE 24-11 Clonal selection.** The clonal selection theory proposes the existence of a large set of lymphocytes, each equipped with its own unique antigen-specific receptor (indicated by different colors). The antigen that shows a fit with the receptor carried by a particular lymphocyte allows that lymphocyte to expand clonally. From a modest number of antigen-specific cells, a large number of cells of the desired specificity (and large amounts of their secreted outputs) may be generated.

(a)

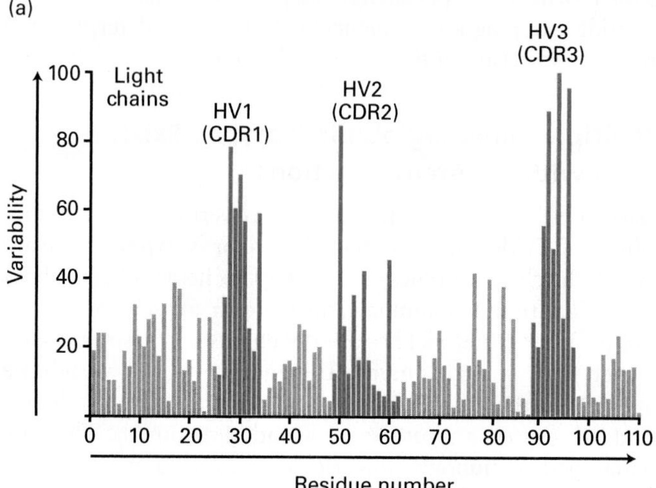

(b)

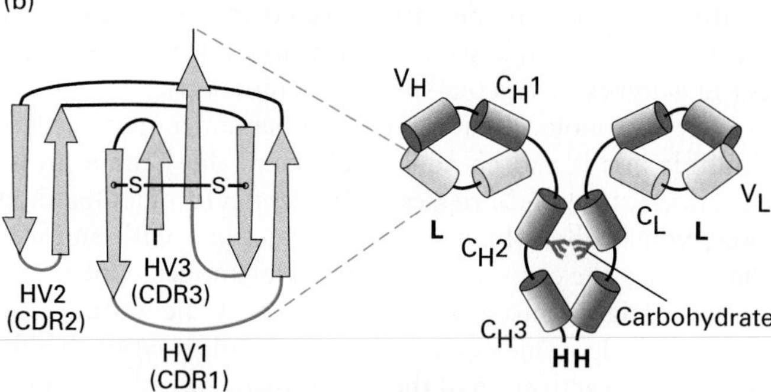

▲ **FIGURE 24-12 Hypervariable regions and the immunoglobulin fold.** (a) Variation in amino acid variability with residue position in Ig light chains. The percentage of variable-region sequences with variant amino acids is plotted for each position in the sequence. Positions for which many different amino acid side chains are present in the data set are assigned a high variability index; those that are invariant among the sequences compared are assigned a value of 0. This analysis reveals three regions of increased variability, hypervariability (HV) regions 1, 2, and 3; these are also called complementarity-determining regions (CDRs). (b) Volume-rendered depiction of $F(ab')_2$ fragment *(right)* and ribbon diagram of a typical Ig light-chain variable domain (V_L) with the positions of the hypervariable regions indicated in red *(left)*. The hypervariable regions are found in the loops that connect the β strands and make contact with antigen. The β strands (rendered as arrows) make up two β sheets and constitute the framework region. Note that each variable and constant domain has a characteristic three-dimensional structure, called the immunoglobulin fold. L = light chain; H = heavy chain; V_H = heavy-chain variable domain; V_L = light-chain variable domain; C_H1, C_H2, C_H3 = heavy-chain constant domains; C_L = light-chain constant domain.

110 amino acids or so. The remainder of the sequence is identical for the different light chains (provided they derive from the identical isotype, κ or λ) and is therefore referred to as the *constant region*, or C_L. From the serum of tumor-bearing individuals, immunoglobulins unique to that individual patient subsequently were purified. Sequencing of the heavy chains from these preparations revealed that the variable residues that distinguish one heavy chain from another were again concentrated in a well-demarcated domain, referred to as the *variable region of the heavy chain,* or V_H.

An alignment of sequences obtained from different homogeneous light-chain preparations showed a non-random pattern of regions of variability, revealing three *hypervariable regions*—HV1, HV2, and HV3—which are sandwiched between what are called framework regions (Figure 24-12a). (Similar alignments for the immunoglobulin heavy-chain sequences also yield hypervariable regions.) In the properly folded three-dimensional structure of immunoglobulins, these hypervariable regions are in close proximity (Figure 24-12b) and make contact with antigen. Thus that portion of an Ig molecule containing the hypervariable regions constitutes the antigen-binding site. For this reason, hypervariable regions are also referred to as *complementarity-determining regions* (CDRs). The difficulty of encoding in the germ line all of the information necessary to generate this enormously diverse antibody repertoire led to suggestions of unique genetic mechanisms to account for this diversity.

Immunoglobulin Domains Have a Characteristic Fold Composed of Two β Sheets Stabilized by a Disulfide Bond

Both the variable and constant domains of immunoglobulins fold into a compact three-dimensional structure composed exclusively of β sheets (see Figure 24-12b). A typical Ig domain contains two β sheets (one with three strands and one with four strands) held together by a disulfide bond. The residues that point inwards are mostly hydrophobic and help stabilize this sandwich structure. Solvent-exposed residues show a greater frequency of polar and charged side chains. The spacing of the cysteine residues that make up the disulfide bond and a small number of strongly conserved residues characterize this evolutionarily ancient structural motif, termed the **immunoglobulin fold.** The basic immunoglobulin fold is found in numerous eukaryotic proteins that are not directly involved in antigen-specific recognition, including the Ig superfamily of cell-adhesion molecules, or IgCAMs (Chapter 19).

The Three-Dimensional Structure of Antibody Molecules Accounts for Their Exquisite Specificity

The three-dimensional structure of immunoglobulins has been solved and the details of how an antibody interacts with an antigen are known at atomic resolution (Figure 24-13; see also Figure 3-19b). The contact area between an antibody and a protein antigen is on the order of 20 × 30 Å and involves mostly interactions that require perfect complementarity. Hydrogen bonds and van der Waals interactions make important contributions to antigen–antibody binding. (For more discussion of the role of molecular complementarity in protein binding and function, see Chapter 2, p. 39.) Antibodies can be elicited not only against proteins, but also against modifications carried by some proteins (e.g., attached oligosaccharide chains or phosphate groups) or even small organic molecules that do not occur in nature. For reasons to be described in Section 24.4, the production of

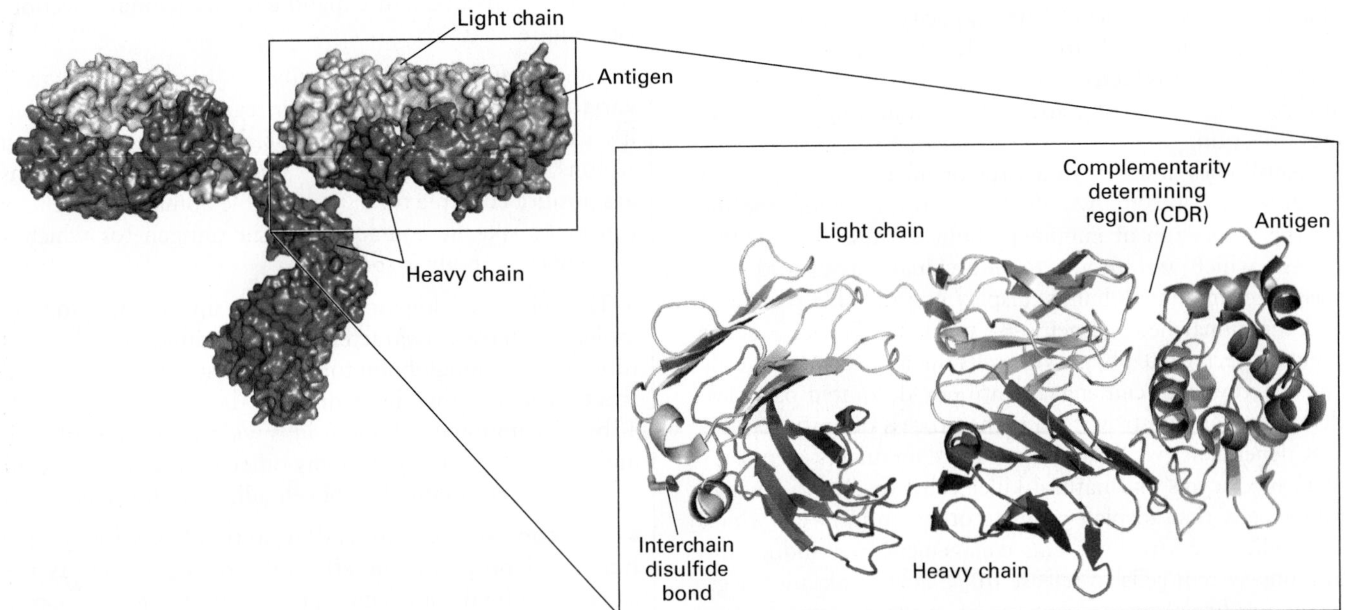

▲ **FIGURE 24-13 Immunoglobulin structure.** This model shows the three-dimensional structure of an immunoglobulin complexed with hen egg-white lysozyme (a protein antigen) as determined by x-ray crystallography. [Based on E. A. Padlan et al., 1989 *PNAS* **86:** 5938.]

antibodies specific for small nonprotein antigens requires conjugation of the antigen to carrier proteins, but the antibodies themselves can recognize these smaller antigens as such. The smaller the antigen, often the deeper it lies buried in the antigen-binding site of the antibody. The hypervariable regions of the heavy and light chains make the most extensive contacts with the antigen to which the antibody binds, with the third hypervariable regions making particularly significant contributions.

The region on an antigen where it makes contact with the corresponding antibody is called an **epitope.** A protein antigen usually contains multiple epitopes, which often are exposed loops or surfaces on the protein and thus accessible to antibody molecules. Each homogeneous antibody preparation, derived from a clonal population of B cells, recognizes a single molecularly defined epitope on the corresponding antigen.

In order to solve the structure of an antibody complexed to its cognate epitope on an antigen, it is important to have a source of homogeneous immunoglobulin and the antigen in pure form. Homogeneous immunoglobulins can be obtained from B-cell tumors (malignant monoclonal expansions of immunoglobulin-secreting B cells), but in that case the antigen for which the antibody is specific is not known. The breakthrough essential for generating homogeneous antibody preparations suitable for structural analysis was the development of techniques to obtain **monoclonal antibody** produced by **hybridomas** by use of a special selection medium (see Chapter 9, pp. 400–402).

An Immunoglobulin's Constant Region Determines Its Functional Properties

Antibodies recognize antigen via their variable regions, but their constant regions determine many of the functional properties of antibodies. An important functional property of antibodies is their neutralizing capacity. By binding to epitopes on the surface of virus particles or bacteria, antibodies may block a productive interaction between a pathogen and receptors on host cells, thereby inhibiting (neutralizing) infection.

Antibodies attached to a virus or microbial surface can be recognized directly by cells that express receptors specific for the Fc portion of immunoglobulins. These *Fc receptors* (FcRs), which are specific for individual classes and subclasses of immunoglobulins, display considerable structural and functional heterogeneity. By means of FcR-dependent events, specialized phagocytic cells such as dendritic cells and macrophages can engage antibody-decorated particles, then ingest and destroy them in the process of opsonization. FcR-dependent events also allow some immune-system cells (e.g., monocytes and natural killer cells) to directly engage target cells that display viral or other antigens to which antibodies are attached. This engagement may induce the immune-system cells to release toxic small molecules (e.g., oxygen radicals) or the contents of cytotoxic granules, including perforins and granzymes. These proteins can attach themselves to the surface of the engaged target cell, inflict membrane damage, and so kill the target (see Figure 24-5). This process, called *antibody-dependent cell-mediated cytotoxicity,* illustrates how cells of the innate immune system interact with, and benefit from, the products of the adaptive immune response.

Depending on the immunoglobulin isotype, antigen–antibody (immune) complexes can initiate the classical pathway of complement activation (see Figure 24-4). IgM and IgG3 are particularly good at complement activation, but all IgG classes can in principle activate complement, whereas IgA and IgE are unable to do so.

KEY CONCEPTS OF SECTION 24.2

Immunoglobulins: Structure and Function

- Most immunoglobulins (antibodies) are composed of two identical heavy (H) chains and two light (L) chains, with each chain containing a variable (V) region and constant (C) region. Proteolytic fragmentation yields monovalent F(ab) and bivalent $F(ab')_2$ fragments, which contain variable-region domains and retain antigen-binding capability (see Figure 24-8). The Fc portion contains constant-region domains and determines effector functions.

- Immunoglobulins are divided into classes based on the constant regions of the heavy chains they carry (see Figure 24-9). In mammals there are five major classes: IgM, IgD, IgG, IgA, IgE; the corresponding heavy chains are referred to as μ, δ γ, α, and ε. There are two major classes of light chain, κ and λ, again characterized by the attributes of their constant regions.

- Each individual B lymphocyte expresses an immunoglobulin of unique sequence and is therefore uniquely specific for a particular antigen. Upon recognition of antigen, only a B lymphocyte that bears a receptor specific for it will be activated and expand clonally (clonal selection) (see Figure 24-11).

- The antigen specificity of antibodies is conferred by their variable domains, which contain regions of high variability, called hypervariable or complementarity-determining regions (see Figure 24-12a). These hypervariable regions are positioned at the tip of the variable domain, where they can make specific contacts with the antigen for which a particular antibody is specific.

- The repeating domains that make up immunoglobulin molecules have a characteristic three-dimensional structure, the immunoglobulin fold: It consists of two β pleated sheets held together by a disulfide bond (see Figure 24-12b). The immunoglobulin fold is widespread in evolution and is found in many proteins other than antibodies, including an important class of cell-adhesion molecules.

- The constant regions endow antibodies with unique functional properties or effector functions, such as the capacity to bind complement, the ability to be transported across epithelia, or the ability to interact with receptors specific for the Fc portion of immunoglobulins.

24.3 Generation of Antibody Diversity and B-Cell Development

Pathogens have short replication times, are quite diverse in their genetic makeup, and evolve quickly, generating even more antigenic variation. An adequate defense must thus be capable of mounting an equally diverse response. Antibodies fulfill this role. B cells, which are responsible for antibody production, make use of a unique mechanism by which the genetic information required for synthesis of immunoglobulin heavy and light chains is stitched together from separate DNA sequence elements, or Ig gene segments, to create a functional transcriptional unit. The act of recombination that combines Ig gene segments itself dramatically expands the variability in sequence precisely where these genetic elements are joined together. This mechanism of generating a diverse array of antibodies is fundamentally different from meiotic recombination, which occurs only in germ cells, and from alternative splicing of exons (Chapter 8). Because this recombination mechanism occurs in somatic cells but not in germ cells, it is known as *somatic gene rearrangement* or *somatic recombination.*

A Functional Light-Chain Gene Requires Assembly of V and J Gene Segments

Immunoglobulin genes encoding intact immunoglobulins do not exist already assembled in the genome, ready for expression. Instead, the required gene segments are brought together and assembled in the course of B-cell development (Figure 24-14). Although the rearrangement of heavy-chain genes precedes the rearrangement of light-chain genes, we discuss light-chain genes first because of their less complex organization.

The immunoglobulin light chains are encoded by clusters of V gene segments, followed at some distance downstream by a single C segment. Each V gene segment carries its own promoter sequence and encodes the bulk of the light-chain variable region, although a small piece of the nucleotide sequence encoding the light-chain variable region is missing from the V gene segment. This missing portion is provided by one of the multiple J segments located between the V segments and the single C segment in the unrearranged κ light-chain locus (see Figure 24-14a). In the course of B-cell development, commitment to a particular V gene segment—a random process—results in its juxtaposition with one of the J segments, again a random choice, forming an exon encoding the entire light-chain variable region (V_L). The act of recombination not only generates an intact and functional light-chain gene, but also places the promoter sequence of the rearranged gene within controlling distance of enhancer elements required for its transcription. Only a rearranged light-chain gene is transcribed.

Recombination Signal Sequences Detailed sequence analysis of the light-chain and heavy-chain loci revealed a conserved sequence element at the 3′ end of each V gene segment. This conserved element, called a *recombination signal sequence (RSS),* is composed of heptamer and nonamer sequences separated by a 23-bp spacer. At the 5′ end of each

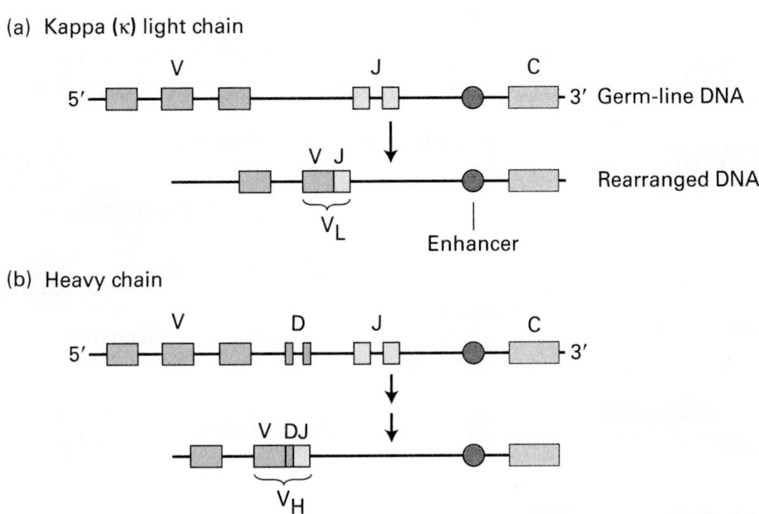

▲ **FIGURE 24-14 Overview of somatic gene rearrangement in immunoglobulin DNA.** The stem cells that give rise to B cells contain multiple gene segments encoding portions of immunoglobulin heavy and light chains. During development of a B cell, somatic recombination of these gene segments yields functional light-chain genes (a) and heavy-chain genes (b). Each V gene segment carries its own promoter. Rearrangement brings an enhancer close enough to the combined VJ sequence to activate transcription. The light-chain variable region (V_L) is encoded by two joined gene segments, and the heavy-chain variable region (V_H) is encoded by three joined segments. Note that the chromosomal regions encoding immunoglobulins contain many more V, D, and J segments than shown. Also, the κ light-chain locus contains a single constant (C) segment, as shown, but the heavy-chain locus contains several distinct C segments (not shown) corresponding to the immunoglobulin isotypes.

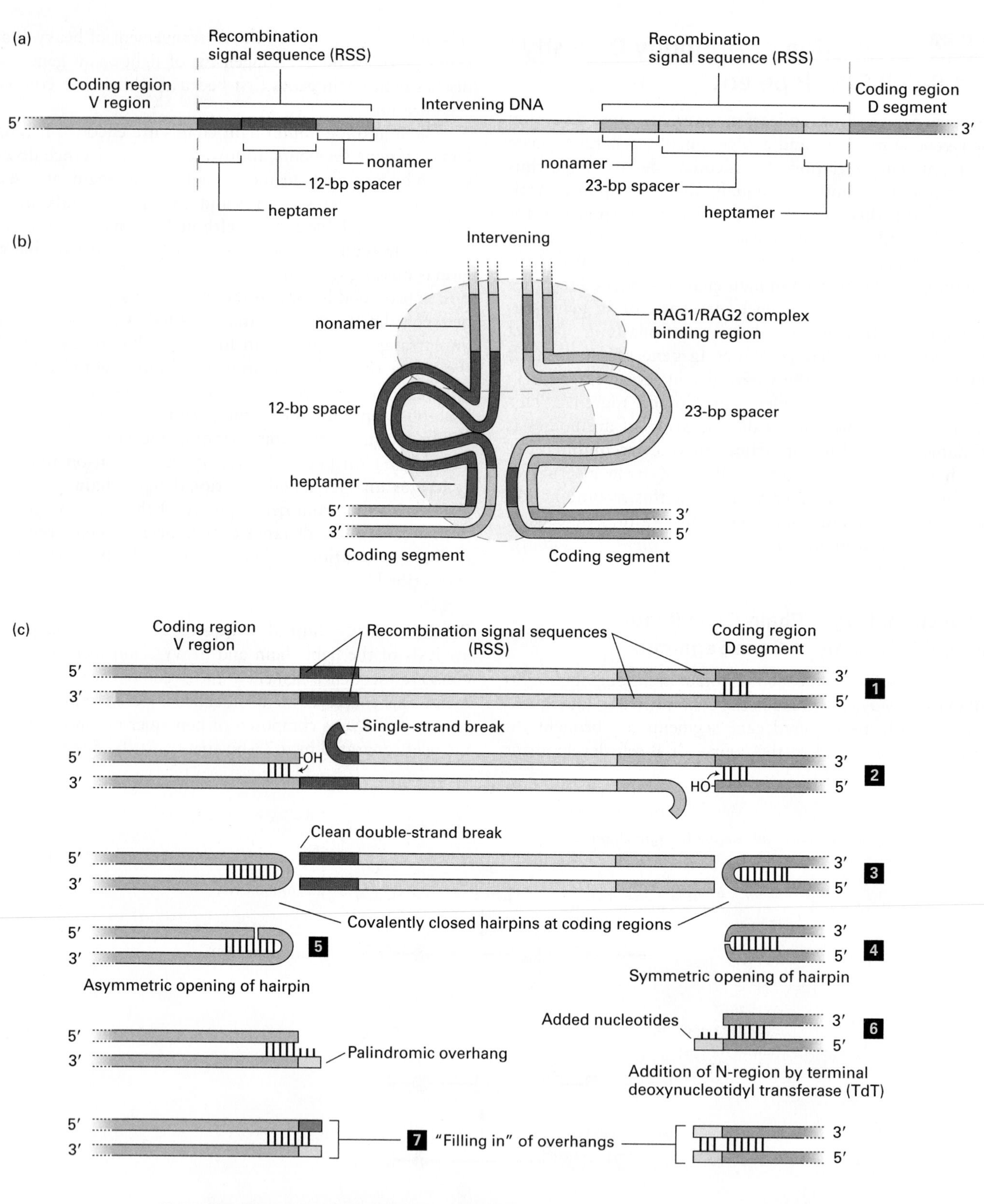

J element, there is a similarly conserved RSS that contains a 12-bp spacer (Figure 24-15a). The 12- and 23-bp spacers separate the conserved heptamer and nonamer sequences by one and two turns of the DNA helix, respectively.

Somatic recombination is catalyzed by the RAG1 and RAG2 recombinases, which are expressed only in lymphocytes. Juxtaposition of the two gene segments to be joined is stabilized by the RAG1/RAG2 complex (Figure 24-15b).

◀ **FIGURE 24-15 Mechanism of rearrangement of immunoglobulin gene segments via deletional joining.** This example depicts joining of a V segment and a D segment. D segments are present in the heavy-chain, but not light-chain, locus (see Figure 24-14b.) (a) Location of the DNA elements involved in somatic recombination of immunoglobulin gene segments. At the 3′ end of all V gene segments is a conserved recombination signal sequence (RSS) composed of a heptamer, a 12-bp spacer, and a nonamer. Each of the D segments with which a V can recombine possess at their 5′ end a similar RSS with a 23-bp spacer. The nonamer and heptamer sequences at the 5′ end of D are complementary and antiparallel to those found at the 3′ end of each V when read on the same (top) strand. (b) Hypothetical model of how two coding regions to be joined may be arranged spatially, stabilized by the RAG1 and RAG2 recombinase complex. Both strands of the DNA are shown. (c) Events in deletional joining of V and D coding regions. The germ-line DNA (**1**) is folded, bringing the segments to be joined close together and the RAG1/RAG2 complex makes single-stranded cuts at the boundaries between the coding sequences and RSSs (**2**). The free 3′ –OH groups attack the complementary strands, creating a covalently closed hairpin at each coding end and a clean double-stranded break at each boundary with a RSS (**3**). The hairpins are opened, either symmetrically (**4**), as shown for the J segment, or asymmetrically (**5**), as shown for the V segment. Terminal deoxynucleotidyl transferase adds nucleotides in a template-independent manner to symmetrically opened hairpins (**6**, *right*), generating an overhang (yellow) of unpaired nucleotides of random sequence; asymmetric opening automatically creates a palindromic overhang (**6**, *left*). The unpaired overhangs at the ends of both the V and D coding regions are filled in by DNA polymerase (**7**) or may be excised by an exonuclease. DNA ligase IV joins the two segments generated from the V and D coding regions (**8**). Rearrangements of light-chain V and J segments occur by the same mechanism except that N-region addition does not take place. See text for additional discussion.

The recombinases then make a single-stranded cut at the exact boundary of each coding sequence and its adjacent RSS. Only gene segments that possess heptamer-nonamer RSSs with spacers of different length can engage in this type of rearrangement (the so-called 12/23 bp spacer rule). Each newly created –OH group then executes a nucleophilic attack on the complementary strand, creating a covalently closed hairpin for each of the two coding ends, and double-strand breaks at the ends of the RSSs. Protein complexes that include the Ku70 and Ku80 proteins hold this complex together, so that the ends about to be joined remain in close proximity. The RSS ends are then covalently joined without loss or addition of nucleotides, creating a circular reaction product containing the intervening DNA, which is lost from the locus altogether. The hairpin ends of the coding segments undergoing recombination then are opened and finally joined as depicted in Figure 24-15c, completing the recombination process.

The recombination mechanism just described, called *deletional joining,* occurs when the V gene segment involved has the same transcriptional orientation as the other gene segments at the light-chain locus. Some V gene segments, however, have the opposite transcriptional orientation. These are joined to J segments by a mechanism, termed *inversional joining,* in which the V segment is inverted and the intervening DNA and RSSs are not lost from the locus.

Defects in the synthesis of RAG proteins obliterates the possibility of somatic gene rearrangements. As described below, the rearrangement process is essential for B-cell development; consequently, RAG deficiency leads to the complete absence of B cells. People with defects in RAG gene function suffer from severe immunodeficiency.

Junctional Imprecision In addition to the sequence variability created by the random selection of V and J gene segments to join, processing of the intermediates created in the course of recombination provide additional means for expanding the variability of immunoglobulin sequences. This additional variability is created at the junction of the segments to be joined. The opening of the hairpins at the coding ends is a key step in this process: this opening may occur symmetrically or asymmetrically (see Figure 24-15, **4** and **5**). The protein Artemis, whose function requires the catalytic subunit of DNA-dependent protein kinase, carries out opening of the hairpins.

If opening of a hairpin is asymmetric, a short, single-stranded palindrome sequence is generated. Filling in of this overhang by DNA polymerase results in the addition of several nucleotides, called *P-nucleotides,* that were not part of the original coding region of the gene segment in question. Alternatively, the overhang may be removed by exonucleolytic attack, resulting in the removal of nucleotides from the original coding region. These possibilities apply equally to the V and the J coding regions. Symmetric opening of a hairpin retains all the original coding information. However, even if the hairpin is opened symmetrically, the ends of the DNA molecule tend to breathe, creating short single-stranded sequences, which also may be attacked exonucleolytically, resulting in removal of nucleotides.

Once the hairpins have been opened and the coding ends processed, the ends are ligated together by DNA ligase IV and XRCC4, generating a functional light-chain gene. Inherent in the rearrangement process is *junctional imprecision* resulting in part from the addition and loss of nucleotides at the coding joints. Whenever a V and a J segment recombine, the sequence and reading frame of the VJ product cannot be predicted. Only one in three recombination reactions results in a reading frame that is compatible with light-chain synthesis.

Light-chain diversity therefore arises not only from the combinatorial usage of V and J gene segments, but also from junctional imprecision. Inspection of the three-dimensional structure of the light chain shows that the highly diverse joint, generated as a consequence of junctional imprecision, forms part of a loop—hypervariable region 3 (HV3)—that projects into the antigen-binding site and makes contact with antigen (see Figure 24-12b).

Rearrangement of the Heavy-Chain Locus Involves V, D, and J Gene Segments

The organization of the heavy-chain locus is more complex than that of the κ light-chain locus. The heavy-chain locus contains not only a large tandem array of V gene segments (each equipped with its own promoter) and multiple J elements, but also multiple D (diversity) segments (see Figure 24-14b). Somatic recombination of a V, D, and J segment generates a rearranged sequence encoding the heavy-chain variable region (V_H).

At the 3′ end of each V gene segment in heavy-chain DNA, there are conserved heptamer and nonamer sequences separated by spacer DNA, similar to the recombination signal sequences (RSSs) in light-chain DNA. These RSSs are also found in complementary and antiparallel configuration at the 5′ end and the 3′ end of each D segment (see Figure 24-15(a)). The J segments are similarly equipped at their 5′ end with the requisite RSS. The spacer lengths in these RSSs are such that D segments can join to J segments, and V segments to already rearranged DJ segments. However, neither direct V to J nor D to D joining is allowed in view of the 12/23 heptamer-nonamer rule. Heavy-chain rearrangements proceed via the same mechanisms described above for light-chain rearrangements.

In the course of B-cell development, the heavy-chain locus always rearranges first, starting with D-J rearrangements. D-J rearrangement is followed by V-DJ rearrangement (Figure 24-16). In the course of D-J and V-DJ rearrangements, terminal

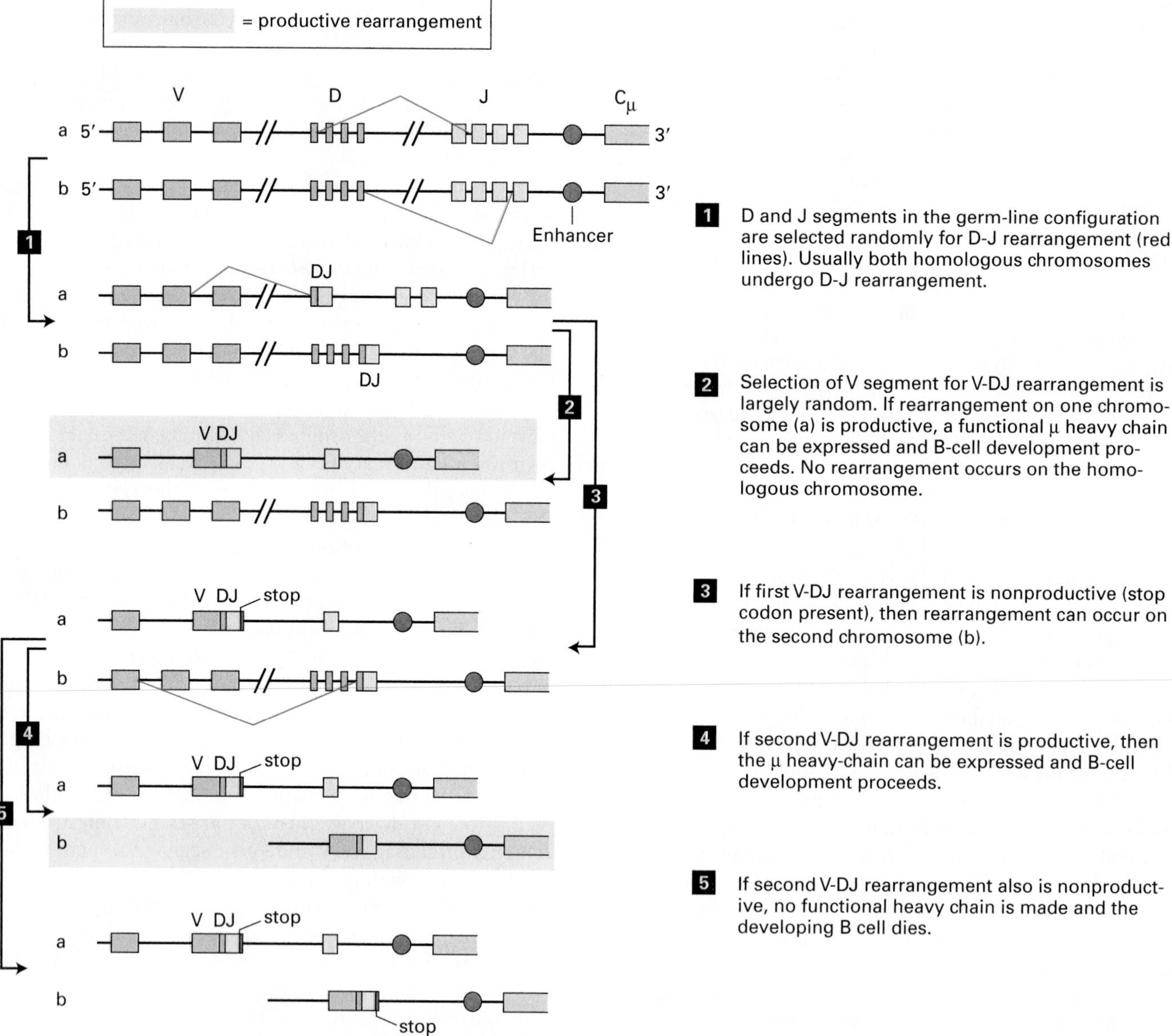

= productive rearrangement
V
D
J
Cμ
a 5′
3′
b 5′
3′
Enhancer
1
D and J segments in the germ-line configuration are selected randomly for D-J rearrangement (red lines). Usually both homologous chromosomes undergo D-J rearrangement.
DJ
a
b
DJ
2
Selection of V segment for V-DJ rearrangement is largely random. If rearrangement on one chromosome (a) is productive, a functional μ heavy chain can be expressed and B-cell development proceeds. No rearrangement occurs on the homologous chromosome.
V DJ
a
b
3
If first V-DJ rearrangement is nonproductive (stop codon present), then rearrangement can occur on the second chromosome (b).
V DJ
stop
a
b
4
If second V-DJ rearrangement is productive, then the μ heavy-chain can be expressed and B-cell development proceeds.
V DJ
stop
a
b
5
If second V-DJ rearrangement also is nonproductive, no functional heavy chain is made and the developing B cell dies.
V DJ
stop
a
b
stop

▲ **FIGURE 24-16 Somatic recombination of the heavy-chain locus.** The immunoglobulin heavy-chain locus includes multiple V, D, and J gene segments that must be rearranged to form a V-region coding sequence. The two copies of the heavy-chain locus present on homologous chromosomes (a and b) are depicted. D to J rearrangements occur first (**1**), on one or on both chromosomes carrying the heavy-chain locus. A V segment then rearranges with the newly rearranged DJ segment on one chromosome. If this rearrangement is productive (**2**), further rearrangement ceases, and B-cell development proceeds. If the first V to DJ recombination is nonproductive (i.e., introduces an inappropriate stop codon), the second chromosome may rearrange (**3**). If the second recombination is productive (**4**), then B-cell development proceeds. However, if the second V to DJ recombination also is nonproductive (**5**), the developing B cell dies.

deoxynucleotidyltransferase (TdT) may add nucleotides to free 3′ OH ends of DNA in a template-independent fashion. Up to a dozen or so nucleotides, called the *N-region,* may be added, generating additional sequence diversity at the junctions whenever D-J and V-DJ rearrangements occur (see Figure 24-15, step **6**). Only one in three rearrangements yields the proper reading frame for the rearranged VDJ sequence. If the rearrangement yields a sequence encoding a functional protein, it is called productive. Although the heavy-chain locus is present on two homologous chromosomes, only one productive rearrangement is permitted, as discussed below.

An enhancer located downstream of the cluster of J segments and upstream of the μ constant-region segment activates transcription from the promoter at the 5′ end of the rearranged VDJ sequence (see Figure 24-16). Splicing of the primary transcript produced from the rearranged heavy-chain gene generates a functional mRNA encoding the μ heavy chain. For both immunoglobulin heavy- and light-chain genes, somatic recombination places the promoters upstream of the V segments within functional reach of the enhancers necessary to allow transcription, so that only rearranged VJ and VDJ sequences, and not the V segments that remain in the germ-line configuration, are transcribed.

Somatic Hypermutation Allows the Generation and Selection of Antibodies with Improved Affinities

In addition to the diversity created by somatic recombination and junctional imprecision, antigen-activated B cells can undergo *somatic hypermutation.* Upon receipt of proper additional signals, most of which are provided by T cells, expression of activation-induced deaminase (AID) is turned on. This enzyme deaminates cytosine residues to uracil. When a B cell that carries this lesion replicates, it may place an adenine on the complementary strand, thus generating a G-to-A transition (see Figure 4-35). Alternatively, the uracil may be excised by DNA glycosylase to yield an abasic site. These abasic sites, when copied, give rise to possible transitions as well as a transversion, unless the nucleotide opposite the gap is chosen to be the original G that paired with the cytosine target. Mutations thus accumulate with every successive round of B-cell division, yielding numerous mutations in the rearranged VJ and VDJ segments. Many of these mutations are deleterious, in that they reduce the affinity of the encoded antibody for antigen, but some improve the encoded antibody's affinity for antigen. B cells carrying affinity-increasing mutations have a selective advantage when they compete for the limited amount of antigen that evokes clonal selection (see Figure 24-11). The net result is generation of a B-cell population whose antibodies, as a rule, show a higher affinity for the antigen.

In the course of an immune response or upon repeated immunization, the antibody response exhibits *affinity maturation,* an increase in the average affinity of antibodies for antigen, as the result of somatic hypermutation. Antibodies produced during this phase of the immune response display affinities for antigen in the nanomolar (or better) range. For reasons that are not understood, the activity of activation-induced deaminase is focused mostly on rearranged VJ and VDJ segments, and this targeting may therefore require active transcription. The entire process of somatic hypermutation is strictly antigen dependent and shows an absolute requirement for interactions between the B cell and certain T cells.

B-Cell Development Requires Input from a Pre-B Cell Receptor

As we have seen, B cells destined to make immunoglobulins must rearrange the necessary gene segments to assemble a functional heavy- and light-chain gene. These rearrangements occur in a carefully ordered sequence during development of a B cell, starting with heavy-chain rearrangements. Moreover, the rearranged heavy-chain is first used to build a membrane-bound receptor that executes a cell fate decision necessary to drive further B-cell development (and antibody synthesis).

Successful rearrangement of V, D, and J segments in the heavy-chain locus allows synthesis of a μ chain. B cells at this stage of development are called *pre-B cells,* as they have not yet completed assembly of a functional light-chain gene and therefore cannot engage in antigen recognition. The newly rearranged heavy-chain gene encodes the μ polypeptide, which becomes part of a signaling receptor whose expression is essential for B-cell development to proceed in orderly fashion. The μ chain made at this stage of B cell development is a membrane-bound version. Following engagement with antigen, the B cell switches to producing soluble, secreted immunoglobulins from the same transcription unit, as we describe below.

In pre-B cells, newly made μ chains form a complex with so-called surrogate light chains, composed of two subunits, λ5 and VpreB (Figure 24-17). The μ chain itself possesses no cytoplasmic tail and is therefore incapable of recruiting cytoplasmic components for the purpose of signal transduction. Instead, early B cells express two auxiliary transmembrane proteins, called Igα and Igβ, each of which carries in its cytoplasmic tail an immunoreceptor tyrosine-based activation motif, or ITAM. The entire complex including Igα and Igβ constitutes the *pre-B cell receptor (pre-BCR).* Engagement of this receptor by suitable signals results in recruitment and activation of a Src family tyrosine kinase, which phosphorylates tyrosine residues in the ITAMs. In their phosphorylated form, ITAMs recruit other molecules essential for signal transduction (see below). Because no functional light chains are yet part of this receptor, it is incapable of antigen recognition.

The pre-B cell receptor has several important functions. First, it shuts off expression of the RAG recombinases, so rearrangement of the other (allelic) heavy-chain locus cannot occur. This phenomenon, called *allelic exclusion,* ensures that only one of the two available copies of the heavy-chain locus will be rearranged and thus expressed. Second, because of the association of the pre-B cell receptor with Igα and Igβ, the receptor becomes a functional signal-transduction unit. The pre-BCR also initiates cell proliferation and so expands the numbers of those B cells that have undergone productive D-J and V-DJ recombination.

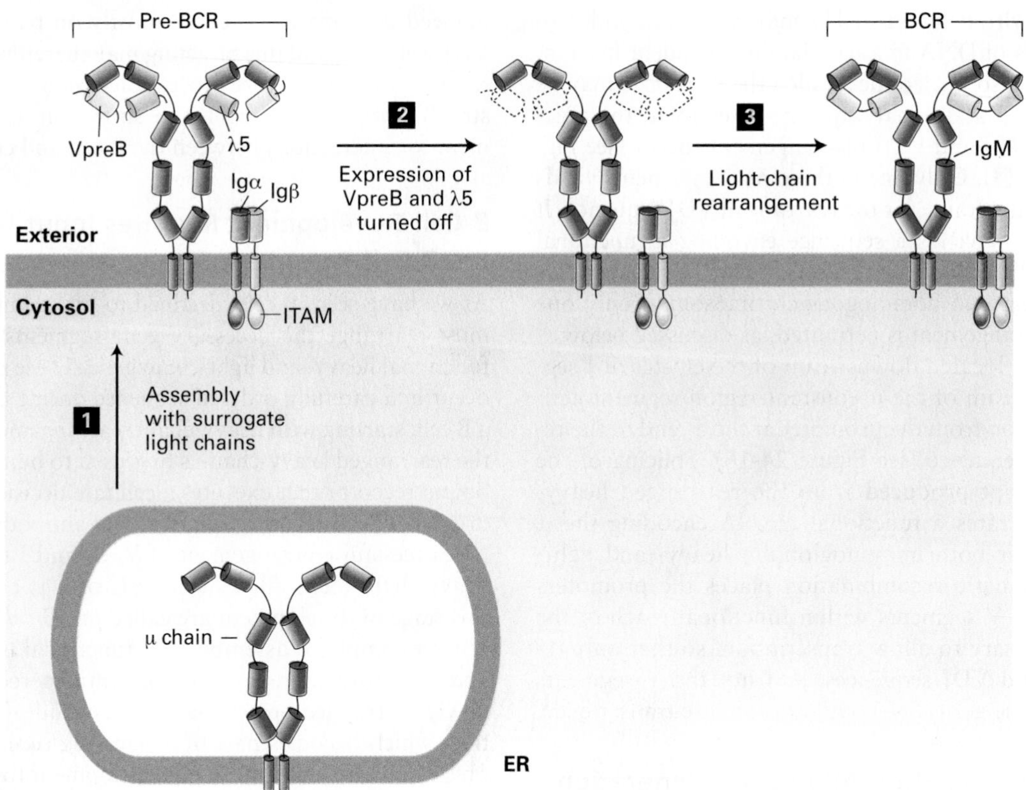

▲ **FIGURE 24-17 Structure of the pre-B cell receptor and its role in B-cell development.** Successful rearrangement of V, D, and J heavy-chain gene segments allows synthesis of membrane-bound μ heavy chains in the endoplasmic reticulum (ER) of a pre-B cell. At this stage, no light-chain gene rearrangement has occurred. Newly made μ chains assemble with surrogate light chains, composed of λ5 and VpreB, to yield the pre-B cell receptor, pre-BCR (**1**). This receptor drives proliferation of those B cells that carry it. It also suppresses rearrangement of the heavy-chain locus on the other chromosome and so mediates allelic exclusion. In the course of proliferation, the synthesis of λ5 and VpreB is shut off (**2**), resulting in "dilution" of the available surrogate light chains and reduced expression of the pre-BCR. As a result, rearrangement of the light-chain loci can proceed (**3**). If this rearrangement is productive, the B cell can synthesize light chains and complete assembly of the B-cell receptor (BCR), comprising a membrane-bound IgM and associated Igα and Igβ. The B cell is now responsive to antigen-specific stimulation.

In the course of this expansion, expression of the surrogate light-chain subunits, VpreB and λ5, is turned off. The progressive dilution of VpreB and λ5 with every successive cell division allows reinitiation of expression of the RAG enzymes, which now target the κ or λ light-chain locus for recombination. A productive V-J rearrangement also shuts off rearrangement of the allelic locus (allelic exclusion). Upon completion of a successful V-J light-chain rearrangement, the B cell can make both μ heavy chains and κ or λ light chains, and assemble them into a functional B-cell receptor (BCR), which can recognize antigen (see Figure 24-17).

Once a B cell expresses a complete BCR on its cell surface, all subsequent steps in B-cell activation and differentiation involve recognition of the antigen for which that BCR is specific. The BCR not only plays a role in driving B-cell proliferation upon successful encounter with antigen, but also functions as a device for ingesting antigen, an essential step that allows the B cell to process the acquired antigen and convert it into a signal that sends out a call for assistance by T lymphocytes. This antigen-presentation function of B cells is described in later sections.

During an Adaptive Response, B Cells Switch from Making Membrane-Bound Ig to Making Secreted Ig

As just described, the **B-cell receptor (BCR)**, a membrane-bound IgM, provides a B cell with the ability to recognize a particular antigen, an event that triggers clonal selection and proliferation of that B cell, thus increasing the number of B cells specific for the antigen (see Figure 24-11). However, key functions of immunoglobulins, such as neutralization of antigen or killing of bacteria, require that these products be released by the B cell, so that they may accumulate in the extracellular environment and act at a distance from the site where they were produced.

The choice between the synthesis of membrane-bound versus secreted immunoglobulin is made during processing of the heavy-chain primary transcript. As shown in Figure 24-18, the μ locus contains two exons (TM1 and TM2) that together encode a C-terminal domain that anchors IgM in the plasma membrane. One polyadenylation site is found upstream of these exons; a second polyadenylation site is present downstream. If the downstream poly(A) site is chosen,

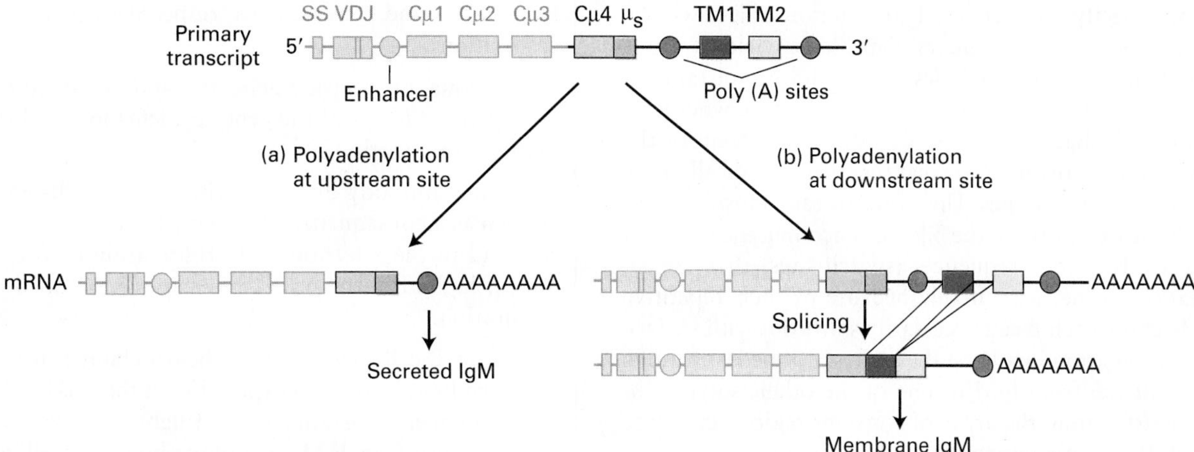

▲ **FIGURE 24-18 Synthesis of secreted and membrane IgM.** The organization of the μ heavy-chain primary transcript is shown at the top: $C_\mu 4$ is the exon encoding the fourth μ constant-region domain; μ_s is a coding sequence unique for secreted IgM; TM1 and TM2 are exons that specify the transmembrane domain of the μ chain. Whether secreted or membrane-bound IgM is made depends on which poly(A) site is selected during processing of the primary transcript. (a) If the upstream poly(A) site is used, the resulting mRNA includes the entire $C_\mu 4$ exon and specifies the secreted form of the μ chain. (b) If the downstream poly(A) site is used, a splice donor site in the Cμ4 exon allows splicing to the transmembrane exons, yielding a mRNA that encodes the membrane-bound form of the μ chain. Similar mechanisms generate secreted and membrane-bound forms of other Ig isotypes. SS = signal sequence.

then further processing yields a mRNA encoding the membrane-bound form of μ. (As described above, this choice is necessary for formation of the B-cell receptor, which includes membrane-bound IgM.) If the upstream poly(A) site is chosen, processing yields the secreted version of the μ chain. Similar arrangements are found for the other Ig constant-region gene segments, each of which can specify either a membrane-bound or a secreted heavy chain.

In the course of B-cell differentiation, the B cell acquires the capacity to switch from the synthesis of exclusively membrane immunoglobulin to the synthesis of secreted immunoglobulin. Terminally differentiated B cells, called *plasma cells,* are devoted almost exclusively to the synthesis of secreted antibodies (see Figure 24-6). Plasma cells synthesize and secrete several thousand antibody molecules per second. It is this ramped up production of secreted antibodies that underlies the effectiveness of the adaptive immune response in eliminating pathogens. The protective value of antibodies is proportional to the concentration at which they are present in the circulation. Indeed, circulating antibody levels are often used as the key parameter to determine whether vaccination against a particular pathogen has been successful. The ability of plasma cells to establish adequate antibody levels is a function of their ability to secrete large amounts of immunoglobulins and so requires a massive expansion of the endoplasmic reticulum, a hallmark of plasma cells.

B Cells Can Switch the Isotype of Immunoglobulin They Make

In the immunoglobulin heavy-chain locus, the exons that encode the μ chain lie immediately downstream of the rearranged VDJ exon (Figure 24-19, *top*). This is followed by

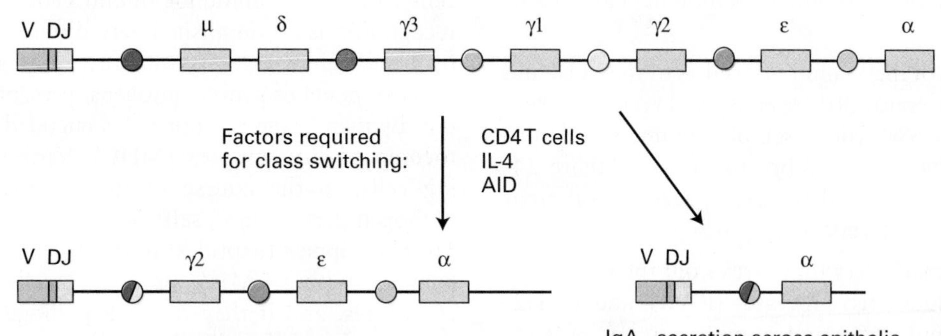

▲ **FIGURE 24-19 Class switch recombination in the immunoglobulin heavy-chain locus.** Class switch recombination involves switch sites, which are repetitive sequences (colored circles) upstream of the heavy-chain constant-region genes. Recombination requires activation-induced deaminase (AID), assistance by T cells, and cytokines (e.g., IL-4) produced by certain T cells. Recombination eliminates the segment of DNA between the switch site upstream of μ exons and the constant region to which switching occurs. Class switching generates antibody molecules with the same specificity for antigen as that of the IgM-bearing B cell that mounted the original response, but with different heavy-chain constant-regions and therefore different effector functions.

exons that specify the δ chain. Transcription of a newly rearranged immunoglobulin heavy-chain locus yields a single primary transcript that includes the μ and δ constant regions. Splicing of this large transcript determines whether a μ chain or a δ chain will be produced. Downstream of the μ/δ combination are the exons that together encode all of the other heavy-chain isotypes. Upstream of each cluster of exons (with the exception of the δ locus) encoding the different isotypes are repetitive sequences (switch sites) that are recombination prone, presumably because of their repetitive nature. Because each B cell necessarily starts out with surface IgM, recombination involving these sites, if it occurs, results in a *class switch* from IgM to one of the other isotypes located downstream in the array of constant-region genes (see Figure 24-19). The intervening DNA is deleted.

In the course of its differentiation, a B cell can switch sequentially. Importantly, the light chain is not affected by this process, nor is the rearranged VDJ segment with which the B cell started out on this pathway. Class switch recombination thus generates antibodies with different constant regions, but of identical antigenic specificity. Each immunoglobulin isotype is characterized by its own unique constant region. As discussed previously, these constant regions determine the functional properties of the various isotypes. Class switch recombination is absolutely dependent on the activity of activation-induced deaminase (AID) and the presence of antigen and T cells. Somatic hypermutation and class switch recombination occur concurrently, and their combined effect allows fine tuning of the adaptive immune response with respect to the affinity of the antibodies produced and the effector functions called for.

KEY CONCEPTS OF SECTION 24.3

Generation of Antibody Diversity and B-Cell Development

- Functional antibody-encoding genes are generated by somatic rearrangement of multiple DNA segments at the heavy-chain and light-chain loci. These rearrangements involve V and J segments for immunoglobulin light chains, and V, D, and J segments for immunoglobulin heavy chains (see Figure 24-14).
- Rearrangement of the V and J, as well as of the V, D, and J gene segments is controlled by conserved recombination signal sequences (RSSs), composed of heptamers and nonamers separated by 12- or 23-bp spacers (see Figure 24-15). Only those segments that have spacers of different length can rearrange successfully.
- The molecular machinery that carries out the rearrangement process includes recombinases (RAG1 and RAG2) made only by lymphocytes and numerous other proteins that participate in nonhomologous end joining of DNA molecules in other cell types as well.
- Antibody diversity is created by the random selection of Ig gene segments to be recombined (combinatorial joining) and by the ability of the heavy and light chains produced from rearranged Ig genes to associate with many different light chains and heavy chains, respectively (combinatorial association).
- Junctional imprecision generates additional antibody diversity at the joints of the gene segments joined during somatic recombination.
- Further antibody diversity arises after B cells encounter antigen as a consequence of somatic hypermutation, which can lead to the selection and proliferation of B cells producing high-affinity antibodies, a process termed affinity maturation.
- During B-cell development, heavy-chain genes are rearranged first, leading to expression of the pre-B cell receptor. Subsequent rearrangement of light-chain genes results in assembly of an IgM membrane-bound B-cell receptor (see Figure 24-17).
- Only one of the allelic copies of the heavy-chain locus and of the light-chain locus is rearranged (allelic exclusion), ensuring that a B cell expresses Ig with a single antigenic specificity.
- Polyadenylation of different poly(A) sites in an Ig primary transcript determines whether the membrane-bound or secreted form of an antibody is produced (see Figure 24-18).
- During an immune response, class switching allows B cells to adjust the effector functions of the immunoglobulins produced but retain their specificity for antigen (Figure 24-19).

24.4 The MHC and Antigen Presentation

Antibodies can recognize antigen without the involvement of any third-party molecules; the presence of antigen and antibody is sufficient for their interaction. Although antibodies contribute to the elimination of bacterial and viral pathogens, it is often necessary to destroy also the infected cells that serve as a source of new virus particles. This task is carried out by T cells with cytotoxic activity. These *cytotoxic T cells* make use of antigen-specific receptors whose genes are generated by mechanisms analogous to those used by B cells to generate immunoglobulin genes. However, antigen recognition is accomplished very differently by T cells than by B cells. The antigen-specific receptors on T cells recognize short snippets of protein antigens, presented to these receptors by membrane glycoproteins encoded by the **major histocompatibility complex (MHC).** Various antigen-presenting cells, in the course of their normal activity, digest pathogen-derived (and self) proteins and then "post" these protein snippets (peptides) to their cell surface in a physical complex with an MHC protein. T cells can inspect these complexes, and if they detect a pathogen-derived peptide, the T cells take appropriate action, which may include killing the cell that carries the MHC–peptide complex.

MHC proteins, which commonly are called **MHC molecules,** also facilitate communication between T cells and B cells. B cells do not usually engage in production of secreted antibodies unless they receive assistance from another subset of T cells, called helper T cells. These T cells also use antigen-specific receptors to recognize MHC-peptide complexes. In this section, we describe the MHC and the proteins it encodes, and then examine how these MHC molecules are involved in antigen recognition.

The MHC Determines the Ability of Two Unrelated Individuals of the Same Species to Accept or Reject Grafts

The major histocompatibility complex was discovered, as its name implies, as the genetic locus that controls acceptance or rejection of grafts. At a time when tissue culture had not yet been developed to the stage where tumor-derived cell lines could be propagated in the laboratory, investigators relied on serial passage in vivo of tumor tissue. It was quickly observed that a tumor that arose spontaneously in one inbred strain of mice could be propagated successfully in the strain in which it arose, but not in a genetically distinct line of mice. Genetic analysis soon showed that a single major locus was responsible for this behavior. Similarly, transplantation of healthy skin was feasible within the same strain of mice, but not when the recipient was of a genetically distinct background. Genetic analysis of transplant rejection likewise identified a single major locus that controlled acceptance or rejection, which is an immune reaction. As we now know, all vertebrates that possess an adaptive immune system have a genetic region that corresponds to the major histocompatibility complex as originally defined in the mouse.

An important step in discovering the functions of the MHC was development of mice strains *congenic* for the MHC. Congenic strains are genetically identical except for the locus or genetic region of interest. Figure 24-20 outlines

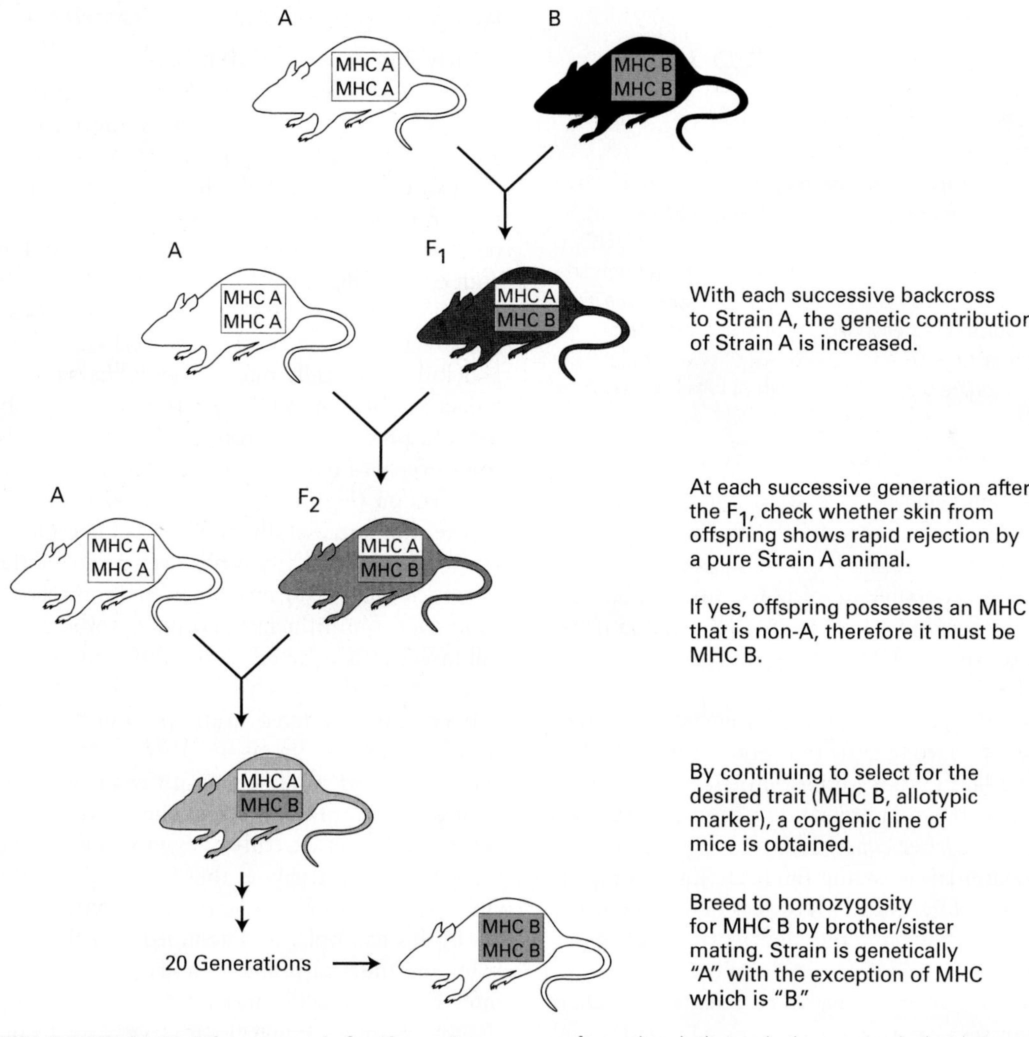
A
B
MHC A
MHC A
MHC B
MHC B
A
F_1
MHC A
MHC A
MHC A
MHC B
With each successive backcross to Strain A, the genetic contribution of Strain A is increased.
A
F_2
MHC A
MHC A
MHC A
MHC B
At each successive generation after the F_1, check whether skin from offspring shows rapid rejection by a pure Strain A animal.
If yes, offspring possesses an MHC that is non-A, therefore it must be MHC B.
MHC A
MHC B
By continuing to select for the desired trait (MHC B, allotypic marker), a congenic line of mice is obtained.
20 Generations
MHC B
MHC B
Breed to homozygosity for MHC B by brother/sister mating. Strain is genetically "A" with the exception of MHC which is "B."

▲ EXPERIMENTAL FIGURE 24-20 **Mice congenic for the major histocompatibility (MHC) are generated by crossing two histo-incompatible strains.** Strain A and strain B, which reject each other's grafts, are said to be histo-incompatible and differ at their MHC. The (A × B) F_1 progeny accept grafts from either parental strain. By backcrossing F_1 mice to one of the parental strains (e.g., strain A) for many generations, the contribution of strain A to the genetic material of the resulting offspring will increase. Breeding is performed such that only those animals that have retained the MHC of strain B are used for breeding. The presence of the strain B MHC is assessed by performing a skin graft onto a strain A recipient. Only if the B-type MHC is present will inbred mice of strain A reject the graft. By performing such backcrosses and assays for the B-type MHC for 20 or more generations, a strain of mice is obtained that is essentially A in its genetic make up, yet retains the MHC of strain B. These mice are said to be congenic for the MHC.

(a) Mouse MHC (H-2 complex)

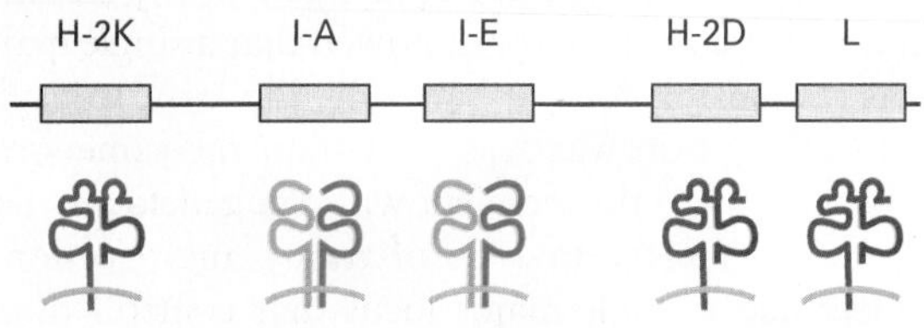

(b) Human MHC (HLA complex)

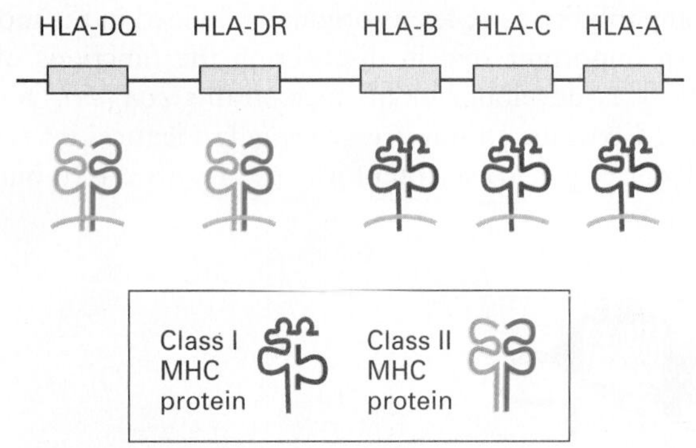

▲ **FIGURE 24-21 Organization of the major histocompatibility complex in mice and humans.** The major loci are depicted with schematic diagrams of their encoded proteins below. Class I MHC proteins are composed of a MHC-encoded single-pass transmembrane glycoprotein in noncovalent association with a small subunit, called β2-microglobulin, which is not encoded in the MHC and is not membrane bound. Class II MHC proteins consist of two nonidential single-pass transmembrane glycoproteins, both of which are encoded by the MHC.

how mice strains congenic for the MHC can be generated. Congenic strains are essential tools for assigning complex immunological functions to a particular locus, such as the MHC. As long as it is possible to select for a particular phenotypic trait in the form of an allelic marker (e.g., graft rejection in the case of the MHC), congenic strains may be produced for other loci.

In the mouse, the genetic region that encodes the antigens responsible for a strong graft rejection is called the *H-2 complex* (Figure 24-21a). The initial characterization of the MHC was followed by an appreciation of the genetic complexity of this region. After coarse mapping by standard genetic means (recombination within the MHC), the complete nucleotide sequence of the entire MHC was determined. The typical mammalian MHC contains dozens of genes, many encoding proteins of immunological relevance.

In humans, the discovery of the MHC relied on the characterization of antisera produced in patients who underwent multiple blood transfusions: Antigens expressed on the surface of the genetically nonidentical donor cells provoked an immune response in the recipient. The predominant target antigens recognized by these antisera are encoded by the human MHC, a genetic region also referred to as the *HLA complex* (Figure 24-21b). All vertebrate MHCs encode a highly homologous set of proteins, although the details of organization and gene content show considerable variation between species.

The human fetus may also be considered a graft: The fetus shares only half of its genetic material with the mother, the other half being contributed by the father. Antigens encoded by this paternal contribution may differ sufficiently from their maternal counterparts to elicit an immune response in the mother. In the course of pregnancy, fetal cells that slough off into the maternal circulation stimulate the maternal immune system to mount an antibody response against these paternal antigens. The antibodies recognize structures encoded by the human MHC. The fetus itself is spared rejection because of the specialized organization of the placenta, which prevents initiation of an immune response by the mother against fetal tissue.

The Killing Activity of Cytotoxic T Cells Is Antigen Specific and MHC Restricted

Clearly the function of MHC molecules is not to prevent the exchange of surgical grafts. MHC molecules play an essential role in the recognition of virus-infected cells by cytotoxic T cells; these cells also are called cytotoxic T lymphocytes (CTLs). In virus-infected cells, MHC molecules interact with protein fragments derived from viral pathogens and display these on the cell surface where cytotoxic T cells, charged with eliminating the infection, can recognize them.

Mice that have recovered from a particular virus infection are a ready source of cytotoxic T cells that can recognize and kill target cells infected with the same virus. The radioactive chromium (^{51}Cr) release assay can be used to detect the presence of cytotoxic T cells in single-cell suspensions prepared from the spleen of an animal that has cleared an infection (Figure 24-22a). If T cells are obtained from a mouse that successfully cleared an infection with influenza virus, cytotoxic activity is observed against influenza-infected target cells, but not against uninfected controls (Figure 24-22b). Moreover, the influenza-specific cytotoxic T cells will not kill target cells infected with a different virus, such as vesicular stomatitis virus. Cytotoxic T cells can even discriminate between closely related strains of influenza virus, and can do so with pinpoint precision: Differences of a single amino acid in the viral antigen may suffice to avoid recognition and killing by cytotoxic T cells. These experiments show that cytotoxic T cells are truly antigen-specific and do not simply recognize some attribute that is shared by all virus-infected cells, regardless of the identity of the virus.

In this example, it is assumed that the T cells harvested from an influenza-immune mouse are assayed on influenza-infected target cells derived from the identical strain of mouse (strain a). However, if target cells from a completely unrelated strain of mouse (strain b) are infected with the same strain of influenza and used as targets, the cytotoxic T cells from the strain a mouse are unable to kill the infected strain b target cells (see Figure 24-22b, **3** vs. **4**). It is therefore not sufficient that the antigen (an influenza-derived protein) is present; recognition by cytotoxic T cells is restricted by strain-specific elements. By making use of MHC congenic

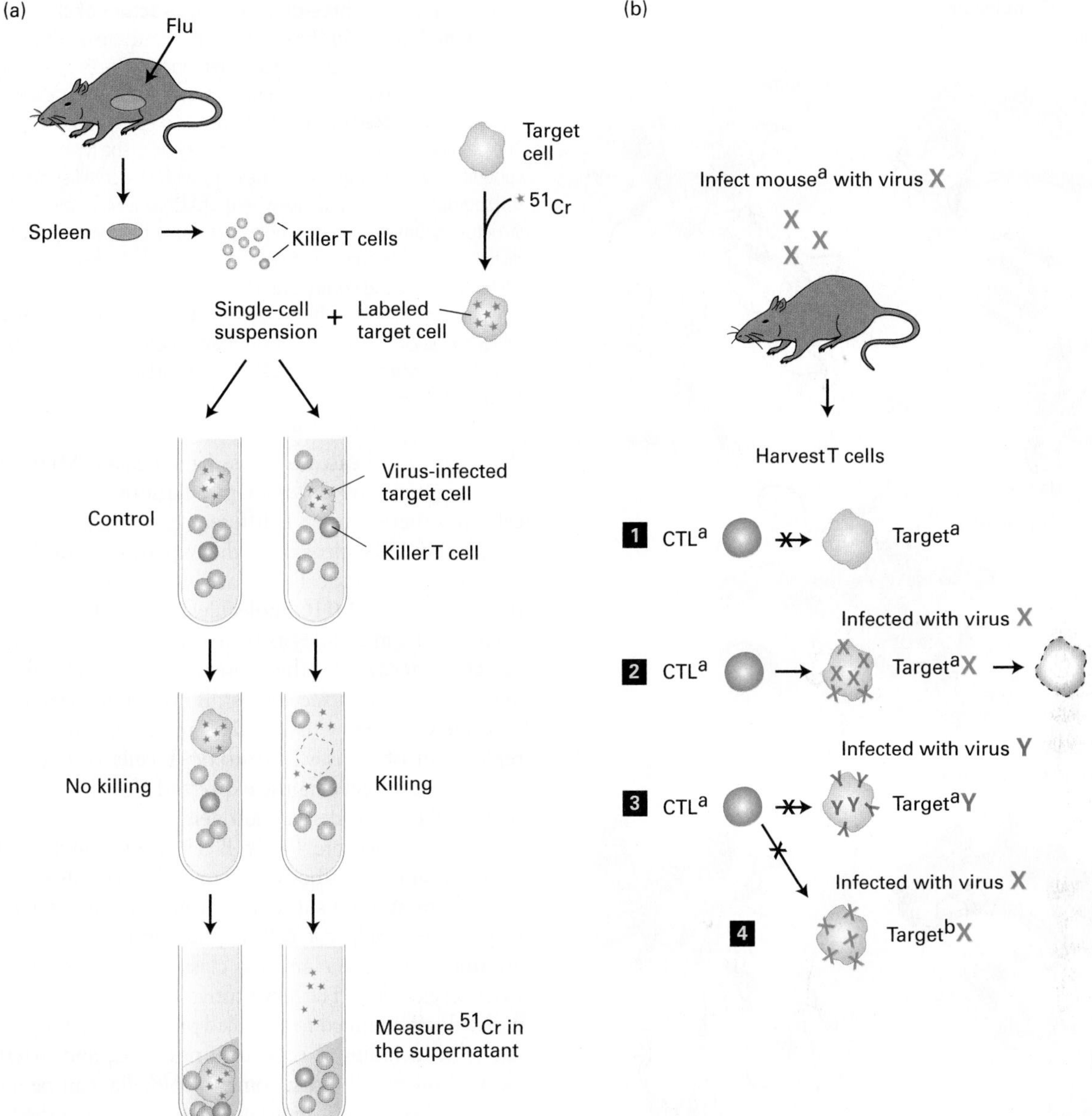

▲ EXPERIMENTAL FIGURE 24-22 **Chromium (^{51}Cr) release assay allows the direct demonstration of the cytotoxicity and specificity of cytotoxic T cells in a heterogeneous population of cells.** (a) A suspension of spleen cells, containing cytotoxic (killer) T cells is prepared from mice that have been exposed to a particular virus (e.g., influenza virus) and have cleared the infection. Target cells obtained from the same strain are infected with the identical virus or left uninfected. After infection, cellular proteins are labeled nonspecifically by incubation of the target-cell suspension with ^{51}Cr. Upon incubation of radiolabeled target cells with the suspension of T cells, killing of target cells results in release of the ^{51}Cr-labeled proteins. Uninfected target cells are not killed and retain their radioactive contents. Lysis of cells by cytotoxic T cells can therefore be readily detected and quantitated by measuring the radioactivity released into the supernatant. (b) Cytotoxic T lymphocytes (CTLs) harvested from mice that have been infected with virus X can be tested against various target cells to determine the specificity of CTL-mediated killing. CTLs capable of lysing virus X-infected target cells (2) cannot kill uninfected cells (1) or cells infected with a different virus, Y (3). When these CTLs are tested on virus X-infected targets from a strain that carries an altogether different MHC type (b), again no killing is observed (4). Cytotoxic T-cell activity is thus virus specific and restricted by the MHC.

strains, the genes that encode these restricting elements were mapped to the MHC. Thus cytotoxic T cells from one mouse strain immune to influenza will kill influenza-infected target cells from another strain only if the two strains match at the MHC for the relevant MHC molecules. This phenomenon is therefore known as *MHC restriction.*

T Cells with Different Functional Properties Are Guided by Two Distinct Classes of MHC Molecules

The MHC encodes two types of glycoproteins essential for immune recognition, commonly called *class I* and *class II*

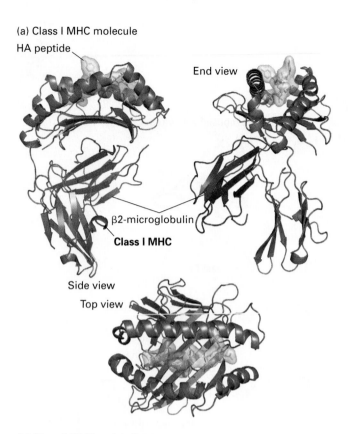

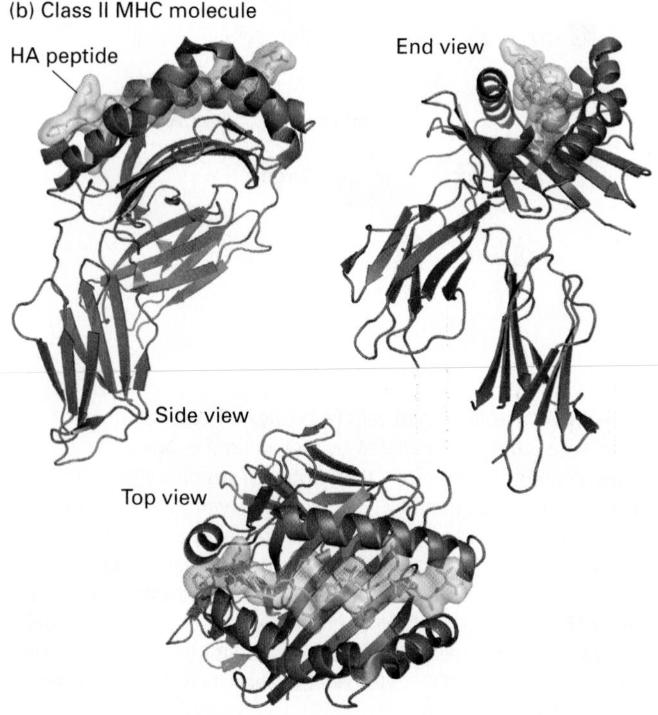

◄ **FIGURE 24-23 Three-dimensional structure of class I and class II MHC molecules.** (a) Shown here is the structure of a class I MHC molecule with bound peptide as determined by x-ray crystallography. The portion of a class I MHC molecule that binds peptide consists of a β sheet composed of eight β strands and flanked by two α helices. The peptide-binding cleft is formed entirely from the MHC-encoded large subunit, which associates noncovalently with the small subunit (β2-microglobulin) encoded elsewhere. (b) Class II MHC molecules are structurally similar to class I molecules, but with several important distinctions. Both the α and β subunits of class II MHC molecules are MHC encoded and contribute to formation of the peptide-binding cleft. The peptide-binding cleft of class II MHC molecules accommodates a wider range of peptide sizes than that of class I molecules. [Part (a) based on D. N. Garboczi, 1996 *Nature* **384:**134, Part (b) based on J. Hennecke et al., 2000, *EMBO* **19:** 5611.]

MHC molecules. A comparison of the genetic maps of the mouse and human MHCs shows the presence of several class I MHC genes and several class II MHC genes, even though their arrangement shows variation between the different species (see Figure 24-21). In addition to the class I and class II MHC molecules, the MHC encodes key components of the antigen-processing and presentation machinery. Finally, the typical vertebrate MHC also encodes key components of the complement cascade. Class I and Class II MHC molecules are recognized by different populations of immune system cells and therefore serve different functions.

As should be clear from the experiments outlined in Figure 24-22, cytotoxic T cells are guided in the recognition of their targets by MHC molecules. These T cells mostly use class I MHC molecules as their restriction elements, and also are characterized by the presence of the CD8 glycoprotein marker on their surface. Most, if not all, nucleated cells constitutively express class I MHC molecules and can support replication of viruses. Cytotoxic T cells recognize and kill the infected targets via the expressed class I MHC molecules that display virus-derived antigen.

As mentioned previously, B cells do not undergo final differentiation into antibody-secreting plasma cells without assistance from another subset of T cells, the *helper T cells* (or T helper cells). Helper T cells express on their surface the CD4 glycoprotein marker and use class II MHC molecules as restriction elements. The constitutive expression of class II MHC molecules is confined to so-called *professional* antigen-presenting cells, including B cells, dendritic cells, and macrophages. (Several other cell types, some in epithelia, can be induced to express class II MHC molecules, but we will not discuss these.)

The two major groups of functionally distinct T lymphocytes—cytotoxic T cells and helper T cells—can thus be distinguished based on the unique profile of membrane proteins displayed at the cell surface and by the MHC molecules used as restriction elements:

- Cytotoxic T cells: CD8 marker; class I MHC restricted
- Helper T cells: CD4 marker; class II MHC restricted

Both CD4 and CD8 belong to the immunoglobulin (Ig) superfamily of proteins, which all include one or more Ig domains. The B-cell and T-cell receptors, polymeric IgA receptor, and many cell-adhesion molecules (Chapter 19) also belong to the Ig superfamily. The molecular basis for the strict correlation between expression of CD8 and utilization of class I MHC molecules or between expression of CD4 and utilization of class II MHC molecules as the restriction element will become evident once the structure and mode of action of MHC molecules has been described.

MHC Molecules Bind Peptide Antigens and Interact with the T-Cell Receptor

Both class I and class II MHC molecules are highly *polymorphic;* that is, many allelic variants exist among individuals of the same species. Both classes of MHC molecules also are structurally similar in many respects as are their interactions with peptides and the T-cell receptor (Figure 24-23).

Class I MHC Molecules Class I MHC molecules belong to the Ig superfamily and consist of two polypeptides. The larger subunit is a type I membrane glycoprotein (see Figure 13-10) encoded by the MHC. The smaller β2-microglobulin subunit is not encoded by the MHC and corresponds in structure to a free Ig domain. Originally purified from human leukocytes by digestion with papain, which releases the extracellular portion of the class I MHC molecules in intact form, these proteins are now produced by recombinant DNA technology procedures and have become important tools for the detection of antigen-specific T cells.

Implicit in the notion that MHC molecules are the targets of graft rejection is their structural variation, attributable to inherited variation (genetic polymorphism): If a recipient rejects a graft, the recipient's immune system must be capable of distinguishing unique features of the donor MHC molecules present on the graft. In fact, the genes encoded by the MHC are among the most polymorphic currently known, with over 2000 distinct allelic products identified in humans. The class I MHC molecules in humans are encoded by the HLA-A, HLA-B and HLA-C loci (see Figure 24-21). Each unique allele is designated by a numerical suffix for that particular locus (e.g., HLA-A2 and HLA-A28 represent two distinct HLA-A allelic products). In the mouse, the class I MHC molecules are encoded by the H-2K and H-2D loci. A superscript is used to denote the identity of the allelic products (e.g., H-$2K^b$ and H-$2K^k$ represent two allelic variants of the H-2K locus product).

The three-dimensional structure of class I MHC molecules reveals two membrane-proximal Ig-like domains. These domains support an eight-stranded β-pleated sheet topped by two α helices. Jointly the β sheet and the helices create a cleft, closed at both ends, in which a peptide binds (see Figure 24-23a). The mode of peptide-binding by a class I MHC molecule requires a peptide of rather fixed length, usually 8 to 10 amino acids, so that the ends of the peptide can be tucked into pockets that accommodate the charged amino and carboxyl groups at the termini. Further, the peptide is anchored into the peptide-binding cleft by means of a small number of amino acid side chains, each of which is accommodated by a pocket in the MHC molecule that neatly fits that particular amino acid residue (Figure 24-24a). On average, two such "specificity pockets" must be filled correctly to allow stable peptide binding. One of these pockets often involves the very C-terminal residue; a second specificity pocket accommodates a residue more central to the peptide. In this manner, a given MHC molecule can accommodate a large number of peptides of diverse sequence, as long as the "anchor" requirements are fulfilled.

The polymorphic residues that distinguish one allelic MHC molecule from another are mostly located in and around the peptide-binding cleft. These residues therefore determine the architecture of the peptide-binding pocket and hence the specificity of peptide binding. Further, these polymorphic residues affect the surface of the MHC molecule and hence the points of contact with the T-cell receptor. A T-cell receptor designed to interact with one particular class I MHC allele will therefore, as a rule, not interact with unrelated MHC molecules because of their different surface architecture (Figure 24-24b). The CD8 marker functions as

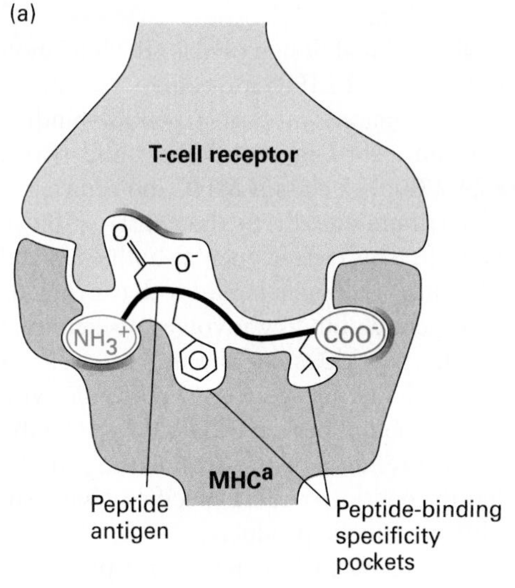

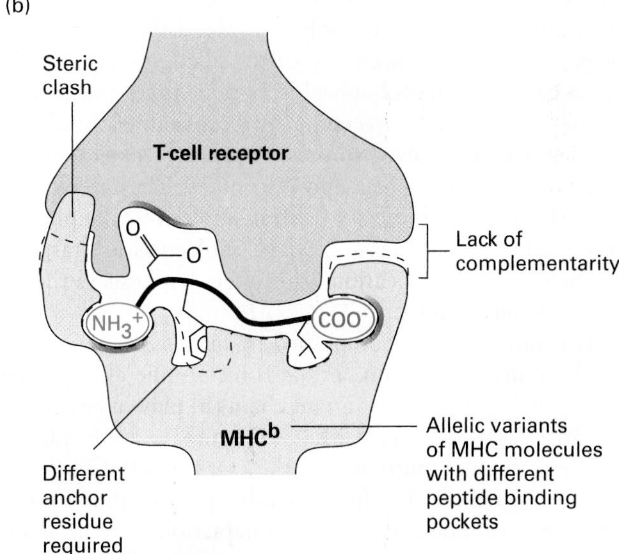

▲ **FIGURE 24-24 Peptide binding and MHC restriction.** (a) Peptides that bind to class I molecules are on average 8–10 residues in length, require proper accommodation of the termini, and include two or three residues that are conserved (anchor residues). Positions in class I molecules that distinguish one allele from another (polymorphic residues) occur in and around the peptide-binding cleft. The polymorphic residues in the MHC affect both the specificity of peptide binding and the interactions with T-cell receptors. Successful "recognition" of a peptide antigen–MHC complex by a T-cell receptor requires a good fit among the receptor, peptide, and MHC molecule. (b) Steric clash and a lack of complementarity between anchor residues and the MHC molecule prevent proper binding. T-cell receptors are thus restricted to specific peptide-MHC products.

a co-receptor, binding to conserved portions of the class I MHC molecules. The presence of CD8 thus "sets" the restriction specificity of any mature T cell that bears it.

Class II MHC Molecules The two subunits (α and β) of class II MHC molecules are both type I membrane glycoproteins and belong to the Ig superfamily. The typical mammalian MHC contains several loci that encode class II MHC molecules (see Figure 24-21). Like the large subunit of class I molecules, both the α and β subunits of class II molecules show genetic polymorphism.

The basic three-dimensional design of class II MHC molecules resembles that of class I MHC molecules: Two membrane-proximal Ig-like domains support a peptide-binding portion composed of an eight-stranded β sheet and two α helices (see Figure 24-23b). For class II MHC molecules, the α and β subunits contribute equally to the construction of the peptide-binding cleft. This cleft is open at both ends and thus supports the binding of longer peptides that protrude from it. The mode of peptide binding involves pockets that accommodate specific peptide side chains, as well as contacts between side chains of the MHC molecule with main-chain atoms of the bound peptide. Class II MHC polymorphisms mainly affect residues in and around the peptide-binding cleft, so that the peptide-binding specificity will usually differ among different allelic products.

A T-cell receptor designed to interact with a particular class II MHC molecule will not, as a rule, interact with a different allelic molecule, not only because of the difference in the peptide-binding specificity of the allelic molecules, but also because of the polymorphisms that affect the contact residues with the T-cell receptor. As discussed below, class II MHC molecules evolved to present peptides generated predominantly in endosomes and lysosomes. The interactions between a peptide and class II MHC molecule take place in these organelles, and class II MHC molecules are targeted specifically to those locations after their synthesis in the endoplasmic reticulum.

This targeting is accomplished by means of a chaperone called the invariant chain, a type II membrane glycoprotein (see Figure 13-10). The invariant chain (Ii) plays a key role in the early stages of class II MHC biosynthesis by forming a trimeric structure onto which the class II MHC $\alpha\beta$ heterodimers assemble. The final assembly product thus consists of nine polypeptides: $(\alpha\beta\text{Ii})_3$. The interaction between Ii and the $\alpha\beta$ heterodimer involves a stretch of Ii called the CLIP segment, which occupies the class II MHC peptide-binding cleft. Once the $(\alpha\beta\text{Ii})_3$ complex is assembled, the complex enters the secretory pathway and is diverted to endosomes and lysosomes at the *trans*-Golgi network (see Figure 14-1). The signals responsible for this diversion are carried by the Ii cytoplasmic tail and do not obviously conform to the endosomal targeting or retrieval signals commonly found on lysosomal membrane proteins. Some of the $(\alpha\beta\text{Ii})_3$ complexes are directed straight to the cell surface from which they may be internalized, but the vast majority end up in late endosomes.

As we saw for class I MHC molecules and their CD8 co-receptor, the CD4 co-receptor recognizes conserved features on class II MHC molecules. Any mature T cell that bears the CD4 co-receptor uses class II MHC molecules for antigen recognition.

Antigen Presentation Is the Process by Which Protein Fragments Are Complexed with MHC Products and Posted to the Cell Surface

The process by which foreign materials enter the immune system is the key step that determines the eventual outcome of a response. A successful adaptive immune response, which includes the production of antibodies and the generation of helper and cytotoxic T cells, cannot unfold without the involvement of professional antigen-presenting cells. Professional antigen-presenting cells include dendritic cells and macrophages, both bone marrow-derived, and B cells. It is these cells that acquire antigen, process it, and then display it in a form that can be recognized by T cells. The pathway by which antigen is converted into a form suitable for T-cell recognition is referred to as *antigen processing and presentation*.

The class I MHC pathway focuses predominantly on presentation of proteins synthesized by the cell itself, and the class II MHC pathway is centered on materials acquired from outside the antigen-presenting cell. This distinction, however, is by no means absolute. Together, the class I and class II pathways of antigen processing and presentation sample all of the compartments that need to be surveyed for the presence of pathogens.

Antigen processing and presentation in both the class I and II pathways may be divided into six discrete steps that are useful in comparing the two pathways: (1) acquisition of antigen, (2) tagging the antigen for destruction, (3) proteolysis, (4) delivery of peptides to MHC molecules, (5) binding of peptide to a MHC molecule, and (6) display of the peptide-loaded MHC molecule on the cell surface. In the next two sections, we describe the molecular details of each pathway.

Class I MHC Pathway Presents Cytosolic Antigens

Figure 24-25 summarizes the six steps in the class I MHC pathway using a virus-infected cell as an example. The following discussion describes the events that occur during each step:

1 *Acquisition of Antigen:* In the case of a virus infection, acquisition of antigen is usually synonymous with the infected state. Viruses rely on host protein synthesis to generate the building blocks for new virions. This includes the synthesis of viral cytosolic proteins as well as membrane proteins. Protein synthesis, unlike DNA replication, is an error-prone process, with 10–30 percent of newly initiated polypeptide chains being terminated prematurely or suffering from errors (misincorporation of amino acids, frame shifts, improper or delayed folding). These mistakes in protein synthesis affect both the host cell's own proteins and those specified by viral genomes. Such error-containing proteins must be rapidly removed so as not to clog up the

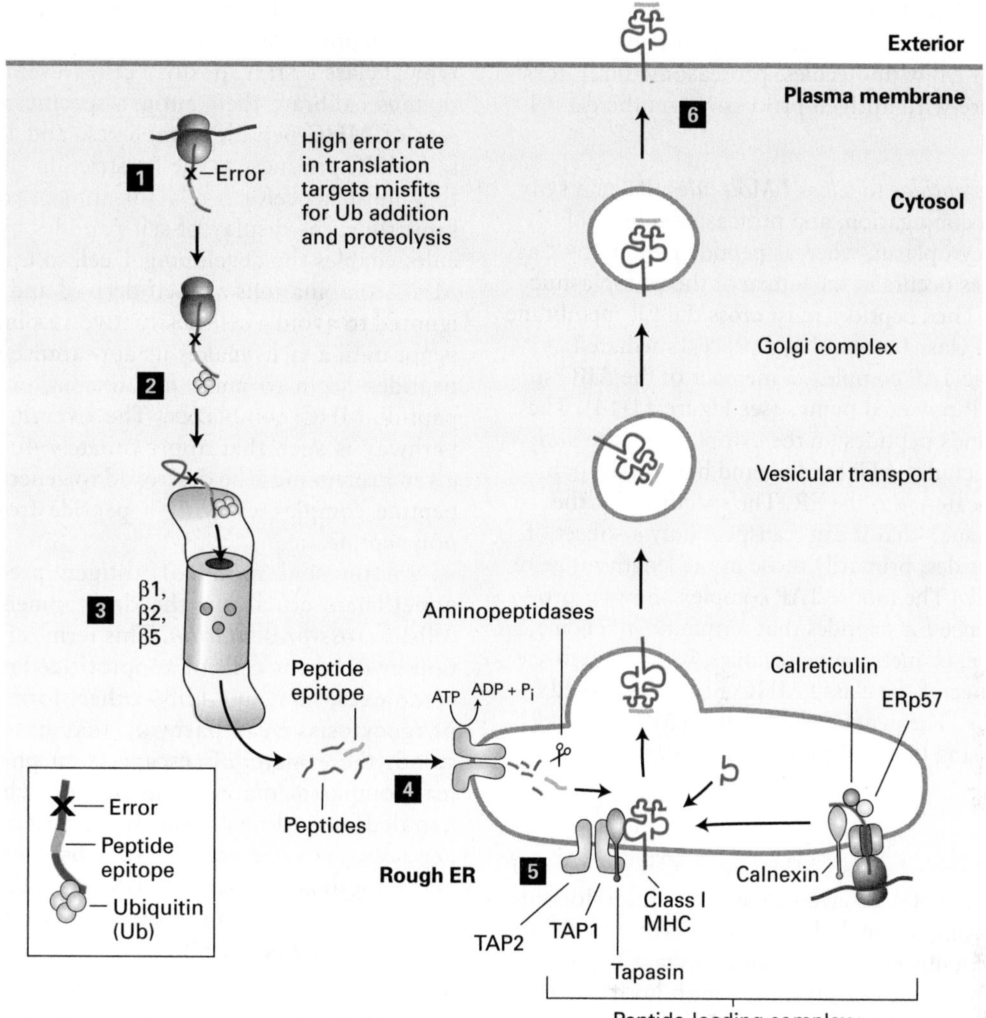

▲ **FIGURE 24-25 Class I MHC pathway of antigen processing and presentation.** Step **1**: Acquisition of antigen is synonymous with the production of proteins with errors (premature termination, misincorporation). Step **2**: Misfolded proteins are targeted for degradation through conjugation with ubiquitin. Step **3**: Proteolysis is carried out by the proteasome. In cells exposed to interferon γ, the catalytically active β subunits of the proteasome are replaced by interferon-induced immune-specific β subunits. Step **4**: Peptides are delivered to the interior of the endoplasmic reticulum (ER) via the dimeric TAP peptide transporter. Step **5**: Peptide is loaded onto newly made class I MHC molecules within the peptide-loading complex. Step **6**: The fully assembled class I MHC–peptide complex is transported to the cell surface via the secretory pathway. See text for details.

cytoplasm or engage partner proteins in nonproductive interactions. The rate of cytosolic proteolysis must be matched to the rate at which mistakes in protein synthesis occur. These rapidly degraded proteins are an important source of antigen peptides destined for presentation by class I MHC molecules. With the exception of cross-presentation (discussed below), the class I MHC pathway results in the formation of peptide-MHC complexes in which the peptides are derived from proteins synthesized by the class I MHC–bearing cell itself.

2 *Tagging the Antigen for Destruction:* For the most part, the ubiquitin conjugation system is responsible for tagging a protein for destruction (see Chapter 3, p. 88). Ubiquitin conjugation is tightly regulated.

3 *Proteolysis:* Ubiquitin-conjugated proteins are destroyed by proteasomal proteolysis. The proteasome is a highly processive protease that engages its substrates and, without the release of intermediates, yields final digestion products, peptides in the size range of 3–20 amino acids (see Figure 3-29). During the course of an inflammatory response and in response to interferon γ, the three catalytically active β subunits (β1, β2, β5) of the proteasome can be replaced by three immune-specific subunits: β1i, β2i and β5i. The β1i, β2i and β5i subunits are encoded in the MHC. The net result of this replacement is the generation of the *immunoproteasome,* the output of which is matched to the requirements for peptide binding by class I MHC molecules. The immunoproteasome adjusts the average length of the peptides produced, as well as the sites at which cleavage occurs. Given the central role of the proteasome in the generation of peptides presented by class I MHC molecules, proteasome inhibitors potently interfere with antigen processing via the class I MHC pathway.

4 *Delivery of Peptides to Class I Molecules:* Protein synthesis, ubiquitin conjugation, and proteasomal proteolysis all occur in the cytoplasm, whereas peptide binding by class I MHC molecules occurs in the lumen of the endoplasmic reticulum (ER). Thus peptides must cross the ER membrane to gain access to class I molecules, a process mediated by the heterodimeric TAP complex, a member of the **ABC superfamily** of ATP-powered pumps (see Figure 11-14). The TAP complex binds peptides on the cytoplasmic face and, in a cycle that includes ATP binding and hydrolysis, peptides are translocated into the ER. The specificity of the TAP complex is such that it can transport only a subset of all cytosolic peptides, primarily those in the length range of 5–10 amino acids. The mouse TAP complex shows a pronounced preference for peptides that terminate in leucine, valine, isoleucine, or methionine residues, which match the binding preference of the class I MHC molecules served by the TAP complex. The genes encoding the TAP1 and TAP2 subunits composing the TAP complex are located in the MHC.

5 *Binding of Peptides to Class I Molecules:* Within the ER, newly synthesized class I MHC molecules are part of a multiprotein complex referred to as the peptide-loading complex. This complex includes two chaperones (calnexin and calreticulin) and the oxidoreductase Erp57. Another chaperone (tapasin) interacts with both the TAP complex and the class I MHC molecule about to receive peptide. The physical proximity of TAP and the class I MHC molecule is maintained by tapasin. Once peptide loading has occurred, a conformational change releases the loaded class I MHC molecule from the peptide-loading complex.

6 *Display of Class I MHC–Peptide Complexes at the Cell Surface:* Once peptide loading is complete, the class I MHC–peptide complex is released from the peptide-loading complex and enters the constitutive secretory pathway (see Figure 14-1). Class I MHC molecules, depending on the species and allelic identity, contain between one and three N-linked oligosaccharides, which receive extensive modifications in the Golgi complex. Transfer from the Golgi to the cell surface is rapid and completes the biosynthetic pathway of a class I MHC–peptide complex.

The entire sequence of events in the class I pathway occurs constitutively in all nucleated cells, which express class I MHC molecules and the other required proteins or can be induced to do so. In the absence of a virus infection, protein synthesis and proteolysis continuously generate a stream of peptides that are loaded onto class I MHC molecules. Healthy, normal cells therefore display on their surface a representative selection of peptides derived from host proteins. There may be several thousand distinct MHC-peptide combinations displayed at the surface of a typical class I MHC positive cell. Developing T cells in the thymus calibrate their antigen-specific receptors on these sets of MHC-peptide complexes, and learn to recognize self-MHC products as the "restriction elements" on which they must henceforth rely for antigen recognition. At the same time, the display of self-peptides by self-MHC molecules enables the developing T cell to learn which peptide-MHC combinations are self-derived and must therefore be ignored to avoid a self-destructive autoimmune reaction. It is not until a virus makes its appearance that virus-derived peptides begin to make a contribution to the display of peptide-MHC complexes. The overall efficiency of this pathway is such that approximately 4000 molecules of a given protein must be destroyed to generate a single MHC-peptide complex carrying a peptide from that particular polypeptide.

An unusual mode of antigen presentation that is nonetheless crucial in the development of cytotoxic T cells is *cross-presentation.* This term refers to the acquisition by dendritic cells of apoptotic cell remnants, immune complexes, and possibly other forms of antigen by phagocytosis. By a pathway that has yet to be understood, these materials escape from phagosomal/endosomal compartments into the cytosol, where they are then handled according to the steps described above. Only dendritic cells are capable of cross-presentation, and so allow the loading of class I MHC molecules complexed with peptides that derive from cells other than the antigen-presenting cell itself.

Class II MHC Pathway Presents Antigens Delivered to the Endocytic Pathway

Although class I MHC and class II MHC molecules show a striking structural resemblance, the manner in which the two classes acquire peptide and their function in immune recognition differ greatly. Whereas the primary function of class I MHC molecules is to guide CD8-bearing cytotoxic T cells to their target cells, class II MHC molecules serve to guide CD4-bearing helper T cells to the cells with which they interact, primarily professional antigen-presenting cells.

As noted previously, class II MHC molecules are expressed primarily by professional antigen-presenting cells: dendritic cells and macrophages, which are phagocytic, and B cells, which are not. Hence, the class II MHC pathway of antigen processing and presentation generally occurs only in these cells. The steps in this pathway are depicted in Figure 24-26 and described below:

1 *Acquisition of Antigen:* In the class II MHC pathway, antigen is acquired by pinocytosis, phagocytosis, or receptor-mediated endocytosis. Pinocytosis, which is rather nonspecific, involves the delivery, by a process of membrane invagination and fission, of a volume of extracellular fluid and the molecules dissolved therein. **Phagocytosis,** the

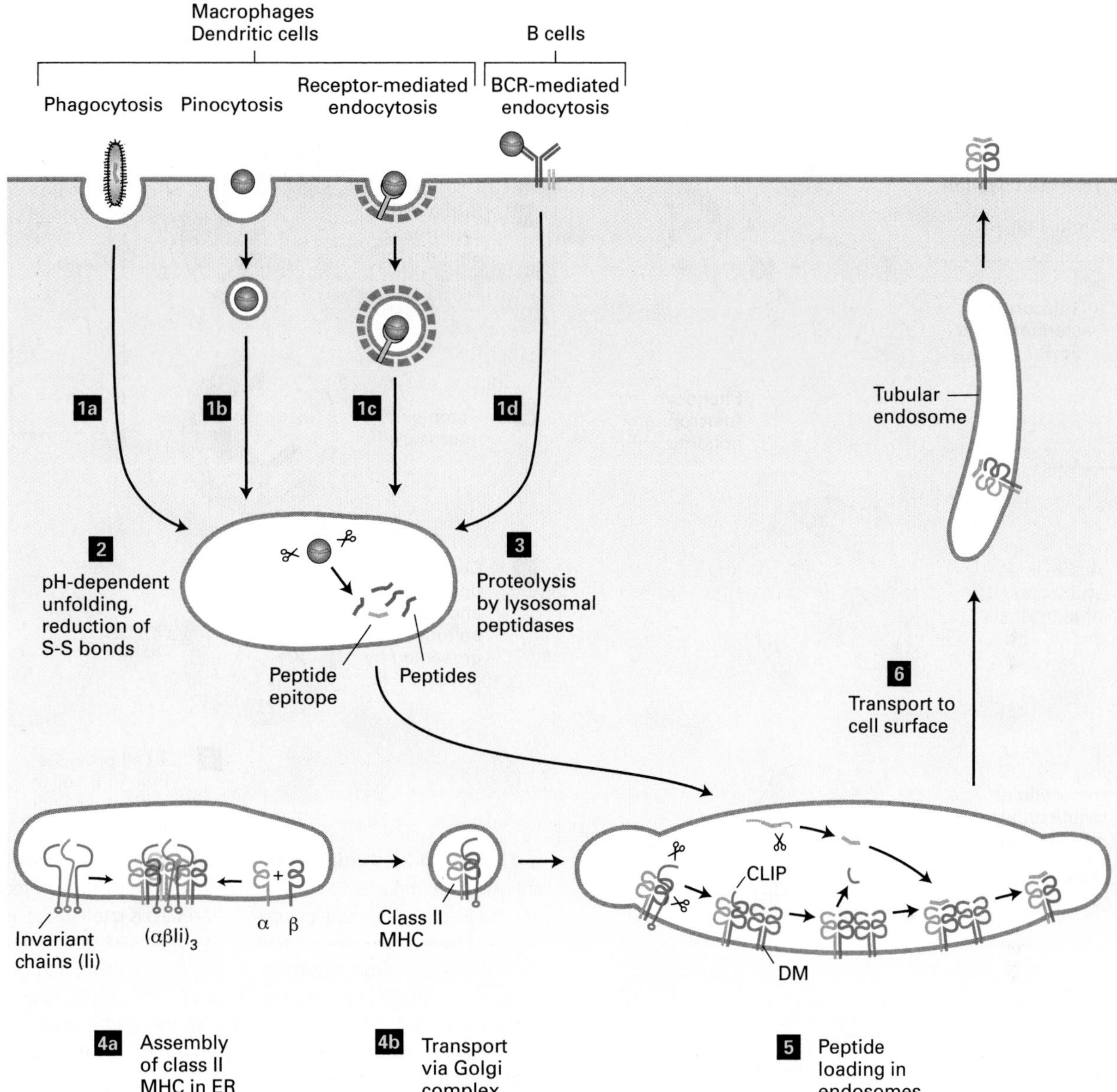

▲ **FIGURE 24-26 Class II MHC pathway of antigen processing and presentation.** Step **1**: Particulate antigens are acquired by phagocytosis, and nonparticulate antigens by pinocytosis or endocytosis. Step **2**: Exposure of antigen to the low pH and reducing environment of endosomes and lysosomes prepares the antigen for proteolysis. Step **3**: The antigen is broken down by various proteases in endosomal and lysosomal compartments. Step **4**: Class II MHC molecules, assembled in the ER from their subunits, are delivered to endosomal/lysosomal compartments by means of signals contained in the associated invariant (Ii) chain. This delivery targets late endosomes, lysosomes, and early endosomes, ensuring that class II MHC molecules are exposed to the products of proteolytic breakdown of antigen along the entire endocytic pathway. Step **5**: Peptide loading is accomplished with the assistance of DM, a class II MHC-like chaperone protein. Step **6**: Display of peptide loaded Class II MHC molecules at the cell surface. See text for details.

ingestion of particulate materials, such as bacteria, viruses, and remnants of dead cells, involves extensive remodeling of the actin-based cytoskeleton to accommodate the incoming particle. Although phagocytosis may be initiated by specific receptor-ligand interactions, these are not always required: Even latex particles can be ingested very efficiently by macrophages. In the process of opsonization, pathogens decorated by antibodies and certain complement components are targeted to macrophages and dendritic cells, recognized by cell-surface receptors for complement components or the Fc portion of immunoglobulins, and then phagocytosed (Figure 24-27). Macrophages and dendritic cells also express several types of nonspecific receptors (e.g., Toll-like receptors, scavenger receptors) that recognize molecular patterns on nonparticulate antigens; the bound antigen is internalized by **receptor-mediated endocytosis.** B cells, which are not phagocytic, also can acquire antigen by receptor-mediated endocytosis using their

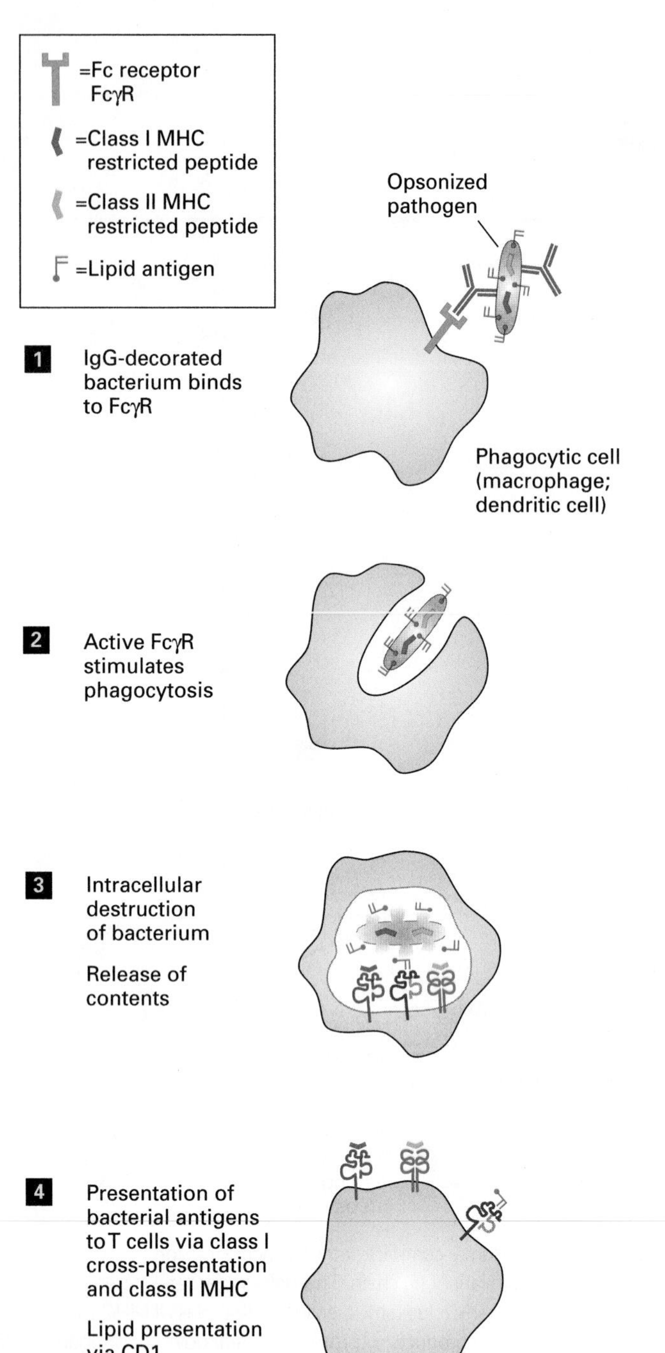
=Fc receptor FcγR
=Class I MHC restricted peptide
=Class II MHC restricted peptide
=Lipid antigen
Opsonized pathogen
1 IgG-decorated bacterium binds to FcγR
Phagocytic cell (macrophage; dendritic cell)
2 Active FcγR stimulates phagocytosis
3 Intracellular destruction of bacterium
Release of contents
4 Presentation of bacterial antigens to T cells via class I cross-presentation and class II MHC
Lipid presentation via CD1

▲ **FIGURE 24-27 Presentation of opsonized antigen by phagocytic cells.** By means of Fc receptors such as FcγR displayed on their cell surface, specialized phagocytic cells such as macrophages or dendritic cells can bind and ingest pathogens that have been decorated with antibodies (opsonization). After digestion of the phagocytosed particle (e.g., immune complex, bacterium, virus), some of the peptides produced, including fragments of the pathogen (orange) are loaded onto class II MHC molecules. Class II MHC–peptide complexes displayed at the surface allow activation of T cells specific for these MHC–peptide combinations. Lipid antigens are delivered to the class I MHC-like molecule CD1, whose binding site is specialized to accommodate lipids. Certain pathogen-derived peptides (purple) may be delivered to class I MHC products by means of cross-presentation. The mechanisms that underlie cross-presentation remain to be clarified.

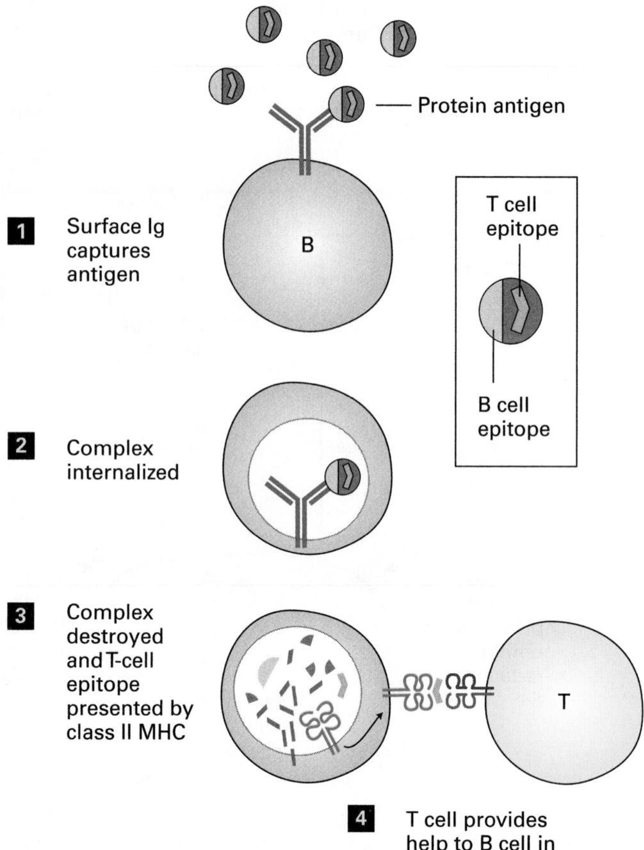
Protein antigen
1 Surface Ig captures antigen
B
T cell epitope
B cell epitope
2 Complex internalized
3 Complex destroyed and T-cell epitope presented by class II MHC
T
4 T cell provides help to B cell in antigen-specific fashion

▲ **FIGURE 24-28 Antigen presentation by B cells.** B cells bind antigen, even if present at low concentration, to their B-cell receptors, or surface Ig. The immune complex that results is internalized and then delivered to endosomal/lysosomal compartments where it is destroyed. Peptides liberated from the immune complex, including fragments of the protein antigen, are displayed as class II MHC–peptide complexes at the cell surface. CD4 T cells specific for the displayed complex can now provide help to the B cell. This help is MHC restricted and antigen specific.

antigen-specific B-cell receptors (surface immunoglobulin) (Figure 24-28). Finally, cytosolic antigens may enter the class II MHC pathway via autophagy (see Figure 14-35). After formation of the autophagic cup, an autophagic vesicle is formed. These vesicles are of a size that can accommodate damaged organelles, and sizable quantities of cytoplasm may be encapsulated in the process. The resulting autophagosomes are destined to fuse with lysosomes, where the contents of the autophagosome then become available for digestion by lysosomal proteases.

2 *Tagging Antigen for Destruction:* Proteolysis is required to convert intact protein antigen into peptides of a size suitable for binding to Class II MHC molecules. Protein antigens are tagged for degradation by progressive unfolding, brought about by the drop in pH as proteins progress along the endocytic pathway. The pH of the extracellular environment is around pH 7.2, and that in early endosomes between pH 6.5 and 5.5; in late endosomes and lysosomes the pH may drop to pH 4.5. ATP-powered V-class proton pumps in the endosomal and lysosomal membranes are responsible for this acidification (see Figure 11-9). Proteins stable at neutral pH tend to unfold when exposed to extremes of pH, through rupture of hydrogen bonds and destabilization of salt bridges. Further, the environment in the endosomal/lysosomal compartments is reducing, with lysosomes attaining a concentration of reducing equivalents in the millimolar range. Reduction of disulfide bonds that stabilize many extracellular proteins is catalyzed also by a thioreductase inducible by exposure to IFNγ. The combined action of low pH and reducing environment prepares the antigens for proteolysis.

3 *Proteolysis:* Degradation of proteins in the class II MHC pathway is carried out by a large set of lysosomal proteases, collectively referred to as cathepsins, which are either cysteine or aspartyl proteases. Other proteases, such as asparagine-specific endoprotease, also may contribute to proteolysis. A wide range of peptide fragments is produced, including some that can bind to class II MHC molecules. The lysosomal proteases operate optimally at the acidic pH within lysosomes. Consequently, agents that inhibit the activity of V-class proton pumps interfere with antigen processing, as do inhibitors of lysosomal proteases.

4 *Delivery of Peptides to Class II Molecules:* Recall that most class II MHC molecules synthesized in the endoplasmic reticulum are directed to late endosomes. Because the peptides generated by proteolysis reside in the same topological space as the class II MHC molecules themselves, the delivery of peptide to class II MHC molecules does not require traversal of a membrane. The sole requirement, therefore, is to allow peptides and class II MHC molecules to meet. This is accomplished in the course of biosynthesis by a sorting step that is dependent on the invariant chain (Ii) and ensures delivery of the class II $(\alpha\beta\text{Ii})_3$ complex to endosomal compartments.

5 *Binding of Peptides to Class II Molecules:* The $(\alpha\beta\text{Ii})_3$ complex delivered to endosomal compartments is incapable of binding peptide because the peptide-binding cleft in the class II molecule is occupied by Ii. The same proteases that act on internalized antigens and degrade them into peptides act also on the (αβIi)3 complex, resulting in removal of Ii with the exception of a small portion called the CLIP segment. Because it is firmly lodged in the class II MHC peptide-binding cleft, CLIP is resistant to proteolytic attack. The class II MHC molecules themselves are also resistant to unfolding and proteolytic attack under the conditions that prevail in the endocytic pathway. The CLIP segment is removed through interaction of the αβCLIP complex with a chaperone, DM. Although the DM protein is MHC-encoded and structurally very similar to class II molecules, it is unable to bind peptides. Newly formed class II MHC–peptide complexes are themselves susceptible to further "editing" by DM, until the class II molecule acquires a peptide so stably bound that it cannot be dislodged by interaction of the complex with DM. The resulting class II MHC–peptide complexes are extremely stable, with estimated half lives in excess of 24 hours.

6 *Display of Class II MHC –Peptide Complexes at the Cell Surface:* The newly generated class II MHC–peptide complexes are localized mostly in late endosomal compartments, which include multivesicular endosomes (or bodies) (see Figure 14-33). Recruitment of the internal vesicles of the multivesicular bodies to the delimiting membrane expands their surface area; by a process of tubulation along tracks of microtubules, these compartments elongate and ultimately deliver class II MHC–peptide complexes to the surface by membrane fusion. These events are tightly regulated: Tubulation and delivery of class II MHC molecules to the surface are enhanced in dendritic cells and macrophages following their activation in response to signals, such as bacterial lipopolysaccharide (LPS), which is detected by their Toll-like receptors.

For professional antigen-presenting cells, the above steps are constitutive—happening all the time—but they can be modulated by exposure to microbial agents and cytokines.

KEY CONCEPTS OF SECTION 24.4

The MHC and Antigen Presentation

- The MHC, discovered as the genetic region responsible for acceptance or rejection of grafts, encodes two major classes of type I membrane glycoproteins, class I and class II MHC molecules. These proteins are highly polymorphic, occurring in many allelic variations in an outbred population (see Figure 24-21).

- The function of MHC products is to bind peptides and display them to antigen-specific receptors on T cells. Class I MHC molecules are found on most nucleated cells, whereas the expression of class II MHC molecules is confined largely to professional antigen-presenting cells such as dendritic cells, macrophages, and B cells.

- The organization and structure of class I and class II MHC molecules is similar and includes a region specialized for binding a wide variety of different peptides (see Figure 24-23).

- Different allelic variants of MHC molecules bind different sets of peptides, because the differences that distinguish one allele from another include residues that define the architecture of the peptide-binding cleft (see Figure 24-24). Allelic residues also include residues contacted by the T-cell receptor for antigen. Thus different allelic variants of an MHC molecule, even if they were to bind the identical peptide, will not usually react with a T-cell receptor designed to interact with only one of the allelic MHC molecules. This phenomenon is called MHC restriction.

- Class I and class II MHC molecules sample different intracellular compartments for peptide: Class I molecules sample predominantly cytosolic materials, whereas class II molecules sample extracellular materials internalized by phagocytosis, pinocytosis, or receptor-mediated endocytosis.

- The process by which protein antigens are acquired, processed into peptides and converted into surface-displayed MHC-peptide complexes is referred to as antigen processing and presentation. This process operates continuously in cells that express the relevant MHC molecules, yet can be modulated in the course of an immune response.

- Antigen processing and presentation can be divided into six discrete steps: (1) acquisition of antigen, (2) tagging the antigen for destruction; (3) proteolysis; (4) delivery of peptides to MHC molecules; (5) binding of peptide to the MHC molecule; and (6) display of the peptide-loaded MHC molecules on the cell surface (see Figures 24-25 and 24-26).

24.5 T Cells, T-Cell Receptors, and T-Cell Development

T lymphocytes recognize antigen through the lens of MHC molecules. The receptors entrusted with this task are structurally related to immunoglobulins. To generate these antigen-specific receptors, T cells rearrange the genes encoding the T-cell receptor (TCR) subunits by mechanisms of somatic recombination identical to those used by B cells to rearrange immunoglobulin genes. The development of T cells is also strictly dependent on successful completion of the somatic gene rearrangements that yield the TCR subunits. We shall describe the subunits that mediate antigen-specific recognition, how they pair up with membrane glycoproteins essential for signal transduction, and how these complexes recognize MHC-peptide combinations.

As pointed out in the preceding section, T cells recognize antigens only together with the polymorphic MHC molecules that happen to be present in that host. In the course of T-cell development, T cells must "learn" the identity of these self MHC molecules and receive instructions about which MHC-peptide combinations to ignore, so as to avoid potentially catastrophic reactivity of newly generated T cells with the host's own tissues (i.e., autoimmunity). We shall describe how T cells develop and introduce the major classes of T cells according to their function.

The Structure of the T-Cell Receptor Resembles the F(ab) Portion of an Immunoglobulin

Much as B cells utilize the B-cell receptor to recognize antigen and then transduce signals that lead to their clonal expansion, so T cells use their **T-cell receptors.** T cells that have been activated through engagement of these antigen-specific receptors proliferate and acquire the capacity to kill antigen-bearing target cells or to secrete cytokines that will assist B cells in their differentiation. The T-cell receptor for antigen recognizes MHC molecules complexed with the appropriate peptides.

The structure of the T-cell receptor was elucidated based on two assumptions. The first was the notion that the variability of T-cell receptors, given their ability to distinguish between closely related antigens, would be considerable. On the assumption that each T cell possesses a clonally distributed antigen-specific receptor, it should be possible to raise antibodies against the features unique to a T-cell receptor expressed by a particular T-cell clone (clonotypic antibodies). Such antibodies should react uniquely with cells from the same clone that served as the source of the T cell receptor and not with other T cells. Such antibodies were used successfully to arrive at a first description of the T-cell receptor. A second assumption was the need for somatic rearrangement to generate the requisite number of antigen-specific T-cell receptors. cDNA clones encoding the putative TCR subunits, when used as probes, should reveal whether somatic rearrangement does in fact occur. A comparison of the organization of a TCR locus in germ-line configuration and in mature T cells should show evidence of somatic rearrangement.

The immunochemical approach, through use of clonotypic antibodies and other T cell–specific monoclonal antibodies, yielded the structure of the T-cell receptor (Figure 24-29). It is composed of two glycoprotein subunits, each of which is encoded by a somatically rearranged gene. The receptors are composed of either an $\alpha\beta$ pair, or a $\gamma\delta$ pair. The structure of these subunits shows structural similarity to the F(ab) portion of an immunoglobulin: at the N-terminal end there is a variable domain, followed by a constant-region domain and a transmembrane segment. The cytoplasmic tails of the TCR subunits are too short to allow recruitment of cytosolic factors that assist in signal transduction. Instead, the TCR associates with the CD3 complex, a set of membrane proteins composed of γ, δ, ε, and ζ chains. (The TCR γ and δ subunits are not to be confused with the similarly designated subunits of the CD3 complex.) The ε chain forms a noncovalent dimer with the γ on the δ chain to yield $\delta\varepsilon$ and $\gamma\varepsilon$ complexes. The external domain of the CD3 subunits is homologous to immunoglobulin domains, and the cytoplasmic domain in each contains an ITAM domain, through which adapter molecules may be recruited upon phosphorylation of tyrosine residues in the ITAMs. The ζ chain is integrated into the CD3-TCR complex as a disulfide-bonded homodimer, and each ζ chain contains three ITAM motifs.

TCR Genes Are Rearranged in a Manner Similar to Immunoglobulin Genes

Virtually all antigen-specific receptors generated by somatic recombination contain a subunit that is the product of V-D-J recombination (e.g., Ig heavy chain; TCR β chain) and a subunit produced by V-J recombination (e.g., Ig light chain; TCR α chain). The mechanism of V-D-J and V-J recombination

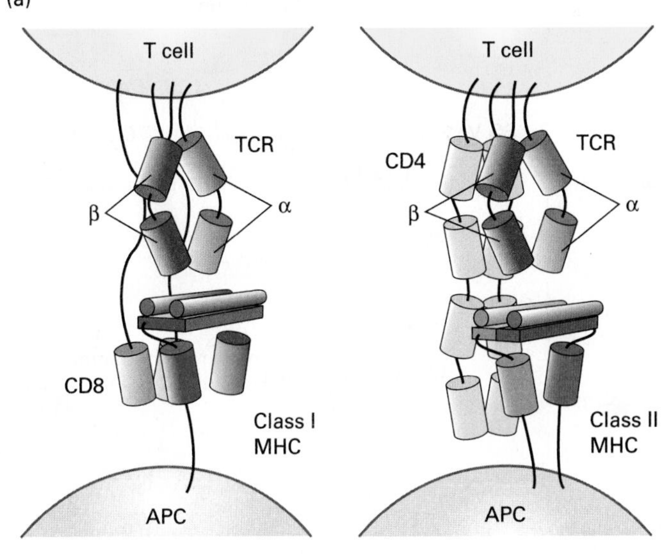

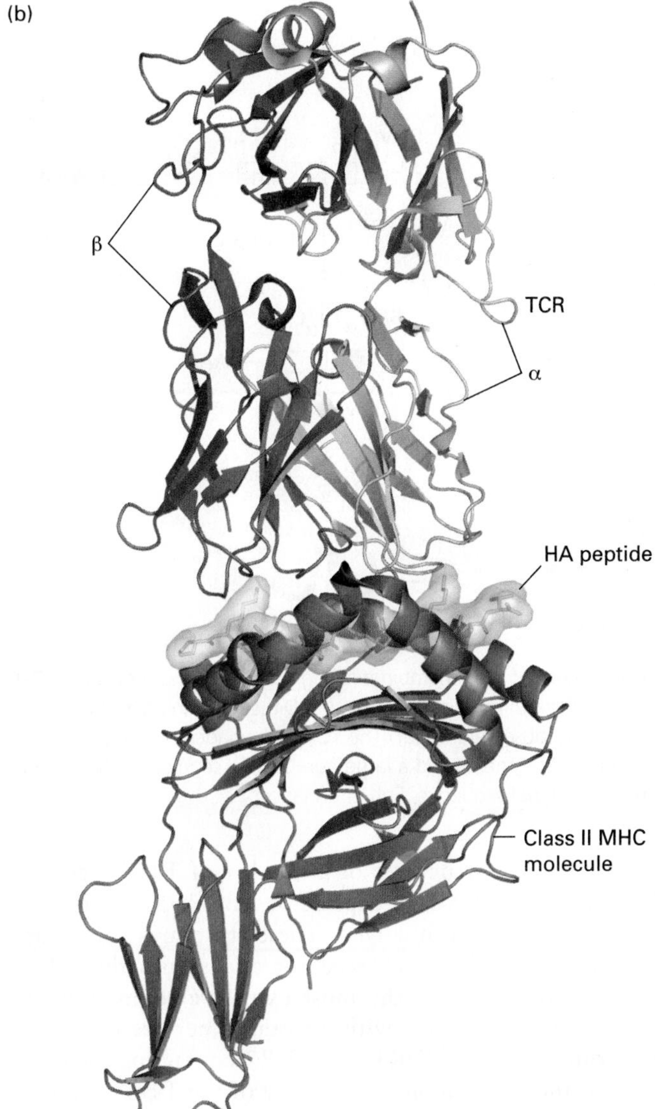

◀ **FIGURE 24-29 Structure of the T-cell receptor and its co-receptors.** (a) The T-cell receptor (TCR) for antigen is composed of two chains, the α and β subunits, which are produced by V-J and V-D-J recombination, respectively. The αβ subunits must associate with the CD3 complex (see Figure 24-31) to allow the transduction of signals. The formation of a full TCRαβ–CD3 complex is required for surface expression. The T-cell receptor further associates with the co-receptors CD8 (light blue) and CD4 (light green), which allow interaction with conserved features of class I MHC and class II MHC molecules, respectively, on antigen-presenting cells. (b) Structure of the T-cell receptor bound to a class II MHC–peptide complex as determined by x-ray crystallography. [Part (b) based on J. Hennecke, 2000, *E MBO* **19:** 5611.]

for the TCR loci is identical to that described for immunoglobulin genes, and requires all the component proteins composing the nonhomologous end-joining machinery: RAG-1, RAG-2, Ku70, Ku80, DNA-PK catalytic subunit, XRCC4, DNA ligase IV, and Artemis. There is an absolute requirement for the recombination signal sequences (RSSs), and recombination obeys the 12/23 bp spacer rule (Figure 24-30).

A number of noteworthy features characterize the organization and rearrangement of the TCR loci. First, the organization of the recombination signals is such that D-to-D rearrangements are allowed. Second, the enzyme terminal deoxynucleotidyltransferase (TdT) is active at the time the TCR genes rearrange, and therefore N nucleotides can be present in all rearranged TCR genes. Third, in the human and mouse, the TCR δ locus is embedded within the TCR α locus. This organization results in complete excision of the interposed δ locus, and so a choice of the TCR α locus for rearrangement precludes utilization of the δ locus, now lost by deletion. T cells that express the αβ receptor and those that express the γδ receptor are considered separate lineages with distinct functions. Among the T cells expressing γδ receptors are some capable of recognizing the CD1 molecule, which is specialized in the presentation of lipid antigens. The γδ T cells are programmed to hone in on distinct anatomical sites (e.g., the epithelium lining the genital tract, the skin) and likely play a role in host defense against pathogens commonly found at these sites.

Deficiencies in the key components of the recombination apparatus, such as the RAG recombinases, preclude rearrangement of TCR genes. As we have seen for B cells, and as will be described below for T cells, development of lymphocytes is strictly dependent on the rearrangement of the antigen-receptor genes. A RAG deficiency thus prevents both B- and T-cell development.

T-Cell Receptors Are Very Diverse with Many of Their Variable Residues Encoded in the Junctions between V, D, and J Gene Segments

The diversity created by somatic rearrangement of TCR genes is enormous, estimated to exceed 10^{10} unique receptors.

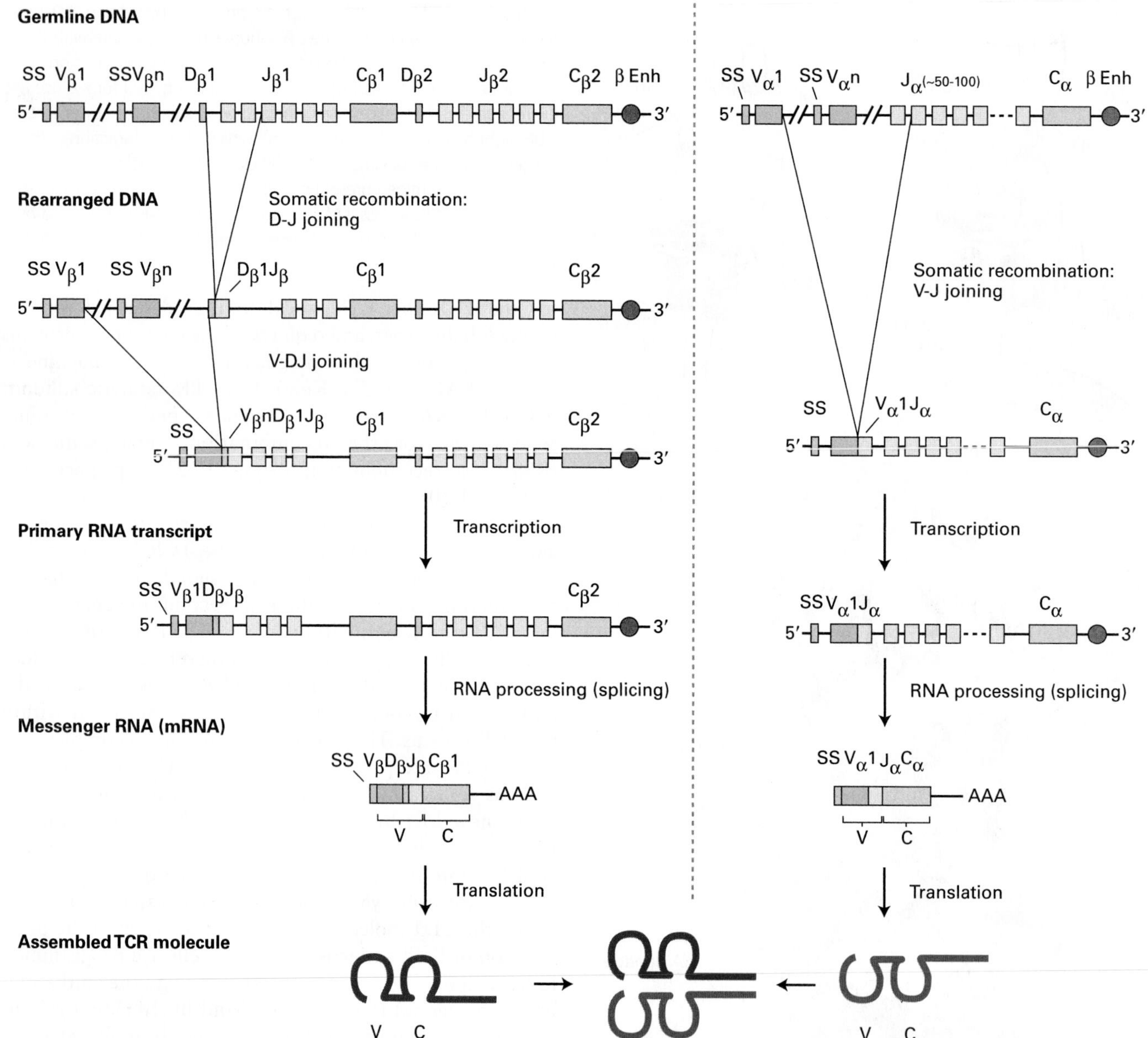

▲ **FIGURE 24-30 Organization and recombination of TCR loci.** The organization of TCR loci is in principle similar to that of immunoglobulin loci (see Figure 24-14). (*Left*) The TCR β-chain locus includes a cluster of V segments, a cluster of D segments, and several J segments, downstream of which are two constant regions. The arrangement of the recombination signals is such that not only is D-J joining allowed, but also V-DJ joining. Direct V-J joining in the TCRβ locus is not observed. (*Right*) The TCR α-chain locus is composed of a cluster of V segments and a large number of J segments. SS = exon encoding signal sequence; Enh = enhancer.

Combinatorial usage of different V, D, and J gene segments makes an important contribution, as do the mechanisms of junctional imprecision and N-nucleotide addition already discussed. The net result is a degree of variability in the V regions that matches that of the CDR3 of immunoglobulins (see Figure 24-12). Unlike immunoglobulin genes, the TCR genes do not undergo somatic hypermutation, and thus T-cell receptors exhibit nothing equivalent to the affinity maturation of antibodies during the course of an immune response.

The crystal structures of a number of T-cell receptors bound to class I MHC–peptide or class II MHC–peptide complexes have been solved. These structures show variability in how the T-cell receptor docks with the MHC-peptide complex, but the most extensive contacts in the CDR3 region are made with the central peptide portion of the complex, with CDR1 and CDR2 contacting the α helices of the MHC molecules. Many of the T-cell receptors for which a structure has been solved sit down diagonally across the peptide-binding portion of the MHC-peptide complex. As a result, the T-cell receptor makes extensive contacts with the peptide cargo as well as with the α helices of the MHC molecule to which it binds. The positions at which allelic MHC molecules differ from one another frequently involve residues that directly contact the T-cell receptor, thus precluding tight binding of the "wrong" allele.

Amino acid differences that distinguish one MHC allele from an other also affect the architecture of the peptide-binding cleft. Even if the MHC residues that interact directly with the T-cell receptor were shared by two MHC allelic molecules, their peptide-binding specificity is likely to differ because of amino acid differences in the peptide-binding cleft. Consequently, the TCR contact residues provided by bound peptide and essential for stable interaction with a T-cell receptor would be absent from the "wrong" MHC-peptide combination. A productive interaction with the T cell receptor is then unlikely to occur.

Signaling via Antigen-Specific Receptors Triggers Proliferation and Differentiation of T and B Cells

Both B-cell and T-cell receptors for antigen transduce signals by means of proteins associated with the antigen-specific portions of the receptor (i.e., Ig heavy and light chains for the BCR; α and β chains for the TCR). The cytosolic portions of the antigen-specific receptors themselves are very short, do not protrude much beyond the cytosolic leaflet of the plasma membrane, and are incapable of recruitment of downstream signaling molecules. Instead, as discussed previously, the antigen-specific receptors on T and B cells associate with auxiliary subunits that contain ITAMs (immunoreceptor tyrosine based activation motifs). Engagement of antigen-specific receptors by ligand initiates a series of receptor-proximal events: kinase activation, modification of ITAMs, and subsequent recruitment of adapter molecules that serve as scaffolds for recruitment of yet other downstream signaling molecules.

As outlined in Figure 24-31, engagement of antigen-specific receptors activates Src family tyrosine kinases (e.g., Lck in CD4 T cells; Lyn and Fyn in B cells). These kinases are found in close proximity to or physically associated with the antigen receptor. The active Src kinases phosphorylate the ITAMs in the antigen receptors' auxiliary subunits. In their phosphorylated forms, these ITAMs recruit and activate non-Src family tyrosine kinases (ZAP-70 in T cells, Syk in B cells) as well as other adapter molecules. Such recruitment and activation involves phosphoinositide-specific phospholipase C_γ and PI-3 kinases. Subsequent downstream events parallel those discussed in Chapter 16 for signaling from receptor tyrosine kinases. Ultimately signaling via antigen-specific receptors initiates transcription programs that determine the fate of the activated lymphocyte: proliferation and differentiation.

T cells depend critically on the cytokine interleukin 2 (IL-2) for clonal expansion. Following antigen stimulation of a T cell, one of the first genes to be turned on is that for IL-2. The T cell responds to its own initial burst of IL-2 and proceeds to make more IL-2, an example of autocrine stimulation and part of a positive feedback loop. An important transcription factor required for the induction of IL-2 synthesis is the NF-AT protein (nuclear factor of activated T cells). This protein is sequestered in the cytoplasm in phosphorylated form and cannot enter the nucleus unless it is dephosphorylated first. The phosphatase responsible is calcineurin, a Ca^{2+}-activated enzyme. The rise in cytosolic Ca^{2+} leading to activation of calcineurin results from mobilization of ER-resident Ca^{2+} stores triggered by hydrolysis of PIP_2 and the concomitant generation of IP_3 (see Figure 15-30, steps **2**–**4**).

The immunosuppressant drug cyclosporine inhibits calcineurin activity through formation of a cyclosporine-cyclophilin complex, which binds and inhibits calcineurin. If dephosphorylation of NF-AT is suppressed, NF-AT cannot enter the nucleus and participate in the up-regulation of transcription of the IL-2 gene. This precludes expansion of antigen-stimulated T cells and so leads to immunosuppression, arguably the single most important intervention that contributes to successful organ transplantation. Although the success of transplantation varies with the organ used, the availability of strong immunosuppressants such as cyclosporine has expanded enormously the possibilities of clinical transplantation. ▮

T Cells Capable of Recognizing MHC Molecules Develop Through a Process of Positive and Negative Selection

The rearrangement of the gene segments that are assembled into a functional T-cell receptor is a stochastic event, completed on the part of the T cell without any prior knowledge of the MHC molecules with which these T-cell receptors must ultimately interact. Similar to somatic recombination of Ig heavy-chain loci in B cells, the first TCR gene segments to rearrange are the TCRβ D and J elements, followed by joining of a V segment to the newly recombined DJ. At this stage of T-cell development, productive rearrangement allows the synthesis of the TCR β chain, which is incorporated into the pre-TCR through association with the pre-T α subunit. This pre-TCR fulfills a function strictly analogous to that of the pre-BCR in B-cell development: It allows expansion of pre-T cells that successfully underwent rearrangement, and it imposes allelic exclusion to ensure that, as a rule, a single functional TCRβ subunit is generated for a given T cell and its descendants. After the expansion phase of pre-T cells is complete, rearrangement of the TCRα locus is initiated, ultimately leading to the generation of T cells with a fully assembled TCR αβ receptor.

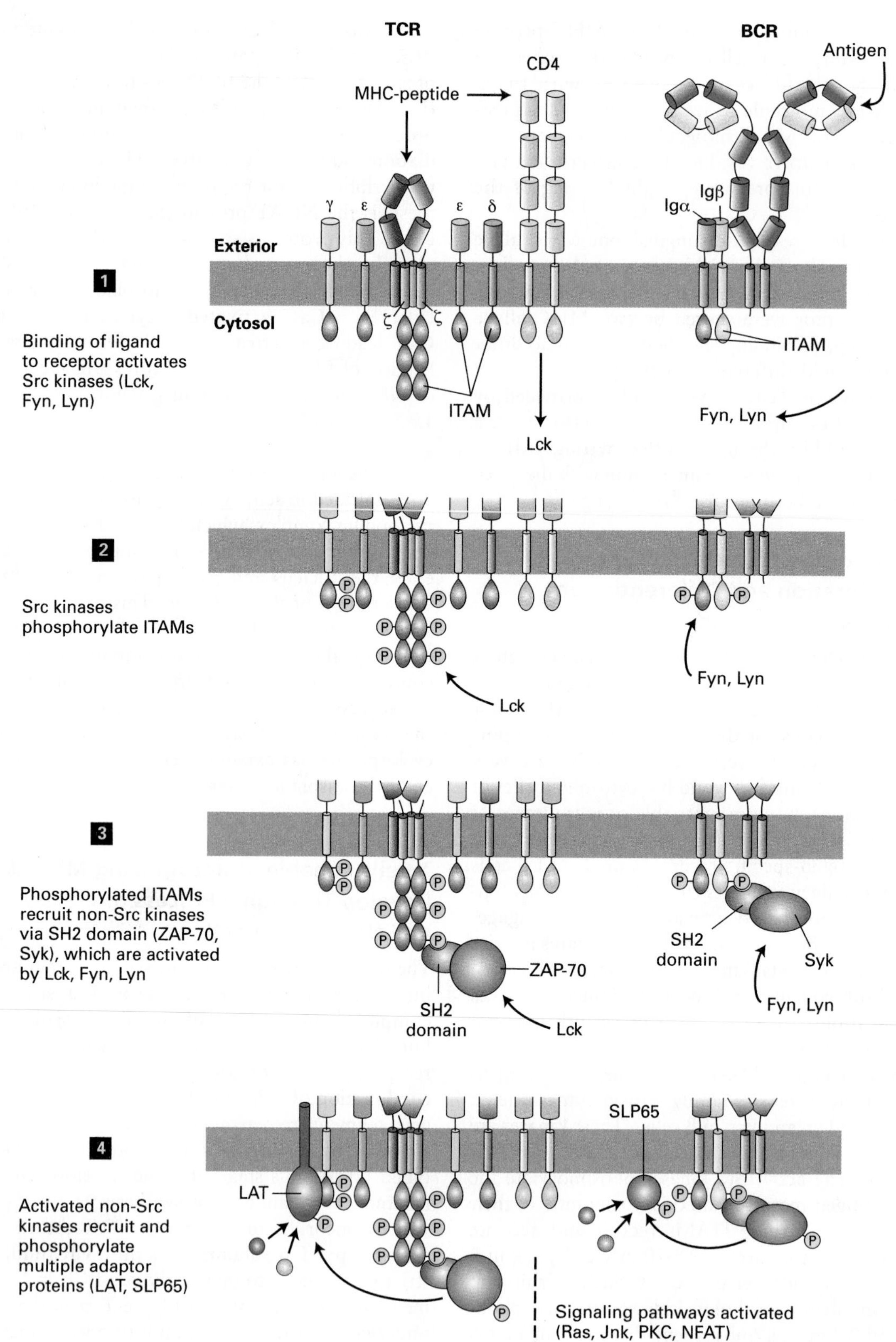
TCR
BCR
CD4
Antigen
MHC-peptide
γ
ε
ε
δ
Igβ
Igα
Exterior
Cytosol
ζ
ζ
ITAM
ITAM
Lck
Fyn, Lyn
1
Binding of ligand to receptor activates Src kinases (Lck, Fyn, Lyn)
2
Src kinases phosphorylate ITAMs
Lck
Fyn, Lyn
3
Phosphorylated ITAMs recruit non-Src kinases via SH2 domain (ZAP-70, Syk), which are activated by Lck, Fyn, Lyn
ZAP-70
SH2 domain
Lck
SH2 domain
Syk
Fyn, Lyn
4
Activated non-Src kinases recruit and phosphorylate multiple adaptor proteins (LAT, SLP65)
LAT
SLP65
Signaling pathways activated (Ras, Jnk, PKC, NFAT)

▲ **FIGURE 24-31 Signal transduction from the T-cell receptor (TCR) and B-cell receptor (BCR).** The signal-transduction pathways used by the antigen-specific receptors of T cells (*left*) and B cells (*right*) are conceptually similar. The initial stages are depicted in this figure; downstream signaling events lead to changes in gene expression that result in proliferation and differentiation of the antigen-stimulated lymphocytes. See text for further discussion.

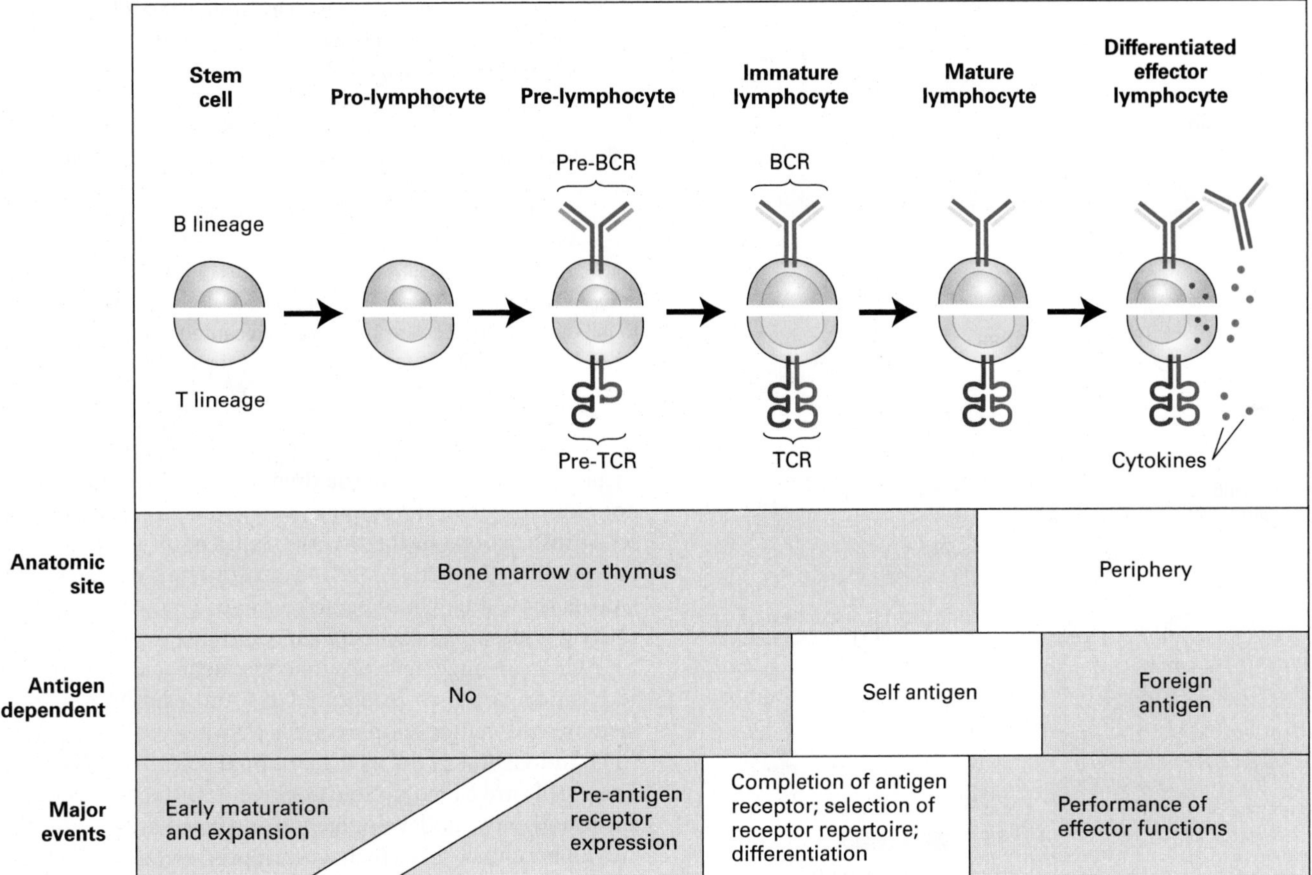

▲ **FIGURE 24-32 Comparison of T-cell and B-cell development.** Cell fate decisions are executed by receptors composed of either the newly rearranged μ chain (pre-BCR) or the newly rearranged β chain (pre-TCR). The pre-B and pre-T cell receptors serve similar functions: expansion of cells that successfully underwent rearrangement and allelic exclusion. This phase of lymphocyte development does not require antigen-specific recognition. Both the pre-BCR and pre-TCR include subunits unique to each receptor and absent from the antigen-specific receptors found on mature lymphocytes: VpreB and $\lambda 5$ (orange, green) for the pre-BCR; pre-Tα (blue) for the pre-TCR. Upon completion of the expansion phase, expression begins of the gene encoding the remaining subunit of the antigen-specific receptor: Ig light chain (light blue) for the BCR; TCR α chain (light red) for the TCR. Lymphocyte development and differentiation occur at distinct anatomical sites, and only fully assembled antigen-specific receptors (BCR, TCR) recognize antigen. Mature lymphocytes are strictly dependent on antigen recognition for their activation.

Figure 24-32 illustrates the analogous steps in the development of T and B cells.

How is the newly emerging repertoire of T cells shaped so that a productive interaction with self-MHC molecules can occur? The random nature of the gene rearrangement process and the enormous variability engendered as a consequence produces a set of T-cell receptors, the vast majority of which cannot productively interact with the host MHC products. Recall that antigen processing and presentation are constitutive processes, and that in the thymus, where these selection events take place, all self-MHC molecules are necessarily occupied with peptides derived from self-proteins. These combinations of self-peptides complexed to class I and class II MHC molecules constitute the template on which the initial set of T-cell receptors must be calibrated.

T cells born with receptors that cannot engage self-MHC are useless. Such T cells will fail to perceive survival signals via the newly generated T-cell receptors and die. At the other extreme are T cells endowed with receptors that show a perfect fit for self-MHC complexed with a particular self-peptide. These T cells, if allowed to leave the thymus and take up residence in peripheral lymphoid organs, would be by definition self-reactive and could cause autoimmunity. If the number of such self-peptide MHC combinations surpasses a threshold sufficient to allow triggering of the T-cell receptor, then these T cells are instructed to die by apoptosis. This is called negative selection and serves to purge the T-cell repertoire of overtly self-reactive T cells. Any T cell equipped with a receptor that perceives signals sufficiently strong to allow survival, but below the threshold that would impel the T cell to initiate apoptosis, is positively selected. The heterogeneity of peptide-MHC complexes displayed on the surface of T cells undergoing selection makes it highly probable that the T-cell receptor interprets signals in additive fashion: The summation of binding energies of the different MHC-self-peptide combinations on display determine the outcome of selection. This is called the *avidity model of T-cell selection.*

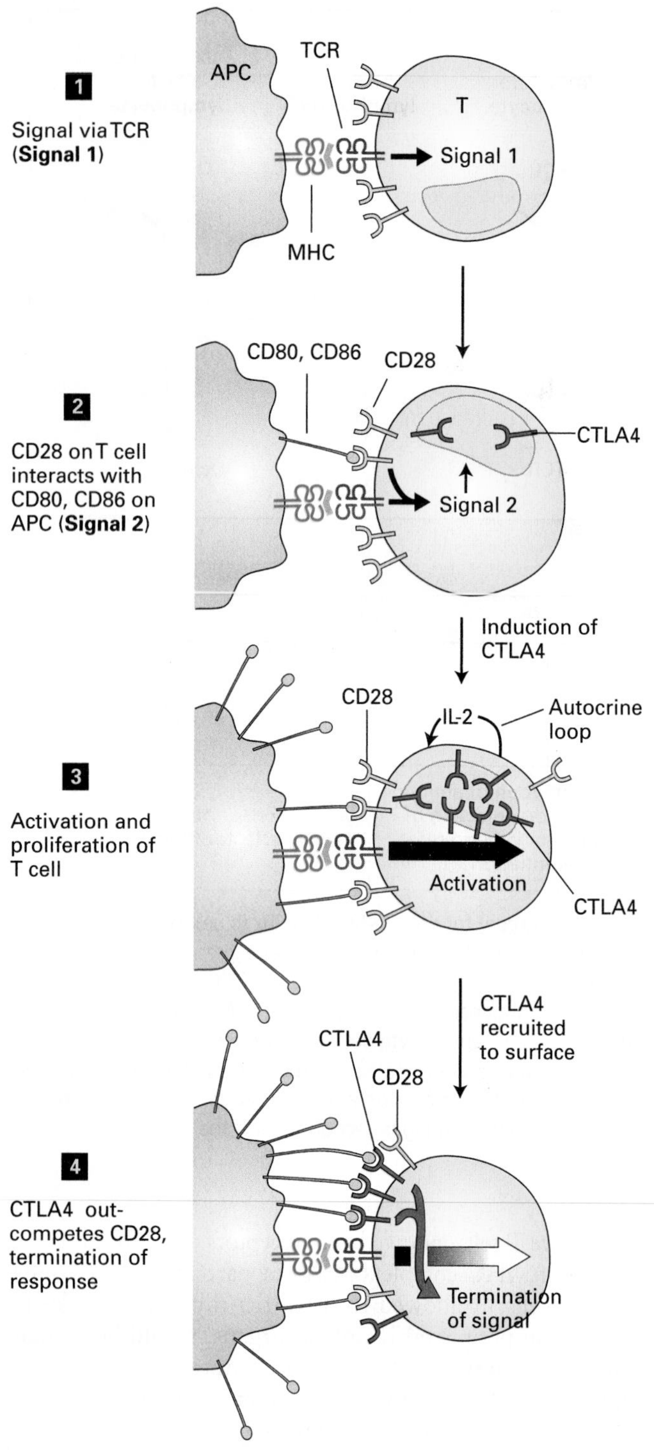

◀ **FIGURE 24-33 Signals involved in T-cell activation and termination.** The two-signal model of T-cell activation involves recognition of an MHC–peptide complex by the T-cell receptor, which constitutes Signal 1 (**1**), along with recognition of costimulatory molecules (CD80, CD86) on the surface of the antigen-presenting cell, which constitutes Signal 2 (**2**). If costimulation is not provided, the newly engaged T cell becomes unresponsive (anergic). The provision of both Signal 1 via the T-cell receptor and Signal 2 via engagement of CD80 and CD86 by C28 allows full activation. Full activation, in turn, leads to increased expression of CTLA4 (**3**). After moving to the T-cell surface, CTLA4 binds CD80 and CD86, leading to inhibition of the T-cell response (**4**). Because the affinity of CTLA4 for CD80 and CD86 is greater than that of the CD28, T-cell activation eventually is terminated.

T cells are killed off by apoptosis only if the appropriate self-antigen is represented adequately in the thymus as MHC-peptide complexes. Proteins that are expressed in tissue-specific fashion, such as insulin in the β cells of the pancreas or the components of the myelin sheath in the nervous system, do not obviously fit this category. However, a factor called AIRE (autoimmune regulator) allows expression in a subset of epithelial cells of such tissue-specific antigens. How AIRE accomplishes this is not known, but it is widely suspected of directly regulating the transcription of the relevant genes. Defects in AIRE lead to a failure to express these tissue-specific antigens in the thymus. In individuals who do not express AIRE, developing T cells fail to receive the full set of instructions in the thymus that lead to elimination of potentially self-reactive T cells. As a consequence, these individuals show a bewildering array of autoimmune responses, causing widespread tissue damage and disease.

TCR rearrangement occurs coincident with the gradual acquisition of the co-receptors CD4 and CD8. A key intermediate in T-cell development is a thymocyte that expresses CD4 and CD8, as well as a functional TCR-CD3 complex. These cells are called double positive (CD4CD8+) cells and are found only as developmental intermediates in the thymus. The choice of which co-receptor (CD4 or CD8) to express determines whether a T cell will recognize class I or class II MHC molecules. How a newborn CD4CD8+ cell is instructed to become a CD8 (class I MHC restricted) T cell or CD4 (class II MHC restricted) T cell is not entirely settled.

T Cells Require Two Types of Signal for Full Activation

All T cells require a signal via their antigen-specific receptor, the TCR, for activation, but this is not sufficient: the T cell further needs costimulatory signals. To perceive costimulatory signals, T cells carry on their surface several different receptors, of which the CD28 molecule is the best-known example. CD28 interacts with CD80 and CD86 on the antigen-presenting cells with which the T cell interacts. Their expression is up-regulated when these antigen-presenting cells have themselves received the proper stimulatory signals, for example by engagement of their Toll-like receptors (TLRs). The signals delivered via CD28 synergize with signals that emanate from the engaged T-cell receptor, all of which are required for full activation (Figure 24-33).

T cells, once activated, also express receptors that provide an attenuating or inhibitory signal upon recognition of these very same costimulatory molecules. The CTLA4 protein, whose expression on T cells is induced only upon activation, competes with CD28 for binding of CD80 and CD86. Because the affinity of CTLA4 for the CD80 and C86 proteins is higher than that of CD28, the inhibitory signals provided through CTLA4 will ultimately dominate the stimulatory signals that derive from engagement of CD28. Costimulatory molecules can thus be stimulatory or inhibitory and provide an important means of controlling the activation status and duration of a T-cell response.

Cytotoxic T Cells Carry the CD8 Co-receptor and Are Specialized for Killing

As we've seen, cytotoxic T cells, also called cytolytic T lymphocytes (CTLs), generally carry the CD8 glycoprotein marker and are restricted in their recognition by class I MHC molecules. They kill target cells that display the appropriate MHC-peptide combinations and do so with exquisite sensitivity: A single MHC-peptide complex suffices to allow a properly primed CD8 T cell to kill the target cell that bears it.

The mechanism of killing by CTLs involves two classes of proteins that act synergistically, *perforins* and *granzymes* (Figure 24-34). Perforins, which exhibit homology to the terminal components of the complement cascade composing the membrane attack complex, form pores of up to 20 nm in membranes to which they attach. The destruction of an intact permeability barrier, which leads to loss of electrolytes and other small solutes, contributes to cell death. Granzymes are delivered to and are presumed to enter the target cell, likely via the pores generated by perforin. Granzymes are serine proteases that activate effector **caspases** and so propel the target on a path of programmed cell death (apoptosis). Perforins and granzymes are packaged into cytotoxic granules, stored inside the cytotoxic T cell. Upon binding of the T-cell receptor to antigen, the cytotoxic granules and their contents are released into the cleft formed between the cytotoxic T cell and target cell. How the T cell avoids being killed upon release of granzymes and perforins into this synaptic cleft is still not known. Natural killer cells also exert cytotoxic activity and likewise rely on perforins and granzymes to kill their targets (see Figure 24-5).

T Cells Produce an Array of Cytokines That Provide Signals to Other Immune Cells

Many lymphocytes and nonlymphoid cells in lymphoid tissue produce cytokines. These small secreted proteins instruct lymphocytes what to do by binding to specific receptors on

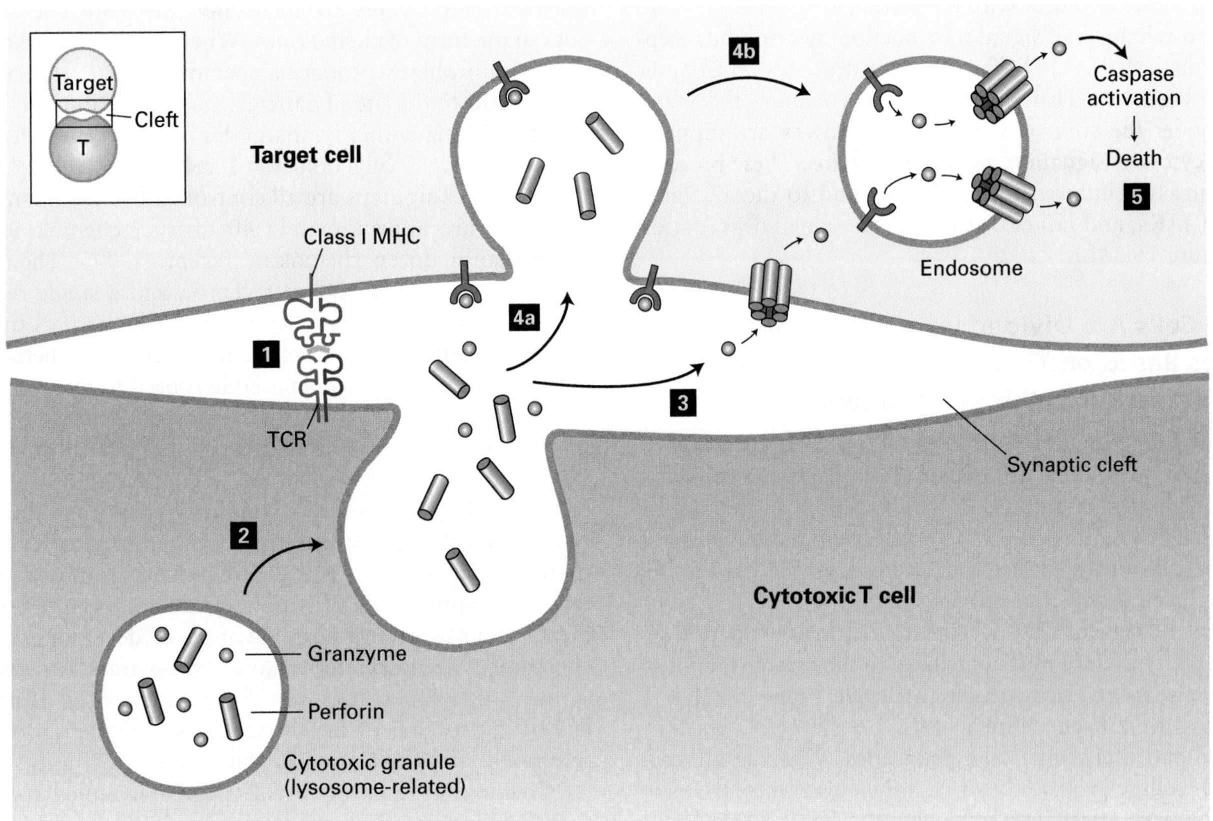

▲ **FIGURE 24-34 Perforin- and granzyme-mediated cell killing by cytotoxic T cells.** Upon recognition of the target cell (1), cytotoxic T cells form tight antigen-specific contacts with target cells. Tight contact results in the formation of a synaptic cleft, into which the contents of cytotoxic granules are released (2). The content of these granules includes perforins and granzymes. Perforins form pores in the membranes onto which they adsorb, and granzymes are serine proteases that enter through the perforin pores (3). Perforins are believed to act not only at the surface of the target cells, but also at the surface of endosomal compartments of target cells after the perforin molecules have been internalized from the cell surface (4). Once in the cytoplasm, the granzymes activate caspases, which initiate programmed cell death (5).

the lymphocyte surface and initiating a transcriptional program that allows the lymphocyte to either proliferate or differentiate into an effector cell ready to exert cytotoxic activity (CD8 T cells), helper activity (CD4 T cells), or antibody-secreting activity (B cells). Cytokines that are produced by or act primarily on leukocytes are called **interleukins;** at least 27 interkeukins have been identified and molecularly characterized. Structurally related interleukins are recognized by cognate receptors with structural similarities, the interleukin-2 receptor being a particularly well-characterized example. Interleukin 2 (IL-2) was identified as T-cell growth factor and is one of the first cytokines produced when T cells are stimulated. IL-2 acts as an autocrine (self-acting) growth factor and drives clonal expansion of activated T cells.

Interleukin 4 (IL-4), which is produced by CD4 T cells specialized in providing help to B cells (discussed below) induces activated B cells to proliferate and to undergo class switch recombination and somatic hypermutation. Interleukin 7 (IL-7), produced by stromal cells in the bone marrow, is essential for development of T and B cells from committed precursors. Both IL-7 and IL-15 play a role in the maintenance of T cells in the form of *memory cells,* which are antigen-experienced cells that may be called upon when reexposure to antigen occurs. These memory T cells then rapidly proliferate and deal with the intruder.

The mechanism of signal transduction by cytokine receptors through the JAK/STAT pathway is described in Chapter 16 (see Figure 16-1b for quick review). Among the many genes under the control of the STAT pathway are suppressors of cytokine signaling, or SOCS proteins. These proteins are themselves induced by cytokines, bind to the activated form of JAKs, and target them for proteasomal degradation (see Figure 16-14b).

CD4 T Cells Are Divided into Three Major Classes Based on Their Cytokine Production and Expression of Surface Markers

The major function of CD4 T cells is to provide assistance to B cells and guide their differentiation into plasma cells that secrete high-affinity antibodies. This function, carried out by helper T cells, requires both the production and secretion of cytokines, as well as direct contact between the CD4 T cell and the B cell to which it provides help.

A second type of CD4 T cell has as its major function secreting the cytokines that contribute to the establishment of an inflammatory environment. Multiple types of CD4 T cells have thus been defined based on the cytokines they produce and their functional properties. Whereas all activated T cells can produce IL-2, other cytokines are produced by particular CD4 T cell subsets. These CD4 T cells are classified as T_H1 cells, characterized by the production of interferon γ and TNF, and T_H2 cells, characterized by the production of IL-4 and IL-10. T_H1 cells, through production of IFNγ, can activate macrophages and stimulate an inflammatory response. Referred to also as *inflammatory T cells,* T_H1 cells nonetheless play an important role in antibody production, notably facilitating the production of complement-fixing antibodies such as IgG1 and IgG3. T_H2 cells, through production of IL-4, play an important role in B-cell responses that involve class switch recombination to the IgG1 and IgE isotypes (discussed below). The combination of cytokines produced by CD4 T cells, and the interaction between the CD40 protein induced on activated T cells with CD40 ligand (CD40L) on B cells is responsible for induction of activation-induced deaminase, and so prepares the B cell for class switch recombination and somatic hypermutation.

A recently discovered subset of CD4 T cells are called *regulatory T cells.* They can attenuate immune responses by exerting a suppressive effect on cytokine production by other T cells. These regulatory T cells restrain the activity of potentially self-reactive T cells and are important in maintaining tolerance (the absence of an immune response to self antigens).

Leukocytes Move in Response to Chemotactic Cues Provided by Chemokines

Interleukins tell lymphocytes what to do by eliciting a transcriptional program that allows lymphocytes to acquire specialized effector functions. On the other hand, chemokines tell leukocytes where to go. Many cells emit chemotactic cues in the form of chemokines. When tissue damage occurs, resident fibroblasts produce a chemokine, IL-8, that attracts neutrophils to the site of damage. The regulation of lymphocyte trafficking within lymph nodes is essential for dendritic cells to attract T cells, and for T cells and B cells to meet. These trafficking steps are all controlled by chemokines.

There are approximately 40 distinct chemokines and more than a dozen chemokine receptors. One chemokine may bind to more than one receptor, and a single receptor can bind several different chemokines. This creates the possibility of generating a combinatorial code of chemotactic cues of great complexity. This code is used to allow the navigation of leukocytes from where they are generated, in the bone marrow, into the bloodstream for transport to their target destination.

Some chemokines direct lymphocytes to leave the circulation and take up residence in lymphoid organs. These migrations contribute to the population of lymphoid organs with the required sets of lymphocytes. Because these movements occur as part of normal lymphoid development, such chemokines are also referred to as *homeostatic chemokines.* Those chemokines that serve the purpose of recruiting leukocytes to sites of inflammation and tissue damage are referred to as *inflammatory chemokines.*

Chemokine receptors are G protein–coupled receptors that function as an essential step in the regulation of cell adhesion and cell migration. Leukocytes that travel through blood vessels do so at high speed and are exposed to high hydrodynamic shear forces. For a leukocyte to traverse the endothelium and take up residence in a lymph node or seek out a site of infection in tissue, it must first slow down, a process that requires interactions of surface receptors called selectins with their ligands, which are mostly carbohydrate in nature. If chemokines are found in proximity, adsorbed to the extracellular matrix, and if the leukocyte possesses a receptor for that chemokine, activation of the chemokine receptor elicits a signal that allows integrins carried by the leukocyte to undergo a conformational change. This change results in an increase in affinity of the integrin for its ligand and causes firm arrest of the leukocyte. The leukocyte may now exit the blood vessel by a process known as *extravasation* (see Figure 19-36).

KEY CONCEPTS OF SECTION 24.5

T Cells, T-Cell Receptors, and T-Cell Development

- The antigen-specific T-cell receptors are dimeric proteins consisting of α and β subunits or γ and δ subunits. T cells occur in at least two major classes based on the expression of the glycoprotein co-receptors CD4 and CD8 (see Figure 24-29).
- Cells that use class I MHC molecules as restriction elements carry CD8; those that use class II MHC molecules carry CD4. These classes of T cells are functionally distinct: CD8 T cells are cytotoxic T cells; CD4 T cells provide help to B cells and are an important source of cytokines.
- Genes encoding the TCR subunits are generated by somatic recombination of V and J segments (α chain) and of V, D, and J segments (β chain); their rearrangement obeys the same rules as those defined for rearrangement of Ig genes in B cells (see Figure 24-30). Rearrangement of TCR genes occurs in the thymus and only in those cells destined to become T lymphocytes.
- A complete T-cell receptor includes the accessory CD3 complex, which is required for signal transduction. Each subunit of the CD3 complex carries in its cytoplasmic tail one or three ITAM domains; when phosphorylated, these ITAMs recruit accessory molecules involved in signal transduction (see Figure 24-31).
- In the course of T-cell development, the TCR β locus rearranges first, encoding a functional β subunit that is incorporated in the pre-TCR, which also includes a specialized pre-TCR α subunit encoded by an unrearranged gene (see Figure 24-32). Similar to the pre-BCR, the pre-TCR mediates allelic exclusion and proliferation of those cells that successfully underwent TCRβ rearrangement.
- T cells destined to become CD8 T cells must interact with class I MHC molecules in the course of their development; those that will become CD4 T cells must interact with class II MHC molecules. Developing T cells that fail to recognize any MHC molecules at all die for lack of survival signals. T cells that interact too strongly with peptide-MHC complexes encountered during development are instructed to die (negative selection); those that have intermediate affinity for peptide-MHC complexes are allowed to mature (positive selection) and are exported from the thymus to the periphery.
- T cells are instructed where to go (cell migration) through chemotactic signals in the form of chemokines. Receptors for chemokines are G protein–coupled receptors that show some promiscuity in terms of their binding of chemokines. The complexity of the chemokine–chemokine receptor family allows precise regulation of leukocyte trafficking, both within lymphoid organs and in the periphery.

24.6 Collaboration of Immune-System Cells in the Adaptive Response

An effective adaptive immune response requires the presence of B cells, T cells, and antigen-presenting cells. For B cells to execute class switch recombination and hypermutation—prerequisites for production of high-affinity antibodies—they require help from activated T cells. These T cells, in turn, can only be activated by professional antigen-presenting cells such as dendritic cells, which detect invaders using their Toll-like receptors (TLRs). The interplay between components of innate and adaptive immunity is therefore a very important aspect of adaptive immunity. In this section we describe how these various elements are activated and how the relevant cell types interact.

Toll-Like Receptors Perceive a Variety of Pathogen-Derived Macromolecular Patterns

An important part of innate defenses is the ability to immediately detect the presence of a microbial invader and respond to it. This response includes direct elimination of the invader, but it also prepares the mammalian host for a proper adaptive immune response, in particular through activation of professional antigen-presenting cells. Antigen-presenting cells are positioned throughout epithelia (airways, gastrointestinal tract, genital tract) where contact with pathogens is most likely to occur. In the skin a network of dendritic cells called *Langerhans cells* makes it virtually impossible for a pathogen that breaches this barrier to avoid contact with these professional antigen-presenting cells. Dendritic cells and other professional antigen-presenting cells detect the presence of bacteria and viruses through members of the Toll-like receptor (TLR) family. These proteins are named after the *Drosophila* protein Toll because of the structural and functional homology between them. *Drosophila* Toll was discovered because of its important role in dorsoventral patterning in the fruit fly, but related receptors are now recognized as capable of triggering an innate immune response in insects as well as in vertebrates.

TLR Structure Toll itself and all Toll-like receptors possess an extracellular domain composed of *leucine-rich repeats*. These repeats form a sickle-shaped extracellular domain believed to be involved in ligand recognition. The cytoplasmic portion of Toll-like receptors contains a domain responsible for the recruitment of adapter proteins to enable signal transduction. The signaling pathways engaged by the Toll-like receptors share many of the components (and outcomes) as those used by receptors for the cytokine IL-1 (Figure 24-35).

The *Drosophila* Toll protein interacts with its ligand, Spaetzle, itself the product of a proteolytic conversion initiated by components of the cell wall of fungi that prey on *Drosophila*. In the fly, activation of Toll unleashes a signaling cascade that ultimately controls the transcription of genes that encode antimicrobial peptides. The activated receptor at the surface communicates with the transcriptional apparatus by means of a series of adapter proteins that activate downstream kinases interposed between the Toll-like receptor and the transcription factors that are activated by signaling from them. A key step is the ubiquitin-dependent proteasomal degradation of the Cactus protein. Its removal allows the protein Dif to enter the nucleus and initiate transcription. This pathway is highly homologous in its operation and structural composition to the NF-κB pathway in mammals (see Figure 16-35).

Diversity of TLRs There are approximately a dozen mammalian Toll-like receptors that can be activated by various microbial products. These receptors are expressed by a variety of cell types, but their function is crucial for the activation of dendritic cells and macrophages. Neutrophils also express Toll-like receptors. The microbial products recognized by Toll-like receptors include macromolecules found in the cell envelope of bacteria such as lipopolysaccharides (LPS), flagellins (subunits of bacterial flagella), and bacterial lipopeptides. Although direct binding of these macromolecules to Toll-like receptors remains to be shown directly, their presence is sensed by distinct receptors: for example, TLR4 for lipopolysaccharide; heterodimeric of TLR1/2 TLR2/6 for lipopeptides; and TLR5 for flagellin. Recognition of all bacterial envelope components occurs at the cell surface.

A second set of Toll-like receptors—TLR3, TLR7, and TLR9—sense the presence of pathogen-derived nucleic acids. They do so not at the cell surface, but rather within the endosomal compartments where these receptors reside. Mammalian DNA is methylated at many CpG dinucleotides, whereas microbial DNA generally lacks this modification. TLR9 is activated by unmethylated, CpG-containing microbial DNA. Similarly, double-stranded RNA molecules present in some virus-infected cells lead to activation of TLR3. Finally, TLR7 responds to the presence of certain single-stranded RNAs. Thus the full set of mammalian TLRs allows the recognition of a variety of macromolecules that are diagnostic for the presence of bacterial, viral, and fungal pathogens.

TLR Signal Cascade As shown in Figure 24-35, engagement of mammalian Toll-like receptors leads to recruitment of the adapter protein MyD88, which in turn allows the binding and activation of IRAK (interleukin 1 receptor-associated kinase). After IRAK phosphorylates TNF-receptor associated factor 6 (TRAF6), several downstream kinases come into play, leading to release of active NF-κB, a transcription factor, for translocation from the cytoplasm to the nucleus,

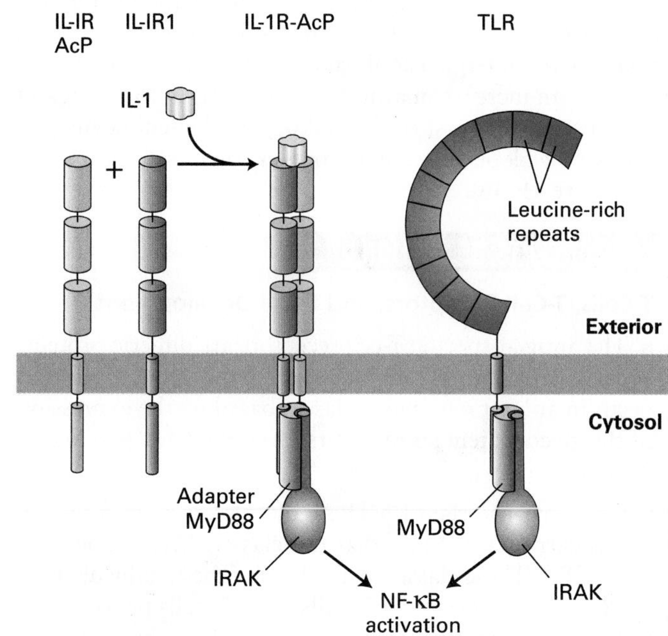

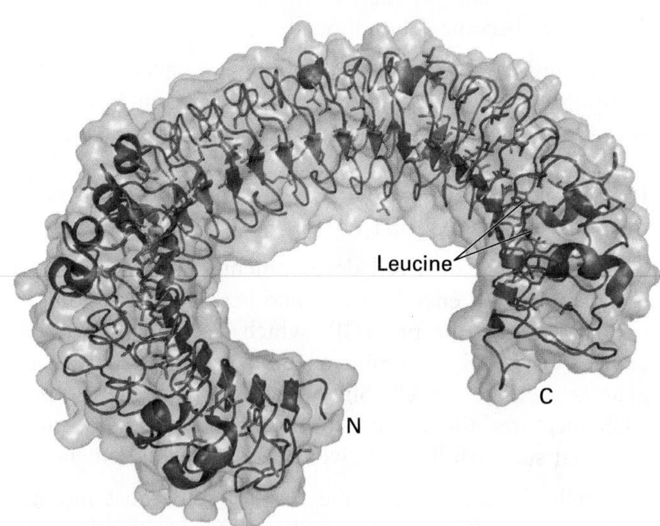

▲ **FIGURE 24-35 Toll-like receptors.** (a) Comparison of signal transduction from the IL-1 receptor and Toll-like receptor (TLR). Interleukin 1 (IL-1) binds to its receptor by co-engagement of the IL-1R accessory protein (IL-1R AcP). The adapter protein MyD88 is recruited to the activated receptor and in turn allows activation, by the interleukin-1 receptor associated kinase (IRAK), of the NF-κB pathway (see Figure 16-35). Toll-like receptors, though structurally quite distinct from the IL-1R at their ligand-binding domains, also recruit MyD88 and IRAK to activate the NF-κB pathway. (b) Structure of the human TLR3 extracellular domain structure. [Part (b) adapted from J. Choe et al., 2005, *Science* **309**:581.]

where NF-κB activates various target genes (see Figure 16-35). These target genes include those encoding IL-1 and IL-6, which contribute to inflammation, as well as the genes for tumor necrosis factor (TNF) and IL-12. Expression of type I interferons, small proteins with antiviral effects, is also turned on in response to TLR signaling.

Cell responses to TLR signaling are quite diverse. For antigen-presenting cells, these responses include not only production of cytokines but also the up-regulation of costimulatory molecules, the surface proteins important for full activation of naive T cells. TLR signaling allows dendritic cells to migrate from where they encounter a pathogen to the draining lymph nodes, where they may interact with naive lymphocytes. Not all activated Toll-like receptors evoke the identical response. Each activated TLR controls production of a particular set of cytokines by dendritic cells. For each engaged TLR, the combination of surface proteins and the cytokine profile induced by TLR engagement thus creates a unique phenotype for the activated dendritic cell. The identity of the microbe encountered sets the pattern of the TLRs that will be activated. These, in turn, shape the differentiation pathways of activated dendritic cells in terms of the cytokines produced, the surface molecules displayed, and the chemotactic cues to which the dendritic cells respond. The mode of activation of a dendritic cell and the cytokines it produces in response create a unique microenvironment in which T cells differentiate. Here they acquire the functional characteristics required to fight the infectious agent that led to engagement of the TLRs in the first place.

Engagement of Toll-Like Receptors Leads to Activation of Antigen-Presenting Cells

Professional antigen-presenting cells engage in continuous endocytosis, and in the absence of pathogens, they display at their surface class I and class II MHC molecules loaded with peptides derived from self-proteins. In the presence of pathogens, the Toll-like receptors on these cells are activated, inducing the antigen-presenting cells to become motile: They detach from the surrounding substratum and start to migrate in the direction of the draining lymph node, where the directional cues are provided by chemokines. An activated dendritic cell, for example, reduces its rate of antigen acquisition, up-regulates the activity of endosomal/lysosomal proteases, and increases the transfer of class II MHC–peptide complexes from the loading compartments to the cell surface. Finally, activated professional antigen-presenting cells upregulate expression of the costimulatory molecules CD80 and CD86, which will allow them to activate T cells more effectively. The initial contact of a professional antigen-presenting cell with a pathogen thus results in its migration to the draining lymph node in a state that is fully capable of activating a naive T cell. Antigen is displayed in the form of peptide-MHC complexes, costimulatory molecules are abundantly present, and cytokines are produced that assist in setting up the proper differentiation program for the T cells to be activated.

Antigen-laden dendritic cells engage antigen-specific T cells, which respond by proliferating and differentiating. The cytokines produced in the course of this priming reaction determine whether a CD4 T cell will polarize towards an inflammatory or a helper cell phenotype. If engagement occurs via class I MHC molecules, a CD8 T cell may develop from a precursor cytotoxic T cell to a fully active cytotoxic T cell. The activated T cells are motile and move through the lymph node, in preparation for their encounter with B cells, or in order to contact the vasculature, enter the circulation, and leave the lymph node to execute effector functions elsewhere in the body.

Production of High-Affinity Antibodies Requires Collaboration Between B and T Cells

In order to generate high-affinity antibodies, which are better at binding antigen and at neutralizing pathogens, B cells require assistance from T cells. Thus, B-cell activation requires a source of antigen to engage the B-cell receptor and the presence of activated antigen-specific T cells. Soluble antigen reaches the lymph node through the afferent lymphatics (see Figure 24-6). Bacterial growth is accompanied by the release of microbial products that can serve as antigens. If the infection is accompanied by local tissue destruction, activation of the complement cascade results in bacterial killing and concomitant release of bacterial proteins, which are also delivered via the lymphatics to the draining lymph node. Complement-modified antigens are superior in the activation of B cells through engagement of complement receptors on B cells, which serve as co-receptors for the B-cell receptor. B cells that acquire antigen via their B-cell receptors internalize the immune complex and process it for presentation via the class II MHC pathway. Antigen-experienced B cells thus convert the BCR-acquired antigen into a call for T-cell help in the form of a class II MHC–peptide complex (Figure 24-36). Note that the epitope recognized by the B-cell receptor may be quite distinct from the peptide ultimately displayed on the surface associated with a class II MHC molecule. As long as the B-cell epitope and the class II-presented peptide—a T-cell epitope—are physically linked, successful B-cell differentiation can be initiated.

The concept of linked recognition explains why there is a minimum size to, for example, peptides to be used successfully as antigens in elicitation of a high-affinity antibody response. Such peptides must fulfill several criteria: They must contain the epitope seen by the B-cell receptor, survive endocytosis and proteolysis, and bind to the allelic class II MHC molecules available in order to be presented as a class II MHC–peptide complex, which serves as a call for T-cell help. For this reason, synthetic peptides used to elicit antibodies are conjugated to carrier proteins to improve their immunogenicity. Only through recognition of this class II MHC–peptide complex via its T-cell receptor can T cells provide the help necessary for the B cell to run its complete course of differentiation.

This concept applies equally to B cells capable of recognizing particular modifications on proteins or peptides. Antibodies that recognize the phosphorylated form of a kinase are commonly raised by immunization of experimental animals

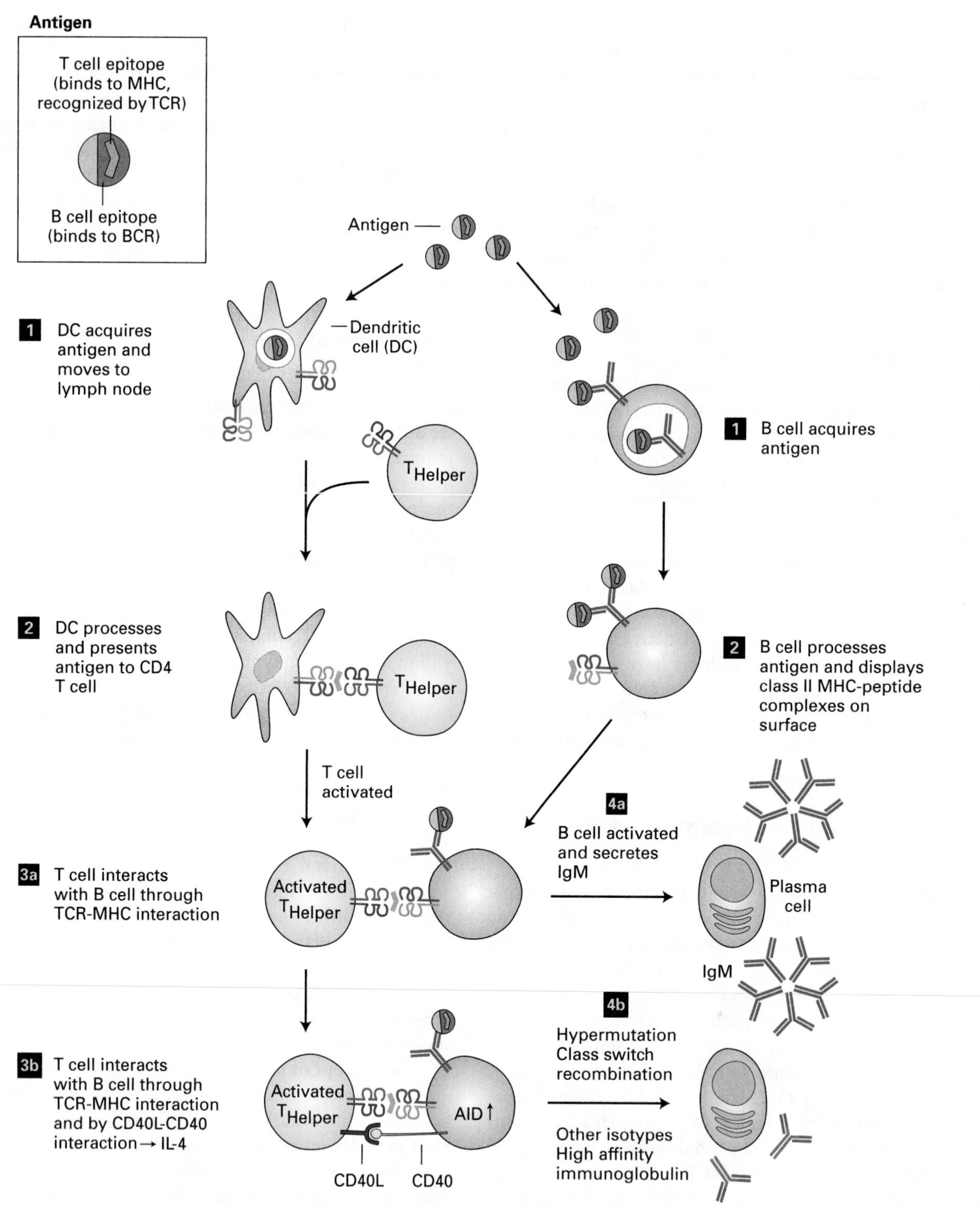

▲ **FIGURE 24-36 Collaboration between T and B cells required to initiate the production of antibodies.** (*Left*) Activation of T cells by means of antigen-loaded dendritic cells (DCs). (*Right*) Antigen acquisition by and subsequent activation of B cells. Step **1**: Professional antigen-presenting cells (DCs, B cells) acquire antigen. Step **2**: Antigen is internalized, processed, and presented to T cells. T-cell activation occurs when dendritic cells present antigen to T cells. Step **3a**: Activated T cells engage antigen-experienced B cells through peptide-MHC complexes displayed on the surface of the B cell. Step **3b**: T cells that are persistently activated initiate expression of the CD40 ligand (CD40L), a prerequisite for B cells becoming fully activated and turning on the enzymatic machinery (activation-induced deaminase) to initiate class switch recombination and somatic hypermutation. Step **4a**: A B cell that receives the appropriate instructions form CD4 helper T cells becomes an IgM-secreting plasma cell. Step **4b**: A B cell that receives signals from activated CD4 helper T cells in the form of CD40–CD40L interactions and the appropriate cytokines can class switch to other immunoglobulin isotypes and engage in somatic hypermutation.

with the phosphorylated peptide in question, conjugated to a carrier protein. An appropriately specific B cell recognizes the phosphorylated site on the peptide of interest, internalizes the complex between phosphorylated peptide and carrier, and generates a complex set of peptides by endosomal proteolysis of the carrier protein. Among these peptides there should be at least one that can bind to the class II MHC molecules carried by that B cell. If properly displayed at the surface of the B cell, this class II MHC–peptide complex becomes the call for T-cell help, provided by CD4 T cells equipped with receptors capable of recognizing the complex of class II MHC molecule and carrier-derived peptide.

The T cell identifies, via its T-cell receptor, an antigen-experienced B cell by means of the class II MHC–peptide complexes displayed by the B cell. The B cell also displays costimulatory molecules and receptors for cytokines produced by the activated T cell (e.g., IL-4). These B cells then proliferate. Some of them differentiate into plasma cells; others are set aside and become memory B cells. The first wave of antibodies produced is always IgM. Class switching to other isotypes and somatic hypermutation (necessary for the generation and selection of high-affinity antibodies) requires the persistence of antigen or repeated exposure to antigen. In addition to cytokines, B cells require cell-cell contacts to initiate somatic hypermutation and class switch recombination. These contacts involve CD40 protein on B cells and CD40L on T cells. These proteins are members of the TNF–TNF receptor family.

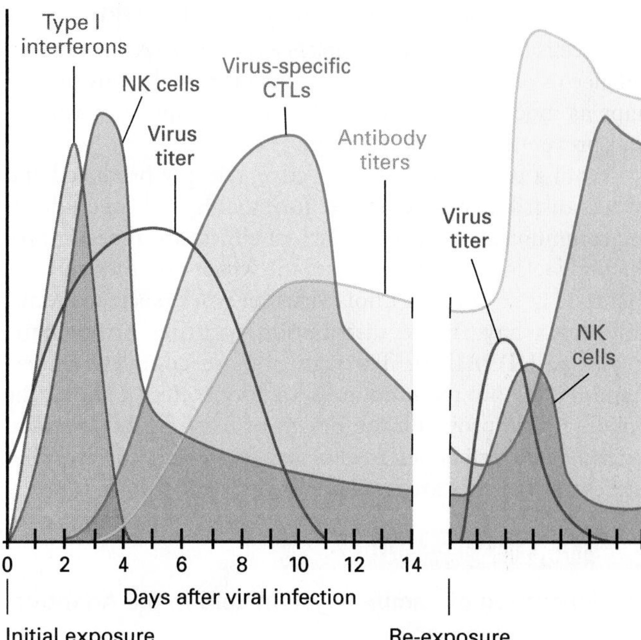

▲ **FIGURE 24-37 Time course of a viral infection.** The initial antiviral response, seen when the numbers of infectious particles rises, includes activation of natural killer (NK) cells and production of type I interferons. These responses are part of innate defenses. The production of antibodies, as well as the activation of cytotoxic T cells (CTLs) follows, and eventually clears the infection. Re-exposure to the same virus leads to more rapid and more pronounced production of antibodies and to more rapid activation of cytotoxic T cells. A successful vaccine induces an immune response similar in some respects to that following initial exposure to a pathogen, but without causing significant symptoms of disease. If a vaccinated person subsequently is exposed to that pathogen, the adaptive immune system is primed to respond quickly and strongly.

Vaccines Elicit Protective Immunity Against a Variety of Pathogens

Arguably the most important application of immunological principles are vaccines. Vaccines are materials designed to be innocuous but that can elicit an immune response for the purpose of providing protection against a challenge with the virulent version of that pathogen (Figure 24-37). It is not always known why such vaccines are as successful as they are, but in many cases, the ability to raise antibodies that can neutralize the pathogen (viruses) or that show microbicidal effects (bacteria) are good indicators of successful vaccination.

Several strategies can lead to a successful vaccine. Vaccines may be composed of live attenuated variants of more virulent pathogens. Serial passage in tissue culture or from animal to animal will often lead to *attenuation,* the molecular basis for which is not well understood. The attenuated version of the pathogen causes a mild form of the disease or causes no symptoms at all. However, by recruiting all the component parts of the adaptive immune system, such live attenuated vaccines can elicit protective levels of antibodies. These antibody levels may wane with advancing age, and repeated immunizations (booster injections) are often required to maintain full protection. Live attenuated vaccines are in use against flu, measles, mumps, and tuberculosis. In the latter case, an attenuated strain of the mycobacterium that causes the diseases is used (Bacille Calmette-Guerin; BCG). Although live attenuated poliovirus was used as a vaccine until recently, its use was discontinued because the risk of re-emergence of more virulent strains of the poliovirus outweighed the benefit. Currently, killed poliovirus is used as the vaccine of choice.

Vaccines based on the cowpox virus, a close relative of the human pathogen variola responsible for smallpox, have been used successfully to completely eradicate smallpox, the first such example of the elimination of an infectious disease. Attempts to achieve a similar feat for polio are nearing completion.

The other major type of vaccine is called a subunit vaccine. Rather than using live attenuated strains of a virulent bacterium or virus, only one of its components is used to elicit immunity. In certain cases this is sufficient to afford lasting protection against a challenge with the live, virulent source of the antigen used for vaccination. This approach has been successful in preventing infections with the hepatitis B virus. The commonly used flu vaccines are composed mostly of the envelope proteins neuraminidase and hemagglutinin (see Figure 3-10); these vaccines elicit neutralizing antibodies. For the human papillomavirus HPV 16, a serotype that causes cervical cancer, virus-like particles are generated composed of the virus capsid proteins, but devoid of its genetic material; these virus-like particles are noninfectious, yet in many respects mimic the intact virion. The HPV vaccine now licensed for use in humans is expected to reduce the incidence of cervical cancer in susceptible populations by perhaps as much as 80 percent, the first example of a vaccine that prevents this particular type of cancer.

From a public health perspective, cheaply produced and widely distributed vaccines are formidable tools in eradicating communicable diseases. Current efforts are aimed at producing vaccines against diseases for which no other suitable therapies are available (Ebola virus) or where socioeconomic conditions have made distribution of drugs problematic (malaria, HIV/AIDS). Through a more complete understanding of how the immune system operates, it should be possible to improve on the design of existing vaccines and extend these principles to diseases for which currently no successful vaccines are available. ▮

KEY CONCEPTS OF SECTION 24.6

Collaboration of Immune-System Cells in the Adaptive Immune Response

- Antigen-presenting cells such as dendritic cells require activation by means of signals delivered to their Toll-like receptors. These receptors are broadly specific for macromolecules produced by bacteria and viruses. Engagement of Toll-like receptors activates the NF-κB signaling pathway, whose outputs include the synthesis of inflammatory cytokines (see Figure 24-35).

- Upon activation, dendritic cells become migratory and move to lymph nodes, ready for their encounter with T cells. Activation of dendritic cells also increases their display of MHC-peptide complexes and expression of costimulatory molecules required for initiation of a T-cell response.

- B cells require assistance of activated T cells to execute their full differentiation program to become plasma cells. Antigen-specific help is provided to B cells by activated T cells, which recognize class II MHC–peptide complexes on the surface of B cells. These B cells generate the relevant MHC-peptide complexes by internalizing antigen via BCR-mediated endocytosis, followed by antigen processing and presentation via the class II MHC pathway (see Figure 24-36).

- In addition to cytokines produced by activated T cells, B cells require cell-cell contact to initiate somatic hypermutation and class switch recombination. This involves CD40 on B cells and CD40L on T cells.

- Important applications of the immunological concept of collaboration between T and B cells include vaccines. The most common forms of vaccines are live attenuated viruses or bacteria, which can evoke a protective immune response without causing pathology, and subunit based vaccines.

Perspectives for the Future

Several areas immunological research promise to have a major impact. The generation of new vaccines that are both safe and effective remains a goal of enormous practical and societal significance: HIV infection, drug-resistant tuberculosis, and malaria are three examples of major killers, each responsible for millions of deaths annually, for which no successful vaccines are currently available. Research must incorporate cell biological as well as immunological concepts to solve this unmet need. Even though some of the most successful vaccines were developed in the absence of detailed immunological knowledge, the current regulatory climate demands a more detailed understanding of the composition of a successful vaccine and why such vaccines work.

Lymphocytes are among the few cell types that can be studied as primary cells that continue to perform cell- and tissue-type specific behavior in tissue culture. This trait makes lymphocytes an attractive model in which to study details of signal transduction, to explore interactions between well-defined, different cell types under defined laboratory conditions, and to accurately measure and model the responses evoked by specific stimuli.

The ability of lymphocytes to create antigen-specific receptors of nearly limitless variability comes at a price: the generation of self-reactive receptors, which is the major contributing factor to autoimmune disease. Vertebrates have several mechanisms for keeping such self-reactive lymphocytes in check, but none of them is fool-proof. We must understand how tolerance to self-antigens is generated, maintained, and ultimately broken at the onset of autoimmune disease. This understanding should help define new strategies for manipulating and controlling self-reactive lymphocytes to prevent or treat autoimmune diseases (e.g., type I diabetes, multiple sclerosis, and arthritis).

With advances in understanding of stem cells and how to use stem cells for therapy in the treatment of diseases (e.g., Parkinson's disease, muscular dystrophy, and spinal cord injuries), transplantation of heterologous stem cells (i.e., derived from an individual other than the patient) is one of several options. The immune system's ability to recognize as foreign such transplanted stem-cell derivatives is a factor that will limit the use of heterologous stem cells. Consequently, new means to suppress responses to these transplants, or to induce tolerance to them, are important goals.

More refined genetic tools continue to allow identification of additional lymphocyte subsets, often with distinct functions. Improved classification schemes will help in understanding lymphocyte function, and so open the door to selective manipulation of lymphocyte function for therapeutic gain.

Our understanding of the interplay between innate and adaptive immune responses, and the many ways in which pathogens interfere in both innate and adaptive immunity, will benefit from our increased understanding of genome sequences and genome structures, both host and pathogen. The use of genetically modified pathogenic organisms as probes for host immune function will help our understanding of basic immunology. This is a rapidly expanding area of basic cell biological research.

Key Terms

affinity maturation 1073
antigen 1055
antigen processing and presentation 1082
autoimmunity 1055
B cells 1056
B-cell receptor (BCR) 1074
chemokines 1061
clonal selection theory 1066
complement 1059
cytokines 1060
cytotoxic T cell 1076
dendritic cells 1059
epitope 1068
Fc receptor 1068
helper T cells 1080
immunoglobulins 1063
inflammation 1061
interleukins 1096
isotypes 1065
junctional imprecision 1071
lymphocytes 1057
macrophages 1059
major histocompatibility complex (MHC) 1076
memory cells 1096
natural killer (NK) cells 1060
neutrophils 1061
opsonization 1060
plasma cells 1073
primary lymphoid organs 1057
secondary lymphoid organs 1058
somatic recombination 1069
T cells 1056
T-cell receptor (TCR) 1088
Toll-like receptors (TLRs) 1059
transcytosis 1065

Review the Concepts

1. Describe the ways in which each of the following pathogens can disarm or manipulate their host's immune system to their own advantage: (a) pathogenic strains of *Staphylococcus aureus* and (b) enveloped viruses.

2. Trace the course of leukocytes as they perform their functions throughout the body.

3. Identify the major mechanical and chemical defenses that protect internal tissues from microbial attack.

4. Compare the classical pathway of complement activation to the alternative pathway.

5. What evidence led Emil von Behring to discover antibodies and the complement system in 1905?

6. What is opsonization? What is the role of antibodies in this process?

7. In B cells, what mechanism insures that only rearranged V genes are transcribed?

8. What prevents further rearrangement of immunoglobulin heavy-chain gene segments in a pre-B cell once a productive heavy-chain rearrangement has occurred?

9. How do antibodies undergo a class switch from producing IgM antibodies to any of the other isotypes?

10. What biochemical mechanism underlies affinity maturation of the antibody response?

11. Compare the structures of class I and class II MHC molecules. What kinds of cells express each class of MHC molecule? What are their functions?

12. Describe the six steps in antigen processing and presentation via the class I MHC restricted pathway.

13. Describe the six steps in antigen processing and presentation via the class II MHC restricted pathway.

14. What prevents self-reactive T cells from leaving the thymus?

15. Explain why T-cell mediated autoimmune diseases are associated with particular alleles of class II MHC genes.

16. How are antigen-presenting cells and helper T cells involved in B-cell activation?

Analyze the Data

To understand how ovalbumin (a protein from chicken eggs) and other foreign antigens in the cytoplasm of a cell are presented for immunosurveillance, ovalbumin can be introduced by electroporation into the cytoplasm of mouse primary B lymphocytes. The ovalbumin is cleaved, and one of its cleavage products is a peptide with the sequence SIINFEKL (single-letter abbreviations). When these B cells are mixed with a clonal T-cell population that specifically recognizes SIINFEKL in the context of MHC molecules, the T cells become stimulated. The graphs below show the secretion of IL-2 after these T cells were mixed with B cells that had been treated in various ways.

Graph A: B cells were electroporated either with ovalbumin (solid line) or with a control protein (dashed line) not recognized by this clonal T-cell population.

Graph B: B cells were first incubated with an inhibitor of lysosomal cysteine proteases (solid line) or an inhibitor of the proteasome (dashed line) and then electroporated with ovalbumin.

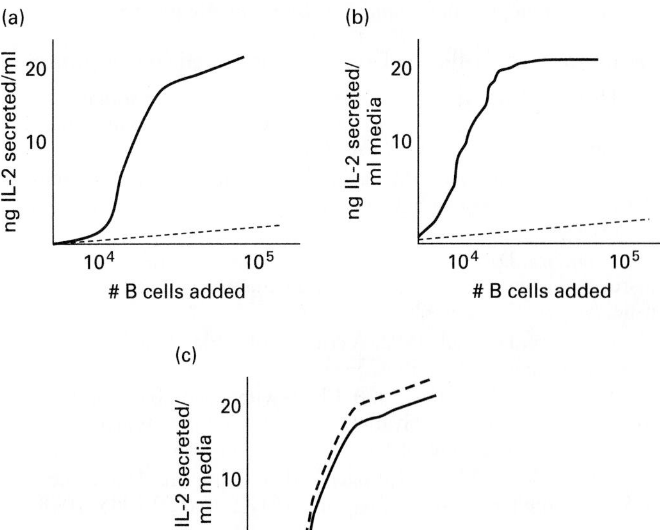

Graph C: B cells were exposed to a high concentration of the SIINFEKL peptide and then immediately fixed [killed] (dotted curve) with formaldehyde or incubated for two hours in the presence (dashed curve) or absence (solid curve) of the proteasome inhibitor and then fixed with formaldehyde.

a. Under what conditions is IL-2 secreted? Is it likely that the T cells or the B cells are secreting IL-2?

b. What information is derived from the use of the lysosomal and the proteasome inhibitors? Is the ovalbumin peptide more likely to be presented complexed with class I MHC or class II MHC molecules? What is the most likely pathway by which ovalbumin in the cytoplasm of the B cell is presented for recognition by the appropriate T cells?

c. Why do you think that the presence or absence of the proteasome inhibitor in experiment C had no effect on the amount of IL-2 secreted, whereas the presence of the inhibitor in experiment B shows a marked effect?

References

Overview of Host Defenses

Akira, S., K. Kiyoshi Takeda, and T. Kaisho. 2001. Toll-like receptors: critical proteins linking innate and acquired immunity. *Nature Immunol.* **2:**675–680.

Heyman, B. 2000. Regulation of antibody responses via antibodies, complement, and Fc receptors. *Ann. Rev. Immunol.* **18:**709–737.

Lemaitre, B., and J. Hoffmann. 2007. The host defense of *Drosophila melanogaster. Ann. Rev. Immunol.* (in press).

von Behring, E., and S. Kitasato. 1890. The mechanism of diphtheria immunity and tetanus immunity in animals. Reprinted in *Mol. Immunol.,* 1991, **28**(12):1317, 1319–1320.

Immunoglobulins: Structure and Function

Williams, A. F., and A. N. Barclay. 1988. The immunoglobulin superfamily—domains for cell-surface recognition. *Ann. Rev. Immunol.* **6:**381–405.

Amzel, L. M., and R. J. Poljak. 1979. Three-dimensional structure of immunoglobulins. *Ann. Rev. Biochem.* **48:**961–997.

Generation of Antibody Diversity and B-Cell Development

Hozumi, N., and S. Tonegawa. 1976. Evidence for somatic rearrangement of immunoglobulin genes coding for variable and constant regions. *Proc. Nat'l. Acad. Sci. USA* 73:3628–3632.

Jung, D., et al. 2006. Mechanism and control of V(D)J recombination at the immunoglobulin heavy chain locus. *Ann. Rev. Immunol.* **24:**541–570.

Kitamura, D., et al. 1991. A B cell-deficient mouse by targeted disruption of the membrane exon of the immunoglobulin mu chain gene. *Nature* **350:**423–426.

Kitamura, D., et al. 1992. A critical role of lambda 5 protein in B cell development. *Cell* **69:**823–31.

Muramatsu, M., et al. 2000. Class switch recombination and hypermutation require activation-induced cytidine deaminase (AID), a potential RNA editing enzyme. *Cell* **102:**553–563.

Schatz, D. G., M. A. Oettinger, and D. Baltimore. 1989. The V(D)J recombination activating gene, RAG-1. *Cell* **59:**1035–1048.

The MHC and Antigen Presentation

Bjorkman, P. J., et al. 1987. Structure of the human class I histocompatibility antigen, HLA-A2. *Nature* **329:**506–512.

Brown, J. H., et al. 1993. Three-dimensional structure of the human class II histocompatibility antigen HLA-DR1. *Nature* **364:**33–39.

Neefjes, J. J., et al. 1990. The biosynthetic pathway of MHC class II but not class I molecules intersects the endocytic route. *Cell* **61:**171–183.

Peters, P. J., et al. 1991. Segregation of MHC class II molecules from MHC class I molecules in the Golgi complex for transport to lysosomal compartments. *Nature* **349:**669–676.

Rock, K. L., et al. 1994. Inhibitors of the proteasome block the degradation of most cell proteins and the generation of peptides presented on MHC class I molecules. *Cell* **78:**761–771.

Rudolph, M. G., R. L. Stanfield, and I. A. Wilson IA. 2006. How TCRs bind MHCs, peptides, and coreceptors. *Ann. Rev. Immunol.* **24:**419–466.

Townsend, A. R., et al. 1984. Cytotoxic T cell recognition of the influenza nucleoprotein and hemagglutinin expressed in transfected mouse L cells. *Cell* **39:**13–25.

Zinkernagel, R. M., and P. C. Doherty. 1974. Restriction of in vitro T cell-mediated cytotoxicity in lymphocytic choriomeningitis within a syngeneic or semiallogeneic system. *Nature* **248:**701–702.

T Cells, T-Cell Receptors, and T-Cell Development

Miller, J. F. 1961. Immunological function of the thymus. *Lancet* 30(2):748–749.

Kisielow, P., et al. 1988. Tolerance in T-cell-receptor transgenic mice involves deletion of nonmature CD4+8+ thymocytes. *Nature* **333:**742–746.

Dembic, Z., et al. 1986. Transfer of specificity by murine alpha and beta T-cell receptor genes. *Nature* **320:**232–238.

Shinkai, Y., et al. 1993. Restoration of T cell development in RAG-2-deficient mice by functional TCR transgenes. *Science* **259:**822–825.

Mombaerts, P., et al. 1992. RAG-1-deficient mice have no mature B and T lymphocytes. *Cell* **68:**869–877.

Lenschow, D. J., T. L. Walunas, and J. A. Bluestone. 1996. CD28/B7 system of T cell costimulation. *Ann. Rev. Immunol.* **14:**233–258.

Sharpe, A. H., and A. K. Abbas. 2006. T-cell costimulation: biology, therapeutic potential, and challenges. *N. Engl. J. Med.* **355:**973–975.

Collaboration of Immune-System Cells in the Adaptive Immune Response

20 years of HIV science. 2003. *Nature Med.* **9:**803–843. A collection of opinion pieces on the prospects for an AIDS vaccine.

Banchereau, J. 2002. The long arm of the immune system. *Sci. Am.* **287:**52–59.

Chang, M-H., et al. for The Taiwan Childhood Hepatoma Study Group. 1997. Universal hepatitis B vaccination in Taiwan and the incidence of hepatocellular carcinoma in children. *N. Engl. J. Med.* **336:**1855–1859.

Cytokines Online Pathfinder Encyclopedia. http://www.copewithcytokines.de/

Jego, G., et al. 2003. Plasmacytoid dendritic cells induce plasma cell differentiation through type I interferon and interleukin 6. *Immunity* **19:**225–234.

Rajewsky, K., et al. 1969. The requirement of more than one antigenic determinant for immunogenicity. *J. Exp. Med.* **129:**1131–1143.

Koutsky, L. A., et al. 2002. A controlled trial of a human papillomavirus type 16 vaccine. *N. Engl. J. Med.* **347:**1645–1651.

Plotkin, S. A., and W. A. Orenstein. 2003. *Vaccines,* 4th ed. Saunders.

Smith, Jane S. 1990. *Patenting the Sun: Polio and the Salk Vaccine.* Wm Morrow and Co.

Steinman, R. M., and H. Hemmi. 2006. Dendritic cells: translating innate to adaptive immunity. *Curr. Topics Microbiol. Immunol.* **311:**17–58.

Steinman, R. M., and Z. A. Cohn. 2007. Identification of a novel cell type in peripheral lymphoid organs of mice. I. Morphology, quantitation, tissue distribution. *J. Immunol.* **178:**5–25.

CLASSIC EXPERIMENT 24.1

TWO GENES BECOME ONE: SOMATIC REARRANGEMENT OF IMMUNOGLOBULIN GENES

N. Hozumi and S. Tonegawa, 1976, *Proc. Nat'l. Acad. Sci. USA* **73**:3629

For decades, immunologists wondered how the body could generate the multitude of pathogen-fighting immunoglobulins, called antibodies, needed to ward off the vast array of different bacteria and viruses encountered in a lifetime. Clearly, these protective proteins, like all proteins, somehow were encoded in the genome. But the enormous number of different antibodies potentially produced by the immune system made it unlikely that individual immunoglobulin (Ig) genes encoded all the possible antibodies an individual might need. In studies beginning in the early 1970s, Susumu Tonegawa, a molecular biologist, laid the foundation for solving the mystery of how antibody diversity is generated.

Background

Research on the structure of Ig molecules provided some clues about the generation of antibody diversity. First, it was shown that an Ig molecule is composed of four polypeptide chains: two identical heavy (H) chains and two identical light (L) chains. Some researchers proposed that antibody diversity resulted from different combinations of heavy and light chains. Although somewhat reducing the number of genes needed, this hypothesis still required that a large portion of the genome be devoted to Ig genes. Protein chemists then sequenced several Ig light and heavy chains. They found that the C-terminal regions of different light chains were very similar and thus were termed the constant (C) region, whereas the N-terminal regions were highly variable and thus were termed the variable (V) region. The sequences of different heavy chains exhibited a similar pattern. These findings suggested that the genome contains a small number of C genes and a much larger group of V genes.

In 1965, W. Dryer and J. Bennett proposed that two separate genes, one V gene and one C gene, encode each heavy chain and each light chain. Although this proposal seemed logical, it violated the well-documented principle that each gene encodes a single polypeptide. To avoid this objection, Dryer and Bennett suggested that a V and C gene somehow were rearranged in the genome to form a single gene, which then was transcribed and translated into a single polypeptide, either a heavy or light chain. Indirect support for this model came from DNA hybridization studies showing that only a small number of genes encoded Ig constant regions. However, until more powerful techniques for analyzing genes came on the scene, a definitive test of the novel two-gene model was not possible.

The Experiment

Tonegawa realized that if immunoglobulin genes underwent rearrangement, then the V and C genes were most likely located at different points in the genome. The discovery of restriction endonucleases, enzymes that cleave DNA at specific sites, had allowed some bacterial genes to be mapped. However, because mammalian genomes are much more complex, similar mapping of the genes encoding V and C regions was not technically feasible. Instead, drawing on newly developed molecular biology techniques, Tonegawa devised another approach for determining whether the V and C regions were encoded by two separate genes. He reasoned that if rearrangement of the V and C genes occurs, it must happen during differentiation of Ig-secreting B cells from embryonic cells. Furthermore, if rearrangement occurs, there should be detectable differences between unrearranged germ-line DNA from embryonic cells and the DNA from Ig-secreting B cells. Thus, he set out to see if such differences existed using a combination of restriction-enzyme digestion and RNA–DNA hybridization to detect the DNA fragments.

He began by isolating genomic DNA from mouse embryos and from mouse B cells. To simplify the analysis, he used a line of B-cell tumor cells, all of which produce the same type of antibody. The genomic DNA was then digested with the restriction enzyme Bam*HI*, which recognizes a sequence that occurs relatively rarely in mammalian genomes. Thus, the DNA was broken into many large fragments. He then separated these DNA fragments by agarose gel electrophoresis, which separates biomolecules on the basis of charge and size. Since all DNA carries an overall negative charge, the fragments were separated based on their size. Next, he cut the gel into small slices and isolated the DNA from each slice. Now, Tonegawa had many fractions of DNA pieces of various sizes. He then could analyze these DNA fractions to determine if the V and C genes resided on the same fragment in both B cells and embryonic cells.

To perform this analysis, Tonegawa first isolated from the B-cell tumor cells the mRNA encoding the major type of Ig light chain, called κ. Since a RNA is complementary to one strand of the DNA from which it is transcribed, it can hybridize with this strand forming a RNA–DNA hybrid. By radioactively labeling the entire κ mRNA, Tonegawa produced a probe for detecting which of the separated DNA fragments contained the κ chain gene. He then isolated the 3′ end of the

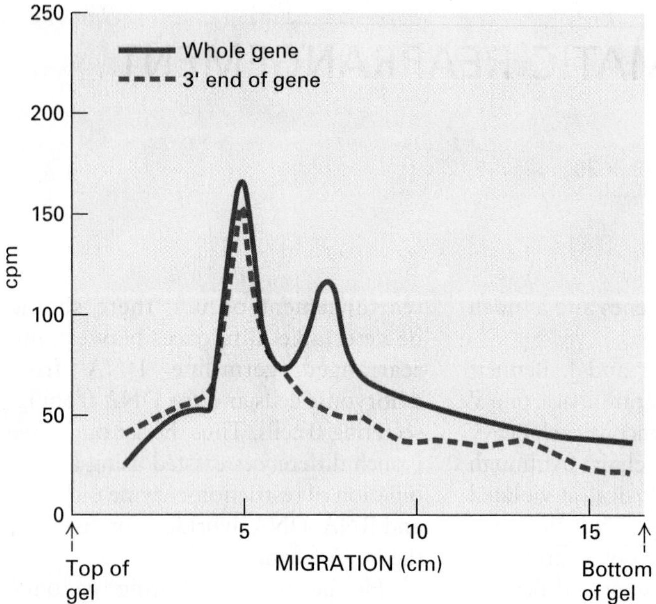

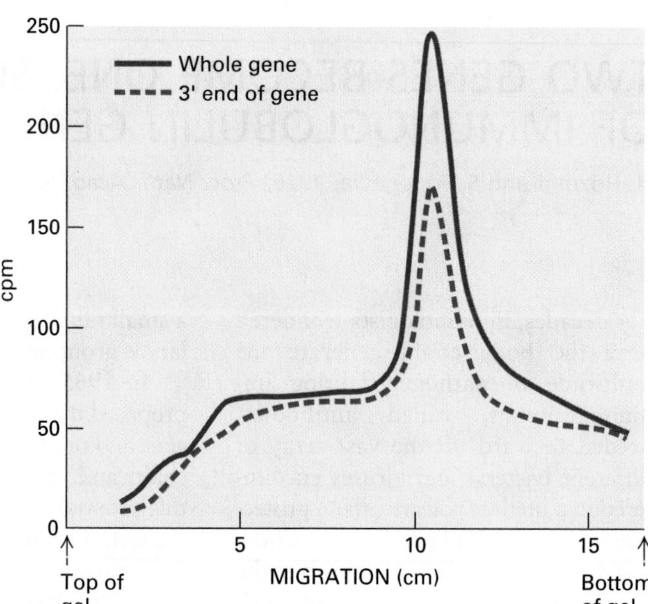

▲ **FIGURE 1 Experimental results showing that the genes encoding the variable (V) and constant (C) regions of κ light chains are rearranged during development of B cells.** These curves depict the hybridization of labeled RNA probes, specific for the entire κ gene (V + C) and for the 3′ end that encodes the C region, to fractions of digested embryonic or B-cell DNA separated by agarose gel electrophoresis. [Adapted from N. Hozumi and S. Tonegawa, 1976, *Proc. Nat'l. Acad. Sci. USA* 73:3629.]

κ mRNA and labeled it, yielding a second probe that would detect only the DNA sequences encoding the constant region of the κ chain. With these probes in hand—one specific for the combined V + C gene and one specific for C alone—Tonegawa was ready to compare the DNA fragments obtained from B cells and embryonic cells.

He first denatured the DNA in each of the fractions into single strands and then added one or the other labeled probe. He found that the C-specific probe hybridized to different fractions derived from embryonic and B-cell DNA (Figure 1). Even more telling, the full-length RNA probe hybridized to two *different* fractions of the embryonic DNA, suggesting that the V and C genes are not connected and that a cleavage site for Bam*HI* lies between them. Tonegawa concluded that during the formation of B cells, separate genes encoding the V and C regions are rearranged into a single DNA sequence encoding the entire κ light chain (Figure 2).

Discussion

The generation of antibody diversity was a problem awaiting development of powerful molecular techniques to answer it. Tonegawa went on to clone V-region genes and prove that the rearrangement must occur somatically. These findings impacted genetics as well as immunology. Where once it was believed that every cell in the body contained the same genetic information, it became clear that some cells take that information and alter it to suit other purposes. In addition to somatic rearrangement, Ig genes undergo a variety of other alterations that allow the immune system to create the diverse repertoire of antibodies necessary to react to any invading organism. Our current understanding of these mechanisms rests on the foundation of Tonegawa's fundamental discovery. For this work, he received the Nobel Prize for Physiology and Medicine in 1987.

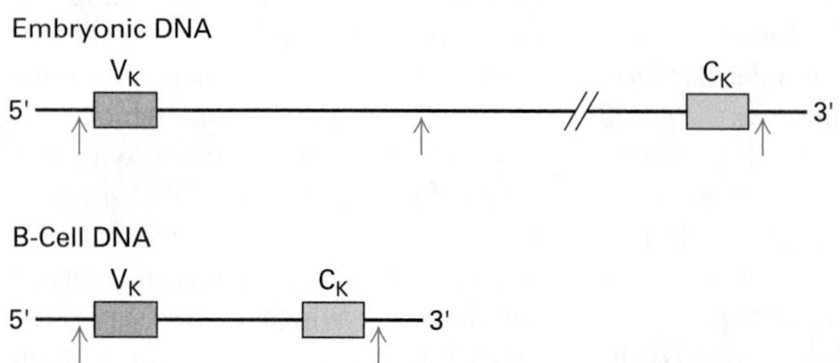

◀ **FIGURE 2 Schematic diagram of κ light-chain DNA in embryonic cells and B cells that is consistent with Tonegawa's results.** In embryonic cells, cleavage with the Bam*HI* restriction enzymes (red arrows) produces two different sized fragments, one containing the V gene and one containing the C gene. In B cells, the DNA is rearranged so that the V and C genes are adjacent, with no intervening cleavage site. Bam*HI* digestion thus yields one fragment that contains both the V and C genes.

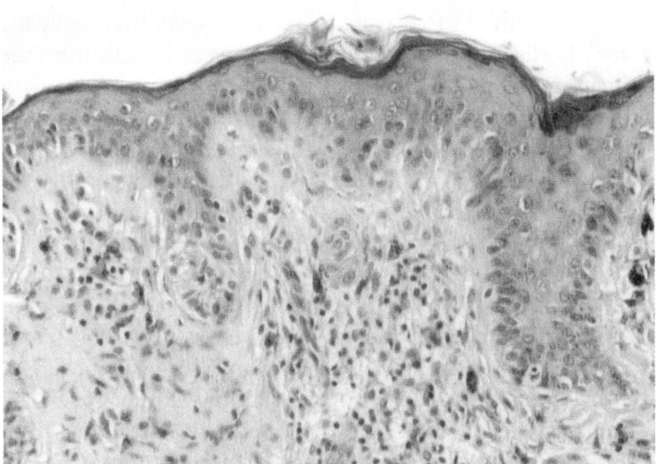

Melanoma tumor in human skin, stained with hematoxylin and eosin. Underlying the relatively normal epithelium at the top are invasive melanoma cells, many with small dark nuclei and some producing brown melanotic inclusions. Melanoma tumors are malignant, moving from their original sites in the skin to seed new tumors in a variety of tissues; these metastases can be fatal. Worldwide each year, about 48,000 people die from melanoma and 160,000 new cases are detected. The only treatment is early detection and excision. (Courtesy of Anthony Oro, Sam Dadras, and Matthew Scott.)

CHAPTER

25

CANCER

Cancer causes about one-fifth of the deaths in the United States each year. Worldwide, between 100 and 350 of each 100,000 people die of cancer each year. Cancer is due to failures of the mechanisms that usually control the growth and proliferation of cells. During normal development and throughout adult life, intricate genetic control systems regulate the balance between cell birth and death in response to growth signals, growth-inhibiting signals, and death signals. Cell birth and death rates determine adult body size, and the rate of growth in reaching that size. In some adult tissues, cell proliferation occurs continuously as a constant tissue-renewal strategy. Intestinal epithelial cells, for instance, live for just a few days before they die and are replaced; certain white blood cells are replaced as rapidly, and skin cells commonly survive for only 2–4 weeks before being shed. The cells in many adult tissues, however, normally do not proliferate except during healing processes. Such stable cells (e.g., hepatocytes, heart muscle cells, neurons) can remain functional for long periods or even for the entire lifetime of an organism. Cancer occurs when the mechanisms that maintain these normal growth rates malfunction to cause excess cell division.

The losses of cellular regulation that give rise to most or all cases of cancer are due to genetic damage that is often accompanied by influences of tumor-promoting chemicals, hormones, and sometimes viruses (Figure 25-1). Mutations in two broad classes of genes have been implicated in the onset of cancer: **proto-oncogenes** and **tumor-suppressor genes.** Proto-oncogenes normally promote cell growth; they are changed into **oncogenes** by mutations that cause the gene to be excessively active in growth promotion. Either increased gene expression or production of a hyperactive product will do it. Tumor-suppressor genes normally restrain growth, so mutations that inactivate them allow inappropriate cell division. A third, more specialized, class of genes called **caretaker genes** are also often linked to cancer. Caretaker genes normally protect the integrity of the genome; when they are inactivated, cells acquire additional mutations at an increased rate—including mutations that damage growth control and lead to cancer. Many of the genes in these three classes encode proteins that help regulate cell birth (i.e., entry into and progression through the cell cycle) or cell death by **apoptosis;** others encode proteins that participate in repairing damaged DNA.

Cancer commonly results from mutations that arise during a lifetime's exposure to **carcinogens,** substances encountered in the environment, which include certain chemicals and

OUTLINE

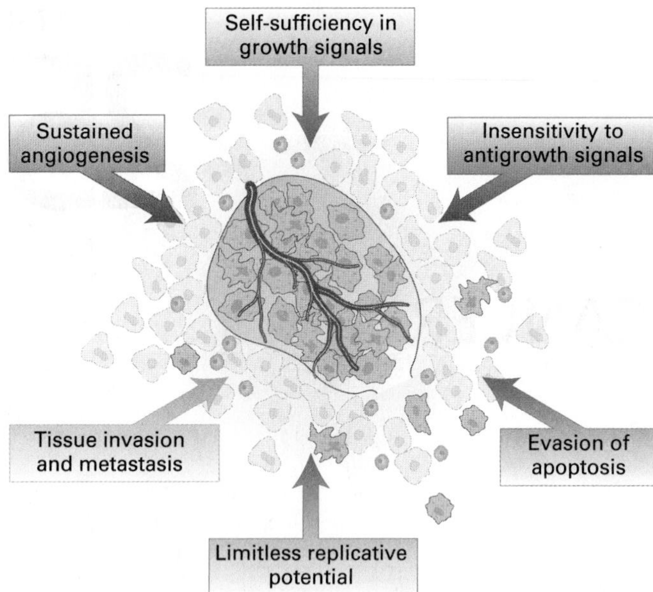

▲ **FIGURE 25-1 Overview of changes in cells that cause cancer.** During carcinogenesis, six fundamental cellular properties are altered, as shown here in this tumor growing within normal tissue, to give rise to the complete, most destructive cancer phenotype. Less dangerous tumors arise when only some of these changes occur. In this chapter we examine the genetic changes that result in these altered cellular properties. [Adapted from D. Hanahan and R. A. Weinberg, 2000, *Cell* **100**:57.]

ultraviolet radiation. Most cancer cells lack one or more DNA repair systems, which may explain the large number of mutations they accumulate. Although DNA-repair enzymes do not directly inhibit cell proliferation, cells that have lost the ability to repair errors, gaps, or broken ends in DNA accumulate mutations in many genes, including those that are critical in controlling cell growth and proliferation. Thus loss-of-function mutations in caretaker genes such as the genes encoding DNA-repair enzymes prevent cells from correcting mutations that inactivate tumor-suppressor genes or activate oncogenes.

Cancer-causing mutations occur mostly in somatic cells, not in the germ-line cells, and somatic cell mutations are not passed on to the next generation. In contrast, certain inherited mutations, which are carried in the germ line, increase the probability that cancer will occur at some time. In a destructive partnership, somatic mutations can combine with inherited mutations to cause cancer.

Thus the cancer-forming process, called *oncogenesis* or *tumorigenesis,* is an interplay between genetics and the environment. Most cancers arise after genes are altered by carcinogens or by errors in the copying and repair of genes. Even if the genetic damage occurs only in one somatic cell, division of this cell will transmit the damage to the daughter cells, giving rise to a **clone** of altered cells. Rarely, however, does mutation in a single gene lead to the onset of cancer. More typically, a series of mutations in multiple genes creates a progressively more rapidly proliferating cell type that escapes normal growth restraints, creating an opportunity for additional mutations. The cells also acquire other properties that give them an advantage, such as the ability to escape from normal epithelia and stimulate the growth of vasculature to obtain oxygen. Eventually the clone of cells grows into a **tumor.** In some cases cells from the primary tumor migrate to new sites where they form secondary tumors, a process termed **metastasis.** Most cancer deaths are due to invasive, fast-growing metastasized tumors.

Metastasis is a complex process with many steps. Invasion of new tissues is nonrandom, depending on the nature of both the metastasizing cell and the invaded tissue. Metastasis is facilitated if the tumor cells produce growth factors and angiogenesis factors (inducers of blood vessel growth). Motile, invasive cells are most dangerous. Tissues under attack, such as bone, blood vessels, and liver, are most vulnerable if they produce growth factors and readily grow new vasculature, since these help to support the invaders. They are more resistant if they produce (1) anti-proliferative factors that restrain tumor cell division, (2) inhibitors of proteolytic enzymes; these block cancer cell proteases that are used to penetrate tissue, and (3) anti-angiogenesis factors that stop tumor cells from spurring the growth of blood vessels.

Research on the genetic foundations of a particular type of cancer often begins by identifying one or more genes that are mutationally altered in tumor cells. Subsequently it is important to learn whether an altered gene is a contributing cause for the tumor, or an irrelevant side event. Such investigations usually employ multiple approaches: epidemiological comparisons of the frequency with which the genetic change is associated with a type of tumor, tests of the growth properties of cells in culture that have the particular mutation, and the construction of mouse models of the disease to see the effects of mutating a putative cancer-related gene. A more sophisticated analysis is possible when the altered gene is known to encode a component of a particular molecular pathway (e.g., an intracellular signaling pathway). In this case it is possible to alter other components of the same pathway and see whether the same type of cancer arises.

Time plays an important role in cancer. Many years may be required to accumulate the multiple mutations that are required to form a tumor, so most cancers develop later in life. The requirement for multiple mutations also lowers the frequency of cancer compared with what it would be if tumorigenesis were triggered by a single mutation. However, huge numbers of cells are, in essence, mutagenized and tested for altered growth during our lifetimes, a powerful selection in favor of cells which, in this case, we do not want. Cells that proliferate quickly become more abundant, undergo further genetic changes, and can become progressively more dangerous. Furthermore, cancer occurs most frequently after the age of reproduction, and therefore plays a lesser role in reproductive success. So cancer is common, in part reflecting increasingly long human lives but also perhaps reflecting the lack of evolutionary selection against the disease.

In this chapter, we first introduce the properties of tumor cells and describe the multistep process of oncogenesis. Next, we consider the general types of genetic changes that lead to the unique characteristics of cancer cells, and the interplay between somatic and inherited mutations. The following sections examine in detail how mutations affecting both growth-promoting and growth-inhibiting processes can result in excess cell proliferation. We conclude the chapter with a discussion of the role of carcinogens and how a breakdown in DNA repair mechanisms, by the loss of caretaker genes or the activation of a telomerase enzyme that allows more cell divisions, can lead to oncogenesis.

25.1 Tumor Cells and the Onset of Cancer

Before examining in detail the genetic basis of cancer, we consider the properties of tumor cells that distinguish them from normal cells and the general process of tumorigenesis. The change from a normal cell into a cancer cell commonly involves multiple steps, each one adding properties that make cells more likely to grow into a tumor. The genetic changes that underlie oncogenesis alter several fundamental properties of cells, allowing cells to evade normal growth controls and ultimately conferring the full cancer phenotype (see Figure 25-1). Cancer cells acquire a drive to proliferate that does not require an external inducing signal. They fail to sense signals that restrict cell division and continue to live when they should die. They often change their attachment to surrounding cells or the extracellular matrix, breaking loose to move away and spread the tumor. A cancer cell may, up to a point, resemble a particular type of normal, rapidly dividing cell, but the cancer cell and its progeny will be immortal. Tumors are characteristically *hypoxic* (oxygen-starved), so to grow to more than a small size, tumors must obtain a blood supply. They often do so by signaling to induce the growth of blood vessels into the tumor. As cancer progresses, tumors become an abnormal organ, increasingly well adapted to growth and invasion of surrounding tissues.

Normal animal cells are often classified according to their embryonic tissue of origin, and the naming of tumors has followed suit. Malignant tumors are classified as *carcinomas* if they derive from epithelia such as endoderm (gut epithelium) or ectoderm (skin and neural epithelia), and *sarcomas* if they derive from mesoderm (muscle, blood, and connective tissue precursors). Carcinomas are by far the most common type of malignant tumor (more than 90 percent). The *leukemias*, a class of sarcomas, grow as individual cells in the blood, whereas most other tumors are solid masses. (The name *leukemia* is derived from the Latin for "white blood": the massive proliferation of leukemic cells can cause a patient's blood to appear milky.) The *lymphomas,* another type of malignant sarcoma, are solid tumors of lymphocytes and plasma cells. Malignant brain tumors can be derived from neural cells, and *glioblastomas* are tumors of the glial cells that constitute the major cell type in the brain.

Metastatic Tumor Cells Are Invasive and Can Spread

Tumors arise with great frequency, especially in older individuals, but most pose little risk to their host because they are localized, and of small size. We call such tumors **benign;** an example is warts, a benign skin tumor. The cells composing benign tumors closely resemble, and may function like, normal cells. The cell-adhesion molecules that hold tissues together keep benign tumor cells, like normal cells, localized to the tissues where they originate. A fibrous capsule usually delineates the extent of a benign tumor and makes it an easy target for a surgeon. Benign tumors become serious medical problems only if their sheer bulk interferes with normal functions or if they secrete excess amounts of biologically active substances like hormones. Acromegaly, the overgrowth of head, hands, and feet, for example, can occur when a benign pituitary tumor causes overproduction of growth hormone.

In contrast, cells composing a **malignant** tumor, or **cancer** (Figure 25-2), usually grow and divide more rapidly than normal, and fail to die at the normal rate (e.g., chronic lymphocytic leukemia, a tumor of white blood cells). A key characteristic of malignant cells is their ability to invade nearby tissue, spreading and seeding additional tumors while the cells

(a)

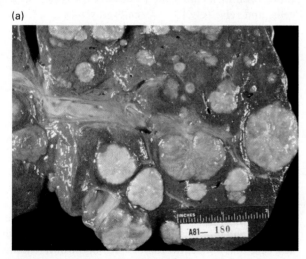

(b)

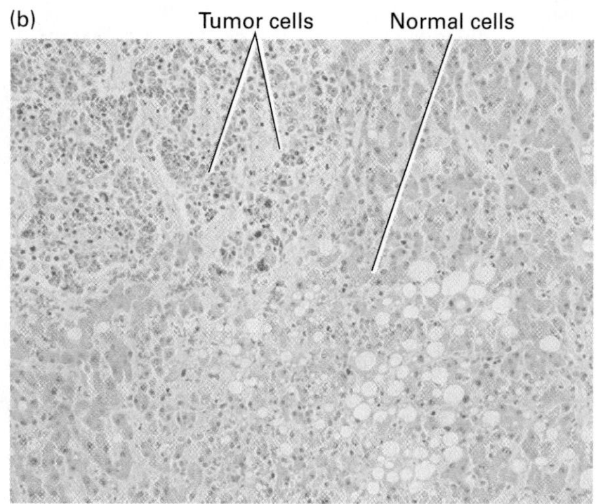

▲ **FIGURE 25-2 Gross and microscopic views of a tumor invading normal liver tissue.** (a) The gross morphology of a human liver in which a metastatic lung tumor is growing. The white protrusions on the surface of the liver are the tumor masses. (b) A light micrograph of a section of the tumor in (a) showing areas of small, dark-staining tumor cells invading a region of larger, light-staining, normal liver cells. [Courtesy of J. Braun.]

continue to proliferate. Some malignant tumors, such as those in the ovary or breast, remain localized and encapsulated, at least for a time. When these tumors progress, the cells invade surrounding tissues, and undergo metastasis (Figure 25-3a). Most malignant cells eventually acquire the ability to metastasize. Thus the major characteristics that differentiate metastatic (or malignant) tumors from benign ones are their abilities to invade nearby tissues and spread to distant ones.

Normal cells are restricted to their place in an organ or tissue by cell-cell adhesion and by physical barriers such as the *basement membrane,* which underlies layers of epithelial cells and also surrounds the endothelial cells of blood vessels (Chapter 19). Cancer cells have a complex relation to the extracellular matrix and basement membrane. They can penetrate the matrix using a cell protrusion called an "invadopodium," formed by the localized assembly of an actin cytoskeleton (Figure 25-3b). The cells must degrade the basement membrane to penetrate it and metastasize, but in some cases cells may migrate along the membrane. Many tumor cells secrete a protein (plasminogen activator) that converts the serum protein plasminogen to the active protease plasmin. Increased plasmin activity, together with other protease activities, promotes metastasis by digesting the basement membrane, thus allowing its penetration by tumor cells (Figure 25-3b). As the basement membrane disintegrates, some tumor cells will enter the blood, but fewer than 1 in 10,000 cells that escape the primary tumor survive to colonize another tissue and form a secondary, metastatic tumor. In addition to escaping the original tumor and entering the blood, cells that will seed new tumors must then adhere to an endothelial cell lining a capillary and migrate across or through it into the underlying tissue. The multiple crossings of tissue layers that underlie malignancy often involve new or variant surface proteins made by malignant cells (Figure 25-3c).

(a)

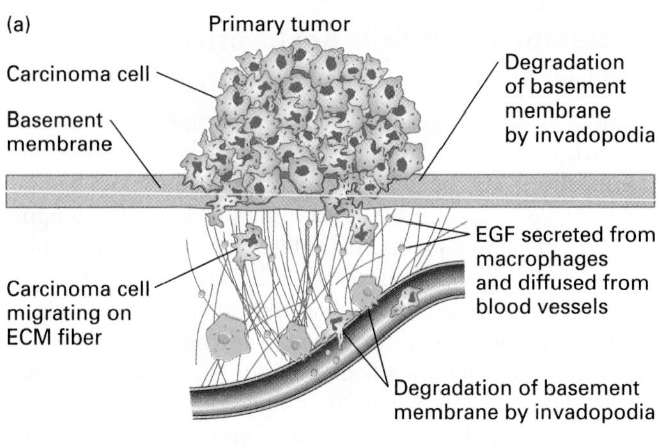

(b)

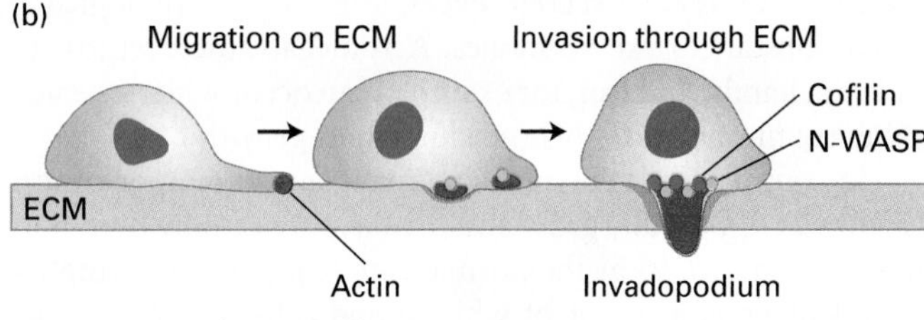

(c) Components of invadopodia

Actin regulation: Cortactin, N-WASP, Arp2/3 complex, WIP, cofilin, tafin
Signaling/adaptors: Cdc42, Nck1, p190RhoGAP, Src, FAK, PKCμ, AMAP1
Adhesion: Integrins, vinculin, paxillin, tensin
Proteases: MMP2, MT1-MMP, seprase, invadolysin
Membrane remodeling: Dynamin-2, Art6, synaptojanin-2

▲ **FIGURE 25-3 Metastasis.** (a) First steps in metastasis, using breast carcinoma cells as an example. Cancer cells leave the main tumor and attack the basement membrane, using extracellular matrix (ECM) fibers to reach the blood vessels. The cancer cells can be attracted by signals such as epidermal growth factor (EGF), which can be secreted by macrophages (yellow). At the blood vessels they penetrate the layer of endothelial cells that forms the vessel walls, and enter the bloodstream. (b) Carcinoma cells penetrate the extracellular matrix and blood vessel wall by extending "invadopodia," which produce matrix metalloproteases and other proteases to open up a path. (c) Proteins used by invadopodia. N-WASP (Chapter 17) controls assembly of actin into filaments that form the internal skeleton of the invadopodia. One hope for fighting cancer is to develop drugs that target components of the invasion and metastasis process without blocking normal essential functions. [Adapted from Yamaguchi et al., 2005, *Curr. Opin. Cell Biol.* **17:**559.]

In addition to important changes in cell-surface proteins, drastic changes occur in the cytoskeleton during tumor-cell formation and metastasis. These alterations can result from changes in the expression of genes encoding Rho and other small GTPases that regulate the actin cytoskeleton (Chapter 17). For instance, tumor cells have been found to overexpress the *RhoC* gene, which encodes a protein that promotes actin/myosin contraction in cells, and this increased RhoC activity stimulates metastasis.

Cancer cells can often be distinguished from normal cells by microscopic examination. They are usually less well differentiated than normal cells or benign tumor cells. In a specific tissue, malignant cells usually exhibit the characteristics of rapidly growing cells, that is, a high nucleus-to-cytoplasm ratio, prominent nucleoli, an increased frequency of mitotic cells, and relatively little specialized structure. The presence of invading cells in an otherwise normal tissue section is used to diagnose a malignancy.

Cancers Usually Originate in Proliferating Cells

In order for most oncogenic mutations to induce cancer, they must occur in dividing cells so that the mutation is passed on to many progeny cells. When such mutations occur in nondividing cells (e.g., neurons and muscle cells), they generally do not induce cancer, which is why tumors of muscle and nerve cells are rare in adults. Nonetheless, cancer can occur in tissues composed mainly of nondividing differentiated cells such as erythrocytes and most white blood cells, absorptive cells that line the small intestine, and keratinized cells that form the skin. The cells that initiate the tumors are not the differentiated cells, but rather their precursor cells. Fully differentiated cells usually do not divide. As they die or wear out, they are continually replaced by proliferation and differentiation of **stem cells,** and these cells are capable of transforming into tumor cells.

In Chapter 21, we learned that stem cells both perpetuate themselves and give rise to differentiating cells that can regenerate a particular tissue for the life of an organism (Figure 21-2). For instance, many differentiated blood cells have short life spans and are continually replenished from hematopoietic (blood-forming) stem cells in the bone marrow (Figure 21-15). Populations of stem cells in the intestine, liver, skin, bone, and other tissues likewise give rise to all or many of the cell types in these tissues, replacing aged and dead cells, by pathways analogous to hematopoiesis in bone marrow. Similarly within a tumor there may be only certain cells with the ability to divide uncontrollably and generate new tumors; such cells are *tumor stem cells,* discussed below.

Because stem cells can divide continually over the life of an organism, oncogenic mutations in their DNA can accumulate, eventually transforming them into cancer cells. Cells that have acquired these mutations have an abnormal proliferative capacity and generally cannot undergo normal processes of differentiation. Many oncogenic mutations, such as ones that prevent apoptosis or generate an inappropriate growth-promoting signal, also can occur in more differentiated, but still replicating, progenitor cells. Such mutations in hematopoietic progenitor cells can lead to various types of leukemia.

Cancer Stem Cells Can Be a Minority Population

Tumors may arise from normal stem cells. In addition, at least some types of tumor appear to have their own stem cells; that is, these are the only tumor cells capable of seeding a new tumor. The concept is that tumor cells are diverse, and among them some cells will cease dividing while others can continue cancerous growth. The latter, of course, are the most dangerous and the most important to destroy with anticancer treatments. Like normal stem cells, tumor stem cells divide to produce two daughters: another tumor stem cell and a cell with more limited replicative potential.

One way to identify cancer stem cells is to purify different classes of cells from a tumor that carry different surface markers, usually using a fluorescent-activated cell sorter (FACS, see Chapter 9). Transplantation tests, usually with mice, reveal which classes of cell have the ability to seed a new tumor and which do not. For example, a few hundred human brain tumor cells with an antigen called *CD133* on the surface are potent in starting new tumors in immunologically deprived mice, whereas many thousands of non-CD133 tumor cells were unable to seed tumors. Similar findings have been made for human multiple myeloma, where fewer than 5 percent of the cells lack a marker called *CD138*, and these cells have considerably greater ability than the rest of the cells to start tumor growth. Surface proteins on human breast cancer cells also allowed identification of a minority class of cells that carried with them most of the dangerous potential to form tumors. These findings suggest that identifying tumor stem cells and then targeting those cells specifically with drugs or antibodies may be a more effective cancer therapy than targeting bulk tumor cells. It is not yet clear how many types of tumors have stem cells that differ from the majority of cells. ▮

The results with cancer stem cells highlight three important points. First, tumors are not made up of uniform cells, even if they originated from a single initiating cell. Second, a minority of the cells may be the really dangerous ones. Third, tumor cells may grow faster or slower depending on whether they find themselves in a particular environment. Just as stem cells can be kept in a dividing, nondifferentiating state by virtue of occupying a suitable niche (Chapter 21), tumor stem cells may behave according to their surroundings. Some neighboring cells may be conducive to tumor cell or tumor stem cell growth more than others. Figure 25-4 depicts normal stem-cell dependence on a niche (Figure 25-4a) and four scenarios in which cancer cells can originate and expand. Changes in niche cells, for example by mutation, can create a larger environment in which stem cells will reproduce themselves, in the worst case becoming cancer stem cells (Figure 25-4b). Alternatively, the prospective cancer stem cells can mutate and adjust to surviving and proliferating in a niche that is not normally their own (Figure 25-4c). In another scenario, stem cells become self-sufficient for growth, no longer requiring sustaining signals from niche cells (Figure 25-4d). Lastly, progenitor cells that normally have the ability to reproduce themselves for a period but not indefinitely become mutated in a way that makes them into cancer cells with large replicative potential (Figure 25-4e). In each case many of the cells that make up the tumor are not themselves able to divide indefinitely or seed new tumors.

The idea that *tumor microenvironment* matters extends to the importance of the extracellular matrix and to one of the most common environments for a tumor cell: inflammatory cells. Cancers frequently arise at sites of injury or infection. Immune cells migrate to sites of injury, growth factors are produced there to promote healing, and the extracellular matrix is reconstructed. All of these local tissue properties may contribute to the establishment and growth of a tumor.

Tumor Growth Requires Formation of New Blood Vessels

Tumors, whether primary or secondary, require recruitment of new blood vessels in order to grow to a large mass. In the absence of a blood supply, a tumor can grow into a mass of about 10^6 cells, roughly a sphere 2 mm in diameter. At this point, division of cells on the outside of the tumor mass is balanced by death of those in the center owing to an inadequate supply of nutrients. Such tumors, unless they secrete hormones, cause few problems. However, most tumors induce the formation of new blood vessels that invade the tumor and nourish it, a process called *angiogenesis.* This complex process requires several discrete steps: degradation of the basement membrane that surrounds a nearby capillary, migration of endothelial cells lining the capillary into the tumor, division of these endothelial cells, and formation of a new basement membrane around the newly elongated capillary.

Many tumors produce growth factors that stimulate angiogenesis; other tumors somehow induce surrounding normal cells to synthesize and secrete such factors. Basic fibroblast growth factor (bFGF), transforming growth factor α (TGFα), and vascular endothelial growth factor (VEGF),

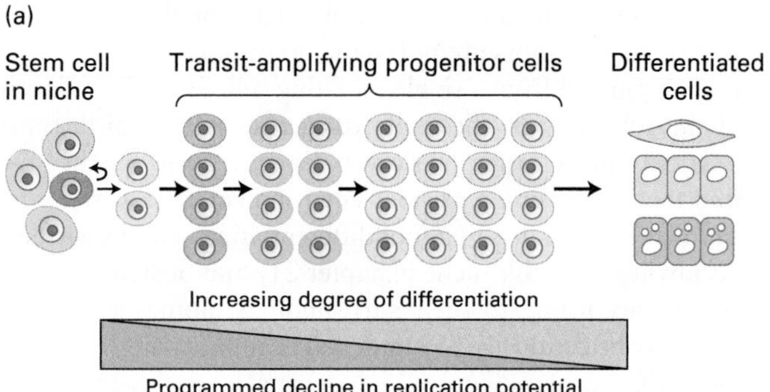

(a)
Stem cell in niche
Transit-amplifying progenitor cells
Differentiated cells
Increasing degree of differentiation
Programmed decline in replication potential
The dependence of normal stem cells on the niche limits their expansion.

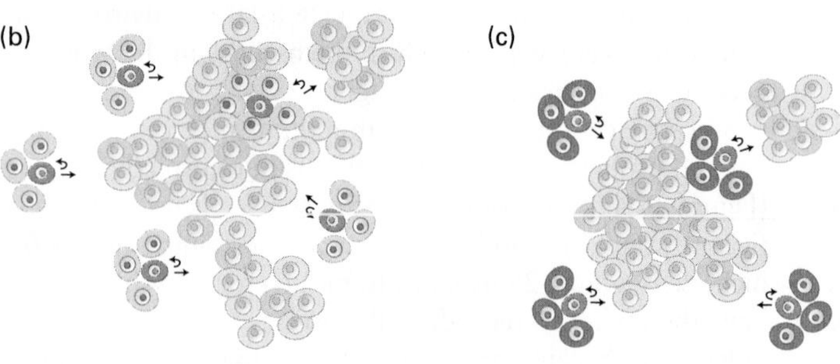

(b)
(c)
Expansion of the normal stem cell niche permits the expansion of cancer stem cells that arose from normal stem cells.
Cancer stem cells that arose from normal stem cells adapt to a different niche allowing their expansion.

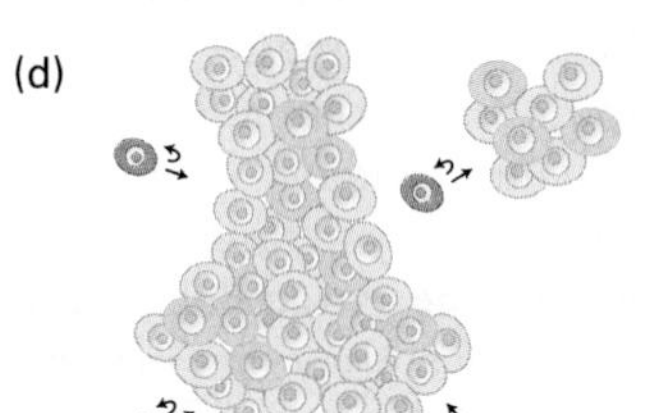

(d)
Cancer stem cells that arose from normal stem cells become niche-independent, and self-renewal is cell-autonomous.

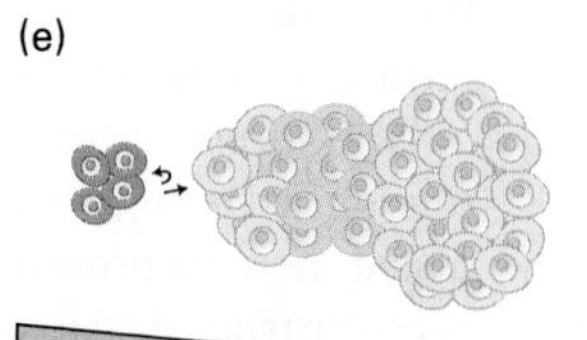

(e)
Shift in the programmed declined in replication potential
Cancer stem cells arising from a progenitor cell.

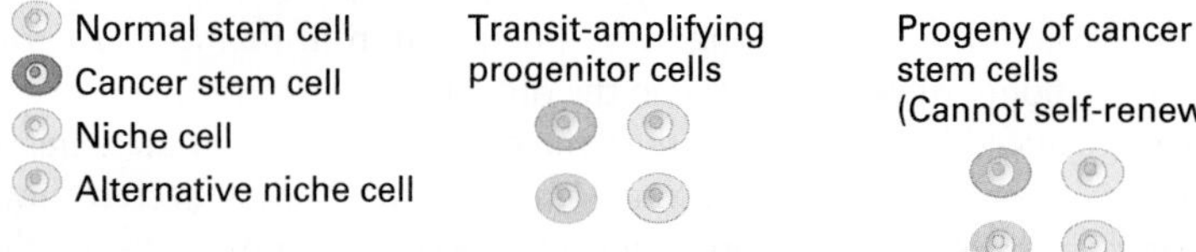

Normal stem cell
Cancer stem cell
Niche cell
Alternative niche cell
Transit-amplifying progenitor cells
Progeny of cancer stem cells (Cannot self-renew)

◀ **FIGURE 25-4 Possible origins of cancer stem cells.** (a) Interactions between normal stem cells and supporting niche cells. In normal tissues, the niche cells (light green) provide signals enabling normal stem cells (blue) to reproduce themselves (curved arrow), while at the same time producing cells that go on to proliferate and then differentiate (pink). These progenitor cells (pink, orange, yellow) do not receive a signal from the niche, so they do not remain stem cells. With each cell division, the proliferation capacity of the progenitor cells declines. Panels (b), (c), (d), and (e) represent possible ways in which tumors may arise from normal stem cells by circumventing niche-dependent restrictions on stem-cell expansion. (b) Expansion of the stem-cell niche (pale green) allows a corresponding expansion of cancer stem cells (purple) that have arisen from normal stem cells. The expansion of niche cells may be driven by genetic alterations in the niche cells. (c) Alterations in cancer stem cells (purple) enable them to be sustained by alternative niche cells (dark green) to provide them with self-renewal signals. This mechanism could result in the invasion of local tissues or may facilitate cancer stem-cell growth at sites of metastasis. The resultant tumors are a mixture of cancer stem cells and their nontumorigenic progeny. (d) Genetic or epigenetic alterations in cancer stem cells enable them to become niche-independent such that they undergo cell-autonomous self-renewal. The resulting tumors containing self-renewing stem cells and their nontumorigenic progeny. (e) Mutations may arise that give progenitor cells the stem-cell property of self-renewal. This loss of growth control can lead to tumor formation. [Adapted from Clarke and Fuller, 2006, *Cell* **124:**1111.]

which are secreted by many tumors, all have angiogenic properties. New blood vessels nourish the growing tumor, allowing it to increase in size and thus increase the probability that additional harmful mutations will occur. The presence of an adjacent blood vessel also facilitates the process of metastasis.

In humans there are five VEGF genes and three VEGF-receptor-protein genes. VEGF expression can be induced by hypoxia, starvation of cells for oxygen that occurs when $[pO_2] < 7$ mmHg. The VEGF receptors, which are tyrosine kinases, regulate different aspects of blood vessel growth such as endothelial (blood vessel wall) cell survival and growth, cell migration, and vessel wall permeability.

The hypoxia signal is mediated by hypoxia-inducible factor (HIF-1), a transcription factor that is activated in conditions of low oxygen and then binds to and induces transcription of the VEGF gene and about 30 other genes, many of which can affect the probability of tumor growth. Among these are glycolytic enzymes such as lactate dehydrogenase; thus HIF-1 may also help tumor cells to adapt to low oxygen by turning to glycolysis rather than oxidative phosphorylation for ATP generation. HIF-1 activity is controlled, in turn, by an oxygen sensor

(a)

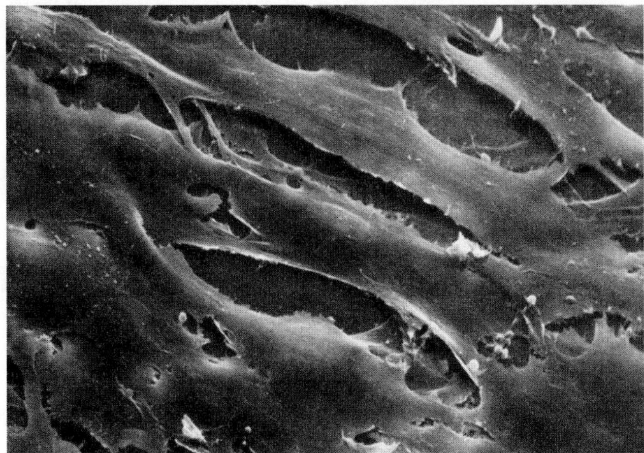

(b)

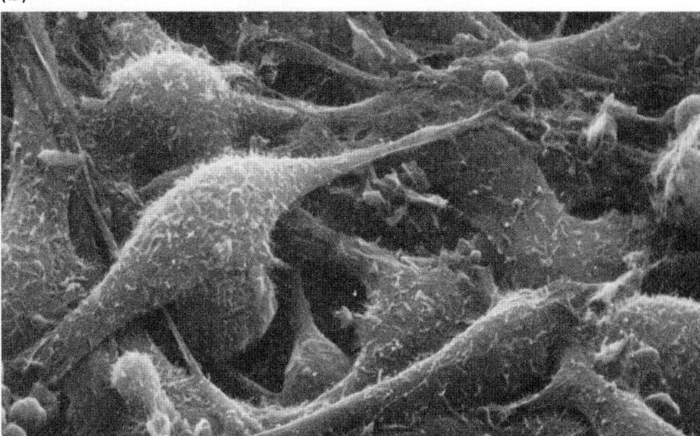

▲ EXPERIMENTAL FIGURE 25-5 **Scanning electron micrographs reveal the organizational and morphological differences between normal and transformed 3T3 cells.** (a) Normal 3T3 cells are elongated and are aligned and closely packed in an orderly fashion. (b) 3T3 cells transformed by an oncogene encoded by Rous sarcoma virus are rounded and covered with small hairlike processes and bulbous projections. The transformed cells that grow have lost the side-by-side organization of the normal cells and grow one atop the other. These transformed cells have many of the same properties as malignant cells. Similar changes are seen in cells transfected with DNA from human cancers containing the *ras*^{D} oncogene. [Courtesy of L.-B. Chen.]

composed of a prolyl hydroxylase that is active in normal O_2 levels but inactive when deprived of O_2. Hydroxylation of HIF-1 causes ubiquitination and degradation of the transcription factor, a process that is blocked when O_2 is low.

Several natural proteins that inhibit angiogenesis (e.g., angiogenin and endostatin) and antagonists of the VEGF receptor have excited much interest as potential therapeutic agents. Although new blood vessels are constantly forming during embryonic development, few form normally in adults except after injury. Thus specific inhibitors of angiogenesis might be effective against many kinds of tumors and have few adverse side effects. New drugs developed according to this rationale are showing promise. Drugs that inhibit the VEGF receptor tyrosine kinase activity are being tested. One recently approved is a monoclonal antibody against VEGF itself, which has been effective against metastatic colon cancer. One of the challenges in such therapies is to avoid an immune response directed against the monoclonal antibody. This is done by engineering changes in the mouse monoclonal antibody so it more closely matches human antibody sequences. Antibody and other anti-metastatic cancer treatments are typically successful in extending survival times for periods of months. It is hoped that combinations of treatments as well as new treatments will move closer to the real goal, complete cures. ▮

Specific Mutations Transform Cultured Cells into Tumor Cells

The morphology and growth properties of tumor cells clearly differ from those of their normal counterparts; some of these differences are also evident when cells are cultured. That mutations cause these differences was conclusively established by transfection experiments with a line of cultured mouse fibroblasts called *3T3 cells*. These cells normally grow only when attached to the plastic surface of a culture dish and are maintained at a low cell density. Because 3T3 cells stop growing when they contact other cells, they eventually form a monolayer of well-ordered cells that have stopped proliferating and are in the quiescent G_0 phase of the cell cycle (Figure 25-5a).

When DNA from human bladder cancer cells is transfected into cultured 3T3 cells, about one cell in a million incorporates a particular segment of the exogenous DNA that causes a distinctive phenotypic change. The progeny of the affected cell are more rounded and less adherent to one another and to the dish than are the normal surrounding cells, forming a three-dimensional cluster of cells (a focus) that can be recognized under the microscope (Figure 25-5b). Such cells, which continue to grow when the normal cells have become quiescent, have undergone oncogenic **transformation.** The transformed cells have properties similar to those of malignant tumor cells, including changes in cell morphology, ability to grow unattached to an extracellular matrix, reduced requirement for growth factors, secretion of plasminogen activator, and loss of actin microfilaments.

Figure 25-6 outlines the procedure for transforming 3T3 cells with DNA from a human bladder cancer and cloning the specific DNA segment that causes transformation. It was remarkable to find a small piece of DNA with this capability; had more than one piece been needed, the experiment would have failed. Subsequent studies showed that the cloned segment included a mutant version of the cellular *ras* gene, where the glycine normally found in position 12 is replaced with a valine. This mutant was designated *ras*^{D}, where the "D" stands for dominant. The mutation is genetically **dominant** because the active protein has an effect even in the presence of the other, normal *ras* allele. Normal **Ras protein,** which participates in many intracellular signal-transduction pathways activated by growth factors (Chapter 16), cycles between an inactive, "off" state with bound GDP and an active, "on" state with bound GTP. The mutated Ras^D

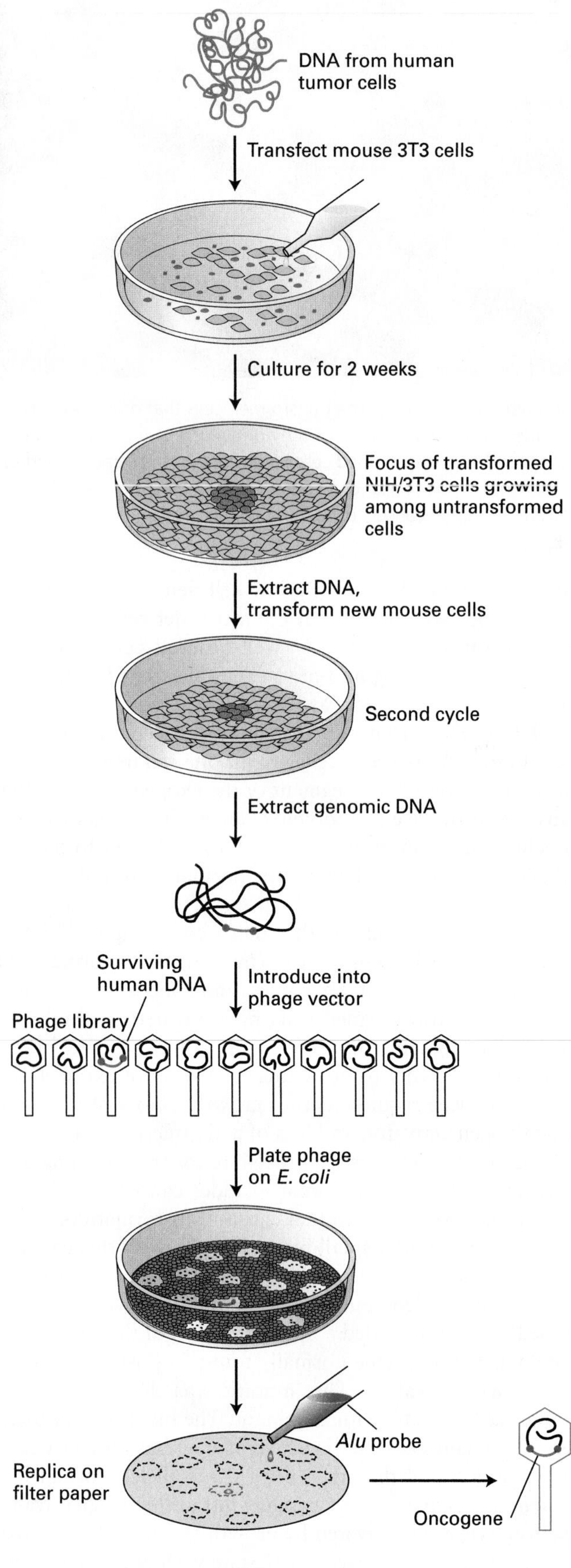

◀ EXPERIMENTAL FIGURE 25-6 **Transformation of mouse cells with DNA from a human cancer cell permits identification and molecular cloning of the *ras*^{D} oncogene.** Addition of DNA from a human bladder cancer to a culture of mouse 3T3 cells causes about one cell in a million to divide abnormally and form a focus, or clone, of transformed cells. To clone the oncogene responsible for transformation, advantage is taken of the fact that most human genes have nearby repetitive DNA sequences called *Alu* sequences. DNA from the initial focus of transformed mouse cells is isolated, and the oncogene is separated from adventitious human DNA by secondary transfer to mouse cells. The adventitious DNA is human DNA that has no effect on cell transformation, but just happened to end up in a cell that also contains the active oncogene. The total DNA from a secondary transfected mouse cell is then cloned into bacteriophage λ; only the phage that receives human DNA hybridizes with an *Alu* probe. The hybridizing phage should contain part of or all the transforming oncogene. This expected result can be proved by showing either that the phage DNA can transform cells (if the oncogene has been completely cloned) or that the cloned piece of DNA is always present in cells transformed by DNA transfer from the original bladder cancer donor cell.

protein hydrolyzes bound GTP very slowly and therefore accumulates in the active state, sending a growth-promoting signal to the nucleus even in the absence of the hormones normally required to activate its signaling function.

The production and **constitutive** activation of Ras^D protein is not sufficient to cause transformation of normal cells in a primary (fresh) culture of human, rat, or mouse fibroblasts. Unlike cells in a primary culture, however, cultured 3T3 cells have undergone loss-of-function mutations in the *p19ARF* or *p53* genes, which are regulators of the cell cycle. Such cells can grow for an unlimited time in culture if periodically diluted and supplied with nutrients, which normal cells cannot do (see Figure 9-31b). These immortal 3T3 cells are transformed into full-blown tumor cells only when they produce a constitutively active Ras protein or other oncoproteins. For this reason, transfection with the *ras*^{D} gene can transform 3T3 cells, but not normal cultured primary fibroblast cells, into tumor cells.

A mutant *ras* gene is found in most human colon, bladder, pancreatic, and other cancers, but not in normal human DNA; thus it must arise as the result of a somatic mutation in one of the tumor progenitor cells. Any gene, such as *ras*^{D}, that encodes a protein capable of transforming cells in culture or inducing cancer in animals is referred to as an oncogene. An oncogene arises from a normal cellular gene, a proto-oncogene, such as *ras*. The oncogenes carried by viruses that cause tumors in animals are often derived from protooncogenes that were hijacked from the host genome and altered to be oncogenic. When this was first discovered, it was startling to find that these dangerous viruses were turning the animal's own genes against them.

A Multi-hit Model of Cancer Induction Is Supported by Several Lines of Evidence

As noted earlier and illustrated by the oncogenic transformation of 3T3 cells, multiple mutations usually are required to convert a normal body cell into a malignant one. According to this "multi-hit" model, evolutionary (or "survival of the fittest") cancers arise by a process of clonal selection not unlike the selection of individual animals in a large population. Here is the scenario, which may or may not apply to all cancers: A mutation in one cell, perhaps a stem cell, would give it a slight growth advantage. One of the progeny cells would then undergo a second mutation that would allow its descendants to grow more uncontrollably and form a small benign tumor; a third mutation in a cell within this tumor would allow it to outgrow the others and overcome constraints imposed by the tumor microenvironment, and its progeny would form a mass of cells, each of which would have these three mutations. An additional mutation in one of these cells would allow its progeny to escape into the blood and establish daughter colonies at other sites, the hallmark of metastatic cancer. This model makes two easily testable predictions.

First, all the cells in a given tumor should contain at least some genetic alterations in common. Systematic analysis of cells from individual human tumors supports the prediction that all the cells are derived from a single progenitor. Recall that during the fetal life of a human female each cell inactivates one of the two X chromosomes. A woman is a genetic mosaic; half the cells have one X inactivated, and the remainder have the other X inactivated. If a tumor did not arise from a single progenitor, it would be composed of a mix of cells with one or the other X inactivated. In fact, the cells from a woman's tumor have the same inactive X chromosome. Different tumors can be composed of cells with either the maternal or the paternal X inactive. Second, cancer incidence should increase with age because it can take decades for the required multiple mutations to occur. Assuming that the rate of mutation is roughly constant during a lifetime, then the incidence of most types of cancer would be independent of age if only one mutation were required to convert a normal cell into a malignant one. In fact current estimates suggest 5–6 "hits," or mutations, must accumulate as the most dangerous cancer cells emerge. As the data in Figure 25-7 show, the incidence of many types of human cancer does indeed increase drastically with age.

More direct evidence that multiple mutations are required for tumor induction comes from transgenic mice. A variety of combinations of oncogenes can cooperate in causing cancer. For example, mice have been made that carry either the mutant ras^{V12} dominant oncogene (one version of ras^{D}) or the c-*myc* proto-oncogene, in each case under control of a mammary cell–specific promoter/enhancer from a retrovirus. The promoter is induced by endogenous hormone levels and tissue-specific regulators, leading to overexpression of c-*myc* or ras^{V12} in breast tissue. Heightened transcription of c-*myc* mimics previously identified oncogenic mutations that turn up c-*myc* transcription, converting the proto-oncogene into an oncogene. By itself, the c-*myc* transgene causes tumors only after 100 days, and then in only a few mice; clearly only a minute fraction of the mammary cells that overproduce the Myc protein become malignant. Similarly, production of the mutant Ras^{V12} protein

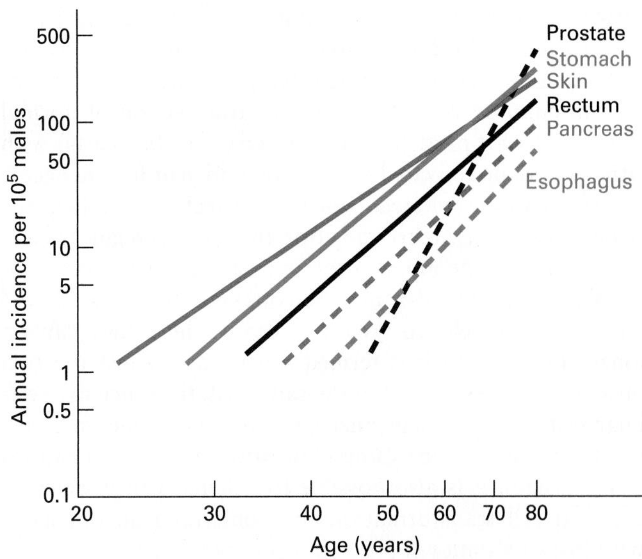

▲ EXPERIMENTAL FIGURE 25-7 **The incidence of human cancers increases as a function of age.** The marked increase in the incidence with age is consistent with the multi-hit model of cancer induction. Note that the logarithm of annual incidence is plotted versus the logarithm of age. [From B. Vogelstein and K. Kinzler, 1993, *Trends Genet.* **9**:101.]

alone causes tumors earlier but still slowly and with about 50 percent efficiency over 150 days. When the c-*myc* and ras^{V12} transgenics are crossed, however, such that all mammary cells produce both Myc and Ras^{V12}, tumors arise much more rapidly and all animals succumb to cancer (Figure 25-8). Such experiments emphasize the synergistic effects of multiple oncogenes. They also suggest that the long latency of tumor

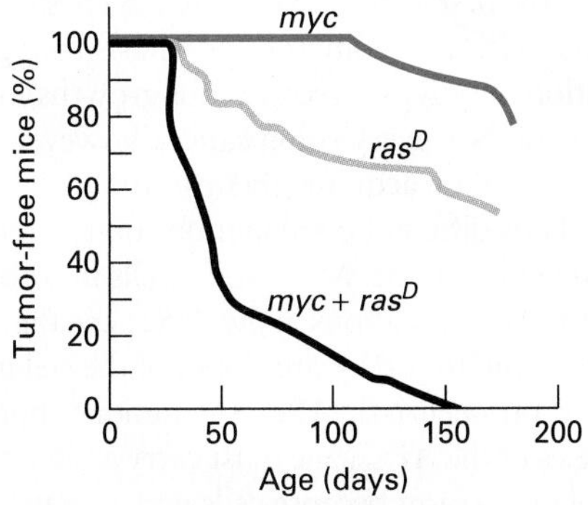

▲ EXPERIMENTAL FIGURE 25-8 **The kinetics of tumor appearance in female mice carrying either one or two oncogenic transgenes shows the cooperative nature of multiple mutations in cancer induction.** Each of the transgenes was driven by the mouse mammary tumor virus (MMTV) breast-specific promoter. The hormonal stimulation associated with pregnancy activates the MMTV promoter and hence the overexpression of the transgenes. The graph shows the time course of tumorigenesis in mice carrying either *myc* or ras^D transgenes as well as in the progeny of a cross of *myc* carriers with ras^D carriers that contain both transgenes. The results clearly demonstrate the cooperative effects of multiple mutations in cancer induction. [See E. Sinn et al., 1987, *Cell* **49**:465.]

formation, even in the double-transgenic mice, is due to the need to acquire additional somatic mutations.

Similar cooperative effects between oncogenes can be seen in cultured cells. For example, transfection of normal fibroblasts (*not* partially transformed 3T3 fibroblasts) with either c-*myc* or activated ras^D is not sufficient for oncogenic transformation, whereas when transfected together, the two genes cooperate to transform the cells. Deregulated levels of c-*myc* alone induce proliferation but also sensitize fibroblasts to apoptosis, and overexpression of activated ras^D alone induces cells to go into a state where they can no longer divide, which is termed *senescence*. When the two oncogenes are expressed in the same cell, these negative cellular responses are neutralized and the cells undergo transformation. While the examples mentioned are combinations of oncogenes, it is also possible to enhance cancer rates or cultured cell transformations by combining an oncogene with loss of a tumor-suppressor gene.

Successive Oncogenic Mutations Can Be Traced in Colon Cancers

Studies on colon cancer provide the most compelling evidence to date for the multi-hit model of cancer induction. Surgeons can obtain fairly pure samples of many human cancers, but since the tumor is observed only at one time, its exact stage of progression cannot be easily determined. An exception is colon cancer, which evolves through distinct, well-characterized morphological stages. These intermediate stages—polyps, benign adenomas, and carcinomas—can be isolated by a surgeon, allowing mutations that occur in each of the morphological stages to be identified. Numerous studies show that colon cancer arises from a series of mutations that commonly occur in a well-defined order, providing strong support for the multi-hit model (Figure 25-9).

Invariably the first step in colon carcinogenesis involves loss of a functional *APC* (*a*denomatous *p*olyposis *c*oli) gene, resulting in formation of polyps (precancerous growths) on the inside of the colon wall. Not every colon cancer, however, acquires all the later mutations or acquires them in the order depicted in Figure 25-9. Thus different combinations of mutations may result in the same phenotype. Most of the cells in a polyp contain the same one or two mutations in the *APC* gene that result in its loss or inactivation; thus they are clones of the cell in which the original mutation occurred. *APC* is a tumor-suppressor gene, and both alleles of the *APC* gene must carry an inactivating mutation for polyps to form because cells with one wild-type *APC* gene express enough APC protein to function normally. Like most tumor-suppressor genes, *APC* encodes a protein that inhibits the progression of certain types of cells through the cell cycle. The APC protein does so by preventing the Wnt signal-transduction pathway (Chapter 16) from activating expression of proto-oncogenes including the c-*myc* gene. The absence of functional APC protein thus leads to inappropriate production of Myc, a transcription factor that induces expression of many genes required for the transition from the G_1 to the S phase of the cell cycle. Cells homozygous for *APC* mutations proliferate at a rate higher than normal and form polyps.

If one of the cells in a polyp undergoes another mutation, this time an activating mutation of the *ras* gene, its progeny divide in an even more uncontrolled fashion, forming a larger adenoma (see Figure 25-9). Mutational loss of a particular chromosomal region (the relevant gene is not yet known), followed by inactivation of the *p53* gene, results in the gradual loss of normal regulation and the consequent formation of a malignant carcinoma. While the four "hits" listed here are certainly crucial parts of the picture, there may be additional contributing genetic events.

About half of all human tumors carry mutations in the p53 tumor-suppressor gene, which encodes a key stress-response transcriptional regulator. Activity of *p53* causes cells in trouble to go into cell-cycle arrest or apoptosis. A critical feature of cell-cycle control is the G_1 checkpoint, which prevents cells with damaged DNA from entering S phase (Figure 20-35, **4a**). The **p53 protein** is a sensor essential for the checkpoint control that arrests cells with damaged DNA in G_1. The arrest can be temporary, allowing correction of cell defects, or permanent, resulting in cell senescence. The p53 protein has several functions, and which of them is most relevant to its tumor-suppressing function remains unclear. For example, a mutant p53 protein that is unable to activate transcription can nonetheless suppress tumor formation. Virtually all p53 mutations abolish its ability to bind to specific DNA sequences and activate gene expression.

DNA from different human colon carcinomas generally contains mutations in all four genes—loss-of-function mutations in the tumor suppressors *APC* and *p53*, and in the as-yet mysterious gene, and an activating (gain-of-function) mutation in the dominant oncogene *K-ras* (one of the *ras* family of genes)—establishing that multiple mutations in the same cell are needed for the cancer to form. Some of these mutations appear to confer growth advantages at an early stage of tumor development, whereas other mutations promote the later stages, including invasion and metastasis, which are required for the malignant phenotype. The number of mutations needed for colon cancer progression may at first seem surprising, seemingly an effective barrier to tumorigenesis. Our genomes, however, are under constant assault. Recent estimates indicate that sporadically arising polyps have about 11,000 genetic alterations in each cell, though very likely only a few of these are relevant to oncogenesis. Genetic instability is a hallmark of cancer cells.

Colon carcinoma provides an excellent example of the multi-hit mode of cancer. The degree to which this model applies to cancer generally is only now being learned, but it is clear that many types of cancer each involve multiple mutations.

DNA Microarray Analysis of Expression Patterns Can Reveal Subtle Differences Between Tumor Cells

Traditionally the properties of tumor and normal cells have been assessed by staining and microscopy. The prognosis for many tumors could be determined, within certain limits,

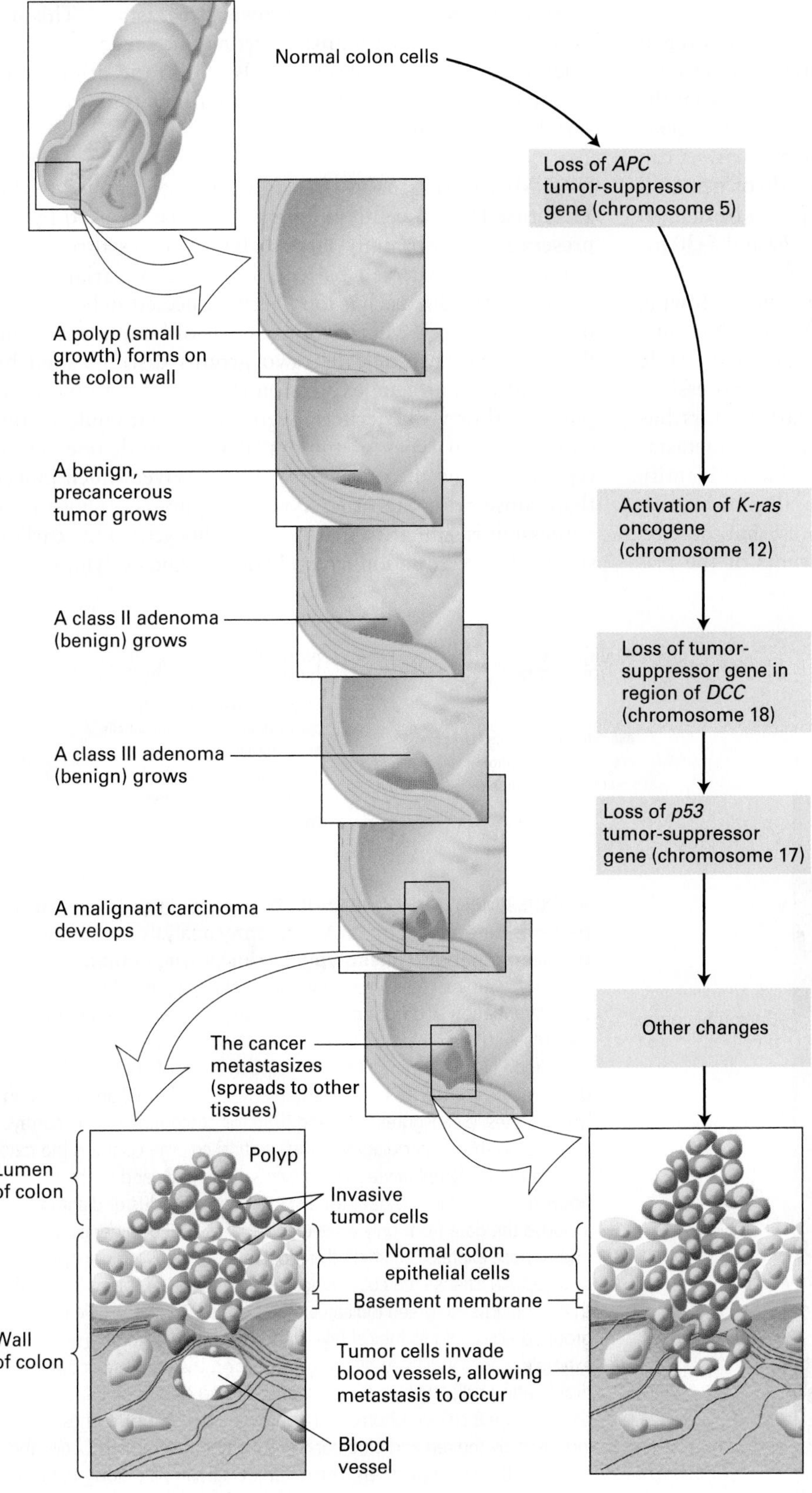

Normal colon cells
Loss of *APC* tumor-suppressor gene (chromosome 5)
A polyp (small growth) forms on the colon wall
A benign, precancerous tumor grows
Activation of *K-ras* oncogene (chromosome 12)
A class II adenoma (benign) grows
Loss of tumor-suppressor gene in region of *DCC* (chromosome 18)
A class III adenoma (benign) grows
Loss of *p53* tumor-suppressor gene (chromosome 17)
A malignant carcinoma develops
Other changes
The cancer metastasizes (spreads to other tissues)
Lumen of colon
Polyp
Invasive tumor cells
Normal colon epithelial cells
Basement membrane
Wall of colon
Tumor cells invade blood vessels, allowing metastasis to occur
Blood vessel

◀ **FIGURE 25-9 The development and metastasis of human colorectal cancer and its genetic basis.** A mutation in the APC tumor-suppressor gene in a single epithelial cell causes the cell to divide, although surrounding cells do not, forming a mass of localized benign tumor cells, or polyp. Subsequent mutations leading to expression of a constitutively active Ras protein and loss of two tumor-suppressor genes—an unidentified gene on chromosome 18 and *p53*—generate a malignant cell carrying all four mutations. The loss of the tumor-suppressor gene on chromosome 18 arises from a deletion that includes the *DCC* (deleted in colorectal cancer) gene. It is not clear that this particular gene is in fact the important gene with respect to cancer, but something in that vicinity serves as a tumor suppressor. This cell continues to divide, and the progeny invade the basement membrane that surrounds the tissue. Some tumor cells spread into blood vessels that will distribute them to other sites in the body. Additional mutations permit the tumor cells to exit from the blood vessels and proliferate at distant sites; a patient with such a tumor is said to have cancer. [Adapted from B. Vogelstein and K. Kinzler, 1993, *Trends Genet.* **9:**101.]

from their histology. However the appearance of cells alone has limited information content, and better ways to discern the properties of cells are desirable both to understand tumorigenesis and to arrive at meaningful and accurate decisions about prognosis and therapy.

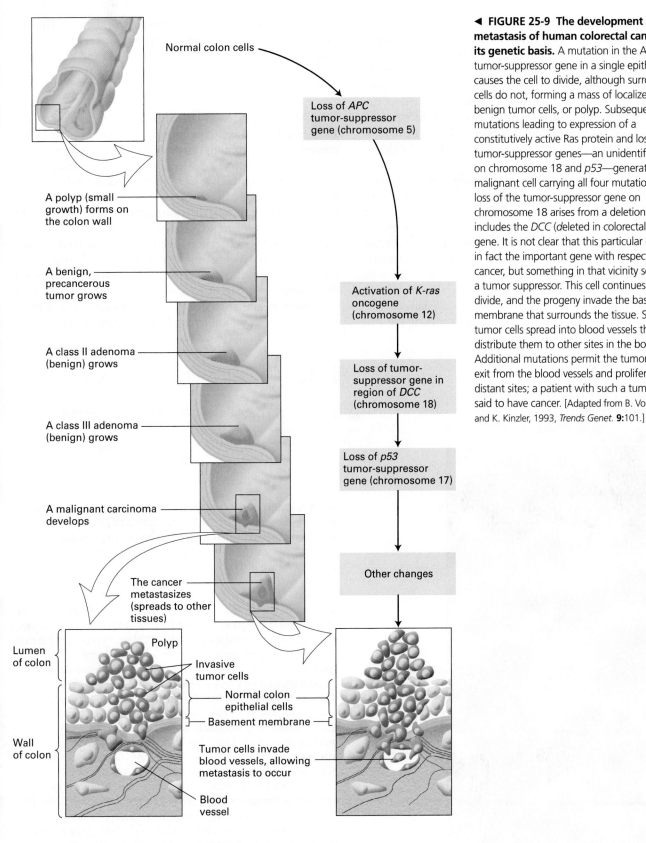

As we've seen, genetic studies can identify the single initiating mutation or series of mutations that cause transformation of normal cells into tumor cells, as in the case of colon cancer. After these initial events, however, the cells of a tumor undergo a cascade of changes reflecting the interplay between the initiating events and signals from outside. As a result, tumor cells can become quite different, even if they arise from the same initiating mutation or mutations. Although these differences may not be recognized from the appearance of cells, they can be detected from the cells' patterns of gene expression. **DNA microarray** analysis can determine the expression of tens of thousands of genes simultaneously, permitting complex phenotypes to be defined at the molecular genetic level. (See Figures 5-29 and 5-30 for an explanation of this technique.)

The advent of DNA microarray technology is allowing more detailed examination of tumor properties. Not surprisingly, primary tumors can often be distinguishable from metastatic tumors by the pattern of gene expression. More interestingly, a subset of solid primary tumors has been found to have characteristics more typical of metastatic tumors, suggesting that it may be possible to identify primary tumors that have a greater probability of becoming metastatic. This also raises the possibility that, at least for some types of cancer, the initiating events of the primary tumor may set a course toward metastasis. This hypothesis can be distinguished from the scenario of the emergence of a rare subset of cells within the primary tumor acquiring a series of further mutations that are necessary for metastasis.

Microarray analysis recently has been applied to diffuse large B cell lymphoma, a disease marked by the presence of abnormally large B lymphocytes throughout lymph nodes. Affected patients have highly variable outcomes, so the disease has long been suspected to be, in fact, multiple diseases. Microarray analysis of lymphomas from different patients revealed two groups distinguished by their patterns of gene expression (Figure 25-10). No morphological or visible criteria were found that could distinguish the two types of tumors. Patients with one tumor type defined by the microarray data survived much longer than those with the other type. Lymphomas whose gene expression is similar to that of B lymphocytes in the earliest stages of differentiation have a better prognosis; lymphomas

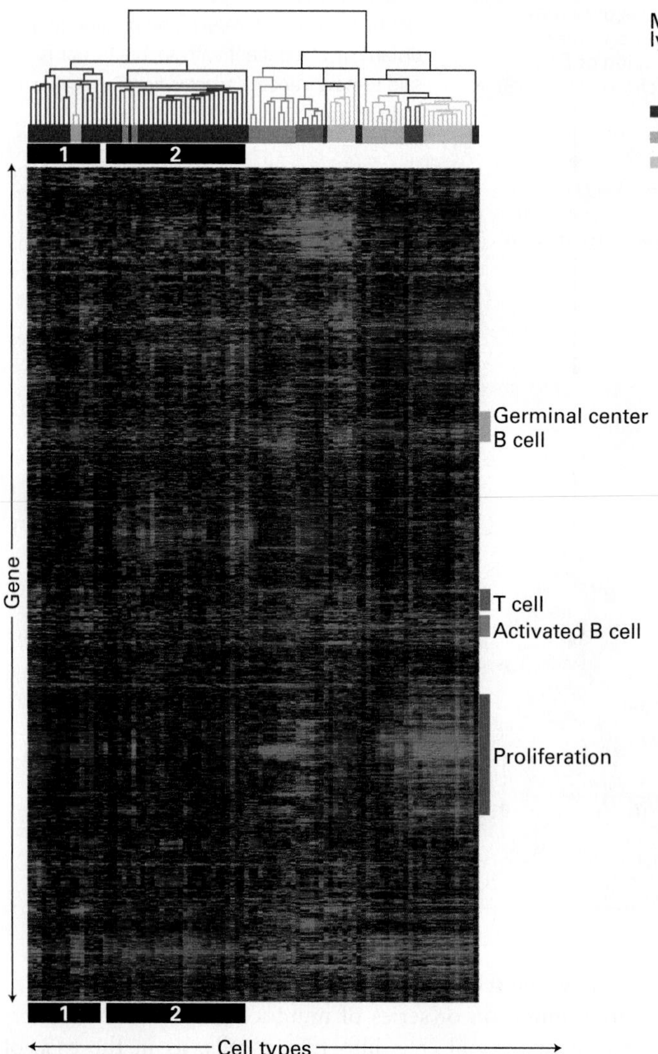

◀ EXPERIMENTAL FIGURE 25-10 **Differences in gene expression patterns determined by DNA microarray analysis can distinguish between otherwise phenotypically similar lymphomas.** Samples of mRNA were extracted from normal lymphocytes at different stages of differentiation and from malignant lymphocytes obtained from patients with three types of lymphoma. DNA microarray analysis of the extracted RNA determined the transcription levels of about 18,000 genes in each of the 96 experimental samples of normal and malignant lymphocytes. (See Figures 5-29 and 5-30 for description of microarray analysis.) In this cluster diagram, each vertical column contains the data for a particular lymphocyte sample (see sample key), and each horizontal row contains data for a single gene. The cluster diagram includes the data from only a selected set of the 18,000 genes, the ones whose expression differs the most in the various lymphocyte samples. An intense red color indicates a high level of transcription of that gene; intense green indicates the opposite. The genes were grouped according to their similar patterns of hybridization. For example, genes indicated by the green bar on the right are active in proliferating cells, such as transformed cultured cells (pink bar at top) or diffuse large B cell lymphoma cells (dark blue bar at top). Genes indicated by the red and blue bars on the right are active in T cells and activated B cells, respectively. The different cell samples along the top of the diagram also were grouped according to their similar expression patterns. The resulting dendrogram (tree diagram) shows that the samples from patients with diffuse large B cell lymphoma (dark blue samples) fall into two groups (black bars 1 and 2). One group is similar to relatively undifferentiated B lymphocytes in germinal centers (orange samples); the other is similar to more differentiated B cells (gray samples). [From A. A. Alizadeh et al., 2000, *Nature* **403**:505.]

whose gene expression is closer to that of more differentiated B lymphocytes have a worse prognosis. Similar analyses of the gene expression patterns, or "signatures," of other tumors are likely to improve classification and diagnosis, allowing informed decisions about treatments, and also to provide insights into the properties of tumor cells. ∎

KEY CONCEPTS OF SECTION 25.1

Tumor Cells and the Onset of Cancer

- Cancer is a fundamental aberration in cellular behavior, touching on many aspects of molecular cell biology. Most cell types of the body can give rise to malignant tumor (cancer) cells.
- Cancer cells can multiply in the absence of at least some of the growth-promoting factors required for proliferation of normal cells and are resistant to signals that normally program cell death (apoptosis).
- Most oncogenic mutations occur in somatic cells and are not carried in the germ-line DNA.
- Cancer cells sometimes invade surrounding tissues, often breaking through the basement membranes that define the boundaries of tissues and spreading through the body to establish secondary areas of growth, a process called *metastasis*. Metastatic tumors often secrete proteases, which degrade the surrounding extracellular matrix (Figure 25-3).
- Cancer cells may arise from stem cells and often arise from proliferating cells. Cancer cells bear more resemblance to dividing precursor cells than to more mature differentiated cell types.
- Both primary and secondary tumors require angiogenesis, the formation of new blood vessels, in order to grow to a large mass.
- Certain cultured cells transfected with tumor-cell DNA undergo transformation (see Figure 25-6). Such transformed cells share certain properties with tumor cells.
- The multi-hit model, which proposes that multiple mutations are needed to cause cancer, is consistent with the genetic homogeneity of cells from a given tumor, the observed increase in the incidence of human cancers with advancing age, and the cooperative effect of oncogenic transgenes and tumor-suppressor gene mutations on tumor formation in mice.
- Colon cancer develops through distinct morphological stages that commonly are associated with mutations in specific tumor-suppressor genes and proto-oncogenes (see Figure 25-9).
- DNA microarray analysis can identify differences in gene expression between types of tumor cells that are indistinguishable by traditional criteria. Some tumor cells appear to be related to specific types of normal cells at certain stages of development based on similarities in their expression patterns.
- Within tumors, subsets of cells can be identified with distinctive gene expression characteristics. These differences may allow identification of dangerous tumor stem cells among a mass of cells with limited replicative or metastatic potential.

25.2 The Genetic Basis of Cancer

As we have noted, mutations in three broad classes of genes—proto-oncogenes (e.g., *ras*), tumor-suppressor genes (e.g., *APC*), and caretaker genes—play key roles in cancer induction. These genes encode many kinds of proteins that help control cell growth and proliferation (Figure 25-11). Virtually all human tumors have inactivating mutations in genes whose products normally act at various cell-cycle **checkpoints** to stop a cell's progress through the cell cycle if a previous step has occurred incorrectly or if DNA has been damaged. For example, most cancers have inactivating mutations in the genes coding for one or more proteins that normally restrict progression through the G_1 stage of the cell cycle, or activating mutations in genes coding for proteins that drive the cells through the cell cycle. Likewise, a constitutively active Ras or other activated signal-transduction protein is found in several kinds of human tumor that have different origins. Thus malignancy and the intricate processes for controlling the cell cycle discussed in Chapter 20 are two faces of the same coin. In the series of events leading to growth of a tumor, oncogenes combine with tumor-suppressor mutations to give rise to the full spectrum of tumor-cell properties described in the previous section (see Figure 25-9).

In this section, we consider the general types of mutations that are oncogenic and see how certain viruses can cause cancer. We also explain why some inherited mutations increase the risk for particular cancers and consider the relation between cancer and developmentally important genes.

Gain-of-Function Mutations Convert Proto-oncogenes into Oncogenes

Recall that an oncogene is any gene that encodes a protein able to transform cells in culture, usually in combination with other cell alterations, or to induce cancer in animals. Of the many known oncogenes, all but a few are derived from normal cellular genes (i.e., proto-oncogenes) whose wild-type products promote cell proliferation or other features important for cancer. For example, the *ras* gene discussed previously is a proto-oncogene that encodes an intracellular signal-transduction protein; the mutant ras^D gene derived from *ras* is an oncogene, whose encoded protein provides an excessive or uncontrolled growth-promoting signal. Other proto-oncogenes encode growth-promoting signal molecules and their receptors, anti-apoptotic (cell-survival) proteins, and some transcription factors.

Conversion, or activation, of a proto-oncogene into an oncogene generally involves a *gain-of-function* mutation. At

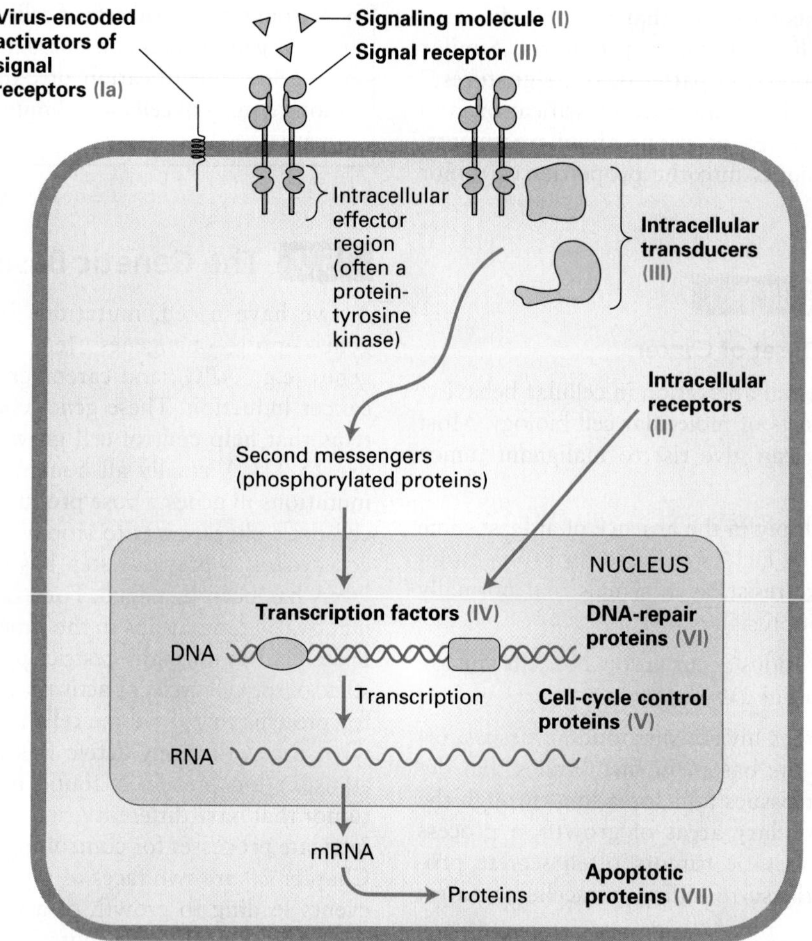

▲ **FIGURE 25-11 Seven types of proteins that participate in controlling cell growth and proliferation.** Cancer can result from expression of mutant forms of these proteins. Mutations changing the structure or expression of proteins that normally promote cell growth generally give rise to dominantly active oncogenes. Many, but not all, extracellular signaling molecules (I), signal receptors (II), signal-transduction proteins (III), and transcription factors (IV) are in this category. Cell-cycle control proteins (V) that function to restrain cell proliferation and DNA-repair proteins (VI) are encoded by tumor-suppressor genes. Mutations in these genes act recessively, greatly increasing the probability that the mutant cells will become tumor cells or that mutations will occur in other classes. Apoptotic proteins (VII) include tumor suppressors that promote apoptosis and oncoproteins that promote cell survival. Virus-encoded proteins that activate signal receptors (Ia) also can induce cancer.

least four mechanisms can produce oncogenes from the corresponding proto-oncogenes:

1. *Point mutation* (i.e., change in a single base pair) in a proto-oncogene that results in a hyperactive or constitutively active protein product

2. *Chromosomal translocation* that fuses two genes together to produce a hybrid gene encoding a chimeric protein whose activity, unlike that of the parent proteins, often is constitutive

3. *Chromosomal translocation* that brings a growth-regulatory gene under the control of a different promoter that causes inappropriate expression of the gene

4. *Amplification* (i.e., abnormal DNA replication) of a DNA segment including a proto-oncogene, so that numerous copies exist, leading to overproduction of the encoded protein

An oncogene formed by either of the first two mechanisms encodes an "oncoprotein" that differs from the normal protein encoded by the corresponding proto-oncogene. In contrast, the other two mechanisms generate oncogenes whose protein products are identical with the normal proteins; their oncogenic effect is due to production at higher-than-normal levels or in cells where they normally are not produced. The localized amplification of DNA to produce as many as 100 copies of a given region (usually a region spanning hundreds of kilobases) is a common genetic change seen in tumors. Normally such an event would be repaired or the cell would stop cycling owing to checkpoint control, so such lesions imply a DNA-repair (caretaker) defect of some kind. This anomaly may take either of two forms: the duplicated DNA may be tandemly organized at a single site on a chromosome, or it may exist as small, independent mini-chromosome-like structures. The former case leads to a homogeneously staining region (HSR) that is visible in the light microscope at the site of the

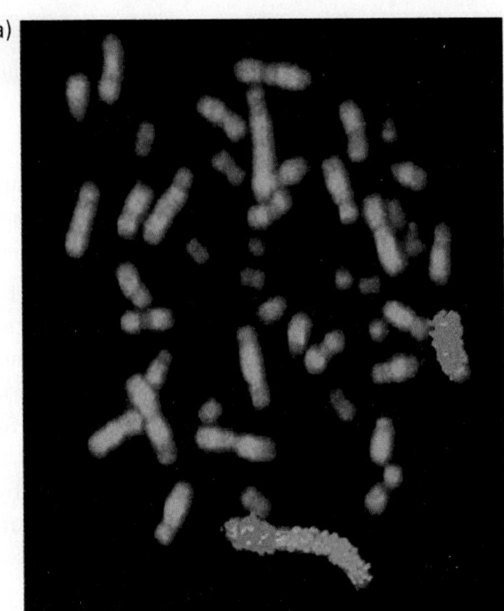

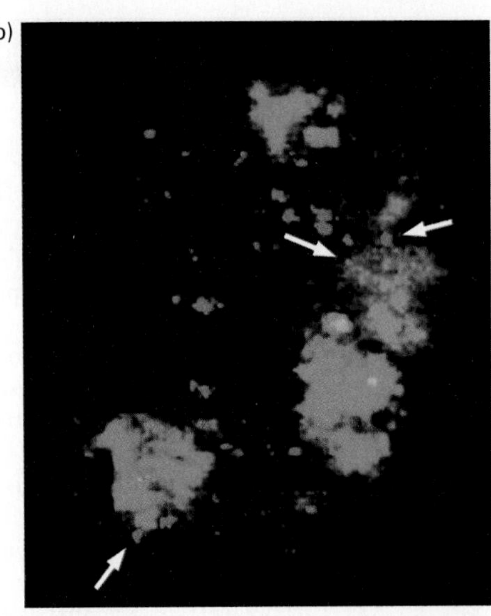

▲ **EXPERIMENTAL FIGURE 25-12 DNA amplifications in stained chromosomes take two forms, visible under the light microscope.** (a) Homogeneously staining regions (HSRs) in a human chromosome from a neuroblastoma cell. The chromosomes are uniformly stained with a blue dye so that all can be seen. Specific DNA sequences were detected using fluorescent in situ hybridization (FISH) in which fluorescently labeled DNA clones are hybridized to denatured DNA in the chromosomes. The chromosome 4 pair is marked (red) by in situ hybridization with a large DNA clone containing the N-*myc* oncogene. On one of the chromosome 4's an HSR is visible (green) after staining for a sequence enriched in the HSR. (b) Optical sections through nuclei from a human neuroblastoma cell that contain so-called double-minute chromosomes. The normal chromosomes are the green and blue structures; the double-minute chromosomes are the many small red dots. Arrows indicate double-minutes associated with the surface or interior of the normal chromosomes. [From I. Solovei et al., 2000, *Genes Chromosomes Cancer* **29:**297–308, figs. 4 and 17.]

amplification; the latter case causes extra "minute" chromosomes, separate from the normal chromosomes that pepper a stained chromosomal preparation (Figure 25-12).

Gene amplification may involve a small number of genes, such as the N-*myc* gene and its neighbor *DDX1*, which are amplified in neuroblastoma, or a chromosome region containing many genes. It can be difficult to determine which genes are amplified, a first step in determining which gene caused the tumor. DNA microarrays offer a powerful approach for finding amplified regions of chromosomes. Rather than look at gene expression, the application of microarrays described previously here and in Chapter 5, these experiments involve looking for abnormally abundant DNA sequences. Genomic DNA from cancer cells is used to probe arrays containing fragments of genomic DNA, and spots with amplified DNA give stronger signals than control spots. Among the amplified genes, the strongest candidates for the relevant ones can be identified by also measuring gene expression. A breast carcinoma cell line, with four known amplified chromosome regions, was screened for amplified genes, and the expression levels of those genes were also studied on microarrays. Fifty genes were found to be amplified, but only five were also highly expressed. These five were better candidates for being new oncogenes, since amplified genes that are not highly expressed are less likely to contribute to tumor growth. ■

However they arise, the gain-of-function mutations that convert proto-oncogenes to oncogenes are genetically dominant; that is, mutation in only one of the two alleles is sufficient for induction of cancer. See Table 25-1 for a comparison of different classes of cancer-related genes.

Cancer-Causing Viruses Contain Oncogenes or Activate Cellular Proto-oncogenes

Pioneering studies by Peyton Rous beginning in 1911 led to the initial recognition that a virus could cause cancer when injected into a suitable host animal. Many years later molecular biologists showed that Rous sarcoma virus (RSV) is a **retrovirus** whose RNA genome is reverse-transcribed into DNA, which is incorporated into the host-cell genome (see Figure 4-49). In addition to the "normal" genes present in all retroviruses, oncogenic transforming viruses like RSV contain an oncogene: in the case of RSV, the v-*src* gene. Subsequent studies with mutant forms of RSV demonstrated that only the v-*src* gene, not the other viral genes, was required for cancer induction.

In the late 1970s, scientists were surprised to find that normal cells from chickens and other species contain a gene that is closely related to the RSV v-*src* gene. This normal cellular gene, a proto-oncogene, commonly is distinguished from the viral gene by the prefix "c" (c-*src*). RSV and other acutely transforming viruses are thought to have arisen by incorporating, or transducing, a normal cellular proto-oncogene into their genome. Subsequent mutation in the transduced gene then converted it into a dominantly acting oncogene, which can induce cell transformation even in the presence of the normal c-*src* proto-oncogene. Such viruses

TABLE 25-1 Classes of Genes Implicated in the Onset of Cancer

	NORMAL FUNCTION OF GENES	EXAMPLES OF GENE PRODUCTS	EFFECT OF MUTATION	GENETIC PROPERTIES OF MUTANT GENE	ORIGIN OF MUTATIONS
Proto-oncogenes	Promote cell survival or proliferation	Anti-apoptotic proteins, components of signaling and signal transduction pathways that result in proliferation, transcription factors	Gain-of-function mutations allow unregulated cell proliferation and survival	Mutations are genetically dominant	Arise by point mutation, chromosomal translocation, amplification
Tumor-suppressor genes	Inhibit cell survival or proliferation	Apoptosis-promoting proteins, inhibitors of cell-cycle progression, checkpoint-control proteins that assess DNA/chromosomal damage, components of signal pathways that restrain cell proliferation	Loss-of-function mutations allow unregulated cell proliferation and survival	Mutations are genetically recessive	Arise by deletion, point mutation, methylation
Caretaker genes	Repair or prevent DNA damage	DNA-repair enzymes	Loss-of-function mutations allow mutations to accumulate	Mutations are genetically recessive	Arise by deletion, point mutation, methylation

are called *transducing retroviruses* because their genomes contain an oncogene derived from a transduced cellular proto-oncogene.

Because its genome carries the potent v-*src* oncogene, the transducing RSV induces tumors within days. In contrast, most oncogenic retroviruses induce cancer only after a period of months or years. The genomes of these *slow-acting retroviruses,* which are weakly transforming, differ from those of transducing viruses in one crucial respect: they lack an oncogene. All slow-acting, or "long-latency," retroviruses appear to cause cancer by integrating into the host-cell DNA near a cellular proto-oncogene and activating its expression. The long terminal repeat (LTR) sequences in integrated retroviral DNA can act as an enhancer or promoter for a nearby cellular gene, thereby stimulating its transcription. For example, in the cells from tumors caused by avian leukosis virus (ALV), the retroviral DNA is inserted near the c-*myc* gene. These cells overproduce c-Myc protein; as noted earlier, overproduction of c-Myc causes abnormally rapid proliferation of cells. Slow-acting viruses act slowly for two reasons: integration near a cellular proto-oncogene (e.g., c-*myc*) is a random, rare event, and additional mutations have to occur before a full-fledged tumor becomes evident.

In natural bird and mouse populations, slow-acting retroviruses are much more common than oncogene-containing retroviruses such as Rous sarcoma virus. Thus, insertional proto-oncogene activation is probably the major mechanism by which retroviruses cause cancer. Although the only retrovirus known to cause human tumors is human T cell leukemia/lymphoma virus (HTLV), the huge investment in studying retroviruses as a model for human cancer paid off both in the discovery of cellular oncogenes and in the sophisticated understanding of retroviruses, which later accelerated progress on the HIV virus that causes AIDS.

A few DNA viruses also are oncogenic. Unlike most DNA viruses that infect animal cells (see Chapter 4), oncogenic DNA viruses integrate their viral DNA into the host-cell genome. The viral DNA contains one or more oncogenes, which permanently transform infected cells. For example, many warts and other benign tumors of epithelial cells are caused by the DNA-containing human papillomaviruses (HPV). A medically much more serious outcome of HPV infection is cervical cancer, the third most common type of cancer in women after lung and breast cancer; we shall examine HPV oncoproteins later in the chapter. The Pap smear, which is used to sample the cervical tissue and screen for possible cancers, is thought to have reduced the death rate by about 70 percent. However thousands of people still die from cervical cancer each year, and some of these deaths would have been prevented had screening been done. Fortunately, not all HPV infections lead to cancer. Unlike retroviral oncogenes, which are derived from normal cellular genes and have no function for the virus except to allow their proliferation in tumors, the known oncogenes of DNA viruses are integral parts of the viral genome and are required for viral replication. As discussed later, the oncoproteins expressed from integrated viral DNA in infected cells act in various ways to stimulate cell growth and proliferation.

Loss-of-Function Mutations in Tumor-Suppressor Genes Are Oncogenic

Tumor-suppressor genes generally encode proteins that in one way or another inhibit cell proliferation. *Loss-of-function* mutations in one or more of these "brakes" contribute to the development of many cancers. Prominent among the classes of proteins encoded by tumor-suppressor genes are these five:

1. Intracellular proteins that regulate, or inhibit progression through, a specific stage of the cell cycle (e.g., p16 and Rb for G_1)

2. Receptors or signal transducers for secreted hormones or developmental signals that inhibit cell proliferation (e.g., TGFβ, the hedgehog receptor patched)

3. Checkpoint-control proteins that arrest the cell cycle if DNA is damaged or if chromosomes are abnormal (e.g., p53)

4. Proteins that promote apoptosis

5. Enzymes that participate in DNA repair, encoded by caretaker genes

Since generally one copy of a tumor-suppressor gene suffices to control cell proliferation, *both* alleles of a tumor-suppressor gene must be lost or inactivated in order to promote tumor development. Thus oncogenic loss-of-function mutations in tumor-suppressor genes are often genetically **recessive** (Table 25-1). In this context "recessive" means that if there is even one working gene, producing about half the usual amount of protein product, tumor formation will be prevented. With some genes, half the amount of product is not enough, in which case the loss of just one of the two genes can lead to cancer. This kind of gene is *haplo-insufficient*. The loss of one copy of the gene is decisive for the final phenotype, so this type of mutation is dominant. It is useful to remember, then, two processes by which cancer genes can be dominant: (1) loss of one copy of a tumor-suppressor gene, resulting in insufficient product to control growth, and (2) activation of a gene or protein that causes growth even in the presence of one normal allele, i.e., a dominant oncogene (as was described in the previous section). In many cancers, tumor-suppressor genes have deletions or point mutations that prevent production of any protein or lead to production of a nonfunctional protein. Another mechanism for inactivating tumor-suppressor genes is methylation of cytosine residues in the promoter or other control elements, which inhibits their transcription. Such methylation is commonly found in nontranscribed regions of DNA (see Chapter 7).

Inherited Mutations in Tumor-Suppressor Genes Increase Cancer Risk

Individuals with inherited mutations in tumor-suppressor genes have a hereditary predisposition for certain cancers. Such individuals generally inherit a germ-line mutation in one allele of the gene; somatic mutation of the second allele facilitates tumor progression. A classic case is retinoblastoma, which is caused by loss of function of *RB*, the first tumor-suppressor gene to be identified. As we discuss later, the protein encoded by *RB* helps regulate progress through the cell cycle.

Hereditary versus Sporadic Retinoblastoma Children with hereditary retinoblastoma inherit a single defective copy of the RB gene, sometimes seen as a small deletion on one of the copies of chromosome 13. The children develop several retinal tumors early in life and generally in both eyes. The deletion or mutation of the normal *RB* gene on the other chromosome is an essential step in tumor formation, giving rise to a cell that produces no functional Rb protein (Figure 25-13). Individuals with sporadic retinoblastoma, in contrast, inherit two normal *RB* alleles, each of which has undergone a loss-of-function somatic mutation in a single retinal cell. Because losing two copies of the *RB* gene is far less likely than losing one, sporadic retinoblastoma is rare and usually affects only one eye. However, RB mutation contributes to many tumors, and while the hereditary

(a) Hereditary retinoblastoma

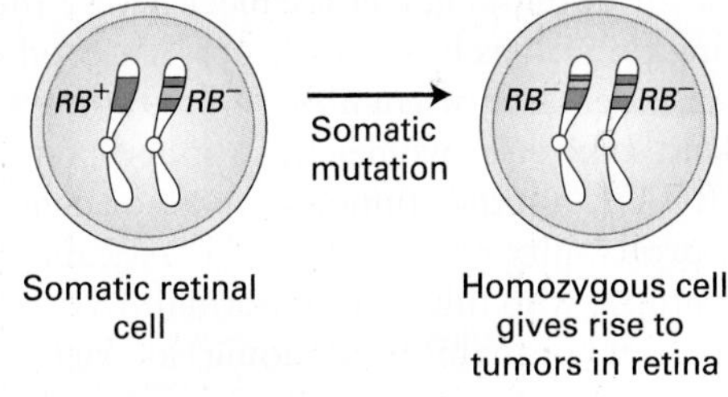

(b) Sporadic retinoblastoma

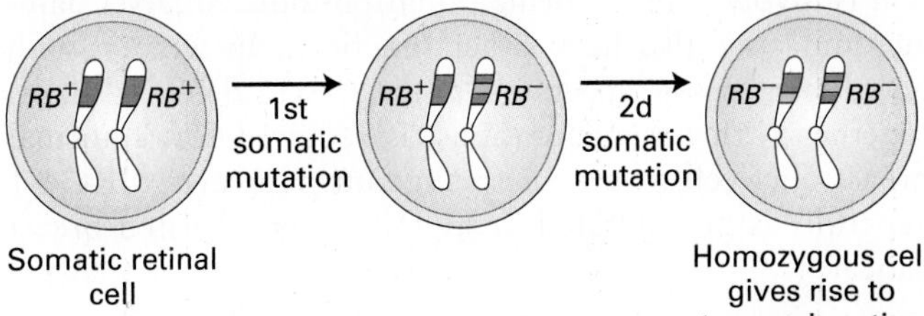

▲ **FIGURE 25-13 Role of spontaneous somatic mutation in retinoblastoma.** This disease is marked by retinal tumors that arise from cells carrying two mutant RB^- alleles. (a) In hereditary (familial) retinoblastoma, a child inherits a normal RB^+ allele from one parent and a mutant RB^- allele from the other parent. A single mutation in a heterozygous somatic retinal cell that inactivates the normal allele will produce a cell lacking any *Rb* gene function due to two mutations. (b) In sporadic retinoblastoma, a child inherits two normal RB^+ alleles. Two separate somatic mutations in a particular retinal cell or its progeny are required to produce a homozygous RB^-/RB^- cell.

retinoblastoma cases number about 100 per year in the U.S., about 100,000 other cancer cases each year involve RB mutations acquired postconception.

If retinal tumors are removed before they become malignant, children with hereditary retinoblastoma often survive until adulthood and produce children. Because their germ cells contain one normal and one mutant *RB* allele, these individuals will, on average, pass on the mutant allele to half their children and the normal allele to the other half. Children who inherit the normal allele are normal if their other parent has two normal *RB* alleles. However, those who inherit the mutant allele have the same enhanced predisposition to develop retinal tumors as their affected parent, even though they inherit a normal *RB* allele from their other, normal parent. Thus the *tendency* to develop retinoblastoma is inherited as a dominant trait; one mutant copy is sufficient to predispose a person to develop the cancer. As we shall see shortly, many human tumors (not just retinal tumors) contain mutant *RB* alleles or mutations affecting other components of the same pathway; most of these arise as the result of somatic mutations.

Inherited Forms of Colon and Breast Cancer Similar hereditary predisposition for other cancers has been associated with inherited mutations in other tumor-suppressor genes. For example, individuals who inherit a germ-line mutation in one *APC* allele develop thousands of precancerous intestinal polyps (see Figure 25-9). Since there is a high probability that one or more of these polyps will progress to malignancy, such individuals have a greatly increased risk for developing colon cancer before the age of 50. Screening for polyps by colonoscopy is a good idea for people 50 or older, even when no *APC* mutation is known to be present. Likewise, women who inherit one mutant allele of *BRCA-1*, another tumor-suppressor gene, have a 60 percent probability of developing breast cancer by age 50, whereas those who inherit two normal *BRCA-1* alleles have a 2 percent probability of doing so. Heterozygous *BRCA-1* mutations also increase the lifetime risk of ovarian cancer from 2 percent to 15–40 percent. BRCA-1 protein is involved in repairing radiation-induced DNA damage and may also have other functions. In women with hereditary breast cancer, loss of the second *BRCA-1* allele, together with other mutations, is required for a normal breast duct cell to become malignant. However, *BRCA-1* generally is not mutated in sporadic, noninherited breast cancer.

Loss of Heterozygosity Clearly, then, we can inherit a propensity to cancer by receiving a damaged allele of a tumor-suppressor gene from one of our parents; that is, we are heterozygous for the mutation. That in itself typically will not cause cancer; since the remaining normal allele prevents aberrant growth, the cancer is recessive. Subsequent loss or inactivation of the normal allele in a somatic cell, referred to as *loss of heterozygosity (LOH),* is usually how a cancer develops. One common mechanism for LOH involves missegregation during mitosis of the chromosomes bearing the affected tumor-suppressor gene (Figure 25-14a). This process, also referred to as *nondisjunction,* is caused by failure of the spindle-assembly checkpoint, which normally prevents a metaphase cell with an abnormal mitotic spindle from completing mitosis (see Figure 20-35, **2**). Another possible mechanism for LOH is mitotic recombination between a chromatid bearing the wild-type allele and a homologous chromatid bearing a mutant allele. As illustrated in Figure 25-14b, subsequent chromosome segregation can generate a daughter cell that is homozygous for the mutant tumor-suppressor allele. A third mechanism is the deletion or mutation of the normal copy of the tumor-suppressor gene; such a deletion can encompass a large chromosomal region and need not be a precise deletion of just the tumor-suppressor gene.

Estimates vary, but hereditary cancers, i.e., cancers that arise due in part to an inherited version of a gene, are thought to constitute about 10 percent of human cancers. Further work tracing contributions of human genes seems likely to increase the percentage. It is important to remember, however, that the inherited germ-line mutation alone is not sufficient to cause tumor development. In all cases, not only must the inherited normal tumor-suppressor allele be lost or inactivated, but mutations affecting other genes also are necessary for cancer to develop. Thus a person with a recessive tumor-suppressor gene mutation can be exceptionally susceptible to environmental mutagens such as radiation.

Aberrations in Signaling Pathways That Control Development Are Associated with Many Cancers

During normal development secreted signals such as Wnt, TGFβ, and Hedgehog (Hh) are frequently used to direct cells to particular developmental fates, which may include the property of rapid mitosis. The effects of such signals must be regulated so that growth is limited to the right time and place. Among the mechanisms available for reining in the effects of powerful developmental signals are inducible intracellular antagonists, receptor blockers, and competing signals (Chapter 22). Mutations that prevent such restraining mechanisms from operating are likely to be oncogenic, causing inappropriate or cancerous growth.

Hh signaling, which is used repeatedly during development to control cell fates, is a good example of a signaling pathway implicated in cancer induction. In the skin and cerebellum, one of the human Hh proteins, Sonic hedgehog, stimulates cell division by binding to and inactivating a membrane protein called Patched1 (Ptc1) (see Figure 16-34). Loss-of-function mutations in *ptc1* permit cell proliferation in the absence of an Hh signal; thus *ptc1* is a tumor-suppressor gene. People who inherit a single working copy of *ptc1* have a propensity to develop skin and brain cancer; either can occur when the remaining allele is damaged. Other people can get these diseases, too, if they suffer loss of both copies of the gene. Thus there are both familial (inherited) and sporadic (noninherited) cases of these diseases, just as in retinoblastoma. Mutations in other genes in the Hh signaling pathway are also associated with cancer.

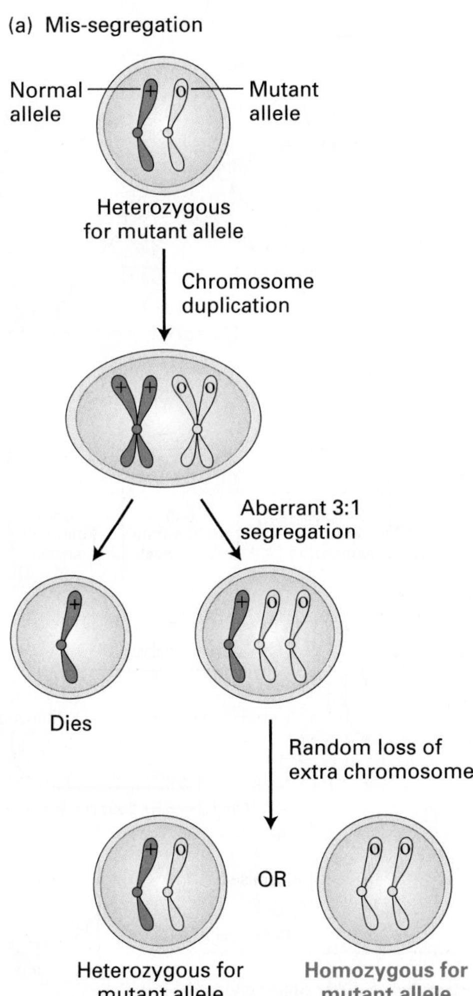

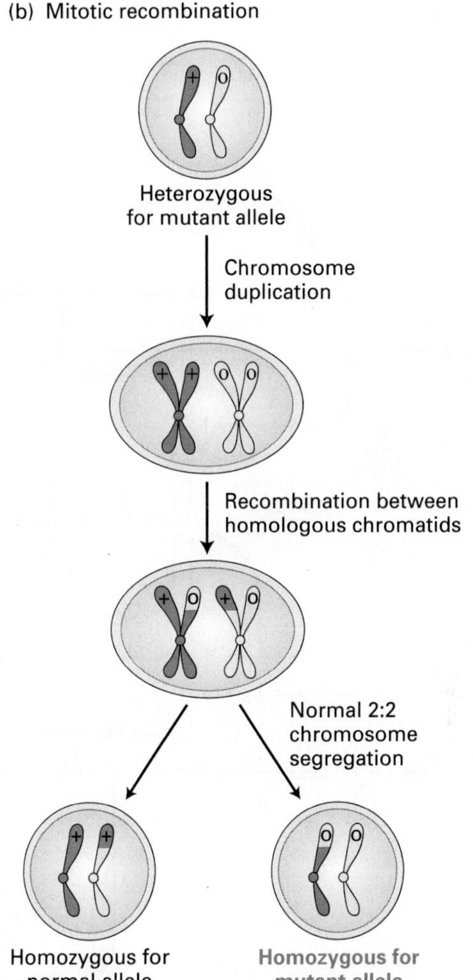

▲ **FIGURE 25-14 Two mechanisms for loss of heterozygosity (LOH) of tumor-suppressor genes.** A cell containing one normal and one mutant allele of a tumor-suppressor gene is generally phenotypically normal. (a) If formation of the mitotic spindle is defective, then the duplicated chromosomes bearing the normal and mutant alleles may segregate in an aberrant 3:1 ratio. A daughter cell that receives three chromosomes of a type will generally lose one, restoring the normal 2*n* chromosome number. Sometimes the resultant cell will contain one normal and one mutant allele, but sometimes it will be homozygous for the mutant allele. Note that such aneuploidy (abnormal chromosome constitution) is generally damaging or lethal to relatively undifferentiated cells that have to develop into the many complex structures of an organism, but can often be tolerated in clones of cells that have limited fates and duties. (b) Mitotic recombination between a chromosome with a wild-type and a mutant allele, followed by chromosome segregation, can produce a cell that contains two copies of the mutant allele.

Some such mutations create oncogenes that turn on Hh target genes inappropriately; others are recessive mutations that affect negative regulators like Ptc1. As is the case for a number of other tumor-suppressor genes, complete loss of Ptc1 function would lead to early fetal death, since it is needed for development, so it is only the tumor cells that are homozygous *ptc1/ptc1*.

Many of the signaling pathways described in other chapters play roles in controlling embryonic development and cell proliferation in adult tissues. In recent years mutations affecting components of most of these signaling pathways have been linked to cancer (Figure 25-15). Indeed, once one gene in a developmental pathway has been linked to a type of human cancer, knowledge of the pathway gleaned from model organisms like worms, flies, or mice allows focused investigations of the possible involvement of additional pathway genes in other cases of the cancer. For example, *APC*, the critical first gene mutated on the path to colon carcinoma, is now known to be part of the Wnt signaling pathway (see Chapter 16), which led to the discovery of the involvement of *β-catenin* mutations in colon cancer. Mutations in tumor-suppressor developmental genes promote tumor formation in tissues where the affected gene normally helps restrain growth. Thus the tumor-suppressor gene mutations will not cause cancers in tissues where the primary role of the developmental regulator is to control cell fate—what type of cell develops—but not cell division. Mutations in developmental proto-oncogenes may induce tumor formation in tissues where an affected gene normally promotes cell proliferation or in another tissue where the gene has become aberrantly active.

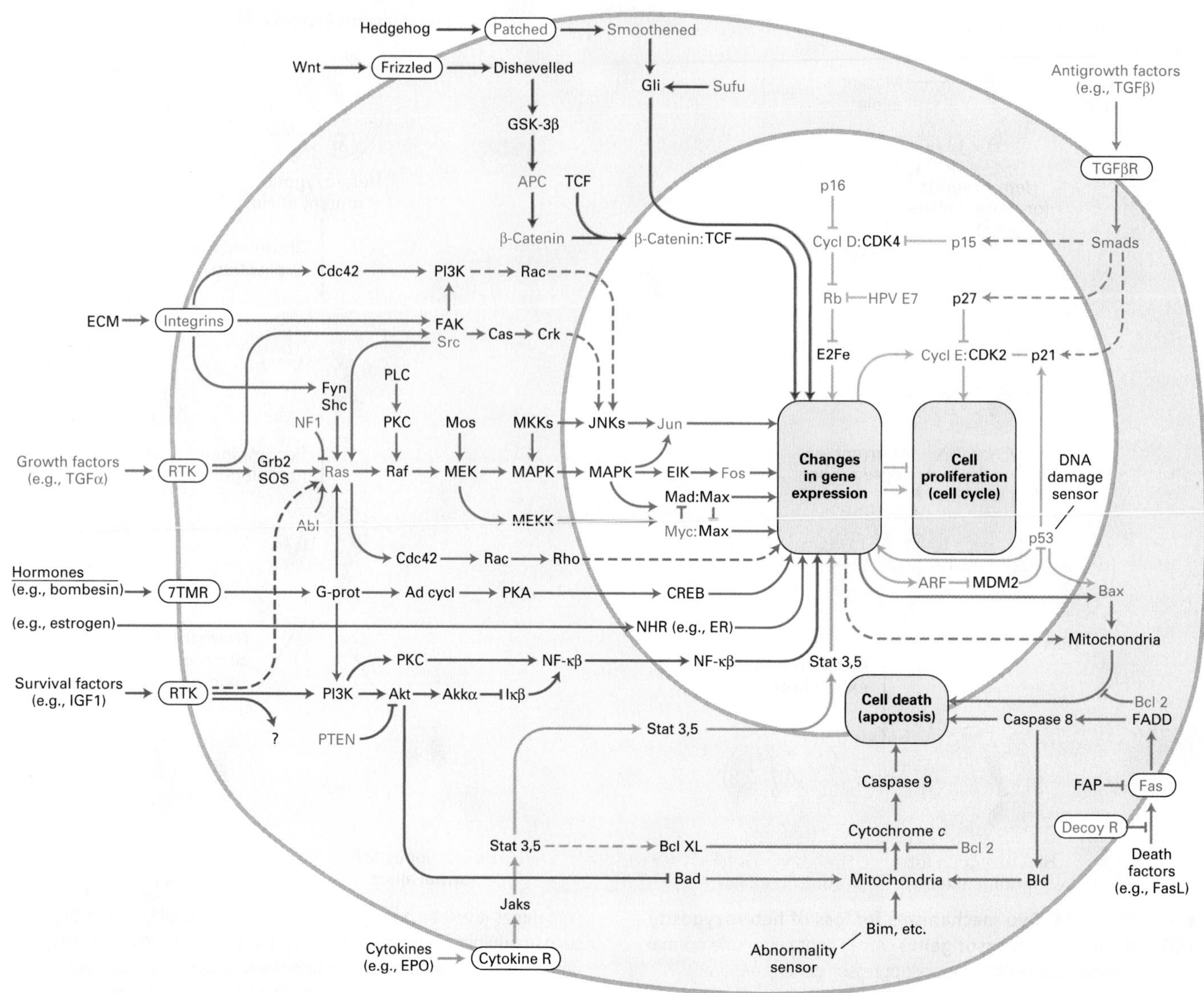

▲ **FIGURE 25-15 Cell circuitry that is affected by cancer-causing mutations.** Growth control and the cell cycle, the heart of cancer, are influenced by many types of signals, and the external inputs become integrated as the cell makes its decision about whether to divide or to continue dividing. Developmental pathways give cells their identity, and along with that identity often comes a commitment to proliferate or not. Genes known to be mutated in cancer cells are shown in red. Less firmly established pathways are indicated with dashed lines. [From D. Hanahan and R. A. Weinberg, 2000, *Cell* **100**:57.]

KEY CONCEPTS OF SECTION 25.2

The Genetic Basis of Cancer

- Dominant gain-of-function mutations in proto-oncogenes and recessive loss-of-function mutations in tumor-suppressor genes are oncogenic.
- Among the proteins encoded by proto-oncogenes are growth-promoting signaling proteins and their receptors, signal-transduction proteins, transcription factors, and apoptotic proteins (see Figure 25-11).
- An activating mutation of one of the two alleles of a proto-oncogene converts it to an oncogene. This can occur by point mutation, gene amplification, gene translocation, and overexpression.
- The first human oncogene to be identified encodes a constitutively active form of Ras, a signal-transduction protein. This oncogene was isolated from a human bladder carcinoma (see Figure 25-6).
- Slow-acting retroviruses can cause cancer by integrating near a proto-oncogene in such a way that transcription of the cellular gene is activated continuously and inappropriately.
- Tumor-suppressor genes encode proteins that directly or indirectly slow progression through the cell cycle, such as checkpoint-control proteins that arrest the cell cycle, components of growth-inhibiting signaling pathways, pro-apoptotic proteins, regulators of kinase activities, and DNA-repair enzymes.

- The first tumor-suppressor gene to be recognized, *RB*, is mutated in retinoblastoma and many other tumors; some component of the RB pathway is altered in most or all tumors.
- Inheritance of a single mutant allele of *RB* greatly increases the probability that a specific kind of cancer will develop, as is the case for many other tumor-suppressor genes (e.g., *APC* and *BRCA-1*).
- In individuals born heterozygous for a tumor-suppressor gene, a somatic cell can undergo loss of heterozygosity (LOH) by mitotic recombination, chromosome mis-segregation, mutation, gene silencing, or deletion (see Figure 25-14).
- Mutations in caretaker genes are genetically recessive, since having even one functional copy is usually sufficient to prevent serious DNA damage. However a single copy leaves the carrier subject to possible LOH.
- Many genes that regulate normal developmental processes encode proteins that function in various signaling pathways (see Figure 25-15). Their normal roles in regulating where and when growth occurs are reflected in the character of the tumors that arise when the genes are mutated.

25.3 Oncogenic Mutations in Growth-Promoting Proteins

Genes encoding each class of cell regulatory protein depicted in Figure 25-11 have been identified as proto-oncogenes or tumor-suppressor genes. In this section we examine in more detail how mutations that result in the unregulated, constitutive activity of certain proteins or in their overproduction promote cell proliferation and transformation, thereby contributing to carcinogenesis. In each case we see how a rare cell that has undergone a very particular sort of mutation becomes abundant owing to its uncontrolled proliferation.

Oncogenic Receptors Can Promote Proliferation in the Absence of External Growth Factors

Hyperactivation of a growth-inducing signaling protein, due to an alteration of the protein, might seem a likely mechanism of cancer, but in fact this rarely occurs. Only one such naturally occurring oncogene, *sis*, has been discovered. The *sis* oncogene, which encodes an altered form of platelet-derived growth factor (PDGF), can aberrantly stimulate proliferation of cells that normally express the PDGF receptor. A more common event is that cells begin to produce an unaltered growth factor that acts upon the cell that produces it. This is called **autocrine** stimulation.

In contrast, oncogenes encoding cell-surface receptors that transduce growth-promoting signals have been associated with several types of cancer. The receptors for many such growth factors have intrinsic protein-tyrosine kinase activity in their cytosolic domains, an activity that is quiescent until activated. Ligand binding to the external domains of these **receptor tyrosine kinases (RTKs)** leads to their dimerization and activation of their kinase activity, initiating an intracellular signaling pathway that ultimately promotes proliferation.

In some cases, a point mutation changes a normal RTK into one that dimerizes and is constitutively active in the absence of ligand. For instance, a single point mutation converts the normal human EGF receptor 2 (Her2) receptor into the Neu oncoprotein ("neu" for its first known role, in neuroblastoma), which is an initiator of certain mouse cancers (Figure 25-16, *left*). Similarly, human tumors called *multiple endocrine neoplasia type 2* produce a constitutively active dimeric glia-derived neurotrophic factor (GDNF) receptor that results from a point mutation in the extracellular domain. GDNF receptor is a protein-tyrosine kinase, so the constitutively active form excessively phosphorylates its downstream target proteins. In other cases, deletion of much of the extracellular ligand-binding domain produces a constitutively active oncogenic receptor. For example, deletion of the

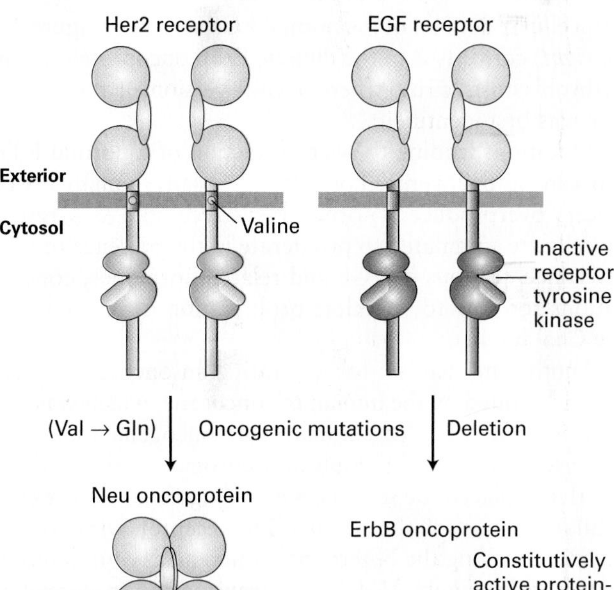

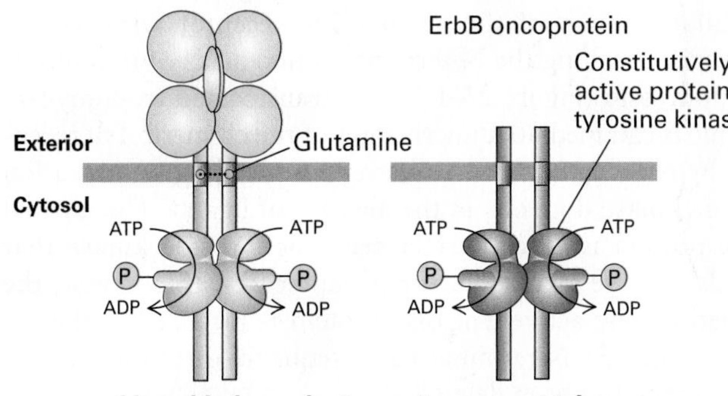

▲ **FIGURE 25-16 Effects of oncogenic mutations in proto-oncogenes that encode cell-surface receptors.** *Left:* A mutation that alters a single amino acid (valine to glutamine) in the transmembrane region of the Her2 receptor causes dimerization of the receptor, even in the absence of the normal EGF-related ligand, making the oncoprotein Neu a constitutively active kinase. *Right:* A deletion that causes loss of the extracellular ligand-binding domain in the EGF receptor leads, for unknown reasons, to constitutive activation of the kinase activity of the resulting oncoprotein ErbB.

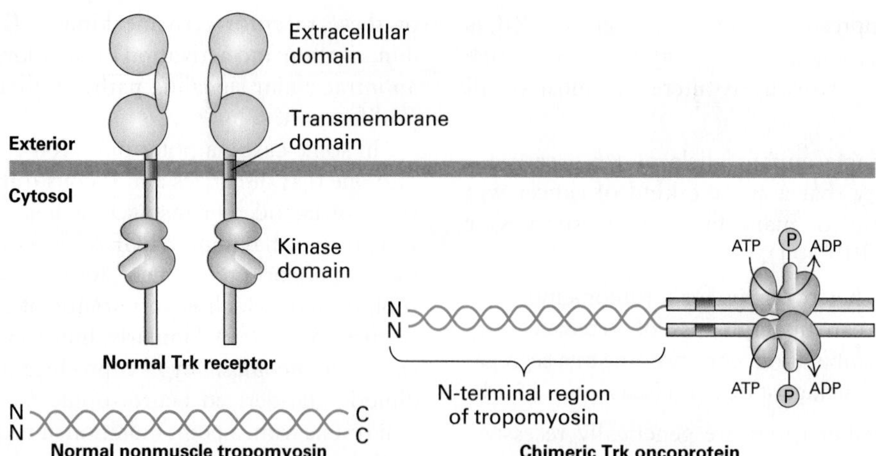

▲ **FIGURE 25-17 Domain structures of normal tropomyosin, the normal Trk receptor, and chimeric Trk oncoprotein.** A chromosomal translocation results in replacement of most of the extracellular domain of the normal human Trk protein, a receptor tyrosine kinase, with the N-terminal domain of nonmuscle tropomyosin. Dimerized by the tropomyosin segment, the Trk oncoprotein kinase is constitutively active. Unlike the normal Trk, which is localized to the plasma membrane, the Trk oncoprotein is found in the cytosol. [See F. Coulier et al., 1989, *Mol. Cell Biol.* **9:**15.]

extracellular domain of the normal EGF receptor (Figure 25-16, *right*) converts it to the dimeric ErbB oncoprotein (from erythroblastosis virus, where a viral version of the altered gene was first identified).

Mutations leading to overproduction of a normal RTK also can be oncogenic. For instance, many human breast cancers overproduce a normal Her2 receptor. As a result, the cells are stimulated to proliferate in the presence of very low concentrations of EGF and related hormones, concentrations too low to stimulate proliferation of normal cells (see Chapter 16).

Another mechanism for generating an oncogenic receptor is illustrated by the human *trk* oncogene, which was isolated from a colon carcinoma. This oncogene encodes a chimeric protein as the result of a chromosomal translocation that replaced the sequences encoding most of the extracellular domain of the normal Trk receptor with the sequences encoding the N-terminal amino acids of nonmuscle tropomyosin (Figure 25-17). The translocated tropomyosin segment can mediate dimerization of the chimeric Trk receptor by forming a coiled-coil structure, leading to activation of the kinase domains in the absence of ligand. The normal Trk protein is a cell-surface receptor tyrosine kinase that binds a nerve growth factor (Chapter 21). In contrast, the constitutively active Trk oncoprotein is localized in the cytosol, since the N-terminal signal sequence directing it to the membrane has been deleted.

Viral Activators of Growth-Factor Receptors Act as Oncoproteins

Viruses use their own tricks to cause cancer, presumably to increase the production of virus from infected cancer cells. For example, a retrovirus called *spleen focus-forming virus (SFFV)* induces erythroleukemia (a tumor of erythroid progenitors) in adult mice by manipulating a normal developmental signal. The proliferation, survival, and differentiation of erythroid progenitors into mature red cells absolutely require erythropoietin (Epo) and the corresponding Epo receptor (see Figure 16-6). A mutant SFFV envelope glycoprotein, termed *gp55*, is responsible for the oncogenic effect of the virus. Although gp55 cannot function as a normal retrovirus envelope protein in virus budding and infection, it has acquired the remarkable ability to bind to and activate Epo receptors in the same cell (Figure 25-18). By inappropriately and continuously stimulating the proliferation of

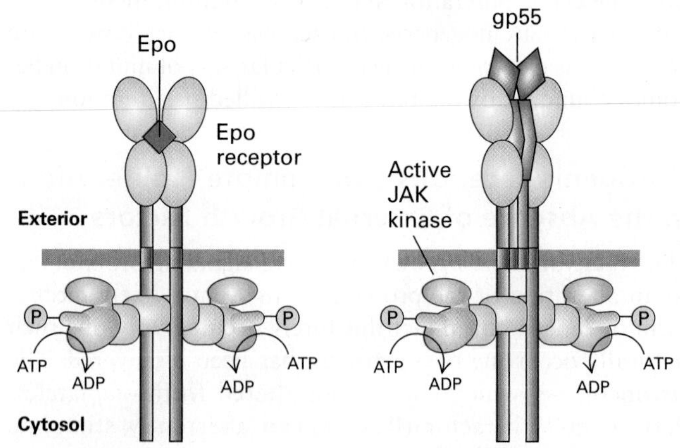

▲ **FIGURE 25-18 Activation of the erythropoietin (Epo) receptor by the natural ligand, Epo, or a viral oncoprotein.** Binding of Epo dimerizes the receptor and induces formation of erythrocytes from erythroid progenitor cells. Normally cancers occur when progenitor cells infected by the spleen focus-forming virus produce the Epo receptor and viral gp55, both localized to the plasma membrane. The transmembrane domains of dimeric gp55 specifically bind the Epo receptor, dimerizing and activating the receptor in the absence of Epo. [See S. N. Constantinescu et al., 1999, *EMBO J.* **18:**3334.]

erythroid progenitors, gp55 induces formation of excessive numbers of erythrocytes. Malignant clones of erythroid progenitors emerge several weeks after SFFV infection as a result of further mutations in these aberrantly proliferating cells.

Another example of this phenomenon is provided by human papillomavirus (HPV), the sexually transmitted DNA virus that causes cervical cancer and genital warts. A papillomavirus protein designated *E5*, which contains only 44 amino acids, spans the plasma membrane and forms a dimer or trimer. Each E5 polypeptide can form a stable complex with one endogenous receptor for PDGF, thereby aggregating two or more PDGF receptors within the plane of the plasma membrane. This mimics hormone-mediated receptor dimerization, causing sustained receptor activation and eventually cell transformation. As we will see later, the HPV genome also codes for several other proteins that act to inhibit tumor-suppressor genes, further contributing to cell transformation. Recently a vaccine against an HPV capsid protein (L1) has been found to protect against cervical cancer caused by at least some types of virus.

Many Oncogenes Encode Constitutively Active Signal-Transduction Proteins

A large number of oncogenes are derived from proto-oncogenes whose encoded proteins aid in transducing signals from an activated receptor to a cellular target. We describe several examples of such oncogenes; each is expressed in many types of tumor cells.

Ras Pathway Components Among the best-studied oncogenes in this category are the ras^D genes, which were the first nonviral oncogenes to be recognized. Any one of a number of changes in Ras protein can lead to its uncontrolled and therefore dominant activity. In particular, if a point mutation substitutes any amino acid for glycine at position 12 in the Ras sequence, the normal protein is converted into a constitutively active oncoprotein. This simple mutation reduces the protein's GTPase activity, thus maintaining Ras in the active GTP-bound state. Constitutively active Ras oncoproteins are produced by many types of human tumors, including bladder, colon, mammary, skin, and lung carcinomas, neuroblastomas, and leukemias.

As we saw in Chapter 16, Ras is a key component in transducing signals from activated receptors to a cascade of protein kinases. In the first part of this pathway, a signal from an activated RTK is carried via two adapter proteins to Ras, converting it to the active GTP-bound form (see Figure 16-20). In the second part of the pathway, activated Ras transmits the signal via two intermediate protein kinases to MAP kinase. The activated MAP kinase then phosphorylates a number of transcription factors that induce synthesis of important cell-cycle and differentiation-specific proteins (see Figure 16-25). Activating Ras mutations short-circuit the first part of this pathway, making upstream activation triggered by ligand binding to the receptor unnecessary. Oncogenes encoding other altered components of the RTK/Ras/MAP kinase pathway also have been identified.

Constitutive Ras activation can also arise from a recessive loss-of-function mutation in a GTPase-accelerating protein (GAP). The normal GAP function is to accelerate hydrolysis of GTP and the conversion of active GTP-bound Ras to inactive GDP-bound Ras (see Figure 3-32). The loss of GAP leads to sustained Ras activation of downstream signal-transduction proteins. For example, neurofibromatosis, a benign tumor of the sheath cells that surround nerves, is caused by loss of both alleles of *NF1,* which encodes a Ras GAP-type protein. Individuals with neurofibromatosis have inherited a single mutant *NF1* allele; subsequent somatic mutation in the other allele leads to formation of neurofibromas. Thus *NF1,* like *RB,* is a tumor-suppressor gene, and neurofibromatosis, like hereditary retinoblastoma, is inherited as an autosomal dominant trait.

Src Protein Kinase Several oncogenes encode cytosolic protein kinases that normally transduce signals in a variety of intracellular signaling pathways. Indeed the first oncogene to be discovered, v-*src* from Rous sarcoma retrovirus, encodes a constitutively active protein-tyrosine kinase. At least eight mammalian proto-oncogenes encode a family of nonreceptor tyrosine kinases related to the v-Src protein. In addition to a catalytic domain, these kinases contain SH2 and SH3 protein-protein interaction domains. The kinase activity of cellular Src and related proteins normally is inactivated by phosphorylation of the tyrosine residue at position 527, which is six residues from the C-terminus (Figure 25-19a, b). Hydrolysis of phosphotyrosine 527 by a specific phosphatase enzyme normally activates c-Src. Tyrosine 527 is often missing or altered in Src oncoproteins that have constitutive kinase activity; that is, they do not require activation by a phosphatase (Figure 25-19b).

Abl Protein Kinase One oncogene encoding a cytosolic nonreceptor protein kinase is generated by a chromosome translocation that fuses a part of the c-*abl* gene, which encodes a tyrosine kinase, with part of the *bcr* gene, whose function is unknown (Figure 25-20a). One consequence of the fusion is that a hybrid protein is produced, and it has odd and dangerous properties. The normal c-Abl protein promotes branching of filamentous actin and extension of cell processes, so it may function primarily to control the cytoskeleton and cell shape. The chimeric oncoproteins encoded by the *bcr-abl* oncogene form a tetramer that exhibits unregulated and continuous Abl kinase activity (Figure 25-20b). (This is similar to dimerization and activation of the chimeric Trk oncoprotein shown in Figure 25-17.) Bcr-Abl can phosphorylate and thereby activate many intracellular signal-transduction proteins, and at least some of these proteins are not normal substrates of Abl. For instance, Bcr-Abl can activate JAK2 kinase and STAT5 transcription factor, which normally are activated by binding of growth factors (e.g., erythropoietin) to cell-surface receptors (see Figure 16-12). Bcr-Abl can also provide a docking site for

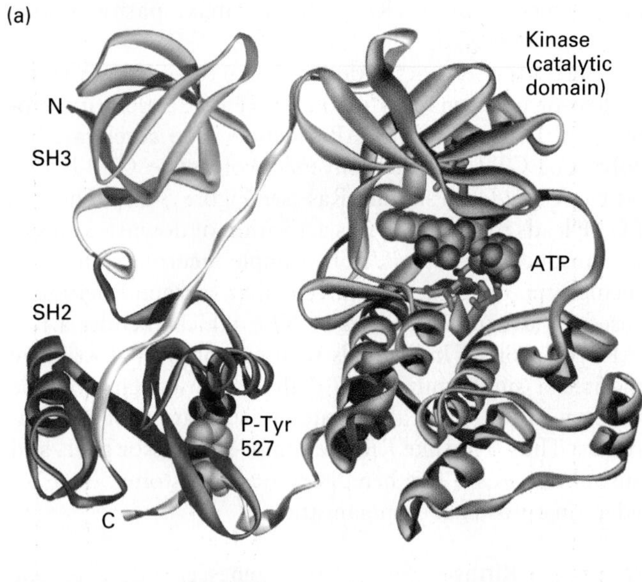

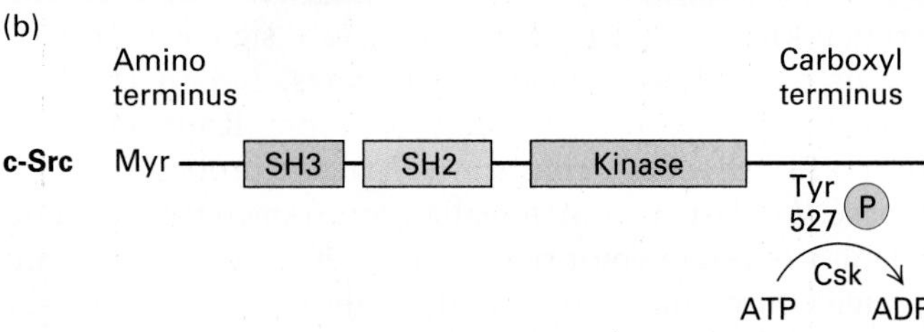

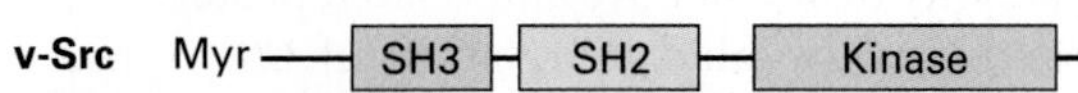

▲ **FIGURE 25-19 Structure of Src tyrosine kinases and activation by an oncogenic mutation.** (a) Three-dimensional structure of Hck, one of several Src kinases in mammals. Binding of phosphotyrosine 527 (P-Tyr 527) to the SH2 domain induces conformational strains in the SH3 and kinase domains, distorting the kinase active site so it is catalytically inactive. The kinase activity of cellular Src proteins is normally activated by removing the phosphate on tyrosine 527. (b) Domain structure of c-Src and v-Src. Phosphorylation of tyrosine 527 by Csk, another cellular tyrosine kinase, inactivates the Src kinase activity. The transforming v-Src oncoprotein encoded by Rous sarcoma virus is missing the C-terminal 18 amino acids including tyrosine 527 and thus is constitutively active. [Part (a) from F. Sicheri et al., 1997, *Nature* **385:**602; see also T. Pawson, 1997, *Nature* **385:**582, and W. Xu et al., 1997, *Nature* **385:**595.]

signal-transduction proteins through the Bcr part, thus potentially stimulating signal transduction.

The chromosomal translocation that forms *bcr-abl* generates the diagnostic *Philadelphia chromosome,* discovered in 1960 (see Figure 25-20a). If this translocation occurs in a hematopoietic cell in the bone marrow, the activity of the chimeric *bcr-abl* oncogene results in the initial phase of human chronic myelogenous leukemia (CML), characterized by an expansion in the number of white blood cells. A second loss-of-function mutation in a cell carrying *bcr-abl* (e.g., in *p53* or *RB*) leads to acute leukemia, which often kills the patient. The CML chromosome translocation was only the first of a long series of distinctive, or "signature," chromosome translocations linked to particular forms of leukemia (Figure 25-21). Many of the gene fusions involve genes encoding transcriptional regulators, in particular, transcriptional regulators of Hox genes (Chapter 22). Each case presents an opportunity for greater understanding of the disease, earlier diagnosis, and new therapies. In the case of CML, that second step to successful therapy has already been taken.

After a painstaking search, an inhibitor of Abl kinase named imatinib (Gleevec) was identified as a possible treatment for CML in the early 1990s. Imatinib, which binds directly to the Abl kinase active site and inhibits kinase activity, is highly lethal to CML cells while sparing normal cells (see Figure 25-20c). After clinical trials showing imatinib is remarkably effective in treating CML despite some side effects, it was approved by the FDA in 2001, the first cancer drug targeted to a signal-transduction protein unique to tumor cells. Imatinib inhibits several other tyrosine kinases that are implicated in different cancers and has been successful in trials for treating these diseases, including forms of gastrointestinal tumors, as well. There are 96 tyrosine kinases encoded in the human genome, so drugs related to imatinib may be useful in controlling the activities of all these proteins. One ongoing challenge is that tumor cells can evolve to be resistant to imatinib and other such drugs, necessitating the invention of alternative drugs. ■

Inappropriate Production of Nuclear Transcription Factors Can Induce Transformation

Oncogenic mutations that create oncogenes or damage tumor-suppressor genes eventually cause changes in gene expression. Experimentally this can be measured by comparing the amounts of different mRNAs produced in normal versus tumor cells. As discussed earlier, we can now measure such differences in the expression of thousands of genes with DNA microarrays (see Figure 25-10).

Since the most direct effect on gene expression is exerted by transcription factors, it is not surprising that many oncogenes encode transcription factors. Two examples are *jun* and *fos,* which initially were identified in transforming retroviruses and later found to be overexpressed in some human tumors. The c-*jun* and c-*fos* proto-oncogenes encode proteins that sometimes associate to form a heterodimeric transcription factor, called *AP1,* that binds to a sequence found in promoters and enhancers of many genes (see Figure 7-29 and Chapter 16). Both Fos and Jun also can act independently as transcription factors. They function as oncoproteins by activating transcription of key genes that encode growth-promoting proteins or by inhibiting transcription of growth-repressing genes.

Many nuclear proto-oncogene proteins are produced when normal cells are stimulated to grow, indicating their direct role in growth control. For example, PDGF treatment of quiescent 3T3 cells induces an ≈50-fold increase in the production of c-Fos and c-Myc, the normal products of the

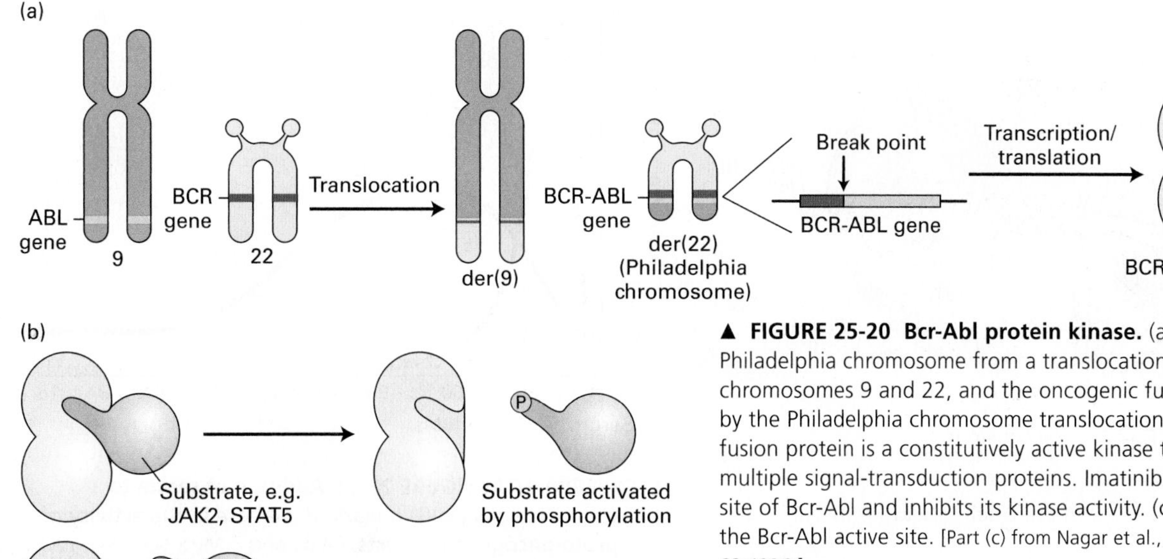

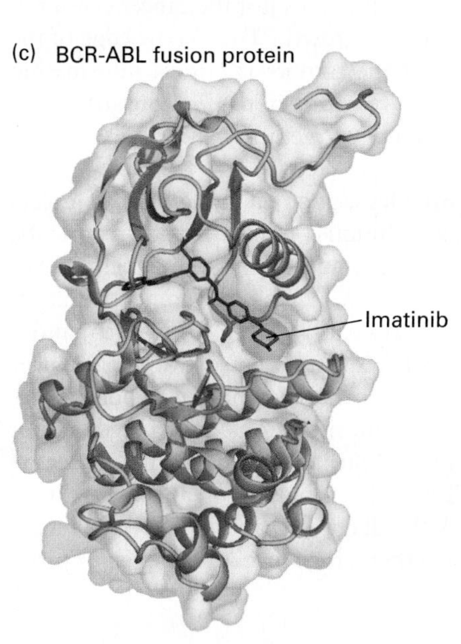

▲ **FIGURE 25-20 Bcr-Abl protein kinase.** (a) Origin of the Philadelphia chromosome from a translocation of the tips of chromosomes 9 and 22, and the oncogenic fusion protein formed by the Philadelphia chromosome translocation. (b) The Bcr-Abl fusion protein is a constitutively active kinase that phosphorylates multiple signal-transduction proteins. Imatinib binds to the active site of Bcr-Abl and inhibits its kinase activity. (c) Imatinib bound to the Bcr-Abl active site. [Part (c) from Nagar et al., 2002, *Cancer Research* **62:**4236.]

fos and *myc* proto-oncogenes. Initially there is a transient rise of c-Fos and later a more prolonged rise of c-Myc (Figure 25-22). The levels of both proteins decline within a few hours, a regulatory effect that may, in normal cells, help to avoid cancer. As discussed in Chapter 20, c-Fos and c-Myc stimulate transcription of genes encoding proteins that promote progression through the G_1 phase of the cell cycle and the G_1-to-S transition. In tumors, the oncogenic forms of these or other transcription factors are frequently expressed at high and unregulated levels.

In normal cells, c-Fos and c-Myc mRNAs and the proteins they encode are intrinsically unstable, leading to their rapid loss after the genes are induced. Some of the changes that turn c-*fos* from a normal gene to an oncogene involve genetic deletions of the sequences that make the Fos mRNA and protein short-lived. Conversion of the c-*myc* proto-oncogene into an oncogene can occur by different mechanisms. In cells of the human tumor known as *Burkitt's lymphoma,* the c-*myc* gene is translocated to a site near the heavy-chain antibody genes, which are normally active in antibody-producing white blood cells (Figure 25-23). The c-*myc* translocation is a rare aberration of the normal DNA rearrangements that occur during maturation of antibody-producing cells. The translocated *myc* gene, now regulated by the antibody gene enhancer, is continually expressed, causing the cell to become cancerous. Localized amplification of a segment of DNA containing the *myc* gene, which occurs in several human tumors, also causes inappropriately high production of the otherwise normal Myc protein.

The *c-myc* gene encodes a basic helix-loop-helix zipper protein that acts as part of a set of interacting proteins that can dimerize in various combinations, bind to DNA, and coordinately regulate the transcription of target genes. Other members of the protein set include Mad, Max, and Mnt. Max can heterodimerize with Myc, Mad, and Mnt. Myc-Max dimers regulate genes that control proliferation, such as cyclins. Mad proteins inhibit Myc proteins, which has led to an interest in using Mad proteins, or drugs that stimulate Mad proteins, to rein in excessive Myc activity that contributes to tumor formation. Myc protein complexes affect transcription by recruiting chromatin-modifying complexes containing histone acetyltransferases (which usually stimulate transcription, see Chapter 7) to Myc target genes. Mad and Mnt work with the Sin3 co-repressor protein to bring in histone deacetylases that help to block transcription. Together, all these proteins form a regulatory network that employs protein-protein association, variations in DNA binding, and transcriptional regulation to control cell proliferation.

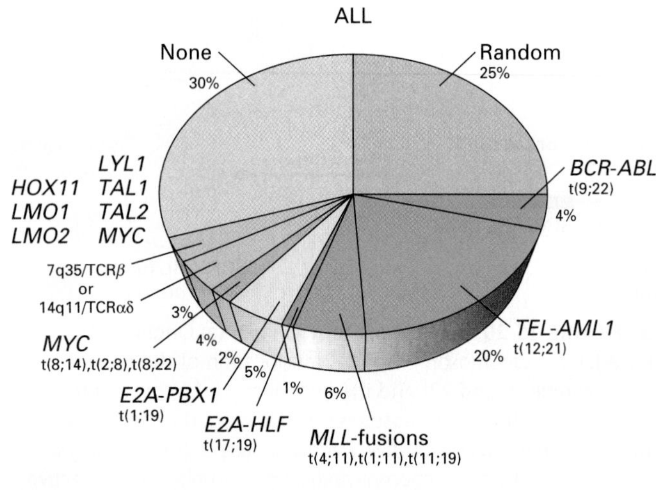

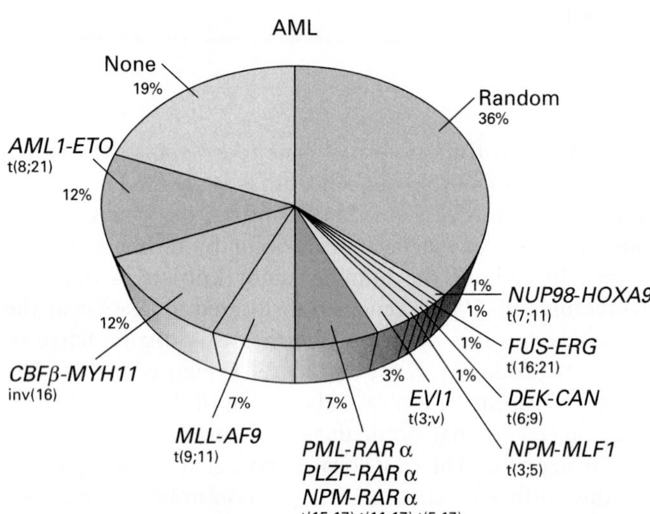

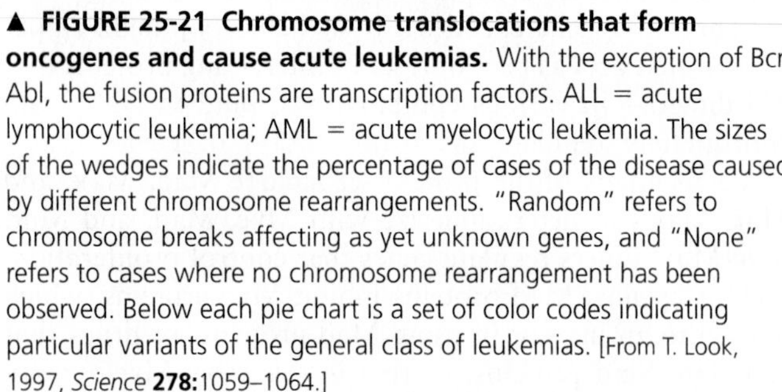

▲ **FIGURE 25-21 Chromosome translocations that form oncogenes and cause acute leukemias.** With the exception of Bcr-Abl, the fusion proteins are transcription factors. ALL = acute lymphocytic leukemia; AML = acute myelocytic leukemia. The sizes of the wedges indicate the percentage of cases of the disease caused by different chromosome rearrangements. "Random" refers to chromosome breaks affecting as yet unknown genes, and "None" refers to cases where no chromosome rearrangement has been observed. Below each pie chart is a set of color codes indicating particular variants of the general class of leukemias. [From T. Look, 1997, *Science* **278:**1059–1064.]

Molecular Cell Biology Is Changing How Cancer Is Treated

The imatinib story described earlier illustrates how genetics—the discovery of the Philadelphia chromosome and the critical oncogene it creates—together with biochemistry—discovery of the molecular action of the Abl protein—can lead to a powerful new therapy. In general, each difference between cancer cells and normal cells provides a new opportunity to identify a specific drug or treatment that kills just the cancer cells or at least stops their uncontrolled growth. Thus knowledge of the molecular cell biology of the tumors is critical information that can be exploited by researchers to develop anticancer treatments that more precisely target just cancer cells.

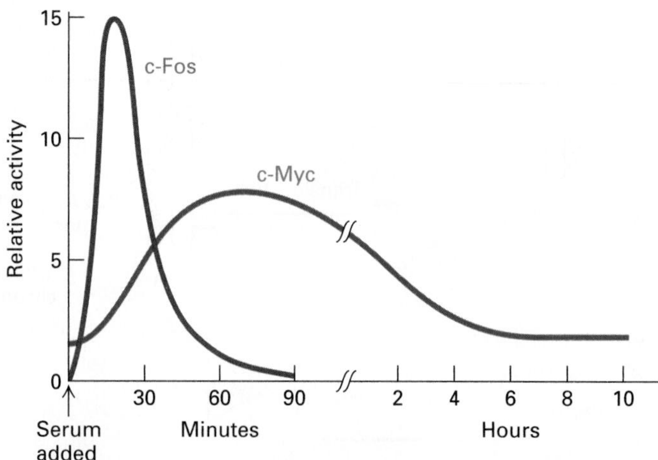

▲ EXPERIMENTAL FIGURE 25-22 **Addition of serum to quiescent 3T3 cells yields a marked increase in the activity of two proto-oncogene products, c-Fos and c-Myc.** Serum contains factors like platelet-derived growth factor (PDGF) that stimulate the growth of quiescent cells. One of the earliest effects of growth factors is to induce expression of *c-fos* and *c-myc,* whose encoded proteins are transcription factors. [See M. E. Greenberg and E. B. Ziff, 1984, *Nature* **311:**433.]

Breast cancer provides a good example of how molecular cell biology techniques have affected treatments, both curative and palliative. Until the rise in the incidence of lung cancer, resulting from an increase in women smokers, breast cancer was the most deadly cancer for women, and it remains the second most frequent cause of women's cancer deaths. The cause of breast cancer is unknown, but the frequency is increased if certain mutations are carried. Breast cancers are often diagnosed during routine mammogram (x-ray) examinations. Typically, a biopsy of a 1–2-cm tissue mass is taken to check the diagnosis, and is tested with antibodies to determine whether a high level of estrogen or prog-

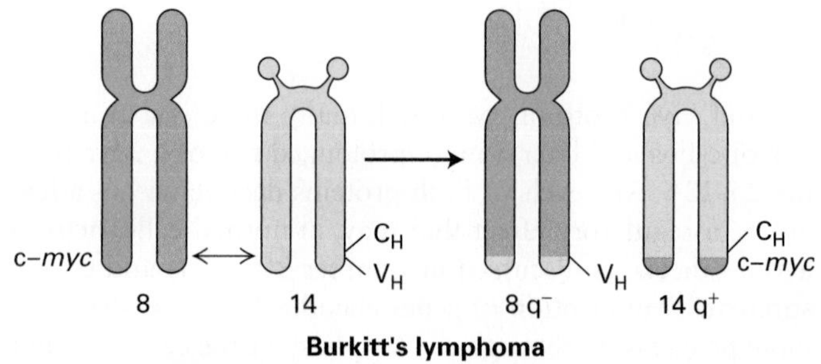

▲ **FIGURE 25-23 Chromosomal translocation in Burkitt's lymphoma.** As a result of a translocation between chromosomes 8 and 14, the *c-myc* gene is placed adjacent to the gene for part of the antibody heavy chain (C_H), leading to overproduction of the Myc transcription factor in lymphocytes and hence their growth into a lymphoma.

esterone receptor is present. These steroid receptors are capable of stimulating tumor growth, and are sometimes expressed at high levels in breast cancer cells. If either receptor is present, it is exploited in the treatment. A drug called tamoxifen, which inhibits the estrogen receptor, can be used to deprive the tumor cells of a growth-stimulating hormone. The biopsy is also tested for amplification of the proto-oncogene *HER2/NEU,* which as we have seen, encodes EGF receptor 2. A monoclonal antibody specific for Her2 has been a strikingly successful new treatment for the subset of breast cancers that overproduce Her2. Her2 antibody injected into the blood recognizes Her2 and causes it to be internalized, selectively killing the cancer cells without any apparent effect on normal breast (and other) cells that produce moderate amounts of Her2.

Breast cancer is treated with a combination of surgery, radiation therapy, and chemotherapy. The first step is surgical resection (removal) of the tumor and examination of lymph nodes for evidence of metastatic tumor, which is the major adverse prognostic factor. The subsequent treatment involves 6 weeks of radiation and 8 weeks of chemotherapy with three different types of agents. These harsh treatments are designed to kill the dividing cancer cells; however, they also cause a variety of side effects including suppression of blood cell production, hair loss, nausea, and neuropathy. This can reduce the strength of the immune system, risking infection, and cause weakness due to poor oxygen supply. To help, patients are given the growth factor G-CSF (Chapter 16) to promote neutrophil formation (a type of white blood cell that fights bacterial and fungal infections) and erythropoietin (Epo; Chapter 16) to stimulate red blood cell formation. Despite all this treatment, an average-risk woman (60 years old, 2-cm tumor, 1 positive lymph node) has a 30–40 percent 10-year risk of succumbing to her cancer. This risk can be reduced by 10–15 percent by hormone-blocking treatment such as tamoxifen, exploiting the molecular data that shows a hormone receptor present on the cancer cells. Mortality is reduced another 5–10 percent by treatment with antibodies against the Her2/Neu oncoprotein. Thus molecular biology is having a huge impact on breast cancer victim survival rates, though still far less than one would like. ∎

The vision for the future of medicine is that the need for radiation and chemotherapy and perhaps even surgery will be substantially reduced, thus reducing toxicity and collateral damage. Increased knowledge of the molecular cell biology of cancer will allow drugs to be administered that are, first, more effective and harmless and, second, tuned to the particular properties of an individual's tumor cells. In breast tumor treatment we can already see progress in this direction.

KEY CONCEPTS OF SECTION 25.3

Oncogenic Mutations in Growth-Promoting Proteins

- Mutations or chromosomal translocations that permit RTKs for growth factors to dimerize in the absence of their normal ligands lead to constitutive receptor activity (see Figures 25-16 and 25-17). Such activation ultimately induces changes in gene expression that can transform cells. Overproduction of growth-factor receptors can have the same effect and lead to abnormal cell proliferation.

- Certain virus-encoded proteins can bind to and activate host-cell receptors for growth factors, thereby stimulating cell proliferation in the absence of normal signals.

- Most tumor cells produce constitutively active forms of one or more intracellular signal-transduction proteins, causing growth-promoting signaling in the absence of normal growth factors.

- A single point mutation in Ras, a key transducing protein in many signaling pathways that promote cell proliferation and differentiation, reduces its GTPase activity, thereby maintaining it in an activated state.

- The activity of Src, a cytosolic signal-transducing protein-tyrosine kinase, normally is regulated by reversible phosphorylation and dephosphorylation of a tyrosine residue near the C-terminus (see Figure 25-19). The unregulated activity of Src oncoproteins that lack this tyrosine promotes abnormal proliferation of many cells.

- The Philadelphia chromosome results from a chromosomal translocation that produces the chimeric *bcr-abl* oncogene. The unregulated Abl kinase activity of the Bcr-Abl oncoprotein is responsible for its oncogenic effect. An Abl kinase inhibitor (imatinib, or Gleevec) is effective in treating chronic myelogenous leukemia (CML) and may work against a few other cancers that are driven by related kinases (see Figure 25-20).

- Many other cases of leukemia are also due to chromosome rearrangements that create new fusion genes and fusion proteins. A rare chromosome breakage and rejoining event in a blood cell that creates the new protein causes uncontrolled growth. The cell grows into a clone of mutant cells.

- Inappropriate production of nuclear transcription factors such as Fos, Jun, and Myc can induce transformation. In Burkitt's lymphoma cells, c-*myc* is translocated close to an antibody gene, leading to overproduction of c-Myc (see Figure 25-23).

- Precise molecular analysis of early tumors is allowing the use of highly targeted drug treatments that are likely to be optimal for a particular case of a particular tumor. These refinements have allowed substantial reduction in breast cancer mortality. There are good prospects for still better targeting of treatments, based on continuing increases in the understanding of tumor-cell regulation.

- The advent of molecular techniques for characterizing individual tumors is allowing the application of drugs and antibody treatments that target the properties of a particular tumor. This permits more effective treatment of individual patients and reduces the use of drugs or antibodies that will be ineffective and possibly toxic.

Focus Animation: TGFβ Signaling Pathway 

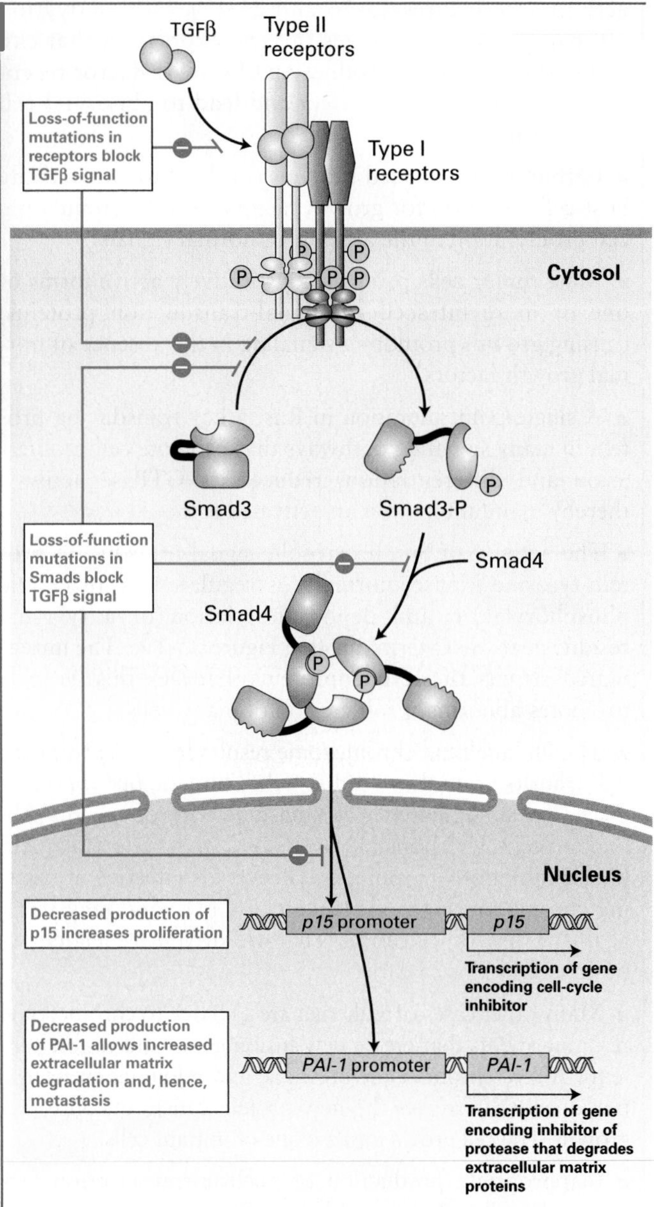

▲ **FIGURE 25-24 Effect of loss of TGFβ signaling.** Binding of TGFβ, an antigrowth factor, causes activation of Smad transcription factors. In the absence of effective TGFβ signaling owing to either a receptor mutation or a SMAD mutation, cell proliferation and invasion of the surrounding extracellular matrix (ECM) increase. [See X. Hua et al., 1998, *Genes & Dev.* **12:**3084.]

25.4 Mutations Causing Loss of Growth-Inhibiting and Cell-Cycle Controls

Normal growth and development depends on a finely tuned, highly regulated balance between growth-promoting and growth-inhibiting pathways. Mutations that disrupt this balance can lead to cancer. Most of the mutations discussed in the previous section cause inappropriate activity of growth-promoting pathways. Just as critical are mutations that decrease the activity of growth-inhibiting pathways when they are needed.

For example, transforming growth factor β (TGFβ), despite its name, inhibits proliferation of many cell types, including most epithelial and immune system cells. Binding of TGFβ to its receptor induces activation of cytosolic Smad transcription factors (see Figure 16-4). After translocating to the nucleus Smads can promote expression of the gene encoding p15, an inhibitor of cyclin-dependent kinase 4 (CDK4), which causes cells to arrest in G_1. TGFβ signaling also promotes expression of genes encoding extracellular matrix proteins and plasminogen activator inhibitor 1 (PAI-1), which reduces the plasmin-catalyzed degradation of the matrix. Loss-of-function mutations in either TGFβ receptors or in Smads thus promote cell proliferation and probably contribute to the invasiveness and metastasis of tumor cells (Figure 25-24). Such mutations have in fact been found in a variety of human cancers. For example, deletion of the *Smad4* gene occurs in many human pancreatic cancers; retinoblastoma and colon cancer cells lack functional TGFβ receptors and therefore are unresponsive to TGFβ growth inhibition.

The complex mechanisms for regulating the eukaryotic cell cycle are prime targets for oncogenic mutations. Both positive- and negative-acting proteins precisely control the entry of cells into and their progression through the cell cycle, which consists of four main phases: G_1, S, G_2, and mitosis (see Figure 20-34). This regulatory system assures the proper coordination of cellular growth during G_1 and G_2, DNA synthesis during the S phase, and chromosome segregation and cell division during mitosis. In addition, cells that have sustained damage to their DNA normally are arrested before their DNA is replicated or in G_2 before chromosome segregation. This arrest allows time for the DNA damage to be repaired; alternatively, the arrested cells are directed either to commit suicide via programmed cell death, or at least not to divide. The whole cell-cycle control system functions to prevent cells from becoming cancerous. As might be expected, mutations in this system often lead to abnormal development or contribute to cancer.

Mutations That Promote Unregulated Passage from G_1 to S Phase Are Oncogenic

Once a cell progresses past a certain point in late G_1, called the **restriction point**, it becomes irreversibly committed to entering the S phase and replicating its DNA (see Figure 20-32). D-type **cyclins, cyclin-dependent kinases (CDKs)**, and the Rb protein are all elements of the control system that regulate passage through the restriction point.

The expression of D-type cyclin genes is induced by many extracellular growth factors, or **mitogens.** These cyclins assemble with their partners CDK4 and CDK6 to generate catalytically active cyclin-CDK complexes, whose kinase activity promotes progression past the restriction point. Mitogen withdrawal prior to passage through the restriction point leads to accumulation of p16. Like p15 mentioned above, p16 binds specifically to CDK4 and CDK6, thereby inhibiting their kinase activity and causing G_1 arrest. Under normal circumstances, phosphorylation of Rb protein is initiated midway through G_1 by active cyclin D–CDK4 and cyclin D–CDK6 complexes. Unphosphorylated Rb binds to and sequesters E2F transcription factors in the cytoplasm. E2F transcription factors stimulate transcription of genes encoding proteins required for DNA synthesis. Rb phosphorylation is completed by other cyclin-CDK complexes in late G_1, allowing release and activation of E2F transcription factors and $G_1 \rightarrow S$ progression. The complete phosphorylation of Rb and its disassociation from E2F irreversibly commits the cell to DNA synthesis. Most tumors contain an oncogenic mutation that causes overproduction or loss of one of the components of this pathway such that the cells are propelled into the S phase in the absence of the proper extracellular growth signals (Figure 25-25; see also Figure 20-33).

Elevated levels of cyclin D1, one of the three D-type cyclins, for example, are found in many human cancers. In certain tumors of antibody-producing B lymphocytes, for instance, the *cyclin D1* gene is translocated such that its transcription is under control of an antibody-gene enhancer, causing elevated cyclin D1 production throughout the cell cycle, irrespective of extracellular signals. (This phenomenon is analogous to the c-*myc* translocation in Burkitt's lymphoma cells discussed earlier.) That cyclin D1 can function as an oncoprotein was shown by studies with transgenic mice in which the *cyclin D1* gene was placed under control of an enhancer specific for mammary ductal cells. Initially the ductal cells underwent hyperproliferation, and eventually breast tumors developed in these transgenic mice. Amplification of the *cyclin D1* gene and concomitant overproduction of the cyclin D1 protein is common in human breast cancer; the extra cyclin D1 helps to drive cells through the cell cycle.

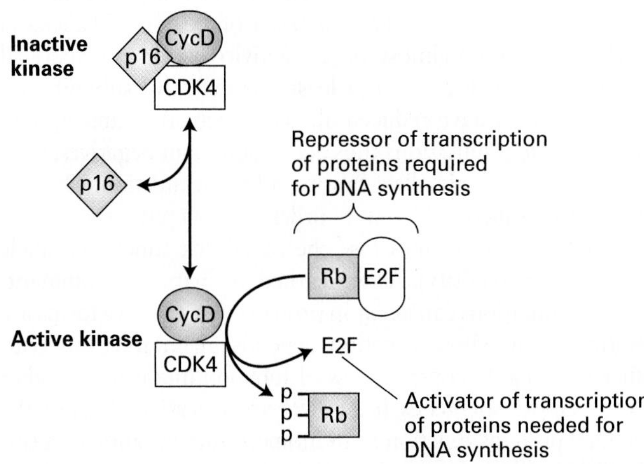

▲ **FIGURE 25-25 Restriction point control.** Unphosphorylated Rb protein binds transcription factors collectively called *E2F* and thereby prevents E2F-mediated transcriptional activation of many genes whose products are required for DNA synthesis (e.g., DNA polymerase). The kinase activity of cyclin D–CDK4 phosphorylates Rb, thereby activating E2F; this kinase activity is inhibited by p16. Overproduction of cyclin D, a positive regulator, or loss of the negative regulators p16 and Rb commonly occurs in human cancers.

The proteins that function as cyclin-CDK inhibitors play an important role in regulating the cell cycle (Chapter 20). In particular, loss-of-function mutations that prevent p16 from inhibiting cyclin D–CDK4/6 kinase activity are common in several human cancers. As Figure 25-25 makes clear, loss of p16 mimics overproduction of cyclin D1, leading to Rb hyperphosphorylation and release of active E2F transcription factor. Thus p16 normally acts as a tumor suppressor. Although the *p16* tumor-suppressor gene is deleted in some human cancers, in others the *p16* sequence is normal. In some of these latter cancers (e.g., lung cancer) the *p16* gene, or genes encoding other functionally related proteins, is inactivated by hypermethylation of its promoter region, which prevents transcription. What promotes this change in the methylation of *p16* is not known, but it prevents production of this important cell-cycle control protein.

We've seen already that inactivating mutations in both *RB* alleles lead to childhood retinoblastoma, a relatively rare type of cancer. However, loss of *RB* gene function also is found in more common cancers that arise later in life (e.g., carcinomas of lung, breast, and bladder). These tissues, unlike retinal tissue, most likely produce other proteins (e.g., p107 and p130, both structurally related to p53) whose function is redundant with that of Rb, and thus loss of Rb is not so critical for preventing cancer. One way or another, Rb functions are eventually shut down. In addition to inactivating mutations, Rb function can be eliminated by the binding of an inhibitory protein, designated E7, that is encoded by human papillomavirus (HPV), another nasty viral trick to create virus-producing tissue. At present, this is known to occur only in cervical cancer.

Tumors with inactivating mutations in Rb generally produce normal levels of cyclin D1 and functional p16 protein. Conversely, tumor cells that overproduce cyclin D1 or have lost p16 function generally retain wild-type Rb. Thus loss of only one component of this regulatory system for controlling passage through the restriction point is all that is necessary to subvert normal growth control and set the stage for cancer.

Loss-of-Function Mutations Affecting Chromatin-Remodeling Proteins Contribute to Tumors

We have seen how mutations can undermine growth control by inactivating tumor-suppressor genes. However, these types of genes can also be silenced by repressive chromatin structures. In recent years the importance of *chromatin-remodeling complexes,* such as the SWI/SNF complex, in transcriptional control has become increasingly clear. These large and diverse multiprotein complexes have at their core an ATP-dependent helicase and often control histone modification and chromatin-remodeling (Chapter 7). By causing changes in the positions or structures of nucleosomes, SWI/SNF complexes make genes accessible or inaccessible to DNA-binding proteins that control transcription. If a gene is normally activated or repressed by SWI/SNF-mediated chromatin changes, mutations in the genes encoding the SWI or SNF proteins will cause changes in expression of the target gene.

Our knowledge of the target genes regulated by SWI/SNF and other such complexes is incomplete, but the targets evidently include some growth-regulating genes. For example, studies with transgenic mice suggest that SWI/SNF plays a role in repressing the *E2F* genes, thereby inhibiting progression through the cell cycle. The relationship between the genes that encode SWI/SNF proteins and the *E2F* gene was discovered in genetic experiments with flies. Transgenic flies were constructed to overexpress *E2F*, which resulted in mild growth defects. A search for mutations that increase the effect of the *E2F* overexpression in these flies identified three components of the SWI/SNF complex. That loss of function of these genes increases the proliferative effects of E2F indicates that SWI/SNF normally counteracts the function of the E2F transcription factor. Thus loss of SWI/SNF function, just like loss of Rb function, can lead to overgrowth and perhaps cancer. Indeed, in mice, Rb protein recruits SWI/SNF proteins to repress transcription of the *E2F* gene. Rb represses genes via this effect on E2F and by recruitment of histone deacetylases and histone methyltransferases.

Recent evidence from humans and mice has strongly implicated the *SNF5* gene in cancer. In humans, inactivating somatic *SNF5* mutations cause rhabdoid tumors, which most commonly form in the kidney, and an inherited (familial) disposition to form brain and other tumors. In mice, 15–30 percent of $snf5^-/snf5^+$ heterozygotes develop rhabdoid tumors, and in all cases the tumor cells have lost the remaining functional allele. Since nothing was known of the mechanism involved, microarray studies were used to discover regulatory changes in the tumors. These studies showed the gene-expression similarity of mouse and human tumors, and that the loss of SNF5 leads to heightened expression of cell cycle genes including many regulated by E2F. Subsequent genetic experiments with double mutant $snf5^-/snf5^-$ $p53^-/p53^-$ mice showed synergism between the genes in tumor formation. In these mice the loss of Snf5 will lead to derepression of the *E2F* gene, and the loss of p53 will cause the E2F to be fully active in promoting the $G_1 \rightarrow S$ cell cycle transition.

With chromatin-remodeling complexes involved in so many aspects of transcriptional control, it is expected that SWI/SNF and similar complexes will be linked to many cancers. In humans, mutations in *Brg1,* which encodes the SWI/SNF catalytic subunit, have been found in prostate, lung, and breast tumors. Components of the SWI/SNF complex also have been found to associate with BRCA-1, the nuclear protein that helps suppress human breast cancer. BRCA-1 is involved in the repair of double-strand DNA breaks (as discussed in Chapter 4) and in transcriptional control, so the SWI/SNF complex may assist BRCA-1 in these functions.

Loss of p53 Abolishes the DNA-Damage Checkpoint

Cells with functional p53 become arrested in G_1 when exposed to DNA-damaging irradiation, whereas cells lacking functional p53 do not. Unlike other cell-cycle proteins, p53 is present at very low levels in normal cells because it is extremely unstable and rapidly degraded. Mice lacking p53 are largely viable and healthy, except for a predisposition to develop multiple types of tumors. In normal mice the amount of *p53* protein is heightened, a post-transcriptional response, only in stressful situations, such as ultraviolet or γ irradiation, heat, and low oxygen. DNA damage by γ irradiation or by other stresses somehow leads to the activation of ATM or ATR, serine kinases that phosphorylate and thereby stabilize p53, leading to a marked increase in its concentration (Figure 25-26). The stabilized p53 activates transcription of the gene encoding $p21^{CIP}$, which binds to and inhibits mammalian G_1 cyclin-CDK complexes, and a plethora of other genes. As a result, cells with damaged DNA are arrested in G_1, allowing time for DNA repair by mechanisms discussed in Chapter 4, or the cells permanently arrest, i.e., become senescent. In addition to its role in the $G_1 \rightarrow S$ transition, p53-induced production of $p21^{CIP}$ results in the inhibition of CDK1, which in turn blocks the cyclin B–CDK1 complex that is required for entry into mitosis, thus causing cells to arrest in G_2 (Figure 20-35, **4d**). Directly or indirectly, p53 also represses expression of the genes encoding cyclin B and topoisomerase II, which also are required for the $G_2 \rightarrow$mitosis transition. Thus if DNA is damaged following its replication, p53-induced G_2 arrest will prevent its transmission to daughter cells.

As mentioned earlier, loss-of-function mutations in the *p53* gene occur in more than 50 percent of human cancers. When the p53 G_1 checkpoint control does not operate properly, damaged DNA can replicate, perpetuating mutations and DNA rearrangements that are passed on to daughter cells, contributing to the likelihood of transformation into metastatic cells. Interestingly, however, homozygous $p53^-/p53^-$ cells have no discernable defect in G_2 arrest, so there seem to be other ways to activate $p21^{CIP}$ or different inhibitors of G_2 progression.

The active form of p53 is a tetramer of four identical subunits. A missense point mutation in one of the two *p53* alleles in a cell can abrogate almost all p53 activity because virtually all the oligomers will contain at least one defective subunit, and such oligomers have reduced ability to activate transcription. Oncogenic *p53* mutations thus act as **dominant negatives,** with mutations in a single allele causing a loss of function. The loss of function is incomplete, so in order to grow more rapidly, tumor cells still sometimes lose the remaining functional allele (loss of heterozygosity). As we learned in Chapter 5, dominant-negative mutations can occur in proteins whose active forms are multimeric or whose function depends on interactions with other proteins. In contrast, loss-of-function mutations in other tumor-suppressor genes (e.g., *RB*) are recessive because the encoded proteins function as monomers and mutation of a single allele has little functional consequence.

Under stressful conditions, the ATM kinase also phosphorylates and thus activates Chk2, a protein kinase that phosphorylates the protein phosphatase Cdc25A, marking it for ubiquitin-mediated destruction. This phosphatase normally removes the inhibitory phosphate from CDK2, a prerequisite

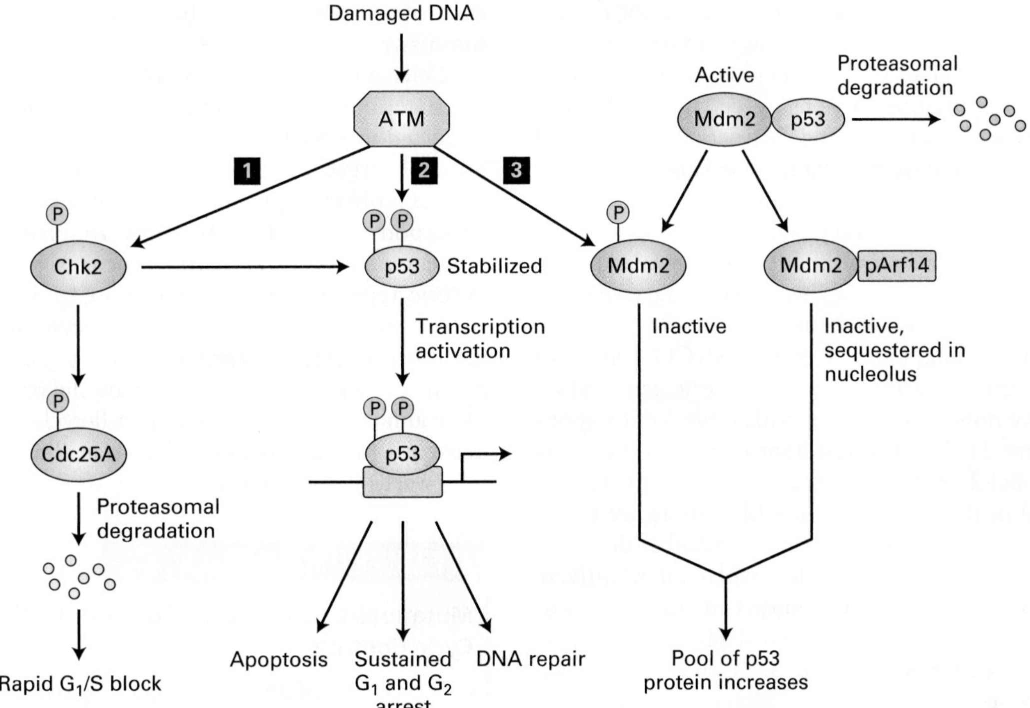

▲ **FIGURE 25-26 G_1 arrest in response to DNA damage.** The kinase activity of ATM is activated in response to DNA damage due to various stresses (e.g., UV irradiation, heat). Activated ATM then triggers three pathways leading to arrest in G_1: **1** Chk2 is phosphorylated and, in turn, phosphorylates Cdc25A, thereby marking it for degradation and blocking its role in CDK2 activation. **2** In a second pathway, phosphorylation of p53 stabilizes it, permitting p53-activated expression of genes encoding proteins that cause arrest in G_1 and in some cases G_2, promote apoptosis, or participate in DNA repair. **3** The third pathway is another way of controlling the pool of p53. The Mdm2 protein in its active form can form a complex with p53, causing p53 ubiquitination and subsequent degradation by the proteasome. ATM phosphorylates Mdm2 to inactivate it, causing increased stabilization of p53. In addition, Mdm2 levels are controlled by pArf14 (pArf19 in mice), which binds Mdm2 and sequesters it in the nucleolus, where it cannot access p53. The human Mdm2 gene is frequently amplified in sarcomas, which presumably causes excessive inactivation of p53. See the text for a discussion.

for cells to enter the S phase. Decreased levels of Cdc25A thus block progression into and through the S phase (see Figure 25-26 and Figure 20-35, **4b**). Thus, loss-of-function mutations in the *ATM* or *Chk2* genes have some of the effects of *p53* mutations.

The activity of p53 normally is kept low by a protein called *Mdm2*. When Mdm2 is bound to p53, it inhibits the transcription-activating ability of p53 and catalyzes the addition of ubiquitin molecules, thus targeting p53 for proteasomal degradation. Phosphorylation of p53 by ATM or ATR displaces bound Mdm2 from p53, thereby stabilizing it. Because the *Mdm2* gene is itself transcriptionally activated by p53, Mdm2 functions in an autoregulatory feedback loop with p53, perhaps normally preventing excess p53 function. The *Mdm2* gene is amplified in many sarcomas and other human tumors that contain a normal *p53* gene. Even though functional p53 is produced by such tumor cells, the elevated Mdm2 levels reduce the p53 concentration enough to abolish the p53-induced G_1 arrest in response to irradiation.

The activity of p53 also is inhibited by a human papillomavirus (HPV) protein called *E6*. HPV encodes three proteins that contribute to its ability to induce stable transformation and mitosis in a variety of cultured cells. Two of these—E6 and E7—bind to and inhibit the p53 and Rb tumor suppressors, respectively. Acting together, E6 and E7 are sufficient to induce transformation in the absence of mutations in cell regulatory proteins. Recall that the HPV E5 protein, which causes sustained activation of the PDGF receptor, enhances proliferation of the transformed cells.

The activity of p53 is not limited to inducing cell-cycle arrest. In addition, this multipurpose tumor suppressor stimulates production of pro-apoptotic proteins and DNA-repair enzymes (see Figure 25-26). Senescence and apoptosis may in fact be the most important means through which p53 prevents tumor growth.

Apoptotic Genes Can Function as Proto-oncogenes or Tumor-Suppressor Genes

During normal development many cells are designated for **programmed cell death,** also known as apoptosis (see Chapter 21). Many abnormalities, including errors in mitosis,

DNA damage, and an abnormal excess of cells not needed for development of a working organ, also can trigger apoptosis. In some cases, cell death appears to be the default situation, with signals required to ensure cell survival. Cells can receive instructions to live and instructions to die, and a complex regulatory system integrates the various kinds of information.

If cells do not die when they should and instead keep proliferating, a tumor may form. For example, chronic lymphoblastic leukemia (CLL) occurs because cells survive when they should be dying. The cells accumulate slowly, and most are not actively dividing, but they do not die. CLL cells have chromosomal translocations that activate a gene called *bcl-2*, which we now know to be a critical blocker of apoptosis (see Figure 21-37). The resultant inappropriate overproduction of Bcl-2 protein prevents normal apoptosis and allows survival of these tumor cells. CLL tumors are therefore attributable to a failure of cell death. Another dozen or so proto-oncogenes that are normally involved in negatively regulating apoptosis have been mutated to become oncogenes. Overproduction of their encoded proteins prevents apoptosis even when it is needed to stop cancer cells from growing.

Conversely, genes whose protein products stimulate apoptosis behave as tumor suppressors. An example is the *PTEN* gene discussed in Chapter 16. The phosphatase encoded by this gene dephosphorylates phosphatidylinositol 3,4,5-trisphosphate, a second messenger that functions in activation of protein kinase B (see Figure 16-30). Cells lacking PTEN phosphatase have elevated levels of phosphatidylinositol 3,4,5-trisphosphate and active protein kinase B, which promotes cell survival and prevents apoptosis by several pathways. Thus PTEN acts as a pro-apoptotic tumor suppressor by decreasing the anti-apoptotic effect of protein kinase B.

The most common pro-apoptotic tumor-suppressor gene implicated in human cancers is *p53*. Among the genes activated by p53 are several encoding pro-apoptotic proteins such as Bax (see Figure 21-41). When most cells suffer extensive DNA damage or numerous other stresses such as growth-factor deprivation or hypoxia, the p53-induced expression of pro-apoptotic proteins leads to their quick demise (see Figure 25-26). While this may seem like a drastic response to DNA damage, it prevents proliferation of cells that are likely to accumulate multiple mutations. When p53 function is lost, apoptosis cannot be induced and the accumulation of mutations required for cancer to develop becomes more likely.

Failure of Cell-Cycle Checkpoints Often Leads to Aneuploidy in Tumor Cells

It has long been known that chromosomal abnormalities abound in tumor cells. We have already encountered several examples of oncogenes that are formed by translocation, amplification, or both (e.g., c-*myc, bcr-abl, bcl-2,* and *cyclin D1*). Another chromosomal abnormality characteristic of nearly all tumor cells is **aneuploidy,** the presence of an aberrant number of chromosomes—generally too many.

Cells with abnormal numbers of chromosomes form when certain cell-cycle checkpoints are nonfunctional. As discussed in Chapter 20, the unreplicated-DNA checkpoint normally prevents entry into mitosis unless all chromosomes have completely replicated their DNA; the spindle-assembly checkpoint prevents entry into anaphase unless all the replicated chromosomes attach properly to the metaphase mitotic apparatus; and the chromosome-segregation checkpoint prevents exit from mitosis and cytokinesis if the chromosomes segregate improperly (see Figure 20-35, **1**–**3**). As advances are made in identifying the proteins that detect these abnormalities and mediate cell-cycle arrest, the molecular basis for the functional defects leading to aneuploidy in tumor cells will become clearer.

KEY CONCEPTS OF SECTION 25.4

Mutations Causing Loss of Growth-Inhibiting and Cell-Cycle Controls

- Loss of signaling by TGFβ, a negative growth regulator, promotes cell proliferation and development of malignancy (see Figure 25-24).

- Overexpression of the proto-oncogene encoding cyclin D1 or loss of the tumor-suppressor genes encoding p16 and Rb can cause inappropriate, unregulated passage through the restriction point in late G_1. Such abnormalities are always seen in human tumors.

- Mutations affecting the SWI/SNF chromatin-remodeling complex, which participates in transcriptional control, are associated with a variety of tumors. In some cases, interaction of the SWI/SNF complex with a nuclear tumor-suppressor protein may have a repressing effect on gene expression.

- The p53 protein is a multifunctional tumor suppressor that promotes arrest in G_1 and G_2, apoptosis, and DNA repair in response to damaged DNA (see Figure 25-26).

- Loss-of-function mutations in the *p53* gene occur in more than 50 percent of human cancers. Overproduction of Mdm2, a protein that normally inhibits the activity of p53, occurs in several cancers (e.g., sarcomas) that express normal p53 protein. Thus in one way or another, the p53 stress-response pathway is inactivated to allow tumor growth.

- Human papillomavirus (HPV) encodes three oncogenic proteins: E6 (inhibits p53), E7 (inhibits Rb), and E5 (activates the PDGF receptor).

- Overproduction of anti-apoptotic proteins (e.g., Bcl-2) can lead to inappropriate cell survival and is associated with chronic lymphoblastic leukemia (CLL) and other cancers. Loss of proteins that promote apoptosis (e.g., p53 transcription factor and PTEN phosphatase) have a similar oncogenic effect.

- Genomic instability is a hallmark of cancer. Many types of DNA damage are observed in tumors. Most human tumor cells are aneuploid, containing an abnormal number of chromosomes (usually too many). Failure of cell-cycle checkpoints that normally detect unreplicated DNA, improper spindle assembly, or mis-segregation of chromosomes permits aneuploid cells to arise.

25.5 Carcinogens and Caretaker Genes in Cancer

Carcinogens, which can be natural or produced by humans, are chemicals that cause cancer. Carcinogens cause mutations that reduce the function of tumor-suppressor genes, create oncogenes from proto-oncogenes, or damage DNA repair systems. As our previous discussion has shown, alterations in DNA that lead to malfunction of tumor-suppressor proteins and oncoproteins are the underlying cause of most cancers. These oncogenic mutations in key growth and cell-cycle regulatory genes include insertions, deletions, and base substitutions, as well as chromosomal amplifications and translocations. In addition, damage to caretaker genes comprising the DNA-repair systems (Chapter 4) leads to an increased mutation rate. Of the many mutations that accumulate, some affect cell-cycle regulators, and the cells bearing these may become cancerous. Furthermore, some DNA-repair mechanisms themselves are error-prone (see Figure 4-40). Those "repairs" also contribute to oncogenesis. The inability of tumor cells to maintain genomic integrity leads to formation of a heterogeneous population of malignant cells. For this reason, chemotherapy directed toward a single gene or even a group of genes is likely to be ineffective in wiping out all malignant cells. This problem adds to the interest in therapies that interfere with the blood supply to tumors or that in other ways act upon multiple types of tumor cells.

Stem cells (see Chapter 21), with their capability of carrying out many cell divisions, are a frequent source of cancer cells. Normal dividing cells usually employ several mechanisms to prevent accumulation of detrimental mutations that could lead to cancer. One form of protection against mutation for stem cells is their relatively low rate of division, which reduces the possibility of DNA damage incurred during DNA replication and mitosis. Furthermore, the progeny of stem cells do not have the ability to divide indefinitely. After several rounds of division they exit the cell cycle, reducing the possibility of mutation-induced misregulation of cell division associated with dangerous tumors. Also, if multiple mutations are required for a tumor to grow, attract a blood supply, invade neighboring tissues, and metastasize, a low rate of replication combined with the normal low rate of mutations (10^{-9}) provides further shielding from cancer. However, these safeguards can be overcome if a powerful mutagen reaches the cells, or if DNA repair is compromised and the mutation rate rises. When cells with stem cell-like growth properties are mutagenized by environmental poisons and unable to efficiently repair the damage, cancer can occur.

In this section, we explore the ways in which various carcinogens act on DNA to induce cancer, as well as how mutations in caretaker genes contribute to cancer by compromising the cell's ability to repair DNA damage. We end the chapter with a discussion of a special enzyme called telomerase, another guardian of the genome that protects against DNA damage, and its role in cancer cells.

Carcinogens Induce Cancer by Damaging DNA

The ability of chemical and physical carcinogens to induce cancer is due to the DNA damage that they cause as well as the errors introduced into DNA during the cells' efforts to repair this damage. Thus carcinogens also are **mutagens.** The strongest evidence that carcinogens act as mutagens comes from the observation that cellular DNA altered by exposure of cells to carcinogens can change cultured cells, such as 3T3 cells, or cells implanted in mice into fast-growing cancer-like cells (see Figure 25-6). The mutagenic effect of carcinogens is roughly proportional to their ability to transform cells and induce cancer in animal models.

Although substances identified as chemical carcinogens have a broad range of structures with no obvious unifying features, they can be classified into two general categories. *Direct-acting carcinogens,* of which there are only a few, are mainly reactive electrophiles (compounds that seek out and react with electron-rich centers in other compounds). By chemically reacting with nitrogen and oxygen atoms in DNA, these compounds can modify bases in DNA so as to distort the normal pattern of base pairing. If these modified nucleotides are not repaired, they allow an incorrect nucleotide to be incorporated during replication. This type of carcinogen includes ethylmethane sulfonate (EMS), dimethyl sulfate (DMS), and nitrogen mustards.

In contrast, *indirect-acting carcinogens* generally are unreactive, often water-insoluble compounds that can act as potent cancer inducers only after the introduction of electrophilic centers. In animals, *cytochrome P-450 enzymes* are located in the endoplasmic reticulum of most cells and at especially high levels in liver cells. P-450 enzymes normally function to add electrophilic centers, such as OH groups, to nonpolar foreign chemicals, such as certain insecticides and therapeutic drugs, in order to solubilize them so that they can be excreted from the body. However, P-450 enzymes also can turn otherwise harmless chemicals into carcinogens. Indeed, most chemical carcinogens have little mutagenic effect until they have been modified by cellular enzymes.

Some Carcinogens Have Been Linked to Specific Cancers

In the earliest days of cancer awareness it became clear that at least some cancers are due to environmental poisons. For example in 1775 it was reported that the exposure of chimney sweeps to soot caused scrotal cancer, and in 1791 the use of snuff (tobacco) was reported to be associated with nasal cancer. Environmental chemicals were originally associated with cancer through experimental studies in animals. The classic experiment is to repeatedly paint a test substance on the back of a mouse and look for development of both local and systemic tumors in the animal. Such assays led to the purification of a pure chemical carcinogen from coal tar in 1933, benzo(*a*)pyrene. The role of radiation in

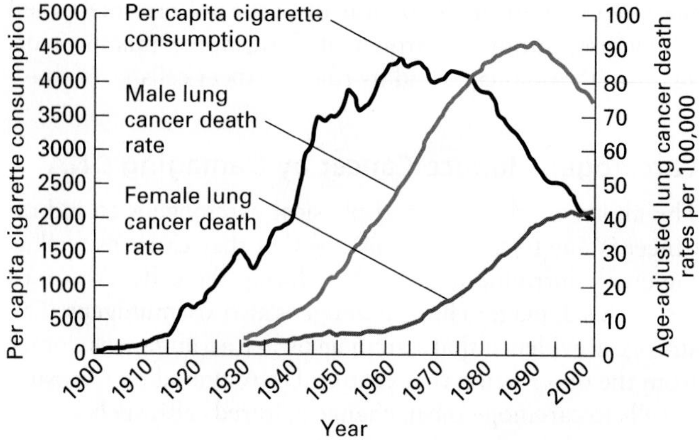

▲ **FIGURE 25-27 Chemical carcinogenesis by tobacco smoke.** Cigarette smoking provides a clear example of a deadly form of chemical carcinogenesis. The rates of lung cancer follow the rates of smoking, with about a 30-year lag. Women began to smoke in large numbers starting in the 1960s, and starting in the 1990s lung cancer passed breast cancer as the leading cause of women's cancer deaths. At the same time a gradual decrease in smoking rates among men starting in the 1960s has been reflected in a decrease in their lung cancer rate. [From the American Cancer Society.]

damaging chromosomes was first demonstrated in the 1920s using x-irradiated *Drosophila*. Later the ability of radiation to cause human cancer, especially leukemia, was dramatically shown by the increased rates of leukemia among survivors of the atomic bombs dropped in World War II (ionizing radiation), and more recently by the increase in melanoma (skin cancer) in individuals exposed to too much sunlight (UV radiation).

Although chemical carcinogens are believed to be risk factors for many human cancers, a direct linkage to specific cancers has been established only in a few cases, the most important being lung cancer and the other cancers (laryngeal, pharynx, stomach, liver, pancreas, bladder, cervix, and more) that are associated with smoking. Epidemiological studies (Figure 25-27) first indicated that cigarette smoking was the major cause of lung cancer, but the reason was unclear until the discovery that about 60 percent of human lung cancers contain inactivating mutations in the *p53* gene. The chemical *benzo*(a)*pyrene,* found in cigarette smoke as well as coal tar, undergoes metabolic activation in the lung (Figure 25-28) to a potent mutagen that mainly causes conversion of guanine (G) to thymine (T) bases, a transversion mutation. When applied to cultured bronchial epithelial cells, activated benzo(*a*)pyrene induces many mutations, including inactivating mutations at codons 175, 248, and 273 of the *p53* gene. These same positions, all within the protein's DNA-binding domain, are major mutational hot spots in human lung cancer. In fact the nature of mutations in *p53* (and other cancer-related genes) gives clues as to the origin of the cancer. The G to T transversions caused by benzo(*a*)pyrene, for example, are present in the *p53* genes of about one-third of smokers' lung tumors. That type of mutation is relatively rare among the *p53* mutations found in

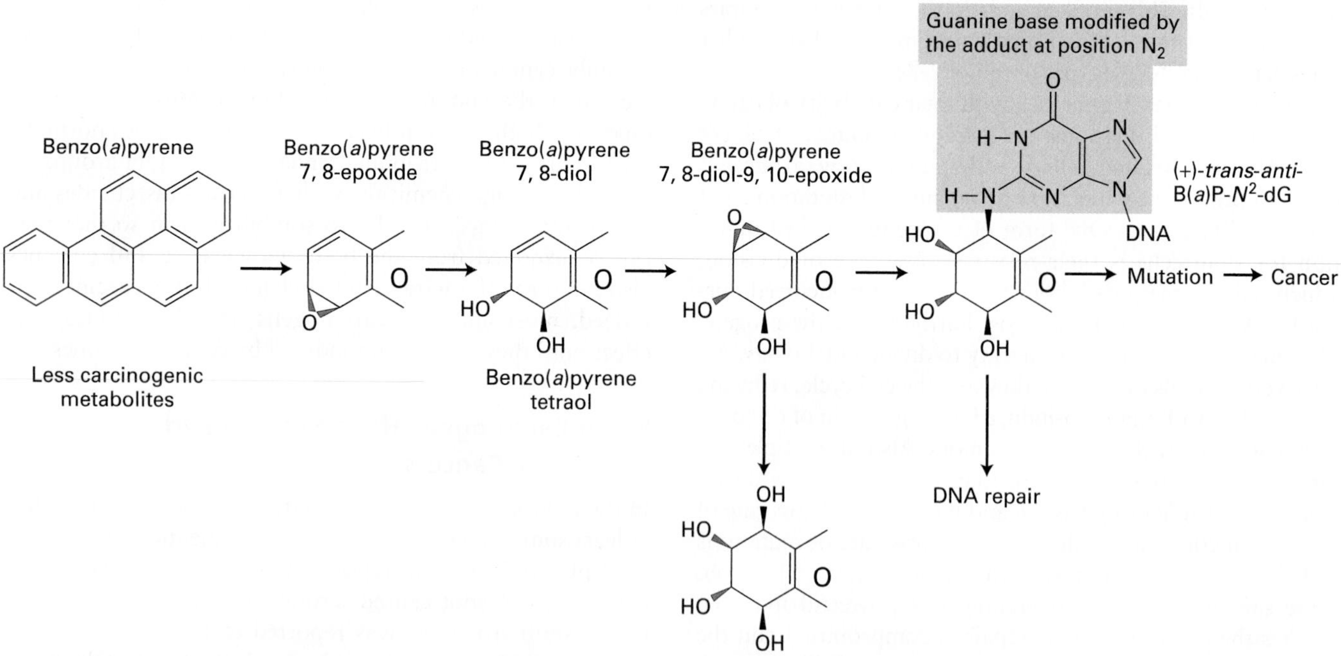

Guanine base modified by the adduct at position N_2
Benzo(*a*)pyrene
Benzo(*a*)pyrene 7, 8-epoxide
Benzo(*a*)pyrene 7, 8-diol
Benzo(*a*)pyrene 7, 8-diol-9, 10-epoxide
(+)-*trans-anti-*B(*a*)P-N^2-dG
DNA
Mutation → Cancer
Less carcinogenic metabolites
Benzo(*a*)pyrene tetraol
DNA repair

▲ **FIGURE 25-28 Enzymatic processing of benzo(a)pyrene to a more potent mutagen and carcinogen.** Liver enzymes, particularly P-450 enzymes, modify benzo(a)pyrene in a series of reactions, producing 7,8-diol-9,10-epoxide, a highly potent mutagenic species that reacts with DNA primarily at the N_2 atom of a guanine base in DNA. The resulting adduct, (+)-*trans-anti*-B(*a*)P-N^2-dG, causes polymerase to insert an A rather than a C opposite the modified G base. Next time the DNA is replicated, a T will be inserted opposite the A, and the mutation will be complete. Horizontal arrows indicate alterations toward greater potency, while vertical arrows indicate changes in the direction of reduced toxicity. The large 'O' symbol represents the rest of the mult-ring structure shown in the complete benzo(a)pyrene molecule at the left. [Adapted from E. L. Loechler, 2001, *Encyclopedia of Life Science.*]

other types of tumors. The carcinogen leaves its footprint. Thus, there is a strong correlation between one defined chemical carcinogen in cigarette smoke and human cancer. It is likely that other chemicals in cigarette smoke induce mutations in other genes, since it contains more than 60 carcinogens. Similarly, asbestos exposure is clearly linked to mesothelioma, a type of epithelial cancer.

Lung cancer is not the only major human cancer for which a clear-cut risk factor has been identified. *Aflatoxin*, a fungal metabolite found in moldy grains, induces liver cancer (Figure 25-29a). After chemical modification by liver enzymes, aflatoxin becomes linked to G residues in DNA and induces G-to-T transversions. Aflatoxin also causes a mutation in the *p53* gene. Furthermore, cooking meat at high temperatures causes chemical reactions that form *heterocyclic amines* (HCAs), which are potent mutagens and cause colon and breast carcinomas in animal models. HCAs react with deoxyguanosine bases to form mutagenic adducts (Figure 25-29b). Exposure to other chemicals has been correlated with minor cancers. Hard evidence concerning dietary and environmental risk factors that would help us avoid other common cancers (e.g., breast, colon, and prostate cancer, leukemias) is generally lacking.

Loss of DNA-Repair Systems Can Lead to Cancer

Even without any external carcinogen or mutagen, normal processes generate a large amount of DNA damage. The damage is due to depurination reactions, to alkylation reactions, and to the generation of reactive species such as oxygen radicals, all of which alter DNA. It has been estimated that in every cell, more than 20,000 alterations to the DNA occur each day from just reactive oxygen species and depurination. Thus DNA repair is a crucial defense system.

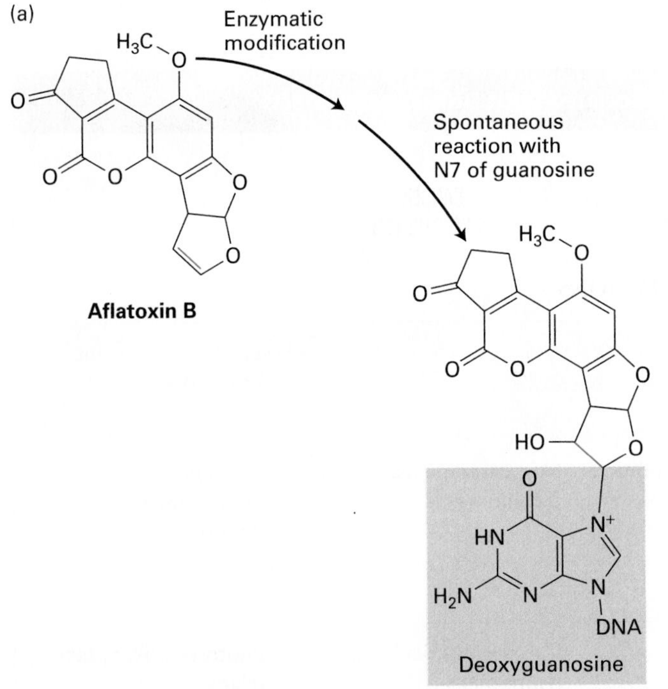

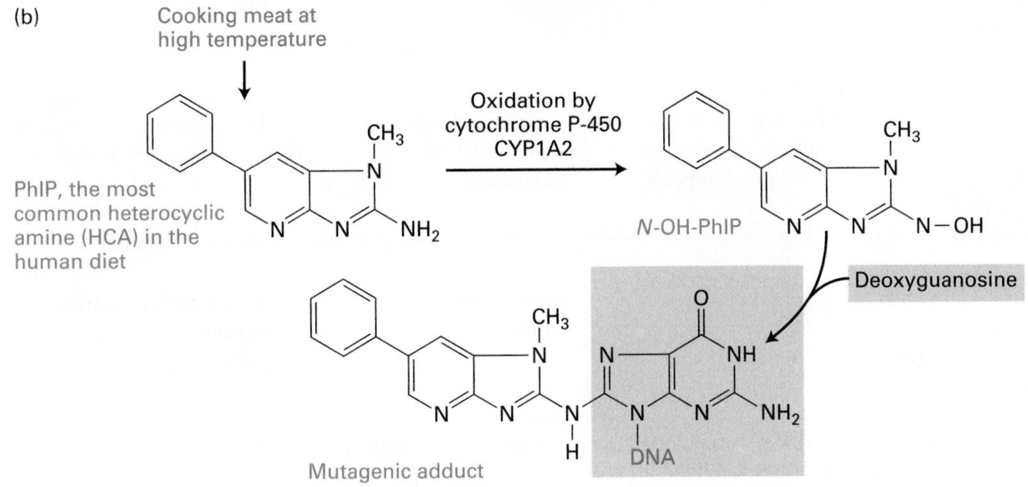

(a)

Enzymatic modification

Aflatoxin B

Spontaneous reaction with N7 of guanosine

Deoxyguanosine

(b)

Cooking meat at high temperature

PhIP, the most common heterocyclic amine (HCA) in the human diet

Oxidation by cytochrome P-450 CYP1A2

N-OH-PhIP

Deoxyguanosine

Mutagenic adduct

◀ **FIGURE 25-29 Actions of two chemical carcinogens.** (a) Like all indirect-acting carcinogens, aflatoxin must undergo enzyme-catalyzed modification before it can react with DNA. In aflatoxin the colored double bond reacts with an oxygen atom, enabling it to react chemically with the N-7 atom of a guanine in DNA, forming a large bulky molecule that causes DNA polymerase to insert an A rather than a C opposite the modified G base during replication. This compound mutates the *p53* tumor-suppressor gene, causing G to T transversions, and is a known risk factor for human cancer. (b) Chemical reactions that occur in human food raised to high temperature, especially red meat, create about sixteen different types of heterocyclic amines (HCAs) from precursors such as creatine and amino acids. The HCA shown here, PhIP, is the most common one in the human diet. Cytochrome P-450 enzymes convert it to a very chemically reactive form, which reacts with a guanine base in DNA to form a DNA adduct that is mutagenic. PhIP causes breast and colon carcinomas in rodents and may be involved in human prostate cancer. Although the P-450 conversions occur mainly in liver cells, the HCAs can migrate to other tissues.

The normal role of caretaker genes is to prevent or repair DNA damage. Loss of the high-fidelity DNA-repair systems that are described in Chapter 4 correlates with increased risk for cancer. For example, humans who inherit mutations in genes that encode a crucial mismatch-repair or excision-repair protein have an enormously increased probability of developing certain cancers (Table 25-2). Without proper DNA repair, people with xeroderma pigmentosum (XP) or hereditary nonpolyposis colorectal cancer (HNPCC) have a propensity to accumulate mutations in many other genes, including those that are critical in controlling cell growth and proliferation. XP causes affected people to develop skin cancer at about a thousand times the normal rate. Seven of the eight known XP genes encode components of the excision-repair machinery, and in the absence of this repair mechanism, genes that control the cell cycle or otherwise regulate cell growth and death become mutated. HNPCC genes encode components of the mismatch-repair system, and mutations in them account for a few percent of the cases of colon cancer. The cancers progress from benign polyps to full-fledged tumors much more rapidly than usual, presumably because the initial cancer cells are undergoing continuous mismatch mutagenesis without repair.

One gene frequently mutated in colon cancers because of the absence of mismatch repair encodes the type II receptor for TGFβ (see Figure 25-24 and Chapter 16). The gene encoding this receptor contains a sequence of 10 adenines in a row. Because of "slippage" of DNA polymerase during replication, this sequence often undergoes mutation to a sequence containing 9 or 11 adenines. If the mutation is not fixed by the mismatch-repair system, the resultant frameshift in the protein-coding sequence abolishes production of the normal receptor protein. As noted earlier, such inactivating mutations make cells resistant to growth inhibition by TGFβ, thereby contributing to the unregulated growth characteristic of these tumors. This

TABLE 25-2 Some Human Hereditary Diseases and Cancers Associated with DNA-Repair Defects

DISEASE	DNA-REPAIR SYSTEM AFFECTED	SENSITIVITY	CANCER SUSCEPTIBILITY	SYMPTOMS
PREVENTION OF POINT MUTATIONS, INSERTIONS, AND DELETIONS				
Hereditary nonpolyposis colorectal cancer	DNA mismatch repair	UV irradiation, chemical mutagens	Colon, ovary	Early development of tumors
Xeroderma pigmentosum	Nucleotide excision repair	UV irradiation, point mutations	Skin carcinomas, melanomas	Skin and eye photosensitivity, keratoses
REPAIR OF DOUBLE-STRAND BREAKS				
Bloom's syndrome	Repair of double-strand breaks by homologous recombination	Mild alkylating agents	Carcinomas, leukemias, lymphomas	Photosensitivity, facial telangiectases, chromosome alterations
Fanconi anemia	Repair of double-strand breaks by homologous recombination	DNA cross-linking agents, reactive oxidant chemicals	Acute myeloid leukemia, squamous-cell carcinomas	Developmental abnormalities including infertility and deformities of the skeleton; anemia
Hereditary breast cancer, BRCA-1 and BRCA-2 deficiency	Repair of double-strand breaks by homologous recombination		Breast and ovarian cancer	Breast and ovarian cancer

SOURCES: Modified from A. Kornberg and T. Baker, 1992, *DNA Replication,* 2d ed., W. H. Freeman and Company, p. 788; J. Hoeijmakers, 2001, *Nature* **411**:366; and L. Thompson and D. Schild, 2002, *Mutation Res.* **509**:49.

finding attests to the importance of mismatch repair in correcting genetic damage that might otherwise lead to uncontrolled cell proliferation.

All DNA-repair mechanisms utilize a family of DNA polymerases to correct DNA damage. Nine of these polymerases, including one called *DNA polymerase β*, are capable of using templates that contain DNA adducts and other chemical modifications, even missing bases. These are called *lesion-bypass* DNA polymerases. Each member of the polymerase family has distinct capabilities to cope with particular types of DNA lesions. Presumably such polymerases are tolerated because often any repair is better than none. They are the polymerases of last resort, the ones used when more conventional and accurate polymerases are unable to perform, and they carry out a mutagenic replication process. DNA Pol β does not proofread and is overexpressed in certain tumors, perhaps because it is needed at high levels for cells to be able to divide at all in the face of a growing burden of mutations. Error-prone repair systems are thought to mediate much of, if not all, the carcinogenic effects of chemicals and radiation, since it is only after the repair that a heritable mutation exists. There is growing evidence that mutations in DNA Pol β are associated with tumors. When 189 tumors were examined, 58 had mutations in the DNA Pol β gene, and most of these mutations were found neither in normal tissue from the same patient nor in the normal spectrum of mutations found in different people. Expressing two of the mutant polymerase forms in mouse cells caused them to grow with a transformed appearance, with focus formation and anchorage independence.

Double-strand breaks are particularly severe lesions because incorrect rejoining of double strands of DNA can lead to gross chromosomal rearrangements and translocations such as those that produce a hybrid gene or bring a growth-regulatory gene under the control of a different promoter. Often the repair of such damage depends on using the homologous chromosome as a guide (Chapter 4). The B and T cells of the immune system are particularly susceptible to DNA rearrangements caused by double-strand breaks created during rearrangement of their immunoglobulin or T cell receptor genes, explaining the frequent involvement of these gene loci in leukemias and lymphomas. *BRCA-1* and *BRCA-2*, genes implicated in human breast and ovarian cancers, encode important components of DNA-break repair systems. Cells lacking one of the BRCA functions are unable to repair DNA where the homologous chromosome is providing the template for repair.

Telomerase Expression Contributes to Immortalization of Cancer Cells

Telomeres, the physical ends of linear chromosomes, consist of tandem arrays of a short DNA sequence, TTAGGG in vertebrates. Telomeres provide the solution to the end-replication problem—the inability of DNA polymerases to completely replicate the end of a double-stranded DNA molecule (Chapter 6). *Telomerase,* a reverse transcriptase that contains an RNA template, repeatedly adds TTAGGG repeats to chromosome ends to lengthen or maintain the 3- to 20-kb regions of repeats that decorate the ends of human chromosomes (see Figure 6-49). Embryonic cells, germ-line cells, and stem cells produce telomerase, but most human somatic cells produce only a low level of telomerase as they enter S phase. As a result of having only modest telomerase function, their telomeres shorten with each cell cycle. Complete loss of telomeres leads to end-to-end chromosome fusions (Figure 25-30) and cell death. Extensive shortening of telomeres is recognized by the cell as a kind of DNA damage, with consequent stabilization and activation of p53 protein, leading to p53-triggered apoptosis.

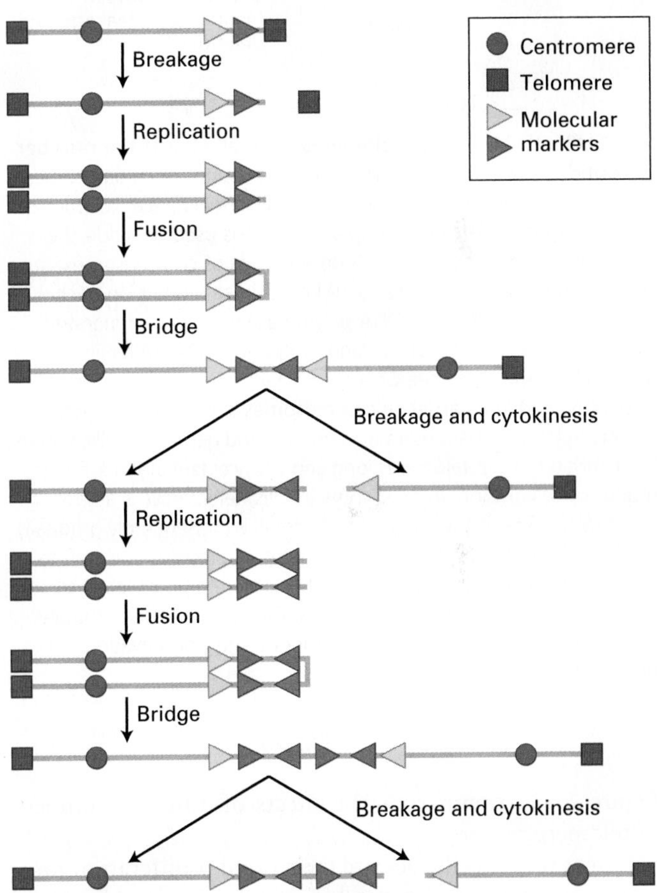

▲ **FIGURE 25-30 Chromosomal fusion and damage from loss of telomeres.** The absence of a telomere from the end of a chromosome triggers a breakage-fusion-bridge (B-F-B) cycle, with consequent major damage to chromosome structures. The loss of a telomere, by breakage or due to the absence of telomerase, allows a chromatid to fuse with its sister chromatid during mitosis. A bridge is formed, and when the centromeres separate at anaphase, the two chromatids are joined together and must break. In each daughter cell, again chromosomes lack telomeres, so the cycle repeats. The sites of the breaks are usually within 1 megabase of the telomere, and each daughter cell will have either terminal deletions or an amplification of terminal genes. Telomeres are often lost even in cancer cells that have ample telomerase, so multiple mechanisms can trigger the B-F-B cycle. [Adapted from S. M. Bailey and J. P. Murnane, 2006, *Nucl. Acid Res.* **34:**2408.]

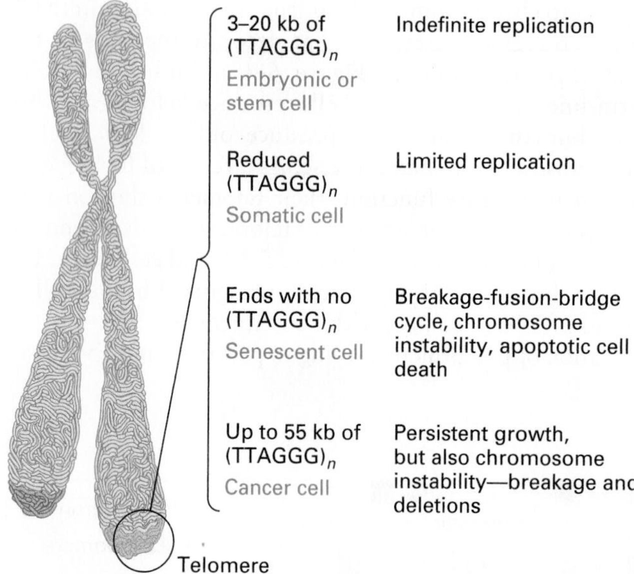

▲ **FIGURE 25-31 Loss of telomeres normally limits the number of rounds of cell division.** Replication of the ends of chromosomes, the telomeres, requires a special enzyme called *telomerase*. Telomerase carries with it a short RNA template, which is used to guide the assembly of repeats of TTAGGG on the ends of chromosomes. Human embryonic cells have 8–10 kb of these repeats on each chromosome end. Because DNA polymerase requires a primer and there is none at the end of the lagging strand, it cannot replicate DNA fully at the ends (see Chapter 6). Telomerase is therefore necessary, and in its absence chromosomes shrink during each mitosis. Telomerase is present in stem cells and germ-line cells, where it is needed to keep telomeres long and allow essentially indefinite rounds of cell division. In most somatic cells, telomerase is at low or very low levels, and the length of telomeres is consequently gradually eroded. The length of the telomeres therefore provides one limit to replicative capacity. Complete loss of telomere repeats leads to triggering of DNA repair and apoptosis. Cancer cells often produce telomerase, allowing them to divide indefinitely and evade programmed death.

Figure 25-31 summarizes the effects of different numbers of telomere repeats.

Most tumor cells, despite their rapid proliferation rate, overcome this fate by producing telomerase. Many researchers believe that telomerase expression is essential for a tumor cell to become immortal, and specific inhibitors of telomerase have been suggested as cancer therapeutic agents. Introduction of telomerase-producing transgenes into cultured human cells that otherwise lack the enzyme can extend their lifespan by more than 20 doublings while maintaining telomere length. Conversely, treating human tumor cells (HeLa cells) in culture with anti-sense RNA against telomerase caused them to cease growth in about four weeks. Dominant-negative telomerases, such as those carrying a modified RNA template, can interfere with cancer cell growth. For example, when such a mutant (Figure 25-32) was expressed in prostate or breast cancer cells in culture, the cells became apoptotic. Furthermore, human breast tumor cells carrying the mutant telomerase, transplanted into a mouse, grew at lower rates.

The prognosis for neuroblastoma, a pediatric tumor of the peripheral nervous system, can be ascertained by assessing the level of telomerase activity in tumor cells. High levels of telomerase predict a poor response to therapy, while low levels predict a good response. The level of *N-myc* protein, a transcription factor that regulates expression of telomerase, is also predictive for the outcomes of these tumors. Tumors that have amplified the *N-myc* gene to high copy number have a worse prognosis. These results show the practical importance of understanding the circuitry of telomere synthesis and its regulation. ■

Genetic approaches have further improved our understanding of the role of telomerase in cancer. The initial finding that mice homozygous for a deletion of the RNA subunit of telomerase are viable and fertile was surprising. However, after four to six generations defects began to appear in the telomerase-null mice as their very long telomeres (40–60 kb) became significantly shorter. The defects included depletion of tissues that require high rates of cell division, like skin and intestine, and infertility.

When treated with carcinogens, telomerase-null mice develop tumors less readily than normal mice do. For example, skin papillomas induced by a combination of chemical carcinogens occur 20 times less frequently in mice lacking a functional telomerase than in normal mice, presumably because p53-triggered apoptosis is induced in response to the ever shortening telomeres of cells that have begun to divide. However, if both telomerase and p53 are absent, there is an increased rate of epithelial tumors such as squamous-cell carcinoma, colon, and breast cancer. Mice with an *APC* mutation normally develop colon tumors, and these too are reduced if the mice lack telomerase. Some other tumors are less affected by loss of telomerase. These studies demonstrate the relevance of telomerase for unbridled cell division and make the enzyme a possible target for chemotherapy.

KEY CONCEPTS OF SECTION 25.5

Carcinogens and Caretaker Genes in Cancer

- Changes in the DNA sequence result from copying errors and the effects of various physical and chemical agents, or carcinogens. All carcinogens are mutagens; that is, they alter one or more nucleotides in DNA.

- Indirect-acting carcinogens, the most common type, must be activated before they can damage DNA. In animals, metabolic activation occurs via the cytochrome P-450 system, a pathway generally used by cells to rid themselves of

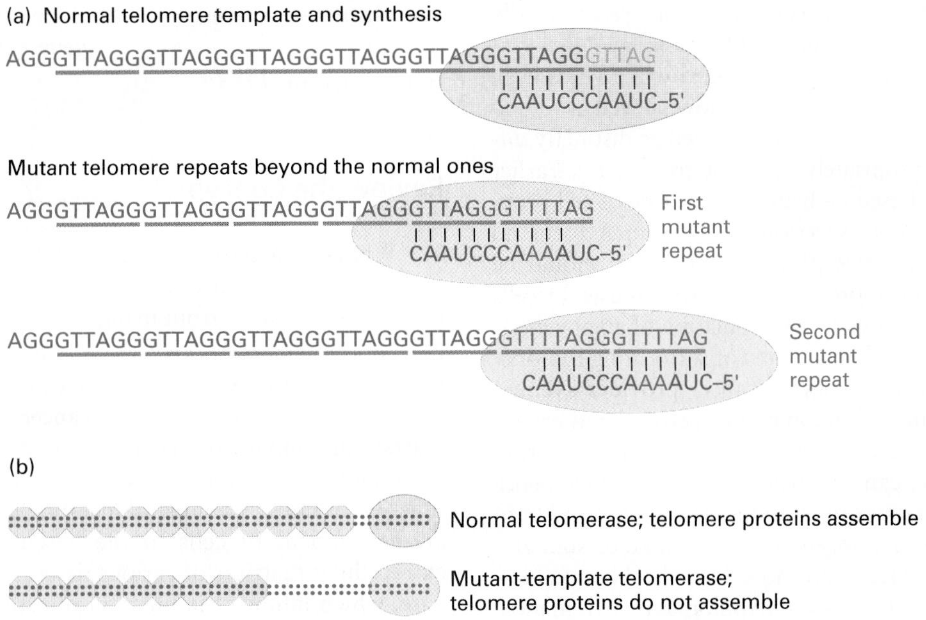

(a) Normal telomere template and synthesis

AGGGTTAGGGTTAGGGTTAGGGTTAGGGTTAGGGTTAGGGTTAG
CAAUCCCAAUC–5'

Mutant telomere repeats beyond the normal ones

AGGGTTAGGGTTAGGGTTAGGGTTAGGGTTAGGGTTTTAG
CAAUCCCAAAAUC–5'
First mutant repeat

AGGGTTAGGGTTAGGGTTAGGGTTAGGGTTAGGGTTTTAGGGTTTTAG
CAAUCCCAAAAUC–5'
Second mutant repeat

(b)

Normal telomerase; telomere proteins assemble

Mutant-template telomerase; telomere proteins do not assemble

▲ **EXPERIMENTAL FIGURE 25-32 Telomerase can be manipulated to inhibit cancer cell growth.** An engineered telomerase that contains a defective RNA template (a mutant-template telomerase) was used to interfere with normal enzyme function. (a) *Top:* Normal telomerase (gray) is shown located on one strand of DNA at the end of a telomere. The dark blue sequence is the RNA template contained within telomerase. The normal telomeric sequence is a series of repeats of the 6-mer GTTAGG (green). Synthesis proceeds to the right. The latest repeat sequence added is in light green. *Bottom:* The mutant-template telomerase (pink) has an altered RNA template with two added A bases (shown in red). When expressed in cells, this mutant telomerase catalyzes the addition of 8-mer repeats with two additional dT bases (red). Two rounds of priming and synthesis are shown. (b) The mutant telomerase acts like a dominant-negative because proteins (yellow) that bind and protect normal telomere sequences are unable to assemble on the altered telomeric sequences synthesized by the mutant enzyme. The consequence is destabilization of the chromosomes, effective uncapping of the telomere, cell sensing of DNA damage, and apoptosis. When expressed in human breast or prostate cancer cells in culture, or in breast cancer cells implanted into a mouse, the mutant-template telomerase blocked cell division and increased apoptosis. This approach was effective with even low levels of the mutant telomerase. [Adapted from M. M. Kim et al., 2001, *Proc. Nat'l Acad. Sci. USA* **98:**7982.]

noxious foreign chemicals. Direct-acting carcinogens such as EMS and DMS require no such cellular modifications in order to damage DNA.

- Benzo(*a*)pyrene, a component of cigarette smoke, causes inactivating mutations in the *p53* gene, thus contributing to the initiation of human lung tumors.
- Caretaker genes encode enzymes that repair DNA, otherwise maintain the integrity of the chromosomes, or promote the death of cells when DNA damage does occur. Mutations in caretaker genes allow the survival of cells that should die, and a continuing mutagenesis of the genome that can lead to damaged cell-cycle control and hence cancer.
- Inherited defects in DNA-repair processes found in certain human diseases are associated with an increased susceptibility for certain cancers (see Table 25-2).
- Cancer cells, like germ cells and stem cells but unlike most differentiated cells, produce telomerase, which prevents shortening of chromosomes during DNA replication and may contribute to their immortalization. The absence of telomerase is associated with resistance to generation of certain tumors, owing to the protective p53 response.

Perspectives for the Future

The recognition that cancer is fundamentally a genetic disease has opened enormous new opportunities for preventing and treating the disease. Carcinogens can now be assessed for their effects on known steps in cell-cycle control. Genetic defects in the checkpoint controls for detecting damaged DNA and in the systems for repairing it can be readily recognized and used to explore the mechanisms of cancer. The multiple changes that must occur for a cell to grow into a dangerous tumor present multiple opportunities for intervention. Identifying mutated genes associated with cancer points directly to proteins at which drugs can be targeted.

Diagnostic medicine is being transformed by our newfound ability to monitor large numbers of cell characteristics. The traditional methods of assessing possible tumor cells, mainly microscopy of stained cells, will be augmented or replaced by techniques for measuring the expression of tens of thousands of genes, focusing particularly on genes whose activities are identified as powerful indicators of the cell's growth properties and the patient's prognosis. Currently, DNA microarray analysis permits measurement of gene transcription. In the future, techniques for systematically measuring protein production, modification, and localization, all important measures of cell states, will give us even more refined portraits of cells. Tumors now viewed as identical or very similar will instead be recognized as distinctly different and given appropriately different treatments. Earlier detection of tumors, based on better monitoring of cell properties, should allow more successful treatment. A focus on that particularly destructive process, metastasis, should be successful in identifying more of the mechanisms used by cells to migrate, attach, and invade. Manipulation of angiogenesis continues to look hopeful as a means of suffocating tumors.

The molecular cell biology of cancer provides avenues for new therapies, but prevention remains crucial and preferable to therapy. Avoidance of obvious carcinogens, in particular cigarette smoke, can significantly reduce the incidence of lung cancer and perhaps other kinds, as well. Beyond minimizing exposure to carcinogens such as smoke or sunlight, certain specific approaches are now feasible. New knowledge of the involvement of human papillomavirus 16 in most cases of cervical cancer led to the development and Federal Drug Administration approval of a cancer vaccine that in tests prevented three-quarters of all cervical cancers. Antibodies against cell-surface markers that distinguish cancer cells are a source of great hope, especially after successes with the clinical use of monoclonal antibodies against human EGF receptor 2 (Her2), a protein involved in some cases of human breast cancer. Further steps must involve medicine and science. Understanding the cell biology of cancer is a critical first step toward prevention and cure, but the next steps are hard. The success with Gleevec (imatinib) against leukemia is exceptional; many cancers remain difficult to treat and cause enormous suffering. Since *cancer* is a term for a group of highly diverse diseases, interventions that are successful for one type may not be useful for others. Despite these daunting realities, we are beginning to reap the benefits of decades of research exploring the molecular biology of the cell. We hope that many of the readers of this book will help to overcome the obstacles that remain.

Key Terms

aflatoxin *1141*
aneuploidy *1138*
angiogenesis *1111*
benign *1109*
benzo(*a*)pyrine *1140*
Burkitt's lymphoma *1131*
carcinogens *1107*
carcinomas *1109*
caretaker genes *1107*
leukemias *1109*
loss of heterozygosity (LOH) *1124*
malignant *1109*
metastasis *1110*
mutagens *1139*
nondisjunction *1124*
oncogene *1114*
p53 protein *1116*
Philadelphia chromosome *1130*
proto-oncogene *1107*
sarcomas *1109*
slow-acting retroviruses *1122*
SWI/SNF complexes *1135*
transducing retroviruses *1122*
transformation *1113*
tumor-suppressor gene *1107*

Review the Concepts

1. What characteristics distinguish benign from malignant tumors? With respect to gene mutations, what distinguishes benign colon polyps from malignant colon carcinoma?

2. Ninety percent of cancer deaths are caused by metastatic rather than primary tumors. Define *metastasis*. Explain the rationale for the following new cancer treatments: (a) batimastat, an inhibitor of matrix metalloproteinases and of the plasminogen activator receptor, (b) antibodies that block the function of integrins, integral membrane proteins that mediate attachment of cells to the basement membrane and extracellular matrices of various tissues, and (c) bisphosphonate, which inhibits the function of osteoclasts. Osteoclasts are cells that digest bones, for example, in remodeling during growth or healing. They can be recruited to a bone-digesting task by signals from other cells.

3. Because of oxygen and nutrient requirements, cells in a tissue must reside within 100 mm of a blood vessel. Based on this information, explain why many malignant tumors often possess gain-of-function mutations in one of the following genes: *bFGF, TGFα,* and *VEGF.*

4. Ras oncogenes have been identified in human tumor cells using cell transformation assays with a line of cultured mouse fibroblasts called *3T3 cells.* How does the 3T3 cell transformation assay work? Why can transfection with a ras gene transform 3T3 cells but not normal cultured primary fibroblast cells?

5. What hypothesis explains the observations that incidence of human cancers increases exponentially with age? Give an example of data that confirms the hypothesis.

6. Distinguish between proto-oncogenes and tumor-suppressor genes. To become cancer-promoting, do proto-oncogenes and tumor-suppressor genes undergo gain-of-function or loss-of-function mutations? Classify the following genes as proto-oncogenes or tumor-suppressor genes: *p53, ras, Bcl-2, jun,* and *p16.*

7. Explain how DNA microarray analysis can distinguish between phenotypically similar lymphomas.

8. Hereditary retinoblastoma generally affects children in both eyes, while spontaneous retinoblastoma usually occurs during adulthood only in one eye. Explain the genetic basis for the epidemiological distinction between these two forms of retinoblastoma. Explain the apparent paradox: loss-of-function mutations in tumor-suppressor genes act recessively, yet hereditary retinoblastoma is inherited as an autosomal dominant.

9. Explain the concept of loss of heterozygosity (LOH). Why do most cancer cells exhibit LOH of one or more

genes? How does failure of the spindle-assembly checkpoint lead to loss of heterozygosity?

10. Many malignant tumors are characterized by the activation of one or more growth-factor receptors. What is the catalytic activity associated with transmembrane growth-factor receptors such as the EGF receptor? Describe how the following events lead to activation of the relevant growth-factor receptor: (a) expression of the viral protein gp55, (b) translocation that replaces the extracellular domain of the Trk receptor with the N-terminal region of tropomyosin, (c) point mutation that converts a valine to glutamine within the transmembrane region of the Her2 receptor.

11. Describe the common signal-transduction event that is perturbed by cancer-promoting mutations in the genes encoding Ras and NF-1. Why are mutations in Ras more commonly found in cancers than mutations in NF-1?

12. What is the structural distinction between the proteins encoded by c-*src* and v-*src*? How does this difference render v-*src* oncogenic?

13. Describe the mutational event that produces the *myc* oncogene in Burkitt's lymphoma. Why does the particular mechanism for generating oncogenic *myc* result in a lymphoma rather than another type of cancer? Describe another mechanism for generating oncogenic *myc*.

14. Pancreatic cancers often possess loss-of-function mutations in the gene that encodes the Smad4 protein. How does this mutation promote the loss of growth inhibition and highly metastatic phenotype of pancreatic tumors?

15. Several strains of human papilloma virus (HPV) can cause cervical cancer. These pathogenic strains produce three proteins that contribute to host-cell transformation. What are these three viral proteins? Describe how each interacts with its target host protein.

16. Loss of p53 function occurs in the majority of human tumors. Name two ways in which loss of p53 function contributes to a malignant phenotype. Explain how benzo(*a*)pyrene can cause loss of p53 function.

17. Which human cell types possess telomerase activity? What characteristic of cancer is promoted by expression of telomerase? What concerns does this pose for medical therapies involving stem cells?

Analyze the Data

Cigarette smoking is a major risk factor in the development of non-small cell lung cancer (NSCLC), which accounts for 80 percent of all lung cancers. NSCLC is characterized by poor prognosis and resistance to chemotherapy. To understand if nicotine affects this resistance, lung cancer cell lines were treated with the chemotherapeutic drugs gemcitabine, Cisplatin, and Taxol, in the presence and absence of nicotine (see Dasgupta et al., 2006, Proc. Nat'l Acad. Sci. USA 103:6332–6337).

a. Three different NSCLC cells lines, A549, NCI-H23, and H1299, were examined by the TUNEL assay, which detects cells that are undergoing apoptosis (programmed cell death). The cells were either untreated or treated with one of the three chemotherapeutic drugs in the presence or absence of nicotine. The following data were obtained. Why are these chemotherapeutic drugs potentially useful for the treatment of lung cancer, and how does nicotine affect their potential usefulness?

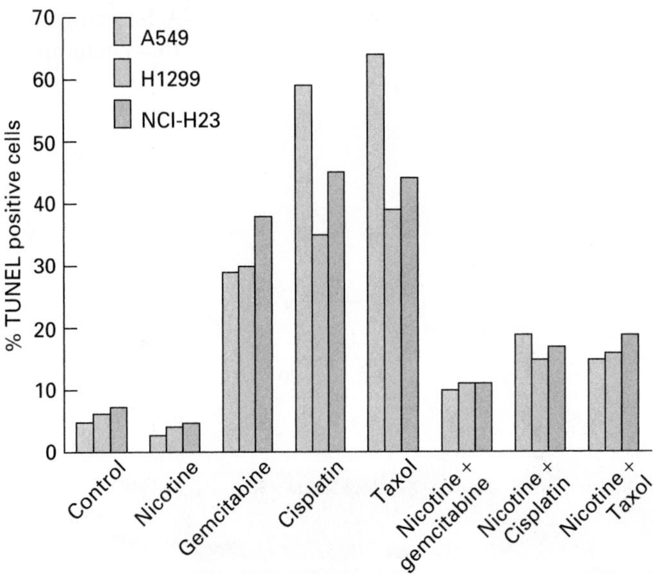

b. A549 cells were incubated with the drugs as in part (a), and in an additional chemotherapeutic drug, camptothecin, in the presence or absence of nicotine. The cells were lysed, extracts were run on SDS gels, and then the gels were blotted and probed with antibodies against the indicated proteins. PARP is a protein that is cleaved during apoptosis. What can you deduce from these data about the effects of nicotine? What is the purpose of assessing the levels of p53, p21, and actin?

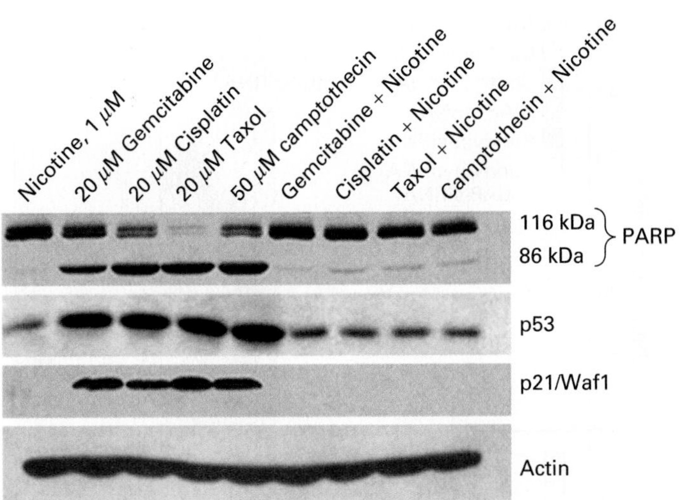

c. Survivin and XIAP are both members of the inhibitor of apoptosis (IAP) family, proteins that protect cells against apoptosis. A549 cells were treated with the chemotherapeutic drugs in the presence or absence of nicotine and presence or absence of LY294002, an inhibitor of PI-3 kinase (see Chapter 16). The levels of PARP, XIAP, survivin, and actin were assessed, as shown in the Western blots below. The graph shows the results of a TUNEL assay in which A549 cells have been transfected with the indicated siRNAs. "None" indicates the amount of cell death with no transfected RNA. Control-siRNAs 1 and 2 are irrelevant RNAs added to control for nonspecific effects of having an interfering RNA entering the cells. What do these studies suggest about the mechanism of nicotine action?

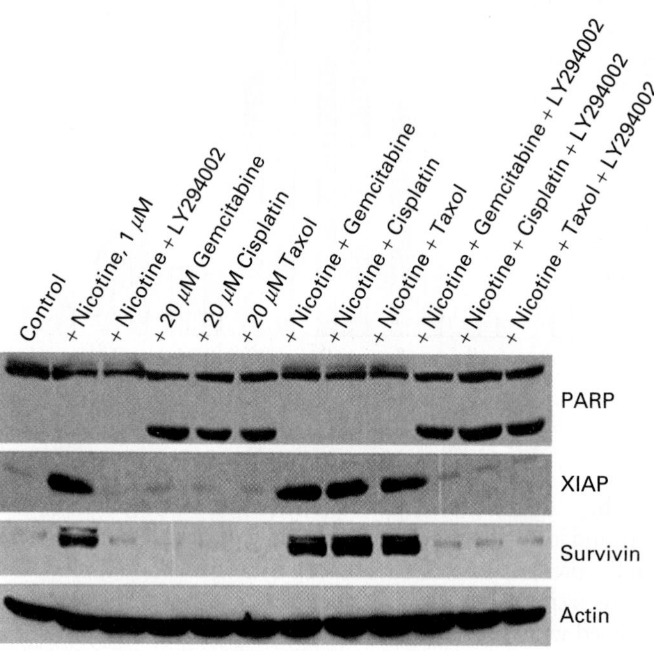

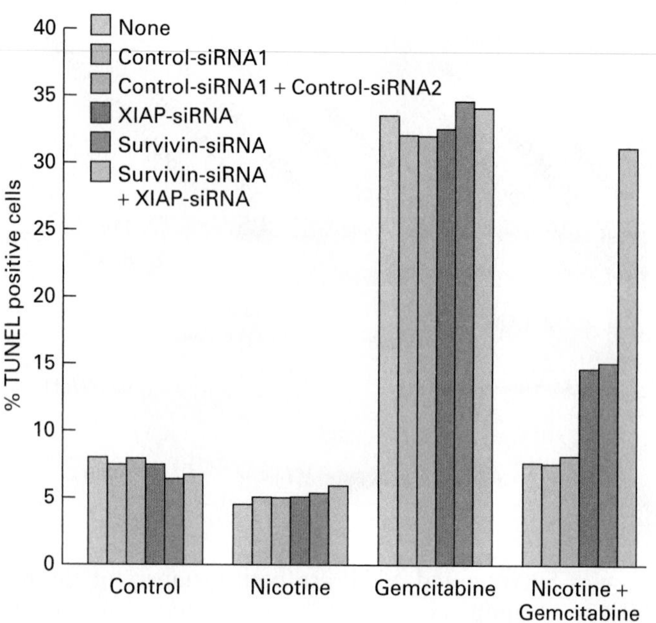

d. Can you provide a biological explanation for why the prognosis is poorer for patients who smoke during chemotherapy compared with those who quit smoking prior to it?

References

Introduction

Weinberg, R. A. 2006. *The Biology of Cancer*. Garland Science.

Tumor Cells and the Onset of Cancer

Al-Hajj, M., et al. 2003. Prospective identification of tumorigenic breast cancer cells. *Proc. Nat'l Acad. Sci. USA* **100**:3983–3988. Erratum in *Proc. Nat'l Acad. Sci. USA* (2003) **100**:6890.

Clarke, M. F. 2004. At the root of brain cancer. *Nature* **432**:281–282.

Clarke, M. F., and M. Fuller. 2006. Stem cells and cancer: two faces of Eve. *Cell* **124**:1111–1115.

Coussens, L. M., and Z. Werb. 2002. Inflammation and cancer. *Nature* **420**:860–867.

Egeblad, M., L. E. Littlepage, and Z. Werb. 2005. The fibroblastic coconspirator in cancer progression. *Cold Spring Harbor Symp. Quant. Biol.* **70**:383–388.

Fidler, I. J. 2002. The pathogenesis of cancer metastasis: the "seed and soil" hypothesis revisited. *Nature Rev. Cancer* **3**:1–6.

Folkman, J. 2002. Role of angiogenesis in tumor growth and metastasis. *Semin. Oncol.* **29**:15–18.

Hanahan, D., and R. A. Weinberg. 2000. The hallmarks of cancer. *Cell* **100**:57–70.

Huber, M. A., N. Kraut, and H. Beug. 2005. Molecular requirements for epithelial-mesenchymal transition during tumor progression. *Curr. Opin. Cell Biol.* **17**:548–558.

Jain, M., et al. 2002. Sustained loss of a neoplastic phenotype by brief inactivation of MYC. *Science* **297**:102–104.

Kinzler, K. W., and B. Vogelstein. 1996. Lessons from hereditary colorectal cancer. *Cell* **87**:159–170.

Klein, G. 1998. Foulds' dangerous idea revisited: the multistep development of tumors 40 years later. *Adv. Cancer Res.* **72**:1–23.

Ludwig, T. 2005. Local proteolytic activity in tumor cell invasion and metastasis. *BioEssays* **27**:1181–1191.

Matsui, W., et al. 2004. Characterization of clonogenic multiple myeloma cells. *Blood* **103**:2332–2336.

Olsson, A. Y., and C. S. Cooper. 2005. The molecular basis of prostate cancer. *Brit. J. Hosp. Med. (Lond.)* **66**:612–616.

Rafii, S., et al. 2002. Vascular and haematopoietic stem cells: novel targets for anti-angiogenesis therapy? *Nature Rev. Cancer* **2**:826–835.

Ramaswamy, S., et al. 2003. A molecular signature of metastasis in primary solid tumors. *Nature Genet.* **33**:49–54.

Trusolino, L., and P. M. Comoglio. 2002. Scatter-factor and semaphorin receptors: cell signalling for invasive growth. *Nature Rev. Cancer* **2**:289–300.

Yancopoulos, G., M. Klagsburn, and J. Folkman. 1998. Vasculogenesis, angiogenesis, and growth factors: ephrins enter the fray at the border. *Cell* **93**:661–664.

Zakarija, A., and G. Soff. 2005. Update on angiogenesis inhibitors. *Curr Opin. Oncol.* **17**:578–583.

The Genetic Basis of Cancer

Bertucci, F., et al. 2004. Gene expression profiling of colon cancer by DNA microarrays and correlation with histoclinical parameters. *Oncogene* **23**:1377–1391.

Clark, J., et al. 2002. Identification of amplified and expressed genes in breast cancer by comparative hybridization onto microarrays of randomly selected cDNA clones. *Genes Chrom. Cancer* **34**:104–114.

Classon, M., and E. Harlow. 2002. The retinoblastoma tumour suppressor in development and cancer. *Nature Rev. Cancer* **2**:910–917.

Fogarty, M. P., J. D. Kessler, and R. J. Wechsler-Reya. 2005. Morphing into cancer: the role of developmental signaling pathways in brain tumor formation. *J. Neurobiol.* **64**:458–475.

Grisendi, S., and P. P. Pandolfi. 2005. Two decades of cancer genetics: from specificity to pleiotropic networks. *Cold Spring Harbor Symp. Quant. Biol.* **70**:83–91.

Lau, J., H. Kawahira, and M. Hebrok. 2006. Hedgehog signaling in pancreas development and disease. *Cell Mol. Life Sci.* **63**:642–652.

Moon, R. T., et al. 2002. The promise and perils of Wnt signaling through beta-catenin. *Science* **296**:1644–1646.

Nevins, J. R. 2001. The Rb/E2F pathway and cancer. *Hum. Mol. Genet.* **10**:699–703.

Polakis, P. 2000. Wnt signaling and cancer. *Genes Devel.* **14**:1837–1851.

Pollack, J. R., et al. 2002. Microarray analysis reveals a direct role of DNA copy number alteration in the transcriptional program of human breast tumors. *Proc. Nat'l Acad. Sci. USA* **99**:12963–12968.

Sasaki, T., et al. 2000. Colorectal carcinomas in mice lacking the catalytic subunit of PI(3)Kgamma. *Nature* **406**:897–902.

Scambia, G., S. Lovergine, and V. Masciullo. 2006. RB family members as predictive and prognostic factors in human cancer. *Oncogene* **25**:5302–5308.

Sherr, C. J., and F. McCormick. 2002. The RB and p53 pathways in cancer. *Cancer Cell* **2**:103–112.

van't Veer, L. J., et al. 2002. Gene expression profiling predicts clinical outcome of breast cancer. *Nature* **415**:530–536.

West, M., et al. 2001. Predicting the clinical status of human breast cancer by using gene expression profiles. *Proc. Nat'l Acad. Sci. USA* **98**:11462–11467.

Oncogenic Mutations in Growth-Promoting Proteins

Bachman, K. E., and B. H. Park. 2005. Dual nature of TGF-beta signaling: tumor suppressor vs. tumor promoter. *Curr. Opin. Oncol.* **17**:49–54.

Beachy, P. A., S. S. Karhadkar, and D. M. Berman. 2004. Tissue repair and stem cell renewal in carcinogenesis. *Nature* **432**:324–331.

Capdeville, R., et al. 2002. Glivec (STI571, imatinib), a rationally developed, targeted anticancer drug. *Nature Rev. Drug Discov.* **1**:493–502.

Downward, J. 2003. Targeting RAS signalling pathways in cancer therapy. *Nature Rev. Cancer* **3**:11–22.

Gregorieff, A., and H. Clevers. 2005. Wnt signaling in the intestinal epithelium: from endoderm to cancer. *Genes Dev.* **19**:877–890.

Rowley, J. D. 2001. Chromosome translocations: dangerous liaisons revisited. *Nature Rev. Cancer* **1**:245–250.

Sahai, E., and C. J. Marshall. 2002. RHO-GTPases and cancer. *Nature Rev. Cancer* **2**:133–142.

Shaulian, E., and M. Karin. 2002. AP-1 as a regulator of cell life and death. *Nature Cell Biol.* **4**:E131–E136.

Shawver, L. K., D. Slamon, and A. Ullrich. 2002. Smart drugs: tyrosine kinase inhibitors in cancer therapy. *Cancer Cell* **1**:117–123.

Mutations Causing Loss of Growth-Inhibiting and Cell-Cycle Controls

Bardeesy, N., et al. 2006. Both p16(Ink4a) and the p19(Arf)-p53 pathway constrain progression of pancreatic adenocarcinoma in the mouse. *Proc. Nat'l Acad. Sci. USA* **103**:5947–5952.

Chau, B. N., and J. Y. Wang. 2003. Coordinated regulation of life and death by RB. *Nature Rev. Cancer* **3**:130–138.

Lane, D. P. 2005. Exploiting the p53 pathway for the diagnosis and therapy of human cancer. *Cold Spring Harbor Symp. Quant. Biol.* **70**:489–497.

Malumbres, M., and M. Barbacid. 2001. To cycle or not to cycle: a critical decision in cancer. *Nature Rev. Cancer* **1**:222–231.

Mooi, W. J., and D. S. Peeper. 2006. Oncogene-induced cell senescence—halting on the road to cancer. *N. Eng. J. Med.* **355**:1037–1046.

Planas-Silva, M. D., and R. A. Weinberg. 1997. The restriction point and control of cell proliferation. *Curr. Opin. Cell Biol.* **9**:768–772.

Spike, B. T., and K. F. Macleod. 2005. The Rb tumor suppressor in stress responses and hematopoietic homeostasis. *Cell Cycle* **4**:42–45.

Zhang, L., et al. 2000. Role of BAX in the apoptotic response to anticancer agents. *Science* **290**:989–992.

Carcinogens and Caretakers in Cancer

Armanios, M., and C. W. Greider. 2005. Telomerase and cancer stem cells. *Cold Spring Harbor Symp. Quant. Biol.* **70**:205–208.

Bailey, S. M., and J. P. Murnane. 2006. Telomeres, chromosome instability and cancer. *Nucl. Acids Res.* **34**:2408–2417.

Batty, D., and R. Wood. 2000. Damage recognition in nucleotide excision repair of DNA. *Gene* **241**:193–204.

Blackburn, E. H. 2005. Telomerase and Cancer: Kirk A. Landon—AACR prize for basic cancer research lecture. *Mol. Cancer Res.* **3**:477–482.

Cleaver, J. E. 2005. Cancer in xeroderma pigmentosum and related disorders of DNA repair. *Nature. Rev. Cancer* **5**:564–573.

D'Andrea, A., and M. Grompe. 2003. The Fanconi anemia/BRCA pathway. *Nature Rev. Cancer* **3**:23–34.

Flores-Rozas, H., and R. Kolodner. 2000. Links between replication, recombination and genome instability in eukaryotes. *Trends Biochem. Sci.* **25**:196–200.

Friedberg, E. 2003. DNA damage and repair. *Nature* **421**:436–440.

Hoeijmakers, J. 2001. Genome maintenance mechanisms for preventing cancer. *Nature* **411**:366–374.

Jiricny, J. 2006. The multifaceted mismatch-repair system. *Nature. Rev. Mol. Cell Biol.* **7**:335–346.

Ju, Z., and K. L. Rudolph. 2006. Telomeres and telomerase in cancer stem cells. *Eur. J. Cancer* **42**:1197–1203.

Kitagawa, R., and M. B. Kastan. 2005. The ATM-dependent DNA damage signaling pathway. *Cold Spring Harbor Symp. Quant. Biol.* **70**:99–109.

Loechler, E. L. 2002. Environmental carcinogens and mutagens. In *Encyclopedia of Life Sciences*. Nature Publishing.

Muller, A., and R. Fishel. 2002. Mismatch repair and the hereditary non-polyposis colorectal cancer syndrome (HNPCC). *Cancer Invest.* **20**:102–109.

O'Driscoll, M., and P. A. Jeggo. 2006. The role of double-strand break repair—insights from human genetics. *Nature Rev. Genet.* **7**:45–54.

Schärer, O. 2003. Chemistry and biology of DNA repair. *Angewandte Chemie* **42**:2946–2974.

Somasundaram, K. 2002. Breast cancer gene 1 (BRCA-1): role in cell cycle regulation and DNA repair—perhaps through transcription. *J. Cell Biochem.* **88**:1084–1091.

Sweasy, J. B., T. Lang, and D. DiMaio. 2006. Is base excision repair a tumor suppressor mechanism? *Cell Cycle* **5**:250–259.

Thompson, L., and D. Schild. 2002. Recombinational DNA repair and human disease. *Mut. Res.* **509**:49–78.

van Gant, D., J. Hoeijmakers, and R. Kanaar. 2001. Chromosomal stability and the DNA double-stranded break connection. *Nature Rev. Genet.* **2**:196–205.

Wogan, G. N., et al. 2004. Environmental and chemical carcinogenesis. *Semin. Cancer Biol.* **14**:473–486.

GLOSSARY

Boldfaced terms within a definition are also defined in this glossary.
Figures and tables that illustrate defined terms are noted in parentheses.

AAA ATPase family A group of proteins that couple hydrolysis of ATP with large molecular movements usually associated with unfolding of protein substrates or the disassembly of multisubunit protein complexes.

ABC superfamily A large group of integral membrane proteins that often function as ATP-powered **membrane transport proteins** to move diverse molecules (e.g., phospholipids, cholesterol, sugars, ions, peptides) across cellular membranes. (Figure 11-14)

acetylcholine (ACh) Neurotransmitter that functions at vertebrate neuromuscular junctions and at various neuron–neuron synapses in the brain and peripheral nervous system. (Figure 23-19)

acetyl CoA Small, water-soluble metabolite comprising an acetyl group linked to coenzyme A (CoA). The acetyl group is transferred to citrate in the **citric acid cycle** and is used as a carbon source in the synthesis of fatty acids, steroids, and other molecules. (Figure 12-9)

acid Any compound that can donate a proton (H^+). The carboxyl and phosphate groups are the primary acidic groups in biological macromolecules.

actin Abundant structural protein in eukaryotic cells that interacts with many other proteins. The monomeric globular form (*G-actin*) polymerizes to form actin filaments (*F-actin*). In muscle cells, F-actin interacts with **myosin** during contraction. See also **microfilament.** (Figure 17-5)

action potential Rapid, transient, all-or-none electrical activity propagated in the plasma membrane of excitable cells (e.g., neurons and muscle cells) as the result of the selective opening and closing of voltage-gated Na^+ and K^+ channels. (Figures 23-3 and 23-9)

activation energy The input of energy required to (overcome the barrier to) initiate a chemical reaction. By reducing the activation energy, an **enzyme** increases the rate of a reaction. (Figure 2-30)

activator Specific **transcription factor** that stimulates transcription.

active site Specific region of an enzyme that binds a **substrate** molecule(s) and promotes a chemical change in the bound substrate. (Figure 3-21)

active transport Protein-mediated movement of an ion or small molecule across a membrane against its concentration gradient or electrochemical gradient driven by the coupled hydrolysis of ATP. (Figure 11-3, **[1]**; Table 11-1)

adenosine triphosphate (ATP) See **ATP.**

adenylyl cyclase One of several enzymes that is activated by binding of certain ligands to their cell-surface receptors and catalyzes formation of **cyclic AMP (cAMP)** from ATP; also called *adenylate cyclase.* (Figures 15-21 and 15-22)

adhesion receptor Protein in the plasma membrane of animal cells that binds components of the **extracellular matrix,** thereby mediating cell–matrix adhesion. The **integrins** are major adhesion receptors. (Figure 19-1, **[5]**)

aerobic Referring to a cell, organism, or metabolic process that utilizes gaseous oxygen (O_2) or that can grow in the presence of O_2.

aerobic oxidation Oxygen-requiring metabolism of sugars and fatty acids to CO_2 and H_2O coupled to the synthesis of ATP.

agonist A molecule, often synthetic, that mimics the biological function of a natural molecule (e.g., a hormone).

allele One of two or more alternative forms of a gene. Diploid cells contain two alleles of each gene, located at the corresponding site (locus) on **homologous chromosomes.**

allosteric Referring to proteins and cellular processes that are regulated by **allostery.**

allostery Change in the tertiary and/or quaternary structure of a protein induced by binding of a small molecule to a specific regulatory site, causing a change in the protein's activity.

alpha (α) carbon atom (C_α) In amino acids, the central carbon atom that is bonded to four different chemical groups (except in glycine) including the **side chain,** or R group. (Figure 2-4)

alpha (α) helix Common protein **secondary structure** in which the linear sequence of amino acids is folded into a right-handed spiral stabilized by hydrogen bonds between carboxyl and amide groups in the backbone. (Figure 3-4)

alternative splicing Process by which the exons of one pre-mRNA are spliced together in different combinations, generating two or more different mature mRNAs from a single pre-mRNA. (Figure 4-16)

amino acid An organic compound containing at least one amino group and one carboxyl group. In the amino acids that are the **monomers** for building proteins, an amino group and carboxyl group are linked to a central carbon atom, the α carbon, to which a variable side chain is attached. (Figures 2-4 and 2-14)

aminoacyl-tRNA Activated form of an amino acid, used in protein synthesis, consisting of an amino acid linked via a high-energy ester bond to the 3′-hydroxyl group of a **tRNA** molecule. (Figure 4-19)

amphipathic Referring to a molecule or structure that has both a **hydrophobic** and a **hydrophilic** part.

anaerobic Referring to a cell, organism, or metabolic process that functions in the absence of gaseous oxygen (O_2).

anaphase Mitotic stage during which the sister **chromatids** (or duplicated homologues in meiosis I) separate and move apart (segregate) toward the spindle poles. (Figure 18-34)

anchoring junctions Specialized regions on the cell surface containing **cell-adhesion molecules** or **adhesion receptors;** include *adherens junctions* and *desmosomes,* which mediate cell–cell adhesion, and *hemidesmosomes,* which mediate cell–matrix adhesion. (Figures 19-12 and 19-14)

aneuploidy Any deviation from the normal **diploid** number of chromosomes in which extra copies of one or more chromosomes are present or one of the normal copies is missing.

anion A negatively charged ion.

antagonist A molecule, often synthetic, that blocks the biological function of a natural molecule (e.g., hormone).

antibody A protein (immunoglobulin), normally produced in response to an **antigen,** that interacts with a particular site (**epitope**) on the same antigen and facilitates its clearance from the body. (Figure 3-19)

anticodon Sequence of three nucleotides in a tRNA that is complementary to a **codon** in an mRNA. During protein synthesis, base pairing between a codon and anticodon aligns the tRNA carrying the corresponding amino acid for addition to the growing polypeptide chain. (Figure 4-20)

antigen Any material (usually foreign) that elicits an immune response. For B cells, an antigen elicits formation of antibody that specifically binds the same antigen; for T cells, an antigen elicits a proliferative response, followed by production of **cytokines** or the activation of cytotoxic activity.

antigen-presenting cell (APC) Any cell that can digest an antigen into small peptides and display the peptides in association with class II MHC molecules on the cell surface where they can be recognized by T cells. *Professional* APCs (dendritic cells, macrophages, and B cells) constitutively express class II MHC molecules. (Figures 24-27 and 24-28)

antiport A type of **cotransport** in which a membrane protein (*antiporter*) transports two different molecules or ions across a cell membrane in opposite directions. See also **symport.** (Figure 11-3, [3C])

apical Referring to the tip (apex) of a cell, an organ, or other body structure. In the case of epithelial cells, the apical surface is exposed to the exterior of the body or to an internal open space (e.g., intestinal lumen, duct). (Figure 19-8)

apoptosis A genetically regulated process, occurring in specific tissues during development and disease, by which a cell destroys itself; marked by the breakdown of most cell components and a series of well-defined morphological changes; also called *programmed cell death.* See also **caspases.** (Figures 1-19, 21-33, and 21-41a)

apoptosome Large, disc-shaped heptamer of mammalian Apaf-1, a protein that assembles in response to apoptosis signals and serves as an activation machine for initiator and effector **caspases.** (Figure 21-40)

aquaporins A family of **membrane transport proteins** that allow water and a few other small uncharged molecules, such as glycerol, to cross biomembranes. (Figure 11-8)

archaea Class of **prokaryotes** that constitutes one of the three distinct evolutionary lineages of modern-day organisms; also called *archaebacteria* and *archaeans.* In some respects, archaeans are more similar to **eukaryotes** than to **bacteria** (eubacteria). (Figure 1-3)

associated constant (K_a) See **equilibrium constant.**

aster Structure composed of microtubules (astral fibers) that radiate outward from a **centrosome** during mitosis. (Figure 18-36)

asymmetric carbon atom A carbon atom bonded to four different atoms or chemical groups; also called *chiral carbon atom.* The bonds can be arranged in two different ways, producing **stereoisomers** that are mirror images of each other. (Figure 2-4)

asymmetric cell division Any cell division in which the two daughter cells receive the same genes but otherwise inherit different components (e.g., mRNAs, proteins) from the parental cell. (Figures 21-26 and 21-28)

ATP (adenosine 5′-triphosphate) A nucleotide that is the most important molecule for capturing and transferring **free energy** in cells. Hydrolysis of each of the two **phosphoanhydride bonds** in ATP releases a large amount of free energy that can be used to drive energy-requiring cellular processes. (Figure 2-31)

ATPase One of a large group of enzymes that catalyze hydrolysis of **ATP** to yield ADP and inorganic phosphate with release of free energy. See also **Na^+/K^+ ATPase** and **ATP-powered pump.**

ATP-powered pump Any transmembrane protein that has ATPase activity and couples hydrolysis of ATP to the active transport of an ion or small molecule across a biomembrane against its electrochemical gradient; often simply called *pump.* (Figure 11-9)

ATP synthase Multimeric protein complex, bound to inner mitochondrial membranes, thylakoid membranes of chloroplasts, and the bacterial plasma membrane, that catalyzes synthesis of ATP during oxidative phosphorylation and photosynthesis; also called *F_0F_1 complex.* (Figure 12-24)

autocrine Referring to signaling mechanism in which a cell produces a signaling molecule (e.g., growth factor) and then binds and responds to it.

autoradiography Technique for visualizing radioactive molecules in a sample (e.g., a tissue section or electrophoretic gel) by exposing a photographic film (emulsion) or two-dimensional electronic detector to the sample. The exposed film is called an *autoradiogram* or *autoradiograph.*

autosome Any chromosome other than a sex chromosome.

axon Long process extending from the cell body of a neuron that is capable of conducting an electric impulse (**action potential**), generated at the junction with the cell body, toward its distal, branching end (the axon terminus). (Figure 23-2)

axonal transport Motor protein–mediated transport of organelles and vesicles along microtubules in axons of nerve cells. *Anterograde* transport occurs from cell body toward axon terminal); *retrograde* transport, from axon terminal toward cell body. (Figures 18-17 and 18-18)

axoneme Bundle of **microtubules** and associated proteins present in **cilia** and **flagella** and responsible for their structure and movement. (Figure 18-29)

bacteria Class of **prokaryotes** that constitutes one of the three distinct evolutionary lineages of modern-day organisms; also called *eubacteria.* Phylogenetically distinct from **archaea** and **eukaryotes.** (Figure 1-3)

bacteriophage (phage) Any virus that infects bacterial cells. Some phages are widely used as **vectors** in **DNA cloning.**

basal See **basolateral.**

basal body Structure at the base of a **cilium** or **flagellum** from which microtubules forming the **axoneme** assemble; structurally similar to a **centriole.** (Figure 18-29)

basal lamina (pl. basal laminae) A thin sheet-like network of extracellular-matrix components that underlies most animal epithelia and other organized groups of cells (e.g., muscle), separating them from connective tissue or other cells. (Figures 19-19 and 19-20)

base Any compound, often containing nitrogen, that can accept a proton (H^+) from an acid. Also, commonly used to denote the **purines** and **pyrimidines** in DNA and RNA.

base pair Association of two complementary **nucleotides** in a DNA or RNA molecule stabilized by hydrogen bonding between their base components. Adenine pairs with thymine or uracil (A·T, A·U) and guanine pairs with cytosine (G·C). (Figure 4-3b)

basic helix-loop-helix See **helix-loop-helix, basic.**

basolateral Referring to the base (basal) and side (lateral) of a polarized cell, organ, or other body structure. In the case of epithelial cells, the basolateral surface abuts adjacent cells and the underlying **basal lamina.** (Figure 19-8)

B cell A lymphocyte that matures in the bone marrow and expresses antigen-specific receptors (membrane-bound **immunoglobulin**). After interacting with antigen, a B cell proliferates and differentiates into **antibody**-secreting *plasma cells.*

B-cell receptor Complex composed of an antigen-specific membrane-bound immunoglobulin molecule and associated signal-transducing Igα and Igβ chains. (Figure 24-17)

benign Referring to a tumor containing cells that closely resemble normal cells. Benign tumors stay in the tissue where they originate but can be harmful due to continued growth. See also **malignant.**

beta (β) sheet A flat **secondary structure** in proteins that is created by hydrogen bonding between the backbone atoms in two different polypeptide chains or segments of a single folded chain. (Figure 3-5)

beta (β) turn A short U-shaped **secondary structure** in proteins. (Figure 3-6)

BLAST A widely used computer program for comparing the amino acid sequence of a protein with the sequences of known proteins stored in databases. BLAST searches can provide clues about the structure, function, and evolution of newly discovered proteins.

blastocyst Stage of mammalian embryo composed of ≈64 cells that have separated into two cell types—*trophectoderm,* which will form extra-embryonic tissues, and the *inner cell mass,* which gives rise to the embryo proper; stage that implants in the uterine wall and corresponds to the *blastula* of other animal embryos. (Figure 22-1)

buffer A solution of the acid (HA) and base (A^-) form of a compound that undergoes little change in pH when small quantities of strong acid or base are added at pH values near the compound's pK_a.

cadherins A family of dimeric **cell-adhesion molecules** that aggregate in adherens junctions and desmosomes and mediate Ca^{2+}-dependent cell–cell homophilic interactions. (Figure 19-2)

calmodulin A small cytosolic regulatory protein that binds four Ca^{2+} ions. The Ca^{2+}/calmodulin complex binds to many proteins, thereby activating or inhibiting them. (Figure 3-31)

calorie A unit of heat (thermal energy). One calorie is the amount of heat needed to raise the temperature of 1 gram of water by 1 °C. The *kilocalorie* (kcal) commonly is used to indicate the energy content of foods and changes in the **free energy** of a system.

Calvin cycle The major metabolic pathway that fixes CO_2 into carbohydrates during photosynthesis; also called *carbon fixation.* It is indirectly dependent on light but can occur both in the dark and light. (Figure 12-44)

cancer General term denoting any of various malignant tumors, whose cells grow and divide more rapidly than normal, invade surrounding tissue, and sometimes spread (metastasize) to other sites.

capsid The outer proteinaceous coat of a **virus,** formed by multiple copies of one or more protein subunits and enclosing the viral nucleic acid.

carbohydrate General term for certain polyhydroxyaldehydes, polyhydroxyketones, or compounds derived from these usually having the formula $(CH_2O)_n$. Primary type of compound used for storing and supplying energy in animal cells. (Figure 2-18)

carbon fixation See **Calvin cycle.**

carcinogen Any chemical or physical agent that can cause cancer when cells or organisms are exposed to it.

caretaker gene Any gene whose encoded protein helps protect the integrity of the genome by participating in the repair of damaged DNA. Loss of function of a caretaker gene leads to increased mutation rates and promotes carcinogenesis.

caspases A class of vertebrate protein-degrading enzymes (proteases) that function in **apoptosis** and work in a cascade with each type activating the next. (Figures 21-37 and 21-38)

catabolism Cellular degradation of complex molecules to simpler ones usually accompanied by the release of energy. *Anabolism* is the reverse process in which energy is used to synthesize complex molecules from simpler ones.

catalyst A substance that increases the rate of a chemical reaction without undergoing a permanent change in its structure. Enzymes are proteins with catalytic activity, and ribozymes are RNAs that can function as catalysts. (Figure 3-20)

cation A positively charged ion.

cDNA (complementary DNA) DNA molecule copied from an mRNA molecule by **reverse transcriptase** and therefore lacking the **introns** present in the DNA of the genome.

cell-adhesion molecules (CAMs) Proteins in the plasma membrane of cells that bind similar proteins on other cells, thereby mediating cell–cell adhesion. Four major classes of CAMs include the **cadherins, IgCAMs, integrins,** and **selectins.** (Figures 19-1 and 19-2)

cell cycle Ordered sequence of events in which a eukaryotic cell duplicates its chromosomes and divides into two. The cell cycle normally consists of four phases: G_1 before DNA synthesis occurs; S when DNA replication occurs; G_2 after DNA synthesis; and M when **cell division** occurs, yielding two daughter cells. Under certain conditions, cells exit the cell cycle during G_1 and remain in the G_0 state as nondividing cells. (Figures 1-17 and 20-1)

cell division Separation of a cell into two daughter cells. In higher eukaryotes, it involves division of the nucleus (**mitosis**) and of the cytoplasm (**cytokinesis**); mitosis often is used to refer to both nuclear and cytoplasmic division.

cell junctions Specialized regions on the cell surface through which cells are joined to each other or to the extracellular matrix. (Figure 19-9; Table 19-2)

cell line A population of cultured cells, of plant or animal origin, that has undergone a genetic change allowing the cells to grow indefinitely. (Figure 9-31b)

cell strain A population of cultured cells, of plant or animal origin, that has a finite life span and eventually dies, commonly after 25–50 generations. (Figure 9-31a)

cellulose A structural polysaccharide made of glucose units linked together by β(1 → 4) **glycosidic bonds.** It forms long microfibrils, which are the major component of the **cell wall** in plants.

cell wall A specialized, rigid extracellular matrix that lies next to the plasma membrane, protecting a cell and maintaining its shape; prominent in most fungi, plants, and prokaryotes, but absent in most multicellular animals. (Figure 19-37)

centriole Either of two cylindrical structures within the **centrosome** of animal cells and containing nine sets of triplet microtubules; structurally similar to a **basal body.** (Figure 18-6)

centromere DNA sequence required for proper **segregation** of chromosomes during mitosis and meiosis; the region of mitotic chromosomes where the **kinetochore** forms and that appears constricted. (Figures 6-40 and 6-46b)

centrosome (cell center) Structure located near the nucleus of animal cells that is the primary **microtubule-organizing center (MTOC)**; it contains a pair of **centrioles** embedded in a protein matrix and duplicates before mitosis, with each centrosome becoming a spindle pole. (Figures 18-6 and 18-35)

chaperone Collective term for two types of proteins—*molecular chaperones* and *chaperonins*—that prevent misfolding of a target protein or actively facilitate proper folding of an incompletely folded target protein, respectively. (Figures 3-16 and 3-17)

chaperonin See **chaperone.**

checkpoint Any of several points in the eukaryotic **cell cycle** at which progression of a cell to the next stage can be halted until conditions are suitable. (Figure 20-35)

chemical equilibrium The state of a chemical reaction in which the concentration of all products and reactants is constant because the rates of the forward and reverse reactions are equal.

chemical potential energy The energy stored in the bonds connecting atoms in molecules.

chemiosmosis Process whereby an electrochemical proton gradient (pH plus electric potential) across a membrane is used to drive an energy-requiring process such as ATP synthesis; also called *chemiosmotic coupling.* See **proton-motive force.** (Figure 12-2)

chemokine Any of numerous small, secreted proteins that function as chemotatic cues for leukocytes.

chemotaxis Movement of a cell or organism toward or away from certain chemicals.

chimera (1) An animal or tissue composed of elements derived from genetically distinct individuals; a hybrid. (2) A protein molecule containing segments derived from different proteins.

chlorophylls A group of light-absorbing porphyrin pigments that are critical in **photosynthesis.** (Figure 12-31)

chloroplast A specialized organelle in plant cells that is surrounded by a double membrane and contains internal chlorophyll-containing membranes (**thylakoids**) where the light-absorbing reactions of photosynthesis occur. (Figure 12-29)

cholesterol A lipid containing the four-ring steroid structure with a hydroxyl group on one ring; a component of many eukaryotic membranes and the precursor of steroid hormones, bile acids, and vitamin D. (Figure 10-5c)

chromatid One copy of a replicated chromosome, formed during the S phase of the cell cycle, that is joined at the **centromere** to the other copy; also called *sister chromatid.* During mitosis, the two chromatids separate, each becoming a chromosome of one of the two daughter cells. (Figure 6-40)

chromatin Complex of DNA, histones, and nonhistone proteins from which eukaryotic chromosomes are formed. Condensation of chromatin during mitosis yields the visible **metaphase** chromosomes. (Figures 6-28 and 6-30)

chromatography, liquid Group of biochemical techniques for separating mixtures of molecules (e.g., different proteins) based on their mass (*gel filtration* chromatography), charge (*ion-exchange* chromatography), or ability to bind specifically to other molecules (*affinity* chromatography). (Figure 3-37)

chromosome In eukaryotes, the structural unit of the genetic material consisting of a single, linear double-stranded DNA molecule and associated proteins. In most prokaryotes, a single, circular double-stranded DNA molecule constitutes the bulk of the genetic material. See also **chromatin** and **karyotype.**

cilium (pl. cilia) Short, membrane-enclosed structure extending from the surface of eukaryotic cells and containing a core bundle of **microtubules.** Cilia usually occur in groups and beat rhythmically to move a cell (e.g., single-celled organism) or to move small particles or fluid along a surface (e.g., trachea cells). See also **axoneme** and **flagellum.**

cisterna (pl. cisternae) Flattened membrane-bounded compartment, as found in the Golgi complex and endoplasmic reticulum.

citric acid cycle A set of nine coupled reactions occurring in the matrix of the **mitochondrion** in which acetyl groups are oxidized, generating CO_2 and reduced intermediates used to produce ATP; also called *Krebs cycle* and *tricarboxylic acid (TCA) cycle.* (Figure 12-10)

clathrin A fibrous protein that with the aid of assembly proteins polymerizes into a lattice-like network at specific regions on the cytosolic side of a membrane, thereby forming a clathrin-coated pit that buds off to form a vesicle. (Figure 14-18; Table 14-18)

cleavage In embryogenesis, the series of rapid cell divisions that occurs following fertilization and with little cell growth, producing progressively smaller cells; culminates in formation of the blastocyst in mammals or blastula in other animals. Also used as a synonym for the hydrolysis of molecules. (Figures 22-1 and 22-8)

cleavage/polyadenylation complex Large, multiprotein complex that catalyzes the cleavage of **pre-mRNA** at a 3′ poly(A) site and the initial addition of adenylate (A) residues to form the poly(A) tail. (Figure 8-15)

clone (1) A population of genetically identical cells, viruses, or organisms descended from a common ancestor. (2) Multiple identical copies of a gene or DNA fragment generated and maintained via **DNA cloning.**

cochlea Snail-shaped structure containing the **organ of Corti,** the sound-sensing part of the inner ear. (Figure 23-30)

codon Sequence of three nucleotides in DNA or mRNA that specifies a particular amino acid during protein synthesis; also called *triplet.* Of the 64 possible codons, three are stop codons, which do not specify amino acids and cause termination of synthesis. (Table 4-1)

coiled coil A protein **structural motif** marked by amphipathic α helical regions that can self-associate to form stable, rodlike structures in proteins; commonly found in fibrous proteins and certain transcription factors. (Figure 3-9a)

collagen A triple-helical glycoprotein rich in glycine and proline that is a major component of the **extracellular matrix** and connective tissues. The numerous subtypes differ in their tissue distribution and the extracellular components and cell-surface proteins with which they associate. (Figure 19-22; Table 19-4)

complement A group of constitutive serum proteins that bind directly to microbial or fungal surfaces, thereby activating a proteolytic cascade that culminates in formation of the cytolytic *membrane attack complex.* (Figure 24-4).

complementary (1) Referring to two nucleic acid sequences or strands that can form perfect **base pairs** with each other. (2) Describing regions on two interacting molecules (e.g., an enzyme and its substrate) that fit together in a lock-and-key fashion.

complementary DNA (cDNA) See **cDNA.**

complementation See **genetic complementation** and **functional complementation.**

concentration gradient In cell biology, a difference in the concentration of a substance in different regions of a cell or embryo or on different sides of a cellular membrane.

conformation The precise shape of a protein or other macromolecule in three dimensions resulting from the spatial location of the atoms in the molecule. (Figure 3-8)

connexins A family of transmembrane proteins that form **gap junctions** in vertebrates. (Figure 19-18)

constitutive Referring to the continuous production or activity of a cellular molecule or the continuous operation of a cellular process (e.g., constitutive secretion) that is not regulated by internal or external signals.

contractile bundles Bundles of **actin** and **myosin** in nonmuscle cells that function in cell adhesion (e.g., *stress fibers*) or cell movement (*contractile ring* in dividing cells).

COPI A class of proteins that coat transport vesicles in the **secretory pathway.** COPI-coated vesicles move proteins from the Golgi to the endoplasmic reticulum and from later to earlier Golgi cisternae. (Table 14-1)

COPII A class of proteins that coat transport vesicles in the **secretory pathway.** COPII-coated vesicles move proteins from the endoplasmic reticulum to the Golgi. (Table 14-1)

cotranslational translocation Simultaneous transport of a secretory protein into the endoplasmic reticulum as the nascent protein is still bound to the ribosome and being elongated. (Figure 13-6)

cotransport Protein-mediated movement of an ion or small molecule across a membrane against a concentration gradient driven by coupling to movement of a second molecule down its concentration gradient in the same (**symport**) or opposite (**antiport**) direction. (Figure 11-3, [3B, C]; Table 11-1)

covalent bond Stable chemical force that holds the atoms in molecules together by sharing of one or more pairs of electrons. See also **noncovalent interaction.** (Figures 2-2 and 2-6)

cross-exon recognition complex Large assembly including RNA-binding SR proteins and other components that helps delineate exons in the **pre-mRNAs** of higher eukaryotes and assure correct **RNA splicing.** (Figure 8-13)

crossing over Exchange of genetic material between maternal and paternal **chromatids** during **meiosis** to produce recombined chromosomes. See also **recombination.** (Figure 5-10)

cyclic AMP (cAMP) A **second messenger,** produced in response to hormonal stimulation of certain G protein–coupled receptors, that activates **protein kinase A.** (Figure 15-9; Table 15-2)

cyclic GMP (cGMP) A **second messenger** that opens cation channels in rod cells and activates protein kinase G in vascular smooth muscle and other cells. (Figures 15-9, 15-18, and 15-31)

cyclin Any of several related proteins whose concentrations rise and fall during the course of the eukaryotic cell cycle. Cyclins form complexes with **cyclin-dependent kinases,** thereby activating and determining the substrate specificity of these enzymes.

cyclin-dependent kinase (CDK) A protein kinase that is catalytically active only when bound to a cyclin. Various cyclin−CDK complexes trigger progression through different stages of the eukaryotic cell cycle by phosphorylating specific target proteins. (Figure 20-32)

cytochromes A group of colored, heme-containing proteins some of which function as **electron carriers** during cellular respiration and photosynthesis. (Figure 12-14a)

cytokine Any of numerous small, secreted proteins (e.g., erythropoietin, G-CSF, interferons, interleukins) that bind to cell-surface receptors on blood and immune-system cells to trigger their differentiation or proliferation.

cytokine receptor Member of major class of cell-surface signaling receptors including those for erythropoietin, growth hormone, interleukins, and interferons. Ligand binding leads to activation of cytosolic JAK kinases associated with the receptor, thereby initiating intracellular signaling pathways. (Figures 16-8 and 16-12)

cytokinesis The division of the cytoplasm following **mitosis** to generate two daughter cells, each with a nucleus and cytoplasmic organelles. (Figure 17-34)

cytoplasm Viscous contents of a cell that are contained within the plasma membrane but, in eukaryotic cells, outside the nucleus.

cytoskeleton Network of fibrous elements, consisting primarily of **microtubules, microfilaments,** and **intermediate filaments,** found in the cytoplasm of eukaryotic cells. The cytoskeleton provides organization and structural support for the cell and permits directed movement of organelles, chromosomes, and the cell itself. (Figures 17-1, 17-2, and 18-1)

cytosol Unstructured aqueous phase of the cytoplasm excluding organelles, membranes, and insoluble cytoskeletal components.

cytosolic face The face of a cell membrane directed toward the cytosol. (Figure 10-8)

DAG See **diacylglycerol.**

dalton Unit of molecular mass approximately equal to the mass of a hydrogen atom (1.66×10^{-24} g).

denaturation Drastic alteration in the **conformation** of a protein or nucleic acid due to disruption of various noncovalent interactions caused by heating or exposure to certain chemicals; usually results in loss of biological function.

dendrite Process extending from the cell body of a neuron that is relatively short and typically branched and receives signals from **axons** of other neurons. (Figure 23-2)

dendritic cells Phagocytic professional **antigen-presenting cells** that reside in various tissues and can detect broad patterns of pathogen markers via their **Toll-like receptors.** After internalizing antigen at a site of tissue injury or infection, they migrate to lymph nodes and initiate activation of **T cells.** (Figure 24-6)

deoxyribonucleic acid See **DNA.**

depolarization Decrease in the cytosolic-face negative electric potential that normally exists across the plasma membrane of a cell at rest, resulting in a less inside-negative or an inside-positive **membrane potential.**

determination In embryogenesis, a change in a cell that commits the cell to a particular developmental pathway (cell fate).

development Overall process involving growth, **differentiation,** and organization by which a fertilized egg gives rise to an adult plant or animal, including the formation, growth, polarization, and movements of individual cell types, tissues, and organs.

diacylglycerol (DAG) Membrane-bound **second messenger** that can be produced by cleavage of **phosphoinositides** in response to stimulation of certain cell-surface receptors. (Figures 15-9 and 15-29)

differential gene expression The expression of different sets of genes by cells with the same genotype, resulting in production of specific sets of proteins characteristic of a particular stage of development or particular differentiated cell type. (Figure 1-24)

differentiation Process usually involving changes in gene expression by which a precursor cell becomes a distinct specialized cell type.

diploid Referring to an organism or cell having two full sets of **homologous chromosomes** and hence two copies (**alleles**) of each gene or genetic locus. Somatic cells contain the diploid number of chromosomes ($2n$) characteristic of a species. See also **haploid.**

disaccharide A small carbohydrate (sugar) composed of two monosaccharides covalently joined by a **glycosidic bond.** (Figure 2-19)

dissociation constant (K_d) See **equilibrium constant.**

disulfide bond (—S—S—) A common covalent linkage between the sulfur atoms on two cysteine residues in different polypeptides or in different parts of the same polypeptide.

DNA (deoxyribonucleic acid) Long linear polymer, composed of four kinds of deoxyribose **nucleotides,** that is the carrier of genetic information. See also **double helix, DNA** (Figure 4-3)

DNA cloning Recombinant DNA technique in which specific cDNAs or fragments of genomic DNA are inserted into a cloning **vector,** which then is incorporated into cultured host cells and maintained during growth of the host cells; also called *gene cloning.* (Figure 5-14)

DNA library Collection of cloned DNA molecules consisting of fragments of the entire genome (*genomic library*) or of DNA copies of all the mRNAs produced by a cell type (*cDNA library*) inserted into a suitable cloning **vector.**

DNA ligase An enzyme that links together the 3′ end of one DNA fragment with the 5′ end of another, forming a continuous strand.

DNA microarray An ordered set of thousands of different nucleotide sequences arrayed on a microscope slide or other solid surface; can be used to determine patterns of gene expression in different cell types or in a particular cell type at different developmental stages or under different conditions. (Figures 5-29 and 5-30)

DNA polymerase An enzyme that copies one strand of DNA (the template strand) to make the complementary strand, forming a new double-stranded DNA molecule. All DNA polymerases add deoxyribonucleotides one at a time in the $5' \rightarrow 3'$ direction to the 3′ end of a short preexisting primer strand of DNA or RNA.

domain Distinct regions of a protein's three-dimensional structure. A *functional* domain exhibits a particular activity characteristic of the protein; a *structural* domain is ≈40 or more amino acids in length, arranged in a distinct secondary or tertiary structure; a *topological* domain has a distinctive spatial relationship to the rest of the protein.

dominant In genetics, referring to that allele of a gene expressed in the **phenotype** of a heterozygote; the nonexpressed allele is **recessive;** also refers to the phenotype associated with a dominant allele. Mutations that produce dominant alleles generally result in a gain of function. (Figure 5-2)

dominant negative In genetics, an allele that acts in a **dominant** manner but produces an effect similar to a loss of function; generally is an allele encoding a mutant protein that blocks the function of the normal protein by binding either to it or to a protein **upstream** or **downstream** of it in a pathway.

double helix, DNA The most common three-dimensional structure for cellular DNA in which the two polynucleotide strands are antiparallel and wound around each other with complementary bases hydrogen-bonded. (Figure 4-3)

downstream (1) For a gene, the direction RNA polymerase moves during transcription, which is toward the end of the template DNA strand with a 5′-hydroxyl group. Nucleotides downstream from the +1 position (the first transcribe a nucleotide) are designated +2, +3, etc. (2) Events that occur later in a cascade of steps (e.g., signaling pathway). See also **upstream.**

dyneins A class of **motor proteins** that use the energy released by ATP hydrolysis to move toward the (−) end of **microtubules.** Dyneins can transport vesicles and organelles, are responsible for the movement of cilia and flagella, and play a role in chromosome movement during mitosis. (Figures 18-24 and 18-25)

ectoderm Outermost of the three primary cell layers of the animal embryo; gives rise to epidermal tissues, the nervous system, and external sense organs. See also **endoderm** and **mesoderm.** (Figures 21-3 and 22-11)

EF hand A type of helix-loop-helix **structural motif** that occurs in many Ca^{2+}-binding proteins such as **calmodulin.** (Figure 3-9b)

electric potential The energy associated with the separation of positive and negative charges. An electric potential is maintained across the plasma membrane of nearly all cells.

electrochemical gradient The driving force that determines the energetically favorable direction of transport of an ion (or charged molecule) across a membrane. It represents the combined influence of the ion's **concentration gradient** across the membrane and the **membrane potential.**

electron carrier Any molecule or atom that accepts electrons from donor molecules and transfers them to acceptor molecules in coupled **oxidation** and **reduction** reactions. (Table 12-2)

electron transport Flow of electrons via a series of electron carriers from reduced electron donors (e.g., NADH) to O_2 in the inner mitochondrial membrane, or from H_2O to $NADP^+$ in the thylakoid membrane of plant chloroplasts. (Figures 12-18 and 12-30)

electron transport chain Set of four large multiprotein complexes in the inner mitochondrial membrane plus diffusible cytochrome *c* and coenzyme Q through which electrons flow from reduced electron donors (e.g., NADH) to O_2. Each member of the chain contains one or more bound **electron carriers.** (Figure 12-16)

electrophoresis Any of several techniques for separating macromolecules based on their migration in a gel or other medium subjected to a strong electric field. (Figure 3-35)

elongation factor (EF) One of a group of nonribosomal proteins required for continued **translation** of mRNA (protein synthesis) following initiation. (Figure 4-25)

embryogenesis Early development of an individual from a fertilized egg (zygote).

embryonic stem (ES) cells A line of cultured cells derived from very early embryos that can differentiate into a wide range of cell types either in vitro or after reinsertion into a host embryo. (Figure 21-7)

endergonic Referring to reactions and processes that have a positive ΔG and thus require an input of **free energy** in order to proceed; opposite of **exergonic.**

endocrine Referring to signaling mechanism in which target cells bind and respond to a **hormone** released into the blood by distant specialized secretory cells usually present in a gland (e.g., pituitary or thyroid gland).

endocytic pathway Cellular pathway involving **receptor-mediated endocytocis** that internalizes extracellular materials too large to be imported by membrane transport proteins and to remove receptor proteins from the cell surface as a way to down-regulate their activity. (Figure 14-29)

endocytosis General term for uptake of extracellular material by invagination of the plasma membrane; includes **receptor-mediated endocytosis, phagocytosis,** and pinocytosis.

endoderm Innermost of the three primary cell layers of the animal embryo; gives rise to the gut and most of the respiratory tract. See **ectoderm** and **mesoderm.** (Figures 21-3 and 22-11)

endoplasmic reticulum (ER) Network of interconnected membranous structures within the cytoplasm of eukaryotic cells contiguous with the outer nuclear envelope. The *rough ER,* which is associated with **ribosomes,** functions in the synthesis and processing of secreted and membrane proteins; the *smooth ER,* which lacks ribosomes, functions in lipid synthesis. (Figure 9-1)

endosome One of two types of membrane-bounded compartments: *early* endosomes (or endocytic vesicles), which bud off from the plasma membrane during receptor-mediated endocytosis, and *late* endosomes, which have an acidic internal pH and function in sorting of proteins to **lysosomes.** (Figures 14-1 and 14-29)

endosymbiont Bacterium that resides inside a eukaryotic cell in a mutually beneficial partnership. According to the endosymbiont hypothesis, both mitochondria and chloroplasts evolved from endosymbionts. (Figure 6-20)

endothermic Referring to reactions and processes that have a positive change in **enthalpy,** ΔH, and thus must absorb heat in order to proceed; opposite of **exothermic.**

enhancer A regulatory sequence in eukaryotic DNA that may be located at a great distance from the gene it controls or even within the coding sequence. Binding of specific proteins to an enhancer modulates the rate of transcription of the associated gene. (Figure 7-16)

enhancesome Large nucleoprotein complex that assembles from transcription factors (activators and repressors) as they bind cooperatively to their binding sites in an **enhancer** with the assistance of DNA-bending proteins. (Figure 7-30)

enthalpy (*H*) Heat; in a chemical reaction, the enthalpy of the reactants or products is equal to their total bond energies.

entropy (*S*) A measure of the degree of disorder or randomness in a system; the higher the entropy, the greater the disorder.

envelope See **nuclear envelope** or **viral envelope.**

enzyme A protein that catalyzes a particular chemical reaction involving a specific **substrate** or small number of related substrates.

Eph Cell-surface receptor for an **ephrin;** also called *Eph receptor.*

ephrins A family of membrane-bound signaling proteins that participate in cell–cell interactions involved in regulating the growth of axons so that they make the proper connections during development of the nervous system.

epidermal growth factor (EGF) One of a family of secreted signaling proteins (the *EGF family*) that is used in the development of most tissues in most or all animals. EGF signals are bound by **receptor tyrosine kinases.** Mutations in EGF signal-transduction components are implicated in human cancer, including brain cancer. See **HER family.**

epigenetic Referring to a process that affects the expression of specific genes and is inherited by daughter cells, but does not involve a change in DNA sequence.

epinephrine A catecholamine secreted by the adrenal gland and some neurons in response to stress; also called *adrenaline.* It functions as both a hormone and neurotransmitter, mediating "fight or flight" responses (e.g., increased blood glucose levels and heart rate).

epithelium (pl. epithelia) Sheetlike covering, composed of one or more layers of tightly adhering cells, on external and internal body surfaces. (Figure 19-8)

epitope The part of an antigen molecule that binds to an antigen-specific receptor on B or T cells or to antibody. Large protein antigens usually possess multiple epitopes that bind to antibodies of different specificity.

equilibrium constant (K) Ratio of forward and reverse rate constants for a reaction. For a binding reaction, $A + B \rightleftarrows AB$, the association constant (K_a) equals K, and the dissociation constant (K_d) equals $1/K$.

erythropoietin (Epo) A **cytokine** that triggers production of red blood cells by inducing the proliferation and differentiation of erythroid progenitor cells in the bone marrow. (Figures 16-6 and 21-5)

euchromatin Less condensed portions of **chromatin** present in interphase chromosomes; includes most transcriptionally active regions. See also **heterochromatin.** (Figure 6-33a)

eukaryotes Class of organisms, composed of one or more cells containing a membrane-enclosed nucleus and organelles, that constitutes one of the three distinct evolutionary lineages of modern-day organisms; also called *eukarya*. Includes all organisms except **viruses** and **prokaryotes.** (Figure 1-3)

excision-repair system, DNA One of several mechanisms for repairing DNA damage due to spontaneous depurination or deamination or exposure to **carcinogens.** These repair systems normally operate with a high degree of fidelity and their loss is associated with increased risk for certain cancers.

exergonic Referring to reactions and processes that have a negative ΔG and thus release **free energy** as they proceed; opposite of **endergonic.**

exocytosis Release of intracellular molecules (e.g., hormones, matrix proteins) contained within a membrane-bounded vesicle by fusion of the vesicle with the plasma membrane of a cell.

exon Segment of a eukaryotic gene (or of its **primary transcript**) that reaches the cytoplasm as part of a mature mRNA, rRNA, or tRNA molecule. See also **intron.**

exon shuffling Evolutionary process for creating new genes (i.e., new combinations of exons) from preexisting ones by recombination between introns of two separate genes or by transposition of mobile DNA elements. (Figures 6-18 and 6-19)

exoplasmic face The face of a cell membrane directed away from the cytosol. (Figure 10-8)

exosome Large exonuclease-containing complex that degrades spliced out introns and improperly processed pre-mRNAs in the nucleus or mRNAs with shortened poly(A) tails in the cytoplasm. (Figure 8-1)

exothermic Referring to reactions and processes that have a negative change in **enthalpy,** ΔH, and thus release heat as they proceed; opposite of **endothermic.**

exportin Protein that binds a "cargo" protein in the nucleus and with the aid of Ran (a member of the **GTPase superfamily**) transports the cargo through a nuclear pore complex to the cytoplasm. See also **importin.** (Figure 13-36)

expression vector A modified **plasmid** or virus that carries a gene or cDNA into a suitable host cell and there directs synthesis of the encoded protein; used to screen a DNA library for a gene of interest or to produce large amounts of a protein from its cloned gene (Figures 5-31 and 5-32)

extracellular matrix (ECM) A complex interdigitating meshwork of proteins and polysaccharides secreted by cells into the spaces between them. It provides structural support in tissues and can affect the development and biochemical functions of cells. (Table 19-1)

F_0F_1 complex See **ATP synthase.**

facilitated transport Protein-aided transport of an ion or small molecule across a cell membrane down its concentration gradient at a rate greater than that obtained by **simple diffusion;** also called *facilitated diffusion.* (Table 11-1)

FACS See **fluorescence-activated cell sorter.**

FAD (flavin adenine dinucleotide) A small organic molecule that functions as an electron carrier by accepting two electrons from a donor molecule and two H^+ from the solution. (Figure 2-33b)

fatty acid Any long hydrocarbon chain that has a carboxyl group at one end; a major source of energy during metabolism and a precursor for synthesis of phospholipids, triglycerides, and cholesteryl esters. (Figure 2-21; Table 2-4)

fibroblast A common type of connective tissue cell that secretes **collagen** and other components of the **extracellular matrix;** migrates and proliferates during wound healing and in tissue culture.

fibronectin An abundant **multiadhesive matrix protein** that occurs in numerous isoforms, generated by alternative splicing, in various cell types. Binds many other components of the extracellular matrix and to integrin adhesion receptors. (Figure 19-30)

FISH See **fluorescence in situ hybridization.**

flagellum (pl. flagella) Long locomotory structure (usually one per cell) extending from the surface of some eukaryotic cells (e.g., sperm), whose whiplike bending propels the cell through a fluid medium. Bacterial flagella are smaller and much simpler structures. See also **axoneme** and **cilium.** (Figure 18-30)

flavin adenine dinucleotide See **FAD.**

flippase Protein that facilitates the movement of membrane lipids from one leaflet to the other leaflet of a phospholipid bilayer. (Figure 11-15)

fluorescence-activated cell sorter (FACS) An instrument that can detect one or a few cells from thousands of other cells and sort them based on differences in their fluorescence. (Figure 9-28)

fluorescence in situ hybridization (FISH) Any of several related techniques for detecting specific DNA or RNA sequences in cells and tissues by treating samples with fluorescent **probes** that hybridize to the sequence of interest and observing the samples by fluorescence microscopy.

fluorescent staining General technique for visualizing cellular components by treating cells or tissues with a fluorescent dye–labeled agent (e.g., antibody) that binds specifically to a component of interest and observing the sample by fluorescence microscopy.

free energy (G) A measure of the potential energy of a system, which is a function of the **enthalpy** (H) and **entropy** (S).

free-energy change (ΔG) The difference in the total free energy of the product molecules and starting molecules (reactants) in a chemical reaction. A large negative value of ΔG indicates that a reaction (or other process) has a strong tendency to occur.

functional complementation Procedure for screening a DNA library to identify the wild-type gene that restores the function of a defective gene in a particular mutant. (Figure 5-18)

G_0, G_1, G_2 phase See **cell cycle.**

gamete Specialized **haploid** cell (in animals either a sperm or an egg) produced by **meiosis** of precursor **germ cells;** in sexual reproduction, union of a sperm and an egg initiates the development of a new individual.

gap genes In *Drosophila,* a group of genes that are activated in the early embryo by transcription factors produced from maternal mRNAs in the zygote; all encode transcription factors that function in early patterning along the anterioposterior axis. (Figure 22-27a, b)

gap junction Protein-lined channel connecting the cytoplasms of adjacent animal cells that allows passage of ions and small molecules between the cells. See also **plasmodesmata.** (Figure 19-18)

gastrulation Process in early animal embryos in which cells of the blastocyst invaginate, giving rise to the three **germ layers,** ectoderm, mesoderm, and endoderm. (Figure 22-11)

gene Physical and functional unit of heredity, which carries information from one generation to the next. In molecular terms, it is the entire DNA sequence—including **exons, introns,** and **transcription-control regions**—necessary for production of a functional polypeptide or RNA. See also **transcription unit.**

gene control All of the mechanisms involved in regulating **gene expression.** Most common is regulation of transcription, although mechanisms influencing the processing, stabilization, and translation of mRNAs help control expression of some genes.

gene expression Overall process by which the information encoded in a gene is converted into an observable **phenotype** (most commonly production of a protein).

gene family Set of genes that arose by duplication of a common ancestral gene and subsequent divergence due to small changes in the nucleotide sequence. (Figure 6-26)

genetic code The set of rules whereby nucleotide triplets (**codons**) in DNA or RNA specify amino acids in proteins. (Table 4-1)

genetic complementation Restoration of a wild-type function in diploid heterozygous cells generated from haploid cells, each of which carries a mutation in a different gene whose encoded protein is required for the same biochemical pathway. Complementation analysis can determine if recessive mutations in two mutants with the same mutant phenotype are in the same or different genes. (Figure 5-7)

genetic mapping Determination of the relative position of genes on a chromosome.

genetic markers Alleles associated with an easily detectable **phenotype** that are used experimentally to identify or select for a linked gene, a chromosome, a cell, or an individual. See also **molecular markers, DNA-based.**

genome Total genetic information carried by a cell or organism.

genomic imprinting Process that occurs during development of gametes involving chromatin modifications so that only certain genes can subsequently be expressed. Because different genes are imprinted in male and female gametes, the phenotypic expression of certain genes is determined by whether a particular allele is inherited from the female or male parent.

genomics Comparative analyses of the complete genomic sequences from different organisms and determination of global patterns of gene expression; used to assess evolutionary relations among species and to predict the number and general types of RNAs produced by an organism.

genotype Entire genetic constitution of an individual cell or organism, usually with emphasis on the particular alleles at one or more specific loci.

germ cell In sexually reproducing organisms, any cell that can potentially contribute to the formation of offspring including gametes and their immature precursors; also called *germ-line cell.* See also **somatic cell.**

germ layers Three primary cell layers—ectoderm, endoderm, and mesoderm—formed during gastrulation of animal embryos that give rise to distinct tissues and organs. (Figure 21-3)

germ line Lineage of germ cells, which give rise to **gametes** and thus participate in formation of the next generation of organisms; also, the genetic material transmitted from one generation to the next through the gametes.

germ-line cell See **germ cell.**

glia Supporting cells of nervous tissue that, unlike neurons, do not conduct electrical impulses; also called *glial cells.* Of the four types, *Schwann cells* and *oligodendrocytes* produce **myelin sheaths;** *astrocytes* function in **synapse** formation; and *microglia* make **trophic factors** and serve in immune responses. (Figure 23-14)

glucagon A peptide hormone produced in the cells of pancreatic islets that triggers the conversion of glycogen to glucose by the liver; acts with **insulin** to control blood glucose levels.

glucose Six-carbon monosaccharide (sugar) that is the primary metabolic fuel in most cells. The large glucose polymers, glycogen and starch, are used to store energy in animal cells and plant cells, respectively.

GLUT proteins A family of transmembrane proteins, containing 12 membrane-spanning α helices, that transport glucose (and a few other sugars) across cell membranes down its concentration gradient. (Figure 11-5)

glycogen A very long, branched polysaccharide, composed exclusively of glucose units, that is the primary storage carbohydrate in animals; found primarily in liver and muscle cells.

glycolipid Any lipid to which a short carbohydrate chain is covalently linked; commonly found in the plasma membrane.

glycolysis Metabolic pathway in which sugars are degraded anaerobically to lactate or pyruvate in the cytosol with the production of ATP. (Figure 12-3)

glycoprotein Any protein to which one or more oligosaccharide chains are covalently linked. Most secreted proteins and many membrane proteins are glycoproteins.

glycosaminoglycan (GAG) A long, linear, highly charged polymer of a repeating disaccharides in which many residues often are sulfated. GAGs are major components of the extracellular matrix, usually as components of **proteoglycans.** (Figure 19-26)

glycosidic bond The covalent linkage between two monosaccharide residues formed when a carbon atom in one sugar reacts with a hydroxyl group on a second sugar with the net release of a water molecule (dehydration). (Figure 2-13)

G protein, monomeric (small). See **GTPase superfamily.**

G protein, trimeric (large) Any of numerous heterotrimeric GTP-binding switch proteins that function in intracellular signaling pathways; usually activated by ligand binding to a coupled seven-spanning receptor on the cell surface. See also **GTPase superfamily.** (Table 15-1)

G protein–coupled receptor (GPCR) Member of a large class of cell-surface signaling receptors, including those for epinephrine, glucagon, and yeast mating factors. All GPCRs contain seven transmembrane α helices. Ligand binding leads to activation of a coupled trimeric G protein, thereby initiating intracellular signaling pathways. (Figures 15-10 and 15-13)

Golgi complex Stacks of flattened, interconnected membrane-bounded compartments (cisternae) in eukaryotic cells that function in processing and sorting of proteins and lipids destined for other cellular compartments or for secretion; also called *Golgi apparatus.* (Figure 9-6)

growth cone Bulbous enlargement, composed of cell membrane extensions, at the leading, growing end of an **axon;** functions as a moving sensory guidance structure. (Figures 23-37 and 23-38)

growth factor An extracellular polypeptide molecule that binds to a cell-surface receptor, triggering an intracellular signaling pathway generally leading to cell proliferation.

GTPase superfamily Group of intracellular switch proteins that cycle between an inactive state with bound GDP and an active state with bound GTP. Includes the G_α subunit of trimeric (large) G proteins, monomeric (small) G proteins (e.g., **Ras,** Rab, Ran, and Rac), and certain **elongation factors** used in protein synthesis. See also **G protein, trimeric (large).** (Figure 3-32)

haploid Referring to an organism or cell having only one member of each pair of **homologous chromosomes** and hence only one copy (**allele**) of each gene or genetic locus. Gametes and bacterial cells are haploid. See also **diploid.**

Hedgehog (Hh) A family of secreted signaling proteins that are important regulators of the development of most tissues and organs in diverse animal species. Mutations in Hh signal-transduction components are implicated in human cancer and birth defects. The receptor is the Patched transmembrane protein. (Figures 16-33 and 16-34)

helicase (1) Any enzyme that moves along a DNA duplex using the energy released by ATP hydrolysis to separate (unwind) the two strands; required for DNA replication. (2) Activity of certain initiation factors that can unwind the secondary structures in mRNA during initiation of translation.

helix-loop-helix, basic (bHLH) A conserved DNA-binding **structural motif,** consisting of two α helices connected by a short loop, that is found in many dimeric eukaryotic transcription factors. (Figure 7-26b)

HER family Group of receptors, belonging to the **receptor tyrosine kinase (RTK)** class, that bind to members of the epidermal growth factor (EGF) family of signaling molecules in humans. Overexpression of HER2 protein is associated with some breast cancers. (Figure 16-18)

heterochromatin Regions of **chromatin** that remain highly condensed and transcriptionally inactive during interphase. (Figure 6-33a)

heterozygous Referring to a diploid cell or organism having two different **alleles** of a particular gene.

hexose A six-carbon **monosaccharide.**

high-energy bond Covalent bond that releases a large amount of energy when hydrolyzed under the usual intracellular conditions. Examples include the phosphoanhydride bonds in ATP, thioester bond in acetyl CoA, and various phosphate ester bonds.

histone One of several small, highly conserved basic proteins, found in the **chromatin** of all eukaryotic cells, that associate with DNA in the **nucleosome.** (Figure 6-29)

homeodomain Conserved DNA-binding **structural motif** (a helix-turn-helix) found in many developmentally important transcription factors.

homeosis Transformation of one body part into another arising from mutation in or misexpression of certain developmentally critical genes. (Figure 22-31)

homologous chromosome One of the two copies of each morphologic type of chromosome present in a **diploid** cell; also called **homolog.** Each homolog is derived from a different parent.

homologous recombination See **recombination.**

homologs Maternal and paternal copies of each morphologic type of chromosome present in a diploid cell; also called *homologues.*

homology Similarity in characteristics (e.g., protein and nucleic acid sequences or the structure of an organ) that reflects a common evolutionary origin. Proteins or genes that exhibit homology are said to be homologous and sometimes are called homologs. In contrast, *analogy* is a similarity in structure or function that does not reflect a common evolutionary origin.

homozygous Referring to a diploid cell or organism having two identical **alleles** of a particular gene.

hormone Generally, any extracellular substance that induces specific responses in target cells; specifically, those signaling molecules that circulate in the blood and mediate **endocrine** signaling.

Hox genes Group of developmentally important genes that encode homeodomain-containing transcription factors and help determine the body plan in animals. Mutations in Hox genes often cause **homeosis.** (Figure 22-32).

hyaluronan A large, highly hydrated **glycosaminoglycan (GAG)** that is a major component of the extracellular matrix; also called *hyaluronic acid* and *hyaluronate.* It imparts stiffness and resilience as well as a lubricating quality to many types of connective tissue. (Figure 19-26a)

hybridization, nucleic acid Association of two **complementary** nucleic acid strands to form double-stranded molecules, which can contain two DNA strands, two RNA strands, or one DNA and one RNA strand. Used experimentally in various ways to detect specific DNA or RNA sequences.

hybridoma A **clone** of hybrid cells that are immortal and produce **monoclonal antibody;** formed by fusion of a normal antibody-producing B cell with a myeloma cell. (Figure 9-35)

hydrocarbon Any compound containing only carbon and hydrogen atoms.

hydrogen bond A **noncovalent interaction** between an atom (commonly oxygen or nitrogen) carrying a partial negative charge and a hydrogen atom carrying a partial positive charge. Important in stabilizing the conformation of proteins and in formation of **base pairs** between nucleic acid strands. (Figure 2-8)

hydrophilic Interacting effectively with water. See also **polar.**

hydrophobic Not interacting effectively with water; in general, poorly soluble or insoluble in water. See also **nonpolar.**

hydrophobic effect The tendency of nonpolar molecules or parts of molecules to associate with each other in aqueous solution so as to minimize their direct interactions with water; commonly called a *hydrophobic interaction* or *bond.* (Figure 2-11)

hyperpolarization Increase in the magnitude of the cytosolic-face negative electric potential that normally exists across the plasma membrane of a cell at rest, resulting in a more negative **membrane potential.**

hypertonic Referring to an external solution whose solute concentration is high enough to cause water to move out of cells due to **osmosis.**

hypotonic Referring to an external solution whose solute concentration is low enough to cause water to move into cells due to **osmosis.**

IgCAMs A family of **cell-adhesion molecules** that contain multiple immunoglobulin (Ig) domains and mediate Ca^{2+}-independent cell–cell interactions. IgCAMs are produced in a variety of tissues and are components of **tight junctions.** (Figure 19-2)

immunity State of being resistant (immune) against the harmful effects of exposure to pathogens either in the form of *innate* responses, which develop within minutes to hours but are relatively nonspecific, or *adaptive* responses, which take several days to develop fully but are highly specific. (Figure 24-1)

immunoglobulin (Ig) Any of the serum proteins, produced by fully differentiated **B cells,** that can function as antibodies; also occur in membrane-bound form as part of the **B-cell receptor.** Immunoglobulins are divided into five main classes (*isotypes*) that exhibit distinct functional properties. See also **antibody.** (Figures 24-8 and 24-9)

immunoglobulin (Ig) fold Evolutionarily ancient structural motif found in antibodies, the T-cell receptor, and numerous other eukaryotic proteins not directly involved in antigen-specific recognition; also called *Ig domain.* (Figure 24-12b)

importin Protein that binds a "cargo" protein in the cytoplasm and transports the cargo through a nuclear pore complex into the nucleus. Return of importin to the cytoplasm requires the aid of Ran (a member of the **GTPase superfamily**) See also **exportin.** Figure 13-35)

induction (1) In embryogenesis, a change in the developmental fate of one cell or tissue caused by signals from another cell or tissue or by direct contact. (2) In metabolism, an increase in the synthesis of an enzyme or series of enzymes mediated by a specific molecule (inducer).

inflammation Localized response to injury or infection that leads to the activation of immune-system cells and their recruitment to the affected site; marked by the four classical signs of redness, swelling, heat, and pain. (Figure 24-6)

initiation factor (IF) One of a group of nonribosomal proteins that promote the proper association of ribosomes and mRNA and are required for initiation of **translation** (protein synthesis). (Figure 4-24)

inositol 1,4,5-trisphosphate (IP_3) Intracellular **second messenger** produced by cleavage of the membrane lipid phosphatidylinositol 4,5-bisphosphate in response to stimulation of certain cell-surface receptors. IP_3, which triggers release of Ca^{2+} stored in the endoplasmic reticulum, is one of several biologically active **phosphoinositides.** (Figure 15-9; Table 15-3)

in situ hybridization Any technique for detecting specific DNA or RNA sequences in cells and tissues by treating samples with single-stranded RNA or DNA **probes** that hybridize to the sequence of interest. (Figure 5-28)

insulator DNA sequence that prevents transcriptional **enhancers** on one side of the insulator from influencing transcription of a gene on the other side of the insulator, thereby preventing inappropriate interactions between the control elements of neighboring genes.

insulin A protein hormone produced in the β cells of the pancreatic islets that stimulates uptake of glucose into muscle and fat cells; acts with **glucagon** to help regulate blood glucose levels. Insulin also functions as a growth factor for many cells.

integral membrane protein Any protein that contains one or more hydrophobic segments embedded within the core of the **phospholipid bilayer;** also called *transmembrane protein.* (Figure 13-10)

integrins A large family of heterodimeric transmembrane proteins that function as adhesion receptors, promoting cell–matrix adhesion, or as cell-adhesion molecules, promoting cell–cell adhesion. (Table 19-3)

interferons (IFNs) Small group of cytokines that bind to cell-surface receptors on target cells inducing changes in gene expression that lead to an antiviral state or other cellular responses important in immune responses.

interleukins (ILs) Large group of cytokines, some released in response to inflammation, that promote proliferation and functioning of T cells and antibody-producing B cells of the immune system.

intermediate filament Cytoskeletal fiber (10 nm in diameter) formed by polymerization of related, but tissue-specific, subunit proteins including **keratins, lamins,** and **neurofilaments.** (Figure 18-45; Table 18-1)

interphase Long period of the cell cycle, including the G_1, S, and G_2 phases, between one M (mitotic) phase and the next. (Figures 1-17 and 20-1)

intron Part of a **primary transcript** (or the DNA encoding it) that is removed by splicing during RNA processing and is not included in the mature, functional mRNA, rRNA, or tRNA.

in vitro Denoting a reaction or process taking place in an isolated cell-free extract; sometimes used to distinguish cells growing in culture from those in an organism.

in vivo Denoting a reaction or process that occurs in an intact cell or organism.

ionic interaction A **noncovalent interaction** between a positively charged ion (cation) and negatively charged ion (anion); commonly called *ionic bond.*

IP_3 See **inositol 1,4,5-trisphosphate.**

isoelectric point (pI) The pH of a solution at which a dissolved protein or other potentially charged molecule has a net charge of zero and therefore does not move in an electric field. (Figure 3-36)

isoform One of several forms of the same protein whose amino acid sequences differ slightly and whose general activities are similar. Isoforms may be encoded by different genes or by a single gene whose primary transcript undergoes **alternative splicing.**

isotonic Referring to a solution whose solute concentration is such that it causes no net movement of water in or out of cells.

karyopherin One of a family of nuclear transport proteins that functions as an **importin, exportin,** or occasionally both. Each karyopherin binds to a specific signal sequence in cargo proteins moving in or out of the nucleus.

karyotype Number, sizes, and shapes of the entire set of **metaphase** chromosomes of a eukaryotic cell. (Chapter 6 opening figure)

keratins A group of **intermediate filament** proteins found in epithelial cells that assemble into heteropolymeric filaments. (Figure 18-46)

kilocalorie (kcal) See **calorie.**

kinase An enzyme that transfers the terminal (γ) phosphate group from ATP to a substrate. Protein kinases, which phosphorylate specific serine, threonine, or tyrosine residues, play a critical role in regulating the activity of many cellular proteins. See also **phosphatases.** (Figure 3-33)

kinesins A class of **motor proteins** that use energy released by ATP hydrolysis to move toward the (+) end of a **microtubule.** Kinesins can transport vesicles and organelles and play a role in chromosome movement during mitosis. (Figures 18-19 through 18-21).

kinetic energy Energy of movement, such as the motion of molecules.

kinetochore A multilayer protein structure located at or near the **centromere** of each mitotic chromosome from which microtubules extend toward the spindle poles of the cell; plays an active role in movement of chromosomes toward the poles during anaphase. (Figure 18-39)

K_m A parameter that describes the affinity of an enzyme for its substrate and equals the substrate concentration that yields the half-maximal reaction rate; also called the *Michaelis constant.* A similar parameter describes the affinity of a transport protein for the transported molecule or the affinity of a receptor for its ligand. (Figure 3-22)

knockdown, siRNA Technique for experimentally inhibiting translation of a specific mRNA by use of **siRNA;** useful for reducing the activity of a protein, particularly in organisms that are not amenable to classical genetic methods for isolating loss-of-function mutants.

knockout, gene Selective inactivation of a specific gene by replacing it with a nonfunctional (disrupted) allele in an otherwise normal organism.

lagging strand One of the two daughter DNA strands formed at a **replication fork** as short, discontinuous segments (Okazaki fragments), which are synthesized in the $5' \rightarrow 3'$ direction and later joined. See also **leading strand.** (Figure 4-30)

laminin Large heterotrimeric **multiadhesive matrix protein** that is found in all **basal lamina.** (Figure 19-21)

lamins A group of **intermediate filament** proteins that form a fibrous network, the **nuclear lamina,** on the inner surface of the nuclear envelope.

lateral See **basolateral.**

lateral inhibition Important signal-mediated developmental process that results in adjacent equivalent or near-equivalent cells assuming different fates. (Figure 22-41)

leading strand One of the two daughter DNA strands formed at a **replication fork** by continuous synthesis in the $5' \rightarrow 3'$ direction. The direction of leading-strand synthesis is the same as movement of the replication fork. See also **lagging strand.** (Figure 4-30)

lectin Any protein that binds tightly to specific sugars. Lectins assist in the proper folding of some glycoproteins in the endoplasmic reticulum, and can be used in affinity chromatography to purify glycoproteins or as reagents to detect them in situ.

leucine zipper A type of coiled-coil **structural motif** composed of two α helices that form specific homo- or heterodimers; common motif in many eukaryotic transcription factors. See **coiled coil.** (Figures 7-26a and 3-9)

ligand Any molecule, other than an enzyme **substrate,** that binds tightly and specifically to a macromolecule, usually a protein, forming a macromolecule–ligand complex.

LINEs (long interspersed elements) Class of **retrotransposons,** $\approx$6 kb long, that are particularly abundant in mammals and constitute about 21 percent of total human DNA. (Figure 6-16).

linkage In genetics, the tendency of two different loci on the same chromosome to be inherited together. The closer two loci are, the lower the frequency of **recombination** between them and the greater their linkage.

lipid Any organic molecule that is poorly soluble or virtually insoluble in water but is soluble in nonpolar organic solvents. Major classes include **fatty acids, phospholipids, steroids,** and **triglycerides.**

lipid-anchored membrane protein Any protein that is tethered to a cellular membrane by one or more covalently attached lipid groups, which are embedded in the phospholipid bilayer. (Figure 10-19)

lipid raft Microdomain in the plasma membrane that is enriched in cholesterol, sphingomyelin, and certain proteins.

lipoprotein Any large, water-soluble protein and lipid complex that functions in mass transfer of lipids throughout the body. See also **low-density lipoprotein (LDL).**

liposome Artificial spherical **phospholipid bilayer** structure with an aqueous interior that forms in vitro from phospholipids and may contain membrane proteins. (Figure 10-6c)

locus (pl. **loci**) In genetics, the specific site of a gene on a chromosome. All the **alleles** of a particular gene occupy the same locus.

long terminal repeats (LTRs) Direct repeat sequences, containing up to 600 base pairs, that flank the coding region of integrated retroviral DNA and viral **retrotransposons.**

low-density lipoprotein (LDL) A class of **lipoprotein,** containing apolipoprotein B-100, that is a primary transporter of cholesterol in the form of cholesteryl esters between tissues, especially to the liver. (Figure 14-27)

lumen The space within a tubular structure (e.g., a vessel or the intestine) or the interior volume of a membrane-bounded compartment within a cell.

lymphocytes Two classes of white blood cells that can recognize foreign molecules (**antigens**) and mediate immune responses. B lymphocytes (B cells) are responsible for production of antibodies; T lymphocytes (T cells) are responsible for destroying virus- and bacteria-infected cells, foreign cells, and cancer cells.

lysis Destruction of a cell by rupture of the plasma membrane and release of the contents.

lysogeny Phenomenon in which the DNA of a bacterial virus (bacteriophage) is incorporated into the host-cell genome and replicated along with the bacterial DNA but is not expressed. Subsequent activation leads to formation of new viral particles, eventually causing lysis of the cell.

lysosome Small organelle that has an internal pH of 4–5, contains hydrolytic enzymes, and functions in degradation of materials internalized by endocytosis and of cellular components in autophagy. (Figure 9-2)

M (mitotic) phase See **cell cycle.**

macromolecule Any large, usually polymeric molecule (e.g., a protein, nucleic acid, polysaccharide) with a molecular mass greater than a few thousand daltons.

macrophages Phagocytic leukocytes that can detect broad patterns of pathogen markers via **Toll-like receptors.** They function as professional **antigen-presenting cells** and are a major source of **cytokines.**

major histocompatibility complex (MHC) Set of adjacent genes that encode class I and class II **MHC molecules** and other proteins required for antigen presentation, as well as some complement proteins; called the *H-2 complex* in mice and the *HLA complex* in humans. (Figure 24-21)

malignant Referring to a tumor or tumor cells that can invade surrounding normal tissue and/or undergo **metastasis.** See also **benign.**

MAP kinase Any of a family of protein kinases that are activated in response to cell stimulation by many different growth factors and that mediate cellular responses by phosphorylating specific transcription factors and other target proteins. (Figures 16-26 and 16-27)

maximal velocity See V_{max}.

mechanosensor Any of several types of sensory structures that are embedded in various tissues and respond to touch, the positions and movements of the limbs and head, pain, and temperature.

mediator A very large multiprotein complex that forms a molecular bridge between transcriptional activators bound to an **enhancer** and to RNA polymerase II bound at a **promoter;** functions as a co-activator in stimulating transcription. (Figures 7-41 and 7-42)

meiosis In eukaryotes, a special type of cell division that occurs during maturation of germ cells; comprises two successive nuclear and cellular divisions with only one round of DNA replication. Results in production of four genetically nonequivalent haploid cells (**gametes**) from an initial diploid cell. (Figure 5-3)

membrane potential Electric potential difference, expressed in volts, across a membrane due to the slight excess of positive ions (cations) on one side and negative ions (anions) on the other. (Figures 11-17 and 11-18)

membrane transport protein Collective term for any integral membrane protein that mediates movement of one or more specific ions or small molecules across a cellular membrane regardless of the transport mechanism. (Figure 11-3)

meristem Organized group of undifferentiated, dividing cells that are maintained at the tips of growing shoots and roots in plants. All the adult structures arise from meristems.

mesenchyme Immature embryonic connective tissue, composed of loosely organized and loosely attached cells, derived from either the **mesoderm** or **ectoderm** in animals.

mesoderm The middle of the three primary cell layers of the animal embryo, lying between the **ectoderm** and **endoderm;** gives rise to the notochord, connective tissue, muscle, blood, and other tissues. (Figures 21-3 and 22-11)

messenger RNA See **mRNA.**

metaphase Stage of mitosis at which condensed chromosomes are aligned equidistant between the poles of the mitotic spindle, but have not yet started to segregate toward the spindle poles. (Figure 18-34)

metastasis Spread of cancer cells from their site of origin and establishment of areas of secondary growth.

MHC See **major histocompatibility complex.**

MHC molecules Glycoproteins that display peptides, derived from foreign (and self) proteins, on the surface of cells and are required for antigen presentation to **T cells.** *Class I* molecules are expressed constitutively by nearly all nucleated cells; *class II* molecules, by professional antigen-presenting cells. (Figures 24-23 and 24-24)

micelle A water-soluble spherical aggregate of phospholipids or other amphipathic molecules that form spontaneously in aqueous solution. (Figure 10-6c)

Michaelis constant See K_m.

microfilament Cytoskeletal fiber (≈7 nm in diameter) that is formed by polymerization of monomeric globular (G) **actin;** also called *actin filament.* Microfilaments play an important role in muscle contraction, cytokinesis, cell movement, and other cellular functions and structures. (Figure 17-4)

micro RNA See **miRNA.**

microtubule Cytoskeletal fiber (≈25 nm in diameter) that is formed by polymerization of α, β-**tubulin** monomers and exhibits structural and functional polarity. Microtubules are important components of cilia, flagella, the mitotic spindle, and other cellular structures. (Figures 18-2 and 18-3)

microtubule-associated protein (MAP) Any protein that binds to microtubules and regulates their stability. (Figures 18-14 and 18-15)

microtubule-organizing center See **MTOC.**

microvillus (pl. **microvilli**) Small, membrane-covered projection on the surface of an animal cell containing a core of actin filaments. Numerous microvilli are present on the absorptive surface of intestinal epithelial cells, increasing the surface area for transport of nutrients. (Figures 17-4 and 19-9)

miRNA (micro RNA) Any of numerous small, endogenous cellular RNAs, 20–30 nucleotides long, that are processed from double-stranded regions of hairpin secondary structures in long precursor RNAs. A single strand of the mature miRNA associates with several proteins to form an RNA-induced silencing complex (**RISC**) that inhibits translation of a target mRNA to which the miRNA hybridizes imperfectly. Several miRNAs must hybridize to a single mRNA to inhibit its translation. See also **siRNA.** (Figures 8-25a and 8-26)

mitochondrion (pl. **mitochondria**) Large organelle that is surrounded by two phospholipid bilayer membranes, contains DNA, and carries out **oxidative phosphorylation,** thereby producing most of the ATP in eukaryotic cells. (Figures 9-8 and 12-6)

mitogen Any extracellular molecule, such as a growth factor, that promotes cell proliferation.

mitosis In eukaryotic cells, the process whereby the nucleus divides, producing two genetically equivalent daughter nuclei with the diploid number of chromosomes. See also **cytokinesis** and **meiosis.** (Figure 18-34)

mitosis-promoting factor. See **MPF.**

mitotic spindle A specialized temporary structure, present in eukaryotic cells during mitosis, that captures the chromosomes and then pushes and pulls them to opposite sides of the dividing cell; also called *mitotic apparatus.* (Figure 18-36)

mobile DNA element See **transposable DNA element.**

molecular chaperone See **chaperone.**

molecular complementarity Lock-and-key kind of fit between the shapes, charges, hydrophobicity, and/or other physical properties of two molecules or portions thereof that allow formation of multiple **noncovalent interactions** between them at close range. (Figure 2-12)

molecular markers, DNA-based DNA sequences that vary among individuals (*DNA polymorphisms*) of the same species and are useful in genetic linkage studies; includes **RFLPs.**

monoclonal antibody Antibody produced by the progeny of a single B cell and thus a homogeneous protein that recognizes a single antigen (epitope). It can be produced experimentally by use of a **hybridoma.** (Figure 9-35)

monomer Any small molecule that can be linked chemically with others of the same type to form a **polymer.** Examples include amino acids, nucleotides, and monosaccharides.

monosaccharide Any simple sugar with the formula $(CH_2O)_n$ where $n = 3–7$.

morphogen A signaling molecule that specifies different cell fates during development as a function of its concentration. (Figure 22-13b)

motif, structural In proteins, combination of secondary and tertiary structure that often is generated by a distinctive primary amino acid sequence; also called *structural fold.* A structural motif usually is indicative of a particular three-dimensional architecture and often is associated with a specific functional property.

motor protein Any member of a special class of mechanochemical enzymes that use energy from ATP hydrolysis to generate either linear or rotary motion; also called *molecular motor.* See also **dyneins, kinesins,** and **myosins.**

MPF (mitosis-promoting factor) A heterodimeric protein, composed of a mitotic **cyclin** and **cyclin-dependent kinase (CDK),** that triggers entrance of a cell into mitosis by phosphorylating multiple specific proteins.

mRNA (messenger RNA) Any RNA that specifies the order of amino acids in a protein (i.e., the primary structure). It is produced by **transcription** of DNA by RNA polymerase. In eukaryotes, the initial RNA product (primary transcript) undergoes processing to yield functional mRNA. See also **translation.** (Figure 4-15)

mRNP-exporter A heterodimeric protein that binds to mRNA-containing ribonucleoprotein particles (mRNPs) and directs their export from the nucleus to cytoplasm by interacting transiently with **nucleoporins** in the nuclear pore complex. (Figure 8-22)

MTOC (microtubule-organizing center) General term for any structure (e.g., centrosome, spindle pole, basal body) that organizes microtubules in cells. (Figure 18-5)

multiadhesive matrix proteins Group of long flexible proteins that bind to other components of the **extracellular matrix** and to cell-surface receptors, thereby crosslinking matrix components to the cell membrane. Examples include **laminin,** a major component of the basal lamina, and **fibronectin,** present in many tissues.

multimeric For proteins, containing several polypeptide chains (or subunits).

mutagen A chemical or physical agent that induces mutations.

mutation In genetics, a permanent, heritable change in the nucleotide sequence of a chromosome, usually in a single gene; commonly causes an alteration in the function of the gene product.

myelin sheath Stacked specialized cell membrane that forms an insulating layer around vertebrate **axons** and increases the speed of impulse conduction. (Figure 23-15)

myofibril Long, slender structures within cytoplasm of muscle cells consisting of a regular repeating array of **sarcomeres** composed of thick (**myosin**) filaments and thin (**actin**) filaments. (Figure 17-29)

myosins A class of **motor proteins** that have actin-stimulated ATPase activity. Myosins move along actin **microfilaments** during muscle contraction and cytokinesis and also mediate vesicle translocation. (Figure 17-20)

NAD^+ (nicotinamide adenine dinucleotide) A small organic molecule that functions as an electron carrier by accepting two electrons from a donor molecule and one H^+ from the solution. (Figure 2-33a)

$NADP^+$ (nicotinamide adenine dinucleotide phosphate) Phosphorylated form of **NAD^+** that is used extensively as an electron carrier in biosynthetic pathways and during photosynthesis.

Na^+/K^+ ATPase A P-class ATP-powered pump that couples hydrolysis of one ATP molecule to export of Na^+ ions and import of K^+ ions; is largely responsible for maintaining the normal intracellular concentrations of Na^+ (low) and K^+ (high) in animal cells; commonly called *Na^+/K^+ pump.* (Figure 11-12)

natural killer (NK) cells Components of the innate immune system that nonspecifically detect and kill virus-infected cells and tumor cells. (Figure 24-5)

necrosis Cell death resulting from tissue damage or other pathology; usually marked by swelling and bursting of cells with release of their contents. Contrast with **apoptosis.**

neurofilaments (NFs) A group of **intermediate filament** proteins, found only in neurons, that contribute to axonal structure and rate of transmission of action potentials down axons. (Figure 18-2b)

neuron (nerve cell) Any of the impulse-conducting cells of the nervous system. A typical neuron contains a cell body; multiple short, branched processes (**dendrites**); and one long process (**axon**). (Figures 23-1 and 23-2)

neurotransmitter Extracellular signaling molecule that is released by the presynaptic neuron at a chemical **synapse** and relays the signal to the postsynaptic cell. The response elicited by a neurotransmitter, either excitatory or inhibitory, is determined by its receptor on the postsynaptic cell. (Figures 23-19 and 23-20)

neurotrophins Family of structurally and functionally related **trophic factors** that bind to receptors called Trks and are required for survival of neurons; include nerve growth factor (NGF) and brain-derived neurotrophic factor (BDNF).

neurulation Formation of the neural tube by infolding of the neural plate, the portion of the ectoderm that develops into neural structures, in vertebrate embryos. (Figures 22-38 and 22-39)

neutrophils Phagocytic leukocytes that are attracted to sites of tissue damage and migrate into the tissue. Once activated, neutrophils secrete various chemokines, cytokines, bacteria-destroying enzymes (e.g., lysozyme), and other products that contribute to **inflammation** and help clear invading pathogens.

nicotinamide adenine dinucleotide See **NAD^+.**

nicotinamide adenine dinucleotide phosphate See **$NADP^+$.**

***N*-linked oligosaccharide** A branched oligosaccharide chain attached to the side-chain amino group of an asparagine residue in a glycoprotein. See also **O-linked oligosaccharide.**

nociceptor Mechanosensor that responds to pain associated with injury to body tissues caused by mechanical trauma, heat, electricity, or toxic chemicals.

noncovalent interaction Any relatively weak chemical interaction that does not involve an intimate sharing of electrons. (Figures 2-6 and 2-12)

nonpolar Referring to a molecule or structure that lacks any net electric charge or asymmetric distribution of positive and negative charges. Nonpolar molecules generally are less soluble in water than polar molecules and are often water insoluble.

Northern blotting Technique for detecting specific RNAs separated by electrophoresis by hybridization to a labeled DNA **probe.** See also **Southern blotting.** (Figure 5-27)

nuclear body Roughly spherical, functionally specialized region in the nucleus, containing specific proteins and RNAs; many function in the assembly of ribonucleoprotein (RNP) complexes. The most prominent type is the **nucleolus.**

nuclear envelope Double-membrane structure surrounding the nucleus; the outer membrane is continuous with the endoplasmic reticulum and the two membranes are perforated by **nuclear pore complexes.** (Figure 9-1)

nuclear lamina Fibrous network on the inner surface of the nuclear envelope composed of lamin intermediate filaments. (Figure 20-16)

nuclear pore complex (NPC) Large, multiprotein structure, composed largely of nucleoporins, that extends across the nuclear envelope. Ions and small molecules freely diffuse through NPCs; large proteins and ribonucleoprotein particles are selectively transported through NPCs with the aid of soluble proteins. (Figure 13-32)

nuclear receptor Member of a class of intracellular receptors that bind lipid-soluble molecules (e.g., steroid hormones), forming ligand–receptor complexes that activate transcription; also called *steroid receptor superfamily.* (Figure 7-50)

nucleic acid A polymer of **nucleotides** linked by **phosphodiester bonds.** DNA and RNA are the primary nucleic acids in cells.

nucleocapsid A viral **capsid** plus the enclosed nucleic acid.

nucleolus Large structure in the nucleus of eukaryotic cells where rRNA synthesis and processing occurs and ribosome subunits are assembled. (Figure 6-33a)

nucleoporins Large group of proteins that make up the **nuclear pore complex.** One class (FG-nucleoporins) participates in nuclear import and export.

nucleoside A small molecule composed of a **purine** or **pyrimidine** base linked to a pentose (either ribose or deoxyribose). (Table 2-3)

nucleosome Structural unit of **chromatin** consisting of a disk-shaped core of **histone** proteins around which a 147-bp segment of DNA is wrapped. (Figure 6-29)

nucleotide A **nucleoside** with one or more phosphate groups linked via an ester bond to the sugar moiety, generally to the 5′ carbon atom. DNA and RNA are polymers of nucleotides containing deoxyribose and ribose, respectively. (Figure 2-16 and Table 2-3)

nucleus Large membrane-bounded organelle in eukaryotic cells that contains DNA organized into chromosomes; synthesis and processing of RNA and ribosome assembly occur in the nucleus.

Okazaki fragments Short (<1000 bases), single-stranded DNA fragments that are formed during synthesis of the **lagging strand** in DNA replication and are rapidly joined by DNA ligase to form a continuous DNA strand. (Figure 4-30)

oligopeptide A small to medium-sized linear polymer composed of amino acids connected by peptide bonds. The terms *peptide* and *oligopeptide* are often used interchangeably.

O-linked oligosaccharide Oligosaccharide chain that is attached to the side-chain hydroxyl group in a serine or threonine residue in a glycoprotein. See also **N-linked oligosaccharides.**

oncogene A gene whose product is involved either in transforming cells in culture or in inducing cancer in animals. Generally is a mutant form of a normal gene (**proto-oncogene**) for a protein involved in the control of cell growth or division. (Figure 25-11)

oncoprotein A protein encoded by an **oncogene** that causes abnormal cell proliferation; may be a mutant unregulated form of a normal protein, or a normal protein that is produced in excess or in the wrong time or place in an organism.

open reading frame (ORF) Region of sequenced DNA that is not interrupted by stop codons in one of the triplet reading frames. An ORF that begins with a start codon and extends for at least 100 codons has a high probability of encoding a protein.

operator Short DNA sequence in a bacterial or bacteriophage genome that binds a repressor protein and controls transcription of an adjacent gene. (Figure 7-2)

operon In bacterial DNA, a cluster of contiguous genes transcribed from one **promoter** that gives rise to an mRNA containing coding sequences for multiple proteins. (Figure 4-13a)

organelle Any membrane-limited subcellular structure found in eukaryotic cells. (Figures 1-2b and 9-1)

organ of Corti Acoustic sensory structure housed within the cochlea of the inner ear and composed of hair cells that transduce sound-generated mechanical movement into electrical impulses; the body's microphone. (Figure 23-30)

osmosis Net movement of water across a semipermeable membrane (permeable to water but not to solute) from a solution of lesser to one of greater solute concentration. (Figure 11-6)

oxidation Loss of electrons from an atom or molecule as occurs when a hydrogen atom is removed from a molecule or oxygen is added; opposite of **reduction.**

oxidation potential The voltage change when an atom or molecule loses an electron; a measure of the tendency of a molecule to loose an electron. For a given oxidation reaction, the oxidation potential has the same magnitude but opposite sign as the **reduction potential** for the reverse (reduction) reaction.

oxidative phosphorylation The phosphorylation of ADP to form ATP driven by the transfer of electrons to oxygen (O_2) in bacteria and mitochondria. Involves generation of a **proton-motive force** during electron transport and its subsequent use to power ATP synthesis.

p53 protein The product of a **tumor-suppressor gene** that plays a critical role in the arrest of cells with damaged DNA. Inactivating mutations in the *p53* gene are found in many human cancers. (Figure 25-26)

pair-rule genes In *Drosophila,* a group of genes expressed in alternating stripes along the anterioposterior axis in the early embryo. All encode transcription factors and function, along with **gap genes** and **segment-polarity genes,** in determining the body segments in flies. (Figure 22-27c)

paracrine Referring to signaling mechanism in which a target cell responds to a signaling molecule (e.g., growth factor, neurotransmitter) that is produced by a nearby cell(s) and reaches the target by diffusion.

patch clamping Technique for determining ion flow through a single ion channel or across the membrane of an entire cell by use of a micropipette whose tip is applied to a small patch of the cell membrane. (Figure 11-21)

pattern formation Process of organizing the cells, organs, and tissues of a developing embryo into well-ordered spatial patterns, like the bones of a hand or the color pattern on a butterfly wing.

P body Dense cytoplasmic domain, containing no ribosomes or translation factors, that functions in repression of translation and degradation of associated mRNAs; also called *cytoplasmic RNA-processing body.*

PCR (polymerase chain reaction) Technique for amplifying a specific DNA segment in a complex mixture by multiple cycles of DNA synthesis from short oligonucleotide primers followed by brief heat treatment to separate the complementary strands. (Figure 5-23)

pentose A five-carbon **monosaccharide.** The pentoses ribose and deoxyribose are present in RNA and DNA, respectively. (Figure 2-16)

peptide A small linear polymer composed of amino acids connected by peptide bonds. The terms *peptide* and *oligopeptide* are often used interchangeably. See also **polypeptide.**

peptide bond The covalent amide linkage between amino acids formed between the amino group of one amino acid and the carboxyl group of another with the net release of a water molecule (dehydration). (Figure 2-13)

peripheral membrane protein Any protein that associates with the cytosolic or exoplasmic face of a membrane but does not enter the hydrophobic core of the phospholipid bilayer. See also **integral membrane protein.** (Figure 10-1)

perlecan A large multidomain **proteoglycan** component of the extracellular matrix (ECM) that binds to many ECM components, cell-surface molecules, and growth factors; a major component of the **basal lamina.**

peroxisome Small organelle that contains enzymes for degrading fatty acids and amino acids by reactions that generate hydrogen peroxide, which is converted to water and oxygen by catalase.

pH A measure of the acidity or alkalinity of a solution defined as the negative logarithm of the hydrogen ion concentration in moles per liter: $pH = -\log [H^+]$. Neutrality is equivalent to a pH of 7; values below this are acidic and those above are alkaline.

phagocyte Any cell that can ingest and destroy pathogens and other particulate antigens. The primary phagocytes are neutrophils, macrophages, and dendritic cells.

phagocytosis Process by which relatively large particles (e.g., bacterial cells) are internalized by certain eukaryotic cells in a process that involves extensive remodeling of the actin cytoskeleton; distinct from **receptor-mediated endocytosis.** (Figure 9-2)

phenotype The detectable physical and physiological characteristics of a cell or organism determined by its **genotype;** also, the specific trait associated with a particular **allele.**

pheromone A signaling molecule released by an individual that can alter the behavior or gene expression of other individuals of the same species. The yeast α and **a** mating-type factors are well-studied examples.

phosphatase An enzyme that removes a phosphate group from a substrate by hydrolysis. Phosphoprotein phosphatases act with protein kinases to control the activity of many cellular proteins. (Figure 3-33)

phosphoanhydride bond A type of **high-energy bond** formed between two phosphate groups, such as the γ and β phosphates and the β and α phosphates in ATP. (Figure 2-31)

phosphodiester bond Chemical linkage between adjacent nucleotides in DNA and RNA; consists of two phosphoester bonds, one on the 5′ side of the phosphate and another on the 3′ side. (Figure 4-2)

phosphoglycerides Amphipathic derivatives of glycerol 3-phosphate that generally consist of two hydrophobic fatty acyl chains esterified to the hydroxyl groups in glycerol and a polar head group attached to the phosphate; the most abundant lipids in biomembranes. (Figures 2-20 and 10-5a)

phosphoinositides A group of membrane-bound lipids containing phosphorylated inositol derivatives; some function as **second messengers** in several signal-transduction pathways. (Figures 15-29 and 16-29)

phospholipase One of several enzymes that cleave various bonds in the hydrophilic end of **phospholipids.** (Figures 10-14)

phospholipase C (PLC) A membrane-associated phospholipase, activated by either $G_{\alpha q}$ or $G_{\alpha o}$, that cleaves the membrane lipid phosphatidylinositol 4,5-bisphosphate to generate two second messengers, **DAG** and **IP$_3$.** (Figures 15-29 and 15-30)

phospholipid The major class of lipids present in biomembranes, including **phosphoglycerides** and **sphingolipids.** (Figures 10-5a, b and 2-20)

phospholipid bilayer A two-layer, sheetlike structure in which the polar head groups of phospholipids are exposed to the aqueous media on either side, and the nonpolar fatty acyl chains are in the center; the foundation for all biomembranes. (Figure 10-6a, b)

photoelectron transport Light-driven electron transport that generates a charge separation across the **thylakoid** membrane that drives subsequent events in photosynthesis. (Figure 12-33)

photorespiration A reaction pathway that competes with CO_2 fixation (**Calvin cycle**) by consuming ATP and generating CO_2, thus reducing the efficiency of photosynthesis. (Figure 12-45)

photosynthesis Complex series of reactions occurring in some bacteria and in plant **chloroplasts** in which light energy is used to generate carbohydrates from CO_2, usually with the consumption of H_2O and evolution of O_2.

photosystems Multiprotein complexes, present in all photosynthetic organisms, that consist of light-harvesting complexes containing **chlorophylls** and a reaction center where **photoelectron transport** occurs. (Figure 12-42)

phragmoplast In plants, a temporary structure, formed during telophase, whose membranes become the plasma membranes of the daughter cells and whose contents develop into the new cell wall between them. (Figure 18-43)

pI See **isoelectric point.**

plakins A family of proteins that help attach **intermediate filaments** to other structures.

plaque assay Technique for determining the number of infectious viral particles in a sample by culturing a diluted sample on a layer of susceptible host cells and then counting the clear areas of lysed cells (plaques) that develop. (Figure 4-45)

plasma membrane The membrane surrounding a cell that separates the cell from its external environment; consists of a **phospholipid bilayer** and associated membrane lipids and proteins. (Figures 10-1 and 10-2)

plasmid Small, circular extrachromosomal DNA molecule capable of autonomous replication in a cell; commonly used as a **vector** in **DNA cloning.**

plasmodesmata (sing. **plasmodesma**) Tubelike cell junctions that interconnect the cytoplasms of adjacent plant cells and are functionally analogous to **gap junctions** in animal cells. (Figure 19-38)

point mutation Change of a single nucleotide in DNA, especially in a region coding for protein; can result in formation of a codon specifying a different amino acid or a stop codon. Addition or deletion of a single nucleotide will cause a shift in the **reading frame.**

polar Referring to a molecule or structure with a net electric charge or asymmetric distribution of positive and negative charges. Polar molecules are usually soluble in water.

polarity In cell biology, the presence of functional and/or structural differences in distinct regions of a cell or cellular component.

polarized In cell biology, referring to any cell or subcellular structure marked by functional and structural asymmetries.

polymer Any large molecule composed of multiple identical or similar units (**monomers**) linked by covalent bonds. (Figure 2-13)

polymerase chain reaction See **PCR.**

polypeptide Linear polymer of amino acids connected by peptide bonds, usually containing 20 or more residues. See also **protein.**

polyribosome A complex containing several **ribosomes,** all translating a single messenger RNA; also called *polysome.* (Figure 4-28)

polysaccharide Linear or branched polymer of **monosaccharides** linked by glycosidic bonds and usually containing more than 15 residues. Those with fewer than 15 residues are often called *oligosaccharides.*

polytene chromosome Enlarged chromosome composed of many parallel copies of itself formed by multiple cycles of DNA replication without chromosomal separation; found in the salivary glands and some other tissues of *Drosophila* and other dipteran insects. (Figures 6-44 and 6-45)

polyunsaturated Referring to a compound (e.g., fatty acid) in which two or more of the carbon–carbon bonds are double or triple bonds.

porins Class of trimeric transmembrane proteins through which small water-soluble molecules can cross the membrane; present in outer mitochondrial and chloroplast membranes and in the outer membrane of gram-negative bacteria (Figure 10-18)

potential energy Stored energy. In biological systems, the primary forms of potential energy are chemical bonds, concentration gradients, and electric potentials across cellular membranes.

pre-mRNA Precursor messenger RNA; the **primary transcript** and intermediates in RNA processing. (Figures 4-15 and 8-2)

pre-rRNA Large precursor ribosomal RNA that is synthesized in the nucleolus of eukaryotic cells and processed to yield three of the four RNAs present in ribosomes. (Figures 8-34 and 8-35)

primary structure In proteins, the linear arrangement (sequence) of amino acids within a polypeptide chain.

primary transcript In eukaryotes, the initial RNA product, containing **introns** and **exons,** produced by transcription of DNA. Many primary transcripts must undergo RNA processing to form the physiologically active RNA species.

primase A specialized RNA polymerase that synthesizes short stretches of RNA used as primers for DNA synthesis (Figure 4-31).

primer A short nucleic acid sequence containing a free 3′-hydroxyl group that forms **base pairs** with a complementary template strand and functions as the starting point for addition of nucleotides to copy the template strand.

probe Defined RNA or DNA fragment, radioactively or chemically labeled, that is used to detect specific nucleic acid sequences by hybridization.

programmed cell death See **apoptosis.**

prokaryotes Class of organisms, including the **bacteria** (eubacteria) and **archaea,** that lack a true membrane-limited nucleus and other organelles. See also **eukaryotes.** (Figure 1-3)

prometaphase Second stage in mitosis during which the nuclear envelope and nuclear lamina break down, and microtubules assembled from the spindle poles "capture" chromosome pairs at specialized structures called **kinetochores.** (Figure 18-34)

promoter DNA sequence that determines the site of **transcription** initiation for a RNA polymerase. (Figure 4-11)

promoter-proximal element Any regulatory sequence in eukaryotic DNA that is located within ≈200 base pairs of the transcription start site. Transcription of many genes is controlled by multiple promoter-proximal elements. (Figure 7-16)

prophase Earliest stage in mitosis during which the chromosomes condense, the duplicated centrosomes separate to become the spindle poles, and the mitotic spindle begins to form. (Figure 18-34)

protease Any enzyme that cleaves one or more peptide bonds in target proteins.

proteasome Large multifunctional protease complex in the cytosol that degrades intracellular proteins marked for destruction by attachment of multiple **ubiquitin** molecules. (Figure 3-29)

protein A macromolecule composed of one or more linear **polypeptide** chains and folded into a characteristic three-dimensional shape (**conformation**) in its native, biologically active state.

protein family Set of homologous proteins encoded by a **gene family.**

protein kinase A (PKA) Cytosolic enzyme that is activated by **cyclic AMP (cAMP)** and functions to phosphorylate and thus regulate the activity of numerous cellular proteins; also called *cAMP-dependent protein kinase.* (Figure 15-23)

protein kinase B (PKB) Cytosolic enzyme that is recruited to the plasma membrane by signal-induced **phosphoinositides** and subsequently activated; also called *Akt.* (Figure 16-30)

protein kinase C (PKC) Cytosolic enzyme that is recruited to the plasma membrane in response to signal-induced rise in cytosolic Ca^{2+} level and is activated by membrane-bound **diacylglycerol (DAG).** (Figure 15-30)

proteoglycans A group of glycoproteins (e.g., perlecan and aggrecan) that contain a core protein to which is attached one or more **glycosaminoglycan (GAG)** chains. They are found in nearly all animal extracellular matrices, and some are integral membrane proteins. (Figure 19-29)

proteome The entire complement of proteins produced by a cell.

proteomics The systematic study of the amounts, modifications, interactions, localization, and functions of all or subsets of proteins at the whole-organism, tissue, cellular, and subcellular levels.

proton Common term for a hydrogen ion (H^+).

proton-motive force The energy equivalent of the proton (H^+) concentration gradient and electric potential gradient across a membrane; used to drive ATP synthesis by **ATP synthase,** transport of molecules against their concentration gradient, and movement of bacterial flagella. (Figure 12-2)

proto-oncogene A normal cellular gene that encodes a protein usually involved in regulation of cell growth or differentiation and that can be mutated into a cancer-promoting **oncogene,** either by changing the protein-coding segment or by altering its expression. (Figure 25-11)

provirus The DNA of an animal virus that is integrated into a host-cell genome; during replication of the cell, the proviral DNA is replicated and appears in both daughter cells. Activation of proviral DNA leads to production and release of progeny virions.

pseudogene DNA sequence that is similar to that of a functional gene but does not encode a functional product; probably arose by sequence drift of duplicated genes.

pulse-chase A type of experiment in which a radioactive small molecule is added to a cell for a brief period (the pulse) and then is replaced with an excess of the unlabeled form of the same small molecule (the chase). Used to detect changes in the cellular location of a molecule or its metabolic fate over time. (Figure 3-39)

pump See **ATP-powered pump.**

purines A class of nitrogenous compounds containing two fused heterocyclic rings. Two purines, adenine (A) and guanine (G), are base components of nucleotides found in DNA and RNA. See also **base pair.** (Figure 2-17)

pyrimidines A class of nitrogenous compounds containing one heterocyclic ring. Two pyrimidines, cytosine (C) and thymine (T), are base components of nucleotides found in DNA; in RNA, uracil (U) replaces thymine. See also **base pair.** (Figure 2-17)

quaternary structure The number and relative positions of the polypeptide chains in multimeric (multisubunit) proteins. (Figure 3-10b)

radioisotope Unstable form of an atom that emits radiation as it decays. Several radioisotopes are commonly used experimentally as labels in biological molecules. (Table 3-1)

Ras protein A monomeric member of the **GTPase superfamily** of switch proteins that is tethered to the plasma membrane by a lipid anchor and functions in intracellular signaling pathways; activated by ligand binding to **receptor tyrosine kinases** and some other cell-surface receptors. (Figures 16-20 and 16-24)

rate constant A constant that relates the concentrations of reactants to the rate of a chemical reaction.

reading frame The sequence of nucleotide triplets (**codons**) that runs from a specific translation start codon in an mRNA to a stop codon. Some mRNAs can be translated into different polypeptides by reading in two different reading frames. (Figure 4-18)

receptor Any protein that specifically binds another molecule to mediate cell–cell signaling, adhesion, endocytosis, or other cellular process. Commonly denotes a protein located in the plasma membrane, cytosol, or nucleus that binds a specific extracellular molecule (**ligand**), which often induces a conformational change in the receptor, thereby initiating a cellular response. See also **adhesion receptor** and **nuclear receptor.** (Figures 15-1 and 16-1)

receptor-mediated endocytosis Uptake of extracellular materials bound to specific cell-surface receptors by invagination of the plasma membrane to form a small membrane-bounded vesicle (early endosome). (Figure 14-29)

receptor tyrosine kinase (RTK) Member of a large class of cell-surface receptors, usually with a single transmembrane domain, including those for insulin and many growth factors. Ligand binding activates tyrosine-specific protein kinase activity in the receptor's cytosolic domain, thereby initiating intracellular signaling pathways. (Figures 16-16 and 16-17)

recessive In genetics, referring to that allele of a gene that is not expressed in the phenotype when the **dominant** allele is present; also refers to the phenotype of an individual (homozygote) carrying two recessive alleles. Mutations that produce recessive alleles generally result in a loss of the gene's function. (Figure 5-2)

recombinant DNA Any DNA molecule formed in vitro by joining DNA fragments from different sources.

recombination Any process in which chromosomes or DNA molecules are cleaved and the fragments are rejoined to give new combinations. Homologous recombination occurs during meiosis, giving rise to **crossing over** of homologous chromosomes. Homologous recombination and nonhomologous recombination (i.e., between chromosomes of different morphologic type) also occur during several DNA-repair mechanisms and can be carried out in vitro with purified DNA and enzymes. (Figure 5-10)

redox reaction An oxidation–reduction reaction in which one or more electrons are transferred from one reactant to another.

reduction Gain of electrons by an atom or molecule as occurs when a hydrogen atom is added to a molecule or oxygen is removed. The opposite of **oxidation.**

reduction potential (*E*) The voltage change when an atom or molecule gains an electron; a measure of the tendency of a molecule to gain an electron. For a given reduction reaction, *E* has the same magnitude but opposite sign as the **oxidation potential** for the reverse (oxidation) reaction.

release factor (RF) One of two types of nonribosomal proteins that recognize stop codons in mRNA and promote release of the completed polypeptide chain, thereby terminating **translation** (protein synthesis). (Figure 4-27)

replication fork Y-shaped region in double-stranded DNA at which the two strands are separated and replicated during DNA synthesis; also called *growing fork.* (Figure 4-30)

replication origin Unique DNA segments present in an organism's genome at which DNA replication begins. Eukaryotic chromosomes contain multiple origins, whereas bacterial chromosomes and plasmids usually contain just one.

reporter gene A gene encoding a protein that is easily assayed (e.g., β-galactosidase, luciferase). Reporter genes are used in various types of experiments to indicate activation of the promoter to which it is linked.

repressor Specific **transcription factor** that inhibits transcription.

residue General term for the repeating units in a polymer that remain after covalent linkage of the monomeric precursors.

resolution The minimum distance between two objects that can be distinguished by an optical apparatus; also called *resolving power.*

respiratory chain See **electron transport chain.**

respiratory control Dependence of mitochondrial oxidation of NADH and $FADH_2$ on the supply of ADP and P_i for ATP synthesis.

resting K^+ channels Nongated K^+ ion channels in the plasma membrane that, in conjunction with the high cytosolic K^+ concentration produced by the **Na^+/K^+ ATPase,** are primarily responsible for generating the inside-negative resting **membrane potential** in animal cells.

restriction enzyme Any enzyme that recognizes and cleaves a specific short sequence, the *restriction site,* in double-stranded DNA molecules; used extensively to produce **recombinant DNA** in vitro; also called *restriction endonuclease.* (Figure 5-11; Table 5-1)

restriction fragment A defined DNA fragment resulting from cleavage with a particular **restriction enzyme.** These fragments are used in the production of recombinant DNA molecules and DNA cloning.

restriction fragment length polymorphisms See **RFLPs.**

restriction point The point in late G_1 of the cell cycle at which mammalian cells become committed to entering the S phase and completing the cycle even in the absence of growth factors; functionally equivalent to START in yeast.

retinotectal maps Corresponding maps of visual information, one in the retina created by incoming light and one in the visual part of the brain (the tectum) created by retinal ganglion cells carrying information from eye to brain. The map in the brain is like the map in the eye. (Figure 23-40)

retrotransposon Type of eukaryotic **transposable DNA element** whose movement in the genome is mediated by an RNA intermediate and involves a reverse transcription step. See also **transposon.** (Figure 6-8b)

retrovirus Type of eukaryotic virus containing an RNA genome that replicates in cells by first making a DNA copy of the RNA. This viral DNA is inserted into cellular chromosomal DNA, forming a **provirus,** and gives rise to further genomic RNA as well as the mRNAs for viral proteins. (Figure 4-49)

reverse transcriptase Enzyme found in retroviruses that catalyzes a complex reaction in which a double-stranded DNA is synthesized from a single-stranded RNA template. (Figure 6-14)

RFLPs (restriction fragment length polymorphisms) Differences among individuals in the sequence of genomic DNA that create or destroy sites recognized by particular **restriction enzymes.** Are one of several types of sequence differences between individuals that can serve as DNA-based molecular markers in human **linkage** studies. (Figure 5-36)

ribonucleic acid (RNA) See **RNA.**

ribonucleoprotein (RNP) complex General term for any complex composed of proteins and RNA. Most RNA molecules are present in the cell in the form of RNPs.

ribosome A large complex comprising several different **rRNA** molecules and more than 50 proteins, organized into a large subunit and small subunit; the engine of **translation** (protein synthesis). (Figures 4-22 and 4-23)

ribosomal RNA See **rRNA.**

ribozyme An RNA molecule with catalytic activity. Ribozymes function in RNA splicing and protein synthesis.

ribulose 1,5-bisphosphate carboxylase Enzyme located in chloroplasts that catalyzes the first reaction in the **Calvin cycle,** the addition of CO_2 to a five-carbon sugar (ribulose 1,5-bisphosphate) to form two molecules of 3-phosphoglycerate; also called *rubisco.* (Figure 12-43)

RISC See **RNA-induced silencing complex.**

RNA (ribonucleic acid) Linear, single-stranded polymer, composed of ribose **nucleotides.** mRNA, rRNA, and tRNA play different roles in protein synthesis; a variety of small RNAs play roles in controlling the stability and translation of mRNAs, and in controlling chromatin structure and transcription.(Figure 4-17)

RNA editing Unusual type of RNA processing in which the sequence of a pre-mRNA is altered.

RNA-induced silencing complex (RISC) Large multiprotein complex associated with a short single-stranded RNA (**siRNA** or **miRNA**) that mediates degradation or translational repression of a complementary or near-complementary mRNA.

RNA interference (RNAi) Functional inactivation of a specific gene by a corresponding double-stranded RNA that induces either inhibition of translation or degradation of the complementary single-stranded mRNA encoded by the gene but not that of mRNAs with a different sequence. (Figures 5-45)

RNA polymerase An enzyme that copies one strand of DNA (the *template* strand) to make the **complementary** RNA strand using as substrates ribonucleoside triphosphates. (Figure 4-11)

RNA splicing A process that results in removal of **introns** and joining of **exons** in pre-mRNAs. See also **spliceosome.** (Figures 8-8 and 8-9)

rRNA (ribosomal RNA) Any one of several large RNA molecules that are structural and functional components of **ribosomes.** Often designated by their sedimentation coefficient: 28S, 18S, 5.8S, and 5S rRNA in higher eukaryotes. (Figure 4-22)

rubisco See **ribulose 1,5-bisphosphate carboxylase.**

S (synthesis) phase See **cell cycle.**

sarcomere Repeating structural unit of striated (skeletal) muscle composed of organized, overlapping thin (**actin**) filaments and thick (**myosin**) filaments and extending from one Z disk to an adjacent one; shortens during contraction. (Figures 17-29 and 17-30)

sarcoplasmic reticulum Network of membranes in cytoplasm of a muscle cell that sequesters Ca^{2+} ions; release of stored Ca^{2+} induced by muscle stimulation triggers contraction. (Figure 17-32)

satellite DNA See **simple-sequence DNA.**

saturated Referring to a compound (e.g., fatty acid) in which all the carbon–carbon bonds are single bonds.

second messenger A small intracellular molecule (e.g., cAMP, cGMP, Ca^{2+}, DAG, and IP_3) whose concentration increases (or decreases) in response to binding of an extracellular signal and that functions in **signal transduction.** (Figure 15-9)

secondary structure In proteins, local folding of a polypeptide chain into regular structures including the α helix, β sheet, and β turns.

secretory pathway Cellular pathway for synthesizing and sorting soluble and membrane proteins localized to the endoplasmic reticulum, Golgi, and lysosomes; plasma membrane proteins; and proteins eventually secreted from the cell. (Figure 14-1)

secretory vesicle Small membrane-bound vesicle derived from the ***trans*-Golgi network** containing molecules destined to be released from the cell.

segment-polarity genes In *Drosophila,* a group of genes encoding components of signaling systems that influence cell fates and the polarity of cytoskeletons along the anterioposterior axis in the early embryo.

segregation The process that distributes an equal complement of chromosomes to daughter cells during mitosis and meiosis.

selectins A family of **cell-adhesion molecules** that mediate Ca^{2+}-dependent interactions with specific oligosaccharide moieties in glycoproteins and glycolipids on the surface of adjacent cells or in extracellular glycoproteins. (Figures 19-2 and 19-36)

shuttle vector Plasmid vector capable of propagation in two different hosts. (Figure 5-17)

side chain In amino acids, the variable substituent group attached to the **alpha (α) carbon atom** that largely determines the particular properties of each amino acid; also called *R group.* (Figure 2-14)

signaling molecule General term for any extracellular or intracellular molecule involved in mediating the response of a cell to its external environment or to other cells.

signal-recognition particle (SRP) A cytosolic ribonucleoprotein particle that binds to the ER **signal sequence** in a nascent secretory protein and delivers the nascent chain/ribosome complex to the ER membrane where synthesis of the protein and translocation into the ER are completed. (Figure 13-5)

signal sequence A relatively short amino acid sequence within a protein that directs the protein to a specific location within the cell; also called *signal peptide* and *uptake-targeting sequence.* (Table 13-1)

signal transduction Conversion of a signal from one physical or chemical form into another. In cell biology commonly refers to the sequential process initiated by binding of an extracellular signal to a receptor and culminating in one or more specific cellular responses.

silencer A sequence in eukaryotic DNA that promotes formation of condensed chromatin structures in a localized region, thereby blocking access of proteins required for transcription of genes within several hundred base pairs of the silencer; also called *silencer sequence.*

simple diffusion Net movement of a molecule across a membrane down its concentration gradient at a rate proportional to the gradient and the permeability of the membrane; also called *passive diffusion.*

simple-sequence DNA Short, tandemly repeated sequences that are found at **centromeres** and **telomeres** as well as at other chromosomal locations and are not transcribed; also called *satellite DNA.*

SINEs (short interspersed elements) Class of **retrotransposons,** 100–400 bp long, that constitute ≈13 percent of total human DNA. In humans, *Alu* elements account for about two-thirds of all SINEs.

siRNA A small, double-stranded RNA, 21–23 nucleotides long with two single-stranded nucleotides at each end. A single strand of the siRNA associates with several proteins to form an RNA-induced silencing complex (**RISC**) that cleaves target RNAs to which the siRNA base pairs perfectly; variously called *short* or *small interfering RNA* and *small inhibitory RNA.* siRNAs can be designed to experimentally inhibit expression of specific genes. See also **miRNA.** (Figure 8-25b)

Smads Class of transcription factors that are activated by phosphorylation following binding of members of the **transforming growth factor β (TGFβ)** family of signaling molecules to their cell-surface receptors. (Figure 16-4)

SMC proteins Structural maintenance of chromosome proteins; a small family of nonhistone chromatin proteins that are critical for maintaining the morphological structure of chromosomes and their proper segregation during mitosis. Members of this family include *condensins,* which help condense chromosomes during mitosis, and *cohesins,* which link sister chromatids until their separation in anaphase. Bacterial SMC proteins function in the proper segregation of bacterial chromosomes to daughter cells. (Figure 6-38 and 20-21)

SNAREs Cytosolic and integral membrane proteins that promote fusion of vesicles with target membranes. Interaction of v-SNAREs on a vesicle with cognate t-SNAREs on a target membrane forms very stable complexes, bringing the vesicle and target membranes into close apposition. (Figure 14-10)

snoRNA (small nucleolar RNA) A type of small, stable RNA that functions in rRNA processing and base modification in the nucleolus.

snRNA (small nuclear RNA) One of several small, stable RNAs localized to the nucleus. Five snRNAs are components of the **spliceosome** and function in splicing of pre-mRNA. (Figures 8-9 and 8-11)

somatic cell Any plant or animal cell other than a **germ cell.**

somatic cell nuclear transfer (SCNT) Procedure for generating specific cell types in culture starting from embryonic or adult stem cells.

sorting signal A relatively short amino acid sequence within a protein that directs the protein to particular transport vesicles as they bud from a donor membrane in the secretory or endocytic pathway. (Table 14-2)

Southern blotting Technique for detecting specific DNA sequences separated by electrophoresis by hybridization to a labeled nucleic acid **probe.** (Figure 5-26)

Spemann organizer Signaling center on the dorsal side of early amphibian embryos that functions in anterior-posterior and dorsal-ventral patterning. (Figures 22-12 and 22-14).

SPF (S phase–promoting factor) A heterodimeric protein, composed of a G_1 **cyclin** and **cyclin-dependent kinase (CDK),** that triggers entrance of a eukaryotic cell into the S phase of the cell cycle by phosphorylating specific proteins.

sphingolipid Major group of membrane lipids, derived from sphingosine, that contain two long hydrocarbon chains and either a phosphorylated head group (sphingomyelin) or carbohydrate head group (cerebrosides, gangliosides). (Figure 10-5b)

spliceosome Large ribonucleoprotein complex that assembles on a pre-mRNA and carries out **RNA splicing.** (Figure 8-11)

SRE-binding proteins (SREBPs) Cholesterol-dependent transcription factors, localized in the ER membrane, that are activated in response to low cellular cholesterol levels and then stimulate expression of genes encoding proteins involved in cholesterol synthesis and import as well as synthesis of other lipids. (Figure 16-38)

starch A very long, branched polysaccharide, composed exclusively of glucose units, that is the primary storage carbohydrate in plant cells. (Figure 12-28)

STAT Signal Transduction and Activation of Transcription. Class of transcription factors that are activated in the cytosol following ligand binding to **cytokine receptors.** (Figure 16-12)

steady state In cellular metabolic pathways, the condition when the rate of formation and rate of consumption of a substance are equal, so that its concentration remains constant. (Figure 2-23)

stem cell A self-renewing cell that can divide *symmetrically* to give rise to two daughter cells whose developmental potential is identical to the parental stem cell or *asymmetrically* to generate daughter cells with different developmental potentials (Figure 21-2)

stereocilia (sing. **stereocilium**) Nonmotile filament-filled protrusions of hair cells in the **organ of Corti** that are moved by sound-induced vibrations, triggering depolarization in the axons associated with each hair cell. (Figures 23-30 and 23-31)

stereoisomers Two compounds with identical molecular formulas whose atoms are linked in the same order but in different spatial arrangements. In *optical isomers,* designated D and L, the atoms bonded to an **asymmetric carbon atom** are arranged in a mirror-image fashion. *Geometric isomers* include the *cis* and *trans* forms of molecules containing a double bond.

steroids A group of four-ring hydrocarbons including cholesterol and related compounds. Many important hormones (e.g., estrogen and progesterone) are steroids. *Sterols* are steroids containing one or more hydroxyl groups. (Figure 10-5c)

substrate Molecule that undergoes a charge in a reaction catalyzed by an enzyme.

substrate-level phosphorylation Formation of ATP from ADP and P_i catalyzed by cytosolic enzymes in reactions that do not depend on a proton-motive force or molecular oxygen.

sulfhydryl group (–SH) A substituent group present in the amino acid cysteine and other molecules consisting of a hydrogen atom covalently bonded to a sulfur atom; also called a *thiol group.*

suppressor mutation A mutation that reverses the phenotypic effect of a second mutation. Suppressor mutations are frequently used to identify genes encoding interacting proteins. (Figure 5-9a)

symport A type of **cotransport** in which a membrane protein (*symporter*) transports two different molecules or ions across a cell membrane in the *same* direction. See also **antiport.** (Figure 11-3, [3B])

synapse Specialized region between an axon terminal of a neuron and an adjacent neuron or other excitable cell (e.g., muscle cell) across which impulses are transmitted. At a *chemical* synapse, the impulse is conducted by a **neurotransmitter;** at an *electric* synapse, impulse transmission occurs via **gap junctions** connecting the pre- and postsynaptic cells. (Figure 23-4)

syncytium A multinucleated mass of cytoplasm enclosed by a single plasma membrane.

syndecans A class of cell-surface **proteoglycans** that function in cell–matrix adhesion, interact with the cytoskeleton, and may bind external signals, thereby participating in cell–cell signaling.

synteny Occurrence of genes in the same order on a chromosome in two or more different species.

synthetic lethal mutation A mutation that increases the phenotypic effect of another mutation in the same or a related gene. (Figure 5-9b, c)

TATA box A conserved sequence in the **promoter** of many eukaryotic protein-coding genes where the transcription-initiation complex assembles. (Figure 7-12)

T cell A lymphocyte that matures in the thymus and expresses antigen-specific receptors that bind antigenic peptides complexed to **MHC molecules.** There are two major classes: *cytotoxic* T cells (CD8 surface marker, class I MHC restricted, kill virus-infected and tumor cells), and *helper* T cells (CD4 marker, class II MHC restricted, produce cytokines, required for activation of B cells). (Figures 24-34 and 24-36)

T-cell receptor A heterodimeric antigen-binding transmembrane protein containing variable and constant regions and associated with the signal-transducing multimeric CD3 complex. (Figure 24-29)

telomere Region at each end of a eukaryotic chromosome containing multiple tandem repeats of a short telomeric (TEL) sequence. Telomeres are required for proper chromosome **segregation** and are replicated by a special process that prevents shortening of chromosomes during DNA replication. (Figure 6-49)

telophase Final mitotic stage during which the nuclear envelope re-forms around the two sets of separated chromosomes, the chromosomes decondense, and division of the cytoplasm (cytokinesis) is completed. (Figure 18-34)

temperature-sensitive (ts) mutation A mutation that produces a wild-type phenotype at one temperature (the permissive temperature) but a mutant phenotype at another temperature (the nonpermissive temperature). This type of mutation is especially useful in identification of genes essential for life. (Figure 5-6)

tertiary structure In proteins, overall three-dimensional form of a polypeptide chain, which is stabilized by multiple noncovalent interactions between side chains. (Figure 3-10a)

thylakoids Flattened membranous sacs in a chloroplast that can be arranged in stacks and contain the photosynthetic pigments and **photosystems.** (Figure 12-29)

tight junction A type of cell–cell junction between the plasma membranes of adjacent epithelial cells that prevents diffusion of macromolecules and many small molecules and ions in the spaces between cells and diffusion of membrane components between the apical and basolateral regions of the plasma membrane. (Figure 19-15)

Toll-like receptor (TLR) Member of a class of cell-surface and intracellular receptors that recognize a variety of microbial products. Ligand binding initiates a signaling pathway that induces various responses depending on the cell type. (Figure 24-35)

topogenic sequences Segments within a protein whose sequence, number, and arrangement direct the insertion and orientation of various classes of transmembrane proteins into the endoplasmic reticulum membrane. (Figure 13-13)

transcription Process in which one strand of a DNA molecule is used as a template for synthesis of a **complementary** RNA by RNA polymerase. (Figures 4-10 and 4-11)

transcription-control region Collective term for all the DNA regulatory sequences that regulate transcription of a particular gene.

transcription factor (TF) General term for any protein, other than RNA polymerase, required to initiate or regulate transcription in eukaryotic cells. *General* factors, required for transcription of all genes, participate in formation of the transcription-preinitiation complex near the start site. *Specific* factors stimulate (**activators**) or inhibit (**repressors**) transcription of particular genes by binding to their regulatory sequences.

transcription unit A region in DNA, bounded by an initiation (start) site and termination site, that is transcribed into a single **primary transcript.**

transcytosis Mechanism for transporting certain substances across an epithelial sheet that combines **receptor-mediated endocytosis** and **exocytosis.** (Figures 14-25 and 24-10)

transfection Experimental introduction of foreign DNA into cells in culture, usually followed by expression of genes in the introduced DNA. (Figure 5-32)

transfer RNA See **tRNA.**

transformation (1) Permanent, heritable alteration in a cell resulting from the uptake and incorporation of a foreign DNA into the host-cell genome; also called *stable transfection.* (2) Conversion of a "normal" mammalian cell into a cell with cancer-like properties usually induced by treatment with a virus or other cancer-causing agent.

transforming growth factor beta (TGFβ) A family of secreted signaling proteins that are used in the development of most tissues in most or all animals. Members of the TGFβ family more often inhibit growth than stimulate it. Mutations in TGFβ signal-transduction components are implicated in human cancer, including breast cancer. (Figures 16-3 and 16-4)

transgene A cloned gene that is introduced and stably incorporated into a plant or animal and is passed on to successive generations.

transgenic Referring to any plant or animal carrying a **transgene.**

***trans*-Golgi network (TGN)** Complex network of membranes and vesicles that serves as a major branch point in the **secretory pathway.** Vesicles budding from this most-distal Golgi compartment carry membrane and soluble proteins to the cell surface or to lysosomes. (Figures 14-1 and 14-17)

transition state State of the reactants during a chemical reaction when the system is at its highest energy level; also called the *transition state intermediate.*

translation The **ribosome**-mediated assembly of a polypeptide whose amino acid sequence is specified by the nucleotide sequence in an mRNA. (Figure 4-17)

translocon Multiprotein complex in the membrane of the rough endoplasmic reticulum through which a nascent secretory protein enters the ER lumen as it is being synthesized. (Figure 13-7)

transmembrane protein See **integral membrane protein.**

transport protein See **membrane transport protein.**

transport vesicle A small membrane-bounded compartment that carries soluble and membrane "cargo" proteins in the forward or reverse direction in the **secretory pathway.** Vesicles form by budding off from the donor organelle and release their contents by fusion with the target membrane.

transposable DNA element Any DNA sequence that is not present in the same chromosomal location in all individuals of a species and can move to a new position by **transposition;** also called *mobile DNA element* and *interspersed repeat.* (Table 6-1)

transposition Movement of a **transposable DNA element** within the genome; occurs by a cut-and-paste mechanism or copy-and-paste mechanism depending on the type of element. (Figure 6-8)

transposon, DNA A **transposable DNA element** present in prokaryotes and eukaryotes that moves in the genome by a mechanism involving DNA synthesis and transposition. See also **retrotransposon.** (Figures 6-9 and 6-10)

triacylglycerol See **triglyceride.**

triglyceride Major form in which fatty acids are stored and transported in animals; consists of three fatty acyl chains esterified to a glycerol molecule. (p. 491)

tRNA (transfer RNA) A group of small RNA molecules that function as amino acid donors during protein synthesis. Each tRNA becomes covalently linked to a particular amino acid, forming an **aminoacyl-tRNA.** (Figures 4-19 and 4-20)

trophic factor Any of numerous signaling proteins required for the survival of cells in multicellular organisms; in the absence of such signals, cells often undergo "suicide" by **apoptosis.**

tubulin A family of globular cytoskeletal proteins that polymerize to form the cylindrical wall of **microtubules.** (Figure 18-3)

tumor A mass of cells, generally derived from a single cell, that arises due to loss of the normal regulators of cell growth; may be **benign** or **malignant.**

tumor-suppressor gene Any gene whose encoded protein directly or indirectly inhibits progression through the cell cycle and in which a loss-of-function mutation is oncogenic. Inheritance of a single mutant allele of many tumor-suppressor genes (e.g., *RB, APC,* and *BRCA1*) greatly increases the risk for developing colorectal and other types of cancer. (Figures 25-9 and 25-11)

ubiquitin A small protein that can be covalently linked to other intracellular proteins, thereby tagging these proteins for degradation by the **proteasome,** sorting to the lysosome, or alteration in the function of the target protein. (Figure 3-29)

uncoupler Any natural substance (e.g., the protein thermogenin) or chemical agent (e.g., 2,4-dinitrophenol) that dissipates the **proton-motive force** across the inner mitochondrial membrane or thylakoid membrane of chloroplasts, thereby inhibiting ATP synthesis.

uniport A type of transport in which a membrane protein (*uniporter*) mediates movement of a small molecule across a membrane down its concentration gradient via **facilitated transport.** The glucose transporters (**GLUT proteins**) are well-studied examples of uniporters. (Figure 11-3, [3A])

unsaturated Referring to a compound (e.g., fatty acid) in which one of the carbon–carbon bonds is a double or triple bond.

upstream (1) For a gene, the direction opposite to that in which RNA polymerase moves during transcription. Nucleotides upstream from the +1 position (the first transcribed nucleotide) are designated −1, −2, etc. (2) Events that occur earlier in a cascade of steps (e.g., signaling pathway). See also **downstream.**

upstream activating sequence (UAS) Any protein-binding regulatory sequence in the DNA of yeast and other simple eukaryotes that is necessary for maximal gene expression; equivalent to an enhancer or promoter-proximal element in higher eukaryotes. (Figure 7-16)

vaccine An innocuous preparation derived from a pathogen and designed to elicit an immune response in order to provide immunity against a future challenge by a virulent form of the same pathogen.

van der Waals interaction A weak **noncovalent interaction** due to small, transient asymmetric electron distributions around atoms (dipoles). (Figure 2-10)

vector In cell biology, an autonomously replicating genetic element used to carry a cDNA or fragment of genomic DNA into a host cell for the purpose of gene cloning. Commonly used vectors are bacterial plasmids and modified bacteriophage genomes. See also **expression vector** and **shuttle vector.** (Figure 5-13)

viral envelope A phospholipid bilayer forming the outer covering of some viruses (e.g., influenza and rabies viruses); is derived by budding from a host-cell membrane and contains virus-encoded glycoproteins. (Figure 4-47)

virion An individual viral particle.

virus A small intracellular parasite, consisting of nucleic acid (RNA or DNA) enclosed in a protein coat, that can replicate only in a susceptible host cell; widely used in cell biology research. (Figure 4-44)

V_{max} Parameter that describes the maximal velocity of an enzyme-catalyzed reaction or other process such as protein-mediated transport of molecules across a membrane. (Figures 3-22 and 11-4)

Western blotting Technique in which proteins separated by electrophoresis are attached to a nitrocellulose or other membrane, and specific proteins then are detected by use of labeled antibodies; also called *immunoblotting.* (Figure 3-38)

wild type Normal, nonmutant form of a gene, protein, cell, or organism.

Wnt A family of secreted signaling proteins used in the development of most tissues in most or all animals. Mutations in Wnt signal-transduction components are implicated in human cancer, especially colon cancer. Receptors are Frizzled-class proteins with seven transmembrane segments. (Figure 16-32)

x-ray crystallography Commonly used technique for determining the three-dimensional structure of macromolecules (particularly proteins and nucleic acids) by passing x-rays through a crystal of the purified molecules and analyzing the diffraction pattern of discrete spots that results. (Figure 3-42)

zinc finger Several related DNA-binding **structural motifs** composed of secondary structures folded around a zinc ion; present in numerous eukaryotic transcription factors. (Figures 3-9b and 7-25)

zygote A fertilized egg; diploid cell resulting from fusion of a male and female **gamete.**

INDEX

Note: Page numbers followed by f indicate figures; those followed by t indicate tables.

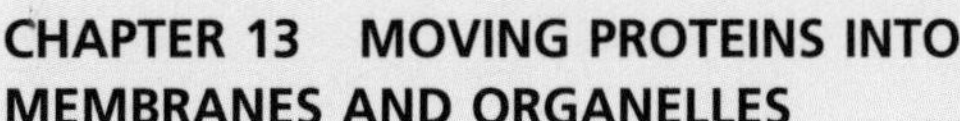

CHAPTER 13 MOVING PROTEINS INTO MEMBRANES AND ORGANELLES

Overview Animation: Protein Sorting (Figure 13-1)

Video: Three-Dimensional Model of a Protein Translocation Channel (Section 13.1)

Podcast: Structure and Function of the Signal Recognition Particle in Protein Translocation into the ER (Figure 13-5)

Focus Animation: Synthesis of Secreted and Membrane-Bound Proteins (Figure 13-6)

CHAPTER 14 VESICULAR TRAFFIC, SECRETION, AND ENDOCYTOSIS

Overview Animation: Protein Secretion (Figure 14-1)

Video: Transport of VSVG-GFP Through the Secretory Pathway (Figure 14-2)

Video: KDEL Receptor Trafficking (Figure 14-13)

Video: 3-D Model of a Golgi Complex (Figure 14-15)

Video: Birth of a Clathrin Coat (Figure 14-18)

Video: Segregation of Apical and Basolateral Cargo in the Golgi of Live Cells (Figure 14-25)

Podcast: HIV Budding from the Plasma Membrane (Figure 14-34)

CHAPTER 15 SIGNALING I: SIGNAL TRANSDUCTION AND SHORT-TERM CELLULAR EFFECTS

Overview Animation: Extracellular Signaling (Figure 15-1)

Focus Animation: Second Messengers in Signaling Pathways (Figure 15-9)

Video: Chemotaxis of a Single *Dictyostelium* Cell to the Chemoattractant cAMP (Section 15.4)

Overview Animation: Extracellular Signaling (Figure 15-13)

Podcast: Activation of G Proteins Measured by Fluorescence Resonance Energy Transfer (FRET) (Figure 15-14)

Overview Animation: Extracellular Signaling (Figure 15-20)

Focus Animation: Second Messengers in Signaling Pathways (Figure 15-30)

Technique Animation: Reporter Constructs (Figure 15-34)

CHAPTER 16 SIGNALING II: SIGNALING PATHWAYS THAT CONTROL GENE EXPRESSION

Focus Animation: TGF Signaling Pathway (Figure 16-4)

Overview Animation: Extracellular Signaling (Figure 16-27)

Podcast: Scaffold Proteins in Yeast MAP Kinase Cascades (Figure 16-28)

Video: Protein Dynamics in Response to cAMP Stimulation of a *Dictyostelium* Cell (Figure 16-30)

Video: Localized Activation of PI-3 Kinase in a Chemotactic *Dictyostelium* Cell (Section 16.5)

Video: Chemotaxis of a Single *Dictyostelium* Cell to the Chemoattractant cAMP (Section 16.5)

Overview Animation: Extracellular Signaling (Figure 16-31)

CHAPTER 17 CELL ORGANIZATION AND MOVEMENT I: MICROFILAMENTS

Focus Animation: Actin Polymerization (Figure 17-7)

Focus Animation: Elongation of Actin Filament by Forming FH2 Dimer (Figure 17-13)

Video: Direct Observation of Actin Filament Branching Mediated by Activated Arp2/3 (Figure 17-15)

Video: In Vivo Assembly of Actin Tails in *Listeria*-Infected Bacteria (Figure 17-17)

Technique Animation: In Vitro Motility Myosin Assay (Figure 17-21)

Focus Animation: Myosin-Actin Cross-Bridge Cycle (Figure 17-24)

Podcast: Myosin Movement Against Actin Filaments (Figure 17-24)

Overview Animation: Movement of Multiple Cargos by Myosin-V in Yeast (Figure 17-36)

Video: Mechanisms of Fish Keratinocyte Migration (Figure 17-38)

Video: Actin Filaments in the Lamellipodium of a Fish Keratinocyte (Figure 17-38)

Video: Actin Dynamics in a Migrating Fibroblast (Figure 17-39)

Video: Chemotaxis of a Single *Dictyostelium* Cell to cAMP (Figure 17-45)

Video: Microtubule and Actin Movements in Migrating Cells (Section 17.7)

CHAPTER 18 CELL ORGANIZATION AND MOVEMENT II: MICROTUBULES AND INTERMEDIATE FILAMENTS

Video: Cytoplasmic Assembly of Microtubules in Cultured Cells (Figure 18-11)

Video: In Vivo Visualization of Microtubule Ends by GFP-EB1 (Figure 18-15)